International Thermal Spray Conference & Exposition (ITSC 2017)

Dusseldorf, Germany
7 - 9 June 2017

Volume 1 of 2

ISBN: 978-1-5108-5822-0

Printed from e-media with permission by:

Curran Associates, Inc.
57 Morehouse Lane
Red Hook, NY 12571

Some format issues inherent in the e-media version may also appear in this print version.

Printed by Curran Associates, Inc. (2018)

For permission requests, please contact DVS Media GmbH
at the address below.

DVS Media GmbH
P. O. Box 10 19 65
40010 Dusseldorf
Germany

Phone: 49 (0)211 / 1591-0
Fax: 49 (0)211 / 1591-150

media@dvs-hg.de

Additional copies of this publication are available from:

Curran Associates, Inc.
57 Morehouse Lane
Red Hook, NY 12571 USA
Phone: 845-758-0400
Fax: 845-758-2633
Email: curran@proceedings.com
Web: www.proceedings.com

TABLE OF CONTENTS

VOLUME 1

AVIATION & POWER GENERATION INDUSTRY II

CHARACTERIZATION & TESTING METHODS I

HVAF SPRAYING

TRIBOLOGICAL COATINGS

AVIATION & POWER GENERATION INDUSTRY III

LASER CLADDING & PTA

PRE- & POST-TREATMENT

EQUIPMENT / CONSUMABLES & POWDERS, WIRES, SUSPENSIONS II

AVIATION & POWER GENERATION INDUSTRY IV

WEAR & CORROSION PROTECTION

HVOF SPRAYING

POWER GENERATION - STEAM

ELECTRONICS & SENSORIC

CORROSION PROTECTION

SUSPENSION SPRAYING III

COLD GAS SPRAYING III

AVIATION & POWER GENERATION INDUSTRY VI

APPLICATIONS – MARITIM INDUSTRY

APPLICATIONS – MEDICAL INDUSTRY

APPLICATIONS – METALS PROCESSING

APPLICATIONS – POWER GENERATION – FUEL CELLS & SOLAR

APPLICATIONS – POWER GENERATION - STEAM

EQUIPMENT / CONSUMABLES – PROCESS DIAGNOSTICS, SENSORS & CONTROLS

PROPERTIES – CERAMICS COATINGS

PROPERTIES – CORROSION PROTECTION

PROPERTIES – METAL COATINGS

PROPERTIES – POLYMER COATINGS

PROPERTIES – PRE- & POST-TREATMENT

PROPERTIES – PROPERTIES CHARACTERIZATION & TESTING METHODS

PROPERTIES – TRIBOLOGICAL COATINGS

PROPERTIES – WEAR PROTECTION

THERMAL SPRAY PROCESSES – ADDITIVE MANUFACTURING

THERMAL SPRAY PROCESSES – ARC SPRAYING & LASER SPRAYING

THERMAL SPRAY PROCESSES – COLD GAS SPRAYING

THERMAL SPRAY PROCESSES – HVOF/HVAF SPRAYING

THERMAL SPRAY PROCESSES – LASER CLADDING

THERMAL SPRAY PROCESSES – MODELING & SIMULATION

THERMAL SPRAY PROCESSES – NEW PROCESSES

THERMAL SPRAY PROCESSES – PLASMA SPRAYING

THERMAL SPRAY PROCESSES – PTA – PLASMA TRANFERRED ARC

THERMAL SPRAY PROCESSES – SUSPENSION SPRAYING

DVS-BERICHTE

International Thermal Spray

ITSC 2017

Conference & Exposition

June 7 – 9, 2017, Düsseldorf / Germany

Conference Proceedings

Organizers:

- DVS – German Welding Society
- ASM International – Thermal Spray Society (TSS)
- IIW – International Institute of Welding

Supporting Partners:

Media Sponsors:

DVS MEDIA

International Thermal Spray

ITSC 2017

Conference & Exposition

Proceedings of the Conference
in Düsseldorf / Germany
on June 7 – 9, 2017

Organized by

DVS – German Welding Society
ASM International – Thermal Spray Society (TSS)
IIW – International Institute of Welding

Supporting Societies:

Media Sponsors:

Bibliographic information published by the Deutsche Nationalbibliothek
The Deutsche Nationalbibliothek lists this publication in the Deutsche Nationalbibliografie; detailed bibliographic data are available in the Internet at http://dnb.dnb.de.

DVS-Berichte Volume 336

The conference contributions are published in form of manuscripts.

Print: rewi druckhaus, Reiner Winters GmbH, Wissen

Preface

Düsseldorf is the capital city of the German state of North Rhine-Westphalia and one of the most populous cities in Germany. The river Rhine, its "Alt Beer" and its carnival are only three reasons to come over. The city with its international business and financial centre is renowned for its fashion and trade fairs. Düsseldorf is headquarters to one Fortune Global 500 and two DAX companies. Messe Düsseldorf (Düsseldorf Trade Fairs) organizes nearly one fifth of premier trade shows.

Düsseldorf is known for its academy of fine arts (Joseph Beuys, Emanuel Leutze, August Macke, Gerhard Richter, Sigmar Polke and Andreas Gursky), its pioneering influence on electronic/experimental music and its Japanese community.

In this extraordinary surrounding of Düsseldorf ITSC will take place in the CCD Congress Center Düsseldorf, Stadthalle (Messe Düsseldorf).

Beside the three-day practice-oriented ITSC technical conference program further highlights, e.g. the Session "Young Professionals" will be offered. ITSC 2017 is again supporting young talents becoming part of the worldwide thermal spray family.

The "Industrial Forum" is another characteristic of ITSC. All practically interested experts will be addressed to get information in this forum.

The Three-Day Exposition is a permanent part of the event. If you are looking for new products and services, have questions or need answers, then ITSC 2017 is the right place for you.

ITSC exhibitors are experts who can offer attendees a wealth of problem-solving information and cost-saving opportunities. You will find information about equipment for thermal spraying, re-search and specialist institutes, applied research, and the latest innovations conveniently located in one big forum.

The organizers look forward to welcoming you in Düsseldorf / Germany!

Düsseldorf, May 2017

DVS – German Welding Society

Dr.-Ing. Roland Boecking
General Manager

Dipl.-Ing. Jens Jerzembeck
Head of Research and Technology

TIME SCHEDULE ITSC 2017 CONFERENCE AND EXPOSITION (Session Overview)

12:00 – 18:00 ITSC 2017 Exposition, CCD Congress Center Düsseldorf, Stadthalle, Hall X, Foyers

Wednesday, June 7, 2017						
Time	Hall Y	Hall 12	Hall 14	Hall 26	Hall 27	Hall 28
09:00	Opening: ■ Welcome ■ Plenary Lecture					
10:30	Coffee Break, CCD Congress Center Düsseldorf, Stadthalle, Hall X, Foyers					
10:50	Plasma Spraying I	Aviation & Power Generation Industry I	Medical Industry I	Process Diagnostics, Sensors & Controls	Ceramic Coatings	Cold Gas Spraying I
12:30	Lunch Break, CCD Congress Center Düsseldorf, Stadthalle, Hall X, Foyers **Poster Session, CCD Congress Center Düsseldorf, Stadthalle, Hall X, Foyers**					
14:00	Maritim Industry & Off-Shore Technologies	Power Generation – Fuel Cell	Medical Industry II	Wear Protection I	Suspension Spraying I	Equipment / Consumables & Powders, Wires, Suspensions I
15:20	Coffee Break, CCD Congress Center Düsseldorf, Stadthalle, Hall X, Foyers					
15:40	Session "Young Professionals" with Awards Presentation					
17:00	Awards Presentation	ITSC Best Paper Awards | René Wasserman Prize | Oerlikon Metco Young Professional Award				
	End of ITSC 2017 Conference Program					
18:00	**Exhibitor Reception, CCD Congress Center Düsseldorf, Stadthalle, Hall X, Foyers** **Poster Session, CCD Congress Center Düsseldorf, Stadthalle, Hall X, Foyers**					

09:00 – 18:00 ITSC 2017 Exposition, CCD Congress Center Düsseldorf, Stadthalle, Hall X, Foyers

Thursday, June 8, 2017					
Time	Hall 12	Hall 14	Hall 26	Hall 27	Hall 28
09:00	Automotive Industry	Industrial Forum I	Wear Protection II	Suspension Spraying II	Modeling & Simulation I
10:40	Coffee Break, CCD Congress Center Düsseldorf, Stadthalle, Hall X, Foyers				
11:00	Aviation & Power Generation Industry II	Industrial Forum II	Characterization & Testing Methods I	HVAF Spraying	Tribological Coatings
12:40	Lunch Break, CCD Congress Center Düsseldorf, Stadthalle, Hall X, Foyers **Poster Session, CCD Congress Center Düsseldorf, Stadthalle, Hall X**				
13:40	Aviation & Power Generation Industry III	Industrial Forum III	Laser Cladding & PTA	Pre- & Post-Treatment	Equipment / Consumalbles & Powders, Wires, Suspen-sions II
15:20	Coffee Break, CCD Congress Center Düsseldorf, Stadthalle, Hall X, Foyers				
15:40	Aviation & Power Generation Industry IV	Industrial Forum IV	Wear & Corrosion Protection	HVOF Spraying	Cold Gas Spraying II
17:20	End of ITSC 2017 Conference Program				
19:30	Awards Presentation	**ITSC 2017 Networking Event with Awards Presentation, ESPRIT ARENA (Düsseldorf's Football Arena)** TSS Hall of Fame | TSS President's Award | JTST Best Paper Award			

09:00 – 16:00 ITSC 2017 Exposition, CCD Congress Center Düsseldorf, Stadthalle, Hall X, Foyers

Friday, June 9, 2017					
Time	Hall 12	Hall 14	Hall 26	Hall 27	Hall 28
09:00	Aviation & Power Generation Industry V	Industrial Forum V	Characterization & Testing Methods II	Metal Coatings	Modeling & Simulation II
10:40	Coffee Break, CCD Congress Center Düsseldorf, Stadthalle, Hall X, Foyers				
11:00	Power Generation – Steam	Electronics & Sensoric	Corrosion Protection	Suspension Spraying III	Cold Gas Spraying III
12:40	Lunch Break, CCD Congress Center Düsseldorf, Stadthalle, Hall X, Foyers **Poster Session, CCD Congress Center Düsseldorf, Stadthalle, Hall X**				
13:40	Aviation & Power Generation Industry VI	Additive Manufacturing	Amorphous Coatings	Plasma Spraying II	Cold Gas Spraying IV
15:20	End of ITSC 2017 Conference Program				

PENTA: A novel high spray rate APS gun

T. Schläfer, A. Wank, C. Schmengler, K. Müller-Roden, GTV Verschleiss-Schutz GmbH, Luckenbach / D

Multi-electrode APS has proven significant advantages with regard to spray rates, deposit efficiency and component life time. So far, three-cathode and three-anode spray guns have been established successfully in several industrial sectors where high spray rates are mandatory. Based on the successful three-anode plasma spray gun DELTA, the five-anode APS gun PENTA was developed to further increase productivity. With its gross power of up to 125 kW it allows very high spray rates and thus significantly reduced coating times. This paper focusses on the development of high spray rate parameter settings for producing oxide ceramic coatings. All coatings will be investigated with regard to their microstructures. Furthermore, economical benefits of the five-anode technology will be highlighted.

1 Introduction

For decades single anode-single cathode D.C. plasma torches are state of the art when producing oxide ceramic coatings.

Conventional D.C. plasma spray guns such as 9MB, F4 (Oerlikon Metco, Wohlen, CH) or F6 (GTV Verschleißschutz GmbH, Luckenbach, D) are characterized by a rod shaped cathode and a hollow, water cooled concentric anode. They allow input powers of up to 55 kW at high currents of up to 1,000 A whereupon the arc voltage typically is below 80 V. With such guns spray rates of 30-60 g/min are industrially established when producing oxide ceramic coatings.

Such plasma guns are characterized by substantial plasma jet fluctuations which cause a non-uniform treatment of the powder feedstock which is radially injected into the plasma plume some millimeters downstream the nozzle exit thus lowering coating quality and deposition efficiency. [1,2]

The reason for the plasma jet fluctuations lies within the axial movement of the anodic arc attachment. Azimuthal movement on the other hand is essential for the operation of such gun type in order to limit erosion caused by high current densities in the anodic arc attachment. [2]

2 Multi Electrode APS Guns

Over the last decade equipment manufacturers and researchers have followed several approaches in order to overcome the disadvantages of conventional 1 cathode/1 anode plasma spray guns and to increase the maximum power input and hence spray rate and productivity.

Very promising approaches lie within cascaded arc channels/nozzles as well as in dividing the single electric arc into several arcs.

With the first of the two approaches, the cascaded arc channel, several water cooled hollow cylindrical segments, so-called neutrodes, which are electrically insulated against each other and the electrodes are installed between the anode and the cathode thus strongly limiting axial arc attachment movement Furthermore, by using a cascaded arc channel the distance between cathode and anode can be increased leading to higher arc voltages. Consequently the input power can be increased without increasing the arc current.

The second improvement of plasma spray guns results from dividing the electric arc into three or more thus reducing the electric power being transferred through the arc attachment(s) and the thermal load at the arc attachments significantly. This measure can be utilized, to increase the input power of the spray gun, too.

The combination of both design principles in one spray gun promises high spray rates in combination with high deposition efficiency, excellent coating quality and improved component lifetime.

Over the last decade two types of such spray guns have entered the market, both can be traced back to the work of Prof. Klaus Landes of the University of Armed Forces (Neubiberg, D):

2.1 3 Cathodes/1 Anode Spray Gun

The TriplexPro-210™ gun (Oerlikon Metco, Wohlen, CH) and its predecessors are characterized by three cathodes, a cascaded arc channel and a hollow cylindrical anode. Azimuthal fixation of the anodic arc roots is ensured by a vortex flow of the plasma forming gases, which are typically Ar/N_2 or Ar/He mixtures. Due to vortex stabilization the optimum area of powder injection changes azimuthally with adjusted gas flow rates and current. The gun is capable of an input power of 65 kW (90 kW in so-called high-velocity mode), chromia coatings can be produced at a spray rate of 90 g/min [3].

2.2 1 Cathode/3 Anodes Spray Gun: Delta

With the Delta gun (GTV Verschleißschutz GmbH, Luckenbach, D), which is also characterized by an cascaded arc channel, the opposite approach was followed: The design is based on one single cathode and three anode segments, which are electrically insulated against each other and are individually connected to three independent units of a common power supply. During operation the single electric arc is divided into three close to the anode segments as depicted schematically in Figure 1.

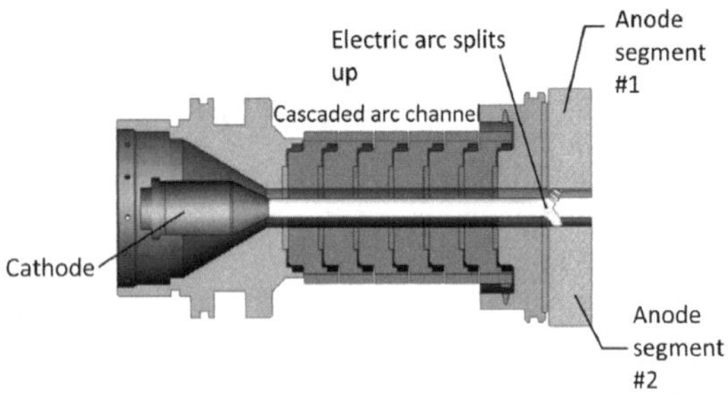

Figure 1: Schematic of the principle of operation of the Delta gun

The use of three anode segments instead of one hollow cylindrical nozzle allows the azimuthal fixation of the anodic roots without the necessity of a vortex flow of the plasma forming gases.

Furthermore the positions of the arc attachments are not dependent on the spray parameters. The Delta gun can be operated at input powers of up to 75 kW while using a Ar/H_2 plasma forming gas mixture. With high spray rates of e.g. 120 g/min for chromia, the Delta gun outperforms conventional plasma guns with regard to the spray rate by a factor of ≥ 2 qualifying the gun particularly for applications in which large areas need to be coated (e.g. paper & printing industry) or where extremely thick coatings are required as in the production of PVD sputter targets [4]. Figure 2 depicts a micrograph of Chromia coating produced with the Delta gun at a spray rate of 120 g/min.

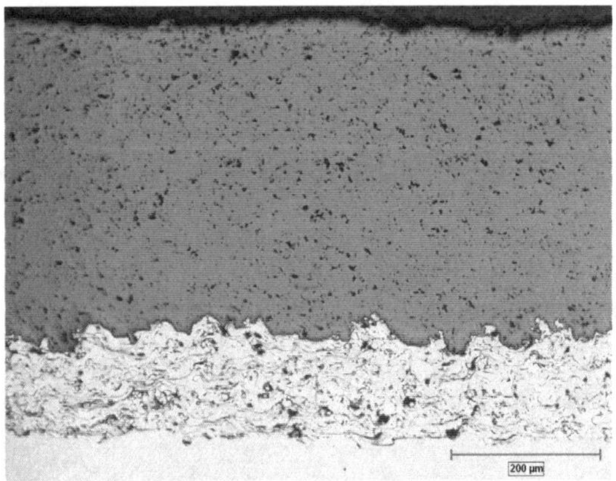

Figure 2: Micrograph of a Delta sprayed chromia coating (I = 420 A, P_{el} = 71 kW, powder feed rate = 120 g/min, powder GTV 41.06.0S, -30 + 10 µm), DE = 50.5 %, hardness = 1,440 HV0.3)

The positive effect of the design principle on plasma fluctuations, voltage fluctuations, respectively, is given in Figure 3 in direct comparison to a standard 1 cathode/1 anode plasma gun for spray parameters which are utilized for spraying NiCr bond coats. With the 1 cathode/1 anode gun an average voltage of 69 V (at 600 A) and fluctuations of up to ±28 V were observed. The absolute fluctuations observed for the Delta gun operating at a much lower current of 350 A reach only values half of those observed for abovementioned spray gun. At the same time the average voltage corresponds to 2.4 times the value observed for the 1 cathode/1 anode gun (166 V). Expressed in percentage terms the fluctuations reach values of 40 % (1 cathode/1 anode) and 9 % (Delta); power fluctuations are in the same range.

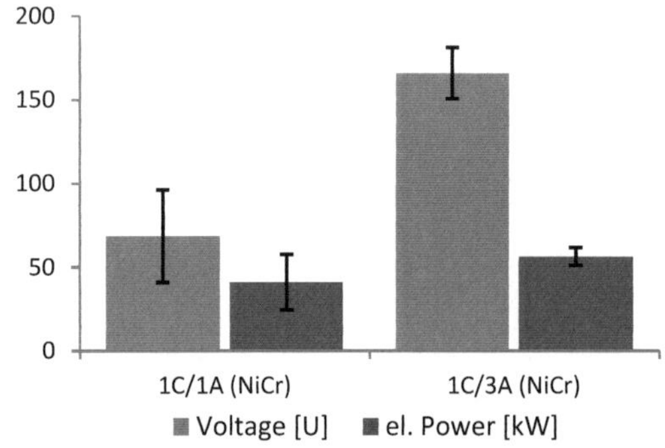

Figure 3: Fluctuations in voltage and electric power for a 1 cathode/1 anode (F6) gun and for the 1 cathode/3 anode APS gun Delta when adjusting typical parameters for spraying NiCr 80/20 powder (measurement times: F6: 2 ms, Delta: 10 ms). All measurements were carried out by Dr.-Ing. Stephan Zimmermann, University of Armed Forces, Neubiberg, D.

2.3 1 Cathode/5 Anodes Spray Gun: Penta

The Penta gun (Figure 4) is based on the same principle of operation as the Delta. The arc emitting from the single cathode is divided in to 5 arcs near the nozzle exit. The arc channel is cascaded. Instead of 3 anode segments the gun possesses 5 pieces. Its layout allows an input power of up to 125 kW. Powder is injected radially into the plasma plume just behind the anode segments via 5 powder ports. Just as the Delta gun the Penta operates with Ar and H_2 as plasma forming gases. Purging gas is injected just behind the anode segments at the nozzle exit in order to provide cooling of the anode and extending anode life time.

Figure 4: Penta gun (top), spraying alumina at a feed rate of 300 g/min onto a roller (bottom)

3 Experimental

So far, coatings development for this new gun has been successfully carried out for alumina powders to produce electrical insulation coatings and some special materials used in the thin film coatings industry.

Within the frame of this work, the capability of the Penta gun to produce chromia coatings at spray rates of 200 g/min is to be investigated. In focus of the work is the coating quality but also the productivity.

Powder used is a fused and crushed chromia with a nominal particle size distribution of -45+15 µm (GTV 40.06.1).

Spray parameters development was carried out using round flat samples with a diameter of 40 mm and a thickness of 10 mm made from C45E steel. In order to achieve a high surface velocity and hence a low thickness per pass all samples were mounted to a rotating specimen holder. Rotating at a revolution speed of 159 min^{-1} the surface velocity was adjusted to 200 m/min. The track offset was set to 8 mm. 30 passes were sprayed in all trials aiming for a coating thickness of 300 µm. Prior to coating the surfaces to be coated were grit blasted, no bond coat was applied. As a result from preliminary investigations, a stand off distance of 150 mm was adjusted. Parameters to be investigated are the flow rates of the plasma forming gases argon and hydrogen as well as the current. The argon flow rate was varied between 45 and 65 slpm, the argon purging gas flow rate of 40 slpm did remain constant with all trials. The fraction of hydrogen was adjusted to values of 9 and 19 % of the total plasma forming gas flow. A neutrode stack with an ID of 7 mm did prove right in preliminary investigations with ceramic powders of abovementioned particle size distribution when aiming for dense coatings, other ID´s of 11 mm and 12 mm are possible, too. Table 1 summarizes the parameter variations.

Table 1: Parameter variations for spraying chromia (GTV 40.06.1, f&c, -45+15µm) at a feed rate of 200 g/min, carrier gas flow rate: 6 slpm

	Center	Variation ±
Ar [slpm]	55	10
H_2 [Vol.-%]	14	5
I [A]	720	25

Trial no.	Ar [slpm]	H_2 [Vol.-%]	*H_2 [slpm]*	I [A]
1	45	9 %	*4,5*	695
2	65	9 %	*6,4*	695
3	45	19 %	*10,6*	695
4	65	19 %	*15,2*	695
5	45	9 %	*4,5*	745
6	65	9 %	*6,4*	745
7	45	19 %	*10,6*	745
8	65	19 %	*15,2*	745

From all coated coupons micrographs were produced and analyzed with regard to apparent coating porosity by digital image analysis (Clemex VISION Lite, Clemex Technologies, inc., Canada). Given values represent the average of 10 measurements. Coating thickness was determined using a light optical microscope and the imaging software DigiVID 8.05 (Mikrovid GmbH, Germany), 20 measurements were carried out for each sample. Vickers hardness of the coatings was determined utilizing a load force of 0.1 kg (Microhardnesstester FM-300, Future-Tech Corp., Japan), here 10 measurements were carried out for each sample. The average roughness Rz of all coatings was determined by means of the profile method. So far or one of the most promising parameter settings from table 1 the deposition efficiency was determined in accordance to ISO 17836:2004.

4 Results

Table 2 gives an overview of achieved results. It could be shown that the coating hardness HV0.1 lies within the typical range of chromia coatings with an average value of 1,414 HV0.1. The highest hardness of 1,512 HV0.1 was observed for the coating produced with parameter setting 6 (high current, high Ar flow rate, low fraction of H_2) while the lowest apparent porosity was observed for trial 4 with 1.8 %. The average porosity within these series of tests was approximately 4 %, with a maximum value of 7 % when spraying at low current, low Ar flow rate and low fraction of hydrogen leading to the lowest input power of 83 kW.

The smallest fraction of porosity was determined for the parameters settings 4 and 7 with values of 1.8 % and 1.9 %, respectively, at which the coating produced in trial 7 exhibits the second highest hardness value (1,474 HV0.1), high thickness (333 µm) and high coating weight (2.1 g). The deposition efficiency was determined to be 50.3 %. Figure 5 depicts a micrograph of the coating produced with this parameter setting.

Table 2: Min. /max. values reached within the frame of the investigations – none of the produced coatings exhibited visible cracks

	Thickness [µm]	Apparent Porosity [%]	HV0.1	Rz [µm]	$m_{coating}$ [g]
min	283	1,8	1350	22	1,8
max	339	7,0	1512	26	2,1

Figure 5: Micrographs of the coating sprayed with parameter setting 7

5 Economical Aspects

One of the main aims for developing the Penta gun was the further improvement of productivity and thus reduction of coating costs. Figure 6 depicts a first approximation of the potential the Penta exhibits in direct comparison to the industrially well established high-spray rate Delta gun. Here, the reduction of coating time T and the consumption of argon, hydrogen as well as electric energy E are given. Basis for these calculation are parameter settings which are industrially established (Delta) or settings resulting from current RnD work (Penta). Calculations were carried out for processing two powders: chromia 99.7% (GTV 40.06.1) and alumina (GTV 40.05.1). Net coating time refers to the time the spray guns is actually coating the part - other times, such as time required for stabilization or time required for reversing the gun - are not taken into account.

It can be seen that the Penta gun leads to a significant reduction of coating time compared to the Delta. When spraying chromia with the Delta gun at a feed rate of 200 g/min the net coating time is reduced by 40 % (Delta: 30 min per m^2 and 300 µm coating thickness vs 17 min). With alumina and a feed rate of 300 g/min the reduction is 70 % (13.2 min vs 4 min).

Besides the reduction of coating time a positive effect of the Penta on the consumption of plasma forming gases (Ar and H_2) and other gases (purging gas and powder carrier gas; both Ar) is clearly visible. E.g. when spraying alumina with the Penta at a spray rate of 300 g/min, the consumptions of Ar and H_2 are reduced by 35 %, 49 % respectively. In this case also the consumption of electric energy input into the gun is halved.

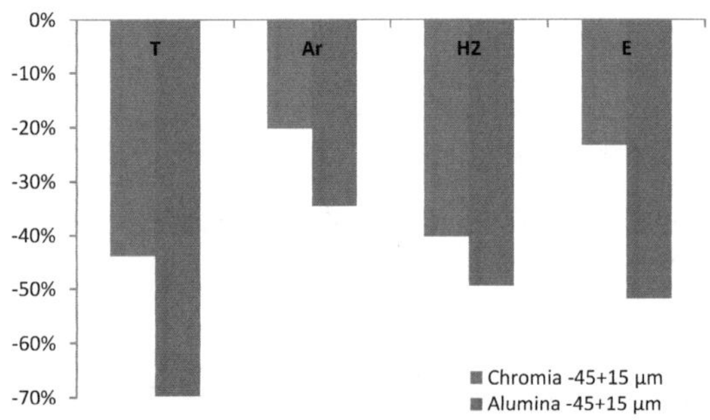

Figure 6: Reduction of net coating time (T), reduction of consumption of argon (Ar), hydrogen (H_2) and electrical energy input E into the plasma guns when spraying chromia GTV 40.06.1 or alumina GTV 40.05.1 (-45+15 µm, f&c). Spray rates: Delta: 120 g/min (both powders), Penta/chromia: 200 g/min, Penta/alumina: 300 g/min

5 Summary and Outlook

Within the frame of this work it was shown that the novel Penta 1 cathode/5 anodes spray gun is suitable to produce chromia coatings which at least meet the state of the art with regard to coating quality at a very high spray rate of 200 g/min which correlates to the quadruple of the value that is industrially established for the 1 cathode/1 anode spray gun F6.

Presented results indicate that the limits of the gun have not been reached yet: future work will focus on a further increase of the spray rate – up to 300 g/min appear realistic without noteworthy losses in coating quality. Furthermore, spraying of chromia powders exhibiting smaller particles sizes (e.g. -30+10 µm, -25+5 µm) will be subject to future investigations in order to further improve coating quality.

With regard to productivity and economical aspects it was shown that the Penta gun can lead to significant advantages even over the industrially well established high spray rate Delta gun.

6 Literature

[1] Dzulko, M., G. Forster G., Landes K. D., Zierhut J., Nassenstein K.: Plasma Torch Developments, Proceedings of the 2005 International Thermal Spray Conference, Basel

[2] Fauchais, P., Heberlein, J., Boulos, M.: Thermal Spray Fundamentals – Form Powder to Part, kindle edition, Springer, 2014, ISBN 10: 0387283196

[3] oerlikon metco: Spray your way! – TriplexPro™-helim-free spray processing: online, https://www.oerlikon.com/ecomaXL/files/metco/oerlikon_Flyer_TriplexPro_He-Free_EN1.pdf&download=1

[4] Stahr C.C.: Der DELTA Plasmabrenner im industriellen Einsatz, Proceedings 5. GTV-Kolloquium „Thermisches Spritzen“, 2012, ISSN 1610-0530

Novel Approaches for thermal spray deposition of fully dense ceramic and metal coatings with well bonded lamellae

Chang-Jiu Li, Shu-Wei Yao, Jia-Jia Tian, Cheng-Xin Li, Guan-Jun Yang, Xiao-Tao Luo, Ying-Kang Wei
State Key Laboratory for Mechanical Behavior of Materials, School of Materials Science and Engineering, Xi'an Jiaotong University, Xi'an/China /E-mail: licj@mail.xjtu.edu.cn

It has been well accepted that thermal spray coatings present a typical lamellar structure with limited lamellar interface bonding. It has been a great challenge to deposit a fully dense coating with fully bonded lamellae. In this report, three different novel approaches are introduced to deposit fully dense ceramic coatings and metal alloy coatings. With the deposition of a specific ceramic coating, it was found that there exists an intrinsic bonding temperature corresponding to the glass transient temperature of spray material. A chemical bonding is formed at the interface upon splatting of a molten ceramic droplet, as far as the maximum interface temperature between the spreading splat and the solid splat reaches over the intrinsic bonding temperature. Moreover, it will be presented that a simple critical bonding temperature in a linear relation with the melting point of coating materials can be utilized to deposit fully dense ceramic coatings by controlling the deposition temperature. Furthermore, with metal alloy coatings, a self-bonding mechanism is proposed utilizing the ultra-high temperature molten droplet for dense coating with fully bonded lamellae. Using specially designed core-shell structured powders, it was demonstrated that a bulk-like metal coating is deposited by creating ultra-high temperature molten droplet. It will be found that such coatings present excellent properties and performance comparable to bulk materials. Moreover, it will be shown that, for ductile metal alloys, the solution-impermeable dense metal coatings can be deposited by using the novel in-situ shot-peening assisted cold spraying.

1 Introduction

Thermal spray has become a well-established coating process due to continuous development in the last century [1]. It is also regarded as a key and environmentally friendly technology to design and modify the properties of surfaces and the characteristics of components. The process has been widely used in many industrial sectors, including transport, energy, materials extraction and processing, biomedical and electronic applications [2]. The survey showed that thermal spray takes up about two-third market of advanced ceramic coating production in northern America [3]. Therefore, the thermal spray process has become indispensible in many industrial fields.

The properties and performance of thermal spray coatings are dependent on the microstructure of coatings [4]. It has been well known that thermal spray coating usually presents a lamellar structure. The main features of the lamellar structure are characterized by certain amount of porosity and limited lamellae interface bonding. Porosity in a thermal spray coating can be adjusted by spray powder design and spray parameters to fulfill the porosity requirements for applications, such as abradable coatings [5-6], thermal barrier coatings [7], and biomedical coatings [8]. As for most wear and corrosion protection applications, thermal spray processes are usually optimized to obtain a dense coating with well bonded lamellae, since a decreased porosity and increased lamellar bonding are associated to increase coating performances [2].

Despite of widely accepted industrial applications, the excellent performance potential of coating materials has not been tapped out yet. It has been known for a long time that the limited interface bonding dominates coating properties and thus wear and corrosion performances [4]. For the coatings deposited following routine approach, a limited inter-lamellae bonding of less than one-third total interface area is present [9]. The unbonded interfaces in the coating acts as pre-cracks, which are activated to propagation along interfaces under thermal-mechanical loading [10] or dynamic loading, such as particle impact in erosion condition, bubble collapse under cavitation, and repeated loading in fretting wear [2]. The propagation of cracks results in crack-linkage, lamellae debonding, and subsequently lamellae spalling, which accelerates the wear of coatings. Therefore, the performance of thermal spray coatings under dynamic loading is generally lower than bulk material of the same kind. It was found that erosion wear resistance of plasma spray Al_2O_3 coating is proportional to lamellar interface bonding [11]. In addition, the investigations showed that, although the HVOF-sprayed Stellite-6 coating presents much better abrasive wear resistance than the ASTM SA213-T-11 steel as a substrate, its erosion rate at 90° impact is higher than the substrate by a factor of about 7 [12-13]. Therefore, the performance of thermal spray coating under dynamic loading may be significant lower than that expected from coating materials. Moreover, there are data illustrating that the spray-fused nickel-based self-fluxing alloy (Ni60) presents a better abrasive wear performance that the plasma-sprayed coating by a factor of about five [14]. Thus, the abrasive wear performance of thermal spray coating can be significantly lower than the same bulk of coating material under severe loading conditions. On the other hand, the limited interfaces link together to constitute connected tunnels for corrosive materials to penetrate through the whole coating [2]. Therefore, the as-sprayed coating cannot generally provide full protection of substrate from the attacking of corrosive substances.

It has been a great challenge to develop thermal spray coatings with full or well bonded lamellae and fully dense enough to provide complete protection from corrosion at the as-sprayed condition. In this review paper, three approaches recently developed in Xi'an Jiaotong University to deposit different types of coatings with special microstructure features towards to the above mentioned challenge will be presented and discussed.

2 Deposition of ceramic coatings with sufficiently bonded lamellae

Previous systematical investigations into the lamellar bonding of plasma sprayed Al_2O_3 coatings revealed that, the particle temperature dominates the formation of the bonding at the interface between lamellae during the spreading of molten ceramic droplet [4,9]. Regarding to the fact that the interface bonding formation is determined by the solid/melt interface temperature, therefore, it is natural to consider that the lamellar bonding can be enhanced by increasing the deposition temperature, although the increase of deposition temperature by preheating substrate may induce cracks in the coatings. Actually, cracking phenomena at preheated conditions have been utilized to produce DVC (dense vertical cracking) thermal barrier coatings [15,16, 17]. It was confirmed that increasing deposition temperature can significantly increase the inter-lamellae bonding in plasma spray both YSZ coatings [18] and Al_2O_3 coatings [19]. The key questions are at what temperature molten ceramic splat can fully bond to ceramic substrate and how the bonding forms.

The dependency of bonding formation on the deposition temperature has been investigated through splat formation test using YSZ splat/YSZ substrate and Al_2O_3 splat/Al_2O_3 substrate. It was revealed that there exists a critical deposition temperature for the ceramic droplet with a temperature near its melting point [9,20]. Recently, it was revealed that the critical condition corresponds to a critical solid/melt interface temperature, being equal to the glass transient temperature of ceramic materials [21]. In fact, when a molten material is supercooled to its glass transient temperature, the further cooling leads to the dramatic increase of melt viscosity and surface tension in the magnitude of orders [22,23]. Thus, it was considered that, when the melt is supercooled near to glass transient temperature, the mobility of molecules in the melt becomes so low that the molecules cannot coordinate with the molecules on the substrate surface to form effective chemical bonds. Based on the above finding, an intrinsic critical bonding temperature, being equal to the glass transient temperature, is defined.

Although more detailed explanation is required to understand the physics and chemistry phenomena involved in the bonding formation, the systematical investigation into the interface bonding formation showed that the critical bonding temperature is almost linearly proportional to the melting point of spray material (Fig.1) [21]. As a result, deposition temperature is divided into the bonding formation region and unbonding region. Therefore, regarding to the critical bonding temperature of 150 °C for TiO_2, when a molten TiO_2 splat is deposited on the identical ceramic substrate surface at the deposition temperature of 115 °C, lower than the critical temperature, no bonding is formed (Fig.2a, b). On the other hand, when the TiO_2 splat is deposited at a deposition temperature of 200 °C, higher than the critical bonding temperature, the splat is completely bonded to the identical substrate (Fig.2e,f). Therefore, based on the concept of critical bonding temperature and the relationship between the critical bonding temperature and melting point of spray materials, the lamellar bonding state in a ceramic coating can be adjusted by deposition in different deposition regions. For example, a TiO_2 coating of sufficiently bonded lamellae was deposited in the deposition temperature region higher than the critical bonding temperature, although some cracks and pores are present in the coating, while a coating of limited inter-lamellae bonding was prepared in the deposition temperature region lower than the critical bonding temperature (Fig.3). Although on the basis of the critical bonding temperature concept, the no lamellar interface bonding should be observed (Fig.1), the deposited ceramic coatings present usually a limited bonding with a mean bonding ratio less than one-third. This fact is explained by the intrinsic critical bonding temperature concept based on the splat cooling condition [24]. Moreover, it should be pointed out that the relationship shown in Fig. 1 is based on the conditions that the temperature of ceramic molten droplet is near its melting point. Therefore, the critical deposition temperature would decrease with the increase of the overheating degree of molten droplet.

Since the performance of plasma spray coating is dependent on specific application service requirements, diverse applications require the coatings of different properties and subsequently different microstructures. Typical cases are for thermal barrier coatings and erosion resistant coatings. The former needs a coating of lower thermal conductivity for better thermal resistance and subsequently a coating of limited lamellar interface bonding. On contrary, the later needs a coating of higher toughness for higher cracking resistance and dense coating of well bonded lamellae. Therefore, single layer coating or multilayered coatings with different microstructures can be designed to give the best performances [9].

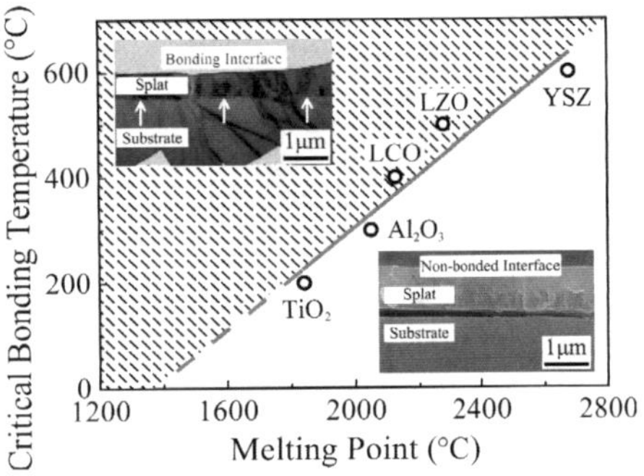

Fig.1 Relationship between the critical bonding temperature and the melting point of spray materials. In the shadow region up-left corner deposited splats bond to the substrate of identical compositions.

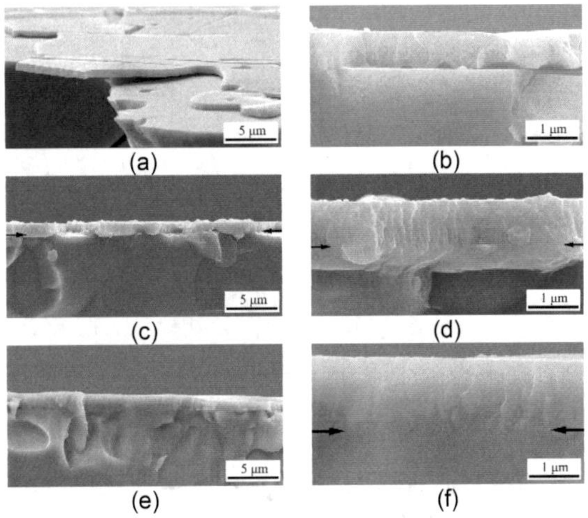

Fig.2 Typical TiO_2 splats deposited on TiO_2 substrate at different temperatures: (a) and (b) 115 °C; (c) and (d) 150 °C; (e) and (f) 200 °C. The critical bonding temperature of TiO_2 splat is 150 °C.

3 Deposition of bulk-like metal alloy coatings by shell-core structured powder particle design

When metal alloy coatings are deposited in an inert atmosphere, the above mentioned approach for ceramic coatings can be applied to enhance lamellar interface bonding formation. However, when metal alloy coating deposition is carried out in ambient atmosphere, it may not be feasible to form metallurgical bonding based on the critical temperature concept approach, since inevitably oxide scale covers over splat surface. Thereafter, how to form sufficiently metallurgically bonded lamellae in the as-sprayed metal coating without post-spray re-melting treatment is another challenge.

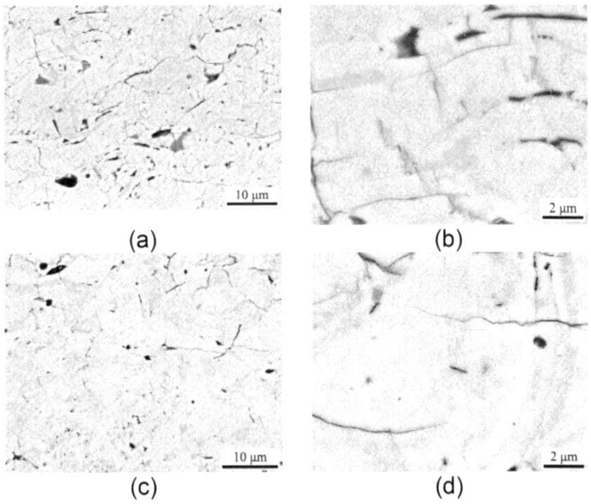

Fig.3 Cross-sectional morphologies of typical TiO_2 coatings deposited at room temperature (a, b) and 200 °C (c, d)

It has been well known that impact of molten droplet of high melting point material with high enthalpy can cause the surface melting of a low melting point substrate [25], leading to the formation of metallurgical bonding at splat/substrate interface. Based on such fact, it is supposed that if one can raise the temperature of a Ni-based alloy or Fe-based droplet to a temperature over Mo melting point, forming an ultra-high temperature (UHT) droplet, its impact causes surface layer melting of the same Ni-based or Fe-based substrate. Thus, during successive deposition of splats in coating formation, the self-bonding mechanism can be effected.

To realize the self-bonding mechanism by creating UHT droplets, a shell-core composite powder design was proposed [26]. A nickel-based alloy or iron-based alloy powders are clad by a layer of refractory metal such as Mo. Using plasma jet, the powder particles can be heated to a temperature over Mo melting point. Since particle temperature of about Mo melting point is close to the boiling points of common alloying elements in Ni-based or Fe-based alloy, such droplet is referred to as the UHT droplet. Moreover, since the total Mo content in powder particle is limited, alloying to NiCr by Mo does not increase the melting point of the coating material. Accordingly, as shown in Fig.4, which presents Mo-NiCr splat deposited on In-738 Ni-based super alloy surface, the deposition of such alloy droplets by plasma spraying induces the surface melting of substrate with the comparable compositions to droplet, and creates metallurgical bonding at the interface. The successive deposition of UHT droplets contributes effective metallurgical bonding throughout all lamellar interfaces by impact of UHT droplets induced melting. Fig.5 shows a typical cross-sectional microstructure of Mo-NiCr coating deposited in the above way. It is clear that the coating presents a dense microstructure of well bonded lamellae.

The erosion behavior is used to evaluate the coating performance [27]. Fig. 6 shows the erosion rates of plasma sprayed Mo-NiCr coating at two impact angles of 30° and 90° in comparison of the data for plasma-sprayed NiCr coating without Mo cladding and IN-738 bulk. It can be found that Mo-NiCr self-bonding coating presents comparable erosion behavior to IN-738 alloy, not only the erosion rates being the same but also dependency of erosion with erodent impact angle. With common plasma sprayed metal alloy coatings, due to limited interface bonding, the erosion rate at high impact angle of near 90 ° is higher than low impact angle of around 30 °. Although material itself is ductile, erosion of the coating presents one similar to brittle material as observed for plasma-sprayed NiCr in Fig. 6. However, erosion of Mo-NiCr coating presents the characteristic of ductile material as IN-738 alloy. This fact indicates that plasma-sprayed Mo-NiCr coating behaviors as IN-738 alloy. Thus, it is evident that by using shell-core structured powder design plasma spraying of bulk-like Ni-based coating becomes possible.

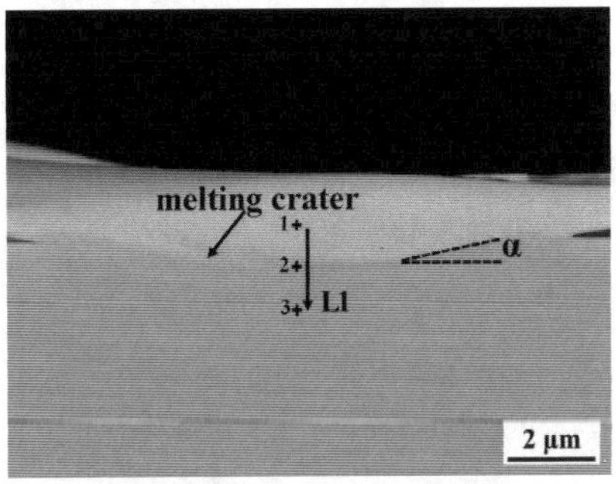

Fig.4 Cross section of Mo-NiCr splat deposited on polished flat IN738 surface showing crater generated by impact-induced melting

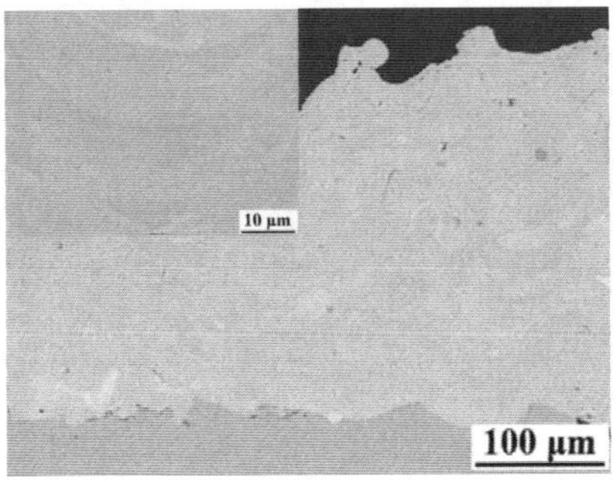

Fig.5 Cross sectional microstructure of Mo-NiCr coating plasma-sprayed using shell-core Mo-clad NiCr powder particles

4 Deposition of fully dense corrosion resistant coating by in-situ shot-peening assisted cold spraying

Cold spray coatings are deposited through intensive plastic deformation caused by the high velocity solid particle impact. The deformation degree of both impacting particles and impacted substrate (or previously deposited particles) determines coating density and inter-particle bonding. A great effort has been made to deposit dense metal coating by cold spraying. Increasing particle velocity and temperature are two common approaches to enhance the deformation of approaching particles and previously deposited particles. Usually, this is implemented by increasing the pressure and temperature of accelerating gas, and preheating powder particles as well. For materials of excellent plasticity, a dense coating without through-pores has been deposited with properly optimized spray conditions.

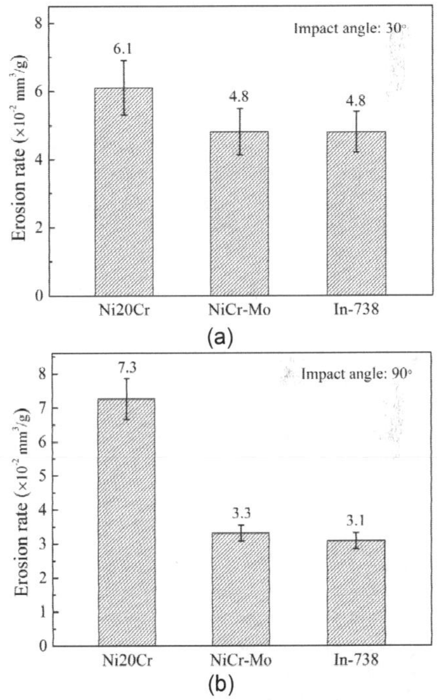

Fig.6 Erosion rates of the NiCr-Mo coating in comparison of Ni20Cr coating and In-738 alloy bulk at the impact angles of 30° (a) and 90° (b).

It has been reported that gas-tight Tantalum coating deposited by using high pressure and high temperature along with preheating of powders presents excellent corrosion resistance [28]. Usually, due to the limited accelerating effect of N_2 gas, use of He is a common option to increase particle velocity. Thus, when He gas is used as processing gas at much higher temperature and pressure, it was demonstrated that pure copper coating of much high cohesive strength of about 250 MPa has been acquired [29]. This value accounts of about one-half ultimate strength of rolled bulk copper. It was reported that the cavitation resistance of cold-sprayed $CuAl_{10}Fe_5Ni_5$ bronze is much lower than casted bulk [30]. Despite of high gas cost, non-uniform deformation of spray particles limits the coating density and bonding formation at particle interfaces even using He. Those facts may imply that the particle interface bonding is limited since the cavitation erosion is sensitive to the lamellar interface bonding. Therefore, how to increase the inter-particle bonding by enhancing plastic deformation and subsequent cold welding effect is still a large challenge for cold spraying.

Based on the densifying mechanism by repeated tamping by spray particle stream one after another during cold spraying [31], the additional more intensive in-situ tamping was introduced as the novel process of cold spraying assisted by in-situ shot peening [32-33]. Through optimizing shot peening loading, it was found that highly dense alloy coatings including Ti-based alloy, Ni-based and Al-based alloy can be deposited by using N_2 gas. Fig.7 shows the cross-sectional microstructure of fully dense Al coating deposited by cold spraying assisted by in-situ peening. The tensile test using the method reported in the literature yielded an ultimate strength of 110 MPa for pure Al coating. In order to evaluate the corrosion performance of Al coating, the coating was deposited on Mg alloy substrate. The electrochemical polarization test revealed that the Al-coated Mg presents the same behavior as Al itself, whereas Mg alloy coated with porous Al coating by conventional cold spraying presents a comparable behavior as Mg alloy itself (Fig.8). The long term immersion corrosion test in 3.5%NaCl aqueous solution up to for one moth proved the feasibility of the coating as cathode protection.

5 Summary

Although conventional thermal spray coatings have been widely and successfully applied in different industrial fields, the potential of excellent wear and corrosion resistance of coating materials is only utilized in a very limited extent due to the lamellar structure features. Limited lamellar interface bonding in thermal spray coatings is the most critical factor dominating the use of spray material performance potential. By understanding the dominant factors for lamellar bonding formation, the effective approaches have been developed for ceramic coatings and metal alloy coatings. Based on the critical bonding temperature concept and the relationship between the critical bonding temperature and the melting point of spray material, the ceramic coating with different bonding ratios can be deposited by simply controlling deposition temperature. Thus, it is possible to design single layer or multiple layered ceramic coatings with different microstructures to fulfill the complicated performance requirements of applications.

Furthermore, a self-bonding mechanism is proposed for ultra-high temperature molten droplet to prepare a dense metal alloy coating with fully bonded lamellae. Using specially designed core-shell structured powders, it was demonstrated that a bulk-like metal coating is deposited by creating ultra-high temperature molten droplet. It was found that such coatings present excellent properties and performance comparable to bulk materials. Moreover, it was shown that for ductile metal alloys, the solution-impermeable dense metal coatings can be deposited by using the novel in-situ shot-peening assisted cold spraying for fully corrosion protection.

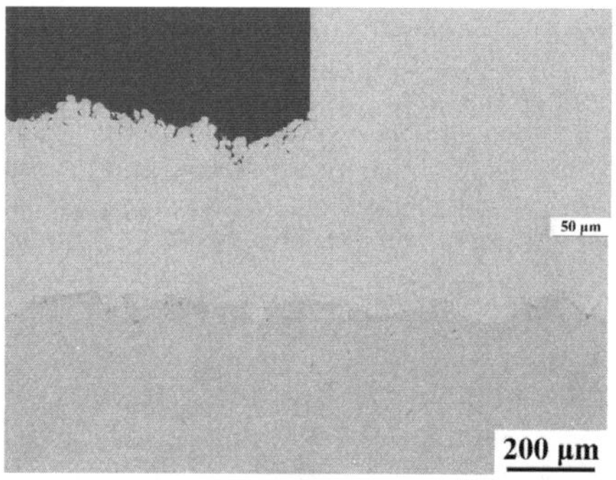

Fig.7 Cross-sectional microstructure of Al coating cold-sprayed assisted by in-situ shot-peening

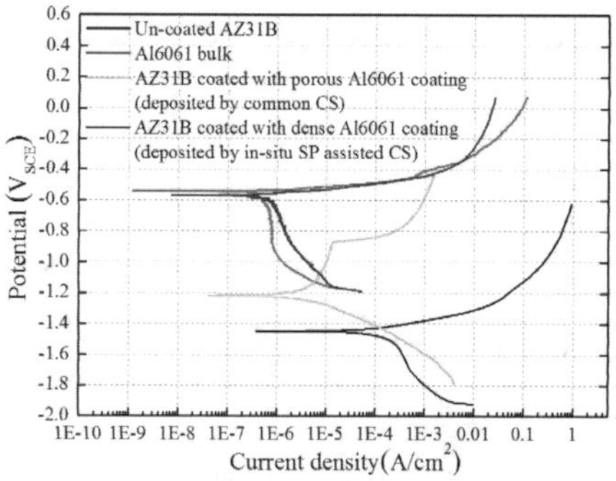

Fig.8 Comparison of dynamic anode polarization behavior of Al6061 alloy coated AZ31B Mg alloy, cold-sprayed assisted by in-situ shot-peening, with those of conventional Al6061 alloy coated Mg alloy, bare ZA31B alloy and Al6061 alloy.

Acknowledgments

The present project is financially supported by National Nature Science Foundation of China (No. 51474171, 51171144).

References

[1] P.Fauchais, J. Heberlein, M. Boulos, Thermal Spray Fundamentals, Springer, 2014.

[2] A. Vardelle, C, Moreau, et al., The 2016 thermal spray roadmap, J. Therm. Spray Technol., 25 (2016), pp.1376-1440.

[3] A. McWilliams, High-Performance Ceramic Coatings: Markets and Technologies, BCC Res. 1 (2016).

[4] C.-J. Li and A. Ohmori, Relationship between the structure and properties of thermally sprayed deposits, J. Therm. Spray Technol., 11 (2002), pp. 365-74.

[5] E. Irissou, A Dadouche, R.S. Lima, Tribological characterization of plasma sprayed CoNoCrAlY-BN abradabe coatings, J. Therm. Spray Technol., 23 (2014), pp. 252-261

[6] C.-J. Li, J. Zou, H.-B. Huo, J.-T. Yao, G.-J. Yang, Microstructure and properties of porous abradable Alumina coatings flame-sprayed with semi-molten particles, J. Therm. Spray Technol., 25 (2016), pp. 264-272.

[7] C.U. Hardwicke, Y.C Lau, Advances in thermal spray coatings for gas turbines and energy generation: A Review, J. Therm. Spray Technol., 22 (2013), pp.564-576.

[8] R.B. Heimann, Plasma-sprayed hydroxylapatite-based coatings: chemical, mechanical, microstructural, and biomedical properties, J. Therm. Spray Technol., 25 (2016), pp. 827-850.

[9] C.-J. Li, G.-J. Yang, C.-X. Li, Development of the particle interface bonding in thermal spray coatings: A Review, J. Therm. Spray Technol., 22 (2013), pp.192-206.

[10] C.-J. Li, H. Dong, H. Ding, G.-J. Yang, C.-X. Li, The correlation of the TBC lifetimes in burner cycling test with thermal gradient and furnace isothermal cycling test by TGO effects, J. Therm Spray Tech., 2017, 26, p.378-387

[11] C.-J. Li, G.-J. Yang, A. Ohmori, Relationship between particle erosion and lamellar microstructure for plasma sprayed alumina coatings, Wear, 2006, 260, p.1166-1172.

[12] H.S. Sidhu, B.S. Sidhu, S. Prakash, Performance of HVOF sprayed NiCr and Stellite-6 coatings under Pin on Disc wear testing, Mater. Sic. Forum, 701 (2012), pp.21-29.

[13] H.S. Sidhu B.S. Sidhu, S. Prakash, Solid particle erosion of HVOF sprayed NiCr and Stellite-6 coatings, Surf. Coat. Technol., 202 (2007), pp. 232-238.

[14] J. Rodriguez, A. Martin, R. Fernandez, J.E. Fernandez, An experimental study of the wear performance of NiCrBSi thermal spray coatings, Wear, 255 (2003), pp. 950-955.

[15] D.M. Gray, Y.C. Lau, C.A. Johnson, M.P. Borom, W.A. Nelson, Thermal barrier coatings having an improved columnar microstructure, US patent, 1998, 5830586t

[16] H.B. Guo, S. Kuroda, H. Murakami, Segmented thermal barrier coatings produced by atmospheric plasma spraying hollow powders, Thin Solid Films, 506-507 (2006), pp.136-139.

[17] A. Tricoire, M. Vardelle, P. Fauchais, Limoges, F. Braillard, A. Malie, Chatellerault, P. Bengtsson, Corbeil-Essone, New concepts for plasma sprayed zirconia TBCs for aeronautical applications, in: E. Lugscheider (Ed.), Thermal Spray Connects: Explore its Surfacing Potential, Proc. 2005 Int. Therm. Spray Conf., Basil, Switzerland, 2005, pp. 924-928.

[18] Y.Z. Xing, C.J. Li, Q. Zhang, C.X. Li, G.J. Yang, Influence of Microstructure on the Ionic Conductivity of Plasma-Sprayed Yttria-Stabilized Zirconia Deposits, J. Am. Ceram. Soc., 91 (2008), pp. 3931-3936.

[19] S. Hao, C.-J. Li, G.-J. Yang, Influence of deposition temperature on the microstructures and properties of plasma-sprayed Al_2O_3 coatings, J. Therm. Spray Technol., 20 (2011), pp.160-169

[20] G.-J. Yang, C.-X. Li, S. Hao, Y.-Z. Xing, E.-J. Yang, C.-J. Li, Critical bonding temperature for the splat bonding formation during plasma spraying of ceramic materials, Surf. Coat. Technol., 235 (2013), pp. 841-847.

[21] S.-W. Yao, C.-J. Li, J.-J. Tian, G.-J. Yang, C.-X. Li, Conditions and mechanisms for the bonding of a molten ceramic droplet to a substrate after high-speed impact, ACTA Mater., 119 (2016), pp. 9-25.

[22] W.H. Wang, The nature and properties of amorphous matter, Prog. Phys., 33 (2013), pp. 177-347.

[23] A. Cavagna, Supercooled liquids for pedestrians, Phys. Report, 476 (2009), pp. 51-124.

[24] S.-W. Yao, J.-J. Tian, G.-J. Yang, C.-X. Li, X.-T. Luo, C.-J. Li, Understanding the formation of the limited inter-lamellar bonding in thermal spray ceramic coatings based on the intrinsic bonding temperature concept, J. Therm. Spray Technol., 25 (2016), pp.1617-1630.

[25] L. Li, X.Y. Wang, G. Wei, A. Vaidya, H. Zhang, S. Sampath, Substrate melting during thermal spray splat quenching, Thin Solid Films, 468 (2004), pp. 113-119.

[26] J.-J. Tian, S.-W. Yao, X.-T. Luo, C.-X. Li, C.-J. Li, An Effective approach for creating metallurgical self-bonding in plasma-spraying of NiCr-Mo coating by designing shell-core-structured powders, ACTA Mater., 110 (2016), pp.19-30.

[27] J.-J. Tian, S.-W. Yao, X.-T. Luo, C.-X. Li, C.-J. Li, Realization of comparable erosion resistance of NiCr-Mo coating to similar bulk by plasma spraying of shell-core-structured powders with improving lamellar interface bonding, Proc. 2017 Int. Therm. Spray Conf., 2017, DVS.

[28] H. Koivuluoto, G. Bolelli, L. Lusvarghi, F. Casadei, P. Vuoristo, Corrosion resistance of cold-sprayed Ta coatings in very aggressive conditions, Surf. Coat. Technol., 205 (2010), pp. 1103-1107.

[29] R. Huang, H. Fukanuma, Study of the influence of particle velocity on adhesive strength of cold spray deposits, J. Therm. Spray Technol., 21 (2011), pp. 541-549.

[30] S. Krebs, F. Gartner, and T. Klassen, Cold spraying of Cu-Al-Bronze for cavitation protection in marine environments, J. Therm. Spray Technol., 24 (2015), pp. 126-135.

[31] C.-J. Li, W.-Y. Li, Deposition characteristics of titanium coating in cold spraying, Surf. Coat. Technol., 167 (2003), pp. 278-283.
[32] X.-T. Luo, Y.-K. Wei, Y. Wang, and C.-J. Li, Microstructure and mechanical property of Ti and Ti6Al4V prepared by an in-situ shot peening assisted cold spraying, Mater. Des., 85 (2015), pp. 527-533.
[33] Y.-K. Wei, X.-T. Luo, C.-J. Li, C.-X. Li, Optimization of in-situ shot-peening-assisted cold spraying parameters for full corrosion protection of Mg alloy by fully dense Al-based alloy coating, J. Therm. Spray Technol., 26 (2017), pp.173-183.

Porosity characterization and its effect on thermal properties of APS sprayed alumina coatings

W. Tillmann, O. Khalil*, M. Abdulgader. Dortmund/ D.

Abstract

In the present work, three different APS alumina coatings were fabricated using three fused and crushed alumina powders of different particle size fine, medium and coarse. The influence of the particle size on thermal properties and micro-structural features of the produced coating were investigated by thermal insulation test and detailed image analysis technique, respectively. The analyzed micro-structural features include the total porosity, pore size (fine, medium, and large) and cracks. All types of cracks were considered in calculations as voids and were evaluated according to their sizes as pores. All spray parameters except the particle size were fixed throughout the spraying process. The results revealed that the fine starting powder has produced the densest coating with the lowest total porosity and that the total porosity increases with an increasing particle size. This was expected as powders of smaller particle size will reach a higher in-flight temperature and velocity than powders of bigger particle sizes as long as the same spray parameters are applied. However, a detailed image analysis investigation on the three produced coatings showed that the fraction of fine pores and cracks versus the total porosity is substantially higher in coatings produced by using fine starting powders than those produced using medium and coarse powders. In this work, a connection between the thermal insulation and the porosity fraction, which includes fine pores and cracks, was revealed.

1 Introduction

Thermal spraying method is a coating process that adds additional thermal and mechanical properties to base substrate materials. In this process, a thick coating is applied to bulk materials substrates which enable them to operate beyond their capabilities in an uncoated state. Ceramics, in general, are particularly suitable for thermal spray processes, with plasma spraying being the most appropriate process due to high temperatures that the plasma jet provides [1]. Atmospheric Plasma sprayed (APS) alumina ceramic coatings are very good electrical and thermal insulators. They also improve the resistance of steel against corrosion, wear and erosion. For these reasons, ceramic coatings have been widely investigated [2,3,4,5]. These coatings are being used by the automotive and aerospace industries, where internal components normally operate in severe environments including extreme temperatures [6].
During the spraying process, molten alumina particles are heated and accelerated by the plasma jet towards a prepared substrate surface. If these accelerated particles are melted, they impact on the substrate, flattened and form so-called splats. Splats are the main structural unit in APS coatings [1].
The successive deposition overruns define the final thickness and compose the layered structure of the whole coating. In addition to the fully melted particles, APS coatings consist of partially melted particles as well as un-melted particles.
The described deposition mechanism leads to a coating with micro structural defects such as pores, delaminations, and micro-cracks. Also, oxide films around the molten particles (droplets), together with un-melted particles, form inclusions in the coating.
Fine pores arising from entrapped gas below the molten droplets are of spherical shape and homogeneously distributed. The coarse pores, however, are formed due to incomplete filling caused by both semi-molten and un-melted particles and they are non- homogeneously distributed [7].
Oxides, foreign inclusions, as well as un-molten particles are considered to be detrimental to the coating properties; at the same time, the micro-structural features including pores, delaminations, and cracks generated during spraying process improve the thermal insulation of the coating and increase its strain tolerance as well. Hence, the degree of the particle's melting state whilst in-flight affects largely the level of cohesion, porosity, and the ultimate coating's properties [1,8].
Thermal stresses and rapid solidification of in-flight particles are disadvantages related to the APS ceramic coatings. Thermal stresses motivate micro-cracks to initiate and then propagate within the coating, which limits the coating thickness.
In a wet environment, micro-structural defects such as pores and cracks may cause local corrosion in the substrate [6], however, the presence of such defects with defined levels enhances the thermal insulation of the coating as well as increases its thermal shock resistance [1]. Un-molten particles exist in the coating when large particles do not have enough time to be fully melted. Only the outer surface of such particles is melted as they impact on the substrate while the inside of the particles is still un-melted [9].
APS alumina coatings have a lamellar microstructure, and defects such as pores and cracks exist generally within/between the lamellae to form an interconnected network. These bi-modular features dramatically influence the mechanical behavior and electrical/thermal insulation of the coating [10,11].
Besides other process parameters, such as the spray power and spray distance, the starting powder particle size has a high influence on the particle's in-flight temperature and velocity. The in-flight particle conditions in turn, determine the particle melting state,

impact behavior, and the ultimately produced microstructure of the coating and its properties. [12].
Many researchers have referred to the fact that total porosity plays only a secondary role while the size, shape, and spatial distribution of the pores play the primary role in determining the properties of a coating [13,14].
The present work is based on the study of APS alumina-coating micro-structures using three feedstock powders of different particle sizes that are bonded to steel substrates. The influence of the particle size on the in-flight particle temperature and velocity as well as the coating's microstructural features and the properties of the coating were investigated. The microstructural features of APS alumina coatings including its total porosity, pore area, and crack network were analyzed quantitatively by means of an image processing tool. The work focuses on the accurate measurement of porosity fractions based on a clear criterion that distinguishes each porosity component. Moreover, coating properties such as the micro-hardness and thermal insulation were evaluated and related to the quantified microstructural features.

2 Materials and Processes:

2.1 Samples preparation and deposition process:

Three sets of steel substrates with nominal dimensions with a diameter of 40 mm and a thickness of 5 mm were grit blasted using alumina (EFK-14) on one side. This procedure was followed by an ultrasonic cleaning for the APS deposition. Three commercial alumina powders (fused and crushed Al_2O_3) of different particle sizes; fine powder, Amperit 740.000 (5-22 µm), medium powder, Amdry 6062 (25-45 µm), and coarse powder, Amperit 740.002 (45-90 µm), H.C.Stark, Germany, were deposited on the steel substrates. The median size (Vol. 50%) of these powders were 14.00 µm, 35.88 µm, and 77.7 µm, respectively.

Fig. 1. SEM images showing the morphology of the: (a)fine, (b)medium, and (c)coarse alumina powders.

SEM images of the utilized powders and their particle size distribution are shown in **Fig. 1**. An Orlikon Metco F4-MB gun, with a nozzle with a 6mm internal diameter and a single powder feeder was used for this study. All spray parameters were kept constant for the three powders in order to investigate the influence of the starting powder's particle size on the microstructural features of the coatings. The APS process parameters are given in **Tab.1**.

Tab. 1. APS process parameters for alumina powders

Parameter	Value
Current [A]	600
Voltage [V]	74 ± 3
Argon [l.min^{-1}]	41
Hydrogen [l.min-1]	12
Power carrier gas[NLPM]	3.4
Nozzle diameter [mm]	6
Spray distance [mm]	120
Step height[mm]	4

2.2 Porosity Analysis

In order to quantitatively evaluate the porosity of the produced depositions; cross-sectional metallographic samples were prepared for the investigations. A series of SEM images with different magnification factors for each cross-section were taken by a Field Emission Scanning Electron Microscope (Joel, JSM-7001F FESEM). For the porosity measurements, ten images along the cross-section were taken randomly, central to the thickness, for each cross-section. The images were magnified with a magnification factor of 1000x, which enables the detection of 0.1 µm-sized features. The image brightness was reduced while the contrast was increased to facilitate the image analysis software to accurately detect any porosity. Additionally, a very low accelerating voltage (1.0 kV) and a low probe current (8-10 µA) were used so that the contrast between the pores and coating's matrix was enhanced. Additionally, back-scattering electrons feature were used to enhance the quality of the obtained SEM images. An image analysis was performed using the Image J1.49r software (provided by the National Institutes of Health, USA) on SEM images at 1000x. Image filtering was applied to remove noises followed by image segmentation by thresholding, which produced binary images; the zero-threshold level (black) was assigned to the pores and one level (white) to the alumina solid state. Subsequently, the pores were measured and categorized according to three area ranges: Fine (0-1 μm^2), medium (1-10 μm^2), and large (>10 μm^2). Moreover, the medium pore size range (1-10 μm^2) for the three coatings was further classified in terms of the circularity (cir.<0.5). Circularity is a measuring unit of pore spheroids. The more a pore becomes spherical, the circularity approaches unity and when pores become more elongated or narrow, circularity decreases. Circularity is defined by:

$$4\pi(A/p2) \qquad (1)$$

where "*A*" is the pore area and "*p*" is the pore perimeter. Various shapes of pores and cracks of different sizes in a cross-sectional SEM image for one coating are identified in **Fig. 2**. The process of pore classification is shown in **Fig. 3**.

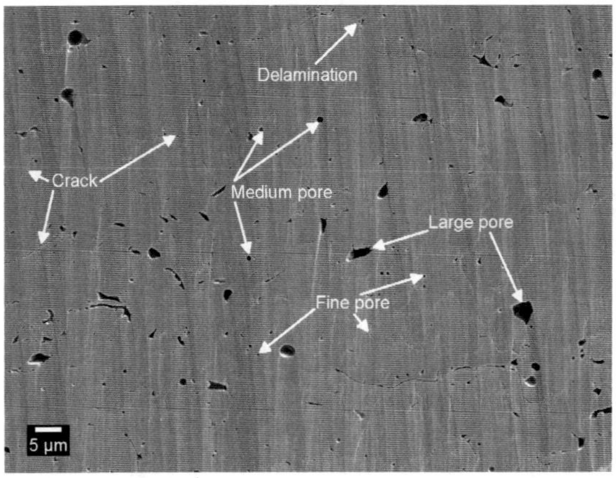

Fig. 2. Cross-sectional SEM image showing the shapes and relative sizes of pores and cracks.

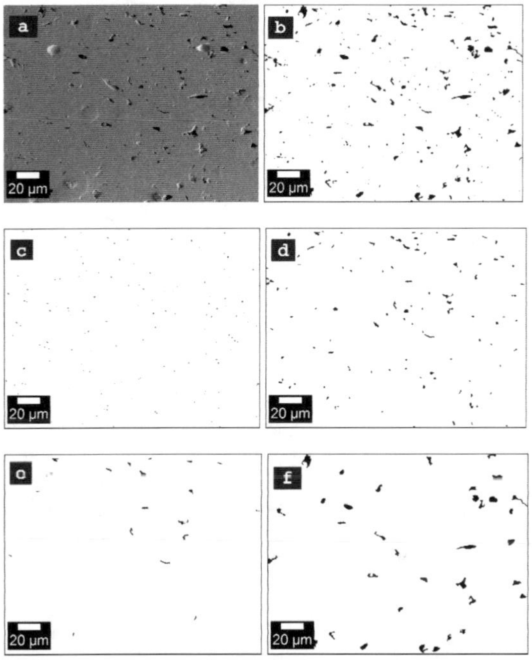

Fig. 3. Pore classification process: a) SEM gray image, b) total porosity (binary image), c) fine-pore, d)medium-pore, e) medium-pore with circularity <0.5, f)large-pore class.

2.2 Characterization

A thermal spray sensor (AccuraSpray G3, Tecnar Co., Canada) was used to measure the temperature and velocity of in-flight particles prior the process of samples deposition. The system was positioned so that measurements are taken exactly at the spray distance listed in **Tab. 1**.

One sample of each coating was used in the insulation test where the coating surface was subjected to an increasing heat source (oven) up to 900°C with a rate of 15°C per minute, and a holding time of 60 minutes. The temperature on both sides of the three coatings was simultaneously measured by means of two thermocouples.

Microhardness measurements were carried out at room temperature on polished cross-sections that represent the three as-sprayed coatings, using a Vickers microindenter (Leco M-400, Germany). The indentations were performed with a load of 2.94 N for a dwell time of 15s. The standard spacing was used between the indentations (at least three times the diagonal) to ensure that no further stresses were produced by the interaction between the consecutive indentations.

3. Results and discussion

3.1 Image processing analysis

SEM images of as-sprayed coatings were used to analyze the porosity of the produced coatings for the three powders.

The porosity level variation in the three coatings is shown in **Fig. 4**. It can be observed that fine pores with an almost spherical shape are uniformly distributed along the coating of the fine-powder. As the particle size increases with medium and coarse powders, more pores of larger sizes are visible and the porosity distribution becomes non-uniform. The image analysis performed on SEM images of coating's cross sections showed that the average total porosity of fine-, medium-, and coarse-powder coatings are 0.93%, 2.47%, and 5.55% of coating area, respectively. In general, the total porosity levels for all coatings increases when increasing the feedstock powder particle size. This increase is mainly due to the simultaneous increase of both the number and size of large pores, whereas fine and medium pores decrease.

As all spray parameters were kept constant for the three powders, fine-powder particles were exposed to a higher in-flight temperature and velocity (2620 °C, 415 m/s in average) than medium- and coarse-powder particles (2530 °C, 365 m/s and 2390 °C, 330 m/s in average, respectively).

Also, thickness measurements of the three coatings revealed that coarse-powder produce thicker coatings than medium- and fine-powders (avg. thicknesses were 668µm, 475µm and 180µm, respectively).

The porosity of the three coatings was evaluated quantitatively and classified into three area ranges. Subsequently, the average porosity of each pore range was calculated. The porosity evaluation was based on ten measurements of ten different SEM images for each coating sample. Given the average pore range porosity and average total porosity for each sample, the percentage of each pore range versus the total porosity was calculated. Additionally, the medium pore range (1-10µm) for each sample was also classified according to the pore's circularity.

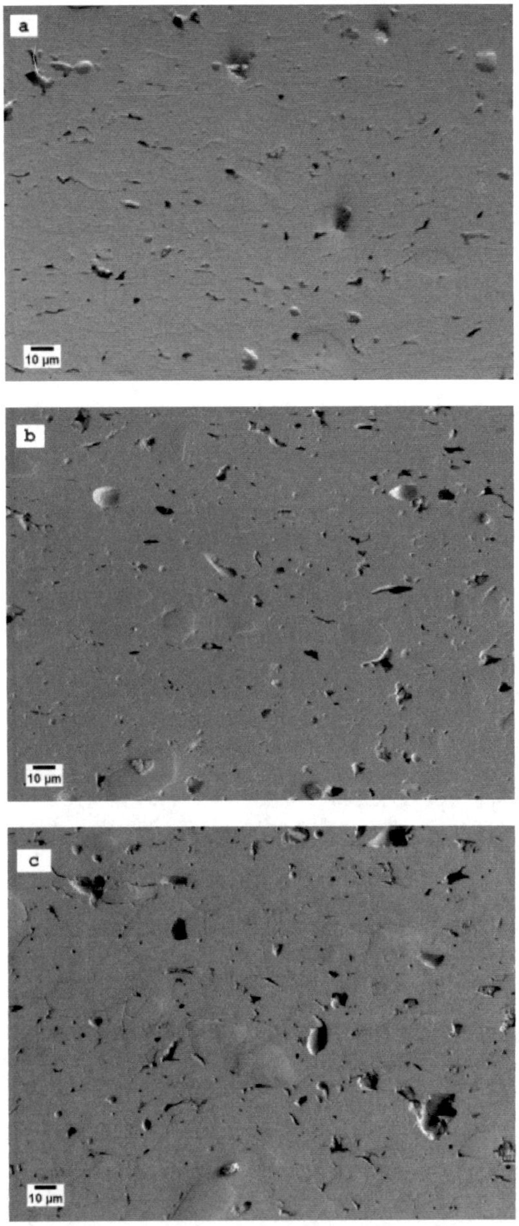

Fig. 4. Cross-sectional SEM micrographs of as-sprayed alumina coatings indicating the coating porosity and cracks deposited with: a) fine-powder, b) medium-powder, and c) coarse-powder.

For the medium pore range, the shape of the pores is crucial with regard to the thermal insulation property as pores with a low circularity (circularity<0.5) highly contribute to the thermal insulation property in comparison to pores with a high circularity (circularity>0.5). These measurements are visualized in **Fig. 5**. It can be seen that fine-powder coating possesses a higher percentage of fine pores and medium pore ranges (34.4%, 62.3%, respectively) than medium-powder and coarse-powder coatings (15%, 45%, and 6%, 27%, respectively). While more than 60% of the medium pore range for fine-powder coatings exhibit a circularity of <0.5, medium-powder and coarse-powder coatings exhibit less than half (48.2% and 48%, respectively) of medium pore range with the same circularity.

In contrast, coarse-powder coatings possess a higher percentage of large pore ranges (67%) than medium-powder and fine-powder coatings (40% and 3.3%, respectively).

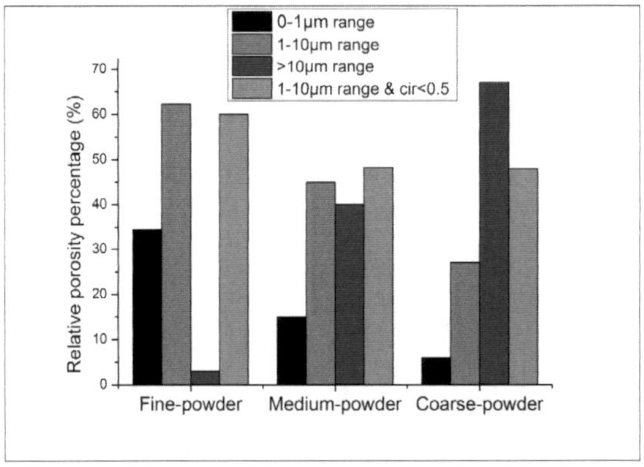

Fig. 5. Relative pore range contribution out of total porosity.

Microhardness tests showed the effect of the coating porosity on microhardness values of the three coatings. **Tab. 2** lists the average microhardness values of the three coatings along with the deviations of these measurements. In comparison to medium- and coarse-powder coatings, fine-powder coatings have the highest microhardness value due to their denser microstructure and lower total porosity. It is noticeable that the deviation value associated with the average microhardness value for the coarse-powder coating is higher than that of the medium- and fine-powder coatings. This can be attributed to the non-uniform distribution of large- and medium-range pores along the coating. In contrast, fine-powder coatings possess the lowest microhardness value due to the relative low pore-size distribution and the very low fraction of large-range pores (3.3%).

Tab. 2. Average microhardness measurements and associated deviations.

Coating of	Avg. microhardness (HV)	Deviatio n
Fine-powder	944	16.2
Medium-powder	881	32.3
Coarse-powder	829	71.4

2.2 Thermal insulation

The thermal insulation of the three coatings is shown in **Fig. 6**. It can be seen that the fine-powder coating

possesses a comparable thermal insulation value when compared to the medium- and coarse-powder coatings although its total porosity is much lower (0.93%) than the total porosities of medium- and coarse-powder coatings, (2.47% and 5.55%, respectively).
As the porosity of fine-powder coatings consists mainly from pores and cracks of the fine- and medium-range (96.7%), it can be concluded that these ranges of pore sizes and shapes contribute largely to the thermal insulation property and play the primary role in determining thermal insulation value of coatings. The high thermal insulation exhibited by fine-powder coatings can be related to the greater number of interfaces between the pores and the matrix of the coating [14]. Also, the relative high thermal insulation of fine-powder coatings can be related to the higher fine pore density (number of fine pores per unit area), as it can be noticed in SEM images in **Fig. 4**.
Additionally, large pores and medium pores of circularity of >0.5 turn to be of low significance and plaz only a secondary role in determining the thermal insulation property of coatings, although their presence in medium- and coarse-powder coatings increases greatly the overall porosity.

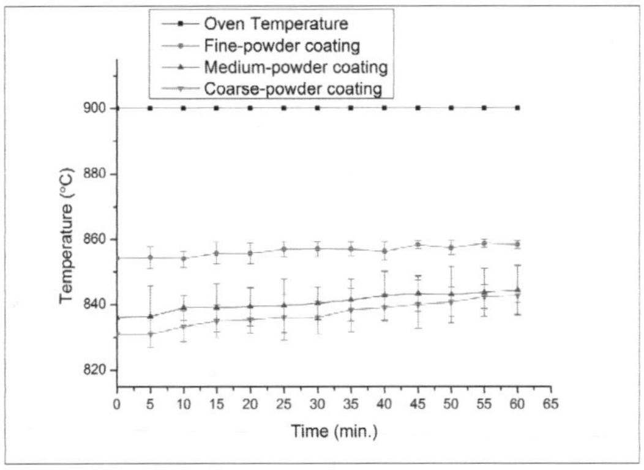

Fig. 6. Thermal insulation of the three as-sprayed coatings.

4 Conclusions

In this study, it was shown that the particle size of the feedstock starting powder has a fundamental effect on the porosity level as well as pore-size distribution, pore shape, and distribution of the overall porosity which highly influence the ultimate thermal coating properties such as the thermal insulation and strain tolerance. According to the obtained results, the following conclusions can be drawn:
1. The porosity of coatings produced by fine-powders (i.e. fine particle size) exhibit a more uniform distribution than the porosity of coatings produced using medium- and coarse-powders. The pore size ranges from few tenth nanometers to approximately 10um. More than 94% of this range highly contributes to the coating's thermal insulation property due to the uniform distribution, pore size, and pore shape.
2. As the starting powder's particle size increases, the percentage of fine and medium pore ranges (0-1 & 1-10um) versus the total porosity decreases, while the percentage of large pore ranges (>10um) versus the total porosity increases. The increment in large pores greatly increases the total porosity with a low contribution to the coating's thermal insulation.
3. APS coatings fabricated by fine-powders have higher micro hardness values than coatings fabricated by medium- and coarse-powders due to their denser coatings. At the same time, fine-powder coatings exhibit thermal insulation value comparable to thermal insulations of coatings produced by medium- and coarse-powders despite of higher total porosity and coating thichness owned by both medium- and coarse-powder coatings. This is due to the higher contribution of the "effective porosity" in fine-powder coatings represented by fine pores and cracks and to their homogeneous distribution within the coatings.

Acknowledgements:

The authors thank the German Federation of Industrial Research Associations (AiF) and the Forschungskuratorium Maschinenbau e.V. (FKM) for financially supporting the project 18627 N "Entwicklung von Dünnschichtthermoelementen zur Inline-Schmelzetemperaturmessung am Beispiel der Flachfolienherstellung".

References

[1] J.R. Davis, Handbook of Thermal Spray Technology, ASM International Materials Park, Las-Vegas, USA, 2009.
[2] Pavitra, B., Padture P. and V. Alexandre: Improved interfacial mechanical properties of Al2O3 – 13 wt% TiO2 plasma-sprayed coatings derived from nanocrystalline powders, *ActaMaterialia*, Vol. 51 (2003), pp. 2959-2970.
[3] Singh, V., Sil, A., and R. Jayaganthan: A study on sliding and erosive wear behaviour of atmospheric plasma sprayed conventional and nanostructured alumina coatings, Materials and design, Vol. 32, No. 2 (2011), pp. 584-591.
[4] Tingaud,O., Bertrand, P., and G. Bertrand: Microstructure and tribological behavior of Suspension plasma sprayed Al2O3 and Al2O3–YSZ composite coatings, Surface & coating technology, Vol. 205, No. 4 (2010), pp. 1004- 1008.
[5] Li, C., Yang, G., and A. Ohmori: Relationship between particle erosion and lamellar microstructure for plasma-sprayed alumina coatings, Wear, Vol. 260, No. 2 (2006), pp. 1166-1172.
[6] Morks, M., Cole, I., Corrigan, C., and A. Kobayashi: Electrochemical characterization of plasma sprayed alumina coatings, Journal of surface engineered materials and advanced technology, Vol.1, No.3 (2011)
[7] Girolamo, G., Brentari, A., Blasi, C., and E. Serra: Microstructure and mechanical properties of

plasma sprayed alumina-based coatings, Ceramic international, 40 (2014).
[8] Schlichting, K., Padture, N., and P. Klemens: Thermal conductivity of dense and porous yttria stabilized zirconia, Journal of materials science, vol. 36, No. 12 (2001), pp. 3003–3010.
[9] Hurevich, V., Smurov, I., and L. Pawlowski: Theoretical study of the powder behavior of porous particles in a flame during plasma spraying, *Surface coatings technoogyl, (*2002**)**, pp. *151–152*, 370–376.
[10] Pawlowski L. et al: The relationship between structure and dielectric properties in plasma sprayed alumina coating, Surface & coatings technology, Vol. 35, (1988), pp. 285-298.
[11] Beauvais, S. et al: Influence of defect orientation on electrical insulating properties of plasma-sprayed alumina coatings, Journal of electroceramics, Vol. 15, (2005), pp. 65-74.
[12] Friis, M., Persson, C., and J. Wigren: Influence of particle in-flight characteristics on the microstructure of atmospheric plasma sprayed yttria stabilized ZrO2, *Surface coating technology*, Vol141, No. 2-3 (2001), p. 115-127
[13] Sevostianov,I., and M. Kachanov: Plasma sprayed Ceramic Coatings; Anisotropic elastic and conductive properties in relation to the microstructure; cross-property correlations, M*aterials scince engineering*, Vol 297, No.1-2 (2001), pp. 235-243
[14] Ravichandran, K., An, K., Dutton, R., and S. Semiatin: Thermal conductivity of plasma-sprayed monolithic and multilayer coatings of alumina and yttria-stabilized zirconia, C*eramic socity*, Vol 82, No.3 (1999), p.673-682

New Investigations of the Arc Instabilities in a One-Cathode-One-Anode Plasma Generator via special Wavelet-Analysis

Stephan Zimmermann, Maximilian Mudra, Oliver Bornschlegl, Silva Schein, Günter Haderer, Jose Marques and Jochen Schein Neubiberg / D

Plasma generators are frequently used in a wide field of applications. These applications include thermal coating, plasma-welding, preprocessing of surfaces, sterilization and propulsion systems. The industrial use of plasma generators demands a high level of accuracy and reproducibility which has to be monitored throughout the process. Rather than analyzing results of e.g. a coating process (by electron microscopic) one prefers to have an online assessment of the plasma generator. The plasma generated is defined by various parameters (e.g., temperature, conductivity...) which can be measured with suitable measuring systems. Determining all plasma parameters in one setup is very costly and requires a big setup. In industrial facilities it is often suitable to measure one or two physical quantities and derive the plasma parameters from these measurements. This paper focuses on the evaluation of the electric current consumed by the plasma generator and the sound emitted by the whole system. A quick analyzation of these signals – that may result in an online assessment –was accomplished by using the wavelet transform. Multiple experiments were performed to gain reliable and significant data (behavior of the arc/plasma).

1 Introduction

Commonly the Fourier Transform is being used to analyze time-dependent data. The Fourier Transform is based on the orthogonal function space of trigonometric functions (sin(ωt); cos(ωt)). Therefore it is especially useful when handling periodic signals. The discrete wavelet transform of a function is another way to display the function's information. The discrete wavelet transformation or wavelet series development is a tool for displaying signals. It follows the Fourier series development of the design specification [8]:

$$f = \sum_{m \geq m_0} \sum_{n \in \mathbb{Z}} v_{m,n} \psi_{m,n}$$

Here, the $v_{m,n}$ are the wavelet coefficients, and $\Psi_{m,n}$ is the wavelet function. The wavelet coefficients contain the same information after the transformation signal. As with the Fourier transform, the signal is thus only represented differently. In addition to the frequency information, the wavelet transformation also contains information about the temporal sequence of the signal. It is to be understood as a compromise between perfect frequency resolution and perfect temporal resolution. Non-periodic signals and impulses can be represented particularly well with the wavelet transformation.

2 (Discrete) Wavelet Transformation

First, the wavelet function $\Psi_{m,n}$ is to be considered. The following properties are expected from this [1-2, 4-5, 8]:

1. A representation is possible for a sufficiently large class of functions; the analysis (calculation of the coefficients $v_{m,n}$) and synthesis (reconstruction)
2. The basic functions are well localized, so only on a limited time range substantially different from zero
3. The basic functions are well localized in the frequency domain, ie their Fourier transforms are well localized

A wavelet function satisfying the criteria (1), (2) and (3) is the Haar's mother wavelet, **Fig. 1**.

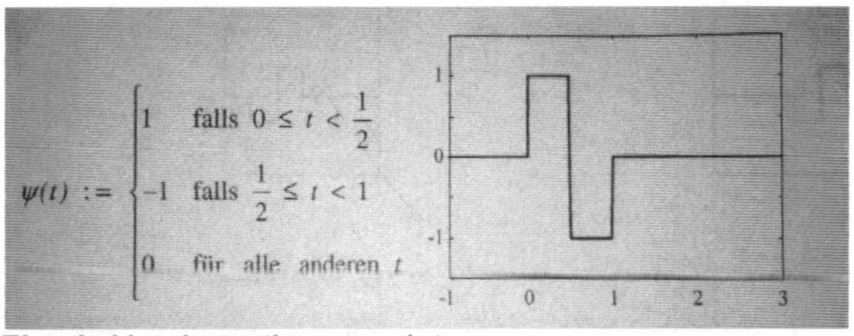

Fig. 1: Haar's mother wavelet

In order to access the general wavelet function from the mother wavelet, the following scaling and displacement of the mother wavelet is used:

$$\psi_{m,n} = 2^{(-m/2)} \psi(2^{-m}t - n)$$

By selecting the prefactor $2^{(-m/2)}$, the wavelet function is normalized. The indices m and n have the following physical meanings:

- m is the scale or scale, comparable to frequency steps
- n is the translation, the position around which the wavelet extends

The Haar's mother-wavelet can be seen as a counterpart to the Fourier-base, since it is localized in the best possible time in the frequency domain (3) in the case of a poor localization (2). Criterion (1) is also met, since the wavelet coefficients are determined.

The temporal localization and the "simple shape" of the mother wavelet result in two simple integrals, which can be solved numerically efficiently. The result is a multiscale representation of the original signal. The magnitude of the coefficients $v_{m,n}$ shows how strongly the signal changes at the local time interval n in the scale range (frequency range) m. A low scale corresponds to a high frequency range. Thus, there are restrictions for the indices m and n without having a high information loss. The choice of m_{max} and m_{min} is thereby dependent on the desired approximation accuracy. If the restriction is less than m, a loss of information is accepted. However, this is of no relevance to the application, since in this paper are only interested in certain scales m, which are intended to correspond to a certain form of arc movement. In the paper a restriction on 4 levels is used. The index n is dependent on the current m and the signal length. The signal is vector shape and extends over a certain time range. The index n is selected, so that $\Psi_{m,n}$ is within the time interval of f, **Fig. 2**.

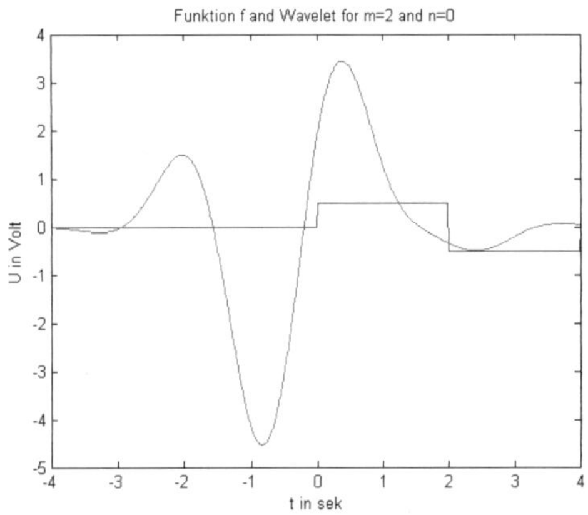

Fig. 2: Example for a Wavelet Transformation (m=2, n=0)

3 Experimental Setup

For the experiments, a classical one-cathode single-anode plasma generator (F4/F6) was used. In this generator the anodic arc root point [4-7] is moving depending on the parameter, **Fig. 3**.

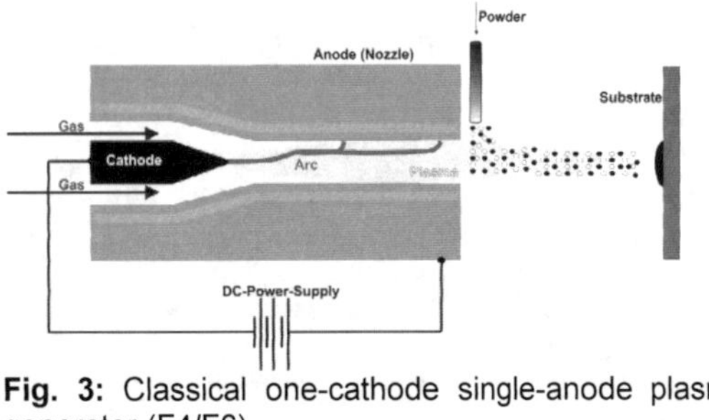

Fig. 3: Classical one-cathode single-anode plasma generator (F4/F6)

Several experiments were carried out with a duration of 4 × 1 h each. Various settings of the current, the gas mixture and the nozzle diameter (or the degree of wear) were chosen, **Table 1**.

Parameter oft the experiments			
Parameter_Nr.	nozzle	current(A)	Ar (SLPM)/ N_2 (SLPM)
I	6mm	350	40/6
II	6mm	500	40/6
III	6mm	150	40/6
IV	8mm new	350	40/6
V	8mm old	350	40/6

Table 1: Parameter of the experiments with the classical one-cathode single-anode plasma generator (F4/F6)

Futhermore all parameters are investiged with the following conditions:

- Gas mixture 40 SLPM Argon / 6 SLPM nitrogen was used
- 84 measurement points were recorded per parameter, one every 3 minutes
- Current measurements with Chauvin Arnoux AC-DC Current Clamp PA 22 (1mV/A) [3]
- Voltage measurements with Contec-9000 Differential Probe (1/100)
- Data processing with LeCroy Waverunner LT364 (500 kS/s)
- Pauses of 45 min were made after 1 h process operation

The measured signals were analyzed with Matlab. Also a fast Fourier transformation was performed to obtain comparability with previous results. It has been shown that the choice of different wavelet types has no effect on the qualitative result [1,5].

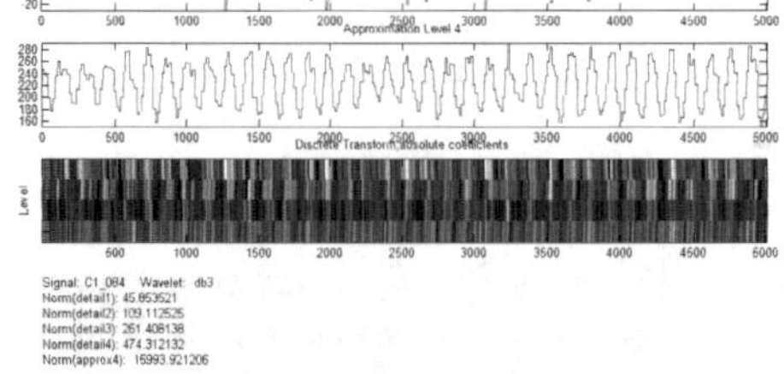

Fig. 4: Individual wavelet coefficients (*d1*, *d2*, *d3*, *d4* and *a4*) of an plasma parameter

Daubeschies wavelets were used here (*db3 Matlab*). The individual wavelet coefficients (*d1*, *d2*, *d3*, *d4* and a4) have now been converted to a measured variable with the 2-standard to assign a measured variable (standard (*d1*)) to a measurement time, **Fig. 4**. These were summarized and possible tendencies were made visible in graphs. The norm values of the detail vectors are plotted over the sequence number of the measurements.

4 Results and Discussion

The various parameters show a clear tendency for the *a4* values. These points represent the low-frequency signal components. At the ignition timing, the *a4* values are higher than in the steady-state operation, but rapidly decrease shortly after the ignition. This indicates a relationship to the temperature of the generator. The *d1* components show a similar behavior. Shortly after the first ignition, the values are increased, then decrease and then follow a downward trend. Only in parameter III can this downward trend not be observed, which suggests a correlation of the *d1* parameter with the current intensity. Both a change in the current and a change in the nozzle diameter have an effect. In the case of increased wear (parameter V), the characteristic frequency of the generator appears to increase. In the next picture the wavelet coefficients (*d1* and *a4*) for different parameters and process time are shown [1], **Fig. 5-8.**

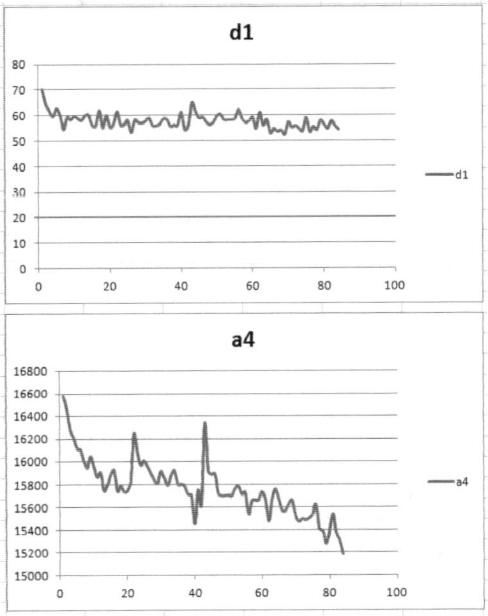

Fig. 5: Wavelet coefficients (*d1* and *a4*) of plasma parameter I

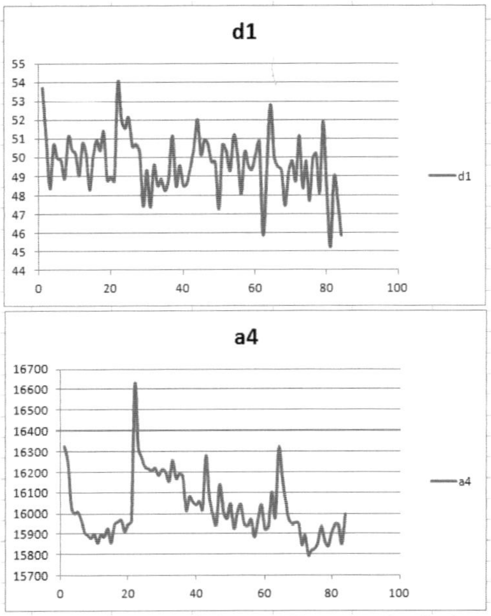

Fig. 6: Wavelet coefficients (*d1* and *a4*) of plasma parameter II

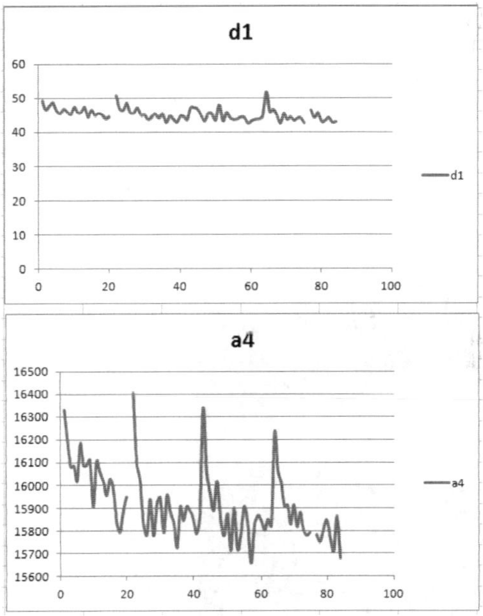

Fig. 7: Wavelet coefficients (*d1* and *a4*) of plasma parameter IV

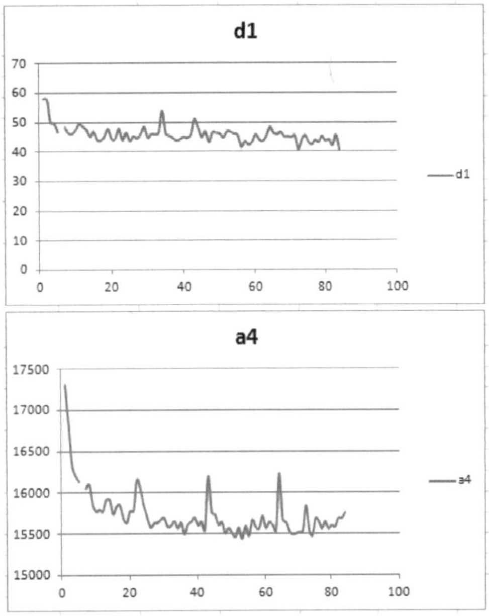

Fig. 8: Wavelet coefficients (*d1* and *a4*) of plasma parameter V

After 4h the cathodes and anodes were recorded under the light image microscope in order to determine the "destruction". These results were compared to the wavelet coefficients and calculated fourier coefficients. In **Fig 9** the "destruction" of parameter II is depicted as one example for this investigation and in **Table 2** the individual characteristic arc root displacement frequencies are shown [1].

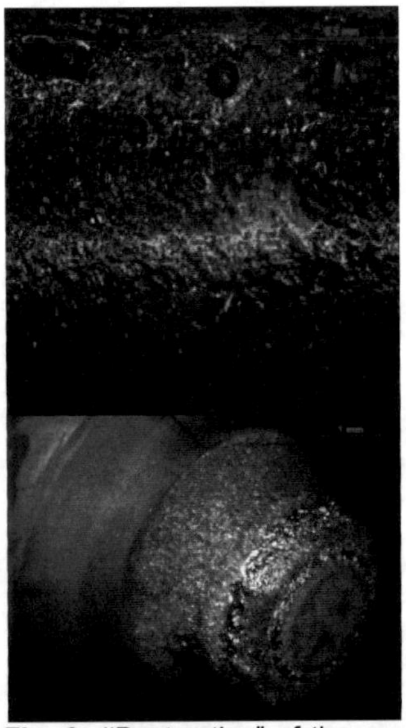

Fig. 9: "Destruction" of the anode and cathode after 4h process operation with parameter II

characteristic arc root displacement frequencies	
Parameter I	3113 Hz
Parameter II	3500 Hz
Parameter III	2044 Hz
Parameter IV	2200 Hz
Parameter V	2300 Hz

Table 2: Characteristic arc root displacement frequencies for argon-nitrogen plasma parameters

5 Outlook

In subsequent investigations, the noise of the wavelet coefficients should be determined at a measurement time. For this purpose, the measuring interval must be reduced. Assuming that the behavior of the torch does not change fundamentally within 1 minute, the normalized wavelet coefficients should not show any great fluctuations. This makes it possible to underline the usability of the wavelet transformation in practice. It must also be examined whether more steps lead to a clearer result in the wavelet transformation. The ratio between the wavelet coefficients could also provide information about the frequency composition of the signal. Futhermore in future measurements the gas mixture must and will be changed to hydrogen parameters.

6 Acknowledgements

The authors gratefully acknowledge the financial support of the AiF/DVS within the project IGF 18.963N / DVS 02.099.

7 Literature

[1] M. Mudra, Analyse von Schall-, Strom- und Spannungsverläufen bei DC-Plasmageneratoren mittels Wavelet-Transformation, 2015

[2] W. Bäni, Wavelets: Eine Einführung für Ingenieure, 2002.

[3] http://www.chauvin-arnoux.com/sites/default/files/D00NTY69_10.PDF

[4] J.-L. Marqués, M. Fuchs, F. Dschung, S. Lange, J. Zierhut und J. Schein, Fluctuations in one-cathode plasma torches: characterization and modelling

[5] Y. Meyer, Wavelets and Operators, 1992

[6] S. Schmettlach, Akustische Emissionsanalyse von Thermischen Spritzprozessen, 2014

[7] S. Zimmermann, K. Kreher, P. Weiland, L. Schnauber, J.L. Marqués, K. Landes und J. Schein, Entwicklungen im Bereich der DC Plasmageneratoren und deren mögliche Applikationsfelder, 2012

[8] P.L. Fauchais, J.V.R. Heberlein , M. Boulos : Thermal Spray Fundamentals From Powder to Part, ISBN 978-0-387-28319-7

Practical Aspects of Suspension Plasma Spray for Thermal Barrier Coating on Potential Gas Turbine Components

Xinqing Ma, Peter Ruggiero, Düsseldorf / D

Suspension Plasma Spraying (SPS) process has attracted extensive efforts and interests to produce fine-structured and functional coatings. In particular, thermal barrier coating (TBC) applied by SPS process gains increasing interest due to its potential for superior thermal protection of gas turbine hot-section components as compared to its conventional APS-TBC and even EB-PVD TBC. Unique columnar architecture and nano- and sub-micron sized grains in a SPS-TBC demonstrate some advantages of thermal shock durability, low thermal conductivity and high-temperature sintering resistance. This work will look into some practical aspects of SPS process for TBC application before it becomes a reliable industry method. The spray capability and applicability of SPS to achieve uniformity thickness and microstructure on curved susbtrates was evaluated in designed spray trials to simulate industrial parts with complex configurations. The performances of SPS-TBCs in erosion, free falling ballistic impact and indentational loading tests were tested and evalauted to simulate SPS-TBC performances in turbine service conditions. The behaviors of SPS-TBCs in those tests were correlated to key test factors including grit incident angles, impact object sizes, indentation head sharpes and coating surface curvatures. Finally, a turbine blade was coated and sectioned to verify SPS sprayability in multiple critical sections. The SPS trials and test results demonstrate that SPS is promising for innovative TBCs, but some challenges need to be addressed before it becomes an economic and reliable industrial process, especially for gas turbine components.

1 Introduction

Traditionally, thermal spray processes use solid materials in a form of powder, wire or rod as feedstock. The earlier research and development demonstrated that liquid-state suspensions or solution precursors could be used as feedstock in thermal spray processes, including air plasma, combustion flame and high velocity oxy-fuel spraying (HVOF) [1-6]. The emerging methods so-called suspension plasma spray (SPS) and solution precursor plasma spray (SPPS) were investigated intensively during last decade, mostly for thermal barrier coatings (TBCs) [7-10]. TBCs are widely used in gas turbines to provide thermal insulation for critical hot-section components such as blades, vanes and fuel combustion assemblies during turbine operation [11-12].

Currently, TBCs are applied by either APS process or EB-PVD process. APS-TBCs have the merits of low thermal conductivity, low cost and wide applicability, in contrast, EB-PVD TBCs have superior resistance to thermal shock and erosion specifically on rotating blades. However, EB-PVD TBCs are limited to rotating blades for aero-turbines due to high capital cost, production cost and EB-PVD chamber availability.

SPS or SPPS process can be economically set up to produce TBCs by retrofitting with minor modifications of conventional APS systems which provide sufficient energy and heat, and thus become an economic process and widely available. Meanwhile, the new SPS or SPPS-formed TBCs offer superior combined coating properties such as low conductivity, adequate high thermal shock and good erosion resistance as EB-PVD TBCs. The improvements of TBC properties applied by SPS or SPPS process relative to conventional APS process are attributed to their unique nanometer or submicron-sized grains, unique morphologies and columnar or vertically cracked microstructures as well [13-15]. The fine, uniform and unique features of TBC microstructures differing from those of APS-formed TBCs, are believed to improve coating performance under high temperature oxidation and thermal cycling conditions in gas turbine engines.

While SPS-based thermal spray techniques have attracted attention of research and development community, the interest of industry for various applications also is increasing. For example, one turbine manufactures have worked on developing SPS process for TBC on its engine components [16]. However, it is important and necessary to understand, develop and evaluate liquid feedstock methods and resultant coatings from the viewpoint of practical aspects of the SPS processes for industrial applications, in particular, TBCs on critical turbine engine components. The nature of SPS process using liquid-based slurry creates some challenges and or concerns about slurry handling, environmental safety, deposition rate, process efficiency, coating performance, evaluation practices, qualification specifications and process applicability as applied to complex configuration parts before the method can be adapt by industries [17, 18].

So far there is not much published work and information to directly address these challenges to date. Therefore, this work will develop experimental methodology to investigate the process potential, capability and applicability of SPS process for TBCs on potential turbine components by addressing the challenges and the concerns related to SPS practical aspects. The aim is to provide better understanding of SPS process from industry expectation and practice, and further overcome the process disadvantages and technical barrier to become a viable and acceptable production TBC process.

2 Experimental Details

2.1 Suspension plasma spray

Suspension slurry for all TBC coatings was made from ZrO2–8 wt.% Y2O3 powder suspended and dispersed in ethanol. 8YSZ solid content in the slurry is about 25 wt% (Treibacher Industrie AG, Austria), with particle size D50 about 0.7 micron. The suspension feedstock was agitated prior to spraying. TBC coatings were fabricated from these suspensions at one of Curtiss-Wright Surface Technology facilities with a setup of Oerlikon-Metco Sinplex plasma spray system equipped with a liquid flow-controled feeding unit. The nominal feeding rate was set at 30-35 g/min. The primary spray parameters to be investigated include plasma current, primary Ar gas flowrate, secondary H_2 gas flow and spray distance. Air cooling jets were used to control substrate temperature during spray trials. Panels and bars (6.35, 12.7 and 25.4 mm in diameters) made of Inconel 718 were used as substrates for test specimens. All the specimens were coated with a NiCrAlY bondcoat by APS, at a typical thickness 200-250 microns. Typical thickness of YSZ topcoat is in the range of 200 to 300 microns. SPS sprayability and applicability was analyzed based on deposition rate and process efficiency, coverage and microstructure.

2.2 Test and evaluation

Microstructural evaluation and characterization of the SPS-TBCs were performed using optical microscope and SEM. Physical property tests included microhardness HV300 primarily. Performance tests included erosion testing and impact testing, Erosion test was conducted per an internal test procedure with the conditions of 50 mm distance, an Al2O3 grit size of 240 mesh and incident angle between 20 and 90 degrees (normal to the surface of a test specimen). The change of coating thickness before and after erosion test was measured, erosion resistance was calculated with the formula: the ratio of time of blast exposure to the loss in coating thickness.

Ballistic impact resistance of the SPS-TBCs applied on the bars with a diameter of 6.35, 12.7 or 25.4 mm, were tested by free falling of 10-mm diameter steel pin with a tip of 10-mm diameter ZrO2 ball and total weight of 62.6 grams. The height of free falling is approximately 4.67 m. The impact velocity and impact energy is estimated about 9.45 m/s and 2.8 J separately. After impact testing, the surface morphologies of the SPS-TBCs at impact areas were examined under a stereomicroscope to check impact area for damages. Regular APS-TBCs were also tested for comparison purpose.

The fracture behaviors of the SPS-TBCs were tested using Rockwell indentation techniques under a load of 100 Kg. Both HRB with a steel sphere tip and HRC with a diamond spheroconical tip were used for fracture testing. After indentation testing, the surface morphologies of the SPS-TBCs at impact areas were examined under a stereomicroscope to check impact area and any coating damages.

2.3 Process applicability

To analyze SPS process capability for full coverage and thickness uniformity on complex configurations with curved surfaces such as turbine blades, process simulation testing was designed and conducted by spraying TBCs onto the bar substrates with 3 different diameters by manipulating gun movement in a pattern of straight line (up and down) or a pattern of ladder. All substrates are stationary without rotation. As a result, coating thickness profile was plotted with the angle between spray direction and coating location. Finally, an actual turbine blade was coated with a selected SPS parameter, and the thickness and microstructure of the resultant TBCs were examined in multiple critical locations including leading and trailing edges, concave and convex airfoils, radiant areas and platforms.

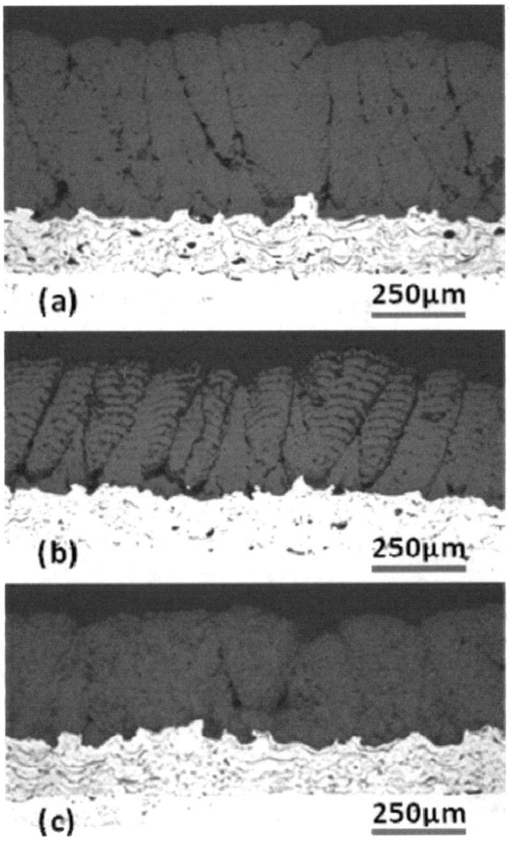

Fig. 1 Columnar microstructures of 8YSZ-TBC applied by SPS. (a) Dense columnar microstructure obtained at a high plasma power; (b) Dense and laminated microstructure; and (c) Porous columnar microstructure at a low plasma power and or a relative longer spray distance.

3 Results and Discussion

3.1 Microstructure of SPS-TBC

Typical microstructures of SPS-TBCs among the spray trials are selectively shown in Fig. 1. The SPS-YSZ topcoat have typically columnar structures with a dense (Fig. 1a), porous (Fig. 1c) or dense and layered (Fig. 1b) microstructure within individual columns. The oriented boundaries between the columns indicate the direction for the columns' growth. As shown in other studies, the top view of a columnar SPS-YSZ shows a morphology of cauliflower-like network [18, 19]. The growth of the columns is affected significantly both in columnar size and orientation by the surface profile of the underneath bondcoat, and spray angle as well. The large columns preferentially nucleated on the peaks of the rough bondcoat. The SPS-TBC on a HVOF bondcoat with a low roughness or a polished bondcoat shows more straight, vertical and even column spaced microstructure [20]. The density of the columns seems to depend on plasma condition and spray distance in the spray trials. A high plasma power and suitable short spray distance can generally produce a denser TBC. In addition, the density of TBC will be reduced if a high liquid flowrate is used for SPS spraying. The mechanism for SPS-TBC formation are described [13, 21].

The deposition rate of the SPS-process is plotted with spray distance (SD) in Fig. 2. The deposition rate falls into 5 zones within the range of spray distances selected in the SPS spray trials. In not workable zone, the deposition rate starts to reduce dramatically because the short SD less than 50 mm mostly attributing to the insufficient resident time for melting of flying particles in plasma torch. Additionally, the short spray distance will create a situation of the torch too close to the substrate. Then issues associated

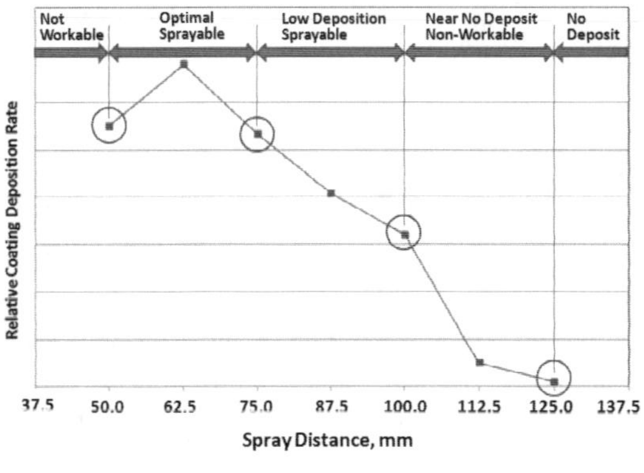

Fig. 2 Plot of the dependence of deposition rate on spray distance for the SPS spray trials.

with over heating or a degraded surface can be produced. The optimal sprayable distance is between 50 mm to 75 mm. The resultant coating should have a uniform, dense and columnar structure with desirable deposition rate and hardness. This optimal zone provides the appropriate heating and acceleration condition for the slurry feedstock in considering the distance is accordance with the length of the plasma torch. With increasing SD > 75 mm, the deposition rate reduced nearly linearly from low deposition sprayable zone, near non-deposition/non workable zone to no deposition zone at a spray distance of 125 mm. In low deposition sprayable zone, normally a coating with much porous microstructure is obtained. In no deposition zone, the particles are cooled down and slow down too much due to the long spray distance out of the length of plasma torch.

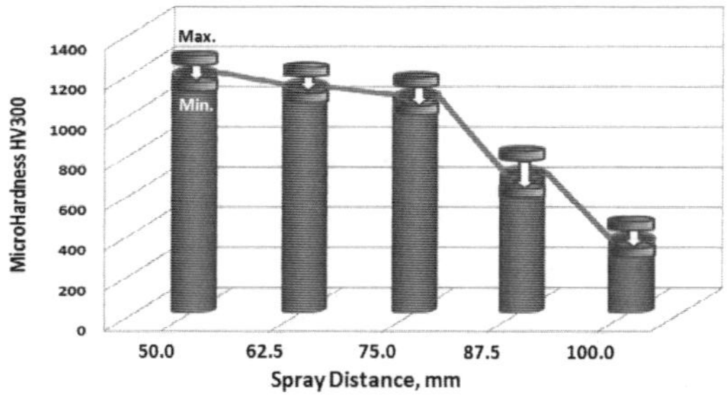

Fig. 3 Plot of TBC microhardness HV300 via spray distance used in SPS spray trials.

The relation between spray distance and microhardness HV300 of the SPS-TBCs is given in Fig. 3. The coating microhardness shows a general trend of reduction with prolonging spray distance from 50 to 100 mm. The average value of microhardness is in the range from 1183 at a SD of 50 mm to 356 at a SD of 100 mm. In the optimal sprayable zone with a SD of 50 to 75 mm, the maximum HV can be achieved, i.e. in the range of 1076 to 1183. After the SD extends to > 75 mm in the low deposition zone, the microhardness HV drops significantly to 356 at a SD of 100 mm due to the increase in coating porosity. It is believed that the SPS-TBC with high hardness is favourable for improving erosion resistance relative to a porous APS-TBC typically with a low HV below 400. By comparing with that of the dense APS-TBC with vertical cracks or a TBC by EB-PVD, the microhardness HV of the optimal SPS-TBC is equivalent or even a higher value can be attainable. The high coating hardness for a dense SPS-TBC should result from the used slurry feedstock which contains primary fine particles at sizes of nanometers and submicrometers, and sequentially their well-melting and high velocity impacting to build up a strong cohesive fine-structured coating during SPS process with the optimal condition.

From the viewpoint of an industrial process, SPS process has a relatively narrow spray window in spray distance, and thus which will limit its sprayabality and applicability to a part with complex configurations and specially curved surfaces and deep pockets. Also, the effort for robot programming will be more time intensive and require highly dimensional precision to achieve uniform coating thickness and microstructure.

3.2 Process applicability

Turbine engine parts like turbine blades will require a coating uniformity in thickness and microstructure, as well as full coverage in designed coating areas including airfoils and platform. To simulate SPS applicability to a curved surface like leading or trailing edge or airfoil of a blade, experimental method was designed by applying SPS-TBC onto a cylindrical substrate with a diameter of 6.35, 12.7 or 25.4 mm. SPS-TBCs were sprayed onto the bar surface with two spray path patterns which are typically used for applying a TBC to turbine parts by thermal spray industry, i.e. straight line path pattern and "ladder" path pattern. The profiles of normalized coating thickness were plotted with radial angle for both spray patterns on the cylindrical substrates, as shown in Fig. 4. For spray pattern – straight line path, this simulation testing is to estimate coating roundness capability and

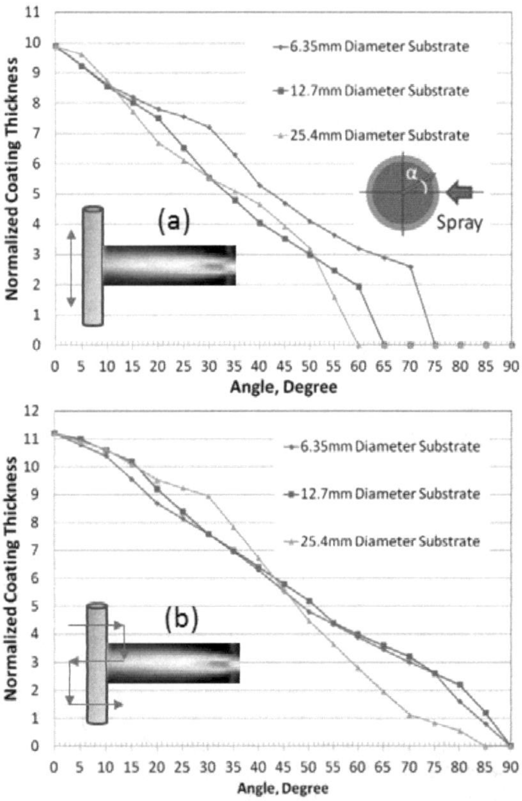

Fig. 4 Normalized thickness profiles of SPS-TBCs coated on cylindrical substrates with 3 different diameters varied with radial angles, SD at 75 mm. (a) Straight line spray pattern; (b) Ladder spray pattern

coverage over a radius. There is a cutoff angle for SPS-TBCs' coverage on the substrates with different diameters. The cutoff angle is about 60, 65 and 75 degrees for the diameters of 25.4, 12.7 and 6.35 mm, separately. The cutoff angle can be used for characterize the roundness capability of SPS process by coordinating to the radius of a curved surface. For example, with a radius of 3.2 mm, coating coverage will be within 75 degrees, not full radial surface. The thickness profile indicates SPS-TBC on 6.35-mm diameter specimen is more uniform compared to the other two larger diameter specimens. This is a desirable merit to obtain a more even thickness TBC coating layer on a sharp corners or edges, for example, leading edge of a turbine blade. The SPS-TBCs on two large diameter substrates have very similar coating thickness profiles, indicating less sensitivity to the surface curvature if the diameter/radius exceed a certain value, at least 12.7 mm in this study. By considering the relative small SD in SPS process, SPS process demonstrates a good capability for coating coverage around a small radius.

For spray pattern – ladder path, this simulation testing is to estimate coating uniformity over a curved surface. As shown in Fig. 2b, coating coverage is improved in full range of 0 to 90 degrees without a cut-off angle. Also the fact that three specimens show very similar thickness profile, indicates the ladder path pattern eliminates the effect of surface curvature on coating thickness profile and creates full coverage as well. By spraying a part with a ladder pattern, SPS process is capable of producing a TBC with improved uniformity and coverage over an entire surface with different curvatures. This capability of SPS process is very important to become an industrial process for turbine components. The near linear relation between coating thickness and radial angle for those specimens can be useful for creating robot programs to achieve a uniform thickness profile on a turbine component.

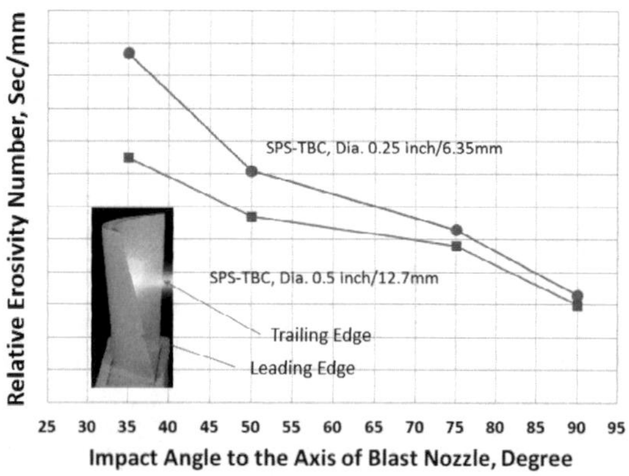

Fig. 5 Erosion resistance of SPS-TBCs applied on cylindrical substrates with diameters of 6.35 mm and 12.7 mm at different incident angles.

3.3 Erosion resistance

One major concern for a regular APS-TBC is its low erosion resistance. Due to porous structure and resultant low hardness, APS-TBCs typically have poor erosion resistance compared to that of EB-PVD or dense APS-TBCs [20]. If SPS-TBCs are considered as economic alternative to expensive EB-PVDTBCs for turbine components, improved erosion resistance should be demonstrated. In our previous comparison study of erosion resistance at an incident angle of 20 degrees, the SPS-TBC with a microhardness above 1,000 demonstrated similar erosivity number to those of dense APS-TBCs or EB-PVD TBCs [18]. To further simulate and evaluate the effect of surface curvature on erosion behaviour, SPS-TBCs were applied to cylindrical substrates with diameters of 6.35 and 12.7 mm, which simulate the radius of trailing and leading edges of a turbine blade separately. The relative erosivity number vs incident angle is given in Fig. 5. First, both specimens show reduced erosivity number with increasing angle from 35 to 90 degrees. In another word, the TBC coatings were eroded more severe at a high angle than at a low angle. At a high incident angle between 75 to 90 degrees, the curvature radius seems to have little effect on erosivity number. However, at a low medium incident angle, the erosivity number for 6.35mm-dia. specimen trends to become bigger than 12.7mm-dia. specimen while the angle reduces from 75 to 35 degrees. It is assumed that the erosion results may result from the differences in the microstructures (such as density, columnar sizes and orientation) between the two specimens.

It is worthy to mentioned that the columnar grained microstructure in SPS-TBCs may be considered as a key and sensitive factor to influence their erosion behaviors. Similarly, it was found that erosion resistance was affected by columnar or multi-layered structured in SPS-formed ceramic coatings [22]. Other factors such as nano-grains and morphologies of TBCs may also have contributed to the behaviors of the SPS-TBCs as described in other TBCs studies [23-26]. Therefore, SPS-TBCs are proved to have comparable erosion resistance to EB-PVD TBCs, and they are less sensitive to the surface curvatures within certain incident angles. Those merits will be beneficial for TBC survival on curved surfaces such as in leading edge of turbine blades. But then again, their erosion behaviors show they are prone to suffer move severe erosion at certain high incident angles.

3.4 Ballistic impact resistance

TBC coated turbine components may be damaged by foreign objects, known as "FOD" [27-29], and can cause disastrous consequences in terms of the performance of critically loaded components, such as the HP turbine blades. In this work, free-falling impact testing were designed and performed aiming to evaluate the foreign object damage to SPS-TBCs by ballistic impact. To better simulate a case of turbine blade, test specimens were prepared by applying SPS-TBCs on different diametered cylindrical substrates, representing leading edge, trailing edge, airfoil and platform, separately. The object with a tip of ZrO_2 ball impacted on the SPS-TBCs at 9.5 m/s and with an impact energy of about 2.8 J.
The coating surfaces of both SPS-TBCs and APS-TBCs were examined for impacted morphologies and damages under a stereomicroscope. As partially shown in Fig. 6, the tested specimens show different degree of damages. For SPS or APS-TBCs with a large diameter of 25.4 mm, impacting create visible large size impact marks or dents (~6 mm diameter), but without obvious coating damages like cracking or spalling. For SPS or APS-TBCs with a medium

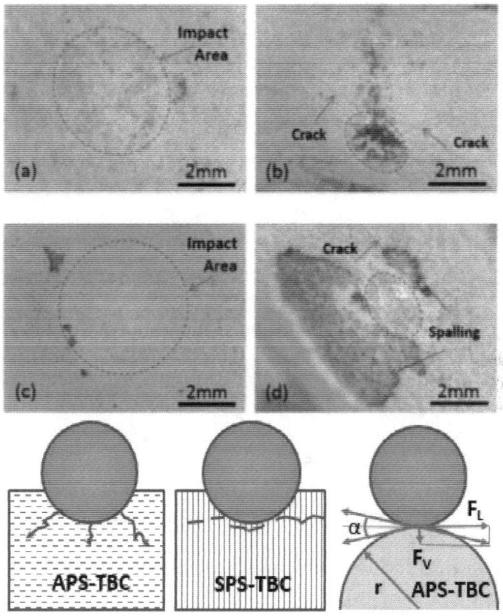

Fig. 6 Morphologies of SPS and APS-TBCs after free falling impact tests and schematics of failure modes under impact forces. (a) APS-TBC with a diameter of 25.4 mm; (b) APS-TBC, 6.35 mm dia.; (c) SPS-TBC, 25.4 mm dia.; (d) SPS-TBC, 6.35 mm dia.

diameter of 12.7 mm, the sizes of impact marks become smaller, accompanying with fewer cracks, but no coating delamination. For SPS or APS-TBCs with a small diameter of 6.35 mm, the sizes of impact marks becomes much less (~2 mm diameter), accompanying with long cracks, and even delamination of SPS-TBC. The results revealed that the risk and chance for impact failure increases with increasing surface curvature in the areas of small edges or corners such as leading edge and radiant areas on a turbine blade.

The experimental results can be interpreted by two factors at least. (i) Mechanical force and pressure factor. During the impact tests, the impact force was constant but the impact pressure varied due to the different impact areas. Based on the sizes of the impact marks on coating surfaces, it is roughly estimated that the impact pressure on the larger 25.4mm sample is about at 10 times lower than that on the larger 6.35 mm one. With the increase in impact pressure, the chance for coatings to fracture and or delamination will rise significantly while the impact intensity exceeds the coatings' cohesive and or adhesive strengths. (ii) Microstructural factor. As schematically explicated in Fig. 6, ASP-TBC and SPS-TBC has a typical laminated and a columnar microstructure, separately. In the case of impacting on a small-radius coated substrate, the contact angle α is larger than on a large-radius substrate. With the bigger impact angle, there is a relatively large lateral force (F_L). Due to the lateral force, the individual columns in the SPS-TBC are more prone to be laterally fractured than the laminated structure in the APS-TBC. In the APS-TBC, contrastly, the impact can induce radially cracking readily. Thus, SPS-TBC tends to peel or spall under a high intensity impact due to the mechanism of lateral breaking of the columns.

Particles or debris sizes between 2 μm and 20 μm may impact inclined surfaces, where significant local aerodynamic effects modify particle impact conditions, and thus these sized particles cause erosion to the trailing edge of turbine aerofoils [30]. Large particles (>40 μm) move relatively slowly in the gas stream and can be hit by the rotating turbine hardware damaging the leading edges. The impact simulation testing proved that TBCs on the small-radius edges (ex. leading edge) are prone to be damaged and failed by peeling or spalling. Relatively, the low-radius areas (ex. airfoils) should have much less damages.

3.5 Indentation fracture behaviour

In above section, the FOD testing was simulated with the impact of an object with a larger sphere tip (10 mm in diameter). However, in real turbine service, typical particles are in the ranges of less than 1 mm and are irregular in shapes such as sharp edge. Thus, the flying particles can impact, embed or cut into a coating. To simulate this type of FOD, two experimental tests were designed and conducted on both SPS and APS-TBCs. Test #1 was performed using HRB indentation tester under a load of 100 Kg, in which a steel ball tip of about 1.5 mm diameter was used. Test #2 was performed using HRC indentation tester under a load of 100 Kg, in which a sharp diamond tip as to simulate a sharp edge penetration.

The results of tests #1 and #2 are summarized in Fig. 7. Similarly, the coatings on the substrates with smaller diameters show more severe coating damages, either cracking and or spalling. However, the HRB indentation tests with smaller sphere tip seem to cause more damages than the impact testing using a large ball. It can be understood, just because the small tip provides a test condition of more stress/force concentration locally in those coatings. In comparison with APS-TBCs, the SPS-TBCs suffered more distinguished coating damages mainly spalling. As discussed in above section 3.4, the spallation of the SPS-TBCs can be explained by columns' breaking under a lateral force during the penetration of indentation head into the coating. Single architectured APS-TBCs seems to be more resistance to an indentation load.
Test #2 was conducted using HRC indentation tester on the same specimens as the test #1. The Test #2

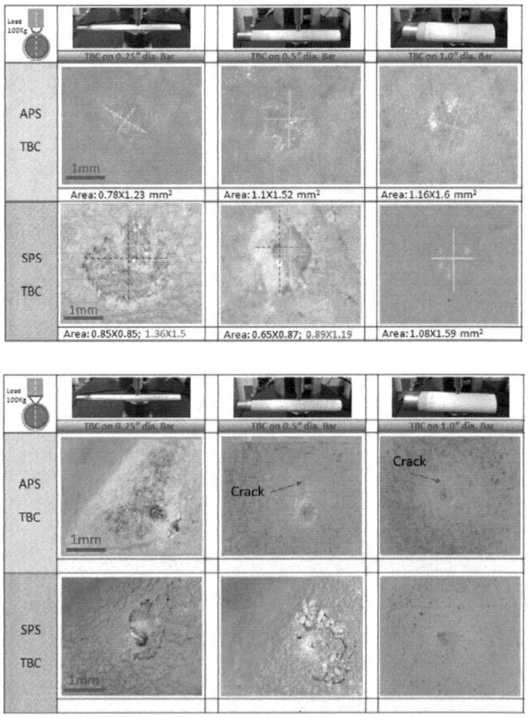

Fig. 7 Test setup and surface observation of the test specimens after HRB and HRC indentation tests on ASP and SPS-TBCs applied on cylindrical substrates. Top: Test #1, HRB indentation test; Bottom: Test #2, HRC indentation test

obtains quite similar results as the test #1, i.e. more damages occurred on the smallest diameter specimens for both types of TBCs. However, several differences are observed: (i) Smaller indented marks compared to those in test #1. It should be attributed mostly to the physical dimension and spheroconical shape of the tester tip; (ii) The tip seems to penetrate deeply in the specimens due to the sharp tip used, thus both types of TBCs have spalled damages; (iii) On the smallest dia. specimens, APS-TBC shows larger area spallation than SPS-TBC, and (iv) the damages show a different pattern, typically, circular cracking and buckling around the indentation in the coatings. Thus, the specimens suffered a progressive loading and resultant coating cracking predominantly. (v) From the cracking in the coatings, SPS-TBCs exhibit better fracture toughness to resist coating cracking on the two larger diameter specimens.

To summarize the indentation tests, the tests #1 and #2 are able to simulate the FOD and related coating toughness more closely. In addition, test #2 can represent the condition of sharp particles' penetration and embedding in a TBC coating. Meanwhile, test #2 is more suitable for testing and comparing fracture toughness among different types of TBC specimens compared to test #1. The coating areas with a small radius such as leading edge or other corners of a turbine components will be riskier to be damaged and cause failure. Although SPS-TBC demonstrated higher fracture toughness, but its columnar architecture may not be benefit considerably to resist high intensity impact or penetration of flying objects. It is necessary to understand better what is optimal grain structure of a SPS-TBC for better and more reliable impact resistance, in addition to thermal strain tolerance.

The columnar structured TBC is produced in all the examined areas, despite that there are some variances in columns' sizes and orientations. Within the columns, there is quite uniform density. There is a well-bonded interface between APS bondcoat and SPS topcoat. It is noticed that the 8YSZ topcoat is not dense as the optimal one, mostly because a longer spray distance was selected as to conveniently manipulate spray gun for better coating coverage and uniform thickness across all the blade surface. The columns' sizes are quite depending on the roughness of bondcoat, instead of the surface curvatures at different locations. The growth orientation of the columns is associated with primary spray angle and secondary bondcoat roughness. Therefore, the control on SPS-TBC's microstructure needs to include bondcoat roughness profile, spray angle and spray

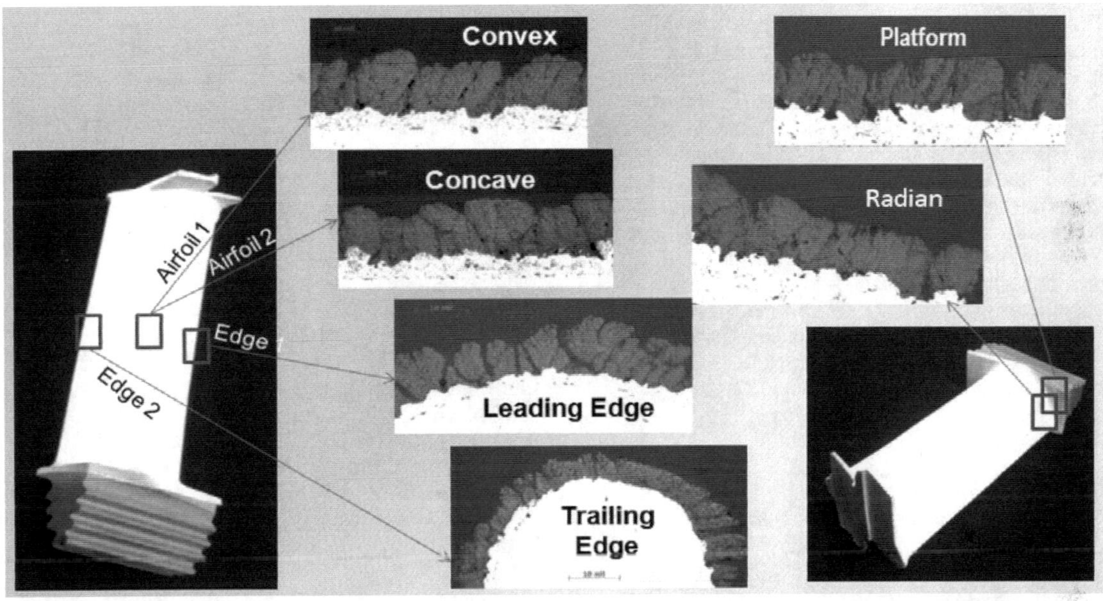

Fig. 8 The cross-section views of microstructures of SPS-TBC coated gas turbine blade in the selected locations of airfoils, leading edge, trailing edge, radiant area and platform.

parameters at least.

3.6 SPS sprayability on gas turbine blade

To further verify spray capability of SPS process, a turbine blade was sprayed with one selected spray parameters at a relative long spray distance. Then, the coated blade was sectioned to examine coating thickness and microstructures in represented and critical areas, including airfoils (convex and concave), leading edge, trailing edge, radiant (near bottom platform) and platform. In Fig. 8, the cross-section images of the sectioned blade coated with SPS-TBC are shown. In general, a uniform thickness profiles as demonstrated in the studies of spray capability and applicability, can be achievable on the blade's leading and trailing edges, as well as radian area between airfoil and platform.

Based on the initial results of SPS spraying on a turbine blade and other simulated specimens, SPS process has a sprayability within a narrow spray distance relative to conventional APS process. For a single airfoil blade, it is quite feasible to produce an acceptable topcoat with an effort of controlled bondcoat and robot programming. However, for a turbine component with complex configurations such as double or triple airfoil vanes, SPS process will face the challenge to offer the extended distance required for an access to the areas between two airfoils. Other challenges may include how to describe, characterize and specify the coating microstructural features for laboratory evaluations. The limitations of current SPS process about liquid injection and spray efficiency are discussed also [17]. Although many challenges associated with SPS process for TBC applications on turbine engine components, it is believed that most of the technical obstacles can be resolved if we gain better understanding of the process and its capability and disadvantages. SPS-TBCs show unique and superior properties, good economy and some Sprayability and applicability. Hence, it is hopeful with further development SPS-TBCs will become a practical and reliable industrial process with on-going research, development and engineering efforts.

4 Summary and Conclusion

Suspension plasma spray (SPS) was investigated for producing finely structured thermal barrier coatings. Columnar structured 8YSZ coatings were prepared with varied density and laminated architecture in spray trials. The spray capability and applicability of SPS was characterized by deposition rate, process efficiency, thickness distribution and microstructure. Several experimental methods were developed with the aim to simulate and evaluate the practical aspects and potential of the SPS process for coating industrial gas turbine components such as turbine blades and vanes. The test results of physical properties, erosion resistance, ballistic impact resistance, fracture toughness and coating a turbine blade, indicate that SPS is promising for innovative TBCs. However, some challenges need to be addressed before it becomes an economic and reliable industrial process needed, mostly associated with low deposition rate, short spray distance and microstructure sensitive to bondcoat roughness.

5 Literature

[1] G. Chow, L. K. Kithara, T. D. Xiao, P. R. Strutt, C. W. Strock and R. A. Zatorski: US Patent 6,447,848 (2002).

[2] F. Gitzhofer, E. Bouyer, M. I. Boulos: US Patent 5,609, 921 (1997).

[3] E. H. Jordan, L. Xie, M. Gell, N.P. Padture, B. Cetegan, A. Ozturk, X. Ma and Others: J. Thermal Spray Technol. 13 (2004), Issue 1, pp. 57/65.

[4] L. Pawlowski: Surf. Coat. Technol. 203 (2009), pp. 2807/2829.

[5] X. Ma, J. Roth, D.W. Gandy, and G.J. Frederick: J. Thermal spray Technol. 15 (2006), Issue 4, pp. 670/675.

[6] P. Fauchais and G. Montavon: J. Thermal Spray Technol. 19 (2010), Issue 1-2, pp. 226/239.

[7] Lech Pawlowski: Surf. Coat. Technol. 202 (2008), pp. 4318/4328.

[8] L. Xie, X. Ma, A. Ozturk, E. H. Jordan, N. P. Padture, B. M. Cetegen, D. Xiao, M. Gell: Surf. Coat. Technol. 183 (2004), pp. 51/61.

[9] K. VanEvery, M. Krane and R. W. Trice: Surf. Coat. Technol. 206 (2012), pp. 2464/2473.

[10] X. Ma, M. Stephen, R. Jeffery, D. Xiao, and B. M. Cetegen: US Patent 7,112,758 (2006)).

[11] R.A. Miller: J. Therm. Spray Technol. 6 (1997) Issue 1, pp. 35/42.

[12] D. R. Clarke, M. Oechsner and N. P. Padture: MRS Bulletin 37 (2012), Issue 10, pp. 891/941.
SPS structure

[13] K. VanEvery, M. Krane, R. W. Trice, H. Wang, W. Porter, M. Besser, D. Scordelet, J. Iiavsky and J. Almer: J. Thermal Spray Technol. 20 (2011), Issue 4, pp. 817/828.

[14] R. Siegert, J.E. Doring, J.L. Marques, R. Vassen, D. Sebold and D. Stover: Proceedings of the International Thermal Spray Conference (2005), C. Berndt and E. Lugsheider, Ed., ASM International.

[15] M. Gell, L. Xie, X. Ma, E. H. Jordan and N. P. Padture: Surf. Coat. Technol. 177-178 (2004), pp. 97/102.

[16] L. S. Leblanc, J. A. Ruud, K. P. McEvoy, and A. J. Kulkarni: US Patent 2014/0178641 (2014).

[17] P. Fauchais, M. Vardelle and S. Goutier: J. Thermal Spray Technol. 24 (2015), Issue 7, pp. 1120/1129.

[18] X. Ma and P. Ruggiero: Surface Protection for Enhanced Materials Performance: Science, Technology and Application, MS&T (2016).

[19] Z. Tang: Thermal Spray of Suspensions & Solution Symposium, ASM International, (2015).

[20] X. Ma and P. Ruggiero: NATO Symposium RTO AV-187/RSY-28, (2011), pp. 11-1/11-13.

[21] J. O. Berghaus, S. Bouaricha, J.G. Legoux, and C. Moreau: Proceedings of the International Thermal Spray Conference, C. Berndt and E. Lugsheider, Ed., ASM International, (2005)

[22] S. Mahade, N. Gurry, S. Bjorklund, N. Markocsan, P. Nylen and R. Vaben: Proceedings of the International Thermal Spray Conference (2016), pp.314/347.

[23] R.G. Wellmana, M.J. Deakinb, J.R. Nicholls: Tribology International 38 (2005), Issue 9, pp. 798/804.

[24] Z. Qu, X. Cheng, J. Wu, R. He, Y. Pei and D. Fang: Surf. Coat. Technol. 299 (2016), pp.129/134.

[25] J. Kitamura, K. Fujimori, T. Wada, J. Liu, B. and Callen: Proceedings of the International Thermal Spray Conference (2016), pp.348/352.

[26] X. Ma, F. Wu, J. Roth, M. Gell, E. H. Jordan: Surf. Coat. Technol. 201 (2006), Issue 7, pp. 4447/4452.

[27] J. R. Nicolls and R. G. Wellman: The Control and Reduction of wear in Military Platforms, NATO RTO-MP-AVT-109 (2003), pp. 20-1/20-30.

[28] F. Cernuschi, L. Lorenzoni, S. Capelli, C. Guardamagna, M. Karger, R. Vaßen, K. Niessen, N. Markocsan, J. Menuey and C. Giolli: Wear 271 (2011), Issues 11–12, pp. 2909/2918.

[29] M. W. Crowell, T. A. Schaedler, B. H. Hazel, D. G. Konitzer, R. M. McMeeking, A. G. Evans: International Journal of Impact Engineering 48 (2012), pp. 116/124.

[30] W. Tabakoff, A. Hamed and B. Beacher: Wear 86 (1983), pp.73/88.

Effect of spray parameters on the microstructure and porosity content of gadolinium zirconate TBCs deposited by suspension plasma spray

Satyapal Mahade [a*], Dapeng Zhou [b], Robert Vassen [b]

[a] Department of Engineering Science, University West, Sweden
[b] Institute of Energy and Climate Research (IEK-1), Forschungszentrum Jülich GmbH, Germany

Abstract: Gadolinium zirconate (GZ) is considered as a promising top coat candidate for high temperature (>1200°C) thermal barrier coating (TBC) applications. Suspension plasma spray (SPS) technique has shown the capability to generate a wide range of microstructures which includes the more desirable columnar microstructure. In this study, GZ single layer TBCs were deposited by axial SPS process. The variable parameters include the standoff distance, solid load content of the suspension and input power. The cross section and top surface of the as sprayed TBCs were analyzed by SEM. The phase content in the as sprayed TBCs was analyzed by XRD. The porosity content of the as sprayed TBCs was measured using image analysis. In the SEM analysis, it was observed that a lower solid load content in the suspension favoured the formation of a columnar microstructure. Additionally, at lower solid load content, increase in standoff distance resulted in columnar microstructure with high porosity content in the TBC. However, with higher solid content suspension and alteration of input power, only a dense vertical cracked microstructure can be obtained.

1 Introduction

Thermal barrier coatings (TBCs) enable the gas turbine engine to operate at elevated temperatures by insulating the underlying metallic components from high heat loads. The desire to achieve a higher engine efficiency has led to an increase in the turbine inlet temperatures [1]. 7-8wt.% Yttria Stabilized Zirconia (YSZ) is the widely used top coat material for TBC applications due to its excellent properties. However, with the rise in operating temperatures (>1200°C), YSZ is reported to be susceptible to CMAS (Calcium Magnesium Alumino Silicates) attack and undesirable phase transformations at such high temperatures which reduce the TBC durability [2] [3] [4] [5] [6] [7]. Due to the limitations of YSZ, new ceramic top coat materials have been proposed and investigated in the past decade [8] [9] [10]. Among the new top coat materials, pyrochlores of rare earth zirconate based compositions (lanthanum zirconate and gadolinium zirconate) present attractive properties such as low thermal conductivity and excellent phase stability compared to YSZ [11] [12] [13] [14]. However, lanthanum zirconate (LZ) based TBCs have shown to be prone to loss in stoichiometry during their processing by plasma route [15]. On the other hand, GZ based TBCs have shown to retain their stoichiometry during plasma spraying [16]. Additionally, GZ based TBCs have shown to be resistant to CMAS attack [17].

The conventional TBC processing routes include electron beam physical Vapor deposition (EB-PVD) and atmospheric plasma spray (APS). Both the techniques impart unique microstructural features such as; columnar microstructure in the EB-PVD processed TBCs whereas splat like morphology in the APS processed TBCs. Columnar microstructures improve the strain tolerance of the TBCs and thus lead to increase in thermal cyclic life. In addition, columnar microstructures have shown to be more erosion resistant than the splat microstructured TBCs [18] [19]. However, EB-PVD is a relatively slow deposition process compared to the APS process. Therefore, a process which can deposit columnar microstructured TBCs using the plasma route (faster deposition rates) would be highly beneficial. Suspension plasma spray (SPS) is one such technique which has shown the capability to deposit columnar microstructured TBCs using liquid feedstock (nano sized powder suspended in a solvent) [20] [21] .

The key to durable TBCs are microstructural features such as columns, porosity etc. as these features influence the functional performance of the TBC (thermal cyclic fatigue, thermal conductivity, erosion resistance etc). To the author's knowledge, so far, no study has been conducted on the influence of SPS process parameters on the microstructure of GZ based TBCs. However, S. Mahade et al. reported the functional performance (thermal cyclic life, erosion resistance, thermal conductivity, hot corrosion resistance) of SPS deposited GZ based TBCs [22] [23] [24]. This work aims at investigating the effect of different SPS process parameters such as the input power, standoff distance (SoD) and solid load content in the suspension on the microstructure and porosity content of GZ single layer TBCs.

2 Experimental Details

2.1 Specimen Preparation

Stainless steel substrates of 25 x 25 mm dimensions were grit blasted using powdered alumina of 0.42-0.6 mm grit size until a surface roughness of 3-4 on Ra (μm) scale was obtained. Bond coat of NiCoCrAlY composition (AMDRY-9954) was deposited on the grit blasted steel substrates using the high velocity oxy fuel process (DJ-2600, Oerlikon Metco). The roughness of the as sprayed HVOF bond coat was about 10.4 μm in Ra. It has been shown that surface roughness influences the column formation in suspension plasma sprayed TBCs [20].

2.2 Suspension Preparation

$Gd_2Zr_2O_7$ (GZ) powder used in this work was synthesized by solid state reaction using stoichiometric amounts of Gd_2O_3 and Zr_2O powder. For making stable $Gd_2Zr_2O_7$ suspension, 300 g of GZ powder was dispersed in 700 g ethanol along with 4.5 g Polyethylenimine (PEI) as the dispersant (Polysciences Inc., Warrington, USA), and 900 g ZrO_2 milling balls (d=3 mm, Sigmund Lindner GmbH, Germany). The mixture was later milled on the roller bank (120 min^{-1}) for 24 h. Finally, the obtained suspension was diluted with ethanol into two different solid load content suspensions, i.e.10 wt.% and 5 wt.% respectively. The milled GZ powder particle size in the suspension was measured using dynamic light scattering particle size analyzer LB-550 (HORIBA Instruments Inc.) where the median particle size D50 was found to be approximately 577 nm.

2.3 Top Coat Deposition

Prior to the top coat deposition, the bond coated steel substrates were ultrasonically cleaned in ethanol to obtain a dirt free surface. Axial III plasma spray gun (Mettech Corp., Vancouver, Canada) was used to deposit the GZ based TBCs; and a suspension feeder developed by Forschungszentrum Jülich GmbH (Germany) was used. The feeding rate of the suspension was set to 30 g/min. Six variations of the top coat were deposited in this work. The parameters used for depositing each variation of the TBC are shown in Table. I.

Table I. Spray parameters used for depositing the TBCs

Sample No.	Power (kW)	SoD (mm)	Solid load (wt.%)
DAZ62	67.2	70	10
DAZ64	79.8	70	10
DAZ66	92.4	70	10
DAZ68	105.5	70	10
DAZ74	105.5	70	5
DAZ76	105.5	100	5

2.4 Coating Characterization

The as sprayed TBCs were cold mounted using a low viscosity epoxy resin and later sectioned using slow speed cutting machine. The sectioned samples were again cold mounted and polished until a scratch free mirror finish surface was obtained. Gold sputtering of the polished samples was carried out and analysed by SEM (Hitachi TM 3000, Japan) in back scattered electron mode.
Porosity measurement of the as sprayed TBCs was performed by image analysis technique using the Image J software. Two different magnifications, 400 X and 4000X, were chosen. Twenty five different SEM micrographs at each magnification were considered and the mean values along with the standard deviation were reported.

XRD analysis of the top surface of as sprayed TBCs was performed using the Bruker AXS GmbH, Germany at IEK-1, Forschungszentrum Jülich GmbH. A slow scan rate of 0.02°/step and a count time of 0.75s/step in the 2 theta range of 20° to 70° was used. The peaks obtained were labelled using the PDF standard.

3 Results and Discussion

3.1 Microstructural analysis

In the cross sectional SEM micrographs of DAZ62 sample deposited with low power (67 kW), horizontal cracks were observed, according to Fig.1(a). These cracks in a TBC could be beneficial to lower the thermal conductivity. However, they act as a weak link in the TBC, leading to early TBC spallation during thermal cycling. The horizontal cracks observed in this case could be due to the poor bonding between the successive splats. The thickness of the TBC was found to be approximately 300±10 µm. In the top view SEM micrograph, column heads which appear round in shape, were observed, as shown in Fig. 1(b). However, the column density observed in the top view SEM micrograph was lower compared to the previous finding [22]. It seems that the microstructure obtained with this set of spray parameter is not fully columnar and it resembles a dense vertically cracked microstructure with the presence of few columns and horizontal cracks.

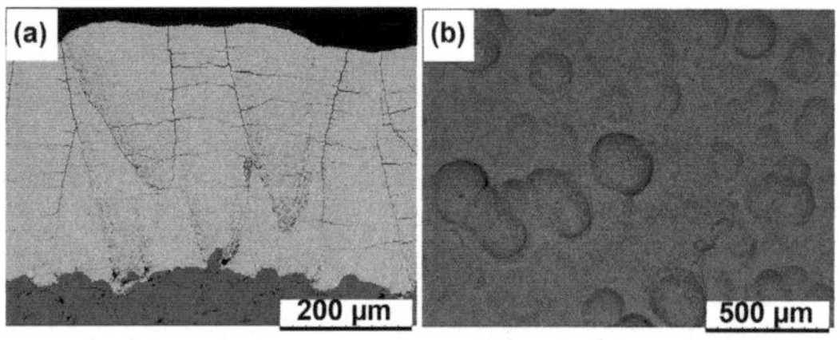

Fig.1 SEM micrographs of DAZ62 sample a) Cross section b) Top surface view

The cross sectional SEM micrograph of DAZ64 samples deposited with a slightly higher power (79kW) showed a microstructure with reduced number of horizontal cracks than DAZ62 sample, according to Fig. 2(a). The top view SEM micrograph showed a microstructure similar to sample DAZ62 with a few column heads, as seen in Fig 2(b).

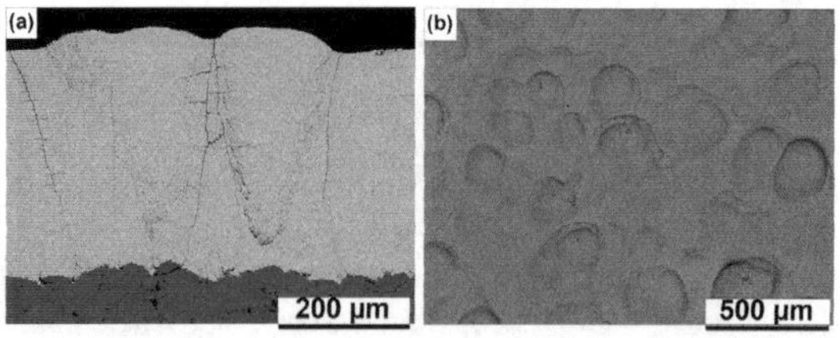

Fig.2 SEM micrographs of DAZ64 sample a) Cross section b) Top surface view

In the case of DAZ66 sample, the horizontal cracks seem to reappear with an increase in input power (92 kW), as seen in Fig. 3(a). The top view SEM micrograph looked similar to the DAZ62 sample with a low number column heads, as seen in Fig 3(b).

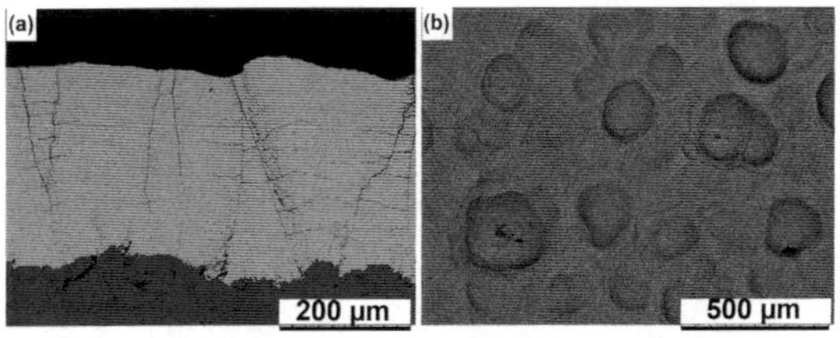

Fig.3 SEM micrographs of DAZ66 sample a) Cross section b) Top surface view

With further increase in input power (105kW), the cross sectional SEM micrograph showed a lower number of horizontal cracks than the DAZ62 sample, according to Fig. 4(a).It seems that a higher spray input power leads to improvement in bonding between the splats due to the increased droplet melting which allows the droplet to be hotter while it reaches the substrate.

Increase in spray input power in the range of 67 to 105 kW could not generate a columnar microstructure at higher solid load content. However, a high spray (105kW) power was beneficial in minimizing the horizontal cracks. Therefore, the next set of spray parameters were varied by keeping the spray input power at 105kW.

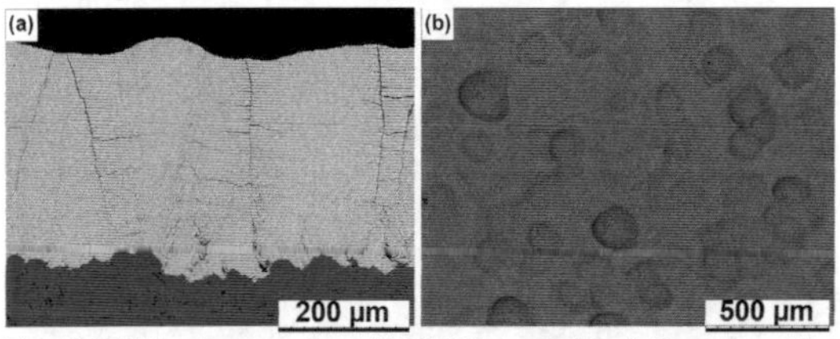

Fig. 4 SEM micrographs of DAZ68 sample a) Cross section b) Top surface view

At high input power (105kW), a lower solid load content (5 wt.%) in the suspension for sample DAZ74 led to a higher column density in the microstructure compared to the DAZ68 sample, as shown in Fig.5 (a). Additionally, the top view microstructure showed an increase in column heads which resembled a cauliflower microstructure reported previously [21], according to Fig. 5(b).

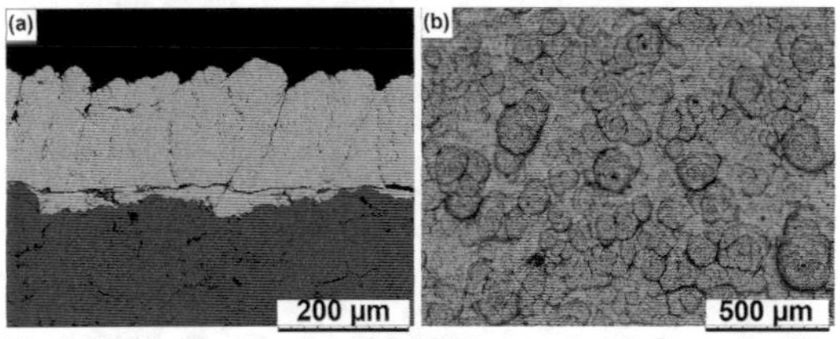

Fig.5 SEM micrographs of DAZ74 sample a) Cross section b) Top surface view

The reason for achieving a columnar microstructure with 5 wt.% solid load compared to the 10 wt.% could be due to better atomization of the droplet. Further studies need to be done to understand the difference in microstructure due to lower solid load content.

At high input power (105kW) and lower solid load content (5wt.%), an increase in the standoff distance (100 mm) in the case of DAZ76 sample led to a columnar microstructure with high porosity content compared to DAZ74 sample, according to Fig. 6(a). The reason for increase in porosity due to increase in standoff distance is not yet clearly understood and further studies need to be made to gain an insight on the inflight characteristics with varying standoff distance. Additionally, the top view SEM micrograph resembled a cauliflower microstructure, as seen in Fig. 6(b).

Fig.6 SEM micrographs of DAZ76 sample a) Cross section b) Top surface view

3.2 Porosity measurement

Porosity measurements were made using SEM micrographs at two different magnifications (400X and 4000X). Low magnification (400 X) SEM micrographs take into account the porosity contribution due to the column gaps or vertical cracks in the TBC whereas the high magnification (4000X) micrographs help in taking into account the porosity contribution due to fine scale porosity. Among the deposited TBCs, sample no: DAZ74 showed a higher porosity content, according to Fig. 7. It seems that a lower solid load content in the suspension along with higher standoff distance leads to the formation of a porous microstructure compared to TBCs deposited with a high solid load content in suspension and low standoff distance. A high porosity in the TBC is beneficial to lower the thermal conductivity and elastic modulus of the coatings.

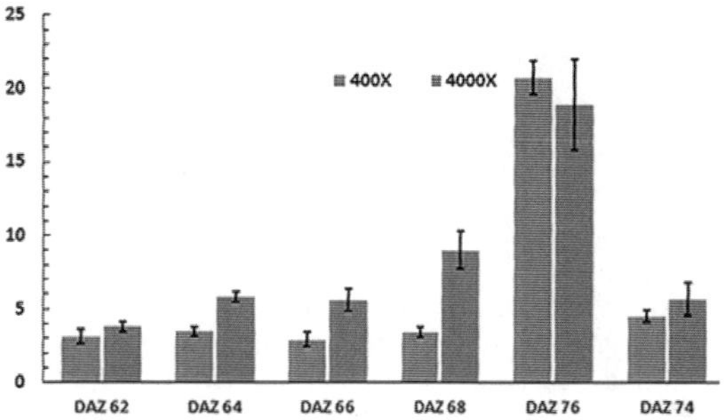

Fig.7 Porosity content of as sprayed TBCs measured using image analysis technique

3.3 XRD analysis

The peaks obtained by XRD were labelled using PDF standard. In all the deposited TBCs, a cubic defect fluorite phase of gadolinium zirconate was observed, as shown in Fig.8. A similar defect fluorite phase of GZ was reported by plasma spraying [21]. Defect fluorite phase of GZ is a disordered variation where the cations and oxygen vacancies occupy random positions in the cubic crystal structure. In SPS processed GZ based TBCs, it seems that a variation in input power, solid content in the suspension and standoff distance does not lead to an alteration in the phase composition.

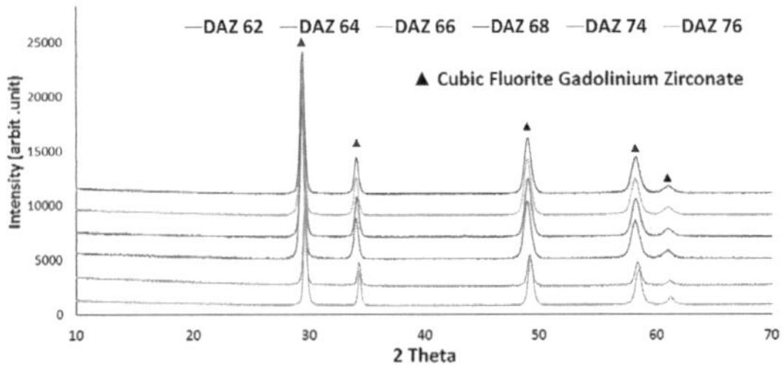

Fig.8 XRD analysis of the top surface of as sprayed TBCs

4 Conclusion

In this work, the effect of three different SPS process parameters, namely; spray input power, solid load content in the suspension and standoff distance, on the microstructure and porosity content of the deposited TBCs were investigated. It was shown that:

- Spray input power alteration at higher solid load content did not favour a fully columnar microstructure formation.
- Lower solid load content in the suspension resulted in columnar microstructure.
- Increase in standoff distance resulted in higher porosity content in the TBC.

Acknowledgements

The authors would like to thank Mr. Karl-Heinz Rauwald and Mr. Frank Kurze from IEK-1, Forschungszentrum Jülich, Germany for their help in spraying the samples. The authors would also like to thank Dr. Yoo Jun Sohn from IEK-1, Forschungszentrum Jülich, for performing the XRD measurements.

References

[1] R. A. Miller, "Thermal barrier coatings for aircraft engines: history and directions," *J. Therm. Spray Technol.*, vol. 6, no. 1, pp. 35–42, Mar. 1997.

[2] R. Vaßen, M. O. Jarligo, T. Steinke, D. E. Mack, and D. Stöver, "Overview on advanced thermal barrier coatings," *Surf. Coat. Technol.*, vol. 205, no. 4, pp. 938–942, Nov. 2010.

[3] S. Krämer, J. Yang, C. G. Levi, and C. A. Johnson, "Thermochemical Interaction of Thermal Barrier Coatings with Molten CaO–MgO–Al_2O_3–SiO_2 (CMAS) Deposits," *J. Am. Ceram. Soc.*, vol. 89, no. 10, pp. 3167–3175, Oct. 2006.

[4] C. Mercer, S. Faulhaber, A. G. Evans, and R. Darolia, "A delamination mechanism for thermal barrier coatings subject to calcium–magnesium–alumino-silicate (CMAS) infiltration," *Acta Mater.*, vol. 53, no. 4, pp. 1029–1039, Feb. 2005.

[5] J. R. Brandon and R. Taylor, "Phase stability of zirconia-based thermal barrier coatings part I. Zirconia-yttria alloys," *Surf. Coat. Technol.*, vol. 46, no. 1, pp. 75–90, May 1991.

[6] J. R. Brandon and R. Taylor, "Phase stability of zirconia-based thermal barrier coatings part II. Zirconia-ceria alloys," *Surf. Coat. Technol.*, vol. 46, no. 1, pp. 91–101, May 1991.

[7] J. M. Drexler, A. L. Ortiz, and N. P. Padture, "Composition effects of thermal barrier coating ceramics on their interaction with molten Ca–Mg–Al–silicate (CMAS) glass," *Acta Mater.*, vol. 60, no. 15, pp. 5437–5447, Sep. 2012.

[8] D. Stöver, G. Pracht, H. Lehmann, M. Dietrich, J.-E. Döring, and R. Vaßen, "New material concepts for the next generation of plasma-sprayed thermal barrier coatings," *J. Therm. Spray Technol.*, vol. 13, no. 1, pp. 76–83, Mar. 2004.

[9] R. Vassen, X. Cao, F. Tietz, D. Basu, and D. Stöver, "Zirconates as New Materials for Thermal Barrier Coatings," *J. Am. Ceram. Soc.*, vol. 83, no. 8, pp. 2023–2028, Aug. 2000.

[10] G. Mauer, M. O. Jarligo, D. E. Mack, and R. Vaßen, "Plasma-Sprayed Thermal Barrier Coatings: New Materials, Processing Issues, and Solutions," *J. Therm. Spray Technol.*, vol. 22, pp. 646–658, Jun. 2013.

[11] G. Suresh, G. Seenivasan, M. V. Krishnaiah, and P. Srirama Murti, "Investigation of the thermal conductivity of selected compounds of gadolinium and lanthanum," *J. Nucl. Mater.*, vol. 249, no. 2, pp. 259–261, Oct. 1997.

[12] S. Mahade, N. Curry, S. Björklund, N. Markocsan, P. Nylén, and R. Vaßen, "Functional performance of $Gd_2Zr_2O_7$/YSZ multi-layered thermal barrier coatings deposited by suspension plasma spray," *Surf. Coat. Technol.*

[13] S. Mahade, N. Curry, S. Björklund, N. Markocsan, and P. Nylén, "Thermal conductivity and thermal cyclic fatigue of multilayered $Gd_2Zr_2O_7$/YSZ thermal barrier coatings processed by suspension plasma spray," *Surf. Coat. Technol.*, vol. 283, pp. 329–336, Dec. 2015.

[14] G. Moskal *et al.*, "Characterization of microstructure and thermal properties of $Gd_2Zr_2O_7$-type thermal barrier coating," *J. Eur. Ceram. Soc.*, vol. 32, no. 9, pp. 2025–2034, Jul. 2012.

[15] X. Q. Cao, R. Vassen, W. Jungen, S. Schwartz, F. Tietz, and D. Stöver, "Thermal Stability of Lanthanum Zirconate Plasma-Sprayed Coating," *J. Am. Ceram. Soc.*, vol. 84, no. 9, pp. 2086–2090, Sep. 2001.

[16] E. Bakan, D. E. Mack, G. Mauer, and R. Vaßen, "Gadolinium Zirconate/YSZ Thermal Barrier Coatings: Plasma Spraying, Microstructure, and Thermal Cycling Behavior," *J. Am. Ceram. Soc.*, vol. 97, no. 12, pp. 4045–4051, Dec. 2014.

[17] S. Krämer, J. Yang, and C. G. Levi, "Infiltration-Inhibiting Reaction of Gadolinium Zirconate Thermal Barrier Coatings with CMAS Melts," *J. Am. Ceram. Soc.*, vol. 91, no. 2, pp. 576–583, Feb. 2008.

[18] J. R. Nicholls, M. J. Deakin, and D. S. Rickerby, "A comparison between the erosion behaviour of thermal spray and electron beam physical vapour deposition thermal barrier coatings," *Wear*, vol. 233–235, pp. 352–361, Dec. 1999.

[19] R. G. Wellman and J. R. Nicholls, "A review of the erosion of thermal barrier coatings," *J. Phys. Appl. Phys.*, vol. 40, no. 16, p. R293, 2007.

[20] K. VanEvery *et al.*, "Column Formation in Suspension Plasma-Sprayed Coatings and Resultant Thermal Properties," *J. Therm. Spray Technol.*, vol. 20, no. 4, pp. 817–828, Mar. 2011.

[21] S. Mahade, R. Li, N. Curry, S. Björklund, N. Markocsan, and P. Nylén, "Isothermal Oxidation Behavior of $Gd_2Zr_2O_7$/YSZ Multilayered Thermal Barrier Coatings," *Int. J. Appl. Ceram. Technol.*, vol. 13, no. 3, pp. 443–450, May 2016.

[22] S. Mahade, N. Curry, S. Björklund, N. Markocsan, P. Nylén, and R. Vaßen, "Erosion Performance of Gadolinium Zirconate-Based Thermal Barrier Coatings Processed by Suspension Plasma Spray," *J. Therm. Spray Technol.*, pp. 1–8, Dec. 2016.

[23] K. P. Jonnalagadda *et al.*, "Hot Corrosion Mechanism in Multi-Layer Suspension Plasma Sprayed $Gd_2Zr_2O_7$ /YSZ Thermal Barrier Coatings in the Presence of V2O5 + Na2SO4," *J. Therm. Spray Technol.*, vol. 26, no. 1–2, pp. 140–149, Jan. 2017.

[24] S. Mahade, N. Curry, S. Björklund, N. Markocsan, and P. Nylén, "Failure analysis of $Gd_2Zr_2O_7$/YSZ multi-layered thermal barrier coatings subjected to thermal cyclic fatigue," *J. Alloys Compd.*, vol. 689, pp. 1011–1019, Dec. 2016.

High Feed Rate Plasma Spraying of YSZ from Various Suspensions

Radek Musalek, Jan Medricky, Jiri Kotlan, Tomas Tesar, Frantisek Lukac, Pavel Ctibor, Ksenia Illkova, Tomas Chraska, Prague / CZ

In this paper, yttria-stabilized zirconia (YSZ) coatings were prepared by plasma spraying of ready-to-spray suspensions provided by three different manufacturers. High-enthalpy hybrid water-argon plasma torch WSP-H 500 was successfully used for deposition of coatings with porous and columnar microstructure consisting of tetragonal non-transformable phase. Sensitivity of the deposition process to variation of deposition conditions was also evaluated by the change of suspension injection point position. Slight differences in the microstructures of the deposited coatings (in particular character of porosity and mutual bonding of the microsplats) were reflected in slight but measurable differences in hardness and wear resistance of the coatings indicating changes in the coating cohesion. Tensile adhesion/cohesion strength of the coatings was found to be in the range of 9 to 15 MPa. High coating porosity desirable for low thermal conductivity combined with high suspension feed rate (from about 100 to 120 ml/min in this study) makes the WSP-H coatings promising for further development for example in thermal barrier coatings applications.

1 Introduction

In the field of thermal spraying, Yttria-Stabilized Zirconia (YSZ) is one of the most commonly sprayed materials having numerous applications, in particular in thermal barrier coatings (TBCs) or solid oxide fuel cells (SOFCs). Recent development in the suspension/solution plasma spraying allowed deposition of novel types of YSZ coatings with very fine microstructures. For example, porous and columnar microstructure may result in combination of low thermal conductivity and improved strain tolerance which promises a major benefit for preparation of the next generation of TBCs. However, economical spraying of liquid feedstocks demands a high plasma enthalpy in order to achieve considerable feed rate and deposition efficiency [1]–[11].

Hybrid water stabilized plasma (WSP-H) torch provides such a high-enthalpy plasma suited for spraying with feed rates of more than 100 ml of suspension per minute while maintaining high deposition efficiency [12]–[14]. Aim of this study was to compare YSZ coatings deposited from suspensions provided by three different manufacturers and to see what differences in the coating microstructures and properties may be expected even though the nominal formulation (including chemical composition, solid load content and particle granulometry) was identical or very similar.

Distance of the suspension injection point from the plasma torch nozzle ("feeding distance") was also varied in order to observe sensitivity of the deposition process to the variation of this parameter, which is in the case of WSP-H commonly individually optimized for various materials in order to tune the deposition conditions.

2 Materials and Methods

2.1 Coating deposition

Three ready-to-spray YSZ suspensions produced by Treibacher Industrie AG (Austria), Oerlikon Metco (USA), and Innovnano (Portugal) were used for spraying. All three suspensions were ethanol-based with solid load concentration of 25 wt.%. Further details on suspensions are listed in **Table 1**.

Particle size distribution of Treibacher and Innovnano suspensions as provided by manufacturers are listed in **Table 2**. Particle size distribution of Oerlikon suspension was measured by Mastersizer 3000 (Malvern, UK). Viscosity of the suspensions was measured by DV2TLV viscometer (Brookfield, USA) with coaxial cylinder-cylinder geometry at 100 RPM speed of the spindle.

For deposition, high-enthalpy hybrid water-argon plasma torch WSP-H 500 (ProjectSoft HK, Czech Republic) was used. Torch was operated at 500 A (~150 kW) and argon flow 15 slpm. Stand-off (spraying) distance was set to 100 mm. Torch was mounted to a programmable robotic arm. Suspension was injected into the plasma jet by pneumatic feeding system through calibrated Ø0.35 mm nozzle. The feeding distance (FD), i.e. distance of injection point from the torch nozzle, was set to 30 or 20 mm, and liquid injection angle was 65 degrees downstream, **Fig. 1**. Injection pressure was set by the help of shadowgraphy system SprayCam (Control Vision Inc, USA).

Grit-blasted AISI 304 substrates (20x30x2.5 mm) were mounted to the rotating carousel with cooling by compressed air. Sample temperature was monitored by thermocouple attached to the back-side of the substrates.

Prior to the deposition, substrates were preheated by brief exposure to the plasma jet. Deposition consisted of 15 individual deposition cycles, each consisting of 3 up-and-down torch strokes in front of the samples and followed by cooling period. Relative speed between the plasma torch and the substrate surface was 650 mm/sec.

Table 1: YSZ suspensions used for spraying.

#	Manufacturer	Designation	Y_2O_3 content
A	Treibacher Ind.	YSZ in Ethanol	8.3 wt.%
B	Oerlikon Metco	METCO 6608	7.0 wt.%
C	Innovnano	4YSZSE	7.7 wt.%

Table 2: Suspensions characteristics.

#	*Particle size [µm]*			*Viscosity*
	Dv 10	*Dv 50*	*Dv 90*	*[mPa.s]*
A	0.23*	0.58*	1.38*	1.6
B	0.09#	0.43#	0.97#	1.2
C	0.23*	0.43*	0.99*	1.5

** provided by manufacturers.*
 # measured by Malvern Mastersizer 3000.

Fig. 1. WSP-H plasma jet without and with activated suspension injection. Experiment A20. In-scale overlay shows position of shadowgraphy area from Fig. 2.

2.2 Coating characterization

Free-surfaces and cross-sections prepared by standard metallographic procedure were observed by SEM microscope EVO MA 15 (Carl Zeiss, Germany). Thickness of the deposits was evaluated by means of image analysis using AxioVision (Carl Zeiss Microscopy, Germany). Porosity was also evaluated from SEM micrographs taken at 300x, 500x, 1000x, and 3000x nominal magnifications using ImageJ 1.49v (NIH, USA) software.
Phase composition of the deposits was evaluated by XRD analysis using D8 Discover (Bruker AXS, Germany) diffractometer.
Microhardness of the coatings was evaluated from 12 Vickers indents imposed on polished cross-sections by 300 g load using Nexus 4504 (Innovatest, Netherlands) hardness tester.
Tensile adhesion/cohesion strength of the coatings was evaluated according to ASTM C633-13 standard. The coated coupons were glued between two grit-blasted cylinders (Ø18 mm) by FM 1000 adhesive.
Wear resistance of the deposits was evaluated by Slurry Abrasion Response (SAR) test according to ASTM G75 standard using 50:50 wt.% mixture of alumina grit (mesh 240) and water.

3 Results and Discussion

Atomization of suspensions in the plasma jet as observed by shadowgraphy is illustrated in **Fig. 2**. Apparent behavior of all suspensions was comparable regardless to the feeding distance. All suspensions showed intense fragmentation accompanied with evaporation of solvent, leading to the reduction of the droplet size. Shadowgraphy was also used for the optimization of the injection pressure, which was increased until the mean droplet trajectory was identical with the plasma jet axis. Injection feed rate was in the range of about 100 to 120 ml/min depending on the injection pressure and suspension viscosity, **Table 3**.

Table 3: Deposition settings.

Coating	Suspension	FD [mm]	FP [bar]	FR [ml/min]
A30	A	30	3	98
A20		20	4	112
B30	B	30	3	112
B20		20	4	120
C30	C	30	3	102
C20		20	3	101

FD = Feeding Distance
FP = Feeding Pressure
FR = Feed Rate

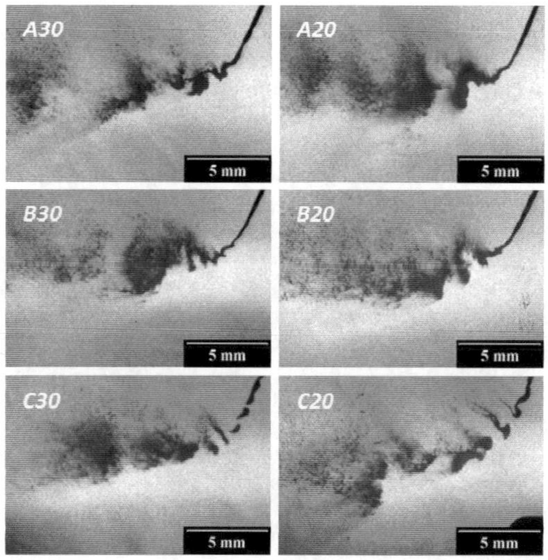

Fig. 2. Atomization of suspensions in the plasma jet for 30 mm (left) and 20 mm (right) feeding distance.

From the sample temperature profile in **Fig. 3**, it is apparent that after the preheating to approximately 350 °C, steady deposition conditions were achieved. Despite intensive air-cooling, sample temperature was increased by several hundred degrees during each deposition cycle, which had to be therefore followed by the cooling period in order to protect the low-mass substrates from overheating. The rapid increase of the temperature during the deposition phases was caused by the combined effect of using high-enthalpy plasma torch at relatively low stand-off distance and injection of combustible ethanol solvent providing additional heat to the spraying process.

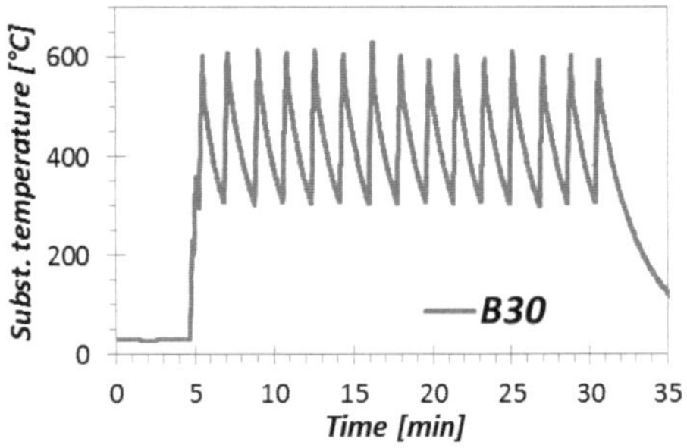

Fig. 3. Record of the substrate temperature.

Fig. 4. Coating B20 deposited on the substrates.

Fig. 4 shows that the coatings uniformly covering the substrates were deposited. All deposits showed cauliflower-like microstructure on the free-surface, **Fig. 5**.

Columnar microstructure with intercolumnar gaps also characteristic for suspension sprayed coatings was observed on the coatings cross-sections, **Fig. 6**. Columns typically originated from the surface asperities of the grit-blasted substrates and grew in the direction of local material deposition confirming the coating build-up model as suggested in [15]. Between the columnar gaps, the microstructure consisted of regions of well-sintered flattened microsplats intermingled with regions of fine porosity, **Fig. 7**.

The coatings deposited from Treibacher and Oerlikon suspensions reached comparable thickness, **Table 4**. Coating A20 showed tendency to form moderate interpass porosity. Slightly higher value of deposition efficiency (in terms of thickness increase per deposition cycle) of B30 and B20 coatings may be explained by their higher porosity leading to their lower density and therefore accelerated build-up. However, in the case of spraying with Innovnano suspension, high coating porosity was accompanied with lower deposition efficiency. The decrease in the deposition efficiency could be a result of disturbed laminarity of the suspension injection (**Fig. 2**) occurring probably due to the wear of the injection nozzle. Such disturbances might have led to the premature segmentation of the laminar suspension stream which did not penetrate so effectively into the plasma jet.

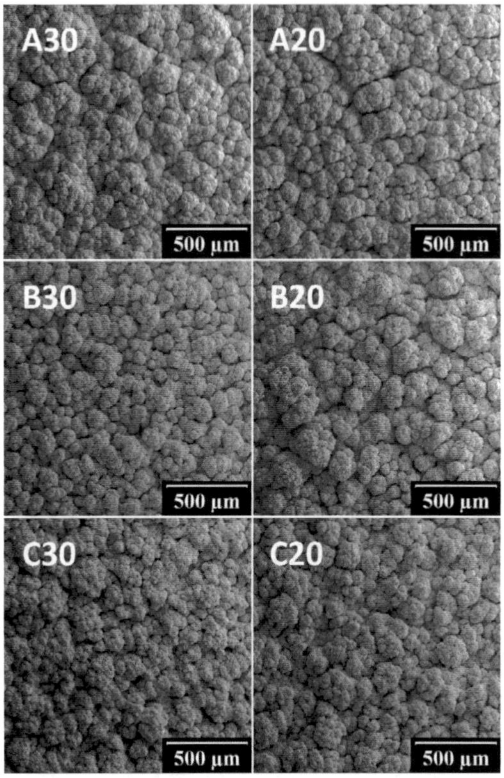

Fig. 5. Free-surfaces of the deposits. SEM.

Table 4: Coatings microstructural characteristics.

Coating	Thickness [μm]	IA porosity [%]	DE [μm/cycle]
A30	168 ± 12	25.0 ± 0.6	11.2 ± 0.8
A20	182 ± 24	24.1 ± 2.2	12.2 ± 1.6
B30	195 ± 38	37.7 ± 1.9	13.0 ± 2.5
B20	207 ± 29	29.2 ± 2.0	13.8 ± 1.9
C30	138 ± 22	31.0 ± 4.8	9.2 ± 1.5
C20	114 ± 34	38.4 ± 2.3	7.6 ± 2.3

IA = Image Analysis
DE = Deposition Efficiency

Table 5: Coatings mechanical properties.

Coating	Hardness [-]	SAR [m/mm^3]	TAT [MPa]
A30	419 ± 66	17.9 ± 2.8	15.4 ± 1.3
A20	403 ± 76	16.5 ± 1.8	11.4 ± 3.5
B30	258 ± 47	13.6 ± 1.8	8.9 ± 3.7
B20	239 ± 36	22.0 ± 3.6	9.7 ± 1.8
C30	171 ± 39	6.8 ± 0.3	12.6 ± 5.2
C20	163 ± 33	4.7 ± 0.1	14.0 ± 3.7

SAR = Slurry Abrasion Response
TAT = Tensile Adhesion/Cohesion Strength

The most compact microstructure of A30 and A20 coatings was reflected in their high hardness and decent abrasion resistance, **Table 5**. A more porous B30 and B20 coatings showed significantly reduced hardness probably due to the collapse of internal voids under the indenter [16]. However, mutual bonding of the splats within these coatings was not seriously compromised judging from a relatively high slurry abrasion resistance comparable to the Treibacher coatings. Contrary, the loose microstructure of both Innovnano coatings led to both low hardness and low slurry abrasion resistance.

Even though evaluation of tensile adhesion/cohesion strength of the coatings with so low thickness and high open porosity is somehow questionable [17], adhesion/cohesion test was carried out. During the visual inspection of fracture surfaces, penetration of the adhesive into the coating was not observed. It may be therefore concluded that a relatively high cohesion strength (in the range of 9 to 15 MPa) is representative for the YSZ coatings deposited by WSP-H torch. All tested coatings failed due to the loss of cohesion meaning that their adhesion to the substrate is even higher.

XRD patterns of the deposits showed that all coatings consisted of tetragonal YSZ phase only, **Fig. 8**. The ratio of cell parameters $c/a\sqrt{2}$ was for all coatings smaller than 1.010 indicating that the coating consisted of non-transformable t' phase [18].

Due to the low ratio of the coating thickness to the substrate thickness, values of thermal conductivity of the coatings were highly sensitive to the sample geometry. Reliable method for evaluation of representative thermal conductivity of deposited coatings is therefore still to be developed, however, preliminary results achieved with laser flash apparatus LFA 1000 (Linseis, Germany) already show that the conductivity values are significantly smaller than 1.0 W/mK and comparable to the values characteristic for low-thermal conductivity TBC coatings deposited by other plasma spraying techniques [8], [19].

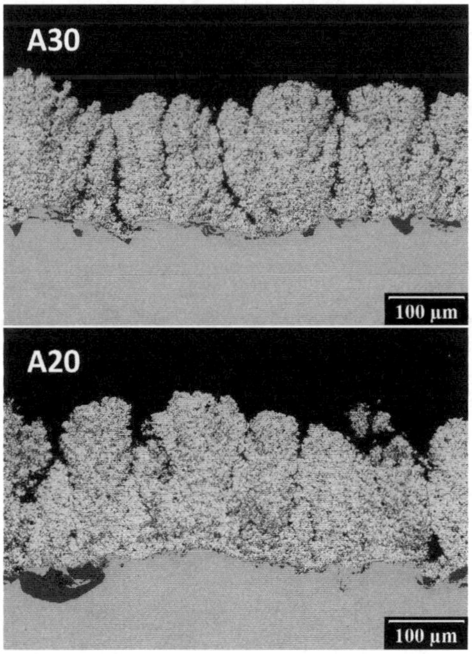

a)

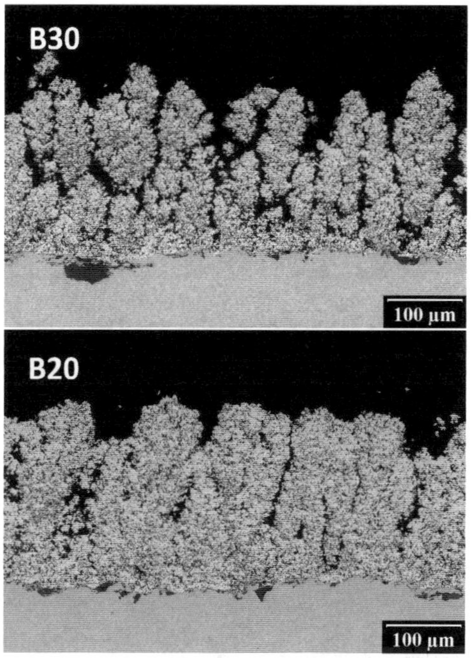

b)

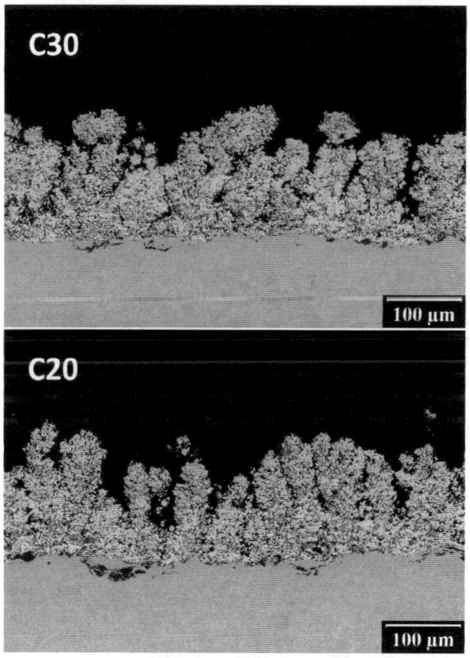

c)

Fig. 6. Cross-sections of the deposits. SEM.

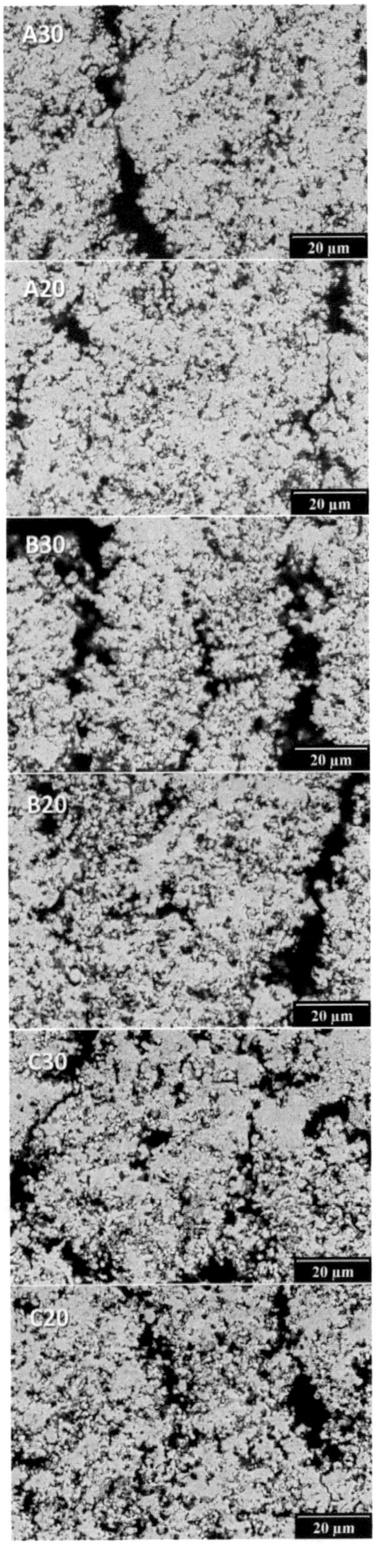

Fig. 7. Details of cross-sections. SEM.

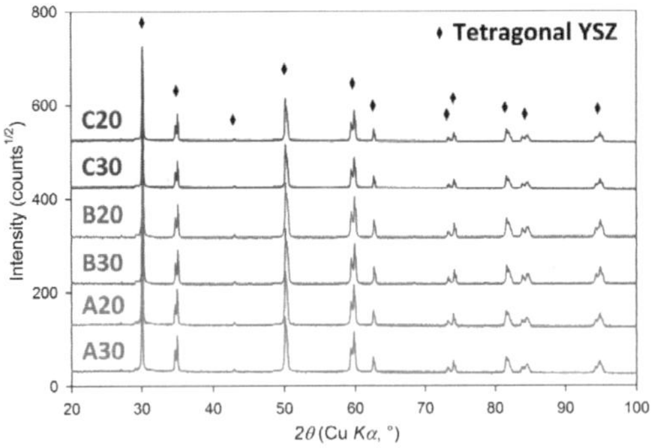

Fig. 8. XRD patterns of the deposits.

4 Conclusions

In this study, YSZ deposition was attempted from three different suspensions provided by three different manufacturers. Results show that the deposition of tetragonal YSZ with columnar microstructure may be achieved with all three suspensions. Slight variations in the coatings morphology are promising for further optimization of the spraying parameters in order to optimize the coating performance for various applications.

Moreover, due to the high achievable feed rate combined with low consumption of plasma-forming medium, WSP-H deposition of YSZ from suspensions opens a new window of thermal spray technology for cost-effective deposition of TBC coatings in applications where protection of large-areas and/or high coating thickness are desirable.

6 Acknowledgement

Financial support provided by Czech Science Foundation through grant No. GA15-12145S (Physical aspects of plasma spray deposition from liquid feedstock) is gratefully acknowledged.

7 Literature

[1] Fauchais, P., Vardelle, M., Goutier, S. and Vardelle, A.: Key Challenges and Opportunities in Suspension and Solution Plasma Spraying. Plasma Chemistry and Plasma Processing 35 (2014), Issue 3, pp. 511/525.

[2] Vassen, R., Kaner, H., Mauer, G. and Stover, D.: Suspension plasma spraying: Process characteristics and applications. Journal of Thermal Spray Technology 19 (2010), Issue 1-2, pp. 219/225.

[3] Pawlowski, L.: Suspension and solution thermal spray coatings. Surface and Coatings Technology 203 (2009), Issue 19, pp. 2807/2829.

[4] Fauchais, P., Vardelle, M., Vardelle, A. and Goutier, S.: What Do We Know, What are the Current Limitations of Suspension Plasma

Spraying? Journal of Thermal Spray Technology 24 (2015), Issue 7, pp. 1120/1129.
[5] Meillot, E., Vert, R., Caruyer, C., Damiani, D. and Vardelle, M.: Manufacturing nanostructured YSZ coatings by suspension plasma spraying (SPS): Effect of injection parameters. Journal of Physics D: Applied Physics 44 (2011), Issue 19, 194008.
[6] Killinger, A., Gadow, R., Mauer, G., Guignard, A., Vassen, R. and Stöver, D.: Review of new developments in suspension and solution precursor thermal spray processes. Journal of Thermal Spray Technology 20 (2011), Issue 4, pp. 677/695.
[7] Fauchais, P., Montavon, G., Lima, R. S. and Marple, B. R.: Engineering a new class of thermal spray nano-based microstructures from agglomerated nanostructured particles, suspensions and solutions: An invited review. Journal of Physics D: Applied Physics 44 (2011), Issue 9, 093001.
[8] Curry, N., VanEvery, K., Snyder, T., Susnjar, J. and Bjorklund, S.: Performance Testing of Suspension Plasma Sprayed Thermal Barrier Coatings Produced with Varied Suspension Parameters. Coatings 5 (2015), Issue 3, pp. 338/356.
[9] Curry, N., Markocsan, N., Li, X.-H., Tricoire, A. and Dorfman, M.: Next generation thermal barrier coatings for the gas turbine industry. Journal of Thermal Spray Technology 20 (2011), Issue 1-2, pp. 108/115.
[10] Ganvir, A., Curry, N., Bjorklund, S., Markocsan, N. and Nylen, P.: Characterization of Microstructure and Thermal Properties of YSZ Coatings Obtained by Axial Suspension Plasma Spraying (ASPS). Journal of Thermal Spray Technology 24 (2015), Issue 7, pp. 1195/1204.
[11] Sokolowski, P., Kozerski, S., Pawlowski, L. and Ambroziak, A.: The key process parameters influencing formation of columnar microstructure in suspension plasma sprayed zirconia coatings. Surface and Coatings Technology 260 (2014), pp. 97/106.
[12] Musalek, R., Medricky, J., Tesar, T., Kotlan, J., Pala, Z., Lukac, F., Chraska, T. and Curry, N.: Suspensions Plasma Spraying of Ceramics with Hybrid Water-Stabilized Plasma Technology. Journal of Thermal Spray Technology 26 (2016), Issue 1-2, pp. 37/46.
[13] Kotlan, J., Pala, Z., Musalek, R. and Ctibor, P.: On reactive suspension plasma spraying of calcium titanate. Ceramics International 42 (2016), Issue 3, pp. 4607/4615.
[14] Musalek, R., Bertolissi, G., Medricky, J., Kotlan, J., Pala, Z. and Curry, N.: Feasibility of suspension spraying of yttria-stabilized zirconia with water-stabilized plasma torch. Surface and Coatings Technology 268 (2015), pp. 58/62.
[15] VanEvery, K., Krane, M. J. M., Trice, R. W., Wang, H., Porter, W., Besser, M., Sordelet, D., Ilavsky, J. and Almer, J.: Column formation in suspension plasma-sprayed coatings and resultant thermal properties. Journal of Thermal Spray Technology 20 (2011), Issue 4, pp. 817/828.
[16] Nohava, J., Musalek, R., Matejicek, J. and Vilemova, M.: A contribution to understanding the results of instrumented indentation on thermal spray coatings - Case study on Al2O3 and stainless steel. Surface and Coatings Technology 240 (2014), pp. 243/249.
[17] Musalek, R., Pejchal, V., Vilemová, M. and Matejicek, J.: Multiple-approach evaluation of WSP coatings adhesion/cohesion strength. In Journal of Thermal Spray Technology 22 (2013), Issue 2-3, pp. 221/232.
[18] Viazzi, C., Bonino, J.-P., Ansart, F. and Barnabe, A.: Structural study of metastable tetragonal YSZ powders produced via a sol-gel route. Journal of Alloys and Compounds 452 (2008), Issue 2, pp. 377/383.
[19] Latka, L., Cattini, A., Pawlowski, L., Valette, S., Pateyron, B., Lecompte, J.-P., Kumar, R. and Denoirjean, A.: Thermal diffusivity and conductivity of yttria stabilized zirconia coatings obtained by suspension plasma spraying. Surface and Coatings Technology 208 (2012), pp. 87/91.

Sporicidal efficacy of thermal spray copper alloy coating with varying degrees of roughness

R. Shafaghi, L. Pershin, M. Riguette, J. Mostaghimi, Düsseldorf / D

Abstract
This study was focused on the biocidal efficacy on spores of copper alloy sheet and copper alloy coating at two different surface topographies. Endospores can remain viable in a dormant state for centuries. Our work compares the effectiveness in killing endospores of copper alloy coating and sheet metal.
A twin-wire arc spray system was used for coating of stainless steel coupons. The feedstock was CuNiZn wire, the coating thickness was 400 µm. The copper alloy sheet metal had the same composition and is registered as antimicrobial by Environmental Protection Agency (US). Uncoated stainless steel coupons were used as controls in all experiments.
The surface was polished to two roughness levels Ra=3.5 µm, and Ra=0.1 µm.The surface topography was analysed by a stylus profilolemeter and 3D image analysis. EDS and FIB were used to characterize the elemental composition and structure flower-like nanostructures and endospores.
The results obtained in this study indicated that changes in Ra values of 0.1 and 3.5 µm had no significant impact on the biocidal activity of sheet metal and the coating on *E. coli, S. epidermidis* and *B. subtilils*. The coating was as effective as the EPA certified sheet metal in the destruction of vegetative cells within 5 minutes. This study indicates that degradation of B. subtilis endospore begins within 2 hours after exposure to the coating. By day seven, only extensively degraded endospores and nanostructures were visible on the both surfaces. Our results show that thermal spray copper alloy coatings were as effective as certified antimicrobial sheet metal in the destruction of endospores within hours however; the coating was more effective in killing after one week of exposure.

1 Introduction

Despite stringent cleaning protocols in healthcare facilities, the spread of microbial pathogens remains a major issue in part because microbes have evolved diverse survival strategies in response to environmental stress, toxins, and antibiotics. Another strategy for overcoming environmental stress is the formation of spores. Spores consist of a core with a copy of bacteria's DNA, the core is surrounded by a lipid membrane and a thick proteinaceous coat (peptidoglycan).In addition to being highly stable and resistant to heat and bactericidal drugs, spores can remain viable in a dormant, metabolically inert state for centuries.
Many surfaces in hospitals are made of stainless steel due to their durability, aesthetic appeal, and chemical inertness. An alternative approach is to take advantage of the biocidal activity of metals such as silver, zinc, and copper. Copper alloys are the most promising because of their availability, better corrosion resistance than zinc, and cost-effectiveness in comparison to silver. Several studies have demonstrated that spores are inactivated when exposed to copper alloys with greater than 60% copper content (Weaver et al. 2008b). However, thin copper coatings would be more cost effective and technologically flexible in comparison with the sheet metal.
During twin wire arc spraying the feedstock wire is subject of a number of transformations such as melting, atomization, oxidation and rapid solidification. As a result it forms different microstructures compared to rolled sheet metal. The microstructure of the substrate, its porosity, grain size surface morphology and structure defects could affect its biocidal properties (Ivanova et al., 2009). In this study, we compared the biocidal effectiveness of sprayed copper alloy coating and EPA certified sheet metal of similar composition and surface roughness.

2 Materials and methods

Twin-wire arc spray system HD 456 (Thermion, Silverdale, WA, US) was used for deposition of the copper alloy onto 25 mm stainless steel discs. The feedstock Cu-Ni-Zn alloy wire was sprayed to a coating thickness of 400 µm onto grit blasted coupons using operational parameters presented in Table 1.

Spray parameters Table 1

Arc current, A	200
Voltage, V	29
Wire feed rate, m/min	4.8
Spray distance, mm	150

The sheet metal coupons were cut out of C75200 copper alloy registered as antimicrobial by Environmental Protection Agency (EPA, US). Uncoated stainless steel coupons were used as controls. All coupons were polished to achieve a comparable surface roughness. The surface roughness was measured using a stylus profiler (Alpha-Step D-120, KLA-Tencor, USA).The surface composition and topography analysis was performed by Hitachi TM3000 SEM equipped with an EDS (Quantax 70 Bruker, D). The surface topographies were analysed with a 3D-image Viewer of TM3000. To compare surface energy of coating and sheet metal the contact angle measurements were performed using distilled water. Equilibrium angle was measured on left and right side of the drop using ImageJ software from National Institute of Health (USA).

The coupons were washed and then rinsed in sterile de-ionized water, sterilized in 70% ethanol, air dried and placed in sterile petri dishes.

B. subtilis spores were generated according to (Setlow 1996). The bacteria were incubated on agar plates at 37°C then the colonies were suspended in 0.05 % NaCl and heated to 100 °C to kill cells. The spores were separated by centrifugations and re-suspended in distilled water. The suspension of spores (25 µl) was placed on the coupons and incubated for 2 hours, 1 day, 3 days, or 7 days at room temperature. Spore survival was assessed according to (Silva et al. 2013). The spores from coupons were rinsed with water and the suspension was heated to 70 °C for 30 min, and cooled on ice for 15 min. After that the spores were activated in germination buffer and spread on agar plates and incubated at 37°C then viable colonies were counted.

For SEM analyses spores were prepared by fixation in 3% glutaraldehyde and phosphate buffered saline (PBS). Coupons than were sputter-coated with osmium tetroxide (OsO_4). After drying the coupons were sputter coated with gold-palladium and observed under SEM. Additionally, samples were sputtered with tungsten in preparation for Focused Ion Beam (FIB) sectioning (FIB-SEM nanoDUE'T NB5000, Hitachi).

3 Results

3.1 Surface evaluation

It was critical for these studies that the resultant surface topography as well as the composition of the thermal sprayed copper alloy was similar to that of the sheet metal. The compositions of initial wire, the deposited coating and the sheet metal are presented in Table 2.

Table 2

Element	Wire [wt.%]	Coating [wt.%]	Metal C75200 [wt.%]
Cu	68.28	69.13	64.92
Ni	18.21	20.48	17.52
Zn	13.51	8.05	15.60
O_2		2.34	1.96

As the EDX results show copper concentration is comparable in the coating and the sheet metal. However oxidation during spraying resulted in increased oxygen from 0 to 2.34% and reduced in Zn content from 13.51 to 8.05 % in the coating in comparison with the initial wire composition.

As a result of the coating formation from molten metal drops and their consequent high speed solidification introduces micro scale inhomogeneity in elements distribution on the coating surface. As it presented in Fig.1 a splat boundaries have higher concentration of copper.

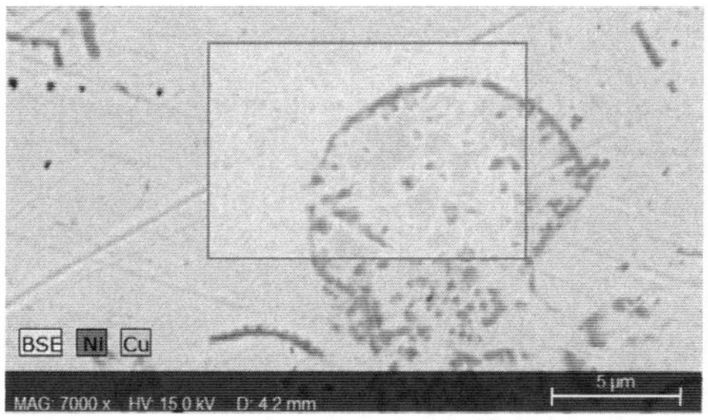

Figure 1 Variations in copper distribution on the coating surface

The untreated deposited coating had an average roughness value (Ra) about 20-30 µm. A positive feeling is associated with a surface roughness Ra between 2-4 µm (Barnes et al., 2004). Based on this the coupons were polished to produce two Ra values of 3.5 µm and 0.1 µm to compare the impact of a 35-fold difference in surface topography on biocidal activity.

The surface hydrophobicity-hydrophilicity is affected by its topography, which also has a major impact on bacterial adhesion and motility. We compared the wetting properties of the polished to Ra 0.1 µm sheet metal and the coating by measuring contact angle (Θ) of a drop of distilled water. Contact angles, 94.4° and 91.1°were observed for the sheet metal and the coating respectively (Fig. 2), indicating that the surfaces had comparable adhesive properties. Hence, differences in biocidal activity would not be expected to be due to differences in their surface energy.

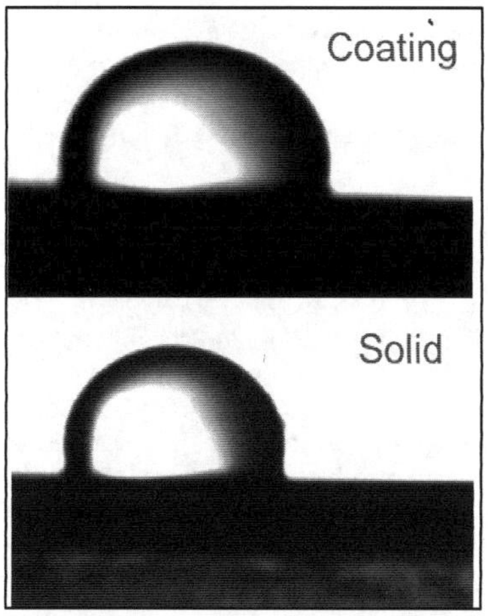

Figure 2 Water drop in equilibrium on polished coating and metal surfaces

Greater than 95% of the vegetative gram-positive, rod-shaped *E. coli*, *S.epidermidis* and *B. Subtilis* were killed at room temperature within 5 minutes by exposure to the sheet metal and the coating with surface Ra values of 3.5 µm and 0.1 µm. Control stainless steel coupons had no significant impact on cell viability. For example Figure 3 presents data on *B. Subtilis* survival on the coatings and sheet metal with two surface roughnesses.

In light that differences in surface roughness within the studied range had no effect on overall biocidal activity of bacterial cells, all subsequent experiments were conducted using coupons with a surface Ra value of 3.5 µm.

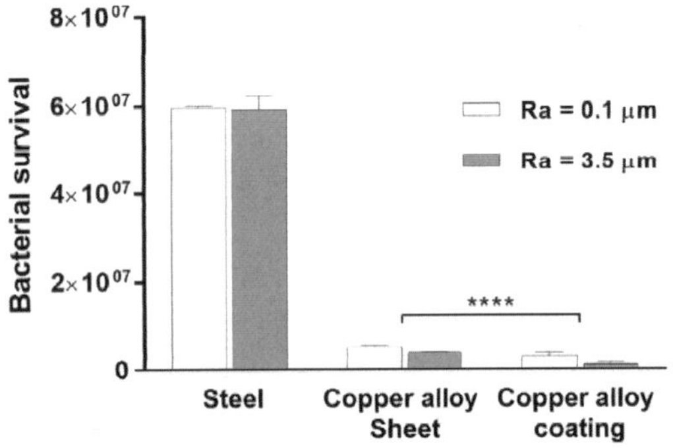

Figure 3 Biocidal activities of copper alloy sheet and thermal spray coating for Ra values of 0.1 and 3.5 µm. for *Bacillus subtilis.*

3.2 Sporocidal activity of sheet metal and coating

Bacteria contact killing occurs when copper ions are released from surfaces, leading to rapid bacterial cell degradation (Mathews et al). On the contrary spores are metabolically inactive, non-reproductive derivatives of bacterial cells that are resistant to environmental factors such as desiccation, heat, and hospital disinfectants.

Thus we sought to determine the effects of copper alloy exposure also on spore viability. Spore formation was induced by nutrient depletion of *B. subtilis* cells. The spore formation was confirmed using Schaeffer-Fulton staining, revealing that approximately 85-90 % of the colonies were composed of spores (Humphreys 2011).

Spore preparations were incubated at 100°C for 10 minutes to kill residual bacteria before seeding them onto the coupons. All coupons were washed with saline after two hours to retrieve spores from the surface. Germination buffer was added to all washes to evaluate to total surviving spores. Large number of colonies was observed with washes from the stainless steel coupons, indicating that the spores maintained a high level of viability on this surface. In contrast, washes from the coupons of copper alloy coating or sheet metal revealed an 85-90% decrease in the number of colonies relative to stainless steel (Fig. 4)

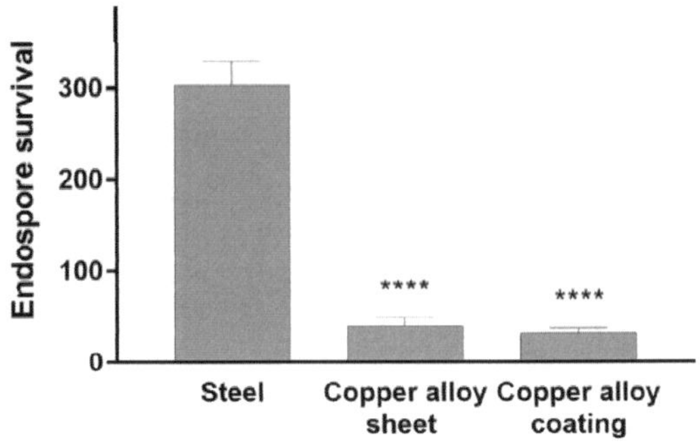

Figure 4 Comparisons of *B. subtilis* spores killing effectiveness of stainless steel, copper alloy sheet and **coating (Ra 3.5µm)**

Exposure of spores to stainless steel control coupons for 2 hours, 1, 3 or 7 days had no visible impact on their morphology as it shown on SEM images (Fig. 5). Hoewer, on the copper alloy surfaces (sheet metal and coating) some deformed spores appeared as early as 2 hours after exposure (Fig. 5 b-1). By day one of exposure, small nanoflower structures were visible (Fig 5 b2, c2).

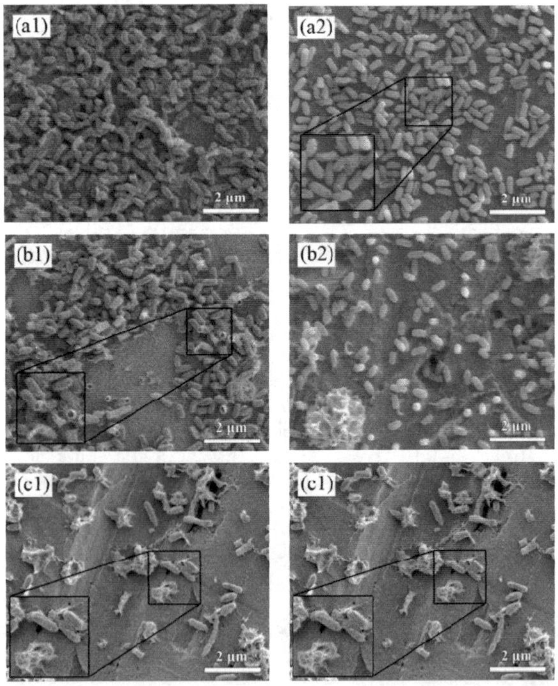

Figure 5 *B. subtilis* spore deposited on stainless steel (a1-2), copper alloy metal sheet (b1-2) and thermal spray copper coating (c1-2) and incubate for 2 hours (a1-c1), 1day (a2-c2).

Studies indicated that the negative charge of the spore coat is a barrier to germination. It is therefore plausible that copper ion react with the negatively charged coat molecules. interaction of bound copper ions with dissolved oxygen can generate hydrogen peroxide and reactive oxygen species that could trigger the destruction of the protective endospore coat and internal structures (Cross et al., 2003).

3.3 Nanostructre formation

Petal-like nanostructures were generated by an initial binding of copper ions to a protein core, triggering the formation of copper phosphate crystals. It is therefore probable that proteins derived during spore degradation serve as nucleation material (Somturk et al. 2015).

A subset of the nanoflowers had randomly oriented petal-like nanostructures different morphologies. Spores appeared more severely deformed on the coating surface than those exposed to the copper sheet. The nanoflowers increased in size and numbers with exposure time and no intact spores were present by day 7 (Fig.6).

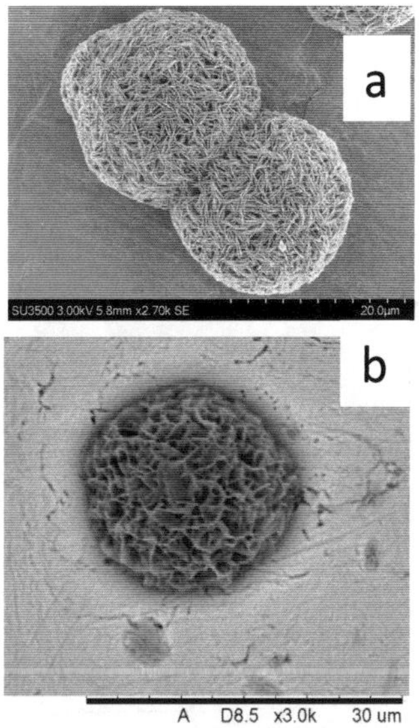

Figure 6 *B. subtilis* spores degradation and transformation into nanoflowers on sheet metal (a) and coating (b)

The nanoflowers composition presented in Figure 7. EDX analysis revealed that on stainless steel spores composition was 14% carbon, 2.9% oxygen, 3.6% phosphorous and only trace levels of copper (0.2%). Composition of nanoflowers on copper alloy sheet showed 48.2% copper, 26.2% oxygen with a 3.8% in phosphorous compared. The nanoflowers on the coating presented a similar elemental composition 40.2% copper, 20.6% oxygen, and 8.7% phosphorous. Collectively, the data are consistent with other studies, indicating nanoflowers are hybrid organic-inorganic nanostructures.

The generation of nanoflowers happens in three steps, in the first step the initial binding of the amine groups of the proteins to copper ions trigger the formation of copper phosphate crystals (Somturk et al. 2015). Repeated nucleation-crystal formation generates petal-like structures. Despite the distinct surface topographies, for the nanoflower formed on the sheet metal and coating (Fig.7) focused ion beam sectioning of mature nanoflowers reveals similar inner structure (Fig. 8) for all of them. The cores of all the mature nanoflowers appeared composed a dense network of irregular petal-like structures of variable sizes, orientation and separation.

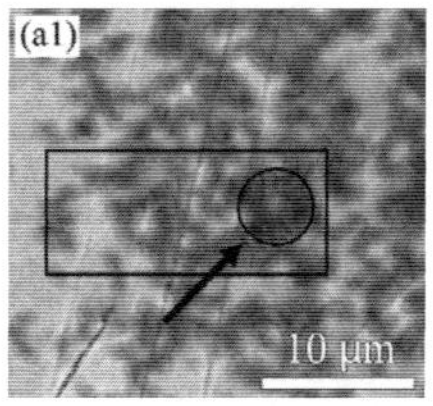

Mass percent	Stainless steel
Carbon (%)	14 ± 0.4
Copper (%)	0.2 ± 0.1
Phosphorus (%)	4 ± 0.3
Oxygen (%)	3 ± 0.3

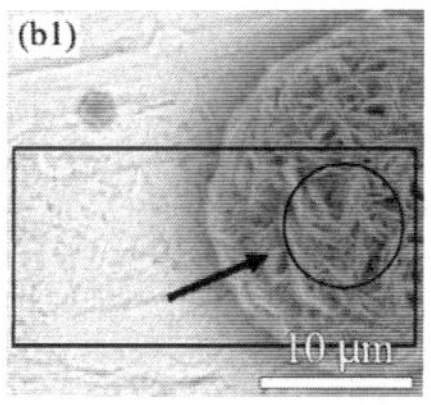

Mass percent	Sheet metal
Carbon (%)	7 ± 0.3
Copper (%)	48 ± 0.4
Phosphorus (%)	14 ± 0.2
Oxygen (%)	26 ± 0.2

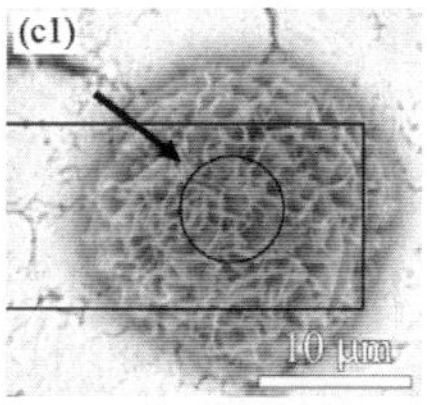

Mass percent	Thermal Spray
Carbon (%)	17 ± 0.3
Copper (%)	40 ± 0.7
Phosphorus (%)	9 ± 0.2
Oxygen (%)	21 ± 0.5

Figure 7 Compositional EDX analysis of endospores and nanoflowers on stainless steel (a1), copper alloy sheet metal (b1) and the coating (c1).

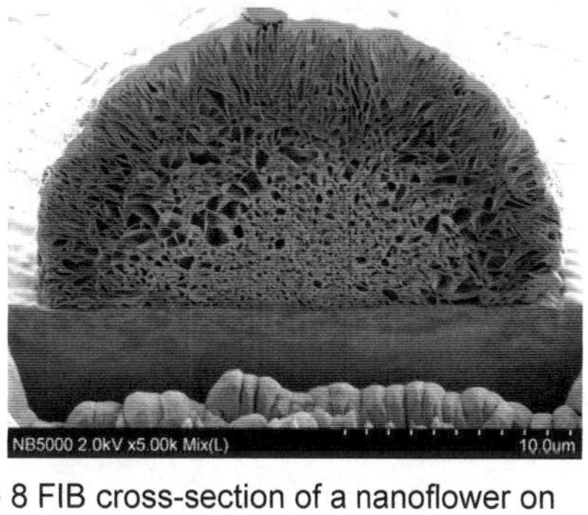

Figure 8 FIB cross-section of a nanoflower on the coating

Conclusions

Our study indicates that thermal spray copper alloy coatings are as effective as copper alloy sheet metal in the destruction of bacteria and spores within hours, thus representing a viable strategy for decreasing the health hazards in hospital settings.

B. subtilis spores degradation was observed after 2 hours of exposure to the copper alloy coating and after 7 days exposure all spores were destroed.

The surface roughness in a range of R_a 0.1 to 3.5 µm doesn't affects the biosidal activities of the Cu-Ni-Zn alloy.

Destraction of spores initiates formation of crystal nanostructures called nanoflowers. It was observed that on the coating and the sheet metal the surface morphology of the nanoflowers are different. However they have similar inner structure.

References

Barnes, C., Childs, T., Henson, B. & Southee, C. 2004. Surface finish and touch—a case study in a new human factors tribology. Wear, 257, 740-750.

Cross, J., Currier, R., Torraco, D., Vanderberg, L., Wagner, G. & Gladen, P. 2003. Killing of Bacillus spores by aqueous dissolved oxygen, ascorbic acid, and copper ions. Applied and environmental microbiology, 69, 2245-2252

Humphreys, P.N. 2011. Testing standards for sporicides. J. Hosp. Infect. 77: 193-198

Ivanova, E. P., Truong, V. K., Wang, J. Y., Berndt, C. C., Jones, R. T., Yusuf, I. I., Peake, I., Schmidt, H. W., Fluke, C. & Barnes, D. 2009. Impact of nanoscale roughness of titanium thin film surfaces on bacterial retention. Langmuir, 26, 1973-1982.

Mathews, S., et al., Contact killing of bacteria on copper is suppressed if bacterial-metal contact is prevented and is induced on iron by copper ions.
Applied and environmental microbiology, 2013. 79(8): p. 2605-2611.

Somturk, B., Hancer, M., Ocsoy, I., Ozdemir, N. 2015. Synthesis of copper ion incorporated horseradish peroxidase-based hybrid nanoflowers for enhanced catalytic activity and stability. Dalton Trans. 44: 13845-13852 doi: 10.1039/c5dt01250c.

Setlow, P. 2006. Spores of Bacillus subtilis: their resistance to and killing by radiation, heat and chemicals. Journal of applied microbiology, 101, 514-525.

Transpiring thermally sprayed alumina layers with integrated fluid flow tubes

M. Rodriguez Diaz[1], K. Möhwald[1], N. Loftfield[2], M. Kästner[2], E. Reithmeier[2], S. Knigge[3], B. Glasmacher[3], H.J. Maier[1]

[1] Institut für Werkstoffkunde, Leibniz Universität Hannover, 30823 Garbsen, Hannover
[2] Institut für Mess- und Regelungstechnik, Leibniz Universität Hannover, 30823 Garbsen, Hannover
[3] Institut für Mehrphasenprozesse, Leibniz Universität Hannover, 30823 Garbsen, Hannover

Osteoarthritis in the hip or knee is one of the most common diseases in industrialized countries. The implantation of an endoprosthesis as a joint replacement represents the most effective way to treat serious pathological changes in these joints. The lifetime of an endoprosthesis can be shortened by aseptic inflammation and osteolysis. The main cause for the aseptic inflammations, osteolysis, and thus, the failure of the endoprostheses are abrasion particles of the acetabular cup inlays that are caused by the tribological load of the prosthesis. This research project aims at developing coatings with enhanced tribological behaviour for endoprostheses by an active hydrodynamic lubrication of the joint with synovial fluid. In addition, biocompatibility, as well as the increase of the strength under static and cyclic loading need be realized. In the current approach, a deterministic fluid flow tube structure is formed in a thermally sprayed alumina layer by the introduction of a leachable placeholder. This tube structure allows the transverse transport of the synovial fluid through the alumina layer. Furthermore the synovial fluid can be transported into the lubricating gap of the sliding surfaces by leaving the alumina layer through the porous surface. First results will be presented and the ramifications in correlation to applications will be discussed.

1 Introduction

The replacement of a hip or knee by an endoprostheses is based on the premise of a long lifetime of the implant, whereby modern implants have a maximum life span of approx. 15 years [1]. This becomes obvious when considering the loading scenarios and the materials used for the components of the implant. While the mechanically strongly loaded femoral stem is often manufactured from the high-strength titanium alloy TiAl6V4, for the femoral head corrosion-resistant steels, cobalt-chromium alloys (CoCr) or particularly wear-resistant ceramics like alumina (Al_2O_3) are suitable. Despite the low coefficient of friction of the polished interfaces, a substantial number of microscopic abrasive particles are produced during service of the prosthesis. These particles lead to a foreign body reaction of the surrounding host tissue [2]. In addition to shortening the lifetime of the prostheses, there is also the risk of aseptic inflammations as well as the breakdown of the bone, which can lead in serious cases to the complete failure of the prosthesis [3, 4, 5]. Both will result in costly and risky revision operations. In order to counteract this, a significant improvement in the tribological behaviour of coatings for endoprostheses is needed. The objective of the present study was to develop an active hydrodynamic lubrication of the joint, while ensuring biocompatibility and excellent mechanical properties. With the present approach, a deterministic fluid flow tube structure is formed in a thermally sprayed alumina coating by the introduction of a leachable placeholder. The introduced tube structure is intended to provide the transverse transport of the body-borne synovial fluid along the coating. Furthermore the fluid is discharged into the lubricating gap via the open-porosity of the coating. In addition to the enhancement of the tribological behaviour, the introduction of anti-inflammatory substances via the tube structure is also conceivable, **Fig 1**.

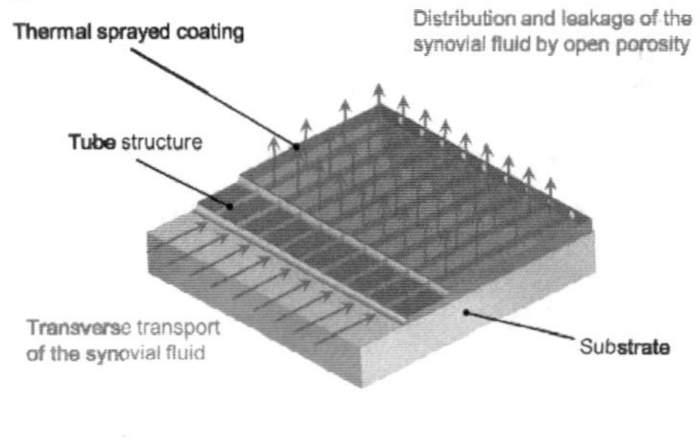

Fig. 1: Schematic diagram of an active lubrication by thermally sprayed transpiring coatings

2 Experimental

The porous implant coating was produced by atmospheric plasma spraying (APS). X5CrNiMo was used as the substrate, and Al_2O_3 was used as the coating material. In a first attempt, nylon threads, wetted with an adhesive, were used to produce the fluid-flowable tube structure, **Fig 2**.

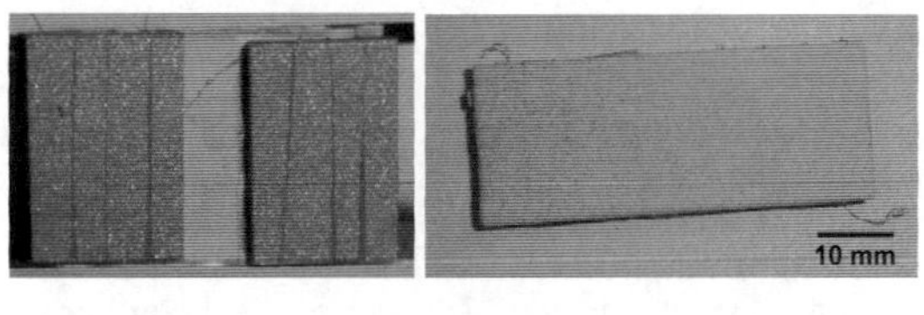

Fig. 2: Nylon threads as placeholders, wetted with adhesive (left) and after the APS process (right)

When using nylon threads, the wetting with an adhesive was indispensable since, on the one hand, a fixation of the threads on the sample was needed, and, on the other hand, the nylon threads themselves were not able to withstand the thermal and kinetic load during the coating process.

After successful coating by the APS process, the nylon threads were removed from the coating. The actual removal of the threads was carried out by a pyrolytic process at 450 °C, with a slow heating of 90 min and a holding and cooling down time of 240 min each. A metallographic cross-section of a representative coating system obtained by this method is shown in **Fig 3**.

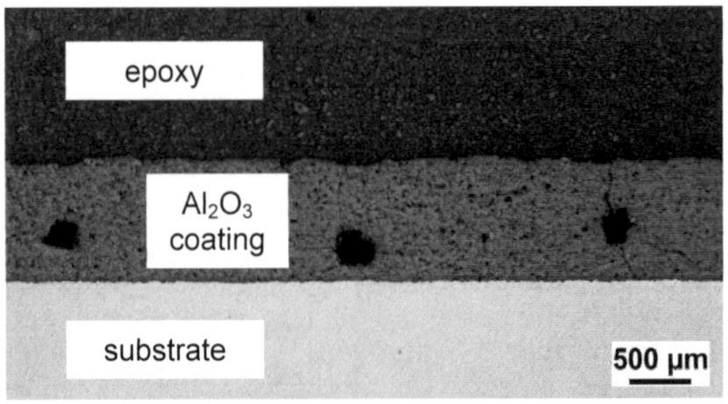

Fig. 3: Cross section of an Al_2O_3 coating with pyrolyzed placeholders

3 Results and discussion

In order to test the fluid permeability of the tube structures introduced into the alumina coating, a sample was immersed in an ink solution. It was observed that most of the tubes were permeable. The ink solution was absorbed solely by the capillary force of the tubes and could reach the surface in the region of the tubes and leaked out there. Within three minutes, the ink was also distributed in the vicinity of the tubes, **Fig 4**.

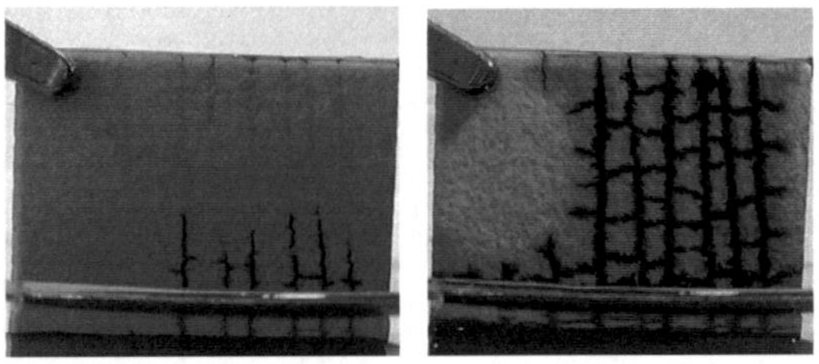

Fig. 4: Permeability test in ink solution: after 5 (left) and 180 (right) seconds

The leakage of the ink not only in the region of the tubes, but also in their vicinity by capillary forces indicates that the reversed path is in principle possible as well. Considering the viscosity of the body-borne synovial fluid, which is higher in comparison to the ink solution, fluid should be less fast.

For a more precise assessment of the internal structure, the tubes as well as the embedded placeholders, were sectioned, ground and polished. The kinetic energies of the impinging particles results in a roughening of the placeholder, **Fig 5**.

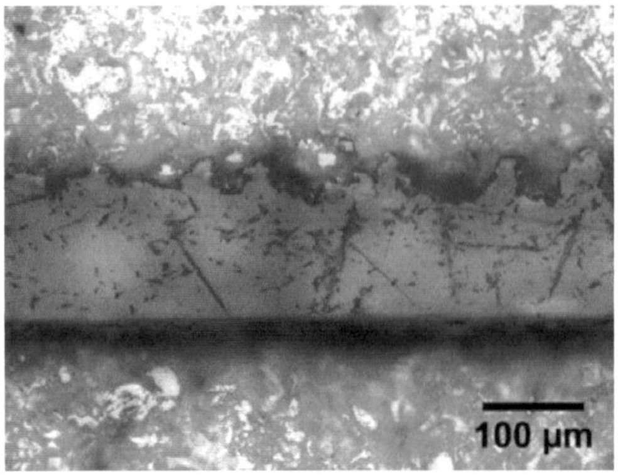

Fig. 5: Cross section of a tube prior to removal of the nylon thread

This results in a relatively high roughness within the tubes after the placeholders had been removed. The determined average roughness of the tube structure was S_a = 19.2 µm. As seen in **Fig. 5**, roughness can vary substantial along the top tube wall and locally changes in tube diameter of up to 80 µm could be detected. For a more precise characterization of the microgeometry of the coatings, an optical analysis was carried out using a VK-X210 laser scanning confocal microscope.

The analysis of the porosity revealed that above the introduced tube structure the porosity increased approx. by 10% compared to the regions remote from the tubes. This increase in porosity can be attributed to due to the changed impact angle of the sprayed particles. Furthermore, the analysis also disclosed a significant crack formation around the tubes, **Fig 6**.

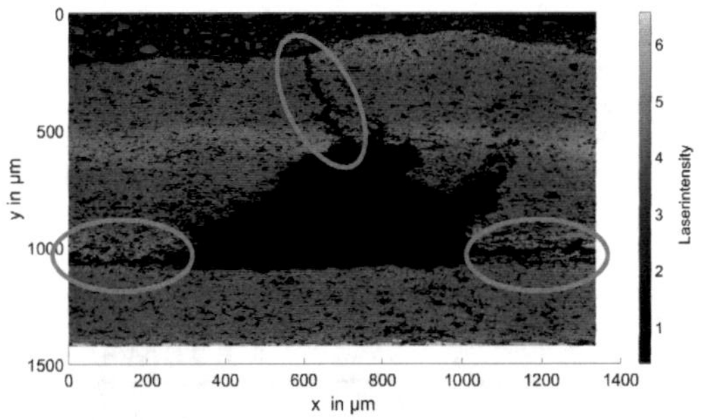

Fig. 6: Laser scanning image of a pyrolyzed sample

Starting from the tubes, the cracks always proceeded in the vertical direction, as well as along the interface between the coating and the substrate. The cracking is due to the different thermal expansion coefficients of the materials used. During the pyrolysis process this difference results in high thermal stresses and finally crack formation. Although the vertical crack formation would have the advantage that transport of the synovial fluid to the coating surface would be eased, a considerable impairment of the coating adhesion is to be expected. In particular the crack formation at the interface is critical.

Although the results obtained with nylon threads demonstrated the principle feasibility of the approach, the difference in thermal expansion coefficient turned out to be a major roadblock. Thus, metallic placeholder were tested. Specifically, cold gas-sprayed metallic placeholders proved to be particularly promising, as were more resistant to thermal load during the APS process.
The cold gas spraying also provided for excellent bonding to the substrate. Moreover, the differences in thermal expansion became less and the metallic placeholders could be etched without the need for an extra thermal process. This approach avoided crack formation around the placeholders. The wedge-shaped geometry of the placeholder resulted from the use of a slit mask during the cold gas spray process, **Fig. 7**.

Fig. 7: Copper placeholders applied by cold gas spraying

After the APS process, the metallic placeholder was anodically removed from the Al_2O_3 coating in a sulfuric acid electrolyte at a cell voltage of 600 mV. The analysis of the coating porosity in this system revealed that the porosity increased in the coating near the tube sides by 28% compared to the areas remote from the tubes.
In the present study, copper was used as placeholder for the sake of ease of processing. As, copper lacks biocompatibility, the copper needs to fully removed. However, using 3D X-ray microscopy it turned out, that in this process the removal of the placeholder represents a major challenge. Some of the copper still blocks the tube, **Fig 8**.

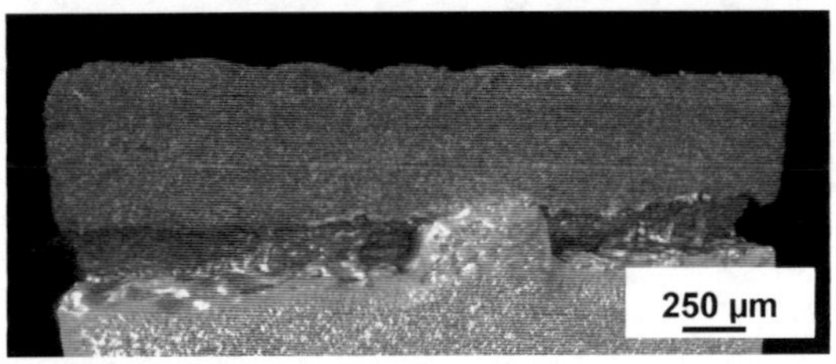

Fig. 8: 3D X-ray microscopic image showing a longitudinal section through the tube

Currently it is only possible to produce 2 to 3 mm long tubes that are free of any copper residues. Thus, other metallic materials such as Mg or Fe will be employed in future studies.

4 Summary and outlook

The results obtained in the present indicate the feasibility of the production of thermally sprayed Al_2O_3 coatings with integrated tube systems. It was successfully demonstrated the system is capable of transporting fluids to the surface of the coating. Moreover, cold gas spraying is a promising approach to deposit metallic placeholders that are firmly bonded to the substrate. In addition, the use of metallic placeholders minimizes the risk of spallation of the coating during removal of the placeholders.
The results obtained so far also suggest transpiring coatings could have applications outside the biomedical field. For example, the coatings could be useful lubrication of technical bearing surfaces, **Fig 9**.

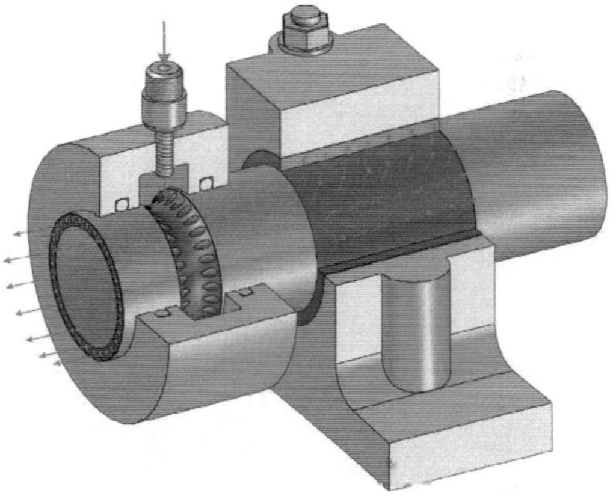

Fig. 9: Schematic representation of the lubrication of plain bearing surfaces

5 Acknowledgment

Financial support of this study by the Dr. Jürgen und Irmgard Ulderup Stiftung is gratefully acknowledged.

6 Literature

[1] Rinio M. and T. Schneider: Fortschritt in der Therapie der Hüftarthrose: Moderne Behandlungs-alternativen bel welt verbreiteten Hüftproblemen, Zeitung am Samstag (2017) Orthopädische Gelenk-Klinik, https://gelenk-klinik.de

[2] Zolotarevova E. and G. Entlicher, E. Pavlova, M. Slouf, D. Pokorny, F. Vesely, J. Gallo, A. Sosna: Distribution of polyethylene wear particles and bone fragments in periprosthetic tissue around total hip joint

replacements. Acta Biomater. 6 (2010), pp. 3595–3600.

[3] Utzschneider S. and A. Paulus, J.C. Datz, C. Schroeder, B. Sievers, B. Wegener, V. Jansson: Influence of design and bearing material on polyethylene wear particle generation in total knee replacement. Acta Biomater. 5 (2009), pp. 2495-2502.

[4] Wilches L.V. and J.A. Uribe, A. Toro: Wear of materials used for artificial joints in total hip replacements. Wear 265 (2008), pp.143–149.

[5] Amstutz H.C. and P. Grigoris: Metal on Metal Bearings in Hip Arthroplasty. Clinical Orthopaedics and Related Research, Number 3298 (1992), pp. 11-34.

Experimental Analysis of Impact Behavior of Ultra-High Molecular Weight Polyethylene-Nano Ceramics Composite Particles by Isolated Particle Deposition Method Using Downstream Injection Cold Spray Technique

Kesavan Ravi[1,2,3], Kazuhiro Ogawa[1,3], Olivier Lame[2], Jean-Yves Cavaillé[3]
1) Fracture and Reliability Research Institute, Tohoku University
2) Material Engineering and Science lab, MATEIS, INSA Lyon - CNRS
3) ELyTMaX, UMI 3757, CNRS, Tohoku University and Université de Lyon, International Joint Unit, Tohoku University, Room#503, MaSC Building, Katahira 2-1-1, Aoba-Ku, Sendai, 980-8577, Japan

Recent advances has made possible to obtain Ultra High Molecular Weight Polyethylene (UHMWPE) coatings by cold spray technique with nano-ceramic additives with the feedstock. However, the exact role of nano particles is largely not understood. In this work, isolated depositions of UHMWPE particles with 0%, 2%, 4%, 10% of fumed nano alumina (FNA) on Al surface were performed at different gas/particle temperatures. Particle velocities and particle temperatures were controlled by varying the carrier gas pressure and temperature. The impact behavior of UHMWPE was analyzed using SEM, FIB and high-speed camera. Increase in gas temperature and percentage of FNA showed a significant variation in the deposition volume. FIB analysis showed that successful depositions were influenced by degree of deformation of particle. Further, addition of FNA helped in deposition of particles that have required a lesser degree of deformation. Finally, high speed camera showed that particles are moving at an incidence velocity of 180-200m/s and rebound velocity of 40-50m/s. This suggests that particles lose a significant amount of their kinetic energy during the impact.

1 Introduction

Ultra High Molecular Weight Polyethylene (UHWMPE) is a semi-crystalline high molecular weight polymer used in the applications demanding high wear resistance, high impact resistance and high cavitation erosion resistance [1][2][3]. The semi-crystalline nature of UHMWPE is characterized by the presence of crystal network and the entanglement network in its polymeric chains. At temperature higher than its melting point, UHMWPE exhibits a highly viscous rubber-like behaviour which makes it possible to be processed by powder metallurgy technique like compression sintering [3]. But, the powder sintering suffers some disadvantages like lack of portability, inability to be applied to large-scale surfaces, inability to be integrated with other material surfaces etc. Subsequently, there have been some serious efforts to mitigate these issues by using cold spray as a processing/coating technique.

Cold spray is a high speed impact coating technique where fine particles (5-100μm) are impacted onto a target surface at high velocities (200-1000m/s) to cause deposition [4]. The parameters, especially particle temperature and particle velocity, essential for deposition, vary largely depending upon the type of materials and type of target surfaces [5]. In the case of UHMWPE, the particle surface, temperature and velocity were controlled to obtain coherent coatings on different surfaces [6][7]. The addition of hydrophilic polar nano-alumina particles like fumed nano alumina helped in better deposition of UHMWPE. But, the exact deposition mechanism and the role of nano-ceramic particles are yet unclear [7][8].

Hence, main idea behind this research work is to understand the behavior of UHWMPE particles in conjunction with nano-ceramic particles during its high strain rate deformation. This research work aims to analyse the deposition process by a technique of isolated particle deposition (IPD) method where particles are deposited on a suitable target surface in an isolated manner and the individual particle impact behavior is characterized and analyzed. The study then proceeds to identify the fundamental parameters for UHMWPE deposition mechanism.

2 Experimental

2.1 Powder

2.1.1 UHMWPE

The UHMWPE used in this study is GUR 4170 obtained from Ticona (Oberhausen, Germany) in the form of powder with an average size distribution of 60 μm. The molecular mass obtained from the viscosity measurements is 10.5 Mg/mol.

2.1.2 Fumed Nano Alumina

Fumed nano alumina (FNA), used in this study as an additive to UHMWPE, was obtained from Evonik under the product name of AEROXIDE® Alu 130. These FNA particles have a primary particle size of approximately 13nm and agglomerate size of 90nm.

2.2 Substrate

2.2.1 Aluminum

The target surface/substrate used in this study is 1100 grade Aluminum.

2.3 Feedstock Preparation

In this study, the feedstock was prepared by sieving the as-received UHMWPE feedstock powder to a size distribution of +45-63μm. Further, 0%, 2%, 4% and

10% by weight of FNA was mixed with the sieved feedstock by US mixing for 30 min in an ethanol bath, and dried before use.

2.4 Isolated Particle Deposition Method

Isolated Particle Deposition (IPD) experiments were carried out using a downstream injection low-pressure cold spray system. It is a conventional low-pressure cold spray system available under the name Dymet 403K (Obninsk Centre for Powder Spraying, Russia). An additional accessory equipped to this system is a **Fig. 1** represents a schematic of the downstream injection cold spray apparatus used in the experiment. Pressurized air (0.2-0.8MPa) was supplied from a compressor, meets the powder particles after passing through converging-diverging section (delaval region) of the nozzle. Gas temperature of the pressurized air can be varied from room temperature to 500°C. A vibratory feeder was used to feed the powder particles at a steady controllable rate. The powder particles finally go through the designed diverging nozzle before impacting the target surface.

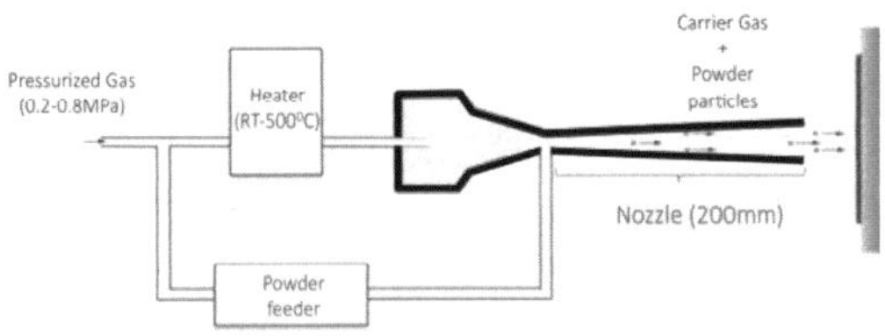

Fig. 1 Schematic of a downstream injection low-pressure cold spray system.

To control the standoff distance between the nozzle and substrate surface and also to control the traverse parameters, the cold spray unit was mounted on a XYZ stage and manipulated by a personal computer.
For the purpose of isolated deposition of powder particles, firstly the nozzle was held right angles to the target surface with the end of the nozzle at a standoff distance of 5mm from the target. The traverse velocity of the nozzle scanning the surface and the feed rate of the powder particles were controlled to achieve the deposition in this manner.

2.5 High Speed Camera Setup

A FASTCAM SA 1.1 model 675-M2 (high speed camera) was setup to calculate the flight-velocity of particles and to observe the spray process during cold-spray. The frame rate and shutter speed were adjusted to 40000 fps and 1/551000 s for an optimal view ability of the spray process.

2.6 Morphological Studies

The morphological investigations were performed using a field emission scanning electron microscope (FE-SEM, Hitachi SU-70), operated at 20 kV. A thin Platinum coating was applied prior to SEM observation by ion sputtering.

Further, the deposited particles were micro-machined along a plane normal to the particle-substrate interface plane to reveal the cross section of the particle and to understand the degree of deformation using Focused Ion Beam technique (FIB). FIB, in this study, uses a finely focused beam of gallium ions that operated at low beam currents for imaging and high beam currents for site specific sectioning of deposited particles in a precise controlled manner.

3 Results and Discussion

3.1 High Speed Camera Measurements

The high-speed camera was used to observe the deposition process of 0% and 4% UHMWPE-FNA composite particles. The two carrier gas temperatures were used in this process, Room Temperature (R.T.) and 380°C. Gas pressure was adjusted to 0.4MPa for both sets of experiments. **Fig. 2** shows the variations in the particle velocities at R.T. and 380°C gas temperatures. It was observed that the nano-particle addition to the feedstock did not change the velocity of the particle and the respective incident velocities at R.T with 0% and 4% FNA were 188±10 m/s and 184±10 m/s. Further, the incident particle velocities at the gas temperature of 380°C was recorded to be 212±10 m/s and 196±10 m/s for 0% and 4% FNA respectively. Rebound velocities for the entire scenario tested were around 40±10 m/s. The loss in KE is predominantly used in the deformation of the particle on impact and heat energy losses.

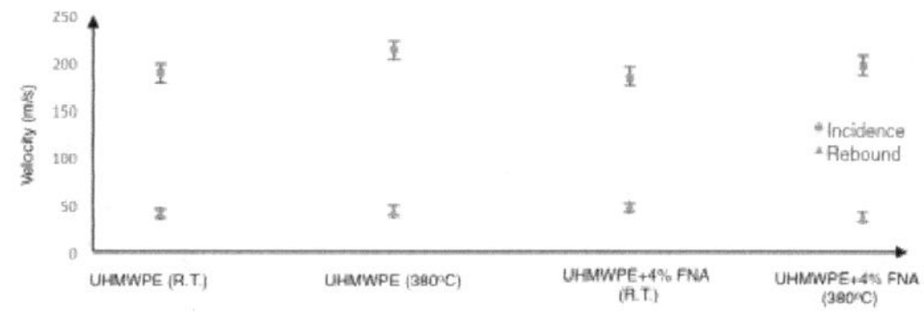

Fig. 2 Distribution of incident velocities and rebound velocities for 0% and 4% UHMWPE-FNA composite particles at gas temperatures corresponding to R.T. and 380°C.

3.2 Effect of Gas Temperature/Particle Temperature

3.2.1 Deposition Behavior

To analyze the effect of particle temperature on its deposition behavior, the temperature of the carrier gas was varied as follows: T1(190°C), T2(275°C), T3(380°C), T4(460°C) and T5(500°C); and the volume of deposition were analyzed for spray deposition of 0%, 2%, 4% and 10% UHMWPE-FNA composite particles of +45-63μm particle size distribution.
As the purpose of this set of experiments was to independently analyze the effect of particle temperature, the gas pressure (which monitors the particle velocity) was kept constant at 0.4MPa. The deposited surface show the presence of deposited particles smaller than the feedstock particle

distribution as well as particle agglomeration at different gas temperature-%FNA sets. The volume of deposited particles on a unit mm^2 surface area of substrate surface, were calculated from SEM images of the deposited surface and plotted. Further, the deposition data was normalized for a unit feed rate. Hence, the deposition data is denoted by $\mu m^3/mm^2/(gm/min)$

Linear approximation on the deposition volume data was done to identify the trend line as shown in **Fig. 3**. Without the nano-particles on the surface, UHMWPE showed little or no deposition at lower gas temperatures. The deposited particles were mostly a few particles of size around 10 μm. At, temperature T5 there was a slight increase in the deposition volume observed which could be attributed to a fluid like behavior of UHMWPE at high temperatures. As the %FNA was increased, there was a detectable increase in the volume of deposition. Further, the slope of the evolution of volume of deposition with temperature increased with an increase in the %FNA.

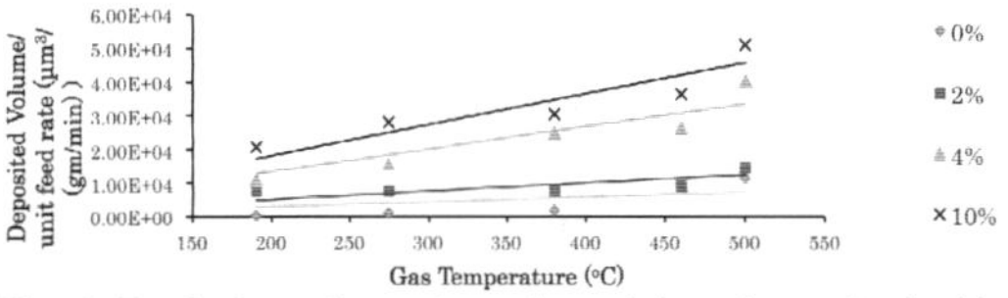

Fig. 3 Evolution of number of particles deposited with increase in gas temperature

It is interesting to note that even though the feedstock particles were of the distribution [45-63μm], but, depending upon the temperature and %FNA, size range of depositions varied from as small as 10μm to as large as 100μm. During the sieving of UHMWPE powder to [45-63μm] particle size distribution, it is possible that some trace amounts of particles smaller than 45μm particles end up in the feedstock. The SEM images of the depositions were analyzed by ImageJ where the area covered by the deposited particle on the surface was mapped when viewed perpendicular to the deposited surface. It was observed that increase in the temperature allowed larger particles to deposit on the substrate surface. For the depositions with UHMWPE-FNA feedstock, the presence of FNA on the surface allowed a larger size distribution of deposits for the same temperature when compared with depositions of UHMWPE without nano-particles. Further, when the high %FNA concentration is coupled with high temperature, the deposited particles were observed to be agglomerated (deposits as large as 100 μm in diameter).

3.2.2 FIB Analysis

The deposited particles were cut along the particle-substrate cross section to evaluate their degree of deformation after impact when deposited under the above conditions. To express the degree of deformation, aspect ratio of the before-impacted (initial) and after-impacted (deformed) particles were calculated. Aspect ratio (L1/L2) of the initial particle is approximately equal to 1 as they are spherical in shape and that of a deformed particle will be less than 1 depending upon the extent of deformation. A higher value of L1/L2 means a lower degree of deformation and vice-versa. Thus, the aspect ratio of the after-impacted particles were plotted for 0%, 2%, 4% and 10% UHMWPE-FNA composite particles at the temperatures T1, T3 and T5 as shown in **Fig. 4**. Aspect ratio of UHMWPE (0% FNA) deposits at T1 and T3 were difficult to be recorded as they were extremely small in size and the energy of FIB ions were too large to make precise cuts along the cross section of the particle. At the same time, at temperatures T1 and T3, deposited particles with FNA on the surface showed a slight deformation compared to the feedstock particles. At T5, the deposits of UHMWPE were heavily deformed. It was observed that in the presence of FNA at T5, there were additional particles increased with a lower degree of deformation. Hence, the average L1/L2 ratio of deposits at >2% FNA was higher than the one with 0% FNA. FNA was observed to allow particles with lower degree of deformation (low contact surface area) to get stuck to substrate surface.

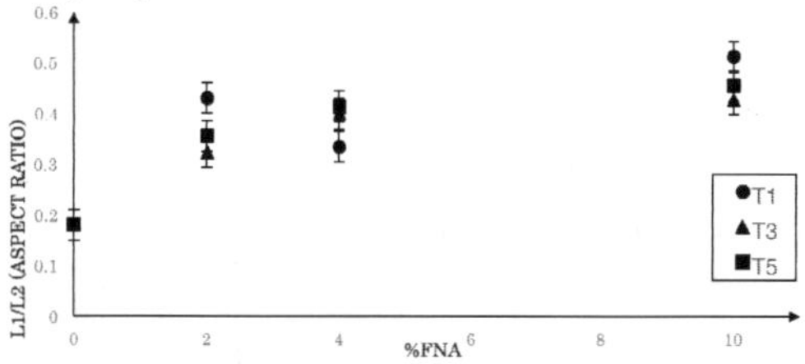

Fig. 4 Degree of deformation of the deposited particles at different %FNA and gas temperatures (T1 (190°C), T3 (380°C) and T5 (500°C).

4 Deposition Mechanism

4.1 Effect of temperature on elastic strain energy

For UHMWPE, when loaded, energy is consumed in deforming in the form of plastic energy, which is utilized to permanently deform, and elastic energy, which is stored in the crystalline lamellae and gets released when the load is removed. When UHMWPE particle impacts the substrate surface, energy in the form of elastic and plastic energy is utilized in deforming the particle to a certain degree of deformation. Once the impact load is released, the elastic strain energy is released and the material has an elastic recovery which is exhibited in the form of rebound energy. Hence, after impact, net non-zero rebound energy or the elastic strain energy released is the deciding factor for a successful deposition or not.

Fig. 5 shows the variation in the elastic energy per unit volume of UHMWPE with respect to specimen temperature. Clearly as seen from the graph, elastic energy/unit volume decreases as the temperature is increased. At higher strain rates, literatures suggest an increase in the Young's modulus and tensile strength values [8]. Although it is very difficult to exactly predict the influence of strain-rate, it is possible to predict a possible shift in the trendline as UHMWPE, on the other hand, shows an increasing elastic energy with the increase in the strain rate. Hence, an illustrative shift in the curve can be predicted as shown in **Fig. 5**.

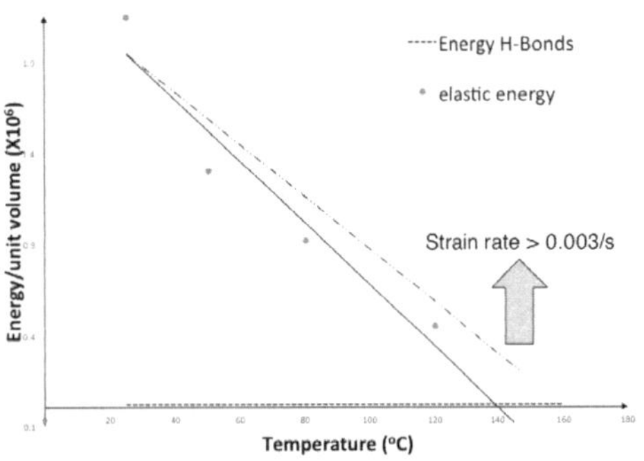

Fig. 5 Variation in elastic energy per unit volume of UHMWPE and an illustrative shift in the trend with the increase in strain-rate.

4.2 Effect of FNA

It was interesting to note that FNA on the surface of UHMWPE particles played a very important role in increasing the volume of deposition to a great extent. The reason for such a phenomena is investigated in this section. From the high-speed camera data, it was clear that impact time of UHMWPE particles is of the order ~10^{-4}s. Hence, firstly, certain effects like recrystallization, induced nucleation by FNA (if any) can be considered highly unlikely as their kinetics are too slow within the time frame of impact. Most likely phenomenon that is possible in this particular time-frame is adhesion via H-bonds. Fumed nano alumina (ALU 130 in this case) is known to have hydroxyl groups on its surface [9]. Weston et. al. estimates 8 hydroxyl groups/nm^2 on the surface of ALU130.

Energy stored by hydrogen bonds are estimated to be around 1-5kcal/mol or 4184-20920J/mol. Considering, a spherical 52μm in diameter particle deforming to a hemisphere of the same volume, total energy stored by these hydrogen bonds is

$$4\times10^{-10}J < E_{H-Bond} < 2\times10^{-9}J$$

Hence, if the rebound energy of the particle is lower than or equal to the energy stored by the hydrogen bonds, the particle will stick to the surface. For UHMWPE surface without the FNA (or H-bonds), the elastic energy/unit volume has to be almost equal to zero to obtain a successful deposition. But, in the presence of FNA, the deposition window is increased as the H-bonds can overcome the non-zero net rebound energy by a value of $\approx 1.6\times10^{4}J$ by helping in "pinning" incoming particles with a elastic energy lower than this value. This value depends upon the contact area furnished by the particle upon deformation. That is, if the 52μm particle upon impact can furnish a surface larger surface area than $\approx 7.2\times10^{-9}J$ as in the calculation above, it will have more H-bonds available for adhesion. Hence the deposition of particles with higher rebound energy/unit volume can adhere on to substrate if it can furnish a sufficiently large contact area for total energy stored by H-bonds to overcome its rebound energy.

Now, H-bonds can be assumed as a bridge bond between UHMWPE and Al surface. H-bonds can bond with electropositive elements like Al quiet easily [10]. But the bond between FNA particles and UHMWPE is highly unlikely via H-bonds as UHMWPE is made of predominantly non-polar C-C covalent bonds. Hence, the FNA-UHMWPE should be visualized as a more mechanical bond. As the particle impacts the substrate, the FNA particles can be visualized as piercing into the UHMWPE interface creating a mechanical interlock and at the same time establishing H-bonds with the electropositive Al surface.

This explanation is consistent with the experimental results where it can be seen that the volume of deposition of UHMWPE-FNA is higher than that of UHMWPE particle with lower %FNA at the same temperature. Hence, the reason for an increased deposition volume depending upon the temperature of particle and %FNA on the surface can be attributed to the "pinning" effect of hydrogen bonds.

5. Conclusions

The present study has attempted to understand the fundamental mechanism of cold-sprayed UHMWPE-FNA particle deposition by identifying the parameters responsible for adhesion and their individual causative effects. The following conclusions can be drawn from the study

1. Major parameters that influence significantly the UHMWPE deposition efficiency is the particle temperature and velocity, elastic strain energy and the amount of FNA (or density of H-bonds) between is impacting UHMWPE/metal interface.
2. Elastic energy/rebound energy of UHMWPE decreases at increasing gas temperature.
3. Deposition or rebound of UHMWPE particles on Al surface depends upon whether UHMWPE particles after impact can furnish a contact area with an interfacial bond stronger than elastic strain energy of the particle.
4. H-bonds on the FNA surface provide sufficiently strong extra bonds at the contact surface to help UHMWPE deposit at higher temperatures, which was otherwise not possible.

Acknowledgements

This work was partly supported by the JSPS Core-to-Core Program, A. Advanced Research Networks"International research core on smart layered materials and structures for energy saving".

References

[1] S.M. Kurtz, The UHMWPE Handbook: Ultra-High Molecular Weight Polyethylene in Total Joint Replacement, Academic Press, New York, 2004.

[2] D.C. Prevorsek, Y.D. Kwon, and H.B. Chin, Analysis of the Temperature Rise in the Projectile and Extended Chain Polyethylene Fiber Composite Armor During Ballistic Impact and Penetration, Polym. Eng. Sci.(1994), 34(2), p 141-152.

[3] Deplancke, T., Lame, O., Rousset, F., Aguili, I., Seguela, R. and Vigier, G. Diffusion versus cocrystallization of very long polymer chains at interfaces: experimental study of sintering of UHMWPE nascent powder.Macromolecules, (2013) 47(1), pp.197-207.

[4] McCune, R. C., W. R. Donlon, E. L. Cartwright, A. N.Papyrin, E.F.Rybicki and J.R.Shadley. Proceedings of the 1996 National Thermal Spray Conference, (1996) 7-11.

[5] Dykhuizen, R. C., M. F. Smith, D. L. Gilmore, R. A. Neiser, X. Jiang, and S. Sampath. "Impact of high velocity cold spray particles." Journal of Thermal Spray Technology 8, no. 4 (1999): 559-564.

[6] Ravi, K., Ichikawa, Y., Deplancke, T., Ogawa, K., Lame, O. and Cavaille, J.Y. Development of Ultra-High Molecular Weight Polyethylene (UHMWPE) Coating by Cold Spray Technique. Journal of Thermal Spray Technology,(2015) 24(6), pp.1015-1025.

[7] Ravi, K., Ichikawa, Y., Ogawa, K., Deplancke, T., Lame, O. and Cavaille, J.Y. Mechanistic Study and Characterization of Cold-Sprayed Ultra-High Molecular Weight Polyethylene-Nano-ceramic Composite Coating. Journal of Thermal Spray Technology, (2016) 25(1-2), pp.160-169.

[8] Russell, B. P., Karthikeyan, K., Deshpande, V. S., & Fleck, N. A. The high strain rate response of ultra high molecular-weight polyethylene: from fibre to laminate. International Journal of Impact Engineering, (2013) 60, 1-9.

[9] Weston, J. S., Venkataramani, D., Aichele, C. P., Grady, B. P., Harwell, J., & Resasco, D. Pseudosolid, Shear-Thinning gel formation in binary dispersions of metal oxide nanoparticles at low volume fractions. Langmuir, (2014), 30(49), 14982-14990.

[10] Kipling, J. J., & Wright, E. H. M. 683. The adsorption of stearic acid from solution by oxide adsorbents. Journal of the Chemical Society (Resumed), (1964), 3535-3540.

The arc movement as a source of process instability for single-cathode-anode systems – a new method for direct investigation

Georg Thomas, Technical University of Berlin, Department of Coating Technology, IWF, Berlin/D
Sebastian Mihm, Medicoat AG, Mägenwil/CH
Marius Limburg, Technical University of Berlin, Department of Coating Technology, IWF, Berlin/D
Heiko Gruner, Medicoat AG, Mägenwil/CH

The almost uncontrollable arc movement in a single-cathode-anode system limits the efficiency and process stability of this global used DC plasma spraying technique. A successful approach to stabilize the arc anode attachment and thus to increase process efficiency was the replacement of the convergent-cylindrical nozzle geometry by a convergent-divergent nozzle geometry with Laval contour. However, the advantages of this nozzle design were connected to an increased anode erosion caused by an excessive heat input from the anode attachment of the arc. Prerequisite for an optimization of the arc movement is its technical measurement and the dedicated correlation out of it with the plasma jet and the trajectory of injected powder particles. For this purpose Technical University of Berlin developed a new optical analysis method based on high-speed video recording to examine the arc movement, its position and intensity. The software based data together with the simultaneous measured voltage of the corresponding DC-arc clearly show for example the dependence of the arc movement from the position of the cathode tip in relation to center axis of the anode, impressively demonstrated for a new three-zone anode (convergent, conical and divergent). Thus a new tool is given to the developer of new anode geometries for increasing process efficiency and the lifetime of the single-cathode-anode system.

1 Introduction

The functional principle of DC plasma spraying is based on the energy transfer from an arc, generated between cathode and anode, to process gases which pass by [1]. The geometrical arrangement of centered cathode surrounded by the anode, functioning as nozzle at the same time, fixes the arc on the cathode tip and allows flexible movements of the arc anode attachment point on the anode surface. Due to the hydrodynamic forces from the gas flow and the Lorentz force from the arc itself, the anode attachment performs axial and rotational movements on the anode surface [1-8], leading to electrical power fluctuations and consequently to process instabilities by varying plasma, spray stream and coating properties [9-11]. Starting from the convergent-cylindrical anode geometry, the convergent-divergent geometry (Laval contur) was developed to realize process stabilization by minimizing the plasma power fluctuations [12]. Focused on the controlled limitation of the axial arc movement (evaluated by arc voltage fluctuation measurements), the rotatory movement of the anode attachment was not investigated due to missing measuring system. However, the property of the Laval contoured anode to improve the process stability and efficienybased on the axial fixation of the anode attachment close to the smallest cross-section (transition divergent-convergent part) leads in operation to a parameter depending increase of the anode erosion [13]. Therfore the convergent-divergent anode geometry (VMT_1.0) was consequently improved with the aim of minimizing anode erosion. Focus of the development was an increased freedom of movement of the anode attachment to spread the heat input over a larger anode surface and to avoid local overheating. During this development, it was shown that the position of local anode erosion is connected to the smallest cross section of the nozzle (for Laval nozzles). The limited anode surface erosion and the typically increased values of the arc voltage amplitude for convergent-cylindrical nozzles indicate an increased axial movement of the anode attachment in this type of nozzle.

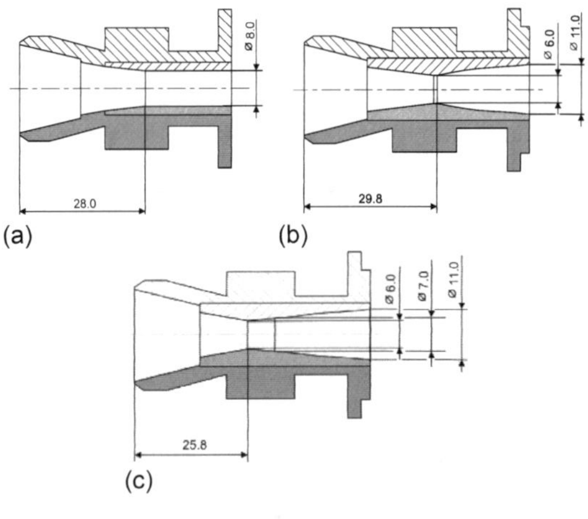

Fig. 1. Cross-section of (a) a standard nozzle, (b) the VMT_1.0 nozzle and c) the VMT_4.0 nozzle, dimensions in mm [13]

The development of the 3-zone anode (VMT_4.0) correlatesthe systematic movement of anode attachment with the plasma jet-shaping [13] [14]. The freedom and limitation of movement of the anode attachment could be improved by a given diffusor effect. Nevertheless, a minimal variations (+- 5%) should be tolerated in deposition efficiency (DE) at constant spray parameters. The predominant movements of the arc in the anode nozzle could only be interpreted limited and indirectly. In the first assumption, this fluctuation can be explained by a statistically not uniformly distributed angular position of the anode attachment responsible for an asymmetrical plasma jet, and therefore for many process instabilities through plasma spraying with a single cathode-anode-system (also with other nozzle designs).

In addition to the analysis of the arc voltage and current, focus on the optical investigation of the resulting plasma jet and the particle spray stream with Spraywatch (Fa. Oseir), Accuraspray (Fa. TECNAR) or NIR-Sensor (Fa. IfU Diagnostic Systems GmbH) [15], explain the limited accessibility of the arc. The arc as the source of energy is the basis for the DC plasma spray process. Therefore the direct investigation of the axial and tangential arc movement in the plasma generator should be successfully implemented, for the development of a commercially available measuring system [1-8]. Due to the lack of high-capacity measurement and analysis techniques (image resolution, recording speed, processed data volume), the arc movements could only restricted and for a discrete state analyzed. The optical recording of the arc in the anode has mostly been triggered by characteristic values of the arc voltage, thus permitting only limited insight into the operation of the arc movement in the anode nozzle. Especially the development of new anode nozzles with a modified inner geometry requires the direct examination of the arc and its movement in the plasma generator. The analysis of the arc movement in the anode nozzle would allow a reduction of all related processes (e.g. plasma jet generation, spray stream generation) of DC plasma spraying.

2 Motivation

During the coating production with fixed parameters some variation of the deposition efficieny of atmospheric plasma spraying (DE) could only be unsatisfied explained. Responsible can be an existing plasma jet asymmetry due to a non-concentric arrangement of anode and cathode as the results of investigations shown (**Fig. 2**). With increased decentration of the cathode tip to a defined direction, a decrease of mean arc voltage and deposition efficiency is observed (**Fig. 3**). The high sensitivity of the arc generation and it`s movement is significantly influenced by geometric change of the relative position of the cathode within anode.

Above mentioned, the investigation of plasma jet or spray stream, does not provide necessary information about the causes of process disturbances such as plasma jet asymmetry or an inefficient powder consumption. Therefore the accuracy and process stability in thermal plasma spraying is thus possible only to a limited extent. Spray spots and deposit analysis remain still the source of data to control and document the ongoing plasma spraying process. On this foundation, the decision for a process release or process termination is made in industrial applications. Also for the adaptation and development of new plasma generators and their components (e.g. gas distributor ring, anode and cathode geometry, positioning anode / cathode), an unexpected problem arises which affects the predictability and the influence of the arc movement.

Due to the direct effects of the arc movement to the process stability during thermal plasma spraying, the development of an analytical method is vital which allows direct observation and evaluation of the axial and rotational arc movement in the plasma generator.

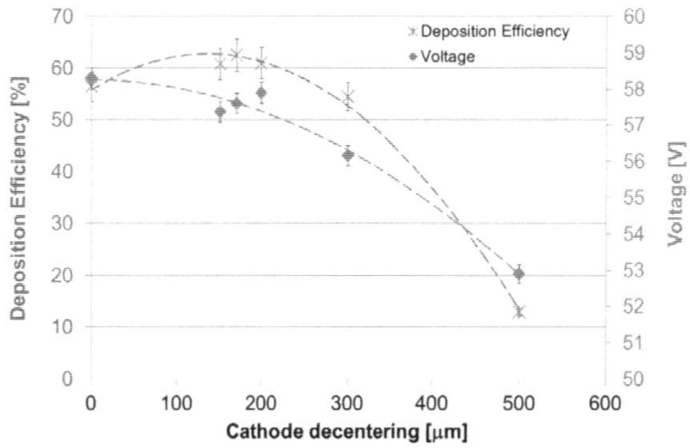

Fig. 2. Influence of cathode tip decentering on DE and voltage (30 kW, plasma gas flow Ar/H_2 30/7.5 slpm)

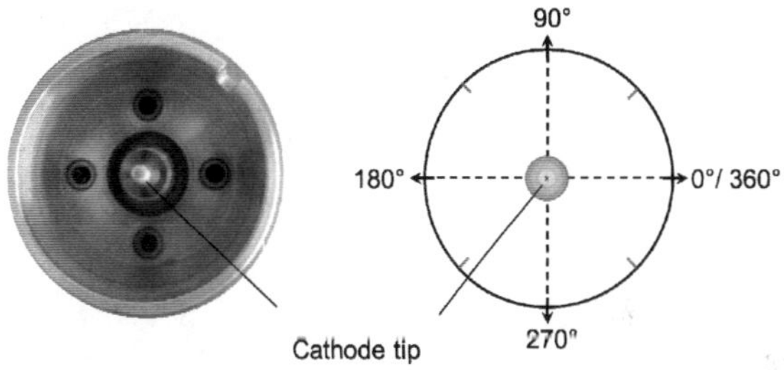

Fig. 3. Front view plasma generator(left), Position cathode tip and arc orientation (right)

The present work is based on the time-synchronous recording of arc movement using a high-speed camera and the corresponding arc voltage. By software analysis a direct correlation of the variables mentioned exact allocation of the arc orientation (position of the anode attachment point on the anode surface is now available). The dependence of the arc motion on the selected plasma parameters, the anode geometry and the cathode position can thus be analyzed. The analysis of the arc movement in the anode nozzle offers a remarkable potential for a very efficientprocess control during plasma spraying. At the same time the optimization of the plasma generator is possible.

2 Experimental Procedure

Taking into account to realize an industrial usage, a simple and robust test setup was chosen. As already tested ([6] [16] [17]), an optical measuring section with a 90 ° deflection (mirror) for frontal observation of the anode outlet is installed. The arc movement in the nozzle is recorded by a high-speed camera with a neutral density (ND) filter. The values of the arc voltage (direct measurement on the plasma generator) are recorded synchronously. Subsequently, the software processing for the evaluation and interpretation of all data is carried out.

2.1 Measurement Setup

As plasma generator a single cathode-anode system MC60 (equivalent to F4) from Medicoat is used, which is widely used in industrial and medical applications. The experiment setup is shown in **Fig. 4**. The frontal observation of the nozzle exit by mirror system the complete arc formation in the nozzle can be studied without further masking.

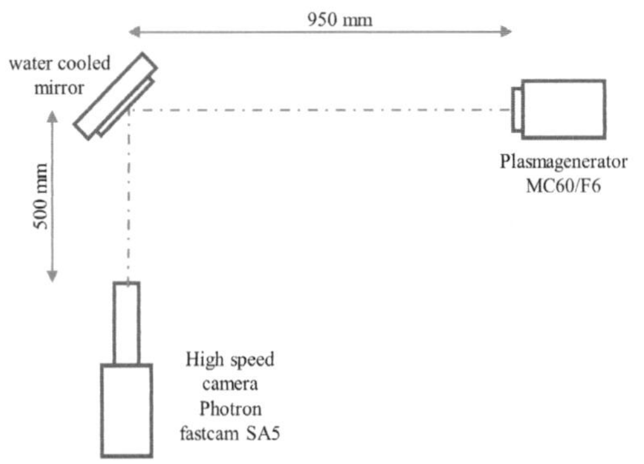

Fig. 4. Measurement setup for the direct optical investigagtion of the nozzle exit and the arc (top view)

For the eyperimental work, the following tools with corresponding parameters are used:

- DC-voltage by a PC measuring card (PC-Scope T512) from IMTEC Co., sampling rate = 1 µs, measuring time = 32 ms (32,000 measuring points)
- Camera lens with neutral density filter and Nikon lens "AF NIKKOR 70-300mm", selected aperture 22-16 (depending on the total image brightness), focal length = 300 mm. The pictured image area is approximately 13 mm x 22 mm with a resulting resolution of around 365 dpi.
- The arc movement is recorded using a high-speed camera (Fastcam SA5) from Photron Co. Recording of a gray scale image (0-255 values), Resolution = 192 x 320 pixel, Image frame rate = 100000 fps, without shutter, Exposure time = 1 / frame rate (10 µs)

The recording of the image data and voltage values takes place synchronously, so voltage values can be unambiguously assigned to each image. A single frame of the obtained video is shown in **Fig. 5**.

2.2 Software Evaluation

The data are registered using MATLAB software from MathWorks Co., which offers interesting possibilities to automatically evaluate large amounts of data. For the analysis, a video with 3200 images is recorded, to which 32000 voltage values can be assigned. In order to ensure a comparability of the image data, automatic image centering and filtering by the self-developed MATLAB-based software is carried out in advance. The subsequent image analysis based on the evaluation of the image brightness and its change detects the anode center axis, the arc orientation as well as the value of the arc voltage (data voltage measurement) with respect to the examined brightness range. In several steps, the image raw data is converted into a filtered image, which produces a circular representation of the arc expansion **(Fig. 6**). The calculated center of area of an image and its orientation is defined in this work as the orientation of the arc.

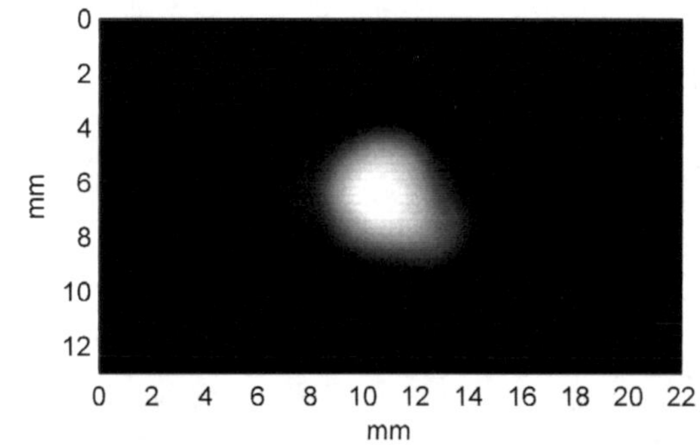

Fig. 5. Original video (single frame), standard nozzle, centered cathode (set up see **Tab. 1**)

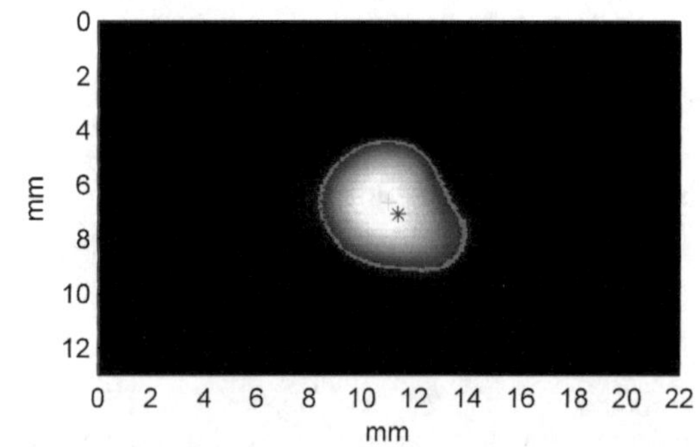

Fig. 6. Filtered Video, Dimension of arc (red line) and center of Area for single frame (blue asterisk) related to the anode center line (green plus sign), Standard nozzle, centered cathode (set up see **Tab. 2**)

The data analysis provides a frequency distribution of the arc orientation and an analysis of the arc voltage frequency spectrum (obtained via Fast Fourier Transformation-FFT). The position-resolved representation of the arc orientation in conjunction with the corresponding mean voltage, enables a three-dimensional representation of the arc orientation in the anode nozzle.

2.3 Plasma Parameter

Based on the developed standard useof the VMT_4.0 anode [14] (see Fig.1), the following parameter set up was investigated (**Tab. 1**).

Tab.1. Plasma parameter setup

Argon [slpm]	30
Hydrogen [slpm]	7.5
Current [A]	300-600
Electrical power [kW]	min 11- max 42
APS-process	

Furthermore, experiments with centered and decentered cathode were carried out for each type of anode. The decentering is adjusted with a centering clock, which has an accuracy of + - 10 µm. **Tab. 2** summarizes the used data.

Tab. 2. investigated set ups

Nozzle	Standard nozzle 8mm	VMT1.0	VMT4.0
Distance of cathode tip to anode center line [µm]	0 (centered)		
	200 (decentered, orientation = 0/360°)	200 (decentered, orientation = 270°)	
Current [A]	300; 400;450;500;550;600		

3 Results

The arc geometry could be determined for all chosen parameters and anodes. The arrangement is tolerant to alignment errors (angularity, distance) and quick to install (10min). Sequential combination of process control and coating process is possible. Various anode geometries were successfully analyzed. The influence of the different brightness values of the recording is as far as possible compensated by the software. For better evaluability, the use of the entire brightness spectrum (0-255 LSB) is advantageous but not necessary. Reproducible results are obtained for tests with the same setup (anode type, cathode alignment, parameters). Variated process parameters are clearly reflected in the evaluated data.

3.1 Voltage analyse

The evaluation of the arc movement shows a statistical uniform distribution over 360° on the anode surface. An arc-shaped movement of the arc takes place in the direction of the nozzle exit, until it comes to a new ignition near the cathode. In this case, the flow of the working gas with tangential velocity components decisively determines the movement of the arc. The investigations of the anodes mentioned with parameters given in **Tab. 1** are shown below by way of example with the parameters gas flow Ar/H_2 30/7.5 slpm and I = 500 A. In the centered state of the cathode, the voltage fluctuations for the VMT_1.0 anode are more restricted compared to the Standard nozzle (**Fig. 7, Fig 8**).

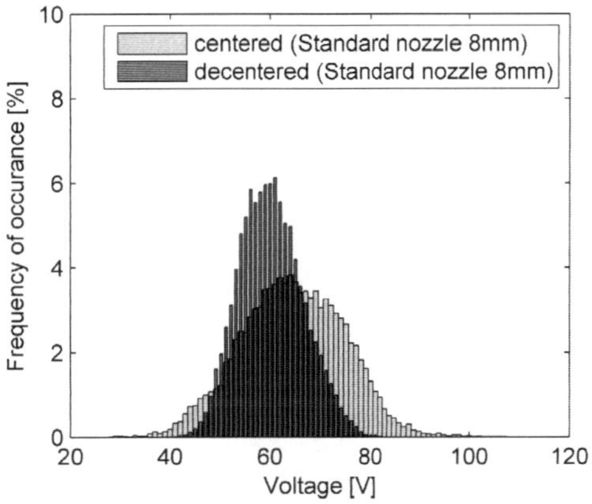

Fig. 7. Standard nozzle 8 mm, frequency of occurence of arc voltage (centered: U_{mean}=64,4 V, U_{min}=27,9 V, max 106,9 V; decentered: U_{mean}=60,0 V, U_{min}=39,1 V, U_{max}= 82,3 V)

The frequency spectrum is showing the absence of the 3.5 kHz fluctuations (**Fig. 9**). The influence of the decentered cathode on the arc voltage can be seen for the standard and VMT_1.0 anode in a reduction of the average arc voltage, while the spread width is shrinking at the same time (**Fig 8**). This effect is more pronounced for the standard anode. The FFT analysis of the arc voltage show a complete suppression of the fluctuations at 3.5 kHz (**Fig. 10**). The reduced spread width and the higher arc voltage attained with a decentered cathode, indicate a reduced axial movement with intensified re-ignition of the arc in the vicinity of the cathode. Consequently, the thermal energy of the anode attachment point can only be dissipated via a reduced anode surface.

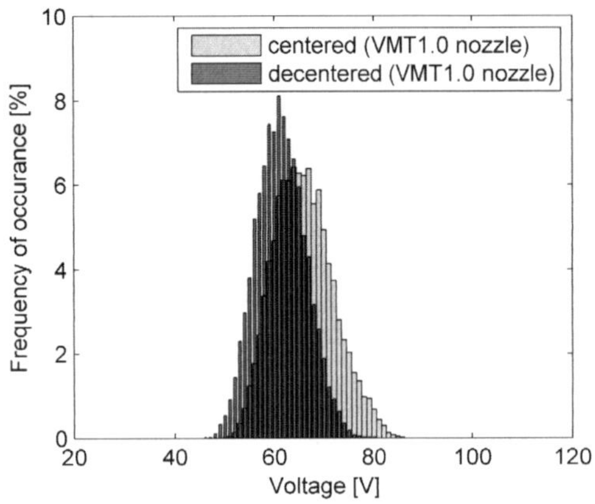

Fig. 8. VMT_1.0 nozzle, frequency of occurence of arc voltage (centered: U_{mean}=65,9 V, U_{min}=50,1 V, U_{max}=85,7 V; decentered: U_{mean}=61,3 V, U_{min}=45,7 V, U_{max}=80,4 V)

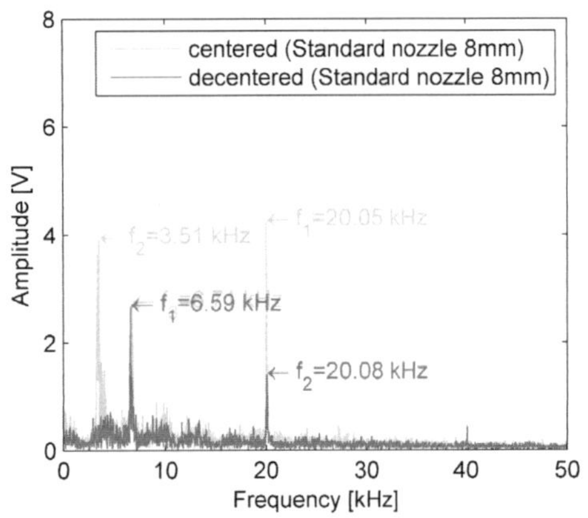

Fig. 9. Standard nozzle 8 mm, single sided amplitude spectrum of arc voltage

For the VMT_1.0 anode, the FFT analysis of the arc voltage shows no differences in the frequency spectrum. In the dominant frequencies, however, a significant amplitude decrease (4 to 2 V) occurs for the range 6.9-7 kHz when using a decentered cathode.

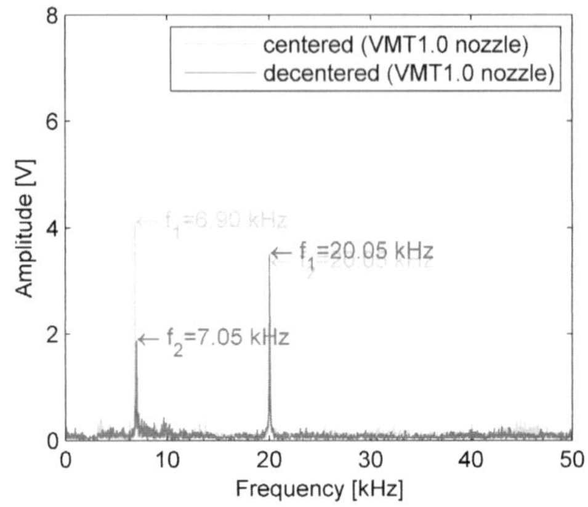

Fig. 10. VMT_1.0 nozzle, single sided amplitude spectrum of arc voltage

The evaluation of the circular movement of the arc results in an approximatively statistical uniform distributed orientation of the center of area of the arc position for standard and VMT1.0 anodes. The occurring fluctuations in the alignment can be classified as non-significant compared to the decentered nozzle (**Fig. 11, 12**).

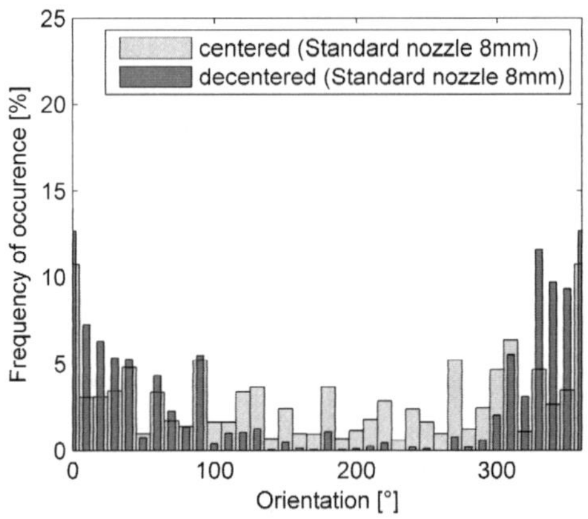

Fig. 11. Standard nozzle 8 mm, angular orientation of centre of area (plasma) to anode center line

If the cathode is decentered, a shift in the center of area and thus of the arc can be observed in the direction of the smallest distance between the anode surface and the cathode tip. Thus, the arc is oriented in agreement with the decentration defined in **Tab. 2.** The VMT_1.0 anode seems to react more sensitive to decentering than the Standard anode (**Fig. 11, Fig. 12**), which in combination with the restricted axial anode attachment movement (**Fig. 8**) can result in overheating and thus anode damage (consider the different angles of decentereing for nozzle types).

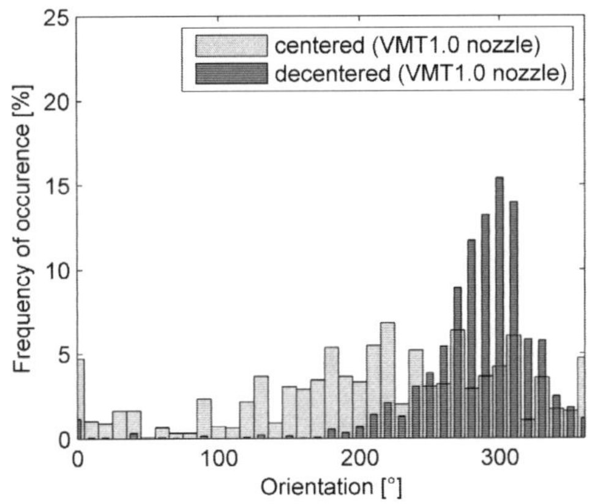

Fig. 12. VMT_1.0 nozzle, angular orientation of centre of area (plasma) to anode center line

The objective of the VMT_4.0 anode was to increase the axial freedom of movement of the arc (anode attachment point) so that the anode-side heat flow can be dissipated via a larger anode surface. As already stated in the spread width of the arc voltage is increased in comparison to the VM_T1.0 anode, but remains behind the Standard anode in the maximum values (**Fig. 13**). The decrease of the mean arc voltage and the narrowing of the spreading width (the arc voltage) are only rudimentary in comparison to the Standard and VMT_1.0 anode. The frequency analysis of the arc voltage shows no significant differences, only a slight amplitude shift can be seen over the entire frequency spectrum (**Fig 14**).

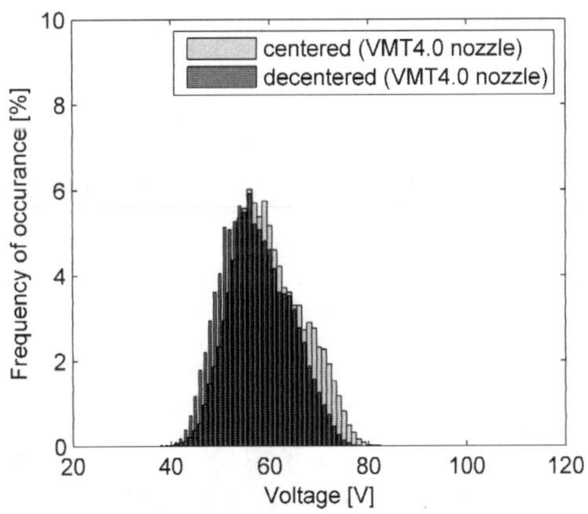

Fig. 13. VMT4.0 nozzle, Frequency of occurence of arc voltage (centered: U_{mean}=59,8 V, U_{min}=39,9 V, U_{max}=81,9 V; decentered: U_{mean}=57,1 V, U_{min}=38,0 V, U_{max}=78,0 V)

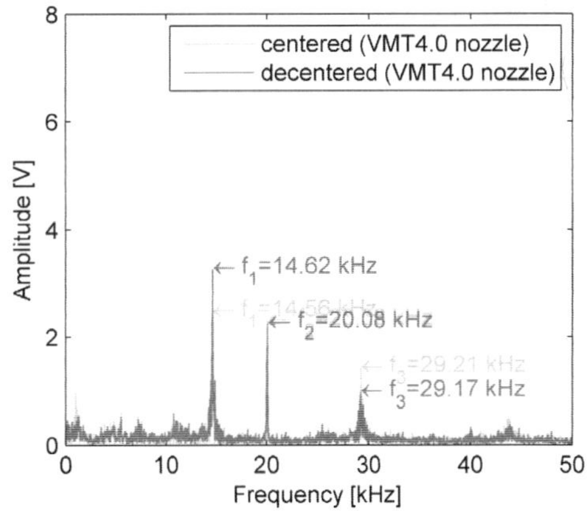

Fig. 14. VMT_4.0 nozzle, single sided amplitude spectrum of arc voltage

The evaluation of the image analysis shows for the centered cathode state a homogeneous distribution of the orientation of the center of area and thus a nearly statistically uniform distributed orientation of the arc. If a decentering of the cathode occurs, this can be clearly observed in a non-uniform distribution of the frequency of the center of area orientation, which corresponds to a displacement of the arc in the direction of decentering (smallest distance between the anode surface and the cathode tip). However, the preferred orientation is not as much defined as for the VMT_1.0 anode, which suggests an increased tangential mobility of the arc. The axial and tangential fixing of the arc when decentering the cathode is significantly reduced in comparison with the VMT_1.0 anode, which can counteract anode erosion.

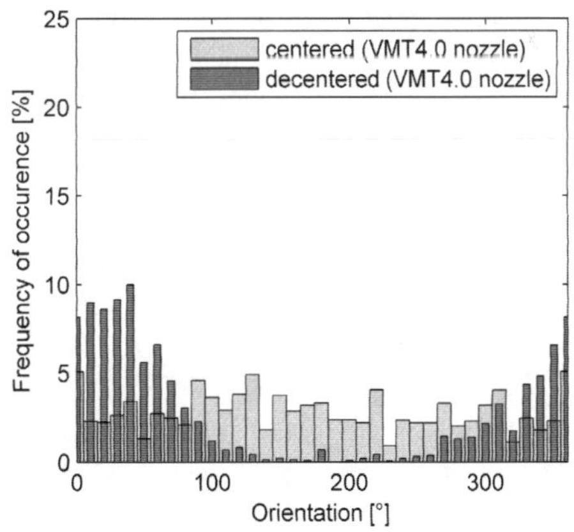

Fig. 15. VMT_4.0 nozzle, angular orientation of centre of area (plasma) to anode center line

Based on the determined data (arc voltage and arc orientation), it is possible to abstract a spatial representation of the arc orientation. The example of the VMT_4.0 anode shown in **Fig. 16** (centered) and

Fig. 17 (decentered) is visualizing the distribution of the arc orientation and the corresponding arc voltage.

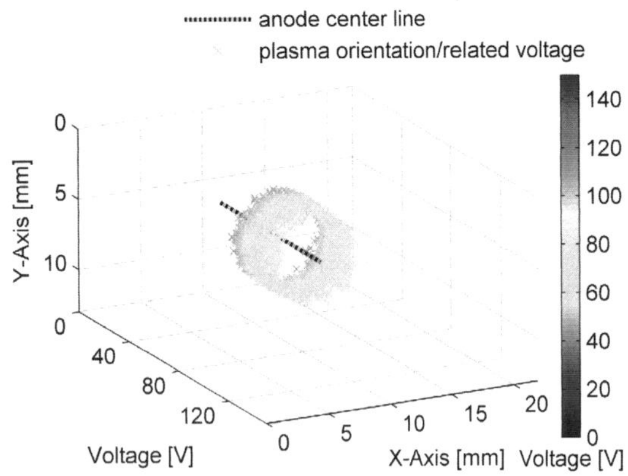

Fig. 16. Centered cathode and VMT_4.0 nozzle, angular orientation of arc (plasma) to anode center line

The detectable asymmetry of the arc orientation with the decentered cathode coincides with the results, which leads to a drop in the DE and thus process fluctuations. The results presented are only a small part of the scope of the investigation. However, on this basis, a positive conclusion of the measuring system for the direct examination of the arc movement in the anode nozzle can be drawn. Furthermore, the sensitivity of the direct current plasma spraying to the relative position of anode and cathode is shown. Experiments with a cathode decentration of 100µm already lead to a detectable deviation from the statistically uniform distributed orientation of the arc, which ultimately results in an asymmetry of the plasma jet / spray stream.

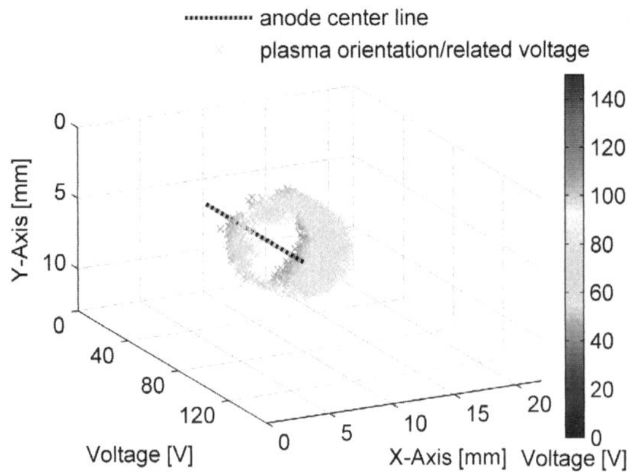

Fig. 17. Decentered cathode and VMT_4.0 nozzle, angular orientation of arc (plasma) to anode center line

3 Conclusion

A measuring system for practical application was developed and successfully used for the detection of the arc movement in the plasma generator. It can be used as an in-situ measuring instrument for process development / process control in the future. The detection of the arc orientation and associated arc voltage values deepen the understanding of the plasma generation in the anode nozzle and show causes for process fluctuations in DC plasma spraying. On the basis of the current state of the data evaluation, this publication also refers to a movement of the arc / center of area. However, it is assumed that the arc movement is mainly defined by the movement of the anode attachment. For a clear statement which is valid for all types of anodes, a further complete analysis of the generated data is required. The orientation of the anode attachment on the anode surface does not necessarily have to correspond to the orientation of the arc, as first investigations show.
Contrary to the idea of a continuous rotational movement of the arc in the anode, a sequence of axial movement in direction of the nozzle outlet with only a tangential component could be observed. A fully circulating movement of a stationary arc (anode attachment) is not present in any parameter setup. The cyclical generation of an arc with variable orientation as a function of the anode geometry produces a discrete, uniform distribution without a preferential direction of arc attachment for an ideal concentric arrangement of the anode and the cathode. The decentering of the cathode tip from the anode center axis leads to a preferential orientation of the arc in the direction of the minimum distance anode surface cathode tip. A direct dependence of the anode life time on the axial and tangential arc movement is given. The influence on the arc movement by decentering of the cathode tip, the anode- and gas distributor ring geometry is possible. The fundamental consideration of the arc movement in the anode

nozzle has an impact on the new conception and design of relevant components (anodes, cathode, gas distributor ring) of a single-anode cathode plasma generator. An eccentric arrangement of the cathode and the anode can be realized by decentering the cathode tip relative to the anode center axis or by an asymmetrical anode inner contour. The significance of the direct current arc and its movement (orientation) during the DC plasma spraying process must be interpreted as a superior function. Thus, the control of the arc position by the above-mentioned methods offer new possibilities for process control in the case of direct current plasma spraying.

4 Literaure

1] Nutsch, G.: Die Anwendung des thermischen Plasmas in der Werkstofftechnik. Habilitationsschrift (1995), Technische Universität Ilmenau

[2] Duan, Z., Heberlein, J.: Anode Boundary Layer Effects in Plasma Spray Torches. Proceedings of the 1st International Thermal Spray Conference (2000), p. 1-7

[3] Duan, Z., Heberlein, J.: Arc Instabilities in a Plasma Spray Torch. Journal of Thermal Spray Technology, ASM International, H. 11 (1) (2002), p. 44-51

[4] Duan, Z.: Investigations of Plasmas Instabilities in a Spray Torch. Ph.D. Thesis (2000), University of Minnesota

[5] Dorier, J.-L., Hollenstein, Ch., Salito, A., Loch, M., Barbezat, G.: Influence of External Parameters on Arc Fluctuations in a F4 DC Plasma Torch Used for Thermal Spraying. 1st International Thermal Spray Conference (2000), ASM International, p. 37-43

[6] Dorier, J.-L., Gindrat, M., Hollenstein, C., Salito, A., Loch, M., Barbezat, G.: Time-Resolved Imaging of Anodic Arc Root Behavior During Fluctuations of a DC Plasma Spraying Torch. IEEE Transactions on Plasma Science (2001), Vol. 29, No. 3

[7] Dorier, J.-L., Hollenstein, Ch., Salito, A., Loch, M., Barbezat, G.: Influence of External Parameters on Arc Fluctuations in a F4 DC Plasma Torch Used for Thermal Spraying. 1st International Thermal Spray Conference (2000), ASM International, p. 37-43

[8] Trellers, J. P., Heberlein, J. V. R.: Simulation Results of Arc Behavior in Different Plasma Spray Torches. Journal of Thermal Spray Technology (2006), Vol. 15(4), p. 563-569

[9] Bisson, J.F., Gauthier, B. , Moreau, C.: Effect of Plasma Fluctuations on In-Flight Particle Parameters. Journal of Thermal Spray Technology (2003), ASM International, Vol. 12 (1), p. 38-43

[10] Bisson, J.F., Moreau, C.: Effect of Direct-Current Plasma Fluctuations on In-Flight Particle Parameters. Journal of Thermal Spray Technology (2003), ASM International, Vol. 12 (2), p. 258-264

[11] Goutier, S., Nogues-Delbos, E., Vardelle, M., Fauchais, P.: Particle temperature fluctuations in plasma spraying. Journal of Thermal Spray Technology (2008), Vol. 17 (5-6), p. 895-901

[12] Schwenk, A.: Entwicklung und Erprobung neuartiger Düsen für das atmosphärische Plasmaspritzen. Dissertation (2005), TU Chemnitz(D)

[13] Mihm, S., Thomas, G., Kuhn, R.: Three-zone anode geometry - The disengagement of classic nozzle design for atmospheric plasma spraying. International Thermal Spray Conference ITSC 2015 (2015), Long Beach, USA

[14] Mihm, S., Duda, T., Gruner, H., Thomas, G., Dzur, D.: Method and Process Development of Advanced Atmospheric Plasma Spraying for Thermal Barrier Coatings. Journal of Thermal Spray Technology (2012), Springer Verlag, Vol. 21, p. 400-408

[15] Mauer, G., Vaßen, R., Stöver, D.: Plasma and Particle Temperature Measurements in Thermal Spray Approaches and Applications. Journal of Thermal Spray Technology (2011), Springer Verlag, Vol 20 (3), p. 391-406

[16] He-Ji, H.; Wen-Xia, P. & Cheng-Kang, W.: Arc Root Motions in an Argon-Hydrogen Direct-Current Plasma Torch at Reduced Pressure. Chinese Physics Letters (2008), Vol. 25, No. 11, 4058

[17] Gindrat, M.: Etude du comportement dynamique d'une torche à plasma DC par imagerie rapide et modulation du courant d'arc. Masterthesis (2000), École polytechnique fédérale de Lausanne(CH)

Investigation of the Usefulness of Particle Jet Monitoring in a Production Environment

Luc Leblanc, Ryan Hinckley, Christopher Spinicci, Malta, NY / USA

In the thermal spray process, particle state at impact is among the key factors influencing the quality, characteristics and properties of the deposit formed. Measuring and eventually controlling in-flight particle jet characteristics can help ensure the repeatability of desired coating properties. Moreover, monitoring particle jet can lead to cost savings by allowing replacement of consumable parts on a need-basis rather than on a preventive-basis. However, successful use of in-flight particle sensors in a production environment requires developing an implementation strategy that minimizes impact on coating cycle time, limits operator intervention and reduces analysis time, while still generating useful data. The usefulness of in-flight particle sensor in a high yield production environment is investigated. A preliminary study was first conducted to define a strategy to implement the data acquisition process using an ensemble in-flight particle monitoring system that allows real-time measurements of average particle temperature and velocity and overall particle jet profile. Data was collected to capture representative torch and particle jet behaviors over the life of several consumable hardware sets. Evolution of the particle jet parameters over time is compared to that of some of the process parameters, as well as to the deposition efficiency and one selected characteristic of the coating deposited. Some trends are identified, and potential benefits and drawbacks of using in-flight particle monitoring in a high yield production environment are highlighted.

1 Introduction

It is of primary importance to ensure constant quality of coatings as well as the reproducibility of their characteristics in a production environment, especially when rejection of a part can lead to significant cost or impact on delivery schedule. It is widely recognized that in-flight characteristics of thermally sprayed particles such as their temperature and velocity play a key role in the deposit formation dynamics and microstructure evolution. As such, in-flight particle sensors have successfully been used to accelerate coating development and establish acceptable process windows [1-16]. Other parameters such as substrate temperature, stresses generation during spraying, and heat flux from the hot gases also play an important role in the coating formation and must be controlled [13].

For some demanding coating applications, controlling the particle jet can be of prime importance to minimize yield issues, and the use of sensors to monitor and help control the process can be critical and easily justified. However, for high yield applications, the use of particle sensors is not vital in terms of decreasing the rejection rate since measures are already in place to insure deposition of coatings that meet the quality requirements are produced most of the time. Moreover, only a fraction of the number of parts rejected can generally be attributed to the coating process itself. Other factors such as masking, operator error, equipment malfunction or others contribute to the rejection of parts [17]. Therefore, in thermal spray processes where high yields are already established, the value of using in-flight particle sensors resides more in trying to increase part throughput in order to optimize equipment usage. This can be accomplished by understanding process limits, variations and drifting such as the ones due to electrode wear, feeding issues, or the use of different feedstock material lots. This understanding can indeed eventually lead to the definition of a higher throughput process at lower cost by generating tighter and/or flexible process specifications within (or outside) the already accepted window for a still better control of the coating characteristics and properties. In turn, this could potentially be done at increased and maintained deposition efficiency while possibly extending the life of the gun consumable parts, which are oftentimes replaced on a preventive basis. Monitoring the general behavior of the process to capture and eventually correct any anomalies is an additional benefit.

The successful integration of in-flight particle sensors in a high yield process requires developing an implementation strategy that minimizes impact on coating cycle time and throughput, limits operator intervention and distraction, and reduces data analysis time while still generating useful data.

In this contribution, the usefulness of using an in-flight particle sensor in a high-yield, high volume production environment with plasma spraying of ceramic materials is investigate. Although the proprietary nature of the coating and process used in this study prevents full disclosure of specific process conditions and materials used, general trends are reported and discussed.

The development of a strategy and its implementation for data collection and extraction is discussed first. The evolution of some of the key process and coating parameters as well as that of particle jet parameters for several sets of gun hardware is then presented and discussed. Some trends are identified and potential benefits and drawbacks of using in-flight particle monitoring in a high yield production environment are highlighted.

2 Experimental

2.1 Process, Material and Data Acquisition

Results presented in this contribution were generated using a single injector feeding a feedstock ceramic material radially, and vertically from the top in the plasma plume. All spray conditions were nominally kept the same throughout the whole study, and no attempts were made to "correct" for the process drifts typically associated with normal wear of the gun components. For example, feedstock feed rate and plasma spray parameters such as plasma gas, arc current and carrier gas flow rate were maintained to their nominal condition. Similarly, deposition conditions were nominally the same. For example, the number of pre-heat passes, feedrate stabilization time and gun-to-part relative motion were kept the same.

Gun hardware sets, such as anodes, cathodes and feedstock material injectors were all replaced at the same time when deemed necessary (end of life). Process, particles jet and coating data was collected as a function of hardware operating time for 10 hardware sets. Two different feedstock material lots (nominally the same) were sprayed, i.e. lot A was used with the first three hardware sets, and lot B was introduced with hardware set #4 and was used for the rest of this work. As anticipated, no statistical differences were noted between the results generated with the two different lots, as both were very similar based on their reported characteristics. For instance, size distribution and chemical compositions were within a few percentage points for both lots.

Before starting this investigation, gun hardware sets were replaced preventively after a predetermined maximum number of operating hours, or earlier when signs of significant drifting were suspected from the process and coating data collected, or based on operator input. This was also the case for the first hardware set used in this study. However, as data was being collected for the second hardware set, it appeared that the process seemed stable enough to warrant extending the pre-established limit for hardware preventive replacement by an extra 33%.

In-flight particle data including particle state and jet intensity profile, as well as process data (only gun power measured at the gun is reported here) were collected at the beginning of each coating run, after a pre-determined stabilization time. The plume intensity profile was also measured at the end of each coating run. Particle jet characterization is detailed in the next section.

Coating deposition efficiency was calculated the same way for each run, by dividing the part weight gain from the deposited coating with the calculated mass of material directed at it. The relative error on the calculated deposition efficiency was estimated to be less than 3%.

The coating from each run was also quantitatively characterized for its key characteristics per standard quality control procedures in place. Only one is reported here. For this contribution, we will call it "Characteristic A". "Characteristic A" is bound by an upper limit, above which the coated part is to be rejected. The process yield is calculated using the amount of rejected parts versus the amount of parts sprayed.

2.2 Particle Jet Characterization

Successful use of in-flight particle sensors in a production environment requires developing an implementation strategy that minimizes impact on coating cycle time, limits operator intervention and distraction, and reduces data analysis time while still generating useful data. Many authors have discussed the use in-flight particle sensors in thermal spray for both research and production use [1-16]. Ensemble particle diagnostic systems are typically more suitable for production environment since in general the data acquisition time is short enough to minimize disruptions to the coating process while collecting a relatively small and manageable number of data which does not require lengthy analysis, and thus enables rapid feedback. However, the data collected generally consists of averaged values over relatively large measurement volumes such that granularity and statistical value of the data is relatively poor, and does not allow for detailed analysis and understanding of the process and its variability.

An Accuraspray-g3 diagnostic system (TECNAR Automation Ltd., St-Bruno, QC, Canada) was used in this work [18]. Average particle velocities are obtained from cross-correlation of signals which are recorded at two locations in close proximity. The average particle temperatures are determined by two-color pyrometry. The other component of this system consists of a CCD camera enabling the analysis of the particle jet profile intensity appearance (position, width, distribution, intensity) along a line perpendicular to the particle jet axis at a selected stand-off distance.

For in-flight particle state, the Accuraspray-g3 provides ensemble average data representing the particle characteristics in a measurement volume of approximately 3.2 mm diameter x 25 mm depth-of-field. In other words, the measurement volume for average particle velocity and temperature can be pictured as a small cylinder extending across and covering the whole particle jet in the field-of-view direction, since in the case studied here the particle jet section is less than 25 mm at the measuring location. On the other hand, the particle jet section is bigger than the 3.2 mm diameter of the measurement volume, therefore the probed section does not capture the whole jet in the direction perpendicular to the field of view (up and down in our configuration). Hence one must carefully choose where to measure in-flight particle characteristics into the particle jet. One approach consists in measuring the data at a given fixed position relative to the gun axis. This approach is simple to implement, however little valuable in terms of providing useful in-flight particle state data since it does not account for the variations in the position of the particle jet with time, which is typically a moving target with gun hardware wear when radial injection is used. Thus, this approach can lead to probing a different population of particles, and potentially not be representative of the whole jet, depending on gun hardware wear. Interpretation of such data can be difficult.

Another approach, which is the one that was adopted in this study, consists in acquiring data at a location more directly related to the particle jet such as at the maximum particle jet profile intensity, which is in principle representative of a more significant fraction of the particles in the jet.

A preliminary study was conducted to determine specific acquisition parameters for the Accuraspray system such as stand-off distance, sampling parameters, shutter speed, signal amplification and sampling time to minimize acquisition time, with minimal operator intervention and minimal processing time for raw data analysis (offline). These parameters were kept the same throughout this work.

The collected or calculated particle jet data includes the total intensity and maximum intensity of the plume profile, the width of the profile at 70% of its maximum intensity, the maximum plume intensity location relative to its location at time zero (how much the plume is moving with hardware wear), and the average particle temperature and velocity at the maximum plume intensity location.

It was determined that the average particle characteristics were consistently measured within 0.6mm of the actual location of the maximum profile intensity. A sensitivity study performed on the actual process showed that this corresponds to a variation of less than 3% in average particle velocity or temperature.

3 Results

3.1 Evolution of Process and Coating Parameters Over Time

Gun power, deposition efficiency and coating "Characteristic A" are reported as a function cumulated operating time for each of the 10 hardware sets in **Fig. 1** (same time scale for all graphs, set identification number in chronological order, and same Y axis scale for the 10 hardware sets). One data point was collected per coating run. A few references lines are also shown to help identify trends. Significant variations or deviations are indicated by arrows. Vertical reference lines on the gun power graphs (top) indicate the maximum operating time at which the hardware sets were preventively replaced before initiating this study.

For most hardware sets, gun power (top) shows an indication of a quick initial change that can likely be associated with a break-in period, followed by a stable or slow and typically upward progression, especially towards the end of the hardware life. Two reference lines separated by a ~1% change in gun power are shown for indication purpose.

Hardware set #5 showed a different behavior compared to the other ones, with a relatively rapid gun power increase over time. This hardware set had to be replaced prematurely after substantial and unusual damages were observed by the operator. Hardware set #9 showed a sudden and significant drop in power before it was replaced. A careful investigation of that hardware set did not reveal what could have caused

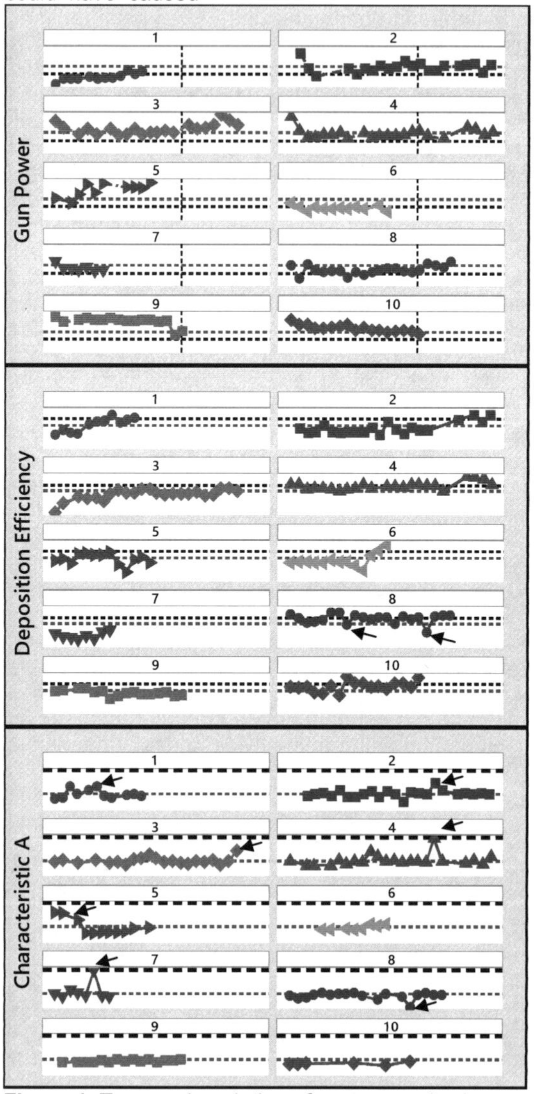

Figure 1. Temporal evolution of gun power (top), deposition efficiency (middle) and "Characteristic A" (bottom) for 10 different hardware sets.

the drop in power, which still remains unexplained. It is worth noting that gun power for that set was significantly higher compared to that measured for the other sets, before it dropped sharply.

Similar to gun power, deposition efficiency results (middle graph, **Fig. 1**) tend to indicate a stable or slight upward trend with time, especially towards the end of the hardware life.

Two reference lines separated by a ~3% change are shown for indication purpose (3% relative change is considered significant for deposition efficiency). Deposition efficiency became somewhat unstable for the last 4 to 5 runs for hardware sets #5 and #6, whereas it was consistently lower than average for set

#7. This hardware set actually had to be replaced prematurely following a gun meltdown.

Also, feedstock material feeding issues were recorded by the operator in two occasions during the use of hardware set #8, which resulted in lower than average calculated deposition efficiency, as pointed out by arrows on the graph (**Fig. 1**, middle). It is worth noting that in cases where feeding problems occurred, erroneous estimation of the quantity of material sprayed also likely occurred, and consequently calculation of the actual deposition efficiency was erroneous.

"Characteristic A" is plotted in the bottom graph of **Fig.1**. The upper limit, above which the part is rejected, is also shown. In general, Characteristic A showed very little variation around its average for all the hardware sets, confirming the stability of the process. The yield was near 99% during the period when this study was conducted, with 2 occurrences where "Characteristic A" was at the upper limit, namely once each for hardware sets #4 and #7 for which parts had to be rejected. In one case the high value of "Characteristic A" could be attributed to an auxiliary equipment malfunction, while in the second case it could be traced back to an operator error (involving the same equipment).

In a few other instances as highlighted by arrows in the graphs, "Characteristic A" was still within the acceptable limit, but deviated from its expected value. For the cases that occurred with hardware set #1, it was determined that the gage measuring "Characteristic A" was unstable. The problem was fixed and did not re-occur afterward. Two occurrences (set #2 and #8) could be linked to feedstock material feeding issues. No clear cause could be identified for the higher values measured at the end of the life hardware set #3. Hardware set #5 also showed non-typical behavior, as was the case for gun power and deposition efficiency. As explained before, that set suffered premature and severe degradation.

3.2 Evolution of Particle State Over Time

Average particle temperature and velocity at the maximum plume intensity location are reported as a function cumulated operating time for each of the 10 hardware sets in **Fig. 2**. Two reference lines separated by a ~3% change are shown on each graph for indication purpose.

Particle temperature and velocity showed relatively little or slow progression over time for most of the hardware sets studied. However, set #1 and #9 stand out from the others, showing a different behavior. For set #1, average particle temperature consistently increased with operating time, while average velocity decreased. A potential explanation could be that at lower velocity the particle dwell time in the plume is longer, hence the higher temperature. This trend is also observed, although to a smaller extent, for hardware set # 4 and #7. In-flight particle state evolution was different from all the other sets for hardware set #9. Particle temperature was initially relatively stable, but significantly higher compared to that measured for all the other sets. However, roughly half-way through, it started decreasing at a relatively high rate. Particle velocity steadily decreased from time zero for the same hardware set. As pointed out earlier gun power for that set was significantly higher compared to the others as well, which could potentially help explain why it behaved differently in terms of particle state. This set was replaced following the realization that both particle temperature and gun power were dropping quickly.

Particle velocity for sets #2 and #6 was generally significantly higher compared to that of the other sets. In one specific instance as indicated by an arrow in the associated graph for set #2, particle velocity was extremely low compared to the expected value. A note from the operator recorded on the runsheet for that particular run indicated that feedstock material "spitting" was observed initially but vanished quickly and before coating deposition started. This condition was thus captured by the in-flight particle sensor.

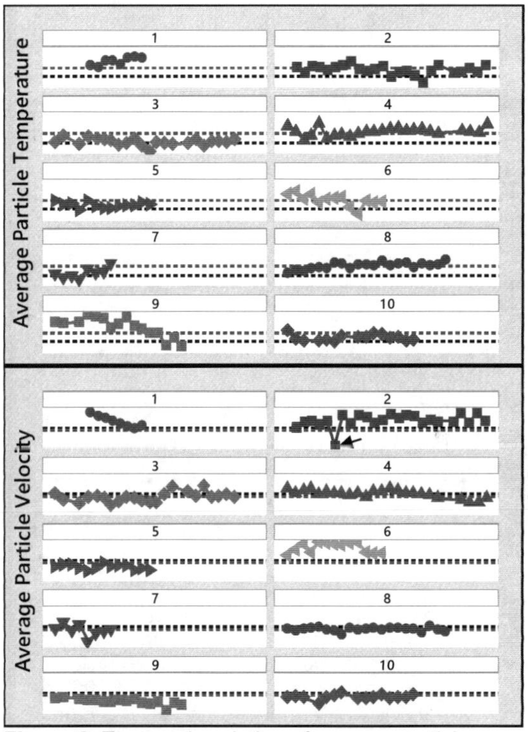

Figure 2. Temporal evolution of average particle temperature (top) and velocity (bottom) for 10 different hardware sets.

3.3 Evolution of Jet Profile Over Time

Location of the maximum particle jet profile intensity relative to its location at time zero is shown in **Fig. 3**, in addition to the total intensity and width of the plume profile as a function cumulated time for each of the 10 hardware sets. The maximum intensity of the plume profiles was also recorded, however not reported here since it showed trends very similar to that of the total intensity of the same profiles.

The maximum particle jet profile intensity relative location (top) generally showed a stable behavior, or a

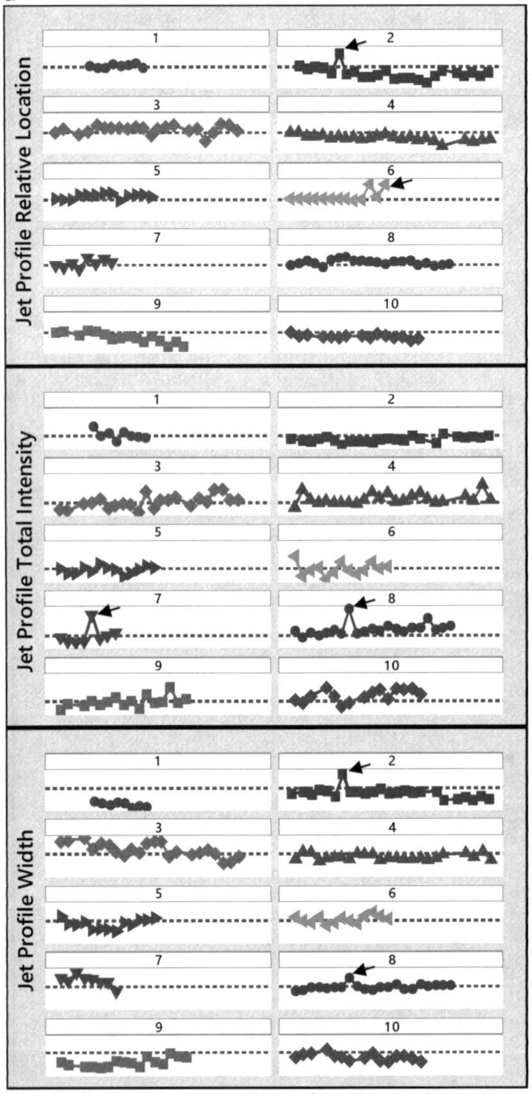

Figure 3. Temporal evolution of particle jet intensity profile. Maximum particle jet profile intensity location relative to its location at time zero (top), total intensity (middle) and width (bottom) of the plume profile.

slow and typically downward progression with time. In other words, the particle jet was fairly consistently located at the same position with regards to the plasma gun axis, and typically moved by less than 1° over the life of the hardware sets.

Hardware set #6 showed an unexpected variation at the end of its life, similar to what was observed for in-flight particle state and deposition efficiency for the same set. That set was replaced following these observations. Hardware set #9 showed a more pronounced downward slope compared to that of the other sets.

The total intensity of the particle jet profile (middle) generally shows a relatively stable or slow upward trend, similar to that observed for gun power and deposition efficiency (see **Fig. 1**). The total intensity significantly deviated from the average value in two occasions, once for set #7 and once for set #8. A lower particle velocity than normal and an increase in particle temperature was also observed for the occurrence noticed in set #7. No indication of abnormal behavior was recorded for gun or coating parameters, suggesting the process reverted to its normal condition rather quickly. Particle jet profile data collected after that specific run revealed a profile intensity in line with the average value, confirming that the process had returned to its normal state during or before coating deposition. In the case of the deviation observed for set #8, it was confirmed by the operator that a feedstock material feeding issues occurred during that run, which would explain why the particle jet profile looked different. A significant drop in deposition efficiency was also recorded for that run.

Except for hardware set #9, the particle jet profile width (bottom) generally shows a relatively stable or slow downward progression indicating that the profile tends to narrow down with time. The profiles actually narrowed by as much as 16% in some case (set #3 for example), indicating a sizable change in the particle jet profile shape. Interestingly and as noted before, it was also observed that the total (and peak) profile intensity tend to increase with time. The narrowing down of the profile, combined with its increased intensity with time suggests a transformation to a more concentrated and possibly hotter particle jet with time, potentially leading to the increase in deposition efficiency trend observed in **Fig.1**.

Again, hardware set #9 behaved differently compared to the other ones, with a widening of the particle jet with time (as opposed to narrowing), which was also accompanied with a decrease of particle temperature and velocity (**Fig. 2**). It was also observed that deposition efficiency for that set decreased with time (**Fig. 1**), which also suggests that the profile shape can be an indicator of how the fed material will deposit to form a coating.

4 Discussion

4.1 General Trends

Some general trends observed in this study are summarized in **Table 1**.

Table 1. Summary of general observations

Parameter	*Typical Evolution with Time*
Gun Power	Quick initial change followed by slow upward trend
Deposition Efficiency	Slow upward trend
"Characteristic A"	Stable
Particle	Slow progression, no general

Temperature	trend
Particle Velocity	Slow progression, no general trend
Jet Profile Position	Slow downward trend
Jet Profile Intensity	Slow upward trend
Jet Profile Width	Slow downward trend

Gun power, deposition efficiency and particle jet profile intensity all tend to increase with hardware operating time. The jet intensity profile narrows down, while its location slowly shifts down with time. No clear trend could be identified for average in-flight particle temperature and velocity. "Characteristic A" was stable for the most of this study, indicating its low sensitivity to typical process variations and hence, the resulting high yield for this coating application.

These observations suggest a correlation between deposition efficiency and the particle jet profile density, defined here as its intensity divided by its width. For instance, a narrow, more concentrated particle jet will result in a higher density number.

This correlation is confirmed in **Fig. 4**, where deposition efficiency is plotted as a function of the calculated profile density for all the data collected in this study, with the few exceptions of those where feeding problems were detected during the run, for which deposition efficiency calculations was inaccurate as mentioned before. It is worth noting that the difference between the highest and smallest data point values in this graph (**Fig. 4**) corresponds to a rather large variation of close to 30% in relative deposition efficiency. Although the scatter is fairly large in the data, a clear trend showing higher deposition efficiency with higher profile density is observed.

This finding could have some practical implications as one could envision using the profile density to estimate deposition efficiency for a given run, which could potentially lead to a better control of the coating thickness, since deposition rate is proportional to deposition efficiency. This result could also help guide development of modified spray conditions for higher and maintained deposition efficiency throughout the life of the gun hardware, which would translate into cost saving and shorter cycle times.

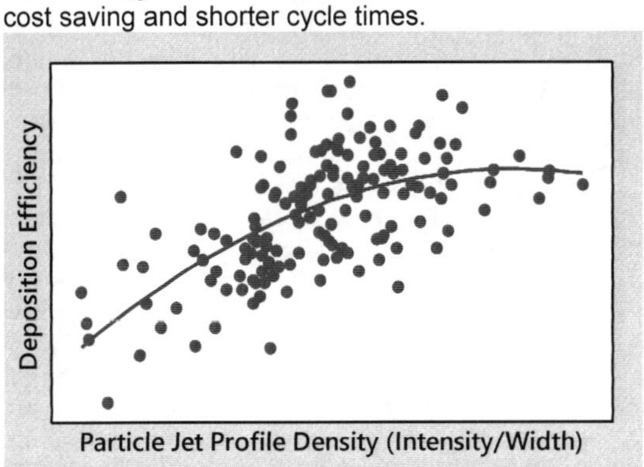

Figure 4. Deposition efficiency as a function of the calculated particle jet profile density.

Deposition efficiency is reported as a function of average in-flight particle temperature and velocity in **Fig. 5**. No clear correlation could be found between the deposition efficiency and in-flight particle data, although a potentially weak upward trend might exist with average particle temperature. This work thus confirms that the data collected with ensemble sensors such as the Accuraspray-g3 does not allow detailed analysis and understanding of the process variability, as a lot of the information is lost in the averaging process [13,19]. Therefore, the particle state can hardly be used in a predictive way, for example to forecast deposition efficiency. Nevertheless, in some cases it could potentially serve as an indicator of strong process drift or abnormal behavior, as illustrated by the rapid decrease of average particle temperature with operating time for hardware set #9 (top graph, **Fig. 2**), This was also detected by monitoring the particle jet intensity profile.

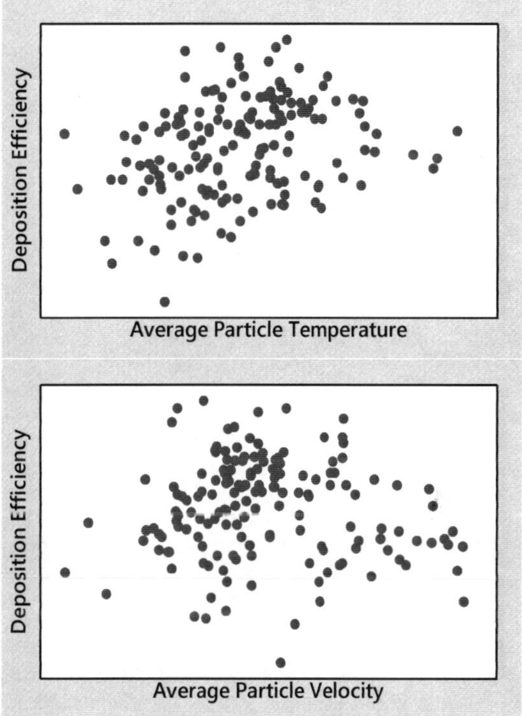

Figure 5. Deposition efficiency as a function of average particle temperature (top) and velocity (bottom).

4.2 Ability to Detect Process Deviations

In a production context, the ability to detect a condition that will potentially lead to substantial coating deviation is important so that it can preventively be corrected to avoid part rejection or sub-standard quality coatings. This work allowed to evaluate the usefulness of an in-flight particle sensor for two cases of the scenario described above. In one case, an unforeseen event occurs, such as irregularities in feeding the material to the process. In the other case, one must decide if the gun hardware has reached the end of its useful life, and needs to be replaced.

A list of unanticipated events or observations compiled during this study is presented in **Table 2**. Two columns are also included identifying if the unanticipated events were captured or not by the operator and/or the sensor.

Table 2. List of unanticipated events with columns indicating if they were captured or not by the operator or sensor.

Set#	Event	Operator?	Sensor?
8	Feeding (2)	Y	Y
2	"Spitting"	Y	Y
8	Feeding	Y	N
3,6	DE variation	N	Y
7	Unidentified	N	Y
2,4	DE variation	N	N
3,5	Characteristic A	N	N
4,7	Parts rejected	(Y)	N

The first three events listed in **Table 2** were related to feeding issues, and most of them were captured by both the operator and the sensor, although in one instance the sensor did not show indication of deviation. It is very likely that in that case the feeding issue occurred after the particle jet was monitored, and hence was not captured by the sensor. In two occasions (sets #3 and #6), a significant variation in deposition efficiency was recorded where the operator did not notice anything unusual with the process, whereas the sensor pointed to a deviation. In two similar cases where deposition efficiency varied considerably (set #2 and #4), neither the operator nor the sensor found any sign of variation. There was one instance (set #7) where the sensor detected an unidentified anomaly that the operator did not. It had no consequence on the coating process and likely reverted to its normal condition rather quickly and before the coating was applied. In two other instances (set #3 and #5), "Characteristic A" deviated from its normal value but was still within it specified limits. In both cases neither the operator nor the sensor noticed any significant change in the process. Finally, parts were rejected in two occasions (set #4 and #7), once due to an auxiliary equipment malfunction, and the other from an operator error, both of which could have been captured by the operator, but were not.

In summary, neither the operator nor the sensor alone could have captured all the unanticipated events. However, the use of the sensor combined with the observations of the operator could have allowed to detect most of the unforeseen events.

It is worth noting that since the in-flight particle state was monitored only before (and after) deposition of the coating on the parts, it could not detect any variation of the process during the actual coating operation. Monitoring the process during that time could potentially generate some more useful information, however likely with added complexity (sensor moving with the gun) and/or more impact on cycle time.

One of the goals of this study was to evaluate if the hardware life could be extended based on the collection of process, coating and in-flight particle jet data. Before that, gun hardware sets were replaced preventively after a predetermined maximum number of operating hours, or earlier when signs of drifting were suspected from the process and coating data collected. The maximum allowable number of operating hours had been established based on experience, general trends from traditional process parameters, and overall concerns that the coating performance could potentially deteriorate if the hardware was used longer, even if the measured coating characteristics where within the specification limits. Testing extended hardware life without additional and quantitative data was considered too risky. However, it was decided very early in this work that the collection of in-flight particle data added the level of confidence needed to use the hardware sets past the limit originally pre-established. Therefore, a new limit was established for preventive replacement of the gun hardware, which was arbitrarily set at roughly 33% longer than the original one.

The type (preventive or early) and reason for why the hardware sets were replaced are summarizes in **Table 3.**

Table 3. Reason for hardware set replacement

Set#	Type	Reason for change
1	Preventive	Prior criteria
2	Preventive	Reached new time limit
3	Preventive	Reached new time limit
4	Preventive	Reached new time limit
5	Early	Gun power, hardware damage
6	Early	DE up, jet profile trend
7	Early	Gun melt-down
8	Early	Unjustified (Feeding issues)
9	Early	Gun power drop, particle state
10	Early	Unjustified

Set #1 was replaced as per the criteria established before the start of this study. Set #2, #3 and #4 were replaced preventively at the new established time limit, while set #5 and #7 were replaced early, after damages were observed to the gun hardware. Set #6 and #9 were replaced early following observations of concerning behavior for process, coating and in-flight particle state parameters. Set #8 and #10 were replaced prematurely for unjustified reasons, and could more than likely have been used longer.

In summary, six of the ten hardware sets tested in this study reached or exceeded the originally established life limit (see **Fig. 1**, top (gun power) where vertical reference lines indicate the maximum operating time at which the hardware sets were preventively replaced before initiating this study). Moreover, an assessment revealed that on average the lifespan of set #2 to set #10 (9 sets total) was more than 30% longer than that of the 20 sets that had been used prior to the start of this study (data not shown). Adding in-flight particle monitoring to our process data collection increased our confidence level, and eventually led to an increase in gun hardware life of roughly 30%.

5 Conclusion

Implementation of in-flight particle monitoring in a production environment requires significant effort and commitment to determine a strategy that minimizes impact on cycle time with limited operator intervention while providing useful data. Subsequent careful analysis and interpretation of the collected data is also essential, and can be time consuming.

It was shown that adding in-flight particle monitoring to collection of process and coating data, along with the observations from the operator, can allow early detection of most of the unanticipated events, and significantly extend the life of the gun hardware without affecting the coating quality or process yield. However, ensemble particle jet monitoring does not allow for detailed analysis and understanding of the process and its variability. Nevertheless, it permits to acquire some understanding of the expected process evolution, as well as the identification of some general trends that may find practical implications. For instance, it could lead to the development of better process control for shorter coating deposition cycle with smaller variability in coating characteristics, that can translate into cost savings and more consistent or improved coating quality and performances.

6 Literature

[1] C. Moreau, P. Gougeon, M. Lamontagne, V. Lacasse, G. Vaudreuil, and P. Cielo, On-Line Control of the Plasma Spraying Process by Monitoring the Temperature, Velocity and Trajectory of In-Flight Particles, Thermal Spray Industrial Applications, C.C. Berndt and S. Sampath, Ed., ASM International, Materials Park, OH, 1994, p 431-437

[2] J. Vattulainen, E. Hämäläinen, R. Hernberg, P. Vuoristo, and T. Mäntylä, Novel Method for In-Flight Particle Temperature and Velocity Measurements in Plasma Spraying Using a Single CCD Camera, J. Therm. Spray Technol., 2001, 10(1), p 94-104

[3] V. Srinivasan, A. Vaidya, T. Streibl, M. Friis, and S. Sampath, On the Reproducibility of Air Plasma Spray Process and Control of Particle State, J. Therm. Spray Technol., 2006, 15(4), p 739-743

[4] S. Sampath, V. Srinivasan, A. Valarezo, A. Vaidya, and T. Streibl, Sensing, Control, and In Situ Measurement of Coating Properties: An Integrated Approach Toward Establishing Process-Property Correlations, *J. Therm. Spray Technol.*, 2009, 18(2), p 243-255

[5] M. Prystay, P. Gougeon, and C. Moreau, Correlation Between Particle Temperature and Velocity and the Structure of Plasma Sprayed Zirconia Coatings, Thermal Spray: Practical Solutions for Engineering Problems, C.C. Berndt, Ed., ASM International, Materials Park, OH, 1996, p 517-523

[6] M. Vardelle, A. Vardelle, A.C. Leger, P. Fauchais, and D. Gobin, Influence of the Particle Parameters at Impact on Splat Formation and Solidification in Plasma Spraying Processes, J. Therm. Spray Technol., 1995, 4(1), p 50-58

[7] P. Fauchais, M. Vardelle, A. Vardelle, and L. Bianchi, Plasma Spray: Study of the Coating Generation, Ceram. Int., 1996, 22, p 295-303

[8] P. Fauchais, M. Vardelle, A. Vardelle, L. Bianchi, and A.C. Leger, Parameters Controlling the Generation and Properties of Plasma Sprayed Zirconia Coatings, Plasma. Chem. Plasma Process., 1996, 16, p S99-S126

[9] P. Wang, S.C.M. Yu, and H.W. Ng, Correlation of Plasma Sprayed Coating Deposition Efficiency with Volume Flux Measurements by Phase Doppler Anemometry (PDA), Plasma. Chem. Plasma Process., 2007, 27, p 311-336

[10] J.-G. Legoux, B. Arsenault, L. Leblanc, V. Bouyer, and C. Moreau, Evaluation of Four High Velocity Thermal Spray Guns Using WC-10% Co-4% Cr Cermets, J. Therm. Spray Technol., 2002, 11(1), p 86-94

[11] L. Leblanc and C. Moreau, The Long-Term Stability of Plasma Spraying, J. Therm. Spray Technol., 2002, 11(3), p 380-386

[12] C.-J. Li, T. Wu, C.-X. Li, and B. Sun, Effect of Spray Particle Trajectory on the Measurement Signal of Particle Parameters Based on Thermal Radiation, J. Therm. Spray Technol., 2003, 12(1), p 80-94

[13] P. Fauchais and M. Vardelle, Sensors in Spray Processes, J. Therm. Spray Technol., 2010, 19(4), p 668-694

[14] G. Mauer, R. Vaßen, D. Stöver, Plasma and Particle Temperature Measurements in Thermal Spray: Approaches and Applications, J. Therm. Spray Technol., 2011, 20(3), p 391-406

[15] P. Fauchais, M. Vardelle and A. Vardelle, Reliability of plasma-sprayed coatings: monitoring the plasma spray process and improving the quality of coatings, Journal of Physics D: Applied Physics, 46 (2013) 224016 (16pp).

[16] Jose Colmenares-Angulo, Kentaro Shinoda, Travis Wentz, Wei Zhang, Yang Tan, Sanjay Sampath, On the Response of Different Particle State Sensors to Deliberate Process Variations, J. Therm. Spray Technol., 2011, 20(5), p 1035-1048.

[17] Wigren J and Johansson J 2011 Improving reliability of thermal spray coatings in production environment Presented at 3rd Plasma Round Table (Johannesburg, South Africa, 31 October–4 November 2011)

[18] J.F. Bisson, M. Lamontagne, C. Moreau, L. Pouliet, J. Blain, and F. Nadeau, Ensemble In-Flight Particle Diagnostics Under Thermal Spray Conditions, Thermal Spray 2001: New Surfaces for a New Millennium, C. Berndt, K.A. Khor, and E.F. Lugsheider, Ed., ASM International, Singapore, 2001, p 705-714

[19] D. Wroblewski Email author , G. Reimann, M. Tuttle, D. Radgowski, M. Cannamela, S. N. Basu, M. Gevelber, Sensor Issues and Requirements for Developing Real-Time Control for Plasma Spray Deposition, J. Therm. Spray Technol., 2010, 19(4), p 723-735

In Situ Acoustic Monitoring of Thermal Spray Process Using High-Frequency Impulse Measurements

W. Tillmann; F. Walther; W. Luo; M. Haack; J. Nellesen; M. Knyazeva; Dortmund / D

In order to guarantee their protective function, thermal spray coatings must be free from cracks, which expose the substrate surface to e.g. corrosive media. Cracks in thermal spray coatings are usually formed because of tensile residual stresses. Most commonly, the crack occurrence is determined after the thermal spraying process by examination of metallographic cross-sections of the coating. Recent efforts focus on *in situ* monitoring of crack formation by means of acoustic emission analysis. However, the acoustic signals related to crack propagation can be absorbed by the noise of the thermal spraying process.
In this work, a high-frequency impulse measurement technique was applied to separate different acoustic sources by visualizing the characteristic signal of crack formation via quasi-real-time Fourier analysis. The investigations were carried out on a twin wire arc spraying process, utilizing FeCrBSi as a coating material. The impact of the process parameters on the acoustic emission spectrum was studied. Acoustic emission analysis enables to obtain global and integral information on the formed cracks. The coating morphology as well as coating defects were inspected using light microscopy on metallographic cross-sections. Additionally, the resulting crack patterns were imaged in 3D by means of X-ray micro-tomography.

1 Introduction

Thermal spray coatings are widely applied to protect components from wear, corrosion, and heat. The coating functionality is significantly influenced by defects e.g. cracks, which are predominantly caused by process-induced residual stresses [1, 2].
A prevailing practice for quality assurance in the thermal spray industry is the optical examination of sample coatings by light or scanning electron microscopy. Applications for a *non*-destructive detection of defective coatings include e.g. dye penetrant inspection or the verification of electrical isolation by high voltage puncture tests. These techniques are commonly limited to inline inspections of readily coated components. Although there are some commercially available solutions for an in-process monitoring of in-flight particle properties, these techniques are not widely used by the industry and do not provide direct information on the structural integrity of the applied coating [3]. Therefore, new approaches for online monitoring of coating depositions are pursued e.g. by the measurement of acoustic emissions (AE) generated during crack formation [4].
A brittle cracking of a mechanically loaded solid body is evoked due to its natural behavior to reduce its energetic state by release of stored elastic energy. In the course of the crack formation, part of the released elastic energy is transmitted through the solid medium as an acoustic wave. This structure borne noise causes micro deflections at the free surface of a body and can be received by e.g. piezoelectric sensors. Hence, the acoustic signal is a carrier of information concerning the structural degradation of a solid body.
The application of AE analysis to evaluate thermal spray coatings predominantly refers to the characterization of coating failure due to mechanical loading *after* processing [5-8]. Acoustic emission is mainly utilized to detect micro- and macro-crack formations caused by a lower ductility of brittle coating materials when compared to the subjacent substrate, which results in uneven stress distributions. Investigated load cases include bending [5], tribological [6], indentation [7] and tensile tests [5, 8]. Similar mechanical stresses are induced by differences in the thermal expansion coefficient during thermal loading [9], in particular the cooling of coated parts after thermal spray processes [10].
Nishinoiri et al. used AE techniques based on non-contact measurements by means of laser-interferometry to study the effect of process parameters on crack formation in NiCrAlY-Al_2O_3 thermal barrier coatings after atmospheric plasma spraying [4]. Utilizing the Fourier analysis, two different types of acoustic events could be distinguished, which the authors associated with spalling/delamination (low frequency, large amplitude) and cohesive failure/cracking (high frequency, small amplitude), respectively.
Cracking can also occur during the coating process, because of high quenching stresses [11, 12]. Hence, an *in situ* AE measurement might be a viable concept for an online quality assurance in thermal spraying. However, the use of AE analyses for *in situ* monitoring during thermal spray processes is challenging due to the superposition of process noises and the thermal loads of the measurement equipment.

2 Experimental procedure

2.1 Thermal spaying

Due to the high potential at the field of wear and corrosion protection as well as low material costs, arc-sprayed FeCrBSi coatings are widely used in industrial applications. However, a large number of defects can appear in FeCrBSi coatings due to some unstable coating conditions, even though the process parameters were optimized previously. For this research, cored FeCrBSi wires (AS812, Durum, Germany) were employed as spray material. The coatings were deposited on quadratic specimens (100 × 100 mm²) made of medium-carbon steel (Mat.

AISI 1045, DIN C45) using the twin wire arc spray system Arc PPT-350 (Sulzer Metco, Switzerland).
An undesirable coating of the mounting device might lead to spalling and acoustic emissions not related to the actual specimen. In order to prevent a coating deposition on the clamping device, a metal plate (bezel) with a quadrate aperture (90 × 90 mm²) was set up in front of the specimen (**Fig. 1**). The bezel was acoustically uncoupled from the rest of the experimental setup by rubber paddings. To avoid an overspray, the plate was painted with anti-bond mask (Super Spray Mask, Sulzer Metco, Switzerland). Four parameter sets were selected based on a preliminary parameter study. Each parameter set was expected to result in different states of micro damages in the FeCrBSi coating. Three specimens were coated for each parameter set and the acoustic emission was measured during the coating process. Every specimen was coated with 2 overruns of the spray gun, designated as layer 1 (*L1*) and layer 2 (*L2*).

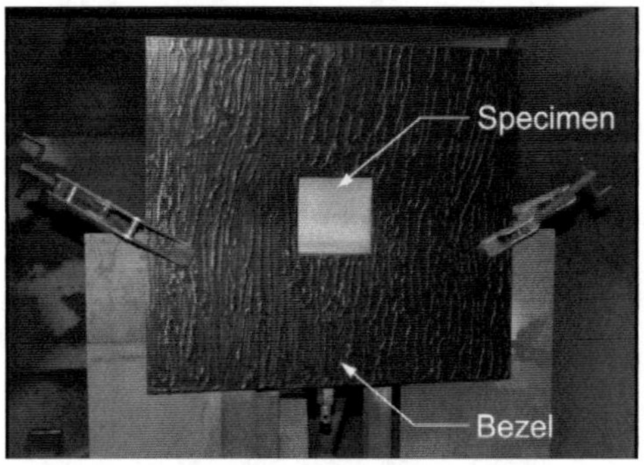

Fig. 1. Metal plate with a rectangular bezel in front of the specimen

2.2 Acoustic emission measurement

Coating deposition was monitored by a dedicated AE measurement device (HFIM: High-Frequency-Impulse-Measurement, Optimizer 4D, Qass, Germany). A piezoelectric sensor was attached to the back of the carbon steel specimens with a set screw. To prevent a loss of the piezo effect and damaging the sensor, a water-cooled dissipator was mounted between the sensor and the surface of the specimens, applying a temperature-resistant grease as a couplant. The water flow was adjusted prior to thermal spraying in order to counteract interfering acoustic signals from the turbulent flow. After the attachment of the sensor assembly to the specimen, the measurement system was set to execute an internal self-test to assure reproducible test conditions. The piezoelectric sensor temporally acts as an actuator to create a reference signal. The echo of this pulse is recorded and analyzed to obtain information on the reproducibility of the sensor-sample coupling.
The amplification, the sampling rate etc. were adjusted based on preliminary research. The data acquisition was triggered via the SPS interface by the controller of the linear motion system. The acquisition started 2 seconds prior to the actual coating process with another 160 seconds of tracking to monitor the crack formation in the cooling phase.

2.3 Signal analysis

The HFIM device conducts quasi-real time Fourier analysis of the acoustic signal. Multiple stages of signal processing had to be applied to provide information on the acoustic emissions in the secondary processing time interval t_s as well as during the cooling phase and to distinguish between the process noise and the acoustic events related to material reactions e.g. crack developments. The processing procedure was implemented as follows:

1. A quasi-real time Fourier analysis is conducted by the measurement system for process visualization of the acoustic energy $E^*(t,f)$ according to **Fig. 3a**.
2. Based on a predefined frequency mask, the process noise is identified and subtracted from the data stream (**Fig. 3b**).
3. The evolution of the acoustic energy $E(t)$ is computed by the measurement system based on the filtered dataset.
4. The data is loaded in an external SciLab program for further evaluation. Peaks associated with the passing of the thermal spray gun are recognized as process related noises E_p and subtracted from the dataset (**Fig. 5a**).
5. The remaining acoustic energy E_m is defined as the material reaction which contains information on crack formations, spalling etc. The accumulated acoustic energy $E_{a,p}$ and $E_{a,m}$ are computed and serve as characteristic variables for the process evaluation (**Fig. 4**).

2.4 Metallographic analysis

After the thermal spraying process, polished cross-sections of the coatings were investigated using a light microscope (BX51M, Olympus, Japan), thus compiling the coating morphology. The coating porosity was determined on micrographs with a magnification of 200 via the commercially available analyzing software (Stream Motion, Olympus, Japan).

3 Results and discussion

3.1 Parameterization of the thermal spray process

In order to create crack-free FeCrBSi coatings as a reference for in situ monitoring tests, a partial-factor 2^{5-1} experimental plan was utilized for the parameter optimization of the arc spraying process. In the design of experiments (DOE), the current, voltage, stand-off distance, and pressures of the primary and secondary gases were varied in three stages each, cf. **Tab. 1**. Using the DOE, 16 experiments and 3 central runs were performed.
The hardness, thickness, and porosity of the coatings were determined and evaluated after the thermal

spraying process. Optically identifying of the porosity by distinguishing the gray values, (micro-) cracks are also included in the characteristic variable. A variation of the porosity of the nineteen specimens from the partial-factor 2^{5-1} experimental plan is given in **Fig 2**. Because of the lowest porosity value, the parameter set *P4* can be recognized to result in qualitatively acceptable coatings with a low intrinsic damage. Accordingly, the parameter sets *P2*, *P3*, and *P12* were selected to create coatings with varying amounts of defects. In the context of this paper, the parameter sets are consecutively renamed *A*, *B*, *C*, *D* (**Tab. 2**).

Tab. 1 DOE: Variation of process parameters

	Current [A]	Voltage [V]	Distance [mm]	Primary [bar]	Secondary [bar]
1	180.0	28.0	100.0	6.0	3.0
2	200.0	30.0	110.0	7.5	4.0
3	220.0	32.0	120.0	9.0	5.0

Tab. 2 Spraying parameter sets

	Current [A]	Voltage [V]	Distance [mm]	Primary [bar]	Secondary [bar]
A (P2)	220.0	28.0	120.0	9.0	3.0
B (P3)	200.0	30.0	110.0	7.5	4.0
C (P4)	180.0	28.0	120.0	9.0	5.0
D (P12)	220.0	28.0	100.0	6.0	3.0

Because of the relatively cold arc spray flame (current: 180 A, voltage: 28 V), high atomization gas pressure (primary: 9 bar; secondary: 5 bar), and a long stand-off distance of 120 mm in parameter set *C*, fine and fast spray particles with a relatively low temperature are generated. These favorable features result in low residual stresses [12, 13]. On the contrary, an increase of the current and/or voltage (*A*, *B*, *C*) exalts the particle temperature; a decrease of the automation gas pressures (*B*, *D*) increases the particle size and decreases the particle velocity. A decrease of the stand-off distance (*B*, *D*) increases the coating thickness and the substrate temperature. These properties lead to high residual tensile stresses in coatings, which can cause cracking or spalling [14].

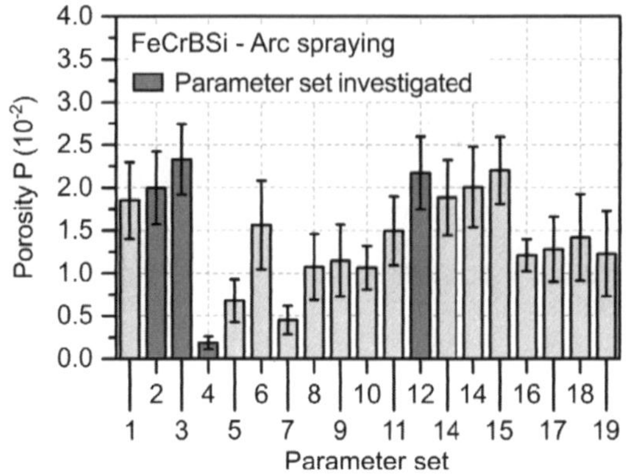

Fig. 2. Evaluation of the porosity of the arc-sprayed FeCrBSi coatings according the DOE

3.2 Acoustic emission analysis

Current signals from a piezoelectric sensor were digitized with a sampling rate f_s = 1.6 MHz; the amplification was adjusted to avoid an overdrive and subsequent loss of information. A Fourier analysis was conducted using the Hann window and 8x oversampling. In **Fig. 3a**, a section of the process landscape is depicted. The ambient noise was subtracted from the measurement data in post-processing. A corresponding frequency mask was defined by an integral over the first second of the measured data, which only comprises of the noise of the running spray gun prior to the first passing. The filtered signal *E(f,t)* (**Fig. 3b**) was converted back into the time-energy domain *E(t)* and exported for further evaluation with the external software application written in SciLab. For an appropriate representation within the scope of this paper, the data points were exported as CSV files and visualized with the scientific software Origin 9.1. In **Fig. 4** the energy *E(t)* as a function of the time interval *t*, assigned to the actual coating process (dotted line at $t \approx$ 86 sec) is exemplarily illustrated. Major periodic peaks are associated with the passing

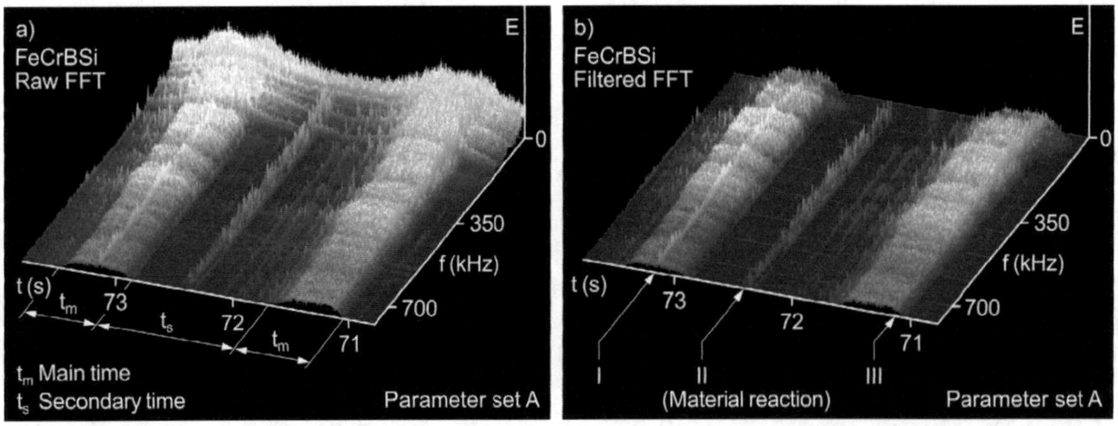

Fig. 3 Visualization of the acoustic emission $E^*(t,f)$: a) FFT as derived in online measurements; b) Filtered FFT – same time interval as in **Fig. 3a**

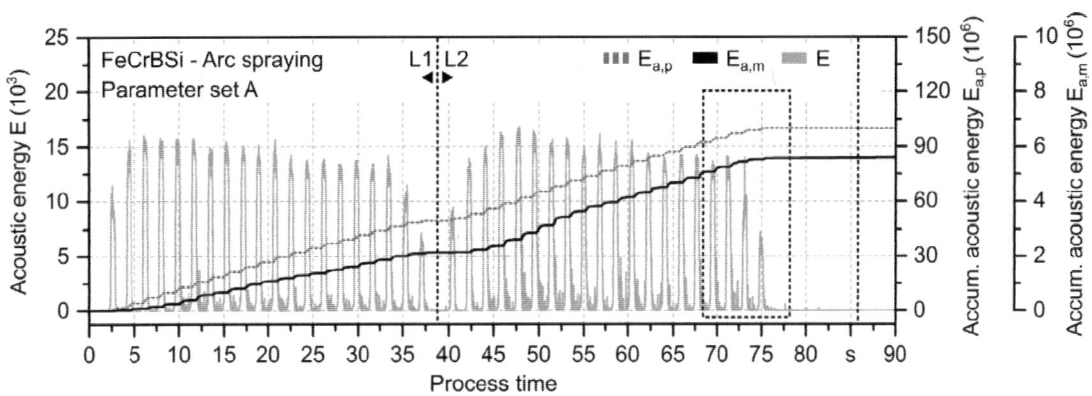

Fig. 4. Evolution of the acoustic energy components in the thermal spray process – exemplified by an experiment based on parameter set A

of the thermal spray gun (main time interval t_m, see **Fig. 3a**), while minor events at falling edges and between the peaks (secondary time interval t_s, see **Fig. 3a**) are attributed to the material reactions e.g. crack initiation. Examples of acoustic events, likely related to a material reaction, are marked in **Fig. 3b** (labelled I, II, III), see also **Fig. 5a**. Additional experiments, which are not further discussed in this paper, were conducted using low alloy steel as spray material, resulting in different acoustic emission with basically no minor peaks in the secondary time interval t_s.

For a quantitative evaluation of the process parameters, a rule for the distinction of energy components related to the coating process $E_p(t)$ and to the material reaction $E_m(t)$ in secondary time intervals t_s had to be established. The present approach comprises of a peak-finding algorithm, based on signal smoothing and threshold levels for amplitudes and a temporal extent. At the transition between the main time interval t_m and the secondary time interval t_s, material reaction-related minor peaks protrude from falling edges of major peaks, cf. **Fig. 5a**. As visible in **Fig. 4** such emission is building up over time and is generally stronger in the second layer *L2*. In order to include parts of this information in the computation of $E_p(t)$, a linear extrapolation of the falling edge is implemented (**Fig. 5a**). This approach seems to be reasonable as visible with the first 3 peaks in **Fig. 4**. At this point in time, not much coating is deposited and therefore no defect-related acoustic emission is to be expected due to low residual stresses.

The process energy $E_p(t)$ is subtracted from the raw signal $E(t)$ to derive the material reaction $E_m(t)$ as shown in **Fig. 5a.** The time interval depicted in **Fig. 5a** is marked in **Fig. 4** by a dotted line. The accumulation of the energy components $E_{a,p}$ and $E_{a,m}$ is calculated according to **Fig. 4** (note the different scaling for both components). The accumulated energy components $E_{a,p}$ and $E_{a,m}$ were adjusted to the impact of the process time, although the feed rate was not varied in this case. Finally, the accumulated value $E_{a,m}$ was divided by $E_{a,p}$ to derive a normalized value $E_{a,n}$ to (at least partly) counteract damping effects based on diverging coupling conditions.

The normalized accumulated acoustic energy $E_{a,n}$ as well as the respective arithmetic mean $\bar{E}$ are depicted in **Fig. 5b** as a function of the investigated parameter sets. The results indicate that more energy is dissipated in t_s for parameter set *D* than for the other 3 parameter sets *A*, *B*, and *C*, whereby parameter set *C* features the lowest values for $E_{a,n}$.

a)

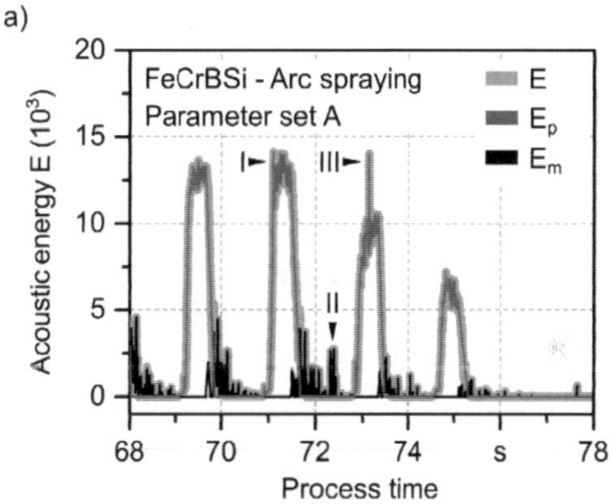

b)

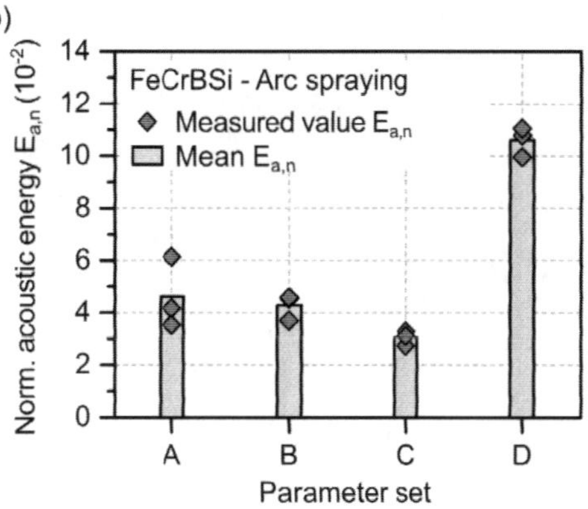

Fig. 5. a) Components of acoustic energy *E* - principle of distinction; b) Normalized

accumulated acoustic energy $E_{a,n}$ - dependence on parameter set

The difference between parameter sets *A* and *B* is not distinctive, but the values clearly differ from parameter sets *C* and *D*. The scatter in parameter set *A* is high; the range R_A is approx. 26.04E-3, which is approx. 56.31 % of the mean value $\bar{E}_A \approx 46.24$E-3. The scatter in parameter sets *B*, *C*, and *D* is considerably lower. The range R_B is approx. 8.74E-3 (≈ 20.43% of $\bar{E}_B \approx$ 42.78E-3), the range R_C is approx. 5.58E-3 (≈ 18.28% of $\bar{E}_C \approx$ 30.52E-3), and R_D is approx. 10.99E-3 (≈ 10.37% of $\bar{E}_D \approx$ 105.97E-3).

In **Fig. 6**, the process-related amount of the accumulated energy component $E_{a,p}$ as well as the material reaction $E_{a,m}$ are depicted for each parameter set. The values for $E_{a,p}$ do not display the same (resp. inverse, as $E_{a,p}$ is the denominator) quantitative trend as illustrated in **Fig. 5b**, whereas the component $E_{a,m}$ does. Therefore, it is concluded that the results presented in **Fig. 5b** can primary attributed to differences in the acoustic emission $E_m(t)$ in secondary time intervals t_s and hence, are a possible indicator for an intrinsic damage developed during quenching.

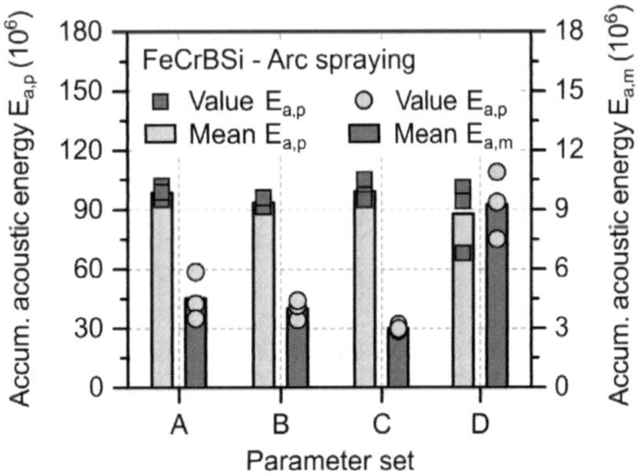

Fig. 6 Process-related accumulated acoustic energy $E_{a,p}$ – dependent on parameter set

As a result of AE analysis, the parameter set *C* is expected to have the least amount of defects, which corresponds with the results of the optical evaluation summarized in **Fig. 2**, although a pronounced manifestation of damages for the parameter set *D* as compared to parameter sets *A* and *B* is not confirmed. At this point, it has to be noted that a statement on the overall damage state is difficult by means of metallographic analyses alone, as the evaluation is limited to certain planes of dissection and may not be representative for the entire coating.

3.3 Metallographic analysis

The structure of the two coatings, which were deposited by using the parameter sets *C* and *D*, are illustrated in **Fig. 7**. Both coatings show a laminar structure, which is a typical structure of arc sprayed coatings. Because of the relative cool spray parameters, coating *C* has a dense coating structure and a tight bind to the substrate. In contrast, there are a large number of cracks in coating *D*. These cracks are not only parallel but also perpendicular to the coating surface. The parallel cracks are the binding defects between the splats, caused by the low particle velocity, which corresponds with the low atomization gas pressure (primary and secondary gas). These cracks are actually crack-shaped pores, which cannot be determined by the AE measurement. The perpendicular cracks are the breaking locations of the splats under high tensile residual stresses, which are caused by the "hot" parameter with a current of 220 A and a voltage of 28 V in parameter set *D*. The appearance of such perpendicular cracks can generate acoustic emissions, which can be determined via the AE measurement. The corresponding peaks appear in the curve after the Fourier transformation. Furthermore, if the tensile stress is high enough, the coating can be torn off from the substrate. A local spalling was clearly recognizable in **Fig. 7b**. This spalling can also causes a high peak in the AE measurement.

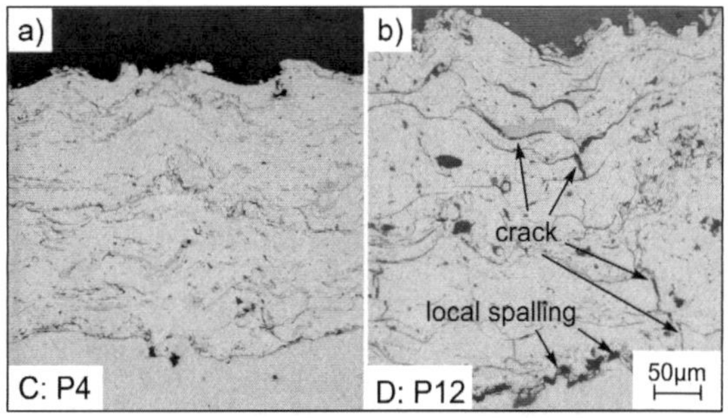

Fig. 7. Coating structure of two FeCrBSi coatings, deposited by using parameter set *C* and *D*.

3.4 Tomographic analysis

In order to inspect the quality of the coatings of the specimens C and D, tiny nearly square shaped specimens were cut from the coated sheets. These specimens exhibit a cross-sectional area of roughly 2 × 2 mm² and a height of approximately 10 mm. Using an industrial µCT scanner, which is equipped with a micro-focus reflection tube and a flat-panel detector, the specimens were tomographically imaged. Thus, the parallel plane microstructure below the coating surface can be studied in 3D. Radioscopic images of the specimens were acquired for 1,200 angular positions of the sample using X-rays with a maximum photon energy E_{ph} = 150 keV. The specimen was projected onto the detector plane with a magnification of 50. Tomograms, which reveal the microstructure of the coating and the interface between the coating and the substrate in 3D, were reconstructed from the set of projections. The tomograms contain roughly 600 × 200 × 1,600 cubic voxels which exhibit an edge length of 4 µm. This value approximately matches the achieved spatial resolution.

In **Fig. 8** exemplary longitudinal sections, which are extracted from the 3D tomograms in a plane underneath the coating surface, are depicted. The qualitative visual inspection of the 3D tomograms suggests that less vertical cracks were formed in specimen *C* than in specimen *D*. It can be recognized that the cracks form patterns or networks.

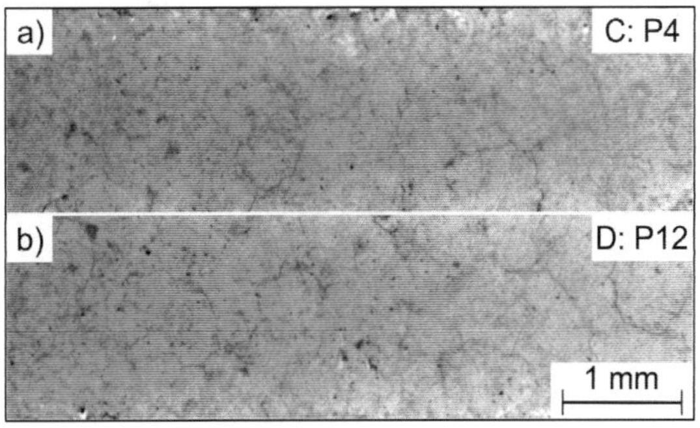

Fig. 8 Top view of specimens C and D, generated via a computer tomography measurement

4 Conclusions

Utilizing a bezel, a rapid alternation between the coating deposition and cooling phases could be achieved in an arc spraying process. This allowed to promptly record acoustic information on the FeCrBSi reaction for every overrun. The evaluation of the filtered AE signal yielded a normalized accumulated acoustic energy as a possible indicator for the damage development in thermal spray coatings after deposition. The indication of damage by the AE analysis correlates qualitatively with optical metallographic analyses, although the rank order could not be entirely validated. The scatter may be as well influenced by the measurement / coupling as well as by the reproducibility of the coating process itself.

5 Acknowledgment

The results of this paper were obtained within the scope of the IGF funded research project No. 18.710 "Non-destructive in-situ monitoring for the optimization of coating microstructure in plasma-, arc- and HVOF-based thermal spray processes – OptiMorph". We thank the research association "Welding and Related Procedures" of the DVS and the AiF e.V. for their financial support.

6 References

[1] Bolelli, G. Cannillo, V. Lusvarghi, L. and Tiziano Manfredini: Wear behaviour of thermally sprayed ceramic oxide coatings. Wear 261 (2006), Issue11-12, pp. 1298/315.

[2] López Cantera, E. and B. G. Mellor: Fracture toughness and crack morphologies in eroded WC-Co-Cr thermally sprayed coatings. Materials Letters 37 (1998), Issue4-5, pp. 201/10.

[3] Sampath, S. Srinivasan, V. Valarezo, A. Vaidya, A. and T. Streibl: Sensing, control, and in situ measurement of coating properties: an integrated approach toward establishing process-property correlations. Journal of Thermal Spray Technology 18 (2009), Issue 2, pp. 243/55.

[4] Nishinoiri, S. Enoki, M. and K. Tomita: In situ monitoring of microfracture during plasma spray coating by laser AE technique. Science and Technology of Advanced Materials 4 (2003), Issue 6, pp. 623/31.

[5] Ma, X.Q. Cho, S. and M. Takemoto: Acoustic emission source analysis of plasma sprayed thermal barrier coatings during four-point bend tests. Surface and Coatings Technology 139 (2001), Issue 1, pp. 55/62.

[6] Zhang, Z.Q. Li, G.L. Wang, H.D. Xu, B.S. Piao, Z.Y. and Li Na Zhu: Investigation of rolling contact fatigue damage process of the coating by acoustics emission and vibration signals. Tribology International 47 (2012), pp. 25/31.

[7] Faisal, N.H. Steel, J.A. Ahmed, R. and R. L. Reuben: The use of acoustic emission to characterize fracture behavior during Vickers indentation of HVOF thermally sprayed WC-Co coatings. Journal of Thermal Spray Technology 18 (2009), Issue 4, pp. 525.

[8] Yao, W.B. Dai, C.Y. Mao, W.G. Lu, C. Yang, L. and Y. C. Zhou: Acoustic emission analysis on tensile failure of air plasma-sprayed thermal barrier coatings. Surface and Coatings Technology 206 (2012), Issue 18, pp. 3803/7.

[9] Fu, L. Khor, K.A. Ng, H.W. and T. N. Teo: Non-destructive evaluation of plasma sprayed functionally graded thermal barrier coatings. Surface and Coatings Technology 130 (2000), Issue 2, pp. 233/9.

[10] Schlichting, K. Padture, N. Jordan, E. and M. Gell: Failure modes in plasma-sprayed thermal barrier coatings. Materials Science and Engineering: A 342 (2003), Issue1-2, pp. 120/30.

[11] Kuroda, S. and T. W. Clyne: The quenching stress in thermally sprayed coatings. Thin Solid Films 200 (1991), Issue 1, pp. 49/66.

[12] Sampath, S. Jiang, X. Matejicek, J. Prchlik, L. Kulkarni, A. and A. Vaidya: Role of thermal spray processing method on the microstructure, residual stress and properties of coatings, An integrated study for Ni–5 wt.%Al bond coats. Materials Science and Engineering: A 364 (2004), Issue1-2, pp. 216/31.

[13] Pourmousa, A. Mostaghimi, J. Abedini, A. and S. Chandra: Particle Size Distribution in a Wire-Arc Spraying System. Journal of Thermal Spray Technology 14 (2005), Issue 4, pp. 502/10.

[14] Newbery, A.P. Grant, P.S. and R. A. Neiser: The velocity and temperature of steel droplets

during electric arc spraying. Surface and Coatings Technology 195 (2005), Issue 1, pp. 91/101.

Epitaxial growth during the rapid solidification of plasma-sprayed molten TiO_2 splat

Shu-Wei Yao, Chang-Jiu Li*, Guan-Jun Yang, Cheng-Xin Li
State Key Laboratory for Mechanical Behavior of Materials, School of Materials Science and Engineering, Xi'an Jiaotong University, Xi'an/China / E-mail: licj@mail.xjtu.edu.cn

Epitaxial grain growth during the rapid solidification of molten TiO_2 in plasma spraying was studied. The crystallographic structure of the TiO_2 splats deposited on rutile and α-Al_2O_3 substrates at 150, 300 and 500 °C was characterized by high resolution transmission electron microscopy and electron back scattering diffraction. The results reveal that homo-epitaxial and hetero-epitaxial TiO_2 splats can be formed at the deposition temperature of 500 °C. Epitaxial growth is significantly influenced by the crystal orientation. It is easier to form an epitaxial TiO_2 splat with a <001> orientation in the direction perpendicular to the substrate surface. In order to explain the formation of epitaxial splat during plasma spraying, a competition mechanism between heterogeneous nucleation and epitaxial growth was proposed. It was indicated that the face (001) of rutile crystal exhibits the largest growth velocity, which is conducive to form an epitaxial splat for the melt with a largest undercooling degree. In addition, the effect of deposition temperature and crystalline orientation on the epitaxy was simulated. The simulation results are in agreement with the experimental observations.

1 Introduction

Plasma spray coatings are built up through the successive deposition of individual fully molten or/and semi-molten particles [1, 2]. Following the high-speed impact, the molten droplets rapidly spread and solidify into thin splats. During plasma spraying, the mean cooling rate of ceramic splat ranges from 10^4 to 10^5 K/s [3]; for metal splat, the mean cooling rate is as high as 10^6~10^8 K/s. The maximum cooling rate at the early stage of cooling process would be much higher than the average value. Owing to the rapid cooling and solidification processes, the splats and coatings always exhibit metastable phases or even amorphous phase [1, 2, 4]. In addition, the polycrystalline splats contain fine elongated grains, with the longitudinal axis perpendicular to the substrate [4-6]. Therefore, it was regarded as impossible to deposit epitaxial splats or coatings by plasma spraying.

Generally, plasma-sprayed ceramic coatings exhibit a typical lamellar structure with numerous non-bonded interface [7]. Gray et al. [8] developed a continuous columnar structure of 8YSZ (8 mol% Y_2O_3-ZrO_2) coating by controlling the deposition temperature. Observing from the transmission electron microscope (TEM) image, Sampath et al. [9] found a columnar grain extending from one splat into an adjacent splat in the continuous-columnar-structured YSZ coatings. However, there also exist non-epitaxial grains and non-bonded lamellar interface just beside the epitaxial one. Thus, both heterogeneous nucleation and epitaxial growth would take place during the deposition of continuous-columnar-structured coatings. In addition, Chraska et al. [5] revealed that epitaxial growth of columnar grains occurs in the subsequent splat, when two YSZ splats were successively deposited on the stainless steel substrate at about 450 °C. Yang et al. [4, 10] investigated the effect of deposition temperature on the crystal structure of plasma-sprayed Al_2O_3 and 8YSZ splats. They found that epitaxial Al_2O_3 splats can be formed on α-Al_2O_3 substrate at 900 °C. In addition, epitaxial growth takes place in the 8YSZ splats deposited on the cubic 8YSZ and the tetragonal 3YSZ substrates at 900 and 1200 °C, respectively.

It is a very interesting phenomenon that epitaxial growth occurs during a rapid solidification process. In addition, epitaxial growth is conducive to the design of coating microstructure and the improvement of coating properties. For example, the formation of coherent or semi-coherent interface due to epitaxy can significantly improve the interlamellar bonding strength and electrical conductivity. However, the epitaxial growth during plasma spraying has not been widely investigated. Gray et al. [8] considered that the deposit surface can be locally melted by a subsequent molten particle at a high deposition temperature, and then columnar directional solidification takes place in the incoming particle from the grains of the adjacent underlying layer. Wei et al. [11] and Zhang et al. [12] suggested that epitaxial growth might be possible when substrate melting takes place. However, as revealed by Li et al. [13] and Zhang et al. [14], substrate melting would not happen unless the melting point of sprayed material is much higher than the substrate material. Thus, the epitaxial growth in plasma-sprayed YSZ coatings does not result from local melting. On the other hand, epitaxial growth would not happen even when the substrate is locally melted by a molten particle with higher melting point. This is because there exist many factors influencing epitaxial growth, such as lattice mismatch. For example, Houben et al. [15] revealed that the steel substrate was locally melted due to the impact of Mo droplet. However, in this case, the substrate melting contributes to form an intermetallic layer instead of an epitaxial Mo splat. Thus, the mechanism of epitaxial growth during a rapid solidification process, such as plasma spraying, needs to be further discussed.

Titanium dioxide (TiO_2) is one of the most widely investigated materials owing to its excellent physical and chemical properties and broad applications [16-18]. Although polycrystalline and amorphous TiO_2 films have been widely used in functional devices [18, 19], the epitaxial TiO_2 film still attracts much interest because of its obviously superior properties compared with polycrystalline and amorphous films [16, 20, 21]. Up to now, high quality hetero-epitaxial and homo-epitaxial TiO_2 films have been successfully prepared by various processes [16, 17, 20, 22, 23], such as metal-organic chemical vapor deposition (MOCVD), and atomic layer deposition (ALD). However, it is worth mentioning that, although the above processes can fabricate high-quality epitaxial TiO_2 films, the expensive costs of processes [24], the low growth rate of film [17] and the special requirement for film deposition [22, 25] result in that these processes cannot be widely used in various fields.

Due to the excellent properties, thermal-sprayed TiO_2 coatings have been widely used [19, 24]. Under the conventional deposition conditions, the TiO_2 coatings exhibit both anatase and rutile phases with a nanocrystal structure. Since superior properties may be explored by epitaxial TiO_2 coatings, it is worth to study the epitaxial growth of TiO_2 splat during plasma spraying. In this study, single TiO_2 splats were deposited on rutile TiO_2 and α-Al_2O_3 substrates, and characterized by TEM and electron-backscattered diffraction (EBSD). The effects of deposition temperature and crystal orientation on the epitaxial growth of TiO_2 splat were revealed. Furthermore, the mechanism of epitaxial growth during rapid solidification was theoretically discussed based on the nucleation and crystal growth theories.

2 Experimental

In this study, the fuse-crushed TiO_2 powder (Sulzer-Metco, USA) with a particle size range from 10 to 45 μm was used to deposit single splats on the (110) and (001) surfaces of rutile TiO_2, polycrystalline rutile TiO_2, the $(10\bar{1}0)$ and $(1\bar{1}02)$ surfaces of α-Al_2O_3, respectively. All the substrates were mirror-polished.

A commercial plasma spray system in the 80 kW class (Jiujiang, China) was utilized to deposit TiO_2 splat. The spray parameters employed to fully melt TiO_2 particles were displayed in Table 1. As shown in Table 1, three different deposition temperatures were chosen to deposit TiO_2 splat, which was precisely controlled by a copper heating stage [26]. In order to avoid the splashing of TiO_2 splat at the deposition temperature of 150 °C, the substrates were first preheated to 250 °C and then cooled down to the preset temperature. As revealed in our previous study, a regular disk-shaped splat can be formed at these conditions [26].

Table 1 Spray conditions for TiO_2 splats

Parameters	Values
Arc power (kW)	36
Arc current (A)	520
Arc voltage (V)	70
Primary plasma gas (Ar/slpm)	45
Secondary plasma gas (H_2/slpm)	4.5
Powder feed gas (N_2/slpm)	3
Spray distance (mm)	80
Traverse speed of torch (mm/s)	1200
Deposition temperature (°C)	150, 300, 500

In this study, the microstructures of TiO_2 splat were characterized by EBSD (Oxford, UK). The EBSD measurement was performed on splat surface at the acceleration voltage of 15 kV. In addition, selected splat samples were carefully prepared utilizing the focus ion beam (FIB) technique (TESCAN, CZE) and examined using a high resolution TEM (JEOL Ltd, Japan), in order to clearly reveal the crystalline structure of splat.

3 Results and Discussion

3.1 Effect of deposition temperature on the homo-epitaxial growth of TiO_2 splats

Figure 1(a) shows a cross sectional bright field (BF) image of a TiO_2 splat deposited on the polycrystalline TiO_2 substrate at 150 °C. The TiO_2 splat is well bonded with the substrate. As shown in Fig 1(a), the top of splat exhibits a typical columnar grain microstructure. The columnar grains with a width of 100~400 nm are perpendicular to the substrate surface. There exists a transition layer with a thickness of about 150 nm at the bottom of splat. The crystalline structure of TiO_2 splat was identified by selected area electron diffraction (SAED) analysis. The SAED patterns taken from the transition layer indicates a rutile structure. In addition, the SAED pattern suggests that the transition layer is composed of nano-crystalline grains. Thus, it was concluded that epitaxial growth of TiO_2 splat cannot take place at the deposition temperature of 150 °C.

The crystal structure of TiO_2 splats deposited on TiO_2 substrate at 300 °C was characterized by TEM. As shown in Fig. 2(a), columnar grains with an average size of about 200 nm are formed on the substrate surface. At the top of splat, grain coarsening takes places. This may be caused by the release of solidification latent heat, which would decrease the solidification rate. The SAED pattern obtained from region B indicates that the TiO_2 splat exhibits a rutile structure. However, the orientation of columnar grain in splat is different from that of substrate. Thus, in this case, heterogeneous nucleation occurs on the substrate surface and then crystal nuclei grow up quickly in the opposite direction of heat transfer to from columnar grain structure.

The microstructure of TiO_2 splat deposited at 300 °C was further measured f by EBSD analysis, Fig. 3(a) shows the orientation map of TiO_2 splat, in which the splat is distinguished from the substrate by a circle. In the orientation map, sub-grain boundaries and grain boundaries are indicated by black and bold black lines, respectively. As shown in Fig. 3(a), the TiO_2 splat mainly consists of coarse grains with a diameter of tens micrometers, which is comparable to the grains size of substrate.

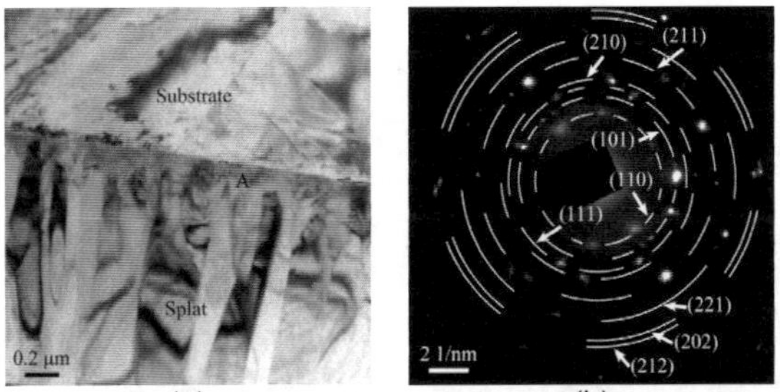

(a) (b)

Fig. 1 (a) Cross-sectional BF-TEM image of a TiO_2 splat deposited on polycrystalline rutile TiO_2 substrate at 150 °C, (b) SAED pattern corresponding to region 'A'

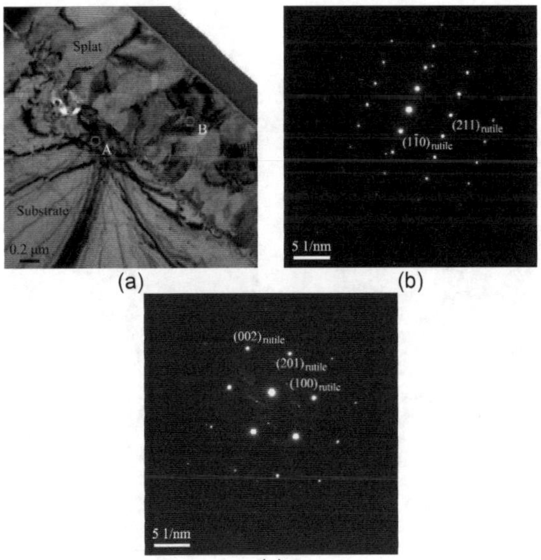

(a) (b)

(c)

Fig. 2 (a) Cross-sectional BF-TEM image of a TiO_2 splat deposited on rutile TiO_2 substrate at 300 °C. SAED pattern corresponding to (b) region 'A' and (c) region 'B'

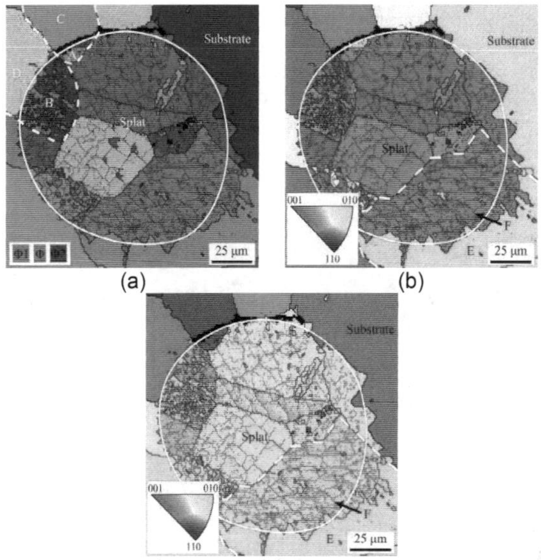

(a) (b)

(c)

Fig. 3 (a) Orientation map, (b) IPF-Y orientation map and (c) IPF-Z orientation map of a typical TiO_2 splat deposited on the polycrystalline TiO_2 structure at 300 °C

As shown in Fig. 3(a), most of the substrate grain boundaries are continuous at the periphery of splat. According to the continuity of grain boundaries, the orientation of underlying grains at the splat periphery can be inferred from the adjacent uncoated regions. Thus, as shown in Fig. 3(a), few regions exhibit the same orientation with the underlying substrate. In addition, it was confirmed that many grains of TiO_2 splats present the same orientation to the underlying grains in one direction, i.e. direction X, Y or Z, at the deposition temperature of 300 °C. This fact suggest that substrate still has some effect on the crystal structure of splat, even though epitaxy does not take place at the deposition temperature of 300 °C, which may be the main reason for the TiO_2 splat presenting a comparable grain size with the substrate.

Figure 4(a) shows the orientation map of a TiO_2 splat deposited on (110) surface of rutile TiO_2 at 500 °C. It can be found that most of regions in the splat present the same orientation with the substrate except some narrow regions at the periphery. The similar phenomenon was observed in the TiO_2 splat deposited on the (001) surface of rutile TiO_2 at 500 °C, as shown in Fig. 4(b). Thus, it was confirmed that epitaxial TiO_2 splat are formed on the low-index rutile substrate.

Figure 5(a) shows the orientation map of a typical TiO_2 splat prepared on the polycrystalline TiO_2 substrate at 500 °C. It was observed that most of the regions presents the same orientation with the substrate except some narrow periphery regions. The epitaxy of TiO_2

splat was confirmed by TEM. As shown in Fig. 5(c), the columnar grain microstructure disappears in this splat. The SAED pattern obtained from the splat/substrate interface illustrates that both the splat and the substrate exhibit the rutile structure, and the orientation of splat is same to the substrate. Thus, it was concluded that epitaxial growth can takes places when a molten TiO_2 splat is deposited on the TiO_2 substrate at 500 °C.

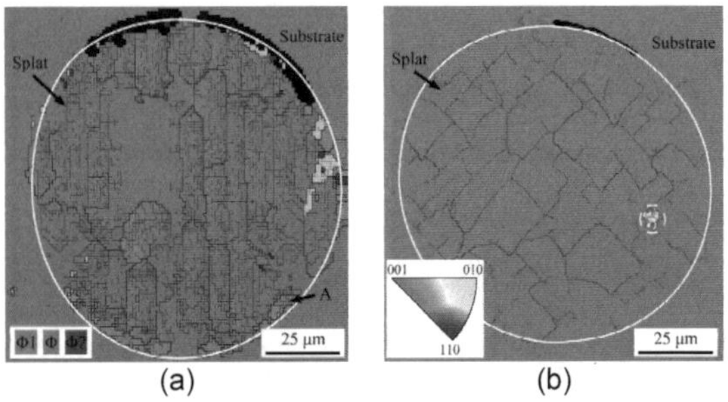

Fig. 4 (a) Orientation map of a typical TiO_2 splat deposited on the (110) surface of rutile TiO_2 at 500 °C, (b) IPF-Z orientation map of a typical TiO_2 splat deposited on the (001) surface of rutile TiO_2 at 500 °C

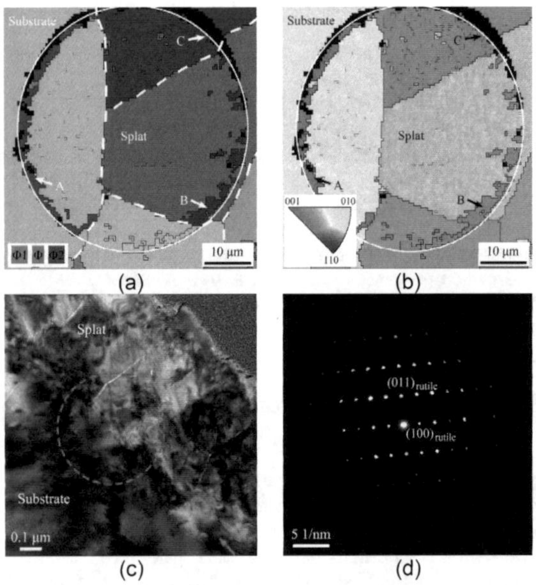

Fig. 5 (a) Orientation map, (b) IPF-Z orientation map and (c) BF-TEM image of a typical TiO_2 splat deposited on the polycrystalline TiO_2 substrate at 500 °C, (d) SAED pattern corresponding to the region med by the circle in Fig. 5(c)

3.2 Hetero-epitaxial growth of TiO_2 splats deposited on α-Al_2O_3 substrates at 500 °C

Figure 6(a) shows a cross-section BF-TEM image of a typical TiO_2 splat deposited on the (10$\bar{1}$0) surface of α-Al_2O_3 at 500 °C. It can be found that the TiO_2 splat is well bonded to the substrate. As shown in Fig. 6(a), a single crystal TiO_2 splat was obtained on the α-Al_2O_3 substrate. The crystalline structure of TiO_2 splat was identified by SAED analysis, which indicates a rutile structure. The epitaxy of the TiO_2 splat deposited on the (10$\bar{1}$0) α-Al_2O_3 surface was further confirmed by EBSD, as shown in Fig. 7(a).

Based on the pole figures, the crystal relationship between rutile TiO_2 splat and α-Al_2O_3 substrate is $(001)[100]_{rutile}//(10\bar{1}0)[2\bar{1}\bar{1}0]_{sapphire}$. As shown in Fig. 7(b), epitaxy was also observed in the TiO_2 splat deposited on the (1$\bar{1}$02) surface of α-Al_2O_3 at 500 °C with the crystallographic relationship of $(011)[100]_{rutile}//(1\bar{1}02)[1\bar{2}10]_{sapphire}$.

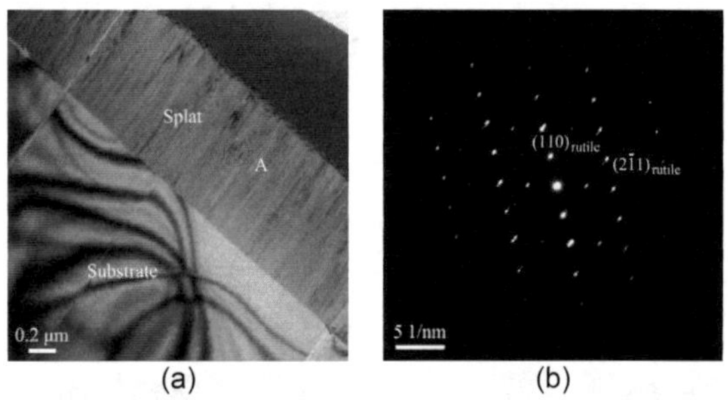

Fig. 6 (a) Cross-sectional BF-TEM image of a TiO_2 splat deposited on (10$\bar{1}$0) surface of α-Al_2O_3 substrate at 500 °C; (b) SAED pattern corresponding to region 'A'

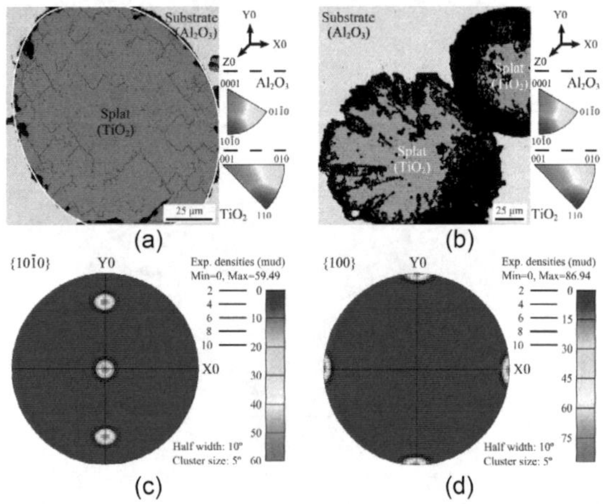

Fig. 7 IPF-Z orientation map of a TiO_2 splat deposited on (a) the (10$\bar{1}$0) α-Al_2O_3 and (b) the (1$\bar{1}$02) α-Al_2O_3 substrate at 500 °C. (c) {10$\bar{1}$0} pole figure of (10$\bar{1}$0) α-

Al_2O_3 surface, (d) {100} pole figure of the TiO_2 splat shown in Fig. 7(a)

3.3 Effect of crystalline orientation on the homo-epitaxial growth of TiO_2 splats

Generally, the spreading and solidification of molten splat are considered as two independent processes. However, as shown in Figs 4(a), 4(b) and 5(a), the periphery of splat tends to exhibit a non-epitaxial structure at 500 °C. It was considered that the molten splat would be cooled down during the rapid spreading process, which contributes to promote the solidification rate at splat periphery. Thus, it would be more difficult for epitaxy to take place at the periphery than in the center of splat.

Observed from Fig. 5(a), it was found that epitaxy occurs only at the periphery region 'C'. As shown in Fig. 5(b), both the epitaxial and the non-epitaxial grains present an orientation closed to <001> in direction Z. This fact suggests that the crystalline orientation may have effect on the epitaxial growth. In addition, as shown in Figs. 7(a) and (b), the TiO_2 splat deposited on the $(1\bar{1}02)$ surface of α-Al_2O_3 has a lower resolution rate than that deposited on the $(10\bar{1}0)$ surface. According to the pole figures, it was found that the TiO_2 splat deposited on $(10\bar{1}0)$ surface exhibits a <001> orientation in direction Z, whereas the splat on the $(1\bar{1}02)$ surface presents an orientation close to <025> in direction Z. These results suggest that it is easier for the hetero-epitaxial rutile grains with <001> orientation in direction Z to form on the α-Al_2O_3 substrate.

In order to reveal the influence of orientation on epitaxy, the grains at the periphery of TiO_2 splats, which were deposited at 500 °C, were characterized by EBSD. Figs. 8(a) and (b) show the orientation distributions of grains in direction X and Y, respectively. It was found that both the epitaxial and the non-epitaxial grains exhibit random orientations in direction X and Y. However, as show in Fig. 8(c), the orientations of non-epitaxial grains are focused on <001> in direction Z. In addition, most of the orientations of epitaxial grains also concentrate on <001> in direction Z, as shown in Fig. 8(d), although the orientations distribution of epitaxial grains is wider than the non-epitaxial grains. Thus, it was indicated that the orientation closed to <001> in direction Z are conducive to epitaxy. On the other hand, the non-epitaxial regions exhibit a <001> fiber texture in direction Z, which is related to the preferred growth of TiO_2 splat. Thus, it is confirmed that the orientation has an effect on the epitaxial growth of TiO_2 splat.

3.4 Simulation of epitaxial growth for plasma-sprayed TiO2 splats

It was confirmed that epitaxy of TiO_2 splat takes place on the rutile TiO_2 and α-Al_2O_3 substrate at the deposition temperature of 500 °C. Generally, epitaxial films are prepared by a series of processes close to equilibrium level [16, 17, 21-23, 25], in which epitaxy is mainly a thermodynamic control process [17, 20, 21]. However, plasma spray is a non-equilibrium process, during which the rapid solidification of molten splat takes place on a very short time scale. Generally, the solidification of melt involves two processes, i.e. nucleation and crystal growth processes. However, epitaxial growth is a special crystallization process in which liquid atoms condense on the substrate surface without the formation of nuclei. It was considered that epitaxial growth is in competition with the crystal nucleation when the molten splat is cooled down below its melting point. Thus, in order to form an epitaxial splat, epitaxial grains are required to grow up faster so that there is no enough time for nuclei to form on a cold substrate surface.

In this study, the time-dependent nucleation theory developed by Shao et al. was applied to simulate the rapid solidification of molten splat, which was described more detail in reference [27]. On the other hand, epitaxial growth is a special crystal growth process, during which the modified growth velocity is expressed in the following form:

$$V = \frac{\beta_{hkl} D}{a_0}[1-\exp(\frac{\Delta H_m \cdot \Delta T}{RT \cdot T_m})] \qquad (1)$$

where β_{hkl} represents the effect of orientation on the growth velocity. More details can be found in reference [28-30].

According to the crystal growth theory, the thickness of epitaxial film is required to exceed the height of critical nucleus before the nucleus forms, in order to form a perfect epitaxial splat. For simplification, a relative growth thickness (h_r) was proposed, which is equal to the ratio between the thickness of epitaxial film and the height of critical nucleus. As shown in Fig. 9, only the h_r corresponding to face {001} is slightly above 1, when critical nucleus is formed at the deposition temperature of 150 °C. When the deposition temperature increases to 300°C, the evolutions of h_r and nuclei number suggest that the epitaxial TiO_2 splat can be formed on some crystalline faces with a larger relative growth velocity. Further by increasing the deposition temperature to 500°C, the h_r corresponding to all the orientations can reach to 1 before the nucleus forms, which suggests that the epitaxial splat can be well formed on the polycrystalline rutile substrate. These results show that the calculations are in good agreement with the experimental observations.

4 Conclusions

In this study, it was revealed that the epitaxial growth of TiO_2 splat can take place during plasma spraying. The epitaxial growth is significantly influenced by the deposition temperatures. Only when the temperature increases to 500 °C, the homo-epitaxial and hetero-epitaxial TiO_2 splats can well grow on the rutile TiO_2 and α-Al_2O_3 substrates. The crystal orientation also has an important effect on the epitaxial growth. It is easier to form an epitaxial TiO_2 splat with a <001> orientation in the direction perpendicular to the substrate surface. In addition, a competition mechanism between the heterogeneous nucleation and the epitaxial growth was proposed in order to explain the phenomenon of epitaxial growth. It was revealed that the epitaxial growth is related to the undercooling degree of the melt and the crystal growth velocity.

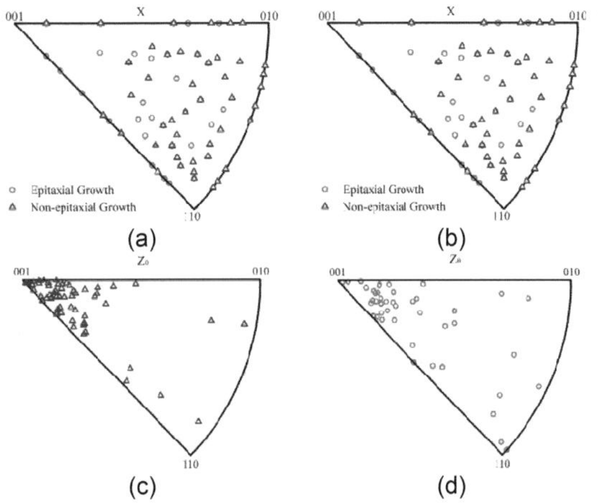

Fig. 8 Orientation distributions of periphery grains in the TiO_2 splats deposited at 500 °C in direction (a) X and (b) Y. Orientation distributions of periphery grains in the TiO_2 splats prepared at 500 °C in direction Z: (c) non-epitaxial grains and (d) epitaxial grains

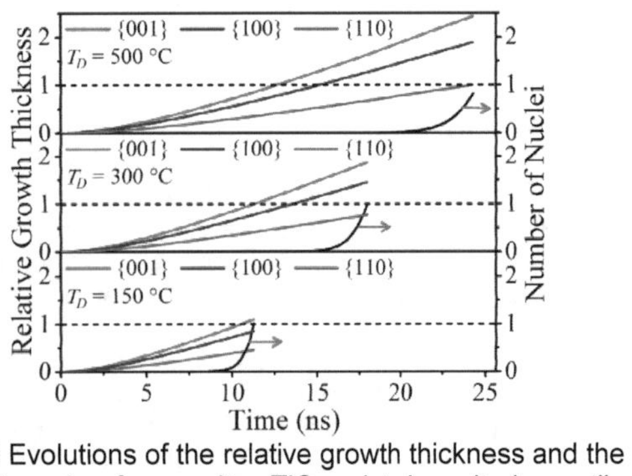

Fig. 9 Evolutions of the relative growth thickness and the nuclei number for a molten TiO_2 splat deposited on rutile substrates at 150, 300 and 500 °C

Acknowledgments

The present project is financially supported by National Nature Science Foundation of China (No. 51171144).

References

[1] L. Pawlowski, John Wiley & Sons, 2008.
[2] J.R. Davis, ASM international, 2004.
[3] L. Bianchi, A. Denoirjean, F. Blein, et al., Thin Solid Films, 299 (1997) 125-135.
[4] E.J. Yang, X.T. Luo, G.J. Yang, et al., Ceram. Int., 42 (2016) 853-860.
[5] T. Chraska, H. King, Thin Solid Films, 397 (2001) 40-48.
[6] S. Safai, H. Herman, Thin Solid Films, 45 (1977) 295-307.
[7] A. Ohmori, C.-J. Li, Thin Solid Films, 201 (1991) 241-252.
[8] D.M. Gray, Y.C. Lau, C.A. Johnson, et al., US, 1998.
[9] S. Sampath, X.Y. Jiang, J. Matejicek, et al, Mater. Sci. Eng., A, 272 (1999) 181-188.
[10] E.-J. Yang, X.-T. Luo, G.-J. Yang, et al., Surf, Coat. Technol. 274 (2015) 37-43.
[11] G. Wei. Proc. Int. Therm. Spray Conf. 2007.
[12] W. Zhang, G.H. Wei, H. Zhang, et al., Plasma Chem. Plasma Phys. 27 (2007) 717-736.
[13] L. Li, X.Y. Wang, G. Wei, et al., Thin Solid Films, 468 (2004) 113-119.
[14] H. Zhang, X.Y. Wang, L.L. Zheng, et al., Int. J Heat Mass Transf., 44 (2001) 4579-4592.
[15] J. M. Houben, G.G. Liempd, Proc. 10th Int. Therm. Spray Conf., DVS German Weld. Res. Ins. (1983) 66-71.
[16] J. Aarik, A. Aidla, H. Mandar, et al., J Cryst. Growth, 242 (2002) 189-198.
[17] S.A. Chambers, Surf. Sci. Rep., 39 (2000) 105-180.
[18] J. Jacimovic, R. Gaal, A. Magrez, et al., Appl. Phys. Lett, 102 (2013).
[19] M. Bozorgtabar, M. Rahimipour, M. Salehi., Surf. Eng. 26 (2010) 422-427.
[20] O.Y. Gorbenko, S.V. Samoilenkov, I.E. Graboy, et al., Chem. Mater., 14 (2002) 4026-4043.
[21] E.S. Ilton, T.C. Droubay, A.M. Chaka, et al., Surf. Sci., 632 (2015) 185-194.
[22] S. Chen, M.G. Mason, H.J. Gysling, et al., J Vac. Sci. Technol. A, 11 (1993) 2419-2429.
[23] M.D. McDaniel, A. Posadas, T.Q. Ngo, et al., J Vac. Sci. Technol. B, 30 (2012).
[24] C. Lee, H. Choi, C. Lee, H. Kim, Surf. Coat. Technol., 173 (2003) 192-200.
[25] Y.J. Kim, S. Thevuthasan, T. Droubay, et al., Appl. Phys. Lett. 84 (2004) 3531-3533.
[26] S.-W. Yao, C.-J. Li, J.-J. Tian, et al., Acta Mater., 119 (2016) 9-25.
[27] G. Shao, P. Tsakiropoulos, Acta Metal. Mater. 42 (1994) 2937-2942.
[28] C.G. Levi, V. Jayaram, J.J. Valencia, et al., J Mater. Res., 3 (1988) 969-983.
[29] W. Weihua, Prog. Phys. 33 (2013) 177-351.
[30] S.-W. Yao, C.-J. Li, (Unpublished Results).

Broken characteristics of the oxide film in cold spray

Yingchun XIE, Chaoyue CHEN*, Marie-Pierre PLANCHE, Rija-nirina RAOELISON, Christophe VERDY, Hanlin LIAO, ICB UMR 6303, CNRS, Univ. Bourgogne Franche-Comté, UTBM, F-90010 Belfort, France
*E-mail : chaoyue.chen@utbm.fr

Abstract: In cold spraying, oxide-free interface is an important factor for fresh metal bonding between particles and substrate, which determines the bonding strength and final coating quality. In this study, a well-designed experiment was performed to examine the deformation behaviour of the oxide film on copper alloy particle surface after deposition. The experiment results show that partial oxide film could be disrupted during the high-speed impact. However, most of the oxide films were found to remain intact after particle deposition, which limited the exposure of oxide free interface. The presence of oxide film at the interfaces between deposited particles and substrate seriously affected the metallurgical bonding. Besides, substrate material is found to have a strong influence on the deformation behaviour and final state of the oxide film. The study also demonstrated that the bonding mode between deposited particle and substrate strongly depends on the type of substrate.

Keywords: cold spray, oxide films, metallurgical bonding, high speed impact.

1 Introduction

Cold spray as a relatively new coating technique has been developed for more than 30 years. In this process, solid powders are accelerated in supersonic gas jet through a convergent-divergent nozzle. When the particles impact onto the substrate at a very high velocity, the particles undergo plastic deformation and adhere to the substrate to form the coating. Metals, polymers, composite materials and even ceramics can be deposited by cold spray [1]. Compared with conventional thermal spray, the powder used for cold spray remains in the solid state without melting during the deposition process. Thus, some defects in the thermal spray coating, such as serious oxidation, thermal residual stress and phase transformation, can be avoided in the cold sprayed coating. Besides, the coating growth is almost unlimited for most metals, which allows cold spray to be used as an additive manufacturing technique for fabricating bulk materials [1, 2].

The understanding of the bonding mechanism is essential for further development of cold spray technology. Over the past few decades, principles of the bonding mechanism have been studied by many experiments and numerical models. Currently, mechanical bonding and metallurgical bonding are considered to be the main bonding mechanisms during the formation of cold sprayed coating [3, 4]. Mechanical bonding as a major bonding mode between the coating and the substrate presents non-chemical reaction. As mechanical interlocking is a typical mechanical bonding, the hard particle can be trapped by the relative soft substrate. As another bonding mechanism, metallurgical bonding is formed due to the chemical reaction occurring at coating/substrate interfaces. Metallurgical bonding can provide high bonding strength and has been proved by some researchers through the evidence of intermetallic or amorphous phase formed at the interfacial region and the dimple-like feature appearing at the fracture surface. However, the actual mechanism for metallurgical bonding during cold spray process is still not well understood. The most prevailing hypothesis is that the particle and substrate materials undergo an extensive localized plastic deformation during the high velocity impact, which can disrupt thin oxide films on the surface of the material during the high velocity impact. The resulting oxide-free surface can provide an intimate contact between the particles and the substrate, then the metallurgical bonding occurs at particle/substrate interface under high local pressure [3, 5-7]. These studies of the metallurgical bonding mechanism are mainly focused on the formation of adiabatic shear instabilities (ASI)[1, 8]. It is suggested that the metallurgical bonding results from the nano-scale chemical reaction at the interface due to the interfacial adiabatic shear instability at high velocity impact. When the shear instability occurs, thermal softening is dominant over work hardening, thus metals behave like a viscous material and form the outward jet feature due to extrusion. The formation of this kind of jetting could be helpful in cleaning the cracked oxide film attached on the metal surfaces, and create fresh metal surfaces for the formation of high-strength metallurgical bonding [5]. Obviously, the oxide-free surface is of great importance for the metallurgical bonding between particle and substrate. The area of oxide-free surface mainly depends on the broken state of the oxide film on the surface of the particle. Although many studies suggested that the native oxide film on the metal surface can be disrupted and pushed out from the interface by the outward metal jet, some researchers observed the debris of oxide films at the coating/substrate interface [9-12]. It means that the debris of oxide films cannot be entirely cleaned up from the coating/substrate interface, some of which will remain at the interface after particle deposition. According to the TEM observation results [9-12], it is clearly seen that the broken oxide film has been included in cold sprayed coating. Yin also proved this hypothesis using simulation method [5], which suggested that the cracked oxides films can be removed in the high velocity impact.

Although TEM has been used to observe the oxide film fragments in cold sprayed coating in many studies

[9-12], the results obtained by TEM cannot fully reveal the oxide-free interface and the actual state of metallurgical bonding due to the limitations of the observation scale. The cleaning effect of the outward metal jet on the oxide film is only a conjecture and simulation result, and there is no direct experimental evidence for the moment. The true condition of the oxide-free interface in cold spray deposition is very important to understand the mechanism of metallurgical bonding and to further optimize the properties of cold sprayed coating. Therefore, it is reasonable to design experiments to characterize the broken characteristics of the oxide film in cold spray deposition, and further understand the reasons for the formation of metallurgical bonding in cold spraying. In this work, in order to deal with the above-mentioned problems, Cu alloy particles with a suitable oxide film thickness were deposited onto substrates with different hardness: Al, Cu and stainless steel (SS). A systematic investigation was conducted on the broken characteristics of the oxide film and true bonding state at the coating/substrate interface in cold spraying.

2 Materials and Experiments

In this study, the Cu alloy powders (LERMPS, FRANCE) with spherical morphology was selected as the feedstock. Fig. 1 (a) displays the spherical morphology of this powder by SEM. As shown in Fig. 1(b), the size distribution that was measured by the laser diffraction particle size analyzer (Mastersizer 2000, MALVERN Instruments, UK) presents a mean particle diameter of 51.1 µm. Substrates of aluminum, copper and stainless steel (SS) after polished were used for individual particle deposition.

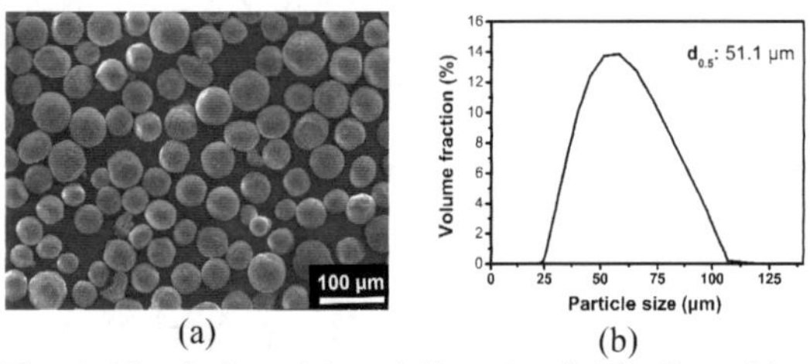

Fig. 1. Morphology (a) and diameter distributions (b) of the Cu alloy powder

In this work, a CS system (CGT K3000) with MOC nozzle (type 24 of CGT) was used to deposit single particle. High pressure compressed air were used as powder carrier gas and propelling gas. The pressure and temperature of the compressed air were about 3.0 MPa and 500°C, respectively. The standoff distance between the exit of the nozzle and the sample surface was held at 30 mm. A high gun traverse speed of 500 mm/s was used to obtain individual deposited particle. In order to observe bonding features at the particle/substrate interface, the deposited particles were detached from the different substrates. Meanwhile, A scanning electron microscope (SEM) equipped with an energy-dispersive spectroscopy (EDS) unit (JSM5800LV, JEOL, Japan) was used to characterize the deposited particle morphology.

3 Results and Discussions

Figure 2 shows the cross-sectional micrographs of a single Cu alloy particle deposited on the substrate of Al, Cu, and SS. For the Al substrate (Fig.2 a), both the Cu alloy particle and Al substrate experienced intensive deformation. The deposited Cu alloy particles was deeply embedded in the Al substrate and enveloped by the jet of substrate, so as to achieve the mechanical interlocking between the particles and the substrate. In the case of Cu alloy particle deposited on Cu substrate, as shown in the Fig. 1 (b), the Cu alloy particle and the Cu substrate can be hardly distinguished. And it was found that the interface between the similar materials was blurred through traditional SEM observation. When the Cu alloy particle deposited on the SS substrate (Fig. 1 c), the particles have undergone severe deformation. The substrate deforms to a limited extent and no metal jet coming from the substrate can be found. This relatively small deformation of substrate provides few opportunity for the formation of mechanical interlocking and metallurgical bonding even though the particle undergoes great plastic deformation and becomes much flattened. By observing the interface between particles and different substrates via SEM (Fig. 1 a-c), it seems that the Cu alloy particle was intimately bonded to the substrate.

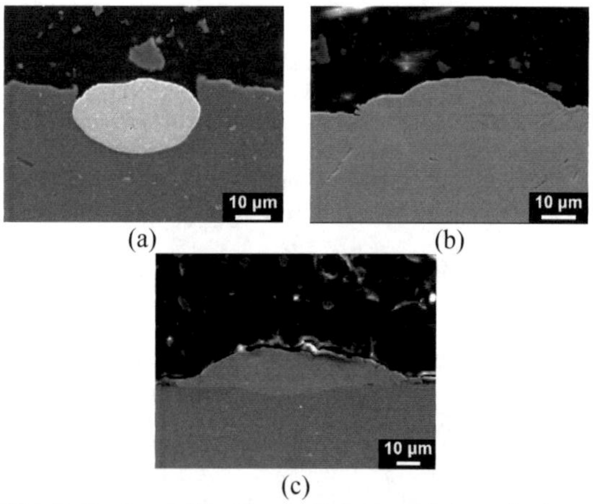

Fig. 2. Cross-sectional morphologies of deposited particle on different substrates (a) Al, (b) Cu and (c) SS

By observing the surface morphologies of the deposited Cu alloy particle on different substrates (Fig. 3), we first characterized the broken behavior of the oxide film in cold spray deposition. For all cases, the oxide film on the surface of the particles can be found to be broken. Severe plastic deformation at the edge of particle caused serious break-up of oxide film, which is a necessary condition for fresh metal exposure. On the contrary, at the center region of the sprayed particle, the degree of fragmentation of the oxide film is relatively smaller. As observed in Fig 3 (i), although the oxide film was broken into nanoscale fragments in the case of SS substrate, only a lot of micro cracks appeared on the surface of the particle , which could not form an effective fresh metal surface at the center region on the surface of deposited Cu alloy particle. In the cases of substrates Al and Cu shown in Fig 3 (g, h), the oxide film at the center region of the sprayed particle did not even break up. It is worth mentioning that, for different substrates, dissimilar crushing behaviors of oxide film are observed. With the increase of substrate hardness, the crushing of the oxide film on the surface of sprayed particle became more intensive (Fig. 3 d-f). As seen in Fig 3 (d), the oxide film were broken into larger pieces, the size of the oxide film fragments was large, reaching the micron level. For the Cu substrate (Fig. 3 e), the oxide film on the particle surface only forms some microcracks. Bare fresh metal surface was only found at the edge of particle surface deposited on the SS substrate (Fig. 3 f), which facilitates the formation of metallurgical bonding with subsequent particle impacts. In order to quantify the element content at the fresh metal surface of deposited particle, EDS scanning was conducted at the points marked in Fig. 3 (f), and the results are given in Table 1. No signs of oxygen were found at the points 1 and 3, which means that fresh metal area was exposed with the breakdown of the oxide film at the high-speed velocity impact.

Table 1 Atomic content of the elements at different positions of deposited particle surface measured by EDS.

	O	Cu	Fe	Cr
1	0.00	94.62	4.31	1.07
2	9.17	86.27	3.63	0.93
3	0.00	94.28	4.56	1.15
4	9.13	86.49	3.39	1.00

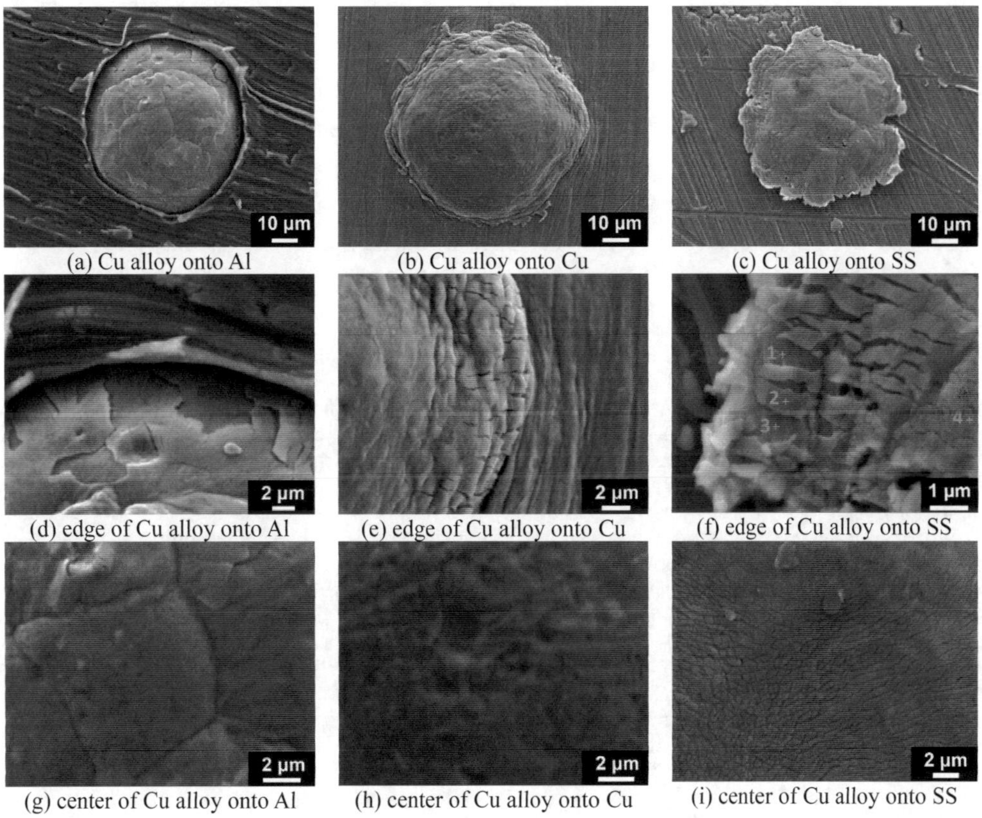

Fig. 3. Surface morphologies of deposited particle on the substrate (a) Al, (b) Cu and (c) Steel, and the EDS results on the surface of deposited particle.

The traditional study of oxide film fragmentation distribution is mainly focused on the observation of cross-sectional microstructure through TEM [9-12]. Although the presence of the oxide film can be detected, the distribution of oxide debris is still unclear due to the limitations of the detection scale. In order to directly observe the interfacial bonding between particle/substrate, the deposited Cu alloy particles were detached from different substrates. The contact surface morphology at deposited particles detached from different substrates are given in Fig. 4 (a-c). It can be found that the bonding surface of Cu alloy particle detached from Al and Cu substrates was similar that both particles experienced less deformation and no metallurgical bonding was formed (Fig. 4 a, b). However, in the case of SS substrate (Fig. 4 c), Cu alloy particle experienced larger deformation compared with the ones on Al and Cu substrates. Larger size of material jet could be observed in the case of hard SS substrate and no oxide debris were found at the periphery of deposited particle which was extruded by the viscous metal jet. At the same time, some oxide debris remains at the center-bottom region. This phenomenon is consistent with the theory of oxide film cleaning effect, which suggests that the oxide film fragments could be removed by the outward viscous metal jet. Moreover, the dimple-like fractures on the fractured bonding surface was found, which is widely considered as the sign of strong metallurgical bonding (Fig. 4 f).

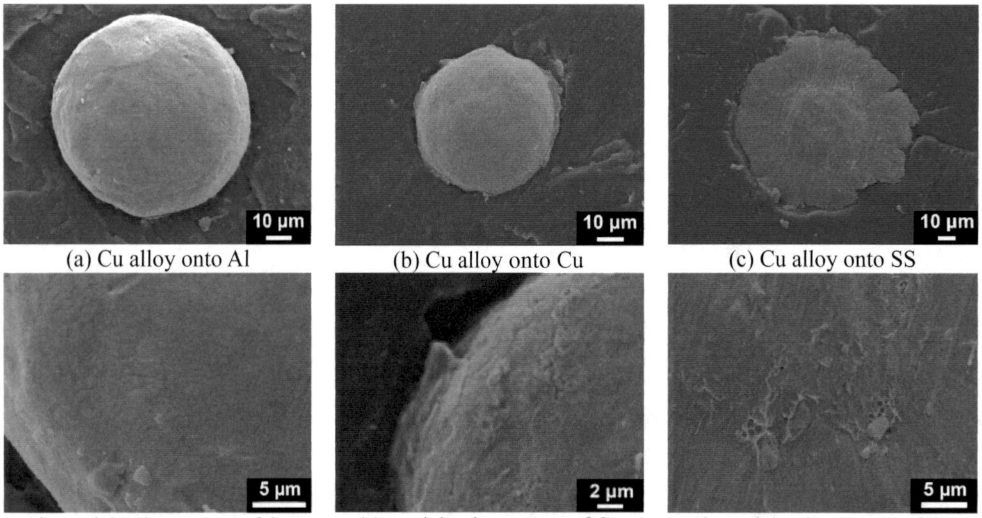

(a) Cu alloy onto Al (b) Cu alloy onto Cu (c) Cu alloy onto SS

(d) partial enlargement of Cu alloy onto Al (e) partial enlargement of Cu alloy onto Cu (f) partial enlargement of Cu alloy onto SS

Fig. 2. Bonding surface morphologies of deposited Cu alloy particle on the substrate (a) Al, (b) Cu and (c) Steel, and the EDS scan result on the bonding surface.

According to the related reports [1, 5-7, 13], it has been known that oxide-free interface is of great importance for the formation of metallurgical bonding. In order to induce metallurgical bonding, the surface oxide films should be broken up or removed to allow the exposure of oxide-free surface. As the substrate hardness increases, strong plastic deformation leads to severe compression and tensile bonding surface of deposited particle. It could be considered that the metal can flow along the interface under high shear stress during particle impact, and a large number of oxide-free surface can be exposed bare out. Thus the formation of metallurgical bonding occurs, which leads to the successful deposition of the Cu alloy particles successfully on SS substrate.

According to the above discussion, the oxide-free area on deposited particle surface caused by strong plastic deformation is the key factor in formation of metallurgical bonding at particle/substrate interface. Based on the most prevailing metallurgical bonding hypothesis [8, 14, 15], the common understanding suggests that the high-temperature at the interface caused by adiabatic shear instability is the reason for the formation of metallurgical bonding. However, the experimental findings in Fig. 4 (c, f) suggest that the metallurgical bonding only occurred around the central area, rather than a wide range of oxide-free areas which are mainly located on the back of the jet. Therefore, it can be considered that adiabatic shear instability is conducive to produce a large area of oxide-free area film, and provides the necessary conditions for the further formation of metallurgical bonding, but it is not the direct cause of the formation of metallurgical bonding. In our previous study [7], it was proposed that the localized high contact pressure is the dominant fact for the formation of metallurgical bonding under the intimate metal-to-metal contact. Thus, the dimple-like features are not available at the whole oxide-free area on the back of the jet.

However, for the substrates of Al and Cu (Fig. 4 a, b), the particle deformation is limited and no effective tangential deformation is formed. So even if the oxide film has been broken in the high velocity impact, the oxide-free surface of metal is not effectively exposed (Fig. 4 d, e). The broken pieces of oxide film directly hinder the formation of metallurgical bonding. Hence, the deposition of individual particles onto Al and Cu substrate depends entirely on mechanical interlock.

The breaking-up and extrusion of the oxide films are critical to the formation of metallurgical bonding during cold spraying. With relatively soft substrates (Al, Cu), more kinetic energy is essential to break up and extrude the oxide film, and thus a higher particle velocity is necessary for the formation of metallurgical bonding.

4 Conclusion

In this work, well-designed experiments were made to investigate the broken and distribution characteristics of the remaining oxides film at the interface between particle and different substrates in cold spray. By depositing single Cu alloy particles onto substrates of Al, Cu and SS, the crushing behavior of the oxide film in cold spray deposition was revealed for the first time. The experiment results show that, the oxide film on the surface of deposited Cu alloy particle suffered different kinds of fragmentation after high speed impact onto substrates with different hardness. The evident oxide-free surface was only found in the case of Cu alloy/SS due to severe deformation of the deposited particles. However, there is no effective oxide-free area exposed on the surface of deposited particle in the case of Al and Cu substrates, although the oxide film has been disrupted during the high velocity impact. Moreover, the metallurgical bonding was only found around the central area on the bonding surface of Cu alloy particle deposited on the steel substrate. It can be concluded that substrate hardness has a significant influence on the deformation behavior of Cu alloy particle. The deformation behavior of the particles further affects the broken characteristics of the oxide film on the surface of particle, thus determining the exposure status of oxide-free surface, which is a necessary condition for the formation of metallurgical bonding at interface between particle and substrate. This fact creates different bonding modes between deposited Cu alloy particle and different substrates. For the cases of Al and Cu substrates, mechanical interlocking is the main bonding mechanism for the successful deposition of particle, and for stainless steel substrates, metallurgical bonding is critical for the formation of coatings.

Acknowledgments

The authors would like to acknowledge the support by China Scholarship Council (CSC) and Marie Curie FP7-IPACT-268696(EU).

References

[1]. Assadi, H. and Gärtner, F. Other: Bonding mechanism in cold gas spraying. Acta Materialia 51 (2003), Issue 15, pp. 4379-94.

[2]. Alkhimov, A.P. and Kosarev, V.F. Other: A Method Of Cold Gas-Dynamic Spraying. Dokl Akad Nauk Sssr+ 315 (1990), Issue 5, pp. 1062-65.

[3]. Grujicic, M. and Saylor, J.R. Other: Computational analysis of the interfacial bonding between feed-powder particles and the substrate in the cold-gas dynamic-spray process. Applied Surface Science 219 (2003), Issue 3-4, pp. 211-27.

[4]. Xie, Y. and Planche, M.-P. Other: Effect of Substrate Preheating on Adhesive Strength of SS 316L Cold Spray Coatings. Journal of Thermal Spray Technology 25 (2015), Issue 1-2, pp. 123-30.

[5]. Yin, S. and Wang, X. Other: Deformation behavior of the oxide film on the surface of cold sprayed powder particle. Applied Surface Science 259 (2012), Issue, pp. 294-300.

[6]. Xie, Y. and Yin, S. Other: New insights into the coating/substrate interfacial bonding mechanism in cold spray. Scripta Materialia 125 (2016), Issue, pp. 1-4.

[7]. Chen, C. and Xie, Y. Other: Evaluation of the interfacial bonding between particles and substrate in angular cold spray. Materials Letters 173 (2016), Issue, pp. 76-79.

[8]. Bae, G. and Xiong, Y. Other: General aspects of interface bonding in kinetic sprayed coatings. Acta Materialia 56 (2008), Issue 17, pp. 4858-68.

[9]. Borchers, C. and Gärtner, F. Other: Microstructural bonding features of cold sprayed face centered cubic metals. Journal of Applied Physics 96 (2004), Issue 8, pp. 4288.

[10]. McCune, R.C. and Donlon, W.T. Other: Characterization of Copper Layers Produced by Cold Gas-Dynamic Spraying. Journal of Thermal Spray Technology 9 (2000), Issue 1, pp. 80.

[11]. Balani, K. and Agarwal, A. Other: Transmission electron microscopy of cold sprayed 1100 aluminum coating. Scripta Materialia 53 (2005), Issue 7, pp. 845-50.

[12]. Li, C.-J. and Li, W.-Y. Other: Formation of metastable phases in cold-sprayed soft metallic deposit. Surface and Coatings Technology 198 (2005), Issue 1-3, pp. 469-73.

[13]. Assadi, H. and Kreye, H. Other: Cold spraying – A materials perspective. Acta Materialia (2016), Issue

[14]. Bae, G. and Jang, J.-i. Other: Correlation of particle impact conditions with bonding, nanocrystal formation and mechanical properties in kinetic sprayed nickel. Acta Materialia 60 (2012), Issue 8, pp. 3524-35.

[15]. Yin, S. and Wang, X. Other: Deposition behavior of thermally softened copper particles in cold spraying. Acta Materialia 61 (2013), Issue 14, pp. 5105-18.

Essential factors influencing the bonding strength of cold sprayed aluminum coatings on ceramic substrates

R. Drehmann[1], T. Grund[1], T. Lampke[1], B. Wielage[1], C. Wüstefeld[2], M. Motylenko[2], D. Rafaja[2]

[1] Institute of Materials Science and Engineering, Chemnitz University of Technology / D

[2] Institute of Materials Science, Technische Universität Bergakademie Freiberg / D

Abstract

The present work summarizes the most important results of a research project dealing with the comprehensive investigation of the bonding mechanisms between cold sprayed Al coatings and various poly- and monocrystalline ceramic substrates (Al_2O_3, AlN, Si_3N_4, SiC, MgF_2). Due to their exceptional combination of properties, metallized ceramics are gaining more and more importance for a wide variety of applications, especially in electronic engineering. Cold spraying provides a quick, flexible and cost-effective one-step process to apply metallic coatings on ceramic surfaces. However, since most of the existing cold spray-related publications focus on metallic substrates, only very little is known about the bonding mechanisms acting between cold sprayed metals and ceramic substrates. In this paper, the essential factors influencing the bonding strength in such composites are identified. Besides mechanical tensile strength testing, a thorough analysis of the coatings and especially the metal/ceramic interfaces was conducted by means of HRTEM, FFT, STEM, EDX, EELS, XRD and EBSD. The influence of substrate material, substrate temperature and particle size is evaluated with regard to the observed bonding behavior. The results suggest that, apart from mechanical interlocking, the adhesion of cold sprayed metallic coatings on ceramics is based on a complex interplay of different mechanisms such as quasi-adiabatic shearing, static recrystallization as well as heteroepitaxial growth.

1 Introduction

The deposition of metallic coatings on ceramic surfaces is the precondition for various practical applications, especially in electronic engineering. Metal/ceramic compounds with good interfacial bonding are for instance needed in insulated-gate bipolar transistors (IGBTs), metal-oxide-semiconductor field-effect transistors (MOSFETs) and similar conductive structures on electrically isolating materials [1–4]. However, due to the significant differences in the properties of substrate and coating material, the metallization of ceramics is often still a challenge. Thus, the conventional processes to produce metallized ceramics such as Direct Copper Bonding, Active Metal Brazing, CVD, PVD or W/Mn metallization are usually energy-intensive, time-consuming and/or procedurally complex [5].

A comparatively quick and flexible one-step process to produce metallic coatings on ceramic surfaces is cold spraying. Cold spraying, which was invented in the middle of the 1980s, has meanwhile become an established method in the field of thermal spraying [6, 7]. It is a high-kinetic, low-temperature coating process, which means that the powder particles are usually accelerated to supersonic velocities while remaining well below their melting temperature. As long as the powder material exhibits sufficient ductility, it is possible to deposit dense and oxide-free coatings with bulk-like material properties.

Most of the existing cold spray-related publications focus on the coating of metallic substrates, which is the main reason that only very little is known about the bonding mechanisms between cold sprayed metals and ceramic substrates [8]. Usually mechanical clamping is considered to be the main, if not the only bonding mechanism in cold sprayed coatings. However, the application of well-adhering metallic coatings on atomically smooth, single-crystalline ceramic substrates provide evidence that there have to be additional bonding mechanisms taking effect at metal/ceramic interfaces. It is assumed that under certain conditions, the lattice of the interface-near metallic coating material adapts to the lattice of the ceramic surface by overcoming a certain lattice misfit – a phenomenon which is known as heteroepitaxy [9–11].

The atomic mobility in the coating material, which is necessary for heteroepitaxial growth, is most likely facilitated by the formation of quasiadiabatic shear bands during the impact of the powder particle on the substrate. According to numerical analyses, the transformation of the particles' kinetic energy into thermal energy upon impact results in a local temperature rise just below the melting point of the metallic material [12, 13]. Experimental observations of several authors even indicate that local melting of the interface-near particle material can occur [14–17].

The main goal of the work presented here was the investigation of cold sprayed metal coatings on ceramic substrates with respect to interfacial bonding mechanisms and the identification of various factors that have an influence on the bonding behaviour. Therefore, five different ceramic substrate materials (Al_2O_3, AlN, Si_3N_4, SiC, MgF_2) were coated with Al by means of cold spraying. In addition to the substrate material, also the substrate temperature, heat treatment regime and powder fraction were varied throughout the experiments. The coating characterization included tensile bond strength tests, microstructural analyses as well as the high-resolution examination of metal/ceramic interfaces.

2 Experimental work

For the investigations, spherical Al99.7 powder from TLS GmbH & Co. Spezialpulver KG (Bitterfeld, Germany) in three different powder fractions (−25+5 µm, −45+25 µm, −63+45 µm) was used. The polycrystalline Al_2O_3, AlN, Si_3N_4 and SiC substrates were ceramic plates from CeramTec GmbH (Plochingen, Germany) with a size of 138 x 190.5 mm^2. The thickness of the plates was 0.32 mm for Si_3N_4, 0.63 mm for Al_2O_3 (Rubalit® 708 HP) and AlN (Alunit®), and 5.5 mm for SiC (Rocar SiG® SiSiC). The polycrystalline MgF_2 substrates from U-Optics Technology, Inc. (Beijing, China) were delivered as round disks with a diameter of 82 mm and a thickness of 2 mm. The surface roughness of the substrates was R_a = 0.29 µm for Al_2O_3 and AlN, R_a = 0.47 µm for Si_3N_4 and R_a = 5.5 µm for SiC, as determined by the tactile profile method according to DIN EN ISO 3274. The MgF_2 disks were the only substrates that had to be corundum-blasted (R_a after blasting 10.6 µm) in order to achieve a coating deposition. All the other substrates were left in their as-delivered state, apart from a surface cleaning with acetone.

In addition to the described polycrystalline substrates, Al coatings were deposited on atomically smooth single-crystalline Al_2O_3, AlN and MgF_2 substrates. The 10 x 10 mm^2 α-Al_2O_3 (sapphire) substrates from CrysTec GmbH (Berlin, Germany) had a surface roughness of R_a < 0.3 nm and a thickness of 530 µm. Substrates with two different surface orientations – (001) and $(1\bar{1}2)$ – were used. The (001)-oriented AlN substrates from CrystAl-N GmbH (Fürth, Germany) were processed in two variants – with Al-polar (R_a < 0.5 nm) and N-polar surfaces (R_a < 1.7 nm). Furthermore, (001)-, (100)- and (110)-oriented MgF_2 single crystals from VM-TIM GmbH (Jena, Germany) were used for the investigations.

The Al coatings were deposited by using a CGT Kinetics 3000 cold spray system with a tungsten carbide Laval type nozzle (Type 27 TC MOC) in conjunction with a high-pressure powder feeder Praxair 1264. The spraying parameters are summarized in **Tab. 1**. In order to investigate the influence of the substrate temperature on the coating adhesion, a self-constructed sample holder with four heating elements was utilized. Three different substrate temperatures were adjusted: room temperature (no heating), 150 °C and 300 °C.

Tab. 1. Cold spray process parameters.

Parameter	Value
Process gas flow (N_2)	1000 slpm
Powder feeder gas flow (N_2)	70 slpm
Gas stagnation pressure (N_2)	2.8 MPa
Gas temperature	350 °C
Rotation velocity powder feeder	0.4 min^{-1}
Stand-off distance	30 mm
Traversal velocity	0.2–0.3 m/s
Line spacing	2 mm

Besides the substrate heating during the coating process, another method of supplying additional thermal energy to the metal/ceramic interface was realized by a heat treatment of cold sprayed Al coatings. The samples were heat treated in an argon-flushed furnace Linn High Therm VMK 39 S. The heat treatment temperature (250, 300, 350 °C) and duration (30, 60, 90 min) were varied. The tensile strength tests revealed that the subsequent heat treatment of samples which were already heated during the spray process, did not result in a significant effect on the adhesion behavior [10]. Therefore, all results of heat treated samples presented here refer to Al coatings which were deposited at room temperature (RT).

The adhesion of the coatings to the substrates was quantified by tensile bond strength measurements according to DIN EN 582. Three samples (Ø 25 mm) of each parameter set were tested, epoxy resin HTK Ultrabond® 100 was used as adhesive. The tests were carried out on a Zwick/Roell UPM 1475 universal testing machine with a drawing speed of 1 mm/min.

The microstructural investigation started with scanning electron microscopical (SEM) analyses of the coatings, accompanied by electron backscatter diffraction (EBSD) of selected samples. Orientation relationships between the coating and substrate material were determined by means of Fast Fourier Transformation (FFT) of high-resolution transmission electron micrographs (HRTEM). A 200 kV analytical HRTEM JEM 2200 FS from JEOL Ltd. (Tokyo, Japan) equipped with an ultra-high-resolution objective lens (C_s = 0.5 mm) and in-column energy filter was used for these investigations. The elemental composition of the interface-near coating and substrate material was determined by energy-dispersive X-ray spectroscopy (EDX) and electron energy loss spectroscopy (EELS) line scans across the interface.

3 Results and discussion

Especially for Al coatings on Al_2O_3 and AlN, the tensile

bond strength tests revealed a pronounced dependency of the bonding strength on the substrate temperature, **Fig. 1**. This observation was also made for the MgF_2 substrates where no coating deposition could be achieved on samples at RT or 150 °C, but only on substrates which were heated up to 300 °C. By contrast, the coatings on Si_3N_4 and SiC did not exhibit a distinct correlation of the tensile bond strength with the substrate temperature. The increase in tensile bond strength for the coatings on Al_2O_3, AlN and MgF_2 substrates is assumed to be caused by an improved atomic mobility in the interface-near coating material due to the higher substrate temperature. As pointed out before, a sufficient atomic mobility is considered to be a necessary precondition for heteroepitaxial growth [11].

The samples, which were heat treated after the deposition of the Al coatings, exhibited a very similar behavior as regards the dependency of tensile bond strength on heat treatment temperature, **Fig. 2**. Again, the coatings on Al_2O_3, AlN and, to a lesser extent, also SiC showed an increase in tensile bond strength with rising heat treatment temperature. The coatings on Si_3N_4, however, demonstrated a slightly contrary behavior. SEM images revealed large disbonded areas in the Al/Si_3N_4 interface [11]. Among the used ceramic substrates, Si_3N_4 has the lowest coefficient of thermal expansion (CTE) ($\approx 2.5 \cdot 10^{-6}\ K^{-1}$). The large difference to the CTE of Al ($23 \cdot 10^{-6}\ K^{-1}$) could be a reason for the low tensile bond strength values at increased temperatures.

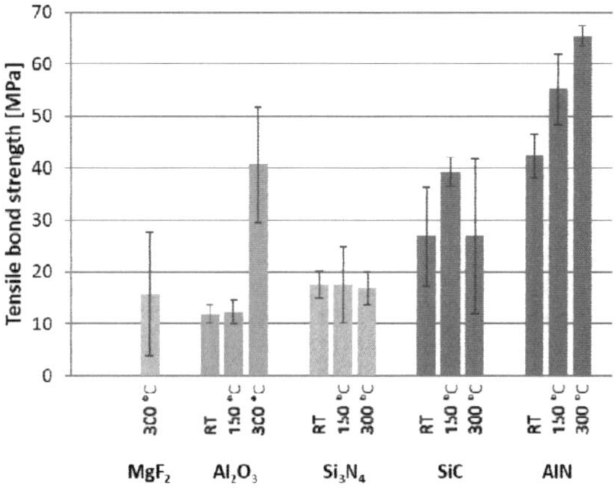

Fig. 1. Influence of substrate temperature on tensile bond strength of cold sprayed Al coatings (powder fraction –45+25 µm) on various polycrystalline ceramic substrates.

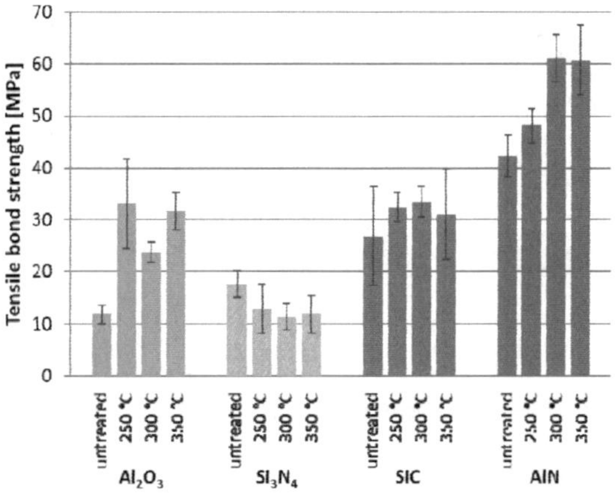

Fig. 2. Influence of heat treatment temperature on tensile bond strength of cold sprayed Al coatings (powder fraction –45+25 µm) on various polycrystalline ceramic substrates (holding time 90 min).

The microstructural characterization of the coatings by means of EBSD revealed fine-grained areas at the substrate/coating interface as well as particle/particle interfaces in an Al coating cold sprayed on AlN at RT, **Fig. 3** (left side). This observation can be explained by mechanically induced grain fining, as known from various high-strain rate deformation processes [18, 19], which can be followed by static recrystallization,

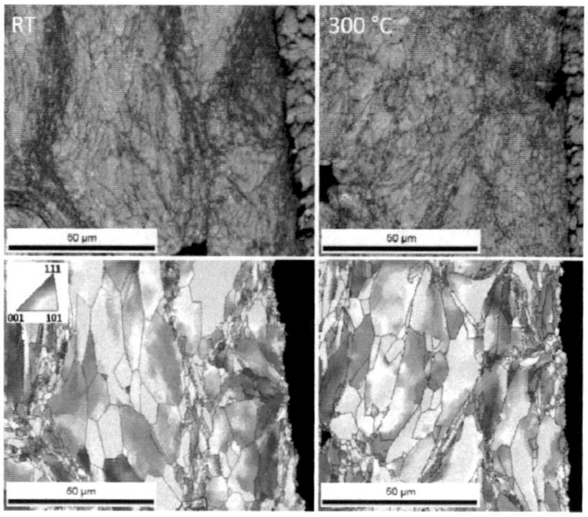

Fig. 3. EBSD patterns (top) and orientation maps (bottom) of two Al coatings (powder fraction –63+45 µm) cold sprayed on polycrystalline AlN at different substrate temperatures (RT, 300 °C).

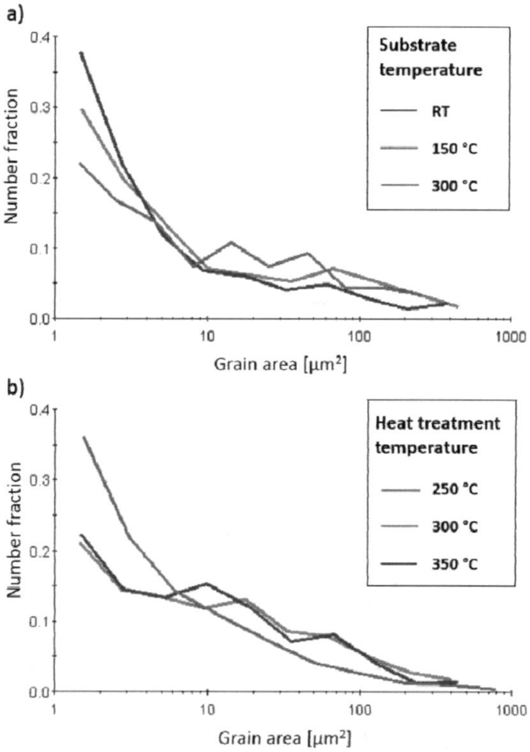

Fig. 4. Number fraction of grain area as determined by EBSD for Al coatings (powder fraction –63+45 µm) on AlN (a) deposited at three different substrate temperatures, (b) heat treated at three different temperatures.

as long as the recrystallization temperature of the coating material is reached. It is assumed that the substrate heating supports the static recrystallization in areas of the coating material, where the necessary minimum degree of deformation is reached, i.e. particularly at the substrate/coating interface and the particle/particle interfaces. These considerations are confirmed by the EBSD pattern of the Al coating deposited at 300 °C substrate temperature, **Fig. 3** (right side), and the changes in number fraction of the grain size as a function of substrate temperature, **Fig. 4a**. With rising substrate temperature, a decrease in the number fraction of small grains (< 10 µm²) could be observed, which was accompanied by an increasing number of larger grains (10–100 µm²). Again, with respect to an increase in heat treatment temperature, the behavior of heat-treated Al coatings is very similar, **Fig. 4b**.

The assessment of **Figs. 1 and 2** already indicates that the tensile bond strength is dependent on the substrate material. The comparison with various material properties of the ceramics, such as ionicity, thermal expansion behavior and hardness, did not show a clear correlation with the measured tensile bond strength values. However, the tensile bond strength seems to correlate with the heat conductivity of the substrates, **Fig. 5a**. This somewhat unexpected result can be explained by considering the thermal effusivity e of the used ceramics, which is linked to the heat conductivity λ, the density ρ and the specific heat capacity c_p by equation (1):

$$e = \sqrt{\lambda \cdot \rho \cdot c_p} \qquad (1)$$

Thermal effusivity is a measure of a material's ability to exchange thermal energy with its surroundings. Furthermore, it can be used to calculate the contact temperature T_m between two materials with the temperatures T_1 and T_2:

$$T_m = \frac{e_1 \cdot T_1 + e_2 \cdot T_2}{e_1 + e_2} \qquad (2)$$

Equation (2) is only valid for semi-infinite bodies in perfect thermal contact with constant initial temperatures – requirements which are usually not met at the particle/substrate contact in cold sprayed coatings. Nonetheless, it is justified to use the equation in comparative considerations as a first approximation.

The high thermal conductivity and thermal effusivity of AlN results in a low contact temperature, **Fig. 5b**. The lower the contact temperature, the smaller the temperature window which has to be passed by the interface-near coating material until it has cooled down to RT. Besides the difference in the coefficients of thermal expansion (CTE) of substrate and coating material, the width of this temperature window influences the residual stress in the coating. Based on the difference to the CTE of Al, which is large for all the investigated substrate materials (**Tab. 2**), the coatings on MgF_2 and Al_2O_3 should exhibit the lowest tensile residual stress and therefore a better bonding. However, the decisive influence as regards the bond strength between coating and substrate seems to be the contact temperature, since tensile bond strength increases with decreasing contact temperature, **Fig. 5b**. This can be attributed to a lower tensile residual stress in the interface-near coating material, as confirmed by GAXRD measurements which are not presented here.

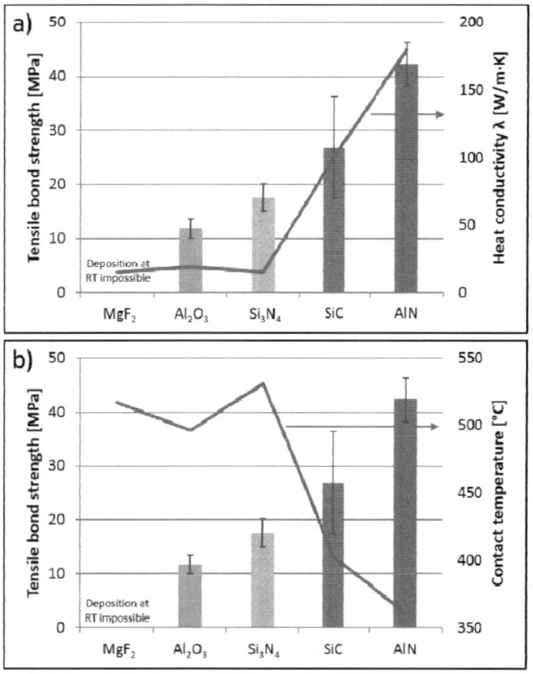

Fig. 5. Correlation of the tensile bond strength of cold sprayed Al coatings (powder fraction –45+25 µm, substrate temperature RT, no heat treatment) with (a) the substrates' heat conductivities and (b) the estimated respective contact temperatures between particle and substrate.

Tab. 2. Coefficients of thermal expansion (CTE) at RT of the coating and substrate materials [10^{-6} K^{-1}].

Al	MgF_2	Al_2O_3	AlN	SiC	Si_3N_4
23.1	8.9	7.0	4.5	4.0	2.5

Moreover, it became apparent that also the powder fraction has an influence on the bonding strength, **Fig. 6**. There are several approaches which could explain the lower tensile bond strength values of the coatings deposited with the smaller powder fractions. Firstly, due to the bigger surface-to-volume ratio of smaller particles, there is a greater influence of the brittle oxide skin on the Al particles, impeding the plastic deformation and thereby the formation of quasiadiabatic shear bands upon impact. Secondly, since all powder fractions were sprayed with the same process parameters, the smaller particles have a significantly higher impact velocity (but less kinetic energy) in comparison to the bigger ones, thus deforming with a higher strain rate. Metallic materials are characterized by an increase in flow stress with increasing strain rate, leading to a lower ductility. The strain rate sensitivity in face-centered cubic metals like Al is very low at low and medium strain rates, but increases rapidly at strain rates above 10^4 s^{-1} [20]. In cold sprayed particles, the strain rates upon impact can locally reach 10^9 s^{-1} [21], which means that the higher impact velocity of the smaller particles will result in a significantly higher flow stress in the deformation zone. Consequently, the lower deformability of the smaller particles might hinder or even prevent the formation of quasiadiabatic shear bands.

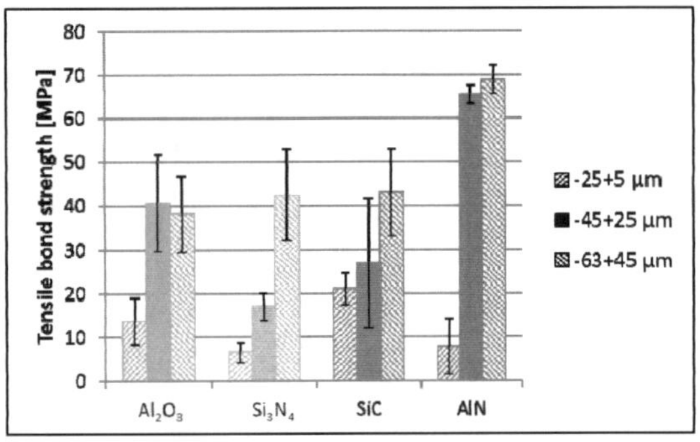

Fig. 6. Influence of powder fraction on the tensile bond strength of cold sprayed Al coatings on various ceramics (substrate temperature 300 °C, no heat treatment).

Furthermore, with regard to the smallest powder fraction –25+5 µm, it is conceivable that these particles experience a significant deceleration when passing the bow shock in front of the substrate, leading to a lower impact velocity and consequently an inferior bonding. In addition, for very small particles, the influence of heat conductivity becomes so large in relation to the particle volume that the deformation in the shear bands can no longer be considered to be quasiadiabatic, resulting in a less pronounced softening of the particle material [8].

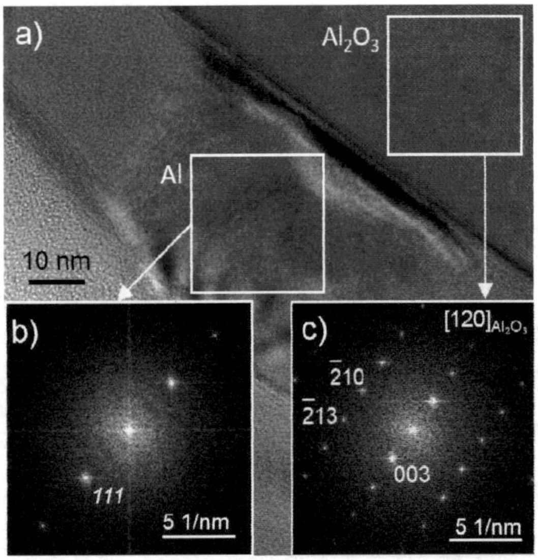

Fig. 7. HRTEM image of a cold sprayed Al coating on (001)-oriented sapphire (a) and the corresponding FFTs from adjacent areas in the Al coating (b) and in the substrate (c).

Besides the mechanical characterization of the coatings, the Al/ceramic interface areas were analysed by means of various microstructural characterization methods. FFT of HRTEM images of cold sprayed Al coatings on (001)-oriented sapphire single crystals indicated heteroepitaxial growth of the coating material. Two examples are shown in **Fig. 7** and **Fig. 9**. In one sample area, the Al is growing with its (111) plane nearly parallel to the sapphire's (001) plane, **Fig. 7**.

The parallel alignment of the planes $(111)_{Al}$ and $(001)_{Al_2O_3}$ was also observed in the model system of an Al film deposited on (001)-oriented sapphire by magnetron sputtering [11]. The orientation relationship (OR) found in the model system is illustrated in **Fig. 8**.

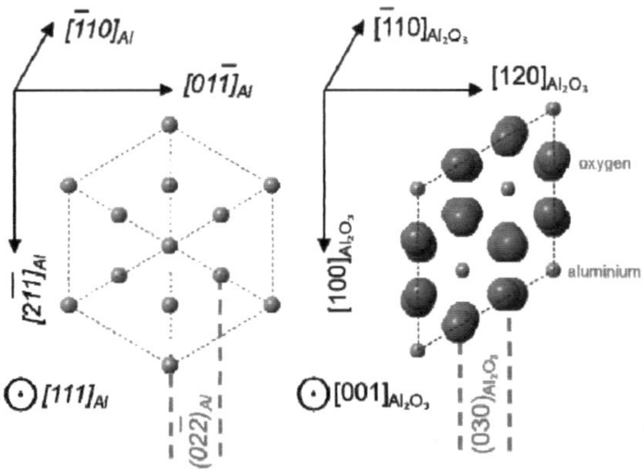

Fig. 8. Orientation relationship of Al and Al_2O_3 observed in the Al-film deposited by magnetron sputtering onto (001)-oriented sapphire.

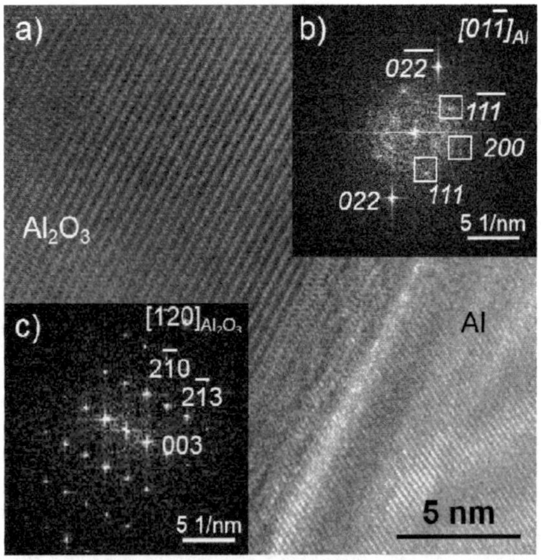

Fig. 9. HRTEM image of cold sprayed Al coating on (001)-oriented sapphire (a) and the corresponding FFTs from adjacent areas in the Al coating (b) and in the substrate (c). Weak reflections in Al are highlighted by squares.

Another sample area of the cold sprayed Al coating on (001)-oriented sapphire showed a further OR with $[120]_{Al_2O_3}$ || $[01\bar{1}]_{Al}$ and $(21\bar{3})_{Al_2O_3}$ || $(1\bar{1}\bar{1})_{Al}$, **Fig. 9**. The directions $[120]_{Al_2O_3}$ and $[01\bar{1}]_{Al}$ correspond to the plane normals of $(030)_{Al_2O_3}$ and $(02\bar{2})_{Al}$, respectively. Both planes are perpendicular to the substrate surface and have a lattice misfit of only 4.2%. The parallel alignment of the upright lattice planes $(030)_{Al_2O_3}$ and $(02\bar{2})_{Al}$ was already observed in the model system, **Fig. 8**, which indicates that also the cold sprayed Al coating prefers a good conformity in the interplanar spacings in order to stick to the Al_2O_3 substrate.

This affects also the bond strength of the cold sprayed Al coating deposited onto sapphire single crystals, as proven by tensile bond strength measurements of cold sprayed Al coatings on (001)- and $(1\bar{1}2)$-oriented sapphire substrates (Ø 25 mm, thickness 500 µm), **Fig. 10**. First of all, taking into consideration that the sapphire substrates are atomically smooth, the average tensile bond strength test result of 16 MPa for the Al coating on the (001)-oriented sapphire is remarkable, as it is solely caused by chemico-physical interface interactions. The superior bond strength of the cold gas sprayed Al coating on the (001)-oriented sapphire substrate as compared to the $(1\bar{1}2)$-oriented sapphire substrate correlates with the better match of Al to the (001) orientation, which could be confirmed by interfacial investigations of magnetron sputtered Al films on single crystals with the two mentioned surface orientations. Furthermore, the comparison with the Al coating on the comparatively rough, polycrystalline Al_2O_3 substrate shows that mechanical clamping definitely has an influence on the bonding strength, **Fig. 10**.

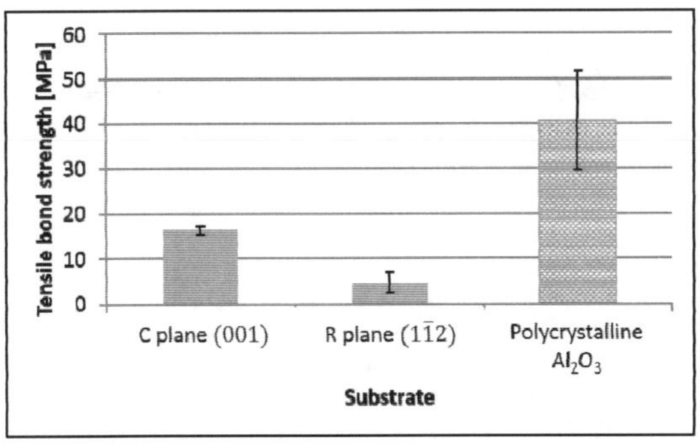

Fig. 10. Influence of the lattice orientation of single-crystalline sapphire substrates on the tensile bond strength of cold sprayed Al coatings and comparison with respective Al coatings on polycrystalline Al_2O_3 (substrate temperature 300 °C, powder fraction −45+25 µm).

The interface characterization and determination of orientation relationships of cold gas sprayed Al coatings on heated AlN substrates (300 °C) (space group $P6_3mc$) also indicated the occurrence of heteroepitaxy, as described in detail in Ref. [22]. One example is shown in **Fig. 11**. The analysis of the local OR revealed $[100]_{AlN}$ || $[101]_{Al}$ and $(011)_{AlN}$ || $(\bar{1}\bar{1}1)_{Al}$. The directions $[100]_{AlN}$ and $[101]_{Al}$ are the plane normals to $(2\bar{1}0)_{AlN}$ and $(202)_{Al}$, respectively. Both planes are perpendicular to the Al/AlN interface, have a lattice misfit of -7.9 % and belong to the orientation type $\{220\}_{Al}$ || $\{110\}_{AlN}$. The parallel alignment of the planes $\{220\}_{Al}$ and $\{110\}_{AlN}$ being perpendicular to the Al/AlN interface was observed for all investigated AlN substrates (polycrystalline / (001)-oriented N-polar single crystal / (001)-oriented Al-polar single crystal) as well as in the model system of an Al film deposited onto (001)-oriented AlN by magnetron sputtering [22].

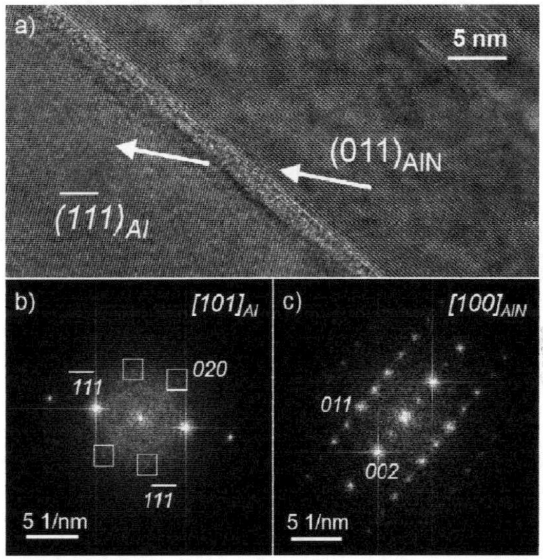

Fig. 11. HRTEM image of a cold sprayed Al coating on (001)-oriented Al-polar AlN single crystal (a) and the corresponding FFTs from the Al coating (b) and the substrate (c). The arrows in (a) indicate the plane normal of parallel lattice planes.

In addition to the parallel alignment of the upright lattice planes $\{220\}_{Al}$ and $\{110\}_{AlN}$ at least one additional pair of parallel lattice planes being perpendicular or inclined to the Al/AlN interface could be identified in all cold sprayed Al coatings. In the example given in **Fig. 11** the second set of parallel lattice planes is $(011)_{AlN}$ || $(\bar{1}\bar{1}1)_{Al}$ which is inclined by 62° to the Al/AlN interface. The lattice mismatch is only -1.3 %. The interface characterization in the cold sprayed Al coatings on AlN substrate indicated that the heteroepitaxial growth is mainly driven by a good conformity in the interplanar spacings, i.e. a small lattice mismatch, rather than by a conformity in the symmetry operations of substrate and coating material [22].

High-resolution analyses of the interface area between cold sprayed Al and Al-polar AlN single-crystal revealed the existence of a transition zone of about 2 nm thickness between substrate and coating material, **Fig. 12**. By means of EDX/EELS line scans across the interface, an increased oxygen concentration in this transition zone was detected [22]. This corresponds with XPS investigations of several authors [23, 24] that provided evidence for the existence of thin films consisting of Al_2O_3, AlOOH and $Al(OH)_3$ on the surface of AlN samples. It is not yet been clarified if and to what extent the presence of oxygen influences the bonding strength between Al and AlN, since similar transition zones were also found in Al/SiC and Al/MgF_2 samples.

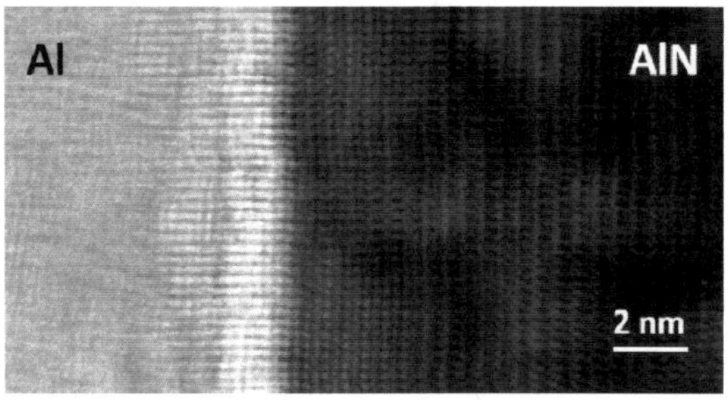

Fig. 12. HR-STEM brightfield image of the transition zone between a cold sprayed Al coating and an Al-polar AlN single crystal.

Despite the difficulties in depositing cold sprayed Al coatings on the very brittle MgF_2 substrates, it was possible to prepare a few samples for TEM investigations. The results, which will be published in the near future, again indicate heteroepitaxial growth between coating and substrate material.

4 Summary

By means of cold spraying, Al coatings were deposited on various poly- and single-crystalline ceramic substrates (Al_2O_3, AlN, Si_3N_4, SiC, MgF_2). Tensile bond strength tests revealed a clear dependency of the bonding strength on additional thermal energy input (substrate heating or subsequent heat treatment), substrate material and powder fraction. One decisive factor is the atomic mobility in the interface-near coating material, as it determines the probability of heteroepitaxial growth. It is assumed that the temperature rise in the interfacial quasiadiabatic shear bands ensures a sufficient atomic mobility. Furthermore, static recrystallization of the deformation-induced ultra fine-grained microstructure leads to a reordering of the atoms in the Al grains adjacent to the ceramic substrate resulting in orientation relationships with a small lattice misfit between the coating and the substrate material, as proven by high-resolution interface investigations. In addition, it was observed that the tensile bond strength increases with the heat conductivity and thermal effusivity of the substrate material. This can be explained by the low coating/substrate contact temperature of good heat conductors such as AlN, leading to a lower tensile residual stress in the coating material near the interface. Thus it can be concluded that the adhesion between cold sprayed Al coatings and ceramic substrates is based on a complex interplay of microscopic and macroscopic influences.

5 Acknowledgement

The authors gratefully thank the German Research Foundation (DFG) for financing the presented work under the project numbers WI688/95-2 and RA1050/15-2. M. Motylenko thanks the DFG for financial support of the subproject A05, which is a part of the Collaborative Research Centre SFB 920 "Multi-Functional Filters for Metal Melt Filtration – A Contribution towards Zero Defect Materials".

6 Literature

[1] Wielage, B; Grund, T.; Rupprecht, C.; Kümmel, S.: New Method for Producing Power Electronic Circuit Boards by Cold-Gas Spraying and Investigation of Adhesion Mechanisms. Surf. Coat. Technol. 205 (2010), Issue 4, pp. 1115–1118.

[2] Donner, K.-R.; Gärtner, F.; Klassen, T.: Metallization of Thin Al_2O_3 Layers in Power Electronics Using Cold Gas Spraying. J. Therm. Spray Technol. 20 (2011), Issue 1–2, pp. 299–306.

[3] King, P.C.; Zahiri, S.H.; Jahedi, M.; Friend, J.: Cold Spray Electroding of Piezoelectric Ceramic. Mater. Forum 31 (2007), pp. 116–119.

[4] Marx, S.; Paul, A.; Köhler, A.; Hüttl, G.: Cold Spraying: Innovative Layers for New Applications. J. Therm. Spray Technol. 15 (2006), Issue 2, pp. 177–183.

[5] Tillmann, W.: Fügen, Technische Keramik (ed. W. Kollenberg), Vulkan-Verlag Essen, 2004, pp. 445–457.

[6] Papyrin, A.; Kosarev, V.; Klinkov, S.; Alkhimov, A.; Fomin, V.: Cold Spray Technology (ed. A. Papyrin), Elsevier Amsterdam, 2007.

[7] Fauchais, P.; Montavon, G.: Thermal and Cold Spray: Recent Developments. Key Eng. Mater. 384 (2008), pp. 1–59.

[8] Assadi, H.; Kreye, H.; Gärtner, F.; Klassen, T.: Cold spraying – A materials perspective. Acta Mater. 116 (2016), pp. 382–407.

[9] Rafaja, D.; Schucknecht, T.; Klemm, V.; Paul, A.; Berek, H.: Microstructural characterisation of titanium coatings deposited using cold gas spraying on Al_2O_3 substrates. Surf. Coat. Technol. 203 (2009), pp. 3206–3213.

[10] Drehmann, R.; Grund, T.; Lampke, T.; Wielage, B.; Manygoats, K.; Schucknecht, T.; Rafaja, D.: Splat formation and adhesion mechanisms of cold-sprayed Al coatings on Al_2O_3 substrates. J. Therm. Spray Technol. 23 (2014), Issue 1–2, pp. 68–75.

[11] Drehmann, R.; Grund, T.; Lampke, T.; Wielage, B.; Manygoats, K.; Schucknecht, T.; Rafaja, D.: Interface characterization and bonding mechanisms of cold gas-sprayed Al coatings on ceramic substrates. J. Therm. Spray Technol. 24 (2015), Issue 1, pp. 92–99.

[12] Assadi, H.; Gärtner, F.; Stoltenhoff, T.; Kreye, H.: Bonding Mechanism in Cold Gas Spraying. Acta Mater. 51 (2003), Issue 15, pp. 4379–4394.

[13] Grujicic, M.; Zhao, C.L.; DeRosset, W.S.; Helfritch, D.: Adiabatic Shear Instability Based Mech-

anism for Particles/Substrate Bonding in the Cold-Gas Dynamic-Spray Process. Mater. Des. 25 (2004), Issue 8, pp. 681-688.

[14] Li, C.-J.; Li, W.-Y.; Wang, Y.-Y.: Formation of Metastable Phases in Cold-Sprayed Soft Metallic Deposit. Surf. Coat. Technol. 198 (2005), Special Issue 1-3, pp. 469–473.

[15] Wank, A.; Wielage, B.; Podlesak, H.; Grund, T.: High-Resolution Microstructural Investigations of Interfaces Between Light Metal Alloy Substrates and Cold Gas-Sprayed Coatings, J. Therm. Spray Technol. 15 (2006), Issue 2, pp. 280–283.

[16] Vlcek, J.; Gimeno, L.; Huber, H.; Lugscheider, E.: A Systematic Approach to Material Eligibility for the Cold-Spray Process. J. Therm. Spray Technol. 14 (2005), Issue 1, pp. 125–133.

[17] King, P.C.; Zahiri, S.H.; Jahedi, M.: Focused Ion Beam Micro-Dissection of Cold-Sprayed Particles. Acta Mater. 56 (2008), Issue 19, pp. 5617–5626.

[18] Hines, J.A.; Vecchio, K.S.: Recrystallization kinetics within adiabatic shear bands. Acta Mater. 45 (1997), Issue 2, pp. 635–649.

[19] Hughes, D.A.; Hansen, N.: High angle boundaries formed by grain subdivision mechanisms. Acta Mater. 45 (1997), Issue 9, pp. 3871–3886.

[20] Klopp, R.W.; Clifton, R.J.; Shawki, T.G.: Pressure-shear impact and the dynamic viscoplastic response of metals. Mech. Mater. 4 (1985), Issues 3–4, pp. 375–385.

[21] Schmidt, T.; Gärtner, F.; Assadi, H.; Kreye, H.: Development of a generalized parameter window for cold spray deposition. Acta Mater. 54 (2006), pp. 729–742.

[22] Wüstefeld, C.; Rafaja, D.; Motylenko, M.; Ullrich, C.; Drehmann, R.; Grund, T.; Lampke, T.; Wielage, B.: Local heteroepitaxy as an adhesion mechanism in aluminium coatings cold gas sprayed on AlN substrates. Acta Mater. 128 (2017), pp. 418–427.

[23] King, S.W.; Barnak, J.P.; Bremser, M.D.; Tracy, K.M.; Ronning, C.; Davis, R.F.; Nemanich, R.J.: Cleaning of AlN and GaN surfaces. J. Appl. Phys. 84 (1998), Issue 9, pp. 5248–5260.

[24] Dalmau, R.; Collazo, R.; Mita, S.; Sitar, Z.: X-ray photoelectron spectroscopy characterization of aluminum nitride surface oxides: Thermal and hydrothermal evolution. J. Electron. Mater. 36 (2007), Issue 4, pp. 414–419.

Metallization of various polymers by cold spray

Hanqing Che, Phuong Vo, Stephen Yue, Montreal/CDN

Abstract: Previous results at McGill University have shown that metallic coatings can be successfully cold sprayed onto polymeric substrates. This paper studies the cold sprayability of various metal powders on different polymeric substrates. Five different substrates were used, including carbon fibre reinforced polymer (CFRP), acrylonitrile butadiene styrene (ABS), polyether ether ketone (PEEK), polyethylenimine (PEI); mild steel was also used as a bench mark substrate. The CFRP used in this work has a thermosetting matrix, and the ABS, PEEK, and PEI are all thermoplastic polymers, with different glass transition temperatures as well as a number of distinct mechanical properties. Three metal powders, tin, copper and iron, were cold sprayed with both a low-pressure system and a high-pressure system at various conditions. In general, cold spray on the thermoplastic polymers rendered more positive results than the thermosetting polymers, due to the local thermal softening mechanism in the thermoplastics. Thick copper coatings were successfully deposited on PEEK and PEI.

1 Introduction

Polymers and polymer matrix composites are increasingly used but their low electrical conductivity has limited their application. It has been reported that cold spray of metal powders onto the carbon fibre reinforced polymer (CFRP) substrates is viable [1]. This paper presents the effect of polymeric substrates on the cold sprayability of various metal powders. Five different substrates were used, including CFRP, ABS (acrylonitrile butadiene styrene), PEEK (polyether ether ketone), PEI (polyethylenimine); mild steel was also used as a bench mark substrate. It was reported by researchers that thermoplastics and thermosets render different responses during cold spray of metallic powders [2, 3]. The CFRP used in this work has a thermosetting matrix, and the ABS, PEEK, and PEI are all thermoplastic polymers, with different glass transition temperatures as well as a number of distinct mechanical properties. The glass transition temperatures, T_g, is a temperature region over which the reversible transition between a hard state ($T < T_g$) and a soft state ($T > T_g$) occurs in an amorphous polymer or in amorphous regions of a partially crystalline polymer [4].

PEEK and PEI are both high-performance polymers that have been widely used in a wide range of applications, especially those that involve high-temperature exposure. PEEK is a semi-crystalline thermoplastic that offers excellent mechanical properties, chemical and wear resistance, high continuous-use temperature (300°C for short-time and 250°C for long-time exposure), low smoke/toxic gas emission etc. [5, 6]. PEI, on the other hand, is an amorphous polymer with good mechanical properties. It is a cheaper alternative to PEEK despite offering a lower continuous performance temperature of 170°C [5]. ABS is a common engineering plastic, which normally cannot be used at elevated temperatures (below 70°C [5]). It has excellent impact resistance, stiffness and toughness, and its main advantage is its low cost [7].

Three single-component powders, tin, copper and iron, were used in this work. Copper and tin were used in our previous CFRP experiments; iron was chosen because its critical velocity, v_{crit}, is reported to be higher than 600 m/s (without heating) [8], higher than those of copper (460-500 m/s) and tin (below 200 m/s), and this enables an investigation over a range of v_{crit}.

2 Experimental methodology

Three single-component powders, copper, tin and iron, were used in this work. The average particle size was measured with a Horiba LA-920 Laser Scattering Particle Size Distribution Analyser, and the results are listed in Table 1. Also listed in Table 1 are the Vickers hardness measurements of the three powders. The scanning electron microscope images and cross-section optical micrographs of the feedstock powders are shown in Fig. 1. It can be seen that the iron powder has a mixed morphology (i.e. a combination of spherical and irregular) and the particles are larger than the other two powders.

Table 1. Feedstock powders used in this work.

Powder	Morphology	Supplier	D_{avg}	Hardness
Cu	Relatively spherical	Plasma Giken	29 µm	55 HV
Sn	Relatively spherical	CenterLine, SST	17 µm	11 HV
Fe	Mixed	Quebec Metal Powders	35 µm	129 HV

The substrates used in this work were CFRP (supplied by Bombardier Aerospace, Montreal, Canada), commercially available ABS, PEEK, PEI and 1020 mild steel. For the cold spray experiments, sheet sections of dimensions 7 x 5 cm were used as the substrates. Prior to the cold spray experiments, the CFRP sections were degreased with acetone, the ABS, PEEK and PEI sections were degreased with ethyl alcohol, and the mild steel sections were degreased with acetone and grit blasted with 24 grit alumina.

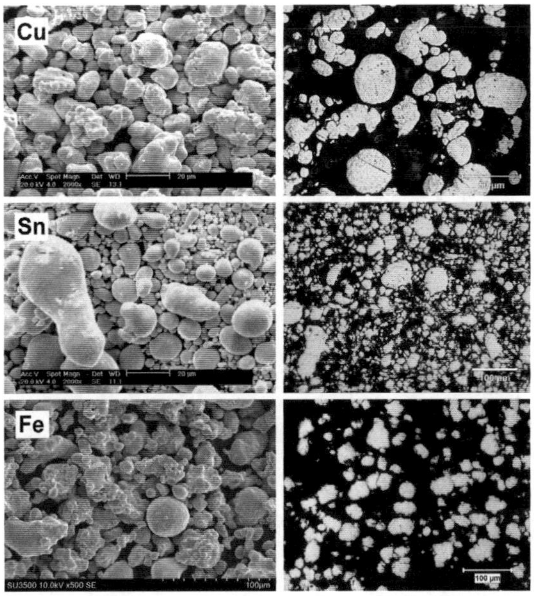

Fig.1. SEM images (left) and cross-sectional optical micrographs (right) of the feedstock powders (images for iron are at lower magnification).

Cold spray experiments were carried out at the McGill-NRC cold spray facility at National Research Council Canada, Boucherville. Both a low-pressure CenterLine SST system and a high-pressure Plasma Giken system were used to investigate the cold sprayability of the powders onto the substrates at a wide range of particle velocities. It should be noted that when directly comparing the results from two different cold spray systems, there are a number of complications resulting from the difference between the two different systems (e.g. differences in geometry of the nozzle, powder injection mode, location of the thermocouple, etc.). This work assumes that for the same gas temperature, the influence of the system was negligible. Nitrogen was selected as the carrier gas for both systems, and the above-mentioned three powders were sprayed at various conditions, which are shown in detail in Table 2. Tin was cold sprayed at 200°C to avoid melting, copper was cold sprayed at 425°C and iron was cold sprayed at both 200°C and 425°C. Other process parameters for the CenterLine system were basically those that resulted in successful deposition in our previous work [1]. For the Plasma Giken system, the 25 $mm \cdot s^{-1}$ gun travel speed is relatively low for this system, but was kept the same as in the CenterLine system for direct comparison purpose.

Table 2. Process parameters for cold spray.

Low-pressure Centerline system					
Powder	**Temperature** °C	**Pressure** psi (MPa)	**SoD** mm	**GTS** $mm \cdot s^{-1}$	**Feeding** $g \cdot min^{-1}$
Sn	200	75-200 (0.52-1.38)	18	25	10
Fe	200	75-200 (0.52-1.38)	18	25	16
Fe	425	50- 200 (0.34-1.38)	18	25	16
Cu	425	50-200 (0.34-1.38)	18	25	11
High-pressure Plasma Giken system					
Powder	**Temperature** °C	**Pressure** MPa	**SoD** mm	**GTS** $mm \cdot s^{-1}$	**Feeding** $g \cdot min^{-1}$
Fe	200	2- 4	40	25	16
Fe	425	2-4.9	40	25	17
Cu	425	2-4.9	40	25	19

3 Results

3.1 Tin

The measured deposition efficiency (DE) results of tin powder cold sprayed at 200°C on five substrates at various gas pressures are presented in Fig. 2. The DE results on the steel substrate can be considered as benchmarks. At 0.5 MPa, no deposition was achieved on steel; at a higher gas pressure of 1.0 MPa (150 psi), deposition started on steel. The DE results on steel increased significantly at gas pressures of 1.2 MPa (175 psi) but did not show further improvement at 1.4 MPa (200 psi).

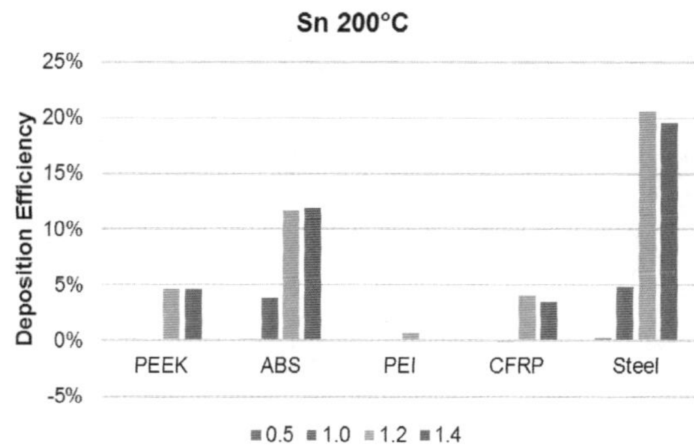

Fig. 2. Deposition efficiency of tin at various pressures at 200°C.

For ABS, the deposition behavior was similar to that on steel substrates, deposition started at 1.0 MPa and increased at 1.2 and 1.4 MPa. However, DE on ABS at each condition was lower than that on steel at corresponding conditions. On CFRP and PEEK substrates, similar DE results were obtained, although one is thermoplastic and the other is a composite with thermosetting matrix. For CFRP, there was slight erosion of the substrate at 0.5 and 1.0 MPa since the measured DE values were slightly negative; whereas no weight change was found on PEEK substrates at these two conditions. At gas pressures of 1.2 MPa (175 psi) and 1.4 MPa (200 psi), deposition was achieved on CFRP and PEEK, but the DE values for both substrates were much smaller than those on steel. No deposition was achieved on PEI at most conditions, with a maximum DE of less than 1 per cent at 1.2 MPa (175 psi).

The cross-sectional microstructures of the tin coatings cold sprayed at 200°C and 1.4 MPa on ABS and PEEK, which showed the best deposition behavior, are presented in Fig. 3. It can be seen that continuous tin coatings were deposited on both thermoplastics, but the coating on ABS is non-uniform but wavy, signifying local material removal (erosion) took place.

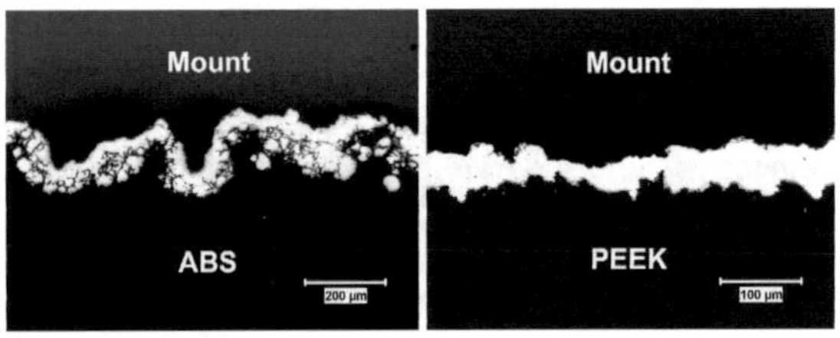

Fig. 3. Optical micrographs showing the cross-sections of the tin coatings cold sprayed at 200°C and 1.4 MPa on ABS (left) and PEEK (right).

3.2 Iron

At 200°C, for all substrates at all gas pressures performed, no effective deposition can be achieved, with the maximum DE being lower than 1 per cent. This indicates that the particle velocities were below v_{crit} at most conditions at 200°C. As for the polymeric substrates, the CFRP with a thermosetting matrix showed significant erosion with increasing gas pressure.

For iron cold sprayed at 425°C, the measured DE results are presented in Fig. 4. It can be seen that on steel substrates, deposition did not begin until 2.0 MPa, at pressures above which DE increased with increasing gas pressure. This suggests that the particle velocity at 2.0 MPa is near/just above the v_{crit} for iron at this temperature. However, the magnitudes of DE for iron on steel were relatively low, even at 4.9 MPa, DE was only 25 per cent.

Two polymeric substrates, the CFRP and ABS, showed negative DE values, which signifies substrate erosion. For CFRP, no deposition can be achieved and DE was negative for all gas pressures. At a gas pressure of 3 MPa or higher, full-thickness removal/erosion of the CFRP substrate occurred, causing fracture of the substrates, as indicated by the patterned bars in Fig. 4. ABS, the thermoplastic with the lowest T_g (105°C) and operating limit (80°C), showed low DE (< 2%) at low pressures (1.4 MPa and lower) and negative DE (at 2 and 3 MPa) and full-thickness erosion (4 and 4.9 MPa) at high pressures. Some deposition, probably particle embedding, was achieved on ABS at pressures (0.3 to 1.4 MPa), lower than the onset deposition pressure on steel (2.0 MPa).

No significant deposition was achieved on either

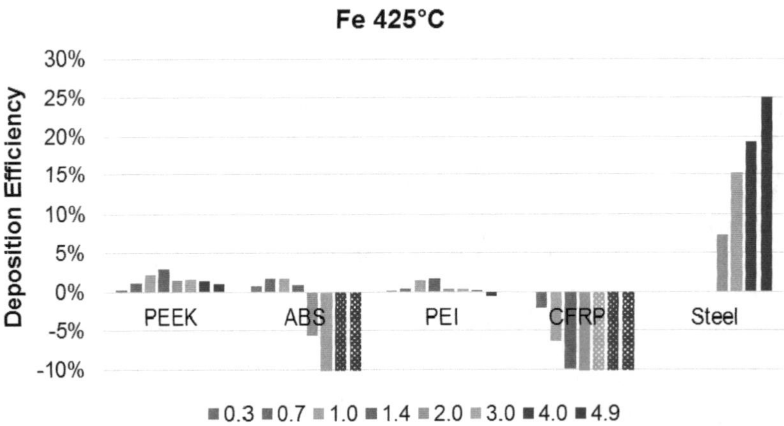

Fig. 4. Deposition efficiency measurements of iron at various pressures at 425°C.

PEEK or PEI, with a maximum DE below 3 and 2 per cent, respectively. DE on PEEK first increased and then decreased with increasing gas pressure, and the same trend was observed on PEI, for which DE became negative at 4.9 MPa.

It should be noted that at low pressures (below 2.0 MPa), the deposition of iron was achieved on the thermoplastics, but not achieved on steel, and this indicates that the deposition was through particle embedding. This can be seen from the cross-sectional micrographs in Fig. 5.

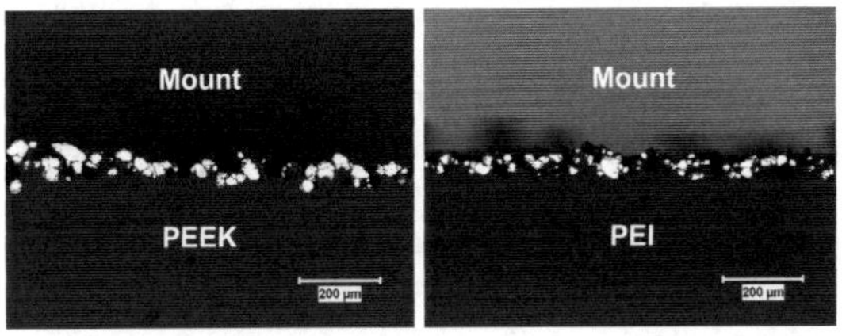

Fig. 5. Cross-sectional optical micrographs of PEEK (left) and PEI (right) after cold spray of iron at 425°C and 1.4 MPa.

3.3 Copper

The DE results for copper cold sprayed at 425°C are presented in Fig. 6. It can be seen that the deposition of copper on the steel substrate started at a gas pressure of 1.0 MPa, and DE increased with increasing gas pressure. At gas pressures higher than 2.0 MPa, DE was not successfully measured since the deposited coating kept falling off the substrate during cold spray, indicating low strength bonding with the substrate, probably caused by effects due to increasing coating thickness (e.g. thermal stress). For CFRP, significant erosion was again observed, but the magnitudes of material loss were lower than those with the iron powder. ABS also showed deposition behavior similar to that in Fig. 4, low DE (<5%) at low pressures (1.4 MPa and lower) and erosion at high pressures (2 MPa and higher).

For PEEK and PEI, both materials showed an increasing trend of DE with increasing gas pressure. They exhibited low DE (<5%) at low pressures till 1.4 MPa, although DE of copper on steel started to increase significantly at a lower gas pressure, 1.0 MPa. With the high-pressure system, DE on PEEK increased to a comparable level as on the steel substrate at 2.0 MPa, and continued increasing at higher pressures. As for PEI substrate, DE also increased largely starting from 2.0 MPa, but DE on PEI was lower than that on steel and PEEK substrates under the same conditions. The DE result on PEI at 4 MPa was not successfully recorded due to the detachment of the coating during the process.

Fig. 7 shows the cross-sectional optical micrographs of PEEK (left) and ABS (right) after cold spray of copper at 425°C and 2.0 MPa. It can be seen clearly that thick copper coating was successfully deposited on PEEK, whereas only some particle embedment was achieved on ABS. In brief, considerable deposition of copper can be achieved on steel, PEEK and PEI substrates when cold spraying at 425°C with the high-pressure Plasma Giken system.

4 Discussion

The results of cold spray of various metal powders onto different substrates showed that different substrates rendered different responses during cold spray. For example, cold spray of copper at 425°C at 2.0 MPa was successful on PEEK, PEI and steel substrates, but failed on ABS and CFRP. Therefore, the substrate played a strong role in determining the cold sprayability of one metal powder.

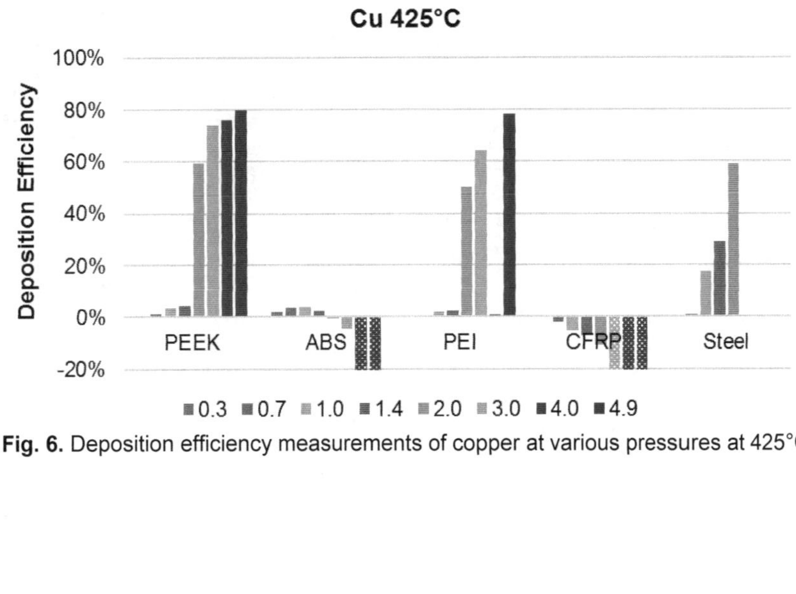

Fig. 6. Deposition efficiency measurements of copper at various pressures at 425°C.

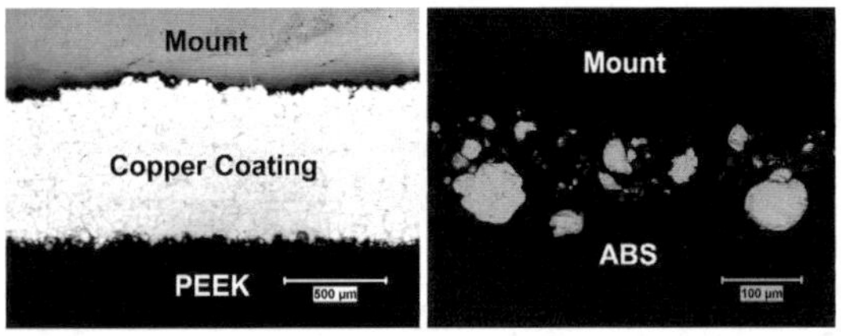

Fig. 7. Cross-sectional optical micrographs of PEEK (left) and ABS (right) after cold spray of copper at 425°C and 2.0 MPa.

It has been mentioned by researchers that cold spray of metals on the thermoplastic polymers are more successful than on the thermosetting polymers, due to the local softening mechanism in the thermoplastics [2, 3]. In this work, it was found that cold spray on the CFRP with a thermosetting matrix was unsuccessful at most conditions, and brittle fracture in the substrate surface and erosion of the substrate were observed (reported in [1]). For the thermoplastics, more positive results were obtained: on PEEK the tin coatings and thick copper coatings were achieved, on PEI the thick copper coatings were achieved, and on ABS the tin coatings were achieved.

The success with the thermoplastic substrates in this work, similarly to those reported by other researchers [3], can be attributed to the local softening of the thermoplastics. Namely, when cold spraying at a temperature higher than the T_g of the thermoplastic, the polymer surface in contact with the metal particles softens upon impact, so that the metal particles penetrate the polymer, and mechanically interlock with the polymer substrate upon cooling. A typical characteristic of this mechanism is the waviness at the coating/substrate interface, which was observed frequently in this work.

At temperatures below T_g, the polymers are in their glassy strate and are relatively hard and brittle, so the particle embedding and mechanical interlocking are relatively difficult at low impact velocities since the local thermal softening is absent. (e.g. the failure in cold spray of tin at 200°C on PEI, whose glass transition temperature is 215°C). Whereas significant erosion of the substrate may occur at high impact velocities due to the brittleness.

On the other hand, if the temperature is much higher than T_g, the mechanical properties of the thermoplastic may largely deteriorate, deposition on the thermoplastics is again difficult and erosion of the substrate may occur, since the polymer loses its strength. An example for this is the cold spray of copper at 425°C on ABS, whose T_g is 105°C. In contrast, the success with PEEK and PEI is due to their well-known capacity to retain their good mechanical properties at relatively high temperatures, especially when compared with ABS. The temperature dependence of the polymer properties is, therefore, important to cold spray of metals on polymer substrates.

5 Summary and conclusions

Three metal powders, tin, copper and iron, were cold sprayed onto five different substrates, including three thermoplastics, PEEK, PEI, and ABS, one polymeric composite with a thermosetting matrix, CFRP, and one metal, mild steel as the benchmark. The cold spray campaign was performed with both a low-pressure system and a high-pressure system to examine a wide range of particle velocities. In general, cold spray on the thermoplastic polymers rendered more positive results than the thermosetting polymers, due to the local thermal softening mechanism in the thermoplastics. Thick copper coatings were successfully deposited on PEEK and PEI at 425°C.

Acknowledgement

The authors wish to acknowledge the financial support of the Consortium for Research and Innovation in Aerospace in Quebec and the Natural Sciences and Engineering Research Council of Canada. The industrial partners, Bombardier Aerospace, Bell Helicopter, 3M Canada, and the collaborating universities, École Polytechnique de Montréal and Université du Québec à Montréal are gratefully acknowledged. National Research Council Canada, Boucherville is acknowledged for its contribution to the cold spray experiments.

References

[1] Che H, Vo P, Yue S: Metallization of carbon fibre reinforced polymers by cold spray. Surface and Coatings Technology 313 (2017), pp. 236/47.

[2] Ganesan A, Yamada M, Fukumoto M: Cold spray coating deposition mechanism on the thermoplastic and thermosetting polymer substrates. Journal of Thermal Spray Technology 22 (2013), pp.1275/82.

[3] Gardon M, Latorre A, Torrell M, Dosta S, Fernández J, Guilemany JM: Cold gas spray titanium coatings onto a biocompatible polymer. Materials Letters 106 (2013), pp. 97/99.

[4] Wikipedia: Glass transition. Wikipedia, The Free Encyclopedia. 2016. https://en.wikipedia.org/w/index.php?title=Glass_transition&oldid=713982756. (accessed 26-April-2016).

[5] Hough M, Dolbey R: The plastics compendium: key properties and sources: Rapra Technology; 1995.

[6] Ashby MF, Johnson K: Materials and design: the art and science of material selection in product design: Butterworth-Heinemann; 2002.

[7] Rosato D: Designing with plastics and composites: a handbook: Springer US; 2013.

[8] Schmidt T, Gärtner F, Assadi H, Kreye H: Development of a generalized parameter window for cold spray deposition. Acta Materialia 54 (2006), pp.729/42.

A case study of repair work on ship engine valve stems

P.T. Nielsen, D.B. Bangsgaard, K.G. Soerensen, FORCE Technology Denmark, P. Pallesen, MarineShaft

The degradation mechanisms of machine parts are well-known facts, usually resulting in replacement of components. However, in some cases, for instance, parts for large ship engines, repair will often seem more attractive. Time-wise, the vessel's down-time will be reduced and the investment will be lower.

This paper describes repair work performed on valve stems from a 52-year-old ship engine. Due to this long time in operation, the stems were damaged by wear and were no longer functional. The requirements for the stems were scarcely documented stating only a few specific requirements; therefore a short study of the possibilities concerning rebuild material and application method was initiated.

This paper illuminates the manufacturing process used; twin wire arc spraying with Monel as the rebuild material. Monel is chosen to meet the requirements for hardness and the subsequent high precision machining. In addition, the mechanical and metallurgical properties of the rebuild coating are examined in regard to adhesion, micro-structure and hardness to interpret and support the functionality.

1 Introduction

The industry is guarantied at least one fact about its machinery – it will eventually fail as a result of degradation mechanisms. A lot of effort is generally successfully put into prolonging the time to failiure and replacements of components. However, in some cases, this extended life can cause several problematic issues. Firstly, a replacement component might not be available anymore. This leaves only possibilities such as manufacturing of a new component or repair of the old. Secondly, the technical documentation on the specific component might be limited, which will force the problem owner to rethink the functionality and concept of the component. Usually none of these options are especially attractive due to the expected work load, lack of experience and economic aspects.

Glancing at the refurbishment business, it is obvious they have a huge experience-based knowledge bank. This is built on solving specific industrial application problems spanning from sectors such as offshore constructions and power plants to pharmaceutical and food production industry.

In this case of repairing valve stems from a 52-year-old ship engine, a similar case was not identified. This combined with few, but imperative, requirements related to the properties of the valve steel, desirable functionality, characteristics of the repair material and a narrow timeframe, a technically satisfactory solution was sought.

Leaving out alternatives or irrelevant methods and materials, this paper will describe the actual repair motivation resulting in one material and one process supported by mechanical and metallurgical studies of the selected repair material.

2 Methods and approach

2.1 Repair technology

The stems are approximately 1.8 m long and have a diameter of 140 mm, and the nature of the valve stems does not allow any form of heat resulting in metallurgical changes to the substrate steel. Thus traditional weld overlay processes are disqualified. The damaged area caused by wear mechanism reaches a deph of 1 mm, hence 2 mm on the diameter.

The repair process should be able to apply a fairly high amount of material not inducinge substrate changes related to mechanical and metallurgical properties. The process used was a twin wire arc process assisted by a x-y robotic gun setup and a rotating stem manipulater as shown in **Fig. 1**. Specifically, a METCO 4RC system was used [1].

Fig. 1. Valve stem repair setup using the twin wire arc spraying process.

2.2 Materials

The properties of the applied surface was defined by a updated, but limited, technical specification from the original manufacturer. The hardness should be between 125-160 HB. This relative soft material pointed at an already commercially available material such as Monel, a Nickel-Copper alloy with approximately 70 % Ni and 30 % Cu, as shown in table 1. [2]

Table 1. Valued characteristics of Monel

Characteristics	Typical Data
Bond strength	27.5-34.5 MPa
Hardness	72-84 HRB
Finishing	Machine/grind

By conversion of the Rockwell hardness to Brinell it correlates well with the values stated in the technical specification.

The repair coating should also be suitable for precision post-machining. The bond strength was seen as an important quality parameter, which could be evaluated by adhesion test of the repair coating.

The fact that Monel in wire form is excellent for twin wire arc spraying was the main reason for choosing this process.

2.3 Repair work and tests

5 valve stems were damaged. All of these were pre-machined 2 mm under defined technical drawing diameter. The stems were cleaned using a mild alkali with surface pacifier followed by standard sandblasting with 0.5-1 mm korund particles.

The spray repair had the following settings:

- 200 A
- 30 V
- Standard pressurised shop air
- 65 psi primary
- 20 psi secondary
- Stem rotary: 18 m/min
- Gun transvers: 5/rot

2 plates, 50 mm by 50 mm, were prepared for metallurgical examination and samples for bond testing according to standard were prepared [3].

The valve stems were coated 3/10 mm above final diameter in consideration of the post-machining to final measurements.

3 Results and discussion

3.1 General mechanical properties

The essential mechanical property relevant for testing is the bond strength. If the repair coating does not adhere to the valve stem, it is useless. An adhesion test in accordance with ASTM C633-01 was performed on two test specimens.The test specimens failed at 68 MPa and 74 MPa, thus with an average adhesion strength of 71 MPa. Comparing with the typical properties of Monel in Table 1, the bond strength of the present repair coating is much higher, thus the adhesion of the repair coating to the valve stem should be sufficient. Failure of the test specimens occurred in the bonding agent, as shown in Figure 2, so the actual bond strength of the repair coating may be even higher.

Fig. 2. The test specimen with the repair coating after adhesion test. Failure occurred in the bonding agent.

3.2 Metallurgical observations

A metallurgical cross section was prepared from a test specimen with the repair coating. Visual examination revealed a relatively uniform layer, as shown in Figure 3.

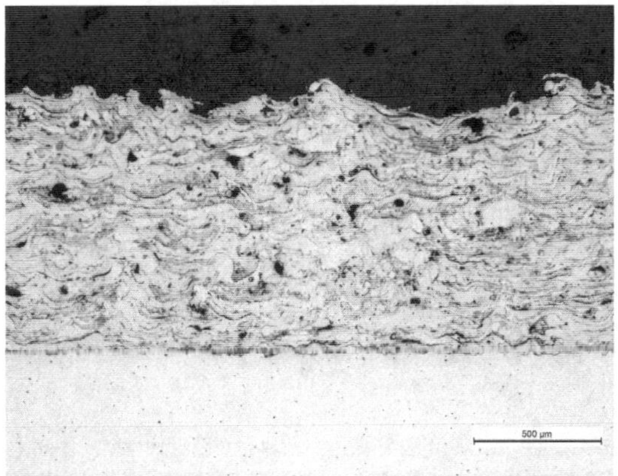

Fig. 3. The repair coating as sprayed onto a test specimen, 50 X magnification.

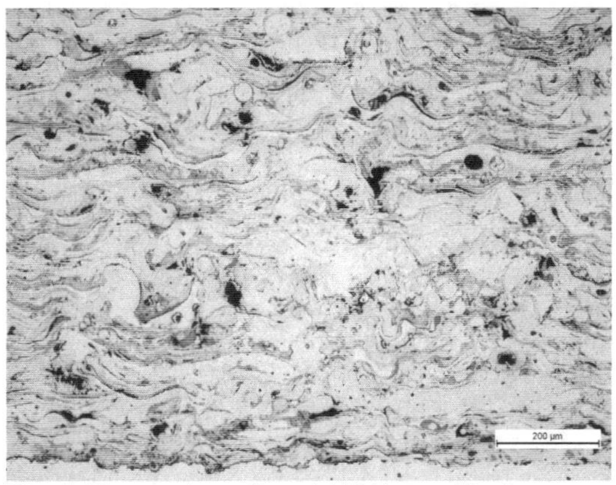

Fig. 4. The repair coating at 100 X magnification.

A magnified representative image is shown in Figure 4. Porosities are present and the magnified image in Figure 5 reveals interlayer lack of fusion, unmelted particles from the process and oxides. This, however, is normal for this process, and as the adhesion test showed, the performance of the repair coating is superior to the typical data for Monel.
Further, porosities in a surface increase the tribological properties if the machinery is run with lubrication.

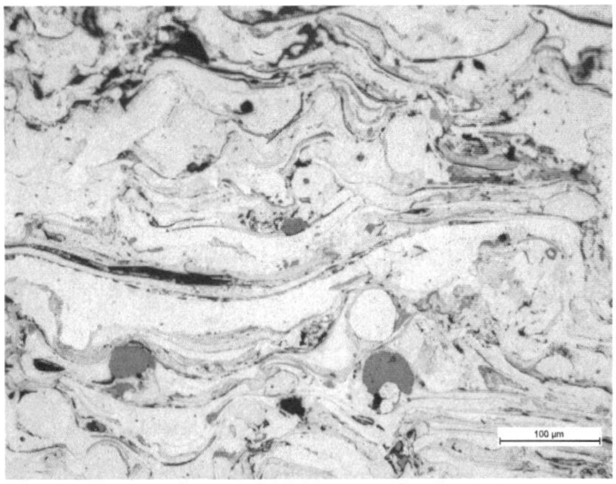

Fig. 5. The coating at 200 X magnification.

In general, a higher hardness increases the resistance to abrasive wear and thus the hardness is an important parameter. Hardness testing was performed on the test specimen with the repair coating. The initial rough surface was ground to simulate the post-machining on the actual valve stem. Hardness testing on the initial rough surface would result in a lower hardness than on the actual post-machined valve stem. The hardness was measured by the standard HRB method and an average value of 69 HRB was measured with very little scatter. This is slightly below the typical values of 72-84 HRB for Monel shown in Table 1. The repair coating has a high level of porosities and was not machined. Thus, the post-machined valve stem may still meet the requirements of the technical specification.

3.3 Interpretation of results

The motivation for re-manufacturing is based on time, cost, and the possibility to optimize functional properties. In this case, down time and cost have been important issues as long as the material properties and their specificatoins are met.

Making new stems in this case could possibly avoid unnessary down time due to factors such as:

- Interpretation of original specifications
- Availability of stem material
- Availability of post-treatments
- Fitting issues in the used engine

These factors would indeed also increase the repair costs. By choosing re-manufacturing the cost would definitely be lowered, as this process only requires three simple steps:

- Pre-machining
- Stem build-up
- Post-machining to final measurements

This maneuver is traditionally a fraction of the cost of replacing with new parts. In this case it would probably be even lower.

The stem build up method, would result in unaffected substrate material with respect to heat transfer, meaning the substrate would not experience any form of metallurgical changes of structure or mechanical properties related to heat treatment.

The typically use of Monel is for corrosion protection purposes. The used build-up material was chosen primarily in regard to its hardness properties, but also due to the sliding and lubricating properties as these are expected to be improved compared to the substrate material.

The typical reference bonding values was lower than the intended use. To enhance the bonding, adjustment and caution should be applied for the following production components:

- Cleaning
- Blasting
- Spraying

The spraying and blasting parameters did not deviate significantly from the recommended settings and procedures. The cleaning process was based on an alkali pacifier that has the ability to clean the surface in depth. This would result in a clean surface, which is non reactive with the environment, i.e. resulting in surface oxidation.
A typical cleaning process does not necessarily remove all impurities. By removing impurities from pores, defects or traces from machining, the bonding quality will not be influenced during the spraying process.

4 Summary and conclusions

The present case focuses on the repair work of valve stems in a 52-year-old ship engine. The goal was to establish and perform a reasonable repair method based on few but imperative specification's. 5 valve stems were repaired using the twin wire arc process with Monel as the re-build material, with the following results;

- A possible way of repairing valve stems

- A measured average bond strength of 71 MPa, indicating an improvement of 100% in comparison to expected values.
- A hardness of 69 HRB was achieved, which is slightly lower than the specification and might relate to porosity content.

5 Outlook

This study has highlighted that conventional materials and re-manufacturing can contribute to increased material functionality. This is not only limited to the marine sector, but has a wide scope for other applications across various sectors, which would reduce production time and lower costs. It would be advantageous to draw experience from previous knowledge in other areas and implement this into new applications.

6 Literature

[1] OERLICON METCO, www.oerlicon.com/metco

[2] MSDS Nickel-Copper (Monel), www.oerlicon.com/metco

[3] ASTM C633-01

Partial repair of thermally sprayed and sealed corrosion protection – Organic coating material or Thermal Spraying?

Wilhelm T., Klesen C., Maghet T., Duisburg / D

For many decades thermally sprayed corrosion protection systems on the basis of ZnAl or Al carry out their service for structures in coastal areas, the offshore sector, as CUI (Corrosion Protection under insulation) or anywhere where the properties of thermally sprayed corrosion protection systems bring important advantages in terms of durability. A thermally sprayed corrosion protection system is about to protect the structure 25 to 30 years against corrosion. During this time it may be damaged due to factors like construction work, improper handling or simple aging.
There are many standards and regulations, which describe the initial design of thermal spray systems, however they remain silent regarding repair. In particular, a mending of partial regions is hardly described. Specific repair instructions are rare and if present, they differ from one another. Overall there is a lack of knowledge of the proper procedures for partially repairing thermal spray coating systems.
This project was concerned with tangible corrosion-technical issues of the coating repairs: How does the critical overlap area perform? Have organic coatings benefits? To what extent does a renewed damage affect the lifetime? The aim of the study was to develop practice-relevant instructions for the repair of thermally sprayed duplex systems.

1 Introduction

Damaged, thermal sprayed (TS) corrosion protection coatings, made of ZnAl and aluminum, offer cathodic protection under certain conditions, but are repaired when larger defects occur. Reasons for the defects are e.g. Mistakes during coating, damage during building activity, or normal corrosive process during lifetime. There are various ways of repairing these damages. The operator of a building or a machine and the executing coater are interested in the factors of efficiency, the environmental impact (offshore), the achievable service life and the behavior in case of renewed damage. Standards and regulations [S1, S2] generally recommend a complete renewal of the entire coating system, which is not usually done [1]. The following variables can be described as having an influence on the corrosion behavior of thermally sprayed coating systems during repair:

- **Repair by organic coating materials versus thermal spraying**
  For economic reasons, the coater is striving to achieve the greatest possible benefit with the smallest effort. When using organic coating materials, a large part of the effort compared to thermal spraying can be dispensed with. Contrary to this is the usually lower lifetime of the organic coating system.

- **Power-Tools versus Grit Blasting**
  Abrasive Blasting of steel structures is more complex than repairing small areas by power tools such as angle grinders or bristle blasters [5]. However, a defined roughness of the substrate surface is essential. A compromise between acceptable roughness and low effort is optimal.

- **Overlap width**
  Recommendations on overlapping width between the repaired and the original layer range from 10 mm (avoidance of crevice corrosion) to approximately 76 mm (3") . A larger overlap means only a relatively small amount of extracting. Important is the long-lasting corrosion protection.

The factors of efficiency and profitability have to be calculated by the user for each individual case, but the question is whether the more cost-effective variants offer comparable corrosion protection. Therefore, the "lifetime of the coating system" under corrosive conditions, in particular in the event of renewed damage to the repaired coating system, was defined as a comparison number in this project. The three influences described above were examined on their influence on the corrosion behavior of the coating systems.

2 Experimental

In order to answer the questions above, samples were realistically and practically coated, aged in the salt spray test (NSS), repaired and repositioned in the NSS. Between these steps and to the end, the properties of the coating systems were examined and documented. The different repaired coating systems were finally evaluated comparatively.
The following parameters were taken into account in the test matrix:

- Spraying material ZnAl15 & Al99
- Surface preparation method of repair by blasting & bristle blaster (power tool)
- Repair coating made of TS and organic coating
- Overlap widths of 10 mm & 76 mm
- Re-damage

The initial coating of the steel samples was carried out on the basis of a surface preparation Sa3 according to DIN EN ISO 8501-1, 50 µm < Rz > 80 µm by abrasive blasting (corundum). According to DIN EN ISO 2063

[S1] and NORSOK M-501 [S2], 200 µm of ZnAl15 or Al99 were applied by arc spraying (OSU G20 / 2).

Table 1: Spraying parameters of the examination

Parameters	Transformer Step Switch	Current [A]	Atomiser-pressure [bar]	Gap contact tubes [mm]
ZnAl15	5	72	5	4,0
Al99	6	125	5	4,0

Subsequently, the thermally sprayed samples were sealed with a 2K epoxy + corrosion protection pigments. According to NORSOK M-501 [S2], the objective was to apply only as much sealer as the porous layer can absorb and, if possible, not exceeds a dry film thickness (DFT) of 40 µm.

The layer systems were initially stored for 380 h in the NSS in order to achieve a defined aging. The hour number 380 h was determined in preliminary tests. For this purpose, comparable layer systems were used after eight years in offshore operation (North Sea) (**Fig.1**). Compared to newly formed coating systems, **Fig. 1** shows increasing porosity and initiating sealer flaking.

Original areas of the samples were saved. In each test step, surface roughness, layer thickness and adhesive strength were determined. The aged samples were subjected to high pressure washing prior to the repair in order to remove coarse salt residues. Subsequently, half of the layer system was removed by blasting or Bristle Blasting and at the same time the surface was prepared for the repair layer. The transition between the substrate and the old layer was as soft as possible, so that no abrupt edge was formed.

The repair coating was carried out either by repeated thermal spraying or by three coats each of 125 µm DFT. The nominal repair layer thicknesses were therefore again 200 µm for thermal spraying and 375 µm for the organic coating according to the manufacturer's instructions. A Bresle test before the repair coating gave a fairly high salt concentration of 150 mg/m^2. The partially-repaired layer systems were scribed and brought to the NSS for further 1000 h. The premature termination criterion was the marked occurrence of red rust (rust degree Ri5 according to DIN EN ISO 4628-3).

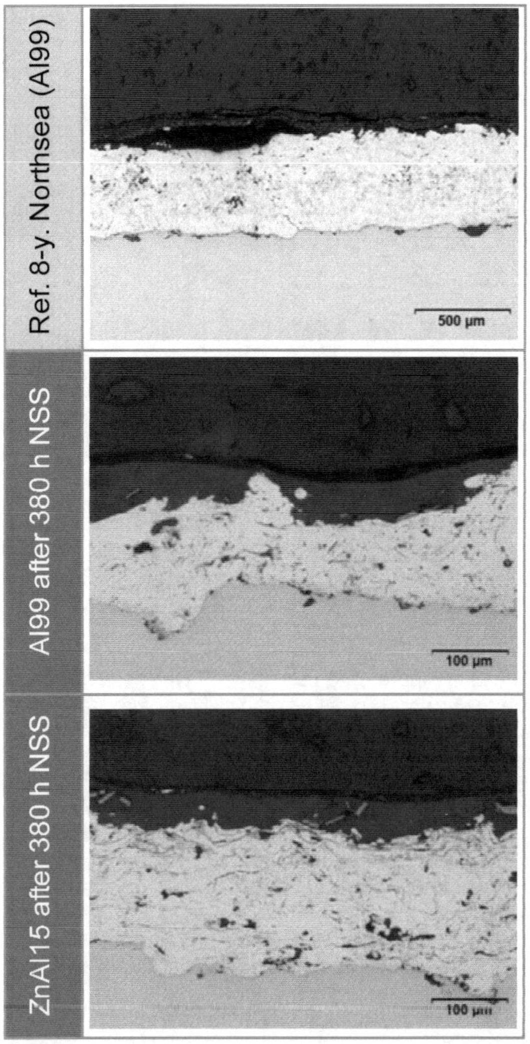

Figure 1: Comparison of the coating appearance after 8 years offshore use or 380 h NSS

Finally, the samples were evaluated visually:

- Optical assessment of degree of blistering according to DIN EN ISO 4628-2
- Optical evaluation of rusting according to DIN EN ISO 4628-3
- Percentage covered by white rust
- Damage around a scribe according to DIN EN ISO 4628-8

3 Results and Discussion

The latest draft of the standard [S1] and [S3] and [4] recommend the use of liquid organic coating materials. They are applied as corrosion protection, but as described, characterised by a shorter service life in a corrosion environment C5-M and C5-I compared to a thermally sprayed coating. In particular, the behavior in case of renewed damage and the possibility of a gap formation between the original and the repair layer could be shown in the work as a not in-depth research.

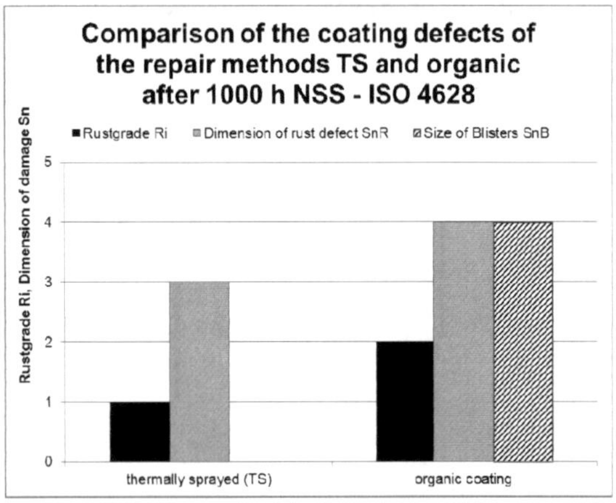

Figure 2: Comparison of coating defects of the repair methods TS and organic after 1000 h NSS

Fig. 2 compares the determined layer defects of the thermally sprayed and organically repaired samples after a 1000 hour salt spray test. The rust level of the organically repaired samples is higher with "2" than with the thermally sprayed samples "1". A corrosion creep in the form of blister formation was evident. It can be assumed that the cathodic protective action of the organic coating is poorer and the organic layer tends to creep rather than the thermally sprayed layer. Also, the size of the rust damage in the purely organically repaired systems is greater with the value "4". It is striking that no blisters are visible in the thermally sprayed systems.

Since the layers were scribed in a defined manner, the cathodic protection of the two coating methods could also be compared. The samples, which had been thermally sprayed with ZnAl15/ Al99, showed a very good remote protection effect in the area of the scribe in the strong electrolyte as described in [3]. Here practically no red rust was visible. The degree "1" determined was shown on a punctiform basis on the entire sample surface and not in the area of the scribe. The organically coated samples showed a creep in the form of blisters after approximately 600 h salt spray test, since this layer system could not protect itself by means of a remote protection effect.

In the area of the scribe, the organically repaired samples show poorer results in terms of lifetime. The organic layers protect due to a barrier effect and offer no cathodic protection. In the event of a defect in the barrier, the electrolyte has unimpeded access to the substrate. The following corrosion infiltrates the layer and corrosion-promoting conditions prevail as a result of the cleavage condition.

Surface preparation parameters are given by the three most common standards for corrosion protection by thermal spraying [S1], [S2] and [S3]. It is common to all of them that they require a surface roughness of Rz between 50 µm and 85 µm. Since the blasting during repairs is associated with a high effort and thus a high cost, the Bristle Blaster would be an attractive alternative because it is less expensive and therefore more economical in use.

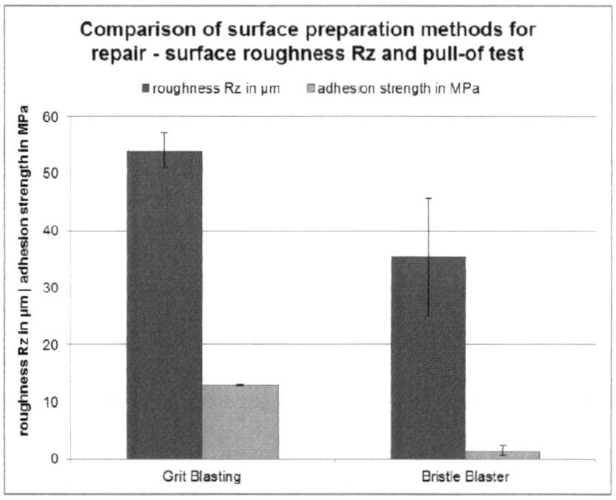

Figure 3: Comparison of surface preparation methods grit blasting vs. Bristle Blaster (Power Tool) regarding surface roughness and adhesion strength

Fig. 3 shows that a roughness Rz of 54.0 µm is achieved on average by blasting with corundum for the repair coating (sample 1, completely removed region). The roughness measurements on the transition region showed slightly higher values (not shown). Here, at reduced pressure, this transition area was "swept". The adhesion strengths of the repair layer determined on the newly produced area showed good values (13.0 MPa) (**Fig. 3**). Blasting is a suitable surface preparation process for repair. Similar roughness and adhesion values are obtained as with the initial coating.

The substrate roughness Rz generated by the Bristle Blaster device was approximately 35.4 µm on average, which is clearly below the minimum value of Rz = 50 µm. Accordingly, a tensile strength of the layer could hardly be detected with 1.4 MPa. According to [S2] at least 7 MPa would be necessary. A thermal spray layer can only adhere to [2] if a minimum roughness prevails. The main adhesion mechanism according to [2] at the phase boundary is, apart from also acting mechanisms such as adhesion and micro-welding, mechanical clamping. It can be assumed that the layer did not have a sufficient roughness or a favorable surface profile for thermal spraying layers. The particles could not cramp. The layer system also dissolved from the surface after approximately 96 h in the salt spray test and showed blister formation of the size "5" according to

DIN EN ISO 4628-2 and layer delamination on the entire surface.
The application of the Bristle Blaster device is not found to be suitable for the repair of thermally sprayed layer systems made of ZnAl15 under the selected conditions since a surface profile which is unfavorable for thermal spraying layers is produced so that clamping is not possible. Sweeping, i.e. lower-pressure blasting and "feathering" (chamfering of the edges) as described in [S3], showed good results.

The sources [3] and the DVS-leaflet [S4] praise the good remote protection effect of thermally sprayed coating systems based on Zn and ZnAl. In strong electrolytes, aluminum should also have this effect. All specimens prepared in this study were scribed, so that the remote protection effect and the behavior could be evaluated in case of renewed damage. The samples first coated and repaired by ZnAl showed no red rusting from the substrate over the entire test period of 1000 h of the salt spray test. The sample prepared by the Bristle Blaster delaminated from the scribe, and also in areas not affected by a scratch. In this case, the adhesive strength as a whole was restricted. As described in [3], the samples coated with ZnAl formed white rust ($Zn(OH)_2$) due to the permanently moist climate:

$$Zn \xrightarrow{O_2} ZnO$$

$$ZnO \xrightarrow{H_2O} Zn(OH)_2$$

As shown in Fig. 4, this resulted in a coverage of 28% to 33% with white rust of the sample area after 1000 h of NSS. After peeling off the white rust, the undercut sealer (**Fig. 4 (1)**) appeared. The sealing layer is delaminated to the same percentage. This results in limited protection, which impairs the service life.

The samples repaired by Al99 also showed no red rust or delamination around the area of the injury. This is due to the very low corrosion potential of aluminum in relation to steel [3]. The values for the rust degree are small with "1" and a size of "3". The samples repaired by thermal spraying with Al99 showed no blistering, showed very good adhesion strengths and good layer appearance after 1000 hours salt spray testing.
The organic layers showed blistering around the scribe of size "4".

Fig. 5 shows exemplified the transition between the original thermally sprayed layer and the repair coating with the same spray material (top) and an organic coating system (bottom). Both surfaces were prepared by blasting.

1	ZnAl Grit Blast	**W**	33	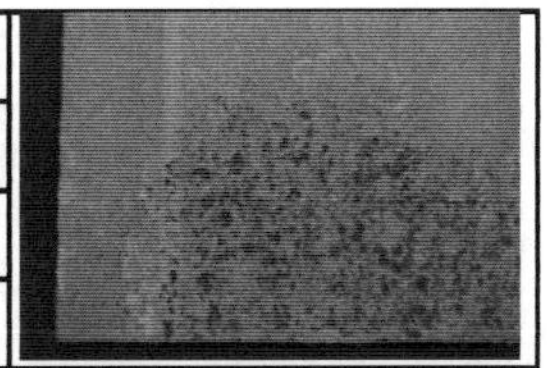
		R	0	
		G	0	
		B	0	
2	ZnAl Bristle Blast	**W**	28	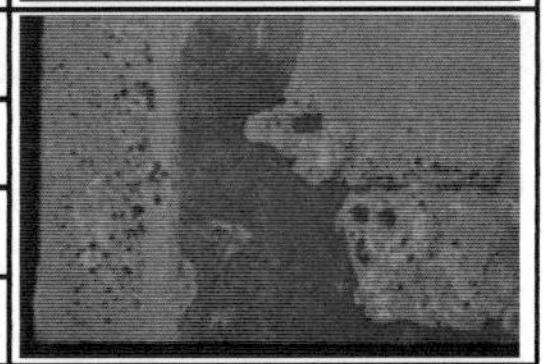
		R	1	
		G	3	
		B	5	
3	Al 10mm overlap	**W**	-	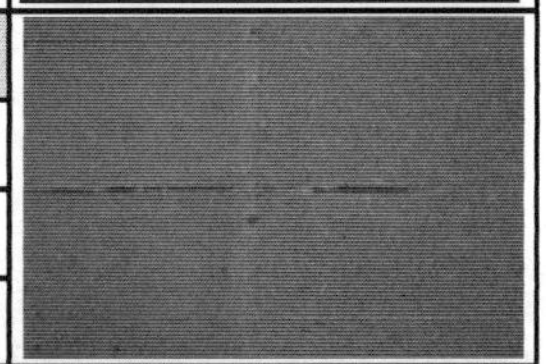
		R	1	
		G	3	
		B	0	
4	Al 76mm overlap	**W**	-	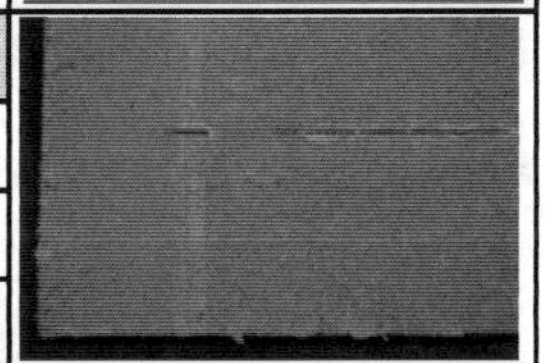
		R	1	
		G	3	
		B	0	
5	Organic 10mm overlap	**W**	-	
		R	2	
		G	4	
		B	4	
6	Organic 76mm overlap	**W**	-	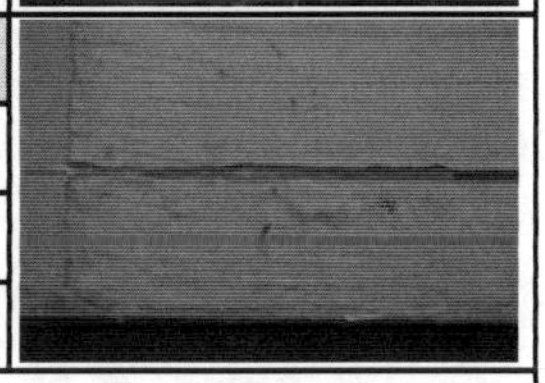
		R	2	
		G	4	
		B	4	
W = White Rust in %, **R** = Rustgrade Ri, **G** = Dimension of Damage Sn, **B** = Blister grade (all acc. DIN EN ISO 4628)				

Figure 4: Comparison of all repaired coating systems after 1000 NSS, washed and scaled of all residues

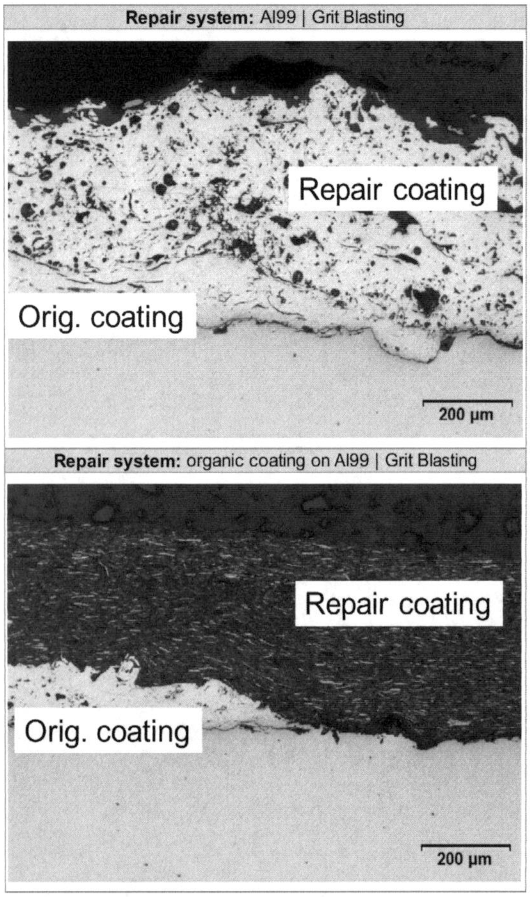

Figure 5: Comparison of the transition between original coating and repair coating (thermal spraying and organic coating)

If a layer system based on Al99 has to be repaired in part and there is the risk of a new injury, the repair by repeated thermal spraying shows very good results in relation to the corrosion behavior of the coating system. The values for the red rust formation, the size of the damage and the blister formation are lower compared to the organic repair concept. In particular, in the event of a renewed damage, the organic systems show a lower service life.
The systems thermally sprayed and repaired by ZnAl showed, if prepared by blasting, good resistance to red rust and bubble formation.

4 Conclusion

The comparative investigation of the repair coating methods thermal spraying against organic coating has shown that thermal spray repair provides a longer service life. The organic systems tended to undercut the coating (blister size Sn = 4 after 1000 h NSS) in the event of renewed damage. The thermally sprayed systems showed a good remote protection effect. Both methods showed no red rust. The assumed gap conditions at the transition between the original layer and the newly applied layer could not be recognized. A close transition was shown at the phase boundary.

For a suitable surface preparation of the partial repair, the blasting with corundum was compared with the Bristle Blaster. The roughness produced by the Bristle Blaster was found to be too low with Rz = 35.4 µm. The thermal spray layer did not adhere to the prepared surface. The reduced pressure blasting methods and the chamfering of the edges for surface preparation of the original layer showed good adhesive strengths of the repair layer and a longer life of the corrosion protection layer than by the bristle blasting.

The investigation of the influence of the overlapping width between the original and the repaired layer showed no significant influence of the overlapping width on the life of the layer system. The tested overlap widths of 10 mm and 76 mm resulted in the same degree of rust and damage for the organically repaired systems (Ri = 2, Sn = 4) and for thermal sprayed systems (Ri = 1, Sn = 3). However, it is to be noted that the organically repaired layers tend to blister from a scribe in the area of overlap. Therefore, it is recommended that in the case of a repair coat a tendentially smaller overlap of about 10 mm should be selected.

Especially in the case of renewed damage, corrosion protection systems repaired by thermal spraying show the longer service life. Due to their pronounced cathodic protection and due to their low corrosion potential compared to steel, they protect the substrate from corrosion. In particular, the systems based on Al99, which were partially repaired by corundum blasting and thermal spraying, showed the greatest durability in the study.

5 Literature

[1] Holcomb, G.R.; Covino, B.S.; Cramer, S.D.; Bullard, S.J.: Thermal-Spray Coatings for Coastal Infrastructure
Albany Research Center, U.S. Department of Energy, 1997

[2] Mathesius H.; Krömmer W.: Praxis des thermischen Spritzens – Anleitung für das Fachpersonal
DVS Media, 2009, ISBN 978-3-87155-552-7

[3] Büteführ, M.: Einfluss des Aluminiumgehaltes gespritzter Zinküberzüge auf den Korrosionsschutz von Stahl
Dissertation, MPA Stuttgart, Schriftenreihe Heft 91, 2006, ISBN 3-936846-62-6

[4] United States Navy, David Taylor Research Center: National Steel and Shipbuilding Company Procedure Handbook for Shipboard Thermal Sprayed Coating Applications
San Diego California, 1992

[5] Stango, R.; Khullar, P.: Fundamentals of Bristle Blasting Process for removing corrosive layer, NACE Int. Corrosion Conference and EXPO, 2009

Standards:

[S1] DIN EN ISO 2063:2005-05
Thermisches Spritzen - Metallische und andere anorganische Schichten - Zink, Aluminium und ihre Legierungen

[S2] NORSOK Standard M-501, Rev.6 February 2012
Surface preparation and protective coating

[S3] ANSI/AWS C2.18-93R
Guide for the Protection of Steel with Thermal Sprayed Coatings of Aluminium and Zinc and their Alloys and Composites

[S4] DVS 2302:2011-09
Korrosionsschutz von Stählen und Gusseisenwerkstoffen durch thermisch gespritzte Überzüge

Improvements of Coating Properties and Residual Stress States in Arc Sprayed Aluminium Bronze Coatings by Using an Alternative Gas Mixture

M. Hauer*[1], S. Krebs*[2], W. Krömmer*[3], K.-M. Henkel*[1]; *[1]Rostock, *[2]Hamburg, *[3]Unterschleißheim / D

The highly cavitation erosion resistant propeller alloys CuAl9Ni5Fe4Mn and CuMn13Al8Fe3Ni2 were arc sprayed with different traverse speeds by using a mixture of nitrogen and 2 % of hydrogen as atomising gas. Residual stresses were measured by the modified hole-drilling method using ESPI. Microstructural, chemical and mechanical analyses were realised to examine adhesive and cohesive properties. Additionally, the cavitation erosion behaviour was investigated. In comparison to coatings sprayed with pressurised air, the results of the study show superior coating qualities with regard to microstructure, cavitation erosion resistance and residual stresses.

1 Introduction

Arc spraying is a simple and cost-effective thermal spray technology, which is determined by a few key factors [1, 2]. Arc voltage, current and gas flow, pressure and type mainly affect coating quality [2, 3]. Besides, residual stresses in the coatings are strongly influenced by these parameters and the spray kinematics [3, 4]. Since arc spraying is already established for corrosion protection of large structures, the process is also considered for restoration of eroded ship rudders [1, 5]. Cavitation erosion, the formation and collapse of vapor bubbles caused by pressure oscillations, has a great impact on these damages [1, 6]. Typically, compressed air is used as atomising gas, which on the one hand results in partly large oxides in the coatings [2]. On the other hand, there is a significant melting loss of alloying elements, having a high oxygen affinity, during the spray process [2, 4]. This negatively affects cavitation erosion resistance of e.g. aluminium bronzes [4]. Furthermore, tensile residual stresses near the coating surfaces, mainly induced by high quenching stresses, contribute to this material loss [4]. Recent process developments revealed that use of different atomising gases improves coating quality [2]. Gases like nitrogen and its mixtures with hydrogen create a less reactive atmosphere and hence diminish the temperature of the particles in the spray jet. Therefore, less oxidation occurs [2] and residual stresses might be reduced due to less quenching stresses. Thus, a mixture of nitrogen and hydrogen is used for the following experiments. The results of the study are compared to former results using pressurised air, see [4].

2 Arc Spray Experiments

Arc spraying experiments were performed by using a Sulzer Metco Smart Arc (Oerlikon Metco Europe GmbH, Kelsterbach, DE). The traverse speed of the robot was varied from 333, 500 to 666 mm/s, while the spraying parameters remained identically ($U = 28$ V, $I = 180$ A, $p_{N2+H2} = 4$ bar, SOD = 100 mm). A mixture of nitrogen and 2 % of hydrogen was used as atomising gas. Due to a necessary minimum coating thickness of 300 µm, the number of layers for Ni-Al-Bronze was raised as compared to previous experiments [4], **Table 1**. Materials used for spray experiments were the cavitation erosion resistant [7] Ni-Al-Bronze CuAl9Ni5Fe4Mn (Lincoln Electric LNM CuAl8Ni6, Ø 1.6 mm, Ratingen, DE) and Mn-Al-Bronze CuMn13Al8Fe3Ni2 (Bedra Bercoweld A300, Ø 1.6 mm, Heuchelheim, DE), see also [4]. The substrates were grit-blasted and of type S275 mild steel (corresponding to DIN EN 10025-2) with dimensions of ca. 65 mm x 50 mm x 5 mm.

3 Results

3.1 Deposition Efficiency

The deposition efficiency (DE) was determined in accordance with DIN EN ISO 17836. It rises for Ni-Al-Bronze with increasing traverse speeds, while the Mn-Al-Bronze coatings reveal a slight reduction, **Table 1**. The effect of atomising gas N_2+H_2 results in an overall increase of the DE as compared to the coatings sprayed with air, revealing one slight drop for the highest traverse speed specimen of Mn-Al-Bronze in **Table 1**.

Table 1 Properties of N_2+H_2 – sprayed coatings compared to coatings sprayed with air [4]. Symbols indicate an increase (⇧), a decrease (⇩) or a stagnation (~)

Traverse Speed	Number of layers	Deposition Efficiency (DE) in %	Defects in total in %	Specific electrical conductivity in MS/m
Ni-Al-Bronze				
333 mm/s	9 (⇧)	73.5 (⇩)	4.9 (⇩)	9.0 (⇩)
SD	...	...	± 1.1	± 0.3
500 mm/s	10 (~)	86.3 (~)	9.4 (⇩)	15.6 (⇧)
SD	...	± 2.0	± 0.8	± 0.8
666 mm/s	20 (⇧)	89.8 (⇧)	6.7 (⇩)	7.1 (⇩)
SD	...	± 1.1	± 1.9	± 0.5
Mn-Al-Bronze				
333 mm/s	8 (~)	85.1 (⇧)	4.7 (⇩)	5.5 (⇩)
SD	...	± 2.4	± 2.6	± 0.2
500 mm/s	10 (~)	83.8 (⇧)	3.0 (⇩)	9.5 (⇧)
SD	...	± 3.5	± 0.9	± 0.6
666 mm/s	15 (~)	82.0 (⇩)	10.5 (⇧)	6.8 (⇩)
SD	...	± 1.3	± 2.1	± 0.5

3.2 Adhesion and Mechanical Properties

The influence of the atomising gas on coating adhesion depending on the traverse speed variation was investigated according to DIN EN 582 by using a static uniaxial testing machine (Zwick Roell Z50, Ulm, DE, three specimens each), **Fig. 1**.

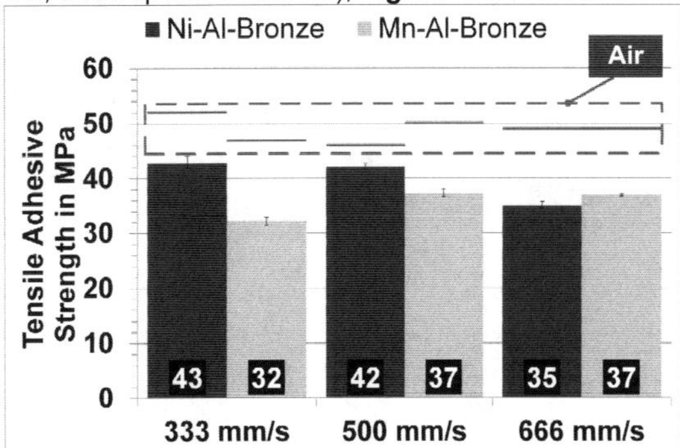

Fig. 1. Average tensile adhesive strength of the N_2+H_2 – sprayed coatings. Grey lines indicate the values of air – sprayed coatings [4].

It is visible, that the tensile adhesive strength is influenced by traverse speed in the arc spray process, **Fig. 1**. There is a tendency of rising values for the Mn-Al-Bronze with increasing traverse speeds, while the Ni-Al-Bronze coatings reveal an opposite trend, **Fig. 1**. Moreover, all values are within a range of about 10 MPa. In comparison to coatings sprayed with air in [4], the use of N_2+H_2 as atomising gas results in strongly reduced tensile adhesive strength values for all specimen. The Mn-Al-Bronze coatings (average decrease of - 28 %) are more affected by the change of the atomising gas than the Ni-Al-Bronze coatings (average decrease of - 18 %).
Hardness measurements on the surface of the coatings were carried out by using a Wolpert 432 SVD hardness testing machine (Wilson Wolpert Instruments, Aachen, DE; five measurements each; applied load of 9.8 N), **Fig. 2**.

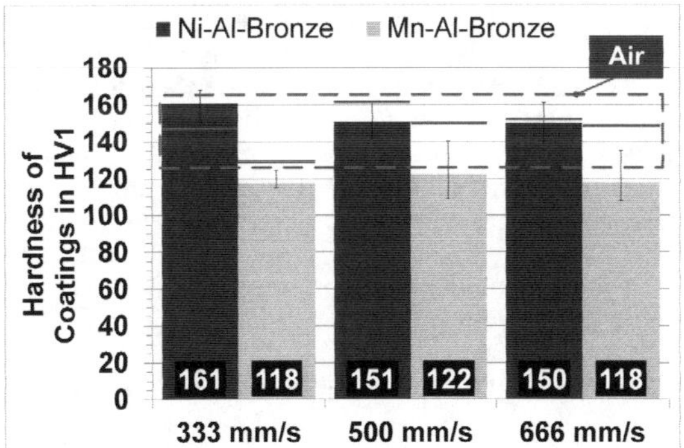

Fig. 2. Average hardness of the N2+H2 – sprayed coatings. Grey lines indicate the values of air – sprayed coatings [4].

The hardness values of the Mn-Al-Bronze coatings exhibit no significant differences and are far lower than for Ni-Al-Bronze coatings, **Fig. 2**. The Ni-Al-Bronze coatings also display a constant level of hardness for the traverse speed specimens with 500 mm/s and 666 mm/s, while only the slowest traverse speed specimen reveals a slight increase. The usage of N_2+H_2 seems to influence solely the Mn-Al-Bronze coatings, since the hardness decreases about 16 % as compared to coatings sprayed with air [4]. In contrast, the hardness of Ni-Al-Bronze is only slightly affected (average growth of 1 %).

3.3 Representative Microstructural Features

Further information about coating quality were obtained by microstructural analyses using an optical microscope (OM) of type Leica DM6000M (Leica Microsystems GmbH, Wetzlar, DE) and the software tool ImageAccess (Imagic Bildverarbeitung AG, Glattbrugg, CH). Specimens for metallographic examinations were hot-mounted via ATM Opal 410 Hot Mounting Press (ATM GmbH, Mammelzen, DE) and incrementally grinded and polished (6 µm, 3 µm suspensions, finally oxide polish). Klemm III colour etchant (stock solution: sodium-thiosulphate in water; etching solution: potassium pyrosulphite) was chosen for enhancing contrasts of the coating microstructures and identification of different phases. Moreover, energy dispersive x-ray spectrometry (EDS) was performed by using a set-up of Bruker QUANTAX (Bruker Nano GmbH, Karlsruhe, DE).

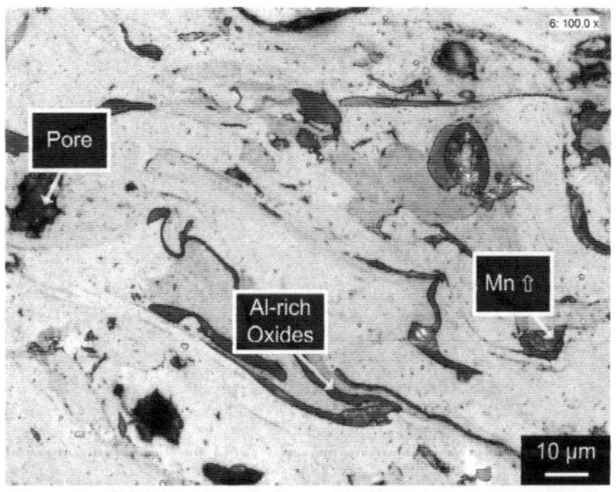

Fig. 3. OM image of coating microstructure; exemplarily for 666 mm/s specimen of N_2+H_2 – sprayed Mn-Al-Bronze

The OM image in **Fig. 3** displays the typical microstructure of the colour etched coatings, supplemented by EDS analyses of the same areas. The focus on Cu and the alloying elements Al, Ni, Fe and Mn reveals only minor local composition variations, which are marked in **Fig. 3**. Partially large oxides are primarily formed by Al and appear dark grey. Enhanced contents of Mn in Mn-Al-Bronze are marked in **Fig. 3** and can be identified having another dark grey tone. These areas are characterised as β-phase. Compared to the coatings sprayed by using pressurised air in [4], the β-phase appears only eutectoid and not martensitic. Additionally, this phase is detectable to a far lesser extent in the N_2+H_2 – sprayed coatings. The major share of the coatings appears in brighter grey tones, which can be identified as α-phase. The microstructures of the Ni-Al-Bronze coatings are even more homogeneous than for the Mn-Al-Bronze coatings and contain no ordered β-phase. In general, all of the alloying elements (most important Al) are more evenly distributed in the matrix of the N_2+H_2 – sprayed coatings in comparison with the air-sprayed coatings [4]. Thus, the enhanced alloying element content might lead to less oxides, an enhanced ductility and higher cavitation erosion resistance of the alloys [7].

3.4 Quantitative Microstructal Analyses and Electrical Conductivity

The software tool ImageAccess was used to identify the overall defect content (examining five pictures per specimen, magnification of 200x). The results of these analyses are shown in **Table 1**.

The Ni-Al-Bronze coatings show highest defect content for specimen of 500 mm/s traverse speed, while Mn-Al-Bronze coatings reach the highest defect value at the highest traverse speed. During the quantitative analyses it became clear, that the Mn-Al-Bronze coatings show a more inhomogeneous pore and oxide distribution, which is evident in higher standard deviations compared to Ni-Al-Bronze coatings. In comparison to the air – sprayed coatings in [4], the N_2+H_2 – sprayed coatings reveal significantly fewer defects, except for one specimen with a slight increase. The Ni-Al-Bronze coatings benefit more from the usage of N_2+H_2 as atomising gas than Mn-Al-Bronze coatings, since the average defect content decreases more severely [4].

The specific electrical conductivity was measured by four-terminal sensing using a Loresta GX MCP-T700 system (Mitsubishi Chemical Analytech Co. LTD, Kanagawa, JP) with a PSP measuring head for small specimens (five measuring points, measurement time of 10 s, current of 1 A). The specimens were tested in the as-sprayed state.

The results of specific electrical conductivity measurements show varying values for both materials, **Table 1**. For Ni-Al-Bronze, the values are in a wider regime than for Mn-Al-Bronze. Overall, the coatings show a far lower degree of variation in comparison to the air-sprayed coatings in [4]. In addition, the Ni-Al-Bronze coatings are more affected by the change of atomising gas than Mn-Al-Bronze coatings.

Table 2 Chemical composition of N_2+H_2 – sprayed coatings compared to coatings sprayed with air [4]. Symbols indicate an increase (⇧), a decrease(⇩) or a stagnation (~).

Type	Traverse Speed	Chemical composition and standard deviation in wt%			
		Al	Ni	Fe	Mn
Wire	Ni-Al-Bronze	9.0	5.0	4.0	2.5
Coatings	333 mm/s	10.42 (⇧)	4.70 (⇩)	2.93 (⇩)	0.90 (⇧)
	SD	± 0.58	± 0.20	± 0.11	± 0.02
	500 mm/s	10.61 (⇧)	4.65 (⇩)	2.85 (⇩)	0.90 (⇧)
	SD	± 0.36	± 0.33	± 0.11	± 0.03
	666 mm/s	10.55 (⇧)	4.80 (⇩)	2.91 (⇩)	0.90 (⇧)
	SD	± 0.76	± 0.08	± 0.20	± 0.02
Wire	Mn-Al-Bronze	7.2	2.3	2.3	11.6
Coatings	333 mm/s	7.92 (~)	2.46 (⇩)	2.20 (⇩)	9.31 (⇧)
	SD	± 0.09	± 0.06	± 0.12	± 0.21
	500 mm/s	7.95 (~)	2.53 (⇩)	2.18 (⇩)	9.55 (⇧)
	SD	± 0.29	± 0.12	± 0.04	± 0.66
	666 mm/s	8.04 (⇧)	2.75 (⇩)	2.36 (~)	10.48 (⇧)
	SD	± 0.24	± 0.08	± 0.23	± 0.92

Pure Cu coatings fabricated by conventional arc spraying are reported to reach approx. 10 MS/m [8]. Except for one specimen, all N_2+H_2 – sprayed coatings are below the reported value for pure arc sprayed Cu. In contrast, the comparable air – sprayed coatings in [4] attain more results above this reference value.

3.5 Chemical Analyses

In order to examine differences in the chemical composition, the coatings are examined via optical emission spectrometry (OES) using Spectromaxx (Spectro Analytical Instruments GmbH, Kleve, DE, three measurements for each specimen).

Additionally, the results are opposed to the chemical composition of the wires according to the manufacturer specifications.

The OES results in Table 2 reveal almost similar contents of alloying elements for all traverse velocities for both materials, except for Mn with a slight increase at the highest traverse speed for Mn-Al-Bronze coatings. In comparison to the manufacturer specifications of Ni-Al-Bronze wire, the contents of Al are higher, while other alloying element contents are lower. For Mn-Al-Bronze, the fractions of Al and Ni are higher than specified, whereas Fe and Mn contents are reduced. In comparison to coatings in [4], the change of atomising gas to N_2+H_2 results in a higher Al and Mn content in the coatings for both bronzes. However, the Ni-Al-Bronze coatings are affected to a far greater extent.

3.6 Cavitation Erosion Properties

The cavitation erosion behaviour is considered to be a measure of coating quality. Thus, indirect cavitation erosion tests in compliance with [6] were executed for a period of 120 min using an UIP1000 (Hielscher Ultrasonics GmbH, Teltow, DE) with a frequency of 19.51 kHz, a peak-to-peak amplitude of 50 µm and a distance from the sonotrode tip to the specimen of 0.5 mm. Since asperities can serve as initial cracks, plane surfaces needed to be ensured. Hence, the specimens were gradually grinded and polished up to

3 µm. Subsequently, the cumulative mass loss of the specimens (three specimens each) was converted to the cumulative thickness loss, which is also designated as the mean depth of erosion (MDE) [6]. This allows comparison of materials with different densities. Another significant parameter of cavitation erosion resistance is the maximum erosion rate (MER). It corresponds to the maximum slope of the MDE curve and is computed by an in-house MATLAB routine.
Both materials reveal different trends, **Fig. 4**. The Ni-Al-Bronze coatings are characterised by an almost linear progress over the test period, resulting in a maximum MDE of up to 16 µm. On the other hand, Mn-Al-Bronze coatings exhibit a linear growth to a time of approx. 60 min and afterwards a more asymptotic course with a maximum MDE of 19 µm, **Fig. 4**. When comparing both materials, the Ni-Al-Bronze coatings obtain lower and more regular material losses as well as erosion rates.
Using N_2+H_2 as atomising gas affects the coating properties of both spray materials. This is especially evident, when comparing the average MDE of each material after 120 min to the air – sprayed coatings in [4]. The MDE values are reduced by 71 % for Ni-Al-Bronze and by 56 % for Mn-Al-Bronze. It is obvious, that the use of N_2+H_2 impacts Ni-Al-Bronze coatings to a greater extent than the Mn-Al-Bronze coatings, **Fig. 4**. As a result, Ni-Al-Bronze coatings are considered more cavitation erosion resistant, which differs from the observations in [4].

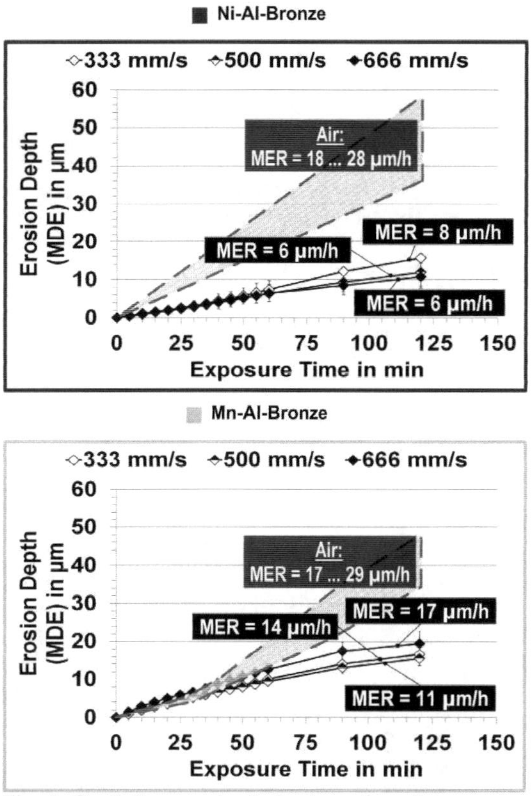

Fig. 4 Cavitation erosion behaviour of coatings sprayed using N_2+H_2. Grey areas indicate the values of air – sprayed coatings [4].

3.7 Residual Stress Analyses

Modified incremental hole drilling method based on electronic speckle pattern interferometry (ESPI) was used to identify the residual stress state inside the coatings. The ESPI technique is well documented in literature and will not be further explained [9, 10]. In combination with the elastic material parameters (Young's modulus, Poisson ratio), the stress values can be calculated by the software [9, 10].
The Stresstech PRISM system (Stresstech GmbH, Rennerod, DE) was utilised for residual stress analyses. The holes were machined in steps of 50 µm using high speed steel end mills (2 fluted, TiN coated, Ø 0.8 mm and 1.6 mm) up to a profile depth of 450 µm. The inner integration radius was 2.5 times the hole diameter and the outer integration radius 4.5, respectively. The Poisson ratios used for calculation were 0.3 for Ni-Al-Bronze and 0.34 for Mn-Al-Bronze [7].
The Young's modulus, which is mandatory for stress calculations, was determined by using a nano indentation machine NHT (CSM Instruments, Freiburg, D) with five measurements for each coating. The measurements of the Young's moduli differ significantly from bulk material [7]. For cast Ni-Al-Bronze, the values lie within a range of 115 GPa to 121 GPa, while the coatings show average values of

102 GPa to 91 GPa and 96 GPa with increasing traverse speed. For Mn-Al-Bronze coatings, the average moduli vary from 72 GPa to 69 GPa and 78 GPa with rising traverse speed, while cast Mn-Al-Bronze alloys have values of 117 GPa. Thus, in comparison to the air – sprayed coatings in [4] both coating materials show a general decrease in moduli by the use of N_2+H_2 as atomising gas, except for one specimen. The Mn-Al-Bronze coatings reveal a much sharper loss in elasticity than Ni-Al-Bronze coatings.

In general, the residual stress curves of Ni-Al-Bronze show a linear decreasing progress of residual tensile stresses to a depth of at least 300 µm, **Fig. 5**. The highest values can be observed near the surface, reaching up to a maximum of about 510 MPa. Compressive stresses appear only for a traverse speed of 500 mm/s. The specimens with the highest traverse speed also show the highest amount of residual stress over the entire drilling depth and are followed directly by the lowest traverse speed specimens.

Likewise, the Mn-Al-Bronze coatings show a linear decreasing progress of the tensile stresses to a depth of at least 350 µm, **Fig. 5**. However, the stresses vary to a lesser extent than Ni-Al-Bronze coatings. This is evident for stress depth profiles of 500 mm/s and 666 mm/s specimens, which are almost identical in progress and reveal lower corresponding values than the curve of the 333 mm/s specimens. The highest stress values occur for the lowest traverse speed. Those stresses reach up to 313 MPa near the surface and remain tensile throughout the depth profile.

The comparison of both materials reveals smaller tensile stresses of corresponding curves for Mn-Al-Bronze, **Fig. 5**.

This is also confirmed by the numerically calculated integral below the tensile area, which is always lower compared to Ni-Al-Bronze, **Fig. 6**. For both materials, a decrease can be observed from 333 mm/s to 500 mm/s specimens. Interestingly, this reduction of the tensile areas correspond to the increasing traverse speeds, since the number of layers is only slightly changed, **Fig. 6**. Additionally, a significant rise regarding the number of layers also correlates with an increase in the integral, which is evident for the change from 500 mm/s to 666 mm/s for both materials, **Fig. 6**. Comparing both materials, a stronger growth can be seen for Ni-Al-Bronze, **Fig. 6**.

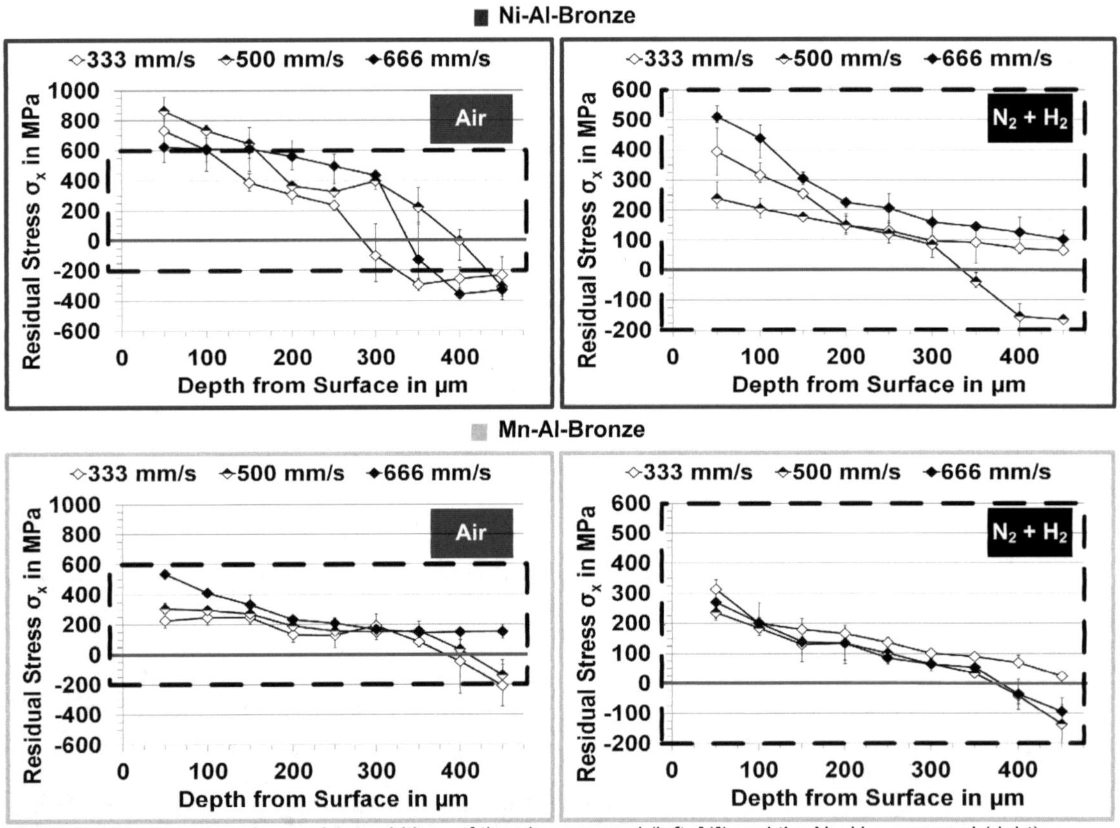

Fig. 5. Residual stresses in specimen width σ_x of the air – sprayed (left, [4]) and the N_2+H_2 – sprayed (right) coatings. Ni-Al-Bronze is displayed on top, while Mn-Al-Bronze is presented below. A different scaling of the ordinate axis was chosen for the N_2+H_2 diagrams (right).

The effect of N_2+H_2 as atomising gas is tremendously visible when comparing the maximum residual stresses with the air – sprayed coatings, **Fig. 5**.

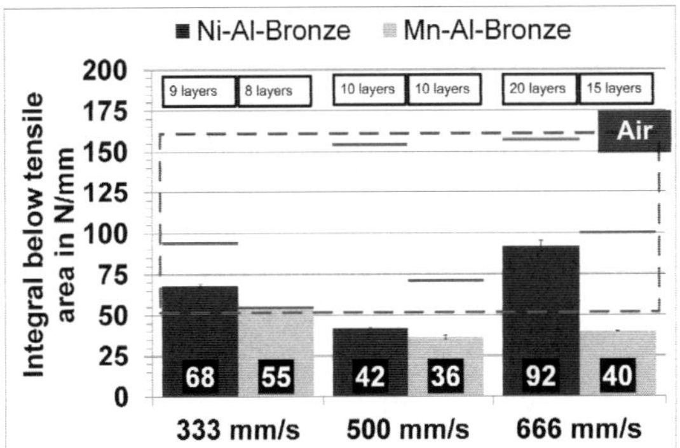

Fig. 6. Average integral below tensile area of stress curves. Calculated by trapezoidal numerical integration. Grey lines indicate the values of air – sprayed coatings [4].

Even near the coating surface, the residual stress curves of the N_2+H_2 – sprayed coatings show lower initial values, except for one specimen. The Ni-Al-Bronze coatings are far more affected by the change of atomising gas than Mn-Al-Bronze, which is also proven when considering the integral below the tensile area, **Fig. 6**. However, higher numbers of layers for 333 mm/s and 666 mm/s of Ni-Al-Bronze specimens need to be considered for evaluation too.

4 Discussion

First of all, the spray process itself is improved, which is evident in higher deposition efficiency, **Table 1**. This can be attributed to the shroud effect [2], resulting in larger particles due to less melting loss and oxidation of the particles in comparison to air-sprayed coatings in [4]. The latter is also confirmed by reduced overall defects, **Table 1**, and enhanced contents of highly oxygen affine alloying elements Al and Mn, cf. [4] and **Table 2**. The lower specific electrical conductivity further verifies this. Even small amounts of Al in the matrix drastically reduce conductivity in aluminium bronze materials, [7]. Bronze coatings sprayed with air show much higher conductivities, which correspond to less Al in the matrix and are caused by a far higher Al-based oxide content, [4]. Hence, considering the reported literature [8], a rising conductivity up to roughly the threshold of pure arc sprayed Cu tends to correlate with a lower defect content, while the trend is vice versa above this value [4].

In contrast to the aforementioned positive effects, the tensile adhesive strength and Young's moduli decrease for both materials, while Mn-Al-Bronze is far more affected. Furthermore, solely the Mn-Al-Bronze coatings reveal a notable reduction of hardness due to the change of atomising gas. This probably originates from a decrease of the retained martensitic β-phase, see section **3.3**, caused by lower cooling rates in the coating process compared to the usage of air. The lower temperatures in the coating process by using N_2+H_2 also likely induce the reduction of tensile adhesive strength and possibly elasticity, especially for Mn-Al-Bronze. However, the tensile adhesive strength values are still sufficient enough for bonding. Considering the cavitation erosion resistance, the loss in elasticity seems to be important too. In comparison to the air-sprayed coatings in [4], the Ni-Al-Bronze coatings reveal less loss of elasticity and a far better resistance against cavitation erosion damages than Mn-Al-Bronze, **Fig. 4**. Nevertheless, for both materials a massively improved cavitation erosion behaviour can be observed in comparison to air – sprayed coatings in [4], which is likely a primary consequence of the reduced defect content, **Table 2**. Moreover, the Ni-Al-Bronze coatings benefit more from the usage of N_2+H_2 regarding the main alloying element Al, **Table 2**. This further enhances cavitation erosion resistance.

Finally, the residual stresses were drastically reduced by the usage of N_2+H_2, **Fig. 5** and **Fig. 6**. It might be reasonably assumed, that lower substrate temperatures and cooling rates mainly contribute to this fact. Additionally, the reduced tensile stresses improve cavitation erosion behaviour. Still, in comparison to the coatings in [4] it must be noted, that the change of atomising gas reduces the dominance of the residual stresses regarding coating properties, cf. **Fig. 6** and **Fig. 4**. Hence, the influence of traverse speed is also lowered. The number of layers seems to be important only once it is increased excessively, which is different from the observation in [4]. Nevertheless, the differing number of layers of two N_2+H_2 – sprayed Ni-Al-Bronze specimens might affect these results and needs to be considered for future investigations.

5 Summary and Outlook

In the present work, the influence of nitrogen and 2 % of hydrogen as atomising gas in the arc spray process was examined. It was found that aluminium bronze coatings sprayed with this mixture show superior coating qualities regarding defect content, melting loss of alloying elements, cavitation erosion resistance and residual stresses in comparison to conventionally sprayed coatings. In contrast, the adhesive strength, elasticity and partially hardness is reduced. Overall, Ni-Al-Bronze coatings benefited more from the change of atomising gas than Mn-Al-Bronze coatings.

Further investigations will focus on determining the change of the temperature regime. On the other hand an optimisation of heat transfer by improved spray patterns will be considered.

Acknowledgements The authors would like to thank all co-workers involved in the study, namely listed in alphabetical order: R. Arndt, A. Herhaus, D. Krüger, N. Németh, A. Novikov, B. Ripsch, M. Schulze and S. Wachtmann.

The research project (No. 18449 BG) of the research community Center of Maritime Technologies (CMT), Bramfelder Straße 164, 22305 Hamburg has been funded by the AiF within the program for sponsorship by Industrial Joint Research (IGF) of the German Federal Ministry of Economic Affairs and Energy based on an enactment of the German parliament. This support is greatly acknowledged.

6 Literature

[1] Kim, J.-H. and M.-H. Lee: A Study on Cavitation Erosion and Corrosion Behavior of Al-, Zn-, Cu-, and Fe-Based Coatings Prepared by Arc Spraying. J. Therm. Spray Technol. 19 (2010), Issue 6, pp. 1224/1230.

[2] Lang, F. and W. Kroemmer: Economic and ecological benefits of using gas mixtures for arc spraying. Proc. 10th HVOF Colloquium (2015), pp. 79/88.

[3] Chi, J.; Zhang, A.; Xie, S. and C. Jin: Process Optimization and Residual Stress Measurement for Arc Spraying Rapid Tooling. Acad. J. Xi'an Jiaotong Univ. 48 (2014), Issue 12, pp. 126/130, in Chinese.

[4] Hauer, M.; Henkel, K. M.; Krebs, S. and W. Kroemmer: Study of Traverse Speed Effects on Residual Stress State and Cavitation Erosion Behavior of Arc-Sprayed Aluminum Bronze Coatings. J. Therm. Spray Technol. 26 (2017), Issue 1-2, pp. 217/228.

[5] Barth, A.: Application areas of arc spraying. Presentation within the seminar „Modern Coating Technologies“. Dortmund, Germany, 05.11.2014, in German.

[6] American Society of Testing and Materials: Standard Test Method For Cavitation Erosion Using Vibratory Apparatus, G32. Annual Book of ASTM Standards (2010), pp. 98/116.

[7] Meigh, H.J.: Cast and Wrought Aluminium Bronzes: Properties, Processes and Structure.
Copper Development Association / IOM Communications, London (2000).

[8] Stoltenhoff, T., Borchers, C.; Gärtner, F. and H. Kreye: Microstructures and key properties of cold-sprayed and thermally sprayed copper coatings. Surf. Coat. Technol. 200 (2006), Issue 16, pp. 4947/4960.

[9] Steinzig, M. and E. Ponslet: Residual Stress Measurement using the Hole Drilling Method and Laser Speckle Interferometry: Part I. Exp. Tech. 27 (2003), Issue 3, pp. 43/46

[10] Schajer, G. S. and M. Steinzig: Full-Field Calculation of Hole-Drilling Residual Stresses from ESPI Data. Exp. Mech. 45 (2005), Issue 6, pp.526/532.

Experimental and Numerical Study of the effect of Gas-shrouded Plasma Spraying on Cathode Coating of Alkaline Electrolysis Cells

T. Liu and A. Ansar (German Aerospace Center), Düsseldorf / D

Dual layer electrode coating for alkaline water electrolysis was prepared by plasma spraying. For improving performance this work aims at reducing the oxide in electrode coating. Regarding the necessity of obtaining high specific area, atmospheric plasma spray was employed under protection of argon which was used as shrouding gas. Composite cathode was established on Ni-coated perforated steel sheet with crushed and gas atomized Nickel-based alloy powders. The dual-layer structure was a composite of 5 layers of NiAl at the bottom and 10 layers of NiAlMo as the top layer. Microstructure and morphology were studied by scanning electron microscope (SEM). Element content was estimated by energy dispersive spectrometer (EDS). Enthalpy probe was introduced for measuring plasma temperature and velocity as well as gas composition. Numerical calculation was carried out with same condition for better understanding the shrouding effect. The results showed moderate protection by using of arranged gas shrouding. Overall the dual layer region, Oxygen was decreased by 0.3% from 3.46% to 3.15%. With gas shrouding coating exhibited similar element contents as coating sprayed by VPS. However, no obvious difference was observed in microstructure and morphology with or without gas shrouding.

1 Introduction

As a clean energy carrier, hydrogen has drawn a lot of interests in industry and in academic research. Especially, the global warming effect brings challenge in environment and society in recent years. As one of the most promising approaches to produce hydrogen alkaline water electrolysis (AWE) exhibits surprising potential of taking advantage of the renewable energy as well as the capacity to be integrated in large-scale energy storage [1]–[4]. It has been intensively investigated to reduce the cost and decrease the degradation of the electrodes as low as possible throughout the development [1], [5]–[7]. The state-of-art cathode is based on nickel alloy which shows low over-potential and moderate current density for hydrogen evolution reaction (HER) [7]–[9]. Recently, a lot of works have been reported focusing on the process to prepare cathode with reasonable performance. Kjartansdóttir et al [3] tested a physical vapour deposited (PVD) Al/Ni electrode with over potential of -1000 mV at 500 mA/cm2. Manazoğlu et al [10] reported over-potential around -200 mV at current density of 100 mA/cm2 on an electroplated Ni/Mo electrode. While Schiller et al [11], [12] presented much better performance with an over-potential as low as -90 mV at 500 mA/cm2 current density based on VPS coated NiAlMo electrode..

2 Methodology and installations

2.1 Experimental setups

2.1.1 Plasma spray installation

This work was carried out in a chamber which was able to be controlled in vacuum state or in atmospheric state by a vacuum pump/ventilation pump (see **Fig. 1**). The pressure of the chamber could be controlled either as low as 6000 Pa for VPS or be maintained at 1 atm. (9.4×10^4 Pa) for APS. Inside the chamber, a water-cooled F4-VB torch was installed on an X-Y movement (see **Fig. 1 (b)**). The feedstock was perpendicularly fed to the plasma jet through an injector with 1.8 mm internal diameter being installed 3 mm upstream to the nozzle exit. A gas shrouding part with 12 arranged shrouding nozzles was mounted to the torch and to prevent any unfavourable entering path of the ambient gas as indicated by P1 in **Fig. 2(a)**. The ground-face of the shrouding part was aligned with the exit of the torch that the ambient gas would be well entrained to the plasma jet through the path P2 in **Fig. 2(a)**. The gas used for shrouding was argon with net pressure as high as 3.76×10^5 Pa; the input argon was delivered to the shrouding nozzles by two entrances as indicated in **Fig. 2(b)** and **(c)**. In **Fig. 2(d)** the simplified nozzle was presented to be a V-like notch with an expanding angle of 30 °.

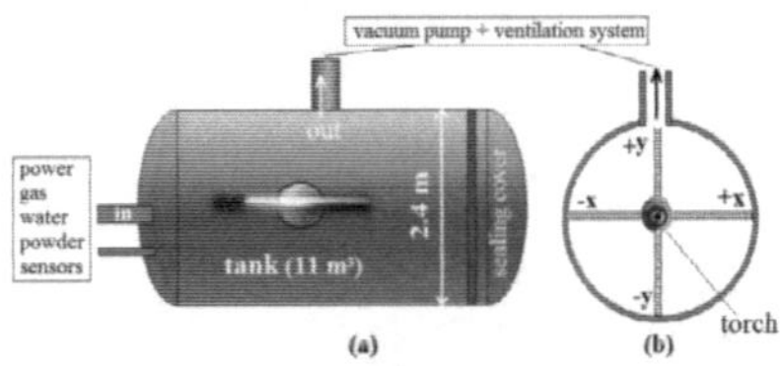

Fig. 1. the plasma spray installation with vacuum pump and ventilation pump

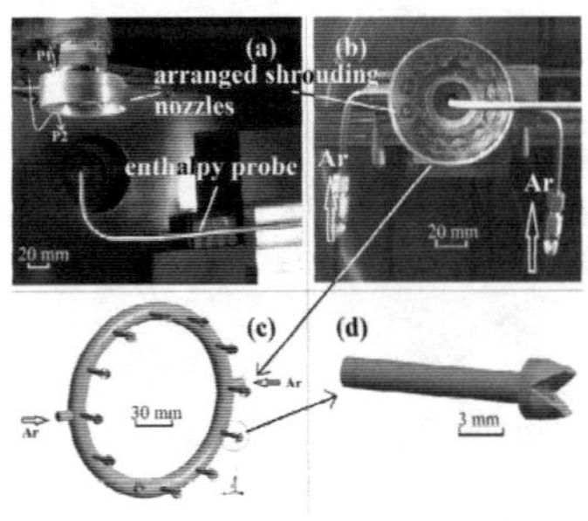

Fig. 2 installations of the arranged shrouding nozzles and the enthalpy probe: (a) top view; (b) front view; (c) model of the arranged nozzles with the gas delivering channel; (d) model of a single shrouding nozzle simplified for simulation

2.1.2 Enthalpy probe

A lot of works have been reported to successfully use enthalpy probe to diagnose plasma jet with reasonable precision [13]–[17].In this work the enthalpy probe system from TEKNA (Québec, Canada) was employed to detect the plasma gas composition, the plasma temperature and the plasma velocity. The probe used was ENT 476 with orifice area of 1.27 mm^2 at the tip. The probe tip was placed on the centreline of the torch aperture and perpendicularly pointed to the centre of the torch. Instead of moving the probe the torch was precisely moved to adjust the measuring position for depicting the distribution of the gas composition as well as the temperature and the velocity of the plasma. The zero point was set as the centre of the plasma which was found by slowly moving the torch. During the measurement, the sampling rate was adjusted to achieve a good isokinetic of the sample gas for reducing error.

2.1.3 Feedstocks

The feedstock used in this work was commercially supplied NiAl (56/44 wt %) as the bond layer and NiAlMo (39/44/17 wt %) as the top layer. The size distribution of the used powders was displayed in **Fig. 3**. The granularity test also indicated that the d_{50} was 15.43 µm and 16.67 µm for NiAl and NiAlMo respectively. Both of the two powders were sphered to improve the flow-ability and thus to diminish the fluctuation in the powder feeding according to our prior work [18]. The morphology of the two powders was shown in **Fig. 4**.

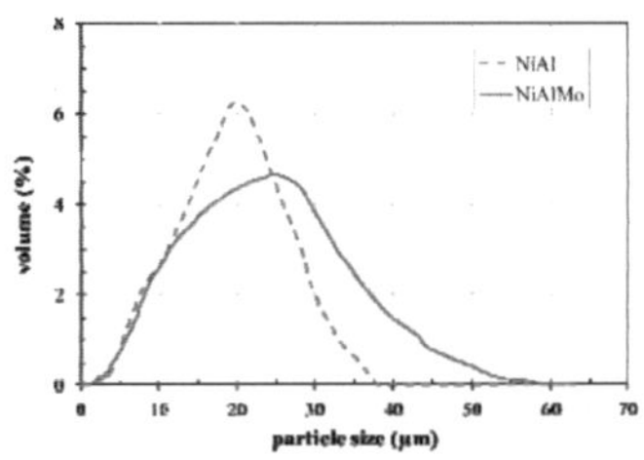

Fig. 3 the granularity of NiAl and NiAlMo

Fig. 4 the morphology of NiAl and NiAlMo

2.1.4 Samples preparation and procedure

Three samples were prepared with different pressure and different shrouding state (see **Tab. 1**). Sample 1 was sprayed in VPS (6000 Pa); sample 2 was obtained in APS (1 atm.) when the plasma jet was being shrouded by compressed argon with net pressure of 3.76x10^5 Pa (4 atm.); sample 3 was also prepared in APS but without plasma jet shrouding. The substrates used were stainless steel sheets with size of ϕ32 mm x 0.5 mm on which a pure nickel layer was pre-coated with thickness about 36±6 µm. The substrates were punched to have open pores with diameter of 2 mm which resulted in an effective area of 555 mm^2 as indicated in **Fig. 5**. After being sandblasted, for each sample, 5 layers of NiAl were firstly sprayed and then 10 layers of NiAlMo were continuously deposited on the surface.

Tab. 1 operating parameters and shrouding state

No.	I (A)	Ar (SLM)	He (SLM)	H_2 (SLM)	pressure (Pa)	shrouding	distance (mm)
1	420	45	10	4	6x10^3	0	80
2	420	45	10	4	9.4x10^5	1	80
3	420	45	10	4	9.4x10^5	0	80

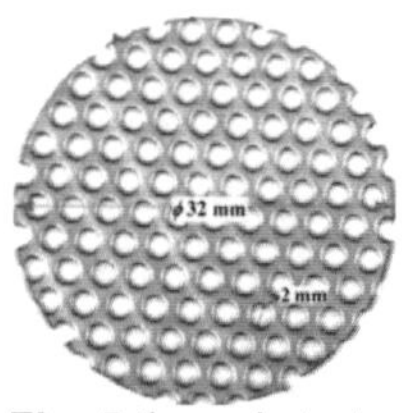

Fig. 5 the substrate used in this work

2.2 Mathematical model

2.2.1 Simulation assumption and governing equations

For verifying the effect of the gas shrouding on the plasma jet, numerical simulations were carried out focusing on the plasma jet behaviour as well as the ambient gas. Thus, the reaction between electrons and ions, such as charge transfer, was not mentioned. The magnetic force on particles and the current density on electrodes were also not interesting for this work. The simulation work was based on the following assumption: the plasma is in equilibrium; the profile of temperature and velocity of the plasma jet is in the form of Gaussian distribution; the plasma gas is a mixture of Ar and H_2; all solid parts are rigid; the heat absorbing effect of the solid part was not considered; the entrained air was a mixture of 79% N_2 and 21% O_2; the properties of the ambient gas was constant with value determined according to the instant temperature.

Two governing equations were used [19]:
The conservation of mass:

$$\nabla \cdot (\rho \boldsymbol{U}) = 0 \qquad (1)$$

The conversation of momentum: $\nabla \cdot (\rho \boldsymbol{U}\boldsymbol{U}) + \nabla p - \nabla \vec{\tau} = 0$ (2)

The conservation of energy:

$$\nabla \cdot (\rho \boldsymbol{U} h) - \alpha \nabla^2 T = 0 \qquad (3)$$

Where ρ denotes the density, kg/m^3; $\boldsymbol{U}$ gives the vector of velocity, m/s; p means the pressure, Pa; τ indicates the stress tensor, Pa; T presents the temperature, K; h gives the enthalpy, J/kg; α is the thermal conductivity, W/m^{2}k.

For closing the equations, the stress tensor τ in equation 2 is given as follow:

$$\vec{\tau} = \mu \left[(\nabla \boldsymbol{U} + \nabla \boldsymbol{U}^T) - \frac{2}{3} \nabla \cdot \boldsymbol{U} I \right] - pI \qquad (4)$$

Where μ gives the viscosity, Pa·s; and I presents the unit tensor calculated as product of $A_{ik}A_{jk}$ which are two matrices automatically produced during the iteration [20].

Standard k-ε model was employed for calculating the turbulence kinetic energy k and the rate of the dissipation ε according to the following equations [21]:

$$\nabla \cdot (\rho k \boldsymbol{U}) = \nabla \cdot \left[\left(\mu + \frac{\mu_t}{\sigma_k} \right) \nabla k \right] + G_k + G_b - \rho \varepsilon - Y_M \qquad (5)$$

$$\nabla \cdot (\rho \varepsilon \boldsymbol{U}) = \nabla \cdot \left[\left(\mu + \frac{\mu_t}{\sigma_\varepsilon} \right) \nabla \varepsilon \right] + C_1 \frac{\varepsilon}{k} (G_k + C_3 G_b) - C_2 \rho \frac{\varepsilon^2}{k} \qquad (6)$$

Where G_k presents the generation of the turbulence kinetic energy from the mean velocity gradients; G_b presents the turbulence kinetic energy generated by buoyancy; Y_m gives the dedication of the fluctuating dilatation in the compressible turbulence to the dissipation rate; C_1, C_2 and C_3 are constants which are also able to be defined by user; μ_t denotes the turbulent viscosity which can be calculated as product of $0.09\rho k^2/\varepsilon$; δ_k and δ_e are the turbulent Prandtl numbers for k and ε with default value of 1 and 1.3 respectively [20].

2.2.2 Computational domain and conditions

The simulation in this work was focused on the behaviour of plasma jet and the effect of the gas shrouding. Thus the plasma was simplified as hot gas with given temperature and velocity entering the computational domain through the plasma inlet as indicated in **Fig. 6**. The nozzle for plasma gas inlet was created on the base of the F4-VB torch with adiabatic wall condition. The computational domain was mainly composed of the plasma inlet, the shrouding system and a cylinder with size of ϕ 120 mm x 200 mm for the progress of the off gas.

For evaluating the entrainment of ambient gas to the plasma jet, gas composition at boundaries was defined as the volume fraction of corresponding component. That the plasma inlet was defined with 100% plasma gas and 0% air; the shrouding gas inlet was defined with 100% argon; the opening was defined with 90% air and 10% plasma gas; and the outlet was defined with 50% air and 50% plasma gas. The temperature of the ambient gas at the opening and the outlet was defined to be 320 K; while the shrouding gas was set at 300 K. The distributions of the inlet temperature for the plasma gas were given as function of the coordinate x and y as follows:

$$T(x, y) = 15000 \times e^{-2.33 \times 10^5 (x^2 + y^2)} \qquad (7)$$

However, the plasma inlet was defined as a mass inlet with flowrate of 0.00122183 kg/s which was calculated according to the total flow rate of plasma gas and its density. Moreover, the plasma gas was simplified as a mixture of 90% argon and 10% hydrogen with a total flow rate of 50 NL/min. Therefore, the plasma gas properties were considered as 90% from argon and 10% from hydrogen.

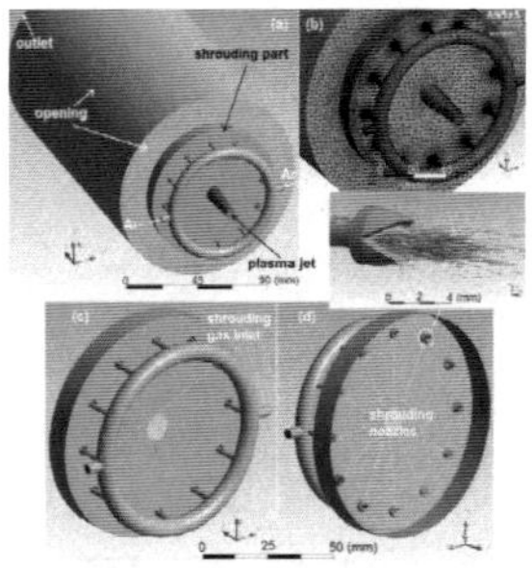

Fig. 6 model of simulation: (a) the mesh; (b) the model with 12-arranged nozzles for shrouding; (c) illustration of the result display and (d) a single nozzle for shrouding

3 Results and discussions

3.1 Effect of shrouding gas on plasma jet

The plasma jet was measured by the enthalpy probe at the centre which was determined by adjusting the torch to find the maximal value of the temperature and the velocity and the density (TVD). Then it was able to obtain plasma TVD with different standing on the centreline by moving back the probe following the spraying direction (Z). In addition, in simulation case, the plasma TVD was able to be directly red out on the centreline. Consequently, the plots of the measured results and the simulated results were presented in **Fig. 7** with lines presenting the simulation results (T_sim_SH0, T_sim_SH4…) and with discrete markers indicating the normal experimental results (T_exp_SH0, T_exp_SH4…). In the legend, "SH0" signified that no shrouding gas was used and "SH4" presented the use of shrouding gas with pressure of 4x10^5 Pa. Accordingly, two aspects were clearly found that firstly the gas shrouding made the plasma jet different and secondly the simulation fit the experiment well. That the shrouding gas brought cooling effect and started to slightly reduce the plasma temperature at 70 mm on the spraying direction; while the plasma density was found to become higher. However, it is hard to point out how the plasma velocity had been changed, except that with shrouding gas the plasma velocity was found to be more instable. Similarly, the simulation also revealed the cooling effect of the shrouding gas on the plasma jet; while the velocities were found to be more or less the same.

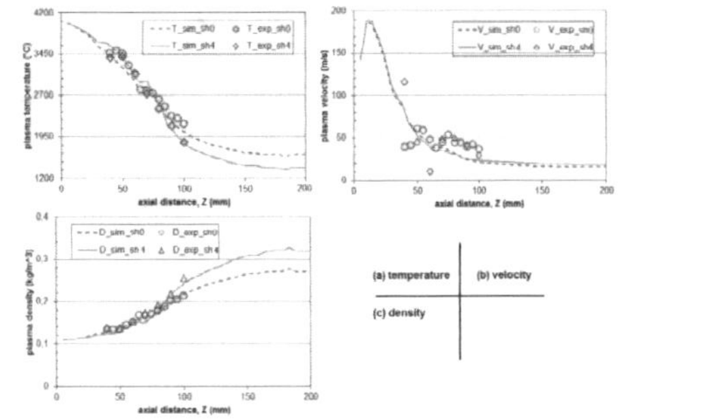

Fig. 7 measured and simulated characteristics of the plasma jet: temperature (a), velocity (b) and density (c)

The gas shrouding brought significate effect to the plasma gas composition as well as the surrounding gas. As shown in **Fig. 8**, the axial and radial distributions of O_2-N_2-Ar without gas shrouding were depicted with dashed lines and void markers; the distributions with gas shrouding were plot with solid lines and filled markers. It can be found that with gas shrouding the axial region between 70 mm and 100 mm was well protected to have higher argon fraction and lower oxygen fraction (see **Fig. 8 (a)**). That at the spraying distance (80 mm) the oxygen fraction was decreased from 3.2% to 0.9%; while the argon fraction was increased from 96.8% to 99.1%; and no N_2 was detected. Radially, the argon-rich region around the plasma jet was obviously expended by the shrouding gas with edge being extended from Y=-4 mm to Y=-10 mm (see **Fig. 8 (b)**). Without gas shrouding, the distributions of the volume fraction for argon and nitrogen were found to have a crossover at about Y=-7 mm; while with gas shrouding, the crossover was predicted at about Y=-13 mm.

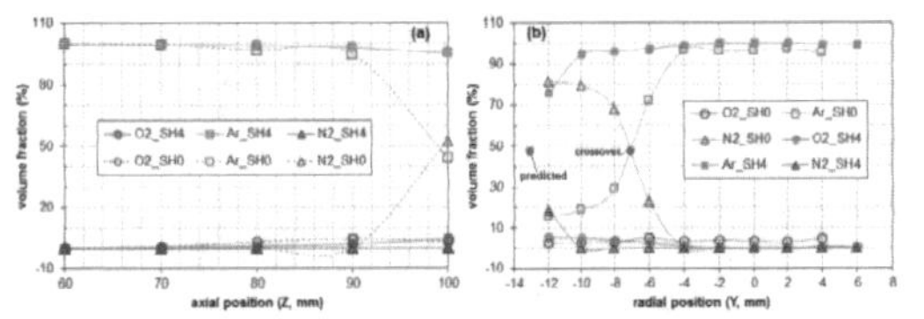

Fig. 8 measured gas composition along axial position (Y=0 mm) (a) and against the radial displacement (Z=80 mm) (b)

Moreover, the shrouding effect was also confirmed by the simulation. As displayed in **Fig. 9 (a)** and **(b)**, with gas shrouding the gas distributions were evidently modified. That in axial direction (see **Fig. 9 (a)**), the region between 50 mm and 170 mm of the centreline was protected with more argon while with less air (O_2 and N_2). At Z=80 mm, the calculated volume fraction of oxygen reduced from 7.5% to about 2.4%; and nitrogen was restricted with volume fraction as low as 9% which was only one third than that without gas shrouding; while the fraction of argon was increased from 57% to 83%. It is clear from **Fig. 9 (a)** that without gas shrouding Ar_SH0 intersected N_2_SH0 at about 100 mm on the centreline while with gas shrouding the crossover was pushed away from the torch exit at about 150 mm. In parallel, the simulation also gave reasonable results about the radial distribution of gas that the shrouding gas was able to optimize the spraying process by shielding the plasma jet from the air. However, due to the simplification of the calculating models, gap might exist between the simulation and the experiment. At least it gave possibility to justify the effect of the gas shrouding on the plasma jet.

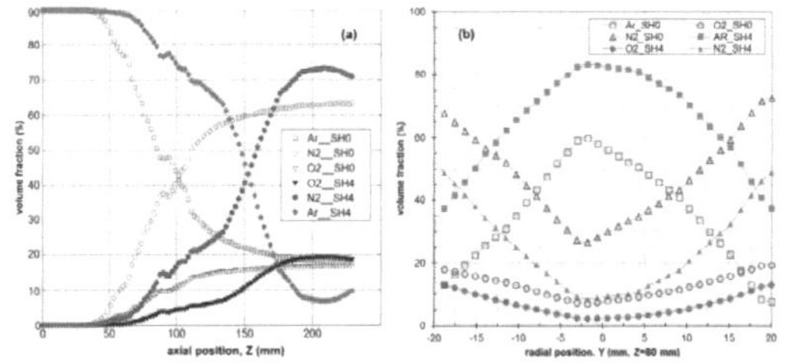

Fig. 9 the axial and radial distribution of Ar_O_2_N_2

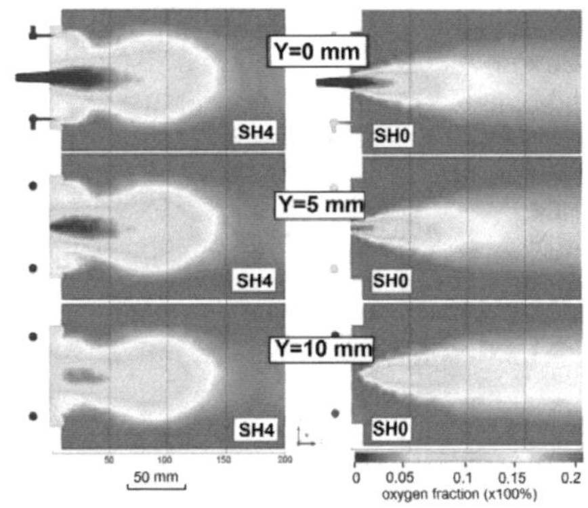

Fig. 10 the oxygen distribution on the selected planes

Additionally, simulation provided visible evidence about the shrouding effect. As shown in **Fig. 10**, the oxygen distributions on three planes, which were defined as Y=0 mm, Y=5 mm and Y=10 mm, indicated that with gas shrouding the plasma jet was found to become radially wider and axially shorter with more region being protected from oxygen. Moreover, as depicted in **Fig. 10**, the shrouding gas released by the shrouding nozzles was simultaneously drawn toward the plasma jet due to the low pressure around the plasma jet. Therefore, a gas shield was formed around the plasma jet, and yet it is impossible to completely eliminate the entrainment of cold ambient gas due to the interval between the arranged nozzles (see **Fig. 6**). The shrouding part with arranged nozzles was designed as a compromise between the shrouding effect and the cooling effect. The stronger the shrouding effect is, the lower the temperature becomes. For example, with annular shrouding part, more shrouding gas will be required to obtain comparable gas shield as that of the arranged nozzles; moreover, the requirement of the shrouding pressure is necessary to be much higher to retain enough momentum of the shrouding gas for withstanding the entrainment of the ambient gas. Consequently, the plasma jet is well protected by the strong gas shield while the plasma temperature will be unavoidably reduced which is detrimental to the coating property. Therefore, the module with 12-arranged-nozzles was finally employed in this work and it showed expected protecting effect to the plasma jet.

3.2 Effect of shrouding gas on coatings characteristics

For estimating the effect of gas shrouding on coating characteristics, microstructure and element composition of the coatings have been analysed by means of SEM and EDS. As mentioned above, the substrate was pre-treated to have a pure nickel layer for protecting the substrate and afterwards, NiAl coating was directly sprayed as a transition between the pure nickel layer and the NiAlMo layer according to the mass percent of nickel. Generally, the electrode of AWE will be activated to have high specific area by etching out Al before being put in service. That means the fraction of etchable Al in the as-sprayed coatings gives significant influence on the electrode performance. Therefore, by applying different spraying conditions, the performance of the resultant coatings will be different. As displayed in **Fig. 11**, sample 1, which was sprayed in vacuum, presented a structure obviously denser and thicker than sample 2 and sample 3. While sample 2 and sample 3 exhibited similar structures with comparable pores being deeply trapped in NiAl layer. It is general accepted that big pores in the electrode coatings may bring unfavourable defect after long term working but its importance to the electrode performance is indistinct comparing with that of the nanostructure produced by etching. Therefore, the electrode performance will be mainly determined by the element composition of the coatings.

As a result of the gas shrouding, sample 2 presented a moderate mass percent of oxygen with a reduction of 0.31% comparing to the value of 3.46% in sample 3 (see **Tab. 2**). While, the content of Al in sample 3 was found to be lower than that of sample 2. That meat after activation sample 2 might have more Nano pores than sample 3 and thus lead to a higher performance to hydrogen evolution. In addition, the content of Ni in sample 3 was found to be as high as 51.56% which was slightly higher than that of sample 1 and sample 2. However, in all cases the content of Mo was almost the same that meant during the spraying Mo was not oxidized. Moreover, regarding the content of Al in the two feedstocks (NiAl 56/44 and NiAlMo 39/44/17), it was able to estimate that the global content of Al in the as-sprayed coatings was about 44%. And thus it was able to evaluate the loss of Al during the spraying for each sample, 5.38% in sample, 6.7% in sample 2 and 8.9% in sample 3. The cause of the loss was mainly due to the vaporization of the in-flight particles and the deposition efficiency. Owing to the relative low boiling point, Al or Al-rich phase was easier to vaporize than Ni and Mo. Therefore, with higher oxide content and lower Al, sample 3 was supposed to have the lowest electrode performance. Executing gas shrouding brought slight reduction of oxide in the coating and it was helpful for retaining Al.

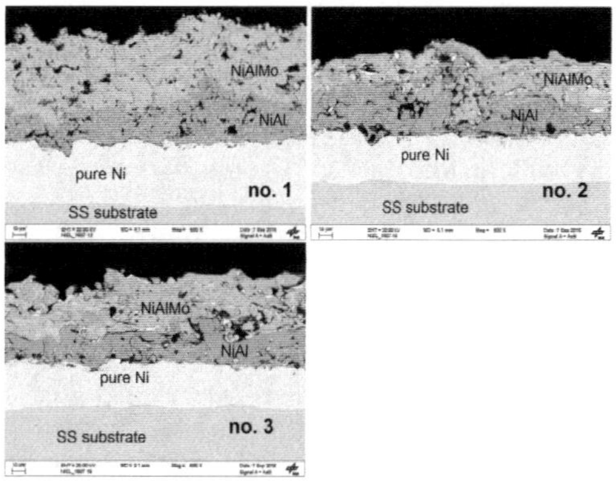

Fig. 11 microstructure of coatings in different cases

Tab. 2 mass percent of elements in different cases

No.	Mass percent (%)			
	O	Al	Ni	Mo
1	1.8	38.62	48.94	10.64
2	3.15	37.30	48.83	10.72
3	3.46	35.11	51.56	9.87

4 Conclusions

As a conclusion, in this work gas shrouding was carried out through a shrouding part with 12 arranged nozzles to protect the plasma jet. Both results of experiment and simulation showed that the shrouding gas was able to restrain the entrainment of oxygen to the plasma jet and retrain the oxygen fraction as low as 2.4% on the centreline at the spraying distance of 80 mm. The plasma jet was surrounded by the shrouding gas and forced to become radially wider and axially shorter which were experimentally detected and successfully numerically predicted. Significant effects of gas shrouding on coating characteristics have been watched to retain Al in the coatings while reducing the oxide. The use of gas shrouding should be optimized to achieve gas shield with enough momentum while without bringing too much reduction to the plasma temperature.

Thermally sprayed porous copper coatings for capillary transport of liquids

Chen Feng, Subramaniam Yugeswaran and Sanjeev Chandra

Cetre for Advanced Coating Technologies, Department of Mechanical and Industrial Engineering, University of Toronto, Canada, M5S 3G8

Porous copper coatings, which act as wicks for liquid transport, were fabricated using a flame spraying process. Copper and aluminum powders were fed independently into the spray torch and deposited on copper substrates to form a composite coating. The aluminum was subsequently removed using chemical leaching leaving a porous copper coating behind. Varying the feed rate of aluminum powder allowed the coating porosity to be controlled. Channels to enhance liquid flow were made in some of the porous copper coatings by placing pieces of aluminum wire mesh on the copper substrate before spraying. During spraying the sprayed powders passed through the mesh opening and created pyramid shaped arrays on the substrates. The groove width was controlled by using different wire mesh sizes. Coatings were made with porosity varying from 2 to 44 %, and groove width ranging from 0.16 to 0.53 mm. The capillary performance of the coatings was evaluated experimentally by measuring the rate-of-rise of ethanol in the coatings. The rate of rise increased with coating porosity, and decreased with groove width.

1 Introduction

Porous metal wicks, which have excellent mass and heat transfer properties, are used in a wide variety of applications including fuel cells, heat exchangers, and heat pipes [1]. Capillary forces in heat pipes wicks drive the working fluid to the hot region where it evaporates and carries heat away. Woven wire meshes and sintered powders are commonly used as wick structures, and are typically spot welded or sintered to the inner surfaces of heat pipes [2]. However, these methods require the use of a mandrel to preserve the shapes of the wicks and keep them pressed against the skins during bonding. If the bonding is not complete the contact area between the wicks and the skins is reduced, subsequently diminishing the overall performance of the heat pipes [3]. Besides these conventional processing techniques, other experimental techniques developed to fabricate porous wicks on surfaces, include electrodeposition [4], ion etching [5], and using pore former that are removed by heating [6], but these methods are difficult to scale up for mass production. An industrially viable method that simultaneously creates porous layers and bonds them to metal skins would be very useful for many applications.

In the present work flame spraying, a common industrial process, was employed to produce porous copper coatings that can be used in wicking applications. During the spraying, copper and aluminum powders were both injected into the flame and sprayed simultaneously on copper substrates. Channels were created in some of the coatings by placing pieces of aluminum wire mesh on the copper substrate before spraying, a technique that has previously been used to create finned surfaces [7]. Due to the shading effect of the mesh wires pyramid shaped structures were deposited on the substrates with open channels between them. After spraying the coatings were chemically leached using sodium hydroxide solution to remove aluminum and create a porous microstructure. In this manner porous copper coating with arteries running through them were produced on copper surfaces.

The microstructures of the porous copper coatings were examined using scanning electron microscope (SEM) images and image analysis software. Coating porosity, thickness, groove width and groove opening sizes were measured for the coating samples. In addition rate-of-rise experiments were conducted to investigate capillary performance of the porous copper coating by standing coated copper strips vertically with one end immersed in a pool of ethanol. The rate of capillary rise of ethanol was measured using an infra-red (IR) camera. The effects of porosity and groove size on capillary performance were studied.

2 Materials and experimental procedures

2.1 Fabrication of artery porous copper coatings

Fig. 1. Experimental setup of co-injection flame spraying technique.

The flame spray system, shown in **Fig. 1**, was developed to simultaneously inject aluminum and copper powders into the flame and spray them onto copper strips. An external powder injector was attached in front of the flame spray torch (Flame Spray Technologies, P5II, Duiven) to introduce aluminum powder into the copper powder being sprayed. The location of the tip of the injector was specified by measuring the distances d_1 and d_2 from the front of the spray torch nozzle, as marked in **Fig. 1**. The torch was mounted on a computer-controlled, six-axis robot to produce a uniform coating. A hydrogen-oxygen mixture was used as fuel for the spray torch.

The typical spray parameters used are given in **Table 1**. The copper powder (99.9%, Alfa Aesar) had a size distribution of 53-88 µm while the aluminum powder (99.9 %, Praxair Surface Technologies) had a size distribution of 35-63 µm. Pure copper powder was used to make dense coatings and copper mixed with aluminum to make porous coatings.

Table 1. Typical co-injection flame spraying operating parameters.

	Cu coating	Cu-Al coating
Fuel flow rates (slpm)		
Oxygen	28	28
Hydrogen	56	56
Gas pressure (Psi)	50	50
Cu feed rate (gpm)	24.4	24.4
Al feed rate (gpm)	-	1.4 and 5.7
d_1, d_2 distance (mm)	-	25, 12
Spray distance (mm)	100	100
Number of passes	20	20

The external injector was connected to a powder feeder containing aluminum powder and nitrogen gas was used as the carrier gas to carry the powder into the flame jet. The position of the injector at the exit of torch nozzle was optimized through numerous trial runs to achieve a narrow spray footprint with copper and aluminum splats well-mixed in the coating.

Fifteen copper strips, 15 mm × 80 mm in size, were coated five at a time to make three sets of samples. Before each spray trial four copper strips were masked with aluminum woven meshes of varying mesh count of 12×12, 16×16, 20×20 and 30×30 number of cells per square inch (McMaster-Carr, #9227T58, #9227T61, #9227T62, #9227T63, USA), while one copper strip was left uncovered. The mesh specifications are presented in **Table 2**. The five strips were clamped next to each other on a holder and sprayed simultaneously. The copper powder feeding rate was kept constant at 24.4 g/min in all three experiments, while the aluminum powder feeding rate was set to 0, 1.4 and 5.7 g/min, respectively, in the three different trials. During spraying powder particles passed through the mesh opening and deposited on the copper strips to form pyramid shaped structures. After spraying the meshes were lifted off the strips, leaving behind a grid of channels on the coating surface.

Table 2. Specifications of woven wire cloth.

Mesh count (ncsi)	Wire diameter (mm)	Opening size (mm)	Spacing between cells (mm)
12X12	0.6	1.5	2.1
16X16	0.5	1.1	1.6
20X20	0.4	0.9	1.3
30X30	0.3	0.6	0.9

After spraying, the composite Cu-Al coatings were chemically leached with 30% sodium hydroxide solution (NaOH) at 100°C to remove aluminum splats from the coating microstructure. During leaching aqueous NaOH reacted with aluminum to produce sodium aluminate ($NaAlO_2$) by the following reaction mechanism. The resultant $NaAlO_2$ is soluble in water and leaves the coating microstructure through the open porous networks.

$$2Al + 2NaOH + 6H_2O \rightarrow 2NaAl(OH)_4 + 3H_2 \quad (1)$$
$$NaAl(OH)_4 \rightarrow NaAlO_2 + 2H_2O \quad (2)$$

The higher the amount of aluminum in the original coating, the greater the final porosity. The copper strips with both solid and porous coatings are shown in **Fig. 2**.

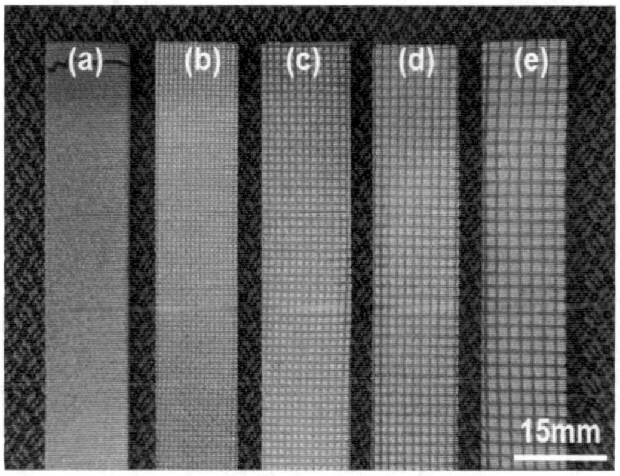

Fig. 2. A photographic image of flame sprayed plain and artery porous Cu coatings.

2.2 Rate-of-rise measurements: apparatus and procedure

Capillary flow of liquid through flame sprayed copper coating samples was investigated by photographing the ethanol rising in the porous copper coatings using an IR camera for imaging. A schematic of the experimental apparatus is shown in **Fig. 3**.The copper strips were suspended vertically from a holder by a wire and dipped into an ethanol reservoir so that the liquid spontaneously rose up in the wicks due to capillary pressure. During the experiment, the amount of ethanol absorbed by the coatings was very small compared with the volume of ethanol in the reservoir, so the level of ethanol in the reservoir was assumed to be constant. An infrared camera (FLIR, ThermaCAM SC 5000, USA) was used to view the rise of liquid in the coating. The lower temperature of the liquid was visible as a change in infrared emission from the surface of the coating. The camera was directly connected to a lab computer, and Altair thermal imaging software was used for analysis. The temperature measuring range of this camera is -20 to 3000°C with an accuracy of ± 1%. The video was taken at a frame rate of 10 frames per second.

Before the experiment, the fifteen copper wick strips were cleaned with acetone and dried overnight at room temperature. The experiments were conducted in ambient condition, and capillary rise in each strip was recorded for five minutes.

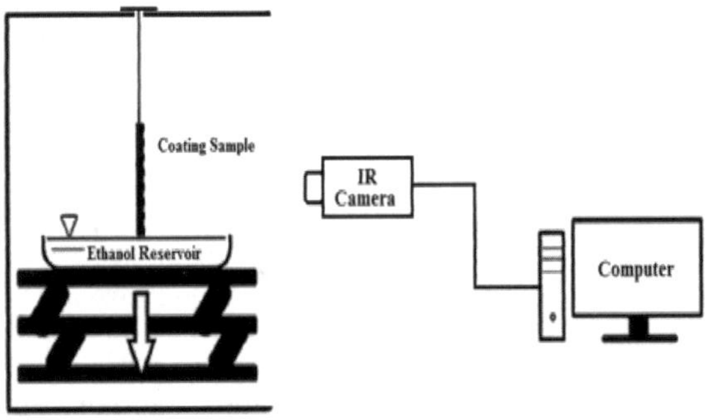

Fig. 3. A schematic of ethanol rate-of-rise test of flame sprayed copper sample strips.

3 Results and discussions

3.1 Microstructure of the porous copper coatings

Microstructures and phase formations of the coatings and the feedstock powders were examined by scanning electron microscope (HITACHI, TM3000, USA) and x-ray diffraction (Rigaku, MiniFlex^{TM}, USA) analysis respectively. Image analysis software (ImageJ) was applied to the SEM images to determine the porosity of the coating microstructure. Cross-sectional microstructures of three plain coatings (with no channels on the coating surface) are shown in **Fig. 4**. As a result of varying flow rates of the aluminum powder, the porosity and formed pore networks are different in the three coating samples. The SEM images show that the porosity increases as the flow rate of the aluminum powders is raised, and more interconnected networks are formed in the highly porous coatings.

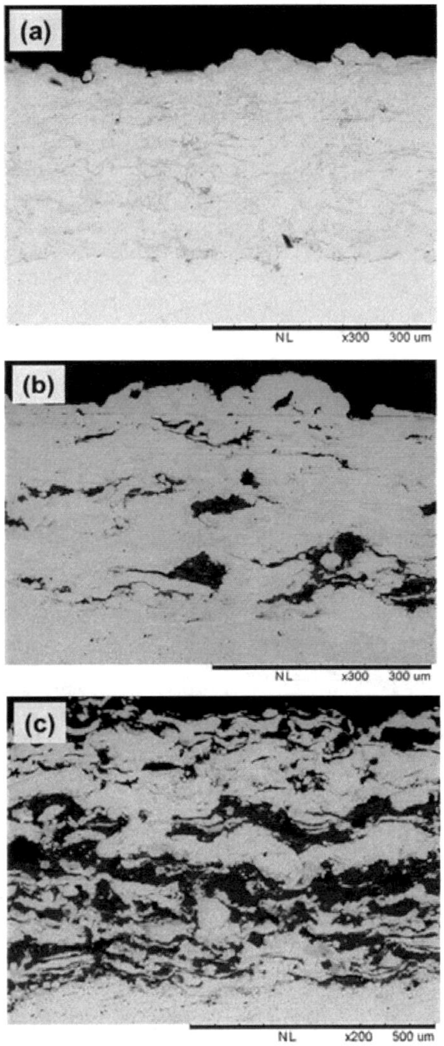

Fig. 4. Cross- section SEM images of plain Cu coating microstructures: (a) dense (2%); (b) low porosity (14%); (c) high porosity (44%).

The shape of the pyramidal structures formed by spraying through a wire mesh is, shown in **Fig. 5a** while a cross-section view through them is shown in **Fig. 5b**. The widths of the grooves between the structures are measured both at the top (W_1) and the bottom (W_2) of the channel. The groove opening (ψ) is defined as the ratio of the channel volume to the bulk volume of the coatings. Measured values of these properties are listed in **Table 3** for all the samples prepared. The results show that the bottom groove width is always smaller than the wire diameter of the mesh. The samples are labeled as being either *Dense* (*D*) for 2% porosity, *Low Porosity* (*L*) for 14% porosity, and *High Porosity* (*H*) for 44% porosity, followed by a number 12, 16, 20 or 30, representing the 12×12, 16×16, 20×20 and 30×30 meshes that were used to create channels.

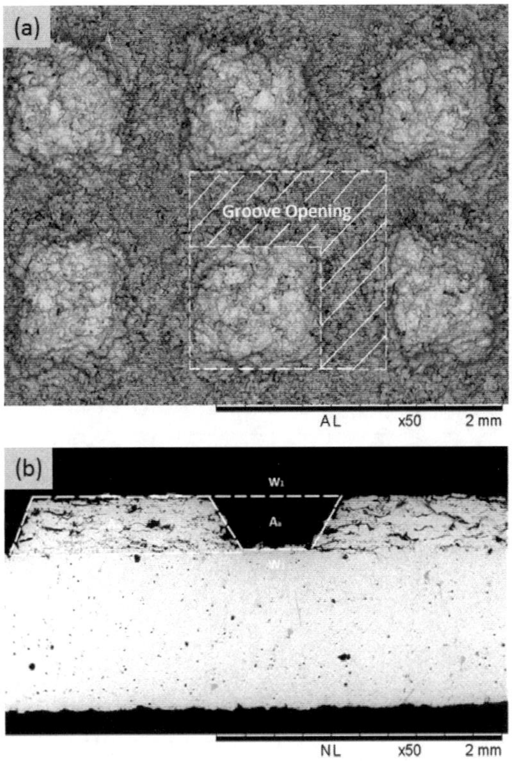

Fig. 5. (a) Surface SEM image of the truncated-pyramidal coating piles. The groove opening (φ) is defined as the ratio of channel volume to bulk volume of the coating. (b) Cross-section view of the adjacent pyramid piles and artery. The groove widths (W_1/W_2) are indicated.

Table 3 Specifications of dense and porous copper coating samples.

Samples	Pattern (ncsi)	Groove widths W_1/W_2, (mm)	Groove opening φ, (%)	Coating thickness x, (um)	Coating porosity ε, (%)
D	N/A	N/A	N/A	280	
D12	12X12	0.88/0.53	57	250	
D16	16X16	0.70/0.43	65	280	2
D20	20X20	0.66/0.35	68	270	
D30	30X30	0.54/0.16	71	220	
L	N/A	N/A	N/A	320	
L12	12X12	0.90/0.51	57	320	
L16	16X16	0.83/0.38	65	300	14
L20	20X20	0.66/0.35	69	260	
L30	30X30	0.49/0.20	71	220	
H	N/A	N/A	N/A	510	
H12	12X12	1.00/0.51	63	510	
H16	16X16	0.90/0.40	72	500	44
H20	20X20	0.76/0.38	65	300	
H30	30X30	0.52/0.33	80	250	

3.2 Effects of groove size and porosity on capillary performance

Capillary rise of ethanol in four dense (2%) copper coatings with arteries in them at four different times are shown, in **Fig. 6.** The IR images show the temperature distributions of the coating surface as a color gradient. As ethanol flowing through the pores and channels on the copper coatings, the region filled with ethanol has a lower temperature than that the dry regions due to the cooling effect of evaporation of ethanol. The dark blue portion of the surface represents the lower-temperature areas on the coatings that are filled with ethanol. The leading edge of the ethanol in the coatings was determined by finding the interface of two contrasting colors in the image.

The images show that the ethanol rises on four coating samples over time. The rate of capillary rise gradually increases from left to right in the samples as the groove width reduces. The driving force acting on the ethanol to transport it upwards on the samples is the capillary force that is generated by both the coatings and the channels. Since the four samples have the same coating porosity, the difference in the capillary rise height is attributed to the difference in channel width. The capillary pressure is inversely proportional to the groove size for a rising liquid column [8].

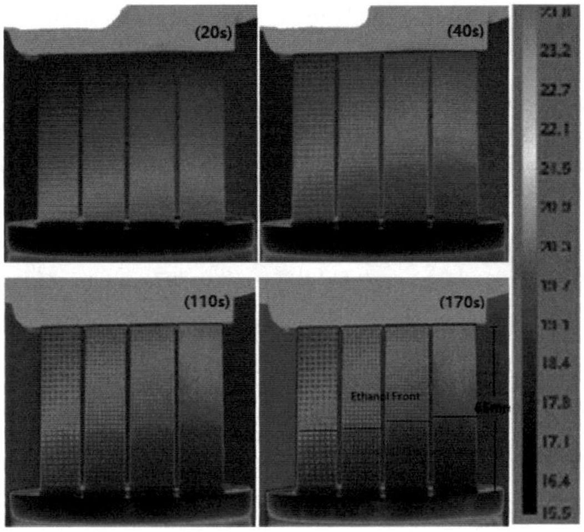

Fig. 6. Consecutive IR thermal images showing ethanol rises within arteries in dense copper coating strips. From left to right D12, D16, D20 and D30. The ethanol leading edges are marked in the last figure at 170 s.

The results of capillary rise height as a function of time of dense (2%) copper coatings, both with and without channels, are presented, in **Fig. 7.** The four copper coatings with channels running through them show faster capillary rise than the plain coating, with D30 having the highest capillary height, followed in order by D20, D16, D12 and D. At 170 s, the capillary heights of D30, D20, D16, D12 and D are 27, 26, 24, 22 and 14 mm, respectively. The maximum capillary height of D30 is almost two times that of the plain coating. The results indicate that the grooves improve the capillary rise of ethanol in the coatings, and the enhancement varies inversely with the groove size of the coating.

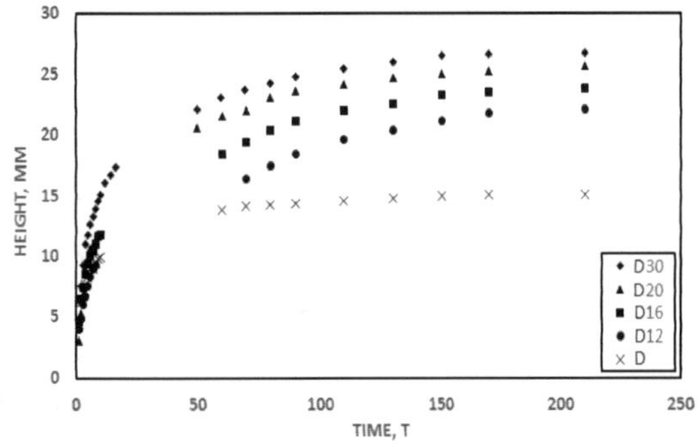

Fig. 7. Capillary height of ethanol in dense copper coating strips that has the same coating porosity.

The coating porosity also has an effect on the capillary rise height in the porous copper coatings. The capillary heights over time of the samples D30, L30 and H30, which had the same groove width but different porosities, are presented in **Fig. 8.** The results indicate that the capillary height increases with time, and the maximum capillary height increases with coating porosity. The maximum capillary height increases only slightly as the porosity increases from 2% to 14%, but by a much more significant amount as porosity increases from 14% to 44%. For example, the capillary heights of D30, L30 and H30 at 210 s are 27, 29 and 45 mm, respectively. As seen in **Fig 4**, the coating microstructures of these three samples are very different. At porosities of both of 2% and 14%, relatively few connected open pores are formed in the coating matrix; However, coatings with a porosity of 44% have many more connected networks of open pores which provide continuous flow paths for the ethanol to propagate.

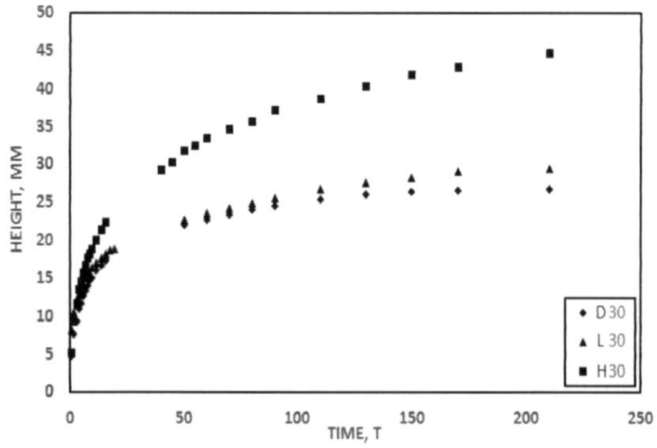

Fig. 8. Capillary height of ethanol in the same size of arteries featured copper coatings that has various coating porosity.

4 Conclusions

Porous copper coatings with arteries running through them were created using a flame spraying technique to simultaneously deposit copper and aluminum powders on copper strips and then chemically leaching away the aluminum. The porosity of the copper coating was increased by increasing the flow rates of aluminum powder. The capillary performance of the porous copper strips was investigated by measuring the rate of rise of ethanol using an infra-red camera. Increasing the porosity and reducing the groove size enhanced the rate of rise of the ethanol on the coatings.

References

[1] Holley, B. and Faghri, A.: Permeability and effective pore radius measurements for heat pipe and fuel cell applications, Applied Thermal Engineering 26 (2006) 448-462.

[2] Reay, D. and Kew, P.: Heat pipes: theory, design and applications. Fifth Edition.
[3] Singh, R., Nguyen, T., and Mochizuki, M.: Capillary evaporator development and qualification for loop heat pipes, Applied Thermal Engineering 63 (2014) 406-418
[4] Gheitaghy, A.M., Saffari, H. and Mohebbi, M.: Investigation pool boiling heat transfer in U-shaped mesochannel with electrodeposited porous coating, Experimental thermal and fluid science 76 (2016) 87-97
[5] Chen, Y.K., Melvin, L.S., Rodriguez, S., Bell, D. and Weislogel, M.M.: Capillary driven flow in micro scale surface structures, Microelectronic Engineering 86 (2009) 1317-1320.
[6] Kim, J. H.: Enhancement of pool boiling heat transfer using thermally conductive microporous coating techniques, PhD Thesis, University of Texas at Arlington, TX, 2006.
[7] Dupuis, P., Cormier, Y., Corbeil, A. and Jodoin, B.: Performance of cold sprayed near-net pyramidal shaped arrays under forced convection, Conv-14: International Symposium on Convective Heat & Mass Transfer, 2014.
[8] Washburn, E. W.: The dynamics of capillary flow, Physical Review, Vol. 17, no. 3, pp. 273, 1921.

A New Technology for Spraying Advanced Low-Temperature (300~600 °C) Solid Oxide Fuel Cells

Kang Yuan[1], Yueguang Yu[1], Xiaoliang Lu[1], Xiaojuan Ji[1], Bin Zhu[2]
[1]*Beijing General Research Institute of Mining and Metallurgy, Beijing 100160, China*
[2]*Department of Chemical Engineering and Technology, Royal Institute of Technology (KTH), S-100 44 Stockholm, Sweden*

Abstract. Solid oxide fuel cell (SOFC) has been developed for a hundred year and met a great challenge on material design and marketing. In recent years, new SOFC materials are dug up to achieve high energy-output performance at lower working temperature (300~600 °C), namely low-temperature SOFC (LTSOFC). In this study, Ni-Co-Al-Li oxide (NCAL) was used for making dense, thin and uniform coatings on grooved bipolar electrode substrate for LTSOFC. Low-pressure plasma spray (LPPS) technology was applied to manufacture the NCAL coatings. The performance of a fuel cell package using the coated bipolars was tested between 350 and 600 °C, showing 6~8 W power output with 4 single fuel cells (active area of 25 cm^2). The LPPS technology is believed to be one of the ultimate ways for manufacturing the thin film/coatings for SOFC applications in future.

1 Introduction

Due to its high energy transfer efficiency (> 60%), solid oxide fuel cell (SOFC) is a potential candidate for power generation and clean energy application [1,2]. Various materials have been studied and new techniques have been attempted for SOFC development in past decades. Too-high working temperature (800~1000 ºC) is the bottleneck which limits the commercial application of the SOFC. Recently, some new SOFC materials using semiconductor-ionic composites show high performance even at 300~600 ºC, which are called low-temperature SOFC (LTSOFC) or single-layer fuel cell (SLFC) [3-6]. For the LTSOFC, multi phases and abundant grain/phase boundaries are highly recommended for high ionic conductivity of the fuel cells [5,7,8].

The nowaday approach to make the LTSOFC layers (electrode and electrolyte) in lab was powder metallurgy, e.g. via powder sintering of nano-composites. The product yield and the quality by such approach was, however, very low and not suitable for a commercial use, so new efficient technologies should be explored. For fuel cells, since dense and thin layers are highly required, advanced coating technologies can be the ways for the manufacture.

In this paper, we introduce low-pressure plasma spray (LPPS) technology to make coatings for LTSOFC. LPPS has been successfully applied into the converntional high-temperature SOFC (YSZ) [9,10], so applying LPPS for LTSOFC is imperative. Some previous work has been done by making dense NCAL coatings for electrode and electrolyte layers for LTSOFC [11]. This paper mainly studies the process control for solving thermal-induced substrate deformation and the influence of the process paramters on the microstructure of the coatings.

2 Experimentals

A commercial $Ni_{0.8}Co_{0.15}Al_{0.05}LiO_{2-\delta}$ (NCAL) powder, with D_{50} of 13 um, was used to manufacture coatings on steel samples or bipolar plates (60*60 mm^2). The powder had typical $Li(Ni_{0.9}Co_{0.1})O_2$ phase (PDF No. 70-4311) as shown in **Fig. 1a.** $Li(Ni,Co)O_2$ is a common cathode material for Li-ion batteries and could be a good candidate for electrode coatings. The macro appearance of the powder is given in **Fig.1b**.

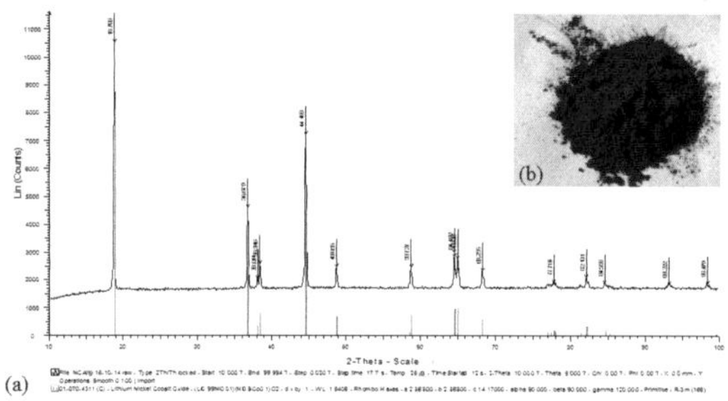

Fig. 1. (a) XRD spectrum of NCAL powder showing $Li(Ni_{0.9}Co_{0.1})O_2$ structure. (b) The naked-eye-looking morphology of the NCAL powder.

The NCAL coatings were manufactured in Sulzer Metco LPPS-TF system. The upright combustion chamber (around 3 meters high) of the LPPS system allowed to form a long, wide and stable plasma stream for uniform coating manufacture. The vacuum degree of the combustion was set at 1.5 mbar, and the plasma stream was widen to 200~400 mm in diameter [12]. Before the spray, the workpieces were cleaned, sand-blasted, and pre-warmed to get an active surface. Then, NCAL powder was fed into the plasma and uniformly sprayed onto the workpieces with a sway of the plasma gun. The parameters used in the LPPS spray for different coating microstructures are shown in **Table 1**. The powder feeding rate was about 25 g/min. The surface temperature of the workpieces was in-situ monitored by an infrared pyrometer.

Table 1. LPPS process parameters.

Process	Spray Distance (SD), mm	Spray plasma Power (SP), kW
1#	700	68
2#	900	50
3#	900	68

The NCAL-coated bipolar plates were used for an electric property measurement in an electrochemical workstation (Gamry Reference 3000) with testing temperature between 350 and 600 °C. More testing details were presented in our previous paper [11]. The microstructures, chemical composition and phase identification of the samples were analyzed via scanning electron microscope (SEM), energy-dispersive X-ray spectrometer (EDX) and X-ray diffraction meter (XRD) with Cu-K_α radiation (λ=1.54060 Å).

3 Results and discussion

Under SEM mode, NCAL powder showed spherical shape and typical agglomeration morphology; the agglomerates were in nano/sub-micro scale (**Fig. 1**). The powder showed good flowability in LPPS feeders and pipelines with carrier gas under pressure, so was directly used for coating spray without additional processing.

Fig. 2. SEM images showing (a) the morphology of NCAL powder and (b) agglomeration structure in the powder.

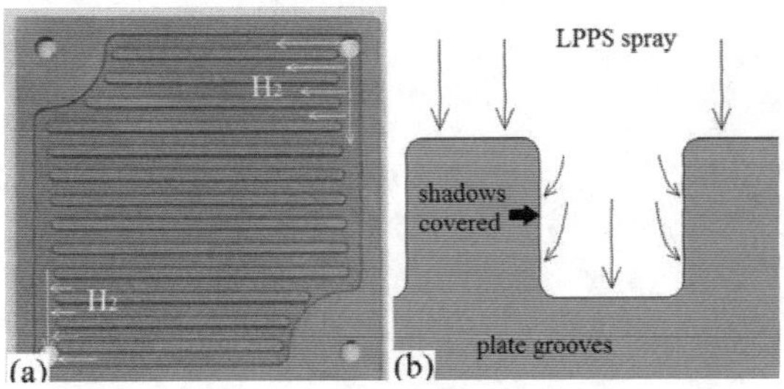

Fig. 3. (a) The bipolar electrode plate showing the gas flow grooves. (b) Schematic illustrating the LPPS spray on the plate grooves which also covers wall shadows.

The bipolar electrode plates, as shown in **Fig. 3a**, had complex surfaces as designed for gas flow (H_2 or air/O_2). **Fig. 3b** schematically shows the none-line-of–sight spray by LPPS. LPPS coatings could cover all the surfaces including the shadow walls of the grooves and corners.

Under the low pressure (~1.5 mbar), the plasma stream expanded to more than 2 m long and 200~400 mm in diameter. The core heating zone in the stream with feeding powder was about 100 mm, as seen in **Fig. 4a**, which can cover a whole workpiece (60*60 mm). The sway of the plasma gun could achieve the coating manufacture on two and more workpieces. The coatings, for a thickness of 50 µm, can be made in the plasma in 30 sec. The "turbulence" occurring at the backside of the fixture was caused by hydromechanics with dynamic pressure heterogeneity around the components. An as-sprayed bipolar plate is given in **Fig. 4b**.

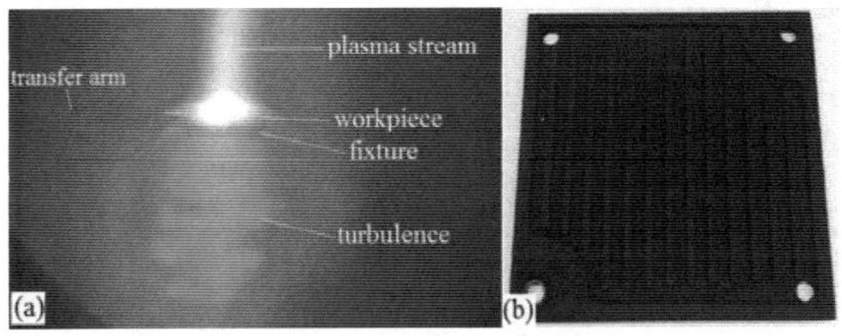

Fig. 4. (a) The working status of coating spray in LPPS chamber. (b) A NCAL-coated bipolar plate.

The LPPS spray was firstly carried out on some sample pieces (without grooves) with the same size of electrode plates (60*60 mm). The samples must be fixed on the fixture and the first trial was done by fixing two holes simultaneously with the two sides of the sample stuck, as shown in **Fig. 5a**. Such fixing caused a high hunching-up deformation of the sample due to thermo-induced expansion of the substrate in plasma. Other trails (**Fig. 5b**) were then performed by removing unnecessary fixing points, and finally the sample bending was successfully avoided with only one hole fixed (**Fig. 5c** and **d**). In the spraying trails, different spray distance (SD) and plasma spray power (SP) were tried. As the sample bending did not cause dramatic coating spallation (only a local coating peeling occurred in the first trail in **Fig. 5a**) and the bending problem was solved, the LPPS spray was believed to have made qualified coatings. **Fig. 5e-g** present the corresponding surface temperature during the plasma pre-heating and the coating spray processes. As shown, the plasma heating brought a quick temperature increase, and the sample temperature could be up to 800 °C during the spray (**Fig. 5g**).

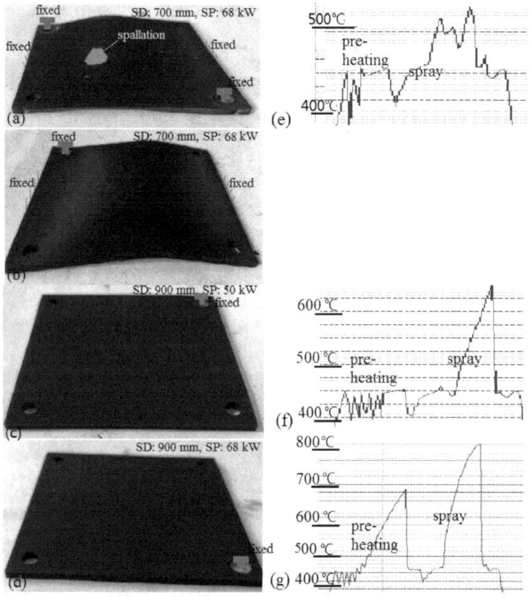

Fig. 5. The sample pieces in different LPPS spraying conditions (a-d) and the corresponding surface temperature of the samples (e-g). Images (a-d) also show different fixing ways of the samples in the spray.

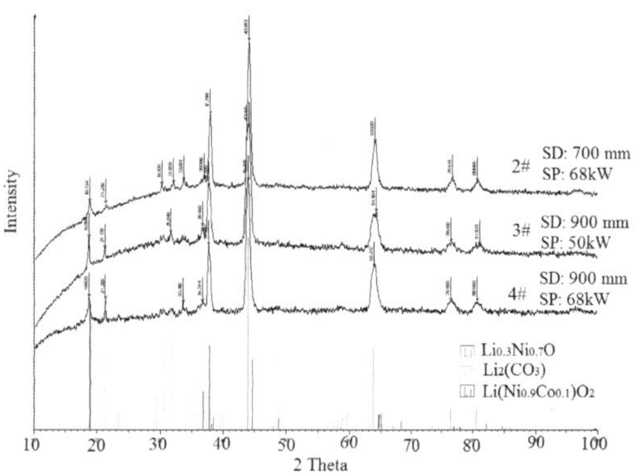

Fig. 6. XRD spectra of NCAL coatings using different spraying parameters. Three main phases were identified: blue lines for $Li_{0.3}Ni_{0.7}O$, green lines for $Li_2(CO_3)$, and red lines for $Li(Ni_{0.9}Co_{0.1})O_2$.

The XRD results of the NCAL coatings are presented in **Fig. 6**. Three phases were identified: cubic lithium nickel oxide $Li_{0.3}Ni_{0.7}O$ (PDF No. 75-0543), monoclinic lithium carbonate $Li_2(CO_3)$ (PDF No. 87-0729), and rhomb-h-axes lithium nickel cobalt oxide $Li(Ni_{0.9}Co_{0.1})O_2$ (PDF No. 70-4311). The variety of the LPPS parameters caused no significant difference on the phase composition. Compared to the NCAL powder, $Li_{0.3}Ni_{0.7}O$ and $Li_2(CO_3)$ were the new formed phases. $Li_{0.3}Ni_{0.7}O$ was formed probably due to high-temperature phase transformation from $Li(Ni_{0.9}Co_{0.1})O_2$; while the formation of $Li_2(CO_3)$ could be due to a reaction with CO_2 [13]; the sourse of carbon was so far unclear yet.

The surface morphology of the NCAL coatings under SEM is shown in **Fig. 7**. The usage of spray distance 900 mm and spray power 68 kW in LPPS (**Fig. 7c**) seems to give the highest sintering extent of the coating as the splats were coarser and the porosity was lower than the coatings made by using the other spray parameters. The high density was beneficial for the gas proofness of the coatings.

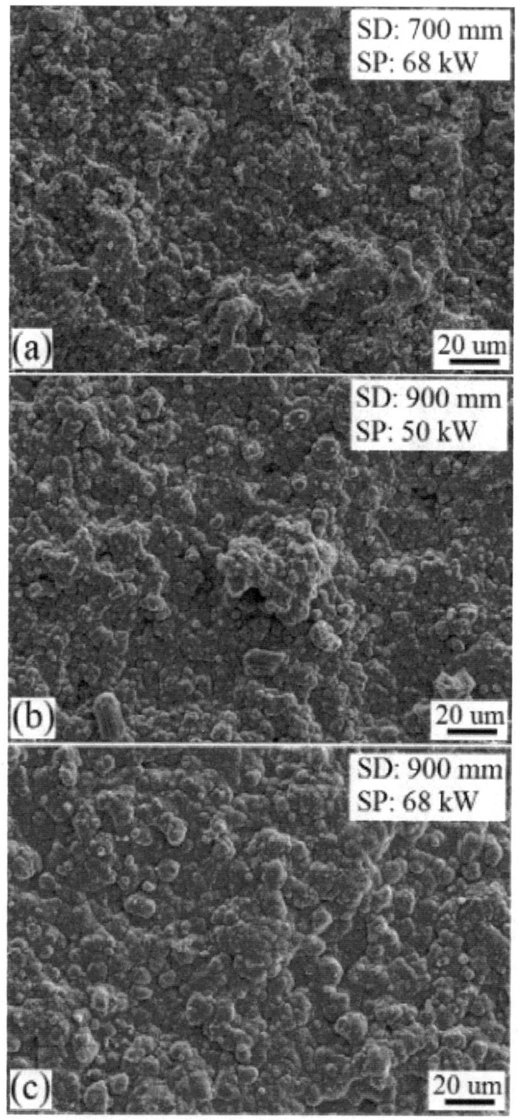

Fig. 7. The surface morphology of NCAL coatings made by LPPS using different spraying processes.

Fig. 8 compares the microstructures of the coatings (higher magnification). "Nodule"-like morphology was shown when spraying distance of 700 mm was used; while nano crystalline structures were formed when using spray distance of 900 mm. Using plasma power of 50 kW, “flake“ structure was formed in the nancyrstals. With a higher spray power (68 kW), “ridge“-like morphology was obtained. The occurrence of the nanocrystallization was due to a recrystallization process as the powders were molten in the plasma stream and then cooled down by hitting to the cold substrate.

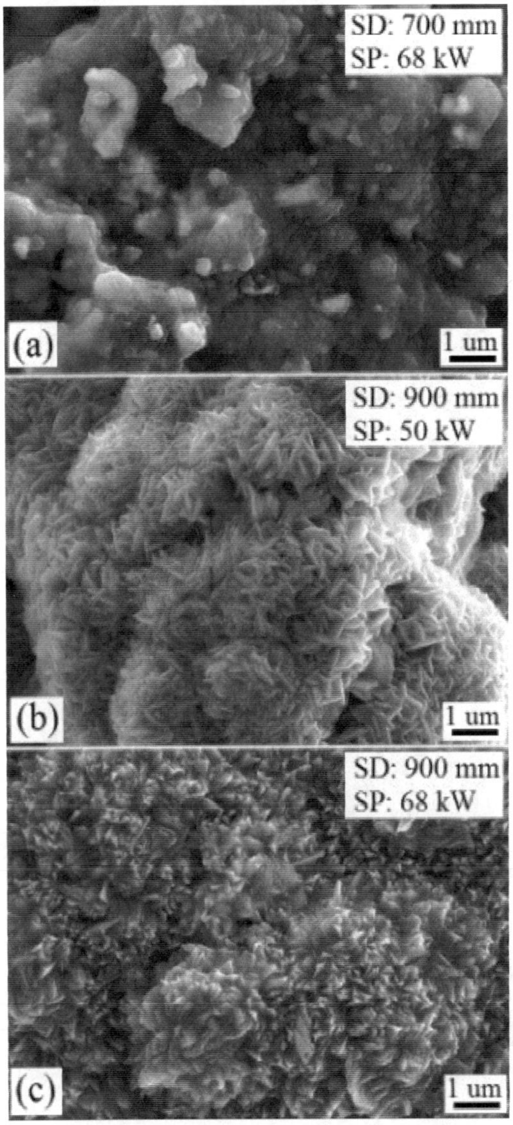

Fig. 8. The surface morphology of NCAL coatings made by LPPS using different spraying processes (magnified images in **Fig. 7**).

The cross sections further demonstrated that the NCAL coatings were tight (**Fig. 9**). Some pores can be found locally but they were closed. The melting of the particles can help the bondings in the coatings. As shown in **Fig. 10**, such melting-bonding of the powders would give a high gas proofness for the coatings.

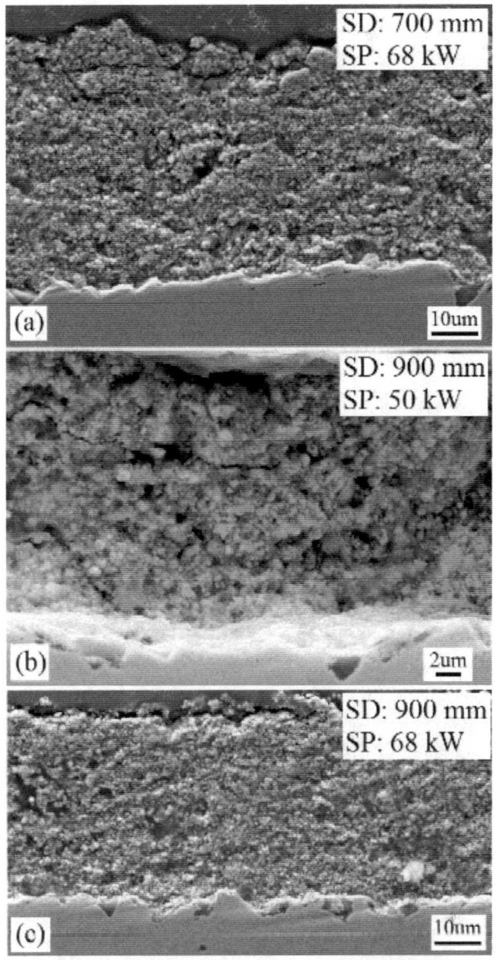

Fig. 9. The cross section morphology of NCAL coatings made by LPPS using different spraying processes.

EDX mapping composition analyses showed no macro heterogeneous distribution of the elements such as carbon, nickel and cobalt. While if tracking the carbon and oxygen content by comparing “in the particle“ and “at particle boundary“ (**Fig. 11**), carbon was likely richer in the molten part of the powders (i.e. at particle boundaries). That indicated the new phases, i.e. $Li_{0.3}Ni_{0.7}O$ and $Li_2(CO_3)$, could be formed from original $Li(Ni_{0.9}Co_{0.1})O_2$ phase during the melting and cooling processes in coating process. If so, the nano crystals may be a mixtures of those phases. Since an efficient electrolyte material of LTSOFC requires multi phases, to form p-n junction, and multiple interfaces so called nanocomposites, to provide fast paths of ion diffusion, the LPPS can make coatings functioning as semiconductor-ionic fuel cells.

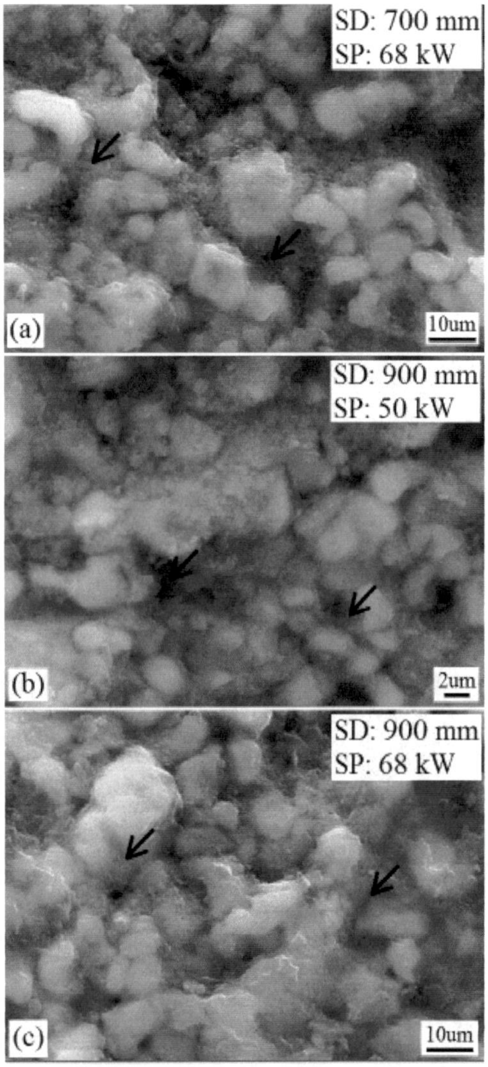

Fig. 10. The cross section morphology of NCAL coatings made by LPPS using different spraying processes (magnified images in **Fig. 9**). Arrows mark the molten particle boundaries.

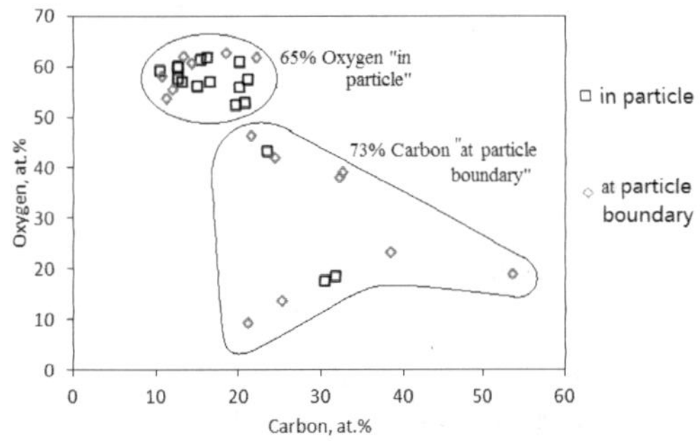

Fig. 11. EDX elemental content (oxygen and carbon) in particles or at particle boundaries in the NCAL coatings.

Fig. 12 schematically shows the structure of a fuel cell unit and a cell package for electric power generation. As shown in **Fig. 12a**, the sprayed NCAL coatings were contacted with a fuel cell layer for collecting electrons from fuel cell layer. In the gas flow grooves, H_2 and air/O_2 were pumped. At a temperature of 300~600 °C, ions moved through the cell layers and electrons in electrode plates, and then electricity can be generated (**Fig. 12b**). By applying the NCAL coatings on bipolar electrodes, the fuel cell package could produce 6~8 W power with 4 single fuel cells (active area of 25 cm^2). The NCAL coatings showed good conductivity (**Fig. 13**), indicating that the LPPS-made coatings worked well for the LTSOFC.

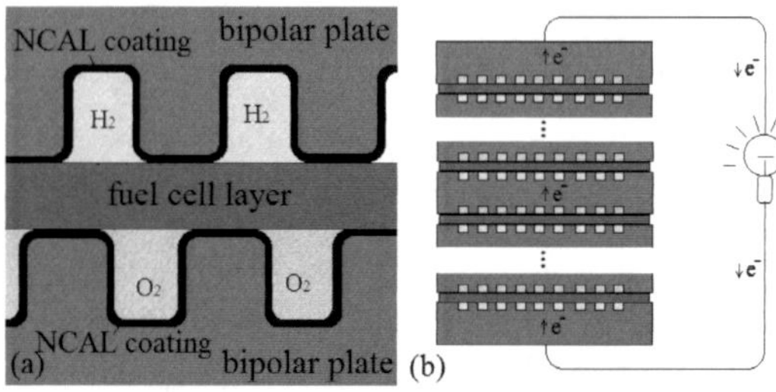

Fig. 12. Schematic drawing to show (a) the structure of the LTSOFC unit and (b) power generation by a fuel cell package.

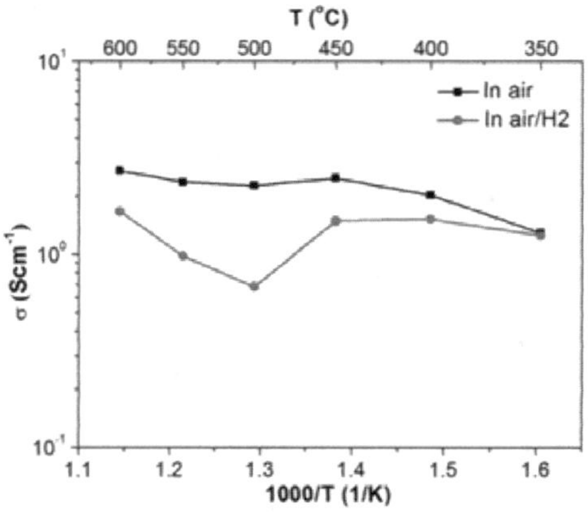

Fig. 13. Arrhenius curve showing the conductivity of the NCAL coatings on bipolar plates at different temperatures [11].

4 Conclusions

In this study, low-pressure plasma spray (LPPS) technology was applied for manufacturing Ni-Co-Al-Li oxide (NCAL) electrode coatings for low-temperature solid oxide fuel cell (LTSOFC). The results demonstrated that the LPPS technique can quick make continuous, uniform and dense coatings. The problem of thermo-induced substrate deformation was solved via freeing the uneccessary fixing points of the workpieces. The spraying distance and plasma power were changed for making different microstructures of the coatings. Under some parameters, nanocrystallization occurred. The existence of the multi phases and the formation of the nanocrystals in the coatings may give a possibility to in-situ form nanocomposite electrolyte for semiconductor-ionic coating application for LTSOFCs via LPPS technique.

5 Acknowledgement

This work was supported by the fundings from the project in Beijing General Research Institute of Mining and Metallurgy (Grant No. YJ-2016-04, and Grant No. QC-02). The authors would also like to thank Hubei University who performed the electric property tests on fuel cells with the coatings.

6 Literature

1. Steele, B.C.H.: Fuel-cell technology - Running on natural gas. Nature 400 (1999), Issue 6745, pp. 619/21.
2. Atkinson, A. Barnett, S. Gorte, R.J. Irvine, J.T.S. McEvoy, A.J. Mogensen, M. Singhal, S.C. Vohs, J.: Advanced anodes for high-temperature fuel cells. Nature Materials 3 (2004), Issue 1, pp. 17/27.
3. Zhu, B. Raza, R. Abbas, G. Singh, M.: An Electrolyte-Free Fuel Cell Constructed from One Homogenous Layer with Mixed Conductivity. Advanced Functional Materials 21 (2011), Issue 13, pp. 2465/69.
4. Zhu, B. Ma, Y. Wang, X.D. Raza, R. Qin, H.Y. Fan, L.D.: A fuel cell with a single component functioning simultaneously as the electrodes and electrolyte. Electrochemistry Communications 13 (2011), Issue 3, pp. 225/27.
5. Zhu, B. Fan, L.D. Lund, P.: Breakthrough fuel cell technology using ceria-based multi-functional nanocomposites. Applied Energy 106 (2013), pp. 163/75.
6. Zhu, B. Fan, L.D. He, Y.J. Zhao, Y.F. Wang, H.: A commercial lithium battery LiMn-oxide for fuel cell applications. Materials Letters 126 (2014), pp. 85/88.
7. Wang, X.D. Ma, Y. Li, S.H. Kashyout, A.H. Zhu, B. Muhammed, M.: Ceria-based nanocomposite with simultaneous proton and oxygen ion conductivity for low-temperature solid oxide fuel cells. Journal of Power Sources 196 (2011), Issue 5, pp. 2754/58.
8. Dong, W.J. Yaqub, A. Janjua, N.K. Raza, R. Afzal, M. Zhu, B: All in One Multifunctional Perovskite Material for Next Generation SOFC. Electrochimica Acta 193 (2016), 225/30.
9. Gindrat, M. Damani, R.: LPPS Hybrid Technologies for Emerging Energy Applications-Recent Developments. International Thermal Spray Conference, 2011 (Hamburg, Germany).
10. Vardelle, A. Moreau, C. Akedo, J. Shrafizadeh, H.: The 2016 Thermal Spray Roadmap. Journal of Thermal Spray Technology 25 (2016). Issue 8, pp. 1376/1440.
11. Yuan, K. Zhu, J. Dong, W. Yu, Y. Lu, X. Ji, X. Wang, J.: Applying Low-Pressure Plasma Spray Technology for Low-Temperature SOFC. International Journal of Hydrogen Energy (submitted), (2017).
12. von Niessen, K. Gindrat, M.: Plasma Spray-PVD: A New Thermal Spray Process to Deposit Out of the Vapor Phase. Journal of Thermal Spray Technology 20 (2011), Issue(4), pp. 736/43.
13. Liu, H.S. Yang, Y. Zhang, J.J.: Reaction mechanism and kinetics of lithium ion battery cathode material LiNiO2 with CO2. Journal of Power Sources 173 (2007), Issue 1, pp. 556/61.

Microstructure and Mechanical Properties of Cold Sprayed Titanium Coatings

W. Żórawski, Kielce /PL, J.Madry, Mielec / PL, J. Sienicki, Mielec / PL, M. Makrenek, Kielce / PL, A. Góral, Cracow / PL, S. Kowalski, Kielce / PL

The work concerns a study of the properties of cold sprayed Ti coatings. This material is an attractive choice for many applications because it exhibits high strength-to-weight ratios, very good oxidation resistance, corrosion resistance and biocompatibility. Cold spraying is applied to deposit Ti coatings and elements as additive manufacturing process, however it needs higher critical velocity for deposition than other, more ductile metals. Nowadays nitrogen as cheap gas is used as working gas in cold spray process, however application helium as accelerating gas allows to obtain elements with higher strength. It allows to understand the mechanism of cohesion between sprayed particles. In carried out experiment Ti powder with angular shape was applied in the cold spraying process. The coatings were sprayed by means of Impact Innovations 5/8 system with nitrogen and addition of helium onto 7075 Al alloy. The investigations revealed that the cold sprayed Ti coatings with addition of helium as working gas exhibit better mechanical properties, lower porosity and roughness.

1 Introduction

Cold spraying is a high kinetic energy coating process, where fine powder particles are propelled by a supersonic gas jet towards a substrate. The ability of a cold sprayed particle to undergo plastic deformation and adiabatic shear is a function of its mechanical (e.g. yield strength) and thermal (e.g. melting point) properties of the powder material as well as parameters such as the deposition temperature, particle velocity and particle size. Low temperatures also aid in retaining the original powder chemistry and phases in the coating, with only changes due deformation and cold working [1-4]. Nowadays the range of applied materials include metals and metal-ceramic composites. Titanium and its alloys are applied in many branches of industry due to their exceptional properties, such as high strength-to- weight ratio, very good corrosion resistance and biocompability. On the other hand, titanium has a high affinity to oxygen which makes them much production costs due to need for a controlled atmosphere [5-8]. Cold spraying technique as low temperature process, allows deposition of titanium without any oxidation which represent significant progress in its deposition, compared with the existing methods. This technique allows decreasing material input, elimination of mould and heating cost and reduces expenses for additional machining of components, creating the possibility of wider application of titanium in different branches of industry. The deposition of cold sprayed coatings is directly associated with exceeding of critical velocity by spraying particles of a powder. This velocity depends on many factors as the type of gas and its temperature and pressure, distance between the gun and substrate, powder size, powder feeding rate, properties of substrate material and its surface topography. In order to achieve the desired coating properties, optimisation of these parameters must be carried out, which gives restricted area of factors called "window of deposition". Cold spraying with parameters outside this area leads to increased porosity of coatings and deterioration of their properties.The fundamental in cold spraying is the choice of process gas which directly influences microstructure, density and bond strength of coatings.

In this work, titanium powder was deposited onto 7075 Al aluminium-alloy substrate by cold spraying using nitrogen and helium in different ratio. The aim of this study was to examine the impact of gas mixture on the microstructure, porosity and mechanical properties of cold sprayed Ti coatings.

2 Experimental

Titanium coatings were cold sprayed with various mixtures of nitrogen and helium, designated as: Ti-c1 – 100% nitrogen,Ti-c2 – 50% nitrogen + 50% helium, Ti-c3 – 20% nitrogen + 80% helium. The cold spraying system Impact Innovations 5/8 was applied to deposition of coatings. Commercialy avaible titanium powder was applied as coating material. Parameters of the cold spray process of Ti coatings were presented in Table 1. Morphology of powder feedstocks, as well microstructure of sprayed coatings were investigated both on the surface and cross section of coatings-substrate systems using scanning electron microscopy (SEM; Jeol JSM-7100 F; E-SEM FEI XL 30). X-ray diffraction phase analysis was performed using Bruker D8 Discover (Co Kα radiation) and program Diffrac.Eva 3.0 with crystallographic database PDF-4+. The micromechanical testing of coatings was carried out with the use of nanoindentation technique (Nanovea) with a Berkovitz indenter (the Olivier and Pharr methodology). Forty nine readings were taken for each coating. The geometry of samples surface before cold spraying and "as sprayed" coatings were analysed by means of a Talysurf CCI-Lite non-contact 3D profiler. Porosity was evaluated according to ASTM-E2109.

Table 1. Parameters of cold spraying of Ti coatings.

Parameter	Value
Pressure, MPa	4,0
Temperature, ^{o}C	800
Powder output, feeder scale	4
Robot arm speed, mm/s	400
Number of layers	4

3 Results and discussion

3.1. Characterization of Ti powder

Titanium grains morphology, sizes and their cross-section are presented in Figs. 1a and 1b, respectively. Ti powder grains had irregular and angular shape what is connected with production method of the powder. Great surface area per unit mass made powder grains more susceptible for oxidation. The cross-section of Ti grains (Fig. 1b) showed dense structure and did not reveal any internal porosity. Mean diameter of titanium powder d_{50} = 35.40 μm.

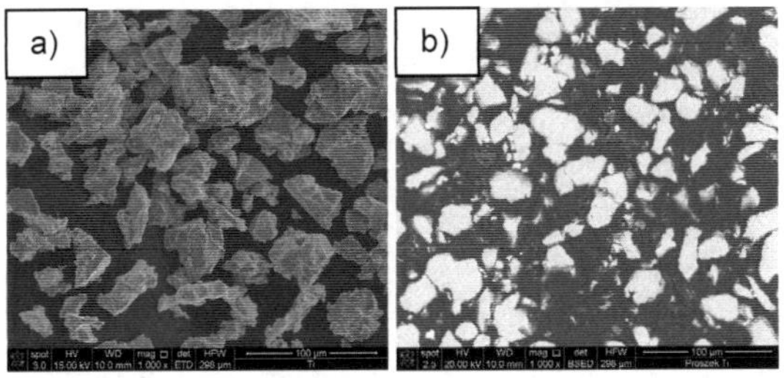

Fig. 1. Ti powder morphology: a) grains, b) cross section.

3.2. Characterization of cold sprayed Ti coatings

SEM examinations of the cold sprayed Ti coating surfaces showed that morphology of powder has considerable effect on the obtained surface microstructure sprayed with nitrogen (Figs. 2a and 2b) and sprayed with mixture 20% nitrogen + 80% helium (Figs. 2c and 2d). The microstructure of top coating comprised of adjoining plastically deformed titanium particles which formed surface with high roughness (Table 2) independently of applied process gas. Between deformed Ti particles small voids were found indicating negligible porosity of coatings.

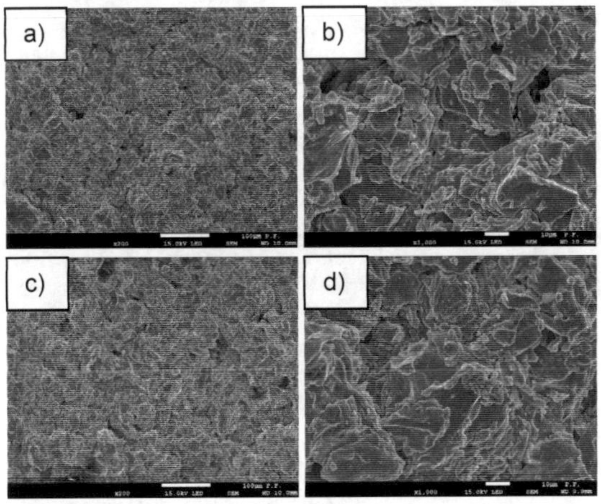

Fig. 2. Surface morphology of cold sprayed: a) Ti-c1 coating, 200x, b) Ti-c1 coating, 1000x, c) Ti-c3 coating, 200x, d) Ti-c3 coating, 1000x.

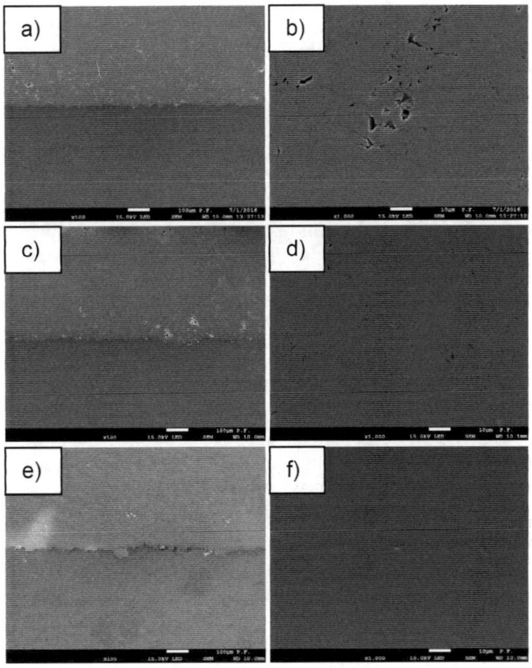

Fig. 3. Cross-section of cold sprayed: a) Ti-c1 coating, 100x, b) Ti-c1 coating, 1000x, c) Ti-c2 coating, 100x, d) Ti-c2 coating, 1000x, c) Ti-c3 coating, 100x, d) Ti-c3 coating, 1000x.

Figs. 3a, c and d present the cross-sections of dense cold sprayed titanium coatings where the interface showed very good bonding between the coating and substrate. At higher magnification (Figs. 3b,d,f) interfaces between sprayed grains of powder were apparently seen. Moreover, observations (Fig. 3a,b) revealed that coating sprayed with nitrogen contain areas with increased porosity. Presence of helium caused decreased porosity in the coating sprayed with mixture 20% nitrogen + 80% helium in comparison with the coating sprayed with nitrogen.

XRD analysis (Fig. 4) showed that titanium powder and sprayed coatings did not contain additional phases in the form of oxides or nitrides which can be formed in high temperature of spraing stream [9,10].

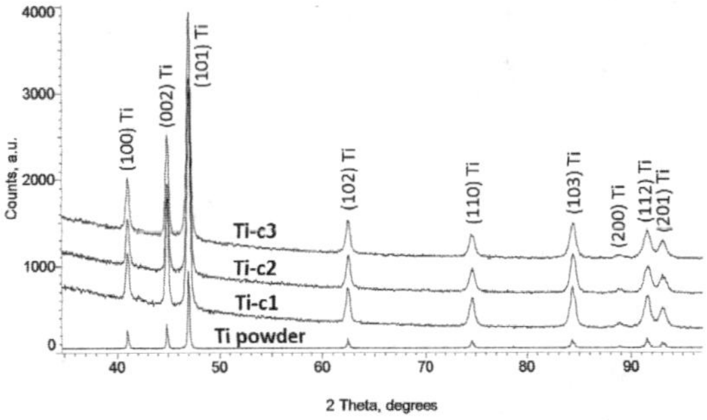

Fig. 4. XRD patterns of Ti powder and cold sprayed Ti-c1, Ti-c2 and Ti-c3 coatings.

Analysis of the sprayed coatings geometry (Fig. 5) showed that all coatings were characterized by a high surface roughness, which is the result of wide particle size distribution of titanium powder and angular shape of its grains. The parameters of surface geometry of cold sprayed coatings are presented in Table 2. The highest arithmetic mean of the surface height Sa of Ti-c1 coating cold sprayed with nitrogen was slighty higher than that showed for the other two coatings. The increased content of helium resulted in nearly 10% decrease of coating roughness. For the mean squared surface height, Sq similar relationship has occured. A negative value of asymmetry of the surface, Ssk indicated that Ti-c3 coating surface was flattened and means that the greater part of the distributon function amplitude was above the average. Steep irregularities and defects were described by kurtosis Sku. A kurtosis of approximately 3 indicated that the distribution of the profile ordinate corresponded to a normal distribution of the mean value. The maximum peak Sp, the maximum valley Sv and the maximum height Sz showed a decrease in value with increasing content of helium [11,12].

Table 2. Surface geometry parameters of cold sprayed Ti coatings according to ISO 25178.

Parameter	Surface of coating, µm		
	Ti-c1	Ti-c2	Ti-c3
S_a	18.0	16.7	16.4
S_q	22.0	20.8	20.6
S_{sk}	0.13	0.05	-0.04
S_{ku}	2.6	3.1	3.0
S_p	83.8	71.5	70.6
S_v	83.4	82.9	79.2
S_z	167.2	154.4	149.8

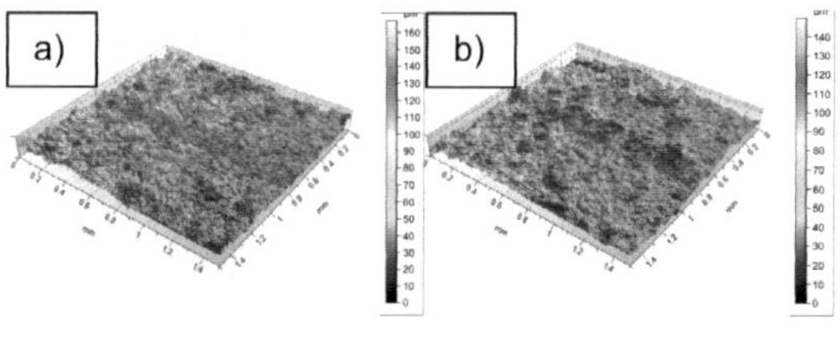

Fig. 5. Surface geometry of cold sprayed coating: a) Ti-c1, b) Ti-c3.

The Olivier and Pharr methodology was applied to measure hardness and Young's modulus of the investigated microstructures. Results were obtained on the basis of analyses using a square grid of 49 nanoindentation tests on the sample of coatings. The distributions of the nanomechanical properties were as then plotted as histogram distributions and maps of surface in which each hardness and Young's modulus result was shown on the charts with the same dimensions and locations, as in the investigated areas. Histograms and probability distributions of the hardness and Young's modulus of all cold sprayed titanium coatings are shown in Fig. 6.

Nanoindentation tests showed the distribution of the mechanical properties (hardness and Young's modulus) related to the different areas of titanium coatings. Mechanical propertis and porosity of cold spayed titanium coatings are presented in Table 3.

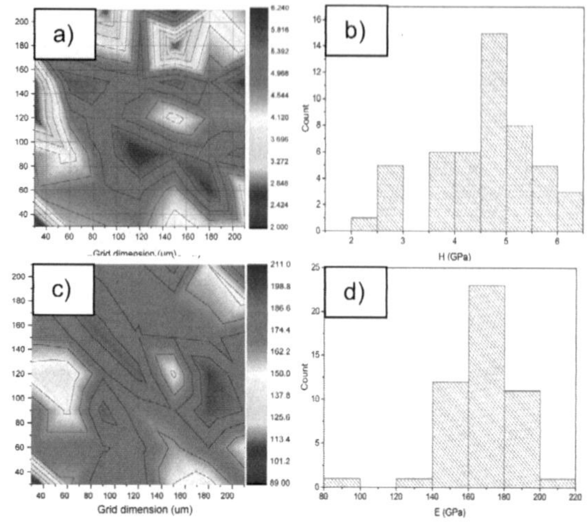

Fig. 6. Distribution of Ti-c3 coating: a) hardness map, b) hardness histogram and probability, c) Young's modulus map, d) Young's modulus histogram and probability.

Table 3. Properties of cold sprayed Ti coatings.

Param.	Ti coating		
	Ti-c1	Ti-c2	Ti-c3
H, GPa	4.17±1.0	4.32±1.11	4.58±0.98
E, GPa	136±21.9	150.9±27.5	169.7±19.8
Porosity, %	1.04±0.21	1.00 ±0.13	0.86±0.12

4 Conclusion

The titanium coatings were sprayed with nitrogen and addition of helium with angular titanium powder onto 7075 Al alloy. Increase of content of helium in process gas caused increase of hardness and Young's modulus of cold sprayed titanium coatings. The parameters of surface geometry of titanium coating decreased with all parameters.

Acknoweledgments

This paper is based on research funded by the National Centre for Research and Development as part of project INNOLOT "Additive Manufacturing Processes and Hybrid Operations Research for Innovative Aircraft Technology Development".

5 Literature

[1] A. Papyrin, Cold spray technology, Adv. Mater. Process 159 (2001) 49-51.

[2] R.C. Dykhuizen, M.F. Smith, D.L. Gilmore, R.A. Neiser, X. Jiang, S. Sampath, Impact of high velocity cold spray particles, J Therm Spray Technol 8 (1999) 559-564.

[3] J. Vlcek, L. Gimeno, H. Huber, E. Lugscheider, A systematic approach to material

eligibility for the cold-spray process, J Thermal Spray Tech 14 (2005) 125-133.

[4] M. Grujicic, J.R. Saylor, D.E. Beasley, W.S. DeRosset, D. Helfritch, Computational analysis of the interfacial bonding between feed-powder particles and the substrate in the cold-gas dynamic-spray process, Appl. Surf. Sci. 219 (2003) 211-227.

[5] T. Marrocco, D.G. McCartney, P.H. Shipway, A.J. Sturgeon, Production of Titanium Deposits by Cold-Gas Dynamic Spray: Numerical Modeling and Experimental Characterization, J Thermal Spray Tech 15 (2006) 263-272.

[6] Shuo Yin, Pengjiang He, Hanlin Liao, Xiaofang Wang, Deposition Features of Ti Coating Using Irregular Powders in Cold Spray, J Thermal Spray Tech 23 (2014) 984-990.

[7] N. Cinca, J. M. Rebled, S. Estrade, F. Peiro, J. Fernandes, J.M. Guilemany, Influence of the particle morphology on the Cold Gas Spray deposition behaviour of titanium on aluminium light alloys, J Alloy Comp., 554 (2013) 89-96.

[8] N.W. Khun, A.W.Y. Tan, K.J.W. Bi, E. Liu, Effects of working gas on wear and corrosion resistancesof cold sprayed Ti-6Al-4V coatings, Surf Coat Tech 302(2016) 1-12.

[9] T. Novoselova, P. Fox, R. Morgan, W. O'Neill, Experimental study of titanium/aluminium deposits produced by cold gas dynamics spray, Surf Coat Tech 200(2006) 2775-2783.

[10] Sadhen H. Zahiri, Christian I. Antonio, Mahnaz Jahedi, Elimination of porosity in directly fabricated titanium via cold gas dynamic spraying, J Mater Proc Tech 209(2009) 922-929.

[11] S. Adamczak, D. Janecki, K. Stępień: Cylindricity measurement by the V-block method; Theoretical and practical problems, *Measurement*, 2011, 44(1), p 164-173.

[12] Adamczak S., Janecki D., Stępień K.: Qualitative and quantitative evaluation of the accuracy of the V-block method of cylindricity measurements, Precision Engineering, vol. 34/3 (2010) pp. 619-626.

The influence of spray parameters on the characteristics of hydroxyapatite in-flight particles, splats and coatings during micro-plasma spraying

X.M. Liu, Y.M. Wang, Z. Zhou, G.H. Wang, Z. Tan, D.Y. He

Phase composition and microstructure of hydroxyapatite (HA) significantly affects the biological and mechanical properties of final hydroxyapatite (HA) coating. In the present study, HA coatings were deposited on Ti-6Al-4V by micro-plasma spraying (MPS) using different spray parameters. The influence of spray parameters on the composition and microstructure of the coatings were investigated. To understand the formation mechanism of HA coatings, the in-flight particles and splats were examined as well. The morphologies of coatings surface, cross-sections, initial powder, in-flight particles and splats were characterized by scanning electron microscopy (SEM). X-ray diffraction (XRD) was employed to analyze the phase composition. Three typical HA coatings were fabricated. The results indicated that the coating composition and microstructure were tightly related to the melting state of in-flight particles. And this was influenced by the spraying parameters. The formation mechanisms of these coatings were discussed.

1 Introduction

Hydroxyapatite (HA) coating has been used as surface coatings on metallic implants because it's close to the chemical composition of teeth and bone [1,2], which can stimulate bone growth at the bone implant interface [3]. Atmosphere plasma spraying (APS) has been widely used to deposit HA coatings on metallic implants surfaces [4,5]. It was reported that high temperature of plasma jet can lead to secondary phases with higher solubility such as amorphous calcium phosphates (ACP), HA decomposition phases, which are not good for the long-term reliability of the implants [6]. Therefore, HA coating with high crystallinity and purity is pursued for long-term reliability purpose. In vitro studies showed that HA coating with high crystallinity and purity show biochemical stability in the simulated body fluid [6]. In order to improve the reliability of implants, highly oriented hydroxyapatite coatings were obtained through radio-frequency thermal plasma spraying (TPS) method and micro-plasma spring(MPS). These may be to improve the mechanical properties and bioactivity [7,8]. However, the formation mechanisms of the preferential orientation during coating formation are still not clear, and the relationship of MPS process and coating structure is still not well understood. It is well known that the melting of in-flight particles and the flattened particle shape significantly affects the structure and properties of HA coating [9]. Therefore, the study on the relationship between particle characteristics and phase composition is essential to understand the phase composition and microstructure formation mechanism of the hydroxyapatite coatings during MPS.

Compared with other spraying equipment, MPS have the feature that HA coating of high crystallinity with strong preferred orientation can easily obtained by changing the spray parameter of MPS equipment. In the present study, HA coatings were prepared by different spraying parameter to investigate the influence of spray parameters on the phase, microstructure and the formation mechanism of the coatings.

2 Experimental procedure

2.1 Powders and substrate

Hydroxyapatite powders, supplied by Medicoat, Switzerland, which was used as spraying materials. Ti-6Al-4V plates of Φ15mm ×2.5mm were used as the substrate. Prior to spraying, the substrate surface was grit blasted with brown alumina to roughen and clean the surface.

2.2 Micro-plasma spraying

The HA powders were sprayed by a micro-plasma spraying (WDP-1,China). Argon was used as a carrier gas for MPS. The spray parameters for MPS are shown in table 1. The corresponding in-flight particles were collected by liquid nitrogen. The splats were deposited on a polished substrate.

Tab 1 Parameters and notions for micro-plasma spraying

	P1	P2	P3
Current(A)	20	40	40
Voltage(V)	25	25	25
Flow rate of Argon(L/min)	1.3	1.3	1.3
Spraying distance(mm)	60	60	110

2.3 Characterization methods

The morphologies of initial powder, in-flight particles, splats and coatings were characterized by scanning electron microscopy (SEM, SU8020). The polished and polished-etched (dilute hydrochloric acid solution) cross-section of coatings were both prepared before observation. Coating porosity was estimated with ImageJ. Phase composition of as-sprayed coatings and sprayed particles was qualitatively determined through X-ray diffraction (XRD, D8 Advance) analysis with Cu K_{α} radiation source. The temperature of substrate was tested by infrared thermometer.

3 Results

3.1 Surface morphology of HA coatings

Surface morphologies of HA coatings are shown in Fig. 1. The surface of P1 coating consists of partially melted particles that indicate a low degree of formation and melted particles that impact on substrate to form splats (Fig. 1a, b). A large amount of pores formed between the partially melted particles (Fig. 1a, b). The morphology of the P2 coating was dominated by splatted particles and a small amount of little spherical particles (2~10μm) (Fig. 1b, c). This coating contains relatively fewer pores. Comparing coating sprayed with P2 condition, P3 coating with longer SOD is non-uniform that contains lots of spherical particles (5~25μm) and smaller splatted splats as well as the quantity of pores increase in P3 coating (Fig. 1e, f). In addition, cracks were observed in all coatings. These cracks often seen on the larger splats.

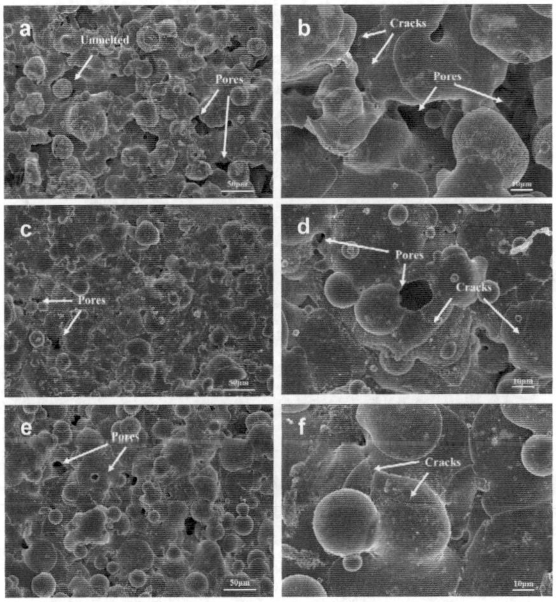

Fig. 1. Surface morphology of the HA coatings: (a, b) P1 coating, (c, d) P2 coating, (e, f) P3 coating.

3.2 Cross-section microstructure of coatings

The cross-section morphology of the coating sprayed under different parameters is shown in Fig. 2. In the polished cross-section of P1 coating, a large amount of pores (~20%) were distributed throughout the coating matrix (Fig. 2a). Spherical particles and no splat boundary were observed in this coating. On the contrary, both P2 and P3 coating (Fig. 2c, e) exhibits lower amounts of pores (<4%) that induce a dense and smooth structure without obvious spherical particles and a splats' boundary. Chemically etched cross-sectional morphology of P1 coating contained many unmelted and partially melted particles (Fig. 2b), which corresponded to the spherical particles that were observed in the polished cross-section of P1 coating (Fig. 2a). P2 coating shows typical lamellar structure with columnar grains after chemically etching (Fig. 2d). This columnar grains grew along deposition direction and through the whole splats. No columnar grains grew across the boundary of the splats. Chemical etched cross-section of the P3 coating without obvious columnar grains. In addition, there are resolidification spherical particles were observed in this coating, however were not observed in the P1 and the P2 coating. In this conditions, fine droplets were solidified into spherical particles, due to the temperature decrease along with increasing SOD. It was possible that these spherical particles stabilized in acid solution.

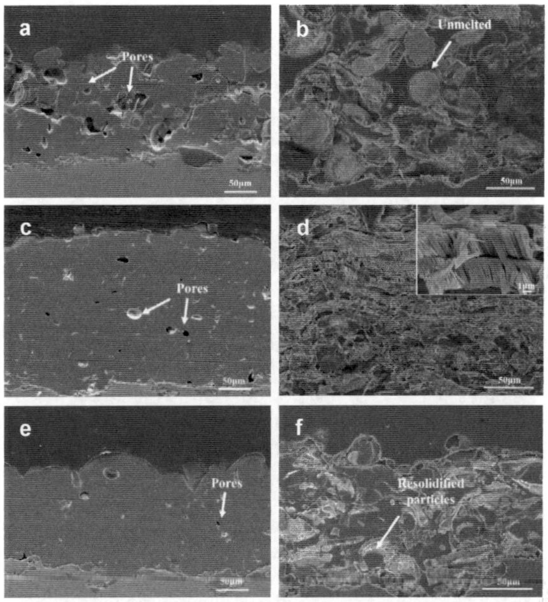

Fig. 2. Cross-section morphology of HA coating: (a) as-sprayed cross-section P1, (b) as-sprayed P2 coating, (c) as-sprayed P3 coating, (d) chemically etched P1 coating, (e) chemically etched P2 coating; (f) chemically etched P3 coating.

3.3 Phase composition of coatings

Figure 3 shows the XRD patterns of the investigated HA coatings. The XRD spectrum of raw powder shows that all diffraction peaks and relative intensities are assigned to standard HA powder pattern (JCPDS 09-432), indicating that the raw powder exhibits random crystal orientation (Fig. 3a). In contrast to HA raw powders, the decomposition phases, such as tetracalcium phosphate (TTCP), alpha-tricalcium phosphate (α-TCP), beta-tricalcium phosphate (β-TCP) and amorphous calcium phosphate (ACP) were observed in HA coatings. The phase CaO can not be identified in the patterns. The XRD analysis reveals that P1 coating (Fig. 3b) consists of 93.3% HA, 3.5% TTCP, and 3.2% α-TCP. P2 coating's intensity of peak (002) increased compared with P1 coating (Fig. 3c). The HA content is 86.7% and accompanied by 13.3% α-TCP. However, TTCP and ACP were not seen from XRD analysis. In addition, P3 coating (Fig. 3d) contains 40.0% HA, 16.4% α-TCP, 2.6% β-TCP and 41% ACP. No TTCP was observed in this coating.

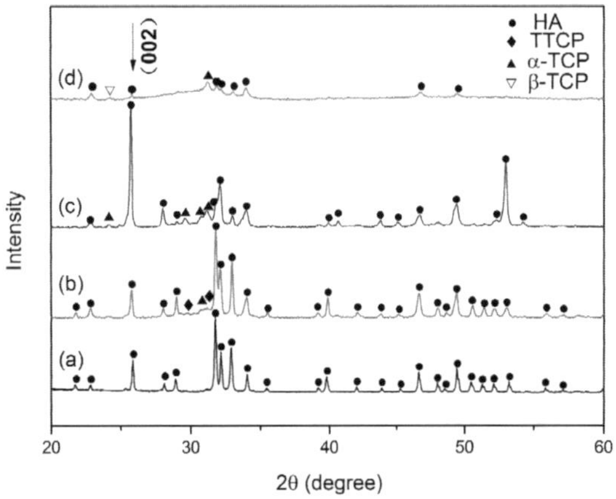

Fig. 3. XRD pattern of as-sprayed HA coatings: (a) Raw powder, (b) P1 coating, (c) P2 coating, (d) P3 coating.

3.4 In-flight particles characteristics

The morphology of raw powders and in-flight particles are show in Fig. 4. Raw powders exhibited a spherical shape with high porosity and inside grains with rodlike (Fig. 4a,b). There are three molten state observed in the in-flight particles: 1)unmelted particles, 2)partially melted particles, 3) fully melted particles. Unmelted and partially melted features were observed in-flight particles made by parameter P1 (Fig. 4c). The resolidification of melted part induces a densified structure and the unmelted part shows a porous structure. Therefor, the boundary between the melted section and un-melted section was formed in the cross-section of partially melted particles (Fig. 4d). Fully melted state can be found with the increase of spraying power (Fig. 4e). In these particles, the unmelted or partially melted particles were rarely observed (Fig. 4e). Meanwhile, pores were observed from the cross-section of the fully melted particles (Fig. 4f). These in-flight particles made by parameter P3 exhibited a lot of fully melted and a little amount of unmelted or partially melted particles(Fig. 3g,h).

Figure. 5 shows that in-flight particles were obtained in experience P1, P2 and P3. It can be seen that the powders sprayed with different parameter reveal different particles size distribution. In-flight particles size distribution was found to be different for raw powder size distribution. In-flight particles made by parameter P1 that includes most coarser particles (>40μm) and fine particles (<20μm) were observed in diagram (Fig. 5b). The particles sprayed using parameter P2 contain more small particles (30~35μm). Fine particles (<20μm) were also not observed in these particles (Fig. 5c). On the contrary, very fine particles (<15μm) were formed in condition that the powder were sprayed in experience P3 (Fig. 5d). Fewer coarse particles (>40μm) are found under this parameter.

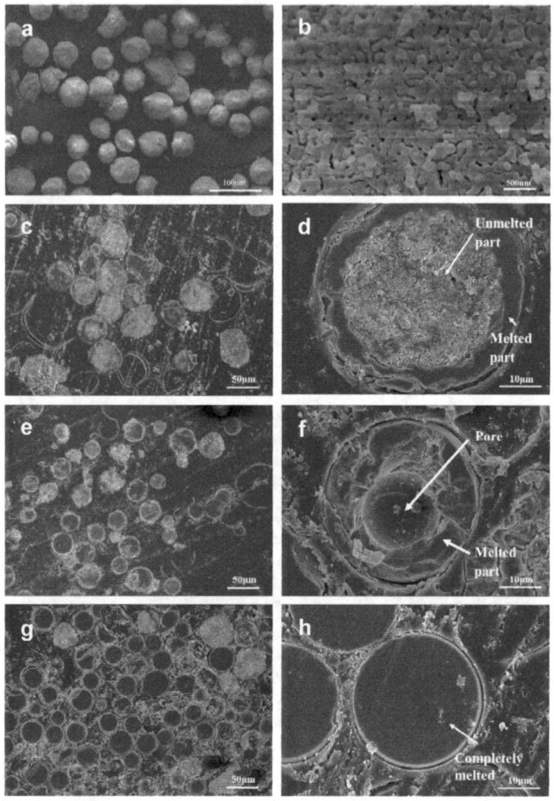

Fig. 4. SEM morphology of raw powders and sprayed particles: (a, b) raw powders surface, (c, d) cross-section of sprayed particles made by parameter P1, (e, f) cross-section of sprayed particles made by parameter P2, (g, h) cross-section of sprayed particles made by parameter P3.

3.5 Morphology of splats

Figure 6 shows the morphology of splats deposited by different conditions. The hemispherical shape splats and a small amount of smaller disk shape splats with some irregularity at the rim were observed from the splats prepared by parameter P1 (Fig. 6a). The diameter of the splat ranges from 8~57μm and the height ranges from 5~40μm. On the contrary, a large number of larger disk shape splats with some irregularity at the rim were observed from the splats prepared by parameter P2 (Fig. 6b). The diameter of the splat is in range of 10~100μm and the height

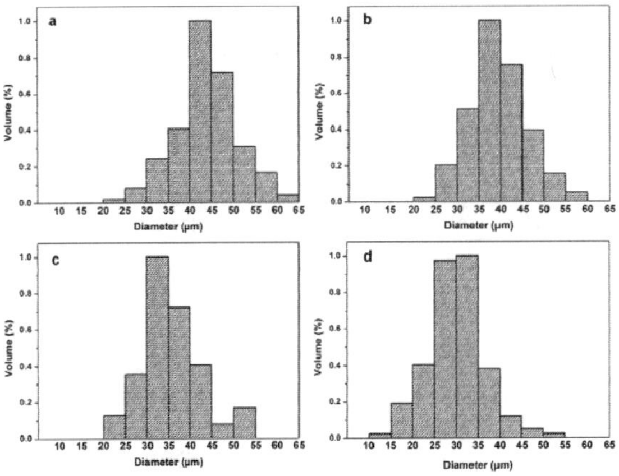

Fig. 5. In-flight particle size distribution: (a) raw powders,(b) made by parameter P1, (c) made by parameter P2, (d) made by parameter P3.

ranges from 1~10μm. Meanwhile, open pores were observed in the surface of the disk shape splats, the size of the pores is revealed to be approximately 1~15μm. An increase in spray distance resulted in a change in splat shape from the disk shape with some irregularity at the rim to the disk shape with splashing (Fig. 6c). In addition, there are some hemispherical shape and thick dish shape splats. The diameter of the splat is in range of 24~146μm and the height ranges from 2~16μm.

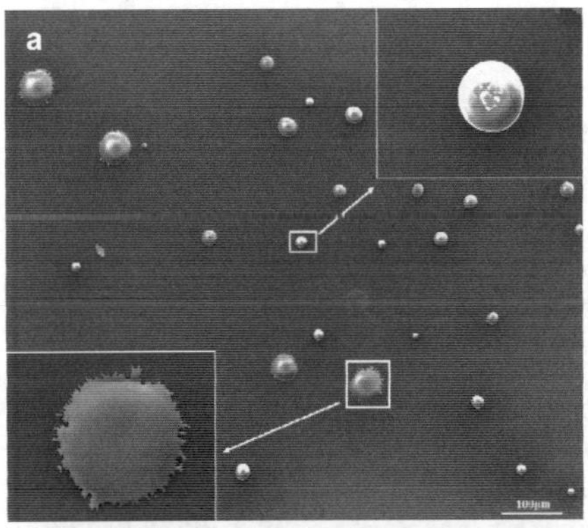

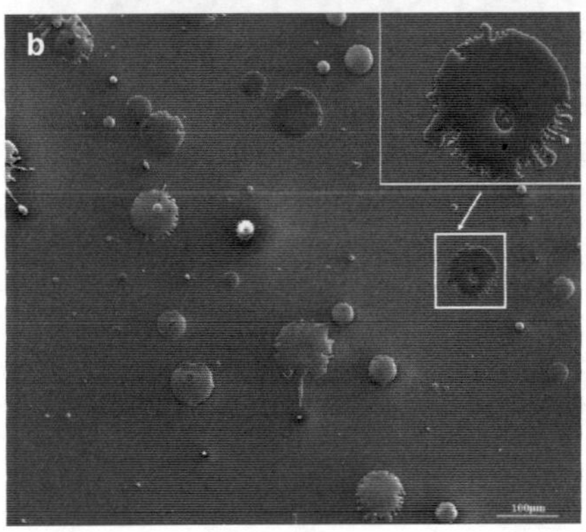

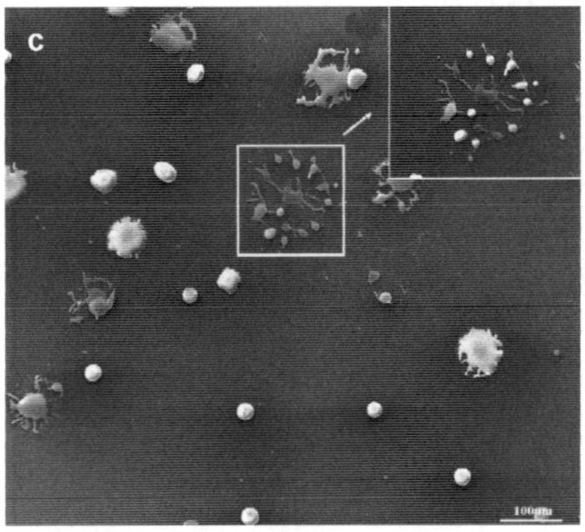

Fig.6. Morphologies of as-sprayed HA splats: (a) made by parameter P1, (b) made by parameter P2, (c) made by parameter P3.

4 Discussion

It is known that the difference in the spray parameters leaded to different phase compositions and microstructures of HA coatings. To understand the process of coating formation, it is worthy to investigate the in-flight particles and splats. The phase composition of coating depended on the in-flight particles state and the solidification rate of the droplet. The morphology of the P1 coating was dominated by unmelted and partially melted particles. This was because that the lower plasma power leaded to a lot of partially melted particles (Fig. 4c). As well, the crystalline HA phase was dominated by the retention of original HA phase in the sprayed particles. The relatively poor bonding between partially melted or unmelted particles resulted in a porous microstructure in the coating cross-section (Fig. 2a). When the spraying power increased, the temperatures of the plasma flame and the particle increased [10], and therefore more HA powders were fully melted within the plasma. These fully melted particles with a relatively high velocity impinged on the substrate and enabled the particles to flatten better on the substrate surface to form disk shape splat, and it resulted in a lamellar structure in the coating cross section (Fig. 2c). The P2 coating with high plasma power transferred more heat to the substrate, and the substrate temperature was increased to 200°C to decreased the cooling rate of splat to improve more recrystallization from. The HA phase of P2 coating was dominated by the recrystallized HA. These collected particles at 60mm underwent high melting and subsequently deposited onto heated substrate, further developed well-flattened splats (Fig.6b). The (002) crystal orientation is of the lowest surface energy among all orientations in crystalline HA [11]. HA films tend to grow along the c-axis with a (002) orientation to minimized the surface energy [11]. In our case, the heat flux of the flattened splats was perpendicular to the substrate surface, hence promoting the (002) crystal growth along deposition direction during MPS. The heat input of plasma flame and follow-up droplet could keep the temperature of coating surface higher than 300℃. Therefore, the splat solidification occurred at slow cooling rate and facilitate the columnar grains formation. The columnar structure might have a beneficial effect on biocompatibility osteogenic differentiation of implants [12]. Moreover, pores within HA coating were mainly caused by the existence of gas (Fig. 2c). This can be inherited from in-flight particles (Fig. 4f). Although some pores within in-flight particles could be filled by the follow-up droplet (Fig. 2c).

As the SOD increased, more in-flight particles experienced melting and dehydroxylation due to the longer duration within plasma. Furthermore, so the temperature of substrate was about 40℃ when the SOD was of 110mm. The lower local temperature also leads to the high cooling rate of the coating. The cooling rate of corresponding splats was higher than that sprayed at short SOD. Thereby, fully melted HA particles was rapidly cooled as they impinged onto substrate or successive coating. This leaded to the formation of large fraction of ACP within HA coating. The P3 coating surface contained more fine spherical particles compared with P1 and P2 coating (Fig. 5d). However, the cross-section of P3 coating exhibited a compact structure with lower porosity than others. This was because that the fully melted particles overlapped each other.

5 Conclusion

In this study, three kinds of HA coatings with different phase composition and microstructure were obtained by MPS. Which includes: 1) a porous coating with high crystalline HA content (93.3%) (P1). It contained partially melted and unmelted particles. 2) a columnar structured HA coating with high crystalline HA content (86.7%) (P2). 3) A HA coating with high fraction of ACP content (40%) and correspondingly, low fraction of crystalline HA (40%) (P3). The microstructure of the coatings was related to the melting state of the in-flight particles. The in-flight particles were not sufficiently melted during the spraying process of P1. However, more adequate melting was acquired by in-flight particles during P2 compared to that during P1. Accompanying by the low cooling rate, columnar structured coating was hence observed. At last, sufficient melting of in-flight particles and the relative high cooling rate leaded to the formation of ACP within P3 coating.

6 References

[1] ParkJ. Bioceramics:Properity, Characterization, and Applications. New York: Springer Science + business Media, LLC;2008.

[2] Locardi B, Pazzaglia UE, Gabbi C, Profilo B (1993) Thermal behaviour of hydroxyapatite intended for medical applications. Biomaterials 1993, Issue 6, pp. 437–441.

[3] Dorozhkin SV. Calcium orthophosphate-based biocomposites and hybrid biomaterials. J Mater Sci 2009, Issue 9, pp. 2343–87.

[4] V. Deram , C. Minichiello , R.-N. Vannier ,A. Le Maguer , L. Pawlowski, D. Murano. Microstructural characterizations of plasma sprayed hydroxyapatite coatings. Surface and Coatings Technology 166(2003), Issue 2-3, pp. 153–159.

[5] Limin Sun, Christopher C. Berndt, Clare P. Grey. Phase, structure and microstructural investigation of plasma sprayed hydroxyapatite coatings. Materials Science and Engineering A 360 (2003), Issue 1-2, pp. 70-84.

[6] Sahar Vahabzadeh, Mangal Roy, Amit Bandyopadhyay, Susmita Bose. Phase stability and biological property evaluation of plasma sprayed hydroxyapatite coatings for orthopedic and dental applications. Acta Biomaterialia 17 (2015),Vol 17 pp. 47–55.

[7] M.Inagaki,T.Kameyama. Phase transformation of plasma-sprayed hydroxyapatite coating with preferred crystalline orientation. Biomaterials 28 (2007), Issue 19, pp. 2923-2931.

[8] Yiming Wang,Tingting Fan,Zheng Zhou,Dingyong He. Hydroxyapatite coating with strong (002) crystallographic texture deposited by micro-plasma spraying. Materials Letters,2016, Vol 185. pp. 484-487.

[9] Saeed Saber-Samandari, Karlis A. Gross. Nano-indentation on armorphous calcium phosphate splats: Effect of droplet size on mechanical properties. Journal of the mechanical behavior of biomedical materials 16 (2012) pp. 29-37.

[10] Jan Cizek, Khiam Aik Khor. Role of in-flight temperature and velocity of powder particles on plasma sprayed. Surface & Coatings Technology 206 (2012), Issue 8-9, pp. 2181-2191.

[11] M.R.T. Filgueiras, D. Mkhonto, N.H. de Leeuw, Computer simulations of the adsorption of citric acid at hydroxyapatite surfaces, J. Cryst. Growth 294 (2006), Issue,1, pp. 60-68.

[12] Wei Chen, BoTian, YongLei, YongLei, Zhen-An Zhu, Ya-Ping Guo. Hydroxyapatite coatings with oriented nanoplate and nanorod arrays: Fabrication, morphology, cytocompatibility and osteogenic differentiation. Materials Science and Engineering C 67 (2016), Vol 67, pp. 395–408.

7 Acknowledgments

This work is financially financed by National Natural Science Foundation of China (Project No. 51471010)

The protection against spray wear using pseudoalloys applied with the method of spraying

R. Winkelmann, Senftenberg / D, H. Pokhmurska, Chemnitz / D

An important parameter that affects the protection from wear and tear, is the geometric structure of the abrasive causing the destruction. The equivalent diameter of the particles of the primary influence on protection from wear and tear must be greater than the width of the particle of the crushed abrasive. Here also belongs the requirement for the distribution of the particles inside the matrix was more uniform and dense. The use of solid materials gives this possibility only in the certain circumstances, therefore, it is necessary to use pseudoalloys. Pseudoalloys can be obtained by means of thermomechanical processing of the conventional layers deposited by means of spraying. The particles of the solid material charged with kinetic energy are applied to the heated layers, and they penetrate into inside. Under the influence of the heat energy inside the layers suitable for this, the additional reaction between the substrate and the layer takes place, thus improving the adhesion of the layers. The appropriate regulations were developed, and the reproducible stratification appeared caused by spraying, that is pseudoalloys. A positive consequence of the inclusion of the particles of a solid ceramic material into the spayed layer is the emergence of the current when compressed. The result of the research of the parameters are the optimal performance indexes, optimization criterion served as an indicator of the intensity of the wear. Manufactured pseudoalloys underwent checking in terms of spray wear. The study of spray wear was carried out at the temperatures up to 750 ° C and speed load amounted to 37.5 m/s. The silica sand fractions + 150-212 µm was used as an abrasive. As the covering material the alloys based on aluminum were used. In the places of transition to the substrate the multiphase alloys of the system Fe-Al are formed. It is advisable to use layers with a high iron content to prevent the loose and destruction.

1 Aim

For the application of layers occupying a large area, the spraying methods are very effective. The users' requirements are satisfied better and better concerning high layers' resistance to corrosion and/or wear and, which is constantly growing. The main comments relate to the low adhesion of the layer to the substrate, and also often with the limited possibilities of applying with the method of pseudoalloys spraying with the contents of the large solid materials. Improving adhesion with regard to formation of phases inside the metal was a part of a series of publications (for example, /1/ and /2/). With the subsequent heat treatment of the sprayed layers on the basis of aluminum the phase system Fe-Al are produced; their properties ensure a high level of adhesion layer to the substrate, as well as significantly reduce the intensity of wear as the layers that do not stimulate the formation of phases of the system Fe-Al. The purpose of the research that formed the basis of this article and were carried out on the basis of this information, was to create pseudoalloys with high adhesion strength with substrate. Pseudoalloys are especially known for their role in the systems of protection against the abrasive wear. In the field of welding it is worth to mention the pseudoalloys consisting of matrix based on nickel, embedded tungsten carbides (WxCy). Solid substances containing tungsten carbides provide a high level of wear resistance of these layers. The layers of this type are different from the layers containing tungsten carbides, which is applied with the method of high-speed flame gas-spraying (HVOF), firstly with the faction and the content level of the solid materials in the matrix. It is known that the size of the particles of a solid material and their distribution inside the sprayed layer significantly influences the intensity of wear, since there is the dependency on the geometrical structure of the abrasive, causing destruction. Inside the layers of the system of WC-Co applied with the method of high-speed gas-flame spraying (HVOF), the solid material is processed, the diameter of the particles of which averages 2.6 µm. Due to the high content of the solid materials, these layers ensure a high level of protection against abrasive wear and tear, if the fraction of abrasives, the influence of which the layers are exposed, approximately is the same as the fraction of particles of solid materials composed in the layers. The larger is the faction of the abrasives exposing the layers, the greater must be the fraction of the particles of a solid material that contains the layers. The studies that were intended to include the composition of the layers of particles of a solid material to a larger faction (20 microns)/3/still had success only under the certain conditions. It is known that during the application of the method of high-speed gas-flame spraying the decrease of the size of the particles was observed, as well as the loss of solid material at 65-80% and reducing the effectiveness of the stratification of 24% /3/. The following objectives: a) application in the spaying with a sufficient amount (> 10% of volume) of the particles of solid material with a larger faction (> 100 microns), b) ensuring lasting adhesion of particles of a solid material with a matrix c) reducing the impact on the layers. It is worth to determine the durability of the developed layers.

2 Experimental design and substances

The task of the layers' development with the content of solid materials blended with the substrate by means of metal, was performed during the process, which consisted of several stages. Traditionally, after the preliminary treatment of the substrate the three layers spaying was applied. Then the inclusion of the particles of solid material under the influence of heat and kinetic energy, the further mechanical treatment and covering with the spayed coating. For the preliminary treatment of the substrate the sand blasting machine is used, where the settings listed in **Tab. 1** were set.

Tab 1. Spray parameters

Al_2O_3/ZrO_2	F16 (1000 + 1400 microns)
Air pressure	7 bar
Diameter of the nozzle opening	4 mm
Fluctuations speed	9.43 mm/s
Distance of the nozzle	13 mm
Installation angle	45°

The original material was the structural quality steel S 235 JR. Sample size: 240*80*4 mm³. For the application of the layers the machine was used for electrical metallization, **Fig. 1**.

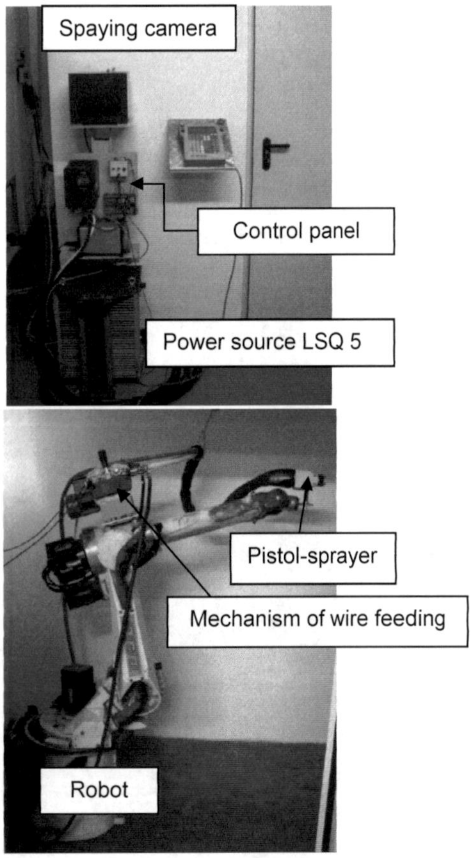

Fig. 1. Machine for electrical metallization

To create the layers based on aluminum the substances are used like AlSi5 and AlSi3Mn in the form of wire thickness in 1.6 mm. As the alloys based on iron the substances of type FeCr22Al5 and 80MnSi are used in the form of wire thickness in 1.6 mm. For further the processing the device is used, **Fig. 2**. The sample that should interact with a solid substance (SiC).

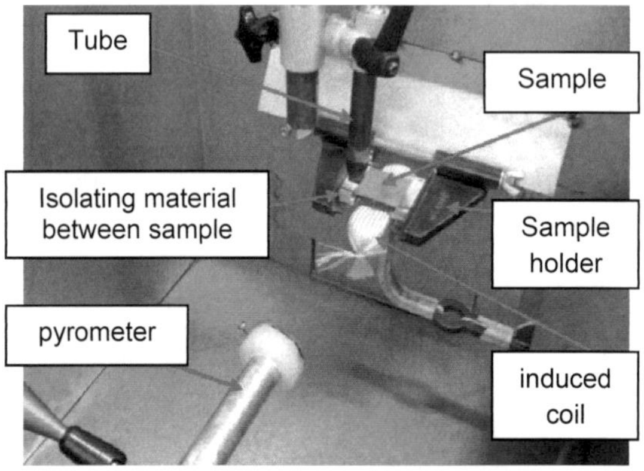

Fig. 2. Device for further processing

heated with induction method. For determining the temperature of the sample set with calibrated pyrometer. At the same time, it is used for adjusting the generator. In normal sand blasting machine, the solid substance is expulsed on the surface of the sample through a tube. The tube has a nozzle of a defined size. The distance to the sample is also determined. For a uniform distribution of solid material on the surface of the sample the tube should be moved as if you inflict the ornament like "meander". The amount of solid material in the sample, the temperature of the sample and the size of the particles of a solid material can be changed.
To assess the level of layers' durability, the researches spray wear is conducted. For this you can use the test stand, **Fig. 3**. It consists of three stoves each of which is equipped with a separate control system: stoves for heating the rotor (air) for heating of carrier environment and for the heating of the sample. The movement of the abrasive is accelerated under the influence of its own weight and action of the injector, and it is sprayed on the surface of the sample through a nozzle with a specified diameter. **Fig. 4** shows the placement of the sample relatively to the nozzle. The nozzle is placed at a certain distances and under a certain angle to the sample. If the angle is 45 °, you can expect the maximum load and correspondingly the maximum wear of the metal.

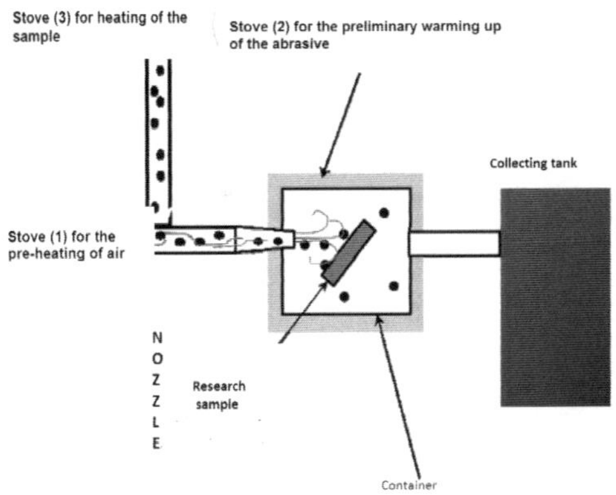

Fig. 3. Test stand to assess the resistance of spray wear, top scheme, bottom figure - order of fulfillment

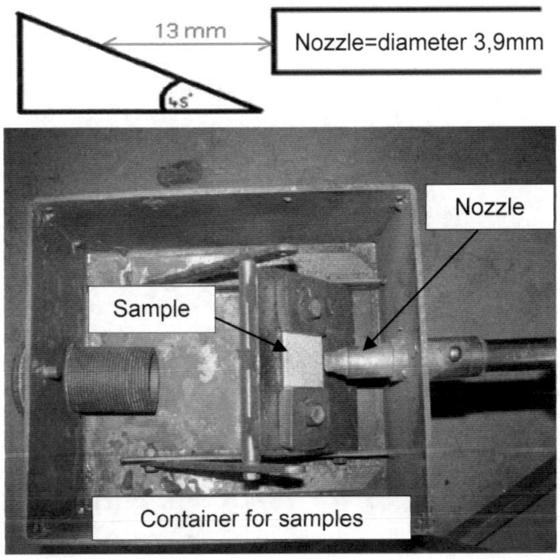

Fig. 4. Samples camera

4 Methodology of the study

Pretreatment of the sample was carried out directly before applying the coating. The result of the three-layer coating is shown in **Fig. 5**. After coating the test material was divided into parts with a size 40 * 40 mm^2.

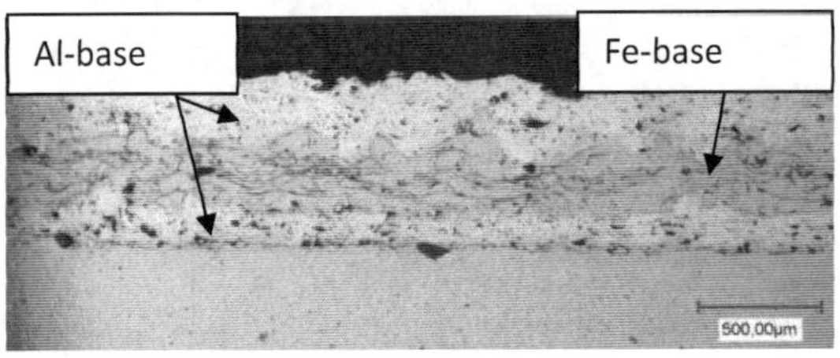

Fig. 5. Structure of material after coating

The thickness of individual layers was about 200 microns. The results of the sealing of solid substances layers are listed in **Tab. 2**. Solid substance SiC was used because of its high hardness and roughness.

Tab. 2. Options for layer

Size of grain of SiC	F70/F120/F180
Abrasive consumption	0.10...0.22 g/mm²
Air pressure	0.5 bar
Temperature	550°C/650°C/750°C
Crystal lattice openings	3.9 mm
Distance between the openings of lattice	11 mm
Grinding angle	90°

As a result of heat treatment, between Al and Fe the intermetal structures are formed. During the mechanical treatment the sealing of layer occurs.

Tab. 3. Additional processing

Grain size of Al_2O_3/ZrO_2	F80
Abrasive consumption	150 g
Air pressure	3 bar
Crystal lattice openings	3.9 mm
Fluctuations speed	12.42 mm/s
Distance between the openings of lattice	11 mm
Grinding angle	90°

Following this procedure, the samples were covered by basic Al-coating with a thickness of 100 microns. After the coating the samples were tested for grinding.

5 Results

The surface roughness was determined according to the standard DIN EN ISO 8503. All investigated samples had average values of Ra = 11 microns and Rz = 120 microns. Coverage options are listed in the **Tab. 4**.

Tab. 4. Options of coverage

Material	FeCr22Al5 / 80MnSi	AlSi5 / AlSi3Mn
Voltage [V]	30.1	28.7
Current [A]	261	98
Air pressure [bar]	5	5
Thrust (progression) [m/min]	6.6	3.6

Determining of the appropriate parameters for the compaction of solid material SiC is conducted by determined by the mass of inflicted substances. The results are presented in **Fig. 6**. See coarser grains

(250 - 180 microns) that are better suited than smaller ones. The covering should be at a temperature of 650° C to 750 ° C.
The measurement of the mechanical treatment, which was carried out before the sealing and the fixing of a sample of the solid substances are listed in the **Tab. 3** and shown in **Fig. 7**. Thus, the prepared samples were tested for erasing. The results of the tests are presented in **Fig. 8**.

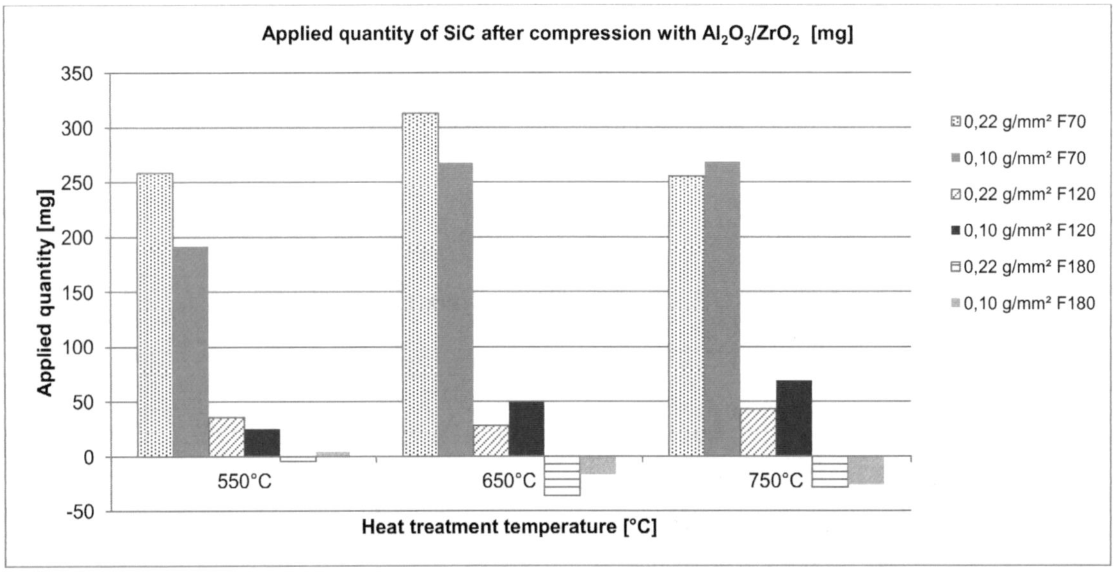

Fig. 6. Sealing solid material

Before processing

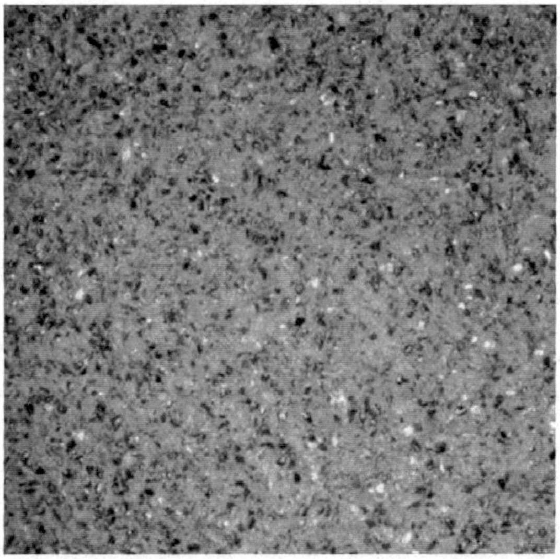

After the thermo-mechanical processing

Fig. 7. Results of thermo-mechanical processing and compression

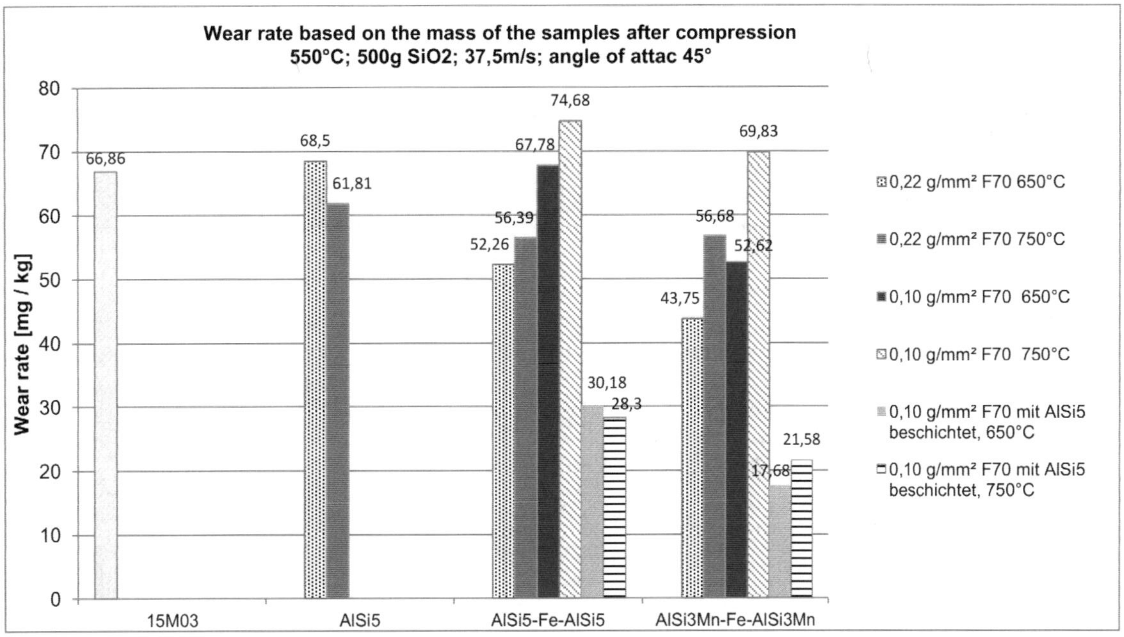

Fig. 8. Test results on erasing

Thanks for erasing test, the following optimal options were found for the covering: AlSi3Mn/80MnSi/AlSi3Mn; 0.1 g/mm^2 at a temperature of 650°C; AlSi3Mn

The analysis of the claimed sample surface is presented in **Fig. 9**.
SiC-particle, the primary protection from wear and tear.

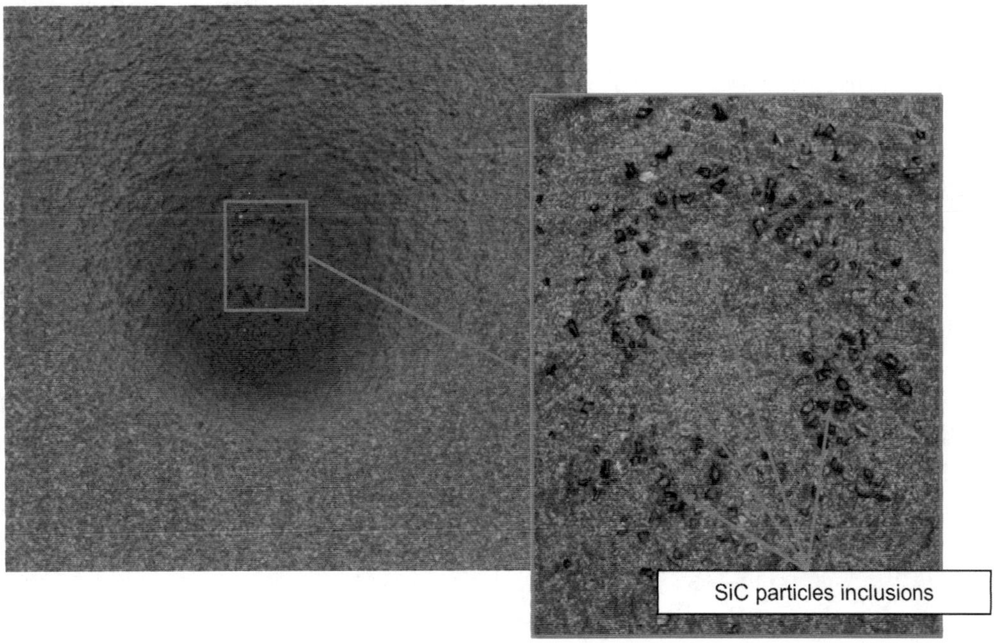

Fig. 9. Point of erase

6 Conclusions

The inclusions of heavy, rough particles in coatings was successful. This was developed by multi-step process of coating and optimal settings. As a criterion of optimization here comes the indicator of erasing. Structure of Fe-Al-compounds is a good basis to reduce adhesion and wear rate, compared with a raw layer. The effect of saved hard particles was provided.

7 Literature

[1] Winkelmann, R.; Pokhmurska, H.: Behavior and grinding of the protective coatings under thermo-abrasive loads. Publ.: Conference on 15-th technical colloquium, on materials and technical applications, 20.09. -21 09.2012 in Chemnitz, vol. 47, Chemnitz, 2012

[2] Winkelmann, R.; Pokhmurska, H.: The development and evaluation of Fe-Al-x layers' wear. Publ.: Conference on 17-th technical colloquium, on materials and technical applications, 11.09. -12 09.2014 in Chemnitz, vol. 47, Chemnitz, 2012

[3] Röttger, A.: The development of new abrasive Fe-based layers. Free access diser. A, University Bochum, 2011

Cavitation Erosion Characteristics of HVOF Thermal Sprayed WC-cermet Coatings

A. Kanno, K. Takagi, M. Arai, Akashi / J

Cavitation erosion (CE) damage, which occurs in the main parts (made of high chromium cast steel) of hydroelectric power generation machine, is one of the serious problems. It is expected that life time of those parts would be prolonged if the suitable CE-resistant coating is applied on the surface of the cast steel. In this study, WC-cermet coatings (WC-CoCr and WC-Cr_3C_2-Ni), which were fabricated by high-velocity oxygen-fuel (HVOF) thermal spraying process, was interested in protecting CE attack to the cast steel. To clarify CE property of the WC-cermet coatings, the ultrasonic vibration tests were conducted, and the amount of volume loss characterized as CE damage was measured. The microstructure and the fracture toughness, which was evaluated by the indentation test method, of the coatings were related with the CE damage. As the results obtained in this study, the fragment which was spalled from the surface after CE test was almost flake-like shape, and its size was from 2µm to 50µm. SEM observation indicated that this fragment included both WC particle and metal binder, which means that WC particle and metal binder was still strongly bonded together. It was also confirmed that the amount of volume loss could relate directly with the fracture toughness K_{IC} rather than Vickers hardness. It was considered that CE damage was progressed into the depth by throwing out the fragment originated from micro crack initiation. Thus, it was required that the CE resistance of the developed coatings could be labelled through the fracture toughness.

1 Introduction

Damage of impeller of a hydraulic power generation machine and the water pump parts caused by cavitation erosion is one of the serious problems [1]. Cavitation erosion behavior causes massive damage to the material surface because of fatigue fracture or brittle fracture from repeated imploding pressure of cavitation bubbles on the surface [2], [3], [4].

High chromium cast steel such as 27%Chromium cast steel (JIS G 3463, from now on 27%Cr steel) is selected as the base material of water pump parts and impeller for the cavitation erosion resistance. By applying erosion resistant thermal spray coating on 27%Cr steel, a lifetime of these parts are expected to be prolonged [5], [6]. In this study, WC cermet coatings sprayed by high-velocity oxygen-fuel (HVOF) spraying and 27%Cr steel were evaluated by cavitation erosion (CE) test using magnetostrictive cavitation test equipment. Fracture morphology of cavitation eroded 27%Cr steel and WC-cermet coating surfaces were analyzed by surface observations with SEM and also the volume loss was measured. Also, it was reported that CE resistance is proportional to the value of the indentation fracture (IF) toughness [7]. Hardness and IF toughness were measured in the coatings by Vickers hardness test, and CE resistance of each coatings was evaluated. The relationship between erosion resistance of each particle and the coating was evaluated by magnetostrictive cavitation tester.

Though HVOF coating exhibited superior CE resistance compared with 27%Cr steel, HVOF is more suitable for the application of thermal spray coating to actual pump parts. To develop a coating with excellent CE resistance, fracture behavior of thermal spray coating by CE was studied.

2 Experimental Procedures

2.1 Test Specimen

The coating material was WC-cermet powder, and the substrate material of specimens for CE test was made of SUS304 steel base material, which is 16mm in diameter. The coating method was HVOF spraying process, and the reference specimen was 27%Cr steel plate. Table 1 shows the chemical composition of material powder and 27%Cr steel plate. The spray conditions by HVOF process is shown in Table 2. And Fig.1 concludes the specification of the material and coating. By polishing, specimens for CE test should be with smooth surface, and surface roughness should be less or equal to 0.8µm.

Table1. *Chemical composition of WC-cermet powder and 27%Cr steel and by Fluorescent X-ray analysis (mass%).*

Spray Powder	Chemical Composition (mass%)				
WC-20mass%Cr_3C_2-7mass%Ni	W	Cr	Ni	C	Fe
	Bal.	18.0	7.2	7.2	0.1
WC-10mass%Co-4mass%Cr	W	Co	Cr	C	Fe
	Bal.	10.3	4.0	5.9	0.02

Reference specimen	Chemical Composition (mass%)				
27mass%Cr cast steel	Fe	Cr	Ni	Mo	Si
	Bal.	27.5	5.2	2.1	0.5

***Table2.** Spray conditions by HVOF process*

Spray condition	Standard	High power
Spray process	HVOF	
Spray gun	TAFA / JP-5000HP	
Fuel + supporting gas	Kerosene + Oxygen gas	
Nozzle length	8 inch	
Combustion pressure	0.7 MPa	0.8 MPa
Spray distance	400 mm	

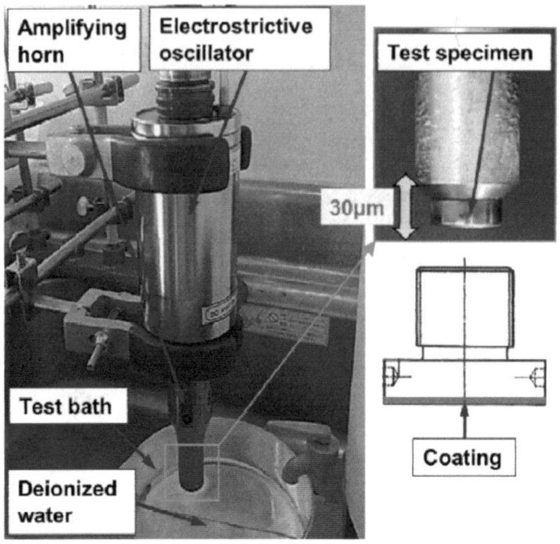

***Fig.2.** Outview of testing apparatus and details of test specimen.*

2.2 Cavitation Erosion Test Method

CE test was performed by electrostrictive cavitation test equipment. Overview of equipment is shown in Fig.2. The test specimen was set on the tip of amplifying horn. Moreover, the test was conducted at a frequency of 20 kHz, and at horn amplitude of 30μm in ion-exchanged water under controlled temperature. The test method is specified in ASTM G32-10 [8]. To evaluate erosion resistance, volume loss per unit time was monitored. Eroded surfaces of coatings were observed by SEM.

2.3 Evaluation of Mechanical Properties

The hardness of coatings was measured by micro Vickers hardness test (JIS-Z2244, load: 300gf, load time: 15sec.) and IF toughness was measured by Vickers hardness test (IF test, load: 1kgf=9.8N, load time: 15sec.) to investigate at the macro level of coatings. Crack length was measured from optical image as it is shown in Fig.3. Crack direction is perpendicular to the spray direction and lies in plane with the elongated side of the splat. To calculate the fracture toughness (as K_{IC}) of the coatings, the Lawn and Fuller equation that is shown in bellow was used [9], [10];

Lawn and Fuller equation

$$K_{IC} = 0.0515 * P/c^{1/2}$$

K_{IC} : Fracture Toughness (MPa*m$^{1/2}$)
P : Load = 9.807N
c : ½ Crack lengths

By these methods, mechanical properties such as hardness and IF toughness of coating can be observed with being influenced by particle boundaries.

Sample	TP1	TP2	TP3	TP4	TP5	TP6	TP7	Fe27Cr
Material compo-sition	$WC-Cr_3C_2-Ni$		WC-CoCr					27%Cr Cast steel
WC Grain size	3μm	1μm	3μm		1μm	0.6μm	0.1μm	
Powder	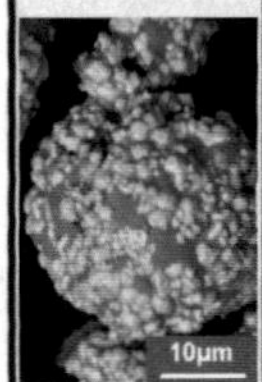	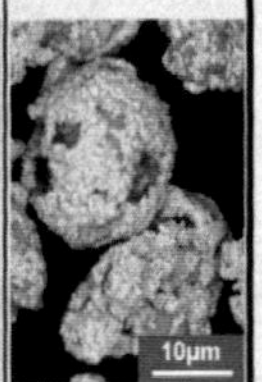	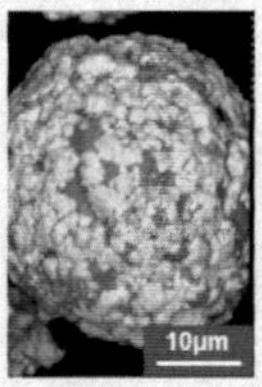		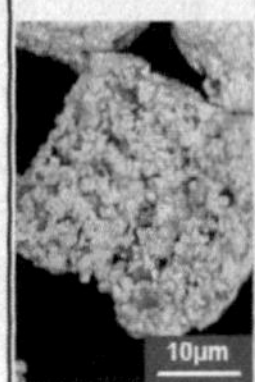	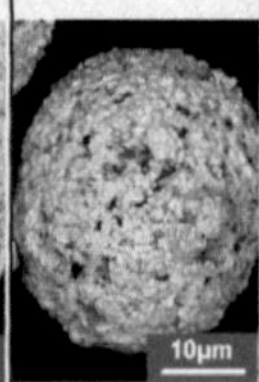		
Spray condition	Standard	Standard	High power	Standard	Standard	Standard	Standard	

***Fig.1.** Specification of the specimens*

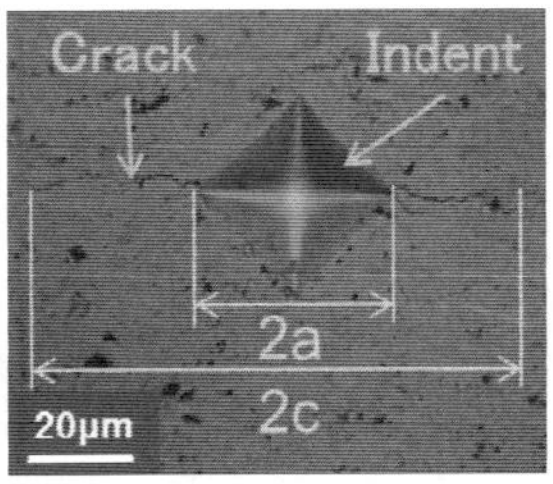

Fig.3. *Measurement of the crack length by IF test.*

3. Results and Discussion

3.1 The result of Cavitation Erosion Test

Figure 4 shows the volume loss per unit time of the WC-cermet coatings sprayed by HVOF and 27%Cr steel after cavitation exposure time of 120 minutes. Compared to 27%Cr steel, WC-cermet coating showed about two third to one fifth amount of volume loss rate. As compare with WC-Cr_3C_2-Ni, WC-CoCr showed higher CE resistance. Moreover, the TP3 to TP7 of WC-CoCr coating showed the relatively high performance of CE resistance. Especially, TP6 (WC grain size ; 0.6μm) and TP3(coated by High power condition) showed excellent performance for CE resistance in particular.

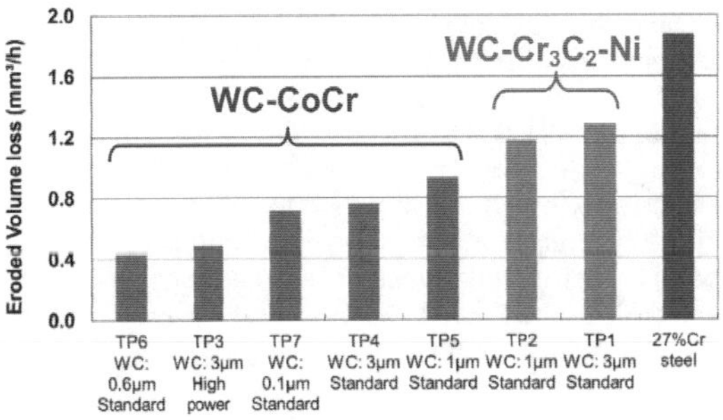

Fig.4. *CE test results of WC-cermet coatings and 27%Cr steel*

3.2 Microstructure of WC-cermet coatings

Figure 5 shows SEM surface images of the TP6 and TP5 after CE test. The surface of TP6 remains the eroded surface where particles locally dropped out (point of the arrow) were observed after CE test. On the other hand, the surface of TP5 showed heavy damage. The area of the flat surface, which is expected as the initial surface by polishing, was quite a few.

Figure 6 (b), (c), (d) shows the SEM images of some fragments which were spalled from the TP4 coating surface after CE test. Most of the case, the shape of fragments was flake-like, and the sizes of them was from 2μm to 50μm. SEM observation indicated that these fragments included both element that are WC particle and metal binder. Therefore, it is assumed that metallic phase keeps the potential as binder. In these observations, it is inferred about CE fracture morphology that fragment with WC grain particle and metal binder are peeled off from the coating surface, and erosion was progressed.

From these observation results, internal CE resistance of a particle was studied by investigating internal mechanical properties of a particle.

TP6 WC: 0.6μm Standard	TP5 WC: 1μm Standard
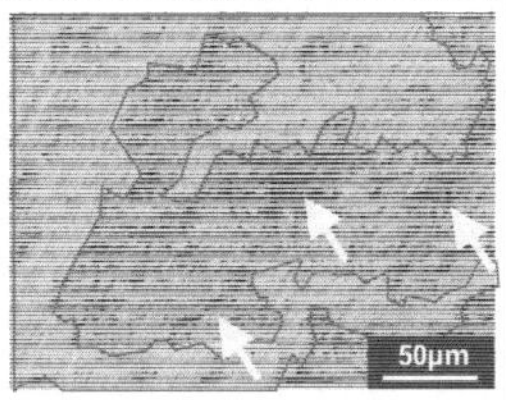 	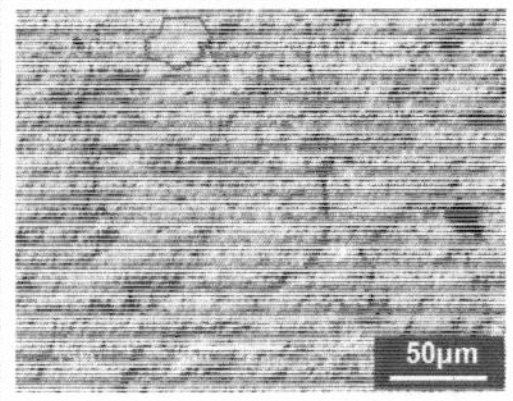

Fig.5. *SEM image of coating surface after CE test.*

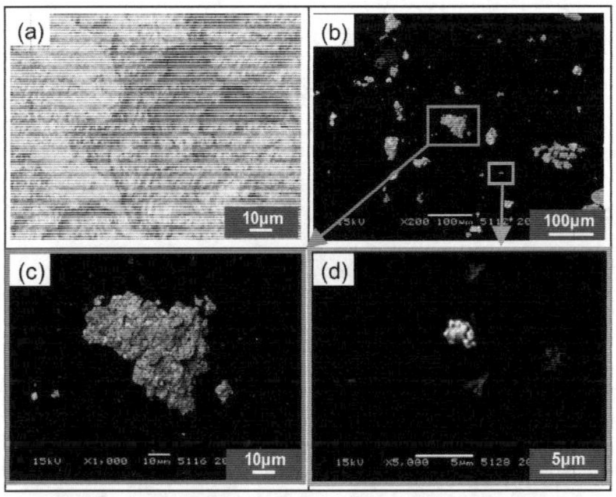

Fig.6. *SEM image of the eroded coating surface and fragment which was spalled from the TP4 coating surface after CE test.*
(a)eroded coting surface (b)spalled particles (c)flake-like particles (d)fine particles

3.3 Mechanical properties of WC coatings

3.3.1 The result of hardness test

Figure 7 shows the relationship between Hardness and CE test results of WC coatings. As compare with WC-Cr_3C_2-Ni, WC-CoCr coating showed higher hardness.

Generally, the grain size of WC has significant effect on the hardness of the cemented tungsten carbide alloy. According to decreasing of the grain size of WC, increasing of hardness of the WC-cermet was reported [11]. However, no correlation between the hardness of WC-CoCr coating and WC grain size was found in this result. WC-CoCr coating showed same tendency.

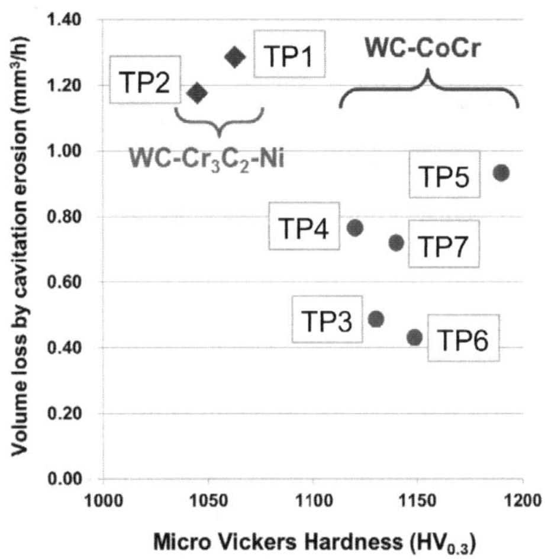

Fig.7. *Relationship between Hardness and CE test results*

3.3.2 Fracture toughness testing result by IF test

Figure 8 shows the result of fracture toughness by IF test. Green bar (TP1, TP2) shows the WC-Cr_3C_2-Ni coatings, and Blue bar (TP3 to TP7) shows the WC-CoCr coatings. As compare with WC-Cr_3C_2-Ni, the WC-CoCr coating shows relatively high fracture toughness that excepts TP5.

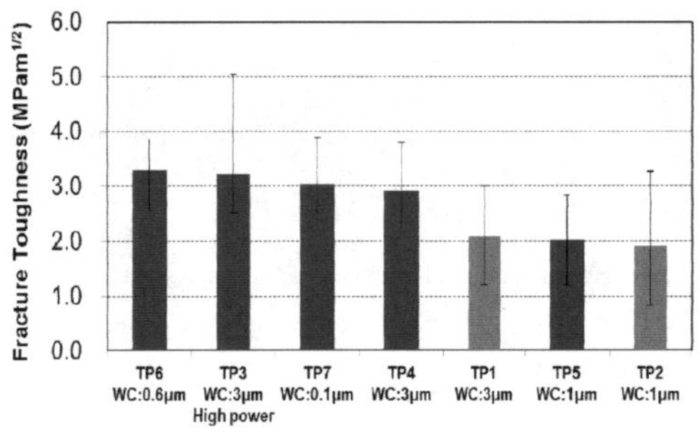

Fig.8. *The result of IF test: TP1 and 2 made by WC-Cr_3C_2-Ni, and TP3 to 7 made by WC-CoCr.*

Figure 9 shows the relationship between CE test results and fracture toughness (K_{IC}). According to increasing of fracture toughness WC-cermet coatings, cavitation erosion resistance was improved. It seems that cavitation erosion resistance and K_{IC} of the WC-cermet coating has correlation.

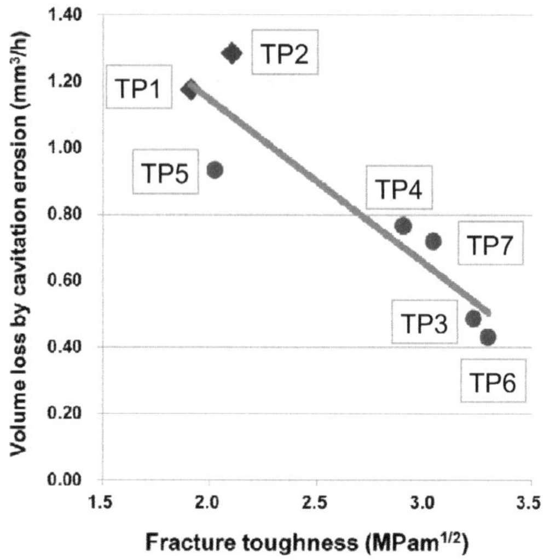

Fig.9. *Relationship between Indentation fracture toughness (K_{IC}) and CE test results*

3.3.3 Summarize of mechanical properties of WC coatings

The mechanical properties of WC-CoCr coating showed higher values compared to WC-Cr_3C_2-Ni coating. From comparison of WC-CoCr coating and WC-Cr_3C_2-Ni coating, hardness of WC-CoCr coating is higher than that of WC-Cr_3C_2-Ni.

Considering about WC-CoCr coatings (TP3, 4, 6, 7), they have high IF toughness. On the other hand, IF toughness of TP5 is lower than that of other WC-CoCr coatings. Because of high hardness of TP5, it performed brittle characteristic. These results indicated that both of deposited particle cohesive strength and IF toughness in WC-CoCr coating were considered to be stronger than that of WC-Cr_3C_2-Ni due to indistinct gap at particle interface.

Fracture process of spray coating by CE was assumed to be as follows (As shown Fig.10).

Firstly, cavitation collapse pressure was applied to the surface of the coating, and then a crack progresses along a weak part in the coating like particle interface or pores (Fig.10 (a)). Second, fragment of particles drop out from spray deposit, and then fracture progresses (Fig.10 (b)). It means that pores and boundaries have weak parts of cohesion strength. So, it is considered that reduction of pores and an increase of inner particle cohesion strength, which are enhanced CE resistance.

WC-CoCr coating (TP3 and TP6 in particular) could be an excellent coating for cavitation erosion resistance because of its strong interparticle cohesive strength can prevent cracks which initiate the process of fracture.

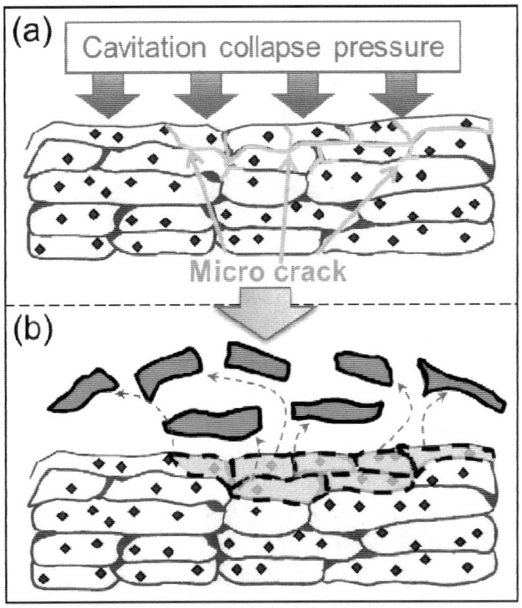

Fig.10 *Fracture behavior of WC cermet-coatings by CE.*
(a) The crack progress along the grain boundaries from many originations
(b) Particles drop off from the pore and the between deposited particles

4. Conclusion

In this study, CE properties and fracture morphology of 27%Cr steel and thermal spray coatings were investigated. Microstructure of WC-cermet coatings sprayed by HVOF, and their erosion resistant properties were evaluated. It's indicated following conclusions.

1. CE resistance of the WC-CoCr coatings are 1.5 to 5 times as much as that of 27%Cr cast steel.

2. CE resistance of WC-CoCr coatings are three times higher than WC-Cr_3C_2-Ni coatings.

3. CE resistance and IF toughness (K_{IC}) of WC-cermet coatings shows a direct correlation.

4. Increasing of IF toughness of the WC-cermet coating improves CE resistance.

Properties of thermal spray coatings were influenced by interparticle cohesive strength. Moreover, erosion was progressed by dropped out particles from a part with weak cohesive strength.

Development of HVOF coating with excellent erosion resistance based on the results of this study is focused.

Reference

[1] S.Hattori: The 7th Pacific Symposium on Flow Visualization and Image Processing (2009)

[2] Y.Kato, The Cavitation (1969), pp.189/204

[3] K.Ogino et al., Journal of High Temperature Society, Vol.30, No.5 (2004), pp.270/278

[4] Japan Society of Corrosion Engineering, Hand book of Corrosion Engineering (2000), p.95

[5] K.Tani et al., Journal of High Temperature Society, Vol.35, No.6 (2009), pp.301/307

[6] M.Adachi et al., Journal of High Temperature Society, Vol.28, No.2 (2002), pp.73/77

[7] Y.Kato, The Cavitation (1969), pp.224/227

[8] ASTM Designation G32-10, Standard Test Method for Cavitation Erosion Using Vibratory Apparatus

[9] E.Lopes Cantera, B.G.Mellor, Materials Letters 37 (1998) pp.201/210

[10] S. De Palo et al., Thermal spray surface engineering applies (2000) pp.245/250

[11] S.Kuroda at al., Journal of High Temperature Society, Vol.36, No.6 (2010) pp.199/209

SPS COATING MICROSTRUCTURE CONTROLLED BY THE SURFACE TOPOGRAPHY USING LASER TEXTURING

R. Kromer[1*], P. Sokołowski[2, 3], R.T. Candidato[2], S. Costil[1], L. Pawłowski[2]

[1] *ICB-LERMPS UMR 6303, Univ. Bourgogne Franche-Comté, UTBM, F-90010 Belfort, France*

[2] *SPCTS, UMR CNRS 7315, University of Limoges, France*

[3] *Faculty of Mechanical Engineering, Wroclaw University of Science and Engineering, Wroclaw, Poland*

Contact: robin.kromer@gmail.com

Abstract

Thermal barrier coatings (TBC's) being produced at present either by atmospheric plasma spraying (APS) or electron beam physical vapor deposition (EB PVD) are widely used in the hot-temperature sections of turbines to provide thermal and corrosion protections. An emerging technology of suspension plasma spraying has become interesting for the manufacturing of thermal barrier coatings thanks to the useful microstructure including columns similarly to the EB-PVD deposits associated with considerably lower price of production. A narrow window of optimal suspension plasma spraying (SPS) parameters remain an outstanding problem in creating the favorable microstructure. The recent studies demonstrated that the substrate roughness may play an important role in reaching columnar growth of the coatings. This study presents a follow up by showing how the substrate topography obtained by laser surface texturing may be controlled to create regular columnar structure thanks to. The laser generated peaks disposed regularly on the surface can promote columnar structure growth. The formulated suspensions were sprayed onto superalloy substrates coated with powder plasma sprayed bond coats. Optimized previously, plasma spray parameters were selected to generate columnar structures and to find out the influence of the suspension behavior on coating microstructures. The results indicate that columnar SPS coating microstructure can be controlled by optimizing the laser treatment parameters. The control of surface topography may be an important factor to improve the performances of TBC-SPS coatings.

Keywords: suspension plasma spraying, thermal barrier coatings, laser surface texturing, coating formation, microstructure control

1 Introduction

The aircraft industry intends to increase the Thermal Barrier Coating (TBC) systems [1]. A key objective is to enhance the insulating power of TBC ceramic layers and its adhesion over thermomechanical stresses [2]. Until now, TBC's are mainly composed of Yttria-Stabilized Zirconia (YSZ) as thermally insulating coating and are commonly applied using the Electron Beam Physical Vapor Deposition (EB-PVD) process [3]. One of the most important features of EB-PVD is its ability to produce coatings with columnar microstructure that offers high thermal compliance leading to high lifetime of TBC's [4]. Moreover, the columnar structure results in an anisotropic heat conduction that lead to relatively high thermal conductivity of zirconia coatings (1.5 $W \cdot m^{-1} \cdot K^{-1}$). Unfortunately, the high cost of the EB-PVD process causes that the method is used mostly for coating the critical parts of gas turbines [5]. This drives also the application and development of other coatings and processing methods.

Atmospheric Plasma Spraying (APS) is also widely used to produce TBC's for aero engines, particularly for combustion chambers [6]. It involves feedstock powder particles injected into a plasma jet by a carrier gas and spread over the metal substrate as splats [7]. The lamellar microstructure resulting from the stacking of successive splats offers porous and highly insulating TBC's. In the same way, this specific microstructure is also more sensitive to thermal expansion mismatch at the topcoat/bondcoat interface affecting [8]. It was found already that the substrate surface treatment prior to the coating deposition enhances bond strength due to the anchoring mechanisms [9].

By using a liquid feedstock, resulting in e.g. Suspension Plasma Spraying (SPS) process, the injection of fine powders is possible. The deposition of submicrometric and nanometric coatings can be achieved. As described in literature [10], SPS allows also the production of microstructures by APS, so called columnar-like microstructures [11]. The most accepted coating build-up mechanism theories incriminates the smallest particles to explain the cone shape of SPS columnar structures [12]. Based on the observation of cauliflower column-type coatings, it is assumed that small satellite droplets, having a significant velocity component parallel to the substrate, impact side of asperities leading to the growth of conical defects [13]. VanEvery et al. have presented that for such small droplets or particles, their deposition trajectory would result in the formation of columnar structures on surface asperities. These columnar microstructures have been identified as promising for the enhancement of TBC's (prolonged life-time and limited thermal conductivity).

The present studies, in the research field, focus strongly on the production and analysis of ceramic topcoats having different microstructures, mainly columnar-like structures. In this work, in accordance to the mechanism of column formation, it is proposed that columns size, morphology and density (per area) is influenced strongly by the substrate surface topography. Therefore, a potential to control the structure of the SPS topcoat has been developed. Laser surface texturing has been used to adapt the surface, on which the coating is deposited. This process was found to be able to produce specific pattern arrangements. This work presents initial spraying of SPS coatings on stainless steel substrates characterized by various topographies. Then, the methodology of laser pre-treatment has been applied for APS bond coats in order to produce new kind of Thermal Barrier Coating systems. The coating microstructure evaluation has been studied depending on the surface preparation by laser to promote columnar structure and to create mechanical anchoring.

2 Methods and materials

2.1 Design of spray experiments

The samples deposition was divided into two distinct stages. The first one (Exp. 1) consisted of YSZ coatings production directly on the laser pre-treated 304L stainless-steel substrates. The aim was to observe the coating formation when different patterns of laser treatment were applied. Different suspensions, in terms of the solid content and the kind of solvent, were used. The second stage of coatings preparation (Exp. 2) included the preparation of complete TBC system. It consisted of SPS YSZ topcoat produced on MCrAlY bond coat manufactured using the APS method. Then, the bond coats were subjected to laser texturing prior to spraying of ceramic topcoat. The metallic interlayers were sand-blasted, grinded or raw (as-sprayed) when submitted to laser pre-treatment. The laser texturing patterns of bond-coats (Exp. 2) were selected based on the preliminary results (Exp. 1).

2.2 Laser surface pre-treatment

Laser surface texturing is a technology easy to implement in industry and can be adapted to different applications. The laser surface texturing (LST) was performed using a pulsed fiber laser device (Laseo, Ylia M20, Quantel, France) with wavelength 1.064 µm, pulse duration 100 ns, maximum power 20W and frequencies varying from 20 up to 100 kHz. The used laser beam had Gaussian energy distribution profile. Different substrate topographies were created using laser surface texturing (Fig.1). Hole and grove patterns were made on the surface by laser ablation and liquid matter ejection during the laser-matter interactions.

The laser pre-treatment was performed to create two kinds of patterns holes and groves. The variable parameters were: (i) hole diameter or groove width and (ii) distance between patterns. The laser texturing was carried out on two substrates:

- on stainless steel substrates, as detailed in the Table 1, five point patterns and six linear patterns depending on their features were made, and shown in Fig.1.

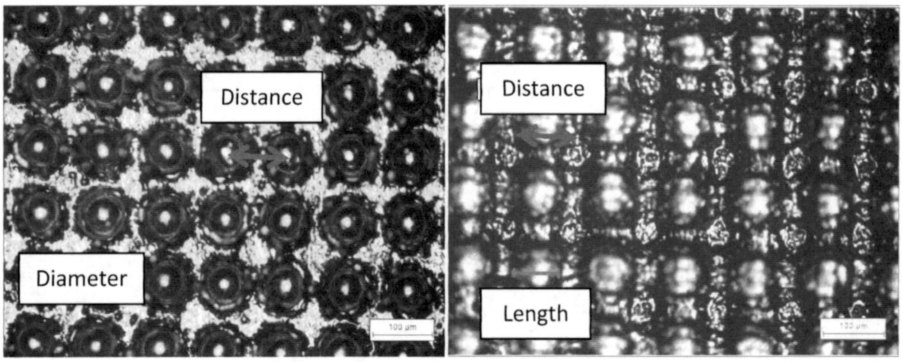

Fig. 1 Examples of laser texturing on stainless steel substrate

- on MCrAlY bond coats for TBC applications. As presented Table 1, several surface preparations have been considered due to laser-matter interaction, hence the metallic interlayers were sand-blasted, grinded or given in as-sprayed state for the laser pre-treatments. The most promising patterns were recreated on the MCrAlY coating selected basing on the preliminary tests. This way, specific patterns were made in each case (Fig. 2).

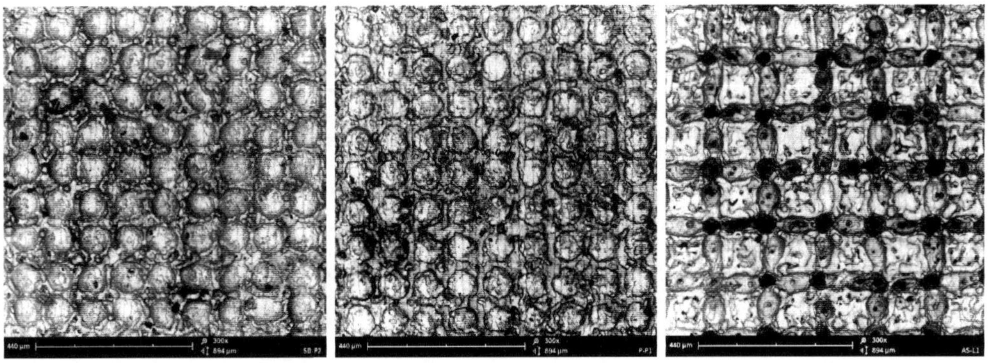

Fig. 2 Top view of bond coats after laser pre-treatment: (a) sand-blasted with point pattern, (b) grinded with point pattern and (c) as-sprayed with linear pattern

Table 1 The features of laser texturing in Exp. 1 (1-5 is point pattern and 6-11 is linear pattern)

	LST Run (LT)	Pattern	Diameter [µm]	Length [µm]	Depth [µm]	Distance [µm]
Exp. 1	1	Hole	55	-	35	100
	2		60	-	45	100
	3		60	-	35	100
	4		75	-	35	100
	5		65	-	50	100
	6	Groove	-	70	30	120
	7		-	60	30	120
	8		-	50	30	130
	9		-	60	30	100
	10		-	60	30	60
	11		-	60	30	130
Exp. 2	21	Hole	80	-	30	100
	22		72	-	30	100
	23		62	-	30	100
	24	Groove	-	60	30	120

2.3 *Spraying conditions*

The powder used for the suspension preparation and the deposition of the ceramic topcoats was 8 mol.% Y_2O_3 + ZrO_2 (TZ-8YS, Tosoh, Japan). This is a commercially available and sub-micrometer sized powder (d_{v50}=0.54 µm; Dynamic Light Scattering, Zetasizer Nano-ZS, Malvern Instruments, UK). The methodology of suspension preparation and stabilization was presented in our previous publications [14]. The metallic bond coats were produced by using NiCrAlY powder (Amdry 963, Oerlikon Metco, Switzerland) with a main powder particle size d_{v50}= 73 µm (Dynamic Light Scattering, Partica LA-950V2, Horiba, Japan). The features of suspension preparation were summarized in Table 2. The solid concentration was ranging from 5 to 20 wt.%. The pure ethanol as well as 50:50 water/ethanol based suspensions were tested. For each kind of substrate preparation four suspensions were tested in Exp. 1. The above means also that the Exp. 1 resulted in the deposition of 44 different samples, depending on feedstock preparation and substrate topography. Then, for Exp. 2, one suspension was selected for each LST pattern. The result of Exp. 2 was the deposition of 12 different TBC's with the spray process P1, P2, P3 and P4 for LT24, LT21, LT22 and LT23 respectively.

Table 2 Details of suspensions preparation used in Exp. 1 and Exp. 2

Suspension preparation (P)	Suspension characteristics		Exp. 1	Exp. 2
	Solid content [wt.%]	Solvent	LST	LST
1	20	water/ethanol	1 to 11	21
2	10	ethanol	1 to 11	22
3	10	water/ethanol	1 to 11	23
4	5	water/ethanol	1 to 11	24

The APS and SPS experiments were carried out using SG-100 plasma torch (Praxair S.T., Indianapolis, IN, USA) installed on the arm of 6-axis industrial robot (ABB, Zürich, Switzerland)). The feedstock material was transported by rotary powder feeder in case of APS and in-house made pneumatic injection system when spraying suspensions. The spraying conditions are collected in the Table 3.

Table 3 Details of APS and SPS parameters

	APS	SPS
Plasma gas composition	Ar+H_2	Ar+H_2
Working gas flow rate	45+5 slpm	45+ 5 slpm
Carrier gas	Argon (7 slpm)	-
Electric power	40 kW	40 kW
Stand-off distance	120 mm	40 mm
Scan speed	400 mm/s	500 mm/s
Feed rate	~25 g/min	~30 g/min
Type of injection	Outside of the torch (0.5 mm internal diameter of injector) radially	Inside the torch (0.3 mm internal diameter of injector) radially

2.4 Microstructure analysis

The microstructure of coatings was analyzed by using Scanning Electron Microscopy (SEM, Philips XL30, The Netherlands). For an observation of coatings' cross-sections, the samples were sectioned using a diamond cutting blade and mounted in low viscosity epoxy based resin using a vacuum impregnation technique. These samples were subsequently polished using a well-established method for TBC's specimens.

The topography of substrates submitted to laser texturing and of coatings was also investigated by SEM (Phenom G2 pro, Eindhoven, The Netherlands). In order to determine the dependences between suspension preparation, substrate topography and coating formation the quantitative analysis of coatings was carried out. The density of columns defined as: columns/mm^2 and the width of columns were calculated based on 3D views of coatings' topographies obtained by using Shape From Shading (SFS) technique. The methodology of topography analysis is shown in Fig. 3.

Fig. 3 Three step quantitative analysis of ceramic topcoats: (a) surface image capture, (b) counting of columns and (c) measuring the width of columns

3 Results

3.1 Hole patterns

In the hole patterns, the holes as well as plateaus can be distinguished. This influenced the formation of very different coating microstructures. Fig. 5a shows a columnar structure that was formed on the plateaus. The peaks were favorable to columnar structure. However, between the substrate peaks (inside the holes) the dense and lamellar structure can be seen. So, large irregularities inside the coating were hence observed. Fig. 5b shows a similar behavior but two peaks can be deduced on the plateau which induced two columns formation. Those peaks were spaced apart by 30 µm. The micro-irregularities enhanced the formation of columns.

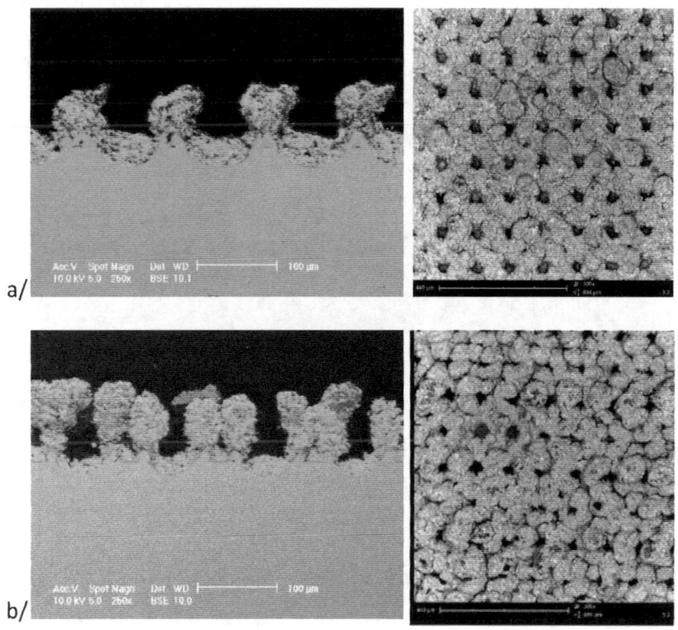

Fig. 5: The examples cross-section (left side) and surfaces (right side) of coatings deposited in Exp. 1 (hole pattern): (a) LT1P4 and (b) LT4P4

As presented, the hole pattern allowed the formation of columnar-like coatings in Exp. 1. The density of columns and their width were analyzed in order to select the pattern shape favorable to the growth-up of columns. The number of columns and width of the single column is presented in (Fig. 6). The density of patterns was equal to 100 per mm². The columns were directly link to this values. Each peak corresponds to one column, as seen on the microstructure analysis. However, one or two textured promoted two columns structures on two peaks but it is not deduced from image analysis. The column structures were joined on the top. Then, the column widths were equivalent for the different structures, but it can be deduced an effect of the process parameters. The suspension characteristics enhance columnar type structure and the hole geometry increased the phenomena. For instance, P3 process created columns link to the hole shape. Larger hole diameter involved smaller plateaus so smaller columns. However, the number of columns does not influence the irregularities.

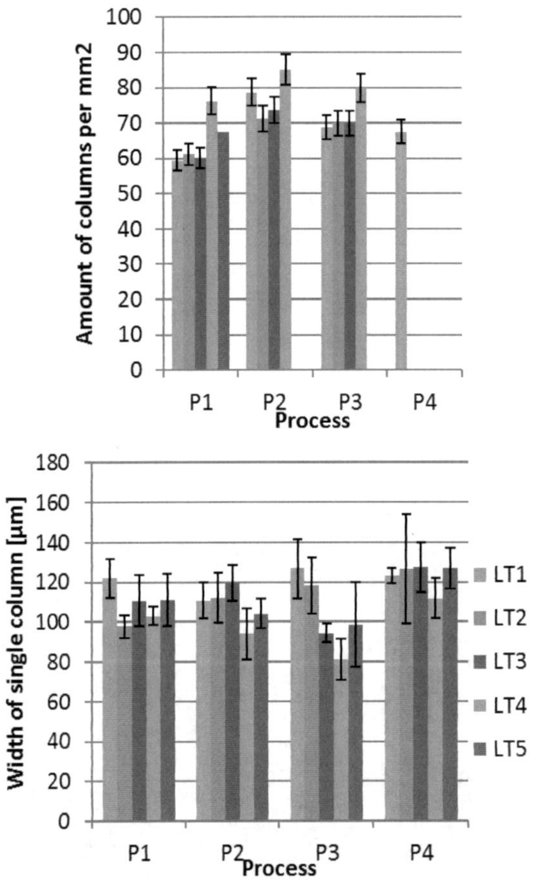

Fig. 6: Number of columns per 0.8mm² and width of single columns for hole patterned stainless steel substrate for the four coating processes

The irregularities can be observed Fig. 7. The columns were built on peaks and leave space between them as large as hole diameter. The plateau are the key induced columns structures. The hole diameter had to be equal to 40 µm and to be spaced to 60 µm to induced columns structure without irregularities. The laser treatments had to be remade with another lens to create those topographies.

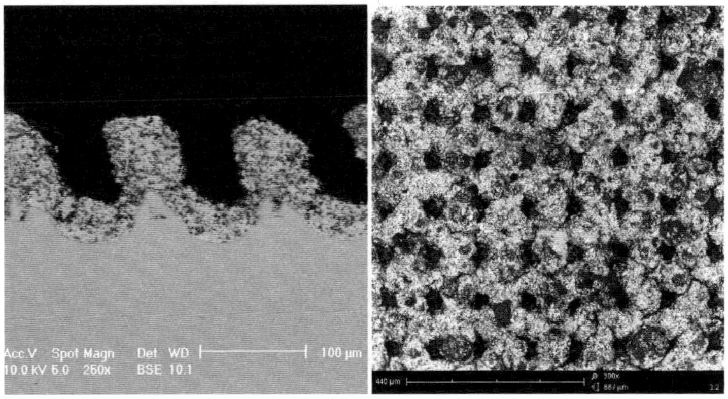

Fig. 7: The best coating basing on the quantitative analysis (LT4P2)

3.2 Groove patterns

The microstructure of coatings deposited onto the substrates with groove LST patterns have been also analyzed (Fig. 8). The different density and width of columns have been observed depending on the pattern features and spray process parameters. First, the process P4 worked on the different topographies for columnar coatings. The other permitted only lamellar structures for LT8, LT10 and LT11. Process parameters involved consequently different columnar structures.

a/ b/

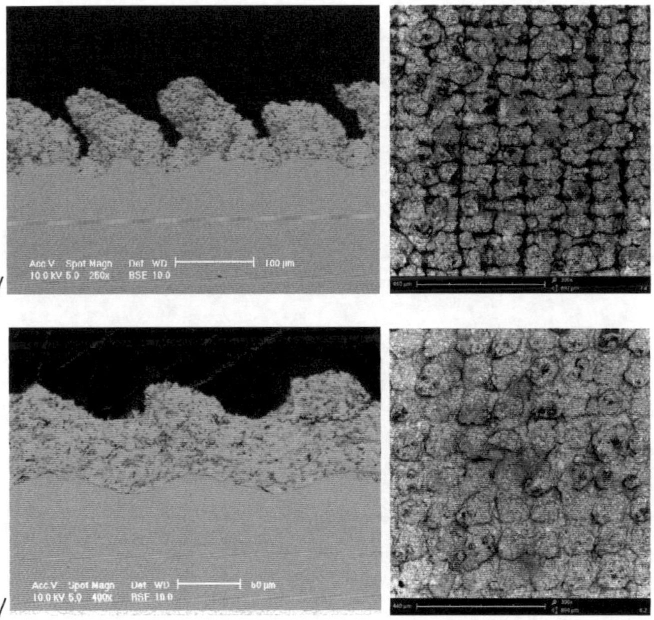

Fig. 8 The examples cross-section (left side) and surfaces (right side) of coatings deposited in Exp. 1 (groove pattern): (a) LT7P4 and (b) LT6P1

Then, the pattern densities had a major influence. The columns built up on peaks on plateaus (as observed previously). The small distances between grooves (LT10) enabled a great number of columns per mm^2 with small columns to be formed. Besides, the process parameter influences have been deduced between P3 and P4. The columns joined or did not respectively corresponding to the coating growth mechanisms. Besides, for a same groove density, columnar grown on the plateaus for P1, P2 and P3. The differences were link to the column growth mechanism types. However, the P4 enabled smaller columns and enabled a higher columns number for the same pattern densities with larger plateau (LT9).

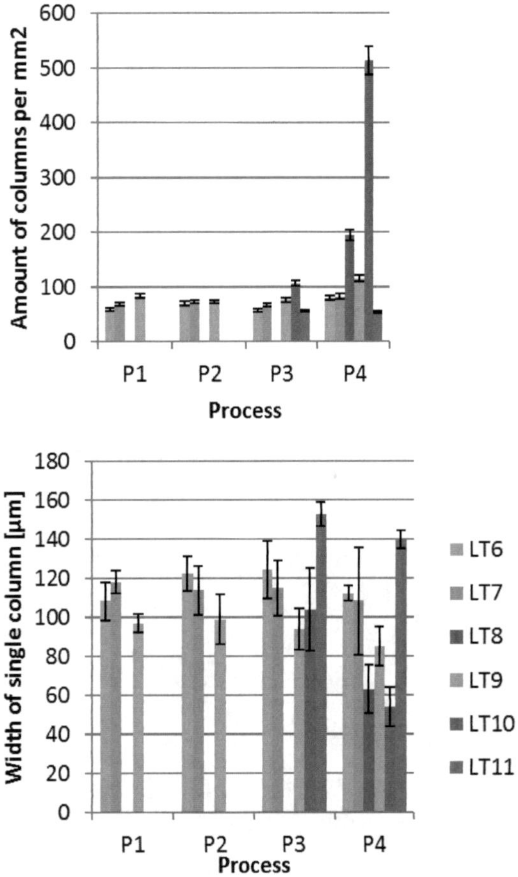

Fig. 9: Amount of column per mm² and width of single columns for groove patterned stainless steel substrate for the four coating processes

As described previously, the much more uniform and dense columnar structures were observed in case of substrates with grooves patterns (Fig. 10). But still some irregularities can be distinguished on the coating free surface. In addition, large pores can also be deduced above grooves. It was link to the columnar growth mechanisms. The grooves need to be smaller in depth. This peaks density seems to correspond to the best configuration.

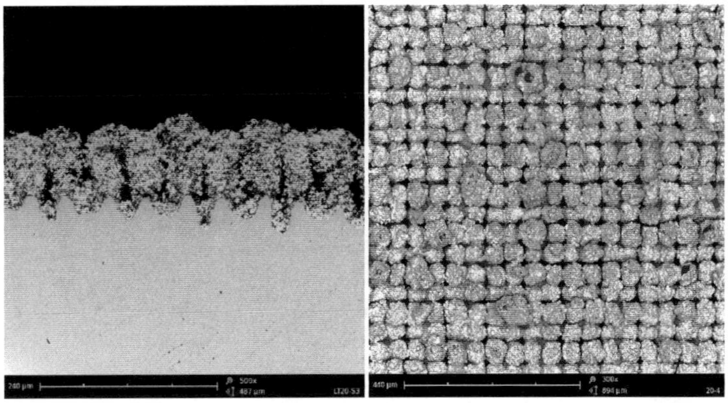

Fig. 10: The best coating basing on the quantitative analysis (LT10P4)

3.3 *Full TBC with the textured bond-coat*

Laser surface texturing has been applied on MCrAlY bond coat for TBC applications. The bond coat was textured in as-sprayed, sand-blasted and polished state. As expected, laser matter interactions can be changed with the roughness so the laser parameters were changed to provide similar patterns. The objective was also to see the influence of the previous roughness on the coating buildup even a likely structured topography. First, the columns analysis has been performed (Fig. 11).

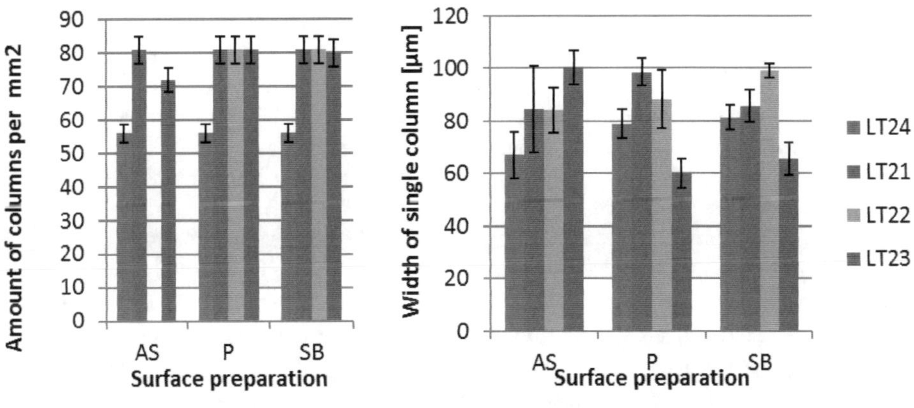

Fig. 11: Amount of column per mm² and width of single columns for groove patterned bond coat (with three previous topographies: as sprayed, sand-blasted and polished)

First, the number of columns was similar with the previous experiment 1 study. Also the top-view structures were likely, as seen Fig. 12. The micrometric structured surface enabled columnar structures for TBC applications. Optimization had to be done. Besides, the roughness and preparation of bond coat prior to laser texturing seems to does not induce different columnar growth mechanisms. The pattern morphologies were in the range of 50 µm with peaks of 10 µm. However, the as-sprayed or sand-blasted surface roughness created peaks of 2-3 µm. The plasma flow was deviated by the patterns enhancing column growth on the plateaus. In addition, the column density and width were nearly equal to the Exp. 1 results. Optimization strategy can then be deduced as presented earlier. Specific pattern density and shape had to be correlated with the columnar growth rate. Further investigation with numerical modeling could be performed to evaluate the effects of the pattern shape on plasma flow and particle deposition.

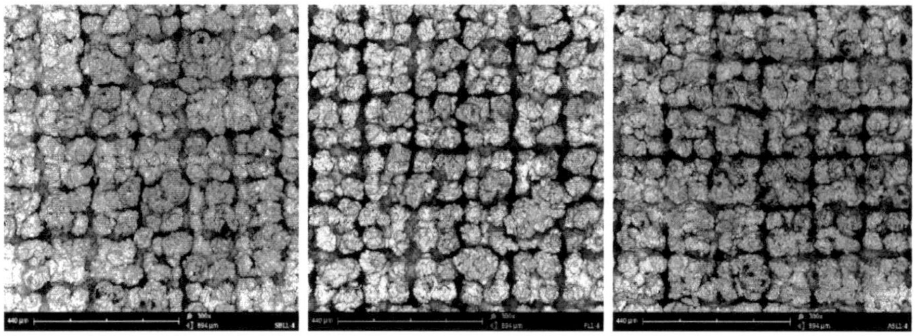

Fig. 12: The top view of best TBC's basing on the quantitative analysis (SBL1 = PL1 = ASL1 in terms of columns density; starting from the left)

4 Discussions

4.1 Laser texturing

The patterns were created by vaporization and matter ejection phenomena. The recast material has been induced by the recoil pressure due to plasma formation in the patterns (Fig. 4). The calculations of the energy balance, melt surface temperature, keyhole/cutting front propagation velocity and melt-ejection velocity could be performed using the physical model earlier proposed by Semak et al. [SEMAK]. The results would show that:

- the recoil pressure can play a significant role in ejection of the melt from the interaction zone even for low melt surface temperatures close to the melting point;
- high-velocity melt flow is generated in the interaction zone for beam intensities typically used in laser cutting/welding.

The matter ejection can be controlled depending on the material properties and the laser intensity. The pressure can be significant or low to eject the liquid matter or to create a ring, respectively. The coating structure could hence be optimized. This behavior is true for hole patterns, but the laser texturing strategy play also a role for groove patterns. The pulses induced matter ejection on a specific zone. Scan speed induced an over-lap of this zone which create a laser-matter interaction on the previous melted matter. A linear groove creates a linear recast matter on the edges. Then, perpendicular groves have the same texturation mechanisms. On each crosses, four peaks can be created on the plateaus. Those peaks can also be deduced by laser intensity and scan speeds for a specific material, as such columnar structures.

It was observed that one sample consists often (but not always) of two different kinds of morphologies (Fig. 5). This was specific for hole patterns when too big spaces between neighboring holes occurred but similar behavior could be deduced for grooves. The objective is to provide cavities to obtain tangential plasma flow for columnar buildup. The recast material generated peaks which enables those structures. An optimization of the laser strategy is then required create specific pattern morphology to provide significant peaks around. So the patterns need to be likely the column sizes, so it depends on the process parameters.

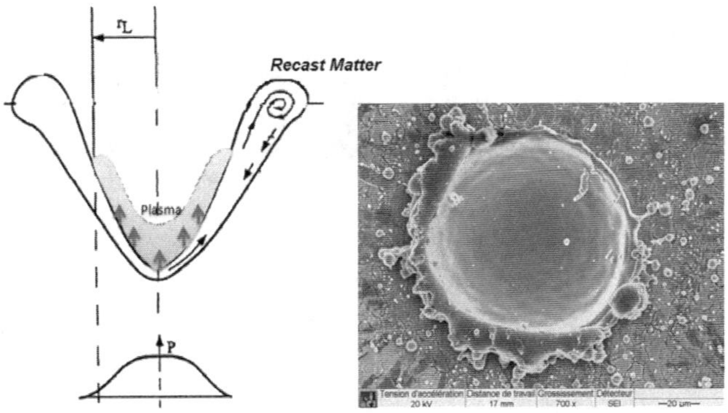

Fig. 4: The scheme of the laser-matter interaction creating the pattern (a) and the top-view of textured 304L stainless steel substrate (hole pattern) (b)

4.2 Coating build-up influenced by the substrate preparation

Suspension Plasma Spraying is a technology that uses very fine feedstocks. The surface topography should be designed in reference to the size of powder particles. Three type of spray deposition can be defined (REF VanEvery):

- Type 1; YSZ droplets move with the plasma. The parallel velocity component of these droplets dominates the normal component. However, as the plasma flows across the substrate, the inertia difference between the plasma and small YSZ droplets cause some of the entrained droplets to be unable to follow the more sudden directional changes produced as the plasma moves around surface asperities. Type 1 results shows droplet impacts on the sides of surface asperities, as shown in Fig. 13a. Initially, deposits formed on substrate asperities row both laterally and vertically. During this time, the lateral growth of taller deposits may engulf shorter deposits.
- Type 2; spray deposition is proposed as the droplet retains a velocity directed primarily perpendicular to the substrate, yet the substrate parallel component still affects deposition. In this case, the droplets are more massive than those depositing according to the type 1 mechanism, and they separate from the plasma before completing the substrate impingement turn. Consequently, deposition in regions between surface asperities increases with 2SD. However, the effects of the impinging plasma drag cause the droplets to follow trajectories that result in surface asperities shadowing portions of the substrate downstream from the asperity, which is illustrated in Fig. 13b. Early in the coating process, the shadowing occurring with 2SD will create variations in the growth rate.
- Type 3; spray deposition is proposed in which plasma drag forces do not affect the deposition characteristics, meaning the droplets are more massive. As a result, deposition is characterized by lamellar structure on the entire substrate surface (Fig. 13c) during each plasma gun pass, i.e., droplets do not preferentially impact on surface asperities, and no shadowing occurs from these asperities.

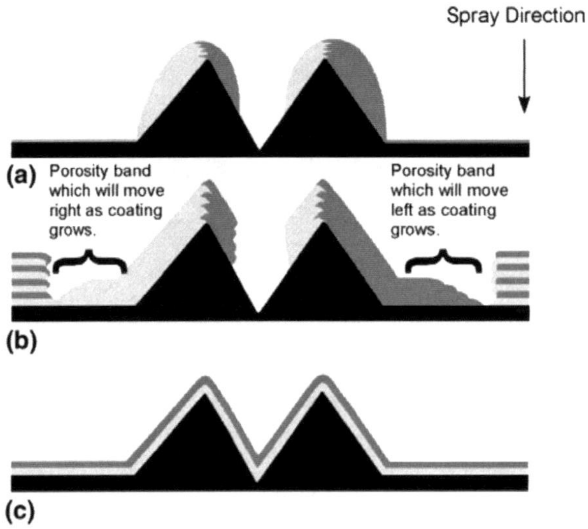

Fig. 13: The schematic representation of the deposition characteristics occurring on and away from substrate surface asperities during a) type 1, (b) type 2, and (c) type 3. The spray direction is the same in each figure. The black regions denote the substrate. The light ray regions represent material deposited by droplets having a substrate parallel velocity component directed from left to right, and the dark gray regions represent material deposited by droplets having a substrate parallel velocity component directed from right to left (REF VanEvery)

In this study it was observed that the laser treatment was to "rough" comparing to the size of powder particles. Fig. 5 shows that columns were formed on very little irregularities of the substrate. The columnar structures were the objective to decrease thermal conductivity and enhance life-span for TBC applications. The conductivity would be link to the extern temperature and the columnar structures would accommodate the thermal stresses. The patterns and spray parameters need to be in accordance to growth such structure. The spray Exp. 1 allowed obtaining coatings with very different microstructures. The patterns enabled flow to go in the pattern. As proposed, a type 2 structure can be assumed for the different spray process. Particle deposition were in and around the patterns.

It was also noted that one sample consists often (but not always) of two different kinds of morphology (Fig. 14). This was observed mainly for hole patterns when too big spaces between neighboring holes occurred. The laser strategy need also be optimize to obtain length between holes minimize (near 60 µm). The squared matrix strategy need to be changed to obtain regular distances for a specific pattern density.

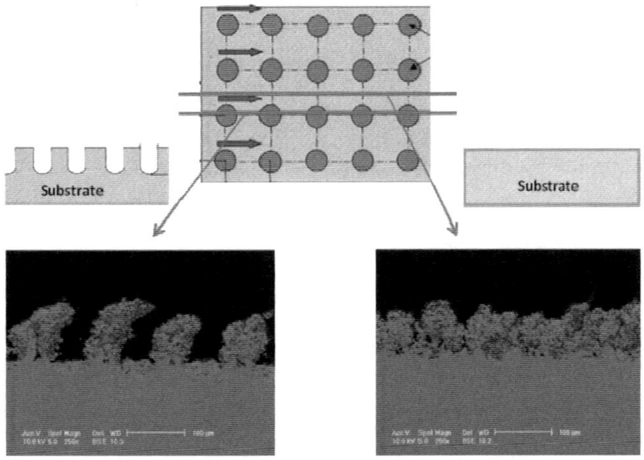

Fig. 14: The dual nature of coatings (LT1S4)

5 Conclusion

The laser surface texturing (LST) was tested as a new method of substrate preparation suitable for Thermal Barrier Coatings (TBC's) deposition by Suspension Plasma Spraying (SPS). The LST technology was tested first for zirconia coating on stainless steel substrate. Different substrate topographies were created by the laser pre-treatment, namely hole and groove patterns. The patterns were characterized by various hole diameters or groove widths and distances between patterns. The analysis of coatings showed that different coating microstructures were obtained depending on the substrate and suspension preparation. When the LST pre-treatment was too rough then the coatings were characterized by dense and lamellar microstructures. Smooth and very dense patterns on the substrate surface allowed obtaining of columnar-like coatings. Basing on the quantitative analysis of the coating free-surfaces, the highest columns density (columns per area) was greater than $500/mm^2$.

Afterwards the trials to prepare full TBC systems were carried out (substrate + MCrAlY bond coat + zirconia top coat). After APS spraying of metallic interlayers, they were subjected to the LST pretreatment. The laser treatments were made on as-sprayed, grinded or grit-blasted state. It was found that all coatings were suitable for LST and all bond coats were successfully treated prior to the top coat spraying. The three layer systems were much more difficult to be controlled. But the columnar-like coatings were also achieved. The highest density of columns was greater than $150/mm^2$ and similar to the density analysis evaluated on stainless-steel with the same patterns.

Basing on this study it can be said that laser surface texturing can be an interesting technology for producing of coatings with designed microstructure and controlled topography. This kind of substrate preparation can be useful for the manufacturing of Thermal Barrier Coatings and many other kinds of thermally sprayed coatings as well. The LST technology should be still optimized in a way to find best substrate topography by considering the plasma flow, particle deposition and coating buildup. The further work will be focused on further LST experiments as well as testing the performance (isothermal oxidation, thermal shock resistance) of TBC's deposited on laser pre-treated bond-coats.

6 Acknowledgements

The authors gratefully acknowledge the ANR for financial assistance in the ARCOLE (12-BS09-0009) project. The work of Pawel Sokolowski was financed by the grant 2014/15/N/ST8/02660 financed by the National Science Centre (Poland). We are also grateful to Ms. Monika Michalak for her help with the microscopic analysis of coatings.

References

[1] R. Eriksson, "Thermal Barrier Coatings: Durability Assessment and Life Prediction," 2013.

[2] M. Bartsch, B. Baufeld, S. Dalkilic, L. Chernova, and M. Heinzelmann, "Fatigue cracks in a thermal barrier coating system on a superalloy in multiaxial thermomechanical testing," *Int. J. Fatigue*, vol. 30, no. 2, pp. 211–218, Feb. 2008.

[3] C. Courcier, V. Maurel, L. Rémy, S. Quilici, I. Rouzou, and A. Phelippeau, "Interfacial damage based life model for EB-PVD thermal barrier coating," *Surf. Coat. Technol.*, vol. 205, no. 13–14, pp. 3763–3773, Mar. 2011.

[4] P. G. Klemens and M. Gell, "Thermal conductivity of thermal barrier coatings," *Mater. Sci. Eng. A*, vol. 245, no. 2, pp. 143–149, 1998.

[5] A. Cahit, K. Ogawa, A. Turk, and I. Ozdemir, "Thermal Shock and Cycling Behavior of Thermal Barrier Coatings (TBCs) Used in Gas Turbines," in *Progress in Gas Turbine Performance*, E. Benini, Ed. InTech, 2013.

[6] N. Goubot, J. Dehrs, and F. Pierrelatte, "Comparison of TBC coatings obtained by conventional and high velocity aps," *DVS-ASM*, pp. 556–560, 1999.

[7] R. Vassen, J. Dorning, and M. Dietrich, "Article - R.Vassen - Recent development in the field of aps tbc.pdf," *Int. Gas Turbine Congr.*, no. 129, pp. 1–9, 2003.

[8] W. R. Chen, X. Wu, and D. Dudzinski, "Influence of Thermal Cycle Frequency on the TGO Growth and Cracking Behaviors of an APS-TBC," *J. Therm. Spray Technol.*, vol. 21, no. 6, pp. 1294–1299, Oct. 2012.

[9] R. Kromer *et al.*, "Laser Surface Patterning to enhance adhesion of Plasma Sprayed Coatings Surface and Coatings Technology."

[10] O. Tingaud *et al.*, "Suspension plasma spraying of zirconia coatings: process and coating structure," in *18th International Symposium on Plasma Chemistry*, 2007, p. 4–pages.

[11] H. Kassner, R. Siegert, D. Hathiramani, R. Vassen, and D. Stoever, "Application of Suspension Plasma Spraying (SPS) for Manufacture of Ceramic Coatings," *J. Therm. Spray Technol.*, vol. 17, no. 1, pp. 115–123, Mar. 2008.

[12] P. Sokołowski, L. Pawłowski, D. Dietrich, T. Lampke, and D. Jech, "Advanced Microscopic Study of Suspension Plasma-Sprayed Zirconia Coatings with Different Microstructures," *J. Therm. Spray Technol.*, vol. 25, no. 1–2, pp. 94–104, Jan. 2016.

[13] N. Curry, K. VanEvery, T. Snyder, J. Susnjar, and S. Bjorklund, "Performance Testing of Suspension Plasma Sprayed Thermal Barrier Coatings Produced with Varied Suspension Parameters," *Coatings*, vol. 5, no. 3, pp. 338–356, Jul. 2015.

[14] P. Sokołowski, S. Kozerski, L. Pawłowski, and A. Ambroziak, "The key process parameters influencing formation of columnar microstructure in suspension plasma sprayed zirconia coatings," *Surf. Coat. Technol.*, vol. 260, pp. 97–106, Dec. 2014.

Characteristics of Dense Al_2O_3 Coating Prepared by Suspension Plasma Spraying

H. Ibe, T. Masuda, K. Sato, N. Kato, Fujimi Incorporated, Gifu / Japan

Sintering ceramics have been widely used in industries which require electrical and mechanical properties. Thermal sprayed ceramics coatings are also applied for the industries, however the coating which has micron size pores are limited their applications due to inferior electrical and mechanical properties compared with sintering bulk. To expand thermal sprayed ceramics coating applications, dense coatings prepared by suspension plasma spraying are widely studied. Dense Al_2O_3 coatings are applicable to fabricating equipment for electronics devices, such as ESC. There are no reports regarding electric properties of plasma sprayed dense Al_2O_3 coating with different spray conditions. In this study to achieve a electric properties of dense Al_2O_3 coating, spray parameters such as plasma power, gas flow rate and spray distance are investigated. Suspension materials prepared with three microns Al_2O_3 powder are sprayed by high power suspension plasma spraying system. Spray conditions, plasma power, gas flow rate and stand off affect the coating density, christal phase, mechanical and electric properties. Mechanism of coating formation by plasma spraying with fine powder suspensions will be discussed based on the findings. Al_2O_3 coatings obtained by the plasma spraying is applied for application to application utilizing the electrical insulation properties of such electronics devise manufacturing equipment components is proceeding.

1 Introduction

Ceramic materials including bulk and various coating technologies are used in various applications according to the properties of the constituent ceramics. Aluminiunm oxide (Al_2O_3) are used as protective coatings for various industries because of its exhibits abrasion resistance, corrosion resistance, lower heat conduction and high electric insulation.

In these years, semiconductor device or liquid crystal panel production process require high electric insulation apparatuses for higher productivities.

Al_2O_3 are known as high technical value material and used in applications which inquire high breakdown voltage and electric insulation. Al_2O_3 sintered-bulk shows almost intrinsic properties of materials themselves which is better than ceramic coatings by atomospheric plasma spraying (APS). For example, APS Al_2O_3 coatings are contained some defects than bulk and have lower hardness of around1000HV compared with bulk of around 1800HV [1].

There are many techniques to obtain Al_2O_3 coatings such as AD, PVD, CVD for dense coatings which have similar density to the sintered-bulk. However, these coating technologies are limited in applications because of their applicable condition, for exalmle, size limit of substrate or thickness upper limit.

On the other hand, thermal spraying has the advantage for its applicable range and thickness compared to AD, CVD, PVD coating technologies. This advantage results ceramic thermal spraying coating has been increasingly applied in the field of semiconductor and displaymanufacturing equipment.

To meet the increasing demands of high breakdown voltage, increasing coating thickness is an effective approach. In this regard, thermal spraying is advantageous as a method for the formation of an Al_2O_3 coating having excellent breakdown voltage because of higher thickness coating is more easily achieved [2]. However a traditional thermally sprayed coating which contains a relatively high amount of porosity is inferior to bulk of Al_2O_3 in electric strength.

Thermal spraying process using with fine particles is effective to deposite a dense Al_2O_3 coating. On other hands the crystal structure of the Al_2O_3 coating with fine particles include γ phases which affect the breakdown voltage characteristics [3]. It was reported that the suspension plasma spraying(SPS) method with higher ratio of α phase in the coating is being developed for a dense thermal spray coating[4]. It is considered that thermal sprayed coating characteristics are decided by material size and spray conditions as a factor which has great influence for the structure.

In this study, Al_2O_3 coatings by SPS have been prepared with several pattern of the plasma power, gas flow rate and spray distance. Microstructure, hardness, crystal structure and electric properties of the coatings was investigated.

Influence of coating density and crystal structure for dielectric strength and electrical resistivity was studied.

2 Experimental

<u>Al_2O_3 Suspension</u>

In this study, DTS-SA28-W30 which was Al_2O_3 suspension product manufactured by FUJIMI INCORPORATED, JAPAN was sprayed for inrestigation. Fused and crushed Al_2O_3 powder with the α-Al_2O_3 (purity >99.8%) wad used as the raw material. Scanning electron microscopic (SEM) images of the powders are shown in Fig. 1(a) .

The particle size of Al_2O_3 powder which was classified precisely seemed a uniform daimeter.

Purified water was used as solvent and mixed with 30wt% concentration Al_2O_3 powder and dispersant additives by stirring mixer.

The particle size distribution in suspensions ranged from 0.8 to 7.0μm (from $Dv_{3\%}$ to $Dv_{97\%}$) and the average particle size ($Dv_{50\%}$) was about 3.2μm (Fig. 1b).

The inspection values of Al_2O_3 suspension such as specific gravity, pH and viscosity around 1.5 mPas were around 1.3, around 4.6 and around 1.5mPa·s, respctively.

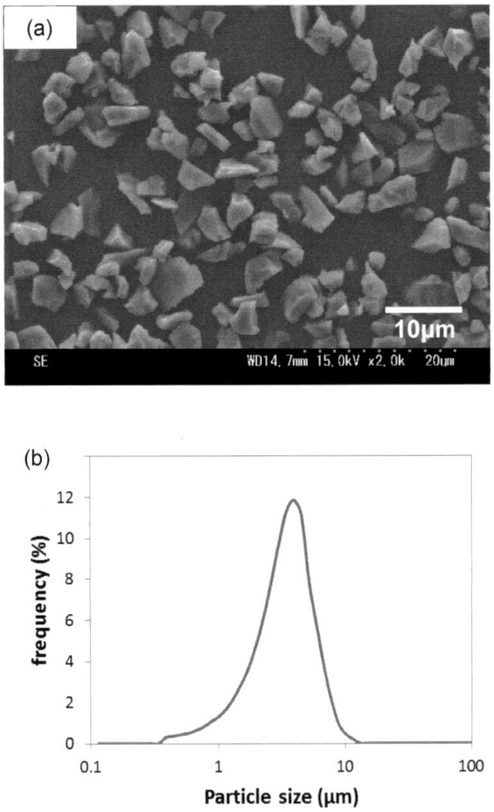

Fig.1. (a) SEM images of Al2O3 alumina powders (b) Cumulative volume distribution of the Al_2O_3 particle in suspension

Plasma Spraying
High power plasma spraying system, 100HE (Progressive Surface) were used for coating preparation. Spray conditions of plasma power, gas flow rate and spray distance (SD) was investigated respectively. Other spray parameters traverse speed, pitch and others were fixed.
Gas flow rate was changed by total gas flow rate and same ratio of three plasma gases, Ar, N2 and H2 were applied. Spray conditions are summarized on Table 1.
Deposition rate was calculated from the coating thickness after 50 passes spraying
Aluminum alloy (A6061) substrate, 50 × 70 × 10t mm, were sand blasted by #220 alumina grit before plasma spraying.
Base of Al_2O_3 suspension and spray conditions were decided by considering previous study of Y_2O_3 SPS[4].

Table 1: Spray Condition

Spray gun	100HE (Progressice Surface)	
Plasma Power	(kW)	90~105
Total Gas Flow Rate	(slm)	240~320
$Ar_2 : N_2 : H_2$ ratio	(-)	56 : 22 : 22
Slurry feed rate	(ml/min)	38
Stand off	(mm)	70~105
Traverse speed	(mm/sec)	1500
Pitch	(mm)	4
Number of pass	(-)	50

Coating Evaluations
Microstructural investigation of the coatings was analysed by cross-sectional SEM. Coating hardness was evaluated by micro Vickers hardness tester. Crystal structure of the coatings and powders were analysed by X-ray diffraction (XRD).
Dielectric strength was measured at room temperature by TOS9201 (KIKUSUI, Japan) based on JIS C2141. An electrode with 25 mm in diameter was placed on the specimen followed by applying direct current (DC) voltage with a ramp rate of 100 V/s (max. 5 kV) in the air until breakdown. Dielectric strength was calculated by dividing breakdown voltage by coating thickness.

3 Results and discussion

Influence of plasma power and spray distance
The results of coating hardness, dielectric strength and depsitionrate on several conditions of plasma power and spray distance are shown in Fig.2~4. In these results, gas flow is fixed with 320slm.
Deposition rate tends to be high when plasma power is high. In adittion, deposition rate of 105kW average was close to three times higher than that of 90kW.
SD for depositon rate is less than 1um/pass in each plasma power conditions. Compared with plasma power efficiency, the effect of SD was much small.
Hardness result shows similar tendency with deposition rate. It is considered that higher plasma power, in other words, a lot of heat quantity is effective for particle melting and hardness.
The best result of dielectric strength was obtained when plasma power and SD were 95kW and 75mm, respectively. The result of 35kV/mm was superior than conventional Al2O3 coating [3].On same spray distance, the lowest result was plasma power of 105kW. In this study, both of high mechanical and high erectric properties were not achived.

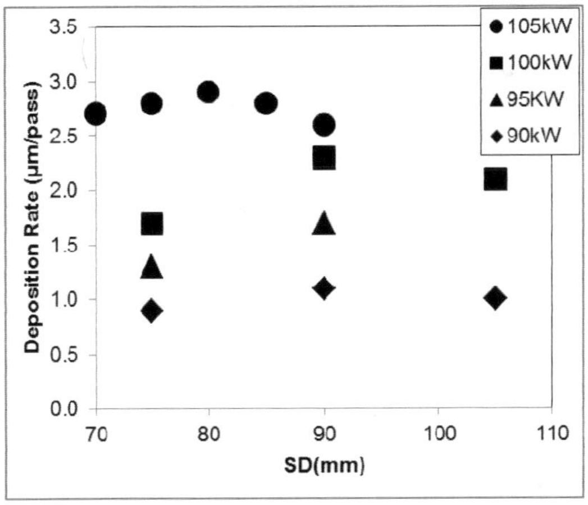

Fig.2. Deposition Rate as a function of spray distance (SD) at gas flow rate of 320slm

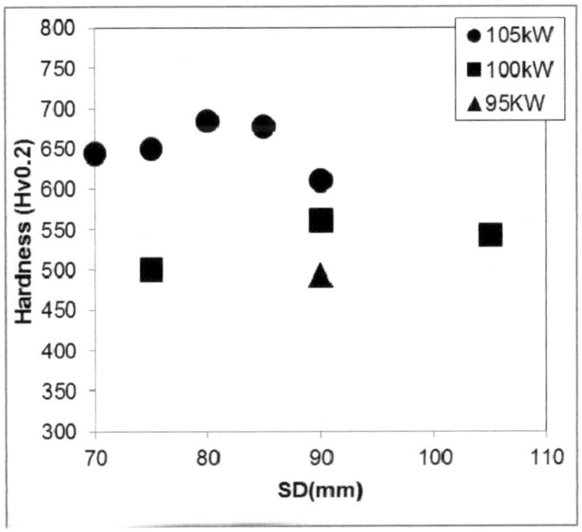

Fig.3. Coating hardness as a function of spray distance (SD) at gas flow rate of 320slm

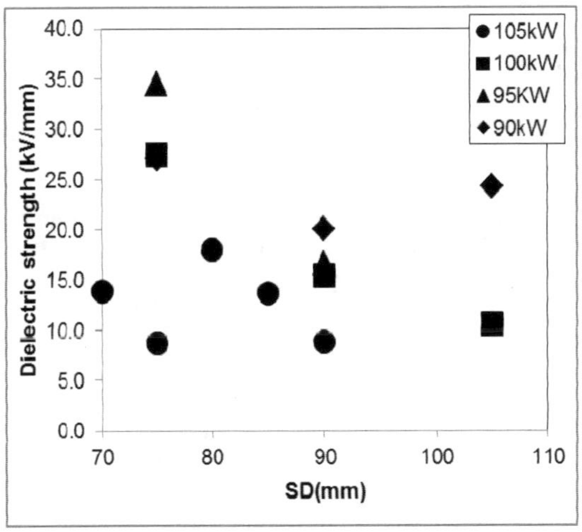

Fig.4. Dielectric strength as a function of spray distance (SD) at gas flow rate of 320slm

Influence of gas flow rate and spray distance

Results of deposition rate, hardness and dielectric strength on several gas flow rate and spray distance are shown in Fig.5~7. In these results, plasma power is fixed with 105kW. Influence of gas flow rate for deposition and coating property was similar to the case of study of plasma power. High flow rate leads high deposition rate, high hardness and low dielectric strength.

It seems that infligtht particle tempertuer of high flow rate and high power condition.

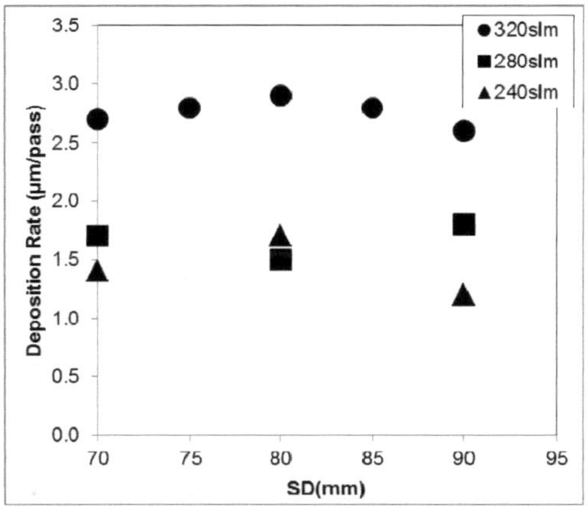

Fig.5. Deposition Rate as a function of spray distance (SD) at plasma power of 105kW

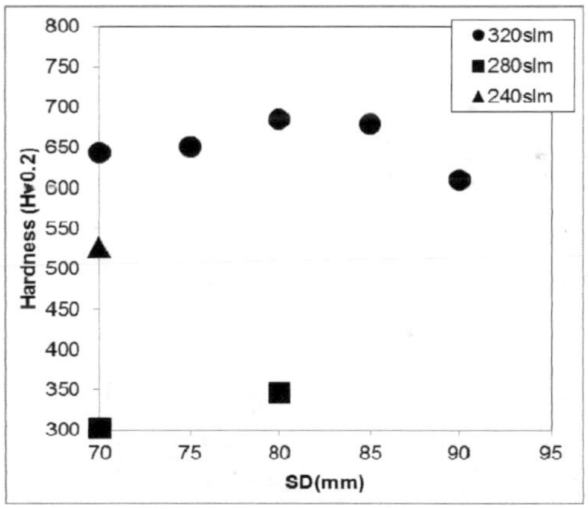

Fig.6. Coating hardness as a function of spray distance (SD) at plasma power of 105kW

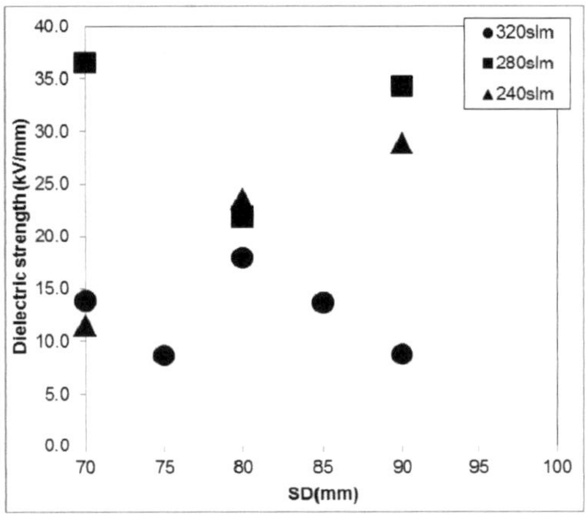

Fig.7. Dielectric strength as a function of spray distance (SD) at plasma power of 105kW

Microstructures

In this studies, cross sectional images of several coatings were observed. The structures of coatings were much different when plasma power or gas flow rate was changed. Cross-sectional SEM images with different plasma power are shown in Fig.8.

Lower plasma power coating tends to be higher porosity. Coarse pores and defects on particle interface were observed at 90kW which is lowest plasma power in this study. High plasma power reduced both of size and number of micro pores and defects on particle interface.

This result is corresponding to the low deposition rate and low hardness with low plasma power which is seemd to be insufficient to melt particles.

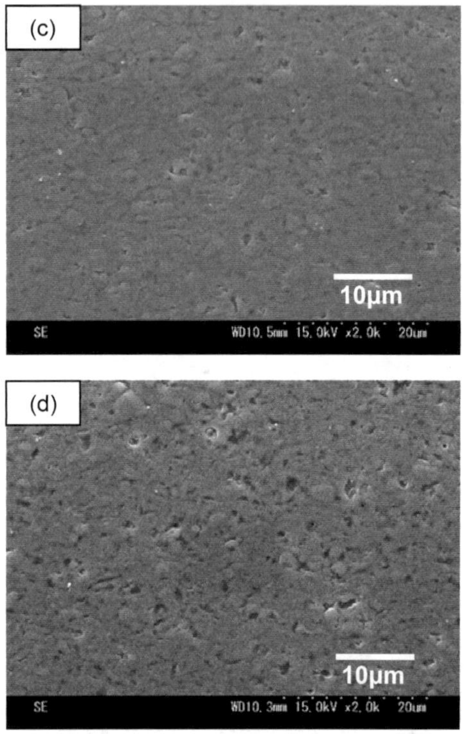

Fig.8. Cross-sectional SEM images of Al_2O_3 coatings at SD 90mm, gas flow rate 320slm:
(a) 105kW, (b) 100kW, (c) 95kW, (d) 90kW

Crystal structure

XRD patterns of coatings which recorded the highest and lowest dielectric strength are shown in Fig.9.

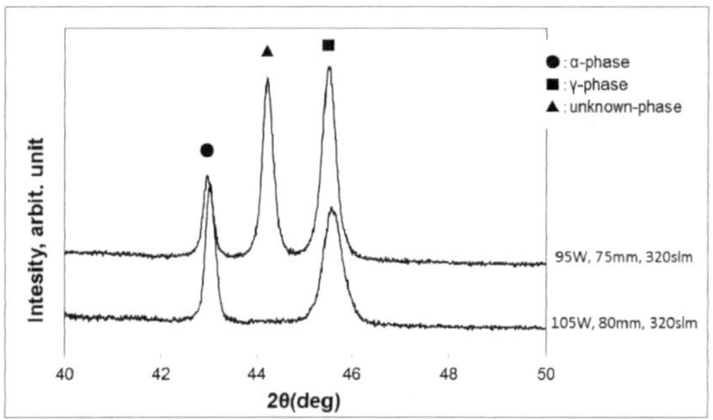

Fig.9. XRD patterns of the highest dielectric strength coating (at 95kW, 75mm, 320slm) and the lowest dielectric strength coating (at 105kW, 80mm, 320slm)

The coating which obtained the highest dielectric strength coating (95kW, 75mm, 320slm) is considered to contain three types of crystal phase which were the metastable γ-Al_2O_3 phase (JCPDS card no. 10-0425) as the main phase, the satable α-Al_2O_3 (JCPDS card no. 46-1212) phase, and the unknown phase. There is no peak pattern in JCPDS card to fit peak between α-phase and γ-phase. It is expected that the peak is Al of substrate or θ-phase Al_2O_3[5].

In the lowest dielectric strength coating (105kW, 80mm, 320slm) showed two types, the α-phase was the main crystalline phase.

Generally, ratio of α-phase in Al_2O_3 sprayed coating is important factor for coating property. Peak ratio of α-Al2O3 phase in the lowest and the highest dielectric strength coatings were 53% and 23%, respectively. This tendency is considered to be remarkable compared with previous study.[2]

Sprayed material was consist of pure α-phase. γ-phase, metastable phase were appeared due to rapid cooling process of liquid phase A_2O_3 onto substrate. APS coating includes higher ratio of γ-phase in the coating [3].

Mechanism of unknown phase appearance is under investigation. However heating and cooling process would be key parameter for the mechanism.

Comparison between dielectric strength and unknown phase peak ratio which is calculated by XRD pattern of each coatings are shown in Fig.10.

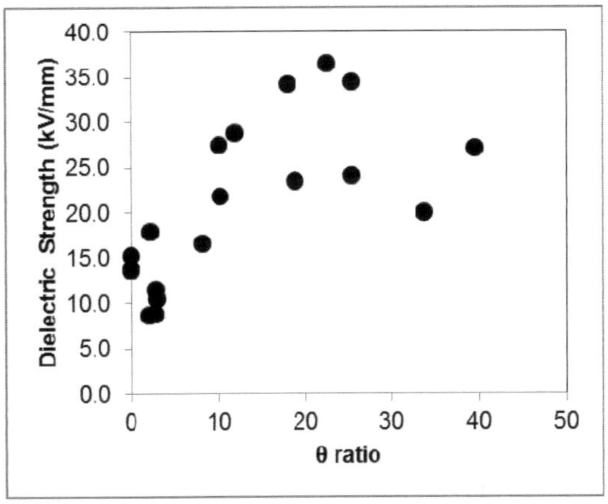

Fig.10. Dielectric strength as a function of θ phase ratio

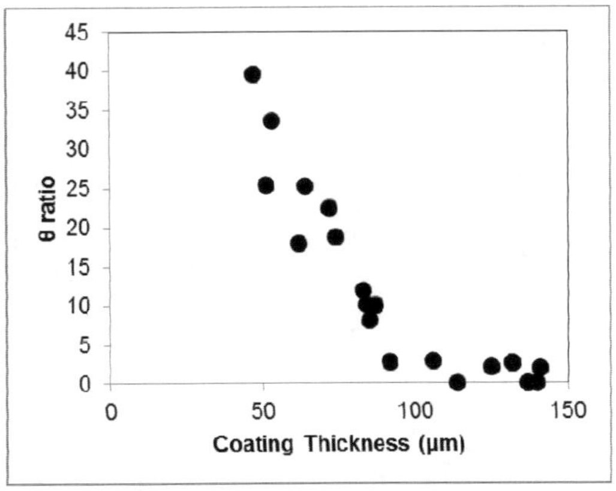

Fig.11. Dielectric strength as a function of coating thickness

Comparison between unknown peak ratio and coating thickness are shown in Fig.11. Higher unknown phase ratio coating were shown thinner coating. And thicker coating showed lower ratio of unknown phase.

From these tendencies, it is expected that the total thermal energy or thermal insulation between substrate and top of the coating during the spraying related with θ phase appearance.

It is considered that the dielectric strength is influenced by unknown phase from this results.

For improving the valu of dielecteric strength, it is considered that it is important to elucidate unexplained mechanism of appearing unknown phase.

4 Conclusions

Al_2O_3 coatings prepared with several SPS conditions were investigated on microstructure, hardness, crystal structure and electric properties. The influence of thermal spraying conditions on deposition mechanical and electrical property of the coating was cleared with following details.

Deposition rate and hardness showed a similar tendency in plasma power and hardness, but the dielectric strength tended to be different.

The influence of spray distance on deposition rate and hardness was smaller than plasma power and gas flow rate.

The dielectric strength showed high at low plasma power or gas flow rate.

Coatings which show the high dielectric strength value tend to thin thickness. Moreover, the thin thickness coating contained three types of crystal phase which was α-phase, γ-phase and unknown phase.

References

[1] R. Westergård, L.C. Erickson, N. Axén, H.M. Hawthorne,S Hogmak, The erosion and abrasion characteristics of alumina coatings plasma sprayed under different spraying conditions, Tribology Internationa,l 1998, 31, p271.

[2] T. Tosaki, Y. Kobayashi, K. Ohta, J. Kitamura: Comparable Study of Electrical and Mechanical Properties in Plasma Sprayed Alumina Coatings, ITSC2013 proc.

[3] F-L Toma, L-M Berger, S. Scheitz, S. Langner, C. Ro¨del, A. Potthoff, V. Sauchuk, and M. Kusnezoff: Comparison of the Microstructural Characteristics and Electrical Properties of Thermally Sprayed Al_2O_3 Coatings from Aqueous Suspensions and Feedstock Powders, Journal of Thermal Spray Technology, Volume 21(3-4) June 2012, p484.

[4] J. Kiatamura, Z. Tang, H. Mizuno, K. Sato, and A. Burgess: Structural, Mechanical and Erosion Properties of Yttrium Oxide Coatings by axial Suspension Plasma Spraying for Electronics applications, ITSC2010 proc., p171.

[5] G. Heintze, S. Uematsu: Plasma sprayed γ-Al_2O_3 and α-Al_2O_3 coatings, Report of Ship Research Institute, JAPAN, 1991, p157.

Influence of Feedstock Powder Modification by Heat Treatments on the Properties of APS-sprayed Al_2O_3-40%TiO_2 Coatings

L.-M. Berger, Dresden / D, R. Vaßen, Y.J. Sohn, Jülich / D

Aluminum titanate (Al_2TiO_5) is a congruently melting compound in the binary Al_2O_3-TiO_2 system, which decomposes below 1200 °C. Its properties (e.g. thermal conductivity, CTE) differ significantly from those of Al_2O_3 and TiO_2. Thus it is of special interest to study the stability of Al_2TiO_5 in the spray process and its influence on the coating properties. A commercial fused and crushed Al_2O_3-40%TiO_2 powder, which was found to be substoichiometric, was selected as the feedstock material for the experimental work, as the composition is close to stoichiometric Al_2TiO_5. Part of that powder was heat-treated in air at 1150° and 1500°C in order to vary the phase composition, while not influencing the particle size distribution and processability. The powders were analyzed by thermal analysis, XRD and FESEM including metallographically prepared cross sections. A powder having Al_2TiO_5 as the main phase was not possible to be prepared due to inhomogeneous distribution of Al and Ti in the original powder. Plasma spraying was performed with a TriplexPro-210 (Oerlikon Metco) using Ar-H_2 and Ar-He plasma gas mixtures with 41 and 48 kW plasma power. Coatings were studied by XRD, SEM of metallographically prepared cross sections, and microhardness HV1. Moreover, the results show a clear influence of the Al_2TiO_5 content in the feedstock powder on the phase composition of the coatings.

1 Introduction

Thermally sprayed coatings of the Al_2O_3-TiO_2 system are used for more than fifty years and have a high technical importance in a big variety of tribological, mechanical, electrical, corrosive, and high-temperature applications. In addition to their unique properties, materials of this system are environmentally friendly and cost-efficient, as both aluminium and titanium are aboundingly found in the earth crust.
However, depending on the technology of feedstock powder preparation and on the spray process conditions a number of so far uncontrolled changes in the chemical and phase composition occur.
According to the Al_2O_3-TiO_2 phase diagram after Pena und de Aza [1] shown in **Fig. 1**, there exist the equimolar compound Al_2TiO_5 (Al_2O_3·TiO_2, mineralogical name: tialite), which can easily be synthesized from alumina and titania and is congruently melting. Thus, it is well applicable for processing by thermal spraying, but decomposes below 1200 °C into these single oxides.
Moreover, there are significant differences between the physical properties between the single oxides Al_2O_3 and TiO_2 and the compound Al_2TiO_5, such as thermal conductivity and coefficient of thermal expansion. Thus, a significant influence on the coating properties by the existence of Al_2TiO_5 can be expected. As shown in recent summaries [2,3] there are contradictory informations in the literature due to its instability influenced by the different existing feedstock powders, spray processes and spray conditions. The appearance of substoichiometry is another factor influencing the coating properties but poorly investigated so far. Usually, coatings of the Al_2O_3-TiO_2 system are prepared by different processes of atmospheric plasma spraying (APS). The use of hydrogen as plasma gas component can create or increase the substoichiometry of the coating arriving from substoichiometric feedstock powders. Based on the current knowledge the affinity of oxygen loss depends on the TiO_2-content of the composition. Oxygen loss of Al_2TiO_5 leads to the formation $Al_{2-x}Ti_{1+x}O_5$ [4-6].

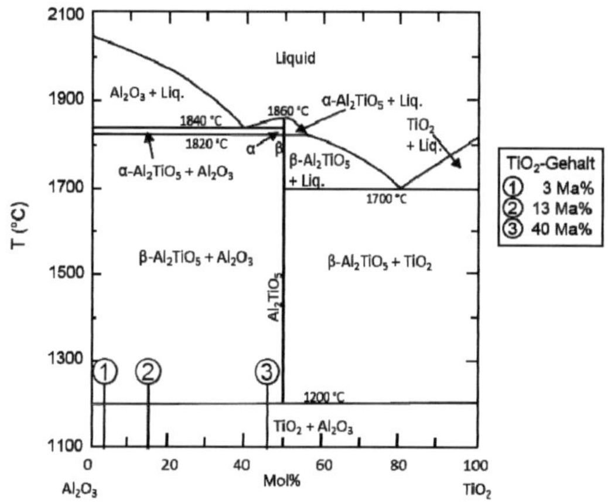

Fig. 1. The Al_2O_3-TiO_2 phase diagram after Pena and de Aza [1] with indication of the binary standard compositions for thermal spray coating preparation.

Since more than fifty years three commercially available standard compositions containing 3, 13 und 40 mass-% TiO_2 are used for coating deposition. They are all situated in the partial Al_2O_3-Al_2TiO_5 system (see **Fig. 1**). The content of Al_2TiO_5 in these feedstock powders is not specified. For Al_2O_3-40%TiO_2–coatings with Al_2TiO_5 as the main phase Young's modulus values higher than for sintered Al_2TiO_5-ceramics were found [7].
The main aim of this study was to investigate the stability of Al_2TiO_5 during the spray process starting from feedstock powders with its different content. For this purpose a commercial jointly fused and crushed Al_2O_3-40%TiO_2 feedstock powder was selected and characterized. The oxygen content and the phase composition of this powder were modified by heat treatments without influencing powder morphology.

2 Experimental

2.1 Feedstock powder

A typical commercial jointly fused and crushed Al_2O_3-40%TiO_2 (-45 +20 µm) feedstock powder was selected for the experiments and designated as powder A in the following. Such jointly fused powders represent one type available on the market [4-7], while in other commercial products particles of separately fused oxides are just mechanically blended [8,9]. The behavior of the feedstock powder in different streaming atmospheres (air, argon, hydrogen) at high temperatures was studied by thermogravimetric experiments (STA 429 and STA 409, Netzsch, Germany) from room temperature up to 1470°C using a heating rate of 5 K/min.
After preliminary tests for the determination of optimum heat treatment conditions, portions of the initial powder were heat-treated at 1150 °C for 4 h in air in order to prepare a modified powder consisting of Al_2O_3 and TiO_2 only. Another portion was heat-treated at 1500 °C for 4h a conventional sintering furnace in air in order to prepare a powder consisting of Al_2TiO_5 as the main phase and a small content Al_2O_3 according to phase diagram (see **Fig. 1**). In the following, these modified powders are designated as powder B and powder C, respectively. For powder C the cooling rate in the temperature range from 1500°C to 900°C was about 25 K/min. It was intended not to change the morphology of the powder particles and particle size distribution by the heat-treatments.
The morphology and metallographically prepared cross sections of the initial and the heat-treated powders were studied by FESEM (NVISION40, Carl Zeiss GmbH, Germany). The phase composition was studied by X-ray diffraction (D4 Endeavor, Bruker AXS GmbH, Karlsruhe, Germany). The X-ray data were collected using Cu Kα radiation in the range of 10° to 80° in 2θ, with a 0.02° step size and 0.75 seconds per step time. Particle size distribution was determined by static laser light scattering (Horiba LA 950 V2).

2.2 Coating deposition

Spray experiments were performed with a TriplexPro-210 three cathode plasma torch (nozzle diameter 9 mm) mounted on a six-axis robot in a Multicoat atmospheric plasma spray facility (Oerlikon Metco, Wohlen, Switzerland). Coatings from the three powders were sprayed with Ar/H_2 (46/4 slpm) and Ar/He (46/4 slpm) plasma gas mixtures and identical parameter sets at two plasma power levels (about 41 kW and 48 kW, respectively) on mild steel substrates of dimension 50·50 mm² which were grit-blasted immediately before spraying and without using a bondcoat. The spray distance was 150 mm, the surface speed of the gun 500 mm/s. Coating thicknesses were in the range 260-740 µm, resulting from 5-10 passes.

2.3 Coating characterisation

The metallographically prepared cross sections were studied by SEM (Zeiss Ultra55). Phase composition of the coatings was determined by X-ray diffraction using the same equipment and conditions as described for the powders. Hardness of the coatings (HV1) was measured in the coating cross sections using Vickers indents with a load of 9.81 N with a Duramin - A300 (Struers).

3 Results and discussion

3.1 Feedstock powder characterisation

3.1.1 Initial powder

The thermogravimetric curves of the initial powder A in different atmospheres are shown in **Fig. 2**. The measurement in air confirmed the deficiency of oxygen, as an mass gain of about 1.7 % was observed at 1400 °C. In a hydrogen atmosphere further mass loss of about 0.6 % at 1400 °C was observed due to reduction, while in argon no mass change was found.

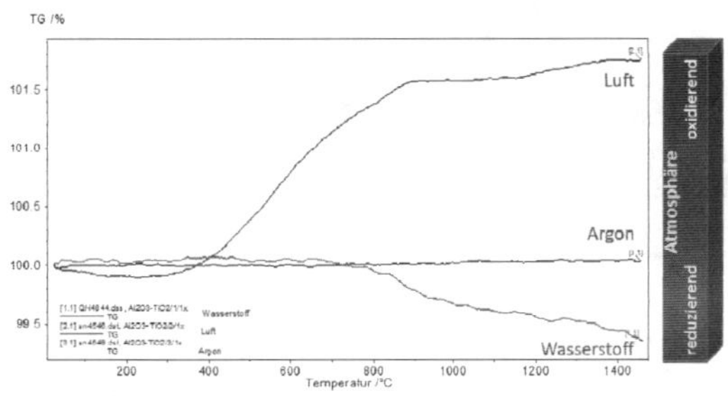

Fig. 2. Thermogravimetric (TG) analysis of the initial feedstock powder in streaming air, argon and hydrogen.

The images of the initial powder A and its particle cross section are shown in **Fig. 3**. The image shows the typical appearance of irregular shaped particles of a fused and crushed feedstock powder. **Figure 4** shows the X-ray diffraction pattern of the initial and heat treted powders together. As in earlier publications [4-6] α-Al_2O_3 was detected in the powder together with $Al_{2-x}Ti_{1+x}O_5$ with lattice parameters close to α-Ti_3O_5 (card 01-089-4733). In the cross section (**Fig. 3b**) occasionally particles with cracks are observed. The particles showed different gray-scales, indicating locally different contents of aluminium and titanium, which were confirmed by detailed investigation by EDX. In some cases such areas of inhomogeneity appeared even in one particle, or particles contained Al_2O_3 only.

Fig. 3. Morphology (a) and cross-section (b) of the initial fused and crushed feedstock powder.

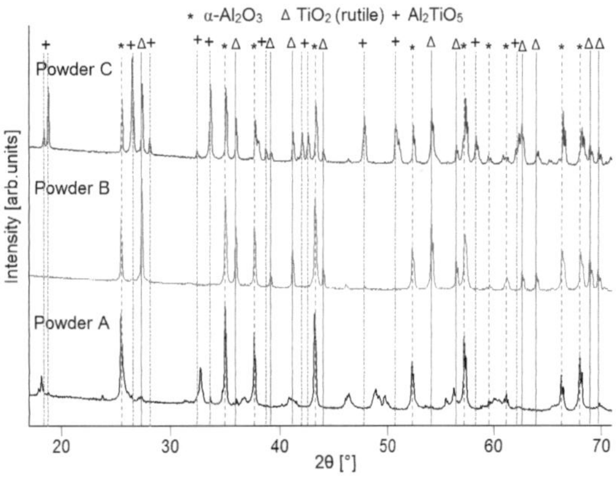

Fig. 4. X-ray diffraction pattern of the initial powder A and after heat treatments at 1150 °C and 1500 °C.

3.1.1 Heat-treated powders

The images of the powder and the particle cross section after heat treatment at 1150 °C (4 h), designated as powder B are shown in **Fig. 5**. The X-ray diffraction pattern is shown in **Fig. 4**. The morphology of powder B in **Fig. 5a** is similar to that of the initial powder A in **Fig. 3a**, the measurement by laser light scattering confirmed that this heat treatment did not influence the particle size distribution and the diffraction pattern of the powder confirms that it consists of Al_2O_3 (46-1212) and TiO_2 (21-1276, rutile) only (**Fig. 4**). However, the inhomogeneity of the distribution of the metallic elements became still more obvious. Particles or areas, were both metallic elements were present show now a very fine grained substructure. Pure Al_2O_3 particles and areas did not change their structure (**Fig. 5b**).

Fig. 5. Morphology (a) and cross-section (b) of the powder after heat-treatment at 1150 °C / 4h.

The images of the powder C and the particle cross section after heat treatment at 1500 °C (4 h) are shown in **Fig. 6**. The X-ray diffraction pattern of this powder is shown in **Fig. 4**. Due to the inhomogeneities in the original powder A besides the main phase Al_2TiO_5 significant residues of Al_2O_3 und TiO_2 (rutile) were detected (**Fig. 6**). The lattice parameter of the Al_2TiO_5 corresponds to that of the standard (41-0258). The Al_2TiO_5 in this powder has the smallest unit cell volume. Due to the high heat treatment temperature some changes in the particle morphology were found. However, as confirmed by laser light scattering the particle size distribution and the processability were not significantly influenced. The existence of very different grayscales in the FESEM image of the particle cross section (**Fig. 7b**), is a result of the inhomogeneity of the distribution of the metallic elements and the solid state reactions during the heat treatment. Different from powder B less fine-grained substructures were found. A large number of cracks in the powder particles can be observed as well.

Fig. 6. Morphology (a) and cross-section (b) of the powder after heat-treatment at 1500 °C / 4h.

There is no work in the literature known to the authors, which describes the homogeneity of distribution of aluminium and titanium in fused and crushed feedstock powders. Thus, the homogeneity of the distribution of the metallic elements has to be checked for a number production lots. In the case of a fused and crushed TiO_2-10%Cr_2O_3 powder no inhomogeneity of distribution of chromium was observed [10].

3.2 Coating characterisation

Figures 7a-c show the FESEM images of the coating cross sections, deposited by using Ar/H_2 plasma with 48 kW plasma power from the powders A, B and C, respectively. All coatings show a low porosity and the typical wavy structure. The explanation of the distribution of the metallic elements and the phases require still further work. The microstructural appearance of the coatings from the other spray conditions (41 kW plasma power, Ar/He plasma) show rather smaller differences.

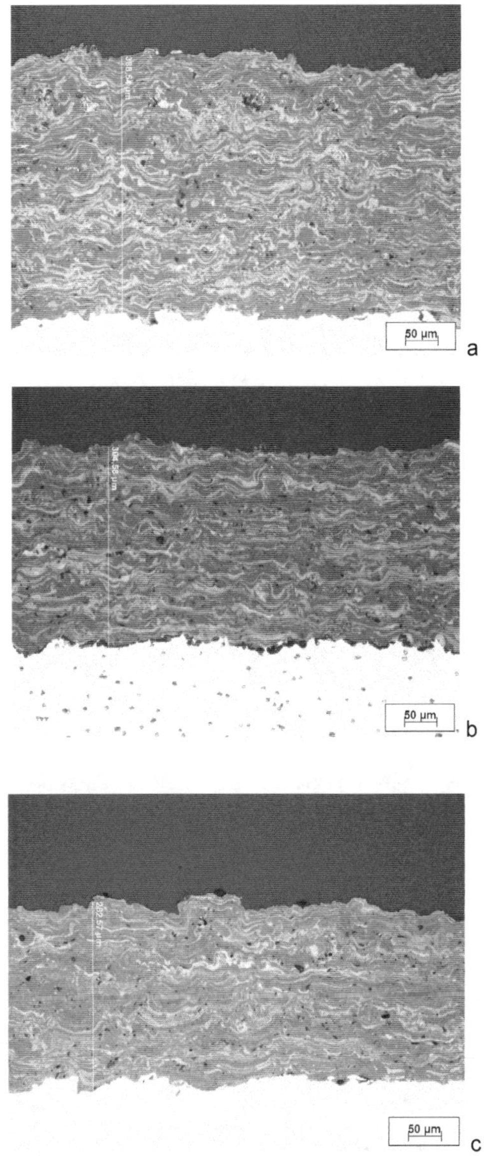

Fig. 7. Coatings sprayed with Ar/H_2 plasma and with 48 kW plasma power using the powder in the as-delivered state (powder A, a), after heat treatment at 1150 °C (powder B, b), and after heat treatment at 1500 °C (powder C, c).

The X-ray diffraction patterns of the coatings sprayed with 48 kW plasma power and the different plasma gases are compiled together with the corresponding powders A, B and C in **Figs. 8-10**, respectively. The α-Al_2O_3 in powder A (**Fig. 8**) has been largely transformed into γ-Al_2O_3 in both coatings, except some residue, which is larger in the case of the Ar/He plasma. The peaks of $Al_{2-x}Ti_{1+x}O_5$ in the coating are broader than those in the powder, but also shifted to higher angles. Substoichiometric titania appears as a new phase in the coatings.

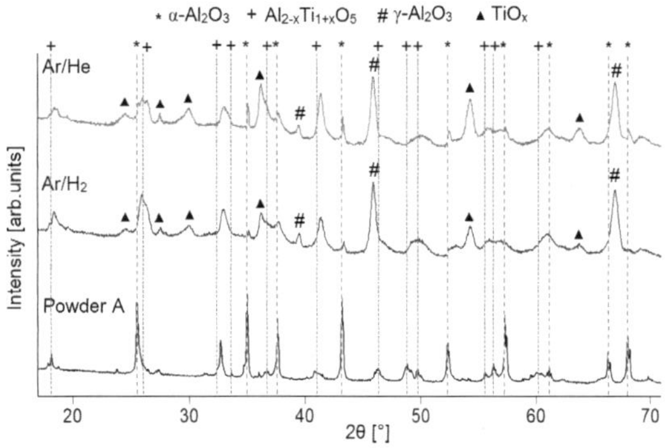

Fig. 8. X-ray diffraction pattern of the initial powder A and the coatings sprayed with Ar/H_2 and Ar/He plasma gases (plasma power 48 kW).

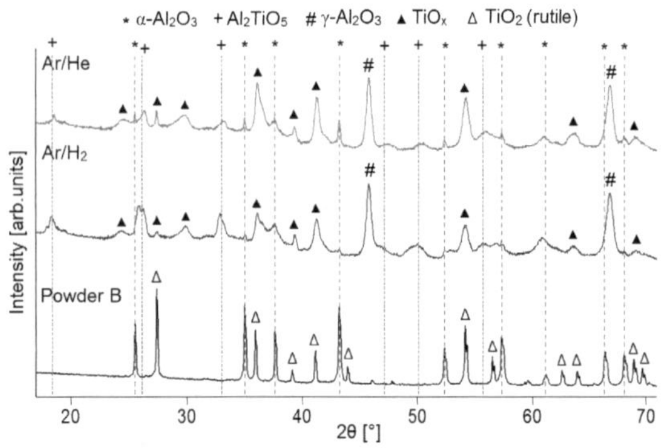

Fig. 9. X-ray diffraction pattern of the powder after heat-treatment at 1150 °C / 4h and the coatings sprayed thereof.

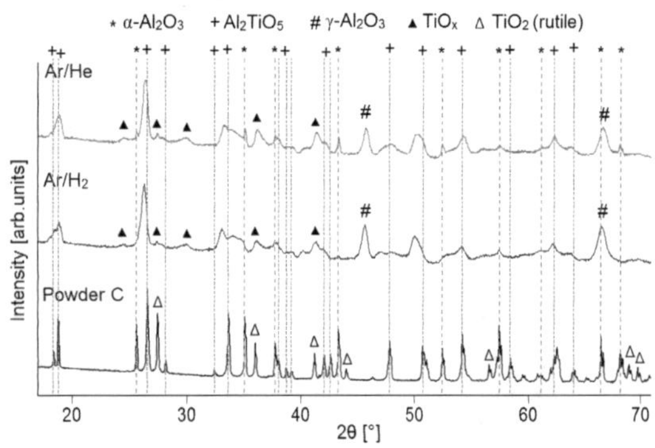

Fig. 10. X-ray diffraction pattern of the powder after heat-treatment at 1500 °C / 4h and the coatings sprayed thereof.

The γ-Al_2O_3 and substoichiometric titania are the main phases in the coatings prepared from powder B (**Fig. 9**). Except some residue of α-Al_2O_3, it is also interesting to mention that Al_2TiO_5 appeared in this coating. Its mechanism of formation is currently not known and has to be studied in detail in upcoming experimental series. Also remarkably, the TiO_2 in the powder has lost significant amounts of oxygen not only by the use of Ar/H_2 plasma gas mixture, but also for Ar/He.

In the coatings from powder C (**Fig. 10**) Al_2TiO_5 is the main phase, with a lattice parameter close to the standard. Again, most of the α-Al_2O_3 was transformed to γ-Al_2O_3 and the TiO_2 in the powder has lost significant amounts of oxygen. In both plasma gas mixtures.
The new binary compound $Al_6Ti_2O_{13}$ ($3Al_2O_3{\cdot}2TiO_2$), reported in the literature [11,12], was not observed in our study, different from another study on thermal spray feedstock powders and coatings of the Al_2O_3-TiO_2 system [13].

Figure 11 shows the Vickers hardness values HV1 of the coatings including the corresponding standard deviations. All hardness values were in the range of about 800-100 HV1. There is some tendency that the hardness values of the coatings sprayed Ar/H_2 plasma are higher than those obtained with Ar/He, but this is not valid without exception.

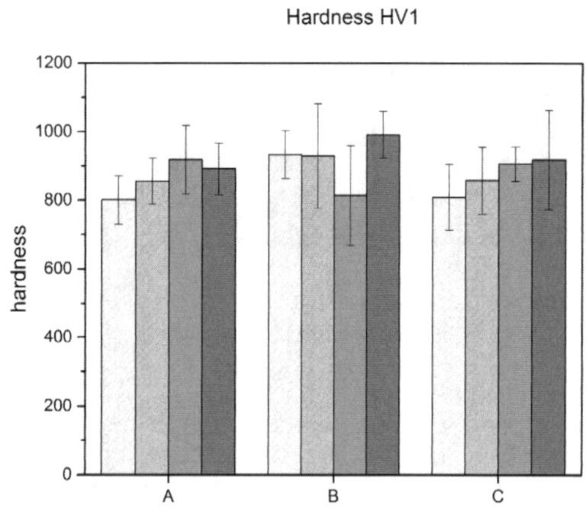

Fig. 11. Vickers hardness HV1 and their standard deviation of the coatings from powders A, B and C. Description of the colors in the columns: bright yellow: Ar/He plasma, 41 kW; dark yellow: Ar/He plasma, 48 kW; bright green: Ar/H_2 plasma, 41 kW; dark green: Ar/H_2 plasma 48 kW.

Despite the different loads applied, all coatings show hardness values, which are in a good agreement with values of plasma sprayed coatings reported earlier, e.g. [4, 6, 14].

4 Summary and Outlook

The oxygen content and the phase composition of the selected fused and crushed Al_2O_3-40%TiO_2 feedstock powder was successfully modified by heat treatments in air without significant changes of particle size distribution and processability by APS. Only the heat treatment temperature of 1500 °C lead to some changes in the powder morphology. As expected, this feedstock powder of jointly fused alumina and titania, was found to consist of alumina and a substoichiometric $Al_{2-x}Ti_{1+x}O_5$. However, both metallic elements were inhomogeneously distributed in the

powder particles, even particles consisting of alumina only existed. Although the heat treatment at 1150 °C resulted in a feedstock powder consisting of Al_2O_3 and TiO_2 only, their inhomogeneous distribution resulted from the original powder. This inhomogeneity prevented the preparation of a powder with a very large content of Al_2TiO_5 by such a heat treatment.
All three powders could be easily sprayed to dense coatings using a TriplexPro 210 plasma gun with Ar/H_2 and Ar/He plasma gas mixtures with plasma power of 41 kW or 48 kW, respectively . All coating hardness values were in the range of about 800-1000 HV1. However, significant differences in the phase compositions were found. The typical phase transformation $\alpha \rightarrow \gamma$ Al_2O_3 during spraying was observed, together with the formation of substoiciometric titania. In the coating sprayed from the Al_2O_3/TiO_2 powder (powder B), Al_2TiO_5 was detected. The lines of this phase are easily shifted, obviously due to changes in the oxygen content. However, the differences in the phase composition due to the different plasma gases were smaller than expected.
In future work, the homogeneity of the distribution of the metallic elements has to be checked for a number production lots. The distribution of the Al_2TiO_5 in the coatings and the changes of the stoichiometry in the powder and the coatings have to be studied in detail.

5 Acknowledgements

Colleagues from Fraunhofer IKTS are thanked for execution of feedstock powder preparation by heat treatment and FESEM studies of the powders. The support of Karl-Heinz Rauwald (FZ Jülich), who performed the spraying runs, is also gratefully acknowledged.

6 Literature

[1] Pena, P. and de Aza, S. The System ZrO_2-Al_2O_3-TiO_2, Ceramica (Florence), 33 (3), **1980**, 23-30.
[2] Berger L.-M., Toma, F.-L., Scheitz, S., Trache, R., Börner, T., Thermisch gespritzte Schichten im System Al_2O_3-Cr_2O_3-TiO_2 - ein Update, Mat.-wiss. Werkstofftechn. 45(6), **2014**, 465-475.
[3] Berger L.-M., Tribology of Thermally Sprayed Coatings in the Al_2O_3-Cr_2O_3-TiO_2 System, in: Thermal Sprayed Coatings and Their Tribological Performances, IGI Global, **2015**, 227-267.
[4] Berger, L.-M., Stahr, C.C., Toma, F.-L., Saaro, S., Herrmann, M., Deska, D., Michael, G., Thiele S., Korrosion thermisch gespritzter oxidkeramischer Schichten, Thermal Spray Bull., 2 (1), **2009**, 40-56.
[5] Toma, F.-L., Stahr, C.C., Berger, L.-M., Saaro, S., Herrmann, M., Michael, G., Corrosion Resistance of APS- and HVOF-Sprayed Coatings in the Al_2O_3-TiO_2 System, J. Therm. Spray Technol. 19(1-2), **2010,** 137–147.
[6] Herrmann, M., Toma, F.-L., Berger, L.-M., Kaiser, G., Stahr, C.C., Comparative Study of the Corrosion Resistance of Thermally Sprayed Ceramic Coatings and Their Bulk Ceramic Counterparts, J. Eur. Ceram. Soc. 34(2), **2014,** 493–504.
[7] Trache, R., Berger, L.-M., Leyens, C., Comparison of Alumina-Titania Coatings Deposited by Plasma and HVOF Spray Techniques, in: Proc. Int. Therm. Spray Conf., May 21-23, 2014, Barcelona, Spain, Düsseldorf DVS Media, **2014**, 368-372.
[8] Niemi, K., Abrasion Wear Characteristics of Thermally Sprayed Alumina Based Coatings, Doctoral Thesis, Tampere University of Technology, Publication 820, Tampere, **2009**, 117.
[9] Ananthapadmanabhan, P.V., Thiyagarajan, T.K., Sreekumar, K.P., Satpute, R.U., Venkatramani, N., Ramachandran, K., Co-Spraying of Alumina-Titania: Correlation of Coating Composition and Properties with Particle Behaviour in the Plasma Jet, Surf. Coat. Technol., 168 (2-3), **2003**, 231–240.
[10] Trache R., Berger L.-M., Toma F.-L., Stahr S., Lima R.S., Marple B.R., Electrical Resistivity of Thermally Sprayed Cr_2O_3-TiO_2 Coatings, Int. Therm. Spray Conf. ITSC 2011, Conf. Proc., September 27-29, 2011, Hamburg, Germany, DVS-Berichte Vol. 276, Düsseldorf, DVS-Verlag GmbH, **2011**, CD, p. 1030-1035.
[11] Norberg, S.T., Hoffmann, S., Yoshimura, M., Ishizawa, N., $Al_6Ti_2O_{13}$, a New Phase in the Al_2O_3-TiO_2 System, Acta Cryst., C61 (X), **2005**, i35-38.
[12] Berger, M.-H., Sayir, A., Directional Solidification of Al_2O_3-$Al_6Ti_2O_{13}$ System, J. Eur. Ceram. Soc. 28(12), **2008,** 2411–2419.
[13] Vargas, F., Ageorges, H., Fournier, P., Fauchais, P., López, M.E., Mechanical and tribological performance of Al_2O_3-TiO_2 coatings elaborated by flame and plasma spraying, Surf. Coat. Technol., 205(4), **2010**, 1132-1136.
[14] Niemi, K., Vuoristo, P, and Mäntylä, T., Properties of Alumina-Based Coatings Deposited by Plasma Spray and Detonation Gun Spray Processes, J. Therm. Spray Technol. 3(2), **1994,** 199–203.

Dry-sliding wear of laser clad nickel aluminide / chromium carbide coatings against cast iron

Sven Bengtsson[a], Senad Dizdar[a], and Karin Gong[b]
[a)] Höganäs /S/, [b)] Göteborg /S/

The wear of piston rings in large marine two-stroke diesel engines is a major maintenance cost. Applying coatings with good oxidation, corrosion resistance and high temperature strength, can lower the total maintenance cost. In the past nickel aluminide with chromium carbide have been applied to pistons by thermal spraying. Using laser cladding a suitable microstructure can be formed while at the same time avoiding cracks and bonding issues. In this report powders and coatings were manufactured in order to be able to investigate the dry-sliding wear behavior. Material with three levels of carbides was atomized. Wear test samples were manufactured by laser cladding. The dry sliding wear-mechanism maps are generated by using block on ring test setup where coated blocks slide against cast iron rings. All alloys exhibited regions of plasticity-dominated wear and oxidational wear with a transition region in-between. The carbide-containing alloys showed lower friction and wear in comparison to the carbide free nickel aluminide alloy.

1. Introduction

The requirements on modern marine diesel engines are becoming higher and higher. The requirement for high power to volume ratio is higher than ever, combined with the ability to run long periods at reduced power and the need to burn low-sulfur fuels [1] in order to reduce environmental pollution [2]. Furthermore, the need for longer service intervals and higher availability is also strong.

Accordingly, piston rings of marine diesel engines are getting increased severity of the working conditions. Sulphur and its compounds are known to act as anti-wear and extreme pressure additives to lubricating oils. These additives forms thin friction and wear reducing boundary layers-films able to separate the contacting surfaces in boundary and mixed lubrication regimes when full oil film lubrication cannot be established or can fail [3]. Reduction in sulfur amount in the diesel fuel, reduces significantly the sulfurs ability to form the protective boundary layers. .

The risks for severe wear of piston rings are highest near the piston top dead center position when high gas pressure, high temperature and low sliding speed do not allow establishing of full oil lubricating film. Further down the piston stroke, the sliding speed increases, and the temperature drops which is beneficial for establishment of full oil film lubrication regime and leads to minimal risks for wear of the piston rings..

Piston rings for large two-stroke marine diesel engines are subjected to wear and corrosion during operation limiting the usable service interval. By extending the life time of the piston rings the service intervals can be increased and total maintenance costs decreased. This improvement can be achieved by using coatings on the piston rings [4–8]. Hard chrome plating was one of the first methods used to improve wear performance of the piston ring [4]. Other widespread methods are plasma spray and high velocity oxy-fuel spray (HVOF) coatings [9]. Other friction reducing materials, such as $MoSi_2$ has also been used [10]. In some cases the thermal spray methods, such as HVOF, cannot achieve enough strong bonding to the substrate or inside the deposited layer. This can result in increased wear rates [9].

An interesting material combination is nickel aluminide [11,12] combined with chromium carbide [13]. This material combination has been applied as a coating using plasma spray and high velocity oxy-fuel coating processes [5,6,10]. The intermetallic Ni_3Al-based alloys are promising candidates for developing high-temperature and high-strength materials in wear-resistant applications [11,12,14]. This group of alloys exhibit two attractive engineering properties their increasing yield strength with increasing temperature up to about 650°C -750 °C and the ability to resist oxidation and corrosion at high temperature [15].

The reinforcing phase, chromium carbide exists in different forms; Cr_3C_2, M_7C_3, $M_{23}C_6$ where M is Cr, Fe, or possibly other metals if available. Typically, the hardness is highest for the Cr_3C_2 carbide followed by M_7C_3, and $M_{23}C_6$. If Cr_3C_2 carbide particles is mixed with nickel aluminide and subsequently welded, the carbide will dissolve quickly in the melt and re-precipitate on solidification. However, depending on amount of Cr_3C_2 added to the mix it will precipitate as one or two of the possible carbides. If the total concentration of carbon is, low it will precipitate as $M_{23}C_6$ and if it is high, it will form Cr_3C_2. However, it is also possible to form graphite if the carbon concentration is too high. A complicating factor is that chromium is a required alloying element in the nickel aluminide matrix. In the present concentration regime for the matrix two phases can form NiAl and Ni_3Al. The amount of NiAl should be kept at low levels since the NiAl is brittle at room temperature and exhibit moderate strength and hardness at high temperature. However, it can be used to buffer the composition and to allow impurities to accumulate in this phase rather than precipitate on grain boundaries since the Ni_3Al does not dissolve typical impurities at all.

Alloys were atomized in the pilot scale atomizer at Höganäs AB, Sweden. The compositions were selected based on the discussion above and the guidelines in previous work [16]. Three compositions are included in this investigation, based on amount of carbide in the alloy. Without carbides the general hardness is expected to be too low for sufficient wear performance, while with too much carbide the weldability will be insufficient, with extensive cracking as result. By this approach it was expected to have a good characterization of the wear behavior of this alloy system in terms of tribological behavior.

The powder is used for laser cladding of wear test blocks for block on ring adhesive dry testing. The rings are made of a cast iron similar to cylinder liner material for large marine diesel engines. The wear tests are performed in dry – unlubricated conditions, at two load and five sliding velocity levels in order to simulate the variety of speeds that the piston ring encounter during its cycle.

The aim of this work is to investigate-rank the dry-sliding friction and wear of nickel aluminide alloys with chromium carbide in a model wear test designed to rank adhesive wear of materials aimed for the piston ring / cylinder liner application.

2. Experimental

The materials were atomized in the Pilot scale gas atomizer and sieved to 53 – 150 micrometer sieve cut. The chemical composition is found in Table 1. The atomized powders were basically spherical with few satellites or other geometrical distortions.

Table 1. Chemical composition of the tested powders.

Code	Composition (wt.%)						
	C	B	Ni	Fe	Cr	Al	Other
A	2.5	0.03	bal		30,4	9,5	
B	6.3	0.18	bal	5,6	43,5	4,6	0.3Mn, 0.3Ti
C	-	0.04	bal	11,1	8,4	10,5	

The laser cladding was performed using an IPG 6 kW fiber coupled diode laser with a Coax 8 powder feed nozzle and a 5 mm round spot. The determination of the process window was typically performed using two to three laser travel speeds a powder feed rate designed to give approximately 1 mm thick single track coating and a variation of laser power in order to reach a modest dilution (target 2-10%) without significant interface or inter track porosity. The dilution is defined as the area fraction of melted substrate divided by the area sum of melted substrate and coating. In some cases pre-heating of the substrate was required in order to avoid welding cracks. In most cases the samples were allowed to cool in vermiculite until the temperature is below 50 °C. Once suitable parameters are found using the single track samples wear test plates are clad using several adjacent tracks with a defined overlap between tracks. The clad samples were examined visually for defects and by liquid penetrant testing for surface cracks. The porosity and the dilution was examined by light optical microscopy of the as polished surface on cross sections of single track coatings. The interface and inter track porosity is also examined on the wear test samples. A more detailed metallographic investigation was performed using a Hitachi SU6600 FEGSEM scanning electron microscope equipped with a Bruker energy dispersive spectrometer.

In Fig. 1 an example of defining the process window for the laser cladding process is given. Here it can be seen at laser power above 3 kW no interface porosity is found, thus ensuring good fusion. At a laser power of 4 kW the dilution is 5%. Acceptable limit of interface porosity is below 0.1% while dilution should be between 2 and 5%. The process window for material A is a laser power of 3.5-4.0 kW with laser travel speed of 16 mm/s and a powder feed rate of 30 g/s. The B material with its higher amount of carbides exhibit a more narrow process window and it is only near 3,5 kW laser power acceptable results can be obtained, while the carbide free C have a slightly lower optimum range at 3,0-3,5 kW.

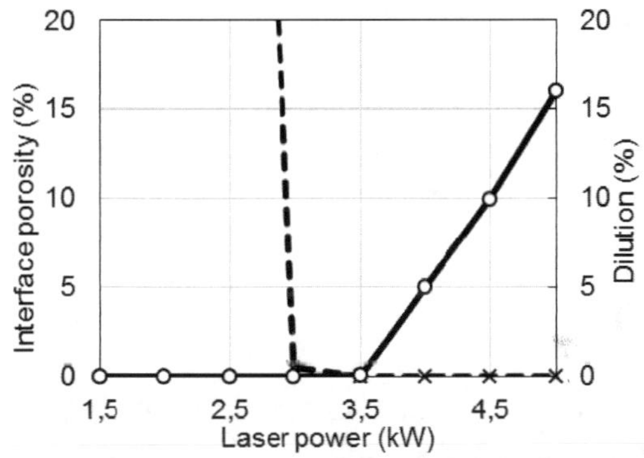

Fig. 1. Process window for material A. Combined diagram of Interface porosity (dashed line) and dilution (solid line). Laser travel speed is 16 mm/s and the powder feed rate is 30 g/min.

The parameters for the cladding of wear samples of material A were selected inside the process window found for the single track clads. A laser power of 3,75 kW, laser travel speed of 16 mm/s and a powder feed rate of 30 g/min was selected. However, using these conditions cracks were found after cladding. In order to decrease the risk for cracks it was decided to pre-heat the plate before cladding to 200 °C. At the same time the laser power was decreased to 3.4 kW in order to avoid too high dilution. The cladding was performed as patches of approximately 30 mm x 100 mm. Three or four such patches fitted onto one plate. The temperature of the plate was measured before the cladding of each patch. If the temperature was above 300 °C the cladding was delayed until, the temperature had dropped below 250 °C.

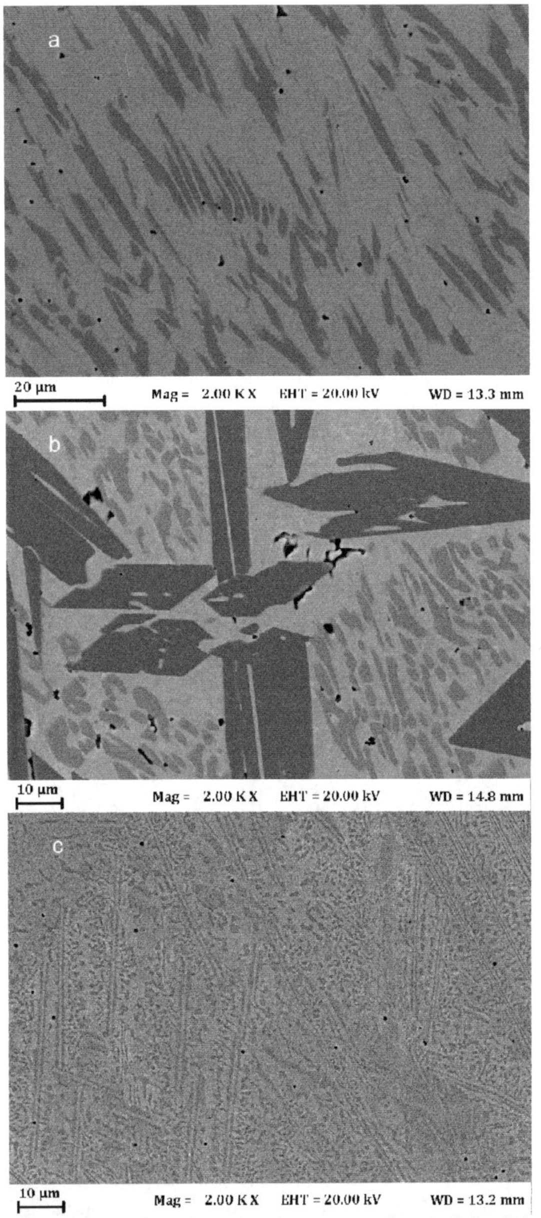

Fig. 2. Scanning electron microscopy of materials a) A, b) B, and c) C. Backscattered electron image.

For the material B it turned out to be necessary to place a buffer layer (Material C) followed by pre-heating to 300 °C in order to completely avoid cracks. Both layers were clad using a laser power of 3,0 kW. Material C did not require any pre-heating and was clad at a laser power of 3,0 kW. A laser travel speed of 16 mm/s and a powder feed rate of 30 g/s was maintained for all these materials.

In the microstructure of the A material carbides can be found as the dark phase in Fig. 2a. The matrix consists of two phases, presumably Ni_3Al and NiAl. Furthermore, small black spots can be found, possibly graphite. The hardness is 570 HV5. The B material with its higher carbon and chromium content exhibits two types of carbides. The darker carbide is presumably Cr_3C_2 while the lighter phase is probably M_7C_3. The matrix also consists of two phases even if it is not evident in Fig. 2b. Furthermore, black spots can be seen that are possibly porosity or carbon/graphite. The carbide free material, C exhibit a very fine structure with only two phases, again most likely Ni_3Al and NiAl. The hardness is 900 HV5. The microstructure of material C consists of two phases, also in this case most likely Ni_3Al and NiAl. It solidified in a cell structure or as dendrites depending on the position in the clad. The hardness is 360HV5.

The piston ring / cylinder liner wear contact can be modelled for wear testing in different ways and at different complexity levels. Investigation presented in this paper focuses on a basic ranking of dry sliding friction and wear of three novel prototype materials for piston rings. Therefore, a model wear test of lowest complexity level with so called primitive body geometries is to be used. All other aspects of the piston ring to cylinder liner contact except material and dry sliding are to be intentionally excluded. Selection of test sliding velocities and normal loads will illustrate wear mechanisms under failed oil film lubrication and slight overloads. The test sliding velocities are to be 0.045, 0.13, 0.42, 1.1 and 2.9 m/s. The normal loads are to be 6 and 52 N corresponding max. Herztian contact pressures of 170 respectively 340 MPa. The five test sliding velocities and the two load levels means that wear mapping technique is to be used. As Bowden [] commented it I 1950', the wear mapping is a very opportune way avoid risks with localized presentation of wear data.

The block on ring setup was selected as the most suitable tribological test here for two main reasons: an appropriate scale of contact stresses and wear track as well as convenience of the specimen manufacture. Dry sliding wear applications usually do not belong to highly loaded/concentrated tribological contacts, and the relatively low contact pressure levels in block-on-ring setups will provide a convenient size of wear track. In addition, laser cladding technology can, without high complexity, deposit plain steel bars, from which rectangular test blocks are manufactured using as common chip machining and finishing by plain grinding..

However, in the present case, it was decided to use Tarkalloy-C, a grey cast iron used for cylinder liners in the marine diesel engines that are the target of this investigation. The rings were crowned in order to accommodate any misalignment present in the contact.

In order to be able to compare the performance of the nickel aluminide / chromium carbide materials to relevant reference materials tests were also performed using blocks of nickel aluminide without carbide (C), cast iron taken from piston ring segments. The tested materials are summarized in Table 2.

Table 2. Material systems used in the dry-sliding adhesive wear test.

Test	Block	Ring
A	Nickel aluminide / 25% Chromium carbide	Cast Iron (Tarkalloy-C)
B	Nickel aluminide / 50% Chromium carbide	
C	Nickel aluminide	

The wear tests were performed on a multi-purpose friction and wear tester – tribometer (Phoenix Tribology TE53). Fig. 3 illustrates the block on ring tribometric test setup. A test block press the block cladded face onto a rotated crowned counter ring. The normal load on the block (F_n) comes from a specimen holder and a load arm assembly with dead weights. A linear displacement transducer (LVDT) registers the total linear wear and a load cell measures the friction force (F_f) in the block on ring contact. Compendx software controls all tribometer sub-systems, log target and acquired values of all test parameters and save in a data file.

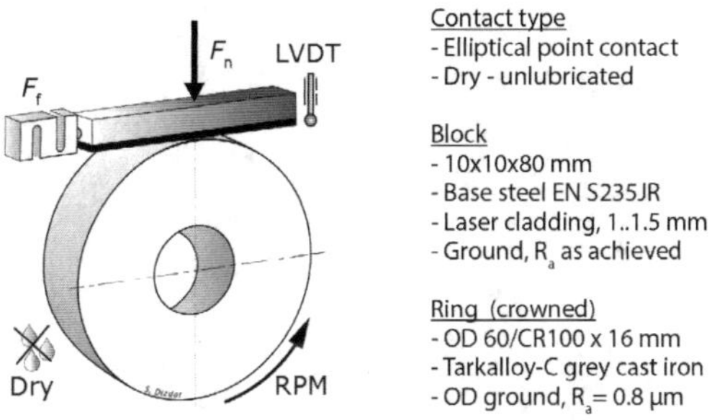

Fig. 3. Block on ring tribometric test setup.

3. Results

The main result are wear mechanism maps shown in Fig. 4. The wear mechanisms were decided based on friction coefficient, wear rate and appearance of the wear surface. – All three wear mechanism maps show wear mechanism regions in principal agreement with those for dry sliding steel to steel contacts [17]. Plasticity dominated wear prevails at low to moderate sliding velocities while oxidational wear occur at relatively high sliding velocities. However, there are substantial differences in friction and wear behavior between the tested materials.

In Test A (material A) exhibit good friction and wear performances, for both 6 and 52 N test normal loads and all five test sliding velocities, see Fig. 4. Friction coefficient is about 0.4-0.6 for the plasticity dominated wear region and drops to about 0.4-0.45 in the oxidational wear region. Wear rate is at 10^{-6} mm^3/Nm level for normal load of 6 N, for both wear regions. At 52 N in normal load, the wear rate is at 10^{-5} mm^3/Nm level in plasticity dominated wear region and drops to 10^{-6} mm^3/Nm in oxidational wear region.

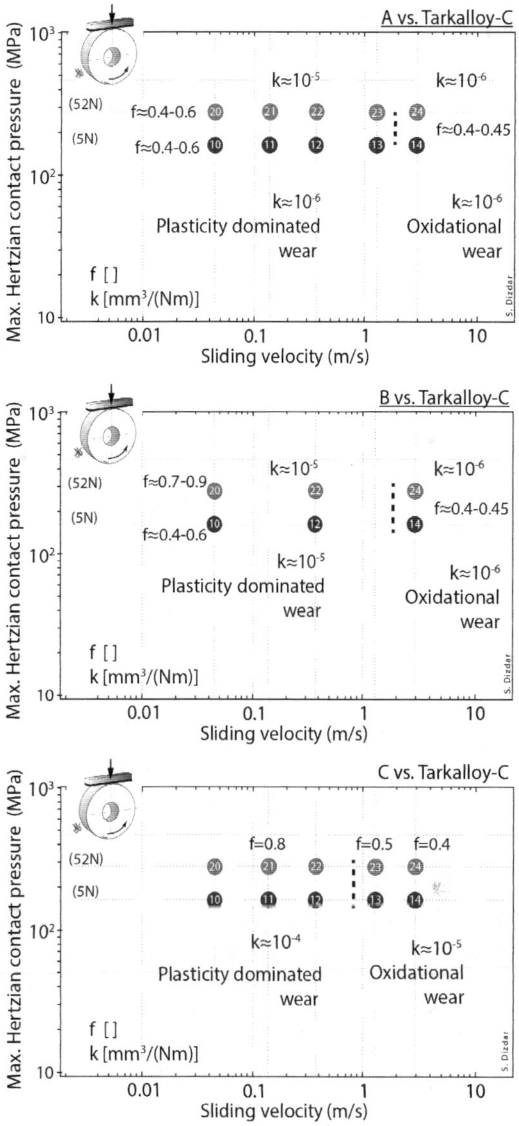

Fig. 4. Wear maps for the tested materials.

Test B (material B) was run as a reduced map including sliding velocities 0.045, 0.37 and 2.9 m/s and it showed a somewhat different behavior, see Fig. 4. Already in the as-ground state the surface is different in that it seems like particles have fallen off, effectively forming small craters, see Fig. 5. However, the size of these craters is quite small about 10 µm. At the lower load of 6 N the coefficient of friction was fairly constant near 0.5 for all three sliding velocities. However, at the higher load the coefficient of friction continuously increased from the start of the test at about 0.3 to about 0.9 at 0.045 mm/s sliding velocity. At the intermediate sliding velocity (0.37 m/s) the behavior was similar, except that the maximum coefficient of friction was 0.7. It also showed some evidence of material falling off, possibly after formation of a heavily worked surface zone. At the highest sliding velocity the oxidational wear mechanism exhibit basically the same coefficient of friction and wear behavior as at the lower load.

In test C (material C) the friction coefficient is high (0.7-0.9) for high (52 N) loads up to a sliding velocity of 0.37 m/s, the region were plasticity dominated wear dominates, see Fig. 4. At low (6 N) loads the friction coefficient is low (0.4-0.6). At 1.1 m/s the coefficient of friction drops to about 0.5-0.6 at the higher load and remains at 0.4-0.5 for the lower load. At the highest sliding speed at 2.9 m/s the coefficient of friction dropped further to 0.4-0.5 at the higher load and to 0.4 for the lower load.

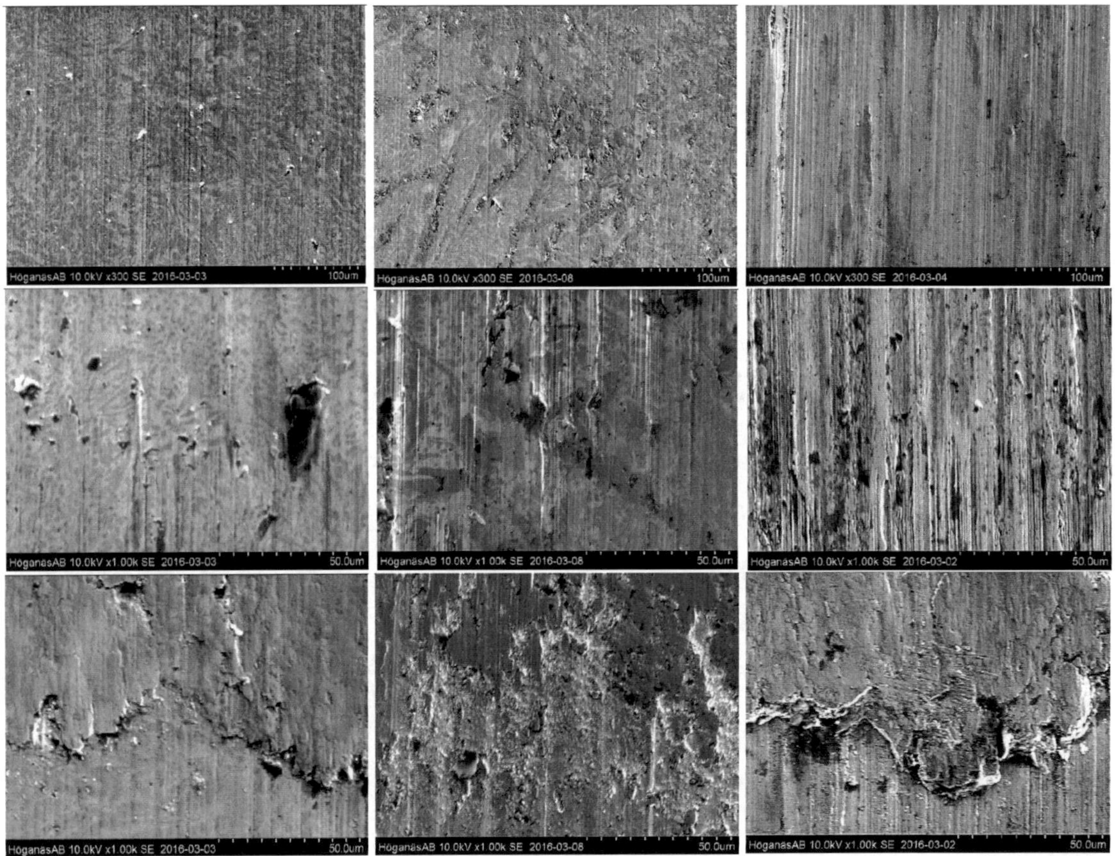

Fig. 5. Left column: A, Middle column: B, Right column: CTop row: As ground surfaces. Middle row: Surfaces after test at sliding speed 0.37 m/s and 6 N normal load. Lower row: Surfaces after test at sliding speed 2.9 m/s and 52 N normal load.

4. Discussion

The carbide free material, C, exhibited very good laser cladding performance. The processing window was relatively wide and no cracks were formed in any of the coupons produced. This indicates some degree of ductility as was expected from literature [14] at this hardness (370 HV_5).

Laser cladding of the powder with the highest carbide content, B, proved to be quite difficult without forming cracks. Pre-heating alone did not provide a solution to avoiding cracks in all cases. The wear samples used in the investigation required both pre-heating and the use of a buffer layer using the carbide free nickel aluminide powder (C). The volume fraction of hard phase is about 50% for this material and most of it is precipitated in the melt. A consequence of this is that the melt pool will have very poor fluidity and that there is a risk for solidification porosity. The observed black spots may be porosity, Fig. 2c. A further consequence is that the accommodation of the residual stresses due to the large temperature drop will be limited since it is only 50% of the material, the matrix, which can plastically deform. At room temperature this is manifested in the very high hardness (900 HV_5) exhibited by this material.

Material A is a good compromise between B and C. It has approximately half of the volume fraction carbide found in B. It gives a significant hardness increase over C, 600 compared to 370 HV_5, but still have a sufficient amount of matrix phase (~75%) to accommodate the residual stresses formed. When cladding using this material, the tendency to form cracks was much less prominent. Depending on the coating size it was sometimes possible to perform the cladding operation without pre-heating the substrate.

In the wear testing all the tested materials exhibit a plasticity dominated wear mechanism for sliding speeds from 0.045 m/s up to about 1.1 m/s. At higher sliding speeds the wear mechanism is dominated by oxidative wear.

Nickel aluminide material without the carbide (C) exhibits poor behavior in the plasticity dominated region with very high coefficient of friction, 0.7-0.9, and a high wear rate of 10^{-4} mm^3/Nm while performance recovers somewhat in the oxidative wear dominated region were coefficient of friction drops to about 0.4 and wear rate to 10^{-5} mm^3/Nm [17].

However, the introduction of the carbide into the nickel aluminide system provides several advantages. The most significant, for the intended application, is the improvement in wear behavior at low speed. Here apparently the carbides prevent adhesive junctions so that coefficient of friction can remain relatively low and that the wear rate for the carbide containing material is lower at all speeds and loads tested.

The carbide containing material (A) exhibits a coefficient of friction of 0.4-0.6, and a wear rate of 10^{-6} mm^3/Nm in the plasticity dominated region and 0.4-0.45 respective 10^{-6} mm^3/Nm in the oxidative wear dominated region.

The material with a higher carbide content (B) performs well at the low load conditions, but at high loads and low speeds, the coefficient of friction is 0.7-0.9. The expected better wear performance may be offset by the fact that it appears that material is falling out forming small craters on the wear surface, Fig. 5. Carbides falling out will create craters and also scratch the surface resulting in increased surface roughness and increased wear rate.

The piston ring to cylinder liner contact has, as previously mentioned, a risk for severe wear in the piston top dead center position where full film lubrication cannot be established due to low sliding velocity and other factors. Here, the wear resistance is provided by friction and wear reducing surface phases. Carbides in nickel aluminide alloy A and B have been showed here to reduce both friction and wear in dry sliding conditions at low sliding velocities in comparison to carbide free nickel aluminide alloy C.

5. Conclusions

The nickel aluminide alloy A with 25% carbides has been showed here to reduce both friction and wear in dry sliding conditions at low sliding velocities in comparison to carbide free nickel aluminide alloy C.

6. References

[1] Woodyard, D. Exhaust Emissions and Control. *Pounder's Mar. Diesel Engines Gas Turbines* 61–86 (2009). doi:10.1016/B978-0-7506-8984-7.00003-5

[2] Liu, Y. Q., Du, J. Z., Tan, P. Q. & Du, A. M. Progress on Denitration and Desulfuration Technologies for Marine Diesel Engines. *Appl. Mech. Mater.* **713–715,** 51–56 (2015).

[3] Cheng, H. S. in *ASM Handbook, Vol 18* 88–97 (ASM International, 1992).

[4] Brenneke, A. M. *Chrome plated piston rings design and application. SAE Technical Paper 490113* (1949). doi:10.4271/490113.

[5] Rastegar, F. & Richardson, D. E. Alternative to chrome: HVOF cermet coatings for high horse power diesel engines. *Surf. Coatings Technol.* **90,** 156–163 (1997).

[6] Aram, M. Thermal Spryaing of a Piston Ring. *US 2005/0073107 A1* **1,** (2005).

[7] Obara, R., Takiguchi, K. & Hosotsubo, Y. Piston rings and thermal spray coating used thereiin, and method for manufacturing thereof. *US 7291384 B2* **2,** (2007).

[8] Hwang, J.-H., Han, M.-S., Kim, D.-Y. & Youn, J.-G. Tribological Behavior of Plasma Spray Coatings for Marine Diesel Engine Piston Ring and Cylinder Liner. *J. Mater. Eng. Perform.* **15,** 328–335 (2006).

[9] Picas, J. A., Forn, A. & Matthäus, G. HVOF coatings as an alternative to hard chrome for pistons and valves. *Wear* **261,** 477–484 (2006).

[10] Yan, J., He, Z., Wang, Y., Qiu, J. & Wang, Y. Microstructure and Wear Resistance of Plasma-Sprayed Molybdenum Coating Reinforced by MoSi2 Particles. *J. Therm. Spray Technol.* **25,** 1322–1329 (2016).

[11] Stoloff, N. S., Liu, C. T. & Deevi, S. C. Emerging applications of intermetallics. *Intermetallics* **8,** 1313–1320 (2000).

[12] Yang, J.-M., Kao, W. H. & Liu, C. T. Development of Nickel aluminide matrix composites. *Mater. Sci. Eng. A* **107,** 81–91 (1989).

[13] Samuelsson, P., Gong, K., Li, C. & Aram, M. Piston ring having wear resistant composition. *US7052018B2* (2006).

[14] Deevi, S. C. & Sikka, V. K. Nickel and iron aluminides: an overview on properties, processing, and applications. *Intermetallics* **4,** 357–375 (1996).

[15] Velon, A. Oxidation of Fe- and Ni-aluminides in dry and humid atmospheres. *Dissertation: Chalmers University of Technology* (Chalmers University of Technology, 2001).

[16] Bengtsson, S. New powder composition and use thereof. *WO 2016/146735 A1* (2016).

[17] Lim, S. C. & Ashby, M. F. Overview no. 55 Wear-Mechanism maps. *Acta Metall.* **35,** 1–24 (1987).

Twin Wire Arc Torch with Optimized Flow Parameters

Amirsaman Farrokhpanah, Larry Pershin, Javad Mostaghimi
Centre for Advanced Coating Technologies (CACT), University of Toronto, Toronto, ON, Canada

Abstract

Wire arc is an effective and affordable type of coating process. Detailed understanding of it helps improving design and optimization of currently available guns. In wire arc process, an arc forming between two charged wires generates enough heat to melt the metallic wires. Blow of air at high velocity over this bath of molten metal leads to atomization. The formed droplets are then accelerated towards the substrate where they impact and solidify during the coating process. The current work is a follow up to our earlier numerical study [1] where numerical simulation of wire arc spraying using ANSYS® FLUENT® revealed flow circulations inside the gun can affect flow pattern and contribute to energy dissipation. In this study, an attempt has been made to validate and measure this matter using experimental procedures and to seek potential ways of improvement.

Introduction

In coating, the surface is covered with a material with the goal of protection or creation of new more advanced material. One of the most important type of coatings is thermal spray process due to high deposition ratio compared to other methods [2]. In this process, the surface is coated from impact and solidification of molten droplets at high velocities. The droplets can reach the molten state either from combustion or using an electric arc.

Wire arc is a popular form of thermal spray with many industrial applications since it is very cost effective. The quality of the coatings generated by this method is acceptable compared to other methods [3], especially with the operation cost of it being nine times lower than other methods [4].

Some studies have performed investigation of the wire arc torches. Hale *et al* [5] studied wire arc using a 2D parabolic laminar flow model. Atomization process however was neglected. Kelkar *et al* [6] has gathered several numerical models into one study, taking supersonic flow inside the nozzle in calculations using compressible flow models. Resulting flow parameters are then applied to breakup of a 2D turbulent jet. Hussary *et al.* [7] has also studied the primary atomization of the molten wire tips.

The current work is a follow-up to an earlier numerical study of ours on the wire arc process. The numerical findings from that study is used to propose design optimizations for the wire arc torch geometry.

Numerical Motivation

ANSYS® FLUENT® software has been used to solve Navier Stokes and energy equations for flow inside of a twin wire arc torch shown in Figure 1. The geometry used in the numerical model has been demonstrated in Figure 2. Turbulence is taken into account using Realizable k-ε with Enhanced near wall treatments.

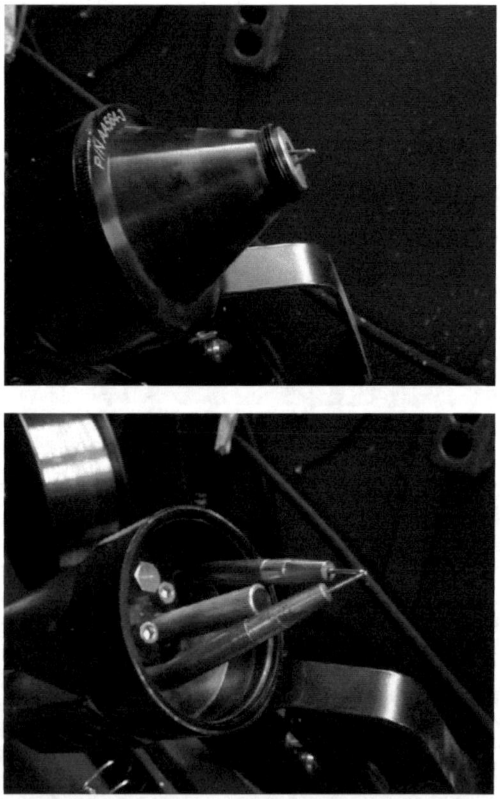

Figure 1: Wire arc torch that has been modeled here. Image on the top shows the arc with closed case. Image on the bottom shows the same arc with its lid opened [1].

The primary atomization has not been modeled and instead, particles in the form of droplets are injected into the flow in the place where the wires meet. The diameter of these droplets is taken from DPV measurements.

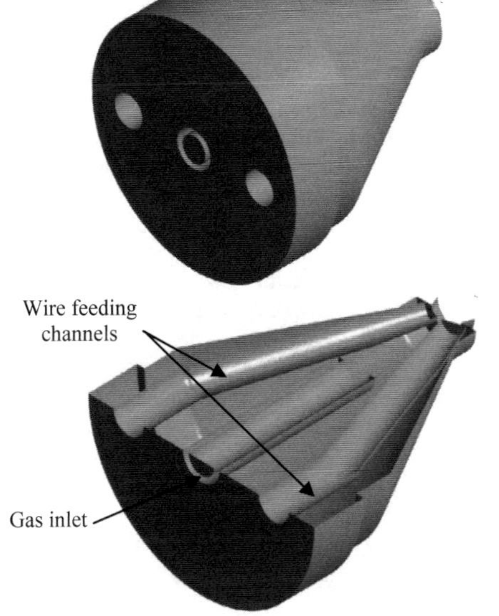

Figure 2: Geometry of the modeled torched on top. Geometry clipped to show details inside on bottom [1].

The arc itself has also been approximated as a heat source, similar to Abkenar [3]. The total input energy of torch from current and voltage are used for this source. Velocity profile for the wire arc torch at a sample working condition obtained from numerical simulation has been shown in Figure 3.

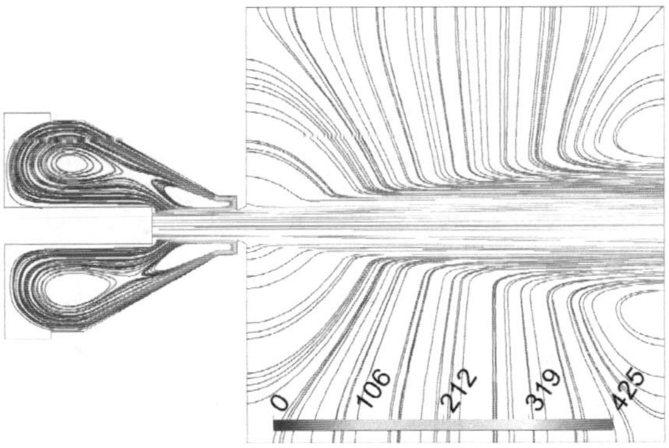

Figure 3: Velocity streamlines colored by velocity magnitude [1].

The unwanted flow circulation inside the torch contributes to loss of energy. Numerical simulations have shown that removing circulations before inlet, Turbulent Kinetic Energy, Eddy Viscosity, and Turbulence Eddy Dissipation inside the entire fluid domain are reduced by 3.46, 1.12, and 9.96 percent, respectively, while the maximum velocity inside domain has increased 4m/s [1].

Prototype Design

From the inner topology of the torch, a volume needs to be filled while accounting for the inner structures, i.e., the path needed for air to pass through the nozzle and also the channels directing the wires to the torch. Consequently, there will be spaces between the wires and air channels that will remain open. These effects are neglected at this moment. The process now requires design of this filling. This filling needs to be thermally resistance, at least near to the tip of the nozzle where the arc and radiation of heat exist. Moreover, the design should be applicable to current available torches without the need for a complete overhaul. The installation should be straight forward. It should also facilitate easy maintenance, for instance, ability to be easily removed for cleaning purposes.

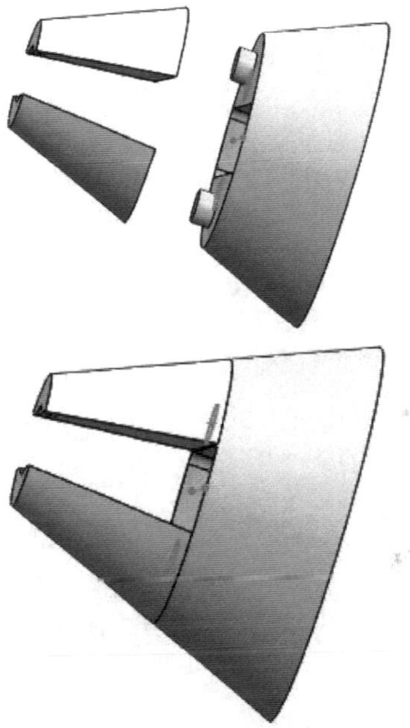

Figure 4. Easy to insert and assemble prototype for filling the void.

After iteration through many possible designs, the one shown in Figure 4 has been chosen as the final prototype and for further experimental testing. The lower half of this prototype (colored yellow in figure) is constructed from M3 crystal using 3D printing. This is done for convenience and leads to a lightweight finish. As the part is sits in the back casing of the torch, it is not exposed to harsh mechanical and thermal stresses. This part also has two cylindrical pins to support the additions in the front. These two parts (colored white in the figure), are machined from Calcium Silicate insulation material. This material ensures temperature resistance and will also help removing flow circulations near the nozzle exit. This design can be easily inserted into the torch and removed at the time of cleaning.

Prototype Testing

Effects of inserting the prototype into the torch is evaluated by performing two sets of experiments here. Several pieces of Stainless steel are coated with the wire arc torch using also Stainless steel wires. The coatings are done with and without the torch insertion. Primary visual inspection of the samples collected, indicates a narrower coating footprint resulting from insertion of the prototype. Adding the filling into the torch leads to removing inner circulations dissipating energy. The energy in the flow, now will be able to exit the nozzle carrying more energy and less initial turbulence. Hence, as expected, the angle of the diverging flow outside the torch has been decreased.

The coated substrates obtained have also examined closely under a scanning electron microscope (SEM). For this purpose, samples were cut utilizing a diamond tipped blade at 3,500RMP with a feed rate of 1.5mm/min, and then polished in epoxy, using 320, 600, and 800 grit sandpapers and then silica fluid has been applied manually.

SEM images obtained, like the one demonstrated in Figure 5 show the porosity in the final coating has been nearly 23% lowered with the porotype insertion. Measurements of the oxidation levels also suggest a nearly 13% improvement with the utilization of the prototype.

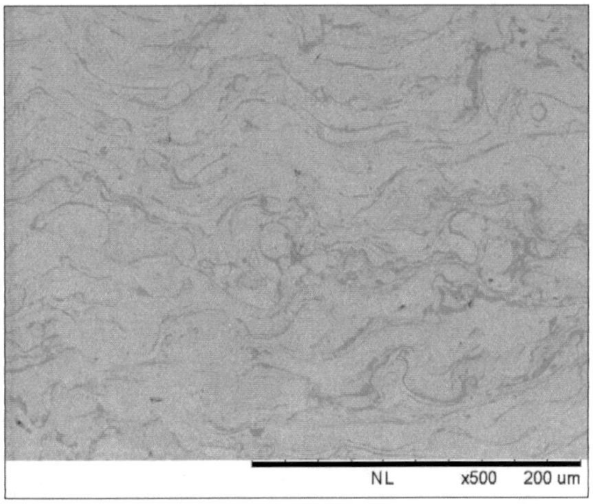

Figure 5. A sample of SEM results of the coating cross-section.

Conclusions

Wire arc spray, is an efficient form of thermal spraying. In the current study, findings from numerical models are used to improve the geometry of the torch. The main issue noted is the circulations and turbulence effects generated in the flow outside the torch which could be reduced by making improvements to inner geometry. In this manner, a prototype from partly ceramic insulation materials is built and tested to prove that quality of coating can be enhanced by filling the void inside the torch.

Acknowledgement

The authors wish to thank Wei Hung Kao, Kevin Hsu, Sahil Nagpal, and Emil Larsson for their assistance in the manufacturing process.

References

[1] A. Farrokhpanah, J. Naffrechoux, L. Pershin and J. Mostaghimi, "Investigation of Flow Parameters in Wire Arc Thermal Spraying using CFD Simulation," in *Thermal Spray 2015: Proceedings of the International Thermal Spray Conference*, Long Beach, California, USA, 2015.

[2] M. Pasandideh-Fard, "Droplet impact and solidification in a thermal spray process," 1998.

[3] A. P. Abkenar, Wire-arc Spraying System: Particle Production, Transport, and Deposition, 2007: University of Toronto.

[4] G. Montavon, S. Sampath, C. C. Berndt, H. Herman and C. Coddet, "Effects of the spray angle on splat morphology during thermal spraying," *Surface and Coatings Technology,* vol. 91, no. 1-2, pp. 107-115, 1997.

[5] D. L. Hale, A. Hodum, G. Irons, L. B. Lundberg, W. Kratochvil, D. J. Varacalle, G. C. Wilson and V. Zanchuck, "An SDE Study of Twin-wire Electric Arc Sprayed Nickel-Aluminum Coatings," in *Proceedings of 8th National Thermal Spray Conference*, Houston, TX, 1995.

[6] M. Kelkar and J. Heberlein, "Wire-Arc Spray Modeling," *Plasma Chemistry and Plasma Processing,* vol. 22, no. 1, pp. 1-25, 2002.

[7] N. A. Hussary and J. V. R. Heberlein, "Primary Breakup of Metal in theWire Arc Spray Process," *Thermal Spray 2003: Advancing the Science and Applying the Technology (ASM International),* pp. 1023-1032, 2003.

Lamellar bonding enhanced NiCr-Mo coating with comparable erosion resistance to similar bulk by plasma-spraying shell-core-structured powders

Jia-Jia Tian, Ya-Ting Wang, Cheng-Xin Li, Guan-Jun Yang and Chang-Jiu Li*
State Key Laboratory for Mechanical Behavior of Materials, School of Materials Science and Engineering, Xi'an Jiaotong University, Xi'an, Shaanxi 710049, China/E-mail: Licj@mail.xjtu.edu.cn

NiCr-Mo composite coating was prepared by plasma spraying of shell-core-structured NiCr-Mo powders. The morphologies of the NiCr-Mo powders and microstructure of the corresponding NiCr-Mo coating were characterized by SEM. Furthermore, the erosion behavior of the NiCr-Mo coating at impact angles of both 30° and 90° was investigated, and was further compared with that of the Ni20Cr coating and the In-738 alloy bulk. Results showed that fully-dense and homogenous NiCr-Mo coating with excellent interface bonding and no pure Mo inclusions was obtained. Furthermore, the erosion test results showed that the erosion rate of the optimized NiCr-Mo coating is lower than that of NiCr coating at both impact angles. Moreover, the NiCr-Mo coating presented excellent erosion resistance which was comparable as that of In-738 alloy bulk, attributing to the fully-dense microstructure and metallurgical interface bonding within coating.

1 Introduction

High performance coatings are always required to possess diverse properties such as wear resistance, corrosion resistance, chemical stability and excellent chemical properties to satisfy a host of industrial applications. Plasma sprayed nickel based alloy coatings are widely used to protect the turbine and boiler tubes from corrosion or high-temperature oxidation and so on [1]. With such applications, these coatings are simultaneously confronted with erosion damage, which is a major life-limiting issue that may result in degradation of engineering devices. Erosion is a general phenomenon presenting in many industrial applications, such as jet engines, gas turbine and so on [2]. Accordingly, enhancing the erosion resistance is a necessary approach to increase the service life of the thermal spray metallic coatings.

It has been reported that the erosion of the bulk materials is influenced by fluid flow conditions and properties of the erodent particles and target [3,4]. Researches over the last decades demonstrated that the erosion of thermal spray coatings is also influenced by those factors mentioned above in a similar way as that of bulk materials [5-7]. Moreover, it has also been confirmed by many researchers that the microstructure and morphology of the coatings have a significant effect on the erosion behavior of the thermal spray coatings [6,8,9]. Particularly, it is believed that the inter-lamellar interface bonding of the coatings is a key factor to control the erosion resistance of the thermally spray coatings. Li demonstrated that the erosion of the plasma sprayed alumina coatings was inversely proportional to the lamellar bonding ratio [6]. Furthermore, it was found that good inter-lamellar bonding and low porosity could result in excellent erosion resistance of thermal spray thermal barrier coatings and protective coatings [5,8,10]. Moreover, the erosion resistance of FeAl and NiAl intermetallic coatings is also improved with the improved inter-lamellar bonding strength [11,12]. Essentially, it is explained as follow: since the thermal spray coatings exhibit typical lamellar structure, the non-bonded inter-lamellar interfaces within coating act as lateral pre-cracks. When the coating is subjected to impact by erodents, these pre-cracks tend to propagate along the bonded fraction of the lamellar interface. The propagation of the cracks throughout whole interfaces will finally result in the spalling of lamella. Consequently, the erosion of the coating occurs through the spalling of lamellas one by one [6,11]. Hence, it is the consensus that the inter-lamellar interface bonding has a determining effect on the erosion behavior of thermal spray coatings, and it is clear that good erosion resistance of coatings is associated with good inter-lamellar bonding strength and low porosity. Therefore, enhancing the inter-lamellar bonding is a necessary approach to improve the erosion resistance of the plasma sprayed metallic coatings.

However, according to previous reports, plasma sprayed coatings exhibit a typical lamellar structure with very limited lamellar bonding ratio. It has been confirmed that, with the variation of spray parameters, the maximum lamellar bonding ratio of plasma sprayed nickel based alloy is less than one third, which is similar to that of plasma sprayed ceramic coatings when they are deposited at ambient temperature [13,14]. Therefore, many investigations have focused on improving interface bonding of coatings [15-17]. It has been recognized that the interface temperature between flattening droplet and pre-deposited particles is a critical factor to control the interface bonding formation. In this way, two basic approaches are responsible for enhancing interface bonding. The one is to increase the deposition temperature, which is the substrate surface temperature before particles are deposited. The other is to enhance the in-flight particle temperature prior to impact onto substrate. Experimental investigations [15,18] showed that the deposition temperature has a positive effect on improving the inter-lamellar bonding of plasma sprayed ceramic coatings and the interface bonding increases with the increase of the deposition temperature. Moreover, as reported by Pershin et al. [17] that the deposition temperature has a significant effect on interface bonding of plasma sprayed metallic coatings in a similar way as that of ceramic coatings. Nevertheless, although the adhesive strength of plasma sprayed nickel coating was increased when the deposition temperature was increased, no metallurgical bonding was formed between inter-lamella due to the oxide films inclusion formed between lamella when the deposition temperature was increased. Therefore, in the case of the plasma sprayed metallic coatings, taking account of the fact that oxide film, that forms readily at the surface of pre-deposited splats with the increase of the deposition temperature, will act as a barrier of formation of metallurgical bonding between lamella, it is not a suitable way to improve the inter-lamellar bonding of plasma sprayed metallic coatings by increasing the deposition temperature. In this way, enhancing the temperature or enthalpy of impinging particles is a necessary approach to improve interface bonding of thermal spray metallic coatings.

It has been known for several decades that metallurgical bonding will be formed at the interface through localized substrate melting when a molten particle with higher melting point impacts onto a substrate surface with lower melting point, for example, Mo, W and Ta molten particles impact on steel substrate, and Ni, Cr particles on Al substrate [19-21]. Therefore, in our previous article [22], we proposed a novel shell-core structured powder method, which was designed by cladding spherical Ni20Cr powder with refractory Mo to raise the particle temperature high enough to cause impact melting and consequently metallurgical bonding at interface. Results showed that with the Mo cladding treatment, the NiCr-20Mo coating presented much dense microstructure with improved interface bonding. Moreover, the localized substrate melting and resultant metallurgical self-bonding at the interface were formed due to the much enhanced particle temperature, which was also confirmed by numerical simulation. However, as pointed out in Ref [22], some Mo inclusions were observed in NiCr-Mo coating. These Mo inclusions may be introduced by separation of Mo from the NiCr core during in-flight process, which could act as a barrier for the formation of metallurgical bonding at these interfaces and hinder the further improvement of bonding strength.

In this study, in order to limit the separation of the shell Mo from the core NiCr and ensure the cladding integrity of shell layer during in-flight process, heat treatment process was employed for the ball-milled NiCr-Mo powders. Moreover, the coating prepared by plasma spraying of heat-treated NiCr-Mo powders were characterized, and furthermore the erosion behavior of this coating was investigated and compared with that of Ni20Cr coating and In-738 alloy bulk.

2 Experimental procedures

2.1 Powder Preparation

The shell-core structured powders were prepared by mechanical alloying process with the same process parameters as that used in [22]. Then the ball-milled NiCr-20Mo powders were heat-treated in an inert argon atmosphere. The heat treatment was conducted at 1000 °C for 30 minutes after parameter optimization.

2.2 Preparation and characterization of Coatings

The optimized NiCr-20Mo coating was prepared by plasma spray of the heat-treated mechanical-alloyed NiCr-Mo powders with the same process parameters as that used in Ref. [22]. Coatings were characterized with the same methods as that in Ref. [22].

2.3 Erosion Test

Erosion test was conducted by a blast erosion tester [23]. The optimized NiCr-Mo coating, Ni20Cr coating and In-738 alloy bulk samples were all of dimensions of 40 mm × 60 mm ×3 mm to ensure all erosive particles impacting on the coating surface. Angular alumina particles with an average grain size of 250 μm were used as erodents. Compressed air was used as an accelerating gas to accelerate erosive particles. The pressure of the compressed air was fixed at 0.35 MPa. The diameter and length of the nozzle for acceleration of the abrasives by the compressed air were 3.6 mm and 22 mm, respectively. Erosion tests at impact angles of both 30° and 90° were carried out to investigate the erosion behavior at different erosion conditions. The distance from the nozzle exit to the coating surface was set at 100 mm. The erosion test was carried out successively with 20 g abrasives at one test at a constant feeding rate and the sample weight loss was measured after each test.

3. Results

3.1 Microstructure of the Shell-Core-Structured NiCr-Mo Powder

3.1.1 Microstructure of the ball-milled NiCr-Mo powder

The cross sectional microstructures and surface morphologies of the shell-core structured NiCr-20Mo powder made by mechanical alloying are shown in Fig. 1. The outer white contrast area and the inner grey contrast area in Fig. 1a and 1b were identified as Mo and NiCr, respectively. It can be seen from Fig. 1a and 1c that the whole surface of the core NiCr particle was clad by Mo shell homogenously. However, from the cross-sectional picture at a high magnification, as shown in Fig. 1b, it can be found that there still existed some cracks and gaps both within Mo cladding layer and at the interface between Mo and NiCr. This fact implied that the Mo cladding layer may be not fully dense, and cladding Mo and NiCr is not in fully tight and intimate contact. This was consistent with the surface morphology, as shown in Fig. 1d, that the contact within

Mo was kind of loose and gapped. On one hand, the unintimate contact between Mo and NiCr may lead to the separation of Mo from NiCr during in-flight process, which will result in the Mo inclusions dispersing in coating, and furthermore the NiCr particle will not be heated to a high temperature due to the absence of Mo around NiCr. On the other hand, the less dense Mo cladding and loose contact between the shell layer and the core will hinder the effective thermal conduction from Mo to NiCr, which will affect the significant temperature raise for the NiCr particle.

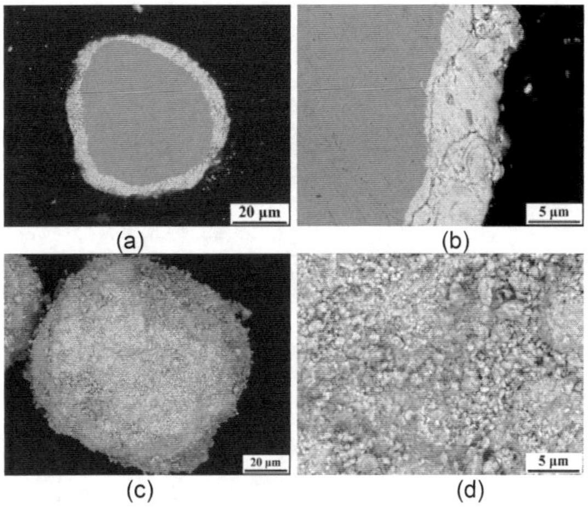

Fig. 1. Cross sectional microstructures (a), (b) and surface morphologies (c), (d) of the ball-milled NiCr-20Mo powder.

3.1.2 Microstructure of the heat-treated ball-milled NiCr-Mo powder

After the optimization of heat treatment parameters, 1000 °C for 30 minutes was chosen as the heat treatment condition for the mechanical-alloyed shell-core NiCr-Mo powders. As can be seen from the cross-section in Fig. 2a and 2b and the powder surface in Fig. 2d and 2e, after heat treatment, the outer Mo cladding layer became dense and the interface between the outer Mo layer and the core NiCr became in intimate contact, and the cracks and pores observed in the as mechanical-alloyed NiCr-Mo powders were disappeared. Moreover, a middle grey contrast band was observed at the interface between the outer Mo layer and the core NiCr. It is obviously found from Fig. 2b and 2c that there exists a diffusion layer with a thickness of about 1 µm consisting of Mo, Ni and Cr elements at the interface of the shell Mo layer and the core NiCr. As a result, after heat treatment at 1000 °C for 30 minutes, not only the outer Mo layer was sintered densely and also Mo and NiCr were bonded together chemically by a diffusion layer, which could suppress the separation of Mo cladding from NiCr particle before impact.

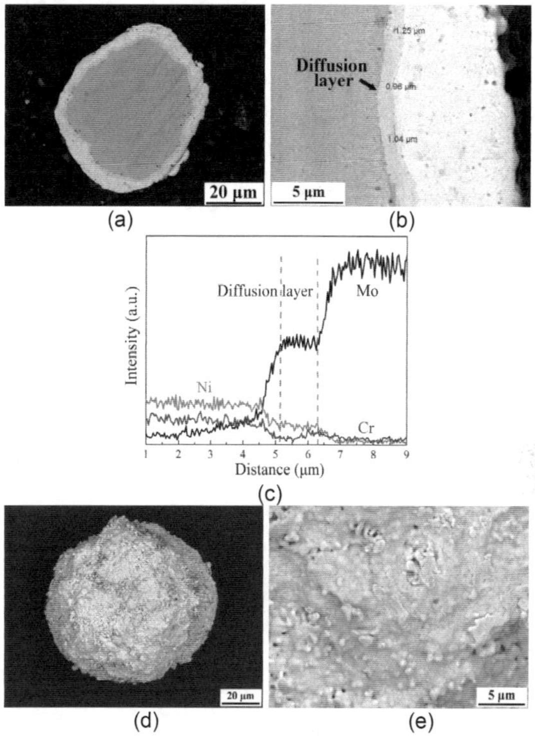

Fig. 2. Cross sectional microstructures (a), (b) and surface morphologies (d), (e) of the ball-milled NiCr-20Mo powder after heat treatment, and (c) is the line-scan result of L1 in (b).

3.2 Characterization of the NiCr-Mo coating using the heat-treated ball-milled NiCr-Mo powders

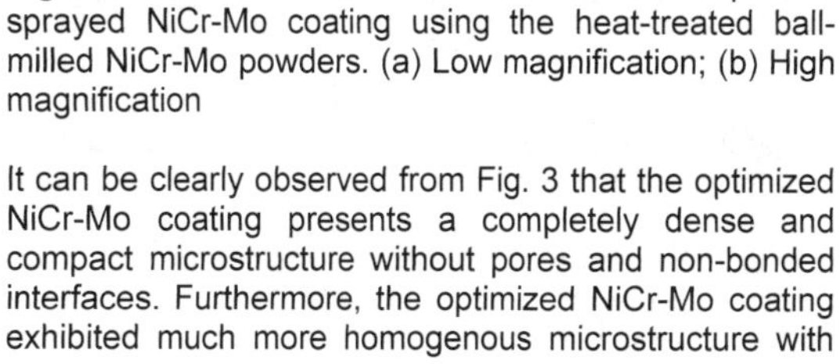

Fig. 3. Cross-sectional microstructure of the plasma sprayed NiCr-Mo coating using the heat-treated ball-milled NiCr-Mo powders. (a) Low magnification; (b) High magnification

It can be clearly observed from Fig. 3 that the optimized NiCr-Mo coating presents a completely dense and compact microstructure without pores and non-bonded interfaces. Furthermore, the optimized NiCr-Mo coating exhibited much more homogenous microstructure with almost no pure Mo inclusions, which is attributed to the formation of a metallurgical bonding layer between Mo cladding layer and core NiCr through heat treatment of the ball-milled powders. Since there is no Mo inclusions acting as a barrier for impact-induced melting of later deposited splats over it, localized substrate melting and consequently the metallurgical bonding are formed at the interfaces layer over layer in the optimized NiCr-Mo coating. This can be evidently confirmed by the well-bonded inter-lamellar interfaces observed in Fig. 3b that the inter-lamellar interfaces are indistinct and even indistinguishable. Accordingly, it can be concluded that with a modified ball-milled NiCr-Mo powders by heat treatment, the coating presented much denser and homogenous microstructure without evident non-bonded interfaces and Mo inclusions, which indicated further improved interface bonding within coating attributing to the formation of metallurgical bonding at interfaces.

3.3 Erosion behavior of the optimized NiCr-Mo coating

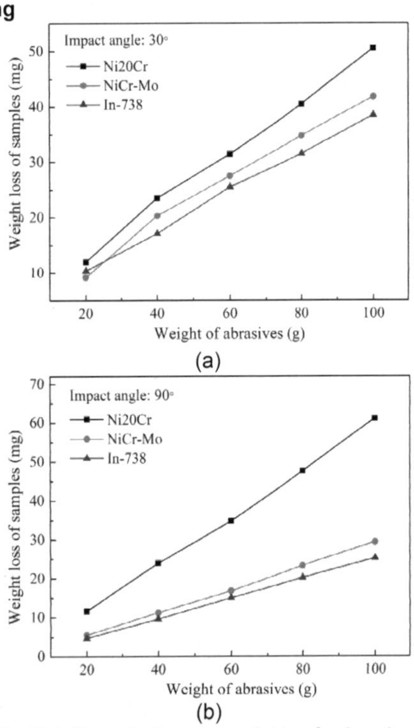

Fig. 4. Relations between weight of abrasives and weight loss of the NiCr-Mo coating, Ni20Cr coating and In-738 alloy bulk during erosion tests at the impact angles of 30° (a) and 90° (b).

The erosion behavior of the optimized NiCr-Mo coating was investigated. It was found from Fig. 4 that the coating weight loss almost increased linearly with the increase of the abrasive weight at both impact angles. At impact angle of 30°, it was observed from Fig. 5a that the erosion rates of the three samples increased with the following sequence: In-738 bulk < NiCr-Mo coating < Ni20Cr coating. On the other hand, at impact angle of 90°, as shown in Fig. 5b, the erosion rate of the optimized NiCr-Mo was similar as that of the In-738 alloy bulk, which was much smaller than that of the Ni20Cr coating.

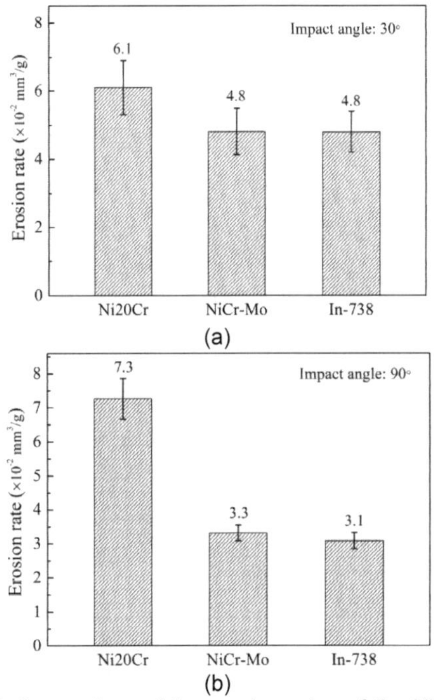

Fig. 5. Comparison of the erosion rates of the NiCr-Mo coating, Ni20Cr coating and In-738 alloy bulk during erosion tests at the impact angles of 30° (a) and 90° (b).

These erosion rates were then calculated and converted to volume erosion rates, which could eliminate the influence of the density differences on erosion samples, as shown in Fig. 5. It is obvious that at impact angle of 30°, the Ni20Cr coating exhibited the highest erosion rate of 6.1×10^{-2} mm^3/g, and the NiCr-Mo coating and the In-738 alloy bulk presented the same volume erosion rate of 4.8×10^{-2} mm^3/g. Moreover, at an impact angle of 90°, the volume erosion rate of the Ni20Cr coating was 7.3×10^{-2} mm^3/g, which was more than two times of that of the NiCr-Mo coating and the In-738 alloy bulk, as the latter two samples had considerable erosion rates of 3.3×10^{-2} mm^3/g and 3.1×10^{-2} mm^3/g, respectively. Therefore, it can be concluded that the erosion resistance of the optimized NiCr-Mo coating is comparable with that of the In-738 alloy bulk at impact angles of both 30° and 90°, which has a great improvement than that of the Ni20Cr coating. As reported in previous publications [6,7,9,11,12], the high erosion resistance which is comparable to that of the nickel based alloy bulk of the optimized NiCr-Mo coating is attributed to the significantly improved metallurgical inter-lamellar bonding within coating.

4. Discussion

4.1 Erosion mechanisms of NiCr-Mo coating and NiCr coating at an impact angle of 30°

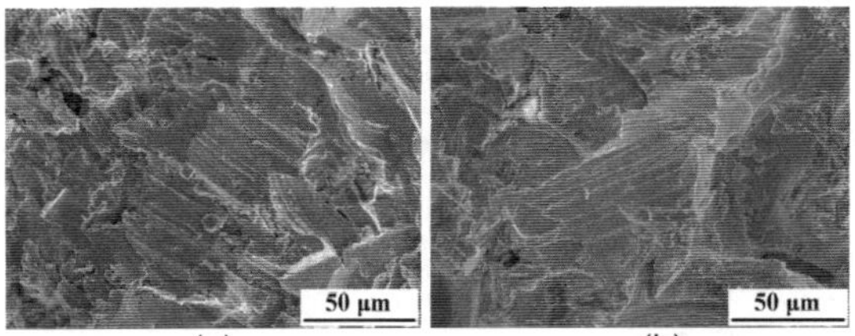

(a) (b)

Fig. 6. Surface morphologies of the (a) Ni20Cr coating and (b) NiCr-Mo coating after erosion at an impact angle of 30°.

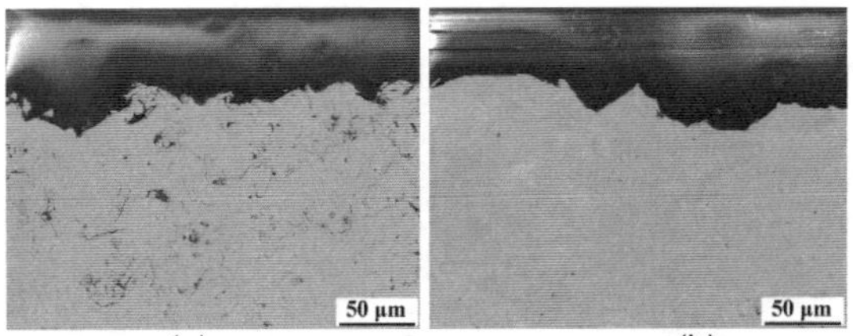

(a) (b)

Fig. 7. Cross-sectional microstructures of the (a) Ni20Cr coating and (b) NiCr-Mo coating after erosion at an impact angle of 30°.

As can be seen from Fig. 6 that many narrow ploughs were observed on the surface of both coatings which were formed through the cutting of the angular alumina erodent. Furthermore, the cross-sectional microstructures of the Ni20Cr coating and NiCr-Mo coating after erosion at an impact angle of 30° are shown in Fig. 7. The coatings only presented the same microstructure features as that observed in the as-sprayed coatings and no cracks were present on the cross-section of both coatings. Therefore, it is obvious that the dominant erosion mechanisms for both the optimized NiCr-Mo coating and the Ni20Cr coating at an impact angle of 30° are microcutting and ploughing.

4.2 Erosion mechanisms of NiCr-Mo coating and NiCr coating at an impact angle of 90°

It is obviously found from Fig. 8 that for both the Ni20Cr coating and the NiCr-Mo coating, many craters and indentation were observed, which showed evidence of plastic deformation. Moreover, it was found that these craters had platelet-like edges, which were produced by the impact extrusion that protrudes outward over adjacent material. In this way, it is obvious that a forging-extrusion mechanism, which can be also referred to as the platelet formation and removal mechanism, is responsible for a portion of mass loss that leads to the erosion of coatings at an impact angle of 90° [24].

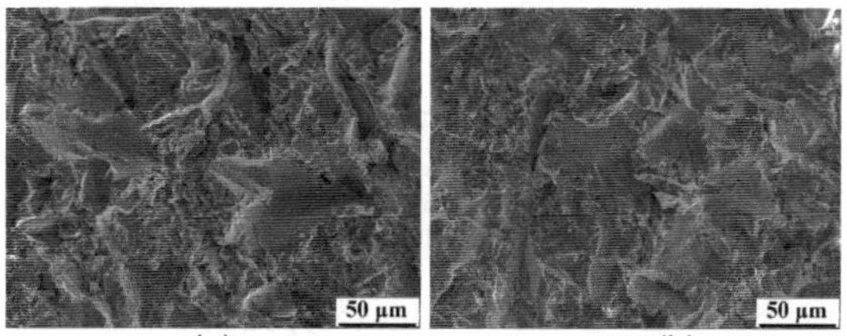

(a) (b)

Fig. 8. Surface morphologies of the (a) Ni20Cr coating and (b) NiCr-Mo coating after erosion at an impact angle of 90°.

(a) (b)

Fig. 9. Cross-sectional microstructures of the (a) Ni20Cr coating and (b) NiCr-Mo coating after erosion at an impact angle of 90°.

In the case of the Ni20Cr coating, as shown in Fig. 9a, large cracks along lamellar interfaces were obviously observed near the coating surface after the coating was subjected to erosion. When these cracks link up together through the whole lamellar interface, erosion and mass loss of coatings occur. Therefore, it can be concluded that owing to the very limited lamellar interface bonding within the Ni20Cr coating, erosion of the Ni20Cr coating occurs very easily through spalling off of splats. On the contrast, for the optimized NiCr-Mo coating, as can be seen in Fig. 9b, completely dense microstructure with sufficiently metallurgically bonded lamellae was obtained. Thus, no pre-cracks existing in the form of non-bonded interfaces are present in the coating. Consequently, much improved erosion resistance of the NiCr-Mo coating comparable to In-738 nickel-based alloy is obtained.

Therefore, for the NiCr-Mo coating, the dominant erosion mechanism at an impact angle of 90° is platelet formation and removal mechanism, which is same as that of the In-738 alloy bulk. On the other hand, in the case of the plasma sprayed Ni20Cr coating, a combined erosion mechanism is responsible for the mass loss of the coating. The dominant one is crack propagation along weakly bonded lamellar interface and the subsequent spalling off of splats, and the other is platelet formation and removal mechanism [25].

5. Conclusions

The optimized NiCr-Mo coating were prepared by plasma spraying of the heat-treated shell-core structured NiCr-Mo powders. Results showed that after heat treatment of the mechanical alloyed NiCr-Mo powder at 1000 °C for 30 minutes, the outer Mo layer and the core NiCr was metallurgically bonded together and Mo cladding layer was sintered compactly. The optimized NiCr-Mo coating exhibited completely dense and homogenous microstructure with no unbonded lamellar interfaces. Consequently, excellent interface bonding was obtained with the NiCr-Mo coating attributing to the more readily realized impact melting effect and subsequent metallurgical bonding at interfaces. The erosion test results showed that the optimized NiCr-Mo coating exhibited excellent erosion resistance, which is comparable with that of the In-738 alloy bulk. The erosion rate of the NiCr-Mo coating is lower than NiCr coating at both impact angles. The results indicate that through the novel shell-core structured powder design, the NiCr-Mo coating presents fully-dense microstructure and erosion behavior comparable with bulk materials attributing to the metallurgical interface bonding at lamellar interfaces.

Acknowledgments

The present project is partially supported by National Natural Science Foundation of China (Granted No.: 51474171).

References

[1] Zeng Z, Kuroda S, Era H. Comparison of oxidation behavior of Ni–20Cr alloy and Ni-base self-fluxing alloy during air plasma spraying. Surface and Coatings Technology, 2009, 204 (1-2): 69-77.

[2] Padavala R. Failure mechanisms in impact erosion of ductile materials. Morgantown, West Virginia: West Virginia University, 2004.

[3] Finnie I. Some observations on the erosion of ductile metals. Wear, 1972, 19 (1): 81-90.

[4] Ninham A. The effect of mechanical properties on erosion. Wear, 1988, 121 (3): 307-324.

[5] Levy AV, Wang B. Erosion of hard material coating systems. Wear, 1988, 121 (3): 325-346.

[6] Li C-J, Yang G-J, Ohmori A. Relationship between particle erosion and lamellar microstructure for plasma-sprayed alumina coatings. Wear, 2006, 260 (11–12): 1166-1172.

[7] Lathabai S, Ottmüller M, Fernandez I. Solid particle erosion behaviour of thermal sprayed ceramic, metallic and polymer coatings. Wear, 1998, 221 (2): 93-108.

[8] Levy AV. The erosion-corrosion behavior of protective coatings. Surface & Coatings Technology, 1988, 36 (1–2): 387-406.

[9] Yang GJ, Wang HT. Effect of annealing on the microstructure and erosion performance of cold-sprayed FeAl intermetallic coatings. Surface & Coatings Technology, 2011, 205 (23): 5502-5509.

[10] Davis AG, Boone DH, Levy AV. Erosion of ceramic thermal barrier coatings. Wear, 1986, 110 (2): 101-116.

[11] Hearley JA, Little JA, Sturgeon AJ. The erosion behaviour of NiAl intermetallic coatings produced by high velocity oxy-fuel thermal spraying. Wear, 1999, 233 (99): 328-333.

[12] Wang Y, Yan M. The effect of CeO_2 on the erosion and abrasive wear of thermal sprayed FeAl intermetallic alloy coatings. Wear, 2006, 261 (11–12): 1201-1207.

[13] Li C-J, Yang G-J, Li C-X. Development of Particle Interface Bonding in Thermal Spray Coatings: A Review. Journal of Thermal Spray Technology, 2012, 22 (2-3): 192-206.

[14] Ohmori A, Li C-J. Quantitative characterization of the structure of plasma-sprayed Al_2O_3 coating by using copper electroplating. Thin Solid Films, 1991, 201 (2): 241-252.

[15] Xing Y-Z, Li C-J, Zhang Q, et al. Influence of Microstructure on the Ionic Conductivity of Plasma-Sprayed Yttria-Stabilized Zirconia Deposits. Journal of the American Ceramic Society, 2008, 91 (12): 3931-3936.

[16] Sampath S, Jiang XY, Matejicek J, et al. Substrate temperature effects on splat formation, microstructure development and properties of plasma sprayed coatings Part I: Case study for partially stabilized zirconia. Materials Science & Engineering A, 1999, 272 (1): 181-188.

[17] Pershin V, Lufitha M, Chandra S, et al. Effect of substrate temperature on adhesion strength of plasma-sprayed nickel coatings. Journal of Thermal Spray Technology, 12 (3): 370-376.

[18] Hao S, Li C-J, Yang G-J. Influence of Deposition Temperature on the Microstructures and Properties of Plasma-Sprayed Al_2O_3 Coatings. Journal of Thermal Spray Technology, 2011, 20 (1-2): 160-169.

[19] Kitahara S, Hasui A. A study of the bonding mechanism of sprayed coatings. Journal of Vacuum Science and Technology, 1974, 11 (4): 747.

[20] Li L, Wang XY, Wei G, et al. Substrate melting during thermal spray splat quenching. Thin Solid Films, 2004, 468 (1–2): 113-119.

[21] Li C-J, Li CX, Yang GJ, et al. Examination of substrate surface melting-induced splashing during splat formation in plasma spraying. Journal of Thermal Spray Technology, 2006, 15 (4): 717-724.

[22] Tian J-J, Yao S-W, Luo X-T, et al. An effective approach for creating metallurgical self-bonding in plasma-spraying of NiCr-Mo coating by designing shell-core-structured powders. Acta Materialia, 2016, 110: 19-30.

[23] Ji GC, Li C-J, Wang YY, et al. Erosion performance of HVOF-Sprayed Cr_3C_2-NiCr coatings. Journal of Thermal Spray Technology, 2007, 16 (4): 557-565.

[24] Wang BQ, Geng GQ, Levy AV. Erosion-

corrosion of thermal spray coatings. Surface & Coatings Technology, 1990, s 43–44: 859-874.

[25] Kingswell R, Rickerby DS, Bull SJ, et al. Erosive wear of thermally sprayed tungsten coatings. Thin Solid Films, 1991, 198 (1–2): 139-148.

Development of wear-resistant high-entropy alloy coatings produced by thermal spray technology

M. Löbel, T. Lindner, T. Lampke
Institute of Materials Science and Engineering, Chemnitz University of Technology

C. Kohrt
Rhein-Ruhr Beschichtungs-Service GmbH

The alloying concept of High-Entropy Alloys (HEA) has attracted much scientific interest due to an interesting combination of properties. Previous investigations have shown that high hardness and strength, comparable to bulk metallic glasses, can be achieved. Furthermore, HEAs show distinct ductility and good high-temperature resistance. First investigations on tribological properties are indicating high wear resistance. Previous investigations of the alloy system AlCoCrFeNiTi in bulk state have shown promising properties. Therefore, the alloy AlCoCrFeNiTi with equimolar composition was selected for transferring bulk properties to thermally sprayed coatings. The focus of this contribution is on studying tribological properties of thermally sprayed HEA coatings to enlarge the field of possible applications.
Feedstock material production was carried out by high-energy ball milling (HEM) and inert gas atomization. Subsequently, coatings were deposited by Atmospheric Plasma Spray (APS). Tribological properties of the coatings under different wear regimes were investigated in ball-on-disk wear tests, oscillating wear tests and scratch tests. The tribological properties are compared with a conventional hard chrome plating and correlated with microstructure.

1 Introduction

The concept of HEA was introduced by J.W. Yeh et al. [1]. This new concept comprises an alloy composition with at least 5 elements each with a share of 5 – 35 at.%. Investigations have shown the formation of simple solid solutions with face centered cubic (FCC) and body centered cubic (BCC) structures due to the stabilization by high entropy effect. Contrary to first expectations, no brittle, intermetallic or complex phases were formed [1-2]. Independent investigations by Cantor et al. revealed a single phase face centered cubic structure in the equimolar alloy FeCrMnNiCo [3]. Further investigations of HEAs showed the formation of a multiphase structure consisting of several phases for most alloy system. For alloys which do not consist of one simple solid solution, the term compositionally complex alloys (CCA) was introduced [4].
In this paper, the equimolar alloy AlCoCrFeNiTi is investigated. Microstructural studies of the alloy system showed a dendritic structure with two BCC solid solutions [5]. Mechanical investigations in compression tests revealed promising properties, including a high fracture strength of 2.58 GPa and plastic strain of 8.8 %. In this particular alloy system, these properties were attributed to solid solution strengthening by large Ti atoms [5]. The focus of subsequent studies were microstructure and mechanical properties. Two alloying elements showed strong influence on alloy properties: Titanium and Aluminium. Aluminium is a promoter of BCC phases. For alloys with high Al content, mainly strong and brittle BCC phase are formed. Adding large Titanium atoms causes strengthening by solid solution hardening. But for alloys with high amounts, intermetallic precipitates were formed, causing embrittlement [5, 6].
First investigations on tribological properties of the CCA system AlCoCrFeNiTi showed good wear resistance, also in comparison with conventional steel (SUJ2, SKH51), which makes it promising for further investigations [7].
The wear resistance of thermally sprayed HEAs and CCAs have only been briefly studied yet [8].
The aim of this study is to apply the alloy AlCoCrFeNiTi for surface engineering to produce wear resistant coatings.

2 Experiments

In this study, the alloy AlCoCrFeNiTi with equimolar composition is investigated. Two different feedstock material production routes are investigated. The first approach to produce thermal spray feedstock was the application of high-energy ball milling (HEM). In a first step, a powder blend was produced. Therefore, pure elemental powders were utilized, except Fe. Fe was added as pre-alloy FeCr13. Cyclic milling was carried out in a Zoz Simoloyer CM08 mill with a rotor speed of 400-700 rpm and a powder mass of 0.8 kg. 8 kg of steel balls (100Cr6) with a diameter of 5 mm were used as grinding medium. Stearic acid was added as a process control agent. The process was carried out under air.
Another approach for the production of thermal spray feedstock was inert gas atomization, where Argon

was used as a process gas. The particle fraction with D_{90}< 100 µm was used for thermal spray.

Both feedstock materials were investigated regarding their powder size distribution by Laser diffraction analysis (Cilas 930). Furthermore, metallographic cross sections were prepared and investigated in Scanning Electron Microscope / SEM (Leo 1455VP) equipped with a backscattered electron detector (BSD) to examine the structure, homogeneity and particle morphology. The chemical composition was investigated by energy dispersive X-Ray spectroscopy / EDX (EDAX Genesis) and phase analysis was conducted by X-Ray diffraction / XRD (D8 Discover / Bruker AXS) using Co-K_α-radiation.

Before coating, steel flat substrates (S235) were prepared by corundum blasting (grain size: 425-600 µm; pressure: 3.5 bar; distance: 100 mm; angle: 60°). The coating process was carried out with the APS system GTV F6, applying the parameters displayed in Table 1.

Table 1. APS parameters.

	HEM powder	atomized powder
Ar	38 l/min	54 l/min
H_2	12.5 l/min	9 l/min
I	600 A	530 A
spray distance	120 mm	120 mm
speed	70 m/min	100 m/min
offset	4 mm	4 mm

The microstructure and phase composition of the coatings was investigated by analyzing cross sections in SEM (BSD). Furthermore, the chemical composition was determined by EDX. Wear behavior under different tribological conditions was tested in ball-on-disk (Tetra Basalt Tester), oscillating wear (Wazau SVT 40) and scratch tests (CSM Revetest-RST). The parameters are shown in Table 2.

Table 2. Wear test parameters.

ball-on-disk test		***oscillating wear test***		***scratch test***	
force	20 N	**force**	26 N	**force**	1-200 N
radius	5 mm	**frequency**	40 Hz	**speed**	2.5 mm/min
speed	96 rpm	**time**	900 s	**length**	5 mm
cycles	15916	**amplitude**	0.5 mm	**tip**	Truncated diamond Cone r: 200 µm
Ø Al_2O_3	6 mm	**Ø Al_2O_3**	10 mm		

Resulting wear marks were investigated with the Laser Scanning Microscope (LSM) Keyence VK-X200.

3 Results and Discussions

Feedstock material particle size distribution was investigated by laser diffraction analysis. For HEM feedstock material, the analysis was not possible due to the irregular particle shape. For the atomized powder, a d_{50} value of 42.9 µm and a d_{90} value of 109.9 µm was revealed. Phase analysis by XRD was carried out for both feedstock materials. Resulting diffractograms are displayed in Fig. 1.

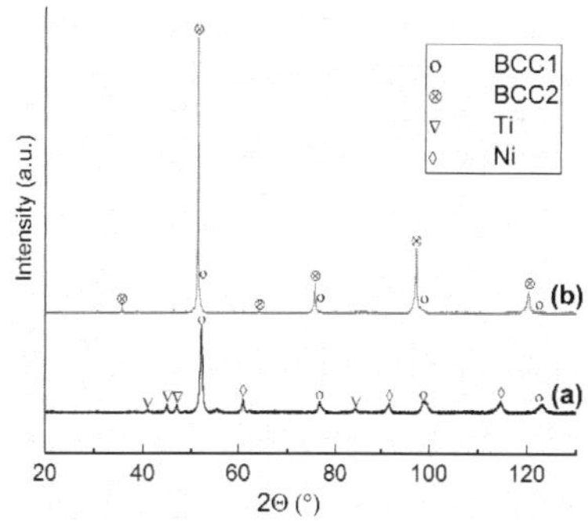

Fig 1. XRD diffractograms of (a) HEM and (b) atomized powder feedstock.

Peaks of elemental Ti and Ni can be seen in the diffraction pattern of the HEM powder, showing that no homogeneous and fully alloyed powder was obtained with this production method. Furthermore peaks appear which can be referred to a BCC phase (BCC1). The diffraction pattern of the atomized powder exhibits high intensity peaks, which can be attributed to another BBC phase (BCC2). Diffraction patterns also match with the BCC1 phase.

Metallographic cross sections of both feedstock materials are displayed in Fig. 2.

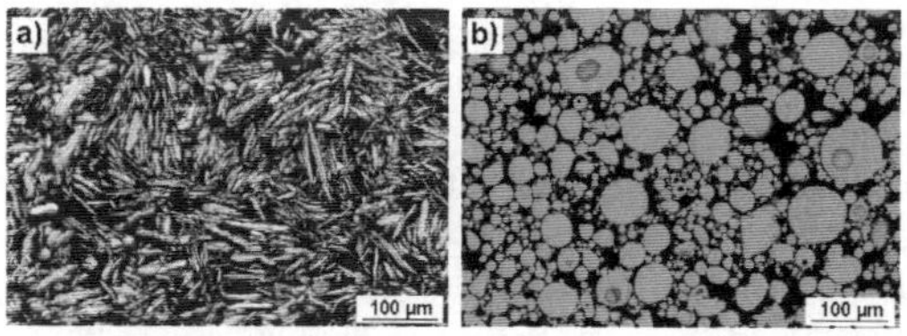

Fig 2. Cross-sections of (a) HEM powder and (b) atomized powder feedstock.

In the cross-section of the HEM powder, plate-like and heterogeneous particles can be seen. The atomized powder consists of more spherical and homogeneous particles.

The chemical composition of both feedstock materials was investigated by EDX. Results are shown in Table 3.

Table 3. Chemical composition of feedstock material.

	Al [at.%]	Co [at.%]	Cr [at.%]	Fe [at.%]	Ni [at.%]	Ti [at.%]
Nominal	*16.7*	*16.7*	*16.7*	*16.7*	*16.7*	*16.7*
HEM powder	24.3	19.8	15.7	13.7	13.8	12.7
atomized powder	18.3	16.0	16.9	16.7	16.1	16.0

For the HEM powder, a distinct deviation between measured and nominal chemical composition exists. In contrast, the atomized powder is in good accordance with the nominal composition.

After the coating process, phase analysis of the resulting coatings was carried out by XRD. The resulting diffractograms are shown in Fig. 3.

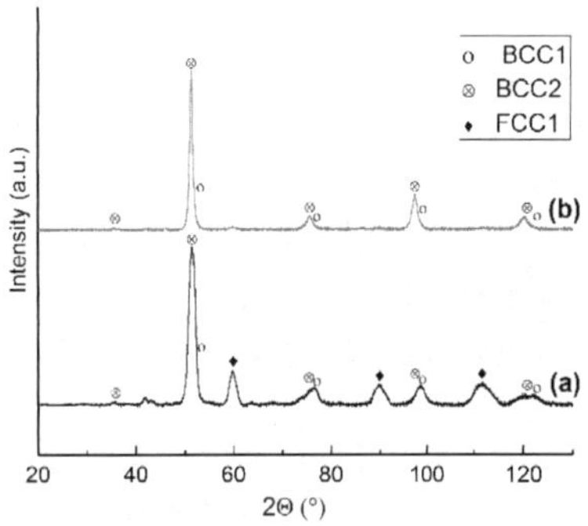

Fig 3. XRD patterns of coatings produced with a) HEM and b) atomized powder feedstock.

For the coatings deposited with HEM powder, feedstock diffraction peaks appear, which can be attributed to two BCC phases. These phases are identical to the phases BCC1 and BCC2 in the feedstock material. In contrast to the HEM powder feedstock, no peaks of elemental Ti and Ni appear. Therefore a FCC phase is formed. The diffraction pattern of the coating produced with atomized powder feedstock shows the same peaks like the feedstock material, comprising two BCC phases.

Subsequently, the coating microstructure was investigated in SEM. Cross- sections of the coatings are displayed in Fig. 4.

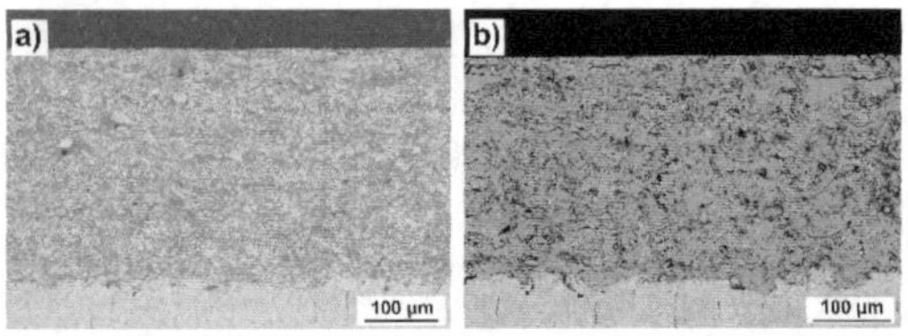

Fig. 4. SEM (BSD) images of coating cross sections produced with a) HEM and b) atomized powder feedstock.

The coating produced with HEM powder feedstock exhibits an inhomogeneous microstructure. Distinct contrast between different phase areas can be seen with BSD contrast, indicating strong differences in chemical compositions of the phase areas. In the cross section of the coating produced with atomized powder feedstock, no distinct contrast between single phase areas is observed, indicating a more homogeneous state. However, black oxide lamella and pores also exist.

The chemical composition of both coatings has been investigated by EDX and is shown in Table 4.

Table 4. Chemical composition of APS coatings produced with HEM and atomized powder feedstock.

	Al [at.%]	Co [at.%]	Cr [at.%]	Fe [at.%]	Ni [at.%]	Ti [at.%]
Nominal	*16.7*	*16.7*	*16.7*	*16.7*	*16.7*	*16.7*
HEM Powder	16.1	16.9	11.2	16.1	16.9	22.8
Atomized powder	18.7	16.3	15.2	16.3	16.5	17.0

The chemical composition of the coating produced with HEM powder feedstock distinctly deviates from the nominal composition, especially the Cr and Ti content. Furthermore, the chemical composition is also markedly changed in comparison to the feedstock material due to vaporization during the coating process. In contrast, the chemical composition of the coating produced with atomized powder feedstock is in better accordance with nominal composition. Vaporization of single elements has been reduced by a more homogeneous state of the feedstock material.

After microstructural investigations, the wear behaviour under different tribological conditions was studied. Results are displayed in Fig 5.

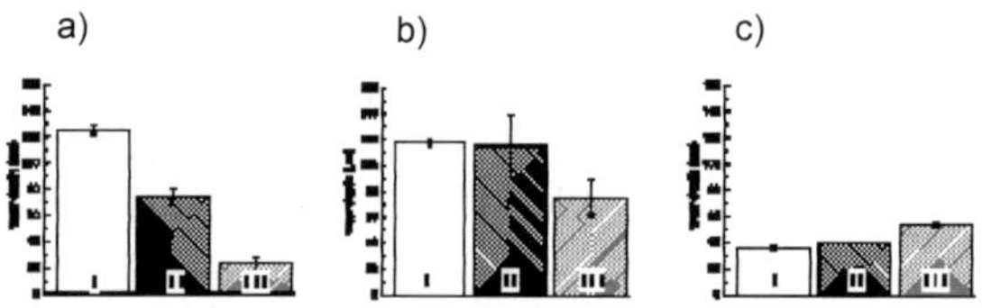

Fig 5. Wear test results of a) ball-on-disk b) oscillating wear and c) scratch test of hard chrome plating (I) and coatings produced with HEM (II) and atomized powder feedstock (III).

For the reference hard chrome plating, the highest wear depth was measured in ball-on-disk test. By applying thermal spray with CCAs, a reduction of wear depth can be achieved. For the coating produced with atomized powder feedstock, the lowest wear depth and highest wear resistance was measured.

Oscillating wear test results reveal similar wear depth for hard chrome plating and APS coating produced with HEM powder feedstock. However the coating produced with HEM powder feedstock exhibits a high standard deviation, which is caused by the inhomogeneous state of the coating. In contrast, the wear depth could be distinctly reduced by applying atomized powder feedstock due to the more homogeneous microstructure and the absence of soft FCC phase.

Results of the scratch test reveal the lowest wear depth for the hard chrome plating. Higher wear depths were measured for both APS coatings. The highest wear depth was measured for the coating produced with atomized powder feedstock. The comparatively bad wear behaviour of CCA coatings produced with APS in scratch test have to be investigated in more detail. The wear behaviour is probably also influenced by the appearance of pores and oxides, which might be reduced by further optimization of thermal spray parameters.

4 Conclusions

The CCA AlCoCrFeNiTi with equimolar composition has been applied for thermal spray technology in this study. In a first step, feedstock powder was produced by the processes of HEM and inert gas atomization. The process of inert gas atomization proved to be most suitable for the production of homogeneous and spherical powder. Phase analysis revealed two BCC phases in this powder. Both feedstock powders were processed by APS. For the coating produced with atomized powder feedstock, a comparatively more homogeneous microstructure occurs. The investigation of wear behaviour in ball-on-disk and oscillating wear tests revealed superior wear resistance for the coating produced with atomized powder feedstock. In these tests, the wear resistance was also superior to the hard chrome plating. However in scratch tests, the coating produced with atomized powder feedstock showed lower wear resistance than the hard chrome plating. This behaviour might also be influenced by the presence of pores and oxides, which can be reduced by further optimization of thermal spray parameters.

5 Acknowledgements

The authors gratefully acknowledge the Arbeitsgemeinschaft Industrieller Forschungs-vereinigungen "Otto von Guericke" e.V. (AiF) for support of this work (AiF-No. KF2152612CK4) with funds from the German Federal Ministry for Economic Affairs and Energy. Furthermore the authors would like to thank Marc Pügner for conducting the XRD measurements and Ilka Günther for conducting the particle size analysis.

6 Literature

[1] Yeh, J.-W.; Chen, S.-K.; Gan, J.-Y.; Lin, S.-J.; Chin, T.-T.; Shun, T.-T.; Tsau, C.-H. and Chang, S.-Y.: Formation of Simple Crystal Structures in Cu-Co-Ni-Cr-Al-Fe-Ti-V Alloys with Multiprincipal Metallic Elements, Metall. Mater. Trans. A (2004), Vol. 35A, pp. 2533-2536.

[2] Yeh, J.-W.; Chen, Y.-L.; Lin, S.-J. and Chen S.-K.: High-Entropy Alloys – A New Era of Exploitation, Mater. Sci. Forum (2007), Vol. 560, pp. 1-9.

[3] Cantor, B.; Chang, I. T. H.; Knight, P. and Vincent, A. J. B.: Microstructural development in equiatomic multicomponent alloys, Mater. Sci. Eng. A (2004), 375-377, pp. 213-218.

[4] Miracle, D. B.: Critical Assessment 14: High entropy alloys and their development as structural materials, Mater. Sci. Tech. Ser (2015), Vol. 31, pp. 1142-1147.

[5] Zhou, Y. J.; Zhang, Y.; Wang, Y. L. and Chen, G. L.: Solid solution alloys of $AlCoCrFeNiTi_x$ with excellent room-temperature mechanical properties, Appl. Phys. Lett. (2007), 90, 181904.

[6] Zhang, K.B.; Fu, Z.Y.; Zhang, J. Y.; Wang, W.M.; Wang, H.; Wang, Y.C.; Zhang, Q.J. and Shi, J.: Microstructure and mechanical properties of $CoCrFeNiTiAl_x$ high-entropy alloys, Mater. Sci. Eng. A (2009), 508, pp. 214-219.

[7] Chuang, M.-H.; Tsai, M.-H.; Wang, W.-R.; Lin, S.-J. and Yeh, J.-W.: Microstructure and wear behavior of $Al_xCo_{1.5}CrFeNi_{1.5}Ti_y$ high-entropy alloys, Acta Mater. (2011), Vol. 59, pp. 6308-6317

[8] Huang, P.-K.; Yeh, J.-W.; Shun, T.-T. and Chen, S.-K.: Multi-Principal-Element Alloys with Improved Oxidation and Wear Resistance for Thermal Spray Coating, Adv. Eng. Mater. (2004), No. 1-2, pp. 74-78.

Thermal Spraying of Thin Metallic Coatings

F. Trenkle, R. Koehler, M. Winkelmann, F. Wuest, J. Luth, S. Hartmann
obz innovation gmbh, Bad Krozingen, Germany
info@obz-innovation.de ; +49 7633 90899 0

Metallic coatings can be produced with thermal and cold gas spraying processes very easy. But are required, for economical or technological reasons, coating thicknesses below 50 µm the tasks become more challenging for those well established spray processes. obz innovation has developed specific spray processes, which even allows producing metallic coatings below 20 µm. Such coatings are very interesting for example for cold gas sprayed silver coatings, which have a high purity and a good electrical conductivity. Thinner coatings with such valuable materials directly gain economical advantages, often also compared with commonly used thin film processes.

1 Introduction

Electroplating is popular as a standard process for producing thin metallic coatings. But for some applications this technique brings along technological drawbacks concerning the usable materials as well as economical disadvantages. Economic issues play a decisive role, particularly related to higher coating thicknesses (>10 µm), compared to electroplating, and smaller surfaces related to the whole component size. Electroplating process times of such applications is often too long. Furthermore, a big component with small surfaces to coat, requires a complex masking to seal it against fluids. Otherwise the material consumption would rise because of plating the non-masked surfaces. Thermal and especially cold gas spraying could be an interesting alternative to these applications, however, they usually have a minimum coating thickness of about 50 µm. There is a technological-economical gap under those 50 µm, what means between 10 and 40 µm. Often, this gap can't be satisfied neither from the established thin film processes, nor conventional thermal spraying.

2 Preliminary tests to optimize the coating parameters

The economical disadvantages of an unneeded thick coating are shown most clearly on expensive materials. Therefore, in this study silver coatings were developed using cold gas spraying on a Kinetiks 4000 cold gas system. For this purpose, the powder feed rate and the amount of gas, the injection of the powder into the gun and the geometry of the nozzle were optimized. In addition to that, based on tests with different temperatures and pressures, an optimized process window was found for the silver coating. Figure 1 illustrates a cold gas silver coating with a higher coating thickness and suitably optimized process parameters. It can be realized that cold sprayed silver coatings are homogeneously coated and no pores can be seen. Related to electroplating, a comparably good coating quality seems to be possible.

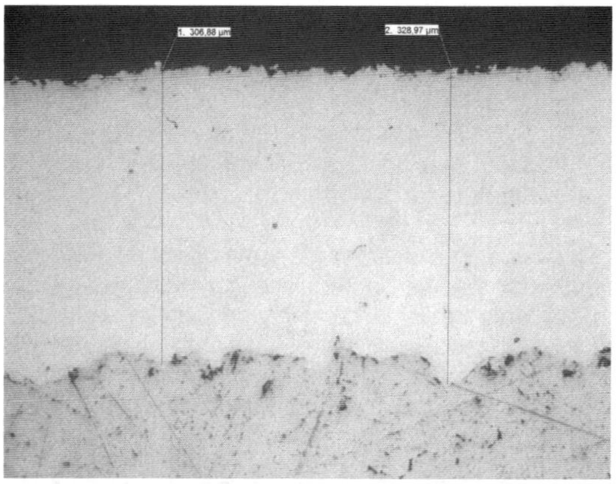

Fig. 1: Cross section of a 300 µm thick cold gas silver coating with optimized spray parameters. A very good coating quality and nearly no pores can be seen.

3 Influence of substrate- and coating roughness

A boundary condition of thin coatings is usually a reproducible minimum material thickness of e.g. 10 µm, which means independent on the surface roughness (max. / min. peaks) of the coating. Especially in case of thin coatings is an optimally adjusted roughness of the sandblasting and the surface of the coating required. Otherwise, the roughness peaks of the sandblasting or the "valleys" of the surface roughness reduce the minimum material thickness or even generate defects. This issue can be easily understood by means of the areas marked with circles in the micrograph in Figure 2. The Roughness of sandblasting is too high for the combination of the used powder fraction and spray parameters. In order to meet the criterion of the minimum material thickness, such a process would increase the consumption up to a multiple of the actually required coating material. On thin coatings it is therefore particularly important from an economic point of view to do an optimization of the roughness.

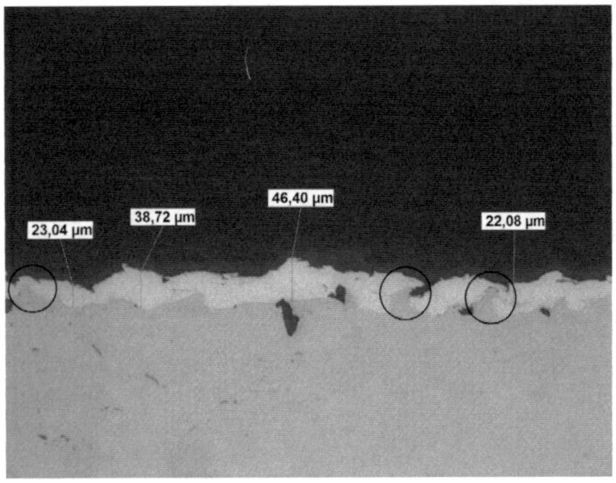

Fig. 2. Without process optimization, especially of the surface roughness and sandblasting, hardly any economically reasonable coating in the range of 10-40 µm can be produced.

4 Optimizing of the minimal coating thickness

As explained in the last section, the optimization of the sandblasting is a prerequisite for minimizing the coating thickness. In combination with an adapted powder grain size, this process optimization is the key for producing the minimum coating thicknesses. It allows the lowest possible material consumption, thus the best possible process economy.

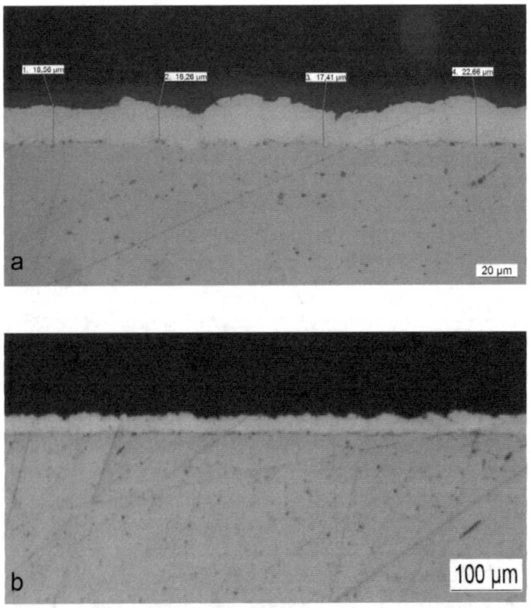

Fig. 3. A microscopic detail (a) of the optimized silver coating illustrates the good bonding. The average coating thickness is about 18 µm and the minimum material thickness is about 13 µm. The overview (b) showns an uniform coating above the copper material.

The objective of a minimized average coating thickness, while achieving a minimum material thickness was evaluated in further tests. For this purpose, the movement speed of the gun was varied. The powder flow was kept constant so that the material thickness depends primarily on the travel speed of the gun, for optimally adjusted sandblasting. The optimized silver cold gas coating is shown in Figure 3. Figure 3a shows a close-up view of a coating with an average coating thickness of approximately 18 µm. The average coating thickness is essential for the total material consumption and is therefore also relevant for economic aspects. The silver coating is thus located in the lower third of the coating thickness range of 10-40 µm as mentioned in the introduction. The spots with a minimal coating thickness in the microscopic image show about 13 µm thickness. The described objective of a minimum material thickness of 10 µm thus has been reached with sufficient certainty. A further reduction of the average coating thickness by approximately two micrometres would be conceivable according to this.

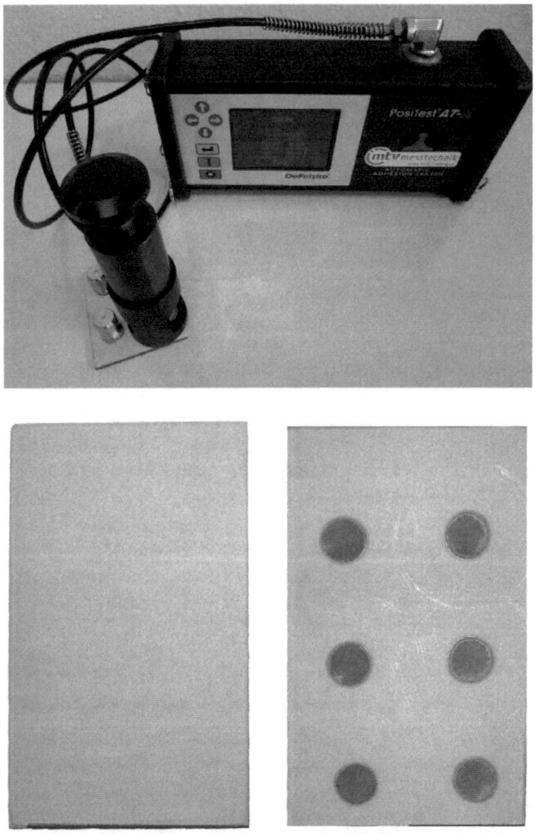

Fig. 4. Adhesive tensile strength tester (top) for testing "as sprayed" coatings directly on a part. Silver coating before (down left) and after (down right) the adhesive tensile strength test.

Overall, the results show the possibility of reaching a difference of approx. 5 to 6 µm between minimum material thickness and average coating thickness, in a serial production process. This allows to set customer and application specific the design as well as calculations of the coating costs. Furthermore, Figure 3a shows the bonding between silver and copper bas material, which looks visually well.

The good bonding is also confirmed by the tests of the adhesive tensile strength, determined with a measuring instrument from DeFelsko (PosiTest AT-A). The adhesive tensile samples were stuck directly onto the silver coated copper plate, cf. Figure 4 (top). The adhesive tensile strength is 31,2 MPa, averaged over five tests, with a standard deviation of 3,5 MPa. Overall, the thin silver coating shows very good properties. It demonstrates that for specific applications, thermal and / or cold gas spraying are generating alternatives to commonly used thin film processes like electroplating.

5 Conclusion

Thin metallic coatings are widespread used in various applications, like corrosion protection, electrically conductive contacts, circuit paths or areas e.g. for heaters, medical technology and many more. Electroplating, as a standard processes for producing thin metallic coatings, provides no practical possibility of coating deposition for some metals or alloys, e.g. Titanium. In addition, a galvanic coating is quite cost-intensive for parts which have to be coated only partially. For higher coating thicknesses > 10 µm, electroplating is less efficient due to the long process times. Thermal and cold gas technology can cover these applications, corresponding to the shown results. With the example of silver cold gas coatings, obz innovation has shown that coatings can be carried out on the basis of conventional equipment down to a region of 10 - 20 µm. These coatings can have an economic and / or technologic advantage in corresponding applications, compared to electroplating. In addition, this technology makes it possible to produce almost any metal or metallic alloy as a coating.

6 Acknowledgment

Supported by:

Federal Ministry
for Economic Affairs
and Energy

on the basis of a decision
by the German Bundestag

Cold Gas Spraying of lead-free bearing bronzes

S. Theimer*, P. Brethack*, F. Gärtner*, T. Klassen*, Hamburg / D

*Department of Mechanical Engineering, Helmut Schmidt University, University of the Federal Armed Forces Hamburg, Hamburg, Germany

In current process techniques to manufacture sliding bearings, bronze components are soldered to the respective parts, which is cost and energy intensive. Apart from that, so far most bearing materials still contain lead, which in new applications is omitted by EU law to avoid associated health risks. The present study aims to offer solutions for both by using cold gas spraying as additive manufacturing technique for processing bearings directly onto steel parts and by applying that to new lead-free bronze alloys.
A lead-free bronze alloy was processed as powders by gas atomization and classified to optimum sizes for cold spraying. During cold spraying, the process gas pressures and temperatures as well as the substrate temperature were varied with the aim to study influences by impact conditions and effective surface temperature on particle deformation and bonding. Respective coatings show low porosity, high hardness and high electrical conductivities. With properties similar to that of bulk cast material respectively manufactured parts should meet the requirements for new bearing applications.

1 Introduction

Bronze and brass parts serve in manifold tribological applications. As first approach, the project aims to meet the requirements concering loads on hydraulic pump parts. The major challenge is to withstand pressures up to 50 MPa as applied to the control plate of such system [1,2]. Up to now, bronze parts for sliding surfaces are cast and soldered to the respective steel construction. Associated process steps are time and energy intensive. Moreover, most tribological bronzes still contain lead to guarantee needed emergency operating features. For avoiding health risks, the European Union prohibited the usage of lead [3, 4].

The present study aims to offer solutions for both problems by using cold spraying as additive manufacturing technique for processing bearings directly onto steel parts and by applying that to new lead-free bronze alloys.

In cold spraying, the powder temperature is always below the melting temperature of the respective spray material. By that, oxidation and undesired phase transformation can be avoided, in contrast to thermal spraying [5] techniques. In general, cold spraying can generate dense and high strength coatings if the spray process is well controlled and adjusted [6 - 12]. By increasing the applied process gas temperature and pressure, cold spraying was developed to a well-established coating technique [13] as well as an additive manufacturing method [14].

2 Methods

2.1 Calculations and powder requirements

Particle impact conditions and critical conditions for bonding were calculated by the KSS ® software package from Kinetic Spray Solutions (Buchholz, Germany).

For different spray parameter sets regarding process gas temperature and pressure, as well as spray distance and nozzle geometry, impact conditions (particle velocity v_p and temperature T_p) were calculated by computational fluid dynamics (CFD) by considering material density ρ and heat capacity c_p, as well as powder sizes. The critical velocities for bonding were calculated on basis of powder properties, in the first instance data concerning ultimate tensile strength UTS, melting temperature T_m, density ρ and specific c_p as available in literature for bulk materials. These first calculations are mainly used to gain general information on parameter selection [7] and to define the most suitable particle size distribution.
As spray material, CuSn6Ni6Zn2-bronze was gas atomized by using the EIGA-methode [15], and classified to the needed size range. Based on measured powder strengths, the calculations were refined, providing more precis information on expected coating qualities. Following the approach by Assadi et al. [7], the dimensionless coating quality parameter η was calculated as the ratio between the impact velocity (v_p) and the critical velocity (v_{crit}) of the particle. For easier comparsion of different spray parameter sets, the calculations refer to the mean diameter (39 µm) of the given particle size distribution.

2.2 Powder analyses

Scanning Electron Microscopy

Scanning electron microscopy (SEM) was performed by using a Quanta 650 instruments from FEI (Brno, Czech Republic) at an acceleration voltage of 20 kV.

Measurements of Particle Size Distribution

Particle size distribution was analysed by laser scattering using a LA-910 from Horiba Ltd. (Kyoto, Japan).

Particle Compression

Particle strength was analysed by the particle compression method established by H. Assadi et al [16] using a modified nanoindenter NHT² from CSM Instruments (Gras, Austria).

2.3 Cold Spraying

For cold spraying, the HSU prototype of the later commercialized CGT-Kinetics 8000 spray system was used. As powder feeder, commercial equipment from CGT (Ampfing, Germany) was used. The experiments were performed with a water-cooled D51 nozzle and with powder injection 180 mm up-stream the nozzle throat for sufficient preheating by using nitrogen as process gas. The spray distance was set to 60 mm. The kinematics of the process was fixed to a traverse velocity of 250 m/s, a line-distance of 2 mm and a spray layer overlap of 1 mm. The powder feed rate was kept to 3 rpm, corresponding to about 10 cm^3/min. Spraying was performed onto steel 1.0503 substrates. Flat sheets in dimensions of 50 mm x 70 mm x 4 mm were used for microstructural and conductivity analyses. Cylinders a diameter of 25 mm and a height of 25 mm were used for bond stength tests.
The deposition efficiency (DE) was determined as the ratio between the mass gain of the coating and the powder mass flow, considering the process kinematics and substrate geometry.
The used material for all substrates was steel 1.0503.

Optical Microscopy

For analysing the coating microstructure and porosity, an optical microscope (OM) type DMRM with analyse software Axio-Vision ® from Leica (Wetzlar, Germany) was used. Quantitative analyses of porosity was performed according to ASTM-Standard E2109-1 using the software AxioVision from Zeiss (Jena, Germany).

Hardness measurements

Vickers Hardness was measured on cross sections as HV0.3 under a load of 2.942 N according to ASTM E384-10 standard, using an indentation tester type miniload from Zeiss (Jena, Germany).

Adhesion testing

The bond strength was measured according to DIN EN 10002 part 1 by using a tensile testing machine Z100 from Zwick/ Roell (Ulm, Germany). The coated abutting face of the specimen (Ø 25 mm x 25 mm) was bonded to counterbodies by using Ultrabond 100 glue from HTK (Hamburg, Germany).

Electrical conductivity

The electrical conductivity was determined according to the standard ASTM E1004 by using a Sigmascope® SMP10-HF and a measuring sensor ES40HF from Fischer GmbH (Sindelfingen, Germany) at a frequency of 1250kHz. The conductivity serves as simple and direct measure of coating quality [17].

3 Results and Discussion

3.1 Impact Conditions

For easier comparison, the results from the calculations are presented in terms of the generalised coating quality parameter η to distinguish influences of gas temperature and gas pressure on coating performance. Figure 1 shows the coating quality parameter as function of particle size or cold spraying with T_{gas} = 800°C and p_{gas} = 4 MPa.

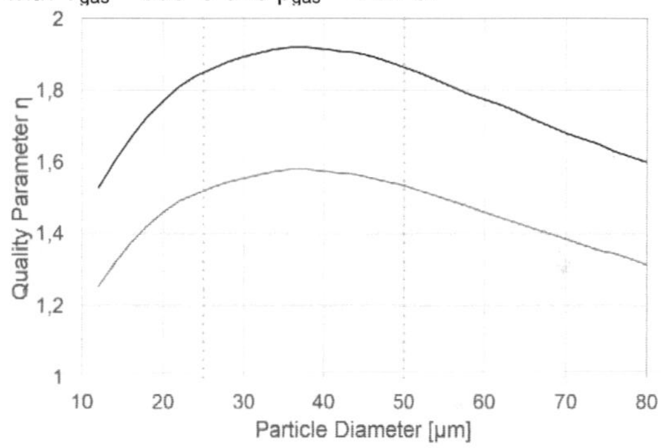

Fig. 1 Quality parameter η for a particle diameter range from 10 µm to 80 µm. The graph distinguishes the quality parameter based on bulk material strength η_{bulk} (black line) and the quality parameter based on powder material strength η_{pow} (grey line), calculated for a gas temperature of 800 °C and a gas pressure of 4 MPa.

The graphs in Figure 1 show a broad maxmium in η at a particle size of about 36 µm. With respect to needed high η-values, good coating qualities are expected for particle diameters between 25 µm and 50 µm. The impact of too small particles suffers from higher critical velocities and the deceleration in front of the substrate. Too large particels by their high momentum will not reach the needed velocities for sufficient good bonding quality.
With respect to parameter settings, it might be noted here that according to calculations showed an increase in η of 0.1 can be reached either by an

increase of gas temperature of 75 °C or by an increase of gas pressure of 1 MPa. Thus, process tuning by increased gas temperature seems to be easier than by increased gas pressure.

3.2 Powder Morphologies

Four different powders batches (delivered by TLS-Technik, Bitterfeld, Germany) were investigated with respect to particle morphology, particle size distribution and particle strength. The most suitable powder lot had an ideal spherical particle morphology, as shown in Figure 2, and a particle size distribution of d10 = 26 µm, d50 = 39 µm and d90 = 59 µm. Despite being slightly larger than ideal, the powder well fulfils the criteria for cold spraying of bronzes.

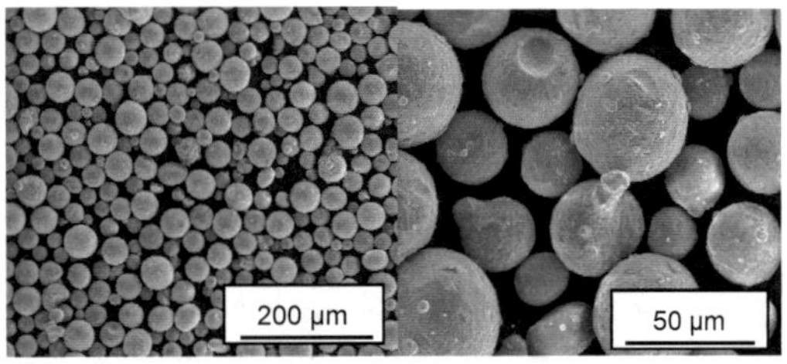

Fig. 2. Morphologies of the gas atomized bronze CuSn6Ni6Zn2 powder feedstock under different magnifications.

According to the compression tests, the powder particles have an ultimate tensile strength of 712 MPa, being significantly higher than that of the bulk material (418 MPa). Obviously, by rapid solidification, the atomisation process resulted in microstructures with low deformability. For a realistic description of critical velocties, subsequently, the real powder strength was taken to calculate the coating quality parameter (compare Figure 1).
Table 1 summarizes expected η-values for the different spray conditions. Despite the high powder strength, η values ≥ 1.5 can be reached under well tuned spray conditions, which are needed to attain dense coatings and good coating quality [7].

3.3 Cold spray efficiency

Based on the calculations, the experimental cold spray parameters were set to the range shown in table 1. For the spray experiments, two different lots of powder were used.

Table 1 Parameters for cold spraying of bronze CuSn6Ni6Zn2 coatings and associated impact conditions and η values based on powder data for a mean particle diameter of 39 µm. In addition, the experimentally observed deposition efficiencies are displayed. The last column indicates, which powder lot was used in the spray experiments.

p_{gas}	T_{gas}	T_{sub}	v_p	T_p	η_{pow}	DE	lot
[bar]	[°C]	[°C]	[m/s]	[°C]		[%]	
40	600	20	608	369	1,29	97	1
60	600	20	652	369	1,38	86	1
40	700	20	628	442	1,42	100	1
40	750	20	638	478	1,50	53	2
40	750	200	638	478	1,50	81	2
40	800	200	647	515	1,58	83	2

For improving the bond strength, two sets of experiments were performed with substrate preheating. Following literature [18], the substrate temperature was set to 200 °C in these experiments.
Experimentally the increase of η was mainly obtained by increasing the process gas temperature constantly from 600 °C up to 800 °C. Well-tuned η values are generally confirmed by high deposition efficiencies in a range of 80 to 100 %.
The use of a second powder lot causes a decrease of the DE, which is probably due to slightly larger particle sizes and thus lower impact velocities.

3.4 Coating microstructure, hardness and porosity

The coatings were optimized regarding their microstructure, porosity and spray parameter sets.
Figure 3 gives shows typical microstructures of cold sprayed CuSn6Ni6Zn2 coatings. The coating in Figure 3 a was sprayed with an η-value of 1.29 on a non-preheated substrate. The micrograph reveals some pores and a crack in the interface between the substrate and the coating.

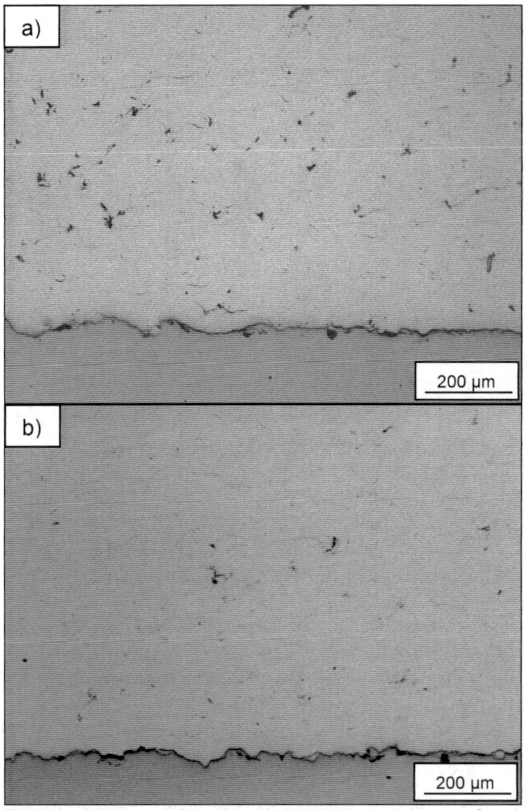

Fig. 3. Micrographs showing cross sections of cold sprayed CuSn6Ni6Zn2 coatings on steel 1.0503 substrate using p_{gas} = 4 MPa and gas temperatures of (a) T_{gas} = 600°C, (b) T_{gas} = 800°C. (a) was sprayed on a non-prehated substrate, for (b), the substrate temperature was set to 200 °C.

Being sprayed with a η-value of 1.58 on a preheated substrate, the second coating in Figure 3 b shows a very dense microstructure with negligiable porosity. In contrast to the overall good coating appearance, the interface between substrate and coating is decorated by some cracks.

Figure 4 shows the different coating porosities obtained under varied spray conditions. All porosities are well below 1 %, indicating overall good coating integrety. By increasing process gas temperature and thus the η-values, coating porosity can be reduced to about 0.2 %. One additional influence is given by the process gas pressure, as shown by the decrease of porosity from 0.9 % of the coating sprayed with an η-values of 1.29 to 0.5 % for the one sprayed under η = 1.38. By higher pressure, particles are faster, and deform to higher extend to fill gaps.

Moreover, substrate preheating could further reduce porosity. Following the argumentation from Assadi et al. [17], better deformabilty of already adhering spray layers by higher surface temperature could enhance bonding and thus overall coating performance. Reached values of less than 0.2 % are close to optimum.

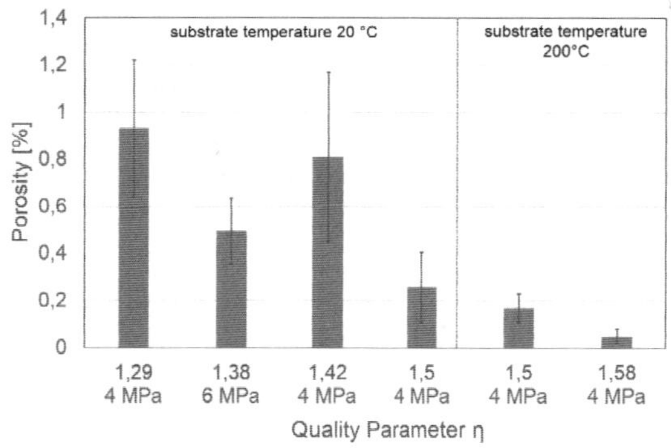

Fig. 4. Relation between the coating porosity and the quality parameter η.

Figure 5 summarizes the coating hardness as obtained under the different spray conditions. Interestingly, the coating hardness varies only over a rather small range from 275 to 288 HV 0.3. This indicates, that differences at observed low porosity contents only have marginal influence on hardness. Moreover, different work hardening effects seem to play no prominent role in the macroscopic scale analysed by hardness testing, as well as thermal recrystallization, which could occur under substrate heating . The results also show, that the coating hardness exceededs that of the bulk material by a factor 2.5. This not not surprising, considering the rather high strength of the powder feedstock and work hardening effects during deposition [5].

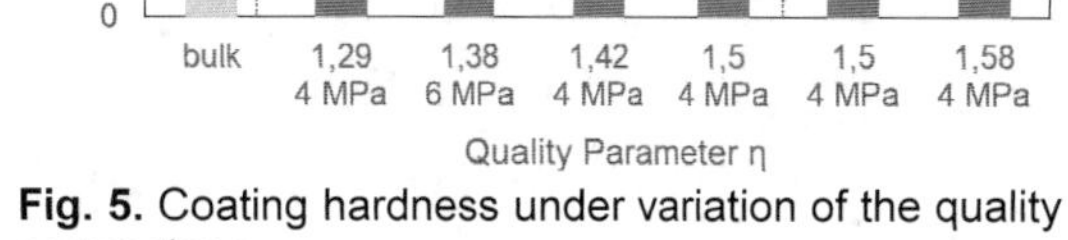

Fig. 5. Coating hardness under variation of the quality parameter η.

Adhesion

The coating bond strengths are summurized in Figure 6. Most of the cold sprayed coatings reached a bond strength of about 20 MPa or even below. By that, thresholds for applications are not reached yet. By preheating the substrate, the bond strength can be improved by in average about 50 %. Nevertheless, further investigations are needed to enhance the bond strength to minimum 30 MPa.

One explanation for the comparably low bond strength could be given by the high amount of internal stress in the coating induced by the cold gas process. This is indicated by thicker coatings trendwise showing lower adhesion, and the larger scatter of individual data sets, to some extend being influenced by secondary effects. In summary, improving adhesion is on of the main topics for further investigations.

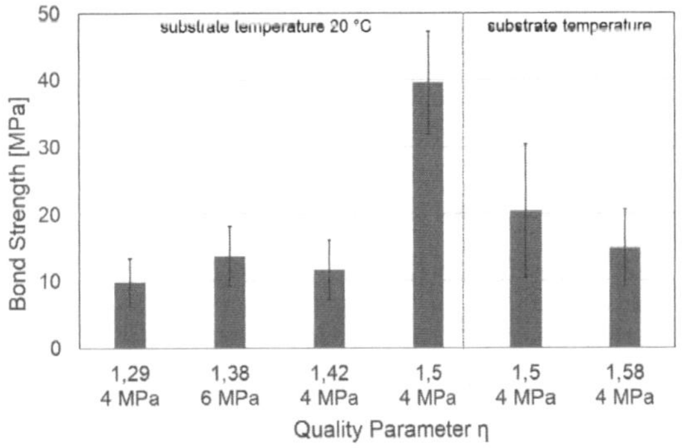

Fig. 6. Relation between the coating bond strength the quality parameter η.

Electrical Conductivity

Figure 7 summarizes the results on electrical conductivities of the bronze coatings under the variation of spray conditions. The cold sprayed coatings achieved an electrical conductivity of minimum 88 % as compared to that of the bulk material, and an increase with η values. An electrical conductivity > 80 as compared to bulk can be taken as measure for good coating integrity, and small amounts of non bonded interfaces being present within the coating [17].
The samples sprayed with a process gas temperature of 750°C and a process gas pressure of 4 MPa reached an electrical conductivity of 7.0 MS/m and exceed the electrical conductivity of the bulk material by 5 %. Such rather high conductivity could be explained by rather low amounts of scattering centres for electrons, here possibly due to the powder containing less solute atoms by vaporization of Zn during the atomization process and less internal phase boundaries by the lack of precipitation formation during rapid solidification.

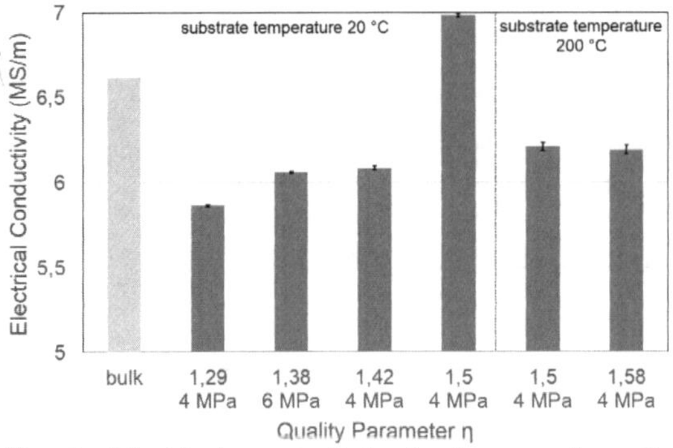

Fig. 7. Electrical conductivity for increased quality parameter η.

4 Summary and Conclusion

The present study shows that lead-free bronze coatings for high stress loads can successfully applied to steel 1.0503 substrates. Obtained coating microstructures are dense with negliable porosity the measured electrical conductivity of the coatings indicates close to bulk material cohesion and coating quality. Despite the overall rather good quating quality, the bond strength of the coating reaches in maximum 20 MPa and need further improvement. First experiments by substrate preheating already showed promising results. A further increase of bond strength is expected for adjusting higher substrate temperatures, for apply a Nickel bond coat between the substrate and the bronze coating or tuning the spray kinematic. Also, heat treatments of the coating could further improve the adhesion, together with other coating properties.

5 Acknowledgments

This research and development project is funded by the German Federal Ministry of Education and Research (BMBF) within the "SME – Innovative: Research for Production" programe (funding number 02P14K560 – 4) and managed by the Project Management Agency Karlsruhe (PTKA), which is greatly ackknowledged. The authors like to thank in alphabetical order Thomas Breckwoldt, Norbert Németh, Camilla Schulze, Matthias Schulze and Uwe Wagener for technical support, as well as Hamid Assadi for fruitful discussions concerning the work of this study.

6 Literature

[1] https://www.boschrexroth.com/de/de/unternehmen/presse/press-detail-1-113280, date: 2017-02-20

[2] https://www.boschrexroth.com/ics/cat/?language=de&id=&cat=Mobile-Hydraulics-Catalog&m=DE&u=si&o=Desktop&p=g261584,g261586,g261587,g261589,p662660, date: 2017-02-20

[3] Environment Pollutant Report 2015, Green Cross and Pure Earth, 2015

[4] Directive 2011/65/EU of the European Parliament and of the Council, (2011)

[5] T. Stoltenhoff, C. Borchers, F. Gärtner and H. Kreye. Microstructures and key properties of cold-sprayed and thermally sprayed copper coatings, Surface and Coating Technoloy 200 (2006), pp. 4947/4960

[6] K. Binder, Cold Spraying of Fatigue Resistant Ti Coatings, Ph.D. Thesis, Helmut-Schmidt-University, 2012

[7] H. Assadi, T. Schmidt, H. Richter, J.O. Kliemann, K. Binder, F. Gärtner, T. Klassen and H. Kreye, On Parameter Selection in Cold Spraying. J. Thermal Spray Technology (2011), pp. 1161/1176

[8] T. Schmidt, F. Gärtner, and H. Kreye. New Developments in Cold Spraying Based on Higher Gas and Particle Temperatures, J. Thermal Spray Technology (2006), pp. 488/494

[9] H. Assadi, F. Gärtner, T. Stoltenhoff and H. Kreye, Bonding Mechanism in Cold Gas Spraying. Acta Materialia 51 (2003), pp. 4379/4394

[10] T. Schmidt, H. Assadi, F. Gärtner, H. Richter, T. Stoltenhoff, H. Kreye and T. Klassen. From Particle Acceleration to Impact and Bonding in Cold Spraying. JTST 18 (2009), pp. 794/808

[11] T. Schmidt, F. Gärtner, H. Assadi and H. Kreye. Development of a generalized parameter window for cold spray deposition, Acta Materialia 54 (2006), pp. 729/742

[12] H. Assadi, H. Richter, F. Gärtner, T. Schmidt, J.-O. Kliemann, K. Binder, T. Klassen, H. Kreye. Applications of parameter selection maps in cold spraying, Proceedings ITSC (2011), pp. 100/104

[13] T. Schmidt, F. Gärtner and H. Kreye. New Developments in Cold Spray Based on Higher Gas- and Particle Temperatures, Thermal Spray 2006: Science, Innovation and Application (2006), s2_2-12003

[14] T. Klassen, F. Gärtner, H. Assadi, M. Villa-Vidaller, K. Binder, K. Onizawa, A. List, K.-R. Ernst, S. Krebs, H.Gutzmann. Kinetisches Spritzen zur Funktionalisierung von Oberflächen und für die generative Bauteilfertigung, Proceedings Hagener Symposium in der Pulvermetallurgie 33 (2014), pp. 97-116

[15] M. Homann and N. Ludwig, German Patent DE 4102 101 A1 (1991)

[16] H. Assadi , I. Irkhin, H. Gutzmann, F. Gärtner, M. Schulze, M. Villa Vidaller, T. Klassen, Determination of plastic constitutive properties of microparticles through single particle compression, Advanced Powder Technology (2015), pp. 1544/1554

[17] Z. Arabgol, M. Villa-Vidaller, H. Assadi, F. Gärtner, T. Klassen. Influence of thermal properties and temperature of substrate on the quality of cold-sprayed deposits, Acta Materialia (2017), pp. 287-301

[18] S. Krebs, F. Gärtner, T. Klassen, Optimizing Cavitation Resistance – An Approach by Cold Spraying of Serveral Bronze Materials, Proceedings ITSC (2015), pp. 1098/1104

Nozzle-internal particle velocity measurements and loading effect on particle acceleration inside a Cold Spray nozzle

M. Meyer, F. Caruso, *R. Lupoi, Trinity College Dublin, The University of Dublin, Department of Mechanical and Manufacturing Engineering, Parsons Building, Dublin 2, Ireland *Contact email: lupoir@tcd.ie

The advantages of the solid state deposition process Cold Spray (CS) over conventional spray technologies go hand in hand with the requirement of high and well-predictable particle velocities. The acceleration of particles primarily takes place within the CS-nozzle while measurements of their velocity are conducted downstream of its exit. Despite their essential value, these observations are limited, in that only the result of the acceleration can be evaluated, not the actual driving mechanisms themselves. Previous work has indicated that there is no conclusive understanding of these mechanisms, especially in cases of increasing particle loading. This study therefore presents a transparent rectangular CS-nozzle design (made out of quartz) for a low stagnation pressure regime. A novelty to the field of thermal spray is the first report of particle in-flight measurements within the CS-nozzle using Particle Tracking Velocimetry (PTV) at varying particle loadings and pressure levels. It is found that particle velocities in the jet decrease with increasing particulate loading as the momentum exchange of the gas is enhanced, while in the subsonic flow region, the average velocity level increases due to particle-particle interactions with shallower axial velocity profiles. This effect is aggravated for higher working pressures, as energetic collisions cause increasing losses, depending on the number density of particles. This study forms the basis for a comprehensive nozzle-internal analysis.

1 Introduction

The coating manufacturing process Cold Spray (CS) makes use of high speed impact and deposition of particles from a feedstock powder onto a substrate. The impact velocities are obtained from a driving process gas flow through a supersonic nozzle, which enables very low temperature levels compared to other coating technologies and therefore a solid state deposition. This minimizes disadvantages of melting and hence provides a possibility to coat oxidisation-sensitive materials and material combinations with different melting temperatures [1][2] opening a wide field of interesting applications. For deposition to succeed, a material-specific critical impact velocity must be crossed, which makes the particle acceleration a critical aspect of the technology [3][4]. Both experimental and numerical studies on the particle-laden gas nozzle flow were conducted in the past decades, identifying the main parameters for particulate acceleration. Primarily, the gas stagnation pressure and temperature [5][6], the gas species [7][8], powder injection conditions [9][10] and particle material and size [11][12][13], as well as shape [14][15] are important. An important part of this progress was the use of optical measurement techniques: Schlieren photography was frequently used to visualise flow features of the gas phase. The quantification of the velocity and particle motion however required other, non-intrusive velocity measurement techniques, of which several were employed over the years: laser-two-focus (L2F) [16], doppler picture velocimetry (DVP) [17], and lastly Particle image velocimetry (PIV) or similar techniques, in particular Particle Tracking Velocimetry (PTV). The latter techniques could be used to record instantaneous particle velocity distributions throughout the field of measurement [18][19][20]. Depending on the particle feed rate and the relative gas consumption, the discrete phase loading can increasingly affect the acceleration process. Although Pardhasaradhi et al. [21] found that the effect is negligible at low mass fractions of the discrete phase, results demonstrated by Samareh et al. [22] indicated that the gas flow structures change and the particle speed reduces as the particle mass fraction increases. With respect to engineering applications, a study by Lupoi [23] showed that a series of experimental observations could not be explained by computational techniques that ignore phase coupling effects, which were improved by Meyer & Lupoi [24] when inter-phase momentum-exchange was included. The authors also published a validation case of most prevalent computational techniques with attention to this aspect [25], stating that the models can reproduce the trend of decreasing velocities with feed rate, nevertheless fail to provide reliable quantitative results, suggesting that neglected effects such as particle-particle interactions and the particle volume fraction in fact contribute noticeably to the particle velocity level. Meyer et al. [26] recently extended their work on the matter by an experimental investigation that begins to identify driving mechanisms for the velocity drop and ties the susceptibility of high-density materials predominantly to mass loading and such of low-density materials rather to volume fraction. It remains unknown what in detail leads to this interplay, and consequently conclusive understanding of the phase coupling effects still has to be achieved.

The benefit of higher particle feed rates is without doubt faster processing times in manufacturing applications. Moreover, a cost analysis of the CS process by Stier [27] provides evidence that it is important to understand the mass loading effect also on an economical level, as it enables the optimisation of gas and powder consumption without loss of deposition efficiency.

The acceleration of particles and the potential causes for coupling effects primarily take place within the CS-nozzle, while the optical measurements to date had to be conducted downstream of its exit. Despite their essential value, these observations are limited as only the result of the acceleration can be evaluated, not the actual driving mechanisms. Therefore, this work aims to investigate the particle velocity not only in the jet of a CS system, but also within the internal channel of the nozzle itself. A transparent quartz nozzle with rectangular cross-section and a de-Laval shape, characteristic for CS nozzles, was designed for this purpose. PTV was used to measure Stellite-21 particle velocities at varying low stagnation pressure and feeding conditions. This study is the very first attempt of a quantitative measurement of a nozzle-internal flow, and therefore represents a very distinct novelty to the field. Establishing such techniques would not only give insight into the mechanisms of the particle acceleration, but also enable direct studies on the injection process.

2 Methods

2.1 Cold Spray system, nozzle and materials

Fig. 1a) illustrates a schematic of the experimental arrangement, including PIV and the CS process. The nitrogen gas flow was fed from a high pressure gas supply line, regulated to different settings of constant pressure between 3 and 8bar-g at the nozzle inlet. The gas line was split into two lines, a main line, which was connected to the nozzle head, and a powder feeder line, both comprising flow meters and pressure gauges. The wheel type powder feeder was set to varying wheel speeds and included a load cell to measure the mass feed rate of powder. The range of pressure and wheel speed settings can be found in Table 1. Downstream of the powder feeder, the gas–solid mixture then merged with the main flow in the nozzle head with a central injection position. **Fig. 1**b) shows the measurement set-up as arranged in the facilities. Details of the imaging procedure can be found in sections to follow, however the core piece of the setup is the transparent quartz nozzle. An illustration of it is presented in **Fig. 1**c). This in-house nozzle-design has rectangular cross-section in order to provide a wall contour only in one dimension and a flat nozzle wall in the other dimension, through which the image can be recorded. The de-Laval converging/diverging shape had a full length of 150mm and an expansion ratio of 3.72. As the design point for this expansion ratio, a pressure level of 10bar-g was chosen, leading to a slightly over-expanded flow condition under such circumstances. Due to beginning design-related leakage at this pressure, the nozzle operation was limited to 8bar. The low stagnation pressure regime was set as the target because the stability of the nozzle is a limiting factor due to its material. Based on pre-tests, quartz was chosen, since it provides the best possible transparency and the least image distortions, while it is hard enough to withstand erosion at moderate speed levels. As can be seen in **Fig. 1**c), the measurements were conducted in two locations for all settings. Firstly, the low-speed converging section down to the nozzle throat was investigated, since this part is characterised by a relatively large particle number density as compared to the high- speed section and therefore is the most important candidate for regions with strong interactions.

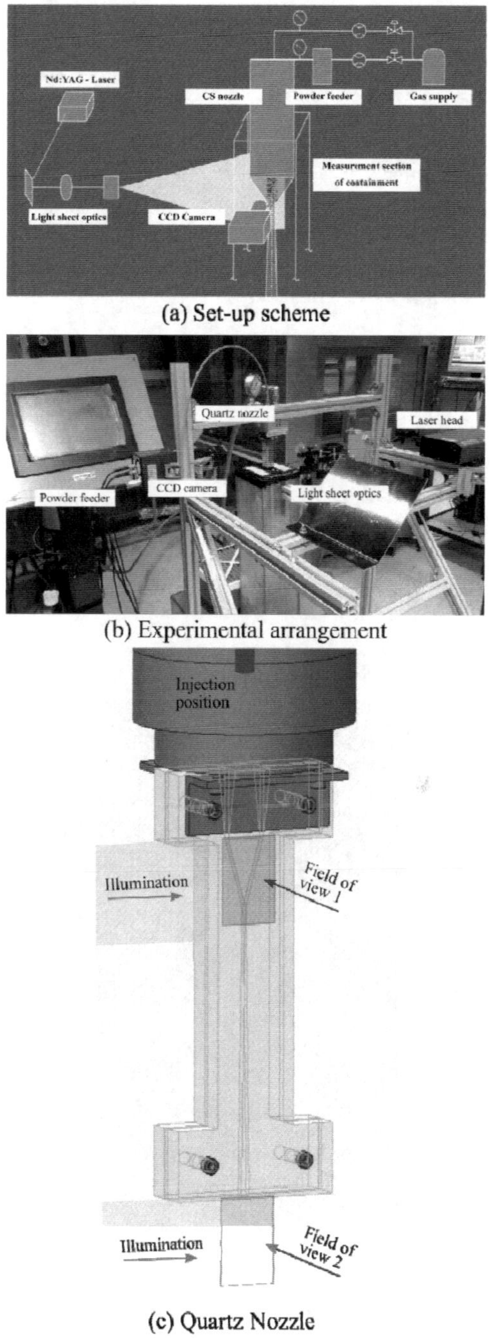

Fig. 1. a) Scheme of experimental set-up ,b) picture of the measurement system in the laboratory, c) illustration of transparent quartz nozzle.

Secondly, the measurement area just downstream of the nozzle exit, downstream to 50mm standoff distance was analysed in order to have a more complete picture of the outcome of the internal effects. The nozzle sprayed into the measurement section of an enclosed area, which ensured undisturbed flow in absence of a substrate.

Feedstock	Stellite-21			8440 kg/m^3
Stagnation temperature	293K			
Pressures at nozzle inlet	3.4	5.1	6.2	8.0
Powder feeder wheel speeds	5%	10%	15%	20%
Locations of measurement	Converging section			Jet (nozzle exit)

Table 1. Parameter and spray conditions.

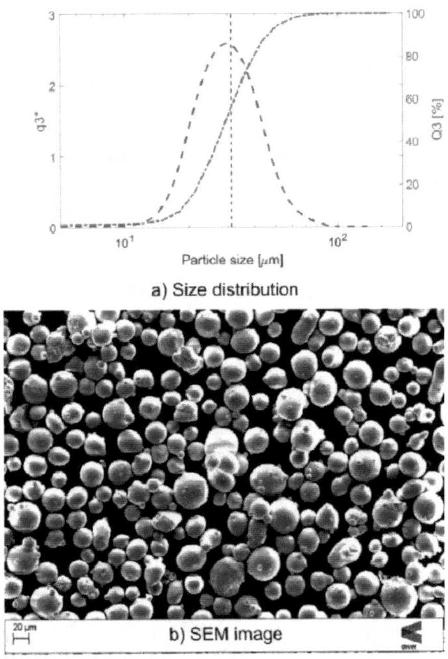

Fig. 2. Measured CS powder, Stellite-21: a) size distribution, b) SEM image.

Fig. 2 shows details of the used powder Stellite-21, a cobalt–chrome alloy. It was used as the spray material representing a relevant material in CS applications, and mostly because it offers advantages both for the injection stability and the optical imaging behaviour inside and outside the nozzle. For the measurement technique, it is essential to gather images at well-defined flow and feeding conditions, such that a stable, continuous flow is even more critical than during the coating application process. Moreover, the illumination as described below depends on the particle size and shape, as well as the reflectivity of the surface. Stellite-21 is a favourable candidate in this respect. As can be seen in **Fig. 2**, it has a standard size distribution between 10 to 60 µm at a Volumetric Mean Diameter (VMD) is approximately 30 µm and a nice spherical particle structure.

2.2 Measurement set-up and testing

Particle Tracking Velocimetry was used to optically measure the particle velocity within the jet. They were illuminated with two subsequent laser pulses formed to a light sheet in the plane of measurement. A camera system captured two images of the scattered light respectively with varying pulse separation time depending on the conditions. These images were processed by a cross–correlation algorithm, deducing the particle displacement, which corresponds to velocity information by knowledge of the pulse separation. A Nd:YAG Laser with a wavelength of 532nm and 6ns pulse duration was used as the light source at 4Hz repetition rate. The pulse separation time was adjusted between 1 and 5µs depending on the velocity level, aiming for an approximate displacement of 10px, allowing for a prolonged dynamic range towards the lower end. The light sheet was formed by a telescope of spherical lenses for the sheet thickness of approximately 1mm and a cylindrical lens for the sheet width of 60mm respectively, illuminating the plane of symmetry in and downstream of the nozzle. A monochromic camera with a resolution of 1280x1024 pixels of 6.7µm pixel size delivered the consecutive image pairs The camera lens aperture was set to f/8.0 in order to minimise lens aberrations. The observed particle image size was found between 2 and 5px. The depth of field was approximately 9mm accordingly. The pre–processing of images included a standard background subtraction of an averaged image from 100 samples in absence of particles and high–pass filtering of the raw data to reduce stationary image features and low–frequency background variations with a kernel width of 5px. The image processing involved a two-step approach, in which a PIV step was used as a predictor for the PTV algorithm, identifying single particle vectors. Along with the hardware, the algorithms were produced by *LaVision* as part of *DaVis v7.2*.

A multitude of sources contributes to the measurement error in PIV and PTV, of which the most important are optical uncertainties (e.g. lens aberrations and calibration errors) and algorithm-related errors (rms displacement error and a bias or "peak-locking" error). From target plane images, we estimate a maximal calibration uncertainty of 0.03px for a 10px displacement. Based on the primary peak ratio in the correlation plane of the measurement images, the rms error could be estimated in the order of 1.3%. This value was conservatively corrected according to previous thorough analysis of the algorithm [28], which results in an rms error of 0.18px or 1.8% on the average displacement. Lastly, the used algorithm for particle tracking was shown to have exceptionally low bias error [28], hence this source was not considered further. In addition, repeatability of the test cases was assured and the respective standard deviation, as well as the deduced PIV errors were included in the reported errors bars of the measurements given in sections to follow.

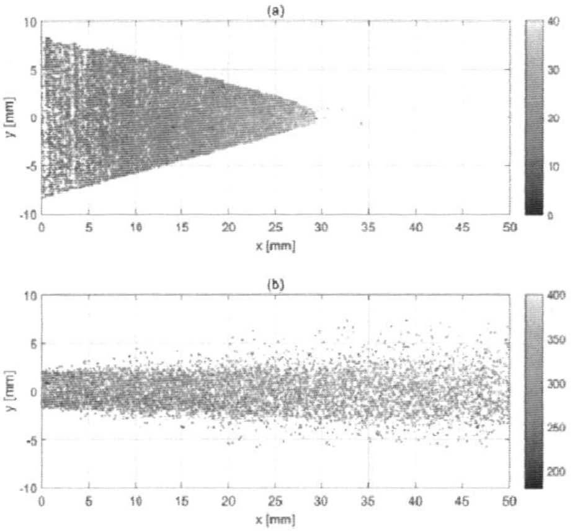

Fig. 3. Scatter plots of particle velocity at 8bar and 140gmin^{-1} (a) in the converging section of nozzle and (b) in the jet.

3 Results and Discussion

3.1 Particle distribution and velocity profile

In order to compare the phase-coupling effect, it is useful to start by exploring the particle velocity distribution under varying conditions. **Fig. 3** shows a comparison of the particle distributions at 8bar in the planes of measurement (a) for the converging section of the nozzle and (b) for the jet downstream of the exit, coloured by the velocity magnitude. This illustration gives a good impression of the velocity change within the nozzle just upstream of the nozzle restriction. Measurement of the velocity further downstream within the nozzle was not feasible due significant change of the optical properties of the quartz surfaces. While the particles on average accelerate up to 30-40m/s, individual particles are much faster than the average throughout almost all the field. At the nozzle exit, the particle velocity spans between 200 and 400m/s. It does not change significantly in the streamwise direction in the jet. As expected, the strongest acceleration must occur just downstream of the nozzle throat. Nevertheless, the observed internal region offers clues about the condition of particles in the more dense region just upstream of this point. It is plausible to say, that during the peak acceleration in the diverging section, effects of interaction between particles drasticly reduce due to their increasingly large average distance.

A comparison of the axial velocity distribution within this dense flow region is shown in **Fig. 4**, comprising results at (a) 3.4 and (b) 8.0bar, each at the lowest and highest feeder wheel speed setting. Each data point represents a moving average including whiskers for the standard deviation of the velocity values. Interestingly, the average velocity at low pressures is almost very similar for both loadings. The statistical distribution is, however, somewhat wider, which suggests that the higher loading results in a stronger mixing and a more unordered flow. At higher pressures, this effect is aggravated drastically, which can be seen by the much larger standard deviation for the high feed rate. More momentum within the flow results in an increase of particle-particle interactions. In fact, the particle velocities are sufficiently more randomly enhanced, that the average velocity lies above the low loading case. At the same time, the axial profile flattens slightly, since the random motion is strong throughout the converging section. It should be mentioned that the study also showed up to a 5-fold increase in the standard deviation of the particle trajectory orientation.

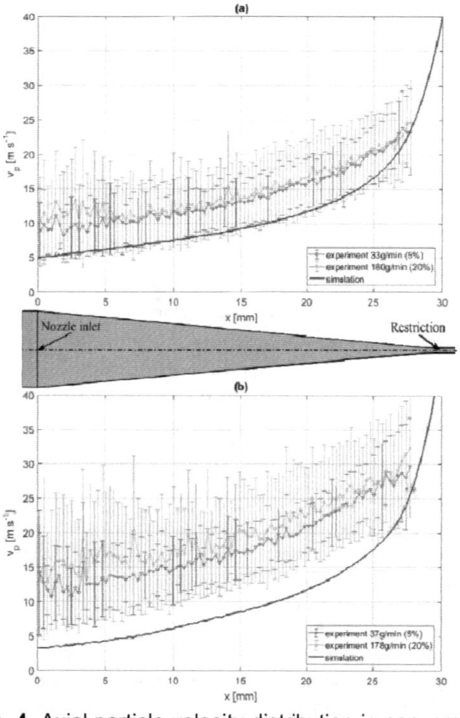

Fig. 4. Axial particle velocity distribution in converging section for (a) 3.4bar and (b) 8.0bar at 5% and 20% feeder wheel speed (ca. 30 and 170g/min), compared to simulation data.

The figure also shows the average velocity profile of particles from a computational model. Fluent v16.0 was employed to model the gas phase in a 3D-steady-state RANS with a realizable k-ε-turbulence model. A density-based solver in an implicit formulation along with a ROE-FDS flux splitting and second-order upwind schemes was used as the discretization method. The particle model is a Lagrangian discrete phase model, which computes the particle trajectory based on a high-Mach number drag law, and accounts for the particle size distribution. This model delivers signifiacantly lower particle velocities than measured, while the rate of increase is in fairly good agreement. It does not account for particle-particle interactions or the volume fraction, which gives an explanation for the mismatch.

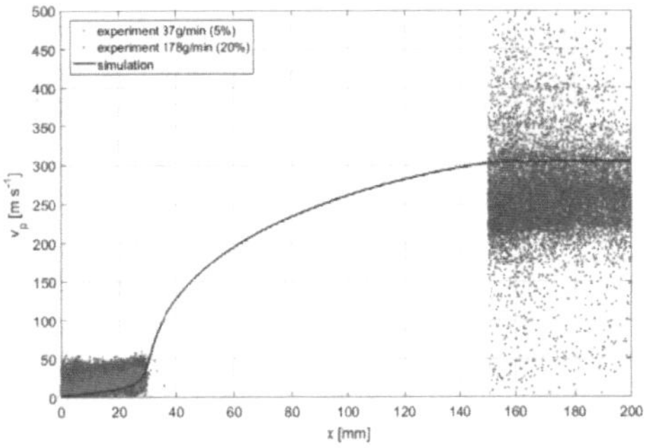

Fig. 5. Axial particle velocity profile at 8bar (experimental and CFD) and increasing particle feed rate.

Conclusively, particle motion within the low-velocity section of the nozzle significantly depends on the particle packing density and is characterised not only by gas particle, but also by particle-particle interactions. This makes the process dependant on both volume and mass fraction of the discrete phase. Downstream of the internal measurement section, the gas expansion leads to a rapid particle acceleration, that increases the average distance between particles. In the diverging section, the primary force on particles is therefore the drag force. In **Fig. 5** the regions 0-30mm and 150-200mm are populated with velocity PTV data points; for comparison a simulation curve is superimposed and also covers the 30-150mm region where there is no experimental data. The figure presents the 8bar case, demonstrating clearly that the increasing feed rate reduces the final velocity level, despite a higher velocity in the throat (by PTV results). The reason lies in the losses that characterise the phase interactions and is discussed below. A comparison of the simulation and experiment shows an underestimation of the mean particle trend in the converging part and an overestimation of the velocity level in the jet. As a sidenote, it should be mentioned that, since the particle velocity upon impact is cruicial for the resulting coating in a process, the simulation tool is not a conservative approach for process design.

3.1 Velocity trends and loading mechanisms

On average, the particle velocity increases in the converging section due to random, collision-driven motion, and decreases as a result of the energy and momentum lost from the gas phase. Hence, the particle behaviour at the inlet and outlet regions of the nozzle is opposite. **Fig. 6** quantifies this connection in a comparison of the mean velocity developments in (a) the converging section and (b) the jet for the different measured pressure levels in dependence of the particle mass loading Z (the fraction of particle mass and gas mass flux), which characterises the momentum exchange between phases according to previous findings [26]. As expected, the overall velocity level increases with pressure throughout the field. As noted above, it can be seen again that the average velocity increases with mass loading in the converging section, and the reverse is true in the jet. Most importantly, the gradients of the shown regression lines differ depending on the pressure. If these gradients are normalised by each velocity level, in order to make a comparison meaningful, we can investigate the susceptibility of the material to mass loading. The same can be done for the volume fraction F (the fraction of particle and an estimated gas volume flux, here with reference to conditions in the restriction). These normalised gradients are denoted $Gr_{Z,n}$ and $Gr_{F,n}$ and represented in **Fig. 7**.

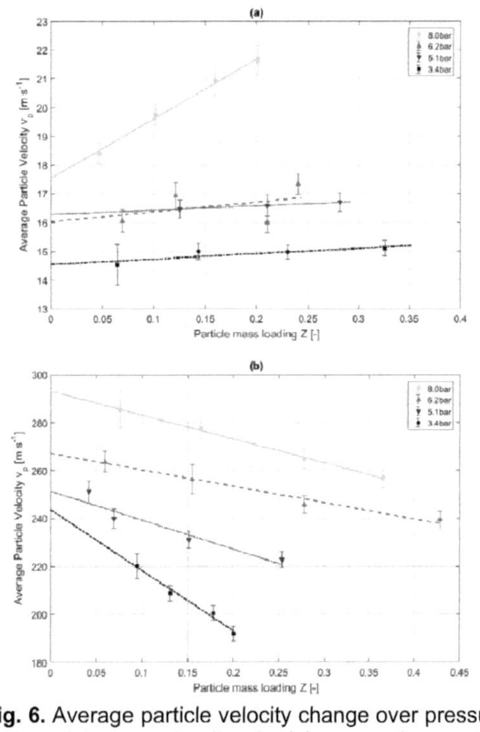

Fig. 6. Average particle velocity change over pressure and particle mass loading in (a) converging section and (b) in the jet.

Focusing on the gradient with mass loading $Gr_{Z,n}$, it should be explained first that the positive gradients for the converging section represent a gain in particle momentum as an outcome of strong mixing, particle collisions and a resulting randomised motion. At low pressures, this gradient is small, as the collisional rate is still low. With higher pressure, more momentum (and hence energy) is gained by particles as collisions become stronger. The negative gradients represent the jet behaviour and stand for a loss of particle momentum (and energy) due to mass loading. At low pressures, this effect is strong, as a the gas flux is rather weak and a unit increment of mass loading can withdraw a significant protion of energy from the gas flux during the acceleration in the nozzle. At higher pressures, the gas flow is much more energetic, and consequently, a smaller part of this access momentum flux is transferred to the particle phase. At the same time, despite stronger collisional losses in the dense flow region, the stronger velocity increase at higher pressure also goes hand in hand with a smaller loss of velocity in the jet, which makes these opposing trends very plausible.

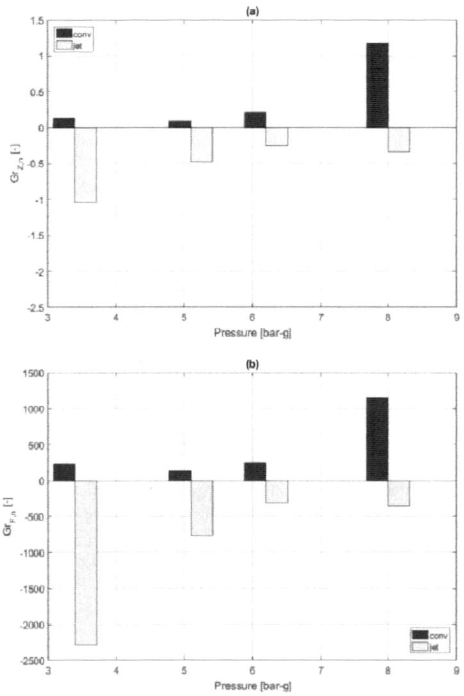

Fig. 7. Pressure dependencte of particle velocity change with (a) mass loading and (b) volume fraction.

Fig. 7b) displays the gradients of same velocity drop with respect to volume fraction. Although the trends remain the same as explained above, the relative change of the magnitude of the gradients has interesting implications. The difference between mass and volume parameters Z and F mainly goes back to the gas density. While the mass is important for the momentum exchange, the volume fraction is directly connected to the displacement of fluid and the geometric distance of particles. This means a high volume fraction can play an important role in the dense flow region. Comparing the two subfigures, we can see that the density significantly scales up the low-pressure gradients with respect to the high-pressure gradients. In a volume fraction-driven loss mechanism, one would expect that a unit increment of volume fraction produces a similar increment of loss or gain for all pressure levels. In other words, in such a mechanism, it should not be gas density-dependent how much engery and momentum is exchanged between phases. In this case, however, particularly in the jet, the low pressure increments of losses are more strongly dependent on volume fraction than on mass loading which suggests that density plays a significant role. This implies that the observed effects are overall predominantly momentum-driven. This is particularly creditable under the consideration a high density material. On the other hand, an increasing dependency could be inferred for the higher pressure, where the trend in the jet begins to reverse and the gradient with respect to volume fraction remains almost constant over a large pressure change. In combination with the strongest enhancement of the particle dynamics in this region, the rising importance of the volume fraction appears logical.

4 Conclusions

The acceleration of particles and the potential causes for coupling effects in Cold Spray primarily take place within the nozzle, while the optical measurements of particle velocity to date had to be conducted downstream of its exit. This limitation to the outcome of the acceleration was firstly overcome in the present study in order to start understanding the driving mechanisms for complex particle-gas interactions. Therefore, this work investigated the particle velocity within the converging section of a 10-bar quartz nozzle and in the free jet. PTV was employed to measure Stellite-21 particle velocities at varying low stagnation pressures and increasing feeding conditions. The study presented an increasing tendency to random motion in the converging section due to higher feed, connected to particle collisions, which is strongly aggravated for higher pressures. Nevertheless, the velocity in the jet decreases with increasing loading as more momentum is withdrawn from the gas phase during the overall particle flight. Because of the stronger momentum flux and the higher momentum particles emerging from the low-speed section, the average drop in the jet is weaker for high pressures. An investigation of the strength of the velocity drops with loading and volume fraction was presented as a susceptibility analysis. In this, it was found that the overall interaction is rather momentum-driven at low-pressure scenarios, whereas the high-pressure case becomes more influenced by the volume fraction, in connection with the strong nozzle-internal particle collisions.
Future investigations are required to extend the region of measurement, as a portion of the nozzle flow still remains implicit.
This study presented the very first attempt to measure particle motion within a CS nozzle. This novelty can be further developed and herein offers opportunities to study the character of particle acceleration and, moreover, enables direct observations of the injection process.

Acknowledgements
The authors wish to express their gratitude to FP7 Marie Curie Actions (Grant 333663) for the financial support.

5 Literature

[1] A. Alkhimov, V. F. Kosarev, and A. Papyrin, "Method of Cold-Gas Dynamic Spraying," *Dokl. Acad. Nauk SSSR*, vol. 315, no. 5, pp. 1062–1065, 1990.

[2] A. Papyrin, "COLD SPRAY TECHNOLOGY," *Adv. Mater. Process.*, vol. 159, no. 9, pp. 49–51, 2001.

[3] H. Assadi, F. Gärtner, T. Stoltenhoff, and H. Kreye, "Bonding mechanism in cold gas spraying," *Acta Mater.*, vol. 51, no. 15, pp. 4379–4394, 2003.

[4] M. Grujicic, C. L. Zhao, C. Tong, W. S. DeRosset, and D. Helfritch, "Analysis of the impact velocity of powder particles in the cold-gas dynamic-spray process," *Mater. Sci. Eng.*, vol. 368, no. 1–2, pp. 222–230, 2004.

[5] T. Han, Z. Zhao, B. a. Gillispie, and J. R. Smith, "Effects of Spray Conditions on Coating Formation by the Kinetic Spray Process," *J. Therm. Spray Technol.*, vol. 14, no. 3, pp. 373–383, 2005.

[6] E. H. Kwon, S. H. Cho, J. W. Han, C. H. Lee, and H. J. Kim, "Particle behavior in supersonic flow during the cold spray process," *Met. Mater. Int.*, vol. 11, no. 5, pp. 377–381, 2005.

[7] T. C. Jen, L. Li, W. Cui, Q. Chen, and X. Zhang, "Numerical investigations on cold gas dynamic spray process with nano- and microsize particles," *Int. J. Heat Mass Transf.*, vol. 48, no. 21–22, pp. 4384–4396, 2005.

[8] H. Katanoda, T. Matsuoka, and K. Matsuo, "Experimental study on shock wave structures in constant-area passage of cold spray nozzle," *J. Therm. Sci.*, vol. 16, no. 1, pp. 40–45, 2007.

[9] S. Yin, Q. Liu, H. Liao, and X.-F. Wang, "Effect of injection pressure on particle acceleration, dispersion and deposition in cold spray," *Comput. Mater. Sci.*, vol. 90, pp. 7–15, Jul. 2014.

[10] S. Yin, X. Suo, H. Liao, Z. Guo, and X. Wang, "Significant influence of carrier gas temperature during the cold spray process," *Surf. Eng.*, vol. 30, no. 6, pp. 443–451, 2014.

[11] F. Gärtner, T. Stoltenhoff, T. Schmidt, and H. Kreye, "The Cold Spray Process and Its Potential for Industrial Applications," *J. Therm. Spray Technol.*, vol. 15, no. 2, pp. 223–232, 2006.

[12] A. Sova, A. Okunkova, S. Grigoriev, and I. Smurov, "Velocity of the Particles Accelerated by a Cold Spray Micronozzle: Experimental Measurements and Numerical Simulation," *J. Therm. Spray Technol.*, vol. 22, no. 1, pp. 75–80, Nov. 2012.

[13] S. Li, B. Muddle, M. Jahedi, and J. Soria, "A Numerical Investigation of the Cold Spray Process Using Underexpanded and Overexpanded Jets," *J. Therm. Spray Technol.*, vol. 21, no. 1, pp. 108–120, Oct. 2011.

[14] X.-F. Wang, S. Yin, and B. P. Xu, "Effect of cold spray particle conditions and optimal standoff distance on impact velocity," *J. Dalian Univ. Technol.*, vol. 51, no. 4, pp. 498–504, 2011.

[15] W.-Y. Li and C.-J. Li, "Optimization of spray conditions in cold spraying based on numerical analysis of particle velocity," *Trans. Nonferrous Met. Soc. China (English Ed.*, vol. 14, no. 2, pp. 43–48, 2004.

[16] D. L. Gilmore, R. C. Dykhuizen, R. A. Neiser, T. J. Roemer, and M. F. Smith, "Particle Velocity and Deposition Efficiency in the Cold Spray Process," *J. Therm. Spray Technol.*, vol. 8, no. 4, pp. 576–582, 1999.

[17] H. Fukanuma, N. Ohno, B. Sun, and R. Huang, "In-flight particle velocity measurements with DPV-2000 in cold spray," *Surf. Coatings Technol.*, vol. 201, no. 5, pp. 1935–1941, 2006.

[18] B. Jodoin, F. Raletz, and M. Vardelle, "Cold spray modeling and validation using an optical diagnostic method," *Surf. Coatings Technol.*, vol. 200, no. 14–15, pp. 4424–4432, 2006.

[19] F. Raletz, M. Vardelle, and G. Ezo'o, "Critical particle velocity under cold spray conditions," *Surf. Coatings Technol.*, vol. 201, no. 5, pp. 1942–1947, 2006.

[20] X.-J. Ning, Q.-S. Wang, Z. Ma, and H.-J. Kim, "Numerical Study of In-flight Particle Parameters in Low-Pressure Cold Spray Process," *J. Therm. Spray Technol.*, vol. 19, no. 6, pp. 1211–1217, Sep. 2010.

[21] S. P. Pardhasaradhi, V. Venkatachalapathy, S. V Joshi, and S. Govindan, "Optical Diagnostics Study of Gas Particle Transport Phenomena in Cold Gas Dynamic Spraying and Comparison with Model Predictions," *J. Therm. Spray Technol.*, vol. 17, no. 4, pp. 551–563, Oct. 2008.

[22] B. Samareh, O. Stier, V. Lüthen, and A. Dolatabadi, "Assessment of CFD Modeling via Flow Visualization in Cold Spray Process," *J. Therm. Spray Technol.*, vol. 18, no. 5–6, pp. 934–943, Aug. 2009.

[23] R. Lupoi, "Current design and performance of cold spray nozzles: experimental and numerical observations on deposition efficiency and particle velocity," *Surf. Eng.*, vol. 30, no. 5, pp. 316–322, Jun. 2014.

[24] M. Meyer and R. Lupoi, "An analysis of the particulate flow in cold spray nozzles," *Mech. Sci.*, vol. 6, no. 2, pp. 127–136, 2015.

[25] M. Meyer, S. Yin, K. A. Mcdonnell, O. Stier, and R. Lupoi, "Feed rate effect on particulate acceleration in Cold Spray under low stagnation pressure conditions," *Surf. Coatings Technol.*, vol. 304, pp. 237–245, 2016.

[26] M. Meyer, S. Yin, and R. Lupoi, "Particle in-flight velocity and dispersion measurements at increasing particle feed rates in Cold Spray," *J. Therm. Spray Technol.*, vol. 16, no. 1–2, pp. 60–70, 2017.

[27] O. Stier, "Fundamental cost analysis of cold spray," *J. Therm. Spray Technol.*, vol. 23, no. 1–2, pp. 131–139, 2014.

[28] M. Stanislas, K. Okamoto, C. J. Kähler, and J. Westerweel, "Main results of the Second International PIV Challenge," *Exp. Fluids*, vol. 39, no. 2, pp. 170–191, 2005.

New Mechanical Roughening Processes as Surface Preparation of Engine Cylinder Bores Before Thermal Spraying

A. Roffe, M. Morgan-Hague, Autocraft Drivetrain Solutions Ltd.; M. Waiblinger, Gehring; G. Sanuk, M. Joschko, Gühring; M. Kesting, Sturm Maschinen- und Anlagenbau GmbH; T. Birkner, MAG IAS GmbH; K. Over, M. Silk, C. Verpoort, Ford - Werke GmbH; R. Wilton, Ford Motor Company, UK; Düsseldorf / D

Several surface preparation techniques are being used like grit blasting, HP water jet roughening as well as mechanical roughening for the preparation of Aluminum cylinder bore surfaces before a thermal spray coating can be applied. However, in case of spray-repaired CI cast iron engine blocks the conventional mechanical roughening processes - using cutting inserts with small dovetail-undercut geometry - are not applicable due to the high hardness and high material toughness. Therefore such CI engine blocks are bored oversize in order to remove the bore wear damage and subsequently this rough-machined surface is coated by a NiAl-bond coating material in order to provide sufficient bond strength for the functional top coating material.

In this paper it will be demonstrated that the 2-step spray-repair process can be replaced by a single-step process by using a new diamond-roll-roughening method. This process leads to significant higher bond strength values than the conventional process, including the bond coating material. PAT Adhesion test results as well as microstructural cross sections of coated cylinder bores will be presented. The principle of the roll-roughening process is outlined. In addition it will be shown that different mechanical roughening methods can be combined to obtain high bond strength values for spray-repaired aluminum blocks which require a higher coating thickness to compensate for the depth of the original mechanical bore activation.

1 Introduction

Thermal spray repair offers an excellent process to salvage damaged, worn-out automotive engine blocks: the cylinder bores are bored over size until the wear damage zone is completely removed. Such machining operation can be done without cooling fluid in dry condition, so that grey cast iron blocks can be coated directly after this machining step. A coating of 0.3 – 0.5mm thickness is technically and economically feasible. In order to guarantee sufficient bond strength of the spray coating, a bond coating of NiCr or NiAl material is used. After spraying, the blocks can be honed back to the original honing diameter. However, special attention is required to adjust the honing morphology according to the typical porosity level of the spray process used, whether it is powder or wire spray technology.

However, the question is whether an alternative surface preparation and surface activation/ roughening process can be developed, so that the bond strength of the coating can be improved in order to increase the robustness of the process. On the other hand it might be possible to eliminate the required process step of spraying the bond coating, thus leading to shorter process time and lower material spray cost.

Thermal spraying is now increasingly used as well for Aluminum engine blocks for the wear protection of cylinder bores. During the German BMBF "NanoMobile" R&D Project (2005-09) a so called "dovetail-undercut" mechanical surface roughening method was developed at the IWF Institute, University of Braunschweig [1]. This technology was then further developed between Gühring and Honsel leading to a serial production PCD Poly-Crystalline-Diamond cutting tool insert that is used since June 2009 in serial production [2,3]. In Figure 1 a cross section through such a spray coating is shown.

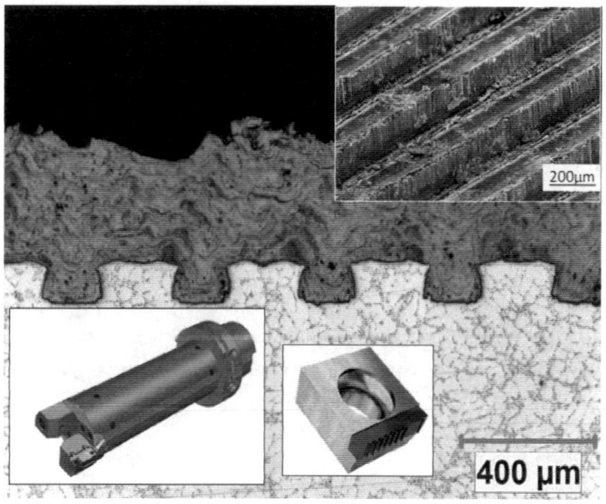

Figure 1: Cross section of a LP Sand Cast AlSi7Mg Al-block roughened with PCD cutting tool Gühring Generation 9.2. and SEM picture after adhesion testing.

The surface roughening was carried out by the patented Gühring Generation 9.2 PCD cutting tool insert. High PAT Adhesion values of above 50MPa can be obtained because the entrapped spray material will deform the edges of the undercut section during the tensile test. The plastic deformation at the profile is shown in the enclosed SEM picture. Such a PCD tool can be used to roughen up to 5.000 cylinder bores depending on the level of casting porosity and oxide inclusions. However, after a given service interval, such spray coated engine blocks might require a rejuvenation process by spray-repair and by the application of a new coating. In that case the old coating has to be removed by rough-honing or milling; then the bores are fine-bored and mechanically roughened again. However, in that case a significantly thicker coating has to be applied, than for the original blocks. Thicker coatings lead to longer spray times, higher temperature increase inside the bores and higher loading of the coatings. Therefore it might be required to improve the surface roughening process in order to allow for improved heat transfer and for additional interlocking of the coating on top of the machined pillars. The patented processes from Ford describe the machining of open rectangular profiles, which are deformed in a second operation to generate the dovetail-undercut-profile, either by using a wiper [4], or by a rolling operation [5]. The concept of using rolls with different grit sizes of diamond particles was now developed between Gehring and Gühring to a serial production process and first test results will be shown. In addition, the combination of different roughening processes will be studied as well.

2 Experimental

2.1 Diamond roll-roughening of Cast Iron Engine Blocks

The new roll-roughening tool developed by Gehring [6], called the **GRP2** (**G**ehring **R**oughening **P**rocess) is shown in Figure 2: a roll with Ni-bonded diamonds is placed inside a rotating tool holder. The roll is spring loaded to compensate misalignment of the pre-machining cylinder bore as well as the stiffness of the spindle bearings. Inside the tool holder 2 rolls can be used: the first roll has a very large grit size of D181 µm diamonds, whereas the second roll is using a finer grit size of D76µm. The large grit will generate a large open indentation into the material, whereas the finer grit passing over such an indented surface will deform the flat surface zone along the crater rim and thereby deflecting the material so that an undercut is generated. Such an undercut is believed to be required in order to obtain higher bond strength values compared to just a milled / turned surface area.

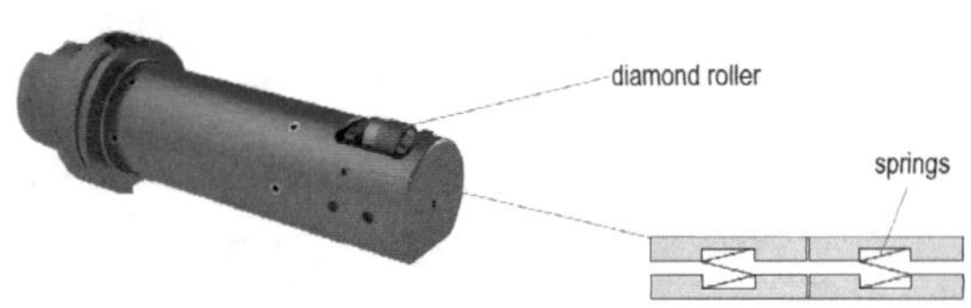

Figure 2: GRP2 roughening tool holder with 2 rolls with different diamond grit sizes.

In Figure 3 a roughened grey cast iron block is shown. It is possible to start with the roughening operation below the head deck so that the non-damaged section of the bore above the top ring reversal point does not have to be sprayed. The load of the springs can be adjusted from 50 N to 1100 N so that the diamonds can be pushed even into hard grey cast iron material. In order to avoid plastic deformation of the roll itself it is made from W-carbide or hardened steel material. The tool holder with the 2 diamond rolls rotates in the center of the bore. A feed-rate of 0.1mm/revolution is used at 1000 rpm.

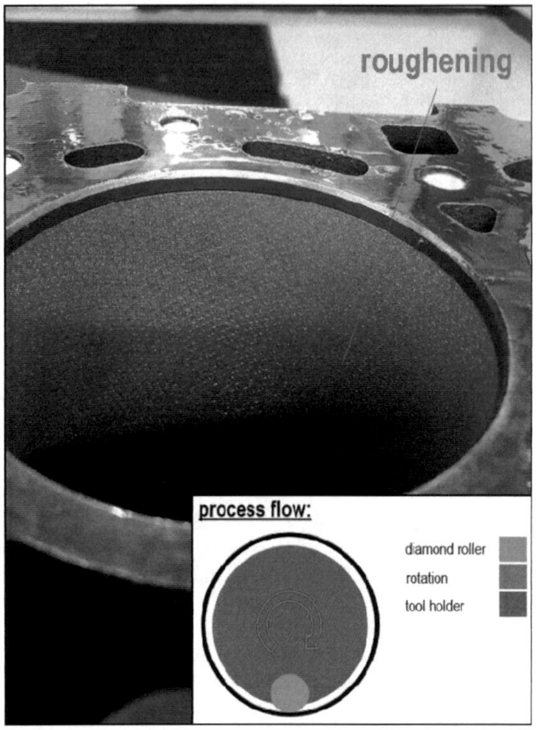

Figure 3: Roll-roughened cylinder bore of grey cast iron diesel block.

Cast iron cylinder blocks were bored oversize followed by different surface preparation steps and different spray coatings (Table 1) applied by the GTV-Ford "RSW" wire spray method at Autocraft Drivetrain Solutions Ltd, UK.

Table 1: Test matrix

Cylinder block no.:	Cylinder bore no.:	Coating system	Activation option	Cutting parameter
4	001	with NiAl	D181	f = 0,1 mm/z n = 1000 1/min
4	002	without NiAl	D181	f = 0,1 mm/z n = 1000 1/min
4	003	with NiAl	D181 + D76	f = 0,1 mm/z n = 1000 1/min
4	004	without NiAl	D181 + D76	f = 0,1 mm/z n = 1000 1/min

The rolling operation resulted in different surface roughness values: cylinder bores 1 / 2 showed Ra = 7.99 / Rz = 53.06µm whereas the double-rolled cylinder bore 3 / 4 showed values of Ra = 3.65 / Rz = 26.79µm. The surface roughness of a cylinder bore surface after the removal of the damaged wear zone is typically in the range of Ra = 2.42 / Rz = 18.02 µm.

After spraying the PAT adhesion bond strength is measured with an 8.16mm dolly diameter. A value of 20 – 35MPa is typical for bored CI surfaces using a NiAl bond coating and 0.8%C-Steel spray wire. In Figure 4 a section of a spray coated cylinder bore is shown, where a ring is machined around the dolly location, which is cut completely through the coating that only the adhesion between substrate and coating is measured under tensile test conditions. The dolly is fixed and spring loaded, while the adhesive FM1000 foil is sintered during oven baking operation at 180°C/2h.

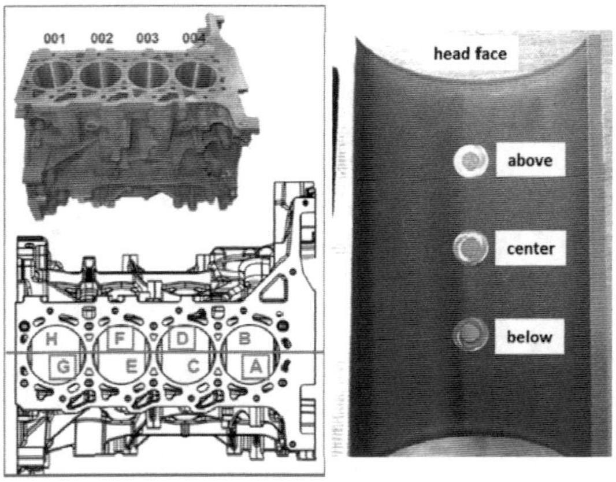

Figure 4: Segment from CI diesel 2.2L diesel block with RSW spray-repair coating being prepared for PAT testing, near head deck, middle of the bore and near crankcase.

2.2. Combination of Roughening Processes

In the GRP1 Process from Gehring [7], a PCD cutting tool with an open rectangular shape is used to generate a threaded groove from head deck to crankcase. In case, however, that a so called "selected-bore-area" concept is preferred with a 4-5mm portion of Aluminum material above the spray coating, then those open rectangular grooves can also be machined by a circular milling tool. In the second step, these pillars are then deformed by a rolling operation with a roll, coated with sharp diamond particles. This step is then leading to a similar undercut in the cross section, as it is used in the GEN 9.2 standard operation. In addition this roll-deformation operation is increasing the surface roughness on top of the pillars and hence particle bonding and heat transfer.

As an alternative the **MSGRP** (**M**AG **S**TURM **G**ÜHRING **R**OUGHENING **P**ROCESS) could be used, which is a combination of the threading operation with the GEN 9.2 PCD cutting tool with a dovetail-undercut profile and a superimposed axial grooving operation: a knife-edge roller-blade is generating V-shaped axial grooves from head deck down to the crankcase. For prototype testing a tool with just six roller blades was used, however, for serial production a special dedicated tool can be made for the desired bore diameter so that the entire cylinder bore can be axially grooved in just four single passes through the bore, i.e. in less than 5 seconds operating time.

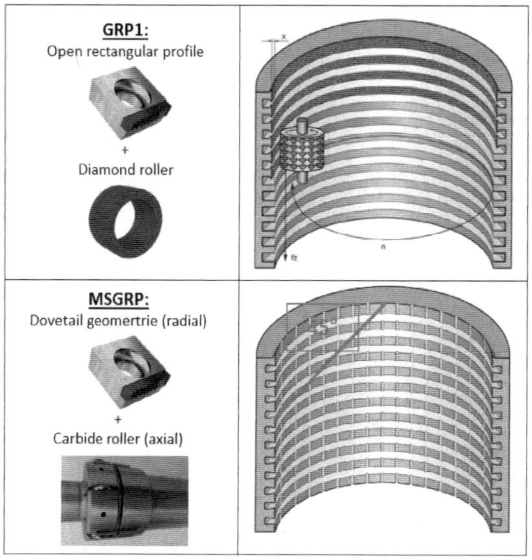

Diamond roller		
Diameter (Ø)	84.3	mm
Number of cutting edges (z)	1	
Cutting speed (v_c)	529	m/min
Feed rate (v_f)	400	mm/min
Rotation (n)	2000	min^{-1}
Feed (fz)	0,2	mm/z
Radial force	78,2	N

Carbide roller		
Diameter (Ø)	84.3	mm
Number of cutting edges (z)	6	
Cutting speed (v_c)	/	m/min
Feed rate (v_f)	50000	mm/min
Rotation (n)	/	min^{-1}
Feed (fz)	/	mm/z
Radial force	240	N per edge

Roughening insert (open rectangular + dovetail geomertie)		
Diameter (Ø)	84.3	mm
Number of cutting edges (z)	1	
Cutting speed (v_c)	600	m/min
Feed rate (v_f)	906	mm/min
Rotation (n)	2265	min^{-1}
Feed (fz)	0,4	mm/z

Figure 5: Schematic presentation of the GRP1 and MSGRP Processes: threading plus different rolling operations. The location of the microstructural cross section under a 45° angle is indicated by the red line.

For machining tests with these 2 processes, liners made from spray-compacted PM Aluminum "DISPAL S-260" material (Erbslöh Aluminium GmbH) were used. This material has a very homogeneous deformation behavior, without any casting porosity and oxides. Liners with a bore diameter of 84mm and 120mm length were clamped in a horizontal CNC system for the machining tests (Figure 6).

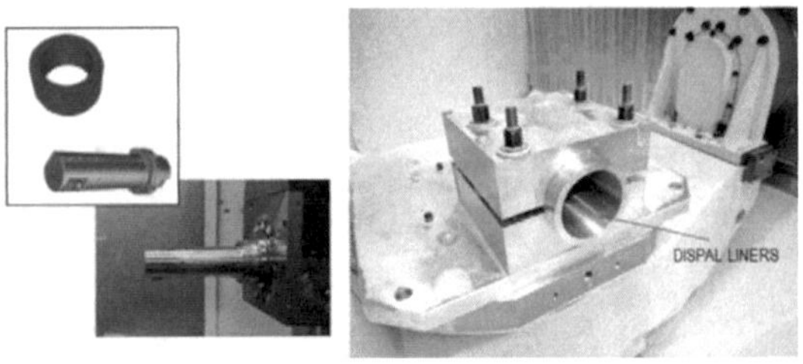

Figure 6: Rolling tool in horizontal machining center MAG SPECHT 600L with a mounted test Al-liner.

Besides these processes GRP1 and MSGRP two additional trails were made:

> Trial 1: a combination of open rectangular grooves plus axial V-shaped grooves.
>
> Trial 2: axial V-shaped grooves followed by a GRP1 roll-roughening step. In the GRP1 process the surface is roll-roughened by a D181 grit size leading to large indentations.

See schematic presentation of these trials in Figure 7.

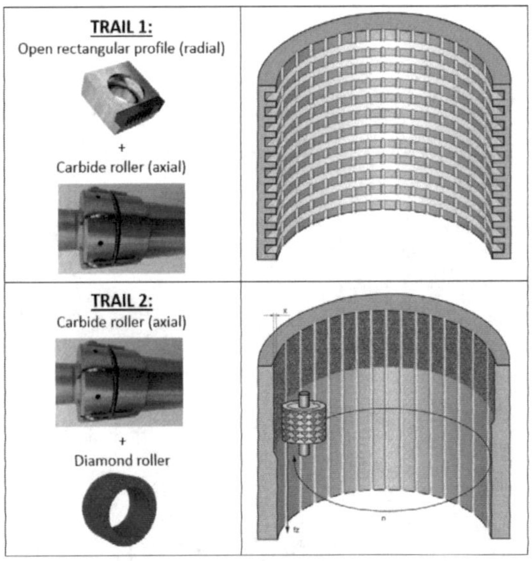

Figure 7: Schematic presentation of Trial 1 / 2.

Those roughened liners were coated in a Sturm ACC System with RSW Fe-0.3%C-Steel coatings, using the spray parameters from Figure 8.

RSW-Coating		
Diameter (Ø)	84,3	mm
Maschine	STURM ACCS-TWS	
Wire Material:	FeC0,3	
Current	150	A
Wire-Feedrate	10,5	m/min
Doublestrokes	6	
Coating-Thickness (avg.)	300	µm
Coating-Time	28,28	s

Figure 8: Set-up for RSW-spraying of Al-Dispal test liners.

3 Results

The PAT tests were carried out at 2 facilities: at Gehring and at IOT, University Aachen. These PAT results of spray repaired cast iron diesel blocks in Figure 9 are indicating that the roll-roughening is achieving similar bond strength values without the use of NiAl-bond coating material as with NiAl-coating.

The cross section of a cylinder bore with NiAl + top coating is shown in Figure 10. The surface roughening profile, generated by the diamond grit is clearly visible. The photos are taken through the coated zone as well as through the dolly test location after the pull-test:

It is obvious that the delamination will take place between the NiAl-coating and the top coating and that the interface bond coating / top coating is the weakest point. The adhesion between the top coating and this roughened CI substrate material seems to be higher than between NiAl and top coating. Because cylinder bore 2 and 3, with and without bond coating are showing similar bond strength values, it can be concluded that the bond coating could be eliminated and that a rolling tool with two diamond roller (D181 +D76 grit size) is sufficient to obtain a bond strength of above 40MPa.

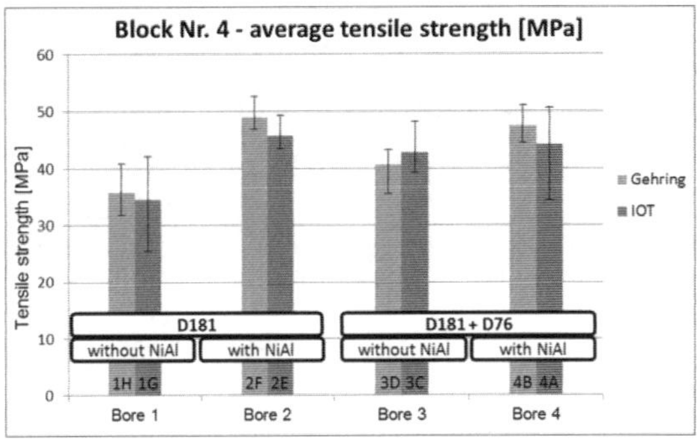

Figure 9: PAT bond strength values in all 4 cylinder bores.

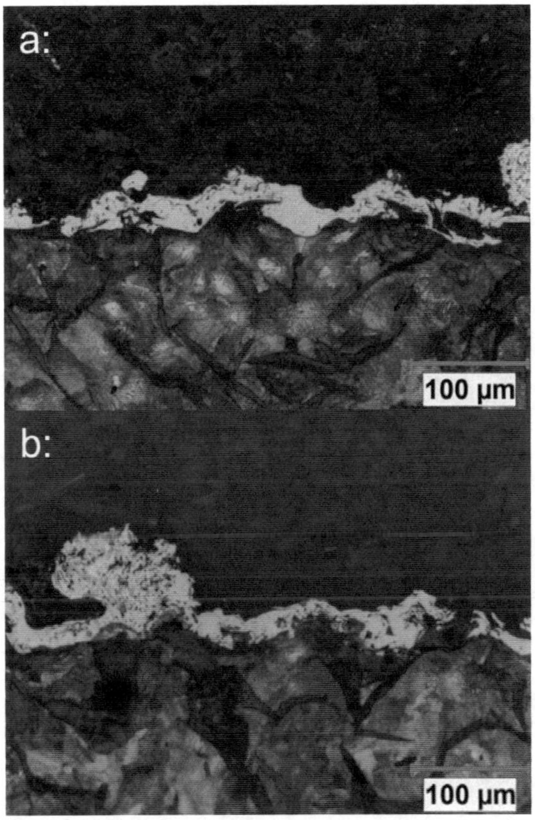

Figure 10: Cross sections of grey cast iron block
a: Cross section through coated bore
b: Cross section through the dolly location after pull test.

The coated Aluminum liners were inspected by light and scanning electron microscopy. In Fig. 11 a cross section of a coated liner, roughened by process GRP1 is shown. In contrast to the smooth surface on the pillars (Figure 1) a rough surface is generated.

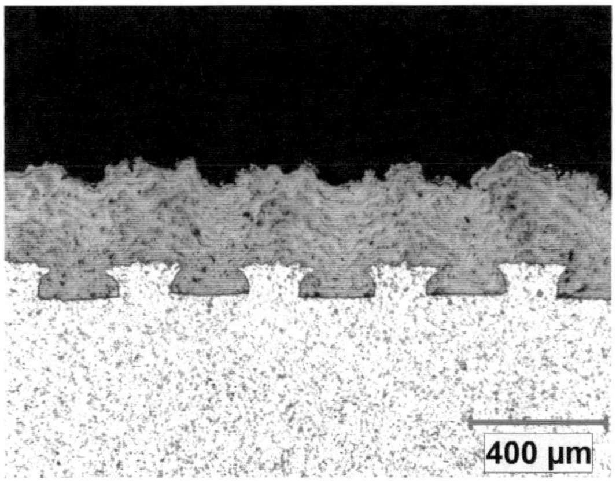

Figure 11: Cross section of RSW coated liner, roughened by the GRP1 Process.

These coated liners were tested by the PAT test: The 8.16mm dolly was glued to the as-sprayed surface and after pull-off, the sample was inspected again in a

Figure 12: Pulled-off RSW spray coating containing fractured Al-material from the deformed dovetail undercut edges. The SEM picture shows that the profile edges are completely removed.

cross section, Fig. 12. It is clear that those sharp, plastically deformed dovetail-undercut edges were fractured and are entrapped inside the removed spray coating material. Therefore only moderate PAT adhesion values can be expected.

In order to visualize the surface morphology of a liner roughened by the MSGRP process, a sample was cut from the liner under a 45° angle. (see Figures. 5/13)

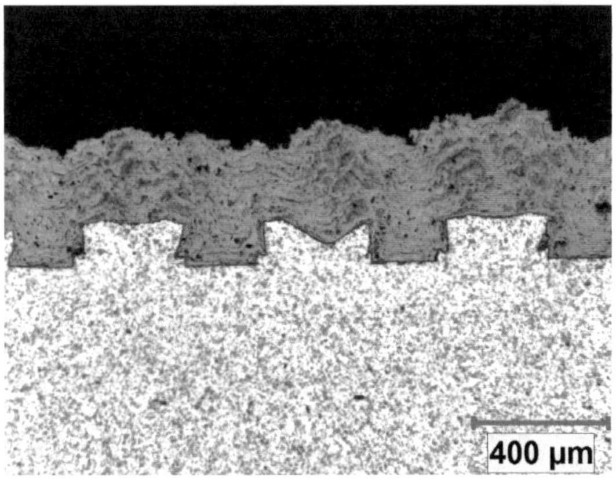

Figure 13: Cross section under 45° of an RSW coated liner, roughened by the MSGRP process

In addition it should be pointed out, that the head deck chamfer of the cylinders bores do not show any coating break-out caused by an remaining open profile. Instead the coating is held in place by the MSGRP honeycomb-type roughness profile.

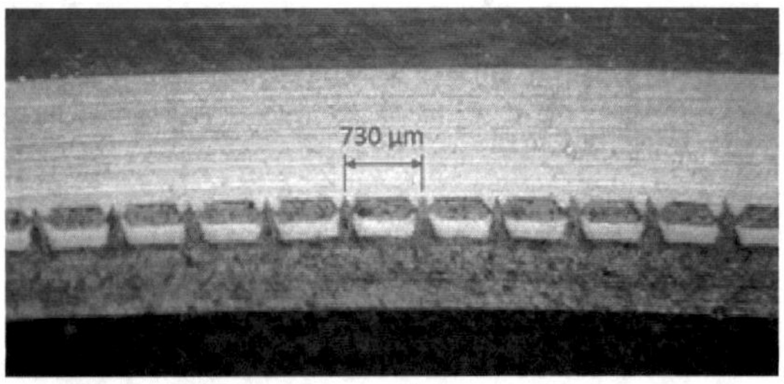

Figure 14: Chamfer at head deck through Aluminium and RSW spray coating

For comparison, the cross sections of Trials 1 / 2 are shown in Fig. 15 a/b

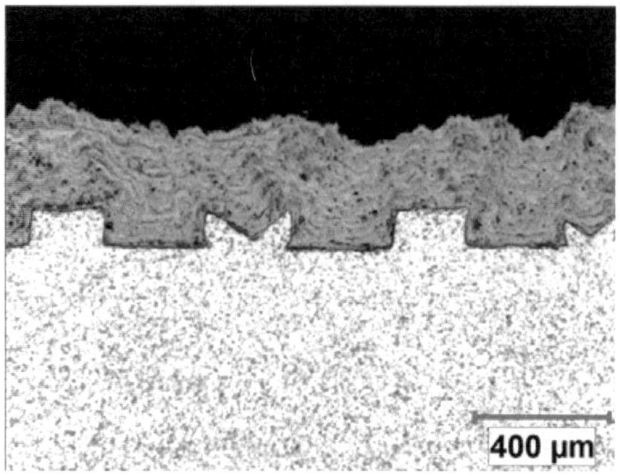

Figure 15: a) Rectangular shaped groves plus MSRP

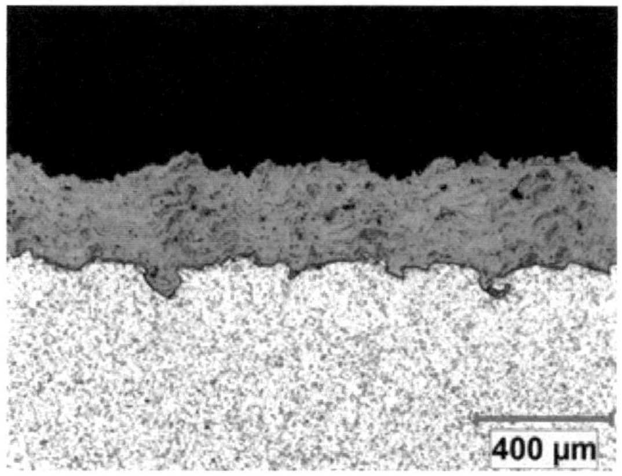

Figure 15: b) MSRP followed by diamond roll roughening.

The results of the PAT tests (Fig. 16) indicate, that only the 2 processes **GRP1** and **MSGRP** have a potential for high bond strength of spray coatings in Al-engine bores.

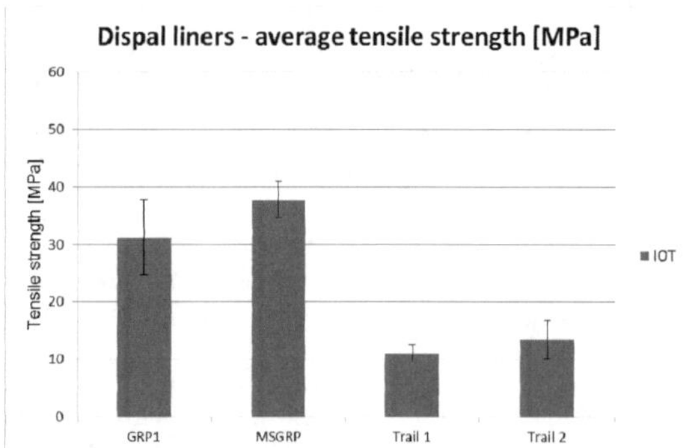

Figure 16: PAT Adhesion test results. (IOT, Aachen University).

The surface at the location of the PAT dolly was inspected after the pull test by using a Keyence microscope and a SEM, Figure 17.

Figure 17: Surface of liner at the location of the pulled-off PAT dolly. (SEM Zeiss, Sigma VP/ Keyence digital microscope, VHX-5000).

4 Summary and Conclusion

The Gehring GRP2 roll-roughening process can increase the surface roughness for spray-repaired grey cast iron blocks, leading to improved bond strength. In case of Al-engine blocks the combination of rectangular open profiles plus plastic deformation / simultaneous surface roll-roughening of the pillar surface is leading to good adhesion of the spray coating on the rough pillar surface. This bonding is very strong so that the weakest point is now the plastically deformed pillar surface material itself. Due to the brittle material behavior this plastically deformed zone becomes the weakest point and the Al-material is pulled-off from the substrate, the coating is no longer entrapped under the dovetail groves. This behavior seems to explain, why the PAT bond strength is not as high as is measured typically on cast Al-material.

Higher bond strength can be obtained by combining the GEN9.2 PCD surface threading and the new MSRP axial roller-blade cutting: The spray droplets are fixed by the dovetail undercut whereas the axial grooves prevent any movement of the coating in circumferential direction. Especially for smaller bores with rapid heating-up of the Al-block such a behavior is important so that the coating cannot peel-off from the substrate during large expansion of the bores due to increasing temperature.

Trails 1 and 2 with the axial grooving process do not lead to good adhesion in the combination with an open rectangular threading or with a roll roughening process. Therefore it can be concluded that an undercut geometry is required for optimum adhesion bevavior.

5 Literature

[1] DE 10 2006 045275 B3: Coated product producing method, involves introducing undercut into surface of base plate i.e. cylindrical hollow body, with cutting tool before coating of product, where undercut is aligned, oriented and opposite to another undercut, Publication Date 2008-02-07.

[2] WO 2010/015229 A1, Method and Tool for Producing a Surface Having a Predetermined Roughness, Publication Date Feb. 11, 2010.

[3] Gand, B., H.J. Salenbauch; AluThinFer-Zylinderlaufflächen – Verbesserung des Kraftstoff-verbrauchs und des Verschleißverhaltens für Motoren mit niedriger und hoher thermo - mechanischer Beanspruchung; VDI Berichte 2109, „Zylinderlaufbahn, Kolben, Pleuel", Baden-Baden, June 7-8, 2010, page 159.

[4] US 2013/8726874 B2, Cylinder Bore with Selective Surface Treatment and Method for the Production Thereof, Publication Date Nov. 7, 2013.

[5] EP 2267178 A2, Method for Keying Metal Surfaces, Publication Date Dec. 29, 2010.

[6] US 2012/0317790 A1, Tool and Method for Mechanical Roughening, Publication Date Dec. 20, 2012.

[7] WO 2016/074862 A1, Method and Device for Roughening a Substrate Surface, Publication Date May 19, 2016.

Development of Novel Fe-Based Coating Systems for Internal Combustion Engines

K. Bobzin, M. Öte, T. Königstein, RWTH Aachen University, Aachen / D
K. Dröder, H. W. Hoffmeister, G. Mahlfeld, Technical University of Braunschweig, Braunschweig / D
T. Schläfer, GTV Verschleißschutz GmbH, Luckenbach / D

Nowadays combustion engines are the most common way to impel vehicles. Thereby losses occur, due to cooling, exhaust gas and friction. Modern engines roughly dissipate 8% of the chemical energy stored in the fuel because of friction in different tribological systems. The highest potentials for optimisation can be found in the tribological system of inner surface of combustion chamber and piston ring. Besides friction, corrosive stress of inner surface of combustion chamber increases e.g. due to the utilization of auxiliary systems such as Exhaust-Gas-Recovery. In order to save energy, reduce emissions and enhance the lifetime of combustion engines innovative coating material systems need to be developed, especially for inner surface of combustion chamber.

This study focuses on the development of innovative iron based materials for combustion chamber application using Plasma Transferred Wire Arc (PTWA) and Rotating Single Wire Arc (RSW) technologies. In order to improve the wear and corrosion resistance boron and chromium are added into the feedstock material. After deposition, different honing topographies are manufactured in order to evaluate their influence on the tribological behavior. Furthermore, electro-chemical corrosion tests are conducted by using an electrolyte simulating the exhaust gas concentrate. In conclusion an optimised coating material deposited by PTWA and RSW and improved surface topographies can be combined.

1. Introduction

Due to the strict requirements with regard to emissions the demand for high performance materials steadily increases in all industry sectors. A prominent sector is represented by the automotive industry. In this field different approaches such as lightweight concepts are investigated regarding applicability e.g. fibre reinforced plastics. Further strategies aim for the optimization of tribological systems in order to reduce wear and friction. In combustion engines, numerous tribological systems can be identified such as gear wheels, clutches, chain drives, bearings or valves [1]. In this context, surface modification, such as structuring and application of coatings have proven to be key approaches in order to reduce friction and wear [2-7].

The tribological system between cylinder inner surface and piston ring represents a critical example due to high amount of energy losses. 50 % of all friction losses occurring in the combustion engines can be attributed to this system showing the importance of optimization [8]. The deposition of a s = 150 µm thick coating on the inner surface of the combustion chamber replaces liners made of cast iron, contributing to a further reduction of total weight [9]. In addition, thermal management can be influenced by adjusting thermal conductivity of the coating material. It was shown, that thermal management can influence the performance of combustion engines significantly [10]. Apart from mechanical load, corrosion effects might also be significant for the functionality and efficiency of combustion engines. These are getting more significant nowadays due to Exhaust-Gas-Recovery (EGR) systems or increased use of biofuels. In the future, due to further increase in peak pressures and extensive use of EGR systems, the development of high performance, wear and corrosion resistant coating materials will be required.

The aim of this study is to introduce an innovative coating material which exhibits improved wear resistance and corrosion behaviour compared to established systems such as 0.8 % C-steel. This will be demonstrated by means of the machinability by honing and model tests, such as a reciprocating sliding test and an electrochemical corrosion test.

2. Test Methods

All coatings were deposited on Al6060 cylinders with inner diameter of d = 80 mm, outer diameter of D = 100 mm and height of h = 120 mm. Prior to coating, the cylinders were mechanically roughened using a specially designed tool developed by Institute of Machine Tools and Production Technology (IWF) Braunschweig in order to manufacture a dovetail shaped contour [11].

Subsequently a reference coating of 0.8 % C-steel was deposited by thermal spraying technology using Plasma Transferred Wire Arc (PTWA) and the successor Rotating Single Wire (RSW) spraying systems, both from GTV Verschleißschutz GmbH (Luckenbach, Germany). Furthermore, a novel iron-based feedstock material of the alloy FeCrBMn was deposited by RSW. In the following, for the sake of simplicity, the reference alloy will be referred as FeC 0.8 % and the novel iron based alloy as FeCr. PTWA and RSW have similar torch designs. Main difference between these processes can be found in the ignition sequence. During ignition for PTWA, the arc is initially ignited between an anode and the cathode. Anode itself forms the nozzle of the torch. Due to plasma gas volume flow, the arc extends towards the positively charged wire, closing the electric circuit. In case of

RSW, the arc is directly ignited on the wire. This reduces wear of the nozzle. For both processes following parameters were kept constant: current (I) = 85 A, argon feed rate ($\dot{V}_{Ar}$) = 70 SLPM, hydrogen feed rate ($\dot{V}_{H2}$) = 10 SLPM, air feed rate ($\dot{V}_{Air}$) = 1,150 SLPM, surface velocity (v_s) = 150 m/min, axial feed were a = 3 mm/rpm. Number of double torch passes (N) and wire feed rate (v_W) were changed and can be obtained in Table 1.

Table 1 Process parameters for PTWA and RSW

Number	C1	C2	C3
Coating technology	PTWA	RSW	RSW
Feedstock material	FeC 0.8 %	FeC 0.8 %	FeCr
v_W [m/min]	6.3	6.3	4,3
N [-]	7	10	12

All coatings were characterized metallographically with regard to their microstructure and bonding quality using Scanning Electron Microscopy (SEM) SEM LEO 1530 VP Gemini by Carl Zeiss AG (Oberkochen, Germany). Furthermore, hardnesses were measured by means of indentation method using a Vickers pyramid and the indentation machine Micromet 1 from Bühler (Esslingen am Neckar, Germany). The tests were executed on the one hand on the cross-sections and on the other hand on the honed surfaces. For both test setups, measurements were conducted with a load of m = 300 g and dwell time of t_d = 10 s. An average hardness value was calculated taking n = 10 measurements as basis.

Honing as the manufacturing process for finishing cylinder bores has a high influence on the tribological properties. To investigate new coating materials it, is necessary to develop a productive honing processes for each honing topography. Therefore, the honing machine Solohone LV1 from Gehring (Ostfildern, Germany) and the cooling fluid Honilo 980 by Castrol was used. All cylinder liners were honed in a four-step honing procedure with a cutting speed of v_c = 55 m/min and a honing angle of α = 39° to allow a direct comparison. The tool contained eight diamond honing stones with a length of 80 mm and a width of 3 mm. For the different surface topographies special grain sizes, shown in Table 2, were combined with optimal infeed rates. Between every manufacturing step the surface characteristics have been measured with a Hommel-Etamic T1000 from Jenoptik (Jena, Germany) and the material removal rate ($Q_w{'}$) was calculated from the pre- and post-process bore diameters measured with an inside micrometer.

Table 2 Grain sizes for the surface topographies

Honing step	Topography		
	Reference	Mirror	Plateau
1	D107	D181	D181
2	D064	D064	D107
3	D025	D015	D064
4	D007	D007	D007

During operation of combustion engines, highest pressure and thereby highest load on the surface occurs immediately after ignition at top position. Therefore, this study focused on the tribological investigation considering the top position. All coatings were tested with regard to their wear and friction behaviour using a reciprocating sliding test, SVT 1000-P from Wazau GmbH (Berlin, Germany). For the model test segments of cylinder liners and nitrided piston rings were used to simulate the tribological system. In order to reduce the test duration for wear tests the mechanical load on the cylinder liner was increased by selecting a smaller piston ring diameter of d_r = 72 mm compared to the inner diameter of the cylinder liners d_L = 81 mm. For wear tests all coatings were post treated identically by machining a reference honing topography. The test duration was set to t = 12 h, normal load to F_N = 200 N, sample temperature to T_S = 130 °C, frequency to f = 20 Hz and stroke to h = 4 mm. A low data acquisition frequency of f_{Scan} = 1 Hz was chosen to reduce the data volume. A standard oil 0W-40 of Castrol was used for lubrication. Wear tracks were analysed by confocal laser scanning microscopy (CLSM) using VK-X210 by Keyence (Tokyo, Japan). In order to investigate the friction coefficient the test parameters were adjusted as follows: t = 4 h, f = 10 Hz, h = 10 mm and F_N = 100 N. The data acquisition frequency corresponded to f_{Scan} = 100 Hz recording 10 measurement points per cycle ensuring high resolution. The temperature and lubricant were kept constant. Besides reference honing, the influence of further honing topographies on the friction coefficient was investigated for the novel FeCr coating.

All coatings were investigated with regard to the corrosion resistance by means of electrochemical test using the potentiostat Reference 600+ of Gamry Instruments (Warminseter, USA). Therefore, segments of top position were aligned to the measurement cell using an adapter. As electrolyte a mixture of different acids was used, simulating the exhaust gas [13]. The composition for V = 1,000 ml basic solution can be found below;

- 958.0 ml distilled water
- 14.0 ml methanoic acid
- 15.0 ml ethanoic acid
- 1.0 ml hydrochloric acid
- 11.0 ml nitric acid
- 1.0 sulphuric acid

ΔV_{BS} = 12.5 ml of this basic solution is blended with V_W = 1,000 ml distilled water and adjusted if necessary until pH-value equals 2.3. For each coating material, the top position was tested. For the measurement a normalized surface area of A = 1 cm^2 was considered. The measurements were conducted using a potential sweep rate of V_U = 0.5 mV/s. After the test micrographs of the surface and polished cross-sections were captured by SEM.

3. Results & Discussion

In the following subchapters, the results of this study will be presented. It is organized according to the methodology including the coating characterization, honing process and the results of tribological and corrosion testing.

3.1. Coating characterization

After coating deposition and honing the cross-sections of the coatings were evaluated regarding their microstructure and bonding quality. It can be observed that all coatings have a lamellar coating build-up, see Fig. 1. Furthermore, all coatings exhibit a degree of darker areas representing oxides. Although same feedstock material, same process parameters, and very similar deposition technologies were used, different oxidation degrees of coatings C1 and C2 can be observed, see Fig 1a and b. Comparing both references, the coating deposited by PTWA (C1) exhibits a higher degree of oxides than the RSW coating (C2). This could be an indicator for a dependency between coating technology and oxidation degree. However, comparing the geometry of the roughening profile, differences especially with regard to the crones can be observed. The profile in Fig 1a shows a sharper tip, while the profile in Fig. 1b has a smooth shape. This is a result of different tool alignment during roughening process. It is assumed, that oxides are mainly formed at positions exhibiting high temperatures. The surface temperature itself is related to the heat transfer rate which is a function of material properties and geometry. Due to missing surrounding materials edges cannot transfer as much heat into the bulk material as for example smooth surfaces. As a consequence edges, for example of the crones of the roughening profile, exhibit higher temperatures. It is assumed, that this higher surface temperature at the edges is jointly responsible for the formation of oxides. Up to now the mechanism behind the oxidation formation of PTWA and RSW coatings is not entirely understood. As a consequence of higher degree of oxides in C1, a higher average hardness value of 420±58 HV0.3 compared to 346±44 HV0.3 of the coating C2 was measured by indentation on the cross-sections. Both reference coatings show fully filled-up profiles, which is an indicator for good bonding. Coating thickness of both reference coatings amount to s = 220 µm measured from top crown position to the honed surface. In Fig. 1c the cross-sections of the FeCr coating can be seen, exhibiting a significantly lower content of oxides, compared to both references. This can be attributed to the different material composition of the feedstock material, especially to the addition of chromium. The hardness amounts to 353±34 HV0.3. The dovetail profile is fully filled-up with the coating material indicating good boding properties. The coating thickness amounts to s = 100 µm and is significantly thinner compared to both reference coatings.

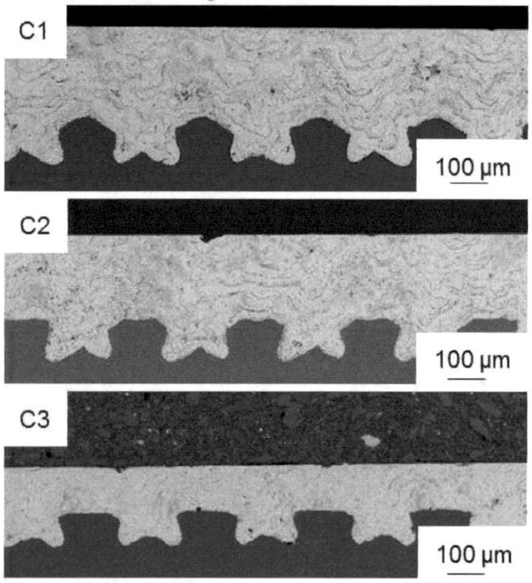

Fig. 1 SEM micrographs of polished cross- sections C1 (a); C2 (b) and C3 (c)

3.2. Honing

To compare the thermal spraying technologies the cylinder liners coated with the reference material FeC 0.8 % have been honed with the same infeed force. Consistent with the lower coating hardness, a higher material removal rate was achieved for the RSW coating, because less coating hardness increases the grain intrusion. This is proven by comparing Rz values after the third honing step. With a grain size of D025, an Rz value of 5.63 µm for the RSW and 4.98 µm for the PTWA coating has been measured. After last honing step with a D007 grain, the porosity at the surface is higher for FeC 0.8 % reference coating deposited by PTWA compared to that deposited by RSW, as shown in Fig. 2. This can be attributed to higher content of brittle oxides in the PTWA coating, breaking out during honing process.

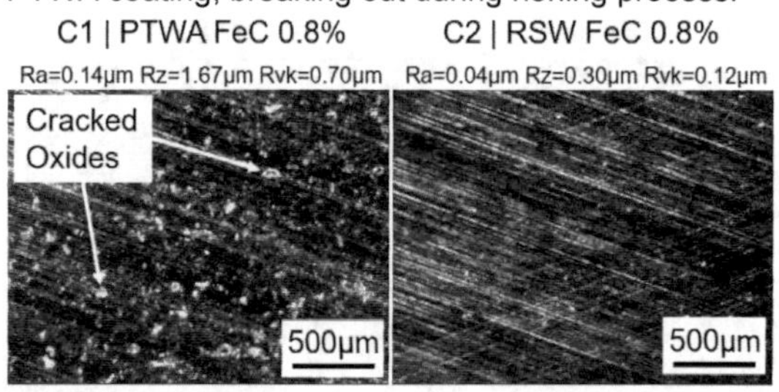

Fig. 2 Final surface topography of PTWA and RSW coating

For the RSW technology almost no tool wear could be quantified during the experiments. This can be traced back to the optimal combination between bond hardness and grain type in the honing stone. Due to the higher hardness and a higher degree of oxides a

G-Factor, ratio between the removed material and the tool wear, of 38 was measured for the PTWA coating.

After running an identical process for the new FeCr material an Rz = 3.59 µm for the D025 grain and a material removal rate Q_w = 30.52 mm³/s was measured. Compared with the reference coatings exhibiting a Q_w = 41.55 mm³/s for the RSW and the Q_w = 18.66 mm³/s for the PTWA technology, the combination of the new RSW technology and the FeCr feedstock materials leads to the desired high machinability rates. With a G-Factor of 74 compared to 38 for the PTWA FeC 0.8% the novel FeCr coating caused less tool wear combined with a higher productivity.

Following the direct comparison, the capability to hone different surface topographies needs to be proven. Therefore, the special strategies in the grain size steps for each topography were tested. Pictures of the final surfaces and the surface characteristics are shown in Fig. 3. For further process optimisations the function between the grain size and Rz is demonstrated.

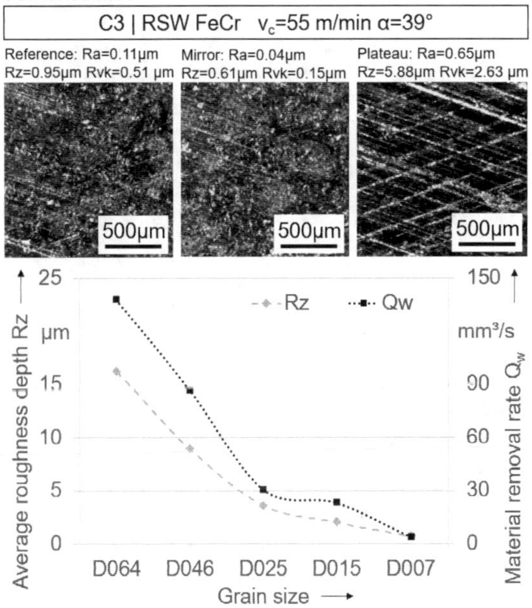

Fig. 3 Final surface topographies for the new iron-based coating material; Surface characteristics depending on grain size

In conclusion the novel FeCr material for cylinder liner application does not set special requirements regarding the manufacturing by honing. With common diamond honing stones, higher material removal rates and G-Factors compared to the reference PTWA FeC 0.8% coating can be achieved. In terms of machinability of special surface topographies, the new coating system is in no way inferior to established systems, as proven in comparison with FeC 0.8 % C-steel.

3.3. Tribological testing

In the following chapter the results of tribological tests will be explained by evaluating the depth and width of the wear tracks of the coatings and the counter bodies. Furthermore, the friction coefficients for the different honing topographies in combination with the FeCr feedstock material will be compared with each other.

The wear tracks on both FeC 0.8 % reference coatings show similar appearances. Both wear track widths amount to w = 3.7 mm. The average depth of PTWA coating amounts to d_{C1} = 2 µm and that of the RSW coating to d_{C2} = 5 µm. The differences correlate with hardness values measured on the cross-sections. In addition, the hardnesses of honed surfaces were measured. PTWA coating (C1) exhibits 453±38 HV0.3 and RSW coating (C2) exhibits 357±37 HV0.3. As already shown, PTWA coating is significantly harder, which was explained by higher degree of oxides observed in the cross-sections. The hardness values measured on the honed surfaced are about 5 % higher compared to the hardness values measured on the cross sections. This can be attributed to the honing process and strain hardening due to stretched lattices. The width of wear track of the FeCr coating amounts to w_{C3} = 3.6 mm and thereby comparable to those of the reference wear tracks. The depth of the wear track amounts to d_{C3} = 7 µm. In this case measured hardness of the honed surface corresponds to 409±25 HV0.3.

For the evaluation of the frictional behaviour, different testing parameters were used. After the test the average friction coefficients, running-in behaviours and trends of single friction cycles were evaluated. FeC 0.8 % reference coatings exhibit an average friction coefficient of $\mu_{A.C1}$ = 0,151 and $\mu_{A.C2}$ = 0,154 respectively. The running-in times do not show any differences, both amount to t_{C1} = t_{C2} = 20 s. The average friction coefficient of the FeCr coating exhibits $\mu_{A.C3}$ = 0.155 and running-in time amounts to t_{C3} = 20 s, both matching to the results of the references. Subsequently, the influence of different honing topographies was evaluated using the FeCr coating material. The average friction coefficients correspond to $\mu_{A.C3}$ = 0.157 for plateau honing and $\mu_{A.C3}$ = 0.149 for mirror honing. No significant differences between the running-in times were observed. In direct comparison of single cycles it can be observed, that each cycle correspond to the targeted cycle time of τ = 0.1 s, see Fig. 4. Furthermore, different friction coefficients can be observed corresponding to different friction regimes such as static friction at turning points and mixed friction in between these points. Maximum velocity of v_{max} = 0.314 m/s at the midpoint of the track borders was calculated. Due to the fact that v_{max} is comparatively small, it is assumed that the hydrodynamic operation condition is not reached [8]. It is assumed that more pronounced differences in friction coefficients will be obtained with increasing relative velocity enabling hydrodynamic operation conditions.

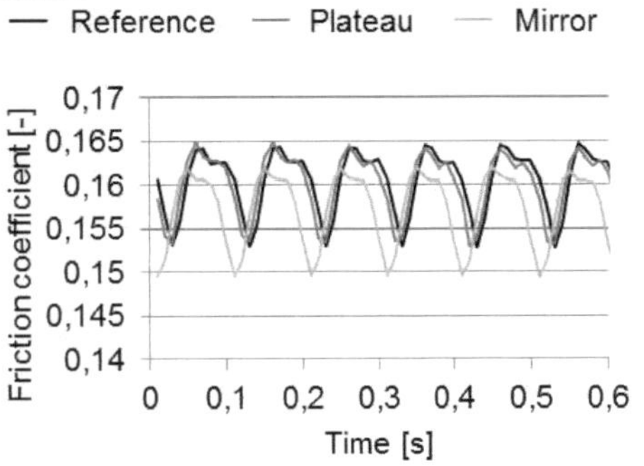

Fig. 4 Friction coefficient of C3 and different honing topographies

3.4. Corrosions behaviour

In the following chapter the results of corrosion tests will be explained by evaluating current density curves and micrographs of the top view and cross-sections after the test. Comparing the current density curves of both FeC 0.8 % references, no significant difference between PTWA and RSW deposition technology can be identified, see Fig. 5. In fact, both curves have a high degree of congruency. The corrosion current densities $i_{corr} = 56 \cdot 10^{-3}$ mA/cm² and resting potentials $E_h = -580$ mV are almost identical. In comparison to the reference, the current density curve of the FeCr coating exhibits a more positive resting potential $E_h = -495$ mV. This indicates a higher corrosions resistance. The current density of the FeCr coating amounts to $i_{corr} = 54 \cdot 10^{-3}$ mA/cm² and thereby is in the same range compared to the references. For high polarisation above $E_h > 300$ mV current density approaches to the values of the substrate. It is assumed that due to small coating thickness electrolyte gains access to the substrate. In that case mixed current density of substrate and coating is measured.

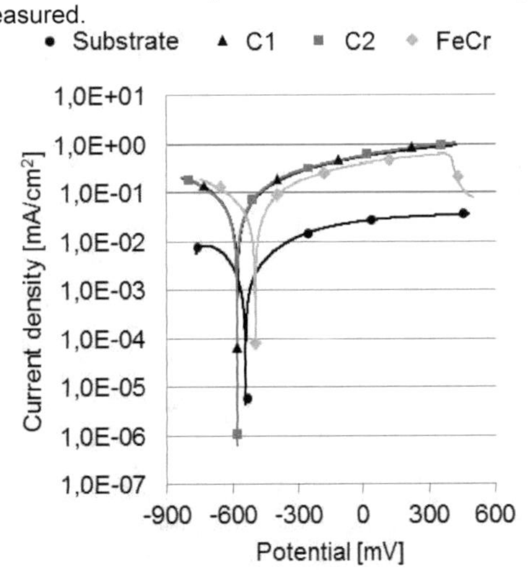

Fig. 5 Current density curves of reference coatings and new iron-based material

SEM micrographs of the tested surfaces after exposition to the electrolyte show significant changes in both FeC 0.8 % references indicating severe chemical reactions of the coating material with the electrolyte, see Fig. 6a. This micrograph confirms major corrosion attack especially along the lamellar interface and oxidized areas. High magnification also confirms corrosion attack on the iron-based matrix, resulting in a pitted microstructure. SEM micrographs of the FeCr coating proofs a significant improved corrosion resistance of the iron-based matrix, see Fig. 6b. Even traces of honing process can be recognized. In that case, corrosion was observed along the interlamellar interfaces and oxidized areas. No attack of the steel matrix was recognized, which can be attributed to the higher chromium content in the coating.

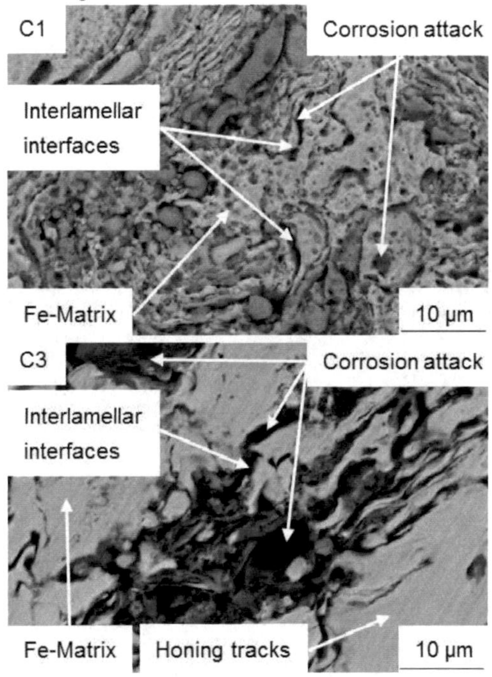

Fig. 6 SEM micrographs of coated surface after corrosion test of C1 and C3

Investigations on the polished cross-sections confirm these results. In Fig. 7a and b, the pitted surface of the FeC 0.8 % coating can be confirmed. The FeCr coating does not show any corrosion attacks of the iron matrix, see Fig. 7c. Corrosion paths can mainly be identified along the oxides. Nevertheless, for high polarisation the current density curve in Fig. 5 decreases and approaches to the current density curve of the substrate. This indicates connection of the electrolyte with substrate presumably along lamella interfaces until the substrate is reached. It must be noted, that the coating thickness of the FeCr material amounts less than half of the coating thicknesses of the references. Further increase of coating thickness shall already solve the issue of breaking through.

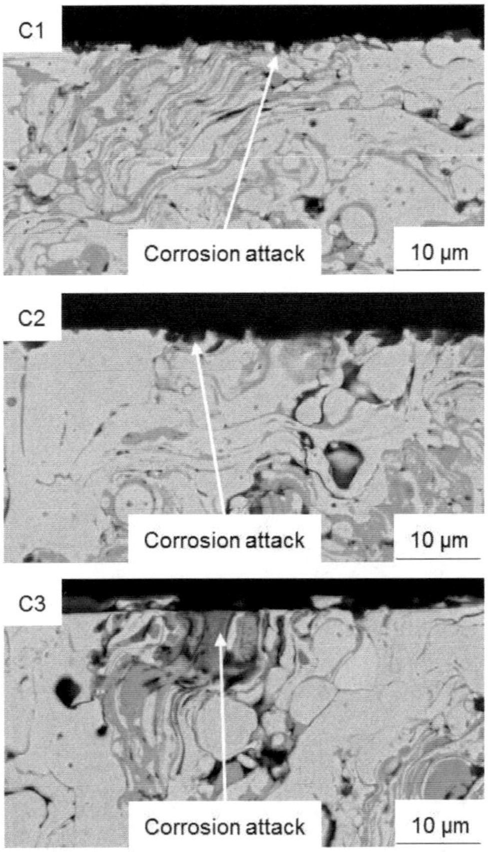

Fig. 7 SEM micrographs of polished cross-sections after corrosion test of C1, C2 and C3

4. Conclusion

A novel iron-based feedstock material was developed and investigated regarding coating quality. Results were compared to the established references FeC 0.8 %. Special focus has been payed on the machinability, tribological performances and corrosion behaviour.

It was shown, that the reference coatings FeC 0.8 % exhibit different degrees of oxides. It is assumed, that this phenomenon is either related to the spraying technology and/or to the different geometry of the machined roughening profile. This will be investigated further in the future. It was shown that the wear behaviour of the FeC 0.8 % coatings correlate with the hardness values measured on the cross-section and surface. With respect to tribological behaviour, no significant differences between the FeC 0.8 % references and the FeCr coating were observed. Electrochemical corrosion tests indicate improved corrosion resistance of the FeCr coating compared to the FeC 0.8 % references. Furthermore, promising G-Factors and Q' values were measured, proofing good machinability of the FeCr coating. In conclusion, a coating system exhibiting superior corrosion resistance and machinability compared to the references has been developed.

References

[1] Bobzin, K.; Öte, M.; Linke, T. F. and Königstein, T. Verschleiß im Antriebsstrang: Oberflächentechnische Herausforderungen, Beschichtungskonzepte und Prüfmethoden, Proceedings GTV-Kolloquium 2014: 6. GTV-Kolloquium Thermisches Spritzen 6 (2014), pp. 60-74

[2] Bobzin, K.; Öte, M. and Königstein, T., Entwicklung neuartiger Eisenbasiswerkstoffe zur Herstellung reibverlustmindernder Zylinderinnenbeschichtungen Proceedings gytt 2016: 4. Györer Tribologie Tagung (2016), pp. 83-93

[3] Bobzin, K.; Ernst, F. B. G.; Richardt, K. R. M.; Schläfer, T.; Verpoort, C. and Flores, G. Thermal spraying of cylinder bores with the plasma transferred wire arc process, Surface & Coatings Technology 202, (2008), Issue 18, pp. 4438/43

[4] Bobzin, K.; Ernst, F. B. G.; Zwick, J. B.; Schläfer, T.; Cook, D.; Nassenstein, K.; Schwenk, A.; Schreiber, F.; Wenz, T.; Flores, G. and Hahn, M., Coating bores of light metal engine blocks with a nanocomposite material using the plasma transferred wire arc thermal spray process, Journal of Thermal Spray Technology 17 (2008), Issue 3, pp. 344-351

[5] Schramm, L.; Verpoort, C.; Schwenk, A.; Bobzin, K.; Bagcivan, N.; Schläfer, T.; Theiß, S. and Yilmaz, K., Friction improvement of new generations of engines by PTWA cylinder bore coating and new piston rings Proceedings ITSC 2011: International Thermal Spray Conference (2011), pp. 473-478

[6] Bobzin, K.; Brögelmann, T.; Stahl, K.; Stemplinger, J.-P.; Mayer, J.; Hinterstoißer and M. Influence of wetting and thermophysical properties of diamond-like carbon coatings on the frictional behavior in automobile gearboxes under elasto-hydrodynamic lubrication, Surface & Coatings Technology 284 (2015), pp. 290-301

[7] Djoufack, M. H.; May, U.; Bagcivan, N.; Brögelmann, T. and Bobzin, K., Tribological Evaluation of Hydrogenated DLC in Diesel Lubricated Model Test, Surface & Coatings Technology 258 (2014), pp. 381-391

[8] Wong, V. W. and Tung, S. C., Overview of automotive engine friction and reduction trends– Effects of surface, material, and lubricant-additive technologies, Friction 4 (2016), pp. 1-28

[9] Gand, B.; Wörfel, A.;Mörz, C.and Saule, H., Integration of a Thermal Spray Process into the GROB Prodution Line for Passenger Car Aluminum Engine Block of Combustion Engines into Mass Production Proceedings ITSC 2016: International Thermal Spray Conference (2016) pp. 206-209

[10] Kosaka H.; Wakisaka Y.; Nomura Y.; Hotta Y.; Koike M. and Nakakita, K., Concept of "Temperature Swing Heat Insulation" in Combustion Chamber Walls, and Appropriate Thermo-Physical Properties for Heat Insulation Coat, SAE International Journal of Engines 6 (2013), Issue 1, pp. 142-149

[11] Hoffmeister, H.-W.; Schnell, C., New Mechanical Pretreatment Process to Prepare Aluminum Crankcases for Thermal Spraying, 9th international conference of the european society for precision engineering and nanotechnology, 2009, pp. 74-77

[12] Hoffmeister, H.-W.; Grosse, T.; Gerdes, A., Investigation of the influence of different process setting parameters on the surface formation at honing of thermally sprayed layers, 5th CIRP Conference on High Performance Cutting 2012, Procedia 1, 388–393

[13] Schramm, L.,Development of Robust Cylinder Bore Coating Process and Investigation of Different Feedstock Materials (2016), Shaker

Acknowledgement

The work in this paper is performed in the context of the research project "Entwicklung eines Innenbeschichtungsprozesses zur Herstellung korrosionsbeständiger und reibverlustmindernder Eisenbasisschichten für Zylinderlaufbahnen" (KF2757705SU4). This project is funded by the German Federal Ministry of Economics and Technology (BMWi) according to a decision of the German Federal Parliament.

ZIM Zentrales Innovationsprogramm Mittelstand

Supported by:

Federal Ministry for Economic Affairs and Energy

on the basis of a decision by the German Bundestag

Effects of carbon contents and gas type on hardness and wear resistance of ferrous coating fabricated by twin wire arc spray process

Jaewon Lee, Jaeick Kim, Changhee Lee*
Hanyang University, Haengdang 1-dong, Seongdong-gu, Seoul, Korea

In automotive industry, thermal spray process is used to reduce engine weight by replacing cast iron liners inserted in cylinder bores. Especially, twin wire arc spray is one of widely used thermal spray processes with inexpensive cost and high deposition rate. In this study, two kinds of wire materials, low carbon steel (0.07 wt.%C) and high carbon steel (0.80 wt.%C) were deposited by twin wire arc spray process using two kinds of process gas (i.e., compressed air and nitrogen) in order to elucidate effects of carbon contents of ferrous coating and process gas type on the hardness and wear resistance of coating. In case of hardness, low carbon steel coatings had higher hardness when air was used as process gas whereas high carbon steel coatings had higher hardness when nitrogen was used, which was caused by the counter effects of carbon loss and oxide formation. The results of sliding wear test in lubricated condition indicated that coatings with higher hardness have better wear resistance and oxides improve wear resistance by playing a role as solid lubricant. The main wear mechanism was splat delamination induced by inter-splat crack, and traces of other wear behaviours such as splat tip fracture and abrasive wear were also observed.

1 Introduction

As a method of weight reduction in vehicles, heavy cast iron engine blocks have been replaced by light aluminium alloy engine blocks. With regard to this, grey cast iron liners are inserted in the cylinder bores since aluminium has insufficient wear resistance to endure harsh wear conditions in cylinder bores. However, such method caused a reduction of fuel efficiency due to cylinder bore transformation induced by thermal expansion coefficient difference between aluminium and iron liners and heavy weight of grey cast iron.

In order to solve this problem, there have been numerous researches to substitute grey cast iron liner to thin ferrous coating layer using several thermal spray processes. Especially, twin wire arc spraying is accepted as a proper process for this due to its low running costs and high deposition efficiency which is suitable for high volume production. Also, compared to thermal spray processes using powder materials (i.e. plasma spraying and HVOF), ferrous wire materials have merits in inexpensive cost and ease of alloy design for desirable properties. Above this, oxides formed in coating layer during the deposition process can play a role of self-lubricants in improvement of wear resistance [1]-[4].

In this study, six kinds of ferrous coatings were fabricated on aluminium substrate via twin wire arc spray process. The purpose of the study is to investigate how carbon contents in carbon steel wire and type of process gas affect the wear properties of coating layer especially focused on hardness and oxide contents with microstructure.

2 Experimental Methods

2.1 Materials and Spray Parameters

Coating samples were prepared by twin wire arc spray system (350PP Electric Arc Spray System SmartArc, Sulzer Metco) using two kinds of ferrous wires and gases. In detail, low carbon steel wire (0.08 wt.%C, Metco 8230, Oerlikon Metco) and high carbon steel wire (0.80 wt.%C, Sprasteel 80, Oerikon Metco) were used as coating materials. Medium carbon steel wire was substituted by mounting low carbon steel wire and high carbon steel wire on each side of wire feeding towers in the spray equipment. Air and nitrogen were used as process gas. Other process parameters such as current, voltage, spray distance and gas pressure were same in all samples as follows: 100 A, 30V, 100mm, and 3.8 bar.

Totally 6 types of coating samples were fabricated by mixing different wires with different process gases: (1) "Metco 8230" wires with air, (2) Mixing "Metco 8230" wire and "Sprasteel 80" wire with air, (3) "Sprasteel 80" wires with air, and (4), (5), and (6) followed the same procedure with wires but the process gas was nitrogen.

For convenience, fabricated coatings were referred to as "Metco/Air", "Mixing/Air", "Spra/Air", "Metco/N_2", "Mixing/N_2", and "Spra/N_2". The designations of coatings are shown in **Table 1**.

Coating specimen	Wire	Atomizing gas
Metco/Air	Metco 8230	Air
Mixing/Air	Metco 8230 + Sprasteel 80	Air
Spra/Air	Sprasteel 80	Air
Metco/N_2	Metco 8230	N_2
Mixing/N_2	Metco 8230 + Sprasteel 80	N_2
Spra/N_2	Sprasteel 80	N_2

Table 1. Six kinds of coating specimen

2.2 Wear Tests

Wear tests were conducted by reciprocating wear tester (TE77AUTO, Pint & Partners) under lubricated condition. The engine oil 5W30 was used as a lubricating oil. The counter pins were made of S45C, 4 mm in diameter. Micro-hardness of the pins was over 700 HV. Normal loads and frequencies applied on the pins were changing from weak wear condition (low load and high frequency) to harsh condition (high load and low frequency). Normal loads varied from 100 N to 600 N, frequencies were from 12.5 Hz to 2.5 Hz and with 10.8 mm stroke.

2.3 Characterization

Carbon sulfer analyser (CS600, Leco, US) was used to compare the carbon loss difference. Vickers microhardness tester (HMV-2 ASSY, SHIMADZU, Japan) was used to measure micro-hardness of overall coating surfaces, splats and oxides. The applied load - time for measuring hardness of coating was: 300 g - 10 s, and that of splats and oxides was 100 g - 10 s. Coating samples were polished and observed using optical microscope (OM) (BX60M, OLYMPUS, USA) and field emission scaning electron microscope (FE-SEM) (SIGMA, Carl Zeiss, Germany). Surface profiler (DektakXT Stylus Profiler) was used to measure the average wear depth of worn surfaces. Wear volume loss was calculated in accordance with following Eq. (1) and (2):

Worn surface area (A) = $1/4\ \pi D^2 + S \times D$ (mm^2) (1)

Wear volume loss = A x average wear depth (mm^3)(2)

where D is diameter of counter pin and S is stroke.

3 Results and Discussion

3.1 Effects of atomizing gas type on hardness of coating

Microhardness of each coating surface is shown in **Fig. 1**. Hardness of both coatings using air and nitrogen increased as carbon contents of wire increased. However, increase of hardness in coating using air was smaller than that of nitrogen case. Also, hardness of Metco/Air was higher than that of Metco/N_2, while hardness of Spra/Air was lower than that of Spra/N_2. The difference in aspect of hardness increment between the coatings using air and nitrogen is attributed to two counter effects of oxidation phenomenon during the deposition process: (1) hardness reducing factor: decrease of hardness in individual Fe splats due to the carbon loss, (2) hardness raising factor: creation of hard oxides.

Table 2 shows the difference of carbon contents caused by process gas type. The carbon contents contained in the coatings using nitrogen were compared to that of air case. Also, according to **Fig.2.**, it is clear that amount of oxides was much higher in the coating using air relative to that of the nitrogen case. The oxides (grey areas) are widely distributed on the surface of air coating (Fig. 2(a)), but it is hard to find oxides in the surface of coating using nitrogen (Fig. 2(b)). The primary reason of such difference of carbon contents and amount of oxides was process gas type. It has been already reported repeatedly that oxidation of coating layer is reduced when the nitrogen is used as a process gas [4, 5]

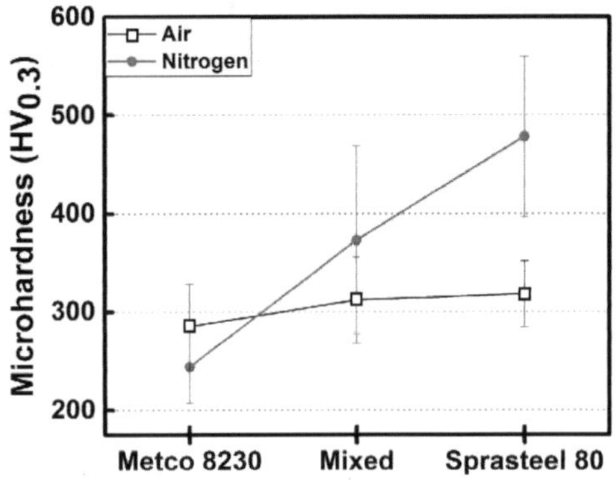

Fig. 1. Hardness of coatings

Regarding this, difference in aspect of hardness increment can be explained by dominance of two counter effects owing to oxidation. In case of low carbon steel coating, hardness of Metco/Air was higher than that of Metco/N_2 because hard oxides formed much in Metco/Air coating. On the other hand, for the case of high carbon steel coating, the hardness of Spra/N_2 was higher than that of Spra/Air because of dominant effect of higher carbon contents in the coatings relative to the lack of hard oxides. Thus, larger increase of hardness in the coating using nitrogen with increasing carbon contents is due to preservation of carbon in the coatings.

Material	Carbon contents after coating (wt.%C)	
	Air	Nitrogen
Metco 8230	0.04	0.07
Mixed	0.15	0.34
Sprasteel 80	0.28	0.53

Table 2. Carbon contents of coatings by atomizing gas type

3.2 Wear resistance

Wear volume loss after sliding wear test under lubricating condition is shown in **Fig. 3**. Wear volume loss decreased as the carbon contents of coating increased in both of gas types (air and nitrogen). Such aspects of wear volume loss corresponded to the increase in hardness.

The better wear resistance of coating with higher hardness also can be contributed to smaller hardness difference between oxides and splat, especially in the case of air, since there is more chance for hard oxides to be detached from the coating when the hardness difference between oxides and splat is larger [6]. The hardness difference between splat matrix and oxides in the coating using air is shown in **Table 3**. The hardness difference is much smaller in the coating with higher carbon contents.

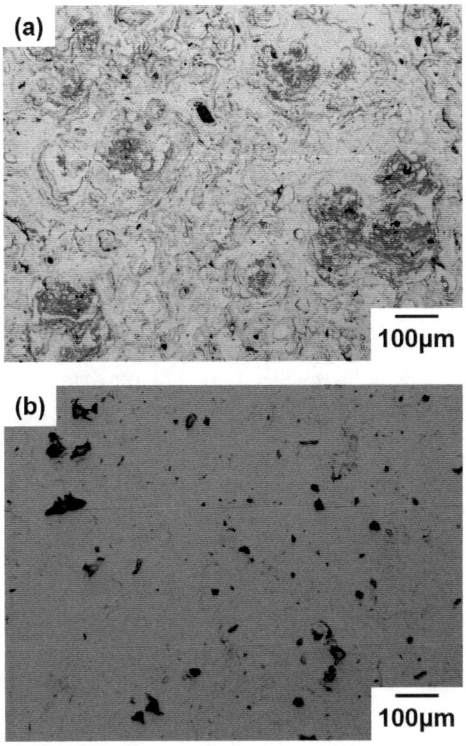

Fig. 2. OM image of coating surface: (a) Metco/Air, and (b) Metco/N_2

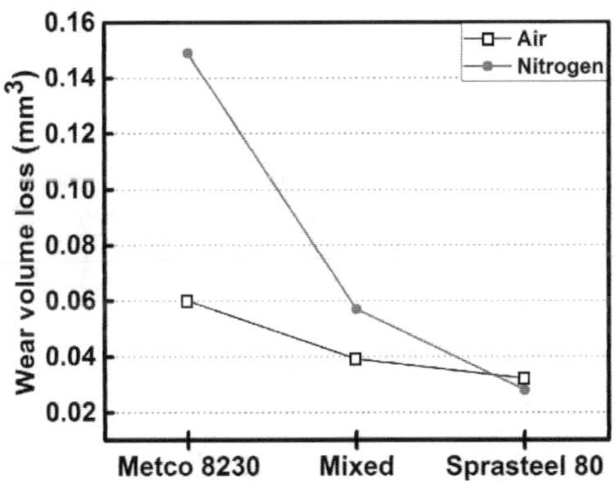

Fig. 3. Wear volume loss of coatings

Coating specimen	Microhardness ($HV_{0.1}$)	
	Splat	Oxide
Metco/Air	271	362
Mixing/Air	279	331
Spra/Air	294	354

Table 3. Hardness difference between splat and oxides in coating using air

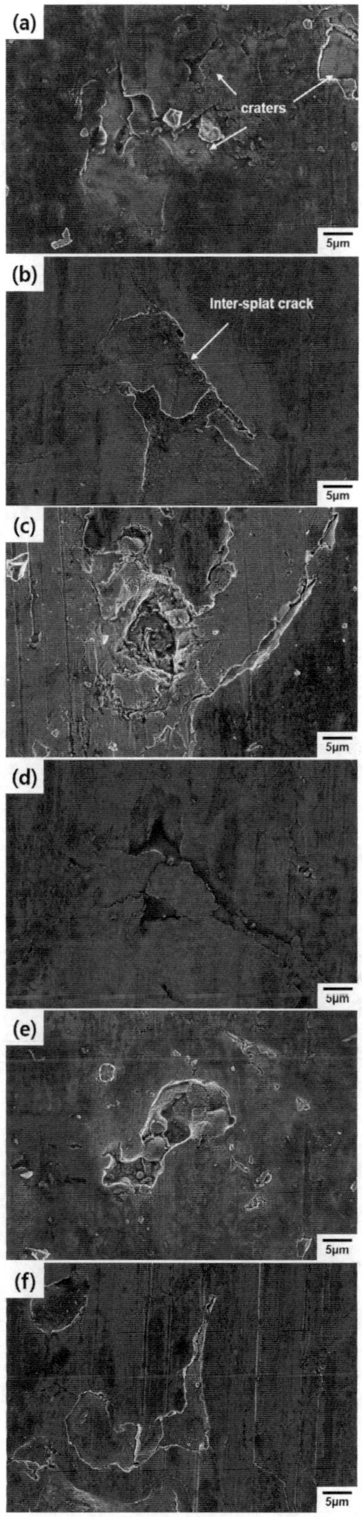

Fig. 4. SEM image of worn surface of each coating (a) Metco/Air, (b) Metco/N_2, (c) Mixing/Air, (d) Mixing/N_2, (e) Spra/Air, (f) Spra/N_2

Fig. 3 also showed that overall wear volume loss is smaller in the case of air compared to the nitrogen case, although the hardness of Mixing/N_2 was higher than that of Mixing/Air. This is probably because some of oxides such as FeO (wustite) and Fe_3O_4 (magnetite) played a role as solid lubricants, which can contribute to improved wear resistance of the coatings [1].

3.3 Wear mechanisms

The main wear mechanism of the coatings was pull-out of splats (splat delamination). The splat delamination easily occurs when splat-splat or splat-oxides boundaries are weak with concentration of stress on the boundaries. With respect to this, lots of observed inter-splat cracks and craters by splat pull-out means that splat-splat or splat-oxides boundaries were much weaker than other regions for the ferrous coatings in this research. Inter-splat cracks and craters in worn surface of each coatings are shown in **Fig. 4**. Besides, the evidence of splat tip fracture and abrasive wear (deep grooves caused by ploughing of hard oxides or the pin) is shown in **Fig.5**. [7, 8].

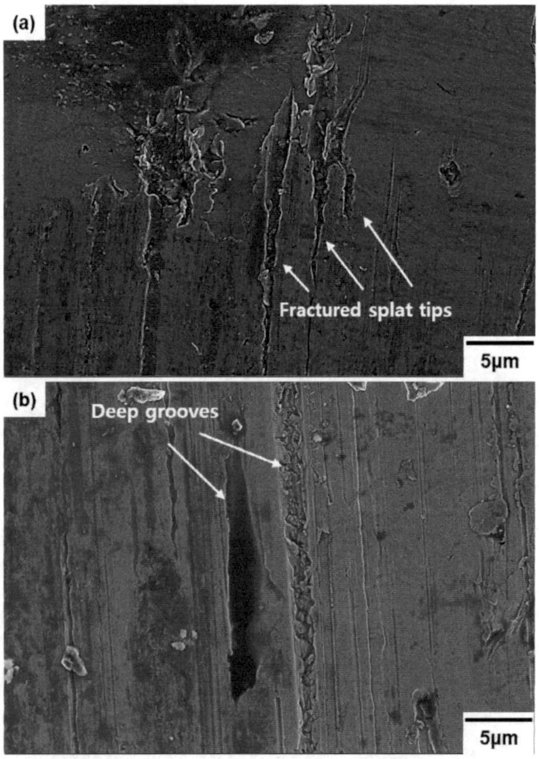

Fig. 5. SEM image of worn surface showing other features of worn surface (a) splat tip fracture and (b) deep grooves formed by abrasion wear

4 Conclusions

Arc-sprayed coatings using different ferrous wires and process gases were fabricated on aluminium substrate. The hardness of the coatings. The result showed that the hardness of low carbon steel (Metco 8230) coating was higher for the case of air due to dominant effect of hard oxides. On the other hand, the hardness of high carbon steel (Sprasteel 80) coating was higher for the case of nitrogen assisted by the dominance of the effect of preserved carbon contents.

In terms of wear, wear resistance was improved as the coating hardness increased. For the coating using air, the wear resistance was better than the case of nitrogen, since the oxides like FeO (Wustite) and Fe_3O_4 (magnetite) serves as solid lubricant.

Finally, it was confirmed that the major wear mechanism was splat delamination induced by inter-splat crack. The splat delamination resulted in splat pull-out and large wear volume loss. Besides, traces of splat tip fracture and abrasive wear were observed.

5 Literature

[1] B. Gerard, Application of thermal spraying in the automobile industry, Surf. Coat. Technol., 201 (2006), p.2028

[2] B. Gerard, Advanced thermal spray technology and coating for lightweight engine block for the automotive industry, Surf. Coat. Technol., 200 (2005), 1990-1993

[3] B. Hwang, J. Ahn, S. Lee, Correlation of Microstructure and Wear Resistance of Ferrous Coatings Fabricated by Atmospheric Plasma Spraying, Metall. Mater. Trans. 33A (2002) 2933

[4] D. Kumar, Q. Murtaza, R. C. Singh, Sliding wear behaviour of aluminium alloy coating prepared by two-wire electric arc spray process, Int. J. Adv. Manuf. Technol., (2016) Vol.85, Issue 1, pp 237-252

[5] A.p. Newbery, P.S. Grant, Oxidation during electric arc spray forming of steel, J.Mater. Proc. Technol. 178 (2006) 259-269

[6] B. Hwang, S. Lee, J. Ahn, Effect of oxides on wear resistance and surface roughness of ferrous coated layer fabricated by atmospheric plasma spraying, Material Science Engineering, Vol. 335 (2002) 268-280

[7] S. Usmani, S. Sampath, Time-Dependent Friction Response of Plasma-Sprayed Molybdenum, Wear, 1999, 225--229(2), p 1131-1140

[8] A. Vencl, Tribological Behavior of Ferrous-Based APS Coatings Under Dry Sliding Conditions, J. Therm Spray Tech, (2015) 24, pp671-682

Significance of personnel qualification and quality management for thermal spray plants

H. Cramer, R. Huber, A. Ohliger-Volmer; Gesellschaft Schweißtechnik International – GSI mbH, Niederlassung SLV München, München / D

Well-trained personnel is decisive for a successful company. International business relations are increasingly demanding internationally recognised qualification and certification measures. The requirements are stringent particularly in sensitive fields, e.g. aviation and rail vehicles.
No matter whether in aircraft engines, power station engineering or the printing and paper industries, numerous requirements on coatings are solved with the aid of thermal spraying. However, in order to be able to master what the language in standards calls a "special process" (i.e. a process whose final result can be checked by means of non-destructive testing to a limited extent only), there is a need for calibrated and monitored installations and parameters as well as for correspondingly experienced, trained and certified personnel.
Frequently, ETSSs, ETSs as well as the qualification tests according to ISO 14918 are already integral constituents of contracts relating to thermal spraying (requirement of the German Railways according to DIN 27201-10).
The article provides detailed information about training and qualification as well as about the certification of plants and personnel in the field of thermal spraying. Not only the significance of a quality management system but also the particular advantages of the GTS certification are highlighted in this respect.

1 Introduction

The deployment of qualified coordination personnel and qualified sprayers is one of the most important prerequisites for the manufacture of high-quality products. Sprayed coatings must withstand particular loads. Failures of thermally sprayed components threaten material damage or, at worst, even personal injury. Spraying technology companies employ metalworkers, installation or structural mechanics, fitters, industrial mechanics, tinsmiths or even lateral entrants unfamiliar with the subject. These training measures do not lead to any qualifications as thermal sprayers or spraying coordinators at all or not comprehensively enough. Further education is necessary. Training and further education measures can be completed at SLV Munich which regularly carries out courses to become "European Thermal Sprayers" and "European Thermal Spraying Specialists". SLV Munich has more than 20 years of experience in the training and certification of spraying personnel

2 Thermal spray plants need employees with background knowliedge

- **Two recognised training measures**

The fact that a plant can only achieve and maintain a certain quality of its products depends on the qualification of the personnel to a great extent. In particular, small and medium-sized thermal spraying plants need all-round employees who know a lot about many fields. An occupational training and subsequent further education in spraying technology can meet these needs. Therefore, it is more than recommendable to provide thermal spray personnel fundamental training and constant further education.
In the case of thermal spraying, the spray personnel can choose between two recognised training levels. These guidelines of the European Federation for Welding, Joining and Cutting (EWF) are standardised all over Europe. The application for the adoption of the EWF guidelines for ETSSs and ETSs as IIW/EWF guidelines was submitted and approved in 2016.
For thermal spraying, a comprehensive set of standards for assuring the quality of thermally sprayed products exists together with the product-related application standards. EN and ISO standards define the qualification requirements for thermal spray personnel which are fulfilled by the EWF guidelines (Fig. 1).

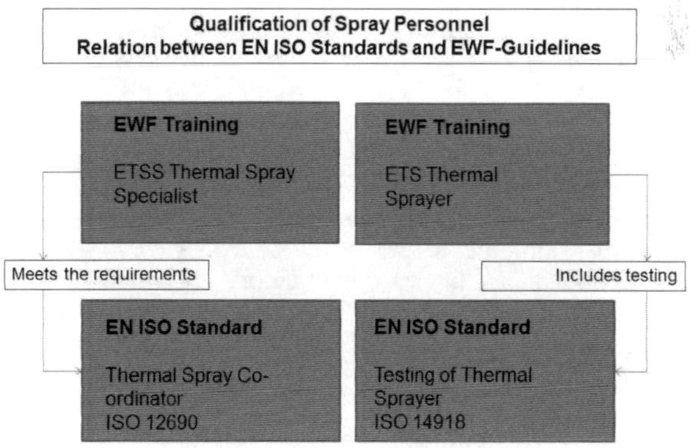

Fig. 1: Connection between the requirements on TS personnel qualification according to EN standards and the training according to EWF guidelines

2.1 Education and training of thermal spray personnel

- **Courses**

The training concept for the field of thermal spraying which has been implemented at SLV Munich for years encompasses the following courses, portrayed on Fig. 2:

- Thermal spraying specialist
  - (ETSS) **E**uropean **T**hermal **S**praying **S**pecialist
- Thermal sprayer
  - (ETS) **E**uropean **T**hermal **S**prayer

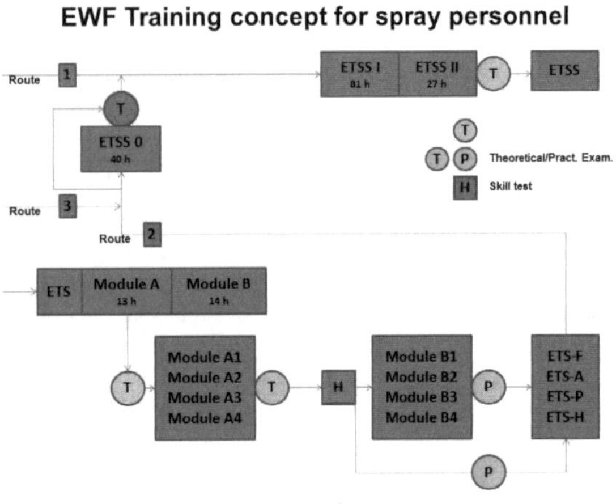

Fig. 2: Training levels for thermal spraying

The system has a permeable structure so that a thermal sprayer can obtain further qualifications.

2.2 ETSS (European Thermal Spraying Specialist)

The highest training and qualification level until now has been the training to become a spraying specialist (ETSS). The personnel trained and certified according to this EWF guideline can perform diverse responsible tasks in the plant, e.g. as a responsible spraying coordinator who is demanded according to ISO 14922-2 and ISO 14922-3 (Chapters 3.1/3.3) and also according to the GTS set of guidelines (Chapter 3.2).

- **ETSS-course – who may attend**

According to the EWF guideline, the following people can take part in the ETSS training:

- Anybody who has a qualification as a foreman in skilled trades or industry or anybody who is a technologist or an engineer and, in each case, has two years of occupation-related experience.
- Anybody who has an ETS qualification and at least two years of occupation-related experience and has passed the entrance qualification test.
- Anybody who can prove at least five years of occupational experience or a qualification as a skilled worker in the metalworking trade with at least three years of experience in activities which are at least related to thermal spraying technology and has passed the entrance qualification test.

On the basis of the ETSS Part 0 script, the participants independently acquire the basic knowledge needed for the ETSS Part 0 entrance qualification test.

- **ETSS-lessons – contents and duration**

With 108 lessons in total, the EWF training offers very comprehensive training. The three-week course imparts very well-founded and practice-related expert knowledge as main focal points. In part, the teaching is provided by external specialists from industry and research.
The personnel trained and certified according to this EWF guideline can perform diverse responsible tasks in the plant, e.g. as a responsible spraying coordinator (ISO 12690) who is demanded according to ISO 14922-2 and ISO 14922-3 and also according to the GTS set of guidelines.

- **ETSS – validity period of the diploma**

After passing the final theoretical examination, the participant receives a "European Thermal Spraying Specialist" diploma which is valid for his/her lifetime [2].

	Main course ETSS I Theoretical Basis	85h
1.	Spray processes: Indroduction and basic knowledge of processes, environmental aspects	20h
2.	Materials and bonding mechanismens	13h
3.	Thermals spraying, procedure, health and safety	15h
4.	Results and documentation	15h
5.	Engineering application and qualitymanagement	18h
	Examination	4h
	Main course ETSS II Practical training/demonstration	27h
1.	Practical demonstration of all processes	7h
2.	Practical instruction to all spray processes	14h
3.	Practical demonstration: post treatment	3h
4.	Testing in workshop and laboratory	3h

Fig. 3: ETSS-course – contence and duration

2.3 ETS (European Thermal Sprayer)

The spraying plant needs thermal sprayers who have background knowledge to be better at recognising connections relevant to quality.

- **ETS-course – who may attend**

Normal physical and mental abilities are sufficient as an admission prerequisite for the participation in the ETS training course.
The participant must have an oral and written command of the course language to such an extent that he/she can follow the course and the instructions in the practical training and take part in the theoretical examination without any problems. Over the years, experience has shown that participants in ETS courses should possess basic skills in metalworking. Otherwise, corresponding practical basic training ("Module B 0") is recommended.
Corresponding special training in advance has proven to be suitable for participants who cannot indicate any or only little practical experience in the thermal spray process the participant intends to be qualified in.

- **ETS-training – contents and duration**

With 13 lessons, "Module A 0 - Theoretical basics" encompasses not only principles of the basics (Fig. 4) of thermal spraying, procedural sequence, occupational health and safety, coating properties and adhesion mechanisms, materials, pretreatment and preparation, execution of thermal spraying, maintenance and calibration, post-treatment as well as measurement and testing but also handling and storage of the sprayed components, mechanisation and automation as well as environmental aspects and questions relating to safety at work.

	ETS Education and Training		
Module A	(Common part) Theoretical Basis	All spraying processes	13h
Module B	(Common part) Practical basis	All spraying processes	14h
	Intermediate examination		2h
Module A1	Specific theoretical education	Flame	1h
Module A2	Specific theoretical education	Arc	1h
Module A3	Specific theoretical education	Plasma	2h
Module A4	Specific theoretical education	HVOF	1h
A1/2/3/4	Final theoretical examination		
Module B1	Specific pracical training	Flame	*
Module B2	Specific practical training	Arc	*
Module B3	Specific practical traning	Plasma	*
Module B4	Specific practical training	HVOF	*
B1/2/3/4	Final practical test		

* Individual duration

Fig. 4: ETS teaching - Contents, duration and sequence

In "Module B 0 - Practical basic training" for participants without any practical experience in thermal spraying, demonstrations and practical exercises are carried out in a spraying workshop in 14 lessons.
The theoretical intermediate qualification test is followed by the procedure-specific theoretical (Modules A 1 - 4) and practical (Modules B 1 - 4) training.
The duration of the practical training is dependent on the already existing experience of the sprayer and his/her ability. The qualification test pieces correspond to those in ISO 14918 entitled "Qualification testing of thermal sprayers".

- **ETS – validity period of the diplomas and certificates**

After passing the training and the qualification test, the participant receives:

- an ETS certificate which complies with the EWF guideline and is valid for his/her lifetime

and, at the same time,

- a certificate which relates to his/her passed qualification test(s) according to ISO 14918 and is valid for three years [3].

3. Thermal Spray plants and quality - certification measures for plants and TS personnel

In the wake of globalisation, the market is increasingly demanding proof of the quality of products or of the working method of a company. Companies must work according to stipulated standards when stringent safety requirements are set on a manufactured product. Manufacturers have themselves certified in their own interests in order to emphasise their own demands on the quality of their products and to possess proof for any third parties such as customers or authorities.

3.1 Relevant standards - Overview

Quality management must not be considered as a necessary evil. Fig. 5 shows what factors are relevant to quality in the spraying plant and to what standards reference can be made.

It is important that the plant's own QM system becomes a functioning control loop which concentrates on the essentials and does not allow anything superfluous to arise at all or, if necessary, rejects it once again.

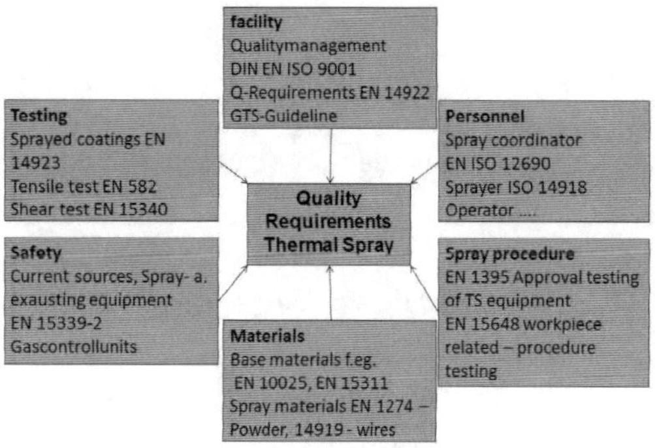

Fig. 5: Factors relevant in the spraying plant so that quality requirements are satisfied

3.2 The correct understanding of quality

The correct understanding of quality includes not only the implementation of requirements from standards and sets of rules but also the incorporation, motivation and self-responsibility of all the employees. The registration and evaluation of customer complaints yield most information about the condition of the plant and whether its own QM system is functioning (Figs. 6 and 7).

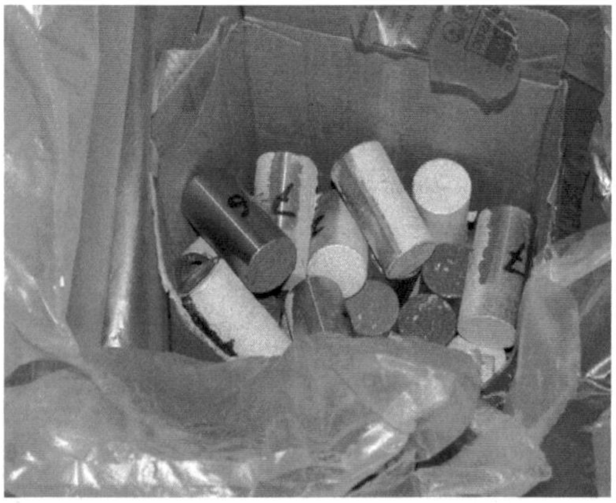

Fig. 6: Qualification test specimens packed in an incorrect and negligent way because of inadequate knowledge on the part of the personnel or because of a lack of quality awareness

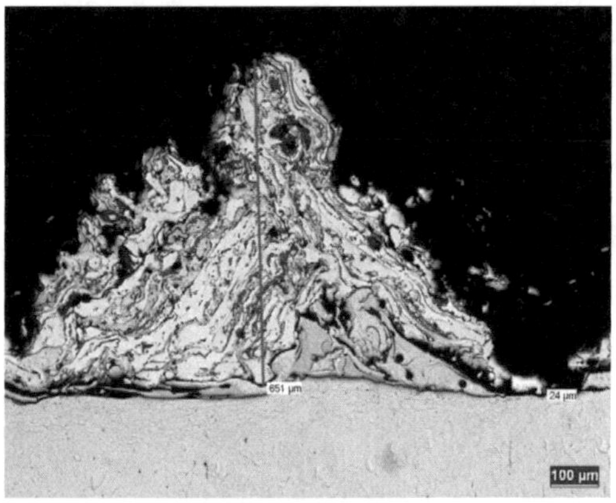

Fig. 7: Front face of a test specimen - sprayed in a wrong way - submitted for the tensile adhesive strength test according to EN 582

3.3 Certification of the special process of "thermal spraying"

If a firm is certified according to ISO 9001, this means that preventive management tasks are proven and all the process sequences in the company are documented.

- **Certification according to ISO 14922**

The functioning capacity of the plant facilities and also that of the in-house sequences can be confirmed by the proof of a QM system which is specially oriented to thermal spraying. The quality requirements which relate to spraying technology and are set on a fabrication plant dealing with spraying technology are stipulated comprehensively in the series of DIN EN ISO 14922 standards. The requirements extend from the offer phase via the conclusion of a contract right up to design, fabrication, qualified spraying personnel and the testing of spraying technology products. This includes not only the maintenance of installations and measuring equipment but also the documentation for quality assurance and the satisfaction of the product requirements.
A certain scope of time and money is unavoidable in order to bring the sequences, the responsibilities and the testing steps into a manual and into procedural and working instructions. However, the setting-up of a new QM system or the optimisation of the already existing QM system also offers the chance to save a lot of time and money because only in this way is it possible to reliably recognise any weak spots and the costs of mistakes and to stipulate any measures for required improvements.
The certification according to DIN EN ISO 14922 is frequently carried out by auditors from SLV Munich in combination with the GTS QM certification and is valid for three years.

- **GTS certification**

One functional QM tool which has been established for years is checking and certification according to the guidelines of the Association of Thermal Sprayers (GTS) which is an association of representatives of industry, skilled trades as well as research and development, has been established for over 20 years and has set itself the objective, amongst others, of assuring the quality of thermally sprayed products.

- **for personnel, process and, if necessary, product inspection**

The GTS QM guideline unites QM elements in ISO 9001 with procedure qualification tests and the theoretical and practical qualification tests for spraying technology personnel. During the GTS audit, expert discussions are also held with the TS personnel taking the qualification test, the thermal spraying of the test pieces is observed and these are assessed.
Full members are firms which industrially apply the thermal spraying in industry or skilled trades. According to the statutes, these members have undertaken to audit the quality of their installation technology with regard to reproducible coatings and the qualification of their personnel as sprayers and spraying coordinators and to have them certified according to the GTS certificate guideline (GTSPA003).
For the GTS certificate, a re-audit must be carried out every three years. On behalf of GTS, SLV Munich

carries out the GTS certification on its own according to stipulated sequences.

• GTS-Product certification

Moreover, thermal spray plants can voluntarily decide in favour of additional product or coating certification. Beforehand, the plant to be certified stipulates the properties which are to be tested and assessed in an independent, accredited test laboratory. The specimens or the components are thermally sprayed under the supervision of the auditor - another confidence-building QM measure which is used for advertising purposes.

3.4 Current standards define the deployment and qualification testing of thermal spray personnel

Thermal spraying is multifaceted. There is a need for comprehensive knowledge about connections and interactions in thermal spraying as well as for many years of practical experience in thermal spraying.

• Thermal Spray co-ordinator

Spraying co-ordinators are deployed in some function or other in every plant.

- **Tasks**

DIN EN ISO 12690 entitled "Thermal spray coordination - Tasks and responsibilities" stipulates the quality-related responsibilities and the tasks, including the coordination of the spraying technology activities.

Every activity may be linked with a number of tasks and responsibilities, e.g.

- the stipulation of specifications and their preparation
- the coordination
- checking of tasks and areas relating to fabrication by means of spraying technology.

- **Who may be entitled as a spray coordinator**

The European Thermal Spraying Specialist (ETSS) or people with comparable knowledge and experience are demanded and recognised as authorised spraying coordination personnel in various standards.

• Qualification testing of thermal sprayers

With the (admission) qualification test for thermal sprayers according to ISO 14918, the thermal spray plant receives proof that spraying work is carried out by trained/qualified and certified personnel. The qualification test according to ISO 14918 affirms to the thermal sprayer that he/she has the current knowledge and ability for the spraying procedure in which he/she is active and for which he/she has taken the qualification test.

- **ISO 14918 – the way to be tested**

Not only the practical qualification test but also the checking of the expert knowledge in the spraying procedure which is the subject of the qualification test in each case are carried out during the first qualification test. Here, the inquiries primarily relate to the knowledge about installation technology, the spraying process including spraying parameters, base materials and spraying consumables, testing methods for thermally sprayed coatings as well as safety at work and accident prevention. The practical qualification test encompasses the evaluation of blasted surfaces, masking procedures, the operation of the installation, the application of the coating and safety at work.

- **Evaluation**

The sprayed test pieces are subjected to visual inspection before and after their machining. Depending on the material and the procedure, tensile adhesive strength tests, metallography and microhardness tests are also performed in order to evaluate the spraying results [4].

- **ISO 14918 - Certificate**

As far as the qualification test is concerned, distinctions are made between manual and mechanised thermal spraying as well as between the various spraying procedures. For information purposes, the material group with which he/she has sprayed the qualification test piece is noted on the certificate, Fig. 8.

European Scheme for the Education and Qualification of Welding and Joining Personnel

EUROPEAN WELDING FEDERATION

EOTC AGREEMENT GROUP
for Welding, Joining, Cutting and Surfacing

Having met the education and training requirements of
EWF Guideline 'European Thermal Sprayer' and
by examination having satisfied the requirements of the
Governing Board of the Authorised National Body

Name: Mario Mustermann
Date of birth: 01.04.1972

is hereby awarded the diploma of

EUROPEAN THERMAL SPRAYER
• Plasma spraying • Flame spraying • Arc spraying

Date: 22.02.2002 Diploma Nr: D-S-80636-EWF507-020222-0210075

Chairman, Head

EWF Authorised National Body: Deutscher Verband für Schweißen und verwandte Verfahren e.V.
Training School: Schweißtechnische Lehr- und Versuchsanstalt SLV München, NL der GSI mbH

Scheme Administration: EOTC AG Scheme Control: EWF

Fig. 8: ISO 14918 - Certificate

- **Period of Validity**

The (admission) qualification test is valid for three years in so far as the sprayer performs spraying tasks regularly in the working field of his/her qualification test.

So that the admission qualification test remains valid, the spraying coordinator responsible in the TS plant

must document the activities of the thermal sprayers in question.
The qualification test loses its validity if a thermal sprayer has not carried out any thermal spraying work for more than six months. Then, the thermal sprayer must take another qualification test.

4. Conclusion

In the future, customer demands vis-à-vis coating businesses will continue to increase. In addition to application standards for spray materials, spray equipment and methods for testing thermally sprayed coatings, standards have also been devised by the technical committee CEN TC 240 “Thermal spraying and thermally sprayed coatings” which provide job shops with tools to produce according to quality requirements. They also help to establish a QM system for spray technology.
Modern joining and surface technologies, especially welding and thermal spraying, are characterized by manual skills, engineering knowledge and innovative processes.
Without a doubt, consistently implementing and applying the standards and guidelines described above will visibly give the field of thermal spraying the necessary impetus forfuture applications.

3 Literature

/1/ Rolf Huber, Annegret Ohliger-Volmer: Training and Certification in thermal Spraying: Experiences and Future Trends, HVOF Colloquium 2015 Erding

/2/ Training course: Thermal Spraying Specialist „EUROPEAN THERMAL SPRAYING SPECIALIST (ETSS)“)“ according to guideline DVS®-EWF 1188

/3/ Training course: Thermal Sprayer „EUROPEAN THERMAL SPRAYER (ETS)“ according to guideline DVS®-EWF 1197

/4/ DIN EN ISO 14918: Tesing of Thermal Sprayers

Efffect of ball milling of Al_2O_3 particle deposition behavior on polycarbonate substrate using vacuum kinetic spray

Hansol Kwon, Hyungkwon Park, Changhee Lee*
Hanyang university, Haengdang 1-dong, Seongdong-gu, Seoul, Korea

In this study, ball-milled Al_2O_3 powder was used as a feedstock material for vacuum kinetic spray to deposit hard ceramic Al_2O_3 coating on a relatively soft polycarbonate substrate. Microstructural and X-ray diffraction analysis of powders and coatings were performed. The results shows that the ball-milled powder has more unstable state than the primary powder. Compared to primary Al_2O_3 coating the crystallite size and coating thickness of ball-milled Al_2O_3 coating are smaller and thicker, respectively. Since the ball-milled Al_2O_3 particles are more easily fragmented during the VKS coating process, it is possible to deposit hard Al_2O_3 coating on the soft polycarbonate substrate.

1 Introduction

Vacuum kinetic spray (VKS), also called aerosol deposition method, is a coating process which can fabricate thin and thick film [1, 2]. In this process, submicron-size particles are accelerated by a pressure difference between two chambers, and ejected through a various type of nozzles. The detailed process have been described in previous research [3]. Due to low cost, various availiable coating material, room temperature process and high deposition rate, VKS coating have been used in various industrial field such as energy harvesting, biomaterial, fuel cell and sensor [4-7]. Polymer materials have some advantages such as flexible, low cost, easy of process, light weight and transparent [8, 9]. Because these various properties of polymeric materials that other materials usually can not have are unique, thereby enabling advanced applications in the above-mentioned fields.

In VKS coating, fragmentation and deformation of the particle take an important role in the deposition of coating layer [10, 11]. However, due to the different mechanical properties, it is challenging to successfully fabricate the hard ceramic coating on the soft polymer, there has been almost no case in which a hard ceramic materials is deposited on soft polymer.

Ball milling process is a technique for processing powders using mechanical energy. The microstructure of the powder can be controlled by the energy generated by the collision of the grinding medium and the powder [12]. Many studies have been conducted to control the microstructure of hard Al_2O_3 particles using a ball milling process [13-15]. These research reported that the size of the raw powder could be reduced through the ball milling process. Balaz et al indicated that the reduction of particle size by ball milling is caused by impact, shear attrition or compression force [16]. In such a complicated stress situation, failure occurs in a part of the brittle particles, and the size of the particles becomes small. A reduction in particle size can be evidence of particle damage and destruction. Thus, It can be proposed that ball milling can make hard Al_2O_3 particles more susceptible to fracture and deformation during VKS coating.

In this study, a planetary ball milling process was used for successful deposition of hard ceramic Al_2O_3 particle on soft polymer polycarbonate substrate. Microstructural and phase characterization of powders and coatings were carried out.

2 Experimental procedure

Commercially used α-Al_2O_3 powder was chosen as a feedstock powder. To facilitate aerosolization, the powder was dried using an oven at about 100°C. A planetray ball milling process was conducted to analyse the deposition behaviour of microstructural control of powder using planetray mill machine (Pulverisette 5, FRITSCH). An Al_2O_3 ball with a diameter of 10 mm was selected as the grinding medium to pevent contamination of the milled powder and used in a weight ratio of about 10 : 1 with the Al_2O_3 feedstock powder. An air atmosphere was used because Al_2O_3, a ceramic material is unlikely to react with airborne materials during milling process, and ethanol was added for smooth flow of feedstock powder. The milling speed was set at 200 rpm and the pause time was added for 10 min after each 60 min milling time. The total milling time was set to 300 min. The ball-milled powder was dried at 100°C in a vauum dryer to totally remove moisture.

Polycarbonate with a size of 75 x 25 x 2 mm^3 was used as substrate material. Polycarbonate substrate was cleaned with enthanol using an ultrasonic for 5 min to remove the contaminants on the surface and totally dried. The slit type nozzle with orifice size 5.0 x 0.4 mm^2 was used in this study. Nitrogen gas was chosen as a process gas for deposition of aerosolized Al_2O_3 particles. The process gas pressure was set at 0.6 MPa and gas flow rate was controlled to 10 L/min. The VKS coating experiments were begun when the deposition chamber pressure reached at 6.0 x 10^{-2} torr. The working distance, stage traverse speed and pass number was fixed to 10 mm, 1 mm/sec and 10, respectively. The VKS coating paramters are summarized in **Table 1**.

The morphology of the powders (as-recieved, ball-milled) and the surface of fabricated coatings were observed using field emission scanning electron microscopy (FE-SEM, SIGMA, Carl Zeiss). The crystal structure of the powder and coating were characterized for phase analysis using X-ray diffraction (XRD, New D8-Advance, Bruker-AXS). The specimen for observation of the cross-sectional microstructure of the coatings was prepared using focused ion beam system (FIB, FB-2100, Hitachi).

VKS coating parameters	Value
Nozzle type	Slit nozzle
Process gas	N_2
Process gas pressure, MPa	0.6
Process gas flow rate, L/min	10
Deposition chamber pressure, Torr (Starting deposition pressure)	6.0 x 10^{-2}
Working distance, mm	10
Stage traverse speed, mm/sec	1
Pass number	10

Table 1. VKS coating parameters

Field emission transmission electron microscopy (FE-TEM, Tecnai G^2 F30 S-Twin, FEI) operating at 300 kV was used to analyse the microstructure of powder and the cross-sectional microstructure of the coatings. The Al_2O_3 powder size, coating thickness and crystallite size of the coating layer were measured using image analysis software (Image-pro plus, Media cybernetics). The FE-TEM iamages were analysed using TEM image analysis software (DigitalMicrograph, Gatan).

3 Results and discussion

3.1 Powder characteristics

Fig. 1(a) and (b) shows the morphology of as-received Al_2O_3 powder and ball-milled Al_2O_3 powder, respectively. As-received Al_2O_3 powder (hereinafter referred to as AR) have a irregular shape and facet surface morphology. Agglomeration of the powder also was observed in most of the sample in **Fig. 1(a)**. Ball-milled Al_2O_3 powder (reffered to as BM5h) also have a irregular and facet morphologies. However, Some debris-like particles which are much smaller than the primary particles was observed to remain on the surface of the powders in **Fig. 1(b)**. The particle size distribution of the AR and BM5h powder was shown in **Fig. 1(c)**. The particle size distribution graph of BM5h powder was shifted toward left compared to that of AR powder. The average particle size of BM5h powder was measured to 326 nm which is relatively smaller value than that of AR power, 414 nm. Through these differences, It can be carefully expected that the state of the Al_2O_3 powder could be changed somewhat due to the ball milling process.

Fig. 2 shows the TEM images of AR, BM5h particles for more precise microsturctural analysis. The AR particle has a clean and gentle surface state in **Fig. 2(a)**. The BM5h powder, however, have a rough and unstable surface condition and have some debris on the surface of the powder (marked by red arrow) in Fig.

Fig. 1. FE-SEM morphologies of (a) AR, (b) BM5h powders, and (c) particle size distribution of powders.

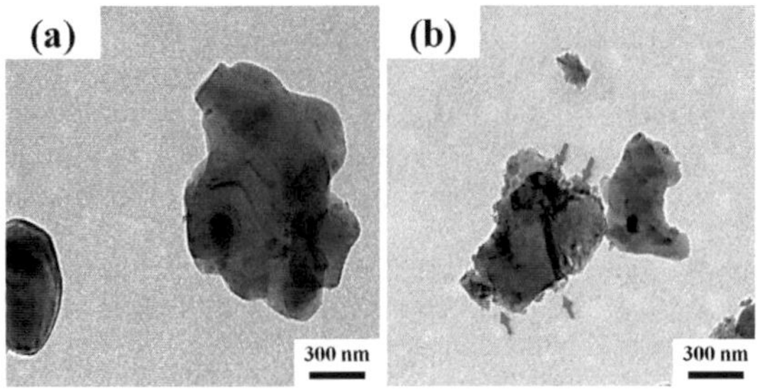

Fig. 2. TEM images of (a) AR and (b) BM5h powder

2(b). It is considered that the microstructure of the Al_2O_3 powder changes due to the force applied during the ball milling process.

The XRD pattern of as-received and ball-milled powders are represented in **Fig. 3(a)**. The main peaks of the AR, BM5h powders were almost identical. It can be confirmed that phase change was not occur and the contamination of Al_2O_3 powder also didn't occur during the ball milling process. For quantitative comparative analysis, the Intensities of base line and four strongest peaks shown as arrows are summarized in **Fig. 3(b)**. The intensity of the base line was the average of the intensity from 20 to 25°. The results show that the intensity of base line of BM5h

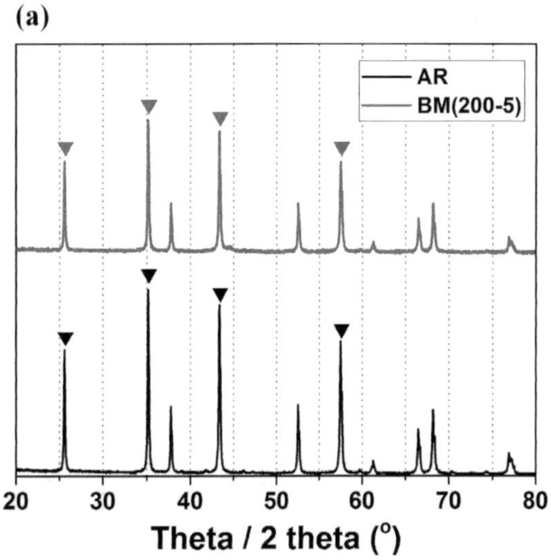

(b)

Peak intensity \ Peak list	Base line (20-25°)	25.57°	35.14°	43.34°	57.48°
AR	137	2768	3686	3325	2304
BM5h	245	2019	2730	2395	1687

Fig. 3. (a) XRD patterns and (b) peak intensity of AR, BM5h powder

powder is larger than that of AR powder, and the intensity of all four strongest main peaks of BM5h powder are smaller than that of AR powder. It can be thought that the combination of the increase in baseline intensity and the decrease in main peak intensity in the xrd pattern are the result of reduced degree of crystallinity [13, 17]. Therefore, It can be carefully expected that the uniformity of crystallinity was inhibited and lattice defects occurred on the local region of the BM5h particle.

Based on the above mentioned results, It can be postulated that microstructural changes of Al_2O_3 powder are induced by ball milling process. A schematic image of the hypothesis for the microstructural change of Al_2O_3 powder is shown in **Fig. 4**. During ball milling, the feedstock Al_2O_3 particles undergo collision with another adjacent Al_2O_3 particle or grinding medium [12]. Thus, the particles are subjected to compressive or shear stresses from the surface where the collision occurs as shown in Fig. 4(a). Due to this stress, deformation or fracture of the surface of the particles could occur as a reaction to the stress. Several possible results of powder microstructure are illustrated in Fig. 4(b). Four possible cases can be considered; (i) There will be a portion where the crystallinity of the atom is somewhat reduced due to the continuous impact on the surface of the powder. Changes in the base line and main peak intensity observed in the XRD pattern may have originated from the surface of these powders. (ii) A crack may be generated from a region where surface stress is concentrated. Such a crack can be a starting point at which the particle can be more easily broken down. (iii) As crack propagates further, a certain area on the surface can be broken and become a debris. These debris have already been confirmed by TEM observation and are evidence of particle demage. (iv) Crack propagation also can result in bulk failure which a relatively large amount of area is destroyed. In summary, It can be thought that the BM5h particle exists as an unstable state containing more defects than the AR particle.

3.2 Coating microstructure

The cross-sectional TEM images of the AR coating on polycarbonate substrate is shown in **Fig. 5(a).** It can be seen that the Al_2O_3 coating and polycarbonate substrate was distinguishable. The average thickness and standard deviation of the coating were measured

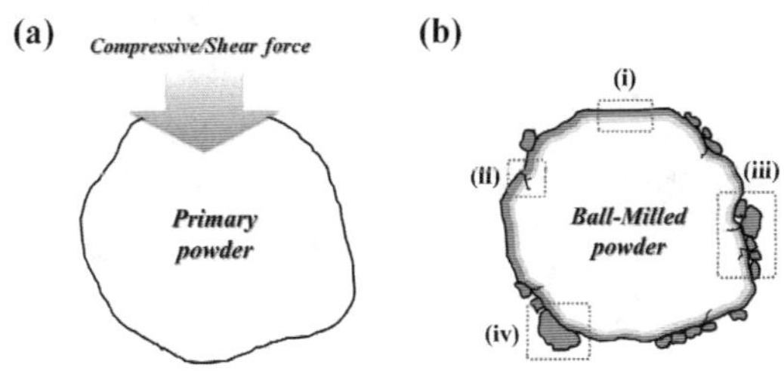

Fig 4. Schematic images of microstructures of primary powder and ball-milled powder

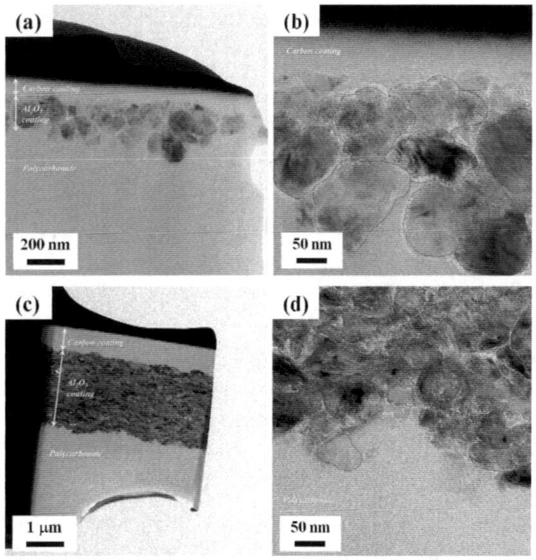

Fig. 5. Cross-sectional TEM images of (a, b) AR coating and (c, d) BM5h coating

by 230 and 46 nm, repectively. Inside the AR coating layer, the shape of the particles is obvious enough to be distinguished. It seems that the particles which are sectioned with red dot line constitue the entire coating layer as shown in **Fig. 5(b)**. Since the particle shape is almost similar to that of the initial powder, it seems that the fracture hardly occurred and particle embedded in polycarbonate substrate maintained almost initial state during deposition. The cross-sectional TEM images of the BM5h coating on polycarbonate substrate is shown in Fig. 5(c). It can be stated that the coating and substrate has continuous and clear interface. The average thickness and standard deviation of the coating were measured by 2031 and 97nm, respectively. The average thickness of the BM5h coating was drastically increased compared with that of the AR coating. Dark/white stripes that were not observed in the AR coating layer were observed in the BM5h coating layer as a whole. It is presumed that this microstructure is caused by the loading induced during the VKS coating process [18]. The high magnification TEM image of the coating/substrate interface region was shown in Fig. 5(d). Only crystallites larger than the size that can be distinguished were sectioned as red dotted lines. There were relatively many regions where it was difficult to clearly distinguish the boundary between crystallites. It means that the bonding between adjacent crystallite is more intimate.

The measured value of crystallite size is shown in **Fig. 6**. In case of AR coating, the average, standard deviation, maximum and minimum values of the crystallite size were 99, 42, 202 and 35 nm, respectively. This average size is smaller than the primary AR particle size. there may be there is a difference in the deposition behavior depending on the size of the particle or the fragmentation phenomena during deposition. However, this is not

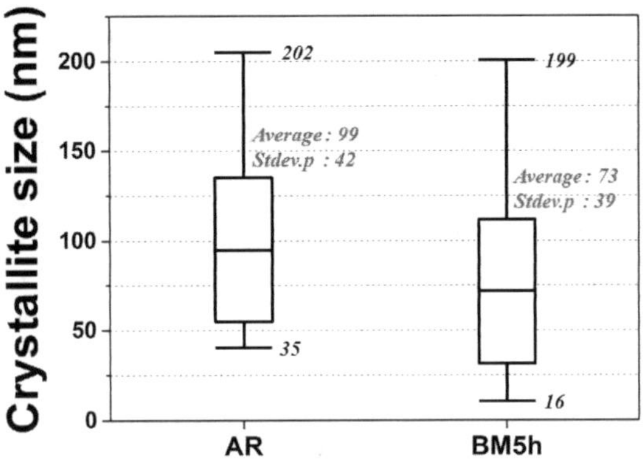

Fig. 6. Crystallite size of AR and BM5h coating

clear, so further analysis is required. In case of BM5h coating, the average, standard deviation, maximum and minimum values of the crystallite size were 73, 39, 199 and 16 nm. It can be seen that the average crystallite size of BM5h coating decreases compared to AR coating. Given that the uncountable crystallites are excluded, the average size is expected to show even greater differences. It is interesting that the maximum values of the AR and BM5h coatings are similar. It means that particles which are not fragmented are also contained in the BM5h coating layer.

Fig. 7(a) represents the surface FE-SEM morphology of AR coating. The area indicated by the red arrow demonstrated as the surface of the polycarbonate substrate. Some damaged regions of the surface were observed. It can be expected that the impact between rapidly accelerated Al_2O_3 particles and substrate cause the result. The embedded particles which is sectioned to yellow dotted line were also observed. These surface morphologies of polymer surface impacted by hard particle have been reported by previous research [19, 20]. It is considered that these particles didn't undergo fragmentation during deposition but only be embedded on the surface of the substrate. These particles are thought to be firstly deposited particle. Because there is no successful subsequent deposition of particles on these particles,at any reason, the satisfactory fabrication of Al_2O_3 coating is impossible. The surface FE-SEM image of BM5h coating is shown in **Fig. 7(b)**. The fine crushed particles and flattened surface morphology indicated by red arrows are observed on the coating surface. This microstructure shows a part of the

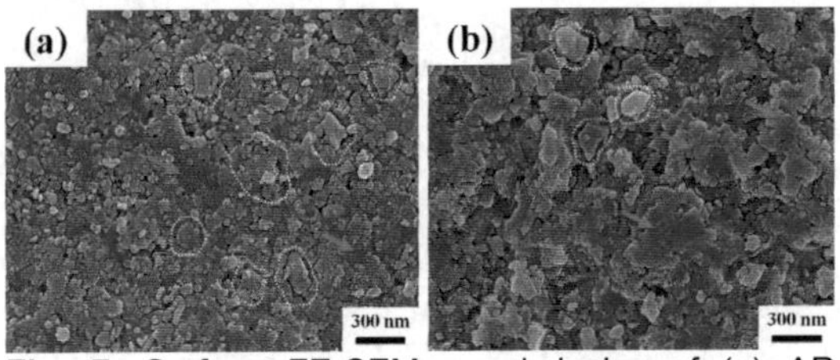

Fig. 7. Surface FE-SEM morphologies of (a) AR coating and (b) BM5h coating

deposition process. For the successful deposition of Al_2O_3 layer using VKS coating, It is necessary to fragment well the deposited particles. The finely fragmented particles undergo consolidation, which is made more dense coating by the collision of the subsequent particles. Due to the combination of the fragmentation and consolidation phenomena occurring at the same time, it is possible to fabricate the successful coating. In case of BM5h, this phenomena seems to have occurred and the reduction of the crystallite size is also evidence. Thus, An unstable state of Al_2O_3 particle can be made by ball milling process, and the weakened particles can be easily fragmented than primary particles. This makes it possible to deposit hard Al_2O_3 coating on soft polycarbonate substrate.

4 Conclusion

In this present study, ball-milled Al_2O_3 powder was deposited on polymeric polycarbonate substrate using VKS coating. Through microstructural observation and XRD pattern analysis of the powders, it was confirmed that the BM5h powder was more unstable state than the AR powder. Compared to AR coating the crystallite size and coating thickness of BM5h coating are smaller and thicker, respectively. Since the BM5h particles are more easily fragmented during the VKS coating, it is possible to deposit Al_2O_3 coating layer on the polycarbonate substrate.

5 Literature

[1] Akedo, J. and Lebedev, M. and Iwata, A. and Ogiso, H. and Nakano, S. : Aerosol deposition method (ADM) for nano-crystal ceramics coating without firing. MRS Proceedings (2003), pp. W7. 10/U8. 10.

[2] Akedo, J. : Aerosol deposition method for fabrication of nano crystal ceramic layer. Materials Science Forum (2004), pp. 43/48.

[3] Akedo, J. : Aerosol Deposition of Ceramic Thick Films at Room Temperature: Densification Mechanism of Ceramic Layers. The American Ceramic Society 89 (2006). 1834/39.

[4] Lee, B. and Lin, S. and Wu, W. and Wang, X. and Chang, P. and Lee, C. : Piezoelectric MEMS generators fabricated with an aerosol deposition PZT thin film. Journal of Micromechanics and Microengineering 19 (2009), Issue 6.

[5] Yang, G. and Li, C. and Fan, S. and Gao, J. : Influence of pore structure on ion diffusion property in porous TiO_2 coating and photovoltaic performance of dye-sensitized solar cells. Surface and Coatings Technology 205 (2011), Issue 10, pp. 3205/10.

[6] Lee, S. and Hahn, B. and Kang, T. and Lee, M. and Choi, J. and Kim, M. and Kim, S. : Hydroxyapatite and collagen combination-coated dental implants display better bone formation in the peri-implant area than the same combination plus bone morphogenetic protein-2–coated implants, hydroxyapatite only coated implants, and uncoated implants. Journal of Oral and Maxillofacial Surgery 72 (2014), Issue 1, pp. 53/60.

[7] Sahner, K. and Kaspar, M. and Moos, R. : Assessment of the novel aerosol deposition method for room temperature preparation of metal oxide gas sensor films. Sensors and Actuators B: Chemical 139 (2009), Issue 2, pp. 394/99.
[8] Young, R. and Lovell, P. : Introduction to polymers. (2011).
[9] Brandrup, J. and Immergut, E. and Grulke, E. and Abe, A. and Bloch, D. : Polymer handbook. (1989).
[10] Park, H. and Kim, J. and Lee, C. : Dynamic fragmentation process and fragment microstructure evolution of alumina particles in a vacuum kinetic spraying system. Scripta Materialia 108 (2015), pp. 72/75.
[11] Park, H. and Kwon, J. and Lee, I. and Lee, C. : Shock-induced plasticity and fragmentation phenomena during alumina deposition in the vacuum kinetic spraying process. Scripta Materialia 100 (2015), pp. 44/47.
[12] Suryanarayana, C. : Mechanical alloying and milling, Progress in materials science 46 (2001), Issue 1, pp. 1/184.
[13] Tomasi, R. and Rabelo, A. and Chinelatto, A. and Reis, L. and Fo, B. and Walter, J. : Characterization of high-energy milled alumina powders. Cerâmica 44 (1998), Issue 289, pp. 166/70.
[14] Eskandari, A. and Aminzare, M. and Aboutalebi, S. and Sadrnezhaad, S. : Effect of high energy ball milling on compressibility and sintering behavior of alumina nanoparticles. Ceramics International 38 (2012), Issue 4, pp. 2627/32.
[15] Chauruka, S. and Hassanpour, A. and Brydson, R. and Roberts, K. and Ghadiri, M. and Stitt, H. : Effect of mill type on the size reduction and phase transformation of gamma alumina. Chemical Engineering Science 134 (2015), pp. 774/83.
[16] Baláž, P. and Achimovičová, M. and Baláž, M. and Billik, P. and Cherkezova-Zheleva, Z. and Criado, J. and Delogu, F and Dutková, E. and Gaffet, E. and Gotor, F. : Hallmarks of mechanochemistry: from nanoparticles to technology. Chemical Society Reviews 42 (2013), Issue 18, pp. 7571/637.
[17] Mahmoud, A. and Wasly, H. and Doheim, M. : Studies of crystallite size an d lattice strain in Al-Al_2O_3 powders produced by high-energy mechanical milling, Jounral of Engineering Science 42 (2014), pp. 1430/39.
[18] Park, H. and Heo, J. and Cao, F. and Kwon, J. and Kang, K. and Bae, G. and Lee, C. : Deposition Behavior and Microstructural Features of Vacuum Kinetic Sprayed Aluminum Nitride. Journal of thermal spray technology 22 (2013, Issue 6, pp. 882/91.
[19] Mathias, P. and Wu, W. and Goretta, K. and Routbort, J. and Groppi, D. and Karasek, K. : Solid particle erosion of a graphite-fiber-reinforced bismaleimide polymer composite. Wear 135 (1989), Issue 1, pp. 161/69.
[20] Friedrich, K. : Erosive wear of polymer surfaces by steel ball blasting. Journal of materials science 21 (1986), Issue 9, pp. 3317/32.

Understanding the influence of micro- and sub-micro structural features on the mechanical properties of HVO/AF sprayed WC-CoCr cermets

M. Parco*, I. Fagoaga, G. Barykin, C. Vaquero, Fundación Tecnalia Research & Innovation, IRUN, Spain
A. Chuvilin, CIC Nanogune - San Sebastian / IKERBASQUE - Bilbao, Spain

HVOF processes represent the state of art for the spray deposition of wear and corrosion resistant coatings since their supersonic gas velocities in combination with moderate flame temperatures allow the deposition of optimal coatings with very high bond strengths, fine surface finishes and low oxide levels. However, new generation coating materials (fine powders), stringent quality requirements and the high productivity demanded by the industry, push the HVOF technologies to their limits. Recently, a novel air-oxygen controlled high velocity combustion process has been development by Tecnalia. The system operates within the supersonic regime using a broad range of fuel/oxidant ratios thanks to the use of air-oxygen mixtures and a carefully optimized gun design. Extremely low flame temperatures can be achieved while keeping a supersonic flow of combustion products, thus allowing the solid state deposition of almost all industrially relevant metal alloys with superior deposit qualities. In this work, a systematic investigation of the influence of the powder particle size and gun configuration on resulting coating microstructural features has been performed. For comparison, two fine structured commercially available WC10Co4Cr powders with different particle size distributions have been investigated. The coating structure has been characterized with by high resolution SEM cross-section imaging and X-ray diffraction analysis. Resulting coatings are characterized by highly dense structures, a high reterion of the primary carbides, average microhardness of up 1885 HV0.3 and fracture toughness varying between 3 and 7 $MPa.m^{-1/2}$ depending on the powder particle size distribution and the process conditions used.

1 Introduction

Because of their high hardness, high wear-resistance and good fracture resistance, hardmetal coatings are attractive and successful composite materials that have been widely used in many industrial applications. In particular, High Velocity Oxy-Fuel (HVOF) spray coatings based on tungsten carbide (WC) have been introduced some years ago as an alternative to electrolytic hard chrome (EHC). Such HVOF sprayed coatings have supposed a clear improvement from a technical and environmental perspective. In applications where the coatings will be used in an erosive-corrosive environment, HVOF WC-CoCr coatings are frequently recommended and are, in fact, the current industrial standard as an alternative to EHC [1].

HVOF sprayed WC-Co coatings are formed by the successive deposition of a stream of solid–liquid two-phase droplets. However, the thermal cycling of WC-Co spray particles up to a temperature over the melting point of the binder phase during thermal spraying leads to the decarburization of carbide, which significantly influences wear performance of the resultant coating. The decarburization of carbide occurs through the dissolution of carbide into molten binder cobalt alloy, thermal decomposition of WC, oxidation of dicarbide and rebounding of large carbide particles within droplet. As a result, the phases present in thermally sprayed WC-Co coatings may include main phase WC carbide, certain content of W_2C (or WC_{1-x}), minor phases of M_6C (Co_3W_3C, Co_2W_4C) and W, trace of Co, and even $M_{12}C$ (Co_6W_6C). Experimental results have shown that the thermal decarburization of WC is limited to the thermal decomposition of WC to W_2C and carbon. The formation of metallic tungsten in the coating results from the oxidation of W_2C. Therefore, the effective suppression of thermal decomposition of WC becomes effective to suppress the formation of tungsten. However, in order to reach a sufficient deformability of the cermet particles upon high velocity impact, it is necessary to heat the feedstock just to melting of the binder phase through spray condition control [2,**[3]**].

For this reason, there is a growing interest within the industry in finding thermal spray processes able to operate under "colder" conditions and/or with higher gas velocities. Introduced around ten yeas ago, the so-called warm spray process is able control the temperature of the propellant gas in a range approximately from 2300 to 1000 K by injecting room temperature inert gas into the combustion gas jet of an HVOF, so that many powder materials can be deposited in thermally softened state at high impact velocity [4]. More recently, High-Velocity Air-Fuel (HVAF) processes and even cold spraying have challenged HVOF deposited WC-CoCr coatings. HVAF processes, for instance, have proved to be a viable technology with many practical advantages over conventional HVOF systems [5]. Activated Combustion HVAF (AC-HVAF) sprayed WCCoCr coatings were found to outperform the wear resistance of their HVOF counterparts sprayed with conventional HVOF systems like JP5000 and DJ2600, performing even 20-30 times better than electroplated hard chrome and spray-fused coatings [6]. Moreover, liquid fuelled HVOF systems can be modified by the addition of an inert gas (N_2) for cooling purpose, together with the reduction of the effective flame power by changing the design of the combustion chamber, like the so-called nanoHVOF® technology [7].

Comparing with kerosene fuelled HVOF systems (~200kw), gas fuelled HVOF systems work at lower

* *Corresponding author: M. Parco, Industry and Transport Division, TECNALIA, Mikeletegi Pasealekua 2, 20009 San Sebastián-Donostia, Spain.*
E-mail address:maria.parco@tecnalia.com

power levels (~100kW), which in combination with the implementation of low temperature fuel gases, as natural gas, and the substitution of O_2 by air, can further reduce the flame temperature while keeping a stable performance in the supersonic regimen. In the practice, the oxidant mixture can be adjusted from pure O_2 to full compressed air, leading to a wide range of energetic conditions in the flame.

In line with this research trend, the thermal spray group from TECNALIA has been woring on the developed a High-Velocity Oxy/Air-Fuel (HVO/AF) system able to operate within a wide range of process conditions. Low gas temperatures are essentially reached by using natural gas as fuel and by injecting compressed air and oxygen to the combustion mixture. In previous reports, the potential of the system for the development of super hard cermet coatings using different types of conventional and fine WC-Co(Cr) powders was showed [8]. In this work, a systematic investigation of the influence of the powder particle and process parameters on the resulting coating structural features at the sub-micron has been performed.

2 Process description

In this work, a novel High-Velocity Oxy/Air-Fuel (HVO/AF) system has been implemented. This system has been under development during the last 10 years at TECNALIA and has proved to be a cost effective thermal spray technology for the deposition of high quality metal [9] and cermet base coatings [8], **Fig. 1**.

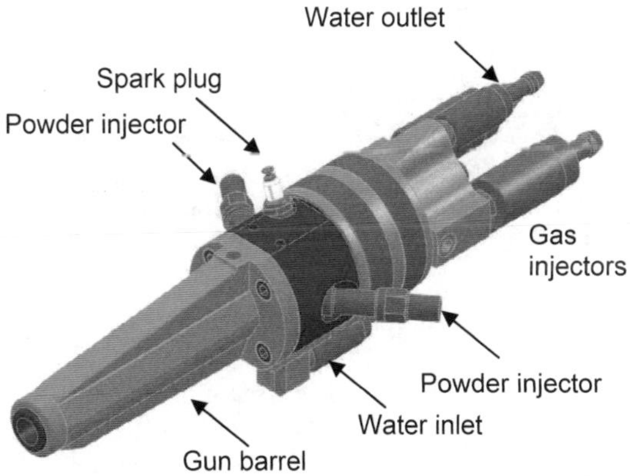

Fig. 1. Overview of the HVO/AF gun (proprietary technology of TECNALIA).

Basically, thanks to the use of air-oxygen mixtures and a carefully designed combustion chamber and gas injection module, the system is able to operate within the supersonic regime using a broad range of fuel/oxidant ratios. This fact allows the operation of the system with extremely low flame temperatures while keeping a supersonic flow of combustion products. Moreover, following a modular design, further critical gun components as the gun barrel and the powder injection module, can be easily exchanged to tailor the gun configuration for the processing of materials with very different physical natures, thus allowing the deposition of almost all industrially relevant metal base and cermet materials with superior deposit qualities. The system is operated using natural gas as fuel. Among the investigated coating materials, the temperature of the flame and thus the process conditions are mainly adjusted by varying the natural gas flow rate between 150 and 250 NLPM. The feedstock powders are injected radially using two feeding lines, allowing the processing of up to 6 kg/h of powder (for tungsten carbide). The whole gun is water cooled.

3 Experimental procedure

3.1 Feedstock powders

As feedstock materials, two commertially available WCCo(Cr) powders from H.C. Starck were employed: Amperit® 554.071 (size distribution betweeb 5–25 µm) and Amperit® 554.074 (size distribution between 15–45 µm). The 554 family consists of blocky tungsten carbide particles with 10 % cobalt and 4 % chromium (sintered and crushed powder). The tungsten carbide is well distributed in the cobalt-chromium matrix and has a size of less than 2 µm (see **Fig. 2**).

Fig. 2. Overview (a) and cross sectional micrograph of the Amperit® 554.074 powder.

3.2 Coating deposition

Coatings were sprayed on flat stainless steel substrates (410) in size of 60 x 30 x 5 mm. Prior to the coating deposition, substrates were degreased and grit blasted with electro corundum at a pressure of 4 bar.

Three different conical expansion nozzles with 70, 100 and 150 mm in length were investigated to assess the influence of the resulting coating properties.

Process parameters have been optimized in previoius works [8]. **Table 1** summarizes the implemented process parameters. The feedstock powders are injected radially using two feeding lines. A relative gun velocity of 1.4 m/s was used in all cases.

The deposition efficiency (DE) has been calculated as a ratio of the mass of the coating deposited on a flat mild steel substrate over the amount of feedstock material fed into the system during time spend to deposit the coating. The mass of the coating has been calculated by weighting the sample before and after the coating deposition.

3.3 Coating characterization

X-ray diffraction (XRD) measurements were carried out on a Bruker D8 Advance diffractometer operating at 40 kV and 40 mA using a Cu radiation (Kα= 1,54184 Å). The integral intensities of the (100) peak of WC (I_{WC}) and of the (101) peak of W_2C (I_{W2C}) were assessed by pattern fitting using pseudo-Voigt functions and the index of carbide retention (I), quantifying the retention of the WC phase against the formation of W_2C, was computed as $I = I_{WC} / (I_{WC} + I_{W2C})$, according to the definition in [10].

On the one side, coatings were cut, mounted, ground, and polished and cross sections were examined by optical microscopy (OM). On the other side, for detailed inspection by Scanning Electron Microscopy (SEM), part of the coating was removed (see **Fig. 3**) by means of the Focused Ion Beam (FIB) technique with the aim of exposing the real undisturbed structure of the coating. In this way, undesired modifications of the coating structure through the common sectioning, grinding and polishing steps are avoided.

For this purpose, a dual beam SEM-FIB system (Helios 450S FIB) was implemented. This system is equipped with both electron and ion beam columns, allowing for simultaneous cutting of the sample by ion and imaging by electron beam. In order to avoid the deterioration of the surface of the coating by the tails of Ga beam, a thin Pd layer was deposited on the top of the working surface.

Fig. 3. Overview of cross sections prepared by ion beam for high resolution SEM inspection: a) Section ablated from the top of the coating; b) section cut from the side.

The microhardness of the coatings was measured on polished cross sections by Vickers indentation according to the standard ASTM E-384-10. Measurements were carried out using a Vickers indenter (Struers Duramin A-300 micrometer) at three different loads of 300 gf (≈3 N), 500 gf (≈5 N) and

Table 1. Optimized process parameters

Ref.	Feedstock	Natural gas [NLPM]	O_2 [NLPM]	Air [NLPM]	Carrier gas [NLPM]	Feeding rate [g/min]	Standoff distance [mm]	Nozzle length [mm]
F1	Amperit 554.071	250	500	1100	70	~37	150	70
F2	Amperit 554.071	250	500	1100	70	~37	150	100
F3	Amperit 554.071	250	500	1100	70	~37	150	150
C1	Amperit 554.074	250	500	1100	70	~47	150	70
C2	Amperit 554.074	250	500	1100	70	~47	150	100
C3	Amperit 554.074	250	500	1100	70	~47	150	150

1Kgf (≈10 N), with a dwell time of 15 s. The reported hardness values are the average of 10 measurements.

Oxygen contents were analyzed by a using LECO TCH-600 oxygen analyser. Analyses involved combustion of 0.03 – 0.10 g of the crushed coating and measurement of combustion gases by infrared absorption. The reported hardness values are the average of 2 measurements.

Indentation fracture toughness was also assessed by performing indentations into the polished cross-sections of the coatings using a Vickers-shaped indenter. A peak load of 5 kgf (≈49 N, load period 10 s) was employed in order to induce crack propagation from the indent corners. Fracture toughness was determined by measuring crack lengths on optical micrographs (200× magnification). Only cracks parallel to the substrate starting at the left and right tip of indents, respectively, were considered according to the method described in [11] , **Fig. 3**. The critical stress intensity (K_{IC}) was calculated according to Shetty et al. based on the Palmqvist crack model:

$$K_{IC} = 0.0889(HW)^{\frac{1}{2}} \quad (1)$$

Where H is the hardness and W is the crack resistance parameter (also quoted as Palmqvist toughness). W is calculated based on the following equation:

$$W = \frac{(P - P_0)}{4\bar{a}} \quad (2)$$

Where $\bar{a}$ is the mean radial crack length, P is the indentation load and P_0 is the threshold indentation load for inducing a crack.

Fig. 3. Microhardness indentation pattern, where a: 100 µm and b: 50 µm.

4 Results and discussion

Fig.4 - **Fig. 5**. Show some representative cross sectional micrographs of the coatings produced under conditions summarized in **Table 1**. It is well known that during flight spray particles melt to different extent depending on their size and dwell times.Coatings produced with the fine WC10Co4Cr powder (Amperit 554.071) exhibit macroscopically dense micro-structures with very low porosity detectable at low magnifications (**Fig. 4a, c and d**). Nevertheless, appreciable differences in porosity are observed in the cross sectional micrographs taken at greater magnifications on the samples prepared by FIB milling from the various deposits (**Fig. 4b, d and f**).

In sample F1, in deposition of which the particle temperature has been lowest (sample produced with the 70 mm long nozzle), the resulting coating features a dense microstructure and a large rentention of the primary carbides, nevertheless with the presence of some volume defects (pores) either in the interface between the particles and inside the particles self. This residual porosity might be attributed to a moderate deformation degree of the particles. No oxide films are observed in the interface between the particles. When increasing the nozzle length to 100 mm by constant process parameters, a large amount of primary carbides are still retained unchanged in the coating but larger pores at the interface between the particles (producing black contrast) become visible (**Fig. 4d**). This can be attributed to contraction of liquid phases during solidification on the substrate. In coating F3, which corresponds to the highest particle temperatures reached with the longest gun nozzle (150 mm), the feedstock material is melted to a greater extent. Noteworthy is the massive carbide dissolution witnessed by the loss of small carbides, the spheroidization of bigger carbides, the formation of white W_2C areas around the carbides and even metallic W stripes in the coating. The presence of a bright phase in the matrix is linked to the formation of η-phases M_6C (e.g. Co_3W_3C). Moreover, the formation of micro pores and voids in the interlamerlar zone becomes more evident (**Fig. 4f**).

Coatings produced with the coarse WC10Co4Cr powder (Amperit 554.074) under equivalent conditions (C1, C2 and C3) are shown in **Fig. 5**. The resulting coatings feature also macroscopically dense microstructures with low porosity detectable at these magnifications. Similarly to the samples produced with the fine feedstock powder, the coating produced with the 70 mm long nozzle shows again a very dense microstructure and a large rentention of the primary carbides, with no evidences of oxide film formation in the interface between the particles (**Fig. 5b**). In this case, however, some few micro cracks can be observed between the particles, evidencing a poor interlamelar cohesion.

When using the 100 mm long nozzle, the coating structure becomes apparently denser and the observed micro cracks are smaller and more scattered (**Fig. 5d**). In the coating produced with the longest gun nozzle (C3), the spray particles are melted to a greater extend, as evidenced by the loss of small carbides and the moderate spheroidization of bigger carbides (**Fig. 5f**). The formation of voids points again at the contraction of liquid phases during solidification on the substrate.

The carbon loss and typical metallurgical reactions observed in HVOF sprayed WC-Co(Cr) coatings are perceivable through XRD patterns (**Fig. 6**). For the WCCoCr composition, WC, the η-phase $(Co,Cr)_3W_3C$, and a (Co,Cr,W) alloy binder are expected to be in thermodynamic equilibrium.

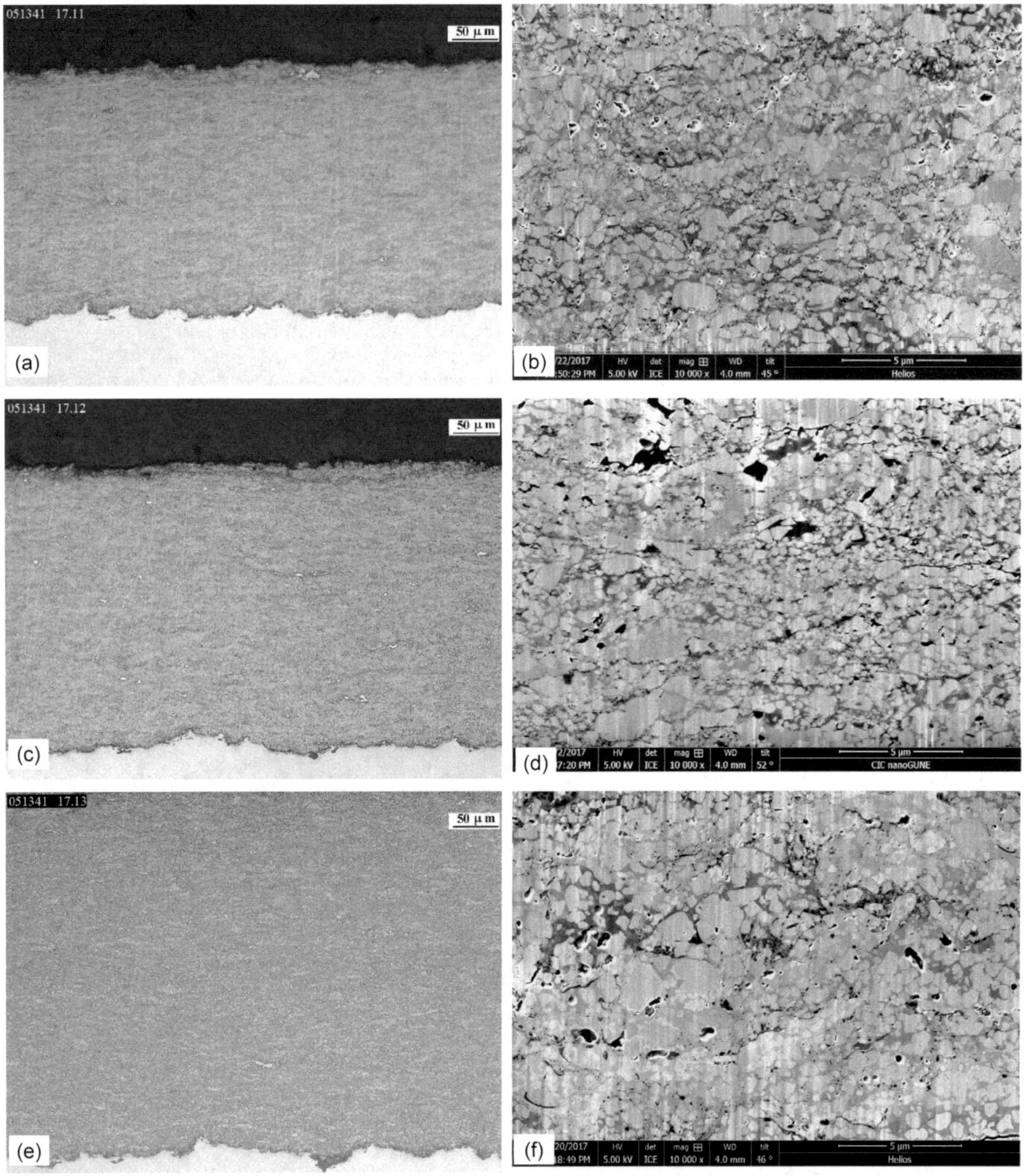

Fig. 4. Cross sectional micrographs of coatings produced with the fine WC10Co4Cr powder (Amperit 554.071) at different process conditions: a) and b) sample ref. F1, c) and d) sample ref. F2; e) and f) sample ref. F3. SEM micrographs shown on the right correspond to the cross sections exposed through the FIB technique (these figures have been filtered in order to decrease the paths left by the tails of Ga beam).

Thus, chromium is contained both in the metallic binder and the hard phase. Basically the chromium addition intensifies the effect of η-phase formation compared to plain WC–Co [1]. **Table 2** summarizes the Index of WC retention calculated from the XRD patterns as well as the measured O content and the average micro harness values (HV0.3). For the coatings produced with the fine WC10Co4Cr powder (554.071), the XRD pattern evidences a large retention of the primary carbide phase (index of WC retention of around 92-93%) when using the short and the intermediate nozzle length (samples referred as F1 and F2). The coating produced with 150 mm long nozzle (F3) undergoes a significant decarburization as result of the higher particle temperature reached inside the nozzle.

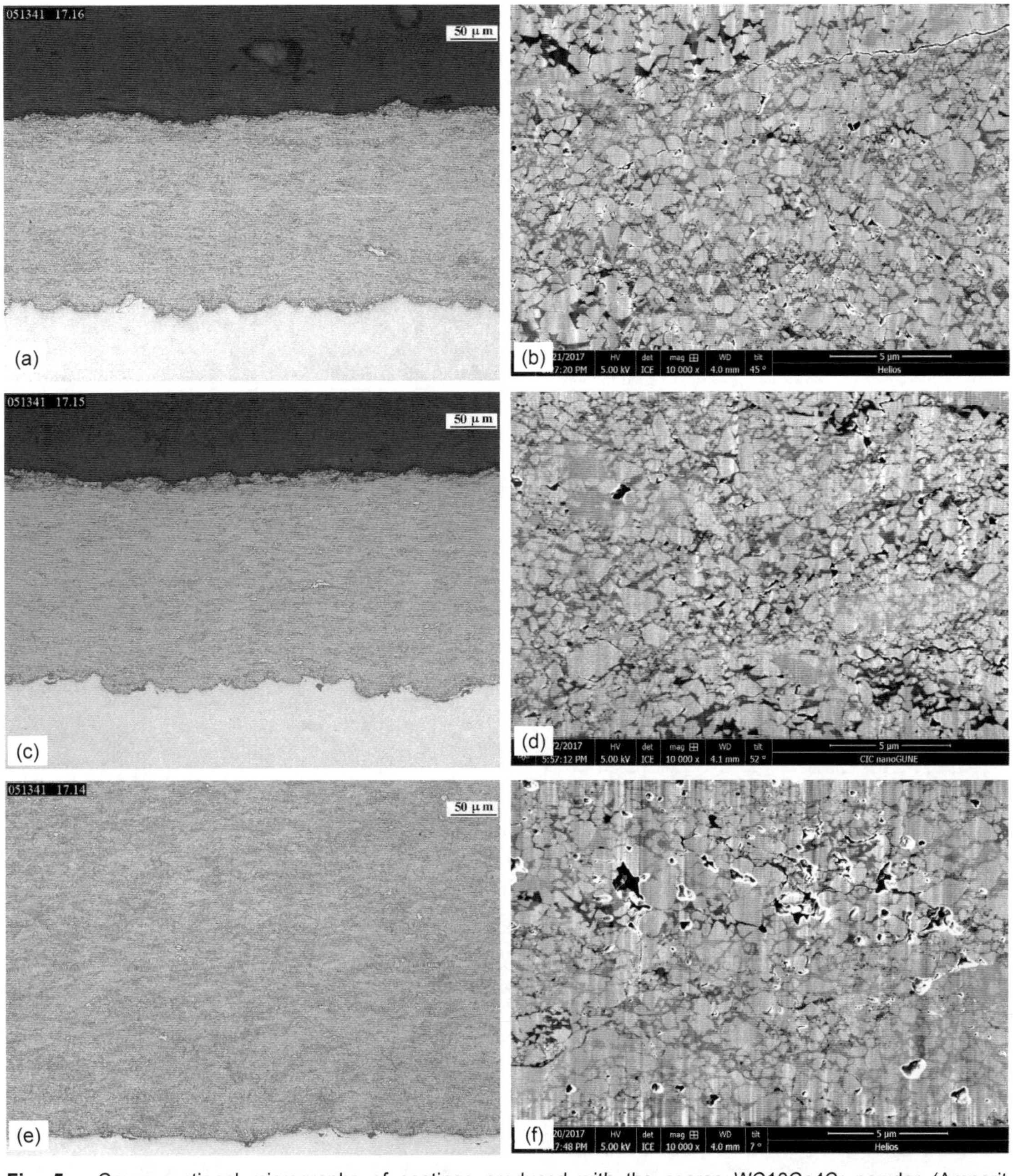

Fig. 5. . Cross sectional micrographs of coatings produced with the coarse WC10Co4Cr powder (Amperit 554.074) at different process conditions: a) and b) sample ref. C1, c) and d) sample ref. C2; e) and f) sample ref. C3. Micrographs shown on the right correspond to the cross sections exposed through the FIB technique (these figures have been also filtered in order to decrease the paths left by the tails of Ga beam).

Nevertheless, the resulting Index of WC retention is moderate in comparison to values reported for WC10Co4Cr coatings produced with conventional gas-/liquid-fuelled HVOF systems [5]. Moreover, the XRD pattern evidences the formation of amorphous and/or nanostructured phases around the main peak of the M_6C phase in coating F3.

Coatings produced with the coarse WC10Co4Cr powder (554.071) exhibit in gerenal a higher retention of the WC phase. Assuming the XRD technique has a detection threshold of about 2%, the index of WC retention is ≥98% for those coatings produced with gun nozzle having 70 or 100 mm in length (samples ref. C1 and C2, respectively). The coating produced with 150 mm long nozzle (C3) undergoes a higher decarburization as result of the higher particle temperature reached inside the nozzle. With an index of WC retention over 95%, the decomposition of the feedsctock material is in any case moderate.

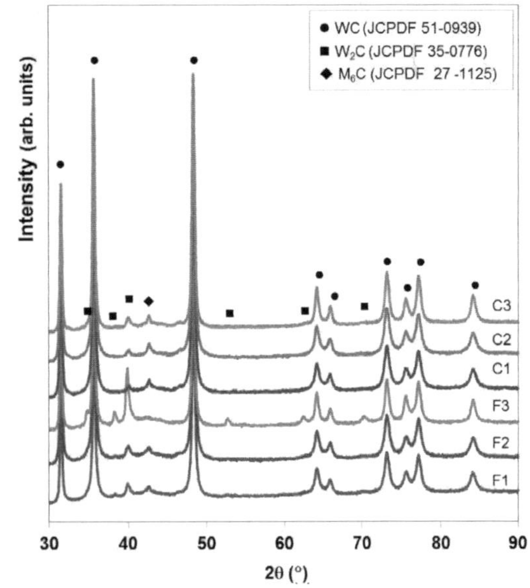

Fig. 6. XRD patterns of all WC-Co(Cr) coatings.

Table 2. Measured properties for WC-Co(Cr) coatings

Ref.	Index of WC retention, %	O Content %	Hardness HV0.3
F1	>92	0.87	1637 ± 53
F2	>93	0.63	1544 ± 128
F3	>75	0.83	1597 ± 83
C1	>98	0.51	1885 ± 80
C2	>98	0.40	1849 ± 111
C3	>95	0.42	1444 ± 155

On the other side, oxygen content in the coatings was found to decrease with the prolonged nozzle lengths. This is expected, since the higher is the temperature the more carbon is dissolved to the molten droplets and the more effective reducing agent carbon is based on the Ellingham diagram [13].

Similar behavior has also been found by Varis et al., when they observed that oxygen content in the coating was found to decrease with increasing flame temperature, even if having higher oxygen parcial pressures under the investigated process conditions. They concluded that carbon-oxygen reaction in WC-CoCr spraying plays a decisive role in the oxidation of metallic material in the coating during the spraying, suggesting that low temperature oxygen-rich flames can be used without the risk of marked loss of properties [8].

On the other side, because of the pronounced risk of chromium oxidation in WC-CoCr compositions, it is likely that a prolonged dwel time in the gun barrel may favour the oxidation of chromium in the CoCr matrix, thus explaining the higer oxide content in the coating produced with the fine powder and the longest gun nozzle (coating F3). As it could be expected, the measured O contents evidence the larger oxidation of the feedstock material for the coatings produced with the fine powder (554.071) in comparison to these produced with the coarse powder (554.074). It has been extensively documented that on small size particles, the whole particle can be affected from melting and decarburization, making the control of the particle temperatures an issue.

For comparison purposes, the different microhardness values are plotted in parallel for all investigated coatings in **Fig. 7**. Coatings produced with the fine WC10Co4Cr powder feature comparable mean microhardness values when using a 300g test load. Neverthelss, when increasing the test load to 500g, remarkably higher values are obtained for the coating produced with short gun nozzle (coating F1).

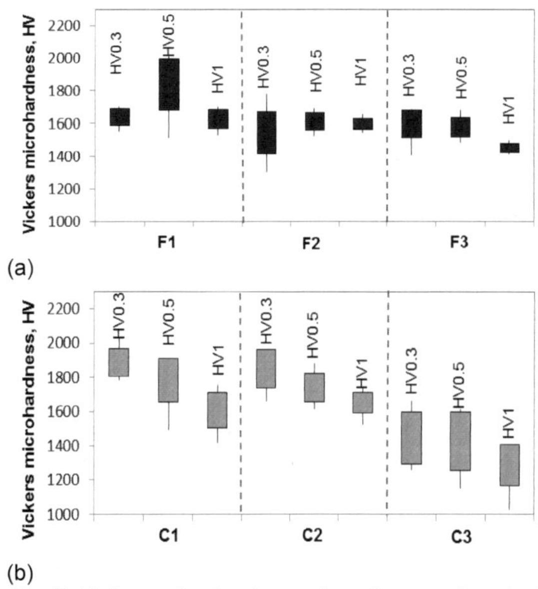

Fig. 7. Vickers microhardness of coatings produced with the (a) fine powder (554.071) and (b) coarse powder (555.074) using the different gun nozzle lengths.

From the one side, the coating is definitively harder, the higher dispersion of the values indicates however the presence of more heterogeneities through the coating cross section, which based on the observed microstructural features (**Fig. 4b**) could be attributed to the presence of microcracks. With an increased melting degree (coating F2 and F3), the coating hardness slightly decreases as well as the scattering of the values when using higher test loads, thus evidencing a real softening of the coating. It is worth mentioning that no cracks were observed around the indentation mark when using a test load of 1Kgf (HV1). In the case of the coatings produced with the coarse powder (**Fig.7b**), there is a more clear correlation of the resulting microhardness with the gun nozzle length and thus with the corresponding dwell time within the flame. Until a nozzle length of 100 mm, the hardness of the resulting coating remains essentially the same. Nevertheless, by a prolonged interaction with the flame (i.e. gun nozzle with 150 mm in length), such coarse particles are melted and/or softened to a longer extent without a significant decomposition of the feedstock material, thus leading to a softer metal matrix in the resulting coating in comparison to coatings obtained with the fine particles. Some preliminary estudies have showed that under the investigated process conditions, the in-flight velocity of the coarse particles might be influenced to a greater extent by the nozzle length in comparison to the coarse particles, thus further explaining the observed tendencies. This effect must in any case be further investigated and will be the focus of future experimental works.

Fig. 8 collects the calculated indentation fracture toughness (K_{IC}) for the different coatings. In correlation to the tendencies observed on the hardness values, coatings produced with the fine powder show only a low variation in the cross sectional fracture toughness. Only the coating produced with the intermediate nozzle length (F2) shows a moderate increase of the thoughness, what could be explained by the denser coating structure (less volume defects) as observed in **Fig. 4d**, in combination with a softer matrix (less embrittled thorugh the carbide dissolution effect).

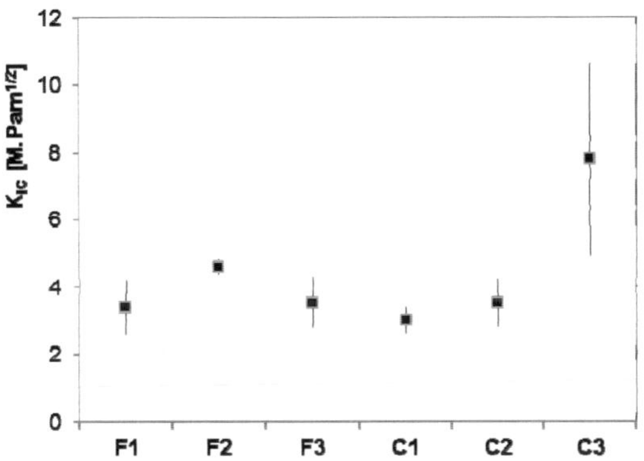

Fig. 8. Indentation fracture toughness (K_{IC}) values of different coatings produced [the standard deviation is shown for each average value].

Similarly, coatings produced with the coarse powder show a tight correlation with the microhardness, having a moderate increase of the thoughness when processing the material with the intermediate nozzle length (100 mm) in comparison to the short nozzle (70 mm), and a sharp increase when using the longest nozzle (150 mm). It is worth mentioning that even if the values have a relatively large scattering, the thoughness of the coating produced with the coarse particles and the longest gun nozzle feature a remarkdably high toughness in comparison to the values reported elsewhere.

In previous studies it was found that the cracks initiating from the edge of the Vickers indenter tip follows the path of lamella boundaries. According to Varis et al. [3], the increased particle temperature improves the lamella bonding and makes more difficult for crack to grow until the toughness of the lamella boundary region has decreased excessively due to carbide dissolution effects. Fracture toughness, which is measured from the cross section of the coating, was found to be therefore connected to the properties of the lamella interfaces.

It explains the comparably lower fracture toughness observed by the coatings produced with the colder particles (short barrel lengths, 70 and 100 mm) in comparison to the more melted coatings, deposited with the longest barrel of 150 mm. Additionaly, the suppression of microcraks at the interface between the particles as a consequence of the higher melting degree of the particles (when comparing in Fig. 5b, d and f), could explain the higher resistance to crack propagation in the coating.

5 Conclusions

Based on two commercially available WC10Co4Cr powders from the Amperit 554 family (sintered and crushed particles with medium sized primary carbides of less than 2 μm), a systematic investigation of the influence of the powder particle size on the resulting microstructure and ascertained mechanical properties of coatings deposited with an experimental HVOAF system trom TECNALIA has been performed.

Different gun nozzle lengths have been investigated while keeping other relevant process parameters constant, in an attempt to assess the influence of selected gun configurations on the resulting coating properties. Collected results showed that the melting degree of the feedstock powders and the resulting coating microstructure and mechanical robustness can be significantly influenced by varying the nozzle length, thus allowing the selection of the most suitable gun configuration for each feedstock material.

An intermediate gun nozzle length of 100 mm seems to be the most suitable option for processing such fine powders (particle size between 5 and 25 μm), leading to highly dense coating structures, a high reterion of the primary carbides (index of WC retention of more than 93%) and average microhardnesses of around 1500-1600 HV0.3 in combination with a good indentation fracture toughness (~4.6 MPa.m$^{-1/2}$).

On the other side, for the processing of the more conventionally sized (coarse) WC10Co4Cr powder, the longest gun nozzle with 150 mm was found to be a good approach for achieving coatings featuring similary dense coating structures, a high primary carbide retention (>95%), high microhardness (~1444 HV0.3) and extremely high thoughness (~7.8 MPa*m$^{1/2}$). Depending on the expected service conditions, an interesting option would be to tailor the gun configuration to reach extremely high hard layers (~1850 HV0.3) with superior carbide retentions (>98%), nevertheless at the expenses of a moderate indentation thoughness (~3.5 MPa*m$^{1/2}$).

Some preliminary estudies have showed that under the investigated process conditions, the in-flight velocity of the coarse particles might be influenced to a greater extent by the nozzle length in comparison to the coarse particles, further explaining the observed tendencies. Further details and analysis of the in-flight particles properties as well as their effect on the coating properties will be needed in order to gain a deeped understanding of the influence of the different process parameters on the resulting coating quality.
Therefore, increased lamella cohesion may explain the increased capacity of coating to bear evolving stress.

6 Literature

[1] L.-M. Berger. Application of hardmetals as thermal spray coatings. Int. Journal of Refractory Metals and Hard Materials 49 (2015), pp. 350–364.

[2] Chang-Jiu, L. and G.-J. Yang. Relationships between feedstock structure, particle parameter, coating deposition, microstructure and properties for thermally sprayed conventional and nanostructured WC–Co. Int. Journal of Refractory Metals and Hard Materials, 39 (2013), pp. 2-17.

[3] T. Varis et al. Formation Mechanisms, Structure, and Properties of HVOF-Sprayed WC-CoCr Coatings: An Approach toward Process Maps. J. Therm. Spray Technol. Volume 23(6) August 2014, pp.1009 - 1018.

[4] Seiji Kuroda, Makoto Watanabe, KeeHyun Kim, and Hiroshi Katanoda. Current Status and Future Prospects of Warm Spray Technology. J. Therm. Spray Technol.Volume 20(4) June 2011, pp. 653 - 676.

[5] G. Bolelli et al. Tribology of HVOF- and HVAF-sprayed WC–10Co4Cr hardmetal coatings: A comparative assessment. Surface & Coatings Technology 265 (2015) 125–144.

[6] A. Verstak and V. Baranovski: AC-HVAF Sprayed Tungsten Carbide: Properties and Applications. Building on 100 Years of Success: Proceedings of the 2006 International Thermal Spray Conference, B.R. Marple, M.M. Hyland, Y.C. Lau, R.S. Lima and J. Voyer, Eds., May 15-18, 2006 (Seattle, WA, USA), ASM International, Materials Park, OH, 2006.[

[7] V. Wesling, R. Reiter, L. Lau and G. Matthäus. nanoHVOF – Process technology for and properties of a new high density protective coating. Mat.-wiss. u. Werkstofftech. 2014, 45, No. 6. (In german).

[8] M. Parco, I. Fagoaga, G. Barykin and C. Vaquero, High velocity spray deposition of WC cermets by an air-oxygen controlled combustion process, ITSC 2016: Thermal Spray: Fostering a sustainable world for a better life!, May 10-12, 2016 (Shanghai, R.P. China).

[9] M. Parco, G. Barykin and I. Fagoaga. HVOF spraying of metal alloys with Natural Gas and Oxygen-Air mixtures. Thermal Spay Bulletin 7 (2014) No.1. Specialist article.

[10] C. Bartuli, T. Valente, F. Cipri, E. Bemporad and M. Tului. Parametric Study of an HVOF Process for the Deposition of Nanostructured WC-Co Coatings. J. Therm. Spray Technol. 14 (2) (2005) 187–195.

[11] M.M. Lima et al. Toughness evaluation of HVOF WC /Co coatings using non-linear regression analysis. Materials Science and Engineering A357 (2003), pp. 337-345.

[12] A. Ghabchi et al. Behavior of HVOF WC-10Co4Cr Coatings with Different Carbide Size in Fine and coarse Particle Abrasion. J. Therm. Spray Technol. Volume 19 (1-2) January 2010, pp. 368 – 377.

[13] D.R. Gaskell, Introduction to the Thermodynamics of Materials, 5th ed., Taylor & Francis, New York, 2008.

Effect of different shroud principles on the performance of a NiTi coating produced by means of twin-wire arc spraying (TWAS) process

W. Tillmann[a], M. Abdulgader*[a], D. Bezerra[a], M. Pohl[b], J. Gembaczka[b], [a)]Dortmund, [b)]Bochum / D

The super-elasticity behavior of a NiTi-shape memory alloy (SMA) is very promising regarding cavitation resistance. The need of high vacuum conditions by thermal spraying processes, to avoid oxidation, has always been and still is the main obstacle for the widespread of NiTi as a coating material. This work deals with studying the effect of the different shroud concepts on the obtained oxide content and the phases of the obtained twin wire arc sprayed (TWAS) coatings. The concepts include the use of argon as a shield in gas shroud (GS) as well as the use of an extended air cap attachment as a massive shroud (MS). The use of MS-concept led to a significant decrease in oxide content and therefore was selected to spray pre-alloyed NiTi-SMA wires. The standoff distance between the MS-outlet and the substrate surface shows also an effect on the obtained phases and thus on the behavior of the obtained coatings. At lower standoff distance a pseudo-elastic behavior was obtained and therefore a higher cavitation and wear resistance. The use of argon as atomization and shield gas with a massive shroud could be a cost-effective alternative for vacuum process in case of spraying NiTi-SMA pre-alloyed feedstock materials.

1 Introduction

Oxidation of the vital compounds in a feedstock is a major drawback by the conventional thermal spraying techniques. Since most of the spraying operations are performed in atmospheric environment, interstitial compounds are emerged more likely between atmospheric gases and the heated feedstock material. Reactivity of the feedstock material at elevated temperatures as well as the presence of the gases mentioned above are the main factors controlling the progress of these reactions. Subsequent buildup of stable oxide layers, which is initiated after coating deposition, is the reason for the higher corrosion resistance of Nickel-Titanium-Shape Memory Alloy (NiTi-SMA) [1-9] or pure Titanium [3, 10]. In addition, NiTi-SMA has a very good resistance against wear, in this case cavitation, which would make it as an exceptional choice as a coating material. The oxidation during the spraying process is detrimental for shape memory effect, the corrosion behavior, cohesion and adhesion of thermally spray coating [1,2, 4-9, 11,12]. In addition, the oxidation changes the ratio between the components that build these alloys. This effect becomes more obvious if the desired feature (e.g. pseudo-elasticity) is assigned to a predefined stoichiometric ratio. Any changes in the composition may shift or even eliminate the desired behavior [1, 4-9, 12].

Nickel-Titanium-alloy with a specified stoichiometric ratio range of Ni to Ti is characterized by its pseudo-elasticity and shape memory effect. Pseudo-elasticity of this alloy has very promising properties to reduce or even eliminate the risk of the cavitation in hydrodynamic systems [1, 5, 6, 8-9, 12]. This effect is correlated to the excellent damping behavior of pseudo-elastic alloys [12], see

Fig. 1 [14]. The basis of the shape memory effect is a martensitic phase transformation, in which the high temperature phase austenite converts into the low temperature phase martensite.

This phase transformation in NiTi is a very sensitive system in which slight fluctuations in the alloy composition, the element distribution or even small amounts of impurities (e.g O, N and C) can already cause significant changes in the ratio between Ni and Ti and thus to a shifting in material behavior. The phase transformation depends strongly on the Ni content, which determines the temperature range of the transformation [15]. The Ti in NiTi-SMA is very affine to interstitial elements like oxygen, nitrogen and carbon and tends to the form oxides, nitrides or carbides. These interstitial compounds are rich in titanium [15,16]. The consumed Ti ratio shifts the NiTi-SMA to higher Ni concentrations and lower transformation temperatures [16].

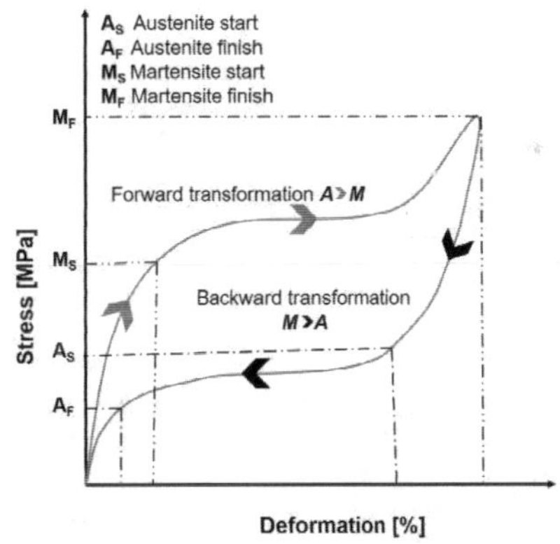

Fig. 1. Pseudo-elastic behavior of NiTi-SMA [edited from [14]]

The excellent corrosion and cavitation resistance behavior of the NiTi-SMA makes it an exceptional choice as a coating material. However, the high reactivity of Ti, as an essential part of this compound, represents a big challenge facing the conventional thermal spraying techniques to produce coatings with the desired pseudo-elastic behavior. This issue is impeded by the fact that in almost all thermal spraying techniques the feedstock material is heated to temperatures above their melting points, which enhances intensely the chemical reactions between feedstock and surrounding atmosphere. Therefore, depositing of NiTi-SMA as protective coatings take place almost exclusively in controlled environment like in Physical Vapor Deposition (PVD) [17-19] or vacuum plasma spraying (VPS) [1, 2, 12, 14, 20] or low pressure arc spraying LPAS [1, 10]. Nevertheless, there have been some attempts made to use conventional atmospheric plasma spraying (APS) and high velocity oxygen fuel (HVOF) to apply pseudo-elastic coating out of NiTi-SMA feedstock material [15].

Different types of a protective shroud were used as a low-cost alternative for the spraying in controlled low-pressure environment [21-31]. However, they all have one goal in common: eliminating the feedstock oxidation during spraying. Gas as protective shield along with massive shroud were applied for the different spraying techniques like HVOF [21, 27-28], APS [22-24, 31] and twin wire arc spraying (TWAS) [11, 26, 30]. The basic idea behind the shroud concept is to separate the spraying plume from the surrounding atmosphere. Argon is used as gas shroud and atomization gas to exclude any potential oxidation. Ward et. al in [21] used an argon shroud to produce a graphene oxide (GO) reinforcement in NiAl APS-sprayed coating. They proclaim in their work that the use of argon shroud leads to the retention of GO in the NiAl-coating. Low-oxygen (<2%) and high corrosion resistance were obtained by using inert gas shroud system was attached to HVOF-spraying gun [27-28].

It is clear that the shroud concept for controlling the oxide content in the obtained coating is very useful and cost effective compared to low pressure spraying techniques. In addition, wires as feedstock material are more promising in case of spraying compounds with reactive material like Ti [10]. The powder feedstock has a higher specific surface area compared to wires, which means a higher potential for oxidation. The contamination (oxidation) of the powder feedstock remains the produced coating even by using inert gases or controlled low-pressure equipment.

TWAS-process is the most economical coating process regarding the utilization of energy to melt a feedstock material [11]. It is also characterized by the highest variety in feedstock among all other thermal spraying techniques by using cored wires. In addition, the high deposition rate of TWAs-process enables a significant decrease in coating cost. However, the generated coatings show a higher porosity and lower adhesion strength compared to APS and HVOF. The use of wires as feedstock, due to their lower specific surface area and lower oxidation potential compared to powder contra part, makes the development of protective shroud for TWAS-process very attractive. The investigation need in order to explore the potential of shroud protection in TWAS-process is emphasized by the fact that this process is difficult to control in low-pressure conditions [11]. However, the heat intensity during arc-ignition is very high, which may result in burn of vital compounds of the feedstock material or undesired phase transformation processes.

Therefore, this work focusses on the investigation of different shroud concepts on the oxide formation by spraying NiTi-SMA. Initial tests are made by simultaneous spraying of nickel and titanium solid wires using the different shrouds and atomization gases. The tests to be conducted are intended to select the proper shroud concept. The selection is based on the obtained oxide content in the deposited coatings. Consequently, the spraying of NiTi-SMA wires is performed and thereby the effect of process parameters standoff distance (STOD) is investigated. Differential scanning calorimetry (DSC), X-ray diffraction (XRD) analyses are conducted for coatings and wires. The wear and friction test (Pin-On Disc) is performed only for the coatings. The focus of this study is to investigate and modify the effective shroud concept for spraying of NiTi-SMA. Thus, to the best of our knowledge, the development of proper shroud concepts to produce NiTi-SMA coatings with pseudo-elastic behavior is not yet thoroughly documented. The findings could provide an important breakthrough in spraying of NiTi-SMA for hydrodynamic systems and components.

2 Materials and experimental

2.1 Materials

Smart Arc 350 PPG (Oerlikon Metco) was used to explore the effect of the different shroud concepts on obtained NiTi-SMA coatings. The individual components of the NiTi-SMA namely pure Ni (06T-Nickel, 99.6 wt.% Ni, 0.00 wt.% O) and Ti (14T – Titanium, 99.8 wt.% Ti, 0.08 wt.% O) were used as feedstock to investigate the effect of shroud concept on oxide content in the obtained coatings. The wires are 1.6 mm in diameter and were supplied by PRAXAIR, Germany. Subsequently, pre-alloyed equiatomic NiTi-wires were used to produce a coating with pseudo-elastic behavior. The company Nitinol-Shop-Austria delivered super-elastic NiTi wire (Ø = 1.6 mm ± 0.010 mm, As <15 ° C, length: 500 m). With these wires, a detailed study regarding the effect of standoff distance on the preservation of the pseudo-elastic behavior and thus the cavitation resistance was carried out.

The following four independent adjustable process parameters were individually changed: primary atomization gas pressure PG (from 0.4 MPa to 0.6 MPa), secondary atomization gas pressure SG (from 0.2 MPa to 0.4 MPa), arc voltage U (from 30 V to 34 V), and arc current I (from 200A to 260A). The higher ionization energy of Argon, as atomization gas, does not allow the lower adjusting values for voltage and current.

2.2 Experimental setups

The spraying of pre-alloyed NiTi-wires is very sensitive to the adjustable voltage and current, since the two setting variables determine the thermal energy in the arc, and to the presence of interstitial elements (O, N, and C) in the surrounding atmosphere. Therefore, the development and modification of an effective shroud concept were implemented in two stages. The first attempt was aimed to investigate the effect of the different shroud concepts on the obtained oxidation. For that reason, simultaneous spraying of pure Ni as anode and pure Ti as cathode was performed. In this attempt, only oxidation test was investigated. The oxide content of the obtained coatings was qualitatively determined by XRD-measurements.

Fig. 2 shows the different shroud concepts developed for this study. Conventional secondary gas cap, also referred to as conventional gas shrouds (C-GS), are used as a reference for the oxidation measurements. The C-GS attachment has 10 secondary gas (SG) outlet holes with discharge angle Θ = 30°. Furthermore, a modified secondary gas air cap attachment (Mo-GS) was used. It is provided by two sets of SG-outlet holes, which are arranged in two concentric rings with outlet holes are provided for this (see

Fig. 2a). The outer one is occupied by 6 holes and the inner one with 4 holes. In addition, the outer ring is slightly angled to the spray jet axis so that the emerging gas flow is expected to provide improved shielding.

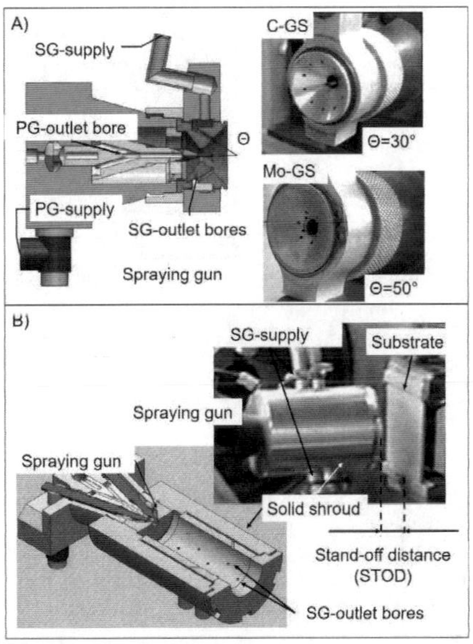

Fig. 2. Shroud concepts for twin wire arc spraying (TWAS), a) gas shroud, b) massive shroud

The other shroud concept is based on the solid shield principle (MS). In this concept the length of the air cap attachment is significantly increased so that the entire spray jet is covered and ends directly at substrate surface. The distance between MS-end and the substrate surface, i.e. standoff distance (STOD), was kept at 15 mm during the first attempt and varied (15, 30, 60 mm) by spraying pre-alloyed NiTi. This way, the length of MS-air cap attachment, in addition to Argon as atomization and protecting gas, provides additional shielding.

Fig. 2b shows the modified air cap MS. It consists of a massive outer jacket and inner part with spirally arranged bores. A gap, into which the secondary gas is fed, is located between the inner and outer parts. The fact that the attachment ends directly at the substrate is also intended to rule out oxide formation during particle flattening and layer formation.

The selection of the proper shroud concept is based on the comparison of the obtained oxide content and the proper shroud concept is the one that produced a coating with lower oxide content.

Therefore, the settings of voltage and current were kept constant at 30 V and 200 A respectively for all standoff distances studied. The standoff distance between the shroud outlet and substrate surface was modified between 15 mm, 30 mm and 60 mm.

The sprayed NiTi layers were tested for their pseudo-elasticity and cavitation erosion resistance. For the detection of the pseudo-elasticity, the phase transformation temperatures were determined by means of differential calorimetry analyzes (DSC) with a device of the type DSC 204 F1 Phoenix from Netzsch in a temperature range from -150 to 150 ° C. When the temperature range passed through at a rate of 10 K / min., a holding time of 3 minutes at the minimum and maximum temperature occurs. The cavitation erosion resistance was tested by means of cavitation tests according to ASTM G32-10 in distilled water at room temperature, which were evaluated using mass loss curves.

The wear resistance of the applied coatings was carried out by means of Pin-on-Disc tribometer (type: CH-2000, CSEM, Switzerland) with a contact force of 10 N, a relative speed of 400 mm/s, and 30,000 rotations. Ceramic ($Al2O3$) Ø 6 mm balls was used to perform the wear and friction tests at tracks with diameters of 18, 24, or 30 mm. The volume of the wear track was determined using a 3D-Profilometer (Type: Infinite- Focus, Alicona Imaging GmbH, Austria). The obtained wear and friction coefficients of the different STODs were compared to each other.

Additionally, SEM and XRD observations of the coating obtained and the wires as delivered were conducted to determine the phases in the feedstock and the obtained coatings.

3 Results

3.1 Oxidation

The results show that the shielding concepts have a direct influence on the oxide content in the Ni-Ti layers. The measurements were carried out using a scanning electron microscope (SEM). The oxide content was measured in several places (in each case 5 sites per sample) by means of energy-dispersive X-ray spectroscopy (EDX), and the average was selected as representative for the measurements. The determined oxide values serve as a qualitative comparison between the different designs of the examined air caps. In the Ni-Ti layers obtained, the oxide content varies depending on the shielding concept used and the nature of the atomizing gas. It is shown that the increased number of outlet holes in the secondary gas cap leads to better shielding of the beam (Mo-GS).

Tab. 1 shows the oxide contents in the layer for the different shielding principles. The use of the GS concepts as well as the MS concept shows positive effects on the reduction of the oxide content in the obtained coatings. Using compressed air as the atomizing gas, the Mo-GS concept provides the lowest oxide contents (5.1 wt.%) in the obtained coating. As expected, the use of argon as an atomizing and shielding gas leads to a further increase in the efficiency of the shroud concept and reduces the oxide formation by all shroud concepts (**Tab. 1** Ar).

Tab. 1. Oxide content of the obtained coatings by the different shroud concepts (A/compressed Air, Ar/Argon)

Concept	MS		C-GS		Mo-GS	
(wt. %)	A	Ar	A	Ar	A	Ar
O	6.3	1.7	11.3	6.9	5.1	3.9
Ni	66.1	52.1	59.4	45.7	63.4	63.4
Ti	27.6	46.2	29.3	47.4	31.5	32.7

Furthermore, the Ti content in **Tab. 1** tends to vary significantly with increasing oxide content. The reason for this is that titanium is much more reactive than nickel at elevated temperatures. Elements with a small atomic radius (for example, oxygen, nitrogen and carbon) can very quickly form interstitial compounds with Ti. Therefore, the oxide formation, measured as a percentage (EDX), reduces the proportion of the titanium.

The lowest oxide contents (1.7 wt.%) were measured for Ni-Ti layers, which were generated with a MS-shield (

Fig. 2. b). In this way, the air oxygen is effectively shielded out of the spray plume and therefore MS is selected for the spraying of pre-alloyed NiTi-SMA wires.

3.2 Spraying of pre-alloyed NiTi

The MS-shroud concept with the lowest oxide content was chosen to perform the spraying of NiTi pre-alloyed wires. The higher ionization energy of the argon limits the variation in voltage and current settings. The attempt was to set the voltage at its lowest possible setting, which allows for a stable arc ignition. This value is also dependent on the adjusted PG-pressure and the current. By the implementation of the experiments a conclusion was made that at higher PG-pressure a higher voltage is required. Therefore, the spraying parameters were fixed at (U = 30 V, I = 220 A, PG = 0.4 MPa and SG less than 0.1 MPa) to study the effect of the STODs by MS-shroud on the oxidation behavior. The SG was set at the lowest value because it was used only for the cooling the inner part of the MS-shroud.

Before layers of pre-alloyed, pseudo-elastic NiTi-wires were sprayed, the wires were first examined by means of DSC analysis. The DSC curve of the wire used in

Fig. 3 shows a complete austenitic-martensitic phase transformation. The phase conversion into the austenite is completed as desired at room temperature. A two-step phase transformation caused by the R-Phase (further martensitic transformation), which is often found in fine-grained structures (e.g. wires), can be observed [12, 16-19].

The DSC curve of the layer sprayed from this wire (

Fig. 3) does not show the desired complete austenitic-martensitic transformation. Here, a two-step transformation into the austenite takes place. The first part of the phase transformation occurs at room temperature as desired, while the second part is shifted towards higher temperatures. This indicates a partial non-uniform distribution of nickel and titanium in the layer since a smaller proportion of nickel leads to phase transformation at higher temperatures [15].

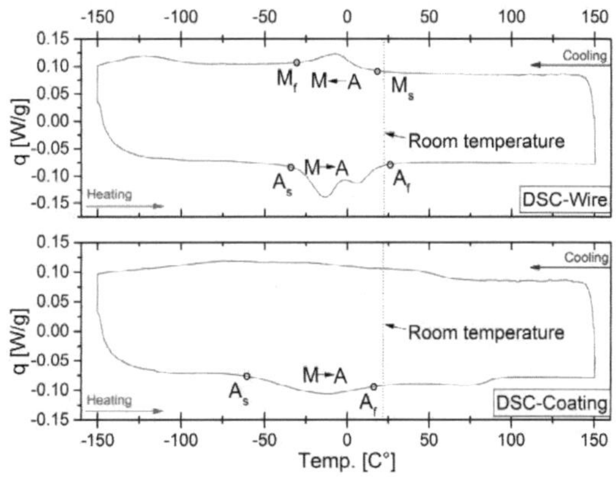

Fig. 3. Differential scanning calorimetry (DSC) graphs showing pseudo-elasticity of pre-alloyed NiTi-wires and deposited coatings

3.3 Phase composition of the obtained coatings

The XRD at the beamline BL9 of the synchrotron light source DELTA (TU Dortmund University, Germany) was used to define the phase compositions of the obtained coatings, which is produced by pre-alloyed NiTi-wires and at different STODs. The angle of incidence was set to 0.2 degrees and an image plate detector (Pilatus 100K) was utilized for photon detection. The obtained 2Θ patterns were converted afterwards for the phase analysis to a wavelength of λ = 1.5406 Å (CuKα-line).

The XRD patterns in

Fig. 4 shows sharps peaks with high intensities, which indicates a well crystallized NiTi-coating. The pattern for the obtained coatings (STODs-15, -30 and -60) is similar. At the STOD-60 a slight shift in the peak position s revealed in

Fig. 4. A strong and well-defined peak can be see for all coatings at Θ between 42.5° and 43°, which indicates the presence of NiTi-B2 (austenite), a rich Ni phase (Ni3Ti) and Ni3Ti3O. The oxidation consumes the Ti (Ti impoverishment) by the coating obtained and leads to enriching of the Ni in the remained compound and thus the initiation of Ni3Ti phases. Diffraction patterns of the NiTi-austenite B2 as well as NiTi-martensite phases were also observed at Θ 62° and at 78°. The peaks are not sharp, which indicates the performance of amorphous phases. The presence of oxide (TiO, and Ni3TiO) are always associated with the coexistence of Ni enriched phases.

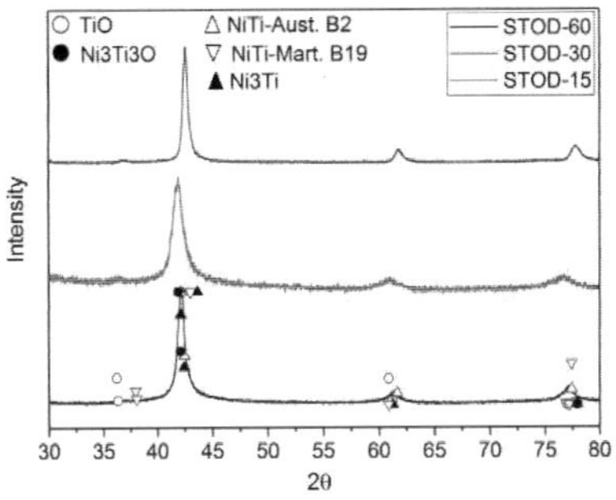

Fig. 4. XRD patterns of the obtained coatings

The SEM-images indicate a homogenous distribution of Ni and Ti in particular at STOD-15, which may indicate dispersal NiTi in the obtained coating. It becomes more inhomogeneous as the standoff distance increases.

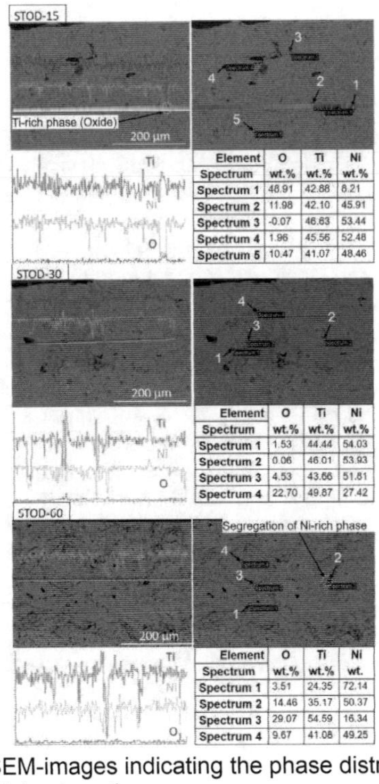

Element / Spectrum	O wt.%	Ti wt.%	Ni wt.%
Spectrum 1	48.91	42.88	8.21
Spectrum 2	11.98	42.10	45.91
Spectrum 3	-0.07	46.63	53.44
Spectrum 4	1.96	45.56	52.48
Spectrum 5	10.47	41.07	48.46

Element / Spectrum	O wt.%	Ti wt.%	Ni wt.%
Spectrum 1	1.53	44.44	54.03
Spectrum 2	0.06	46.01	53.93
Spectrum 3	4.53	43.66	51.81
Spectrum 4	22.70	49.87	27.42

Element / Spectrum	O wt.%	Ti wt.%	Ni wt.
Spectrum 1	3.51	24.35	72.14
Spectrum 2	14.46	35.17	50.37
Spectrum 3	29.07	54.59	16.34
Spectrum 4	9.67	41.08	49.25

Fig. 5. SEM-images indicating the phase distribution in the obtained coating

The images reveal also phases with higher Ni content as well oxide phases with higher Ti content. Formation of Ti-oxides leads in general to the initiation of Ni-rich phases, which may lead to a shifting in phase transformation temperature or even losing the transformation behavior and thus the pseudo-elastic character of NiTi-coating.

3.4 Cavitation resistance

To investigate the influence of the spray distance, samples were coated, polished and then subjected to a cavitation test at three different distances (15 mm, 30 mm, 60 mm). After 90 minutes of cavitation (**Fig. 6**), it is found that a spray distance of 15 mm has the least mass loss. This could be related to the homogenous distribution of the Ni and Ti in the obtained coating, which shows some indications of pseudo-elasticity as can be seen in DSC-graph in **Fig. 3**. The larger the distance with which the layer was applied, the greater the mass loss after 90 minutes.

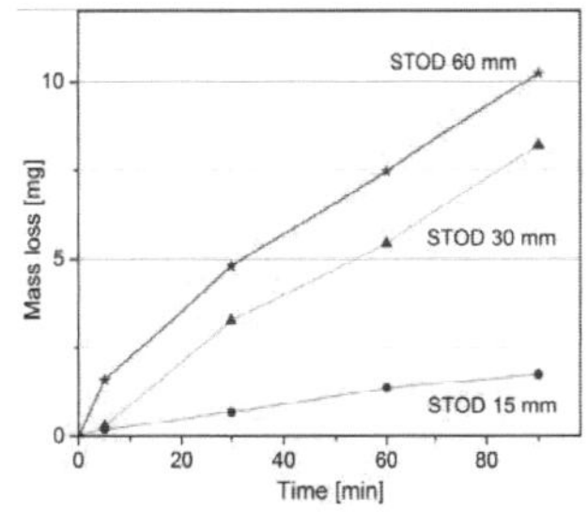

Fig. 6. Cavitation curves of the obtained coatings

3.4 Wear resistance and friction coefficient

The wear and friction behavior of the obtained coatings were determined by Pin-on-disc tests. The surfaces of the obtained coatings were ground and polished to eliminate the effect of the different roughness.

Fig. 7 shows that the coating produced at standoff distance 15 mm reveals the highest wear resistance and lowest friction.

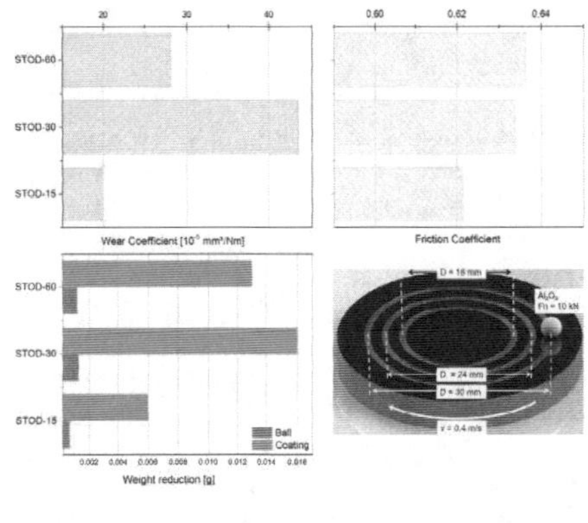

Fig. 7. Wear and friction coefficient of the obtained coatings

A clear tendency is obvious in case of friction, which is higher when increasing the standoff distance. The lower friction coefficient in case of STOD-15 confirms the lower weight reduction of the test specimen and the wear counterpart. In case of wear coefficient there is no clear tendency. At STOD-30 the highest wear losses were recorded. One possible reason is the remained coating thickness after grounding and polishing, with 298 µm was the lowest compared to 423 µm for STOD-15 and 312 for STOD-60.
The wear track of STOD-15 is shallow and shows a homogenous distribution of the Ni and Ti at the coating surface, as can be seen in
Fig. 8. Accumulation of Al was also recorded in the investigated track. The wear tracks at STOD-30 and -60 are wider than in case of STOD-15.

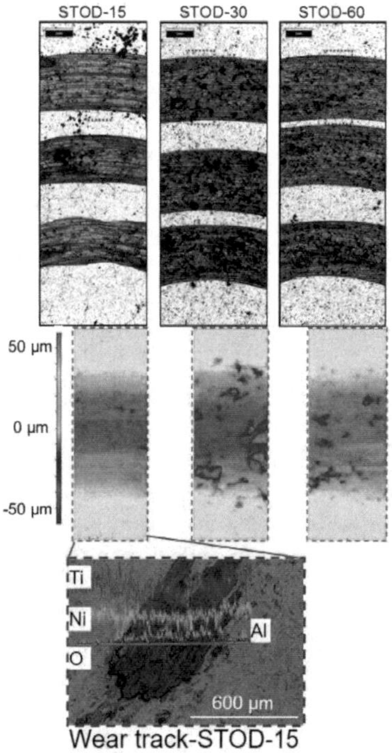

Fig. 8. Wear tracks of the coatings obtained at 18, 24 and 30 mm

4 Conclusion

It has been shown that the atomization gas used as well as the selected shroud concept has a significant influence on the phase changes in the sprayed Ti-rich wire feedstock materials. The oxidation of sprayed particles, in particular NiTi coatings, can be specifically reduced. A reduction of 11.3 wt.% to 1.7 wt.% Oxygen in the NiTi layer structure can be achieved with a massive shroud (MS) in combination with argon as the atomizing gas. In this way, semi equiatomic NiTi layers can be produced without the use of the expensive vacuum modification, which may also limit the component that can be sprayed. The investigations of the pseudo-elastic behavior as well as the cavitation resistance show very promising results with regard to using MS in spraying pre-alloyed NiTi-SMA feedstock. The distance between the MS-outlet and the substrate surface plays a crucial role on the preserved pseudo-elasticity of the sprayed coatings of pre-alloyed NiTi-wires. Based on the findings, the MS-concept will be a subject of further optimization, in an ongoing research DVS-financed study, to further reduce the oxidation and preserve the shape memory and cavitation protection properties of the NiTi-SMA wire feedstock.

Acknowledgment

The results were obtained within the scope of the IGF funded research project No. 18.788 with the topic "Kavitationsschutzschichten aus pseudoelastischen Nickel-Titan-Legierungen hergestellt durch modifizierte Lichtbogenspritzprozesse", Engl. “cavitation protection layers made of pseudo-elastic nickel-titanium alloys sprayed by modified twin wire arc spraying process”. We would like to thank the Research Association welding and related procedures of the DVS and the AiF e.V. for the financial support.

5 Literature

[1] Halter, K., Sickinger, A., Zysset, L. and S. Siegmann: Low pressure wire arc and vacuum plasma spraying of NiTi shape memory alloys. ITSC 2003, Olendo USA, (2003), pp. 589/95.
[2] Guilemany, J.M., Cinca, N., Dosta, S.and A.V. Benedetti: Corrosion behaviour of thermal sprayed nitinol coatings, Corrosion Science 51 (2009), pp. 171/80.
[3] Fernandez, J., Isalgue, A., Cinca, N., Dosta, S. and J. Ventayol: Wear and corrosion of metal-matrix (stainless steel or NiTi)-TiC coatings, Physics Procedia 10 (2010), pp. 77/80.
[4] Zhu, L., Trepanier, C., Fino, J.M. and A.R. Pelton: Oxidation of NiTiNol and its Effect on Corrosion Resistance, Medical device materials: proceedings, OH USA (2004), pp. 156/61.
[5] Li, D.Y.: Development of novel wear-resistant materials TiNi-based pseudo-elastic tribomaterials, Materials and Design 21 (2000), pp. 551/55.
[6] Lin, H.C., Liao, H.M., Hee, J.L., Chen, K.C. and K.M. Lin: Wear characteristics of TiNi shape memory alloys, Metallurgical and Materials Transactions A 28 (1997) 1871/77.
[7] Zhang, T. and D.Y. Li: An experimental study on the erosion behavior of pseudo-elastic TiNi alloy in dry sand and in aggressive media, Materials Science and Engineering A 293 (2000), pp. 208/14.
[8] Liu, W., Zheng, Y.G., Liu, C.S. and Z.M. Yao: Cavitation erosion characteristics of a NiTi alloy, Metallurgical and Materials Transactions A 35 (2004), pp. 356/62.
[9] Wu, S.K., Lin, H.C. and C.H. Yeh: A comparison of the cavitation erosion resistance of TiNi alloys, Wear 244 (2000), pp. 85/93.
[10] Steffens, H.O., and E. Ertürk: Low pressure arc spraying of reactive materials. Thin Solid Films, 12 (1984), pp. 143/50.

[11] Wang, X., Heberlein, J., Pfender, E. and W. Gerberich: Effect of nozzle configuration, gas pressure, and gas type on coating properties in wire arc spray. Journal of Thermal Spray Technology 8 (1999), Issue 4; pp. 565/75.
[12] Bitzer, M., Rauhut, N., Mauer, G., Bram, M., Vaßen, R., Buchkremer, H.-P., Stöver, D. and M. Pohl: Cavitation-resistant NiTi coatings produced by low-pressure plasma spraying (LPPS), Wear 328-329 (2015), pp. 369/77.
[13] Belakacem, M., Hamma, Z., Hamid, D. and B. Mourad: Three-dimensional modeling of pseudo-elastic effect of the shape memory alloys, Journal of the Serbian society for computational mechanics 9 (2015), issue 1, pp 1/14.
[14] Cinca, N., Isalgué, A., Fernández, J., and J.M. Guilemany: Structure characterization and wear performance of NiTi thermal sprayed coatings. Smart Materials and Structures 19 (2010), issue 8, pp. 085011.
[15] Frenzel, J., George, E.P., Dlouhy, A., Somsen, C., Wagner, M.F.X. and G. Eggeler: Influence of Ni on martensitic phase transformations in NiTi shape memory alloys, Acta Mater.58 (2010), pp. 3444/58.
[16] Khalil-Allafi, J., Dlouhy, A. and G. Eggeler, Ni4Ti3-precipitation during aging of NiTi shape memory alloys and its influence on martensitic phase transformations, Acta Mater. 50 (2002), 4255/74.
[17] Tillmann, W. and S. Momeni: Tribological performance of near equiatomic and Ti-rich NiTi shape memory alloy thin films, Acta Materialia 92 (2015), pp. 189/96.
[18] Tillmann, W., and S. Momeni: Investigation of the self-healing sliding wear characteristics of NiTi-based PVD coatings on tool steel, Wear 368-369 (2016), pp. 53/59.
[19] Tillmann, W., Momeni, S., and M. Pohl: Composite cavitation resistant PVD coatings based on NiTi thin films, Materials and Design 110 (2016), pp. 830/38
[20] Sickinger, A. and J.P. Teter: Superelastic shape memory alloy for seal applications produced by low pressure thermal spray technology, Materials Research Society Symposium – Proceedings 604 (2000), pp. 135/40.
21] D.W Parker: HVOF low oxide coatings - An alternative to LPPS, Annual Aerospace/Airline Plating and Metal Finishing Forum and Exposition, SAE Technical Papers, (1993), USA, Code 90640
[22] Planche M.P., Liao H. and C. Coddet: Oxidation control in atmospheric plasma spraying coating. Surface & Coatings Technology 202 (2007), pp. 69/76.
[23] Ward, D., Gupta, A., Saraf, S., Zhang, C., Sakthivel, T.S., Barkam, S., Agarwal, A., and S. Seal: Functional NiAl-graphene oxide composite as a model coating for aerospace component repair Carbon 105 (2016), pp. 529/43.
[24] Zhou, H., Gabbitas, B., Mathews, S. and D. Zhang, Titanium and titanium alloy coatings for corrosion protection, Proceedings of the 12th World Conference on Titanium, 3 (2012), pp. 1906/10.
[25] Liu, T., Zheng, L. and H. Zhang: Effect of Solid Shield on Coating Properties in Atmospheric Plasma Spray Process Journal of Thermal Spray Technology, 25 (2016), issue (8), pp. 1502/15.
[26] Rinaldi, C., Ferravante, L., Uberti Paci, F. and P. Bianchi: Extensive Characterization of Dense and Corrosion Resistant Coatings Produced by Improved Shrouded Techniques, Proceedings of the International Thermal Spray Conference (2001), pp. 1185/93.
[27] Wielage, B., Paczkowski, G. and C. Rupprecht: Gas-dynamic improvement of HVOF systems - Development aspects and applications, Proceedings of the International Thermal Spray Conference (2009), pp. 709/13.
[28] Dolatabadi, A., Pershin, V. and J. Mostaghimi: New attachment for controlling gas flow in the HVOF process Journal of Thermal Spray Technology 14 (2005), issue (1), pp. 91/99.
[29] Kawakita, J., Kuroda, S., Fukushima, T. and T. Kodama: Development of dense corrosion resistant coatings by an improved HVOF spraying process Science and Technology of Advanced Materials 4 (2003), issue (4), pp. 281/89.
[30] Jamais, S., Creaven, M. and C. Rinaldi: Calculated and measured oxygen contents in the gas flow of a shrouded thermal spray process Proceedings of the International Thermal Spray Conference (2001), pp. 903/10.
[31] D.W. Parker: HVOF low oxide coatings - An alternative to LPPS, Annual Aerospace/Airline Plating and Metal Finishing Forum and Exposition, SAE Technical Papers, (1993), USA, Code 90640

Microstructural characteristics and performances of Cr_2O_3 and Cr_2O_3-15%TiO_2 S-HVOF-coatings obtained from water-based suspensions

F.-L. Toma[1], N. Kulissa[1], S. Langner[1], A. Potthoff[2], M. Barbosa[1,3], C. Leyens[1,3]
[1)] Fraunhofer Institute IWS, Dresden / D
[2)] Fraunhofer Institute IKTS, Dresden / D
[3)] Technische Universität Dresden / D

Cr_2O_3 coatings and Cr_2O_3-based binary coatings offer high hardness, excellent sliding wear performance and corrosion resistance. Therefore, they are widely applied in the paper and textile industry, as well as for pumps and mechanical sealing systems. Compared to the conventional spray processes, the technology of Suspension-HVOF (high velocity oxy-fuel flame spraying of suspensions, S-HVOF) using submicron-scaled raw materials allows the production of dense, finely-structured coatings with smooth surfaces and improved mechanical properties.
This work investigates the microstructure and performance of Cr_2O_3 and Cr_2O_3-15wt.%TiO_2 coatings obtained by S-HVOF with water-based suspensions. For the development of the binary composition two routes were used to produce ready-to-spray suspensions: i) dispersion of an appropriate alloyed material in the solvent and ii) mixture of two stable suspensions in the desired ratio. In order to evaluate the potential of suspension spraying over the conventional APS and HVOF processes, the mechanical properties and the corrosion resistance of the S-HVOF Cr_2O_3 and $85Cr_2O_3/15TiO_2$ coatings were compared with the coatings produced from spray powders. From the experimental results it was observed that the suspension sprayed coatings showed in most of cases denser microstructure, good mechanical properties and superior corrosion performances.

1 Introduction

Chromium(III)-oxide Cr_2O_3 is one of the most important oxide ceramics for producing thermally sprayed coatings. Thanks to their high hardness, excellent sliding wear performance, corrosion resistance, as well as good engraving laser properties, they are widely applied in the paper and textile industry, as well as for pumps and mechanical sealing systems [1-5]. The state of the art is given by plasma-sprayed Cr_2O_3 coatings, which are mainly produced by spraying fused and crushed Cr_2O_3 powders. In order to avoid or to reduce common problems like low deposition rates, high coating brittleness or the presence of metallic chromium and possible formation of Cr(VI) during the thermal spraying process, mixed crystals containing Al_2O_3 or TiO_2 are used. As a result, the occurrence of Cr(VI) decreases, but hardness and wear resistance of coatings are also reduced [6].

In comparison to conventional spraying processes which use powders as feedstock, suspension spraying provides significant benefits. It is based on the direct use of commercially available fine-dispersed powders and their dispersion in an aqueous and/or organic solvent. Coating thicknesses in a wide range from ca. 10 µm up to several hundred µm, as well as reduced coating roughness can thereby be achieved. At the same time the hardness, corrosion and wear resistance of these coatings are comparable to, or even exceed, those of conventionally sprayed coatings [7-10].

Only few papers concerning suspension-sprayed Cr_2O_3 coatings have been published [10-13]. Alcohol or alcohol-water based suspensions with a solid content of 10 to 20 % were used to produce Cr_2O_3 coatings by suspension high velocity oxy-fuel spraying (S-HVOF or HVSFS) [10-12]. Powders with particle sizes below 1 µm were used as feedstock, and a deposition efficiency (DE) between 12 to 18 % was achieved. By applying an $85Cr_2O_3/15TiO_2$ suspension, the DE could be increased up to 20 % [12].

We have previously presented the first results on the development of Cr_2O_3 S-HVOF coatings starting from aqueous suspensions [13]. In this work we have extended these studies by developing two binary 25 wt.% solid $Cr_2O_3/15TiO_2$ suspensions using two different production routes. These were processed via S-HVOF and the resulting coating properties (microstructure, phase composition, microhardness, E-Modulus, corrosion resistance and wear resistance performances) were evaluated and compared to that of conventional powder coatings.

2 Materials and experimental procedure

2.1 Suspension development

The development of a suspension as feedstock for thermal spraying requires the use of appropriate raw materials. In order to ensure a stable spraying process, colloid-chemical stable suspensions with low viscosity, high stability and pH > 4 to prevent corrosion of hardware are required [8, 9]. For the development of binary suspensions knowledge about particle-particle interaction as well as particle-liquid interaction are needed [14-16].

Commercially available raw materials were used to produce aqueous suspensions for spraying [17]: Cr_2O_3 99 (Ceram GmbH, Germany) with average particle size $x_{50;3}$ of 0.8 µm, TiO_2 (Ceram GmbH) with $x_{50;3}$ of 1.7 µm and mixed-crystal 85%Cr_2O_3/15%TiO_2 powder (Ceram GmbH) with $x_{50;3}$ of 4,5 µm.

	suspension type A		suspension type B
Raw materials	Cr_2O_3 99	TiO_2 99	Cr_2O_3/TiO_2 85/15

Suspension	25 wt.% in water + additive	25 wt.% in water + additive	25 wt.% in water + additive
	Mixing of stable single suspension at a ratio 85/15		

Fig. 1. Development of binary suspensions with identical composition using two routes ($85Cr_2O_3/15TiO_2$): suspension type A und suspension type B.

For the development of binary suspensions with 25 wt.% total solid contents, two procedures were used (**Fig. 1**) [17]:
A) Selection of single oxide powders – dispersion of the single raw materials and preparation of stable corresponding suspensions – mixture of single-oxide suspensions → suspension type A;
B) Selection of a mixed-crystal raw material – dispersion → suspension type B.

The viscosity of the suspensions was determined by using a rheometer MCR 301 (Anton Paar, A). The Zeta potential was measured with the Zeta Probe (Colloidal Dynamics, USA).

For comparative purposes fused & crushed spray powders Cr_2O_3 (+10-45 µm) and $Cr_2O_3/15TiO_2$ (+5-25 µm and +10-45 µm), both from Ceram GmbH were used to produce conventional Cr_2O_3-based coatings.

2.2 Coating development

For the feeding of the suspensions, specific hardware components are needed. At the Fraunhofer IWS, an industrial suitable three pressurized-vessels suspension feeder, which can also be used for the simultaneous feeding of two different suspensions was develop and used in this work. The feeder enables the continuous monitoring of the suspension feed rate and pressure during the spray process. The S-HVOF spraying trials were carried out with a Top gun (GTV Verschleißschutz GmbH, Germany) using ethylene as fuel. Injectors with diameters of 0.25 mm were developed for the axial injection of the suspension as a compact jet directly into the combustion chamber. The long-term stability of the S-HVOF process was considered during the coatings development.

For powder-based coatings, atmospheric plasma spraying (APS) using a F6 (GTV) plasma gun and Ar/H_2 plasma gas mixture was used to obtain Cr_2O_3 and Cr_2O_3/TiO_2 coatings; HVOF process with Top Gun was used to spray binary Cr_2O_3/TiO_2 coatings. Steel substrates, previously cleaned and grit-blasted with corundum, were used as substrates. During the spraying, the substrates were cooled using air jets systems.

2.3 Coating characterization

Coating microstructure, porosity in cross-section, Vickers microhardness, E-Modulus and phase composition were investigated. The E-Modulus was determined using the non-destructive laser acoustic wave method LAwave® developed at Fraunhofer IWS [18].

The corrosion tests of the sprayed coatings were carried out in a 1N H_2SO_4 solution at 85 °C in a thermostat up to 300 h exposure time, using the procedure described elsewhere [19]. Coatings with thickness of about 150 ± 20 µm were sprayed onto high corrosion resistance 1.4462 duplex steel (X2CrNiMoN22-5-3) coupons with a diameter of 25 mm, without and with a 50 µm thick APS-CoNiCrAlY bondcoat. The corrosion resistance was evaluated by determining the mass loss (mg/cm^2) of the coatings during exposure.

The sliding wear resistance of the sprayed coatings was investigated with an oscillating test ring (SR4 wear test machine, Optimol Instruments, Germany) using a ball-on-flat geometry at room temperature and 40 to 60 % relative air humidity. The test conditions were: oscilatting frequency of 10 Hz with a stroke length of 5 mm, normal force of 25 N, an average sliding speed of 0.1 m/s and a test duration of 1 h (sliding distance: 360 mm). S-HVOF and conventional coatings were coated on disc samples (1.4462, 24 mm-Ø). The tests were applied on polished (Ra < 0.05 µm, Rz < 0.5 µm) samples. As counterparts 100Cr6 steel and sintered Al_2O_3 balls (100 mm-Ø) were used. The sliding tests were performed unlubricated and lubricated using an synthetical oil (PAO). The wear resistance was evaluated by the estimation of the coating volume wear rate (mm^3/Nm) and coefficient of friction, as well as by the observation of the wear tracks using SEM microscopy.

3. Results and Discussion

3.1. Suspension characterization

In a suspension with two solid systems, it has to be taken into account that every solid interacts differently with the aqueous ambience. Moreover during the development of a binary suspension it should be ensured that the different type of particles stay in a dispersed state, which means that agglomeration does not occure. The particle agglomeration is avoided when the repulsion forces between the particles ist stronger than the attraction. The repulsion is achieved by electrostatic interaction, which is characterized by high zeta potential values. The zeta potential strongly depends of the raw material, the pH value and the presence of additives in the liquid.

3.1.1. Suspension type A

When dispersing Cr_2O_3 in deionized water a pH value of 6 was measured and the Zeta potential is slightly positive (+7 mV) (**Fig. 2a**). For the dispersion of TiO_2 a pH value of 3 and a Zeta potential of -12 mV were measured. Taking into consideration the dependence of zeta potential with pH value [20], agglomeration processes due to interaction of particle surfaces were to be expected. In order to avoid this phenomena, the single suspensions were stabilized by an appropriate dosage of an additive (LOPON 885) which interacts both with the Cr_2O_3 and the TiO_2 particles' surfaces.

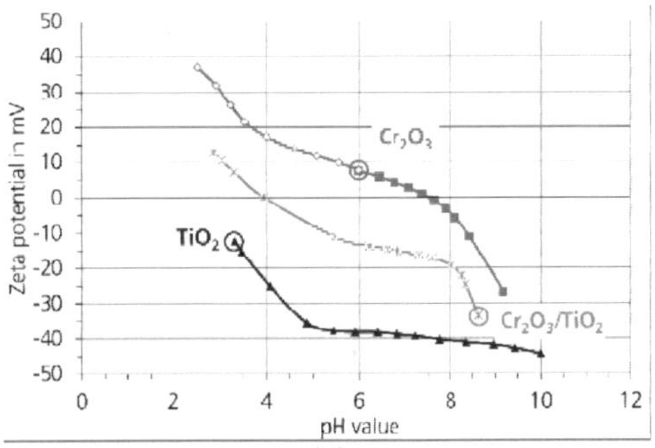

(a)

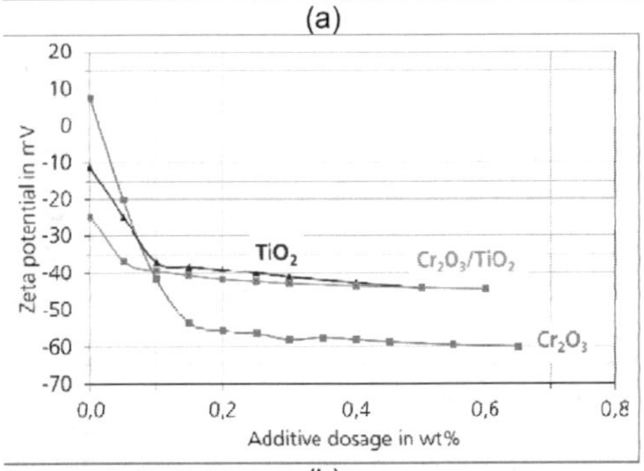

(b)

Fig. 2. Zeta-potential curves of single raw powders function of : (a) pH value; (b) additive content.

The single suspensions stabilized with LOPON showed high negative Zeta potential values (**Fig. 2b**); their mixing in the composition 85wt.%Cr_2O_3 – 15wt.%TiO_2 resulted in the suspension type A.

3.1.1. Suspension type B

The mixed-crystal $85Cr_2O_3/15TiO_2$ raw powder showed a completely different behaviour in comparison to the single components. In this case a pH value of 8.7 was measured (**Fig. 2a**). The Zeta potential value was negative and strong enough to avoid agglomeration; but for comparison purposes, the suspension type B the mixed-crystal was also stabilized with the additive LOPON (**Fig. 2b**).

Low viscosities of < 1.4 mPa·s at shear rates up to 1000 s^{-1} for both types of binary suspensions were measured which is favourable for good feeding behaviour and processability.

3.2. Coating characterization

The microstructures of S-HVOF Cr_2O_3 and $Cr_2O_3/15TiO_2$ coatings are shown in SEM micrographs of **Fig. 3**. The microstructures of conventional APS and HVOF Cr_2O_3-based coatings are shown in the micrographs of **Fig. 4**.

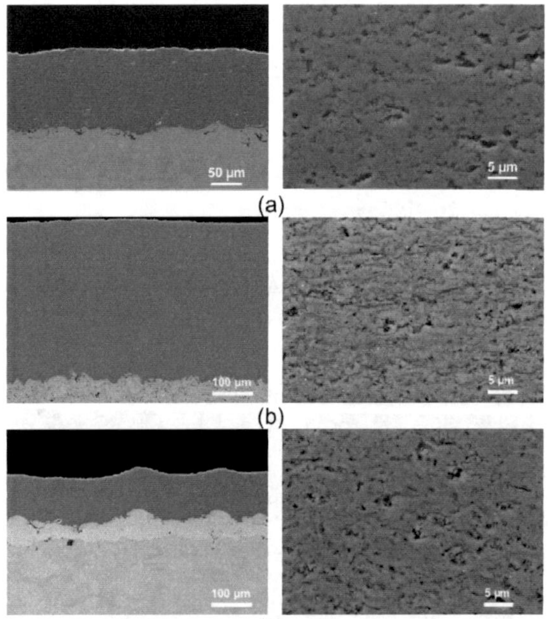

(a)
(b)
(c)

Fig. 3. SEM micrographs of S-HVOF coatings starting from different suspensions: (a) Cr_2O_3 suspension; (b) $Cr_2O_3/15TiO_2$ suspension type A; (c) $Cr_2O_3/15TiO_2$ suspension type B.

By the S-HVOF-spraying of aqueous suspensions with 25 wt.% solid content dense Cr_2O_3 coatings were obtained (**Fig. 3a**); the porosity estimated by image analysis using SEM-micrographs at a magnification of 200x was about 2.5 ± 0.5%.

(a)

(b) (c)

Fig. 4. Optical micrographs of: (a) APS Cr_2O_3 coating; (b) $Cr_2O_3/15TiO_2$ APS coating; (c) $Cr_2O_3/15TiO_2$ HVOF coating.

The typical crack-like structure of conventional APS Cr_2O_3 coatings (**Fig. 4a**) was not observed in the S-HVOF Cr_2O_3 coating (**Fig. 3a**). Vickers hardness of 1400 ± 150 HV0.3 was measured for S-HVOF coatings, while the hardness of APS Cr_2O_3 coatings was 1150 ± 50 HV0.3. The E-Modulus measured by LAwave® for S-HVOF coatings was about 151 ± 20 GPa; the values for APS coatings were around 75 ± 2 GPa. Deposition efficiency (after DIN EN ISO 17836) of 30 – 35 % was achieved for Cr_2O_3 coatings obtained with the S-HVOF process. The S-HVOF coatings are characterized by a smooth surface with an average roughness Ra of 1.5 ± 0.5 µm.

Higher deposition efficiency up to 45% could be obtained by spraying $85Cr_2O_3/15TiO_2$ binary suspensions and dense coatings were produced by S-HVOF (**Fig. 3b,c**). No significant difference was observed between the microstructures for coating manufactured using the suspensions type A and type B. Moreover, the bimodal microstructure as observed in the APS $Cr_2O_3/15TiO_2$ coatings (**Fig 4b**) could not be observed for the S-HVOF coatings. The porosity was estimated at 2.9 ± 0.5%. The mixed-crystal phase $(Cr_{0.88}Ti_{0.12})_2O_3$, eskolaite phase ($Cr_2O_3$) and traces of TiO_2-rutile were detected by XRD analysis in both S-HVOF coating systems. Hardness values of 1200 ± 100 HV0.3 were measured for S-HVOF binary coatings; these values are comparable to those measured for the HVOF $Cr_2O_3/15TiO_2$ coatings (1150 ± 50 HV0,3) and higher than those of APS $Cr_2O_3/15TiO_2$ coatings 890 ± 90 HV0,3). The E-Modulus of $Cr_2O_3/15TiO_2$ S-HVOF coatings sprayed from the different suspensions were comparable and values of 163 ± 30 GPa were measured; E-Modulus values of 56 GPa and 146 GPa were measured for binary APS and HVOF coatings, respectively.

No singnificant differences in the microstructure and hardness of the binary coatings produced using suspension type A and B were noticed. Thus, for the advanced comparison between $Cr_2O_3/15TiO_2$ S-HVOF coatings and conventional powder based coatings only coatings from suspension type A were tested.

3.3 Corrosion resistance: S-HVOF coatings vs. conventional coatings

Only a small mass loss (0.1 mg/cm²) was observed during the exposure of duplex substrate in acidic media (**Fig. 5a**); thereby the corrosion resistance of the coatings alone can be assessed.

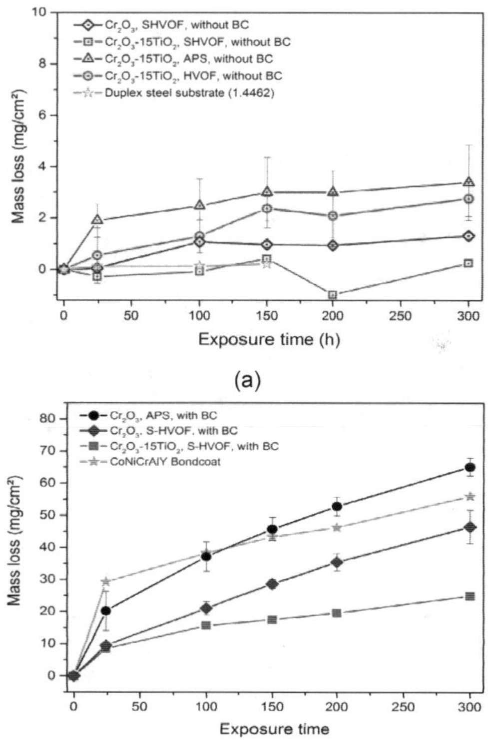

(a)

(b)

Fig. 5. Comparison of mass losses of spray coatings during exposure to 1N H_2SO_4 at 85 °C: (a) coatings without bondcoat; (b) coatings with bondcoat. Cr_2O_3-$15TiO_2$ S-HVOF coatings were sprayed with the binary suspension type A. The mass loss of the 1.4462 substrate is given for comparative purpose.

For the coated samples without bondcoat, small mass losses were measured after 300 hours of exposure, with suspension coatings showing a slightly higher corrosion resistance (**Fig. 5a**). Higher mass losses were obtained for the coatings sprayed with bondcoat, with the highest corrosion rate for APS coatings **(Fig. 5b)**.

CoNiCrAlY bondcoat alone exhibited also a very high mass loss. In the case of coatings sprayed onto a bondcoat, the mass loss measured is a total value comprising both the corrosion of coating and the bondcoat. This is an indicator, that the coatings are not completely dense and that the porosity is an open porosity.

S-HVOF coatings were characterized by higher corrosion resistance than the conventionally sprayed coatings. The addition of TiO_2 to Cr_2O_3 improved the corrosion resistance of the coatings.

Traces of Ni-, Co-elements coming from the corroded bondcoat could be identify by EDX-Analysis on top-surface of the coatings after exposure (**Fig 6**). Additionally, Ni-sulfate could be identified in the XRD-difractogramm of corroded coatings. The presence of bondcoat elements on the top-surface of corroded coatings can be correlated to the microstructure of the spray coatings.

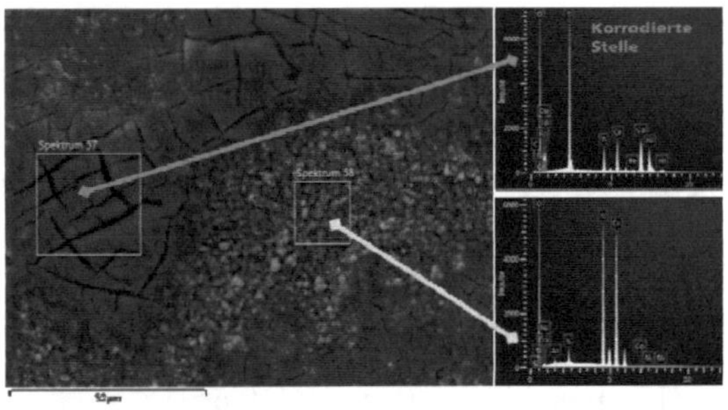

Fig. 6. EDX-analysis of a corroded $Cr_2O_3/15TiO_2$ after 300 h exposure. Coating with bondcoat.

APS coatings are characterized by a network of micro-fissures and open porosity through which the corrosion media can penetrate, reach and corrode the bondcoat. S-HVOF coatings have a denser structure but there are still some open porosity which enables the corrosion of the bondcoat.

3.4. Sliding wear resistance: S-HVOF coatings vs. conventional coatings

The results of oscilatting sliding wear tests (coating volume wear rate and coefficient of friction) are summarized in **Fig. 7**. In unlubricated conditions and in presence of steel counterpart, the coefficients of friction were about 0.60 for S-HVOF Cr_2O_3 and $Cr_2O_3/15TiO_2$ coatings and 0.71-0.72 for the conventional APS and HVOF coatings (**Fig. 7a**).

A barely quantifiable coating volume wear rate could be measured for the sliding tests carried out with the metallic counterpart without lubricant. By microscopic examination of the wear tracks, a plastic deformation and an abrasive wear of the steel ball were visible. In addition, material transfer from the steel ball to the coating surface was observed (**Fig. 8**).

In lubricated conditions and in the presence of an Al_2O_3 ball, the values for the coefficient of friction were of about 0.14 - 0.16, slightly lower for S-HVOF-Cr_2O_3 coatings. In presence of Al_2O_3 counterpart but unlubricated conditions, coefficients of friction of about 0.56 for APS Cr_2O_3 coatings and about 0.78 for S-HVOF Cr_2O_3 coatings were obtained. For the binary coating systems coefficient of friction of 0.85 for the S-HVOF $Cr_2O_3/15TiO_2$ coatings and up to 0.89 for the conventional APS and HVOF $Cr_2O_3/15TiO_2$ coatings were determined (**Fig. 7a**).

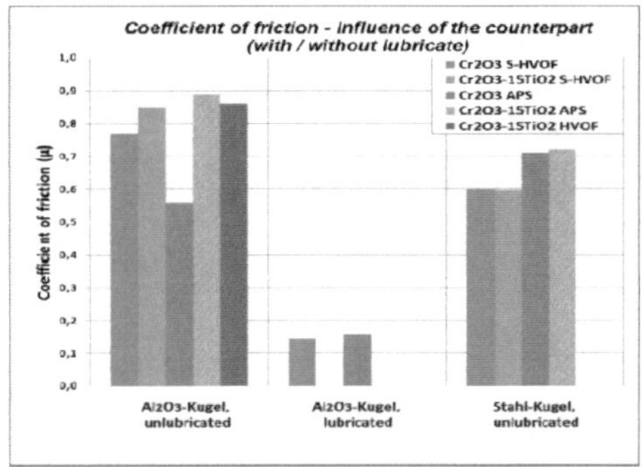

(a)

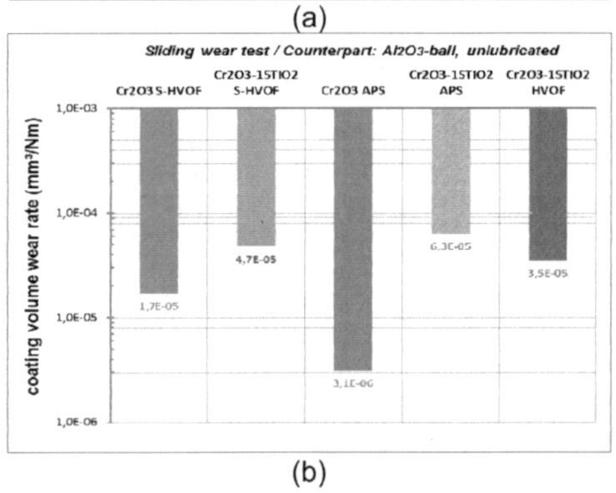

(b)

Fig. 7. Comparision of the sliding wear behaviour of spray coatings function of spray process, coating system and test conditions (counterpart ball, with and without lubricant): (a) coefficient of friction; (b) coating volume wear rate.

S-HVOF Cr_2O_3 coatings showed slightly lower volume wear rate compared to the APS Cr_2O_3 samples (**Fig. 7b**). For the binary systems Cr_2O_3-15TiO_2, the volume wear coefficients were comparable for both S-HVOF and HVOF coatings. An abrasive wear mechanism of the sprayed coatings was determined with the Al_2O_3 counterpart, with wear tracks on the S-HVOF coatings wider than on the APS coatings. Cr_2O_3 pull-outs and fractures, as well as Cr_2O_3 particles in the wear debris were identified. On the surface of an APS Cr_2O_3 coatings, traces of transferred material from Al_2O_3 ball to coating surface were observed.

The volume wear rates after tests with lubricant could not be quantified for both conventional (APS, HVOF) or S-HVOF coatings since the values were below the measuring range.

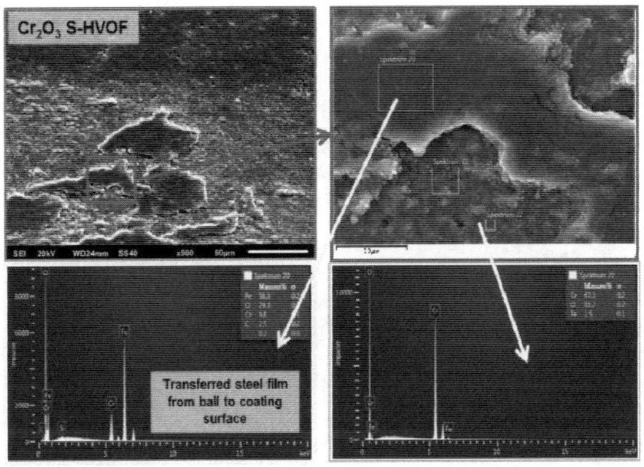

Fig. 8. SEM micrographs on top-surface and EDX-analysis of wear track on an S-HVOF Cr_2O_3 coating after sliding with steel counterpart. Transferred steel film from ball to coating surface was identified.

4 Conclusions

In this work, investigations on the microstructure and performance of Cr_2O_3 and Cr_2O_3-15wt.%TiO_2 coatings obtained by S-HVOF with water-based suspensions were presented. For the development of spray-ready binary suspension two routes were compared: a) dispersion of an appropriate alloyed material in the solvent and b) mixture of two stable suspensions in the desired ratio. Both routes resulted in dense S-HVOF coatings, but the route used to produce suspension type A has a stronger economical interest due to its flexibility in suspension composition and higher availability of single oxide powders.

In order to evaluate the potential of suspension spraying over the conventional APS and HVOF processes, mechanical properties, corrosion resistance in acidic media and sliding wear resistance of the S-HVOF Cr_2O_3 and $85Cr_2O_3/15TiO_2$ coatings were compared with the coatings produced from spray powders. From the experimental results, it was observed that the suspension sprayed coatings showed, in most of cases, denser microstructure, high hardness. The wear and corrosion resistance perfomances of S-HVOF coatings were comparable and even superior to those produced from feedstock spray powders.

Acknowledgements

The results here presented are part of the AiF-Project „Advanced Cr_2O_3-coatings by thermal spraying with suspensions“ (IGF-project Nr.: 18.154B / DVS-Nr.: 02.094). IGF-Project Nr.: 18.154B / DVS-Nr.: 02.094 of Forschungsvereinigung, Forschungsvereinigung Schweißen und verwandte Verfahren e. V. des DVS, Aachener Straße 172, 40223 Düsseldorf was carried out via AiF within the framework of the industrial joint research funding programme (IGF) of the Federal Ministry of Economics and Technology (BMWi). The authors gratefully acknowledge this financial support.

We thank our colleagues from Fraunhofer IWS: Irina Shakhverdova und Beate Wolf für die metallographic preparation of the cross-sections and microstructural analysis, Stefan Makowski and Frank Schaller for sliding wear measurements.

5 Literature

[1] Steffens, H. D., J. Wilden, and D. Haumann: Haftung und Verschleißverhalten plasmagespritzter Keramikschichten für den Einsatz in Schraubenverdichtern (in German). Thermische Spritzkonferenz 1996, DVS 175, pp. 262/265.

[2] Pawlowski, L.: Technology of Thermally Sprayed Anilox Rolls: State of the Art, Problems, and Perspectives. J. Thermal Spray Technol. 5 (1996), Issue 3, pp 317/334.

[3] Bolelli, G., V. Canillo, L. Lusvarghi and T. Manfredini: Wear Behaviour of Thermally Sprayed Ceramic Oxide Coatings. Wear 261 (2006), pp. 1298/1315.

[4] Berger, L.-M., C.C. Stahr, F.-L. Toma, G.C. Stehr und E. Beyer: Ausgewählte Entwicklungstendenzen bei der Herstellung thermisch gespritzter keramischer Schichten. Jahrbuch Oberflächentechnik 63 (2007), pp. 242/266 (in German).

[5] Döring, J-E. and L. De Schoulepnikoff: Preview of the Art of Anilox-Manufacturing – from the Coated Cylinder up to Gravure". Proc Int'l Thermal Spray Conference ITSC 2011, Hamburg, Germany, DVS 276 (2011), pp. 564/571 (CD-Rom Version).

[6] Börner, T., L.-M. Berger, S. Saaro, and S. Thiele: Systematische Untersuchung der Abrasionsbeständigkeit thermisch gespritzter Schichten aus dem Werkstoffsystem Al_2O_3-TiO_2-Cr_2O_3". 13. Werkstofftechnisches Kolloquium, Chemnitz, Band 37 (2010), pp. 178/187 (in German).

[7] Toma, F.-L., L.-M. Berger, C.C. Stahr, T. Naumann and S.Langner: Microstructures and Functional Properties of Suspension-Sprayed Al_2O_3 and TiO_2 Coatings: An Overview". J. Thermal Spray Technol. 19 (2010), Issue 1-2, pp. 262/274.

[8] Berger, L.-M., F.-L. Toma, A. Potthoff: Thermal spraying with suspensions – an economical spray process. Therm. Spray Bulletin 2 (2013), pp. 98/101.

[9] Toma, F.-L., A. Potthoff, L.-M. Berger, and C. Leyens: Demands, Potentials, and Economic Aspects

of Thermal Spraying with Suspensions: A Critical Review. J. Therm. Spray Technol. 24 (2015), Issue 7, pp 1143/1152.
[10] Killinger, A., R. Gadow, G. Mauer, A. Guignard, R. Vaßen, and D. Stöver: Review of New Developments in Suspension and Solution Precursor Thermal Spray Processes. J. Thermal Spray Technol. 20 (2011), Issue 4, pp. 677/695.
[11] Rauch, J., G. Bolelli, A. Killinger, R. Gadow, V. Cannillo and L. Lusvarghi: Advances in High Velocity Suspension Flame Spraying (HVSFS)", Surf. Coat. Technol. 203 (2009), Issue 15, pp. 2131/2138.
[12] Killinger, A., A. Rempp, A. Manzat, P. Müller and R. Gadow: High-velocity oxy-fuel spraying with suspensions consisting of nanoscale and submicronscale oxide powders. Therm. Spray Bulletin 8 (2015), Issue 1, pp. 62/70.
[13] Toma, F.-L., S. Scheitz, R. Trache, S. Langner, C. Leyens, A. Potthoff and K. Oerschlägel: Effect of Feedstock Characteristics and Operating Parameters on the Properties of Cr_2O_3 Coatings Prepared by Suspension-HVOF Spraying. Proc Int'l Thermal Spray Conference ITSC 2015 Long Beach, California USA, 2015 ASM International, pp. 329/334.
[14] Kratzsch, R., A. Potthoff, M. Barbosa, N. Kulissa, O. Kunze and F.-L. Toma: Development and application of binary suspensions in the ternary system Cr_2O_3-TiO_2-Al_2O_3 for S-HVOF spraying. Proc Int'l Thermal Spray Conference ITSC 2017 Düsseldorf, Germany, DVS (2017).
[15] Grasso, D, K. Subramaniam, M. Butkus, K. Strevett and J. Bergendahl: A review of non-DLVO interaction in environmental colloidal systems. Reviews in Envirom. Sci. Biotechnol. 1 (2002), Issue 1, pp. 17/38.
[16] Liang, Y., N. Hilal, P. Langston and V. Starov: Interaction forces between colloidal particles in liquid: Theory ad experiment. Adv. Colloid Interface Sci. 135-135 (2007), pp. 151/166.
[17] Toma, F.-L., S. Langner, N. Kulissa, R. Trache and A. Potthoff: Development of advanced Cr2O3 coatings by thermal spraying with suspensions. Thermal Spray Bulletin 2017, submitted.
[18] Berger, L.-M., D.Schneider, M. M. Barbosa, and R. Puschmann: Laser acoustic surface waves for the non-destructive characterisation of thermally sprayed coatings. Thermal Spray Bulletin 1 (2012), pp. 56/64.
[19] Hermann, M., F.-L. Toma, L.-M. Berger, G. Kaiser and C. C. Stahr: Comparative study of the corrosion resistance of thermally sprayed ceramic coatings and their bulk ceramic counterparts. J. Europ. Ceram Soc. 34 (2014), pp. 493/504.
[20] Večeř, M. and J. Pospíšil: Stability and rheology of aqueous suspensions. Procedia Engineering 42 (2012), pp. 1720/1725.

The effects of substrate preheating temperature and torch power on crystallization and microstructure of ZnO films via SPPS

Z. X. Yu, Y. L. Zhao, M. Moliere, H. L. Liao*, Université de Bourgogne Franche-Comté, CNRS, Lab ICB, UMR 6303, Site UTBM, Belfort/ FR
Z. Q. Feng, W. Z. Wang, Key Lab of Safety Science of Pressurized System, Ministry of Education, East China University of Science and Technology, Shanghai / P.R. China

ZnO films were deposited by solution precursor plasma spray (SPPS) process with different substrate preheating temperatures and torch powers, which were used to study the effects on crystallizations and microstructures. With increasing substrate preheating temperature from 0 °C to 400 °C, ZnO films were always preferential orientation along (002) plane with much higher crystallinity. And more apparent crystallized particles appeared with higher agglomeration degree forming cauliflower-like microstructure under higher preheating temperature. For adjusting hydrogen flow rate, the moderate hydrogen flow rate was the suitable condition for obtaining oriented growth along (002). Besides, all ZnO films under different hydrogen flow rates with a constant preheating temperature as 400 °C were always combined with crystallized particles. Moreover, the increment of torch power makes microstructure becomes denser with less interspace between neighbouring particles. Moreover, it is found that crystallinity and crystallized particles is more dependent on preheating temperature and torch power plays a more important role on densification by two staggered experiments. Taking applications of metal oxides films via SPPS into consideration, choosing moderate substrate preheating temperature and hydrogen flow rate will obtain crystallized particles, unusual preferentially oriented planes and high specific surface area, which is very favourable for optical, electrical, electrochemical properties.

1 Introduction

Due to the feature of industrial-friendly and feasibility, plasmas spray process has been employed in various applications, such as thermal barrier coatings (TBCs), wear resistance, solid oxide fuel cell (SOFC), corrosion resistance, etc. [1]. In a conventional plasma spray process, such as air plasma spray (APS), solid powders are injected into plasma plume as feedings, which forms micrometre-sized molten splats with diameter from a few tens to a few hundreds of micrometres for building coatings. With the development and understanding of TBCs, nanometre-sized splats from finer feedstock particles have been attractive with increasing works during the past 30 years. Solution precursor plasma spray, as a novel thermal spray method, makes it possible for obtain nano-structured coatings through injecting metal salts solutions into plasma plume [2-3]. Additionally, some work discussed the effect of preheating temperature on layer formation, which mainly focus on melting, flatten degree and fragmentation [4].

Based on our previous work, the crystallization process has been observed in ZnO films via SPPS process with crystallized particles instead of common molten splats. As we known, input heat is an important factor on crystallization behaviour. Particularly, in SPPS process, preheating temperature supplies energy for in-suit crystallization on substrate and very hot plasma plume supplies energy for crystallization during flight. Therefore, this work seeks to study effect of input heat on crystallization of ZnO films via SPPS with adjusting preheating temperature and torch power by hydrogen flow rate. The crystallinities, preferentially oriented crystal planes, and microstructures of ZnO films under different spraying conditions were characterized by scanning electron microscope (SEM) and X-ray diffraction (XRD).

2 Experiment and simulation procedures

2.1 Solution precursor preparation

Zinc acetate dehydrate $Zn(CH_3COO)_2{\cdot}2H_2O$ (Alfa Aesar, ACS 98.0–101.0%) was chosen as a precursor for ZnO films with a constant concentration of 0.4M. This salt was dissolved in mixture solvent of water and ethanol.

2.2 Plasma spray conditions

In the deposition procedure, the solution feedstock was injected into the plasma plume via pressurized nitrogen with 6.5 bar through a needle, which keeps solution flow rate of 20 mL/min. A direct current (DC) plasma torch, Metco F4 gun, was attached to a six-axis robotic arm. The axial and radial distances between the needle tip and the centre of torch nozzle as well as the leaning angle of the needle were all fixed by means of a specific lab device. Argon and hydrogen were used as the primary gas and secondary gas, respectively. And the flow rate of argon was 35 L/min. In order to obtain different input energy during flight and on substrate, different hydrogen flow rates and preheating temperatures were employed to study their effects, which were listed in Table 1. These seven different ZnO films were deposited on non-grit blasting commercial aluminium substrate.

2.3 Characterization of the films

The compositions of these films were determined by X-ray diffraction (Bruker AXS D8 focus, Germany) with a cobalt anticathode (λ=1.78897 Å) at 35 kV, 40 mA and by another X-ray diffraction (D/Max-2550 PC) with a copper anticathode (λ=1.5418 Å) at 40kV, 100mA. A rapid scanning speed at 0.1°/s was chosen to determine the phase composition and a slow scanning speed at 0.004°/s was followed to analyse the relative intensities of XRD peaks and preferential orientation degree. Moreover, the surface topography of films was observed by scanning electron microscopy (JEOL, JSM-5800LV).

Table 1 Substrate preheating temperatures and hydrogen flow rates for different ZnO films

Sample Name	Substrate preheating temperature (°C)	Hydrogen flow rate (L/min)
Zn1	0	4
Zn2	200	4
Zn3	400	4
Zn4	400	3
Zn5	400	5
Zn6	600	3
Zn7	0	6

3 Results and discussion

3.1 Effect of substrate preheating temperature on XRD patterns

The rapid scanning XRD patterns of ZnO films deposited with different preheating temperature on alumina substrate were illustrated in Fig. 1. At first, it was confirmed that all films were made up of hexagonal wurtzite ZnO (PDF 75-0756). Due to the thin thickness of the ZnO films, the alumina substrates (PDF 75-0783) were also detected. Through the rapid scanning of XRD patterns, it was shown that when preheating temperature no higher than 400 °C, ZnO films were preferentially oriented along (002) plane. However, for standard PDF card 75-0756, the most intensive crystal plane should be (101) plane. This preferentially oriented growth along the (002) plane of ZnO has also been found in other processes, such as reactive RF magnetron sputtering and ultrasonic spray pyrolysis [5-8]. That was illustrated that that high temperature and high-velocity micro-jets during deposition are more inclined to result in a preferential orientation growth along (002) face [7-8]. It was well-known that very high temperature of plasma and high flight velocity between the torch and the substrate are the features of plasma spray processes comparing with other fabrication methods, which probably accounts for the preferential growth along (002) plane for these three samples. In order to have a further evaluation of this tendency, ZnO films under slow scanning of XRD test were detected between 36° and 44°, as shown in Figure 2. This scanning range includes the top three strong (100), (002) and (101) peaks of ZnO. Furthermore, in order to estimate the degree of oriented growth along the (002) face, a parameters named $P_{(002)}$ was employed as a criterion:

$$P_{(002)} = \frac{I_{m(002)}/I_{m(100)} + I_{m(002)}/I_{m(101)}}{I_{r(002)}/I_{r(100)} + I_{r(002)}/I_{r(101)}}$$

where $I_{m(hkl)}$ is the measured relative intensity of the (hkl) face and $I_{r(hkl)}$ is the intensity as given by the standard reference (PDF 75-0576). And this parameter, $P_{(002)}$, is positively correlated with the degree of c-axis oriented growth of ZnO films via SPPS. From Table 1, it is clear that increasing preheating temperature from 0 °C to 400 °C, the parameter $P_{(002)}$ also rapidly increased from 2.35 to 21.08. However, it should also note that the preferential growth along c-axis of ZnO might be not linearly with preheating temperature or torch power. As mentioned in former work [9], in terms of sol-gel process, upon the temperature of heat-treatment more than 450 °C, the degree of orientation along c-axis even decreased. Similar tendency of the effect of torch power will be pointed out for SPPS process in section 3.3.

Besides, the absolute intensity of these three major peaks of ZnO film went through apparent change as well, which indicates the different degree of crystallinity. Using relative intensity of XRD patterns for evaluating the degree of crystallization has already been used in previous work [10]. Then intensity of (002) plane was chosen as the standard to calculate degree of crystallization of SPPS sprayed ZnO films. From Fig.2, it is obviously that sample Zn3 was with most intensive (002) plane among all samples. Therefore, the sample Zn3 was selected as the reference sample with the relative intensity as 100% to calculate relative intensities of (002) plane for other samples. It reveals that the maximum crystallization degree was achieved with 400 °C preheating temperature, which is the optimized preheating temperature for crystallization among these three samples. Especially, for no-preheating condition, it was found that the relative intensity of sample Zn1 as 16.7% is much lower than samples Zn2 (51.3%) and Zn3 (100%), which indicates very weak crystallinity of Zn1. Moreover, secondary pyrolysis on pre-heated substrate has been reported for SPPS [11-12]. Thus, it is quite possible to realize crystallization on preheated substrates in SPPS simultaneously. And the reason for rapid increment on crystallinity of ZnO film under preheating temperature as 400 °C could be that the preheating temperature as 400 °C is higher than decomposition temperature of zinc acetate dehydrate (about 250 °C), which could supply enough energy for better crystallization from liquid phase than sample Zn1 and Zn2 [13].

3.2 Effect of preheating temperature on microstructures

Fig. 3 shows the surface morphologies of ZnO films with different preheating temperatures. First of all, many crystallized shaped particles (like hexagonal flakes) instead of common molten flatten splats via SPPS have been observed under relative high preheating temperatures, as such 400 °C. Without preheating, some in-suit sintered amorphous particles could be found in Fig. 1(a) and Fig.1 (b), which have been marked with read rectangles. Comparing the sample Zn2 and Zn1, the amorphous particles almost disappeared, which is agreement with XRD patterns of sample Zn2 with higher intensity than sample Zn1. With further increment of preheating temperature up to 400 °C, the degree agglomeration of particles become higher with more porosity between each agglomerated parts. In addition, in Fig. 1(e) and (f), sample Zn3 was combined with more apparent hexagonal-like flakes marked with red circles, which were believed to be obtained through crystallization. Therefore, it also proved that sample Zn3 with higher substrate preheating temperature could be with a better crystallinity. Meanwhile, comparing sample Zn3 with sample Zn1 and Zn2, it also demonstrated that higher preheating temperature promotes degree of agglomeration in Zn3 with larger pores between each agglomerated part. As we know, the rougher and porous surface in Zn3 could be with higher specific surface area than sample Zn1 and Zn2. All in all, from no preheating condition to preheated condition, agglomerated crystallized-like particles with high crystallinity and high specific surface area could be obtained via SPPS.

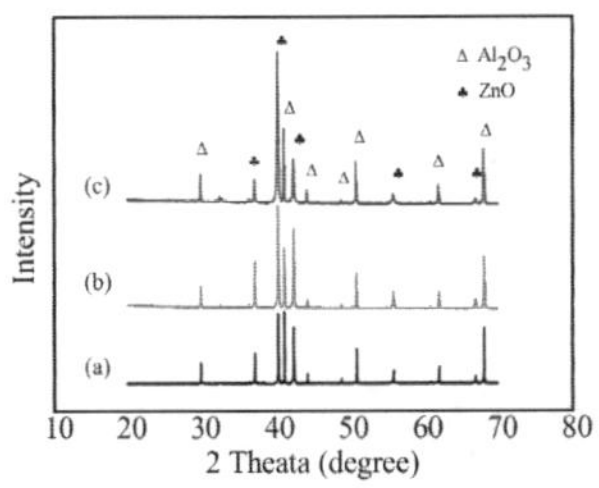

Fig. 1. X-ray diffraction spectra of ZnO films under rapid scanning with different preheating temperature (a) 0 °C, (b) 200 °C, (c) 400 °C.

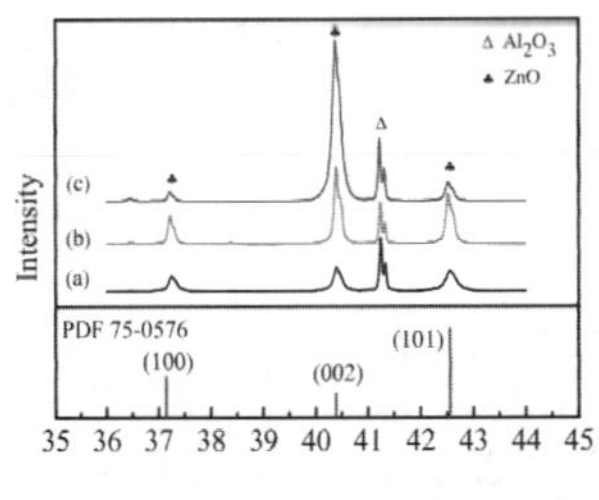

Fig. 2. X-ray diffraction patterns for ZnO under slow scanning with different preheating temperature (a) 0 °C, (b) 200 °C, (c) 400 °C.

Table 2 Degree of preferential orientation growth, relatively intensity of ZnO films with different preheating temperatures

Preheating temperature/ °C	Preferentially orientated crystal plane	$P_{(002)}$	Relative intensity of (002) peak
0	(002)	2.35	16.7%
200	(002)	3.82	51.3%
400	(002)	21.08	100.0%

3.3 Effect of torch power on XRD patterns

The rapid XRD patterns of ZnO films by different torch powers were shown in Fig. 4 (a) and Fig. 4(b), which using copper and cobalt as anticathode, respectively. It is obviously that expect employing a moderate hydrogen flow rate as 4 L/min, other three samples were oriented along (101) crystal planes. In addition, upon the hydrogen flow rate set as 4 L/min, (002) plane is always the most intensive one among all the XRD patterns for ZnO in Fig. 2. By comparison, Fig. 4 demonstrates that preferentially oriented crystal plane might be sensitive to hydrogen flow. In an analogy to sol-gel, in SPPS process, heating procedure during flight is similar preheating step in [14] for evaporation solvent, decomposition and other procedures, and heating procedure on substrate is similar to post-heating step in [14] for further decomposition and crystallization. Similar to the explanation in [14], it was believed that if hydrogen flow rate is too high, the abrupt vaporization, thermal decomposition of the zinc acetate solution may take place simultaneously with crystallization during flight, which might disturbing the oriental crystal growth along (002) plane. On the contrary, when the hydrogen flow rate is too low, too much solvent was left to evaporation and too much solute was left for decomposition on substrate, which could also destroy the preferential orientation growing along (002). And this result is also agree with the previous work about ZnO via sol-gel, which also mentioned an optimized baking temperature to obtain more oriented along (002) ZnO films [15]. Therefore, employing hydrogen flow rate as 4 L/min is the suitable value for making oriental growth along c-axis of ZnO films with 400 °C preheating temperature via SPPS. As mentioned in former work, (002) crystal plane is with highest performance activity for gas sensors and so on [16]. Based on the above result, it is noted that using a moderate hydrogen flow rate in SPPS could be favourable for obtaining preferential orientation along (002) plane of ZnO, which could be interesting for some functional applications.

(e) (f)

Fig.3. Surface morphologies of ZnO films with different preheating temperatures (a, b) 0 °C, (c, d) 200 °C, (e, f) 400 °C.

And choosing intensity of (002) plane as evaluation standard for crystallinity in Section 3.1, from rapid scanning in Fig. 4, the sample Zn3 was with higher degree of crystallinity than sample Zn4 and Zn5. Although relative intensity cannot be quantitatively evaluate by rapid XRD scanning patterns, the relative relationship of intensities of (002) plane of these three ZnO samples could still be compared. It could be found that the crystallinity of sample Zn4 is lowest. The sample Zn4 with hydrogen flow rate as 3 L/min left too much undecomposed or not pyrolysis solution on substrate, which could not be crystallized well on the preheated substrate. Besides, for sample Zn5 with hydrogen flow rate as 5 L/min, higher input energy than sample Zn4 during flight, which perhaps makes less liquid phase left for crystallization on substrate. Therefore, after flight procedure, sample Zn3 with suitable quantity liquid phase is attributed to best crystallization condition on substrate with preheating temperature of 400 °C.

3.4 Effect of torch power on microstructures

The morphologies of ZnO films sprayed by different hydrogen flow rates were displayed in Fig. 5. In comparison with Fig. 3, it is clear that upon keeping the preheating temperature as high as 400 °C, the crystallized particles could always be obtained under three different torch powers. With the lowest plasma power in sample Zn4, some amorphous particles still formed, which were marked with red rectangles. That is due to lowest torch power without enough energy for evaporation, decomposition, crystallization, which is also consistent with its XRD pattern with lowest intensity in Fig. 4. With increment of hydrogen flow rate, those amorphous particles decreased, which is accounted for higher energy during flight for evaporation and decomposition to promote further crystallization of ZnO. Besides, well molten particles under high hydrogen flow rate could be more easily to get together after impacting on substrates with less interspace between connected particles, as shown in Fig. 5 (e) and Fig. 5 (f). Especially, taking hydrogen sample Zn5 as an example, almost no pores were observed with very good connectivity between each particle. Generally speaking, through comparing the microstructures of samples Zn3, Zn4 and Zn5, higher torch power seems to make an apparent densification tendency on ZnO films compared with the condition under relative higher preheating temperature (such as Fig. 3 (c), (d), (e) and (f)). Therefore, it demonstrated that adjusting power torch (such as hydrogen flow) could be a more efficient way to change porosity of films via SPPS than adjusting preheating temperature.

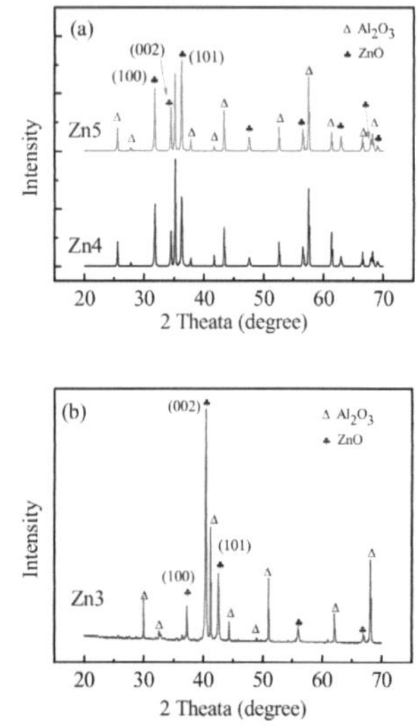

Fig. 4. Rapid XRD scanning using copper (a) and cobalt (b) as anticathode for ZnO films with different hydrogen flow rate 3 L/min (Zn4), 4 L/min (Zn3), 5 L/min (Zn5).

Fig. 5. Surface morphologies of ZnO deposited with different hydrogen rates by SPPS (a, b) 3 L/min, (c, d) 4 L/min, (e, f) 5 L/min.

3.5 Relative roles of substrate preheating temperature and torch power on crystallization and microstructures of ZnO films

Two inversed spraying conditions were used to study the relative roles of substrate preheating temperature and torch power on crystallization and microstructure of SPPS sprayed films. One condition is using highest substrate preheating temperature as 600 °C with lowest hydrogen flow rate as 3 L/min. Another comparative condition is using highest hydrogen flow rate as 6 L/min without substrate preheating temperature. The rapid XRD patterns of these two staggered experiments were shown in Fig. 6. Firstly, both the two ZnO films were preferentially oriented along (002) plane. For better study the relative role of preheating temperature and torch power in crystallinity and preferentially oriented growth, corresponding slow scanning of XRD results were displayed in Fig. 7. As mentioned in Section 3.1, employing the absolute intensities of (002) peak to evaluate the degree of crystallinity of these two samples. There is no doubt that the absolute intensity of (002) peak of sample Zn6 is stronger than sample Zn7, which demonstrated that preheating temperature plays more important role in crystallinity than torch power. This point is also agree with the Fig. 5, which always shows visible crystallized-like particles under different hydrogen flow rate with the same preheating temperature as 400 °C. However, in the Fig. 3, lower preheating temperature with the same hydrogen flow rate resulted in many amorphous particles indicating low crystallinity. In further, it could be inferred that the main process of crystallization of sprayed particles might occur on substrate, since crystallinity and crystallized-like particles are more sensitive to preheating temperature. Besides, based on the value of $P_{(002)}$ for these two samples, higher preheating temperature prefer to grow oriental along (002) plane. Moreover, crystallized-like particles were both observed under these two conditions in Fig. 8. There is no doubt that the shape of crystallized particle in sample Zn6 is more clear than sample Zn7, which could also means a better crystallinity in sample Zn6. And the apparent difference on microstructure between samples Zn6 and Zn7 is the interspaces between neighbouring particles. The interspaces were apparent in Fig. 8 (b) marked with red circles. On the contrary, no pores between connected particles could be observed in Fig. 8 (d). In other words, the porosity of films are mainly depending on torch power rather than preheating temperature of substrates in SPPS process. On one hand, in order to obtain crystallized particles and high crystallinity, preheating temperature should be taken into attention. On the other hand, in order to get denser microstructure, torch power deserves to be more focused on.

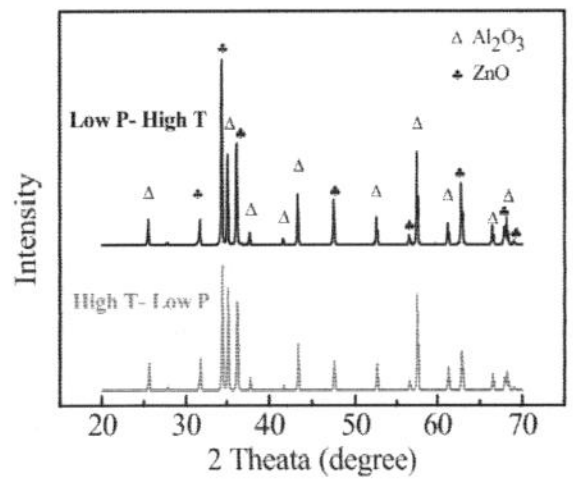

Fig. 6. Rapid scanning x-ray diffraction patterns for (a) Hydrogen flow rate as 3 L/min with the preheating temperature as 600 ° C (sample Zn6) and (b) hydrogen flow rate as 6 L/min without preheating (sample Zn7).

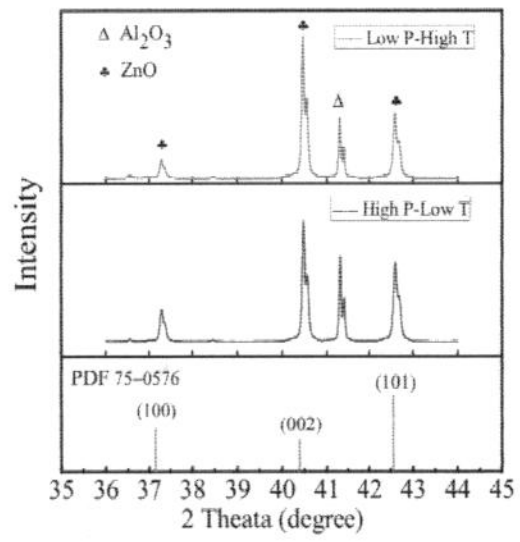

Fig. 7. Slow scanning x-ray diffraction patterns for (a) Hydrogen flow rate as 3 L/min with the preheating temperature as 600 ° C and (b) hydrogen flow rate as 6 L/min without preheating.

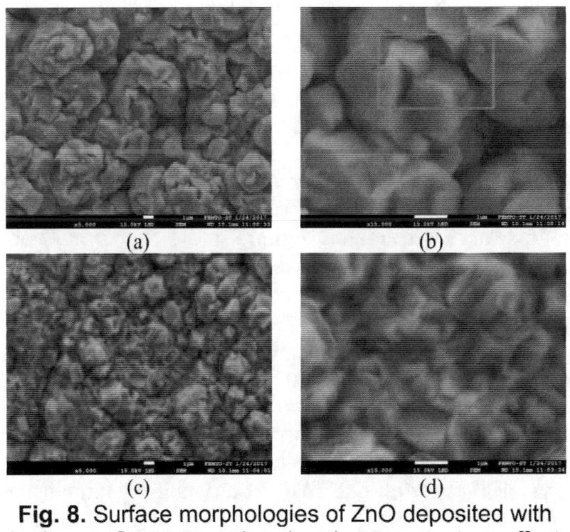

Fig. 8. Surface morphologies of ZnO deposited with hydrogen flow rate and preheating temperature offset collocation by SPPS (a, b) Hydrogen flow rate with 3 L/min and preheating temperature with 600 °C (sample Zn6), (c, d) Hydrogen flow rate with 6 L/min and without preheating temperature (sample Zn7).

Table 3 Degree of preferential orientation growth, crystallite size of ZnO films with staggered preheating temperatures and torch powers

Preheating temperature/ °C	Flow rate hydrogen/ L/min	Preferential orientated crystal plane	Relative intensity of (002) peak	$P_{(002)}$
600	3	(002)	59.4%	8.85
0	6	(002)	49.1%	5.29

4 Conclusion

Solution precursor plasma spray (SPPS) process was employed to deposit ZnO films with different substrate preheating temperatures and torch powers. The effects of these two spraying parameters on crystallization behaviour and microstructures were studied. Increasing preheating temperature from 0 to 400 °C, the degree of preferential orientation growth along (002) plane became more obvious. And the crystallinity of ZnO film via SPPS is also positive with preheating temperature in our research range. Without preheating temperature, crystallinity of ZnO films is very weak (16.7%) compared with preheating conditions (51.3%, 100% for sample Zn2 and Zn4, respectively). Keeping the preheating temperature as 400 °C, the orientation growth along (002) plane was only observed under some suitable torch power (hydrogen flow rate as 4 L/min). And the highest crystallinity is also obtained under moderate hydrogen flow rate, which could be due to suitable quantity of liquid phase left for crystallization. For the microstructure of ZnO films, under higher preheating temperature or larger hydrogen flow rate, crystallized particles became more obvious. Besides, with increment of plasma power, the whole trend of microstructure of ZnO films also becomes denser with better together with well molten particles. In contrast, through augmenting the substrate preheating temperature, it seems induce more agglomerated crystallized shape particles without densification of ZnO films. In addition, the respective roles of preheating temperature and torch power on crystallinity, preferentially oriented crystal plane and microstructure were proven in staggered experiments. It was inferred that preheating temperature is more responsible for crystallization of films and torch power plays main role on densification of films. And the crystallization process might complete on substrate instead of flight procedure. Choosing a certain preheating temperature and moderate hydrogen flow rate is with more possibility to deposit films with higher crystallinity, unusual preferentially oriented crystal and high specific surface area microstructure. This kind metal oxide films are favourable for some optical, electrical, electrochemical applications.

Acknowledgements
The authors gratefully acknowledge finical support from the China Scholarship Council (Grant No. 201504490038).

References

[1] Vardelle A, Moreau C, Akedo J, et al.: The 2016 Thermal Spray Roadmap. Journal of Thermal Spray Technology 25 (2016), Issue 8, pp. 1376/65.
[2] Gell M, Jordan E H, Teicholz M, et al.: Thermal barrier coatings made by the solution precursor plasma spray process. Journal of Thermal Spray Technology 17(2008), Issue 1, pp. 124/12.
[3] Wang W Z, Coyle T, Zhao D: Preparation of Lanthanum Zirconate Coatings by the Solution Precursor Plasma Spray. Journal of Thermal Spray Technology 23 (2014), Issue 5, pp. 827/6.
[4] Michaux P, Montavon G, Grimaud A, et al.: Elaboration of porous NiO/8YSZ layers by several SPS and SPPS routes. Journal of thermal spray technology 19 (2010), Issue 1-2, pp. 317/11.
[5] Zhao J L, Li X M, Zhang S, et al. : Highly (002)-oriented ZnO film grown by ultrasonic spray pyrolysis on ZnO-seeded Si (100) substrate. Journal of materials research, 21 (2006), Issue 09, pp. 2185/6.
[6] Erica P S, Michel C, Gilvan J S, et al. Al-doping effect on the surface morphology of ZnO films grown by reactive RF magnetron sputtering. Materials Sciences and Applications (2013), pp. 761/7.
[7] Dimitrov O, Nesheva D, Blaskov V, et al. Gas sensitive ZnO thin films with desired (002) or (100) orientation obtained by ultrasonic spray pyrolysis. Materials Chemistry and Physics 148(2014), Issue3, pp. 712/8.
[8] Gao X D, Li X M, Yu W D. Structural and morphological evolution of ZnO cluster film prepared by the ultrasonic irradiation assisted solution route. Thin Solid Films 484 (2005), Issue1, pp. 160/5.
[9] Bao D, Gu H, Kuang A.: Sol-gel-derived c-axis oriented ZnO thin films. Thin solid films 321 (1998), Issue1-2, pp. 37/3.
[10] Lo C L, Duh J G, Chiou B S, et al.: Low - Temperature Sintering and Microwave Dielectric Properties of Anorthite - Based Glass - Ceramics. Journal of the American Ceramic Society 85(2002), Issue 9, pp. 2230/6.
[11] Bhatia T, Ozturk A, Xie L, et al.: Mechanisms of ceramic coating deposition in solution-precursor plasma spray. Journal of materials research 17 (2002), Issue 09, pp. 2363/10.
[12] Joshi S V, Sivakumar G, Raghuveer T, et al.: Hybrid plasma-sprayed thermal barrier coatings using powder and solution precursor feedstock. Journal of thermal spray technology 23 (2014), Issue 4, pp. 616/9.
[13] Raoufi, Davood, and Taha Raoufi.: The effect of heat treatment on the physical properties of sol–gel derived ZnO thin films. Applied Surface Science 255 (2009), Issue 11, pp. 5812/6.
[14] Ohyama, Masashi, Hiromitsu Kouzuka, and Toshinobu Yoko.: Sol-gel preparation of ZnO films with extremely preferred orientation along (002) plane from zinc acetate solution. Thin solid films 306 (1997), Issue 1, pp. 78/8.
[15] Guo D, Sato K, Hibino S, et al.: Low-temperature preparation of (002)-oriented ZnO thin

films by sol–gel method. Thin Solid Films, 550 (2014), pp. 250/9.
[16] Han X G, He H Z, Kuang Q, et al.: Controlling morphologies and tuning the related properties of nano/microstructured ZnO crystallites. The Journal of Physical Chemistry C 113(2008), Issue 2, pp. 584/6.

A comparative CFD study of the influence of combustion chamber geometry in the HVSFS process

M. Plachetta, P.Eckert, M. Kroner, A. Killinger, P. Krieg, R. Gadow
Institute for Manufacturing Technologies of Ceramic Components and Composites, University of Stuttgart, Stuttgart/DE

Suspensions have shown a great potential for being employed as the spraying materials in flame spraying processes with the aim of producing thin and dense coatings. The internal axial injection of a suspension within processes like the high-velocity suspension flame spraying (HVSFS) offers the advantage of complete suspension entrainment within the gas stream, which therefore results in enhanced momentum and heat transfer to the particles. Experimental assessment of the achieved particle velocities and temperatures within the combustion chamber is nonetheless practically infeasible. A better insight into the process is attainable through employing computational simulations.
Following a computational fluid dynamic (CFD) modelling approach for HVOF processes, combustion and gas flow turbulences were simulated for different combustion chamber geometries and ethene/oxygen ratios commonly used in the HVSFS process. Simulations were done with the commercial software ANSYS CFX 16.2. To account for the highly turbulent flow characteristics, the k-ε model and the Shear Stress Transport (SST) turbulence model were chosen, employing a eddy-dissipation-model for fuel gas combustion. Second-order upwind discretization was used to enable a good resolution of flow features like shock diamonds. The results of the simulation using different levels of detail of the combustion reaction were compared to experiments employing the modelled combustion chambers and gas flows. Chamber pressure and positions of the shock diamonds were monitored in order to allow a qualitative evaluation of the calculated values.

1 Introduction

Thermally sprayed coatings find use in a wide range of application, like wear resistance, corrosion protection and thermal barriers [1-3]. Generally speaking, thermal spraying uses a particle loaded hot gas jet to form metal, ceramic or metal ceramic coatings. Using different particle sizes and spraying methods have been shown to have significant effects on the coating properties [4]. Notably, nano-sized particles delivered to the process by means of a suspension have resulted in coatings with extraordinary qualities [5]. The high-velocity suspension flame-spraying (HVSFS) process specifically uses an internal injection of the suspension into the combustion chamber of the spray gun, which results in a complete entrainment of the suspension within the gas stream, which leads to an enhanced momentum and heat transfer to the particles. This internal injection leads however to high demands concerning the process stability. The heat and pressure at the injection point increase the risk of a clogging of the injector nozzle and build up of coating material inside the combustion chamber. The actual processes occuring inside the spray gun are nonetheless hard to observe due to firstly the extreme conditions inside the combustion chamber during the process and secondly experimental mesurements are limited to the outside of the nozzle.

To properly assess the factors leading to such process instabilities a computational fluid dynamic (CFD) modelling approach offers itself as a suitable method. This way it is possible to gain insight into factors like local temperatures, gas velocities and flow direction which are major contributing factors to process stability, melting behaviour of the particles and finally coating properties. For example, Shock diamonds are an important feature of the flame, since local conditions lead to turbulence amplification as well as temperature and pressure increase and a velocity reduction after the shock plane. This can result in a dispersion of the particle trajectories, especially of of the very fine particles in the HVSFS process. Consequently, they have to be reproduced correctly by the model [6].

A number of different approaches to model thermal spray processes have been pursued in literature.
Jadidi et al. modeled a premixed combustion of oxygen/propylene in a suspension HVOF process using the eddy dissipation model (EDM) in combination with one-step global reactions. The realizable k-ε turbulence model was used to simulate the gas phase. The suspension was introduced as discrete spherical droplets with a diameter of 120 µm, their atomization calculated with the Taylor Analogy Breakup model [7].
Dongmo et al. used the eddy dissipation model and the k-ε turbulence model to simulate a premixed oxyxgen/propane flame for a variety of combustion chambers and nozzle geometries. Suspension droplets with a an uniform diameter equivalent to the diameter of the injection nozzle were included. The breakup was modeled with the blob method. Postions and number of shock diamonds showed good matches with experimental results, however flame temperatures were greatly overestimated [8].
For an in depth overview refer for example to the extensive review by Dolatabadi et al. [9].

2 Modelling

For modelling, four different converging combustion chambers and parallel sided barrel nozzles for the TopGun thermal spray gun (GTV Verschleißschutz GmbH, Luckenbach, Germany) were selected, as they are commonly used for the HVSFS process.

Table 1: Dimensions of the investigated combustion chambers and nozzles.

	Length of comb	Length of nozzle	Nozzle diameter	Special features
1	22	56	8	-
3	22	113	8	-
3	30	113	8	-
4	19	59	8	Ridge

Table 1 lists the different geometries in the scope of the study. In the following text, the different chambers will be referred to by sets of three numbers relating to their dimensions, where the first number is the length of the combustion chamber, the second the diameter of the nozzle and the third the total length of combustion chamber and nozzle combined.

For the 22-8-135 combustion chamber and nozzle, a study of mesh dependence of the results was conducted. A convergence towards stable values revealed itself with an increasingly finer mesh.

Accordingly, a mesh of $3.5*10^6$ nodes was chosen for most of the subsequent calculations as a trade-off between reliable values and calculation time. At the back of the combustion chamber, 24 inlets for the premixed ethene and oxygen are located in a circular pattern.

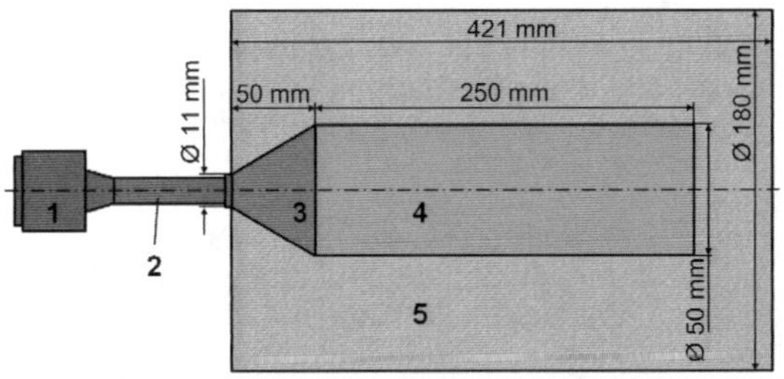

Fig. 1: Geometrie of the fluid domain and overview of different meshing areas.

The combustion chamber itself (1) and the expansion area just outside the nozzel (3) were meshed using unstructured tetrahedral elements to achieve a good mesh quality in their more complex geometrie with element sizes of 0.21 mm and 0.62 mm respectively. The nozzle (2) and the free flow area (4) were meshed using a structured hexahedral elements, with element sizes of 0.27 mm and 0.90 mm respectively, **Fig. 1**.

A number of different gas flow rates was considered, as listed in **Table 2**. Different ethen/oxygen-ratios were investigated. One sub-stoichiometric, stoichiometric and super-stoichiometric compostiotion were chosen respectively. The compostion is given by λ, the oxyygen–fuel equivalence ratio, which is calculated as

$$\lambda = \frac{\dot{V}(O_2)/\dot{V}(C_2H_4)}{\left[\dot{V}(O_2)/\dot{V}(C_2H_4)\right]_{stoich}}$$

with $\dot{V}$ being the volumetric flow rate.

In one set the ethene flow rate was kept constant (at 0.1 m^3/min), in the other set the total gas flow rate (at 0.4 m^3/min). The gases initial temperature was 300 K.

Table 2: different gas flow rates with compositions below, at and above the stoichiometric composition

	Mass flow [g/s]	
λ	Constant ethene flow rate	Constant total gas flow rate
0,8	7,006	8,243
1	8,289	8,289
1,2	9,572	8,324

Modelling turbulence is a challenge in itself due to its three-dimensional, chaotic and instationary nature [10]. The k-ε-turbulence model has been often used in literature to model HVOF processes [9]. The Shear-Stress-Turbulence (SST) model, which combines aspects of the k-ε and k-ω turbulence models, has been used in fewer instances for flame spraying processes, yet it has become an industrial standard for the accurate prediction of flows with strong pressure gradients[11]. Accordingly, both models were compared for their applicability in modelling the processes with the given geometries.

For the highly turbulent, premixed combustion, the eddy dissipation model (EDM) was chosen, which is also a standard model for HVOF processes in literature [9]. The EDM is appropriate for modeling reactions that are controlled by the rate at which fuel and oxidant are mixed, which is the case for combustions with fast reaction kinetics.

Four different combustion reactions of increasing complexity were compared. The choice of reactions and parameters is based on the work of Westbrook and Dryer, which have been shown to deliver good results in a number of publications [12,13,9].

The number of reaction steps included in the model has an enormous effect on the resulting flame temperature. For a more detailed look at the combustion reaction, the effect of dissociation has to be considered. Starting at temperatures of about 2000 to 2500 K, H_2O dissociates into a number of different species. With increasing temperature, the ratio of H_2 and O_2, as well as those of the radicals $H^\bullet$, $O^\bullet$ and $OH^\bullet$ increase and play a more important role in the reaction. The complete inclusion of all the resulting reactants and reactions is outside of the scope of a combustion model, however it should correctly represent the effects of characteristic reactions.

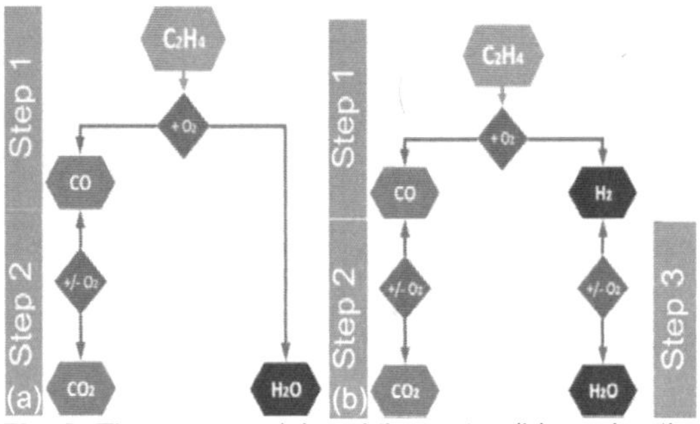

Fig. 2: The two step (a) and three step (b) combustion reaction schemes.

Four different combustion reactions were investitgated: A one step reaction which includes only the reaction of C_2H_4 with O_2 to CO_2 and H_2O (WD1), a two step reaction includes the formation and subsequent oxidation of CO (WD2), a two step reaction including a back reaction in the second step and a modified set of parameters taken from [13] (WD2b), a three step reaction including the formation and oxidation of H_2(WD3), **Fig. 2**.

A further increase in number of reaction steps did not lead to a significant difference in results, rather to a lessened stability of the simulation, which agrees with the results of Oberkampf and Talpallikar [14].

First order upwind discretization showed lower temperatures all over the domain and was not able to model any shock diamonds. It did however show a good stability and convergence during calculation, which is in agreement with literature [15]. It was therefore used to calculate an initial solution for subsequent caculations with a second-order upwind discretization (also called high resolution discretization). This way, a good resolution of flow features like shock diamonds was possible [15].

3 Numerical and Experimental Results

Convergence behaviour of the simulation showed a dependence on the combination of turbulence model and combustion reactions. In general, a good convergence could be reached and the SST turbulence model showed a better convergence and a more turbulent behaviour in the combustion chamber itself. However, in areas of locally overestimated temperatures inside the combustion chamber, residuals remained high.

Using more complex combustion reaction schemes however reduced the stability of the simulation when combined with the SST turbulence model. For those, the more stable k-ε turbulence model was used further on.

Combustion chamber geometry showed a significant effect on the resulting streamlines. The ridge at the outlet of the 19-8-78 combustion chamber in increased turbulent flows, as is the intention of this geometry. The length of the attached nozzle has no significant influence on the temperatures inside the combustion chamber, however a longer nozzle results in a higher gas velocity along the lenght of the nozzle. This is as expected, since a longer nozzle means a longer time of acceleration. The long 30 mm-combustion chamber shows less turbulence and and a rather laminar flow along the longer chamber walls, **Fig. 3**.

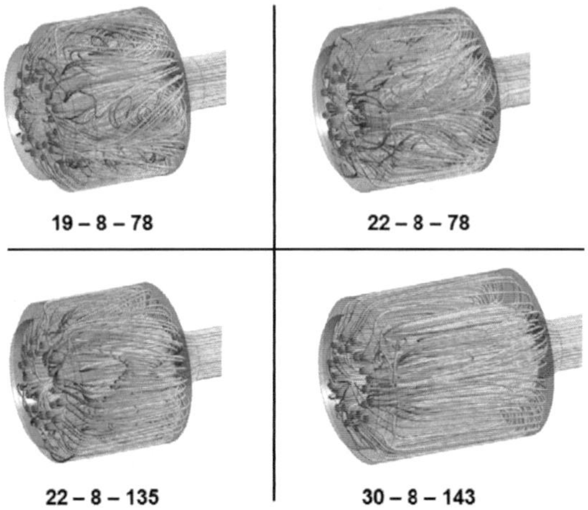

Fig. 3: The streamlines inside the investigated combustion chambers and nozzles.

For a constant ethene volume flow, an ethene surplus leads to an increase in temperature in comparison to stoichiometric conditions, while a surplus of oxygen leads to a reduction in temperature. Absolute pressure changes usually in accordance with temperature, however in the latter case the higher total gas flow rate (refer to **Table 2**) outweighs this effect due to the increased stagnation pressure inside the combustion chamber. The highest pressure is therfore calculated for λ = 1,2.

For a constant gas volume flow, a lower λ goes along with an increase in combustion temperature, the qualitative connection is in agreement with Dongmo et al. [8]. Higher combustion temperatures resulting from a lower λ also leads to an increased gas expansion and subsequently higher pressures and therefore increased acceleration and velocities of the gases in the expansion nozzle.

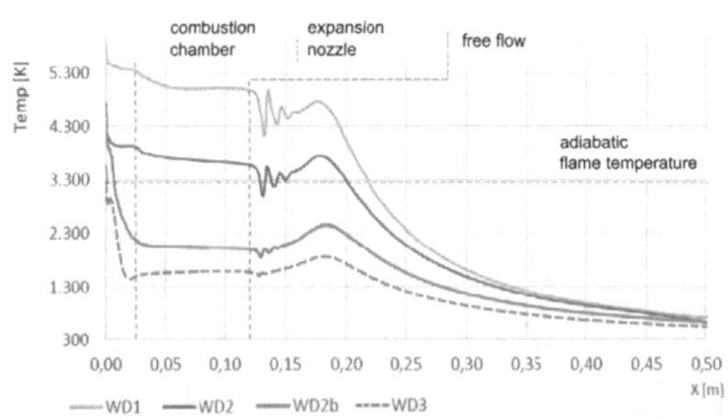

Fig. 4: Temperatures resulting from the four different modeled combustion reactions along the middle axis of the combustion chamber and nozzle.

For ease of comparison, an adiabatic flame temperature of 3300 K is assumed in **Fig. 4**. The adiabatic flame temperature is pressure dependend and varies between 3272 K und 3355 K for the expected pressure between $2*10^5$ Pa und $3.5*10^5$ Pa.
The one step combustion model WD1 showed the highest temperatures overall, at and above 5000 K, well above the adiabatic flame temperature inside the torch. Adding the second reaction step (WD2) reduced temperature significantly but still about 1000 K above adiabatic flame temperature inside the torch.
The optimised reaction scheme WD2b showed another decrease in overall temperature. Temperatures above the adiabatic flame temperature were limited to a small area near the back wall of the combustion chamber. Using the three step reaction scheme WD3, flame temperature quickly dropped below the adiabatic flame temperature and further, staying below 2000 K inside the torch.
Notable is an increase in temperature outside the torch. This is likely to result from local flow characteristics and go along with a decrease in CO. In this area, turbulence is strongly increased by the detaching of the flow from the nozzle walls and the shock diamonds. These turbulences lead to an increase in reaction rate and consequently to an increase in temperature. This phenomenon has been observed in literature before [16, 17].

The pressure in front of the location of the suspension inlet derived with different combustion reactions was compared to values from experimental measurements, **Fig. 5**. For the measurement, a pressure gauge was connected to the suspension inlet in the combustion chamber. A total gas flow rate of 0.4 m^3/min was used. The comparison shows an overestimated pressure for the models WD1 und WD2, resulting from the significantly overestimated temperatures. The combustion reaction WD2b showed values almost identical to the experiment. The WD3 combustion reaction calculated a lower pressure at the measuring point.

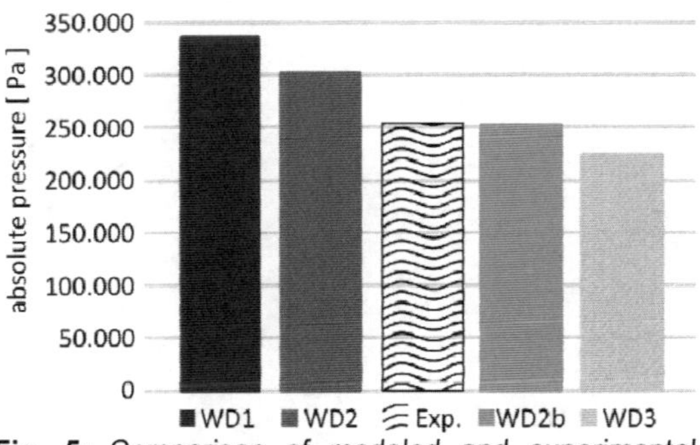

Fig. 5: Comparison of modeled and experimental values of absolute pressures resulting from different combustion reactions.

Another characteristic feature of gas flow with a pressure higher than that of the environment, like it is the case in the investigated setup, are shock diamonds. Since their number and positions depend on the pressure and temperature of the gas jet, they can be used to judge the accuracy of the simulation.
To conduct the validation experiment, the spray torch was run with the stoichiometric gas parameters and a 22-8-135 combustion chamber and nozzle without the injection of a suspension or powder. Photographs of the resulting flame were taken. The centers of the first three shock diamonds were determined and the distance between them was measured. These distances were compared to those derived by the simulation, **Table 3**.
The positions of the shock diamonds mirror the results of the analysis of pressure in the combustion chamber. Model WD2 resulted in the greatest distances between shock diamonds, while distances calculated with WD3 were the lowest. Reaction scheme WD2b shows a good match with experimental values.

Table 3: Positions of the shock diamonds relative to each other as derived from three reaction models and the experiment.

Model	Distance SD1 to SD2 [mm]	Distance SD2 to SD3 [mm]
WD2	9,91	9,70
WD2b	8,34	7,51
WD3	7,16	6,76
Experiment	8,50	7,50

4 Conclusion

In order to to increase understanding for the processes inside the combustion chamber and expansion nozzle during the HVOF process, the combustion and gas dynamics of an ethen-fueled HVOF thermal spray gun were modeled using the commercial CFD software ANSYS CFX 16.2. Experimental studies were conducted to validate the numerical model.

Convergence behaviour of the simulation showed a dependence on the chosen combination of turbulence model and combustion reaction scheme. While a good convergence could be reached with the SST turbulence model, combining it with more complex reaction schemes lead to instabilities in the simulation. For those, the well established k-ε turbulence model was chosen.
The occuring overestimation of combustion temperatures with some reaction schemes can be attributed to the Eddy Dissipation Model, which is well known to overestimate temperatures in fuel gas rich environments assuming a complete combustion [9].
Increasing the complexity of the combustion reaction by adding energy consuming reaction steps lead as expected to a reduction of overall temperatures.
The optimized two step reaction model WD2b showed a good match with experimental measurements of combustion chamber pressure and shock diamond positions.

5 Literature

[1] Di Girolamo, G., Serra, E.: Thermally sprayed nanostructured coatings for anti-wear and TBC applications: State-of-the-art and future perspectives. Teoksessa: Aliofkhazraei, M.: Anti-Abrasive Nanocoatings - Current and Future Applications. Cambridge, UK : Woodhead Publishing Limited (2015) pp. 513–541

[2] Vassen, R., Cao, X., Tietz, F., Basu, D. and Stöver, D.: Zirconates as New Materials for Thermal Barrier Coatings. Journal of the American Ceramic Society, 83 (2000), Issue 8, pp. 2023–2028.

[3] Fauchais, P., Vardelle, A.: Thermal Sprayed Coatings Used Against Corrosion and Corrosive Wear, Advanced Plasma Spray Applications (2012), pp. 14-29

[4] Pawlowski, L.: The Science and Engineering of Thermal Spray Coatings. Second Ed., West Sussex, England : John Wiley & Sons (2008)

[5] Fauchais, P., Montavon, G., Lima, R. S. and Marple, B. R.: Engineering a new class of thermal spray nano-based microstructures from agglomerated nanostructured particles, suspensions and solutions: an invited review. Journal of Physics D, Applied Physics, 44 (2011), Issue 9

[6] Fauchais, P. L., Heberlein, J. V.R., Boulos, M. I.: Thermal Spray Fundamentals. Springer US (2014)

[7] Jadidi, M., Moghtadernejad, S. and Dolatabadi, A.: Numerical Modeling of Suspension HVOF Spray. Journal of Thermal Spray Technology 25 (2016), Issue 3, pp. 451-464

[8] Dongmo, E., Gadow, R., Killinger, A., Wenzelburger, M.: Modeling of Combustion as well as Heat,Mass, and Momentum Transfer During Thermal Spraying by HVOF and HVSFS. Journal of Thermal Spray Technology 18 (2009), Issue 5, pp. 896–908

[9] Jadidi, M., Moghtadernejad, S., Dolatabadi, A.: A Comprehensive Review on Fluid Dynamics and Transport of Suspension/Liquid Droplets and Particles in High-Velocity Oxygen-Fuel (HVOF) Thermal Spray. Coatings 5 (2015), pp. 576-645

[10] Menter, F. R.: Turbulence Modeling for Engineering Flows. Ansys, Inc. (2011)

[11] Menter, F. R., Kuntz, M., R. Langtry: Ten years of industrial experience with the SST Turbulence Model. Turbulence, Heat and Mass Transfer 4 (2003)

[12] Westbrook, C.K., Dryer, F.L.: Simplified Reaction Mechanisms for the Oxidation of Hydrocarbon Fuels in Flames. Combustion Science and Technology 27 (1981), Issue 1-2, pp. 31-43

[13] Westbrook, C. K., Dryer, F. L.: Chemical Kinetic Modeling of Hydrocarbon Combustion. Prog. in Energy and Combust. Science 10 (1984), Issue 1, pp. 1-57

[14] Oberkampf, W.L. and Talpallikar, M.: Analysis of a High-Velocity Oxygen-Fuel (HVOF) Thermal Torch, Part 1: Numerical Formulation. Journal of Thermal Spray Technology 5 (1996), Issue 1, pp.53-61

[15] Lecheler, S.: Numerische Strömungsberechnung (2014), pp. 84-89

[16] Wang, X., Song, Q., Yu, Z.: Numerical Investigation of Combustion and Flow Dynamics in a High Velocity Oxygen-Fuel Thermal Spray Gun Journal of Thermal Spray Technology 25, (2016), Issue 3, pp. 441–450

[17] Li, M., Christofides, P. D.: Modeling and Control of High-Velocity Oxygen-Fuel (HVOF) Thermal Spray: A Tutorial Review. Journal of Thermal Spray Technology 18, (2009), Issue 5-6, pp. 753–768

Interaction between rotary arc and injected particles in a non-transferred DC plasma spray with externally applied magnetic field

H. Saito and T. Fujino, Tsukuba / JPN
H. Takana, Sendai / JPN
J. Mostaghimi, Toronto / CAN

The authors performed a time-dependent, three-dimensional numerical simulation of a non-transferred DC plasma spray with externally applied magnetic fields. Compressible Navier-Stokes equations with MHD source terms and Maxwell's equations were used as the governing equations for plasma flows. In the simulation, two operating conditions, electric currents and strength of externally applied magnetic fields, were parametrically varied in a range of 300 A to 500 A and 0.2 T to 0.8 T, respectively. Numerical results show that the application of strong magnetic fields such as 0.4 T and 0.8 T is recommended for an anode arc rotation leading to elongating an anode lifetime. A voltage variation due to the anode arc rotation shows periodic behavior with a small amplitude, which is expected to be good for plasma spraying processes. Lagrangian approach was used to track injected particles in the plasma jet and the particle temperature and position distributions on a cross section normal to the central axis of spray were studied. Swirl flows induced by the arc rotation hinder the particles from reaching the hot plasma jet. Our numerical results demonstrated that injecting particles in the opposite direction to the swirl flow is an effective way to heat the particles.

1 Introduction

Coating quality of a thermal plasma spray is strongly influenced by arc root movement in a plasma torch, plasma-particle interaction, and molten particle deposit process. Arc root movement is an also important factor for electrode's lifetime.

The application of external magnetic fields has been suggested to control arc root movement [1-3]. Figure 1 shows a schematic view of the plasma torch with a solenoid coil for inducing external magnetic fields, which was developed by Pershin *et al* [1]. The interaction between arc currents and external magnetic fields produces Lorentz force, $\boldsymbol{F} = \boldsymbol{J} \times \boldsymbol{B}$, which rotates the arc root on the anode. Pershin. *et al.* [1] has experimentally reported that the application of external magnetic fields is effective for improving the anode lifetime, and also Hu. *et al.* [2] has experimentally demonstrated the suppression of voltage fluctuation by using external magnetic fields.

No full understanding for the physical mechanism of voltage suppression effect, however, has been obtained yet. To clarify the mechanism, studying the relation between voltage fluctuation and arc's rotation is essential as the first approach. Effects of externally applied magnetic fields on coating quality should be also studied in detail. Coating quality is directly affected by impacting particle properties such as the position, the velocity, and the melting state.

In this work, we examine effects of operating currents and external magnetic fields on time variation of arc voltage by three-dimensional MHD numerical simulations. Trajectories and heating histories of in-flight particles in plasma jets are analysed using the Lagrangian method. Our previous work [4] numerically showed that applying external magnetic fields produces higher enthalpy plasma jets, which carry injected particles with well-molten state. However, swirl flows produced by arc rotational movement tend to hinder the particles from reaching the hot plasma plume. In this work, a suitable particle injection manner

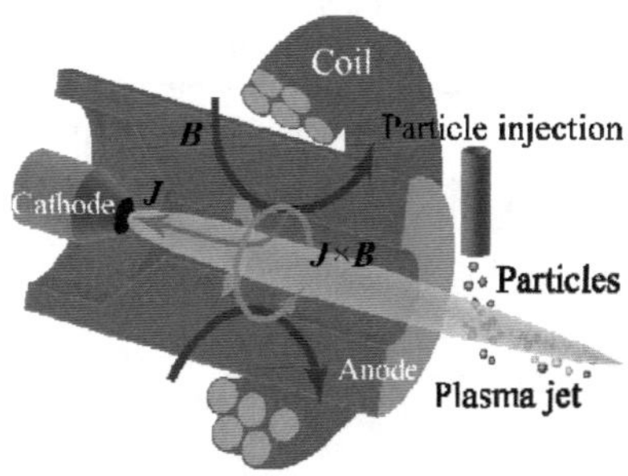

Fig. 1. Schematic view of plasma torch with solenoid coil.

to overcome this problem is suggested, in which the particles are radially injected with a diagonal angle so as to have an azimuthal velocity component opposite to the swirl direction of plasma jets. In this paper, effects of this injection manner on the trajectories and heating histories of the injected particles are also discussed.

2 Description of mathematical model

2.1 Governing equations for plasma flow

The mathematical model for plasma flows is based on a fully coupled treatment of flow fields together with electromagnetic fields. The model is based on the following main assumptions:

- Working gas is pure argon (Ar).
- Plasma flows are compressible, laminar flows in local thermo and chemical equilibrium (LTCE) states.
- Thermodynamic and transport properties depend on the gas pressure and temperature.
- Quasi-neutrality condition holds.
- Ar plasma is optically thin.
- Gravitational effect and Hall effect can be neglected.

Under these assumptions, the governing equations for plasma flows, which are composed of the mass, momentum, and total energy conservation equations, are written as follows:

Mass conservation equation

$$\frac{\partial}{\partial t}\int_V \rho dV + \oint_S \rho(\boldsymbol{u}\cdot\boldsymbol{n})\, dS = 0 \quad (1)$$

Momentum conservation equations

$$\frac{\partial}{\partial t}\int_V \rho\boldsymbol{u} dV + \oint_S \rho\boldsymbol{u}(\boldsymbol{u}\cdot\boldsymbol{n})\, dS$$

$$= \int_V (\boldsymbol{J}\times\boldsymbol{B})dV - \oint_S p\boldsymbol{n}\, dS + \oint_S \bar{\bar{\tau}}\cdot\boldsymbol{n}\, dS \quad (2)$$

Total energy conservation equation

$$\frac{\partial}{\partial t}\int_V \rho E_{flu} dV + \oint_S \rho H_{flu}(\boldsymbol{u}\cdot\boldsymbol{n})\, dS = \oint_S \kappa(\boldsymbol{\nabla}T\cdot\boldsymbol{n})\, dS$$

$$+ \int_V (\boldsymbol{J}\cdot\boldsymbol{E} - \dot{q}_{rad})dV + \oint_S (\bar{\bar{\tau}}\cdot\boldsymbol{u})\boldsymbol{n}\, dS \quad (3)$$

where $\boldsymbol{B}$ is vector of magnetic flux density, $\boldsymbol{E}$ electric field vector, E_{flu} specific total energy, H_{flu} specific total enthalpy, $\boldsymbol{J}$ vector of electric current density, $\boldsymbol{n}$ unit vector of normal direction, p gas pressure, $\dot{q}_{rad}$ radiative energy loss, S area of cell interface, t time, T gas temperature, $\boldsymbol{u}$ gas velocity vector, V cell volume, κ gas thermal conductivity, ρ gas mass density, and $\bar{\bar{\tau}}$ viscous stress tensor. The specific total energy E_{flu} and the specific total enthalpy H_{flu} are defined, respectively, as

$$E_{flu} = e_{flu} + \frac{|\boldsymbol{u}|^2}{2} \quad (4)$$

$$H_{flu} = E_{flu} + \frac{p}{\rho} \quad (5)$$

where e_{flu} is specific internal energy. These conservation equations coupled with the equation of state are discretized using the upwind finite volume method, in which the advection upstream splitting method (AUSM)-DV scheme [5] and the central differential scheme are utilized for evaluating the numerical flux of the convective and diffusion terms, respectively. The time integration is performed by the lower-upper symmetric Gauss-Seidel scheme [6]. Low magnetic Reynolds number MHD approximation is adopted for electromagnetic modeling. The governing equations for electromagnetic fields are written as follows:

Maxwell's equations

$$\boldsymbol{E} = -\boldsymbol{\nabla}\phi \quad (6)$$

$$\boldsymbol{\nabla}\cdot\{\sigma(-\boldsymbol{\nabla}\phi + \boldsymbol{u}\times\boldsymbol{B})\} = 0 \quad (7)$$

$$\boldsymbol{\nabla}^2\boldsymbol{A} = -\mu_0\boldsymbol{J} \quad (8)$$

$$\boldsymbol{B} = \boldsymbol{\nabla}\times\boldsymbol{A} + \boldsymbol{B}_0 \quad (9)$$

Generalized Ohm's law

$$\boldsymbol{J} = \sigma(\boldsymbol{E} + \boldsymbol{u}\times\boldsymbol{B}) \quad (10)$$

where $\boldsymbol{A}$ is vector potential, $\boldsymbol{B}_0$ vector of external magnetic flux density, ϕ electric potential, μ_0 permeability of free space, and σ electrical conductivity. Each of Eqs. (7) and (8) is discretized with the Galerkin finite element method, and then the derived linear equations on the electric potential ϕ and the vector potential $\boldsymbol{A}$ are solved with Bi-CGSTAB2 method [7].

2.2 Governing equations for particle trajectories and heating histories

For modelling the behavior of the in-flight particles, the following assumptions are adopted:

- The shape of a particle is spherical.
- The temperature gradient within a particle is negligible.
- Effects of in-flight particles on plasma flows are negligible.
- Particle-particle interaction is negligible.
- Drag force $\boldsymbol{F}_D$ and thermophoretic force $\boldsymbol{F}_{th}$ are dominant forces working on a particle.

The trajectories and heating histories of in-flight particles are obtained by solving the following equations:

Equations of particle motion [8]

$$\frac{d\boldsymbol{x}_p}{dt} = \boldsymbol{u}_p \quad (11)$$

$$\frac{d\boldsymbol{u}_p}{dt} = \boldsymbol{F}_D + \boldsymbol{F}_{th} \quad (12)$$

$$\boldsymbol{F}_D = \left(\frac{3\rho C_D}{8r_p\rho_p}\right)\cdot|\boldsymbol{u}-\boldsymbol{u}_p|(\boldsymbol{u}-\boldsymbol{u}_p) \quad (13)$$

$$\boldsymbol{F}_{th} = -9\left(\frac{\mu^2}{\rho\rho_p r_p^2}\right)\cdot K_{th}\cdot\left(\frac{1}{T}\nabla T\right) \quad (14)$$

where C_D is drag coefficient, r_p particle radius, $\boldsymbol{u}_p$ particle's velocity vector, $\boldsymbol{x}_p$ particle's position, μ gas dynamic viscosity, and ρ_p particle mass density. The drag coefficient C_D in the drag force $\boldsymbol{F}_D$ and the coefficient K_{th} in the thermophoretic force $\boldsymbol{F}_{th}$ are evaluated according to the Ref. [8].

Particle energy equations

When the in-flight particles are in solid-phase, the particle temperature is calculated by

$$\frac{dT_p}{dt} = \frac{3(q-q_r)}{\rho_p C_{p,s} r_p} \quad (15)$$

where $C_{p,s}$ is solid-phase specific heat of the particle material, q specific heat flux from plasma to the in-flight particles, q_r radiative heat loss from the particle surface, and T_p particle temperature. When the in-flight particles are in melting phase, the particle temperature would keep the melting point. The liquid-phase fraction increment within the *i*th time step Δt_i is calculated by

$$\Delta f_l = \frac{3(q-q_r)}{L_m\rho_p r_p}\Delta t_i \quad (16)$$

where L_m is the melting latent heat. When the in-flight particles are in liquid-phase, the particle temperature is calculated by

$$\frac{dT_p}{dt} = \frac{3(q-q_r)}{\rho_p C_{p,l} r_p} \quad (17)$$

where $C_{p,l}$ is liquid-phase specific heat of the particle material. After the particle temperature reaches its evaporation point, the particle radius reduces with the time as

$$\frac{dr_p}{dt} = -\frac{q-q_r}{\rho_p L_v} \quad (18)$$

where L_v is the evaporation latent heat. The specific heat flux q from plasma to in-flight particles is calculated by

$$q = \beta_1 \beta_2 q_c \qquad (19)$$

where the factor β_1 and β_2 are introduced to correct the Knudsen effect [9] and the particle evaporation effect [10], respectively. q_c is the heat flux when no Knudsen effect and evaporation exist, and it is calculated using the following equations for the Nusselt number defined as $Nu = 2q_c r_p/(S_{h_\infty} - S_{h_w})$:

$$Nu = 2\left[1 + 0.63 Re_\infty Pr_\infty^{0.8} \left(\frac{Pr_w}{Pr_\infty}\right)^{0.42} \left(\frac{\rho_\infty \mu_\infty}{\rho_w \mu_w}\right)^{0.52} C^2\right]^{0.5} \qquad (20)$$

where S_h is heat conduction potential, which is calculated by

$$S_h = \int_{T_0}^{T} \kappa dT \qquad (21)$$

where T_0 is an arbitrary reference temperature, Re Reynolds number, and Pr Prandtl number. The value of T_0 is taken to be 300 K The factor C is calculated by

$$C = \frac{\left[1 - \left(\frac{h_w}{h_\infty}\right)^{1.14}\right]}{\left[1 - \left(\frac{h_w}{h_\infty}\right)^{2}\right]} \qquad (22)$$

The subscripts w and ∞ stand for the particle surface and the gas, respectively. The value of q_r in Eqs. (15)-(18) is calculated by

$$q_r = \epsilon \sigma_{st} T_w^4 \qquad (23)$$

where ϵ is particle surface emissivity and σ_{st} is Stefan-Boltzmann constant. The differential equations of Eqs. (15), (17), and (18) are solved by fourth-order Runge-Kutta method.

2.3 Thermodynamic and transport properties

The chemical composition in plasma under the LTE state is determined by minimizing the Gibbs free energy with RAND method [11]. Thermodynamic properties are directly calculated from the particle number density and the particle partition function. The transport properties are calculated on the basis of Chapman-Enskog approximation [12].

2.4 Computational domain and boundary conditions

The computational domain, which is illustrated in Fig. 2, is composed of the plasma torch region (Fig.2(a)) and the plasma jet region (Fig.2(b)). The two regions are connected at the torch outlet and a substrate is not included in the computational domain. The torch geometry of the plasma torch is based on the plasma torch used in the experiment in Ref. [1]. Operating conditions are shown in Table 1. In this work, a ring-shaped anode is assumed and external magnetic field distribution in the plasma torch is given by a solenoid coil wound about the external surface of the anode-side nozzle wall. The electric potential on the cathode is set to 0 V and the one on the anode is determined so as to satisfy a specific operating current. A more detailed description of the numerical condition is found in Ref. [4].

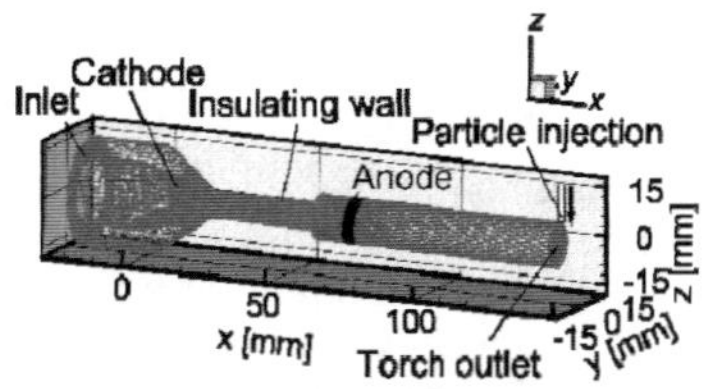

(a) Torch region

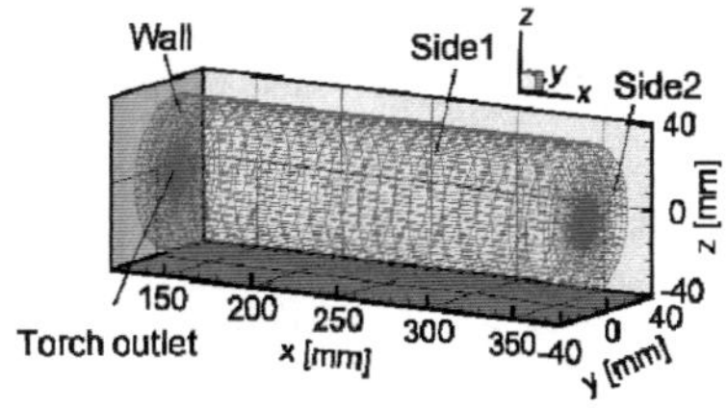

(b) Jet region

Fig. 2. Computational domain of (a) torch region and (b) jet region.

Table. 1. Operating conditions.

Gas	Current [A]	Flow rate [slm]	Externally applied magnetic field [T]
Ar	300	115	0.8
Ar	400	115	0.2
Ar	400	115	0.4
Ar	400	115	0.8
Ar	500	115	0.8

2.5 Particle injection conditions

Powder material specified in this work is ZrO_2 and its physical properties are specified by referring Ref. [13]. The standard Gaussian distribution is given for particle diameter, y and z component of particle injection velocity; please note that the coordinate system used in this work is illustrated in Fig. 2. The mean Initial particle diameter is set to 30 µm and the diameter of particle distributes in the range of about 22-38 µm. The mean injection velocity is (u_{px0}, u_{py0}, u_{pz0}) = (0 m/s, 0 m/s, -10 m/s). The injection point is (x_{p0}, y_{p0}, z_{p0}) = (143.5 mm, 0.0 mm, 6.7 mm).

3 Results and discussion

Figure 3(a) presents the time variation of the voltage between anode and cathode electrodes for different external magnetic fields $\boldsymbol{B}_0$ with operating current I = 400 A. The voltage increases by applying the external magnetic field. Without $\boldsymbol{B}_0$, the anode arc root stays on the same spot and no voltage variation occurs. For the case of max|$\boldsymbol{B}_0$| = 0.2 T, little variation is obtained under the present time scale due to weaker Lorentz force acting on the anode arc. For the case of max|$\boldsymbol{B}_0$| = 0.4 T and 0.8 T, the anode arc keeps rotating and the frequency of voltage fluctuation is almost equivalent to the one of the anode arc rotation on the ring-shaped anode. The result show the arc rotational frequency increases with $\boldsymbol{B}_0$ and the amplitude of the voltage variation is unproportional to the strength of the external magnetic field.

Figure 3(b) presents the time variation of the voltage for different operating currents with max|$\boldsymbol{B}_0$| = 0.8 T. With increasing the operating current, the average arc voltage gets higher. The number of the peak in 3 ms also increases with the velocity of the anode arc rotation. The amplitude of voltage fluctuation is small compared with the total voltage for each current condition. From Figs. 3(a) and 3(b), applying strong external magnetic fields is recommended for the anode arc rotation leading to elongating the anode lifetime. The anode arc rotation shows periodic behavior with a small amplitude, which is expected to be good for plasma spraying processes. It should be noted that in the present work a ring-shaped anode is assumed and voltage variation associated with anode arc movement along the torch axis is not included. In addition, the computed voltage is about twice as high as the experimental one [1]. We expect the assumption of the ring-shaped anode and the overestimation of the voltage will be overcome by introducing a thermally non-equilibrium model for plasma transport properties.

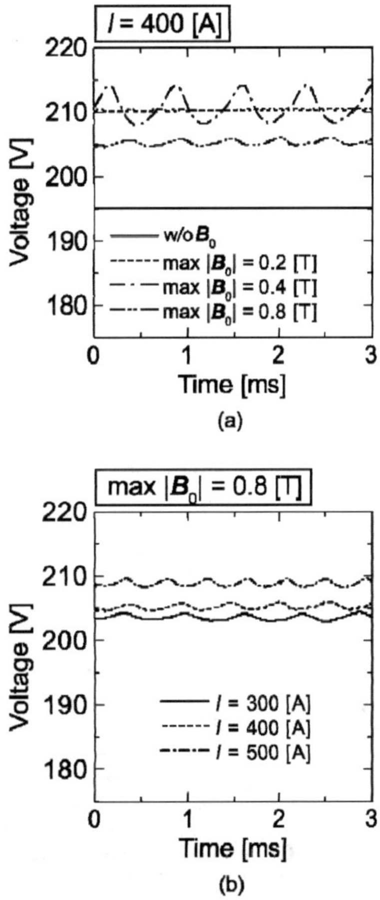

Fig. 3. Voltage time variation for (a) different external magnetic fields (I = 500 A), and (b) different operating currents (max|$\boldsymbol{B}_0$| = 0.8 T).

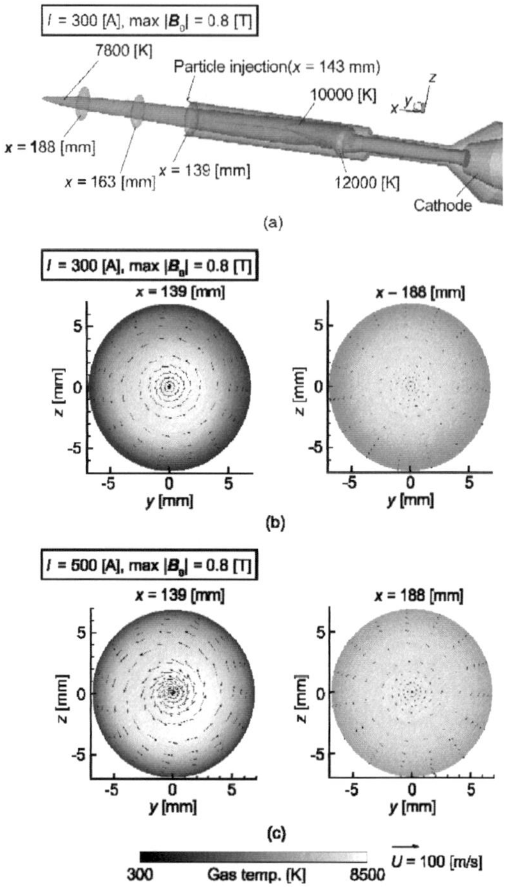

Fig. 4. (a) Iso-surface of plasma jet temperature (I = 300 A, max|$\boldsymbol{B}_0$| = 0.8 T), temperature and gas velocity vector distribution of different axial cross sections for (b) 300 A and (c) 500A (max|$\boldsymbol{B}_0$| = 0.8 T).

The iso-surface of plasma jet temperature and the temperature distribution of different cross sections are shown in Fig. 4. The vectors represent the gas velocity on the y-z plane. The Lorentz force generated by the interaction between the current and the external magnetic field rotates the anode arc in the direction counterclockwise when looking from the downstream. The rotation of the anode arc produces a swirl flow. As can be seen in Figs. 4(b) and 4(c), the temperature and the swirl velocity get higher as the operating current increases. The swirl flow carries particles in the direction counterclockwise just after particle injection near the torch outlet (the cross section at x = 139 mm), while the gas velocity of the swirl component rapidly diminishes downstream (the cross section at x = 188 mm), having little effect on the particle trajectories.

Figure 5 shows the ensemble distribution of the particle position, diameter, temperature, and liquid-phase fraction on y-z plane at 100 mm from the torch exit. The operating current is changed from 300 A to 500 A and the external magnetic field is 0.8 T. The origin corresponds to the axial centreline of the plasma torch where hot plasma jets come out. With increasing the operating current, stronger swirl flow deflects the particle positions in direction of the arc rotation and the particles pass through the cross section with more molten state due to higher joule heating. The results indicate the majority of the particles reaches up to the melting point of ZrO_2 (2950 K). As can be seen in Fig. 5, however, the particles radially injected from the upper edge of the plasma torch fly away from the origin due to the swirl flow and fail to pass through the hot plasma jet. The deflection by the swirl flow should be avoided to heat and accelerate the particles effectively. As explained in Fig. 4, the swirl flow strongly influences on the particle trajectories just after particle injection. If the particle trajectory deflection by swirl flow is reduced at injection point, the operation of a plasma torch with an externally applied magnetic field will be easier to control. The simple idea for the problem is to adjust the particle injection angle. Figure 6 shows the liquid-phase distribution for the different angles of the particle injection with I = 500 A and max$|B_0|$ = 0.8 T. In this simulation, the y-component of mean injection velocity is given for particles to match a specified injection angle.

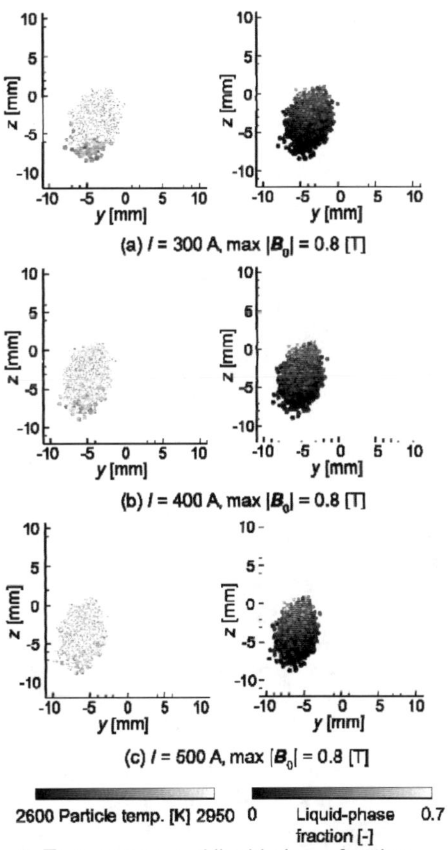

Fig. 5. Temperature and liquid-phase fraction distribution for operating current of (a) 300 A, (b) 400 A, and (c) 500 A (max$|\mathbf{B}_0|$ = 0.8 T, x = 243.5 mm).

The mean injection velocity of z-component is set to -10 m/s. Giving injection angle so as to be opposite to the swirl flow direction can cancel the particle deflection and the injected particles pass through the hot plasma jet.

Figure 7 shows the frequency count of the liquid-phase fraction of the particles injected from different injection angles. The operating current and the external magnetic field is 500 A and 0.8 T, respectively. The results in Fig. 7 show that more particles change into the liquid-phase as the particle injection angle increases. For the case of θ = 30°, the particles fly in the hot plasma plume and pass through the x-plane around the origin as can be seen in Fig. 6. The adjustment of the particle injection angle could be an effective way for heating particles in a plasma spray with an externally applied magnetic field.

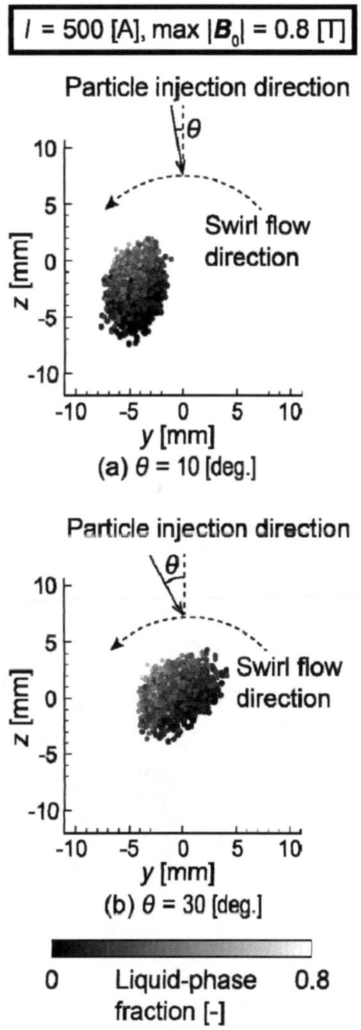

Fig. 6. Liquid-phase fraction for particle injection angle of (a) θ = 10° and (b) θ = 30° (I = 500 A, max$|\mathbf{B}_0|$ = 0.8 T, x = 243.5 mm).

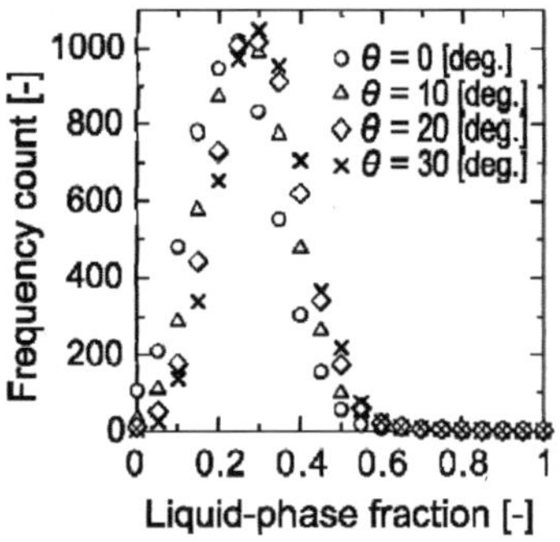

Fig. 7. Frequency count of liquid-phase fraction (I = 500 A, max|$\boldsymbol{B}_0$| = 0.8 T, x = 243.5 mm).

4 Conclusions

The authors performed a time-dependent, three-dimensional MHD numerical simulation of plasma jets generated by a DC plasma torch with externally applied magnetic fields. We also numerically analysed the behavior of in-flight particles in the plasma jet by using the Lagradian approach. The numerical results mainly obtained in this work are summarized as follows:

(1) The frequency of voltage fluctuation is almost equivalent to the rotation frequency of arc's root on the anode.
(2) The swirl velocity of plasma jets increases with arc currents.
(3) The voltage variation due to the anode arc rotation shows periodic behavior with a small amplitude.
(4) Particle injection with a diagonal angle to the vertical direction against the rotation direction of swirling plasma jet is suitable for introducing the particles into a hot core of the swirling plasma jet.

In this work, a ring-shaped anode, not an axially long and cylindrical-shaped anode used by conventional plasma torches, was adopted to artificially set the position of the anode arc root for the axial direction. This artificial manipulation has to be removed to clarify the physical mechanism of the suppression of voltage fluctuation due to rotary arc effects by numerical simulation. In addition, a turbulent model and a thermally non-equilibrium model should be implemented in our computational program, because we predict from Refs. [14,15] that these models could impact on the voltage fluctuation strongly.

Acknowledgement

A part of the work was carried out under the Collaborative Research Project of the Institute of Fluid Science, Tohoku University. The calculations were performed using SGI UV 1000 of super-computing system in the Institute of Fluid Science, Tohoku University.

Literature

[1] Pershin, L., Mitrasinovic, A., and Mostaghimi, J.: Treatment of refractory powders by a novel, high enthalpy dc plasma. Journal of Physics D: Applied Physics 46 (2013), Issue 22, 224019.
[2] Hu, M., Wan, S., Xia, Y., Ren, Z. and Wang, H.: Effect of an external magnetic field on plasma torch discharge fluctuation. Europhysics Letters 102 (2013), Issue 5, 55002.
[3] Park, J. M., Kim, K. S., Hwang, T. H., and Hong, S. H.: Three-dimensional modeling of arc root rotation by external magnetic field in nontransferred thermal plasma torches. IEEE Transactions on plasma science 32 (2004), Issue 2, pp. 479/487.
[4] Saito, H., Nakane, Y., Fujino, T, and Takana, H.: Numerical analysis of in-flight particles in plasma jet with an externally applied magnetic field. Journal of Fluid Science and Technology 11 (2016), Issue 4, 16-00184.
[5] Wada, Y and Liou, M. S.: A flux splitting scheme with high-resolution and robustness for discontinuities. AIAA paper, (1994), 94-0083.
[6] Yoon, S. and Jameson, A.: Lower-upper symmetric-Gauss-Seidel method for the Euler and Navier-Stokes equations. AIAA Journal 26 (1988), Issue 9, pp. 1025/1026.
[7] Gutknecht, M. H.: Variants of BICGSTAB for matrices with complex spectrum. SIAM Journal on Scientific Computing 14 (1993), Issue 5, pp. 1020/1033.
[8] Li, H.-P. and Chen, X.: Three-dimensional modeling of the turbulent plasma jet impinging upon a flat plate and with transverse particle and carrier-gas injection. Plasma Chemistry and Plasma Processing 22, (2002), Issue 1, pp. 27/58.
[9] Chen, X. and Pfender, E.: Effect of the Knudsen number on heat transfer to a particle immersed into a thermal plasma. Plasma Chemistry and Plasma Processing 3 (1983), Issue 1, pp. 97/113.
[10] Chen, X. and Pfender, E.: Heat transfer to a single particle exposed to a thermal plasma. Plasma Chemistry and Plasma Processing 2 (1982), Issue 2, pp. 185/212.
[11] White, W. B., Johnson, S. M. and Dantzig, G. B.: Chemical equilibrium in complex mixtures. The Journal of Chemical Physics 28 (1958), Issue 28, pp. 751/755.
[12] Yos, J. M.: Transport properties of nitrogen, hydrogen, oxygen, and air to 30,000 K. AVCO Tech. Mem. RAD-TM-63-7, ASTIA Doc. AD-435-053, (1963).
[13] Xiong, H.-B., Zheng, L.-L., Sampath, S., Williamson, R. L., and Fincke, J. R.: Three-dimensional simulation of plasma spray: effects of carrier gas flow and particle injection on plasma jet and entrained particle behavior. International Journal of Heat and Mass Transfer 47 (2004), Issue 24, pp. 5189/5200.
[14] Trelles, J. P., Heberlein, J. V. R, and Pfender, E.: Non-equilibrium modelling of arc plasma torches. Journal of Physics D: Applied Physics 40 (2007), Issue 19, pp. 5937/5952.
[15] Trelles, J. P.: Computational study of flow dynamics from a dc arc plasma jet. Journal of Physics D: Applied Physics 46 (2013), Issue 25, 255201.

Numerical investigation of the melting degree of ceramic powder particles during air plasma spraying

K. Bobzin, M. Öte, M. A. Knoch, I. Alkhasli, Aachen / D

Air plasma spraying is a variation of thermal spraying that is used, among others, for the production of thermal barrier and wear resistant coatings. High plasma temperatures enable the processing of ceramic powder particles which have a high melting point and cannot be processed otherwise. Due to their low heat conductance, the ceramic particles are not necessarily fully melted during their flight in the free jet and prior to the impact on the substrate surface. Experimental particle temperature measurements by means of particle diagnostics systems deliver merely the surface temperature of the particles while the melting degree of the ceramic particles remains unknown. Therefore, the temperature field within spherical Al_2O_3 particles is numerically investigated for a commonly used particle size distribution by considering different particle sizes. The model includes a two-way coupled particle-laden free jet model and takes the latent heat of melting and evaporation into account. The effect of the particles size as well as the stand-off distance on the melting degrees of the particles in the given powder size distribution is determined.

1 Introduction

In thermal spraying, heated feedstock particles are accelerated and impacted on the substrate. The melting degree of the particles before their impact plays an important role in the coating deposition behaviour and the resultant microstructure. The internal heat transfer within particles can influence the molten state of the particles. The Biot number is a dimensionless quantity which helps to estimate the relevance of the internal heat transfer within a particle. The assumption of homogenous heating of the particles is considered to be valid for processes where the Biot number is lower than 0.02 [1, 2]. For such cases, the lumped capacitance method is commonly implemented in simulations in order to determine the particle temperatures [3–6]. For the simulation of the air plasma spraying (APS) of ceramic feed stock materials, the validity of this assumption is no longer obvious. For the case considered in the present study the Biot number is well above 0.2 for the most part of the free jet, as shown in **Fig. 1**. Particle diagnostic methods that determine the particle in-flight temperatures rely on the emission from the particle surface and are therefore limited to determining only the surface temperatures [7] while the melting degree of the particle remains unknown. In a simplified approach, the melting rate can be correlated with the heat transfer rate between the plasma gas and the particle and the averaged conductive heat transfer rate within the particle [8, 9]. This method has an advantage of determining the average melting degrees of in-flight particles based on the energy balance without the need to resolve the heat transfer within them. Several studies have also addressed the issue of determining the melting degrees of particles by taking the heat transfer within the injected particles during their flight in the free jet into account [10–14]. Sprayed particles are commonly assumed to have spherical symmetry, which allows the reduction of the heat conduction to a one dimensional model. The present work proposes a similar approach to determine the temperature distributions within the particles, where the heat transfer within the particles is not calculated during the simulation of the free jet but in a segregated fashion.

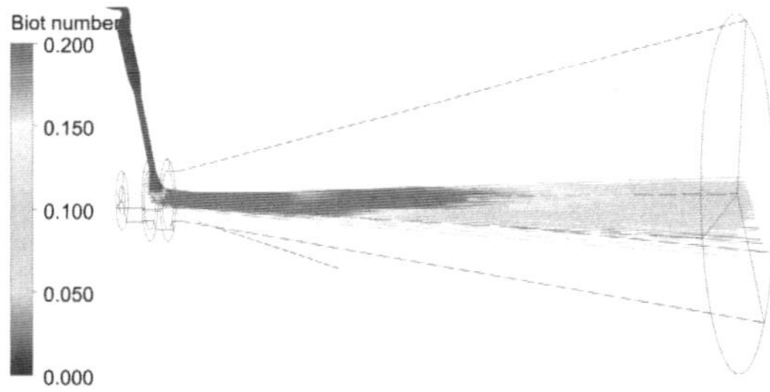

Fig. 1. Particle trajectories in the free jet coloured according to their local Biot numbers.

A plasma jet generated by a three cathode plasma generator, TriplexPro™-210 from Oerlikon Metco, is modelled in this work. An a-priori conducted plasma generator simulation with process parameters of $\dot{V}$ = 50 SLPM volumetric Argon gas flux and I = 500 A electric current is used to determine the temperature and velocity distributions at the outlet of the generator. This information forms the inletting boundary conditions of the free jet simulation. Temperature gradients within Al_2O_3 particles of the sizes between Ø = 22 - 60 µm are investigated for the resultant free jet. The following sections describe the modelling of the free jet and the particle heating. They are then followed by the results and discussion and a summary.

2 Modelling of the plasma generator and the free jet

Due to the multiphysical phenomena and a wide range of time and length scales, the numerical analysis of air plasma spraying is divided into several sub-models which are calculated separately. A detailed description of the multi-model approach for the simulation of APS as well as the modelling of the TriplexPro™-210 plasma generator is given in previous works of the authors [8, 15–19]. A brief description of the plasma generator model and the plasma flow is given in the following paragraph.

The electrical energy is converted into the heat energy in the plasma generator. This is achieved by ionizing the process gas between the electrodes, which increases the conductivity of the gas and allowing the electrical current to flow. The process gasses enter the plasma generator from 24 inclined inlets which cause a swirl of the process gases and are, thus, elongating the plasma arc length. The cathodes and the anode ring at the outlet of the plasma generator are separated with a section of neutrode rings. Three radially placed powder injection units are located near the outlet of the plasma generator. When the process gas exits the nozzle of the plasma generator in plasma state, it can have temperatures T_P > 12,000 K and velocities v_P > 1,000 m/s [19, 20]. Flow characteristics at the nozzle exit of the three cathode plasma generator TriplexPro™-210 exhibit significantly less fluctuations in time than the single cathode plasma generators [17]. The stability of the plasma characteristics enables the use of a steady state simulation of the plasma flow. The fluid model of the plasma, as opposed to the kinetic model, is used to simulate the magneto-hydrodynamic flow phenomena inside the plasma generator. This modelling approach determines the macroscopic parameters that describe the plasma flow by solving the conservation of mass, momentum and energy equations, which are coupled with the Maxwell equations. The forces of the electromagnetism and the fluid flow are coupled by the Lorenz force. The electric heating of the process gas in the plasma generator is defined by the Joule heating. The radiation model is based on the temperature and pressure dependent net emission model coefficients, according to [21, 22]. The influence of the injected particles on the plasma is defined by heat and momentum sources which ensure the particle-gas coupling within the model. The temperature and pressure dependent material properties of the process gas is taken from the literature [23, 24]. The turbulent mixing and heat dissipation of the plasma jet in the environment is taken into account by an established two-equation SST turbulence model [8, 16, 17].

In the present work, the plasma serves as heat source for the particles in the free jet. Particle kinetics are simulated in the free jet using a steady state model, in which each particle trajectory is associated with a particle of a certain size. Despite the steady state nature of the simulation, the particles are assigned a flight time variable according to their location on their trajectory. The particles absorb heat from the plasma depending along their position on their respective flight trajectory and, therefore, depending on their flight time. The heat sources that the particles extract from the plasma along their flight trajectory are coupled with an external solver, which resolves the dynamics of the particle heating. Heat losses due to particle heating are passed on to the free-jet model to ensure the two-way thermal particle-gas coupling. This kind of segregated approach for plasma-particle coupling allows more realistic particle models to be implemented in the future, which could for example consider a particle of an arbitrary shape, its surface roughness, porosity and inhomogeneous chemical composition.

Convective heat exchange and the aerodynamical drag coefficients are described in various ways in the literature [8, 9]. An approach described in [25] is used for the impulse transfer in this work. The particles injected into the plasma jet are heated up and carried by the gas. Except for low mass flow rates, the influence of the particles on the plasma temperatures and velocity is not negligible [26–28]. The influence of the particles is incorporated back into the plasma by coupling the free jet and the particle heating models with heat and momentum sources obtained from the calculation of the particle heating.

3 Modelling of the particle heating

Flight trajectories of the injected particles through the plasma and the free jet are determined using the model described in the previous section. Particle slip velocity and the plasma gas temperature that the particles are experiencing as well as the gas properties at the given temperature determine the heat flux on the particles which serves as the heat source boundary condition for the transient model of the particle heating. The equations of the transient heat transfer including the melting and the evaporation enthalpies, as well as the interaction conditions of the particle with the plasma jet are solved numerically to establish a variation of the temperature gradient within the particle along its flight trajectory as shown schematically in **Fig. 2**.

It should be noted that the two-way coupling between gas and particulate phases is accomplished by taking into account only the lumped capacitance model of the particle heating. The difference in the heat exchange between the gas and the particulate phases has not been corrected according to the newly calculated temperature rise on the surface of the particle due to the inhomogeneous temperature distribution within the particles.

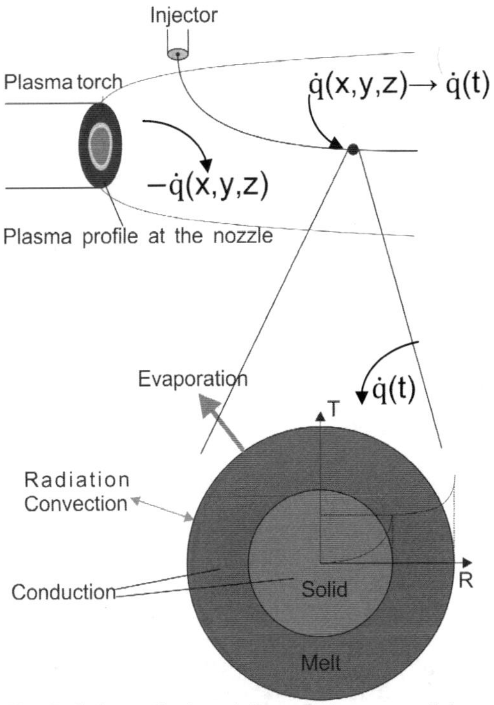

Fig. 2. Schematic description of plasma-particle coupling along the particle flight trajectory

The material properties of the powder material are given in **Table 1**. The following assumptions were made while modelling the particle heating:

- There exists a spherical symmetry of:
  - particle geometry
  - particle temperature distribution
  - local plasma temperature and the velocity distributions along the boundary layer.
- Evaporation on the particle surface doesn't influence the heat exchange between the particle and the plasma.
- Particle size can be influenced by the evaporation.

Table 1. Material properties of the feedstock material

Property, unit	Value
Material	Al_2O_3
Melting temperature, *K*	2,327
Boiling temperature, *K*	3,250
Latent heat of melting, J g^{-1}	700
Latent heat of vaporization, J g^{-1}	19,380
Material density, g cm^{-3}	3.690
Heat capacity, J g^{-1} K^{-1}	NASA polynomials [29]
Thermal conductivity, W m^{-1} K^{-1}	6

4 Results and discussion

The model described above was used to determine the temperature gradients and the melting degrees of the particles with three different particle sizes. Four stand-off distances were chosen for the comparison of the temperature gradients within the particles. **Fig. 3** illustrates the temperature gradients of the particles with Ø = 22, 40 and 60 µm diameter at the stand-off distance of d = 5, 7, 12 and 17 cm. r/R denotes the radial coordinate normalized by the particle radius. The stand-off distances are calculated downstream the plasma generator nozzle.

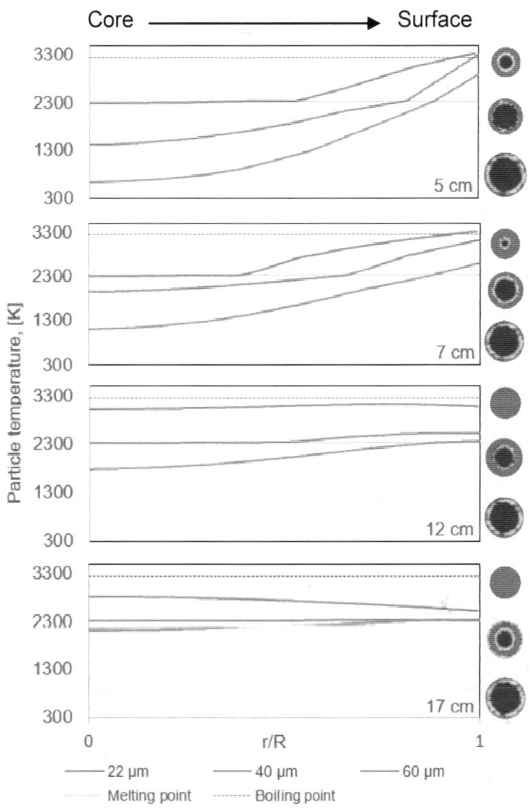

Fig. 3. Temperature gradients of particles of different size and their cross-sections coloured by their melting degrees at different stand-off points.

Highest temperature gradients are observed at the d = 5 cm distance. At d = 12 cm the gradients starting to level out. At d = 17 cm, the cooling at the surface of the Ø = 22 µm particle is visible due to the fact that the particle was heated to a higher temperature than the gas temperature at the given stand-off distance. The melting degrees of the particles are visualised alongside the temperature gradients, with the colour red denoting the liquid phase and the blue the solid phase. As expected, the smallest particle shows the fastest melting; it is completely molten at stand-off distances beyond d = 12 cm. The Ø = 40 µm sized particle shows partial melting throughout the flight path. At the distance of d = 17 cm, it can be seen that a thin layer of solid material has started to form enclosing the inner liquid and solid phases. It should be noted that a thin layer of solidified material is highly instable in reality and would probably break under the aerodynamical forces of the plasma jet and mix with the liquid phase. The Ø = 60 µm particle stays at approximately the same melting state at all investigated stand-off distances, despite the variation of its temperature gradient. For all 4 particles, it is evident that the advancement of the temperature gradient paused at the melting temperature of the material due to the latent heat required for the phase transformation.

5 Summary and outlook

The plasma spraying process has been simulated using sequentially coupled sub-models. The plasma generator model has been coupled with the free jet model which subsequently was coupled with the particle heating model. Melting degrees of different sized particles were determined at different stand-off distances by taking into account the heat transfer and the phase change within the particles. In further studies, more realistic particle morphologies as well as a simplified approach for the determination of the melting degrees of the particles can be considered. The melting degrees obtained with the presented method can be compared to melting degrees of particles obtained without resolving the heat transfer within the particles through a simplified energy balance between gas and particle. Moreover, the heat exchange between the gas and the particulate phases can be corrected by taking into account the temperature rise of the particle surface which has been calculated in the present work to give a more realistic representation of the heat exchange.

6 Acknowledgements

All presented investigations were conducted in the context of the Collaborative Research Centre SFB1120 "Precision Melt Engineering" at RWTH Aachen University and funded by the German Research Foundation (DFG). For the sponsorship and the support we wish to express our sincere gratitude.

7 References

[1] Wei DYC, Farouk B, Apelian D, 19, 1988 Melting metal powder particles in an inductively coupled r.f. plasma torch. *MTB:*213–26. doi:10.1007/BF02654205.

[2] Bourdin E, Fauchais P, Boulos MI, 26, 1983 Transient Heat Conduction under Plasma Conditions. *Int. J. Heat Mass Transf.:*567–82. doi:10.1016/0017-9310(83)90008-X.

[3] Fauchais P, 37, 2004 Understanding plasma spraying. *J. Phys. D: Appl. Phys.:*86–108. doi:10.1088/0022-3727/37/9/R02.

[4] Khelfi D, El-Hadj AA, Aït-Messaoudène N, 2008 Modeling of a 3D plasma thermal spraying and the effect of the particle injection angle. *In Conference Proceeding of Revue des Energies Renouvelables CISM'08:*205–16.

[5] Pasandideh-Fard M, Pershin V, Chandra S, Mostaghimi J, 2002 Splat shapes in a thermal spray coating process: Simulations and experiments. *Journal of Thermal Spray Technology:*206–17. doi:10.1007/BF02648528.

[6] Vardelle M, Vardelle A, Leger AC, Fauchais P, Gobin D, 1995 Influence of particle parameters at impact on splat formation and solidification in plasma spraying processes. *Journal of Thermal Spray Technology:*50–8. doi:10.1007/BF02648528.

[7] Fauchais P, Vardelle M, 2010 Sensors in Spray Processes. *Journal of Thermal Spray Technology:*668–94. doi:10.1007/s11666-009-9435-x.

[8] Öte M, 2016 Understanding Multi-Arc Plasma Spraying. 1. Auflage. Herzogenrath: Shaker ISBN:978-3-8440-4598-7.

[9] Bobzin K, Öte M, 2016 Modelling the plasma-particle interaction in multi-arc plasma spraying. *Journal of Thermal Spray Technology:*279–91. doi:10.1007/s11666-016-0514-5.

[10] Borisov Y, Bushma A, Krivtsun I, 15, 2006 Modeling of motion and heating of powder particles in laser, plasma, and hybrid spraying. *Journal of Thermal Spray Technology:*553–8. doi:10.1361/105996306X146884.

[11] Ahmed I, Bergman TL, 9, 2000 Three-dimensional simulation of thermal plasma spraying of partially molten ceramic agglomerates. *Journal of Thermal Spray Technology:*215–24. doi:10.1361/105996300770349953.

[12] Chen X, Chyou YP, Lee YC, Pfender E, 5, 1985 Heat Transfer to a Particle under Plasma Conditions with Vapor Contamination from the Particle. *Plasma Chemistry and Plasma Processing:*119–41. doi:10.1007/BF00566210.

[13] Dyshlovenko S, Pawlowski L, Pateyron B, Smurov I, Harding JH, 200, 2005 Modelling of plasma particle interactions and coating growth for plasma spraying of hydroxyapatite. *Surface and Coatings:*3757–69. doi:10.1016/j.surfcoat.2005.04.002.

[14] Bobzin K, Öte M, Knoch MA, Alkhasli I, Reisgen U, Mokrov O, et al., 2017 Simulation of the Particle Melting Degree in Air Plasma Spraying. *Conference Proceedings of High Tech Plasma Processes HTPP 14.*

[15] Bobzin K, Öte M, Linke TF, Alkhasli I, Hopmann C, Nikoleizig P, et al., 3, 2015 Development of Simulative Approaches for Precisely Designing the Properties of Plasma Sprayed Coatings for Application in Injection Moulding. *Conference Proceedings of the YIC GACM 2015:*6–10.

[16] Bobzin K, Öte M, Schein J, Zimmermann S, Möhwald K, Lummer C, 25, 2016 Modelling the Plasma Jet in Multi-Arc Plasma Spraying. *Journal*

*of Thermal Spray Technology:*1111–26. doi:10.1007/s11666-016-0438-0.
[17] Bobzin K, Öte M, 25, 2016 Modeling Multi-Arc Spraying Systems. *J Therm Spray Tech:*920–32. doi:10.1007/s11666-016-0407-7.
[18] Bobzin K, Öte M, 2015 A Numerical Investigation: Air Plasma Spraying by means of a Three-Cathode Spraying Torch. *In Proceedings of International Thermal Spray Conference and Exposition (ITSC):*217–22 eISBN: 978-1-62708-093-4.
[19] Bobzin K, Kopp N, Warda T, Schäfer M, Öte M, 2013 A Numerical Investigation: Influence of the Operating Gas On the Flow Characteristics of a Three-Cathode Air Plasma Spraying System. *In Proceedings of International Thermal Spray Conference and Exposition (ITSC):*400–5 ISBN-13: 978-1-62708-021-7.
[20] Bobzin K, 2013 Oberflächentechnik für den Maschienenbau. Germany: Weinheim Wiley-VCH Verlag ISBN:978-3-527-33018-8.
[21] Bauder U, 39, 1968 Radiation from high-pressure plasmas. *Journal of Applied Physics:*148–52. doi:10.1063/1.1655721.
[22] Menart J, Malik S, 35, 2002 Net emission coefficients for argon-iron thermal plasmas. *J. Phys. D: Appl. Phys.:*867–74. doi:10.1088/0022-3727/35/9/306.
[23] Murphy AB, Arundelli CJ, 14, 1994 Transport coefficients of argon, nitrogen, oxygen, argon-nitrogen, and argon-oxygen plasmas. *Plasma Chemistry and Plasma Processing:*451–90. doi:10.1007/BF01570207.
[24] Murphy AB, Tam E, 47, 2014 Thermodynamic properties and transport coefficients of arc lamp plasmas: argon, krypton and xenon. *Journal of Physics D: Applied Physics:*1–10. doi:10.1088/0022-3727/47/29/295202.
[25] Vardelle M, Vardelle A, Fauchais P, Boulos MI, 9, 1983 Plasma-particle momentum and heat transfer: Modeling and measurements. *AIChE Journal:*236–43. doi:10.1002/aic.690290210.
[26] Lamoreaux RH, Hildenbrand DL, Brewer L, 16, 1987 High-Temperature Vaporization Behavior of Oxides II. Oxides of Be, Mg, Ca, Sr, Ba, B, Al, Ga, In, Tl, Si, Ge, Sn, Pb, Zn, Cd, and Hg. *Journal of physical and chemical reference data:*419–43. doi:10.1063/1.555799.
[27] Bobzin K, Öte M, 2017 Numerical Coupling of the Particulate Phase to the Plasma Phase in Modeling of Multi-Arc Plasma Spraying. *Journal of Physics, Conference Series.*
[28] Bobzin K, Öte M, 2017 Influence of Powder Particles on the Plasma Characteristics in Multi-arc Plasma Spraying. *Conference Proceedings of High Tech Plasma Processes HTPP 14.*
[29] Burcat A, Ruscic B, 2005 Third Millennium Ideal Gas and Condensed Phase Thermochemical database for combustion with updates from active thermochemical tables. Lemont: Argonne National Laboratory.

Thermal conductivity and interfacial thermal resistance of thermal barrier coatings

Satoru Takahashi, Tokyo / J, Megumi Akoshima, Tsukuba / J, Akihiro Kanno, Akashi / J, Tatsuo Suidzu, Akashi / J

A method measuring the thermal conductivity and the interfacial thermal resistance of thermal barrier coatings (TBCs) which consist of metallic bond-coats (BCs) and ceramics top-coats (TCs) on superalloys was newly developed. It was based on the areal heat diffusion time method analysing the heat diffusion across multilayers. The developed method was experimentally verified using the BC and the TBC specimens coated by APS. It was found that there were the interfacial thermal resistance not only between the TC and the BC but also between the BC and the substrate. Furthermore, the thermal conductivities of the BC and the TC obtained from the BC and the TBC specimens by this method considering the interfacial thermal resistance were in good agreement with those measured from the free-standing specimen of each coating. Thus, it was confirmed that the newly developed method is effective to evaluate the thermal conductivity and the interfacial thermal resistance of the TBC.

1 Introduction

Thermal barrier coatings (TBCs) are highly advanced material systems, generally applied to surfaces of hot-section components such as combustors, blades, and vanes of power-generation gas turbines in thermal power plants. A function of the TBC is to protect metallic components for extended periods at elevated temperatures by employing thermally insulating materials. The TBC consists of the metallic bond-coat (BC) and the ceramic top-coat (TC) on the superalloy. The thermal barrier effect of the TBC depends strongly on the thermal conductivity of the TC. Therefore, the thermal conductivities of various TCs have been reported [1-3]. In many studies, the free-standing specimens of the TCs were used to determine the thermal conductivity. In the components which the TBC is applied, however, there are the interfaces not only between the TC and the BC but also between the BC and the substrate. Hartmann et al. proposed the thermal diffusivity measurement on two-layered and three-layered systems with the laser flash method, and pointed out the importance of measuring the thermal contact resistance of layered specimens [4]. Accordingly, it is necessary to know the interfacial thermal resistance between the layers as well as the thermophysical properties of each layer in order to understand the internal heat transfer in the TBC. Baba proposed the areal heat diffusion time method [5]. This is a technique to analyse the one-dimensional heat diffusion across multilayers with and without the boundary thermal resistance based on the impulse response method in Laplace transform.

In the present study, the method measuring the thermal conductivity of each layer and the interfacial thermal resistance between the layers in the TBC separately was developed based on the areal heat diffusion time method. The developed method was verified experimentally using multilayer specimens.

2 Method measuring thermal conductivity and interfacial thermal resistance of TBC

2.1 Areal heat diffusion time method for multilayers [5]

The laser flash method is a well-known technique for measuring the thermal diffusivity of solid materials. The front surface of the specimen is pulse-heated using by a uniform laser beam and the temperature change of the rear surface is detected using an infrared radiometer. The areal heat diffusion time is defined as the areal surrounded by the horizontal line at the height of the maximum temperature rise and the temperature rise curve on the rear surface after pulsed heating as shown in **Fig.1**.

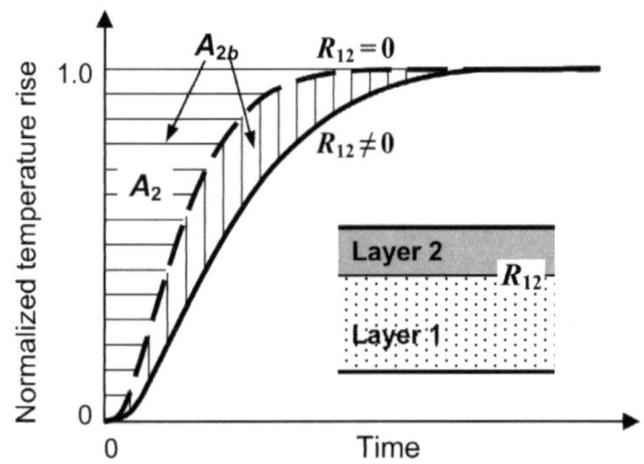

Fig.1. Areal heat diffusion time defined as the area surrounded by the horizontal line at the height of the maximum temperature rise and the temperature rise curve. This shows the case of the two layered model.

2.1.1 Two layered model

The areal heat diffusion time A_2 (**Fig.1**) across the two layered model without the interfacial thermal resistance is expressed as

$$A_2 = \frac{b_1\tau_1^{3/2} + 3b_2\tau_1\tau_2^{1/2} + 3b_1\tau_1^{1/2}\tau_2 + b_2\tau_2^{3/2}}{6(b_1\sqrt{\tau_1} + b_2\sqrt{\tau_2})}, \quad (1)$$

where τ_i is the heat diffusion time and b_i is the thermal effusivity of the i-th layer. From eq. (1), the heat diffusion time of the 2nd layer τ_2 is

$$\tau_2 = \frac{1}{3C_1 + C_2}\left[6(C_1 + C_2)A_2 - (C_1 + 3C_2)\tau_1\right] \quad (2)$$

$$C_i = c_i\rho_i d_i \ ,$$

where c_i is the specific heat capacity, ρ_i is the density and d_i is the thickness of the i-th layer, respectively. Since thermal diffusivity of each layer is defined as

$$\alpha_i = \frac{d_i^2}{\tau_i} \; , \tag{3}$$

α_2 is calculated using eqs. (2) and (3) when A_2 measured from the two layered specimen, α_1 , c_1 , ρ_1 , d_1 , c_2 , ρ_2 and d_2 are known.

Similarly, the areal heat diffusion time A_{2b} (**Fig.1**) across the two layered model with the interfacial thermal resistance R_{12} between the 1st layer and the 2nd layer is expressed as

$$A_{2b} = A_2 + \frac{b_1 b_2 R_{12} \tau_1^{1/2} \tau_2^{1/2}}{b_1\sqrt{\tau_1} + b_2\sqrt{\tau_2}} \; . \tag{4}$$

From eq. (4), the heat diffusion time of the 2nd layer τ_2 is expressed as

$$\tau_2 = \frac{1}{3C_1 + C_2}\begin{bmatrix} 6(C_1 + C_2)A_{2b} - (C_1 + 3C_2)\tau_1 \\ -6C_1C_2R_{12} \end{bmatrix} . \tag{5}$$

A_{2b} can be obtained from the temperature rise curve of the two layered specimen. But, τ_2 and R_{12} are not able to be determined from only this equation separately, because both are unknown.

2.1.2 Three layered model

The areal heat diffusion time A_3 across the three layered model without the interfacial thermal resistance is expressed as

$$A_3 = \frac{\begin{bmatrix} b_1\tau_1^{1/2}\left(\frac{\tau_1}{6} + \frac{\tau_2}{2} + \frac{\tau_3}{2}\right) + b_2\tau_2^{1/2}\left(\frac{\tau_1}{2} + \frac{\tau_2}{6} + \frac{\tau_3}{2}\right) \\ + b_3\tau_3^{1/2}\left(\frac{\tau_1}{2} + \frac{\tau_2}{2} + \frac{\tau_3}{6}\right) + \frac{b_1 b_3}{b_2}\tau_1^{1/2}\tau_2^{1/2}\tau_3^{1/2} \end{bmatrix}}{b_1\sqrt{\tau_1} + b_2\sqrt{\tau_2} + b_3\sqrt{\tau_3}} . \tag{6}$$

From eq. (6), the heat diffusion time of the 3rd layer τ_3 is shown as

$$\tau_3 = \frac{\begin{bmatrix} 6(C_1 + C_2 + C_3)A_3 - (C_1 + 3C_2 + 3C_3)\tau_1 \\ -(3C_1 + C_2 + 3C_3 + 6\frac{C_1C_3}{C_2})\tau_2 \end{bmatrix}}{3C_1 + 3C_2 + C_3} . \tag{7}$$

Since A_3 is obtained from the temperature rise curve of the three layered specimen, α_3 can be calculated using eqs. (3) and (7) when α_1 , c_1 , ρ_1 , d_1 , α_2 , c_2, ρ_2 , d_2, c_3 , ρ_3 and d_3 are known.

Similarly, the areal heat diffusion time A_{3b} across the three layered model with the interfacial thermal resistance is expressed as

$$A_{3b} = A_3 + \frac{\begin{bmatrix} R_{12}b_1b_2\tau_1^{1/2}\tau_2^{1/2} + R_{23}b_2b_3\tau_2^{1/2}\tau_3^{1/2} \\ +(R_{12} + R_{23})b_1b_3\tau_1^{1/2}\tau_3^{1/2} \end{bmatrix}}{b_1\sqrt{\tau_1} + b_2\sqrt{\tau_2} + b_3\sqrt{\tau_3}} \; , \tag{8}$$

where R_{23} is the interfacial thermal resistance between the 2nd layer and the 3rd layer. From eq. (8), the heat diffusion time of the 3rd layer is expressed as

$$\tau_3 = \frac{\begin{bmatrix} 6(C_1 + C_2 + C_3)A_{3b} - (C_1 + 3C_2 + 3C_3)\tau_1 \\ -(3C_1 + C_2 + 3C_3 + 6\frac{C_1C_3}{C_2})\tau_2 \\ -6C_1(C_2 + C_3)R_{12} - 6C_3(C_1 + C_2)R_{23} \end{bmatrix}}{3C_1 + 3C_2 + C_3} . \tag{9}$$

c_i , ρ_i , d_i , A_{3b}, τ_1 , τ_2 are able to be measured. But, τ_3 , R_{12} and R_{23} can not be determined from only eq (9), because these are unknown.

2.2 Method measuring thermal conductivity and interfacial thermal resistance of TBC based on areal heat diffusion time method

2.2.1 BC specimen (Two layered material)

As shown in **Fig.2**, consider two kinds of two layered models only with the different thickness of the 2nd layer. Here, it is assumed that the interfacial thermal resistance R_{12} is the same in both models, and R_{12} and τ_2 are unknown. The thickness ratio of the 2nd layer for both models, k, is defined as

$$k = \frac{d_2'}{d_2} \quad (k \neq 1) \; . \tag{10}$$

From eqs. (3) and (10), the heat diffusion time of the 2nd layer for each model is related as follows:

$$\tau_2' = \frac{d_2'^2}{\alpha_2} = k^2\tau_2 \; . \tag{11}$$

From eqs. (5) and (11), the heat diffusion time of the 2nd layer for each model is expressed as

Model A

$$(3C_1 + C_2)\tau_2 + 6C_1C_2R_{12} = \begin{bmatrix} 6(C_1 + C_2)A_{2b} \\ -(C_1 + 3C_2)\tau_1 \end{bmatrix} \tag{12}$$

Model B

$$k^2(3C_1 + kC_2)\tau_2 + 6kC_1C_2R_{12} = \begin{bmatrix} 6(C_1 + kC_2)A_{2b}' \\ -(C_1 + 3kC_2)\tau_1 \end{bmatrix} \; . \tag{13}$$

From eqs. (12) and (13), the heat diffusion time of the 2nd layer τ_2 and the interfacial thermal resistance R_{12} are expressed as

$$\tau_2 = \frac{6C_1C_2(kL_1 - L_2)}{L_0} \tag{14}$$

$$R_{12} = \frac{(3C_1 + C_2)L_2 - k^2(3C_1 + kC_2)L_1}{L_0} \tag{15}$$

where

$$L_0 = 6kC_1C_2\left[(3C_1 + C_2) - k(3C_1 + kC_2)\right]$$

$$L_1 = 6(C_1 + C_2)A_{2b} - (C_1 + 3C_2)\tau_1$$

$$L_2 = 6(C_1 + kC_2)A_{2b}' - (C_1 + 3kC_2)\tau_1 \; .$$

By substituting τ_2 and d_2 into eq. (3), the thermal diffusivity of the 2nd layer is calculated. Then, the thermal conductivity of each layer is determined as

$$\lambda_i = \alpha_i \; c_i \; \rho_i, \qquad (16)$$

λ_i is the thermal conductivity of the i-th layer. Thus, the thermal diffusivity τ_2 , the thermal conductivity λ_2 and the interfacial thermal resistance R_{12} can be determined respectively by using the two models when c_i , ρ_i , d_i , τ_1 , A_{2b} and A_{2b}' are known.

In the case of the BC specimen, the 1st layer is the substrate and the 2nd layer is the BC. R_{12} indicates the interfacial thermal resistance between the BC and the substrate.

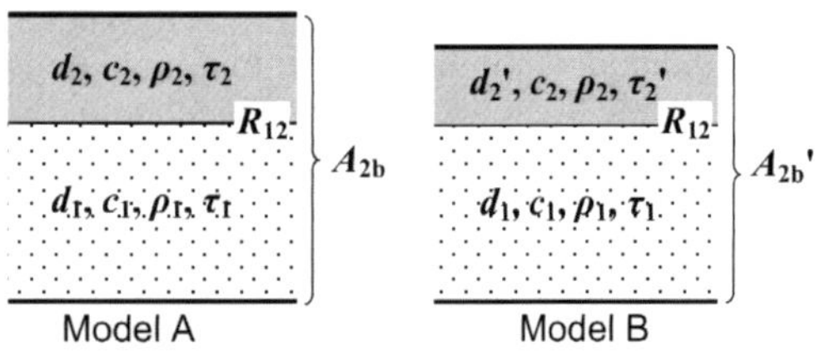

Fig.2. Two kinds of two layered models.

2.2.2 TBC specimen (Three layered material)

Similarly, consider two kinds of three layered models with only the different thickness of the 3rd layer, as shown in **Fig.3**. Here, it is assumed that R_{12} is known, and R_{23} and τ_3 are unknown. The thickness ratio of the 3nd layer for both models, p, is defined as

$$p = \frac{d_2'}{d_2} \quad (p \neq 1) . \qquad (17)$$

From eqs. (3) and (17), the heat diffusion time of the 3rd layer for each model is related as follows:

$$\tau_3' = \frac{d_3'^2}{\alpha_3} = p^2 \tau_3 \; . \qquad (18)$$

From eqs. (9) and (18), the heat diffusion time of the 3rd layer for each model is expressed as

Model C

$$(3C_1 + 3C_2 + C_3)\tau_3 + 6C_3(C_1 + C_2)R_{23} = \begin{bmatrix} 6(C_1 + C_2 + C_3)A_{3b} - (C_1 + 3C_2 + 3C_3)\tau_1 \\ -(3C_1 + C_2 + 3C_3 + 6\frac{C_1C_3}{C_2})\tau_2 - 6C_1(C_2 + C_3)R_{12} \end{bmatrix} \qquad (19)$$

Model D

$$p^2(3C_1 + 3C_2 + pC_3)\tau_3 + 6pC_3(C_1 + C_2)R_{23} = \begin{bmatrix} 6(C_1 + C_2 + pC_3)A_{3b}' - (C_1 + 3C_2 + 3pC_3)\tau_1 \\ -(3C_1 + C_2 + 3pC_3 + 6p\frac{C_1C_3}{C_2})\tau_2 - 6C_1(C_2 + pC_3)R_{12} \end{bmatrix} . \qquad (20)$$

From eqs. (19) and (20), the heat diffusion time of the 3rd layer τ_3 and the interfacial thermal resistance R_{23} are expressed as

$$\tau_3 = \frac{6C_3(C_1 + C_2)(pM_1 - M_2)}{M_0} \qquad (21)$$

$$R_{23} = \frac{(3C_1 + 3C_2 + C_3)M_2 - p^2(3C_1 + 3C_2 + pC_2)M_1}{M_0} \qquad (22)$$

where

$$M_0 = 6pC_3(C_1 + C_2)\begin{bmatrix} (3C_1 + 3C_2 + C_3) \\ - p(3C_1 + 3C_2 + pC_3) \end{bmatrix}$$

$$M_1 = \begin{bmatrix} 6(C_1 + C_2 + C_3)A_{3b} - (C_1 + 3C_2 + 3C_3)\tau_1 \\ -(3C_1 + C_2 + 3C_3 + 6\frac{C_1C_3}{C_2})\tau_2 \\ -6C_1(C_2 + C_3)R_{12} \end{bmatrix}$$

$$M_2 = \begin{bmatrix} 6(C_1 + C_2 + pC_3)A_{3b}' - (C_1 + 3C_2 + 3pC_3)\tau_1 \\ -(3C_1 + C_2 + 3pC_3 + 6p\frac{C_1C_3}{C_2})\tau_2 \\ -6C_1(C_2 + pC_3)R_{12} \end{bmatrix}.$$

By substituting τ_3 and d_3 into eq. (3), the thermal diffusivity of the 3rd layer is calculated. Furthermore, the thermal conductivity of the 3rd layer is determined from eq. (16). Thus, the thermal diffusivity τ_3, the thermal conductivity λ_3 and the interfacial thermal resistance R_{23} can be determined respectively by using the two models when c_i , ρ_i , d_i , τ_1 , τ_2 , A_{3b}, A_{3b}' and R_{12} are known.

In the case of the TBC specimen, the 1st layer is the substrate,the 2nd layer is the BC and the 3rd layer is the TC. R_{23} indicates the interfacial thermal resistance between the TC and the BC.

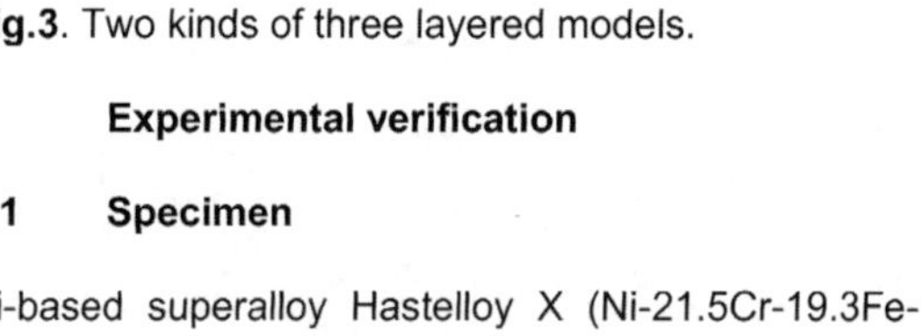

Fig.3. Two kinds of three layered models.

3 Experimental verification

3.1 Specimen

Ni-based superalloy Hastelloy X (Ni-21.5Cr-19.3Fe-8.3Mo-1.3Co-0.7Mn-0.6W-0.3Si, in mass %) was selected as the substrate alloy. Disk specimens, 10 mm in diameter and 1.5 mm in thickness, were machined from the substrate alloy. In this study, the substrate specimen, the BC specimen which was substrate with the BC and the TBC specimen which was substrate with TBC were prepared. The BC and the TC were coated by atmospheric plasma spraying (APS). The powder and the processing condition of each coating are summarized in **Table 1**. The substrate surface was grit-blasted, cleaned and subsequently coated with CoNiCrAlY (Co-32Ni-21Cr-8Al-0.5Y, in mass %, Oerlikon Metco / AMDRY 9954) as the BC. Then, YSZ (8 mass % Y_2O_3-stabilized ZrO_2, Oerlikon Metco / Metco 204NS) was used in TC spraying. The TC and the BC were fabricated under conventional spraying conditions. In the BC specimen, three kinds of specimens with different thickness of the BC were prepared. In the TBC specimen, the thickness of the BC and the TC was 0.30 and 0.50 mm respectively. The TC surface of the TBC specimen and the BC surface of the BC specimen were smoothen by the mechanical polishing. In addition, free-standing specimens of 1mm in thickness of the BC and the TC which were removed mechanically from the substrate were manufactured.

Figure 4 shows a typical cross-sectional microstructure as observed by back scattered electron (BSE) imaging for the TBC specimen. The TC has many micropores and microcracks.

Table 1. Coating powders and processing conditions.

	BC	TC
Powder	CoNiCrAlY	YSZ
Spray system	Oerlikon Metco / F-4	
Power (kW)	38 kW	45 kW
Plasma gas	Ar / He	Ar / He
Atmosphere	In air	In air
Spray distance (mm)	130	120
Thickness (mm)	0.10, 0.20, 0.30	0.50

Fig.4. Typcial cross-sectional microstructure of TBC.

3.2 Measurement of thermal conductivity and interfacial thermal resistance

The thermal diffusivity, the specific heat capacity and the bulk density were measured, and finally the thermal conductivity and the interfacial thermal resistance were determined. The thermal diffusivity was measured using a laser flash method according to ISO 18755 [6].

The measurements were carried out in air at room temperature. Prior to the measurement, the surfaces of the TC were vapor-deposited with a gold, and also were painted with a graphite.

The diffusivity of the BC and the interfacial thermal reistance between the BC and the substrate were evaluated using three kinds of BC specimens with the different thickness of the BC. On the other hand, the diffusivity of the TC and the interfacial thermal resistance between the TC and the BC were evaluated using one TBC specimen, because the surface roughness of the BC as well as the interfacial feature between the TC and the BC differ significantly from specimen to specimen. The TC of the TBC specimen was thinned stepwise by the mechanical polishing and the thermal diffusivity was measured at each stage.

The specific heat capacity was measured according to the DSC method. The bulk density was determined by measuring the mass and the volume. The physical properties are summarized in **Table 2.**

Table 2. Physical properties of substrate and coatings

	Sub.	BC	TC
c, J $kg^{-1} K^{-1}$	441	477	461
ρ, $x10^3$ kg m^{-3}	8.21	6.59	5.41

4 Results and discussions

4.1 Thermal conductivity and interfacial thermal resistacne of BC specimen

Figure 5 shows the thermal diffusivity of the BC. In the figure, the data obtained from the free-standing BC is also shown. Assuming without the interfacial thermal resistance R_{12} between the BC and the substrate, the thermal diffusivity determined from the BC specimen was lower than that of the free-standing BC. The difference between them increased with decreasing the thickness of the BC. Therefore, the interfacial thermal resistance R_{12} was calculated by combining two specimens among three BC specimens with the different thickness of the BC. From the three combinations of the specimens, the interfacial thermal resistance was calculated respectively. The interfacial thermal resistance determined from their average was R_{12} = 1.56 x 10^{-5} [m^2KW^{-1}]. Considering this thermal resistance, the thermal diffusivity of the BC calculated from the BC specimen was in good agreement with that of the free-standing BC. Namely, it was found that there was the interfacial thermal resistance between the BC and the substrate in the as-sprayed BC specimen. **Figure 6** shows the thermal conductivity of the BC. The thermal conductivity showed similar tendency as its thermal diffusivity. It was confirmed that the thermal conductivity and the interfacial thermal resistance of the BC specimen can be measured by using the developed method.

4.2 Thermal conductivity and interfacial thermal resistacne of TBC specimen

Figure 7 shows the thermal diffusivity of the TC. In the figure, the data obtained from the free-standing TC is also shown. Assuming without the interfacial thermal resistance not only R_{12} between the BC and the substrate but also R_{23} between the TC and the BC, the thermal diffusivity of the TC determined from the TBC specimen was lower than that of the free-standing TC. Therefore, the interfacial thermal resistance R_{23} was calculated by combining the results measured in two stages, which the thickness of the TC in the TBC specimen differs. In this study, the measurement was carried out in four stages. R_{12} determined from the BC system was used. The average of the interfacial thermal resistances calculated from the six combinations among the four stages was R_{23} = 7.05 x 10^{-5} [m^2KW^{-1}]. Considering the determined R_{23} and R_{12}, the thermal diffusivity of the TC calculated from the TBC specimen was almost the same as that of the free-standing TC. Namely, it was found that there was the interfacial thermal

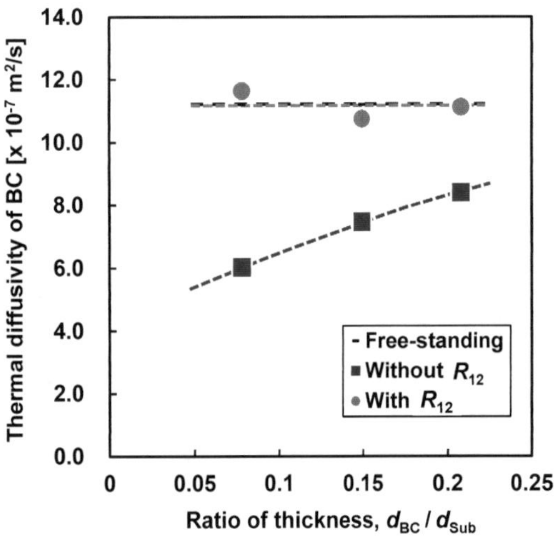

Fig.5. Thermal diffusivities of BC with and without R_{12}, and free-standing BC.

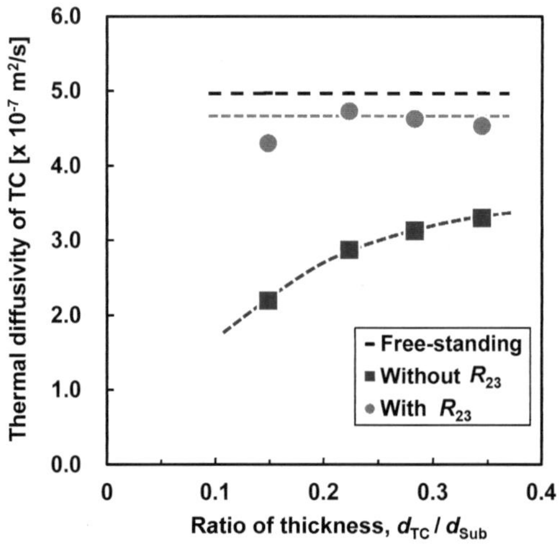

Fig.7. Thermal diffusivities of TC with and without R_{23}, and free-standing TC.

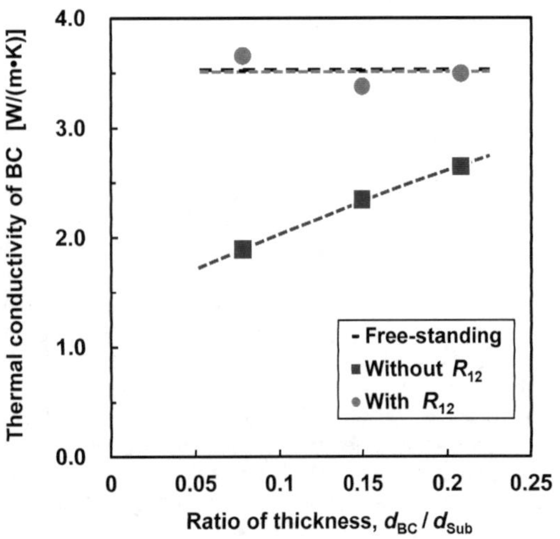

Fig.6. Thermal conductivities of BC with and without R_{12}, and free-standing BC.

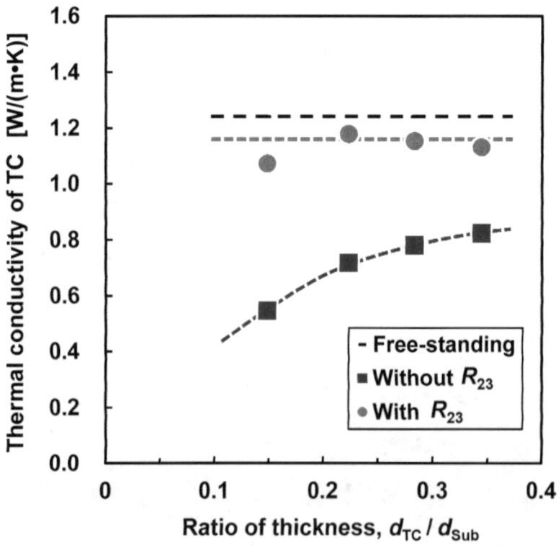

Fig.8. Thermal conductivities of TC with and without R_{23} , and free-standing TC.

resistance between the TC and the BC in the as-sprayed TBC specimen.

Figure 8 shows the the thermal conductivity of the TC. As in the case of the BC specimen, the thermal conductivity of the TC in the TBC specimen exhibited similar tendency as the thermal diffusivity. Namely, it was confirmed that the thermal conductivity and the interfacial thermal resistance of the TBC specimen can be measured by using the developed method.

4.3 Influence of microstructure on interfacial thermal resistance of TBC

From the results in the present study, it was found that there were the interfacial thermal resistance not only between the TC and the BC but also between the BC and the substrate. Furthoremore, the R_{23} (= 7.05 x 10^{-5} [m^2KW^{-1}]) between the TC and the BC was higher than the R_{12} (= 1.56 x 10^{-5} [m^2KW^{-1}]) between the BC and the substrate. The difference of such interfacial thermal resistance was disucussed in association with the microstructure of the TBC.

The cross-sectional microstructures of the BC / Sub interface and the TC / BC interface are shown in **Fig.9**. Since these cross sections were polished by an argon ion beam, pores and microcracks in the BC and the TC can be clearly identified. Furthermore, there were

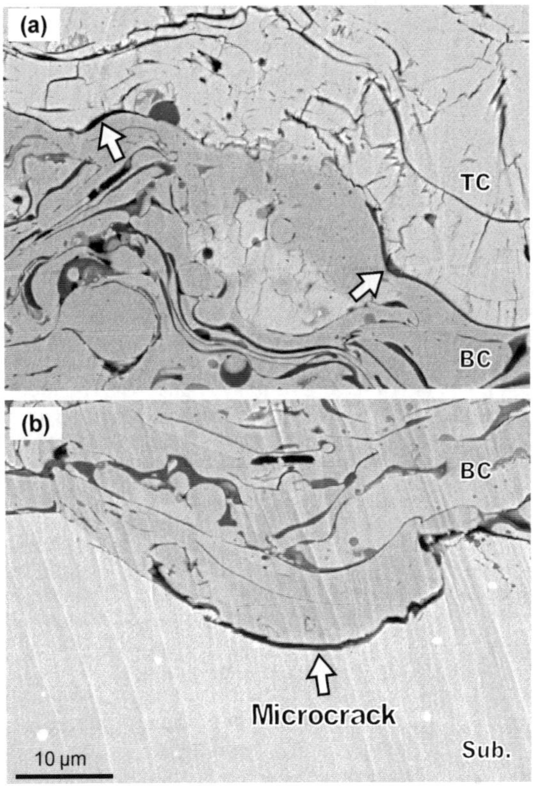

Fig.9. Typical cross-sectional microstructures of interface; (a) TC / BC and (b) BC / substrate. Arrows indicate microcracks at interface.

some microcracks not only between the TC and the BC but also between the BC and the substrate. These microcracks act as air gaps and prevent the heat conduction across the TBC. In addition, the surface roughness on the BC is higher than that on the blasted substrate. Such rough surface will increase the interfacial thermal resistance. Therefore, it is considered that the R_{23} between the TC and the BC is higher than the R_{12} between the BC and the substrate due to the microcracks and the roughness on the interface.

Thus, the developed method is very effective for evaluating the thermal conductivity and the interfacial thermal resistance of the TBC.

5 Conclusions

1) The method measuring the thermal conductivity and the interfacial thermal resistance of the TBC was developed based on the areal heat diffusion time method.
2) In the as-sprayed TBC, there are the interfacial thermal resistance not only between the TC and the BC but also between the BC and the substrate. The former is higher than the latter.
3) The thermal conductivities and the thermal diffusivities of the TC and the BC determined by considering the interfacial thermal resistance almost agree with those of the free-standing TC and BC.

Acknowledgment

This work was supported by JSPS KAKENHI Grant Number 15K06490. A part of this work was supported by Ministry of Economy, Trade and Industry (METI) of Japan.

References

[1] R. E. Tayer: Thermal conductivity determinations of thermal barrier coatings. Mater. Sci. Eng., A245 (1998), pp. 160/167.
[2] W. Chi, S. Sampath, H. Wang: Ambient and high-temperature thermal conductivity of thermal barrier coatings. J. Therm. Spray Technol., 15(4) (2003), pp. 773/778.
[3] N. Markocsan, P. Nylen, J. Wigren, X.-H. Li: Low thermal conductivity coatings for gas turbine appricatoins. J. Therm. Spray Technol., 16(4) (2007), pp. 498/505.
[4] J. Hartmann, O. Nilsson, J. Fricke: Thermal diffusivity measurements on two-layerd and three-layered systems with the laser-flash method. High Temp. High Press., 25 (1993), pp. 403/410.
[5] T. Baba: Analysis of One-dimensional Heat Diffusion after Light Pulse Heating by the Response Function Method. Jpn. J. Appl. Phys., 48 (2009), 05EB04.
[6] ISO 18555:2016 Metallic and other inorganic coaitngs – Determination of thermal conductivity of thermal barrier coatings.

MECHANICAL PROPERTIES OF YTTRIA-STABILISED-ZIRCONIA FOR THERMAL BARRIER COATING SYSTEMS: EFFECTS OF TESTING PROCEDURE AND THERMAL AGING

Pierre PLANQUES[1,2,3], Vanessa VIDAL[2], Philippe LOURS[2], Vincent PROTON[3], Fabrice CRABOS[3], Julitte HUEZ[1], Bernard VIGUIER[1]

[1] CIRIMAT, Université de Toulouse, UPS-INP-ENSIACET-CNRS, Institut Carnot CIRIMAT, 4 allée Emile Monso - BP 44362 - 31030 Toulouse cedex 04 - France
[2] Université de Toulouse; CNRS, Mines Albi, INSA, UPS, ISAE; ICA (Institut Clément Ader); Campus Jarlard, F-81013 Albi - France
[3] SAFRAN HELICOPTER ENGINES, avenue Joseph Szydlowski, 64511 Bordes - France

Cyclic oxidation failure of Atmospheric Plasma Sprayed Thermal Barrier Coatings systems (APS TBCs), commonly used to insulate hot sections in gas turbines, usually results from the spallation of the ceramic top coat. Consequently, in order to predict such spalling phenomena, understanding the mechanisms for cracks initiation and propagation in thermal barrier coatings is of utmost concern for engine-makers. Failure of the TBC is strongly related to the thermal and mechanical properties of each component of the multi-materials system (substrate, bond coat and ceramic) but also to the response of the TBC as a whole. The purpose of the work is to assess the mechanical behaviour of thick TBC using experimental approach for TBC standard lamellar, porous and micro-cracked microstructure (classically obtained through APS coatings).
The experimental characterisation of the mechanical behaviour of the ceramic top coat of the TBC is addressed on specifically designed and prepared free-standing specimens using three points bending (3PB) tests and Small Punch Testing (SPT). The tests are performed on free-standing top coats made of YSZ in the as deposited states and for specimens that undergone isothermal aging at 1100°C for various durations (1h, 10h and 100h). The results of test performed at room temperature using both mechanical testing techniques are compared. This allows to show the evolution of mechanical properties after thermal aging. Tests performed at 850°C in the SPT ring show that the evolution of properties resulting from this aging may be different at room temperature as compare to 850°C.

1 Introduction

Nowadays, APS TBCs commonly used in hot sections of gas turbine engines as thermal insulation systems, are mainly deposited on static parts as combustion chambers or outer air seals (segments or rings) in aircraft engines and on blades and vanes in land-based engines. The most widely used APS TBC system consists in a bilayer system composed by a 150 µm thick MCrAlY (M = Ni and/or Co) bond coat and a 250-1000 µm thick ZrO_2 stabilized with 7–8wt% Y_2O_3 (YSZ) ceramic top coat. The failure of plasma sprayed TBCs upon cumulative thermo-mechanical cycles imposed to the engine is usually induced by the spallation of the ceramic coating. Consequently understanding such spalling phenomena and predicting the life of the thermal barrier coatings is a major issue for engine-makers willing to build and properly implement a relevant model of life to be used for design purpose. The present work proposes a mechanical characterisation of YSZ samples with standard lamellar, porous and micro-cracked microstructure classically obtained for APS coatings.

Because of the large scattering of results reported in the literature [1-4] even for a given and fixed fabrication procedure, obtaining reliable mechanical properties of YSZ is complex, and each TBC system shows its own set of properties (Young's Modulus, tensile strength, fracture toughness,...) depending on its specific microstructure (density, porosity, thickness,...).
Conventional standard uniaxial 3-(or 4-)points bending tests are often employed to evaluate the mechanical properties of brittle materials such as ceramics [5-7], but they show limitations. They are first inapropriate for strength measurements as edge defects can initiate early failure. Furthermore the loading of real components is rather biaxial than uniaxial. Therefore biaxial bending tests of ceramics on disc-shaped specimens are often prefered for many applications [8-11]. They only require easy and straightforward specimen preparation and drastically limit the contribution of edge defects to failure.
In this study, the mechanical behaviour of free-standing TBCs is obtained through three points bending tests and Small Punch Tests.

2 Materials and experimental

Free-standing APS ZrO2–8 wt.% Y_2O_3 thermal barrier coatings (with approximately 12% porosity) were used for the mechanical investigation. The free-standing samples are obtained by spraying the deposit on 5-mm-thick sand blasted steel (42CD4) coupons or disc shaped samples. Following APS deposition of YSZ powders to obtain a dense 1.1mm-thick topcoat, a 1 hour high temperature oxidation treatment is applied to the material system in order to separate YSZ sample from the sacrificial substrate. Isothermal ageing treatments were performed in air in a box-type resistance furnace at 1100°C for 1h, 10h and 100h in order to promote sintering and address its effect on the mechanical properties.

Specimens for 3PB tests were machined in the shape of parallelepipedes with dimensions 25 x 5 x 1.1 mm^3, whereas the SPT samples fare 20 mm diameter and 1.1 mm thick discs.

The 3PB device, is fitted to an INSTRON testing machine with a load cell of 500N. The punch and support rods are made by 5 mm radius steel cylinder and the distance between the two supports is 20 mm,. The deflection of the specimen as loading proceeds upon testing is monitored by a camera with a macro lens allowing a precision of about 3 to 5 µm. The displacement rate of the punch is 0.1 mm/min.

A specific instrumented non-standard SPT setup has been designed, allowing mechanical testing of the materials in natural air in the temperature range 25°C-1000°C, up to 10kN. The sample deflection is measured by a laser micrometer acting as extensometer. The thermal control guaranties a temperature homogeneity in the furnace of 0.5°C between the top and the bottomparts of device.

The disc-shaped specimen (Radius R = 10 mm) is supported and centered by a ring (drawing radius r) and loaded centrally by a punch with spherical punch and then deformed up to failure (**Fig. 1**). The punch radius is R_B = 2.5 mm and a supporting ring radius of Ra = d/2 + r_a = 7 mm were chosen. The resulting stress field is axisymmetrical. The applied load, the punch displacement and the deflection are recorded during the test.

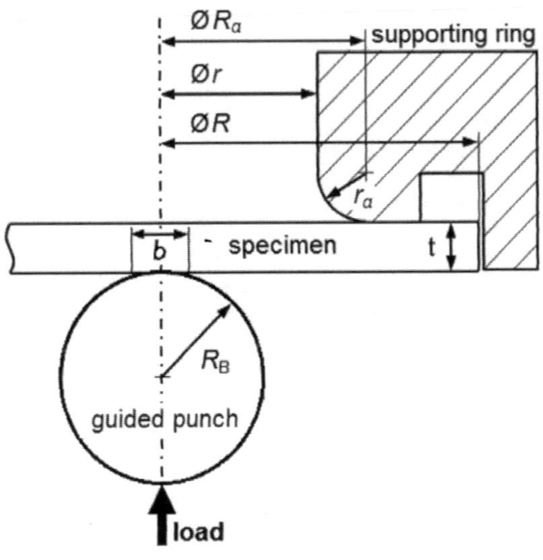

Fig. 1. Small punch bending test details.

The test is conducted in displacement-controlled mode with a constant punch velocity of 0.1 mm/min. Note that for tests at 850°C, a pre-charge of 10 N is applied using load control during the heating phase and temperature stabilisation (total process duration is about 30 mn). For most of the experiments, the test was repeated twice, as summarised in **Table 1**.

Table 1
Numbers of samples and ageing treatments.

	As-sprayed	1h @1100°C	10h @1100°C	100h @1100°C
3PB	2	2	2	2
SPT@RT	2	2	2	1
SPT@850°C	2	2	1	1

3 Results

3.1 3 Points Bending Tests

An example of 3PB sample is shown in **Fig. 2**. All specimens (either as deposited and aged ones) broke in two parts in the middle of the specimen, i.e. close to the punch cylinder. Basically, aging does not change the failure behaviour which remains fully brittle for all specimens.

Fig. 2. Top view of 3PB specimens a) as sprayed– b) 1h@1100°C after failure.

Fig. 3 shows the different load-displacement plots for the various 3PB tests. For all specimens the behaviour seems to be the same, with a monotic increase of the load followed by a sudden rupture. Note that the bending strength increases as a function of the aging time at 1100°C, from 12 N for an as-srayed specimen to 32 N for a 100h aged specimen. The variation of the failure force is roughly continuous, which is quite surprising, considering the sintering kinetics that shows major effect during the ten first hours of treatment [12-13].

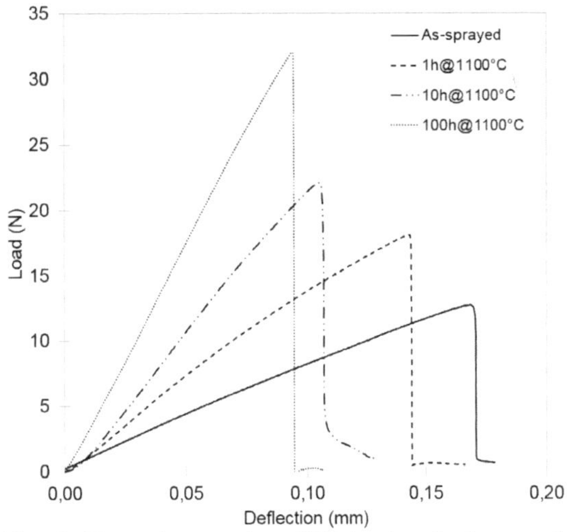

Fig. 3. Room temperature 3PB load – displacement curves following various agings at 1100°C.

Similar conclusion can be drawn for the bending modulus which increases as well continuously with aging time. The growing stiffness of the material, associated with a decrease of the failure deflection of the YSZ, can be attributed to the improvement in inter-splat bonding efficiency and to microcrack healing resulting from sintering [12].

3.2 Small Punch Tests

3.2.1 Room temperature tests

Fig. 4 shows the disc-shape specimen of SPT before and after testing. Similarly to that resulting from 3PB tests, the failure mode is brittle. Almost all the samples exhibit a failure in three parts, joining at the center precisely at the load application point, by the punch. This is typical for a brittle material.

Typical loading curves for all the specimens tested in SPT at room temperature are showed in **Fig. 5**. As for the 3PB, the as-sprayed YSZ specimen exhibits a monotonous increase of the load followed by the rupture, like a brittle behaviour. For all the aged specimens, the initial increase of load is followed by a plateau, corresponding to some energy dispersion looking like a plastic behaviour. This plateau extends for about 0.5 mm then a sharp rupture occurs very similar for all aging times.

As suggested above, and unlike what occurred for 3PB tests, in the case of SPT testing the strength does not evolves continuously with aging duration. Instead, a sharp increase of the SPT strength appears for specimens aged for duration as short as 1 hour, indicating the expected earliness ofthe sintering effects. For 10 hours and 100 hours aging at 1100°C, the tendancy is confirmed but the further increase in the failure load is less pronounced in perfect agreement with previous works on the sintering of YSZ [12-13].

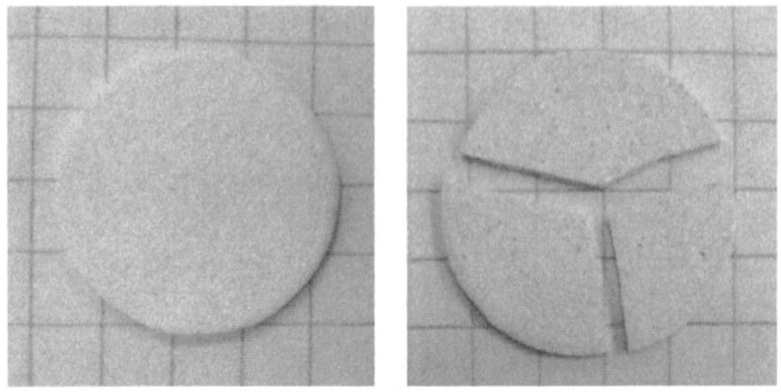

a) b)

Fig. 4. Top view of SPT specimens a) as-sprayed, – b) 10h@1100°C after failure testing.

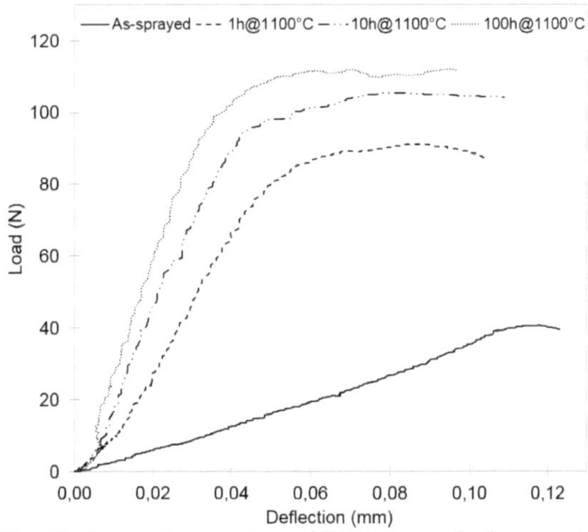

Fig. 5. Room temperature SPT load – displacement curves following various aging at 1100°C.

Classical analysis of SPT loading curve for a ductile metallic alloy, is done by defining different zones, each one corresponding to different deformation phases [14]. Similarly we can define two zones: Zone I corresponds to the elastic bending of the sample, and Zone II describes the plateau mentioned above. The transition between these two zone is characterised by a load level which by analogy to plastic behaviour can be labelled P_y. For the present results, P_y was defined as the vertical projection of the crossing point of the two tangents on the test curve [15] as illustrated in **Fig 6**.
The variation of this parameter P_y for the different specimens is plotted in **Fig. 7**. The rapid evolution of P_y between as deposited and aged specimens is clearly visible (from 40 to 80 N), whereas Py increases only slightly with increasing aging duration from 1h to 100h.

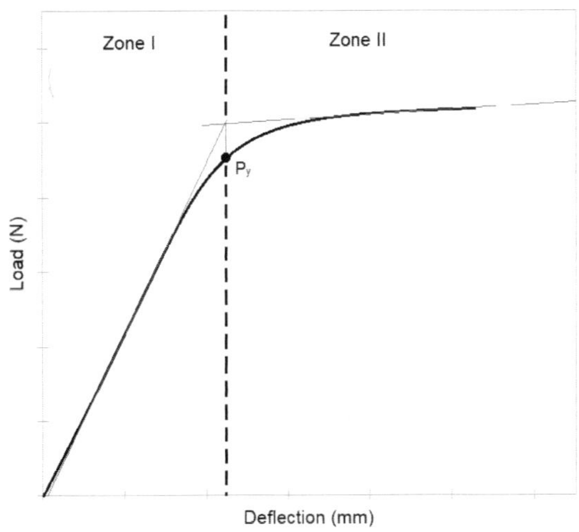

Fig. 6. Typical SPT load-deflection curve with identification of the different curve zones.

3.2.2 High temperature tests (850°C)

For the SPT tests performed at 850°C, the failure of SPT samples is brittle as well, but in this case, the as-sprayed sample exhibits a plateau, resulting from a total deflection of 0.15 mm significantly greater than at room temperature. This plateau ends by the brittle failure of SPT samples into in most cases three pieces as for the tests performed at room temperature. The load to failure is also higher at 850°C than at room temperature, which can not be attributed to sintering effects as the test temperature and the holding time are too low to activate any significant microstructural evolution. Unlikely, for aged YSZ, the values of P_y are slightly smaller for tests at 850°C than for tests at room temperature, and the corresponding deflection decreases as well, down to some 0.075 mm for the 100 hours aged material. The same tendency for both types of specimens (unaged vs aged), however much more pronounced, is observed for the values of the maximum load P_m showing respectively higher and lower values at 850°C than at room temperature for as-sprayed and for aged materials. **Fig. 7** illustrates the evolution of P_y as a function of aging, for both room temperature and high temperature tests.

4 Discussion

In order to better understand the mechanical behaviour and the rupture characteristics of YSZ samples we first try to compare experimental results obtained during 3PB and SPT tests. When doing such a comparison one must keep in mind that sample for both tests have different geometry and identical thickness. A direct comparison of load values or deflection would be meaningless. However it is interesting to notice that the maximum deflection of the as-sprayed specimens divided by the lever length give the same macroscopic shear value (0.17/10 = 0.017 for 3PB and 0.12/7 = 0.017 for SPT). Note this is only true for fully brittle specimen, such coincidence does not hold after aging at 1100°C.

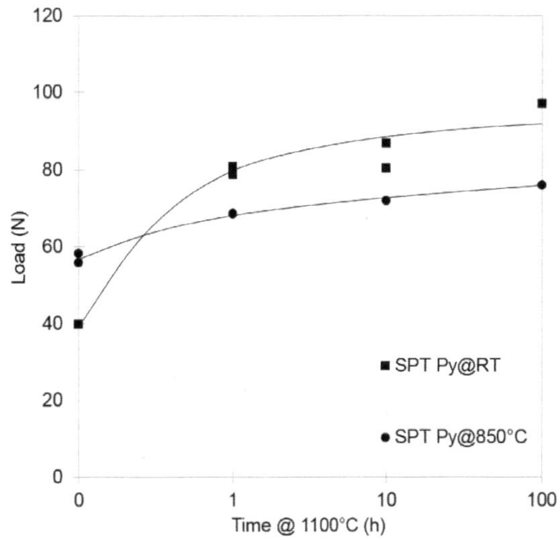

Fig. 7. SPT P_y load after different heat treatments at 1100°C tested at room temperature and 850°C.

As to analyse and compare the 3PB and SPT results, an energetic approach is chosen. For this, the volume energy related to the elastic regime and the volume fracture energy were both derived through an estimation of the area under the load-displacement curves, respectively in the reversible and total elongation domains. Those energies, divided by the effective volume of the specimen (volume between the cylindrical bearings for the 3PB test and volume inside the supporting ring for the SPT test) are then converted into volume energy:

$$Ee\ (resp\ Ef) = \frac{1}{V} \int_{0}^{De\ (resp\ Df)} F.dD$$

where Ee (resp. Ef) is the elastic (resp. fracture) volume energy, F is the force, V is the volume, De (resp Df) is the elastic (resp. failure) deflection.

In the case of 3PB testing, for which only elastic deformation is observed, the fracture energy corresponds to the sole elastic energy. As surface defects and edge effects are likely to provoke randomly and systematically very brittle failure. It can be admitted that the uniaxial loading and subsequent fracture does not satisfactorily represent the whole damage process and mechanisms of the material.
The energy measurement results for the 3PB and the SPT tests are presented in **Fig. 8**. The volume elastic energies for the 3PB are fairly constant, around 12 kJ/m^3. The increase in load with the sintering time

is roughly balanced by the decrease of the deflection, as clearly visible in **Fig. 3**

In the case of SPT, the volume elastic energy shows roughly equivalent values (12 kJ/m^3) as well as similar evolution with the aging time. As shown in **Fig. 5**, very little plastic deformation is observed or the as-sprayed sample. As a consequence, the fracture energy and the elastic energy are assumed to be very close. By contrast for aged specimens, the load plateau induces a clear difference between the elastic and the fracture volume energy. The biaxial stress field in the ceramic disc might be responsible for this non-elastic deflection, resulting from a very likely non occurrence of detrimental edge effects. The fracture energy greatly increases between the as-sprayed and the 1h-aged specimens, from approximately 19 to 42 kJ/m^3. This is related to the occurrence of a pseudoplastic deformation of the aged YSZ samples. A sliding movement between the less bonded splats may be responsible for this non-elastic deformation of the ceramic. This fracture energy keeps rising up to 10h aging, then levels off, remaining nearly constant on a plateau from 10 h to 100h exposure at 1100°C.

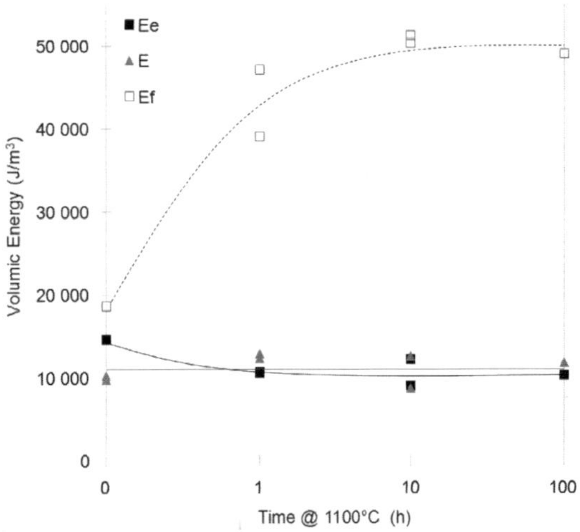

Fig. 8. Volumic energies calculated for 3PB test (E) and SPT (Ee and Ef) after different heat treatments at 1100°C tested at room temperature.

The elastic behaviour of the YSZ samples tested in SPT at 850°C exhibits some similarities as indicated in **Fig. 9**. Indeed, though the elastic energy namely 14 kJ/m^3 is somewhat higher, than that exhibited at room temperature, it is globally independent on the exposure time at 1100°C as well. The test temperature is thus shown to have a slight effect on the elastic energy, which rises as the temperature is increased from room temperature to 850°C. However, the exposure time at high temperature, i.e. 1100°C, activating significant sintering effects, does not strongly modify the elastic energy of deformation of the YSZ, both in 3PB test and SPT.

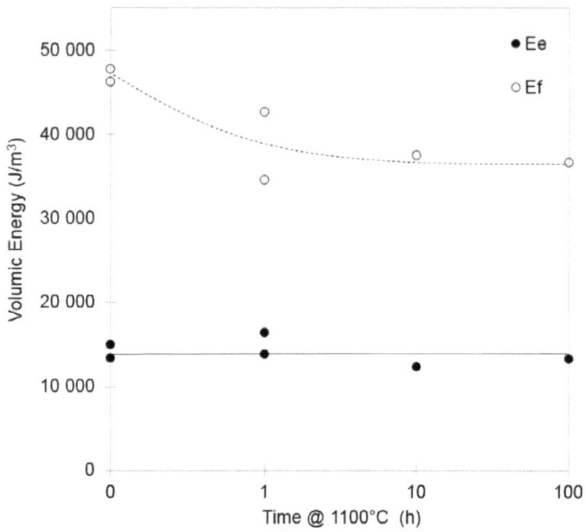

Fig. 9. Volume energies calculated from SPT (Ee and Ef) at 850°C following various agings at 1100°C.

Conversely, a quite large amount of plastic deformation is exhibited by the as-sprayed YSZ when tested at 850°C, indicating a high behavioural difference as compared to the room temperature mechanical response. This significant plastic strain leads to a fracture energy around 47 kJ/m^3, much higher than that measured at room temperature (19 kJ/m^3), which shows no plastic flow at all. Upon aging, for instance after 1h holding at 1100°C, a drop of the volumic fracture energy, from 47 kJ/m^3 to 37 kJ/m^3, is revealed. Longer exposures, 10 hours, then 100 hours, do not result in a further decrease of the energy which remains almost constant (37 kJ/m^3) independently of the progressing sintering of the material..

Unambiguously, these observations raise the question of the balance between the contributive influence of the sintering effects of YSZ and the SPT biaxial loading test temperature. For tests at room temperature, the sintering of the YSZ due to high temperature exposure results in an increase of the ductility, shown by the occurrence of significant plastic deformation, even for exposure as short as 1 hour. For tests conducted at 850°C, the as-sprayed material is rather highly ductile and the aging - whatever the exposure time is - tends to decrease slightly the capacity of the material to plastically deform.

5 Conclusions

Combined 3PB test and SPT were performed on as-sprayed and aged samples of YSZ

The main conclusions that can be drawn are the following:

i) A very good reproductibility of behaviour is observed for as-sprayed specimens (brittle fracture wit the same amount of macroscopic bending before rupture).

ii) For aged YSZ, 3PB tests are inadequate to exhibit all the characteristics of the material behaviour. In particular, the plateau of deformation which induces a much greater energy dissipation at the failure is only visible for aged SPT sample.

The behaviour observed for SPT shows that ageing treatment at 1100°C produces a sintering, which creates or improves the intersplat cohesion.

The high temperature tests (850°C) denote that, even for an as-sprayed specimen, accommodation modes of the deformation can be activated, allowing to obtain some compliance.

References

[1] Kuroda, Y.and Fukaura, K. and Sunada, H. and Izumi, H.: Mechanical properties of injection molded and sintered zirconia/alumina ceramic composites. Journal of the Japan Society of Powder and Powder Metallurgy(Japan) 38 (1999), issue 6, pp.787–790.

[2] Feuerstein, A, Knapp, J., Taylor, T., Ashary, A.,Bolcavage, A. andHitchman, N.:Technical and economical aspects of current thermal barrier coating systems for gas turbine engines by thermal spray and EBPVD: a review. Journal of Thermal Spray Technology 17 (2008), issue 2, pp. 199–213.

[3] Schwingel, D.,Taylor, R.,Haubold, T.,Wigren, J. andGualco, C.:Mechanical and thermophysical properties of thick pysz thermal barrier coatings : correlation with microstructure and spraying parameters. Surface and Coatings Technology 108 (1998), pp. 99–106.

[4] Planques, P., Vidal, V., Lours, P., Proton, V., Crabos, F., Huez, J. and Viguier, B.: Mechanical and Thermo-Physical Properties of Plasma Sprayed Thermal Barrier Coatings: A Literature Survey. Oxidation Of Metals (2017), DOI: 10.1007/s11085-016-9693-1

[5] Kucuk, A., Berndt, C.C., Senturk, U., Lima, R.S. and Lima, C.R.C.: Influence of plasma spray parameters on mechanical properties of yttria stabilized zirconia coatings. I: Four point bend test. Materials Science and Engineering A 284 (2000), pp. 29-40

[6] Yamazaki, Y., Schmidt, A., Scholz, A.: The determination of the delamination resistance in thermal barrier coating system by four-point bending tests. Surface & Coatings Technology 201 (2006), pp. 744-754.

[7] Zhao, Y., Shinmi, A., Zhao, X., Withers, P.J., Van Boxel, S., Markocsan, N., Nylen, P. and Xiao, P.: Investigation of interfacial properties of atmospheric plasma sprayed thermal barrier coatings with four-point bending and computed tomography technique. Surface & Coatings Technology 206 (2012), pp.4922–4929.

[8] Morrell, R., McCormick, N.J., Bevan, J., Lodeiro, M., Margetson, J.: Biaxial Disc Flexure-Modulus and Strength Testing. British Ceramic Transactions 98 (1999), pp. 234–240.

[9] Börger, A., Supancic, P., Danzer, R.: The ball on three balls test for strength testing of brittle discs: stress distribution in the disc. Journal of the European Ceramic Society 22 (2002), pp. 1425–1436.

[10] Rasche, S., Kuna, M.: Improved small punch testing and parameter identification of ductile to brittle materials. International Journal of Pressure Vessels and Piping 125 (2015), pp. 23-34.

[11] Rasche S, Strobl, S., Kuna M, Bermejo, R. and Lube, T.: Determination of strength and fracture toughness of small ceramic discs using the small punch test and the ball on three balls test. Procedia Materials Science 3 (2014), pp. 961–966.

[12] Thompson, J.A., Ji, W., Klocker, T. and Clyne, T.W.: Sintering of the top coat in thermal spray tbc systems under service conditions. Superalloys 2000 (2000), pp. 685–692.

[13] Thompson, J.A. and Clyne, T.W.: The effect of heat treatment on the stiffness of zirconia top coats in plasma-sprayed tbcs. Acta Materialia, 49 (2001), issue 3, pp.1565–1575.

[14] García, T.E., Rodríguez, C.,Belzunce, F.J., Suárez, C.: Estimation of the mechanical propertiesof metallicmaterials by means of the small punch test, Journal of Alloys & Compounds 582 (2014), pp. 708–717.

[15] CEN Workshop Agreement “Small Punch Test Method for Metallic Materials CWA 15627:2007 D/E/F, December 2007.

Porosity measurements on heat treated suspension plasma sprayed YSZ coatings using NMR cryoporometry

J. Ekberg, U. Klement, S. Creci, and L. Nordstierna, Chalmers University of Technology, Gothenburg / SE

Suspension plasma sprayed (SPS) coatings can be produced with fine powder particles and tailor-made porosity. This allows to achieve low thermal conductivity which makes the coatings attractive as e.g. topcoats in thermal barrier coatings (TBCs). Used in gas turbine applications, the TBCs are exposed to high temperatures which leads to alterations of the microstructure. To obtain coatings with optimized properties, possible microstructure alterations like closing of pores and opening of cracks have to be taken into account. Hence, in this study, TBC topcoats consisting of 8 wt.% yttria-stabilized zirconia (YSZ) were heat treated in air at 1150°C and thereafter investigated using scanning electron microscopy, X-ray diffraction, and nuclear magnetic resonance (NMR) cryoporometry. For all investigated samples, the porosity decreased as a result of the heat treatment. The finer pores and cracks disappeared and the larger pores grew slightly and achieved a more distinct shape as the material seemed to become more compact.

1 Introduction

In order to decrease the emissions and prolong the lifetime of gas turbines, the demand for more heat resistant TBCs has increased. The TBCs need to have low thermal conductivity and the ability to handle higher temperature loads in form of strain tolerance (lifetime) as well as high temperature effects such as phase transformations, grain growth and sintering. A TBC consists of a topcoat that acts as heat insulator and a bond coat that promotes the adherence of the topcoat and prevents severe oxidation [1]. During operation of the gas turbine, the bond coat oxidizes forming a thermally grown oxide (TGO) [1].

In yttria-stabilized zirconia (YSZ) coatings, heat transfer takes primarily place by conduction. However, at temperatures above ~1200°C also radiation becomes important [2]. The thermal resistance is improved by phonon scattering at inhomogeneities, e.g. grain boundaries, gas-filled pores and vacancies [3]. The small grain sizes in SPS TBCs are affecting the thermal conductivity in a favourable way by enhancing the amount of grain boundaries that scatter phonons [4-5]. A higher porosity contributes to the reduction of thermal conductivity as the pores are effective scattering sites and the air entrapped in the pores has a lower thermal conductivity than YSZ [6]. Moreover, when zirconia is doped with yttria, zirconium cations (Zr^{4+}) are replaced by yttrium cations (Y^{3+}) creating oxygen vacancies [3], which contribute to phonon scattering and the reduction of thermal conductivity. However, the oxygen vacancies also lead to an increased oxidation of the bond coat.

The preferable amount of yttria in zirconia is 6 - 8 wt.% as it provides a coating with a non-transformable tetragonal (*t´*) structure [7-8]. Important to consider is the slow diffusion-induced phase separation into tetragonal and cubic phase that *t´*-phase undergoes upon cooling from temperatures above 1200°C [9]. Such a phase transformation would lead to a volume change of the topcoat that causes extensive cracking and could result in spallation and failure of the TBC [10].

Hence, an important part in the optimization of the TBC design is the microstructure of the topcoat and how it is affected by the high temperatures occurring during operation of the gas turbine. In this study, TBCs were heat treated at 1150°C and the phases present before and after heat treatment were investigated by use of X-ray diffraction. Scanning electron microscopy was employed for microstructure characterization, and the change in porosity was studied by means of NMR cryoporometry.

1.1 NMR cryoporometry

Nuclear magnetic resonance (NMR) cryoporometry is a technique which has recently been introduced as a suitable technique for measuring open and connected porosity in TBC topcoats made of YSZ [11]. NMR cryoporometry is capable of measuring pore sizes in the range of a few nanometers up to a micrometer depending on the probe liquid properties and the instrumental hardware [12]. The method relies on the Gibbs-Thomson effect, i.e. that the phase transition temperature in smaller pores is lower than the corresponding bulk temperature [13-15]. Experimentally, a porous material is placed in an NMR vial, saturated with a chosen probe liquid and cooled down to complete solidification of the probe material. Afterwards, the temperature is step by step increased and the frozen material confined in small pores will start to melt followed by the larger pores and the bulk. Due to the difference in spin-spin relaxation characteristics of the liquid and solid state of the probe material, NMR spectroscopy can, by the application of a so-called T_2-filter, separate the material's liquid phase from its solid content [16]. A variation of signal intensity from the liquid phase with respect to temperature reflects the pore size distribution and since NMR spectroscopy is quantitative, if handled according to a few guidelines, the total porosity for pore sizes below a certain threshold is recorded. An additional capability with NMR cryoporometry is the possibility to achieve information regarding the pore shape in the studied material, i.e. the YSZ coatings. **Equations 1a** and **1b** are modified Gibbs-Thomson equations that describe the melting and freezing temperature depression, ΔT, of a confined liquid within a pore with radius r.

$$\Delta T_m = T_m - T_{bulk} = -G_1K/(r-\tau) \quad \textbf{(Eq. 1a)}$$
$$\Delta T_f = T_f - T_{bulk} = -G_2K/(r-\tau) \quad \textbf{(Eq. 1b)}$$

where T_m and T_f are the melting and freezing temperatures, respectively, T_{bulk} is the bulk equilibrium temperature, parameter K is relating to the bulk properties of the probe liquid, the constants G_1 and G_2 depend on the geometry of the pores, and τ is the thickness of the omnipresent liquid layer at the surface [13]. By evaluation of the melting and freezing data, the ratio $\Delta T_f/\Delta T_m$ can be assessed and afterwards compared and matched to a corresponding radius, hence providing the geometry.

2 Experimental

2.1 Coating production

The investigated TBCs consist of Hastelloy X substrates, CoNiCrAlY bond coats (produced by High Velocity Air-Fuel thermal spray process) and ceramic topcoats. The topcoats were deposited by axial suspension plasma spraying (ASPS) using an Axial III high power plasma torch. Three different topcoats were produced with a suspension consisting of 8 wt.% YSZ dispersed in ethanol (25 wt.% solid load), i.e. coating A, B and C. The spray parameters were varied and are given elsewhere [17]. The YSZ powder particles have a median particle size of 500 nm as provided by the supplier INNOVNANO (Coimbra, Portugal).

2.2 Heat treatment

The TBCs were heat treated in air at 1150°C. The temperature was chosen to be close to the real temperature in a gas turbine but below the phase transformation temperature of YSZ (1200°C) [9]. The sintering was performed for 50 h and 200 h, respectively, and the samples were cooled down inside the furnace before removal.

2.3 Microstructure investigation

The coatings were analysed in cross section using scanning electron microscope (SEM) in a Philips XL-30 SEM and a Leo 1550 Gemini SEM equipped with a field emission gun, respectively. Prior to the SEM investigations, the samples were sectioned using a diamond cutting blade, embedded into epoxy resin using vacuum impregnation technique, polished and gold coated.

2.4 Phase analysis using XRD

Phase analysis was performed using a D8 Discover X-Ray Diffractometer (XRD) with Cr Kα radiation (λ = 2.29 Å). The diffraction angle (2θ) was varied between 20° and 155° with a step size of 0.05. Match! and TOPAS software was used for identification of phases and quantitative Rietveld analysis, respectively.

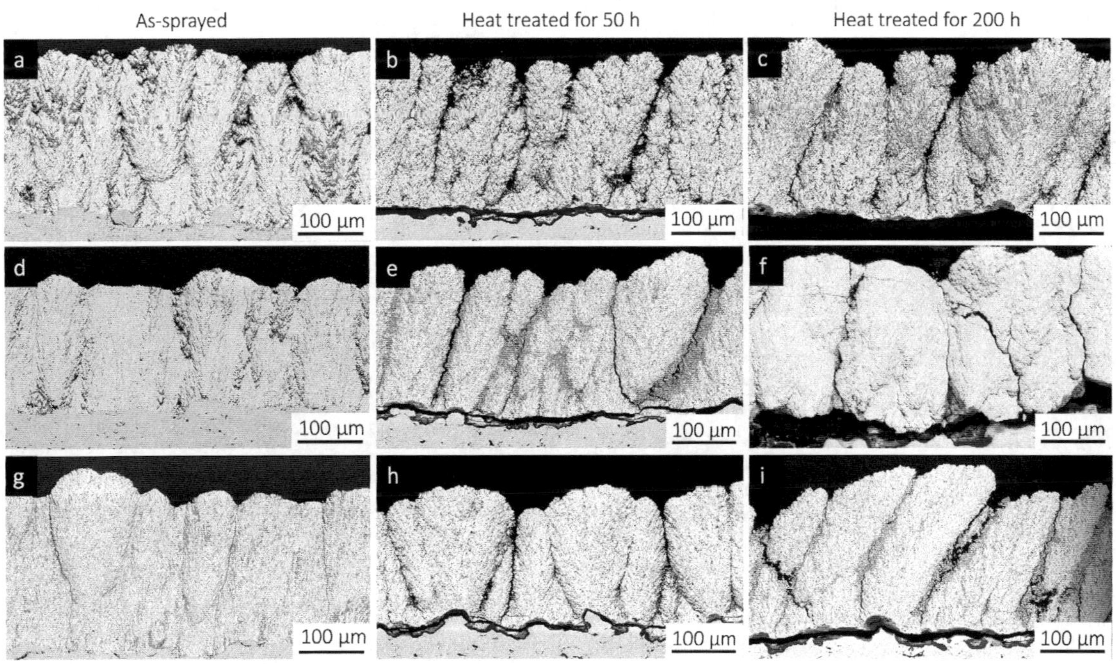

Fig. 1. Cross section micrographs of TBC topcoats in as-sprayed condition and heat treated for 50 and 200 hours, respectively; a - c) Coating A, d - f) Coating B, g - i) Coating C.

2.5 Porosity measurements using NMR cryoporometry

For determining the porosity by use of NMR cryoporometry, the YSZ topcoat was detached from the Hastelloy X substrate by dissolving the metallic bond coat in Aqua Regia (nitro-hydrochloric acid). The topcoats were cleaned repeatedly with deionized water and then freeze-dried to remove absorbed liquid from the pores. The topcoat samples were then saturated with octamethylcyclotetrasiloxane (OMCTS) during 1 h centrifugation at 1400 rpm. OMCTS was chosen because it completely wets the YSZ, facilitates to investigate a wide pore size range and has previously shown good results when used as probe liquid. All samples were cooled down to a temperature of 185 K for complete solidification before insertion into the Bruker Avance 600 spectrometer. For all experiments, the starting temperature was set to 255 K and raised slowly between each NMR acquisition. The NMR integral intensity of the liquid OMCTS signal was recorded at each temperature after sufficiently long time to ensure equilibrium. The bulk density was calculated from the weight of the sample and the sample dimensions measured using an optical microscope. For details about the experiments see Ekberg et al. [11].

3 Results and Discussions

The produced coatings have a columnar or feathery columnar structure, which is beneficial with respect to strain tolerance, **Fig. 1**. Among the investigated topcoats, Coating A appears most porous while Coatings B and C have a more dense structure. After heat treatment, Coatings A and C manifested good adherence, while Coating B showed de-bonding of the topcoat.

3.1 Microstructure investigation

As can be seen in **Fig. 1**, through thickness cracks in the YSZ coatings widened during heat treatment. The cracks had a width of 1 - 3 µm after a heat treatment for 50 h, while they are up to 10 µm wide after 200 h of heat treatment. When the cracks open up, they might cause convective heat flow through the topcoat, which then increases the thermal conductivity unfavourably. At the interface between topcoat and bond coat the formation of TGO can be observed (**Fig. 1**). The TGO is approx. 5 µm thick after heat treatment for 50 h (**Figs. 1 b, e, h**) and achieved a thickness of about 8 µm in topcoats heat treated for 200 h (**Figs. 1 c, f, i**). The denser part of the columnar area of the YSZ coatings are shown at higher magnification, **Fig. 2**. As can be seen, the pores are heterogeneously distributed and have a complex structure as well as a wide size distribution. After 200 h of heat treatment, it can clearly be seen that some of the fine cracks have disappeared. On one hand, this might be favourable for the mechanical properties, however, on the other hand, the lack of finer cracks might reduce the insulative properties. It appears as if many of the smaller pores also disappeared and that the larger pores grew slightly. The shape of the large pores has become more distinct and the material seems to be more compact.

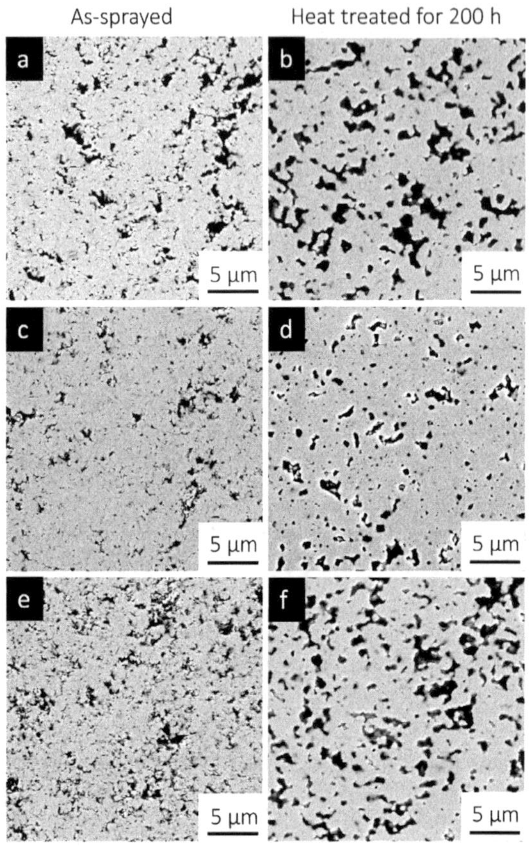

Fig. 2. Cross section micrographs of TBC topcoats in as-sprayed condition and heat treated for 200 hours, respectively; a - b) Coating A, c - d) Coating B, e - f) Coating C.

3.2 XRD for phase characterization

The XRD measurements show that Coating A - C consist of non-transformable tetragonal (*t'*) zirconia as major phase and some monoclinic zirconia. A slight peak broadening is visible after heat treatment when comparing the XRD pattern of the as-sprayed and the heat treated sample (see e.g. Coating A in **Fig. 3**). This peak broadening is not due to phase transformation but likely caused by strains in the crystal lattice. The fractions of monoclinic and tetragonal phases are given in **Table 1**. The calculated amounts are influenced by peak overlap, but it is evident that the phase contents did not change due to heat treatment.

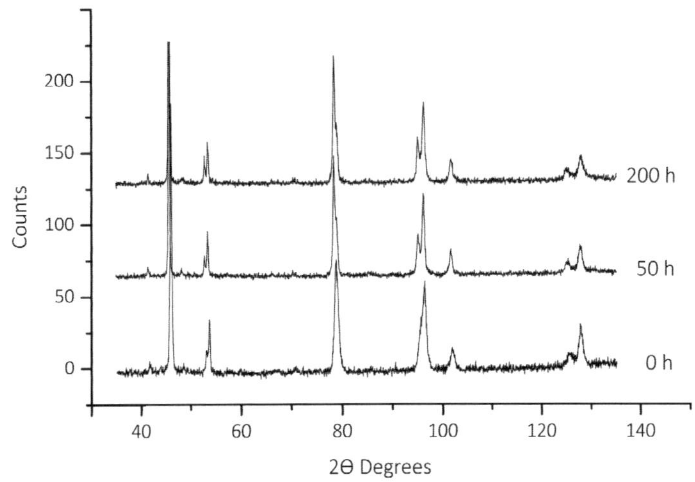

Fig. 3. XRD patterns of Coating A in as-sprayed condition (0 h) and after heat treatments for 50 and 200 h, respectively.

Table 1. XRD phase analysis showing monoclinic and tetragonal zirconia phases present (in wt.%) in Coating A, B and C before and after heat treatment for 50 and 200 hours, respectively.

	As-sprayed		**50 hours**		**200 hours**	
	Monoclinic ZrO_2 [wt.%]	Tetragonal ZrO_2 [wt.%]	Monoclinic ZrO_2 [wt.%]	Tetragonal ZrO_2 [wt.%]	Monoclinic ZrO_2 [wt.%]	Tetragonal ZrO_2 [wt.%]
Coating A	2.9	97.1	2.4	97.6	2.2	97.8
Coating B	1.9	98.1	2.9	97.1	0*	100
Coating C	2.8	97.2	3.7	96.3	2.1	97.9

*The absence of monoclinic phase in Coating B after 200 h of heat treatment may be due to high noise ratio for this sample because of de-bonding of the topcoat causing difficulties with the XRD measurements.

3.3 Porosity measurements using NMR cryoporometry

For the different samples, pore shape and pore size distribution were investigated using NMR cryoporometry. As seen in the SEM images in **Fig. 2**, the pore shapes are rather complex. This has also been shown by Klement et al. [18-19] using 3D X-ray microscopy. Using NMR cryoporometry, the pore shape was approximated and provided for the three topcoats (Coating A - C). An elongated pore shape for the smaller pore sizes and a rather spherical pore shape for the larger pores, both before and after heat treatment. For all three coatings, the change between spherical and elongated pore shape occurred at a pore radius around 150 - 200 nm.

Pore radii (limited by the instrumental hardware and probe liquid properties) between 5 and 500 nm could be measured in the three different coatings. The pore size distribution is presented with total volume of pores of a given radius per mass of material ($ml \cdot g^{-1} \cdot nm^{-1}$) plotted against pore radius (nm), **Fig. 4**. As can be seen, the as-sprayed coatings have a large amount of pores with a radius below 8 nm, around 30 nm and above 80 nm. Moreover, the pore size distributions show that the decrease in porosity during heat treatment of the different coatings is largest for small (r < 50 nm) and large pores (r > 80 nm). The reduction in number of pores smaller than 50 nm can be due to sintering, i.e. that small pores and/or cracks closed during the heat treatment or that they grew in size e.g. by coalescence of pores. Larger pores can also have closed but it is more likely that they grew bigger (e.g. formed through-thickness cracks) and are beyond the measuring range (> 500 nm). Hence, other techniques like image analysis could be used in this case for obtaining information from larger pore sizes.

By integrating the pore size distribution curves, a quantification of pore volume per gram of material can be achieved. When comparing the three as-sprayed coatings, Coating A has the highest porosity per gram YSZ, followed by Coating C and Coating B. The porosity values are 12, 8, and 7 vol.%, respectively, **Table 2**. The value of the total porosity encompasses all pores with a radius up to 200 nm. A value of the total porosity which includes the complete pore size range up to 500 nm pore radius is possible to achieve if the sample material is more densely packed in the NMR vial, i.e. when excess of probe liquid (OMCTS) is avoided. The NMR signal intensity of bulk OMCTS is normally negligible as the volume of probe liquid is rather small compared to the volume of sample material. However, in the current investigation of thin YSZ topcoat flakes, the value for the larger pore sizes (>200 nm) was influenced by the NMR signal intensity from the bulk OMCTS, due to an excess of probe liquid. As a result, Coating C shows values below zero for some size ranges.

During heat treatment of the coatings, the amount of porosity decreased for all coatings as presented in **Fig. 4** and **Table 2**, respectively. For Coating B and C there is no change in porosity between 50 and 200 h of heat treatment and the porosity is affected at an early stage of the heat treatment (i.e. closing of pores and opening of cracks). For Coating A on the other hand, the porosity changed from 5 to 8 vol.% between 50 and 200 h of heat treatment, which needs to be investigated further. Image analysis investigations on these samples are on-going. Both Coating B and C have a similar porosity in as-sprayed state. During heat treatment the porosity decreases more for Coating C, i.e. Coating B demonstrate the smallest change in porosity with a decrease from 7 to 3 vol.%.

Table 2. Total porosity in vol.% for pore radii between 5 and 200 nm in the as-sprayed and heat treated topcoats.

	0 hours	50 hours	200 hours
Coating A	12%	5%	8%
Coating B	7%	3%	3%
Coating C	8%	1%	1%

4 Summary and Conclusions

The NMR cryoporometry investigations on SPS YSZ coatings provided porosity, pore size distribution as well as an assessment of pore shape. The comparison of the as-sprayed and heat treated coatings showed that the porosity decreased during

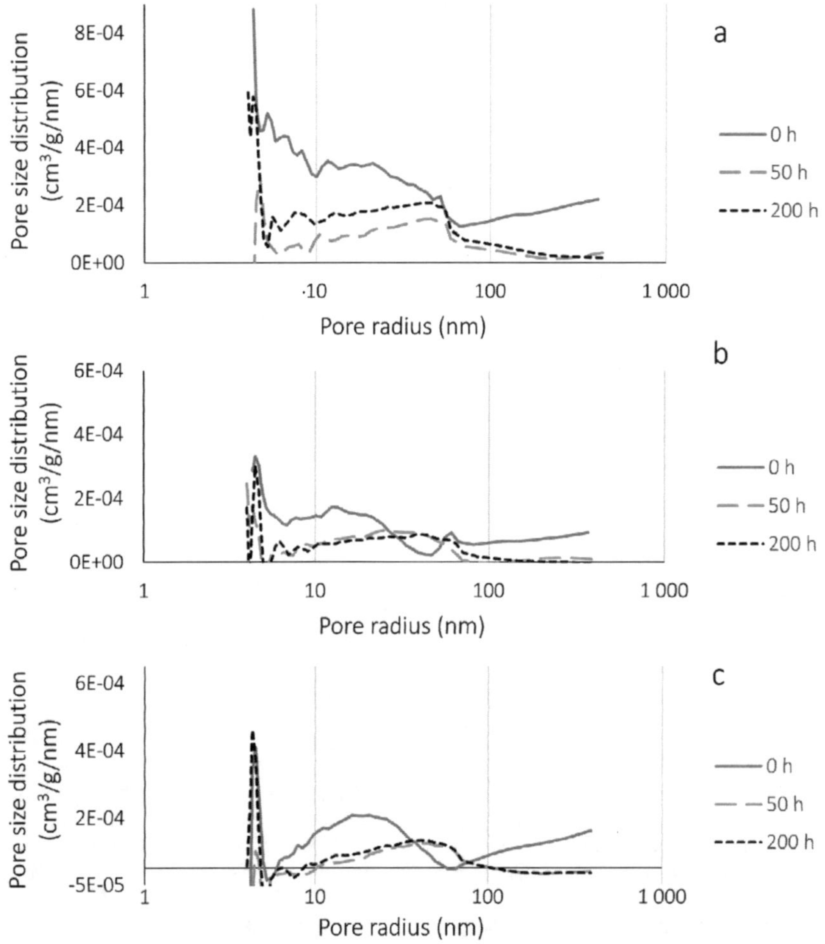

Fig. 4. Pore size distribution in as-sprayed condition and after heat treatment for 50 and 200 h; a) Coating A; b) Coating B; c) Coating C.

heat treatment. It can be concluded that the major change in porosity occured at an early stage of the heat treatment, since the porosity did not change between 50 and 200 h of treatment for two of the coatings, (Coating A is an exception and needs to be investigated further). Pores with a radius of 5 - 500 nm were investigated and in connection with the heat treatments the number of pores with a radius below 50 nm and above 80 nm, respectively, decreased. The reduction of the smaller pore sizes can be due to sintering effects on finer cracks and nanopores, i.e. that small pores and/or cracks closed. Alternatively, they can also have grown in size, e.g. by coalescence of pores. The reduction of the number of large pores is likely caused by growth of the pores and achieving a radius that is beyond the measured range (> 500 nm). For the three topcoats, the assessable pore shape information indicates that the smaller pores are of elongated shape, while the larger pores (>150 - 200 nm) are more confined and spherical.

Failure of the TBCs is likely to occur in connection with the growth of the TGO, causing de-bonding and spallation of the topcoats. For all investigated coatings, the TGO grew to a thickness of 8 µm after 200 h of heat treatment (before, the thickness was negligible). De-bonding of the topcoat occurred for one of the coatings (Coating B) after 200 h of heat treatment.

With the help of porosity investigations of topcoats subjected to operational temperatures information can be attained on changes in porosity and pore size distribution, which are affecting the thermal conductivity.

Acknowledgments

The authors would like to thank S. Björklund and A. Ganvir, University West, Trollhättan, Sweden, for producing the coatings. Thanks to Monica Vilemova, IPP Prague, Czech Republic, for assistance with XRD analysis. The NMR cryoporometry measurements were performed at the Swedish NMR Centre, Göteborg, Sweden. Västra Götalandsregionen is acknowledged for financial support of this work in connection with the PROSAM project.

References

[1] Padture, N.P., Gell, M. and Jordan, E.H. Thermal Barrier Coatings for Gas-Turbine Engine Applications. Science 296 (2002), Issue April, pp. 280/84.

[2] Golosnoy, I.O., Cipitria, A and Clyne, T.W. Heat transfer through plasma-sprayed thermal barrier coatings in gas turbines: A review of recent work. J. Therm. Spray Technol. 18 (2009), Issue 5–6, pp. 809/21.

[3] Subbarao, E.C. Advanced ceramics - an overview. Sadhana 13 (1988), Issue July, pp. 1/11.

[4] Yang, H.S., Bai, G.R., Thompson, L. J. and Eastman, J.A. Interfacial thermal resistance in nanocrystalline yttria-stabilized zirconia. Acta Mater. 50 (2002), Issue 9, pp. 2309/17.

[5] Limarga, A.M., Shian, S., Baram, M. and Clarke, D.R. Effect of high-temperature aging on the thermal conductivity of nanocrystalline tetragonal yttria-stabilized zirconia. Acta Mater. 60 (2012) Issue 15, pp. 5417/24.

[6] Kassner, H., Stuke, A., Rodig, M., Vassen, R., and Stover, D. Influence of Porosity on Thermal Conductivity and Sintering in Suspension Plasma Sprayed Thermal Barrier Coatings. Ceram. Eng. Sci. Proc. 29 (2009), Issue 4, pp.147/58.

[7] Stecura, S. Optimization of the NiCrAl-Y / ZrO2-Y2O3 Thermal Barrier System. NASA-TM 86905, (1985).

[8] Vassen, R., Stuke, A. and Stover, D. Recent developments in the field of thermal barrier coatings. J. Therm. Spray Technol. 18 (2009), Issue 2, pp. 181/86.

[9] Viazzi, C., Bonino, J.-P., Ansart, F. and Barnabé, A. Structural study of metastable tetragonal YSZ powders produced via a sol–gel route. J. Alloys Compd. 452 (2008), Issue 2, pp. 377/83.

[10] Brandon, J.R. and Taylor, R. Phase stability of zirconia-based thermal barrier coatings part I. Zirconia-yttria alloys. Surf. Coatings Technol. 46 (1991), Issue 1, pp. 75/90.

[11] Ekberg, J., Nordstierna, L. and Klement, U. Porosity investigation of yttria-stabilized zirconia topcoats using NMR cryoporometry. Surf. Coatings Technol. (2017), in press. (DOI: 10.1016/j.surfcoat.2017.02.067)

[12] Mitchell, J., Webber, J.B.W. and Strange, J.H. Nuclear magnetic resonance cryoporometry. Phys. Rep. 461 (2008), Issue 1, pp. 1/36.

[13] Boissier, C., Feidt, F. and Nordstierna, L. Study of Pharmaceutical Coatings by Means of NMR Cryoporometry and SEM Image Analysis. J. Pharm. Sci. 101 (2012), Issue 7, pp. 2512/22.

[14] Strange, J., Rahman, M. and Smith, E. Characterization of porous solids by NMR. Phys. Rev. Lett. 71 (1993), Issue 21, pp. 3589/91.

[15] Strange, J.H., Webber, J.B.W. and Schmidt, S.D. Pore size distribution mapping. Magn. Reson. Imaging 14 (1996), Issue 7–8, pp. 803/05.

[16] Lampman, G.M., Pavia, D.L., Kriz, G.S. and Vyvyan, J.R. Spectroscopy 4th ed. Brooks/Cole, (2010).

[17] Ganvir, A. Curry, N., Markocsan, N. Nylen, P., Joshi, S., Vilemova, M. and Pala, Z. Influence of Microstructure on Thermal Properties of Axial Suspension Plasma-Sprayed YSZ Thermal Barrier Coatings, J. Therm. Spray Technol. 25 (2016), Issue 1–2, pp. 202/12.

[18] Klement, U., Ekberg, J. and Ganvir, A. EBSD analysis and assessment of porosity in thermal barrier coatings produced by axial suspension plasma spraying (ASPS). Mater. Sci. Forum. 879 (2017), pp. 972/77.

[19] Klement, U., Ekberg, J. and Kelly, S. 3D analysis of porosity in a ceramic coating using X-ray microscopy. J. Therm. Spray Technol. 26 (2017), Issue 3, pp. 456/63.

Study of the interface strength and residual stresses within plasma sprayed alumina coatings involving LASAT (Laser Shock Adhesion Test)

H. Sapardanis, V. Guipont, F. Borit, A. Debray, A. Koster, V. Maurel Evry /F

The study of both the interface strength and residual stresses within a plasma sprayed ceramic coating is of great interest which main purpose is a better understanding of the mechanical properties of metal/ceramic systems. In this work, experiments involving a LASAT facility (LASAT: Laser Shock adhesion test) were implemented in order to analyse the adhesion and the damaging behaviour (debonding and buckling) of alumina coatings onto Co-based alloy. Similar alumina coatings were deposited using same plasma parameters with various surface preparations: smooth or severe grit blasting, with and without pre-oxidation. The non-destructive analyses (Optical and IR imaging) of the buckled region after LASAT have allowed to compare and discuss the interface strength of the studied coated samples. Further discussion was carried out by analysing the blister, resulting from the release of residual stresses within the coating after LASAT. It was thus evidenced that the residual stress state is a key parameter on resulting adhesive properties. This exploring work suggests using the LASAT method to analyse the adhesion and residual stresses within thermal spray coatings.

1 Introduction

The control of coating's adhesion is a major concern to ascertain the reliability of industrial process like thermal spray or electron beam-physical vapour deposition (EB-PVD). Such mechanical assessment is particularly crucial in case of ceramic layers that fulfil an essential function during the life service while the intrinsic bonding strength is low. Numerous and various adhesion tests (qualitative and quantitative) applied to thermally sprayed coatings can be proposed on as-spayed coatings [1]. In case of thermal barrier coating (TBC), high temperature isothermal or thermal cycling tests can be used to assess the lifetime of the system through the evaluation of the coating's resistance to spallation [2]. Nevertheless, the estimation of adhesion just after manufacturing is still a key issue to be addressed in case of ceramic/metal bonds. The tensile adhesion test involving the ASTM C633 standard with glued specimens is undoubtedly the most widely used test to control an industrial production. However, to overcome some limits of this pull-out test (problem of glue penetration in pores, obvious discrepancy of the results and problem of reproducibility), adhesion tests easier to implement and more reliable for the identification of the interface strength must be envisaged. In this context, the Laser Shock Adhesion Test (LASAT) has been developed in the past decades [3]. This non-contacting method can generate a high-level of tensile stress at the substrate/coating interface by irradiating areas of a few mm² on simple plate samples. With a nanosecond pulsed laser source and for a given target thickness (substrate and coating thicknesses), the input power density P can be varied from about P=0.1 GW/cm² to P=10 GW/cm². Thus, a qualitative assessment can be easily achieved by comparing the laser-induced debonding threshold for similar sample geometries. The 'LASAT threshold' is determined by processing several laser shocks on a same sample with an incremental increase of the laser energy (for a frozen laser spot dimension) in order to find the energy range leading to a first observable debonding. Quantitatively, the shockwave stress shall be calibrated with velocity profiles recorded on the opposite side of the sample during the shock propagation using an optical interferometer. Such experimental signature has to be combined to numerical wave mechanics-based simulations to quantify the stress generated at the interface [4]. Optical interferometer can also be employed as a non-destructive tool to detect a debonding threshold from the typical shape of the velocity profile when debonding occurs [5]. In most cases, for new coated systems, post-mortem metallography is needed to check the presence of the interfacial crack induced by LASAT. By applying laser shock conditions below a selected debonding threshold and employing a non-destructive tool for crack detection, a non-destructive LASAT method has been already studied in case of EB-PVD TBC and could be envisaged for other systems [6]. Therefore, the ongoing developments of the LASAT method to test various industrial coatings are related on the improvement of non-destructive techniques to detect interfacial and within coating damages. These developments are necessary combined and both are very attractive in case of industrial coated parts.

On the other hand, it was established that the presence of one interfacial crack after a laser shock above the debonding threshold could easily be detected by a simple visual inspection of the hydroxyapatite coating top-surface, leading to the so-called 'white spot' analysis. Indeed, it was demonstrated that the diffuse reflection of light in the visible spectrum within the ceramic coating was significantly amplified where a crack was present at the interface. Thus, based on image analysis, brighter areas were proven to be correlated to debonded areas by LASAT. The white spot analysis enabled to compare easily different interface strengths for various hydroxyapatite/Ti-alloy interfaces [5, 7]. Such optical phenomenon in case of damaged interface was also established for yttria stabilized zirconia (YSZ) and pure alumina coatings with thicknesses up to 250 µm [8].
In case of EB-PVD YSZ coating, it was thoroughly checked that the diameters of the white spots after

laser shock were similar to the corresponding interfacial cracks. This investigation associated resin-mounted cross section observations and non-destructive Photo-Luminescence Piezo-Spectroscopy (PLPS) [6]. The EB-PVD YSZ coatings gave the opportunity to develop another testing route, which was no longer based on the first appearance of a crack at the interface but on the measurement of the debonded area diameter (D) and its evolution when the laser power density (P). One of the main advantages of the white spot analysis is the good dimensional reproducibility. For example, tests and measurements have been repeated during 20 months on a same TBC sample. In this specific case, an average diameter (D=3.9 mm) was measured exhibiting a low standard error (±0.2 mm), Figure 1.

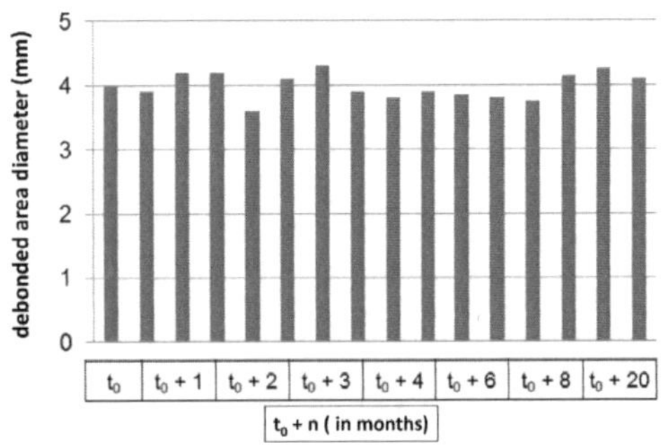

FIG.1: Series of same laser shock on a TBC sample repeated during 20 months (according to [6])

The combination of shockwave-induced interfacial damaging with potential further buckling-driven crack propagation is not considered yet. Therefore, the implementation of a simple cut-off failure criterion with two-dimensional shockwave propagation could lead directly to a debonding threshold if the debonded area can be measured. Assuming a monotonic decreasing of loading stress along the interface in the radial direction, all the debonded region (inside the white spot) is loaded above the debonding threshold while the non-debonded region (outside the white spot) is below. Such approach could be applied in principle with only one white spot but in our studies, it was preferred to analyse series of white spot. Thus, we introduced the so-called 'LASAT-2D curves', D=f(P), as a main experimental result of the LASAT-2D method applied to ceramic coatings. This LASAT-2D method was first applied in case of YSZ/(Ni,Al)Pt/Ni system by EB-PVD for industrial turbine blades [9-12]. In these recent studies, both qualitative and quantitative approaches were investigated to compare the adhesion of various interfaces conditions and its evolution with thermal cycling. Unfortunately, costly industrial samples were not numerous to envisage a more systematic approach of the LASAT-2D method devoted to ceramic deposits.

In this study, we selected a plasma-spayed alumina coating deposited by plasma spray that is easier to obtain. Such ceramic/metal system, without bond-coat, is used as a 'model system' to better investigate the LASAT-2D method and their applications in order to measure adhesion and evaluate residual stress state simultaneously. Three different surface preparations of the substrate were achieved and thermally sprayed with the same ceramic deposit process. The LASAT-2D curves were analysed in order to compare adhesion qualitatively and quantitatively. For this latter, numerical modelling was implemented with a cut-off criterion to estimate the failure stress value for each design of interface. Moreover, the residual stress state within the ceramic layer is investigated through a buckling mechanics analysis and the measurement of the height of the circular blisters.

2 Experimental Methods

2.1 Materials and Coating Process

The substrate material was a cobalt base superalloy Haynes® 188 (see wt. % composition in Table 1). Three types of plate samples with 65×30×3mm^3 dimensions were prepared.

Table 1: HAYNES 188 substrate composition (wt. %)

Co	Ni	Cr	W	Fe	Mn	Si	C	La	B
39	22.0	22.0	14.0	3.0	1.25	0.35	0.10	0.03	0.015

Glass beads blasting (GB) and alumina grit blasting (AG) enabled to obtain two types of surface roughness. A third type of specimens was prepared by pre-oxidation of a grit blasted sample at 1050°C during 3h in air-furnace (AGP). A 65 µm thick ceramic coating was deposited by plasma spraying of pure alumina powder (Oerlikon Metco #105SFP). Main spraying parameters are summarized in Table 2.

Table 2: Plasma parameters

Plasma torch	F4-MB type
Plasma gas	Ar:41 L/min-H_2:14 L/min
Arc current	530 A
Spraying distance	120 mm
Number of pass	1
Robot velocity	300 mm/s
Powder flow rate	32 g/min
Particle size range	3.9 - 31.0 µm

During the spraying operation, a pre-heating step up to 200-250°C was achieved by moving the plasma plume without powder towards the sample. Then, when the temperature reached 150°C, the deposition process was executed. The maximum deposition temperature before final cooling down was measured at 220°C for all samples.

2.2 Laser Shock Experiments

LASAT experimental: LASAT experiments were performed using an Nd-YAG laser source (SAGA 330,

Thales Laser, France) exhibiting a 5.2 ns pulse and a 532 nm radiation. The beam energy can be varied from ~0.1 to 2.0 J. With help of a converging lens, laser irradiated areas, ranging from 2 to 3 mm in diameter, could lead to laser power densities from ~0.1 to 5.0 GW/cm^2. In these experiments, the laser beam is focused on the metallic side with deionized water as a confinement media. The qualitative levels of adhesion were assessed through the drawing of experimental LASAT-2D curves. In case of AG and AGP samples, 2 sets of 14 laser shocks with increasing energy were repeated on a same sample involving a 2.2 mm laser spot diameter. In case of GB sample, two laser spot diameters 2.2 and 3 mm and less number of shock were achieved.

LASAT numerical modelling: Shock wave propagation with elastic assumptions has been modelled to predict stress-time distribution within the whole bilayer system with a special attention paid to the interface. An axi-symetrical model has been performed on Abaqus Explicit for the Finite Element (FE) calculations. The interface was assumed to be flat and perfect representing by a 'Tie' constraint. The input pressure temporal and spatial profiles have been calibrated in a previous study [11]. The material data selected for this finite element analysis are summarized in Table 3. The materials were considered as dense, homogenous and isotropic.

Table 3: Materials data for FE calculations

	Thickness h	Density ρ	E	ν
Alumina γ	65 µm	3.65 g/cm^3	253GPa*	0.24*
Co alloy	3 mm	8.98 g/cm^3	232GPa	0.33

* From ref [13]

2.3 Observations and Analyses

Substrate roughness (S parameters) was measured by optical surface profilometry (Altisurf 500, Altimet, France) on 8×8mm² area. This facility was also used to measure the coating thickness and the blister profile induced by laser shock. Cross sections views were obtained on polished and resin mounted samples using a scanning electron microscope (Nova NanoSEM 450, FEI, The Netherlands). For white spots analysis, optical images of the top-surface were carried out using a macroscope (Makroskop M420, Wild/Leica, Germany). Top-surfaces were also investigated by infrared thermography using an infrared camera with a 320×256 pixels sensor in the wavelength range of 1.9-5.1µm (Titanium serie, Cedip Infrared Systems, France). The IR sensor and focusing distance led a spatial digital resolution of 45 µm/pixel. The coated sample was placed on a heating table with a set point at 50°C. An overview of the thermographic set-up and typical top-surface optical and IR images are given in Figure 2. Image analysis was performed on ImageJ software to measure white spot areas [14].

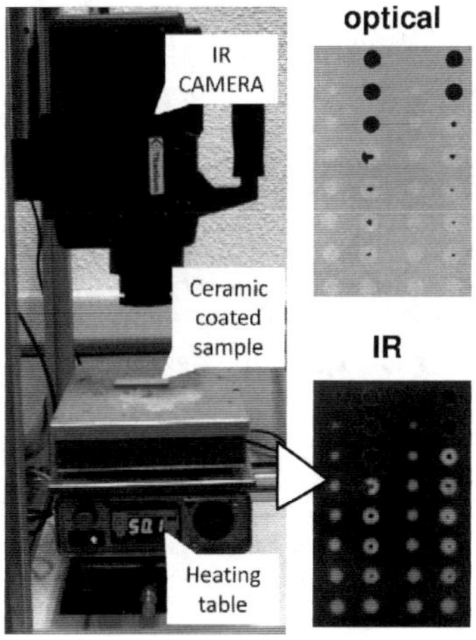

FIG. 2: IR experimental set-up and top-surface views by optical and IR imaging (same sample after LASAT)

3 Results and Discussion

3.1 Coatings and Interfaces

As shown on SEM image in Figure 3, the one-pass alumina coating mainly exhibited globular pores with a typical size around 5 µm in diameter. The spraying of a fine powder led to a dense microstructure with only a few elongated voids or cracks. The presence of resin penetrating inside the whole layer showed that the majority of voids were interconnected. A porosity level of ~ 5% was measured in a previous study [15].

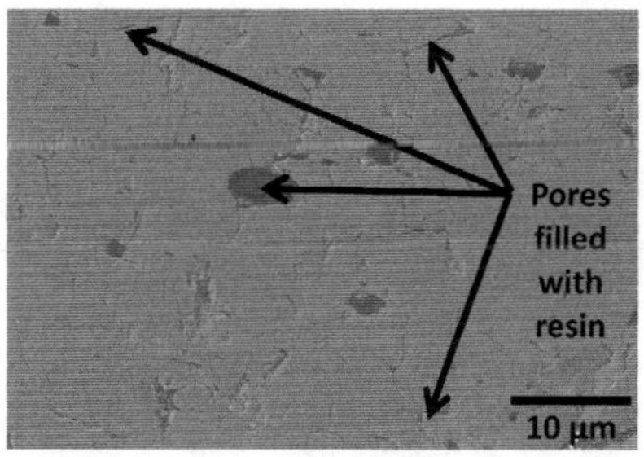

FIG. 3: SEM cross-section view of the alumina coating (Back-Scattered Electrons mode)

Detailed views of the three different ceramic/metal interfaces are given in the Figure 4. These observations and the mean average roughness (Sa) measurement of the substrate, Table 4, highlight the drastic variation of the surface morphology obtained before deposition. A smooth surface was observed in for glass beads blasting (GB) and a much rougher one in case of alumina grit blasting (AG). The oxidation of the AG sample led to an intermediate roughness level.

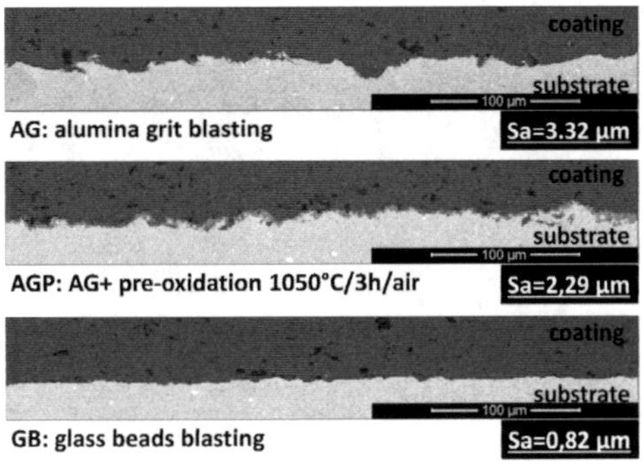

FIG. 4: SEM cross-sections views of the interface of AG, AGP and GB alumina coatings

Further comparative analysis of roughness parameters was achieved and resulting amplitude and hybrid S parameters are summarized in Table 4. In case of alumina grit blasted surfaces, the morphology exhibits an average roughness of 3.32 µm and is skewed upward relatively to the average line (S_{Sk}>0) with a sharp distribution of height (S_{Ku}>3) and pronounced sloping peaks and valleys. Consequently, the corresponding developed area ratio is rather high. Pre-oxidized and glass beads blasting surfaces are characterized by a lower Sa than alumina grit blasted sample. GB sample is found much smoother as obviously evidenced in Figure 4. According to S_{Sk} , S_{Ku} and S_{Δ_q} values in Table 4, the AGP and GB roughness surfaces are more flat, less slopping and more symmetrical as compared to AG sample. It is confirmed by the measurement of developed area ratios that are significantly low in case of AGP and GB samples. Such different morphological features may have a role on adhesion level measured by LASAT.

Table 4: Selection of roughness parameters

S parameters	AG	AGP	GB
S_a, Average roughness (µm)	3.32	2.29	0.82
S_{Sk}, Skewness	-1.11	-.015	0.22
S_{Ku}, Kurtosis	8.48	3.85	3.66
S_{Δ_q}, Root mean square slope	0.86	0.35	0.19
S_{dr}, Developed area ratio (%)	31.5	3.9	1.8

3.2 Assessment of adhesion by LASAT

From white spots non-destructive analysis to IR measurement: In this work, two non-destructive techniques were used to detect circular debonded areas and measure these diameters: IR and optical imaging, typical top-surface observations are given in Figure 2. Before selecting the most accurate, these techniques were compared, result plotted in Figure 5. A standard error of ±0.1 mm was estimated. It is clear that values obtained by IR imaging are always smaller that values obtained by optical imaging. It is worth noting that some discrepancy could occur for diameters below 2.5 mm.

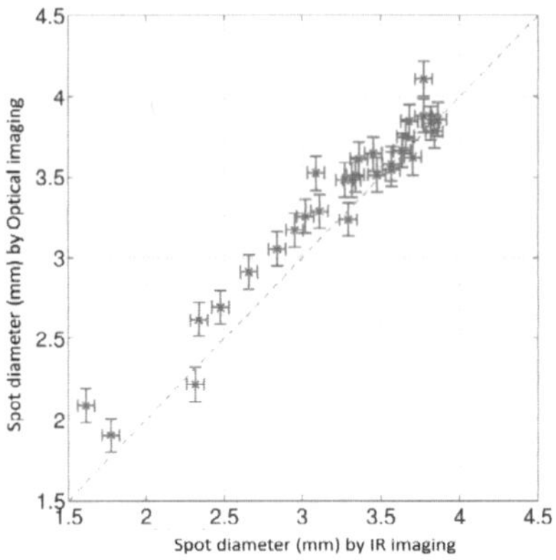

FIG. 5: Comparison of spot diameters measured by optical and IR imaging

In this specific case, cross-sections examination shall be envisaged to choose the most appropriate non-destructive method and to better calibrate the sensitivity of these methods for a 65 µm thick alumina coating. It is worth noting that the crack measurement by metallography is also a tough task. X-ray computed tomography may be an alternative but it will not used in this work. It is also important to underline that small interfacial debonded areas, typically below 1 mm in diameter, are difficult to detect. Nevertheless, optical and IR imaging showed promising results in this dimension range. Even if optical imaging can have high-resolution imaging, it cannot efficient for thick coatings while IR imaging has already shown some good potential to detect a crack throughout a 1 mm thick alumina coating [16]. Moreover, imaging could be drastically improved by increasing the sensor spatial resolution.

LASAT-2D analysis based on IR: The three different coated systems have been submitted to LASAT. Debonded area diameters were measured using IR imaging in order to plot the LASAT-2D curves, Figure 6. In case of AG and AGP samples, numerous laser shocks were achieved and a debonding threshold is evaluated between 0.75 and 0.92 GW/cm². Above the debonding threshold, the debonded area diameter increases continuously with a low level of dispersion as the laser power density increases. The presence debonded areas larger than the laser spot evidences the two-dimensional propagation of the shockwave. From macroscopic observation, induced blisters were also characterized. Above 1.5 GW/cm², laser shock waves lead to local blisters spallation. Below this value LASAT leads to interfacial debonding without further damage of the ceramic blister. In case of GB, the debonded area after LASAT are significantly larger that the other specimens. Therefore, due to sample dimensions, less numerous laser shocks were implemented and only power densities below 1.5 GW/cm² were selected to avoid the damage of the coating. In case of GB, the debonding threshold was below 0.75 GW/cm² and therefore lower than AG and AGP thresholds.

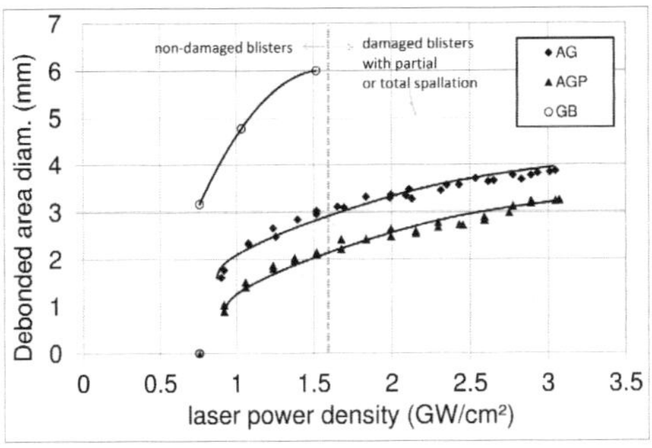

FIG. 6: LASAT-2 D curves ($Ø_{laser}$=2.2mm)

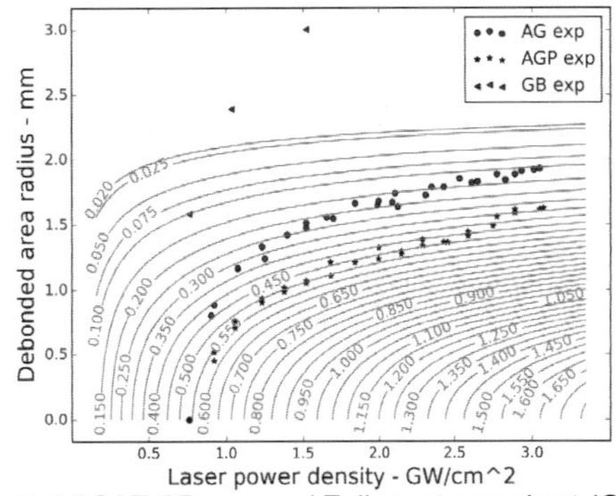

Fig. 7: LASAT-2D curves / Failure stress chart (GPa)

In case of AG and AGP, the evolution of debonded area as a function of the laser power density, the so-called LASAT—2D curves, are observed to be very similar with a quasi constant shift of about 1 mm between the two curves. It proves that the LASAT method applied to ceramic/metal bonds is reliable even with laser conditions leading to coatings spallation. Moreover, it reveales qualitatively that the adhesion of AGP is significantly higher than AG. The higher debonded area for a lower laser power density obtained for the GB sample confirmes that this surface preparation led to a much lower adhesion level. The significant difference between AGP, AG and GB is well illustrated for a laser intensity of 1.5 GW/cm². Indeed, crack diameters of 2, 3 and 6 mm respectively were measured which is consistent with the well-known influence of the roughness level on the mechanical anchoring of thermal spray coatings. Further (and less investigated [7, 17]), physico-chemical effects on adhesion when a 1-3 µm oxide interlayer exists are also evidenced by the LASAT method. It is obvious that the presence of an oxide interlayer can improve significantly the adhesion of the ceramic coating even if the initial roughness is lowered by the thermally-grown oxide layer.

LASAT-2D analysis based on numerical modelling:
Numerical modelling involving a simple cut-off interfacial failure criterion was implemented. A failure stress iso-value chart is plotted and compared to LASAT-2D curves in Figure 7. From such comparison of modelled chart and experimental data, it can be seen that both experimental and simulated shapes are consistent but not fully in agreement. Due to the good reproducibility of LASAT, it is needed first to improve the modelling tool and to investigate a more appropriate failure criterion. Indeed, there is a need to better understand the actual failure mechanism of a ceramic/metal bonds under shockwave. The assumption of instantaneous debonding without delayed crack propagation is probably not satisfactory. Complementary buckling-driven delamination may occur and can influence the final crack size. However, this point does not question the actual difference and ranking between the studied samples.

Blister analysis and evaluation of residual stresses:
The delamination of a thin film may form a blister in order to release pre-existing residual compressive stresses once a critical size of debonded area has been generated [18]. In this study, circular blisters of the porous ceramic coating were observed and the maximum height δ has been measured by 3D profilometry. This amplitude was plotted in function of laser power densities only for undamaged blisters obtained below 1.5 GW/cm² (Figure 8). The three different coated systems exhibited a similar buckling behaviour with distinct typical dimensions of the debonded area and the associated blister. Significant amplitude up to 80 µm is obtained that demonstrated a compressive state of residual stress within the three coatings. It is interesting to further investigate such mechanical behaviour in order to better analyse the mechanics of these coatings exhibiting various adhesion levels.

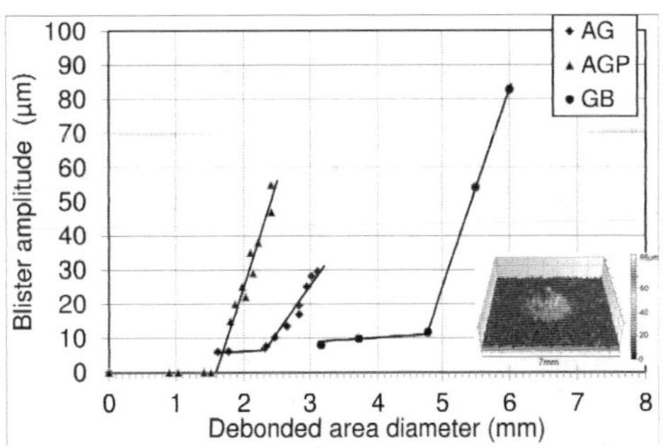

FIG. 8: Blister height and debonded area analysis

Assuming an elastic regime for the residual stress release, the activation of the buckling phenomenon should start above a minimum size of LASAT debonded area. It seems in agreement with the plots in Figure 8. The buckling is initiated and then the blister height increases with the debonded area diameter above the following different limits: AGP: 1.7 mm; AG: 2.3 mm, GB: 4.7 mm. A similar mechanical approach was used in case of SiO_2 thin film on Si(100) substrate by Mc Donald et al. [19]. Circular SiO_2 blisters with height below 1 µm were achieved by a femtosecond laser focused onto the coating and not on the substrate like in our case. Nevertheless, the following equation (1), first established by Hutchinson et al. [18], could be applied here in order to estimate the level of residual stress released with LASAT. For this calculation the Young modulus, *E*, is assumed to be 110 GPa as determined previously by tensile test on free standing alumina deposit [15] and the Poisson coefficient, ν, is 0.24. The results obtained from Figure 8 data are given in Figure 9.

$$\sigma_0 = 1{,}2235\frac{E}{1+\nu^2}\left[c_1\left(\frac{\delta}{D/2}\right)^2+\left(\frac{h}{D/2}\right)^2\right] \quad (1)$$

with $c_1 = 0.2473\,(1+\nu) + 0.2231\,(1-\nu^2)$

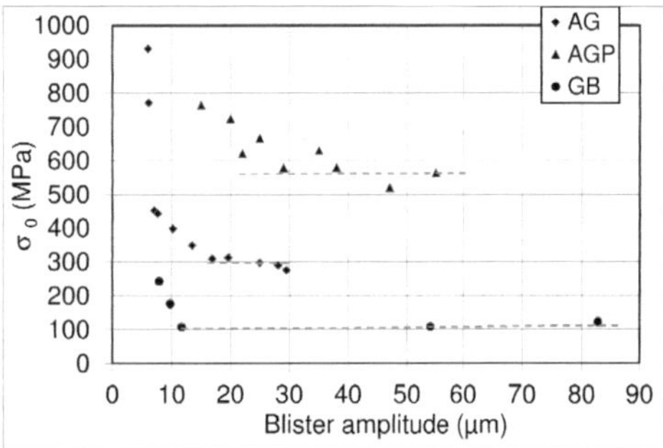

FIG. 9: Residual stress analysis according to Eq. 1

Residual stress levels increase drastically as the blister amplitude increases below 20 μm and seem to be rather constant above this limit. Therefore, typical values can be estimated graphically by drawing an horizontal asymptot: AGP: -570 MPa, AG: -300 MPa, GB: -100 MPa. Beyond the stress values that are probably over estimated, it is interesting to note that these results are consistent with adhesion levels. A higher thermal stress level could be obtained when intrinsic adhesion is also higher. For weaker interfaces, a part of the thermal stresses shall be released through sliding and local debonding effects at the interface during cooling down of the deposition process. Further work is needed to measure the actual Young's Modulus for a complete quantitative mechanical analysis of the ceramic/metal system.

4 Conclusions

In this work LAser Shock Adhesion Test were implemented on alumina coatings onto Co-based alloy. Firstly, according to LASAT-2D curves and white spot analysis by optical or IR imaging, a remarkable continuity of resulting debonded area as a function of the laser power density is observed for each tested condition. Thus, adhesion levels of AGP, AG and GB were easily discriminated. The presence of an oxide layer improved the adhesion of the alumina grit blasted specimen. A lower adhesion was obtained in case of glass beads blasting. The quantitative assessement of the interface strength by mumerical modelling was not satisfactory and further work is needed to investigate the influence of buckling-driven delamination during LASAT. The buckling behaviour of LASAT-induced circular blisters was characterized and completed by a residual stress analysis according to fundamental work of Hutchinson [17]. Residual stress levels are consistent with adhesion level. Such original and integrated experimental analysis of adhesion and residual stresses on a model system involving complementary non-invasive methods could bring new knowledge ceramic/metal systems like TBC. Moreover, the analysis of the propagation under quasi-static loading of a calibrated circular blister introduced by LASAT can be achieved [20]

5 Acknowledgments

Project funded by Carnot M.I.N.E.S Grant n° 30648

6 References

[1] C. K. Lin, C. C. Berndt, J. Therm. Spray. Technol., 1994,3, 75-104

[2] O. Trunova, T. Beck, R. Herzog, R.W. Steinbrech, L. Singheiser, Surf. Coat. Technol., 2008, 202(20), 5027–5032

[3] L. Berthe, M. Arrigoni, M. Boustie, J.P. Cuq Lelandais, C. Broussillou, G. Fabre, M. Jeandin, V. Guipont, M. Nivard, Nondestruc.Test. Eval., 2011, 26, 303-317

[4] A. Kobayashi, A. Jain, V. Gupta, V. Kireev, Vacuum, 2004, 73(3–4), 533–539

[5] S. Boileau, V. Guipont, M. Jeandin, M. Nivard, L. Berthe, J. Jerome, M. Boustie, H. Li, K.A. Khor, , proc. ITSC 2005, Basel, Germany, 2005, DVS Pub. 6 pp.

[6] G. Bégué, G. Fabre, V. Guipont, M. Jeandin, P. Bilhé, J.Y. Guédou, F.Lepoutre, Surf. Coat. Technol, 2013, 237, 305-312

[7] V. Guipont, S. Bansard, M. Jeandin, K.A. Khor, M. Nivard, L. Berthe, J.P. Cuq-Lelandais, M.Boustie, J. Biomed. Mat. Res. A, 2010, 95A, 1096-1104

[8] V. Guipont, Habilitation Dissertation, U. Limoges, France, 2013, p. 101

[9] S. Barradas, F. Bourlier, V. Guipont, M. Jeandin, 2008, French Patent 2926137

[10] J.P. Cuq-Lelandais, PhD thesis, ENSMA, 2010, theses.fr/2010ESMA0017

[11] G. Fabre, PhD thesis, MINES ParisTech, s, 2013, theses.fr/2013ENMP0066

[12] G. Fabre, V. Guipont, M. Jeandin, M. Boustie, J.P. Cuq-Lelandais, L. Berthe, A. Pasquet, J.Y. Guédou, Adv. Mat. Res., 2011, 278, 509–514

[13] M.R. Gallas, G.J. Piermarini, J. Am. Ceram. Soc., 1994, 77(11), 2917–2920

[14] W.S. Rasband, ImageJ, U. S. NIH, Bethesda, Maryland, USA, https://imagej.nih.gov/ij/, 1997-2016

[15] O. Amsellem, PhD thesis, MINES ParisTech, 2008, theses.fr/2008ENMP1550

[16] S. Chaki, P. Marical, S. Panier, G. Bourse, A. Mouftiez , NDT and E Int., 2011, 44, 519-522

[17] S. Valette, A. Denoirjean, P. Lefort, Surf. Coat. Technol., 2008, 202(12), 2603–2611

[18] J.W. Hutchinson, M.D. Thouless, E.G. Liniger, Acta Metall. Mater., 1992, 40(2), 295–308

[19] J.P. McDonald, M.D. Thouless, S. M. Yalisove, J. Mater. Res., 2010, 25(6), 1087–1095

[20] H. Sapardanis, V. Maurel, A. Köster, S. Duvinage, F. Borit, V. Guipont, Surf. Coat. Technol., 2016, 291,430-443

On the determination of delamination toughness in multi-layer thermal barrier coating systems

M. Adam, M. Elsaß, M. Frommherz, A. Scholz, M. Oechsner, Darmstadt / D

A Vickers indentation method has been applied to determine the interfacial fracture toughness of modern multi-layer thermal barrier coatings. The delamination behavior of four types of coating systems will be discussed and compared with results based on modified four-point bending (4PB) tests. The investigated multi-layer coating system consists of a CoNiCrAlY-bond coat applied via low-pressure plasma spray (LPPS) on a nickel-based superalloy and an atmospheric-plasma sprayed (APS) top layer of type gadolinium zirconate (GZO) and yttria-stabilized zirconia (YSZ). A conventional YSZ mono-layer system is used for reference. The effects of GZO and YSZ microstructure were investigated using top coats with low and high porosities for both (multi- and single-layer) coating systems. Isothermal oxidation tests at 1100 °C up to 500 hours were performed to study the interaction between thermal aging and fracture behavior. Investigations of microstructure and sintering behavior show a significant influence of the annealing conditions on fracture toughness. It has been observed, that with increasing annealing time, the stiffness and thus the crack driving force of the GZO layer is increased due to sintering effects and healing of submicron defects. The lower stiffness and higher defect density of GZO seem to be the main reason for the reduced fracture toughness of the YSZ / GZO interface compared to the YSZ / CoNiCrAlY interface. As a result, the delamination of the top coat is observed to shift from the top coat / bond coat interface into the top coat double-layer.

1 Introduction

Thermal barrier coatings (TBCs) have been a major part of stationary gas turbines and jet engines for several decades. To protect the underlying substrate material, mostly nickel or cobalt based single crystal superalloys, TBCs are used in the high-temperature sections. Especially heat shields in combustion chambers and the turbine blades and vanes in the first stages of high-pressure turbines are typically coated with 7 - 8 wt. % yttria-stabilized zirconia (YSZ) using atmospheric plasma spraying (APS) or electron-beam physical vapor deposition (EB-PVD). During long-term operation, YSZ is able to withstand maximum temperatures of approximately 1200 °C [1, 2].

From the Carnot process of ideal gases, it is known, that an increase of efficiency is directly correlated to an increasing upper process temperature. Therefore, gas turbine manufacturers are interested in higher combustion and turbine inlet temperatures, which leads to the requirement of new high-temperature materials as well as material combinations and architectures.

Pyrochlores are one of the most promising material groups for the development of advanced TBCs. Especially gadolinium zirconate (GZO) is predicted to withstand temperatures above the current limit of about 1300 °C caused by its high transition temperature ranging between 1530 and 1550 °C [3-5]. Further advantages compared to YSZ are the lower thermal conductivity ($9 - 11 \cdot 10^{-6}$ / K) and the equivalent lower thermal expansion coefficient (1.2 - 1.8 W / (m·K)) [6-8]. In high-temperature and long-term operation, GZO reacts with the thermally grown oxide (TGO) when applying it directly to the bond coat. Therefore and due to the lower fracture toughness compared to YSZ, a ceramic interlayer of YSZ has to be applied between the bond coat and the GZO top coat.

Often TBCs fail at the interface between bond coat and top coat layer as a result of bond coat oxidation and sintering of ceramic caused by thermal exposure. Subsequently, microcracks nucleate, grow and join, forming a network of macrocracks [2].
For characterization, failure analysis and lifetime prediction, the knowledge about the material describing mechanical properties, especially stiffness and fracture toughness of each layer, as well as their time and temperature dependent evolution is essential.
To determine the fracture toughness of bonded ceramics several measurement techniques have been developed in the past. Particularly bending tests have the advantage to obtain a global fracture toughness. Unfortunately, the preparation of specimens is very time consuming and cost-expensive.

As shown inter alia in [9-12] the fracture toughness of a bi-material interface can be obtained using a standard Vickers hardness tester. Based on the studies by Yamazaki et al. [10], the determined interfacial fracture toughness K_{IFC} is mainly correlated to the crack length c along the interface resulting from the indentation, the residual size of impression a and the stiffness, and the Poisson's ratio of the pairing materials, described by the bi-material constant ε, see Eq. 1. Herein A is a scaling parameter to compare the toughness obtained by indentation with four-point bending results. For ceramic coatings A is calculated to 0.081 [10].

$$K_{IFC} = \frac{A}{\sqrt{2\pi}} \frac{F}{a\sqrt{c}} \cosh(\pi\varepsilon) \qquad \text{Eq. 1}$$

Note that the half-diagonal length of impression should be not more than 20 % of the layer thickness of each pairing material at the investigated interface, to avoid an influence of the layer thickness on fracture toughness [10].

2 Experimental work

A schematic of the investigated mono- and multi-layer TBCs is shown in **Fig. 1**. The substrate material consists of a standard single crystal nickel based superalloy. A CoNiCrAlY-bond coat has been deposited via low-pressure plasma spraying (LPPS). For the application of ceramic top coats, YSZ and YSZ / GZO, atmospheric plasma spraying (APS) have been used.

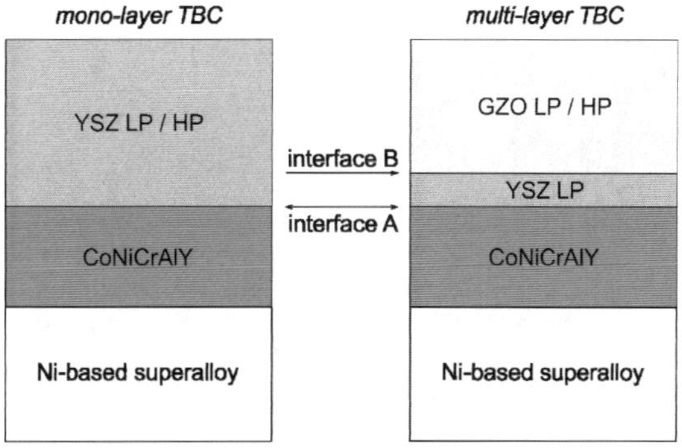

Fig. 1. Schematic of investigated mono- and multi-layer thermal barrier coating systems.

To study the effects of different spraying-process parameters on microstructure and mechanical properties, the ceramic layers were deposited by two coating facilities, marked with G1 and G2. The thickness of YSZ coatings of the mono-layer systems was in the range of about 510 - 520 µm and varied between 110 and 140 µm for the multi-layer systems. The thickness of GZO ranges between 400 and 500 µm. Image analysis delivered porosity levels of 10.2 % (LP: low porosity) and 20.6 % (HP: high porosity) for the YSZ layer. The porosity of GZO was between 8 and 9 % for LP coatings and 12 to 17 % for HP coatings [13].

Tab. 1. Overview of investigated specimens and oxidation tests

ID		Layer system	Annealing conditions		
			Type	Temperature	Time
YSZ LP	G1	mono	iso	1100 °C	0 / 100 / 300/ 500 h
YSZ HP	G1				
YSZ/GZO LP/LP	G1 G2	multi	iso	1100 °C	0 / 100 / 300 / 500 h
YSZ/GZO LP/HP	G1 G2				

iso: isothermal oxidation

Isothermal oxidation tests at 1100 °C up to 500 hours on air have been performed, to investigate the influence of thermal exposure on the delamination behavior, **Tab. 1**.

2.2 Vickers indentation

The fracture toughness along both interfaces (bond coat / YSZ and YSZ / GZO) has been investigated using metallographic specimens of about 15.0 x 5.0 x 5.0 mm (length x height x width), formerly embedded into resin. To remove the resin before testing, all specimens have been annealed at 680 °C for one hour. Both, the heating and cooling process, lasts about 12 hours to prevent the specimens from high temperature gradients and thus the top coat from premature delamination.
It is noticed that all specimens originated from previous modified four-point bending tests (4PB) and only the unstressed sections were used.
In order to achieve thickness-independent results, two indentation load levels have been utilized. Investigations on mono-layer systems at interface position A, see Fig. 1, have been performed with HV5 (F = 49.1 N). At interface position B (only multi-layer systems) load level HV3 (F = 29.4 N) has been used. Thus, the ratio of crack length to impression size could be set to c/a = 3.
A maximum of five impressions per specimen has been set, to inhibit a merging of adjacent cracks.

2.1 Modified four-point bending tests

Frommherz et al. [13] studied the delamination behavior of mono- and multi-layer TBCs using modified four-point bending tests [13-18], performed with an asymmetric setup in conformity to [17].
The measurement procedure consists of two steps. First, a pre-crack is initiated via three-point bending followed by the modified four-point bending test. This practice allows the measurement of crack length independent energy release rates (G_c). Determined G_c-values correspond to a mixed-mode in-plane toughness. The use of stiffeners on top of the ceramic coating applied using an adhesive foil (FM1000), supports the delamination of the ceramic layer.
For further information about the test setup, specimen geometry and preparation see [13].

3 Results and discussion

To compare the results from the two different measurement methods, the energy release rate G_c obtained by modified four-point bending tests has to be converted into an equivalent fracture toughness K using the Irwin correlation (Eq. 2) between both properties [20].

$$Gc = \frac{|K|^2}{E'} = \frac{|K|^2}{E/(1-\nu^2)} \qquad \text{Eq. 2}$$

Based on the derivations in [19] a plain-strain condition for the modified four-point bending tests has been assumed. The Young's modulus for plain-strain can be calculated using the in-plane Young's modulus of the layer in which the delamination crack occurs (YSZ in mono-layer systems and GZO in multi-layer systems) and a Poisson's ratio of ν = 0.15 for both ceramics, **Fig. 2**.

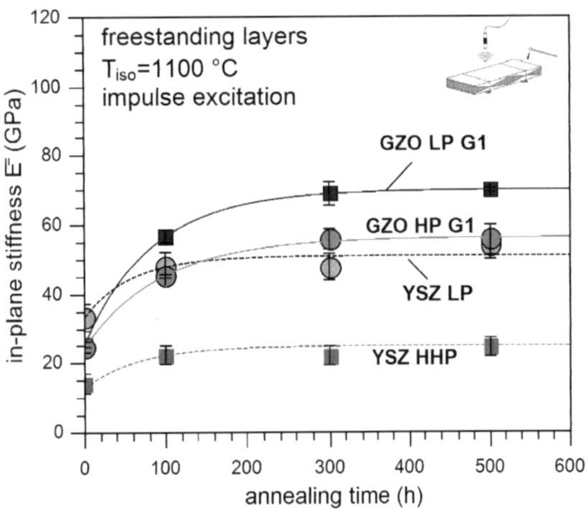

Fig. 2. Development of Young's modulus of YSZ and GZO freestanding layers due to isothermal oxidation after 100, 300 and 500 hours at 1100 °C compared to the as-sprayed condition [13].

3.1 Interfacial fracture toughness

Fig. 3 shows the results from modified four-point bending tests compared to energy release rates determined from Vickers indentations.

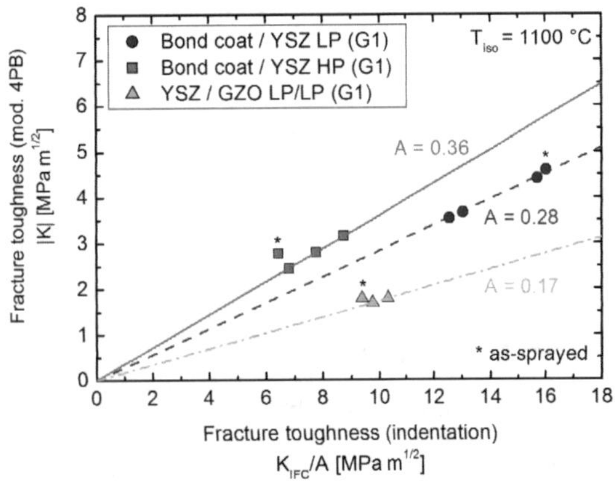

Fig. 3. Comparison of the interfacial fracture toughness K_{IFC} measured by indentation and the calculated |K|-values based on modified four-point bending tests.

The local indentation measurements along the interface are able to represent the global fracture behavior qualitatively. Even a quantitative comparison is possible by fitting the values of K_{IFC} and $|K|$ with a first-order polynomial function. The calculated slope represents a scaling factor *A* for every coating system and is quite similar to the dimensionless constant *A* in Eq. 1. For the investigated YSZ-LP coating, *A* can be determined to 0.28. The slope for the YSZ-HP top coat is calculated to 0.36. A similar relationship has been found for the YSZ / GZO interface in multi-layer systems with a coefficient of *A* = 0.17.

All calculated constants differ from the result presented [10], which can be explained by the use of various four-point bending setups (symmetric versus asymmetric) and the difference in the investigated coating systems and does not detract the applicability of the measurement technique.

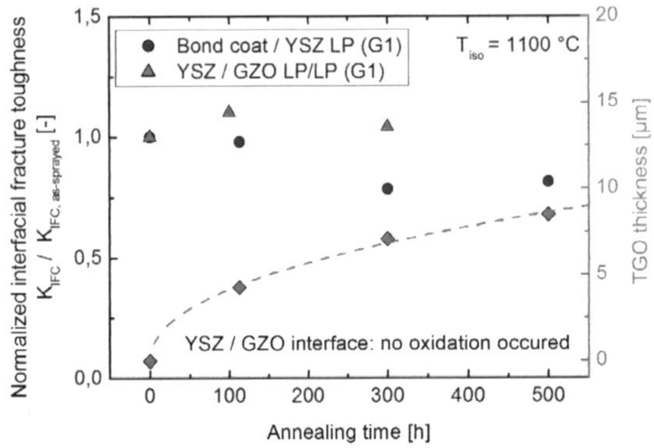

Fig. 4. Normalized interfacial fracture toughness obtained by indentation and TGO thickness as a function of isothermal oxidation time.

As pointed out in **Fig. 4**, the interfacial fracture toughness seems to remain stable within the first 100 hours of annealing. With increasing oxidation time (≥ 300 hours) the toughness is observed to be only about 80 % of the as-sprayed condition. The increasing thickness of TGO leads to a decrease of adhesive strength between bond coat and YSZ. Concurrently, the increasing stiffness of YSZ is suspected to ensure an increase in fracture toughness. With progressing thermal exposure time (> 300 hours), the temporal change of stiffness is going to reach a minimum while TGO grows continuously. Thus, the influence of oxidation on fracture toughness is set to increase, which results in an obvious degradation of the interface.

In contrast to that, no oxidation induced failure mechanism has been identified at the YSZ / GZO interface and the delamination toughness increases with increasing stiffness.

Nonetheless, the role and impact of the mixed-mode crack opening on the determination of the crack length in Eq. 1 have to be considered critically. One should note that the Vickers hardness tester has been located directly between both materials; thereby the crack is initiated directly at the interface and is going to deviate into the ceramic top coat, depending on the mechanical and microstructural characteristics of the pairing materials. As presented in [20], for example, the deviation angle can be described by the mode I and mode II stress intensity factors (K_I and K_{II}) and depends, among others, on the Young's modulus ratio at the bi-material interface, which varies between as-

sprayed and annealed conditions of the represented coating systems.
The crack deflection behavior differs to that, known from the bending tests, in which the crack was initiated via three-point bending near the interface, and propagated almost parallel to the interface during four-point bending. This implies a time-temperature depending alteration of the mixed-mode crack opening, especially in the case of shorter annealing times, when the change of the mechanical properties is at its highest, **Fig. 2, 5**.

Furthermore, the impression size in Eq. 1 is an implicit function of the hardness. During isothermal oxidation, the hardness of the ceramics increases, **Fig. 5**. The change in hardness (*dH/dt*) reaches a maximum within the first 100 hours of oxidation, whereas the hardness decrease of the bond coat is negligible. With increasing annealing time, *dH/dt* is set to descend and Eq. 1 becomes more and more independent from the hardness.

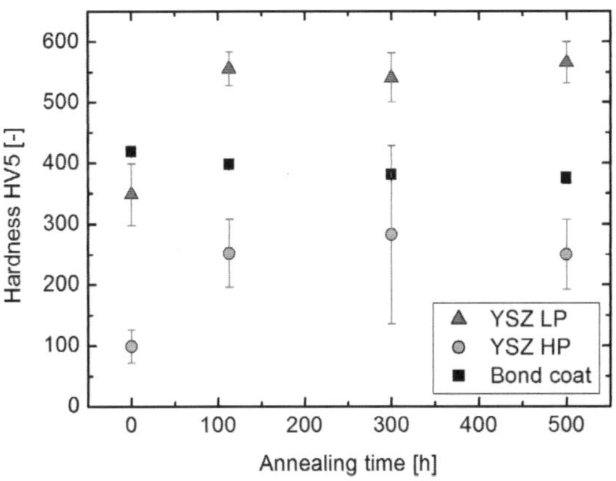

Fig. 5. Vickers hardness of YSZ coatings and bond coat

These observations were supposed to be reasonable for the underestimation of the interfacial fracture toughness, obtained by indentation, in the as-sprayed condition of the presented coating systems, **Fig. 3**.

Further investigations, e.g. finite fracture mechanics analysis has to be performed, in order to study the influence of the variation of mechanical properties (Young's modulus, Poisson's ratio, etc.) on the delamination behavior, see for example [21, 22]. A finite element analysis of the indentation measurements and the asymmetric four-point bending setup is predicted, to deliver reliable information about the present stress-strain states.

3.2 Microstructural investigations

In a multi-layer system of type YSZ / GZO, the interface inside the ceramic top coat or the GZO layer itself seems to be the weak point as indicated by metallographic investigations of test specimens after modified four-point bending tests, **Fig. 6**.

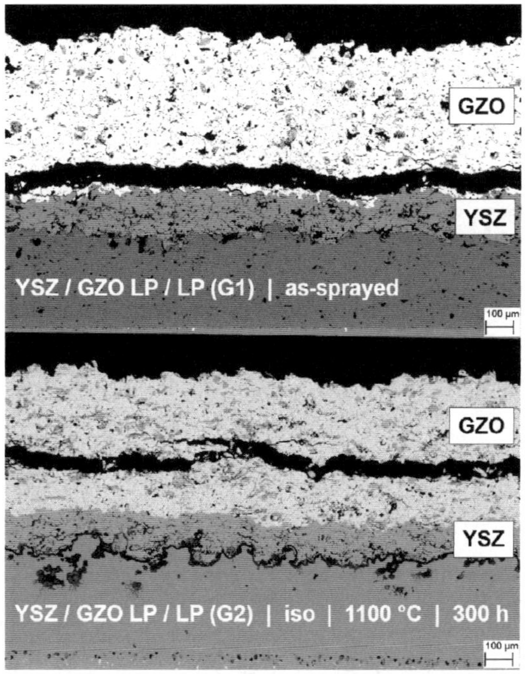

Fig. 6. Cross-sections of the YSZ / GZO LP / LP (G1) coating system in as-sprayed condition and YSZ / GZO LP / LP (G2) after isothermal oxidation. Specimens have been tested in modified 4PB tests.

Dependent on different spray conditions, referred to as G1 and G2, a delamination occurs slightly above the YSZ / GZO interface (interface delamination) or inside the GZO layer (intralayer delamination). The interface delamination inside the ceramic top coat has great similarity to the failure behavior of mono-layer coatings parallel and near the bond coat / YSZ interface, also known as "white failure" [23].

In the as-sprayed condition of the LP system, the energy release rate for delamination of the YSZ / GZO interface is only about ~ 62 % of the bond coat / YSZ interface. For the high-porosity (HP) system, G_c ranges between 54 and 58 % [13].

A variation of the spray parameters in the APS process leads to different microstructures and a higher amount of interlamellar cracks, segmentation cracks and unmolten particles of GZO compared to YSZ, **Fig. 7**. Especially the GZO (G2) coatings seem to exhibit a higher submicron defect density, which is mainly attributed to a higher amount of unmolten particles [24, 25]. Those represent an additional weakness, reducing the adhesive strength inside the upper ceramic layer and are supposed to be one reason for the intralayer delamination. This kind of failure has been observed for the as-sprayed coatings as well.

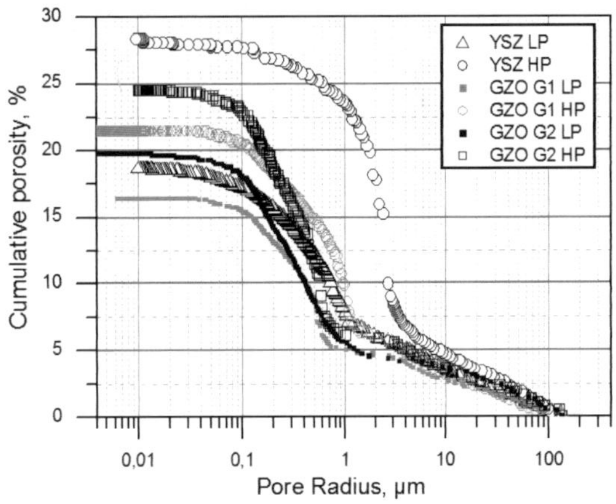

Fig. 7. Distribution of pore size of YSZ and GZO layers (LP and HP) depending on spraying parameters (G1 and G2) [13, 24, 25].

In summary, lower toughness values of YSZ / GZO interface compared to bond coat / YSZ and the higher submicron defect density of GZO are supposed to be reasonable for the delamination switching from the bottom interface into the ceramic top coat layer.

One has to note, that a degradation of both interfaces is influenced by different mechanisms, e.g. growth of TGO at bond coat / YSZ interface as well as sintering and microcracking of YSZ and GZO at both interfaces and inside the ceramic layers, and cannot be compared easily. Additional investigations of the time and temperature dependent, anisotropic mechanical properties and microstructure of YSZ and GZO are necessary; to further characterize the failure mechanisms inside the ceramic multi-layer top coat.

4 Conclusion

The utilization of an indentation technique offers the use of less and smaller specimens for determination of global interfacial fracture toughness with standardized measuring equipment.
It has been demonstrated that interfacial fracture toughness, obtained by a global measurement method, e.g. modified four-point bending tests, can be qualitatively correlated to local measurement results using a Vickers indentation method along the interfaces of mono- and multi-layer thermal barrier coating systems. Furthermore, a linear correlation between both measurement techniques, described by a scaling factor, has been found to vary for different top coat microstructures and interfaces. The relation seems to be independent of the annealing time, except for the as-sprayed condition.
Based on microstructural investigations, it has been observed, that ceramic coatings of type YSZ / GZO fail quite different than YSZ mono-layer coatings. The delamination is shifted from the bond coat / YSZ interface into the ceramic top coat and occurs slightly above the YSZ / GZO interface or inside the GZO layer. This kind of failure shifting is mainly attributed to the lower interfacial fracture toughness of YSZ / GZO interface and the GZO microstructure, depending on the spraying conditions of the APS process.

5 Acknowledgements

This work was supported by the German Research Foundation (Deutsche Forschungsgemeinschaft, DFG) and the German Research Association for Combustion Engines (Forschungsvereinigung für Verbrennungskraftmaschinen e.V., FVV) within the research project "Development of a mechanism-based lifetime model for bi-layer thermal barrier coatings". We thank the German working group "Thermal Barrier Coatings" under the supervision of Dr. Werner Stamm, Siemens AG, for their guidance and the scientific discussions. Besides we are grateful to Siemens AG, ALSTOM (Switzerland) Ltd., MAN Turbo & Diesel SE, Rolls-Royce Deutschland Ltd. & Co. KG and MTU Aero Engines GmbH for the additional financial support for the processing of the coatings and for providing substrate material. The contribution of Lukas Pfeuffer is gratefully acknowledged.

6 Literature

[1] Clarke, D. R., Oechsner, M., and Padture, N. P.: Thermal barrier coatings for more efficient gas-turbine engines. MRS Bulletin 37 (2012). pp. 890-941.

[2] Padture, N. P., Gell, M., and Jordan, E. H.: Thermal barrier coatings for gas-turbine engine applications. Science 296 (2002). pp. 280-284.

[3] Haile, S.M., and Meilicke, S.: The kinetics of ordering in gadolinium zirconate: an unusual oxygen ion conductor, MRS Proc. 398 (1995).

[4] Karaulov, A. G., and Zoz, E. I.: Phase formation in the ZrO2–HfO2–Gd2O3 and ZrO2–HfO2–Yb2O3 systems. Refract. Ind. Ceram. 40 (1999). pp. 479–483.

[5] Lee, Y. H., Sheu, H. S., Deng, J. P., and Kao, H.-C.I.: Preparation and fluorite–pyrochlore phase transformation in Gd2Zr2O7. J. Alloys Compd. 487 (2009). pp. 595–598.

[6] Cao, X. Q., Vaßen, R., and Stöver, D.: Ceramic Materials for Thermal Barrier Coatings. J. Eur. Ceram. Soc., 24 (2004). pp. 1–10.

[7] Vaßen, R., Stuke, A., and Stöver, D.: Recent developments in the field of thermal barrier coatings. J. Therm. Spray Technol., 18 (2009). pp. 181–6.

[8] Stöver, D., Pracht, G., Lehmann, H., Dietrich, M., Döring, J. E., and Vaßen, R.: New material concepts for the next generation of plasma-sprayed thermal barrier coatings. J. Therm. Spray Technol., 13 (2004). pp. 76–83.

[9] Chicot, D., Démarécaux, P., and Lesage, J.: Apparent interface toughness of substrate and coating

couples from indentation tests. Thin Solid Films 283 (1996). pp. 151-157.

[10] Yamazaki, Y., Arai, M., Miyashita, Y., Waki, H., and Suzuki, M.: Determination of interfacial fracture toughness of thermal spray coatings by indentation. Journal of Thermal Spray Technology 22 (2013), pp. 1358-1365.

[11] Yamazaki, Y., Kuga, S., and Yoshida, T.: Evaluation of interfacial strength by an instrumented indentation method and its application to an actual TBC vane. Acta Metall. Sin. 24 (2011). pp. 109-117.

[12] Marot, G., Lesage, J., Démarécaux, P., Hadad, M., Siegmann, S., and Staia, M. H.: Interfacial indentation and shear tests to determine the adhesion of thermal spray coatings. Surface & Coatings Technology 201 (2066), pp. 2080-2085.

[13] Frommherz, M., Scholz, A., Oechsner, M., Bakan, E., and Vaßen, R.: Gadolinium zirconate/YSZ thermal barrier coatings: Mixed-mode interfacial fracture toughness and sintering behavior. Surface & Coatings Technology 286 (2016), pp. 119-128.

[14] Bast, U., and Schumann, E.: Development of novel oxide materials for TBC's. 26th Annu. Conf. Compos. Adv. Ceram. Mater. Struct. B Ceram. Eng. Sci. Proc. John Wiley and Sons, Inc. 2002, pp. 525-532.

[15] Bahr, H.-A., Balke, H., Fett, T., Hofinger, I., Kirchhoff, G., and Munz, D.: Cracks in functionally graded materials, Mater. Sci. Eng. A 362 (2003), pp. 2–16.

[16] Hofinger, I., Oechsner, M., Bahr, H.-A., and Swain, M.: Modified four-point bending specimen for determining the interface fracture energy for thin, brittle layers, Int. J. Fract. 92 (1998), pp. 213–220.

[17] Oechsner, M.: Ein Beitrag zur Lebensdauervorhersage von keramischen Wärmedämmschichten. VDI Verlag, Düsseldorf, 2001.

[18] Yamazaki, Y., Schmidt, A., and Scholz, A.: The determination of the delamination resistance in thermal barrier coating systems by four-point bending tests. Surface & Coatings Technology 201 (2006). pp. 744-754.

[19] Charalambides, P. G., Lund, J., Evans, A. G., and McMeeking, R. M.: A test specimen for determining the fracture resistance of bimaterial interfaces. J. Appl. Mech. 56 (1989). pp. 77–82.

[20] Gross, D., and Seelig, T.: Bruchmechanik – Mit einer Einführung in die Mikromechanik. Vol. 5, Berlin: Springer-Verlag (2011).

[21] He, M.-Y., and Hutchinson, J. W.: Crack deflection at an interface between dissimilar elastic materials. Int. J. Solids Structures 25 (1989). pp. 1053-1067.

[22] Ivanova, J., Nikolova, G., Dineva, P., and Becker, W.: Interface behavior of a bimaterial plate under dynamic loading. J. Eng. Mech. 136 (2010), pp. 1194-1201.

[23] Bose, S., and DeMasi-Marcin, J.: Thermal barrier coating experience in gas turbine engines at Pratt & Whitney. Journal of Thermal Spray Technology 6 (1997), pp. 99-104.

[24] Frommherz, M., Scholz, A., and Oechsner, M.: Fundamental properties of modern double-layer coatings: Delamination and sintering behavior. Turbine Forum, Advanced Coatings for High Temperatures (2016). Unpublished.

[25] Bakan, E., Mack, D. E., Mauer, G., and Vaßen, R.: Gadolinium zirconate/YSZ thermal barrier coatings: Plasma spraying, microstructure, and thermal cycling behavior. J. Am. Ceram. Soc. 97 (2014), Issue 12, pp. 4045-4051.

Microscale adhesion strength evaluation of cold sprayed copper deposit

Y. Ichikawa, R. Tokoro, K. Ogawa, Sendai / JAPAN

Adhesion strength is one of the most important characteristics when discussing the reliability of a cold spray. A focused ion beam (FIB) was used to conduct ultra-micro tensile tests for micro-scale adhesive strength evaluations of high-pressure type cold-sprayed copper deposition on an aluminum substrate.It was also used to determine the essential factor of adhesion strength and the coating formation mechanism. The micro-scale local adhesion strength of cold-sprayed copper deposits on an aluminum substrate was successfully evaluated by FIB microtensile tests. The average local adhesion strength of this deposit was 223 MPa. The variations in adhesion strength between deposit and substrate were smaller than the interfacial strength of cold-sprayed deposits. This was caused by the repeated collision of these subsequent particles.

1 Introduction

Adhesion strength is one of the most important characteristics when discussing the reliability of a surface modified coating, represented by a thermal spray coating and a cold spray. Many test methods have been developed and used to evaluate the adhesion strength between coatings and substrates, and some are mentioned below. The indentation method calculates a stress intensity factor from crack propagation behavior with nanoindentation on thin coatings [1-3]. The shear test uses a single lap joint specimen to determine the shear adhesion strength, in accordance with the ASTM D1002 standard [4-9].
Where a large scatter in the results was observed, the double cantilever beam (DCB) test was used to obtain quantitative data, even when the system under test was very brittle [10-12]. For the tensile adhesive strength test, the test jig and the coatings were fixed with an adhesive glue, and a load was applied in the perpendicular direction to the coating to evaluate the break stress. This was done in accordance with the standards ASTM C633, ASTM F1147, ISO 14916, and EN 582 [13-15].
In any testing method, peeling is caused by an external force at the interface between the coating and the substrate. The mechanically dominant fracture factor that causes peeling depends on the test method, and the parameters to be evaluated are also different. It is therefore very difficult to establish a universal evaluation method, and to understand the adhesion strength controlling factors.
Previous research established a microstrength test method using a focused-ion beam (FIB), which could evaluate the local strength of cold sprayed deposits [16-20]. We attempted to establish an ultra-microtensile adhesive strength evaluation method by applying this FIB microstrength method. Local strength was evaluated by use of this method. It can be expected to know the essential factor of adhesion strength, and coating formation mechanism.
In this study, we aim to establish a local adhesion strength evaluation using cold sprayed copper deposition on the aluminum substrate, and discuss the coating deposition mechanism of the cold spray technique.

2 Experimental

2.1 Cold sprayed copper deposit on aluminum substrate

A pure copper powder was used for the deposition material. The copper powder (HQW-35, Fukuda), was water-atomized spherical-shaped powder, 37 μm in diameter (Fig. 1). The powder was sprayed on a pure aluminum (A1100) substrate using high-pressure type CS equipment (Plasma Giken Co., LTD, PCS-304), using nitrogen gas. The gas temperature was 600 °C, and its pressure 3 MPa. The spray condition is summarized in Table 1. A 15-mm thick deposit was formed on the aluminum substrate (Fig. 2).
In order to conduct a microtensile adhesion test, the deposit was cut and mounted with the epoxy resin. The mounted specimen was cross-sectionally polished and subjected to a scanning electron microscopy (SEM) observation. Fig . 3 shows the cross-sectional observation result of the aluminum deposition interface.

Fig. 1 SEM image of the feedstock copper powder.

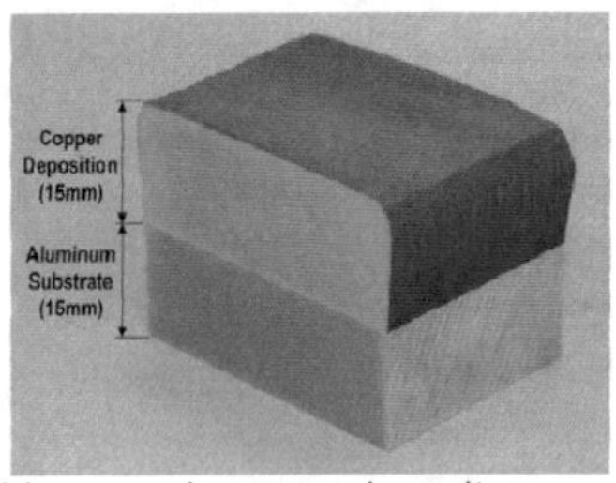

Fig. 2 Cold sprayed copper deposit on an aluminum substrate.

Fig. 3 Cross-sectional observation image of cold sprayed copper deposit on an aluminum substrate.

Table 1 Cold spray conditions.

Facility	Plasma GIken PCS-304
Gas	
Type	Nitrogen
Pressure	3 MPa
Temperature	800°C
Spray Distance	20-25 mm
Traverse speed	20 mm/s
Powder	
Material	Copper (Fukuda Metal Foil & Powder Co., Ltd.)
Mean diameter	37 µm
Process	Water atomized
Substrate	
Material	Aluminum A5052

2.2 FIB microtensile test

In situ microtensile adhesion tests were performed using an FIB system (FB2200, Hitachi). The FIB system was equipped with a microsampling system, consisting of a servo-controlled microsampling probe and a tungsten deposition gun. The microsampling probe could be used for positioning and loading specimens during tensile testing.
As this system cannot measure the applied load directly, a silicon beam was used to perform load measurements. The microtensile specimen, silicon beam, and microprobe were conected by tungsten deposition, and the specimen was loaded by moving the microsampling probe as shown in Fig. 4. The load was measured from the deformation of the silicon beam. The scanning ion microscopic (SIM) video recording monitored the cantilever deformation, and the fracture load was calculated by the difference between the images before testing, and just before fracture occurred.

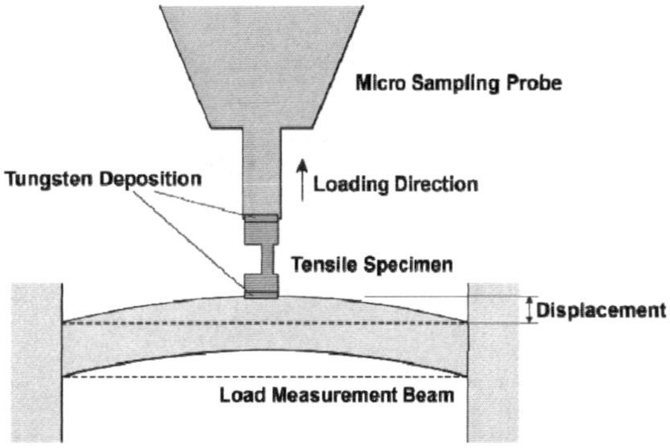

Fig. 4 Schematic diagram of FIB microtensile testing system [16].

2.3 Cantilever design

The load (F) applied to the specimen, was calculated from the displacement (*d*) of the single crystal silicon beam from the initial position,

$$F = Kd \qquad (1)$$

where *K* is the spring constant. It was assumed that the specimen was fixed at the tip of a cantilever, and *K* was described by the equation,

$$K = \frac{Ewt^3}{4l^3} \qquad (2)$$

where *E* was an elastic modulus of a cantilever, and l, w, and t were the length, width, and thickness of the cantilever respectively.
The beam dimensions were determined by the elastic modulus of the beam material, maximum applied load for a specimen, and the maximum displacement of a beam. The maximum displacement had to be within thc FIB-SIM observation area. In this test, the beam material was single crystal silicon, the elastic modulus of which is 170 GPa in the <110> direction, and the SEM observation area was 8 µm x 8 µm. The expected maximum applied load was 1.6 µN, which corresponded to 800 MPa for 2 µm^2 cross-sectional area specimens. The beam that satisfied the above conditions was then designed, as summarized in Table 2.

Table 2 Designed cantilever dimensions.

Material	Single crystal Silicon
Elastic Modulus	170 GPa in <110>
Dimensions	
L	80 µm
T	7 µm
D	3 µm

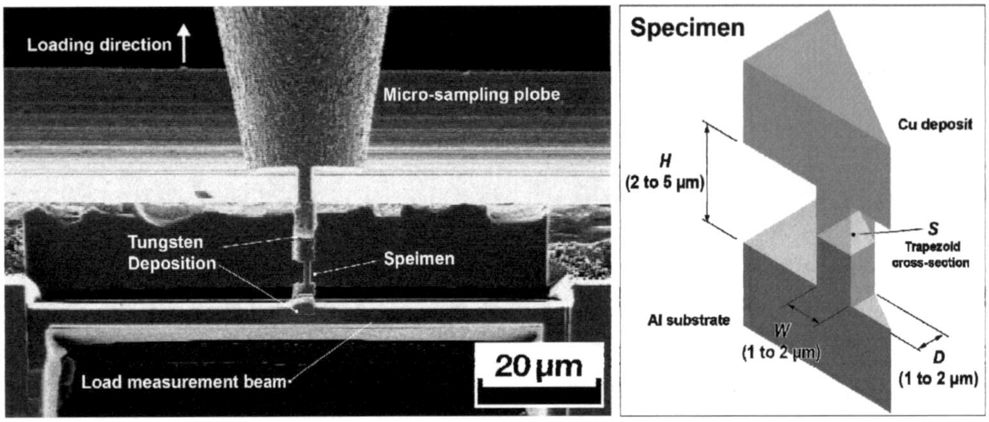

Fig. 5 Fabricated microscale adhesion strength test specimen and load measurement beam.

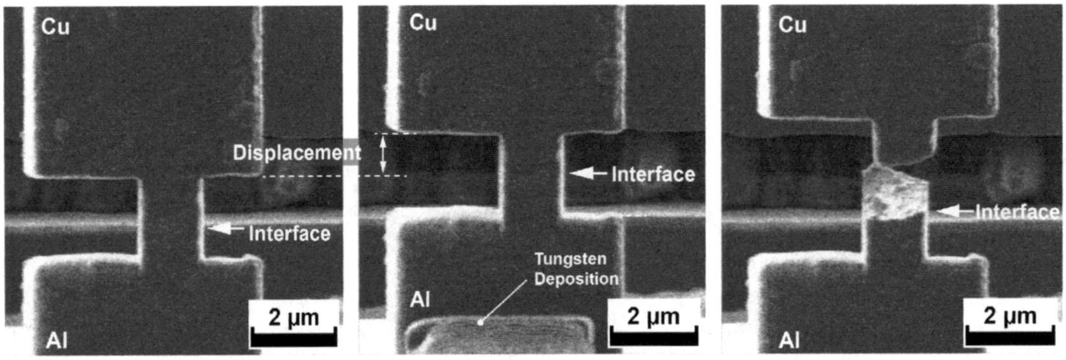

(a) Set-up (befor test) **(b) Just befor fracture** **(c) After tested**

Fig. 6 In-situ SIM snapshots of microadhesion strength test.

2.4 Microspecimens

Three microspecimens were picked-up from the interface between the copper deposition and the aluminum substrate. The SIM image of fabricated load measurement beam and specimen are shown in Fig. 5. The specimens were fabricated using FIB, and their dimensions are summrized in Table 3.

Table 3 Dimensions of fabiricated specimens

Specimen	*D*	*W*	*H*
No. 1	4.33 µm^2	1.32 µm	3.34 µm
No. 2	3.40 µm^2	1.48 µm	1.82 µm
No. 3	3.70 µm^2	1.10 µm	2.96 µm

3 Results and Discussion

Figure 6 shows SIM images captured during the microtensile adhesion testing of the first specimen. When the specimen fractured, the displacement of the load measurement beam was 1.8 µm. The adhesion strength, calculated using the method described above for cross-sectional areas, was 130 MPa. The same experiments were performed on specimens two and second, and third of 135 MPa and 182 MPa were obtained, respectively. The results obtained from these tests are summarized in Table 4.

T Table 4 Summary of microadhesion strength tests

Specimen	Adhesion strength
No. 1	194 MPa
No. 2	273 MPa
No. 3	202 MPa
Average	223 MPa

From these results, it can be seen that the average local adhesion strength was 149 MPa, and the scatter range is small.

In previous studies [16], it was seen that the local cohesion strength of cold spray coating varied, depending on the location. This indicated that the

adhesion state of the particles constituting the deposit was not uniform, and this was also a feature of the mechanical properties of the cold spray deposition.
As can be seen from the results, however, there was little difference in the local adhesion strength of the deposition interface, and its close contact state was more uniform than the inside of the deposits. This was because the deposit constitution near the interface was at the lowest part of the deposition, and the hardening progressed by repeated collisions of succeeding collision particles. It was found that the adhesion strength at the deposition interface increased by repeated collision of these subsequent particles, and uniform strength was obtained.
The fracture surface observation results are shown in Fig. 7. This fracture surface was from the lower half of the specimen. Moreover, the substrate side specimen had been observed by the vertical direction of the fracture surface. From the result of EDX analysis, it can be seen that this fracture surface is composed only of aluminum and the fracture is broken at the substrate and coating interface. Although it was slightly deformed, the lower fracture surface was deformed convexly, and it is thought that only one dimple in the macroscopic fracture test was formed, which showed a mode similar to inter-granular cracking. From this, it could be seen that it had a pseudo-grain boundary structure, in which a good metallic bonding was formed by the cold spray deposition and the substrate interface.

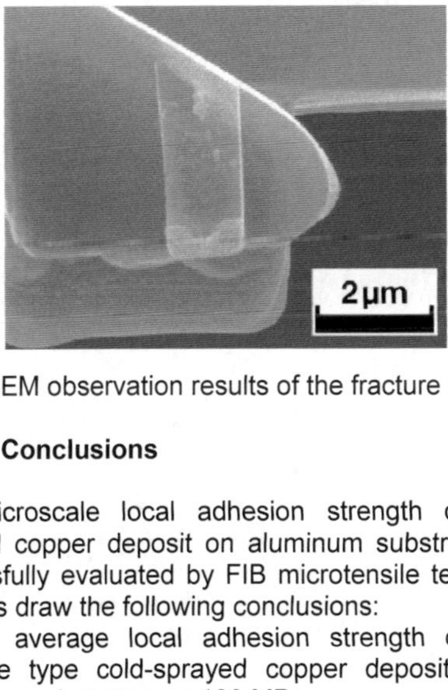

Fig. 7 SEM observation results of the fracture surface.

4 Conclusions

The microscale local adhesion strength of cold-sprayed copper deposit on aluminum substrate was successfully evaluated by FIB microtensile tests. We can thus draw the following conclusions:
1. The average local adhesion strength of high-pressure type cold-sprayed copper deposit on an aluminum substrate was 139 MPa.
2. The variations in adhesion strength between deposit and substrate are less than the interfacial strength of cold sprayed deposits. This was due to the repeated collisions of these subsequent particles.

Acknowledgement

This work was supported by JSPS KAKENHI Grant Number 15H05501.

References

[1] Bhushan, B.: Chemical, mechanical and tribological characterization of ultra-thin and hard amorphous carbon coatings as thin as 3.5 nm: recent developments. Diamond and Related Materials 8 (1999), Issue11, pp.1985–2015,.
[2] Drory, M.D. and Hutchinson, H.W.: Measurement of the adhesion ,of a brittle film on0 a ductile substrate by indentation. Proc. R. Soc. Lond A (1996), pp. 452, 2319.
[3] Vasinonta, A. and Beuth, J.L.: Measurement of interfacial toughness in thermal barrier coating systems by indentation. Engineering Fracture Mechanics 68 (2001), Issue 7, pp. 843-860.
[4] Kitamura, T., Shibutani, Ueno, T.: Crack initiation at free edge of interface between thin films in advanced LSI. Engineering Fracture Mechanics 69 (2002), Issue 12, pp.1289-1299.
[5] Agrawal, D.C. and Raj, R.: Measurement of the ultimate shear strength of a metal-ceramic interface. Acta Metallurgica 37 (1989), Issue 4, pp. 1265-1270.
[6] Leterrier, Y.: Durability of nanosized oxygenbarrier coatings on polymers. Progress in Materials Science 48 (2003), Issue 1, pp. 1-55.
[7] Shieu, F.S. and Shiao, H.: Measurement of the interfacial mechanical properties of a thin ceramic coating on ductile substrates. Thin Solid Films 306 (1997), Issue 1, pp.124-129.
[8] Beuth, J.L.: Cracking of thin bonded films in residual tension. In. J. Solids Structures 29 (1992), Issue 13, pp. 1657-1675.
[9] Era, H., et al.: A Modified shear test for adhesion evaluation of thermal sprayed coating. Materials Science and Engineering: A 251 (1998), pp. 166-172.
[10] Akisanya, A.R. and Fleck, N.A.: Brittle fracture of adhesive joints. International Journal of Fracture 58, , (1992), Issue 2, pp. 93-114.
[11] Piggott, M.R.: Why interface testing by singlefibre methods can be misleading. Composites Science and Technology 57 (1997), Issue 8, pp. 965-974.
[12] Sener, J.Y., Ferracin, T., Caussin,L., Delannay, F.: On the precision of the wedgeopened double cantilever beam method for measuring the debonding toughness of adhesively bonded plates. International Journal of Adhesion & Adhesives 22 (2002), Issue 22, pp.129-137.
[13] Greving, D.J., Shadley, J.R. and Rybicki, E.F.: Effects of coating thickness and residual stresses on the bond strength of ASTM C633- 79 thermal spray coating test specimens. Journal of Thermal Spray Technology 3 (1994), Issue 4, pp. 371-378.
[14] Han, W., Rybicki, E.F. and Shadley, J.R.: Application of fracture mechanics to the interpretation of bond strength data from ASTM Standard C633-79. Journal of Thermal Spray Technology 2 (1993) Issue 3, pp. 235-241.
[15] Mencík, J.: Mechanics of components with treated or coated sufaces. Solid Mechanics and its Applications, ed. G.M.L. Gladwell. Vol. 42 (1995), Kluwer Academic Publisher, P.O. Box17, 3300 AA Dordrecht, Netherlands. 366.

[16] Ichikawa, Y., Tokoro, R. and Ogawa, K.: Micro-Scale Strength Evaluation for Bonding Interface of Cold Sprayed Coatings, Materials Science Forum 879 (2017), pp. 795-800,.
[17] Fujii, K. and Fukuya, K.: Development of micro tensile testing method in an FIB system for evaluating grain boundary strength. Materials Transactions 52 (2011), pp. 20-24.
[18] Fujii, K. and Fukuya, K.: Development of micro tensile testing method in an FIB system for evaluating grain boundary strength. INSS JOURNAL 17 (2010), pp. 172-77.
[19] Ichikawa, Y., Watanabe, Y., Ogawa.K. and Miura, H.: Micro-scale interface strength evaluation of cold sprayed deposit, Thermal Spray 2014: Not Fiction: Thermal Spray the Key Technology in Modern Life! (2014), pp. 707-710.
[20] Ichikawa, Y. and Ogawa, K.: Study on micro-scale strength and structural evaluation for bonding interface of solid state bonded materials. Proceedings of The Japan Society of Mechanical Engineers M&M 2015 Conference (2015), OS1210-319.

Effect of Nozzle Geometry on the Microstructure and Properties of HVAF Sprayed Hard Metal Coatings

V. Matikainen*, H. Koivuluoto, and P. Vuoristo, Tampere University of Technology, Tampere, Finland
J. Schubert, VZÚ Plzeň, Pilsen, Czech Republic
Š. Houdková, University of West Bohemia, Pilsen, Czech Republic

* Corresponding author: ville.matikainen@tut.fi

Thermally sprayed hard metal coatings are the industrial standard solution for numerous demanding applications. Often the performance of thermally sprayed coatings is improved by using finer particle sizes due to improved surface finish and decreased defect sizes. In the aim of utilizing finer particle and primary carbide sizes in thermal spraying of hard metal coatings, several approaches have been studied to control the spray temperature. The most viable solution is to use the modern high velocity air-fuel (HVAF) spray process, which has already proven to produce high quality coatings with dense structures. In HVAF spray process, the particle heating and acceleration can be efficiently controlled by changing the nozzle geometry. In this study, fine WC-10Co4Cr powder (-25+5 µm) was sprayed with three nozzle geometries to investigate their effect on the particle temperature, velocity and coating microstructure. The study demonstrates that the particle melting and resulting W_2C formation can be efficiently controlled by changing the nozzle geometry from cylindrical to convergent-divergent. Moreover, the average particle velocity was increased from 780 to over 900 m/s. This increase in particle velocity significantly improved the coating structure and density while deposition efficiency decreased slightly. Further evaluation was carried out to resolve the effect of particle in-flight parameters on coating structure and cavitation erosion resistance, which was significantly improved with the increasing average particle velocity.

1 Introduction

Thermally sprayed tungsten carbide (WC) based hard metal coatings, especially WC-10Co4Cr, have been extensively studied and used as protective coatings against wear in various applications [1-3]. Gas and liquid fuelled high velocity oxygen-fuel (HVOF) spray processes are commonly used to produce WC-10Co4Cr coatings, while also high velocity air-fuel (HVAF) spraying has been increasingly used in industry and research during the last decade. The HVOF spray process uses kerosene or gaseous, e.g. propane, ethane or hydrogen, fuel and pure oxygen to produce high temperature supersonic gas stream, which accelerates and heats the injected feedstock particles to high velocities. Modern HVAF spray process, on the other hand, uses gaseous fuel, typically propane, and compressed air as the components for the combustion. The lower oxygen concentration reduces the temperature of the combustion but allows the axial injection and efficient heating and acceleration of the particles. The HVAF sprayed WC-Co and WC-10Co4Cr coatings have been studied already for two decades since the early liquid fuelled HVAF systems [4,5], and later with the modern gas fuelled HVAF systems [3]. Several comparative studies between HVOF and HVAF spray processes have been carried out [1,2], but limited attention has been paid on the parameter effects of the HVAF spray process on coating properties and performance [6,7]. Recently Kumar et al. [8] studied the cavitation erosion performance of HVAF sprayed coatings with different nozzle geometries compared to HVOF sprayed coating. In the present work, the effect of nozzle geometry and two spray parameter sets on cavitation erosion resistance are studied. The particle temperatures and velocities are measured with spray sensor and reference coatings are produced with HVOF spray processes. The study provides additional information about the capabilities and limitations of the HVAF spray process compared to the industrial standard HVOF spray processes. In addition, the cavitation erosion resistance of the coatings is compared to AISI 316L stainless steel bulk and the wear debris is analysed after the tests.

2 Experimental

2.1 Coating manufacturing

Agglomerated and sintered WC-10Co4Cr powder (Amperit 558.025: H.C. Starck GmbH, Germany) with a nominal particle size distribution of -25+5 µm was sprayed on S235 structural steel substrates with commercial HVAF spray process (M3, Uniquecoat Technologies LLC, United States). The substrates were grit blasted with mesh 36 Al_2O_3 grit prior to spraying. Two spray parameter sets *High* (H) and *Low* (L), described in **Table 1**, were used for HVAF spraying with each of the three nozzle setups: 4L0 (cylindrical), 4L2 (convergent-divergent) and 4L4 (convergent-divergent). Long combustion chamber and short powder injector (#3) were used for all sprayings. Target coating thickness was 300 µm and compressed air cooling was used during the spraying to maintain the substrate temperature below 200 °C. Two reference coatings were sprayed with gas and liquid fuelled HVOF systems. The gas fuelled HVOF-G coatings were sprayed with DJ2700 (Oerlikon Metco AG, Switzerland) using standard parameters, 230 mm spray distance, 2.6 m/s surface speed and 60 g/min powder feedrate. Liquid fuelled HVOF-L coatings were sprayed with JP-5000 (Praxair Inc., United States) using 360 mm spray distance, 1.6 m/s surface speed, 125 g/min powder feedrate and optimised spray parameters. Coarser powder (Amperit 558.074: H.C. Starck GmbH, Germany) with size distribution of -45+15 µm was used for both HVOF spray processes.

Deposition efficiencies were measured for the HVAF spray process with 4L2 nozzle and *Low* parameter set and for the reference coating sprayed with gaseous fuel HVOF. The measurements were done separately on a cylinder with 180 mm diameter by spraying two passes and measuring the weight of the produced coating compared to the fed powder amount. The DE value was determined as an average of three depositions. DE values for the other coatings sprayed with HVAF and the reference HVOF-L coating were normalised based on the measured results.

Average particle temperatures and velocities were measured with Spraywatch 4S online spray sensor by using 2 and 3 µs exposure times for HVAF and HVOF spray processes, respectively. Camera was located at the corresponding spray distance for each spray process, perpendicular to the spray stream.

Table 1. Spray parameters used for the HVAF spraying of the WC-10Co4Cr powder.

Parameter	Low (L)	High (H)
Air [psi]	108	120
Fuel	Propane	Propane
Fuel1 [psi]	103	113
Fuel2 [psi]	100	100 *)
Spray distance [mm]	300	300
Surface speed [mm/s]	500	500
Increment [mm]	5	5
Powder feedrate [g/min]	200	200

*) 110 psi with 4L4 nozzle.

2.2 Characterisation

Powders, as-sprayed coatings and the wear surfaces of the cavitation erosion specimen were characterised with scanning electron microscope (SEM: XL-30, Philips, Netherlands). Both secondary electron (SE) and backscattered electron (BSE) detectors were used. In addition, x-ray diffractometry (XRD: Empyrean, PANalytical, Netherlands) was used to determine the phase composition of the powder and sprayed coatings (Cu-K_α 1.541874Å, 0.026° step). Peak intensities for WC (35.6°), W_2C (40.0°) and Co_3W_3C (M_6C at 40° and 42.5°) were determined from the XRD patterns. Carbide retention index (CRI) was calculated for each coating following the approach by Bartuli et al. [9] (**Equation 1**) with modifications to reduce the overlapping M_6C peak from the W_2C peak at 40° (**Equation 2**):

$$CRI = \frac{I_{WC}}{I_{WC}+I_{W_2C}+I_W} \quad (1)$$

$$I_{W_2C} = I_{W_2C(40°)} - I_{M_6C(42.5°)} * \frac{I_{powder,M_6C(40°)}}{I_{powder,M_6C(42.5°)}} \quad (2)$$

The intensity of the overlapping M_6C (40°) peak in the as-sprayed coating was reduced from the W_2C peak intensity at 40° by taking into account the ratio of M_6C peaks of the powder at 40° and 42.5°. Carbide retention index values below 1 indicate that carbide dissolution and secondary phase formation has taken place. The lower the number, the more significant is the dissolution and resulting loss of primary WC.

Vickers microhardness tester (MMT-X7, Matsuzawa Co. Ltd, Japan) was used to measure coating hardness values as an average of 10 measurements with 300 gf (2.94 N) load and 10s loading time.

Surface profiles and roughness values (S_a) were measured with InfiniteFocus G5 (Alicona Imaging GmbH, Austria) optical profilometer. Three area profiles (1.62x1.62 mm^2) were analysed for each wear surface and an average S_a value was calculated.

2.3 Cavitation erosion test

Cavitation erosion tests were carried out following the ASTM G32-16 standard by using an ultrasonic transducer (20 kHz) and stationary sample test method with 0.5 mm distance between the replaceable 15.9 mm diameter Ti6Al4V tip and sample surface. Before testing, the surfaces were grinded with SiC papers (P220 -> P1200) and polished with 3 µm diamond suspension. De-ionised water was used for the tests and the temperature was maintained at 25°C by a cooling coil. Samples were ultrasonically cleaned in ethanol, dried and weighed before each 2 hour test cycle. The testing was continued up to 12 hours. The replaceable tip was grinded with P4000 SiC paper every 2 hours to remove the wear marks created during the test. Stainless steel 316L sample cut from 5 mm thick plate was used as the reference material. Vickers hardness of the polished 316L sample surface was 205 HV1.

3 Results and discussion

3.1 Feedstock material

The fine feedstock powder presented typical structure and round morphology of an agglomerated and sintered WC-10Co4Cr material (**Fig. 1**). Carbides were embedded in metal matrix while some of the particles showed larger metallic clusters (arrows in **Fig. 1**) or significant amount of porosity.

Fig. 1. SEM images (BSE) of the fine WC-10Co4Cr feedstock powder a) cross section and b) morphology. Arrows indicate a large metal cluster.

3.2 Spray diagnostics

The three spray processes provided very different particle conditions at the corresponding spray distance where the deposition took place (HVOF-G: 230 mm, HVOF-L: 360 mm, HVAF: 300 mm). The gas fuelled HVOF torch with axial powder feeding produced the highest particle temperature and lowest particle velocity of 1820 °C and 618 m/s in **Table 2**, respectively. As a result of the generally higher combustion chamber pressure in the liquid fuelled HVOF torches, the measured average particle velocity with the same coarse WC-10Co4Cr powder was 794 m/s for HVOF-L. The measured average particle temperature was over 100 °C lower compared to the HVOF-G. This is due to the radial powder injection into the expanding section of the nozzle and longer spray distance. The measured particle temperatures produced by the HVAF spray process were significantly lower between 1310 °C and 1420 °C. The reduced temperature of the process arises from the use of compressed air as the oxidiser for the combustion with propane. The particle velocities, on the other hand, were comparable with HVOF-L when cylindrical nozzle was used and 100 to 150 m/s higher when convergent-divergent nozzles were used. The acceleration of the axially injected particles by the supersonic gas jet in HVAF is efficient and similar velocities and even higher temperatures have been measured with coarse -45+15 µm particle size distributions [7,10].

Table 2. Measured average particle temperatures (*T*) and velocities (*v*), coating microhardness, carbide retention index (*CRI*) and deposition efficiency (*DE*).

Sample*	*v* [m/s]	*T* [°C]	Hardness [HV0.3]	CRI	DE [%]
4L0-L	784	1400	1138 (±71)	0.92	50
4L0-H	777	1370	1098 (±112)	0.92	38
4L2-L	894	1420	1362 (±90)	0.94	42
4L2-H	913	1390	1331 (±107)	0.95	36
4L4-L	889	1310	1383 (±99)	0.97	35
4L4-H	942	1340	1437 (±82)	0.96	32
HVOF-G	618	1820	1292 (±168)	0.84	64
HVOF-L	794	1698	1245 (±103)	0.90	44**

*) L= low parameter, H= high parameter
**) Estimated DE based on the result of HVOF-G, expecting linear effect of feedrate, surface speed and pass spacing.

The highest deposition efficiency was measured for the HVOF-G coating with 64% (in **Table 2**). As the particle velocity was significantly increased with the liquid fuelled HVOF, the deposition efficiency also decreased to an estimated value of 44%. Comparable deposition efficiency was measured for the HVAF spray process with the convergent-divergent 4L2 nozzle. The deposition efficiency was further increased when the cylindrical nozzle was used and, on the other hand, decreased with the 4L4 nozzle. It has to be pointed out that increasing the combustion chamber pressure resulted in decreased deposition efficiency with all nozzle geometries indicating decreased particle deposition and/or increased erosion due to lower particle temperature and higher particle velocity.

3.3 Coating microstructure

The significant differences in the measured particle temperatures suggest some differences also in the particle state, thermal history and produced coating structure. Indeed, the XRD patterns in **Fig. 2** reveal a significant hump between 35° and 45° for both HVOF sprayed coatings. The more pronounced increase in the background intensity for HVOF-G coating arises from the axial injection of the particles into the high temperature combustion, which results also in the highest measured particle temperature. The injection of the particles into high temperature is especially detrimental for fine particles, which produce carbon saturated areas in the matrix with amorphous or nanocrystalline structure due to carbide dissolution [11]. The phenomena is reduced in the case of the HVOF-L coating and the difference can be numerically expressed by the higher carbide retention index of 0.90 compared to the 0.84 for HVOF-G coating (**Table 2**).

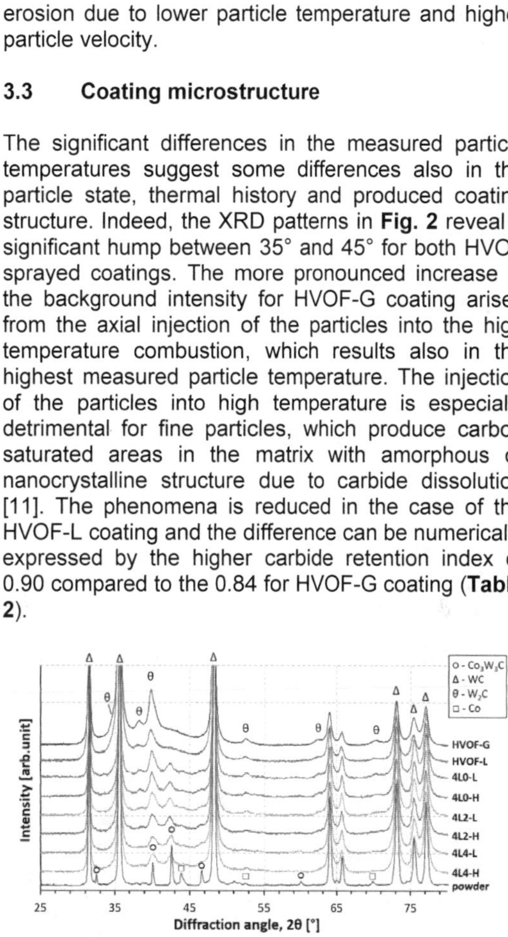

Fig. 2. XRD patterns of the fine powder and as-sprayed coatings.

The HVAF sprayed coatings revealed decreasing content of W_2C phase when moving from cylindrical nozzle (4L0) to convergent-divergent nozzles (4L2 and 4L4). This increased the calculated carbide retention index from 0.92 to 0.97 representing the minimal changes taking place in the original particle structure when near melting temperature deposition with 4L4 takes place. The results confirmed that the thermal history of the particles sprayed with nozzle 4L2 lay between those of nozzles 4L0 and 4L4. The 4L4-L and 4L4-H coatings did not show distinctive W_2C peaks, but when taking into account the intensity of the overlapping M_6C peak it could be concluded that some of the intensity comes from W_2C. Actually, all HVAF sprayed coatings showed the presence of Co_3W_3C originating from the feedstock powder, whereas in the HVOF coatings the peaks were supposedly covered by the increased background intensity.

The SEM micrographs of the HVAF sprayed coatings in **Fig. 3a-d** show homogeneous structures with fine WC particles embedded in the metal matrix. Both presented coatings *4L0-L* (in **Fig. 3a**) and *4L4-H* (in **Fig. 3b**) contain metallic clusters with high Cr concentration shown as dark spots in the BSE images (pointed by arrows). Interestingly the microstructure of the *4L4-H* coating contains higher amount of the dark metallic clusters, while the clusters are not as frequent in the *4L0-L* coating and show flattened morphology (arrow in **Fig. 3b**). This indicates that melting, subsequent mixing of the metallic regions and flattening had taken place during the spraying with 4L0 nozzle. Deposition with 4L4 nozzle, on the other hand, had taken place just below the melting temperature of the metallic alloy preventing such effect. These observations are in line with the measured particle temperatures and observed differences in the XRD patterns and resulting carbide retention indexes.

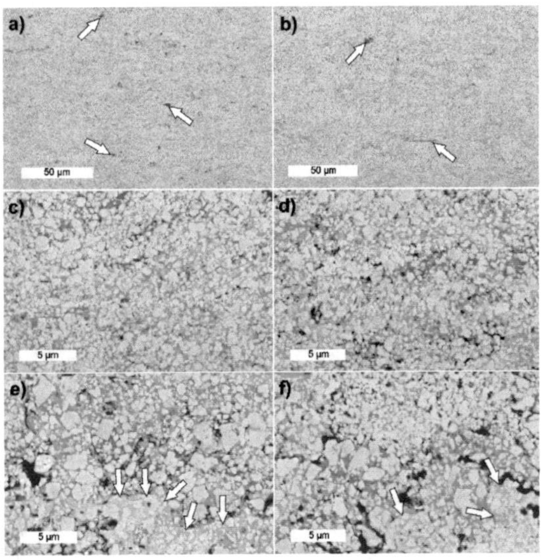

Fig. 3. SEM micrographs (BSE images) of the as-sprayed cross sections of a) 4L4-H, b) 4L0-L, c) 4L4-H, d) 4L0-L, e) HVOF-G and f) HVOF-L coatings. The arrows point metallic clusters (a and b) and carbide dissolution (e and f).

As was expected based on the XRD patterns, the microstructures of the HVOF-G and HVOF-L coatings in **Fig. 3e** and **Fig. 3f** showed signs of significant carbide dissolution. The areas pointed out by arrows show dissolution of numerous fine carbides creating light grey areas into the metal matrix indicating high W content [11]. The areas are less frequent in the HVOF-L coating supported by the lower carbide retention index.

The measured microhardness values presented in **Table 2** ranged from 1100 to over 1400 HV0.3. Lowest hardness values were measured for the coatings sprayed with the cylindrical 4L0 nozzle with the lower particle velocities of 780 m/s compared to the 890-940 m/s that was measured for coatings sprayed with convergent-divergent 4L2 and 4L4 nozzles. An average hardness of over 1200 HV0.3 was measured for both HVOF sprayed coatings. The standard deviation of the measured values for HVOF-G were significantly increased compared to the other coatings suggesting more heterogeneous structure.

3.4 Cavitation erosion resistance

The cavitation erosion test results in **Fig. 4** revealed interesting differences between the HVAF sprayed coatings. The 4L0-H coating had the highest mean erosion rate together with the 4L0-L coating, 4.36 and 3.67 µm/h, respectively. The wear surfaces after 12h testing show high roughness in **Fig. 5a** and **5b**, and many of the produced wear debris particles ranged between 50 and 100 µm in size. Significant improvement was achieved when convergent-divergent 4L2 (**Fig. 5c-d**) and 4L4 (**Fig. 5e-f**) nozzles were used. With the latter, the cavitation erosion rate dropped down to one tenth compared to *4L0-H* and HVOF-G coatings. Also, the wear debris size was significantly decreased with the largest detected particles being 50-60 µm. This resulted in significantly decreased surface roughness, which was further decreased when 4L4 nozzle was used instead of 4L2. The enhanced performance indicates that structural cohesion and fatigue wear resistance of the coatings were improved with the increased particle velocity, complying with the findings of Kumar et al. [8]. While both convergent-divergent nozzles produced highly cavitation erosion resistant coatings, the 4L4-L and 4L4-H coatings experienced 24% to 64% lower wear rates compared to 4L2-L and 4L2-H coatings. As a result of the low wear rates, the HVOF-L coating and the HVAF coatings sprayed with the convergent-divergent nozzles showed areas of the original polished coating surface even after 12h (arrows in **Fig. 5**). Based on the wear surface and produced debris in **Fig. 5** it becomes clear that majority of the material loss happens by the removal of coating pieces that consist of several flattened particles. This highlights the importance of homogeneous coating structure and reduced defect size.

Indeed, heterogeneous structure and brittle areas in the coating can promote the crack propagation, which resulted in high cavitation erosion rate (4.56 µm/h) in the case of the HVOF-G coating (in **Fig. 4**). The size of the wear debris was large compared to the HVAF sprayed coatings and it has a distinctive blocky structure (in **Fig. 5g**). Both the wear surface and wear debris showed signs of brittle fracture with numerous straight fracture surfaces after 6 hours of testing. The removal of the coating in large blocks revealed some poor adhesion between particles, which left smooth flattened splat surfaces visible on the wear surface.

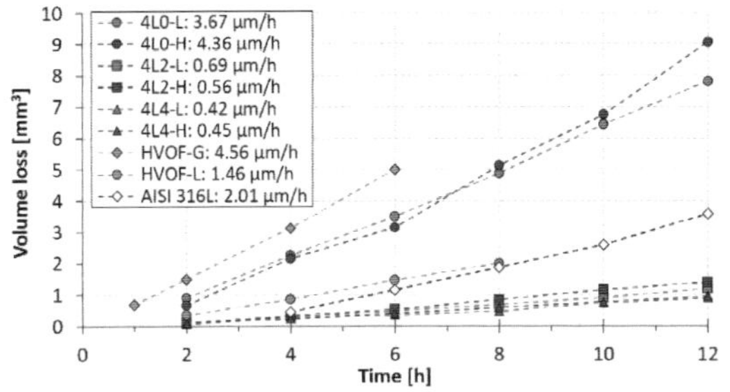

Fig. 4. Cumulative volume losses and calculated mean erosion rates (µm/h) for the tested HVAF and HVOF sprayed coatings and AISI 316L reference sample.

The HVOF-L coating showed significantly improved performance (1.46 µm/h) compared to the HVOF-G coating. The produced wear debris was finer, which resulted also in decreased roughness of the wear surface after 8h of testing. The higher particle velocity and more homogeneous structure compared to the HVOF-G coating highlight the benefits of the high-pressure liquid fuelled HVOF spray process over the gas fuelled HVOF process.

An interesting aspect in wear behaviour of the coatings is their performance compared to ductile stainless steel bulk material. The debris size and roughness of the AISI 316L wear surface (in **Fig. 5i**) were comparable with the 4L4-L and 4L4-H coatings. However, the maximum erosion rate of the stainless steel sample was 340-380% higher compared to the coatings sprayed with the 4L4 nozzle. During the acceleration stage of the cavitation erosion wear the performance of the AISI 316L sample is comparable with the coatings sprayed with 4L2 and 4L4 nozzles. However, after 4 hours the stainless steel sample reaches the maximum erosion rate, while the coating samples maintain their low wear rate. Improved cavitation erosion resistance was also observed on HVOF-L coating. The results demonstrate that the HVAF sprayed coatings have great potential to bring improved protection and performance in e.g. mining, hydro power and marine applications.

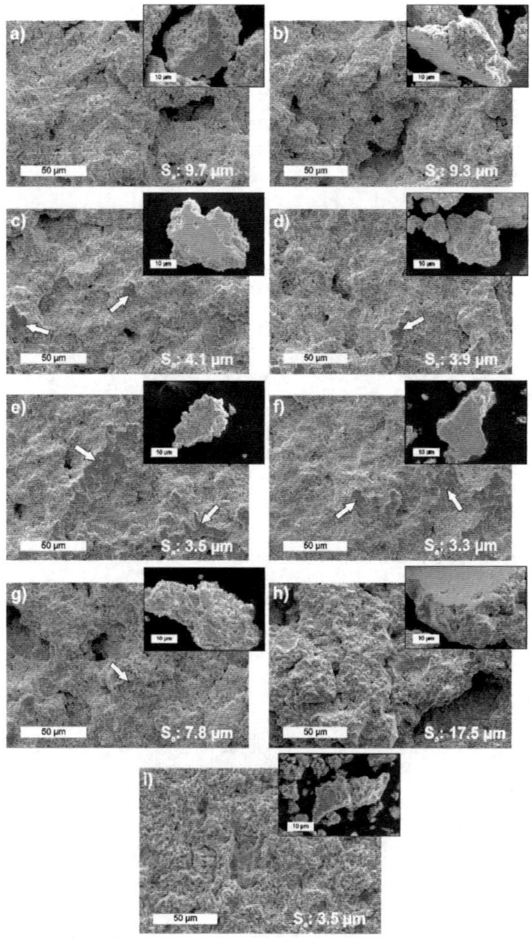

Fig. 5. Wear surfaces of the tested HVAF samples after 12h cavitation erosion test: a) 4L0-L, b) 4L0-H, c) 4L2-L, d) 4L2-H, e) 4L4-L, f) 4L4-H, g) HVOF-G (6h), h) HVOF-L (8h) and AISI 316L reference sample (12h). The arrows point areas of polished unaffected coating surface.

Cavitation erosion takes place as a result of numerous stress pulses generated by the collapsing cavitation bubbles near the surface. As was shown in Section 3.3, the coating structures consisted of fine WC particles bound together by narrow areas of metal matrix, i.e. having low mean free path between the carbides [12]. Such structure does not allow extensive plastic deformation similar to metallic bulk materials when subjected to cavitation erosion produced by the vibratory test method. This is demonstrated in **Fig. 6a**, which shows the progressive cavitation damage of the polished AISI 316L stainless steel surface (1). The damage starts by plastic deformation of the surface without material removal (2). After severe plastic deformation the material removal starts from the grain boundaries (3) and further progresses until maximum erosion rate is reached (4). Thermally sprayed coating structures, on the other hand, contain defects and splat boundaries that provide potential weak spots for crack initiation. Instead of plastic deformation, the HVAF sprayed WC-10Co4Cr coating undergoes crack formation, crack propagation and eventual material removal. The wear process of the polished coating surface (1) in **Fig. 6b** starts by the removal of loose splats or particles on the surface (2). After this the wear rate is controlled by the crack initiation, crack propagation and further material removal (3) until the maximum erosion rate is achieved (4).

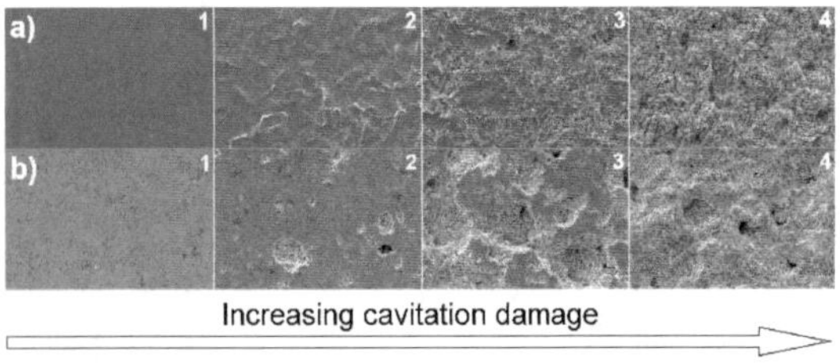

Fig. 6. Increasing cavitation damage from polished surface to maximum erosion rate for a) AISI 316L reference and b) HVAF sprayed WC-10Co4Cr coating.

4 Conclusions

Hard metal coatings were sprayed with HVOF and HVAF spray processes from WC-10Co4Cr feedstock materials. The coatings were characterised and tested for cavitation erosion by using a vibratory apparatus. Following conclusions were made:

- Changing the HVAF nozzle geometry from cylindrical to convergent-divergent increased the particle velocities and decreased the deposition efficiency.
- Heating, melting degree and resulting carbide dissolution, i.e. W_2C formation, were effectively controlled by different nozzle geometries.
- Higher particle velocity increased the HVAF sprayed coating hardness and cavitation erosion resistance.
- Stainless steel reference sample (AISI 316L) experienced 2 to 4 times higher wear rates and HVOF sprayed reference coatings 3 to 10 times higher wear rates compared to the HVAF sprayed coatings.

5 Acknowledgements

The authors would like to thank Mr. Mikko Kylmälahti (Tampere University of Technology) for the assistance with HVAF spraying of the samples and Mr. Leo Hyvärinen (Tampere University of Technology) for carrying out the XRD measurements. The authors gratefully acknowledge Tekes (Finnish Funding Agency for Technology and Innovation), the participating companies of the HYBRIDS research programme, and DIMECC Ltd. for financial support.

6 References

[1] Q. Wang, S. Zhang, Y. Cheng, J. Xiang, X. Zhao, and G. Yang: Wear and corrosion performance of WC-10Co4Cr coatings deposited by different HVOF and HVAF spraying processes. Surf. Coatings Technol. 218 (2013), pp. 127/36.

[2] G. Bolelli, L. M. Berger, T. Börner, H. Koivuluoto, L. Lusvarghi, C. Lyphout, N. Markocsan, V. Matikainen, P. Nylen, P. Sassatelli, R. Trache, and P. Vuoristo: Tribology of HVOF- and HVAF-sprayed WC-10Co4Cr hardmetal coatings: A comparative assessment. Surf. Coatings Technol. 265 (2015), pp. 125/44.

[3] A. Verstak and V. Baranovski: Activated Combustion HVAF Coatings for Protection against Wear and High Temperature Corrosion. In Proceedings of ITSC2003 (2003), pp. 535/41.

[4] L. Jacobs, M. M. Hyland, and M. De Bonte: Comparative Study of WC-Cermet Coatings Sprayed via the HVOF and the HVAF Process. J. Therm. Spray Technol. 7 (1998), pp. 213/18.

[5] K. Akimoto and Y. Horie: Study of HVAF WC-Cermet Coatings. In Proceedings of ITSC'95 (1995), pp. 313/16.

[6] C. Lyphout, S. Björklund, M. Karlsson, M. Runte, G. Reisel, and P. Boccaccio: Screening Design of Supersonic Air Fuel Processing for Hard Metal Coatings. J. Therm. Spray Technol. 220 (2014), pp. 1323/32.

[7] V. Matikainen, K. Khanlari, A. Milanti, H. Koivuluoto, and P. Vuoristo: Spray parameter effect on HVAF sprayed (Fe,Cr)C-30FeNiCrSi hardmetal coatings. In Proceedings of ITSC2016 (2016), pp. 184/89.

[8] R. K. Kumar, M. Kamaraj, S. Seetharamu, T. Pramod, and P. Sampathkumaran: Effect of Spray Particle Velocity on Cavitation Erosion Resistance Characteristics of HVOF and HVAF Processed 86WC-10Co4Cr Hydro Turbine Coatings. J. Therm. Spray Technol. 25 (2016), pp. 1217/30.

[9] C. Bartuli, T. Valente, F. Cipri, E. Bemporad, and M. Tului: Parametric Study of an HVOF Process for the Deposition of Nanostructured WC-Co Coatings. J. Therm. Spray Technol. 14 (2005), pp. 187/95.

[10] I. Hulka, V. A. Serban, M. L. Dan, V. Matikainen, and P. Vuoristo: Corrosion Behavior of WC-FeCrAl Coatings Deposited by HVOF and HVAF Thermal Spraying Methods. Chem. Bull. "Politehnica" Univ. Timisoara 61(2016).

[11] C. Verdon, A. Karimi, and J. Martin: A study of high velocity oxy-fuel thermally sprayed tungsten carbide based coatings. Part 1 : Microstructures. Mat. Sci.&Eng. 246 (1998), pp. 11/24.

[12] C. Lyphout and K. Sato: Optimization of WC-based HVAF-sprayed Coatings for Alternative to Hard Chromium : Role of Carbides Grain Size and Mean Free Path on Coating Wear and Corrosion Properties. In Proceedings of ITSC2014 (2014), pp. 956/61.

Development of HVAF-sprayed novel Fe-based coatings for large area applications

K. Bobzin, M. Öte, M.A. Knoch, J. Sommer, Surface Engineering Institute (IOT), RWTH Aachen University, Aachen/D

One important trend in thermal spraying is the application of novel Fe-based corrosion/wear protection coating systems. A typical field of application for such corrosion and abrasive wear protection coatings are rotary dryers of paper machines. At the moment, these cylinders are coated by wire arc spraying. A disadvantage of the wire arc sprayed coatings is their high thickness, which has a heat-insulation effect, and their high roughness. Therefore, an expensive post production grinding process is necessary in order to achieve the required surface quality. The goal is to develop a HVAF process that enables the production of thin, dense and near net shape corrosion/wear protection coating systems, which significantly reduce the post-production time and costs. In this study, the HVAF coating process and a novel Fe-based feedstock material are investigated. In the first step the Fe-based powder is analysed thermally using differential scanning calorimetry, to investigate the solidification and melting temperature of the feedstock material. Furthermore, the influence of the spraying distance and the powder feed rate on the microstructure and porosity of the resulting coatings is investigated using light microscopy. Furthermore, the deposition efficiency of HVAF coatings is analysed regarding their economic efficiency.

1 Introduction

The novel Fe-based feedstock materials are in focus of research and development activities dealing with the thermally sprayed coating solutions for wear and corrosion protection, due to their favourable economical and ecological properties. High percentages of Cr promote the corrosion behaviour of coatings. Moreover, elements promoting hard phase formation are included in the coating system to increase wear resistance [1, 2]. These novel feedstock materials can be deployed in several application areas, which require high wear and corrosion protection.

A typical example of these applications is the rotary dryers (yankee cylinders) of paper manufacturing machines. At the moment, these cylinders are coated by wire arc spraying (WAS). Although suitable coatings can be produced by WAS, the process-inherent low invest costs and high deposition rates are countered by the high post production costs associated with WAS in case of rotary dryers. Due to the high roughness of WAS coatings, an expensive and time consuming grinding process is certainly necessary to achieve the required surface roughness. A typical as-ground coating thickness corresponds to d_{ct} = 800 - 1200 µm in this application [3, 4]. These thick coatings exhibit unfavourable thermal isolation behaviour, which lead to higher energy expenditure during paper manufacture. Although HVOF sprayed thin WC-CoCr or Cr_3C_2/NiCr 75/25 coatings distinguish themselves with their very high corrosion and abrasive wear resistance, these coatings are economically unattractive for this application due to their higher production costs. In this paper a Cr_3C_2/NiCr 75/25 feedstock material sprayed by means of HVOF was chosen as reference representing a high-performance wear and corrosion protection system.

The HVAF process is a modification of the HVOF process [5] and promises high deposition rates and low particle in-flight temperatures, compared to standard HVOF-processes [6]. The gas flow rates in HVAF are higher than those achievable in HVOF. This can, in turn, enable a higher powder feed rate and consequently higher deposition rates [7]. Compared to conventional HVOF sprayed cemented carbides, the high deposition rates of the HVAF process shall allow the coating of large area components in an acceptable processing time and - in combination with novel Fe-based feedstock materials, - production costs shall be decreased. The lower process temperature is achieved by using air instead of oxygen as oxidant [8]. The high particle velocities and the lower particle temperatures in HVAF process result in dense coatings with a low content of oxides [6]. A number of studies confirm this hypothesis [5,8 – 11]. The variety of different torch designs must, however, not be neglected in this consideration. Thus, no general statement concerning the oxygen content of HVOF and HVAF coating can be made [9]. In [13], WC-CoCr deposited by means of HVAF coatings showed less oxidized phases in comparison to HVOF coatings and the amount of dissolved cemented carbides was lower. This effect is ascribed to the lower combustion temperatures in HVAF process.

The quality of the novel Fe-based corrosion and wear protection coatings which are applied by means of HVAF process; have not been investigated intensively yet. The aim of this study is therefore, to develop a novel cost-effective Fe-based coating system by means of HVAF technology and analyse it, in terms of its wear and corrosion protection capability.

2 Experimental setup and description

The first part of this study describes the coating development of the novel FeCrB/WC-Co HVAF coating. After a basic characterisation of the feedstock material, process parameters which allow the deposition of dense and crack-free HVAF sprayed FeCrB/WC-Co protection coatings are developed. Subsequently, the corrosion and wear protection behaviour of the coatings are determined and compared to a HVOF sprayed reference. The FeCrB/WC-Co feedstock material has a grain size fraction of -45 +11 µm (GTV Verschleißschutz GmbH, Luckenbach, Germany) and the composition is given hereinafter:

- Cr: 20 – 23 wt.-%
- B: 3.5 – 4 wt.-%
- C: 1 – 1.2 wt.-%
- W: 10 – 12 wt.-%
- Co: 1 – 1.5 wt.-%
- Fe-Rest

A high content of Cr was alloyed, to achieve a good corrosion resistance. B was included in the composition to promote the formation of hard phases. WC particles embedded in a Co matrix were added to the material composition to achieve a high wear resistance. As reference feedstock material, Cr_3C_2/NiCr 75/25 with a grain size fraction of -45 +15 µm (GTV Verschleißschutz GmbH, Luckenbach, Germany) was used. Compared to the reference feedstock material, the retail price of the FeCrB/WC-Co feedstock material is remarkably low. The reasons for the more favourable retail price of the FeCrB/WC-Co feedstock material are the lower production and raw material costs. The price of the FeCrB/WC-Co is less than the half retail price of the reference feedstock material.

A scanning electron microscopy (SEM) image of the FeCrB/WC-Co feedstock material is shown in Figure 1. The image shows the Fe-based particles and the WC-Co particles, which consist of WC embedded into a Co-matrix. The liquidus and solidus temperature of the feedstock material were determined in a dynamic differential calorimetry (DSC) analysis using a Setsys Evolution TGA-DTA/DSC system (Setaram Instrumentation, Caluire, France). The samples were heated up to T_{max} = 1400 °C.

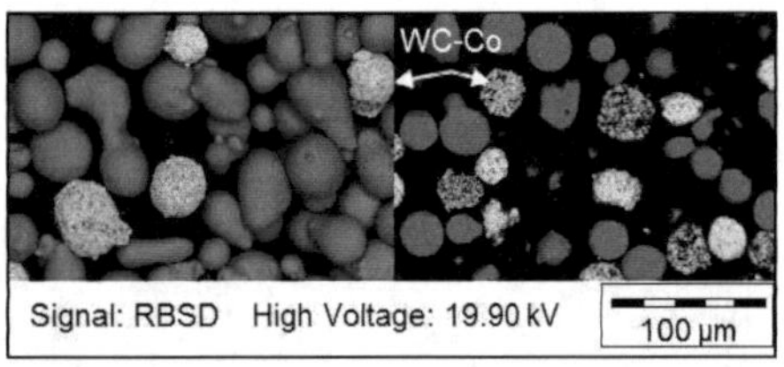

Figure 1: SEM-micrograph of the FeCrB/WC-Co

All coatings were applied on 1.0038 steel substrates with the dimensions of 40 x 50 x 8 mm³. Prior to the coating process, the substrates were roughened using an injector blasting system with corundum. The spraying parameters for the HVAF sprayed feedstock material FeCrB/WC-Co are given in Table 1. These parameters are known to be suitable to produce WC-CoCr coatings by means of HVAF technology [14] and therefore used as a starting point. The coatings were produced using an AK-07 system (Kermetico Inc., Benicia, USA). The thermal conductivity of the FeCrB/WC-Co is much lower than that of the WC-CoCr material. Therefore, the stand-off-distance was varied in order to find the appropriate distance, at which the FeCrB/WC-Co particles exhibit an appropriate melting grade and a crack free coating can be produced. The powder feed rate was set to 40 g/min. An appropriate amount of hydrogen flow was added to increase the heat transfer rate from hot gas flow to the powder particles [6] and to achieve a high melting degree of particles.

The foreword goal of this study was the production of crack free and dense coatings with a low content of oxides. After finding a parameter set for the powder feed rate of 40 g/min, the powder feed rate was increased to 100 g/min. In further studies it is planned to increase the powder feed rate gradually. The powder deposition efficiency (PDE) was investigated, in accordance to DIN EN ISO 17836 [15] and is determined for two selected HVAF parameter sets. A 1.0038 substrate with geometry of 150 x 100 x 8 mm³, which differs from the geometry recommended in the standard, was used.

Table 1: Spraying parameters for the HVAF process

Parameter	Value
Compressed air [bar]	6.3
Propane [bar]	5.6
Nitrogen [SLPM]	18
Hydrogen [SLPM]	35
Stand-off-distance d [mm]	200, 250, 300
Meander distance [mm]	5
Transversal speed [mm/s]	900
Powder Feed rate [g/min]	40, 100
Nozzle [Kermetico-No.]	#3L
Transitions	20

The HVOF sprayed Cr_3C_2/NiCr 75/25 coating was used as reference for the following wear and corrosion tests. This system was used, since it is an industrially established coating system with very good wear and corrosion properties. The spraying parameters for the HVOF sprayed Cr_3C_2/NiCr 75/25 are given in Table 2.

Table 2: Spraying parameters for the HVOF process

Parameter	Value
Oxygene [SLPM]	900
Kerosene [l/h]	26
Oxygene/Fuel ratio λ	1,0
Nitrogene [SLPM]	10
Stand-off distance d [mm]	350
Meander distance [mm]	5
Powder feed rate [g/min]	70
Transversal Speed [mm/s]	1,000
Nozzle [GTV-No.]	780.150-150/12
Transitions	10

A K2 gun (GTV Verschleißschutz GmbH, Luckenbach, Germany) was used to apply the reference coating.

The parameters given in Table 2 are established in the industry and represent a good compromise between coating properties and production efficiency.

Images of the cross sections were taken with a Zeiss Axiophot light microscope. The porosity of the coatings was determined with the evaluation software Image Analysis (both Carl Zeiss AG, Oberkochen, Germany).

For the investigation of the wear properties, a Pin-on-Disc (PoD) tribometer (CSM Instruments SA, Freiburg, Germany) was used. Prior to wear test, samples were ground to a roughness of Ra < 1 µm to adjust the same testing conditions for all samples. An Al_2O_3 ball with a diameter $ø_{cb}$ = 6 mm was used as counter body. An Al_2O_3 counter body was used to conduct an abrasive wear tests and reduce the adhesive influence. Established test parameters for thermally sprayed cemented carbide coatings are given in Table 3. The tests were performed without lubricant. Subsequent to the wear test, the wear coefficient K for different samples was determined using equation (1) [2]. For this purpose, the volumetric wear V and the measured wear track diameter d_t was determined using confocal laser scanning microscopy (VKX 200, Keyence Corporation, Osaka, Japan).

Table 3: PoD test parameters

Parameter	Value
Path diameter d_{pd}	10 mm
Wear track length s	1,000 m
Normal load F	10 N
Rotation speed	100 mm/s
Test temperature	RT

$$K\,[mm^3/Nm] = \frac{V}{F \cdot (s \cdot \frac{d_t}{d_{pd}})} \qquad (1)$$

To analyse the corrosion resistance, polarization tests were performed in a 5 % NaCl-solution at room temperature. A Calomel-(Hg_2Cl_2) electrode was used as reference electrode and the potential scan rate was set to 0.5 mV/s. The rest potentials and the corrosion current densities were investigated using the Tafel slope analysis.

3 Results and discussion

DSC measurement

The results of the DSC-measurement can be seen in Figure 2. A significant increase and decrease of the heat flow at certain temperatures can be observed. These represent the melting and solidification respectively. The liquidus T_l and solidus T_s temperature were observed at T_l = 1,158 °C and T_s = 1,130 °C.

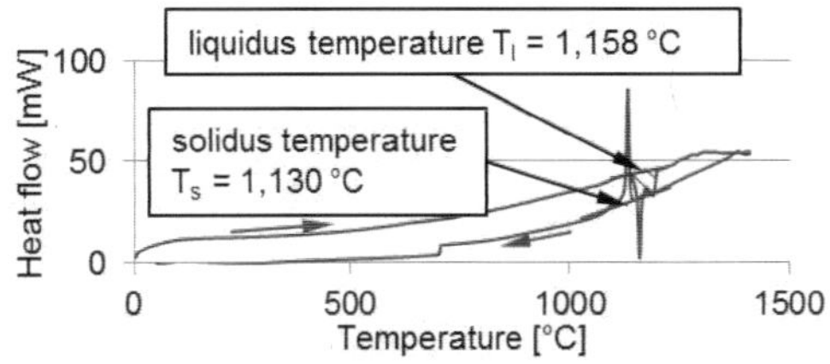

Figure 2: Determination of liquidus and solidus temperature of FeCrB/WC-Co feedstock material

Microstructure

The cross sections of the HVAF coatings are shown in Figure 3. The protruding and darker areas in the cross section are the WC-Co phases, which are embedded into the FeCrB-matrix. For the first 3 images, the same parameters were used with the exception of the stand-off-distance, which was varied between d = 200 mm, d = 250 mm and d = 300 mm. The powder feed rate was kept constant at 40 g/min.

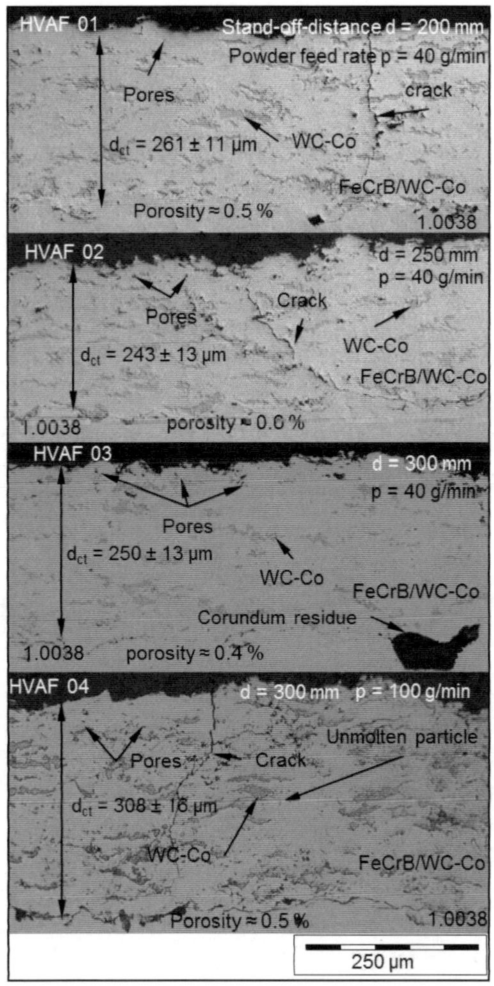

Figure 3: Cross section of the HVAF coatings for the parameter sets HVAF 01, 02, 03, 04

The images show that a good bonding between the substrate and the coating was achieved. Furthermore, the coatings exhibit a uniform WC-Co distribution in case of all stand-off-distances. All HVAF parameters show little or no oxidation in the coating. Typical lamellar structure of thermally sprayed coatings cannot be observed. Pores can be found especially in the near vicinity of coating surface. This is assumed to be due to the missing compaction of the upper layers at the end of the process. As these pores are ground away during post-processing, their influence on the following measurements can be assumed to be negligible. The lowest measured porosity was 0.4%, which is determined for the sample HVAF 03. The porosity for all shown HVAF samples is much lower than 1.5 %, which is a typical value for the HVOF sprayed reference coating Cr_3C_2/NiCr 75/25 [16]. A dense structure can improve the corrosion resistance of these coatings. The coating thickness d_{ct} is in a similar range for all stand-off distances examined at a powder feed rate of p = 40 g/min. The coatings sprayed with the stand-off distances d = 200 and d = 250 mm exhibit, in each case crack formation in the coating. One of the reasons can be the weak spots between the not fully molten particles and the remaining coating. In the cross section of the coating applied with parameter set HVAF 03, no cracks can be observed. Similar to the other HVAF coatings, the coating HVAF 03 exhibits a very high density as well. Therefore, this coating system is investigated further regarding its wear and corrosion behaviour.

Another goal of this study is to determine maximal achievable powder feed rates with HVAF process, providing that coatings exhibit desired wear and corrosion properties. The sample HVAF 04, shown in Figure 3, was a first attempt to analyse the effect of increased powder feed rate on the coating properties. The process parameters, which are not optimised for higher powder feed rates, leads to crack formation in coating, as expected.

Moreover, pores can be observed in the top layers and, in contrast to the other parameter sets, in the lower layers. Unmolten particles can be observed as well. The crack formation, the pores and the unmolten particles can be a result of uneven heating of the particles, because of the high powder feed rate. In further studies, the following strategies will be followed to avoid the crack and pore formation. The hydrogen flow will be increased to increase the heat transfer rate between the process gas and particles. Another strategy will be the increasing of the propane flow rate, the air flow rate and the stand-off distance by keeping the air-to-propane ratio constant. With the last strategy, the combustion temperature and the dwell time of the particles in the process gas can be increased, thereby achieving a uniform melting of the particles at higher powder feed rates.

Powder deposition efficiency

The results of the PDE measurements are shown in Table 4. For the parameter set HVAF 03, a PDE of 70 % and for the parameter set HVAF 04, a PDE of 68 % was measured. The PDE of the HVOF sprayed Cr_3C_2/NiCr 75/25 coating was taken from literature. It can be seen that the PDE of the HVAF coatings is significantly higher compared to the HVOF sprayed Cr_3C_2/NiCr 75/25 coating.

Table 4: Powder Deposition Efficiency

Sample	PDE η_P
HVAF 03	≈ 70 %
HVAF 04	≈ 68 %
Cr_3C_2/NiCr 75/25	≈ 40 % [16]

Corrosion Test

The rest potential U_R and the corrosion current density i_c are shown in Table 5. The corresponding polarization curves of the coatings and the uncoated substrate can be found in Figure 4.

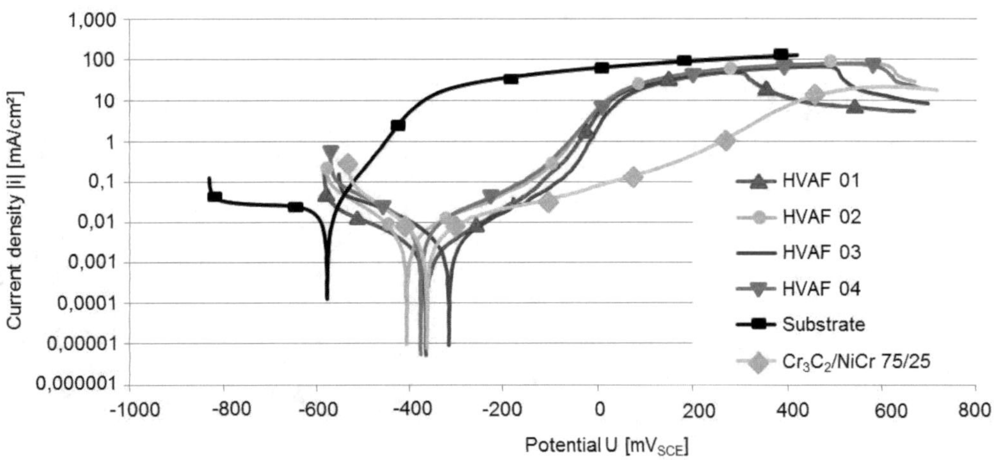

Figure 4: Corrosion current potential curves of the samples in 5 %NaCl-solution at room temperature

At room temperature in a 5 % NaCl-solution the substrate has the lowest corrosion resistance, which results in the lowest rest potential of U_R = -578 mV_{SCE} and the highest corrosion current density of i_c = 0.0273 mA/cm².

Table 5: Rest potential and corrosion current density

Sample	Rest Potential U_R [mV_{SCE}]	Corrosion Current density i_c [mA/cm²]
HVAF 01	-365	0.00263
HVAF 02	-407	0.007240
HVAF 03	-317	0.00569
HVAF 04	-377	0.0113
Cr_3C_2/NiCr 75/25	-363	0.00747
Substrate	-578	0.0273

The highest rest potential of the coatings was measured for the parameter set HVAF 03. This coating has a rest potential of U_R = -317 mV_{SCE}, which means it is the noblest one of all considered coating systems. The rest potential of the parameter HVAF 03 is even higher than that of the Cr_3C_2/NiCr 75/25 reference coating. Significantly lower rest potentials were measured for the samples HVAF 02 and HVAF 04. This result can be traced back to the cracks in these coating systems. Especially, the comparison of the rest potentials of the samples HVAF 03 and HVAF 04 shows the influence of the process parameters on the microstructure and therefore on the corrosion behaviour. An optimisation of the parameter set HVAF 04 can result in a better corrosion behaviour for a parameter set with a powder feed rate p = 100 g/min. A further point that needs to be emphasised is that at small overpotentials the current densities of HVAF 03 and Cr_3C_2/NiCr 75/25 are comparable.

To sum up the result of the corrosion potential curves, it can be stated that for small overpotenatials and in the examined range, the parameter set HVAF 03 achieves a similar or a superior corrosion resistance than that of the HVOF Cr_3C_2/NiCr 75/25 reference. For high temperature applications however, the results might change in favour of the Cr_3C_2/NiCr 75/25 coating.

The cross sections after the corrosion test of the samples HVAF 03 and the HVOF sprayed Cr_3C_2/NiCr 75/25 reference are shown in Figure 5. A corrosion depth of d_{cd} = 81 µm was measured for the reference coating Cr_3C_2/NiCr 75/25 and a d_{cd} = 73 µm for the sample HVAF 03. An interesting point is that the corrosion depth for both samples is within similar range. Sub-corrosion was detected for the sample Cr_3C_2/NiCr 75/25. The images show that the high density of the coating HVAF 03 avoids sub-corrosion. The corrosion tests demonstrated that the cost efficient feedstock material FeCrB/WC-Co can achieve comparable corrosion protection as the Cr_3C_2/NiCr 75/25 coating for potentials U < -150 mV_{SCE}.

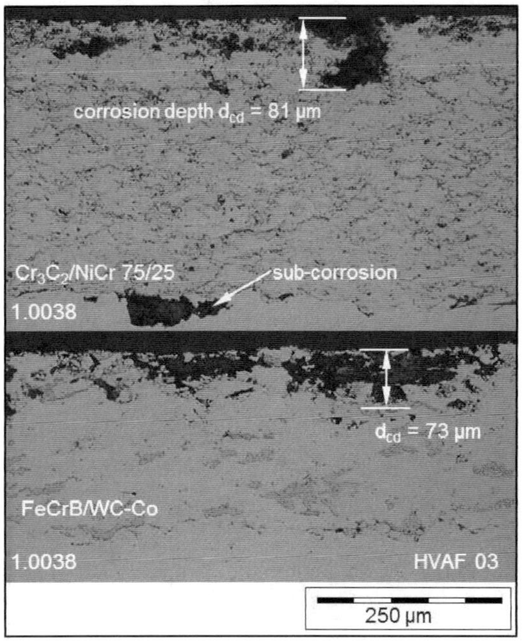

Figure 5: Cross sections of the coatings Cr_3C_2/NiCr 75/25 and HVAF 03 after the polarization test in 5 % NaCl-solution at room temperature

Wear Test

The results of the PoD test are shown in Figure 6. All coatings show significantly less wear than the substrate. Figure 6 a) shows the wear track images of the samples HVAF 03, HVAF 04, Cr_3C_2/NiCr 75/25 and the substrate. The substrate shows the widest wear track and the sample HVAF 03 the thinnest wear track. For the coated and the uncoated samples non-homogenous wear tracks can be observed in each case. All samples exhibit abrasive wear behaviour and splat breakages. The sample HVAF 03 exhibits significantly less outbreaks than the other samples, even compared to the established wear protection coating Cr_3C_2/NiCr 75/25. The cracks which were found in the coating HVAF 04 and the porosity of the sample Cr_3C_2/NiCr 75/25 can promote such breakages of the coating. The porosity and the cracks offer a target for the counter body and a chunk of the coating can break out.

Figure 6 b) shows the profiles of the wear tracks, which confirm the qualitative results of the images shown in Figure 6 a). For the sample HVAF 03, the smallest cross-sectional area of the wear track was measured. It is significantly lower than the wear track cross-sectional area of the Cr_3C_2/NiCr 75/25 reference and the other HVAF samples. Compared to the Cr_3C_2/NiCr 75/25 reference a wider but lower profile can be observed for the sample HVAF 04.

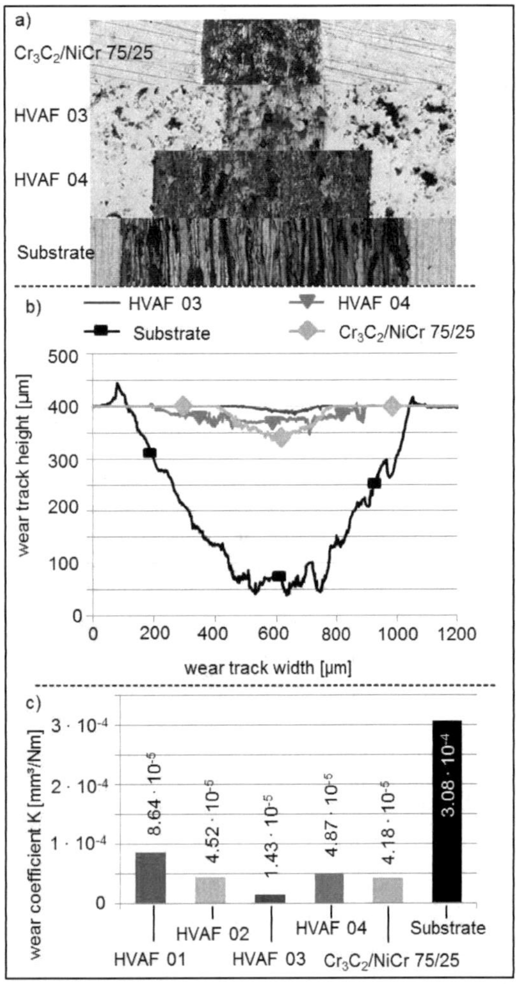

Figure 6: Results of the PoD test: a) wear tracks of the samples, b) profile of the wear tracks and c) wear coefficient K of all samples

In Figure 6 c), the wear coefficients K for all investigated samples are shown. All coatings show a significantly lower wear coefficient than the substrate. The calculated wear coefficients confirm with the results of the previous wear investigations shown in Figure 6 a) and b). The lowest wear coefficient $K = 1.43 \cdot 10^{-5}$ mm³/Nm was measured for the coating HVAF 03. An interesting point is that even the measured wear coefficients of the not optimised parameter sets HVAF 02 and HVAF 04 are comparable to the wear coefficients of the Cr_3C_2/NiCr 75/25 reference. The wear coefficients for these three samples are in a range of $K \approx 4.5 \cdot 10^{-5}$ mm³/Nm. The results of the PoD wear tests show that the novel FeCrB/WC-Co feedstock material exhibits a superior wear protection for optimised HVAF parameters although its cemented carbide content is much lower than that of the Cr_3C_2/NiCr 75/25 reference feedstock material.

4 Summary and outlook

In this study, the successful application of novel FeCrB/WC-Co coatings was demonstrated using a HVAF process. The microstructure and the PDE of these coatings were investigated. Subsequent to this, the wear and corrosion properties were investigated and compared to a HVOF Cr_3C_2/NiCr 75/25 reference. The results of this study can be summed up as followed.

- A dense, crack- and oxidation- free FeCrB/WC-Co coating can be produced using HVAF at a powder feed rate of p = 40 g/min and a stand-off distance of d = 300 mm.
- The PDE of the HVAF sprayed FeCrB/WC-Co was measured at 68 % and is significantly higher than that of HVOF sprayed Cr_3C_2/NiCr 75/25.
- The parameter sets enabling crack-free coatings at powder feed rates p > 100 g/min needs to be developed further.
- Under the used test conditions the novel HVAF sprayed FeCrB/WC-Co feedstock material achieved better wear and corrosion properties then the HVOF reference, although the retail price is significantly lower and its PDE is higher.

The focus in this study was the development of coatings with the novel FeCrB/WC-Co feedstock material using a HVAF process. The following investigations are planned in the future to achieve a more efficient and economic process.

- A parameter study to adjust the melting grade of the particles to enable higher powder feed rates (p ≈ 300 g/min).
- Development of new process parameters with the aim of deposition of a finer grain size fraction of -20 +3 µm. The usage of this grain size fraction can be favourable to produce a more near net shape coating, which would lead to a significant decrease in the post production costs. A finer grain size fraction can also result in denser coatings [17, 18, 19] and a more uniform hard particle distribution, which would improve the wear and corrosion properties.

5 Acknowledgement

This work based on the DFG project BO 1979/45-1 “Ressource efficient production of novel Fe-based coatings for large-area applications by AC-HVAF coating process”. The authors gratefully acknowledge the financial support of the German Research Foundation (DFG).

6 References

[1] Schwenk, A.: HVOF-K2-Spritzen von Einsenbasislegierungen für Anwendungen in der Druckindustrie. Proceedings of 8. HVOF-Kolloquium (2009). Erding.

[2] Osamanda, A., Battenbough, A. and Staines, A.: Maßgeschneiderte HVOF-Beschichtungs-syteme auf Fe-Cr und Fe-Cr-Ni Basis für den Verschleiß- und Korrosionsschutz. Thermal Spray Bulletin 2 (2012), pp. 117/22. DVS Media GmbH, Düsseldorf, Germany.

[3] Perdikaris, C.: Methods of coating yankee dryer drums, European patent application, EP 1158066A1, 2001.

[4] Tenkula, J., Hellman, B., Majava, J.: Yankee Clyinder with plasma sprayed carbide coating, Patent, US005123152A, 1992.

[5] Guo R.Q., Zhang C., Chen Q. and Yang, Y.: Study of structure and corrosion resistance of Fe-based amorphous coatings prepared by HVAF and HVOF. Corrosion Science 7 (2011), Vol. 53, pp. 2351/56. Elsevier, Amsterdam, Netherland.

[6] Verstak, A. and Baranovski, V.: AC-HVAF sprayed tungsten carbide: properties and applications. Proceedings of ITSC 2006, May 2006, pp. 643/648. Seattle, USA.

[7] Salman, A., Gabbitas, B., Cao, P. and Zhang, D.: Tribological properties of Ti(AlO/Al2O3) composite coating by thermal spraying. International Journal of Modern Physics 23 (2009), pp. 1407/12.

[8] Wang, Q., Zhang, S., Cheng, Y. and Xiang, J.: Wear and corrosion performance of WC-10Co4Cr coatings deposited by different HVOF and HVAF spraying processes. Surface & Coating Technology 1 (2013), Vol. 218, pp. 117/123. Elsevier, Amsterdam, Netherland.

[9] Bolelli, G., Berger, L. M., Boerner and T., Koivuluoto, H.: Tribology of HVOF- and HVAF-sprayed WC-10Co4Cr hardmetal coatings: A comparative assessment. Surface & Coating Technology (2015), Vol. 265, pp. 125/44 Elsevier, Amsterdam, Netherland.

[10] Hulka, I., Serban, V. A., Niemi, K. and Vuoristo, P.: Comparison of Structure and Wear Properties of Fine-Structured WC-CoCr Coatings Deposited by HVOF and HVAF Spraying Processes. Solid State Phenomena (2012), Vol. 188, pp. 422/27.

[11] Bolelli, G., Hulka, I., Koivuluoto, H. and Lusvarghi, L.: Properties of WC-FeCrAl coatings manufactured by diffeent high velocity thermal spray processes. Surface & Coating Technology (2014), Vol. 247, pp. 74/89. Elsevier, Amsterdam, Netherland.

[12] Berger, L.-M., Puschmann, R., Spatzier, J. and Matthews, S.: Potential von HVAF-Spritzprozessen. Thermal Spray Bulletin 1 (2013), pp. 16/20. DVS Media GmbH, Düsseldorf, Gemrany.

[13] S. Hashmi: Comprehensive Materials Processing: Thermal Spray Coating Process. Newnes (2014), pp. 233/266. Elsevier, Amsterdam, Netherland.

[14] Bobzin K, Öte M, Linke T F and Malik, K A: Wear and corrosion resistance of Fe-based coatings reinforced by TiC particles for application in hydraulic systems. Journal of Thermal Spray Technology 1 – 2 2016, Vol. 25, pp. 365/74. Springer Science+Bussines Media S.A., Luxembourg

[15] Norm: DIN EN ISO 17836, Thermisches Spritzen – Bestimmung der Auftragrate beim thermischen Spritzen. 2005, Beuth Verlag, Berlin, Germany.

[16] Berger, L.-M., Trache, R., Toma, F.-L., Thiele, S., Norpoth, J. and Janke, L.: Development of cost-effective hardmetal coating solutions for high-temperature applications Part one: Feedstock powders, cost-effectiveness and coating properties. Thermal Spray Bulletin 2 (2015), Vol. 8 pp. 126/36. DVS Media GmbH, Düsseldorf, Germany.

[17] Wielage, B., Wank, A., Pokhmurska, H., Grund, T., Rupprecht, C., Reisel, G. and Friesen, E.: Development and Trends in HVOF spraying technology. Surface & Coating Technology 5 (2006), Vol. 201, pp. 2032/37. Elsevier, Amsterdam, Netherland.

[18] Nebel, J., Tillmann, W., Piotrowski, W. and Laemmerhirt, I.A.: Influence of fine powder feedstock (-10 +2 µm) on the HVOF spraying characteristics, Coating morphology and properties of WC-CoCr 87-10-4. Proceedings of ITSC 2012, May 2012, Houston, USA.

[19] Tillmann, W,. Vogli, E., Baumann, I., Matthaeus, G. and Ostrowski, T.: Influence on the HVOF gas compositionon the thermal ypraying of WC-Co submicro powders (-8 +1 µm) to produce superfine structured cermet coatings. Journal of Thermal Spray Technology 3 2008, Vol 14, pp. 405/13. Springer Science+Bussines Media S.A., Luxembourg.

DEVELOPMENT OF HVAF ID COATING TECHNOLOGY AND MATERIALS

Real world experience

The HVAF process provides a slightly different (to HVOF and other thermal spray processes) and unique combination of thermal and kinetic energy, which presents a technical challenge to material manufacturers who have over the past years optimized their material offerings to the demands and characteristics of plasma, combustion and more recently the various HVOF processes. The recent development of HVAF ID torches, capable of applying the highest quality coatings into dimensions of less than 100mm (4"), has provided a further need to optimize spray materials for the best possible interaction between the spray device and the feedstock in order to meet developing coating specifications from various industries. This paper will discuss the methodologies utilized from the points of view of torch design, material manufacture, spray parameter and procedure optimization, as a means of producing the highest performing coatings to meet specific industry aspirations.

Andrew Verstak - Kermetico, Dr Benno Gries – HC Starck GmbH, Michael Breitsameter – HC Starck GmbH.

6/5/2017

Development of HVAF ID coating technology and materials

Real world experience

The fundamentals of the HVAF process make it uniquely suited to the challenge of producing high quality carbide and metallic coatings in small inside diameters (ID) on a production basis, for use across a broad range of industries.

TORCH DESIGN

In the recently developed Kermetico HVAF guns, the combustion process is organized as a micro-flame combustion on the small orifices of the catalytic ceramic insert inside the combustion chamber. The length of each micro-flame jet is very small. Thus, combustion is completed within a very short distance, in a small combustion chamber. As a result, the stand-off distance during HVAF spraying can be significantly shorter than in similar power HVOF guns where combustion continues beyond the nozzle. This ability to utilize short stand-off distance is a key factor for practical spraying in small internal diameters.

Additionally, the relatively low combustion temperature in the HVAF process (1800-1950°C vs 2500+ °C for HVOF) provides the opportunity to use small spray particle sizes with a significantly reduced risk of the material being thermally degraded or destroyed during the spraying process (through decarburization, oxidation or elemental burn out). The reduction in particle size is necessary when the available length/time for the particle heating and acceleration is limited by the size of the internal diameter of the part to which the coating will be applied, and the reduced energy of the flame to impart heat and acceleration to the individual particle, without which coating quality would likely be severely compromised.

POWDER OPTIMIZATION

The commercial production of these smaller particle size ranges, on a production scale creates a unique set of challenges for the powder manufacturer. HC Starck GmbH have jointly developed a range of gas atomized and agglomerated and sintered/spray dried and sintered materials (WC, CrC and Metallics) that have been optimized for use with HVAF spray devices, including those designed to spray into small ID's (as small as 80mm/3.15").

Coatings were optimized using standard metallography techniques, along with permeability testing as an overall indicator of coating density and through porosity in the coating. In many cases the coatings produced are being used in aqueous corrosive environments, where the coating is required to perform as a barrier for protection of a less corrosive resistant substrate material.

The development of these materials has been focused on several specific aspects:

1. Particle sizing (range and mean) tailored to suit the smaller and lower temperature energy source of the individual HVAF ID torch size and power
2. Primary carbide size as a key parameter when developing such coatings
3. Quality and manufacturing controls to ensure batch to batch and bottle to bottle consistency for use in a production environment
4. Packaging designed with a greater degree of sensitivity to potential ingress of moisture based on the highly hygroscopic nature of finer powders
5. Apparent density and other aspects to ensure best possible deposit efficiencies to ensure commercial viability of the coatings
6. Particle shape/morphology based on the need to have free flowing material for practical feeding from powder hopper to spray torch consistently

Fig 1 and 2 show examples of coating structures produced with new ID torch designs:

Fig 1 – WC/Co/Cr (86/10/4) coating produced for use in 100mm (4") ID, hardness 1300 HV300, porosity <0.5%

Fig 2 – WC/Co/Cr (86/10/4) coating produced for use in 100mm (4") ID, optimized for deposit efficiency, hardness ~1100 HV300, porosity <0.5%

APPLICATIONS

We will herein show a number of examples of coatings and applications that have been recently developed for industrial end users based on the simultaneous optimization of spray process (hardware, parameters, procedures and automation) and powder consumables.

Fig 3 – Coating of ID surface of pump sleeve with WC/Co/Cr coating 200um thick

Fig 4 – Pump sleeves after coating and finishing operation to <1Ra

Fig 5 – Coating of flange on vessel with Ni base superalloy applied with fully automated rotating ID torch, coating thickness of 250um

Fig 6 – Complete vessel after coating of all internal surfaces with 250um of Ni based superalloy, vessel weight more than 1500kg

Fig 7 – 90 degree bends coated for refinery application exposed to erosion and corrosion with WC/Co/Cr, thickness 200um +/- 25um for all internal surfaces

Fig 8 – Round to rectangular ID section coated for refinery application with WC/Co/Cr, coating thickness of 200um +/-25um

The coating performance in all of these applications has been superior (in terms of actual service life, and consistency of service life from part to part and batch to batch) to all previously tested coatings, plating, heat treatment, and weld overlay. In many cases these parts, and others that have been developed, were not practical to coat until the development of HVAF ID hardware and the supporting materials tailored for use with the process. All parts shown are coated on an ongoing basis as required by the end user.

CONCLUSION

The development of HVAF ID spray devices and coatings, used in a production environment, on production applications, in addition to the development of specifically tailored spray materials to support this new technology, has resulted in the growth of new industrial applications previously impossible or impractical to coat with the highest quality of thermal spray coatings available. Previous technologies such as plasma, or wire coatings, while functional, are not of the highest quality, and hence do not provide optimal performance in service. The optimization of spraying hardware, powder materials, along with robust parameters and procedures, has resulted in new opportunities for growth for the thermal spraying industry.

A Study on the Tribological Behavior of Arc Sprayed Vanadium doped Stellite Coatings

W. Tillmann, L. Hagen, D. Duda, Dortmund / D

Due to good performance in abrasive and sliding wear and enhanced oxidation behavior, coatings based on Co-Cr-W alloys are widely used in industrial applications, where the material is exposed to high temperature. Within the scope of this study, a Co-based alloy similar to commercial Stellite® 6, which additionally contains 20.6 wt.% of vanadium, was deposited by Twin Wire Arc Spraying (TWAS). Multi-criteria optimization using statistical design of experiments (DoE) have been carried out in order to produce adequate coatings. The produced coatings have been analyzed with respect to their tribological behavior at elevated temperatures. Dry sliding experiments were performed in the temperature range between 25°C and 750°C. Oxide phases were identified in the investigated temperature range by x-ray diffraction (XRD) using synchrotron radiation. The V-doped Stellite®-based coating possesses a reduced coefficient of friction (COF) of about 0.37 at elevated temperatures (above 650°C), which was significant lower when compared to conventional Stellite® 6 coating that serves as reference. In contrast, both produced coatings feature a similar COF under room temperature. X-ray diffraction reveals the formation of cobalt vanadate and vanadium oxides above 650°C. The formation of vanadium oxides exhibits the ability of self-lubricating behavior, thus leading to enhanced tribological properties.

1 Introduction

The interaction between running surfaces and counterparts, or tool and workpiece material, is mainly determined by their surface properties. Accordingly, a major objective is focused on a reduction of friction losses. The addition of lubricants provides the ability to reduce friction losses or to prevent the emergence of frictional forces. However, under certain conditions, lubricant supply is not possible or economically feasible. Therefore, and because of its potential application at very high temperatures or in aggressive media, in vacuum or, inter alia, for reasons of maintenance, friction-reducing coatings represent a promising candidate to enhance the tribological characteristics of stressed surfaces. For instance, the use of self-lubricous hard coatings for high temperature applications, such as machining, can lead to decreased frictional forces, which in turn contribute to the wear resistance [1]. Thus, the costs of manufacturing can be reduced.

In the field of thin film technology, the positive effects of self-lubricating coatings are already well-known. A reduction in friction has been achieved, in particular, by using titanium- and vanadium-containing coatings [2].

Oxides of transition metals such as vanadium form structures of the Me_nO_{2n-1}-, Me_nO_{3n-1}- or Me_nO_{3n-2}-type with recurring vacancies under thermal load and oxidizing environment [3]. Some of these lubricious oxides generate a series of sub-stoichiometric compounds of so-called Magnéli phases. Magnéli phases show on certain crystallographic planes oxygen defects, which reduce the binding forces between the shear planes, so that an easy shearing of the planes is provided [4]. Accordingly, due to its low decohesion energies G [3] some vanadium oxides such as vanadium pentoxide represent promising candidates for solid lubricant at elevated temperatures. Above its melting point [5] it also acts as solid lubricant with a certain liquid amount [1].

Due to good performance in wear resistance and enhanced oxidation behavior, coatings based on Co-Cr-W alloys are widely used in industrial applications, where the material is exposed to high temperature. Numerous studies focused on the wear and hot corrosion behavior of HVOF sprayed Stellite® 6 coatings [6-8]. Although on the one hand Stellite®-based feedstocks and, on the other hand, arc spraying has long been established, only few studies exist which have been dealing with the investigation on arc sprayed Stellite®-based coatings. Varavallo et al. [9] examined arc sprayed Stellite® 6 coatings with respect to their metallographic characteristics such as microstructure and microhardness as well as adhesion to the substrate. The Phase evolution during arc spraying and tribological characteristics of produced coatings have not been taken into account.

Within the scope of this study, a Co-based alloy similar to commercial Stellite® 6, which additionally contains 20 wt.% of vanadium, was deposited by Twin Wire Arc Spraying (TWAS). Multi-criteria optimization using statistical design of experiments (DoE) have been carried out in order to produce adequate coatings. The tribo-mechanical properties of a vanadium-doped Co-based arc sprayed coating system were examined in dry sliding experiments in the temperature range between 25°C and 750°C. The findings were compared to the behavior of arc sprayed Stellite®-based coatings. Oxide phases were identified in the investigated temperature range by x-ray diffraction (XRD) using synchrotron radiation.

2 Experimental

2.1 Substrate and feedstock material

Round C45 steel specimens (1.0503) with a diameter of 40 mm and a thickness of 4 mm were employed as substrate. To ensure the adhesion between the substrate and the coating, the surfaces of the test specimens were grit blasted with alumina (grain size: 1180 - 1700 µm) using a blasting air pressure of 5 bar, a stand-off distance of 100 mm, and a blasting angle of 45°. Subsequently the specimens were cleaned in an ethanol ultrasonic bath. Prior coating deposition the

samples were preheated in a convection oven to the temperature of about 80°C.
A Co-based cored wire similar to commercial Stellite® 6 which additionally contains 20.6 wt.% of vanadium, serves as feedstock. Within this study, the V-doped Co-based (Co-20.57V-9.8Cr-8.55Fe-1.72W-1.47Ni-0.68Si-0.43C-0.39Mn wt.%) cored wire is referred to as LitV66. The commercial available Durolit 6 (Co-29.7Cr-5.24W-2.64Fe-1.47Ni-1.42Si-1.23C-1.17Mn wt.%) cored wire (Durum Verschleissschutz, Germany) serves as reference. The composition was calculated by the manufacture (Durum Verschleissschutz, Germany) based on the results taken from wet chemical analyses.

2.2 Spray parameter settings

In order to produce adequate coatings, a multi-criteria optimization using statistical design of experiments (DoE) have been carried out. In this study 3 independent variables were subjected to a 3^3 full factorial design with 27 runs. In addition, one center run was added to the design in order to reduce the residual variance of the model. The optimum for the parameters was defined on the basis of multi-criteria optimization using the Derringer's desirability function [10].
With respect to the coating deposition, the Smart Arc 350 PPG spraying system (Oerlikon Metco, Switzerland) was used. Regarding the 3^3 full factorial design, the voltage (U), primary gas pressure (PGP), and stand-off distance (SOD) were selected as independent variables.
Since the atomization behavior gets affected by the electrode phenomena as well as by the flow regimes at the intersection point of the electrode tips, the PGP and U significantly influence the produced coating quality [11, 12]
Moreover, the SOD was taken into account as the distance (inter alia the dwell time) interacts with the in-flight spray particle characteristics, and thus affects the coating properties [13]. The level settings for all the variables were converted to -1 (low), 0 (center), and +1 (high) for the DoE. The current and handling parameters such as the spray angle, gun velocity, track pitch, and amount of overruns were kept on a constant level. The spray parameter settings are summarized in **Tab. 1**.

Tab. 1. Parameter setting for the 3^3 full factorial design

Factor		3^3 full factorial design		
		-1	0	+1
U [V]		28	31	34
SOD [mm]		90	120	150
PGP (compressed air) [MPa]		0.65	0.8	0.95
Handling parameters (constant level)			Spray parameters (constant level)	
Spray angle [°]	90		Current [A]	200
Gun velocity [m/s]	200			
Track pitch [mm]	5			
Overruns	2 passes			

2.3 Coating characterization

The sprayed coatings were examined with respect to their microstructural characteristics, chemical composition and tribological behavior. Diamond grinding discs and polishing cloths with a diamond suspension (steps: 9 µm, 6 µm, 3 µm, and 1 µm) were utilized in order to prepare cross-sections of the coated samples. The microstructure of the coating was analysed by using a field emission scanning electron microscope (FE-SEM) type JSM-7001F (Joel, Germany). The porosity (Po) was determined via image analysis using optical microscopy (microscope BX51M, Olympus, Japan) with the integrated software Axiovision AutoMess at ten different spots across the coating. The Vickers microhardness (MH) was obtained on the cross-section using the hardness testing device M400 (Leco, Michigan) with 300 g loads. The mean roughness depth (Rz) of the test specimens were measured by using the tactile roughness testing device T-1000 (Hommel, Germany). Dry sliding experiments were conducted with the use of the high-temperature ball-on-disk (BOD) tribometer (CSM, Switzerland) at room temperature and elevated temperatures (from 350°C up to 750°C in 100°C steps). An alumina ball with a diameter of 6 mm, and a hardness of 2300 HV 0.3 was used as a counterbody. The velocity was kept at 40 cm/s, and the radius of the circular path was maintained constant at 10 mm. A sliding distance of 200 m, and a load of 5 N without lubricant were applied. After testing, the worn surfaces were examined by means of 3D profilometry using the device Infinite Focus (Alicona, Austria). Taken from the reconstructed surfaces, the volumetric material loss was determined. The wear coefficient k was calculated based on the volumetric material loss (mean value considering four different positions at three samples), the applied normal force and the sliding distance. The phase composition of the different specimens (in accordance to the operating temperature used for the dry sliding experiments) was detected ex-situ via x-ray diffraction (XRD) at the beamline BL9 of the synchrotron light source DELTA [14] (Dortmund, Germany). The photon energy was set to 27 keV (wavelength λ = 0.4592 Å) and a MAR345 image plate detector was used for the detection. The beam size was set to 0.1 x 1.0 mm^2 (v x h) and the angle of incidence was 3°.

3 Results and Discussion

3.1 Statistical design of experiments

Tab. 2 summarizes the level settings with its corresponding results taken from metallographic examination. The statistical combination of the experiments allows to estimate the linear and quadratic effects and also the interactions between the variables on the results.

Tab. 2. Level settings and corresponding results of the 3^3 full factorial design with additional center point run

3^3 Full factorial design				Results (mean values)		
Run no.	U, V	PGP, MPa	SOD, mm	Po, %	MH, HV 0.3	Rz, µm
1	-1	-1	-1	5.2	354	70.5
2	-1	0	-1	3.7	302	70.2
3	-1	+1	-1	3.1	419	65.0
4	-1	-1	0	4.4	367	56.8
5	-1	0	0	3.8	350	61.2
6	-1	+1	0	4.4	379	53.4
7	-1	-1	+1	4.4	334	62.8
8	-1	0	+1	3.6	340	50.1
9	-1	+1	+1	2.9	376	52.6
10	0	-1	-1	2.9	314	79.5
11	0	0	-1	4.1	312	68.5
12	0	+1	-1	3.7	379	57.6
13	0	-1	0	4.5	391	67.3
14 (C)	0	0	0	4.4	411	65.7
15	0	+1	0	3.3	418	55.3
16	0	-1	+1	5.1	375	65.5
17	0	0	+1	4.5	392	61.6
18	0	+1	+1	4.5	367	56.6
19	+1	-1	-1	3.9	394	67.2
20	+1	0	-1	4.0	358	61.0
21	+1	+1	-1	3.7	482	70.6
22	+1	-1	0	4.3	390	64.5
23	+1	0	0	4.8	399	62.1
24	+1	+1	0	4.3	386	57.5
25	+1	-1	+1	5.5	392	67.7
26	+1	0	+1	3.8	390	67.4
27	+1	+1	+1	4.5	404	61.4
28 (C)	0	0	0	3.9	383	56.9

Tab. 3 shows the estimates and the p-values for each response variable (Po, MH, Rz). The independent variables affect the response variable positively or negatively, which is shown by the sign of the estimate. The significance level was set to 10% (α = 0.1) and the significant effects are marked in bold.

Tab. 3. Effect estimates and p-values (α = 0.1) for the 3^3 full factorial design

	Po [%]		MH HV 0.3		Rz [µm]	
	Est.	*p*-Value [$\cdot 10^{-4}$]	Est.	*p*-Value [$\cdot 10^{-4}$]	Est.	*p*-Value [$\cdot 10^{-4}$]
Intercept	4.10	0	377	0	62.80	0
(1) U (L)	0.36	1978	**42**	**97**	**4.10**	**597**
U (Q)	-0.04	8524	-4	7456	1.42	4168
(2) PGP (L)	**-0.69**	**185**	**33**	**330**	**-7.99**	**10**
PGP (Q)	-0.08	7143	**-22**	**888**	-0.23	8942
(3) SOD (L)	**0.48**	**881**	6	6692	**-7.15**	**25**
SOD (Q)	0.15	5213	18	1566	**-4.21**	**243**
1L by 2L	0.37	2694	-4	8318	1.54	5461
1L by 3L	0.54	1151	-4	8389	**6.34**	**207**
2L by 3L	-0.26	4410	-28	1245	-0.20	9357

L = linear, Q = quadratic

With respect to Rz, it can be seen that the PGP and the SOD have a considerable influence. An increase of one of this variable leads to a decreased roughness. The increase of the PGP also leads to a significant decrease of the Po. In contrast to that, the Po increases within a growing SOD. However, the importance of the U, PGP and SOD with regard to their main effects (L), quadratic effects (Q), and their two-way interaction effects will be discussed in detail in a further contribution.

To find an optimum combination of the independent variables the multi-criteria optimization was used. In order to combine the response variables, they must be transformed into in a common scale by using Derringer's desirability function [10]. According to this, the desirability *d* of each response variable gets a value between 0 and 1, corresponding to the type of optimization goal (desirability at a maximum or minimum), **Tab. 4**. The desirability of a response variable takes the value 0 when the response drops to a lower specification limit (LSL). The value 1 is defined when the response variable goes beyond the upper specification limit (USL). Values in between the limits are also assigned to desirabilities, whereas the transformation is linear. Following the desirability transformation, the desirability index is calculated as shown elsewhere [15]. Within this study, the desirability index is calculated, using the software Statistica (Statsoft, Oklahoma), with an even weighting for all criteria (response variables).

Tab. 4. Parameters for desirability transformation

	Po [%]		MH HV 0.3		Rz [µm]	
	Po (min)	*d (Po)*	MH (max)	*d (MH)*	Rz (min)	*d (Rz)*
LSL	3.6	1	330	0	56	1
USL	4.6	0	400	1	68	0

Using this tool, the theoretical optimum with a desirability index of 0.95 is reached by the following spray parameter setting: U = 29 V, SOD = 115 mm, and PGP = 0.95 MPa. The predicted values of the response values with these settings are the following: Po = 3.7%, MH = 397.3 HV 0.3 and Rz = 55.7 µm. Using the optimized spray parameter setting, coatings (verification experiments) were produced and metallographically examined, **Fig. 1**.

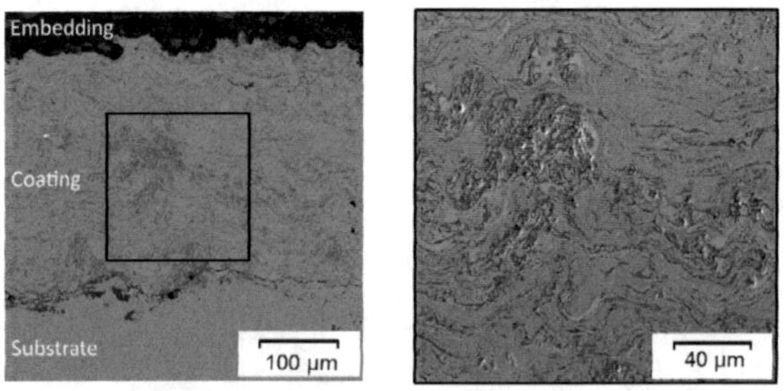

Fig. 1. Cross section images of the V-doped coating taken by SEM

In comparison of predicted and experimentally determined coating properties (Po = 4.1±1.6 %, MH = 534±93 HV0.3 and Rz = 57.4±9.2 µm), it is found that the results of the verification experiments are within the prediction intervals, except for MH. Here, the experiment shows a high deviation in terms of the actual MH compared to the predicted value. Prediction lines along with their 95% prediction intervals can be calculated from models, which were fitted based on the 28 data points from the 3^3 full factorial design with additional center point run.

3.2 Tribological investigation

Fig. 2 shows the wear coefficient k of the V-doped Stellite®-based coating (LitV66) and the reference (Durolit 6) between 25°C (room temperature) and 750°C.

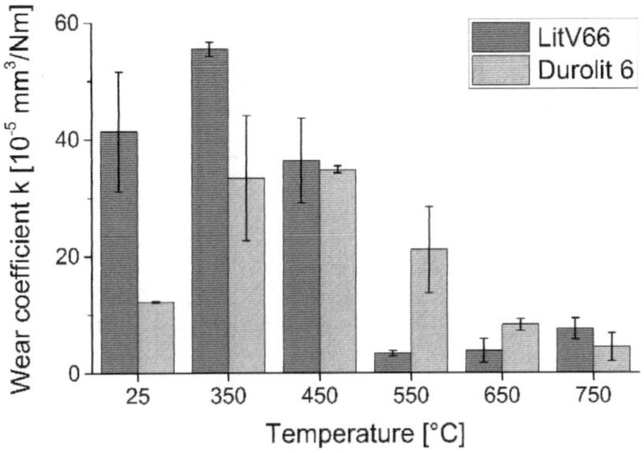

Fig. 2. Wear coefficient k of different specimens depending on the temperature

For the Durolit 6 coating, k increases from k = $12.2 \cdot 10^{-5}$ mm³/ Nm at RT to k = $34.8 \cdot 10^{-5}$ mm³/ Nm at 450°C. Above 450°C k decreases continuously. Thus, the sample possesses a k of $4.4 \cdot 10^{-5}$ mm³/ Nm at 750°C. For the LitV66 coating, k increases from k = $41.4 \cdot 10^{-5}$ mm³/ Nm at RT to k = $55.5 \cdot 10^{-5}$ mm³/ Nm at 350°C. With increasing temperature k decreases. A k of $36.4 \cdot 10^{-5}$ mm³/ Nm at 450°C and $3.4 \cdot 10^{-5}$ mm³/ Nm at 550°C is measured. Above 550°C the coating exhibit a slight increase of k (k = $3.8 \cdot 10^{-5}$ mm³/ Nm at 650°C; k = $7.5 \cdot 10^{-5}$ mm³/ Nm at 750°C). When compared to room temperature, both coatings reveal an enhanced wear resistance, in particular above 550°C. This encourages the assumption that the V-doped Stellite®-based deposit represents a promising candidate as wear resistant coating when exposure to high temperatures.

With rising temperature, a moderate increase of the hardness can be observed (ex-situ measurements) for both sample types, **Tab. 5**.

Tab. 5. Microhardness of different specimens obtained after tribological testing (ex-situ measurements)

	25°C	350°C	450°C	550°C	650°C	750°C
LitV66	603±31	562±46	619±43	680±61	683±46	693±31
Durolit6	585±30	569±27	593±33	638±52	683±76	674±38

Microhardness [HV0.3]

The LitV66 and Durolit 6 coatings feature a microhardness of 603±31 HV0.3 and 585±30 HV0.3, respectively, at the surface without annealing. Above 450°C a continuous increase of microhardness is shown reaching a microhardness of 693±31 HV0.3 (LitV66) and 674±38 HV0.3 (Durolit 6) at 750°C.

Fig. 3 shows the COF of the two coatings from 25°C up to 750°C. The COF of the Durolit 6 coating remains nearly constant between 25°C and 550°C of approx. 0.56 → 0.51. A substantial decrease of the COF was observed for the Durolit 6 coating above 550°C. A significant drop of the COF can be noticed between 550°C and 650°C (0.51 → 0.43). A further increase of temperature leads to a slight increase of the COF to 0.45 at 750°C. As opposed to that, the LitV66 coating exhibits a different behavior. For increasing temperature, a continuously falling COF is observed. Thus, at a temperature of 25°C the coating possesses a COF of 0.55, whereas at 550°C the COF is 0.46. With further increase of temperature, the COF decreases significantly to 0.37.

It was found that at 750°C the V-doped LitV66 coating generates a significant reduction of approx. 24 % of the COF, when compared to the Durolit 6 coating. In contrast, both sample types exhibit a similar COF at 25°C. In comparison between room temperature and 750°C, the COF of the V-doped coating decreases by 34%.

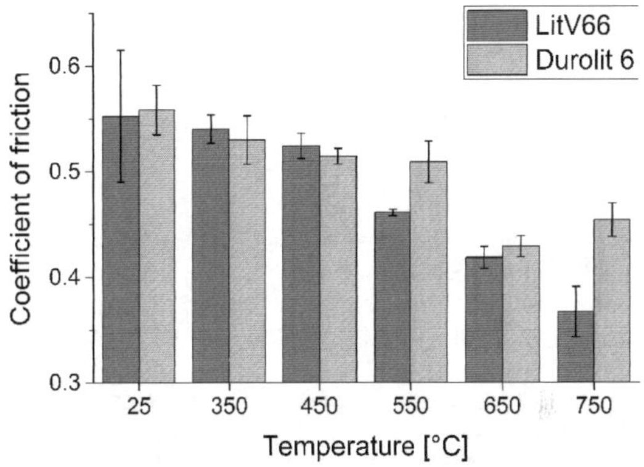

Fig. 3. Coefficient of friction of different specimens depending on the temperature

According to Franz et al. [1], a further reduction in friction can be achieved with V-containing layers at temperatures around 700 °C due to the formation of vandadium oxides such as V_2O_5, which resembles a liquid lubricant above its melting temperature (678°C [5]), although it is reported that his may be associated with a deterioration in wear resistance [1]. These relationships might correspond to the results observed for the temperature range between 550°C and 750°C (**Fig. 2** and **Fig. 3**).

3.3 X-ray diffraction analysis

Fig. 4 shows the diffraction patterns of the V-doped coating (LitV66) and the Stellite®-based coating (Durolit 6) exposed to different temperatures.

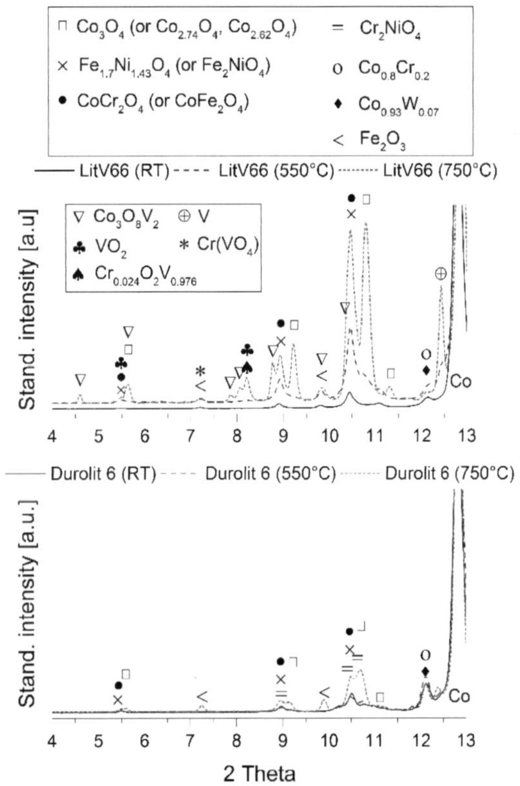

Fig. 4. Diffraction patterns of the V-doped coating (LitV66) and the Stellite®-based coating (Durolit 6) at different temperatures. 2-theta is scaled regarding to a photon energy of 27 keV (wavelength λ = 0.4592 Å)

XRD analyses reveal that the Durolit 6 coating is mainly composed of Co, Co_3O_4 (or $Co_{2.74}O_4$, $Co_{2.62}O_4$), $CoCr_2O_4$ (or $CoFe_2O_4$), $Fe_{1.7}Ni_{1.43}O_4$ (or Fe_2NiO_4), as well as Cr_2NiO_4, $Co_{0.8}Cr_{0.2}$ and $Co_{0.93}W_{0.07}$. The XRD pattern obtained from samples exposure to excessive temperature (sample Durolit 6 (750°C)) confirms that the coating additionally consists of Fe_2O_3 and a significant higher amount of Co_3O_4.

For sample LitV66 (RT), a similar composition is ascertained when compared to Durolit 6. However, XRD patterns reveal that the presence of $Co_{0.8}Cr_{0.2}$ and $Co_{0.93}W_{0.07}$ is less pronounced. Moreover, Cr_2NiO_4 was not detected. The absence of Cr_2NiO_4, and the reduced content of $Co_{0.8}Cr_{0.2}$ and $Co_{0.93}W_{0.07}$ are attributed to the reduced amount of Cr, Ni and W contained in the feedstock (LitV66). The additional peaks in the diagram obtained from sample LitV66 (550°C) and LitV66 (750°C) versus those from sample LitV66 (RT) indicate that new phases have been formed. With respect to the XRD pattern of sample LitV66 (550°C), the coating exhibits a higher amount of Co_3O_4 (or $Co_{2.74}O_4$, $Co_{2.62}O_4$), $CoCr_2O_4$ (or $CoFe_2O_4$), $Fe_{1.7}Ni_{1.43}O_4$ (or Fe_2NiO_4). In contrast, it is found that the sample LitV66 (750°C), which was exposed to a further increase in temperature, additionally consists of a certain amount of $Co_3O_8V_2$ and elementary V, as well as some traces of VO_2 (or $Cr_{0.024}O_2V_{0.976}$) and $Cr(VO_4)$. However, the phase evolution in both arc sprayed sample types is complex and depends on various environmentally conditions (predominantly inherent process characteristics such as constitutional supercooling during rapid solidification of spray particles), which leads to a wide varying non-equilibrium state. Consequently, the determination of phase transformation is a great challenge and not all phases can be identified.

Inman et al. [16] examined the sliding wear behavior of Incoloy MA956 against Stellite® 6 with respect to its tribological characteristics between room temperature and 750°C. As verified by Stellite® 6 versus Stellite® 6 tests, Co-Cr based oxides form wear protective 'glaze' layers at elevated temperatures. It is reported that the material losses due to severe wear became increasingly restricted on raising the temperature from 510°C to 750°C. With regard to the debris, XRD analysis indicated the presence of $CoCr_2O_4$ and Co_3O_4. The authors discussed the free energies of formation for key oxides at 727°C under conditions of static oxidation, indicating the preferential oxidation of Co and Cr from the Stellite® 6.

Sidhu et al. [17] studied the hot corrosion behavior of plasma and laser remelted Stellite® 6 in molten salt (Na_2SO_4–60%V_2O_5) environment at 900 °C under cyclic conditions. XRD analyses revealed that the specimens were prone to form CoO, $CoCr_2O_4$, $NiCr_2O_4$, $NiWO_4$, Cr_2O_3, Fe_2O_3, FeS and $Co_3V_2O_8$. The occurrence of cobalt vanadate has also been depicted by Jones et al. [18] studying the hot corrosion behavior of a CoCrAlY gas turbine blade coating by molten sulfate-vanadate deposits in air environment at 700°C.

The iron present in both sample types, even if both feedstocks contain only a reduced amount of Fe, might have reacted to form iron oxides. Accordingly, a slight amount of Fe_2O_3 can be measured at 750°C. Thermodynamic calculations of the free enthalpy show that the presence of Fe_2O_3 within the temperature range is more likely than other iron oxides such as Fe_3O_4 [19]. Subsequently, Fe_2O_3 is thermodynamically more stable. Studies on the Fe–O system show that Fe_2O_3 should be exist independent from the partial pressure in the temperature range above 550°C up to 750°C [19, 20].

Moreover, both sample types (LitV66 and Durolit 6) show a marginal amount of tungsten carbides (the data is not displayed). It can be stated that both coatings consist of WC_{1-x} under as-sprayed conditions, whereas at elevated temperature some slight amount eta carbides is generated. This is in a good accordance with the findings observed in a study on the microstructural characteristics to assess the phases present in Stellite® 6 deposits using gas tungsten arc welding and plasma transferred arc welding processes. According to this study, Rao et al. [21] reported that for Stellite® 6 deposits also several carbides such as WC, Co_6W_6C, and Co_3W_3C were observed.

Acknowledgments

The contributions of DURUM Verschleissschutz GmbH are gratefully acknowledged for their support in providing the vanadium-doped feedstock. The authors would like to thank the DELTA machine group for providing synchrotron radiation.

Literature

[1] Franz, R. and C. Mitterer: Vanadium containing self-adaptive low-friction hard coatings for high-temperature applications: A review. Surf. Coat. Technol. 228 (2013), pp. 1-13.

[2] Brugnara, R. H.: Hochtemperaturaktive HPPMS-Verschleißschutzschichten durch Bildung reibmindernder Magnéli-Phasen im System (Cr,Al,X)N. Schriftenreihe Oberflächentechnik, dissertation, Technische Hochschule Aachen, Herzogenrath, Shaker, 2016

[3] Bobzin, K.; Bagcivan, N. and M. Ewering: DC-MSIP/HPPMS (Cr, Al, V)N and (Cr, Al, W)N Thin Films for High-Temperature Friction Reduction, Surf. Coat. Technol. 205 (2011), Issues 8-9, pp. 2887-2892.

[4] Zhang, Z. and H. Dong: A State-of-the-art overview: Recent development in low friction and wear-resistant coatings and surfaces for high-temperature forming tools. Manufacturing Review 1 (2014), pp. 1-12.

[5] Wriedt, H.A.: The O-V (Oxygen-Vanadium) System. Bull. Alloy Phase Diagr. 10 (1989), Issue 3, pp. 271-277.

[6] Sidhu, T.S.; Prakash, S. and R.D. Agrawal: Studies of the metallurgical and mechanical properties of high velocity oxy-fuel sprayed Stellite-6 coatings on Ni- and Fe-based superalloys. Surf. Coat. Technol. 201 (2006), Issues 1-2, pp. 273-281.

[7] Jegadeeswaran, N.; Bhat, K.U. and M.R. Ramesh: Oxidation studies on as-received and HVOF sprayed Stellite-6 coating on turbine alloys at 800 °C. Int. J. Sci. Eng. Res. 4 (2013), Issue 6, pp. 214-220.

[8] Kusmoko, A.; Dunne, D. and H. Li: A comparative study for wear resistant of Stellite 6 coatings on nickel alloy substrate produced by laser cladding, HVOF and plasma spraying techniques. Int. J. Curr. Eng. Technol. 4 (2014), Issue 1, pp. 32-36.

[9] Varavallo, R.; Manfrinato, M.D.; Rossino, L.S.; Maluf, O. and F. Camargo: Adhesion of Thermally Sprayed Metallic Coating. Journal of ASTM International 9 (2012), Issue 2, pp. 1-11.

[10] Derringer, G. and R. Suich: Simultaneous Optimization of Several Response Variables. J. Qual. Technol. 12 (1980), pp. 214-219.

[11] Boronenkov, V. and Korobov, Y.: Fundamentals of Arc Spraying: Physical and Chemical Regularities, Springer, Berlin, 2015

[12] Hussary, N.A. and Heberlein, J.V.R.: Effect of System Parameters on Metal Breakup and Particle Formation in the Wire Arc Spray Process, J. Therm. Spray Technol. 16 (2007), pp. 1-13.

[13] Krebs, B.: Konturgenaue Bauteilbeschichtung für den Verschleißschutz mittels atmosphärischen Plasmaspritzens und Lichtbogenspritzens, dissertation, TU Dortmund University, Dortmund, Vulkan-Verl, Essen, 2004

[14] Krywka, C. et al.: The new diffractometer for surface X-ray diffraction at beamline BL9 of DELTA, J. Synchrotron Radiat. 13 (2006), pp. 8-13.

[15] Tillmann, W., Hagen, L., Kokalj, D., Paulus M. and Tolan, M.: A Study on the Tribological Behavior of Vanadium-Doped Arc Sprayed Coatings, J. Therm. Spray Technol. 26 (2017), Issue 3, pp. 503-516.

[16] Inman, I.A.; Rose, S.R. and P.K. Datta: Studies of high temperature sliding wear of metallic dissimilar interfaces II: Incoloy MA956 versus Stellite 6. Tribology International 39 (2006), Issue 11, pp. 1361-1375.

[17] Sidhu, B.S. and S. Prakash: Studies on the behaviour of stellite-6 as plasma sprayed and laser remelted coatings in molten salt environment at 900 °C under cyclic conditions. Journal of Materials Processing Technology 172 (2006), Issue 1, pp. 52–63.

[18] Jones, R.J., Williams, C.E.: Hot corrosion of Co-Cr-Al-Y by molten sulfate-vanadate deposits, Mater. Sci. Eng. 87 (1987), pp. 353-360.

[19] Sundman, B.: An Assessment of the Fe-O System, JPE 12 (1991), Issue 2, pp. 127-140.

[20] Hidayat, T., Shishin, D., Jak, E. and Decterov, S.A.: Thermodynamic Reevaluation of the Fe-O System, Calphad 48 (2015), pp. 131-144.

[21] Rao, K.P., Damodaram, R., Rafi, H.K., Ram, G.J., Reddy, G.M. and Nagalakshmi, R.: Friction surfaced Stellite 6 coatings. Materials Characterization 70 (2012), pp. 111-116.

Practical experiences using HVOF sprayed tungsten carbide coatings in the plastic foil industry

Oliver Brandt (Thun / CH), Kai Möhwald (Witten / D), Wilfried Reimche (Garbsen / D), Oliver Bruchwald (Garbsen / D)

For most HVOF sprayed tungsten carbide coatings powders with identical particle-size distributions and compositions are interchangeable, within the meaning that as long as the manufacturing process is identical other properties are nearly irrelevant. In plastic foil industry often complex strains occur on the surfaces of the used tools. These strains comprises from a chemical attack by means of released process gases due to the thermal conditions and from an abrasive component caused by additives of the plastic. Shaping calender rolls with a high gloss mirror finished surface may show only a minimal wear in such complex strains because otherwise the quality of the foil is affected in a negative way. In this range of applications the practical experience shows that with commercial characterization values such as microhardness and porosity HVOF sprayed tungsten carbide coatings are not sufficiently described. The article describes the effects of different microstructures on the application behaviour of a coating. For that purpose analytical methods such as quantitative phase analysis due to X-ray diffraction by means of Topas-Rietveld-Method were used. Preliminary results of the use of a novel high frequency eddy current test method suggest the potential to allow a local correlation of specific properties of the coatings with their microstructure.

1 Introduction

The plastics industry, especially the manufacture of films, is growing continuous and steady growth, **Fig. 1**.

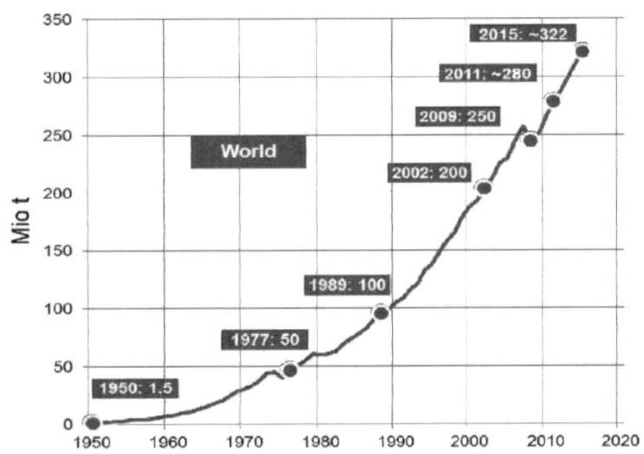

Fig. 1: Groth of market of the plastic foil industry [1]

This interesting field of application opens a whole range of applications in a growth market for coating technology due to the wide variety of different requirements. The applications range from the non-stick coatings, preferably made of plastic materials, to corrosion and wear-resistant coatings. The latter in particular are always used when especially advanced plastics are processed with additives made of abrasives such as titanium oxide or chalk, for example.
Especially in the plastic film industry, very complex stresses are often generated on surface of the tools used. The stresses consist of exposure to chemicals from the process gases released due to the thermal conditions and an abrasive component due to the additives in the plastic. Calender rolls with a high gloss mirror surface are only allowed to exhibit a minimum of wear, even when subject to such complex stresses, because otherwise it will have a negative impact on the quality of the film. This is important because the calender rolls are located near the beginning of the process chain in a film production line, and any defects in a film cannot be corrected anymore in the subsequent production steps and will actually become even more evident. The complexity of a calender line is illustrated in **Fig. 2**.

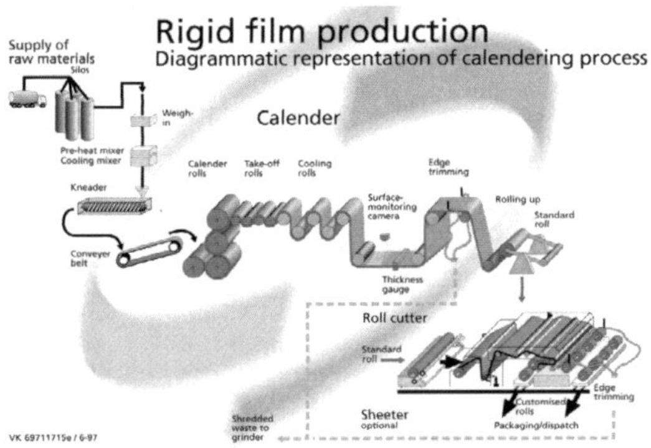

Fig. 2: Schematic diagram of a calender roll plant in plastic foil industry [2]

In conventional systems, the calender rolls are hard chrome plated galvanically and used in the desired surface quality. HVOF sprayed tungsten carbide coatings have proven to be a good alternative in this case since they have a much higher resistance to abrasion as well as even better resistance to mechanical damage.

2 The problem

HVOF sprayed tungsten carbide coatings have been in use in industrial applications since the 1980's. In

this area, layers of tungsten carbide and chromium carbide have become more or less the standard. For these materials, a variety of powders with various particle-size distributions and hard material/binder matrix combinations are available commercially. In most applications, powders with identical particle size distributions and the same composition can be used as desired, which means that no further properties play a role in the manufacturing method [3], [4], [5].
In contrast to galvanically deposited layers of chromium, the tungsten carbide coatings are characterized by a two-phase substrate consisting of the hard material and a soft binder matrix. Machining the coating later on by grinding, polishing, and lapping the surface produces structural differences between the surface layer and the layers of chromium. When both have about the same roughness values, the "gloss" of a tungsten carbide coating looks different. The reason for this is the microtopography of the tungsten carbide coating, which is created when the surface is machined, and especially when polishing. When used on a calender roll, for example, this can have a positive effect on the removal properties of a film. When used in the plastic film industry, a careful analysis of the application in which it will be used is essential for the selection of a coating because chemical interactions with the product, for example a PVC film, play a significant role in addition to the purely mechanical characteristics.
In this complex range of applications, it has been observed that the commercial characterization of a HVOF sprayed tungsten carbide layer using values such as low load hardness or porosity is not an adequate description. According to these observations, tungsten carbide layers with the same values have exhibited completely different behavior in actual applications.

2.1 Film calender boundary conditions

The requirements placed on a calender roll in the plastic film industry are oriented to a great extent on the quality, the material, and the thickness of the film to be produced. In terms of the surface roughness, the gloss level and the machining processes required to achieve it are often prescribed in addition to the roughness values such as Ra, Rz, and Rt. In many cases, the number and size of potential defects on the surface of a roll are defined together with the user. This is crucial when manufacturing very high quality films and requires a great deal of experience. **Table 1** lists examples of typical requirements for a calender roll used in the plastic film industry that were derived from the application of galvanic hard chrome layers. The values contained in the table clearly show the high requirements to be fulfilled.

Tab. 1: Typical basic data of calender rolls

Calender roll diameter (mm)	500 – 800
Barrel length (mm)	1500 - 2800
Weight (to)	5 – 15
Mean roughness index Ra (µm)	up to 0.02
Aver. surface roughness Rz (µm)	0.3 – 0.5
Max. surface roughness Rt (µm)	0.3 – 0.5
Shape/circular runout tolerance	≥ 2 µm
Material	Steel

2.2 Selection of the HVOF process used

For this application the GTV KII method with a typical WCCoCr 86/10/4 powder was selected based on experience regarding the stability of the process and the level of quality of the layers produced. During the development and testing phase, only commercially available powders manufactured through agglomeration and sintering were processed. The typical morphology of such a powder is shown in **Fig. 3**.

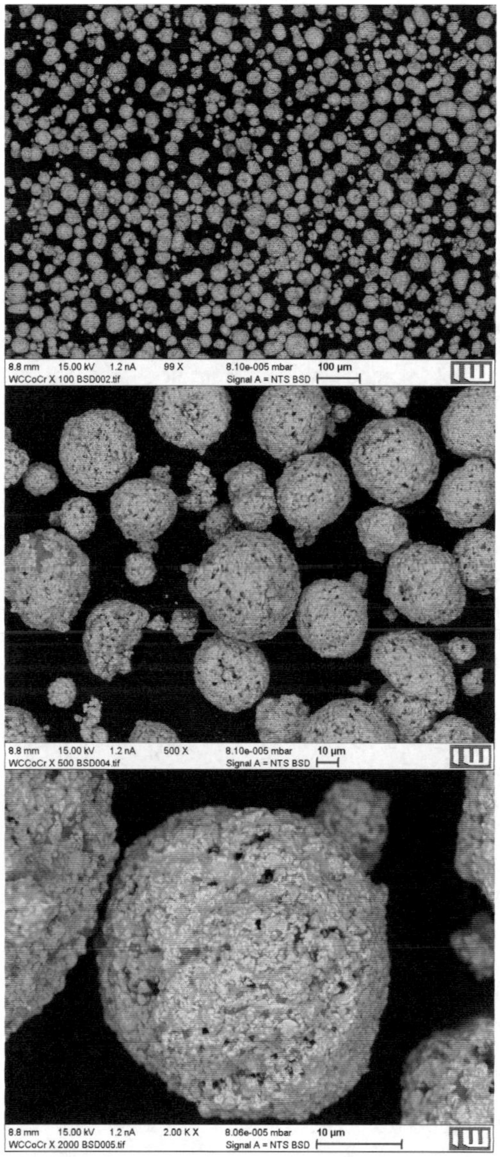

Fig. 3: SEM-view of a typical agglomerated and sintered tungsten carbide thermal spray powder

According to initial experience gained from applications, it was clearly recognized that not all powders exhibited the same behavior in spite of having identical quality characteristics such as the particle size distribution, chemical composition, and bulk density, and that the layers produced from the powders had different properties even though identical process parameters were used.

3 Analysis methods and results

To determine the powder and layer properties, classic metallography using incident light microscopes and scanning electron microscopes as well as quantitative phase analysis using X-ray diffractometry according to the Topas-Rietveld method were applied. This made it possible to detect even the finest differences in the structure of the powder and the layer, and therefore to specify additional requirements for the coating on the roll. In addition to these methods, low load hardness measurements according to HV0.3 [6] were taken on the layers and the hardness indentations were characterized in the scanning electron microscope in terms of crack formation. The individual analysis methods are not presented here in detail. Refer to the relevant literature for descriptions of these methods.
First, three different spray powders of type WCCoCr 86/10/4 with a nominal particle size of -45+15 µm were analyzed. They differed already during the formation of the phase structure in spite of having an identical morphology, manufacturing process, and composition. After the spraying process in a GTV KII HVOF system with identical parameters, the differences were even more apparent and it was seen that the metallic phase of cobalt and chromium was completely converted by the process. Furthermore, a high amount of amorphous phases is produced due to the process conditions, and the production of these phases is promoted even further by a higher sub-sieve spray powder. The **table 2** shows an overview of the results of these analyses.

Tab. 2: Quantitative X-ray diffraction phase analysis (weight percent) of used powders and coatings of them, produced with GTV II gun and identical parameters

	WC	Co-Cr	W_3Co_3C (eta-phase)	W_2C	Amount of amorphous phases
Powder 1	74	18	8	-	-
Powder 2	70	13	17	-	-
Powder 3	79	12	7	-	2
Coating 1	68	-	2	6	24
Coating 2	65	-	-	9	26
Coating 3	53	-	-	11	36

In the coating process ("layer formation"), the metallic binder phase Co-Cr (fused) forms an alloyed and therefore increased amount of melt in a chemical-metallurgical reaction with the WC (solid) and the eta phase W_3Co_3C (solid), from which a small amount of the newly created fused carbide phase W_2C crystallizes and the remaining melt solidified while cooling (quenching) and formed an amorphous phase of W-C-Co-Cr (metallic glass).
A larger amount of sub-sieve powder and/or a larger amount of eta phase W_3Co_3C in the powder in the standard HVOF processes led to an increase in the amount of amorphous phase as well as an increase in the amount of fused carbide phase W_2C in the layer structure.
By optimizing the HVOF process parameters, it is possible to minimize the amount of amorphous phase as well as the amount of fused carbide phase, and therefore maximize the amount of tungsten carbide WC.
All layers analyzed showed no significant differences in terms of hardness and porosity. Nevertheless, the analysis of the Vickers hardness indentations with the scanning electron microscope showed clear differences in crack formation on the edges. The characteristics of the cracks caused by the hardness indentations correlated in this case with the amount of wear determined in the abrasive wear test.

- The lowest amount of wear was found in the samples without crack formation on the hardness indentation.
- Simple cracks on the edge of the hardness indentation led to an average amount of wear.
- Strong radial cracks and cracks on the edges of the pyramid indentation led to a high amount of wear, **Fig. 4**.

Fig. 4: HV0.3 intendation onto HVOF tungsten carbide coating, cracks on the edges of the pyramid indentation led comparatively to a higher amount of wear

4 Summary

Based on the current status of the analyses, the following conclusions can be drawn from the results:
The powder should have the smallest possible amount of sub-sieve particles and an adequately large amount of metallic Co-Cr phase while simultaneously having the smallest possible amount of eta phase.
The spraying process should be performed so that the amount of amorphous phase and the amount of fused carbide phase W_2C is minimized in the structure. Furthermore, a lower, yet sufficiently high amount of eta phase W_3Co_3C should be present, and the amount of WC phase should be as high as possible.
Based on these results, it was possible to successfully produce tungsten carbide layers on some calender systems instead of hard chrome layers. This substitution has proven very effective in actual applications.

5 Outlook

Further analyses will be conducted to completely describe the metallurgical interactions and use this information as a basis for further optimizations. In this context, a non-destructive high frequency eddy current testing method will be developed with the ability to check the quality of fully processed layers on a roll as well as characterize and evaluate the coating results directly, **Fig. 5**. Initial results are already available and show the great potential of this measurement technology.

Fig. 5: High frequency eddy current test of a calender roll

6 Literature

[1] Scheller, K.M.: Der weltweite PVC–Markt, Würzburg, 28.06.2016
[2] Kohlert, C.: internal information, Montabaur 2016
[3] Lampke T.: Ressourceneffizienz durch Oberflächen- und Beschichtungstechnik. In: R. Neugebauer (Hrsg.) Handbuch Ressourceneffiziente Produktion, München, Wien: Hanser, 2014, ISBN 978-3-446-43008-2, 685 – 686
[4] Wielage, B.; Grund, T.; Rupprecht, C.: Thermisches Spritzen – Überblick, Einordnung, Anwendung. ZVO Report, 4 (2007), 36 – 38
[5] Bach, F.-W.; Möhwald, K.; Laarman, A.; Wenz, T.: Moderne Beschichtungsverfahren. Wiley-VCH Verlag GmbH & Co KGaA, Weinheim, ISBN 3-527-30977-2, 131 – 135, 161 – 177
[6] Tillmann, W.; Hussong, B.; Erne, M.; Maier, H. J.: Qualifikation der Poren- und Eindruckhärtemessung an thermisch gespritzten Schichten. Thermal Spray Bulletin 65 (2013), /1/, 30 – 37

Thermally Sprayed Hydrodynamic Main Bearings for Wind Turbines

Kirsten Bobzin, Mehmet Öte, Tim Königstein, Wolfgang Wietheger, Surface Engineering Institute (IOT),
Tim Schröder, Georg Jacobs, Dennis Bosse, Chair for Wind Power Drives (CWD),
RWTH Aachen University, Aachen / D

At present, main bearings in wind turbines are equipped with rolling bearings without exception. Sliding bearings instead can offer a number of advantages, including easier maintenance and extended lifetime. While conventional manufacturing processes for large sliding bearings face their limits regarding processable materials, thermal spraying can provide an effective alternative to meet the requirements by applying coating systems on the bearing surfaces. Within this study a wide range of different feedstock materials based on standard bearing materials and common wear and friction reducing coating systems are investigated. The coatings are tested on tribometers based on the load distribution within the main bearing at critical operating conditions of the wind turbine gained from a validated simulation model. A tribological methodology is developed to investigate the application related properties of the thermally sprayed coatings. The effects of load and geometry of the counter body on the friction and wear behavior of the coatings are investigated using a pin-on-disc and a modified highload ring-on-disc tribometer. The presented results provide a major contribution to the purpose of identifying an appropriate coating system to meet the requirements of slow-moving and highly loaded sliding bearings.

1 Introduction

For the main bearings of wind turbines (WT), rolling bearings are currently used without exception [1]. The failure rate of these rolling bearings is relatively low. Nevertheless, they are co-responsible for high down times since the disassembly of the whole rotor is often necessary for the replacement and a time-consuming use of a crane is required [2, 3]. In terms of replaceability, sliding bearings provide a major advantage regarding the structural design as they can be designed in a segmented way. This allows a fast and economic replaceability without the need of the rotor disassembly.

Main bearings in WT are exposed to a wide variety of different load cases, which lead to special requirements for sliding bearings and their materials. During the start-up process the shaft rotates slowly with adverse impact on the hydrodynamic operation conditions. In addition, the bearing is subjected to high dynamic loads during operation at a comparatively low circumferential speed and even high loads in parked position.

The use of a sliding bearing under these special conditions requires innovative material concepts and production processes. One innovative approach to meet these special challenges is the application of new bearing materials as a coating on the functional surfaces. In this context thermal spraying (TS) offers high flexibility since the properties of the functional surface can be adjusted by the choice and combination of different materials and can be modified according to the application. In contrast to conventional methods, material concepts which cannot be manufactured by metallurgical fusion can be produced by means of TS. For example, solid lubricants can be integrated into the material matrix. In addition, the coating topography can be influenced e.g. by number, size, distribution and orientation of pores, which serve as a lubricant reservoir and thus can improve the tribological properties. TS also allows the application of coatings to large freeform surfaces with consistent coating quality. As a result, the production of large bearings with even complex functional surfaces would be possible.

2 Methodology

In this paper the requirements for the sliding bearing are analyzed and a bearing concept promising easy replaceability and robustness against the load situations in WT is designed. To calculate the surface pressure distribution on each sliding segment a load analysis based on standard load cases is conducted. A recursive design optimization ensures surface pressures in the range of typical sliding bearing applications.

In a comprehensive parameter development, various coating systems were developed to meet the special requirements for a slow-running and highly loaded sliding bearing. Coating systems were subjected to model tests on two different tribometers in regard to friction and wear performance. Creep deformation tests were accomplished under a long term exposure of stress and pulsating tests were carried out in order to analyze the adhesion strength of the coatings.

3 Investigation

In order to develop a material concept, the special requirements concerning the load situations in WT are determined by the CWD. On the basis of the bearing design, coating concepts are developed by the IOT and being examined by both institutes together.

3.1 Requirements analysis

Since the main shaft bearing is subjected to complex loads resulting from the wind impinging on the rotor, it is necessary to conduct a load analysis regarding all important load cases. The IEC 61400 standard provides design requirements to ensure an appropriately engineered WT and specifies boundary conditions such as wind conditions. Based on this, eleven design load cases (DLC) were identified as relevant, since these load conditions represent the highest loads as well as scenarios with a slow rotating shaft. These DLCs provide detailed information such as wind velocity (v_{Wind}), turbulence, direction of the wind and gusts. Within this study the following conditions, as shown in **Fig.1**, were considered.

DLC	Wind Conditions - - IEC61400 - -	v_{Wind}
1.1	Normal Turbulence Model [NTM]	12 m/s
1.1	Normal Turbulence Model [NTM]	25 m/s
1.3	Extr. Coh. Gust w. Dir. Change [ECD]	12 m/s
1.3	Extr. Coh. Gust w. Dir. Change [ECD]	25 m/s
1.4	Normal Wind Profile [NWP]	25 m/s
1.5	Extreme Operating Gust [EOG]	12 m/s
1.5	Extreme Operating Gust [EOG]	25 m/s
3.1	NWP - Start up [NWPs]	12 m/s
5.1	NWP - Emergency shut down [NWPe]	12 m/s
6.1	Extreme Wind Model, Parked [EWM]	52,5 m/s
6.1	Extreme Wind Model, Parked [EWM]	30 m/s

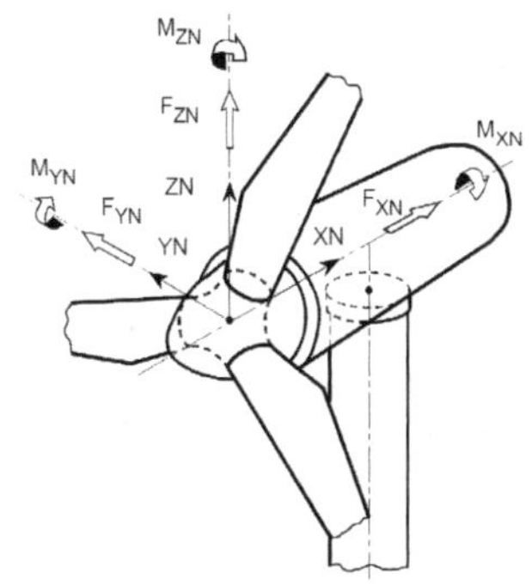

Fig. 1. Calculated wind conditions and considered 6-DOF loads on the rotor shaft flange [4]

Wind fields according to these DLCs were calculated with the aid of the software Turbsim and Matlab. Besides a Mutibody simulation (MBS) model of a 850 kW WT was created with the aid of the simulation tool Simpack. This MBS model contains the tower, the rotor including the hub and the nacelle with actual geometry. The gearbox and the generator are simplified by considering their mass and inertia. By means of NRELs software library AeroDyn the created wind fields were applied on the MBS model, to calculate the 6-dimensional loads on the rotor shaft flange. A co-simulation with Matlab Simulink ensured realistic system behavior by controlling the pitch angle depending on the wind speed. As a result of the calculation all forces and bending moments on the rotor shaft flange from the DLCs, as shown in **Fig. 1** (bottom) concerning all six degrees of freedom (DOF), can be obtained.

The simulation revealed that especially during extraordinary events, like extreme coherent gusts with tilted inflow, the forces and bending moments reach maximum values. Also, which represents additional stress for the bearings, the loads are relatively high during start ups or emergency stops where the rotational speed is comparatively low. Low rotational speeds reduce the load capacity of sliding bearings significantly [5]. **Tab. 1** summarizes the calculation results by showing the highest forces and bending moments on the rotor shaft flange considering all described DLCs in **Fig. 1**.

Tab. 1. Summary of the calculated 6-DOF loads on the rotor shaft flange

abs. Max. Load	DLC	Fx [kN]	Fy [kN]	Fz [kN]	Fyz [kN]	Mx [kNm]	My [kNm]	Mz [kNm]	Myz [kNm]
Fx	NTM 12m/s	127	2	77	77	318	3	21	21
Fy	ECD 25m/s	3	71	84	110	62	25	60	65
Fz	NTM 25m/s	59	3	102	102	341	47	147	154
Fyz	ECD 25m/s	3	71	84	110	62	26	59	64
Mx	NTM 25m/s	68	7	70	70	452	52	36	63
My	ECD 12m/s	34	35	84	91	47	743	240	781
Mz	ECD 12m/s	14	33	75	82	31	462	417	622
Myz	ECD 12m/s	34	37	84	91	50	738	271	786

3.2 Bearing design

Besides the common bearing design with two axial and two radial sliding bearings, the design as a conical sliding bearing offers the opportunity to support axial and radial forces as well as bending moments with just two sliding surfaces needed. Therefore it is suitable to withstand the 6-dimensional loads in the WT with minor space requirement. It also gives prospect to easy maintenance and replaceability because of the opportunity to accomplish the sliding area in segments. **Fig.2** shows the principal design of the considered bearing as a concept in a 850KW WT.

Fig. 2. Conical Sliding Bearing Concept shown in a system test rig for a 850 kW WT

3.3 Pre-dimensioning of the bearing

The calculation of the expected loads on the WTs rotor shaft flange enables the pre-dimensioning of the sliding bearing. Pre-dimensioning standards exist for conventional sliding bearings like radial or axial sliding bearings, whereby the mean surface pressure must not exceed the materials limits. To calculate this mean surface pressure, the associated surface needs to be known, which is for radial sliding bearings the projected surface of the shaft or the direct contact surface for axial sliding bearings. For conical sliding bearings such standards do not exist, but by the reason of the multi axiality of the loads in the WT and the resulting load distribution on the bearing it is possible to calculate the direct surface pressure on each sliding segment similar to the calculation of axial sliding bearings. **Fig. 3** (left).

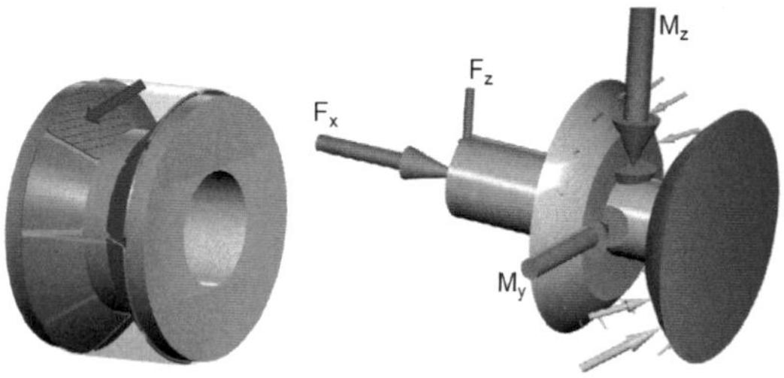

Fig. 3. Load on contact surface (left) and applied loads and resulting forces on bearing surface (right)

In a certain geometry of the bearing the surface of the sliding area is a known quantity, so now the perpendicular forces on each segment resulting from the wind loads have to be calculated in order to determine the surface pressure. This determination was realized in a contact analysis, where a contact point between every segment and the housing was created. These contact points were design to transfer only perpendicular forces. At nominal wind conditions an averaged sliding velocity of v = 1.05 m/s was calculated.

To determine the worst case scenario, three static load scenarios from the calculated shaft flange loads from Tab. 1. were chosen: The load scenario with maximum axial force, the scenario with maximum radial forces and the scenario with maximum bending moments. The applied loads and the resulting perpendicular forces are illustrated in **Fig. 3** (right). In a recursively process the geometry of the bearing, considering the diameter, the angle of sliding surface and the supporting width, was changed and the forces were calculated at each change again. As a result a bearing geometry was determined where the surface pressure at the highest stressed segment is in the range of customary sliding bearing applications. This range typically lies between p = 5 and 25 MPa [5]. With the chosen geometry and the highest perpendicular force working on the segments surface, a surface pressure of p = 7.6 MPa was calculated. To meet the requirements of a slow moving and high loaded sliding bearing new bearing materials are developed by the use of thermal spraying, considering this limit.

3.4 Coating development

Due to the small geometry of the torch and the flexibility of the particle velocities and temperatures which can be achieved, atmospheric plasma spraying (APS) is particularly suitable for the production of large sliding bearing coatings and for generating functional topographies which can be adjusted by process parameters [6]. This allows, for example, the optimization of the emergency running characteristics of the bearing surface, since the resulting pores can specifically improve the tribological properties so that the solid body contact occurring in real operation results in reduced wear. For the cylinder liner of combustion engines, it is already known that the tribological properties in the sliding contact can be positively influenced by the surface structure of thermally sprayed coatings. A reduction in the friction coefficient of 6.8% compared to cylinder liners made of gray cast iron could already be achieved on engine test bench [7].

The aim of the coating development is to ensure a high wear resistance of the tribological system. An additional focus lies on the emergency running and friction characteristics. These objectives should be met by a targeted material composition, coating morphology and topography as well as the manufacturing process in the best possible compromise. On the basis of tribological, surface and material engineering considerations, innovative coating systems have to be developed. Within this study, different copper and aluminum base alloys were developed as "soft" coatings, cobalt base alloys as comparatively "hard" coatings, but also so-called matrix materials, as shown in **Fig. 5**. Hard materials or friction-reducing solid lubricants are added to these matrix materials.

On the basis of standard bearing materials like aluminum and copper based alloys and also common wear and friction reducing coating systems like cobalt base alloys, lead-free feedstock materials were selected. To ensure good bond strength of the coating to the substrate, samples of C10E steel were pretreated by grid blasting with corundum. The coatings were then applied automatically by the atmospheric plasma spraying system TriplexPro™-210 from Oerlikon Metco (Winterthur, Switzerland) with a 9 mm nozzle within a closed spray booth by a robot. For optimal coating results, relevant process parameters such as the magnitude of electric current, gas composition, gas flow rate, powder feed rate, robot velocity, nozzle geometry and the spray distance have been varied. Subsequently a mechanical post-processing of the samples was carried out by means of a cup wheel grinding machine in order to produce a defined surface profile for the tribological investigations.

After an extensive parameter development, four coating variants were successfully manufactured on the basis of various spray materials. The investigated coating variants are listed in **Tab. 2a)** according to their increasing microhardness. The last row represents a reference coating material produced by conventional casting. The variants are assigned to the three different coating concepts, 'soft', 'hard' and matrix. The parameters of the APS process are shown in **Tab. 2b)**.

Tab. 2. Examined coating variants
Coating concept: -1→'soft' -2→'hard' -3→Matrix

a) Material composition, Microhardness, Roughness

Var.	Composition [wt-%]	Hardness [$HV_{0.1}$]	Roughness Ra [µm]
A -1	Al12Si	90±23	0.12
B -1	Cu20Sn1Al	184±26	0.11
C -3	Co25Cr5Al0.27Y1.75Si15hBN	204±137	0.48
D -2	Co28.5Mo8.5Cr2.6Si	465±95	0.05
E -*	Al40Sn *Reference material	35±3	0.12

b) Process parameter

Var.	Current [A]	Ar [nlpm]	N_2 [nlpm]	Feed rate [g/min]	Spray distance [mm]	Surf. speed [m/s]
A -1	360	60	4	50	120	1
B -1	360	60	4	50	140	0.5
C -3	460	60	4	40	120	0.25
D -2	500	60	4	55	100	1

The coatings morphology was characterized by means of light microscopy and scanning electron microscopy (SEM). Based on this, in addition to the coating morphology and the porosity, the quality of the coating adhesion to the substrate can be assessed. In addition, the microhardness was determined for each coating system at ten measuring points in the cross section by the hardness testers Micromet 1 from Bühler (Lake Bluff II, USA).

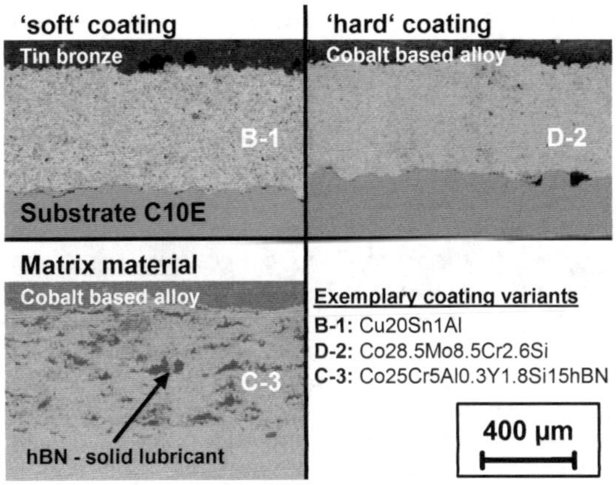

Fig. 5. SEM images of the different coating concepts

3.5 Tribological investigation

A pin-on-disc tribometer from CSM Instruments (Graz, Austria) and a modified highload ring-on-disc tribometer TRM 5000 from WAZAU (Berlin, Germany) were used for tribological analysis. The different material concepts are characterized in terms of friction and wear. The tribological pre-tests were carried out on coated flat samples in an oil bath of polyalkyleneglycol of the viscosity class ISO VG 100 with a temperature of T = 20 °C. The low oil viscosity was intentionally chosen to adjust a small lubricating film thickness. A ball made of 100Cr6 with a diameter of Ø = 6 mm was brought into contact with the coated samples with a normal force of F = 5 N in accordance with GfT Worksheet 7 [8]. This results in hertzian pressures of approx. $p_H \sim 1$ GPa, depending on the material variant. At a sliding speed of v = 0.1 m / s, a total distance of $s_2 = 500$ m was traveled. After a running-in distance of $s_1 = 450$ m the friction coefficient were recorded and the wear volume of the coating and counter body were evaluated at the end of the test. These values are averaged in **Tab. 3a)**.

Tab. 3. Tribological results of the coating variants

a) Pin-on-disc

Var.	μ_{PoD} [-]		$Wear_{Coating}$ [mm³]		$Wear_{CounterBody}$ [mm³]	
A -1	0.042	+	0.8440	–	0.0259	–
B -1	0.045	+	0.0004	+	0.0001	+
C -3	0.050	+	0.0000	+	0.0005	+
D -2	0.093	–	0.0001	+	0.0000	+
E -*	0.031	+	0.0417	–	0.0000	+

b) Ring-on-disc

Var.	μ_1 [-]	μ_2 [-]	μ_3 [-]	μ_4 [-]	μ_5[-]	$Wear_{Coating}$ [mm³]	
A -1	0.110	0.109	0.089	0.087	*Stop	46.0059	–
B -1	0.096	0.078	0.064	0.057	0.046	0.5089	+
C -3	0.099	0.084	0.073	0.068	0.055	0.3420	+
D -2	0.116	0.080	0.069	0.072	0.067	0.0925	+
E -*	0.116	0.066	0.067	0.050	0.049	10.8594	–

In addition to the reference material, the low friction coefficient (μ_{PoD}) of the tin bronze alloy B-1 and the aluminum base alloy A-1 are particularly remarkable. The variant C-3 shown in **Fig. 5** with integrated hexagonal boron nitride as a solid lubricant exhibit lower coefficient of friction in comparison to the cobalt based material without solid lubricant inclusions (D-2).

In addition to the coefficient of friction, wear characteristics are also important to assess the suitability of the coating systems for the application. After conducting the tribological tests, the coatings were analyzed by a confocal laser scanning microscope (CLSM) from Keyence (Osaka, Japan) to examine the material removal in the contact area and at the counter body (ball). The results are listed in **Tab. 3a)**. These show that as the microhardness increases, the wear on the coating and the counter body decreases. The high level of wear of the aluminum based coating A-1 and the reference coating E-* is particularly noticeable, while the copper (B-1) and cobalt based (C-3, D-2) coatings are exhibiting high wear resistance and load carrying capability. Here it must be noted that, in the case of tin alloys the recommended maximum surface pressure is as low as p_{max} = 15 - 20 MPa which is exceeded definitely in the Pin-on-disc test-setup which is used for the preliminary tests [9].

Therefore, in the context of this study, further experiments were conducted to realise the tribological tests under more realistic loading conditions and to provide supplementary information. For these tests a modified highload ring-on-disc tribometer was used, as shown schematically in **Fig. 6a)**. The test were performed with a ring-on-disc configuration with a steel counter ring made of the prospective alloyed steel shaft material 34CrNiMo6 (DIN EN 10083-3: 1.6582) with a surface area of A = 150 mm² under synthetic polyalphaolefine oil of the viscosity class ISO VG 320 at a temperature of T = 60 °C. This oil is a standard transmission oil for WT in order to renounce the need of an additional lubrication circuit.

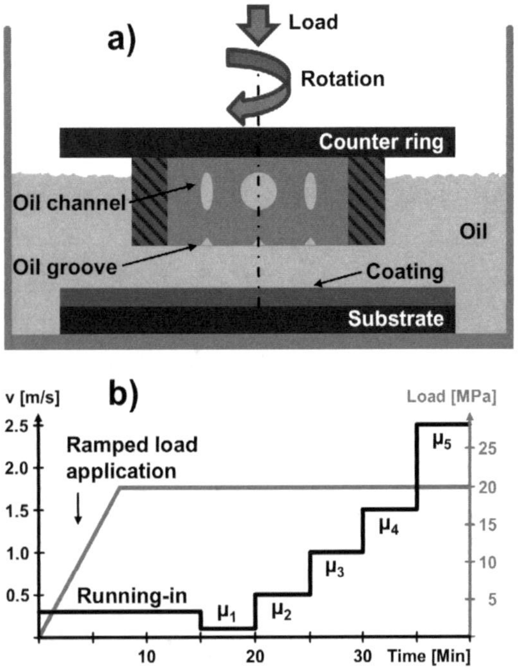

Fig. 6. Sketch of the highload ring-on-disc tribometer (a) and test strategy (b)

The test strategy was developed according to the expected loads and velocities inside the main bearing. The surface pressure was raised from the calculated pressure of p = 7.6 MPa to p = 20 MPa in order to measure the wear resistance of the coating variants and in order to reduce the duration of the test cycle. An initial running-in time was chosen in order to gain temporally constant measured values. Following steps were carried out, delivering five averaged coefficients of friction for five different relative velocities, as illustrated in **Fig. 6b)**:

0: Oil heating 60 °C
1: Running-in at ramped load application of p = 20 MPa at 0.3 m/s - 7.5 min
2: Running-in at 0.3 m/s - 7.5 min
3: Constant velocity 0.1 m/s - 5 min → μ_1
4: Constant velocity 0.5 m/s - 5 min → μ_2
5: Constant velocity 1.0 m/s - 5 min → μ_3
6: Constant velocity 1.5 m/s - 5 min → μ_4
7: Constant velocity 2.5 m/s - 5 min → μ_5
8: End - Wear determination

The positive behaviors of coating variants B-1 and C-3 with respect to the coefficient of friction and wear rate are confirmed by the highload tribometer tests in **Tab. 3b)**. The results are largely consistent with the preliminary pin-on-disc tests, whereby the test had to be stopped due to excessive torque for the rotation.

According to the tribological investigations, the coating variants B-1 and C-3 show the greatest potential for use in a WT. Due to the high hardness compared to the variant A-1, these coating variants promise a higher load bearing capacity. This is also accompanied by a better adaptability of these coating systems to the counter body compared to tougher variants such as D-2. In addition, B-1 and C-3 have a comparatively low coefficient of friction and a low wear volume.

3.6 Creep and pulsating compression test

Due to the fact that in this application of the bearing even high loads in idle are calculated by the simulation model, it is important to prove whether a material tends to creep deformation under a long term exposure of stress. Therefore creep tests were conducted. In this test a cylindrical stamp is used to strain the material with constant compression force, while the sample is heated to T = 120 °C in an oil bath. This temperature is above the expected operation temperature in the bearing application to accelerate the tests. First test results show, that especially the tin and aluminum based coatings tend to creep deformation under these conditions. Coatings based on copper or cobalt instead did not show any deformation within the test time of t = 168 hours. The tests are currently ongoing.

The thermally sprayed materials are not melted to an entire homogeneous alloy during the spraying process. Melted particles impinge on the surface separately resulting in a lamellar coating build-up. Thus it is important to check the cohesion within the coating as well as the adhesion strength between the thermally sprayed coating and the substrate surface. To check the adhesion, pulsating tests with compression force are executed on the thermally sprayed surfaces. Therefore small circular samples with a diameter of d = 25 mm were stressed with pulsating load between p = 10 MPa and p = 180 MPa at f = 55Hz in a hydraulic test bench. After n = 1 to 5 mil. load cycles the samples were dismantled and had to undergo an inspection for coating detachment and cracks in the surface by dye penetrant testing. To the present day, 4 thermally sprayed material variants are investigated and each of them reached the stop criteria of n = 70 mil. load cycles without any coating detachment or crack.

4 Summary

The completed load calculation of the rotor shaft flange for a variety of critical load cases and wind profiles (DLCs) enabled the pre-dimensioning of the conical sliding bearing. A capable geometry as result of this pre-dimensioning keeps the surface pressure in the range of typical sliding bearing applications. The presented material screening tests show a high potential of thermally sprayed sliding bearing materials for use in the main bearing of WT, whereby a comparison to different established bearing materials is necessary for a final evaluation. The creep tests showed that the selected materials and manufacturing processes do not tend to creep. According to the investigations, the tin bronze alloy B-1 and the cobalt based alloy C-3 with integrated hexagonal boron nitride show the greatest potential for use in a WT because of their comparatively low coefficient of friction and low wear volume.

5 Outlook

As a further step a hydrodynamic calculation of the bearing with the objective to optimize the detailed geometry will be conducted. Due to the high loads and the size of the bearing, overall system deformations in the scale of the bearing clearance are expected [10]. This leads to the necessity that the deformation of the bearing housing and the surrounding has to be considered in the detailed calculation.

Additional coating variants will be produced in the future by means of High Velocity Flame Spraying (HVOF), Cold Gas Spraying (CGS) and Wire Arc Spraying (WAS). The coating concept will be also supplemented by multilayer systems which are intended to protect against abrasion, especially by means of tin-based top coatings by embedding abrasive particles and having optimum running properties under emergency conditions. Additionally, the achieved tribological results are compared with the tribological properties of further already established sliding bearing materials.

To verify the capability of the new developed thermally sprayed material coatings for the application in a bearing, 3 to 5 most promising coating variants are going to be tested on a radial sliding bearing test rig. After this, the most suitable coating variant will be chosen to build up a full scale sliding bearing for the application in the 850 kW WT test rig, shown on **Fig.2.** The test rig is capable of testing the novel bearing under real conditions by using a hydraulic load application system which is able to emulate the wind loads.

The potential presented here of novel sliding bearing concepts with innovative coatings is particularly aimed at the requirement for a reliable wind energy. Beyond that, the knowledge acquired in the project on thermally sprayed bearing coatings allows long-term development of high-performance bearing materials which are not yet available, future-oriented and environmentally friendly, which could also be used in other fields of application.

6 Acknowledgements

The presented investigations are part of the joint project "Thermally sprayed coatings of journal main bearing in wind turbines - Wind turbine drivetrain and surface engineering – WEA-GLiTS", promoted by the German Federal Ministry for education and research in the range of the program “Energy and climate fund" (grant ID 03EK3036A).The Authors would like to thank the German Federal Ministry of Research and Education for funding the “WEA-GLiTS” project.

SPONSORED BY THE

Federal Ministry
of Education
and Research

7 Literature

[1] Hau, E: Wind Turbines, Fundamentals, Technologies, Application, Economics, Springer, Berlin, 2013

[2] McMillan, D.; Ault, G. W.: Quantification of Condition Monitoring Benefit for Offshore Wind Turbines in Wind Engineering, Volume 31, No. 4, 2007, p. 267-285

[3] Schröder, T.; Jacobs, G.; Bosse, D.; Bobzin, K.; Öte, M.; Königstein, T.; Wietheger, W.: Thermisch gespritzte Gleitlagerwerkstoffe für die Rotorlagerung von Windenergieanlagen in Ingenieurspiegel, Volume 4, 2016, p.28-29

[4] Guideline for the Certification of Wind Turbines, Germanischer Lloyd, 2010

[5] Jacobs, G.: Maschinengestaltung, Part 2, 2016

[6] Mang, T.; Bobzin, K.; Bartels, T.: Industrial Tribology: Tribosystems, Friction, Wear and Surface Engineering, Lubrication, Wiley, 2010

[7] Bobzin, K.; Ernst, F.; Zwick, J. et al.: Coating Bores of Light Metal Engine Blocks with a Nanocomposite Material using the Plasma Transferred Wire Arc Thermal Spray Process in Journal of Thermal Spray Technology, Volume 17, Issue 3, Sep. 2008, p. 344-351

[8] GfT - Society for Tribology e.V., Arbeitsblatt 7: Tribologie, 2002

[9] VDI (Association of German Engineers) Guideline 2204 Part 1: Design of plain bearings, Basic instructions, 1992

[10] Koch, D.: Dynamische Wälzlagersimulation Einsatz als virtueller Lagerprüfstand, Schaeffler Technologies AG & Co. KG, ATK, 2013

Tribological behaviour of flame-sprayed HDPE/Cu composite coatings

Y.T. Xu, Y. Liu, Z.M. Jia, X.K. Suo, H. Li, Ningbo / CN

Electroless plating was employed for making copper (Cu)-high density polyethylene (HDPE) core-shell particles for following coating deposition by flame spraying. Our previous works already reported large-scale fabrication of HDPE/Cu composite coatings against corrosion, biofouling and ageing for marine applications. In this work, we further investigated tribological behaviours of the HDPE and the HDPE/Cu composite coatings. The structure of the composite coatings was designed via controlling the thickness of the copper shell coated on the composite powder. The result shows that the addition of copper slightly decreased the anti-wear property of the composite coating. The tribology mechanisms of the composite coating and the HDPE coating were discussed.

1 Introduction

Polymer are used in a range of applications because of high specific strength, reducing friction, excellent chemical stability, easy processing [1]. However, the poor thermal stability and mechanical property still limit their applications in many industries [2]. Ceramic, metal and dissimilar polymer are usually added in a polymer matrix to improve these properties [3-5]. Among these kinds of composite materials, metal/polymer composite materials present a promising application due to their combination mechanical and thermal properties of the metal. N. Espallargas et. al fabricated a composite coating containing of polymer capsule and Fe-based alloy matrix, and reported the lubricating property of the coating [6]. J.M. Chen et. al also designed a metal/polymer composite surface using poly-(p)-oxybenzoyl and aluminium bronze, and reported a surface with a low friction coefficient and a high sliding wear resistance [7]. In the previous studies, we also reported a metal/polymer composite with HDPE and copper [8, 9]. The composite coating presents high efficiency antifouling and anti-ageing property with a help of the copper inclusion.

Thermal spraying is one of the process fabricating the metal/polymer composite [10, 11]. In thermal spraying, three composite techniques are usually used to tailor the composite coating. The powders are deposited separately but simultaneously by different guns [12] or a mechanical mixture of different powders is employed [13]. Another method to the deposit composite is through spraying the composite powder, which are usually compounded by mechanical alloying [14], electric plating [15] and electroless plating [16]. However, for the metal/polymer composite, few reports on the composite preparations via powder design are reported.

Therefore, in this study, a novel technique route was introduced to design powder for fabrication the metal/polymer composite coating. The microstructure of the composite powder was designed through controlling the parameters of electroless plating. The sliding wear behaviours of the polymer and the composite coating were characterized, and the tribological mechanisms of the coatings were also discussed.

2 Experiments

2.1 Powder fabrication

HDPE powder (Korea Petrochemical Ind. Co., Ltd) was used as the starting feedstock. The size distribution of the powder ranged from 54 µm to 140 µm with an average size of 85 µm. The HDPE particles were uniformly coated with Cu thin films through electroless plating. The plating was carried out through immersing the particles in an electroless plating bath for 40 min at 50°C. The protocols employed for the powder pre-treatment, including degreasing, coarsening, sensitizing and activating, have been reported previously [8]. The reagents used for the powder pre-treatment were purchased (Sinopharm Chemical Reagent Co., Ltd., China) and used without any further purification. The thickness of the copper shell was controlling by changing the number of plating cycles. Mild steel plates (MS, E235B, 20 x 30 x 2 mm^3) were used as the substrates. Prior to the spraying, the steel substrates were cleaned with acetone, hydrochloric acid and deionized water, and then mechanically coarsened by blasting alumina.

2.2 Coating deposition

The HDPE-Cu powder was used as the feedstock for deposition of HDPE-Cu coatings using flame spraying (CDS 8000, Castolin, UK). Oxygen and acetylene were used as the combustion-supporting gas and the fuel gas with the pressure of 0.55 MPa and 0.075 MPa, respectively. The powder feed rate was set as 60 g/min. The standoff distance from the nozzle exit to the sample surface was 250 mm. The traverse speed of the gun was 200 mm/s.

2.3 Sliding wear test

Friction and wear experiments were carried out employing a ball-on-plate linear wear tester (Rtec, USA). The GCr15 steel ball of 6 mm in diameter was used, and ultrasonically cleaning in alcohol prior to the testing was necessary. Before the friction and wear tests, the coatings were ground, polished, decreased and cleaned. The surface roughness of the HDPE coating and the composite coating were 0.10 um and 0.12 um, respectively. The sliding tests were examined under the conditions: 5 mm reciprocating amplitude, 2 mm/s reciprocating velocity, 420 min sliding time (referring to 504 m sliding distance), 5 N normal load. The temperature and humidity during the test is 12° C and 72%, respectively. The friction coefficient was recorded automatically via the system. The volume loss of the plate was evaluated with a profile tester (Alpha-StepIQ, USA) and the specific wear rate (K, $mm^3/N \cdot m$) is calculated as K = V / L × S, where V is the volume loss (mm3), L is the applied normal load (N) and S is the total sliding distance (m).

2.4 Microstructure characterization

The cross-sectional microstructures and topographical morphologies of the coatings were characterized by field emission scanning electron microscope (FESEM, Quanta FEG 250, USA). Typical wear tracks of the tested specimens were also observed using FESEM. The microhardness of the samples were measured using microhardness tester (HV-1000, JMTT, Shanghai Jiming Measuring Equipment Co., Ltd., China). Ten measurements were carried out per sample in order to get a reliable data.

3 Results and discussions

The microstructures of the composite coatings with different plating cycles were shown in Fig.1. Fig. 1a, 1c, 1e and 1g show the morphology of the composite powder. Under 1 plating cycle condition, there are still some black contrast zones in Fig. 1a, corresponding to the HDPE surface. With the increase of the plating cycle, no black contrast zone is found, suggesting the HDPE particle is totally covered by the copper shell. The microstructures of the composite particles are presented in Fig 1b, 1d, 1f and 1h, in which the increase of the copper shell can also be observed intuitively. The thickness of the copper shell was calculated quantitatively, and the result is shown in Fig.2. Both Fig. 1 and Fig. 2 proves that controlling the relative content of the polymer component and the metal component is feasible via altering the electroless plating cycles.

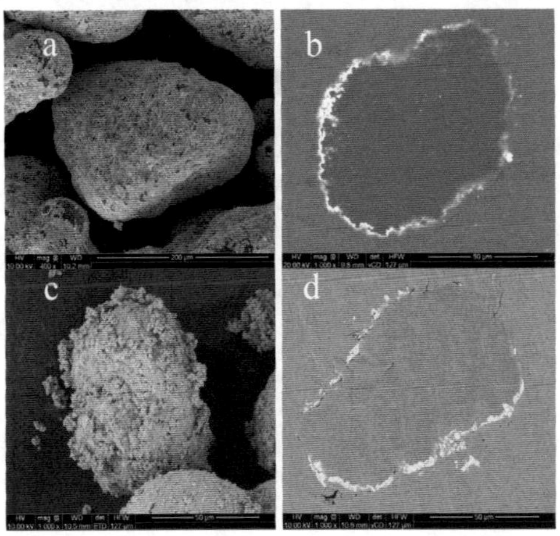

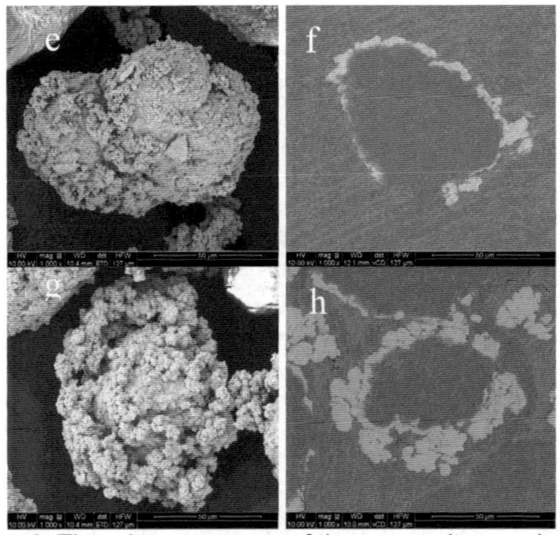

Fig. 1. The microstructures of the composite powders under different plating cycles: (a-b) one cycle, (c-d) three cycles, (e-f) five cycles, (g-h) seven cycles.

The HDPE and HDPE-Cu powders were deposited using flame spraying, ant the microstructures of the coatings are shown in Fig.2. It is found that both the HDPE coating (Fig. 2a) and the HDPE-Cu coating (Fig. 2b) present a dense structure. No crack and pore is found. Similar to the composite powder, there also are two contrast zones for the composite coating. White contrast zones corresponds to the copper element. It is also found that the copper element distributes along the boundary of the HDPE particles. The volume content of the copper element was also measured using image analysing. For the composite coating deposited using the composite powder with 1.1 μm copper shell, 9.57 ± 1.28 vol.% copper element is found. It is believed that the copper content in the composite would increase as the increase of the copper shell thickness. The microstructures of the coatings fabricated using the others composite powders will be investigated later.

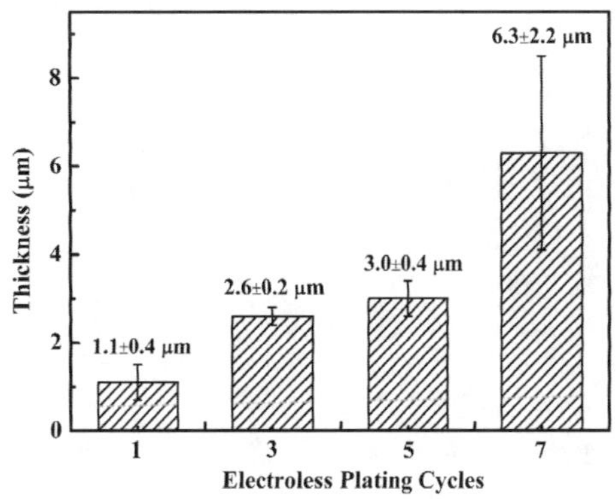

Fig. 2. The evolution of copper shell thickness as a function of plating cycles.

The mechanical properties including microhardness and wear resistance of the HDPE coating and the HDPE-Cu coating were also characterized (Table 1). The result shows that the microhardness of the

HDPE-Cu composite increases. As observed from Fig. 3, copper exists along the polymer particles, which contributes to the increase of the microhardness. Interestingly, the sliding wear resistance did not present significant variation for these two kinds of coatings. The friction coefficient and wear rate of the HDPE-Cu coating presents a slight augment.

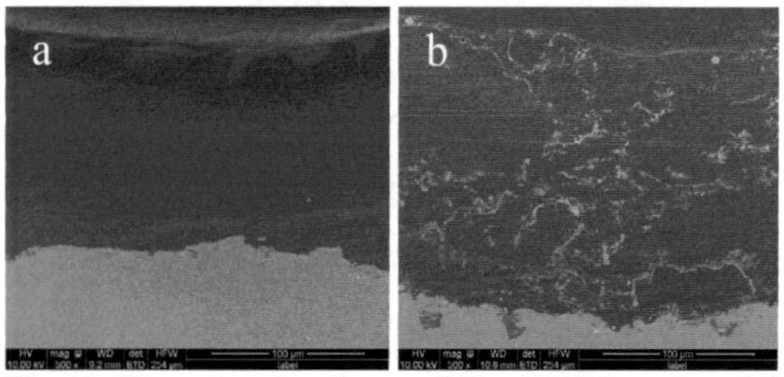
Fig. 3. The microstructures of the HDPE coating (a) and the HDPE/Cu composite coating (b).

Table1 Mechanical properties of the coatings.

	HDPE	HDPE-Cu
Microhardness	6.889±0.123	9.298±0.485
Friction coefficient	0.156±0.020	0.158±0.008
Specific wear rate (10^{-6} $mm^3/N{\cdot}m$)	4.083±2.054	4.392±1.112

The wear morphologies of the HDPE coating and the HDPE-Cu coating were investigated, and the results are shown in Fig.4. Deformation can be found both on the HDPE coating and on the HDPE-Cu coating, as the white arrows marked in Fig. 4a and Fig. 4c.

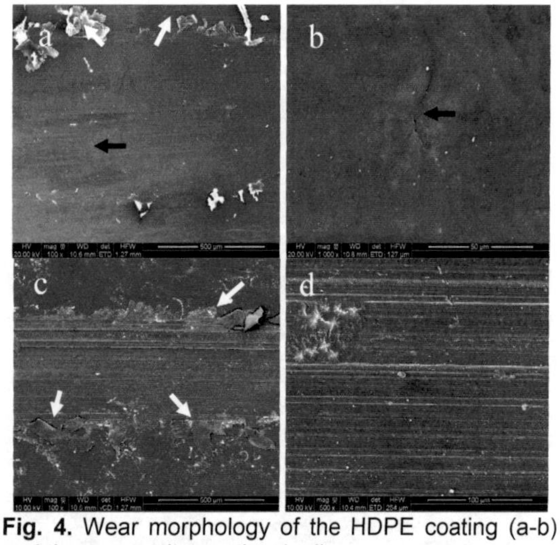
Fig. 4. Wear morphology of the HDPE coating (a-b) and the composite coating (c-d).

Some microcrack can also be found on the surface of the HDPE coating, suggesting that the wear mechanism is mainly fatigue wear. For the HDPE-Cu coating, no microcrack is found, instead of furrows, which is the characterization of abrasive wear. It is reduced that the copper shell broke and participated in wear. It is believe that the copper content may also influence the wear behaviour and mechanism, therefore, the wear behaviour of the composite coatings fabricated using the powders with different copper shell thickness will be investigated later.

4 Conclusions

The HDPE/Cu composite coating was fabricated using flame spraying, and its sliding wear behaviour was investigated. The main conclusions are drawn:

(1) The composite coating is dense, and copper element distributes along the HDPE particles.

(2) Copper inclusion did not alter the sliding wear resistance significantly but change the friction mechanism of the coating. For the HDPE coating and the composite coating, the friction mechanism is fatigue wear and abrasive wear, respectively.

5 References

[1] Wang, H. Liu, Z. Zhu, Y. Meng, Y. Wang, H. and Zhang, X: Facile preparation of superhydrophobic and high oleophobic polymer composite coatings with self-cleaning, heat-resistance and wear-resistance. Journal of Polymer Research 23 (2016), Issue 7, pp. 1/9.

[2] Schadler, L.S. Laut, K.O. Smith, R.W. and Petrovicova, E: Microstructure and mechanical properties of thermally sprayed silica/nylon nanocomposites. Journal of Thermal Spray Technology, 6(1997), Issue 4, pp. 475/85.

[3] Li, Y. Ma, Y. Xie, B. Cao, S. and Wu, Z: Dry friction and wear behavior of flame-sprayed polyamide1010/n-SiO_2 composite coatings. Wear, 262 (2007), Issue 9-10, pp. 1232/38.

[4] Yamane, M. Stolarski,T.A. and Tobe, S: Wear and friction mechanism of PTFE reservoirs embedded into thermal sprayed metallic coatings. Wear, 263 (2007), Issue 7, pp. 1364/74.

[5] Jacobs, O. Mentz, N. Poeppel, A. and Schulte, K: Sliding wear performance of HD-PE reinforced by continuous UHMWPE fibres. Wear, 244 (2000), Issue 1, pp. 20/28.

[6] Espallargas, N. and Armada, S: A new type of self-lubricated thermal spray coatings: Liquid lubricants embedded in a metal matrix. Journal of Thermal Spray Technology, 24 (2015), Issue 1, pp. 222/34.

[7] Chen, J. Yang, J. Zhao, X. An, Y. Hou, G. Chen, J. and Zhou, H: Preparation and properties of poly-(p)-oxybenzoyl/aluminum bronze composite coating by atmosphere plasma spraying. Surface & Coatings Technology, 253 (2014), Issue 9, pp. 261/67.

[8] Jia, Z. Huang, J. Gong, Y. Jin, P. Suo, X. and Li, H: Incorporation of copper enhances the anti-ageing property of flame-sprayed high-density polyethylene coatings. Journal of Thermal Spray Technology, 2017, DOI 10.1007/s11666-017-0525-x

[9] Jia, Z. Liu, Y. Wang, Y.Gong, Y. Jin, P. Suo, X. and Li, H: Flame spray fabrication of polyethylene-Cu composite coatings with enwrapped structures: A new route for constructing antifouling layers. Surface & Coatings Technology, 309 (2017), pp. 872/79.

[10] Niebuhr, D. and Scholl, M: Synthesis and performance of plasma-sprayed polymer/steel coating system. Journal of Thermal Spray Technology, 14 (2005), Issue 4, pp. 487/94.

[11] Niezgoda, S. Gupta, V. Knight, R. Cairncross, R. and Twardowski, T: Effect of reinforcement size on the scratch resistance and crystallinity of HVOF sprayed nylon-11/ceramic composite coatings. Journal of Thermal Spray Technology, 15 (2006), Issue 4, pp. 731/38.

[12] Rempp, A. Killinger, A. and Gadow, R: New approach to ceramic/metal-polymer multilayered coatings for high performance dry sliding applications. Journal of Thermal Spray Technology, 21 (2012), Issue 3-4, pp. 659/67.

[13] Tharajak, J. Palathai, T. and Sombatsompop, N: Morphological and physical properties and friction/wear behavior of h-BN filled PEEK composite coatings. Surface & Coatings Technology, 273 (2015), Issue 1, pp. 20/29.

[14] Jackson, L. Ivosevic, M. Knight, R. and Cairncross, R: Sliding wear properties of HVOF thermally sprayed nylon-11 and nylon-11/ceramic composites on steel. Journal of Thermal Spray Technology, 16 (2007), Issue 5-6, pp. 927/32.

[15] Iacovetta, D. Tam, J. and Erb, U: Synthesis, structure, and properties of superhydrophobic nickel–PTFE nanocomposite coatings made by electrodeposition. Surface & Coatings Technology, , 279 (2015), pp. 134/41.

[16] Narkis, M. Yacubowicz, J. Vaxman, A. and Marmur, A: Electrically conductive composites prepared from polymer particles coated with metals by electroless deposition. Polymer Engineering & Science, 26 (2010), Issue 2, pp. 139/43.

Hybrid powder-suspension Al_2O_3-ZrO_2 coatings by axial plasma spraying: Processing, characteristics & tribological behaviour

S Goel[a], S Björklund[a], U Wiklund[b], SV Joshi[a]
[a] Department of Engineering Science, University West, Trollhättan, Sweden
[b] Materials Science Division, Uppsala University, Uppsala, Sweden

The ability of suspension plasma spraying (SPS) to overcome difficulties associated with feeding of fine (submicron or nano-sized) powders and achieve more refined microstructures than possible in atmospheric plasma spraying (APS) is well established. In recent times, the use of axial injection plasma spray systems has yielded substantial enhancement in deposition rates/efficiencies due to improved thermal exchange between the plasma plume and injected feedstock. The present paper describes utilization of both the above advances in plasma spraying to create various function-dependent coating architectures through simultaneous and/or sequential spraying of hybrid powder-suspension feedstock. A specific variant of such hybrid axial plasma spraying that enables deposition of composite coatings by simultaneous injection of a powder and a suspension is discussed in particular detail. Results obtained using an Al_2O_3-ZrO_2 material system as a case study reveal that composite coatings combining the micron-size features arising from the spray-grade Al_2O_3 powder and submicron or nano-sized features attributable to the ZrO_2 suspension can be conveniently realized. The surface morphology, microstructure, and composition of these coatings, as well as their tribological behaviour determined using scratch and ball-on-disc tests, are presented herein. The utility of this method to develop a wide array of composite coatings is also discussed.

1 Introduction

Atmospheric Plasma Spraying (APS) has been an industrially well-accepted thermal spray technique for developing coatings for diverse applications for the past several years. However, in recent times there has been growing interest in developing more finely structured coatings than possible by conventional thermal spray routes in an effort to obtain improved properties. Early efforts in this direction, involving use of nano- or submicron sized powders to achieve more refined coating microstructures, typically resorted to use of agglomerated feedstock [1] to obviate the practical problem of feeding the fine powders into the plasma plume, thereby adding another processing step and increasing costs. Moreover, the results obtained using nanosized powders were often contradictory [2]. Thus, the above approach was largely uninteresting from an industrial standpoint.

Compared to the above attempts, two emergent variants of the plasma spray technology, namely Solution Precursor Plasma Spraying (SPPS) and Suspension Plasma Spraying (SPS), present elegant liquid-based thermal spray approaches to achieve finely structured coatings [3]. Despite its ability to deposit varied coatings with nano-sized features [4, 5], interest in the SPPS technique has so far been relatively muted because of narrow processing windows as well as due to apprehensions of low throughput, both of which raise concerns regarding its industrial adaptability. On the other hand, the versatility of the SPS technique for producing a wide variety of finely structured advanced coatings [6, 7] has now been demonstrated.

Another relatively recent development in thermal spraying pertains to advent of plasma spray systems that enable axial injection of feedstock [8], thereby permitting improved thermal exchange between the plasma plume and the injected feedstock. In contrast to conventional radial injection of feedstock and concomitant problems of penetration of fine powders into the viscous plasma plume [9], axial injection can enable substantial enhancement in deposition rate / efficiency, particularly in case of more energy demanding solution-based thermal spray processes. The availability of such systems capable of yielding higher throughputs [10] has also contributed to the considerable current excitement in the SPS technology which, unlike SPPS, does not rely on *in situ* formation of powder particles as an additional energy-consuming step for coating formation.

The array of metallic, intermetallic, and ceramic materials that are today readily available in spray-grade powder form from various suppliers is extremely vast. In view of the encouraging results that SPS research has yielded in the past few years, commercial interest in manufacturing suspensions specifically for spraying applications is also now growing. Appreciation of the above, plus the high energy capability of axial feed plasma guns that facilitate intimate energy transfer between the plasma plume and injected feedstock, have motivated the authors to explore hybrid plasma spraying using powder-suspension combinations. One of the authors has previously been part of studies involving similar hybrid processing using powders and solution precursors [3, 11–13]. However, the suspensions, being somewhat more forgiving in energy demand compared to solution precursors, *prima facie* provide a more attractive alternative. Use of a hybrid feedstock comprising a conventional 'coarse' spray-grade powder and a suspension of a 'fine' powder can be a convenient pathway to realizing unique function-dependent coating architectures. For example, sequential feeding of suspension followed by that of a powder (or vice versa) can yield a layered coating, while simultaneous feeding of both can yield a composite coating. In a specific variant of the simultaneous feeding arrangement, the relative feed rates of the two feedstocks can be continuously varied to obtain functionally graded coatings. Demonstrating the ability to generate such different architectures was one of the main goals of the present study. It was decided to investigate co-deposition of dual feedstock in particular detail because of challenges associated with simultaneous feeding of powder and suspension feedstocks. To the best of authors' knowledge, the present study is a first of its kind attempt involving sequential / simultaneous axial injection of powder and suspension feedstock.

2 Experimental

2.1 Coating architectures

Figure 1 illustrates the three different kinds of coating architectures that were sought to be generated during the present study adopting the powder-suspension hybrid plasma spraying approach, namely layered, composite, and functionally graded. The figure also depicts the material systems utilized in each case for demonstrating the above approach.

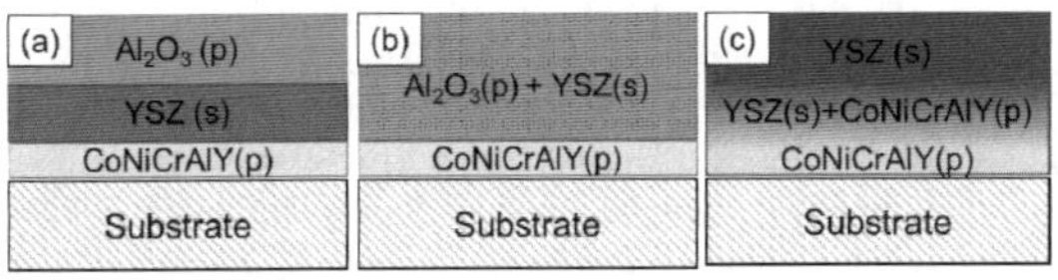

Fig. 1: Schematic illustration of different coating architectures generated using hybrid plasma spraying: (a) layered - by sequential spraying of suspension and powder; (b) composite - by simultaneous spraying of suspension and powder; and (c) functionally graded - by simultaneous spraying of suspension and powder, with continuously manipulated relative feed rates

2.2 Materials

The following feedstock materials were used: (a) 8 wt.% yttria stabilized zirconia (8YSZ) suspension in ethanol with 25wt% solid loading (Treibacher, Austria); (b) Al_2O_3 powder (AMPERIT® 740, H.C. Starck GmbH, Germany); (c) CoNiCrAlY powder (AMPERIT® 410, H.C.Starck GmbH, Germany). Hastelloy X® was the substrate material in case of layered and graded coatings, and Domex 355 steel for composite coatings.

2.3 Substrate preparation

Prior to coating deposition, the substrate surface was degreased and grit blasted with alumina grit of 80 µm size for good coating adhesion. A surface roughness of approximately R_a= 3 µm was obtained after grit blasting. Substrates were pre heated to ~150°C prior to coating deposition.

2.4 Coating deposition

An axial feed capable high power plasma torch (Mettech Axial III, Northwest Mettech Corp., Vancouver, Canada) was used for coating deposition. The above system was integrated with a Nanofeed 350 suspension feeder as well as a powder feed system (Model PF50WL, Uniquecoat, Richmond, USA). An illustrative schematic of the overall feeding arrangement for the powder-suspension hybrid feedstock is shown in Fig. 2.

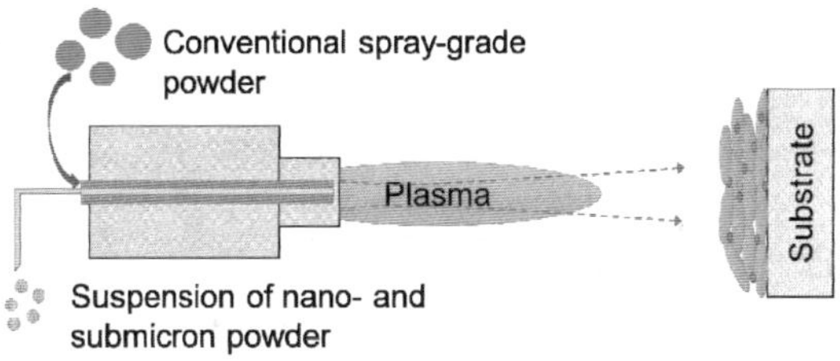

Fig. 2: Arrangement for simultaneous and/or sequential feeding of conventional spray-grade powder and suspension feedstocks

In case of the layered coating sought to be illustrated in the present work, sequential spraying of an 8YSZ suspension and an Al_2O_3 powder was resorted to on a previously CoNiCrAlY bond-coated substrate. A 8YSZ layer was first deposited using a suspension feedstock, followed by a top layer of Al_2O_3 deposited by seamlessly switching to a powder feedstock. Composite coatings were realized by using simultaneous injection of Al_2O_3 powder and 8YSZ suspension. The graded coating architecture was generated by simultaneous deposition of CoNiCrAlY powder and 8YSZ suspension, with the powder feed rate gradually varied at the end of each coating pass from 80 to 0 g/min, while the suspension feed rate was kept constant at 100 ml/min once the suspension feeding was initiated after deposition of the CoNiCrAlY bond coat. However, the concept itself is not limiting in any manner and graded coatings can be achieved by controlling relative feed rates by manipulating feeding of either or both feedstocks.

2.5 Coating characterization

Sample preparation for microstructure investigation involved cutting of coated specimens using an alumina cutting blade and mounting them in a low viscosity epoxy resin to observe the cross sections. After cold mounting the transverse sections using vacuum impregnation, semi-automatic polishing of mounted specimens was carried out using a Buehler PowerPro 5000 (Buehler, USA) system. For cross-sectional microstructure and surface morphology observations, the corresponding specimens were gold sputtered and observed under a Scanning Electron Microscope (SEM) (Model TM3000, Hitachi, Japan). The phase constitution of the composite coatings was analysed using a Siemens D500 X-Ray Diffractometer (XRD) with Cr Kα radiation (λ = 2.29 Å), with the diffraction angle (2θ) varied between 30° and 120°. In addition, Electron Dispersive Spectroscopy (EDS) (EDAX, USA) incorporated in the SEM was also used for elemental analysis of the composite coatings. Scratch testing (CSEM Revetest) and ball-on-disk wear studies (VTT Finland) with a sintered Al_2O_3 ball were also carried out on the composite coatings to assess their tribological behaviour.

3 Results and Discussion

3.1 Realizing different coating architectures

A typical micrograph of the layered coatings deposited by sequentially employing the suspension and powder feedstock is shown in Fig. 3(a). The micrograph clearly reveals the columnar structure in the YSZ layer, typical of axial suspension plasma sprayed (ASPS) YSZ [8], followed by a relatively dense Al_2O_3 top-layer. It also suggests good adherence of the Al_2O_3 top-layer to the underlying YSZ coating. Such a coating architecture could, for example, be attractive for high temperature oxidation/corrosion applications such as in thermal barrier coatings (TBCs) [14]. The denser Al_2O_3 layer can potentially seal the paths provided by the inter-columnar space in ASPS YSZ coatings for ingress of oxidizing or corrosive species, which can be detrimental to coating life and performance, while keeping the columnar structure/porosity in the YSZ coating intact. Other variants of such a layered structure are also possible and practically relevant. For instance, deposition of a SPS coating over an APS layer by Aghasibeig et al. [15] has shown improved performance in electrocatalytic activity of Ni electrode for Hydrogen Evolution Reaction (HER). Such a layered architecture fabricated with seamless transition from powder to suspension feedstock, or vice versa, can also be used to tailor toughness, porosity etc. within the coating.

Figure 3(b) depicts the cross-sectional micrograph of an Al_2O_3-YSZ composite coating generated by simultaneously introducing the Al_2O_3 powder and YSZ suspension into the plasma plume. The fairly uniform distribution of YSZ in the Al_2O_3 'matrix' is clearly evident from the figure, raising prospects of this approach providing a convenient pathway to realize novel composite microstructures involving coarse-fine bimodal features attributable to the powder and suspension feedstocks, respectively. Such a strategy of introducing a fine and discretely distributed hard second phase, solid lubricant etc. can have particularly interesting implications for tribological applications.

In one of its manifestations, the above approach to deposit composite coatings can also be utilized to obtain functionally graded coatings, by suitably manipulating relative feed rates of the constituents. Such an architecture has, for long, been of interest to the TBC community to enhance coating performance but complicated to practically realize. A typical TBC comprises of an oxidation/corrosion resistant MCrAlY bond coat (M=Ni, Co, Fe or its combination) and a YSZ insulating layer. Avoiding a sharp interface between MCrAlY and YSZ is desirable, since elimination of a sharp thermal property mismatch can potentially augment TBC life. The flexibility of the present hybrid approach allows virtually continuous composition variation by changing the relative feed rates at the end of each pass. In contrast, various other reported studies on functionally graded TBCs have involved deposition of successive plasma sprayed layers with pre-blended powder compositions [14, 16], which is relatively more difficult to implement. A typical cross-sectional micrograph of a functionally graded coating obtained by hybrid plasma spray processing, revealing a smooth transition between the bond coat (in this case CoNiCrAlY deposited from a powder feedstock) and the top coat (YSZ deposited from a suspension) with no distinct interface is shown in Fig. 3(c). The above approach is amenable for creation of different functionally graded material systems to suit any specific application. For example, a variety of other interesting functionally graded coating systems, for instance those with graded pore size [17, 18] and/or amount of porosity [19], can also be created.

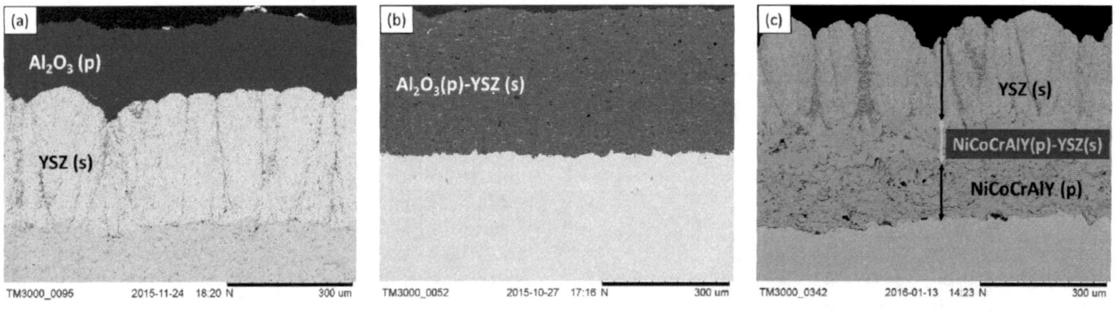

Fig. 3: Typical cross-sectional micrographs of (a) layered, (b) composite, and (c) functionally graded coatings deposited by powder-suspension hybrid plasma spraying

3.2 Characterization of composite coatings

Unlike in the layered architecture example discussed previously, wherein feeding of powder and suspension feedstock is sequential and a corresponding change in process conditions is easily possible, simultaneous injection of hybrid powder-suspension feedstock is considerably more challenging. Foremost, it demands identification of a suitable process window as a fair trade off between operating conditions best suited for spraying either the powder or the suspension individually. In realization of the above, the resulting composite coatings were evaluated in greater detail.

Surface Morphology
The surface morphology of a typical Al_2O_3-YSZ composite coating is shown in Fig. 4. The figure depicts usual splat formation associated with the conventional powder plasma spray process. These 15-30 micron size Al_2O_3 splats co-exist with much finer zirconia-rich splats not exceeding about 5 microns (appearing bright in the figure and separately confirmed by XRD), derived from the simultaneously injected suspension. The multi-scale structure achieved through simultaneous deposition of micron and nano-sized feedstocks is clearly evident.

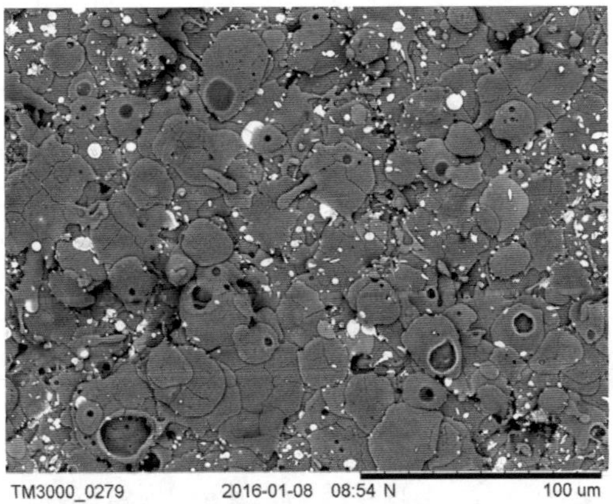

Fig. 4: Surface morphology of a typical Al_2O_3-YSZ composite coating showing multi-scale structure

Microstructure
Microstructural investigation of the transverse section of the composite coatings revealed homogeneous incorporation of YSZ (seen as the brighter phase) in the coating, with typical high-magnification micrographs being depicted in Fig. 5. With variation in YSZ content in the hybrid feedstock from 5wt% to 15wt%, a pronounced difference could be seen in the coating microstructures, with higher density of the brighter Zr-rich regions being notable. The fact that the brighter regions are Zr-rich was confirmed by EDS analysis. It can be inferred from these micrographs that introduction of second phase in the matrix can be successfully controlled during hybrid spraying.

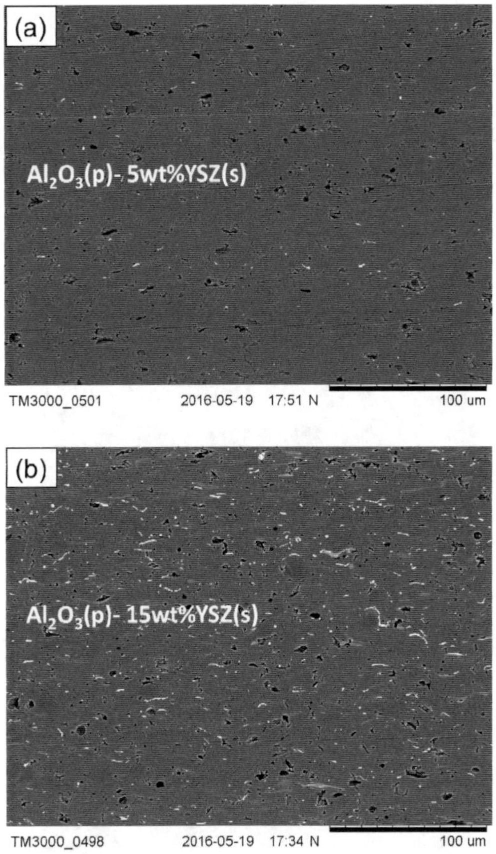

Fig. 5: High magnification micrographs of Al_2O_3-YSZ composite coatings with (a) 5 wt.% YSZ and (b) 15 wt.% YSZ

Phase constitution
The presence of ZrO_2 in the Al_2O_3 matrix was confirmed by XRD spectra of a typical composite coating as shown in Fig. 6. The XRD pattern reveals large peaks corresponding to gamma Al_2O_3 usually observed in plasma sprayed Al_2O_3 [7, 20], as well as peaks corresponding to tetragonal ZrO_2, confirming incorporation of the latter from the suspension feedstock as targeted.

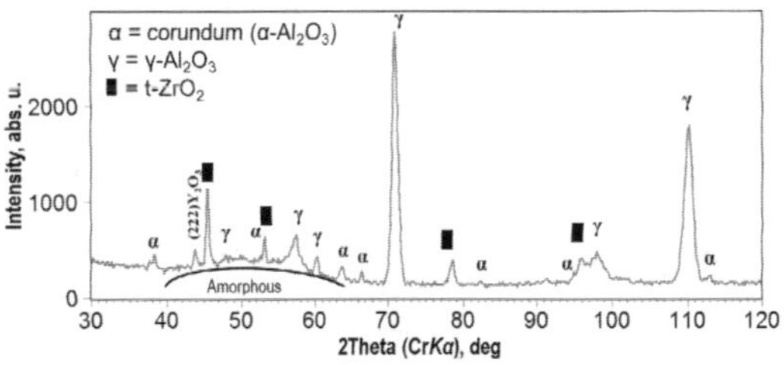

Fig. 6: Typical XRD spectrum of a composite coating showing presence of ZrO_2 in Al_2O_3 matrix

Tribological behaviour

Tribological studies on Al_2O_3-YSZ composite coatings revealed that addition of ZrO_2 as a fine sized second phase led to significantly improved scratch behaviour and wear resistance, compared to coatings produced solely using Al_2O_3 powder. The results are depicted in Fig. 7 and Table 1. It can be concluded from these studies that addition of an appropriate fine sized second phase can potentially enhance functional attributes and can have exciting implications particularly in designing a new generation of high performance coatings for tribological applications.

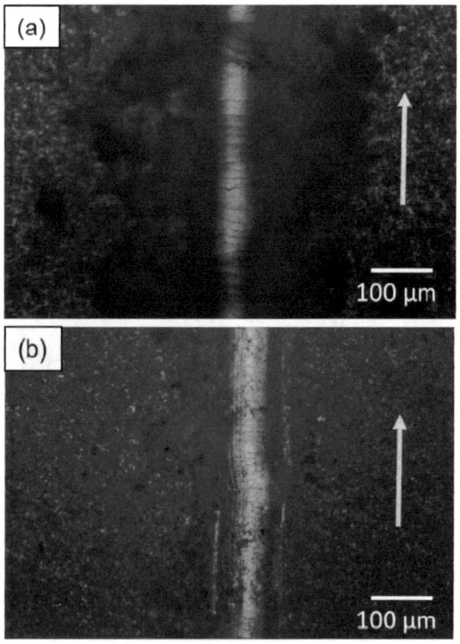

Fig. 7: Scratch tracks in the load range 78-83 N normal load for (a) Al_2O_3-powder, (b) Al_2O_3(p)-YSZ(s) coating. Arrow indicates the direction of the scratch obtained under 10 Nmm^{-1} progressive loading condition

Table 1: Wear rate of coatings during dry sliding test conducted at 10 N normal load for 10,000 cycles

Specimen	Wear rate (10^{-3} mm^3 N^{-1} m^{-1})
Al_2O_3-powder	0.55
Al_2O_3(p)-YSZ(s)	0.35

Although the ability to create the desired architectures has been demonstrated in the present preliminary study, proper process optimization is essential to maximize performance of these coatings for targeted applications. The APS process is known to be governed by a large number of process variables, such as powder size, spray distance etc., with minor changes causing significant variations in properties of the sprayed coatings [21]. Similarly, optimization of SPS coatings involves a detailed parametric study covering particle velocity (controlled by nozzle size, plasma composition, total gas flow rate), suspension flow rate, powder concentration etc. [22, 23]. Thus, finding an optimal set of processing parameters for a given feedstock is essential, and notably challenging when both suspension and powder feedstock injection is involved.

4 Conclusions

The feasibility of sequential/simultaneous injection of conventional spray-grade powder and suspension feedstock axially into a plasma plume has been successfully demonstrated. Such a hybrid approach has shown the potential for being a versatile process for production of diverse function-dependent coating architectures. The ability to realize layered, composite, and functionally graded coatings has been specifically validated in the present work using illustrative case studies. Although process optimization as governed by the coating materials in question is recommended to achieve optimum properties, the results herein amply demonstrate the possibility of identifying a suitable parametric window to accommodate powder and suspension-based feedstocks. Hybrid sprayed composite coatings, which are most challenging to process, have shown good homogeneity and can be potentially extended to spray virtually any material combination, thereby opening new vistas for production of new and superior coating systems not restricted by the commercially available off-the-shelf spray grade powder chemistries. Such hybrid processing can also be extended to radial injection plasma spray systems, particularly when the thermal energy requirement is limited. This is foreseen in cases where low-melting feedstocks are involved or when a discretely distributed high-melting point second phase (e.g. carbides, nitrides or borides) is desired.

Acknowledgment

The assistance of Dr. Zdeněk Pala (The University of Nottingham) for compositional analysis of composite coatings is gratefully acknowledged. The authors are also thankful to Mr. Kenneth Andersson for his guidance during metallographic preparation and to Mr. Esmaeil Sadeghimeresht for assistance in conducting XRD analysis.

5 References

[1] V. P. Singh, A. Sil, and R. Jayaganthan, 'A study on sliding and erosive wear behaviour of atmospheric plasma sprayed conventional and nanostructured alumina coatings', Mater. Des., vol. 32, no. 2, pp. 584–591, Feb. 2011.

[2] R. S. Lima and B. R. Marple, 'Nanostructured YSZ thermal barrier coatings engineered to counteract sintering effects', Mater. Sci. Eng. A, vol. 485, no. 1–2, pp. 182–193, Jun. 2008.

[3] S. V. Joshi and G. Sivakumar, 'Hybrid Processing with Powders and Solutions: A Novel Approach to Deposit Composite Coatings', *J. Therm. Spray Technol.*, vol. 24, no. 7, pp. 1166–1186, Jul. 2015.
[4] E. H. Jordan, L. Xie, M. Gell, N. P. Padture, B. Cetegen, A. Ozturk, X. Ma, J. Roth, T. D. Xiao, and P. E. C. Bryant, 'Superior thermal barrier coatings using solution precursor plasma spray', *J. Therm. Spray Technol.*, vol. 13, no. 1, pp. 57–65, Mar. 2004.
[5] A. Ganvir, N. Curry, S. Govindarajan, and N. Markocsan, 'Characterization of Thermal Barrier Coatings Produced by Various Thermal Spray Techniques Using Solid Powder, Suspension, and Solution Precursor Feedstock Material', *Int. J. Appl. Ceram. Technol.*, vol. 13, no. 2, pp. 324–332, Mar. 2016.
[6] A. Ganvir, N. Curry, N. Markocsan, P. Nylén, S. Joshi, M. Vilemova, and Z. Pala, 'Influence of Microstructure on Thermal Properties of Axial Suspension Plasma-Sprayed YSZ Thermal Barrier Coatings', *J. Therm. Spray Technol.*, vol. 25, no. 1, pp. 202–212, Nov. 2015.
[7] O. Tingaud, P. Bertrand, and G. Bertrand, 'Microstructure and tribological behavior of suspension plasma sprayed Al_2O_3 and Al_2O_3–YSZ composite coatings', *Surf. Coat. Technol.*, vol. 205, no. 4, pp. 1004–1008, Nov. 2010.
[8] A. Ganvir, N. Curry, S. Björklund, N. Markocsan, and P. Nylén, 'Characterization of Microstructure and Thermal Properties of YSZ Coatings Obtained by Axial Suspension Plasma Spraying (ASPS)', *J. Therm. Spray Technol.*, vol. 24, no. 7, pp. 1195–1204, Jun. 2015.
[9] N. Curry, 'Design of Thermal Barrier Coating Systems', (Ph.D. Thesis, University West, 2014), 1-110.
[10] A. Moign, A. Vardelle, N. J. Themelis, and J. G. Legoux, 'Life cycle assessment of using powder and liquid precursors in plasma spraying: The case of yttria-stabilized zirconia', *Surf. Coat. Technol.*, vol. 205, no. 2, pp. 668–673, Oct. 2010.
[11] S. Govindarajan and S. V. Joshi, 'An improved hybrid methodology for producing composite, multi-layered and graded coatings by plasma spraying utilizing powder and solution precursor feedstock', CA2784395 C, 16-Sep-2014.
[12] S. V. Joshi, G. Sivakumar, T. Raghuveer, and R. O. Dusane, 'Hybrid Plasma-Sprayed Thermal Barrier Coatings Using Powder and Solution Precursor Feedstock', *J. Therm. Spray Technol.*, vol. 23, no. 4, pp. 616–624, Feb. 2014.
[13] A. Lohia, G. Sivakumar, M. Ramakrishna, and S. V. Joshi, 'Deposition of Nanocomposite Coatings Employing a Hybrid APS + SPPS Technique', *J. Therm. Spray Technol.*, vol. 23, no. 7, pp. 1054–1064, Feb. 2014.
[14] M. Saremi and Z. Valefi, 'Thermal and mechanical properties of nano-YSZ–Alumina functionally graded coatings deposited by nano-agglomerated powder plasma spraying', *Ceram. Int.*, vol. 40, no. 8, Part B, pp. 13453–13459, Sep. 2014.
[15] M. Aghasibeig, C. Moreau, A. Dolatabadi, and R. Wuthrich, 'Fabrication of nickel electrode coatings by combination of atmospheric and suspension plasma spray processes', *Surf. Coat. Technol.*, vol. 285, pp. 68–76, Jan. 2016.
[16] K. Kokini, J. DeJonge, S. Rangaraj, and B. Beardsley, 'Thermal shock of functionally graded thermal barrier coatings with similar thermal resistance', *Surf. Coat. Technol.*, vol. 154, no. 2–3, pp. 223–231, May 2002.
[17] Y.-C. Yang, 'Investigation of residual stress generation in plasma-sprayed hydroxyapatite coatings with various spraying programs', *Surf. Coat. Technol.*, vol. 205, no. 21–22, pp. 5165–5171, Aug. 2011.
[18] R. H. R. Jr, P. J. Lare, and H. Hahn, 'Surgical implant having a graded porous coating', US4542539 A, 24-Sep-1985.
[19] M. Thieme, K.-P. Wieters, F. Bergner, D. Scharnweber, H. Worch, J. Ndop, T. J. Kim, and W. Grill, 'Titanium powder sintering for preparation of a porous functionally graded material destined for orthopaedic implants', *J. Mater. Sci. Mater. Med.*, vol. 12, no. 3, pp. 225–231, Mar. 2001.
[20] J. O. Berghaus, J.-G. Legoux, C. Moreau, F. Tarasi, and T. Chráska, 'Mechanical and Thermal Transport Properties of Suspension Thermal-Sprayed Alumina-Zirconia Composite Coatings', J. Therm. Spray Technol., vol. 17, no. 1, pp. 91–104, Dec. 2007.
[21] F. Azarmi, T. W. Coyle, and J. Mostaghimi, 'Optimization of Atmospheric Plasma Spray Process Parameters using a Design of Experiment for Alloy 625 coatings', *J. Therm. Spray Technol.*, vol. 17, no. 1, pp. 144–155, Jan. 2008.
[22] D. Waldbillig and O. Kesler, 'Effect of suspension plasma spraying process parameters on YSZ coating microstructure and permeability', *Surf. Coat. Technol.*, vol. 205, no. 23–24, pp. 5483–5492, Sep. 2011.
[23] F. Tarasi, M. Medraj, A. Dolatabadi, J. Oberste-Berghaus, and C. Moreau, 'Effective Parameters in Axial Injection Suspension Plasma Spray Process of Alumina-Zirconia Ceramics', *J. Therm. Spray Technol.*, vol. 17, no. 5–6, pp. 685–691, Nov. 2008.

An Assessment of Thermal Spray Technologies for Deposition of Environmental Barrier Coatings (EBC)

E. Bakan, G. Mauer, R.Vaßen, Forschungszentrum Jülich GmbH, Institute of Energy and Climate Research, Materials Synthesis and Processing (IEK-1), 52425 Jülich, Germany

Environmental barrier coatings (EBC) are currently being investigated to protect ceramic matrix composite (CMC) turbine engine components in water-vapor rich combustion environments. Dense, crack-free, uniform and well-adhered coatings are demanded for this purpose. This paper represents an assessment of different thermal spray techniques for deposition of $Yb_2Si_2O_7$ and silicon (Si) EBC layers. Plasma spraying of refractory silicates is known to be complicated by undesired glass transition due to rapid solidification as well as evaporation of Si-bearing species during spraying. Plasma spraying of low-density Si also requires careful optimizations as it tends to oxidize during spraying, particularly at atmospheric conditions.

Bearing these problems in mind, the $Yb_2Si_2O_7$ coatings were deposited by atmospheric plasma spraying (APS), high-velocity oxygen-fuel spraying (HVOF), and plasma-spray physical vapor deposition (PS-PVD) techniques. As-sprayed microstructure, amorphous content and phase composition of the coatings were analyzed. Based on the findings, the advantages and disadvantages of each method over other techniques are discussed with respect to process parameters and material properties.

1 Introduction

SiC/SiC CMCs offering high-temperature capability, good mechanical strength and thermal shock resistance are considered as the new material of choice for different components of advanced gas turbine engines. By utilizing these materials as a substitution of single crystal superalloys, higher inlet temperatures with less cooling air is aimed to be reached which results in greater efficiency. To that end, environmental barrier coatings (EBCs) are developed for protection of Si-based CMCs because they degrade when exposed to water vapor-rich gasses in high-temperature and pressure combustion environments [1,2]. For that purpose multi-layered EBC systems typically consisting of silicon bond coat and mullite/rare-earth silicate top coat layers were investigated using different manufacturing methods including plasma spray, electron beam physical vapour deposition, magnetron sputtering and sol-gel [3-6]. Common problem observed was the difficulty of obtaining a crystalline and dense silicate layer in the as-deposited state due to high tendency of glass-formation. Such amorphous coatings are not desired because they shrink and crack when they exposed to high temperatures. Therefore in this contribution the results obtained from different thermal spray methods regarding the crystallinity and porosity of $Yb_2Si_2O_7$ coatings will be discussed.

2 Experimental Procedure

2.1 Powder

$Yb_2Si_2O_7$ powder (d10=24µm, d50=38µm, d90=68µm) provided by Oerlikon Metco (US) Inc. (Westbury, New York, USA) was used in this study for the deposition of coatings by APS. Part of this powder was sieved and the particle fractions captured on (d10=35µm, d50=49µm, d90=68µm) and under the sieve (d10=22µm, d50=33µm, d90=48µm) were used for the PS-PVD and HVOF processes, respectively.

2.2 Processing

A MultiCoat system (Oerlikon Metco, Wohlen, Switzerland) with three-cathode TriplexPro 210 (Ø9mm nozzle) and Diamond Jet 2700 (type 2705 nozzle) spray torches mounted on a six-axis robot (IRB 2400, ABB, Switzerland) was employed for APS and HVOF spray experiments, respectively. PS-PVD investigation was performed with an O3CP torch operated by the MultiCoat platform (Oerlikon Metco, Wohlen, Switzerland) in a controlled-atmosphere chamber. Process parameters can be seen in [7]. In each method, prior to spraying the substrate temperatures were increased via plasma/HVOF flame heating as high as possible. This resulted in deposition temperatures of ~150°C, 600°C and 950°C in HVOF, APS and PS-PVD methods, respectively.

2.3 Characterization of deposits

XRD measurements were performed on the as-deposited top coats with the same measurement parameters for the feedstock given above (D4 Endeavor & TOPAS software, Bruker AXS, Germany). The samples were sectioned, polished, and examined with a scanning electron microscope (Carl Zeiss NTS GmbH, Oberkochen, Germany).

3 Results and Discussions

Fig.1 shows the XRD data of the $Yb_2Si_2O_7$ coatings deposited with APS, HVOF and PS-PVD methods. The APS coating clearly reveal an almost fully amorphous phase, while the patterns of HVOF and PS-PVD coatings show sharp, high intensity peaks, which suggest the presence of crystalline phases. This striking difference in the amorphous contents of the coatings can be explained by the deposition temperatures, cooling conditions as well as the melting state of the deposited particles on the substrate. As mentioned above, the highest deposition temperature could be obtained in the PS-PVD chamber. Furthermore, following the deposition, the deposit is post-heated via plasma-heating and this allowed maintaining the deposition temperatures for a longer time. As a result of this procedure highly crystalline $Yb_2Si_2O_7$ coatings could be deposited. In the atmospheric conditions of APS and HVOF, on the other hand, the deposition temperatures were significantly lower. This resulted in a highly amorphous coating in the APS, as molten particles still rapidly cool down to room temperature. However, although the deposition temperature is even lower in the HVOF process, the deposit has higher crystallinity. This is most likely due to the partially molten or non-molten particle deposition in this process, as these particles preserve their crystallinity in the as-sprayed state. Furthermore, as they would serve as nuclei they can even facilitate the further crystallization.

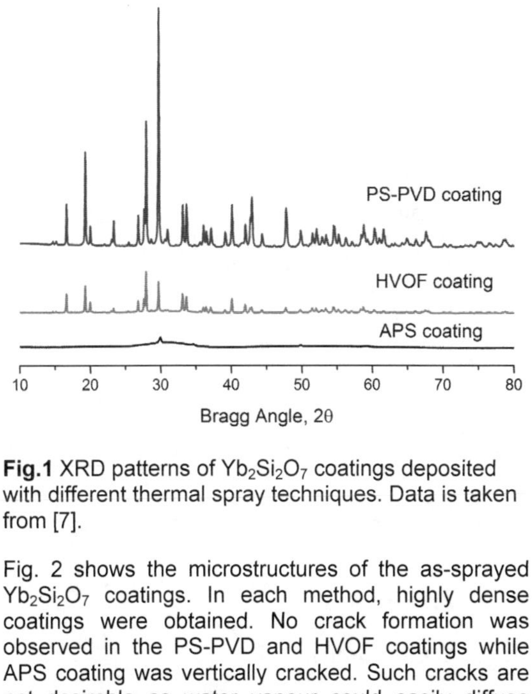

Fig.1 XRD patterns of $Yb_2Si_2O_7$ coatings deposited with different thermal spray techniques. Data is taken from [7].

Fig. 2 shows the microstructures of the as-sprayed $Yb_2Si_2O_7$ coatings. In each method, highly dense coatings were obtained. No crack formation was observed in the PS-PVD and HVOF coatings while APS coating was vertically cracked. Such cracks are not desirable as water vapour could easily diffuse through them. The reason of this different cracking behaviour could be attributed to different crystallinity levels of the coatings or residual stresses assuming no significant difference in their chemistry.

4 Conclusions

Yb2Si2O7 coatings were manufactured with different thermal spray methods. Highly dense coatings with each method could be deposited but the amorphous content of the coatings were found to be highly varying according to deposition temperatures, cooling conditions and melting state of the particles e.g. molten or non-molten. Vertical cracks were observed in the APS coating possibly due to low crystallinity of the coating and rapid quenching conditions.

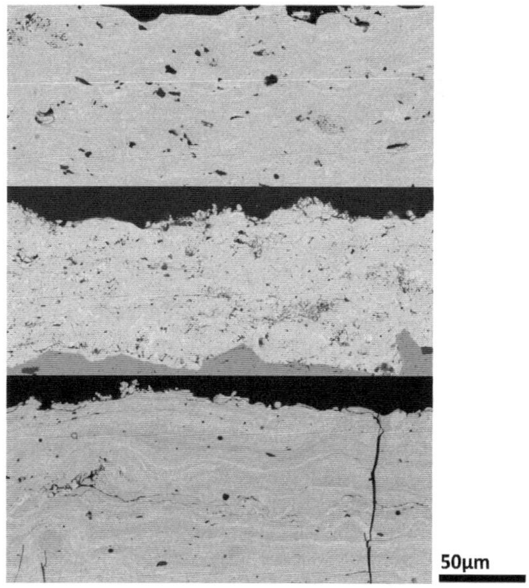

Fig.2 As-sprayed microstructures of PS-PVD, HVOF and APS $Yb_2Si_2O_7$ coatings sorted from top to bottom.

5 References

1. E.J. Opila, R.E. Hann, Paralinear Oxidation of CVD SiC in Water Vapor, *J. Am. Ceram. Soc.*, **80**(1), 197-205 (1997)
2. R.C. Robinson, J.L. Smialek, SiC Recession Caused by SiO_2 Scale Volatility under Combustion Conditions: I, Experimental Results and Empirical Model, *J. Am. Ceram. Soc.*, **82**(7), 1817-1825 (1999)
3. K.N. Lee, D.S. Fox, N.P. Bansal, Rare earth silicate environmental barrier coatings for SiC/SiC composites and Si_3N_4 ceramics, *J. Eur. Ceram. Soc.*, **25**(10), 1705-1715 (2005)
4. B.T. Richards, H.N.G. Wadley, Plasma spray deposition of tri-layer environmental barrier coatings, *J. Eur. Ceram. Soc.*, **34**(12), 3069-3083 (2014)
5. P. Mechnich, Y_2SiO_5 coatings fabricated by RF magnetron sputtering, *Surf. Coat. Technol.*, **237**, 88-94 (2013)
6. D.D. Jayaseelan, S. Ueno, T. Ohji, S. Kanzaki, Sol–gel synthesis and coating of nanocrystalline $Lu_2Si_2O_7$ on Si_3N_4 substrate, *Mater. Chem. Phys.*, **84**(1), 192-195 (2004)
7. E. Bakan, D. Marcano, D. Zhou, Y.J. Sohn, G. Mauer, R. Vaßen, $Yb_2Si_2O_7$ Coatings Deposited by Various Thermal Spray Techniques: A Preliminary Comparative Study *J. Therm. Spray Technol.*, **Submitted Paper**, (2016)

Multilayered suspension plasma sprayed thermal barrier coatings for high temperature gas turbine applications

M. Gupta, N. Markocsan, University West, Trollhättan / SE
R. Rocchio-Heller, J. Liu, Oerlikon Metco, Westbury / USA
X.-H. Li, Siemens Industrial Turbomachinery AB, Finspång / SE
L. Östergren, GKN Aerospace, Trollhättan / SE

Improvement in the performance of thermal barrier coating systems (TBCs) is one of the key objectives for further development of gas turbine applications. The material most commonly used as TBC topcoat is yttria stabilised zirconia (YSZ). However, the usage of YSZ is limited by the operating temperature range which in turn restricts the engine efficiency. Materials such as pyrochlores, perovskites, rare earth garnets, etc. are suitable candidates which could replace YSZ as they exhibit lower thermal conductivity and higher phase stability at elevated temperatures.
The objective of this work was to investigate different multi-layered TBCs consisting of advanced topcoat materials fabricated by Suspension Plasma Spraying (SPS). The investigated topcoat materials were YSZ, dysprosia stabilised zirconia, gadolinium zirconiate, cerium doped YSZ and yttria fully stabilised zirconia. All topcoats were deposited with TriplexPro-210 plasma spray gun and radial injection of suspension. Lifetime of these samples was examined by thermal cyclic fatigue and thermal shock testing. Microstructure analysis of as-sprayed and failed specimens was performed with scanning electron microscope. The failure mechanisms in each case have been discussed in this article. The results show that SPS could be a promising route to produce multilayered TBCs for high temperature applications.

1 Introduction

Development of thermal barrier coatings (TBCs) allowing higher combustion temperatures is of great interest for all gas turbine manufacturers since it results in a more efficient combustion which in turn results in higher fuel efficiency as well as environmentally cleaner emissions. The most commonly used TBC material is yttria stabilised zirconia (YSZ). Even though YSZ has been used extensively for the past thirty years due to its favourable thermal and mechanical properties, its usage in high temperature applications is limited as undesirable phase transformation occurs at higher temperatures. Moreover, at temperatures above 1200°C, the YSZ coatings also undergo sintering and are susceptible to CMAS (Calcium Magnesium Alumino Silicates) attack, both of which may significantly affect the TBC lifetime [1, 2].
New ceramic materials such as the pyrochlores, perovskites, and co-doped zirconia, have been identified as alternative candidates to YSZ for high temperature applications [3, 4] However, these materials usually exhibit a lower toughness and thermal expansion coefficient than YSZ [5] and some of them (e.g. gadolinium zirconate) also have compatibility issues with the bondcoat layer due to oxidation at higher temperatures [6]. In order to overcome these drawbacks, a multi-layer TBC is preferably applied with a first layer of YSZ and top layer made of an advanced TBC material which can result in a significantly enhanced lifetime [7, 8]. In this multi-layer system, the YSZ layer provides a good fracture toughness close to the topcoat-bondcoat interface and high thermal expansion coefficient, while the subsequent material layers provide low sintering, high thermal stability and CMAS attack resistance. Studies done on these materials sprayed by atmospheric plasma spraying (APS) showed that multi-layered thermal barrier coatings show good functional performances while their thermal properties can be well preserved [6, 9].
Columnar microstructures produced by suspension plasma spraying (SPS) are of significant commercial interest as SPS offers a cheaper alternative than the conventionally used Electron Beam – Physical Vapour Deposition (EB-PVD) process, whereas the coatings thermal conductivity and life-time are preserved or even improved [10, 11]. Multilayered TBCs produced by SPS are of high relevance for the gas turbine industry as they can allow higher combustion temperatures and enhanced functional perfomances [12].
The objective of this work was to investigate single layer and double layer topcoat architectures deposited by SPS and to determine their failure mechanism under thermal cyclic conditions.

2 Experimental

2.1 Sample preparation

Two topcoat architectures were investigated in this study – single layer and bilayer. The single layer topcoat architectures which were investigated were 8 wt.% YSZ, and 10 wt.% dysprosia stabilised zirconia, while the bilayer topcoat architectures which were investigated were YSZ/gadolinium zirconate (GZO), and YSZ/cerium doped YSZ (CeYSZ).
The total thickness of topcoats was kept the same as 300 µm for all samples. In case of bilayer topcoats, the first layer deposited was 100 µm in thickness, while the second layer was 200 µm in thickness. All topcoats were sprayed by SPS using the TriplexPro-210 spray gun (Oerlikon Metco) while all suspensions were developed by Oerlikon Metco, Westbury, USA.

The samples along with the suspension number IDs are summarised in Table 1.
The bondcoat material for all samples was NiCoCrAlY (AMDRY 386-2, Oerlikon Metco) while the substrate material was Hastelloy-X. Substrates in dimensions 25.4 mm diameter x 6 mm thickness were used for microstructure analysis and thermal shock testing, while for thermal cyclic fatigue (TCF) testing, substrates in dimensions 50 mm x 30 mm x 6 mm were used.

Table 1. Samples produced in this study. The suspension number IDs are given in the brackets.

Sample ID	Topcoat material	
	1st layer (100 μm)	2nd layer (200 μm)
YSZ	YSZ (*AE10966*)	
DySZ	DySZ (*AE10873*)	
YSZ/GZO	YSZ	GZO (*AE10874*)
YSZ/CeYSZ	YSZ	CeYSZ (*AE10876*)

Preheating was done on all layers of each set of samples before spraying topcoats. Principles from prior multi-layer systems were used while spraying where the first topcoat layer was sprayed dense and the second topcoat layer was sprayed porous [13].

2.2 Microstructure

For metallographic preparation, the samples were first cold mounted in low viscosity epoxy resin, then sectioned using a diamond cutting disc and cold mounted again in high viscosity epoxy resin for polishing. The polishing was performed using a Buehler Power Pro 5000 equipment.
Microstructure of both as-sprayed as well as failed samples after lifetime testing were analysed by scanning electron microscopy (SEM) using a Hitachi TM3000 tabletop equipment.

2.3 Thermal cyclic fatigue testing

TCF testing is performed to analyse the performance of TBCs under long exposures at high temperature. This test allows for significant oxide growth due to bondcoat oxidation. The failure in TCF testing is mainly driven by crack propagation due to swelling of thermally grown oxide (TGO) and mismatch in thermal expansion coefficients of different layers in the TBC system. In addition to this, sintering of topcoat as well as the thermal fatigue induced by thermal cycling also play a significant role in failure.
In this study, TCF testing was performed at Siemens Industrial Turbomachinery, Sweden. The samples were exposed to cycles of firstly isothermally heating up to dwell temperature of 1100°C for one hour, and then cooling down to 100°C in ten minutes with compressed air. The samples were cycled until failure which was determined by 20% spallation of the topcoat. A photograph of the samples was recorded after each cycle to determine the failure.

Two samples of each coating set were tested by TCF testing in this study.

2.4 Thermal shock testing

Thermal shock testing is performed to analyse the performance of TBCs under short exposures at high temperatures and then sudden quenching. This test normally does not allow significant TGO growth due to short exposure times at high temperatures. The failure in thermal shock testing is mainly driven by crack propagation near the topcoat-bondcoat interface due to thermal mismatch between the different layers in the TBC system. Strain tolerance of the topcoat plays a significant role in failure as it determines the resistance of the topcoat against thermal fatigue induced by thermal cycling.
In this study, thermal shock testing was performed at GKN Aerospace, Sweden using their burner rig testing (BRT) equipment which has been developed in-house. During this test, the samples are first exposed to heating at front surface with combustion burners for 75 seconds which results in topcoat surface temperatures up to 1300°C, and then cooled down to around 450°C with compressed air in 75 seconds. Samples are cooled down with compressed air from the back side during the heating cycle which maintains a gradient across the sample. The samples are cycled until failure which is determined by 20% spallation of either complete topcoat or part of the topcoat layer (chipping failure). A short video of the samples is recorded after every four cycles to determine the failure.
In addition to the cycle durations, another major difference between TCF testing and BRT is that while in TCF the samples are exposed to isothermal heating, a thermal gradient is maintained across the sample during BRT.
Four samples of each coating set were tested by BRT in this study.

3 Results and Discussions

3.1 Microstructure

The microstructure images of the samples in as-sprayed condition are shown in **Fig. 1**. The single layer topcoats YSZ and DySZ showed in Fig. 1a and Fig. 1b exhibited a densely packed columnar structure looking similar to the vertically cracked microstructure typically produced by APS. The interpass porosity can be clearly distinguished in the single layer topcoats which although could be beneficial for reducing thermal conductivity of these topcoats, but could also be detrimental to coating lifetime as it could promote crack propagation during thermal cycling. Some branching cracks between the interpass layers can also be observed in these coatings especially close to the column gaps indicating that stress relaxation occurred during spraying probably due to high packing density of the columns with relatively low porosity.
The bilayer topcoats shown in Fig. 1c and Fig. 1d also exhibited a columnar structure with more distinct

column gaps than the single layer coatings. The columns in second layer continued to build on the columns in the first layer in all three bilayer topcoats. It can be observed that the first topcoat layer is denser than the second topcoat layer as intented during spraying. A denser layer is generally sprayed first as it provides resistance against detrimental crack propagation near the topcoat-bondcoat interface during thermal cycling, while the subsequent porous layer provides the necessary strain tolerance required for high lifetime of TBCs [13].

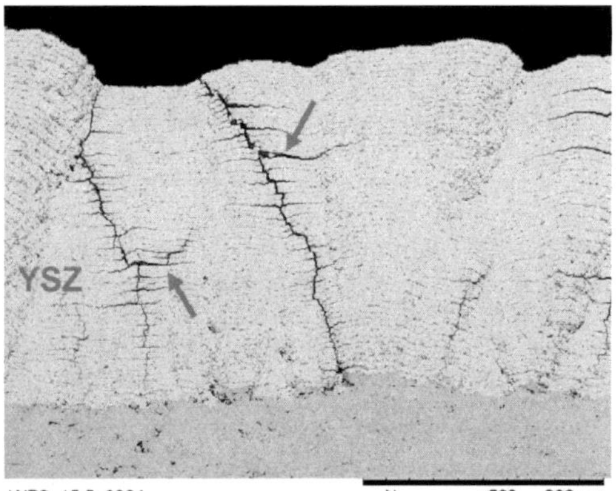

(a)

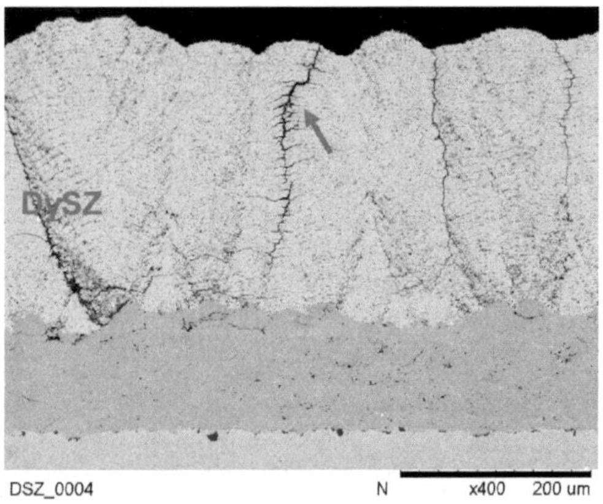

(b)

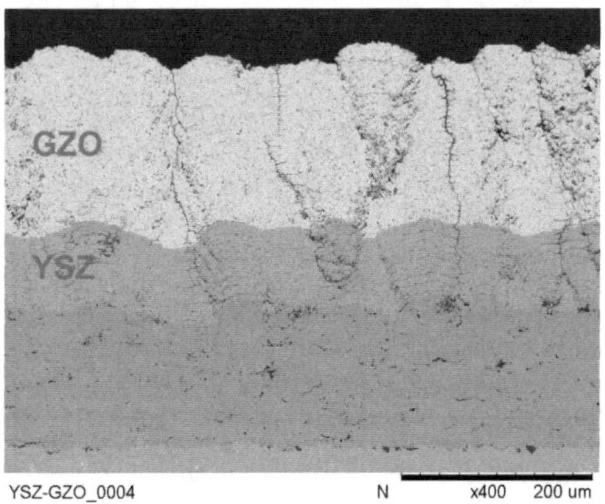

(c)

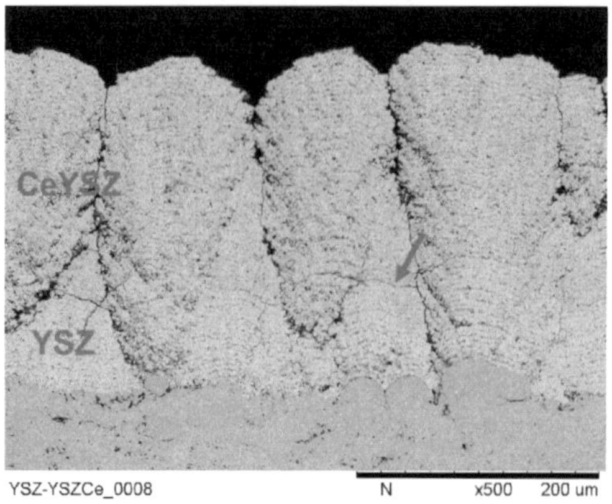

(d)

Fig. 1. Microstructure images of samples in as-sprayed condition (a) YSZ, (b) DySZ, (c) YSZ/GZO, and (d) YSZ/CeYSZ. The arrows indicate cracking between the interpass layers or the two topcoat layers.

The interpass porosity bands were less distinct in the bilayer topcoats than the single layer topcoats. However, some cracking between the interpass layer similar to the single layer topcoats can also be observed in bilayer topcoats. The horizontal cracking is especially pronounced at the interface between the topcoat layer in YSZ/CeYSZ sample indicating low adhesion between the layers. The low adhesion between the layers could be due to different material properties, especially thermal expansion coefficients, and/or due to different porosities of the two layers, which result in cracking during cooling after spraying.

3.2 Thermal cyclic fatigue testing

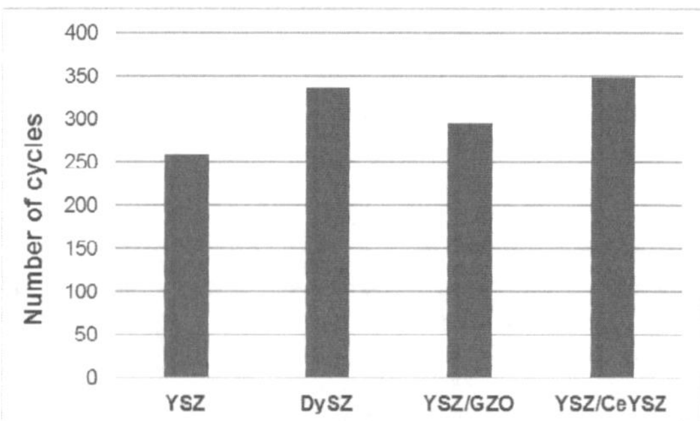

Fig. 2. Thermal cyclic fatigue lifetime results.

The results from TCF testing are shown in **Fig. 2**. Between the single layer topcoats, DySZ showed around 20% higher TCF lifetime than YSZ in TCF testing. This result shows that in addition to providing reduced thermal conductivity, stabilising zirconia with dysprosia instead of yttria would also result in higher TCF lifetime as observed previously for APS TBCs [14]. Between the bilayer topcoats, YSZ/CeYSZ showed a better TCF lifetime than YSZ/GZO which was only slightly below YSZ/CeYSZ.

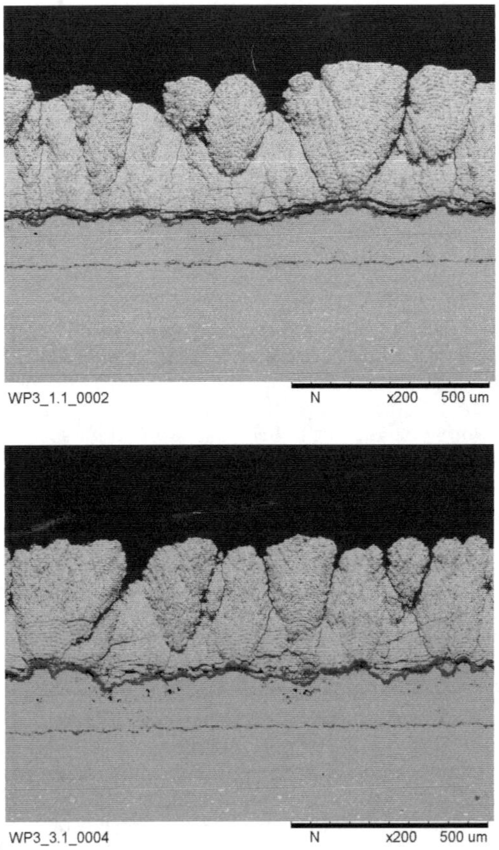

(a)

(b)

Fig. 3. Microstructure images of samples after TCF testing (a) DySZ, and (b) YSZ/CeYSZ.

The microstructure images of DySZ and YSZ/CeYSZ samples after failure in TCF testing are shown in **Fig. 3**. A dark grey layer of TGO with thickness in the range of 6-8 µm can be observed in the coatings which was also observed for all other coatings. It can be clearly noted in Fig. 3 that failure occurred due to cracking near the topcoat-bondcoat interface which was also observed for all other coatings. This indicates that the failure was driven by, as also discussed in section 2.3, crack propagation due to TGO growth and thermal mismatch between different layers during thermal cycling. Both of these phenomena induce high stresses in the topcoat near the topcoat-bondcoat interface resulting in cracking and thus failure [15]. It can be noticed in Fig. 3 that cracking occurred through the interpass porosity bands closest to the topcoat-bondcoat interface in both cases, suggesting clearly that the interpass porosity bands provide a path of least resistance for crack propagation. These results show that formation of interpass porosity should be avoided especially near the topcoat-bondcoat in order to avoid detrimental crack propagation and thus increase TBC lifetime.

Another feature which can be noted in the microstructure images shown in Fig. 3 is that the beta-phase is completely depleted in both cases after the TCF testing. This shows that the bondcoat has reached the end of its lifetime as it could not facilitate the formation of the slow growing alumina scale any longer. It is well known that after complete beta-phase depletion, other oxides such as spinel oxides, chromia, nickel oxide, etc. form rapidly and induce very high stresses resulting in chemical failure of the TBC system [15]. It can be concluded from these observations that both DySZ and YSZ/CeYSZ topcoats survived until the end of bondcoat's chemical lifetime. Therefore, for improvement in TCF lifetime of these TBCs, apart from improvements in topcoat microstructures, either the bondcoat thickness should be increased so as to enlarge the aluminium reservoir in the bondcoat or the bondcoat chemistry should be modified.

3.3 Thermal shock testing

The results from thermal shock testing are shown in **Fig. 4**. Between the single layer topcoats, similar to the TCF lifetime results, DySZ showed around 30% higher average lifetime than YSZ in BRT. This result again shows that dysprosia would be a better stabiliser than YSZ providing both lower thermal conductivity and longer lifetime.

Between the bilayer topcoats, YSZ/GZO showed better BRT lifetime by far as compared to YSZ/CeYSZ which showed even lower average lifetime than the single layer topcoats.

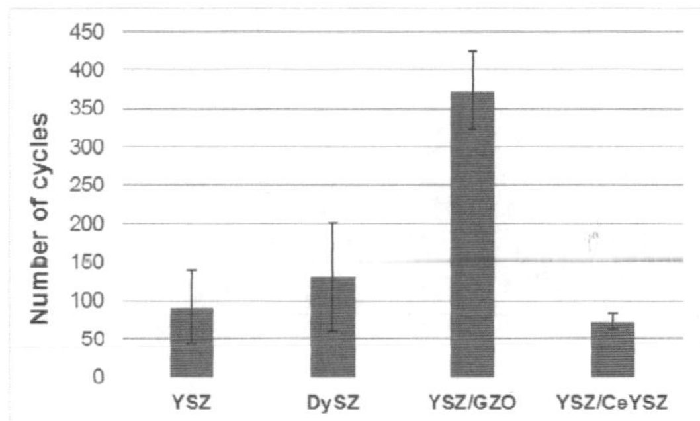

Fig. 4. Thermal shock lifetime results.

The microstructure images of DySZ, YSZ/GZO and YSZ/CeYSZ samples after failure in BRT are shown in **Fig. 5**. The thickness of TGO in case of BRT can be observed to be 1-2 µm in Fig. 5 which was also observed for the rest of the investigated coatings.

It can be observed that failure in single layer DySZ mainly occurred due to cracking in the topcoat near the topcoat-bondcoat interface along with some chipping failure in the topcoat. This failure mode was also observed in the YSZ topcoat. These results show that the failure in single layer topcoats was driven by crack propagation near the topcoat-bondcoat interface due to thermal mismatch stresses induced during thermal cycling.

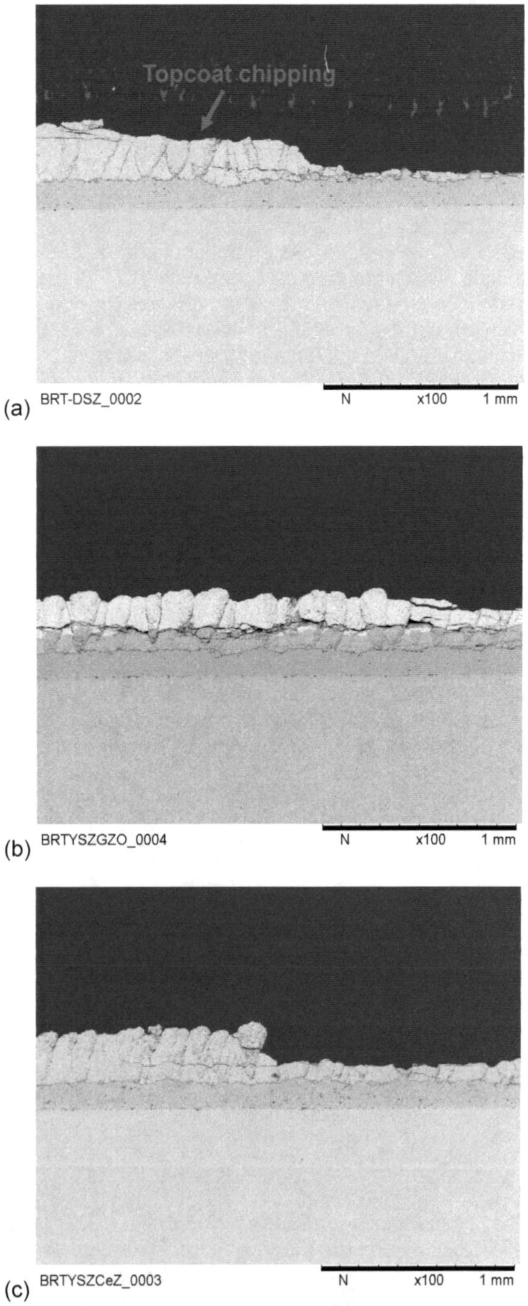

Fig. 5. Microstructure images of samples after BRT (a) DySZ, (b) YSZ/GZO, and (c) YSZ/CeYSZ.

On the other hand, failure in the bilayer topcoats YSZ/GZO and YSZ/CeYSZ occurred mainly in the second topcoat layer near the interface between the two topcoat layers. This failure mode has also been observed in previous work done on SPS multilayered TBCs [12] and is deemed to occur due to the lower fracture toughness of GZO and CeYSZ as compared to YSZ promoting crack propagation in the second topcoat layer near the interface between the two topcoat layers instead of YSZ near the topcoat-bondcoat interface.

The reason for significantly higher BRT lifetime of YSZ/GZO topcoat is deemed to be its comparable thermal expansion coefficient to YSZ [5], phase stability and high sintering resistance at high temperatures [17] during BRT.

4 Summary and conclusions

In this work, single layer and bilayer topcoat architectures deposited by SPS for TBCs were investigated. Lifetime of these samples was determined by thermal cyclic fatigue testing and thermal shock testing. The microstructure of as-sprayed and failed samples was investigated and the failure mechanisms in each case were discussed.

The failure mode in TCF testing was cracking near the topcoat-bondcoat interface in all cases. In BRT, in single layer topcoats, cracking occurred near the topcoat-bondcoat interface while in bilayer topcoats, failure occurred near the interface between the two topcoat layers due to lower fracture toughness of the second topcoat layer.

The lifetime results show that SPS seems to be a promising route to deposit multilayered TBCs for high temperature applications. YSZ/GZO showed the best lifetime results when considering both TCF and BRT. The reason for the good performance of YSZ/GZO topcoat was deemed to be high fracture toughness of YSZ combined with comparable thermal expansion coefficient of GZO, its phase stability and high sintering resistance at high temeperatures.

5 Acknowledgements

The authors would like to acknowledge the Knowledge Foundation for the funding of this research work. Thanks to Stefan Björklund and Kenneth Andersson at University West for the help with spraying and metallographic preparation respectively.

6 References

[1] A. Cipitria, I. O. Golosnoy, and T. W. Clyne: A sintering model for plasma-sprayed zirconia TBCs. Part I: Free-standing coatings. Acta Materialia 57 (2009), Issue 4, pp. 980/92.

[2 J. M. Drexler, A. L. Ortiz, and N. P. Padture: Composition effects of thermal barrier coating ceramics on their interaction with molten Ca–Mg–Al–silicate (CMAS) glass. Acta Materialia 60 (2012), Issue 15, pp. 5437/47.

[3] X. Cao: Application of Rare Earths in Thermal Barrier Coating Materials. Journal of Material Science and Technology 23 (2007), Issue 1, pp. 15/35.

[4] X.Q. Cao, R. Vassen, and D. Stoever: Ceramic materials for thermal barrier coatings. Journal of the European Ceramic Society 24 (2004), Issue 1, pp. 1/10.

[5] R. Vaßen, M.O. Jarligo, T. Steinke, D.E. Mack, and D. Stöver: Overview on advanced thermal barrier coatings. Surface & Coatings Technology 205 (2010), pp. 938/42.

[6] E. Bakan, D.E. Mack, G. Mauer, and R. Vaßen: Gadolinium Zirconate/YSZ Thermal Barrier Coatings: Plasma Spraying, Microstructure, and Thermal Cycling Behavior. Journal of the American Ceramic Society 97 (2014), Issue 12, pp. 4045/51.

[7] X. Zhong, H. Zhao, C. Liu, L. Wang, F. Shao, X. Zhou, S. Tao, and C. Ding: Improvement in thermal shock resistance of gadolinium zirconate coating by addition of nanostructured yttria partially-stabilized zirconia. Ceramics International 41 (2015), Issue 6, pp. 7318/24.

[8] K. S. Lee, D. H. Lee, and T. W. Kim: Microstructure controls in Gadolinium Zirconate/YSZ double layers and their properties. Journal of the Ceramic Society of Japan 122 (2014), Issue 1428, pp. 668/73.

[9] V. Viswanathan, G. Dwivedi, and S. Sampath: Multilayer, Multimaterial Thermal Barrier Coating Systems: Design, Synthesis, and Performance Assessment. Journal of the American Ceramic Society 98 (2015), Issue 6, pp. 1769/77.

[10] A. Ganvir, N. Curry, S. Björklund, N. Markocsan, and P. Nylén: Characterization of Microstructure and Thermal Properties of YSZ Coatings Obtained by Axial Suspension Plasma Spraying (ASPS). Journal of Thermal Spray Technology 24 (2015), Issue 7, pp. 1195/204.

[11] N. Curry, K. VanEvery, T. Snyder and N. Markocsan: Thermal Conductivity Analysis and Lifetime Testing of Suspension Plasma-Sprayed Thermal Barrier Coatings. Coatings 4 (2014), Issue 3, pp. 630/50.

[12] S. Mahade, N. Curry, S. Björklund, N. Markocsan, P. Nylén, and Robert Vaßen: Functional performance of $Gd_2Zr_2O_7$/YSZ multi-layered thermal barrier coatings deposited by suspension plasma spray. Surface & Coatings Technology (2016), DOI: http://dx.doi.org/10.1016/j.surfcoat.2016.12.062.

[13] V. Viswanathan, G. Dwivedi, and S. Sampath: Engineered Multilayer Thermal Barrier Coatings for Enhanced Durability and Functional Performance. Journal of the American Ceramic Society 97 (2014), Issue 9, pp. 2770/78.

[14] N. Curry, N. Markocsan, L. Östergren, X.-H. Li, and M. Dorfman: Evaluation of the Lifetime and Thermal Conductivity of Dysprosia-Stabilized Thermal Barrier Coating Systems. Journal of Thermal Spray Technology 22 (2013), Issue 6, pp. 864/72.

[15] A.G. Evans, D.R. Mumm, J.W. Hutchinson, G.H. Meier, and F.S. Pettit: Mechanisms controlling the durability of thermal barrier coatings. Progress in Materials Science 46 (2001), Issue 5, pp. 505/53.

[16] R. Eriksson, H. Brodin, S. Johansson, L. Östergren, and X.-H. Li: Fractographic and microstructural study of isothermally and cyclically heat treated thermal barrier coatings. Surface and Coatings Technology 243 (2014), pp 82/90.

[17] D. Stöver, G. Pracht, H. Lehmann, M. Dietrich, J-E. Döring, and R. Vaßen: New Material Concepts for the Next Generation of Plasma-Sprayed Thermal Barrier Coatings. Journal of Thermal Spray Technology 13 (2004), Issue 1, pp. 76/83.

Comparing the Microstructures and Properties of YSZ TBCs Manufactured via Air Plasma Spray (APS), Suspension Plasma Spray (SPS) and Finely-dispersed-particle Air Plasma Spray (FAPS)

R. S. Lima, B. M. H. Guerreiro and B. R. Marple - National Research Council of Canada (NRC), Boucherville, QC, Canada

ZrO_2-Y_2O_3 (YSZ) thermal barrier coatings (TBCs) were manufactured via conventional Air Plasma Spray (APS), Suspension Plasma Spray (SPS) and an additional technology hereby termed Finely-dispersed-particle Air Plasma Spray (FAPS). The FAPS processing employs the exact same classification of finely dispersed particles as used in SPS; however, whereas SPS uses a liquid medium, in the case of FAPS the particles are fed conventionally via a carrier gas into the plasma spray torch by using a newly developed powder feeder for fine (suspension-like) particles (NRC patented technology). These finely dispersed YSZ particles consist of irregularly shaped (fluffy-like) agglomerates made from individual nano-sized particles. The conventional APS YSZ TBC was sprayed via a Metco 3MB torch, whereas, both SPS and FAPS YSZ TBCs were sprayed using the Mettech Axial III torch (using the same set of spray parameters). Both SPS and FAPS YSZ TBCs exhibited porous and vertically-cracked microstructures. The conventional APS YSZ TBC microstructure exhibited the traditional lamellar morphology. Elastic modulus, hardness and thermal conductivity values were evaluated for all YSZ TBCs. Microstructures and phase analysis were investigated via SEM and XRD.

1 Introduction

1.1 Thermal Barrier Coatings (TBCs)

Thermal barrier coatings (TBCs) provide thermal insulation from the hot combustion gas stream to the metallic parts in the hot sections of gas turbine engines (e.g., turbine blades and combustion chambers) [1, 2]. A thermally sprayed TBC system typically exhibits a bi-layered structure, with a ceramic top coat and a metallic bond coat (BC) [2]. The ceramic top coat (e.g., ZrO_2-Y_2O_3, a.k.a., YSZ) provides thermal insulation and reduces the heat flow to the turbine metallic part. The metallic BC is an oxidation/corrosion-resistant metallic layer. It protects the underlying component and improves the adhesion of the ceramic top coat on the part.

1.2 TBC Manufacturing via APS

Air plasma spray (APS) is the state-of-the-art process for top coat YSZ TBC (~250-500 μm) manufacturing on the static parts of turbine engines; more specifically, combustion chambers, vanes and afterburners. The lamellar nature of thermal spray coatings (with pores and interfaces perpendicular to the heat flow to the component), reduces the thermal conductivity of the top coat to roughly 50% of the fully dense YSZ material (~2.3 W/mK), i.e., to ~1.0-1.5 W/mK [1, 2]. This is highly desirable because it maximizes the temperature protection of the metallic parts. Moreover, thermal spraying exhibits versatility and low production cost and is, therefore, attractive to the industry. In spite of all these advantages, thermally sprayed YSZ TBCs exhibit drawbacks in performance on the dynamic components of jet engines (turbine blades).

1.3 TBC Manufacturing via EB-PVD

Electron beam physical vapour deposition (EB-PVD) is the state-of-the-art technique to deposit top coat YSZ TBCs (~125-250 μm) on the dynamic parts of turbine jet engines; specifically, on turbine blades. EB-PVD YSZ TBCs exhibit few micron-size width (~2-10 μm) vertical columnar grains (perpendicular to the component surface), which are loosely bonded and separated by vertical crack gaps [1, 2].
This type of coating microstructure is paramount for the rotating turbine blades because they operate at high temperatures and high RPM levels. For example, the Snecma M88 engine that is the propulsion of the Rafale fighter has a maximum turbine inlet temperature of 1580°C [3] reaching 17,000 RPM [4]. The combination of high temperature (from combustion) and centrifugal force (from rotation) generates creep on the blades. The "disconnected columns" provide strain tolerance for the YSZ TBC, thereby accommodating creep and coefficient of thermal expansion (CTE) mismatch stress effects. This is the reason why thermally sprayed TBCs are not employed in jet turbine blades; i.e., they lack the strain tolerance to creep.
However, due to the absence of major horizontal defects perpendicular to heat flow and because the gaps are parallel to the heat flow, the thermal conductivity values of EB-PVD YSZ TBCs are higher than those of APS ones, around ~1.5-2.0 W/mK [1, 2]. Moreover, although EB-PVD processing is paramount for YSZ TBC production on jet turbine blades, due to the need of employing vacuum during manufacturing, it lacks versatility and leads to high cost. Therefore, there is a market-pull driving force to replace this technique.

1.4 TBC Manufacturing via SPS

Suspension plasma spray (SPS) is a thermal spray manufacturing technique, based on APS [5-8], which dates from early 2000s [5]. The main difference between the two techniques resides on the size of the powder particles and how these particles are fed into the torch. In APS, microscopic particles (e.g., 10 to 100 μm range) are fed into the torch by a carrier gas (e.g., Ar or N_2). However, in SPS, finely dispersed particles from nano-to-submicron size are employed. These tiny particles cannot be fed using the regular thermal spray powder feeders, because they would clog and block the powder hoses and tubing, thereby impeding spraying.

In SPS, these nano-to-submicron particles are put in a suspension (i.e. slurry), typically using alcohol (or sometimes water) as the liquid medium. The suspension is pumped into the torch, thereby enabling spraying and coating deposition [5-8].

In SPS, the typical lamellar splat-based APS structure is not present. SPS coatings can exhibit (i) dense and crack-free microstructures, (ii) dense and vertically-cracked microstructures, (iii) porous and vertically-cracked microstructures and (iv) porous and vertical feathery-like microstructures [5-8]. The formation of these different microstructures will be highly dependent on the temperature and velocity of these nano-to-submicron size particles, i.e., the processing map [5, 6 and 8].

Due to the capacity to form vertically growth-oriented microstructures, SPS has the potential to replace EB-PVD in the manufacturing of TBCs for turbine blades [7]. The absence of a high-temperature vacuum chamber during SPS manufacturing, which is required in EB-PVD TBC deposition, would reduce the processing cost and improve versatility. This is a strong market-pull driving force in the jet turbine industry.

1.5 TBC Manufacturing via FAPS

As previously stated, nano-to-submicron size finely dispersed particles cannot be fed into thermal spray torches using regular powder feeders due to clogging problems. To address this, in the mid-2010s the National Research Council of Canada (NRC) developed a powder feeder able to feed nano-to-submicron finely dispersed particles into an APS torch by using a carrier gas (e.g., Ar or N_2) [9]. Therefore, a liquid-based suspension is not necessary to spray suspension-type-sized particles. The nano-to-submicron size particles are sprayed just like regular APS micron-sized particles. This process has been termed Finely-dispersed-particle Air Plasma Spray (FAPS), in order to differentiate it from regular APS and SPS. One major advantage of FAPS over SPS is the fact that no liquid-based carrier is needed. This reduces the costs associated with the feedstock transportation, spray booth installation/maintenance and manufacturing costs. Besides, as alcohol is typically used in suspensions [5-8], flammability issues are not present during FAPS manufacturing.

1.6 Objective

The goal of this study was to compare the microstructure, density/porosity, phase composition and elastic properties of one SPS TBC and one FAPS TBC manufactured using the same finely-dispersed YSZ powder and sprayed with the same set of spray parameters (as-sprayed and heated-treated coatings). An APS YSZ TBC was used as a benchmark.

2 Experimental Procedure

2.1 Feedstock Powders and Suspension

For SPS and FAPS processing, a ZrO_2-14wt%Y_2O_3 (14YSZ) (a.k.a., ZrO_2-8%molY_2O_3) was employed (Inframat 4039OR-8601, Inframat Advanced Materials, Manchester, CT, USA). The same powder lot (#IAM10275YZ8) was employed for SPS and FAPS processing in order to guarantee a fair comparison between the two techniques. According to the manufacturer, this powder was synthesized by wet chemical co-precipitation method [10]. The 14YSZ suspension was manufactured at the NRC. The liquid medium was ethanol-based (dispersant added) and the powder loading in the suspension was 20wt% (14YSZ-20wt%). Before SPS processing, the suspension was ball-milled (Al_2O_3 balls) for 2 h.

A ZrO_2-8wt%Y_2O_3 (8YSZ) conventional powder (Metco 204B-NS, Oerlikon Metco, Westbury, NY, USA) was used to produce the benchmark TBC.

2.2 Thermal Spraying

For SPS and FAPS processing, an axial feeding plasma spray torch was employed (Axial III, Northwest Mettech, North Vancouver, BC, Canada). For SPS, the 14YSZ-20wt% suspension was delivered to the torch by a suspension feeder prototype (Polycontrols Technologies Inc., Brossard, QC, Canada), which employed a precision pump and flowmeters.

For FAPS, the finely dispersed 14YSZ as-received powder was fed into the torch by a carrier gas (N_2) via an ultrafine powder feeder prototype (Polycontrols Technologies Inc., Brossard, QC, Canada), which is an NRC-licenced technology [9].

It is important to highlight that both the 14YSZ-20wt% suspension and the 14YSZ finely dispersed powder were sprayed using exactly the same set of spray parameters, including plasma gas composition and flow, torch power, nozzle and spray distance. Moreover, both 14YSZ TBCs (~180-210 µm thick) were sprayed on a carrousel-mounted bond-coated (~160 µm) Inconel 625 substrate pucks (2.5 cm (1") diam. x 3.2 mm (1/8") thick); using the same tangential speed (i.e., RPM) and torch transverse speed. The only difference in spraying was the feed rate, which was 67 g/min for the 14YSZ-20wt% suspension and 20 g/min (1.2 kg/h) for the 14YSZ finely dispersed powder. The MCrAlY BC for both 14YSZ TBCs was sprayed via an APS torch (Axial III, Northwest Mettech, North Vancouver, BC, Canada). The benchmark TBC (8YSZ top coat and MCrAlY BC) was deposited via an APS torch (Metco 3MB, Oerlikon Metco, Westbury, NY, USA). All spray parameters are considered as intellectual property (IP) of the NRC, and therefore cannot be disclosed.

2.3 Free-standing YSZ and Heat-Treatment

It is important to point out that the microstructures of coatings produced using finely dispersed powders (e.g., in suspensions) are highly affected by the roughness/asperities of the substrate surface [7]. Therefore, for the SPS and FAPS 14YSZ TBC samples, free-standing YSZ coatings were obtained by spraying the YSZ onto MCrAlY bond-coated low carbon steel substrates. This procedure was performed in order to guarantee that the both types of free-standing 14YSZ TBCs would exhibit identical coating microstructures as those described in Section 2.2 (including the BC). The BC and substrate were detached from the 14YSZ top coats via acid dissolution.

The 8YSZ was sprayed directly on low carbon steel substrates, which were dissolved in acid for free-standing coating production.

Subsequently, the 14YSZ and 8YSZ free-standing coatings were dried in a furnace at 150°C for 12 h. In addition, sets of free-standing YSZ coatings were also heat-treated in air at 1300°C for 10 h, for an initial microstructural stabilization.

The density (*d*) values of the as-sprayed and heat-treated YSZ samples were measured via the Archimedes technique: d (g/cm^3) = (m_{dry} / (m_{wet} – $m_{immersed}$)); where *m* is mass in grams. A total of 3 samples per coating were used to generate the density values.

2.4 Thermal Conductivity

The thermal conductivity (*TC*) values of the YSZ TBCs were calculated using the traditional thermal conductivity equation:

K (W/mK) = α (m^2/s) x C_p (kJ/kgK) x d (kg/m^3); where K is *TC*, α is thermal diffusivity (*TD*), C_p is the specific heat of the YSZ and d is the coating density. The TD values were measured via the laser-flash diffusivity method (at 1300°C) and the C_p values at 1300°C were obtained from the literature.

2.5 Microstructural Characterization

The TBC samples were vacuum impregnated in low viscosity epoxy resin prior to cutting and polishing. The microstructures (14YSZ powder and TBCs) were analyzed by scanning electron microscopy (SEM). X-ray diffraction (XRD), using CuKα radiation, was employed to evaluate the phase compositions of the 14YSZ powder and as-sprayed TBCs (14YSZ-20wt% SPS and 14YSZ FAPS).

The porosity values of the YSZ TBCs were estimated based on the ratio between the densities of the coatings in relation to the densities of the respective fully dense bulk materials.

2.6 Elastic Modulus and Hardness Values

The elastic modulus (*E*) and hardness (*H*) values of the as-sprayed and heat-treated (1300°C-10h) YSZ TBCs were measured via instrumented indentation testing (Nano Indenter G200, Agilent Technologies, Santa Clara, CA, USA) on the cross-sections of the coatings. On average, 10 indentation measurements (Berkovich indenter at 50 gf maximum peak load) were used to determine *E* and *H* values, at room temperature (RT).

3 Results and Discussion

3.1 14YSZ Powder Microstructure

Images of the microstructure of the 14YSZ powder can be found in **Fig. 1**. It is possible to observe that the particles are highly dispersed in size (**Fig. 1a**), and some large porous agglomerated particles can exhibit diameters of up to ~100 µm (**Fig. 1b**). However, the external surface of the powder shows that the "fluffy" agglomerates are formed by individual nano-sized particles (**Fig. 1c**).

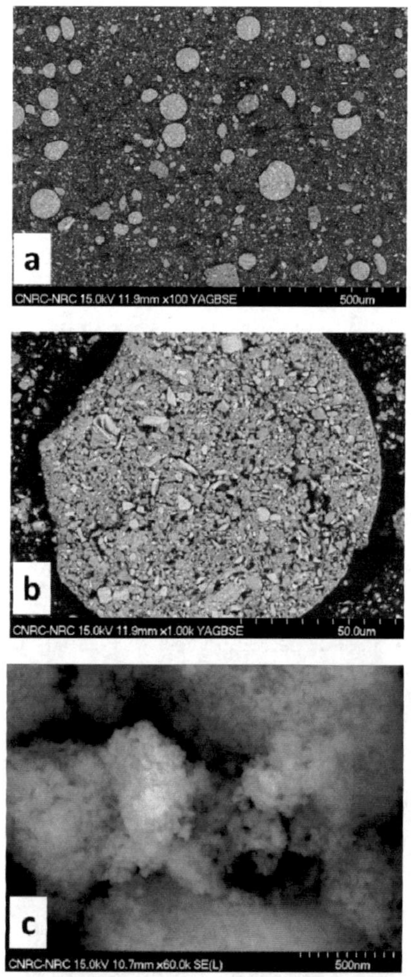

Fig. 1. 14YSZ powder – (a-b) polished cross-section and (c) external surface.

3.2 As-sprayed TBC Microstructure

It is important to highlight that, except for the feed rates, both 14YSZ TBCs were sprayed using the exact same set of spray parameters and torch/substrate relative motion. The maximum 14YSZ coating temperature deposition (pyrometer) was ~400°C for SPS and ~500°C for FAPS. The feed rates for SPS (67 g/min) and FAPS (20 g/min) processing yielded deposition rates of ~6 µm/pass and ~22 µm/pass, respectively. Both 14YSZ TBCs were sprayed on the same type of MCrAlY BC; which exhibited a roughness (R_a) of 5.2 ± 0.7 µm. The R_a of the SPS and FAPS YSZ TBCs were 3.6 ± 0.7 µm and 5.8 ± 0.8 µm; respectively. The polished cross-section microstructure of the as-sprayed 14YSZ-20wt% SPS TBC can be found in **Fig. 2**. It is possible to observe the vertically cracked (**Fig. 2a**) but uniformly porous coating microstructure (**Fig. 2b**).

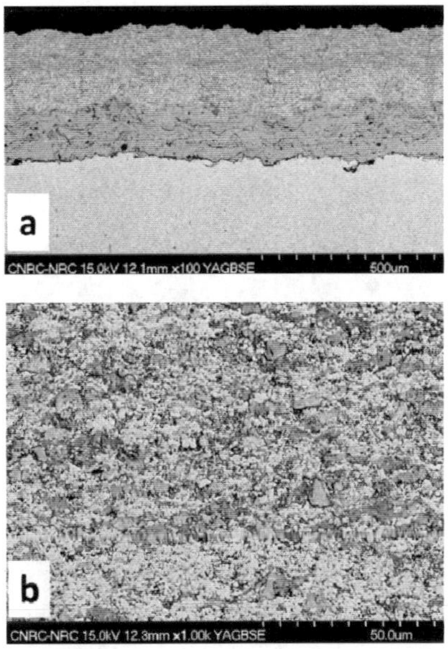

Fig. 2. As-sprayed 14YSZ-20wt% SPS TBC.

The polished cross-section microstructure of the as-sprayed 14YSZ FAPS TBC can be found in **Fig. 3**. The coating is well-adhered on the bond-coated substrate surface. The vertically cracked microstructure is also present; however, the coating is denser and exhibits a bimodal microstructure (**Fig. 3a**). The bimodal microstructure is formed by particles that were fully or highly molten in the spray jet (matrix) and pockets of large partially molten agglomerates that were embedded in the microstructure during spraying, flattening upon impact (**Fig. 3b**). Regarding the embedded particles (**Fig. 3b**), it is possible to notice the similarities with the microstructure of the large agglomerates (**Fig. 1b**). The as-sprayed 14YSZ-20wt% SPS and 14YSZ FAPS YSZ TBCs exhibit ~5 vertical cracks/mm and ~4 vertical cracks/mm, respectively. Therefore, FAPS processing has the potential to create a similar columnar microstructure as that of SPS, but without using suspensions. Columnar vertically cracked SPS YSZ TBCs are being tested in order to try to replace EB-PVD YSZ TBCs in the rotating turbine blades of gas turbine engines [7].

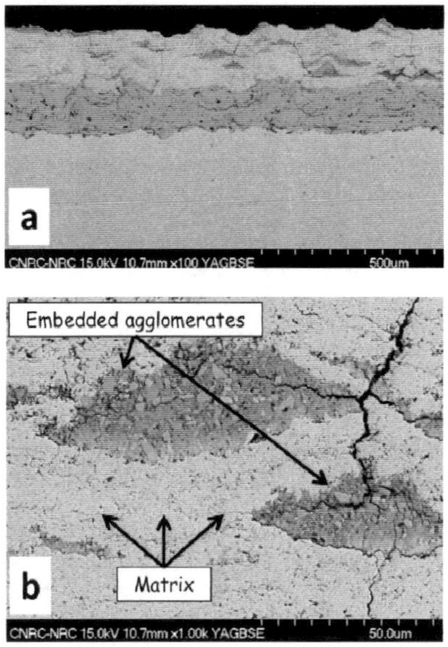

Fig. 3. As-sprayed 14YSZ FAPS TBC.

The overall microstructure of the as-sprayed APS 8YSZ TBC benchmark is shown in **Fig. 4a**. It exhibits the traditional lamellar micro-cracked morphology, with inter-splat and globular pores (**Fig. 4b**).

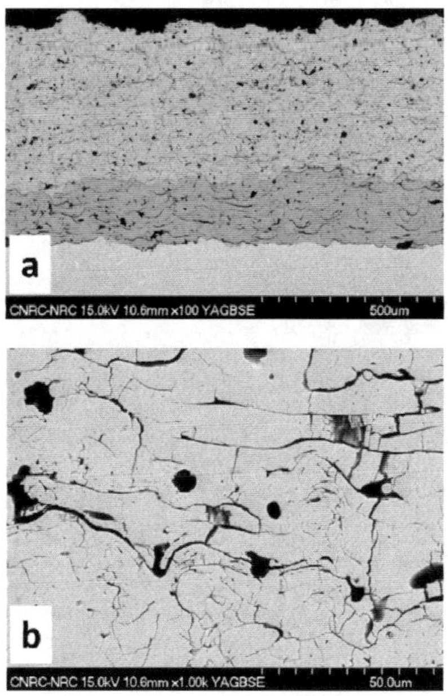

Fig. 4. As-sprayed 8YSZ APS TBC benchmark.

3.3 Phase Composition via XRD

The XRD diffraction patterns of the 14YSZ feedstock powder, and as-sprayed 14YSZ-20wt% SPS and 14YSZ FAPS TBCs are shown in **Fig. 5**. The main phase of the 14YSZ powder is cubic YSZ, with a minor presence of the monoclinic YSZ. For both 14YSZ coatings, all the main peaks of the cubic phase are present and no other phase is detected. This occurs because at 14wt% of Y_2O_3 or higher, ZrO_2 stabilizes in the cubic phase [11]. The results indicate that the processing technique (SPS or FAPS) did not affect the phase composition of the coatings.

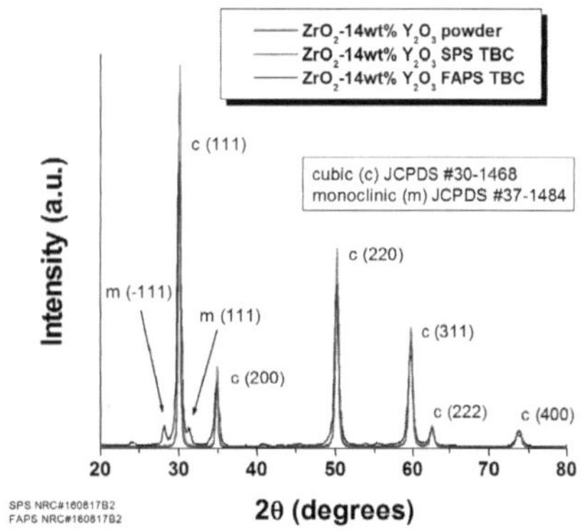

Fig. 5. XRD patterns of 14YSZ powder and TBCs.

3.4 Mechanical and Physical Properties

The *E* and *H* values of the YSZ TBCs are listed in **Table 1**, as well as, density and porosity values. The porosity values were estimated based on the ratio between the coating density (Archimedes technique) with that of a fully dense bulk sample. The density values for dense bulk 14YSZ and 8YSZ have been reported as 6.00 g/cm^3 and as 6.03 g/cm^3, respectively [12].

Table 1 – Mechanical and Physical Properties

Coating	d (g/cm^3) (n=3)	P (%) (n=3)	H (GPa) 50gf (n=10)	E (GPa) 50gf (n=10)
14YSZ-20wt% SPS (as-sprayed)	4.0	33	2.8±0.3	62±4
14YSZ-20wt% SPS (1300°C-10 h)	4.6	23	6.6±1.4	106±16
14YSZ FAPS (as-sprayed)	4.6	23	7.2±3.1	118±31
14YSZ FAPS (1300°C-10 h)	4.8	20	10.4±2.7	148±24
8YSZ APS (as-sprayed)	5.2	14	7.6±1.3	143±15
8YSZ APS (1300°C-10 h)	5.3	12	12.2±1.6	180±24

The *E* and *H* values ranked from the highest to the lowest are found for the APS, FAPS and SPS TBCs; for the as-sprayed and heat-treated samples (**Table 1**). This is directly linked to their porosity levels, which follow the exact opposite trend.

The *E* and *H* values for the SPS and FAPS TBCs are lower than those for the APS TBC benchmark. Consequently, the SPS and FAPS TBCs probably exhibit lower stress levels under strain than those of the APS TBC.

3.5 Heat-treated TBC Microstructures

The cross-sectional microstrucures of all heat-treated YSZ TBCs (1300°C-10h) are shown in **Fig. 6**.

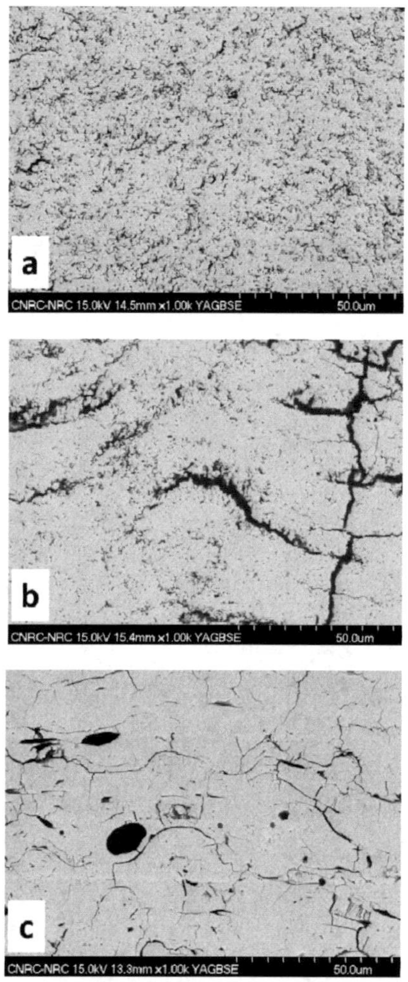

Fig. 6. Heat-treated (1300°C-10h) (a) SPS, (b) FAPS and (c) APS YSZ TBCs.

When comparing the microstrucures at the same magnification (1 kX) of the as-sprayed (**Fig. 2b**) and heat-treated (**Fig. 6a**) 14YSZ-20wt% SPS, it is possible to observe a marked and uniform coating densification after the high temperature exposure. This microstructural effect helps to explain the average increase of ~2X of the *E* and *H* values of the SPS TBC after heat treatment (**Table 1**).
On the other hand, when comparing the microstrucures at the same magnification (1 kX) of the

as-sprayed (**Fig. 3b**) and heat-treated (**Fig. 6b**) 14YSZ FAPS, the densification was less intense. It is possible to observe the formation of horizontal pores or voids after heat treatment. These horizontal pores were formed from the pockets of large partially molten agglomerates that were embedded in the as-sprayed microstructure during spraying (**Fig. 3**). At high temperature, the fine particles in these pockets likely exhibited higher sintering rates than the matrix, thereby collapsing and creating the voids. This is very similar to an event previously reported [13].
The microstructures at the same magnification (1 kX) of the as-sprayed (**Fig. 4b**) and heat-treated (**Fig. 6c**) 8YSZ benchmark TBC, also show a slight densification of the microstructure after heat treatment. The horizontal lamellar interfaces are clearly distinguishable in the coating microstructure, even after the high temperature exposure.

3.6 Thermal Properties

The *TD* and *TC* values of the heat-treated (1300°C-10h) YSZ TBCs, measured at 1300°C, are listed in **Table 2**. The heat treatment at 1300°C-10h was employed to guarantee that the microstrucutres would remain nearly stable prior to *TD* measurement at 1300°C.

Table 2 – Thermal Properties at 1300°C

Coating	*TD* x 10^{-7} (m^2/s) @1300°C (*n=3*)	*TC* (W/mK) @1300°C (*n=3*)
14YSZ-20wt% SPS (1300°C-10 h)	3.5	1.08
14YSZ FAPS (1300°C-10 h)	3.3	1.04
8YSZ APS (1300°C-10 h)	3.0	1.06

The *TD* values ranked from the lowest to the highest are found for the SPS, FAPS and APS TBCs. This is the same trend of the porosity values for the same coatings (**Table 1**). *TD* can be described as a measure of "how fast" heat propagates through the thickness of the coating. One could argue that for a given material the lower the porosity values, the higher the expected *TD* values. It is not the objective of this paper to investigate and explain this "apparent paradox". However, by looking at the microstructures of the as-sprayed and heat-treated TBCs, one can establish a hypothesis. It is possible to observe that the microscopic horizontal interfaces (in the relation to the substrate surface) disappeared for the heat-treated SPS (**Fig. 6a**) and FAPS (**Fig. 6b**) TBCs; i.e., only the pores can be observed. On the other hand, the horizontal inter-lamellar interfaces are still very visible for the heat-treated APS TBC, in addition to the pores (**Fig. 6c**). Therefore, the microscopic horizontal inter-lamellar interfaces may have a more important role than the porosity itself in determining the TD values of thermal spray coatings.
Regarding the *TC* values, they are similar for the APS, FAPS and SPS YSZ TBCs (~1.1 W/mK). Consequently, the *TC* values for SPS and FAPS YSZ TBCs are in the range of those for APS YSZ TBCs (i.e., ~1.0-1.5 W/mK) and below the values reported for EB-PVD YSZ TBCs (i.e., ~1.5-2.0 W/mK) [1, 2]. Therefore, FAPS is a potential candidate to compete with SPS to replace EB-PVD YSZ TBCs (dynamic turbine components). In addition, the premilinary results on (i) microstructure, (ii) mechanical and (iii) thermal properties suggested that FAPS may also have the potential to replace APS processing in the manufacturing of TBCs (static turbine components).

4 Conclusions

SPS-type finely dispersed particles can be thermally sprayed without suspension by using a newly developed powder feeder. The fine particles were fed into an air plasma spray torch via a carrier gas and a well-adhered coating was uniformly deposited over the bond-coated substrate surface (like traditional thermal spraying). This new processing approach was termed Finely-dispersed-particle Air Plasma Spray (FAPS).
14YSZ TBCs were produced via SPS and FAPS and both coatings exhibited porous and vertically-cracked microstructures. Therefore, FAPS processing has the potential to engineer SPS-like coatings without using suspensions. Morever, both SPS and FAPS 14YSZ TBCs exhibited similar *TC* levels to those of the 8YSZ APS TBC benchmark. The *E* and *H* values for the FAPS TBC were found to be between those of SPS and APS TBCs. Therefore, FAPS has the potential to replace EB-PVD and/or APS in TBC manufacturing. Extensive sets of testing need to be performed to validate this hypothesis.

5 Literature

[1] N. P. Padture, M. Gell and E. H. Jordan, *Science*, 2002, **296**, p 280-284
[2] A. Feuerstein, J. Knapp, T. Taylor, A. Ashary, A. Bolcavage and N. Hitchman, *Journal of Thermal Spray Technology*, 2008, **17(2)**, p 199-213
[3] www.snecma.com/military-engines/training-and-combat-aircraft/m88 (January 12, 2016)
[4] www.defense-aerospace.com/articles-view/release/3/10441/rafale-m88-successful-in-carrier-ops-(june-17).html (January 6, 2017)
[5] J. Oberste-Berghaus, S. Bouaricha, J.-G. Legoux and C. Moreau, *Proceedings of the International Thermal Spray Conference 2005 (ITSC 2005)*, E. Lugsheider, Ed., May 2-4, 2005 (Basel, Switzerland), DVS, Dusseldorf, Germany, p 1-7
[6] K. VanEvery, M. J. M. Krane, R. W. Trice, H. Wang, W. Porter, M. Besser, D. Sordelet, J. Ivasky and J. Almer, *Journal of Thermal Spray Technology*, 2011, **20(4)**, p 817-828
[7] N. Curry, K. VanEvery, T. Snyder and N. Markocsan, *Coatings*, 2014, **4**, 630-650; doi: 10.3390/coatings4030630
[8] R. M. Seshadri, G. Dwivedi, V. Viswanathan and S. Sampath, *Journal of Thermal Spray Technology*, 2016, **25(8)**, p 1666-1683
[9] B. Harvey, E. Irissou, J.-G. Legoux, R. S. Lima, J.-C. Tremblay, E. Gagnon and S. Desaulniers, Powder Feeder Method and System, *US Patent 9,505,566 B2*, Nov. 29, 2016
[10] www.advancedmaterials.us/4039OR-8601.htm (October 12, 2016)
[11] K. Muraleedharan, J. Subrahmanyam and S. B. Bhaduri, *Journal of the American Ceramic Society*, 1988, **71(5)**, C-226-C-227
[12] R. P. Ingel and D. Lewis III, *Journal of the American Ceramic Society*, 1986, **69(4)**, p 325-332
[13] R. S. Lima and B. R. Marple, *Journal of Thermal Spray Technology*, 2008, **17(5-6)**, p 846-852

Iron based hardfacing alloys for abrasive and impact wear

B. Maroli, Höganäs AB, Sweden,
S. Dizdar, Höganäs AB, Sweden
S. Bengtsson, Höganäs AB, Sweden

Iron-based hardfacing alloys are widely used to counteract abrasive and impact wear of industrial components in soil, sand and mineral processing applications. These alloys show a high performance to cost ratio as well as a low environmental impact. The wear resistance of the components hardfaced with these alloys depends on achieved coating microstructure i.e. on the alloys chemical composition, the coating method and process parameters selected.

The present work focuses on iron based hardfacing alloys with varying amount of chromium, vanadium, tungsten, molybdenum, boron and carbon deposited by plasma transferred arc (PTA) overlay welding. Weldability, hardness, abrasive and impact wear of the overlays are presented and interpreted through their microstructure. The performance of the iron based overlays is compared with that of nickel-based metal matrix composite coatings with tungsten carbide (MMC) commonly used for hardfacing of parts subjected to severe abrasive wear. The hardness of the iron based overlays investigated ranges between 60 and 65 HRC while abrasive wear is typically below 20 mm^3 (ASTM G65, procedure A). Microstructure consists of different primary precipitated carbides or borides, a martensitic matrix and eutectic structures.

1 Introduction

Industrial parts are replaced at regular intervals due to wear issues. Hardfacing of the wear-exposed region by laser cladding or PTA welding is commonly used to prolong the life of a component and reduce production costs. Both laser cladding and PTA overlay welding ensure strong metallurgical bonding to the substrate combined with limited dilution from the parent metal, typically below 15% and relatively thick and pore free deposits. The low heat input and fast speed of the laser cladding process result in lower dilution from the substrate, when compared to PTA and limited distortion of the coated parts. PTA welding allows instead depositing weld beads having larger width and height when compared to laser cladding.

Selection of the alloy that best solve a specific wear problem is often based on practical experience, on the hardness of the coating or the amount of hard phase particles present. State of the art alloys for PTA overlay welding and laser cladding, normally used to combat abrasive wear, are nickel-based mixes with different amounts of tungsten carbides typically ranging from 40 to 60 wt.%. In applications where both abrasive and impact wear resistance are required iron based grades can be an alternative where high performance to cost ratio is combined with low environmental impact.

Iron based grades for hardfacing of parts exposed to wear are alloyed with different amounts of chromium, molybdenum, vanadium, tungsten, carbon and boron and normally consist of a martensitic matrix reinforced by carbide and/or boride particles. The hard phase particles solidify either as primary precipitates from the melt, as part of eutectic structures or a combination of both. The nature, size and amount of the hard phase particles are critical in determining the wear properties of the final coating.

Nickel-based alloys with tungsten carbides and iron based alloys are used for hardfacing wear plates, parts of earthmoving equipment, cutting shears, mill hammers, digging tools, extrusion screws and ball mills [1].

The present work focuses on iron based hardfacing alloys with varying amount of chromium, vanadium, tungsten, molybdenum, niobium, carbon and boron deposited by PTA overlay welding. Weldability, hardness, abrasive and impact wear of the overlays are presented and interpreted through their microstructure. The performance of the iron based coatings is compared with that of nickel-based coating with tungsten carbides.

2 Experiments

Four different high-speed steel powders were investigated; Höganäs grades M2, M4, A11 and Rockit®701. All powders were gas atomized and sieve cut was 53 to 150µm to fit the requirement of the powder feeder used for the overlay welding. The chemical composition of the investigated grades is summarised in Table 1. The powders were deposited on EN S235JR mild structural steel plates using a commercial PTA unit (Commersald 300I). A welding current of 130A, feed rate of 30 g/min and welding speed of 10 cm/min were selected for coating of the substrate samples. A mixture of argon and 5% hydrogen was used as shielding gas.

Table1. Typical chemistry of investigated grades (wt%)

	C	V	Cr	Mo	W	Nb	Others	Fe
M2	0,8-1,1	1,6-2,2	3,2-4,5	4,5-5,5	5,5-6,8			Bal.
M4	1,1-1,4	3,7-4,2	4,2-4,7	4,2-4,8	5,2-5,7			Bal.
A11	2,4-2,6	9-10	4,7-5,7	1,1-1,5				Bal.
701	<1		<4		<9,5	<1,6	<7	Bal.

The properties of the iron-based overlays were compared to those of a metal matrix composite coating consisting of a Ni-based matrix with tungsten carbides (MMC). A gas atomized Ni-1.5Si-2.9B powder (Höganäs grade 1559-40) with typical hardness 49 HRC, was admixed with 50 wt% macrocrystalline tungsten carbides, WC, (Höganäs grade 4580). Size of the carbide and of the NiSiB powder particles was 53 to 150µm. Similarly, to the iron based grades the mix was deposited using the same PTA unit. Welding current was 90 A, feeding rate was 30 g/min and cladding speed was 11 cm/min. The welding parameters were selected in order to assure good bonding to the substrate, good mixing of the coating material with the parent metal, even distribution of the tungsten carbides and dilution below 15%.

The overlays were investigated for presence of cracks and other surface flaws. They were cleaned (CRC Crick 110) and then coated with a red dye (CRC Crick 120) penetrating into surface defects or cracks through capillary forces. After 10 minutes excess dye was removed from the surface and a white developer (CRC Crick 130) applied. The developer drew the penetrant out of crevices, cracks or other hollow imperfections communicating with the surface.

The overlays were sectioned perpendicular to the coating direction, moulded in bakelite, ground and polished using standard procedures for metallographic sample preparation. The cross section was examined using a light optical microscope (Leica DM 6000) and a FEGSEM (Hitachi FU6600) equipped with a silicon drift detector (SDD) for EDS analysis (Quantax 800 Bruker). The as-polished samples were examined for dilution of the coating by the substrate material and for defects, such as pores, cracks, bonding defects, etc. Dilution was determined geometrically on the cross section of the coating. Prior to measurement, the metallographic moulds were etched in Nital 1% to attack the substrate material and in this way facilitate the detection of the coating. The total coating area ($A_{coating}$ + $A_{substrate}$) and the area of the coating that used to be substrate prior to overlay welding ($A_{substrate}$) were measured by image analysis. Dilution was calculated as shown in Eq. 1.

$$Dilution\% = \frac{A\,substrate}{(A\,coating + A\,substrate)} x100 \qquad \text{[Eq. 1]}$$

Estimation of the area fraction of the microstructure constituents in the overlays was carried out by lineal analysis on oxide polished samples to highlight hard phase particles and eutectic structures. The samples were then etched in Vilella (100 ml 95 % ethyl alcohol, 1 g picric acid, 4 ml concentrated HCl) to reveal the matrix microstructure. Hardness of the microstructure constituents was measured on the metallographic moulds using a hardness tester (Matsuzawa MMT-7). A load of 25 g was used. Hardness HV0,025 was expressed as the average of seven measurements.

Rockwell hardness HRC was measured using Wolpert Universal hardness tester. The coatings were ground, seven hardness indents were performed on the flat surface and the average was calculated.

Low stress abrasive wear testing was performed according to ASTM G65 standard, procedure A [2], by using a commercial multiplex sand/wheel abrasion tribometer (Phoenix tribology TE 65). A schematic of the testing is illustrated in Fig. 1a. Five sample replicas per material were tested.

Impact wear testing was performed by using an inhouse build tribometer. The schematic is shown in Fig. 1b [3]. Standard steel bearing balls drop from pre-defined heights onto the coated test specimen. A data point corresponds to the total number of ball strikes for a pre-defined height i.e. impact energy until occurrence of a first circular crack around the impact dent. This type of model impact wear testing is suitable to rank impact wear resistance of materials exposed to impact overloads at relatively low impact velocities. Operation conditions closest to the modelling in this testing can be exemplified by a first contact of the excavator bucket teeth with the ground; by filling of the excavator buckets by the dig-out material; by forwarding the dig-out material to the truck bed etc. Abrasive wear is removed from this testing in difference to a combined abrasive-impact wear testing [4].

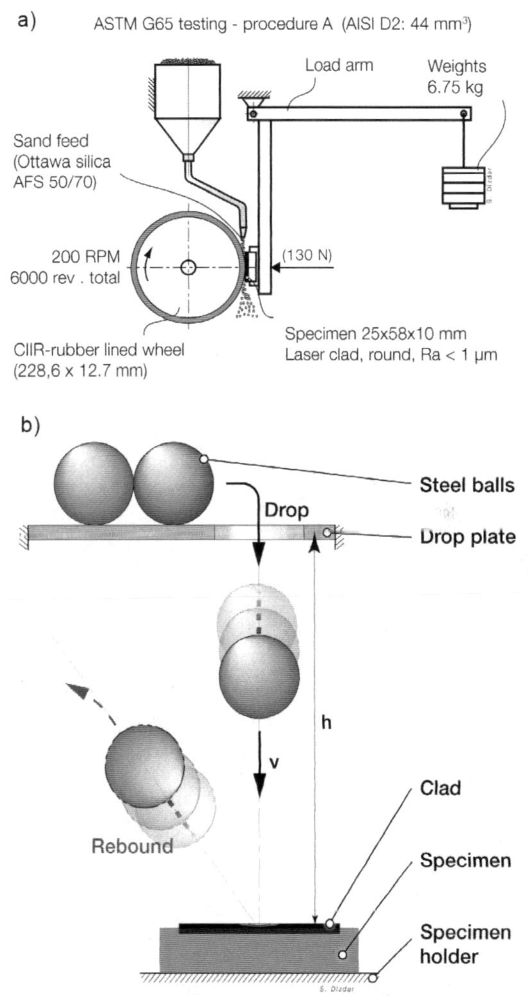

Fig. 1. a) Schematic of abrasive wear testing according to ASTM G65 standard, procedure A. b) Schematic of the impact wear testing setup [3].

3 Results

Overview of the overlays produced with M2, M4, A11 and Rockit®701 is presented in Fig. 2a while the results from the crack investigation by using the liquid penetrant test are shown in Fig. 2b. No defects were observed except for porosity in the stop position illustrated by the arrow in Fig. 2b.

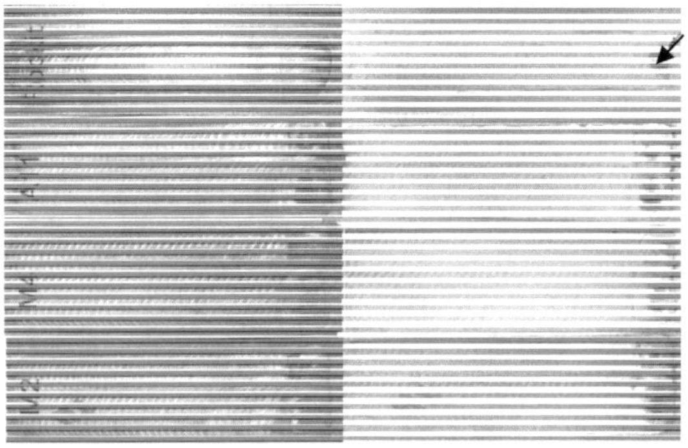

Fig. 2. a) Photograph of PTA overlay welds made of Rockit ®701 (single track), A11, M4 and M2 (two overlapping tracks). b) Liquid penetrant test of samples.

Overview of the coating cross section is shown in Fig. 3. The overlays exhibited good bonding to the substrate with no cracks and few sporadic pores below 100µm. The dilution from the parent metal measured on the cross section of the coating was below 15% for all investigated grades. Dilution was 8% for the M2, 2% for M4 and 5% for A11 and Rockit®701.

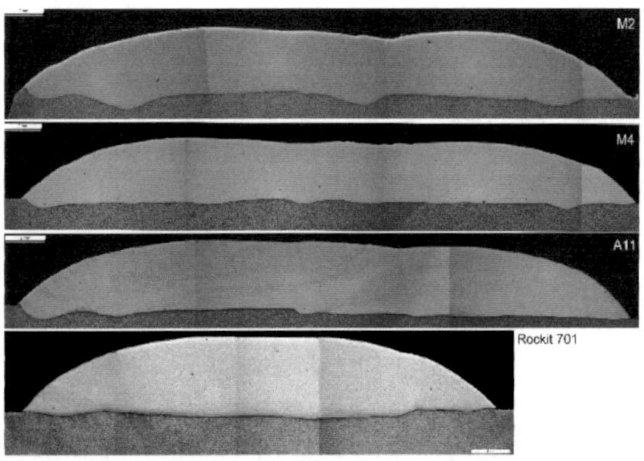

Fig. 3. Cross section of PTA weld. Etching was carried out in Nital 1% to attack only the substrate material and in this way highlight the coating.

Typical microstructure of investigated coatings is illustrated in Fig. 4 and Fig. 5. Analysis by light optical microscopy of M2, M4 and Rockit®701 overlays showed primary dendrites and cells of martensite (M) as well as a eutectic structure between the primary grains (E), Fig. 4. SEM EDS analysis of the M2 and M4 overlays revealed the presence of two types of carbides: a) vanadium rich carbides of a few microns in size located in the grain boundaries (C1); b) molybdenum and tungsten rich eutectic carbides (C2). The microstructure of M2 and M4 overlays was similar except for a somewhat larger amount of vanadium rich carbides in M4. The area fraction of eutectic structure was estimated to 5% for M2 and 8% for M4. The area fraction of the vanadium rich carbides was estimated to be 1% for M2 and 5% for M4. Hardness HV0.025 of the martensitic matrix was similar for both material and corresponding to 802 ± 40 for M2 and 798 ± 30 for M4.

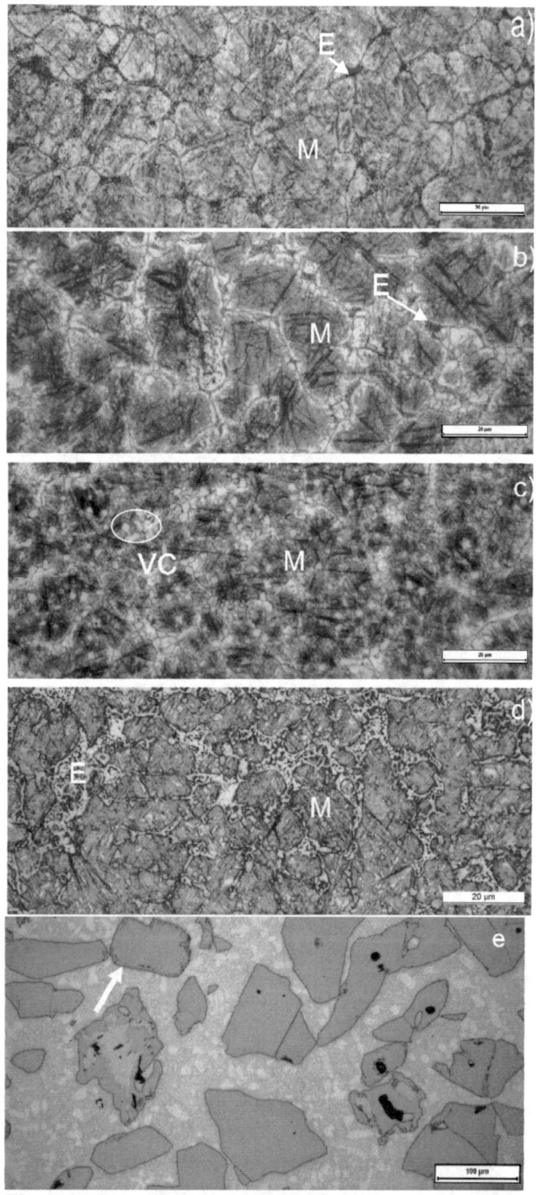

Fig. 4. Light optical micrograph of a cross section of the investigated overlays. a) M2; b) M4; c) A11; d) Rockit ®701 (a-d are etched in Vilella); e) MMC overlay, oxide polished using colloidal SiO_2, lower magnification than for the iron based coatings.

Examination of Rockit® 701 by light optical microscopy showed a martensitic matrix (M) with a hard phase network bewteen the martensitic grains (E), Fig. 4d. SEM EDS analysis showed the presence different primary hard phase particles, Fig. 5d: a) niobium rich carbides (C); b) tungsten rich borides (B); c) Cr rich eutectic carbides (E). Size of the eutectic islands was up to approximately 15 µm, size of the tungsten rich boride areas was up to 10 µm and finally the primary carbides were in the range of 1 micron. The area fraction of hard phase including the eutectic structure was estimated to be 30%, whereof 3% consisted of niobium rich carbides. Hardness HV0.025 of the martensitic matrix was 830±30.

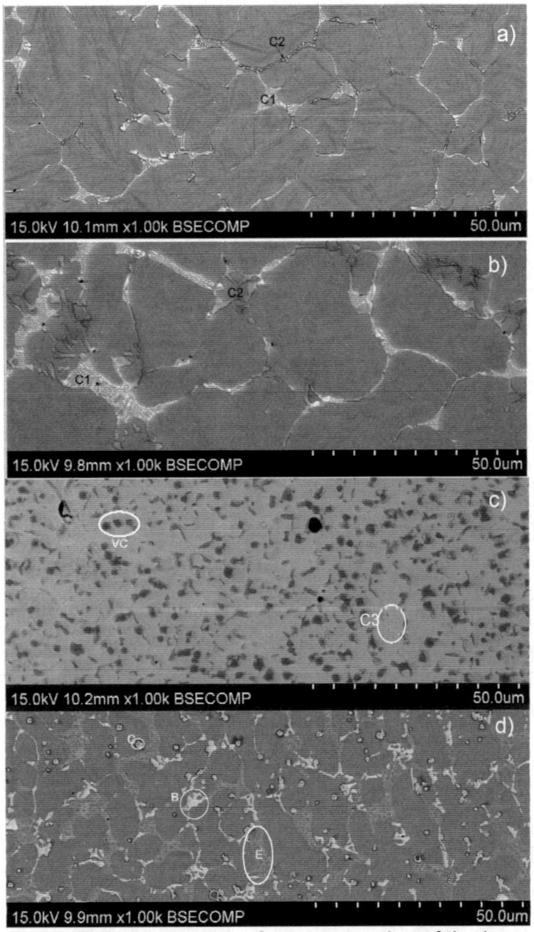

Fig. 5. SEM micrograph of a cross section of the iron based overlays as-oxide polished. a) M2; b) M4; c) A11; d) Rockit® 701.

Microstructure of A11 is illustrated in Fig. 4c and 5c. In A11 solidification started with the formation of primary vanadium rich carbides followed by the formation of primary austenitic dendrites and cells transforming into martensite upon cooling. Very small amounts of chromium and molybdenum rich carbides were observed in the grain boundaries (C3). Some retained austenite was found at the periphery of the martensitic grains (A). The area fraction of the primary vanadium rich carbides was 20%. The vanadium rich carbides were evenly distributed along the coating; they were rounded in shape and roughly 2µm in size.

The microstructure of the MMC overlays consisted of coarse WC particles embedded in a NiSiB matrix consisting of austenitic primary dendrites and mainly Ni-Si-B eutectic structure, Fig. 4e. Dissolution of the carbides during overlay welding was limited and mainly localized in an area of few microns in depth at the periphery of the WC particles, as indicated by the arrow in Fig. 4e. Estimated area fraction of the tungsten carbide particles was 35%.

Hardness HRC and abrasive wear measured according to ASTM G65 – Procedure A, for the investigated materials is presented in Fig. 6a. Rockit®701 showed highest hardness - 65 HRC. The high speed steels M2, M4 and A11 showed similar hardness, 61 HRC for M2 and M4 and 60 HRC for A11. Abrasive wear was lowest, 8 mm^3, for the MMC overlay with 50wt% WC. Rockit®701 and A11 showed similar abrasive wear of 17 mm^3 while M4 showed a wear of 45 mm^3 and finally M2 a wear of >50 mm^3. The impact energy as a function of the number of strikes needed to achieve the first crack in 50% of the tests is shown in Fig. 6b. Rockit®701 overlay showed a higher overload capacity and about ten times longer survival time than the MMC overlay.

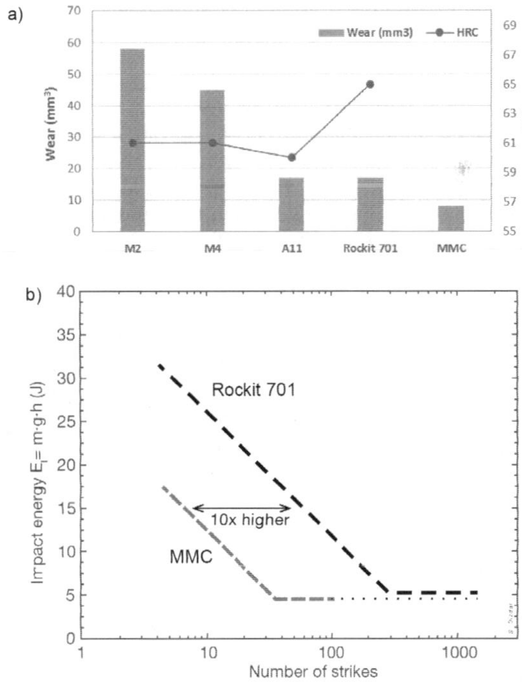

Fig. 6. a) Low stress abrasive wear (ASTM G65, procedure A) and hardness HRC for M2, M4, A11, Rockit® 701 and MMC overlays. b) Impact energy as a function of number of strikes to achieve first crack in 50% of the tests.

SEM analysis of the wear track after abrasive wear testing showed typical ploughing marks – scratches by the silica sand abradent, Fig. 7. The ploughs had a uniform direction in all overlays except for the MMC one. In the MMC overlay, abrasion of the nickel-based matrix around the tungsten carbide particles and rounding of the carbide edges were observed. For the iron-based overlays instead removal of the matrix and hard phase particles was more even. Mapping of the alloying elements in the wear scar by EDS showed that the asperities present (white arrows in Fig 7), were mainly vanadium rich carbides in M4 and A11 and niobium rich as well as tungsten rich hard phase particles in Rockit®701.

5 Discussion

The hardness of the iron-based overlays was, as expected, not in correlation with abrasive wear resistance. Instead, the wear was in correlation with the properties of the hard phase particles formed during solidification from the melt. M2, M4 and A11 overlays showed similar hardness of 60-61 HRC but different wear, 17mm^3 for A11, 40mm^3 for M2 and >50mm^3 for M2. Rockit®701 instead showed highest hardness but it performed similar to A11 in the low stress abrasive wear test. All overlays showed a martensitic matrix with similar hardness. Main difference between the iron-based grades was the type and amount of hard phase particles present in the overlay.

The M4 overlay showed a larger amount of hard phase than M2. Further size of the carbides was somewhat coarser in M4. The higher vanadium content resulted in the formation of a slightly larger amount of vanadium rich carbides in M4. The vanadium rich carbides are most probably hard and abrasive wear resistant MC carbides [5, 6]. Typical Vickers hardness for VC is 2700 [7]. The molybdenum and tungsten rich carbides present in the eutectic structures were most probably relatively soft M_6C or M_2C carbides (1500-2000 HV0,025), less abrasive wear resistant than vanadium rich MC carbides [5, 6]. This was confirmed by analysis of the wear track showing least wear for the vanadium rich carbides. The increased amount of eutectic structure and mainly of vanadium rich carbides substantially increased the abrasive wear resistance of M4 compared to M2, without affecting the coating hardness. M2 and M4 showed lower resistance to abrasive wear than A11 and Rockit®701 due to the smaller volume fraction of hard phase particles combined with the small amount of hard vanadium rich carbides.

The hard phase particles contributing to the abrasive wear resistance of Rockit®701 overlay were mainly the primary niobium rich MC carbides and tungsten rich borides, probably M_2B [5] as confirmed by the analysis of the wear track. Typical Vickers hardness for NbC is 2000 [6], while for W_2B it is 23,7GPa corresponding to 2400HV [8]. The chromium rich carbides present in the eutectic structure were most probably relatively soft $M_{23}C_6$ [5]. Hardness of these carbides is in the range of

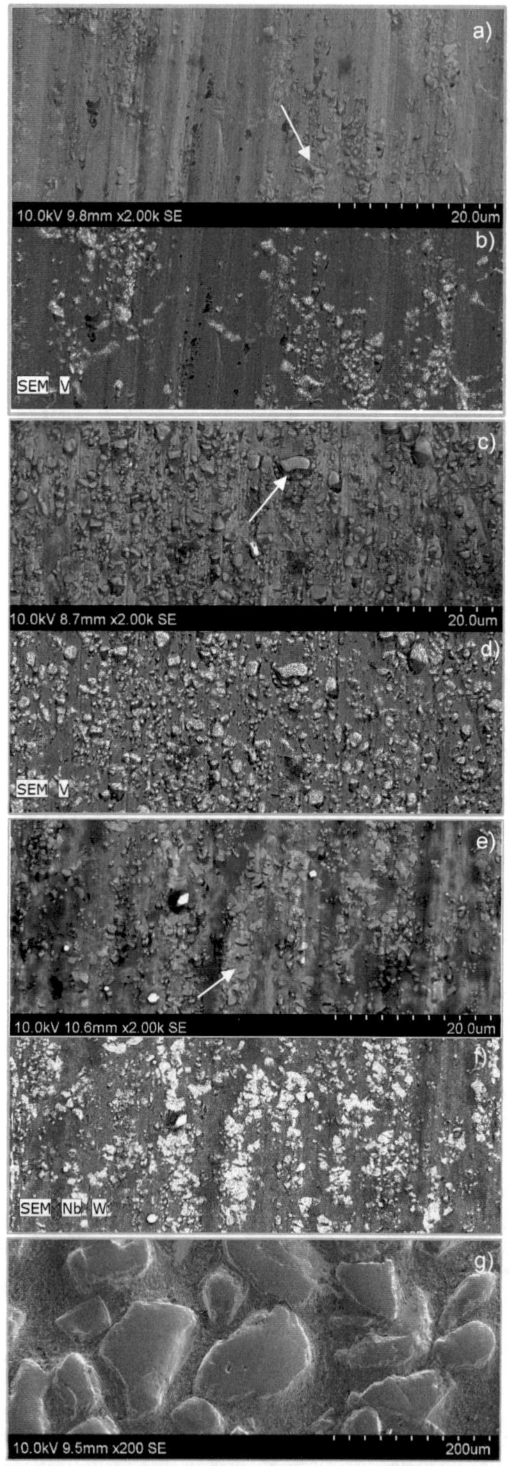

Fig. 7. Morphology of wear scars and elemental mapping of the hard phase particles. a) M4, wear scar morphology; b) M4, elemental mapping of V c) A11 wear scar morphology; d) A11 elemental mapping of V; e) Rockit®701 wear scar morphology; f) Rockit®701 elemental mapping of Nb and W; g) 1559-40 + 50wt% 4580 wear scar morphology

1000HV [6] and very similar to that of the matrix. In A11, the vanadium rich, MC, carbides provided resistance to abrasive wear. These carbides were harder than the hard phase particles formed in Rockit®701. The higher hardness and even distribution of the vanadium rich, carbides, formed in A11, could explain why the wear resistance of A11 was similar to that of Rockit ®701 despite the total volume fraction of hard phase particles was smaller in A11. The lower HRC hardness, measured on A11, when compared to Rockit®701, could be due to the smaller volume fraction of hard phase particles in A11 combined with the presence of retained austenite.

In the abrasive wear test the MMC overlay outperformed the iron based grades showing 50% less wear than A11 and Rockit®701. The area fraction of the tungsten carbide particles in the MMC overlays was 5% larger than in Rockit®701 and 15% larger than in A11. Further, the nickel-based matrix consisted mainly of NiSiB eutectic. Another major difference between the MMC and iron based overlays was the size of the hard phase particles. Size of the tungsten carbide particles in the MMC overlay was 53-150 µm while that of the hard phase particles in the iron-based coatings was in the range of a few microns. Finally, size of the abrading silica sand particles was 212-425 µm. For the MMC overlay, the abrading silica sand particles bridged over the hard phase particles. The matrix wear occurred by ploughing on surface areas where the bridge could not be established due to less dense distribution of the hard phase particles and /or by abrasion caused by oblong silica sand particles. The highest abrasive wear resistance of the MMC overlay was due the larger area fraction of hard phase combined with the larger size of the tungsten carbide particles.

Impact wear resistance of Rockit®701 and MMC overlays differed at high impact energies, such as 10 to 30 J, but appeared to be similar approaching low impact energies such as 5J. It is known that impact loading affects inertia of the microstructure constituents, which "flows" through the deforming region in the impact stress field [10]. Relatively large and heavy hard phase particles, tungsten carbide, more than twice heavier in comparison to nickel based matrix, likely accelerated crack formation in MMC overlay at high impact energies/velocities. Low impact energies, such as 5 J, did not cause crack formations. In contrast, Rockit®701 overlay included much smaller hard phase particles, which as shown, resulted in less cracking.

6 Conclusions

Properties of PTA welded iron based hardfacing grades with different amounts of chromium, vanadium, molybdenum, tungsten, carbon and boron were investigated and compared with those of a NiSiB overlay with 50 wt. % macrocrystalline tungsten carbides.

- The investigated iron based grades performed well during PTA welding and sound clads with neither cracks nor critical porosity or other types of discontinuities were produced.
- Hardness of the investigated iron based overlays is not a measure of their low stress abrasive wear resistance. Type and amount of hard phase particles are the major factors affecting abrasive wear resistance, if size is similar.
- M2, M4 and A11 overlays exhibit similar hardness but abrasive wear is 17mm^3 for A11, 40 mm^3 for M2 and >50 mm^3 for M2. Rockit®701 hardness is 5 HRC units higher than for A11 but the abrasive wear resistance is similar for both grades.
- For low stress abrasion, nickel-based coatings with admixed tungsten carbides outperform the iron-based grades. However, the iron based grade Rockit®701 shows a larger ability to take impact, when compared to a nickel based coating with 50 wt.% WC.

7 Acknowledgements

The author would like to thank Peter Olsérius, Mattias Weibull and Marcus Ewald for support in the preparation and evaluation of the investigated samples.

8 Literature

[1] ASM Handbook on line, Welding, Brazing and Soldering, Vol. 6
[2] ASTM G65: Standard Test Method for Measuring Abrasion Using the Dry Sand/Rubber Wheel Apparatus, 2010.
[3] S. Bengtsson, S. Dizdar, B. Maroli, L. Baguette "New hardfacing material with high impact wear resistance" ITSC 2016, Shanghai
[4] Abrasive impact wear and surface fatigue wear behaviour of Fe–Cr–C PTA overlays, Wear 301(2013) 102-108.
[5] Termo-Calc 2016a, Thermocal Software AB
[6] H. Brandis et al. "Metallurgical aspects of Carbides in High Speed Steels" Thyssen Edelst. Techn. Ber., 1983
[7] A.T. Santhanam, P. Tierney, and J.L. Hunt, Cemented Carbides, Properties and Selection: Nonferrous Alloys and Special-Purpose Materials, Vol 2, ASM Handbook, ASM International, 1990, p 950–977
[8] Crystallography and Engineering Properties of Ceramics, *Engineered Materials Handbook Desk Edition*, ASM International, 1995, p. 922–975
[9] O. Pierson, "Handbook of refractory carbides and nitrides", Noyes Publications, 2009
[10] K.L. Johnson, Contact mechanics, Cambridge university press, Cambridge, 1985.
[11] M. Kirchgassner "Behaviour of iron-based hardfacing alloys under abrasion and impact" Wear 265 (2008) 772–779
[12] Powder mix: 1559-00 / 4580 in "Thermal Surfacing: Powder Choice with Ease", Höganäs AB, 2016

Advanced processes and system technology for high-performance laser cladding

S. Nowotny[1], H. Hillig[1], F. Kubisch[1], F.-L. Toma[1], C. Leyens[1,2]
[1)] Fraunhofer Institute for Material and Beam Technology IWS, Dresden / D
[2)] Technical University of Dresden, Dresden / D

Laser cladding technology is widely used in industry to precisely apply tailored surface coatings, as well as three-dimensional deposits for repair and additive layer-by-layer fabrication of metallic parts. However, the processing of larger components, like tools for oil and gas production, is economically challenging due to the conventionally low deposition rates. Consequently, industry is requesting more powerful technologies that maintain the quality advantages of the laser technology, but also make the process more productive and time effective.
The modern highest power diode lasers offer practical solutions for applying of large-area laser cladding with significantly increased productivity. Using a fiber-coupled diode laser of 20 kW power and the accordingly developed laser cladding heads, real deposition rates of metal alloys, e.g. Inconel 625, could reach 14 kg/h. With the new-developed powder nozzles with rectangular profile of the powder jet allows at a laser power of 20 kW single tracks with 45 mm-width can be produced. Besides the laser source, the processing laser head is the key parameter for a high productivity and efficiency of the whole cladding procedure. The paper presents a new generation of high-performance laser cladding heads with integrated process sensors, which guarantee a stable long-time operation at highest power levels. The deposition rates achieved with this technology are equal or even exceed typical values of the common PTA technique. Current applications are large-area coatings on power plant components, hydraulic cylinders for off-shore equipment, and large metal forming tools for automotive bodies.

1 Introduction

Laser cladding technology is widely used to apply tailored surface coatings as well as three-dimensional deposits for repair and additive layer-by-layer fabrication of metallic parts [1]. The main applications are corrosion, wear or oxidation protection for oil and gas, automotive, aircraft, shipping or heavy industries [2,3].

Compared to other thermal coating technologies such as atmospheric plasma spraying (APS) or high velocity oxy-fuel spraying (HVOF), laser cladding has several advantages: high-precision deposition of the materials, low dilution, high level of adhesive strength, nearly zero porosity and a smooth interface with the substrate, which prevents stress concentration at the clad–substrate interface during operation. These features are beneficial for the long life and high performance.

However, deposition rate is a factor of great significance for competitiveness of a cladding technology. In the past deposition rates of laser cladding were too low (up to 1.6 kg/h at laser power up to 4 kW) and specific costs too high especially for large-area claddings. In our previously work laser laser cladding using a 10 kW high power diode laser showed an improvement in the deposition rates, but there are still limitations.

As a result of the dynamic development of the laser sources, as well as the cladding processes, the conditions have changed right favorable even for the high-performance processes. Nowadays, diode-fiber lasers are available with up to 20 kW, expecting 40 kW in the nearest future. Due to a significant price reduction, which took place simultaneously, large-area laser cladding became possible for surface protection of large and heavily loaded components. Additionally, specially designed system technology like powder nozzles and optics for that high-power level guarantee success in industrial production.

2 Materials and experimental procedure

In the last years many works were dedicated at Fraunhofer IWS to develop modular systems of optic and laser heads with coaxial disposal of laser optics and powder lines (COAXn product line) which are suitable for diode, fiber and disc lasers in a wide range of laser power, from 20 W to 20 kW.

In the present work, maximum deposition rates and process limits have been investigated using a 20 kW high power diode laser (Laserline GmbH Mülheim-Kärlich) and an Inconel 625 powder (-150 + 63 µm, Crapenter Micro-Melt® 625).

Various configurations of powder nozzles developed by Fraunhofer IWS were tested and compared. For this purpose, two different configurations have been used: on one hand a round laser spot of 12.5 mm diameter, combined with a coaxial powder nozzle (COAXpowerline); on the other hand, a rectangular laser spot and the flat-jet powder nozzle COAX11(V3) with up to 50 mm width of the powder stream (**Fig. 1**).

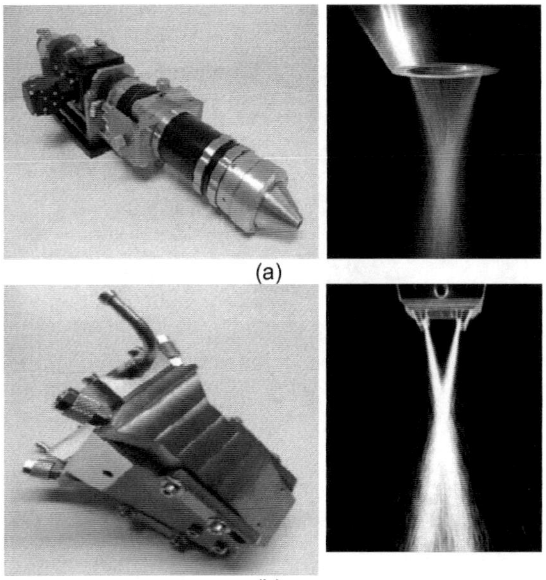

(a)

(b)

Fig. 1. Configurations of powder nozzles developed at Fraunhofer IWS: (a) coaxial powder nozzle (COAXpowerline) with round laser spot; (b) flat-jet powder nozzle (COAX11V3) with rectangular laser spot.

3. Results

3.1. Laser cladding with rectangular laser spot and flat jet nozzle

The investigations show a linear dependence of the cladding rate from the laser power (**Fig. 2**). Thus, a real deposition rate increases almost proportionaly with the laser power and deposition rates up to 14 kg/h Inconel 625 at 20 kW laser power. The coating qualities achieved with 15 kW and 20 kW were comparable to those produced with optimized cladding parameters with 10 kW laser power (**Fig. 3**).

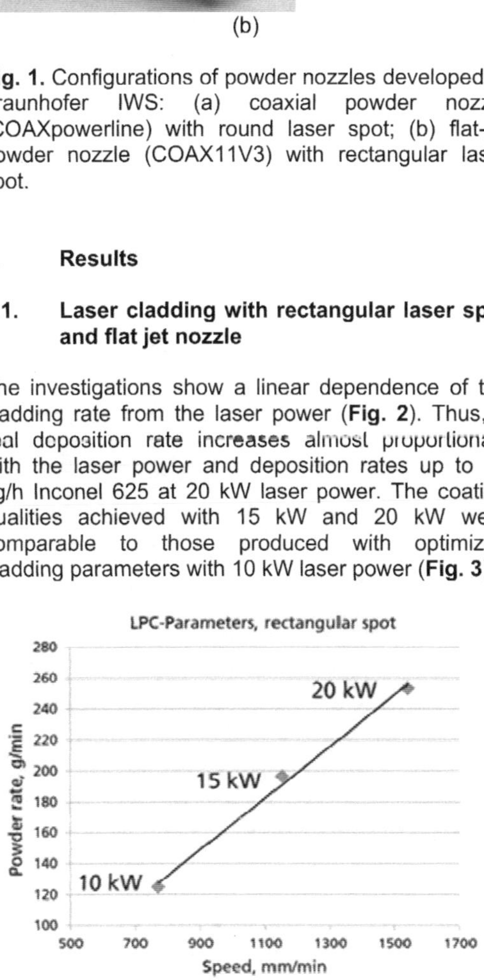

Fig. 2. Coating parameters for rectangular spot and wide stream powder nozzle COAX11(V3).

Fig. 3. Laser cladding tracks of Inconel 625 using the wide stream powder nozzle COAX11(V3) at different laser powers: top-view und cross-sections.

The cladding tracks show a metallurgical bonding with low dilution (iron mixing of about 3 to 5 %). The cross-section microstructure is cracks free with near zero porosity and contains fine-grained solidificated particles. Morever the low degree of overlapping between single tracks allows producting of high area cladding rate.

Under optimised conditions, the single track width up to 45 mm at 20 kW power level were obtained(**Fig. 4**). This is currently the maximum track width ever reached using the laser technology. The technology is applied today for the cladding of very large parts of nuclear power plants [4].

Fig. 4. (a) Picture of the wide stream powder nozzle and 20 kW laser power; (b) cross-section of a large 45 mm single track of coating.

3.2. Laser cladding with round laser spot and coaxial powder nozzle

For cladding applications of more complex parts, free-forms and 3D structures, a coaxial powder nozzle is the right choice. The experiments with round laser spot and annular gap coaxial powder nozzle COAXpowerline were carried out with the technology shown in **Fig. 1.**

In the first step, the optimisation of the parameters set for 10 kW laser power was performed. Then the coating parameters were increased proportionally and stepwise to 20 kW laser power. With 20 kW laser power no cracks on the coating surface were observed (**Fig. 5a**). The metallographic evaluation of the cross-section revealed the presence of hot cracks (**Fig. 5b**), which appear mostly probably because of high cladding speed of 3.9 m/min. Furthermore, some melting/solidification faults were observed in upper coating area. These defects can be explained by too short interaction time for complete melting of powder in the melt pool, or because of a too fast cooling. This shows that the limit of productivity was partly reached here, and further adjustments are necessary. In case of powder materials, which are susceptible for hot cracks, the cladding speed is a limiting parameter.

High deposition rates of coatings of more than 14 kg/h were obtained using COAXpowerline powder nozzle; these values are definitively higher than those found in the literature and are comparable to those measured for the flat-jet nozzle (**Fig. 6**).

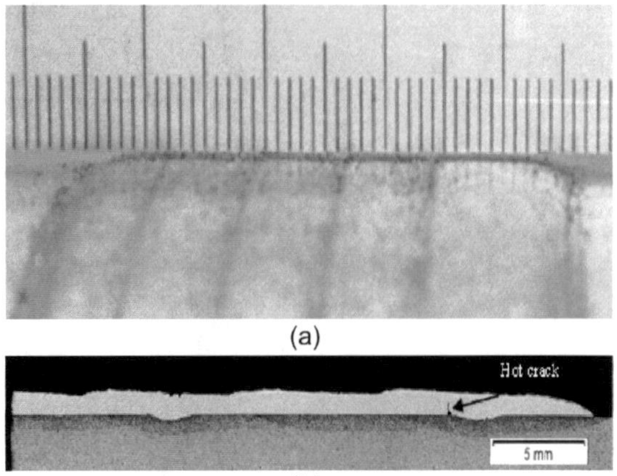

Fig. 5. Top-view (a) and cross-section of Inconel 625, cladded by overlapping of single tracks (b), using round spot and coaxial powder nozzle (COAXpowerline).

The deposition rates achieved with this technology equal or even exceed typical values of the common PTA technique. Anyhow, the coatings show the laser-typical quality in form of low dilution of 4...7%, high accuracy, and good surface quality of the as-cladded layers. For comparison purposes, different laser powers and the hybrid laser cladding technology with induction assistance are shown in **Fig. 6**, too.

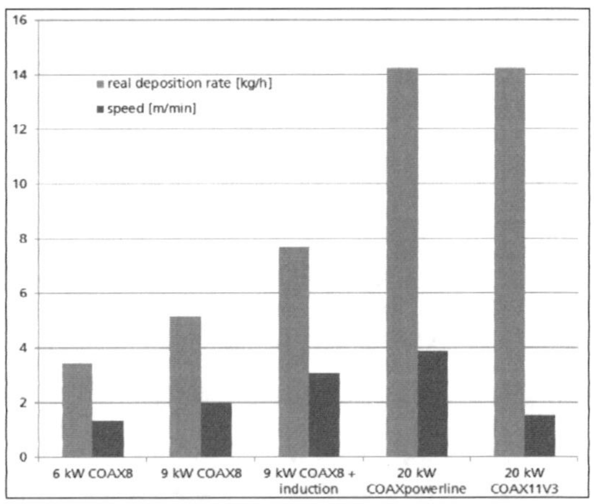

Fig. 6. Deposition rates and cladding speed of Inconel 625 in dependence of laser power; process with/without induction assistance and laser spot shape: COAX8 and COAXpowerline use round laser spot and coaxial ringslit powder nozzle, COAX11V3 use rectangular laser spot and wide jet powder nozzle [5].

Current applications are large-area coatings like hydraulic cylinders for off-shore equipment, or large metal forming tools for automotive bodies.

4 Conclusions

The results described here show the potential of the new laser systems to increase significantly the productivity using high laser power of up to 20 kW. Deposition rates of 14 kg/h Inconel 625 with higher laser speeds and single track width of 45 mm have been achieved. The high quality, which is typical for the laser cladding process, could be guaranteed even at that high power level. However, the tests show that a special selection and adaptation of system technology is necessary. The adaptation of laser spot geometry and suitable powder feeding are especially important.

At laser powers lower than 10kW, the coaxial powder nozzles with round spots are preferred, because of better powder utilisation, omni-directional welding performance, and lower effort for adjustment. At laser powers higher than 10kW, a critical cladding speed for round spots was reached caused by limiting track width. The high speed in this case can lead to coating defects, for instance, hot cracks and melting faults.

Therefore, for that power level, the flat-jet powder nozzle with rectangular laser spot is ultimatively the right solution. Both types of nozzles are ready for practical use and have been already proven in industrial environment.

Acknowledgements

We thank our colleagues from Fraunhofer IWS Irina Shakhverdova und Beate Wolf for the metallographic preparation of the cross-sections and microstructural analysis.

5 Literature

[1] R Vilar: Laser cladding. Journal of Laser Applications 11 (1999) Issue 2, pp. 64/79.

[2] P Goodwin: Laser Cladding of Industrial Components. Paper presented at the ILAS 2015, Kenilworth, UK.

[3] Y.P. Kathuria Some aspects of laser surface cladding in the turbine industry. Surface and Coatings Technology 132 (2000), Issue 23, pp. 262/269.

[4] Baufeld B, and H. Hillig: Laser cladding with 45 mm wide tracks. The International Journal of Advanced Manufacturing Technology 2017, submitted, under Review.

[5] C. Bratt and H. Hillig: Possibilities for High Deposition Rate Cladding: High Power Lasers and Hybrid Techniques. 10th International Laser Symposium and International Symposium "Tailored Joining", 28th February 2014, Dresden, Germany.

High speed steel deposited by pulsed PTA – frequency influence

Rohan, P., Boxanová, M., Zhang, L., Kramar T., Lukac, F. Praha / CZ

The powder of HSS (HSS23, AISI M3:2) was deposited by pulsed-PTA method on to low alloyed steel substrate. The influence of pulsation frequency was evaluated on the surface of deposits and on their cross sections by both light microscope and by Vickers hardness measurement apparatus and extreme properties mapping (XPS). Surfacing parameters at current frequency from 0 to 200Hz were tested during deposition of single weld bead. Dilution and heat affected zone were evaluated and compared for all tested parameters. The presence of retained austenite after deposition was determined by X-ray diffraction. The beads deposited with different frequencies differ in their shape, dilution degree, microhardness and penetration depth. It was found that the microhardness increases with current frequency.

1 Introduction

Plasma transferred arc is widely used as a source of heat both in welding [1], [2] and hardfacing [3], [4], [5], [6], of alloyed metallic materials. This technology can be considered as a development of gas tungsten arc welding technique, while plasma plum is constricted by water cooled copper orifice and tungsten rod is placed inside the torch. The constricted and focused plasma jet is needed in welding processes, however in case of hardfacing the wide and soft plasma plum is required in order to minimalize penetration, resp. dilution [7], [8] of filler material by parent material. Usage of current pulsation is well known in GTAW as a way of increasing bead geometry characteristics [9], [10], [7], [11] but in plasma welding and hardfacing only restricted number of studies can be found [12], [13].
High speed steels (HSS) are commonly known as complex iron-base alloys of carbon, chromium, vanadium, molybdenum, or tungsten, or their combinations and in some cases substantial amounts of cobalt. The carbon and alloy contents are balanced to give high attainable hardening response, high wear resistance [14], high resistance to the softening effect of heat, and good toughness for effective use in industrial cutting operations [15]. There are many fabrication methods of HSS parts preparation such as metallurgy [16], [17], thermal spraying [18], laser cladding [19], etc. However, only few studies follow up PTA deposition of HSS [20]. The aim of this study is to describe influence of plasma current pulsation frequency on different characteristics of HSS deposit such as bead geometry characteristics, hardness, and residual austenite content. Hardness of individual phases was measured by extreme property mapping.

2 Experimental

Single bead of HSS 23 (AISI M3:2) powder (chemical composition - Tab.1) was deposited following the programed procedure on commercially available plasma hardfacing automate PPC 250 R6 (KSK, s.r.o., Czech Republic). This machine is designated for glass industry applications, mainly for hardfacing of glass mold edges.The hardfacing automate can work with plasma current pulse up to 200Hz between 50 and 250 A. Pulsation between 80 and 240 A was used to create an average plasma current at 160A. Continuous plasma current was 160A. Continuous current and six different frequencies (1, 50, 90, 115, 150, 200 Hz) were used in order to deposit one single bead on 5mm thick low alloyed steel plate. Deposits cooled down in open air. Cross sections of each bead were studied by light microscopy and basic bead characteristics were measured. Phase composition was determined from X-ray diffraction (XRD) patterns obtained at room temperature by Co$K\alpha$ parallel beam of 1 mm diameter pointed in the area of interest and collected by 1D LynxEye detector (Feβ filter in primary beam) mounted on Bruker D8 Discover (DE). Subsequent Rietveld refinement [21] of whole powder pattern performed in TOPAS 5 [22] showed strong inherent texture of both ferritic and austenitic phases. Ratio of austenite was calculated from the diffraction peaks integrated intensities. Extreme property mapping (XPM) was performed by Hysitron TI 950 TriboIndenter apparatus (Bruker) on a sample deposited by 200Hz plasma current frequency. Distance between each of 400 (20 x 20) indents was 0,5μm and maximum force was 3mN. Choice of the XPM area was based on previous electron microscopy study [23] in order to cover both hard carbides and matrix of HSS. Microhardness was measured on IndentaMet apparatus (Buehler) as average value from 5 indentations 1mm below free deposit surface in the middle of bead. Alloyed steel DIN 54 SiCr 6 was used as parent material. This steel is commonly used in conditions of impact, vibration, or long-term stress, which require higher tensile strength, limit of elasticity and fatigue strength [12], [13].

Tab. 1: Chemical composition of HSS23 (DEW, Germany)

	C	Si	Mn	Cr	Mo	V	W	Fe
HSS23 No25584	1,30	0,36	0,20	4,05	5,19	3,18	6,07	bal.

3 Results and Disscussion

Metalography – all deposited beads present good metallurgical bond between deposit and parent material. All deposits present mostly similar penetration shape except the bead welded by 1Hz current frequency (Fig.1). Beads deposited by frequency greater than 50Hz present good wettability with parent material. Beads deposited by current frequency up to 90Hz present smaller width than those deposited by frequency higher than 100Hz. Some longitudinal cracks appear in deposits hardfaced by 150Hz and 200Hz. This cracks start from the surface of deposit and propagate perpendicularly to the centre of the bead.

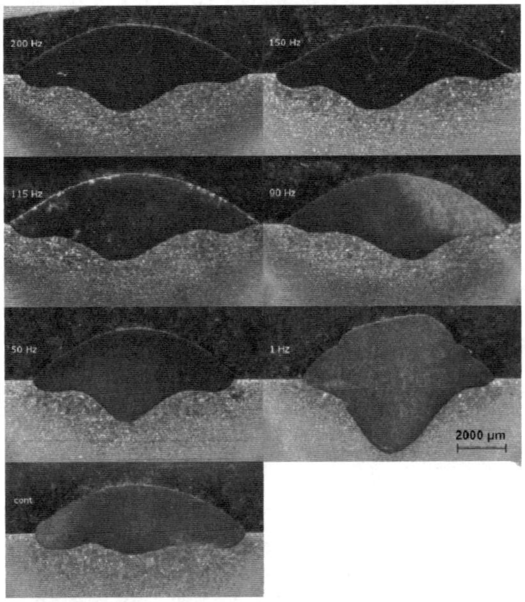

Fig. 1:Cross section of single HSS23 bead

The samples present similar microstructure as cast HSS [24] (Fig.2). This structure is characterised by two types of carbides, thick lamellae and small point like shape in martensitic matrix with some amount of residual austenite.

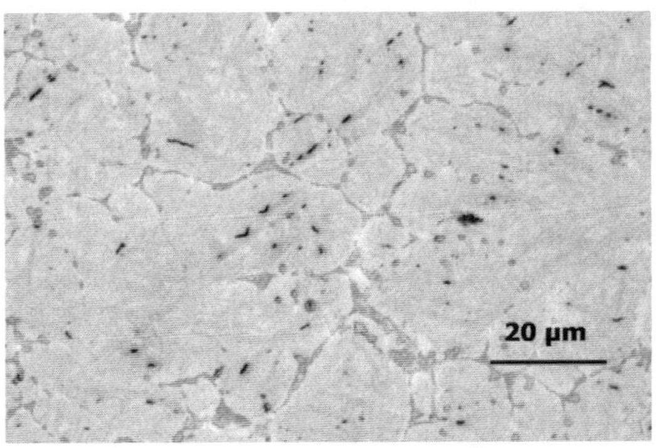

Fig. 2: Microstructure of sample made with current frequency 200Hz, 500x

"200Hz" sample was studied by XPM (Fig. 3,4) and it proved, that the hardest phase in structure is VC, followed by Mo(W)C [25]. Matrix presented only 180 GPa of reduced modulus while its maximum level was 260 GPa. This quite high difference may explain quite high scatter in HV measurement (up to 30HV). Low hardness of matrix corresponds with high content of Fe_γ found by X-Ray diffraction (Fig.5).

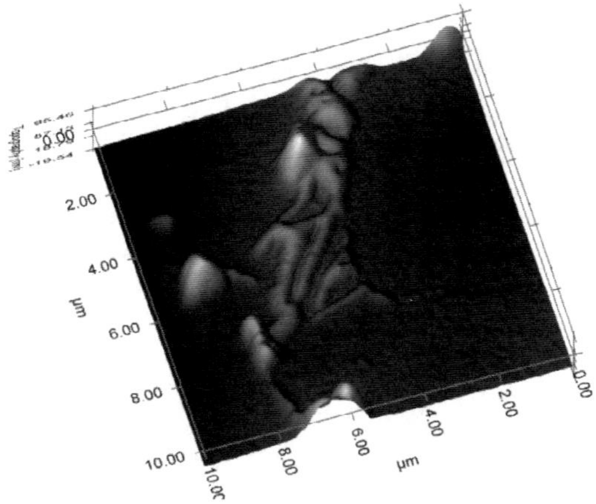

Fig. 3: Topology of area studied by XPM

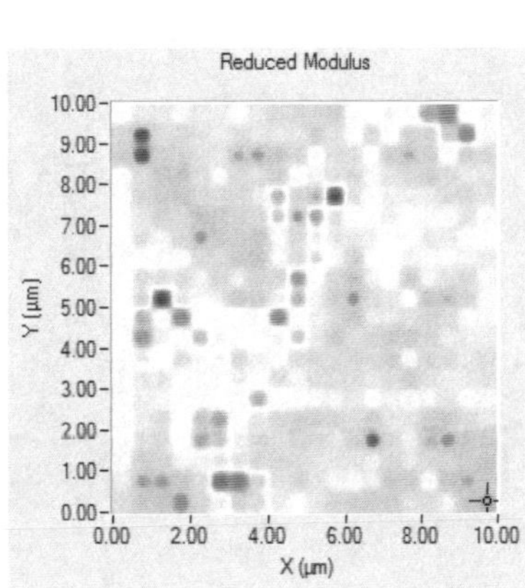

Fig. 4: XPM on sample made with 200Hz current frequency.

The lowest hardness presented the sample made by frequency 90Hz while the hardest sample was deposited by 200Hz (Fig.6). However the deposit cladded by 200Hz frequency also presents high content of residual austenite. This means that hardness of this sample is caused by another reinforcement principle than martensitic. This fact is supported by XPM measurement where very low hardness of matrix was measured. High hardness of "200Hz" deposits is caused by size and distribution of hard particles in its structure.

X-ray diffraction measurement reveals that content of Fe_α depends on current pulsation frequency. The lowest content of residual austenite was found in sample made by 50Hz current pulsation. In contradiction with "200Hz" sample those in the range from 90Hz to 115Hz are the samples where the structure reinforcement can be taken as martensitic because of higher content of residual austenite, which causes their lower hardness.

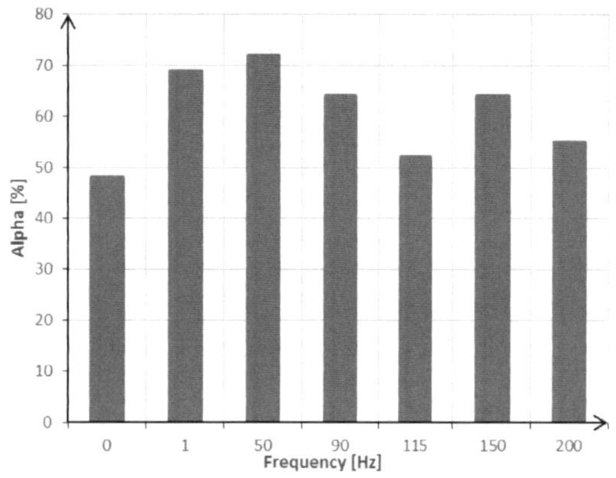

Fig. 3: Content of Fe alpha in bead deposited by pulsed PTA

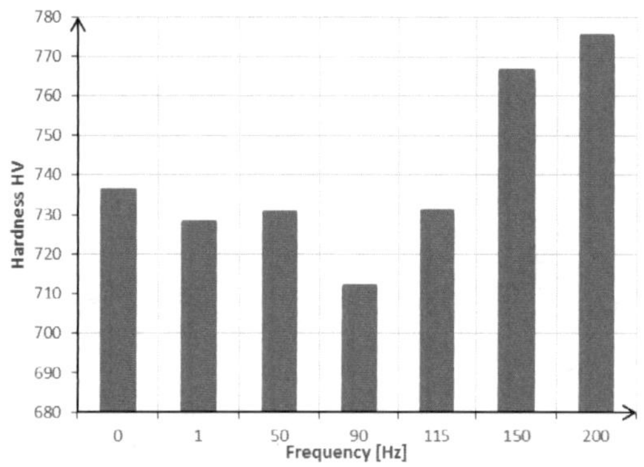

Fig. 4: Hardness of single bead deposited by pulsed PTA, 1mm under surface

4 Conclusions

In this study seven different parameters of PTA hardfacing of HSS were used to investigate the influence of plasma current frequency on structure and properties of deposited single bead on low alloyed steel.

All the samples present good metallurgical bond between deposits and parent material. Wettability of liquid HSS is higher when pulsed plasma is used.

The microstructure of HSS deposited by PTA is similar to cast structure of HSS and consists of hard vanadium and molybdenum/tungsten carbides in martensitic/austenitic matrix.

Frequency of plasma current influences hardness and content of residual austenite in deposits. Hardness decreases up to plasma current frequency of 90Hz followed by significant increasing up to 200Hz. The maximal difference between high and low hardness is more than 60HV.

When hardness of deposits exceeds 730 HV the deposits become too hard and brittle and some cracks appear. Those cracks start from the deposit surface.

Pulsed PTA hardfacing of HSS23 on low alloyed steel can be recommended with frequencies up to 150Hz.

5 Literature

[1] J. C., Metcalfe a Quigley M. B. C. Keyhole stability in plasma arc welding. Welding Journal. 1975, 54(11), 401.

[2] LIU, Zu Ming, ShuangLin CUI, Zhen LUO, ChangZhen ZHANG, ZhengMing WANG a YuChen ZHANG. Plasma arc welding: Process variants and its recent developments of sensing, controlling and modeling: Process variants and its recent developments of sensing, controlling and modeling. Journal of Manufacturing Processes. 2016, 23, 315-327

[3] FAUCHAIS, Pierre a Armelle VARDELLE. Chapter 9 - Thermal Plasmas Surface Treatment Chapter 9 - Thermal Plasmas Surface Treatment. European Materials Research Society Series. Oxford: Elsevier, 2006, s. 311-344.

[4] MELLOR, B.G. 9 - Welding surface treatment methods for protection against wear. Woodhead Publishing Series in Metals and Surface Engineering. Woodhead Publishing, 2006, s. 302-376

[5] Chapter 3 - Weld surfacing. Woodhead Publishing Series in Metals and Surface Engineering. Woodhead Publishing, 1998, s. 42-118.

[6] SIGOLO, Erich, Juliano SOYAMA, Guilherme ZEPON, Claudio Shyinti KIMINAMI, Walter José BOTTA a Claudemiro BOLFARINI. Wear resistant coatings of boron-modified stainless steels deposited by Plasma Transferred Arc. Surface and Coatings Technology. 2016, 302, 255-264

[7] BASHENKO, V. V. a N. A. SOSNIN. Optimization of the plasma arc welding process. Welding Journal. 1988, 67(10), 233-237.

[8] BALASUBRAMANIAN, V, AK LAKSHMINARAYANAN, R VARAHAMOORTHY a S BABU. Application of Response Surface Methodolody to Prediction of Dilution in Plasma Transferred Arc Hardfacing of Stainless Steel on Carbon Steel. Journal of Iron and Steel Research, International. 2009, 16(1), 44-53.

[9] EITNER, R., E., G.,H. MCELHINNEY a E.,L. PRUITT. An investigation of pulsed GTA welding variables. Welding journal. American Welding Society, 1973, 52(9).

[10] MADADI, F., M. SHAMANIAN a F. ASHRAFIZADEH. Effect of pulse current on microstructure and wear resistance of Stellite6/tungsten carbide claddings produced by tungsten inert gas process. Surface and

Coatings Technology. 2011, 205(17–18), 4320-4328.

[11] YU, Shi, Li CHUNKAI, Du LEIMING, Gu YUFEN a Zhu MING. Frequency characteristics of weld pool oscillation in pulsed gas tungsten arc welding. Journal of Manufacturing Processes. 2016, 241, 145-151.

[12] D'OLIVEIRA, A.S.C.M., R.S.C. PAREDES a R.L.C. SANTOS. Pulsed current plasma transferred arc hardfacing. Journal of Materials Processing Technology. 2006, 171(2), 167-174.

[13] CORREA, E.O., S.C. COSTA a J.N. SANTOS. Weldability of iron-based powder metal materials using pulsed plasma arc welding process. Journal of Materials Processing Technology. 2008, 198(1–3), 323-329.

[14] ROHAN, Pavel, Tomáš KRAMÁR a Jaroslav PETR. HSS DEPOSITION BY PTA – FEASIBILITY AND PROPERTIES. Advances in Science and Technology Research Journal. 2016, 10(29), 57-61.

[15] BAYER, AM, and B. A. Becherer. "T, V. HIGH–SPEED TOOL STEELS." 16. BAYER, A.M. a B.A. BECHERER. ASM Handbook. 16. ASM International, ASM World Headquarters 9639 Kinsman Road Materials Park, OH 44073-0002: ASM International, 1989, 51–59.

[16] PARK, Joon Wook, Huo Choon LEE a Sunghak LEE. Composition, microstructure, hardness, and wear properties of high-speed steel rolls. Metallurgical and Materials Transactions A. 1999, 30(2), 399-409.

[17] HWANG, Keun Chul, Sunghak LEE a Hui Choon LEE. Effects of alloying elements on microstructure and fracture properties of cast high speed steel rolls: Part I: Microstructural analysis: Part I. Materials Science and Engineering: A. 1998, 254(1–2), 282-295.

[18] WANG, Hebin, Longgang HOU, Jinxiang ZHANG, Lin LU, Hua CUI a Jishan ZHANG. The secondary precipitates of niobium-alloyed M3:2 high speed steel prepared by spray deposition: 2 high speed steel prepared by spray deposition. Materials Characterization. 2015, 106, 245-254.

[19] WANG, S.-H., J.-Y. CHEN a L. XUE. A study of the abrasive wear behaviour of laser-clad tool steel coatings. Surface and Coatings Technology. 2006, 200(11), 3446-3458.

[20] BOURITHIS, L a G.D PAPADIMITRIOU. Synthesizing a class "M" high speed steel on the surface of a plain steel using the plasma transferred arc (PTA) alloying technique: microstructure and wear properties: microstructure and wear properties. Materials Science and Engineering: A. 2003, 361(1–2), 165-172.

[21] RIETVELD, H.,. Line profiles of neutron powder-diffraction peaks for structure refinement. Acta Crystallographica. International Union of Crystallography, 1967, 22(1), 151-152.

[22] Coelho, A.A. TOPAS: Coelho Software. Version 5. Brisbane, Australia, 2016.

[23] ROHAN, P., T. KRAMÁR a J. PETR. HSS deposition by PTA – feasibility and properties. Advances in Science and Technology Research Journal. 2016, 10(29), 57-61.

[24] VITRY, V., S. NARDONE, J.-P. BREYER, M. SINNAEVE a F. DELAUNOIS. Microstructure of two centrifugal cast high speed steels for hot strip mills applications. Materials & Design. 2012, 34, 372-378.

[25] VITRY, V., S. NARDONE, J.-P. BREYER, M. SINNAEVE a F. DELAUNOIS. Microstructure of two centrifugal cast high speed steels for hot strip mills applications. Materials & Design. 2012, 34, 372-378.

Alloyed Carbides Beyond WC as a New Material Platform for Solving Challenges in Hardfacing

Petr Fiala[1], Fort Saskatchewan / CAN, Rolf Hepp[2], Barchfeld / D, Arkadi Zikin[3], Wohlen / CH

1. Oerlikon Metco (Canada) Inc.; 2. Oerlikon Metco WOKA GmbH (D); 3. Oerlikon Metco AG (CH)

Tungsten carbide in nickel based self-fluxing alloy overlays has been dominating hardfacing applications due to its excellent properties, namely extremely high wear resistance. Nevertheless, there are still applications and limits which tungsten carbide has not conquered. This study focuses on $(TiW)C_{1-x}$ which was deposited with several matrix materials and tested in wear, corrosion and impact resistance and benchmarked against tungsten carbide. Results for several other carbides such as $(NbW)C_{1-x}$, $(VW)C_{1-x}$, NbC_{1-x} and TiC_{1-x} overlays deposited by plasma transferred arc (PTA) and laser cladding (LC) will be presented and discussed. As a result of deposition trials and overlay testing, it was found that better thermodynamic stability of alloyed carbides allows them to be used in an iron based matrix and/or a matrix with a high chromium content, in applications requiring improved corrosion and oxidation resistance, better impact resistance and lower weight.

1 Introduction

The requirements for efficient wear protection solutions in terms of cost and performance in different applications continue to grow. A good example of this customer demand are overlays on ground engaging tools and slurry pipelines in open pit mining. High performance requirements placed on the ground engaging tools and the slurry pipelines require new solutions with regards to effective, protective overlays. They are expected to combine good performance in corrosive as well as in erosive environments. Besides the open pit mining industry, these applications can be also found in the conventional oil and gas industry, power generation, and agricultural industry. Existing well-established hardfacing materials are mostly overlays containing tungsten carbide. Virtually all possible combinations of particle sizes, morphologies, types of tungsten carbide and matrix chemistries have been tried with almost no stone left unturned in attempts to achieve its full potential. The result of this continuous effort is an immensely successful material platform that addresses a plethora of service conditions. However, tungsten carbide also has some disadvantages which in some cases can limit its use. Tungsten carbide tends to suffer from extensive degradation during the weld cladding process. It has inferior thermodynamic stability compared to many other carbides, which prevents it from use in high alloyed matrices; its temperature capability is only up to about 500° C. Also, due to its high weight, it is not suitable for some applications. There is a growing demand, especially in the mining industry, for overlays that provide even more wear resistance, are cheaper and work well in niche applications where tungsten carbide does not fit this criteria. Considering that the limits of the applications of tungsten carbide appear to have been reached, further advances look to other materials which would be able to achieve results beyond those of tungsten carbide. The major roadblock for using other hard phases in hardfacing was the availability of coarse, dense carbides suitable for hardfacing deposition processes. These carbides have become recently available and new overlay compositions containing new alloyed carbides have been tested for PTA and laser cladding hardfacing applications to alleviate some of the limitations associated with tungsten carbide based overlays.

2 Experimental

The powder particle size was determined by screen analysis according to ASTM B214 – Standard Test Method for Sieve Analysis of Granular Metal Powder.
True density of the powder materials and overlays were measured by an UltraPyc1200e Pycnometer (Quantachrome).
A Nikon Epiphot-TME Inverted Microscope was used for all optical light microscopy photos and a JEOL JSM-T330A Scanning Electron Microscope was used for SEM and SEM-BSE photos.
Microhardness HV0.05 was measured on a Leco M-400-G2 Microhardness tester according to ASTM E384 Standard Test Method for Microhardness of Materials.
Wear resistance of overlays was tested according to the Standard Test Method for Measuring Abrasion Using the Dry Sand/Rubber Wheel Apparatus ASTM G65, procedure A. Two tests in the identical spot are used for each sample to account for the inhomogeneity of the microstructure. The first test is disregarded since its purpose is to condition the tested surface. The second test result is reported. The test results are measured as weight loss but since overlays have different densities, the results were converted into volume loss, allowing for direct comparison.
Impact resistance was tested on an impact tester at Alberta Innovates Technology Futures, Alberta, Canada. The principle of the test is for a hammer to be spun inside of a drum as seen in **Fig. 1**, which impacts the surface of two test coupons mounted to the inside of the unit, opposite to each other and perpendicular to the drum's interior surface. The repeated striking of the coupons causes material loss from the overlay. An important aspect of the test is that the hammer only strikes the surface of the coupon and does not drag across it. The rotational

velocity and size of the impact hammer produce an impact of approximately 8J [1]. The test is performed on 3 samples and an average weight loss is calculated. The test measures weight loss in grams but since density of overlays is different due to different carbide and matrix density, these results are converted to volume loss to allow for direct comparison. The test lasts 24 minutes and samples are checked at 4 minute intervals. The arm rotates at 150±5 RPM.

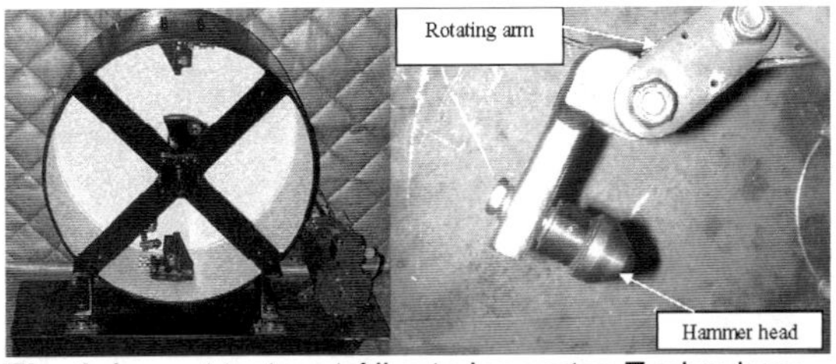

Fig. 1. Impact tester at Alberta Innovates Technology Futures [1]

Electrochemical corrosion testing was performed on a Gamry potentiometer in 0.08% and 3.5% NaCl distilled water solution with the electrolyte's pH adjusted to 8.3 using NaOH, no aeration, and at room temperature. The test water salinity and pH were chosen to approximate service conditions encountered in oil sands mining operations, i.e. 0.08% NaCl and pH 8.3, and in off-shore oil and gas applications, i.e. 3.5% NaCl and pH 8.3. The tested samples were allowed to stabilize for 16 hours before proceeding with corrosion rate measurements. A linear polarization test was run before disturbing the natural equilibrium of the cell or inducing any changes on the overlay surface. A Tafel experiment to measure the Tafel slopes was performed. Then finally a cyclic curve was measured. Generalized corrosion rate estimation was calculated in mils per year (mpy).

3 Results and Discussion

3.1 Carbide Particle Size

Carbide particle size is an important factor for all hardfacing deposition processes. While thermal spray applications usually use carbides agglomerated and sintered with matrix particles and a particle size of individual carbide grains is only a few micrometers in size, powders produced for PTA or laser cladding have typical carbide particle sizes starting at 53 µm and going up to 150 µm or higher. It is also possible to use fine carbides for weld overlays but such carbides need to be first pelletized with a matrix to make coarse, dense agglomerates which have good feeding properties. Such pelletized materials are generally more expensive than blended materials but they also address different application needs. The processes of making coarse tungsten carbide in both the cast eutectoid W_2C/WC form or the stoichiometric WC form are well established. As a result, tungsten carbide is readily available in all forms and sizes up to several hundred micrometers or more.

For most other carbides it is not a simple matter to produce them in particle sizes of 100 µm or more because the most common process is a direct carburization of fine metal powders. There have been attempts to make larger particles by agglomerating and sintering smaller carbides, however, such sintered carbides contain inherent porosity and original grain boundaries are still visible. As a result of this, the agglomerates do not possess the necessary mechanical strength to perform well in demanding wear applications.

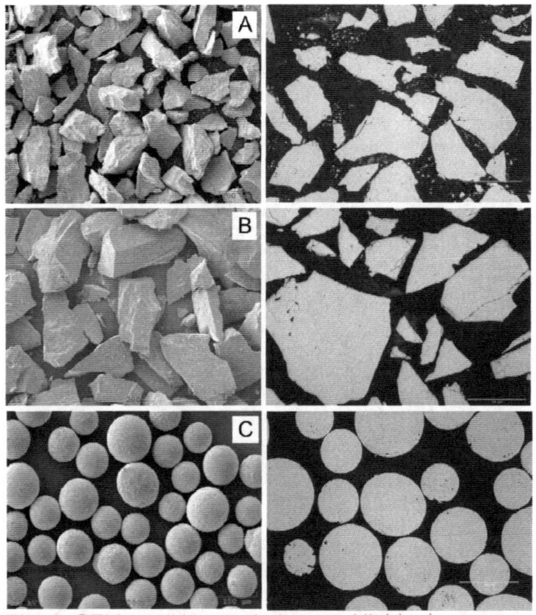

Fig. 2. SEM particle morphology and light microscopy cross-sections, all at 200x magnification, A) $(TiW)C_{1-x}$, B) NbC_{1-x}, C) $(VW)C_{1-x}$

Recently a process has been developed to allow for making various carbides in a coarse and dense particle form. TiC_{1-x}, NbC_{1-x}, and VC_{1-x}, as well as alloyed tungsten carbides such as $(TiW)C_{1-x}$, $(NbW)C_{1-x}$, and $(VW)C_{1-x}$ were produced in particle sizes up to several millimeters. Some examples are shown in **Fig.2**. These carbides are made in their angular morphology and can be also made spherical. The angular morphology carbides are available in very coarse particle sizes up to several millimiters and the spherical carbides can be currently made up to 150 µm. The powders shown in **Fig. 2** are examples of the angular and spherical morphology and of the particle sizes used in the PTA and LC applications. Moreover, the cross-sections show dense carbides with no internal porosity or other features of agglomerated particles.

3.2 Densities

Measured densities of the studied carbides are summarized in **Table 1**. For a typical powder blend of

WC in a 35HRC NiCrBSi self fluxing alloy with density of 8.42 g/cm^3, the theoretical volume content of WC would be 44.7%. In order to obtain the same volume of another carbide in a powder blend it is possible to use less weight of the lower density carbide and as a result to produce lighter overlays. The 3rd column in **Table 1** lists the blend requirement of the carbide content by weight to obtain the same carbide volume as for WC blends. The last column shows overall weight differences of overlays considering the same volume of carbides and the same volume of matrix for all powder blends. Theoretically it is possible to reduce overlay weight by over 40% using TiC_{1-x} containing overlays. This consideration may be very important in machinery with vibratory or reciprocating movement where lower weight allows for thicker overlays and longer service life. Another application benefiting from lower overlay densities would be transportation, where improved fuel consumption is critical.

Table 1. Carbide densities

Carbide	Carbide Density [g/cm^3]	Carbide weight for same volume [wt%]	Overlay weight difference for same volume [%]
W_2C/WC	16.5	63.3	3.3
WC	15.63	60.0	0.0
$(VW)C_{1-x}$	13.14	50.4	-9.6
$(NbW)C_{1-x}$	12.68	48.7	-11.3
$(TiW)C_{1-x}$	10.1	38.8	-21.2
NbC_{1-x}	7.51	28.8	-31.2
TiC_{1-x}	4.84	18.6	-41.4

Note: All using Ni self fluxing alloy 35HRC with density 8.42 g/cm^3

Still another important factor to consider is particle behavior in a molten pool. The bigger the density difference between carbide and matrix material, the bigger the segregation in both a blended powder and an overlay. As for overlays, higher density carbide particles tend to settle and leave a carbide free zone at the top of an overlay and low density carbides have tendency to float and concentrate in the upper part of an overlay. This behavior can be somewhat controlled by deposition parameters and by a particle morphology. For example spherical cast W_2C/WC settles very fast in a molten pool due to a higher density and lower drag force while less dense TiC_{1-x} with a spherical morphology distributes throughout an overlay.

3.3 Microhardness

Vickers microhardness with load of 50g, HV0.05, was measured on cross-sections of carbide particles. The reported values in **Table 2** are averages of 10 measurements, where they are sorted from highest hardness to lowest hardness.

Table 2. Microhardness HV0.05 of select carbides

Carbide	Hardness HV0.05
sph. W_2C/WC	2965
TiC_{1-x}	2963
$(TiW)C_{1-x}$	2818
sph. $(TiW)C_{1-x}$	2775
sph. $(NbW)C_{1-x}$	2550
NbC_{1-x}	2524
W_2C/WC	2457
$(NbW)C_{1-x}$	2448
sph. $(VW)C_{1-x}$	2373
$(VW)C_{1-x}$	2270
WC	2200

Microhardness of all alternative carbides lie between microhardness values of the spheroidized cast W_2C/WC and mono WC.

3.4 Thermodynamic Stability

Chemical reactivity of a substance depends on its thermodynamic stability. The greater the thermodynamic stability of a substance, the lower will be its tendency to undergo a chemical change. An important thermodynamic parameter to determine the stability of a compound is its standard enthalpy of formation. The comparison of the standard enthalpy of formation in kJ/mole of select carbides, is shown in the Ellingham diagram in **Fig. 3**. The results were normalized per mole of a reacting metal.
The more negative the enthalpy of formation, the greater will be the stability of the compound formed and the lower will be its tendency to react in a given environment. As can be seen in **Fig. 3**, WC and W_2C are not very stable in relation to other carbides. W_2C is actually unstable at ambient temperature but is retained in different structural forms depending on quenching and other conditions [3]. The most stable carbides are formed by Hf, Zr, Ti, Ta, and Nb (only select carbides are shown in **Fig. 3**). Every element forming a more stable carbide can replace another element present in an already existing, less stable carbide. This is a known problem for tungsten carbide in that it is sensitive to overheating and readily dissolves in its surrounding matrix; small particles completely dissolve and larger particles dissolve on their surface. Its dissolution is further accelerated by alloying elements which can form more stable carbides. The dissolved tungsten then competes with other alloying elements, particularly Cr that is present in almost all corrosion resistant matrices, but carbon rather forms more stable carbides with Cr or other elements. The more Cr is in the matrix, the more tungsten carbide gets dissolved and the more chromium carbide or complex carbides precipitate. This introduces another problem for corrosion resistant matrices where most alloying elements are depleted to such a level that the matrix ceases to function for the corrosion application it was designed for. Instead, the alloying elements precipitate out of the solution as complex carbides, which often leads to matrix embrittlement, tungsten carbide depletion, and a loss of corrosion resistance.

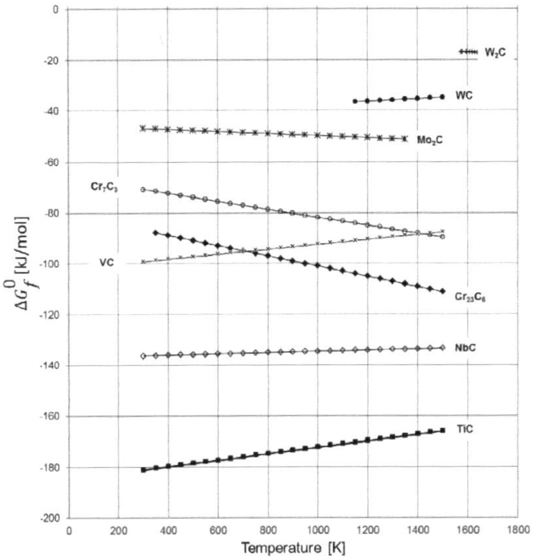

Fig. 3. Temperature dependence of the change in the standard enthalpy of formation ΔG for carbide formation reactions [4].

This is demonstrated in an experiment performed with PTA deposited powder blends containing WC and Ni self-fluxing alloy with 4 different Cr content levels. A WC + 40% Ni self-fluxing alloy with a matrix hardness of 35HRC and size distribution of -150+53 µm was modified by boosting its matrix with more Cr from the baseline of 6.91% to 12.04%, 18.34%, and 24.63%. All powders were PTA welded using a Stellite StarWeld 400A system with an Excalibur torch. Identical parameters were used for all overlays. Overlays were cross-sectioned and submitted for optical light metallography, SEM-BSE, and EDS analysis. SEM-BSE photos are in **Fig. 4** with a magnification of 100x. The microstructure observed on SEM-BSE shows a significant increase in the amount of secondary phases with increasing Cr content. The remaining primary WC content in overlays was determined by image analysis. The results were 32.5%, 30.6%, 27.7%, and 19.7 vol% of the primary tungsten carbide for increasing Cr content in the matrix. The conclusion is that the increasing Cr content promotes the dissolution of primary WC particles. Another interesting observation was that all solidified overlays reached a similar Cr content in the matrix, with only a slight linear increase from 5 to 7 wt% across the whole tested Cr concentration range. There should be enough carbon to completely tie up all available Cr leaving essentially no Cr in the matrix. The activity of Cr likely decreases with its decreasing concentration to the point where Cr at about 5 wt% in a rapidly cooling weld may not form carbides anymore. Quantitatively, the majority of precipitates in all evaluated overlays were of M_3C_2 (M=Cr, W, Fe) and M_7C_3 (M=Cr, W) types. Less frequent were precipitates of the following types: M_6C (M=Ni, Cr, W, Si), M_3C (M=W, Ni, Cr), and M_2C (M=Ni, W, Cr). Borides of chromium and tungsten were not identified due to the used analytical equipment limitations.

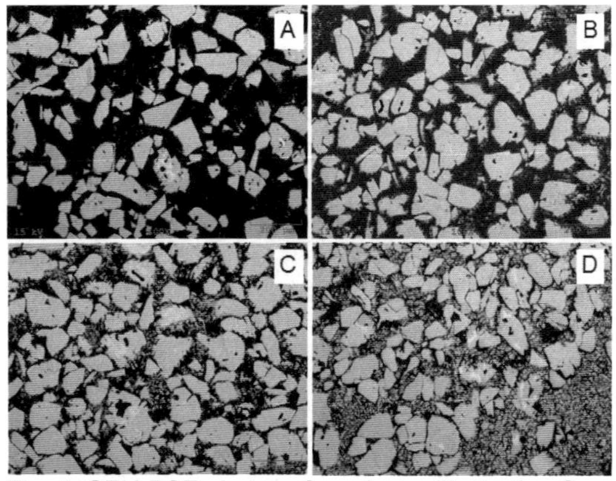

Fig. 4. SEM-BSE photos of overlays with starting Cr content of A) 6.91%, B) 12.04%, C) 18.34%, D) 24.63%

There are several ways to reduce carbide dissolution in a matrix. One of them is to use as low heat input as possible to reduce the thermal degradation of carbides. In most cases this method is not practical because it leads to poor bonding with the substrate and microstructural defects. Another way is to use a matrix that does not contain elements that will form more stable carbides than the primary carbide. In the case of tungsten carbide this limits its use to unalloyed Ni or Co matrices or matrices with a low alloying content. Still another way is to use more stable carbides as primary carbides. These more stable primary carbides have a lower dissolution rate than that of WC in nickel or iron-based highly alloyed matrices, where they maintain their carbide composition rather than losing carbon to the other carbide forming elements. This can be demonstrated on PTA deposited overlays of TiC_{1-x}, NbC_{1-x}, $(TiW)C_{1-x}$, and WC (all -150+53 µm) blended with a MetcoClad 625 matrix (its composition is similar to Inconel 625). Changes in the chemical make-up of these overlay matrices were evaluated by EDS analysis.

TiC_{1-x} did not dissolve or produce almost any precipitates and the matrix chemistry remained almost unchanged with a Cr content of ~21% compared to the original powder's Cr content of 21.4%. As a result of this lack of dissolution, the matrix microhardness remained soft at 305 HV0.3. NbC_{1-x} performed very well, forming only a minimal amount of precipitates and its matrix hardness also remained low at 302 HV0.3. $(TiW)C_{1-x}$ produced significantly less precipitates as compared to WC. The Cr content of the $(TiW)C_{1-x}$ containing overlay decreased on average to 15.5%. The overlay with WC performed the most poorly with its Cr content decreasing to 11% and the matrix hardness increasing to 474 HV0.3.

3.5 Wear Resistance

The G65 test results are summarized in **Fig. 5**. and sorted according to their performance from low to high volume loss. Most of the test results are averages of 2 or more tests. If the same Ni self-fluxing alloy matrix (further in the text as NiSF), with a hardness of about 35 HRC, is used with the different types of carbides it is possible to deduce from the graph that a smaller amount of $(TiW)C_{1-x}$ by weight can be used to achieve comparable wear resistance. This point is illustrated in tests 2 and 3 with 60 vs. 50 wt% of carbide, respectively. Of note in **Fig. 5**, are three overlays of $(TiW)C_{1-x}$ and different types of steel matrices which are only marginally worse than the best performing WC based overlays. It is evident from these results that $(TiW)C_{1-x}$ particles can be deposited in a highly alloyed iron matrix without the catastrophic dissolution of the carbide particles. This is virtually impossible to achieve with tungsten carbide particles.

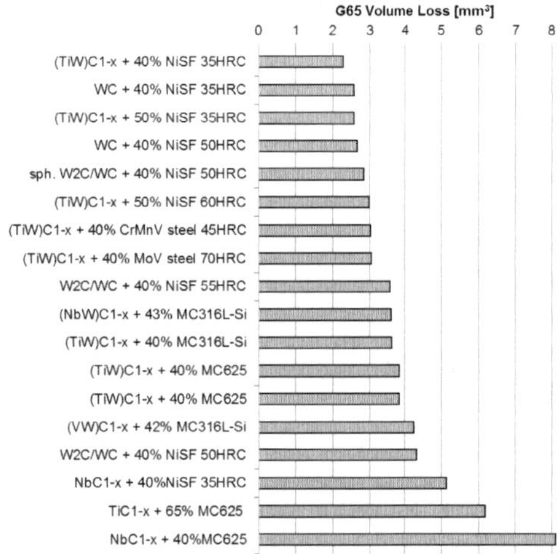

Fig. 5. G65 test results for alternative carbides when compared with WC based overlays.(NiSF=Ni self-fluxing alloy, sph.=spherical, MC=MetcoClad)

More results for other types of carbides in a MetcoClad 625 or MetcoClad 316L-Si matrix (MetcoClad 316L-Si is similar to stainless steel 316L) are also shown. **Fig. 5** demonstrates that carbide/overlay combinations up until the NbC_{1-x}+NiSF alloy 35HRC overlay performed within the industry accepted range which is currently serviced by tungsten carbide overlays. None of the tested overlays, with exception of the tungsten carbide overlays and $(TiW)C_{1-x}$ + MetcoClad 625, were optimized to improve their performance. Further optimization will likely lead to further wear resistance improvements. For example, matrix material should be such that it already provides a very high wear resistance without further precipitation of hard phases as a result of dissolving primary carbides, which stands in contrast with tungsten carbide overlays benefiting from their matrix becoming harder after precipitation of secondary carbides. As the precipitation of secondary carbides happens only to a limited extent with more stable primary carbides, it will be necessary to match suitable matrix materials to each of these stable carbides to achieve comparable or better wear resistance to the tungsten carbide overlays.

3.6 Impact Resistance

All common carbides are brittle at room temperature. At high temperatures (about 1000 °C) they undergo a brittle-to-ductile transition, and at temperatures above the transition temperature they can plastically deform in a manner similar to that for fcc metals [3]. Since the vast majority of industrial applications happen at much lower temperatures, a test for impact resistance tries to identify the less brittle carbides. Because carbides and matrices form a composite material, there may be some unexpected characteristics resulting from the interaction of these components.

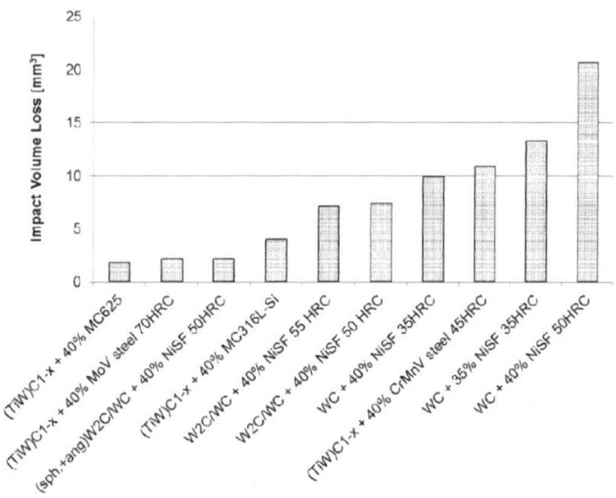

Fig. 6. Impact test results

The lowest impact volume loss was measured for $(TiW)C_{1-x}$ based overlays followed by a spherical cast tungsten carbide overlay, **Fig. 6**. As expected, tungsten carbide's impact resistance decreased from best to worst in this order: spherical cast tungsten carbide, angular cast tungsten carbide, and mono tungsten carbide. In following the same decreasing performance of morphology established by the WC overlays, it is reasonable to expect an improvement of $(TiW)C_{1-x}$ impact resistance by using a spheroidized morphology. Such an experiment will be performed in the future.

3.7 Electrochemical Corrosion Testing

Electrochemical corrosion test results are in **Fig. 7**. A well-established baseline WC+40%NiSF 35HRC overlay was benchmarked against a $(TiW)C_{1-x}$ + 40%MC625 overlay, A516Gr70 steel and the individual overlay components.

The corrosion rate of the $(TiW)C_{1-x}$ + 40%MC625 overlay was very small compared to the WC+40%NiSF baseline. In order to investigate the contribution of the carbides and matrices to these results, these components were tested separately. Please note that WC was not available in a form suitable for electrochemical corrosion testing and as a close approximation, a cast W_2C/WC sample was used instead. The MetcoClad 625 corrosion rate was less than that of the Ni self-fluxing alloy 35HRC but both matrix materials performed very well under the test corrosion conditions. Corrosion rates measured on both carbides were considerably different, with $(TiW)C_{1-x}$ corrosion rate being only a fraction of that of cast W_2C/WC.

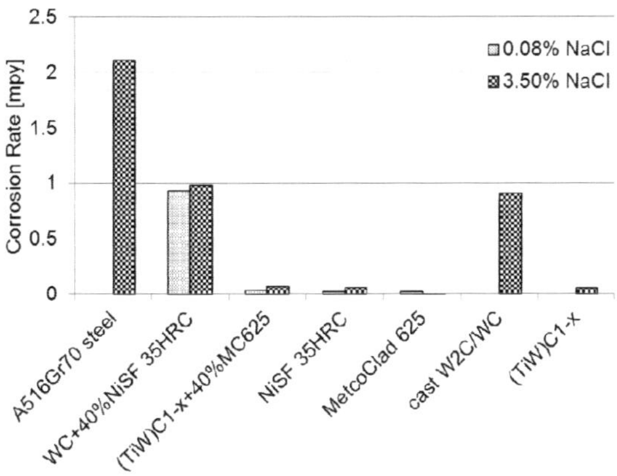

Fig. 7. Generalized corrosion rate by Tafel fit for pH=8.3

The cause of the high corrosion rate on cast W_2C/WC is due to the repeating cycle of carbide oxidation to form WO_3 on its surface, dissolution of WO_3 in water and exposure of fresh carbide surface for new oxidation. While tungsten carbide is not water soluble, WO_3 is and its solubility is considerably increased by the addition of NaCl and even more so by the addition of NaOH [5]. The WO_3 would be also present on the carbide/matrix interface, thus facilitating the removal of WC particles during a wear process if this interface gets weakened. Another aspect that needs to be taken into consideration is that the behavior of carbides can change in the presence of matrix material. For example, in an experiment described in [6], the WO_3 surface layer, on the oxidized tungsten carbide powder, dissolved at pH> 3 with the tungsten concentration increasing linearly with time. However, in the presence of cobalt the dissolution rate was significantly reduced at pH<10. The reduced dissolution rate may be related to the formation of surface complexes containing Co which adsorb onto the oxidized tungsten carbide surface [6].

4 Conclusions

Carbides alternative to tungsten carbide have been produced and are now available in particle sizes up to several hundred micrometers in dense, angular and spherical morphology. They were successfully deposited by plasma transferred arc and laser cladding processes and exhibited promising performance when compared with tungsten carbide based overlays. These carbides can be used in compositions and applications where WC reaches its performance limits. Compared to WC based overlays, TiC_{1-x} based overlays can lower overlay weight by over 40% but at the expense of a lower wear resistance. $(TiW)C_{1-x}$ based overlays can provide weight savings with better impact and corrosion resistance while providing comparable wear resistance. The other carbide MMCs (metal matrix composites) have not been optimized yet for intended end applications but already perform remarkably well. Each of these carbides can become a new material platform to be used with variety of matrix materials addressing specific needs in ways similar to that of tungsten carbide.

5 Acknowledgements

The authors would like to acknowledge Daniel Garcia and Gail Kite for the electrochemical corrosion testing, Brent Bartlett and Carlos Lorena for the G65 testing, Garth Murphy and Wanru Liu for the samples deposition, and Bernard Larouche for carbides analysis. All other Oerlikon Metco staff involved in the experimental work, and preparation of this report, are also gratefully acknowledged.

6 Literature

[1] Fisher, G. et al: Impact Testing of Materials for Oil Sands Processing Applications. NACE International Corrosion 2007 Conference and Expo.
[2] Gronvold et al.: Heat Capacity and Thermodynamic Properties of Ditungsten Carbide, $W2C_{1-x}$, From 10 to 1000K. Thermochimica Acta, 129 (1980) 115-125.
[3] Toth, L. E.: Transition Metal Carbides and Nitrides. Academic Press, New York 1971.
[4] Mazurovsky, V. et al: Carbide Formation During Crystallization Upon Welding. In: Third International Conference on Mathematical Modelling and Computer Modelling and Computer Simulation of Materials Technologies MMT (2004).
[5] Wood, S.: Experimental determination of the hydrothermal solubility and speciation of tungsten at 500°C and 1 kbar, in International Journal of Refractory Metals and Hard Materials, Vol. 14, Issues 5-6, 1996, p. 345-353.
[6] Andersson, K. M. and Bergstrom, L.: Oxidation and dissolution of tungsten carbide powder in water, in International Journal of Refractory Metals & Hard Materials (2000), Issue18, pp. 121-129.
[7] Brookes, K.: Hardmetals and Other Hard Materials. 1998, 3rd Edition, Published by International Carbide Data, East Barnet Hertfordshire, UK.
[8] Pierson, H. O.: Handbook of Refractory Carbides and Nitrides. Noyes Publications, 1996.

Aptitude of different types of carbides for production of durable rough surfaces by laser dispersing

A. Wank, C. Schmengler, K. Müller-Roden, F. Beck, T. Schläfer, GTV Verschleiss-Schutz GmbH, Luckenbach / D

Different types of tungsten carbide materials (fused tungsten carbide, nickel clad fused tungsten carbide, macrocrystalline WC and sintered and crushed WC/Co) are used for laser dispersing of construction steel surfaces. Surface roughness analyses and metallographic evaluation of cross sections concerning efficiency of carbide embedding as well as crack formation tendency are carried out. Generally, all types of tested carbides permit production of rough surfaces with metallurgical bonding to the metallic matrix, but only use of nickel clad fused tungsten carbide permits to prevent crack formation. The effectiveness of silicon and silicon carbide for production of durable rough surfaces on aluminium alloys is investigated. Both silicon and silicon carbide qualify for production of rough surfaces by laser dispersing. While silicon carbide particles show higher hardness, use of silicon does not include danger of embrittlement due to formation of aluminium carbide.

1 Introduction

Surfaces with defined, long-term stable roughness are needed in various applications e.g. for traction control in logistic centres or on moving walkways, pulp production in paper industries, paper conveyors in paper or printing machines, rolls that transport or deflect sheet metal in steel or aluminium mills, fixation of semi-finished parts in moulds or massive forming tools, etc.. In some applications like rolls for transportation of sheet steel [1] and jaws for clamping of cylinders in massive forming processes [2] only relatively small roughness of about R_a 5 µm is needed, which can be covered by HVOF spraying. Other thermal spray processes permit production of rougher coatings, but limited bond strength and / or coating cohesion restrict their use.

Laser based processes are particularly suitable for high precision of surface layer modification or cladding both concerning geometry and microstructure [3]. Among these processes laser cladding and laser dispersing are best suited to achieve surface layers that contain embedded hard materials as a prerequisite for high durability rough surfaces. Thereby laser dispersing results in only small change of the respective components surface contour and is therefore particularly suitable for treatment of already completely machined components. That makes the process especially interesting for job shops that cover various applications with relatively small amount of components.

Potential of hard materials to be used as reinforcing phase depends on the type of base material. Generally, irregular shaped powders show better aptitude to form surfaces with durable roughness peaks. Ideally, reinforcing phases should be wetted easily by the base material´s melt, but only dissolve slowly, and exhibit high fracture toughness rather than high hardness in the first place.

Generally, tungsten carbide tends to dissolve in iron based alloys quickly forming brittle mixed carbides. However, there is a large variety of tungsten carbide based materials with different chemical composition and microstructure as well as particle size distribution commercially available. In contrast vanadium carbide does not tend to dissolve in iron based alloys, especially in vanadium rich alloys, that fast and does not form brittle mixed carbides. But, it is not easily available in a particle size range suitable for laser dispersing processes. This study aims for evaluation of different tungsten carbide based materials concerning their aptitude for laser dispersing with the aim to produce durable rough surfaces.

Additionally, for manufacturing of durable rough surfaces on aluminium alloys the capabilities of silicon carbide and silicon are studied and evaluated.

2 Experimental

Steel dispersing experiments are carried out with high power diode laser type LDF 6000-100 (Laserline GmbH, Mühlheim-Kärlich, Germany) with wavelength between 900 nm and 1,070 nm, 1,500 µm diameter fibre, 80 mm collimation and 250 mm focal length, which results in 4.7 mm focal spot diameter. Feedstock is fed coaxially using powder nozzle PN6625 (GTV Verschleiss-Schutz GmbH, Luckenbach, Germany) with 1.5 mm ID injectors. Argon is used both as laser process gas and carrier gas. S355J2G3 substrates are treated with meander movement of the laser processing head. Properties of four different kinds of tungsten carbide based reinforcing materials are listed in **table 1**. Besides macrocrystalline WC (GTV 10.80.11), fused tungsten carbide (GTV 10.80.4) and sintered and crushed WC/Co compounds (GTV 10.80.60) a chemically nickel clad fused tungsten carbide powder (GTV 80.70.1) is applied. All powders show irregular shape.

Table 1. Properties of applied tungsten carbide based powders

powder	**10.80.11**	**10.80.4**	**10.80.60**	**80.70.1**
material	WC	WC/W_2C	WC/Co 88/12	WC/W_2C Ni clad
particle size [µm]	+75 -180	+30 -75	+75 -150	+10 -53
W [wt.-%]	base	base	base	base
C [wt.-%]	6.21	3.97	5.48	3.72
Co [wt.-%]	-	-	11.85	-
Ni [wt.-%]	-	-	-	8.87

Ranges of investigated process parameters for dispersing of S355J2G3 steel are listed in **table 2**.

Table 2. Process parameters for laser dispersing of S355J2G3 with tungsten carbide based powders

Laser power [W]	900 - 1,500
Laser process gas flow rate [l/min]	20
Carrier gas flow rate [l/min]	4 - 12
Powder feed rate [g/min]	10 - 20
Surface speed	1.0 m/min
Offset between paths	2.0 mm

Laser alloying / dispersing of AlSi10Mg substrates with silicon and silicon carbide is carried out with high power diode laser type LDF 6000-60 (Laserline GmbH, Mühlheim-Kärlich, Germany) with wavelength between 980 nm and 1,060 nm, 1,000 µm diameter fibre, 80 mm collimation and 250 mm focal length, which results in 3.1 mm focal spot diameter. Feedstock is fed coaxially using powder nozzle PN6625 (GTV Verschleiss-Schutz GmbH, Luckenbach, Germany) with 1.0 mm ID injectors. Argon is used both as laser process gas and carrier gas. GTV 70.11.6 and GTV 70.85.2 are used as silicon and silicon carbide feedstock respectively. The silicon powder shows 99.9% purity and a particle size distribution +45 -125 µm, while silicon carbide feedstock shows 99.5% purity and particle size distribution +45 -90 µm. The ranges of investigated process parameters are listed in **table 3**.

Table 3. Process parameters for laser alloying / dispersing of aluminium substrates

Laser power [W]	600 - 1,400
Laser process gas flow rate [l/min]	15 - 25
Carrier gas flow rate [l/min]	3 - 5
Powder feed rate [g/min]	2 - 3
Surface speed	1.0 - 4.0 m/min
Offset between paths	1.5 - 2.0 mm

Cross sections of laser dispersed specimens are prepared and analysed by optical microscopy. Surface roughness is determined by stylus instrument Mitutoyo SJ 201P.

3 Results

3.1 Laser dispersing of S355J2G3 with tungsten carbide based powders

When dispersing S355J2G3 surfaces carrier gas flow rate takes strong influence on topography. **Figures 1** and **2** show the effect at the example of dispersing with nickel clad fused tungsten carbide. Despite large optimal working distance of PN6625 nozzle, i.e. 25 mm, even relatively small carrier gas flow rates of 6 l/min cause displacement of carbide reinforced melt from the centre to the sides and according wavy surface appearance. In contrast 4 l/min carrier gas flow rate permits avoiding melt displacement to the sides with all other parameters kept constant. However, this carrier gas flow rate also represents the lower limit in order to ascertain continuous transportation of relatively heavy tungsten carbide based powders through the feed hoses.

Fig. 1. Surfaces of S355J2G3 dispersed with nickel clad fused tungsten carbide using carrier gas flow rate of 6 l/min (top) and 4 l/min (bottom)

Ratio of powder feed rate and laser power determines content of reinforcing particles as well as microscopic surface roughness of dispersed layers, **figure 3**. Melt pool size increases with laser power and thereby provides an according volume that can take up hard phases. With increasing content of hard phases injected into the molten volume melt flow is hindered, which results in formation of an increasing density of microscopic peaks at the surface of dispersed layers. Both investigated pure carbides, i.e. macrocrystalline WC and fused tungsten carbide, tend to dissolve relatively fast inside the steel melt. Upon cooling high contents of carbon and tungsten cannot remain dissolved inside the metallic matrix and form precipitates, **figure 4**.

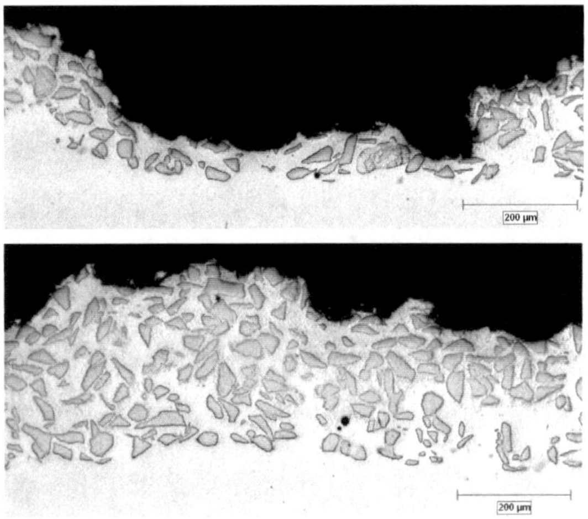

Fig. 2. Cross section of S355J2G3 dispersed with nickel clad fused tungsten carbide using carrier gas flow rate of 6 l/min (top) and 4 l/min (bottom)

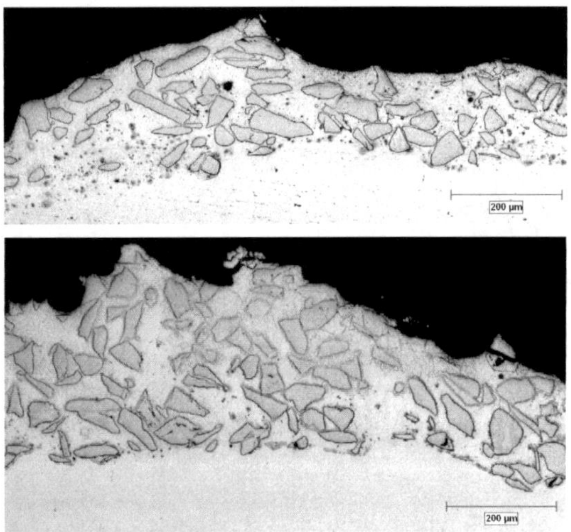

Fig. 3. Cross section of S355J2G3 dispersed with fused tungsten carbide using 10 g/min powder feed rate at 1,000 W laser power (top) and 20 g/min powder feed rate at 1,200 W laser power (bottom)

Observed dissolution inside the steel melt is higher for GTV 10.80.11 compared to GTV 10.80.4, which can be attributed to smaller particle size and accordingly larger specific surface area of the latter. Macrocrystalline WC might also tend to dissolve less in the steel melt, which is reported at least for nickel based matrix alloys [4-5].

Despite their nickel cladding fused tungsten carbides in particularly fine sized GTV 80.70.1 also undergo some dissolution in the steel melt. However, dissolution is less than for GTV 10.80.4.

Also, sintered and crushed WC/Co 88/12 compounds partially dissolve in the steel melt, **figure 5**. Thereby cobalt phase is dissolved in the first place and WC particles with average size of about 2 µm remain mostly intact and become distributed in the cobalt enriched steel matrix. Some WC/Co particles form core-shell structures, which might be due to partial impregnation by steel melt. However, high resolution SEM and EDS were needed to verify that explanation approach.

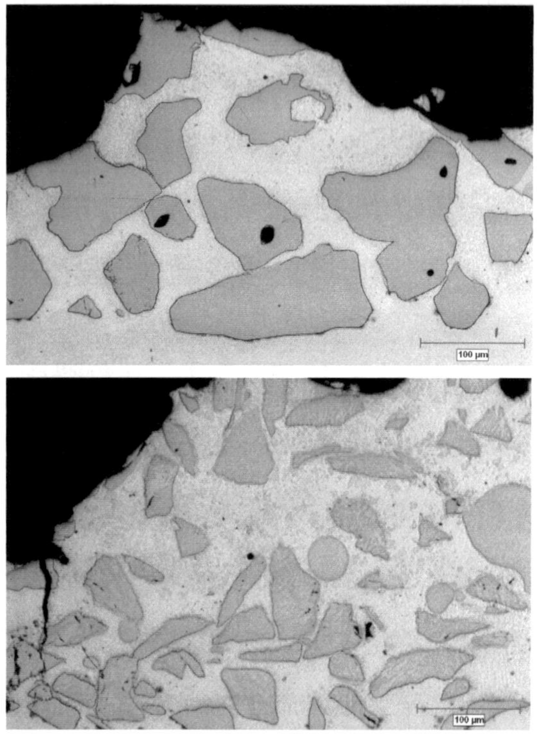

Fig. 4. Precipitates of macrocrystalline (top) and fused (bottom) tungsten carbides that were dissolved in S355J2G3 melt

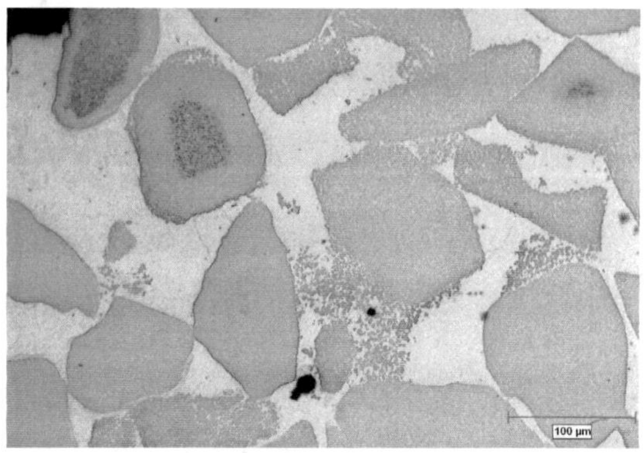

Fig. 5. Partial dissolution of WC/Co compound particles in S355J2G3 melt and partial formation of core-shell structure particles

There is no clear correlation between particle sizes of dispersed tungsten carbide based particles and roughness of surfaces that have been produced in the limits of this study, **table 4**. That can be attributed to the relatively deep melting of the substrate surface that results in thickness of reinforced surface layers that clearly exceeds the size of dispersed tungsten carbide based particles.

Table 4. Roughness of laser dispersed S355J2G3 surfaces

powder	**10.80.11**	**10.80.4**	**10.80.60**	**80.70.1**
R_a [µm]	16 - 26	12 - 24	14 - 33	17 - 23
R_z [µm]	87 - 118	65 - 110	75 - 150	92 - 113

In the limits of the conducted experiments crack formation could not be avoided, when using macrocrystalline WC, fused tungsten carbide or WC/Co compounds, **figure 6**. In case of macrocrystalline WC and fused tungsten carbide dispersed surfaces cracks occur exclusively in places with minimum thickness of the reinforced layer, while cracks in WC/Co compound dispersed surface layers occur in random places. In any case crack orientation is vertical and cracks propagate both through tungsten carbide based particles and metallic matrix. Only for use of nickel clad fused tungsten carbide as reinforcing material crack formation can be avoided completely. However, use of too high laser power levels that cause excessive dissolution of the clad fused tungsten carbide particles can still result in crack formation.

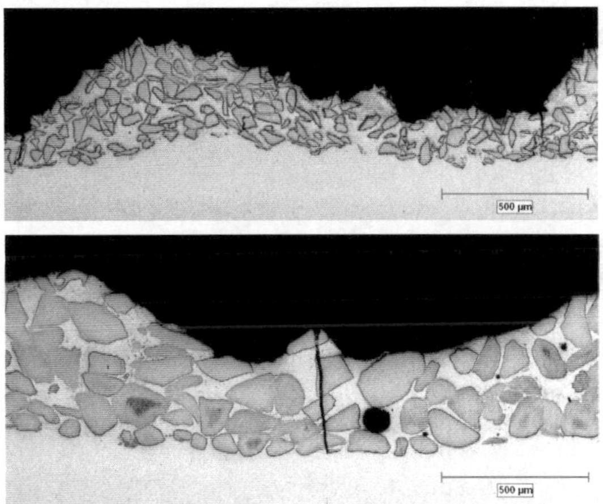

Fig. 6. Crack formation in S355J2G3 surfaces that were laser dispersed with fused tungsten carbide (top) and WC/Co compounds (bottom)

3.2 Laser dispersing of AlSi10Mg

Displacement of reinforced melt from the centre to the sides is observed for dispersing with both silicon and silicon carbide even for use of only 4 l/min carrier gas flow rate, **figure 7**. However, when using even lower carrier gas flow rate stable feeding of powders cannot be ascertained anymore and there is risk of feed line blocking.

Fig. 7. Surfaces of AlSi10Mg alloyed with silicon (top) and dispersed with silicon carbide (bottom)

In the limits of the investigated parameters rough surfaces are produced, but no continuous surface layers with reinforcing phases are formed. Instead reinforcing phases form spikes on a layer of remelted substrate material that shows extremely fine microstructure due to process inherent high cooling rates, **figure 8**.

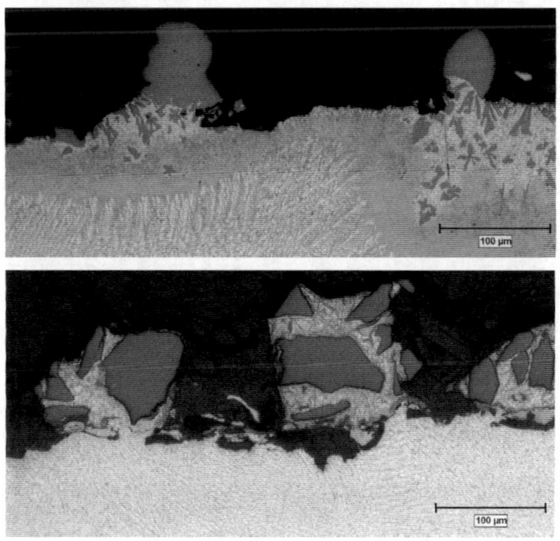

Fig. 8. Surfaces of AlSi10Mg dispersed with silicon (top) and silicon carbide (bottom)

Use of silicon as reinforcing material results in partial dissolution in the aluminium alloy melt and precipitation of primary silicon during cooling process as a result of hypereutectic silicon content in respective areas. The part of silicon particles that is not dissolved forms metallurgically well bonded spikes on the surface of the aluminium alloy.

In contrast silicon carbide is easily wetted by aluminium alloy melt and does not form pure silicon carbide spikes on the surface. A part of silicon carbide dissolves and results in precipitation of needle shaped aluminium carbides during cooling of the accordingly alloyed melt. Silicon carbide reinforced aluminium areas are partially separated from the underlying remelted aluminium alloy by thin oxide bands. Formation of such oxide bands during laser dispersing of aluminium alloys with silicon carbide has been reported for treatment of wrought precipitation hardened high strength aluminium alloys before [6]. In these studies separating oxide bands could be avoided by pre-heating to 150 °C and by adding 50 vol.-% AlSi40 to silicon carbide. The content of silicon in the here investigated aluminium cast alloy is too low to achieve such effect - at least without pre-heating.

Although surfaces with silicon and silicon carbide reinforcement show very different microstructure comparable roughness is observed. Typically R_a between 20 - 25 µm and R_z between 100 - 120 µm is achieved.

4 Summary

Laser dispersing is suitable for manufacturing of rough surfaces with high durability. Both for dispersing of S355J2G3 with different tungsten carbide materials and for dispersing of AlSi10Mg with silicon or silicon carbide roughness R_a exceeding 20 µm and R_z exceeding 100 µm can be achieved.

In the limits of investigated process parameters macrocrystalline tungsten carbide, fused tungsten carbide, sintered and crushed WC/Co as well as nickel clad fused tungsten carbides all formed surface layers with even distribution of reinforcing tungsten carbide based particles in S355J2G3. But, only use of nickel clad fused tungsten carbide permits avoiding crack formation completely.

Dispersing of AlSi10Mg with high hardness silicon carbide is accompanied by metallurgical problems. On the one hand dissolution of silicon carbide results in precipitation of needle shaped aluminium carbides and according embrittlement. On the other hand oxide bands that separate reinforced surface layers from remelted base materials are embedded with according detrimental effect on bonding of the reinforced surface layer. In contrast silicon is metallurgically fully compatible and bonds perfectly to alloyed surface layers that form as a consequence of partial dissolution of silicon in the aluminium alloy melt. Hypereutectic composition of the alloyed surface layer results in precipitation of primary silicon and accordingly strengthens surfaces which support silicon particles that form spikes.

In order to avoid displacement of melt to the sides of seams and accordingly wavy surfaces carrier gas flow rate needs to be minimized, but must still ascertain continuous feed of reinforcing particles. Powder nozzles with large working distance are particularly suitable with respect to this goal.

5 Perspectives

Further investigations are planned concerning dispersing of steel surfaces with vanadium carbide. It is expected to avoid crack formation even more secure than for use of nickel clad fused tungsten carbide due to advantageous re-precipitation behaviour of dissolved vanadium carbide from steel melt during cooling process.

6 Literature

[1] Fukubayashi, H.H.: Present Furnace and Pot Roll Coatings and Future Development. Thermal Spray 2004: Advances in Technology and Application, pp. 125-131

[2] Reisel, G., B. Wielage, U. Mahn, S. Steinhaeuser, D. Weidlich, R. Neugebauer: Friction-promoting layers for the clamping of heavily loaded workpieces with little damage. Proc. THE Coatings in Manufacturing Engineering 2004, Erlangen, Germany, April 5th - 7th, 2004, pp.139-147

[3] Wank, A., B. Wielage, H. Podlesak, K.-J. Matthes, G. Kolbe: Protection of Ti6Al4V Surfaces by Laser Dispersion of Diborides. Journal of Thermal Spray Technology, Vol. 14, No. 1, 2005, pp. 134-140

[4] Jones, M., R.J. Llewellyn: Erosion-corrosion assessment of WC-based PTAW overlays. Corrosion, Vol. 68, 2012, No. 2, pp. 026003-1-026003-14

[5] Jones, M., U. Waag: The influence of carbide dissolution on the erosion-corrosion properties of cast tungsten carbide/Ni-based PTAW overlays. Wear, Vol. 271, 2011, No. 9-10, pp. 1314-1324

[6] Hoenig, T., K.-J. Matthes, B. Wielage, H. Pokhmurska, H. Podlesak, A. Wank: Laserstrahl-dispergieren von SiC in Aluminiumlegierungen zum partiellen Verschleißschutz. Schweissen und Schneiden, Vol. 58, 2006, No. 3, pp. 119-122

Influence of the Surface Structure on the Adhesion Strength of ZnAl2 Coatings on Thermally Instable Polymers Printed by Means of Fused Layer Modelling (FLM)

W. Tillmann, C. Schaak, K. Bleicher, TU Dortmund, Institute of Materials Engineering

Additive manufacturing (AM) techniques give access to completely new manufacturing processes. AM techniques using metals, ceramics, or plastics feedstock are predestined for lightweight construction and for components with complex shapes or internal functions. AM processing with plastics stands out due to the low density of polymers, a good process capability, and low initial costs. The properties of polymer components are extremely dependent on the utilized plastics and the reinforcements, e.g. in the form of fibres. Furthermore, coatings can improve the properties and enhance the possible range of applications for plastics. In the present study, PLA (polylactic acid) was printed utilizing Fused Layer Modeling (FLM). The surfaces of the PLA samples were directly structured with pits with different widths during printing. Subsequently, the surfaces were coated with ZnAl2 by means of Twin Wire Arc Spraying (TWAS). Adhesion tests meeting DIN EN 582 were conducted to measure the adhesion of the coating on the structured plastic surface. The results were compared to the adhesion of ZnAl2 coatings on grit blasted and as-built surfaces. Overall, the surface adhesion was significantly better for the samples with directly structured surfaces. Hence, a direct structuring of the surface during a 3D building process promises to be an outstanding possibility to prepare surfaces prior to coating processes.

1 Introduction

Coating of thermally instable materials such as e.g. polymers, fibre reinforced plastic (FRP), wood, and vulcanized fibre by means of thermal spraying is still a challenge due to the high thermal and kinetic energy used in these coating processes as a high thermal energy leads to the melting or degradation of these substrates. High particle energies can lead to erosion effects on the surface. However, there is a high demand for these materials, because of their low density, good corrosion resistance, and as they are easy to machine. Yet, their mechanical strength and wear resistance are very low when compared to e.g. metals. These problems could be solved by a surface functionalization or metallization using thermal spraying processes.

Gonzalez et al. [1] and Wielage et al. [2] provide a good overview of the process of deposition coatings on polymers by means of thermal spraying. The DVS (German Welding Society) published some basic guidelines on this topic as well [3]. The current publications show that nearly all standard thermal spray techniques were tested to coat polymer materials [1]. Liu et al. [4,5] proved that arc-spraying can be used to coat polymers. Overall primarily, only low melting metals and alloys (e.g. copper, zinc, aluminum) can be deposited due to the low thermal stability of polymers. Hence, a typical application of polymer metallization is the thermal and mechanical protection of polymers. Other applications are the deposition of electrically conductive structures or coatings that prevent an ice accretion on polymers [1]. Furthermore, traditional pre-processing steps, such as grit blasting and/or cleaning in ethanol are not suitable for instable polymer materials [1]. A pre-treatment, however, is pretty important to clean and activate the surface and hence, to guarantee a good adhesion of the coating. Finally, the grit blasting pressure and grit size should be reduced when pre-treating polymers [3]. Other researchers showed that for metal substrates, grit blasting as a prior step could be substituted by geometrically defined surface structures, generated by means of milling [6,7]. In the present research paper, the focus is on the coating of additive manufactured ("3D-printed") polymers by means of thermal spraying. These techniques offer the possibility to generate geometrically defined and thus, well coatable surface structures immediately during printing. Additive manufacturing processes are procedures in which parts are not manufactured in a destructive or removing process but instead by using an additive method. In this sense, elements or layers are assembled until the part or component is completed [8,9].

2 Experimental details

2.1 Substrate

To produce polymer samples, fused layer modelling/manufacturing (FLM) was used. This additive procedure is also referred to as fused deposition modelling (FDM). In this process, a polymer filament is completely melted and applied in lines through an extruder nozzle. This process is repeated layer by layer until the object is completely built up. Figure 1 shows a sketch of the FLM process. The dimension of the cylindrical samples is based on the dimension of adhesion tensile test samples (diameter 40 mm). The top surface was directly structured with different land (highs) and pit (lows) widths during printing. The dimensions of the surface have to be a multiple of the nozzle diameter and the vertical step width (vertical resolution). A sketch of the structured samples is given in figure 2. PLA was used as a polymer material. The samples were printed with a Ultimaker2 (Ultimaker B.V., Netherlands) with a 0.25 mm and 0.4 mm nozzle. The vertical solution was set to 0.1 mm. The small nozzle (0.25 mm) was necessary to print the fine surface structures (figure 2).

Additionally, some samples were printed with a flat top surface. These samples were grit blasted prior to the coating process (EKF 100, p = 2 bar, α = 45 °). The grit blasting pressure was reduced to 2 bar. Higher grit blasting pressures would be used for metals, e.g.

steel. The idea of using the smaller pressure, was to preserve the polymer structure.
Conventional adhesion test samples made of steel (C45 / 1.1191) were grit blasted and coated for further comparisons.

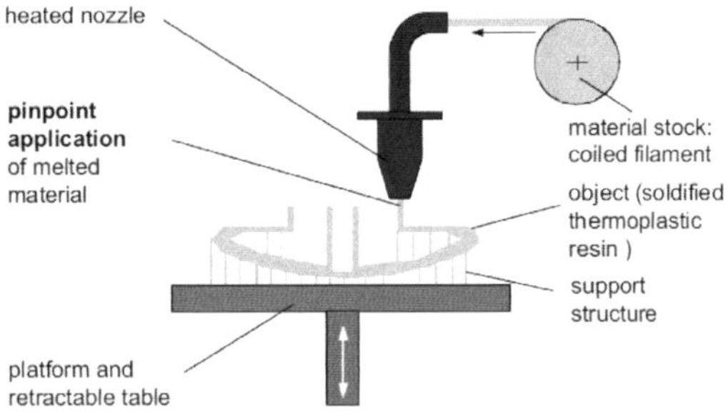

Figure 1: Sketch of the fused layer manufacturing process (taken from VDI 3404 [7]).

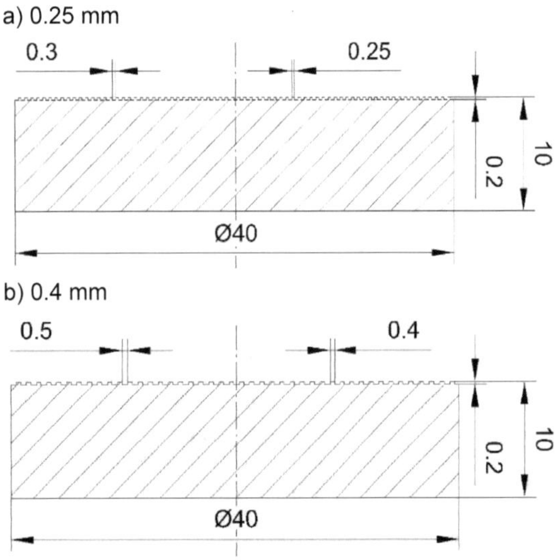

Figure 2: CAD-Image of the printed samples with surface structure

2.2 Coating process

The coating process was conducted with an SmartArc twin wire arc-spraying device (Oerlikon Metco AG, Switzerland). Twin wire arc-spraying (TWAS) is a simple, robust, and cost-effective process. Generally, TWAS exhibits moderate thermal and low kinetic particle energies. A solid zinc-aluminium (ZnAl2) wire with a diameter of 1.6 mm was used as feedstock material. Because of the low melting temperature of the zinc aluminium alloy, this wire can be atomized with low-energy spraying parameters. This should ensure low thermal loads on the plastic substrates. Furthermore, the moving speed of the spray gun was set to the maximum and the number of overruns was set to two times. Both settings should reduce the thermal load. Pretests showed that these parameters are suitable to coat the thermally sensitive polymer material. The arc-spraying process parameters are summarized in table 1.The substrates were cleaned with ethanol prior to the coating process.

Table 1: Arc-spraying parameters used to coat polymer samples

Spray system	TWAS SmartArc PPG
Material (wire)	ZnAl2 (1.6 mm) T_m: 382-407 °C
Parameters	Current I = 220 A Voltage U = 26 V Primary gas: 2.4 bar Overruns: 2 Standoff distance: 220 Distance between lines: 5 mm Speed.: 24,000 mm/min

2.3 Test and analyses methods

The surfaces of the printed substrates were characterized by means of a 3D measurement system (InfiniteFocus-G3, Alicona Imaging GmbH, Grambach, Switzerland). The results were compared to the raw CAD data to identify deviations and to evaluate the printing results.
After the coating process, some samples were prepared metallographically to investigate the coating thickness, the filling of the pits, and the interface. For this purpose, a light-microscope (AxioPhot, Carl Zeiss MicroImaging GmbH, Göttingen, Germany) was used. Most of the coated samples were used for adhesion tensile tests according to DIN EN 582 [10]. For this purpose, the uncoated back of the cylindrical polymer samples (figure 2) were glued to a metal stamp that has a retainer for the tensile test machine (Mohr u. Federhaff AG, Mannheim). Sequentially, the metal stamp with the coated polymer was glued to another metal stamp. Ultrabond 100 (HTK Hamburg GmbH, $\sigma_{max} = 100$ N/mm^2) was used as glue. After curing, the samples were tensile tested. The maximum force (force at failure) was measured. The maximum adhesion strength was calculated by dividing the maximum force by the area of the coated surface. Only samples that show a clear failure in the interface were used for this purpose. The samples were excluded from analyses if the failure occurred in the polymer, because this failure do not generate any information about the adhesion of the coating. Furthermore, the fractured surface was investigated.

3 Results and Discussion

3.1 Surface conditions of the substrate

The CAD data files were the basis for printing the polymer samples. The resolution of the samples after the printing strongly depends on the slice process and the FLM printer parameters. In the slice process, the CAD data or rather the CAD model is sliced into single layers and the moving paths (g-code) are generated for the 3D-printer [13]. The moving paths directly describe the movement of the extrusion nozzle, but also contain futher information such as the extrusion speed and temperature. If wrong slice or printing parameters are used, small or thin features will not accord with the raw CAD data. To check the correlation of the printed parts with the CAD model, the surfaces of the samples were investigated by means of InfiniteFocus-G3. Figure 3 shows 3D images of the samples' surfaces after manufacturing by means of FLM.

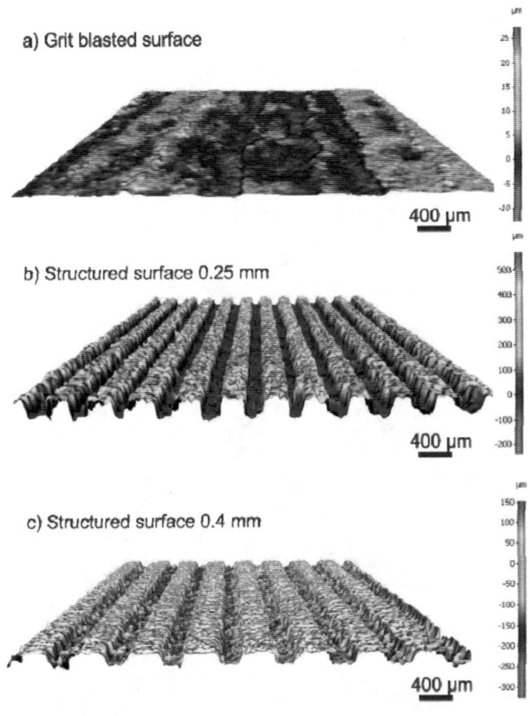

Figure 3: 3D-Image of the real surface structure of printed polymer samples (uncoated)

The images show that the surface is structured as planed (Figure 3b and 3c). The macrostructure of the surface is homogenous and defect free over the complete sample. Additionally, figure 4 shows a line measurement of the primary profile diagonally to the surface structure. The pit depths are close to the desired values (200 µm = 2 layers) and the contour shows sharp sides (flanks). The pits und lands show an inherent microstructure (line shape hollows) in the bottom and at the plateau. This microstructure is elongated into the moving direction of the nozzle during printing. This microstructure is especially well visible in figure 5. A reason why this microstructure is generated is the smearing of the nozzle because of adhesion of the polymer during the printing process [14]. As a consequence, the lands themselves show a micro roughness of Rz = 10.9 µm and Ra = 2.8 µm (measured transversally to the orientation of the lands on a sample with a 0.4 mm surface structure).

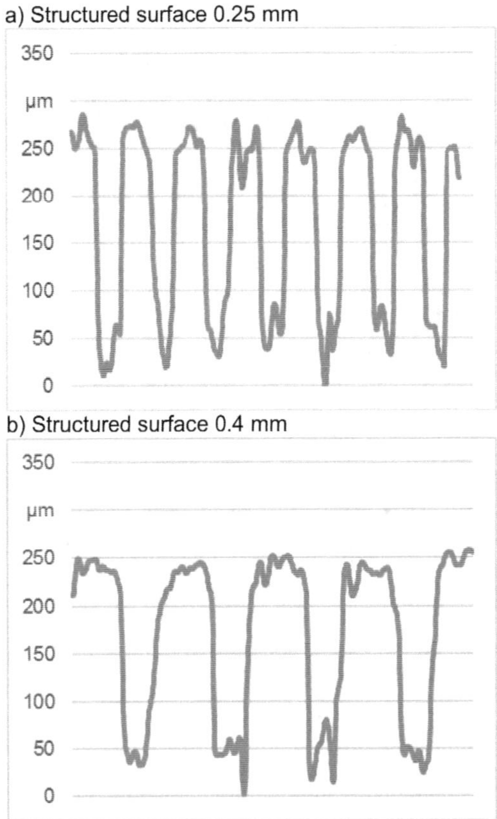

Figure 4: Primary profile along a line scan on the structured surfaces (uncoated)

As comparison, figure 3a shows a flat and grit blasted polymer sample. It has to be mentioned that the grit blasting process with the selected parameters did not damage or destroy the printed polymer. Even this flat samples show some irregular and inhomogeneous smooth pits. Furthermore, this effect is reflected in the roughness values. The roughness values are approximately Rz = 35.9 µm and Ra = 7.6 µm, measured in moving direction of the nozzle. Transverse to the printing direction, the roughness values are Rz = 63.6 and Ra = 11.1 µm. All values are mean values.

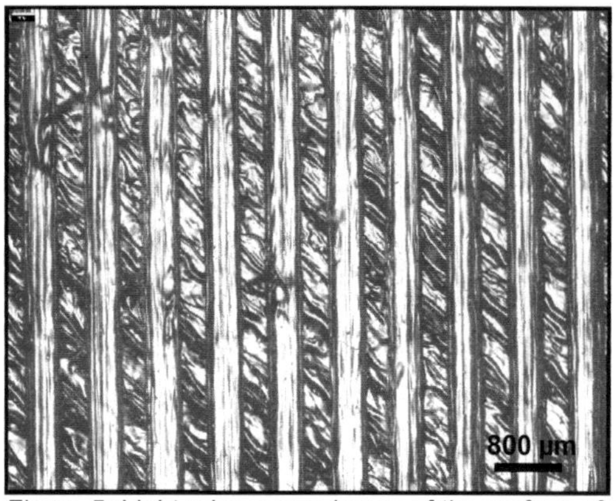

Figure 5: Light-microscope image of the surface of a sample with a 0.25 mm structure

The top surface of FLM samples is almost never really flat due to the extrusion and deposition of several adjacent molten plastic tracks that form a layer. The single layers show a convex curvature after solidification, thus resulting in a small hollow between the individual tracks (figure 3a).
To sum up, the microstructure is strongly influenced by the 3D-printing process, but the macrostructure was as planned. Overall, the samples showed ideal pre-conditions for the coating process.

3.2 Coating morphology and adhesion

After the coating process, some minor delamination were observed at the edge of the samples. One reason for this effect could be thermal stresses caused by the shrinkage of the coating during rapid cooling after spraying. The coefficient of thermal expansion (CTE) strongly differs for both materials (Zn-Al alloys CTE: 22 - 28*10-6 1/K [11]; PLA CTE = 85*10-6 1/K [12]). If samples showed a visible delamination, the samples were excluded from the tensile adhesion tests. Figure 6 shows light-microscope images of cross-sections of differently structured polymer samples, coated with ZnAl2. In all cases the coating was well connected to the polymer without gaps or delamination. In the cross-section image, both printed surface structures (lands and pits) are clearly recognizable (Figure 6b and 6c). With the spray parameters employed, the ZnAl2 coating completely fills the pits and covers the surface structure. Hence, the surface structure is no longer visible after the coating process and a flat surface was obtained (Figure 6b and 6c). Images with higher magnification prove that the surface structure is faultless after spraying (figure 7). Degradation effects of the polymer are not visible in this scale. The average porosity was calculated to 8.37 % with a SD of 0.55 %. Overall, the coating results are very promising.

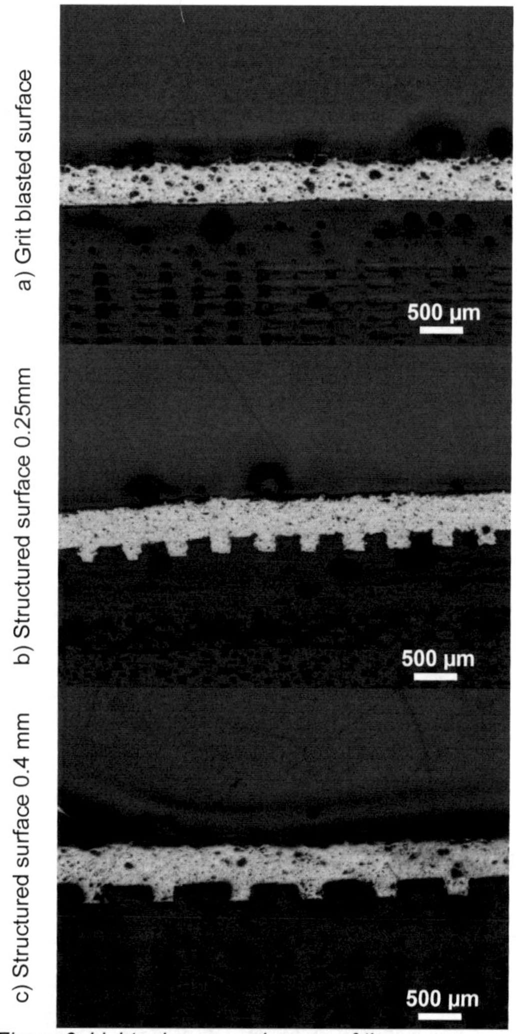

Figure 6: Light-microscope images of the cross-sections of coated polymer samples with different surface structures

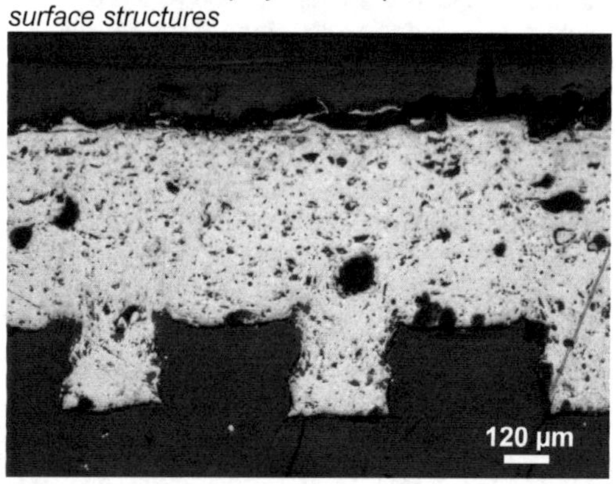

Figure 7: Light-microscope image of the cross-section of a coated polymer sample (0.25 mm structure)

The quantitative results of the tensile adhesion tests are visualized in figure 8. The measurement results of the samples with a 0.25 mm structured surface are not included in the diagram, because the failure always occurred in the polymer and not in the inner coating or the interface. The adhesion of the coating seems to be higher than the cohesion of the printed polymer. The average strength was calculated as 1.61 MPa, with a high standard deviation (SD) of 0.49 MPa. The only possible conclusion is that the adhesion strength of the ZnAl2 coating is higher than this value. One possible reason why especially the samples with a surface structure with a 0.25 mm width, without exception, fail in the substrate could be the printing process itself. To print the small pits, a 0.25 mm nozzle was used. Standard FLM nozzles have a diameter of 0.4 mm. The cohesive strength of the printed polymer seems to be lower for the 0.25 mm nozzle. Reasons for this could be an insufficient melting and connecting of the single, extruded polymer lines. Figure 6b implies that the single polymer layers and lines are not well connected, because of large gaps that are visible in the inside of the substrate. All other samples were printed with a standard 0.4 mm nozzle. These samples showed a better cohesion behavior. Failure mostly occurred in the interface between the ZnAl2 coating and the polymer substrate. Hence, the samples can be used for adhesion strength evaluations.

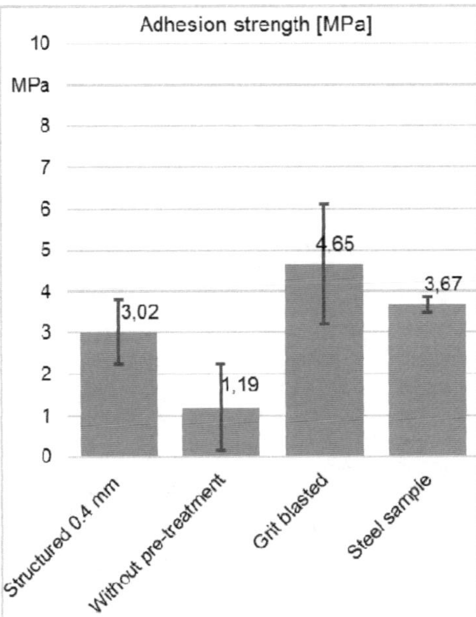

*Figure 8: Results of the tensile adhesion tests. The samples with a 0.25 structured surface are not included because all samples fail in the polymer. The error bars show the confidence interval for ±1*SD*

The ZnAl2 coating showed the lowest adhesion strength on flat polymer samples without a pre-treatment (σ = 1.19 MPa). This behavior was predictable as this surface shows less possibilities to enable mechanical bonding. A better adhesion of the coating was achieved with the 0.4 mm surface structure (σ = 3.02 MPa). For polymer samples, the coating on the grit blasted surface shows the best adhesion (σ = 4.65 MPa). This adhesion strength is even higher than the adhesion strength of the coating on grit blasted steel samples (σ = 3.67 MPa).
One possible reason for the low adhesion on the 0.4 mm structured surface could be the microstructure. Although the structured samples show a well-defined macrostructure (figure 3), the microstructure on the local surfaces on the lands and in pits are comparable to the microstructure of samples without pre-treatment. These samples exhibit an insufficient coating adhesion. Additionally, the microstructure is not isotropic, but orientated in nozzle movement direction (compare figure 5).
Since the grit blasted samples show the best results, the microstructure, respectively the microroughness, seem to be more important for the adhesion strength than the macrostructure. This theory correlates with studies in [7] for HVOF coatings applied on milled surfaces. Furthermore, it has to be mentioned that the adhesion strength values for coated polymer substrates show a large deviation. This large deviation only occurs for polymer substrates. In comparison, the results of the steel samples show a low deviation (figure 8).

4 Summary and Conclusion

Polymer samples were printed by means of FLM. The FLM process enables to structure the surface of the samples directly during printing. Grit blasted samples, samples without pre-treatment, and samples structured with pits and lands were coated with ZnAl2 by means of TWAS. After this process step, the adhesion strength of the coatings was measured. The samples without surface pre-treatment exhibit the worst adhesion strength values. The directly printed surface structure (0.4 mm) improves the adhesion strength when compared to samples without pre-treatment. Nevertheless, the grit blasted polymer samples showed the highest adhesion strength. The samples with a 0.25 mm surface structure always fail inside the polymer substrate.
In summary, the surface structures do not enhance the adhesion strength of the coating above the values of samples that are pre-treated by means of grit blasting for both polymer or steel materials. One possible reason is that the microroughness is more important for adhesion than the macrostructure. But, it is possible that the surface structure is an imporvement for other load directions like for example the shear load. But this has to be shown in future expermiments.
The reflection of all experiments shows that the printed polymer substrate has a strong influence on the results. This becomes apparent in the high deviation of the adhesion strength measurement. For future experiments, the printing process parameters should be investigated in detail to find robust and repeatable parameters.

Acknowledgement

The authors thank the VolkswagenStiftung (A110812) for its financial support.

5 Literature

[1] Gonzalez, R.; Ashrafizadeh, H.; Lopera, A.; Mertiny, P.; McDonald, A.: A Review of Thermal Spray Metallization of Polymer-Based Structures. In: Journal of Thermal Spray Technology 25 (2016) 5, 897-919.
[2] Wielage, B.; Rupprecht, D.; Paczkowski, G.; Winkler, R.: Thermisches Beschichten von Hochleistungspolymeren, Thermal Spray Bulletin 2/2010, S.148-152, ISSN 1866-6248.
[3] Merkblatt DVS 2315 (German Welding Society) - Thermische Spritzschichten auf Bauteilen aus unverstärkten und faserverstärkten Kunststoffen.
[4] Liu, A.; Guo, M.; Zhao, M.: Arc sprayed erosion-resistant coating for carbon fibre reinforced polymer matrix composite substrates. Surface & Coatings Technology 200, 2006, 3073–3077.
[5] Liu, A.; Guo, M.; Gao, J.: Influence of bond coat on shear adhesion strength of erosion and thermal resistant coating for carbon fibre reinforced thermosetting polyimide. Surface & Coatings Technology 201, 2006, 2696–2700.
[6] Nicolaus, M.; Möhwald, K.; Maier, J.; Abrahams, H.; Vogel, F.; Biermann, D.: Geometrically defined surface structures for the positive-locking substrate bonding of thermally sprayed coatings. Thermal Spray Bulletin, 9 (2016) 1, S. 54–59.
[7] Tillmann, W.; Nebel, J.; Schaak, C.: U Haftung HVOF gespritzter WC-CoCr Schichten auf gefrästen, gestrahlten und geschliffenen Oberflächen. In: Wielage, B. (Hg.): Tagungsband zum 13. Werkstofftechnischen Kolloquium. Chemnitz, 2010, S. 120–127.
[8] DIN EN ISO/ASTM 52900:2015: Additive manufacturing – General principles 09.2016.
[9] VDI 3404: Additive fabrication Rapid technologies (rapid prototyping) - Fundamentals, terms and definitions, quality parameters, supply agreements. 12.2009.
[10] DIN EN 582:1993: Thermal spraying — Determination of tensile adhesive strength 01.1994.
[11] Cverna, F.: ASM ready reference. Thermal properties of metals. Materials Park 2002.
[12] http://omnexus.specialchem.com/polymer-properties/properties/coefficient-of-linear-thermal-expansion. SpecialChem SA.
[13] Sabourin, E.; Houser, S. A.; Helge Bøhn, J.: Adaptive slicing using stepwise uniform refinement, Rapid Prototyping Journal. 4 (1996), 20-26.
[14] Ahn, D.; Kweon, J.-H.; Kwon, S.; Song, J.; Lee, S.: Representation of surface roughness in fused deposition modeling. In: Journal of Materials Processing Technology 209 (2009) 15–16, 5593-600.

Microstructural evolution and mechanical property enhancement of a cold sprayed Cu-Zn alloy coating by post-spray friction stir processing

C.J. Huang *, Y.C. Xie, C. Verdy, M.P. Planche, H.L. Liao, G. Montavon, IRTES – LERMPS, ICB UMR 6303, CNRS, Univ. Bourgogne Franche-Comté, UTBM, F-90010 Belfort, France
W.Y. Li, Y. Feng, Shaanxi Key Laboratory of Friction Welding Technologies, Northwestern Polytechnical University, Xi'an 710072, Shaanxi, P.R. China

An innovative hybrid process which combines the two very effective solid-state techniques of cold spraying (CS) and friction stir processing (FSP), was proposed to fabricate a high-strength ultrafine-grained Cu-Zn coating. Results show that the CSed coating had an elongated microstructure with 78.42% of low-angle grain boundaries. Following FSP, there appear ultrafine grains with 90.47% of high-angle grain boundaries and a composition of α, β' and γ phases while the CSed coatings was mainly α. Significant mechanical properties enhancement is achieved, i.e. with the ultimate tensile strength increasing from 87.2 MPa to 257.5 MPa and fracture elongation increasing from 0.17% to 0.81%. The precipitates have a significant effect on the fracture behavior of FSPed coating.

1 Introduction

In the last decade, cold spraying (CS) has been recognized as a promising solid-state coating technique which allows the deposition of oxidation sensitive or phase transformation sensitive materials onto a substrate at relatively low temperatures [1-4]. In addition to depositing coatings, CS has been suggested to be used in additive manufacturing [3] for its low processing temperature, high deposition efficiency and controllable spraying ability and it is possibly the only continuous method to produce thick metal components of metastable powdered materials. However, in CS particles deform plastically during high-velocity impact which usually produces limited porosity and high dislocations in the coating, whereas the relatively inhomogeneous microstructure and poor interface bonding result in low coating adhesion or cohesion [4]. For example, Assadi et al. [4] reported a grain-refined region at the center of the bond zone due to dynamic recrystallization, and other microstructural features of CSed coatings which were affected by severe plastic deformation as the high density of dislocations and other defects, e.g. non-bonded interfaces at the center of impact zones show. Therefore, when deposition conditions are not adequate, metallic or metallurgical bonding forms in parts of the interface, and mechanical interlocking becomes the dominant bonding mechanism producing low strength coatings. Furthermore, Choi et al. [5] reported that the elastic moduli of CSed coatings were significantly lower than those of bulk materials due to hardening. In addition, their plastic behavior is brittle, a cracks underneath indents show [5].

It is known that post-treatment can homogenize the coating microstructure and improve interface bonding through solid-state diffusion [6], therefore improving cohesion between deposited particles with annealing. However, the strength of the annealed coating may be lowered due to coarsening of precipitates and growth of grains [7]. To obtain a coating with ultrafine grains and high strength, another important solid-state technique , . friction stir processing (FSP) [8, 9], is used to modify microstructuresthroughout the coating. In our previous studies [8, 9], a CSed SiCp/Al5056 composite coating was modified with FSP , where results showed a refinement of reinforcements and an improvement of mechanical properties . In order to explore further the post-spray FSP, we focused tothe α/β dual phase transformation system of Cu-Zn alloy, which is rich in nonequilibrium phases [11]. Moreover, the improvement of the microstructure, eliminating multiple defects and enhancing mechanical properties of the CSed Cu-Zn alloy coating were also investigated.

2 Experimental procedures

The feedstock is a commercially irregular Cu60Zn40 alloy powder with an average size of 35.6 µm. A grit-blasted Cu plate was used as the substrate to deposit coatings. A commercial cold spray gun was used with an MOC type nozzle. High-pressure compressed air was used as the driving gas at a pressure of 2.8 MPa at a temperature of 400 °C. The standoff distance from nozzle exit to substrate surface was 30 mm. The CSed coating was then FSPed parallel to spraying direction with a single pass with a commercial FSW machine at a moving speed of 100 mm/min and a tool rotating speed of 1500 rpm. The stir tool was threaded with a root diameter 3.4 mm and length 2.5 mm, having a concave shoulder of 10 mm in diameter. The tool was tilted 2.5° (Z axis). A field emission scanning electron microscope (SEM) equipped with an electron backscatter diffraction (EBSD) system was used for microstructural and phase characterizations. Tensile specimens were prepared according to China National Standard GB/T228-2010 [12]. The room temperature tensile test was performed at a cross-head speed of 1 mm/min using a computer-controlled tensile testing machine. The tensile properties were evaluated using three tensile specimens cut from the same FSPed coating. The fracture features of samples were analyzed with SEM.

3 Results and discussion

As shown in Figs. 1a and 1b, the CSed coating exhibits a flattened structure with an average grain size of 1.9 µm, with some Cu-Zn particles being very elongated and deformed as long strips while some having undeformed zones. This suggests an inhomogeneous deformation of particles during deposition. By comparison, the FSPed coating has a fully recrystallized structure of ultrafine equiaxed grains with an average grain size of 0.6 μm as shown in Figs. 1c and 1d. During FSP, the thermo-mechanical coupled process under the combined actions of solid-state stirring and frictional heating result in full recrystallization similar to severe plastic deformation (SPD) at elevated temperatures [10], which is a promising post-treatment to produce an ultrafine-grained coating for industrial applications. It has been reported that the Cu-Zn alloy with a low stacking-fault energy (SFE) is easier to achieve a smaller grain size with SPD [13-15]. In addition to this, certain defects like large pores form due to insufficient deformation in the CSed coating, which would produce high porosity and low strength of coating. However, a completely stirred and compacted coating with very low porosity is obtained after FSP, which has good metallurgical bond, producing a high strength coating.

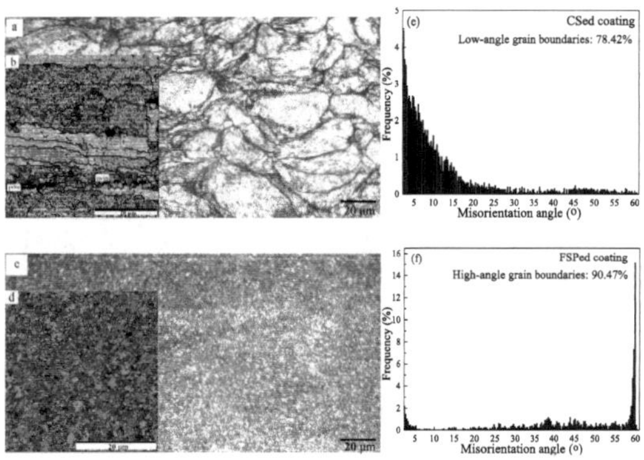

Fig. 1. (a) OM micrograph and (b) EBSD map of CSed coating, and (c) OM micrograph and (d) EBSD map of
FSPed coating, the black, white and yellow lines in EBSD maps corresponding to the HAGBs (θ≥15°), LAGBs (15°>θ>2°), and TBs, respectively. Misorientation angle distribution in (e) CSed coating and (f) FSPed coating.

The grain boundary (GB) map of the CSed coating is shown in Fig. 1b and the corresponding distribution of misorientation angles is shown in Fig. 1e. As it can be seen the CSed coating is characterized by high-angle grain boundaries at 22.58 % (HAGBs) and low-angle grain boundaries (LAGBs) in the remaining 78.42%, which shows that a high density of dislocations was generated after FSP. When a particle with a sufficiently high-velocity impacts upon the substrate or on previously deposited particles [16], a large number of dislocations accumulate near the grain boundary and might be responsible for the HAGBs in the intensively elongated grains [4, 16].
The GB maps of the FSPed coating and the corresponding distribution of misorientation angles are shown in Figs. 1d and 1f, respectively. When considering all GBs misorientation angles which are bigger than 2o, the microstructure of the FSPed coating has a large number of HAGBs (90.47%) (27.49% of Σ3 boundaries) and fewer of LAGBs (9.53%). The interfaces of dislocation cell and/or LAGBs accumulate with FSP and then transform into HAGBs through dislocation rearrangement and/or grain rotation after FSP. As the FSPed coating has experienced complete dynamic recrystallization, HAGBs are directly related to the formation of new grains. Furthermore, a large number of HAGBs is beneficial as it increasesstrain hardening. HAGBs hinder and block dislocations more efficiently, therefore enhancing the strain-hardening capacity and improving tensile strength and ductility [17]. Moreover, reasonable ductility can be obtained with HAGBs sliding and their related activities in ultrafine-grained materials [17, 18]. In addition to these, 67% of HAGBs, shown by yellow boundary lines in Fig. 1d, are responsible for coincidence site lattice (CSL) Σ3 boundaries during recrystallization in FSP. It is stated by Zhao et al. [15] that the lower SFE of Cu-Zn alloy was associated with a decrease in crystallite size and increases in annealing twin densities. As shown in Fig. 1d, there are massive annealing twins during FSP, while fewer annealing twins (marked with the red circle) are in the CSed coating.

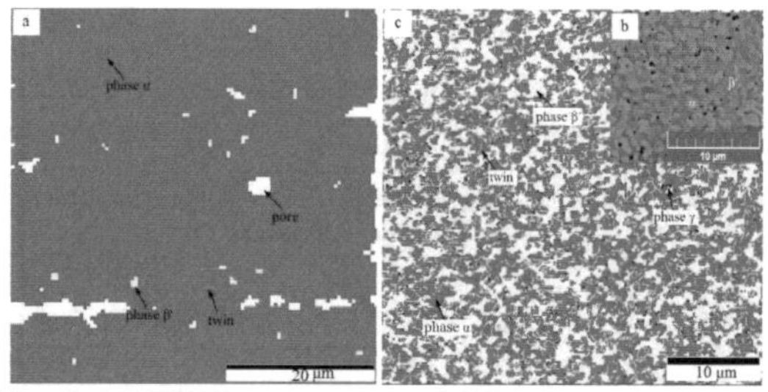

Fig. 2. (a) EBSD phase map of CSed coating, and (b) SEM micrograph and (c) EBSD phase map of FSPed coating. The red, yellow and blue colors corresponding to the phases of α (Cu_3Zn), β′ (CuZn) and γ (Cu_5Zn_8), respectively.

The EBSD phase map of Fig. 2a shows the presence of two solid solutions in the CSed coating, i.e. the majority of α (Cu_3Zn) with an ordered fcc structure (red colour) and few β' intermetallic phases (CuZn) with an ordered bcc structure (yellow colour). The phases of the FSPed coating are marked in the SEM micrographs as shown in Fig. 2b and they are qualitatively examined with elemental analysis. The results of further analysis of the phase distribution are given in Fig. 2c, the equilibrium phase of β' and the nonequilibrium phase of γ (Cu_5Zn_8) with complex cubic structure were produced due to the thermo-mechanical effect during FSP and the statistical analysis of phases present shows a composition of 62.76% of α, 36% of β' and 1.23% of γ in FSP coating. When the coating is under the stirring tool at high temperatures (above 480 °C), β' transforms into β (a disordered bcc structure). But this structure is not retained during quenching and an inverse transformation of β into β' occurs. At the same time, α precipitated from the β. Therefore, the microstructure is referred to as an α/β dual phase with the ultrafine α grains (fcc phase) embedded inside the matrix of β' grains (bcc phase). The β' has a CsCl structure (B2) and which is found to be transformed into two metastable phases of Cu-Zn. One called 9R, a type of martensite, has an orthorhombic nine layer stacking sequence of close packed planes [19]. However, β' is harder and stronger than α, it is brittle at room temperature which affects mechanical properties [19, 20]. Further, the non equilibrium phase of γ forms in these regions having the higher concentration of Zn and this could be the reason for the poor mechanical response of the FSPed coating as it contains γ. When comparing the two EBSD phase maps, it can be seen that annealing twins are uniformly present only in α grains following FSP.

Fig. 3 shows representative stress-strain curves and the ultimate tensile strength (UTS) for CSed and FSPed coating, respectively. Error bars in Fig. 3b identify the total range of test results for the three specimens per condition. A mean UTS value of about 87.2±5.9 MPa was obtained for the CSed coating. The elongation to failure of all three CSed coatings does not exceed 0.2%, which reflects the brittleness of CSed coatings. Following FSP, the mean ultimate strength increases dramatically to 257.5±4.8 MPa, compared to about 381 MPa of the bulk cast Cu-Zn alloy [11]. The different fracture behavior between CSed and FSPed coatings may be related to their microstructures, which have different grain sizes, phase compositions and distribution of defects in the coatings. Therefore, it would confirm that using FSP had a significant effect on the tensile behavior of the CSed coating. The elongations to failure of all FSPed coatings have exceeded 0.8%, which is much lower than that of the base metal (61%). This disadvantage is a crucial limitation of the ultrafine equiaxed FSPed coatings. As Zhang et al. [21] observed high strength is achieved at a loss of ductility due to the inability to accumulate dislocations in small grains as they are saturated with them, and furthermore, an approach to decrease the SFE might improve the ductility as the twinning capability increases, which was proposed in their study [21]. Therefore, in our future studies, the increase of ductility in FSPed coatings would be further improved to produce high performance coatings.

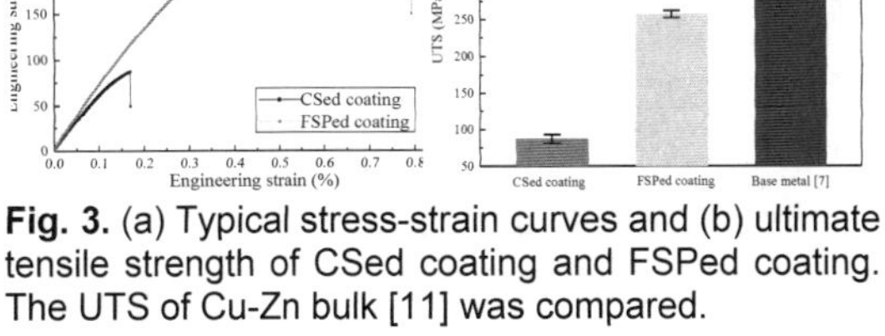

Fig. 3. (a) Typical stress-strain curves and (b) ultimate tensile strength of CSed coating and FSPed coating. The UTS of Cu-Zn bulk [11] was compared.

In order to obtain detailed information on the micro-mechanism of deformation and fracture of both CSed and FSPed coatings, the SEM micrographs of fracture surfaces are shown and analyzed in Fig. 4. The fracture surface of the CSed coating in Fig. 4a shows that non-bonded interfaces are an important microstructural feature in the CSed coating which induces fracture. Crack propagation occurs along the interface of the deposited particles rather than through them. It can be assumed that there is mechanical bonding to a large extent between the interfaces and the mechanical properties of the CSed coatings are limited by mechanical interlocking between deformed particles, which produces lower tensile strength and further demonstrates a brittle type of fracture in CSed coatings. Following FSP, a different dimpled fracture surface is shown in Fig. 4b, and this reveals metallurgical bonding between the deposited particles. Furthermore, fracture surfaces without obvious tearing edges show that the FSPed coating does not have a large elongation to failure due to the precipitates of brittle phases and the FSPed coating has a brittle failure mode. This is the case when the CSed Cu-Zn coating deforms with FSP to increase the strength, however, its ductility would not increase dramatically [22].

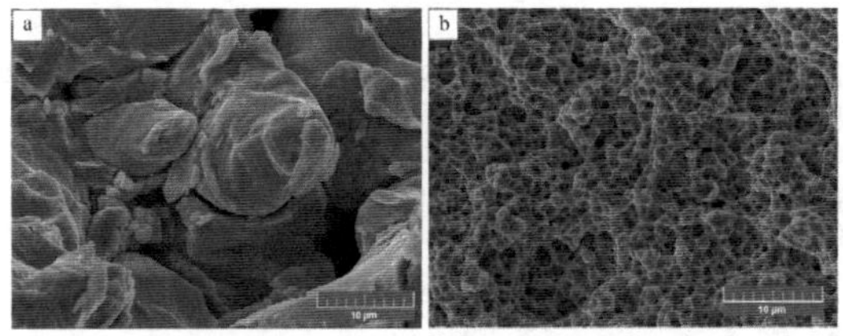

Fig. 4. SEM micrographs of fracture surface of tensile specimens obtained in (a) CSed coating and (b) FSPed coating without substrate.

4 Conclusion

The microstructural evolution and mechanical property enhancement by FSP of a CSed Cu-Zn alloy coating have been investigated. An ultrafine-grained coating with relatively high tensile strength was obtained following FSP. The FSPed coating was highlighted most HAGBs and TBs while LAGBs were dominant in the CSed coating. Phases have transformed from the α phase in the CSed coating to phases α, β' and γ in the FSPed coating. In addition, a significant enhancement of tensile strength in the FSPed coating was observed. However, limited fracture elongation was observed because of precipitates of β' and γ.

Acknowledgments

This work was supported by the financial support from the program of China Scholarship Council and the program of National Natural Science Foundation of China (515774196) and 111 Project (B08040).

5 Literature

[1] Grigoriev, S. and Okunkova, A. Other: Cold Spraying: from process fundamentals towards advanced applications. Surf. Coat. Tech. 268 (2014), pp. 77-84.

[2] Meng, X.M. and Zhang, J.B. Other: Influence of gas temperature on microstructure and properties of cold spray 304SS coating, J. Mater. Sci. Tech. 27 (2011), pp. 809-815.

[3] Pattison, J. and Celotto, S. Other: Cold gas dynamic manufacturing: A non-thermal approach to freeform fabrication, Int. J. Mach. Tools Manuf. 47 (2007), pp. 627-34.

[4] Assadi, H. and Kreye, H. Other: Cold spraying-A materials perspective, Acta Mater. 2016 (116), pp. 382-407.

[5] Choi, W.B. and Li, L. Other: Integrated characterization of cold sprayed aluminum coatings, Acta Mater. 55 (2007), pp. 857-866.

[6] Li, W.Y. and Guo, X.P. Other: Improvement of microstructure and property of cold-sprayed Cu-4at.%Cr-2at.%Nb alloy by heat treatment, Scripta Mater. 55 (2006), pp. 327-330.

[7] Li, W.Y. and Yang, C.L. Other: Effect of vacuum heat treatment on microstructure and microhardness of cold-sprayed TiN particle-reinforced Al alloy-based composites, Mater Des. 32 (2011), pp. 32 388-394.

[8] Huang, C.J. and Li, W.Y. Other: Modification of a cold sprayed SiCp/Al5056 composite coating by friction stir processing, Surf. Coat. Tech. 296 (2016), pp. 69-75.

[9] Huang, C.J. and Li, W.Y. Other: Effect of tool rotation speed on microstructure and microhardness of friction-stir-processed cold-sprayed SiCp/Al5056 composite coating, J. Therm. Spray Tech. 2016 (25), pp. 1357-1364.

[10] Zhang, Z.H. and Li, W.Y. Other: Global anisotropic response of friction stir welded 2024 aluminum sheets, Acta Mater. 92 (2015), pp. 117-125.

[11] Park, H.S. and Kimura, T. Other: Murakamic, Y. Naganod, K. Nakata, M. Ushio, Microstructures and mechanical properties of friction stir welds of 60% Cu-40% Zn copper alloy, Mater. Sci. Eng: A. 371 (2004), pp. 160-169.

[12] GB/T228-2010/ISO 6892: 2009. Tensile test method for the metal materials at room temperature. Standardization Administration of the People's Republic of China; 2010.

[13] Bahmanpour, H. and Youssef, K.M. Other: Deformation twins and related softening behavior in nanocrystalline Cu-30% Zn alloy, Acta Mater. 60 (2012), pp. 3340-3349.

[14] M, X.L. and Zhou, H. Other: Stacking-fault energy effect on zero-strain deformation twinning in nanocrystalline Cu-Zn alloys, Scripta Mater. 109 (2015), pp. 89-93.

[15] Zhao, Y.H. and Horita, Other: Evolution of defect structures during cold rolling of ultrafine-grained Cu and Cu-Zn alloys: Influence of stacking fault energy, Mater. Sci. Eng: A. 474 (2008), pp. 342-347.

[16] Borchers, C. and Gartner, F. Other: Formation of Persistent Dislocation Loops by Ultra-High Stain-Rate Deformation during Cold Spraying, Acta Mater. 53 (2005), pp. 2991-3000.

[17] Zhao, Y.H. and Zhu, Y.T. Other: Preparation of nanostructured materials having improved ductility, Adv. Eng. Mater. 12 (2010), pp 769-788.

[18] Yang, L.Y. and Li, S.M. Other: Twinned dendrite growth during Bridgman solidification, Acta Mater. 97 (2015), pp. 269-281.

[19] Andrade-Gamboa, and Gennari, J. F.C. Other: Stability of Cu-Zn phases under low energy ball mil, Mater. Sci. Eng: A. 447 (2007), pp. 324-331.

[20] Haberkorn, N. and Ahlers, M. Other: Tuning of the martensitic transformation temperature in Cu-Zn thin films by control of zinc vapor pressure during annealing, Scripta Mater. 61 (2009), pp. 821-824.

[21] Zhang, P. and An, X.H. Other: Optimizing strength and ductility of Cu-Zn alloys through severe plastic deformation. Scripta Mater. 67 (2012), pp. 871-874.

[22] Cai, B.Z. and Ma, X.L. Other: Enhanced mechanical properties in Cu-Zn alloys with a gradient structure by surface mechanical attrition treatment at cryogenic temperature, Mater. Sci. Eng: A. 626 (2015), pp. 144-149.

Understanding the infiltration behaviour of sealers into thermally sprayed coatings

Kirsten Bobzin, Mehmet Öte, Martin Andreas Knoch
Surface Engineering Institute, RWTH Aachen University, Germany
Carsten Kunde, Pascal Dettloff
DIAMANT Metallplastic GmbH, Mönchengladbach, Germany

Sealants are widely used in order to enhance the performance of porous thermally sprayed coatings, e.g. for electrical insulation or corrosion protection. In order to accomplish required properties, a good infiltration is necessary. The methods to assess the success of a sealing often rely on determining the infiltration depth by SEM or adding colour pigments to the sealer. In this study, a new approach for assessing the success of a sealing operation is investigated. The underlying assumption is that porous coatings are not gas-tight and by sealing them, the measurable gas flow can be reduced. Therefore, the success of a sealing operation may be assessed by comparing gas flows at defined conditions prior and subsequent to sealing. This hypothesis is investigated by coating special highly porous substrates with a wide range of coating porosities and thicknesses, sealing these coatings, comparing nitrogen flows at a defined pressure prior and subsequent to the sealing operation and correlating the measured changes of nitrogen flow with traditionally assessed infiltration depth and filling degree.

1 Introduction

Porosity is a common occurrence in thermally sprayed coatings. Depending on the process variant, the porosity level c_p can be as low as c_p < 1 % and as high as c_p > 30 %. Although porosity is often associated with flaws, pores can also present a desired or even necessary attribute of a part or coating. In lightweight design for example, pores can be used to reduce weight while maintaining the required stiffness by sophisticated use of the part geometry. Abradable coatings also usually exhibit a certain porosity to enable inter-particle fracture in order to promote the formation of a clean cut. Thermal barrier coatingsrely heavily on pores as well in order to lower the overall thermal conductivity of a coating. In other cases, such as corrosion protection coatings, pores are usually not desired but rather a result of external restrictions. For example, ZnAl15 is often used as a corrosion protection coating in highly corrosive environments. The properties of the feedstock material would allow the use of several process variants, e.g. cold gas spray and wire-arc spraying, for the coating deposition. While coatings produced by cold spraying usually exhibit a very low porosity, which could even be considered to be negligible in case of corrosion protection coatings, the equipment, deposition rates and running costs present a disadvantage compared to wire-arc spraying. Wire-arc sprayed coatings, on the other hand, are almost impossible to be produced with a negligible porosity. However, the low invest, robust process, high deposition rates as well as the possibility to apply coatings on-site are the main reason for choosing wire-arc spraying over cold gas spraying, especially for large structures such as offshore wind-turbines or bridges [2, 7, 11, 12].

One possibility to reduce the effective porosity is the use of sealers. Sealers are materials which are applied after the coating process itself in order to infiltrate the coating and fill the open porosity [11]. Thereby, a hybrid material is created which can take advantage of synergy effects and, thus, exhibit a superior lifetime than the superposition of both materials on their own. The positive effect of sealers on the corrosion resistance of thermally sprayed coatings has been known for many years, e.g. [8]. This effect is, for example, used in so called duplex coatings; a combination of a thermally sprayed cathodic corrosion protection coatings, a sealer and several layers of organic top-coats. Such duplex coatings are often used for the corrosion protection in maritime environments [9].

These apparent advantages of sealers do, however, require a good infiltration, meaning a high infiltration depth combined with a high filling degree of the entire open porosity. Whether a sealing operation was successful or not can be determined with several methods, depending on the coating and sealer material. In case of organic sealers, which are used in most cases, one could prepare a cross section and check whether carbon is present in the pores. For this purpose, energy dispersive spectroscopy (EDS) or an electro-magnetic micro probe analysis (EMPA) can be used. Both methods use an electron beam in order to raise electrons to a higher energy level within the topmost few micrometres of the investigated surface. When these electrons relax into their equilibrium state, radiation is created. This radiation has a specific wavelength depending on the atom from which it originated and thus, carbon based sealers can be differentiated from the coating material. Another possibility is the use of Micro Raman Spectroscopy (MRS). In this case, a laser is used to stimulate the vibration or rotation of chemical bonds via inelastic scattering. The laser photon loses a specific amount of energy to the bond and, therefore, changes its wavelength corresponding to the energy loss. By comparing the initial and reflected wavelength, the lost energy amount can be detected. This energy amount enables, in turn, the detection of chemical bonds, e.g. a carbon-carbon bond. All presented analysis methods have a distinctive disadvantage in common. They do require the preparation of a cross-section as well as access to very expensive equipment. Furthermore, the interpretation of the generated data is difficult and requires highly skilled workers.

Another possibility for detecting the infiltration depth of sealers is the use of fluorescent additives. This method produces easy to understand images and can be very useful for the investigation of fundamental coating-sealer interactions. This method is, however, rather complex with regard to interpreting the results correctly and, therefore, better suited for research purposes than for industrial practice, especially in small companies. For example, the additives may change the sealer's properties and, thus, produce a result which cannot be transferred to the considered system. Furthermore, the infiltration depth can only be detected in a cross section using special microscopes. This, in turn, constitutes ready access to full-fledged metallographic laboratory as well as a skilled metallograph for the preparation. An example for the use of fluorescent additives in sealers is given in **Fig. 1**.

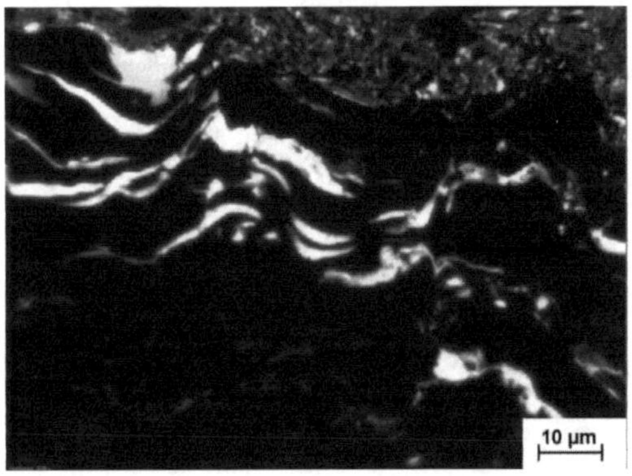

Fig. 1: Wire-arc sprayed and sealed Fe-based coating using a fluoresced sealer [10]

In this paper, a novel approach for assessing the effective infiltration depth of sealers into thermally sprayed coatings will be presented. In contrast to EDS, EMPA or MRS, no expensive equipment is necessary and the analysis gives a quick and easy to understand answer to the question: "to what depth did the sealer seal the pores effectively?" The results of first trials are compared to an EDS analysis in order to verify the applicability of this novel approach.

Although the reliable determination of the infiltration depth of sealers is very important, understanding the mechanisms which enable a deep penetration of sealers into thermally sprayed coatings is equally important. Due to the small diameter of pores and especially the connection between pores, the liquid sealer is draw into the coating by capillary forces. While there are several theories which properties influence the infiltration behaviour in a certain way, the coating's surface energy as well as the sealers surface tension are widely considered to be decisive for infiltration depth [4–6, 8, 10]. Although the interactions on a molecular level, which dictate the infiltration behaviour of sealers, are not the main focus of this paper, they must not be neglected either. Therefore, a quick overview of the measurement methods for these quantities will be given. Furthermore, a measurement technique for assessing the surface energy of porous ZnAl15 coatings will be presented.

2 Experimental setup

In order to detect the infiltration depth of sealers into thermally sprayed coatings, the porosity itself is taken advantage of. The underlying assumption is that open porosity is not gas-tight and, therefore, the gas flux through the coating is a reliable indicator for the occurrence of open porosity. In **Fig. 2**, a schematic representation of the proposed measurement technique is given. Due to the novelty of the presented method, the aim is a reliable assessment of the infiltration depth is defined as the depth to which the sealed coating is completely gas-tight and, therefore, free of open porosity. The flux itself should also be proportional to the porosity level. Since the flux levels

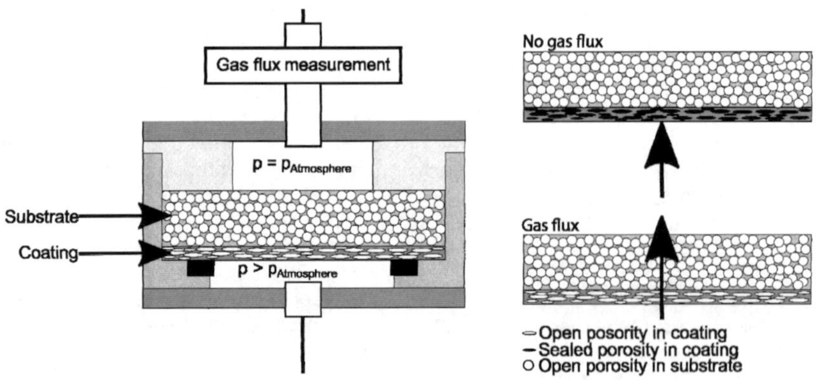

Fig. 2: Schematic representation of the proposed measurement technique, left: measurement setup, right: gas fluxes in case of open porosity and no open porosity in the coating.

depends on many factors, e.g. the dynamic gas properties, pressure, temperature, coating thickness, etc., assessing the porosity level is not considered in this paper. If pressure, temperature and gas composition are kept constant, higher fluxes do, however, indicate a higher porosity.

Although the basic idea is fairly straightforward, measuring gas fluxes through sealed coatings presents a challenge as bulk substrate material is always gas-tight. In order to overcome this severe influence, either the coating can be removed from the substrate after the coating and sealing operation or the substrate itself is changed in such a way that its influence on gas fluxes through the coating-substrate combination may be neglected. The main disadvantage in removing sealed coatings from a substrate is the method of removal may have a huge influence on the sealed coating and, thus, making the subsequent measurement of gas fluxes obsolete as the results do not correspond to the coating properties. This would, for example, be the case if the substrate is dissolved with an acid to which the coating material is not immune, e.g. stainless steel on construction steel.

Therefore, special, sintered steel substrates were used in this study. Due to their high porosity, they do not possess any relevant resistance to gas flows. These substrates were grit-blasted and coated with ZnAl15 using wire-arc spraying with the parameters given in **Tab. 1**. These spraying parameters were chosen to produce coatings with varying thickness and microstructure.

Tab. 1: Spraying parameters

Spray gun	**30/4SF-Push-LD/U2 (Oerlikon Metco, Keltersbach, Germany)**		
Feedstock	ZnAl15 wire, 2,5 mm diameter		
Voltage	20 V		
Current	150 A		
Stand-off distance	a) 50 mm	b) 60 mm	c) 70 mm
Number of passes	a) 3	b) 6	c) 6
Traverse speed	1,000 mm/s		

After the coating process, the coating thickness was measured with a micrometre screw. Subsequently, a sealer was applied to the coatings while paying special attention that the sealer does not come in contact with the substrate in order to eliminate an infiltration of the porous substrate. The sealing operation was performed using the commercially available sealer Dichtol #2407 (Diamant Metallplastic, Mönchengladbach, Germany). This is an epoxy based sealer which consists of two components, resin and hardener. These two components were mixed just prior to the sealing and afterwards heated up to the suggested temperature of 80 °C. The sample itself was also heated up to 80 °C. The increased temperature is used to lower the sealer's viscosity and, thus, improve the infiltration depth. Furthermore, the increased temperature does also shorten the hardening time and, thus, lowering the time required to seal a part. The sample pre-heating is necessary since Dichtol #2407 was developed for the application on hot surfaces immediately after the thermal spray process. Consequently, surface temperatures in the range of $80 < T_{surface} < 120$ [°C] are recommended by the manufacturer.

The gas fluxes were measured before the spraying process, after the spraying process and after the sealing process. Furthermore, an iterative removal of the sealed coating was performed as long as the sealed coatings were completely gas-tight. The gas flux measurements were performed as schematically depicted in **Fig. 2** using a GPT-02 (Kermetico, Benicia, USA). In the context of this study, a measurement scale from 0 to 5.6 ml/min was used. Nitrogen with a pressure p up to p = 7 bar was used during these measurements. Any coating which does not exhibit a measureable gas flux at such a high pressure may safely be considered to be tight against liquids as well, provided that the liquid does not dissolve either the coating material or the sealer. For each spraying parameter set, three coatings were prepared. The parameters, the corresponding gas-flows and coating thicknesses in the as-spayed state are given in **Fig. 3**. These parameters were chosen in order to produce coatings with varying thickness and porosity resulting in different nitrogen flows.

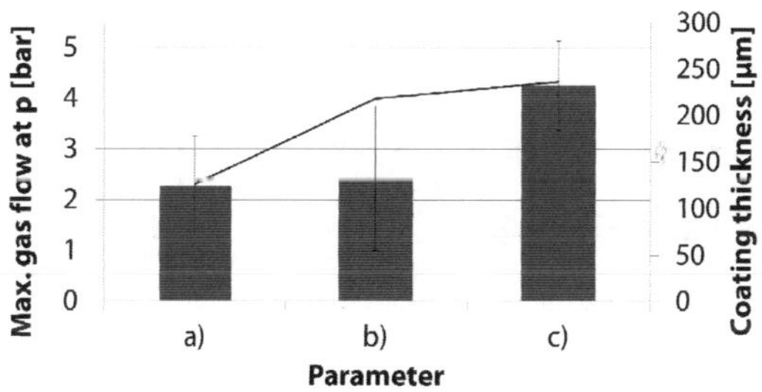

Fig. 3: Coating thickness (line) and pressure required for max. gas flow (5.6 ml/min) of as-sprayed coatings

One sample per parameter was used to establish a suitable method for grinding and polishing the coatings during the iterative coating removal. During each step, the coating thickness is reduced by roughly 10 - 20 µm. This step is essential as wrong grinding parameters can damage the coating material and, thus, induce a gas-flow although the sealed coating is still gas-tight. The second coating was ground and polished until a gas flux was measureable. For the last coating, a different approach was taken. At first, the thin layer of organic material on top of the coating was removed to check whether the sealing operation was successful. Afterwards, a cross-section was prepared without using either an embedding agent or diamond- or oil-based polishing suspensions.

Afterwards, the cross-sections were analysed using EDS in order to measure the infiltration depth. The special preparation method was necessary since polymeric embedding agents, diamond and oil are carbon-based materials and cannot be distinguished from the sealer by EDS. Due to the high reactivity of ZnAl15 with water and the highly porous substrate; a strong oxidation of the coating material during the cross-section preparation cannot be prevented. As the aim of this study is detecting the sealer in pores, preventing carbon-based residue is of utmost importance and oxidation was assumed to be tolerable.

In order to measure the surface tension of sealers, contact angle measurements were conducted on a single hanging drop as well as on Polytetra-fluoroethylene (PTFE). The hanging drop can be used to calculate the total surface tension. By measuring the contact angle on PTFE – a material which has a polar surface energy of 0 and is, therefore, completely disperse, – the total surface tension can be split into a polar and a disperse part. This is important as both parts may play a vital role in the wetting behaviour on various substrates and thus on the capillary forces drawing the sealer into the coatings. The determination of the surface energy of to-be-sealed surfaces is rather challenging as to-be-sealed surfaces do usually exhibit a certain amount of open porosity which causes the sealer to be drawn into the substrate. In this case, the substrate is defined as the solid material on which the contact angle of the sealer is to be measured. Since the sealer is drawn into the coating during the contact angle measurement, either an equivalent surface with no open porosity must be created or the surface energy may be roughly assessed based on the chemical composition and phase composition of the coating material. In this paper, thermally sprayed ZnAl15 coatings are investigated. The surface may be assumed to be a typical metal surface which consists mainly of oxides, hydroxides and a thin film of adsorbed water.

3 Results

Gas flux measurements of the substrate were not possible since even the slightest pressure difference lead to a gas flux beyond the upper limit of the available scales. This clearly shows that the used substrates do not hinder the nitrogen flow in any way and are, therefore, ideal for their intended purpose. After the coating was applied, the three parameter sets resulted in various coating thicknesses and gas flow rates at a defined pressure of p = 7 bar. Results are given in **Fig. 3**. After the sealing operation, all coatings had a thin layer of organic material on top of the sealed coating and no nitrogen-flow could be detected. In a spot check, the pressure was even increased to 8.5 bar without any indication of even a slight gas flow. For parameters a) and b), the sealed samples did not exhibit a measureable gas flow until the iterative grinding process removed the coating all the way to the substrate-coating interface. This indicates a very good infiltration and shows that even a very thin layer of sealed coating can act as an effective barrier. For parameter c), a different behaviour was detected. At first, the sealed coating did not have a measureable gas flux either. Unlike a) and b), a first gas flux was measured before the entire coating was removed, in fact, the remaining coating thickness was roughly 100 µm. Considering the original coating thickness of this particular sample, about 250 µm, an infiltration depth $d_{infiltration}$ = 150 µm was measured.

The EDS measurement proved to be more difficult to understand than expected based on previous trials using a cross fracture on a sealed free standing coating. Although the sealer can easily be detected as the dark phase in the cross fracture, this method is not suited for general use since SEM/EDS analysis requires a smooth surface due to the limited focus range. Therefore, the rougher the surface, the smaller the area which can be investigated simultaneously. Furthermore, voids and edges can induce measurement artefacts during EDS analysis which is necessary in order to assign a certain grey-scale value in the SEM image to a corresponding element or element mix. Therefore, this method is only practical if the investigated surface is sufficiently smooth, which is increasingly hard to achieve with increasing fracture thickness, see **Fig. 4**.

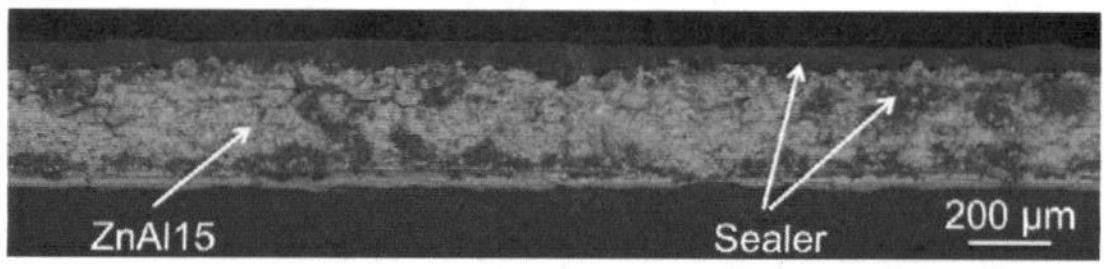

Fig. 4: SEM analysis of a sealed, free-standing ZnAl15 coating using the cross fracture technique

The roughness within a cross fracture is, in general, dependant on several parameters like material properties, fractures thickness and fracture temperature. Due to the remaining coating thickness, combined with a few micrometres of remaining substrate thickness, no acceptable surface roughness within the cross fracture was achieved in pre-trials. Therefore, the preparation method described above was employed to prepare the sample c) for SEM/EDS analysis after the first occurrence of a gas flux. As shown in **Fig. 5**, the entire coating is riddled with dark spots and areas. EDS analysis revealed that the coating was heavily oxidized during the sample preparation.

Oxides formed, as expected, predominantly at the interface between two splats as well as in proximity to the substrate-coating interface. Especially the oxides near the interface can easily be explained by the potential difference between the steel substrate and the ZnAl15 coating. The results reveal, furthermore, that all but one dark spot investigated by EDS point measurements contain a significant amount of silicon. Therefore, these spots are assumed to represent SiC residue which originated from the SiC grinding paper. Considering the low quantitative measurement accuracy of EDS with regard to light elements like oxygen or carbon, only Point 3 exhibits a high enough amount of carbon and a low enough amount of silicon to have any confidence in the presence of sealer in this point. Due to the close proximity of the only detectable occurrence of sealer to the infiltration depth established via the proposed iterative gas flux measurement, this sample gives strong evidence that the proposed method is indeed accurate and applicable for assessing the infiltration depth of sealers into thermally sprayed coatings. It should also be noted that the analysis by means of SEM/EDS can be tricky and difficult to assess, even for highly specialised technicians and material scientists.

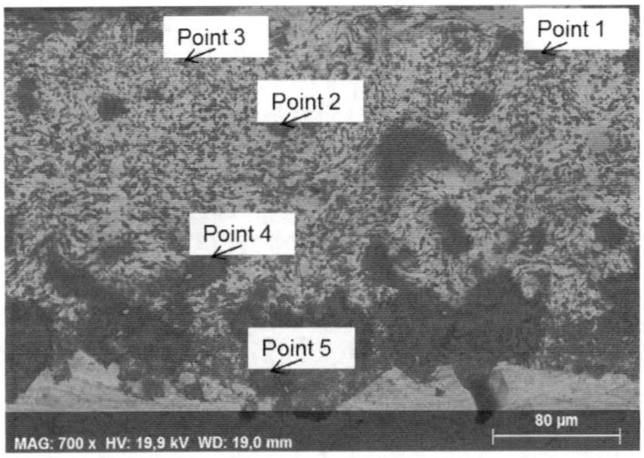

	Al [at. %]	Zn [at. %]	O [at. %]	C [at. %]	Si [at. %]
Point 1	14,3	3,1	26,6	45,1	11,0
Point 2	2,9	1,9	33,5	43,6	18,1
Point 3	30,6	3,7	11,6	52,6	1,4
Point 4	0,8	1,6	22,7	55,1	19,7
Point 5	0,2	1,0	33,7	45,5	19,6

Fig. 5: SEM/EDS analysis of the remaining coating of sample c) after grinding to the first occurrence of a gas flux

In addition to the samples analysed by the gas flux method, one sample per parameter set was analysed by means of SEM/EDS without prior reduction of the coating thickness. Although less oxidation and filled pores could be found throughout the coating for parameter a), EDS revealed high silicon content within these filled pores. Even though the carbon/silicon ratios suggest the presence of at least some sealer, the infiltration depth of this sample could not reliably be assessed. For parameter set b), the high carbon and low silicon content in Point 4 suggest the presence of sealer in the interface, see **Fig. 6**.

This result is in very good agreement with the gas flux method which indicated an infiltration of the entire coating to the substrate-coating interface. Therefore, further evidence for the applicability of this method is given by the investigations of parameter b).

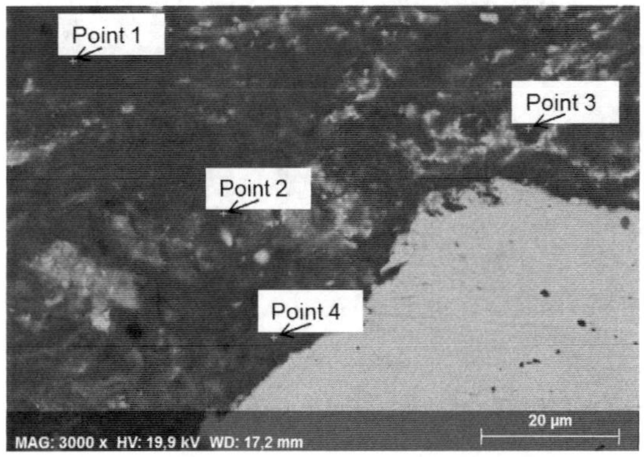

	Al [at. %]	Zn [at. %]	O [at. %]	C [at. %]	Si [at. %]
Point 1	10,6	3,1	30,8	37,5	18,0
Point 2	40,9	13,2	10,4	32,8	2,7
Point 3	43,9	6,0	9,1	39,9	1,1
Point 4	41,5	5,5	11,6	40,5	1,0

Fig. 6: SEM/EDS analysis of the reference coating for parameter b)

For parameter set c), the analysis by means of SEM/EDS proved to be difficult as well. A filled pore could be detected in the proximity of the coating surface, however, the very low porosity of this coating which may be considered gas-tight for most purposes, make the search for pores difficult, even if alternative preparation methods are used which are not suitable for the detection of sealer. Therefore, no reliable detection of the infiltration depth of the sealer was possible for this parameter set either. By analysing the sample which was used to establish a suitable method for the iterative coating removal, the extremely low porosity of this coating is revealed, see **Fig. 7**. Considering that some of the pores are usually contaminated by SiC and can, therefore, not be assessed, the employed SEM/EDS analysis is not particularly suited for this parameter set.

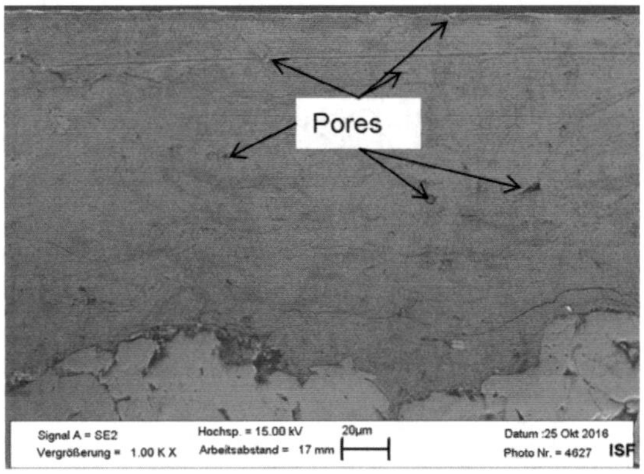

Fig. 7: SEM image of a traditionally prepared coating for parameter c)

4 Conclusions and Discussion

In this paper, a novel approach for measuring the infiltration depth of sealers into thermally sprayed coatings is presented. First results show that this method is, in some cases, in good agreement with a comparative SED/EDS analysis. In some cases, SEM/EDS analysis was not suitable to detect the sealer in the coating. The reason for this was the combination of a substrate-coating combination which is highly susceptible to corrosion, combined with restrictions in the cross section preparation in order to minimize carbon residue in the coatings. For such coatings, a more sophisticated preparation method has to be developed for the SEM/EDS analysis. These results do, however, show that the proposed measurement method is not only cheaper, quicker and easier to interpret but might also be more robust. Furthermore, the analysis by SEM/EDS, or any method used in the state of the art, provides only information about the presence of the sealer at a certain depth. The novel approach presented inhere does, on the other hand, provide information about the effective infiltration depth to which the sealer can effectively seal the coating against gas fluxes at a defined pressure.

Although the first results are very promising, further test are necessary to qualify this test method. For this purpose, further coating systems should be investigated, e.g. a ceramic coating applied by plasma spraying. With regard to ZnAl15 coatings, investigating thicker coatings with a higher porosity could be used for an easier verification by SEM/EDS. Furthermore, the porous substrate has been proven to exhibit a negligible resistance to gas flows. This is, on the one hand, an essential property for the considered test method; on the other hand, this may also have an influence on the infiltration behaviour of a sealer into the coating since the air inside the open porosity can easily be pushed out through the substrate material. If a solid substrate is used, this air cannot be removed so easily and, therefore, might create an increased pressure within the open porosity which could influence the infiltration with the sealer. The importance of the pressure within the open porosity due to e.g. air or vaporous products produced during the hardening of the sealer has already been discussed by Knuuttila et al. [5].

The surface energy of metals is usually assumed to be high. In order to verify this assumption for thermally sprayed ZnAl15, contact angle measurements are necessary. Due to the porosity of such coatings, the test fluids cannot form a stable drop on the surface but are rather sucked into the coating by capillary forces, the exact effect which is employed during the sealing operation. Since the coating itself cannot be analysed this way, an alternative process may be used to produce ZnAl15 samples by rapidly cooling down a ZnAl15 melt without producing any open porosity. One possibility for such a process would be the melt-spinning process. In a melt-spinner, a melt is produced via inductive heating and afterwards pressed through a defined slit and, thus, accelerated onto a cooled, spinning wheel. The whole process takes place in a low vacuum; the pressure is applied by means of inert gas in order to ensure very low oxide content in the melt spinning bands. During the contact with this wheel, the melt solidifies rapidly and cools down rapidly. Typical cooling rates in the order of $10^5 - 10^6$ K/s and, therefore, more or less comparable to the cooling rates of $10^4 - 10^8$ K/s which have been found during the solidification of single splats in thermal spraying [1, 3]. Furthermore, the negligible oxide content of as sprayed ZnAl15 coatings promises a good comparability of ZnAl15 melt-spinner bands with thermally sprayed ZnAl15 coatings. Nonetheless, a thorough analysis of samples produced by both processes with regard to their chemical composition, phase composition, oxide content, etc. would be required to ensure both processes produce similar surfaces and thus comparable surface energies.

5 Acknowledgement

The work in this paper is performed in the context of the research project "Entwicklung eines Verfahrens zur Messung und Qualifizierung der Gaspermeabilität von versiegelten thermisch gespritzten Schichten hinsichtlich der Bewertung ihrer Korrosionsschutz-wirkung" (ZF4059001CK5). This project is funded by the German Federal Ministry of Economics and Technology (BMWi) according to a decision of the German Federal Parliament.

6 Literature

[1] Bobzin, K. 2013. *Oberflächentechnik für den Maschinenbau.*

[2] Champagne, V. K. 2007. *The cold spray materials deposition process. Fundamentals and applications.* Woodhead Publishing in materials. Woodhead, Cambridge, Boca Raton.

[3] Edmund Bühler GmbH. *Materials Science.*

[4] Kabza, K. G., Gestwicki, J. E., and McGrath, J. L. 2000. Contact Angle Goniometry as a Tool for Surface Tension Measurements of Solids, Using Zisman Plot Method. A Physical Chemistry Experiment. *J. Chem. Educ.* 77, 1, 63.
[5] Knuuttila, J., Sorsa, P., and Mäntylä, T. 1999. Sealing of thermal spray coatings by impregnation. *J Therm Spray Tech* 8, 2, 249–257.
[6] Lauth, G. J. and Kowalczyk, J. 2016. *Einführung in die Physik und Chemie der Grenzflächen und Kolloide*. Springer Spektrum, Berlin, Heidelberg.
[7] Lugscheider, E. and Bach, F.-W., Eds. 2002. *Handbuch der thermischen Spritztechnik. Technologien - Werkstoffe - Fertigung*. Fachbuchreihe Schweißtechnik 139. Verl. für Schweißen und Verwandte Verfahren DVS-Verl., Düsseldorf.
[8] Lugscheider, E., Jakiel, P., Messerschmidt, V., and Beckschulte, G. 2013. Subsequent sealing of thermally sprayed coatings to increase corrosion resistance. *Surface Engineering* 10, 1, 46–51.
[9] Norwegian Oil Industry Association. 2012. *NORSOK STANDARD - M-501Surface preparation and protective coating*. Standards Norway, Lysaker, 6.
[10] Ruprecht, C. 2013. *Neue Methoden und Anwendungen des Thermischen Spritzens. Habilationsschrift.*
[11] Tucker, R. C., Ed. 2013. *Thermal spray technology*. ASM International, Materials Park Ohio.
[12] Zhao, D. and Zhang, S., Eds. 2013. *Aerospace materials handbook*. Advances in materials science and engineering. CRC Press, Boca Raton, Fla.

Development and application of binary suspensions in the ternary system Cr_2O_3-TiO_2-Al_2O_3 for S-HVOF spraying

Robert Kratzsch, Annegret Potthoff, Fraunhofer IKTS, Dresden, Germany
Maria Barbosa, Nick Kulissa, Oliver Kunze, Filofteia-Laura Toma, Fraunhofer IWS, Dresden, Germany
Maria Barbosa, Technische Universität Dresden, Germany

Next to ZrO_2, compositions in the system Cr_2O_3-TiO_2-Al_2O_3 are the most used ceramic materials for thermally sprayed coating solutions. Cr_2O_3 coatings present good sliding wear resistance, Al_2O_3 coatings show excellent insulation behaviour and TiO_2 striking corrosion properties. In order to combine these properties, coatings containing more than one oxide are highly interesting.
The conventional spraying process is limited to the availability of binary feedstock powders with defined compositions. The use of suspensions offers the opportunity for tailor-made compositions: within the triangle of Cr_2O_3-TiO_2-Al_2O_3 each mixture of oxides can be created.
The formulation of binary or ternary suspensions is based on the preparation of stable single-oxide suspensions, which will later be mixed to achieve the desired composition. This procedure requires the development of water-based suspensions with suitable behavior, otherwise the ceramic particles and dissolved organic additives may interact and form agglomerates, which are prejudicial to the thermal spray process.
This work presents criteria for the selection of raw materials as well as relevant aspects for the development of binary suspensions to be used in thermal spray. For procedure validation, coatings sprayed using Cr_2O_3-TiO_2 suspensions and S-HVOF will be shortly presented and analysed in terms of their microstructure and microhardness.

1 Introduction

The demand for multifunctional ceramic coatings in the Cr_2O_3-TiO_2-Al_2O_3 system is continuously growing. For applications such as off-shore piston rods, pump sealing seals or hydraulic cylinder rods, the coatings have to connect properties like improved hardness, wear and corrosion resistance. Since this combination of properties cannot be fullfiled by using only one oxide powder, two or more oxide powders may be combined. For the conventional thermal spraying process, the availability of fused feedstock powder composed of two oxides or more is limited, although for industrial applications coatings with a wide spectra of properties are needed. Suspension spraying is the key to satisfy these requirements by offering tailored compositions in the Cr_2O_3-TiO_2-Al_2O_3 system [1–4].
For the preparation of ready-to-spray suspensions, the raw materials have to meet several criteria. These can be separated into physical, chemical, economical and safety attributes [5–7]. Therefore, when choosing the materials to prepare a suspension, one should take the following aspects into consideration: their particle size distribution and their dispersibility into a liquid, as well as crystallinity, chemical composition and purity grade. Furthermore the raw materials should be commercially available and should show a good reproducibility of batches. For the health and safety aspects it is advantageous to use water as dispersing media. The development of binary suspensions needs background knowledge about particle-particle interaction as well as particle-liquid interaction [8–10].
The main goal is to create a colloidal-chemical stable suspension with the following characteristics:

- High stability, i.e. sedimentation rate of particles below 1 mm/h
- High flowability and low viscosity, ideally below 10 mPas, so that the suspension can be fed at constant feed rate, and process stability is guaranteed
- Zeta potential higher than 30 mV (absolute value) in order to avoid particle agglomeration
- pH value between 4 and 10 to prevent the corrosion of hardware components.

In this work, the development of binary aqueous suspensions of tailor-made composition in the Cr_2O_3-TiO_2-Al_2O_3 system is based on the mixing of two stable single-oxide suspensions. Theoretically, the mixing of two sediment-stable suspensions should result in a sediment-stable binary suspension. But challenges, like particle agglomeration due to different surface properties or interactions between chemical reagents (dispersing agents) used for the stabilization of the single-oxides suspensions, could arise [11]. This work presents relevant aspects to take into account upon the development of binary suspensions to be used in thermal spray. For procedure validation, coatings sprayed using the developed suspensions and High Velocity Oxygen Fuel spraying (S-HVOF) will shortly be presented and analysed in terms of their microstructure and microhardness.

2 Materials and Experimental Methods

2.1 Suspensions Formulation

The powders used for the suspension development are TiO_2 R320 (Sachtleben, DE), Cr_2O_3 99 (Ceram mbH, DE) and Al_2O_3 (MZS-1 Martinswerk, DE). The main characteristics of the powders are summarized in Table 1.

Table 1 Properties of selected raw materials

	Al_2O_3	Cr_2O_3	TiO_2
product (provider)	MZS-1 (Martinswerk)	Cr_2O_3 99 (Ceram)	R320 (Sachtleben)
particle size $x_{50,3}$	2.2 µm	0.8 µm	0.4 µm
BET	2.6 m²/g	7.0 m²/g	6.4 m²/g
crystal phase	mostly α-Al_2O_3	eskolaite	mostly rutile

The powder crystallinity plays an important role on the coating properties. Therefore, since a functional suspension should be developed, a powder with the desired crystal phase and high purity should be chosen, in order to achieve coatings with the necessary characteristics [4].
For the powder characterization the Mastersizer 2000 (Malvern Instruments, UK) is used to determine the particle size distribution. The specific surface area was analysed with ASAP2020 MP (Micromeritics, USA).
As previously mentioned the first step consisted in the development of single-oxide suspensions containing Al_2O_3, Cr_2O_3 or TiO_2. The solid content was set to 25 wt.%.The pH value is adjusted using ammonium liquid and acetic acid. In the particular case of Cr_2O_3 a small amount of the dispersing agent LOPON885 had to be used. The general preparation of the suspension is done either with the Dispermat (VMA-Getzmann, DE) or an Ultra-Turrax (IKA-Labortechnik, DE) is used. After a validation of the single-oxide suspensions, the binary suspensions are prepared by mixing the single-oxide suspensions in the required ratios to achieve the desired composition.
The sedimentation rate of the suspension is measured using the LUMiSizer apparatus (LUM GmbH, DE). It is a analytical centrifuge, where the movement of solid-liquid interface is recorded in a transmission profile. The viscosity is determined by using a rheometer MCR 301 (Anton Paar, AT). The zeta potential is measured by Zeta Probe (Colloidal Dynamics, USA) [12–14].

2.2 Suspension Spraying and Coating Characterization

The suspensions were sprayed using an HVOF TopGun (GTV Verschleißschutz mbH, DE) with a nozzle of diameter 8 mm and ethylene as the fuel gas. The suspensions were fed from an industrial-suitable pressurized suspension feeder developed by Fraunhofer IWS and injected axially into the combustion chamber of the HVOF gun. Spray parameters such as ethylene flow rate, spray distance, and suspension flow rate have been varied in order to obtain dense and mechanically stable coatings. Steel substrates previously cleaned and grit-blasted with corundum were used as substrates. During the spraying, the substrates were cooled using air jets systems.

The coating microstructures were examined by optical microscopy on metallographically polished cross sections. The Vickers microhardnesses HV0.3 were measured on the cross-sections using an HP Mikromat 1-HMV tester (Hegewald & Peschke Meß und Prüftechnik GmbH, DE). The rate of coating built-up (coating deposition) was estimated as µm coating thickness deposited per each torch pass.

3 Results and Discussion

3.1 Development and characterization of single-oxide suspensions

Independently of the envised suspension composition, the first goal is the development of colloid-chemical stable and homogeneous suspensions. The particles should be dispersed, which means that agglomeration should not occure. Therefore the particle-particle and particle-liquid interaction in a suspension have to be known. These interactions can be separated into attractive and repulsive. To avoid agglomeration, the repulsion between the particles should be much higher than the attraction. The repulsion is achieved by electrostatic interaction, which is characterized by the zeta potential. The aim is to achieve high absolute values of this parameter (more than 30 mV). Between the range of -30 mV to 30 mV the suspension is not stable, since the attraction between particles is stronger than their repulsion. The zeta potential depends on different factors like the type of solid (slightly soluble, easily soluble), the pH value and the presence of additives (dispersing agents) in the liquid. In a suspension with two solid systems, it has to be taken into account that every solid interacts differently with the aqueous ambience. As result, the surface properties of the solids shift and differ among themselves. Therefore it is important to measure the zeta potential curve according to the pH value. It is necessary that both solid systems have the same charge sign at the desired pH value, otherwise the electrical repulsion turns into an electrical attraction and agglomerates will be observed. The variation of the zeta potential in function of the pH value for the developed single-oxide suspensions is depicted in Figure 1.

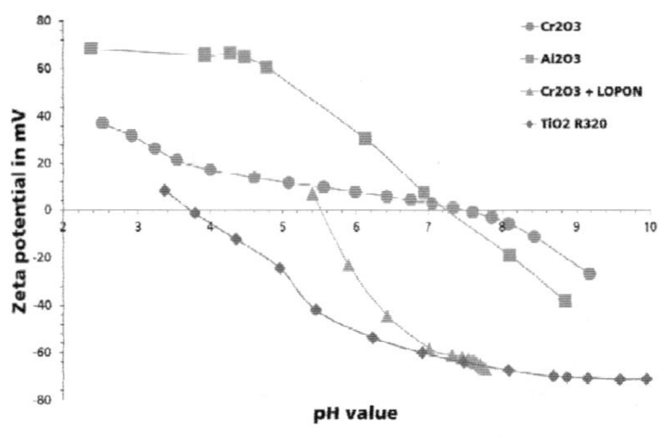

Fig. 1. Zeta potential curves of single-oxide suspensions

Only in basic medium high values of zeta potential were measured for all of the solids so that the electrostatic repulsion between particles can be assured. For Cr_2O_3 the addition of the disperging agent induces a change of the zeta potential, moving it to the desired value range, as it can be noticed in Figure 1. As result of these investigations the pH value for the development of binary systems is set to pH 10.

3.2 Development and characterization of binary Al_2O_3/Cr_2O_3 and Al_2O_3/TiO_2 suspensions

The concept here presented for the development of binary suspensions is the manufacture of stable single-oxide suspensions, which are later mixed to achieve a stable binary suspension. The sedimentation rates of Al_2O_3, Cr_2O_3 and TiO_2 single-oxide suspensions are 1.00 mm/h, 0.22 mm/h and 0.20 mm/h, respectively. The influence of the addition of a second solid system in the sedimentation rate is depicted in Figure 2.

There the sedimentation rates are shown in the dependence on the ratios of Al_2O_3/TiO_2 and Al_2O_3/Cr_2O_3. As it can be seen, the sedimentation rate of the binary suspensions decreases by decreasing the Al_2O_3 content. This can be understood by taking into consideration the sedimentation law of Stokes and that the particle size distributions of single-oxide suspensions are combined in a binary suspension (increased particle-particle interaction). The Cr_2O_3 particles are much smaller than the Al_2O_3 particles and so the "obstruction factor" in sedimentation is increased by the repulsive interactions between Al_2O_3 and Cr_2O_3 or TiO_2 particles.

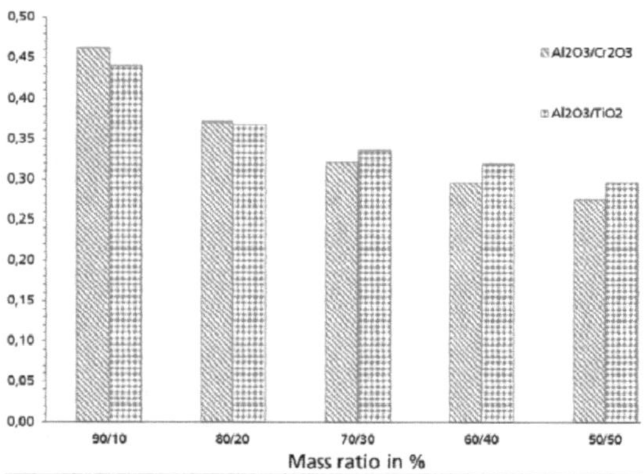

Fig. 2. Sedimentation rate in dependence on mass ratios Al_2O_3/other in %.

After the general investigations on the influence of the different solid systems in a suspension it can be stated, that the change of the particle size distribution and the repulsive interactions between particles leads to an improvement of the sedimentation rate.

Another important parameter is the viscosity of the suspension. For application as thermal spray feedstock, it is important to know if and how the viscosity changes when mixing single-oxide suspensions. The viscosity of a random suspension depends on the particle size distribution, pH value, added chemical reagents and the mass weight of solid [15–18]. The flow curves of the single-oxide and binary Al_2O_3/TiO_2 (50/50, 70/30 and 90/10) suspensions are illustrated in Figure 3.

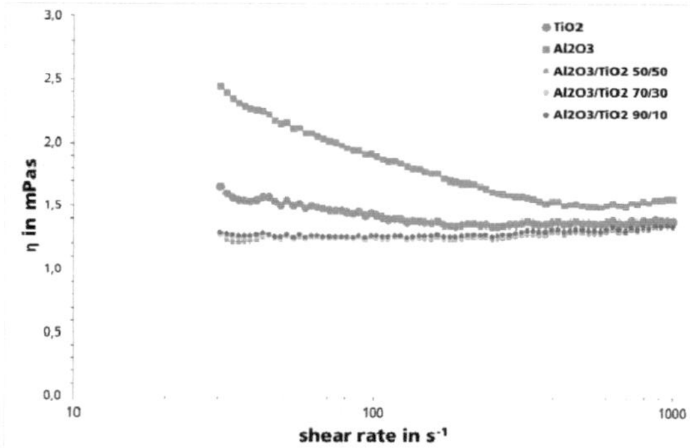

Fig. 3. Comparison of flow curves of single oxide to binary suspensions

It can be recognized that the viscosities of the single-oxide suspensions are higher than that of the binary mixtures. This is due to the flow aid effect of the smaller TiO_2 particles. All suspensions are ready to be sprayed expecting a stable spraying process.

3.3 Development and characterization of binary Cr_2O_3/TiO_2 suspensions

The analysis of suspensions in the binary system Cr_2O_3/TiO_2 ranging between 75-95 % Cr_2O_3 and 5-25 % TiO_2 show that both the sedimentation rate as well as the viscosity are independent of the composition.The sedimentation rate of this system is 0.22 mm/h at any ratios.

Figure 4 shows the flow curves for these binary compositions. As it can be seen with the addition of TiO_2 no significant change in viscosity takes place. All suspensions show a good flowability.

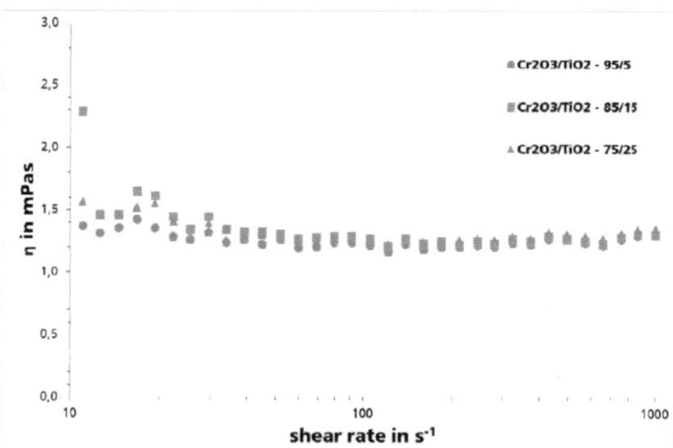

Fig. 4. Flow curves of binary suspensions Cr_2O_3/TiO_2

3.4 Coating microstructures in the Cr_2O_3/TiO_2 binary system

Not only the properties of the suspension determine the properties of the coatings. An appropriate choice of process parameters is important to combine improved coating properties with process stability.

Cr_2O_3 is one of the most challenging materials for thermal spray processing. For powder-based processes, it's sprayability can be significantly improved by the addition of TiO_2. The same improvement in the process stability is expected for the binary system Cr_2O_3/TiO_2. This system was chosen to test the feasibility of the here presented process for the manufacture of binary suspensions as thermal spray feedstock material.

To investigate the influence of the suspension composition on the process stability and coating properties, three different ratios were tested. The ratios are 95%Cr_2O_3/5%TiO_2, 85 %Cr_2O_3/15%TiO_2 and 75%Cr_2O_3/25%TiO_2. The coating evolution for different compositions can be observed in Figure 5.

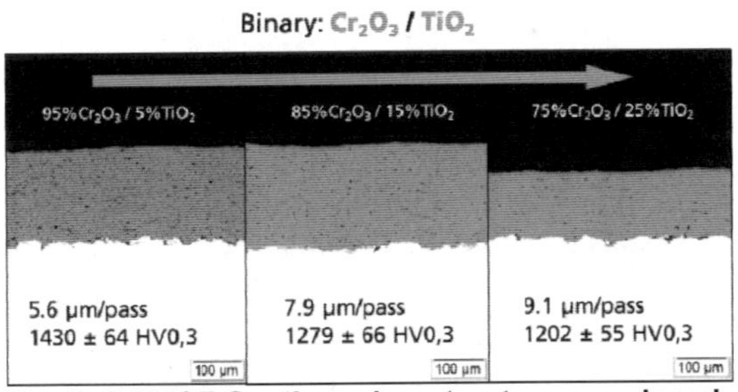

Fig. 5. S-HVOF Coating microstructures produced from tailor-made 25 wt. %-suspensions in the Cr_2O_3/TiO_2 system

Dense structured coatings could be produced. As expected, the coating hardness decreases by increasing the TiO_2 content, since S-HVOF coatings of the single oxides have ca. 1500 HV0.3 [6] [reference]for Cr_2O_3 and ca. 900 HV0.3 for TiO_2. Moreover, the process stability increases tremendously, i.e. not only is clogging significantly reduced, but also there is a high increase in the achieved coating thickness per pass. Parallel studies concerning the wear behavior of these coatings are taking place.

4 Conclusion

In this work a strategy for the development of spray-ready binary suspensions was described. The importance of physical, chemical and safety criteria upon the choice of the powder raw materials was discussed. Particularly important to achieve stable and homogeneous binary suspensions are particle-particle interactions. These can be controlled by having single-oxide suspensions with zeta potentials with the same charge signs and by setting the absolute value of the zeta potential above 30 mV, to assure that particle repulsion is stronger than particle attraction.

Stable and ready-to-spray binary suspensions of different compositions within the systems Al_2O_3/TiO_2, Al_2O_3/Cr_2O_3 and Cr_2O_3/TiO_2 were, by following the here described procedure, successfully produced. To demonstrate the suspension sprayability, the Cr_2O_3/TiO_2 system was chosen. Coatings were produced via S-HVOF and showed a dense and homogenous microstructure. It was observed that with increasing amount of TiO_2 the process stability rises while the coating hardness decreases. The possibility of tailored suspension/coating compositions, allows to find the necessary harmony between the required coating properties and process stability for each application.

5 Acknowledgement

The results presented in this work are part of the research project **(E!9756) "New Cost Effective SUspension ceramic coatings with SUPERior WEAR and insulating properties"** funded by the Bundesministerium für Bildung und Forschung (**BMBF**, DE) under the Programm **EUROSTARS**.

eurostars™

The Eurostars Programme is powered by EUREKA and the European Community

EUREKA

6 Literature

[1] Toma, F.-L., Berger, L.-M., Langner, S., and Naumann, T. 2010. Suspensionsspritzen - das Potential einer neuen Spritztechnologie Suspension spraying - the potential of a new spray technology. *Thermal spray bulletin* 3, 1, 24–29.

[2] Toma, F.-L., Berger, L.-M., Stahr, C., Naumann, T., and Langner, S. 2010. Microstructures and Functional Properties of Suspension-Sprayed Al_2O_3 and TiO_2 Coatings: An Overview. *J Therm Spray Tech* 19, 1-2, 262–274.

[3] Trache, R., Berger, L. M., and Toma, F. L. 2010. Untersuchung der Einflüsse wichtiger Beschichtungsparameter auf die elektrischen Eigenschaften von Cr_2O_3-TiO_2 Schichten. *13. Werkstofftechnisches Kolloquium Chemnitz* (Sep. 2010).

[4] Toma, F.-L., Berger, L.-M., Scheitz, S., Langner, S., Rödel, C., Potthoff, A., Sauchuk, V., and Kusnezoff, M. 2012. Comparison of the Microstructural Characteristics and Electrical Properties of Thermally Sprayed Al_2O_3 Coatings from Aqueous Suspensions and Feedstock Powders. *Journal of Thermal Spray Technology* 21, 3, 480–488.

[5] Berger, L.-M., et al. 2013. “Thermal Spraying with Suspensions: an Economic Spray Process“. *Therm Spray Bull,* Vol. 6, pp. 98-101.

[6] Toma, F.-L., Scheitz, S., Trache, R., Langner, S., Leyens, C., Potthoff, A., and Oelschlägel, K. 2015. Effect of Feedstock Characteristics and Operating Parameters on the Properties of Cr_2O_3 Coatings Prepared by Suspension-HVOF Spraying. Proc Int'l Thermal Spray Conference ITSC 2015, Long Beach, USA, May 2015, ASM , pp. 329-334 (CD-Rom Version)., pp.329-334.

[7] Toma, F.-L., Potthoff, A., Berger, L.-M., and Leyens, C. 2015. Demands, Potentials, and Economic Aspects of Thermal Spraying with Suspensions: A Critical Review. *Journal of Thermal Spray Technology* 24, 7, 1143–1152.

[8] Dörfler, H.-D. 2002. *Grenzflächen und kolloid-disperse Systeme. Physik und Chemie ; mit 88 Tabellen*. Springer, Berlin [u.a.].

[9] Grasso, D., Subramaniam, K., Butkus, M., Strevett, K., and Bergendahl, J. 2002. A review of non-DLVO interactions in environmental colloidal systems. *Reviews in Environmental Science and Biotechnology* 1, 1, 17–38.

[10] Liang, Y., Hilal, N., Langston, P., and Starov, V. 2007. Interaction forces between colloidal particles in liquid: Theory and experiment. *Adv Colloid Interface Sci* 134-135, 151–166.

[11] Connah, M. 2006. Zeta potential, the key to successful suspensions. *American Ceramic Society Bulletin* 85, 7, 38–39.

[12] International Organization for Standardization (Hg.). 2012. *Colloidal systems — Methods for zetapotential determination — Part 1: Electroacoustic and electrokinetic phenomena,* 13099-1.

[13] International Organization for Standardization (Hg.). 2012. *Colloidal systems — Methods for zetapotential determination — Part 2: Optical methods,* 13099-2.

[14] O'Brien, R. W., Cannon, D. W., and Rowlands, W. N. 1995. Electroacoustic Determination of Particle-Size and Zeta-Potential. *Journal of Colloid and Interface Science* 173, 2, 406–418.

[15] Fukuda, Y., Togahi, T., Suzuki, Y., Naito, M., and Kamiya, H. 2001. Influence of additive content of anionic polymer dispersant on dense alumina suspension viscosity. *Chemical Engineering Science* 56, 9, 3005–3010.

[16] Ke-Fa, Z. J.-F. a. L. Z.-Y. a. N. M.-J. a. C. 2009. Dependence of Nanofluid Viscosity on Particle Size and pH Value. *Chinese Physics Letters* 26, 6, 66202.

[17] Nguyen, C. T., Desgranges, F., Galanis, N., Roy, G., Maré, T., Boucher, S., and Angue Mintsa, H. 2008. Viscosity data for Al_2O_3–water nanofluid—hysteresis: is heat transfer enhancement using nanofluids reliable? *International Journal of Thermal Sciences* 47, 2, 103–111.

[18] Nguyen, C. T., Desgranges, F., Roy, G., Galanis, N., Maré, T., Boucher, S., and Angue Mintsa, H. 2007. Temperature and particle-size dependent viscosity data for water-based nanofluids - Hysteresis phenomenon. *International Journal of Heat and Fluid Flow* 28, 6, 1492–1506.

Suspension Plasma Spray YSZ Feedstocks and Delivery System for Improved Spray Distance and Cost Effective Throughput

B.W. Callen[1], Ft. Saskatchewan / CAN. R. Roccio-Heller[2], J. Liu[2], O. Sabouni[2], Westbury / USA

1. Oerlikon Metco (Canada) Inc.; 2. Oerlikon Metco (US) Inc.

Yttria-Stabilized Zirconia (YSZ) suspensions are currently popular in developing strain resistant columnar structured thermal barrier coatings by Suspension Plasma Spray (SPS) as a less costly alternative to conventional EB-PVD. Coatings produced by SPS have a disadvantage of reduced usable spray distance, compared to conventional APS, due to quenching of the plasma by the suspension liquids, which are most commonly alcohol-based. The reduced spray distance can interfere with the coating process for substrates with complex geometries such as turbine blades. This paper shows how spray distance can be increased by using larger suspension particle sizes that are not normally considered for SPS. Such large particle suspensions are shown capable of producing columnar or segmented YSZ coating microstructures that are similar to those produced by submicron particle suspensions, but at longer and more practical spray distances. Another limitation to SPS process technology is the delivery system of feedstock from the point of manufacture to the SPS feed hopper. Current commercial ready-to-use suspensions have limitations involving cost, transportation and storage that effect both the producers and the end-users. An alternative suspension delivery system may be applied to SPS feedstock materials, including current sub-micron and the coarser particle size cuts described herein. Discussed is a pre-formulated dry feedstock that is constituted into fresh suspension by the end-user with locally sourced liquid media and appropriate high-speed mixing equipment. This alternative delivery system for suspensions provides lower cost materials and process flexibility that is particularly suited to commercial scale SPS coating facilities.

1 Introduction

Suspension plasma spray (SPS) materials and process technology is currently in a developmental stage for producing strain tolerant thermal barrier coatings applied to high temperature sections of turbine engines. The economic and practical advantages of utilizing SPS over established higher cost processes such as EB-PVD has been discussed in literature [1]. For SPS it has been demonstrated that the conditions of the coating substrate, plasma spray parameters, suspension feed, and the liquid suspension itself are intimately connected as variables in producing coating microstructures with designed properties [2,3]. As an adaptation of conventional APS technology, SPS has an inherent limitation of reduced spray distance due to quenching of the plasma by the liquid media. Reduced spray distance is a limiting factor for SPS coatings applied to complex geometry components such as turbine blades. Combustible liquid media such as alcohols are commonly used over less volatile non-combustible liquids such as water to partially offset the effect of quenching [4]. This work demonstrates that the problem of spray distance may further be improved through increased suspension particle size. Current suspension feedstock use submicron and nano particle sizes for columnar and segmented microstructured TBC coatings [5,6,7]. Relatively coarse particle suspensions that have particle size distributions above 2 microns are shown in this work to produce viable suspensions that can produce strain tolerant columnar coatings at longer spray distance than smaller particle size suspensions under the same spray conditions. This work compares microstructures, mechanical properties of hardness, tensile strength, erosion resistance, and resistance to thermal cycling of coatings produced by three coarse particle suspensions at three spray distances. Also discussed is an alternative suspension delivery system that may be applied to coarse and submicron-size particle SPS feedstocks that is suitable for commercial applications.

2 Experimental

Suspension powders were produced by milling 8 wt% Yttria stabilized zirconia (8YSZ) from Metco 204-XCL (Oerlikon Metco) down to three particle sizes (Table 1). The powders were formulated into suspensions with composition 25% by weight 8YSZ powder and 75% denatured ethanol with dispersant. The particles were fully dispersed and suspended via jar milling prior to plasma spraying. The three suspensions are named SF, SM, and SC denote relative fine, medium and coarse particle sizes, respectively.

Table 1: Particle size of test suspensions as determined by Microtrac.

Name	D10 (µm)	D50 (µm)	D90 (µm)
SF	0.19	0.55	1.99
SM	0.41	1.75	3.50
SC	0.46	2.68	5.39

Substrates used for this work were Hastelloy X with APS bond coat Ni 22Co 17Cr 12Al 0.5 Hf 0.5Y 0.4Si (Amdry 386-4, Oerlikon Metco) with Ra surface roughness of 8.3 µm. 25.4 mm x 76.2 mm coupons were used for metallographic images and erosion resistance tests. 25.4 mm diameter discs were used for tensile strength, burner rig testing and furnace cycle testing.

Suspension plasma spray top coats were produced using a TriplexPro-210™ torch with 5MPE-SF suspension feeder, both by Oerlikon Metco. Details of the SPS system used are described in other works [7,8,9]. Coatings produced from the three suspensions shared the same spray parameters using Ar/H_2 plasma gas and spray distances of 50 mm, 75 mm and 100 mm. The feed rate of the liquid suspensions was set to 32 g/min. The top coats produced ranged in thickness from 230 µm to 280 µm. The three suspension types and three spray distances were the only variables in preparing the coating samples for this work.

The parameters for furnace cycle testing (FCT) were 10 minutes heat-up to 1135 ^{0}C, steady state for 40 minutes and fan cool-down for 10 minutes. The defined fail-point of the FCT was 10% or greater spallation of the top coat. All FCT coating samples were tested in duplicate, together as a single furnace charge.

The parameters for burner rig testing (BRT) were 20 seconds under burner flame to reach a back face temperature of 1050 ^{0}C, 20 seconds cool-down with compressed air followed by 40 seconds idle at ambient temperature. The eight burner rig samples (one sample per coating condition as per Table 2) were tested together as a single charge on the rig.

3 Results and Discussion

3.1 Coating microstructures

Cross section images of the coatings produced by the three suspensions at three spray distances are shown in **Fig.1** and **Fig. 2**. The fine particle suspension, SF, could only produce coatings at short spray distances of 50 and 75 mm. Under the set conditions for this work, the SF suspension did not produce a viable adherent coating at the longest spray distance of 100 mm.

Table 2: Coatings produced from suspensions.

Type	Coatings produced at spray distance		
	50 mm	75 mm	100 mm
SF	cracked w/ porosity	columnar	no coating*
SM	cracked w/ porosity	intermediate	columnar
SC	cracked w/ porosity	intermediate	columnar

*adherent coating could not be produced

At 50 mm the SF suspension produced a dense cracked coating with localized areas of porosity that quickly transitioned to a lower density columnar coating at 75 mm, **Fig. 1**. The coarser particle suspensions SM and SC produced viable coatings at all three spray distances, **Fig. 2**. The short spray distance of 50 mm produced coatings with comparable appearance featuring cracks with localized areas of porosity near the interface for all three suspensions. At the longest spray distance of 100 mm the SM and SC suspensions produced coatings with porosity and columnar-type structure, with the SC suspension producing a slightly more dense appearing structure compared to the SM. At a spray distance of 75 mm the coating structures for the SM and SC suspensions have intermediate porosity and emergence of dense columns. Similar work by Kitamura [7] reported a structural transition from dense cracked structures to lower density columnar structures by increasing spray distance. The same report, comparing YSZ suspensions with particle size D90 = 2.1 µm to D90=1.1 µm indicated that the larger particles allowed longer spray distances.

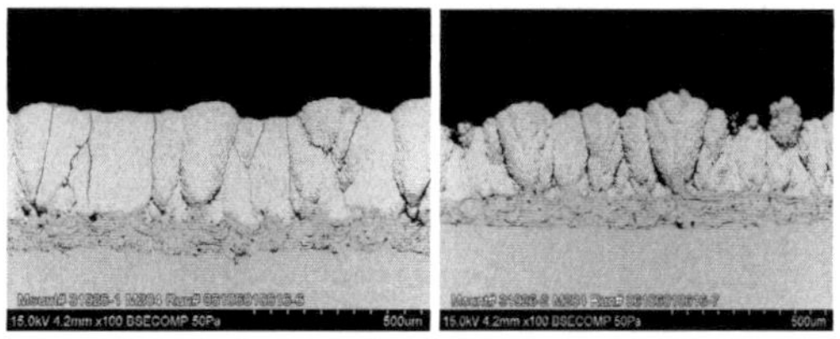

Fig. 1. Coating cross sections produced from SF suspension at spray distance of 50 mm (left) and 75 mm (right).

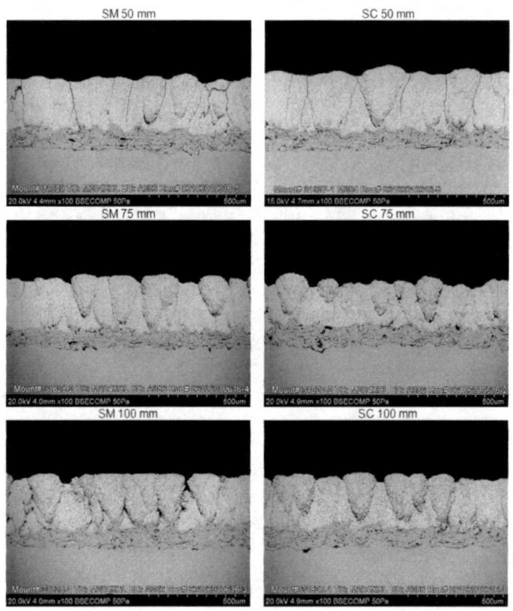

Fig. 2. Coating cross sections produced from SM (left images) and SC (right images) suspensions at spray distances of 50 mm (top images) 75 mm (middle images) and 100 mm (bottom images).

3.2 Coating hardness, tensile strength and erosion resistance

Macro hardness measurements (HR15N ASTM E18) of the coatings are plotted as a function of spray distance in **Fig. 3**. Consistent with the microstructure, the medium and course particle suspensions SM and

SC have comparable hardness profiles that decrease as the coatings densities decrease with increasing spray distance. Similarly, the hardness of the fine particle suspension SF coatings dropped steeply and it was noted that the coating density also dropped sharply between spray distances of 50 mm and 75 mm.

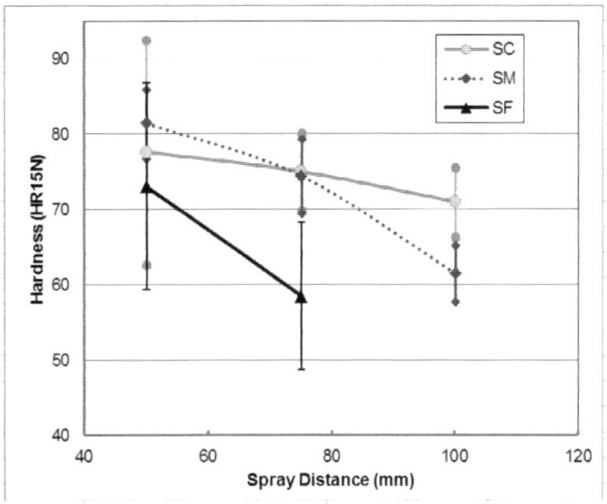

Fig. 3. Macrohardness of test coatings produced by three suspensions.

The tensile test of the coatings (ASTM C633) produced fracture failure points at the interface between the bond coat and the top coat. The fine particle suspension SF coatings had relatively constant tensile strength despite the large differences in microstructures produced by the different spray distances, **Fig. 4**.

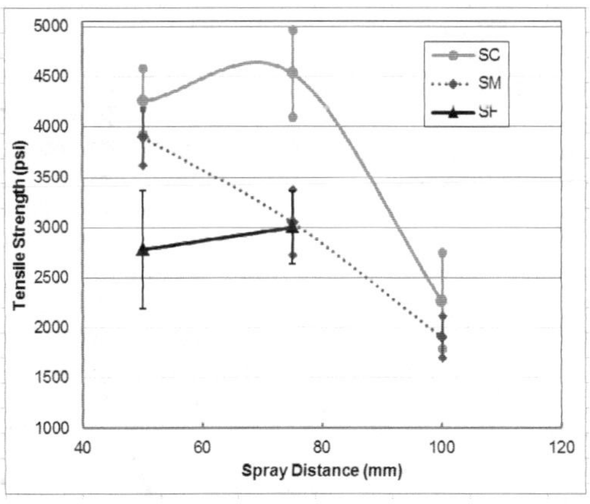

Fig. 4. Tensile strength of test coatings produced by three suspensions.

Similarly, the medium particle size suspension SM produced coatings of similar strength in the 50 – 75 mm spray distance range despite changes in microstructure. At a spray distance of 100 mm the SM and SC coatings sharply dropped in tensile strength. While the longest spray distance produced the weakest coatings, as would be expected, the relative bond strength behavior of the coatings is not easily explained based on the current data.

Erosion resistance (GE specification E50TF121) of the coatings are plotted as a function of spray distance, **Fig. 5**. The erosion resistance follows an overall expected trend of correlation with hardness. Coatings produced by the medium and coarse suspensions, SM and SC, are seen here to exhibit improved erosion resistance over the coatings produced by the fine suspension, SF.

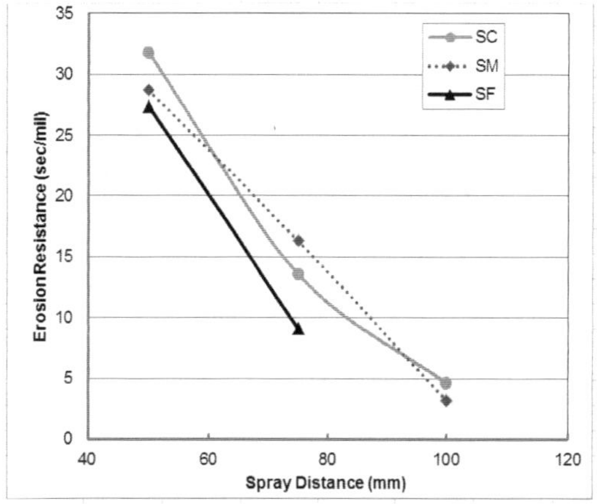

Fig. 5. GE erosion resistance of coatings produced by three suspensions.

3.3 Furnace cycle testing

The coatings with the earliest failures in furnace cycle testing were the dense coatings produced at the shortest spray distance of 50 mm **Fig. 6**. Those dense coatings, which survived between 50 and 380 cycles, had poor repeatability between the sample pairs, a and b, in reaching their failure points. The earliest failures were the coatings produced by the fine suspension, SF, followed marginally by the coatings produced by the SM and SC suspensions, respectively. Longer fail points and improved repeatability between sample pairs is evident for the longer spray distances, with 75 mm appearing to provide similar or slightly longer fail points compared to 100 mm. Coatings produced by the SM suspension provided slightly better furnace cycling performance, compared to those produced by SC, but given the scatter in paired sample repeatability, the improvement may be random. The failure mode of samples in FC testing was chipping of the top coat at the edge of the disk and/or delamination accompanied by cracking of the top coat, **Fig. 7**. It is apparent that the lower density columnar structures produced at longer spray distances provided significantly improved thermal fatigue tolerance in the furnace cycling.

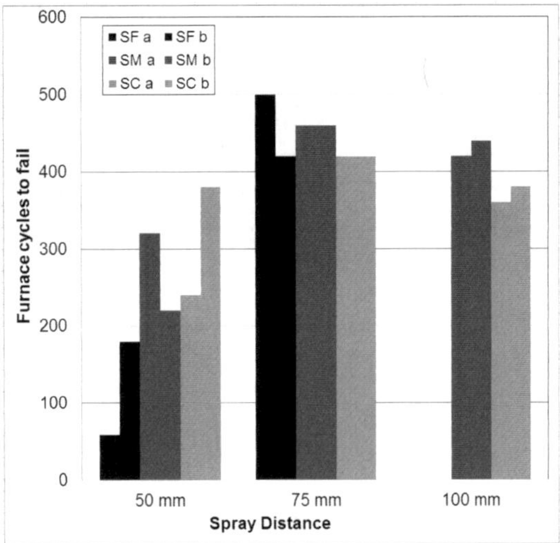

Fig. 6. Furnace cycles to reach >10% spallation of top coats for duplicate test coatings.

Fig. 7. Coatings produced by SC suspension with 100 mm spray distance after failure at 380 cycles in FCT (left) and 5000 cycles in burner rig test (right).

3.4 Burner rig testing

The burner rig test produced similar coating failure profiles to the FC test, **Fig. 8**. For burner rig testing the high density coatings produced at 50 mm spray distance produced very rapid failures, all less than 30 cycles. The remaining samples lasted much longer, between 4905 – 5848 cycles. The burner rig test clearly differentiated the lower density intermediate/columnar structures as having greatly improved thermal fatigue tolerance over the denser cracked structures produced at the shortest spray distance. After the extended period of nearly 4900 cycles with no failures, the remaining five coating samples eventually failed in a relatively narrow test span of 943 cycles. Thus, there was no clear winner among the lower density coatings produced at the 75 mm and 100 mm spray distances. The failure mode for the burner rig samples was radial spalling of the top coat from the middle to the outside of the discs, **Fig. 7.**

The 75 mm spray distance parameter provided the best compromise between all of the tests that were performed. Coating property improvements were demonstrated by FCT and burner rig results of coatings sprayed at longer distances, while still maintaining integrity in hardness, tensile strength, and erosion resistance. Further work would be required to produce coatings with improved erosion resistance while maintaining the long spray distance and thermal fatigue performance.

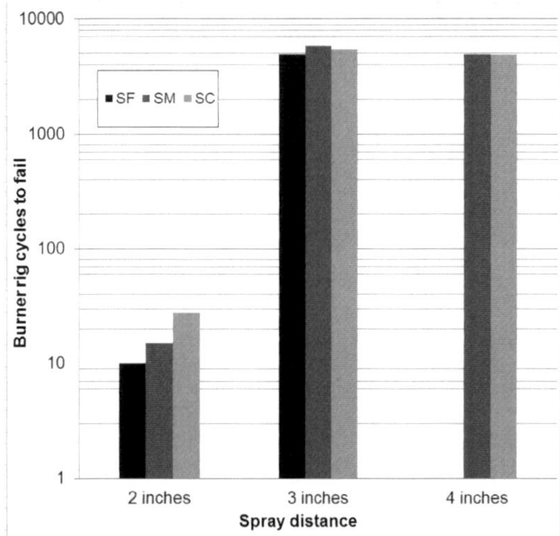

Fig. 8. Burner rig cycles to reach >10% spallation of top coats for duplicate test coatings.

3.5 Alternative delivery system

Options for SPS feedstocks have expanded as the materials and technology are becoming more commercialized. Formerly, for the purpose of research, suspensions were produced by the end-user from individual formulations of powder sources, liquid media, and additives to provide dispersion and stability. In recent years commercial ready-made suspensions have appeared on the market by a number of manufacturers [4]. Current commercial suspensions are produced, shipped and stored in small ready-to-use containers. Such ready-made suspensions, particularly those with flammable liquids such as alcohols, have safety issues and high costs related to specialized packaging, transportation of dangerous goods, storage, handling and shelf life. As a result, there is a desire to provide suspension feedstocks by alternative delivery systems that are more efficient and cost effective, particularly for anticipated commercial scale SPS operations. An alternative suspension delivery system consists of dry loosely agglomerated powder feedstock that is pretreated with dispersant, packaged, transported and stored in the same manner as conventional thermal spray powders. The pre-formulated feedstock is constituted into fresh suspension by the end-user using locally sourced liquid media and appropriate high-speed mixing equipment. This alternative delivery system for suspensions provides lower cost materials and process flexibility that is particularly suited to commercial scale SPS coating projects. Such a delivery system would be suitable for all suspension feedstock particle sizes, including the course particle size range discussed in this paper. The mixing/dispersion process involves inexpensive commercially available high-speed batch mixers, or customized options, that provide sufficient sheer mixing energy to fully disperse the particles. Small batches may alternatively be mixed and dispersed though jar milling, a common practice for lab scale SPS work. An example of a commercial dry pre-formulated SPS feedstock powder to be used for such purposed is 8YSZ Metco 6609 (Oerlikon Metco).

4 Conclusions

Coatings were prepared from ethanol suspensions of 8YSZ with three unconventional coarse particle size cuts. The coatings were produced under set conditions of spray parameters, suspension feed conditions and substrate, with spray distance being the only variable aside from suspension particle size. The coatings were evaluated for mechanical properties and thermal cycling resistance in furnace cycle testing and burner rig testing. The main findings are summarized as follows

1. Coarse powders up to at least D90 = 5 µm are capable of producing viable liquid suspensions, using current suspension feeder equipment to produce microstructured YSZ coatings.

2. Coarse powders with D90 > 2 microns provided a longer practical spray distance, in this case at least 100 mm compared to a finer powder, which did not produce coatings at such distance.

3. The coating microstructures produced by the coarse particles ranged from a relatively dense cracked structure with localized porosity to lower density columnar type structures that are similar to those produced by finer sub micron or nano size powder suspensions.

4. The medium and coarse particle suspensions produced coatings with higher hardness and erosion resistance compared to those produced by the fine particle suspension

5. The low density microstructures produced at long spray distances demonstrated significantly improved thermal fatigue performance compared to the denser cracked coatings produced at short spray distance. The overall best coatings were produced by the coarser suspension, SM and SC, at a spray distance of 75 mm.

5 Acknowledgements

All Oerlikon Metco staff involved in the experimental work, and preparation of this report, are gratefully acknowledged.

6 References

[1] Hardwicke, C.U., Lau, Y.-C., J., Advances in Thermal Spray Coatings for Gas Turbines and Energy Generation: A Review. J. Thermal Spray Technol., Vol. 22(5), (2013) pp. 564/576.

[2] Fauchais, P., Suspension and Solution Plasma or HVOF Spraying. J. Thermal Spray Technol., Vol. 17(1), (2008), pp. 1/3.

[3] Toma, F.-L., Berger, L.-M., Potthoff, A., Demands and economic aspects of thermal spraying with suspensions – an overview. Proc. Int. Thermal Spray Conf. 2014, Barcelona May 21-23, DVS vol. 302, pp. 368/372.

[4] Toma, F.-L., Berger, L.-M., Potthoff, A., Leyens, C., Demands, Potentials, and Economic Aspects of Thermal Spraying with Suspensions: A Critical Review. J. Thermal Spray Technol., Vol. 24(7), (2015) pp. 1143/1152.

[5] Curry, N., VanEvery, K., Snyder, T., Susnjar, J., Bjorklund, S., Performance Testing of Suspension Plasma Sprayed Thermal Barrier Coatings Produced with Varied Suspension Parameters. Coatings, Vol. 5, (2015), pp. 338/356.

[6] Chen, X., Honda, H., Seiji, K., Hiroshi, A., Murakami, H., Watanabe, M., Sakka, Y., Highly Segmented Thermal Barrier Coatings Deposited by Suspension Plasma Spray: Effects of Spray Process on Microstructure. J. Thermal Spray Technol., Vol. 25(8), (2016) pp. 1638/1649.

[7] Kitamura, J., Fujimori, K., Wada, T., Liu, J., Callen, B. Microstructural control on yttria stabilized zirconia coatings by suspension plasma spraying. Proc. Int. Thermal Spray Conf. 2016, Shanghai, May 10-12, DVS vol. 324, pp. 348/352.

[8] Callen, B.W., Gindrat, M., Liu, J., Rauch, J., Suspension Spraying – an emerging coating technology. Sulzer Technical Review, Vol. 2 (2013) pp. 4/8.

[9] Molz, R.J., Cotler, R., Liquid Suspension Spray, New challenges, New Opportunities. Sulzer Technical Review, Vol. 2 (2011) pp. 26/29.

Ultrafine powders for low-power HVOF and HV-APS spraying

Götz Matthäus, Vadim Verlotzki, Dortmund / D

The currently available powders are for conventional high power HVOF or APS. There is an increasing need for low-power HVOF and HV-APS torches for coating temperature-critical parts and inner diameters. Using them for the new applications causes massive restrictions such as poor coating efficiency, much overspray, rough surface or defects in the layer structure.
However, just using ultrafine powders does not solve the problem. The disadvantages in terms of fluidity and oxidation resistance need to be solved, too.
Beside the development of modern HVOF, ID-HVOF and Three-cathode-APS coating systems Thermico pursues a continuous enhancement of customized powders. Due to this steady development, Thermico has realized spraying inner diameters with just 10 – 20 mm stand off by using a new WC carbide powder which is able to absorb enough energy out of the low power flame to build up a very dense high quality coating. The modified WC Co 83 17 powder is available for ID HVOF.
The high quality coating of outer diameters with a HV-APS is another solution which Thermico has successfully developed. Beside this Thermico had the goal to establish a high quality HV-APS coating to prices competitive to HVOF.

1 Introduction

By solving the problem for coating inner diameters by using fine powders you are facing new problems. The most important are the affinity of fine powders to agglomerate and oxidate especially in the powder fractions of -9+1µm. thermico solved the problem of feeding these grain sizes with their powder feeder CPF2 but the problem of the agglomeration is still there. The particles which are fed to the flame are no single particles but particle agglomerates. The smaller the single particle the bigger and more solid are the agglomerates. These are slower and cooler compared to single particles. Another problem is the oxidation of the fine particles. The smaller the particle the bigger the surface. Due to the fact that smaller particles become hotter than bigger ones the mass of oxides is propotionally to third or fourth power of the diameter reduction because of the faster and deeper oxidation of the small particles. The oxidation reduces the coheasion and adheason of the coating because the oxide layer around the particles and between particle and substrate. One solution is the use of with high temperature fluxing agents modified powders.

In the field of high quality outer diameter thermico developed an HV-APS coating for corrosion-, oxidation- and wear resistance up to 900°C and it protects the substrate even against hot sulfuric acid and phosphoric acid. Typical applications are piston rods, calender rolls, valve shafts or extrusion machines. The necessary coatings do not only need to be corrosion resistand but also impermeable against corrosive media. This is the most challenging demand.
The state of the art are hard chrome coatings but they are not dense. Hard chrome and also nearly all plasma sprayed ceramic coatings have a crack network and are porous. For room temperature applications there is the possibility of a polymer sealing but it is very limited. Especially in application with also friction demands like extruder, valve shafts or valves. One coating that combines most of the properties is NiCrBSi but it has an insufficient hardness and friction behaviour.

2 Development

2.1 Modified low-power HVOF powder

The positive effects of silicate glass regarding the coating quality are already proven in some patents and scientific publications [1- 8]. The most important function of glasses as an additive is its use as a high temperature fluxing agent. These modified powders are new and experimental powders and no commercially available powders. It is possible to modify standard powders to achieve an improvement of the coating properies ot these powders in specific applications. The fluxing agents are in very low concentrations less than 1% of the powder weight. The coating properties are not influenced by the fluxing agent. The advantage is in the improvement not only in the cohesion but also in the adhesion of the coating without influencing the other chemical and physical properties. In other words it is the same coating but in better quality!
There is no need to modify every powder available. It is reasonable to modify the powder if the following requirements are fulfilled:

- The procedure works best with porous agglomerated sintered particles. Only in this case the fluxing agent is not only on the surface of the powder but also in the inner of the particle. This is necessary because otherwise it is not possible to realise a fluxing agent content of 0,1 – 1% of the powder weight. Having only the fluxing agents only on the suface of the particle leads to a decreasing of the flowability and a clogging of single particles.
- The fluxing agent needs to be soluble in water to get a homogenious distribution.

Agglomerated and sintered powder is submerged in the water solution. After the mixing the excess water will be removed and the powder dried.
- The fluxing agent needs to be not hygroscopicto prevent the modified powder for clogging.
- The melting point of the fluxing agent needs to be similar to the temperature of the particle during spraying.
- Remaining fluxing agent in the coating are not influencing the coating properties like wear resistance, ductility and chemical resistance.

With regards to the above mentioned requirements only the following powders, fluxing agents and thermal spray technologies are worth to be considered:
- Ultrafine powder with chemical composistions like WCCo, WCCoCr, $WCCr_3C_2Ni$ and Cr_3C_2NiCr. The effect of the fluxing agent is the bigger the smaller the grain size is. Only grain sizes less than 22µm are effective.
- All water soluble silicates especially sodium silicate and potassium silicate are applicable. For these not hygroscopic water soluble silicates is the maximum content of 1% possible. If the silicate is hygroscopic only 0.5% are reasonable.
- In general are the modified powder applicable for all high velocity oxy fuel systems like HVOF, HVAF and Detonation. The limitating thing is that the temperature of the modified powder needs to be high enough to melt the flucing agent. The shorter the spray distance the hotter and slower the particles. Due to this fact the modified powders are more suitable for ID coatings than OD coatings. The maximum effect will be for ID coatings with an ID smaller than 100mm

2.1.1 nanoHVOF® coating sprayed with modified low-power HVOF powder

Starting powder was agglomerated sintered and crushed WC83Co17 powder in the fraction of -3+17µm. Fluxing agent was sodium silicate Na_2O $3{,}5SiO_2$ – 30% solution. This leads to the modified powder WC83Co17 (Na_2O $3{,}5SiO_2$)0,5 in the fraction of -5+25µm. To show the advantages of the modification samples have been sprayed with the same nanoHVOF® system and the same parameters in comparison to the starting powder.
The samples have been sprayed in a tube with an ID of 130mm and a stand off of 50mm. Subjects of the investigation are the surface roughness (**Fig. 1**), microscopy inlcluding hardness testing, SEM (**Fig. 2**) including comparison of Brinell imprints (**Fig. 3**) as an indication of the ductility of the coating and also Pin on Disc Test (**Fig. 4**).

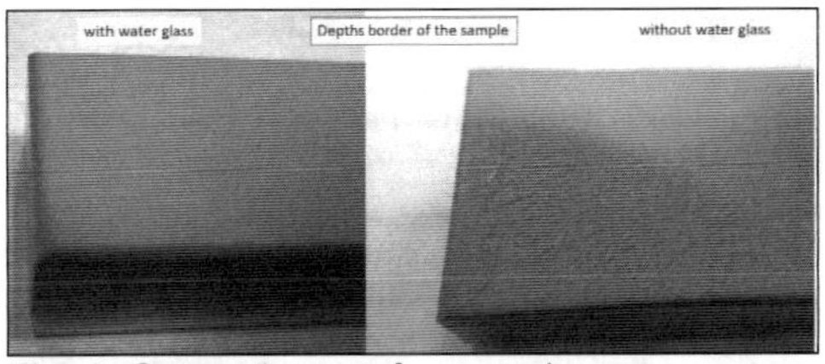

Fig. 1: Comparison surface roughness

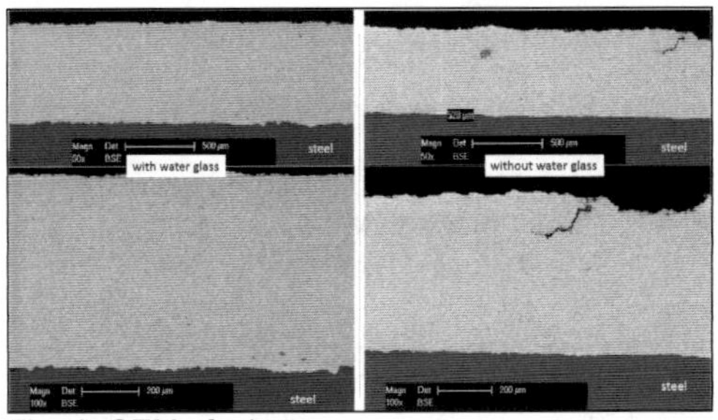

Fig. 2: SEM of micros

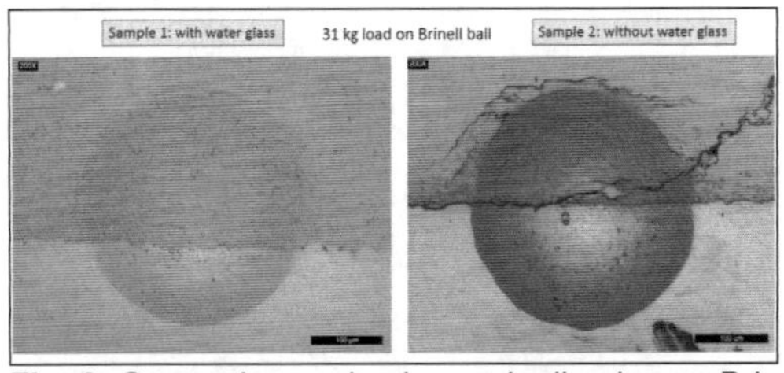

Fig. 3: Comparison cohesion and adhesion on Brinell imprints, load 31 kg

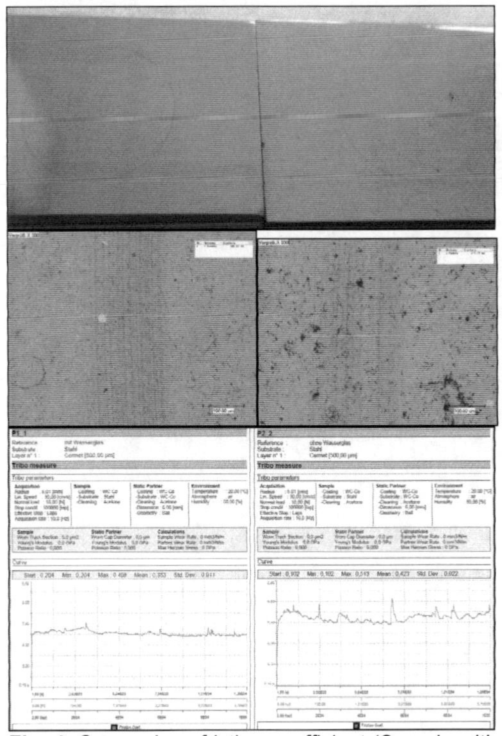
Fig. 4: Comparison friction coefficient (Sample with modified powder on the left side)

2.2 HV APS powder development

The solution is the development of the "ChroSiMol" coating. This type of coating is based on a APS sprayed high temperature resistant chrome based coating [9]. This coating already fulfilled the most important requirement; it is gas tight and has a very good adhesion to the substrate. In contrast to galvanic hardchrome the APS chrome coating has no cracks and is temperature resistant up to 1,000°C. Furthermore is the corrosion resistance of the APS coating better compared to galvanic hardchrome. Unfortuantely the hardness is only about 600HV and the friction behaviour was insufficient.
To increase the friction behaviour, corrosion resistance and wear resistance a part of the chrome has been substituted by molybdenum. Another additive is silicon. This is the base for the ChroSiMol coatings; Cr – Mo – Si. To adjust the thermal expension coefficient and improve the spray behaviour Ni80Cr20 or Co was added.
These separate metallic powders combined with cristobalite in a coarse grain size as a carrier for the fine metallic powders is the ChroSiMol powder. The cristobalite disappears during the APS coating process nearly completely and is only seen in the coating as small intermediate phases. Fig. 5 – 7 show micros of the ChroSiMol coating directly after coating with the Axial III three cathode plasma torch.

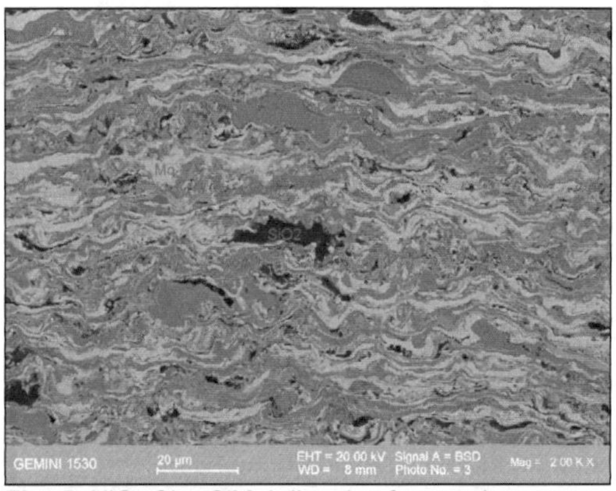

Fig. 5: NiCr-ChroSiMol directly after coating; Magnification 2,000x

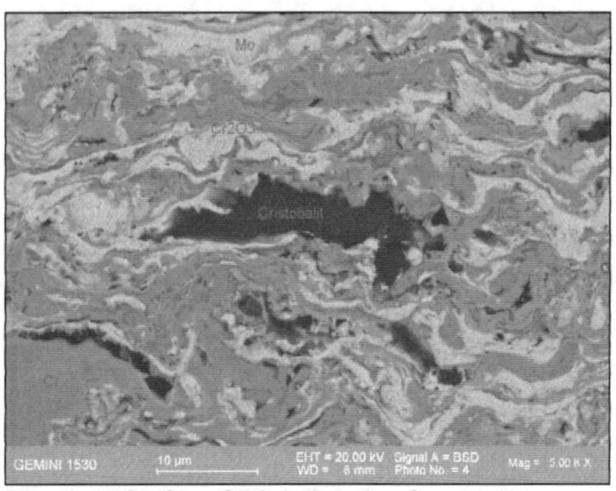

Fig. 6: NiCr-ChroSiMol directly after coating; Magnification 5,000x

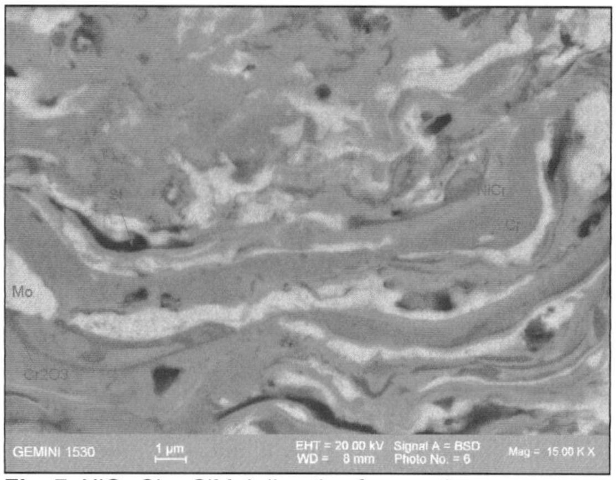

Fig. 7: NiCr-ChroSiMol directly after coating; Magnification 15,000x

It can be seen in the micrographs that the powder components form corresponding phases in the coating. The time the particles are in the plasma during spraying is too short for alloying. The non metallic phases (cristobalite and Cr_2O_3) are below 3%. To obtain a pseudo-alloy the coated part needs to be heat treated. The temperature of the heat treatment is in the range of 750-900°C, duration from 1 hr on. Due to the fact that the coating is already gas tight the heat treatment can be performed in athmospheric conditions. The heat treatment is not only for generating the hard and corrosion resistant alloy but also for the improvement of the adhesion and cohesion of the coating. The bond strength increases up to 250MPa due to the heat treatment. In addition the residual stresses turn to compressive residual stresses.
Figures 8 - 10 are showing the micro after the heat treatment at 900°C for 1 hr in air.

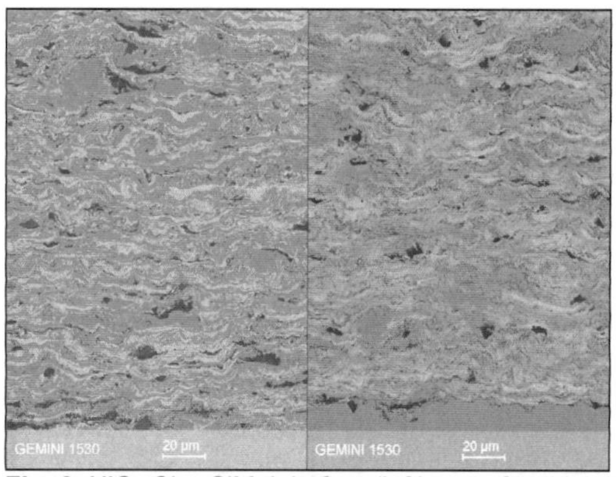

Fig. 8: NiCr-ChroSiMol; before (left) and after (right) heat treatment; Magnification 1,000x

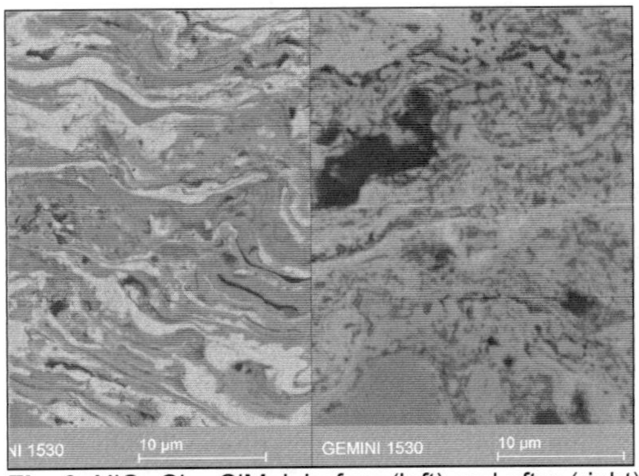

Fig. 9: NiCr-ChroSiMol; before (left) and after (right) heat treatment; Magnification 5,000x

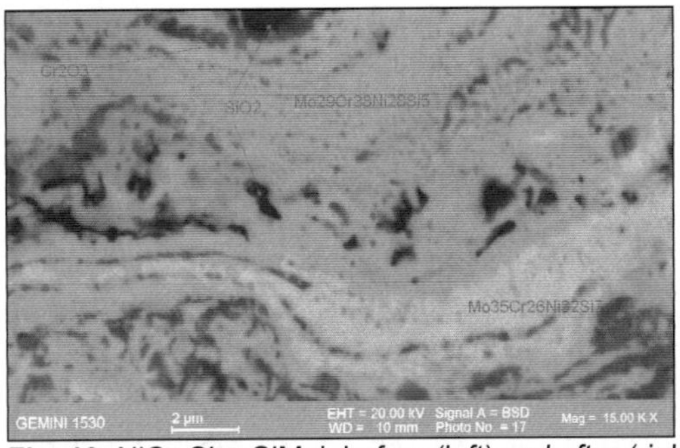

Fig. 10: NiCr-ChroSiMol; before (left) and after (right) heat treatment; Magnification 15,000x

The following table shows the properties of ChroSiMol as well with NiCr as with Co.

Table 1: Properties ChroSiMol

Hardness (RT)	850 - 950 HV0,2
Bending strength (4 point)	300 - 400 MPa
Young´s modulus (RT)	approx. 200 Gpa
Bond strengths	> 250 Mpa
Thermal expension coefficient (RT)	9 - 9,5 x 10^{-6} K^{-1}
Thermal conductivity (RT)	10 - 20 W/mK
Density	7,7 g/cm^3
open Porosity	0
sealed Porosity	< 0,5%

Corrosion resistance was tested in different media and compared to the corrosion resistance of Ti6-4 alloy. Table 2 gives an overview about the corrosion restance.

Table 2: Corrosion resistance NiCr – ChroSiMol

Corrosive Media, Temperature, Duration	ChroSiMol	Ti6-4
10%-H_2SO_4, 60°C, 48h	no corrosion	corrosive damage approx. 0.05 mm
20%-H_2SO_4, 60°C, 48h	no corrosion	corrosive damage approx. 0.02 mm
40%-H_2SO_4, 60°C, 48h	no corrosion	corrosive damage approx. 4 mm
96%-H_2SO_4, 60°C, 48h	no corrosion	corrosive damage approx. 1 mm
200-300 g/L CrO_3 + 2-2,5 g/L H_2SO_4, 20°C, 72h	no corrosion	no corrosion
200-300 g/L CrO_3 + 2-2,5 g/L H_2SO_4, 60°C, 72h	corrosion because of pores	no corrosion
30%-H_3PO_4, 60°C, 48h	no corrosion	corrosive damage approx. 0.1 mm
5%-HCl, 20°C, 48h	no corrosion	no corrosion
Corrosive Media, Temperature, Duration	**ChroSiMol**	**Ti6-4**
5%-HCl, 40°C, 48h	no corrosion	corrosive damage approx. 0.1 mm
10%-HCl, 20°C, 48h	corrosive damage approx. 0.1 mm	corrosive damage approx. 0.5 mm
20%-HCl, 20°C, 1h	corrosive damage approx. 0.1 mm	corrosive damage approx. 1 mm
37%-HCl, 20°C, 0,5h	corrosive damage approx. 0.1 mm	corrosive damage approx. 2 mm
15%-HNO_3, 20°C, 24h	corrosion because of pores	no corrosion
15%-HNO_3, 60°C, 2h	corrosion because of pores	no corrosion
65%-HNO_3, 20°C, 24h	corrosion because of pores	no corrosion
HNO_3(65%) + 3HCl(37%), 20°C, 24h	no corrosion	corrosive damage approx. 0.05 mm
HNO_3(65%) + 3HCl(37%), 40°C, 24h	corrosion because of pores	corrosive damage approx. 0.2 mm

48%-HF, 20°C, 48h	no corrosion	not tested (reaction too fast)
5%-HCl + 10%-HF (1:1), 20°C, 100h	no corrosion	not tested (reaction too fast)
96%-H_2SO_4 + 48%-HF (1:1), 20°C, 48h	no corrosion	not tested (reaction too fast)
10%-H_2SO_4 + 10%-HF (1:1), 20°C, 100h	no corrosion	not tested (reaction too fast)
20%-NaOH, 60°C, 48h	no corrosion	corrosive damage approx. 0.05 mm
3,5%-NaCl, 100°C, air ventilation, 500h	no corrosion	no corrosion

It can be clearly seen that the ChroSiMol coating shows a better resistance against wet chemical corrosion in H_2SO_4 and H_3PO_4.

2.3 Conclusions

2.3.1 Modified low-power HVOF powder sprayed with nanoHVOF®

The use of the modified powder shows clearly a better cohesion of the coating. No defects or spalling of the coating during coating or lab preparation could be seen compared to the reference sample which shows cracks and porous inclusions in the lab samples as well as spalling during preparation.
Furthermore no negative overspray effects appearing during coating of small diameters could be seen on the sample with modified powder. The surface roughness of this sample shows no gradient over the lengths of the sample. A roughness gradient is clearly seen on the reference sample which shows an increasing of defects the deeper the coating inside a pipe.
In general leads the use of the modified powder to an increase of the coating quality for ID applications.

2.3.2 HV APS powder

ChroSiMol can replace hard chrome coatings in nearly all fields as well as in combined corrosive, heat and wear attack it shows improved properties compared to thermal sprayed Cr_3C_2NiCr or WCCoCr and other wear resistant coatings. The densitiy against liquid and gaseous corrosive media and enormous acid resistance makes ChroSiMol out of reach for any porous coatings and even dense and remelted self fluxing coatings. It is also an appropriate coating for high temperature applications in air. Compared to hardchrome the wear resistance is much better and in the same range like the wear resistance of Cr_3C_2NiCr HVOF sprayed coatings. It is even better in the regards of friction combination. Due to the very good adhesion and cohesion as well as the compressive residual stresses the impact strength of ChroSiMol is higher compared to all carbidic HVOF coatings. This makes it to an alternative to the Stellite coatings on seat rings. The costs of the ChroSiMol coating is comparable to conventional Cr_3C_2NiC and WCCoCr HVOF coatings. The supposed disadvantage of the necessary heat treatment of the ChroSiMol coating disappears often because it can be done in one step with the quenching and tempering.

3 Literature

The literature reffered to:

[1] Verlotski, V., Stöver, D., Buchkremer, H.P., Vaßen, R., Wärmedämmende Glas-Metall/Keramik-Schichten. Deutsche Patent Nr.: 19852285, Mai 2000

[2] Verlotski, V., Stöver, D., Buchkremer, H.P., Vaßen, R., Wärmedämmende Glas-Metall/Keramik-Schichten. Eurpean Patent No.: 1141437, Oktober 2003

[3] Verlotski, V., Thermal sprayed protective layer for metallic substrates. European Patent EP2006410-A2, 24.12.2008

[4] Verlotski, V., Thermally sprayed gastight protective layer for metal substrates. US Patent: US 8,784,979 B2, 22.07.2014

[5] Verlotzki, V. Dissertation: Metall-Glas-Wärmedämmschichten für Turbinenschaufeln, Berichte des Forschungszentrums Jülich 3818, ISSN 0944-2952, D294 Diss. Universität Bochum, Oktober 2000

[6] Dietrich, M., Verlotzki, V., Vaßen, R., and Stöver, D., Metal-Glass Based Composites for Novel TBC-Systems. Mat.-wiss. U. Werkstofftech. 32, 669-672 (2001)

[7] Dietrich, M., Verlotzki, V., Vaßen, R., and Stöver, D.,Microstructure and performance of new metal glass composites TBC. 26th annual converence on composites, advanced ceramics, materials and structures, Cocoa Beach FL, 13-18 January 2002

[8] Verlotski, V., Stanglmaier, R., Moormann, G., Mineral-Metal, Multiphase Coatings to Protect Combastion Chamber Components Against Hof-Gas Corrosion and Thermal Loading. Jounal of Engineering for Gas Turbines and Power, October 2011, Vol. 133

[9] Verlotski, V., Process for producing a protective chromium layer, US 20140287261 A1

Nanostructural WC-Co coatings by utilizing novel powder manufacturing route using water soluble raw materials

Marjaana Karhu, Juha Lagerbom, Kimmo Kaunisto, Tomi Suhonen, Tomi Lindroos and Erja Turunen
VTT Technical Research Centre of Finland Ltd

Tungsten carbide -based hard metal coatings are extensively used in demanding industrial applications like for wear protection purposes. Continuously increasing demands set new limits and need for materials with enhanced features. One solution is to improve hard metal properties by nanostructures. Presented study is part of a research where novel and safe route to manufacture nanostructural WC-Co powders starting from water soluble raw materials was developed. In this study powders' workability in thermal spray coatings is studied. WC-12Co powder was manufactured using water soluble raw materials: ammonium metatungstate as a tungsten source, glycine as a carbon source and cobalt acetate as a cobalt source. The powder was manufactured via optimized spray drying and heat treatment method producing a correct phase structure and chemical composition. Experimental powder was sprayed by HVAF-spraying to study its workability and functionality. Morphology, microstructure and properties were analyzed from the experimental nanostructural powder and the HVAF-coatings.

1 Introduction

WC-Co system based hard metal coatings applied by thermal spray processess are extensively used in wear resistant industrial applications. Continuously increasing demands set new limits and need for better materials. One solution is to improve hard metal properties by nanostructures. Generally, the hardness of WC–Co hard metal materials increases with the decrease of WC carbide size since the mean free path between cobalt matrix decreases with the decrease of carbide size [1]. The increase of hardness is expected to increase the wear resistance of WC–Co coatings [2]. The wear resistance of WC–Co hardmetals largely depends on their hardness and is also influenced by their toughness [1,3]

The conventional method for producing micro-sized WC is a solid state synthesis. However, due to the long duration time and high temperature of the solid state process, the manufacturing of nano-sized carbides is limited. A number of techniques are presented in the literature for producing nano-sized tungsten carbide powder, such as mechanical alloying (MA) [4,5], plasma methods [6,7] chemical vapour condensation processing (CVC) [8] and solid state reaction method [9]. The major obstacles in benefitting nanotechnology in tungsten carbide applications are lack of economical and commercial nano-carbide manufacturing methods and inhibition of carbide grain size growth [10]. For these reasons fully nanostructured dense carbide powders are not commercially available. Novel precursor synthesis methods have been developed with water-soluble raw materials and simple heat treatment steps [11,12,13,14].

Thermal decomposition of WC carbide, dissolution of carbide into molten binder and oxidation decarburization, and physical rebounding of large carbide particles upon high velocity impact are the basic degradation mechanisms involved in thermal spray deposition. The former three mechanisms lead to the change of crystallographic structure of WC–Co coatings compared to the starting powders. The later mechanism results in a remarkable change of coating composition and carbide content compared with the starting powders [1]. One essential problem related to thermal spray coatings with nanocarbides is concerned with how to limit the decomposition of WC because the the reduced WC particle size increases the specific surface area of carbide particles. That makes dissolution of carbide into molten cobalt binder matrix even more severe. The decarburization not only reduces the wear-resistant but also leads to the formation of a brittle Co–W–C ternary binder phase [1].

Presented study is part of a research where novel and safe route to manufacture nanostructural WC-Co powders starting from water soluble raw materials is being developed. Handling of solid (dry) nano sized raw materials can be avoided with this novel powder manufacturing method. In the presented study, usability of an experimental nanostructural WC-12Co powder in thermal spray coatings was studied. Powder applicability and functionality was tested with the HVAF (High Velocity Air Fuel) -coating deposition. Particle size and distribution, as well as morphology, microstructure and phase structure of the powder were characterized. The HVAF-coatings' phase structure, microstructure and some mechanical property, hardness value, were analyzed.

2 Materials and methods

2.1 Powder manufacturing

Nanostructural WC-12Co powder was synthesized using a novel powder manufacturing method based on use of water soluble raw materials. Raw materials for the experimental precursor were ammonium metatungstate ($(NH_4)_6H_2W_{12}O_{40} \cdot 3H_2O$), AMT, as a tungsten source, glycine ($C_2H_5NO_2$) as a carbon source and cobalt acetate $Co(C_2H_3O_2)_2$ as a cobalt source. Precursors were dissolved in distilled water by using a magnetic mixer with a heating unit. AMT, glycine and cobalt acetate were mixed with 80°C water in separate vessels and combined after dissolution. Precursors were dried with a industrial scale spray dryer Niro Atomizer. The spray dryer was used to ensure the homogeneity of the mixture and thus avoid separation of the raw materials during water evaporation. Heat treatments of the dried product were carried out under an inert Argon (Ar) gas atmosphere. More details of powder synthesis can be found in [13] and [14]. Nanosized WC-10Co4Cr1Al powder provided by MBN Nanomaterials and submicron WC-12Co powder (WOKA 3110 FC) provided by Oerlicon Metco were used as reference materials.

2.2 Coatings deposition

Coatings were deposited by a HVAF gun M3 (Uniquecoat Technologies LLC, Oilville, United States), using propane as the fuel gas and nitrogen as the carrier gas. HVAF gun was used in order to keep heat input low, compared to the HVOF or plasma spray techniques. Coatings were deposited on steel substrates. Spray parameters are presented in Table 1.

Table 1. HVAF spray parameters.

Parameter	HVAF
Air	7,4 bar
Fuel	209 slpm
Spray distance	250 mm
Surface speed	1.9 m/s

2.3 Characterization techniques

Powders' particle size distributions were determined with laser diffractometry (Malvern Mastersizer). Powders' and coatings cross-sections, powders morphology and carbide size of the powders and final products were analysed by an UltraPLUS Thermal Field Emission Scanning Electron microscope (Carl Zeiss Microscopy GmbH). Qualitative phase and crystalline size analyses were carried out for the powders and coatings by a X-ray diffractometer (XRD, Empyrean, PANalytical B.V., ALMELO, Netherlands) with Cu-Kα radiation source, and analysed using the HighScore Plus software. Powders' apparent densities were measured from free flown powders using a 25 ml container. Hardness values of the coatings were obtained by a Vickers indenter using a 300 g load.

3 Results and discussion

3.1 Powders characterization

Detailed characteristics of experimental nanosized WC-12Co powder and reference powders WC-10Co4Cr1Al provided by MBN Nanomaterial and WC-12Co (WOKA 3110 FC) provided by Oerlicon Metco are presented in Table 2. All the employed powders show similar particle size and distribution being suitable for the HVAF spraying. The major difference between the powders is in apparent density values: experimental powder shows density less than half with respect to that of the reference powders. The difference in densities originates partly from the production method and partly from low sintering temperature (1000 °C) of the experimental powder. Low sintering temperature was used in order to ensure carbide grain growth inhibition.

Table 2. Characterictics of the employed powders.

Powders	Apparent density [g/cm^3]	Particle size distribution d10-d90 [µm]
Experimental	2.3	9.27-32.2
MBN	4.7	4.84-29.7
WOKA 3110 FC	5.0	12.8-32.6

Figure 1 shows XRD patterns of the experimental powder and the reference powders. As identified by X-ray diffraction, all employed powders contain only WC carbide phase and pure cobalt binder phase. Average crystallite sizes according to XRD analysis were found to be 15 nm for nanosized MBN and 30 nm for the experimental powder. Figure 3 presents the experimental powder cross-section SEM-image which shows invidual carbide size ranging from 10 to 100 nm.

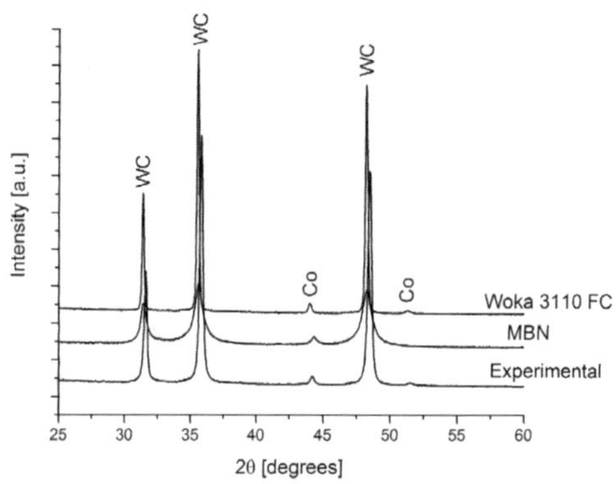

Fig. 1 XRD patterns of the studied powders.

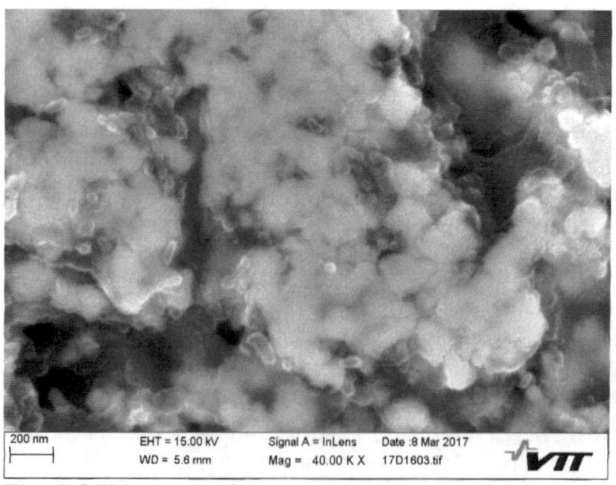

Fig. 2 SEM-image of the experimental powder.

Figure 3 shows morphology SEM-images of the experimental and reference powders (a-c), as well as the cross section images (d-f) of the same powders. It can be seen that the experimental powder shows more spherical shape than the reference powders. This is propably due to difference in the manufacturing method. Reference powders seem to have been heat-treated at high temperature and crushed after the sintering. This procedure provides more uneven shape particles than the mere spray drying. On the other hand, higher porosity of the experimental powder with respect to the reference materials can be observed. Also, the presence of hollow agglomerates in the experimental powder is clearly visible. This is, at least partly, due to the lower sintering temperature and the different type of a production method. Spherical shape is considered beneficial for the thermal spraying in respect of consistent powder flow. Exessivelly porous structure is considered detrimental for WC-Co coating structure.

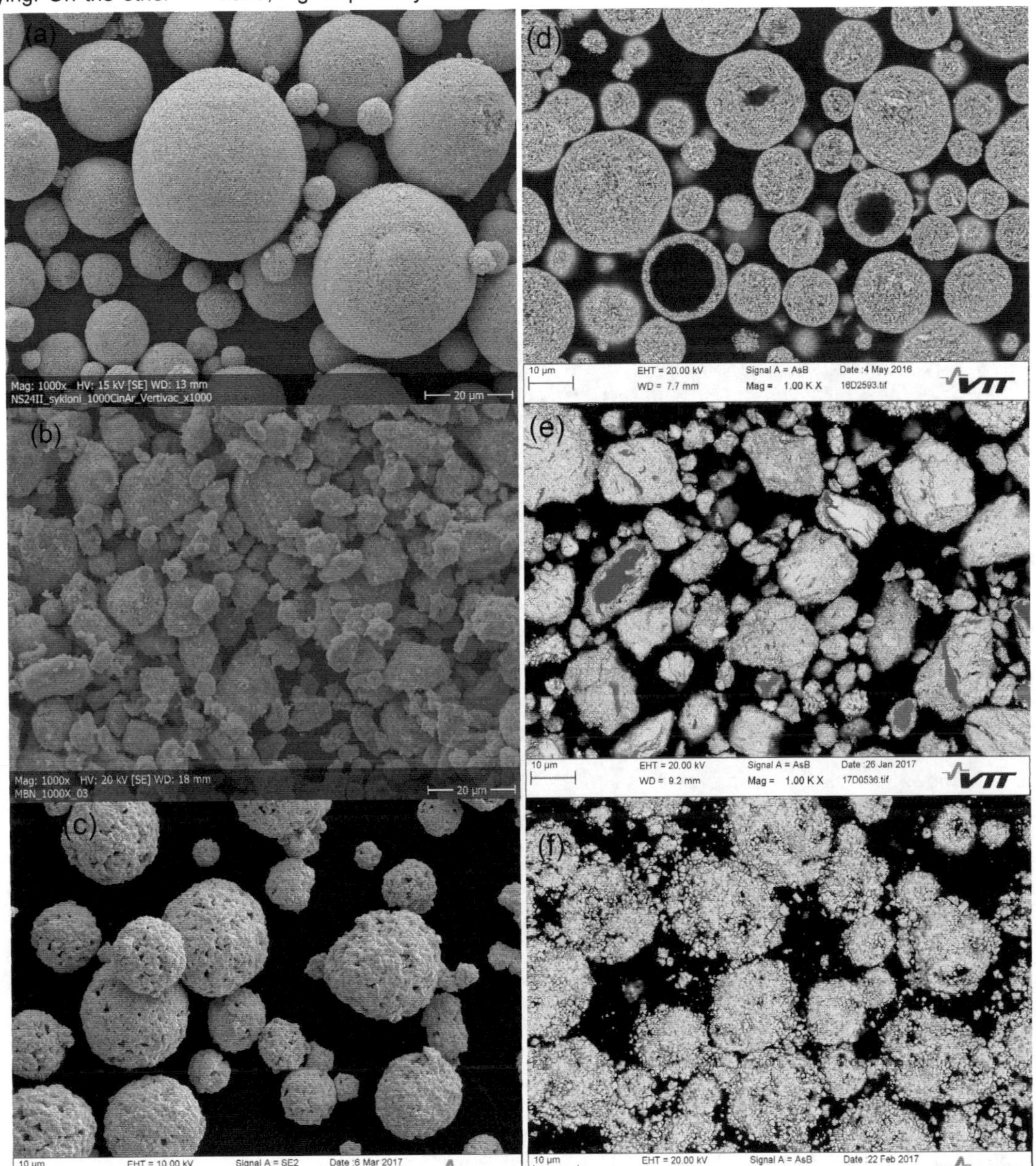

Fig. 3 Morphology SEM-images of the (a) experimental, (b) MBN reference and (c) WOKA 3110 FC powders, and cross-section images of the (d) experimental, (e) MBN reference and (f) WOKA 3110 FC powders.

3.2 Coatings characterization

The SEM figures of the coating microstructures are presented for the experimental (a-b) and MBN reference (c-d) powders in Figure 4. Both coatings resulted in nanocarbide structure. The coatings' porosity is similar at a macro level but the experimental powder shows higher porosity at a microscale.

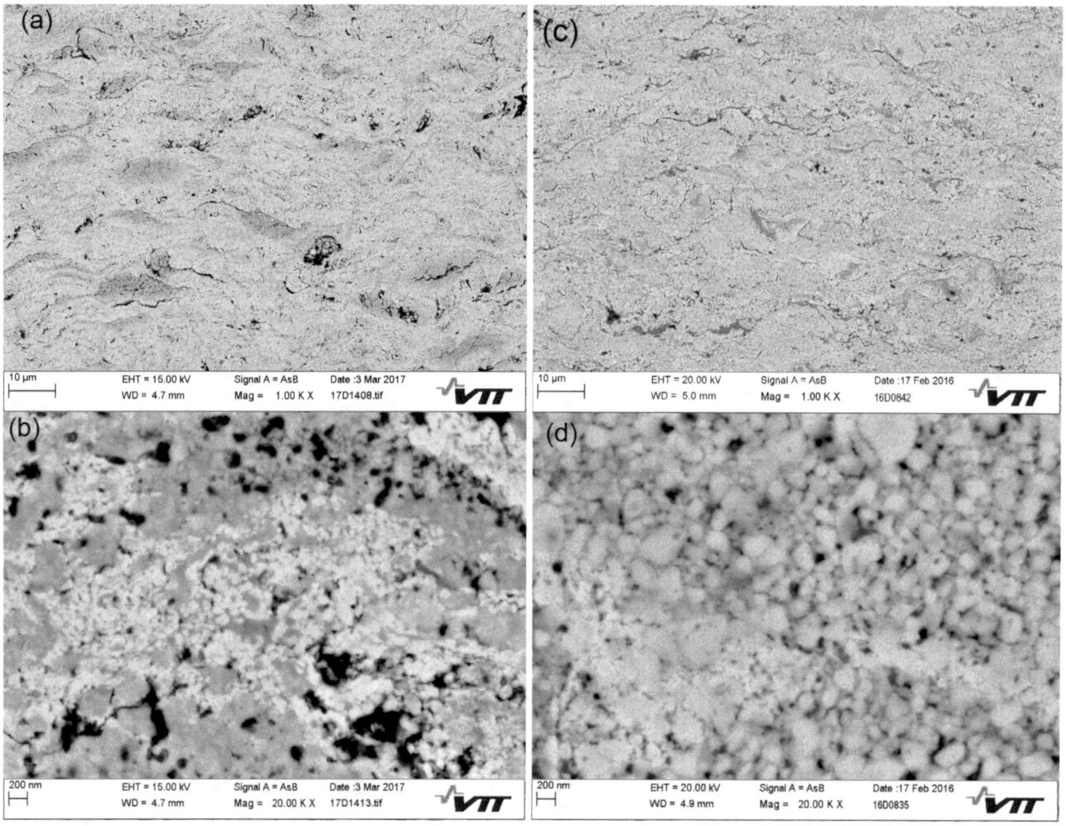

Fig. 4. SEM microstructures of the HVAF-coatings of the experimental (a-b) and reference MBN (c-d) powders .

Figure 5 shows the XRD patterns of the HVAF-sprayed experimental WC-12Co and MBN reference powders to illustrate the phase composition of the coatings. Besides the WC phase, W, Co_3W_3C and W_2C were present in both coatings. Li and Yang (2013) suggest in their study [1] that a broad hump-like peak at 2θ of around 43° is usually observed in the XRD pattern of WC–Co coating as the binder phase, being a Co–W–C ternary alloy, appearing as amourphous phase. Because Co is the main element in the Co-W-C ternary phase, the peak appears at the position corresponding to the fcc structure of cobalt. The significant WC decarburization and dissolution into the matrix phase has occurred during the HVAF-spraying of both powders. Both coatings show WC decomposition to metallic tungsten. Decomposition to metallic tungsten has been more severe in a case of the experimental powder, most probably due to its higher porosity. From the XRD data, the average carbide crystallite size of the coatings were determined to be in nanosize range, corresponding 18 and 30 nm for the MBN reference and experimental powders, respectively.

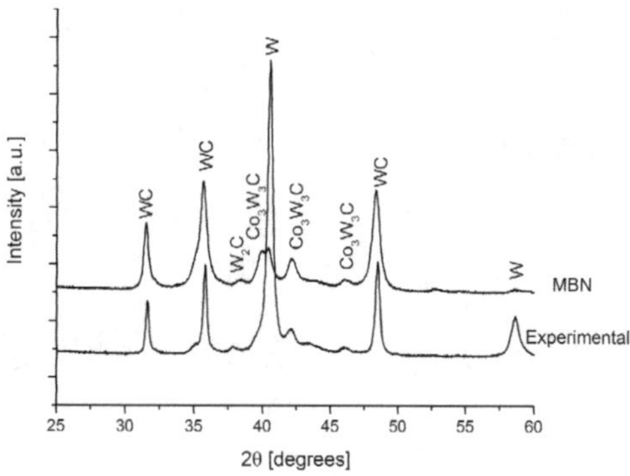

Fig 5. XRD patterns of the HVAF coatings.

In literature [15,16], it was proposed one solution to achieve a dense coating, a multimodal WC–Co powder consisting of both nanostructured and microstructured WC particles. Hardness values of the coatings are presented in Table 3.

Table 3. Hardness values of the coatings.

Powders	Hardness, HV 0.3kg
Experimental	1000±32
MBN	1350±29

4 Conclusions

In the present study, an experimental nanostructural WC-12Co powder was synthesized using a novel powder manufacturing method based on the use of water soluble raw materials. The novel powder manufacturing route produces a powder with the correct WC-Co phase structure and nanosized carbides. Produced powder has more spherical particle shape in comparison to the reference powders that is considered beneficial for the thermal spraying in respect of consistent powder flow. Powder usability and functionality was tested with a HVAF-coating deposition. Microstructure analysis of the coatings revealed nanocarbide structure. The significant WC decarburization and the dissolution into the matrix phase occurred during the HVAF-spraying. In addition, decomposition of WC to metallic tungsten took place during the the HVAF-spraying. To decrease the WC decarburization and decomposition during spraying, the future work concentrates on densification of the porous particle structure by optimizing the spray drying suspension. Also the possibilities of colder deposition techniques will be studied.

Acknowledgment

The authors would like to thank the Academy of Finland for financial support and Ms Hanna-Mari Sinilehto and Mr Jarkko Metsäjoki for participating to experimental work and characterisation.

References

[1] Li, C-J and Yang G-J. Relationships between feedstock structure, particle parameter, coating deposition,microstructure and properties for thermally sprayed conventional and nanostructured WC–Co. Int. Journal of Refractory Metals and Hard Materials 39 (2013), pp 2–17.

[2] Jia K, Fischer TE. Sliding wear of conventional and nanostructured cemented carbides. Wear 203–204 (1997), pp. 310–8

[3] Jia K, Fischer TE, Gallois B. Microstructure, hardness and toughness of nanostructured and conventional WC–Co composites. Nanostruct Mater 10 (1998), pp. 875–94.

[4] Liu, S., Huang, Z.-L., Liu, G.Yang, G.-B. Preparing nano-crystalline rare earth doped WC/Co powder by high energy ball milling, Int. J. Refract. Met. Hard Mater. 24 (2006), pp. 461–464.

[5] Mi, S., Courtney, T.H. Synthesis of WC and WC-Co cermets by mechanical alloying and subsequent hot isostatic pressing, Scr. Mater. 38 (1997), pp. 171–176.

6] Samokhin, A.V., Alekseev, N.V., Kornev, S.A., Sinaiskii, M.A., Blagoveschenskiy, Y.V., Kolesnikov, A.V. Tungsten carbide and vanadium carbide nanopowders synthesis in DC plasma reactor, Plasma Chem. Plasma Process. 33 (2013), pp. 605–616.

[7] Schreuders, C., Leparoux, M., Shin, J.-W., Dvorak, M. and Siegmann, S. Nanosized tungsten carbide powder produced by thermal plasma. International Congress on Particle Technology 2004, Proceedings, Nürnberg, Germany 2004.

[8] Kim, J.C. and Kim, B.K. Synthesis of nanosized tungsten carbide powder by the chemical vapor condensation process, Scr. Mater. 50 (2004), pp. 969–972.

[9] Singh, H. and Pandey, O.P. Direct synthesis of nano crystalline tungsten carbide from scheelite ore by solid state reaction method, Ceram. Int. 39 (2013), pp. 785–790.

[10] Fang, Z.Z., Wang, X., Ryu, T., Hwang, K.S. and Sohn H.Y. Synthesis, sintering, and mechanical properties of nanocrystalline cemented tungsten carbide — a review, Int. J. Refract. Met. Hard Mater. 27 (2009), pp. 288–299

[11] Jin, Y., Liu, D., Li, X. and Yang, R. Synthesis of WC nanopowders from novel precursors, Int. J. Refract. Met. Hard Mater. 29 (2011), pp. 372–375.

[12] Lina, H., Tao, B., Xiong, J., Li, Q. and Li, Y. Tungsten carbide (WC) nanopowders synthesized via novel core–shell structured precursors, Ceram. Int. 39 (2013), pp. 2877–2881.

[13] Kanerva, U., Lagerbom, J. Karhu, M. Kronlöf, A., Laitinen, T. and Turunen, E. Synthesis of nano-WC from water soluble raw materials: Effects of tungsten source and synthesis atmosphere on chemical and phase structure evolution. Int. Journal of Refractory Metals and Hard Materials 50 (2015), pp 65–71.

[14] Kanerva, U., Karhu, M., Lagerbom, J., Kronlöf, A., Honkanen, M., Turunen, M. and Laitinen T. Chemical synthesis of WC–Co from water-soluble precursors: The effect of carbon and cobalt additions to WC synthesis. Int. Journal of Refractory Metals and Hard Materials 56 (2016) pp. 69–75

[15] Skandan G, Yao R, Sadangi R, Kear BH, Qiao Y, Liu L, et al. Multimodal coatings: a new concept in thermal spraying. J Therm Spray Technol 9 (2000), pp. 329–31.

[16]] Skandan G, Yao R, Kear BH, Qiao Y, Liu L, Fischer TE. Multimodal powder: a new class of feedstock material for thermal spraying of hard coatings. Scr Mater 4 (2001), pp.1699–702.

Isothermal oxidation behavior of HVAF-sprayed NiCoCrAlY coatings: Effect of surface treatment

Pimin Zhang*1, Esmaeil Sadeghimeresht[2], Ru Lin Peng[1], Xin-Hai Li[3], Nicolaie Markocsan[2], Shrikant Joshi[2], Sten Johansson[1]
1 Department of Management and Engineering, Linköping University, SE-58183 Linköping, Sweden
2 Department of Engineering Science, University West, SE-46153 Trollhättan, Sweden
3 Siemens Industrial Turbomachinery AB, SE-61283 Finspång, Sweden
*Corresponding author: pimin.zhang@liu.se, esmaeil.sadeghimeresht@hv.se

Abstract: NiCoCrAlY coatings are widely used as bond coats for ceramic thermal barrier coatings (TBCs) and oxidation and corrosion protective overlay coatings in industrial gas turbines. High temperature oxidation behaviour of NiCoCrAlYs has a great influence on the coating performance and lifetime of TBCs. A promising route to decrease the oxidation rate of such coatings is post-coating surface modification which can facilitate formation of a uniform alumina scale with a considerably slower growth rate compared to the as-sprayed coatings. In this work, the effect of surface treatment by means of shot peening and laser surface melting (LSM) on the oxidation resistance of high velocity air-fuel (HVAF) sprayed NiCoCrAlY coatings was studied. Isothermal oxidation was carried out at 1000 ^{0}C for 1000h. Results showed that the rough surface of as-sprayed HVAF sprayed coatings was significantly changed after shot peening and LSM treatment, with a compact and smooth appearance. After the exposure, the oxide scales formed on surface-treated NiCoCrAlY coatings showed different morphology and growth rate compared to those formed on as-sprayed coating surface. The oxidation behaviour of surface treated HVAF-sprayed NiCoCrAlY coatings were revealed and discussed.
Keywords: Surface Treatment; Shot Peening; Laser Surface Melting (LSM); High Velocity Air-Fuel (HVAF); NiCoCrAlY Coating

1 Introduction

Thermal barrier coatings (TBCs) have been widely used to protect metallic components of hot section in advanced gas turbines [1,2]. The use of TBCs provides a temperature reduction of 100-300 ^{o}C to the underlying superalloys, which enables operation of modern gas turbines at temperatures above the melting point of superalloy (~1300 ^{o}C), thereby enhancing the engine performance [3]. The TBC life depends on the quality of a stable, dense, thin, continuous, slow-growing and adherent thermally grown α-Al_2O_3 layer on a MCrAlY (M=Ni and/or Co) bond coat which offers protection against oxidation [4–6]. The oxide scale also acts as a diffusion barrier to suppress formation of other detrimental oxides and improve the durability of the system under oxidation conditions [7]. However, other oxides such as mixed chromia-alumina ($(Cr,Al)_2O_3$), spinel ($(Ni,Co)(Cr,Al)_2O_4$) and nickel oxide (NiO), may form in the thermally grown oxide (TGO). It has been reported that, when spinels form, they act as preferred sites for TBC failure [8]. The spinels are detrimental oxides which may promote crack nucleation during thermal exposure, leading to premature TBC failure [7,9,10]. The formation of a protective alumina scale depends on a number of factors including the surface condition prior to oxidation exposure. Post deposition treatments that modify the surface condition may, therefore, may affect the oxidation behavior of the treated coatings.

Surface mechanical treatments such as shot peening are commonly used for machined components to increase lifetime by introducing compressive stress and strain hardening in a surface layer [11]. When being used as post deposition treatment on NiCoCrAlY coatings, shot peening can improve the oxidation resistance of bond coats by reducing coating porosity and surface roughness, along with increasing the number of paths for diffusion of the protective-scale forming elements [12–14].

Laser surface melting (LSM) has also been observed to have beneficial effects on improving the oxidation resistance of Vacuum Plasma Sprayed (VPS), Cold Sprayed, and High Velocity Oxy-Fuel (HVOF) sprayed MCrAlY coatings [15–17]. The rapid remelting of the surface by high energy density laser irradiation and its subsequent fast solidification after laser irradiation produces a zone below the alloy surface with a fine grained and homogeneous microstructure [15,18,19]. The fine grains of LSM treated NiCoCrAlY coating aids the fast formation of protective α-Al_2O_3 oxide layer, while supressing the formation of other oxides [20].

High velocity air-fuel (HVAF) spray process takes advantages of high kinetic and low thermal energy of the jet stream, which results in dense coating with low oxide content [21]. Although the coating lifetime is enhanced by HVAF, it could be further increased by adding some extra steps such as post deposition treatments. However, the effect of post treatment on the oxidation behaviour of HVAF-sprayed NiCoCrAlY coatings has not been reported. In this study, the oxidation of NiCoCrAlY coating with different surface modification techniques, i.e., shot peening and LSM, was examined.

2 Experimental

Commercially available NiCoCrAlY (AMDRY 386 wt%: 45.1Ni-23.4Co-20.4Cr-10.5Al-0.6Y) powder supplied by Oerlikon Metco was used as feedstock for the deposition of the NiCoCrAlY coating on Hastelloy® X (wt%: 22.1Cr, 17.9Fe, 8.9Mo, 1.6Co,

0.5W, 0.1C, 0.9Mn, 0.8Si and Ni balance) superalloy substrate with size of Φ25.4 x 6 mm. The powder particles size ranged from 15 to 40 μm and the density was around 7.3 g/cm^3. Details of the powder have been described in [22]. A HVAF system (Uniquecoat M3™) was used to spray the NiCoCrAlY coating with the parameters presented in Table 1. Prior to spraying, the substrates were grit blasted using the same HVAF spray gun with F220 alumina powder (mesh 220) to obtain a clean surface with desired roughness.

Table 1 Parameters used for HVAF spraying of NiCoCrAlY coating

Parameter	Value
Fuel	Propane
Nozzle	3L2
Spraying distance (mm)	350
Gun traverse velocity (m/min)	100
Powder feed rate (g/min)	150
Nitrogen carrier gas flow (1/min)	60
Air pressure (MPa)	0.80
Fuel 1 pressure propane (MPa)	0.66
Fuel 2 pressure propane (MPa)	0.70

After deposition, one set of samples was shot peened using spherical glass shots (300 μm in diameter) with 150 to 200% surface coverage and a shot peening intensity of 0.2 to 0.3 mmA. The working pressure was 0.3 MPa and the working distance was 150 mm. The surface roughness was checked using a stylus-based profilometer (Surftest 301, Mitutoyo, Japan). Three measurements in three different directions were performed and the measured values were averaged. The parameter R_a was employed to characterize the surface roughness.

Another set of samples was treated by laser surface melting (LSM) using a 6 kW YAG fiber laser which reproduced a laser spot of 0.8 mm in diameter at the substrate. Processing parameters for all samples were 400 W laser power and 25 mm/s scanning speed. Argon was used as shielding gas with a flow rate of 11.5 L/min.

Three sets of samples including: as-sprayed, shot peened and LSM treated NiCoCrAlY coated specimens were subjected to oxidation at 1000 °C in ambient air for 1000 h. The morphologies of the coating surface and cross sections before and after oxidation test were analysed in a scanning electron microscope (SEM) Hitachi SU70 equipped with energy dispersive X-ray spectrometry (EDS). Phase identification was performed using X-ray diffraction (XRD) with CuK_α radiation in a XPERT-PRO X-ray diffractometer.

3 Results and discussion

3.1 Characterization of surface topography

Fig. 1 shows the surface morphology of as-sprayed, shot-peened and LSM treated NiCoCrAlY coating. A typical as-sprayed NiCoCrAlY coating has a roughness of Ra= 4.7±0.6 μm and is therefore, channel linked by "peak" and "troughs" with lots of portuberances, as shown in Fig. 1(b). The near spherical unmelted particles with similar size as the original powder can be observed on the coating surface. However, after shot peening, the unmelted particles that weakly bonded to the coating surface and protuberances were removed completely, presenting a smooth, compact and uniform surface, as showed in Fig. 1(c) and 1(d). The surface roughness was reduced from 4.7±0.6 μm in the as-sprayed state to 3.8±0.3 μm in shot-peened state.

After LSM treated, as shown in Fig. 1(e), the coating surface was remelted completely, and the entire surface consisted of smooth melting paths with some bulged nodules in the gap between melting paths and few bugled nodules dispersed on the paths. The width of melting path is the same as the laser spot size. The magnified morphology of the melting path in Fig. 1(f) shows a smooth surface with long columnar dendrites and interdendritic porosity. The shrinkage interdendritic microporosity is well known in aluminium casting alloys. The formation of primary dendrites causes a metal shrinkage, then the empty volume of metal has to be compensated by the flow of interdendritic liquid that last solidify. When the liquid flow is unable to cope with the cavities, pores are formed along the interdendritic spaces [23,24].

The XRD patterns of NiCoCrAlY coating of as-spayed, shot-peened and LSM treated are shown in Fig. 2. The as-sprayed NiCoCrAlY coating was mainly composed of β and γ phase. After shot peening treatment, the broadening of γ and β phase peaks were observed indicating refinement of grain size and the atomic-level strain introduced during shot peening process [17,25].

After LSM treatment, β and γ phases and $YAlO_3$ were detected. A layer of Y-rich oxide was observed on bugled nodules due to the selective oxidation of Y on LSM treated NiCoCrAlY coating surface, as shown in Fig. 1(e). The EDS results also confirmed the presence of $YAlO_3$ on the dispersed bulged nodules, see Table 2. It's well accepted that Y-rich oxides in NiCoCrAlY coatings have a significant effect on the growth of alumina and its adherence to bulk bond coat [26]. Thus, the improvement of the TBCs lifetime would be expected [27,28]. Compared with as-sprayed coating, the intensity of γ phase peak in LSM treated coating significantly increased, however β phase content decreased. It can be due to the composition difference between the individual phases for the original and LSM treated coatings. According to [29], atoms need time to arrange themselves such that equilibrium prevails at the solid/liquid interfaces, since the high solidification rates achieved by LSM inhibit the full partition of the alloy elements between the γ and β phases [30], these difference in phase composition and phase fraction within LSM coating are expected. It is worth noting that both β and γ peaks became narrower, which could be owing to that the melting and solidification of LSM treatment on coating reduced the coating defects and residual stresses induced during HVAF spraying process.

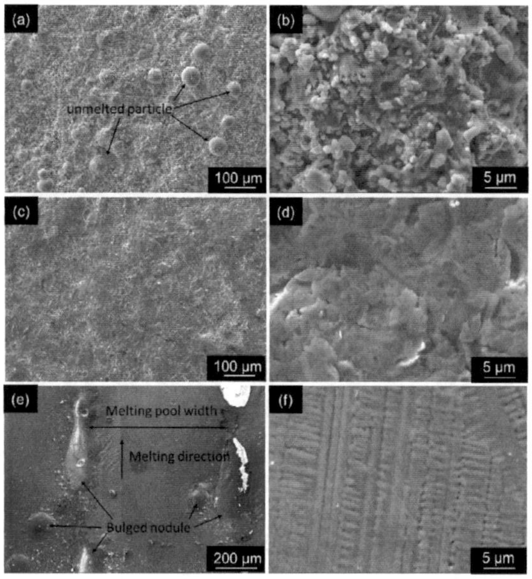

Figure 1. Surface morphology of as-sprayed (a), shot-peened (c) and LSM treated (e) HVAF NiCoCrAlY coatings, (b) higher magnification of (a), (d) higher magnification of (c), (f) higher magnification of (e). Note that (e) is taken at a different magnification to show the full view of one melting path.

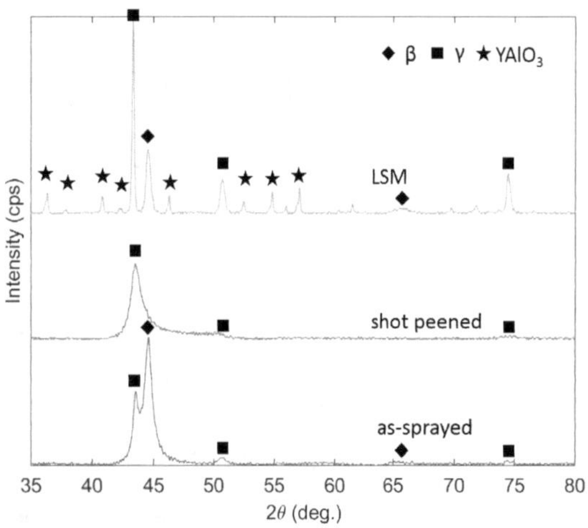

Figure 2. XRD pattern of as-sprayed, shot-peened and LSM treated HVAF NiCoCrAlY coating.

Table 2. EDS results of point A and B in Fig. 1(e), (f).

wt %	Ni	Co	Cr	Al	Y	O
A	-	-	-	23.1	37.5	39.4
B	51	19.8	18.6	10.6	-	-

3.2 Oxidation of coating

3.2.1 Surface morphology

After oxidation test at 1000 ^{0}C in air for 1000 h, differences in oxide morphology between as-sprayed, shot-peened, and LSM treated coating were clearly observed, see Fig. 3. The scale formed on the as-sprayed coating was composed of large spinels protuberance with underlying Al_2O_3 layer and some large unmelted spherical particles, while that of the shot-peened and LSM treated coatings consisted of mainly Al_2O_3 with very little amounts of mixed spinels, which was also confirmed by XRD pattern (Fig 4) and EDS results (Table 3). The significantly rougher structure of as-sprayed coating surface with large surface area promotes Al consumption. Yang et al [31] reported that a good bonding feature of the surface particles to the bulk bond coat increased the diffusion of Al from bulk coating during oxidation, which reduced the formation of spinels. The protuberances with poor bonding feature to bulk coating on as-sprayed coating enhance the formation of spinels, as shown in Fig. 1(a) and (b). Thus, the formation of a large amount of spinels on as-sprayed coating surface is facilitated. Spinels are detrimental oxides; it has been generally believed that spinels, when formed, serve as preferred sites for TBC failure [8]. A rapid increase in local volume [32] accompanying the formation of spinels may promote crack nucleation during thermal exposure, leading to premature TBC failure [7,9,10]. Since no spinels were observed on shot-peened or LSM treated coating surface, the oxide scale on shot-peened and LSM treated coating surface formed a better oxidation barrier, which is expected to improve TBC durability.

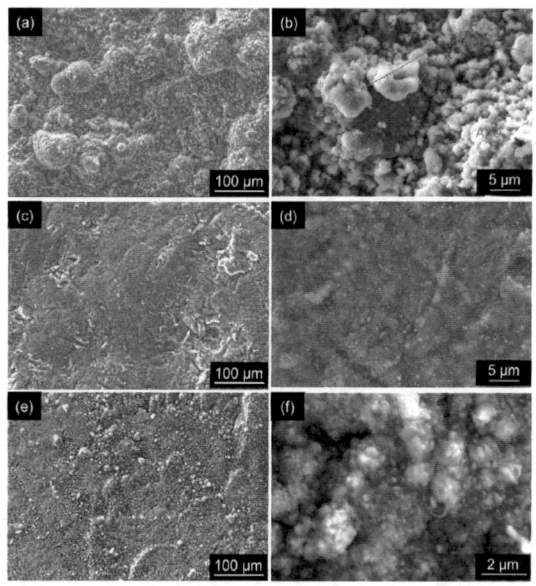

Figure 3. Surface morphology after 1000 h oxidation at 1000 ºC in lab air of as-sprayed (a), shot-peened (c) and LSM treated (e) NiCoCrAlY coatings, (b) higher magnification of (a), (d) higher magnification of (c), (f) higher magnification of (e).

Table 3. EDS results of as-sprayed, shot-peened and LSM treated coatings after 1000 h oxidation at 1000 °C in Fig. 3.

Wt. %		Ni	Co	Cr	Al	O
As-sprayed	Spinel	20.8	11.8	6.1	37	25.9
	Al_2O_3	6.5	3.9	3.2	48.2	37.9
Shot-peened		-	-	-	51.9	48.1
LSM treated		3.6	0.5	2.4	41.4	32.1

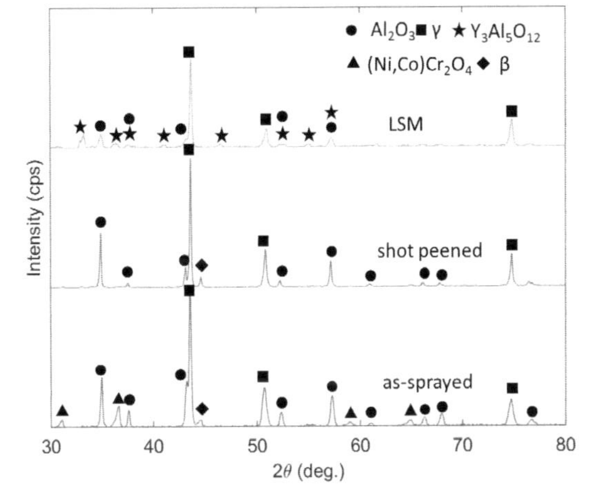

Figure 4. XRD pattern of 1000 h oxidation at 1000 °C of as-sprayed, shot-peened and LSM treated HVAF NiCoCrAlY coating.

3.2.2 Microstructural characterization

Fig .5 shows the cross sectional microstructure of the as-sprayed, shot-peened and LSM treated coating after 1000 h oxidation at 1000 °C. A non-uniform oxide layer grew on the as-sprayed coating surface after oxidation. From the EDS results in Fig. 5(e), the oxide scale formed on as-sprayed coating surface consisted of spinels outer layer and α-Al_2O_3 inner layer, which has also been confirmed by XRD in Fig. 4. Meanwhile, a layer of uniform α-Al_2O_3 formed on shot-peened and melting paths of LSM treated coating surface. The thickness of α-Al_2O_3 layer was measured using image analysis software, see Table 4. Apparently, the oxide scales that formed on shot-peened and LSM treated coatings are thinner than the one grown on top of the as-sprayed coating. Evidently, both shot peening and LSM treatment significantly retard the TGO growth during improved the oxidation behaviour of the HVAF sprayed NiCoCrAlY coating. Furthermore, A lower TGO growth rate can be beneficial for the TBC's durability [8].

In addition, a different oxide layer was observed on the bulged nodules which dispersed on LSM treated coating, see Fig. 5(d) and EDS result of oxide scale in Fig. 6(f). Clearly, the bulged nodule consisted of an outer Y-rich oxide layer and an inner α-Al_2O_3 layer. The distribution of the Y-rich oxide precipitates formed during manufacturing has been proved to affect the growth rate and mechanical stability of the alumina scales during service. Sloof et al. [33] reported that oxide scales with a regularly distributed and larger amount of small Y-oxide inclusions grow faster. The selective oxidation of Y in the outer layer of bulged nodules reduces its availability in the rest of the bond coat. Consequently, the effect of Y on promoting Alumina growth is less profound in such regions of the LSM treated coating. Therefore, the Alumina scale grown on LSM treated coating is thinner than in the as-sprayed and shot-peened coating.

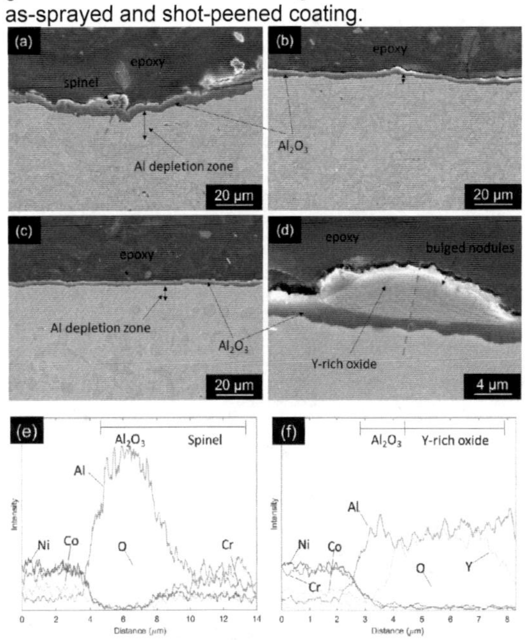

Figure 5. Cross-sectional microstructures of as sprayed (a), shot peened (b), LSM treated (c) coating and high magnified section of Y-rich oxide outer layer on LSM treated coating (d). Distribution of major elements in oxide layer by EDS line profile (e), (f) corresponding to the dot line marked in as-sprayed coating (a) and LSM treated coating (f).

Table 4. Measured thickness of α-Al_2O_3 layer in as-sprayed, shot-peened and LSM treated coating after 1000 h oxidation at 1000 °C using image analysis software.

Sample	Thickness of Al_2O_3 (μm)
As-sprayed	~4
Shot-peened	~3
LSM treated	~2

3.2.3 Spallation of TGO

Thermal stresses developed during cooling from the oxidation temperature may cause local TGO spallation in coatings. γ phase islands surrounded by α-Al_2O_3 were observed on peaks of the as-spayed coating after 1000 h of 1000 °C oxidation test, as shown in Fig. 6. The TGO spallation was mainly due to thermal stress induced by the large coefficient of thermal expansion (CTE) mismatch between TGO and NiCoCrAlY coating upon thermal gradient. The fast grown TGO formed on as-sprayed coating partially spalled off from the peaks while no spallation was observed on either shot-peened or LSM treated coatings. It has been reported that the spallation of TGO depends strongly on its thickness in overlay MCrAlY coating [34,35], the slow growth of TGO on surface treated coatings is beneficial against its spallation. Besides, the roughness of coating surface also affect TGO growth and adherence on coating. Compared to the troughs on coatings, the reduction of the Y incorporation at the peaks in as-sprayed coating may contribute to the frequently observed earlier scale spallation in peaks [26]. Low roughness (3.8±0.3 μm) of shot-peened coating surface and smooth surface of LSM treated surface reduce uneven distribution of Y-Al oxide pegs between peaks and troughs, thus, decrease the risk of TGO spallation from peaks.

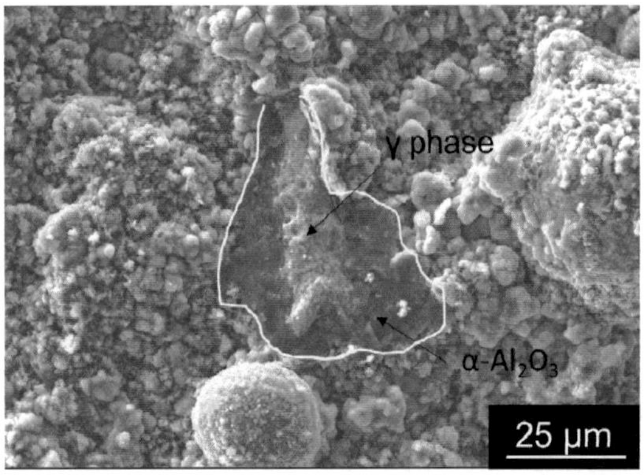

Figure 6. TGO spallation on as-sprayed coating after 1000 h oxidation at 1000 °C. γ phase region marked with dot line, α-Al_2O_3 cross section marked with solid line.

4 Conclusion

NiCoCrAlY coatings were deposited by HVAF spraying and then surface treated with shot peening and laser surface melting respectively. Both shot peening and laser surface melting treatments on as-sprayed coating surface significantly improve the oxidation resistance of the HVAF sprayed NiCoCrAlY coatings.

The results showed that shot peening treatment produces a compact, smooth uniform surface instead of a very rough surface with dispersed unmelted powders as in the as-sprayed coating. The surface roughness of the shot-peened coating was significantly reduced from 4.7±0.6 μm at the as-sprayed state to 3.8±0.3 μm at shot-peened state. After oxidation at 1000 °C for 1000 h, spinels and α-Al_2O_3 mixed oxide formed on as-sprayed coating surface, in contrast, a dense protective α-Al_2O_3 layer formed on the shot-peened coating surface without detrimental oxides. Besides, the redistribution of Y-rich oxides contributes to the slow growth of TGO.

The laser surface melting treated coating surface shows a different morphology: smooth melting paths with some bulged nodules in the gap between melting paths and few bugled nodules dispersed on the paths. After the thermal exposure at 1000 °C for 1000 h, only α-Al_2O_3 were observed on LSM treated surface and the low growth rate of α-Al_2O_3 oxide are due to the early selective oxidation of Y-rich oxide on the bugled nodules after the LSM treatment.

5 Acknowledgement

The author would like to thank the Siemens Industrial Turbomachinery AB (Finspång, Sweden) and Swedish Energy Agency through KME consortium - ELFORSK for their financial support. The research school of Agora Materials and Strategic Faculty Grant SFO-MAT-LiU#2009-00971 and project teams at Linköping University are also acknowledged for their support on this research. Mr. Olle Widman from Curtiss-Wright Surface Technologies AB is appreciately acknowledged for shot peening the coatings.

6 References

[1] R.A. Miller, Current status of thermal barrier coatings — An overview, Surf. Coatings Technol. 30 (1987) 1–11. doi:10.1016/0257-8972(87)90003-X.

[2] X.Q. Cao, R. Vassen, D. Stoever, Ceramic materials for thermal barrier coatings, J. Eur. Ceram. Soc. 24 (2004) 1–10. doi:10.1016/S0955-2219(03)00129-8.

[3] N.P. Padture, M. Gell, E.H. Jordan, Thermal Barrier Coatings for Gas-Turbine Engine Applications, Science (80-.). 296 (2002) 280 LP-284. http://science.sciencemag.org/content/296/5566/280.abstract.

[4] M. Gell, J. Eric, V. Krishnakumar, K. McCarron, B. Barber, Y.-H. Sohn, et al., Bond strength, bond stress and spallation mechanisms of thermal barrier coatings, Surf. Coatings Technol. 120 (1999) 53–60. doi:10.1016/S0257-8972(99)00338-2.

[5] B.A. Pint, I.G. Wright, W.Y. Lee, Y. Zhang, K. Prüßner, K.B. Alexander, Substrate and bond coat compositions: factors affecting alumina scale adhesion, Mater. Sci. Eng. A. 245 (1998) 201–211. doi:10.1016/S0921-5093(97)00851-4.

[6] E.P. Busso, J. Lin, S. Sakurai, M. Nakayama, A mechanistic study of oxidation-induced degradation in a plasma-sprayed thermal barrier coating system.: Part I: model formulation, Acta Mater. 49 (2001) 1515–1528. doi:10.1016/S1359-6454(01)00060-X.

[7] W.R. Chen, X. Wu, B.R. Marple, P.C. Patnaik, The growth and influence of thermally grown oxide in a thermal barrier coating, Surf. Coatings Technol. 201 (2006) 1074–1079. doi:10.1016/j.surfcoat.2006.01.023.

[8] A. Rabiei, A.. Evans, Failure mechanisms associated with the thermally grown oxide in plasma-sprayed thermal barrier coatings, Acta Mater. 48 (2000) 3963–3976. doi:10.1016/S1359-6454(00)00171-3.

[9] W.R. Chen, X. Wu, B.R. Marple, P.C. Patnaik, Oxidation and crack nucleation/growth in an air-plasma-sprayed thermal barrier coating with NiCrAlY bond coat, Surf. Coatings Technol. 197 (2005) 109–115. doi:10.1016/j.surfcoat.2004.06.027.

[10] R. Eriksson, M. Gupta, E. Broitman, K.P. Jonnalagadda, P. Nylén, R. Lin Peng, Stresses and Cracking During Chromia-Spinel-NiO Cluster Formation in TBC Systems, J. Therm. Spray Technol. 24 (2015) 1002–1014. doi:10.1007/s11666-015-0270-y.

[11] S.A. Meguid, G. Shagal, J.C. Stranart, J. Daly, Three-dimensional dynamic finite element analysis of shot-peening induced residual stresses, Finite Elem. Anal. Des. 31 (1999) 179–191. doi:10.1016/S0168-874X(98)00057-2.

[12] Y.. Sohn, J.. Kim, E.. Jordan, M. Gell, Thermal cycling of EB-PVD/MCrAlY thermal barrier coatings: I. Microstructural development and spallation mechanisms, Surf. Coatings Technol. 146–147 (2001) 70–78. doi:10.1016/S0257-8972(01)01478-5.

[13] Y. Li, C.J. Li, Q. Zhang, L.K. Xing, G.J. Yang, Effect of chemical compositions and surface morphologies of MCrAlY coating on its isothermal oxidation behavior, J. Therm. Spray Technol. 20 (2011) 121–131. doi:10.1007/s11666-010-9590-0.

[14] A. Hesnawi, H. Li, Z. Zhou, S. Gong, H. Xu, Isothermal oxidation behaviour of EB-PVD MCrAlY bond coat, Vacuum. 81 (2007) 947–952. doi:10.1016/j.vacuum.2006.03.024.

[15] R. Streiff, M. Pons, P. Mazars, Laser-induced microstructural modifications in a vacuum plasma sprayed NiCoCrAlYTa coating, Surf. Coatings Technol. 32 (1987) 85–95. doi:10.1016/0257-8972(87)90099-5.

[16] J. Cizek, M. Matejkova, J. Kouril, J. Cupera, I. Dlouhy, Potential of New-Generation Electron Beam Technology in Interface Modification of Cold and HVOF Sprayed MCrAlY Bond Coats, Adv. Mater. Sci. Eng. 2016 (2016). doi:10.1155/2016/9070468.

[17] Q. Zhang, C.J. Li, C.X. Li, G.J. Yang, S.C. Lui, Study of oxidation behavior of nanostructured NiCrAlY bond coatings deposited by cold spraying, Surf. Coatings Technol. 202 (2008) 3378–3384. doi:10.1016/j.surfcoat.2007.12.028.

[18] K. Nagarathnam, K. Komvopoulos, Microstructural analysis and oxidation behavior of laser-processed Fe-Cr-Al-Y alloy coatings, Metall. Mater. Trans. A. 27 (1996) 381–390.

[19] M.J. Tobar, J.M. Amado, A. Yáñez, J.C. Pereira, V. Amigó, Laser cladding of MCrAlY coatings on stainless steel, Phys. Procedia. 56 (2014) 276–283. doi:10.1016/j.phpro.2014.08.172.

[20] Y.N. Wu, M. Qin, Z.C. Feng, Y. Liang, C. Sun, F.H. Wang, Improved oxidation resistance of NiCrAlY coatings, 2003. doi:10.1016/S0167-577X(02)01244-2.

[21] R.Q. Guo, C. Zhang, Q. Chen, Y. Yang, N. Li, L. Liu, Study of structure and corrosion resistance of Fe-based amorphous coatings prepared by HVAF and HVOF, Corros. Sci. 53 (2011) 2351–2356. doi:10.1016/j.corsci.2010.12.022.

[22] E. Sadeghimeresht, H. Hooshyar, N. Markocsan, S. Joshi, P. Nylen, Oxidation Behavior of HVAF-Sprayed NiCoCrAlY Coating in H2-H2O Environment, Oxid. Met. 86 (2016) 299–314. doi:10.1007/s11085-016-9637-9.

[23] M.K. Surappa, E. Blank, J.C. Jaquet, Effect of macro-porosity on the strength and ductility of cast Al□7Si□0.3Mg alloy, Scr. Metall. 20 (1986) 1281–1286. doi:10.1016/0036-9748(86)90049-9.

[24] L. Thijs, K. Kempen, J.-P. Kruth, J. Van Humbeeck, Fine-structured aluminium products with controllable texture by selective laser melting of pre-alloyed AlSi10Mg powder, Acta Mater. 61 (2013) 1809–1819. doi:10.1016/j.actamat.2012.11.052.

[25] J. He, J.M. Schoenung, Nanostructured coatings, Mater. Sci. Eng. A. 336 (2002) 274–319. doi:10.1016/S0921-5093(01)01986-4.

[26] A. Gil, V. Shemet, R. Vassen, M. Subanovic, J. Toscano, D. Naumenko, et al., Effect of surface condition on the oxidation behaviour of MCrAlY coatings, Surf. Coatings Technol. 201 (2006) 3824–3828. doi:10.1016/j.surfcoat.2006.07.252.

[27] A. Gil, D. Naumenko, R. Vassen, J. Toscano, M. Subanovic, L. Singheiser, et al., Y-rich oxide distribution in plasma sprayed MCrAlY-coatings studied by SEM with a cathodoluminescence detector and Raman spectroscopy, Surf. Coatings Technol. 204 (2009) 531–538. doi:10.1016/j.surfcoat.2009.08.034.

[28] J. Zou, K. Zhang, C. Dong, Y. Qin, S. Hao, T. Grosdidier, et al., Selective surface purification via crater eruption under pulsed electron beam irradiation Selective surface purification via crater eruption under pulsed electron beam irradiation, Appl. Phys. Lett. 41913 (2009) 50–53. doi:10.1063/1.2234306.

[29] W. Kurz, D.J. Fisher, Fundamentals of solidification, 1986, Trans Tech Publ. Switz. (n.d.).

[30] C. Kwakernaak, T.J. Nijdam, W.G. Sloof, Microstructure Refinement of NiCoCrAlY Alloys by Laser Surface Melting, 37 (2006) 695–703.

[31] G.J. Yang, X.D. Xiang, L.K. Xing, D.J. Li, C.J. Li, C.X. Li, Isothermal oxidation behavior of nicocraltay coating deposited by high velocity air-fuel spraying, J. Therm. Spray Technol. 21 (2012) 391–399. doi:10.1007/s11666-012-9740-7.

[32] C.H. Lee, H.K. Kim, H.S. Choi, H.S. Ahn, Phase transformation and bond coat oxidation behavior of plasma-sprayed zirconia thermal barrier coating, Surf. Coatings Technol. 124 (2000) 1–12. doi:10.1016/S0257-8972(99)00517-4.

[33] W.G. Sloof, T.J. Nijdam, On the high-temperature oxidation of MCrAlY coatings, Int. J. Mater. Res. 100 (2009) 1318–1330.

[34] H.E. Evans, M.P. Taylor, Creep relaxation and the spallation of oxide layers, Surf. Coatings Technol. 94–95 (1997) 27–33. doi:10.1016/S0257-8972(97)00471-4.

[35] H.E. Evans, Stress effects in high temperature oxidation of metals, Int. Mater. Rev. 40 (2013) 1–40. doi:10.1179/imr.1995.40.1.1.

Heat treatment of the thermally sprayed coating system NiCrSi/NiCoCrAlY/Al for repair brazing high pressure turbine blades

M. Nicolaus, K. Möhwald, H. J. Maier, Leibniz Universität Hannover, Garbsen

Depending on the size and type defects of nickel-based alloy turbine blades two procedures are used mainly: cladding and high temperature brazing. The repair brazing of turbine blades is used to regenerate cracks and surface defects and is the focus of this work. In this contribution a two stage hybrid repair brazing process is presented which allows reducing the current process chain for repair brazing turbine blades. In the first stage of this process the filler metal (NiCrSi) then the hot gas corrosion protective coating (NiCoCrAlY) and finally the aluminium are applied in this order by atmospheric plasma spraying. In the second stage of this hybrid technology the applied coating system undergoes a heat treatment in which brazing and aluminising are combined. The temperature-time regime has an influence on the microstructure of the coating which is investigated in this work.

1 Introduction

Components in aero engines and stationary gas turbines like turbine blades (vanes and air foils) as well as compressors are exposed to extreme conditions. Due to high temperatures and pressure and also foreign object impacts such components have different defects. This includes deformation, wear by means of hot gas and CMAS corrosion, impact damages and cracks. The consequences can be fatal und high costs arise. To increase the lifecycle of such components maintenance, repair and overhaul (MRO) are playing an increasingly significant role due to increasing manufacturing costs.

For repairing high pressure turbine blades which are made of nickel-based alloys and which are used in aviation and power plant technology, brazing is established besides welding and cladding [1 – 4]. The sequence is described in [5]:

1. decoating of the damaged turbine blade to reveal the substrate material
2. fluoride ion cleaning (FIC)
3. manual application of the filler metal
4. brazing under high vacuum
5. near net shape machining.

For the protection against hot gas corrosion the brazed turbine blade is coated with a MCrAlY-coating (M =nickel and/or cobalt) which is mainly applied by electron beam vapour deposition (EB-PVD) or vacuum plasma spraying. But other thermal spraying techniques like high velocity oxygen fuel spraying (HVOF) and atmospheric plasma spraying (APS) are used increasingly as well, because of a continuous development due to lower oxide coatings. To raise the oxidation resistance of the MCrAlY-coating (especially the APS process), the brazed and coated turbine blade is aluminised. The formation of the β-phase (NiAl) is responsible for the oxidation resistance [6]. The aluminising is usually made by a CVD-process (chemical vapour deposition) using a pack cementation consisting of aluminium, alumina and an aluminium halide [7]. Another method is to carry out the process under a hydrogen atmosphere using aluminium chloride ($AlCl_3$) as an activator. Aluminium granules represent the "aluminium source" [8]. A further development due to hot gas corrosion protection are the so called smart overlay coatings [9]. These coatings are graded and have an outer aluminised coating, a chromium rich interlayer and the MCrAlY-coating.

In this contribution it is reported about the development of a joining- and coating hybrid process which allows shortening the repair brazing process of turbine blades. The material used in a high pressure turbine blade, starting with the base material followed by the filler metal and finally the MCrAlY-coating, are nickel-based alloys. These alloys are metallurgical compatible and can be widely used together without any problems. This is especially applied to the filler metal alloys (NiCr(B)Si) and the hot gas corrosion protective coating (NiCoCrAlY). These alloys are manufactured powder metallurgical which suggest a common handling by thermal spraying. Beside the application of the hot gas corrosion protective coating, the application of the filler metal by thermal spraying is described [10]. This established the possibility to apply the filler metal and the hot gas corrosion protective coating in a mutual process using the thermal spray technology.

In earlier works [11, 12] the development of a hybrid technology for repair brazing turbine blades is described. The filler metal (Ni-based alloy Ni650) and the MCrAlY-coating were applied on flat specimens (Inconel 718) by thermal spraying. Subsequently the coated specimens were brazed and aluminised in a common process using a pack cementation. With this hybrid technology it is possible to shorten the process chain for repair brazing turbine blades, **Fig. 1.**

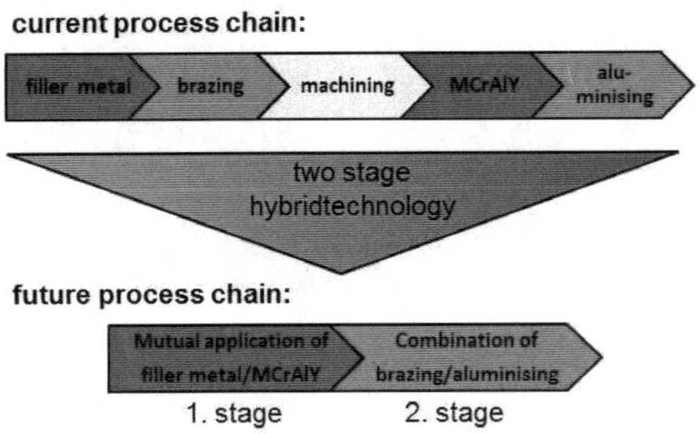

Fig. 1. Process chains for repairing turbine blades

New in this work is that the aluminium is applied by thermal spraying as well. The filler metal (Ni650), then the hot gas corrosion protective coating and finally the aluminium are applied in this order by atmospheric plasma spraying. The coating system is subjected to a common heat treatment and a common brazing and aluminising is carried out. Hereby the separate aluminising process step using neither a pack cementation or the implementation in a hydrogen atmosphere lapses.

2 Experimental methods

2.1 Materials

Inconel 718 flat specimens (40 mm x 30 mm x 1 mm) were used as the base material and were prepared for the coating by corundum blasting using fused alumina (grain size 250 – 355 µm). To remove the residues of the grit the specimens were cleaned in an ultrasonic bath with isopropanol. At first the filler metal (Ni650) was applied with a thickness of approximately 100 µm. Subsequently the application of the hot gas corrosion protective coating was carried out and the thickness was 250 µm. Aluminium was the final layer of the coating system and was approximately 50 µm. The compositions of the used materials are shown in table 1.

Table 1. Composition (in wt.%) of the materials used

Material	**Ni650**	**NiCoCrAlY**	**Inconel 718**
Ni	71	47.5	56.45
Co		23	
Cr	19	17	19
Si	10		
Al		12	0.5
Y		0.5	
Fe			18
Nb + Ta			5.1
Ti			0.95

The liquidus temperature of the filler metal is 1135 °C and the solidus temperature is 1080 °C.
The coating was carried out by atmospheric plasma spraying (APS, Triplex-II torch, Oerlikon Metco) and the deposition parameters are listed in table 2.

Table 2. APS deposition parameters (Triplex-II torch)

Deposition parameter	Ni650, NiCoCrAlY	Al
Ar /L·min^{-1}	35	25
He / L·min^{-1}	25	14
J /A	240	390
Nozzle distance /mm	120	120
Traverse velocity / m·s^{-1}	1.2	1.2
Spray angle /°	90	90
Powder size /µm	–106 + 63	–153 + 53

2.2 Heat treatment

The coated Inconel 718 specimens were heat treated in a heat treatment furnace under high vacuum (approximately 10^{-3} Pa). Three different temperature-time regimes were used:

1. Heating up the specimens to the brazing temperature of 1190 °C with a heat rate of 20 °C/min and the dwell at this temperature was 15 min. After that a free cooling under vacuum was effected. With this heat treatment a simple standard brazing process is carried out.

2. Heating up the specimens to 1190 °C (20 °C/min). After a dwell of 15 min at 1190 °C the temperature was lowered to 1000 °C. At this temperature the dwell was 120 minutes followed by a free cooling. This heat treatment is a typical brazing process with subsequent diffusion annealing and precipitation hardening.

3. Heating up the specimens to 1190 °C and then cooling down to room temperature with a rate of 1 °C/min. This temperature-time regime represents another way of brazing process with diffusion annealing and precipitation hardening.

2.3 Characterisation of the coatings

The microstructures of the brazing/aluminising process were investigated by light optical and electron microscopy. An Axioplan 2 microscope (Zeiss) was used for the light optical microscopic images and field emitter scanning electron microscope (FE-REM SUPRA 40 VP, Zeiss) was used for the determination of the elements in the coatings.

3 Results and discussion

The cross section (SEM backscatter image) of the flat specimen in the as-sprayed state shows the coating with the filler metal (Ni650), the hot gas corrosion-protective layer (NiCoCrAlY) and the aluminium, **Fig. 2**. But the boundary between these materials cannot be clearly detected, **Fig. 2a**. Detailed information about the coating´s microstructure shows the EDX-mapping, **Fig. 2b**.

The coating has a microstructure which is characteristic of atmospheric plasma spraying and has a porosity of about 5%.

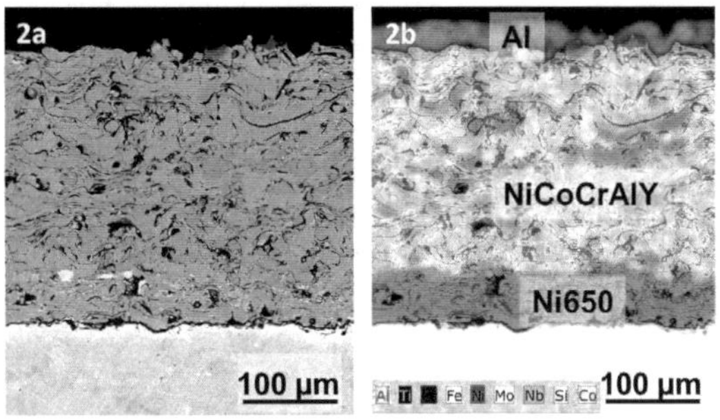

Fig. 2. Cross section of coated flat specimen (as-sprayed)

The result of the coated specimen which is subjected to the first heat treatment as described in chapter 2.2 (heating up to the brazing temperature and a dwell of 15 minutes) can be seen in the cross section, **Fig. 3**.

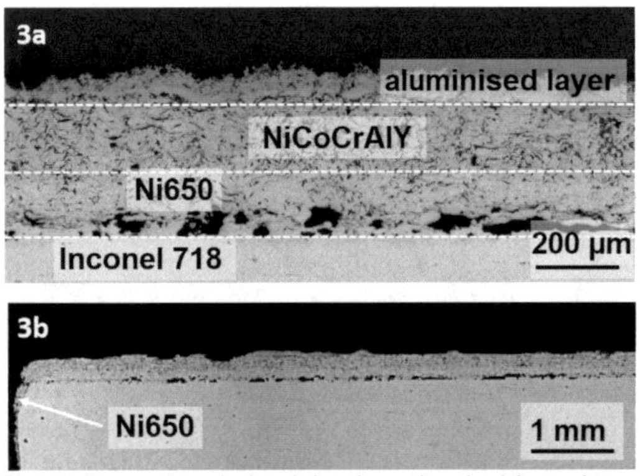

Fig. 3. Brazed and aluminised coating (subjected to heat treatment 1)

The coating can be divided into three areas. Starting from the base material there is the filler metal (Ni650), then the NiCoCrAlY layer which is aluminised on top. It can be seen that the filler metal was molten and that pores or rather cavities are formed in the brazed seam and are distributed evenly over a wide range of the coating, **Fig 3b**. The formation of pores and cavities is caused by infiltration of the filler metal into the porous NiCoCrAlY layer. It can also be observed that the filler metal flowed down the edge of the Inconel 718 specimen. But also diffusion processes due to local concentration differences and segregation effects are responsible for the formation of pores and cavities.

The results are similar to a coated specimen subjected to the second heat treatment, **Fig. 4**. The coating can also be divided into three areas and the formation of pores and cavities in the brazed seam can be observed. But the distribution of the pores/cavities is different to the coating subjected to the first heat treatment.

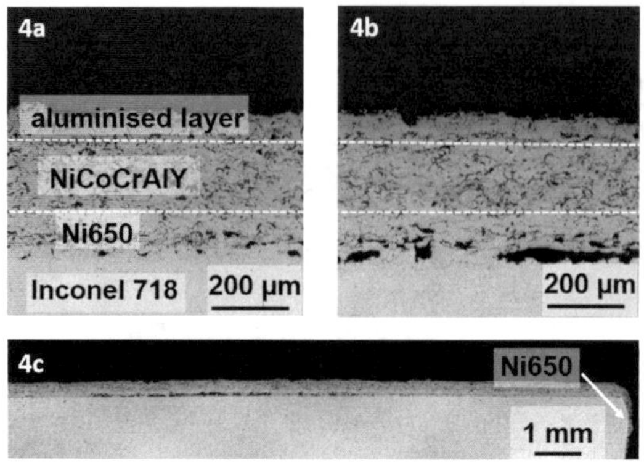

Fig. 4. Brazed and aluminised coating (subjected to heat treatment 2)

In this heat treated coating there are sections which are nearly free of pores, **Fig. 4a** and sections in which pores are formed more frequently, **Fig. 4b**. It can also be seen that the filler metal flowed down the edge of the specimen. This results a non-porous bonding of the filler metal to the MCrAlY-layer in this section. Due to the flow direction of the filler metal to the edge of the specimen and into the MCrAlY-layer there is a significant formation of pores in some sections, respectively, **Fig. 4c**.

The result of the coating which is subjected to the third way of heat treatment is depicted in the light microscopic picture, **Fig. 5.**

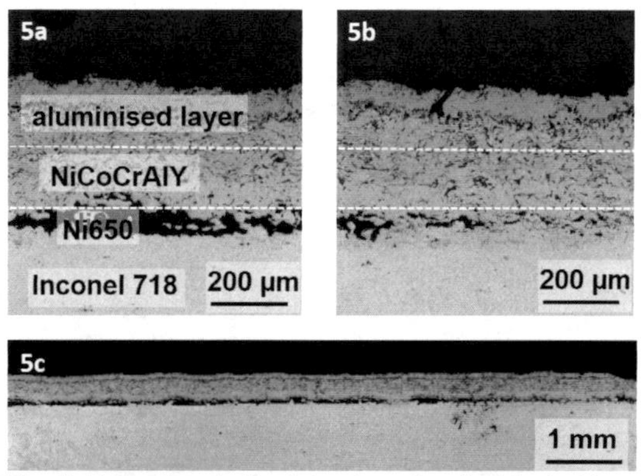

Fig. 5. Brazed and aluminised coating (subjected to heat treatment 3)

Due to this heat treatment pores in the complete coating are formed and there are sections in which large pores are formed in such a way that the coating is delaminated. The formation of the pores over the complete coating is caused by the slow cooling rate of 1 °C/min. The filler metal is in the liquid state for about 55 minutes (form 1190 °C to 1135 °C) and in the two phase region for about 55 minutes as well (from 1135 °C to 1080 °C). However, there is the possibility to reach a metastable state of the liquid filler metal when the liquidus- or even the solidus temperature is fallen below (supercooled liquid). In this time the liquid filler metal can infiltrate into the NiCoCrAlY layer as well as in first and second heat treatments but in this case the amount of infiltrated filler metal is higher.

The results show that different ways of heat treatment have an influence on the microstructure of the coating system. With regard to the distribution of the elements in the coating, EDX-analysis (element mappings) were carried out, **Fig. 6.** In contrast to the coating's microstructure the distribution of the elements is nearly independent of the heat treatment. The EDX-mappings of the coating show that the boundary between the filler metal, the MCrAlY and the aluminium layer can be seen. On top of the coating a continuous aluminium zone can be detected. Below this zone the aluminium concentration is lower than in the upper region due to the composition of the MCrAlY layer and aluminium is found in the filler metal because it can solve in the liquid brazing material. A concentration gradient of aluminium starting from the top all over the coating down to the base material is the result. The concentration gradient of nickel starts from the filler metal to the top of the coating due to the concentration ratios of the filler metal and the MCrAlY. The EDX-mapping of cobalt shows that the concentration is highest in the MCrAlY layer and the boundary between MCrAlY and the filler metal as well as the boundary between MCrAlY and the aluminised layer can be detected. The concentration distribution of nickel and cobalt in the upper aluminised zone is nearly homogenous. Further EDX-analyses show and suggest that the phases NiAl (β-phase), Ni_3Al (γ'-phase) and aluminium-rich cobalt phases like Al_9Co_2 and Al_3Co are formed [13, 14].

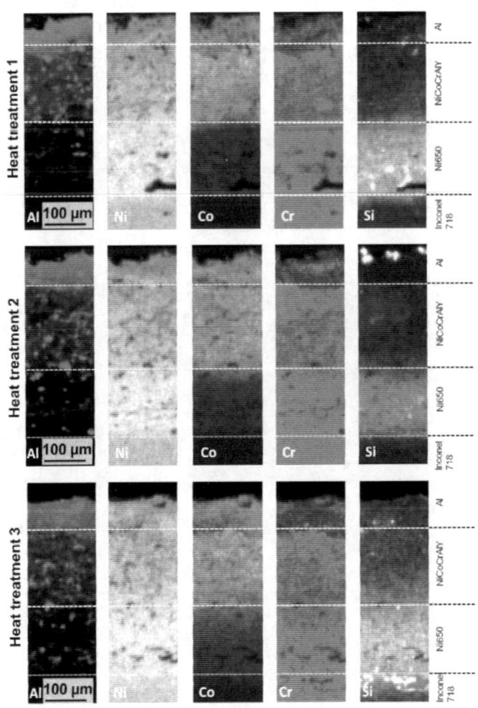

Fig. 6. EDX-mappings of brazed and aluminised coatings

In this upper zone the chromium concentration is much less in the middle than at the boundary to the top as well as to the inner of the coating. The reason is that there is a solution of chromium in a nickel-aluminium-cobalt matrix at higher temperatures and chromium phases precipitate upon cooling, especially at the grain boundaries.

As mentioned above pores and cavities are formed because of diffusion processes due to local concentration differences, the flow direction of the filler metal and segregation effects. Furthermore volumetric contractions can lead to the formation of contractions cavities during the solidification of the filler metal. However, the formation of cavities and pores is problem regarding the mechanical properties. Therefore further investigations must be carried out to understand the mechanism of the formation of cavities and pores.

With this hybrid technology it is also possible to take into account the thermal barrier coating (TBC). Starting from the coating system described here, the TBC (e.g. $ZrO_2 \cdot Y_2O_3$) is finally applied and the following coating system results: filler metal/MCrAlY/Al/TBC. During the subsequent heat treatment the aluminium can improve the hot gas corrosion protection on the one hand. On the other hand the aluminium as an "active element" can improve the bonding of the TBC to the MCrAlY-layer. This coating system has the further advantage to carry out the heat treatment (brazing/aluminising) in a continuous shielding gas furnace. According to the state-of-the-art vacuum-like conditions can be realised in such furnaces with protective gases (e.g. argon doped with monsilane SiH_4), respectively [15, 16]. Carrying out the heat treatment in a continuous shielding gas furnace allows to replace the discontinuous brazing/aluminising process by a continuous process.

4 Conclusion and outlook

The feasibility of a combined brazing- and aluminising process is shown. The filler metal, the hot gas corrosion protective coating and aluminium is applied by thermal spraying to the substrate in this order. This coating system is subjected to a common heat treatment with different temperature-time regimes which have an influence on the coating's microstructure in which pores and cavities are formed. However, the temperature-time regimes have no influence on the distribution of the elements. The formation of nickel-rich and low-cobalt phases can be detected which improve the protection against hot gas corrosion.

This hybrid technology has technical and economic advantages. Due to the application of aluminium by thermal spraying the independent aluminising process can be omitted. The aluminising process is expensive because it requires the embedding of the substrate in a pack cementation or the deposition in an aluminising furnace in a hydrogen atmosphere.

Acknowledgement

The work presented here was supported by the German Research Foundation (DFG) within the scope of the sub-project B1 "Near-net shape turbine blade repair using a joining and coating hybrid process" of the Collaborative Research Centre (SFB 871 "Product Regeneration"). Thanks go to the DFG for their support.

4 Literature

[1] J. McGraw, R. Anton, G. Deventer, A. Burns: Advancements in gas turbine vane repair, Proc. of PWR 2006: ASME 2006 Power Conference, May 2 – 4, 2006 (Atlanta, Georgia), Power Division (2006), pp. 385/89.

[2] X. Huang, W. Miglietti: Wide Gap Braze Repair of Gas Turbine Blades and Vanes - A Review, J. Eng. Gas Turbines Power 134 (2011), Issue 1, pp. 010801/17.

[3] W.A. Demo, S. Ferrigno, S. Budinger, E. Huron: Improving Repair Quality of Turbine Nozzles Using SA650 Braze Alloy, Superalloys 2000, T.M. Pollock, R.D. Kissinger, K. A. Bowman, M. McLean, S. Olson, J.J. Schirra, Ed., TMS The Minerals, Metals & Materials Society (2000), pp. 713/18.

[4] M.B. Henderson, D. Arrell, R. Larsson, M. Heobel, G. Marchant: Nickelbased superalloy welding practices for industrial gas turbine applications, Science and Technology of Welding and Joining 9 (2004), Issue 1, pp. 13/21.

[5] R. Stolle: Conventional and advanced coatings for turbine airfoils, MTU Aero Engines Munich http://www.academia.edu/7789702/Conventional_and _advanced_coatings_for_turbine_airfoils, 2004.

[6] D.B. Miracle: The Physical and Mechanical Properties of NiAl, Overview No. 104, Acta metall. mater. 41 (1993), pp. 649/84.

[7] Z. Zhan, Y. He, L. Li, H. Liu, Y. Dai: Low-temperature formation and oxidation resistance of ultrafine aluminide coatings on Ni-base superalloy, Surface & Coatings Technology 203 (2009), pp. 2337/42.

[8] M. Pytel, M. Góral, A. Nowotnik, J. Sieniawski, M. Drajewicz, M. Ziaja: Heat treatment and CVD aluminizing of Ni-base René 80 superalloy, Journal of Achievements in Materials and Manufacturing Engineering 51 (2012), Issue 1, pp. 30/38.

[9] J.R. Nicholls, N.J. Simms, W.Y. Chan, and H.E. Evans: Smart overlay coatings - concept and practice, Surface and Coatings Technology 149 (2002), pp. 236/44.

[10] Fr.-W. Bach, K. Möhwald, C. Bach, U. Holländer: Thermally sprayed filler metal coatings for high temperature brazing, Thermal Spray 2004: Advances in Technology and Application, on CD-Rom, ASM International, May 10-12, 2004 (Osaka, Japan), ASM International (2004), pp. 1129/31.

[11] M. Nicolaus, K. Möhwald, Fr.-W. Bach, H.J. Maier: Heat treatment of thermal sprayed Ni-base-fillermetal/NiCrAlY-coating systems for repairing turbine blades, Thermal Spray Bulletin 6 (2013), Issue 2, pp. 119/23.

[12] M. Nicolaus, K. Möhwald, H.J. Maier: Combined brazing and alitising process for thermally sprayed Ni-based alloys for the repair of turbine blades, Thermal Spray Bulletin 8 (2015), Issue 1, pp. 56/61.

[13] I. Ansara, N. Dupin, H.L. Lukas, B. Sundman: Thermodynamic assessment of the Al-Ni system, Journal of Alloys and Compounds 247 (1997), pp. 20/30.

[14] A.J. McAlister: The Al-Co (Aluminium-Cobalt) System, Journal of Phase Equilibria 10 (1989), Issue 6, pp. 646/50.

[15] Fr.-W. Bach, K. Möhwald, U. Holländer: Physico-chemical aspects of surface activation during fluxless brazing in shielding-gas furnaces, Key Engineering Materials 438 (2010), pp. 73/80.

[16] Fr.-W. Bach, K. Möhwald, U. Holländer, J. Schaup, C. Roxlau, A. Langohr: Boron and phosphorous free nickel based filler metals for brazing stainless steel in shielding gas furnaces, International Journal of Materials Research 8 (2011), pp. 964/71.

The influence of the coating deposition process on the interdiffusion behavior between nickel-based superalloys and MCrAlY bond coats

M. Elsaß, M. Frommherz, M. Oechsner, Darmstadt / D

In this work, interdiffusion between different nickel-based superalloys and two MCrAlY bond coats, containing different chemical compositions, is investigated. To determine the influence of the coating deposition process, the MCrAlY bond coats were applied using two different spraying processes, high velocity oxygen fuel spraying (HVOF) and low-pressure plasma spraying (LPPS). Of primary interest is the evolution of Kirkendall porosity, which can form at the interface of substrate and bond coat and depends largely on the chemical compositions of the coating and substrate. Experimental evidence suggested also a dependence on the coating deposition process. Formation of porosity at the interface causes a degradation of the bonding strength between substrate and coating, with functional breakdown of the coating system as a worst result. After coating deposition, the samples were annealed at 1050 °C for varying test periods up to 2000 hours. Microstructural and compositional analyses were performed to determine and to evaluate the Kirkendall porosity. The results reveal a strong influence of both the spraying process and the chemical composition. The amount of Kirkendall porosity formed, as well as the location of appearance and the shape, is largely influenced by the coating deposition process. In general, samples with bond coats applied by means of HVOF show accelerated element diffusion. It is hypothesized that recrystallization of the substrate material, as a consequence of the surface treatment prior to coating deposition, is the main root cause for these observations.

1 Introduction

Oxidation and corrosion resistant coatings for very high temperature application in gas turbines are commonly of the MCrAlY type, where M is the base element, Ni or Co. They are either used as overlay coatings or as bond coats in thermal barrier coating systems [1–3]. These type of coatings consist of high amounts of chromium and aluminum to achieve the sufficient resistance to oxidation and hot corrosion to protect the underlying Ni-based substrate. An reactive element, most commonly Y, is added in small amounts to promote the bonding strength of the oxide layer that forms in service [4,5].

MCrAlY-coatings are applied by using thermal spraying processes, such as high velocity oxygen fuel spraying (HVOF) or low-pressure plasma spraying (LPPS) [6]. In thermal spraying processes, melted (or partially melted) particles are sprayed onto a surface, which leads to a characteristic splat structure of the coating. The feedstock (usually powder for MCrAlY-type coatings) is heated by a plasma (LPPS) or a combustion flame (HVOF) [7].

In the LPPS process, the particles reach very high temperatures and high velocities, which results in very dense coatings. Protective gas in a low pressure atmosphere is used to achieve almost no oxidation of the coating during the coating deposition process. The substrate surface is carefully cleaned by grid blasting prior to the coating deposition to ensure sufficient bonding strength [8].

The HVOF process is conducted in an oxygen rich atmosphere. Consequently, HVOF-sprayed coatings contain small amounts of oxides, which are located along the splat boundaries. The particles reach very high velocities and high temperatures that brings about a very dense coating. The substrates surface has to be grid blasted more extensively compared to the LPPS process to achieve the desired bonding strength. This results in a much rougher surface and residual particles at the substrate-coating interface. The main advantage of using the HVOF over the LPPS method is the much lower price [8–10].

In a high-temperature oxidative environment, two main mechanisms are responsible for the degradation of the coating: the formation of a thermally grown oxide (TGO) and interdiffusion between substrate and the coating.

The TGO preferably consists of alpha-alumina, since α-Al_2O_3 forms a stable layer with low growth rate. High amounts of aluminum are necessary to promote the consistent grow of the alumina layer for extensive service time; consequently, the microstructure of MCrAlY alloys consists of Al-rich phases, most commonly β-NiAl. The complete depletion of the β-NiAl phase in the bond coat usually indicates that the remaining amount of aluminum isn't sufficient to form pure alumina. In consequence, large amounts of secondary oxides, e.g. Ni and Co spinel, will form, which can cause the spallation of the overlaying ceramic coating [11,12].

Interdiffusion between MCrAlY coating and Ni-base substrate causes various phenomena such as the dissolution of phases, the precipitation of brittle phases and the formation of Kirkendall porosity. The diffusion of aluminum from coating to substrate is especially problematic, as it will reduce the amount of aluminum that is available for the alumina scale formation. Interdiffusion causes the evolution of a β-phase depleted zone (BDZ) in the MCrAlY coating and a γ'-depleted zone (GDZ) in the Ni-base alloy [11,13].

The formation of Kirkendall porosity is of particular importance. Interdiffusion between materials with different diffusion velocities of main alloying elements causes the accumulation of vacancies at the interface, where the vacancies can coalesce to voids [14]. It is evident, that the amount of Kirkendall porosity formation is mostly dependent on the alloy composition of the involved materials. In experiments, it has also been observed that the coating deposition process has

a high influence on the formation of Kirkendall porosity. Large amounts of Kirkendall porosity may lead to the degradation of the bonding strength between the two materials and result in the failure of the coating system [15].
In this work, samples with four different combination of nickel-based substrate and MCrAlY bond coat have been prepared by both, LPPS and HVOF method each. The effect of the coating deposition process on the different interdiffusion phenomena is investigated and discussed in detail.

2 Experimental work

Cylindrical rods consisting of René 80 and CM 247 with a diameter of 10 mm and 8 mm, respectively, were used as base materials. The Ni-based substrates were coated with two different MCrAlY alloys, referred to as BC1 and BC2; see Table 1 (The table does not account for small amounts of Zr and B in the superalloys).
The differences in composition that are expected to have the biggest impact on the interdiffusion behavior are the amount of chromium, aluminum and Nickel. René 80 is a Ni-based superalloy with higher amounts of chromium and low amounts of aluminum, while CM 247 contains reduced amounts of chromium and higher amounts of aluminum. Both Ni-based alloys contain almost the same amount of nickel.
The BC1 alloy consists of very high amounts of chromium and moderately high amounts of aluminum. BC2 contains only half as much chromium as BC1 and some more aluminum. As a result of the low chromium content, BC2 contains very high amounts of nickel.

Table 1. Elemental composition of the substrates and coating alloys used (wt%)

	Ni	Co	Cr	Al	Ta	Ti	Mo	W	Hf	Y	Si	C
René 80	Bal	9.5	14	3	-	4.8	4	4	-	-	-	0.17
CM 247	Bal	9.2	8.1	5.6	3.2	0.7	0.5	9.5	1.4	-	-	0.07
BC1	Bal		28	8		-	-	-	-	0.6	0.7	-
BC2	Bal		15	11.5		-	-	-	-	0.3	-	-

The coatings were deposited by the LPPS and HVOF methods by an industrial manufacturer. The samples were corundum blasted prior to the coating deposition to clean the substrates' surface and to ensure sufficient bonding of the MCrAlY coatings. The samples received a bonding heat treatment after coating deposition to further enhance the adhesion of the coating. The resulting coating thickness was about 280 µm for the LPPS coatings and 220 µm for the HVOF coatings.
The samples were annealed for 120, 500, 1200 and 2000 hours at 1050 °C.
The specimen were prepared for optical microscope examination to evaluate the inner β-phase depletion zone (BDZ), γ'-depletion zone of the Ni-based alloy (GDZ) and the formation of Kirkendall porosity, Fig. 1.
To obtain comparable data for the evolution of the Kirkendall porosity, the overall pore area near the interface was measured by image analyses and standardized to the length of the examined interface; see Equation (1).

$$\text{Kirkendall porosity} = \frac{\text{Pore area at the interface } [\mu m^2]}{\text{Examined interface length } [\mu m]} \quad (1)$$

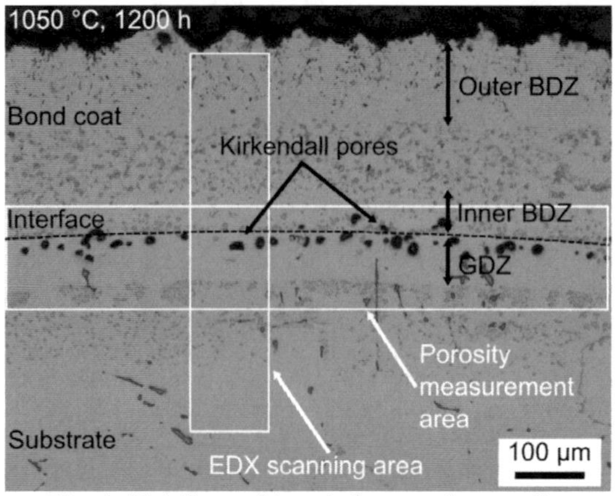

Fig. 1. Metallographic cut of a sample near the interface after annealing for 1200 h at 1050 °C. Important interdiffusion phenomena and the areas for EDX and porosity analysis are marked [16].

To further evaluate the interdiffusion behavior of the alloying elements, EDX (energy dispersive x-ray spectroscopy) element profiles perpendicular to the interface were prepared by performing area scans according to Fig. 1 and averaging the horizontal values. The analysis was done with respect to all elements in Table 1, except C. The measurements also accounted for oxygen and, in the case of HVOF coatings, for nitrogen.

3 Results and discussion

Fig. 2 shows a metallographic cut of the as-sprayed condition (0 hours) near the coating-substrate interface in comparison with the condition after annealing for 2000 hours at 1050 °C for all investigated combinations of coating and substrate and for the LPPS and HVOF version each.
The corresponding measurements of the Kirkendall porosity are shown in Fig. 3.
As expected, only very small amounts porosity has been observed in the as sprayed condition for any of the samples. The presence of small amounts of porosity is believed to be partly due to imperfections from the spraying process and partly due to the formation of Kirkendall porosity in the bonding heat treatment process.

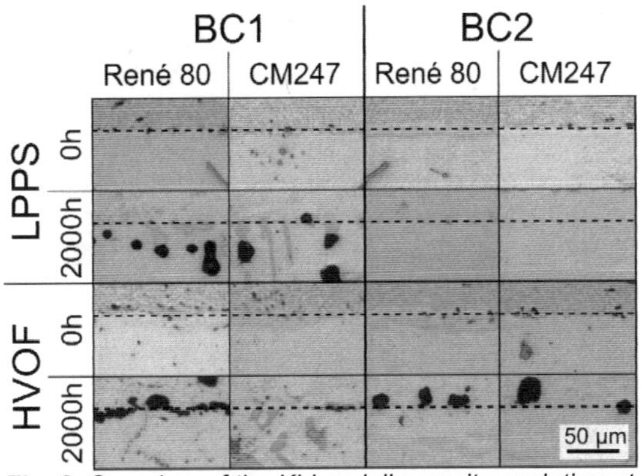

Fig. 2. Overview of the Kirkendall porosity evolution at the interface between substrate (bottom) and bond coat (top). The annealing temperature was 1050 °C. The original interface is marked by the dotted line.

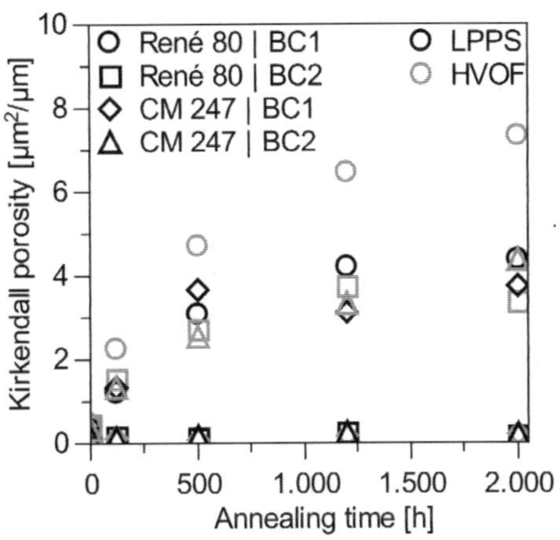

Fig. 3. Measured Kirkendall porosity for all samples after annealing for up to 2000 h at 1050 °C

A couple of observations can be made on the samples in the annealed condition (Fig. 2 and Fig. 3):

I. For the LPPS variants, there is apparent formation of Kirkendall porosity in both BC1 coated samples, while almost no porosity could be observed in the BC2 coated samples.

II. In the LPPS version of the BC1 coated samples, Kirkendall porosity forms on the substrate side of the interface.

III. In all HVOF sprayed samples where porosity was observed, the pores formed directly at the interface or on the coating side of the interface.

IV. The HVOF variant of René 80 coated with BC1 displays an increased amount of Kirkendall porosity.

V. The HVOF variant of René 80 coated with BC2 shows no porosity at all.

VI. In the HVOF variants of the BC2 coated samples, distinct formation of Kirkendall porosity was observed.

Observations **I** and **II** strongly suggests that in the LPPS samples, the diffusion of nickel from the substrate into the coating is the main reason for the formation of Kirkendall porosity. The formation of pores on the substrate side of the interface implies that element diffusion out of the substrate is faster than into the substrate. Nickel is the only element that can diffuse in significant amounts and with reasonable speed. Since the nickel content in BC2 is much higher than in BC1 (see Table 1), it's also reasonable to assume that the driving force for the diffusion of nickel in BC2 samples is much lower, which results in almost no formation of porosity. In addition, uphill diffusion of chromium from coating and substrate has been reported for BC2 samples, which can compensate for some loss of matter [16].

The shift of the location of Kirkendall porosity formation (Observation **III**) indicates that in the HVOF sprayed samples, the diffusion of elements from the coating into the substrate is faster than the diffusion vice versa. Since the coating deposition process should have no impact on the driving force of diffusion, the formation of porosity can't be attributed to the diffusion of nickel in the case of HVOF sprayed coatings. In conclusion, the diffusion of the remaining main alloying elements of the bond coats, aluminum, chromium and cobalt, needs to be the root cause for porosity formation. Additionally, porosity formation preferably initiates at imperfections, such as the oxides along grain boundaries in HVOF coatings. This may be part of the reason for higher amounts of porosity and the accumulation of pores directly at the interface.

The shift of the location of porosity formation (Observation **III**) and the increased amount of porosity in the HVOF version of the BC1 coated René 80 (Observation **IV**) implies that element diffusion from coating to substrate has to be highly accelerated, since it has to compensate for the Ni-diffusion into the coating. In contrast, the formation of very little amounts of porosity in the HVOF-sprayed version of BC1 on CM 247 (**V**) indicates that the Ni-diffusion into the coating is offset by the diffusion of coating elements into the substrate. Another possible explanation can be found in the formation of a β-precipitate zone in the substrate. Faster aluminum diffusion into the substrate could lead to the formation of higher amounts of β-phase and therefore bind nickel atoms.

Observation **VI** is directly related to the conclusions of Observations **II** and **III**. The overall diffusion speed has to be higher from substrate to coating than vice versa in the HVOF variants of the BC2 coated samples. Since the driving force for nickel diffusion is supposed to be low, faster diffusion of the coating elements results in the formation of pores.

The development of the GDZ in the Ni-based substrate and the BDZ in the bond coat can be used as an indicator of the diffusion speed, since the interdiffusion of the precipitation forming elements causes the dissolution of the precipitates.

The formation of the GDZ is largely influenced by the materials combination, while there is almost no influence of the coating deposition process, Fig. 4. The depletion of the γ'-phase is most likely linked to the diffusion of titanium (René 80, CM 247) and tantal (CM 247) into the coating. The coating deposition process seems to have a neglectable amount of influence on the Ti and Ta diffusion. In the BC1 coated René 80 sample, considerable amounts of titanium diffuse into the coating, as reflected by the larger GDZ. The diffusion of titanium and tantal seems to be largely influence by the diffusion of aluminum, as reported for other elements as well [16]. The precipitation of β-phase on the substrate side of the interface in BC1 coated CM 247 sample is believed to be the reason for the even larger GDZ.

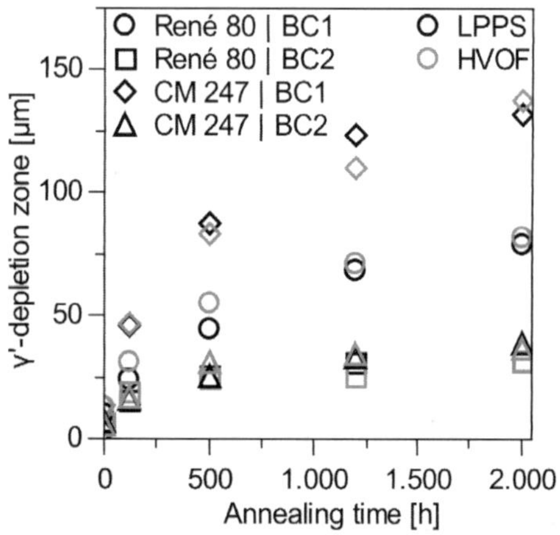

Fig. 4. Measured GDZ thickness for all samples after annealing for up to 2000 h at 1050 °C

In contrast to the GDZ, the development of the BDZ in the bond coat appears to be dependent on the materials combination as well as on the coating deposition process, Fig. 5.

The BDZ has been observed to be much thicker in the René 80 samples, than in the CM 247 samples. Since the dissolution of the β-phase is mainly linked to the diffusion of aluminum, the driving force for Al diffusion in the René 80 samples appears to be higher than in the CM 247 samples. This conclusion is reasonable, since the Al content in CM 247 is much higher than in René 80, Table 1. The higher BDZ thickness in BC2 coated samples is believed to be mainly caused by the higher aluminum content at which the β-phase starts to precipitate [16].

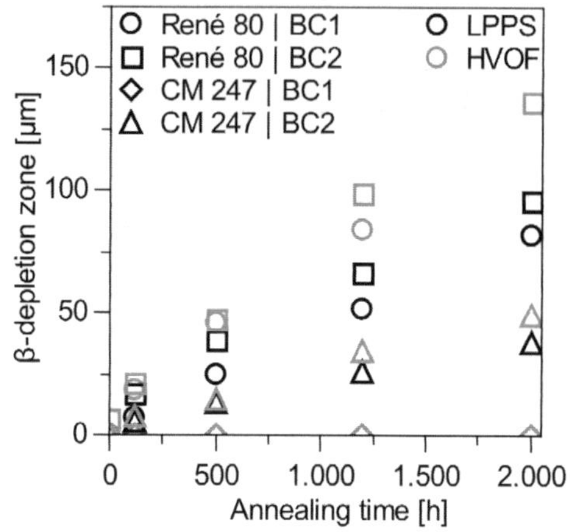

Fig. 5. Measured BDZ thickness for all samples after annealing for up to 2000 h at 1050 °C

In the BC1 coated CM 247 sample, no BDZ was observed. This is due to the fact that a β-phase precipitation zone is forming on the substrate side of the interface. In consequence, the driving force for aluminum diffusion from coating to the substrate is too low for the dissolution of the β-phase in the coating.

In all samples that display the development of a BDZ, the HVOF variants develop much wider BDZ than their LPPS counterparts. It has to be noted, that the coating in the René 80 sample coated with BC1 was completely depleted of β-phase after 2000 h of annealing. The faster β-depletion of the HVOF coatings implies that the aluminum diffusion from coating to substrate in the HVOF sprayed samples is much faster than in the LPPS sprayed samples.

To further investigate the influence of the coating deposition process on the interdiffusion behavior, excessive amounts of EDX area scans have been conducted to obtain element profiles of all diffusing species. The element profiles of Al, Co and Cr after 500 h of annealing at 1050 °C in the René 80 sample coated with BC2 have been chosen to representatively illustrate the differences between LPPS and HVOF variants of the samples, Fig. 6. The diffusion of nickel has been observed to be influenced for only a small amount by the coating deposition process, as expected from the GDZ analysis. The diffusion of the remaining alloying elements of coating and substrate have been investigated as well but are assumed to have very little impact on the formation of Kirkendall porosity.

At first, it has to be noted that a distinct interdiffusion zone is forming in the BC2 coated samples, as reported previously [16].

As stated above, chromium is diffusing uphill, from areas of lower Cr content in the coating and the substrate to an area of higher Cr content at the interface. This process appears to be accelerated in the HVOF sample, as the area of high chromium content (grey solid line) is larger than in the LPPS sample (black solid line).

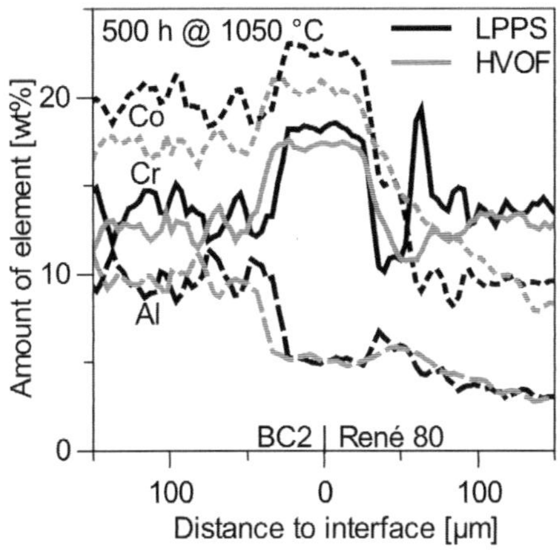

Fig. 6 Element profiles of Al, Co and Cr obtained from EDX area scans near the interface of René 80 and BC2, LPPS and HVOF variants, after 500 h of annealing at 1050 °C

The Al element profile (dashed lines) also reveals an accelerated diffusion of aluminum from the coating into the substrate. The profile shows a sharp transition between high amounts of aluminum in the area of the coating, where the β-phase is stable, and the interdiffusion zone with low amounts of aluminum. It has to be pointed out, that the aluminum content in the area of the coating, where the β-phase is stable doesn't change with ongoing annealing time and doesn't differ between the LPPS and HVOF variant. The distance between the interface and the point where Al content rises is larger in the HVOF sample than it is in the LPPS sample, which means that higher amounts of aluminum have already diffused into the substrate. This correlates to the measured BDZs, Fig. 5.
The diffusion of cobalt (dotted line) is also highly accelerated in the HVOF-sprayed sample. The Co content of the HVOF sample is lower in the coating and higher in the substrate compared to the LPPS sample. From these observations, it can be concluded that the accelerated diffusion of aluminum and cobalt from the coating into the substrate is the main reason for the formation of Kirkendall porosity on the coating side of the interface in HVOF-sprayed coatings.

To investigate the root cause for the accelerated diffusion speed from coating to substrate in the HVOF samples, EBSD measurements of the area near the interface have been conducted. As reported before, it has been found, that in the HVOF samples the surface area of the substrate is recrystallizing, while the LPPS samples show only little signs of recrystallization, Fig. 7. The findings presented before have been confirmed to be correct for the investigated samples in this work as well.
In the EBSD-measurements a certain color is assigned to each grain, depending on its orientation. The uniform color of the substrate in the LPPS sample represents the large scaled grains of the nickel based substrate. In comparison, the recrystallized areas of the substrate in the HVOF sample are evident.

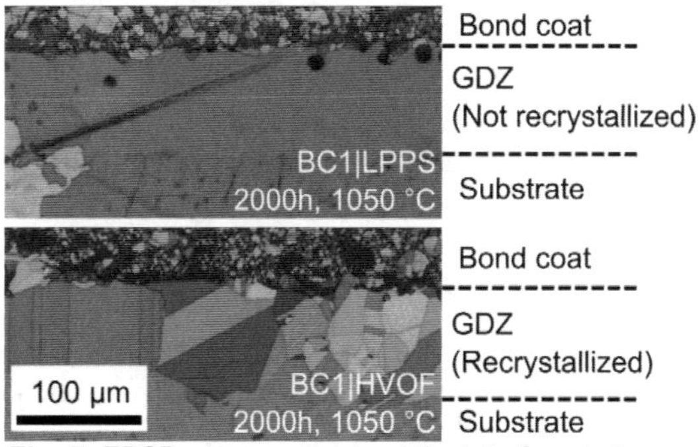

Fig. 7. EBSD measurements at the interface between René 80 substrate and BC1 after 2000 h of annealing at 1050 °C [16]

The formation of large amounts of small sized grains induced plenty of new, lower energy diffusion paths along the grain boundaries and results in the observed accelerated element diffusion from coating to substrate in the HVOF samples.
The recrystallization of the substrate is most likely caused by the surface treatment prior to the coating deposition process. In contrast to the LPPS samples, the HVOF samples received a much more extensive corundum blasting to clean and roughen the surface. As a result, the surface experienced high plastic deformation, which causes the recrystallization of the substrate in the deformed area.

4 Conclusion

In this work, the influence of the coating deposition process on the interdiffusion behavior on the example of different combinations of MCrAlY bond coats and nickel-based superalloy has been examined.
The experimental results show a strong influence of the process used to deposit the coating.
Kirkendall porosity in LPPS sprayed samples is believed to mainly be caused by the diffusion of nickel into the bond coat, as the porosity forms on the substrate side of the interface. In contrast, the pores in the HVOF samples form on the coating side of the interface, which is most likely due to accelerated diffusion of aluminum and cobalt into the substrate.
Samples coated by means of HVOF in general showed higher amounts of Kirkendall porosity and faster aluminum depletion of the coating than the LPPS-sprayed samples. The coating deposition process had no noticeable influence on the diffusion of nickel.
It has been found, that the recrystallization of the substrate as a result of the surface treatment prior to the coating deposition process causes the accelerated element diffusion from coating to substrate in the HVOF-sprayed coatings.

5 Acknowledgments

This work was financially supported by the DFG in the framework of the Excellence Initiative, Darmstadt Graduate School of Excellence Energy Science and Engineering (GSC 1070).

6 Literature

[1] G.W. Goward, Protective coatings – purpose, role, and design, Mater. Sci. Technol. 2 (1986) 194–200.

[2] G.W. Goward, Progress in coatings for gas turbine airfoils, Surf. Coat. Technol. 108–109 (1998) 73–79.

[3] M.J. Pomeroy, Coatings for gas turbine materials and long term stability issues, Mater. Des. 26 (2005) 223–231.

[4] R.J. Christensen, V.K. Tolpygo, D.R. Clarke, The influence of the reactive element yttrium on the stress in alumina scales formed by oxidation, Acta Mater. 45 (1997) 1761–1766.

[5] B. Gudmundsson, B.E. Jacobson, Yttrium oxides in vacuum-plasma-sprayed CoNiCrAlY coatings, Thin Solid Films. 173 (1989) 99–107.

[6] P. Fauchais, A. Vardelle, B. Dussoubs, Quo vadis thermal spraying?, J. Therm. Spray Technol. 10 (2001) 44–66.

[7] S. Safai, H. Herman, Microstructural investigation of plasma-sprayed aluminum coatings, Thin Solid Films. 45 (1977) 295–307.

[8] M. Shibata, S. Kuroda, H. Murakami, M. Ode, M. Watanabe, Y. Sakamoto, Comparison of Microstructure and Oxidation Behavior of CoNiCrAlY Bond Coatings Prepared by Different Thermal Spray Processes, Mater. Trans. 47 (2006) 1638–1642.

[9] E. Lugscheider, C. Herbst, L. Zhao, Parameter studies on high-velocity oxy-fuel spraying of MCrAlY coatings, Surf. Coat. Technol. 108–109 (1998) 16–23.

[10] B. Wielage, A. Wank, H. Pokhmurska, T. Grund, C. Rupprecht, G. Reisel, E. Friesen, Development and trends in HVOF spraying technology, Surf. Coat. Technol. 201 (2006) 2032–2037.

[11] D. Renusch, M. Schorr, M. Schütze, The role that bond coat depletion of aluminum has on the lifetime of APS-TBC under oxidizing conditions, Mater. Corros. 59 (2008) 547–555.

[12] A. Rabiei, A.G. Evans, Failure mechanisms associated with the thermally grown oxide in plasma-sprayed thermal barrier coatings, Acta Mater. 48 (2000) 3963–3976.

[13] K.V. Dahl, J. Hald, A. Horsewell, Interdiffusion between Ni-based superalloy and MCrAlY coating, in: Defect Diffus. Forum, Trans Tech Publ, 2006: pp. 73–78.

[14] A.D. Smigelskas, E.O. Kirkendall, Zinc diffusion in alpha brass, Trans Aime. 171 (1947) 130–142.

[15] J.A. Nesbitt, R.W. Heckel, Modeling degradation and failure of Ni-Cr-Al overlay coatings, Thin Solid Films. 119 (1984) 281–290.

[16] M. Elsaß, M. Frommherz, A. Scholz, M. Oechsner, Interdiffusion in MCrAlY coated nickel-base superalloys, Surf. Coat. Technol. 307, Part A (2016) 565–573.

A comparative study of the wear performance of Stellite 6 hardfacing coatings applied by HVOF and GTAW hot wire cladding onto steel substrates

C.R.C. Lima, M.J.X. Belém, F. Camargo, R. Libardi, São Paulo, BR

Stellite 6 is a widely applied hardfacing alloy with good corrosion and wear resistance. Several different techniques have been used to deposit this alloy, like oxyacetylene, Tungsten Inert Gas welding overlay, CO2 laser cladding and thermal spray methods. This paper evaluates the wear and corrosion resistance of Stellite 6 coatings applied by high velocity oxygen fuel (HVOF) spraying and gas tungsten arc welding (GTAW) hot wire technique. Mechanical properties and microstructural characteristics of the coatings are assessed and compared. The influence of these properties in the wear results is further discussed.

Key words: HVOF; GTAW hot wire cladding; Wear; Erosion; Stellite.

1 Introduction

In Wear is one of the principal causes for degradation and failure of equipment parts in several industries [1]. The erosive environment in some manufacturing processes presents distinctive forms and sometimes is combined with corrosive agents making the search for failure prevention a challenging task. Coatings have become a very important element for equipment manufacture and repair and equipment components can achieve improved performance if properly coated. The recovery of pumps, valves and other parts of industrial machinery using engineered coatings constitutes today one of the most important engineering alternatives [2, 3]. Co-based alloys are one of the selected materials for coating industrial components because of their exceptional wear, corrosion and oxidation resistance in a broad range of temperatures. Among them Stellite® (CoCrW alloy) is the most applied in industrial applications. Different technologies can deposit Stellite coatings onto a component surface, like plasma transferred arc (PTA), GTAW welding, thermal spraying and laser cladding [3-7]. Widely recognized as the main joining process, welding is also an important process for coating application in the production and recovery of parts, equipment and structures. There are a number of different welding processes requiring the selection of that suitable for a given application [8]. Normally elected for the application of coating overlays, gas metal arc welding (GMAW) and gas tungsten arc welding (GTAW) use heat from an electric arc between a naked electrode, fed continuously, and the base metal. The heat melts the electrode and the surface of the base metal to form the melt pool [8]. The GTAW process or Tungsten Inert Gas (TIG) welding process uses a non consumable tungsten electrode for the electric contact with the fed wire. TIG process with the addition of hot-wire is described as a process highly productive, presenting higher deposition rates and lower dilution percentages in comparison with conventional TIG process with cold-wire, making it an optimum alternative for cladding applications [9, 10].

Thermal spraying is a recognized alternative for coating deposition aiming the improvement of wear and corrosion resistance of the coated substrate. Combustion processes, like flame spray (FS) and high-velocity oxygen fuel (HVOF) have proven their ability to deposit coatings for wear protection. The use of the HVOF process for the process of carbide coatings is broadly adopted mainly because of the very high particle velocities and relatively low temperatures that lead to an outstanding protection [11].

Erosion is the dominating wear process in some components of several industries and occurs as a result of erosive particle impacts typical for a given process environment. The erosive wear process occurs when hard particles normally in a liquid flow impact the surface of the object causing wear of the surface which leads to material losses in the external layer. It is generally accepted that the hardness and fracture toughness of the coatings are the most important mechanical properties determining the wear resistance [12]. The hardness of the composites determines the depth of penetration of the hard particles free to roll and slide between two surfaces. Then, an increase in hardness should reduce abrasive wear [13].

This work performs a comparison of the performance in erosive wear tests of Stellite 6 coatings deposited by HVOF spraying and TIG hot wire cladding overlay. The results are compared with the help of microstructural evaluation and literature results. The hardness of the coatings is also discussed.

2 Materials and Methods

Stellite® 6 was used as the coating material for the two different technologies. For TIG hot wire cladding, the material was in the form of wire, 1.2 mm diameter, specification 20401098 KST TUB KESTRALLIT 6 G (Kestra, SP, BR). The material for the HVOF deposition was in powder form, 38-45 µm, specification STELLITE 6 JK576 (Kennametal, Goshen, IN, USA). The substrate was an ASTM 1020 low carbon steel with final dimensions of 80 x 20 x 5 mm. The substrate was cleaned before every coating deposition. Just before spraying, the substrate surface was degreased and grit blasted with aluminum oxide.

The chemical composition of the materials is presented in Table 1. The equipment used for cladding overlay was a TRANSTIG 2200 job G/F Model A 4600 Wels (Fronius, Portage, IN, USA). For HVOF spraying, a JP 5000 HP-HVOF SYSTEM, model 5120 (Praxair –TAFA, Concordia, NH, USA) was used. The coatings were applied following standard commercial conditions. The main parameters used in both processes are shown in Table 2.

Table 1 Chemical composition of feedstock materials

Material	Chemical composition (Wt.%)											
	C	Co	Cr	Fe	Mn	Ni	W	P	Mo	Si	S	Other
HVOF powder	1.07	Bal	26.03	3.35	0.35	0.06	3.27	0.012	0,25	0.53	0.016	0.53
TIG wire	1.20	Bal	28.30	1.60	0.30	2.30	4.70	0.003	0,06	1.20	0.003	0.20

Table 2. Main process parameters for the application of powder and wire materials

Material	Nitrogen (SCFH)		Standoff Distance (mm)	Oxygen (SCFH)	Acetylene (gph)		Barrel (inch)
HVOF Powder	26		380	1950	575		4
TIG Hot-wire	Vs (cm/min)	Ip (A)	Vdp (cm/min)	Idp (A)	Vb (mm)	Ig (A)	Vdg (cm/min)
Layer 1	24	225	175	50	12.5	125	175
Layer 2	20	235	185	50	12.5	135	185

Vs: Substrate rotation velocity Vb: Distance substrate-consumable wire
Vdp: Deposited wire velocity at maximum pulse Vdg: Deposited wire velocity at minimum pulse
Ip: Maximum amperage pulse Idp: Machine Amperage Ig: Minimum amperage pulse

The coated samples were cut, embedded into Bakelite and polished, then ultrasonically cleaned in acetone to remove moisture. Microhardness measurements (H_V) of the coatings were performed by Vickers indentation with a load of 300 g and 15 s dwell time using a Shimadzu model HMV-G microhardness tester. The indentation was performed in the middle plane of the cross section of the coatings in order to minimize the effects of the edge and the interface with the substrate. An average value of 10 microhardness measurements is reported. The roughness of the coatings was measured considering the polished coated surface. The presented results are the average of five replicates performed on each sample with Mitutoyo Surftest 211 digital equipment.

All the samples were machined and polished to 2 µm roughness at the surface to be tested. The erosive wear tests were performed in a lab version of mixture vessel with solid particles in suspension, also defined as a tribometer of mixtures with rotational motion. The method of mixture vessel with erosive particles in suspension is simple in construction, easy to operate and inexpensive, allowing a rapid sorting of different materials by their erosion resistance [14]. Two devices for the assembly of samples were designed to evaluate the influence of the impact angle (α) of the particles in the erosive wear of the material, 30° and 90°, respectively. The influence of the impact velocity (v) of the particles at 3.61 m/s and 9.33 m/s was also evaluated, for which the devices were rotated at 600 rpm and 1550 rpm, respectively. Erosive particles used in the test were quartz sand (SiO2 - 30/40 SE) with a grain size between 300 - 425 µm mixed with water in a concentration of 30% Wt. of erosive particles.

All the tested samples were ultrasonically cleaned in acetone and dried to remove moisture. Then they were weighed on an analytical balance with an accuracy of 0.1 mg. The tests were conducted for a maximum of 6 hours at 1-hour intervals, in which the mass loss of the tested samples was determined to assess the wear performance of the coatings. Two replicates of each test were conducted. After testing, the samples were cleaned again ensuring that all the eroded material had been removed. A new test mixture was used for each of the tested conditions. To evaluate the performance of the coatings, the erosion rate Er (mm^3/kg) was calculated as the lost volume of the material per mass unit of the erosive particles. Micrographic images of the surface and cross section of the coatings before and after testing were obtained by an EVO MA 15 - Carl Zeiss scanning electron microscopy.

3 Results and Discussion

The thickness, microhardness (H_v300) and roughness of the coatings are presented in Table 3. It should be noted that the roughness values obtained for HVOF coatings averaged three times smoother than cladding coatings for similar finishing methods.

Table 3: Mechanical properties of the coatings

Method	Material	Thickness (µm)	Microhardness ($Hv_{0.3}$)	Roughness (Ra - µm)
TIG Cladding	KESTRALLIT 6 G	3800±450	514±15	0,39±0,02
HVOF	STELLITE 6 JK576	750±50	693±34	0,11±0,01

The largest value of microhardness was that of the HVOF sprayed coating (693 Hv), about 34 % higher than that of the TIG cladding. Such difference could be credited to the higher volume fraction of tungsten in the coating as observed in the cross section image of Figure 1. The main alloying element of Stellite alloys is Chromium, which is added for resistance to corrosion and oxidation. It is also the principal carbide former, i.e., most of the carbides in Stellite alloys are Cr-rich. Meanwhile, Cr provides strengthening to the solution matrix. Tungsten and molybdenum provide additional strength to the solution matrix as solute atoms in Stellite alloys [15].

Micrographs of the cross sections of the coatings can be seen in Figure 1. It is evident that the coatings obtained have different microstructures mainly related to the coating process. A lamellar structure results for the thermally sprayed coatings while TIG cladding shows a cast structure, typical of welding processes. It is further observed that the coatings have a dense structure, especially those obtained by cladding, and good adhesion to the substrate. The presence and distribution of tungsten (light phase) can be also observed in the figure and it was confirmed by EDS spectra analysis.

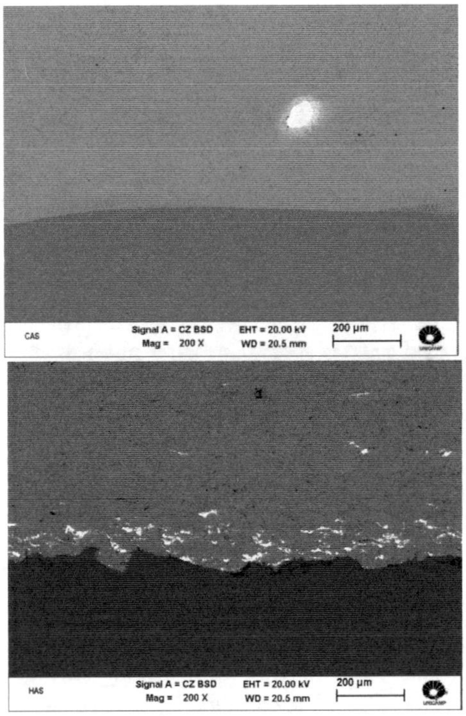

(a) (b)

Fig. 1. SEM (BSD) cross sections of the coatings produced by (a) cladding and (b) HVOF. Light phases refer to the presence of tungsten in the alloy.

The irregular nature of the interface of the HVOF sprayed is common for thermal sprayed samples contrasting with the interface of cladding sample that despite its visual flattening presents a tick region of substrate/coating mixture which is mainly due to the dilution of the substrate at the welding pool zone. For the erosion test evaluation, there is a necessity of performing machining and polishing on the coating surface to reduce error introduced by surface irregularities of the coatings. Figure 2 shows the surface aspect of the coatings after finishing processes just prepared for the erosion tests.

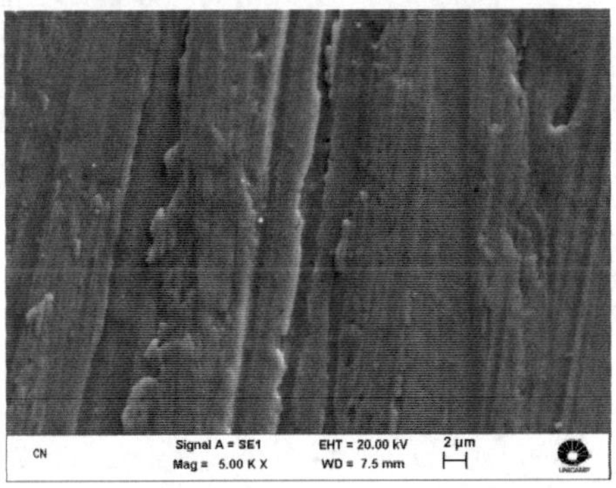

(a)

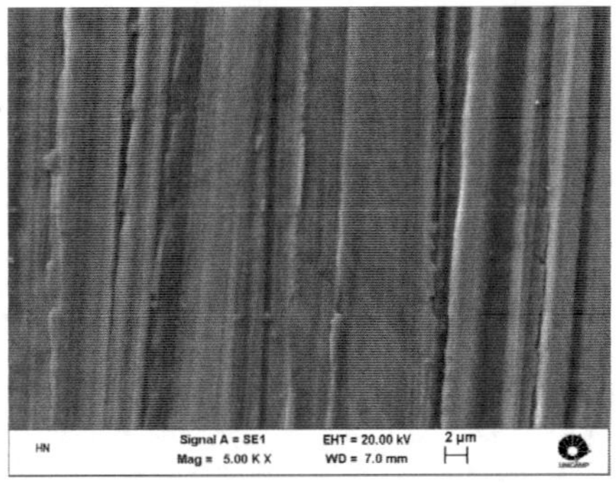

(b)

Fig. 2. Surface SEM ((SE) images of the finished coatings applied by (a) TIG Cladding and (b) HVOF

Figure 3 shows the results of the erosive wear tests for the cladding and thermally sprayed coatings.

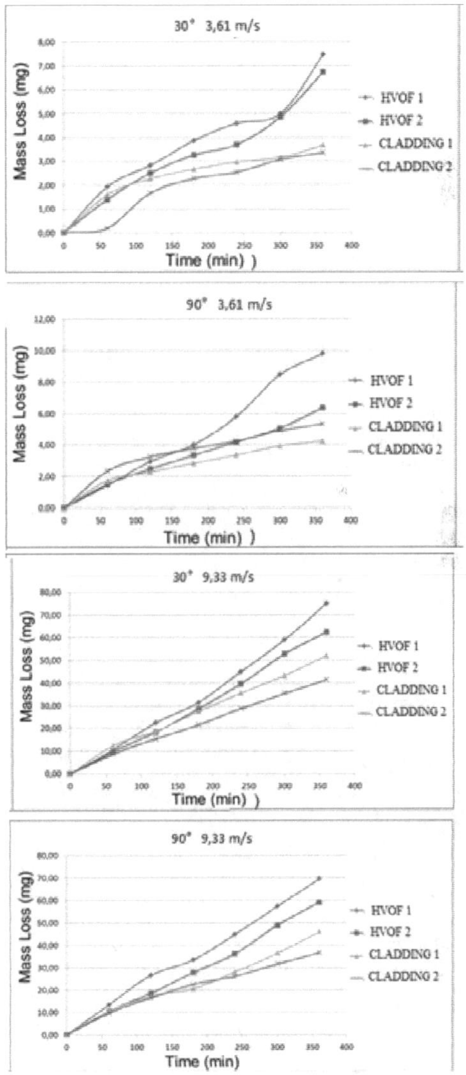

Fig. 3. Erosive wear results for Stellite 6 applied by TIG hot wire cladding and HVOF

The most severe worst condition of testing for the coatings is expected for a particle incidence angle of 90° with the highest particle velocity of 9.33 m/s, mainly observing the results for the two incidence angle at a lower velocity. However, it can be seen from Figure 3 that the highest cumulative wear was for the samples in the condition of 30° and 9.33 m/s, which can be explained by the microstructure of the coatings that are more susceptible to the impact at an acute angle because of their ductile characteristic. Wear of HVOF coatings was at this condition is about 50% higher than cladding ones. It should be observed that at low velocities the wear behavior of the HVOF is comparable to the behavior of the thicker cladding coatings. As the material resistance to erosive wear depends on ductility and hardness, harder surfaces would present a brittle behavior. Then, at the outer surface, micro cutting mechanism is observed, mainly in the lamellar HVOF coatings, combined with plastic deformation; then plastic deformation is induced at the subsurface causing eventual fracture and material removal, as observed by several authors [15, 16].

Figure 4 shows the eroded surfaces of the coatings for the particle impingement angle of 90° and particle velocity of 9.33 m/s. Figure 5 shows enlarged images of the eroded surfaces at the same conditions. Cutting and fracture are mainly observed at the HVOF coating surface. Plastic deformation prevails at the cladding coating surface.

(a)

(b)

Fig. 4. SEM (SE) surface images of the eroded (a) cladding and (b) HVOF coatings at the condition of 90° at 9.33 m/s.

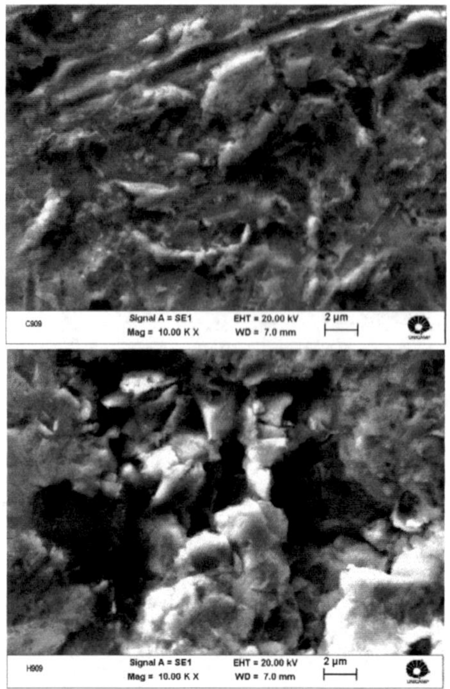

(a)

(b)

Fig. 5. SEM (SE) surface high magnification details of the eroded (a) cladding and (b) HVOF coatings at the condition of 90° at 9.33 m/s. Cutting and fracture are mainly observed at the HVOF coating surface. Plastic deformation prevails at the cladding coating surface.

4 Conclusions

Erosive wear was evaluated for Stellite® 6 coatings obtained with the use of HVOF spraying and TIG hot wire cladding. It can be concluded that:

- Both techniques have performed well for Stellite® 6 deposition resulting in high performance coatings. Cladding coatings show hardness values in the range of 514±15 Hv, about 35% lower than the hardness of HVOF coatings in the range of 693±34 Hv.
- The highest cumulative wear was for the samples in the condition of 30° and 9.33 m/s. Wear of HVOF coatings at this condition is about 50% higher than cladding ones. At low velocities the wear behavior of the HVOF is comparable to the behavior of the thicker cladding coatings.
- Plastic deformation is the principal erosion mechanism for cladding coatings. For HVOF coatings plastic deformation and cutting are combined with some prevalence of microcutting.

Acknowledgements

The authors acknowledge the technical support of Dr. Claudenete Leal from the State University of Campinas – UNICAMP, Brazil.

5 Literature

[1] Pamuk, U., Baydoğan, M., Nilüfer, B. and Çimenoğlu, H.: An energetic approach to abrasive wear of a martensitic stainless steel, Scripta Materialia, 42 (2000), n. 9, 881-885.

[2] Zhao, H.Y., Liu, S.G., Feng, Y.S., Yao, J.P. and Xu, Y.S. Failure Analysis of Desulfurization Slurry Circulating Pump and HVOF Coating Protection Technology Progress. Applied Mechanics and Materials, 320 (2013), 374-382.

[3] Hattori, S., Taruya, K., Kikuta, K. and Tomaru, H. Cavitation Erosion of Silver Plated Coatings Considering Thermodynamic Effect. Wear, 300 (2013) 136-142.

[4] J.C. Shin, J.M. Doh, J.K. Yoon, D.Y. Lee, J.S. Kim, Effect of molybdenum on the microstructure and wear resistance of cobalt-base Stellite hardfacing alloys, Surf. Coat. Technol. 163 (2003) 117–126.

[5] R. Arabi Jeshvaghani, M. Shamanian, M. Jaberzadeh, Enhancement of wear resistance of ductile iron surface alloyed by stellite 6, Mater. Des. 32 (2011) 2028–2033.

[6] A.S.C.M. D'Oliveira, R. Vilar, C.G. Feder, High temperature behaviour of plasma transferred arc and laser Co-based alloy coatings, Appl. Surf. Sci. 201 (2002) 154–160.

[7] P.V. Marques, C.R.C. Lima, Studies of cavitation resistant thermally sprayed and welded coatings, in: C. Moreau, B. Marple (Eds.), Thermal Spray 2003: Advancing the Science & Applying the Technology, Published by ASM International, Materials Park, Ohio, USA 2003, pp. 389–393.

[8] Davis, J.R. Hardfacing, Weld Cladding, and Dissimilar Metal Joining: Welding, Brazing and Soldering, Metals Handbook, 10ª ed. ASM International, Metals Park, Ohio, v.6, 1993

[9] Manz A.F., inventors; Union Carbide Corp, assign. Consumable electrode arcless electric working. United States patent US 3122629A. 1964 Feb 25.

[10] Egerland, S., Zimmer, J., Brunmaier, R., Nussbaumer, R., Posch, G., Rutzinger, B., Advanced Gas Tungsten Arc Weld Surfacing Current Status and Application. Soldagem & Inspeção. 20(2015), n.3, 300-314.

[11] Sobolev, V.; Guilemany, J.M. and Nutting, J. High Velocity Oxy-Fuel Spraying. Maney, London, 2004, 397p.

[12] Houdková, T. and Kašparová, M.Š. Experimental study of indentation fracture toughness in HVOF sprayed hardmetal coatings, Eng Fract Mech 110 (2013) 468–476.

[13] Evans, A.G. and Wilshaw, T.R. (1976) Quasi-static solid particle damage in brittle solids-I. Observations analysis and implications, Acta Metall. 24(10):939-956.

[14] Thakur, L., Arora, N., Jayaganthan, R., Sood, R. An investigation on erosion behavior of HVOF sprayed WC–CoCr coatings Appl. Surf. Sci. 258 (2011) 1225-1234.

[15] J.G.A. Bitter, A study of erosion phenomena. Part I, Wear 6 (1963), n.1, 5–12.

[16] Liu, R., Yao, J., Zhang, Q., Yao, M.X., Collier, R. Effects of molybdenum content on the wear/erosion and corrosion performance of low-carbon Stellite alloys. Materials and Design 78 (2015) 95–106.

Influence of the stand-off distance and spray angle on the coating formation and properties during three-cathode plasma spraying

Kirsten Bobzin[1], Lidong Zhao[1], Mehmet Öte[1], Tim Königstein[1], Matthias Oechsner[2], Marius Siebers[2], Georg Andersohn[2], Jörg Ellermeier[2]

1: Surface Engineering Institute, RWTH Aachen University, Kackertstr. 15, 52072 Aachen, Germany
2: Center for Engineering Materials, State Materials Testing Institute Darmstadt (MPA)
Chair and Institute for Materials Technology (IfW), TU Darmstadt, Grafenstr. 2, 64283 Darmstadt, Germany

Three-cathode plasma spraying is characterized by high process robustness in terms of particle in-flight properties due to high stability of plasma jet. However, during coating parts with complex geometries, process parameters such as stand-off distance and spray angle usually cannot be kept at optimized conditions. In this study, the process robustness during three-cathode plasma spraying of a novel FeCrMnBC alloy was investigated by varying stand-off distance and spray angle to simulate spray conditions for complex geometries. A three-cathode plasma generator TriplexPro™-210 was used to spray FeCrMnBC powder with a fraction of -45 +20 µm onto substrates of cast iron EN-GJL-250. The stand-off distance and the spray angle were varied from d=90 mm to d=110 mm and between ø=90° and ø=60°, respectively, while the other process parameters were kept constant. The results revealed that the reduction of spray angle caused an about 15 % decrease of coating thickness, about 40 % porosity increase and about 15% increase of bond strength. In contrast, the variation of stand-off distance changed only insignificantly the coating thickness, porosity and bond strength. The variation of stand-off distance and spray angle did not result in significant changes in the coating microhardness. The reduction of spray angle resulted in significant worsening of the corrosion protection ability of the coating system.

1 Introduction

Major advantages of Fe-based wear and corrosion resistant coatings are their favourable lower costs and good environmental behavior. Up to now, most of high performance Fe-based coatings have been developed by means of high-velocity-oxygen-fuel (HVOF)-technology [1-7]. High Cr-contents up to X_{cr}=33 wt. % in the coatings provide the desirable corrosion resistance. The precipitated fine borides and carbides in the coatings as well as the amorphous-nanocrystalline composite contents lead to their high wear resistance. An increased sensitivity to cracking was reported in the literature in case of coatings applied by means of air plasma spraying(APS) [7].

In the previous research work [8-10], a novel economical coating system based on FeCr25Mn10BC, which is not prone to above mentioned cracking sensitivity, has been developed using APS spraying. Because this coating system does not contain cost-intensive alloying elements, it is more cost-effective compared with the coating systems as reported in [1-7]. A comparable corrosion resistance to that of 316L stainless steel coating could be confirmed due to the high chromium content of X_{cr}= 25 wt. % [8]. The fine borides in the coatings provide high wear resistance of the coatings. The rig tests in [10] showed that cast iron components could be well protected by this coating system against erosion loads as in pump systems.

Three-cathode plasma generators improve the plasma stability compared to conventional one-cathode plasma generators significantly, leading to more stable and uniform treatment of particles [11-13]. As shown in [8-10], high quality coatings could be well reproduced using TriplexPro™-210 with different FeCr25Mn10BC powders due to the high process robustness in terms of particle in-flight properties under optimized spray conditions. However, during coating deposition on complex geometries, process parameters such as stand-off distance and spray angle usually cannot be kept at optimized conditions as developed in [8-10]. Therefore, the influence of verying stand-off distance and spray angle on the characteristics of the novel FeCr25Mn10BC coating system, which is applied by means of a three-cathode plasma spraying system, will be investigated in this study.

2 Experimental procedure

The same powder, which as developed in [9,10] and exhibit a particle distribution of -45 +20 µm, is used in this study. **Fig.1** shows a SEM-micrograph of this powder. A high number of spheric particles can be recognized. Both ferrite and boride were identified in this powder.

Fig. 1. SEM-micrograph of the spray powder FeCr25Mn10BC -45 +20 µm.

Plates made of grey cast iron EN-GJL-250 were used as substrate samples. After being blasted with corundum, the samples were coated using a three-cathode plasma generator TriplexPro™-210 from Oerlikon Metco Europe GmbH (Kelsterbach, Germany). Four parameter sets were developed to investigate the process robustness. The optimized parameter set used in [9,10] was selected as "Reference". Based thereon, both the stand-off distance and the spray angle were varied, while the other process parameters were kept constant. The four spray parameter sets examined in this study are listed in Table 1.

Table 1: Spray parameters for TriplexPro™-210.

	Reference	Coating No. 1	Coating No. 2	Coating No. 3
Current (A)	360	360	360	360
Ar (SLPM)	70	70	70	70
Distance (mm)	100	**110**	**90**	100
Angle (°)	90	90	90	**60**

The microstructures of the coatings were investigated by means of optical light microscopy with Axiophot from Carl Zeiss Microscopy GmbH (Jena, Germany) and scanning electron microscopy (SEM) with Zeiss LEO 1530 Gemini (Carl Zeiss Microscopy GmbH). Microhardnesses were measured on the cross-sections of the coatings using a Vickers hardness tester of Buehler Ltd (Düsseldorf, Germany) applying a load of F=2,94 N. A microhardness value was given as an average of ten measurements. Coating porosities were determined on cross-sections of the coatings using image analysis system AxioVision of Carl Zeiss Microscopy GmbH. The bond strengths of the coatings were determined using Precision Adhesion Tester (PAT). Scratch tests were carried out using the Scratch-Tester "LSRH-Revetest" of CSM Instruments (Freiburg, Germany) to simulate simple abrasion process. Two loads of F=50 N and F=100 N as well as a scratch length of d=8 mm were applied onto polished surface. The scratch volumes were determined using the confocal laser-scanning microscope (CLSM) VK-X200 from Keyence Deutschland GmbH (Neu-Isenburg, Germany). Salt spray test according to ISO9227 was conducted to evaluate the corrosion resistance of the coatings comparatively.

3 Results and discussion

Fig. 2 shows light optical micrographs of cross-sections of all coatings, giving an overview about the microstructures of the applied coatings. **Fig. 3** presents exemplarily a SEM-micrograph of the coating "Reference" with higher magnification, demonstrating detailed microstructural characteristics. It is recognizable that all coatings exhibited similar lamellar microstructures, indicating that the particles were mostly molten prior to their impact onto the substrate. The particles flattened well on the substrate so that most of lamellae were thinner than 5 µm as shown in **Fig. 3**. The coatings exhibited limited amounts of oxide inclusions.

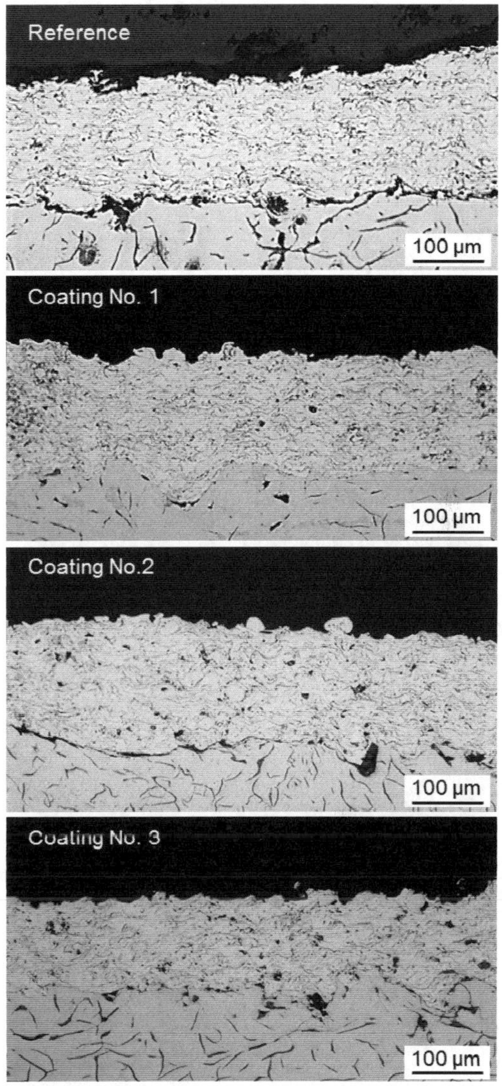

Fig. 2. Light optical micrographs of cross-esctions of all coatings.

It can be seen that the coatings exhibits a dense microstructure. The measured porosity amouts to about p= 2.0%, 1.6%, 2.0% and 2.8% for the coatings "Reference", Coating No.1, Coating No. 2 and Coating No. 3, respectively. The reduction of spray angle led to high porosity compared to the variation of stand-off distance, because the velocity vector vertical to the substrate surface was reduced about 30 % compared to the spray angle of θ=90°. The coating No. 2 exhibited the lowest porosity. It is assumed that the average particle velocity during particle impact was higher due to the shorter stand-off distance.

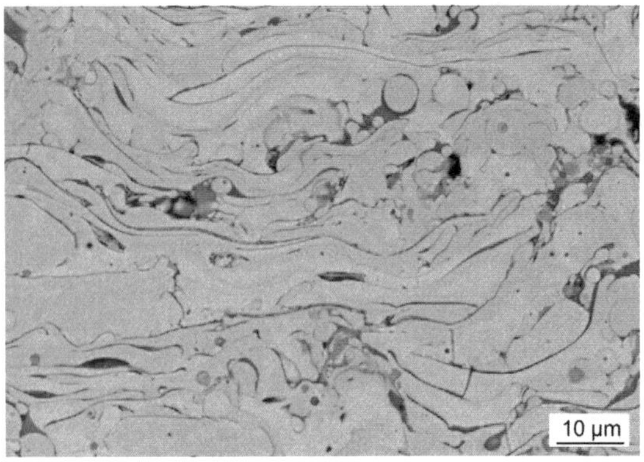

Fig. 3. SEM-micrograph of a cross-section of the coating "Reference".

The variation of spray angle caused about 15 % reduction in the coating thickness, while the variation of stand-off distance led to changes of less than 5 % in the coating thicknesses. The reduced velocity vector vertical to the substrate surface and the additional velocity vector parallel to the substrate surface increases the probability of particles bouncing off when impacting onto the substrate compared to other parameter sets.

Fig. 4 shows the measured microhardness values of the coatings. Generally speaking, both the variation of stand-off distance and spray angle led only to slight changes in the microhardness.

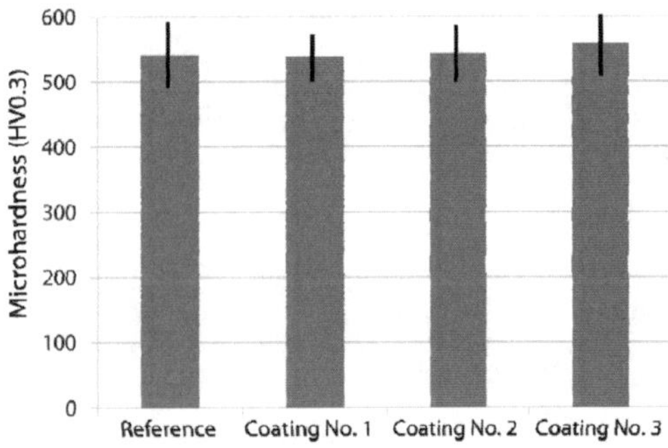

Fig. 4. Measured microhardness values of the coatings.

Fig. 5 shows the determined bond strengths of the coatings. It can be seen that the coatings exhibited bond strength values varying in a range of σ=25-29 MPa. The variation of stand-off distance influenced hardly the bond strength of the coating. In contrast, the bond strength increased more than 10 %, due to the spray angle reduction from θ=90° to θ=60°. This can be mainly attributed to the lower residual stress in the coating inherent to the lower coating thickness of Coating No. 4.

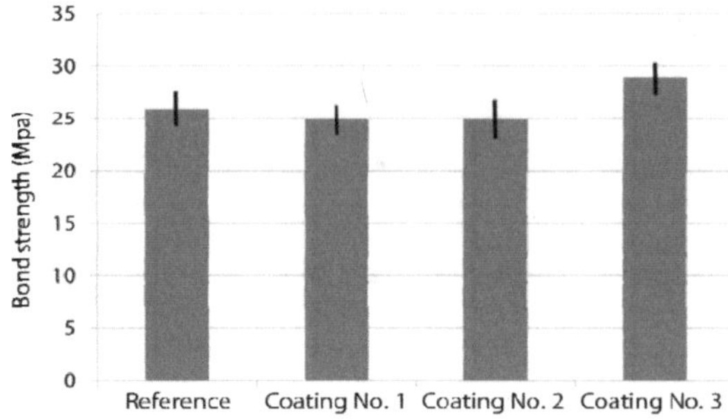

Fig. 5. Determined bond strengthes of the coatings.

The scratch test was carried out to simulate simple abrasion process. **Figs. 6** shows exemplarily the scratch tracks of the coatings under a load of F=50 N. It can be seen that all scratch tracks were very similar. The CLSM-analysis of these sratch tracks showed that the coatings exhibited upward edges, indicating that they were plastically deformed during scratch test.

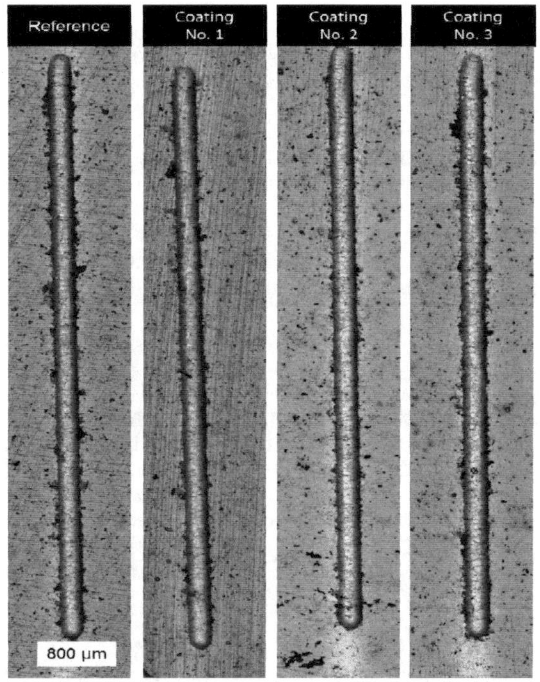

Fig. 5. Sractch tracks of the coatings under a load of F=50 N.

The measured track volumes by CLSM are given in Table 2. It is recognizable that the variation of stand-off distance had slight influence on the scratch resistance of the coating under the load of F=50 N. In contrast, the variation of spray angle caused clearly more increase in track volume amounting to about 35 %. In case of F=100 N, the percental change was significantly lower compared with F=50 N.

Table 2: Volumes of the scratch tracks.

Load (N)	Reference (μm^3)	Coating No. 1 (μm^3)	Coating No. 2 (μm^3)	Coating No. 3 (μm^3)
50	3.1×10^6	3.2×10^6	3.4×10^6	4.2×10^6
100	9.6×10^6	10.4×10^6	10.2×10^6	10.8×10^6

The degradations of the coatings in salt spray test were observed by visual inspection. The coatings were inpected after t=6 h and 30 h. The most severely corroded coating Coating No. 3 was selected for microstructural investigation of the cross section. The states of the coating surfaces in as-sprayed condition, after t=6 h and 30 h test can be seen in **Fig. 6**. It is very evident that after t=6 h test only the coating Coating No. 3 exhibited clearly corrosion products with rust colour on its surface, indicating that this sample was already attacked severely. After t=30 h test, lean corrosion products are visible on the other coatings. Again, Coating No.3 exhibited highest degree of corrosion and show a significant deterioration compared to the state after t=6h. As discussed in [8,9], this novel FeCr25Mn10BC coating system exhibits high corrosion resistance due to its high Cr-content. Therefore, the corrosion products on the coating surface resulted mainly from the corrosion of the substrate material at the coating interface, because the corrosive medium penetrated to the substrate. The microstructural investigation of the Coating No. 3 confirmed the corrosion of the substrate material at the interface as shown in **Fig. 7**. The higher porosity and lower coating thickness in case of Coating No. 3 promoted the penetration of the corrosive medium. The results show that the variation of spray angle influenced significantly the corrosion behavior of the coated sample due to the reduced thickness and increased porosity.

4 Conclusions

This study covers the investigation of the process robustness during spraying of the novel FeCr25Mn10BC alloy by TriplexPro™-210 in terms of complex part geometries by varying the stand-off distance and the spray angle. The results show that under the given test conditions both coating formation and properties are generally influenced more sensitively by the variation of spray angle in comparison with the variation of stand-off distance. The corrosion behavior of the coating was significantly more sensitive to the varying conditions in comparison to the hardness and abrasion resistance.

5 Acknowledgements

This work developed within the scope of IGF-project 17727 N. The IGF-project of VDMA has been, financially supported by the German Ministry of Economic Affairs and Technology via AiF within the frame work of the program for the promotion of joint industrial research and development (IGF). The authors would like to thank all parties involved for funding and support.

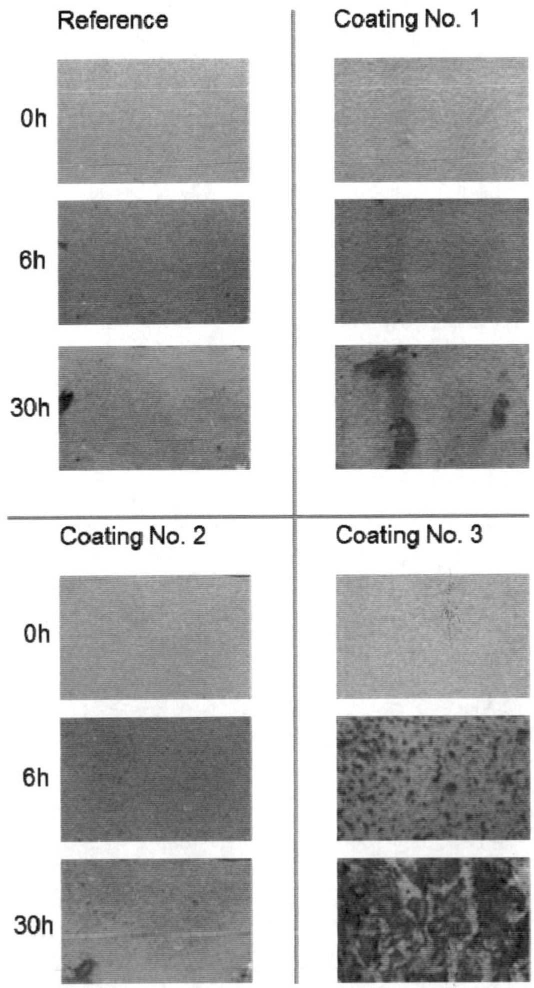

Fig. 6. Pictures of the coatings after different test durations in salt spray test.

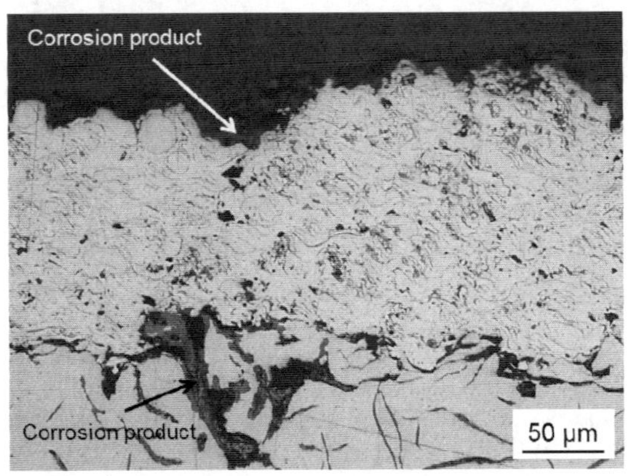

Fig. 7. Micrograph of a cross-section of the coating Coating No. 3 after 30 h test.

6 Literature

[1] A. Milanti, H. Koivuluoto, P. Vuoristo, G. Bolelli, F. Bozza and L. Lusvarghi, Microstrucutral Characteristics and Tribological Behavior of HVOF-Sprayed Novel Fe-Based Alloy Coatings, Coating, 4, 2014, pp. 98-120.

[2] A. Schwenk, HVOF-K2-Spraying of Iron-Based Materials for Applications in Printing Industry. Proc. of 8.Colloquium High Velocity Oxy-Fuel Flame Spraying, Erding, 2009, pp. 191-198

[3] A.M. Osmanda, A.J. Battenbough and A. M. Staines, Fe-Cr and Fe-Cr-Ni Based HVOF Coating Systems Customised for Wear and Corrosion Protection, Thermal Spray Bulletin 5, 2012, pp. 117-122.

[4] R.Q. Guo, C. Zhang, Q. Chen, Y. Yang, N. Li and L. Liu, Study of Structure and Corrosion Resistance of Fe-Based Amorphous Coatings by HVAF and HVOF, Corrosion Science, 53, 2011, pp. 2351-2356.

[5] H.S. Ni, X.H. Liu, X.C. Chang, W.L. Hou, W. Liu and J.Q. Wang, High Performance Amorphous Steel Coating Prepared by HVOF Thermal Spraying, Jornal of Alloys and Compounds, 467, 2009, pp. 163-167.

[6] D. Zois, A. Lekatou, M. Vardaoulias, Preparation and Characterization of Highly Amorphous HVOF Stainless Steel Coatings, Journal of Alloys and Compounds, 504, 2010, pp. 283-287.

[7] V. Varadaraajan, R. K. Guduru and P.S. Mohanty, Synthesis and Microstructural Evolution of Amorphous/Nanocrystalline Steel Coatings by Different Thermal-Spray Processes, Journal of Thermal Spray Technology, 22, 2013, pp. 452-462.

[8] K. Bobzin, L. Zhao, M. Öte, T. Königstein, Novel Fe-based Wear and Corrosion Resistant Coatings by Three-Cathode Plasma Technology, Surface&Coatings Technology, http://dx.doi.org/10.1016/j.surfcoat.2016.08.041

[9] K. Bobzin, L. Zhao, M. Öte, T. Königstein, Highly Economical Fe-based Protective Coating System by Novel Material Design and Full Utilization of the Atomized Powder, Proc. of ITSC 2016, Shanghai, China, 2016, pp. 700-705

[10] K. Bobzin, L. Zhao, M. Öte, T. Königstein, M. Oechsner, M. Siebers, G. Andersohn, J. Ellermeier, Development of a Novel Economical Iron-based Coating to Increase Lifetime of Components of Cast Irons, Part 1, Materials Science & Engineering Technology (accepted)

[11] K. Landes, Plasma Generators for Thermal Plasma Processes, International Journal of Materials Research,102, 2011, pp. 959-963.

[12] K. Bobzin, N. Kopp, T. Warda, M. Schäfer, M. Öte, A Numerical Investigation: Influence of the Operating Gas on the Flow Characteristics of a Three-cahtode Air Plasma Spraying System, Proc. of ITSC 2013, Busan, Korea, 2013, pp. 400-405

[13] K. Bobzin, M. Öte, Modelling Multi-Arc Spraying Systems,JTST, 25(5),2016,pp. 920-932

Impact of the stand-off distance and spray angle on the corrosion, cavitation and erosion behaviour of the thermal spray layers deposited by the three-cathode spraying

Matthias Oechsner[1], Marius Siebers[1], Georg Andersohn[1], Jörg Ellermeier[1], Kirsten Bobzin[2], Lidong Zhao[2], Mehmet Öte[2], Tim Königstein[2],

1: Center for Engineering Materials, State Materials Testing Institute Darmstadt (MPA) Chair and Institute for Materials Technology (IfW), TU Darmstadt, Grafenstr. 2, 64283 Darmstadt, Germany
2: Surface Engineering Institute, RWTH Aachen University, Kackertstr. 15, 52072 Aachen, Germany

Fe-based coatings, such as novel FeCrMnBC alloys, have both economic and ecological advantages compared to other coatings like Ni-based or Co-based coatings. In recent years, high performance Fe-based wear and corrosion resistant coating systems have been developed. Some of them have even been introduced into the market. However, the suitability of the FeCrMnBC alloy as coating for cast iron under complex erosive and corrosive stresses in particle-loaded fluids for pump parts has not been investigated yet. Especially the impact of the process robustness of three-cathode plasma spraying coatings applied with variable process parameters like stand-off distance and spray angle is in the focus of interest. The objective of the present work has been the characterization of novel FeCrMnBC alloys, for the first time deposited via Thermal Spray processes. The corrosion resistances as well as the cavitation and erosion properties were separately evaluated by current density-potential measurements and supersonic cavitation in artificial sea water. Erosion corrosion behavior has been investigated in a pump test rig with 10 wt.-% corundum (Al_2O_3) particles. The results show that the reduction of spray angle and the variation of stand-off distance limit the corrosion and cavitation resistance in different ways. The erosion behavior shows only small variations for the tested parameters. The results reveal that the FeCrMnBC coatings exhibit high process robustness for the chosen parameter variations and a large potential to improve the protection of cast iron even for not optimized conditions.

1 Introduction

For applications in the pump industry the used pumped media varies typically regarding consistency, and promotes in many cases corrosion and wear, leading to failures of pump parts [1,2] during service. In particular marine pumps suffer under rapid flow of particle-loaded seawater in their environments. Therefore pump parts are often exposed to complex mechanical-corrosive stresses, causing material erosion in flow-affected areas [3]. The requirements for a developed and designed superior erosion-corrosion resistant material are good corrosion resistance and excellent erosion resistance in a definite environment [2,4].
Surface coating technologies, such as thermal spray technology has been successfully applied to produce protective coatings to satisfy increasing demands in terms of wear and corrosion protection in different industrial branches. Service run parts reveal that combined erosion-corrosion appears on surfaces and local areas under high loading. One approach of several studies is to apply different coatings to relatively low-cost materials, especially to repair worn or corroded areas [1]. Cast irons are widely used to produce pump components due to their excellent castability, their good machinability and mechanical properties, in addition of their economic advantages compared to other materials, like frequently used stainless steels (e.g. 1.4462).
The high velocity oxygen fuel (HVOF) spray process is a standard process to perform dense coatings (e.g. Ni-based, Co-based and WC reinforced alloys) with excellent wear and corrosions resistance. A disadvantage of the Ni- and Co-containing coatings are the expensive and fluctuating prices on the world market, next to unecological properties [5].
Thermally sprayed Fe-based coatings without Ni or Co offer both, economic and ecological advantages. Several studies have been done to verify anti-wear and corrosion properties of Fe-based coatings [6-9]. The achievable quality of the thermal spray coatings is depending on the chosen process and the process parameters, describing the process robustness. To deposit dense layered structures and low porosity containing coatings, recently the atmospheric plasma sprayed (APS) technique, next to the frequently used HVOF technique, is used [10].
The objective of the current work is to investigate atmospheric plasma sprayed (APS) novel Fe-based coatings as a valid alternative to expensive and conventional thermal sprayed Ni-based coatings on grey cast iron components (EN-GJL-250). The relationships among the microstructure, corrosion, cavitation, and erosion of novel developed thermal spray FeCrMnBC-coatings are studied to gain first insights into the mechanisms. Therefore the influence of variations in stand-off distance as well as spray angle are investigated, with regard to the coating properties. The tests were carried out with four variations of process parameters for the FeCrMnBC-coating and an iron cast substrate, as well as a duplex steal as reference.

2 Experimental procedure

The material used in this work was a newly developed and not yet commercially available $Fe_{rest}Cr_{25}Mn_{10}B_{1.5}C_{0.25}$ (wt.%) alloy with a powder size of -45+20 µm. The coating was prepared using atmospheric plasma spraying (APS) method. Plates of grey cast iron EN-GJL-250 (DIN EN 1561:2012-01) with a size of 100 mm x 100 mm x 8 mm were used as a substrate material. Further a duplex steel 1.4462 (ASTM S31803 (2205)) was used as a reference material. To deposit the investigated Fe-based coatings a three-cathode plasma generator (TriplexPro-210 Oerlikon Metco Europe GmbH, Kelsterbach, Germany) with different spray parameters was used. The details of the thermal spray process are described elsewhere [11,12]. It must be considered that all spray parameters are kept constant, except variations in the stand-off distance and the spray angle. In this case, the distance of 100 mm and the angle of 90° were regarded as initial condition, **Table 1**. All spraying experiments were carried out at ambient atmosphere.

Tab. 1. Different process parameters for coatings

Fe-based thermal spray alloys Powder size: -45+20 µm	**Parameter:** Stand-off distance/ Spray angel	**Coating name**	**Thick-ness of coating**
Fe25Cr10Mn1.5B0.25C	D100 mm/ A90°	D100	≈ 200 µm
Fe25Cr10Mn1.5B0.25C	D90 mm/ A90°	D90	≈ 219 µm
Fe25Cr10Mn1.5B0.25C	D110 mm/ A90°	D110	≈ 152 µm
Fe25Cr10Mn1.5B0.25C	D100 mm/ A60°	A60	≈ 135 µm

By cutting the coated plates and the reference material, coupons of 24 mm in diameter and a thickness of 8 mm were manufactured. The described samples were used for subsequent investigations. Prior to testing, all samples were cleaned in an ultrasonic bath using alkaline solvents and afterwards dried in warm air. In order to simulate the corrosion occurring under service-like conditions (e.g. offshore) the investigations were carried out in artificial sea water (DIN 50905-4:1987-01) as media saturated with air.

The corrosion behavior of the Fe-based coatings, the substrate EN-GJL-250 and the reference material 1.4462 were evaluated by potentiodynamic current density/potential curve measurements. Electrochemical potentiodynamic experiments for evaluation of current density/potential curve were performed with a potentiostat (Gamry Reference 600, Gamry Instruments, USA). A standard three-electrode system in a 1-l glass cell opened to ambient air at 30 °C with artificial sea water as electrolyte was used. The counter electrode (CE) was a platinum gauze, and all potentials were measured relative to a silver/silver chloride (Ag/AgCl) electrode used as reference electrode (RE). The samples were connected as working electrode (WE) with an exposure area of 1 cm^2. After immersing the specimen and prior to starting the potentiodynamic test, the open-circuit potential was measured for 0.5 h, until the system reached stability. The potential for polarization was swept from E = -200 mV to E = 1200 mV relative to the open-circuit potential with a fix ratio of 1000 mV/h. In order to estimate the coating fatigue under superimposed corrosive loads, several cavitation experiments with an ultrasonic cavitation test rig according to ASTM G32-10 standard was carried out [13-15], **Figure 1**.

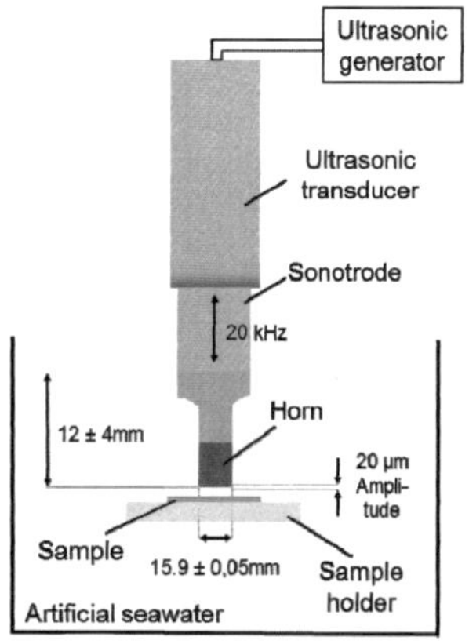

Fig. 1. Schematics of the ultrasonic cavitation test rig with glass vessel and sample holder

The used sonotrode is based on a 20 kHz generator (Telsonic DG-200-02, Switzerland). The samples were held stationary below the vibrating horn with a distance of 0.4 mm and 12 mm below the liquid surface. The vibration frequency was 20 kHz with an amplitude (peak-to-peak) on the horn tip of 20 µm. Experiments took place in artificial sea water at 30 °C. Before and after the test, the samples were ultrasonically cleaned with acetone and dried under airflow. The weight loss was measured via a precision scale (Sartorius RC 250 S, Göttingen, Germany) to evaluate the cumulative mass loss rate (mg/h). Out of three runs of each probe type, the averaged cumulative mass loss rate was determined and plotted versus the testing time. The overall test duration was 5 h. The cavitated surfaces were investigated by scanning electron microscope (SEM - Zeiss CrossBeam 1540 EsB (Carl Zeiss, Jena, Germany)) in order to study the main corrosion and wear mechanisms.

In order to investigate the corrosion and wear on pump components exposed to the corrosive medium, an erosion-corrosion test rig was used [16]. The design consists of a pump impeller, which is driven at distinct speed (drive power 1,1 kW) and rotates in a test vessel with integrated samples in the shell surface. Testing parameters are given in **Table 2**.

Tab. 2. Parameters of the erosion corrosion test rig

Parameter	Value
Solid (Corundum):	Al_2O_3 (-75+43 µm)
Quantity of solid:	100 g l^{-1}
Temperature:	50 °C
Rotation speed:	1400 min^{-1}
Test duration per sample:	200 h

By rotation of the impeller the erosive particle-containing liquid media (10 wt.%) is transported and accelerated to the sample surface. Artificial seawater (DIN 50905-4:1987-01) with corundum was used as erosive media. The resulting flow (v ~ 15.5 m s^{-1}) impinged at an angle of approximately 30° to the sample surface [17]. The erosion loading on the surface of the sample differs in dependence of the impact angle of the particles to the surface. This provides the possibility to investigate increasing erosion degradation on one sample surface [17]. During the particle wear testing the erosive media was exchanged after every 20 h test time. In order to investigate the wear appearances of the surfaces, caused by erosion-corrosion testing, a scanning electron microscopy (SEM) was used.

To validate the potential of the Fe-based thermal spray coatings each time the uncoated substrate material EN-GJL-250 and the reference material 1.4462 were investigated by the previously mentioned methods.

3 Results and discussion

3.1 Corrosion behavior

The corrosion behavior of the coatings/materials are investigated via current/density potential curve measurements (iE) as mentioned above. The polarization curves of the four coatings in comparison with the substrate material and the reference material differed in main curve parts, **Figure 3**.

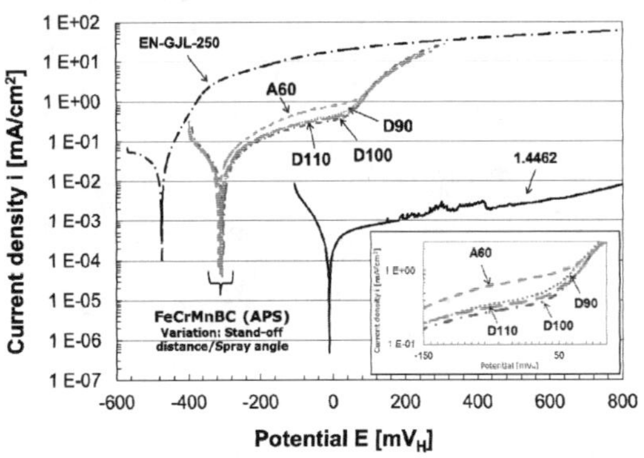

Fig. 3. Polarization curves of parameter variation of Fe25Cr10Mn1,5B0,25C-coatings com-pared with EN-GJL-250 substrate material and 1.4462 reference material (artificial sea water, 30 °C, potential swept rate 1000 mV/h, air saturated media)

For the substrate material EN-GJL-250 in its uncoated state, polarization out of the negative open circuit potential (OCP) (≈ -475 mV_H) in the anodic regime reveals a rapid increase in the current density, leading to immediate general corrosion on the surface. Compared to this, the reference material 1.4462 is characterized by a pronounced passivation regime with a very low passivation current density of 0.0006-0.006 mA/cm^2. The polarization curves of the coatings are located between the values of the substrate and reference material for all varied parameter sets. However, all the thermal spray coatings show more positive and slightly different OCPs (≈ -310 mV_H), indicating an improvement on the corrosion resistance compared to the uncoated substrate. A small passivation region of ≈ 310 mV and a lower current density (0.25-0.45 mA/cm^2) compared to uncoated substrate is measured [18]. This implies a corrosion protection effect, although associated effects of porosity, micro-cracks and oxidation in the coating structure are existent. For the variations regarding the stand-off distance (D90, D100, D110) only marginal differences in between the iE-curves are detectable. Lowering the spray angle from 90° to 60° (A60) increases the current density on the order of $\Delta \approx$ 0.25-0.3 mA/cm^2 (see enlargement in **Figure 3**) and therefore decreases the corrosion resistance.
To investigate the mechanism of corrosion in a first approach, the coatings morphology has been documented after the polarization experiments once the transpassive dissolution started and the first corrosion damages occurred, **Figure 4**. All microstructures are characterized by a high particle flattening ratio with different relative amounts of unmolten particles in the lamellar structure.

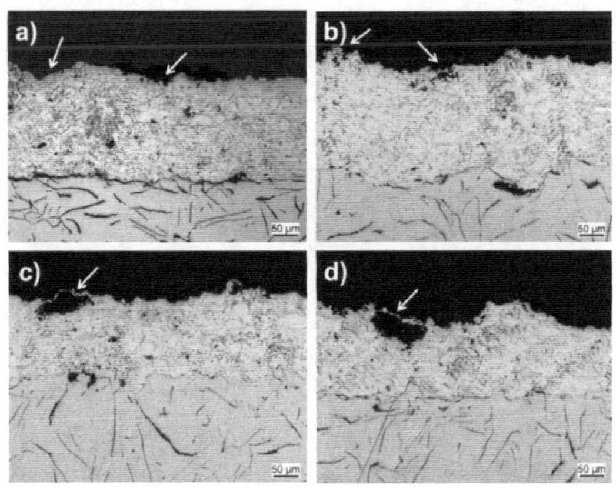

Fig. 4. Metallographic sections of parameter variations of Fe25Cr10Mn1.5B0.25C thermal spray coating after potentiostatic polarization a) D100 b) D90 c) D110 d) A60 (arrows = slight corroded zones)

Galvanic coupling causing corrosion between the coating/substrate interfaces could not be documented for any coating variant. No corrosive attack at the interface (especially on the lamellar graphite of the substrate) with a de-adhesion by the corrosive electrolyte was present. Corrosion phenomena in the coatings, labelled as microgalvanic corrosion, are detected for all parameter variations [19,20]. The microgalvanic corrosion results from the propagation of micro-cracks, oxidation and depends on the level of porosity in the coating, which may cause short-circuit paths for the electrolyte [20,21]. This reduces the corrosion protection capability of the coatings. The variation in stand-off distance (D90, D110) shows slight deviations in corrosion. By shortening the stand-off distance from 100 mm to 90 mm, the relative amount of unmelted particles in the coating increased, causing increased porosity and micro-cracks. An increase in the distance to 110 mm (D110) increased the micorgalvanic corrosion which could be due to increased oxidation of particles. By changing the spray angle from 90° to 60° (A60) the overspray of particles is intensified which results in a thickness decrease. Furthermore, the relative amount of porosity, oxides and micro-cracks is increased, which leads to more pronounced microgalvanic corrosion in comparison to the stand-off distance variation, **Figure 4 d)**. It is observed that defects (micro-cracks, porosity and surface inhomogeneities) are initiation points for the formation of interlamellar corrosion. The resultant microgalvanic corrosion is based on the selective dissolution of the FeCr coating, supporting the findings of other researchers [19,20]. Further studies were necessary to qualify the different corrosion mechanisms, e.g. by immersion tests.

3.2 Cavitation

The cumulative mass loss rate as a function of the testing time for all samples after 5 h cavitation tests show distinct differences between the coatings with their parameter variations and the bulk materials, **Figure 5**.

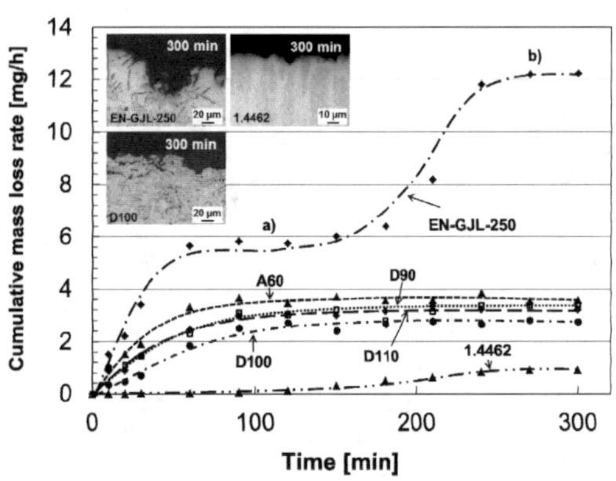

Fig. 5. Cumulative mass loss rate versus testing time of parameter variation of Fe25Cr10Mn1.5B0.25C-coatings compared with EN-GJL-250 substrate material and 1.4462 reference material (artificial sea water, 30 °C)

Generally speaking, the reference material 1.4462 reveals the best cavitation resistance among the tested materials while the EN-GJL-250 showed the worst behavior. Compared to 1.4462 the mass loss of EN-GJL-250 is approximately 13 times higher, and ~ 3.5 times higher compared to the coated specimens. The reference material exhibited the typical incubation stage, where the mass loss rate is zero or negligible small compared to later observed stages. After 6 h of testing the cavitation rate is approx. constant. For the substrate material EN-GJL-250 a steep increase in mass loss rate without incubation is detected. The mechanism can be documented as a time-dependent two-step increased mass loss rate with a) crack growth and eruptions on graphite lamellae in the beginning and b) crack growth and eruptions of the remaining iron matrix. The Fe25Cr10Mn1.5B0.25C-coatings with their different utilized process parameters show only minor deviations and their cumulative mass loss rates vary only slightly. Thus, it is not possible to directly compare the coatings by metrology, only tendencies can be shown. All curves revealed a direct increase in the beginning. Among the coatings, the D100 exhibited the best behavior. Variation in the angle (A60) increases the mass loss rate in small amounts. The examinations of the surfaces in worn state after testing reveal different characteristics of the materials, **Figure 6**.

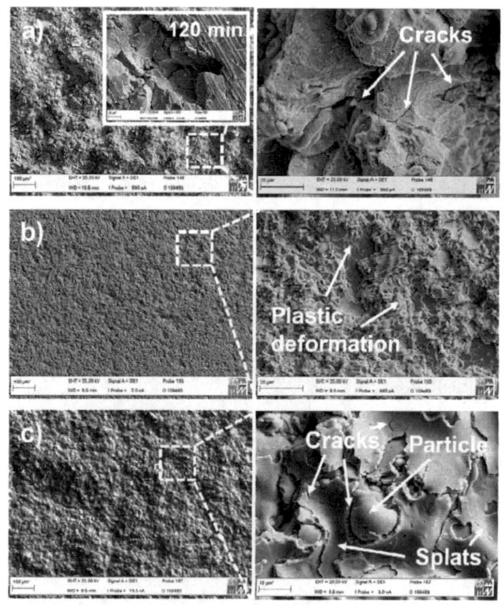

Fig. 6. Worn surfaces after cavitation test of a) substrate material EN-GJL-250 b) reference material 1.4462 c) Fe25Cr10Mn1.5B0.25C-coating D100 (left: survey, right: detail, SEM images)

The distinct surface deterioration of the EN-GJL-250 showing the high cumulative mass loss rate is caused on crack growth and excessive eruptions in the previously described two-step degradation mechanism. After 120 min, eruptions at the lamellae graphite occurred at the surface in the first step, enlargement **Figure 6 a)**. The entire surface was cavitated after 300 min and a brittle fracture of the cast matrix is documented. For the 1.4462 material, a smooth surface with plastic deformation appears, **Figure 6 b)**. The mechanism is based on fatigue delamination due to the accumulation of plastic deformation in distinct local areas. Numerous craters and grooves resulting from localized plastic deformation after 300 min testing time. The worn surfaces of the Fe25Cr10Mn1.5B0.25C-coatings reveal similar degradation mechanism for all parameter variations (D90, D100, D110, A60). A coarse and rough surface with cracks in the lamellar structure, around unmelted particles and pores is documented, **Figure 6 c)**. The wear mechanism seems to be based on delamination at the interface between the splats, due to surface fatigue. Pores and imperfections (unmolten particles) have a detrimental influence. For changing the spray angle (A60) the degradation increases marginal, based on increase in relative amount of porosity, unmolten particles and micro-cracks in the lamellar structure. The shock waves of the cavitation are absorbed by the coating without significant plastic deformation, which lead to the formation and growth of fatigue micro-cracks [14].

3.3 Erosion-corrosion

The degradation of surfaces tested in the erosion-corrosion test rig appears non-homogeneous after 200 h and can be separated in distinct areas, **Figure 7**. From the left side to the middle of the sample regions with mild increasing wear, depending on the material, are documented. At the right side all samples show a differently pronounced erosion track, due to severe erosion conditions based on localized circulating particle swirls.

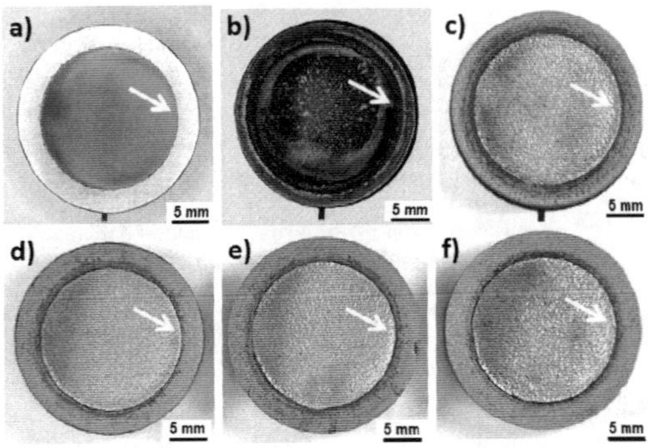

Fig. 7. Light microscope images of a) 1.4462 b) EN-GJL-250; Fe25Cr10Mn1.5B0.25C: c) D100 d) D90 e) D110 f) A60 after 200 h erosion-corrosion testing (arrow: position of erosion track, fringe: origin state)

Concerning the 1.4462 specimens, the erosion reveals small material removal, mainly induced by passive film removal based on cutting from corundum particles and attack of chloride ions. Ductile erosion damage with slight increased surface roughness and a low erosion track (~19 µm) occur, **Figure 8 a)** and **b)**. Contrary to the 1.4462 specimens, the uncoated EN-GJL-250 surface exhibits superimposed eroded and corroded areas with deep grooves and pronounced wear patterns on the entire surface. A deep erosion track with ~1.6 mm is measured, **Figure 8 c)** and **d)**.

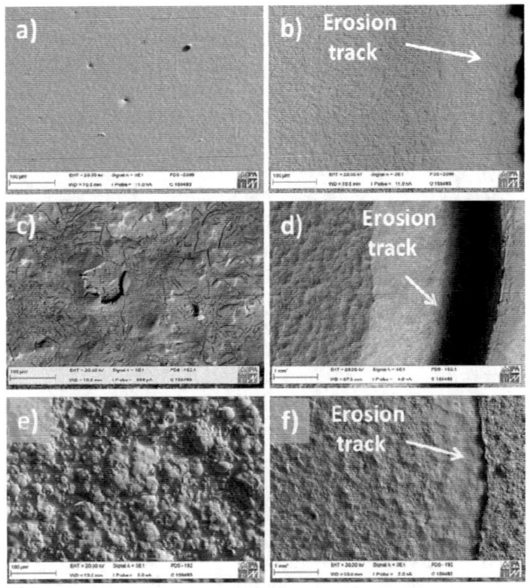

Fig. 8. Degradation of surface after erosion test of: a),b) substrate material EN-GJL-250 c),d) reference material 1.4462 e),f) Fe25Cr10Mn1.5B0.25C-coating D100 (left: area of uniformly directed flow, right: erosion track, SEM images)

Note that all coated EN-GJL-250 parts are protected against erosive particles and no corrosion at the interface appeared. The surfaces of the Fe25Cr10Mn1.5B0.25C-coatings seem to be polished by the particles for the stand-off distance of 100 mm (D100), causing a leveling of roughness peaks, **Figure 8 e)**. Across the entire surface a homogeneous erosive wear with a smaller distinct erosion track (~40 µm) compared to the uncoated substrate EN-GJL-250 is observed. It is suggested that in the erosion track an impact region with some detached fragments of the lamellar coating exhibited, in which a subsequent polishing by the corundum particles occurred. A uniform corrosion behavior without pits or micro-cracks is documented. Furthermore, the erosion-corrosion mechanism for the other investigated coatings with stand-off distance variation (D90, D110) and spray angle variation (A60) exhibit equal constant results, **Figure 7 d)** to **f)**. For all chosen parameter variations regarding the

Fe25Cr10Mn1.5B0.25C-coatings, an improved resistance against surface degradation under complex mechanical-corrosive stress was documented for tests in the erosion corrosion test rig.
However additional investigations on the hypothesis of the mechanisms needed to be done with further experiments, to verify the results.

4 Conclusions

In this study corrosion, cavitation and erosion tests of newly developed thermally sprayed Fe25Cr10Mn1.5B0.25C-coatings were performed. The focus was on parameter variations in the stand-off distance and spray angle, next to characterization of a substrate material EN-GJL-250 and a reference material 1.4462. Comparative investigations to determine the corrosion and wear resistance of the surfaces were carried out.
The main conclusions are:
- Variations within the spray parameters (stand-off distance and spray angle) exhibited slight differences within the respective test methods. Overall a good process stability of the atmospheric plasma spraying (APS) in the selected parameters was documented.
- Polarization curves revealed an increased corrosion resistance of Fe25Cr10Mn1.5B0.25C-coatings compared to uncoated EN-GJL-250, limited by micorgalvanic corrosion in the lamellar structure. Variation in spray angle from 90° to 60° increased the corrosion current density and decreased corrosion resistance. Microgalvanic properties and the microstructure of the coatings are supposed to be dominant properties.
- The resistance to cavitation of all studied Fe25Cr10Mn1.5B0.25C-coatings was lower compared to the reference material 1.4462, attributed to brittle fracture of porosity, weak cohesion between splats and preexistence of micro-cracks in the coating. Compared to uncoated substrate EN-GJL-250 the mass loss rate of the coatings was distinctly smaller.
- By deposition of coatings with different parameter variants, the erosion resistance of EN-GJL-250 was improved in the direction of service-like wear resistance of the reference material 1.4462.
- The Fe25Cr10Mn1,5B0,25C-coatings with better corrosion resistance may be candidates in application for solving the different wear problems of pumps in corrosive particle-containing media.

Acknowledgements

The IGF project (17727/N2) of the mechanical engineering industry association (VDMA) was supported by the german federation of industrial research associations (AiF) within the scope of the program to promote industrial community research and development (IGF) by the Federal Ministry for Economic Affairs and Energy (BMWi) based on a decision by the German Bundestag. The authors would like to thank the VDMA for the opportunity to carry out the project as well as the BMWi and the AiF for the financial support. The project was accompanied by the VDMA research group pumps and systems association. This research group deserves our thanks for the great support.

References

[1] Wang, Y. et al. Slurry erosion-corrosion behaviour of high-velocity oxy-fuel (HVOF) sprayed Fe-based amorphous metallic coatings for marine pump in sand-containing NaCl solutions. Corros. Sci. 53 (2011), Issue 10, pp. 3177/3185.
[2] Wang, Y. et al. Corrosion and erosion-corrosion behaviour of activated combustion high-velocity air fuel sprayed Fe-based amorphous coatings in chloride-containing solutions. Corros. Sci. 98 (2015), pp. 339/353.
[3] Reuschel, A., Merkle, S. & Schmauder, S. Dem Verschleiß von Pumpenlaufrädern auf der Spur. Process 3 (2010), pp. 64/66.
[4] Burstein, G. T., Sasaki, K. The birth of corrosion pits as stimulated by slurry erosion. Corros. Sci. 42 (2000), Issue 2, pp. 841/860.
[5] Milanti, A. et al. Microstructural Characteristics and Tribological Behavior of HVOF-Sprayed Novel Fe-Based Alloy Coatings. Coatings 4 (2014), Issue 1, pp. 98/120.
[6] Pang, S. J., Zhang, T., Asami, K. & Inoue, A. Bulk glassy Fe–Cr–Mo–C–B alloys with high corrosion resistance. Corrosion 44 (2002), pp. 1847/1856
[7] Kumar, S., Kim, J., Kim, H. & Lee, C. Phase dependence of Fe-based bulk metallic glasses on properties of thermal spray coatings. J. Alloys Compd. 475 (2009), Issue 1-2, pp. 10/13.
[8] Yuping, W. et al. Cavitation erosion characteristics of a Fe-Cr-Si-B-Mn coating fabricated by high velocity oxy-fuel (HVOF) thermal spray. Mater. Lett. 61 (2007), Issue 8-9, pp. 1867/1872.
[9] Bolelli, G. et al. Micromechanical properties and sliding wear behaviour of HVOF-sprayed Fe-based alloy coatings. Wear 276–277 (2012), pp. 29/47.
[10] Zhou, Z., Wang, L., He, D. Y., Wang, F. C. & Liu, Y. B. Microstructure and electrochemical behavior of Fe-based amorphous metallic coatings fabricated by atmospheric plasma spraying. J. Therm. Spray Technol. 20 (2011), pp. 344/350.
[11] Bobzin, K. et al. Influence of the stand-off distance and spray angle on the coating formation and and properties during three-cathode plasma spraying. Düsseldorf, ITSC (2017), in press.
[12] Bobzin, K. et al. Entwicklung einer neuartigen wirtschaftlichen, eisenbasierten Beschichtung zu Erhöhung der Lebensdauer von Gussbauteilen unter dem Gesichtspunkt der Korrosionsbeständigkeit. Materwiss. Werksttech. 48 (2017), in press.
[13] Hahn, M. & Fischer, A. Investigation of surface fatigue of thermally sprayed micro- And nanocrystalline cylinder wall coatings by means of cavitation testing. Int. J. Mater. Res. 99 (2008), Issue 10, pp. 1079/1089.
[14] Espitia, L. a. & Toro, a. Cavitation resistance, microstructure and surface topography of materials used for hydraulic components. Tribol. Int. 43 (2010), pp. 2037/2045.
[15] Santa J.F., Espitia L.A., Blanco J.A., Romo S.A., Slurry and cavitation erosion resistance of thermal spray coatings. Wear 267 (2009), pp. 160/167.
[16] Durst, O., Ellermeier, J., Troßmann, T. & Berger, C. Erosions-Korrosions-Untersuchungen an gradierten Multilayer-Chromkarbidschichten. Materwiss. Werksttech. 40 (2009), pp. 756/768.
[17] Depner-Miller, U. et al. Surface & Coatings Technology In fl uence of application technology on the erosion resistance of DLC coatings. Surf. Coat. Technol. 237 (2013), pp. 284/291.
[18] Zhang, C., Guo, R. Q., Yang, Y., Wu, Y. & Liu, L. Influence of the size of spraying powders on the microstructure and corrosion resistance of Fe-based amorphous coating. Electrochim. Acta 56 (2011), Issue 18, pp. 6380/6388.
[19] Verdian, M. M., Raeissi, K. & Salehi, M. Corrosion performance of HVOF and APS thermally sprayed NiTi intermetallic coatings in 3.5% NaCl solution. Corros. Sci. 52 (2010), Issue 3, pp. 1052/1059.
[20] Redjdal, O., Zaid, B., Tabti, M. S., Henda, K. & Lacaze, P. C. Characterization of thermal flame sprayed coatings prepared from FeCr mechanically milled powder. J. Mater. Process. Technol. 213 (2013), Issue 5, pp. 779/790.

[21] Guilemany, J. M., Fernandez, J., Delgado, J., Benedetti, A. V & Climent, F. Effects of thickness coating on the electrochemical behaviour of thermal spray Cr C–NiCr coatings 3 2. Surf. Coatings Technol. 153 (2002), pp. 107/113.

SUBSTRATE PREPARATION BY LASER TEXTURING FOR IMPROVED COATING ADHESION ON AZ91D MAGNESIUM ALLOY

S. S. Aulakh[1], G.Kaushal[2]
[1] Lecturer, Shinas College of Technology, Shinas, Oman
[2]*Assistant Professor*, Mechanical Engineering Section, Yadavindra College of Engineering, Punjabi University, Guru Kashi Campus, Talwandi Sabo, Punjab, India

WC-12 Co powder was thermal sprayed on AZ91D Magnesium alloy using HVOF technique. Laser texturing substrate preparation technique was used to roughen the surface for good coating adhesion. Adhesion strength of the coating was determined using ASTM 633C adhesion strength test for thick coatings. were used to investigate coating microstructure. Coating was characterized for porosity, micro hardness and Scanning electron microscopy and X- ray diffraction techniques. The coating hence prepared, were found to achieve good adhesion strength, micro hardness and low porosity.

Keywords – Laser texturing, substrate preparation, WC-Co, adhesion strength.

1 Introduction

Thermal spraying is the method of producing corrosion, wear and heat resistant coatings for various industrial applications. It is the process of applying layers of metallic and non metallic materials on a surface prepared for the process. In thermal spray processes, the coating material is melted by some sort of thermal energy like, gases, plasma, electric arc etc. The molten particles are impinged on the substrate surface, on impact with substrate surface the particles flattens and produces adhesion to the substrate by interlocking of molten and semi molten particles with asperities of roughened surface [1]. A great number of interactive parameters have to be controlled in order to obtain good coatings and promote substrate adhesion [2]. Coating adhesion with the substrate surface is one of the fundamental factors behind the performance of the coating. To create good conditions to anchor coatings on the substrate, appropriate surface preparations are required [1]. Specific chemical, physical and mechanical characteristics are required for the substrate surface for good coating adhesion. [3]. For good coating adhesion the substrate surface should be free from any adsorbed molecules. The adsorbed molecules evaporate by the heat of molten particles and promote splashing. The substrate wettability has to as high as possible to improve splat spreading [2]. In order to remove organic contaminants from the surface chemical degreasing is used. Degreasing improves chemical behaviour of the surface resulting in good spreading of the particles. But to improve interlocking of the coating particles with surface further surface treatment is required. Conventionally grit blasting has been used to enhance the asperities on the surface to improve surface roughness and mechanical interlocking. The surface roughness, surface temperature, residual stress on surface, cleanliness of the surface, presence of oxide film are some of the important factors which affect the coating adhesion and hence the bond strength and subsequently the coating performance. A number of researcher have experimented with different ways of preparing the substrate surface for thermal spray coating [4]. Grit blasting, laser ablation, laser texturing and dry ice blasting are some of the substrate preparation techniques used by different researchers. Conventional substrate preparation techniques consist of degreasing the surface with some solvents and subsequent grit blasting. Although it is simple and cost effective way of preparing the surface for thermal spraying but it has some limitation [5-6]. Chemical solvents are not good for envvironemnt and pose health hazard to operators and grit blasting cause some damage to the surface and some grits get embedded into the surface, specially for ductile materials like aluminum and magnesium alloys. Grit blasting is very effective for most materials except ductile materials that may be damaged with a risk of micro-cracks on the surface [7]. Alternative substrate preparation processes are being developed to over the problems in surface preparation of ductile materials. Among the new methods dry ice blasting is one of the methods being researched other than laser technology [8].

Advances in laser technology presents new opportunities in the area of substrate preparation. Some of the researchers have already tried to develop laser surface treatments as an alternative to grit blasting process. S. Costil et al [9] studied the effect of flame preheating and laser preheating combined with laser ablation on splat morphology to clean the TA6V surface prior to plasma spraying of NiAl. A flame was used to preheat the surface prior to laser ablation and followed by plasma spraying. The author reported the improvement in splat shapes and reduction in splashing as the temperature of the surface is increased from 25^0C to 360^0C.

Similar results were also achieved with preheating done thorugh laser followed by laser ablation. Combined action of heating and cleaning promotes substrates wettability and adhesion between the particles and substrate. Laser ablation has also been used as surface pre-treatment on Ti-6Al-4V and Al 2017 before plasma spraying of NiAl coating. Surface morphology of laser ablated surface was found to be diferent for Ti-6Al-4V and Al 2017 alloy, whereas craters were found on the surface of Al alloy the titanium alloy had kind of melted surface. The authors corelate this phenomenon on the difference of thermal diffusivity of two mateials. So, thernal diffusivtiy of a material affects the surface morphology of a material after laser treatment. In the same study authors concluded that during laser ablation surface pre treatment change in chemcial affinity of the coating with substrate play a role in improvement of interfacial coating toughness. Laser ablation along with continuous heating of substrate during cold spray has been used for Ti-6Al-4V coating. Smoother surface prepared using laser ablation compared to grit blasting, along with increased plasticity of substrate due to laser heating tend to increase the coating adhesion. Apart from laser ablation, laser texturing has also been used as substrate surface preparation method prior to thermal spray coating. Laser surface texturing is a realtively new surface pretreatment technique that involves ceating precise diameter holes following preset patterns throughout surface. The accuracy of texturing is achieved by controlling laser and scanner parameter that influence the quality of surface to be coated by thermal spraying. This technique produces physico-chemical and mechanical modification on the surface [10].

Lamraoui et al [10] introduced laser surface texturing as a substrate preparation technique. They used pulsed Nd:YAG laser(1.064µm radiation ,120ns) laser with repititive scans to texture AL 2017. The effct of several parameters like scanning rate, distance between laser impacts, frequency of pulses and incident power was studied to optimise the laser texturing process to achieve good NiAl coatings.

Robin Kromer et al. [11] optimised the laser texturing prcoess by optimizing the hole diameter/depth ratio as well as the distance between holes for Ni-Al arc sprayed coating on Al 2017 and AISI 304L stainless steel.

Use of laser technology as surface preparation technique has it own unique selling points like, selective area preparaton, accuracy, better control over surface morphology. A specific laser tool adapted to the material to be treated (in terms of wave length, pulse duration, spot size and pulse frequency) added to a scanner for 3D shape modification can promote mechanical adherence for thick coating elaborated by thermal spraying [11].

Some of the researchers have already successfully used laser ablation and laser texturing techniques for Titanium and aluminium alloys but the effect of laser texturing on surface morphology and coating adherence in case of magnesium alloys has yet to be studied. Further, laser texturing and laser ablation has been used for plasma and cold spray coating techniques but it is yet to be investigated for HVOF thermal spray coating. The aim of the present study is to reaslise laser texturing on magnesium alloy AZ91D using HVOF thermal spray coating and investigate the effect of laser texturing on morphology of AZ91D and bond strength of thermal spray coating done on laser textured surface.

2 Experimenal Procedures

2.1 Material

The substrate material used in this study was magnesium alloy AZ91D (Al 8.4%, Zn 3.2%, Mn 0.04%, Cu 0.03%, Fe 0.02%). AZ91D alloy has been selected for the study because like aluminium alloy, magnesium alloys also sufer from the problem of sub-surface damage and grit inclusion from grit blasting process. Magnesium alloys are increasingly used in automobile parts, aircraft and aerospace industry due to its high specifc strength but at the same time, there are limitation over the use of magnesium alloys becasue of low wear and corrosion resistnace.

The substrate were machined into cylinderical shape Φ25.4 mm and length 38.1 mm as per ASTM633C standard.

2.2 Laser Texturing

Laser texturing has been realised on Nd:YAG Q switch laser system. The laser characteristics used are pulse repitition rate 2KHz, pulse wave duration 100ns and average power 20W. A pattern consisting of matrix of holes at distance of 100µm from each other was used to create texture. Scanner stops the laser beam at the points given in drawing and multiple pulses are generated for given duration to create the holes. Helium gas has been used as shroud gas to protect the surface from oxidation.

2.3 HVOF spray parameters

HVOF thermal spray coating technique was used to coat WC-12%Co on laser textured AZ91D samples. HVOF technique was used for thermal spray coating because in HVOF technique particles strike the laser textured holes at high velocity as well as in plasticized state due to high temperature of spray particles. It is assumed that it will lead to good mechanical anchoring as well as some chemical bonding between particles and substrate. Commercial agglomerated and sintered WC-12Co Sulzer metco powder with particle size -45+15µm was used for coating. WC-12%Co has been choosen for study because of its good wear and corrosion resistance properties. HVOF spray parameters were as given in the table I.

Table I HVOF spray parameters

Gun	HIPOJET 2700
Spray distance	180 mm
Oxygen Flow rate	260 $L.min^{-1}$
LPG flow rate	60 $L.min^{-1}$
Carrier gas- N_2 flow rate	9.4 $L.min^{-1}$
Air flow rate	550 $L.min^{-1}$
Powder feed rate	38 g/min

2.4 Characterization

JSM-6510LV scanning electron microscope was used to study the morphology of laser textured surface and coating. EDX technique was used to study the changes in composition of AZ91D surface affected by laser texturing. Coating adhesion strength was investigated using ASTM C633 thermal spray coating adhesion test. Coating fracture toughness was measured using indentation method using the following equation [12], load of 10 Kg and dwell time of 15 s was used.

$$Kc = 0.079\left(\frac{P}{a^{\frac{3}{2}}}\right) log\left(4.5\frac{a}{c}\right)$$

Where P(N) represents the load applied to the indentor, a (µm) and c(µm) corresponds to half diagonal and crack length of the indentation respectively.

3. Results and discussions

3.1 Laser texture morphology

Magnesium alloy AZ91D samples were laser textured as part of surface treatment to prepare the surface for subsequent HVOF thermal spray coating. The purpose of laser texturing was to create micro holes on the surface. It was assumed that these micro holes will act as anchors for WC-Co powder particles enhancing the adhesion bond strength of the coating. The SEM image (figure 1) shows the overall view of laser textured sample. It can be clearly seen that a matrix of micro holes has been successfully created on the alloy surface. Figure 2 shows the closeup image of texture, the distance between holes is found to be 100 µm and hole diameter is 50 µm. It can be observed from the image that some of the melted alloy from the hole has been redeposited on the periphery of the holes. Figure 3 shows the cross-section of coated surface, from the image it can be observed that depth of the holes is approximately 110 µm and diamter of hole is not same throughout the hole. It is more near the surface and less as the depth increases. The reason behind this effect may be that near the surface laser melted magnesium alloy is ejected out of the hole but as the depth of the hole increases melted alloy get deposted inside the hole itself. Further there appear to be porosity in coating inside the holes. It may be because some of the larger WC-Co particles may get obstructed in the hole resutling in porosity inside the holes. Therefore, for complete filling of holes diamter and depth of the hole need to be optimsed according to the coating particle size. There seem to be good adhesive bonding of the coating with AZ91D substrate as at the interface (figure 3) coating appears to be well adhered.

Fig 1 SEM image of laser textured surface of AZ91D.

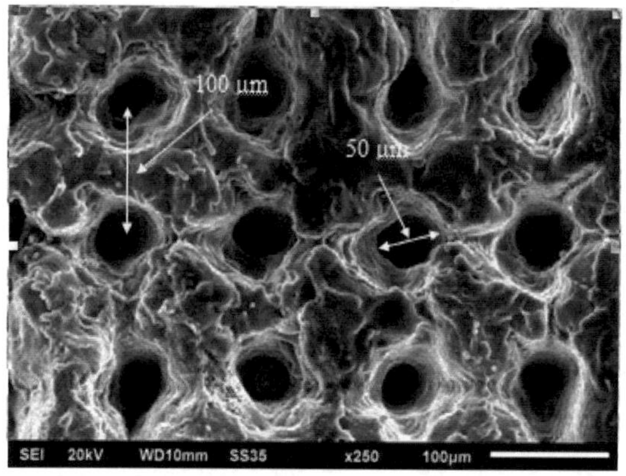

Fig. 2 Surface morphology of laser textured AZ91D showing diamter and distance between holes.

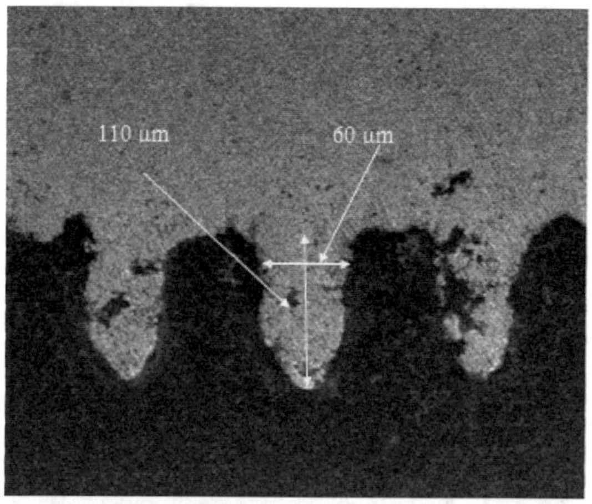

Fig. 3 Cross-sectional view showing coating substrate interface and dimensions of key hole.

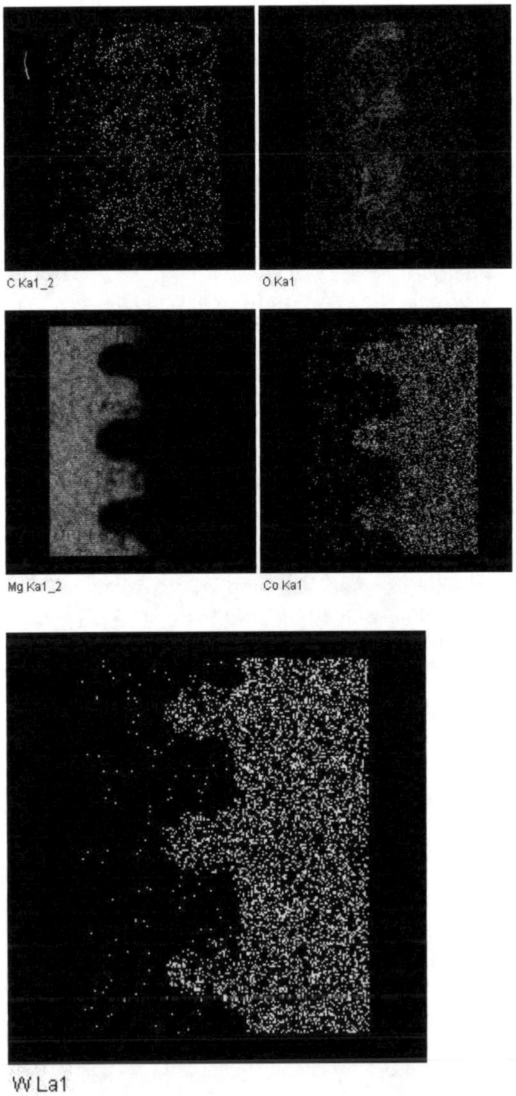

Figure 4 EDX area mapping of cross-section showing distribution of various elements in coating.

Substrate oxidation

To see the effect of laser texturing on oxidation of substrate surface, EDX mapping of cross-section was performed. From the figure 4 it can be clearly observed that relative percentage of oxygen is more at the interface of coating and the substrate and it follows the contours of laser textured micro holes. This phenomenon may be attributed to oxidation of substrate surface due to laser texturing. Oxide formation is supposed to improve the wettability of the surface leading to improved adherence of coating [2]. In the present study laser texturing and HVOF coating were a decoupled process. But by coupling both the process and optimsing the inert gas flow, an optimised thickness of oxide film may be achieved on the substrate surface to achieve better wettability, adherence of coating and corrosion resistance.

Coating Adhesion strength

Performance of any coating depends on the adhesion bond strength of coating with substrate. In the present study, a relatively new method, laser texturing, for preparing the substrate before coating has been investigated. ASTM C633 is the standard test for evaluating coating adhesion strength of thick coatings. For performing the coating adhesion strength test, specimen of diamter 25.4 mm were laser textured and HVOF coated with WC-12%Co coating. Flat top surface of the coated specimen was joined with grit blasted flat surface of another specimen using FM1000 adhesive. Coupled specimen were subjected to tensile force in a universal testing machine until the two specimen separate due to coating failure or pull off.

Coating adhesion strength has been found to be 45.6 MPa and mode of failure was glue failure as is evident from the figure 5. So, well adhered coating wtih high ahdesion bond strength can be achieved using laser texturing substrate preparation technique. Also, during adhesion strength test, glue failed instead of coating, which means that coating adhesion strnength may actually be much higher than the measured value of 45.6 Mpa.

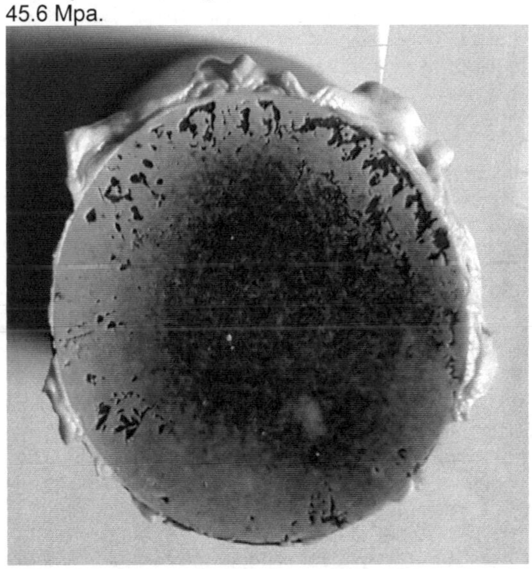

Figure 5 Optical Image of adhesion strength tested sample showing glue failure.

Coating fracture toughness

The purpose of realising WC-Co coating on AZ91D is to be able to use AZ91D for wear resistance applications. Wear resistance of any coating is largely dependent on fracture toughness of the coating [11]. Evans and Wilshaw model [12] was used to measure fracture toughness of the coating. Previously, many of the researchers have used the same model for calulating fracutre toughness. Fracutre toughness of the coating was measured to be 5.90 Mpa $m^{1/2}$. It is in accordance with measured fracture toughness of other studies [13-14].

4. Conclusions

Laser texturing as substrate preparation technique prior to HVOF thermal spray coating has been studied in this paper. The important findings from the research are as follows,

- Nd:YAG Q switch pulse laser can be successfully used to prepare a surface texture on magnesium alloy AZ91D.
- WC-Co coating deposited using HVOF processon the laser textured surface seems to be well adhered at the interface and inside the key holes. Good adhesion strenght of the coating is further confirmed by the reuslts of ASTM C633 adhesion strength test.
- Fracture toughness of the coating is found to be good and in accordance with other studies.

References

[1] R.S.C. Paredes*, S.C. Amico, A.S.C.M. d'Oliveira: The effect of roughness and pre-heating of the substrate on the morphology of aluminium coatings deposited by thermal spraying. Surface & Coatings Technology 200 (2006), pp. 3049 – 3055.

[2] Y. Danlos, S. Costil, H. Liao, C. Coddet: Influence of Ti–6Al–4 V and Al 2017 substrate morphology on Ni–Al coating adhesion — Impacts of laser treatments. Surface & Coatings Technology 205 (2011) 2 pp 2702–2708.

[3] R.S.C. Paredes, S.C. Amico, A.S.C.M. d'Oliveira, Surf. Coat. Technol. 200 (2006), pp. 3049.

[4] S. S. Aulakh, G. Kaushal, Substrate preparation and thermal spray coating performance- a review, Internaltion Thermal Spray coating conference and exposition, Shanghai, May 10-12, 2016, pp 478-483.

[5] Audisio, M. Caillet, A. Galerie, and M. Mazille, Surface Preparation, Surface treatments and Protective layer against the Corrosion, Les editions de physiques, Paris, France, 1987, p 169-174.

[6] V. V. Sobelev, J. M. Guilemany, J. Nutting, and J. R. Miquel, Development of substrate coating adhesion in Thermal Spraying, Int. Mater. Rev., Vol 42 (no. 3) 1997, p 117-135.

[7] C. Coddet, G. Montavon, S. Ayrault-Costil, O. Freneaux, F. Rigolet, G. Barbezat, F. Folio, A. Diard, and P. Wazen, Surface Preparation and Thermal Spray in a single step: The Protal Process – Example of Application for an Aluminium-Base Substrate, Journal of Thermal Spray Technology, vol. 8, no. 2, pp. 235-242, 1999.

[8] Shujuan Dong, Bo Song, Bernard Hansz, Hanlin Liao, Christian Coddet, Improvement of adhesion of plasma-sprayed Al_2O_3 coatings by using dry-ice blasting, Material Letters, Vol 66, 2012, pp 289-291.

[9] S. Costil, H. Liao et al: Influence of Surface Laser Cleaning Combined with Substrate Preheating on the Splat Morphology. Journal of Thermal Spray Technology Volume 14(1) March 2005, pp. 31-39.

[10] A. Lamraoui, S. Costil: Laser surface texturing (LST) treatment before thermal spraying: A new process to improve the substrate-coating adherence. Surface & Coatings Technology 205 (2010), pp. S164–S167.

[11] Robin Kromer, Sohpie Costil et al: Laser Surface Patterning Pre-treatment before thermal spraying- A way to adapt and control the surface topography to the considered materials and stress imposed. Proceedings of International Thermal Spray Conference, May, 2015, pp. 120-128.

[12] A. G. Evans, T. R. Wilshaw, Quasi-Static solid particle damage in brittle solids. Pt. 1. Observations, analysis and implications, Acta Metallic. 24 (1976) 939-956.

[13] A. Lekatou et al: A comparative study on the microstructure and surface property evaluation of coatings produced from nanostructured and conventional WC-Co powders HVOF sprayed on Al7075. Surface and Coatings Technology 276 (2015), pp. 539-556.

[14] Mingxiang Xie et al: Comparative investigation on HVOF sprayed carbide-based coatings. Applied Surface Scinece 273 (2013) 799-805.

Effect of nozzle-length, powder feed-rate and spray distance on HIPOJET® 2700 HVOF sprayed Cr_3C_2-NiCr coatings.

S.Matthews, J. Woo, and J. Daniel, Auckland / NZ

The high wear resistance of Cr_3C_2-NiCr coatings is reliant on the formation of a dense coating containing a high percentage of carbide grains, with minimal carbide degradation. Such coating characteristics are typically achieved through the use of high velocity oxygen fuel (HVOF) spraying. The propane fuelled, manually operated HIPOJET® 2700 HVOF system is one of a suite of smaller sized commercial HVOF systems recommended for smaller job shops. However, few works have characterised the properties of carbide composite coatings produced with this system. In this work a full factorial design of experiment analysis was used to assess the effect of key operating parameters on the quality of Cr_3C_2-NiCr coatings. The combustion parameters (fuel and oxygen flows) were fixed at the manufacturers recommended settings in order to focus on the effect of nozzle length, powder feed rate and, spray distance. The effect of these variables on the porosity/oxide content, carbide content, microhardness, coating thickness, and relative deposit efficiency is discussed.

1 Introduction

Hard, wear resistant, carbide composite thermal spray coatings have become an industry standard to mitigate erosive, abrasive and sliding wear in applications from ambient temperature up to 900°C [1-4]. The high wear resistance is dependent upon the formation of a dense coating, containing a high percentage of carbide grains, with minimal carbide phase degradation. Such coating characteristics are typically achieved through the use of high velocity oxygen fuel (HVOF) spraying. Typical industrial HVOF systems utilise relatively high fuel and oxygen flows and are generally machine or robot mounted in order to achieve accurate and repeatable coating deposition under the high thrust generated by the HVOF torch. The majority of prior research on carbide composite coatings has been based on the use of such systems. The propane fuelled, manually operated HIPOJET® 2700 HVOF system is one of a suite of smaller sized HVOF systems recommended for smaller job shops. Aside from the small size of the components within the torch, the other main difference between this HVOF torch and the larger conventional HVOF torches is that the powder is fed directly through the combustion chamber. In contrast, most current HVOF torches inject the powder downstream of the combustion chamber, near the entrance of the nozzle. Few studies have characterised the effect of process parameters on the coating properties generated with this smaller HVOF system [5-7].

In this work a full factorial design of experiment analysis was used to assess the effect of key operating parameters on the quality of HIPOJET® 2700 HVOF sprayed Cr_3C_2-NiCr coatings. The combustion parameters (fuel and oxygen flows) were fixed at the manufacturers recommended settings in order to focus on the effect of nozzle length (short, air cooled and longer, water cooled nozzles), powder feed rate, and spray distance for an agglomerated and sintered Cr_3C_2-25NiCr powder. The aim of this work was to assess the effect of these variables on the porosity/oxide content, carbide content, microhardness, coating thickness, and relative deposit efficiency, so as to benchmark the results against those from other industrial HVOF systems.

2 Experimental Procedure

All coatings were sprayed by a HIPOJET® 2700M HVOF system (Metallizing Equipment Co PVT Ltd, India) using a Cr_3C_2-25NiCr powder (Amperit 588.074, H C Starck, Germany) of nominal size range of -45/+15µm. The fuel, oxygen and air flow rates were fixed at the supplier's recommended settings, **Table 1**. Trials were performed at powder feed rates of 25g/min and 40g/min. Degreased and grit blasted mild steel samples (100mm x 25mm x 3mm) were mounted onto a spray stand in front of the torch at distances of 150mm and 200mm and sprayed in a raster pattern. The traverse speed was 1m/s and the robot program was repeated 15 times in each trial.

Table 1. Spray parameters

Parameter	Setting
Propane flow (93% propane-7% butane)	Pressure: 6.5 bar Flowmeter setting: 60
Oxygen flow	Pressure: 10 bar Flowmeter setting: 240
Air Flow	Pressure: 4.4 bar
Carrier Gas (nitrogen)	Pressure: 6 bar Flowmeter: 5.5 SLPM
Water cooling flow	10.8 LPM
Traverse Speed	1 m/s

Two torch configurations were used – a "standard" air cooled system and a "hybrid" water cooled system, **Fig. 1a&b**. The air cooled system employed a short, shaped nozzle insert (approximately 40mm in length) of which the narrow diameter nozzle section was approximately 20mm long. The water cooled "hybrid" system employed a much longer shaped nozzle insert (approximately 98mm in length) of which the narrow diameter nozzle section was approximately 70mm in length.

The full factorial experimental design (3^2 factorial) assessed the effect of nozzle length, powder feed rate and spray distance (each at a high and low setting), **Table 2**. The results from the full factorial design were analysed using the statistical software package IBM SPSS Statistics. For each coating property the process variable and variable interactions that were statistically significant at the 0.05 level of probability were identified and discussed in the text.

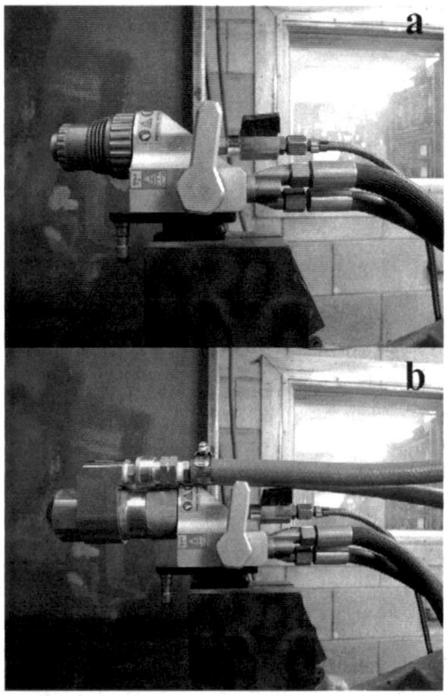

Fig. 1. Images of the “standard” short air cooled nozzle HVOF system (a) and the “hybrid” water cooled nozzle system (b).

Table 2. Spray parameters and coating properties.

Trial	Nozzle length	Spray distance (mm)	Powder feedrate (g/min)
1	Short	150	25
2	Short	150	40
3	Short	200	25
4	Short	200	40
5	Long	150	25
6	Long	150	40
7	Long	200	25
8	Long	200	40

Trial	Porosity+ Oxide (%)	Carbide content (vol%)	Hardness (VHN_{300})	Thickness (µm)	Relative deposit efficiency (µm/(g/min))
1	11.2	54.0	718	177	7.1
2	13.1	52.8	715	261	6.5
3	10.8	52.5	767	160	6.4
4	11.8	55.7	717	215	5.4
5	7.3	32.1	936	249	10.0
6	7.5	37.2	987	377	9.4
7	5.6	32.9	990	186	7.5
8	7.3	36.8	939	296	7.4

Samples from each coating were prepared using standard metallographic procedures. Cross sectional images were taken using back scatter electron (BSE) imaging on a secondary electron microscope (Quanta 200 ESEM, FEI, USA). Image analysis (software package ImageJ [8]) was used to quantify the coating thickness, the porosity+oxide content (magnification 1000X) and the retained carbide content (magnification 2000X) (Note: the brightness and contrast settings were optimised to enhance the differences between the carbides and the Ni alloy matrix. Under these settings the low atomic mass oxide phases appeared very dark and could not be accurately distinguished from the coating porosity. Therefore the oxide and porosity were incorporated together in this analysis). The retained carbide content was calculated as a percentage of the solid coating material i.e. (carbide area %)/(100-(oxide+porosity area %)) x 100. This was in order to remove the effect of the porosity content on the actual retained carbide content. The relative deposit efficiency was calculated from the thickness data by dividing the coating thickness by the powder feed rate. Coating variables were averaged from the analysis of a minimum of five images. The Vickers microhardness was measured under a 300g load for 10 seconds (LECO M400 Hardness Tester, LECO, USA). The data was averaged from a minimum of ten indents.

3 Results and Discussion

3.1 Coating Porosity + Oxide Content

The very dark contrast of the porosity and oxide features meant they were combined together during quantification. However, when observing the coating cross sections it was apparent that the porosity content greatly exceeded that of the oxide content. Therefore, the following discussion regarding the effect of process parameters focus's on their effect on porosity alone. The short nozzle trials exhibited high concentrations of globular and crescent shaped porosity indicative of poor heating of the particles, **Fig. 2a**. The long nozzle trials had a much denser microstructure, formed from highly molten splats, **Fig. 2b**. The porosity in these well molten coatings was dominated by pull-out features and cracking from shrinkage stresses.

The porosity values varied significantly between the two experimental groups, but were similar within each of the groups, **Table 2**. The long nozzle results (5.6-7.5 vol%) were comparable with previous HVOF Cr_3C_2-NiCr coatings presented in the literature, albeit at the higher end of the published range of 1.2-6.3 vol% [2, 4, 9, 10], while the short nozzle coatings were significantly higher (10.8-13.1 vol%).

Over the range of variables considered in these trials the powder feed rate, spray distance and nozzle length all had statistically significant effects on the porosity values. The interaction between all three of these variables was also statistically significant, however, it is difficult to interpret this 3-way interaction effect in a practical context and so emphasis was given to the effect of the individual factors. Higher powder feed rates increased the porosity content. Higher powder feed rates result in less heating per particle and possibly lower average velocities. Such factors would reduce the degree of melting and the extent of deformation upon impact, leading to higher porosity values – particularly for the poorly molten particles in the short nozzle trials.

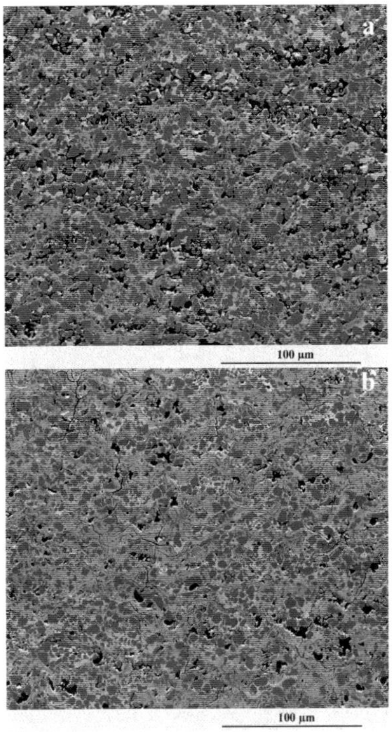

Fig. 2. Representative BSE images of the short nozzle (Trial 2)(a) and long nozzle (Trial 6)(b) coatings.

Longer spray distance reduced the porosity content. Longer spray distances increase the exposure time of the particles to the combustion flame, leading to greater degrees of heating and high velocities – both of which assist in splat spreading. This would reduce the extent of intra-particle porosity and also lead to better splat bonding, which would contribute to greater resistance to "pull-out" porosity formation.

The porosity values were greatly reduced with the use of the long water cooled nozzle compared to the short, air cooled nozzle. The longer nozzle greatly increased the time and efficiency of particle heating and acceleration, leading to much greater degrees of particle melting and splat deformation upon impact as seen in **Fig. 2b**.

3.2 Carbide Concentration

The coating carbide concentrations showed a significant drop from the 79.4 vol% concentration in the starting powder, but were similar within each experimental group. The low degree of particle heating evident in the BSE images of the short nozzle coatings, **Fig. 2a**, was reflected in the relatively high retained carbide content of 52-56 vol%, **Table 2**. In contrast, the greater degree of particle heating in the long nozzle coatings, **Fig. 2b**, generated greater reductions in the carbide content to 32-37 vol%. These values are consistent with those presented in the HVOF spraying of sintered and crushed Cr_3C_2-25NiCr powders [11] at 34.3-41 vol%, but lower than those published for HVOF coatings using agglomerated and sintered powders at 48.5-57.1 vol% [4] and 65.7 vol% [12].

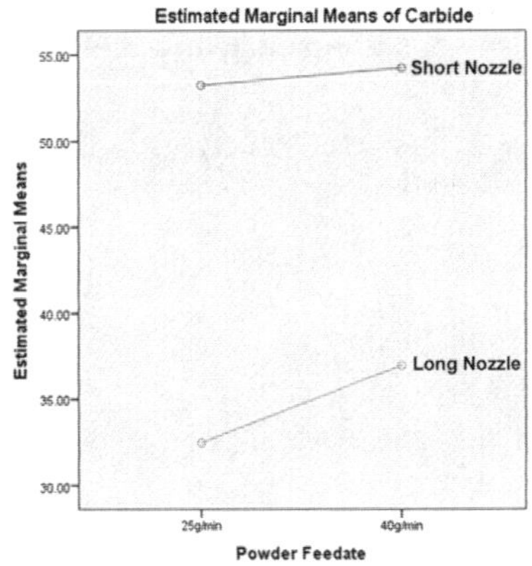

Fig. 3. Carbide concentration - powder feedrate x nozzle length interaction.

Over the range of variables considered in these trials the powder feed rate, nozzle length, powder feed rate x nozzle length interaction, and the three-way powder feed rate x spray distance x nozzle length were all statistically significant. Because the powder feed rate x nozzle length interaction was significant, the individual variables were not analysed independently. From **Fig. 3** it can be seen that the carbide content increased with increasing powder feed rate for both the short and long nozzles, but the effect was more significant for the longer nozzle system. This effect was related to the magnitude of carbide dissolution. The short nozzle generated limited melting of the nickel binder in the particles, as seen in **Fig. 2a**. As a result there was limited carbide dissolution in flight, as well as limited opportunity for carbide rebounding to occur at impact due to the low fluidity of the particles [4, 13]. While more carbide dissolution occurred at the lower powder feed rate, due to the high energy input per particle, the overall degree of particle heating was low, which reduced the significance of the powder feed rate effect. In contrast the particles were well molten in the long nozzle coatings, enabling significant carbide dissolution to occur and accounting for the lower overall retained carbide content values. Under these conditions the high energy input per particle at the lower powder feed rate generated more effective heating, binder melting and, therefore, carbide dissolution. The more fluid nature of the particles at impact would also increase their susceptibility to carbide rebounding [4, 13].

3.3 Microhardness

The porous coatings generated in the short nozzle coatings had the lowest microhardness values of 715-767 VHN_{300}. The greater degrees of heating and carbide dissolution in the long nozzle coatings led to their higher hardness values of 936-990 VHN_{300},

Table 2. The values from these well molten coatings lie at the upper end of the range of values presented in the literature for HVOF Cr_3C_2-NiCr coatings which span the range 873-1092 VHN_{300} [2, 3, 9, 10, 12, 14].

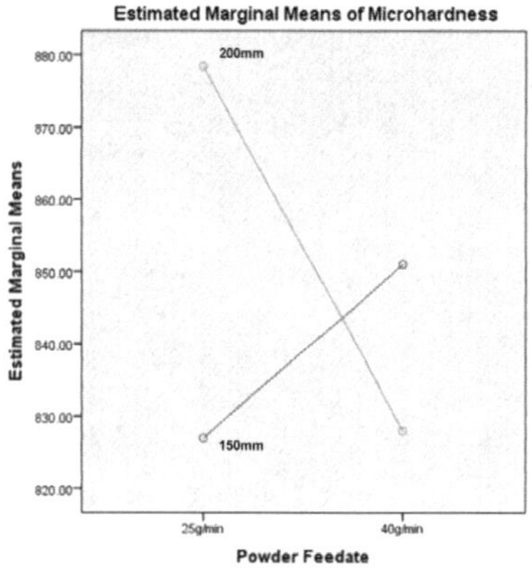

Fig. 4. Microhardness - powder feedrate x spray distance interaction.

Over the range of variables considered in these trials the nozzle length and powder feed rate x spray distance interaction were statistically significant. The longer nozzle generated considerably higher hardness values compared to the short nozzle, due to the greater degree of particle heating. This not only led to lower coating porosity, as discussed above, and greater splat deformation and cohesion, but also a greater degree of carbide dissolution, which combined to generate the high hardness values.

Fig. 4 presents the interaction effect between powder feed rate and spray distance on the coating microhardness. At the low powder feed rate, there was an increase in hardness with increasing spray distance. This is consistent with the arguments presented for both the porosity and carbide percent results above. At the higher powder feed rate there was a reduction in hardness with increasing spray distance. This runs counter to the effect of spray distance discussed in the previous sections. In comparing the raw data for this powder at the higher powder feed rate, it appears that this trend resulted primarily from a reduction in hardness with spray distance in the long nozzle trial. The reason for this is unclear and additional trials are required to determine if this was a consistent result or an experimental anomaly. It should also be noted that this interaction effect had a very low partial Eta squared value of 0.033, meaning that it played only a very minor role in contributing to the variation in microhardness.

3.4 Coating Thickness and Relative Deposit Efficiency

Over the range of variables considered in these trials, the powder feed rate, spray distance, nozzle length and all combinations of these variables had a statistically significant effect on the coating thickness, **Fig. 5**.

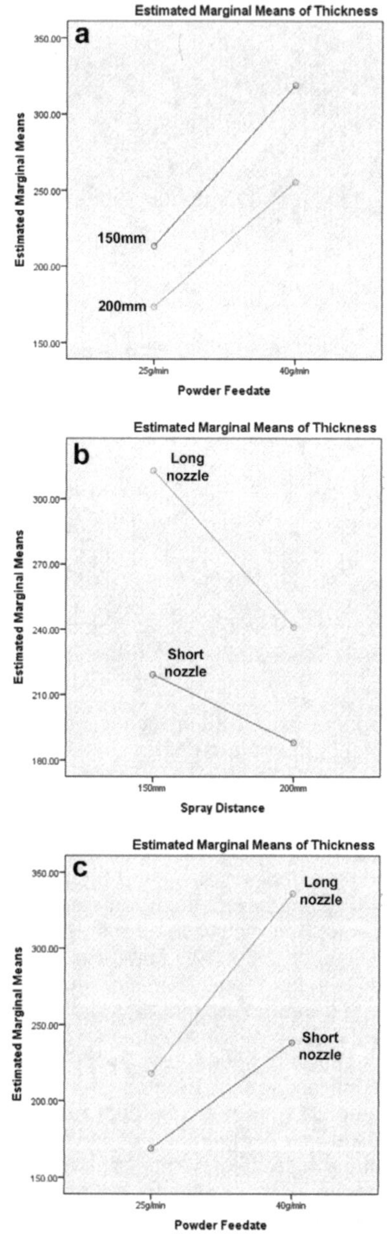

Fig. 5. Thickness - powder feedrate x spray distance interaction (a), spray distance x nozzle length interaction (b) and powder feedrate x nozzle length interaction (c).

With regards to the deposit efficiency, the powder feed rate, spray distance, nozzle length and the powder feed rate x nozzle length and spray distance x nozzle length interactions were all statistically significant, **Fig. 6**. Because the interaction effects were statistically significant, the individual variables were not discussed independently.

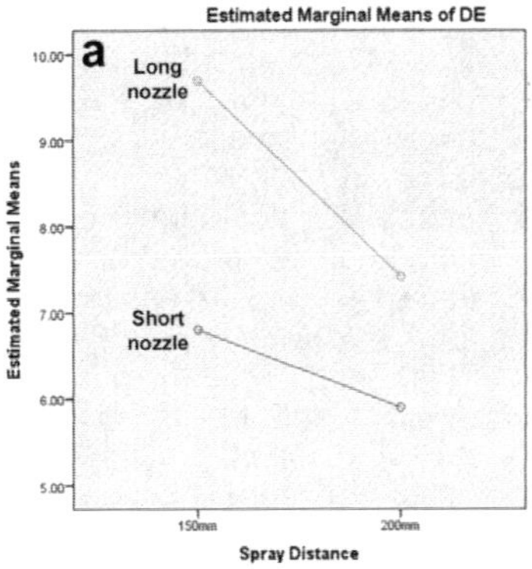

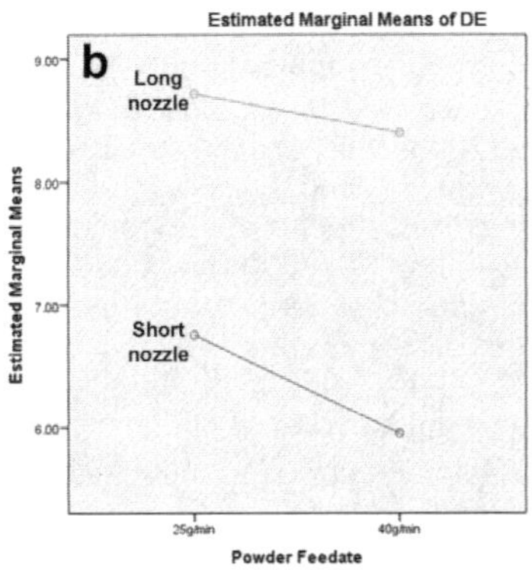

Fig. 6. Relative deposit efficiency - spray distance and nozzle length interaction (a) and powder feedrate x nozzle length interaction (b).

The powder feed rate x spray distance interaction was statistically significant for the coating thickness, but not for the deposit efficiency, **Fig. 5a**. The coating thickness increased with powder feed rate at both spray distances, but the effect was greater at the shorter spray distance. This indicates some form of reduction in deposit efficiency with increasing spray distance at high powder feed rates.

The spray distance x nozzle length interaction was statistically significant and showed the same trends for both the thickness and deposit efficiency, **Fig. 5b** and **Fig. 6a** respectively. The deposit efficiency, and as a result the coating thickness, decreased with increasing spray distance for both nozzle lengths, but the effect was greater for the longer nozzle. The long nozzle trials had a significantly greater deposit efficiency based on the greater degree of particle heating evident in **Fig. 2b**. The reduction in deposit efficiency in this instance is thought to relate to variations in particle temperature and velocity with spray distance – specifically the cooling of particles below the temperature required for them to adhere to the coating. Complicating this interaction is that the porosity decreased with increasing spray distance, which was linked to improved melting of the particles with longer exposure time in flight. A similar effect of spray distance on deposit efficiency was observed in the HVAF spraying of an agglomerated and plasma densified Cr_3C_2-NiCr powder [10]. In that work the response was postulated to result from the particle size distribution. Smaller particles are known to rapidly heat up and accelerate to the maximum values soon after leaving the nozzle. With increasing spray distance they tend to follow the temperature and velocity profiles of the combustion flame, both cooling and slowing down [15, 16]. Larger particles take longer to heat up and reach their peak velocities at longer spray distances. Therefore, at short spray distances the smaller particles would have the highest deposit efficiency, while the larger particles may not yet have achieved high enough temperatures to adhere to the coating. At longer distances the smaller particles may cool and slow down sufficiently to no longer adhere to the coating at impact and simply bounce off. The similarity in trends to that previous work suggests that the variation in thickness and deposit efficiency observed in this work was also particle size dependent. While the trends in **Fig. 5b** and **Fig. 6a** indicate that this mechanism occurs for both nozzle lengths, it was more significant for the longer nozzle. The lower variation in thickness and deposit efficiency with spray distance in the short nozzle trials is thought to result from the poor heating of the powders. The low deposit efficiency meant that the percentage change in this response with spray distance was not as significant in these trials. A further complication is that these short nozzle coatings were formed from blocky partially molten splats which generated a greater thickness for a given number of splat layers, compared to the same number of splat layers formed from thinner, well molten splats in the longer nozzle trials. As a result the depost efficiency based on the coating thickness is higher than it would be if it was instead based on the coating mass, since the porosity of the coating is not considered.

The powder feed rate x nozzle length interaction was statistically significant for both the coating thickness and deposit efficiency, but with contrasting effects **Fig. 5c** and **Fig. 6b** respectively. The coating thickness increased with increasing powder feed rate for both nozzle lengths, but the increase was more significant for the longer nozzle. As noted above, the variation in coating thickness between the short and long nozzle coatings is in effect larger than illustrated if the globular, porous splats are taken into account versus the thinner, well molten splats. However, the deposit efficiency decreased with increasing powder feed rate, with this effect being more significant for the short nozzle. The long nozzle generated much more efficient heating of the particles. This not only led to the higher coating thickness and deposit efficiencies relative to the short nozzle trials, but also enabled a greater capacity to process more powder under the same combustion conditions. This meant that while there was a drop in deposit efficiency for the long nozzle with increasing powder feed rate, it was not as significant as for the short nozzle which did not generate effective heating of the particles, even at the lowest powder feed rate. The poor powder heating with the short nozzle made it particularly sensitive to reductions in deposit efficiency with increasing powder feed rate.

4 Conclusions

In this work a full factorial design of experiment analysis was used to assess the effect of key operating parameters on the quality of Cr_3C_2-NiCr coatings. The key conclusions from this work were:

- The longer water cooled nozzle generated much more effective heating of the particles compared to the short air cooled nozzle across all parameter settings. The short air cooled nozzle generated insufficient heating of the powder particles, leading to very porous coatings
- Higher powder feed rates led to higher porosity, higher carbide contents, higher coating thicknesses, but lower deposit efficiencies. This was linked to the lower degree of heating of the particles at the higher powder feed rate.
- Longer spray distance led to denser coatings, with lower porosity, but lower coating thickness and deposit efficiency. The reduction in deposit efficiency was postulated to result from cooling of the smaller particles below the point required to adhere to the coating with increasing spray distance.

5 Acknowledgements

The authors gratefully acknowledge the support of Metal Spray Suppliers (NZ) Ltd and Holster Engineering Company (2003) Ltd for producing the samples for analysis. We would also like to thank the technical staff within the Department of Chemical and Materials Engineering, The University of Auckland for their support in this work.

6 References

1. Berger, L.-M., Hardmetals as thermal spray coatings. Powder Metallurgy, 2007. 50(3): p. 205-214.
2. Poirier, D., J.G. Legoux, and R.S. Lima, Engineering HVOF-sprayed Cr_3C_2-NiCr coatings: The effect of particle morphology and spraying parameters on the microstructure, properties, and high temperature wear performance. Journal of Thermal Spray Technology, 2013. 22(2-3): p. 280-289.
3. Vernhes, L., et al., HVOF Coating case study for power plant process control ball valve application. Journal of Thermal Spray Technology, 2013. 22(7): p. 1184-1192.
4. Bolelli, G., et al., Sliding and abrasive wear behaviour of HVOF- and HVAF-sprayed Cr_3C_2-NiCr hardmetal coatings. Wear, 2016. 358-359: p. 32-50.
5. Buytoz, S., et al., Microstructural and Wear Characteristics of High Velocity Oxygen Fuel (HVOF) Sprayed NiCrBSi-SiC Composite Coating on SAE 1030 Steel. Arabian Journal for Science and Engineering, 2013. 38(6): p. 1481-1491.
6. Chatha, S.S., H.S. Sidhu, and B.S. Sidhu, High-temperature behavior of a NiCr-coated T91 boiler steel in the platen superheater of coal-fired boiler. Journal of Thermal Spray Technology, 2013. 22(5): p. 838-847.
7. Sidhu, H.S., B.S. Sidhu, and S. Prakash, Wear characteristics of Cr_3C_2-NiCr and WC-Co coatings deposited by LPG fueled HVOF. Tribology International, 2010. 43(5-6): p. 887-890.
8. Rasband, W.S., ImageJ. US National Institute of Health: Bethesda, Maryland, USA. http://imagej.nih.gov/ij/, 1997-2012.
9. Matthews, S., Erosion-Corrosion of Cr_3C_2-NiCr High Velocity Thermal Spray Coatings (PhD Thesis). Department of Chemical and Materials Engineering. 2004, Auckland, New Zealand: The University of Auckland.
10. Matthews, S.J. and M.M. Hyland. Statistical Optimization of HVAF Sprayed Cr_3C_2-NiCr Coatings for Minimizing Decarburization. Proceedings of the International Thermal Spray Conference. (2000), pp 543-549.
11. Ji, G.C., et al., Microstructural characterization and abrasive wear performance of HVOF sprayed Cr_3C_2-NiCr coating. Surface and Coatings Technology, 2006. 200(24): p. 6749-6757.
12. Matthews, S. and L.M. Berger, Long-term compositional/microstructural development of Cr_3C_2-NiCr coatings at 500°C, 700°C and 900°C. International Journal of Refractory Metals and Hard Materials, 2016. 59: p. 1-18.
13. Li, C.J., et al., Dominant effect of carbide rebounding on the carbon loss during high velocity oxy-fuel spraying of Cr_3C_2-NiCr. Thin Solid Films, 2002. 419(1-2): p. 137-143.
14. Matthews, S., M. Hyland, and B. James, Microhardness variation in relation to carbide development in heat treated Cr_3C_2-NiCr thermal spray coatings. Acta Materialia, 2003. 51(14): p. 4267-4277.
15. Li, M. and P.D. Christofides, Multi-scale modeling and analysis of an industrial HVOF thermal spray process. Chemical Engineering Science, 2005. 60(13): p. 3649-3669.
16. Li, M., D. Shi, and P.D. Christofides, Modeling and control of HVOF thermal spray processing of WC-Co coatings. Powder Technology, 2005. 156(2-3): p. 177-194.

New developments in HVOF spraying for internal diameter coatings

J. Gutleber, R. Molz, J. He, C. Weber & J. Colmenares, Westbury, NY/USA

An HVOF extension gun for spraying internal diameters down to 5" (127 mm) has been developed and evaluated. The spray gun utilizes vortex stabilization of the flame which permits operation with a minimal length combustion chamber. Small spray distances and subsequent reduced particle dwell times present a challenge for heating and accelerating powders to achieve typical HVOF coating benchmark properties such as hardness, density and residual stresses. Initial coating characteristics and process/system capabilities are discussed.

1 Introduction

High Velocity Oxygen Fuel (HVOF) thermal spraying is a well-established surface technology for replacing hard-chrome applications where direct line-of-sight is possible [1,2]. Internal Diameter (ID) applications for HVOF of less than 180mm (~7") are a recent development [3]. In contrast, air plasma spray (APS) small diameter extension guns have been available in the market for decades. The challenges for HVOF spraying of internal diameters less than 180 mm are multiple [3]. The standoff distance of traditional HVOF guns tends to be significantly longer than in APS. Given the slower particle velocity and much higher plume temperature of the plasma process, HVOF particle dwell time is maximized by spraying at longer distances before particles begin to decelerate. For example, a typical outer diameter APS standoff distance is 75 to 150 mm from the gun nozzle, while traditional HVOF spray gun designs typically use spray distances of 200 to 400 mm from the nozzle exit. Additional rationale for longer spray distances in HVOF is the significant amount of heat that can be transferred to the substrate. In contrast to plasma guns with plume temperatures in the range of tens of thousands of degrees Celsius, HVOF flames have a maximum temperature of approximately 3000°C. Within the plasma the dissociated ion species recombine quickly and the thermal gradient is very sharp [4], while in HVOF the heat of combustion does not dissipate as quickly and transfers readily to the substrate. Thus, thermal management of the substrate temperature in ID HVOF spraying is an additional challenge due to the limited spray distance options inside a cylindrical component.

The benefits of HVOF coatings are numerous over their plasma sprayed counterparts, including:

- High adhesion
- High density
- Lamellar cohesion
- Residual stress state
- Less compositional degradation

These benefits make ID-HVOF spray capability an attractive technology for wear applications for internal diameters. With carbide materials, the excessive temperatures of plasma torches promote decarburization and solutioning of the carbide phases into the matrix which results in hard but rather brittle coatings [5]. Due to the lower flame temperature of HVOF, these degradation reactions are reduced leading to tougher and more wear resistant coatings. Due to the high particle velocity, HVOF coatings typically exhibit neutral or compressive residual stress which is beneficial for fatigue applications such as aircraft landing gear components. HVOF processing of metallic materials can result in lower oxidation given the lower particle temperatures than plasma spray [6]. For both material classes, the high density due to high impact velocity is also a desirable quality.

A novel approach to ID HVOF spraying is presented concerning the development of a vortex accelerated combustion torch. This variation of the widely used Diamond Jet™ gas fuel spray gun uses hydrogen as fuel and is designed to spray HVOF quality coatings on internal diameters currently down to **125 mm**.

2 Technology overview

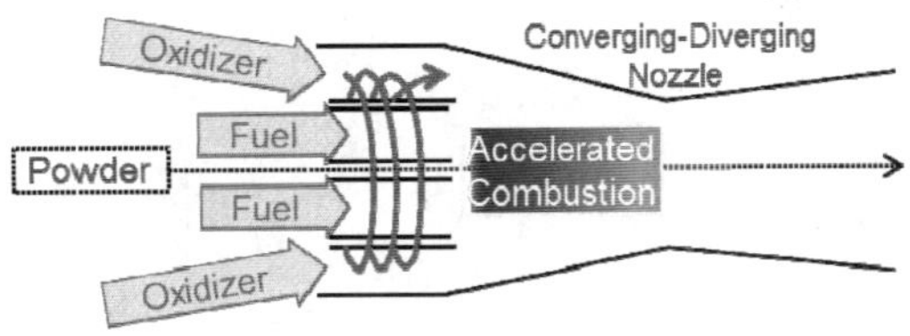

Fig. 1. Schematic of flow directions (rotational for oxidizer) for Diamond Jet™ Vortex ID gun.

The Diamond Jet™ Vortex (DJV) ID is a gas fuel axial powder injection spray torch that accelerates combustion by the use of patented vortex [7] technology, eliminating the need for a combustion chamber and reducing hardware size and complexity. The vortex combustion stabilization principle was devised by James Browning, renowned for the invention of HVOF technology. The basic principle of operation is shown in **Fig. 1**. The oxidizing gas, which can be pure oxygen or a combination of oxygen and air/inert gas is injected in a vortex geometry with a high rotational flow component. The hydrogen fuel gas is injected on a straight forward trajectory. The hydrogen flow is disrupted by the rotational component of the oxidizing flow in the accelerated combustion zone. This straightforward design allows for rapid mixing and combustion as shown in **Fig. 1**, reducing the size and complexity of the gun. Additionally the high rotational component of the oxidizer allows for a shroud effect without the need of an additional shroud gas inlet flow as is utilized in the standard Diamond Jet™ torch. More explicitly, the vortex flow effect limits heat transfer between the flame and the gun component walls due to a boundary layer effect. Additionally, since there is no combustion chamber and no "turning" of the flame within the gun (e.g., vertical to horizontal), heat loss is minimal. These design simplifications result in the Diamond Jet™ Vortex having very low water cooling losses of less than 2 kW.

The accelerated mixing also allows for a wide stoichiometric ratio, ranging from 7 to 28 oxygen/fuel by mass with 8 being stoichiometry for pure hydrogen. Combustion power levels range from 15 to 50 kW. 'Vortex' technology refers to the combustion stabilization method of fuel and oxygen as opposed to a conventional combustion chamber. The HVOF spray gun using vortex design principle, without combustion chamber, has the following advantages:

- Smaller physical gun size
- "Bending of flame" not required within ID (typical with combustion chamber ID guns)
- Lower cooling power requirement
- Axial powder injection (vs radial)

The supersonic vortex stabilized flame has characteristic shock diamonds as shown in **Fig. 2.** Due to limited gun power and spray distance within the ID components, finer powder size distributions are required compared to conventional HVOF powders.

Fig. 2. Image of DJV prototype gun flame with characteristic shock diamonds of supersonic flow

3 Experimental setup

For this work, the DJV gun was adapted to an Oerlikon Metco MultiCoat® spray system with an integrated Twin 120AH series disk powder feeder fitted with an air driven vibrator to assist in feeding the fine powders. Process gases were hydrogen and oxygen for combustion as well as air for shroud/dilution gas and nitrogen for carrier gas. The DJV gun was mounted onto a prototype extension arm to allow for coating of internal diameters. An open platen turntable was used to accommodate exhaust suction through the bottom of the tubes during spraying. Rotation speed of the tube was set to achieve an effective surface speed of 150 m/min. All coatings where sprayed normal (90°) to the ID of a steel tube, where test samples were mounted for spraying and subsequent evaluation and testing. Several material types, carbides and an alloy, were chosen for initial spray trials using this new technology. These materials and corresponding spray parameters are given in **Table 1**, where nlpm is normal liters per minute of gas flow.

Table 1. Materials and spray parameters

Designation	**WOKA 3661**	**AE11346**	**W7211**	**AE11074**
Composition (wt%)	WC-10Co4Cr	WC-10Co4Cr	Cr_3C_2-25(NiCr)	Ni17Cr4Si4Fe3.5B1C
Nominal size (µm)	-20/+5	-20/+5	-20/+5	-20/+5
Hydrogen (nlpm)	135	135	135	135
Oxygen (nlpm)	100	100	100	85
Air (nlpm)	68	68	68	125
Carrier Gas (nlpm)	25	25	25	45
Feed rate (g/m)	20	20	20	20
Spray distance (mm)	50	50	50	60

Experimental powders (other than WOKA 3661) were prepared for the spray trials which are somewhat finer than conventional HVOF size fractions, typically within range of 5-60 microns. Different HVOF guns on the market use specific particle size distributions optimized for each gun [8]. This holds for the DJV gun, where for the work presented here the particle size range was approximately 5-20 microns. Optimization of this particle size range is part of future gun development and is a function of coating material type as well as the diameter of the component to receive the coating.

An Accuraspray G3C sensor system (Tecnar Automation, Canada) was used to measure the average particle temperature and velocity of the WC-CoCr particles in the spray stream. Coatings were sprayed onto AISI 1018 low carbon steel plates and bond samples. Coating characterization included:

- Metallography for microstructure evaluation
- Microhardness (HV300gf, ASTM E384-11)
- Porosity (ASTM E2109-01)
- Bond strength (ASTM C633-13)

In addition, Almen strips (Type N, SAE AMS 2432B) were sprayed to acquire coating residual stress indications for the WC-CoCr coatings. All substrates were grit blasted with 60 grit aluminum oxide to achieve an average surface roughness (Ra) value of 2-3µm prior to coating.

3 Results

In-flight diagnostic measurements of AE11346 WC-CoCr powder are given in **Fig. 3.** Parameter 1 (Par 1) is equivalent to values given in **Table 1**, with the exception of 45 nlpm carrier gas (vs. 25 nlpm). Parameter 2 (Par 2) is simply an increase of 10% in fuel, oxygen and air flow values over Parameter 1 (all else remains equal). From the data given in **Fig. 3**, it is observed that both average particle velocity and temperature reach a maximum at approximately 60 mm spray distance. However, in order to spray an ID of 125 mm and allow the spray gun to fit inside the bore at 90°, a spray distance of 50 mm was used for the carbide materials. Measured particle temperatures were in the range of 1700-2000°C, and particle velocities in range of 700-850 m/s. These values are in line with conventional HVOF inflight particle measurements conducted outside of this work by the authors. The spray stream width was measured at 2.3 mm at 45 mm spray distance, this is typically 5-15 mm for conventional HVOF guns at their respective spray distance again based on measurements conducted outside of this work by the authors.

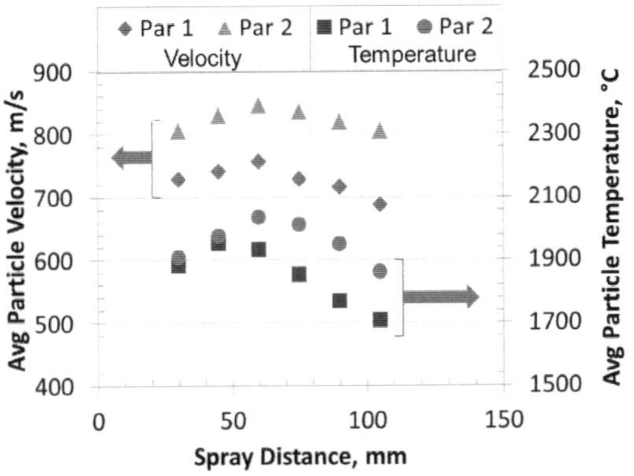

Fig. 3. Measured values of average particle velocity and temperature vs. spray distance for WC-CoCr powder at two parameters

Results of coating trials available for this work were limited and are still work in progress. Initial spray runs used a standard "off the shelf" powder, WOKA 3661, with marginal microstructure results compared to conventional HVOF coatings. The WOKA 3661 coating cross-section is given in **Fig. 4** (top), with:

- Hardness: 960±79 HV300
- Porosity: 3.4±0.2%
- Bond strength: >11 ksi (75.8MPa)

Bond strength tests resulted in failure within the glue (FM1000). The coating microstructure shows micro-porosity at the spray particle boundaries. Primary carbides are well distributed throughout the structure.

AE11346 coating cross-section is given in **Fig. 4** (bottom), with:

- Hardness: 915±79 HV300
- Porosity <1%
- Bond strength: >11 ksi (75.8MPa)

Bond strength failure mode was within the glue (FM1000). The microstructure shows a high degree of lamellar flattening and coalescence with a minimum of distributed porosity observed in the micrographs. Primary carbides are observed to be uniformly distributed, while some metallic chromium or cobalt is observed in discrete islands.

Fig. 4. WC-CoCr coating microstructure: WOKA 3661(top) and AE11346 (bottom)

Almen strips were sprayed for the WC-CoCr coatings, WOKA 3661 and AE11346, to a thickness of 0.127±0.025mm (0.005±0.001") using the methodology of Ref [9]. The WOKA 3661 coating was measured as compressive with a net normalized Almen deflection of +0.104mm (0.0041"), while the AE11346 was significantly more compressive with a net deflection of +0.295mm (0.0116"). This compressive stress indication of AE11346 shows potential for landing gear applications. Future in-flight particle measurements will be used to investigate these results further.

The coating cross-section for W7211 (Cr_3C_2-NiCr) is given in **Fig. 5**, with:

- Hardness: 904±43 HV300
- Porosity: <1%

Bond strength was not measured for this coating though is expected to be in the 75 MPa range. The microstructure shows a favorable degree of lamellar flattening and coalescence with a minimum of distributed porosity observed in the micrographs. Primary carbides appear uniform and well distributed.

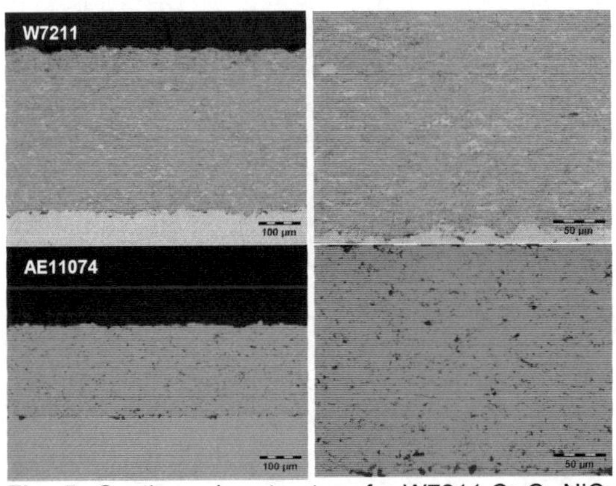

Fig. 5. Coating microstructure for W7211,Cr_2C_3-NiCr (top) and as-sprayed AE11074, Ni-based S/F alloy (bottom)

Spray trials with metallic powders were relegated to AE11074, a nickel-based self-fluxing alloy, at the time of writing. The coating cross-section is given in **Fig. 5**, with:

- Hardness: 707±55 HV300
- Porosity: 1.0±0.1%

Bond strength was not measured for these coatings. A network of small distributed porosity is observed in the micrographs. The degree of lamellar flattening and coalescence could be improved through parameter manipulation during future spray trials.

A summary of the measured coating properties is given in **Table 2**.

Table 2. Summary of ID coating properties

Designation	**WOKA 3661**	**AE11346**	**W7211**	**AE11074**
Composition (wt%)	WC-10Co4Cr	WC-10Co4Cr	Cr_3C_2-25(NiCr)	Ni17Cr4Si4Fe3.5B1C
Porosity (%)	3.4±0.2	<1%	<1%	1.0±0.1
Hardness (HV300)	960±79	915±79	904±43	707±55
Bond strength (MPa)	>75	>75	NA	NA
Almen deflection (mm/0.001")	+0.104 +4.1	+0.295 +11.6	NA	NA

4 Summary and outlook

Oerlikon Metco has developed a prototype ID HVOF gun based on vortex combustion stabilization technology. Initial spray trials with standard and experimental carbide and alloy powders show the vortex stabilized DJV ID HVOF gun can currently deposit near conventional HVOF quality coatings with hardness > 900 HV300, porosity of 1% or less and bond strength in excess of 75 MPa. Additionally, the AE11346 WC-CoCr material demonstrated substantial compressive residual stress using the Almen strip technique demonstrating potential for landing gear applications. Further ID HVOF development will focus on:

- Operating parameter space of the DJV gun
- Feedstock powder characteristics
- Thermal management of component
- Spray rate (higher feed rate)
- Spray angle (gun to target)

This is required to optimize coating properties and performance for ID wear applications.

5 References

[1] Rastegar, F. and Craft, A.E., Piston ring coatings for high horsepower diesel engines, Surface and Coatings Technology (1993), Volume 61, Issue 1, pp.36-42.

[2] Nestler, M. C., Prenzel, G., and Seitz, T. HVOF spraying vs. hard chrome plating coating characteristics and aircraft applications, Thermal Spray 1998: Meeting the Challenges of the 21st Century (1998), pp. 1073-1078.

[3] Matthäus, G., Bobzin, K., Lugscheider, E., and Zwick, J., Coating Properties of HVOF Sprayed Carbide-based Ultrafine Powders, Thermal Spray 2006: Science, Innovation, and Application (2006), pp.673-678

[4] Fauchais, P. Understanding plasma spraying. Journal of Physics D: Applied Physics (2004), *37*(9), R86.

[5] Matthews, S., Hyland, M. and James, B. Microhardness variation in relation to carbide development in heat treated Cr_3C_2–NiCr thermal spray coatings, Acta Materialia (2003), Vol. 51, pp. 4267-4277.

[6] Dulin B., Kushner, B., and Rotolico, A., High Velocity Oxy-Fuel (HVOF) Sprayed Inconel 718 Coatings Internal Diameter, SAE Technical Paper 900954 (1990)

[7] Browning, J.A., U.S. Patent No. US7628606 B1 (2009)

[8] Oerlikon Metco Material Product Data Sheet: DSMTS-0030.11 – 83WC17Co Agglomerated and Sintered Powders, Oerlikon Metco (2016)

[9] Sauer, J. and Sahoo, P., HVOF Process Control Using Almen and Temperature Measurement, IMR Test Labs Publications (2001)

Thermal spraying of pure nickel via HVOF: Effect of fuel and shroud gas variation on particle in-flight characteristics and final coating properties

C. Hambrock, C. Grill, G. Schimo, A.W. Hassel, Linz / AT

In the course of this investigation, thermal spraying with different fuel and shroud gas combinations was investigated in terms of its effect on the in-flight particle properties (temperature, velocity) and on the final coating properties (coating thickness, porosity, oxygen content and corrosion behaviour). Independent on the shroud gas, the particle in-flight temperature and velocity were highest when using ethylene as fuel gas and lowest when using propane. Methylene resulted in intermediate properties. The change in the shroud gas from air to nitrogen generally resulted in lower in-flight particle temperatures and also lower velocity.
The coating properties in terms of porosity and oxygen content directly correlated to the particle in-flight properties. With decreasing velocity and increasing temperature, the porosity and the oxygen content increased, respectively. The corrosion behaviour of the nickel coatings was studied in 0.5 M sulfuric acid media by means of potentiodynamic polarization curves. Good corrosion properties were observed when methylene and air served as fuel gas and shroud gas, respectively.

1 Introduction

There is a wide variety of materials that can be deposited via thermal spraying. Via gas-fueled HVOF thermal spraying, mainly metals and cermets are deposited for obtaining e.g. improved corrosion protection and/or wear resistance of standard metallic substrates [1, 2]. Suitable coatings containing low amounts of oxides and pores are usually obtained by using standard gas-fueled HVOF processes, especially compared to arc, powder or plasma spraying. However, for depositing pure metallic coatings like e.g. pure nickel with such properties recently developed thermal spraying processes, e.g. cold gas or warm spraying, are usually applied .With the help of these processes, increasing in-flight velocity of the particles and/or decreasing in-flight temperatures are obtained [3, 4]. Recently, it has been tried to adapt the standard gas-fueled HVOF thermal spraying equipment, e.g. the combustion chamber or the nozzle design, in order to achieve such improved coating morphologies [5, 6]. However, such equipment is not commercially available. Additionally, it is very time consuming to develope optimized thermal spraying parameters.
In the course of this study it was therefore tried to optimize the thermal spraying parameters. Not only the variation of the fuel gas but also the variation of the shroud gas was taken into account for obtaining better coating properties. A strong influence of the choice of the shroud gas could be revealed, taking into account not only the in-flight particle properties (temperature and velocity) but also the overall coating properties, like the porosity, the oxygen content as well as the corrosion properties.

2 Experimental

Thermal spray coatings of nickel were deposited onto steel substrates by using gas-atomized nickel powder +45-22 µm (supplied by H.C. Starck) after grit-blasting the surface with Tl-grit (0.6 – 1.0 mm).
The powder was sprayed using the gas-fueled HVOF thermal spraying gun DJ2700 with a de Laval nozzle (both supplied by OerlikonMetco) with the spray parameters displayed in Table 1. The powder feed rate of 60 g min^{-1}, the carrier gas flux of 30 FMR (flow meter reading) at a pressure of 12 bar, the spraying distance of 210 mm the gun traverse speed (600 mm min^{-1}) as well as the amount of coating cycles (10 cycles) were chosen similar for all samples. This enabled the relative evaluation of the deposition efficiency.

Table 1: Spraying parameters

Sample	Fuel Gas		Shroud Gas		Oxygen
	Type	Flux / FMR	Type	Flux / FMR	Flux / FMR
P-A	Propane	40	Air	45	20
M-A	Methane	62	Air	45	27
E-A	Ethylene	40	Air	52	28
P-N	Propane	22	Nitrogen	45	20
M-N	Methane	42	Nitrogen	45	27
E-N	Ethylene	30	Nitrogen	58	28

For all spraying parameters, the in-flight particle temperature and velocity was monitored by using the system Spraywatch© (Oseir) [7].
The coatings were analytically evaluated concerning their porosity (Delesse principle [8]), their oxygen content (GDOES depth profiling / Spectruma Analytik GDA 750) and their corrosion behaviour. Corrosion studies were performed using a 3D printed electrochemical cell, which was pressed on the sample giving a circular area of examination (38.5 mm^2). A Ag/AgCl micro-reference electrode filled with 3 M KCl immobilised in agar [9] and a Au counter electrode were positioned inside the cell. Potentiodynamic polarisation curves were recorded using an Ivium CompactStat potentiostat (Ivium Technologies) starting at −250 mV vs. open circuit potential (OCP) and sweeping to the positive direction

with a scan rate of 1 mV s^{-1} until the transpassive regime was reached at about 2 V vs. standard hydrogen electrode (SHE). Via Tafel extrapolation method the corrosion current densites of the nickel coatings were determined.

3 Results and Discussion

3.1 Shroud gas: Air

The in-flight properties of the thermal spray particles – their velocity and their temperature, respectively – at the spraying distance of 210 mm are displayed in **Fig. 1**, directly comparing the different fuel gases ethylene (E), methane (M) and propane (P). The in-flight particle temperature was highest when spraying with ethylene (1770°C) while it was lowest when spraying with propane (1570°C). This is well in accordance with literature describing highest temperature of combustion for thermal spraying with ethylene [8]. Using methane as fuel gas resulted in particle in-flight temperatures comparable to that of propane. Generally, a temperature difference between the highest and lowest particle temperature of 200 °C could be observed. Concerning the in-flight velocity of the particles, the same order of fuel gases was visible, revealing the highest velocity for the particles sprayed with ethylene (700 m s^{-1}) and the lowest for those sprayed with propane (550 m s^{-1}). Methane as fuel gas resulted in an intermediate in-flight velocity of the particles (600 m s^{-1}) which was, compared to the temperature, significantly higher than that of propane. The general difference between the highest and the lowest particle in-flight velocity is 150 m s^{-1}.

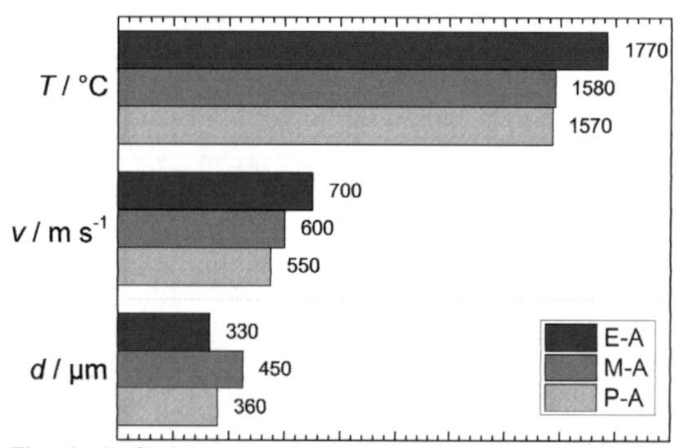

Fig. 1: In-flight properties (temperature and velocity) of the spray particles during thermal spraying with three different fuel gases (ethylene E, methane M and propane P) using the shroud gas air and their corresponding coating thickness obtained during deposition

The thickness of the coatings deposited by using three different fuel gases were evaluated relatively to each other by using similar deposition cycles and measuring the final coating thickness. The highest thickness could be determined for spraying with fuel gas methane (450 µm). Despite the highest in-flight particle temperature and velocity when spraying with ethylene, the coating thickness was lowest for this setup (330 µm). An intermediate coating thickness was obtained when thermal spraying with fuel gas propane (360 µm).

In addition to the particle in-flight properties also the properties of the final nickel coatings were investigated in terms of their porosity, oxygen content and corrosion behaviour. The direct comparison of these values sprayed with the three different fuel gases ethylene, methane and propane and with shroud gas air is displayed in **Fig. 2**.

The porosity of the coatings strongly depended on the fuel gas used for deposition. Using ethylene as the fuel gas resulted in the lowest porosity of 1.8 % while the use of propane resulted in the highest porosity of 3.9 %. The coating sprayed with the fuel gas methane revealed an intermediate porosity of 2.8 %. It is generally known that the in-flight velocity of particles strongly influences the porosity of the coating [8]. With increasing velocity of the particles also a coating with a lower porosity is expected to form. This correlation was determined in this case where thermal spraying with ethylene, revealing the highest in-flight particle velocity. The lowest particle in-flight velocity obtained with fuel gas propane resulted in the highest porosity.

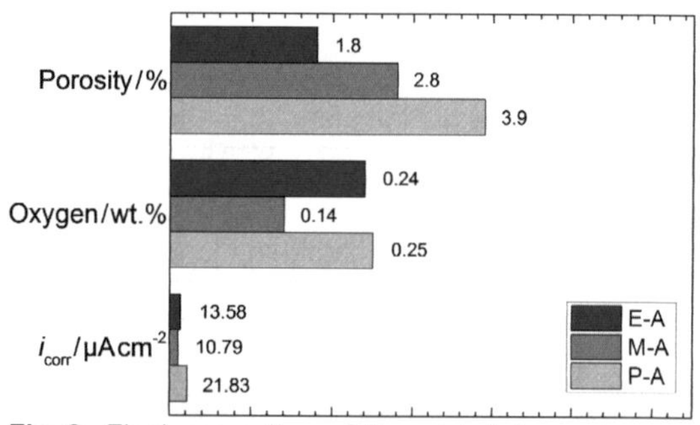

Fig. 2: Final properties of the nickel thermal spray coating revealing the porosity, the oxygen content as well as the corrosion current density i_{corr} for the thermal spray coatings sprayed with three different fuel gases (ethylene E, methane M and propane P) using the shroud gas air A

The oxygen content of the coating using the three different fuel gases ethylene, methane and propane behaved differently. It was expected that with increasing in-flight temperature of the particles also the oxygen content of the coating would increase. This, however, could not directly be revealed. Ethylene as fuel gas with the highest particle in-flight temperature also revealed the highest oxygen content. A comparable oxygen content could also be determined when spraying with propane although here the lowest particle in-flight temperatures were observed. In this case it was expected that in addition to the temperature also the particle velocity influenced the in-flight oxidation phenomena. With its significantly lower particle velocity, the residence time of the particles in the spray plum is higher, resulting in oxidation also at lower particle in-flight temperatures.

Thermal spraying with fuel gas methane showed, compared to the other thermal spraying parameters, a much better combination of particle in-flight temperature and velocity, resulting in the lowest oxygen content measured in the coating.
The corrosion current densities i_{corr} of the samples sprayed with the different fuel gases were also determined (see Figure 2). The sample sprayed with fuel gas propane showed the highest corrosion current of 21.8 µA cm^{-2}. The lowest corrosion current density (10.8 µA cm^{-2}) was observed for the sample for which methane served as fuel gas, corresponding to a low corrosion rate of 11.6 µm y^{-1}. Further analysis of the potentiodynamic polarisation curves (not shown here) gave information about the passivity of the nickel coatings. Full passivity was reached at the passive potential E_p, which was significantly lower for the coating thermally sprayed using methane as fuel gas. The passive potential for sample M-A was 418 mV vs.
SHE.

In contrast to this value the passive potentials of the

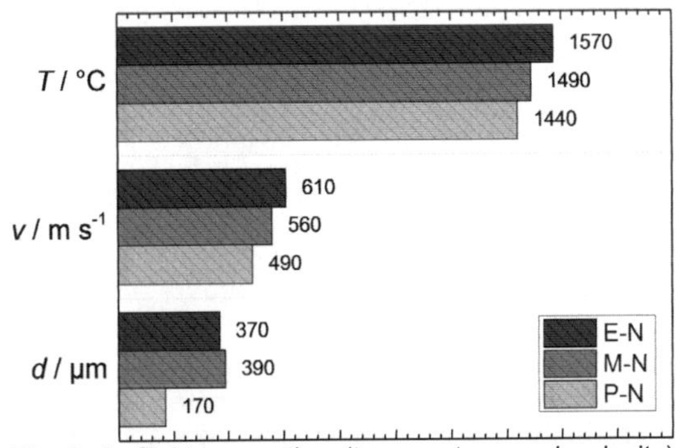

Fig. 3: In-flight properties (temperature and velocity) of the spray particles during thermal spraying with three different fuel gases (ethylene E, methane M and propane P) using the shroud gas nitrogen N

other coatings were all in the range of 600 mV vs. SHE. Furthermore, the transpassive potential E_b of sample M-A was more positive (1.39 V vs. SHE) compared to the coatings, which were fabricated using different fuel gases. As a result, the widest constant passive current region (between the passive potential and the transpassive potential) was obtained for the coating M-A, comprising a potential range of 975 mV. The constant passive current region for the coating sprayed using propane was 730 mV, and 770 mV for spraying with ethylene as fuel gas. In addition, the lowest passive current density i_p was determined for the nickel coating sprayed using methane (0.21 mA cm^{-2}), corresponding to low rates of nickel dissolution in the passive region. The highest passive current density was found for the sample thermally sprayed using propane as fuel gas (1.08 mA cm^{-2}), wheras using ethylene as fuel gas gave an intermediate passive current density of 0.61 mA cm^{-2}. These results show, that the conditions during methane-fueling for HVOF thermal spraying using air as shroud gas improve the passive properties of pure nickel coatings.

3.2 Shroud gas: Nitrogen

Thermal spraying with the three different fuel gases ethylene (E), methane (M) and propane (P) was also performed using the shroud gas nitrogen instead of air. No standard thermal spraying parameters were available for this which is why the standard thermal spraying parameters used for spraying with shroud gas air were adapted. Also for these thermal spraying parameters the in-flight particle properties as well as the final coating properties were determined.
The in-flight particle properties (temperature and velocity) using shroud gas nitrogen and the three different fuel gases are displayed in **Fig. 3**. The use of the shroud gas nitrogen had an influence on the general temperature but not in the already described order. Therefore, the highest in-flight particle temperature was obtained for spraying with fuel gas ethylene (1570 °C) while the lowest temperatures were observed when spraying with propane (1440 °C). It was obvious that this temperature obtained when spraying with fuel gas propane and shroud gas nitrogen was below the melting temperature of pure bulk nickel (1455 °C). Spraying with fuel gas methane resulted in intermediate temperatures of 1490°C. Generally, a temperature difference of 130 °C between the highest and the lowest temperature was observed.
The particle in-flight velocity behaved in the same order like the temperature, revealing the highest particle in-flight velocity when using fuel gas ethylene (610 m s^{-1}) and the lowest when using propane (490 m s^{-1}). Again, also of the velocity of the particles sprayed with methane was intermediate.

The coating thickness obtained for the different fuel gases using similar deposition cycles was highest when spraying with fuel gas methane. The lowest coating thickness was obtained when using the fuel gas propane. This was expected to be directly correlated to the in-flight particle properties. Using fuel gas propane in combination with shroud gas nitrogen resulted in lowest particle in-flight velocities (compared to the other fuel gases) and at the same time also in lowest in-flight particle temperatures which were even below the melting temperature of pure nickel. If the particles are neither heated nor accelerated suffiently, a direct influence on the deposition efficiency of the particles was expected.
The final properties of the themally sprayed coatings were investigated concerning the porosity of the coating as well as the content of oxygen in the coating. Additionally also the corrosion behavior of the coating was determined. The results of these measurements are displayed in **Fig. 4**.
The porosity of the coatings strongly variated comparing the different fuel gases to each other. The lowest porosity was obtained when spraying with ethylene (2.3 %). This was expected because of the

high particle in-flight velocity and temperature during thermal spraying. The opposite was the case when spraying with fuel gas propane. Here, the highest porosity was visible (5.5 %) because in this case, the lowest particle in-flight temperatures and velocities were achieved. Therefore, the particles neither were melted completely nor were impacting fast enough for obtainind a dense coating. Using fuel gas methane resulted in intermediate porosities (3.9 %).

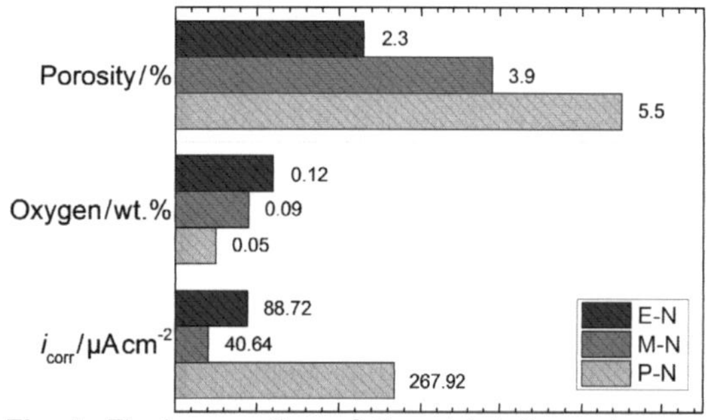

Fig. 4: Final properties of the nickel thermal spray coating revealing the porosity, the oxygen content as well as the corrosion current densities i_{corr} for the thermal spray coatings sprayed with three different fuel gases (ethylene E, methane M and propane P) using the shroud gas air A

Measuring the oxygen content in the coating revealed very low oxygen contents of less than 0.12 wt.% for all coatings. This highest oxygen content of 0.12 wt.% was obtained when spraying with fuel gas ethylene. As a result of the particle in-flight properties, especially of the higher temperature of the particles, stronger oxidation phenomena were expected, e.g. compared to thermal spraying with propane. In the case of propane, the lowest in-flight oxidation could be revealed with only 0.05 wt.%. A low oxidation of the thermal spray powder was expected to occur because the in-flight particle temperature was below the melting temperature of nickel. Using such thermal spraying parameters below the melting temperature of the thermal spray material is one way of obtaining least oxidised thermal spray coatings. However, the methods using such temperatures, e.g. cold spray, are using much higher temperatures so that in addition to the lowest oxidation also lowest porosities can be achieved [10, 11]. The oxygen content for the coating sprayed with methane was again intermediate (0.09 wt.%).

The corrosion current densities i_{corr} of the samples thermally sprayed using different fuel gases were determined from potentiodynamic polarisation curves via Tafel analysis. The i_{corr} values are given in Figure 4. The lowest corrosion current density (40.64 µA cm^{-2}) was found for the nickel coating thermally sprayed using methane as fuel gas. On the other hand, a more than six times higher corrosion current density (267.92 µA cm^{-2}) was determined for the sample for which fuel gas propane was used. Applying ethylene as fuel gas resulted in a corrosion current density of 88.72 µA cm^{-2}. As the potential for recording potentiodynamic polarisation curves was increased further with a constant scan rate of 1 mV s^{-1} until the transpassive region was reached (not shown here), information about the passive behaviour of the nickel coatings could be obtained from the measurements.

The passive potential E_p, indicating completed establishment of the passive layer on the nickel surface, was lowest (480 mV vs. SHE) for the coating thermally sprayed using methane as fuel gas. The highest E_p value of about 600 mV vs. SHE was obtained for the sample where propane served as fuel gas. The coating sprayed with fuel gas ethylene showed an intermediate passive potential of about 550 mV vs. SHE. From the passive potential on the current remained constant until the transpassive region was reached, indicated by the transpassive potential E_b. The potential range of constant passive current was widest for the sample thermally sprayed using methane as fuel gas (900 mV) and lowest for fuel gas propane (600 mV). The constant passive current regime for thermal spraying with ethylene as fuel gas comprised a potential range of 790 mV. Moreover, the lowest passive current density i_p was observed for the sample E-N (1.67 mA cm^{-2}) and the highest i_p was found for the coating sprayed with propane as fuel gas (2.11 mA cm^{-2}). Intermediate passive current density was determined from the potentiodynamic polarisation of the coating thermally sprayed usind fuel gas methane (1.86 mA cm^{-2}).

3.3 Comparison

The different thermal spraying parameters revealed that not only the choice of the fuel gas but also the choice of the shroud gas has a strong effect on the particle in-flight properties as well as on the final coating properties.

The in-flight temperature of the particles sprayed with the above-mentioned thermal spraying parameters was always highest when using fuel gas ethylene, independent on the shroud gas used. However, using the shroud gas nitrogen resulted in a temperature that was 200 °C lower. The influence of the shroud gas on the in-flight particle temperature sprayed with methane and propane was comparable because also here a decrease in particle temperature could be determined when nitrogen was used as shroud gas. However, the influence was lower because temperature differences in the range of around 100 °C could be observed. By changing the shroud gas from air to nitrogen, particle temperatures below the melting temperature of the thermal spray material could be obtained. This is generally important for decreasing the oxygen content in the final coating. It could generally be observed that using shroud gas nitrogen resulted in a decrease of the particle in-flight temperature. This was expected because of literature dealing with shrouding processes during HVOF thermal spraying. At the same time, the effect of the choice of the fuel gas was less important because the temperature of the particles sprayed with shroud gas nitrogen was comparably tight.

The particle in-flight velocity behaved comparably to the particle temperature. Also here, the strongest effect of the variation in shroud gas was observed when spraying with fuel gas ethylene. In this case, the velocity decreased by almost 100 m s^{-1} upon spraying with shroud gas nitrogen. Spraying with fuel gas methane and changing the shroud gas from air to nitrogen resulted in the lowest decrease in particle velocity because the particles were only 40 m s^{-1} slower than before. Propane as the fuel gas resulted in a change of 60 m s^{-1} when changing the shroud gas from air to nitrogen. Generally, the use of nitrogen as the shroud gas for thermal spraying resulted in a decrease of the in-flight velocity of the particles [12].

Correlating the particle in-flight properties to the coating thickness revealed a strong influence of the choice of thermal spray gases. The best coating thicknesses, independent on the variation of the shroud gas, were obtained for the coatings sprayed with fuel gas methane. However, the coating thickness was slightly lower when shroud gas nitrogen was used instead of air (390 µm vs. 450 µm). This was expected to be a result of the slightly lower temperature and velocity of the particles.

The sample sprayed with fuel gas ethylene revealed a higher coating thickness when shroud gas nitrogen was used (330 µm vs. 370 µm). In this case it is expected that the very high temperature and velocity of the particles when spraying with fuel gas ethylene and shroud gas air resulted in well-melted particles that splat well upon impact. However, the impact of the particles is very strong which does not always result in the formation of a coating but in overspray. This would explain the decreased coating thickness when using ethylene-air for thermal spraying. The use of ethylene-nitrogen results in lower temperatures and also less velocity of the particles and therefore the impact is lower resulting in less overspray and therefore a higher coating thickness. The particle in-flight properties of the sample E-N are comparable to that of the sample M-A. However, the sample M-A still resulted in a significantly higher coating thickness as compared to sample E-N. It is therefore expected that the use of air as the shroud gas has a positive effect on the deposition efficiency.

The thermal spray parameters using fuel gas propane and shroud gas nitrogen revealed a significantly lower coating thickness than all other coatings. This was again correlated to the particle in-flight temperature below the melting temperature of nickel as well as to the comparably low particle in-flight velocity. In this case, the melting as well as the impact of the particles is too low for forming a thicker coating.

The final coating properties in terms of porosity, oxygen content as well as corrosion current also reveal a strong influence of the choice of the fuel and shroud gases. The porosity of the coatings sprayed with shroud gas nitrogen was generally higher than compared to the coatings sprayed with shroud gas air. The porosity value was around 20-30 % higher for coatings sprayed with shroud gas nitrogen. This effect is expected to be directly influenced by the particle in-flight properties, especially by the particle velocity. With decreasing velocity of the particles, also a higher porosity is expected [8].

Also the content of oxygen revealed significant differences when comparing the samples sprayed with shroud gas air or nitrogen. Using shroud gas nitrogen instead of air significantly decreased the oxygen content in the coating. Even the highest amount of oxygen determined for these samples spayed with nitrogen was still lower than the lowest oxygen content determined for the samples spayed with shroud gas air. A decrease of 50% of the amount of oxygen for the samples sprayed with fuel gases ethylene and methane could be achieved. Spraying with propane and nitrogen even resulted in the reduction of oxygen content by 80%. Despite this great reduction the overall performance of the coating and the and the spraying process has to be taken into account. The very low oxygen content of the sample sprayed with propane and nitrogen gets along with a high porosity as well as a low deposition efficiency compared to the samples sprayed with other fuel- and shroud gas combinations [13].

Generally, higher corrosion current densities and passive current densities were determined for the nickel coatings thermally sprayed using nitrogen as shroud gas. Although the oxygen content in the coating could be significantly decreased by applying nitrogen as the shroud gas, the porosity was increased at the same time, which might be a result of the lower particle in-flight velocities and temperatures. Despite of many studies attributing improved corrosion resistance to low oxygen content of the coatings [13-15], in this work better corrosion and passivation behaviour was found for thermally sprayed nickel coatings containg relatively more oxygen, which was achieved by using air as shourd gas. These findings suggest, that the combination of relatively more oxygen in the coating (although still very low ≤ 0.25 wt.%) and lower porosity decreases the corrosion current density, the corrosion rate, the passive current density and widens the passive region of the nickel coatings. The poorest behaviour towards corrosion and passivation was found for the nickel sample thermally sprayed using fuel gas propane and shroud gas nitrogen. This could be a result of the low particle in-flight velocity and temperature below the melting temperature of pure nickel, leading to insufficient interlocking of impacted spray particles in the coating. Nevertheless, even when air was used as shroud gas propane-fueled HVOF thermal spraying of nickel showed poorest corrosion performance among the other fuel gases methane and ethylene. HVOF thermal spraying combining methane as fuel gas and air as shroud gas showed superior performance towards corrosion and passivation of pure nickel coatings. In addition the nickel coatings thermally sprayed using fuel gas methane were found to exhibit the lowest corrosion rates compared to the other fuel gases propane and ethylene, independent of applying air or nitrogen as the shroud gas.

4 Conclusions

By using the standard HVOF thermal spraying equipment and varying the combination of fuel and the shroud gas, strong influences on the particle in-flight as well as on the coating properties could be determined.

Using shroud gas nitrogen resulted in generally decreased particle in-flight temperatures and in-flight velocities, independent on the fuel gas used. The general tendency of the different fuel gases, revealing the hottest flame when spraying with ethylene and the coldest with propane, did not vary by changing the shroud gas.

The coating properties revealed, directly compared to spraying with shroud gas air, a strongly decreased oxygen content for the coatings thermally sprayed with shroud gas nitrogen. At the same time, however, also the porosity increased. Both these coating properties are directly correlated to the particle in-flight properties, the decreasing particle temperature and decreasing particle velocity, respectively.

Additionally, the corrosion and passivation behaviour of the thermally sprayed pure nickel coatings was investigated for the different fuel gas and shroud gas combinations. In the case of these measurements, no such direct correlation for the shroud gases could be determined. The lowest corrosion current density was found for combining fuel gas methane with shroud gas air, resulting in a i_{corr} of 10.8 µA cm^{-2}. This, and also the generally decreased corrosion resistance when shroud gas nitrogen was used, resulted in the assumption that the increased porosity had a strong effect on this coating parameter – stronger than the decreased oxygen content in the coating. It is therefore still necessary to improve the thermal spraying parameters in order to obtain denser and therefore more corrosion resistant coatings.

5 Literature

[1] Fauchais, P. and A. Vardelle: Thermal sprayed coatings used against corrosion and corrosive wear, in: Jazi, H.S. (Ed.) Advanced Plasma Spray Application, InTech, 2012.

[2] Chidambaram, D., C.R. Clayton and M.R. Dorfman: Evaluation of the electrochemical behavior of HVOF-sprayed alloy coatings. Surface and Coatings Technology 176 (2004) pp. 307-317.

[3] Koivuluoto, H. and P. Vuoristo: Structural Analysis of Cold-Sprayed Nickel-Based Metallic and Metallic-Ceramic Coatings. Journal of Thermal Spray Technology 19 (2010) pp. 975-989.

[4] Kawakita, J., H. Katanoda, M. Watanabe, K. Yokoyama and S. Kuroda: Warm Spraying: An improved spray process to deposit novel coatings. Surface and Coatings Technology 202 (2008) pp. 4369-4373.

[5] Ishikawa, Y., J. Kawakita, S. Osawa, T. Itsukaichi, Y. Sakamoto, M. Takaya and S. Kuroda: Evaluation of Corrosion and Wear Resistance of Hard Cermet Coatings Sprayed by Using an Improved HVOF Process. Journal of Thermal Spray Technology 14 (2005) pp. 384-390.

[6] Sun, B., H. Fukanuma and N. Ohno: Study on stainless steel 316L coatings sprayed by a novel high pressure HVOF. Surface and Coatings Technology 239 (2014) pp. 58-64.

[7] Oksa, M., E. Turunen, T. Suhonen, T. Varis and S.-P. Hannula: Optimization and Characterization of High Velocity Oxy-fuel Sprayed Coatings: Techniques, Materials, and Applications. Coatings 1 (2011) pp. 17-52.

[8] Fauchais, P.L., J.V.R. Heberlein and M.I. Boulos: Thermal Spray Fundamentals - From Powder to Part, Springer New York, USA, 2014.

[9] Hassel, A.W., K. Fushimi and M. Seo: Electrochemistry Communications 1 (1990) pp. 180-183.

[10] Ajdelsztajn, L., B. Jodoin and J.M. Schoenung: Synthesis and mechanical properties of nanocrystalline Ni coatings produced by cold gas dynamic spraying. Surface and Coatings Technology 201 (2006) pp. 1166-1172.

[11] Schmidt, T., F. Gaertner and H. Kreye: New Developments in Cold Spray Based on Higher Gas and Particle Temperatures. Journal of Thermal Spray Technology 15 (2006) pp. 488-494.

[12] Fukushima, T., H. Yamada, J. Kawakita and S. Kuroda: Correlation between the in-flight conditions of HVOF sprayed alloy particles and the coating structure, in: International Thermal Spray Conference, Essen, Germany, 2002, pp. 912 - 917.

[13] Hanson, T.C. and G.S. Settles: Particle temperature and velocity effects on the porosity and oxidation of an HVOF corrosion-control coating. Journal of Thermal Spray Technology 12 (2003) pp. 403-415.

[14] Kawakita, J., S. Kuroda, T. Fukushima and T. Kodama: Development of dense corrosion resistant coatings by an improved HVOF spraying process. Science and Technology of Advanced Materials 4 (2003) pp. 281-289.

[15] Hofman, R., M.P.W. Vreijling, G.M. Ferrari and J.H.W. de Wit: Electrochemical Methods for Characterisation of Thermal Spray Corrosion Resistant Stainless Steel Coatings. Materials Science Forum 289-292 (1998) pp. 641-654.

Coldgas Coatings with Adjusted Curie Temperatures for Influencing the Magnetic Susceptibility

F. Trenkle, E. Schopp, S. Hartmann
obz innovation gmbh, Bad Krozingen, Germany
info@obz-innovation.de ; +49 7633 90899 0

Abstract

obz innovation presents the latest developments in the field of induction capable coatings, mainly used for induction cookware on aluminum or copper base material. The aim was to develop a new type of "intelligent" induction coating that influences the magnetic susceptibility depending on its temperature. This not only allows to limit the maximum reachable temperature, but also supports the uniform heat distribution over the bottom of the cookware. Thus, this new type of "intelligent" induction coating regulates such important boundary conditions, independently of undesirable or even unsafe settings of an enduser. Finally, this makes cooking easier, healthier and more durable as it prevents damaged PTFE sealings and safer by reducing the danger of fire because of overheating and burning. Additionally, such coatings can be used in further applications, whenever a passively limited or uniform heat distribution is advantageous, for example when the temperatures are too high or in general an electronic control system is for technical or economical reasons not suitable.

1 Introduction

Induction cooking became more and more popular in recent years. Although the advantages of induction cooking are well known for a long time, especially in the last decade prices continuously dropped to a level that allows widespread use in private kitchens. For spreading in industrial kitchens a main reason is the technological progress that was made. Mainly the electronic components and as a result the power transition and switching times have been improved and allow perfect and well controlled cooking of food, compared to gas ovens, but combined with advantages of much lower heating of the workspace and better cleanability of the working area. From cookware side, the progress in the last decade was often focused on design and improvement of anti-adhesive sealings, but less on the induction capable materials. Sealings based on PTFE have now and then the best performance concerning anti-adhesion properties, but have a major drawback since they are not heat resistant for temperatures of >280°C. This can lead first of all to irreversible damage of the sealing and reduced or completely damage anti-adhesion properties. Second, vaporization of PTFE is discussed to be unhealthy, especially repeatedly and in enclosed rooms. A temperature of 280°C is reached in a pan within 2-3 minutes and then the PTFE coating can begin to be irreversibly damaged. Additionally, the danger of even causing fire by coming in contact with flammable material, e.g. oil inside of the cookware, significantly increases with higher temperatures. Thus, a coating that is limiting the maximum temperature is a safety critical function and likewise important for the enduser as well as for the cookware manufacturers. The last ones can guarantee a longer life time of their anti-adhesion sealing, since an overheating of the PTFE will not be possible anymore on induction ovens.

Even more visibility and comfort for the enduser brings along the latest by obz innovation integrated functionality, next to the temperature limitation. The cold gas sprayed coatings also have an immanent function for uniformly distributing the heat over the complete bottom of the cookware, which will be presented in the following chapters.

2 Serial Production of Induction Coatings for Cookware

Aluminum or copper cookware is not heatable on an induction stove, which implies to have an additional ferritic steel bottom, connected to the aluminium or copper base material. For induction, an electric coil induces an alternating magnetic field that induces electrical eddy currents in the cookware bottom. Thus, eddy currents together with reversal magnetism by turning of Weiss domains produces heat in the bottom of induction capable cookware. No direct contact between the cookware and the induction coil is required, nevertheless heat is directly coupled to the area of the induction capable material.
obz is coating cookware with ferritic chromium steel in serial production processes for many years. The coating works as good as other manufacturing technologies, but has advantages from design aspects. Furthermore, the cold sprayed induction coating can be produced with the minimum thickness that is necessary for coupling the complete power into the cookware. This helps to have the best possible heat transition into the cookware, without losses because of unnecessary thick steel material with worse heat conductivity compared to copper or aluminum.
For the production process, the bottom is cleaned and coated with ferritic chromium steel by cold spraying. The cookware is rotating during this coating process, so the rotation velocity has to be fit automatically to

the coating position to receive a homogeneous thickness.

Afterwards the sprayed coating has to be turned. Coating and turning are in respect to cost efficiency optimized to the requirements of serial production in the field of household goods. In figure 1 are shown two pans with cold sprayed induction layer. On the left side the material is as sprayed and rough. On the right side the steel is turned and has its characteristic metallic shine of steel.

The induction coating is corrosion resistant and contains a very good thermal conductivity, compared to convetional induction cookware. Also exists a high scratch and shock resistance, since the coating thickness is after final turning only about 0,5 – 0,6 mm in thickness. Cold spraying causes the ferritic steel particles to micro-weld onto the cookware bottom, so that a very good binding between bottom and coating occurs, which leads to optimal heat transfer.

Fig. 1: Pans with as sprayed cold gas induction coating (left) and after machining (right).

2 Temperature Dependent Influencing of the Magnetic Suceptiblity

To offer additional safety and comfort to the customer, a special induction coating was developed with an immanent temperature limitation and uniform heat distribution (T-LIM coating). In case the user of induction cookware puts the empty cookware to the induction stove and switches on the power, conventional ferritic chromium steel reaches temperatures >300°C fast, within 2-3 minutes. This causes danger of fire and damages the PTFE sealing of the cookware. With the T-LIM coating, the temperature on an induction oven will never rise upon an adjustable limit of e.g. 260°C, which is below the temperature of ca. 280°C where PTFE starts to pyrolyse.

To achieve both of those functionalities, temperature limitation and uniform heat distribution, obz uses its patent applied technology of two (or even more) ferromagnetic materials with different Curie temperatures, combined within one coating. By exceeding the Curie temperature of one of those materials, only this one loses its magnetizability. This leads to a reduced magnetic susceptibility, meaning that less induction power can be coupled into this coating area. So, all this not only happens within the coating as a whole unit, but also within each microscopic area of the coating, which is sketched in figure 2. State 1 illustrates the situation after switching on the oven. The magnetic field mainly heats the material in the center, in the area directly above the induction coil. After reaching the Curie temperature of material 2 in this region, the magnetic susceptibility is reduced in the center and therefore less induction power can be coupled into this area. This is illustrated through the changing color of material 2 from black (=high susceptibility) to light grey (=low susceptibility). When this happened the situation is like shown in state 2, more energy is coupled into the outer regions of the cookware bottom. As in state 2 the center area is cooling down slowly, below the Curie temperature of material 2. Thus, this effect repeatedly happens over the complete bottom within microscopic areas, whenever the temperature in this region is approaching the Curie temperature of material 2. Finally, an uniform heat distribution over the complete cooking area results.

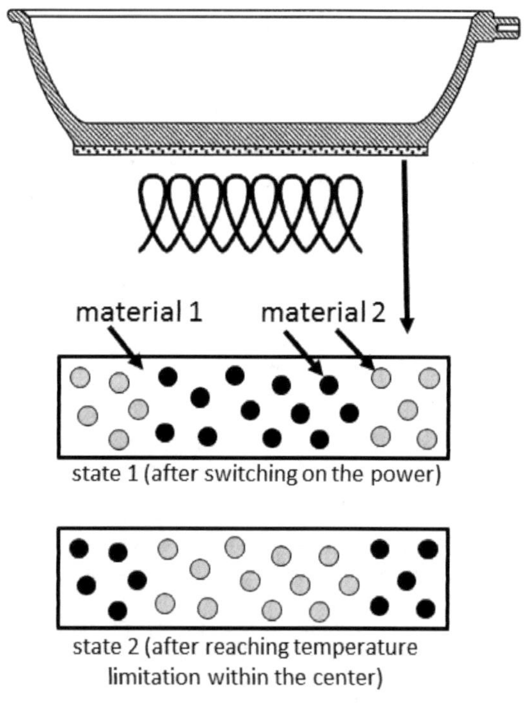

Fig. 2: Material 2 controls with its Curie point the temperature limitation. Furthermore, it cares for a uniform heat distribution over the complete coating area, since the usually not uniformly distributed magnetic field of the induction coil is forced by the alternating magnetic susceptibility (grey = lower and black = higher susceptibility) to the areas of lower temperatures.

3 Results of Changing Suceptibility on Uniform Heat Distribution

As explained in chapter 4, the varying magnetic susceptibility leads to a temperature limitation and a uniform heat distribution, since less power can be coupled into the lower susceptible, higher temperatured microscopic areas of the coating. In figure 3 is presented the difference in a heat image between a conventional induction pan (right) and a pan with T-Lim coating (left). The pan left with obz T-Lim coating cares for much better heat uniformity over the bottom of the cookware and regulates the maximum temperature in the range of about 250°C, while the conventional cookware right not uniformly heats, spreading from the center area that has far above 300°C.

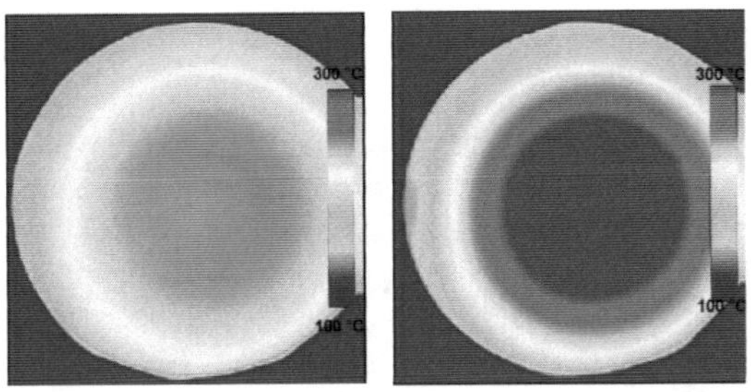

Fig. 3: "Intelligent" induction coating (left) shows in the heat image much better uniformity of the heat distribution compared to conventional induction coating (right).

Furthermore, the effect concerning uniform heat distribution was practically tested by caramelizing sugar. Therefore, the bottom of the pan was uniformly covered with sugar and the power was switched on. The uniform heat distribution of the T-Lim coating cares for melting the sugar all over the bottom of the pan, without burning, as figure 4 shows. Compared to that, a conventional induction pan is melting and caramelizing the sugar in the center area first. And even an aluminum pan with a thicker material strength, that usually supports uniform heat distribution, can not prevent burning sugar in the center – before caramelizing or even melting the sugar in the outside areas. A similar effect can be expected for example by baking pan cakes or all other food that uses a larger area .
One reason for all this is, that the coil of an induction oven is roughly placed in the area between 30% to 60% of the bottom diameter of the cookware. This leads to heating in the center area first, and since conventional induction material has a Curie temperature of >500°C, there is no temperature limitation in the relevant region.

Fig. 4: "Intelligent" T-Lim induction coating (left) cares for temperature limitation and uniform heat distribution, as the equally caramelization process over the complete bottom of the pan shows. Conventional induction cookware (right) shows over- and under heated areas, depending on where the induction coil is placed.

4 Results of Changing Suceptibility on Temperature Limitation and Power Consumption

The temperature scala in figure 3 (left) and figure 5 indicates that the T-Lim coating limits the maximum reachable temperature to ca. 250°C. This is sufficiently below the crictical temperature for damaging the PTFE sealing of the cookware and however more than adequate for roasting meat or similar cooking tasks. Furthermore, this temperature is adjustable to the demands of the cookware manufacturer or the end user.
The function of temperature limitation on cookware is known since about 20 years, but commonly had a major drawback up to now. It only consists of one Curie material and this can lead to problems with induction ovens. The reason is a safety function, that every induction plate must have integreated. It concerns the minimum power that needs to be coupled into the cookware. Below this limit, the induction oven automatically has to switch off its power. Thus, once a single material is heated up towards its Curie point, the power that can be coupled in drops. Finally, it falls below the limit and then it can happen that the induction oven switches completely off the power (see red line in figure 5), so that cooking is not possible any longer. Alternatively, it repeatedly turns on / off / on / off around this power limit, which can cause damage of the electronics and makes an unattractive noise while cooking (see orange line in figure 5).
Compared to this, the newly developed T-Lim coating of obz does not allow the power to drop towards zero, since material 1 (see figure 2) has its Curie point far above material 2 and prevents the power to drop too much. The cookware behaves like the triangular pointed measurment in figure 5 shows, the power does not fall below an adjustable minimum that prevents switching off the induction oven.

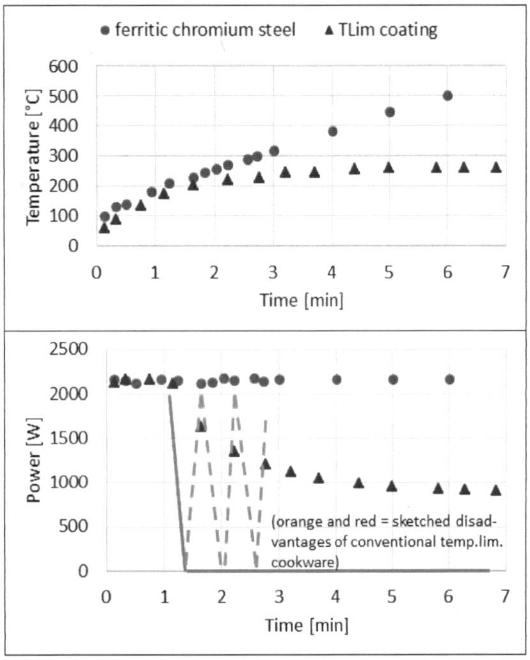

Fig. 5: Temperature and power development for a conventional ferritic chromium cookware and the newly developed T-Lim coating.

5 Functionallity and design of modern cookware

The newly developed T-Lim coating can be combined with mostly every design rules of cookware manufacturers. This means for example that the top sealing in the cooking area and outside can be independently choosen, like PTFE, ceramic sealing or steel inlet. Even the T-Lim coating itself can be combined with design specifications, as figure 6 shows with the "coloduct coating", developed by obz.

Fig. 6: obz developed "coloduct coating" can be combined with standard and T-Lim induction coatings.

6 Summary

T-Lim coating is an induction capable coating, developed by obz innovation, that shows superior functionality compared to state of the art cookware products in the market. It not only has an immanent temperature limitation, but also brings along uniform heat distribution over the cooking area. The latter one especially is advantageous for cooking meals that use most of the bottom area, like pan cakes, caramelizing sugar, fried eggs, and many more. The immanent temperature limitation protects PTFE sealings from degradation and reduces the danger of fire. So, beside the advantges for the cooking process, this is at the same time an aspect of health and safety for the enduser and durability as a quality feature for the cookware manufacturer.

Finally, this patent applied T-Lim coating can be used in other applications, whenever a passively controlled induction coating is useful.

An experimental approach to gain insight into cold gas spraying of ceramics

H. Gutzmann, I.Irkhin, F. Gärtner, T. Klassen, Hamburg / D

In this study, an experimental approach to investigate basic dependencies on impact and bonding of agglomerated ceramic particles in cold spraying is presented. Single impact morphologies of ceramic particles obtained from wipe tests are correlated with data obtained from powder compression experiments with a modified nanoindenter. Different feedstock powders of agglomerated TiO_2-nanoparticles were used and also partially heat treated. The powder shapes and sizes prior and after the compression tests were analyzed by confocal microscopy. The single particle impacts were characterized by SEM. Besides the expected influence of substrate material, substrate temperature, and spray conditions, the deformation and bonding of ceramic particles to metal substrates critically depend on the powder properties. To which degree particles fracture or contribute to layer formation upon the high-energy impact is highly correlated to their individual deformation behaviour in quasi-static compression tests.

1 Introduction

Cold gas spraying of ductile metallic materials is well established today. Manifold applications exist and numerous experiments and simulations led to a comprehensive understanding of the process [1-4]. In cold spraying of metals, bonding is attributed to the generation of adiabatic shear instabilities during particle impact. However, for cold spraying of ceramics, only very little generally accepted knowledge exists up to date. Since the technique can avoid phase transformations and process ceramic material by preserving the crystallographic structure of the powder feedstock [5], it has a high application potential to develop functional coatings, for instance based on photocatalytically active metastable anatase-TiO_2.

In contrast to numerous publications on cold spraying of metals only a couple of papers exist on cold spraying of ceramics [5-16]. In these publications, bonding is attributed to three possible reasons, (i) the generation of adiabatic shear instabilities in metal substrates, (ii) mechanical interlocking or (iii) via formation of chemical bonds. Further experimental investigations and simulations are needed to aquire additional knowledge concerning the impact process of ceramic particles in cold spraying. One important question to answer is for what reason one batch of ceramic powder leads to deposition while another, barely distinguishable batch does not.

The present study aims to shed light to the role of ceramic powder properties in cold spraying in particular. For that, agglomerated anatase TiO_2 powders are characterized with respect to their properties. Spray experiments were performed under wide variation of parameter sets, to gain information about respective influences on particle impact morphologies. For mechanical powder testing, a single particle static compression test method was utilized, that has also been used for the analyses of metallic materials [17].

2 Experimental

2.1 Materials

In this study, two different TiO_2 powders were used as feedstock material; a pure anatase powder Hombikat T40 (Sachtleben Chemie GmbH), and VP Aeroperl P25/20 (Evonik Industries AG) that contains up to 20 % rutile. Sub-batches of the Hombikat powder were also heat treated in air at temperatures in the range of 570°C to 600°C for a duration of 3 h.

Titanium grade 2, stainless steel (EN 1.4301, AISI 304), copper and the aluminium alloy AlMg3 were used as substrate materials, because they possess different hardness, influencing the deceleration and particle deformation upon impact in cold spraying.

2.2 Mechanical powder analyses

The single powder particle compression experiments were conducted with a NHT^2 nanoindenter (CSM Instruments, Peseux, Switzerland). The single particles were separately positioned on a polished WC-Co base plate. The load was applied by a diamond punch that had a circular area with a diameter of 200 µm, **Fig. 1**. By the device, the load is limited to 500 mN. For each powder batch, more than 20 individual particles were tested.

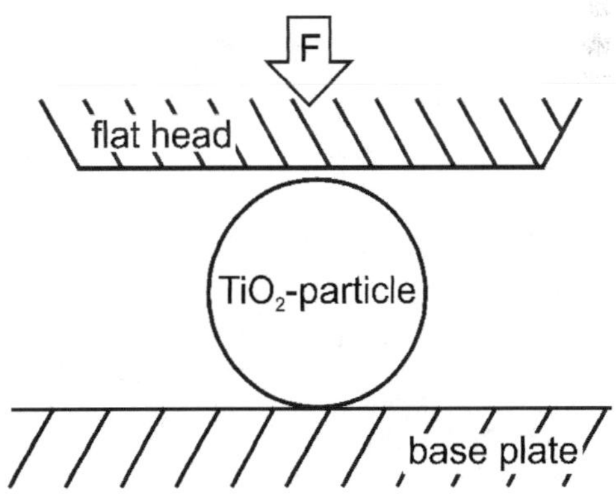

Fig. 1. Schematic side view of the compression experiments on single powder particles.

The separately positioned powder particles were investigated prior and after compression by using a VK-X200 confocal microscope from Keyence, Osaka, Japan to gain information on height and final deformation.

2.3 Sample preparation by cold spraying

As single particle impact morphologies provide valuable information on bonding and particle deformation, respective so called wipe tests were performed. The substrates were ground and polished to a final finish of 3 µm before spraying, to ensure that particle impacts can be easily identified.
The wipe tests were carried out by either moving the substrate perpendicular through the particle beam or the spray gun with a travers velocity ≥ 1.5 m/s.
Sets of experiments were perfomed with heated wipe test substrates. This was done (i) in order to study the effect of the substrate temperature on bonding and (ii) to adjust the substrate temperature to typical values generated during coating production.
All wipe test samples were prepared with the HSU CGT Kinetiks 8000 prototype with the standard type 24 WC-Co nozzle and powder feeder PF4000 from CGT, Ampfing, Germany. Main process parameter settings are summarized in **Table 1**. When different process parameters were used additional information is given in the text.

Table 1 Process parameters used for the preparation of the wipe test samples.

process gas, N_2	800°C / 4 MPa
carrier gas, N_2	6 m^3/h
feed rate	0.5 to 2 U/min ≈1-8 g/min
stand-off distance	60 mm

2.4 Analysis of powder and impact morphologies

Scanning electron microscopy (SEM) was performed to investigate powder morphology, powder microstructure and the wipe test samples in as sprayed condition. The device used was a Quanta 650 from FEI, Netherlands.

3 Results

3.1 Powder characteristics

The agglomerated TiO_2 powder particles have a spherical shape as a result of the production process and appear dense in the SEM micrographs, **Fig. 2**. The applied heat treatments did not change the appearance in the SEM images. It should be noted, that the resolution of the SEM does not supply any information on a conceivably existing nano-porosity.
The compression experiments of single powder particles showed that the behaviour of an individual particle is representative for the whole batch. Three distinct types of force-displacement curves were detected, **Fig. 3**. They are denoted as (i) brittle fracture, (ii) multiple fracture and (iii) quasi-plastic.
The "brittle fracture" characteristic is distinguished by a nearly linear and rather steep increase of the force over displacement that suddenly changes to an almost constant force over a wide range of the total displacement and a final steep increase in force to the maximum value defined as limit for the experiment. At the end, by unloading, the force drops to zero. The final drop by unloading is identical for each powder.

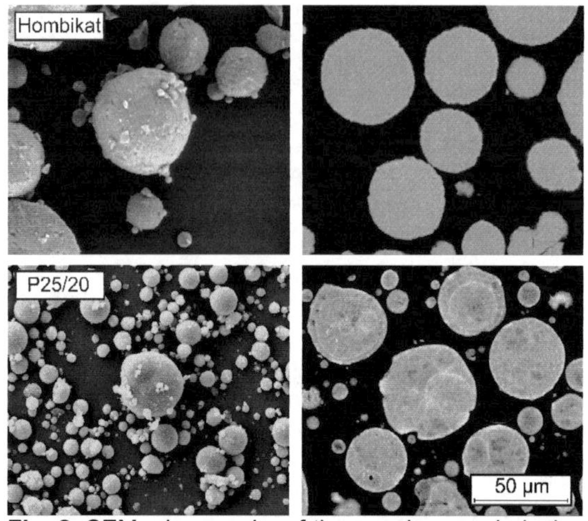

Fig. 2. SEM micrographs of the powder morphologies (left) and microstructure in cross section (right) of the as delivered TiO_2 powders. The agglomerated TiO_2 particles have a spherical shape and appear dense.

In contrast to the "brittle fracture" characteristic, the "quasi-plastic" behaviour is distinguished by a low yield point and a very smooth and steady rise of the force over a wide range of the total displacement, passing into an increasing slope up to the end of the loading section of the individual experiment.

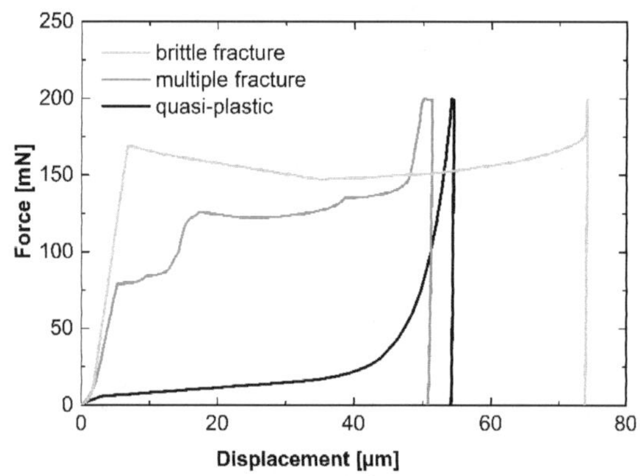

Fig. 3. Typical force-displacement diagrams of three different single TiO_2 particle types under compression. The behaviour of a single particle in each case is characteristic for the whole powder batch. The powder batches shown here are Hombikat (brittle fracture), heat-treated Hombikat (multiple fracture) and P25/20 (quasi-plastic). The maximum load during the experiments was set to 200 mN.

With a steep rise in the beginning of the test and a abrupt change in slope, the "multiple fracture"

characteristic is similar to the "brittle fracture" behaviour. However, during the overall deformation several of such distinct and abrupt changes occur before the maximum load is applied. In the sections between this abrupt changes, features similar to the quasi-plastic behaviour appear, as for instance slight and steady increases of the force. Thus, the "multiple fracture" characteristic appears to be a mixed fracture behaviour of the "brittle fracture" and "quasi-plastic" modes.
Confocal microscopy images reveal that in the case of "brittle fracture" characteristic, numerous particle fragments are generated and also at least partially scatter, **Fig. 4**. In contrast to the brittle fracture, for quasi-plastic deformation a rather compact "cluster" of titania is generated, **Fig. 4**.

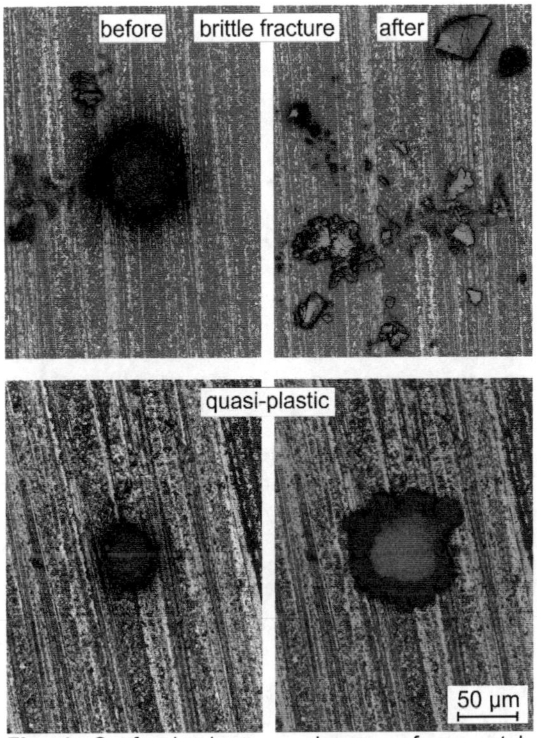

Fig. 4. Confocal microscopy images of separately positioned TiO_2 particles before (left) and after (right) a compression experiment. The notations "brittle fracture" and "quasi-plastic" correspond to those from **Fig. 3**.

3.2 Influence of cold spray conditions and substrate material on deposition

For studying the influence of powder deformation on bonding, wipe tests were performed by using the four different substrate materials and the Hombikat powder in as delivered (brittle fracture) and heat-treated (multiple-fracture) conditions. The spray parameters were varied over a wide range from T_{gas} = 400°C and p_{gas} = 3 MPa to T_{gas} = 1000°C and p_{gas} = 4 MPa.

For all these wipe test experiments on room temperature substrates, no bonding of entire particles could be observed. In most cases, even no remnants of titania were found on the substrates surfaces, only impact craters. Just for the combinations of heat-treated Hombikat (multiple-fracture) and the hard stainless steel and titanium substrates remnants of titania were found for spray conditions equal to or exceeding T_{gas} = 600°C and p_{gas} = 4 MPa.

3.3 Influence of substrate temperature on deposition

The substrate temperature was varied from RT to 300°C in 100°C steps. These experiments were performed with stainless steel substrates and the Hombikat powder in as delivered (brittle fracture) and heat-treated (multiple-fracture) conditions. In addition, spray experiments with the heat-treated Hombikat on AlMg3 substrates were accomplished. It was found, that entire particles of the "brittle fracture" type were only bonded at a substrate temperature above 200°C. Even for higher sunstrate temperatures only around 20 % of the impacting particles could be bonded. For the tempered Hombikat "multiple-fracture" type powder entire particles started to bond at 200°C and up to around 80 % of the particles could be bonded at 300°C. Using AlMg3 as substrate material for wipe tests of the "multiple-fracture" type tempered Hombikat powder, bonding of entire particles started already at 100°C. At a substrate temperature of 200°C, nearly every impacting particle was bonded.

3.4 Influence of powder properties on deposition

For the comparison of the most distinct powder types, wipe test samples were processed under identical conditions by spraying onto heated stainless steel substrates. Typical impact morphologies are shown in **Fig. 5.** It was found, that particles of the "brittle fracture" type usually do not bond. In contrast, particles of the "quasi-plastic" type seem to adhere well, only very few leave empty impact craters. In addition, these TiO_2 particles of the "quasi-plastic" type show substantial flattening under the high velocity impacts.

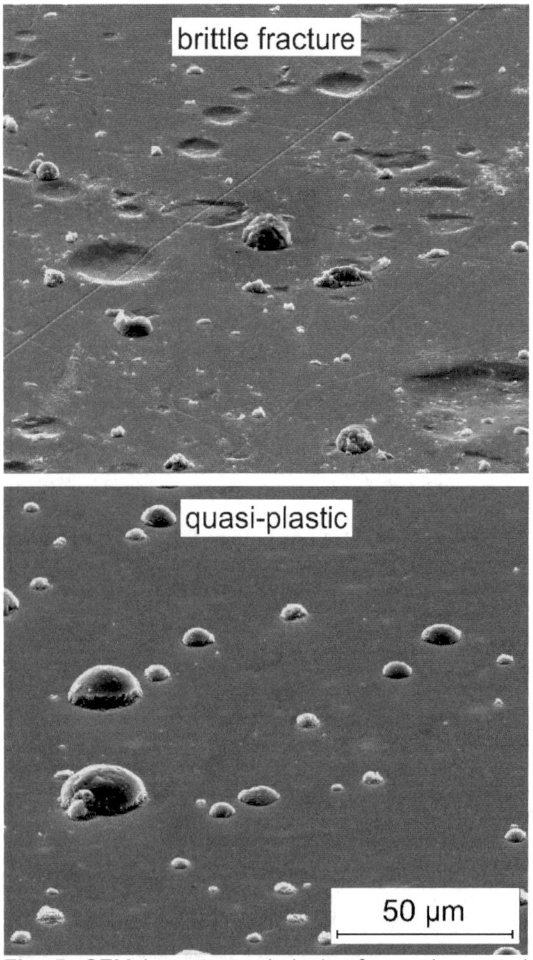

Fig. 5. SEM impact morphologies for agglomerated titania powders of types "brittle fracture" and "quasi-plastic" according to the deformation characteristics given in **Fig. 3**. Using identical N_2 gas temperature of 800°C and pressure of 4 MPa, the wipe tests were sprayed on stainless steel surfaces, heated to 300°C.

4 Discussion

The results show that bonding of the agglomerated TiO_2 particles depends on substrate material, substrate temperature, spray conditions and powder properties. Substrate temperatures of approximately minimum 300°C are necessary for successful bonding of titania particles onto the metal substrates. On the softer substrate materials, more complete TiO_2 particles deposit. On the harder substrate materials, more prominently remnants of titania are observed. In general, it seems that harder spray conditions are more suitable to achieve bonding.

However, independent of the before mentioned influence of process variables, bonding can normally not be achieved or to a much lesser extent if the particles show brittle fracture.

Looking into more detail by comparing **Fig. 3**, **Fig. 4** and **Fig. 5**, the results from the static powder compression experiments seem to agree well to those from the wipe tests. A "quasi-plastic" deformation behaviour under static load promotes bonding in cold spraying while the "brittle fracture" does not. In case of the "brittle fracture" type, the particles undergo catastrophic failure once the accumulated stress by deformation in the particle exceeding certain limits. Numerous particle fragments are generated, scattered and not bonded to the metal substrate. In contrast to this behaviour, the particles of the "quasi-plastic" type are bonded to the metal substrate and even show substantial flattening. Such strain rate independent plastic deformation of ceramics demands for explanations in absence of significant dislocation densities. Conceivably, the quasi-plastic deformation of the TiO_2 particles can be interpreted as progressive formation of large amounts of micro and nano cracks between and within the primary subparticles that already occur at low stresses and in summary enable huge displacements. The significant deformation could explain the flattening of the ceramic particles during impact and respective bonding.

Furthermore, the micro and nano cracks may generate small nanosized fragments with chemically highly reactive surfaces due to unsaturated bonds formed during microcrack formation. Such unsaturated bonds could facilitate rebonding of the nanosized fragments as long as attractive interactions can be ensured. A separation can be excluded since all fragments have a high momentum towards the substrate. An activated surface might be interpreted as contribution to surface energies and respective energy gain under coarsening. However, due to large interaction distances and short time scales, solely surface energy related phenomena may be excluded.

In summary, bonding of entire agglomerated TiO_2 particles in cold spraying on metal substrates is facilitated by a quasi-plastic deformation behaviour of the particles. Whether ceramic powder particles are prone to bonding under the high deformation rates in cold spraying can be derived already from static compressive tests. The needed quasi-plastic characteristic is probably due to micro and nano cracks and the formation of nanosized fragments with highly reactive surfaces, which enable fusion and thus rebonding of the nanosized fragments. The proposed mechanism will conceivably depend on the maximum stress and the rate of stress increase upon impact, thus involving impact energy as well as substrate hardness and critical deceleration. However, since no dislocation mechanisms are involved, strain rate is less important, thus facilitating the observed correlation between quasi-static and high-velocity impact deformation behaviour of particles. A certain sintering stage of the nano-primary particles to agglomerates (processable by cold spraying), between too loose agglomerates and too dense ceramics seems crucial for the desired fracture and fusion behaviour. Similarities to impacting nano-powder clouds in aerosol deposition in vacuum may exist.

5 Conclusions

The present study demonstrates the influence of powder characteristics for deposition in cold spraying of ceramic feedstock material. Static compression experiments of single particles show three distinct particle deformation characteristics that interestingly correlate well to characteristics of single impact phenomena in cold spraying and associated ultra-high deformation rates. Bonding to the metal substrate is facilited by a quasi-plastic deformation behaviour of the ceramic particles. Such behaviour could be attributed to a fracture and fusion mechanism, i.e. formation and subsequent closure of micro and nano cracks.

For agglomerated TiO_2, cold spray conditions of minimum T_{gas} = 800°C and p_{gas} = 4 MPa seem to be required for successful powder deposition. In continuous operation for coating formation, such spray conditions result in substrate temperatures of around 250°C. As demonstrated by the wipe test experiments, substrate temperatures of such magnitude seem to be required to deposit the agglomerated TiO_2 particles.

For cold spraying of ceramics, particular attention needs to be paid to the powder characteristics and spray parameter sets, with a particular view on sufficiently high process gas temperatures.

Acknowledgements

The authors would like to thank the laboratory staff Thomas Breckwoldt, Norbert Németh, Camilla Schulze and Matthias Schulze for their support in the presented work. We are also grateful to the German Research Foundation (Deutsche Forschungsgemeinschaft, DFG) for funding of this work under grant No. KL 1945/1-2.

6 Literature

[1] T. Klassen, F. Gärtner, H. Assadi: High Pressure Cold Spray: Principles and Applications, ed. by C.M. Kay and J. Karthikeyan: ASM International, 2016.

[2] T. Schmidt, H. Assadi, F. Gärtner, H. Richter, T. Stoltenhoff, H. Kreye, T. Klassen: From Particle Acceleration to Impact and Bonding in Cold Spraying. J. Thermal Spray Technol. (2009), 18(5-6), p 794-808.

[3] T. Schmidt, F. Gärtner, H. Assadi, H. Kreye: Development of a Generalized Parameter Window for Cold Spray Deposition. Acta Mater. (2006), 54(3), p 729-742.

[4] H. Assadi, F. Gärtner, T. Stoltenhoff, H. Kreye: Bonding Mechanism in Cold Gas Spraying. Acta Mater. (2003), 51(15), p 4379-4394.

[5] J.-O. Kliemann, H. Gutzmann, F. Gärtner, H. Hübner, C. Borchers and T. Klassen: Formation of Cold-Sprayed Ceramic Titanium Dioxide Layers on Metal Surfaces. J. Thermal Spray Technol. (2010), 20(1-2), p 292-298.

[6] C.-J. Li, G.-J. Yang, X.-C. Huang, W.-Y. Li, A. Ohmori: Formation of TiO_2 photocatalyst through cold spraying. Proceedings of the International Thermal Spray Conference (2004), p 315-319.

[7] H. Y. Lee, Y. H. Yu, Y. C. Lee, Y. P. Hong, K. H. Ko: Interfacial studies between cold-sprayed WO_3, Y_2O_3 films and Si substrate. Applied Surface Science (2004), 227, p 244-249.

[8] H. Y. Lee, Y. H. Yu, Y. C. Lee, Y. P. Hong, K. H. Ko: Thin Film Coatings of WO_3 by Cold Gas Dynamic Spray: A Technical Note. J. Thermal Spray Technol. (2005), 14(2), p 183-186.

[9] J. Vlcek, L. Gimeno, H. Huber, E. Lugscheider: A Systematic Approach to Material Eligibility for the Cold-Spray Process. J. Thermal Spray Technol. (2005), 14(1), p 125-133.

[10] J. Han, S. W. Lee, E. A. Lee, T. Xiong, Z. Bao, H. Du: Photocatalytic Properties of TiO_2 Coatings Prepared by Cold Spray Process. Materials Science Forum (2006), 510-511, p 130-133.

[11] I. Burlacov, J. Jirkovský, L. Kavan, R. Ballhorn, R.B. Heimann: Cold gas dynamic spraying (CGDS) of TiO_2 (anatase) powders onto poly(sulfone) substrates: Microstructural characterisation and photocatalytic efficiency. Journal of Photochemistry and Photobiology A: Chemistry (2007), 187, p 285-292.

[12] G.-J. Yang, C.-J. Li, F. Han, W.-Y. Li, A. Ohmori: Low temperature deposition and characterization of TiO_2 photocatalytic film through cold spray. Applied Surface Science (2008), 254, p 3979-3982.

[13] M. Yamada, Y. Kandori, S. Kazunori, M. Fukumoto: Fabrication of Titanium Dioxide Photocatalyst Coatings by Cold Spray. Journal of Solid Mechanics and Materials Engineering (2009), 3(2), p 210-216.

[14] M. Yamada, H. Isago, H. Nakano, M.Fukumoto: Cold Spraying of TiO_2 Photocatalyst Coating With Nitrogen Process Gas. J. Thermal Spray Technol. (2010), 19(6), p 1218-1223.

[15] M. Yamada, H. Isago, K. Shima, H. Nakano, M. Fukumoto: Deposition of TiO_2 Ceramic Particles on Cold Spray Process. Proceedings of the International Thermal Spray Conference (2010), p 172-176.

[16] H. Gutzmann, S. Freese, F. Gärtner, T. Klassen: Cold Gas Spraying of Ceramics Using the Example of Titanium Dioxide. Proceedings of the International Thermal Spray Conference (2011), p 391-396.

[17] H. Assadi, I. Irkhin, H. Gutzmann, F. Gärtner, M. Schulze, M. VillaVidaller, T. Klassen: Determination of plastic constitutive properties of mircoparticles through single particle compression. Advanced Powder Technology (2015), 26, p. 1544-54.

Spray Pattern and Microstructure of Copper Coatings with the Optimized Rectangular Cross-Section Nozzle by the Computational Fluid Dynamics (CFD) in High-Pressure Cold Spraying

Kazuhiko SAKAKI and Shinji ARAI, Nagano / J

In the cold spray process, cross-sectional shape of the nozzle has a significant effect on spray pattern of coatings. There is a rectangular and a circular cross-sectional shapes on the cold spray nozzle. It has been reported that the rectangular nozzle provide a more uniform particle velocity across the exit of the nozzle. The circular exit nozzle is parabolic in shape. So spray pattern with the rectangular nozzle is wider than that with the circular spray nozzle. The goal of this investigation is to establish a design for the cold spray gun nozzle in order to gain more uniform spray profile of coatings. We have investigated the influence of expansion ratio, nozzle total length and the ratio of nozzle length of divergent section and parallel section of rectangular nozzle on behaviours of gas and particle by the computational fluid dynamics (CFD) in high pressure cold spraying.
In this study, the spray pattern and microstructure of copper coatings with the rectangular cross-section nozzle optimized by CFD analysis in high-pressure cold spraying are reported. The optimized rectangular nozzle provide a more uniform spray profile of coatings. Relatively finer powder tended to deposit on the end of the coating.

1 Introduction

One of the major elements of the cold spray (CS) process [1] is the high-speed gas jet, which is governed by gas dynamics. In gas dynamics, supersonic flows are obtained with convergent-divergent (or de Laval) nozzles. In thermal spraying including cold spraying, the main purpose is to obtain better coating quality [1-16]. Previous studies on thermal spraying show the coating properties are principally determined by the thermal and kinetic energy states of particles upon impact with the substrate.
The nozzle geometry is also important with regard to the cold spray method. In the cold spray method, a coating is formed by exposing a substrate to high-velocity solid-phase particles, which have been accelerated by supersonic gas flow at a temperature much lower than the melting or softening temperature of the feedstock. Cold spray systems employ various kinds of gun nozzle contours, such as convergent-divergent (or de Laval) [1, 4-12, 14-15, 18, 19], convergent-divergent-barrel [8, 15, 18, 19, 20] and convergent-barrel [9, 10] nozzle contours.
And the cross-section of divergent section of the cold spray nozzle has either circle or rectangle. The circular nozzles are one-piece construction, are machined using standard fabrication practices. Rectangle cross-section nozzle is fabricated in two pieces to active the rectangular geometry and weld and machined to final configure ration. Thus, in the commercial CS device, circular section nozzles are often found.
The particle velocity profiles are quite different comparing the circular and rectangular exit geometries in cold spray. The rectangular nozzles provide a more uniform particle velocity across the exit of nozzle. The circular exit is parabolic in shape. Both profiles are characteristic of their respective exit geometries. A particular desired spray profiles may be obtain by changing the nozzle exit geometry [12]. And there is a report [15] that deposition efficacy of copper coatings became maximum with 12 from 11 in of the expansion ratio of circular nozzle for stagnation pressure 3MPa and stagnation temperature 350℃ of nitrogen gas. On the other hand, the expansion ratio of rectangular nozzle is from 1.8 to 4.8 [11], or from 3.8 to 4.0[16]. This expansion ratio is considerably smaller than 11which is the optimum expansion ratio of the circular section nozzle.
The goal of this investigation is to establish a design for the cold spray gun nozzle in order to gain a flatter the spray pattern of coatings. We have investigated the effect of expansion ratio of rectangular nozzle on spray pattern of copper coatings in cold spraying by the computational fluid dynamics (CFD) and experiments [18, 19, 20]. And the particle velocity and deposition efficiency of the copper coatings were dependent on the expansion ratio and geometry of the rectangular nozzle. In the previous report [20], the CFD and experiments investigated the influence of length of divergent and parallel section of the converging-diverging parallel-shaped rectangular cross-section (C- D-Para-Re) nozzle on behaviours of the gas and copper particles in high-pressure cold spray. And we study about the influence of length of divergent section of the converging-diverging shaped rectangular cross-section (C-D-Re) nozzle, to compare with the circular nozzle. As the results, it found that there is an optimal ratio of the length of the diverging section and the parallel section for C-D-Para-Re nozzle.
From the results of CFD, as divergent section became shorter, the gas maximum axial velocity increased. In addition, as divergent section became shorter, gas velocity of the rectangular width direction(Y direction) at the divergent section end increased, too. At a result converging-diverging parallel-shaped nozzle was fabricated on the basis of the CFD, it was able to produce a coating of more flat thermal spray pattern than the converging-diverging nozzle shape.
In this study, distribution of copper particles on the substrate are analysed by CFD. And the experimental results of the spray pattern and microstructure of copper coatings with the rectangular cross-section

nozzle optimized by CFD analysis in high-pressure cold spraying are reported.

2 Numerical simulation of cold spray

There are several techniques that can be used to calculate the gas flow of cold spray systems. Recent analyses have used computational fluid dynamics (CFD) methods to simulate complex phenomenon in two dimensions. In the present paper, Ansys Fluent (version 15.0) commercial software [21] was used to predict the gas and particle flow. A schematic drawing of the cold spray equipment for the present study is shown in **Fig. 1**. **Fig.2** is shown a modelling of an axial-symmetric supersonic nozzle with rectangular cross-section. The nozzle throat diameter d_t is 2 mm (constant), the length of the converging section $L_{con.}$ is 50 mm (constant), the total length of divergent section $L_{div.}$ and the parallel length $L_{para.}$of the nozzle is 200 mm (constant), and ten different ratio of the length of the diverging section and the parallel section for C-D-Para-Re nozzle are used, as shown in **Table 1**. It has been optimized in the previous report [9, 19] that the total length of divergent section $L_{div.}$ and the parallel length $L_{para.}$of the nozzle is 200 mm.

The quarter three-dimensional models were used to reduce the computation cost. The computational domain was partitioned into a number of hexahedral cells. In order to achieve a grid independent solution, a variable number of elements about 140,000 and number nodes of about 55,000 were required, depending on the cross-sectional shape of the nozzles and nozzle outlet peripheral region.

The gas inlet was chosen as pressure inlet with a pressure value of 3.0 MPa [gage] and a temperature value of 623 K with process (nitrogen) gas. The particle velocity of 10 m/s was used in many models, thus in our model all particles are injected randomly at the surface of powder inlet with a constant velocity of 10 m/s at room temperature (300K). The outlet of the computational domain was treated as pressure outlet with an atmosphere pressure (0.0MPa [gage]). A standard no-slip condition was used for the nozzle wall and substrate surface. The heat transfer process between the gas and the wall was not considered. In order to capture the turbulent flow features accurately, the standard K–ε turbulence model available in FLUENT was used to model the turbulent flow in the simulation. The standard wall function was chosen for the near-wall flow treatment.

The trajectory of magnesium particles was simulated. The physical properties of copper particles are shown in **Table 2**. The acceleration of particles was computed using discrete phase modelling (DPM). The heat transfer between the gas and particles had been considered. Particle–particle interactions and the effect of particles on the gas phase were neglected. 1000 particles were used to obtain reliable statistical data in each case. In order to take turbulence effects into account, the Stochastic-Tracking type model implemented in Fluent was used. In this approach, the Discrete Random Walk (DRW) model was used to predict the particle velocity and distribution. The fundamental fluid-dynamic and particle dragging equations employed in the model can be found in the reference [21].

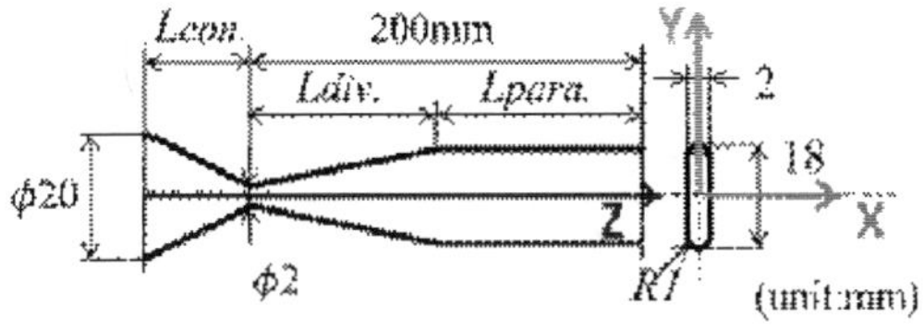

Fig. 1. Schematic cross-section diagram of converging-diverging parallel-shaped rectangular cross-section (C-D-Para-Re) cold spray nozzle used[20].

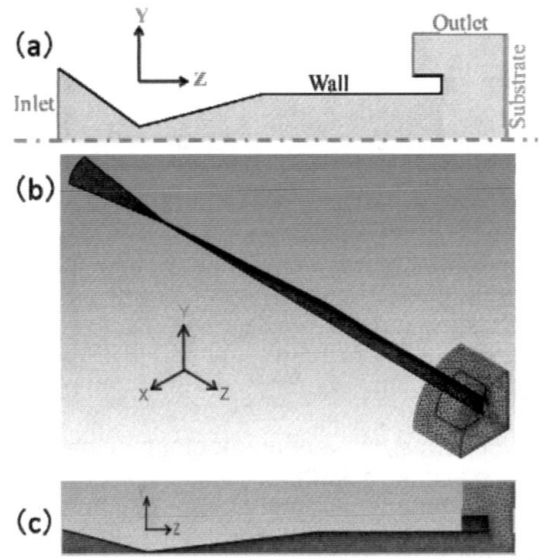

Fig. 2. Modeling of an axial-symmetric supersonic nozzle with rectangular cross-section; (a) schematic of simulation modelm, an example of a computational mesh (b) overall view(outside), (c)overall view(plane of symmetry)[20].

Table 1 Cold splay nozzle dimensions[20] unit: mm

No.	Nozzle Name C-D-Para(Re)-	$L_{con.}$	$L_{div.}$	$L_{para.}$	$L_{div.}:L_{para.}$	$\frac{L_{para}}{L_{div}+L_{para}}$
1	20-180	50	20	180	1 : 9	0.9
2	40-160		40	160	1 : 4	0.8
3	60-140		60	140	3 : 7	0.7
4	80-120		80	120	2 : 3	0.6
5	100-100		100	100	1 : 1	0.5
6	120-80		120	80	3 : 2	0.4
7	140-60		140	60	7 : 3	0.3
8	160-40		160	40	4 : 1	0.2
9	180-20		180	20	9 : 1	0.1
10	200		200	0	10 : 0	0

Table 2 Physical properties of Copper particles.

Diameter (μm)	20
Density (kg/m^3)	8933
Specific heat, C_p (J/(kg· K))	384.5
Thermal conductivity (W/(m· K))	399.7

3 Experimental equipment and methods

A schematic drawing of the cold spray equipment for the present study is shown in **Fig. 1**. Compressed nitrogen gas is introduced to a heater and a powder feeder. The pressurized gas is heated in an electric furnace. Usually, a convergent-divergent (C-D) nozzle is used in a typical cold spray device, and feedstock is injected axially into the main process gas flow at the gun nozzle intake. The cold spray equipment for this study was developed at Shinshu University [3, 4] and is shown in **Fig. 3**.

In this experiment, three kinds of C-D-Para-Re nozzles were used in **Fig.4**. As the results of CFD, C-D-Para-Re-100-100, 140-60, 180-20 nozzle, of which the nozzle throat diameter d_t is 2 mm, the length of the converging section $L_{con.}$ is 50 mm, shown in Table 1 No.5, 7, 9. And C-D-Re-120 nozzle were also used for reference[18].

The gas pressure value in nozzle chamber is 3.0 MPa [gage] and the gas temperature is 623 K. The pressure of the carrier gas is 3.1 [gage] MPa in order to insure the powder feeding.

The Feedstock powder is delivered by a precision metering device called "powder feeder" and typically is injected axially into the center of the high-pressure side of the nozzle intake (axial injection of powder: A/I) . Copper is cold sprayed onto mild steel (JIS SS400) and aluminium alloy substrates using pure copper (Cu-HWQ 20, Fukuda Metal Foil & Power Co., Ltd., Kyoto) with averaged particle diameter of 19.2 μm as the starting material.

Spray pattern of copper coating was measured every 2mm with a dial gauge.

The deposition efficiency was ascertained by measuring the weight gain on substrates for a spray time of 60 s taking into account the known feed-stock powder flow rate.

4 Results and discussion

4.1 Influence of the ration of length between divergent section and parallel section on the gas and Cu particles velocity by CFD

Fig. 5 shows the Influence of the ration of length between divergent section and parallel section on the process (nitrogen) gas velocity on the nozzle Z-axis in nozzle the influence of the ratio of the divergent section of the nozzle and the parallel section length on the distribution of the gas velocity on the nozzle central axis [20]. Gas reaches sonic velocity at the throat section becomes a supersonic accelerated by expansion in a nozzle divergent section, reaches the nozzle end slowly decreases due to pipe friction in the nozzle the nozzle cross-sectional area does not change the parallel section. Nozzle expansion ratio used in the present paper (optimum area ratio of the throat and the exit in nozzle), with respect to the nozzle inlet pressure 3.0MPa [gage], since the gas jet is the proper expansion, of the gas jet at the nozzle outlet and later rapid expansion (under expansion jet)

Fig. 3 Cold spray equipment used.

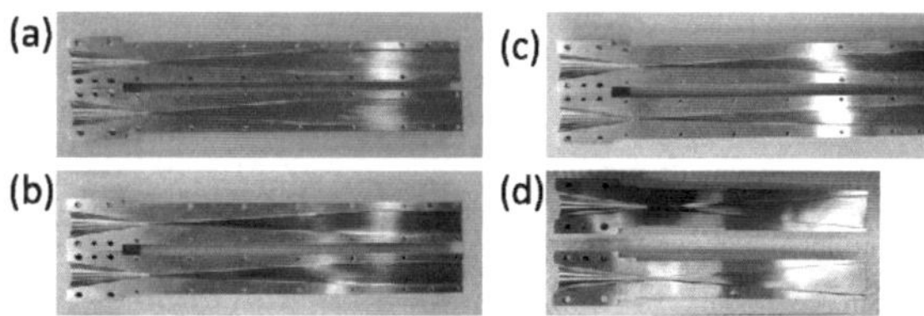

Fig. 4 Cold spray nozzles used: (a) C-D-Para-Re-100-100 nozzle, (b) C-D-Para-Re-140-60 nozzle, (c) C-D-Para-Re-180-20 nozzle, (d) C-D-Re-120 nozzle.

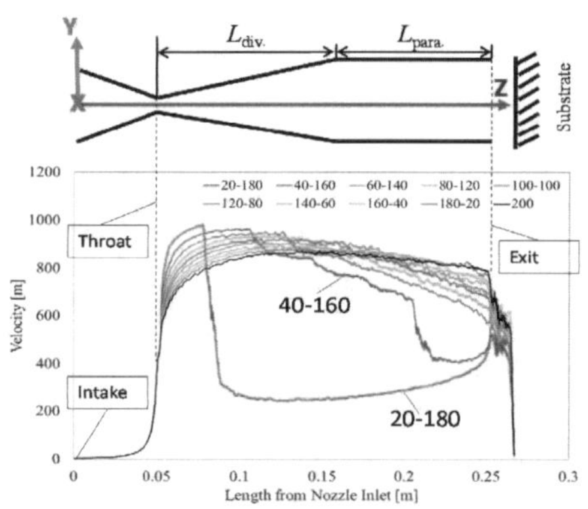

Fig.5 CFD results of influence of the ration of length between divergent section and parallel section on the process (nitrogen) gas axial velocity on the nozzle Z-axis in nozzle. (P_i=3MPa, T_{gi}=350°C).

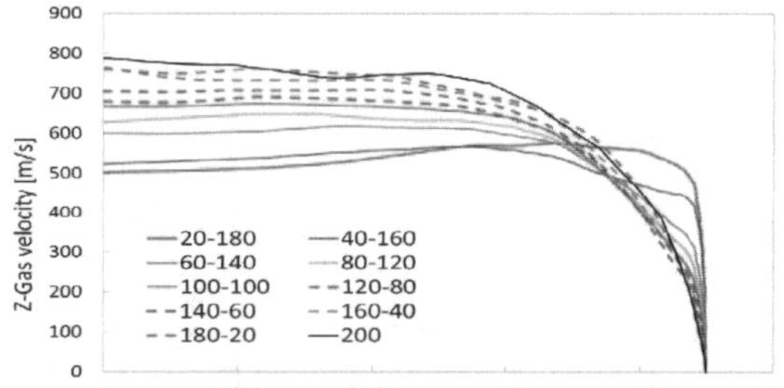

Fig. 6 CFD results of influence of the ration of length between divergent section and parallel section on the process (nitrogen) gas axial velocity distribution at the nozzle exit.
(P_i=3MPa, T_{gi}=350°C).

and contraction (hyperinflation jet) are not. However, since the substrate is in a 15mm away, abruptly decelerated from around 10 mm of the substrate immediately before. Although not shown the pressure is suddenly occurs up plate shock wave. Incidentally, C-D-Para. (Re) -20-180, -40-160 since the abruptly expanding short divergent section, since the flow occurs is peeled from the nozzle inner surface, not shown.

Fig. 6 shows the Influence of the ration of length between divergent section and parallel section on the process (nitrogen) gas axial velocity distribution at the nozzle exit. Providing parallel section of an appropriate length at the exit part of the rectangular nozzle is effective for flattening the axial velocity of the gas at the nozzle exit.

In the previous paper [20], it shows the distribution in the nozzle width direction of the Y direction velocity of the gas flow in the nozzle divergent section end. In CD-Para. (Re) -60-140, has a significant peak compared to other nozzles, Y = fairly early gas of about 1/10 of 100m / s and the axial gas velocity is in the width direction in 8mm It flows, the driving force that attempts to move from the center of the particles in the width direction.

In **Fig 8(a)** and **(b),** CFD results of the ration of length between divergent section and parallel section on particle impact velocity on substrate and number of particle distribution on substrate are shown.

4.2 Experimental results of effect of nozzle shape on the coating spray pattern and microstructure of copper coatings

Photos of spray pattern of copper coatings with 4 types of nozzles are shown in **Fig. 7**. And copper coating spray pattern by experimental are shown in **Fig. 8(c)**.

The boundary of the copper coatings with 3 types of C-R –Para(Re) nozzle in **Fig. 7 (a), (b)** and **(c)** are clear. This is also reflected in the coating thickness distribution of the spray pattern in **Fig 8(c)**, it has spread as the further 19 ㎜ in the width and 80μm in height.

Cross-sectional structure of etched copper coating with C-D-Para(Re)-100-100 nozzle are shown in **Fig. 9**.

No significant difference is observed in the coating structure at the center (in Fig.9(c-1)) and the edge (in Fig.9(c-2)).

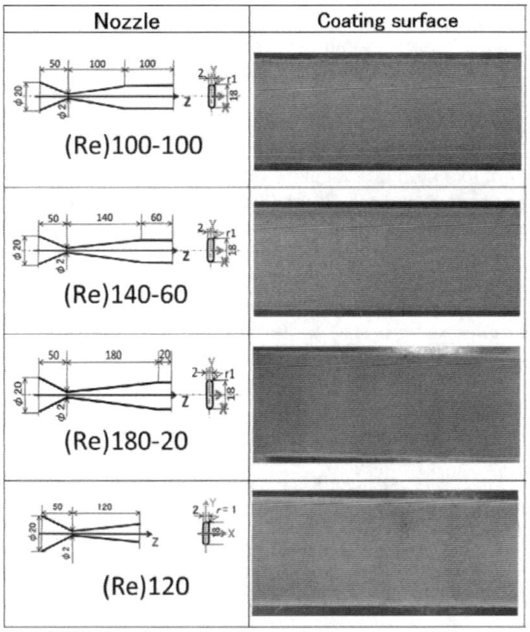

Fig. 7 Photo of spray pattern of copper coatings with (a)C-D-Para(Re)-100-100 nozzle[20], (b)C-D-Para(Re)-140-60 nozzle, (c)C-D(Re)- 180-20 nozzle,and (d)C-D(Re)-120 nozzle [18] (N_2, P_i=3MPa, T_{gi}=350°C).

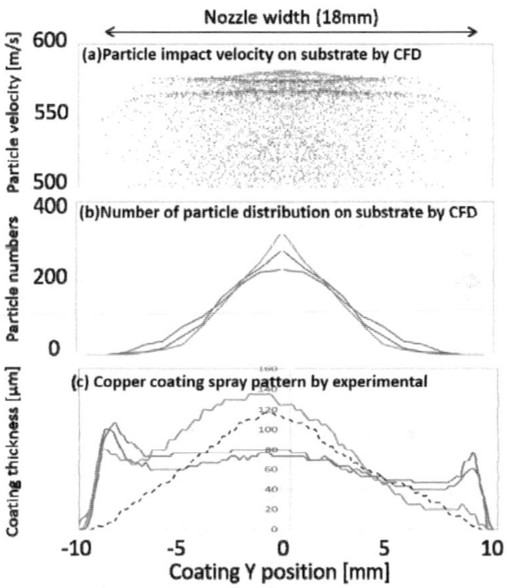

Fig.8 Influence of the ration of length between divergent section and parallel section on (a) particle impact velocity on substrate by CFD, (b) number of particle distribution on substrate by CFD, and (c) copper coating spray pattern by experimental. (P_i=3MPa, T_{gi}=350°C).

5 Summary

A length ratio of the divergent section and the parallel section of the converging-diverging parallel shaped rectangular cross-section nozzles was varied, and with the following results.

(1) From the results of CFD, parallel section is shorter it is, the maximum axial velocity of the gas is as high as 950m / s, it occurs near the end of the divergent section. In addition, in the divergent section end, the gas velocity of the rectangular width direction, still high shorter nozzle of the parallel part, fairly high.

(2) In order to produce a flat coating pattern, the rectangular section nozzle with the converging-

diverging-parallel shape is effective, and an appropriate length ratio of the diverging section and the parallel section is necessary. For example, it is 1: 1, 7: 3, and so on.

ACKNOWLEDGMENTS

Part of this study, Scientific Research Grant Program (Academic Research Grants Fund grant) (Foundation (C) (general): 2015 - 2017 fiscal year) was carried out by the issue number 15K06460.

The authors thank Fukuda Metal Foil & Power Co., Ltd. for supplying the cupper feedstock powder and Plasma Giken Co., Ltd. for lending the powder feeder.

6 Literature

[1] A. P. Alkimov, A. N. Papyrin, V. F. Kosarev and M. M. Shushpanav, US Pat. 5302414(1994)

[2] C. M. Hackette and G. S. Settles: The Influence of Nozzle Design on HVOF Spray Particle Velocity and Temperature", Thermal Spray Science & Technology, C. C. Berndt and S. Sampath, Ed, ASM International, Material Park, OH, (1995), p.p.135/40.

[3] K.Kopiola, J. P .Hirvonen, L. Laas and F.Rossi: The Influence of Nozzle design on HVOF Exit Gas Velocity and Coating Microstructure", J. Thermal Spray Technology, I6 (4), (1997), pp.469/74.

[4] K. Sakaki, Y. Shimizu, Y. Gouda and A. Devasenapath: Effect of nozzle geometry on High Velocity Oxygen-Fuel (HVOF) Thermal spraying process," proceedings of the 15th International Thermal Spray Conference, C .Coddet, ed., ASM International, (1998), pp. 445-450.

[5] K. Sakaki, Y. Shimizu and Y. Gouda, 'Effect of Expansion State of Combustion Gas Jet on High Velocity Oxygen-Fuel Thermal Spraying Process', J. Japan Institute of Metals, 63(2),(1999), 269/76 (in Japanese).

[6] D.L. Gilmore, R.C. Dykhuizen, R.A. Neiser, T.J.Roemer, and M.F.Smith: Particle Velocity and deposition Efficiency in Cold Spray Process," J. Thermal Spray Technology, 8 (4), (1999), pp. 576/82.

[7] K. Sakaki and Y. Shimizu: Effect of Increase in Entrance Length of Gun Nozzle on HVOF and Cold Spray Process," J. of Thermal Spray Technology, 10 (3), (2001), pp.487/96.

[8] T. Stoltenhoff, H. Kreye, and H.J.Richter "An analysis of the cold spray process and its coatings," J. Thermal Spray Technology, 11 (4), (2002), pp. 542/50.

[9] K. Sakaki, N. Huruhashi, K. Tamaki and Y. Shimizu: Effect of nozzle geometry on cold spray process, International Thermal Spray Conference, Lugscheider, E., and Berndt, C.C., ed.,DVS Deutscher Verband für Schweißen, (2002),pp. 385/89.

[10] W-Y. Li, et al.: Optimal design of a convergent-barrel cold spray nozzle by numerical method," Applied Surface Science, 253(2), (2006), pp.708/13.

[11] K.Sakaki, S.Shinkai, N.Ebara and Y.Shimizu: Effect of geometry of the gun nozzle, the increase in the entrance convergent section length and powder injection point on cold sprayed titanium coatings' Materials Science Forum Vols. 534-536, (2006), pp.413/16

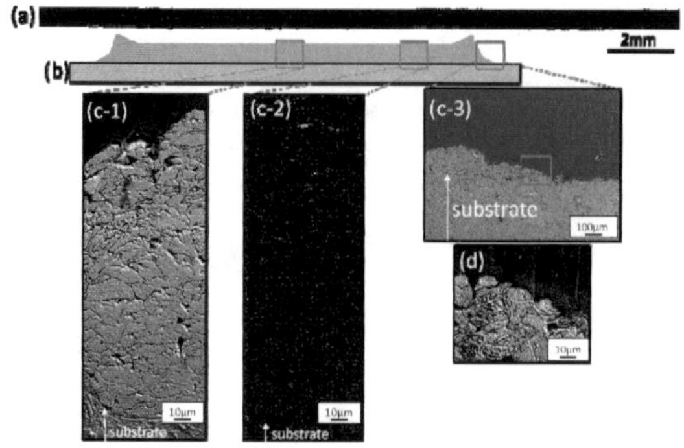

Fig. 9 Cross-sectional structure of etched copper coating with cold spray C-D-Para(Re)-100-100 nozzle : (a) optical micrograph, (b) schematic diagram, (c)SEM image,and (d) enlarged photo (N_2, P_i=3MPa, T_{gi}=350°C).

[12] A. N. Papyrin Ed.: Cold Spray Technology, Elsevier, Cambridge, U.K., (2007), p.119/78.

[13] A. N. Papyrin ed., Cold Spray Technology, Elsevier, Cambridge, U.K., (2007), p.216/20.

[14] V. K. Champagne, ed.: The Cold Spray Materials Deposition Process," Woodhead Publishing Materials, Cambridge, U.K., (2007), pp. 117/26.

[15] W.-Y. Li, H.L. Liao, G. Douchy, C. Coddet: Optimal design of a cold spray nozzle by numerical analysis of particle velocity and experimental validation with 316L stainless steel powder", Materials and Design, 28 (7), (2007), pp.2129/37.

[16] K. Sakaki, K. Takeda, K. Takada, T. Hosono, and Y. Shimizu: Influence of the expansion ratio of the gun nozzle and gas pressure on properties of cold sprayed copper coatings, Proceedings of the 3rd JSME/ASME International Conference on Material and Processing, (2008), MSEC_ICM&P2008-72015.

[17] X.K. Suo, T.K. Liu, W.Y. Li, Q.L. Suo, M.P. Planche, H.L. Liao: Numerical study on the effect of nozzle dimension on particle distribution in cold spraying", Surface & Coatings Technology, Vol.220, (2013), pp. 107/11.

[18] K.Sakaki, T. Akashi, and T.Hosono, "Influence of Cross sectional Shape of Cold Spray Nozzle on Spray pattern of copper coatings", DVD of Proc. of the JSME/ASME 2014 International Conference on Materials and Processing, Detroit (2014), ICMP2014-4961.

[19] S. Arai, K. Sakaki, T. Tagami and T. Tuchida: Optimization of the Shape of Rectangular Cross-Section Nozzle in High-Pressure Cold Spray (Influence of Length of Barrel and Divergent Section of Nozzle), Proc. of 2015 Japan Thermal spray society fall meeting, (2015), pp.5/6(in Japanese).

[20] K. Sakaki and S. Arai: Influence of Length of Divergent and Parallel Section of the Rectangular Cross-Section Nozzle on Behavior of Particles of Copper in High-Pressure Cold Spray, Proc. of International Thermal Spray Conference, (2016) , pp.1035-1039.

[21] ANSYS, Inc. : ANSYS FLUENT Theory guide for Release 15.0, (2015).

Evolution of residual stresses in PS-PVD TBCs with thermal cycling

Shunyan TAO [a], **Jiasheng YANG** [a,b], **Huayu ZHAO** [a], **Jinxing NI** [a], **Xinghua ZHONG** [a], **Yin ZHUANG** [a]
[a]Key Laboratory of Inorganic Coating Materials CAS, Shanghai Institute of Ceramics, Chinese Academy of Sciences, Shanghai 201899, PR China
[b]University of Chinese Academy of Sciences, Beijing 100039, China

Abstract

Cr^{3+} Photoluminescence piezo-spectroscopy (CPLPS) has been developed as a non-destructive inspection technique for the measurement of residual stresses within the thermally grown oxide (TGO) in thermal barrier coatings (TBCs). In this study, plasma spray - physical vapor deposition (PS-PVD) process was used to deposit yttria stabilized zirconia (YSZ) topcoat with quasi-columnar structures. Evolution of the microstructures and residual stress distribution in such kind structured TBCs before and after thermal cycle test on burner rigs were investigated. The accumulated tensile stress in the as-sprayed ceramic topcoat changed to compressive state after 100 cycles, and then gradually increased. In addition, the mapping compressive stresses in the TGO measured through the ceramic topcoat surface decreased rapidly firstly and then essentially maintain at a relatively stable value with further testing. Moreover, the pre-oxidation of the bondcoat could significantly affect the stress distribution in the TGO, in contrast, no obviously influence on the stresses in the YSZ topcoat.

Keywords: Plasma spray - physical vapor deposition (PS-PVD); Thermal barrier coatings (TBCs); Thermal cyclic test; Residual stress; CPLPS

Introduction

Plasma sprayed (PS) thermal barrier coatings (TBCs) have been widely used in a variety of gas turbine and diesel engine applications over the last decades. In the process, the coating material is injected into a plasma jet to be melted, accelerated and deposited onto the substrate surface. When droplets impact the substrate, the material solidifies and builds up in the form of lamellar splats [1, 2]. A new spray process called plasma spray - physical vapor deposition (PS-PVD) was developed as an extension of the low pressure plasma spray process (LPPS) [3-8]. Due to the low chamber pressure, the plasma jet gets broadened and the coating area is enlarged, which makes the deposition profile more uniform and relatively thin and dense ceramic coatings can be obtained. By increasing the power input of the plasma torch, either more spraying powder can be melted resulting in lamellar structures or the powder can be even evaporated so as to deposit feather-like columnar structures [9, 10]. PS-PVD is intended to combine the merits of plasma spraying (low cost and high deposition rate) and electron beam-physical vapor deposition (EB-PVD) to deposit coating with high strain tolerance and high thermal shock resistance [11, 12].

Thermal barrier coatings (TBCs) are multilayer-structured consisting of yttria-stabilized zirconia (YSZ) topcoat, an oxidation-resistant bondcoat and an intermediate alumina protective layer (thermally grown oxide, TGO) between the bondcoat and YSZ as a result of bondcoat oxidation [13-15]. The TBC lifetime often depends on a serious of crack initiation and propagation events which occur near or in the TGO layer [16-19]. It is believed that the driving force for the TBC spallation is related to the accumulation of stored elastic energy and damage due to cyclic straining and oxidation. Therefore, monitoring the residual stress in the TGO can potentially provide critical information regarding the degradation and failure mechanisms of the TBCs. Cr^{3+} photo-stimulated luminescence piezo-spectroscopy (CPLPS) is an optical technique for stress determination in alumina and alumina-containing ceramics [20-23]. Its high spatial resolution and fast acquisition make it a powerful non-destructive evaluation tool for the measurement of residual stress within the TGO (predominantly α-Al_2O_3). The technique has proven to be very effective for such measurements in YSZ thermal barrier coatings fabricated by EB-PVD [24, 25]. The PLPS method might also be suitable for measuring the residual stresses in PS-PVD coatings since the quasi-columnar structure can act as waveguides which limit dispersion of the incident laser beam and reduce attenuate of the optical signals. The piezo-spectroscopic technique is an optical method that utilizes the luminescence from trace Cr^{3+} dopants incorporated into the aluminum oxide formed by oxidation. The spectrum of aluminum oxide consists of two characteristic R-lines, R1 and R2. As the oxidation-induced stress builds up in the TGO, the R1 and R2 lines are shifted to higher frequencies and the R2 frequency shift can be used as a measure of the residual stresses in the TGO scale. Moreover, micro-Raman spectroscopy has been applied to measure residual stress in ZrO_2 and also residual stress of YSZ coatings [26-28]. The spectroscopic technique usually uses laser beam as an excitation source and the method provides local stress information. Therefore, it is possible to measure stress distribution, although the measured area is limited to near the surface.

However, measuring the residual stresses distribution in thermal cycled PS-PVD TBCs have not been systematically investigated up to now. The objective of this work was to assess residual stresses in the topcoat layer and TGO scale formed on TBCs with vacuum plasma sprayed bondcoat by micro Raman spectroscopy and CPLPS, respectively. Based on these results, discussion is made on the residual stress of PS-PVD TBCs.

Experiment procedures

Coating preparation

All ceramic coatings were fabricated by a PS-PVD system (VPS / LPPS-PVD / CVD hybrid system, Oerlikon Metco, Wohlen, Switzerland). The feedstock material used for the coatings was an agglomerated 7 wt.% yttria partially stabilized zirconia powder (7YSZ) produced by Oerlikon Metco (Switzerland) designated as Metco 6700, as shown in Fig. 1. This agglomerated powder consisted of submicron ZrO_2 particles and nano-sized Y_2O_3 particles with the grain size distribution with d_{10}=2 μm, d_{50}=8 μm, and d_{90}=18 μm, respectively. Ni-based superalloy (GH3128, Beijing CISRI-GAONA Materials & Technology Co., Ltd., China) was used as substrates, which were disk shape with a diameter of 30 mm and a thickness of 5 mm. Before the spraying processes, the substrate was sand blasted using alumina powder and then followed by ultrasonically cleaning in ethanol. Prior to spraying, the substrate was transfer arc cleaned and then preheated by the plasma jet and the temperature was controlled by sweep movements of the plasma jet over the substrate. About 100 μm thick CoNiCrAlY bond coat was deposited by vacuum plasma spraying (VPS). Before the deposition of the topcoat, the surface of the bondcoat was polished to be smooth with surface roughness Ra about 0.2 μm with/without pre-oxidation process. During pre-oxidation and the coating deposition process, feeding additional oxygen into the vacuum chamber was used to prevent a loss of oxygen due to the high temperatures in the plasma plume (up to 15000 K) in combination with the low oxygen pressure plasma plume, respectively. As to the ceramic top coat, a ~170 μm YSZ coating with quasi-column structure was applied, which was manufactured with the deposition parameters summarized in Table 1.

Figure 1: SEM images of the Metco 6700 YSZ powder

Burner rig test

Burner rig tests of PS-PVD TBCs were accomplished in an automatic flame shock test facility system (Shanghai Institute of Ceramics, Chinese Academy of Sciences, Shanghai, China). The coatings were heated for 3 min at 1250 ± 15 °C (coating surface temperature). During the heating stage, the backside of the samples was cooled by compressed air to maintain a temperature gradient through the coatings thickness. During the cooling stage, the samples were cooled by compressed air from both sides for 3 min to below 200 °C. Sets of Two samples (Sample 1 and Sample 2) were extracted from the holder for micro Raman spectroscopy and CPLPS experiments after 50, 100, 200, 300 cycles, respectively, as well as after the end of life determined as the time at which the spalled area reached about 10% of the total coating surface.

Table 1: General PS-PVD operating parameters

Parameters	Preheating	Deposition
Current / A	1400	2600
Net power / kW	41	63
Primary plasma gas (Ar) / slpm	50	35
Secondary plasma gas (He) / slpm	110	60
Oxygen / slpm	0 / 2 (*)	2
Carrier gas (Ar) / slpm	6	18
Powder feed rate / g/min	-	5
Spray distance / mm	900	1000

**Note that the parameters of pre-oxidation process are the same as that preheating process except additional introducing oxygen.*

The phase composition of the as-sprayed coating was characterized by X-ray diffractometer (XRD, D/max 2550 V, Rigaku Industrial Corporation, Japan) with filtered Cu Kα (40 kV, 40 mA) radiation at a scan rate of 4°/min in the 2θ range of 20°-80°. The microstructures of the coating samples were examined using a scanning electron microscopy (SEM, S4800, HITACHI, Japan) operated in backscattered electron image mode.

Stress measurement using micro-Raman spectroscopy and CPLPS

The measurement of residual stress in the YSZ topcoat and mapping stress in the TGO layer were carried out using Raman spectroscopy (RS) and Cr^{3+} photo-stimulated luminescence piezo-spectroscopy (CPLPS), respectively. The equipment was a confocal Raman microprobe system (Renishaw Invia Confocal micro-Raman Microscope, Glouchestershire, UK). Micro-Raman spectroscopic measurements were made at a fully controlled temperature (297±1 K) in ambient air. A green laser was used as excitation source of Raman spectroscopy and the beam was incident perpendicular to the surface of the specimens. The typical Raman spectra for the PS-PVD coatings were shown in Fig. 2. The Raman peak around 640 cm^{-1}, which has pronounced intensity and was chosen to calculate the stress distribution in the YSZ topcoat since it has a good signal to noise ratio. In the present PS-PVD TBCs, the stress value, σ(GPa), has been calculated using the following equation:

$$\Delta\nu_{TBC} = \Pi_{TBC} \cdot \sigma_{TBC} \quad (1)$$

where Δυ referred to the frequency shift (cm^{-1}) from the stress-free state, σ_{TBC} was the residual stress in the topcoat, the piezo-spectroscopic coefficient Π_{TBC} was 5.6 cm^{-1}/GPa [27], obtained previously using diamond anvil cell by Liu et al [23, 27]. It was considered to be a tri-axial stress that was measured in the probed volume of the coating. A shift to higher frequency indicated compressive stress.

The measurement of residual stresses in TGO scales were carried out by CPLPS. The excitation source was a green Ar ion laser with a photon wavelength of 532 nm and nominal power of 50 mW. The laser was delivered through an objective lens and focused perpendicular to substrate. Before and after each experiment, the spectrometer was calibrated by taking a spectrum from a strain-free single crystal sapphire sample. In luminescence mapping, measurements were taken on a rectangular grid area. Each spectrum from a particular location on the map grid was assigned to a pair of *x*- and *y*-coordinates using a computer controlled motorised stage. In the mapping a total of 15625 (125×125) spectra were recorded in a grid of a 1 mm×1 mm area with a step of 8 μm. For mapping the stress distribution in the TGO layers, the incident was focused at the specimen surface. Using the piezo spectroscopic coefficients measured by Clarke [20] the relationship between frequency shift and biaxial stress given by:

$$\Delta\nu_{TGO} = \Pi_{TGO} \cdot \sigma_{TGO} \qquad (2)$$

where the piezo spectroscopic tensor Π_{TGO} was 5.07 cm^{-1}/GPa.

Results and discussion

Phase composition of the powder and coatings

The XRD spectrum and Raman spectra of the PS-PVD YSZ coating and primary powder were shown in Fig. 2. The results show that a remarkable phase change took place during the plasma spraying process. All of the strong Raman peaks appeared at 176.39, 188.73, 379.40, 498.88, 535.74 and 574.25 cm^{-1} in the spectra of the primary powder were disappeared after plasma spraying, shown in Fig. 2(b). Among these Raman peaks, 176.39 and 188.73 cm^{-1} were the characteristic peaks of the monoclinic phase of zirconia. Meanwhile, 498.88 and 574.25 cm^{-1} were peculiar to the cubic phase of yttria. The typical Raman bands for the tetragonal phase of yttria stabilized zirconia coatings were 147.06, 255.62, 315.82, 466.53 and 637.02 cm^{-1}, respectively, appeared in the as-sprayed coating. The Raman results show that the PS-PVD coating wholly consisted of tetragonal phase of YSZ (t, t'-ZrO_2). The monoclinic phase of ZrO_2 (m-ZrO_2) and cubic phase of Y_2O_3 (c-Y_2O_3) disappeared after plasma spraying process. The XRD patterns of the YSZ powder and the corresponding as-sprayed coating were exhibited in Fig. 2(a). The original YSZ powder was composed of m-ZrO_2 and c-Y_2O_3; however, the resulting coating consist of t, t'-ZrO_2. No monoclinic bands were observed which confirmed the results obtained by Raman spectra.

Fig. 3 shows low-angle (27-32°) and high-angle (72-76°) regions of XRD patterns, respectively. The low-angle region (27-32°) was considered to distinguish the amount of the monoclinic phase, while the high-angle region (72-76°) was considered to distinguish the relative proportions of t'-, t- and c-zirconia phases. Two zirconia phases could be detected: a high-yttria t'-phase, and a low-yttria t-phase. Their lattice parameters depend on the yttria content. Fig. 3(a) the (1 1 1) peak for (t' + t) zirconia phase was detected, while no $(1\ 1\ \bar{1})$ and (1 1 1) peaks for m-ZrO_2.

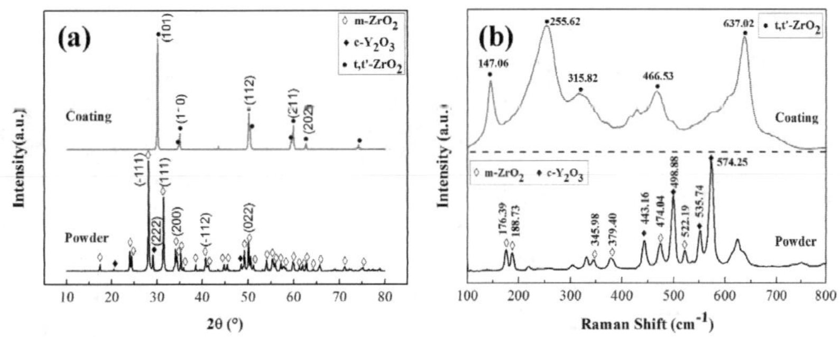

Figure 2. X-ray diffraction spectrum (a) and Raman spectra (b) of the powder and as-sprayed YSZ coating

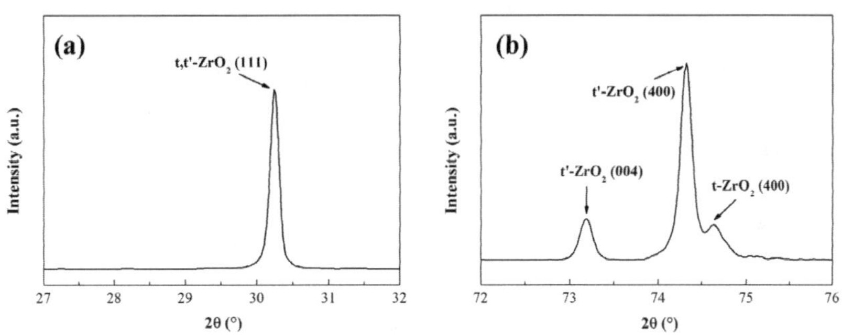

Figure 3. X-ray diffraction patterns of the PS-PVD coatings with 2θ range 27-32° and 72-76°

In Fig. 3(b) the (0 0 4) and (4 0 0) peaks for t'- and t-zirconia phases could be found. The ratio in peak intensities for the t'-zirconia phase was a result of the strong crystallographic texture associated with the columnar microstructure typical of the PVD coatings.

Microstructure characterization of the coatings during burner rig test

Photographs of the PS-PVD YSZ coatings before and after burner rig test were presented in Fig. 4. From Fig. 4(d), it was clear that obvious 'color difference' whereas no apparent coating spalling has occurred at the center of the topcoat even after 300 cycles. Thus, the quasi-columnar microstructure and thereby the improved strain tolerance of the PS-PVD coatings led to a large improvement in durability. It is assumed that a defined pre-oxidation of the bond coat to achieve a thin α-Al_2O_3 layer before TBC deposition can obviously increase the good performance of the PS-PVD TBCs. Such an approach was successfully verified by other authors [3, 11] and the similar process was applied in the present work. Moreover, further thermal cycling test will be carried out in our future research.

Surface SEM images of PS-PVD YSZ coatings before and after burner rig tests were present in Fig. 5. From Fig. 5(a), the surface morphology of the as-sprayed coating was relatively uniform and has been described as a 'cauliflower-like' morphology. The gaps between the columns provide a high strain tolerance and significantly improved thermal-mechanical fatigue performance was yielded in thermal cyclic tests compared with conventional atmospheric plasma spraying (APS) coatings [11]. No apparent spot spallation were observed at the center of the YSZ topcoat after 50 cycles as shown in Fig. 5(b). With the thermal cycling going on, local spot spallations (about the size of several column grains) could be found, shown in Fig. 5(c). Further inspection showed that the segments spallation could be related to a crack bridging and segments pull-out phenomenon. Local spot spallations could effectively improve the thermal shock cycling lifetime of the PS-PVD coatings. However, the thermal insulation performance of the coating would be reduced.

Figure 4. Photographs of PS-PVD YSZ coating (Sample 1) before and after thermal cycling tests (a) as-deposited, (b) 50, (c) 200 and (d) 300 cycles.

Figure 5. Surface SEM images of PS-PVD YSZ coatings before and after burner rig tests (a) as-deposited, (b) 50 and (c) 300 cycles

The real-time detecting temperature difference between the surface of the coatings and the backside of the substrate during burner rig tests was presented in Fig. 6. As can be see, the temperature difference decreased with the number of thermal cycles. Thus the structure of the PS-PVD coatings needed to be further optimized in order to obtain both thermal insulation and excellent thermal shock resistance.

Residual stress in the Topcoat

Fig. 7 shows the variation of position of the Raman peak at around 640 cm^{-1} and the average residual stresses in the topcoat as a function of thermal shock cycles. The measured results show the same trend of stress variation for different samples. Taking the Sample 1 as an example, the Raman peak of the sample after thermal shock for 300 cycles shifted to a higher wavenumber (641.32 ± 0.45 cm^{-1}) compared with the as-deposited one (634.65 ± 0.23 cm^{-1}) and the powder (637.35 ± 0.11 cm^{-1}). According to the formula (1), the conversion of Raman peak frequency shifts measured the in-plane residual stress in the YSZ topcoat were gained. For the as-sprayed coatings, a tensile stress (0.48 ± 0.04 GPa) was present in the topcoat. After thermal shock for 100 cycles, compressive stress was obtained. The average topcoat compressive stress was found to increase with thermal shock cycles and the value was σ_{TBC}= -0.12 ± 0.05, -0.27 ± 0.06, -0.70± 0.08 GPa for the 100, 200 and 300 cycles, respectively. The stress state changed from tensile to compressive after thermal cycling which was mainly due to the different coefficients of thermal expansion between the component of the TBC system ($\alpha_{Topcoat}$=11~13×10^{-6}/°C, $\alpha_{Bondcoat}$=13~16×10^{-6}/°C) [16]. The coating stress relief mode at high temperature was predominantly cracking of the brittle PS-PVD YSZ topcoating. Micro-cracking for ceramic coatings during thermal cycling have been detected using acoustic emission (AE) technique by other authors [31]. The high AE activity shows typical microcracks were induced preferentially during the heating process when the ceramic coating was in a tensile-stress state rather than during the cooling stage. Thus, due to this stress relaxation at high temperature, compressive thermal stress was then again generated within the coating during the cooling procedure and the top coating developed higher compressive stress with further thermal cycling test.

Mapping residual stress distribution in the TGO

Contour map and corresponding histograms of the distribution of TGO stress in the PS-PVD YSZ coating were shown in the Fig. 8. The X and Y axes in Fig. 8(a) represent the absolute positions of the specimen during mapping. Each map covered an area of 1 mm× 1 mm with a step size of about 8 μm. Within the contour maps, the discrete stresses have been linearly interpolated. In the specimen, the compressive residual stresses in the TGO were measured to be within the range of about -3.95 to -1.58 GPa with an average of -2.43 GPa and standard deviation of 0.31 GPa. The wide stress distribution in the TGO observed for the specimens suggest that the stress-relief mechanisms may occur at a localized area.

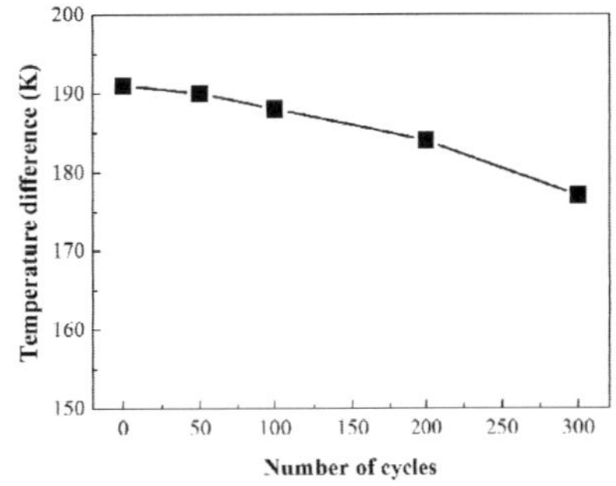

Figure 6. The measured temperature difference between the surface of the PS-PVD YSZ coating and backside of the substrate during burner rig tests

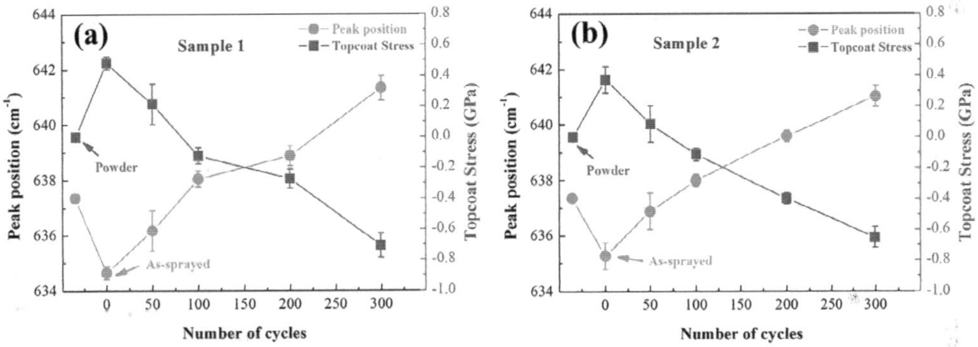

Figure 7. Variation of the residual stresses in the YSZ topcoat as a function of burner rig test (a) Sample 1 and (b) Sample 2.

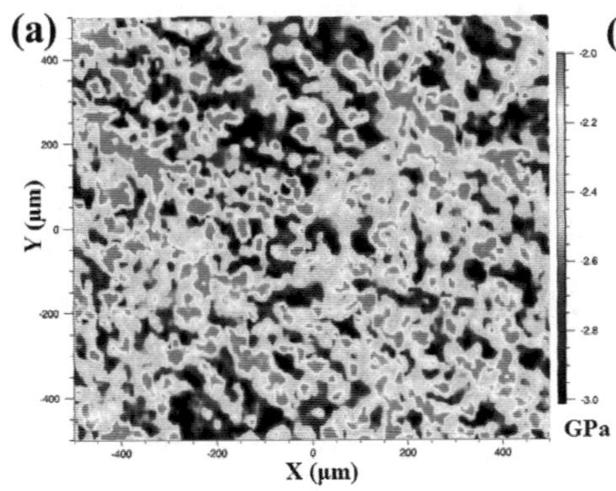

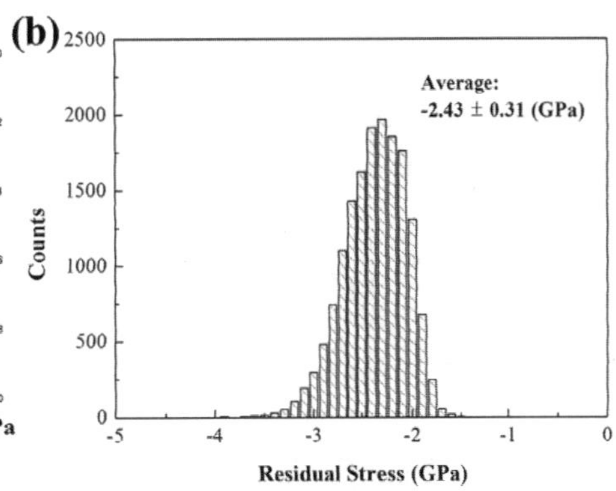

Figure 8. Contour map (a) and corresponding histograms of the distributions of TGO stress on PS-PVD coating specimen. Units of the stress on contour are in GPa.

The TGO/bondcoat interface morphology was mainly the result of surface preparation before TBC deposition. For flat interface, the TGO stress was in a state of compressive. Due to the undulations in the TGO geometry, out-of-plane stress (σ_N) was produced, tension at undulations crests (or at the peaks) and compressive at troughs (or in the valleys). The stress distribution for a sinusoidal film has been calculated by Gong and Clarke [29] using finite-element modeling (FEM). The FEM calculations predicted that this undulation interface would lower the residual stress compared with that measured on a flat surface. Tolpygo and Clarke [30] experimentally investigated the influence of interface roughness on the stress in convoluted-alumina scales formed on FeCrAl alloy and the measured results show a difference in frequency shift of approximately 7-10 cm^{-1} (corresponding to a stress difference of approximately 1.4-2.0 GPa) between the ''peaks'' and the ''valleys'' of the convolutions.

The distribution of residual stress in the TGO layer as a function of thermal shock cycles was presented in Fig. 9. The trend of the change in the distribution of TGO residual stresses for different samples was also in a similar way. The in-plane TGO stresses measured non-destructively through the YSZ topcoat surface varied with the thermal shock cycles, from -2.09 ± 0.13 to -0.11 ± 0.24 GPa for Sample 1 and from -2.77 ± 0.20 to -0.96 ± 0.56 GPa for Sample 2, respectively. Compared with Sample 1, the more compressive TGO stresses generated in another specimen might due to the different pre-oxidation time of the bond coat to achieve a thin α-Al_2O_3 layer before TBC deposition aiming to improve thermal cycling durability of PS-PVD TBC system. From the Fig. 7 and Fig. 9, it was obviously seen that the pre-oxidation process of the bondcoat significantly affect the stress distribution in the TGO layer, while there was no obviously influence on the stress distribution in the YSZ ceramic topcoat.

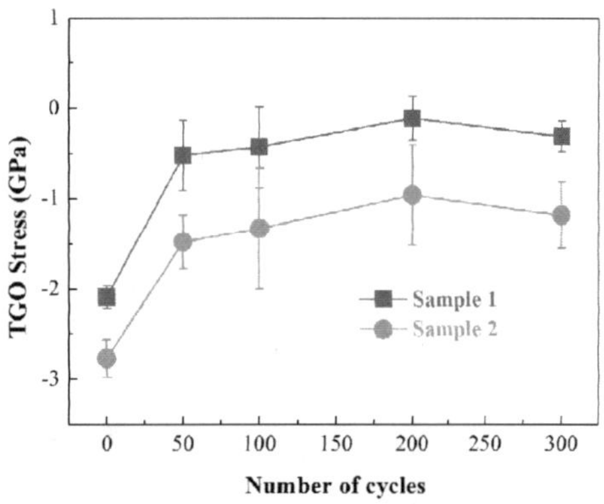

Figure 9. Variation of the residual stresses in the TGO as a function of thermal cycling test

As Sample 2 for example, the value of compressive stresses reached to σ_{TGO}= -1.48 ± 0.30, -0.96 ± 0.50, -1.18 ± 0.37 GPa for the 50, 200 and 300 cycles, respectively. At the beginning of thermal cycling, the measured in-plane TGO compressive stress firstly rapidly decreased and then essentially maintain at a relatively stable value with further testing. The decrease in stress with thermal cycling can arise from several causes. The surface geometry of the bondcoat generally got rougher during cycling due to a process often referred to as rumpling, which in turn allowed the TGO to get longer potentially relieving the compressive stresses in the TGO. There was also an effect of just the oxide geometry on the stress that led to decreasing average stresses with increasing TGO thickness or rough surfaces. In addition, this effect was further increased if the interface undulation increase due to rumpling. Moreover, as thermal cycling further proceeds, the formed Cr_2O_3, spinel and other detrimental oxides that results in increased cracking associated with the TGO and then longer cracks within the topcoat/bondcoat interface. Crack nucleation and propagation then resulting in lower residual stresses. The relative role of these several processes was not yet resolved. Nevertheless, the corresponding failure mechanism of PS-PVD coatings will be further investigated in the future work.

Conclusions

Quasi-columnar structured thermal barrier coatings were fabricated by plasma spray-physical vapor deposition (PS-PVD) process. The residual stresses in the YSZ topcoat and TGO layer during burner rig test have been investigated. The results can be summarized as follows:

(1) The as-sprayed quasi-columnar coatings consisted of t'- and t-zirconia phase and no monoclinic phase was detected. Cr^{3+} photoluminescence piezeospectroscopy technique was effective for mapping residual stress measurements in TGO layer and proved to be indicated as a non-destructive tool for the characterization of the coating quality.

(2) The accumulated tensile stress in the as-sprayed ceramic topcoat changed to compressive state after 100 thermal cycles, and then gradually increased. In addition, the pre-oxidation of the bondcoat have no influence on the stress distribution in the YSZ topcoat.

(3) The compressive stresses in the TGO measured through the ceramic topcoat surface decreased rapidly firstly and then essentially maintain at a relatively stable value with further testing. The decrease in stress with thermal cycling can arise from several causes and further work is needed to resolve this issue.

Acknowledgments

This work was jointly supported by the National natural Science Foundation (NSFC) under the Grant No.51671208, Natural Science Foundation of Shanghai (No.17ZR141Z200) and Laboratory foundation of Chinese Academy of Sciences (Grant No.16S084).

References

[1] Clarke, D. R. *et al.*, "Materials design for the next generation thermal barrier coatings", *Annu. Rev. Mater. Res*, Vol. 33, (2003), pp. 383-417.

[2] Fauchais, P., "Understanding plasma spraying", *J.*

Phys. D: Appl. Phys, (2004), pp. 86-108.
[3] Von Niesen, K. M. *et al.*, "Vapor phase deposition using plasma spray-PVD™", *J. Therm. Spray. Technol*, Vol. 19, No. 1/2 (2010), pp. 502-509.
[4] Hospach, A. *et al.*, "Columnar-structured thermal barrier coatings (TBCs) by thin film low-pressure plasma spraying (LPPS-TF)", *J. Therm. Spray. Technol*, Vol. 20, No. 1/2 (2011), pp. 116-120.
[5] Niessen, K. V. *et al.*, "Plasma sprayed-PVD: a new thermal spray process to deposit out of the vapor phase," *J. Therm. Spray. Technol*, Vol. 20, No. 4 (2011), pp. 736-743.
[6] Hospach, A. *et al.*, "Characteristics of ceramic coatings made by thin film low pressure plasma spraying (LPPS-TF)," *J. Therm. Spray. Technol.*, Vol. 21, No. 3/4 (2012), pp. 435-440.
[7] Goral, M. *et al.*, "PS-PVD deposition of thermal barrier coatings," *Surf. Coat. Technol*, Vol. 237 (2013), pp. 51-55.
[8] Mauer, G. *et al.*, "Novel opportunities for thermal spray by PS-PVD," *Surf. Coat. Technol.*, Vol. 268 (2015), pp. 52-57.
[9] Mauer, G. *et al.*, "Process development and coating characteristics of plasma spray-PVD," *Surf. Coat. Technol.*, Vol. 220 (2013), pp. 219-224.
[10] Mauer, G. *et al.*, "Process conditions and microstructures of ceramic coatings by Gas phase deposition based on plasma spraying," *J. Therm. Spray. Technol.*, Vol. 22, No. 2/3 (2013), pp. 83-89.
[11] Rezanka, S. *et al.*, "Improved thermal cycling durability of thermal barrier coatings manufactured by PS-PVD," *J. Therm. Spray. Technol.*, Vol. 23, No. 1/2 (2012), pp. 182-189.
[12] Gao, L. H. *et al.*, "Microstructure, thermal conductivity and thermal cycling behavior of thermal barrier coatings prepared by plasma spray physical vapor deposition," *Surf. Coat. Technol.*, Vol. 276 (2015), pp. 424-430.
[13] Miller, R. A., "Thermal barrier coatings for aircraft engines: history and directions," *J. Therm. Spray. Technol.*, Vol. 6, No. 1 (1997), pp. 35-42.
[14] Vassen, R. *et al.*, "Recent development in the field of thermal barrier coatings," *J. Therm. Spray. Technol.*, Vol. 18, No. 2 (2009), pp. 181-186.
[15] Clarke D. R. *et al.*, "Materials design for the next generation thermal barrier coatings," *Annu. Rev. Mater. Res.*, Vol. 33 (2003), pp. 383-417.
[16] Rabiei A. *et al.*, "Failure mechanisms associated with the thermally grown oxide in plasma sprayed thermal barrier coatings," *Acta. Mater.*, Vol. 48 (2000), pp. 3963-3976.
[17] Trunova, O. *et al.*, "Damage mechanisms and lifetime behavior of plasma sprayed thermal barrier coating systems for gas turbines-Part I: Experiments, *Surf. Coat. Technol.*," Vol. 202 (2008), pp. 5027-5032.
[18] Shillington E. A. G. *et al.*, "Spalling failure of a thermal barrier coating associated with aluminum depletion in the bond-coat," *Acta. Mater.*, Vol. 47 (1999), pp. 1297-1305.
[19] Schlichting, K. W. *et al.*, "Failure modes in plasma-sprayed thermal barrier coatings," *Mater. Sci. Eng A*, Vol. 342 (2003), pp. 120-130.
[20] Christensen, R. J. *et al.*, "Nondestructive evaluation of the oxidation stresses through thermal barrier coatings using Cr^{3+} piezospectroscopy," *Appl. Phys. Lett.*, Vol. 69, No. 24 (1996), pp. 3754-3756.
[21] Schlichting, K. W. *et al.*, "Application of Cr^{3+} photoluminescence piezo-spectroscopy to plasma-sprayed thermal barrier coatings for residual stress measurement," *Mater. Sci. Eng A.*, Vol. 291 (2000), pp. 68-77.
[22] Liu, D. *et al.*, "Effect of substrate curvature on residual stresses and failure modes of an air plasma sprayed thermal barrier coating system," *J. Euro. Ceram. Soc.*, Vol. 33 (2013), pp. 3345-3357.
[23] Liu, D. *et al.*, "The role of beam dispersion in Raman and photo-stimulated luminescence piezo-spectroscopy of yttria-stabilized zirconia in multi-layered coatings," *Acta. Mater.*, Vol. 61 (2013), pp. 12-21.
[24] Liu, D. *et al.*, "Effect of substrate curvature on the evolution of microstructure and residual stresses in EBPVD-TBC," *J. Euro. Ceram. Soc.*, Vol. 35 (2015), pp. 2563-2575.
[25] Heeg, B. *et al.*, "Damage evolution in thermal barrier coatings with thermal cycling." *J. Am. Ceram. Soc.*, Vol. 94 (2012), pp. 112-119.
[26] Teixeira, V. *et al.*, "Analysis of residual stresses in thermal barrier coatings," *J. Mater. Proc. Technol.*, Vol. 92/93 (1999), pp. 209-216.
[27] Liu, D. *et al.*, "Calibration of Raman spectroscopy in the stress measurement of air-plasma-sprayed yttria-stabilized zirconia," *Appl. Spectrosco.*, Vol. 66, No. 10 (2012), pp. 1204-1209.
[28] Tanaka, M. *et al.*, "Measurement of residual stress in air plasma-sprayed Y_2O_3-ZrO_2 thermal barrier coating system using micro-Raman spectroscopy," *Mater. Sci. Eng A.*, Vol. 419 (2006), pp. 262-268.
[29] Gong, X. Y. *et al.*, "On the measurement of strain in coatings formed on a wrinkled elastic substrate," *Oxid. Met.*, Vol. 46 (1998), pp. 355-376.
[30] Tolpygo, V. K. *et al.*, "Tensile cracking during thermal cycling of alumina films formed by high-temperature oxidation," *Acta. Mater.*, Vol. 47, No. 13 (2010), pp. 3589-3605.
[31] Voyer, J. *et al.*, "Study of the performance of TBC under thermal cycling conditions using an acoustic emission rig," *J. Therm. Spray. Technol.*, Vol. 7, No. 2 (1998), pp. 181-190.

Tribological Performance of NiCrAlYSi-BaF_2/CaF_2-Polyester Coating Deposited by Air Plasma Spray

Q. L. Li, S. Q. Li, Beijing Aeronautical Manufacturing Technology Research Institute, Beijing / CN

NiCrAlYSi-BaF_2/CaF_2-Polyester material was developed for abradable seal of high-pressure compressor in gas turbine. The NiCrAlYSi-BaF_2/CaF_2-Polyester coatings were deposited on Ni-based superalloy substrates with an air plasma spray system. Friction and wear properties of the coatings were investigated in view of friction coefficient and wear morphologies of the coatings. The morphologies of the coatings were evaluated by field emission scanning electron microscopy. The results showed that NiCrAlYSi-7%BaF_2/CaF_2-10%Polyester coating exhibits lower macrohardness, lower friction coefficient and excellent abradability.

1 Introduction

Abradable seal coatings are widely used for clearance reduction in a compressor of a gas turbine engine. The seal coatings are applied onto the inside diameter of stationary components by thermal spray process. During operation of the compressor, a tight clearance is produced by the blades cutting into the seal coatings. Abradable seal coatings cause minimal wear on rotational blades[1]. Ni-graphite, NiCrFeAl-hBN(hexagonal boron nitride), MCrAlY-polyester, MCrAlY-hBN-polyester coatings deposited by thermal spraying are wide used as seal coatings in a compressor of a gas turbine engine[2-4].
Inorganic fluorides (BaF_2, CaF_2) have been successfully employed as solid lubricants. BaF_2 and CaF_2 possess adequate thermophysical properties and stable thermochemistry at high temperatures, as well as low shear strength. BaF_2 and CaF_2 become lubricating above 500°C, when they undergo internal structural changes and a brittle-to-ductile transition. BaF_2, CaF_2 and BaF_2/CaF_2 eutectic have been used in a number of high temperature applications, primarily as a component of self-lubricating composites[5-10]. Before this paper, some compositions of NiCrAlYSi-BaF_2/CaF_2-Polyester coatings had been prepared and tested in our laboratory. In the present experiment, a new composition of NiCrAlYSi-7wt%BaF_2/CaF_2-10wt%Polyester coatings deposited by air plasma spray were investigated. Here BaF_2/CaF_2 eutectic with a composition of 62wt% BaF_2 and 38wt% CaF_2 was chosen as solid lubricant and the polyester was chosen as pore forming material. The coatings exhibit lower macrohardness, lower friction coefficient and excellent abradability against Ni_3Al-based superalloy counterpart. Through strict control of the plasma spraying process parameters, the macrohardness of the coatings after polyester burnout could be controlled between HR45Y 35 and 65, the average friction coefficient coupled with Ni_3Al-based superalloy was below 0.5 at 900°C.

2 Experimental Procedures

2.1 Characteristics of the feedstock

Commercially available NiCrAlYSi alloyed powder, BaF_2/CaF_2 powder and polyester powder were chosen as starting materials.

Field Emission Scanning Electron Microscope (SEM, Hitachi S-4800) was used to evaluate the morphologies of the feedstock. NiCrAlYSi and BaF_2/CaF_2 powders showed spherical shape, as shown in **Fig. 1** and **Fig. 2** respectively. The X-ray diffraction(XRD) pattern of BaF_2/CaF_2 powder is shown in **Fig. 3**.

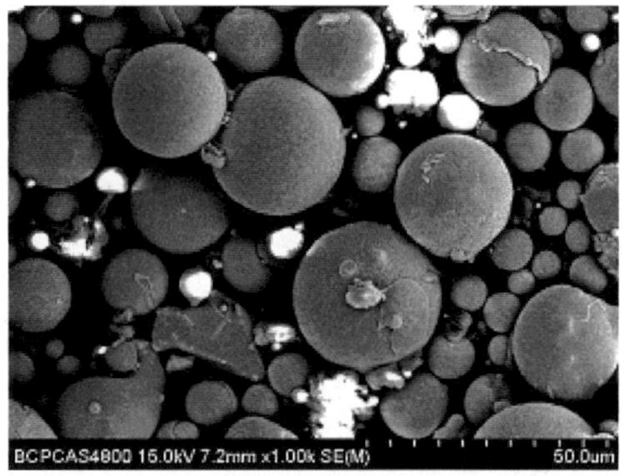

Fig. 1. SEM micrograph of the NiCrAlYSi powder

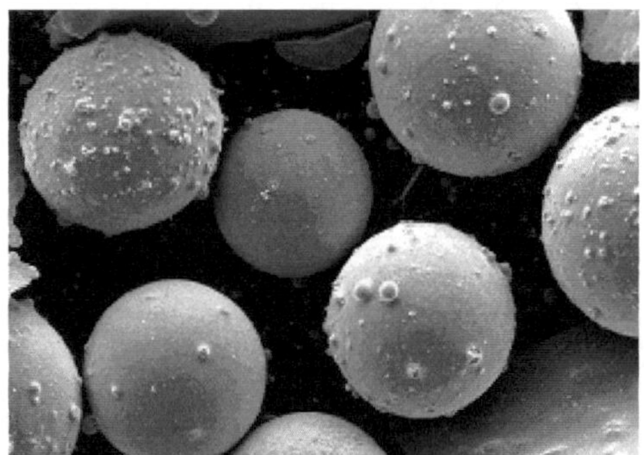

Fig. 2. SEM micrograph of the BaF_2/CaF_2 powder

Commercially available polyester powder was mixed with water, and a binder (polyvinyl alcohol 1788). Then the slurry was spray-dried and agglomerated into spherical micrometer-size granules (50-100μm). The morphology of the agglomerated polyester powder is shown in **Fig. 4**. Commercially available NiCrAlYSi alloyed powder, BaF_2/CaF_2 and the agglomerated polyester powder were mixed in a designed proportion for plasma spray.

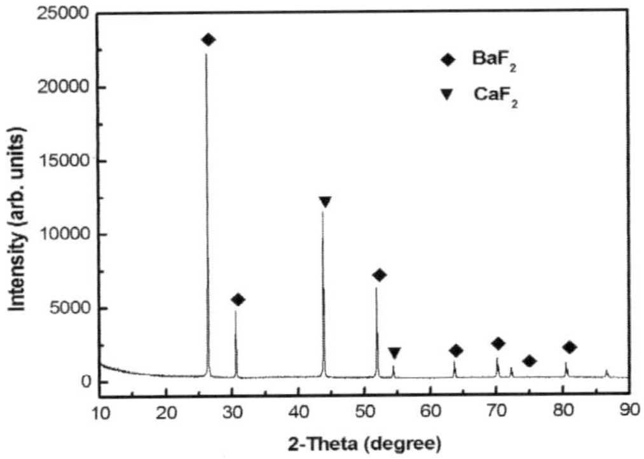

Fig. 3. XRD pattern of BaF_2/CaF_2 powder

Fig. 4. SEM micrograph of the agglomerated polyester powder

2.2 Plasma spraying process

NiCrAlYSi metallic bond coating and NiCrAlYSi-7%BaF_2/CaF_2-10%Polyester top coating were deposited on Ni-based superalloy substrates, Φ25mm×6mm, with A3000 air plasma spray system (Sulzer Metco). Before spraying, Ni-based superalloy specimens were grit blasted with brown corundum to a roughness Ra of about 10μm.

The thickness the bond coating was about 0.2mm, and the thickness of the top coating was about 2.0mm. **Table 1** provides the plasma spray parameters for the coatings under study. **Fig. 5** illustrates the typical microstructural features of as-sprayed NiCrAlYSi-7%BaF_2/CaF_2-10%Polyester coating with a NiCrAlYSi metallic bond coating on a Ni-based superalloy substrate. **Fig. 6** is the magnified image of the metallographic cross section of the coating. The coating exhibited a porous, lamellar microstructure. The metallic area percentage was about 75%, and non-metallic area percentage was about 25%. Non-metallic area includes polyester, BaF_2/CaF_2 and porosity. Polyester would be burned out from the as-sprayed coatings at 500°C for 8 hours to form porosity.

Table 1. Plasma spray parameters, with A3000 plasma spray system

Parameter	bond coating	top coating
Current (A)	500	400
Voltage (V)	70	70
Primary gas flow (Ar, L/min)	60	60
Secondary gas flow (H_2, L/min)	7	7
Carrier gas flow (L/min)	4	4
Powder feed rate (g/min)	40	60
Spray distance (mm)	150	80

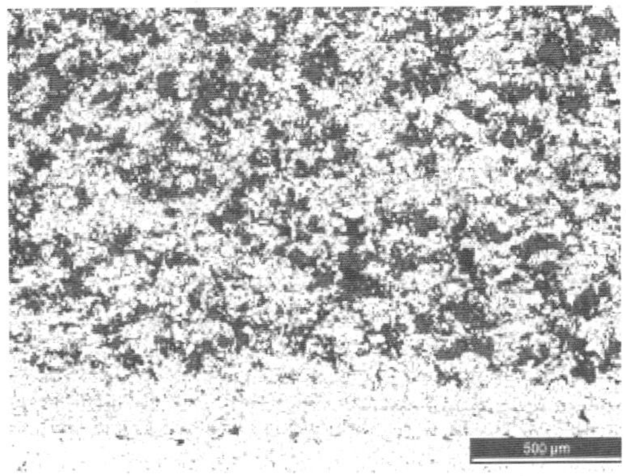

Fig. 5. A typical metallographic cross section of NiCrAlYSi-7%BaF_2/CaF_2-10%Polyester coating

2.3 Macrohardness

Hardness specimens with a thickness of 2 mm NiCrAlYSi-7%BaF_2/CaF_2-10%Polyester coatings were heat treated at 500°C for 8 hours to remove the polyester in the as-sprayed coatings and the coatings were then carefully ground with 180-mesh silicon carbide paper. Macrohardness values were determined in the surface of the coatings on a HR45Y scale with 12.5mm steel ball and 45 kg load (Rockwell Hardness Tester Model TH320, Time Group Inc.). A total of three microhardness measurements were carried out for each specimen. The average hardness of the coatings was HR45Y 59.

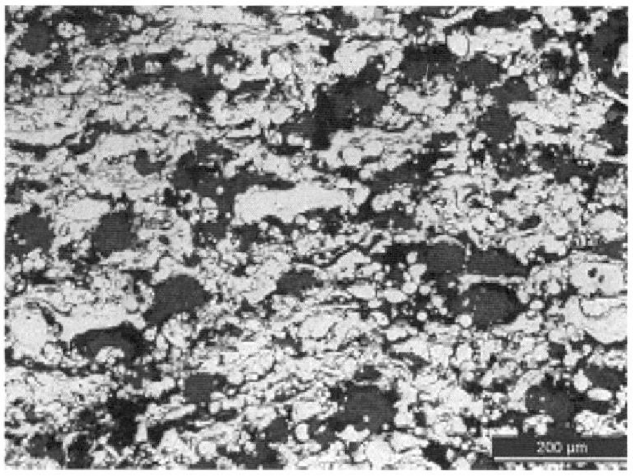

Fig. 6. A magnified image of the metallographic cross section of NiCrAlYSi-7%BaF_2/CaF_2-10%Polyester coating

2.4 Friction and wear

Friction and wear tests were conducted with a friction and wear tester (CETR, UMT-3, USA) in a dry ball-on-disc configuration at 900°C. The disks were NiCrAlYSi-7%BaF_2/CaF_2-10%Polyester coating specimens with a diameter of 35 mm after a heat treatment of 8 hours at 500°C for removing the polyester. The Ni_3Al-based superalloy balls were used as a counterpart with a diameter of 9.7 mm. In the test the ball was pressed against the coating sample at a normal load of 5N, the sliding speed 3m/sec, and the duration time was 20 min. The friction coefficient was monitored by a computer during the test.

3 Results and Discussion

3.1 Tribological properties

Friction coefficient curves for NiCrAlYSi with 7%BaF_2/CaF_2 and 10%Polyester coatings and NiCrAlYSi with 10%Polyester but without BaF_2/CaF_2 coatings against Ni_3Al-based superalloy balls are shown in **Fig. 7** and **Fig. 8** respectively. The mean value of friction coefficient was 0.39 for NiCrAlYSi-7%BaF_2/CaF_2-10%Polyester coatings, and 0.82 for NiCrAlYSi-10%Polyester coatings.

It is seen that the friction coefficient of NiCrAlYSi-7%BaF_2/CaF_2-10%Polyester coating is much lower than that of NiCrAlYSi-10%Polyester coating. It can be drawn that the improvement of BaF_2/CaF_2 to the friction coefficient at the test conditions is obvious for the NiCrAlYSi coating.

The morphologies of the worn surfaces were examined by scanning electron microscope. **Fig. 9** and **Fig. 10** show the typical worn surface SEM morphologies for the counterpart ball and the NiCrAlYSi-7%BaF_2/ CaF_2-10% Polyester coating at 900°C respectively.

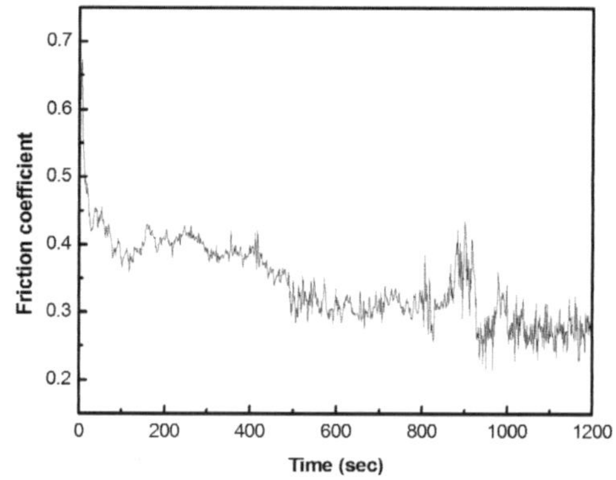

Fig. 7. Curve of friction coefficient vs. the testing time for NiCrAlYSi-7%BaF_2/CaF_2-10%Polyester coating

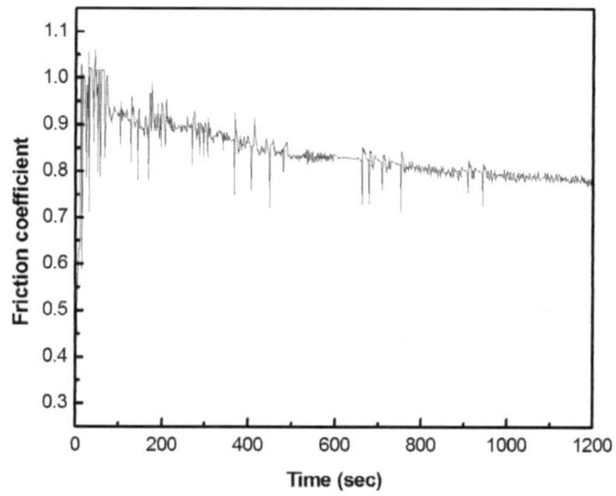

Fig. 8. Curve of friction coefficient vs. the testing time for NiCrAlYSi-10%Polyester coatings

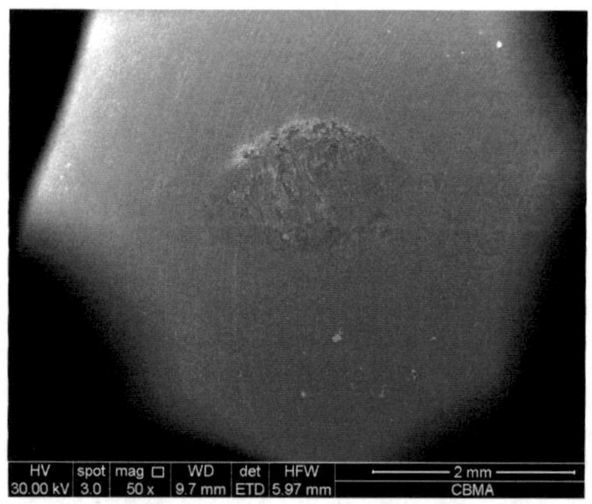

Fig. 9. Typical worn surface SEM morphology SEM morphology for the counterpart ball

Comparing **Fig. 9** with **Fig. 10**, It can be seen that the wearing depth of the coating is much more than that of the counterpart ball. According to the measurements, the wearing depth ratio of the counterpart ball to the NiCrAlYSi-7%BaF_2/CaF_2-10%Polyester coating was about 5%. NiCrAlYSi-

7%BaF_2/CaF_2-10%Polyester coating exhibits excellent abradability.

Fig. 10. Typical worn surface SEM morphology for the NiCrAlYSi-7%BaF_2/ CaF_2-10% Polyester coating

3.2 Wear mechanism analysis

The morphologies of the worn surface of the NiCrAlYSi-7%BaF_2/CaF_2-10%Polyester coating showed evidence of obvious cutting, furrow and some plastic deformation as shown in **Fig. 11** and **Fig. 12**. EDS spectrum of the worn surface of the coating is shown in **Fig. 13** and **Table 2**. Typical three-dimensional contour of the worn surface of the coating is shown in **Fig. 14**. At 900°C, the fluoride was squeezed out and formed lubricating film on the friction surface so it could greatly reduce the friction coefficient of NiCrAlYSi-7%BaF_2/CaF_2-10% Polyester coating. The solid lubricant BaF_2/CaF_2 exhibited excellent self-lubricating property.

Fig. 11. The typical morphology of the worn surface of the coating

Fig. 10 and **Fig. 11** show that adhesion dropping between the friction pairs is not obvious because of BaF_2/CaF_2 lubricating and antifriction properties. Plastic deformation, materials removal and furrows were observed on the worn surface of the coating obviously. So the plastic deformation, cutting action and furrow action are the major wearing mechanism in the wear process of NiCrAlYSi-7%BaF_2/CaF_2-10%Polyester coating against Ni_3Al-based superalloy.

Fig. 12. The typical morphology of the worn surface of the coating

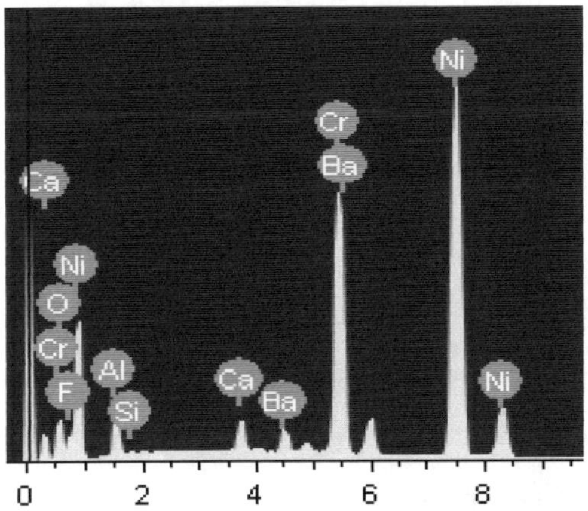

Fig. 13. EDS spectrum of the worn surface of the coating

Table 2. EDS results of the worn coating surface

elements	wt%	at%
O	5.78	14.39
F	11.12	23.33
Al	6.03	8.91
Si	0.39	0.56
Ca	1.57	1.56
Cr	19.70	15.10
Ni	51.60	35.04
Ba	3.81	1.11

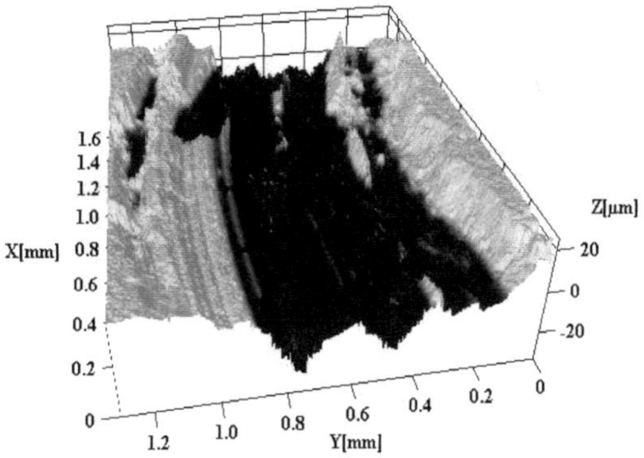

Fig. 14. Typical three-dimensional contour of the worn surface of the coating

4 Conclusions

From the experimental results above, it can be found that NiCrAlYSi-7%BaF_2/CaF_2-10% Polyester coating against Ni_3Al-based superalloy ball exhibited excellent self-lubricating and antifriction properties at high temperature. BaF_2/CaF_2 greatly decreases friction coefficient. The mean value of friction coefficient was 0.39 for NiCrAlYSi-7%BaF_2/CaF_2-10%Polyester coating against Ni_3Al-based superalloy ball at 900°C. The plastic deformation, cutting action and furrow action are the major wearing mechanism in the wear process. NiCrAlYSi-7%BaF_2/CaF_2-10%Polyester coating exhibits excellent abradability and can be explored as a novel prospective candidate material of high temperature abradable seal coating application for advanced gas turbine engines.

5 Literature

[1] F. Ghasripoor, R. K. Schmid, M.R. Dorfman, et al., Development of Advanced Materials for Clearance Control Applications up to 850°C. Advances in Science and Technology (1999), vol.20, pp. 641/50.

[2] T. Steinke, G. Mauer, R. Vaßen, et al., Process design and monitoring for plasma sprayed abradable coatings. Journal of Thermal Spray Technology (2010), 19(4), pp. 756/64.

[3] R. E. Johnston, Mechanical characterisation of AlSi-hBN, NiCrAl-Bentonite, and NiCrAl-Bentonite-hBN freestanding abradable coatings. Surface and Coatings Technology (2011), 205(10), pp. 3268/73.

[4] H. I. Faraoun, T. Grosdidier, J. L. Seichepine, et al., Improvement of thermally sprayed abradable coating by microstructure control. Surface and Coatings Technology (2006), 201(6), pp. 2303/12.

[5] J.H. Ouyang, S. Sasaki, K. Umeda, Low-pressure plasma-sprayed ZrO_2-CaF_2 composite coating for high temperature tribological applications. Surface and Coatings Technology (2001), 137, pp. 21/30.

[6] J.H. Ouyang, S. Sasaki, T. Murakami, K. Umeda, The synergistic effects of CaF_2 and Au lubricants on tribological properties of spark-plasma-sintered $ZrO_2(Y_2O_3)$ matrix composites. Materials Science and Engineering A (2004), 386, pp. 234/43.

[7] J. Deng, L. Liu, X. Yang, et al., Self-lubrication of Al_2O_3/TiC/CaF_2 ceramic composites in sliding wear tests and in machining processes. Materials Design (2007), 28, pp. 757/64.

[8] G.Kim, H. Choi, C. Han, S. Uhm, C. Lee, Characterization of atmospheric plasma spray NiCr–Cr_2O_3–Ag–CaF_2/BaF_2 coatings. Surface and Coatings Technology (2005),195, pp. 107/15.

[9] C. Huang, L. Du, W. Zhang, Preparation and characterization of atmospheric plasma-sprayed NiCr/Cr_3C_2–BaF_2/CaF_2 composite coating. Surface and Coatings Technology (2009), 203, pp. 3058/65.

[10] L. Q. Kong, Q. L. Bi, S. Y. Zhu, et al., Tribological properties of $ZrO_2(Y_2O_3)$-Mo-BaF_2/CaF_2 composites at high temperatures. Tribology International (2012),45, pp.43/49.

Wetting Properties of Ceramic Reinforced Metal Matrix Composites on Varied Roughness Profiles

S. Vijay, B. Roy, BITS Pilani, Pilani, India
N Markocsan, C. Lyphout, University West, Trollhättan

Superhydrophobic surfaces are of great importance in many industrial applications, especially where components are exposed to wet environments and low temperatures.Texturing of surfaces to reach superhydrophobicity can be achieved by thermal spraying technology, which is an attractive coating method as it is cheap, flexible and can employ a large range of feedstock materials. In this study, ceramic reinforced metal matrix composite (WC-CoCr) powders were sprayed using High Velocity Air Fuel method. They were varied based on their powder parameters such as carbide grain size, binder grain size and powder strength. The purpose was to investigate their hydrophobic characteristics and how these are influenced by different roughness profiles. The wetting properties such as contact angle and contact angle hysteresis were first investigated for the as-sprayed coatings. The roughness properties and Hausdorff Dimension were then related to the wetting properties. Aside from as-sprayed coatings, the effect of roughness and inherent wetting characteristics were studied by investigating the coating surface after grit blasting and polishing. Results show that powder parameters can lead to designing surfaces with higher surface roughnesses and thus having higher contact angles. It was also shown that surface composition of cermets has an impact on wettability, with the binder accounting for wetting characteristics and carbides accounting for roughness.

1 Introduction

Over the last decade many studies have been conducted on texturing of solid surfaces to attain superhydrophobicity or water repelling state (1,2,3). Inspiration for these is derived from the lotus leaf, which stays clean even in dirty waters (4). A good measure of wetting behaviour is the contact angle, which is the angle that the tangent of the liquid-gas boundary makes with the solid surface at the three-phase contact line (5). The three phase line is an imaginary curve which represents the meeting of three phases – the liquid in question, air and the solid surface. While the contact angle does provide understanding on wetting abilities of a surface, adhesion of the droplet to the surface is also an important factor. This is described by Contact angle hysteresis (CAH), which is defined as the difference between the static advancing contact angle and the static receding contact angle. The receding contact angle is the angle that the tangent makes with the solid surface just after the three phase line has moved inwards. Likewise, the advancing contact angle is the angle that the tangent makes just after the three phase contact line has moved outwards. Hydrophobic surfaces are those that have a contact angle of greater than 90° and superhydrophobic surfaces are defined as surfaces with contact angles greater than 150° and hysteresis values lower than 10° (6). There are largely two routes to create a hydrophobic surface - by using an inherently hydrophobic material or by imparting hierarchical roughness to the surface. (7)

While there are many methods that can be used to create superhydrophobic surfaces (8), thermal spraying methods are relatively new. Thermal spraying is a flexible and versatile method of producing coatings that can significantly improve the functional and/or physical properties of the components they are applied on (8,9). The thermal spraying technique has been used to obtain superhydrophobic surfaces before (11,12). However, most of them require some surface activation which invariably leads to a decrease in the industrial usability of these surfaces.

Ceramic reinforced metal matrix composites or cermets are commonly sprayed by the method of thermal spraying. They consists of carbides dispersed in a metallic matrix which holds them together. Thus, they consist of a binder phase holding together a carbide phase. One such cermet which is used in this study, WC-CoCr is traditionally known for its wear and erosion properties.

There are two largely accepted models for the water droplet that are used to explain hydrophobic behaviour on rough (i.e. non ideal) surfaces - The Wenzel state and the Cassie-Baxter state. The Wenzel model assumes that the air/solid interface is completely replaced by the liquid/solid interface, i.e. complete wetting (13). The Wenzel wetting relation is given by:

$$\cos\theta^* = r\cos\theta \quad (1)$$

Where r is the roughness parameter, $\theta*$ is the apparent contact angle and θ is the Young's contact angle. This state is characterized by relatively high adhesion to the surface. The Cassie-Baxter model assumes that the droplet is in contact with both air and the solid, i.e. the droplet does not completely 'wet' the surface (14).

Thermal spraying cermets leads to different surface textures being formed which can be quantified by broadly two methods - statistical analysis and fractal analysis. Statistical analysis defines many parameters based on the shape of the surface at a particular magnification. Fractal analysis captures the roughness irrespective of scale of magnification used by one number - the Hausdorff dimension. Both methods have been used in describing the relationship between roughness and wettability. (15)

This study aims to understand wetting behaviour of ceramic reinforced metal matrix composites, specifically WC-CoCr. This is done by performing various surface modifications - such as grit blasting and polishing which are aimed at changing the composition of either binder or carbide at the surface. Suggestions on how to improve as-sprayed roughness by choice of initial powder properties and coating microstructural features are also explored.

2 Experimental Procedure

2.1 Feedstock Materials and Processes

Spherical, agglomerated, and sintered WC-CoCr 86/10/4 powder was obtained from Fujimi Incorporated (Japan). Overall powder nominal size distribution was within the range of 10-30μm. Commercially available feed precursor is chosen based on the size of the primary carbide. In this study, precursor features were varied by three powder parameters - primary Carbide Grain Size (CGS), Binder Grain Size (BGS) and Powder Strength (PS). A full factorial design was engineered at the early stages of the manufacturing route to produce nine sets of precursors based on these powder parameters (Tab1). Domex 355 steel substrate coupons of cylindrical geometry with 6mm thickness were positioned in a rotating carousel. To improve adhesion, substrates were grit blasted prior to deposition using the HVAF-M3 facility at Production Technology Centre, Sweden utilizing mesh 220 (size distribution: 45-75μm) DURALUM fused aluminium oxide abrasive blast medium supplied by Washington Mills. All specimens were sprayed using the same HVAF M3 facility with identical spray parameters. Orientation of the gun was perpendicular for all samples and 15 sequential spray passes were used. Samples were numbered as per the full factorial design with two replicates and KB10 chosen as outside the design cube.

Post-spraying surface alteration was carried out by grit blasting (GB) followed by grinding and polishing. HVAF M3 system was used to grit blast the WC-CoCr surface using mesh 220 (size distribution: 45-75μm) DURALUM fused aluminium oxide abrasive blast medium. Orientation of the spray gun was 45° with carrier gas flow rate was 60l/min. Only one spray pass was used to minimise grit pollution of the surface. Grinding and polishing were carried out using procedures similar to metallographic preparation. Automatic polishing machine BUEHLER EcoMet 300 was used along with 135μm and 65μm diamond grinding disks and appropriate diamond suspension. Polishing was carried out by using 9μm and 3μm spacing diamond polishing clothes until a mirror like finish was obtained.

Sample	CGS (μm)	BGS (μm)	PS (MPa)
KB1-1	2.0	0.5	300
KB1-2	2.0	0.5	300
KB2	0.2	0.5	100
KB3	4.0	0.5	100
KB4	0.2	2.0	100
KB5	4.0	2.0	100
KB6	0.2	0.5	500
KB7	4.0	0.5	500
KB8	0.2	2.0	500
KB9	4.0	2.0	500
KB10	0.2	2	300

Tab1: Powder Parameters and Values

2.2 Characterization Methods

Microstructure Analysis:
Coating microstructure was examined by scanning electron microscopy (SEM) images by using TM3000 Table-Top Microscope (HITACHI). Tungsten filament was used and accelerating voltage was set at 15kV for all images with maximum magnification of x10,000. Plan view images required no special preparation. For cross section view, samples were cut in half and were cold mounted using standard epoxy moulding procedures. Cross sections were then grinded and polished using the same procedure as performed for post-spray surface alteration and repeated until there were no pull-outs. No conductive coating was needed for any sample.

3D areal Profilometry:
3D surface topography was analysed by contactless optical interference profiling method at Toponova AB, Sweden. Surface roughness was evaluated according to roughness standard ISO 25178. A 0.12x0.16mm (magnification:50X) square grid was chosen for interferometry because the length scale is ideal to evaluate hierarchical roughness. 10X magnification measurements were carried out on a square grid of

2.46x1.88 mm. The main elements of a profile are the motifs which are distinguished from each other by segmentation methods using watershed algorithms. Two parameters discussed here are root mean square roughness (Sdq) and skewness of the profile (Ssk).

Wettability Measurements:
Contact angle measurements were taken by the procedure outlined by Drelich (16). Krüss DSA30 was used to take contact angle measurements. Measurements were taken at constant temperature of 21°C. Droplet volume was always chosen to be between 8-30µL to prevent either incorrect measurement or gravity or evaporation effects becoming relevant.

3. Numerical Methods

Fractal Analysis
Hausdorff dimension was found by using the divider method, where a yardstick of a given length is used to measure the length of a curve. Dh can be found by plotting the log-log plot between the length measured by the yardstick by the length of the yardstick. Two dimensional roughness profiles which were generated from interferometry under the 50X magnifications were used to find the Hausdorff dimensions (Dh) along 250 linear profiles with change in x coordinate of 0.00022mm. Different yardsticks were used to capture the roughness profile.

DOE Analysis
Full factorial design was used to understand how the powder parameters alter roughness of the resulting thermally sprayed surface. A full factorial setup composing of 11 experimental runs (including two replicates) was performed in random order. Multi-linear Regression (MLR) was used to relate powder parameters to responses (roughness parameters). Data was filtered and best fit models for each response was chosen. Commercial statistical analysis software - MODDE, MKS Umetrics AB, Sweden was used.

Grit Residues
Image Analysis was performed by taking twenty SEM images (magnification: x100) of plan view of all samples. Grit percentage was found using a custom made routine run on commercial image analysis software APHELION.

4 Results and Discussion

Microstructure Analysis:
Fig1 shows the plan view SEM images of the as-sprayed samples with the highest (KB9) and lowest (KB2) contact angles. It can be seen that the plan view of KB2 (Fig1a) is mostly covered with carbides while KB9 (Fig 1b) has a large share of its plan view covered by binder.

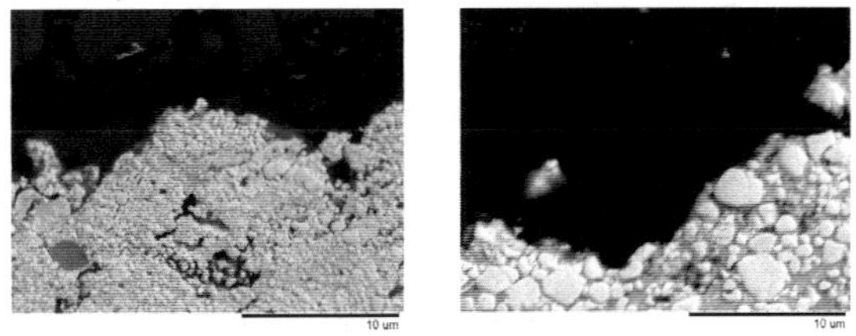

Fig1: Cross Section SEM images: (a) KB2 (b) KB9

Fig2 reveals the post-GB morphology at two different magnifications. The roughness and surface composition of as-sprayed coatings were changed after modification by grit-blasting.

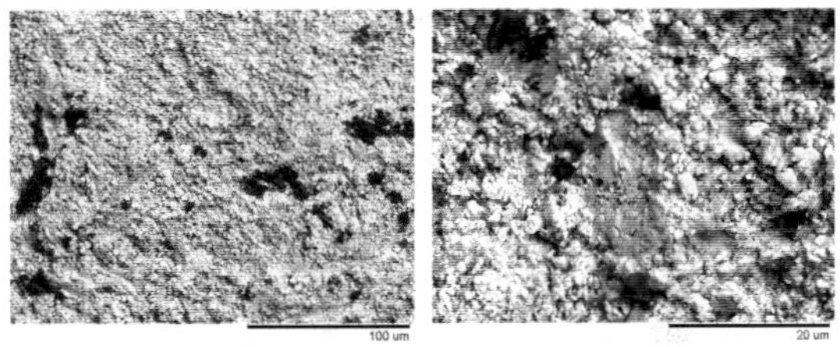

Fig2: KB9 Grit Blasted at (a) x1000 (b) x5000

Fig3 depicts the plan view of the polished surface of KB9 at two different magnifications. Polishing the surface opens up bulk structure, morphology and composition. It can be clearly seen that there is a greater percentage of carbide phase on the surface as compared to as-sprayed samples. The surface was smooth with little or no roughness present.

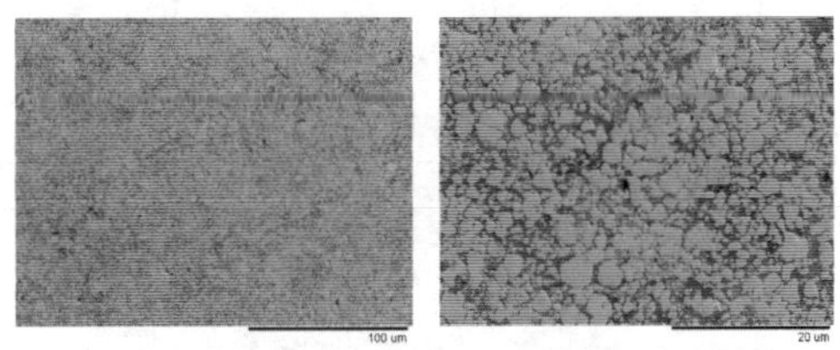

Fig3: Polished Surface at x1000 and x5000

Wettability Analysis:
Advancing contact angles (CA) for samples (as-sprayed, grit blasted and polished) are shown in Fig4. Receding angle measurements were too low to be measured accurately as the pinning effect of the droplet to surface made sure that the drop base diameter never decreased. Furthermore, when the coatings were slowly upturned, the droplet stayed firmly adhered to the sample. Surfaces where CA is also equal to CAH can be said to be either partly or wholly in Wenzel state (17).

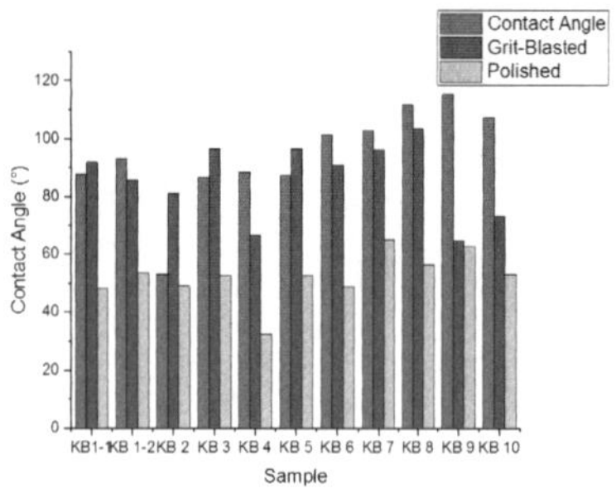

Fig4: Advancing Contact Angles

CA for the as-sprayed samples are in the range of 85-115°. However, KB2 shows a substantially lower CA of 53°, which is a significant departure from from the hydrophobic characteristics of the other as-sprayed samples. GB did not improve the hydrophobicity with either marginal increase or considerable decrease in CA. Polished surfaces (with little or no roughness) show the least contact angles. DOE analysis based on surface roughness was carried out only for as-sprayed samples as they showed the highest CA.

DOE Analysis for As-Sprayed WC-CoCr:

MLR models for each roughness response were analysed. Tab2 shows the relevant model statistics. Fig5 shows that roughness parameter Sdq can be correlated to powder parameters. Higher values of Sdq (1.61) can be obtained by not just high CGS (4μm) but also high PS (500), irrespective of the value of BGS. Thus, conventional studies that rely on just CGS would not be able to predict high values of Sdq (50X). Hence, powder parameters such as PS and BGS are integral to a complete understanding of the resultant roughness profile.

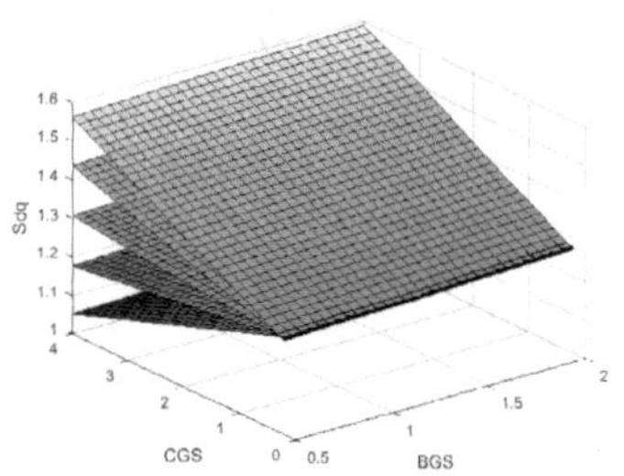

Fig5: Variation of Powder Parameters with Sdq

Theoretically, droplets that are in Wenzel state have a linear relationship between roughness and CA (Eq1). In practice, we must know the roughness parameter to use and the magnification of measurement of the parameter. Mean data fitting on Sdq (50X) shows an upward sloping trend with CA (Fig 6). Thus, high Sdq roughness surfaces can be designed with relevant powder parameters which would lead to higher values of CA. When a similar analysis was repeated for Sdq (10X), more scattered data was observed, with no clear trend. However, 50X is the relevant magnification owing to an in-between micro and submicro length scale of 0.2μm which is ideal for gauging hierarchical roughness.

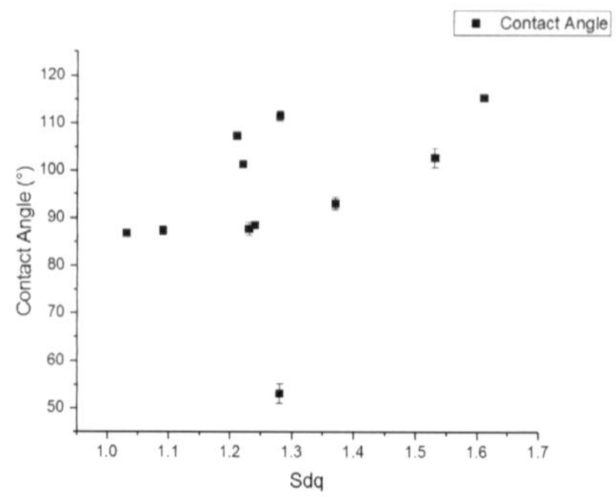

Fig6: Parametric Variation of CA with Sdq (50X)

For the 10X measurement scale, another parameter, Ssk was used. Fig7 describes the variation between Ssk and powder parameters. Again, we see that by either increasing or decreasing just CGS, a wide range of values of Ssk can be obtained. Within each envelope, it is the BGS and PS which dictates the value of Sdq.

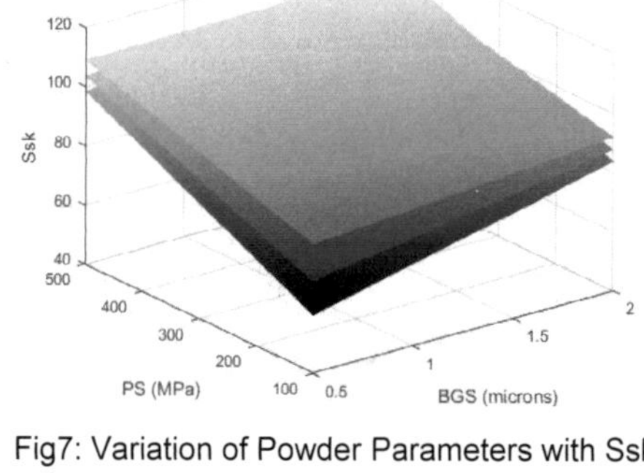

Fig7: Variation of Powder Parameters with Ssk (10X)

Mean data fitting for CA variation with Ssk (10X) shows that a somewhat linear dependence is observed here (Fig8), similar to the case of Sdq (50X). Thus, by choosing appropriate CGS, BGS and PS, higher Ssk values can be obtained, leading to higher CA.

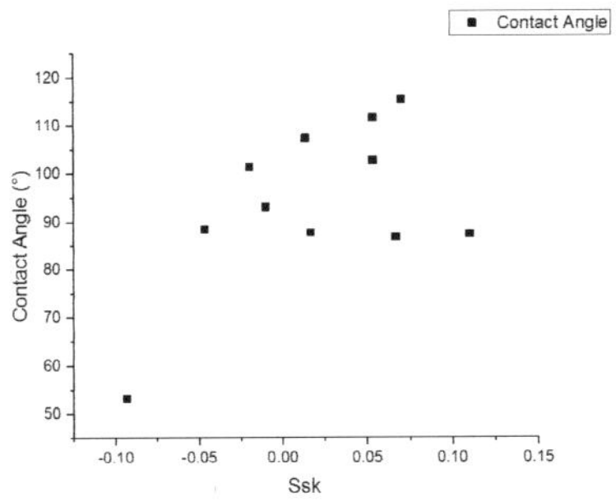

Fig8: Parametric Variation of Ssk (10X) with CA

	Fig5	Fig6
R2	0.94	0.96
Q2	0.84	0.84
Model Validity	0.98	0.92
Reproducibility	0.69	0.90

Tab2: Model Statistics

Fractal analysis of the coating has the advantage of providing an estimate of roughness, in the form of the Hausdorff Dimension (Dh) over all length scales by using measurements from one length scale. Being a measure of how a curve is spread out over an area (18), theoretically Dh should show good relation to wettability. However, as can be seen from Fig9, the CA for shows no relation with Dh.

Grit-blasted Wettability:

There is no across the board variation of the contact angle with grit blasting. While it may decrease for some, it increases for others. It is possible that extraneous factors from the GB might be biasing the experiment. Grit pollution by the aluminum oxide particles was checked to see if it had chemically altered the surface and was the cause for the variation in CA. However, maximum grit percentage on the surface was only 15%.

Polished Wettability:

The intent behind polishing the surface was twofold. One was to ascertain the non-roughness wettability of the samples and the other was to see how surface composition of carbides affected wettability. Due to the lack of roughness, the CA was much lower in the case of

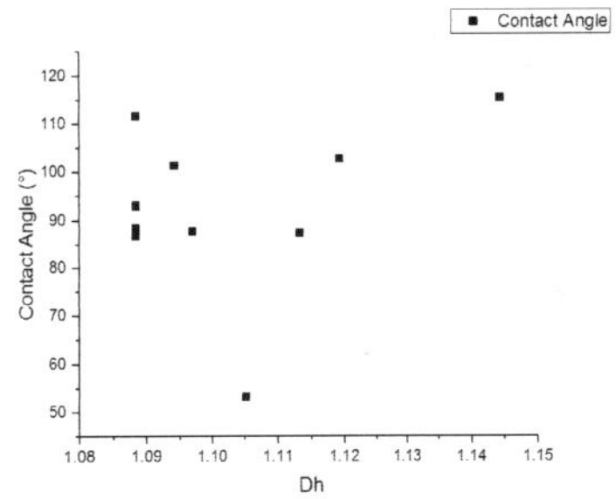

Fig9: Hausdorff Dimension vs CA

polishing as compared to as-sprayed wettability except in the case of KB2, where the CA post polishing returns to roughly the same value as it in the case of as-sprayed. This could mean that KB2 has a larger concentration of carbides on its top surface as compared to other samples.

Fig10: Grit Blasted Top Surface KB2

Overall Analysis of Wetting Behaviour:

There are primarily two factors which influence hydrophobic behaviour of a surface, roughness and surface composition. We have discussed wetting properties of as-sprayed roughnesses. Surface composition of cermets consist of mainly two materials on the surface (carbide and binder). It is possible that the carbide and binder have different roles to play. Consider an example of a flat surface of PTFE. This has a CA of 120°. Now, if we consider a rough surface of PTFE, the CA increases(19). We can consider 120° to be a baseline over which any roughness of the surface will lead to an increase in CA. Tab3 shows the CA for flat CoCr and WC along with average CA for all as-sprayed samples with the exception of KB2. It was seen before, that samples with higher CA seemed to have a larger amount of binder of their top surface. It is thus possible that the binder is the cause of wetting behaviour of the cermet and the carbides provide the roughness, which increases the CA from its baseline i.e. 60°. A notable exception to this is KB2. Due to the small size of carbides and small binder size, the carbides protrude out

Sample	CA
CoCr (20)	~60
WC cutting tool	34.79 ± 5.33
As-sprayed except KB2	98.16 ± 0.94

Tab3: CA of flat and rough surfaces

of the surface more. Hence, CA baseline for this sample is nearer to 35°. Roughness leads to CA being ~53°. By GB, the surfaces loses a large part of the binder and the carbides are exposed. Thus, the CA baseline for all samples is now about 35°, which explains the decrease in CA. Polished samples with little roughness have lower CA, most in the range of the flat surfaces of WC and CoCr. Thus, besides appropriate roughness values, surface composition of the cermet is an important parameter.

4 Conclusion

Through this paper, we have attempted to understand wetting properties of ceramic reinforced metal matrix composites through characterisation of WC-CoCr. A full factorial design was constructed on the basis of the three powder parameters - Carbide Grain Size, Binder Grain Size and Powder Strength. Microstructures show that the samples with the highest contact angles had a large concentration of binder on the surface. An attempt was made to understand how higher values of roughnesses can be attained based on powder parameters. It was found that Sdq for 50X magnification and Ssk for 10X magnification can be altered by varying CGS, BGS and PS. The same roughness parameters were seen to show somewhat linear variation with the advancing contact angles. Surface modifications such as grit blasting and polishing were carried out. Both of these modifications showed that the binder was the cause of the wetting properties of the cermet and the carbides provided the roughness. This forms the basis for using more hydrophobic binders leading to better wetting properties for cermets.

5 References

1. Tuteja, A., Choi, W., Mabry, J. M., McKinley, G. H., & Cohen, R. E. (2008). Robust Omniphobic Surfaces. *Proceedings of the National Academy of Sciences*,

2. Nilsson, M. a, Daniello, R. J., & Rothstein, J. P. (2010). A novel and inexpensive technique for creating superhydrophobic surfaces using Teflon and sandpaper. *Journal of Physics D: Applied Physics*, *43*, 45301.

3. Guo, Z., Liu, W., & Su, B. L. (2011). Superhydrophobic surfaces: From natural to biomimetic to functional. *Journal of Colloid and Interface Science*, *353*(2), 335–355.

4. Barthlott, W., & Neinhuis, C. (1997). Purity of the sacred lotus, or escape from contamination in biological surfaces. *Planta*, *202*(1),

5. Decker, E. L., Frank, B., Suo, Y., & Garoff, S. (1999). Physics of contact angle measurement. *Colloids and Surfaces A: Physicochemical and Engineering Aspects*,

6. Wang, Shutao, and Lei Jiang. "Definition of superhydrophobic states." *Advanced Materials* 19.21 (2007): 3423-3424.

7. Ma, Minglin, and Randal M. Hill. "Superhydrophobic surfaces." *Current opinion in colloid & interface science* 11.4 (2006): 193-202.

8. Li, Xue-Mei, David Reinhoudt, and Mercedes Crego-Calama. "What do we need for a superhydrophobic surface? A review on the recent progress in the preparation of superhydrophobic surfaces." *Chemical Society Reviews* 36.8 (2007): 1350-1368.

9. Abdullah Uzun, I^smet Çevik, Mustafa Akçil, Effects of thermal barrier coating on a turbocharged diesel engine performance, Surface and Coatings Technology, Volumes 116–119, September 1999, Pages 505-507,

10. Miller, R. A. (1997). Thermal barrier coatings for aircraft engines: history and directions. *Journal of Thermal Spray Technology*

11. Xiuyong Chen, Jianhui Yuan, Jing Huang, Kun Ren, Yi Liu, Shaoyang Lu, Hua Li, Large-scale fabrication of superhydrophobic polyurethane/nano-Al2O3 coatings by suspension flame spraying for anti-corrosion applications, Applied Surface Science, Volume 311, 30 August 2014, Pages 864-869, ISSN 0169-4332,

12. H. Teisala, M. Tuominen, M. Aromaa, J.M. Mäkelä, M. Stepien, J.J. Saarinen, M. Toivakka, J. Kuusipalo, Development of superhydrophobic coating on paperboard surface using the Liquid Flame Spray, Surface and Coatings Technology, Volume 205, Issue 2, 15 October 2010, Pages 436-445, ISSN 0257-8972,

13. Wenzel, Robert N. "Surface Roughness and Contact Angle." *The Journal of Physical Chemistry* 53.9 (1949): 1466-1467.

14. Cassie, A. B. D., and S. Baxter. "Wettability of porous surfaces." *Transactions of the Faraday society* 40 (1944): 546-551.

15. Nosonovsky, Michael, and Bharat Bhushan. "Biologically inspired surfaces: broadening the scope of roughness." *Advanced Functional Materials* 18.6 (2008): 843-855.

16. Guidelines to measurements of reproducible contact angles using a sessile-drop technique Jaroslaw DrelichSurface Innovations 2013 1:4, 248-254

17. Wetting and Roughness David Quéré Annual Review of Materials Research, Vol. 38: 71 -99 (Volume publication date August 2008)

18. Falconer, Kenneth. *Fractal geometry: mathematical foundations and applications*. John Wiley & Sons, 2004.

19. Miller, Jan D., et al. "Effect of roughness as determined by atomic force microscopy on the wetting properties of PTFE thin films." *Polymer Engineering & Science* 36.14 (1996): 1849-1855.

20. Woodruff, M. A., Jones, P., Farrar, D., Grant, D. M., & Scotchford, C. A. (2007). Human osteoblast cell spreading and vinculin expression upon biomaterial surfaces. *Journal of Molecular Histology*, *38*(5), 491–499

The effect of components content in fire-proof coating for protection of metal surfaces against open flame in fire conditions is investigated

Ismagilova Renata, TSPC Ltd., Moscow, Russia, Dr. Sergey Baldaev, TSPC Ltd., Moscow, Russia, Dr. Ivan Mazilin, TSPC Ltd., Moscow, Russia, Dr. Lev Baldaev, TSPC Ltd., Moscow, Russia, Fedorova Mariya, TSPC Ltd., Moscow, Russia

High temperature and fire-proof protection of metal structures issue involves formation of dense insulating coatings which possess low thermal diffusivity and capable to withstand high temperatures caused by open flame and other sever conditions. Such coating should be considered as sacrificial as it slowly decomposes during extreme high temperature impact. Such coatings intend significant extension of time required for heating and development of inelastic deformation in metal-based structures increasing service time in sever conditions.

Several modifications of fire-proof coatings composed of organic binders were developed and investigated.Fire-proof coatings were subjected to open flame test on an adapted burner rig. Open flame simulation with 1100°C was carried to estimate coating's protection properties. Investigates coating showed reduction of temperature for 1000°C during 10 minutes. Such results achieved due to spumescent effect of coating's coke layer.

1 Introduction

The most important element of the fire protection system of buildings and structures is the fire protection of building structures. It should ensure the fire resistance of structures to the required level, reduce their fire hazard, prevent the development and spread of the flame. Meeting these requirements reduces the likelihood of deaths and material losses from fires. One of the most effective and affordable ways to impart fire resistance to various materials is the coloration of their fire retardant coatings.

2 Intumescent coatings

Intumescent coatings have been used as passive fire protection for over 20 years to protect structural steel in high-rise buildings to avoid fire-induced structural collapse. An intumescent coating swells on exposure to fire in a controlled manner several times their original thickness, producing a carbonaceous protective char.

The char acts as heat transfer barrier to physically and thermally protect the coated steel structure.

Intumescent coatings can swell upto 100 times on heating (from 1-mm to 10-cm-thick foam). In the event of fire, they can act as passive fire protection for steel which loses half of its load bearing capacity above 500°C limiting structural damage and prolonging evacuation time during fire breakout, preventing the loss of lives and property [1].

Intumescent coatings are increasingly used because they provide a high-quality finish and also have good fire ratings on buildings requiring high fire protection, like stadiums, skyscrapers, multistoried buildings. Most high-rise and steel-framed constructions use intumescent coatings to protect them from the fire. A char with enhanced strength can help to protect the steel structure exposed to heat or fire and remain stable in high wind velocities. Inorganic intumescent coatings based on soluble alkali silicates were widely used in earlier days because of their low cost but they have serious drawbacks in durability and water resistance. Due to their aesthetics and performance, nowadays organic thin film intumescent coatings are becoming more popular. An organic intumescent coating consists of the following ingredients:

- Carbon donor or char former, e.g., Pentaerythritol;
- Acid donor or catalyst, e.g., Ammonium polyphosphate (APP);
- Blowing agent, e.g., Melamine;
- Binder, e.g., Polyvinyl acetate, Epoxy.

Organic intumescent coatings are thin film intumescent coatings that swell considerably when exposed to heat. They have a good quality finish and can also be topcoated if required for outdoor exposure. They have many advantages like ease of application, aesthetic, flexibility but have a limitation of sometimes forming a fluffy char which may dislodge at high wind velocities [2].

3 Organic intumescent coatings

Organic intumescent coatings are thin film intumescent coatings and are based on an acid catalyst, e.g., ammonium polyphosphate (APP), a char former e.g., pentaerythritol (PER) and a blowing agent, e.g., melamine (MEL) in solvent borne or waterborne binders. These kinds of coatings are more weather stable and water resistant than alkali silicate-based coatings. They are favored by architects and designers for passive fire protection of steel-framed structures as they provide a finish that does not detract from the appearance of the exposed steelwork as in the case of cementitious coatings. They are increasingly used today in modern day airports, skyscrapers, sports complexes, shopping malls, hotels and other places, and enabled the architect to create and design elements using the steel itself.
Intumescence mechanism The intumescence process usually takes place in following steps: a mineral acid is released by breaking the acid source which results in dehydration and carbonization of char formers. Further, the blowing agent decomposes to release gases which get constrained in the melted matrix making it swell and forming an insulating multi-cellular protective char layer. This carbonaceous char shield restricts the transfer of heat from the source to the substrate and prevents further degradation of the underlying material (Fig. 1), [3,4].

Fig. 1. Intumescent coatings

The sequence of the intumescence process and reactions are given as follows:
1. Softening and melting of the polymeric binder.
2. Releasing of inorganic acids.
3. Carbonization of char formers (e.g., of polyalcohols).
4. Release of gaseous products by blowing agents (e.g., melamine).
5. Foaming and expansion of the mixture.
6. Crosslinking and solidification of char.
There are three intumescent ingredients: an acid source, a carbon source and a blowing agent. The intumescent coating has to be formulated and optimized so as to form an efficient protective char. The protective and heat barrier properties of the char are largely influenced by its structure. A coating should form a large char volume, thick and continuous inner char structure. The char should be compact, and should have smaller cell size and narrower cell size distribution. The cells should be closed like solid foam. The coating should have other performance properties to withstand the service environment in the long term [5].

4 Influence of binder in intumescent coatings

Various studies have reported on the role of the binder in intumescent formulations. The binders, besides providing typical protective performance, also have to degrade over the same temperature range as other intumescent fire retardant additives in the formulation so as to produce intumescence, leading to a carbonaceous char.
Additionally, it must have a melt viscosity which would not be too low to cause the molten coating to slump leading to dislodging of char during intumescence, nor too high to prevent the expansion of the foam. The polymer melt should be so viscoelastic that the gases released during the intumescence process should remain contained within it, forming a foamed char. Some binder also aids in char formation process. They have synergistic interaction with ammonium polyphosphate and produce thermally more stable chars.
A range of polymeric resins can be used as a binder in intumescent coatings. Selection

of binder also depends upon the service environment to which the coating is exposed and its expected durability. Waterborne acrylics are used mostly in dry, internal locations, solventborne acrylics in internal or sheltered external locations while solvent-less or solventborne epoxy coatings find use in any location where high weatherability and protective performance are anticipated. Thermoplastic binders have good char-forming behavior compared to thermosets. Polyvinyl acetate emulsion is mostly preferred in waterborne while epoxies are in solventborne intumescent formulations. The combination of binders also gives good performance.

5 Other additives, fillers, and fibers in intumescent formulations

Many inorganic fillers and fibers find use in intumescent formulations. They increase char residue of the coatings at high temperatures. Selecting the right combination of flame-retardant fillers also strongly influence the fire protective as well as other properties of the coatings. Fillers such as magnesium hydroxide are reported to increase bonding strength to the metal surface due to its effective interface adhesion [6].
Addition of inorganic fillers results in reinforcement of the char which improves the fire-protective performance, maintaining it for a longer time. Some inorganic fillers react with APP or with its degradation products and leads to the formation of thermally stable phosphate compounds which helps in maintaining char integrity [7]. Inert fillers like zirconium silicate help in the reinforcement of char. It is also reported that its addition results in an increase in residual weight of char and release of less gaseous products[8]. Inorganic compounds e.g., borides, nitrides, or carbides of Titanium, Zirconium, or other metals are chemically inert and have a much higher thermal decomposition temperature. They serve as a filler and stabilize char in intumescent formulations [9]. Glass flake as a modifier in intumescent coatings improves the fire protection and water resistance of the coatings. Coatings modified with glass flake restrict the migration of APP and PER in the presence of water, maintain their fixed ratio and have an excellent intumescent effect and fire protection [10].

6 Experience

Our company has developed two modifications of the moisture-resistant fire-proof coating: Sample №TSPC-1 based on silicone rubber and Sample № TSPC -2 - based on epoxy resin. As an intumescent agent, the best intercalated graphite was chosen. (Fig. 2).

Fig. 2. Coated samples

As a comparison, studies were carried out: Differential Scanning Calorimetry (DSC), thermogravimetric analysis, and a special heat resistance test.
The DSC method was used to determine the parameters of phase transitions. Below are the DSC curves for the samples under study in the temperature range 20-550 °C. (Fig.3, 4). It was established that the melting temperature of sample №TSPC-1 was about 290 ° C, as the temperature was further increased, the destruction of the material began. The melting point of sample №TSPC-2 is about 288 °C.

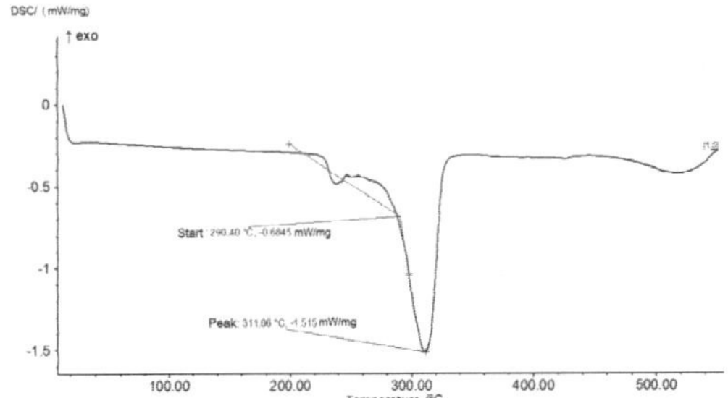

Fig. 3. Graph of the curves for sample №TSPC-1.

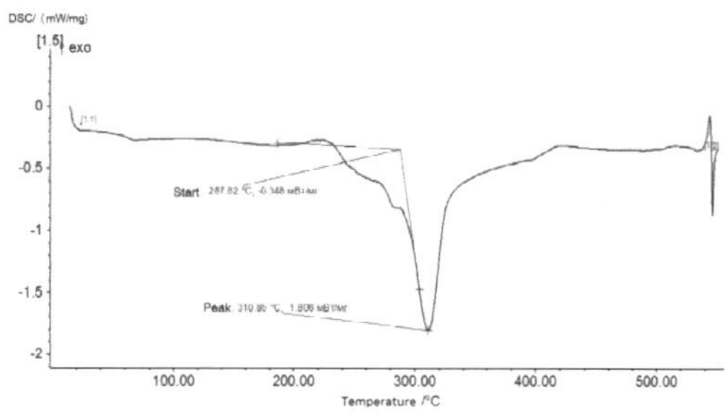

Fig. 4. Graph of the curves for sample №TSPC-2.

Thermogravimetric analysis was carried out, to determine the thermal stability of the samples, they were heated from 25 to 800 ° C in a stream of air. The obtained TG-curves are given below. (Fig.5, 6). For sample №TSPC-1, a noticeable weight loss begins at 255 ° C, the residual mass after heating to 800 ° C is 47%. For sample №TSPC -2, the change in mass begins at 178 ° C and reaches a maximum value of 52%.

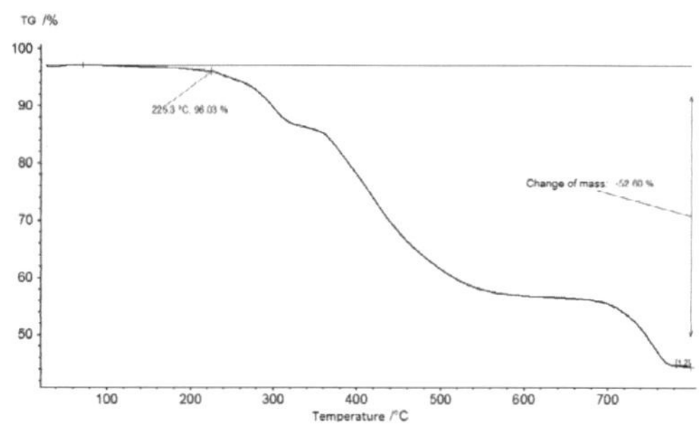

Fig. 5. Graph of the TG-curves for sample №TSPC-1.

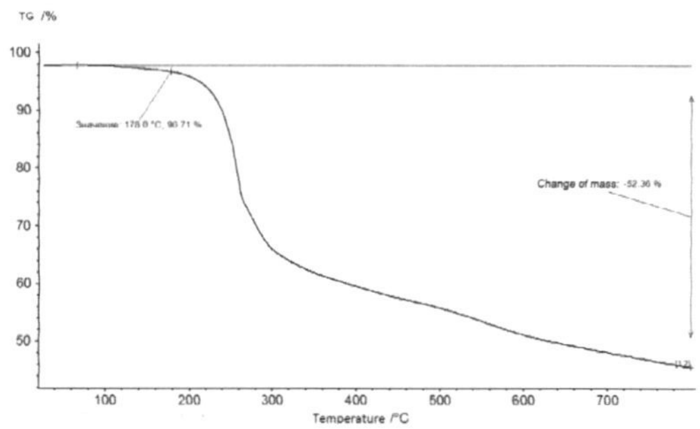

Fig. 6. Graph of the TG-curves for sample №TSPC-2.

Full-scale tests to determine the heat resistance were carried out on a special stand of its own design (**Fig. 7**).

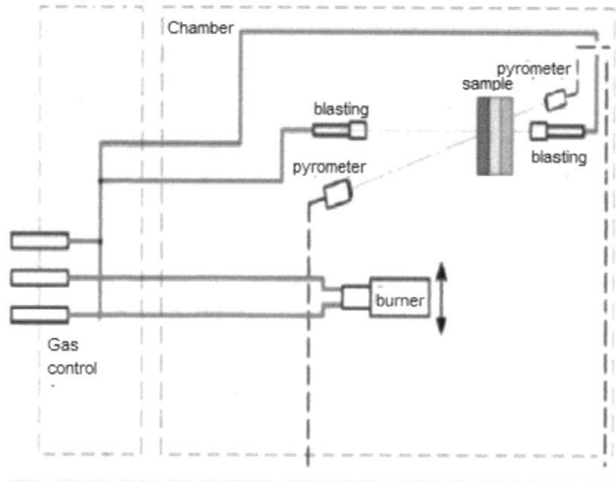

Fig. 7. Scheme of the test stand

Fire resistance tests consist of heating a sample of a given time at a constant flame temperature (1100 °C) until the temperature on the rear wall of the sample reaches 200 ° C. The source of heating is a gas burner, operating on propane mixture and oxygen. To measure the temperature on the front wall of the sample, an IMPAC IPE 140/39 MB18 pyrometer was used (spectral range 3.9 µm, blackness factor 0.32, measurement range + 200 ° C ... + 1800 ° C, error ± 0.5% The calibration certificate NSP 0280867 is valid until 10.2015), and on the back wall of the sample there is a Kelvin pyrometer (spectral range 7.8-14 µm, measuring range + 100 °C ... + 650 °C, error ± 2%).
The graph with curves for the temperature change of the rear wall as a function of time is shown in Fig. 8.

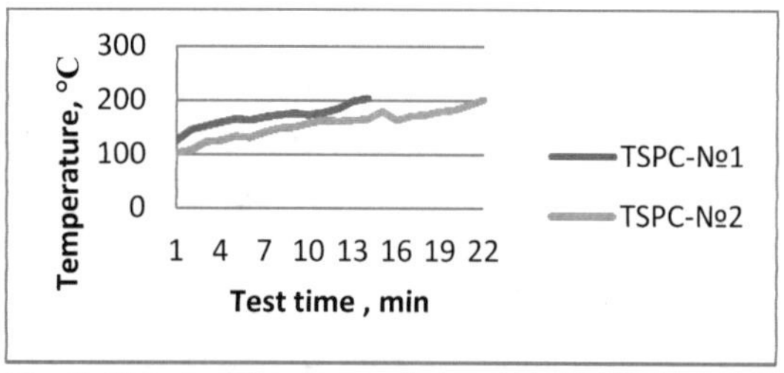

Fig. 8. Graph of the curves for sample №TSPC-1,2.

The temperature on the back wall of the №TSPC-1 sample increases gradually from 125 to 203 °C for 7 minutes when the flame on the coating is applied. At №TSPC-2, the temperature increased from 102 to 201 °C for 11 minutes. The thickness of the coatings before the tests was 2-2.5 mm. During the tests, the thickness of the coating increased significantly to 10 times, but on the substrate remained a layer not subject to degradation.

After carrying out the tests, it is possible to conclude that the coating successfully performs its functions in case of fire and depending on the thickness of the coating, it is possible to vary the time of maintaining the specified substrate temperature on the back side.

7 Conclusion

Intumescent coatings provide passive fire protection with a rating of few minutes to several hours and are consistently used by architects and designers. In addition, they also offer aesthetic appearance and ease of application. They can be formulated for use in cellulosic fire and hydrocarbon fire including jet fire and are effective in prolonging fire-induced structural collapse. Modern day intumescent coatings provide effective passive fire protection in the event of a fire but still have some pitfalls. Lower fire ratings and application involving several layers with long drying times make the use of these coatings difficult on site.
The carbonaceous char formed may have poor cohesive strength and fragility and may dislodge due to mechanical disturbances of falling objects during a fire event.

8 Literatures

[1] Horacek, H, Pieh, S, "The Importance of Intumescent Systems for Fire Protection of Plastic Materials." Polym.Int., 49 (10) 1106–1114 (2000).
[2] Kirbyshire, A, Stewart, A, A Fire Protective Coating. WO2011151649 A1
[3] Hao, J, Chow, WK, "A Brief Review of Intumescent Fire Retardant Coatings." Arch. Sci. Rev., 46 (1) 89–95 (2003)
[4] Camino, G, Costa, L, Martinasso, G, "Intumescent Fire-Retardant Systems." Polym. Degrad. Stab., 23 (4) 359–376(1989).
[5] Camino, G, Costa, L, "Mechanism of Intumescence in Fire Retardant Polymers." Rev. Inorg. Chem., 8 (1–2) 69–100 (1986).
[6] Yew, MC, RamliSulong, NH, Yew, MK, Amalina, MA, Johan, MR, "Influences of Flame-Retardant Fillers on Fire Protection and Mechanical Properties of Intumescent Coatings." Prog. Org. Coat., 78 59–66 (2015).
[7] Duquesne, S, Bachelet, P, Bellayer, S, Bourbigot, S, Mertens, W, "Influence of Inorganic Fillers on the Fire Protection of Intumescent Coatings." J. Fire Sci., 31 (3) 258–275 (2013).
[8] Ullah, S, Ahmad, F, "Effects of Zirconium Silicate Reinforcement on Expandable Graphite Based Intumescent Fire Retardant Coating." Polym. Degrad. Stab., 103 49–62 (2014).
[9] Kobayashi, N, Yoshioki, H, Yoshida, K, Sagawa, K, Ishihara, S, "Intumescent Fire-Resistant Coating, Fire-Resistant Material, and Process for Producing the Fire-Resistant Material." US Patent US5401793A.
[10] Wang, G, Yang, J, "Influences of Glass Flakes on FireProtection and Water Resistance of Waterborne IntumescentFire Resistive Coating for Steel Structure." Prog. Org.Coat., 70 (2–3) 150–156 (2011).

Influence of alloying elements on properties of arc sprayed Fe-C-Cr-B-Al Y coatings from cored wires

Yu. Korobov, V. Shumiakov, S. Nevezhin, M. Filippov, A. Makarov, I. Malygina, R. Savrai, N. Soboleva, G. Tkachuk, Ekaterinburg, Russia

Arc sprayed coatings from Fe-Cr-B-Y cored wires additionally alloyed by Al, Si, C were studied. SEM and XRD analyses, microhardness characterized their morphology, microstructure, and phase composition. The coatings were also evaluated by two-body abrasive wear test, and heat resistance test in air, 700°C, 100 h. Among compared coatings, the Cr12B5Al5Y one with maximum Al and minimum C content showed better indicators of porosity, adhesion strength, wear- and heat resistance comparing to other coatings alloyed with Si or less content of Al. It is resulted from the largest quantity of the carbon enriched eutectic containing reinforcing carboboride phases (Fe, $Cr)_2B$, $Fe_2(C, B)$, and a rational ratio of the alloying elements. As shown, Cr12B5Al5Y coating wear resistance is more than twice as high comparing to the coating from typical wear resistant 150Cr8Ti2Al cored wire. The latter one consists of martensite phase, forming from the metastable austenite during loading, and chromium carbides. Weight loss tests showed that coatings from studied Fe-Cr-B-Y cored wires are of the same heat resistance level as the austenite steels and 1-2 orders of magnitude higher as compared to the pearlite and ferrite-martensitic steels, which are widely used as boiler's materials.

1 Introduction

Repair costs in coal-fired power plants are 12% of the total cost of the generated electricity in Russia [1]. At the same time, about a quarter of the equipment worktop operates at temperatures up to 600°C and comes into contact with aggressive fly ash fractions moving at high velocity in the flue gases [2]. Its wear leads to an unscheduled failure of a number of reliability-critical parts, such as cases and blades of fans and smoke exhausters [3]. Many researches and approaches have been reported for reducing the erosion damage, one of which is thermal spraying coating. Among the thermal spray techniques, arc spraying is widely employed because of its economic advantage and flexibility of operation. To produce Fe- based coatings, wires like FeNiCrB, NiCr, NiCr-WC, FeCrAlB, FeCrAlB-WC are used [4-6]. However, a wide implementation of such high-alloyed feedstock materials is limited because of economic reasons and technological limitations. At the same time, FeCrAl-type alloys are known to be suitable materials for high-temperature applications. Its heat resistance is provided by a surface alumina layer, which forms at high operating temperatures. This oxide layer is characterized by high melting temperature, chemical and thermal stability and low rate of growth [7, 8].

As shown earlier, the arc-sprayed coatings of FeCr6B3Al system demonstrated enhanced resistance to abrasive wear at elevated temperatures up to 600°C. Phase and spectral analyzes showed that the dispersion hardening of such coatings is provided by selections in their boride structure $(FeCr)_2B$, CrN and AlN nitrides, and also $FeAl_3$, Ni_3Al intermetallides [9, 10].

With respect to solid steels, heat resistance rises sharply with increasing content of chromium and/ or silicon. The first is connected with the formation on the surface a firm and refractory chromium oxide or spinel.

At high-temperature exposure, mutual diffusion of iron out of substrate and from atmosphere gases through porous coating is accelerated. Forming iron oxides and aluminum nitrides also negatively effect on heat resistance of chromium-aluminum coatings. Alloying up to 4% silicon results in a formation of oxide layer like SiO_2. It serves as diffusion barrier below alumina layer, which prevents forming of unfavorable oxides like Fe_2O_3 and nitrides like AlN [12, 13]. Maximum heat resistance up to 600 °C is achieved when the content of chromium approaches to 13%, which is higher than that of the above mentioned cored wires. It also grows with the addition of silicon [14].

Addition of yttrium is favorable for protection against gas corrosion and wear by the following reasons. First, Y-containing Fe-Cr-Al alloys showed good oxide adherences on the arc sprayed coatings at high temperature operation under thermal cycling [15, 16]. Second, during arc spraying process an alloying elements burnout takes place. It leads to a decrease of alloying elements concentration reducing the coating heat resistance. Yttrium affinity to oxygen is higher than that of other alloying elements. Therefore, yttrium protects them from oxidation at transition from the feedstock to the coating. Accordingly, the elements save their ability to resist against gas corrosion and wear. In case of Fe-Cr-B-Al cored wire, the addition of yttrium increases the coating's oxidation resistance by 15 % and the wear resistance by 20% comparing to the coatings from the yttrium-free cored wire [17].

Active Arc Spraying (AAS) was applied for coating deposition in the present study. The process differs from typical arc spraying process by the use of propane-air combustion products as a carrier gas instead of compressed air. It increases the particle

velocity, temperature, and forms a reducing atmosphere.
These features can decrease the burnout of spraying metal in flight and the porosity of the coating, increase adhesion strength of the coating [18].
The purpose of this study is to analyze influence of alloying of Fe-Cr-B cored wires by Al, C, Si on the performance of arc sprayed coatings for heat resistant applications up to 600°C.

2 Research Methods

The following parameters of AAS Gun ADM-10 (Ural Welding Institute, Ekaterinburg, Russia) were employed: current 320 A, voltage 34 V, spraying distance 100 mm, moving speed 0.1 m/s, input pressure of gases: air - 0.42 MPa, propane - 0.41 MPa. Fe-based cored wires ∅ 2.0 mm were used. Chemical composition of the coatings from the studied cored wires is seen in **Table 1**. Pointed numbers were accepted in recording data later on.

Table 1 Chemical composition of the coatings from studied cored wires, wt. %

№	Fe	C	Cr	B	Al	Si	Y
1	base	0.6	12.0	5.0	2.0	0	0.1
2	base	0.2	12.0	5.0	0	3.0	0.1
3	base	0.2	12.0	5.0	5.0	0	0.1

The contents of Al, B, Si, and Y in the wires were designated according to developed neural network model [19]. The contents have been optimized by criterion of minimum alloying elements burnout. Calculation results due to model of oxygen input into the metal droplets of Fe-Cr-Al-type at arc spraying [20] were used as the training data. The samples of ASME 1020 steel, with the dimensions of 30x20x3 mm and 10x10x50 mm were used. The coating thickness was 0.4 mm at the 30x20 mm surface and 1.5 mm at 10x10 mm surface correspondingly. The first type of coatings was used for evaluation of the heat resistance and the second one was used for wear test. The amount of oxygen in the coating was measured by melting the coating in inert gas flow by analyzer ON-900 (Eltra, Haan, Germany). The porosity of the coatings was determined by metallographic method on the transverse sections using commercial hardware and software SIAMS 700 (SIAMS, Ekaterinburg, Russia) to estimate the distribution of pores by size, area, and volume. The adhesion strength of the coatings was determined by separation of a conical pin [21]. A schematic view of the test is presented in **Fig. 1.** Here the washer 1 serves as the basis; pin 2 is inserted into its hole so that its end surface is flush with the external plane of the washer. The total surface of the pin and the washer after preparation is coated with the coating № 3. Diameters of the washer and the pin are 25 and 3 mm correspondingly. The test consists of pulling the pin by applying a force (P). The results were averaged for three samples.

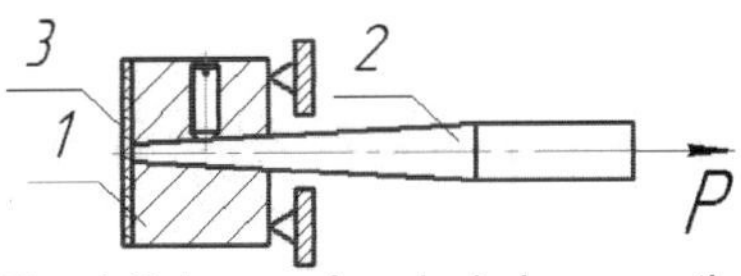

Fig. 1 Scheme of conical pin separation.

Heat resistance of the coatings from these wires was evaluated in accordance with standard test [22]. The gravimetric method was used according to weight-loss measurements of the samples exposed in the air, 700°C, 100 h. To decrease measuring error the 30x20x3 mm samples were deposited by 10 μm galvanic nickel layer. It was removed from 30x20 mm surface to be sprayed. The results were averaged for four samples. Two-body abrasive wear studies were carried out by pin-on-plate reciprocating technique at room temperature. Corundum abrasive paper with grain size up to 150 μm was fixed on the plate and the sprayed sample was fixed in a holder. Test conditions were the following: full friction path 60 m, pin velocity 0.16 m/s, specific load 1 MPa, work surface area 10×10 mm.
As a basis for comparison of the wear resistance, arc sprayed coatings from developed cored wire 150Cr8Ti2Al, belonging to metastable austenite type alloys were used [23]. Alloy of metastable austenite structure is a proper material to withstand various mechanical loads. When the external load is applied to such material, synergetic effects of transformation from austenite to martensite take place. First, rise of martensitic phase share leads to hardness increase. When loading the micro-hardness of 150Cr8Ti2Al coatings is increased from 600 to 880 $HV_{0.1}$. Second, energy of external load, which is applied to the working surface, is dissipated due to strain-induced nucleation of martensitic transformation. As a result, wear resistance under different conditions (e.g. abrasive, hydro and gas-abrasive, erosion, cavitation, adhesion, and fatigue) is improving [24-25]. Metastable austenite type cored wires are used for increasing wear resistance by arc surfacing and by arc spraying [23, 26, 27].
Microhardness measuring was performed by Leica VMHT device (Walter Uhl, Aßlar, Germany) under 50 and 100 g loads on the polished surface of the sample. Microstructure was studied by straight type microscope Altami-3D (Altami, S-Petersburg, Russia). Phase composition on the coating surface was determined by X-ray diffractometer XRD-7000 (SHIMADZU, Kyoto, Japan, Cr-Kα radiation). Chemical composition and structure of the coatings were studied by scanning electron microscope VEGA II XMU (Tescan, Brno, Czech Republic). It was

equipped with a wavelength dispersive X-ray spectrometer INCA Wave 700 (WDX) and energy-INCA Energy 450XT (EDX) microanalyzers. SEM analysis was carried out with Philips microscope.

3 Results and discussion

3.1 Porosity and adhesion strength of the coatings

Coating № 3 is of lowest porosity among studied coatings, **Table 2.**

Table 2 Porosity analysis results

№	Analyzed square, mm^2	Porosity, %	Size of pores, μm		
			Min	Max	Mean
1	1.94	3.5	1.2	24.2	3.2
2	1.94	3.0	1.2	27.7	3.3
3	0.78	2.8	0.5	44.1	6.1

Adhesion strength of the coating № 3 is 1.3-1.5 time as high among studied coatings. Comparing to the data presented in the sites of well-known manufacturers of cored wires (a-c), we see that it is on the same level with the coatings for the same purpose, **Table 3**.

Table 3 Adhesion strength of the coating, MPa

Coating mark	1	2	3	a	b	c
Value	28	32	43	43	46	54

a - PMet 731 (Polymet); b - EuTronic Arc 509 (Castolin Eutectic); c - TAFALOY 26CA (Praxair TAFA)

Reduced porosity and increased adhesive strength of coating № 3 is reasoned by lower degree of oxidation of the sprayed particles and improved interaction in contact "particle-substrate". In turn, yttrium and aluminum influence it. As shown by authors earlier, a strong affinity of yttrium to oxygen prevents burnout of other alloying elements during spraying, which allows keeping their proper content for further operation. Besides formation of complex oxides like $(Fe, Al, Y)_2O_3$ takes place [13]. Additional rise of adhesive strength of coating № 3 may be caused by increased content of aluminum, which is associated with the exothermic reactions (heat is shown per gram of charge):

$3Fe+Al=Fe_3Al+0.75$ kJ/g,

$2/3Fe_2O_3+2Al=4/3Fe+Al_2O_3+3.5$ kJ/g,

$2/3Cr_2O_3+2Al=4/3Cr+Al_2O_3+2.7$ kJ/g.

It leads as to rise of particle's temperature on 200-400 degrees comparing to Al-free wires and to deoxidizing of Fe and Cr [28].

3.2 Microhardness and wear resistance

According to the measurements, the coating № 3 is of the highest microhardness, **Table 4.** For all coatings, the microhardness is of high level through all height of the cross section, **fig. 3**. The microhardness of the coating from 150Cr8Ti2Al cored wire, chosen as the base for comparison, is (600-800)$HV_{0.1}$.

Table 4 Microhardness of the coatings, mean values

№	$HV_{0,05}$	$HV_{0,1}$
1	1158	925
2	1233	904
3	1442	1364

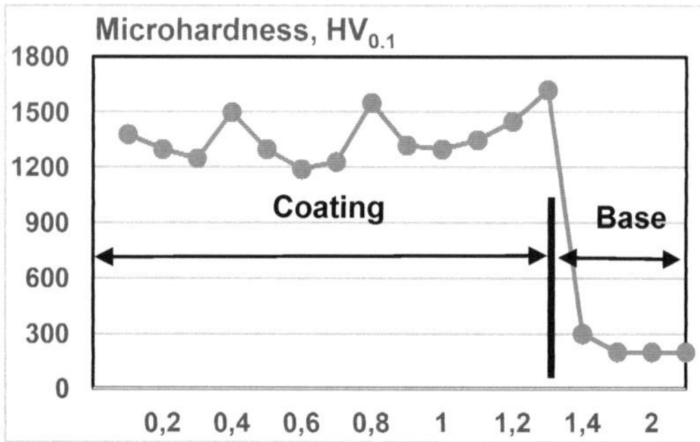

Fig. 3 Microhardness distribution of the coating № 3 along its thickness, mm, from the surface

The coating № 3 is of better wear resistance among studied coatings, **Fig. 4**.

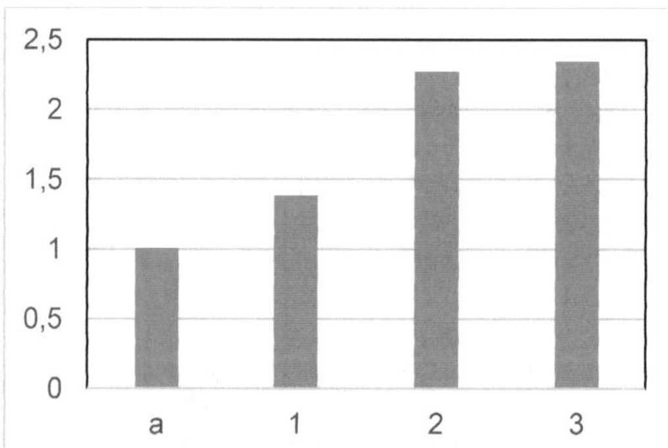

Fig. 4 Relative wear resistance of the coatings from studied cored wires, a - 150Cr8Ti2Al

As shown, wear resistance of the coatings increases at growth of aluminum containing components in the charge due to aluminothermy reactions [29]. It promotes uniform melting of boron containing components and the formation of a dense structure of the coating. Wear resistance of all coatings is higher comparing to the coating from 150Cr8Ti2Al cored wire, chosen as the base for comparison. In case of abrasive wear, apparently, strain induced martensite and chromium carbides in a coating of 150X8Ti2Al type are less effective as strengthening phases than carbo-borides in the Cr-B-Al-Y coating.

3.3 Heat resistance of the coatings

Coatings № 3 showed better heat resistance by the measuring of specific weight loss. Comparison was also done with commercial boiler steels by reference data [30, 31]. It is an order of magnitude smaller than that of steels of pearlite type (12Cr1MoV, like T11) and martensite-ferrite type (1Cr12W2MoV, like T122) and similar to the heat resistance of austenite steel like Cr18Ni12T, Cr23Ni18, **Table 5**.

Table 5 Specific weight loss of the coatings and the commercial boiler steels, g/(m^2·h)

Coatings			
1	2	3	
0.65	0.57	0.54	
Commercial boiler steels [30, 31]			
12Cr1MoV	Cr12W2MoV	Cr18Ni12T	Cr23Ni18
80	10	0,4	0,1

3.4 Phase and structure analysis

Phase and structure analysis was done for the coating № 3, which showed better indicators of porosity, adhesion strength, wear- and heat resistance. According to surface analysis, the coating № 3 consists of iron α-phase (solid solution of Fe-Cr-B, enriched with carbon) with reinforcing phases: borides Fe_2B, $(Fe,Cr)_2B$ and carboborides $(Fe,Cr)_2(B,C)$, **Fig. 5**.

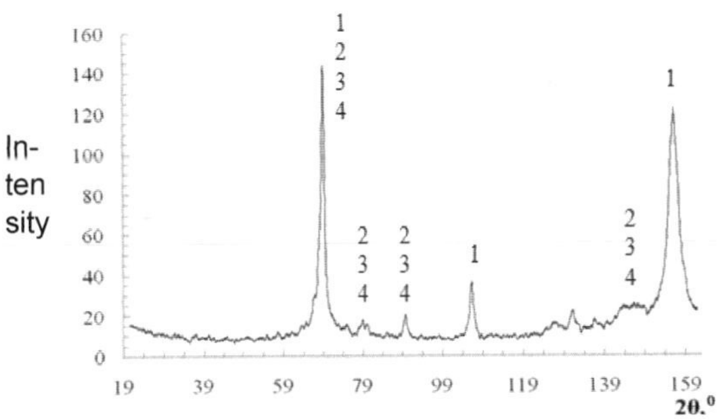

Fig. 5 Diffraction patterns of the coating № 3, surface. 1 – α-Fe, 2 – Fe_2B, 3 – $(Fe,Cr)_2(B,C)$, 4 – $(Fe,Cr)_2B$

Cross section of the coating № 3 is shown in **Fig. 6**. Zones pointed as (A, B, C), **Fig. 6a**, were detected by EDX/WDX microanalysis. The basis (point 1) is a Fe-Cr-B solid solution. The structures indicated by point 2, are enriched by carbon, which substitutes part of the iron and boron. The structure indicated by point 3 represents a carbon-rich eutectic with phases (Fe, Cr) B, and Fe_2 (C, B). Low content of oxygen, (1.3-1.6) wt. %, indicates effective shielding of alloying elements by Y from burnout.

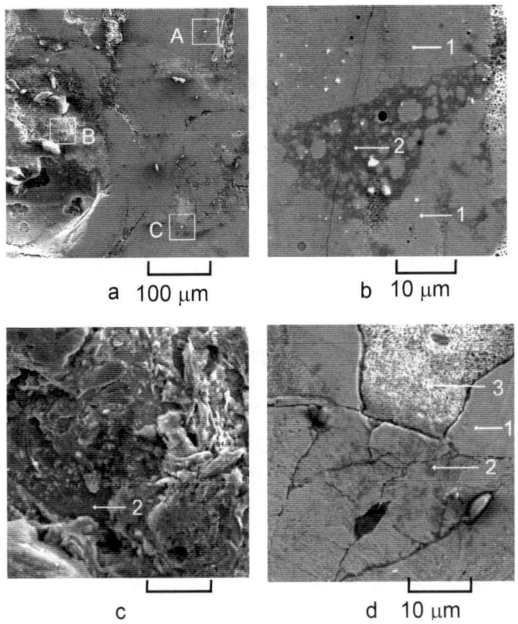

Point	Element content, at. %						
	B	C	Al	Si	Cr	Fe	Y
1	23.2	3.62	0.55	0.99	8.87	62.7	0.04
2	20.5	16.8	0.48	1.01	8.98	52.1	0.03
3	16.6	9.86	0.42	0.39	4.89	64.1	0.07

Fig. 6 Microstructure of the coating № 3 (a) and quantitative elemental analyses of the indicated zones; b – area A, c – area B, d – area C

4 Conclusions

1) It was found that in all coatings the main phase is a solid solution on the basis of α-Fe. A eutectic of variable composition is uniformly distributed in it. The eutectic is enriched by carbon-and contains reinforcing phase $(Fe, Cr)_2$ B and $Fe_2(C, B)$ The greatest quantity of the eutectic is observed in the coating 2Al5Cr12B5Y.

2) Study of performance characteristics of coatings from Fe-Cr-B-Y cored wires additionally alloyed by Al, Si, C was done. It showed that indicators of porosity, adhesion strength, wear- and heat resistance are better in case of the coating like 2Al5Cr12B5Y with maximum Al and minimum C content. Comparing to coatings of lower contents of Al or alloyed by Si instead of Al, porosity is 7-25% less, other indicators are higher: adhesion strength on 25-34%, micro-hardness on 15-20%, heat resistance on 5-20% and wear resistance up to 40 %. Better performance characteristics of that coating are resulted from the largest quantity of the carbon enriched eutectic containing reinforcing carboboride phases $(Fe, Cr)_2B$,

$Fe_2(C, B)$, and a rational ratio of the alloying elements.

3) Heat resistance of the coatings from Fe-Cr-B-Al type core wires was evaluated in air, 700°C, 100 h. The coatings are of the same heat resistance level as the austenite steels like Cr18Ni12T, Cr23Ni18, and 1-2 orders of magnitude higher as compared to the pearlite and ferrite-martensite steels like T11, T122, which are widely used as boilers materials. It is associated with the formation of complex oxides like $(Fe, Al, Y)_2O_3$.

4) The microhardness of the Fe-B-Cr-Y coatings additionally alloyed by Al, Si, C is 1.5-1.8 times as high on average and abrasive wear resistance is more than twice as high compared to arc sprayed wear-resistant coating from 150X8Ti2Al cored wire. In case of abrasive wear, apparently, strain induced martensite and chromium carbides in the 150X8Ti2Al coating are less effective as strengthening phases than complex carbides like $(Fe,Cr)_2(B,C)$ in the Fe-Cr-B-Al-Y coating.

5 Acknowledgements

Authors thank their colleagues from UrFU for assisting in sample preparation. Thanks to L. Chumakova, V. Verkhorubov.

6 Literature

[1] Kakaras, E. Current Situation of Coal fired Power Plants in Russia Federation and the Implementation Options of Clean Coal Technologies. *5th European Conference on Coal Research & Its Applications*, Scotland, 2004, pp. 263-268.

[2] Baranov, P.A. *Prevent of boiler's accidents.* Energoatomizdat, Moscow, 1991, 272 p. (In Russian).

[3] Belyaev, S.A. *Reliability of power engineering equipment.* NLT, Tomsk, 2008. 218 p. (In Russian).

[4] Stack, M., and D. Pena. Particle size effects on the elevated temperature erosion behaviour of Ni-Cr/WC MMC-based coatings. *Surface and Coatings Technology*, 1999, 113: 5-12.

[5] SHS 7170 TWAS. *The NanoSteel Company, Super hard steel. Coating solution technical data.* 2008.

[6] Lai-ma, LUO, LIU Shao-guang, YU Jia, LUO Juan, and LI Jian. Effect of Al content on high temperature erosion properties of arc-sprayed $FeMnCrAl/Cr_3C_2$ coatings. *Trans. Nonferrous Met. Soc. China* 20 (2010), pp 201-206.

[7] Borisova, A.L., I.V. Mits, T.V. Kaida, and A.Sh. Kleiman, Structure and properties of coating based on ferrochrome and ferrochrome aluminum alloys obtained by arc spraying flux-cored wires. *Autom. Welding*, 1995, 6, pp. 3-6. (In Russian).

[8] Sims, Ch.T., N.S. Stoloff, and W.C. Hagel. Superalloys II: High-Temperature Materials for Aerospace and Industrial Power. John Wiley & Sons, New York, 1987, 640 p.

[9] Pokhmurskii, V., M. Student, V. Gozdecki, and H. Pokhmurska. Cored wires of FMI series for arc spraying (Review). *Autom. Welding*, 2011, 9, p. 52-57 (in Russian).

[10] Wielage, B., H. Pokhmurska, M. Student, V. Gvozdeckii, T. Stupnyckyj, and V. Pokhmurskii. Iron-based coatings arc-sprayed with cored wires for applications at elevated temperatures. *Surface & Coatings Technology* 220 (2013), pp 27–35.

[11] Lai, G.Y. *High-Temperature Corrosion and Materials Applications*. ASM International, 2007. 461 p.

[12] Goldstein, M.I., S.V. Grachev, and Yu.M. Veksler. *Special steels*. Metallurgy, Moscow, 1985, 408 p. (In Russian).

[13] Korobov, Yu., M. Filippov, S. Nevezhin, and M. Karabanalov. High-Temperature Oxidation Behavior of Fe-25Cr-5Al Arc Spraying Coatings. *Thermal Spray 2012: Proceedings of the International Thermal Spray Conference, Air, Land, Water, and the Human Body: Thermal Spray Science and Applications,* Basil R. Marple, Arvind Agarwal, Laura Filofteia-Toma, Margaret M. Hyland, Yuk-Chiu Lau, Chang-Jiu Li, Rogerio S. Lima, and André McDonald, Ed., May 21-24, 2012 (Houston, Texas), pp. 729-733.

[14] Ximushin, F.F. *Stainless steels.* Metallurgy, Moscow, 1976, 798 p. (In Russian).

[15] Amano, T. Rare earth application for heat-resisting alloys. *J. of Rare Earth*, 2010, 28 (Spec. Issue), pp. 12-21.

[16] T. Izumi, H., Arikawa, Y. Kojima, J. Hitachi, M. Okada and T. Hisamatsu. Oxidation behavior of bond coats with and without Y addition *Proc. of Int. Thermal Spray Conf. & Exposition ITSC 2011*, Hamburg, September 27-29, 2011. P. 579-582

[17] Korobov, Yu., S. Nevezhin, M. Filippov, A. Makarov, I. Malygina, D. Fantozzi, A. Milanti, H. Koivuluoto, and P. Vuoristo. Study of High Velocity Arc Sprayed heat resistant coatings from FeCrAlBY cored wire. *Thermal Spray: Fostering a sustainable world for a better life: Proceedings from the International Thermal Spray Conference and Exposition ITSC 2016*, May 10–13, 2016, Shanghai, China, 2016, p. 852-856.

[18] Korobov, Yu. S. Efficiency of using activated arc metallization for the deposition of protective coatings. *Welding international*, 19(7), 2005, pp. 580-582.

[19] Korobov, Yu., S. Nevezhin, V. Verkhorubov, G. Rimer, and A. Zhilin. Optimization of alloying of heat resistant core wires for arc spraying by neural network modeling. *Thermal Process Modeling and Computer Simulation*, June 16-19, Orlando, USA, 2014, pp. 322-327.

[20] Boronenkov, V., and Yu. Korobov. *Fundamentals of Arc Spraying. Physic-chemical Regularities.* (Switzerland), Springer International Publishing, 2015, 256 p.

[21] Tyshinski, L.I., A.V. Plokhov, O.A. Tokarev, and V.I. Sindeev. *Methods of study the materials.* Mir: Moscow [in Russian], 2004, 384 p.

[22] *Unified system of corrosion and ageing protec-*

tion. Protection coatings. Methods of heat resistance determination. GOST 9.312, USSR State Committee on Standards, 1989, 11 p. [in Russian].
[23] Korobov, Yu., V. Verkhorubov, S. Nevezhin, M. Filippov, G. A. Tkachuk, A. Makarov, and I. Zabolotskikh. An Influence of Strain-Induced Nucleation of Martensitic Transformations on Tribological Properties of Sprayed and Surfaced Depositions. *Thermal Spray: Fostering a sustainable world for a better life: Proceedings from the International Thermal Spray Conference and Exposition ITSC 2016*, May 10–13, 2016. Shanghai, China, 2016, p. 694-699.
[24] Olson, G. B., and M. A. Cohen, Mechanism for the Strain-Induced Nucleation of Martensitic Transformation, J. of the Less-Common Metals, 28, (1972), pp. 107-118.
[25] Schastlivtsev, V. M., and M. A. Filippov. Role of the Bogachev - Mints Concept of Metastability of Austenite in Choosing Wear-Resistant Materials, *Metal Science & Heat Treatment*. 2005, Vol. 47 Issue 1/2, pp. 3-5 (in Russian).
[26] Kulishenko, B. A., A.N. Balin, and M.A. Filippov. Electrodes for Hardfacing Components Subjected to Abrasive and Impact-Abrasive Effects. *Welding International*, 19 (4), 2005, pp. 326-329.
[27] Pukasiewicz, A.G.M., A. R. Capra, and R. S. C. Paredes, Cavitation Erosion Mechanism in Fe-Mn-Cr-Si-Ni Arc Thermally Sprayed Coatings, *Thermal Spray 2011: Proceedings of the International Thermal Spray Conference*, B.R. Marple, A. Agarwal, M.M. Hyland, Y.-C. Lau, C.-J. Li, R.S. Lima, and A. McDonald, Ed., September 27-29, 2011 (Hamburg, Germany), ASM International, pp. 902-907.
[28] Pokhmurskii, V., M. Student, I. Sidorac, I. Ryabtsev, and Yu. Kuskov. Phase content of arc sprayed Fe-Cr-C-Al coatings. *Autom. Welding*, 2002, 10, p. 20-23 (in Russian).
[29] Borisova, A.L., I.V. Mits, T.V. Kaida, I.Ya. Dzikovich, and V.N. Korzhik. Structure and properties of arc sprayed FeB coatings from cored wires. *Autom. Welding*, 1991, 9 (462), p. 66-68 (In Russian).
[30] Nikitin, V.I. *Calculation of heat resistance of metals*. Metallurgy: Moscow, 1976, 207 p. (In Russian).
[31] Davis, J.R. *ASM Specialty Handbook: Heat-Resistant Materials*. ASM, 1997, 591 p.

Energy efficient production of thermally sprayed coatings through the use of modified twin wire arc spraying processes – Final results of a Bavarian research foundation funded project

A. Atzberger[1], K. Hartz-Behrend[1], J. Schein[1], E. Steiner[2], L. Blau[2], W. Mayr[2], M. Szulc[3], J. Zierhut[3], W. Krömmer[4], F. Lang[4], M. Aumiller[5] and E. Abler[6]

[1]Lab for Plasma Technology (LPT), University of the Federal Armed Forces Munich, Neubiberg, Germany
[2]Electrical Measurement Techniques, University of Applied Sciences Munich, Munich, Germany
[3]Zierhut Messtechnik GmbH, Munich, Germany
[4]Linde AG, Unterschleißheim, Germany
[5]Federal-Mogul Friedberg GmbH, Friedberg, Germany
[6]Abler GmbH & Co. KG, Waltenhofen, Germany

E-mail: alexander.atzberger@unibw.de

The main goal of the project was to investigate the influence of different gas properties, in combination with different spraying parameters, on the wire arc spraying process. For this purpose, investigations with several gas compositions (pure nitrogen as well as nitrogen combined with hydrogen or ethene – compared to compressed air) have been carried out for different spraying materials which are currently being used in today's industries, such as copper, carbon (St0.8) and stainless steels (316L). The preheating of the process gas as well as the variation of the gas pressure up to 1,400 kPa (14 bar) were also subject to research. The resulting coating properties have been analyzed in terms of oxide content, porosity and hardness as well as deposition efficiency and adhesive tensile strength. Additionally, in order to enhance the process stability, a system to detect the cause of cold shuts has been developed.

1 Introduction

Wire arc spraying is one of the oldest thermal spray processes, which is preferably used due to its low operating costs, easy handling and high deposition rates. By being the most cost-efficient of all thermal spray processes, it has a broad range of applications in today's industries, especially in terms of corrosion protection or the production of wear-resistant coatings [1,2].

The main objective of the project was to investigate the wire arc spraying process in order to enhance the process performance in economical terms. Therefore the use of advanced process gas mixtures was one key aspect. It is known that inert gases (compared to compressed air) enhance the coating quality significantly [3-5], thus investigations aiming at suitable industrial applications have been performed. Also the amount of process gas used is an important factor in terms of both coating quality and economic efficiency. Another important topic was to develop a measure in order to detect malfunctions during the ongoing spraying process by activity analysis and surveillance. By being able to detect the cause of cold shuts the overall system performance can be enhanced significantly. The results presented in this paper have been obtained in a research project funded by the Bavarian research foundation (Bayerische Forschungsstiftung). The University of German Federal Armed Forces Munich as well as the University of Applied Sciences Munich were part of the research group university-wise. Linde AG, Federal-Mogul Friedberg GmbH, ABLER GmbH & Co. KG and Zierhut Messtechnik GmbH were part of this group on the corporate side in order to investigate different aspects of the topic under economical process conditions.

2 Process Conditions

Since several industrial partners were part of this research group, different spraying equipment has been used for the investigations. Thus, the components and equipment used for the experiments are listed in every paragraph respectively. The materials used for the investigations (and their chemical composition) are listed in table 1 as well as the process gases.

The materials used are commonly known in today's industries, hence by improving the coating quality of these materials small and midsize companies in Bavaria could gain a competitive advantage.

Table 1: Materials (and their chemical composition) and process gases used in this project

Materials							
Copper	Cu 99.5						
316L	Fe 65	Cr 17	Ni 12	Mo 2,5	Mn 2,0	Si 1,0	C 0,08
St0.8	Fe 98.3	C 0.8	Mn 0.7	Si 0.2			
AlSi12	Al 88.6	Si 11.2	Fe 0.2				
G3Si1	Fe 97.5	Mn 1.5	Si 0.9	C 0.08	P 0.02	S 0.01	

Process gases	
CA	Compressed Air
N_2	Nitrogen
N_2 + 4-Vol.% H_2	Nitrogen-Hydrogen (N_2-H_2)
N_2 + 4-Vol.% C_2H_4	Nitrogen-Ethene (N_2-C_2H_4)

In order to enhance the process performance a combination of nitrogen-hydrogen has been used for the following investigations. A combination of nitrogen-ethene has also been used for first experiments. The idea was to dissociate the fuel gas C_2H_4 to receive "free carbon", which is able to increase the carbon content in the coating and therefore result in a higher hardness. Coatings have been deposited for different spraying materials and gas compositions. Their analysis included the determination of the hardness according to Vickers, the oxide content as well as the deposition efficiency according to DIN EN ISO 17836 (norm), unless stated otherwise (section: preheating of process gas).

3 Results

I. Influence of gas composition variation on different feedstock materials

Material:	Cu, 316L, St0.8 (Ø 1.6 mm)
Gas:	CA, N_2, N_2-H_2
Spray system:	TAFA CoArc
Wfr:	Cu (10.5 m/min), 316L (12 m/min), St0.8 (15 m/min)
Voltage:	Cu (38/39 V), 316L (35 V), St0.8 (35 V)

Coatings of the material and gas combinations listed above have been performed and analyzed in terms of oxide content, porosity and deposition efficiency. As the microsections in Fig. 1 show, the substitution of the process gas (N_2-H_2, right) instead of CA shows a significant change in the coating properties: Oxide enriched areas are eliminated, which results in a more homogenous coating structure as well as in a smoother surface roughness. This reduces grinding time and therefore the costs of the final product.

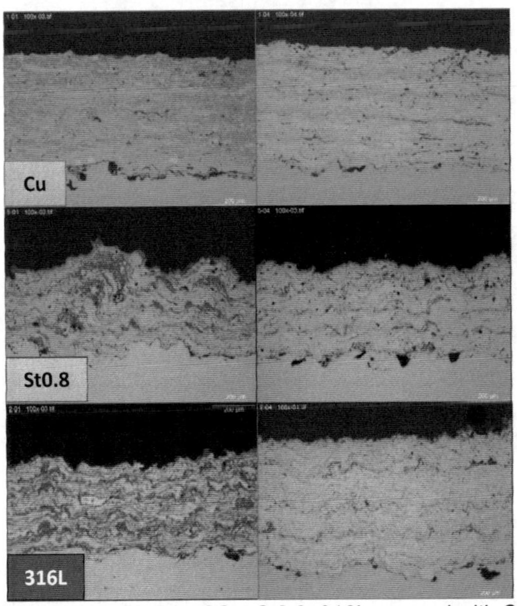

Fig. 1: Micrograph of Cu, St0.8, 316L sprayed with CA (left) und N_2-H_2 (right)

Oxide content

The resistance of a coating against corrosion can be enhanced by a reduction of the oxide content. Providing an inert gas atmosphere in a spray process is a significant tool in minimizing the oxide content for different materials, as displayed in Fig. 2. The spraying distance is another important factor, thus it has been kept constant at 100 mm (for reasons of comparability).

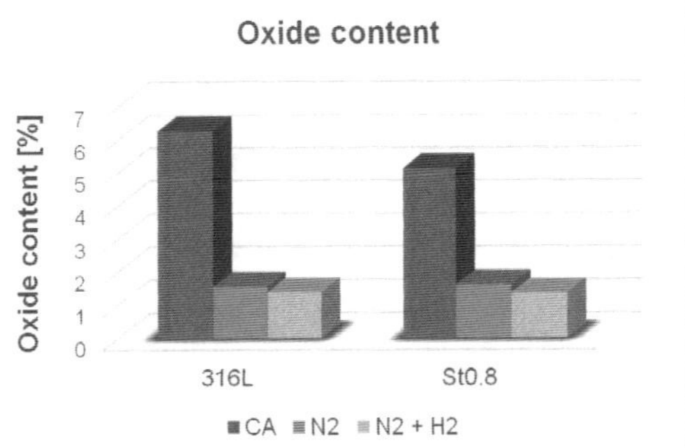

Fig. 2: Oxide content for 316L and St 0.8 for CA, N_2 and N_2-H_2

The oxide content of 6.3 % for 316L when using CA is almost four times higher than for N_2 (1.57 %). By adding a small amount of hydrogen to nitrogen (4% H_2), the oxide content in the resulting coating can be reduced even further. Due to the reaction of the hydrogen molecule with oxygen to H_2O an oxide content of 1.4 % can be reached. Similar results are obtained when St0.8 is used as feedstock, with a huge reduction in the oxide content when CA is substituted by N_2 (and even further for N_2-H_2). No values for copper in terms of the oxide content can be given, yet the coating shows a significant reduction when sprayed with an inert gas.

Hardness

The hardness of coating depends on the oxide content as well as the carbon content. It is commonly known that the substitution of CA with N_2 shows a significant reduction in the oxide content as well as a decrease in carbon burn-off [5,6]. Yet for materials with a low carbon content, such as 316L, the burn-off of carbon is not the key factor regarding the hardness of the coating; the oxide content is the rather decisive. If the carbon content in a base material is relatively high, providing an inert gas atmosphere is important in minimizing the burn-off of the carbon. Another reason for the increase in the hardness is the finer atomization of the molten particles. Smaller particles (with a larger specific surface-to-volume ratio compared to larger ones) tend to oxidize more, especially in standard (non-inert) atmosphere. As displayed in Figure 3, the hardness of the coating depends on the base material. For 316L, a low carbon steel, the substitution of CA with N_2 or N_2-H_2 shows a reduction due to the lower oxide content. However when St0.8 is used, the hardness rises for N_2 (-mixtures) due to the minimized burn-off of carbon. For copper the hardness hardly changes for different process gases and thus does not depend on the oxide or cabon content.

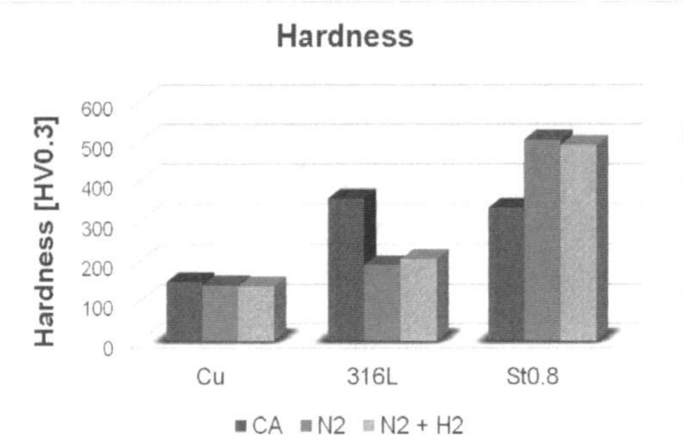

Fig. 3: Hardness of Cu, 316L and St0,8 for CA, N_2 and N_2-H_2

Deposition efficiency

Investigations regarding the deposition efficiency are a crucial factor from an economical point of view. The results, as displayed in Figure 4 show that the substitution of CA with N_2-H_2 does not necessarily improve the deposition efficiency. It can be observed that materials with a low melting point show an increase in deposition efficiency when inert gas (mixtures) are being used. For materials with a higher melting point (TM) hardly any increase can be observed (values for copper, aluminium and G3Si1 are not displayed).

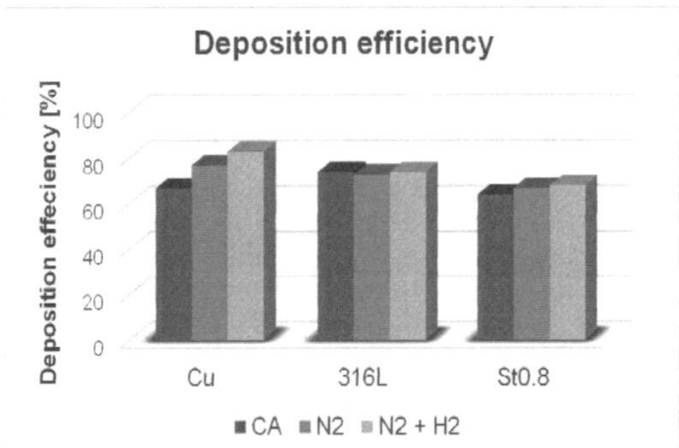

Fig. 4: Deposition efficiency for Cu, 316L and St0.8 for CA, N_2 and N_2-H_2

For materials with a low melting point such as copper (TM = 1,085 °C) and aluminium (TM = 660 °C) an increase in the deposition efficiency of up to 10 – 16 pp (percentage points) when using N_2-H_2 instead of CA can be observed. For G3Si1 and St0.8 (TM = 1,450 °C) a slight increase by 4 pp can be seen for N_2-H_2. When 316L (TM = 1,750 °C), a high melting material, is being used no difference in the resulting deposition efficiency can be observed.

Carbon content

As stated before, the influence of the atomizing gas on the carbon content of a spray coating is generally based on the oxidation of the particles in the atmospheric oxygen. The analysis showed that the carbon content of a coating sprayed with air decreases by 85 % for chrome-nickel steel 316L compared to the feedstock (base material). Fig. 5 illustrates that by

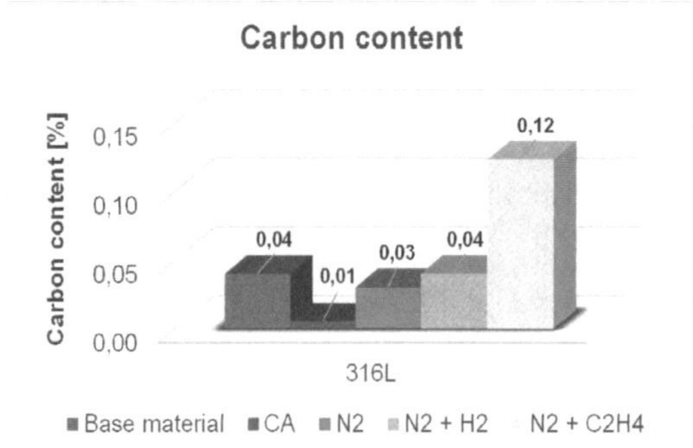

Fig. 5: Carbon content of 316L in the base material vs. when sprayed with different gas compositions

reducing the O_2-content in the vicinity of the spray particles the carbon content can be substantially increased (when using pure nitrogen). The residual content of carbon comes to 75 % of the original feedstock concentration. The addition of hydrogen to the nitrogen in the case of 316L shows that any oxidation of the carbon can practically be prevented. The carbon content of the sprayed coating is now comparable to that of the original feedstock. This can be attributed to the reducing effect of the hydrogen. Hydrogen atoms react with the atmospheric oxygen thus decreasing the number of reaction partners available to the carbon. The addition of the hydrocarbon ethene to the nitrogen increases the carbon content in the spray coating notably. An explanation for the rise in carbon content above the original content of 316L may well be the input of carbon from the fuel gas. The carbon atoms released by dissociation are then bound interstitially into the resulting coating.

II. Preheating of process gases

Material:	G3Si1, 316L (Ø 1.6 mm)
Gas:	N_2 (20 °C and 230 °C)
Spray system:	Experimental spray torch
Wfr:	4 m/min
Voltage:	G3Si1 (29 V), 316L (34 V)

Another approach which has been investigated was to evaluate the influence of spraying gas temperature. The temperature has been preheated to 250 °C by a 17 kW filament heater (resulting in a temperature at the wire tips of 230 °C). Nitrogen was used and coatings have been sprayed, which have then been analyzed in terms of oxide content and porosity. The values for Fig. 6 and 7 have not been obtained according to DIN EN ISO 17836. They have been obtained by an in-house software which is able to analyze images of microsections in order to determine the porosity, oxide content as well as the spray material by means of grey level threshold adjustment. Thus, the values represent the area content (area-%) of the respective pictures. In order to receive reliable data, 12 measurements of different cross-sections have been performed.

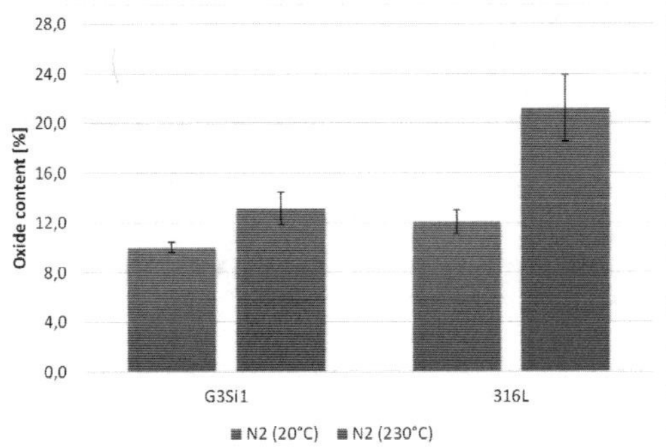

Fig. 6: Measured oxide contents (from micrographs) of cold (blue) and preheated (red) nitrogen gas for G3Si1 and 316L

When taking a look at Figure 6, in which the oxide content has been measured for G3Si1 (left) and 316L (right), a distinct increase for the preheated nitrogen samples can be observed. Moreover, when analyzing the porosity of the coatings (Fig. 7), a decrease can be observed for a higher gas temperature.

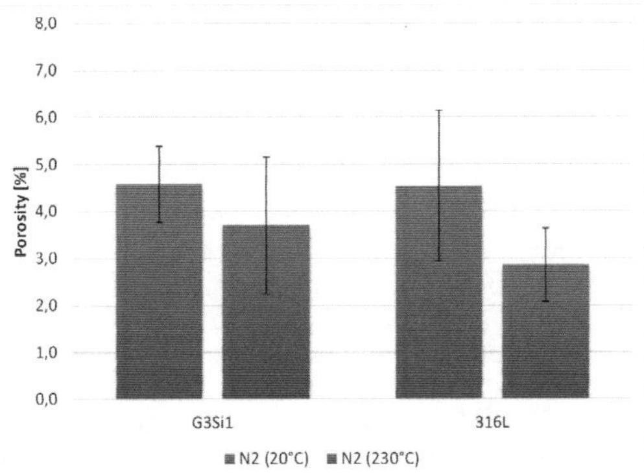

Fig. 7: Measured porosity of cold (blue) and preheated (red) nitrogen gas for G3Si1 and 316L

Besides analyzing the micrograph-images the arc current and voltage traces have also been recorded. With the power source being voltage-controlled a variation in the melting capacity (energy needed to melt a piece of material) is in direct dependence to the current. By preheating the process gas by 210 °C, the energy needed to supply to the wire arc spraying system by the power source is lower.
Increasing the temperature of the process gas does show a change in the resulting coating properties, yet it is very energy-consuming to heat the amount of process gas necessary for wire arc spraying, for which (depending on the applied pressure and the nozzle geometry) the consumption can go up to 100 m³/h (which needs to be kept at a constant high temperature). Keeping the economical perspective in mind no further investigations have been performed with preheated gases. Also the fact that preheating even small amounts of H_2 is a safety-relevant aspect led to that decision.

III. <u>Variation of process gas pressures</u>

Material:	St0.8, AlSi12 (Ø 1.6 mm)
Gas:	CA, N_2
Spray system:	Experimental spray torch
Wfr:	4.7 m/min
Voltage:	St0.8 (36 V), AlSi12 (36 V)

In the following part, the goal was to evaluate the influence of varying process gas pressures (up to 14 bar for CA and 13 bar for N_2) in terms of deposition efficiency, hardness and adhesive tensile strength.

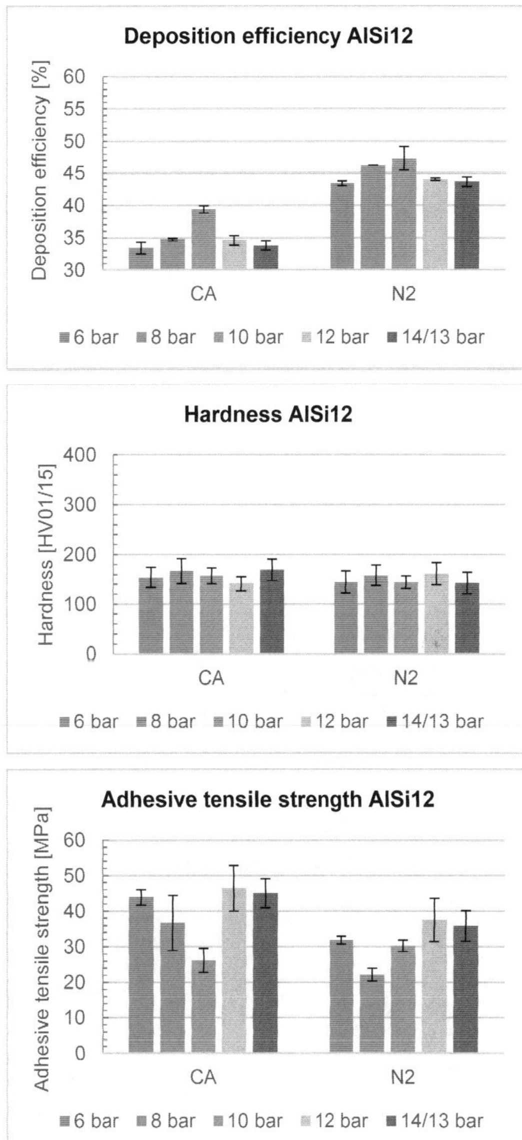

Fig. 8: Deposition efficiency, hardness and adhesive tensile strength of AlSi12 for CA and N_2 at varying gas pressures

The resulting coatings have been sprayed with an in-house wire arc spraying torch by one of the project partners (deLaval gas nozzle, 6 mm diameter in the constriction) for a 1.6 mm wire and pressure of 8 bar. When taking a look at Figure 8 for AlSi12, the highest deposition efficiency can be reached when a pressure of 10 bar is applied for both CA and N_2, with N_2 showing higher values overall. In terms of the hardness no significant differences between the applied gas pressures as well as the gas compositions can be seen. When analyzing the values of the adhesive tensile strength, the highest values have been measured for high pressures. However, a low pressure of 6 bar also shows a reasonable adhesive tensile strength.

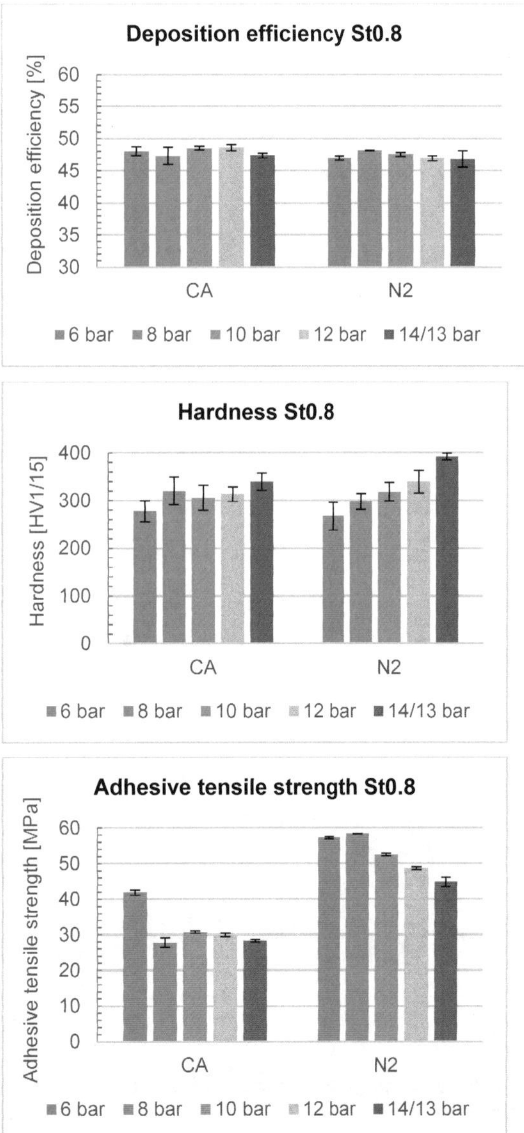

Fig. 9: Deposition efficiency, hardness and adhesive tensile strength of St0.8 for CA and N_2 at different gas pressures

When taking a look at Fig. 9 for St0.8, the values of the deposition efficiency for both process gases range in the same region. Yet when taking a look at the hardness, the highest values can be obtained for a high pressure, especially for N_2. For CA the values equal or greater than 8 bar range in the same region. In terms of the adhesive tensile strength, the highest values for CA and N_2 can be achieved by the lowest pressure.
Summing up, the materials AlSi12 and St0.8 differ significantly from each other for the investigated properties. From an economical point of view, the use of a the process gas as well as the applied pressure depends primarily on the application. AlSi12 shows reasonable values at low pressures in terms of the adhesive tensile strength, with the highest deposition efficiency at 10 bar. By contrast, when longing for a high hardness for St0.8 a higher pressure is better. In terms of the adhesive tensile strength a low pressure is sufficient.

IV. Cold shut analysis and methods for prevention

Besides optimizing coating quality by substituting the process gas another approach is to analyze the cause for cold shuts (unmolten wire pieces in the spray jet) and other instabilities such as wire feed discontinuances. (for AlSi12, Ø 3.2 mm). The contact sleeves are consumables which need to be replaced after a certain amount of time in order to guarantee the quality of a coating. The need for replacement are abrasive wear and so-called micro-arcs eroding the contact sleeves. Micro-arcs can be detected if the voltage drop *dV* between contact sleeve and wire exceeds a certain level. With increasing operating time these micro arcs arise more intensively and may cause a sticking of the wire and hence a cold shut can originate (as explained in detail in [7]). Monitoring the process voltage, current and the voltage drops dV are the basis of the investigation. Measurements are performed by means of a special designed real time measuring equipment featuring the capability to trigger on definable instabilities. In Figure 10 the results are given in a time interval prior to and partially during a cold shut.
In regular operating mode ("random shape" interval) the process voltage and current fluctuate in a random way, caused by the continuous movement of the arc between the electrodes due to the strong flow of the atomizing gas. The voltage drops at the contact sleeves are mostly below 1 V. During the period "similar to periodic shape" (see fig. 10) the current and voltage fluctuations become more similar to a periodic oscillation with a frequency of about 600 Hz. The duration of this interval is mostly longer than 30 ms. At about 0.492 s the "cold shut" time interval starts with a short, indicated by a very low voltage and a current rise up to more than 4000 A. Significant spikes of the dV´s are observed in this interval, which in combination with high current values lead to high electrical power of the micro-arcs and further stimulate the occurrence of the cold shut. The duration of "cold shut" interval is 24 ms (not displayed completely).

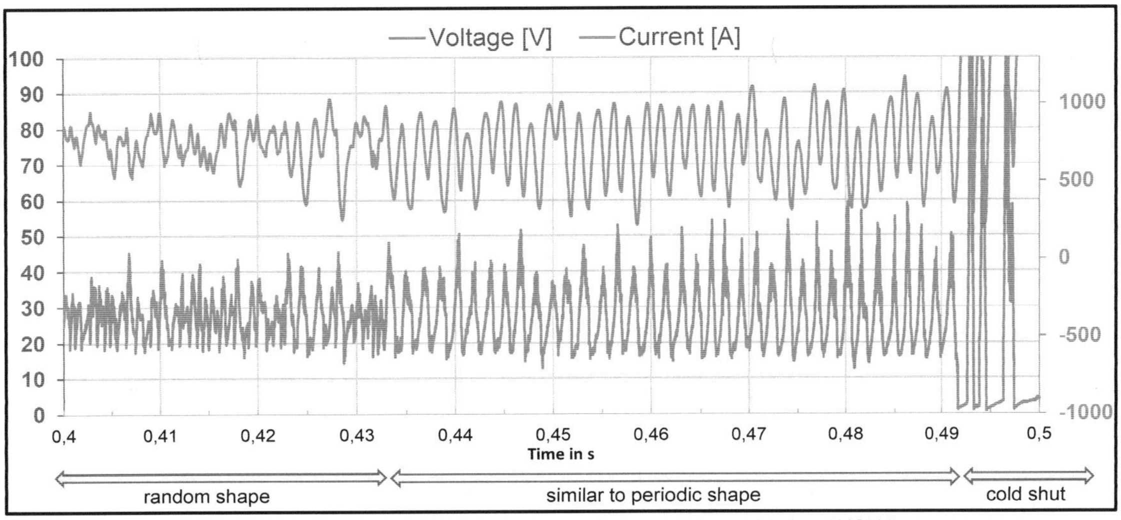

Fig. 10: Process voltage (blue) and current traces (orange) of an occurring cold shut (AlSi12,

In the project two options to avoid cold shuts are considered:

- For fast controllable power supplies it should be possible to detect the "similar to periodic" state in time and modify the process parameters high dynamically to avoid a cold shut before arising. An implementation of such a system by preventing the significant current drop and voltage increase during this period could prevent the expiry of the arc.
- To get an in situ measure of contact sleeve wear, a cost-efficient stand-alone device was developed to sum up (integrate) the total amount of time the voltage drop dV exceeds a selectable level for each contact sleeve. If the values of the integrals for "sleeve replace time" converge, an alert can be established before the value is reached and the risk to produce waste is reduced. The test for convergence is still in progress.

4 Conclusion

With several industrial partners being part of the research team, the main goal of the project was to enhance the competitiveness of the Bavarian industrial partners in the thermal spray environment.
First, the influence of the process gas has been displayed and the coatings have been analyzed in terms of oxide content, porosity and hardness as well as the carbon content. Next, the preheating of the process gas N_2 to 230° C has been compared with coatings sprayed at ambient temperature. Subsequently, different process gas pressures ranging from 6 to 14 bar for both N_2 and CA have been applied for AlSi12 and St0.8. It can be stated that depending on the demanded properties the proper gas pressure has to be chosen. Additionally, a system to detect the cause of cold shuts has been developed.
Different aspects have been investigated in this project, yet some (such as the preheating) do not justify their industrial application.

5 Acknowledgement

The presented results were conducted in the course of the project "Energy efficient production of thermally sprayed coatings through the use of a modified wire arc spraying process" funded by the Bayerische Forschungsstiftung (funding number AZ-1054-12).

6 Literature

[1] Steffens, H.-D.; Babiak, Z.; Wewel, M.: *Recent Developments in Arc Spraying*. IEEE Transactions on Plasma Science, Vol. 18 (1990), pp. 974-979

[2] Pfender, E.: Thermal Plasma Technology: Where Do We Stand and Where Are We Going? Plasma Chemistry and Plasma Processes (1999), 19(1)

[3] Watanabe, T.; Wang, X; Pfender, E.; Heberlein, J.: *Correlations between electrode phenomena and coating properties in wire arc spraying*. Thin Solid Films, Vol. 316 (1998), pp. 169-173

[4] Wang, X.; Heberlein, J.; Pfender, E.; Gerberich, W.: *Effect of nozzle configuration, gas pressure, and gas type on coating properties in wire arc spray*. Journal of Thermal Spray Technology, Vol. 8 (1999), pp. 565-575

[5] Jandin, G.; Liao, H.; Feng, Z. Q.; Coddet, C.: *Correlations between operating conditions, microstructure and mechanical properties of twin wire arc sprayed steel coatings*. Materials Science and Engineering A, Vol. 349 (2003), pp. 298-305

[6] Newberry, A.P.; Grant, P.S.: *Oxidation during electric arc spray forming of steel*. Journal of Materials Processing Technology, Vol. 178 (2006), pp. 259-269

[7] W. Mayr, W.; Steiner, E.; Heilmeier, S.; Blau, L.; Szulc, M.; Atzberger, A.; Hartz-Behrend, K.; Zimmermann, S.; Zierhut, J.; Schein, J.: *Cause study to the occurrence of unmelted wire pieces during wire arc spraying*. Proc. of ITSC (2016), pp. 516 - 526

The effect of Mo substrate oxidation on splat formation

Jun Wang, Chang-Jiu Li
State Key Laboratory for Mechanical Behavior of Materials, School of Materials Science and Engineering, Xi'an Jiaotong University, China / E-mail: licj@mail.xjtu.edu.cn

Disk splats are usually observed when the deposition temperature exceeds the transition temperature, whereas thick oxide layer will reduce the adhesion resulting from high deposition temperature. In present study, single molybdenum splats were sprayed onto polished molybdenum substrates with different preheating processes to clarify the effect of surface oxidation on the splat formation. Three preheating processes included heating the substrate to 350 °C, 550 °C, and cooling the substrate from 550 °C to 350 °C, which were performed in argon atmosphere. The chemistry and compositions of substrate surface was examined by XPS. The cross sections of splats were prepared by focus-ion-beam (FIB), and then characterized by SEM. Nearly disc-shaped splat with small fingers in the periphery was observed on the substrate preheated to 350 °C. Perfect disc-shape splat was deposited at 550 °C. Flower-shaped splat exhibited a central core and discrete periphery detached by some voids on the substrate preheated to 350 °C (cooling down from 550 °C). The results of peeling off splats by carbon tape and morphology of FIB sampled cross-sections indicated that no effective bonding formed in the splat-substrate interface for the substrate ever heated to 550 °C, due to the increasing content of MoO_3 on preheated molybdenum surface.

1 Introduction

Molybdenum and its alloy have characteristics of high melting point, high hardness, low thermal expansion, and good corrosion resistance at high temperature. Coatings of molybdenum and molybdenum alloys are frequently used as targets in automotive parts such as synchronizer, rings, cylinder bores, and piston rings [1]. Molybdenum coatings fabricated by plasma spraying have enhanced resistance to wear and heat [2]. The durability and performances of sprayed coatings are directly related to the typical lamellar microstructure with a limited bonding ratio by the maximum value of 32%.

Recent investigations have shown that the properties of coating are improved by increasing the substrate temperature. The fracture characteristics change gradually from interlamellar to translamellar for molybdenum coatings with increased substrate temperature [3]. Furthermore, Yang et al. [4] proposed that there was a critical bonding temperature for the formation of effective splat interface bonding during plasma spraying of ceramic coatings. Then Yao et al. [5] proposed a model for the bonding mechanism at the melt/solid interface to explain the bonding formation reasonably.

However, the oxide layer formation during the pre-heating process or the following splat deposition is inevitable during thermal spraying of metal coatings. Much attention has been paid to investigate the influence of oxidation on splat formation. Disk splats formed on the substrate with a thin oxide layer, reported by McDonald et al. [6]. Whereas, prolonged heating process of metal substrate surfaces increases splat splashing and reduces adhesion. Syed et al. [7] observed fingered splats formed because of entrapped gases in the porous oxide layer. Our previous results [8] also showed that splat morphology changed from disk shape to finger-like shape with the increase of oxide layer thickness and surface roughness, and confirmed that no effective bonding is formed at the whole interface if the surface roughness exceeds 4 nm at a pre-oxidizing temperature exceeded 800 °C. Thick oxide layer inhibited the effective bonding formation unless surface oxide was deformed or disrupted [9]. Therefore, to deposit Mo coating with good lamellar bonding at a higher deposition temperature, it is necessary to investigate the influence of oxidation on splat formation for molybdenum.

In this paper, the molybdenum particles were sprayed onto the preheated molybdenum substrates by LPPS in an argon atmosphere, and the influences of substrate oxidation on the splat morphology and bonding formation were investigated, to clarify the effect of in-situ oxidation during APS of molybdenum coating surface on the inter-splat bonding evolution.

2 Experimental procedures

Commercial molybdenum powder with a size range from 50 to 60 μm was sprayed onto polished molybdenum substrates to deposit a single splat in an argon atmosphere by low pressure plasma spraying. Substrates were preheated directly from room temperature to the deposition temperature of 350 °C and 550 °C with a heating rate of 100 °C/min. Another specimen cooling down from 550 °C to 350 °C was also compared with the former. All specimens were placed onto a stable copper plate heated to a pre-set deposition temperature by direct resistance heater under the copper plate, and the temperature was measured by a K-type thermocouple fixed into a hole near the surface of the copper plate. For the comparison of splat formation processes with different preheating processes, the same spray conditions were used for splat deposition.

The morphology of splats deposited at various conditions was examined by scanning electron microscope. Surface chemistry was characterized by x-ray photoelectron spectroscopy. For the XPS analyses, a hemispherical spectrometer using the Kα

X-ray line of aluminum was employed. The 3d-doublet peak was fitted with Gaussian peaks corresponding to the known bonding states of molybdenum to investigate the possible molybdenum valence states. Adhesion of splats deposited different substrate conditions was compared in the manner reported by Qu and Gouldstone [10]. Splats were pulled up by carbon tape and the comparison of tapped and untapped regions of splats was used to assess the relative adhesion. To examine the bonding formation at the interface between splat and the substrate by scanning electron microscopy (SEM), the cross sections of splats were prepared by focus ion beam cutting at different regions.

3 Results

3.1 Oxide formation

During the LPPS process, the oxygen partial pressure is about 1 Pa in the argon atmosphere. Regarding the short particle in-flight time, the oxidation of droplet will be neglected. Therefore, the substrate oxidation during the preheating process is dominant. Since the oxidation states of molybdenum presents a wide range from +2 to +6, a variety of oxides exist. The most stable oxides of molybdenum are MoO_2 and MoO_3. MnO_3 is relatively easy to prepare, which is commonly found on the oxidized specimen. The preparation of MoO_2 needs a highly-controlled oxygen partial pressure during the molybdenum oxidation process [11]. Mo_2O_5 is highly unstable, and it has been shown that the insertion of hydrogen, alkali, and alkaline metals between the molybdenum oxide layers of stable MoO_3 allows for the stabilization of pentavalent molybdenum [12].

In general, the thickness and compositions of oxide layer increase with increasing preheating temperature. In this study, polished substrates were preheated to 350 °C and 550 °C in an argon atmosphere. The oxidation state on the pre-oxidized substrates was investigated by X-ray photoelectron spectroscopy (XPS), and compared with that on the un-oxidized substrate.

Fig. 1 shows the XPS spectrum in the Mo 3d binding energy region of substrate surface. The Mo 3d spin-orbit doublet peaks of components on oxidized molybdenum were assigned as: 231.49 eV ($Mo_{5/2}$ 0+), 228.29 eV ($Mo_{3/2}$ 0+), 232.47 eV ($Mo_{5/2}$ 4+), 229.27 eV ($Mo_{3/2}$ 4+), 234.8 eV ($Mo_{5/2}$ 5+), 231.6 eV ($Mo_{3/2}$ 5+), 235.44 eV ($Mo_{5/2}$ 6+) and 232.32 eV ($Mo_{3/2}$ 6+). Curve fitting of XPS spectrum suggested that presence of four spin-orbit doublets corresponding to four species. Fig. 2 presents the percentage concentration of different Mo oxidation states, estimated by integration of the area under the Mo $3_{5/2}$ and Mo $3d_{3/2}$ peaks, respectively. For the polished and un-oxidized substrates, a certain percentage of pure Mo was oxidized into MoO_2 in air. Although the substrates heated in argon atmosphere, the surface was fully coved by molybdenum oxide. With the increase of preheating temperature, a decrease of concentration of Mo(4+) and Mo(5+) was accompanied by an increase in the concentrated of Mo(6+).

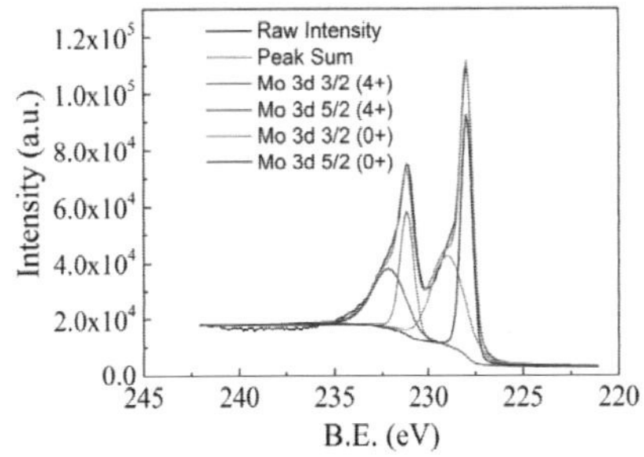

(a) As-polished without preheating

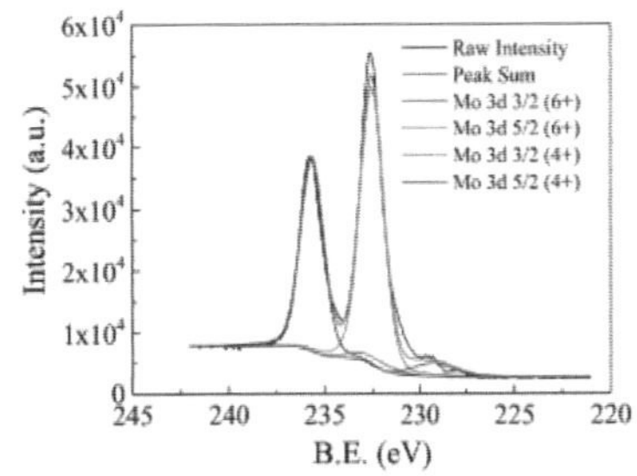

(b) Pre-oxidized at 350 °C

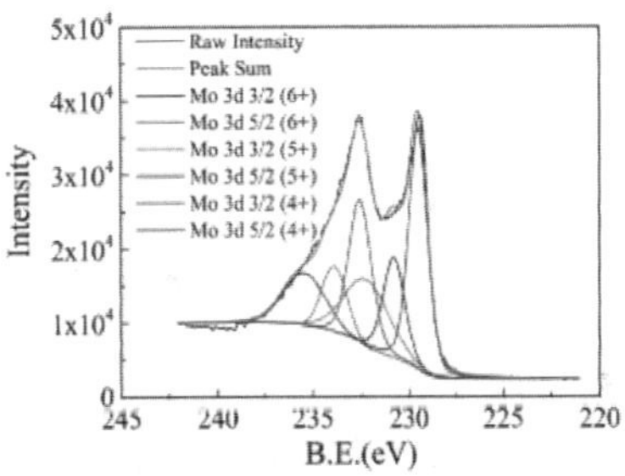

(c) Pre-oxidized at 550 °C

Fig. 1. XPS spectra in the Mo 3d binding energy region of molybdenum substrates experienced different preheating processes.

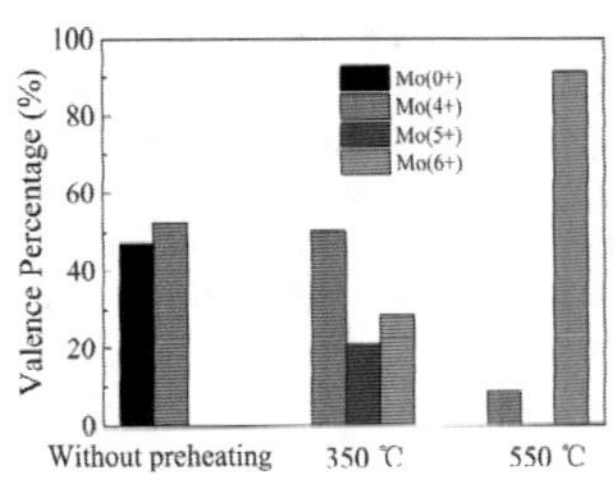

Fig. 2. Percentage concentration of different Mo oxidation states

3.2 Splat morphology

Fig. 3 shows the typical splat morphology at different preheating temperatures. The splat morphology varied with changing the preheating process, presenting nearly disk-shape at 350 °C, perfect disk-shape splat at 550 °C and flower-shaped splat at 350 °C (cooling down from 550 °C). Few fingers were found at the splat periphery for nearly disk-shaped splat. Flower-shaped splats exhibited a central core and discrete periphery detached by some voids.

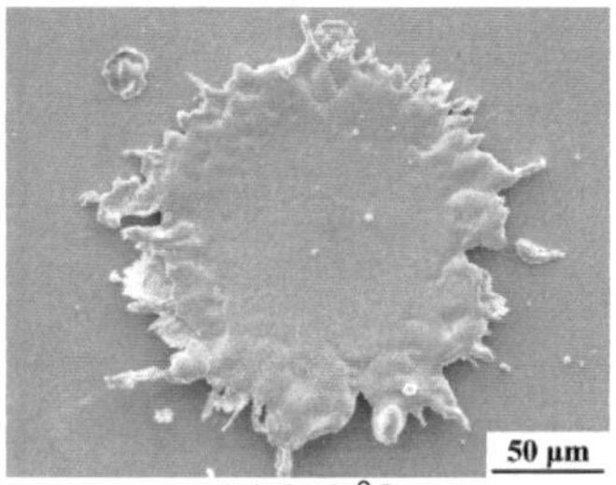

(a) 350 °C

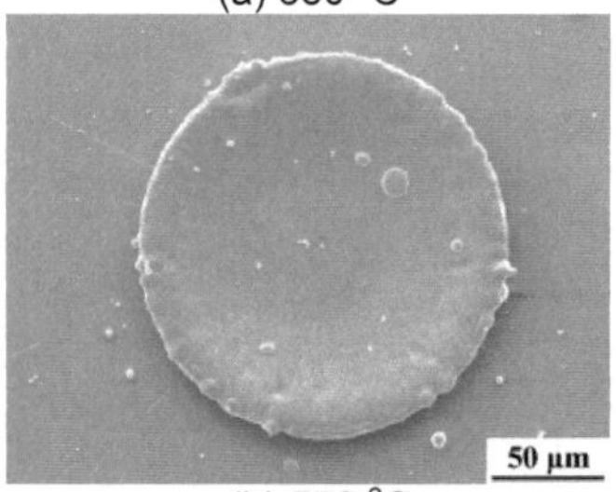

(b) 550 °C

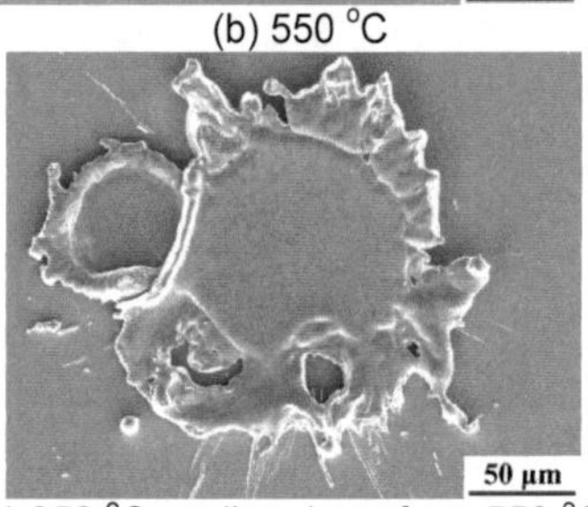

(c) 350 °C cooling down from 550 °C

Fig. 3. Morphology of Mo splats deposited on Mo substrates experienced different preheating processes

3.3 Splat adhesion

Splats were pulled up by carbon tape to assess the relative adhesion. Fig. 3 shows micrographs of splat morphology after pull off test by adhesive tape in comparison of these of the as-deposited ones. From these images it is evident that some splats have been pulled off the substrate due to weak adhesion. Moreover, the three samples showed large difference in pull-off state of splats from the substrate. For the substrate pre-heated to 350 °C very few splats were removed by pull-off test, while for two substrates experiencing pre-heating up to 550 °C nearly all splats have been removed. Results suggest that splat adhesion is much higher on the substrate at a lower preheating temperature.

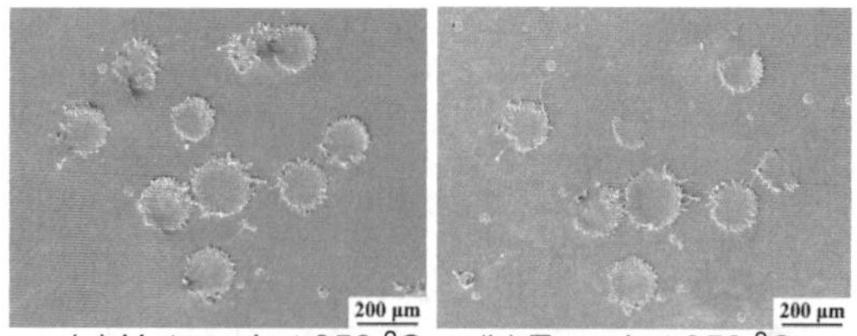

(a) Untaped at 350 °C (b) Taped at 350 °C

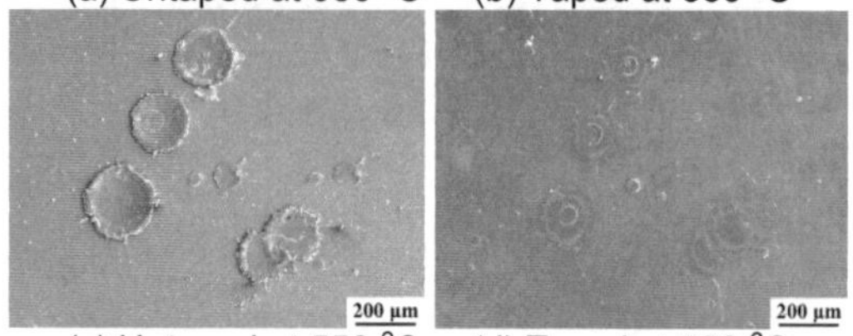

(c) Untaped at 550 °C (d) Taped at 550 °C

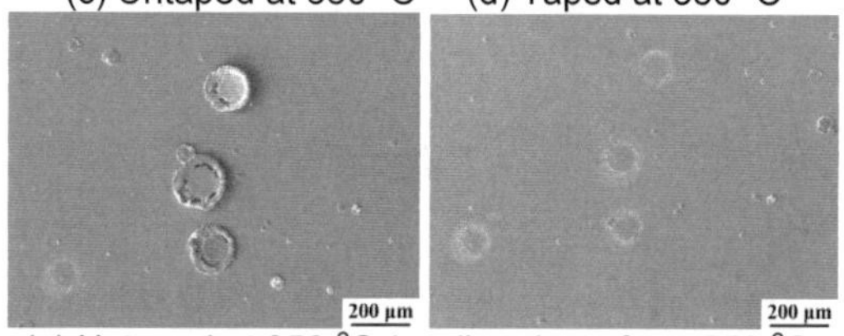

(e) Untaped at 350 °C (cooling down from 550 °C)
(f) Taped at 350 °C (cooling down from 550 °C)

Fig. 4. Micrographs of morphology of Mo splats deposited on the Mo substrates experienced different preheating histories before pull-off test and after test to show splat adhesion

3.4 Splat-substrate interface bonding

To reveal the bonding condition of the interface between the splat and substrate, the cross-sectional morphologies of FIB prepared samples were examined by SEM and shown in Fig. 5 and Fig. 6 for the substrates preheated to 350 °C and 550 °C, respectively. For the splats deposited at 350 °C, effective interface bonding was observed at the interface from the central part of splat marked A to the region marked as B where a fraction of the interface bonded to the substrate as shown in Fig. 5b. The other part of the interface in the region marked as B extended to the periphery of splat as shown in Fig.5c exhibited a distinct boundary between the splat and substrate. On the other hand, a gap was present in the whole splat-substrate interface at a deposition temperature of 550 °C. The results obtained by direct examination of the interface bonding using FIB prepared samples are consistent with that observed for the splat adhesion by adhesives tape pulling off test. All these facts indicate that the interface bonding formation is inhibited if the substrate is ever heated to a temperature of 550 °C.

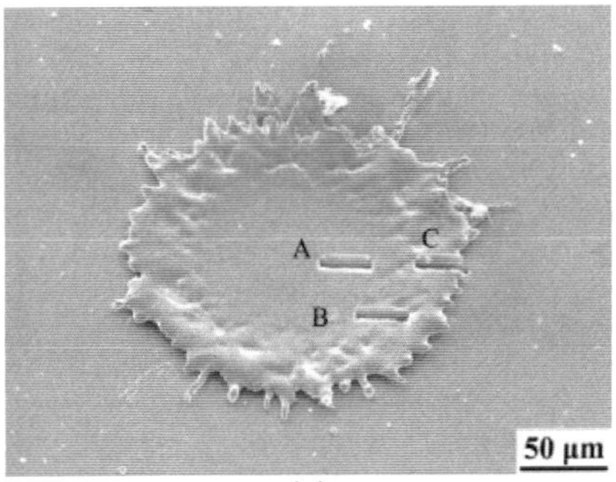

(a)

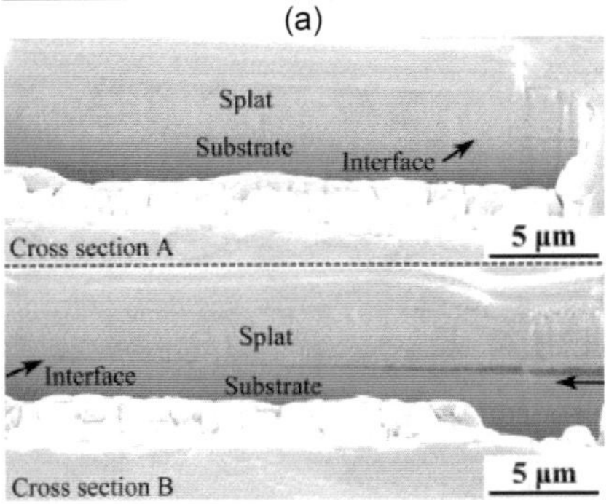

(b)

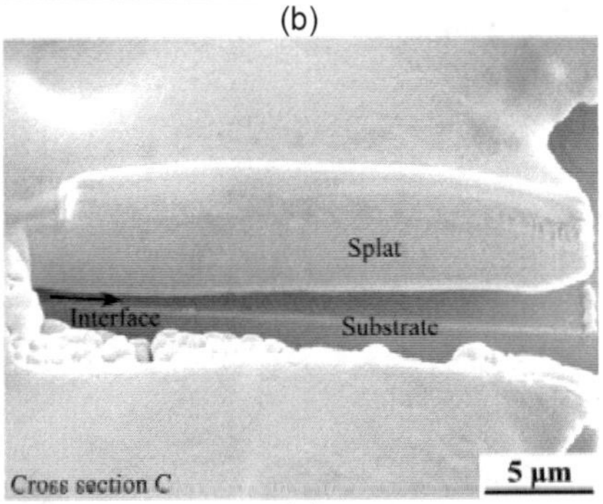

(c)

Fig. 5. FIB sampling positions for typical splat interface at a deposition temperature of 350 °C (a), (b) and (c) presenting the interface at typical regions marked as A, B and C in (a)

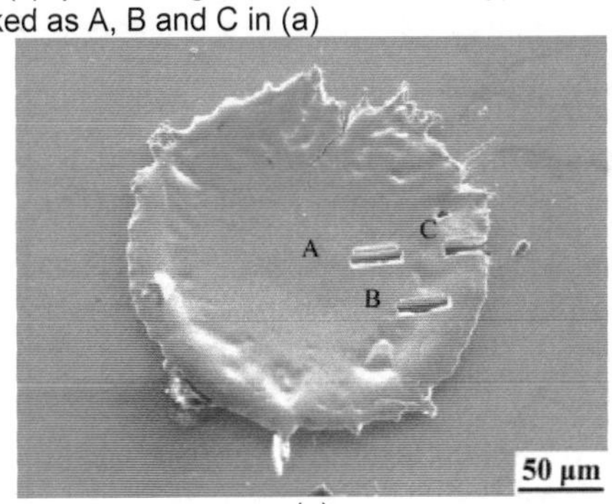

(a)

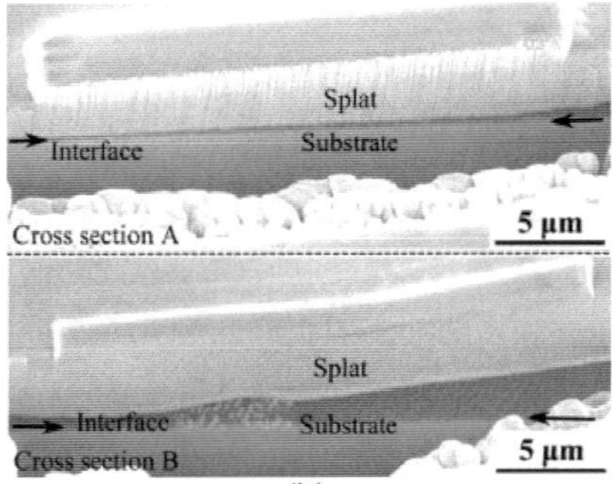

(b)

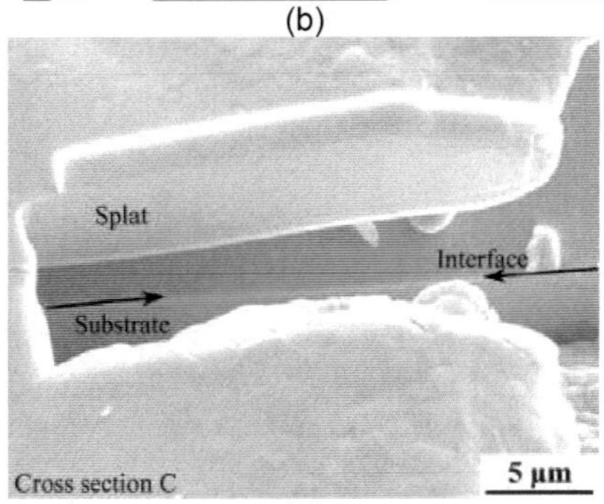

(c)

Fig. 6. FIB sampling positions for typical splat interface at a deposition temperature of 550 °C (a), (b) and (c) presenting the interface at typical regions marked as A, B and C in (a)

4 Discussion

4.1 Effect of substrate oxidation on the interface bonding

The interface bonding formation is related to the interface temperature and contact condition. Recent studies [4,5] showed that increasing the deposition temperature improved the interface bonding.

However, the FIB prepared cross sections showed that the interface bonding formation for Mo splat on Mo substrate is inhibited when the deposition temperature increases from 350 °C to 550 °C. On the other hand, by comparing the splat morphology after pull off test for splat deposited at 350 °C which was cooled down to 350 °C from 550 °C, it is found that the splat adhesion condition deteriorates when the substrate is ever heated to 550 °C. Therefore, it can be considered that the oxidation during the preheating process rather than the deposition temperature has a great influence on interface bonding formation for molybdenum splat.

As discussed in Section 1, the prolonged heating process reduces interface bonding formation because of the altering surface roughness and composition. In this study, the molybdenum was heated in the argon atmosphere with a low temperature and short time; hence a thin layer of oxide forms on the surface. There is no distinct difference for surface morphology observed by SEM. However, a great change in surface chemistry is found on the substrates experienced different preheating processes. The polished pure molybdenum substrate surface is oxidized to form MoO_2, Mo_2O_5 and MoO_3 when it is preheated. With the increase of preheating temperature, the percentage concentration of MoO_3 increases and plays a dominant role on splat formation. The melting point of molybdenum is 2890 K, but the normal boiling point of its trioxide is 1428 K. Despite of the deposition temperature, the MoO_3 volatizes under the impact of droplet with high temperature. Since with the increase of preheating temperature, the Mo surface is covered by more MoO_3. The vitalization of MoO_3 during splatting leads to the formation of a gas layer at the bottom of the splat, which inhibits the direct contact between splat and substrate, and eventually inhibits the interface bonding formation.

4.2 Effect of substrate oxidation on splat morphology

Splat morphology depends on the spreading and solidification process. The splashing phenomenon is commonly observed based on the splat morphology when splats are deposited on a substrate surface in ambient atmosphere. In the last two decades, several hypothesis models [13] have been proposed to explain the transition temperature, across which the splat changes its morphology from a fragmented irregular shape to a regular disk shape.

In this study, it was found that when the deposition temperature for Mo splats exceeds the transition temperature, two types of splashing phenomena are observed at the periphery of the splats deposited on two Mo substrates at 350 °C experiencing two different heating proceeses. One is directly preheated to 350 °C, while the other is cooled down to 350 °C from 550 °C.

The former one shows disk-shaped splat with a few fingers at splat periphery as shown in Fig. 3a. By observing the interface bonding state by FIB prepared samples, it is found that effective bonding forms only in the central zone of splat deposited at 350 °C. The limited bonding surrounding the periphery of splat means poor contact between spread molted Mo and substrate. The poor contact lowers the heat transfer from splat to the substrate and reduces splat cooling rate. The fingering splashing is possibly resulted from the instability of supercooled melt, drived by spreading inertia.

The other one presents a discrete periphery detached by voids as shown in Fig. 3c. As discussed in section 4.1, the volatilization of MoO_3 on the substrate preheated to 550 °C has a great influence on interface bonding formation. Similarly, the volatilization of MoO_3 may affect the splat morphology. The extended heating time for the samples which experienced preheating process to 550 °C and then cooling down to 350 °C leads to the formation of more MoO_3. The rapid heating up of the Mo substrate by spreading Mo droplet causes the volatilization of MoO_3 and forms a gas cushion at the interface between the substrate and spreading molten Mo. Li et al. [14] proposed the evaporated-gas-induced splashing model to explain the splashing phenomena involved gas evolution. By brushing evaporative organic substances on substrates and then removing them by heating, the effect of adsorbates was simulated. Their results showed that regular disk-type splats were formed when the preheating temperature exceeds 50 °C over the boiling point of several different organic substances brushed on substrate surface. The model that included the rapid heating induced evaporation and subsequently evaporation induced upward pressure have reasonably explained the dominant effect of adsorbates on the transition of splat morphology during splatting. Similarly, the poor contact at the interface between Mo substrate and melt due to the gas cushion resulting from the volatilization of MoO_3 lowers the cooling rate of droplet and thus lead to fingering splashing. Moreover, the volatilization of MoO_3 during the splatting process and its concentration may provide an upward pressure to break the thin liquid layer. Therefore, flower-shaped splat was observed on highly oxidized substrate, exhibiting discrete periphery detached by some voids.

5 Conclusion

The effect of molybdenum substrate oxidation on the splat morphology and splat-substrate interface bonding were investigated. Results show that the preheated substrate surface was fully covered by molybdenum oxide, although the substrate was heated in an argon atmosphere. With the increase of preheating temperature, a decrease of concentration of Mo(4+) and Mo(5+) was accompanied by an increase in the concentrated of Mo(6+). The splat on substrate ever heated to 550 °C presents a discrete rims detached by voids. Moreover, by increasing the substrate temperature to 550 °C, the effective interface bonding formation is inhibited. The fact indicates that the volatilization of MoO_3 formed at high temperature evolves as a gas layer during spreading of molten Mo, inhibiting the interface bonding formation and providing upward pressure to break the periphery of splat and inducing splashing.

Acknowledgments

The present project is financially supported by National Nature Science Foundation of China (No. 51474171).

Literature

[1] J. Yan, Z. He, Y. Wang, J. Qiu, and Y. Wang, Microstructure and Wear Resistance of Plasma-Sprayed Molybdenum Coating Reinforced by MoSi2 Particles, Journal of Thermal Spray Technology 25(2016), Issue 7, pp. 1322-1329

[2] J. Ahn, B. Hwang, and S. Lee, Improvement of wear resistance of plasma-sprayed molybdenum

blend coatings, Journal of Thermal Spray Technology 14(2005), Issue 2, pp. 251-257
[3] X. Jiang, J. Matejicek, and S. Sampath, Substrate temperature effects on the splat formation, microstructure development and properties of plasma sprayed coatings: Part II: case study for molybdenum, Materials Science and Engineering: A 272(1999), Issue 1, pp. 189-198
[4] G.-J. Yang, C.-X. Li, S. Hao, Y.-Z. Xing, E.-J. Yang, and C.-J. Li, Critical bonding temperature for the splat bonding formation during plasma spraying of ceramic materials, Surface and Coatings Technology 235(2013), pp. 841-847
[5] S.-W. Yao, C.-J. Li, J.-J. Tian, G.-J. Yang, and C.-X. Li, Conditions and mechanisms for the bonding of a molten ceramic droplet to a substrate after high-speed impact, Acta Materialia 119(2016), pp. 9-25
[6] A. McDonald, C. Moreau, and S. Chandra, Effect of Substrate Oxidation on Spreading of Plasma-sprayed Nickel on Stainless Steel, Surface and Coatings Technology 202(2007), Issue 1, pp. 23-33
[7] A.A. Syed, A. Denoirjean, B. Hannoyer, P. Fauchais, P. Denoirjean, A.A. Khan, and J.C. Labbe, Influence of Substrate Surface Conditions on the Plasma Sprayed Ceramic and Metallic Particles Flattening, Surface and Coatings Technology 200(2005),Issue 7, pp. 2317-2331
[8] J. Wang, C.-J. Li, G.-J. Yang, and C.-X. Li, Effect of Oxidation on the Bonding Formation of Plasma-Sprayed Stainless Steel Splats onto Stainless Steel Substrate, Journal of Thermal Spray Technology 26(2017),Issue 1, pp. 47-59
[9] W. Trompetter, M. Hyland, D. McGrouther, P. Munroe, and A. Markwitz, The Effect of Substrate Surface Oxides on the Bonding of NiCr Alloy Particles HVAF Thermally Sprayed onto Aluminum Substrates, Journal of Thermal Spray Technology 19(2010), Issue 5, pp. 1024-1031
[10] M. Qu, and A. Gouldstone, On the Role of Bubbles in Metallic Splat Nanopores and Adhesion, Journal of Thermal Spray Technology 17(2008), Issue 4, pp. 486-494
[11] R.S. Patil, M.D. Uplane, and P.S. Patil, Electrosynthesis of Electrochromic Molybdenum Oxide Thin Films with Rod-Like Features, International Journal of Electrochemical Science 3(2008), Issue 3, pp. 259-265
[12] N.A. Dhas, and A. Gedanken, Characterization of Sonochemically Prepared Unsupported and Silica-Supported Nanostructured Pentavalent Molybdenum Oxide, The Journal of Physical Chemistry B 101(1997), Issue 46, pp. 9495-9503
[13] M.L. Kun Yang, Kesong Zhou, recent developments in the research of splat formationn process in thermal spraying 2013(2012), pp. 1-14
[14] C.-J. Li, and J.-L. Li, Evaporated-gas-induced Splashing Model for Splat Formation during Plasma Spraying, Surface and Coatings Technology 184(2004), Issue 1, pp. 13-23

Characterization of aluminum based coatings deposited by Very Low Pressure Plasma Spray

X. FAN, M.-P. PLANCHE, G. DARUT, N. Kang, G. MONTAVON
Univ. Bourgogne Franche-Comté, ICB, UTBM, LERMPS, BELFORT / F

The objective of this research is to investigate the changes of the microstructure and mechanical property of aluminum based coatings manufactured by VLPPS along the radial directions of the plasma plume. Aluminum powders were sprayed with a F4-VB low-power plasma gun under a working pressure of 150 Pa. Coatings deposition is studied at different distances from the plasma plume impact. Front of the plasma plume, in-situ reactions between aluminum and substrate elements (such as Fe, Cr, Ni) present in the base metal take places. It mainly forms aluminum based intermetallic Al_3Fe coating according to the XRD. Based on the SEM observation, the packed columnar microstructure mixed with nanometer particles is formed with a majority of pure vapor condensation due to evaporated particles from the plasma jet and/or aluminum coating already made. For different distances relative to the center of plasma plume (i.e. from 10 mm to 110 mm along the radial directions), the deposited coatings exhibit a lamellar binary structure which was formed by the mixed deposition of vapor and molten droplets. The coatings morphologies vary from nearly dense to loose and highly porous. Finally, the hardness of typical coating is investigated. The Al based intermetallic Al_xFe_y coating, on the center of the plasma plume, reached $448HV_{0.025}$, which is much higher than those obtained at other positions.

1 Introduction

Since some years, VLPPS process emerges. It differs from the traditional vacuum thermal spray processes (VPS and LPPS) in that coatings are deposited at pressures typically less than ~500 Pa [1]. The low-pressure atmosphere induces a plasma jet extension larger in both axial and radial directions. It results in considerably improving particle velocities and kinetic impact energies compared to the traditional plasma spray [2]. Under such condition, the deposited material can be formed by very fine molten droplets or a mixture of vapor and droplets, or only vapor phase, according to spraying parameters (plasma power, material, powder size distribution, etc.). In this way, thin dense and defect-free coatings were successfully manufactured [3]. Moreover, VLPPS coatings can be similar in some aspects to coatings produced by alternative coating technologies, such as physical vapor deposition (PVD) or chemical vapor deposition (CVD), excepted to the deposition rates about an order of magnitude higher [1]. So the VLPPS is a very promising manufacturing process which combines the advantages of both traditional plasma spraying technique and vapor deposition process (PVD, CVD) to deposit a wide range of materials (Al, Ti, YSZ, Cu, etc.) [2,4-7]. Mauer *et al.* reports the manufacturing of thin (<50µm) dense Mg-spinel ($MgAl_2O_4$) coatings deposited from droplets state [8]. These coatings have been investigated for different applications such as thermal barrier or, solid oxide fuel cells (SOFC) for example.
This study deals with the manufacturing of VLPPS aluminum coatings to investigate the different morphologies and compositions depending on the radial positions. Aluminum was chosen as a demonstrative material due to its "moderate" vaporization enthalpy (i.e., 38.23 kJ cm^{-3}) compared to copper one (i.e., 55.33 kJ cm^{-3}), cobalt (i.e., 75.03 kJ cm^{-3}), or even tantalum (i.e., 87.18 kJ cm^{-3}) [9]. The injected aluminum powder can be well melt or vaporized under specific spraying conditions, in such a way those molten particles and vapors of Al can be widely distributed throughout the very-low pressure plasma chamber during spraying. As a result, the coating is formed through the solidification of molten droplets, condensation of the vapor phase or a mixed mode (including vapor phase and molten droplets). The objective of this work is first to better understand the characterization and property of the coating along the radial directions of the plasma plume and further research the distribution of vapors compared to molten droplets in the chamber to better control the factors affecting the coating microstructure and property.

2 Experimental protocols

Atomized 1050 grade aluminum powder manufactured at IRTES-LERMPS laboratory was used. Its size distribution is ranged from 8 to 22 µm with an average size diameter (d_{50}) of 13 µm. Stainless steel substrates were employed (diameter 25 x 10mm). In order to obtain a clean surface and improve the adhesion of the coating, the substrates were degreased in ethanol and grit-blasted prior spraying. They were mounted on a bar side by side and spaced by 30 mm.
A vacuum plasma-spray system was employed to deposit coating in this work. A F4-VB torch (50 kW, Oerlikon–Metco) operated at a very low pressure of 150 Pa was used. The torch is mounted on a 6-axis robot (ABB, IRB 1400) located in-house composed of a 12 m^3 of vacuum chamber. The detailed VLPPS operating parameters are listed in **Table 1** and a drawing of the experimental configuration is displayed in the **Fig. 1**. Firstly the vacuum chamber was pumped to the working pressure of 150 mbar, then the substrate number (b) (front of the torch), monitored with a thermocouple, is preheated to about 600°C with the plasma jet. No cooling device is used. The chamber pressure was kept constant for the whole set

Table 1. Plasma spray parameters

parameters	values
pressure (Pa)	150
plasma torch	F4-VB
plasma forming gas ($L.min^{-1}$)	Ar/H_2 (45/10)
arc current intensity (A)	650
voltage (V)	50
carrier gas flow rate ($L.min^{-1}$)	2,5
spraying distance (mm)	900
mass enthalpy ($MJkg^{-1}$)	15.4
powder mass rate ($g.min^{-1}$)	1

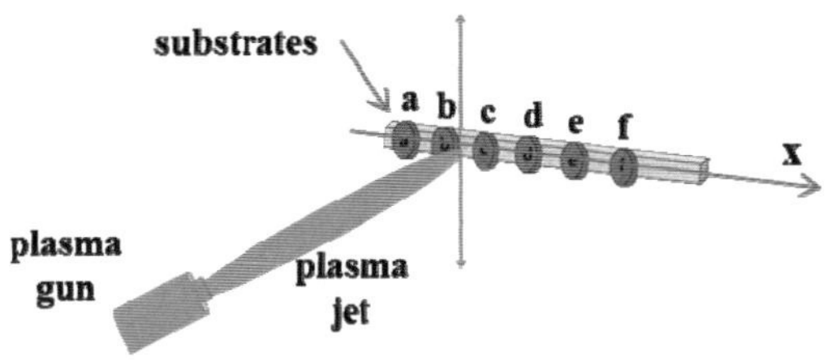

sample	a	b	c	d	e	f
x [mm]	-40	-10	+20	+50	+ 80	+ 110

Fig. 1. VLPPS experimental configuration

of experiments. The kinematic of the torch corresponded to a vertical displacement front of the substrates placed on a sample holder. The impact localization of the plasma has been accurately detected using a laser in order to be front of the sample number b. During the spraying, the real impact has been observed. It appears that the position of the plume passage is between b and c but just near sample b (**Fig. 1**). The diameter of the plume impacting on substrates is estimated to be around 30 mm.

The surface of the coating and chemical etching cross-section were observed by SEM in secondary electron mode (JEOL 7800F FESEM). An X-ray diffractometer (D8 Advance, BRUCKER AXS) equipped with a CoKα anti-cathode (λ_{Co} = 0.179 nm) was employed to determine the phase composition of the coatings (scanning step: 0.02°). The coatings composition and distribution of the elemental chemical species were performed by EDS (BRUCKER SDDX-Flash 6130). The microhardness was measured on the coating cross-section using a digital Vickers device (Leitz, Germany). The load of 25 g and the duration of 15 s were selected and 10 indentations were impressed at random.

3 Results and discussion

3.1 Cross section microstructure

The cross sections of the coatings are presented in **Fig. 2**. First, it appears that the thicknesses of the coatings are different according to their position. Of course, the closer is the coating to the plasma plume impact, the higher is the thickness excepted for the samples (b) and (c) which are front of the jet (b is nearest). In this specific position correlated to the highest thermal flux at the surface of the sample, the temperature is above aluminum melting one (temperature measured in the substrate and not on surface). So the coating is likely always in melting state which prevents the growth. This is confirmed by the different microstructures observed: granular compared to the lamellar of the other coatings. It is founded that vapor phase deposition

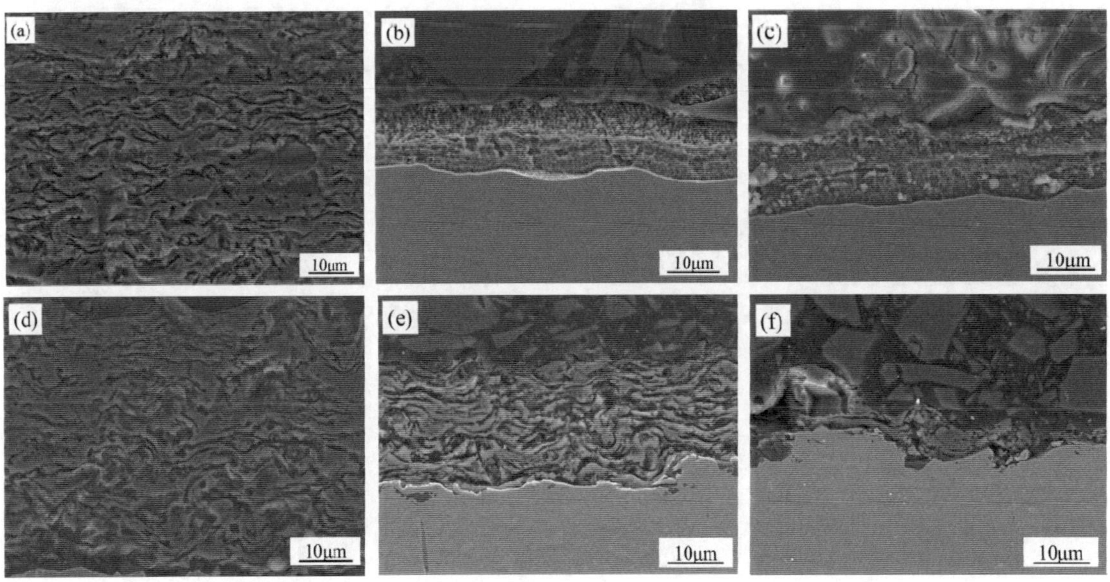

Fig. 2. Cross-section morphologies of the coating deposited along the radial directions of plasma plume center (a) -40mm, (b) -10mm (c) -20mm (d) +50mm (e) +80mm (f) +110mm

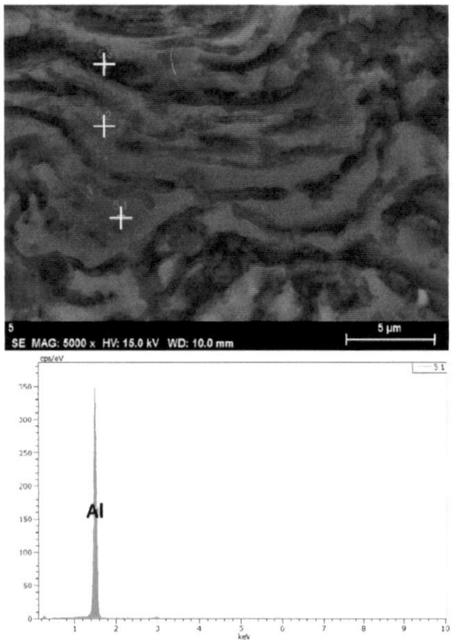

Fig. 3. EDS analysis of coating microstructure out of plasma plume impact

contributes to a column-like grain structure (b), included ball-like particulate in cross-section of the coating (c). So a gas-liquid mixed deposition was consisted by large amount of melted particles and limited vapor phase condenses into the (a), (d),(e), (f) cross-section coatings, even for sample (c) which is not so far from the plume too. In addition, it can been seen the continuous stacking microstructure of splats was formed by melted particles, and limited vapor condensation situated between the splats. **Fig. 3** shows that all the contrast observed are only composed of Al element. Thus, the coatings exhibit a dual microstructure. Besides, the coatings morphologies vary from nearly dense to loose and highly porous due to the coatings exist different vapor condensation fraction and molten droplets condensation fraction. It is obvious that higher vapor condensation fraction was deposited in the coating with the closer distance to the plasma plume from **Fig. 2**.

However, a coating is get far away from the real impact of the jet at 110 mm. So it can supposed that this coating is formed by the ejected particles from the plume before the impact or by those having followed the gas flow after the impact at the surface of substrates. In fact, during the impact of the plume onto the substrates, the flow carries on perpendicularly to the jet and so tangent to the substrate surface. Some particles can follow this latter and hold on asperities.

3.2 Surface microstructure

The surface microstructure of the coatings is presented in **Fig. 4**. For the sample placed next to the plasma plume impact (b), the surface has plenty of nanometer needle-like grain particles and also a few ultra-fine ball-like shape ones formed by the vapor deposition. Front of the plasma plume, the particles with their high-speed impact (molten state) or condensate (vapor state) onto a high temperature substrate. The molten drops nucleate quickly and these nucleates do not have time to grow during the process. Therefore, many refined particles are

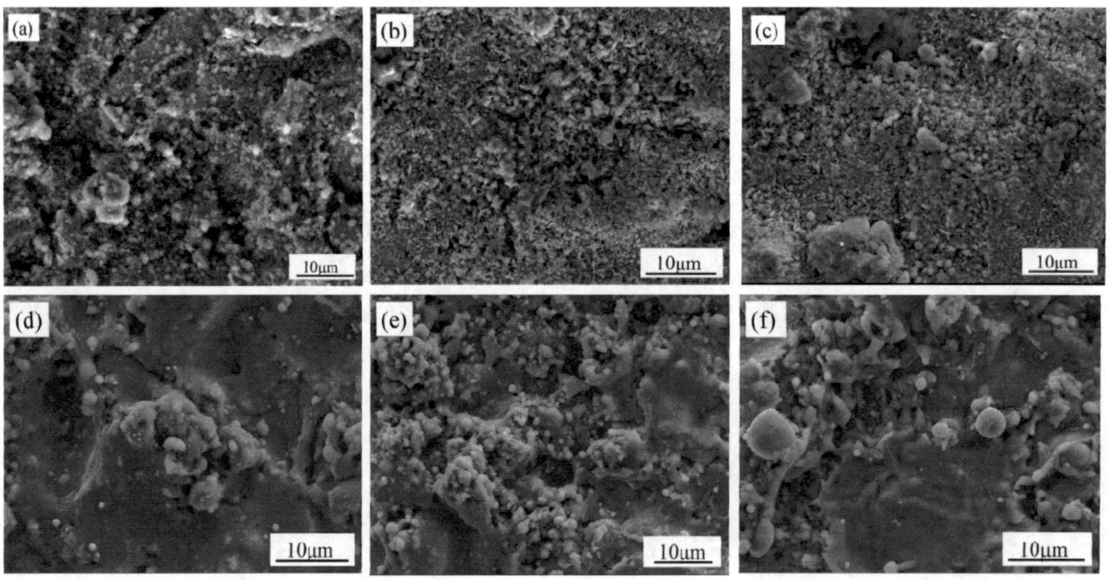

Fig. 4. SEM morphology of the coating deposited along the radial directions of plasma plume centre (a) -40mm, (b) -10mm (c)+20mm, (d)+50mm, (e)+80mm, (f)+110mm

distributed on the surface of the coating. This confirms observations done in the previous part in the **Fig. 2** (b),(c).
If the distance increases compared to impact point (sample (b)), there are higher ball-like shape particles formed by the deposition of vapor but no more needle like grain. In this way, the increased distance modifies the microstructure likely because of the lower surface temperature. For the samples placed outside of the plume impact, the microstructure is significantly different from (b) and (c). The deposited coating is composed of a large number of splats, which resulted from the solidification of molten droplets. This is in correlation with cross sections observed before. In addition, some ball-like shape particles are observed as for sample (c) and some bigger ball-like particles are observed as for example (a). This confirms the cluster deposition or condensation of particles ejected from the plume before the impact.

3.3 Coatings compositions

Elements distribution around the coating/substrate interface has been analysed in order to determine if diffusion phenomenon occurs (**Fig. 5**). For samples (b) and (c), it appears that Fe element is found in all the coating with a mass percentage of 25% wt. (top) to 35% wt. (interface), whereas in the substrate the value is nearly 70% wt. For Al element, the content decreases from 65% wt. (top) to 50%wt. (near interface). So a diffusion takes place all the time during the coating manufacturing inducing an intermetallic coating Al_xFe_y.
For the samples (a) and (d), placed near the plasma plume impact, there is also the presence of the intermetallic coating. The first difference compared to the previous samples (b) and (c) is the smallest thickness of the diffusion, around 2 µm because of the lower thermal flux at the surface. The second main difference is the composition of coating. The percentage of Al reach nearly 100% wt. whereas for (b) or (c), the value is 65% wt. in maximum.

The XRD patterns of the coating are displayed in **Fig. 6**. The coatings placed out of the center of the plasma plume are mainly composed of Al phase. The intensity of Al peak gradually decreases with increasing the distance from the center of the plasma plume. (as shown in the **Fig. 2** (a)(d)(e)(f).The predominant phase of the coating placed front of plasma plume is an intermetallic Al_xFe_y phase (identified as $Al_{13}Fe_4$), and also Al_3Ni_2 and Al_8Cr_5 ones. The peak intensity of the Al phase is very weak (as shown in the **Fig. 4** (b)(c)). This corroborates the EDS element analysis where Fe element is found in high amount in the coating.
It exists obviously differences in coatings phases formed front of the plasma plume and around. During the spray, the temperature of substrates and resulting coatings manufacturing is kept around 600°C (sample (b) front of the torch) without cooling device. But the temperature can reach locally higher temperatures above the melting temperature of aluminum material. So, aluminum in liquid state reacts with the substrate (Fe, Ni, Cr phases) to form the intermetallic phases. The higher temperature of plasma jet in its center is sufficient to cause adequate melting of a substrate surface layer. The in-situ reaction between the Al powder and some elements (such as Fe, Cr, Ni) present in the base metal, leads to the mainly

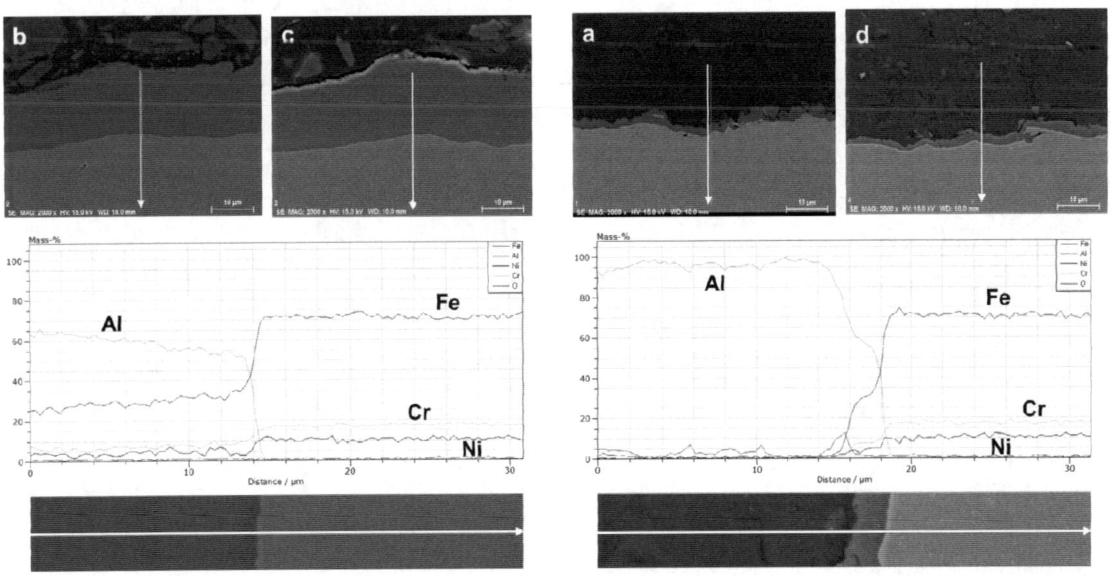

Fig. 5. EDS analysis of elements distribution around coatings/substrate interfaces at different spraying distances to the plasma plume center (a) -40mm, (b) -10mm (c) +20mm, (d) +50mm.

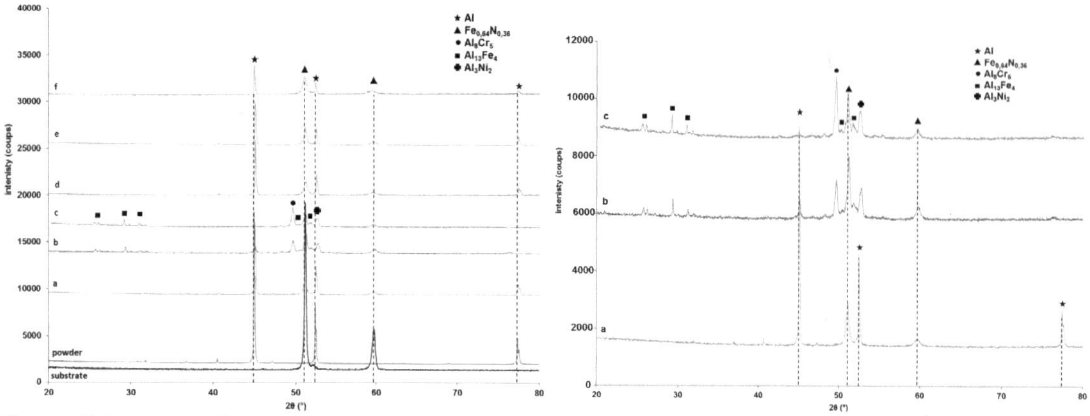

Fig. 6. XRD patterns of substrate, powder and as-sprayed coating at different spraying distances to the plasma plume center (a) -40mm, (b) -10mm (c) +20mm, (d) +50mm, (e) +80mm, (f) +110mm.

intermetallic $Al_{13}Fe_4$ phase. As the temperature is always high, the growth in thickness is difficult due to the aluminum vapor phase in plasma jet that can't condensate at the surface of the substrate.
Because the substrate temperature is strongly dependent of the plasma jet positions, different phases can be formed at different position. For samples (a) and (d), the intermetallic phase disappears because of the lower thermal flux from by plasma jet impact. Only Al element is found. If the distance increases radially compared to plasma center, the Al peak is reduced and substrate one is higher as the thickness is thinner.

3.3 Coatings hardness

Coatings hardness was investigated. It reveals that the hardness of the intermetallic coatings manufactured front of the plasma plume reached $448HV_{0.025}$, that is nearly 8 times higher than the hardness of others coatings outside of the plasma plume ($435HV_{0.025}$),even is more than 2 times higher than the hardness of the Al_3Fe coating by conventionally casting [10]. This is due to the presence of the substrate elements (Fe). At last, the mainly intermetallic $Al_{13}Fe_4$ phase formed and evenly distributed in the coating, playing a role in the dispersion of the hardness of the coating.

4 Conclusion

Aluminum coatings manufacturing by VLPPS thermal spray technology have been studied and more particularly the influence of the substrate position compared to plasma plume impact. Different radial distances have been observed. The uncooled coating front of the plasma plume induces, according to the robot kinematic chosen (substrate temperature front of the torch close to aluminum melting one), the apparition of intermetallic phase $Al_{13}Fe_4$, and also Fe-Ni and Fe-Cr ones. Aluminum melting state reacts with the substrate leading to Fe element diffusion in Al matrix, which increases highly the hardness. However, this temperature prevents the growth of the coating because of the evaporation of aluminum hence the difficulty to condensate properly the vapor of aluminum. All these phenomena induce a typical microstructure (needle shape) of the coating with packed columnar structure mixed with nanometer particles.
For the substrate positions more distant from the plasma plume impact (40mm, 50mm, 80mm, 110mm), the deposited aluminum coatings exhibited a lamellar binary structure formed by melted particles and the vapor condensation layers originating from partially evaporated particles. The coatings vary from nearly dense to loose and highly porous according to the amount of condensated vapor compared to melted phase of aluminum.
In addition, the hardness of the aluminum based inter metallic coating on the center of the plasma plume reach $435HV_{0.025}$ whereas the value is around 48 $HV_{0.025}$ for the others.

References

[1]. K. von Niessen and M. Gindrat, Plasma Spray-PVD: A New Thermal Spray Process to Deposit Out of the Vapor Phase, J. Therm. Spray Technol. (2011), 20(4), pp. 736-743

[2] A. Refke, M. Gindrat, K. von Niessen, R. Damani, LPPS Thin Film: A Hybrid Coating Technology between Thermal Spray and PVD for Functional Thin Coatings and Large Area Applications, Proceedings of the Inter. Therm. Spray Conf., Thermal Spray 2007: Global Coating Solutions, Beijing, (2007), p. 705/10

[3] M. F. Smith, A. C. Hall, J. D. Fleetwood., Very Low Pressure Plasma Spray—A Review of an Emerging Technology in the Thermal Spray Community, Coatings (2011), 1(2), pp.117/32.

[4] A. Hospach, G. Mauer, R. Vaßen, and D. Stöver, Columnar-Structured Thermal Barrier Coatings (TBCs) by Thin Film Low-Pressure Plasma Spraying (LPPS-TF), J. Therm. Spray Technol. (2011), 20(1-2), pp. 116/20

[5]. K. Niessen, M. Gindrat, A. Refke, Vapor phase deposition using plasma spray-PVD™. J Therm Spray Technol.(2010), 19(1–2), pp. 502/9.

[6]. R. Vassen, A. Stuke, D. Stöver, Recent developments in the field of thermal barrier coatings, J. Therm. Spray Technol. (2009), 18, pp. 181/86
[7] B. Vautherin, M.-P. Planche, G. Montavon, F. Lapostolle, A. Quet, L. Bianchi, Study of metallic powder behavior in very low pressure plasma spraying (VLPPS) — Application to the manufacturing of titanium–aluminum coatings, Surf. Coat. Technol. (2015), 275, pp.341/48
[8]. G. Mauer, R. Vaßen, D. Stöver, Thin and Dense Ceramic Coatings by Plasma Spraying at Very Low Pressure, J. Therm. Spray Technol. (2010), 19 (1-2), pp. 495-501
[9] B. Vautherin, M.-P. Planche, R. Bolot, A. Quet, L. Bianchi, G. Montavon, Vapors and Droplets Mixture Deposition of Metallic Coatings by Very Low Pressure Plasma Spraying, J. Therm. Spray Technol. (2013), 23(4), pp. 596-608
[10] E. Karaköse and M. Keskin. Structure and mechanical properties of Al-3Fe rapidly solidified alloy, AMPT Conference Proceedings (2010), 1315(1), pp. 645

Detonation spraying of refractory metals

V. Ulianitsky, I. Batraev, A. Shtertser, D. Dudina / Lavrentyev Institute of Hydrodynamics of SB RAS, Russia
N. Bulina / Institute of Solid State Chemistry and Mechanochemistry SB RAS, Russia
I. Smurov / Université de Lyon, Ecole Nationale d'Ingénieurs de Saint-Etienne (ENISE), Saint-Etienne Cedex 2, France

Due to high melting temperatures and excellent corrosion resistance of refractory metals, they are used for manufacturing parts working under extreme conditions. The formation of refractory metal coatings by thermal spraying is associated with two major challenges: 1) particles of materials having high melting temperatures should be heated to reach a semi-molten or a molten state; 2) oxidation of the metals should be prevented. In this work, the CCDS*2000* detonation spray system was used for obtaining molybdenum and tantalum coatings. The coatings were deposited on steel substrates at O_2/C_2H_2=1.1 and stand-off distances of 20 mm and 100 mm. The calculation of the particle temperatures and velocities were carried out to find the optimal spraying modes for Mo and Ta powders. No oxide phases were found in the coatings obtained by spraying of the Mo powder. In the Ta-based coatings, Ta_2O_5 was found as a second phase. The hardness of the Mo coatings sprayed at 20 mm and 100 mm was 500 HV_{300} and 625 HV_{300}, respectively. The porosity of the Mo coatings was less than 0.5% for both stand-off distances. The hardness of the Ta-based coatings sprayed at 20 mm and 100 mm was 800 HV_{300} and 1000 HV_{300}, respectively. The porosity of these coatings was less than 1% for both stand-off distances. The bond strength of the Mo coatings determined by the pin test method was 92 and 126 MPa and that of the Ta-based coatings was 43 and 77 MPa, for coatings deposited at 20 and 100 mm, respectively.

1 Introduction

Detonation spraying is a promising method of the coating deposition and can be used for a variety of materials [1-8]. Coatings made of refractory metals can be used as protective and corrosion-resistant coatings on parts working under extreme conditions (high temperatures, aggressive media). Coatings present an alternative to full pieces made of these metals and, therefore, can be used to reduce the weight of the parts and their cost.

Thermal spraying of refractory metals is associated with two major challenges: 1) particles of materials having high melting temperatures must be heated to reach a semi-molten or a molten state; 2) oxidation of the metals should be prevented.

Experiments on the plasma spraying of refractory metals have been reported [9-12]. In order to avoid oxidation, spaying had to be conducted in vacuum. In detonation spraying, mixtures with low oxygen contents can be used to create spraying conditions unfavorable for the oxidation of metals. The literature shows no reports on the detonation spraying of molybdenum and tantalum powders. Therefore, the goal of the present work was to evaluate the detonation spraying as a method for obtaining molybdenum and tantalum coatings. The microstructure and phase composition of the coatings were investigated and the porosity, bond strength and hardness of the coatings were measured.

2 Experimental

Experiments on the detonation spraying were carried out using a Computer-Controlled Detonation Spray (CCDS*2000*) system developed in LIH SB RAS [13]. A barrel 1000 mm long having a diameter of 20 mm was used. Spraying was conducted at O_2/C_2H_2=1.1 and an explosive charge (volume fraction of the barrel filled with the O_2+C_2H_2 mixture) of 60%. Nitrogen was used as a carrier gas. The stand-off distances of 20 and 100 mm were used.

The morphology of the powders and the surface morphology and microstructure of the coatings were studied by Scanning Electron Microscopy (SEM) using a MERLIN Compact Microscope (Carl Zeiss, Germany). The SEM images of the powders used in the experiments are shown in Fig. 1.

The average size of the Mo and Ta powders was determined using a LS 13 320 laser diffraction particle size analyzer (Beckman Coulter, USA). The particle size of the Mo powder was measured to be 43 µm and that of the Ta powder was measured to be 47 µm.

The porosity of the coatings was measured on their polished cross-sections using an OLYMPUS Stream Image Analysis Software "Stream Essentials 1.9.1". The optical images of the samples were obtained using an OLYMPUS GX-51 Optical Microscope (Japan).

X-ray diffraction (XRD) patterns of the powders and detonation coatings were recorded by a D8 ADVANCE diffractometer (Bruker AXS, Germany).

The bond strength of the coatings was determined using the pin test method using a conical pin [14]. The reported values are the average of 5 measurements.

Vickers hardness of the coatings was measured on the cross-sections of the samples using a DuraScan 50 hardness tester (EMCO-TEST, Austria) with a load of 300 g. The average value of the hardness was determined from 20 measurements.

3 Results and Discussion

Before conducting the spraying experiments, the particle temperature and velocity calculations were carried out to find the optimal spraying modes for Mo and Ta powders. The calculations were performed using the "LIH" code described in ref. [15].

a

b

c

d

Fig. 1. Morphology of the Mo (a-b) and Ta (c-d) powders.

The calculations showed that at an explosive charge of 60%, the Mo particles with a size ranging from 20 to 45 µm are heated up to the melting point of the metal (T_m(Mo)=2890 K), while their velocities are in the 400–600 m/s range. Mo particles with a size ranging from 5 to 20 µm are heated up to 3000–3300 K and have velocities in the 660–880 m/s range.

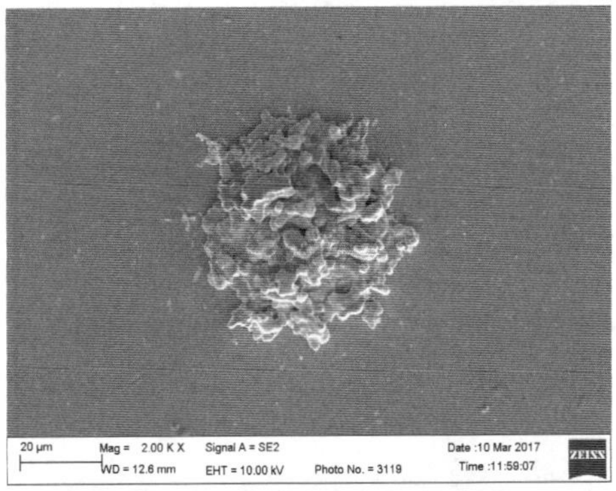

a

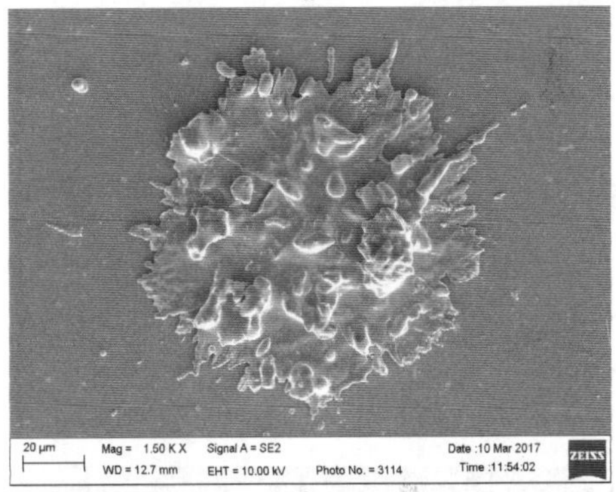

b

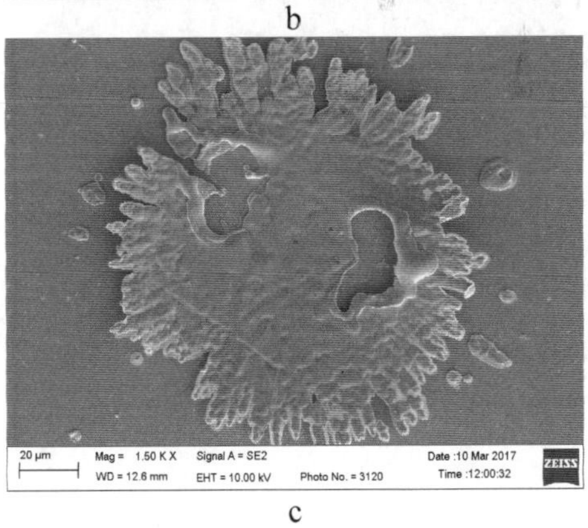

c

Fig. 2. Mo splats obtained at a stand-off distance of 100 mm and an explosive charge of 60%: (a) solid state, (b) semi-molten state, (c) molten state.

At an explosive charge of 60%, the Ta particles with a size ranging from 5 to 30 µm are heated up to a temperature of 2800–3350 K and accelerated up to velocities of 400–770 m/s (T_m(Ta)=3290K). The Ta particles with a size from 30 to 60 µm are heated up to 2100–2800 K and accelerated up to 250–370 m/s.

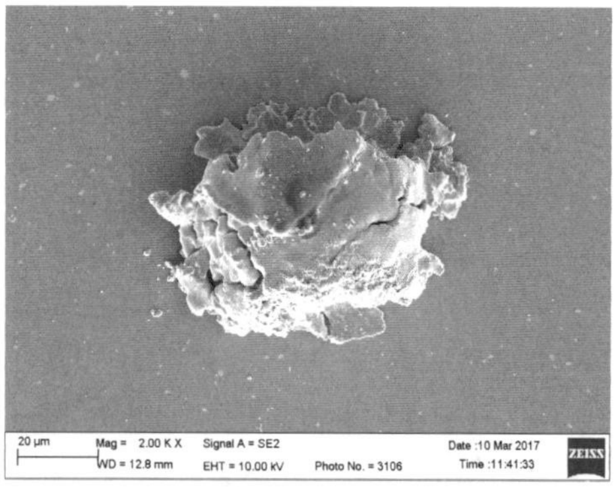

a

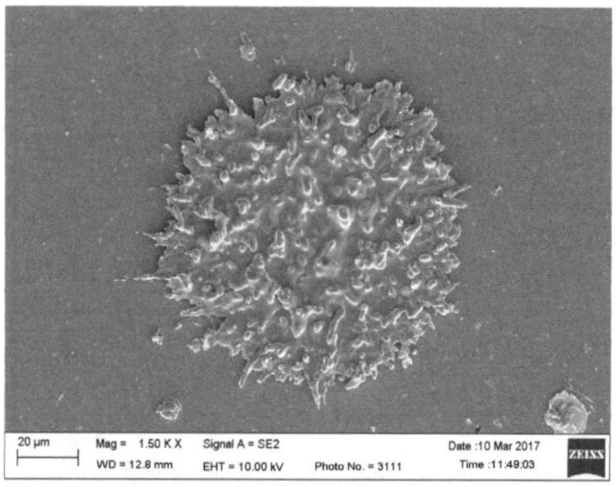

b

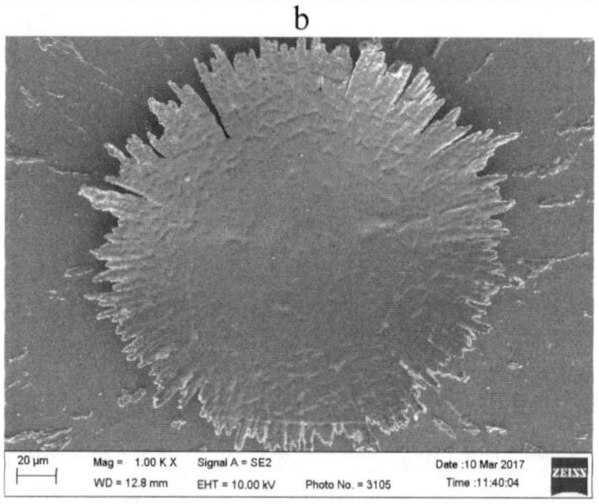

c

Fig. 3. Ta splats obtained at a stand-off distance of 100 mm and an explosive charge of 60%: (a) solid state, (b) semi-molten state, (c) molten state.

a

b

c

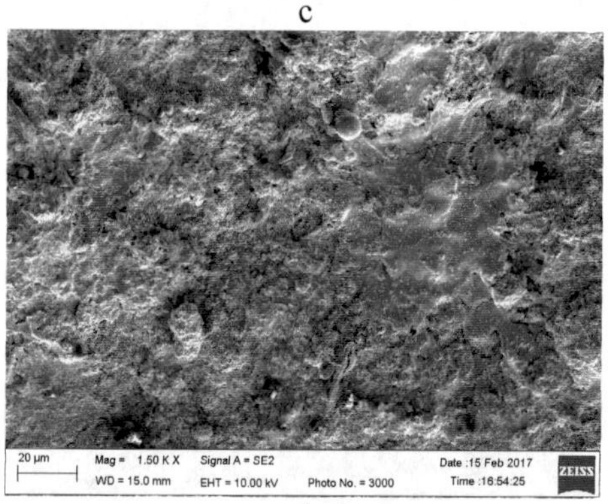

d

Fig. 4. Surface morphology of Mo (a-b) and Ta-based (c-d) coatings obtained at a stand-off distance of 100 mm and an explosive charge of 60%.

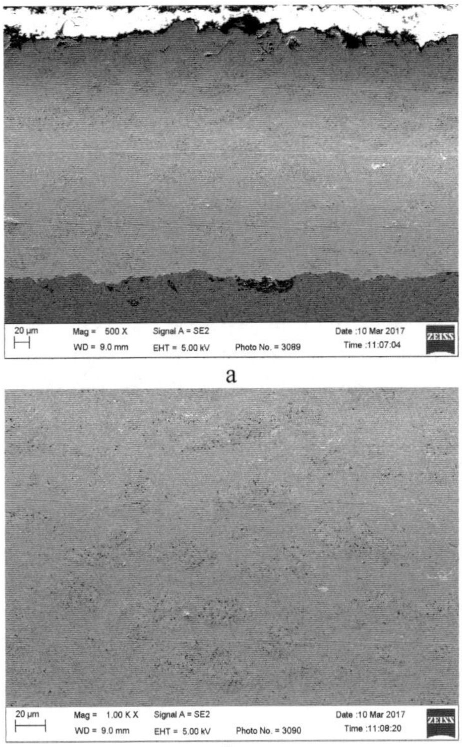

Fig. 5. Cross-section of the Mo coating obtained at a stand-off distance of 100 mm and an explosive charge of 60%: (a) general view; (b) microstructure.

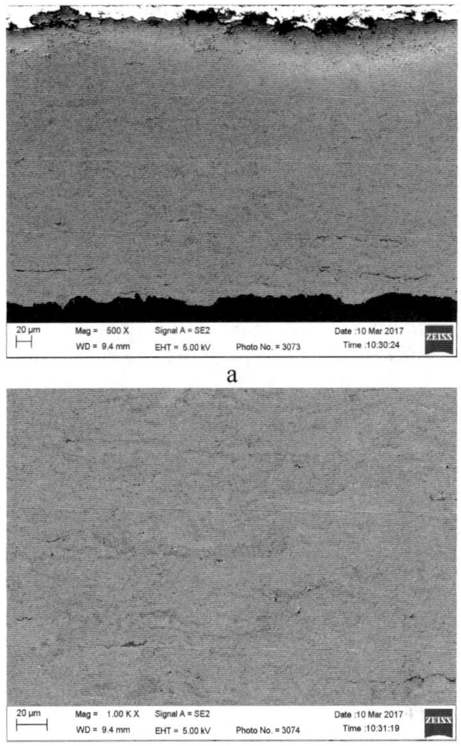

Fig. 6. Cross-section of the Ta-based coating obtained at a stand-off distance of 100 mm and an explosive charge of 60%: (a) general view; (b) microstructure.

The XRD phase analysis showed that no oxidation takes place during detonation spraying of the Mo powder under the selected conditions. The XRD patterns of the coatings obtained by detonation spraying of the Ta powder showed reflections of a second phase – tantalum oxide Ta_2O_5, along with reflections of metallic tantalum (the major phase).

Fig.2-3 show Mo and Ta splats obtained at an explosive charge of 60% and a stand-off distance of 100 mm. The splats formed from the particles of the two metals have similar morphologies and show that the particles deposit on the substrate in the solid, semi-molten and molten states. The surface morphology of the Mo and Ta-based detonation coatings (Fig. 4) confirms these conclusions. The semi-molten state of the particles is preferred for the formation of coatings with a dense structure and good adhesion and cohesion characteristics. After leaving the barrel, the particles can experience additional heating while travelling to the substrate. Taking this into account, the observed splat morphologies can be rationalized.

The general view and the microstructure of the coatings are presented in Fig. 5-6. The porosity of the Mo coatings was less than 0.5% for both stand-off distances used. The porosity of the coatings obtained from the Ta powder was less than 1%.

The hardness of the Mo coatings sprayed at 20 mm and 100 mm was 500 HV_{300} and 625 HV_{300}, respectively. The hardness of the Ta-based coatings sprayed at 20 mm and 100 mm was 800 HV_{300} and 1000 HV_{300}, respectively. The nature of the influence of the stand-off distance on the coating hardness requires further investigation.

The bond strength of the Mo coatings was 92 and 126 MPa, for coatings deposited at stand-off distances of 20 and 100 mm, respectively. The bond strength of the Ta-based coatings was 43 and 77 MPa, for coatings deposited at the same stand-off distances.

The results of the bond strength measurements together with the results of the microstructural studies of the coatings show that the detonation spraying method can produce high-quality low-porosity coatings from Mo and Ta powders.

4 Conclusions

In this work, Mo and Ta-based coatings of low porosity (0.5-1%) were successfully obtained by detonation spraying of Mo and Ta powders using the CCDS2000 detonation spray system. The coatings were deposited on steel substrates at O_2/C_2H_2=1.1.

The temperature and velocity calculations helped determining the spraying conditions for the efficient deposition. No oxide phases were found in the coatings obtained by spraying of the Mo powder. In the Ta-based coatings, Ta_2O_5 was the minor phase. The hardness of the Mo coatings depended on the stand-off distance: the coating sprayed at 20 mm had a hardness of 500 HV_{300}, while the coating sprayed at 100 mm had a hardness of 625 HV_{300}. A similar effect of the stand-off distance on the coating hardness was observed in the case of the Ta-Ta_2O_5 coatings: the coating obtained at 20 mm had a higher hardness (800 HV_{300}) than the coating obtained at 100 mm (1000 HV_{300}). The nature of this effect requires further investigation. The bond strength of the Mo coatings was 92 and 126 MPa and that of the Ta-based coatings was 43 and 77 MPa, for coatings deposited at stand-off distances of 20 and 100 mm, respectively.

5 Literature

[1] L. Singh, V. Chawla, J.S. Grewal, A review on detonation gun sprayed coatings, J. Minerals & Mater. Charact. & Eng. 11 (3) (2012) 243-265.

[2] P.A. Manojkumar, A.S. Gandhi, M. Kamaraj, V. Thomas Paul, N. Kumar, A.K. Tyagi, Role of nanocrystalline feedstock in the tribological behaviour of alumina coatings deposited by detonation gun, Int. J. Refractory Metals & Hard Mater. 35 (2012) 108.

[3] Y. Gao, Z. Hei, X. Xu, G. Xin, Formation of molybdenum boride cermet coating by the detonation spray process, J. Thermal Spray Technol. 10 (2001) 456.

[4] H. Du, W. Hua, J. Liu, J. Gong, C. Sun, L. Wen, Influence of process variables on the qualities of detonation gun sprayed WC–Co coatings, Mater. Sci. Eng. A 408 (2005) 202–210.

[5] M. Kaur, H. Singh, S. Prakash, Surface engineering analysis of detonation-gun sprayed Cr_3C_2–NiCr coating under high-temperature oxidation and oxidation–erosion environments, Surf. & Coat. Technol. 206 (2011) 530–541.

[6] J. K. N. Murthy, D.S. Rao, B. Venkataraman, Effect of grinding on the erosion behaviour of a WC–Co–Cr coating deposited by HVOF and detonation gun spray processes, Wear 249 (2001) 592–600.

[7] A. Shtertser, C. Muders, S. Veselov, S. Zlobin, V. Ulianitsky, X. Jiang, V. Bataev, Computer controlled detonation spraying of WC/Co coatings containing MoS2 solid lubricant, Surf. Coat. Technol. 206 (2012) 4763-4770.

[8] V. Ulianitsky, I. Batraev, D. Dudina, I. Smurov, Enhancing the properties of WC/Co detonation coatings using two-component fuels, Surf. Coat. Technol.(2016), dx.doi.org/10.1016/j.surfcoat.2016.08.008

[9] E. Lugscheider, H. Eschnauer, B. Hauser, D. Jager, Vacuum plasma spraying of tantalum and niobium, J. Vac. Sci. Technol. A 3 (1985) 2469.

[10] R. Kingswell, D. S. Rickerby, S. J. Bull, K. T. Scott, Erosive wear of thermally sprayed tungsten coatings, Thin Solid Films 198 (1991) 139.

[11] T. Kinos, S. L. Chen, P. Siitonen, P. Kettunen, Densification of plasma-sprayed titanium and tantalum, J. Thermal Spray Technol. 5 (1996) 439.

[12] J. Ahn, B. Hwang, S. Lee, Improvement of wear resistance of plasma-sprayed molybdenum blend coatings, J. Thermal Spray Technol. 14 (2005) 251.

[13] V. Ulianitsky, A. Shtertser, S. Zlobin, I. Smurov, Computer-controlled detonation spraying: from process fundamentals toward advanced applications, J. of Thermal Spray Technol. 20 (2011) 791-801.

[14] T. P. Gavrilenko, S.B. Zlobln, Yu.A. Nikolaev, V.Yu. Ulianitsky, About the pin-test for bonding strength measurement, Strengthening Technologies and Coatings. 8 (2007) 54-56.

[15] T. P. Gavrilenko, Yu.A. Nikolaev, V.Yu. Ulianitsky, M.Ch Kim, J.W. Hong, Computational Code for Detonation Spraying Process, ITSC-1998 Proc., Nice, France, 25-29 May, 1998, pp. 1475-1483.

Effect of Electromagnetic Field during Solidification of Ni-Based Alloyed Splats

Recalde O., Castro W., Bejarano M.L., Vargas, M., Valarezo A., Quito / Ecuador

A high electromagnetic field (0.3 T-Teslas) was applied during the solidification of Ni-based alloyed splats. Ni, NiCr, NiCrAl, NiCrBSiFe powders were deposited over steel polished substrates using a flame spray and a plasma spray torch. A strong electromagnet was used to produce sufficient magnetic field to induce effects over the splats during solidification. A remarkable change in splat morphology and chemical segregation was identified specially in NiCrBSiFe and the other alloys. Optical microscopy, surface profilometry, and SEM images revealed changes in the regular cracking trends, splashing, and thickness of the splats. This experimental study discusses the possible explanations for this phenomena. The adherence of the coating is the main property to be analyzed with the goal of improving the mechanical interlocking, and therefore, adhesion by engineering the applied electromagnetic field.

1 Introduction

In this exploratory work, a high electromagnetic field (0.3 T-Teslas) was applied during the rapid solidification of Ni-based alloyed splats deposited by the thermal spray (TS) technique, specifically by plasma and flame spraying. To the extent of the knowledge of the authors, the effect of electromagnetic field during solidification in thermal spray has not been studied before, although it is well known that electromagnetism can be used in solidification processing to improve, for instance: homogeneity in casting by electromagnetic stirring, or deep penetration of a molten pool of metal during welding by in situ electromagnetic stirring.
Several effects over solidification of splats have been studied [1–5], for instance: 1) substrate temperature effect [2,6], where it has been found that higher temperatures allows for the formation of pancake like splats whereas at lower temperatures, the splats splash and fracture easily; 2) adsorbates effect, [1], where it has been found that organic residues burn during solidification, impeding the correct adhesion; 3) gas entrapment [4,7]- where air bubbles hindered the effective adhesion of nickel splats, 4) substrate roughness effects , 5) Hardness of the substrate [8,9]; 6) angle of impact [5]; 7) particles state [3], etc. However, no study has been dedicated to investigate interaction of electromagnetic effects on solidification. The experimental evidence confirm that the time scale of about 4 µs for spreading and 10-40 µs for solidification of the splat are in the frequency and range of action of the electromagnetic field. Also the experimental observations assure an improvement in splat adhesion in smooth surfaces by application of the field. The technological success of this idea is not necessarily valid at this moment; however, the scientific interest justify this study. The results will potentially open a new field of research related to Thermal Spray (TS) processing, towards improving the deposition technique, and modifying the microstructure of the coatings. By allowing modification of particle spreading with the magnetic field, factors such as particle coalescence, splash morphology, coating-substrate interface and intersplats interfaces could be modified at desire. This brings new tools for engineers for applications for coating design.

Preliminary results of the authors will be presented in this paper as to show evidence that the solidification of splats is affected by the electromagnetic field. Deeper understanding of such interaction is necessary as to engineer this principle.

2 Experimental Methods

2.1 Materials

A 3 mm thick plate of low carbon steel was cut into sixteen pieces of 8x9 mm each one to be used as substrate material to collect splats. The selection of this material was based on its electromagnetic properties, and provided its common use as bulk material in TS applications. All substrates were polished starting with a 240 grit until a 15 µm suspension solution of Alumina particles.
Splats were collected from various Ni-alloys by rastering the torch quickly over the substrate (raster speed 1 m/s). A self-fluxing alloy of Ni, Cr, B, Si, Fe (Eutectic 13496, Castolin-Eutectic, WI, USA) by flame spraying. Ni (Metco 56C-NS, NY, USA), NiCr (Metco 4C-NS, NY, USA) and NiCrAl (Metco 443NS, NY, USA) powders were applied through a plasma torch.

2.2 Processing

A homemade electromagnet of 600 coil turns with a steel core was fabricated for the experiments. A power supply of 7 [A] and 24 [v] was used to power the electromagnet. The electromagnet was set up behind the substrate sample to impose an electromagnetic field of about 0.3 [Teslas]. A temperature pyrometer was used to monitor preheating temperature of 170ºC. Preheating was carried out by using the same TS torch. Two TS process were used in this study: flame and plasma spraying. The deposition parameters for flame and plasma spraying are described in, **Table 1**.
Four materials and two processes were used to produced samples of splats and coatings. In all cases, a sample without applied magnetic field was produced for reference. Thus, two samples were produced for every combination of material and process, as presented in, **Table 2**.

Table 1. Spraying parameters for sample preparation.

Parameter	Flame Spray	Plasma Spray
Torch	TeroDyn System 2000-Eutectic-Castodyn	F4-Oerlikon Metco
Air pressure	14.5 psi	-
Acetylene	10.0 psi	-
Oxygen	58.0 psi	-
Argon flow	-	47.5 lpm
Hydrogen flow	-	6 lpm
Current	-	550 A
Spray distance	180 mm	100 mm

Coatings for adhesion tests were produced only for NiCrBSi powder by flame spraying. The samples with and without applied magnetic field were placed one next to the other one, and were produced in the same raster of the torch, with the same temperature history. The electromagnet was placed and powered on right before the spraying and after the preheating. Thus, little heat contribution from the electromagnet is expected.

Table 2. Materials and processes used to produce samples.

Powder	Proces s	Without magnetic field	With magnetic field
Ni-Cr-Br-Si	Flame Spray	Splats / coating	Splats / coating
Ni	Plasma Spray	Splats	splats
Ni-Cr	Plasma Spray	Splats	splats
Ni-Cr-Al	Plasma Spray	Splats	splats

To assure a minimal affectation from the flame spray and plasma spray's parameters over the splat formation during different applications, all the substrates were placed in a row in each process. In other words, the Ni-Cr-Br-Si powder was applied in one single pass over four substrates put in a row: two of them without magnetic field and the others with the presence of a magnetic field.

2.3 Characterization

Splats were analyzed using an optical microscope Nikon MA200, a Scanning Electron Microscope(SEM) Tescan-Mira3, and a ZYGO profilometer. The optical microscopy was the first technique observe changes in splat morphology. Further analysis with a SEM was suggested to explore phases using EDS and elemental mapping.

The adherence of the coating was tested by the pull-off test using an automatic adhesion gauge by Elcometer 510 equipment (MI, USA). Three tests were developed for each material/process system.

3 Results and Discussion

Splats that were deposited under applied electromagnetic field showed a different morphology compared to their similar ones, **Fig. 1**. A large number of NiCrBSiFe splats deposited by flame spray showed a dimple in the center of the splat, **Fig.1b**. Apparently the applied magnetic field makes the splat material to flow towards the substrate surface.

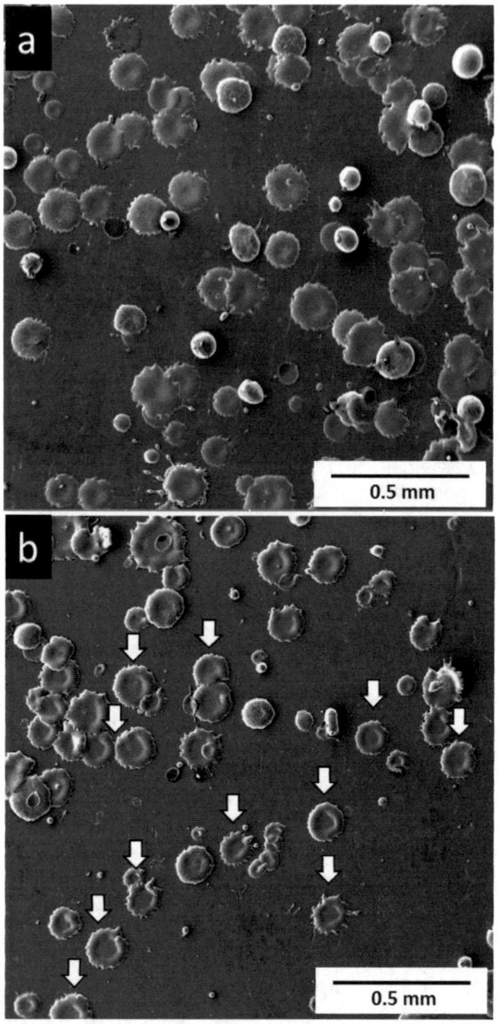

Fig. 1. NiCrBSiFe splats by flame spray: a) Splats deposited without applied electromagnetic field; b) Splats deposited with applied electromagnetic field. Notice the dimples in many of the splats of sample b) marked by white arrows.

Samples in Fig 1a and 1b, were obtained under a similar experimental conditions of preheating (170°C) and along the same raster of the torch which assures constant deposition parameters, and therefore the difference can be attributed to the applied field, only.

The depth of the characteristic dimple was measured by a profilometer to be about 5-10 µm in the case of splats without applied electromagnetic field, whereas, 15-20 µm for splats with applied field, respectively, **Fig. 2a and 2b**.
The material of the splat is pushed towards the substrate and as consecuence a rim is lifted in the surrounding.

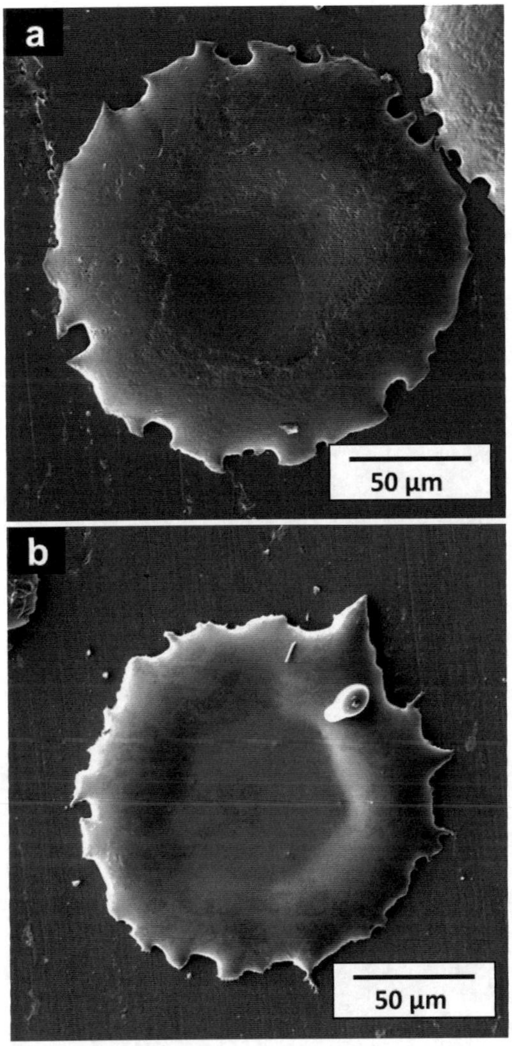

Fig. 2. NiCrBSiFe splats by flame spray at high magnification: typical splat deposited **a)** without applied electromagnetic field; and **b)** with electromagnetic field.

In principle, a magnetic field will induce electric currents in a material conductor, whether solid or fluid, and hence, the interaction of the electrical current and the magnetic field determine the Lorentz force. The force produces convective flow in the conductor. This principle is observed in electromagnetic stirring of the molten pool in welding, and is used in metallurgical processing like aluminium smelting, or continuous casting. [10]
Profilometry images of samples for splats with and without electromagnetic field are presented in, **Fig. 3a, 3b**. The "volcano" shape is pronounced for samples with applied field. In order to compare the shapes of both morphologies, the ratio of spreading and the ratio of splat height to crater depth was analysed for several splats of both conditions. **Fig. 3c**, shows a typical splat for both cases as if they were formed from a round particle of 0.166 mm in. diameter. It can be observed that the application of electromagnetic field reduces significantly the spreading and determines a deeper crater. The mass is concentrated near the center and thus it maybe enhance the adhesion by increased momentum above the interface.

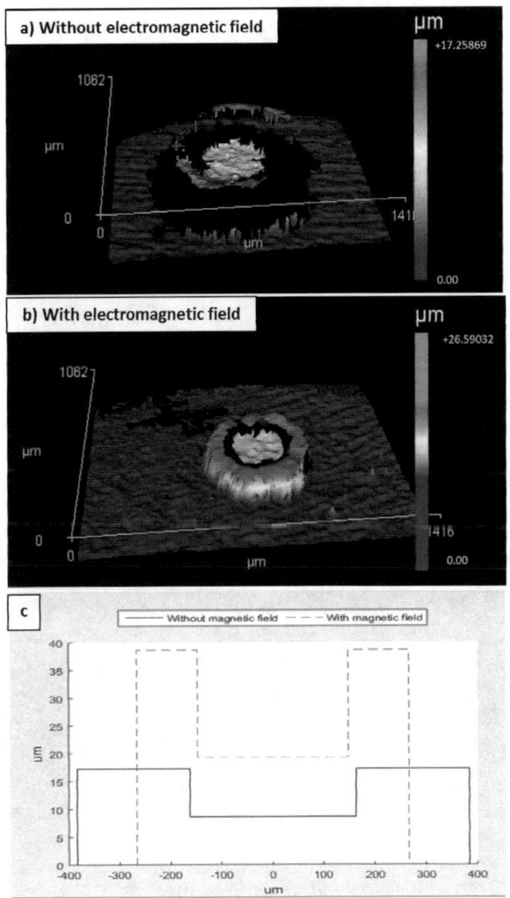

Fig. 3. NiCrBSiFe splats by flame spray by profilometer: typical splat deposited **a)** without applied electromagnetic field; and **b)** with electromagnetic field. **c)** Comparison of the morphology of both types of samples if they were of similar total volume (original splat diameter = 0.166 mm).

Further tests were conducted using plasma spraying and Ni, NiCr, and NiCrAl alloys as feedstock, **Fig. 4**.

Samples sprayed without electromagnetic field showed a higher deposition efficiency in the three materiales, **Fig. 4 a,c,e**. In the samples that were obtained with applied electromagnetic field, the splats were smaller and few, **Fig. 4 b,d,f**. Plasma spraying regularly overheats particles more than any flame process, thus the particle temperature might be considerable higher in comparison to the splats made out of NiCrBSiFe, and their resultant shapes differ from the previous ones. The presence of splashing in the splats without applied field evidence a high overheating of the particles. The fact that samples with applied field showed fewer and smaller splats, and no splashing can be explained by two reasons:

- Particles might have been repelled by the electromagnetic field and the Lorents force, even before they impacted the surface. This effect could have affected more to the slower particles which are regularly the larger ones. Thus, only those particles of sufficient speed could have reached the substrate and got adhere to it.
- Particles might have reached the substrate and bounced back, entirely or in fractions. Only the splats of improved adhesion were left attached to the substrate.

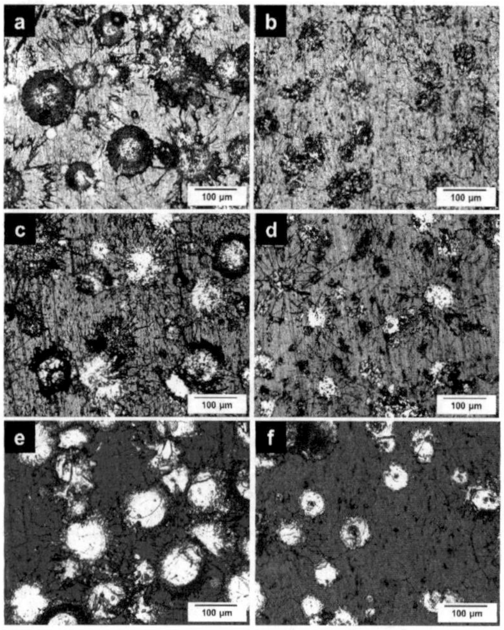

Fig. 4. Optical micrographs of Ni, NiCr, and NiCrAl splats by plasma spraying: a) Ni-without; and b) Ni-with; c) NiCr-without; and d) NiCr-with; e) NiCrAl-without; and f) NiCrAl-with – applied electromagnetic field.

In both cases, the splats deposited with applied electromagnetic field have suffered a selective process allowing only the very-well-adhered splats to be left in the surface. This effect can potentially improve the cohesion and adhesion strength of a coating,as it will be shown later.

Evidence that particles are repelled by the field is illustrated in, **Fig. 5**. NiCrAl splats reach the steel substrate, spread and solidify. However, the branches of the splashing have been removed. The steel substrate was preheated and it formed a thin layer of oxide. The branches of the splat that were removed left marks in the oxide of their presence. These portions are likely to have a poor adhesion due to their direction of spreading, or it could have been that the force produced by the field tore apart the splashes in the in-plane direction.

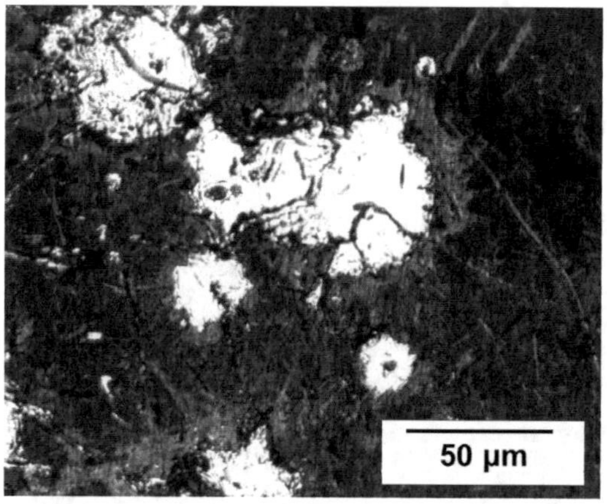

Fig. 5. Optical micrograph of NiCrAl splats by plasma spraying with applied electromagnetic field. The splashing surrounding the splat has been repelled by the field.

Full thickness coatings (0.3 mm) were deposited using NiCrBSiFe and the flame spray process. The results of the adhesion pull-off tests are shown in, **Fig. 6.** The results show an improvement of 25% in adhesion strength. Adhesion strength values are low in absolute magnitude provided that susbtrates were not gritblasted in order to reproduce the substrate condition of the splat collection. However, the overall increase in adhesion strength is significant and important. As discussed earlier, the response may be the result of a selective process of well-bonded splats. As seen in **Fig. 1**, the the splats are brought/pushed towards the substrate, likely improving the bonding of the splats. Additionally, **Fig. 5**, suggests that only the well bonded areas within a single splat are left bonded to the substrate. These contributions certainly can improve the adhesion strength.

When observing the elemental mapping by EDS-SEM, the well-known effect of preferential oxidation of certain elements in Ni-alloys was identified. In NiCr coatings, the Cr oxidized preferentially to form Cr_2O_3 [11], and in NiCrAl [12], the Al oxidized preferentially to form Al_2O_3. No significant effect of phase segregation during the rapid solidification due the applied electromagnetic field could be identified. Literature report this effect in certain materials during solidification [13] but there are no reports available for rapid solidification.

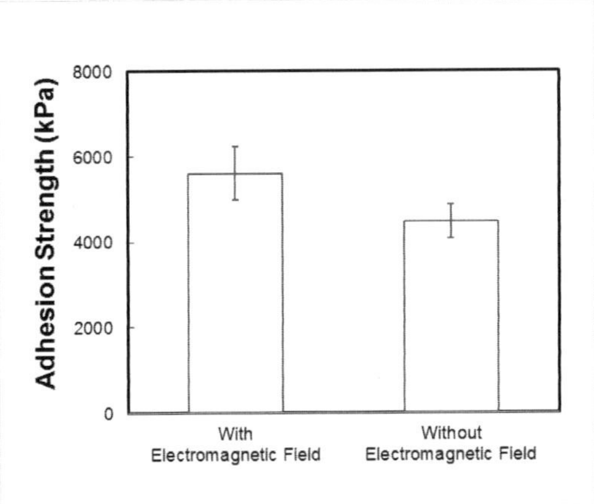

Fig. 6. Adhesion strength of two NiCrBSiFe-coatings of 0.3 mm of thickness, deposited by flame spray, with and without applied electromagnetic field during deposition. The adhesion is measured by the pull-off test.

It has been observed that splats deposited with the applied electromagnetic field present a more uniform distribution of oxygen over the splats, whereas the ones without the applied field show heterogeneous distribution of the oxide, as in **Fig. 7**.

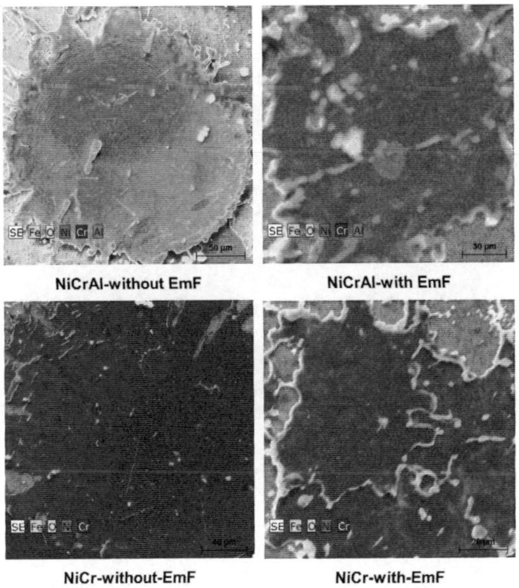

Fig. 7. Elemental mapping of NiCrAl, and NiCr splats by plasma spray, with and without applied electromagnetic field (EmF) during deposition.

It is unlikely that this difference arises from the effect of the applied field, provided that most of the oxidation occurs in-flight and not during splat formation. Instead, it is hypothesized that splats that are oxidized in flight are not found in the batch of applied field as they were repelled due to poor adhesion. This is supported by the fact that branches from the splashing which are regularly found to have been oxidized, are not present in the samples with applied field.
Again, a self-selective process of keeping the best adhered splats in place, makes interesting the use of electromagnetic field during deposition to improve properties like adhesion strength.

4 Conclusions

The application of a strong electromagnetic field during splat formation has shown to have an influence in the splat morphology. The electromagnetic field reduces the spreading and determines deep craters on the top of the splats. The field can selectively repel particles and/or fragments of splats. It is hypothesized that those splats that reach the substrate and deposit are the ones that have got a sufficient degree of adherence to resist to be repelled by the field. As a consequence, the adhesion strength is increased (25% for NiCrBSiFe by flame spray). Phase precipitation during the rapid solidification apparently is not affected by the field.

5 Acknowledgement

The authors would like to thank the collaboration with Stony Brook University and the Center for Thermal Spray Research for sample preparation and characterization, and CENCINAT-Centre for Nanosciences and Nanotechnology at ESPE University for SEM characterization. The authors also thank the grant HUBI-5489 from USFQ-Universidad San Francisco de Quito.

5 References

[1] X. Jiang, Y. Wan, H. Herman, S. Sampath, Role of condensates and adsorbates on substrate surface on fragmentation of impinging molten droplets during thermal spray, Thin Solid Films. 385 (2001) 132–141. doi:10.1016/S0040-6090(01)00769-6.
[2] S. Sampath, X. Jiang, Splat formation and microstructure development during plasma spraying: deposition temperature effects, Mater. Sci. Eng. A. 304–306 (2001) 144–150. doi:10.1016/S0921-5093(00)01464-7.
[3] A. Valarezo, W.B. Choi, W. Chi, A. Gouldstone, S. Sampath, Process Control and Characterization of NiCr Coatings by HVOF-DJ2700 System: A Process Map Approach, J. Therm. Spray Technol. 19 (2010) 852–865. doi:10.1007/s11666-010-9492-1.
[4] M. Qu, J.R. Colmenares, A. Valarezo, A. Gouldstone, Nanoporous surfaces via impact of

molten metallic droplets, Appl. Phys. A Mater. Sci. Process. 96 (2009) 391–397. doi:10.1007/s00339-009-5192-4.
[5] P. Fauchais, M. Fukumoto, a. Vardelle, M. Vardelle, Knowledge Concerning Splat Formation: An Invited Review, J. Therm. Spray Technol. 13 (2004) 337–360. doi:10.1361/10599630419670.
[6] X. Jiang, J. Matejicek, S. Sampath, Substrate temperature effects on the splat formation , microstructure development and properties of plasma sprayed coatings Part II : case study for molybdenum, Coating. 272 (1999) 189–198. doi:10.1016/S0921-5093(99)00461-X.
[7] M. Qu, Y. Wu, V. Srinivasan, A. Gouldstone, Observations of nanoporous foam arising from impact and rapid solidification of molten Ni droplets, Appl. Phys. Lett. 90 (2007) 1–3. doi:10.1063/1.2749183.
[8] W. Trompetter, M. Hyland, D. McGrouther, P. Munroe, a. Markwitz, Effect of Substrate Hardness on Splat Morphology in High-Velocity Thermal Spray Coatings, J. Therm. Spray Technol. 15 (2006) 663–669. doi:10.1361/105996306X147261.
[9] W.J. Trompetter, M. Hyland, P. Munroe, a. Markwitz, Evidence of Mechanical Interlocking of NiCr Particles Thermally Sprayed onto Al Substrates, J. Therm. Spray Technol. 14 (2005) 524–529. doi:10.1361/105996305X76568.
[10] H.K. Moffatt, Electromagnetic stirring, Appunti. (1990) 1336–1343.
[11] A. Valarezo, S. Sampath, An integrated assessment of process-microstructure-property relationships for thermal-sprayed NiCr coatings, J. Therm. Spray Technol. 20 (2011) 1244–1258. doi:10.1007/s11666-011-9665-6.
[12] S. Deshpande, S. Sampath, H. Zhang, Mechanisms of oxidation and its role in microstructural evolution of metallic thermal spray coatings - Case study for Ni-Al, Surf. Coatings Technol. 200 (2006) 5395–5406. doi:10.1016/j.surfcoat.2005.07.072.
[13] H. Conrad, Influence of an electric or magnetic field on the liquid–solid transformation in materials and on the microstructure of the solid, Mater. Sci. Eng. A. 287 (2000) 205–212. doi:10.1016/S0921-5093(00)00777-2.

AUTHOR INDEX

AUTHOR INDEX

AUTHOR INDEX

AUTHOR INDEX

AUTHOR INDEX

AUTHOR INDEX

International Thermal Spray Conference & Exposition (ITSC 2017)

Dusseldorf, Germany
7 - 9 June 2017

Volume 2 of 2

ISBN: 978-1-5108-5822-0

Printed from e-media with permission by:

Curran Associates, Inc.
57 Morehouse Lane
Red Hook, NY 12571

Some format issues inherent in the e-media version may also appear in this print version.

Printed by Curran Associates, Inc. (2018)

For permission requests, please contact DVS Media GmbH
at the address below.

DVS Media GmbH
P. O. Box 10 19 65
40010 Dusseldorf
Germany

Phone: 49 (0)211 / 1591-0
Fax: 49 (0)211 / 1591-150

media@dvs-hg.de

Additional copies of this publication are available from:

Curran Associates, Inc.
57 Morehouse Lane
Red Hook, NY 12571 USA
Phone: 845-758-0400
Fax: 845-758-2633
Email: curran@proceedings.com
Web: www.proceedings.com

TABLE OF CONTENTS

VOLUME 1

AVIATION & POWER GENERATION INDUSTRY II

CHARACTERIZATION & TESTING METHODS I

HVAF SPRAYING

TRIBOLOGICAL COATINGS

AVIATION & POWER GENERATION INDUSTRY III

LASER CLADDING & PTA

COLD GAS SPRAYING II

AVIATION & POWER GENERATION INDUSTRY V

CHARACTERIZATION & TESTING METHODS II

METAL COATINGS

VOLUME 2

MODELING & SIMULATION II

POWER GENERATION - STEAM

ELECTRONICS & SENSORIC

CORROSION PROTECTION

SUSPENSION SPRAYING III

COLD GAS SPRAYING III

AVIATION & POWER GENERATION INDUSTRY VI

ADDITIVE MANUFACTURING

AMORPHOUS COATINGS

PLASMA SPRAYING II

COLD GAS SPRAYING IV

POSTERS

APPLICATIONS – AVIATION INDUSTRY

APPLICATIONS – MARITIM INDUSTRY

APPLICATIONS – MEDICAL INDUSTRY

APPLICATIONS – METALS PROCESSING

APPLICATIONS – POWER GENERATION – FUEL CELLS & SOLAR

APPLICATIONS – POWER GENERATION - STEAM

EQUIPMENT / CONSUMABLES – PROCESS DIAGNOSTICS, SENSORS & CONTROLS

PROPERTIES – CERAMICS COATINGS

PROPERTIES – CORROSION PROTECTION

PROPERTIES – METAL COATINGS

PROPERTIES – POLYMER COATINGS

PROPERTIES – PRE- & POST-TREATMENT

PROPERTIES – PROPERTIES CHARACTERIZATION & TESTING METHODS

PROPERTIES – TRIBOLOGICAL COATINGS

PROPERTIES – WEAR PROTECTION

THERMAL SPRAY PROCESSES – ADDITIVE MANUFACTURING

THERMAL SPRAY PROCESSES – ARC SPRAYING & LASER SPRAYING

THERMAL SPRAY PROCESSES – COLD GAS SPRAYING

THERMAL SPRAY PROCESSES – SUSPENSION SPRAYING

Effect of Substrate Roughness and Thickness on the Gas-Substrate Convective Heat Transfer during Cold Spraying

A. Mahdavi, A. McDonald, Edmonton, CA

In this study, the effect of the substrate roughness and thickness on the heat transfer coefficient of the impinging air jet upon a flat substrate was investigated. A low-pressure cold spraying unit was used to generate a compressed air jet that impinged on a flat substrate. A detailed mathematical model was developed and coupled with experimental data to determine the heat transfer coefficient and surface temperature of the substrate. It was found that increasing the roughness of the substrate enhanced the heat exchange between the impinging air jet and the substrate. As a result, higher surface temperatures on the rough substrate were measured. It was further found that the Nusselt number that was predicted by the model was independent of the thickness of the substrate. The results of the current study were aimed to cover the influential substrate parameters on surface temperature of the substrate that eventually can affect the final quality of the cold-sprayed coatings.

1 Introduction

Cold-gas dynamic spraying is a coating fabrication process in which a supersonic gas flow is produced to propel un-melted metal or alloy powder particles to deposit, with high impact forces, upon a substrate to form a highly adherent coating [1]. Due to the high velocity of the under-expanding gas jet upon the substrate, and the high temperature of the impinging jet, a significant amount of thermal energy is expected to be transferred from the gas jet to the substrate [2]. The high velocity of the impinged gas on the flat substrate, furthermore, ensures that the predominant mode of heat transfer is convection. Therefore, determining the convective heat transfer coefficient will enable determination of the heat transfer rate from the impinging gas jet to the substrate, or previous layers of the deposited coating. It has been shown that the final quality of cold-sprayed coatings can be notably influenced by gas-substrate heat exchange, due to the dependence of the deposition efficiency of the particles on the substrate temperature distribution [3, 4]. Therefore, studying the effects of the cold spray parameters on the heat exchange and surface temperature of the substrate and the coating becomes important. Regarding the influence of cold spray parameters, substrate parameters, such as substrate roughness and thickness, can also affect the heat transfer rate from the gas jet to the substrate.

The effect of patterns on fabricated surfaces on the heat transfer coefficient of an impinging air jet on those surfaces has been studied [5-7]. Chakroun, *et al.* [5] patterned a surface with cubes of 1 mm length, which were distributed uniformly along the substrate to simulate a rough surface. It was shown that the local and average Nusselt number of the impinging air jet on the fabricated rough surface with Reynolds number ranging from 6,500 to 19,000 increased from 8.9% to 28% compared to that on the smooth surface [5]. In a similar study, Beitelmal, *et al.* [6] employed a circular array of protrusions of 0.5 mm base diameter and 0.5 mm high. The results of their study indicated that the Nusselt number over the rough surface increased up to 6% for an impinging air jet with Reynolds number ranging from 9,600 to 38,500. In a recent numerical study, Dobbertean and Rahman [7] calculated the Nusselt number for an impinging air jet with Reynolds number of 500 to 1,000 over a rough surface consisting of indentation depths ranging from 0.125 to 0.5 mm. While these studies have placed emphasis on substrates with patterned surfaces, an investigation of the effect of the surface roughness generated by grit-blasting on the heat transfer coefficient of the impinging air jet, within the domain of application of cold spraying is still lacking.

The effect of the substrate thickness on the heat transfer coefficient and the surface temperature of the substrate was studied. In a numerical study by McDonald, *et al.* [8], the effect of substrate thickness on the maximum surface temperature of the substrate of different materials was studied. It was shown that for all the materials that were studied, the maximum surface temperature decreased, as the thickness of the substrate increased [8]. In the present work, through a semi-empirical analytical study, the effect of the surface roughness and thickness on the heat transfer coefficient and surface temperature during the cold spray process was investigated. The purpose of this study was to emphasize on the substrate parameters on the heat exchange from the impinging air jet on a flat substrate, which can influence the final quality of the cold-sprayed coating.

2 Experimental Procedure

A low-pressure cold-gas dynamic spraying unit (SST Series P, Centerline, Ltd., Windsor, ON, Canada) was utilized to produce a supersonic air jet. The pressure and the temperature of the compressed air jet were set at 660 kPa (96 psig) and 300°C at the cold spray system console, respectively. The nozzle of the cold spray torch was a 140 mm-long converging-diverging de Laval nozzle, with throat and exit diameters of 2.54 mm and 6.3 mm, respectively. A robot (Motoman-HP20, Yaskawa Electric Corp., Waukegan, IL, USA) was employed to control the cold spray nozzle during the experiments. The cold spray nozzle was directed at the center of the front surface of the substrate and was held stationary at a stand-off distance (SOD) of 25 mm.

Aluminium 6061-T6 was used as the substrate. This material is known to have a relatively high thermal conductivity (order of 0.2 W/m-K), which increases heat conduction through the substrate. Thus, this material is expected to achieve steady-state thermal conditions quickly after impingement of the hot air jet on the substrate. The geometry of the substrate was a disk with dimension of 200 mm in diameter. The thickness of the substrate was 1.5, 2.5, or 6.5 mm in this study. In order to prevent the back surface of the substrate from exchanging heat with the ambient air, a thermal insulation material, vitreous aluminosilicate fiber (ISAFORM 2300° L.O., Insulation Specialties of America, Inc., Wanatah, IN, USA), was attached at the back surface of the substrate. The substrate was mounted onto a substrate holder.

In order to study the effect of the substrate roughness on the heat transfer coefficient and surface temperature of the substrate, the cold spray unit was used to grit-blast the aluminium substrate with thickness of 2.5 mm by depositing garnet sand particles (Super Garnet, V.V. Mineral, Tamil Nadu, India).

The transient and steady-state temperatures of the front surface of the substrate were measured by using a calibrated infrared camera (FLIR A65, FLIR Systems, Inc., Nashua, NH, USA). In the case of smooth (polished) substrates, to enable more accurate measurements, the front surface of the substrate was painted black so as to approach conditions of a black body. Accordingly, the emissivity of the substrate was set to 0.96 on the infrared camera. However, for the grit-blasted (roughened) substrate, it was not reasonable to paint the front surface, since the black paint would fill the asperities of the substrate and cause smoothness the rough surface. Therefore, in order to measure the surface temperature of the roughened surface by using the infrared camera, the emissivity of the rough substrate was first evaluated. To this end, an experiment was conducted to measure the temperature-dependent emissivity of the rough aluminium substrate. The substrate holder was fabricated from copper, where two cartridge heaters (50 W Miniature High Temperature Cartridge Heater (D 1/8" ×1 1/4"), McMaster-Carr, Aurora, OH, USA) were embedded inside the copper holder. A temperature controller (CNI8A42, Omega Engineering Inc., Stamford, CT, USA), equipped with a J-type thermocouple, was employed to adjust the temperature of the substrate holder at the desired set-point [9]. The setup was used to heat up the rough substrate. The surface temperature of the substrate was measured by using an external J-type thermocouple. Simultaneously, the infrared camera was directed at the substrate. The emissivity that was indicated on the infrared camera was modified until the temperature measurement on the infrared camera panel was in accord with the measurement of the thermocouple. Figure 3 provides a schematic of the experimental assembly that was used to measure the emissivity of the rough substrate.

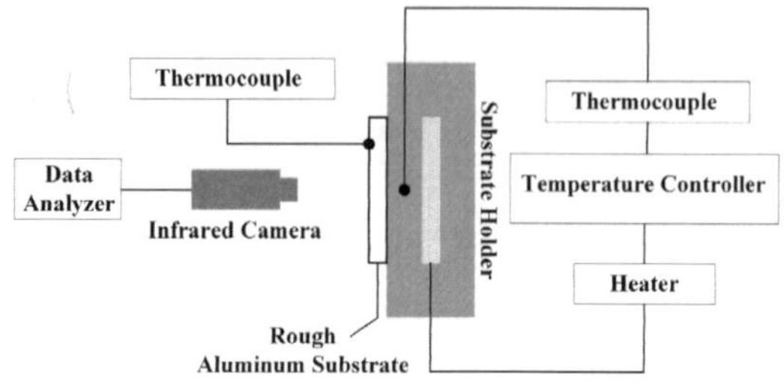

Fig. 1. Schematic of the experimental assembly for measuring the emissivity of the rough aluminium substrate [9].

3 Analytical Model

A significant amount of thermal energy is expected to be exchanged between the under-expanding air jet and the substrate, due to the impingement of the supersonic hot air jet on the flat substrate. The predominant mode of heat transfer between the impinging air jet and the substrate is forced convection; therefore, estimates of the convective heat transfer coefficient (*h*) are required in order to determine the heat flux and the temperature distribution within the substrate. A detailed semi-empirical analytical heat transfer model has been developed to evaluate the radial convective heat transfer coefficient between a flat substrate and an under-expanding supersonic air jet generated by a cold spray system [10]. The model is further capable of estimating the transient temperature distribution within the substrate during the impingement of the air jet. Since analytical determination of the heat transfer coefficient is outside the scope of the current study, a summary of the model is described here. A schematic of the geometry of the model problem is shown in Fig. 2.

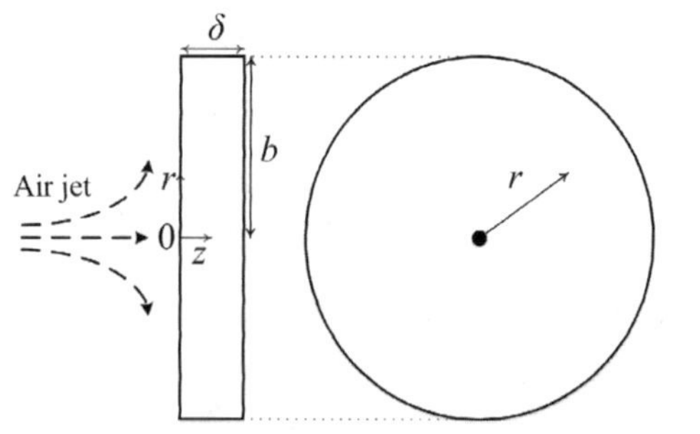

Fig. 2. Schematic of the geometry for the mathematical model [10].

The governing equation of the transient two-dimensional heat conduction model within the substrate is

$$\frac{1}{r}\frac{\partial}{\partial r}\left(r\frac{\partial T}{\partial r}\right)+\frac{\partial^2 T}{\partial z^2}=\frac{1}{\alpha_s}\frac{\partial T}{\partial t}, \quad (1)$$

where α_s is the thermal diffusivity of the substrate. The boundary conditions and the initial condition are

$$T(r=0,z,t)=\text{finite}, \quad (2)$$

$$\frac{\partial T(r=b,z,t)}{\partial r}=0, \quad (3)$$

$$k_s\left[\frac{\partial T(r,z=0,t)}{\partial z}\right]=h(r)\left[T(r,z=0,t)-T_{AW}(r)\right], \quad (4)$$

$$\frac{\partial T(r,z=\delta,t)}{\partial z}=0, \quad (5)$$

$$T(r,z,t=0)=T_0. \quad (6)$$

In Eq. (4), h is the convective heat transfer coefficient between the substrate and the impinging air jet, and T_{AW} represents the adiabatic wall temperature. The thermal conductivity and the initial temperature of the substrate are k_s and T_0, respectively.
A Green's function approach was employed to solve the governing equation of this problem. The final expression for the temperature distribution within the substrate was found to be

$$\begin{aligned}T(r,z,t)=&\sum_{i=1}^{\infty}\sum_{j=1}^{\infty}\left[T_0\frac{\exp\left[-\alpha_s\left(\varsigma_i^2+\beta_j^2\right)t\right]}{N(\varsigma_i)N(\beta_j)}J_0(\varsigma_i r)\right.\\&\times\left[\tan(\beta_j\delta)\sin(\beta_j z)+\cos(\beta_j z)\right]\\&\times\int_{r'=0}^{b}\int_{z'=0}^{\delta}J_0(\varsigma_i r')\left[\tan(\beta_j\delta)\sin(\beta_j z')+\cos(\beta_j z')\right]r'dz'dr'\\&+\left(\frac{\alpha_s}{k_s}\right)\left[\frac{h(r)}{N(\varsigma_i)N(\beta_j)}\right]J_0(\varsigma_i r)\\&\times\left[\tan(\beta_j\delta)\sin(\beta_j z)+\cos(\beta_j z)\right]\\&\times\frac{1}{\alpha_s\left(\varsigma_i^2+\beta_j^2\right)}\left[1-\exp\left(-\alpha_s\left(\varsigma_i^2+\beta_j^2\right)t\right)\right]\\&\left.\times\int_{r'=0}^{b}J_0(\varsigma_i r')T_{AW}(r')r'dr'\right],\end{aligned} \quad (7)$$

where ζ_i and β_j are the eigenvalues in the radial ($0 < r < b$) and the axial ($0 < z < \delta$) coordinates, respectively, and they can be obtained from

$$\varsigma_i J_1(\varsigma_i b)=0, \quad (8)$$

$$\beta_j\tan(\beta_j\delta)-\frac{h(r)}{k_s}=0. \quad (9)$$

In Eq. (7), $N(\zeta_i)$ and $N(\beta_j)$ are the Norms of the differential equations in the radial and axial coordinates, respectively, and can be expressed as

$$N(\varsigma_i)=\int_{r=0}^{b}J_0^2(\varsigma_i r)rdr, \quad (10)$$

$$N(\beta_j)=\int_{z=0}^{\delta}\left[\tan(\beta_j\delta)\sin(\beta_j z)+\cos(\beta_j z)\right]^2dz. \quad (11)$$

Analytical determination of the transient temperature distribution within the flat substrate that was exposed to an under-expanding supersonic air jet of a cold spray system is expressed by Eq. (7). Eqs. (7), and (9) can be used to determine the value of the heat transfer coefficient under various substrate conditions. The shooting method was applied to estimate the point-wise heat transfer coefficient between the substrate and the impinging air jet. A zero-dimensional code in MATLAB (MathWorks, Inc., Natick, MA, USA) was developed to solve the explicit expression for h by using Eqs. (7), (8), and (9). Terms in the infinite series of Eq. (7) were added until the solutions converged to within 1% of the sum.

4 Results and Discussion

4.1 Effect of substrate thickness on heat transfer coefficient and surface temperature

Figure 3 shows the radial variation of the Nusselt number of the impinging under-expanding supersonic air jet on the aluminium surface.

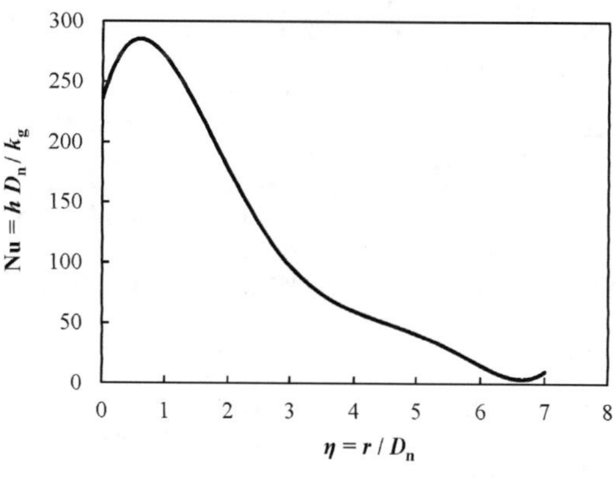

Fig. 3. Curve of the Nusselt number of the impinging air jet as a function of non-dimensional radial distance from the stagnation point.

The Nusslet number, Nu, is defined as

$$\mathrm{Nu} = \frac{hD_\mathrm{n}}{k_\mathrm{g}}, \tag{12}$$

where D_n and k_g are the nozzle diameter and thermal conductivity of the impinging air, respectively. The distance from the stagnation point of the impinging jet was non-dimensionalized as

$$\eta = \frac{r}{D_\mathrm{n}}. \tag{13}$$

The effect of the substrate thickness on the heat transfer coefficient of the impinging air jet and the surface temperature of the substrate was investigated. The surface temperature of the substrate under an impinging air jet may be affected by the thickness of the substrate. It is, however, expected that the heat transfer coefficient of the impinging air jet on the substrate will be independent of the thickness of the substrate, since the heat transfer coefficient mainly depends on geometry of the surface in contact with the fluid, fluid properties, flow regime (that is laminar or turbulent flow), and in some cases, temperature difference between the fluid and the surface [11]. In order to investigate the dependence of the heat transfer coefficient on substrate thickness, the heat transfer coefficient that was estimated for a substrate with thickness of 1.5 mm was substituted into Eq. (7). The non-dimensional surface temperature of the substrate was calculated at non-dimensional Fourier number (Fo = $\alpha_\mathrm{s}\, t / \delta^2$) of 81 and for substrate thicknesses of 2.5 and 6.5 mm. The surface temperature was non-dimensionalized as

$$\theta_\mathrm{s} = \frac{T_\mathrm{s} - T_\infty}{T_\mathrm{g} - T_\infty}, \tag{14}$$

where T_∞ is the ambient air temperature. Figure 4 shows a comparison of the estimated and experimentally-measured non-dimensional surface temperature for different thicknesses. Although the heat transfer coefficient, which was used in the calculations, was measured at δ = 1.5 mm, the results of the semi-empirical analytical model that pertain to the thicker substrates was in close agreement with the experimental results, confirming that the heat transfer coefficient of an impinging air jet upon the substrate was independent of the thickness of the substrate. It was also observed that the surface temperature of the substrate decreased as the radial distance from the stagnation point of the impinging air jet increased and as the thickness of the substrate increased. A similar observation was reported by McDonald, *et al.* [8]. Due to the increase in thickness of the substrate, the volume of the substrate increased; thus, a higher quantity of thermal energy was conducted thorough the substrate. Therefore, the surface temperature of the thicker substrate was lower than that of the substrate with lower thickness [8].

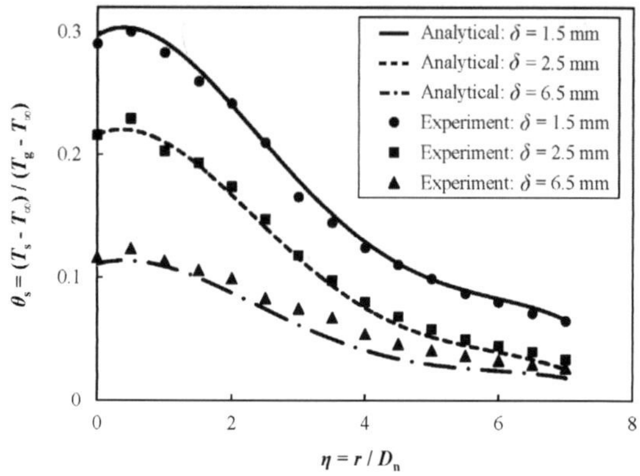

Fig. 4. Comparison of predicted and experimentally-measured non-dimensional surface temperature variation of the substrate at different substrate thicknesses.

4.2 Effect of substrate roughness on heat transfer coefficient and surface temperature

The effect of the surface roughness on the surface temperature of the substrate and the heat transfer coefficient between the substrate and the impinging air jet was investigated. A confocal microscope was used to study and evaluate the topology and roughness parameters of the smooth and roughened substrates within a surface scan area of 1150 × 1150 μm^2 from the front surface of the substrate. Figure 5 shows the surface topology and the peaks and valleys of the asperities of the smooth and rough (grit-blasted) substrates. Three roughness parameters of the substrate, namely average roughness (R_a), skewness (R_sk), and kurtosis (R_ku), were investigated in order to have a better understanding of the results in Fig. 5 [12]. By definition, the average roughness (R_a) is the average of the distance of the asperities from a hypothetical mean plane. Skewness (R_sk) is defined to show the symmetry of the peaks and valleys of the asperities with regards to the hypothetical flat plane. Surfaces with positive R_sk have relatively high peaks, while surfaces with negative R_sk have deep valleys and fairly smoother plateaus [13]. Kurtosis (R_ku) represents the degree of the pointedness of the asperities on the front surface. Surfaces with higher values of R_ku exhibit steep asperities; while, as the R_ku becomes smaller, the surface becomes smoother [13]. Table 1 shows the values of the aforementioned roughness parameters for the smooth and rough substrates.

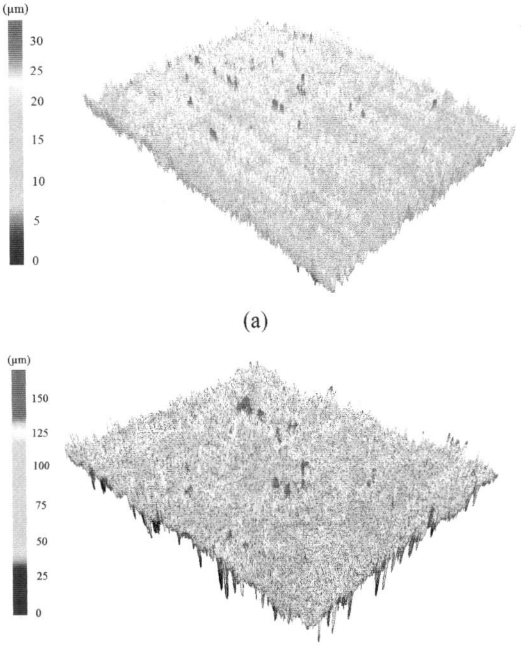

Fig. 5. 3-D representation of the surface topology of (a) smooth, and (b) rough (grit-blasted) aluminium 6061-T6 substrates.

Table 1. Average roughness, skewness, and kurtosis of the smooth and the rough (grit-blasted) aluminium 6061-T6 substrates.

Aluminum 6061-T6 substrate	R_a (μm)	R_{sk}	R_{ku}
Smooth	3.33	0.39	2.51
Rough (grit-blasted)	13.93	-0.59	5.28

According to Fig. 5 and Table 1, grit-blasting significantly increased the average roughness of the substrate and created steep asperities on the surface of the substrate, which led to an increase in value of R_{ku}. The relatively high negative values of skewness of the substrate after grit-blasting show that the grit-blasted surface consisted of deep valleys with smooth plateaus. It is likely that the increased roughness of the substrate would result in an increase in heat transfer coefficient between the impinging air jet and the substrate. Increasing the roughness of the substrate would result in higher magnitudes of the velocity fluctuations within the boundary layer over the substrate [14]. As a result, turbulence intensity in the flow, T_u, would increase, since it is defined as the ratio of the root-mean-square of the fluctuating velocity component and the mean velocity of the flow. The ratio is expressed as [14]

$$T_u = \frac{\left(\overline{u'^2}\right)^{1/2}}{\bar{u}_m}. \tag{15}$$

Higher quantities of turbulence intensity and velocity fluctuation will produce higher shear stress in the turbulent flow over the substrate. Additional shear stress on the substrate will be due to the increased velocity gradients, and the momentum transfer by eddies within the turbulent boundary layer. A fundamental study by Reynolds [15] has shown that the heat transfer between the flow and the substrate is directly proportional to the shear stress on the substrate. The aforementioned theory, known as Reynolds analogy, is expressed as [16]

$$\frac{h}{\rho \bar{u}_m C_P} = \frac{f}{2}, \tag{16}$$

where f is known as the skin friction coefficient and is defined as [16]

$$f = \frac{\tau_W}{0.5 \rho \bar{u}_m^{\,2}}. \tag{17}$$

Since increasing the roughness of the substrate would generate higher skin friction coefficient, resulting in an increase in heat transfer coefficient, this would in turn increase heat exchange between the air and the substrate. As a result of the increased heat exchange between the under-expanding air jet and the substrate, higher surface temperatures of the substrate are expected. In order to evaluate the aforementioned claim, the steady-state surface temperatures of both smooth and rough substrates were measured by using an infrared camera. For the rough aluminium 6061-T6 substrate, the emissivity was measured experimentally and it was found that at temperatures higher than 100°C, the emissivity converged to 0.32. Therefore, this estimated emissivity was used to estimate the steady-state surface temperature of the roughened substrate. Figure 6 shows the comparison of the radial variation of the non-dimensional steady-state surface temperature of the rough substrate with that of the smooth substrate, under the same cold spray process conditions.

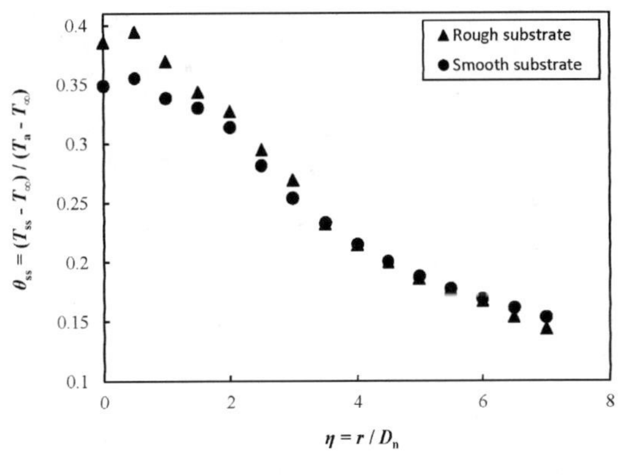

Fig. 6. Comparison of the radial variations of the non-dimensional steady-state surface temperature of the rough and smooth substrate under cold spray air jet impingement.

As is observed in Fig. 6, the steady-state surface temperature of the rough substrate was higher than that of the smooth substrate in the vicinity of the stagnation point of the impinging air jet. In addition to the influence of the heat transfer coefficient, the increased surface area of the rough substrate due to the asperities on the surface (see Fig. 5) may have increased the heat transfer from the air to the surface, thus increasing the surface temperature. This could be explained by Newton's Law of Cooling [11]

$$\dot{Q} = hA_s\Delta T \,, \qquad (18)$$

where A_s and ΔT are the surface area and temperature difference between the under-expanding air jet and the substrate. As the surface area increases, the heat transfer rate ($\dot{Q}$) from the air jet to the substrate increases. According to Table 1, high values of kurtosis yielded steep asperities on the rough surface, which could act as extended surfaces or fins. The areas of the extended surfaces would increase the heat transfer, and as a result, higher surface temperatures are observed on the rough substrate.

5. Conclusion

In this experimental and analytical study, the effect of the substrate roughness and thickness on the heat transfer coefficient between the substrate and the impinging air jet, and also on the surface temperature profile during cold spraying was investigated. A low-pressure cold-gas dynamic spraying unit was used to produce a compressed supersonic air jet that impinged on a flat substrate. A detailed mathematical model was developed and coupled with experimental data to estimate the heat transfer coefficient and the surface temperature of the substrate. The effect of the roughness and thickness of the substrate on the surface temperature of the substrate and the heat transfer coefficient was studied. The heat transfer coefficient of the impinging air jet upon a roughened (grit-blasted) substrate increased, compared to that upon a smooth substrate. It was further found that the heat transfer coefficient is independent from the thickness of the substrate. Moreover, it was shown that as the thickness of the substrate increased, the surface temperature decreased in all radial distances from the stagnation point.

6. References

[1] Kosarev, V.F., Klinkov, S.V., Alkhimov, A.P., and Papyrin, A.N.: On some aspects of gas dynamics of the cold spray process. J. Therm. Spray Technol., Vol. 12 (2), (2003) pp. 265/81.

[2] Limaye, M.D., Vedula, R.P., and Prabhu, S.V.: Local heat transfer distribution on a flat plate impinged by a compressible round air jet. Int. J. Therm. Sci., Vol. 49, (2010) pp. 2157/68.

[3] Legoux, J.G., Irissou, E., and Moreau, C.: Effect of substrate temperature on the formation mechanism of cold-sprayed aluminum, zinc and tin coating. J. Therm. Spray technol., Vol. 16 (5), (2007) pp. 619/26.

[4] Fukumoto, M., Wada, H., Tanabe, K., Yamada, M., Yamaguchi, E., Niwa, A., Sugimoto, M., and Izawa, M.: Effect of substrate temperature on deposition behavior of copper particles on substrate surfaces in the cold spray process, J. Therm. Spray technol., Vol. 16 (5), (2007) pp. 643/50.

[5] Chakroun, W.M., Abdel-Rahman, A.A., and Al-Fahed, S.F.: Heat transfer augmentation for air jet impinged on a rough surface. Appl. Therm. Eng., Vol. 18, (1998) pp. 1225/41.

[6] Beitelmal, A.H., Saad, M.A., and Patel, C.D.: Effects of surface roughness on the average heat transfer of an impinging air jet. Int. Comm. Heat Mass Transfer, Vol. 27 (1), (2000) pp. 1/12.

[7] Cobbertean, M.M., Rahman, M.M.: Numerical analysis of steady state heat transfer for jet impingement on patterned surfaces. Appl. Therm. Eng., Vol. 103, (2016) pp. 481/90.

[8] McDonad, A., Ryabini, A.N., Irissou, E., and Legoux, J.G.: Gas-substrate heat exchange during cold-gas dynamic spraying. J. Therm. Spray Technol., Vol. 22 (2), (2013) pp. 391/397.

[9] Ashrafizadeh, H., Mertiny, P., and McDonald, A.: Evaluation of the effect of temperature on mechanical properties and wear resistance of polyurethane elastomers, Wear, Vol. 368/9, (2016) pp. 26/38.

[10] Mahdavi, A., McDonald, A.: Analytical Evaluation of Temperature distribution within the substrate under an impinging cold spray hot air jet, Thermal Spray 2016: Proceedings of the International Thermal Spray Conference, May 10-12, 2016 (Shanghai, People's Republic of China), ASM International, (2016) pp. 1/5.

[11] Jiji, L.: Heat Convection, Springer-Verlag, Netherlands, (2006).

[12] McDonald, A., Lamontagne, M., Chandra, S., and Moreau, C.: Photographing impact of plasma-sprayed particles on metal substrate. J. Therm. Spray. Technol., Vol. 15 (4), (2006) pp. 708/16.

[13] Chauvy, P., Madore, C., and Landolt, D.: Variable length scale analysis of surface topography: characterization of titanium surfaces for biomedical applications, Surf. Coat. Technol., Vol. 110, (1998) pp. 48/56.

[14] Lyles, L., Disrud, L.A., and Krauss, R.K.: Turbulence intensity as influenced by surface roughness and mean velocity in a wind-tunnel boundary layer. T ASAE, (1971) pp. 285/89.

[15] Reynolds, O.: On the extent and action of the heating surface of steam boilers, Proc. Lit. Phil. Soc., Manchester, Vol. 4 (1874) pp. 7/12.

[16] Shapiro, A.: The dynamics and thermodynamics of compressible fluid flow, vol. II, The Ronald Press Company, New York, (1954).

OpenFOAM Modelling of Particle Heating and Acceleration in Cold Spraying

K.-H. Leitz*, M. O'Sullivan*, A. Plankensteiner*, H. Kestler*, L. S. Sigl*
*Plansee SE, Metallwerk-Plansee-Straße 71, 6600 Reutte, Austria

In cold spraying a powder material is accelerated and heated in the gas flow of a supersonic nozzle to velocities and temperatures that are sufficient to obtain cohesion of the particles to a substrate due to plastic deformation. The deposition efficiency of the powder particles is significantly determined by their velocity and temperature. The particle velocity correlates with the kinetic energy of the particles and thereby with the amount of energy that is converted to plastic deformation and thermal heating. The initial particle temperature significantly influences the mechanical properties of the particle. Velocity and temperature of the particles have nonlinear dependence on the pressure and temperature of the gas at the nozzle entrance. Whereas the particle velocity can easily be measured during the process, the particle temperature is not directly accessible by experimental techniques. Generally information about the particle temperature can be obtained based on theoretical models. In this contribution a simulation model based on the *reactingParcelFoam* solver of OpenFOAM is presented and applied for an analysis of the cold spray process. The model combines a compressible description of the gas flow in the nozzle with a Lagrangian particle tracking. The predictions of the simulation model are verified based on an analytical description of the gas flow, the particle acceleration and heating in the nozzle. Based on experimental data the drag model according to Plessis and Masliyah is identified to be best suited for OpenFOAM modelling particle heating and acceleration in cold spraying.

1 Cold Spraying – Technology Principle and Challenges

In cold spraying a powder coating material is accelerated in the gas flow of a supersonic nozzle to velocities of up to 1000 *m/s*. Additionally the particles are heated by the gas. When the powder hits a substrate to be coated, the kinetic energy of the particles is converted to plastic deformation work. It is generally assumed that the particle adheres to the substrate when thermal softening at the contact interface overcomes strain and strain rate hardening. The velocity at which this occurs is called "critical velocity". **Fig. 1** shows the principle of cold spraying. More detailed descriptions of the cold spray technology can be found in [1 - 6].

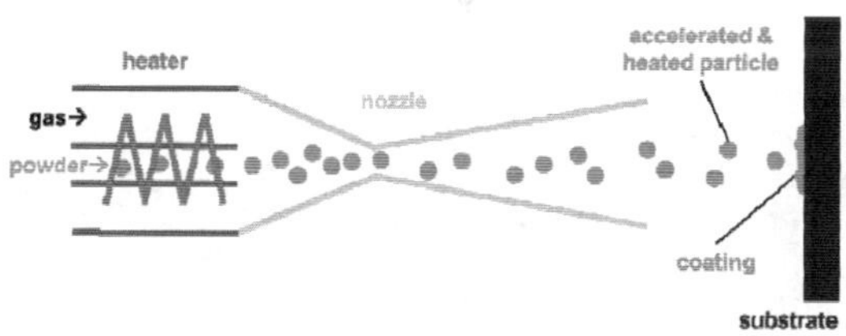

Fig. 1. Cold spraying: technology principle.

Velocity and temperature of the particles in cold spraying are indirectly controlled by the pressure and the temperature of the process gas at the entrance of the nozzle. Whereas there exist various approaches for an experimental measurement of the particle velocity, a measurement of the particle temperature is not trivial, as classical pyrometric approaches do not work for the low particle temperatures of typically below 1300 K in cold spraying [7]. Therefore, the theoretical description and simulation of gas flow, particle heating and acceleration in the cold spray nozzle is of crucial importance, as it allows acquiring information about particle velocity and temperature at certain processing parameters (gas pressure, gas temperature, nozzle geometry and position of particle injection).

Against this background the theory of gas flow, particle heating and acceleration in the cold spray nozzle is described in detail and a numerical simulation model based on the computational fluid dynamics toolbox OpenFOAM is presented. Based on this different models for gas-particle interaction are compared to each other. The predictions of the simulation model are verified based on the described theory. Furthermore, a validation of the simulation model by experimental measurement data on particle velocity is performed. Building on that a suggestion for the choice of the *particleForce* model is made when simulating the cold spray process based on OpenFOAM.

2 Theoretical Description of the Cold Spray Nozzle

2.1 Gas Flow

The gas flow in a cold spray nozzle can be described theoretically based on the Mach number relation [8]:

$$\left(\frac{A}{A^*}\right)^2 = \frac{1}{{M_g}^2}\left[\frac{2}{\gamma+1}\left(1+\frac{\gamma-1}{2}{M_g}^2\right)\right]^{\frac{\gamma+1}{\gamma-1}} \qquad (1)$$

In this equation A is the area of the nozzle cross section and A^* the cross section area at the nozzle throat (see **Fig. 2**). γ is the ratio of specific heats of the gas and M_g the Mach number of the gas flow. Based on equation (1) a Mach number M_g can be assigned to each position along the nozzle axis and the gas temperature T_g, gas velocity u_g, gas pressure p_g and gas density ρ_g can be calculated from the temperature T_{in}, pressure p_{in} and density ρ_{in} at the nozzle entrance based on the following relations [8]:

$$\frac{T_{in}}{T_g} = 1 + \frac{\gamma - 1}{2} {M_g}^2 \qquad (2)$$

$$a_g = \frac{u_g}{M_g} = \sqrt{\gamma R_S T_g} \qquad (3)$$

$$\frac{p_{in}}{p_g} = \left(1 + \frac{\gamma - 1}{2} {M_g}^2\right)^{\frac{1}{\gamma - 1}} \qquad (4)$$

$$\frac{\rho_{in}}{\rho_g} = \left(1 + \frac{\gamma - 1}{2} {M_g}^2\right)^{\frac{1}{\gamma - 1}};\ \rho_{in} = \rho_0 \frac{p_{in}}{p_0} \frac{T_0}{T_{in}} \qquad (5)$$

ρ_0 is the density of the gas at normal pressure p_0 = 1 *bar* and normal temperature T_0 = 293 *K*. a_g is the sound velocity of the gas and R_S its specific gas constant

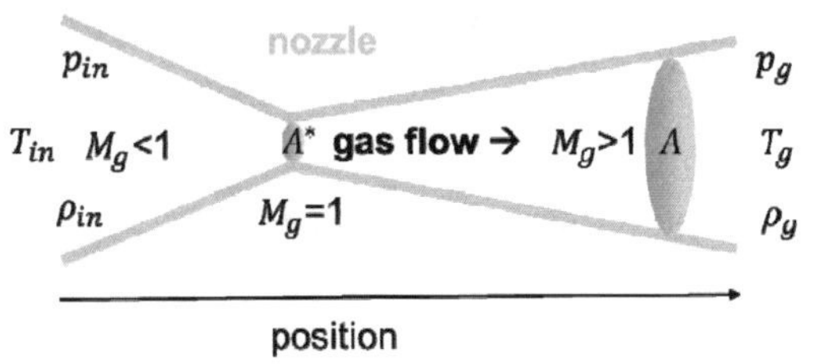

Fig. 2. Gas flow in the cold spray nozzle: theoretical description.

2.2 Particle Heating and Acceleration

The acceleration of a particle of mass m_p in the gas flow of a cold spray nozzle (compare **Fig. 3**) can be described by the following differential equation [4]:

$$m_p \frac{\partial u_p}{\partial t} = \frac{1}{2} C_D(Re_p) \rho_g A_p (u_g - u_p)|u_g - u_p| \qquad (6)$$

The change in particle velocity ∂u_p depends on the difference between gas velocity u_g and particle velocity u_p, the particle cross sectional area A_p, the gas density ρ_g and the drag coefficient $C_D(Re_p)$, that generally is a function of the particle Reynolds number $Re_p = \rho_g d_p |u_g - u_p| / \eta_g$. d_p is the particle diameter and η_g the dynamic viscosity of the gas.

In an analogous way the heating of the particle in the gas flow (compare **Fig. 3**) can be described by [4]:

$$m_p C_{p,p} \frac{\partial T_p}{\partial t} = \frac{\nu \lambda_g}{d_p} O_p (T_g - T_p) \qquad (7)$$

The change in particle temperature ∂T_p depends on the difference between gas temperature T_g and particle temperature T_p, the surface area O_p of the particle and the heat transfer coefficient ν, also called Nusselt number. $C_{p,p}$ is the heat capacity of the particle and λ_g the thermal conductivity of the gas. Based on the gas velocity and temperature profile along the nozzle, that can be calculated as described in section 2.1, equations (6) and (7) allow a calculation of particle velocity u_p and particle temperature T_p by numeric integration.

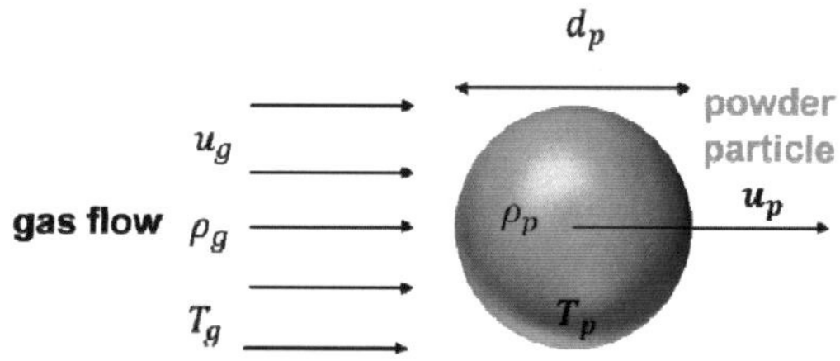

Fig. 3. Heating and acceleration of a particle of density ρ_p and diameter d_p in a gas flow.

3 OpenFOAM Simulation Model for the Cold Spray Nozzle

A more precise approach for analysing particle heating and acceleration in cold spraying are numerical simulations. In contrast to the theoretical models described in the previous section simulation models do not require as severe assumptions regarding geometry and contained physics. Simulations allow to take into account the influence of more sophisticated nozzle geometries and to describe the gas flow outside the nozzle as well as the influence of the substrate. Below a simulation model for particle heating and acceleration in the cold spray nozzle based on the open source finite volume computational fluid dynamics toolbox OpenFOAM shall be described. OpenFOAM specific expressions are written in italics.

The simulation model is based on the *reactingParcelFOAM* solver of OpenFOAM. This transient solver uses the PIMPLE algorithm for the solution of the compressible Navier-Stokes equation and allows the inclusion of Lagrangian tracer particles. The nozzle geometry was meshed using the automatic split hex mesher *snappyHexMesh* of OpenFOAM. In the physical description viscosity and heat capacity were held constant and gravity vertical to the nozzle orientation was considered. The compressible gas flow was described laminar. The particles were inserted with an initial velocity of 5 *m/s* at the entrance of the convergent nozzle section and a certain particle size distribution was taken into account. For the analysis OpenFOAM 2.3.x was applied.

A key issue in modelling the cold spray nozzle is the description of the interaction (force and energy transfer) between gas and particles. In OpenFOAM the force transfer is described by the *particleForce* model and the heat transfer by the *heatTransfer* model. OpenFOAM comprises different drag and particleForce models, respectively. In the *particleForce* model *sphereDrag* the drag coefficient is defined according to Schiller and Naumann [9]:

$$C_D(Re_p)=\begin{cases} 0.424 & Re_p > 1000 \\ \frac{24}{Re_p}\left(1+\frac{1}{6}Re_p{}^{0.66}\right) & Re_p < 1000 \end{cases} \qquad (8)$$

The second *WenYuDrag particleForce* model according to Wen and Yu [10, 11] has a similar form and takes into account the particle volume fraction α_s:

$$C_D(Re_p)=\begin{cases} \frac{0.44}{(1-\alpha_s)^{3.65}} & Re_p > 1000 \\ \frac{24}{(1-\alpha_s)^{3.65}Re_p}\left(1+0.15((1-\alpha_s)Re_p)^{0.687}\right) & Re_p < 1000 \end{cases} \qquad (9)$$

The *ErgunWenYuDrag* particleForce model [10 - 12], also available in OpenFOAM, was not regarded in this analysis, as for the low particle volume fractions α_s in cold spraying of typically 10^{-6}, it is identical to the *WenYuDrag* model.
The third *particleForce* model regarded in this analysis is called *nonSphereDrag.* It can actually handle nonspherical particles. For the spherical particles regarded in this analysis the corresponding drag coefficient has the following form, when neglecting nonspherical contributions [13]:

$$C_D(Re_p) \approx \frac{24}{Re_p}\left(1+0.17Re_p^{0.65}\right) \qquad (10)$$

In the fourth *particleForce* model called *PlessisMasliyhaDrag* according to Plessis and Masliyha [10, 14] the drag coefficient has the following form:

$$C_D(Re_p) = \frac{4(1-\alpha_s)}{3}\left(\tilde{A}\frac{\alpha_s}{1-\alpha_s}+\tilde{B}\cdot Re_p\right) \quad ,(11)$$

with the coefficients

$$\tilde{A} = \frac{26.8(1-\alpha_s)^3}{\alpha_s^{2/3}(1-\alpha_s^{1/3})(1-\alpha_s^{2/3})^2} \qquad (12)$$

and

$$\tilde{B} = \frac{(1-\alpha_s)^2}{(1-\alpha_s^{2/3})^2} \quad .(13)$$

The *PlessisMasliyhaDrag particleForce* model delivers a drag coefficient 2 – 3 times higher than the other models for the Reynolds number range relevant in cold spraying (compare **Fig. 4**).
For the heat transfer between gas and particles in OpenFOAM only one *heatTransfer* model called *RanzMarshall, according to Ranz and Marshall* [15, 16], is available. The heat transfer coefficient contained in this model is the following:

$$\nu = 2 + 0.6Re_p{}^{0.5}Pr^{0.33} \qquad (14)$$

At this $Pr = \eta_g C_{p,g}/\lambda_g$ is the Prandtl number of the gas, with the dynamic viscosity η_g, the heat capacity $C_{p,g}$ and the thermal conductivity λ_g.
Figs. 4 and **5** show the drag and heat transfer coefficients implemented in OpenFOAM for the Reynolds number range relevant at typical cold spray conditions.

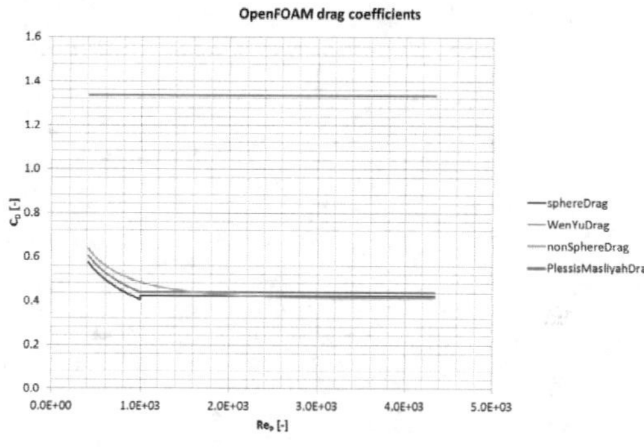

Fig. 4. OpenFOAM drag coefficients.

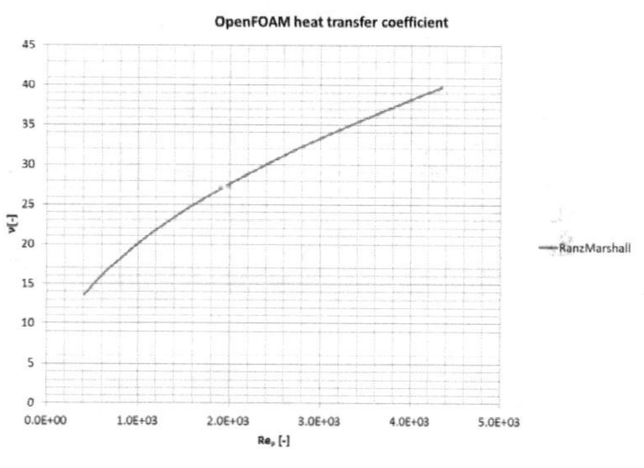

Fig. 5. OpenFOAM heat transfer coefficient.

4 Simulation Results and Discussion

Based on the simulation model heating and acceleration of copper powder in the nitrogen flow of a cold spray nozzle was analysed. In the simulation spraying of a spherical copper powder with a medium particle size of 25 *µm* at three different parameter sets of gas pressure and temperature (32 *bar* / 673 *K*, 40 *bar* / 673 *K* & 40 *bar* / 873 *K*) was regarded and the four different drag models described in section 3 (*sphereDrag*, *WenYuDrag*, *nonSphereDrag* and *PlessisMasliyahDrag*) were compared to each other. The analysed nozzle has a length of 140 *mm*, an exit diameter of 6.4 *mm* and a throat diameter of 2.7 *mm*.

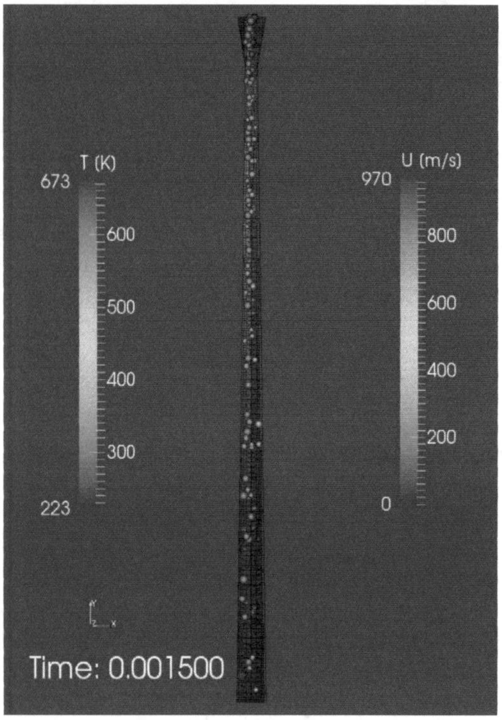

Fig. 6. Simulation of gas flow (nitrogen), particle heating and acceleration in the cold spray nozzle for copper powder ($\bar{d}$ = 25 µm) at a pressure of p_g = 32 *bar* and a temperature of T_g = 673 *K*.

The powder is injected 10 *mm* before the nozzle throat in the convergent section of the nozzle.
The obtained results for velocity and temperature of gas flow and powder particles at a gas pressure of 32 *bar* and a gas temperature of 673 *K* are shown in **Fig. 6**. The left hand side shows the temperature T [*K*] and the right hand side the velocity u [*m/s*] of gas and particles, respectively. The size of the particles is enhanced by a factor of 50. For the shown results the *PlessisMasliyhaDrag particleForce* model was applied.
Fig. 7 shows the evolution of the regarded quantities along the axis of the cold spray nozzle at a gas pressure of 32 *bar* and a gas temperature of 673 *K*. The velocity of the process gas at the exit of the nozzle is 969 *m/s* at a temperature of 222 *K*. Particle velocities and temperatures along the nozzle length after a calculation time of 0.0015 s are plotted as single dots. The width of the particle distribution in combination with the statistics of particle insertion leads to a broadening of particle velocities and temperatures along the nozzle.
The medium exit velocity of the particles is 571 *m/s* ± 42 *m/s* and their medium temperature has a value of 357 *K* ± 10 *K*. The simulations were carried out with the four different drag models available in OpenFOAM. The results for particle velocities and temperatures at the nozzle exit for the three regarded parameter sets are plotted in **Figs. 8** and **9**.
In order to verify the simulation model its predictions were compared to the results of the theoretical description of gas and particle flow in the cold spray nozzle described in section 2. For the theoretical calculation a solid fraction α_s of 10^{-6} was assumed.

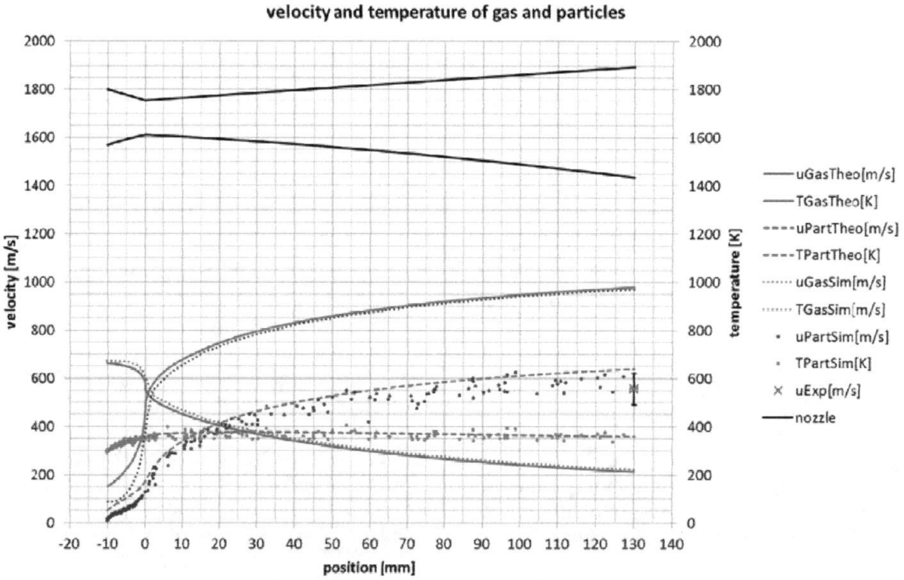

Fig. 7. Velocity and temperature of gas and particle flow on the axis of the cold spray nozzle for copper powder ($\bar{d}$ = 25 µm) at a pressure of p_g = 32 *bar* and a temperature of T_g = 673 *K*: simulation vs. theory & experiment.

Furthermore, for a validation of the predictions of simulation and theory the calculation results for the velocity were also compared to experimental measurement data for particle velocities obtained with an OSEIR SprayWatch system.
Both theoretical and experimental values are plotted in **Figs. 8** and **9**.

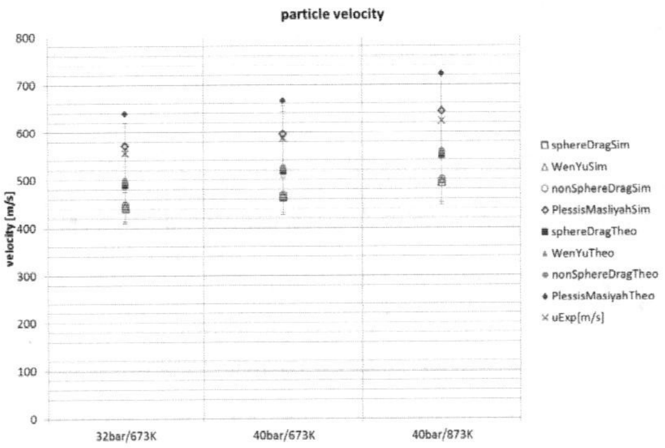

Fig. 8. Particle velocities obtained from simulation, theory and experiment.

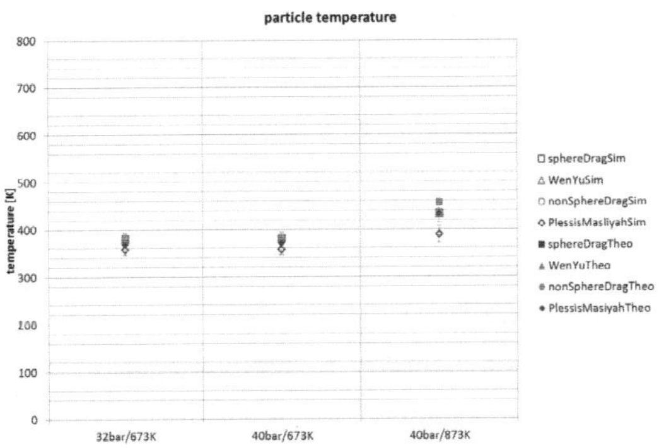

Fig. 9. Particle temperatures obtained from simulation and theory.

A look at the data makes obvious that generally the theoretically calculated particle velocity is slightly higher than that predicted by the simulation model. This deviation can be explained by severe assumptions necessary in the theoretical calculation. The theoretical values for particle temperatures are more or less within the range predicted by the simulation model. The differences between the drag models are less pronounced, as particle heating mainly takes place in the convergent nozzle section and acceleration in the divergent one. When comparing the theoretical predictions of theory and simulation to the experimentally determined particle velocity data, it becomes obvious that for all three parameter sets the *PlessisMasliyahDrag particleForce* model (compare equation (11)) delivers the best prediction of particle velocity. Among the drag models available in OpenFOAM it appears best suited for the simulation of particle heating and acceleration in cold spraying. This result is in accordance to drag model analyses in coupled computational fluid dynamics discrete element modelling of cold spraying [17, 18]. In this the drag model according to Koch and Hill [19], delivering a drag coefficient on the same scale as *PlessisMasliyahDrag* for the particle Reynolds number range relevant in cold spraying, was identified to be well suited for cold spray simulation.

5 Conclusions

The results presented demonstrate, that fluid dynamical modelling in combination with particle tracking based on the *reactingParcelFoam* solver of OpenFOAM can be successfully applied in order to analyse particle heating and acceleration in a cold spray nozzle. The simulation results obtained for cold spraying of copper powder at different spray parameters show good correlation to theoretical estimations for gas and particle temperature respectively velocity for different drag models. The *PlessisMasliyhaDrag particleForce* model of OpenFOAM delivers very good correspondence between simulated and experimentally determined particle velocites for the regarded parameter sets. It appears well suited for an OpenFOAM based modelling of particle heating and acceleration in cold spraying.

7 Literature

[1] A. Papyrin, V. Kosarev, S. Klinkov, A. Alkimov, V. Fomin: Cold Spray Technology. Elsevier, 2007. – ISBN 978-0-008-045155-8
[2] V. K. Champagne: The cold spray materials deposition process – Fundamentals and applications. CRC Press, 2007. – ISBN 978-1-4200-6670-8
[3] T. Stoltenhoff: Kaltgasspritzen von Kupfer – Eine strömungsmechanische und werkstoffkundliche Analyse und Optimierung des Spritzprozesses. Dissertation Universität Hamburg, Shaker Verlag, 2004. – ISBN 3-8322-2302-9
[4] T. Schmidt: Kaltgasspritzen – Eine Analyse des Materialverhaltens beim Partikelaufprall und die daraus abgeleitete Prozessoptimierung. Dissertation Universität Hamburg, Shaker Verlag, 2007. – ISBN 978-3-8322-6399-7
[5] T. Schmidt, H. Assadi, F. Gärtner, H. Richter, T. Stoltenhoff, H. Kreye, T. Klassen: From Particle Acceleration to Impact and Bonding in Cold Spraying. Journal of Thermal Spray Technology 18 (2009), pp. 794-808.
[6] H. Assadi, H. Kreye, F. Gärtner, T. Klassen: Cold spraying – A materials perspective. Acta Materialia 116 (2016), p. 382-407.
[7] P. Fauchais, M. Vardelle: Sensors in Spray Processes. Journal of Thermal Spray Technology (2010), Volume 19(4), pp. 668-694.
[8] J. D. Andersion: Modern Compressible Flow. McGraw-Hill, 1982 – ISBN 0-07-001654

[9] L. Schiller , A. Naumann: Über die grundlegenden Berechnungen bei der Schwerkraftaufbereitung. Zeitschrift des Vereins deutscher Ingenieure, Band 77, Nr. 12 (1933).
[10] J. Lundberg, B. M Halvorsen: A review of some existing drag models describing the interaction between phases in a bubbling fluidized bed. Proc. 49th Cand. Conf. Simulation and Modeling (2008).
[11] C. Y. Wen, Y. H. Yu: Mechanics of Fluidization. Chemical Engineering Progrsss Symposium, Vol. 62, No. 62 (1966), pp. 100-111.
[12] S. Ergun: Fluid Flow Through Packed Columns. Chemical Engineering Progress, Vol. 48, No. 2 (1952), pp. 89-94.
[13] A. Haider, O. Levenspiel: Drag Coefficient and Terminal Velocity of Spherical and Nonspherical Particles. Powder Technology, 58 (1989), pp. 63-70.
[14] J. P. du Plessis, J. H. Masliyah: Mathematical Modelling of Flow Through Consolidated Isotropic Porous Media. Transport in Porous Media, 3 (1988), pp. 145-161.
[15] W. E. Ranz, W. R. Marschall: Evaporation From Drops Part I. Chemical Engineering Progress, Vol. 48, No. 3 (1952), pp. 141-146.
[16] W. E. Ranz, W. R. Marschall: Evaporation From Drops Part II. Chemical Engineering Progress, Vol. 48, No. 4 (1952), pp. 173-180.
[17] K.-H. Leitz, M. O'Sullivan, A. Plankensteiner, T. Lichtenegger, S. Pirker, H. Kestler: CFDEM Modelling of Particle Heating and Acceleration in Cold Spraying. Proceedings of the 19th Plansee Seminar 2017, 29 May – 2 June 2017, Reutte, Austria, (2017) – in press
[18] K.-H. Leitz, M.O'Sullivan, A. Plankensteiner, H. Kestler, L.S. Sigl: Comprehensive Modelling of the Cold Spray Process – from the Heating and Acceleration of the Particle in the Nozzle to its Impact on the Substrate, Proceedings of the Euro PM2017, 1-4 October 2017, Milan, (2017) – submitted
[19] D. L. Koch, R. J. Hill: Inertial Effects in Suspension and Porous-Media Flows. Annu. Rev. Fluid Mech, 33 (2001), pp. 619-647.

Design of Cold Spray Nozzle to Optimize the Particle Velocity by Numerical Simulation

T.P. Han [a], W.Y. Li [a,*], X.P. Guo [b], X.W. Yang [a]
[a]State Key Laboratory of Solidification Processing, Shaanxi Key Laboratory of Friction Welding Technologies, School of Materials Science and Engineering, Northwestern Polytechnical University, Xi'an, P.R. China
[b]Marine Engineering College, Jimei University, Xiamen, P.R. China

Abstract: This study investigated the accelerating behavior of spray particles during cold spraying (CS) by employing a computational fluid dynamics program, FLUENT. Optimization of the dimensions of CS nozzle was conducted to maximize particle velocity. The results show that the expansion ratio, divergent length, particle density and size, operating temperature significantly influence particle acceleration. It is found that the spray particles in nozzles with long divergent length can obtain a relatively higher impact velocity, but too long divergent length will reduce the particle velocity. Besides, the particle impact velocity shows a downward trend with increasing the particle size or density. Hence, the optimal divergent length should increase with the increase of particle mass. Moreover, higher gas temperature leads to a higher particle velocity, but it has no influence on the optimal divergent length.

1 Introduction

Cold spraying（CS） is a relatively new coating technique invented in the mid-1980s at the Institute of Theoretical and Applied Mechanics of the Siberian Division of the Russian Academy of Science in Novosibirsk a[1]. In this process, particles are accelerated by a supersonic jet of compressed gas to a very high velocity (300-1200 m/s) and impact on the substrate at a low temperature, which is below the melting point of the materials. Particles undergo severe plastic deformation and bond to substrate, forming a coating. Cold spray differs significantly from thermal spray because of the high velocity that particles reach and their low temperature. This process does not have disadvantages like oxidation and phase transformations, and it can be used to coat temperature-sensitive materials[1].

The most important factor that determines whether a particle can deposit successfully on the substrate is its velocity prior to impact. It is widely accepted that there exists a critical velocity for successful particle deposition which when exceeded a particle bonds to the substrate [2-4]. In order to get high deposition efficiency, spray particles should be accelerated to a velocity higher than the critical one. Therefore, it is significant to understand and control the particle acceleration behavior.

The gas flow inside and outside the nozzle significantly affects particle acceleration behavior. Although it is possible to measure particle velocity and study gas flow experimentally [5, 6], high cost and time consuming limit the wide application of this method. To solve this problem, analytical and numerical modeling were developed and used in many studies [7]. In published literature, there are two sets of factors that have been investigated to improve particle velocity. The first set is that of operating parameters, such as inlet pressure, gas type and carrier gas conditions [8-11]. It has been found that higher inlet pressure, temperature or the use of helium as a driving gas lead to higher impact velocity. The second set of papers focus on the nozzle geometry. In the early stage, the 1D-isentropic-flow theory was used to design nozzles and always referred to the design of engine tail pipes[12]. Recently, with the rapid development of the computer power, computational fluid dynamics (CFD) technique was popularly used to study influences of nozzle geometry on gas flow and particles accelerate behavior [13-19]. A nozzle optimization design method based on this new technique was developed [13-15]. It has been found that there exists an optimal expansion ratio (expansion ratio is the ratio of the area of the exit to the throat) for a given state where particles can achieve the maximum velocity. However, the design of cold spray nozzle is more than these. For example, the design purpose of cold spray nozzle is vastly different from engine tail pipes which aim at the maximum thrust, while CS wants the maximum particle velocity. The previous design approaches lacked of principles that are exclusively used for cold spray nozzle. In this paper, particle properties, operating temperature, divergent length are chosen as the key concerns to optimize design CS nozzle.

2 Modeling

2.1 Parameters of the model

The geometry of a converging-diverging nozzle used in this model is shown in Fig.1. In this study, the spray nozzle has a circular cross-section and five geometric parameters used in its design included the inlet diameter (D_i), throat diameter (D_t), exit diameter (D_e), upstream length (L_u), and downstream length (L_d). According the previous study [14,18], inlet diameter, throat diameter and convergent length have a little or no effect on the particle velocity. Meanwhile, the throat diameter can influence the gas flow and convergent length can influence particle temperature. Therefore, D_i, D_t and L_u were fixed at 16 mm, 2.7 mm and 30 mm respectively, which are close to previous study[14]. The exit diameter and divergent length were changed to maximize particle velocity. In order to analysis the influence of exit diameter in a wide range, the expansion ratio was changed to change the exit diameter and three expansion ratios (4, 8, 12) have been studied. Ld was altered from 120 mm to 520 mm with an interval of 80 mm.

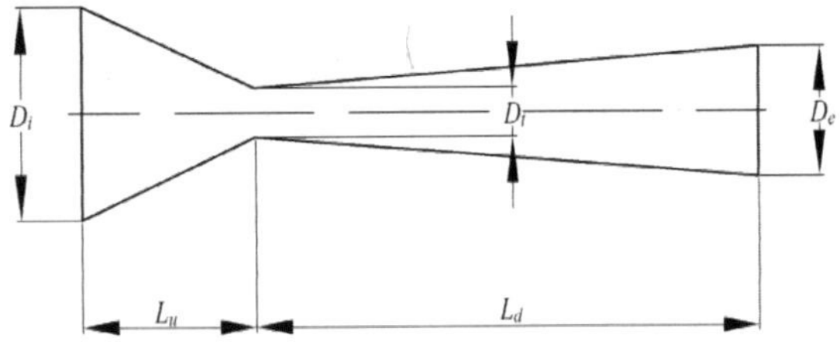

Fig. 1. Nozzle geometry

Air is selected as the driving gas. The inlet pressure was fixed in all cases at 3 MPa and the gas temperature changed from 300 to 800℃. The room temperature is 27℃. Spherical copper, titanium and aluminum particles were chosen as the feedstock. In each case, a single particle at a velocity of 15 m/s was introduced axially into the gas flow. The initial temperature of the particle at the room temperature.

Table 1. Properties of particles used in simulation

	Density (J/kg·K)	Specific heat capacity (J/kg·K)	Particle size (μm)
Copper	8978	381	10,20,30,40,50
Titanium	4850	544	20
Aluminum	2719	871	20

2.2 Numercial simulation method

Numerical simulations were carried out with the commercial software ANSYS-FLUENT 15.6 to determine the flow field of the driving gas inside and outside the nozzle, and subsequently the accelerating and heating of particles in a cold spray stream. As shown in Fig.2, a two-dimensional symmetrical model was employed to reduce computational effort [14]. The model from the nozzle exit was a cylinder of 100 mm in radius and 200 mm in length. A substrate plate with 20 mm in radius and 5 mm thick was placed as in the actual process at a standoff distance of 30 mm from the nozzle exit. A mesh of quadrilateral elements was used to fill the whole model. The mesh at the nozzle throat and exit zone was further refined to keep the calculation precise due to the highly compressible characteristic of the supersonic driving gas.

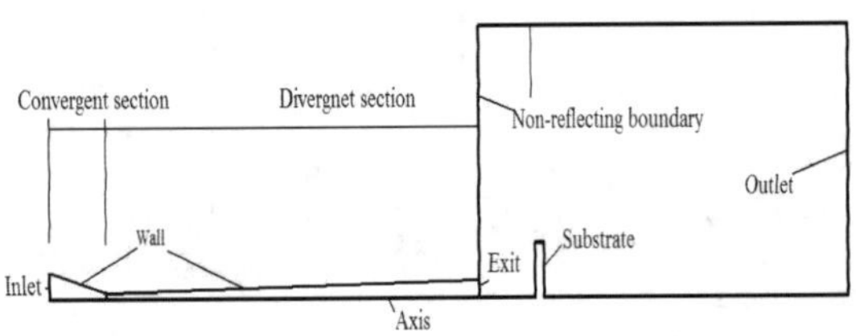

Fig. 2. Computational domain and boundaries of model developed by using FLUENT

The driving gas was considered to be an ideal and compressible one. The particle was taken as spherical and the interaction between the particle and gas was not included in the analysis. The simulation model used was the standard K-ε turbulence model in FLUENT. For near-wall areas, the standard wall function of the software was utilized. Discrete phase modeling (DPM) was chosen to compute the acceleration of the particles. A coupled implicit method was used to solve the flow field and obtain a steady state flow field. The equations of steady compressible flow follow the laws of conservation of mass, momentum and energy. The equations of acceleration of a spherical particle by a compressed gas described in details in [13,14].

3 Results and discussion

3.1 Effect of particle size on optimal nozzle design

Fig.3 shows the effect of Cu particle size on particle velocity for different nozzles dimensions. Copper particles of different sizes (10 μm , 20 μm, 30 μm, 40 μm and 50 μm) were used in each run. It can be seen in Fig.3 that the particle velocity is inversely proportional to particle diameter. This result is consistant with previous study [13-14,19]. It should be noted that the particle size also affects the optimal divergent length. With increasing particle size, the optimal divergent length increases. These facts are because large particles have larger weight and inertia compared to small ones. By increasing weight and inertia, the effect of the driving gas on the particles reduces gradually. Therefore, for a large particle, a long divergent length is required to accelerate it to maximum speed.

It also can be seen from Figs.3(a) and (b) that increasing the divergent length, the particle impact velocity increases at first and decreases after it reaches a maximum value. In Fig.3(c), the particle velocity rises continuously with divergent length. According to the previous literatures [13,14], the compressed gas in a large expansion ratio nozzle yields a higher Mach number than in a small expansion ratio nozzle, and gas with higher Mach number also has higher velocity. Therefore, particles can accelerate continuously with a large expansion ratio nozzle. The Mach number contour of a gas flow for different divergent length nozzles at an expansion ratio of 8 was studied and shown in Fig.4. It can be seen that Mach number of the gas inside the nozzle reduces gradually with the increase of divergent length. This result is consistant with Yin's study [19]. So the velocity of the gas inside nozzle lowers and affects particle impact velocity as a consequence. Over long divergent length can provide a long accelerate channel, but the gas velocity decreased seriously, so the particle velocity decreased.The combinatory effect of gas velocity and divergent length contributes to these results and there exists an optimal state that particle can accelerate to maximum velocity.

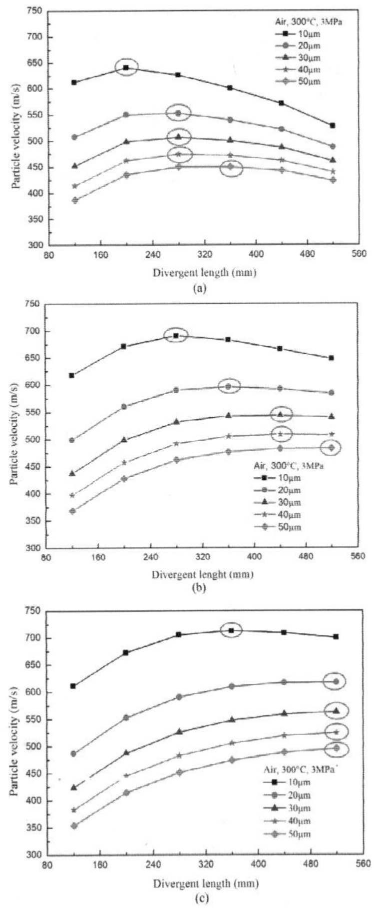

Fig. 3. Effect of Cu particle size on particle velocity for different nozzles dimensions. Expansion ratio (a) 4 (b) 8 (c) 12

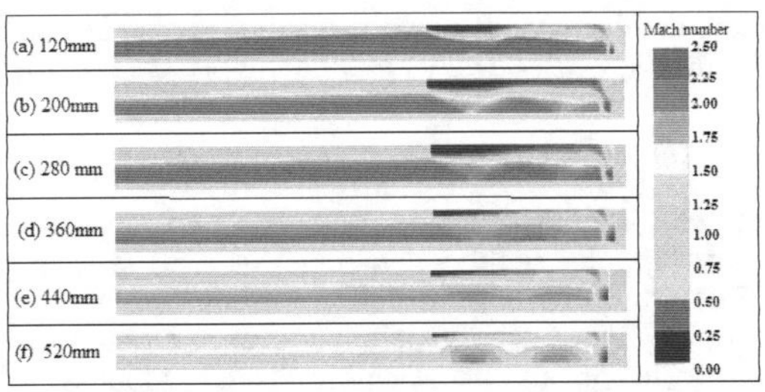

Fig. 4. Effect of divergent length on the Mach number contour for nozzles with an expansion ratio of 8

3.2 Effect of material type on on optimal nozzle design

Fig.5 shows the effect of material types on particle velocity for different nozzles. It is clearly that in each runs that copper particle acclerate to the lowest velocity and aluminum particle acclerate to the highest velocity. In addition, it can be seen from Figs.5(a) and (b) the effect of the divergent length on velocity, for expansion ratios of 4 and 8, was non- linear with a maximum value for particles of all materials but a different level of velocity for each one of them. However, in the case when the expansion ratio was set to 12(Fig.5(c)) velocities of particles of all materials appear to increase with divergent length for all runs reaching a plateau, again at a different level of velocity for each one of them. These observations can be explained when the different densities of particles are taken into account. For copper, which has the largest density among all three materials, it requires a lot of energy to accelerate. So the impact velocity of copper changes slowly in the nozzles with expansion ratios of 4.0 and 8.0 and it requires a longer accelerating channel for the maximum velocity. When nozzle with a long divergent length is employed, the particle velocity is dependent mainly on the gas velocity inside the nozzle. Although driving gas inside nozzles with expansion ratio of 12 reaches a high speed and the longer divergent length provides a longer acceleration time, this gas velocity isn't adequate for copper particle to attain a higher velocity. So particle with larger density require a much higher expansion ratio for a higher acceleration.

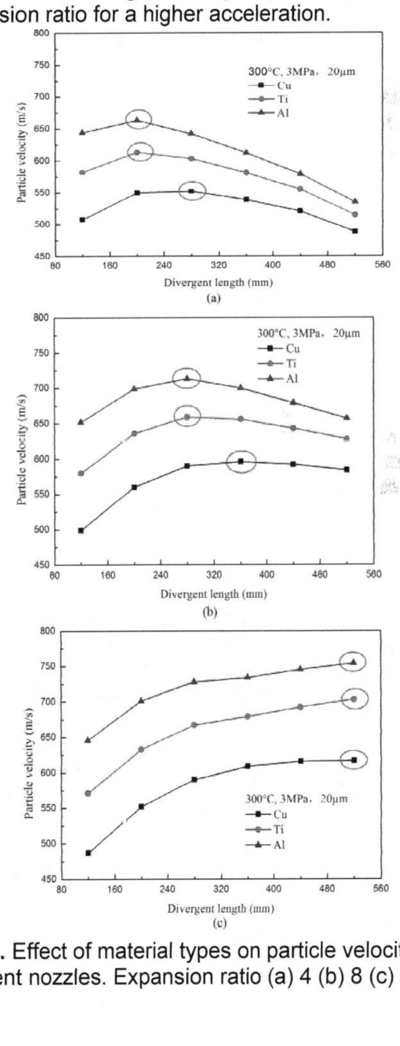

Fig. 5. Effect of material types on particle velocity for different nozzles. Expansion ratio (a) 4 (b) 8 (c) 12

3.3 Effect of temperature on optimal nozzle design

Fig.6 shows the velocity of 20 µm diameter copper particles sprayed in different temperatures. It can be seen that particle sprayed in high temperature can accelerate to a higher velocity. This result is agree with previous study [13] and this is because gas velocity will increase with the increase of temperature. It should be noted that the divergent length where particle attains maximum velocity has no change when temperature increases. It has been said in Section 3.2 there exists an optimal length that particle can accelerate to maximum velocity and the particle velocity is influenced by gas velocity and divergent length. When the temperature increases, the gas velocity increases in all nozzles, so the optimal divergent length has no change and can still provide the best accelerate length.

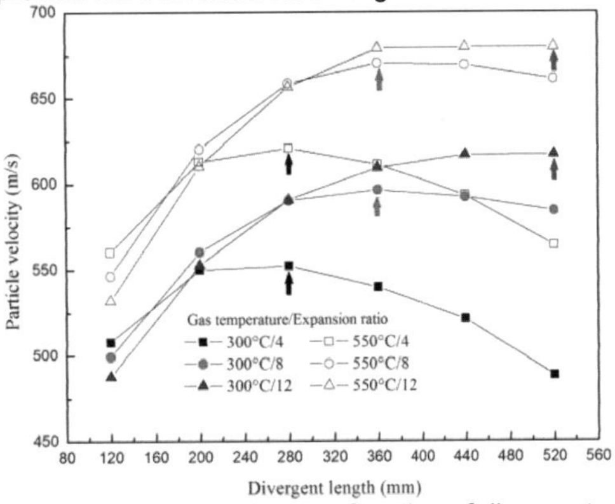

Fig. 6. Particle velocity as a function of divergent length under different gas temperature

3 Conclusion

In this paper, the effect of expansion ratio,divergent length, particle material and temperature on particle velocity in CS was studied using numerical modelling. According to the simulation results, it was found that the divergent length and particle material significantly affect particle velocity. Long divergent length provides a long acceleration channel for particles, but over long divergent length will reduce particle velocity. There exists an optimum relationship between acceleration channel length and gas velocity which maximizes particle velocity. In addition, large values of inertia make particles with large size or high density difficult to accelerate, so the optimal divergent length should be proportional to particle mass. Moreover, with increasing of gas temperature, the velocity of particles increases, but gas temperature has no influence on Mach number and the optimal divergent length.

Refences

[1] Alkimov A P, Kosarev V F, Papyrin A N. A method of "cold" gas-dynamic deposition. Dokl Akad Nauk SSSR 315 (1990), pp. 1062/65.

[2] Assadi H, Gärtner F, Stoltenhoff T, and Kreye H. Bonding mechanism in cold gas spraying. Acta Materialia 51 (2003), Issue 15, pp. 4379/94.

[3] Bae G, Kumar S, Yoon S, Kang K, Na H, Kim H J, and Lee C. Bonding features and associated mechanisms in kinetic sprayed titanium coatings. Acta Materialia 57 (2009), Issue 19, pp. 5654/66.

[4] Assadi H, Schmidt T, Richter H, Kliemann J O, Binder K, Gärtner F, Klassen K, and Richter H . On parameter selection in cold spraying. Journal of Thermal Spray Technology 20 (2011), Issue 6, pp.1161/76.

[5] Wong W, Irissou E, Ryabinin A N, Legoux J G, and Yue S. Influence of helium and nitrogen gases on the properties of cold gas dynamic sprayed pure titanium coatings. Journal of Thermal Spray Technology 20 (2011), Issue 1, pp. 213/26.

[6] Klinkov S V, Kosarev V F, Zaikovskii V N. Influence of flow swirling and exit shape of barrel nozzle on cold spraying. Journal of Thermal Spray Technology 20 (2011), Issue 4, pp. 837/44.

[7] Yin S, Meyer M, Li W, Liao H , and Lupoi R. Gas flow, particle acceleration, and heat transfer in cold spray: A revie. Journal of Thermal Spray Technology 25 (2016), Issue 5, pp. 1/23.

[8] Li S, Muddle B, Jahedi M, and Soria J. A numerical investigation of the cold spray process using underexpanded and overexpanded Jets. Journal of Thermal Spray Technology 21 (2012), Issue 1, pp. 108/20.

[9] Yin S, Liu Q, Liao H, and Wang X. Effect of injection pressure on particle acceleration, dispersion and deposition in cold spray. Computational Materials Science 90 (2014), Issue 1, pp. 7/15.

[10] Yin S, Suo X, Liao H, Guo Z, and Wang. Significant influence of carrier gas temperature during the cold spray process. Surface Engineering 30 (2014), Issue 6, pp. 443/50.

[11] Sova A, Grigoriev S, Kochetkova A, and Smurov. Influence of powder injection point position on efficiency of powder preheating in cold spray: Numerical study. Surface & Coatings Technology 242 (2014), pp. 226/31.

[12] Alkhimov A P, Kosarev V F, Klinkov S V. The features of cold spray nozzle design. Journal of Thermal Spray Technology 10 (2001), Issue 2, pp. 375/81.

[13] Li W Y, Li C J. Optimal design of a novel cold spray gun nozzle at a limited space. Journal of Thermal Spray Technology 14 (2005), Issue 3, pp. 391/96.

[14] Li W Y, Liao H, Douchy G, and Coddet C. Optimal design of a cold spray nozzle by numerical analysis of particle velocity and experimental validation with 316L stainless steel powder. Materials & Design, 28 (2007), Issue 7, pp. 2129/37.

[15] Sova A, Grigoriev S, Smurov I. Velocity of the Particles Accelerated by a Cold Spray Micronozzle: Experimental Measurements and Numerical Simulation. Journal of Thermal Spray Technology 22 (2013), Issue 1, pp. 75/80.

[16] Yin S, Wang X F, Li W Y. Computational analysis of the effect of nozzle cross-section shape on

gas flow and particle acceleration in cold spraying[J]. Surface & Coatings Technology 205 (2011), Issue 8–9, pp. 2970/77.

[17] Suo X K, Liu T K, Li W Y, Suo Q L, Planche M P, and Liao H L. Numerical study on the effect of nozzle dimension on particle distribution in cold spraying[J]. Surface & Coatings Technology 220 (2013), pp. 107/11.

[18] Jung H B, Park J I, Park S H, Kim H J, Lee C H, and Han J W. Effect of the expansion ratio and length ratio on a gas-particle flow in a converging-diverging cold spray nozzle. Metals and Materials International 15 (2009), Issue 6, pp. 967/70.

[19] Yin S, Zhang M, Guo Z, Liao H, and Wnag X. Numerical investigations on the effect of total pressure and nozzle divergent length on the flow character and particle impact velocity in cold spraying. Surface & Coatings Technology 232 (2013), pp. 290/97.

The high temperature wear and oxidation behavior of CrC-based HVOF coatings

Š. Houdková, Z.Česánek, E. Smazalová, Plzeň/CZ

The 3 commercially available CrC-based powders with different kind of matrix (Cr_3C_2-25%NiCr; Cr_3C_2-25%CoNiCrAlY and Cr_3C_2-50%NiCrMoNb) were deposited by HVOF JP-5000 spraying gun, evaluated and compared. The sliding wear resistance, measured at room and elevated (T=600°C) temperature according to ASTM G-133, the influence of heat treatment on the microstructure and properties, as well as the oxidation resistance in hot steam environment (p=24 MPa; T=610°C) were evaluated with respect to their potential application in steam power industry. The surface oxidation and microstructure changes were evaluated by SEM and XRD. The NiCr matrix proved to provide high oxidation and sliding wear resistance of the coating regardless volume content. On the contrary, the Cr_3C_2-25%CoNiCrAlY coating was subjected to massive oxidation of carbide particles. The low cohesive strength and high porosity of Cr_3C_2-25%CoNiCrAlY was identified responsible for its poor oxidation resistance in hot steam environment. The sliding wear resistance was found comparable at room temperature, regardless the matrix composition and content, while at elevated temperatures, the higher volume content of matrix led to higher wear of coating material.

1 Introduction

The power generation industry is one of the most challenging from the materials point of view. The most stressed components are exposed to the combination of high temperature, aggressive environment and mechanical load. To withstand such conditions, surface treatment are often considered as a possible solution. The thermal spray technology, namely High Velocity Oxygen Fuel (HVOF) is technology suitable for applying of hard, wear and oxidation resistant coatings. Among HVOF sprayed wear resistant coatings, applicable in high temperature environment, the CrC-based hardmetals are considered to be the most effective [1]. In the group of CrC-based hardmetals, the Cr_3C_2-25%NiCr coating is the most used. Its microstructure and behavior under various loading conditions were described in detail many times. The influence of heat treatment on the Cr_3C_2-25%NiCr microstructure and related changes of mechanical, wear and corrosion properties were also addressed [2, 3]. The detailed study of Cr_3C_2-25%NiCr behavior was published by Matthews et al [4-7]. While a quite detailed knowledge of as-sprayed and heat treated Cr_3C_2-25%NiCr coatings exists, less is known about the microstructure and behavior at high temperatures.
Some of the studies reported the decrease of wear rate due to the secondary carbides precipitation or formation of oxide tribofilm [8, 9]. In recent study of Bolelli et al. [10], the increase of wear by one order of magnitude compare to the room temperature values was measured at 400°C. The increase of wear is related to thermal softening on the matrix, in accordance with the observation of Matthews [11]. On the other hand, an agreement is generally found in the decrease of COF values, measured during high temperature sliding, for both Al_2O_3 [9,10] or self-mating counterpart material [12].
Considering the high temperature behavior, the oxidation has to be addressed. Aso here, just a few studies are available [5,6]. In [6], the mechanism of inner preferential oxidation of carbides at simulated turbine conditions is proposed. Compare to Cr_3C_2-25%NiCr, low attention was paid up to now to CrC-based materials with alternative matrix such as CoNiCrAlY or NiCrMoNb, which can provide improved corrosion, oxidation or wear resistance.
Picas [13] studied the tribological properties of HVOF as-sprayed and heat treated Cr_3C_2-CoNiCrAlY. Also here, the precipitation of secondary carbide was observed in the vicinity of heat treatment. With increasing temperature, the presence and growth of another phase, identified as Cr_2O_3 not only on the coating surface but also inside, was described. The influence of heat treatment on the wear resistance was not clearly demonstrated. The Cr_3C_2-50%NiCrMoNb spraying powder was introduced into the market recently. Just a few references are available concerning its performance. In [14] the microstructure and phase composition of both the original powder and coating is well described. The Cr_3C_2-50%NiCrMoNb sliding wear behavior at room and elevated temperature was further studied and referred in [9]. Compare to the usual Cr_3C_2-25%NiCr, the Cr_3C_2-50%NiCrMoNb showed higher wear due to the higher content of matrix.
As still low information exist concerning the performance of CrC-based coatings at elevated temperatures, namely concerning the coatings with alternative matrix, further research is needed.
The aim of this study is to evaluate wear properties at room and elevated temperature and oxidation resistance of CrC based coating with typical NiCr matrix, CrC-based coating with highly oxidation resistant CoNiCrAlY and wear and corrosion resistant NiCrMoNb matrix. The oxidation resistance in hot steam environment was tested to reveal the potential of the coatings to serve in the steam power industry.

2. Experimental set-up

The coatings were deposited onto carbon steel substrates and high chrome steel samples. Prior to

the spraying, the substrates were grit blasted by alumina with 0.8-1 mm grain sizes (F22) to achieve roughness of Ra = 8, approximately.
Three commercially available CrC-based powders were used: Cr_3C_2-25%NiCr (Amperit 588.074), Cr_3C_2-25%CoNiCrAlY (Amperit 594.074) and Cr_3C_2-25%NiCrMoNb (Amperit 595.074). The HP/HVOF TAFA JP5000 spraying device was used. The thicknesses of the sprayed coatings were set to reach 400 µm.
The temperatures of annealing procedures was chosen with respect to the intended coatings application operating at approx. 580°C. Hot air annealing of the samples was realized in a muffle furnace at 600°C for 116 hours and cooled in air. Hot steam annealing was realized in the home-made equipment of the laboratory of surface analyses, University of Chemistry and Technology Prague. The temperature of the autoclave was maintain in the range 608-609°C, its inside pressure reached 24 MPa. For hot steam oxidation test, the cylindrical samples were grinded to reach the surface roughness Ra 0.2. The as-sprayed and annealed coatings were then analysed in terms of microstructure, phase composition, microhardness changes and sliding wear behavior. The microhardness was measured on the cross sections of both as-sprayed and annealed coatings using the HV0.3 method. The sliding wear resistance were evaluated using the Ball-on-Flat test, according to ASTM G133. Test parameters were as follows: 25 N Load; steel 100Cr6, 6 mm diameter ball counterpart; 5 Hz oscillating frequency; 10 mm stroke length; 1000 s testing time. The as-sprayed coating was measured at room temperature and at 600°C. The hot air annealed samples were measured at room temperature, to evaluate the influence of previous heat treatment on the wear properties. Prior to testing, the the coatings was polished to the 0.04 ± 0.02 Ra. For each coating, three different measurements were performed. The wear resistance was characterized by the coefficient of wear K [mm^3/Nm], calculated from measured wear tracks profiles.

2. Results

2.1 Feedstock powders microstructure

The feedstock powder cross sections were analysed by BSE (Fig. 1) and EDS. The Cr_3C_2-25%NiCr powder (Fig. 1a) consist of CrC carbides (1), embedded in the NiCr matrix (2). Inside the powder particles, certain amount of porosity is presented, some of the carbides are cracked. According to the XRD analyses the powder consists just from Cr_3C_2 carbides and NiCr matrix. No lower carbides, such as Cr_7C_3 were detected in the feedstock material. On the contrary, in the cross section of Cr_3C_2-25%CoNiCrAlY powder particle (Fig. 1b) more phases can be identified: the dark gray phase (3), corresponding to Cr-based carbide; the light gray phase (4), consisting mainly from Cr, Co and C with small amount of Ni; and light phase (5) with Ni as an dominant element with Co, C and Cr. Correspondingly, the XRD analyses confirmed the presence of both carbides Cr_3C_2 and Cr_7Cr_3, together with the Ni-based phase and low amount of CoNiAl. The high content of C in the matrix, namely in area (4), indicates the high amount of carbide dissolution during powder manufacturing process. Compare to Cr_3C_2-25%NiCr, the Cr_3C_2-50%NiCrMoNb (Fig. 1c) powder particle cross section has higher amount of porosity, but the carbide cracking is rather limited. The carbides seems to have two different composition: the dark gray center (6) surrounded by light gray rim (7). In the light grey rim, the EDS revealed presence of Mo (cca 7 wt %). This observation is in agreement with those referred by Matikainen et al. [14]. In his study, he mentioned the possible solubility of Mo in Cr_7C_3, substituting the Cr atoms, so rather than Cr_7C_3 the $(Cr,Mo,Ni)_7C_3$ carbide should be referred. The XRD analyses confirmed the presence of both type of carbides, $(Cr,Mo,Ni)_7C_3$ being even dominant. In the matrix (8) high amount of C is presented (cca 11 wt %), confirming the dissolution of C during powder manufacturing. Embedded in the matrix, small white precipitates can be found (9), identified as NbC. The NbC presence is also reported in [14].

2.2 As-sprayed coatings microstructure

The microstructure of as sprayed Cr_2C_3-25%NiCr coating is shown in the Figure 2a. Except of carbides (1), the various shades of gray can be seen in the matrix, as a result of carbide dissolution during spraying process. The darker the gray is, the higher the concentration of C was measured by EDS analyses, 20 wt % and 15 wt % for locations (2) and (3) resp. According to Rietveld analyses, more than half of than carbides transfer from Cr_3C_2 to Cr_3C_7. In the original powder, only Cr_2C_3 carbides were detected. All the transformation carried out during spraying indicating the too high temperature of the flame. Dissolution of carbides in the matrix was reported many times [10,14,15].
The as-sprayed Cr_3C_2-25%CoNiCrAlY coating (Fig. 2b) is highly porous. There are numerous cracks and decohesions. The evidence of carbide dissolution in the matrix can be found. The amount of C reaching almost 23 wt. % in the matrix (5). Even if there was significant amount of Cr_7C_3 carbide already in the original powder, more carbides were dissolved during spraying. The carbide dissolution of Cr_3C_2-25%CoNiCrAlY coating was already reported by Picas [13].
Cr_3C_2-50%NiCrMoNb as-sprayed coating (Fig. 2c) contains, similarly to original powder, two types of carbides: the dark grey (7), composed from C and Cr only (Cr_3C_2), and light grey carbide (8), where Mo and Ni are taking part, probably already mentioned $(Mo,Ni,Cr)_7C_3$. In the surrounding matrix (9), more than 25 wt. % of C is dissolved. The NbC precipitates are embedded in the matrix, similarly to the Cr_3C_2-50%NiCrMoNb original powder.

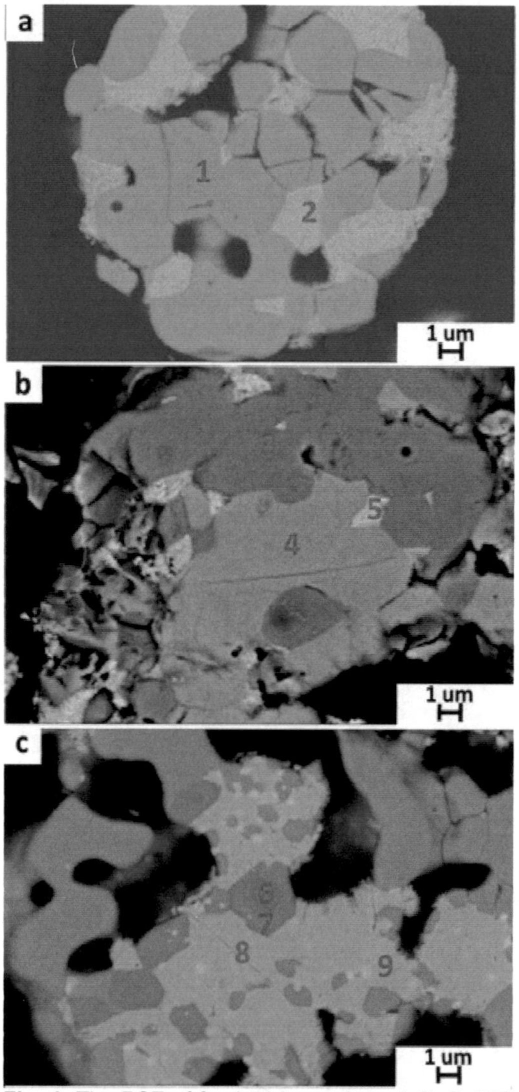

Fig. 1. The BSE of feedstock powders a) Cr_3C_2-NiCr, b) Cr_3C_2-CoNiCrAlY, c) Cr_3C_2-NiCrMoNb

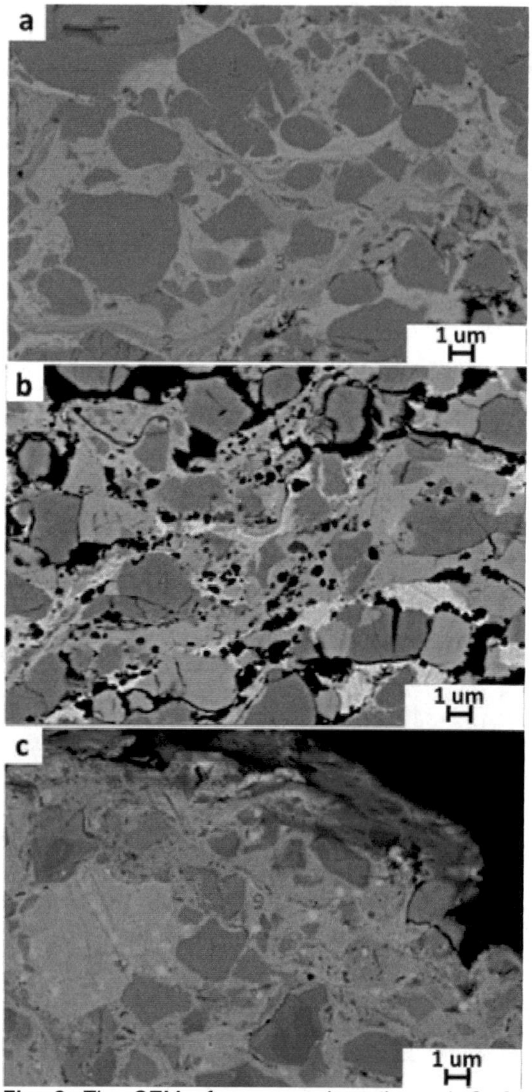

Fig. 2. The SEM of as-sprayed coatings a) Cr_3C_2-NiCr, b) Cr_3C_2-CoNiCrAlY, c) Cr_3C_2-NiCrMoNb

2.1 Air-annealed coatings microstructure

As expected, heat treatment led to a precipitation of secondary carbides from supersaturated matrix of all three kinds of CrC-based coatings. In the Cr_3C_2-25%NiCr microstructure, no Cr_7C_3 carbides were detected. The phase composition return back to Cr_3C_2 carbides in NiCr matrix. On the coatings surface, significant amount of Cr_2O_3 were measured as a result of oxidation in air atmosphere. Nevertheless, no signs of inner oxidation was observed (Fig 3a). In Cr_3C_2-25%CoNiCrAlY air-annealed coating, the secondary carbides precipitation can be also observed (Fig. 3b). Unlike the Cr_3C_2-25%NiCr coating, not all Cr_7C_3 carbides disappeared during the heat treatment, even if its amount was significantly lowered. As a result of oxidation in air, the Cr_3O_3 appeared on the surface as well as in the coating.The EDS analyses confirmed the high amount of oxygen and Cr in the measured areas (1) and (2). The preferential oxidation of carbides, starting from the carbide-matrix boundary, was previously described by Mathews [6]. High level of porosity in Cr_3C_2-25%CoNiCrAlY coating, together with many decohesion between carbides and matrix in as-sprayed coating contributes to creation of free path for oxygen, enhancing the inner oxidation. The inner oxidation of carbides, confirmed by presence of Cr_2O_3 inside the coating microstructure after the 900°C heat-treatment of Cr_3C_2-25%CoNoCrAlY coating, was also reported by Picas [13]. The precipitation of secondary carbides was also observed after heat treatment of Cr_3C_2-50%NiCrMoNb coating in air atmosphere (Fig. 3c). Both types of carbides are present: Cr_3C_2 (3) and $(Mo,Ni,Cr)_7C_3$ (4).

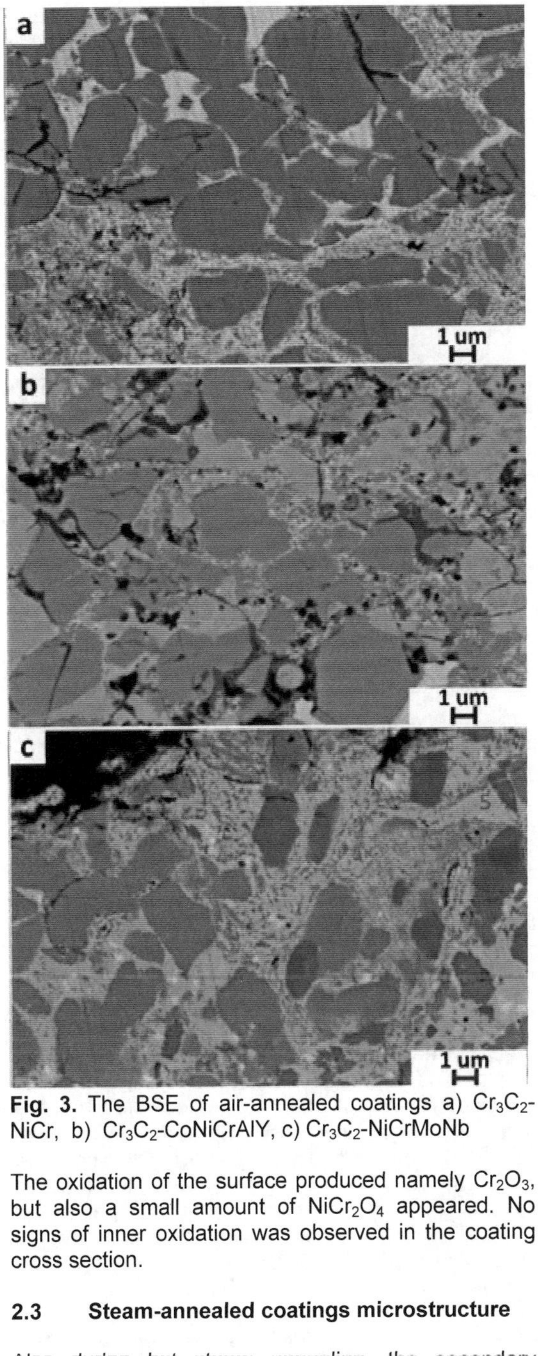

Fig. 3. The BSE of air-annealed coatings a) Cr_3C_2-NiCr, b) Cr_3C_2-CoNiCrAlY, c) Cr_3C_2-NiCrMoNb

The oxidation of the surface produced namely Cr_2O_3, but also a small amount of $NiCr_2O_4$ appeared. No signs of inner oxidation was observed in the coating cross section.

2.3 Steam-annealed coatings microstructure

Also during hot steam annealing, the secondary carbides precipitate from the supersaturated matrix. The similar temperature conditions and time of annealing have led to the carbides morphology similar to those created during annealing in hot air. The change of the environment and increase of pressure influenced namely the inner oxidation, which can be recognized in the steam-annealed Cr_3C_2-25%NiCr coating (Fig. 4a). The EDS analyses confirmed the presence of oxygen and Cr in the area (1) and (2).

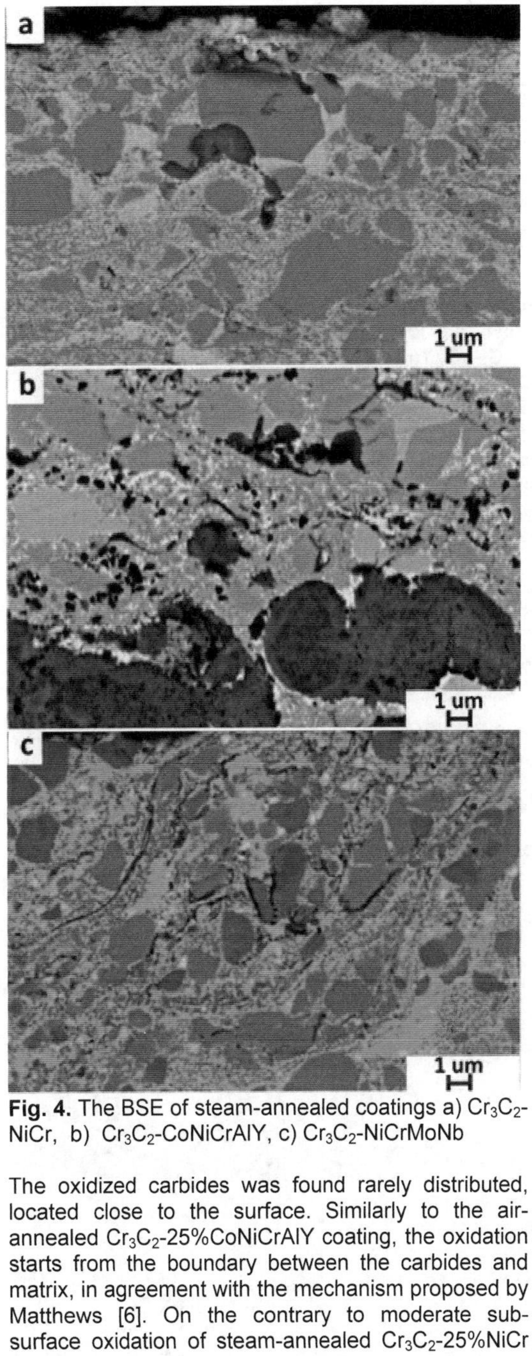

Fig. 4. The BSE of steam-annealed coatings a) Cr_3C_2-NiCr, b) Cr_3C_2-CoNiCrAlY, c) Cr_3C_2-NiCrMoNb

The oxidized carbides was found rarely distributed, located close to the surface. Similarly to the air-annealed Cr_3C_2-25%CoNiCrAlY coating, the oxidation starts from the boundary between the carbides and matrix, in agreement with the mechanism proposed by Matthews [6]. On the contrary to moderate sub-surface oxidation of steam-annealed Cr_3C_2-25%NiCr coating, massive oxidation occurred in Cr_3C_2-25%CoNiCrAlY coating (Fig. 4b). The coating was destroyed through all its thickness. Except the porosity and week boundaries between the carbides and matrix, the boundaries between spraying passes also serve as paths for oxygen to introduce into the coating and enhance further oxidation. The XRD analyses showed that all the lower carbides Cr_7C_3 oxidized, together with the majority of regular Cr_2C_3 carbides. There is no difference observed between air- and steam-annealed Cr_3C_2-50%NiCrMoNb coatings (Fig 4c). The absence of inner oxidation was confirmed by EDS analyses of carbide-matrix boundaries (3) and (4). In both places, the content of Cr and C significantly prevailing the oxygen amount. The dark areas was then identified as a decohesion between carbide and matrix.

2.4 Microhardness

To evaluate the heat-treatment influence, the microhardness HV0.3 was measured on the coatings cross section (Fig. 5). In as-sprayed state, the Cr_3C_2-25%NiCr was the hardest compare to other two CrC-based coating. After heat treatment, the hardness increase can be observed. With respect to the scatter there are no difference between air- and steam-annealed samples. In the literature, the microhardness of Cr_3C_2-25%NiCr was reported many times, in the as-sprayed state [2,7,10] and also after the heat treatment [2,3,4,7] . The average measured values are in the agreement with the results measured in this study. Matthews et al [7] brings the detailed analyses of microhardness evolution in dependence on temperature and time of exposure. For 500°C heat treatment, the microhardness increase was related to precipitation of small carbide grains. The same mechanism can be proposed also for 600°C heat treatment. Contrary to expectations, the hardness of Cr_3C_2-25%CoNiCrAlY is lower compare to Cr_3C_2-25%NiCr. It is caused by the poor microstructure of the coating, containing high amount of porosity and decohesions.

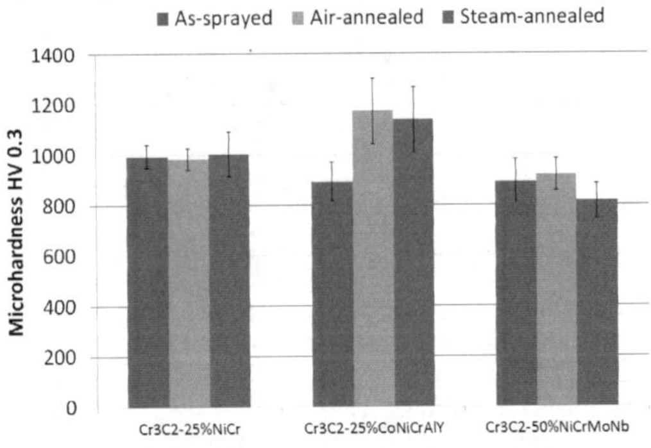

Fig. 5. Microhardness

The annealing procedures increased the microhardness of Cr_3C_2-25%CoNiCrAlY quite significantly. It has to be noted, that the indents was located in the areas of matrix strengthened by secondary carbides, so the influence of oxidation was supressed. Picas [13] referred the Cr_3C_2-25%CoNiCrAlY to have microhardness reaching almost 1050 HV0.3, being further increased by heat treatment up to 1050 HV0.3. Except of precipitation, the oxides development was also mentioned as a reason for strengthening of the coating. The microhardness of Cr_3C_2-50%NiCrMoNb was also lower, but not so dramatically taking into account the higher ratio of matrix content. On the other hand, steam-annealed coating showed the decrease of measured values. The microhardness of Cr_3C_2-50%NiCrMoNb was repoted to be 806 HV0.3 in as-sprayed state and 905 HV0.3 after heat treatment by Matikainen [9]. While the as-sprayed coatings microhardness value is in agreement with measured values, the precipitation strengthening was not recorded in this study.

2.5 Sliding wear resistance

There is just a low difference between the sliding wear resistances of the CrC-based coatings in as-sprayed state (Fig. 6). The average wear coefficient value of Cr_3C_2-50%NiCrMoNb is even lower compare to usual Cr_3C_2-25%NiCr, despite the higher amount of soft matrix. The poor microstructure of Cr_3C_2-25%CoNiCrAlY coating does not led to a significant deterioration of the sliding wear measured at room temperature. Generally, the measured values are in a good agreement with the values reported in literature: the most often referred Cr_3C_2-25%NiCr wear resistance varied between in the range 1.10^{-6} to 1.10^{-5} mm^3/Nm [9,10,12], the Cr_3C_2-25%CoNiCrAlY 7.10^{-6} mm^3/Nm [13], and Cr_3C_2-50%NiCrMoNb 2.10^{-6} mm^3/Nm [9]. Similarly to the low effect of matrix material, ambiguous effect of the heat treatment was observed. Also, no straight correlation with measured hardness was observed. There are several aspects to be taken into account considering the changes of sliding wear resistance, mentioned in the literature [2,3,4,7], including solid solution hardening of supersaturated matrix, precipitation hardening connected with heat treatment, effect of oxidation, which can both increase the hardness, but decrease the toughness and deteriorate the wear resistance, sintering of splat boundary and also the change of inner stress state caused by annealing. As the coatings differ in composition and microstructure, the effects of each aspects can play its role in different extent. The Cr_3C_2-25%NiCr coating showed the increase of hardness connected with decrease of the wear. The heat treatment improved its behavior through the precipitation hardening mechanism. The Cr_3C_2-25%CoNiCrAlY coating increase its hardness not only due to the precipitation or boundary sintering, but also due to the inner oxidation. The oxidation can be responsible for deterioration of the wear resistance despite of the higher hardness [2]. The Cr_3C_2-50%NiCrMoNb coating, containing higher amount of matrix, can decrease its wear resistance due to the higher influence of inner stress relief.

On the contrary, there is a significant difference testing the coatings at elevated temperatures. All the measured coatings increase its wear in the level of magnitude. The same was observed by Bolelli [10] and Matikainen [9]. The highest increase of wear was observed for Cr_3C_2-CoNiCrAlY coating, connected with inner oxidation. The Cr_3C_2-50%NiCrMoNb increase the wear more than the Cr_3C_2-25%NiCr due to the higher amount of soft matrix, prone to plastic deformation at higher temperatures.

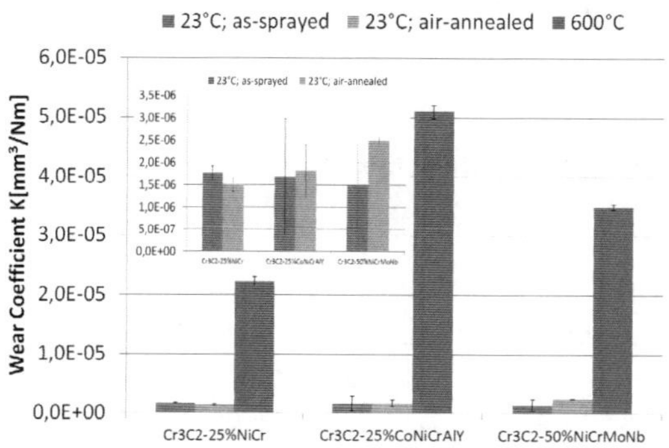

Fig. 6. Wear coefficient of coatings determined after the ASTM G-133 sliding test

3. Conclusions

The heat treatment in air and hot steam environment of three kind of CrC-based coatings showed the potential risk of inner oxidation during service. The Cr_3C_2-25%CoNiCrAlY coatings oxidized massively during steam-heat treatment, namely due to the poor cohesion strength and high porosity. The CrC-based coating with modified matrix composition Cr_3C_2-50%NiCrMoNb showed to be the most oxidation resistant. The sliding wear tests confirmed its ability to withstand the mechanical loading room temperature comparable with usual Cr_3C_2-25%NiCr and with just slightly higher wear at high temperatures. Cr_3C_2-50%NiCrMoNb coating can be used to replace the Cr_3C_2-25%NiCr in hot steam environment.

Acknowledgement
The paper has originated in the framework of the solution of the Technology Agency of the Czech Republic project number TE01020068. The authors would like to thank F. Lukáč from IPP ASCR for providing the XRD analyses and M. Bystrianský from RTI UWB for BSE analyses.

Literature

[1] Berger, L.M.: Application of hardmetals as thermal spray coatings. International Journal of Refractory Metals and Hard Materials 49 (2015) Issue 1, pp.350/64

[2] Guilemany, J.M. et al.: Role of heat treatments in the improvement of the sliding wear properties of Cr3C2–NiCr coatings. Surface and Coatings Technology 157 (2002) pp.207/13

[3] Janka, L., Norpoth, J., Trache R and Berger, L.M.: Influence of heat treatment on the abrasive wear resistance of a Cr3C2-NiCr coating deposited by an ethene-fuelled HVOF spray process. Surface and Coatings Technology 291 (2016) pp.444/51

[4] Matthews, S., Hyland, M and James, B.: Microhardness variation in relation to carbide development in heat treated Cr3C2-NiCr thermal spray coatings. Acta Materialia 51 (2003) Issue 14, pp.4267/77

[5] Matthews, S., James, B. and Hyland, M.: Erosion of oxide scales formed on Cr3C2-NiCr thermal spray coatings. Corrosion Science 50 (2008) Issue 11, pp.3087/94

[6] Matthews, S., James, B. and Hyland, M.: High temperature erosion-oxidation of Cr3C2-NiCr thermal spray coatings under simulated turbine conditions. Corrosion Science 70 (2013) pp.203/11

[7] Matthews, S. and Berger L.M.: Long-term compositional/microstructural development of Cr3C2-NiCr coatings at 500°C, 700°C and 900°C. International Journal of Refractory Metals and Hard Materials 59 (2016) pp.1/18

[8] Bolelli, G. et al.:, Micromechanical properties and sliding wear behaviour of HVOF-sprayed Fe-based alloy coatings. Wear 276–277 (2012) pp.29/47

[9] Matikainen, V. et al, High-temperature sliding wear behaviour of thermally sprayed Cr3C2-based coatings.In 17th Nordic Symopziom on Tribology NORDTRIB 2016, 14th - 17th June 2016, Aulanko, Hämeenlinna, Finland (2016)

[10] Bolelli, G. et al.: Sliding and abrasive wear behaviour of HVOF- and HVAF-sprayed Cr3C2-NiCr hardmetal coatings. Wear 358–359 (2016) pp.32/50

[11] Matthews, S., James, B. and Hyland, M.: High temperature erosion of Cr3C2-NiCr thermal spray coatings - The role of phase microstructure. Surface and Coatings Technology 203 (2009) Issue 9, pp.1144/53

[12] Berger, L.M., Woydt, M. and Saaro, S.: Comparison of self-mated hardmetal coatings under dry sliding conditions up to 600°C. Wear 266 (2009) Issue 3–4, pp.406/16

[13] Picas, J.A. et al.: Microstructural and tribological studies of as-sprayed and heat-treated HVOF Cr3C2-CoNiCrAlY coatings with a CoNiCrAlY bond coat. Surface and Coatings Technology 268 (2015) pp.317/24

[14] Matikainen, V., Koivuluoto, H. and Vuoristo, P.: Microstructural characteristics of different Cr3C2 coating compositions sprayed with HVOF and HVAF processes, in: Sudharsan, T.S., P. Vuoristo, and H. Koivuluoto (Eds.), 28th Interantional Conference Surf. Modif. Technol., Valardocs, 2015: pp. 77/87

[15] Matthews, S., James, B. and Hyland, M.: The role of microstructure in the mechanism of high velocity erosion of Cr3C2-NiCr thermal spray coatings: Part 1 - As-sprayed coatings. Surface and Coatings Technology 203 (2009) Issue 8, pp.1086/93

Amorphous Thermal Spray Coatings to Extend Life of Boiler Tubes

Abstract:

Boiler tube failure is the number one source of forced outages in all coal-fired and biomass0fired power generation plants. It is estimated that plants lose approximately 6% of their power generation annually, due to boiler tube leaks. The major causes for premature tube failure are excessive fireside boiler tube erosion and corrosion caused by impact, abrasive wear, oxidation and molten corrosion of low eutectic alloys.

LiquidMetal Coatings has developed an amorphous metal alloy composition from its unique amorphous coatings family that has a similar thermal expansion coefficient with carbon steel eliminating the risk of spallation and offer extraordinary high erosion and corrosion resistance under severe conditions in boiler tubes.

Amorphous metals are a new class of materials that have a disordered, non-crystalline, glassy structure, that are created when metals or their alloys are either cooled very quickly or because of an alloy combination, bypassing crystallization during solidification. Due to their unique microstructure, amorphous metals combine ultrahigh strength, high hardness and ductility in one single structure. They are more corrosion resistant compared to conventional metals due to the lack of long-range periodicity, related grain boundaries and crystal defects such as dislocations. In addition, they are stronger than crystalline metals and they can sustain larger reversible deformations than crystalline alloys.

Effect of Carbide Dissolution in the Metal Matrix of HVOF and HVAF Sprayed Cr_3C_2-NiCrMoNb Coatings on the Initial Stage of Chlorine High Temperature Corrosion

D. Fantozzi[a], V. Matikainen[a], M. Uusitalo[b], H. Koivuluoto[a], P.Vuoristo[a]

[a] Tampere University of Technology, Laboratory of Materials Science, Tampere / FIN

[b] Valmet Technologies Ltd., Tampere / FIN

Highly corrosion and wear resistant thermally sprayed chromium carbide (Cr_3C_2) based cermets coatings are nowadays a potential highly durable solution to allow traditional fluidised bed combustors (FBC) to be operated with ecological waste and biomass fuels. However, the heat input of thermal spraying processes causes carbide dissolution in the metal binder. This alters the coating structure and forms carbon saturated amorphous and nanocrystalline metastable areas, which can affect the behaviour of the materials under the corrosive chlorides containing environment of the flue gases. This study analyses the effect of carbide dissolution in the metal matrix of MMC coatings and its effect on the onset of chlorine induced high temperature corrosion. Four Cr_3C_2-NiCrMoNb coatings were thermally sprayed with high-velocity air-fuel (HVAF) and high-velocity oxygen-fuel (HVOF) spray processes in order to obtain microstructures with increasing amount of carbide dissolution in the metal matrix. The specimens were heat treated in an inert argon atmosphere at 700°C for 5 hours to induce secondary carbide precipitation. As-sprayed and heat-treated self-standing coatings were covered with KCl and their corrosion resistance was investigated with thermogravimetric analysis (TGA) at 550°C for 4 hours. High carbon dissolution in the metal matrix appeared to be a detrimental factor in the initial stage of corrosion. The microstructural changes induced by the heat treatment hindered the corrosion onset in the coatings. Moreover, an optimal amount of oxides and melting degree seemed beneficial.

1 Introduction

Fossil combustion-based energy production needs to be replaced by renewable sources in order to reduce green house gas emissions. Biomass and municipal solid waste (MSW) are ecological alternatives to fossil fuels able to lead the industry of power generation to a greener new era. It has been estimated that replacing only 10% of the coal combusted worldwide with biomass would reduce the annual emissions by 450 M tonnes of CO_2[1]. However, such fuels have high content of corrosive agents that cause a severe accelerated corrosion mechanism known as chlorine active oxidation[2-5]. Erosion phenomena are also often present, especially in fluidised bed combustors (FBC), and they further increase materials losses. These problems are forcing boilers to be operated at lower steam temperatures to minimise the material degradation. Unfortunately, this results in a drastic drop of energy efficiency and overall performance[6,8]. Currently, there are no materials able to offer adequate erosion-corrosion protection in in the most severe environments. The gap between the industrial need and the available materials causes significant efficiency and performance losses in power plants. Therefore, novel highly erosion-corrosion resistant protective coatings are required.

Thermal spraying is one of the most versatile surface engineering techniques to produce thick (20-1000 µm) coatings. The state-of-the-art materials in erosion-corrosion applications are metal matrix composite (MMC) coatings with Cr_3C_2-NiCr being the most common composition[9]. In this study, a modern Cr_3C_2-NiCrMoNb coating composition was studied. The superalloy matrix is designed to provide higher corrosion protections than the traditional NiCr alloy. However, extremely limited information is available in literature about the thermally sprayed coatings deposited with such composition[10-12] and further knowledge is therefore needed.

Unfortunately, MMC materials experience severe degradation in boilers because of the high reactivity of carbides in presence of chlorine[13,14]. A portion of the carbides in the feedstock material are melted and diluted into the metal matrix forming a metastable C- and Cr-rich alloy during the high temperature thermal spray processing[15,16]. Under exposure to high temperature, enough energy is available to enable the metastable matrix to undergo various microstructural refinements including the precipitation of fine secondary carbide phases and thus, achieving a more stable state[17].

This work aims to verify whether the secondary carbide precipitates and the metastable composition of the matrix have detrimental effect on the high temperature corrosion resistance of thermally sprayed Cr_3C_2-based coatings in chlorine containing environment. In this study, four coatings were sprayed with different melting degrees by varying the spray process and spray parameters. The coatings were prepared with high-velocity air-fuel (HVAF) and high-velocity oxygen-fuel (HVOF) spray processes. HVAF process was used to reach the coldest spray parameters whereas HVOF was used to obtain the hottest ones. The hypothesis is that hotter spray parameters will result in higher carbon dissolution from the carbide hard phase to the matrix. The heat treatment carried out on the as-sprayed coatings induced the precipitation of the dissolved carbon and obtained four coatings with increasing amount of secondary carbide precipitates. The corrosion properties of the as-sprayed and heat-treated coatings were studied with thermogravimetric analyses (TGA).

2 Experimental

Coatings materials and manufacturing parameters are reported in the following chapters together with the experimental procedures.

2.1 Coatings materials

Four coatings of the same composition were deposited with HVOF and HVAF processes from agglomerated and sintered powders. The depositions of the coatings were carried out with M3 HVAF gun (Uniquecoat Technologies LLC) and with Diamond Jet 2700 HVOF gun (DJ2700, Oerlikon Metco). The coating composition was Cr_3C_2-NiCrMoNb in which the hard phase Cr_3C_2 is dispersed in Inconel625-like metal matrix. The coatings were sprayed with two powders with same composition but different particle size distribution. The in-flight particle parameters measured with Spraywatch 4S thermal spray sensor and details of the feedstock powders are listed in Table 1. The coarser particle size was exploited to achieve colder particle temperatures. The process parameters for the deposition of the coatings are reported in Table 2. The nominal alloy composition of Inconel 625 is 20-23 % Cr, 8-10 % Mo, 3.15-4.15 % Nb, max 5 % Fe and with Ni bal. The process parameters and feedstock powders were chosen in order to obtain four different microstructures with increasing degree of carbon dissolution in the matrix. The coatings were marked as HVAF-grade 1, HVAF-grade 2, HVOF-grade 3 and HVOF-grade 4 with carbon dissolution increasing from grade 1 to grade 4, respectively. However, HVAF-grade 2 and HVOF-grade 3 were sprayed with the most optimised process parameters while the parameters used for the deposition of HVAF-grade 1 and HVOF-grade 4 samples were designed with the main purpose of controlling the amount of melting degree and carbon dissolution in the metal matrix.

***Table 1**: Feedstock powder information and inflight particle properties a measured with the SprayWatch 4S thermal spray sensor.*

Code	Powder (H.C. Starck)	Particle size	Process	Particle Temp. [°C]	Particle velocity [m/s]
HVAF Grade 1	Amperit 595.074	-45+15	HVAF M3	1310±3	967±82
HVAF Grade 2	Amperit 595.059	-30+5	HVAF M3	1460±9	949±109
HVOF Grade 3	Amperit 595.074	-45+15	HVOF DJ2700	1770±9	767±40
HVOF Grade 4	Amperit 595.059	-30+5	HVOF DJ2700	1900±5	981±44

***Table 2**: Thermal spray process parameters used for HVAF and HVOF coatings*

Parameter	HVAF grade 1	HVAF grade 2	HVOF grade 3 and 4
Oxygen	-	-	240 slpm
Air	7.4 bar	7.4 bar	383 slpm
Fuel1 (propane)	7.3 bar	7.0 bar	70 slpm
Fuel2 (propane)	6.9 bar	7.3 bar	-
Spray distance	300 mm	300 mm	230 mm

2.2 Heat treatment

The heat treatment was carried out in an in-house built vertical alumina tube furnace under an inert atmosphere (Ar-5 vol.%H_2) at 700°C for a duration of 5 hours with heating rate of 5 °C/min. The samples were cooled to room temperature in inert atmosphere at a cooling rate of 11 °C/min.

2.3 Microstructural characterisation

The microstructures of the coatings were analysed before and after the heat treatment with scanning electron microscope (SEM, Philips XL30) equipped with Energy Dispersive X-ray (EDS) microanalysis. The phase compositions of the as-sprayed coatings were assessed by X-Ray diffractometry (XRD, Empyrean, PANanalytical, Cu-K_α radiation).

2.4 Thermogravimetric analyses

The mass changes of the freestanding coatings were measured with thermogravimetric analyser NETZCH STA 409CD equipped with TASC 414/4 temperature controller and PU 1.851.01 power unit to evaluate their corrosion behaviour. Prior to the test, the coating-substrate interface of the freestanding coating was grinded with P1200 SiC abrasive paper in order to remove possible contaminations from the substrate and grit blasting residues. The heat-treated free-standing coatings were grinded on both sides. The coatings were then cut to small fragments with an approximate size of 1x1 mm each. The total surface of the samples was measured with image analysis with the software "ImageJ" from optical and stereomicroscope images. The TGA measurements were carried out in alumina crucibles in which the coating fragments were mixed with an excess of KCl powder. The weight of the specimens ranged between 10 and 20 mg while the amount of KCl was about 30 mg. The test atmosphere was dry synthetic air with a gas flow rate of 100 ml/min. The temperature program consisted of a heating ramp of 12 °C /min followed by an isothermal step at 550 °C for four hours. The output data consisted of mass change resulting from the oxidation phenomena that took place during the test plotted in function of time. The absolute mass change was normalised by unit of area [mg/cm^2] in order to compare the samples with each other.

3 Results

The initial stage of high temperature corrosion of the heat-treated and as-sprayed coatings is discussed in the following chapters in relation to their microstructural properties.

3.1 Microstructure of coatings

The microstructures of the coating cross-sections were analysed with SEM, Figure 1. All the sprayed coatings presented highly dense microstructures. As expected, the four coatings showed an increasing amount of melting degree with negligible content in HVAF-grade 1 to significant matrix melting in HVOF-grade 4.

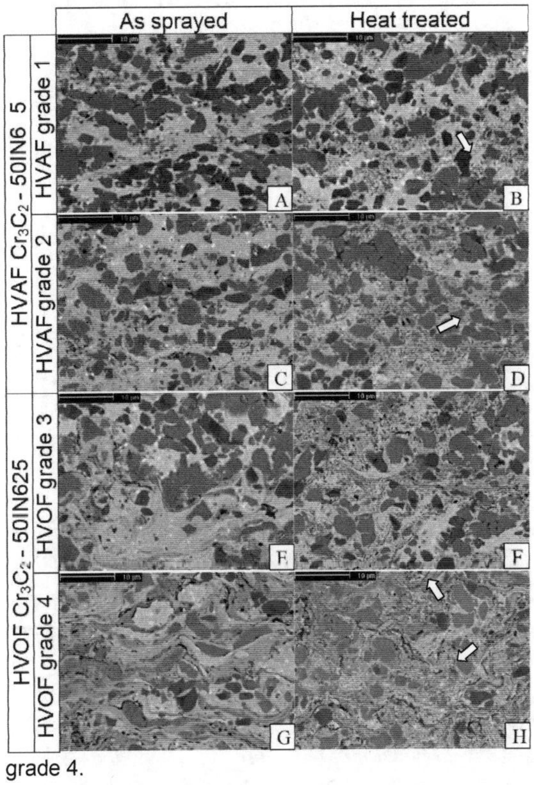

Figure 1: SEM (BSE) images of the cross sections of the as sprayed (A,C, E, G) and heat-treated coatings (B, D, F, H). Areas where no precipitation took place are pointed by arrows.

All the coatings showed mainly two types of carbides in the SEM images taken with BSE detector. The light coloured ones have higher content of Ni and lower content of Cr than the dark coloured ones. The amount of Ni rich carbides seems to be higher in the HVAF-grade 2 and HVOF-grade 4 coatings. The difference is shown also by the XRD patterns in Figure 2 (top) where the Cr_7C_3 peak intensities (at 38.93°) of grades 2 and 4 are increased in relation to the partially overlapping Cr_3C_2 peak (at 38.77°) compared to the XRD patterns of grades 1 and 3. This indicates that more light grey $(Cr,Ni)_7C_3$ carbides are indeed present in the coatings sprayed from the finer particle size distribution. The fine white coloured dots observed in all the coatings are Nb-rich carbides. In HVOF-grade 3 and 4 coatings, the portion of matrix that was melted during the spraying is clearly visible and presents different grey shades in BSE contrast images deriving from the different levels of carbon dissolution.

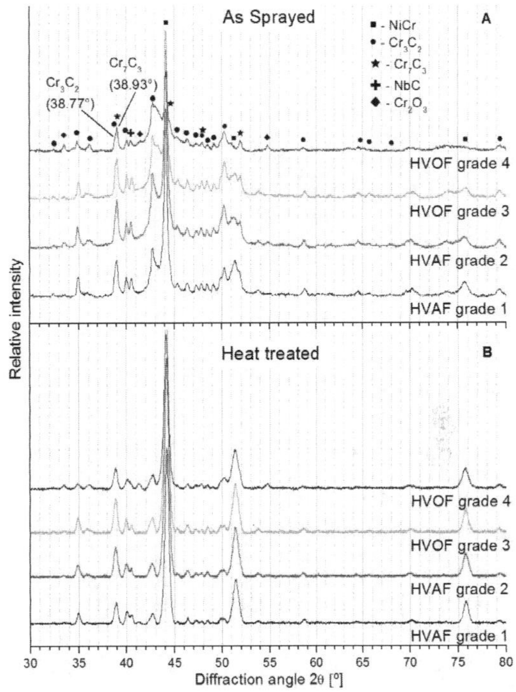

Figure 2: Superimposed XRD patterns of the as-sprayed coatings (A) and heat treated coatings (B).

After heat treatment, fine-grained secondary carbide particles precipitated in the metal binder. Their amount in the coating was incremental from grade 1 to grade 4

as expected. The precipitation seemed to occur especially in the matrix areas that were melted during the spraying and therefore diluted a high amount of carbon from the primary carbide particles. All coatings, especially HVAF-grade 1, contained some areas with little or no precipitate formation in Figure 1 (pointed by arrows).

The phase compositions of the coatings were studied with XRD. The superimposition of the XRD pattern of the as-sprayed coatings is presented in Figure 2.

All as-sprayed coatings presented the same crystallographic phases with the before mentioned differences between the powders. The hard phase is composed of two different chromium carbide compounds, namely Cr_3C_2 and Cr_7C_3. NbC was also detected in all the coatings even though in little amount, as it was also observed by SEM/EDS in Figure 1.

The relatively low intensity of the peaks of the metallic binder denotes higher amount of amorphous and nanocrystalline phase in the metal binder of the as

sprayed coatings[15,18], especially in coating HVOF-grade 4. High content of amorphous phase in coating HVOF-grade 4 was expected because of the higher temperature parameters used in its deposition that induce high carbon dissolution in the metal matrix. The following rapid cooling ultimately resulted in a C saturated partly amorphous metastable alloy.
After heat treatment, all coatings showed marked changes in the microstructure. The main observation is related to the significantly higher intensity and definition of the peaks of the metal matrix. This denotes the transformation of metastable amorphous phases in the alloy into a mostly crystalline phase. The Cr_7C_3 has been reported to transform into Cr_3C_2 during short-term isothermal heat treatments[19], but it was difficult to confirm to take place in the present study. However, slight decrease in the peak intensity of the Cr_7C_3 phase seemed to take place during heat treatment.

3.1 Corrosion behaviour

The early stage corrosion behaviour of the self-standing coatings was measured with thermogravimetric analyses under KCl salt deposit for as-sprayed and heat-treated coatings. The results are presented as mass gain per surface unit during the four hours isothermal step at 550 °C in Figure 3.
The results showed some clear trends. Namely, both heat-treated and as-sprayed coatings showed increased corrosion degradation from grade 2 to grade 4, denoting correlation between declined corrosion resistance and warmer spray parameters. The heat-treated samples, despite their higher content of secondary carbide precipitates, always incurred lower mass gain than that of their as sprayed counterparts.

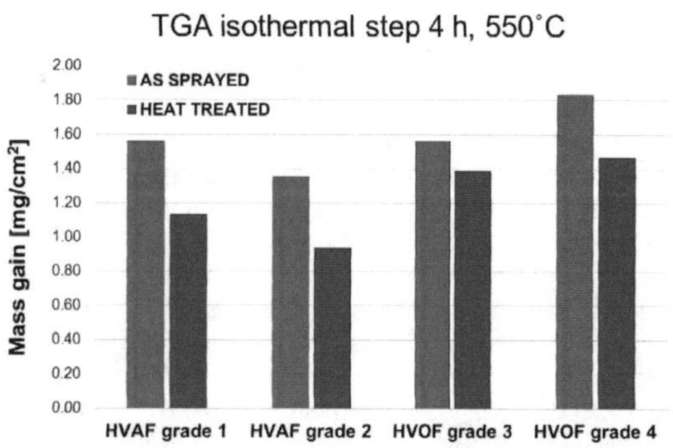

Figure 3: Total mass gain of the as-sprayed and heat-treated self-standing coatings during TGA isothermal step at 550 °C.

The trend observed in the mass gain results might be explained by the content of carbon dilution in the metal matrix, which increases from grade 2 to grade 4 samples. Therefore, rather than the secondary carbide precipitates, the amount of metastable carbon rich matrix seems to play a role in controlling the corrosion rate at this stage. Indeed, during heat treatment, the excess carbon in the metal matrix is expelled in the form of secondary carbide precipitates[15]. The reduction of C content in the metal binder probably increased its corrosion resistance. HVAF-grade 1 represents an exception as it corroded more than HVAF-grade 2 and equally to HVOF-grade 3 which were both deposited with warmer spray parameters. This behaviour, which does not follow the trend of the other coatings, denotes the influence of other variables. HVAF-grade 1 coating present the lowest amount of melting degree and the lowest oxide content of all the coatings. These microstructural features are considered helpful in corrosion protection of thermally sprayed coatings[20,21]. It is possible that there exists an optimal proportion of oxides content, melting degree and carbon dissolution in the matrix that allow reduced corrosion degradation such as in the case of coating HVAF-grade 2.
The mass gain in function of time of all the coatings is presented in Figure 4.

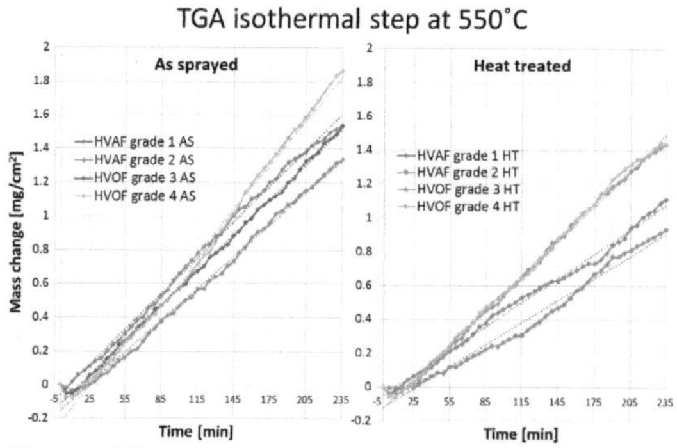

Figure 4: TGA mass change during the four-hour isothermal step at 550°C.

During the four hours at 550 °C, all the samples were subjected to linear corrosion kinetics. Heat treated sample HVAF-grade 2, which outperformed all the other coatings, presented the lowest linear constant and a substantial initial delay before starting to gain mass. The corrosion reactions that are thought to take place during the test involve the first stages of the chlorine active oxidation mechanism. These include first, the reactions between KCl and the oxides embedded in the microstructure and on surface of the coatings as presented in Eq. 1[2].

$$Cr_2O_{3\,(s)} + 4KCl_{(s)} + \frac{5}{2}O_{2\,(g)} \rightarrow 2K_2CrO_{4\,(s)} + 2Cl_{2\,(g)} \qquad \text{Eq.1}$$

This reaction might also explain the higher corrosion rate of the as sprayed coatings as the oxides on their surface (deriving from the spray process) might have contributed in the observed mass gain.
Secondly, further corrosion reactions might have taken place between gaseous Cl_2 evolved from Eq.1 and the metal binder (Eq. 2) or carbides (Eq. 3) based on the following reactions[22]:

$$Cr^0_{(s)} + Cl_{2\,(g)} \rightarrow CrCl_{2\,(s)} \qquad \text{Eq. 2}$$

$$Cr_3C_{2\,(s)} + Cl_{2\,(g)} \rightarrow 3CrCl_{2\,(s)} + 2C_{(s)} \quad \text{Eq. 3}$$

And the subsequent oxidation of the metal chlorides[4]

$$2CrCl_{2\,(g)} + \frac{3}{2}O_{2\,(g)} \rightarrow Cr_2O_{3\,(s)} + 2Cl_{2\,(g)} \quad \text{Eq. 4}$$

The high airflow rate (100 ml/min) employed in the present tests is thought to provide enough O_2 to complete Eq. 4 and therefore no loss of volatile metal chlorides is supposed. However, in the long term, this assumption might change and therefore, longer tests will be carried out in the future to further investigate this aspect. The test procedures will be optimised to evaluate the effects of other microstructural properties and test variables.

3 Conclusions

In this study, the spray parameters of HVOF and HVAF processes were successfully designed in order to obtain four Cr_3C_2-NiCrMoNb coatings with tailored melting degree and carbon dissolution in the metal matrix. The initial stage of chlorine induced high temperature corrosion of HVAF and HVOF self-standing coatings was studied with thermogravimetric analyses.
The results showed that in the initial stage of the corrosion, the actual metal binder composition is probably the rate-controlling factor with the amount of carbon dissolution perhaps being the most detrimental quality on the corrosion resistance of the coatings. However, a certain amount of melting degree and oxides content appear to be beneficial in order to increase corrosion resistance.
Longer test duration are necessary to assess the corrosion mechanism of such coatings in the long-term behaviour.

4 Acknowledgments

The work has been done within the DIMECC HYBRIDS (Hybrid Materials) programme. We gratefully acknowledge the financial support from the Finnish Funding Agency for Innovation (Tekes) and the participating companies. A special thank goes to Mr. Mikko Kylmälahti of Tampere University of Technology, for spraying the HVOF and HVAF coatings.

5 References

[1] Lempp, P., Biomass co-firing. Technology brief E21–January.International Energy Agency-Energy Technology Systems Analysis Programme (IEA-ETSAP), and International Renewable Energy Agency (IRENA) (2013).

[2] Oh, J., M. McNallan, G. Lai, M. Rothman, High temperature corrosion of superalloys in an environment containing both oxygen and chlorine. Metallurgical Transactions A 17 (1986), 6, pp. 1087-94.

[3] Spiegel, M., A. Zahs, H. Grabke, Fundamental aspects of chlorine induced corrosion in power plants. Materials at high temperatures 20 (2003), 2, pp. 153-9.

[4] Zahs, A., M. Spiegel, H.J. Grabke, Chloridation and oxidation of iron, chromium, nickel and their alloys in chloridizing and oxidizing atmospheres at 400–700°C. Corrosion Science 42 (2000), 6, pp. 1093-122.

[5] Cha,S. C., M. Spiegel, Local reactions of KCl particles with iron, nickel and chromium surfaces. Materials and Corrosion 57 (2006), 2, pp. 159-64.
[6] G.Y. Lai, High-Temperature Corrosion and Materials Applications, ASM International, 2007.

[7] Nuss, P., S. Bringezu, K.H. Gardner, Waste-to-materials: the long-term option, Waste to Energy, Springer, 2012, pp. 1-26.

[8] Rukes, B., R. Taud, Status and perspectives of fossil power generation. Energy 29 (2004), 12, pp. 1853-74.

[9] L. Pawlowski, The Science and Engineering of Thermal Spray Coatings, John Wiley & Sons, 2008.

[10] Matikainen, V., H. Koivuluoto and P. Vuoristo. (2014), Microstructural Characteristics of Different Cr3C2 Coating Compositions Sprayed with HVOF and HVAF Processes. Proceeding of the Twenty Eight International Conference on Surface Modification Technologies, Tampere, Finland, June 16-18. Valardocs.

[11] Zimmermann, S. and B. Gries. (2011), New Cermet Powders for HVOF Spraying with Improved Corrosion and Oxidation Resistance for Offshore, Mining and Power Generation Applications. Thermal Spray 2011: Proceedings of the International Thermal Spray Conference (DVS-ASM), Hamburg (DE), 26-29 Sept. 2011.

[12] Zhou, W., K. Zhou, C. Deng, K. Zeng, Y. Li, Hot corrosion behaviour of HVOF-sprayed Cr_3C_2-NiCrMoNbAl coating. Surface and Coatings Technology 309 (2017), pp. 849-59.

[13] Uusitalo, M., R. Backman, L. Berger, P. Vuoristo, T. Mäntylä, Oxidation resistance of carbides in chlorine-containing atmospheres. High Temperature Materials and Processes 21 (2002), 5, pp. 307-20.

[14] Uusitalo, M., P. Vuoristo, T. Mäntylä, Elevated temperature erosion–corrosion of thermal sprayed coatings in chlorine containing environments. Wear 252 (2002), 7, pp. 586-94.

[15] Guilemany, J., J. Nutting, N. Llorcalsern, Microstructural examination of HVOF chromium carbide coatings for high-temperature applications. Journal of Thermal Spray Technology 5 (1996), 4, pp. 483-9.

[16] Matthews, S., M. Hyland, B. James, Long-term carbide development in high-velocity oxygen fuel/high-velocity air

fuel Cr_3C_2-NiCr coatings heat treated at 900° C. Journal of Thermal Spray Technology 13 (2004), 4, pp. 526-36.

[17] Matthews, S., B. James, M. Hyland, The role of microstructure in the mechanism of high velocity erosion of Cr_3C_2–NiCr thermal spray coatings: Part 1—As-sprayed coatings. Surface and Coatings Technology 203 (2009), 8, pp. 1086-93.

[18] Ding, Y., T. Hussain, D. McCartney, High-temperature oxidation of HVOF thermally sprayed NiCr–Cr_3C_2 coatings: microstructure and kinetics. Journal of Materials Science 50 (2015), 20, pp. 6808-21.

[19] Matthews, S., L.-. Berger, Long-term compositional/microstructural development of Cr_3C_2-NiCr coatings at 500 °C, 700 °C and 900 °C. International Journal of Refractory Metals and Hard Materials 59 (2016), pp. 1-18.

[20] Fantozzi, D., V. Matikainen, M. Uusitalo, H. Koivuluoto, P. Vuoristo, Chlorine-induced high temperature corrosion of Inconel 625 sprayed coatings deposited with different thermal spray techniques. Surface and Coatings Technology (2016).

[21] Dobler, K., H. Kreye, R. Schwetzke, Oxidation of stainless steel in the high velocity oxy-fuel process. Journal of Thermal Spray Technology 9 (2000), 3, pp. 407-13.

[22] Nielsen, H., F. Frandsen, K. Dam-Johansen, L. Baxter, The implications of chlorine-associated corrosion on the operation of biomass-fired boilers. Progress in energy and combustion science 26 (2000), 3, pp. 283-98.

Post-plasma-spraying gas nitriding of some metallic coatings on a Fe-based superalloy and their high temperature corrosion behavior

VIKAS CHAWLA

Mechanical Engineering Department, IK Gujral Punjab Technical University (Main Campus), Kapurthala, Punjab-152001, India

Abstract

Ti-Al and Al-Cr metallic coatings were deposited on Superfer800H (Fe-based superalloy) through a plasma spray process. Then the gas nitriding of the coatings was done in the lab and the parameters were optimized after conducting several trials on plasma sprayed coated specimens. Characterization and high temperature corrosion behaviour of coatings after exposure to air and molten salt at 900°C were studied under cyclic conditions. Techniques like XRD, SEM/EDAX and EPMA analysis have been used for characterization of the coatings and to analyze the oxide scale. Both the coatings have successfully protected the substrate and were effective in decreasing the corrosion rate when subjected to cyclic oxidation at 900°C for 50 cycles in air and molten salt. The coatings subjected to cyclic oxidation in air have shown relatively high weight gains in the early cycles of the exposure. Uncoated Superfer800H (Fe-based superalloy) showed very poor resistance to hot corrosion in molten salt environment.

Keywords: *Plasma Spray, Gas Nitriding, Characterization, Corrosion, Oxide Scale*

1. INTRODUCTION

In the last few decades, development of protective coatings has become an active area in the field of materials engineering [1]. Coatings provide a way of extending the limits of use of materials at the upper end of their performance capabilities, by allowing the mechanical properties of the substrate materials to be maintained while protecting them against wear or corrosion [2, 3]. Today, numerous Ni and Fe-based superalloys are used as structural components in various applications such as gas turbines, steam boilers, heat exchangers, aero engines, and industrial springs as they possess superior mechanical properties along with corrosion and oxidation resistance to some extent. These superalloys cannot provide high-temperature corrosion resistance without affecting its mechanical properties as tailoring of microstructural and compositional parameters to achieve the former would have adverse influence on the later [4, 5]. Tiwari and Prakash [6, 7, and 8] have evaluated hot corrosion behavior of superalloys in pure Na_2SO_4, Na_2SO_4–15%V_2O_5 and Na_2SO_4–60%V_2O_5 in the temperature range of 700–900 °C. They observed accelerated corrosion rates for Na_2SO_4–60%V_2O_5 composition, i.e. eutectic with melting point of 500°C. Due to the harsh environments in which they are used, it is necessary to protect them with coatings, against heat, oxidation and corrosion [9, 10]. It is now generally accepted practice to apply coatings to components in fossil fuel energy processes to provide thermal insulation, corrosion and wear resistance, and in chemical process plants or boilers to protect the surface of structural steels against surface degradation processes such as wear, oxidation, corrosion and erosion [11].

In many tribological applications, hard coatings of metal nitrides are now commonly used [12]. The major properties required for such coatings are hardness and wear resistance. However, because of severe operating conditions, there is a need to combine mechanical features with corrosion resistance properties. Recently these coatings are gaining importance for high temperature wear, corrosion and oxidation resistance applications [13, 14]. Adachi and Nakata [15] investigated the process of the coating formation by plasma spraying with Ti powder mixed with Al powder and reported that the adhesive strength of the coating was improved by filling cracks and pores of the Ti compounds layers with the Al phase to form a dense microstructure, and also by hardening the Ti compounds and Al phase with the oxygen and nitrogen contained in the coating. Since the addition of aluminum increases the oxidation resistance of Ti-N coatings, a similar effect on the oxidation resistance of Cr-N coatings is expected [16].

Plasma spray techniques have been reported to be the major coating processing technologies which are used worldwide for coating deposition due to their ease of operation and ability to deposit various coating materials on a substrate alloy [17-19]. This flexibility is probably one of the major reasons for the rapid expansion of this technology [20]. However, the residual porosity in plasma spray coatings allows corrosive liquids to penetrate through them which lead to the debonding or spallation of the coatings associated with the accumulation of corrosion products at the coating/substrate interface. Therefore, the coatings may require post-deposition treatments to improve these properties [8, 21]. An attempt has been made to develop Ti-Al and Al-Cr metallic coatings by plasma spraying process directly onto the Superfer 800H superalloy substrates and then subjected to post-plasma-spraying gas nitriding to modify the surface layer of the coatings.

The present investigation aims to characterize the coatings and to evaluate their cyclic behavior when exposed to high temperature oxidation in air and molten salt (Na_2SO_4–60%V_2O_5) environment. X-Ray diffraction (XRD), scanning electron microscopy (SEM), and energy-dispersive X-ray analysis (EDAX) techniques were used to characterize the coatings and corrosion products after oxidation and hot corrosion.

2. EXPERIMENTAL DETAILS

2.1 Development of coatings

2.1.1 Selection of substrate material and sample preparation

The selection of candidate material for the study has been made after consultation with Mishra Dhatu Nigham Ltd, Hyderabad (India). The selected superalloy named Superfer800H (Fe-based superalloy) is widely used for steam boilers, furnace equipment and piping in the chemical industry, baffle plates/tubes in fertilizer plants. The superalloy was procured in the form of rolled sheets from Mishra Dhatu Nigham Ltd, Hyderabad (India). The actual chemical composition of the selected alloy has been determined with the help of Optical Emission Spectrometer (Thrmo Jarrel Ash, TJA181/81, U.S.A); which is reported in Table 1 along with nominal chemical composition. Specimens with dimensions of approximately 20mm x 15mm x 5mm were cut from the alloy sheet. Polished using emery papers of 220, 400, 600 grit sizes and grit blasted with alumina powders (Grit 45) prior to the deposition of the coatings.

Table 1: **Chemical composition (Wt %) and industrial applications of the alloys used**

Alloy Grade ⇩	Elements ⇨	C	Mn	Si	Cr	Ni	Ti	Al	Fe	Recommended Applications (Manufacturer's Catalogue)
Superfer 800H	Nominal	0.10	1.0	0.6	19.5	30.8	0.44	0.34	Bal.	Steam boilers. Furnace equipment, heat exchanger and piping in chemical industry, reformer, baffle plates/tubes in fertilizers plants
	Actual	0.1	1.5	1.0	21	32	0.3	0.3	Bal.	

2.1.2 Coating materials

Coating powders namely Ti-50Al (50 Ti-50 Al Wt %) and Al-30Cr (70 Al-30 Cr Wt %) were chosen for plasma spray deposition on the substrate. These powders were prepared in laboratory ball mill for 8 hrs to form a uniform mixture from three types of commercially available alloy powders namely Titanium powder with minimum assay 98.0 % (Art.6322) supplied by Loba Chemie, Mumbai, India, Aluminum powder with minimum assay 90.0 % and size 200 mesh (Art. 37064) and Chromium powder with minimum assay 99.0 % and size 60 mesh (Art. 36054) supplied by S.D. Fine Chemical Limited, Mumbai, India.

2.1.3 Formulation of Coatings

A 40 kW Miller Thermal (USA) plasma spray apparatus was used to apply the coatings. Argon was used as the powder carrying and shielding gas. The process parameters, including the spray distance, were constant throughout the coating process and are summarized in Table 2.

Table 2: Parameters of the argon shrouded plasma spray process

Arc Current (A)	750
Arc Voltage (V)	45
Powder Feed Rate (rev/min)	5.2
Spraying Distance (mm)	90-110
Plasma Arc Gas (Argon) Pressure	58 psi
Powder Gas Pressure	60 psi
AVX Gas Pressure	10 psi

2.1.4 Post-plasma-spraying gas nitriding

After reviewing the literature critically; it was decided to do the gas nitriding of the plasma sprayed conventional thick Ti-Al and Al-Cr coatings in the lab. The apparatus used for the gas nitriding process is shown in Fig.1; which consist of a long steel tube placed in a silicon tube furnace as reported by Chawla et al. [5].

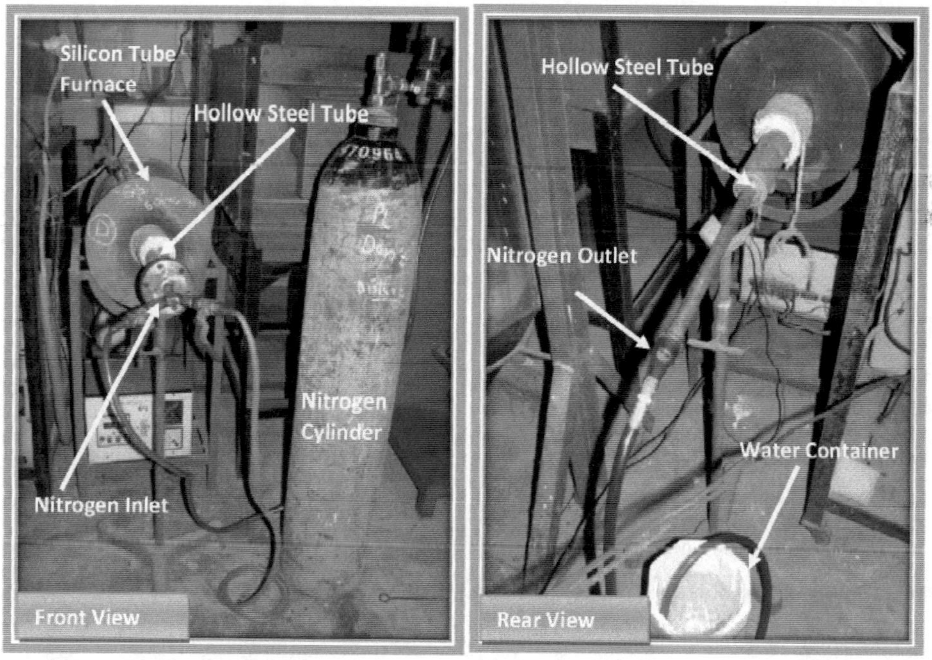

Fig. 1 Front and rear views of the set-up used for gas nitriding of the plasma sprayed Ti-Al and Al-Cr coatings in the laboratory [5]

The steel tube is connected to the nitrogen cylinder for nitrogen supply and the outlet of the nitrogen gas is through water for slow and constant discharge rate. The gas nitriding parameters were optimized after conducting several trials on plasma sprayed coated specimens. The optimized process parameters employed for the gas nitriding are as summarized in Table 3.

Table 3: Parameters for the gas nitriding process

Temperature	650 °C	
Duration	3 hrs	For Ti-Al coating
	5 hrs	For Al-Cr coating
Discharge Rate of nitrogen	0.3 l/min	

2.1.5 Characterization of the Coatings

The as coated specimens were wheel polished and subjected to optical microscopy, XRD and SEM/EDAX analysis to characterize the surface morphology. A Zeiss Axiovert 200 MAT inverted optical microscope fitted with imaging software Zeiss Axio Vision Release 4.1, from Germany was used for the optical microscopy; whereas the XRD analysis was carried out using Bruker AXS D-8 advance diffractometer (Germany) with Cu Kα radiation. The scan rate used was 2°/min and the scan range was from 20° to 120°. Field emission scanning electron microscope (FESEM, FEI, Quanta 200F Company) with EDAX Genesis software attachment (made in Czech Republic) was used for SEM/EDAX analysis. The porosity of the coatings is also a prime importance in the high temperature oxidation studies. The porosity measurements were made with image analyser, having software of Dewinter Materials Plus 1.01 based on ASTM B276. A PMP3 inverted metallurgical microscope was used to obtain the images. Further the bond strength of the gas nitrided plasma sprayed coatings was tested using the ASTM standard C633-01. The surface roughness (Ra) values of the coatings were measured using Surface Roughness Tester (Mitutoyo SJ-201, Japan).

After preliminary characterization, the samples were cross-sectioned, mounted in transoptic powder and subjected to mirror polishing to identify the cross-sectional details. The thickness of some of the as-coated specimens was measured by taking a back scattered electron (BSE) images using a Field emission scanning electron microscope (FE-SEM; FEI Quanta 200F, Made in Czech Republic). Energy Dispersive X-ray Analysis (EDAX) and X-ray mapping were employed while imaging on FE-SEM was carried out to obtain elemental composition at different areas of the coating and surface morphology of the coatings, respectively.

2.2 Oxidation and molten salt corrosion tests

Cyclic oxidation studies were conducted in air and molten salt (Na_2SO_4–60%V_2O_5) environment at 900°C temperature in the laboratory silicon carbide furnace as depicted in Figure 2 [22]; calibrated up to the variation of ± 5°C for 50 cycles. Each cycle consisted of 1 hour heating at 900°C followed by 20 min cooling at room temperature [23, 24]. The aim of cyclic loading is to create severe conditions for testing [22, 25]. The studies were performed for uncoated as well as coated specimens for the purpose of comparison. The uncoated specimens were subjected to mirror polishing, whereas coated specimens were subjected to wheel cloth polishing for 5 min. After washing with acetone, the specimens along with alumina boats were then subjected to tube furnace for oxidation studies in air. Whereas, for oxidation studies in molten salt environment: a salt mixture of Na_2SO_4–60%V_2O_5 dissolved in distilled water was applied with a camel hair brush on preheated samples (250°C) as depicted in Figure 2 [22]. Amount of the salt coating was kept in the range of 3.0 -5.0 mg/cm2. The salt coated specimens as well as the alumina boats were then dried in the oven (Fig.2.d) for 3 hours at 100°C and weighed before being exposed to hot corrosion tests[22, 25]. During oxidation runs, the weight of boats along with the specimens were measured together after visual observations at the end of each cycle with the help of Electronic balance Model CB-120 (Contech, Mumbai, India) with a sensitivity of 1mg.

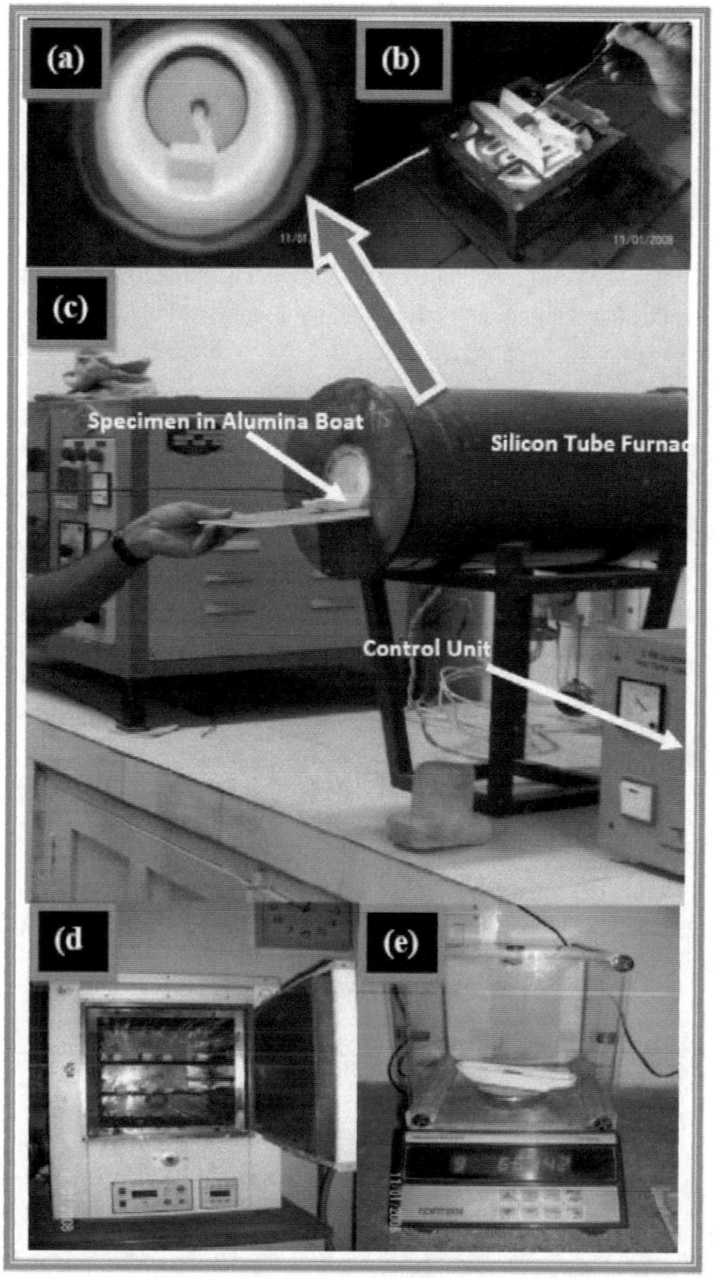

Fig. 2 Experimental set up for high temperature oxidation and hot corrosion studies; (a) Heat Zone inside furnace, (b) Molten salt coating, (c) Silicon tube furnace, (d) Oven for pre heating the specimens, (e) Electronic balance [22]

The spalled scale, if any was also included in the weight change determination. The kinetics of corrosion was determined from the weight change measurement. After the oxidation studies, the exposed specimens were analyzed by XRD (using Bruker AXS D-8 advance diffractometer (Germany) with Cu Kα radiation at the scan rate of 2°/min for 20° to 120°) and SEM-EDAX analysis using FE-SEM (FEI, Quanta 200F). The oxidized specimens were then cut using Buehler's Precision Diamond saw (Model ISOMET 1000, USA make) across the cross-section and mounted for the cross-sectional analysis using SEM/EDAX and elemental X-ray mapping. The kinetics of the cyclic oxidation of coated as well as uncoated specimens was determined using the thermogravimetric analysis and by evaluating the parabolic rate constants. The parabolic rate constant K_p was calculated by a linear least-square algorithm to a function in the form of $(W/A)^2 = K_p\,t$, where W/A is the weight gain per unit surface area (mg/cm^2) and 't" indicates the number of cycles representing the time of exposure.

3. OBSERVATIONS & RESULTS

3.1 Characterization of coatings

The macrographs of the as-coated Superfer 800H superalloy are shown in Fig.3. Visual observations indicate; the surface of the coatings is rough and color appeared as dark grey. XRD diffractograms for TiAlN and AlCrN coated superalloy are depicted in Fig.4 on reduced scale. As indicated by the diffractograms; the main phases identified for the TiAlN coating are Al_2O_3, TiN, Ti_3Al, AlN with small peaks of TiO_2. The phases identified in case of AlCrN coating are CrN, AlN, and Al_2O_3 with minor peaks of Cr_2O_3.

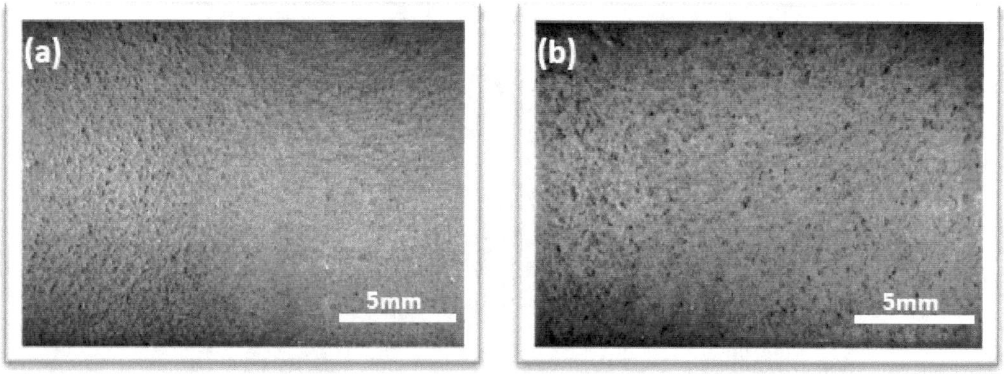

Fig. 3 Surface macrographs for the as coated Superfer 800H superalloy (a) TiAlN coating, (b) AlCrN coating

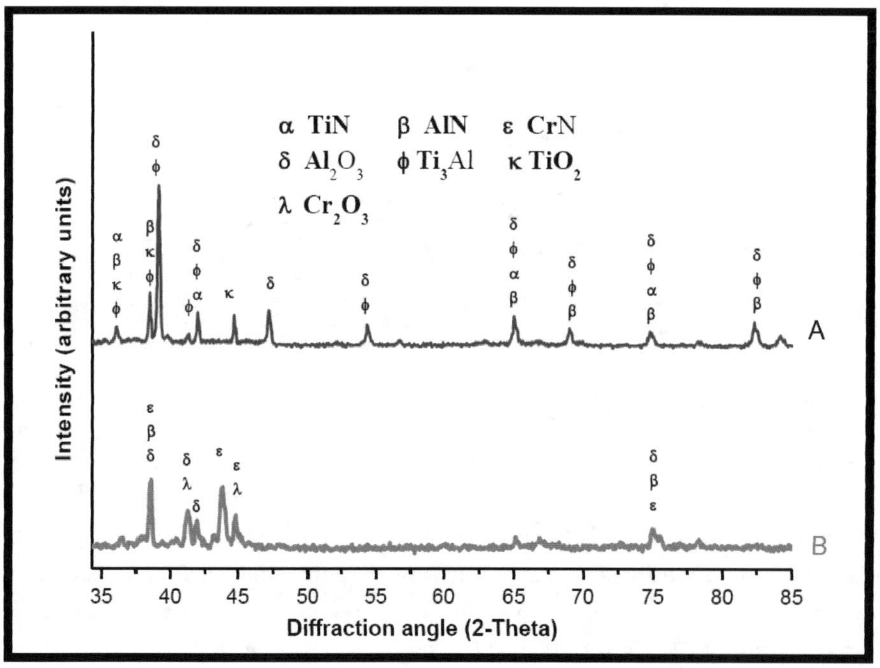

Fig. 4 X-Ray Diffraction pattern of as coated Superfer 800H superalloy (a) TiAlN coating, (b) AlCrN coating

The porosity analysis is of prime importance in high temperature corrosion studies. The dense coatings are supposed to provide very good corrosion resistance as compared to porous coatings. The porosity of as sprayed Ti-Al (1.90-4.20 %) and Al-Cr (2.0-4.10 %) coatings was also evaluated; which is reported in Table.4.

Table 4 Microstructural and mechanical properties of TiAlN and AlCrN coated Superfer 800H superalloy

Coating	Surface Roughness (μm)	Porosity (% age)		Bond Strength (MPa)	Coating Color	Coating Thickness (μm)
		As Sprayed	*After Gas Nitriding*			
TiAlN Coating	10.01-12.48	1.90-4.20	< 0.65	68.74	Dark-grey	140
AlCrN Coating	08.93-11.11	2.00-4.10	< 0.60	54.69	Bluish-grey	126

A considerable decrease in the porosity has been observed after gas nitriding and it was found to be less than 0.65 % (Table.4). The coating surface was very rough due to the presence of melted/partially melted particles and the roughness was found to be in the range of 10.35-15.13 μm and 11.84-15.23 μm respectively. The centre line average (CLA) method was used to obtain the R_a values. Also, the bond strength of the coatings was measured on three specimens as per ASTM standard C633-01. The coatings failed at the substrate-coating interface while remaining attached to the adhesive. Average bond strength of 68.74 MPa and 54.69 MPa was observed in case of TiAlN and AlCrN coatings respectively (Table.4).The microhardness of the coatings has been measured along the cross-section. Figure 5, shows the microhardness profiles along the cross-section of the coatings as a function of distance from the coating-substrate interface.

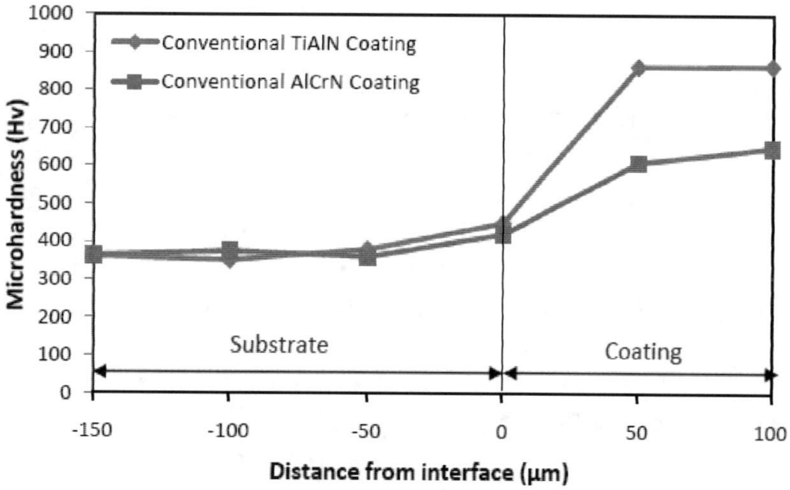

Fig. 5 Microhardness profile across the cross-section for TiAlN and AlCrN coating on Superfer 800H superalloy

The critical mocrohardness value of the substrate Superfer 800H superalloy was found to be in the range 300-400 Hv. The increase in hardness of the substrate at the interface is due to the high-speed impact of the coating particles during the plasma spray process [26]. Hardness value of the order of 900-950 Hv was measured for the TiAlN coating; whereas the AlCrN coating has shown microhardness of the order of 600-650 Hv. The SEM micrographs along with EDAX point analysis reveal the surface morphology of the TiAlN and AlCrN coated Superfer 800H superalloy; are shown in Figure 6. Some indications of unmelted particles can be seen in the coatings. In general microscopic features indicate that the coatings are homogeneous and massive, free from cracks. The EDAX point analysis (Point 1 and 2 in Fig.6) for TiAlN coating indicates the higher concentration of Ti and Al in the coating along with N and oxygen. The presence of Al and Cr along with N and oxygen revealed by EDAX analysis in case of AlCrN coating (Fig.6.c & d); the white contrast region (Point 3) shows higher concentration of Cr as compared to the dark region (Point 4).

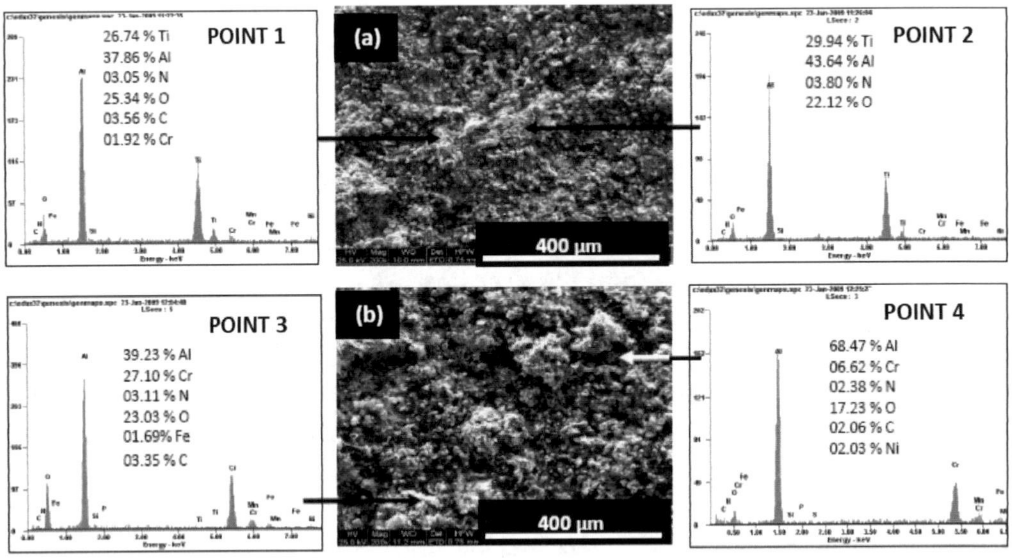

Fig. 6 Surface-scale morphology and EDAX patterns from different spots on as coated Superfer 800H superalloy (a) TiAlN coating, (b) AlCrN coating

Further, BSEI and X-ray mapping analysis for TiAlN and AlCrN coated superalloy is shown in Figure 7. The thickness of the coatings was measured from the BSEI, taken along the cross-section of the mounted samples. The measured average coating thickness values for TiAlN and AlCrN coatings are found to be 140 and 126 µm respectively (reported in Table 4). In case of TiAlN coated superalloy, Figure 7(a): Al and Ti rich coating along with small amount of N and O has been observed. The splats in the coatings are visible. Figure 7 (b) depicts the BSEI and X-ray mapping in case of AlCrN coated Superfer 800H superalloy. The coating is rich in Al and Cr with negligible presence of N and O. The X-ray mapping analysis indicates the presence of Fe in coating only for AlCrN coated superalloy and no sign of diffusion of substrate elements to the coating was observed in case of TiAlN coating.

3.2 Coatings subjected to cyclic oxidation in air and molten salt corrosion in Na_2SO_4–60%V_2O_5 at 900°C

3.2.1 Cyclic oxidation in air

The weight gain per unit area (mg/cm^2) versus time expressed in number of cycles plot for coated and bare Superfer800H superalloy subjected to cyclic oxidation in air at 900°C for 50 cycles, is presented in Figure 8.

It can be inferred from the plots that all the coated and uncoated specimens have shown Initial rapid oxidation rate and the weight gain have shown tendency to reach steady state with the progress of exposure time with some minor deviations. The weight gain for coatings is found to be slightly greater than that of the uncoated superalloy.

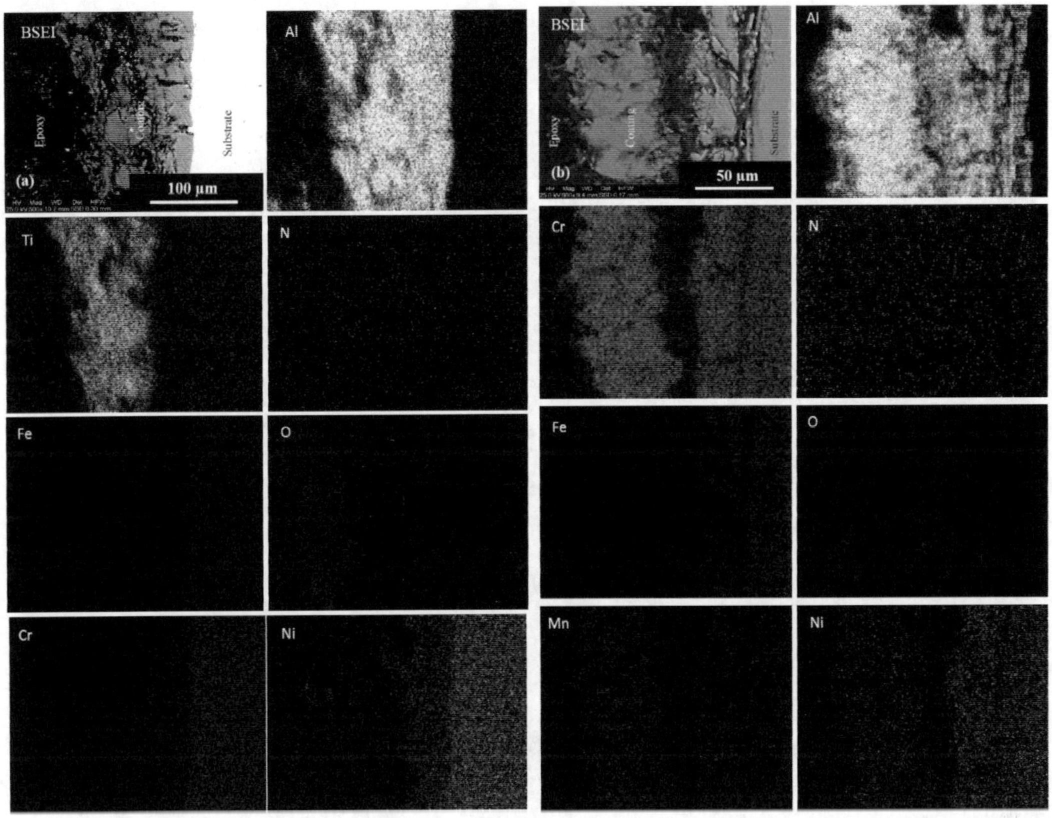

Fig. 7 Composition image (BSEI) and X-ray mapping of the cross-section of as-coated Superfer800H superalloy (a) TiAlN coating, (b) AlCrN coating

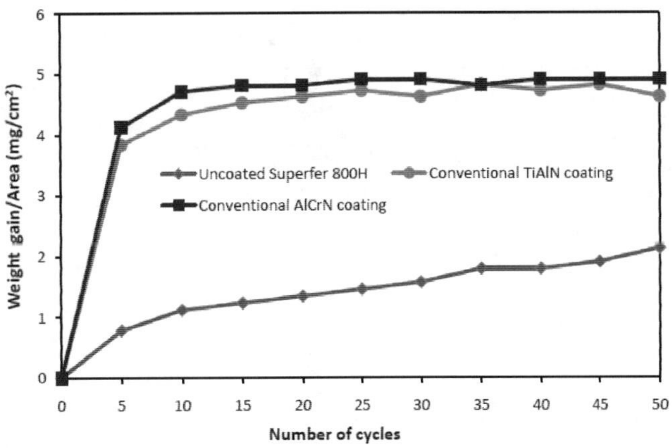

Fig. 8 Weight gain/area *vs* time (number of cycles) for the uncoated and coated Superfer 800H superalloy subjected to cyclic oxidation in air at 900°C for 50 cycles

The cumulative weight gain per unit area for the coated and uncoated Superfer800H superalloy subjected to cyclic oxidation in air at 900°C for 50 cycles is shown in Figure 9. The overall weight gain is highest in case of conventional AlCrN coating and is lowest in case of bare alloy. Figure 10; the (weight gain/area) ^{2} versus number of cycles plot are shown for all the cases to ascertain conformance with the parabolic rate law. All the coated and uncoated Superfer800H superalloy followed the parabolic rate law as evident from the Figure 10. The parabolic rate constants for the bare and coated Superfer 800H superalloy calculated on the basis of 50 cycle's exposure data are shown in Table 5.

Table 5 Parabolic rate constant 'K_p' values of uncoated and coated Superfer800H superalloy subjected to cyclic oxidation in air and in molten salt at 900°C for 50 cycles

Boiler steel substrate / Coating	Oxidation in air K_p x 10^{-12} $gm^2cm^{-4}s^{-1}$	Oxidation in molten salt K_p x 10^{-10} $gm^2cm^{-4}s^{-1}$
Bare Superfer800H superalloy	22.00	33.65
TiAlN coating	79.42	02.60
AlCrN coating	78.25	05.41

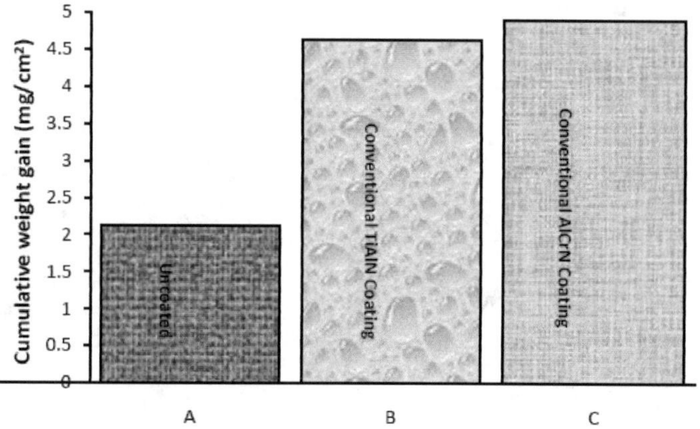

Fig. 9 Column chart showing cumulative weight gain per unit area for the uncoated and coated Superfer 800H superalloy subjected to cyclic oxidation in air at 900°C for 50 cycles: (A) Uncoated Superfer 800H superalloy, (B) Conventional TiAlN coating, (C) Conventional AlCrN coating

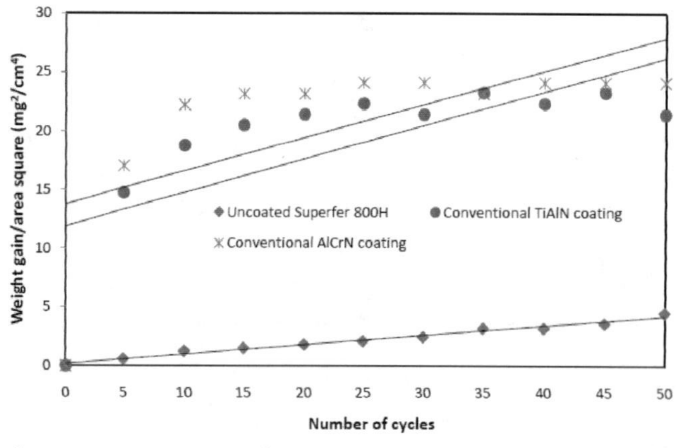

Fig. 10 Weight gain/area square *vs* time (number of cycles) for the uncoated and coated Superfer 800H superalloy subjected to cyclic oxidation in air at 900°C for 50 cycles

3.2.2 Cyclic corrosion in molten salt

Figure 11 depicts the weight gain per unit area (mg/cm^2) versus time expressed in number of cycles plot for coated and bare superalloy subjected to cyclic oxidation in Na_2SO_4-60%V_2O_5 (molten salt) environment at 900°C for 50 cycles. It can be inferred from the plots that the bare superalloy have shown high rate of oxidation as compared to coatings. The coatings have shown relatively higher weight gain in the early cycles of study, followed by tendency to show almost no weight gain with further increase in number of cycles. The cumulative weight gain per unit area for the coated and uncoated superalloy subjected to cyclic oxidation in air at 900°C for 50 cycles is shown in Figure 12. The overall weight gain is highest in case of uncoated superalloy (24.51 mg/cm^2) and is lowest in case of TiAlN coating (8.544 mg/cm^2). Figure 13 shows $(\text{weight gain/area})^2$ versus number of cycles plot for all the cases to ascertain conformance with the parabolic rate law. It can be inferred from Figure 13 that all the coated and uncoated superalloy followed the parabolic rate law. The parabolic rate constants for the bare and coated Superfer800H superalloy calculated on the basis of 50 cycle's exposure data are shown in Table 5.

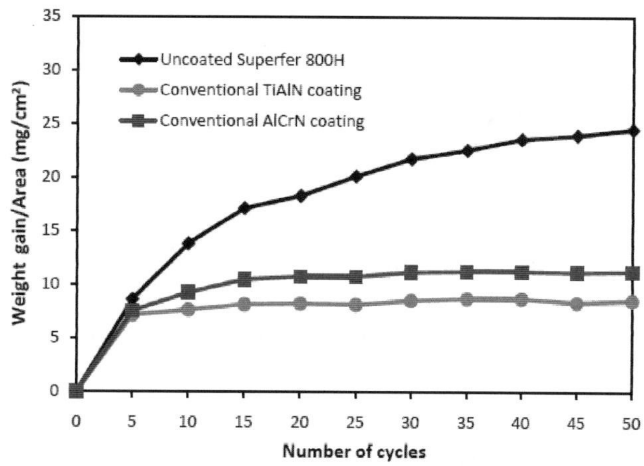

Fig. 11 Weight gain/area *vs* time (number of cycles) for the uncoated and coated Superfer 800H superalloy exposed to molten salt (Na_2SO_4-60%V_2O_5) environment at 900°C for 50 cycles

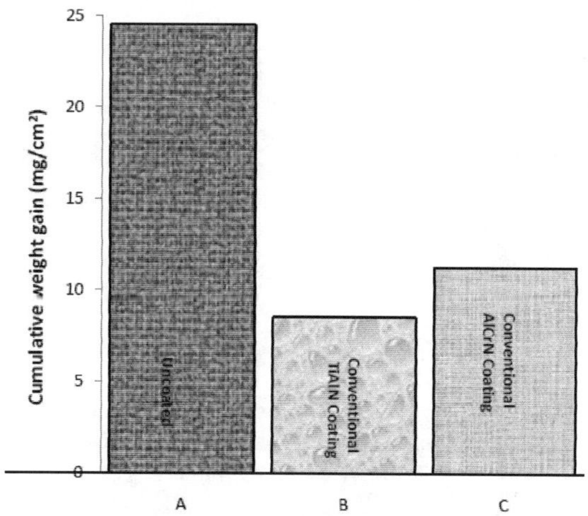

Fig. 12 Bar chart showing cumulative weight gain per unit area for the uncoated and coated Superfer 800H superalloy exposed to molten salt (Na_2SO_4-60%V_2O_5) environment at 900°C for 50 cycles: (A) Uncoated Superfer 800H superalloy, (B) TiAlN coating, (C) AlCrN coating

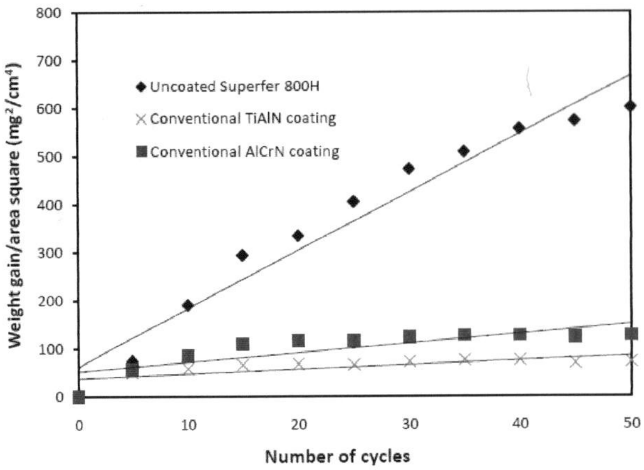

Fig. 13 Weight gain/area square *vs* time (number of cycles) for the uncoated and coated Superfer 800H superalloy exposed to molten salt (Na_2SO_4-60%V_2O_5) environment at 900°C for 50 cycles

3.3 X-Ray diffraction analysis (XRD) of oxide scale

3.3.1 High temperature oxidation in air

XRD diffractograms for bare and coated Superfer800H superalloy subjected to cyclic oxidation in air at 900°C for 50 cycles are depicted in Figure 14 (a & b) on reduced scale. As indicated by the diffractograms, in case of uncoated superfer800H superalloy the oxide phases found are Fe_2O_3, Cr_2O_3, NiO, $NiCr_2O_4$ and $NiFe_2O_4$. Further, the main phases identified for the TiAlN coating are Al_2O_3, Ti_2O_3 and NiO. The oxide phases found in case of AlCrN coating are Al_2O_3 along with NiO.

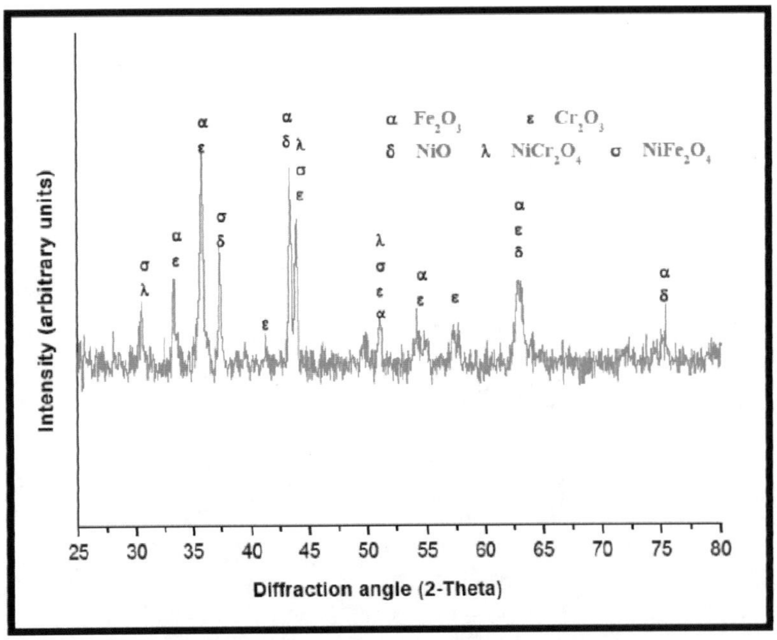

Fig. 14 (a) X-Ray Diffraction pattern of uncoated Superfer 800H superalloy exposed to cyclic oxidation in air at 900°C for 50 cycles

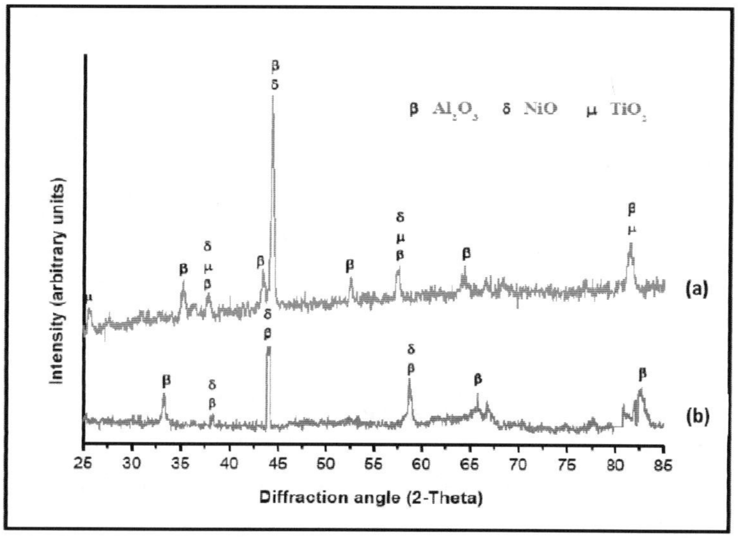

Fig. 14 (b) X-Ray Diffraction pattern of coated Superfer 800H superalloy exposed to cyclic oxidation in air at 900°C for 50 cycles: (a) TiAlN coating, (b) AlCrN coating

3.3.2 High temperature corrosion in molten salt

XRD diffractograms for coated and bare Superfer 800H superalloy subjected to cyclic oxidation in Na_2SO_4-60%V_2O_5 (molten salt) environment at 900°C for 50 cycles are depicted in Figure 15 (a & b) on reduced scale. As indicated by the diffractograms, in case of uncoated Superfer 800H superalloy the oxide phases found are Fe_2O_3, Cr_2O_3, NiO, $NiCr_2O_4$ and $NiFe_2O_4$. The main phases identified for the TiAlN coating are Al_2O_3, Ti_3Al and NiO. The oxide phases found in case of AlCrN coating are Al_2O_3, Fe_2O_3, and Cr_2O_3 along with NiO.

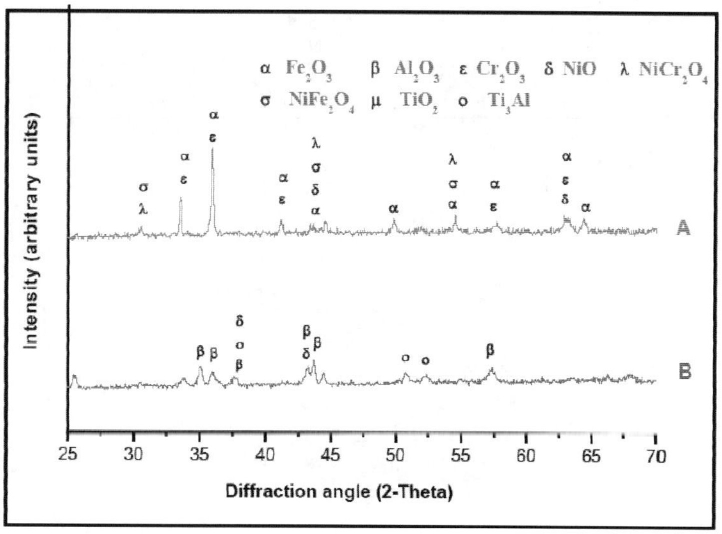

Fig. 15 (a) X-Ray Diffraction pattern of uncoated and coated Superfer 800H superalloy exposed to molten salt (Na_2SO_4-60%V_2O_5) environment at 900°C for 50 cycles: (A) Uncoated superalloy, (B) TiAlN coating

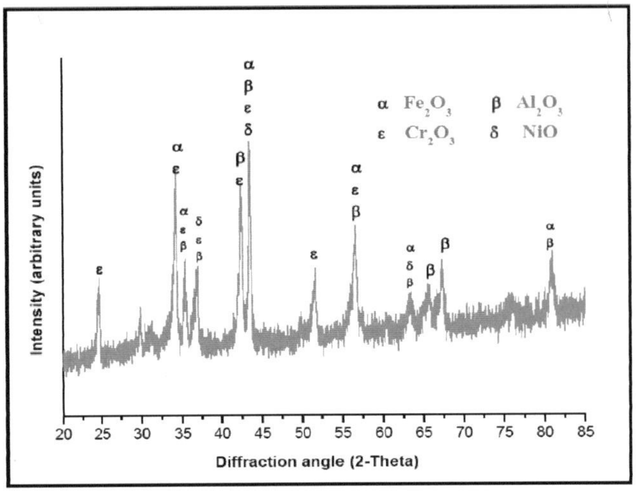

Fig. 15 (b) X-Ray Diffraction pattern of Conventional AlCrN coated Superfer 800H superalloy exposed to molten salt (Na_2SO_4-60%V_2O_5) environment at 900°C for 50 cycles

3.4 Surface morphology of scale

3.4.1 Oxidation in air *at 900°C*

The macrographs for uncoated and coated Superfer800H subjected to cyclic oxidation in air at 900°C for 50 cycles, are shown in Figure 16. For the uncoated superalloy, a dark grey colored scale appeared on the surface right from the 1st cycle, which turned to blackish green colored surface along with a light grey spot at the end of 50th cycle (Fig.16.a).

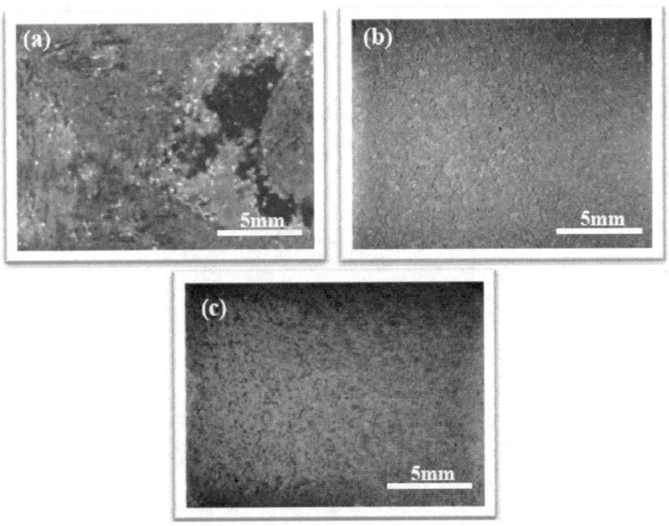

Fig. 16 Surface macrographs of uncoated and coated Superfer 800H superalloy after exposure to cyclic oxidation in air at 900°C for 50 cycles: (a) Uncoated Superfer 800H superalloy, (b) TiAlN coating, (c) AlCrN coating

The fragile porous corrosion products (NiO and Cr_2O_3) were formed on the surface for initial few cycles, there after sticky (adherent) oxide scale formed which lasted up to 50 cycles. A visual observation of TiAlN coated Superfer800H superalloy (Fig.16.b), showed the formation of very light brownish grey color matrix with cream color spots after cyclic oxidation studies. Some amount of superficial spalling was noticed at the end of 3rd cycle, which may be due to the loosely bond particles on the surface of the coating. In case of conventional thick AlCrN coated Superfer800H superalloy, the scale showed no signs of spalling and cracks (Fig.16.c). After 3rd cycle, light bluish grey color was seen which turns to whitish grey colored surface along with a light brown side layer. The scale remains intact with the substrate throughout the study.

SEM micrographs along with EDAX point analysis reveals the surface morphology of the coated and uncoated superalloy subjected to cyclic oxidation in air at 900°C for 50 cycles are shown in Figure 17. The oxide scale for uncoated Superfer800H superalloy indicates the dominance of Fe and O along with some amount of Cr, Mn and Ni (Point 2). EDAX analysis at point 1 on Figure 17(a) i.e. at the portion of the matrix, from which the upper layer has been removed during oxidation studies, shows the presence of Cr, Mn and O as the main elements along with small amount of Fe, Ni and Ti. The surface scale developed on TiAlN coating is with white contrast (Point 3) and black contrast regions (Point 4). The surface appearance is like eroded surface. EDAX analysis at point 3 indicates the presence of Ti, Al and O as the main phases along with very small amount of Fe, whereas the dark region (Point 4 on Fig.17.b) indicates maximum amount of Al (52.76%) and O (38.56%). A homogeneous and continuous surface scale is developed in case of AlCrN coating, which has Al and O as main elements (Fig.17.c).

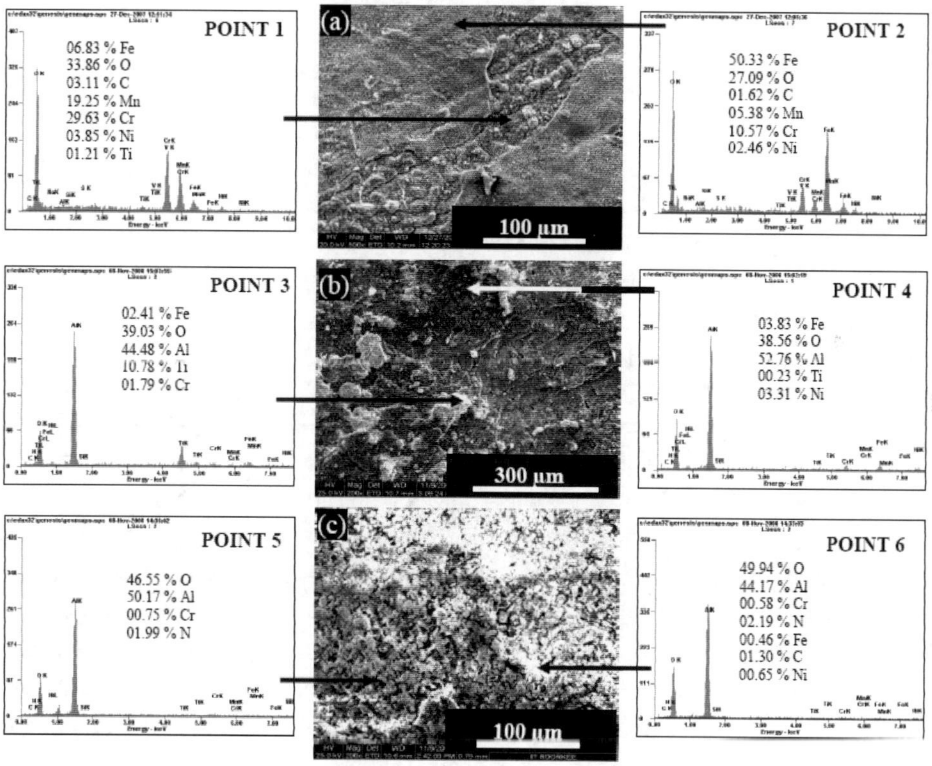

Fig. 17 Surface-scale morphology and EDAX patterns from different spots on uncoated and coated Superfer 800H superalloy after exposure to cyclic oxidation in air at 900°C for 50 cycles : (a) Uncoated Superfer 800H superalloy, (b) TiAlN coating, (c) AlCrN coating

3.4.2 Hot corrosion in molten salt *at 900°C*

The macrographs for uncoated and coated Superfer800H (Fe-based superalloy) subjected to cyclic oxidation in Na_2SO_4-60%V_2O_5 (molten salt) environment at 900°C for 50 cycles, are shown in Figure 18. For the uncoated superalloy, a dark grey colored scale appeared on the surface right from the 1st cycle, which turned to blackish green colored along with a small light grey spot at the end of 50th cycle (Fig.18.a). After 20th cycle some shining dots were also appeared in the scale. A visual observation of TiAlN coated superalloy (Fig.18.b), showed the formation of very light brownish grey color matrix with cream color spots after cyclic oxidation studies in Na_2SO_4-60%V_2O_5 (molten salt) environment. Some amount of superficial spalling was noticed at the end of 2nd cycle, which may be due to the loosely bond particles on the surface of the coating. In case of AlCrN coated superalloy, the scale showed no signs of spalling and cracking (Fig.18.c). The scale remains intact with the substrate throughout the study and the color of the scale is light grey at the end of the study.

Fig. 18 Surface macrographs of uncoated and coated Superfer 800H superalloy exposed to molten salt (Na_2SO_4-60%V_2O_5) environment at 900°C for 50 cycles: (a) Uncoated Superfer 800H superalloy, (b) TiAlN coating, (c) AlCrN coating

SEM micrographs along with EDAX point analysis reveals the surface morphology of the coated and uncoated Superfer 800H superalloy subjected to cyclic oxidation in Na_2SO_4-60%V_2O_5 (molten salt) environment at 900 °C for 50 cycles are shown in Figure 19. The oxide scale for uncoated superalloy indicates the dominance of Fe, Cr and O along with some amount of Mn and Ni (Fig.19.a). The surface scale developed on TiAlN coating is with black (Point 3) and white areas (Point 4 on Fig.19.b). The EDAX point analysis at point 3 indicates the presence of Al (42.29%) and O (39.12%) as the main phases along with small amount of Fe, Ni, Ti and Cr.The white contrast region (Point 4 in Fig.19) shows the presence of Al (28.87%), Fe (11.75%), Ti (6.79%) and O (39.11%) along with Cr (7.51%). A homogeneous and continuous surface scale is developed in case of conventional thick AlCrN coating (Fig.19.c). The EDAX point analysis revealed more concentration of Cr and Fe in the white contrast region (Point 5) as compared to the dark grey region (Point 6), whereas the concentration of Al and O is less.

3.5 Cross-sectional analysis for oxide scale

3.5.1 High temperature oxidation in air *at 900°C*

The oxidized samples were cut across the cross section using Buehler Isomet 1000 precision saw and mounted in transoptic mounting resin and subsequently mirror polished to obtain scanning electron back scattered micrographs and X-ray mapping of different elements for coated and uncoated Superfer800H superalloy. The scale thickness values were measured from SEM back scattered micrographs as shown in Figure 20. The measured average scale thickness values for Superfer800H superalloy, TiAlN and AlCrN coatings are 12, 100 and 120 μm respectively.

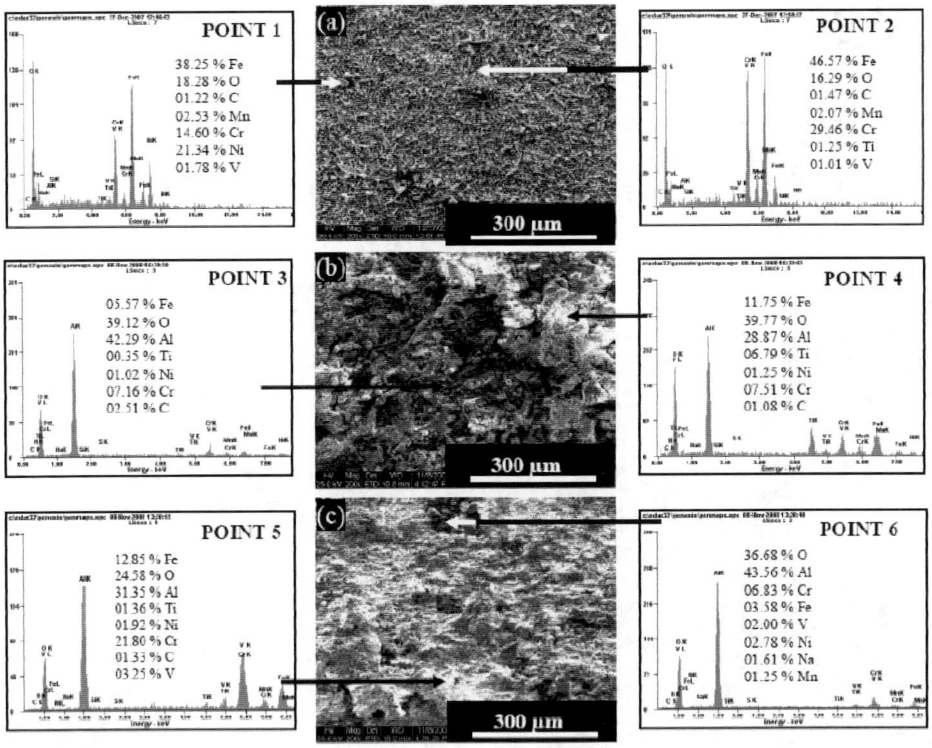

Fig. 19 Surface-scale morphology and EDAX patterns from different spots on uncoated and coated Superfer 800H superalloy exposed to molten salt (Na_2SO_4-60%V_2O_5) environment at 900°C for 50 cycles : (a) Uncoated Superfer 800H superalloy, (b) TiAlN coating, (c) AlCrN coating

Back Scattered Electron Image (BSEI) micrograph and elemental variation across the cross-section for coated and uncoated Superfer800H superalloy subjected to cyclic oxidation in air at 900°C for 50 cycles are shown in Figure 20. In case of uncoated superalloy, EDAX analysis reveals the presence of iron, oxygen, nickel and chromium throughout the scale. The scale is dense and nearly uniform in thickness. At point 3 (Fig. 20.a), the concentration of Cr and O is maximum, which points out the possibility of existence of Cr_2O_3. The concentration of Fe is decreasing as one move away from the alloy-scale interface, whereas Ni has shown increase in the concentration along outward direction. In case of TiAlN coated Superfer800H superalloy, the scale is uniform and adherent as depicted in Figure 20(b). The EDAX point analysis indicates the presence of Fe, Ni, Al, Ti and Cr in the scale with variable amounts. The concentration of oxygen is less except at certain oxygen rich pockets (Point 4) where the concentration of Al is also higher. The AlCrN coated Superfer800H superalloy (Fig.20.c) indicates continuous, thick and adherent scale. EDAX point analysis shows the uniform presence of Fe, Al, Cr, O and Ni throughout the scale.

X-ray mappings for a part of oxide scale of uncoated and coated Superfer800H superalloy oxidized in air at 900°C for 50 cycles are shown in Figure 21. BSEI and X-ray mapping for the oxidized uncoated Superfer 800H shows nickel dominating continuous scale, which contain substantial amounts of iron and oxygen (Fig.21.a). Chromium and oxygen have shown their co-presence in the form of a band at the scale-substrate interface. In case of TiAlN coated superalloy (Fig.21.b), Fe and O are present throughout the scale. Thick band of Ti is present in the top scale and of Ni is present at scale-substrate interface. Oxygen and aluminum rich pockets can be seen in the sub scale region. Figure 21 (c), depicts the BSEI and X-ray mapping in case of AlCrN coating. Aluminum is uniformly distributed in the scale along with substantial amount of Cr, Ni and Fe. The other elements like O, Ti and N have shown negligible presence.

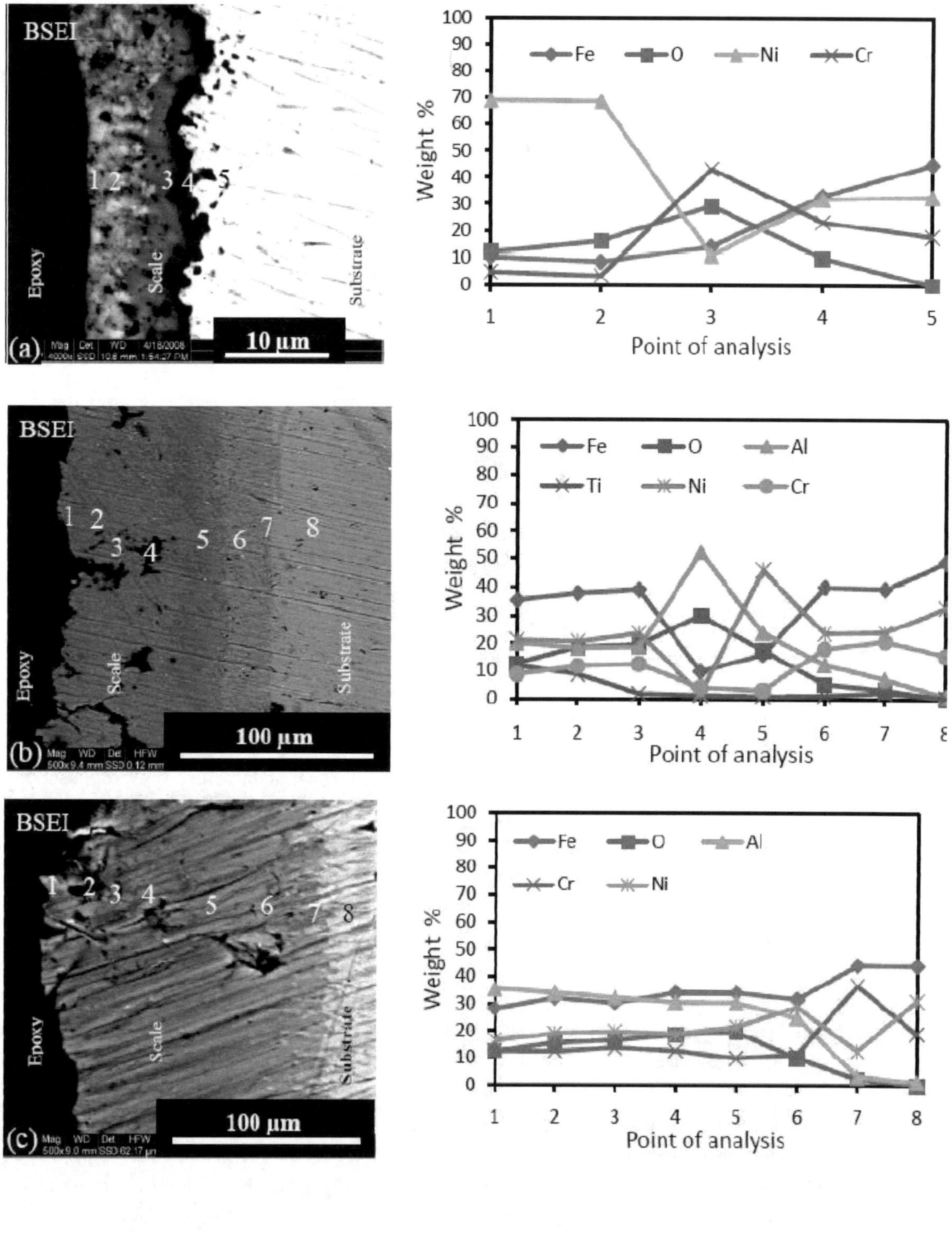

Fig. 20 Oxide scale morphology and variation of elemental composition across the cross- section of the uncoated and coated Superfer 800H superalloy after exposure to cyclic oxidation in air at 900°C for 50 cycles: (a) Uncoated Superfer 800H superalloy (4000 X), (b) TiAlN coating (500 X), (c) AlCrN coating (500 X)

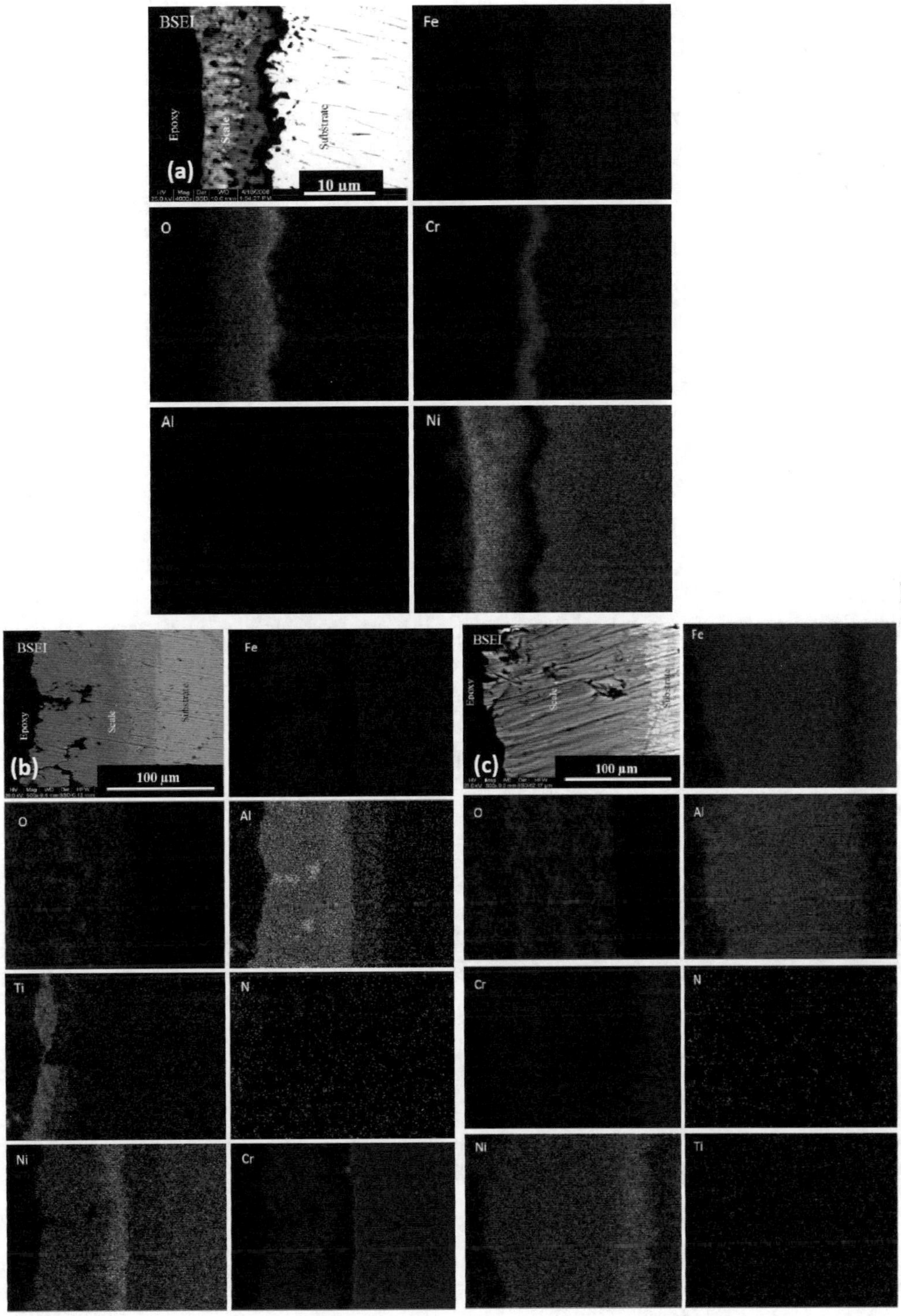

Fig. 21 Composition image (BSEI) and X-ray mapping of the cross-section of uncoated and coated Superfer 800H superalloy after exposure to cyclic oxidation in air at 900°C for 50 cycles (a) Uncoated Superfer 800H superalloy, (b) TiAlN coating, (c) AlCrN coating

3.5.2 Hot corrosion in *Na_2SO_4–60%V_2O_5 at 900°C*

The oxidized samples were cut across the cross section using Buehler Isomet 1000 precision saw and mounted in transoptic mounting resin and subsequently mirror polished to obtain scanning electron back scattered micrographs and X-ray mapping of different elements for coated and uncoated Superfer 800H superalloy. The scale thickness values were measured from SEM back scattered micrographs as shown in Figure 22. The measured average scale thickness values for Superfer 800H superalloy, TiAlN and AlCrN coatings are 84, 108 and 106 µm respectively. Back Scattered Electron Image (BSEI) micrograph and elemental variation across the cross-section for coated and uncoated Superfer 800H superalloy subjected to cyclic oxidation in Na_2SO_4-60%V_2O_5 (molten salt) environment at 900°C for 50 cycles are shown in Figure 22. In case of uncoated Superfer800H superalloy (Fig.22.a), EDAX analysis reveals the presence of iron, oxygen, nickel and chromium throughout the scale. The scale is dense and nearly uniform in thickness. At point 3 and 5 the concentration of Cr and O is higher, which points out the possibility of existence of Cr_2O_3. The amount of Ni is higher at point 4. The concentration of O is varying throughout the scale and is almost zero at point 2 and 4. In case of conventional TiAlN coated Superfer800H superalloy, the scale is uniform and adherent as depicted in Figure 22(b). The EDAX point analysis indicates the presence of Fe, Ni, O, Al, Ti and Cr in the scale with variable amounts. The top scale is rich in Ti, Ni, Al, Fe and O. The concentration of Ti is almost zero in the subscale region. The analysis at point 4 shows increase in the amount of Al, Ni and O with decrease in case of Fe. The conventional thick AlCrN coated Superfer800H superalloy (Figure 22.c) indicates continuous, thick and adherent scale. EDAX point analysis shows the uniform presence of Fe, Al, Cr, O and Ni throughout the scale. The analysis at point 4 revealed increase in concentration of Al, Ni and O with decrease in case of Fe and Cr., whereas point 5 shows increase in concentration of Cr.

X-ray mappings for a part of oxide scale of uncoated and coated Superfer 800H superalloy oxidized in Na_2SO_4-60%V_2O_5 (molten salt) environment at 900 °C for 50 cycles are shown in Figure 23. BSEI and X-ray mapping for the oxidized bare superalloy shows nickel dominating continuous scale, which contain substantial amounts of iron and oxygen (Fig.23.a). Chromium and oxygen have shown their co-presence in the form of some pockets in the scale and at the scale-substrate interface, where iron and nickel is completely missing. In case of TiAlN coated superalloy (Fig.23.b), Fe and O are present throughout the scale. Thick band of Ti is present in the top scale whereas thin band of Ni is present at scale-substrate interface. The sub scale is rich in Al and Fe with substantial amount of O, Ni and Cr. Figure 23 (c), depicts the BSEI and X-ray mapping in case of conventional AlCrN coating. Aluminum is uniformly distributed in the scale along with substantial amount of Cr, Ni and Fe. The other elements like Ti and N have shown negligible presence.

4. DISCUSSION

The plasma spray process provides the possibility of forming Ti-Al and Al-Cr coatings with high temperature and rapid cooling rates simultaneously. The temperature may drop from 15 000 K to almost room temperature over a distance of 10 mm, and the velocity may drop from close to 1000 mys to almost zero over the same distance. The freezing time of small particles (approx. 30 µm) is in the range of 10^{-7} –10^{-6} s [26]. The measured values of porosity (1.90-4.10%) for as sprayed conventional Ti-Al and Al-Cr coatings are almost in close agreement with the findings of Chen et al. [27], Erickson et al. [28], Hidalgo et al. [29, 30], Singh [31] and Sidhu et al. [32, 33] for thermal plasma sprayed coatings. Further, the porosity values found to be less than 0.6% after gas nitriding, which may be because of elimination of microstructural in-homogeneities by filling of pores and voids by nitrogen during gas nitriding. Hardness is the most frequently quoted mechanical property of the coatings [34].

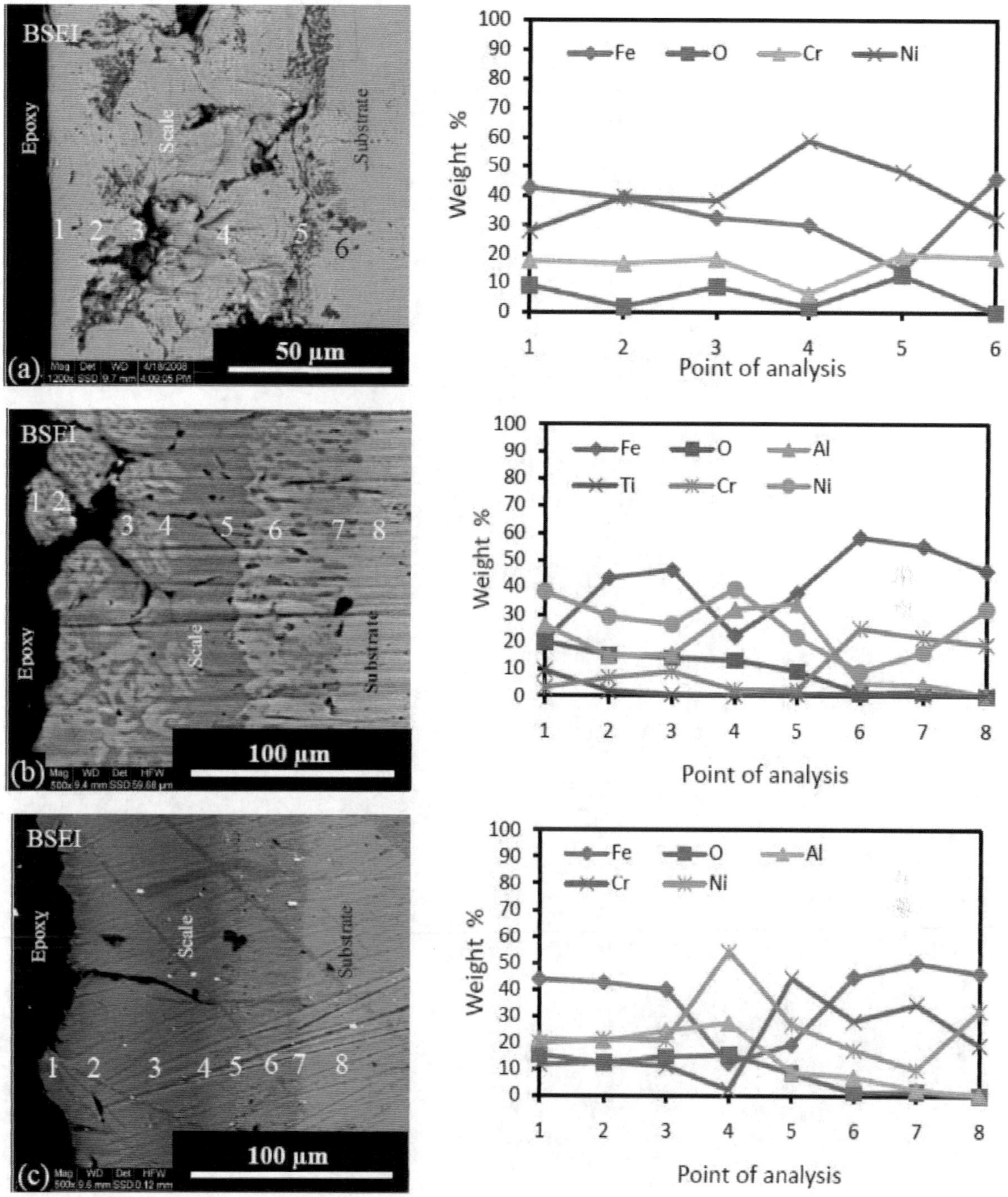

Fig. 22 Oxide scale morphology and variation of elemental composition across the cross- section of the uncoated and coated Superfer800H superalloy exposed to molten salt (Na_2SO_4-60%V_2O_5) environment at 900°C for 50 cycles: (a) Uncoated Superfer800H superalloy (1200 X), (b) TiAlN coating (500 X), (c) AlCrN coating (500 X)

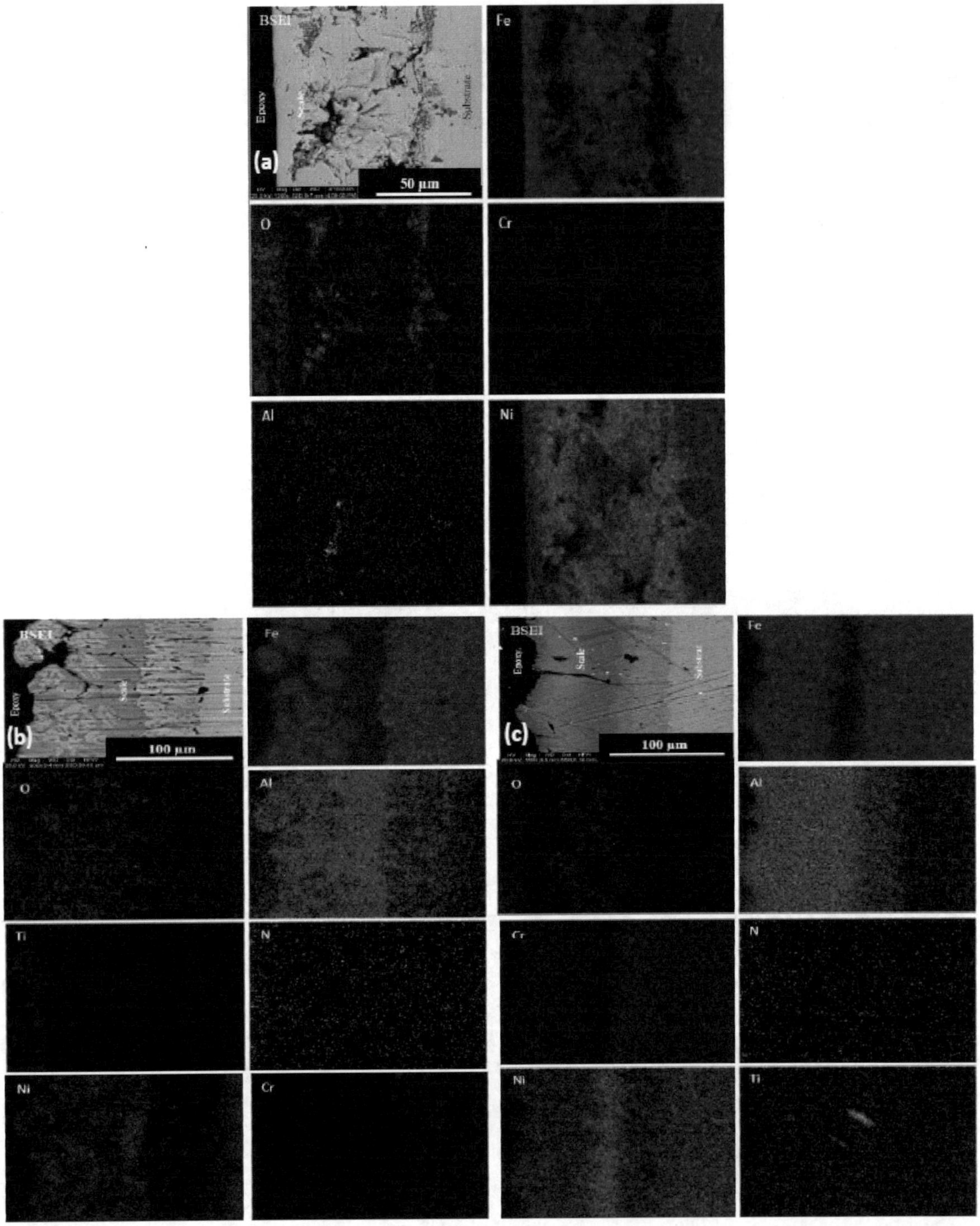

Fig. 23 Composition image (BSEI) and X-ray mapping of the cross-section of uncoated and coated Superfer 800H superalloy exposed to molten salt (Na_2SO_4-60%V_2O_5) environment at 900°C for 50 cycles (a) Uncoated Superfer 800H superalloy, (b) TiAlN coating, (c) AlCrN coating

The observed microhardness values (Fig. 5) for the conventional coatings are within the range of microhardness values reported for plasma coatings by Adachi and Nakata [15], Vuoristo et al. [35], Chen and Hutchings [36] and Westergard et al. [37]. Microhardness plots indicate some increase in the microhardness of substrate steel at the substrate coating interface. The hardening of the substrates as observed in the current study might have occurred due to the high speed impact of the coating particles during plasma spray deposition similar to the findings of Sidhu et al. [33], Hidalgo et al. [29, 30, 38 and 39] and Sidhu et al. [32]. The observed non-uniformity in the hardness values along the thickness of the coatings may be due to the microstructural changes along the cross section of the coatings [40]. Moreover, the microhardness and other properties of the thermal spray coatings are anisotropic because of typical splat structure and directional solidification [34]. Further, Tucker [34] has reported extensive data on the properties of plasma coatings and the values of porosity and microhardness data were found to be similar to those published by him.

The bond strength of the conventional thick TiAlN and AlCrN coatings was measured on three specimens as per ASTM standard C633-01. The coatings failed at the substrate-coating interface while remaining attached to the adhesive. Average bond strength of 68.74 MPa and 54.69 MPa was observed in case of conventional TiAlN and AlCrN coatings respectively. The bond strength in case of conventional TiAlN coatings (68.74 MPa) is almost in good agreement with the results reported by Adachi and Nakata [15].

The main phases identified for the thick TiAlN coating are Al_2O_3, TiN, Ti_3Al, AlN and small peaks of TiO_2 and Fe_2O_3. The phases identified in case of thick AlCrN coating are CrN, AlN, and Al_2O_3 with minor peaks of Cr_2O_3. The presence of metal nitride phases indicates that the gas nitriding process has successfully produced the desired coatings. Further, the phases analyzed are also in agreement with that reported by Adachi and Nakata [15].

Surface EDAX analysis supported the results obtained by XRD (Fig.4) and X-ray mapping analysis (Fig.7). The XRD analysis indicates the presence of the oxide phases in the coatings i.e. Al_2O_3 and TiO_2 in case of TiAlN coating and the presence of Al_2O_3 and Cr_2O_3 in case of AlCrN coatings. The oxides may form due to the in-flight oxidation during spraying process and/or preexisting in the feed material as reported by Bluni and Mardar [41]. The latter reason for the oxides formation in the structure of coatings under study looks to be more relevant as the chances of in-flight oxidation are meager in case of the shrouded plasma spraying. Deshpande et al. [42] proposed that, during in-flight oxidation, a layer of oxide is formed on the molten particle due to chemical reactions between the surface of the liquid phase and oxygen or due to diffusion of oxygen into the liquid. The turbulent mixing of the liquid part of the powder particle during its flight destroys the surface layer of oxides and causes the oxides to be distributed more uniformly through the bulk volume of the particle. However, when temperature of the particle starts dropping during later part of the flight, these oxides tend to solidify and a thin oxide shell would form around the droplet. The oxidation time during thermal spray coating is short typically less than 0.01 s, and can occur in either the solid or molten state. The oxidation of coatings is not always harmful, it is equally important to control and understand the different aspects of oxidation of coatings; therefore, it is important to find an optimum level for oxidation of coatings [5, 43-45].

4.1 High temperature oxidation in air *at 900°C*

The sequence of oxidation rates based on weight gain values for the coated and uncoated Superfer800H superalloy after 50 cycles of oxidation is:

AlCrN coating > TiAlN coating > Bare SF 800H

The plasma sprayed gas nitrided AlCrN and TiAlN coatings subjected to cyclic oxidation in air have shown relatively high weight gains (Fig.8) in the early cycles of the exposure. These initial high oxidation rates in general might partially be attributed to the rapid formation of oxides at the coating splat boundaries and within open pores due to the penetration of the oxidizing species along the splat boundaries/open pores in the early cycles of the study. Once the oxides are formed at places of porosity and splat boundaries, the coating becomes dense and the diffusion of oxidizing species to the internal portions of the coatings gets slowed down and the growth of the oxides becomes limited mainly to the surface of the specimens. This, in turn, will make the weight gain and hence the oxidation rate steady with the further progress of exposure time.

The surface scales for the coatings, have shown excellent spallation resistance against the cyclic oxidation in air, which indicates further effectiveness of the coatings under study as the cycle-oxidation behavior of an alloy is dictated mainly by scale spallation resistance as per the opinion of Stott [46]. Therefore, it can be inferred that the necessary protection has been provided by the coating to the substrate, in spite of the fact that the conventional thick coatings have shown high overall weight gains as compared to that shown by their uncoated and nanostructured coated counterparts.

These XRD results are well supported by the surface EDAX and X-Ray mapping analysis in the present study. In case of uncoated Superfer 800H superalloy, the oxide phases found are Fe_2O_3, Cr_2O_3, NiO, $NiCr_2O_4$ and $NiFe_2O_4$. The results of X-ray mappings analysis; Fig.21 (a) for the oxidised superalloy Superfer 800H are in good agreement with those reported by Polman et al. [47] for a similar superalloy Incoloy 800H. The presence of Cr_2O_3 and NiO in the top scale of bare superalloy is supported by the surface color of these samples after oxidation studies, which is blackish green. Greenish color might be attributed to the presence of both Cr_2O_3 and NiO as reported by Khajavi and Shariat [48].

The oxidation phenomenon for the uncoated Superfer 800H in the light of explanation regarding oxidation put forward by Stott [49] which is identical to the one reported by Singh, Harpreet [50]; was observed as in the initial stage of oxidation process, when the alloy surface adsorb the oxygen molecules, small impinging nuclei of all the thermodynamically stable oxides such as those of Ni, Cr, Fe and Mn develop on the surface of the alloy. These nuclei then coalesce rapidly to give a transient layer of various oxides in a very short time as the temperature of study becomes very high. As the concentration of Fe and Ni is higher in the superalloy Superfer 800H, the amount of nickel and iron oxide in this layer will be the higher, whereas chromium oxide will be the second dominating phase, followed by oxides of Mn. In this transient layer, mixed type oxide such as $NiCr_2O_4$ may also form by reaction of oxides of Ni and Cr. As the oxidation progresses, this transient layer continues to grow with NiO at the top of the scale along with iron oxide, while the thermodynamically favored chromia attempt to establish a complete layer at the base of this transient layer along with MnO. With the progress of exposure, this transient layer continues to grow and sufficient nuclei of the various oxides develop to coalesce to form complete layers. Once a continuous layer Cr_2O_3 is formed at the alloy/scale interface, the rate of oxidation is then controlled by transport of reactants across this layer, which is much slower process than across the initially formed NiO-rich layer. Thereafter, oxidation process enters a steady state. The BSEI and X-ray mapping (Fig.21.a) support the above explained mechanism of oxidation in air for uncoated superfer 800H.

Further, the main phases identified for the TiAlN coating are Al_2O_3, Ti_2O_3 and NiO and the oxide phases found in case of AlCrN coating are Al_2O_3 along with NiO. The SEM-EDAX analysis of the surface and X-ray mapping analysis of the cross section of the corroded samples support the XRD observations in case of the coated samples. Both the coatings are protecting the substrate with protective oxides layers at the surface. Development of some superficial minor cracks near or along the edges of the coated specimens, and subsequent minor spalling may be attributed to the thermal shocks due to differences in the heat expansion coefficients of the oxides, coatings and the substrate [31, 32 and 51-55]. Furthermore, the stress concentration factor at the sharp edges of the specimens might have also contributed to this minor cracking and spalling. It is pertinent to mention here that the edges of the specimens were kept sharp deliberately to add one more severe condition to the accelerated testing of the coatings. Otherwise in actual designs the sharp edges are always avoided to eliminate stress concentration.

The weight change plots (Fig.8 & 10) for the uncoated and coated Superfer 800H superalloy indicated that the oxidation behavior has shown conformance to parabolic rate law. The parabolic kinetic behavior is due to the diffusion controlled mechanism operating at 900°C under cyclic conditions [4]. Small deviation from the parabolic rate law might be due to the cyclic scale growth. The higher weight gain during the first few cycles might be attributed to the rapid formation of oxides at the splat boundaries and within the open pores due to the penetration of the oxidizing species, further the subsequent increase in weight is gradual [56]. Hussain et al. [57] have also established the validity of parabolic rate law for a Fe-based superalloy Incoloy 800H.

It can be inferred based on the present investigation that TiAlN and AlCrN coatings can provide a very good oxidation resistance at high temperature.

4.2 Hot corrosion in Na_2SO_4–60%V_2O_5 at 900°C

The weight gain plot, Figure 11; shows that the weight gained by bare Superfer 800H super alloy increases continuously due to the accelerated oxidation in the molten salt environment, where as the coated superalloy showed better hot corrosion resistance in terms of weight gain. The rapid increase in the weight gain during the initial period of exposure to Na_2SO_4-60%V_2O_5 (molten salt environment) environment at 900°C can be attributed to the possible formation of $NaVO_3$. Kolta et al. [58] proposed that in the temperature range of 900°C, the Na_2SO_4 and V_2O_5 will combine to form $NaVO_3$, as represented by eq.(7.1) having a melting point of 610°C.

$$Na_2SO_4 + V_2O_5 = 2NaVO_3\,(l) + SO_2 + \tfrac{1}{2}\,O_2 \qquad (1)$$

This $NaVO_3$ acts as a catalyst and also serves as an oxygen carrier to the base alloy through the open pores present on the surface, which will lead to the rapid oxidation of the base elements of the substrate to form a protective oxide scale. There may be simultaneous dissolution of protective oxide Cr_2O_3 in the molten salt due to the reaction [59, 60]:

$$Cr_2O_3 + 4NaVO_3 + 3/2O_2 = 2Na_2CrO_4 + 2V_2O_5 \qquad (2)$$

The Na_2CrO_4 gets evaporated as a gas [61]. The rapid increase in the weight gain during the initial period was also reported by Sidhu et al. [8], Harpreet Singh et al.[50], Tiwari and Prakash [7] and Ul-amid [62] during hot corrosion studies. Also, the higher weight gain during the first few cycles can be attributed to the rapid formation of oxides at the splat boundaries and within the open pores due to the penetration of the oxidizing species. Once the oxides are formed at places of porosity and splat boundaries, the coating becomes dense and the diffusion of oxidizing species to the internal portions of the coatings gets slowed down and the growth of the oxides becomes limited mainly to the surface of the specimens. This, in turn, will make the weight gain and hence the oxidation rate steady with the further progress of exposure time [8, 56].

It can be inferred from the weight gain plots that the uncoated Superfer 800H superalloy have shown high rate of oxidation as compared to the coatings. In case of TiAlN and AlCrN coatings; after 5th cycle the weight gain rate becomes almost constant. Thus, it can be think of that these coatings are providing full protection to the substrate alloy in molten salt environment at 900°C. Aluminum [63] and Chromium [50] exhibits higher affinity for oxygen to form Al_2O_3 and Cr_2O_3 during the earlier stages of hot corrosion. The formation of Al_2O_3 protective oxides in case of TiAlN coating and Al_2O_3 and Cr_2O_3 protective oxides in case of AlCrN coating; has also been supported by EDAX analysis of surface (Fig.17. b & c) and cross-section (Fig. 22. b & c), XRD analysis (Fig.15) and X-ray mapping (Fig.23. b & c).

Further, the overall weight gain is highest in case of uncoated superfer 800H superalloy (24.51 mg/cm^2) and is lowest in case of TiAlN coating (8.544 mg/cm^2). The sequence of oxidation rates based on weight gain values for the coated and uncoated superalloy after 50 cycles of oxidation is:

Bare SF > AlCrN Coating> TiAlN Coating

The surface scales for the coatings, have shown excellent spallation resistance against the cyclic oxidation in air, which indicates further effectiveness of the coatings under study as the cycle-oxidation behavior of an alloy is dictated mainly by scale spallation resistance as per the opinion of Stott [46]. The XRD analysis results (Fig.15) are well supported by the surface EDAX and X-Ray mapping analysis in the present study. In case of uncoated Superfer 800H superalloy; the oxide phases found are Fe_2O_3, Cr_2O_3, NiO, $NiCr_2O_4$ and $NiFe_2O_4$. Furthermore, the X-ray mappings, Fig.23 (a) for the oxidized bare superalloy Superfer 800H are in good agreement with those reported by Polman et al. [47] for a similar superalloy Incoloy 800H. Further, the main phases identified for the TiAlN coating are Al_2O_3, Ti_2O_3 and NiO. The oxide phases found in case of AlCrN coating are Al_2O_3, Fe_2O_3 and Cr_2O_3 along with NiO.

It could be seen from the respective X-Ray mapping (Fig.23) analysis that there is no sign of depletion of basic elements from the substrate alloys. This indicates that the coatings have been successful in acting as a reservoir for the formation of protective oxides/spinels, consequently may increase the service life of the substrate alloy. Moreover, it is clear from the back scattered images (Fig.22) of cross-section for the coatings subjected to oxidation for 50 cycles in molten salt environment that the coatings appear to be integral with the substrates. The X-Ray mapping analysis for the TiAlN and AlCrN coatings coating shows that the oxygen has not penetrated into the substrate alloys. In case of uncoated superalloy; some pockets in the scale can be seen where Cr and O are present where Fe and Ni are completely missing (Fig.23.a). Also, the top scale is rich in NI, Fe, Cr and O which is supported by XRD (Fig.15) and EDAX analysis of the surface (Fig.19.a) and cross-section (Fig.22.a). The presence of non protective NiO in the top scale of oxidized bare superalloy allows the penetration of corrosive species through the scale to the substrate. The identification of Fe_2O_3 in the scale (Fig.15) of uncoated superalloy after hot corrosion experiments indicated that non-protective conditions were established when Na_2SO_4-60% V_2O_5 (molten salt) environment was present on the surface. This might be the reason of higher weight gain rate in case of uncoated Superfer 800H superalloy.

Some negligible cracking and minor spalling may be attributed to the thermal shocks due to differences in the heat expansion coefficients of the oxides, coatings and the substrate [26, 51-53, 55 and 64]. The stress concentration factor at the sharp edges of the specimens might have also contributed to this minor cracking and spalling. It is pertinent to mention here that the edges of the specimens were kept sharp deliberately to add one more severe condition to the accelerated testing of the coatings. Otherwise in actual designs the sharp edges are always avoided to eliminate stress concentration. It can be inferred based on the present investigation that TiAlN and AlCrN coatings have provided a very good oxidation resistance in molten salt environment at high temperature.

5. CONCLUSIONS

The plasma spray process provides the possibility of developing Ti-Al and Al-Cr coatings on Fe-based superalloy Superfer 800H. The porosity of the coatings lay in the range of 1.90-4.20 percent for the given process parameters, which is consistent with the range specified in the literature. Further after gas nitriding of the coatings; the porosity was found to be less than 0.65 %. In case of TiAlN coating, the formation of TiN, AlN, Ti_3Al, Al_2O_3 and TiO_2 as the main phases where as in case of AlCrN coating; CrN, AlN, and Al_2O_3 with minor peaks of Cr_2O_3 were the phases confirmed by XRD analysis. Average bond strength of 68.74 MPa and 54.69 MPa Psi was observed in case of TiAlN and AlCrN coatings respectively. The metallurgical bonding was believed to be established after gas nitriding of the plasma sprayed Ti-Al and Al-Cr coatings.

The plasma sprayed gas nitrided TiAlN and AlCrN coatings subjected to cyclic oxidation in air at 900 °C for 50 cycles, exhibited fairly good adherence to the substrate superalloy and developed oxide scale that would protect the substrate from oxidation. The coatings have shown relatively high weight gains in the early cycles of the exposure. These initial high oxidation rates in general might partially be attributed to the rapid formation of oxides at the coating splat boundaries and within open pores due to the penetration of the oxidizing species along the splat boundaries/open pores in the early cycles of the study. The coatings developed a protective scale mainly consisting on Al_2O_3 as revealed by EDAX analysis of the oxidized samples. The X-Ray mapping analysis for the conventional thick TiAlN and AlCrN coatings shows that the oxygen has not penetrated into the substrate alloys and the oxide scale formed is adherent to the substrate in all specimens after hot corrosion studies.

The plasma sprayed gas nitrided TiAlN and AlCrN coatings when subjected to cyclic oxidation in Na_2SO_4-60%V_2O_5 molten salt at 900°C for 50 cycles developed a protective scale mainly consisting on aluminum oxide. The coatings have shown relatively higher weight gain in the early cycles of study, followed by tendency to show gradual weight gain with further increase in number of cycles. In case of uncoated superalloy; the weight gain is highest. The oxidation rate (total weight gain values after 50 cycles) of the coated and uncoated Superfer 800H superalloy follows the sequence as given below:

Bare SF > AlCrN Coating> TiAlN Coating

The formation of Al_2O_3 protective oxides in case of TiAlN coating and Al_2O_3 and Cr_2O_3 protective oxides in case of AlCrN coating; has also been supported by EDAX analysis of surface and cross-section, XRD analysis and X-ray mapping. It can be inferred based on the present investigation that the selected coatings have provided a very good oxidation resistance in molten salt environment at high temperature.

6. REFERENCES

1. **Wang**, Y. and Chen, W., (2003), "Effect of Ceria on the Erosion Resistance of HVOF Thermal Sprayed NiAl Intermetallic Coatings," J. Mat. Sci. Lett., Vol. 22, pp. 845-848
2. **Sidky,** P.S., Hocking, M.G., (1999), "Review of Inorganic Coatings and Coating Processes For Reducing Wear and Corrosion," Brit. Corros. J., Vol. 34, No. 3, pp. 171-183.
3. **Vikas Chawla**, S. Prakash, D. Puri, Buta Singh, "**State Of Art: Plasma Sprayed Nanostructured Coatings: A Review**", Materials Forum, Vol. 32, 2008, 137-143.
4. **Mahesh**, R. A., Jayagandhan, R. and Prakash, S., (2008), "A Study on Hot Corrosion Behaviour of Ni-5AL Coatings on Ni- and Fe-based Superalloys in an Aggressive Environment at 900°C," J. Alloy and Compounds, Vol. 460, No. 1-2, pp. 220-231
5. **Vikas Chawla, "Microstructural characteristics and mechanical properties of nanostructured and conventional TiAlN&AlCrN coatings on ASTM-SA210 Grade A-1 boiler steel",** ISRN Corrosion (Hindawi Publishing Corporation), Vol. 2013, Art. I D 231035, 2013, 1-14.
6. S.N. Tiwari, S. Prakash, Hot corrosion behaviour of an iron-base superalloy in salt environment at elevated temperatures, in: Proceedings of the Symposium on Metals and Materials Research, IIT, Madras, 4–5th July, 1996, pp. 107–117.
7. S.N. Tiwari, S. Prakash, Studies on the hot corrosion behaviour of some superalloys in Na2SO4-V2O5, in: Proceedings of the SOLCEC, Kalpakkam, India, 22–24th January, 1997, Paper C33.

8. **Buta Singh Sidhu and S. Prakash**, (2006), "Studies on the behavior of satellite-6 as plasma sprayed and laser remelted coatings in molten salt environment at 900°C under cyclic conditions," J. Materials Processing Technology, Vol. 172, pp. 52-63.
9. **Vikas Chawla**, S. Prakash, D. Puri, Buta Singh, "**Performance of plasma sprayed nanostructured and conventional coatings**", Journal of the Australian Ceramic Society (ISSN ISSN 0004-881X), Vol. 44, No. 2, 2008, 56-62.
10. **Vikas Chawla**, S. Prakash, D. Puri, Prema G. Gurbuxani' and Buta Singh, **"Hot Corrosion and Erosion problems in coal based power plants in India and possible solutions: a review**", Journal of Minerals and Materials Characterization and Engineering (ISSN 2327-4077 print / 2327-4085 online), Vol. 10, No.4, 2011, 367-385.
11. **Wang**, B. Q., Geng, G. Q. and Levy, A.V. (1992), "Erosion and Erosion-corrosion Behavior of Chromized-siliconized Steel," Surf. Coat. Technol., Vol. 54-55, pp. 529-35.
12. **Bertrand**, G., Mahdjoub, H. and Meunier, C., (2000), "A Study of the Corrosion Behaviour and Protective Quality of Sputtered Chromium Nitride Coatings," Surf. & Coat. Technol., Vol. 126, pp. 199-209.
13. **Kalss**, W., Reiter, A., Derflinger, V., Gey, C. and Endrino, J.L., (2006), "Modern Coatings in High Performance Cutting Applications," Int. J. Ref. Met. Hard Mat., Vol. 24, No. 7, pp. 399-404.
14. **Fujita**, K., (2005), "Research and Development of Oxidation, Wear and Corrosion Resistant Materials at High Temperature by Surface Modification using Ion Processing," Surf. Coat. Tech., Vol. 196, pp. 139-144.
15. **Adachi**, S. and Nakata, K., (2007), "Improvement of Adhesive Strength of Ti-Al Plasma -Sprayed Coatings," Surf. Coat. Technol., Vol. 201, pp. 5617-5620.
16. **Spain**, E., Batista, J.C., Letch, M., Housden, J. and Lerga, B., (2005), "Characterisation and Applications of Cr–Al–N Coatings," Surf. & Coat. Technol., Vol. 200, pp. 1507 – 1513.
17. **DeMasi-Marcin,** J. T. and Gupta, D. K., (1994), "Protective Coatings in the Gas Turbine Engine," Surf. Coat. Technol., Vol. 68-69, pp. 1-9.
18. **Vikas Chawla,** Lakhwinder Singh, and J. S. Grewal,**"A Review on Datonation Gun Sprayed Coatings",** Journal of Minerals and Materials Characterization and Engineering (ISSN 2327-4077 print / 2327-4085 online), Vol. 11, No. 3, 2012, 243-265.
19. **Vikas Chawla** , Satjot Singh Dhillon, Jasmaninder Singh Grewal, **"Hot corrosion and its preventive measures- a review**" Journal of the Australian Ceramic Society (ISSN ISSN 0004-881X), Vol. 47, No. 2, 2011, 01-07.
20. **Vikas Chawla ,** Pawan Deep Singh Grewal and Jasmaninder Singh Grewal, **"High Velocity oxy-fuel coatings-a review"** Journal of the Australian Ceramic Society (ISSN ISSN 0004-881X), Vol. 47, No. 2, 2011, 30-36.
21. **Buta Singh Sidhu, D. Puri, S. Prakash**, (2004), "Characterization of plasma sprayed and laser remelted NiCrAlY bond coats and Ni_3Al coatings on boiler steels," Materials Science & Engineering A, Vol. 368, pp. 149-158.
22. **Vikas Chawla**, Amita Chawla,Buta Singh Sidhu, S. Prakash and D. Puri, **"Performance of nanostructured metal nitride coated T-22 boiler steel in Na_2SO_4–60% V_2O_5 environment at 900°C under cyclic conditions",** Journal of Minerals and Materials Characterization and Engineering (ISSN 2327-4077 print / 2327-4085 online), Vol. 10, No.7, 2011, 583-608.
23. **Vikas Chawla** , Dinesh Gond, D. Puri and S. Prakash, **"High Temperature Corrosion Behaviour of T-91 and T-22 Bare Steel in 75wt.%Na2SO4+25wt.%NaCl Molten Salt Environment at 900°C",**

Journal of Minerals and Materials Characterization and Engineering (ISSN 2327-4077 print / 2327-4085 online), Vol. 9, No. 7, 2010, 593-606.

24. **Vikas Chawla** , Dinesh Gond, D. Puri and S. Prakash, **"Oxidation Studies of T-91 and T-22 Boiler steel in air at 900°C",** Journal of Minerals and Materials Characterization and Engineering (ISSN 2327-4077 print / 2327-4085 online), Vol. 9, No. 8, 2010, 749-761.

25. **Vikas Chawla**, Amita Chawla,Buta Singh Sidhu, S. Prakash and D. Puri, **"Oxidation behavior of nanostructured TiAlN and AlCrN thin coatings on ASTM-SA213-T-22 boiler steel",** Journal of Minerals and Materials Characterization and Engineering (ISSN 2327-4077 print / 2327-4085 online), Vol. 9, No.11, 2010, 1037-1057.

26. **Buta Singh Sidhu, S. Prakash**, (2003), "Evaluation of the corrosion behavior of plasma-sprayed Ni_3Al coatings on steels in oxidation and molten salt environment at 900°C," Surface & Coating Technology, Vol. 166, pp. 89-100.

27. H. C. Chen, Z. Y. Liu, and Y. C. Chuang, "Degradation of plasma-sprayed alumina and zirconia coatings on stainless steel during thermal cycling and hot corrosion," *Thin Solid Films*, vol. 223, no. 1, pp. 56–64, 1993.

28. L. C. Erickson, R. Westerg˚ard, U. Wiklund, N. Ax´en, H. M. Hawthorne, and S. Hogmark, "Cohesion in plasma-sprayed coatings: a comparison between evaluationmethods," *Wear*, vol. 214, no. 1, pp. 30–37, 1998.

29. V. Higuera Hidalgo, F. J. Belzunce Varela, and A. Carrilez Menendez, "Characterization and high temperature behavior of thermal sprayed coatings used in boilers," in *Proceedings of the 15th International Thermal Spray Conference*, pp. 617–621, Nice, france, May 1998.

30. V. H. Hidalgo, F. J. B. Varela, S. P. Martinez, and S. G. Espana, "Characterization and high temperature behaviour of Cr3C2-NiCr plasma sprayed coatings," in *Proceedings of the United Thermal Spray Conference*, pp. 683–686, Germany, 1999.

31. H. Singh, D. Puri, and S. Prakash, "Some studies on hot corrosion performance of plasma sprayed coatings on a Fe based superalloy," *Surface and Coatings Technology*, vol. 192, no. 1, pp. 27–38, 2005.

32. T. S. Sidhu, R. D. Agrawal, and S. Prakash, "Hot corrosion of some superalloys and role of high-velocity oxy-fuel spray coatings: a review," *Surface and Coatings Technology*, vol. 198, no. 1–3, pp. 441–446, 2005.

33. B. S. Sidhu, D. Puri, and S. Prakash, "Characterisations of plasma sprayed and laser remelted NiCrAlY bond coats and Ni_3Al coatings on boiler tube steels," *Materials Science andEngineering A*, vol. 368, no. 1-2, pp. 149–158, 2004.

34. R. C. Tucker Jr., "Advanced thermal spray deposition techniques,"in *Handbook of Deposition Technologies For Films and Coatings*, R. F. Bunshah, Ed., chapter 11, Noyes Publications, Park Ridge, NJ, USA; William Andrew Publishing, New York, NY, USA, 1994.

35. P. Vuoristo, K. Niemi, A. Makela, and T. Mantyla, "Abrasion and erosion wear resistance of Cr3C2-NiCr coatings prepared by plasma, detonation and high-velocity oxyfuel spraying," in *Proceedings of the 7th National Thermal Spray Conference*, pp. 121–126, Boston, Mass, USA, 1994.

36. H. Chen and I. M. Hutchings, "Abrasive wear resistance of plasma-sprayed tungsten carbide-cobalt coatings," *Surface and Coatings Technology*, vol. 107, no. 2-3, pp. 106–114, 1998.

37. R. Westerg˚ard, L. C. Erickson, N. Ax´en, H. M. Hawthorne, and S. Hogmark, "The erosion and abrasion characteristics of alumina coatings plasma sprayed under different spraying conditions," *Tribology International*, vol. 31, no. 5, pp. 271–279,1998.

38. V. Higuera Hidalgo, F. J. Belzunce Varela, and E. Fern´andez Rico, "Erosion wear and mechanical properties of plasmasprayed nickel- and iron-based coatings subjected to service conditions in boilers," *Tribology International*, vol. 30, no. 9, pp.641–649, 1997.
39. V. Higuera Hidalgo, J. Belzunce Varela, J.Mart´ınez De La Calle, and A. CarrilesMen´endez, "Characterisation ofNiCr flame and plasma sprayed coatings for use in high temperature regions of boilers," *Surface Engineering*, vol. 16, no. 2, pp. 137–142, 2000.
40. **Staia,** M. H., Valente, T., Bartuli, C., Lewis, D.B. and Constable, C.P., (2001), "Part I: Characterization of Cr_3C_2-25% NiCr Reactive Plasma Sprayed Coatings Produced at Different Pressures," Surf. Coat. Technol., Vol. 146-147, pp. 553-562.
41. **Bluni,** S.T. and Mardar, A.R., (1996), "Effects of Thermal Spray Coating Composition and Microstructure on Coating Response and Substrate Protection at High Temperatures," Corros., Vol. 52, No. 3, pp. 213-218.
42. S. Deshpande, S. Sampath, and H. Zhang, "Mechanisms of oxidation and its role in microstructural evolution of metallic thermal spray coatings: case study for Ni-Al," *Surface and Coatings Technology*, vol. 200, no. 18-19, pp. 5395–5406, 2006.
43. H. Herman, *Plasma Sprayed Coatings*, Scientific American, 1988.
44. K. Korpiola and P. Vuoristo, "Effect of HVOF gas velocity and fuel to oxygen ratio on the wear properties of tungsten carbide coating," in *Thermal Spray: Practical Solutions For Engineering Problems*, C. C. Bernt, Ed., Cincinnati,Ohio, USA, 1996.
45. J. E. Nerz, B. A. Kushner, and A. J. Rotolico, "Microstructural evaluation of tungsten carbide-cobalt coatings," *Journal of Thermal Spray Technology*, vol. 1, no. 2, pp. 147–152, 1992.
46. **Stott,** F. H., (1992), "Developments in Understanding the Mechanisms of Growth of Protective Scales on High-Temperature Alloys," Mater. Charact., Vol. 28, No. 3, pp. 311-325.
47. **Polman,** E.A., Fransen, T. and Gellings, P.J., (1990), "High-Temperature Corrosion and Mechanical Properties of Protective Scales on Incoloy 800H: The Influence of Preoxidation and Ion Implantation," Oxid. Met., Vol. 33, No. 1-2, pp. 135-155.
48. **Khajavi,** M.R. and Shariat, M.H., (2004), "Failure of First Stage Gas Turbine Blades," Eng. Fail .Anal., Vol.11, pp.589–597.
49. **Stott,** F. H., (1998), "The Role of Oxidation in the Wear of Alloys," Trib. Int., Vol. 31, No. 1-3, pp. 61-71.
50. **Singh**, Harpreet, Puri, D. and Prakash, S., (2005), "Studies of Plasma Spray Coatings on a Fe-base Superalloy, their Structure and High Temperature Oxidation Behaviour," Anti-Corrosion Methods and Materials, Vol. 52, No. 2, pp. 84-95.
51. **Rapp,** R. A. Devan, J. H., Douglass, D. L., Nordine, P. C., Pettit, F. S. and Whittle, D. P., (1981), "High temperature corrosion in energy systems," Mater. Sci. Eng., Vol. 50, No. 1, pp. 1-17.
52. **Evans**, H. E. and Taylor, M. P., (2001), "Diffusion Cells and Chemical Failure of MCrAlY Bond Coats in Thermal Barrier Coating Systems," Oxid. Met., Vol. 55, pp.17-34.
53. **Wang**, B., Gong, J., Wang, A. Y., Sun, C., Huang, R. F., and Wen, L. S., (2002),"Corrosion Behavior of Plasma-sprayed NiCrAlY Coatings on Ni-based Superalloy," Surf. & Coat. Technol., Vol. 149, pp. 70-75.
54. **Niranatlumpong**, P., Ponton, C. B. and Evans, H. E., (2000), "The Failure of Protective Oxides on Plasma-Sprayed NiCrAlY Overlay Coatings," Oxid. Met., Vol. 53, No. 3-4, pp. 241-256.
55. **Liu**, P.S., Liang, K.M. and Gu, S.R., (2001), "High-temperature Oxidation Behavior of Aluminide Coatings on a New Cobalt-bae Superalloy in Air," Corros. Sci., Vol. 43, pp. 1217-1226.

56. **Singh**, H., Prakash, S. and Puri, D., (2007), "Some Observations on the High Temperature Oxidation Behaviour of Plasma Sprayed Ni_3Al Coatings," Mater. Sci. Engg. A, 444 (2007) 242-250.
57. **Hussain,** N., Shahid, K.A., Khan, I.H. and Rahman, S., (1994), "Oxidation of High-Temperature Alloys (Superalloys) at Elevated Temperatures in Air: I," Oxid. Met., Vol. 41, No. 3-4, pp. 251-269.
58. **Kolta,** G. A., Hewaidy, L. F. and Felix, N. S., (1972), "Reactions Between Sodium Sulphate and Vanadium Pentoxide," Thermochim. Acta, Vol. 4, pp. 151-164.
59. **Seierdten**, M and Kofstad, P., (1987), "Effect of SO3 on Vanadate Induced Hot Corrosion," High Temp. Technol., Vol. 5, No. 3, pp. 115-122.
60. **Swaminathan**, J., Raghavan, S. and Iyer, S.R., (1993), "Studies on the Hot Corrosion of Some Nickel-base Superalloys by Vanadium Pentaoxide," T. Indian I. Metals, Vol. 46, No. 3, pp. 175-181.
61. **Fryburg**, G. C., Kohl, F. J., Stearns, C. A. and Fielder, W. L., (1982), "Chemical Reactions Involved in the Initiation of Hot Corrosion of B-1900 and NASA-TRW VIA," J. Electrochem. Soc., Vol. 129, No. 3, pp. 571-85.
62. **Ul-Hamid,** A., (2004), "A TEM Study of the Oxide Scale Development in Ni-Cr-Al Alloys," Corros. Sci., Vol. 46, No. 1, pp. 27-36.
63. **Belzunce,** F.J., Higuera, V. and Poveda, S., (2001), "High Temperature Oxidation of HFPD Thermal-Sprayed MCrAlY Coatings," Mater. Sci. Engg. A, Vol. 297, No. 1-2, pp. 162-167.
64. **Singh**, Harpreet, Puri, D. and Prakash, S., (2004), "High Temperature Oxidation Behaviour of Plasma Sprayed NiCrAlY Coatings on Ni-based Superalloys in Air," Proc. of International Symposium of Research Students on Material Science and Engineering (ISRS), Department of Metallurgical and Materials Engineering, Indian Institute of Technology Madras, Chennai, India, December 20-22, pp. 1-17.

Oxidation behaviour of HVAF-sprayed NiCr coating in moisture-laden environment

E. Sadeghimeresht[a], J. Eklund[b], J. Phother Simon[b], J. Lyske[b], N. Markocsan[a], S. Joshi[a]
a Department of Engineering Science, University West, Trollhättan, 46153, Sweden
b Department of Environmental Inorganic Chemistry, Chalmers University of Technology, 412 96 Gothenburg, Sweden

Abstract: Reducing CO_2 emissions from power generation plants is intimately related to enhancing their thermal efficiency, which can be achieved by increasing the temperature/pressure of steam. However, any increase in steam temperature is inevitably accompanied by accelerated oxidation of boiler components. The use of renewable fuels such as biomass increases the problem by introducing a number of corrosive compounds into the boiler environment, resulting in more rapid degradation of components. Although thermal sprayed coatings are techno-commercially attractive solutions for augmenting the durability of degradation-prone boiler components and are already used, further improvements in their performance are continuously sought. High-velocity air fuel (HVAF) coatings are promising in this context. In the present work, isothermal oxidation behavior of a candidate HVAF-sprayed Ni21Cr was studied in N_2 + 5% O_2 + 20% H_2O at 600°C for 168h. The oxide scale growth mechanisms were studied by BIB/SEM/EDX to evaluate the effectiveness of the coatings. It was found that the water vapor effect is insignificant due to the Cr reservoir in the Ni21Cr coating, which yielded enhanced oxidation protection by forming nano-scale Cr_2O_3.

1 Introduction

High temperature corrosion of load-bearing components such as superheater tubes in biomass-fired boilers is critical as it significantly limits their functionality and leads to substantial economic losses due to the long downtimes required for service and overhaul [1], [2]. In order to enhance the durability of such components, and thereby increase the boiler efficiency, a promising route is applying a dense and protective coating that could last longer in the boiler's corrosive environment [3]. The chemistry and microstructure of such coatings can be designed to provide superior oxidation resistance [4].

Ni-based materials alloyed with Cr [5] are commonly used as coatings to protect components against oxidation by forming protective oxide scales of Cr_2O_3 [6]. However, the oxidation attack in boilers is particularly severe in the presence of water vapor (H_2O (g)), where some undesirable reactions (for instance, formation of a volatile, unstable and non-protective $CrO_2(OH)_2$(g)) deplete the oxide in Cr and destroy the inherent protective properties of the oxide [7], [8]. It is worth noting that several mechanisms have been suggested to determine the role of H_2O (g), however the exact effect of water vapor in enhancing the degradation is still under debate [9].

It is well-known that the microstructure of the coatings should be dense, defect-free, and have a sufficient reservoir of protective scale-forming elements such as Cr to perform as an efficient and durable barrier in a corrosive environment [3]. Prior coating efforts have focused on the oxidation behavior of Ni-based coatings deposited by commonly used thermal spraying techniques such as atmospheric plasma spraying (APS) and high-velocity oxy fuel (HVOF) [4], [10]. However, the coating features, associated with the above techniques, such as inter-lamellar boundaries, pores and *in-situ* formed oxides [11], lead to the formation of a discontinuous oxide scale which provide motivation to seek even more efficient protective barriers.

High-velocity air fuel (HVAF) spraying technique is being increasingly employed to address the above-mentioned challenges and produce coatings with better protective characteristics [12]. In HVAF, air and a fuel (usually propane) are mixed in the combustion chamber of the spray gun. The feedstock material is axially injected into the chamber by a carrier gas (e.g. N_2). Due to the thermal energy produced by combustion, the feedstock particles are heated up to a temperature below 1800 °C [13], [14] and accelerated towards the substrate to velocities of around 700-1200 m/s [15]. While the supersonic spray particle velocity helps in obtaining very dense coatings, the low particle temperature and dwell time ensures that there is almost no *in-situ* oxide pick-up or elemental depletion during spraying [16]. Therefore, the protective scale-forming elements such as Cr are not "used" during the spraying process but preserved for oxidation protection. The few studies performed so far on HVAF Ni-based coatings show that the coatings present excellent oxidation behavior in high-temperature environments [17], [18]. However, the oxidation mechanisms in HVAF Ni-based coatings need to be studied further in terms of comprehensive investigations of both thermodynamic and kinetic aspects.

In order to address the effects of water vapor in chromia-forming coatings, the oxidation behavior of the NiCr coating thermally sprayed by HVAF on substrate of 16Mo3 steel (a commonly used material for corrosion-prone power plant components) was investigated in N_2 + 5% O_2 + 20% H_2O at 600°C for 168h using BIB/SEB and EDS techniques. This environment was selected in order to get a better understanding of the complex corrosion reactions that occur during field exposures such as in biomass boilers in a more simplified and well-controlled environment.

2 Experimental Details

2.1 Materials

2.1.1 Substrate and reference materials

A commercially available low carbon steel -16Mo3 (in wt%; 0.01Cr- 0.3Mo- 0.5Mn- 0.3Si- 0.15C- Bal. Fe)- was used as the substrate material. The alloy was sourced in form of rods with 25-mm diameter.

2.1.2 Feedstock powders

A commercially available gas-atomized powder supplied by H.C.Starck GmbH, i.e. NiCr (in wt%; 21.3Cr- 0.1O, Bal. Ni) was used to spray onto the 16Mo3 substrate. Fig. 1 illustrates the particle size distribution, and spherical morphology with very few satellite particles.

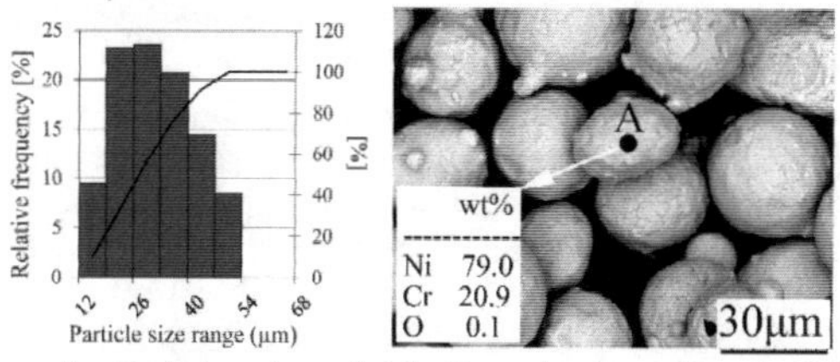

Fig. 1. Particle size distribution, back-scattered SEM morphology, and EDS point analysis of the NiCr powder.

2.2. HVAF spraying

A HVAF gun (Uniquecoat M3[TM], Oilville, VA, USA) was utilized to produce the coatings. A 16Mo3 rod (length=500 mm) was fixed in a horizontal rotating mandrel for applying the coating on the cylindrical surface. After spraying, the rod was cut into pieces of 5 mm length and coatings were applied on the flat tips of the cylinder so that the specimens were coated on all sides. The substrates were grit blasted prior to spraying for cleaning and roughing the surfaces. Several combinations of thermal spray parameters were initially tried and the resulting coatings characterized to determine spray conditions that yielded least porosity.

The coating was manufactured using 3L2G nozzle with air flow rate of 0.8 MPa and pressures of both fuel 1 (propane) and fuel 2 (propane) of around 0.7MPa. The N_2-carrier gas pressure and feed rate was set to 0.4MPa and 150 g/min, respectively. The spray distance was 300 mm. The coating was sprayed to a thickness of around 250 µm.

All surfaces of the coating were polished with 0.2 µm SiC suspension to a roughness of R_a< 0.1 µm prior to oxidation exposure in order to skip the complex effect of the coating's surface roughness on the oxidation behavior.

2.3 Oxidation test

The exposure was performed in a tube furnace fitted with a 44 mm i.e. silica-glass tube. The experiments were carried out in N_2 + 5% O_2 + 20% H_2O, and the flow rate was 3 cm/s. The exposures were carried out at ambient pressure. The water-vapor content was regulated by setting the dew point of the gas to 60,4 °C using a reflux condenser. The flow rate of the dry gas was measured by a BIOS DC2 Flow Calibrator. The samples were placed in separate alumina boats, one sample in each. The alumina boats were positioned on an alumina tray in the furnace with the samples parallel to the flow direction. Three samples were exposed isothermally at 600 °C (±3 °C) for 168h.

2.4. Characterization of coatings

Broad ion beam (BIB) milling was used to prepare wide cross sections in the millimeter range using a Hitachi E-3500 broad ion beam instrument, equipped with an argon ion gun, which was operated at 6 kV for sputtering. Before each milling, a thin gold film was applied on the sample surface and, then, a low speed saw was used to cut the samples. Morphology of both as-sprayed and exposed coatings was characterized using a QUANTA 200 FEG scanning electron microscope (SEM) equipped with X-ray energy dispersive spectroscopy (EDS).

The porosity content of the coatings was determined by image analysis (IA) technique using ImageJ software [19] by converting the micrographs of the coatings (taken from Olympus BX60M optical microscope) into binary images, and quantifying the percentages based on the gray scale contrast [14].

Micro-hardness measurements were performed on the polished cross-sections of the coatings according to ASTM E384 standard with a Vickers indenter (Shimadzu, HMV-2, Tokyo, Japan) using a load of 300 g and dwell time of 15 s [20]. The hardness of each coating was determined as an average of 10 measurements performed on the sample.

The surface roughness (R_a) of the coatings was measured using a stylus-based profilometer (Surftest 301, Mitutoyo, Japan). Three measurements in three different directions were performed and the measured values were averaged.

3 Results & Discussion

3.1 As-sprayed coating microstructure

Fig. 2 shows the as-sprayed microstructure of the NiCr coating sprayed by HVAF. The values of surface roughness (R_a), microhardness and porosity of the coating were measured as <0.1 µm, 391±19 $HV_{0.3}$, 0.1±0.02 vol.% respectively. In general, the sample showed a well-bonded interface (between the coating and the substrate). The coating showed a relatively dense structure, with presence of pores at splat boundaries. Some semi-molten particles could be also identified from the micrograph (Fig. 2 b). The oxygen pick-up was indicated by the EDS analysis. Typically, oxides are formed *in situ* due to the powder particles being exposed to high temperature in ambient air environment during the spraying process [21] and/or because of pre-existing oxygen in the feedstock material [22]. The influence of the former is minimal in the investigated sample, as the dwell time in the HVAF process is too short and the temperature (<1800 °C) is lower compared to other thermal spray processes, e.g., HVOF (>2200 °C) and APS (>5000 °C) [14]. The NiCr coating produced in this study had less porosity and better adherence (as evident from the well-bonded interface without the presence of defects) to the substrate than the coatings produced with the other spraying techniques such as HVOF [11] and APS [23]. The higher values of microhardness of the coatings compared to values reported in the literature for the HVOF and APS coatings [11], [23] with identical composition may be attributed to the less porosity of the HVAF coatings. The presence of O and Al at the coating/substrate interface was attributed to the embedded alumina grits used for grit blasting in the substrate.

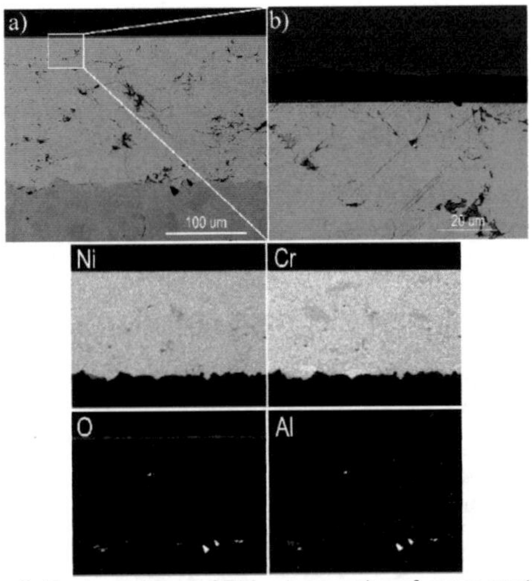

Fig. 2. Back-scattered SEM micrographs of cross-sections of as-sprayed NiCr coating and EDS elemental mapping analysis, a) low magnification b) high magnification.

3.2 Kinetic studies of the NiCr coating

The mass gain values for the three exposed NiCr coatings exposed at 600 °C for 168 h is given in Table I. A relatively low mass gain was observed in the coating, which is most probably due to the protective oxide scale-forming ability of the coating. The weight gain is related to the oxide thickness of chromia scale. Assuming the scale to be dense, pure chromia, the thickness of the oxide scale, X (µm), is given by eq. 1;

$$X = \frac{M_{Cr_2O_3} \times W_O}{3M_O \times \rho_{Cr_2O_3}} \quad (1)$$

Where W_O is the total amount of oxygen present in a protective chromia scale, M_O is atomic weight for oxygen, M_{Cr2O3} the molecular weight of Cr_2O_3, and ρ the density of Cr_2O_3 (5.21 g/cm^3). The calculated values of the thickness of the oxide scale is also given in Table I.

Table I. Mass gain values for the exposed NiCr coating exposed at 600 °C for 168 h

Sample	Mass gain (mg/cm^2)	Thickness of chromia scale (µm)
NiCr 1	0.1709±0.0115	1.039±0.070
NiCr 2	0.1636±0.0115	0.994±0.070
NiCr 3	0.1511±0.0351	0.918±0.213

The values indicate that the NiCr coating was successful in reducing the oxidation rate of the 16Mo3 substrate. Although the oxidation study was performed in only one exposure period (168h), it is well-known that a high oxidation rate could be reported at the beginning of exposure that could be due to the discontinuity in oxide films (Cr_2O_3) in the coating [24]. On increasing the exposure time, the Cr_2O_3 scale forms gradually and laterally becomes continuous, thereby isolating the coating from the oxidation environment at high temperature. It was reported that the major weight gain of the coatings during the early stages of the exposure might be associated with the rapid formation of oxides at the inter-lamellar boundaries or/and within interconnected pores due to the high diffusion of oxygen [25], [26]. Once a continuous oxide scale forms, inward diffusion of oxygen into the coating slows down and the growth of the oxides become restricted only to the coatings surface [11]. This leads the weight gain to become constant in due course.

3.3 Microstructural characterization of exposed NiCr coating

Fig. 3 shows the cross-sectional SEM micrographs and EDS elemental mapping of the coating exposed in N_2 + 5% O_2 + 20% H_2O at 600 °C for 168h.

The surface color of the coating was not found to change over the exposure by showing a shiny appearance. After 168 h of exposure at 600 °C, no sign of coating spallation was observed. As the coating had a Cr content of ≈21 wt%, a thin and protective oxide scale of chromia was observed as also reported in the literature [24]. The fact that the scale neither spalls nor grows is amply indicative of its protective nature.

A factor which can lead to enhanced degradation of coatings is the interdiffusion of elements at the coating/substrate interface [27]. It has been reported in prior studies that such interdiffusion can lead to depletion of important elements, such as Cr from the coating, in addition to the formation of brittle intermetallic phases, which can lead to early coating delamination in service conditions [28]. As seen in Fig. 4, no significant depletion of Cr was observed at the coating sides adjacent to the substrate. Figs. 3 and 4 show that the inter-lamellar boundaries could serve as fast diffusion paths for oxygen to diffuse and cause the internal oxidation of the coating. No sign of oxidation attached was also observed at the coating/substrate interface. The formation of the protective and very thin oxide scale on the NiCr can be seen in Fig. 5. The high temperature corrosion resistance of the coating is provided by the formation of a compact and continuous layer of chromia over the coating. It seems that at a Cr level of around 21wt%, the depletion in Cr is relatively small and breakaway is avoided [29]. The difference between the thickness of the scale observed in Fig. 5 (≈200 nm) and the expected thickness in Table I (≈1 µm) might be attributed to the formation of volatile Cr species such as $CrO_2(OH)_2(g)$ which was not considered in eq. (1).

Due to the formation of oxides within the coating, a potential mechanism to explain the role of water vapor might be that H_2O leads hydrogen to dissolve into the coating's defects such as pores through lamellar boundaries. The dissolved hydrogen is supposed to enhance oxygen diffusivity in the pores, thereby enhancing internal oxidation of the coating [30]. However, more experiments need to be performed to support this mechanism. Moreover, increased hydrogen content in the coating in connection to the water vapor effect has to be studied. Finally, an increased diffusivity of oxygen in the coating in the presence of dissolved hydrogen need to be verified by further investigations.

To sum up, the NiCr coating was observed to be successful in maintaining its continuous surface contact with the substrate over the period of exposure. Moreover, it was clear from the cross-sectional EDS that oxygen activity was not substantial enough to penetrate through the coatings and reach the substrate. This highlights the beneficial protective capability of the dense HVAF coatings, since the substrate did not suffer any oxidation. Furthermore, the oxide scales formed over the coatings showed high spallation resistance, which is desirable to ensure effectiveness of the coatings as the oxidation behavior of an alloy is defined mainly by scale spallation resistance [31]. Therefore, the HVAF NiCr coating, with chemistry promoting formation of a protective oxide scale, could be useful in imparting oxidation resistance in the test conditions that simulated an oxidizing environment.

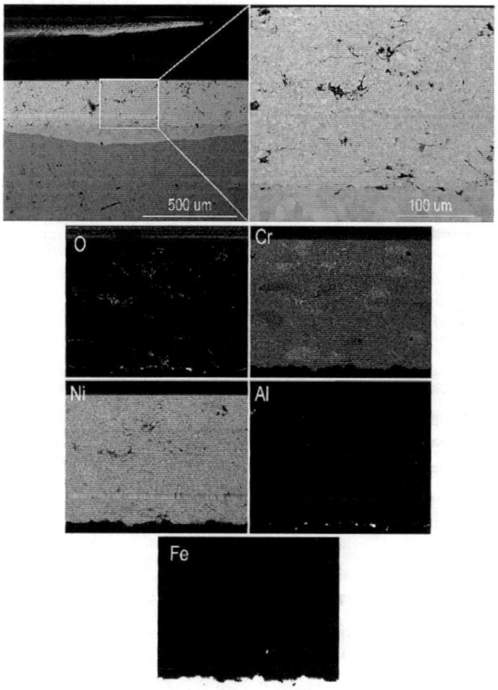

Fig. 3. Cross-sectional SEM micrograph and EDS elemental mapping analysis of NiCr coating oxidized in N_2 + 5% O_2 + 20% H_2O at 600 °C for 168 h.

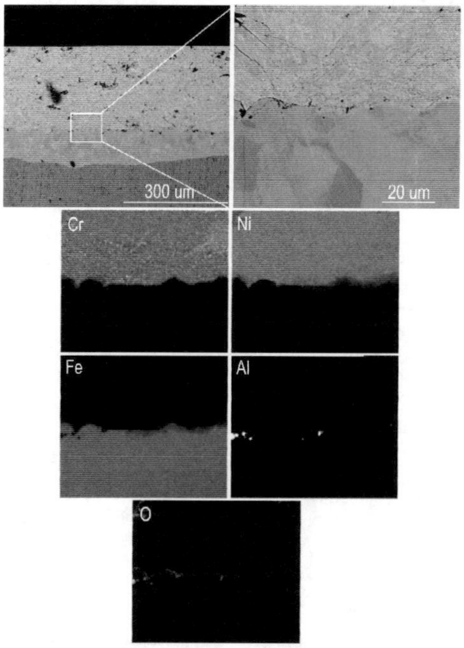

Fig. 4. Cross-sectional SEM micrograph and EDS elemental mapping analysis of NiCr coating/substrate interface oxidized in N_2 + 5% O_2 + 20% H_2O at 600 °C for 168 h.

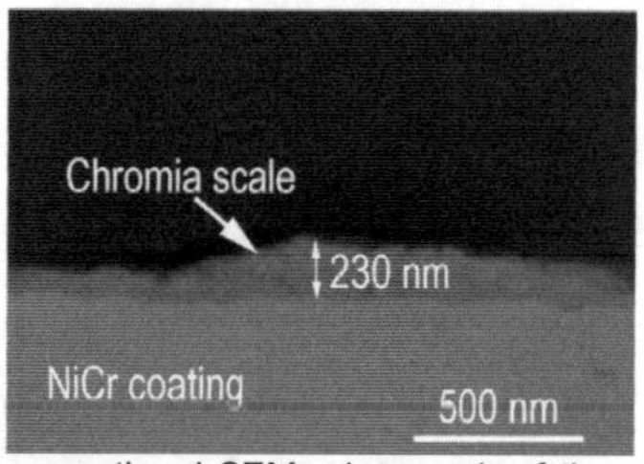

Fig. 5. Cross-sectional SEM micrograph of the oxide scale formed over the NiCr coating oxidized in N_2 + 5% O_2 + 20% H_2O at 600 °C for 168 h.

4 Conclusions

The aim of this work was to study the oxidation behavior of the HVAF-sprayed NiCr coating in N_2 + 5% O_2 +20% H_2O at 600 °C for 168 h.

1) A thin chromia scale was observed on the NiCr coating alloy after 168 h of reaction.

2) No sign of breakaway oxidation or internal oxidation of Cr was observed in the presence of water vapor.

3) The water vapor effect was insignificant in such a high pO_2 environment due to the large Cr reservoir, and the continued growth of Cr_2O_3.

3) The NiCr coating found to be successful in maintaining the integrity with their respective substrates throughout the exposure to the environment of the study, without any significant spalling of the oxide scales.

4) The HVAF-sprayed NiCr coating may be recommended as a proper chemistry to be used for boilers (as well as other similar high-temperature

applications) as far as oxidation resistance in water vapor is concerned.

5) Further studies need to be performed on the coatings to verify the proposed oxidation mechanism in the presence of water vapor. Investigations in other environments is required to assess more accurately the coatings performance in boiler applications.

Acknowledgements

The authors would like to thank Mr. Kenneth Andersson for his valuable advice, Mr. Jonas Olsson, and Mr. Stefan Björklund for spraying the coatings.

References

[1] K. Szymański, A. Hernas, G. Moskal, and H. Myalska, "Thermally sprayed coatings resistant to erosion and corrosion for power plant boilers - A review," *Surf. Coat. Technol.*, vol. 268, pp. 153–164, Apr. 2015.

[2] S. C. Okoro, M. Montgomery, F. J. Frandsen, and K. Pantleon, "Effect of Water Vapor on High-Temperature Corrosion under Conditions Mimicking Biomass Firing," *Energy Fuels*, vol. 29, no. 9, pp. 5802–5815, Sep. 2015.

[3] E. Sadeghimeresht, H. Hooshyar, N. Markocsan, S. Joshi, and P. Nylén, "Oxidation Behavior of HVAF-Sprayed NiCoCrAlY Coating in H2–H2O Environment," *Oxid. Met.*, pp. 1–16, Jul. 2016.

[4] M. Oksa, P. Auerkari, J. Salonen, and T. Varis, "Nickel-based HVOF coatings promoting high temperature corrosion resistance of biomass-fired power plant boilers," *Fuel Process. Technol.*, vol. 125, pp. 236–245, Sep. 2014.

[5] A. J. López, M. Proy, V. Utrilla, E. Otero, and J. Rams, "High-temperature corrosion behavior of Ni–50Cr coating deposited by high velocity oxygen–fuel technique on low alloy ferritic steel," *Mater. Des.*, vol. 59, pp. 94–102, Jul. 2014.

[6] M. Kumar, H. Singh, and N. Singh, "Study of Ni-20Cr Coatings for High Temperature Applications – A Review," *Arch. Metall. Mater.*, vol. 58, no. 2, pp. 523–528, 2013.

[7] H. Asteman, J.-E. Svensson, and L.-G. Johansson, "Oxidation of 310 steel in H2O/O2 mixtures at 600 °C: the effect of water-vapour-enhanced chromium evaporation," *Corros. Sci.*, vol. 44, no. 11, pp. 2635–2649, Nov. 2002.

[8] X. Peng, J. Yan, Y. Zhou, and F. Wang, "Effect of grain refinement on the resistance of 304 stainless steel to breakaway oxidation in wet air," *Acta Mater.*, vol. 53, no. 19, pp. 5079–5088, Nov. 2005.

[9] H. Hooshyar, T. Jonsson, J. Hall, J.-E. Svensson, L. G. Johansson, and J. Liske, "The Effect of H2 and H2O on the Oxidation of 304L-Stainless Steel at 600 °C: General Behaviour (Part I)," *Oxid. Met.*, pp. 1–22, Nov. 2015.

[10] M. Oksa, J. Metsäjoki, and J. Kärki, "Thermal Spray Coatings for High-Temperature Corrosion Protection in Biomass Co-Fired Boilers," *J. Therm. Spray Technol.*, vol. 24, no. 1–2, pp. 194–205, Sep. 2014.

[11] B. Song, Z. Pala, K. T. Voisey, and T. Hussain, "Gas and liquid-fuelled HVOF spraying of Ni50Cr coating: Microstructure and high temperature oxidation," *Surf. Coat. Technol.*, 2016.

[12] A. A. Verstak and G. Kusinski, "High-Velocity Air-Fuel Spraying and Its Applications in Oil and Gas Industry," in *International Thermal Spray Conference and Exposition (ITSC) 2012 American Society for Metals*, 2012.

[13] Z. Zeng, N. Sakoda, T. Tajiri, and S. Kuroda, "Structure and corrosion behavior of 316L stainless steel coatings formed by HVAF spraying with and without sealing," *Surf. Coat. Technol.*, vol. 203, no. 3–4, pp. 284–290, Nov. 2008.

[14] E. Sadeghimeresht, N. Markocsan, P. Nylén, and S. Björklund, "Corrosion performance of bi-layer Ni/Cr2C3–NiCr HVAF thermal spray coating," *Appl. Surf. Sci.*, vol. 369, pp. 470–481, Apr. 2016.

[15] E. Sadeghimeresht, N. Markocsan, P. Nylén, and S. Björklund, "Corrosion performance of bi-layer Ni/Cr2C3–NiCr HVAF thermal spray coating," *Appl. Surf. Sci.*, vol. 369, pp. 470–481, Apr. 2016.

[16] E. Sadeghimeresht, N. Markocsan, and P. Nylén, "Microstructural and electrochemical characterization of Ni-based bi-layer coatings produced by the HVAF process," *Surf. Coat. Technol.*, vol. 304, pp. 606–619, Oct. 2016.

[17] G.-J. Yang, X.-D. Xiang, L.-K. Xing, D.-J. Li, C.-J. Li, and C.-X. Li, "Isothermal Oxidation Behavior of NiCoCrAlTaY Coating Deposited by High Velocity Air-Fuel Spraying," *J. Therm. Spray Technol.*, vol. 21, no. 3–4, pp. 391–399, Jan. 2012.

[18] A. P. Wang, Z. M. Wang, J. Zhang, and J. Q. Wang, "Deposition of HVAF-sprayed Ni-based amorphous metallic coatings," *J. Alloys Compd.*, vol. 440, no. 1–2, pp. 225–228, Aug. 2007.

[19] C. A. Schneider, W. S. Rasband, and K. W. Eliceiri, "NIH Image to ImageJ: 25 years of image analysis," *Nat. Methods*, vol. 9, no. 7, pp. 671–675, Jul. 2012.

[20] "ASTM E384-11e1, Standard Test Method for Knoop and Vickers Hardness of Materials, ASTM International, West Conshohocken, PA, 2011, www.astm.org." .

[21] M. H. Enayati, F. Karimzadeh, M. Tavoosi, B. Movahedi, and A. Tahvilian, "Nanocrystalline NiAl Coating Prepared by HVOF Thermal Spraying," *J. Therm. Spray Technol.*, vol. 20, no. 3, pp. 440–446, Oct. 2010.

[22] S. T. Bluni and A. R. Marder, "Effects of Thermal Spray Coating Composition and Microstructure on Coating Response and Substrate Protection at High Temperatures," *CORROSION*, vol. 52, no. 3, pp. 213–218, Mar. 1996.

[23] T. Hussain, T. Dudziak, N. J. Simms, and J. R. Nicholls, "Fireside corrosion behavior of HVOF and plasma-sprayed coatings in advanced coal/biomass co-fired power plants," *J. Therm. Spray Technol.*, vol. 22, no. 5, pp. 797–807, 2013.

[24] L. Intiso, L.-G. Johansson, S. Canovic, S. Bellini, J.-E. Svensson, and M. Halvarsson, "Oxidation Behaviour of Sanicro 25 (42Fe22Cr25NiWCuNbN) in O2/H2O Mixture at 600 °C," *Oxid. Met.*, vol. 77, no. 5–6, pp. 209–235, Jan. 2012.

[25] H. Choi, B. Yoon, H. Kim, and C. Lee, "Isothermal oxidation of air plasma spray NiCrAlY bond coatings," *Surf. Coat. Technol.*, vol. 150, no. 2–3, pp. 297–308, Feb. 2002.

[26] P. Niranatlumpong, C. B. Ponton, and H. E. Evans, "The failure of protective oxides on plasma-sprayed NiCrAlY overlay coatings," *Oxid. Met.*, vol. 53, no. 3–4, pp. 241–258, 2000.

[27] F. Otsubo, H. Era, and K. Kishitake, "Interface reaction between nickel-base self-fluxing alloy coating and steel substrate," *J. Therm. Spray Technol.*, vol. 9, no. 2, pp. 259–263, Jun. 2000.

[28] S. Brossard, P. R. Munroe, A. T. T. Tran, and M. M. Hyland, "Study of the microstructure of NiCr splats plasma sprayed on to stainless steel substrates by TEM," *Surf. Coat. Technol.*, vol. 204, no. 9–10, pp. 1608–1615, Jan. 2010.

[29] N. K. Othman, N. Othman, J. Zhang, and D. J. Young, "Effects of water vapour on isothermal oxidation of chromia-forming alloys in Ar/O2 and Ar/H2 atmospheres," *Corros. Sci.*, vol. 51, no. 12, pp. 3039–3049, Dec. 2009.

[30] W. J. Quadakkers, J. Żurek, and M. Hänsel, "Effect of water vapor on high-temperature oxidation of FeCr alloys," *JOM*, vol. 61, no. 7, pp. 44–50, Jul. 2009.

[31] H. Singh, D. Puri, S. Prakash, and R. Maiti, "Characterization of oxide scales to evaluate high temperature oxidation behavior of Ni–20Cr coated superalloys," *Mater. Sci. Eng. A*, vol. 464, pp. 110–116, 2007.

Demonstrating Usability of Thermally Sprayed Coatings for Mechatronics and Power Electronics Applications

J. Luth[1], R. Trache[2 3], F.-L. Toma[3], S. Hartmann[1], F. Trenkle[1]

1 obz innovation GmbH, Bad Krozingen, D
2 Technische Universität Dresden, IfWW, Dresden, D
3 Fraunhofer IWS, Dresden, D

In order to meet the increased requirements for power electronics in the automotive sector, an effective utilization of difficult installation spaces is necessary. A new production concept to realize this 3D integration of electronic circuit boards directly on components is the combination of thermal spraying and cold gas spraying to create multilayer-coating systems consisting of conducting and insulating coatings. In this study two- and tree-dimensional demonstrators were developed, showing the flexible use of thermal spraying in mechatronics and power electronics. In contrast to past studies on this construction concept, the main focus of this study was on the optimization of the ceramic insulting coatings and bond strength of the metallization. The ceramic coatings showed a dielectric strength and electrical resistance, which was suitable for most applications. Additional post treatment improved the electrical resistance in humid conditions. Already 150 µm thick electrical insulation layers showed a breakdown voltage of more than 5 kV AC and a specific electrical resistance of $5 \cdot 10^{11}$ Ω.m.

1 Introduction

Electronic components become more and more important for modern automotive technology. Recent introduction of mass production of hybrid vehicles with integrated electric drive shows the importance of high power automotive electronics. Even in conventionally powered vehicles extensive power electronic and mechatronic system devices are currently in development or already state of the art.
In order to meet the increased requirements for power electronics in the automotive sector, an effective utilization of difficult installation spaces is necessary. 3D integration is a key technology with which the advantages of fast switching semiconductors can only be fully utilized [1].
A new production concept is the production of power electronic circuit boards with a combination of thermal spraying and cold gas spraying. This circuit board can be sprayed directly on the substrate (e.g. a cooling plate made of aluminum alloy). This makes the construction more compact and improves improves heat dissipation significantly. The components in this concept consist of a system of different coating materials: base plate - primer coating - electrically insulation ceramic coating - metallization.
In contrast to past studies on this construction concept [2], the main focus of this study was on the optimization of the ceramic insulting coatings and bond strength of the metallization. State of the art material for insulating coatings is aluminum oxide. It has excellent insulating properties, good thermal conductivity at relatively low material costs. Al_2O_3 is usually applied using atmospheric plasma thermal spray technology (APS). In contrast to sintered Al_2O_3 it can happen that thermally sprayed Al_2O_3 has worse insulation properties and lower thermal conductivity. The reasons are alpha-gamma phase transformation taking place during the spray process and the special coating microstructure consisting of solidiyed splats and porosity in between them [3].

Thus, an approach to improve electrical insulation properties and thermal conductivity for thermal sprayed ceramic coatings, is the use of high velocity flame spraying (HVOF). Compared to APS, HVOF spraying has the following advantages: Lower porosity of Al_2O_3 coatings and a potentially lower phase transformation rate of γ - Al_2O_3 to α - Al_2O_3 [4-7].
The approach shown in the following is to create HVOF sprayed Al_2O_3 coatings with additions of TiO_2 or MgO. The applicability of the developed coatings in electronics is finally shown by two different demonstrators.

2 Experimentation

2.1 Optimization of the Insulation Coating

Samples of AlMg3 (100 x 60 x 5 mm³) received a 50 micron, NiCr 80/20 primer HVOF coating followed by a coating with the powders shown in **Tab. 1**. HVOF coatings were sprayed with a Top Gun system (8-mm nozzle, GTV mbH) using ethylene as a fuel gas. The coating thickness was varied in three steps: 100 µm, 150 µm and 200 µm.

Tab. 1. Spraying powders for electrical insulating coatings with varying chemical composition, suppliers and purity.

Label	Material	Suppl.	Total amount of contamination [%]
A1	Al_2O_3	A	0,2
S5	Al_2O_3 – MgO 72/28	H	
S6	Al_2O_3 – MgO 72/28	I	0,4
T4	Al_2O_3 – TiO_2 97/3	F	0,6

2.2 Basic coating characterization

Deposition efficiency was examined according to DIN EN ISO 17836. The coating microstructures were examined by optical microscopy on metallographically polished cross sections. The phase compositions of powders and sprayed coatings were evaluated by x-ray diffraction using a D8 (Advance Bruker AXS) diffractometer. The volume content of the crystalline α-Al_2O_3 (C_α^{Al2O3}) was detected using the following equation:

$$C_\alpha^{Al2O3} = \frac{A_\alpha(113)}{A_\alpha(113) + 0.89A_\gamma(400)}$$

where *A* is the integral area of the respective peak [7].

2.3 Investigation of Electrical Insulating Properties

In order to measure the electrical resistance of the oxide ceramic coatings, a cold gas sprayed aluminum coating was applied on the coating surface. It provides a good electrical contact and reduces the influence of the surface roughness. The cold sprayed contacts were applied in an arrangement based on IEC 60093. A circular plate electrode of 26 mm diameter with protective ring electrode was used. Prior to the measurement of electrical resistance, the samples were stored at room temperature (22 °C) for 48 hours at different relative air humidities (RH). The measurement was then performed, according to IEC 60093, at the same climate conditions and after 60 seconds of current flow using an using an Advantest R8340 Ultra-high Resistance Meter (Advantest Corporation, Tokyo,Japan). In contrast to IEC 60093 a direct current (DC) of 200 V (instead of 100 V) was used because these conditions are closer to the requirements in power electronics.
Dielectric strength was measured with a procedure based on DIN EN 60243 using an A/C testing instrument from Hippe OHG at 50 Hz current frequency. In order electrically contact the ceramic coating surface, a cold gas sprayed aluminum contact was applied. Each metallized measurement pad has dimensions of 10 x 10 mm².
For further series of measurements, the samples were subjected to a post-treatment process. Resistivity and dielectric strength was determined in the same manner as described for the not post-treated samples.
The thermal shock resistance of insultion coatings was detreminded by measuring the dielectric strength before and after a thermal cycle test, which was based on the requirements in the automotive sector: 50 cycles between +20°C and +200°C with 30 min hold time.

2.4 Demonstrator construction and evaluation

For the demonstrators cold gas sprayed coatings were applied on top of the ceramic coating that performed best, powder T4 in **Tab. 1**. Commercial aluminium (99,97 wt.% purity) and copper (99,95 wt.% purity) powders with appropriate particle sizes for cold gas processes were used as the material feedstock for coating deposition. The design of the demonstrators is described in section 3.3 and 3.4 in length.
The adhesion strength of the multilayer coatings was examinded according to DIN EN 582. The electric conductivity was determined according to DIN EN 2004-1 and ASTM E 1004 using a SIGMASCOPE SMP 10.

3 Results and Discussion

3.1 Optimization of the Insulation coating

An important voltage class for converters with insulated-gate bipolar transistor *IGBT* is 600 V. This voltage class requires a resistance against 1200 V until safety is achieved. As the insulation layers must remain as thin as possible to ensure adequate thermal conductivity, the dielectric strength of 100 µm thick coatings has to be > 12kV / mm. The dielectric strength E_D of the tested coatings is between 20 kV / mm and 30 kV / mm, **Fig. 2**. Therefore the requiremens for E_D are met.
The specific resistivity R_{spec} behaves largely independently of the dielectric strength. This behaviour has already been observed in a previous study [8]. R_{spec} is between 10^7 Ω·m und 10^8 Ω·m, **Fig. 3**. The material for the coatings influences the electrical insulation properties most. Al_2O_3-$3TiO_2$ (powder T4) has both the best dielectric strength E_D and the highest specific resistivity R_{spec}. With this powder, values of E_D = 30 kV / mm and $R_{spec} = 2 \cdot 10^8$ Ω·m were achieved. One reason why coatings with powder T4 have the best insulation properties can be their higer content of α - Al_2O_3 C_α^{Al2O3}. In these coatings C_α^{Al2O3} is 20 % and is thus twice as high as for coatings with pure Al_2O_3 (powder A1). A coating microstructure that is more dense compared to A1 cannot be determined by means of cross-sections, **Fig. 1**.

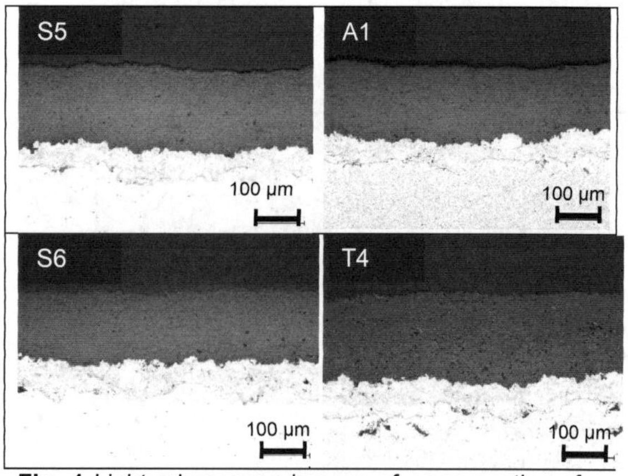

Fig. 1 Light microscope images of cross sections for four different oxide coatings (A1, S5, S6 and T4).

For all coatings, the deposition efficiency η and the thermal shock resistance were determined. After 50 cycles between 20 °C and 200 °C no change in the coating structure could be detected. Also no influence on the dielectric strength was evident. This indicates that the coating is suitable for the use in the automotive field. The deposition efficiency was 61 % for A1, 57 % for T4 and 40 % for both S5 and S6.These values are comparable to conventional ceramic coating processes.

In the following section, the influence of the layer thickness on the insulation properties is discussed. A higher layer thickness leads, as expected, to a higher resistance R and a higher breakdown voltage U_D. Corresponding intensive quantities R_{spec} and E_D decrease slightly, **Fig. 2** and **Fig. 3**. The fact that the resistance and the breakdown volatage do not increase linearly with the thickness is due to the increase of material defects with increasing layer thickness and has already been observed frequently [7].

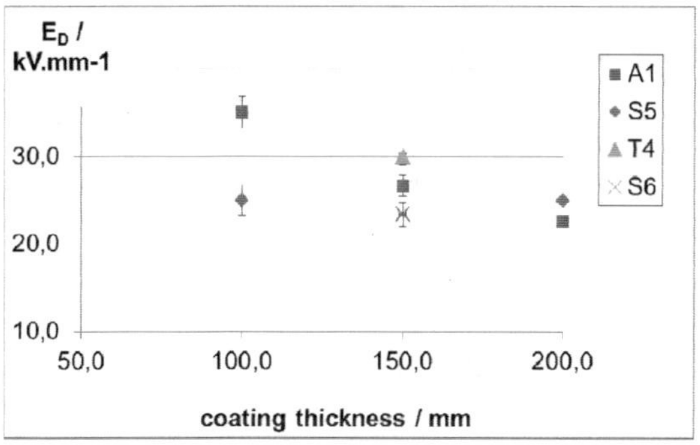

Fig. 2. Dielectric strengh E_D and depending on coating thickness for four different oxide coatings (A1, S5, S6 and T4). The error bars indicate the standard error of the mean value for a confidence interval of 95% (N = 5).

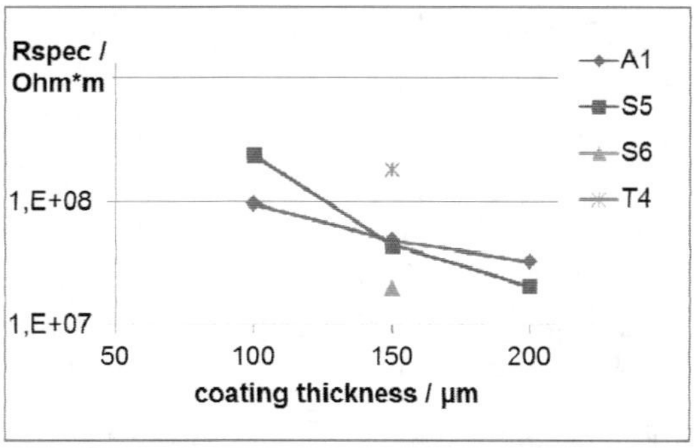

Fig. 3. Specific electrical resistance R_{spec} depending on coating thickness for four different oxide coatings (A1, S5, S6 and T4).

3.2 Need for post treatment

To avoid high leakage currents especially in applications for power electronics, resistivity of at least 10^8 Ω·m is necessary. While the values for E_D already meet the requirements for use in the power electronics sector, a post treatment must be applied to achieve a sufficiently high R_{spec}.

The final post treatment of the coating raises the quality of the electrical insulation of all tested coatings to a similar level, **Fig. 4.** The dielectric strength can be increased from 30 kV / mm to 40 kV / mm and the specific resistance increases from 10^8 Ω·m to $5 \cdot 10^{11}$ Ω·m. The coating material has almost no influence on the insulation properties any more. So the process becomes more and more stable and the influence of a higher humidity is then only little, **Fig. 4**.

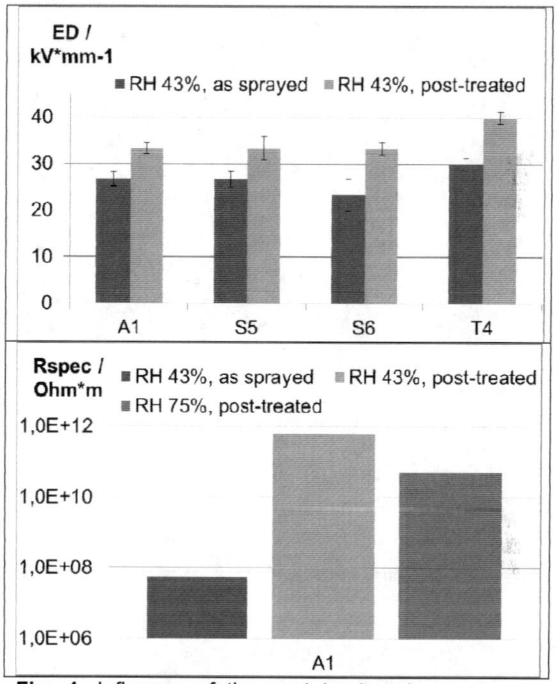

Fig. 4. Influence of the post treatment process on dielectric strength (above, for coating A1, S5, S6 and T4) and specific resistance (below, only for coating A1) at different relative humidities RH. E_D values show error bars of the standard deviation (N = 5), R_{spec} was only measured once.

3.3 Flat Demonstrator

Since the main application area of the coatings is expected to be in the area of (power) electronics and mechatronics, the two demonstrators present the design possibilities for the previously developed coating process. The lid of a hollow cylinder made of aluminum was chosen for the flat-type demonstrator. The layer system consists of a 50 µm thick NiCr 80-20 bondcoat layer and a 150 µm thick insulation layer made of the material T4. The post treated insulation layer appears gray in **Fig. 5**. On it is a coating of aluminum cirucits. Copper was applied at the positions where the SMD components should be soldered in a further step. In addition, it was shown that the adhesion strength of the multi-layer system made of bondcoat layer, ceramic, aluminum and copper is sufficiently large for applications in electrical engineering. It was tested to 32 ± 9 MPa.

The SMD components (LEDs, resistance components and pressure switches) could finally be soldered on the copper cold gas metalization. After a post treatment process of the component, the functional capability could be proven, as **Fig. 5** shows.

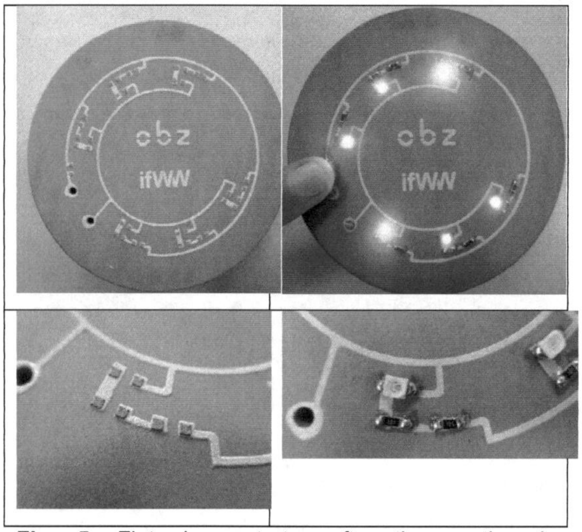

Fig. 5. Flat demonstrator after the coating is completed (left) and fully functional with soldered SMD components (right); enlarged picture sections (below).

3.4 3D Demonstrator

For the 3D demonstrator a 50 mm thick aluminium tube (Ø 80 mm x 200 mm) was coated with bondcoat, insulation ceramics (T4) and metallization, **Fig. 6**. The demonstrator thus shows the transferability of the multilayer coating process to more complex components. Even without mechanical finishing, high edge sharpening and a minimum structural width of 1 mm were realized. In addition, a direct coating with copper without an intermediate layer made of aluminum was possible. As a result, the conductivity of the metal layer could be increased to 80 % of the International Annealed Copper Standard *IACS*.

Fig. 6. 3D Demonstrator with aluminium (left) and with direct copper metalization (right).

4 Conclusion

The applicability of a multilayer layer system in mechatronics and power electronics has been demonstrated. The weak point of previous development was the insulation layer and the bond strength of the metalization. Both was improved, and now meets the requirements in terms of dielectric strength, specific resistance and bond strength of the metalization. The dielectric strength of this layer was measured to be 40 kV / mm and the specific resistance was $5 \cdot 10^{11}$ Ω.m at a relative humidity of 43%. Thermal resistance was verified for at least 50 cycles between room temperature and + 200 ° C. The bondstrength of the metalization was measured to be 32 ± 9 MPa.

For the best found insulation layer Al_2O_3-3TiO_2 (powder T4) should be used. A post treatment process ensures process stability and the resistance remains high enough even in humid environments.

5 Acknowledgment

Supported by:

Federal Ministry
for Economic Affairs
and Energy

on the basis of a decision
by the German Bundestag

6 Literature

[1] P. Friedrichs, J. Milan, T. Harder, N. Kaminski, L. Lorenz, L. F Schindele and P. Ward: CPE Position Paper on Next Generation Power Electronics based on Wide Bandgap Devices – Challenges and Opportunities for Europe (2016).

[2] K.-R. Donner, F. Gärtner, T. Klassen: Metallization of thin Al2O3 layers in power electronics using cold gas spraying in Journal of Thermal Spray Technology 20 (1-2) (2011), S. 299 ff.

[3] L. Pawlowski, The Science and Engineering of Thermal Spray Coatings, John Wiley and Sons (2008)

[4] P. Vuoristo, K. Niemi, V. Matikainen, L. Hyvärinen, H. Koivuluoto, L.-M. Berger, S. Scheitz und I. Shakherdova: Structure and Properties of HVOF and Plasma Sprayed Ceramic Alumina-Chromia Coatings Deposited from Fused and Crushed Powders in Proc. Int. Thermal Spray Conference (2013), Busam, Republic of Korea

[5] L.-M. Berger, C. Stahr, F.-L. Toma, G. Stehr und E. Beyer, Ausgewählte Entwicklungstendenzen bei der Herstellung thermisch gespritzter keramischer Schichten in Jahrbuch Oberflächentechnik, Bd. 63 (2007), Bad Saulgau: R. Suchentrunk, pp. 71-84.

[6] A. K. Sharma, K. Vijayakumar und R. Krishnamurthy: Evaluation of Plasma Deposited Alumina-Titania Ceramic Composite Coatings in Proc. Int. Thermal Spray Conference (2012), Essen

[7] F-L. Toma, S. Scheitz, L.-M. Berger, V. Sauchunk, M. Kusnezoff and S. Thiele: Comparative Study on the Electrical Properties and Charactereristics of Thermally Sprayed Alumina and Spinel Coatings in Journal of Thermal Spray Technology, Volume 20 (2010), pp. 195-204.

[8] J. Luth, M. Cichosz, M. Lehmann, S. Hartmann, F. Trenkle: HVOF-sprayed insulation coatings for applications in Thermal Spray Bulletin 8 (2015) NO. 1.

Transfer of Wire Arc Sprayed Metal Coatings onto Plastic Parts

K. Bobzin, M. Öte, M.A. Knoch, X. Liao, Surface Engineering Institute (IOT), Aachen / D
Ch. Hopmann, P. Ochotta, Institute of Plastic Processing (IKV), Aachen / D

By means of In-Mold-Metal-Spraying (IMMS), wire arc sprayed metal coatings are transferred onto plastic parts during the injection molding process for the efficient production of metallized plastic parts. One potential field of application of IMMS parts are electrical applications such as electrically conductive tracks or electromagnetic shielding. In the current study, the properties of the transferred coatings, especially the electrical resistivity, are determined. Different feedstock materials are used for the application of the coatings. In the first investigation, pressurized air is used as atomizing gas for wire arc spraying. In contrary to Zn coatings, Cu coatings applied with pressurized air have a significantly higher electrical resistivity in comparison to massive copper. One possible reason for this is the oxidation of the Cu particles during the spraying process. Therefore, N_2 and a mixture of N_2 and H_2 are used as atomizing gas to reduce the oxidation of particles. Consequently, the electrical resistivity of IMMS parts can be significantly reduced. Furthermore, spraying distance, current and pressure of the atomizing gas are varied to investigate the influence of these process parameters on the coating properties.

1 Introduction

The materials known under the term "plastic" are a relatively young material group. The introduction of the synthesizing possibilities of plastic by Hermann Staudinger at the beginning of the 1920s lead to the increased economic importance of plastic since the 1930s [1]. Since then, the use of plastic has increased considerably. In 1983, the global volumetric consumption of plastic exceeded that of iron with $1.25 \cdot 10^8$ m^3 [2]. Today, plastic are used in almost every industrial branch. Although plastic have been widely used, the growing complexity of operating conditions and requirements on functionalities leads to increased challenges. One example for that is integrating electrical functionality in plastic parts for electrical and electronic applications due to the increasing requirement on the compactivity of parts. To achieve this, plastic parts can be produced by injection molding of electrically conductive plastic melts, which contain electrically conductive additives [3]. However, the high filler content reduces the flowability of the plastic melts and limits the design freedom of parts. Furthermore, defined conducting tracks cannot be fabricated by this method. Another way of integrating electrical functionality in plastic parts is the use of metal/plastic hybrid parts. Prefabricated metal parts, such as films and grids, can be placed into injection mold and over molded by means of the in-mold assembly process [4]. In this case, the geometrical freedom of the parts is limited by the elasticity of the used metal. Furthermore, the fabrication of the metal parts often requires time consuming steps, e.g. milling, drilling and bending. The direct application of metallic coatings onto the surface of the plastic parts is also feasible. Different coating methods can be used, such as electroplating, physical vapor deposition (PVD) and thermal spraying. Although coatings with high quality can be fabricated by these methods, they still have some disadvantages. Elaborate preparation works and the use of special plastics are required for the electroplating process. The part size is limited by the size of the vacuum chamber of coating equipment in case of PVD. The direct application of thermally sprayed coatings onto plastic parts has also been studied [5-6]. A possible damage of parts with thin walls by the high temperature process gas and particles, in combination with the high particle velocities, must be considered.

To overcome the above-mentioned problems, a new method, the In-Mold-Metal-Spraying (IMMS) has been developed at Surface Engineering Institute (IOT) and Institute of Plastic Processing (IKV) of RWTH-Aachen University. In this process, the metal coatings are firstly applied onto the surface of the mold insert for injection molding, which acts as a carrier body for the coatings. By pre-adjusting of the coating bond strength to the carrier body, using suitable surface pre-treatment for the mold insert, the coatings can be transferred to the plastic parts during the injection molding process. After demolding, plastic parts with metal coatings on defined area of the part surface can be fabricated. IMMS enables an efficient production of metal/plastic parts due to the shortening of the process chain. Because the part dimension is defined by the mold, no further remachining of the parts or coating surfaces is necessary. The details of the process have been introduced elsewhere [7-10]. In the current study, the focus is the investigation of the coating properties with respect to the electrical resistivity of the coatings.

2 Experimental setup

At the current stage of development, the coatings are applied onto steel sheets made of the steel 1.0037 (AISI 1015), which are afterwards placed into the mold cavity for coating transfer. In this way, more parts can be fabricated at the same time for research purpose without using multiple expensive mold inserts. In the current study, the focus is the investigation of the electrical resistivity of coatings directly after the coating application. In this way, the influence of the coating process on the electrical resistivity of the coatings can be analysed without superposition by the effect of the injection molding process. To characterise the electrical resistivity of the coatings, a mask with the pattern in the shape of a conducting track is firstly fixed onto the carrier body. After the thermal spray process, coatings in the form of conducting track can be obtained. The coatings are carefully detached from the carrier bodies for further investigation, see Fig. 1.

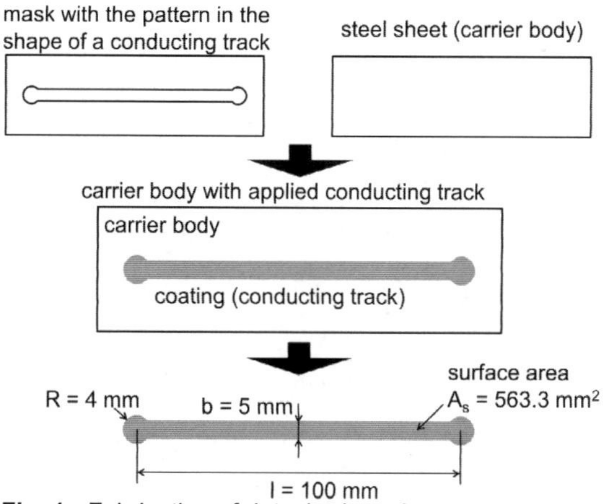

Fig. 1: Fabrication of detached coatings as conducting tracks

The electrical resistivity is a material constant that is dependent on the chemical composition, micro structure of the materials. It can be calculated by the following formula:

$$\rho = R \cdot A_c / l \quad (1)$$

with ρ electrical resistivity
R electrical resistance
A_c cross-sectional area of the specimen
l length of the specimen

The electrical resistance of the coatings is measured by 4-wire sensing to eliminate the lead and contact resistance. The length of the coatings is assumed to be the distance between the two contact points of a conducting track as indicated in Fig. 1. The cross-sectional area can be calculated from the width and thickness of the coatings. Due to the roughness and the inhomogeneity of the coatings in as-sprayed status, the determination of a representative, average coating thickness may require an analysis of a large amount of cross-sections. Therefore, in the current study, the weight of the coatings is measured. After that, A_c can be calculated as followed:

$$A_c = b \cdot m / (\rho_c \cdot A_s) \quad (2)$$

with b width
m weight
ρ_c density of feedstock material
A_s surface area of the specimen as indicated in Fig. 1

Although the density of the feedstock material is not equal to the density of the coatings due to oxides and porosity in the coatings, this assumption enables an approximation of the thickness and a comparison between coatings applied with different process parameters. By substituting A_c with equation 2 in equation 1, the electrical resistivity can be obtained:

$$\rho = R \cdot b \cdot m / (\rho_c \cdot A_s \cdot l) \quad (3)$$

3 Results and discussion

In the current study, Zn coatings are applied with different atomization gases (pressurized air and N_2) while the other process parameters were kept constant. The same procedure is carried out for Cu coatings but with three kinds of atomization gas (pressurized air, N_2 and a mixture of N_2 and H_2) to investigate the effect of the atomizing gases. A 2^3-experimental matrix with central point has also been conducted for Zn coatings by varying the process parameters: current, stand-off distance and pressure.

3.1 The effect of the atomizing gas

The first series of experiments is designed to investigate the effect of atomizing gas on the electrical resistivity of the coatings. Therefore, other process parameters, the current I_A, the stand-off distance d_s, the pressure p, the voltage V and the traverse velocity v, are held constant for each feedstock material. These process parameters are listed in Table 1a. Two kinds of atomizing gases have been used for the application of Zn coatings while one additional atomization gas, a mixture of N_2 with 5 Vol.% H_2 is used for Cu coatings as listed in Table 1b.

Table 1:

a) Constant process parameters for the application of Zn and Cu coatings

Feedstock Material	I_A/A	d_s/mm	p/bar	V/V	v/(mm/s)
Zn	130	120	3.5	20	500
Cu	120	120	3.5	30	500

b) Different atomizing gases

Feedstock Material	Option 1	Option 2	Option 3
Zn	Pressurized air	N_2	-
Cu	Pressurized air	N_2	N_2 with 5 Vol.% H_2

The experiment series contains 5 experiment points. Four repetitions have been carried out for each experiment point. After detaching the coatings from the carrier bodies, the electrical resistivity is determined as described in Section 2. The results as well as the electrical resistivity of massive Zn and Cu [11] are shown in Fig. 2.

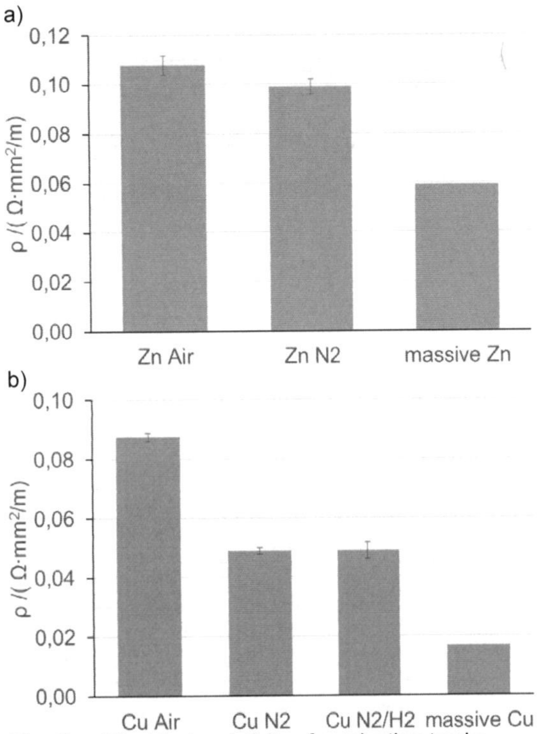

Fig. 2: Electrical resistivity of conducting tracks sprayed with different atomizing gases in comparison with that of massive material: a) Zn, b) Cu

To investigate the significance of the differences between the electrical resistivity of coatings sprayed with different atomizing gases, a one-way analysis of variance has been performed on the results. The difference between the means of electrical resistivity of the two kinds of Zn coatings has a 95% confidence interval of (0.0029, 0.0149) $\Omega \cdot mm^2/m$. Since the 95% confidence interval doesn't contain zero, the difference can be considered as significant. Both kinds of Zn coatings have higher electrical resistivity than that of massive Zn. It can be assumed that the higher electrical resistivity results from the porosity, cracks and oxides inside the coatings. Bobzin et al. have reported in [12] that in wire arc sprayed ZnAl15 coatings no oxides can be detected, which means that the amount of the oxides in the coatings are too low for detection. A similar effect may also be expected in wire arc sprayed Zn coatings which would result in a very low oxide content in the Zn coatings. Nevertheless, a thin oxide film between the Zn particles can form an interface that is sufficient to increase the electrical resistivity of Zn coatings in comparison to massive Zn. By using N_2 instead of pressurized air as atomizing gas the amount of oxides in the coatings cannot be significantly reduced since there are already few oxides inside the Zn coatings sprayed with pressurized air.

In contrast, there is a much bigger difference between the electrical resistivity of Cu coatings sprayed with pressurized air and N_2. Furthermore, the difference between electrical resistivity of Cu coatings sprayed with pressurized air and massive Cu ($\rho_{Cu,air} \approx 5\rho_{Cu,massive}$) is also significantly higher than between Zn coatings and massive Zn ($\rho_{Zn,air} \approx 2\rho_{Zn,massive}$). This suggests that there is more severe oxidation in the Cu coatings than the Zn coatings. By using N_2 as atomizing gas, which functions also as a protecting gas for the spray particles, the oxidation can be significantly reduced which leads to the decrease of the coating electrical resistivity. The difference between electrical resistivity of Cu coatings sprayed with N_2 and a mixture of N_2 and H_2 has a 95% confidence interval of (-0.0037, 0.0037) $\Omega \cdot mm^2/m$. Therefore, the difference is not significant which means the use of the mixture of N_2 and H_2 as atomization gas hasn't further improve the electrical resistivity of Cu coatings compared to using pure N_2.

3.2 The effect of other process parameters

A 2^3-experimental matrix has been conducted to investigate the effect of the current I_A, the stand-off distance d_s and the pressure p on the electrical resistivity of the sprayed coatings. In this experiment, Zn is used as feedstock material and N_2 is used as atomizing gas. The other process parameters, voltage and traverse velocity, are held constant as indicated in Table 1a. In the design of the experiment, for each factor, the current (factor A), the stand-off distance (factor B) and the pressure (factor C), a low level ("-1") and a high level ("1") are chosen to form a 2^3 full factorial design, see Table 2. In addition, a centre point labelled level "0" in Table 2 is used for each factor has also been included in the experimental matrix. As reference, all the experimental points, except the centre point, will be called corner points hereafter.

Table 2: Setup of the 2^3 full factorial design with center point

Factor	Level	Value
A/A	-1	100
	0	130
	1	160
B/mm	-1	90
	0	120
	1	150
C/bar	-1	2.4
	0	3.5
	1	4.6

The corner points with factor setup at level "-1" or "1" are carried out once. The centre point has been replicated four times. This way, the effect of each factor (main effect: A, B and C) and the interaction between the factors (interaction effect: AB; AC, BC and ABC) can be investigated. The 12 experiments and their setup as well as the measured electrical resistivity of the applied coatings are listed in Table 3.

Table 3: Experiment points and results

Nr.	A	B	C	AB	AC	BC	ABC	ρ ($\Omega \cdot mm^2/m$)
1	-1	-1	-1	1	1	1	-1	0.0973
2	1	-1	-1	-1	-1	1	1	0.0927
3	-1	1	-1	-1	1	-1	1	0.1084
4	1	1	-1	1	-1	-1	-1	0.0997
5	-1	-1	1	1	-1	-1	1	0.1007
6	1	-1	1	-1	1	-1	-1	0.1029
7	-1	1	1	-1	-1	1	-1	0.1128
8	1	1	1	1	1	1	1	0.1026
9	0	0	0	0	0	0	0	0.1031
10	0	0	0	0	0	0	0	0.0964
11	0	0	0	0	0	0	0	0.0986
12	0	0	0	0	0	0	0	0.0975

The analysis of variance is also conducted on these results. As the first step, the contrast, which is the sum of the multiplication of the column "ρ" and the column of the to be calculated effect in Table 3, should be determined, e.g. for main effect A:

$$\text{Contrast}_A = \sum_{i=1}^{8} \rho_i \cdot A_i \quad (4)$$

To check the significance of each effect and interaction effect, the sum of squares (SS) or the mean square (MS) is needed. The determination of the two values is shown in followed equations:

$$SS = (\text{Contrast})^2/2^3 \quad (5)$$
$$MS = SS/DF \quad (6)$$

DF is the degree of freedom of an effect. In the current study, each effect has one degree of freedom because there are only two levels for each factor. The F-value of an effect, which is the indicator of the significance of an effect, is the mean square of an effect divided by the mean square of error. Since the corner points are carried out only once, the experimental matrix in the current study is a unreplicated 2^3 full factorial design. In this situation, the mean square of error can be determined by two methods. When some of the interaction effects can be assumed to be negligible, the mean square of them can be used as an estimation of the error. If a centre point with multiple replication is present, the standard deviation of the results of the centre point can be used as the mean square of error. The second option is chosen in the current study. The F-values of each effects in comparison to the threshold F-value, F_{limit}, with 90% confidence level for the significance of the effects are shown in Fig. 3. An effect with a F-value higher than F_{limit} can be considered as significant effect. F_{limit} is determined by referring to the percentage points of F distribution in [13]:

$$F_{limit} = F_{\alpha,v1,v2} = F_{0.1,1,3} \quad (7)$$

with α 1-90% (confidence level)
v1 degree of freedom of the effect
v2 degree of freedom of the error

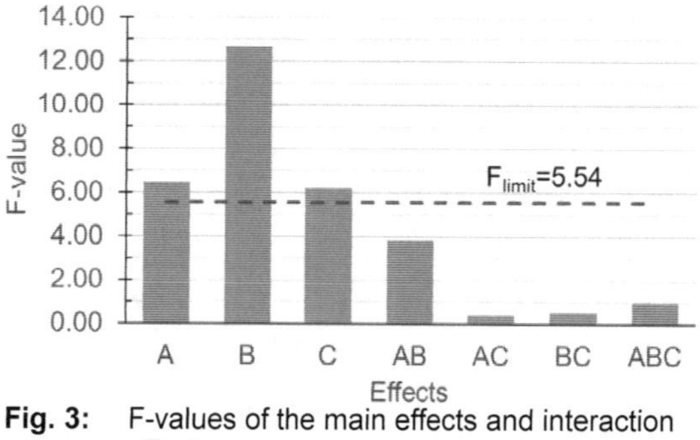

Fig. 3: F-values of the main effects and interaction effects

Based on the data in current study, the 3 main effects are significant. All interaction effects are insignificant, as shown in Fig. 3. The interaction plots in Fig. 4 confirm the conclusion that all second order interactions are insignificant. In each plot, two factors or parameters are observed. A data point in an interaction plot refers to a certain combination of factor levels of the two observed factors in the plot. The value of data points is the mean of electrical resistivity of coatings applied using the referred factor level combination. If the "-1" and "1" lines in an interaction plot are parallel or almost parallel to each other, the effect of a factor hasn't been significantly changed by changing the setup of the other factor in the plot. Therefore, the interaction effect between the two factors should be considered as small. The interaction plots for the stand-off distance and pressure as well as for the current and the pressure corresponds to the above-mentioned situation, see Fig. 4a and 4b. In Fig. 4c, the two lines are not parallel. But they both have a negative slope and do not intersect with each other. This means that there is a slight interaction effect between the current (A) and the stand-off distance (B). This finding is consistent with the result of the F-value analysis in Fig. 3, in which the F-value of the interaction effect AB is close to F_{limit} but still smaller than it. In general, a system is dominated by some of the main effects and lower order interactions. Therefore, it is not unusual that the interaction effect ABC has a low F-value.

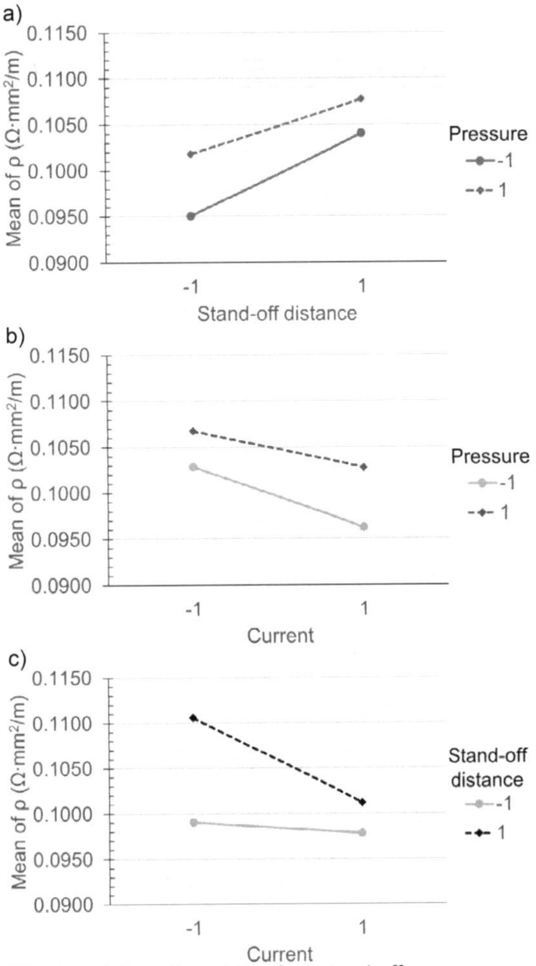

Fig. 4: Interaction plots of: a) stand-off distance*pressure; b) current*pressure; c) current*stand-off distance

The main effect plots are shown in Fig. 5. The value of a data point in a main effect plot is the average electrical resistivity of coatings applied with the overserved factor set at the corresponding level. It can be concluded that the electrical resistivity of the coatings increases with decreasing current and increasing stand-off distance and pressure. The decrease of current and increase of stand-off distance and pressure can all lead to a same effect, which is a better atomization. Lower current means lower wire feed rate which leads to lower amount of to be atomized molten zinc per unit of time. The melt can, therefore, be better atomized resulting in finer and more particles. By increasing the stand-off distance, the chance of secondary or tertiary atomization of the particles respectively is higher which also leads to higher particle count. Higher pressure also results in better atomization effect and, therefore, more particles. By increasing the particle amount, there are more interfaces inside the coatings. Pores, cracks and oxides can form at these interfaces. Therefore, the electrical resistivity of these coatings is higher.

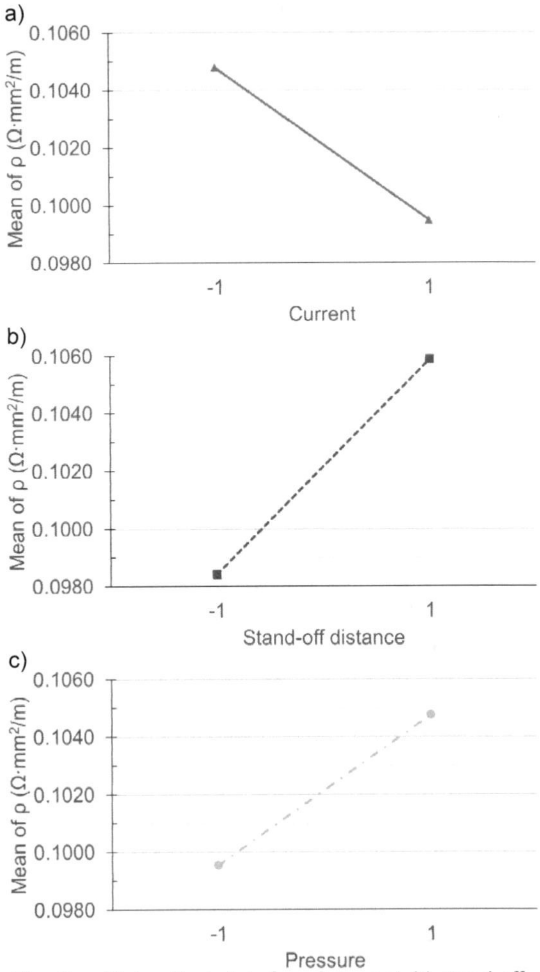

Fig. 5: Main effect plots for: a) current; b) stand-off distance; c) pressure

In addition to the effect on the atomization, the increase of stand-off distance reduces the protecting effect of N_2 against particle oxidation at the end of the free jet. It promotes the formation of oxides at the surface of particles and, therefore, increases the resulting coating electrical resistivity.

4 Conclusion and outlook

In the current study, the influence of different process parameters on the electrical resistivity of wire arc sprayed coatings has been investigated. The use of N_2 instead of pressurized air as atomizing gas leads to a slight decrease of the electrical resistivity in case of Zn coatings. However, it has a more significant effect on Cu coatings. The reason for that might be the different extent of oxide inclusion in Zn and Cu coatings, when pressurized air is used as atomizing gas. A third atomizing gas, a mixture of N_2 and H_2, has also been used for the application of Cu coatings. In this case, no significant change of the electrical resistivity of coatings in comparison to that of coatings sprayed with N_2 can be overserved. It must be noted that, adding H_2 into the atomizing gas may however influence other properties of the coatings, e.g. the bond strength of the coating to the carrier body.

By conducting a 2^3 experiment and the following analysis of the results the effects of the current, the stand-off distance and the pressure on the electrical resistivity of the coatings could be investigated. The results show that the interaction effects of the three factors are not significant. In contrast to that, the three main effects are identified as significant. The electrical resistivity of the coatings increases with decreasing current and increasing stand-off distance and pressure. The reason might be a better atomization which leads to fine particles and increases the amount of interfaces inside the coatings.

In the current study, the density of the feedstock material is used to calculate the cross-sectional area of the coatings which is needed for the determination of the electrical resistivity. Although it enables an approximate estimation of the cross-sectional area, a better solution is a direct measurement of the volume of the coatings. One possible method is the measurement using a pycnometer. The sample volume should be however as large as possible for an accurate measurement. For the corner points, only one conducting track has been applied. Therefore, the sample volume is not large enough for the measurement using a pycnometer. To verify the observation in current study, larger carrier bodies and masks should be used to apply more conducting tracks with same process parameters at the same time. This way, a larger sample volume can be provided for the pycnometer measurement so that the cross-sectional area of the conducting tracks can be determined more accurately.

5 Acknowledgements

The authors would like to thank the German Research Foundation DFG for the support of the depicted research within the Cluster of Excellence "Integrative Production Technology for High-Wage Countries".

6 Literature

[1] Hopmann, C. and Michaeli, W.: Einführung in die Kunststoffverarbeitung. Carl Hanser Verlag (2015), München, pp. 2
[2] Bonten, C.: Kunststofftechnik – Einführung und Grundlagen, Carl Hanser Verlag (2014), München, pp. 2
[3] Pflug, G.: Polymer MWNT Composites: Shielding Plastic Housing. Kunststoffe international (2008), Issue 2, pp. 94-98
[4] Paul, H., Luke, M. and Henning, F.: Polymer Metal Hybrid Structures - Experimental Investigations of Deformation and Failure Behavior. Journal of Plastic Technology 10 (2014), Issue 4, pp. 118-141
[5] Liu, A., Guo, M., Zhao, M. and Hu, M.: Arc Sprayed Erosion-Resistant Coating for Carbon Fiber Reinforced Polymer Matrix Composite Substrates. Surface and Coating Technology 200 (2006), Issue 9, pp 3073-3077
[6] Robitaille, F., Yandouze, M., Hind, S., Jodoin, B.: Metallic Coating of Aerospace Carbon/Epoxy Composites by the Pulsed Gas Dynamic Spraying Process. Surface and Coating Technology 203 (2009), Issue 19, pp 2954-2960
[7] Bobzin, K., Öte, M., Linke, T.F., Liao, X., Hopmann, Ch. and Ochotta, P.: Metallisierung von Kunststoffoberflächen – ein neues Verfahren: „In-Mold-Metal-Spraying (IMMS)". Jahrbuch Oberflächentechnik 71 (2015), pp. 67-73.
[8] Bobzin, K., Öte, M., Linke, T.F., Liao, X., Hopmann, Ch. and Ochotta, P.: Application of Thermal Spraying for the Manufacturing of Metal/plastic components. Thermal Spray Bulletin 67 (2015), Issue 1, pp. 28-31
[9] Hopmann, Ch., Bobzin, K., Schoeldgen, R., Wunderle, J., Linke, T.F. and Ochotta, P.: IMKS and IMMS: Two Methods for the Production of Plastic Parts Featuring Metallic Areas. Journal of Polymer Engineering 36 (2016), Issue 6, pp. 549-556
[10] Bobzin, K., Öte, M., Knoch, M.A., Liao, X., Hopmann, Ch. and Ochotta, P.: Investigation of Influencing Factors on the Transfer of Wire Arc Sprayed Zn Coatings for the Metallisation of Plastic Parts, Proceeding of International Thermal Spray Conference 2016, Shanghai, pp. 24-29
[11] Weast, R.C.: CRC Hand Book of Chemistry and Physics 48th Edition, The Chemical Rubber Co. (1968), Cleveland, pp. F-118
[12] Bobzin, K., Öte, M., Knoch, M.A.: Characterisation of Thermally Sprayed ZnAl15 Corrosion Protection Coatiqungs for Offshore-Wind-Turbines, Proceeding of International Thermal Spray Conference 2016, Shanghai, pp. 179-183
[13] Montgomery, D.C.: Design and Analysis of Experiments, John Wiley & Sons, Inc. (2012), Hoboken, pp. 689

Liquid flame spray fabrication of WO_3-graphene nanocomposite coatings for gas-sensing applications

Yi Liu, Jing Huang, Yongfeng Gong, Hua Li*
Key Laboratory of Marine Materials and Related Technologies, Ningbo Institute of Materials Technology and Engineering, Chinese Academy of Sciences, Ningbo 315201, China. *: lihua@nimte.ac.cn.

There have been increasing demands for adequate gas sensors to monitor O_3, a respiratory irritant gas associated with a spectrum of adverse health events. Here we report film construction by liquid flame spray route and characterization of nanostructured WO_3-reduced graphene oxide (rGO) composites and their gas-sensing activities to O_3. The starting feedstock was prepared from WCl_6 and rGO for pyrolysis synthesis by flame spray. Nanosized WO_3 grains exhibited oriented nucleation on rGO flakes and rGO retained intact nano-structural features after the spraying. Constrained grain growth of WO_3 was realized in the rGO-containing films with 60-70 nm size as compared to ~220 nm in the pure WO_3 film. The WO_3-rGO film sensors showed quicker response to O_3 and faster recovery than the rGO-free WO_3 film sensors. Addition of rGO in 1.0wt.% or 3.0wt.% in the films caused significantly reduced effective working temperature of the film sensors from ~250°C to ~150°C. These results might shed some light on liquid flame spray fabrication of novel functional nanocomposites for gas-sensing applications.

1 Introduction

In recent years, ozone generator-related air purifiers like ionic and electrostatic air cleaners with ozone as a byproduct have been developed for normal daily use, medical therapy and industrial applications [1,2]. However, ozone gas is harmful to human beings even under low concentration levels like several parts per million (ppm). In many cases there have been increasing demands for ozone sensors for possibly controlling the gas emission. Among the existing gas sensors, nanostructured semiconducting metal oxides were successfully developed for sensing a variety of gases and vapors.

WO_3 is an important n-type metal oxide semiconductor with oxygen vacancies, which act as donors and exhibit high sensitivity particularly to ozone[3,4]. Extremely high carrier mobility of graphene at room temperature indicates that graphene is a promising candidate for gas sensing operated at low temperature. In particular, graphene is a p-type semiconductor in nature, the potential barrier at the n-WO_3/p-graphene interfaces could be modified by introduction of graphene. Challenges persist yet pertaining to developing appropriate technical routes for making the semiconductor films for their long-term functional services.

As one of the most extensively applied surface coating techniques, thermal spray already showed promises in producing WO_3 films with exciting gas-sensing performances [5,6]. In this work, we fabricated WO_3-reduced graphene oxide (rGO) nanocomposite films by liquid flame spray. Graphene was selected as the additives with an aim of achieving remarkably enhanced sensing capability of WO_3 films. Spray pyrolysis was made during the flame spray processing for synthesizing WO_3 and microstructure of the films was characterized. The ozone-sensing performances of the films were examined and elucidated, giving clear insight into flame spray fabrication of the novel nanocomposites for gas-sensing applications.

2 Materials and methods

2.1 Fabrication of WO_3-rGO films

Tungsten chloride (WCl_6) was used for pyrolysis synthesis of tungsten trioxide (WO_3) by flame spraying. The preparation procedure is schematically shown in **Fig. 1 a**. Gold electrodes were printed by screen printing method and sintered on alumina substrate. Then the WO_3-rGO films were deposited by liquid flame spray (**Fig. 1 b**). During the deposition, the precursor was fed vertically into the flame. WO_3 films with/without addition of rGO platelets were liquid flame sprayed onto Al_2O_3 substrates. The liquid flame deposition was carried out using the CDS 8000 system (Castolin, UK). Oxygen and acetylene were used as the combustion-supporting gas and fuel gas with the pressure of 0.7 MPa and 0.1 MPa, respectively. The composite feedstock feeding rate was 40 ml/min and the spray distance was 200 mm.

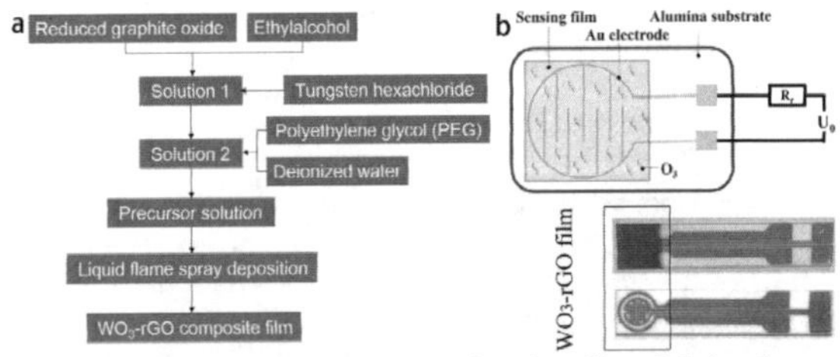

Fig. 1. Processing flow chart for the film fabrication (a), and schematic depiction of the basic circuit for the WO_3-rGO coated sensor to measure its sensing response to O_3 (b).

2.2 Characterization of microstructure

Microstructure of the powder and the films was characterized by transmission electron microscopy (FEI Tecnai F20, the Netherlands) and field emission scanning electron microscopy (FEI Quanta FEG250, the Netherlands). Phase composition of the samples was analyzed by X-ray diffraction (D8 Advance, Bruker AXS, Germany). rGO dispersed in WO_3 matrix was characterized by Raman spectroscopy (Renishaw inVia Reflex, UK). The Diamond TG/DTA system (PerkinElmer, USA) was used to carry out the thermogravimetric analysis (TGA) of the samples in air with the heating rate of 10°C/min.

2.3 Assessment of O_3 sensing properties

Gas-sensing capability of the coated sensors was examined by measuring their sensitivity and response to ozone gas. The concentration of the ozone gas in the testing system was adjusted from 0.5 to 10.0 ppm and calibrated by an ozone detector. The sensors were heated to different working temperatures of 100°C to 300°C. The sensor responses for the gas (O_3 in air) were measured, which were determined by: $S = \frac{R_{gas}-R_{air}}{R_{air}}$, where R_{gas} and R_{air} denote the resistance of the as-deposited film in the environment of gas mixture and air, respectively. The response performances of the sensors was measured by the transient time to the steady-state resistance upon their exposure to O_3 gas. Response time and recovery time were defined as the time for the sensors to attain 90% change due to the chemical adsorption.

3 Results and discussion

TEM image of the precursor synthesized via the chemical approach without addition of rGO shows well-dispersed nanosized WO_3 particles (**Fig. 2 a**). The tungsten oxide grains have the size of 8 to 50 nm. XRD pattern of tungsten oxide precursor matches with $WO_3 \cdot H_2O$ (**Fig. 2 b**). As the amount of water in mixed solvent increases, mole fraction of oxygen in the tungsten oxide system also increases, which is likely due to the fact that water is the source of oxygen for the solvothermal synthesis [7]. Room-temperature dried products of the precursor show no trace of WO_3, indicating clearly that heat source is essential for WO_3 synthesis from the solution precursor and in fact monoclinic WO_3 was successfully synthesized during the spraying.

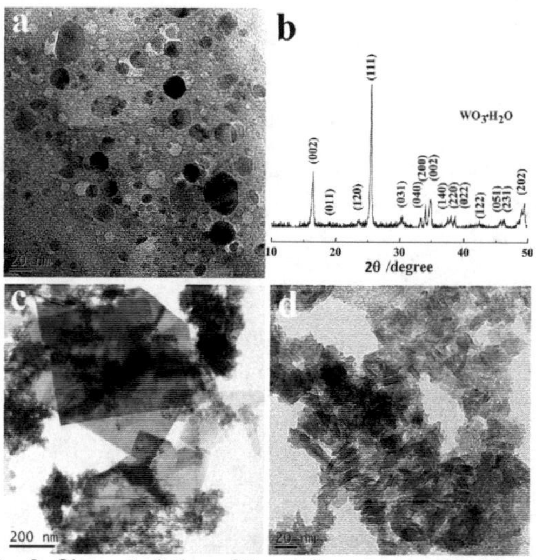

Fig. 2. Characteristics of the WO_3 precursor (a, b) and the WO_3-rGO precursor (c, d), (a) TEM image of the pure WO_3 precursor showing $WO_3 \cdot H_2O$ grains, (b) XRD pattern of the WO_3 precursor showing sole component of $WO_3 \cdot H_2O$, (c) TEM image of the WO_3-rGO precursor showing wrinkled paper-like morphology of rGO and adhered $WO_3 \cdot H_2O$ nano grains on rGO flakes, and (d) enlarged TEM image of the rod-like $WO_3 \cdot H_2O$ nano grains in the rGO-containing precursor.

The as-received rGO shows wrinkled paper-like morphology. TEM observation revealed that $WO_3 \cdot H_2O$ nanograins intimately attach to rGO flakes (**Fig. 2 c**). Close view demonstrates rod-like particles in the rGO-containing precursor (**Fig. 2 d**). $WO_3 \cdot H_2O$ grains have the size of 20 nm in length and 5 nm in diameter, thereby implying preferred crystallographic orientation of WO_3 growth induced by rGO sheets.

During the liquid flame spraying of the WO_3-rGO precursor, liquid droplets injected into the flame are subjected to a maximum temperature of 2600°C and are accelerated to about 160 ms^{-1}. Thermal analyses were therefore carried out to examine the behaviors of WO_3 phase during high temperature processing in air atmosphere (**Fig. 3**). DTA and TG measurements were made in static air from room temperature to 1200°C. It is noted that there is an obvious weight loss (as indicated by the TG curve, **Fig. 3a**) and a strong exothermic peak (as suggested by the DTA curve, Fig. 3 b) from 80°C to 110°C, referring to release of solvent, ethyl alcohol molecules and water molecules in this case. Weight loss of the surfactant PEG is not detected due to its low content. In addition, dehydration of $WO_3 \cdot H_2O$ is suggested with the formation of WO_3 below the temperature of 200°C [7]. The small exothermic DTA peak at 698.7°C (**Fig. 3b**) with undetectable weight loss (**Fig. 3a**) is presumably attributed to combustion of few rGO nansheets. The peak appeared at ~910°C likely refers to phase transformation of WO_3 [8].

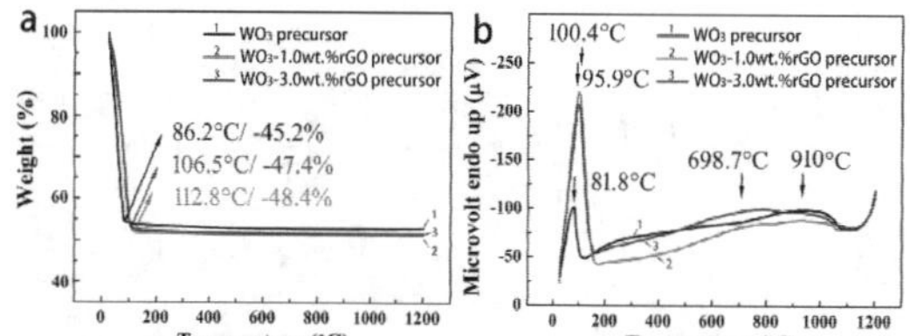

Fig. 3. TG curves (a) and TGA curves of the WO_3 and the WO_3-rGO precursors.

Nanostructured WO_3 and WO_3-rGO films were successfully deposited on alumina substrates by the liquid flame spray processing. Raman and XRD detection reveals the presence of monoclinic WO_3 in the films without other phases relating to tungsten oxide (**Fig. 4**), indicating complete synthesis of WO_3 during the spraying. In addition, it is noted that rGO is retained after the film deposition (**Fig. 4b**). rGO is suggested by the G band at 1578 cm^{-1} on the Raman curve (**Fig. 4b**).

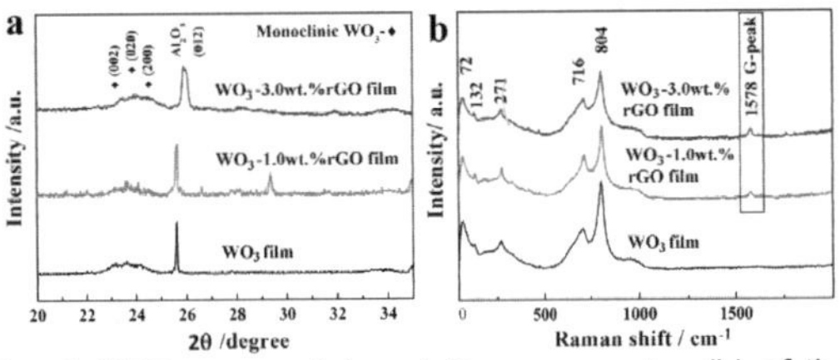

Fig. 4. XRD curves (a) and Raman spectra (b) of the samples (the Al_2O_3 XRD peaks are detected from the substrate).

Fractured cross-sections show unique film thickness (**Fig. 5 a**), and it is easily tailorable by changing deposition time. It is noted, however, that addition of rGO reduced the deposition efficiency of the precursors for the film fabrication (**Fig. 5 c-1, b-1 versus a-1**). The films show the thickness of 5 μm, 0.9 μm and 0.6 μm for the pure WO_3, the WO_3-1.0wt.%rGO and the WO_3-3.0wt.%rGO films. It is established that film thickness is critical in deciding diffusivity and reactivity of target gas molecules. Electrical resistance and sensor response of the sensor films to certain gases decreased monotonically with increasing film thickness, while some gases responded almost independently of the film thickness [9]. In this case, thickness, microstructural characteristics, and chemistry of the films might synergistically play roles in regulating their gas-sensing performances. In addition, FESEM views from top surfaces of the films (**Fig. 5 a-2, b-2 and c-2**) suggest remarkably different microstructural features for the films with/without addition of rGO. The pure WO_3 film exhibits more porous network with large pores of ~4 μm (**Fig. 5 a-2, a-3**), while the rGO-containing WO_3 films present meso-/micro- porous topographical feature. rGO sheets are uniformly dispersed in WO_3 matrix (**Fig. 5 b, c**).

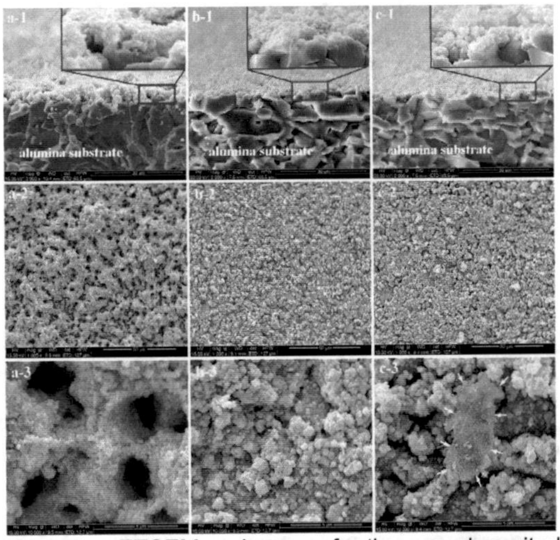

Fig. 5. FESEM views of the as-deposited nanostructured films, (a) the pure WO_3 film, (b) the WO_3-1.0 wt.% rGO film, and (c) the WO_3-3.0 wt.% rGO film. (-1: cross-sectional view, -2: surface view, and -3: further magnified surface view). The red circles highlight typical rGO existing on the top layer of the films, and the yellow arrows point to rGO sheet).

For macropores with pore size larger than 50 nm, gas transport occurs mainly by molecular diffusion (confinement by the pore boundaries is less significant), whereas in micropores with pore size less than 2 nm, surface diffusion becomes predominant. Transport without external pressure can be described by Knudsen diffusion and the diffusion coefficient $D_k = 4r/3\sqrt{(2RT/\pi M}$ [10], where R is the universal gas constant, M is molecular weight of the gas molecule, and r is pore radius. In theory, larger pores would result in better diffusivity of the target gas, giving rise to higher gas sensitivity of the films.

In fact, gas diffusion is a complex issue involving the influence of various structural and textural parameters of sensing films. Grain size and specific surface area of the films are predominately the key factors responsible for their gas sensing performances. For the rGO-containing films, a large number of tiny particles are clearly recognized, which mainly grow on rGO sheets. The particles show a size of ~60-70 nm (**Fig. 6b**), and the pure WO_3 film shows the particle size of ~220 nm (**Fig. 6a**). The structural features suggest intact state of rGO during the spraying and WO_3 particles opt to adhere on rGO sheets during the film deposition. A schematic depiction is proposed to illustrate the WO_3 formation superimposing on the surface of rGO sheets (**Fig. 6c**).

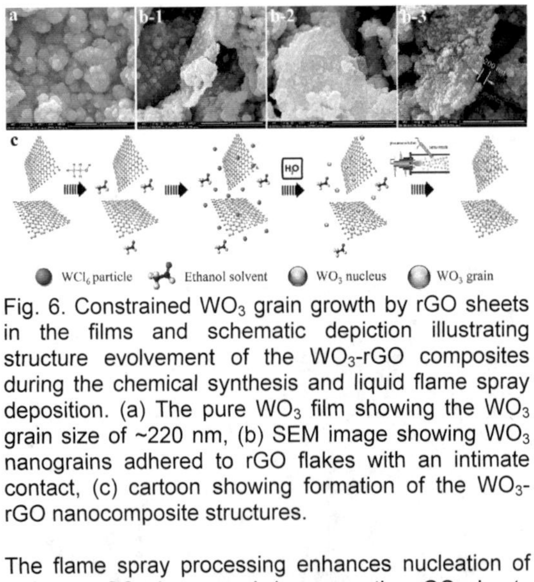

Fig. 6. Constrained WO_3 grain growth by rGO sheets in the films and schematic depiction illustrating structure evolvement of the WO_3-rGO composites during the chemical synthesis and liquid flame spray deposition. (a) The pure WO_3 film showing the WO_3 grain size of ~220 nm, (b) SEM image showing WO_3 nanograins adhered to rGO flakes with an intimate contact, (c) cartoon showing formation of the WO_3-rGO nanocomposite structures.

The flame spray processing enhances nucleation of WO_3 on rGO sheets and, however, the rGO sheets constrain fast growth of the adhered WO_3 grains (60-70 nm versus 220 nm). This could be explained by the inhibited grain growth along at least one direction, that is, the direction perpendicular to the interface between the adhered WO_3 grain and rGO sheet. These structural features alone might already affect the gas-sensing performances of WO_3. Theoretically, the changes in conductance of sensing materials occur at the exterior regions of their grains. Beyond the depletion layer, the interior parts of the grains do not contribute to gas response [11]. As a consequence, grain size affects the sensor performance quite substantially. Smaller crystal size gives rise to greater sensitivity of the bulk film [11,12]. Our results suggest that the addition of rGO increases the specific surface area of WO_3-based films, which could in turn provides more adsorption sites for gas reaction.

To gain clear insight into the impact of the WO_3-rGO composite structure on gas sensing performances, response of the films to O_3 was examined (**Fig. 7**). As one of the crucial factors that affect the sensing performances, the operation temperature was altered to inspect the response of the films to O_3. Results show that the responses are enhanced and reach the maximums and then decline rapidly with increase of the operation temperature (**Fig. 7a**). Surprisingly, it is realized that addition of rGO platelets altered the temperature at which point maximum response is attained. The temperature changed from ~250°C for the rGO-free WO_3 films to ~150°C for the rGO-containing WO_3 films. Medium to high working temperature (typically in the range of 200-500°C) has been the major hurdle for applications of semiconductor-based sensors [3,13,14], since the markedly high operation temperature triggers many obstacle problems, e.g. undesired grain growth [44,45], elongated response time [15], collapsed pore structure, noticeable power consumption, and others, which could either shorten service duration of the sensors or reduce their sensitivity. It is exciting to note that incorporation of rGO effectively reduces the working temperature of the sensor. Compared with previously reported semiconductor-based sensors for ozone detection [15,16], the WO_3-rGO films fabricated by the liquid flame spray route show much lower optimum operation temperature (T_{opt}) and higher sensitivity. The rationale for the necessarily required temperature lies in the kinetics of both the reactions of target gas molecules with surface-adsorbed oxygen and the replacement of the oxygen from gas molecules. In addition, the working temperature is also essential for facilitating gas diffusion through the sensing layer. The tendency between operation temperature and sensor response is commonly observed and this is believed to be resulted from the competition between slow kinetics at low temperature and enhanced desorption at high temperature [13,10]. For the WO_3-based sensor films, upon their exposure to the oxidizing O_3 gas, the gas molecules can easily adsorb on the active sites of the films and capture the electrons from WO_3, in turn enhancing the electron-depletion for increased electrical resistance [3]. For the synthesized rGO existing in the WO_3-rGO films, there have been considerable structural disorders in graphene lattice. Due to well retained functional groups of rGO in the films, the interaction of gas with rGO is anticipated at the defects [17,18]. In O_3 atmosphere, carbon atoms could be removed from graphene lattice and vacancies are formed which could further enhance the sensing performances [17]. In addition, graphene has large surface area with respect to its volume due to its unique two-dimensional structural features. For the rGO-containing films, the large surface area can presumably facilitate O_3 adsorption and diffusion on active surfaces. Moreover, the superior electrical property of rGO and potential barrier (Schottky barrier) at the WO_3/rGO interface could contribute to improved conductivity and specific capture and migration of electrons from WO_3 to rGO, leading to a better sensing behavior [19].

Gas sensing assessment revealed significantly enhanced sensitivity of the WO_3 sensors to ozone by the addition of rGO (**Fig. 7b**). The response of the sensor operated at 150°C is almost proportional to ozone concentration from 0.5-10 ppm. Results show that higher content of rGO in the films gives rise to better sensing response and the WO_3-3.0wt.%rGO sensor has almost 4 times sensing response more than the pure WO_3 sensor. It has been reported WO_3 nanorods/graphene nanocomposites performed high-efficiency visible-light-driven photocatalysis and NO_2 gas sensing [19]. Our finding is consistent with the existing results that graphene promotes the sensing performances of the semiconductor sensors.

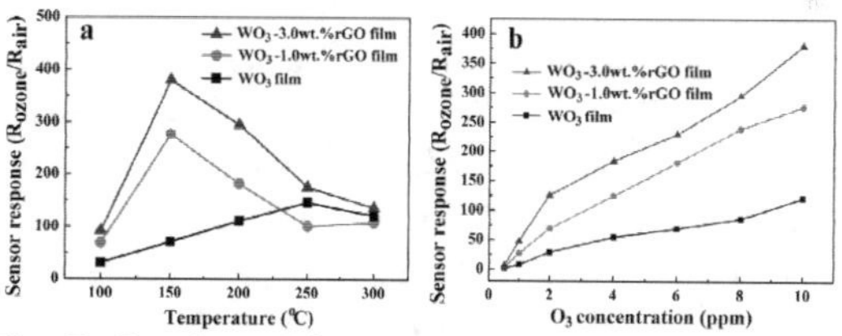

Fig. 7. Sensing performances of the films to O_3, (a) sensitivity versus operation temperature (O_3: 10 ppm), and (b) sensitivity versus O_3 concentration tested at 150°C.

In addition, the responses and the recovery responses of the sensors based on the WO_3 and the WO_3-rGO films were also examined under 10 ppm O_3 tested at 150°C (**Fig. 8**). The resistance of the WO_3 and WO_3-rGO sensors can be recovered to the initial values after they were put back to air. The response time and the recovery time of the WO_3 and WO_3-1.0wt.%rGO sensors are 64.6 s and 40.6 s, 26.9 s and 29.1 s, respectively. The WO_3-3.0wt.%rGO sensor shows the values of 17.1 s and 32.7 s, respectively, which are remarkably shorter than the values exhibited by other semiconductor-based sensors like ZnO-rGO sensor to NO_2 [14]. Compared to that of pure WO_3 sensor, the higher sensitivity and quicker response of the rGO-containing WO_3 sensors could partly due to the smaller grain size and higher surface area, which could enhance the adsorption capability for O_3 gas. The mechanisms as to how the WO_3/rGO interfaces affect the sensing performances are to be further explored.

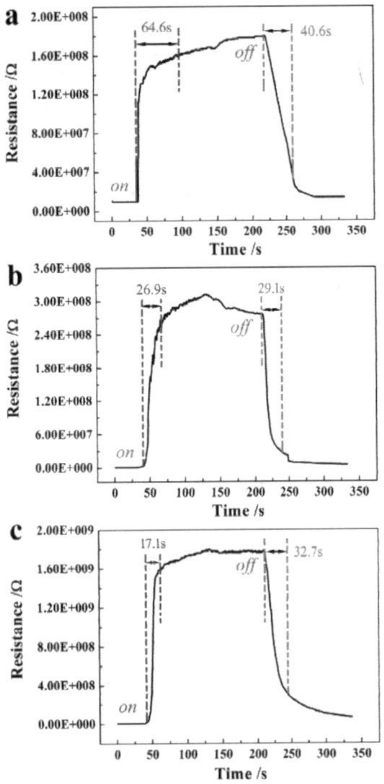

Fig. 8. Response and recovery curves of the WO_3 sensor (a), the WO_3-1.0wt.%rGO sensor (b), and the WO_3-3.0wt.%rGO sensor (c).

4 Conclusions

Nano-porous WO_3-rGO films have been successfully fabricated by liquid flame spraying WCl_6 and rGO precursor. The films showed tailorable thickness and WO_3 grains opted to nucleate on and grew along rGO nanosheets during the spraying. Significantly constrained growth of WO_3 grains was realized in the nanocomposite films and the structural features of rGO were retained after the spraying. Addition of rGO significantly enhanced the sensitivity of the WO_3 films to O_3 and reduced the effective working temperature of the sensors from ~250 °C to ~150 °C. The results would give insight into selection and preparation of new sensing nanomaterials and their film fabrication by liquid flame spray technical route for gas-sensing applications.

5 References

[1] Singer, B.C., Coleman, B.K., Destaillats, H., Hodgson, A.T., Lunden, M.M., Weschler, C.J., and Nazaroff, W.W.: Indoor secondary pollutants from cleaning product and air freshener use in the presence of ozone. Atmos. Environ. 40 (2006), Issue 35, pp. 6696/10.

[2] Britigan, N., Ahmad, A. and Sergey, A.N.: Quantification of ozone levels in indoor environments generated by ionization and ozonolysis air purifiers. J. Air Waste Manage. 56 (2006), Issue 5, pp. 601/10.

[3] Guérin, J., Aguir, K., Bendahan, M. and Lambert-Mauriat, C.: Thermal modelling of a WO_3 ozone sensor response. Sensor. Actuat. B 104 (2005) , Issue 35, pp. 289/93.

[4] Guerin, J., Aguir, K. and Bendahan, M.: Modeling of the conduction in a WO_3 thin film as ozone sensor. Sensor. Actuat. B 119 (2006) , Issue 1, pp. 327-334.

[5] Li, X.L., Lou, T.J., Sun, X.M. and Li, Y.D.: Highly sensitive WO_3 hollow-sphere gas sensors. Inorg. Chem. 43 (2004) , Issue 17, pp. 5442/49.

[6] Wu, Q.F., Huang, J. and Li, H.: Deposition of porous nano-WO_3 coatings with tunable grain shapes by liquid plasma spraying for gas-sensing applications. Mater. Lett. 141 (2015), pp. 100/03.

[7] Choi, H.G., Jung, Y.H. and Kim, D.K.: Solvothermal synthesis of tungsten oxide nanorod/nanowire/ nanosheet. J. Am. Ceram. Soc., 88 (2005), Issue 6, 1684/86.

[8] Salje, E. and Viswanathan, K.: Physical properties and phase transitions in WO_3. Acta Cryst. A31 (1975) , Issue 3, pp. 356/59.

[9] Sakai, G., Baik, N.S., miura, N. and Yamazoe, N.: Gas sensing properties of tin oxide thin film fabricated from hydrothermally treated nanoparticles. Sensor. Actuat. B 77 (2001) , Issue 1, pp. 116/21.

[10] Sakai, G., Matsunaga, N., Shimanoe, K. and Yamazoe, N.: Theory of gas-diffusion controlled sensitivity for thin film semiconductor gas sensor. Sensor. Actuat. B 80 (2001) , Issue 2, pp. 125/31.

[11] Rothschild, A. and Komem, Y.: The effect of grain size on the sensitivity of nanocrystalline metal-oxide gas sensors. J. Appl. Phys. 95 (2004) , Issue 11, pp. 6374/80.

[12] Xu, C.N., Tamaki, J., Miura, N. and Yamazoe, N.: Grain size effects on gas sensitivity of porous SnO_2-based elements. Sensor. Actuat. B 3 (1991) , Issue 2, pp. 147/55.

[13] Wang, C., Yin, L., Zhang, L., Xiang, D. and Gao, R.: Metal oxide gas sensors: sensitivity and influencing factors. Sensors 10 (2010) , Issue 3, pp. 2088/06.

[14] Liu, S., Yu, B., Zhang, H., Fei, T. and Zhang, T.: Enhancing NO_2 gas sensing performances at room temperature based on reduced graphene oxide-ZnO nanoparticles hybrids. Sensor. Actuat. B 202 (2014), pp. 272/78.

[15] Hattori, A., Tachibana, H., Yoshiike, N. and Yoshida, A.: Ozone sensor made by dip coating method. Sensor. Actuat. B 77 (1999) , Issue 2, pp. 120/25.

[16] Catto, A.C., da Silva, L.F., Ribeiro, C., Bernardini, S., Aguir, K., Longo, E. and Mastelaro, V.R.: An easy method of preparing ozone gas sensors based on ZnO nanorods. Rsc Adv. 5 (2015) , Issue 25, pp. 19528/33.

[17] Cheng, Y.C., Kaloni, T.P., Zhu, Z.Y. and Schwingenschlögl, U.: Oxidation of graphene in ozone under ultraviolet light. Appl. Phys. Lett. 101 (2012) , Issue 7, pp. 073110.

[18] Lee, G., Lee, B., Kim, J. and Cho, K.: Ozone Adsorption on graphene: Ab initio study and experimental validation. J. Phys. Chem. C 113 (2009) , Issue 32, pp. 14225/29.

[19] An, X.Q., Yu, J.C., Wang, Y., Hu, Y.M., Yu, X.L. and Zhang, G.J.: WO_3 nanorods/graphene nanocomposites for high-efficiency visible-light-driven photocatalysis and NO_2 gas sensing. J. Mater. Chem. 22 (2012) , Issue 17, pp. 8525/31.

Enhanced corrosion resistance of magnesium alloys by transplantation of thermally sprayed coatings

M. Rodriguez Diaz[1], P. Knödler[1], M. Otten[1], K. Möhwald[1], D. Freiburg[2], P. Kersting[2], D. Biermann[2],

[1] Institut für Werkstoffkunde, Leibniz Universität Hannover, 30823 Garbsen, Hannover, Germany
[2] Institut für Spanende Fertigung, Technische Universität Dortmund, 44227 Dortmund, Germany

Despite their excellent specific mechanical properties magnesium-based alloys are not widely used in the industry due to their high affinity to oxygen. Given the need for lightweight design, there are increasing efforts to replace high density materials by magnesium. One way to cope with the high oxygen affinity of magnesium is the use of thermally sprayed anti-corrosion coatings. However, conventional thermal coating processes have various process-related limitations. A case in point is coating of complex geometries and internal coatings with small diameters that often can not be realized by conventional processes. Due to the changed process order some of the limitations of conventional coating methods can be resolved by the transplantation of thermally sprayed coatings. This method is a composite casting process for the coating of die cast components, where the thermally sprayed coating is applied to the corresponding area of the mold prior to the casting process.
The aim of this study is to compare the effectiveness of transplanted thermally sprayed coatings with corrosion protection properties to conventional coatings deposited by thermal spraying and to discuss the ramifications with respect to industrial applications.

1 Introduction

The developments of recent years have led to an increasing demand for lightweight materials. These materials should not only feature low density, but also exhibit competitive chemical and mechanical properties as compared to the widely used conventional structural materials. Despite its outstanding lightweight properties magnesium has hardly attracted attention in the past, due to its high affinity for oxygen in comparison to other materials [1]. Recent research, however, has shown the potential to compensate the disadvantage of the high reactivity and to exploit the positive properties of the magnesium. An already standardized process is the application of wear and corrosion protection coatings by thermal spraying. However, the conventional thermal coating methods have substantial process-related limitations. In particular, coatings of complex component geometries or boreholes with small diameters often can not be realized or are associated with increased finishing work on the corresponding surfaces [2, 3,].
Knödler et al. have recently demonstrated that the process of transplantation of thermally sprayed coatings has the potential to circumvent these limitations [4, 5]. Transplantation is a composite casting process for the coating of diecast components in which the coating is applied before the actual component is produced. Subsequently it is transferred to the surface of the casting during the casting process. In the course of the diecasting process the melt infiltrates the porous coating and thus creates a form-locked connection. As a result of the changed process chain, some of the process limitations of the conventional thermal spraying can thus be avoided. However, detailed knowledge whether the peculiarities of the transplantation process are also suitable to ensure a satisfactory corrosion protection is still limited. The objective of this study was therefore to generate transplanted coatings with corrosion protection properties that meet industrial requirements. Their effectiveness in comparison to conventional thermally sprayed coatings was analysed and evaluated. Specifically, CuAl corrosion protection coatings were applied to magnesium die-castings made of the AZ91 magnesium alloy using both, conventional coating deposition and coating transplantation. Their corrosion protection properties as a function of infiltration depth were characterized and compared with conventionally applied coatings.

2 Experimental procedures

The knowledge gained in previous experiments on transplantation of coatings indicates that surface roughness provided by standard finishing milling processes is not sufficient to ensure adequate adhesion on flat specimens while gritblasting processes do not ensure a destruction-free demolding [4]. Therefore an appropriate structuring process had to be integrated into the process chain. In this study a structure was introduced into the flat coating surface of the mold insert by means of high-feed milling with a toroidal cutter of 10 mm diameter, 2 mm corner radius, at a cutting speed of 110 m/min and a rotational speed of 3500 min^{-1}, **Fig. 1**.

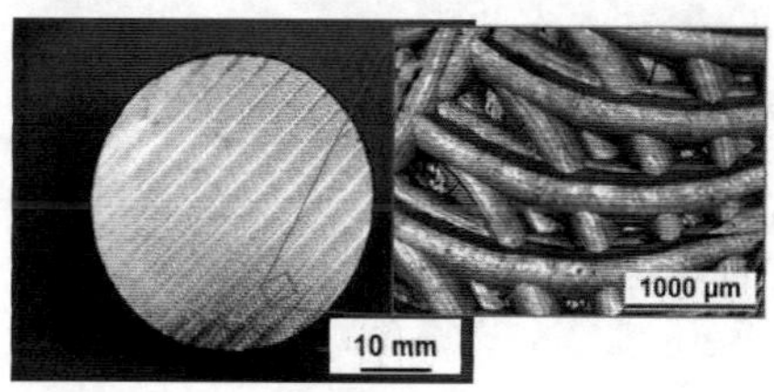

Fig. 1. High-feed-milling structure used on the mold to provide for tailored adhesion of the coatings

Subsequently, CuAl coatings were deposited onto the structured mold inserts by an atmospheric plasma spraying process. In addition to homogeneous single layer (sl) coatings (coatings #1, #2), multilayer (ml) systems consisting of several different coatings with regard to their porosities were applied (coatings #3, #4). The idea behind the application of multilayer systems was to tailor the infiltration depth and, thus, allow for control of the functional layer thickness. In order to vary the coating porosities, two different CuAl fractionations were used: a coarse (c) fractionation with -106 +45 µm and a fine (f) fractionation with -53 +5 µm. In order to compare the properties of transplanted coatings with conventional (con) coatings, the magnesium alloy was also coated conventionally with the coarser CuAl powder (coating #5). The coating systems studed are listed in **Tab. 1**.

Tab. 1. Coated CuAl-systems employed in the present, see main text for details

Coating no.	1	2	3	4	5
Material	CuAl	CuAl	CuAl	CuAl	CuAl
Coating type	sl	sl	ml	ml	Con
Fraction	c	f	f c	c f c	C
Morphology (schematical)					
Coating thickness, µm	400	400	400	400	400
Proportionate coating thickness, µm	400	400	250 150	100 150 150	400

In the actual prozess the coatings 1 – 4 were transplanted onto AZ91 die-cast panels in a cold chamber die-casting process, **Fig. 2**.

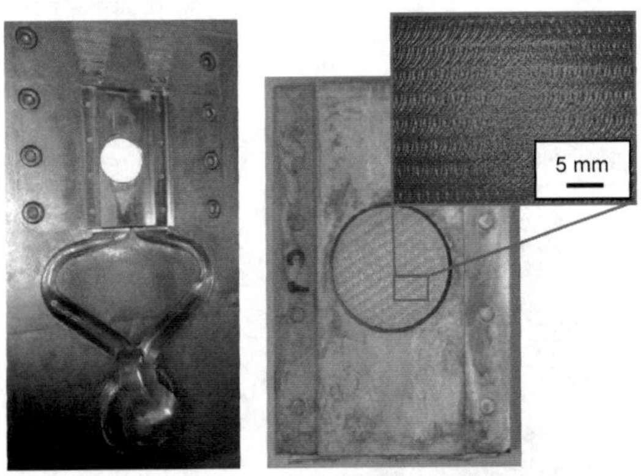

Fig. 2. AZ91 die-cast panel with transplanted coating

The infiltration depths were measured using metallographic cross-sections, while the corrosion protection properties of the coatings on the AZ91 components were determined by electrochemical current density-potential measurements, using a standard three-electrode arrangement. The electrolyte was a 3 wt.-% NaCl solution at 21 °C. As counter and reference electrode a platinum and a silver-silver chloride electrode were used respectively. The reference electrode had a potential of +197.6 mV against the standard hydrogen electrode. The voltage range in the tests was between -1600 mV and 1000 mV, whereby the potential change rates used during measurements of the density potential curves (DPC) were 10 mV / s, 5 mV / s and 1 mV / s.

3 Results

3.1 Porosities and infiltration depths

By varying the powder fractionations and the coating parameters, it was possible to influence the coating porosity. The optical analysis revealed that the pores of the coarser CuAl fractionation were relatively large, spherical and approximately uniformly distributed over the entire coating. On the other hand, the pores of the finer CuAl fractionation were smaller and had a lamellar form. Based on the metallographic analysis of the coatings, the infiltration depth of the melt and the porosities could be correlated, **Fig. 3**.

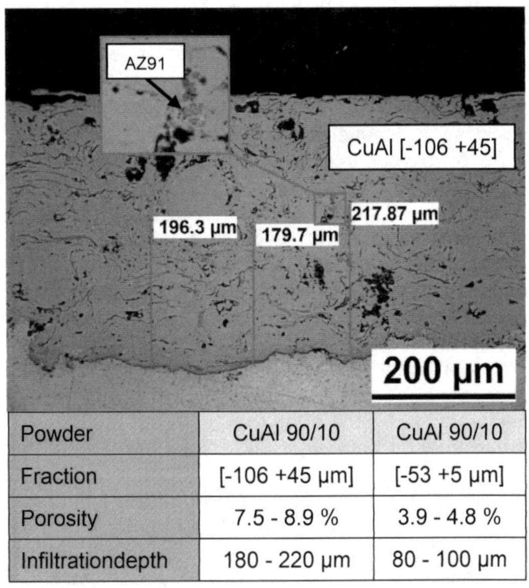

Powder	CuAl 90/10	CuAl 90/10
Fraction	[-106 +45 µm]	[-53 +5 µm]
Porosity	7.5 - 8.9 %	3.9 - 4.8 %
Infiltrationdepth	180 - 220 µm	80 - 100 µm

Fig. 3. Infiltration depth as a function of coating porosity

Starting from a total coating thickness of 400 µm, the functional layer changes its properties with increasing infiltration depth. Specifically this has a negative effect on the corrosion protection properties of the coating [6]. In order provide for enhance corrosion protection, the control of infiltration depth by deposition of multilayer coatings appeared as viable approach. In fact the analysis of the metallographic cross-sections revealed that the infiltration can be completely inhibited at the specific interfaces in multi-coatings layers, cf. **Fig. 4**.

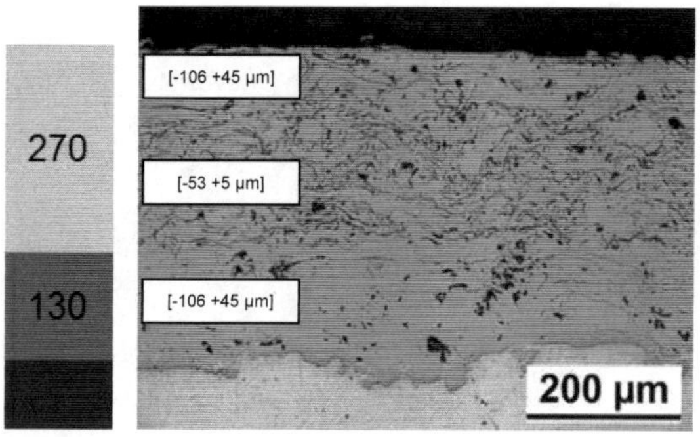

Fig. 4. Infiltration of a CuAl tri-layer system

As summarized, the infiltration depth and the functional layer thickness can be adjusted over a wide range using this approach, **Tab. 2.**

Tab. 2. Infiltration depth and resulting functional layer thickness, where coatings #3 and #4 represent multilayer systems

Coating no.	1	2	3	4
Max. infiltration depth, µm	220	100	150	130
Min. functional layer thickness, µm	180	300	250	270

A change in the coating porosity and morphology leads to an influence on the infiltration depth. However, it should be noted that the infiltration of the coatings does not change in a linear fashion with the porosities. It has been found that the infiltration capacity depends primarily on the prevailing surface energies, the spatial expansions of the pores, the interactions between the surfaces of the pores and the melt, and the capillary effects. The wetting behaviour between the partners, as well as the oxide content of the layer, have also a substancial influence on the infiltration capacity of a melt.

3.2 Corrosion test

For an overview of the corrosion protection effect of thermally sprayed coatings, the DPC of an uncoated AZ91 sample is presented along with the one obtained on the transplanted coating #1, **Fig. 5**.

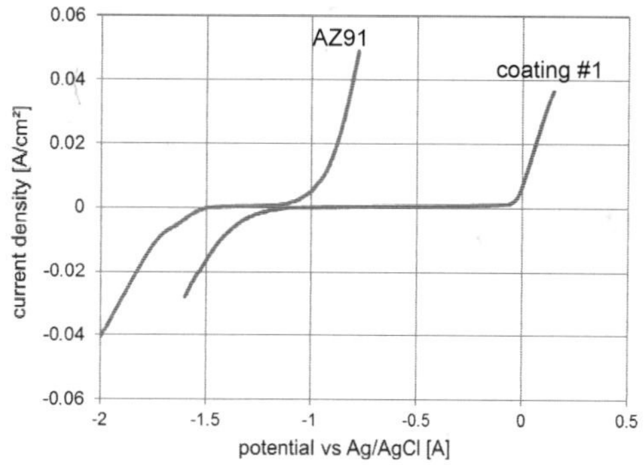

Fig. 5. DPCs of AZ91 and CuAl coating

The DPCs of the other coatings are not displayed for the sake of clarity, since the differences are not apparent at this scale. For the DSCs shown in Fig.5 voltage scan rate was 5 mV/s. By graphical evaluation of the recorded DPC, a corrosion potential of U_K = -1.49 V and a breakdown potential of U_D = 1.1 V were obtained for the AZ91 alloy. As the experiments were startet at a negative potential sunstancial amount of hydrogen gas was formed during the initial part of the scan. Part oft he hydrogen can also react to magnesium hydride and magnesium hydroxide is formed by the contact with water. After reaching the breakdown potential U_D is reached, the metal dissolves rapidly with initial formation of local corrosion holes.

The free corrosion potentials and breakthrough potentials of the transplanted and conventionally applied CuAl coatings as determined from the recorded DPCs are summarized, **Tab. 3**.

Tab. 3. Corrosion and breakdown potentials

Coating no.	1	2	3	4	5
U_K, V	-1.07	-0.74	-1.03	-0.93	-0.87
U_D, V	-0.025	0.002	-0.008	-0.002	-0.03

The starting potential U_S of all coatings was -1.6 V. All the DSCs for the coatings showed only small variations between similary deposited coatings, cf. **Tab 3**. The scan rate of the evaluated DPCs is 5 mV / s. As compared to the AZ91 alloy there is a strongest attenuation of the metal dissolution. Furthermore, the evaluation of the DPCs reveals that, compared to the conventionally applied coatings, all DPCs of transplanted coatings had a flatter curve progression and thus a slower metal dissolution. The anodic branches of the DPC of the conventionally applied coating #5 and of the morphologically identical coating #1 featured nearly identical characteristics. The passive current densities of all CuAl layers differ, in contrast to their transpassive current densities, not significantly, whereas the tri-layer coating 4 has the highest passive current density. On the other hand, the conventionally applied coating #5 has the lowest passive current density. On the basis of the DPCs, an influence of the porosities on the corrosion protection effect can be recognized: the coatings with a higher proportion of finer powder show a breakthrough potential shifted in the anodic direction. In this case, the DPC shifts are a function of the denser layer. Thus, the coating #2, which consists entirely of the finer powder fractionation, has the highest breakdown potential of all the investigated coatings, while the coatings from the coarser powder, and thus a higher porosity, show the strongest anodic reaction. However, the evaluation of the DPCs also shows a hazard potential of the investigated coatings: the CuAl-DPCs are clearly shiftet in relation to the AZ91-DPCs. Although this is an indication of the corrosion protection potential of the coatings, it also bears witness to the potential danger of contact corrosion.

4 Conclusion

In addition to the targeted control of the infiltration depth, the corrosion protection potential of transplated coatings could be demonstrated. It was thus found that the corrosion protection properties of a coating depends on the coating material and morphology. No restrictions with regard to the corrosion protection effect are to be expected by changing the process sequence, due to the transplantation process.

5 Literature

[1] Kainer, K. U. and H. Dieringa: Technologische Eigenschaften und Potential von Magnesiumlegierungen. Gießereirundschau 56, 7/8 (2009), pp. 114/119.

[2] Flores G. and A. Schwenk, C. Schnell: Process chain for the manufacture of thermally coated cylinder bores, VDI Berichte 2109 (2010), pp 191/202.

[3] Krebs, B.: Konturgenaue Bauteilbeschichtung für den Verschleißschutz mittels atmosphärischen Plasmaspritzens und Lichtbogenspritzens. Werkstofftechnologische Schriftenreihe Bd. 4. Vulkan-Verlag, (2011)

[4] Knödler, P. and M. Otten, K. Möhwald, H. J. Maier, D. Freiburg, A. Peuker, D. Biermann: Transplantation von thermisch gespritzten Schichten, Thermal Spray Bulletin 8, (2015), pp. 50/55.

[5] Bauer, A.F.: Transplant Coated Aluminum Cylinder Bores – Physical Properties of this New Protective Coating in Comparison to Known Processes, Society of Automotive Engineers - Summer Meeting (1961)

[6] Carmen Parco Camacaro, M.: Thermische Beschichtung von Leichtbaulegierungen aus Magnesium zum Verschleiß- und Korrosionsschutz, Schriftenreihe Oberflächentechnik 2, dissertation Aachen (2005)

Fe-based powder alloys deposited by HVOF and HVAF spraying – a salt spray test ranking

S. Dizdar, Höganäs AB, S-263 83 Höganäs, Sweden,

Three Fe-based powder alloys, Höganäs Fe SP529, Fe SP586 and 6AB, have been deposited by HVOF and HVAF spraying onto mild steel plates. The sprayed samples were first ground and then shot peened using glass shot in order to seal the surface interconnected pores and other surface imperfections. The samples as ground and ground/glass shot peened were tested by salt spray (fog) exposition for 238 h according to ASTM B117/ISO 9227. FeSP586, HVOF and HVAF sprayed and glass shot peening samples achieved surface sealing enough to pass the test with appearance rating R_A = 9 according to ISO 10289. All other samples achieved moderate to excessive pitting and/or moderate to excessive staining types of corrosion defects.

1 Introduction

HVOF and HVAF thermal spray techniques achieve high-density carbide and metal coatings for applications in aerospace, pulp and paper, process and other industries [1]. Normally, the coatings achieve pore content less than 5%. Compared to flame sprayed/fused coatings with a common pore content 10 to 20%, this is a large improvement. However, surface related component performances such as fatigue, wear and corrosion, are significantly reduced by presence of surface interconnected pores and other surface flaws such as poorly bonded particles and splats in the last-sprayed coating layer.

Elimination or substantial reduction of the surface interconnected pores and other surface flaws, i.e. surface sealing can be achieved by techniques such as fusing, selective surface re-melting, chip machining - finishing techniques, shot peening, surface rolling and sealing [1]. Choice of a suitable surface sealing technique depends on a component's particularities such as overall size, geometry and material. The present investigation will focus on shot peening as a potential process for surface sealing of HVOF/HVAF sprayed components.

Shot peening (SP), also called blasting, is a cold working mechanical process which plastically deforms a component's metallic surface, produces a compressive residual stress layer and improve surface related performances of the component. It entails repeated impacting of a surface with mostly round shot particles made of metal, glass, or ceramic [2]. Shot peening finds many applications, from surface cleaning from rust using steel or alumina shot, surface roughening prior to thermal spraying by using steel or alumina shot, to a gentle glass shot peeing for improvement of the surface aesthetic appearance by imparting matt finish. The aesthetic peening is the gentlest peening, it causes only a very shallow plastic deformation of the surface but change the surface topography. Due to these reasons, it is interesting to understand if the glass shot peening (GSP) could seal surface of thin porous HVOF/HVAF sprayed Fe-based coatings.

A testing method which could be useful to investigate how the glass shot peening seals the HVOF/HVAF sprayed surface is salt spray (fog) testing. The testing was first standardised by ASTM in 1939 [3] and is today widely used corrosion testing technique for stainless steels, paintings, coatings and other applications. Accordingly, there are many variants of salt spray testing developed by for example automotive and paint companies. However, the salt spray testing, as all other testing methods, must be used with understanding. International stainless steel forum reviews benefits and drawbacks with the salt spray testing [4]. Briefly summarized, the testing is suitable for evaluation of a component surface homogeneity depending of the surface manufacturing/finishing technology as well as leakage in mechanical assemblies. The testing causes for sure surface corrosion, often pitting, but is not accelerated enough for purpose of development of new stainless steel and corrosion resistant materials. For this purpose, there are other, more accelerated corrosion testing standards and methods.

For the present investigation, the salt spray test will be used to evaluate quality of the sprayed coating surface i.e. presence of interconnected surface pores and flaws, cracks possibly interconnected with the base metal, coverage of chromium oxide passive layer and other surface imperfection which affect corrosion nucleation and growth. The preferred sites for corrosion nucleation and growth were surface defects such as pores and cracks [5, 6].

The aim of investigation presented in this paper is to perform a salt spray test ranking of samples made by three HVOF and HVAF spraying by using Höganäs Fe-based alloys, FeSP529, FeSP586 and 6AB, in the as ground and ground/GSP conditions. The ranking is to be performed according to ASTM B117 / ISO 9227 and evaluated according to ISO 10289 [7]. A deeper analysis of the corrosion types is to be left for future investigations.

2 Method

Shot peening (SP) is performed in well-sealed chamber (**Figure 1**). Peening shot is picked-up by pressurized air by using simple Venturi-type pump from a sand collector in the chamber bottom. The pressurized sand flow proceed through a hose until a peening gun with ceramic nozzle. The beam from the gun hits the sample surface with a target area coverage, often multiply of 100%. The peened surface changes its light reflection which is quite easy to notify for a first peening coverage but not for the successive ones. Properly performed peening results in plasticised, orange peel like surface with compressive residual stresses close to the surface. Intensity of shot peening can also be specified by Almen intensity [2].

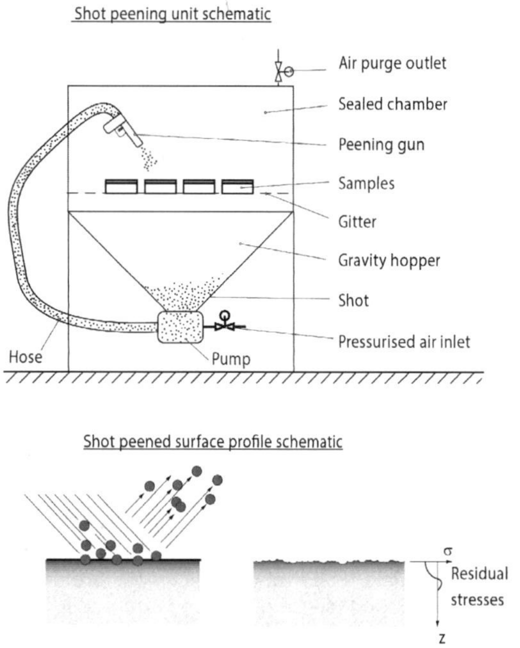

Figure 1. Schematic illustration of shot-peening process.

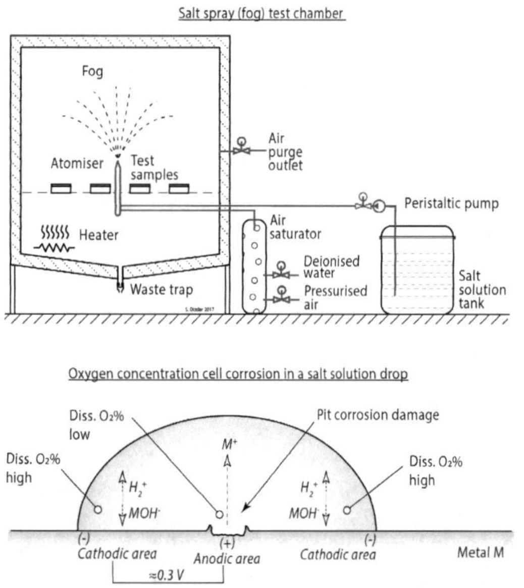

Figure 2. Schematic of salt spray (fog) testing

Salt spray (fog) testing according to ASTM B117 (eq. to ISO 9227) can be briefly described as it follows. The test specimens are placed in an enclosed climate chamber (**Figure 2**) and exposed to a continuous indirect spray of neutral (pH 6.5 to 7.2) 5±1 mass % salt-water solution. The sprayed solution drops down onto the specimens at a rate of 1.0 to 2.0 ml / 80 cm^2 / h under chamber temperature of 35±2°C. The test duration is variable, but normally multiplication of 24 h. The test is also referred to as a neutral salt spray or NSS test. The test theoretical background is oxygen concentration cell corrosion in a salt solution drop [9].

Rating of the salt spray test results can be based on methodology presented in ISO 10289 including ASTM dot charts. However, the investigator have to take care about judgment of the results. The reasons are complexity of the corrosion process and common corrosion types and modelling of the accelerated testing.

3 Experiment

The iron-based powder grades used for HVOF/HVAF spraying in the present investigation are Höganäs FeSP629, FeSP586 and 6AB. The HVOF/HVAF deposition was performed at the same time as for the samples in investigation reported in [12,13,14]. Fe-based thermal surfacing powder alloys can offer various degrees of resistance to wear and corrosion for applications exposed to low to moderately high temperatures. [9, 10]. Table 1 lists the typical chemical composition for the three Fe-based powder alloys with HVOF/HVAF sieve cuts, while Table 2 lists outcome of Thermocalc simulations showing the present phases. In all cases except one austenite is present as the ductile matrix phase while carbides and borides are the reinforcing phases. FeSP529 exhibits a very high amount of hard phases with 66.2 at.% carbides and borides with the remainder mainly austenite and some ferrite. FeSP586 is more ductile with 60,7 at.% ductile phase and the remainder carbides and borides. In 6AB the ductile phase is ferrite with some austenite, the hard phase is a mixture of carbides and borides. Some graphite may be precipitated in this material.

Test specimens were manufactured using HVOF/HVAF spraying onto alumina shot peened 8 mm thick plane blanks made of structural steel EN S355 MC, followed by water jet cutting to 25x58 mm coupons and plane grinding in a knife grinding machine. The quality of sprayed samples is confirmed with XRD analysis in **Figure 4**, microstructure cross cut of the samples is shown in **Figure 3**, and surface roughness listed in Table 3. The microstructure in **Figure 3**, shows typical voids and oxide stringers for HVOF Fe SP529 and 6AB samples. HVAF samples of the same powder show the same features but to a lower extent. Samples of Fe SP586 powder, show the best overall microstructure with lowest content of voids and oxide stringers. XRD analysis in **Figure 4** shows matrix

The samples were glass shot peened in a commercial peening unit with a 120 l chamber, 7 bar workshop air and a peening gun with 5 mm ceramic nozzle. The shot were glass balls with sieve cut 0.25 to 0.42 mm. The samples were peened with 200% coverage, cleaned and painted with an alkyd oil based wear and weather resistant paint for metals (Quick Bengalack) to corrosion protect the faces with base metal and sharp edges. A sample face mid-section of ca. 12 x 25 mm of the coated 25x58 mm was left for the testing.

Table 1. Typical composition of Fe-based powder grades

Designation		FeSP529	FeSP586	6AB
Elemental composition (wt. %)	C	1.9	0.7	2.1
	Mo	5.8	-	-
	Cu	1.5	-	-
	Ni	4.2	12.3	5.9
	V	0.2		
	Fe (bal)	(70.7)	(52.3)	(70.9)
	Mn	0.2	-	0.3
	Cr	9.7	30.6	14.4
	Si	2.1	0.6	3
	B	3.7	3.5	3.4
Sieve cut: HVOF-DJH, -45+20µm, HVAF-M3, -36+20µm				

Table 2. Outcome of Thermocalc simulation, at.%

	FeSP529	FeSP586	6AB
Austenite	26,1	60,7	6,8
Ferrite	6,2		28,5
M_7C_3	21,9	7,4	24,8
MoB	3,2		
Cr_5B_3		3,2	
CrB	0,7	27,4	5,3
M_2B	40,4		33,5
Graphite			0,5
Total	98,5	98,7	99,4

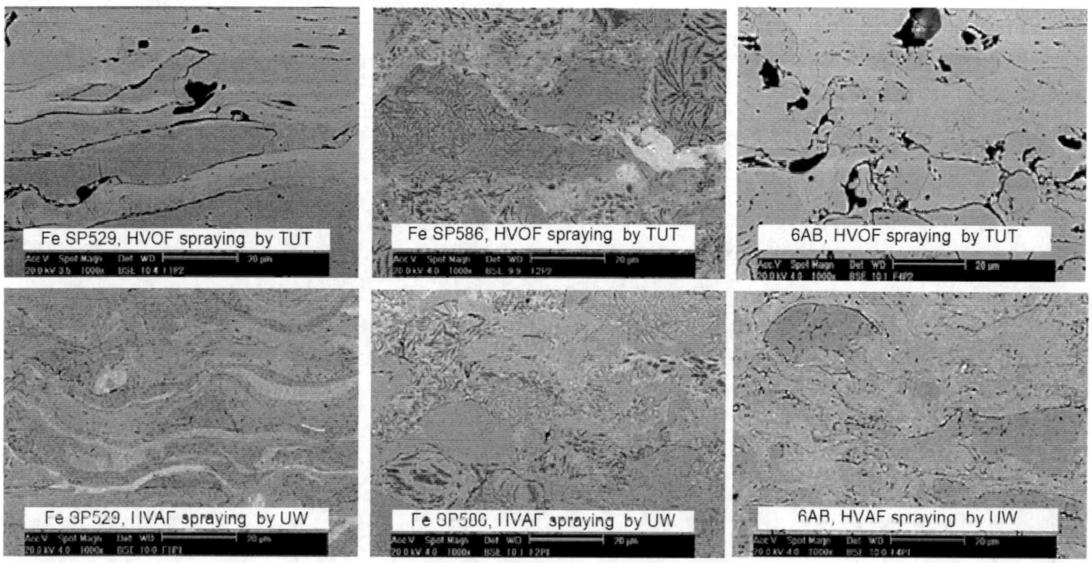

Figure 3. Typical SEM BSE microstructure cross cut photographs of test samples. HVOF spraying courtesy of Tampere University of Technology, Finland, (TUT) and HVAF spraying courtesy of University West, Sweden, (UW)[12]

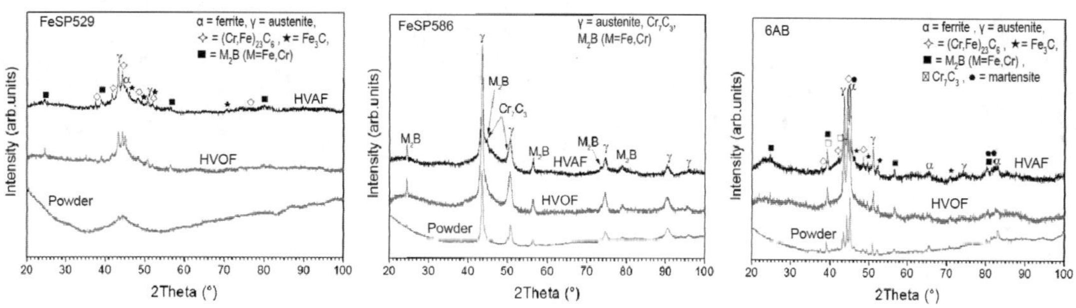

Figure 4. XRD analysis of the powder and sprayed specimens [].

4 Results

The test samples were exposed to salt spray (fog) conditions according to ASTM B117/ISO9227 for 235 hours or almost 10 days. Table 3 shows a chronological description of the appearance the test samples.

Figure 5 shows appearance of the selected test samples after 235 h while Figure 6 shows SEM photos and surface roughness profile of the samples.

The most evident result is moderate to excessive pitting and staining on all samples except two FeSP586

GSP ones. FeSP529, HVOF and FeSP529, HVAF, GSP samples achieved pitting but no respective some rust. Except for FeSP529, HVOF, GSP; all shot peened samples achieve less number and size of the pits.

Table 3. Chronological description of the appearance the test samples exposed to salt spray testing in a corrosion test chamber (Ascott CC450XP) according to ASTM B117/ISO 9227. Classification of the coating deterioration according ISO 10289.

Sample	Surface appearance for respective control stops		
	67 h / ≈3 days	118 h / ≈5 days	235 h / ≈10 days
Fe SP529, HVOF	M pitting, m staining	M pitting, m staining	M pitting, m staining
Fe SP529, HVOF, GSP	M pitting, x staining	X pitting, x staining	X pitting, x staining
Fe SP529, HVAF	X staining	X staining	M pitting, x staining
Fe SP529, HVAF, GSP	S pitting, s staining, s dulling	S pitting, m staining, s dulling	M pitting, m staining
Fe SP586, HVOF	S pitting, m staining, m dulling	M pitting, m staining, m dulling	M pitting, x staining
Fe SP586, HVOF, GSP	S dulling	S dulling	S dulling / R_A = 9
Fe SP586; HVAF	M pitting, x staining	M pitting, x staining	X pitting, x staining
Fe SP586, HVAF, GSP	Vs pitting, s dulling	Vs pitting, s dulling	Vs pitting, s dulling / R_A = 9
6AB; HVOF	X pitting, m staining	X pitting, m staining	X pitting, x staining
6AB, HVOF, GSP	S pitting, m staining	S pitting, x staining	S pitting, x staining
6AB; HVAF	Pitting, some rust	M pitting, m staining	M pitting, x staining
6AB, HVAF, GSP	M pitting, x staining	M pitting, x staining	M pitting, x staining
EHC	M pitting, s dulling	M pitting, x staining	M pitting, x staining
ISO 10289: vs. = very slight amount, s = slight amount; m = moderate amount, x = excessive amount			

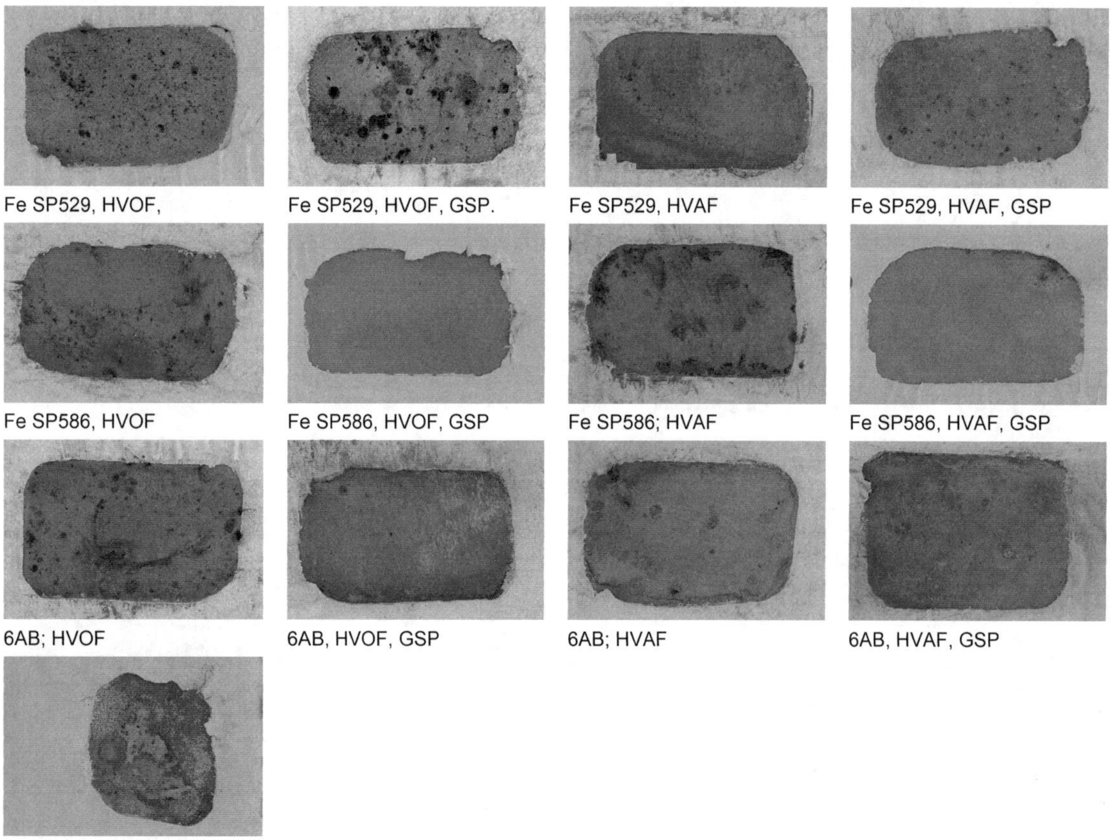

Fe SP529, HVOF, | Fe SP529, HVOF, GSP. | Fe SP529, HVAF | Fe SP529, HVAF, GSP

Fe SP586, HVOF | Fe SP586, HVOF, GSP | Fe SP586; HVAF | Fe SP586, HVAF, GSP

6AB; HVOF | 6AB, HVOF, GSP | 6AB; HVAF | 6AB, HVAF, GSP

EHC plating reference

Figure 5. Appearance of the test samples after 235 h ASTM B117 salt spray (fog) testing.

FeSP586, HVOF, Ground, 238 h

FeSP586, HVAF, Ground, 238 h

FeSP586, HVOF, Ground, GSP, 238 h

FeSP586, HVAF, Ground, GSP, 238 h

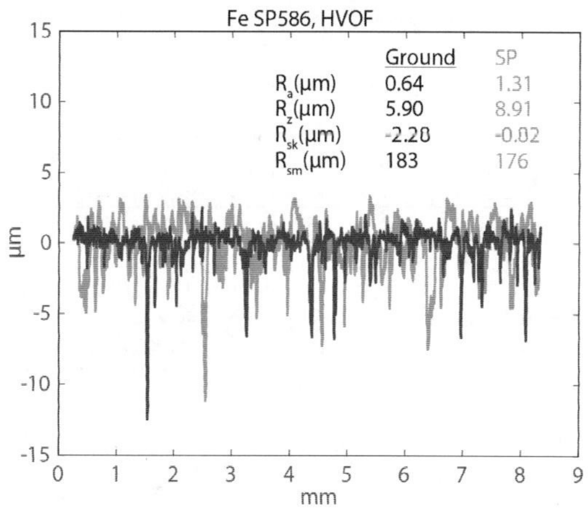

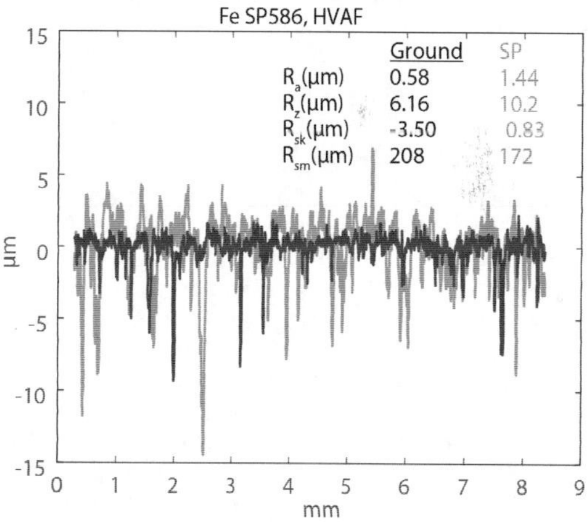

Surface profiler (Talysurf Intra), 60 mm probe with R=2 µm stylus, ISO 4287 - 2D roughness paramaters, cut-off 0.8 mm.

Figure 6. SEM photos and roughness profiles of the sample surfaces after 238 h ASTM B117 salt spray testing.

Chronologically, the pitting corrosion increases with the exposition time. Staining coverage increases also but some samples such as Fe SP529, HVOF, GSP; Fe SP586; HVAF and 6AB, HVAF, GSP reach excessive staining before 67 h of the exposition time.

Reference EHC plated samples achieved moderate pitting, and excessive staining after 67 h.

GSP is evidently a strong factor affecting pitting and staining. Figure 6 show surface roughness profile and some selected 2D surface roughness parameters for samples FeSP586, HVOF/HVAF as ground and GSP. The ground surface has 0.6 µm in average roughness R_a, 6 µm in ten-point-height R_z, 200 µm in mean spacing R_{sm} but -2,28 respective -3,5 in skewness R_{sk} as a measure of the asymmetry of the profile amplitude density curve. That negative skewness shows over-representation of the surface pores in the ground samples. In contrast, the GSP surface profile has -0.8 in skewness what places the GSP surface closer to a neutral surface with zero skewness and equal representation of surface pores / walleys and surface ridges. With other words, number of surface pores/valleys is much less in comparison to the ground surface.

A complemental description of the ground and GSP surfaces can be seen in Figure 6. The SEM photos of ground FeSP586 HVOF/HVAF surfaces show surface fragment with presence of both grinding marks, surface pores and staining. In contrast, photos of FeSP586 HVOF GSP surface show orange peel like surface with roundish multiple indentation marks after glass shot. The FeSP586 HVAF GSP surface has the same appearance as the FeSP586 HVOF ones, but the darker partition represent dulling mentioned in Table 3 and seen in **Figure 5**. Also, the dulled partition does not differ in surface roughness in comparison to the non-dulled one. The likely explanation is formation of a very thin chromium oxide protecting layer.

Another strong factor is material microstructure of the HVOF/HVAF coatings. Alloy FeSp586 with austenitic work harden able matrix, lowest content of hard phase, lowest carbon and highest both chromium and nickel content achieves highest corrosion resistance in the investigation. A local pitting on the HVAF, GSP sample is likely caused by uneven GSP coverage.

FeSP529 alloy is the alloy close to metallic glass alloys see smooth curve in Figure 4, with austenitic-ferritic matrix but highest content of hard phase. Pitting occurred clearly on all samples, but for the HVOF samples rust occurred intensively on the shot peened ones but not on the ground ones. For HVAF samples, the opposite occurred. 6AB samples achieved pitting and staining with minor variation.

5 Discussion

The combination of austenitic matrix with work hardening ability and GSP resulted in effective pore sealing which prevented pitting and staining in the salt spray test for 238 hours. This illustrates the importance of pore sealing for HVOF/HVAF sprayed components. GSP is demonstrated here as a suitable pore sealing technology. An optimisation of pore sealing together with a successive polishing/honing can lead to HVOF/HVAF components suitable for marine and other applications with salt exposition.

6 Conclusions

Three Fe-based powder alloys, Höganäs Fe SP529, Fe SP586 and 6AB, have been deposited by HVOF and HVAF spraying onto mild steel plates.

- FeSP586, HVOF and HVAF sprayed and glass shot peened samples achieved surface sealing enough to pass the ASTM B117 test for 238 with appearance rating R_A=9 according to ISO 10289.

- The FeSP529 samples show different behavior in comparison to other samples sprayed with Fe-based grades. The reason can be closeness to metallic glass microstructure.

- The 6A samples showed moderate to excessice staining for all conditions

7 Acknowledgements

The author acknowledges Thermocalc calculations by Dr. Sven Bengtsson and extensive laboratory work by Mr. Henrik Holmqvist.

8 Literature

1. P.L. Fauchais, J.V.R. Heberlein, M.I. Boulos, *Thermal spray fundamentals – From powder to part*, Springer, Berlin,2014.
2. A. Momber, *Blast cleaning technology*, Springer, Berlin, 2008,
3. ASTM B117, Standard Practice for Operating Salt Spray (Fog) Apparatus; ASTM Int., West Conshohocken, USA
4. The salt spray test and its use in ranking stainless steels – The test and its limits, ISSF, May 2008, Brussels, Belgium.
5. Sadeghimeresht, E, Markocsan, N., Nylén, P., Microstructural characteristics and corrosion behaviour of HVAF- and HVOF-sprayed Fe-based coatings, *Surf. Coat. Techn.*, In press.
6. Beardsley, M B, Iron-Based Amorphous Coatings Produced by HVOF Thermal Spray Processing-Coating Structure and Properties, *Metall. Mater. Trans.*, **40A** (2009) 1-20.
7. ISO 10289, Methods for corrosion testing of metallic and other inorganic coatings on metallic substrates -- Rating of test specimens and manufactured articles subjected to corrosion tests, ISO, Geneva, Switzerland.
8. McCafferty, E., *Introduction to corrosion science*, Springer, Berlin, 2010
9. Kotecki, D.J., Ogborn, J.S., Abrasion Resistance of Iron-Based Hardfacing Alloys, *Weld. Res. Suppl.*, **74** (8) 995, pp. 265/278
10. EN 14700: Welding consumables - Welding consumables for hard facing, August 2005.
11. R. González, M. Cadenas, R. Fernández, J.L. Cortizo, E. Rodríguez, Wear behaviour of flame sprayed NiCrBSi coating remelted by flame or by laser, *Wear* **262** (2007) 301–307.
12. Dizdar, S, Lyphout, C., Milanti, A., Koivuluoto H., Vuoristo P., Fe-based Powder Alloys Deposited by HVOF and HVAF for Abrasive Wear Applications, *Int. Therm. Spray Conf.*, 21-23 May 2014, Barcelona, Spain
13. Dizdar, S., Kumar, M., Fe-based Powder Alloys Deposited by HVOF and HVAF for Applications Exposed to Solid Particle Erosion, *Int. Therm. Spray Conf.*, 11-14 May 2015, Long Beach CA, USA.
14. Dizdar, S. Fe-based Powder Alloys Deposited by HVOF and HVAF for Sliding Wear Applications, *Int. Therm. Spray Conf.*, 10-12 May 2016, Shanghai, P.R. of China

Comparative Study of the Corrosion and Cavitation Resistance of HVOF and HVAF FeCrMnSiNi and FeCrMnSiB Coatings

L. L. Silveira[1]*, A. G. M. Pukasiewicz[1], S.Björklund[2], P. Nylén[2], A. J. Zara[3]
*luciana.llsilveira@gmail.com
[1] Post-graduate Program in Mechanical Engineering, Federal University of Technology - Parana, Ponta Grossa, PR, Brazil.
[2] University West, Trollhättan, Sweden
[3] State University of Ponta Grossa, Ponta Grossa, PR, Brazil

Cavitation and corrosion on hydrodynamic components and systems reduces the operational efficiency. The use of wear resistant coatings have been studied as a solution to the problem of corrosion and cavitation in the industrial environment. Thermal spray processes are recognized as excellent technique to deposit coatings. The high velocity oxy-fuel process (HVOF) can produce high density and bond strength coatings. High velocity air-fuel process (HVAF) is an alternative process, shown to be superior regarding corrosion protection and production costs. HVAF can deposit coating with shorter dwell time and lower temperature, producing coating with lower oxide content. This paper presents the use of HVOF and HVAF process to deposit FeCrMnSiNi and FeCrMnSiB coatings, studying the resistance against corrosion and cavitation in comparison to 316L HVOF coating. Microstructure was analyzed by XRD, microscopic means and mechanical testing. Cavitation and corrosion behavior of the coatings were also studied comparatively. HVAF coatings presented lower porosity and oxide levels, as well as higher hardness values, compared with the coatings deposited by HVOF process. The HVAF process presented better cavitation resistance than HVOF coatings. The FeCrMnSiNi HVAF coating had the best corrosion protection performance between the developed alloys.

Key-words: Protective coatings. Thermal Spray. Fe-based alloy. Boron.

1 Introduction

The corrosion process combined with the cavitation process, are very common processes of material degradation in hydrodynamic systems, resulting in mechanical and economical losses. Cavitation erosion is related to mechanical damages, while corrosion is a electrochemical process. Both effects lead to stress concentration, plastic deformation, crack initiation and growth and material deterioration [1-4].

To decrease component degradation, thermal spray has been used to create coatings with high resistance to corrosion and cavitation processes. High performance alloys create coatings with good mechanical and chemical properties. Ni-based and Fe-based alloys are known to have good corrosion protection properties, and have been used by several studies as replacement coatings for different substrate materials. Fe-based alloys have superior cost advantages, as well as being environment friendly [5-7]. Addition of alloying elements such as Cr and Ni increases corrosion resistance of the Fe-based alloys, as well as their mechanical properties [6-9]. Boron addition promotes the formation of hard fine-grained precipitates [10,11], increasing coating hardness. High Si additions can increase the corrosion resistance by forming a SiO2 passive film, which occurs from the diffusion of Si to the surface of the coating [12].

The high-velocity-oxyfuel (HVOF) process produces coatings with high density and adhesion, but can present some disadvantages concerning corrosion resistance and residual stress formation [13-16]. The high velocity air-fuel (HVAF) is an alternative process for the deposition of Fe-based coatings. Operation costs are decreased by the use of compressed air instead of pure oxygen, producing high quality coatings with reduced oxidation of the sprayed particles [17,18]. HVAF sprayed coatings have shown superior corrosion resistance compared to HVOF coatings, due to lower oxidation levels [19, 20].

This paper studies the influence of different high velocity thermal spray process on the properties, morphology and resistance to corrosion and cavitation process of FeCrMnSiNi and FeCrMnSiB coatings. Properties, cavitation resistance and corrosion behavior of these alloys were compared to the commercial steel Diamalloy 1003 (316L), deposited by HVOF process.

2 Materials and Methods

The Fe-based FeCrMnSiNi and FeCrMnSiB alloy's powder, patent pending (Table 1) was prepared by gas atomization with argon (Clausthal University of Technology, Clausthal-Zellerfeld, Germany). The alloys were coated by the thermal spray processes HVOF and HVAF on SAE 1020 carbon steel plates, machined to final dimensions of 35 x 75 x 8 mm (L x W x H). The surface was prepared by abrasive blasting to produce a clean surface with adequate roughness. A Sulzer-Metco Diamond Jet 2700 commercial system with gas fuel (propylene) were used for the HVOF deposition by Revesteel, Pinhais, PR, Brazil, Table 2. A UNIQUECOAT M3 system with propane fuel was used for the HVAF deposition by University West in Trollhättan, Table 2.

Table 1. Chemical composition of the studied alloys, wt. %.

Alloy	Fe	C	Cr	Mn	Ni	Si	B	Mo
FeCrMnSiNi				FeCrMnSiNi				
FeCrMnSiB				FeCrMnSiB				
316L	Bal.	0.03	17	-	12.0	2.3	-	2.5

Table 2. HVOF and HVAF parameters

Parameter	HVOF	HVAF
Gun type	DJ-2700	M3
Oxygen flow rate, SLPM	265.4	--
Air flow rate, SLPM	317.0	--
Air pressure, bar	75	113
Carrier gas, SLPM	60	50
Propylene flow rate, SLPM	71.4	--
Propane flow rate, SLPM	--	(1) 97.0, (2) 95.0
Powder feeding rate, g/min	45.0	90.0
Spray distance, mm	200.0	300.0

Characterization of the coatings was performed by optical microscopy (OM, Zeiss A2.m) and scanning electron microscopy (SEM, TESCAN VEGA 3). For chemical composition analysis, energy dispersive spectroscopy (EDS, Oxford) measurements have been conducted. Using the cross-sectional images, porosity and oxide levels, as well as the presence of other defects, such as non-deformed particles, have been evaluated by image analysis (with Axio Vision software). ASTM E2109-01 Standard [21] was used to determine the porosity and oxide area fraction of the coatings. Mechanical properties were measured by Vickers microindentation hardness testing (Shimadzu HMV-G), following ASTM E 384-99 Standard [22] with a 300 gf of load and 15 seconds of indentation time.

Cavitation erosion test was performed with vibratory ultrasonic cavitation equipment (ASTM G32-03 Standard) [23], using indirect method. A peak-to-peak amplitude of 50 μm and frequency of 20 kHz, in a 19.5 mm of diameter horn, were used, and the distilled water was maintained at 20 ± 1 °C. The weight loss of the samples was measured during cavitation erosion tests. The cavitation erosion rate (CER) of the materials was determined after 90 minutes of cavitation test.

Potentiodynamic polarization resistance measurements were carried out in a Autolab Potentiostat, in accordance to ASTM G59-97 and G102-89 Standard [24, 25], with intend to determine corrosion current density, polarization resistance and corrosion rate of the coatings in a 3.5 wt.% NaCl solution at 25 °C. For corrosion tests, samples with 0.5 cm of diameter were used as working electrode. A calomel reference electrode and a platinum counter electrode were used in the tests.

3. Results and Discussion

3.1 Microstructure and Mechanical Properties of the Coatings

Figure 1 and 2 show the cross-sectional micrographs of HVAF and HVOF FeCrMnSiNi and FeCrMnSiB, respectively.

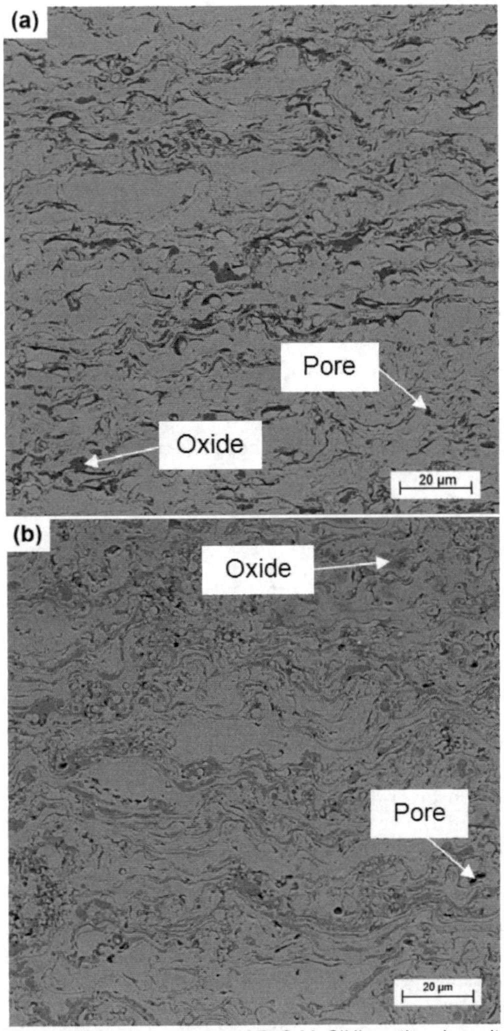

Fig. 1. SEM micrographs of FeCrMnSiNi coating deposited by (a) HVAF and (b) HVOF.

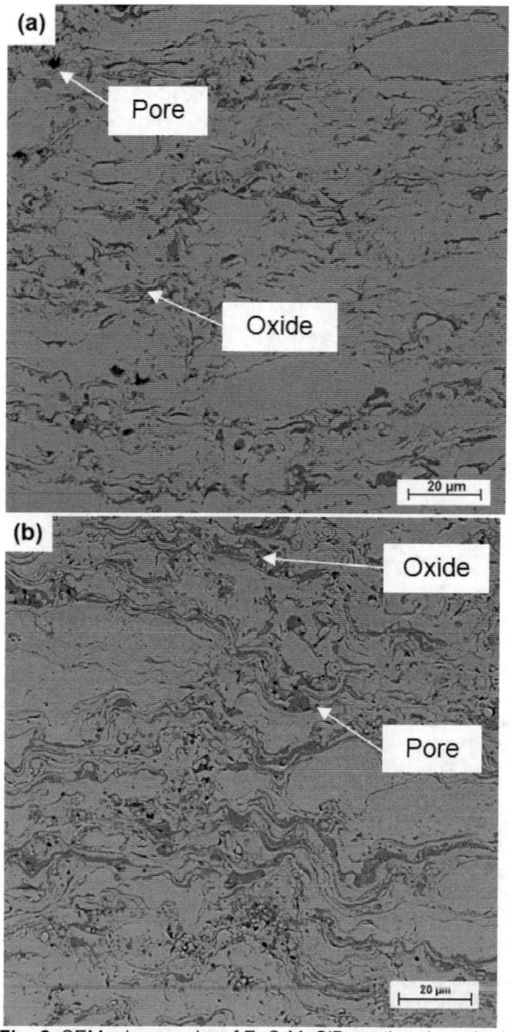

Fig. 2. SEM micrographs of FeCrMnSiB coating deposited by (a) HVAF and (b) HVOF.

The microstructure of the HVAF and HVOF coatings presented pores and oxides, as shown in Figures 1 and 2. Splats in HVAF coating showed a more deformed and more elongated structure as in the HVOF coatings, independently of the alloy. All alloys presented lower levels of porosity than 316L, with porosity inferior to 0.5% (Table 4). Low content of pores in these coatings suggest a more dense and homogeneous coatings than the 316L commercial alloy. The wetting improvement with Boron and Nickel addition was observed previously [26]. The improvement of the splats deformation can be occurred due to melting temperature reduction with Boron addition in iron alloys [27], and due to oxide thickness reduction with Nickel addition [28]. It is expected a higher corrosion resistance of the FeMnCrSiNi coating in comparison to that of the FeMnCrSiB due its Ni content.

HVOF coatings presented higher degree of oxidation with HVAF (Table 4), which can be related to the larger oxygen content due to its use as fuel for this process and higher dwell time of particles in the HVOF process. The HVAF process has lower flame temperatures compared to the HVOF process (due to the use of air instead of oxygen) leading to a lower level of oxidation, due to the limitation of particle thermal alteration [29-31].

The weight ratios of the elements of the sprayed material have slightly changed after the spraying process. Boron levels are not detectable by EDS analysis. A sufficient thickness of the coatings is essential to a effective corrosion protection [6], as the AISI 1020 substrate is less noble and has a lower resistance to corrosion as the sprayed material. The coatings presented a thickness of 200-300 µm after metallographic preparation.

Table 3. Calculated pore and oxide content and microhardness of the studied materials

Material	Porosity content (% area)	Oxide content (% area)	Vickers microhardness (HV)
FeCrMnSiNi HVAF	0.09 ± 0.03	18.6 ± 1.5	487 ± 26
FeCrMnSiNi HVOF	0.14 ± 0.09	27.9 ± 7.0	386 ± 25
FeCrMnSiB HVAF	0.14 ± 0.07	16.1 ± 1.7	535 ± 44
FeCrMnSiB HVOF	0.10 ± 0.10	29.5 ± 3.9	527 ± 33
316L HVOF	0.43 ± 0.27	12.6 ± 3.5	251 ± 38

The microhardness values of the HVAF coatings were higher than that of the HVOF coatings, this behavior was also observed in the literature [32]. This result can be correlated with the oxide content in the coatings, as for micro-hardness testing both inter-lamellar properties and lamellar cohesion have influence in the results [15, 33-34], and lower levels of oxide lead to a higher interlocking of particles and also higher mechanical properties. Additionally to the oxide content, HVAF process present higher particle velocities [32], which generates a stronger peening effect, increasing cold work and dislocation densities in the coating, resulting in higher hardness of the HVAF coating [35]. Both FeCrMnSiB coatings presented higher hardness values in comparison to FeCrMnSiNi, as expected by the B content in this alloy [27, 36-37].

3.2 Cavitation Resistance of the Coatings

The results of the cavitation tests for HVOF and HVAF coatings are listed on Table 5. These results were obtained from the cumulative weight loss curves (Fig. 3). As expected all coatings presented weight loss, even from the beginning of cavitation test. It was not observed incubation stage on the HVAF and HVOF coatings [38]. Thermally sprayed coatings have porosity, oxide and lamellae structure, which provides preferential locations for weight loss. The cavitation damage appears to be initiated near porosity, as observed by previous work [32] and has increased weight loss in oxide sites which present a lower adherence on the substrate.

The HVAF coatings presented lower erosion rates than the HVOF coatings, 1.97 mg/h compared to 3.76 mg/h for the FeCrMnSiNi alloy and 3.79 mg/h compared to 5.68 mg/h for the FeCrMnSiB alloy. The better cavitation resistance of the FeCrMnSiNi alloy is probable related to its Ni content, which for the FeCrMnSiB alloy was replaced with Boron, which did not have the same improvement in cavitation resistance. The 316L HVOF coating presented the highest erosion rate, with 9.44 mg/h. The HVAF has higher particle velocities compared to the HVOF process, which causes increase in the cohesive properties of sprayed layers [32], providing better cavitation resistance. The consequent peening effect and cold work of particles due to the higher velocities also have influence on the cavitation resistance of the coatings, by increasing the anchorage of particles and consequently improving the coating resistance to the cavitation damage.

Table 4. Cavitation erosion rate of specimens tested (from 90 to 240 minutes of test)

Sample	Cavitation erosion rate (mg/h)	Total mass loss (mg)
FeCrMnSiNi HVAF	1.97	15.0
FeCrMnSiNi HVOF	3.76	22.0
FeCrMnSiB HVAF	3.79	28.1
FeCrMnSiB HVOF	5.68	32.1
316L HVOF	9.44	47.9

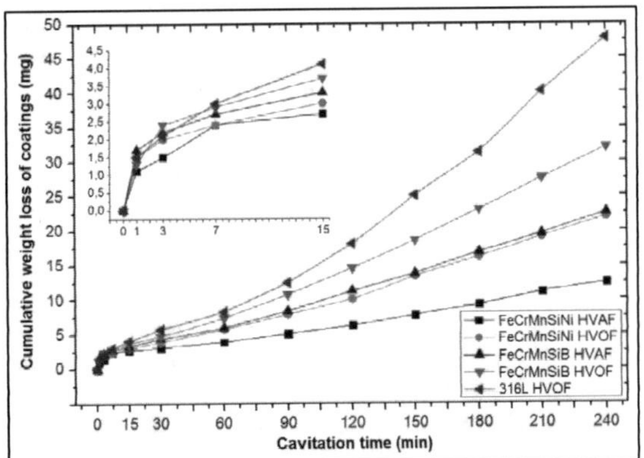

Fig. 3. Cumulative weight loss of the studied alloys after ultrasonic cavitation tests

3.2 Corrosion Properties of the Coatings

The corrosion properties of the coatings were measured by potentiodynamic polarization technique. The results are shown in Table 5, which include corrosion potential (E_{corr}), corrosion current density (i_{corr}), polarization resistance (R_p), anodic (b_a) and cathodic (b_c) Tafel slopes. The corrosion potential values of the HVAF coatings, -0.194 V for FeCrMnSiNi and -0.326 V for FeCrMnSiB, are inferior to the HVOF ones, -0.354 V and -0.435 V, respectively, and -0.369 V for the 316L, showing greater difference between these spraying techniques than between the studied alloys.

Regarding the studied alloys, higher values of polarization resistance were found to the FeCrMnSiNi samples, when compared to the same deposition method. These results indicate that the Ni content in the FeCrMnSiNi had an important influence on the corrosion resistance of the coating. The FeCrMnSiNi HVAF sample presented highest value of polarization resistance, 1047.35 kΩ cm^2, which resulted in a very low corrosion current density, 0.127 μA/cm^2. HVOF FeCrMnSiB sample presented the lowest polarization resistance (15.04 kΩ cm_2) and the highest corrosion current density (11.285 μA/cm^2). 316L sample presented better results than other HVOF samples (65.47 kΩ cm_2 and 3.015 μA/cm^2) but worse than the HVAF ones. E. Sadeghimeresht et al. [20] studied a HVAF 316L sample, obtaining worse results than the ones of this study (3.2 kΩ cm_2 and 7.7 μA/cm^2). The lowest values of corrosion current density obtained by the HVAF coatings indicate that this deposition method lead to the formation of a more stable and protective superficial layer in the coatings.

The annual corrosion rate of the studied coatings verify the previous statements, with FeCrMnSiNi HVAF coating presented a very low value of 0.0014 mm/year, followed by the FeCrMnSiB HVAF coating, with 0.0199 mm/year. The HVOF coatings FeCrMnSiNi and FeCrMnSiB presented higher values, 0.0595 mm/year and 0.118 mm/year, respectively. Commercial 316L presented an intermediate value of 0.0368 mm/year.

Compared to the HVOF coatings, HVAF coatings displayed better corrosion resistance, as indicated by the lower results of annual corrosion rate and higher corrosion potential. This behavior can be related to the lower oxide content of the HVAF coatings, as similar behavior was studied by Guo et al., 2011 [39], which observed that higher metal oxide content (mainly Fe-oxides) between the partially melted particles in the HVOF coating increase the corrosion process of the coating. These metal oxides are found to be favorable sites for microgalvanic and/or microcrevice corrosion [39, 40], facilitating the penetration of chloride ions, causing a larger inner corrosion of the coatings.

Table 5. Corrosion potential (Ecorr), corrosion current density (icorr) and annual corrosion rate (CR) of specimens tested.

Sample	i_{corr} (μA/cm^2)	E_{corr} (V *vs.* SCE)	b_a (V/dec)	b_c (V/dec)	R_p (kΩ cm^2)	CR (mm/year)
FeCrMnSiNi HVAF	0.127±0.02	-0.194±0.03	0.216±0.006	0.085±0.002	1047.35±198	0.0014±0.0002
FeCrMnSiNi HVOF	5.385±0.83	-0.354±0,04	0.203±0.052	0.129±0.019	32.03±4.57	0.0595±0.0091
FeCrMnSiB HVAF	1.885±0.73	-0.326±0.015	0.216±0.009	0.111±0.008	92.12±32.35	0.0199±0.0076
FeCrMnSiB HVOF	11.285±2.75	-0.435 ± 0,009	0.128±0.034	0.184±0.02	15.04±0.94	0.118±0.0312
316L HVOF	3.015±0.615	-0.369±0.012	0.189±0.065	0.140±0.005	65.47±14.4	0.0368±0.0074

Such behavior was observed in the present study, as the HVOF coatings presented higher levels of oxides, darker regions of the coatings (Fig. 1 and 2), as observed in Table 4, 27.9% area and 29.5% area, compared to the HVAF coatings, Table 4, 18.6% area and 16.1% area for the FeCrMnSiNi and FeCrMnSiB, respectively. The commercial 316 L HVOF coating low oxide content (12.6% area) compared to the other HVOF coatings alloys, which can contribute to the higher corrosion resistance of this coating in comparision to the other HVOF coatings. The intention of future work would be the study of the resistance of HVAF thermally sprayed 316L coatings.

4 Conclusions

The cavitation and corrosion behavior of HVOF and HVAF thermally sprayed FeCrMnSiNi and FeCrMnSiB coatings was studied using ultrasonic cavitation and eletrochemical measurements. The results were compared to HVOF thermally sprayed 316L coating. The major conclusions can be summarized as follows:

1. The microstructure of the FeCrMnSiNi and FeCrMnSiB coatings presented lower levels of porosity than the 316L commercial alloy. Levels of oxidation varied with the spraying process. HVOF coatings presented higher levels of oxide, due to the use of oxygen as fuel and higher dwell time for particles for this process.
2. Mechanical properties of the coatings were influenced by the oxide content, with higher values of hardness in the coatings with lower levels of oxide. Additionally, higher velocities of the HVAF process lead to a stronger peening effect and cold work of the coatings, with increase of the mechanical properties.
3. Lower cavitation erosion rates were presented by the HVAF coatings, which can be related to better cohesive properties of the coatings due to higher spraying velocities, additionally to the peening and cold work effect. Ni content of the FeCrMnSiNi alloy improved the resistance of the coatings, compared to the FeCrMnSiB alloy. 316L HVOF coating presented lower cavitation resistance compared to the other alloys.
4. HVAF coatings presented lower corrosion potential values and corrosion current density, which indicates, along with lower annual corrosion rates, a better corrosion resistance. Such behavior is related to lower oxide content, since metal oxides are found to be favorable sites for microgalvanic and/or microcrevice corrosion. Ni content influenced the corrosion properties, since the FeCrMnSiNi alloy had superior corrosion resistance properties, in comparison to the FeCrMnSiB alloy. The 316L coating, with the lowest oxide content, presented better corrosion resistance than the other HVOF coatings.

References

[1] C. T. Kwok, F. T. Cheng, H. C. Man: Synergistic effect of cavitation erosion and corrosion of various engineering alloys in 3.5% NaCl solution. Material Science and Engineering A 290 (2000), pp. 145/54.

[2] A. Neville, B. A. B. McDougall: Erosion- and cavitation-corrosion of titanium and its alloys. Wear 250 (2001), pp. 726/35.

[3] D. G. Li, J. D. Wang, D. R. Chen, P. Liang: Ultrasonic cavitation erosion of Ti in 0.35% NaCl solution with bubbling oxygen and nitrogen. Ultrasonics Sonochemistry 26 (2015), pp. 99/110.

[4] I. Mitelea, E. Dimian, I. Bordeasu, C. Craciunescu: Ultrasonic cavitation erosion of gas nitrided Ti-6Al-4V alloys. Ultrasonics Sonochemistry 21 (2015), pp. 1544/48.

[5] R. Q. Guo, C. Zhang, Y. Yang, Y. Peng, L. Liu: Corrosion and wear resistance of a Fe-based amorphous coating in underground environment. Intermetallics 30 (2012), pp. 94/9.

[6] K. Bobzin, M. Öte, T. F. Linke, K. M. Malik: Wear and Corrosion Resistance of Fe-Based Coatings Reinforced by TiC Particles for Application in Hydraulic Systems. Journal of Thermal Spray Technology 25, (2016), pp. 365/74.

[7] A. Milanti, H. Koivuluoto, P. Vuoristo: Influence of the Spray Gun Type on Microstructure and Properties of HVAF Sprayed Fe-Based Corrosion Resistant Coatings. Journal of Thermal Spray Technology 24 (2015), pp. 1312/22.

[8] C. Suryanarayana, A. Inoue: Iron-based bulk metallic glasses. International Materials Reviews 58 (2013), pp. 131/66.

[9] A. Milanti, H. Koivuluoto, P. Vuoristo, G. Bolelli, F. Bozza, L. Lusvarghi: Microstructural Characteristics and Tribological Behavior of HVOF-Sprayed Novel Fe-Based Alloy Coatings, Coatings, 2014, 4, p 98-120

[10] S. Kumar, J. Kim, H. Kim, C. Lee: Phase dependence of Fe-based bulk metallic glasses on properties of thermal spray coatings. Journal of Alloys and Compounds 475 (2009), pp. L9-L12.

[11] K. Chokethawai, D.G. McCartney, P.H. Shipway: Microstructure evolution and thermal stability of an Fe-based amorphous alloy powder and thermally sprayed coatings. Journal of Alloys and Compounds 480 (2009), pp. 351/59.

[12] C. A. C. Souza, D. V. Ribeiro, C. S. Kiminami: Corrosion resistance of Fe-Cr-based amorphous alloys: An overview. Journal of Non-Crystalline Solids 442 (2016), pp. 56/66.

[13] Z. Zhou, L. Wang, F. C. Wang, H. F. Zhang, Y. B. Liu, S. H. Xu: Formation and corrosion behavior of Fe-based amorphous metallic coatings by HVOF thermal spraying. Surface & Coatings Technology 204 (2009), pp. 563/70.

[14] B. Rajasekaran, G. Mauer, R. Vaßen, A. Röttger, S.Weber, W. Theisen: Development of cold work tool steel based-MMC coating using HVOF spraying and its HIP densification behaviour. Surface & Coatings Technology 204 (2010), pp. 3858/63.

[15] G. Bolelli, B. Bonferroni, J. Laurila, L. Lusvarghi, A. Milanti, K. Niemi, P. Vuoristo: Micromechanical properties and sliding wear behaviour of HVOF-sprayed Fe-based alloy coatings. Wear 276-277 (2012), pp. 29/47.

[16] Z. B. Zheng, Y. G. Zheng, W. H. Sun, J. Q. Wang: Erosion–corrosion of HVOF-sprayed Fe-based amorphous metallic coating under impingement by a sand-containing NaCl solution. Corrosion Science 76 (2013), pp. 337/47.
[17] S. Liu, D. Sun, Z. Fan, H. Y. Yu, H. M. Meng: The influence of HVAF powder feedstock characteristics on the sliding wear behaviour of WC–NiCr coatings,. Surface & Coatings Technology 202 (2008), pp. 4893/900
[18] Q. Wang, S. Zhang, Y. Cheng, J. Xiang, X. Zhao, G. Yang: Wear and corrosion performance of WC–10Co4Cr coatings deposited by different HVOF and HVAF spraying processes. Surface & Coatings Technology 218 (2013), pp. 127/36.
[19] R.Q. Guo, C. Zhang, Q. Chen, Y. Yang, N. Li and L. Liu: Study of structure and corrosion resistance of Fe-based amorphous coatings prepared by HVAF and HVOF. Corrosion Science 53 (2011), pp. 2351/56.
[20] E. Sadeghimeresht, N. Markocsan and P. Nylén,: A Comparative Study of Corrosion Resistance for HVAF-Sprayed Fe- and Co-Based Coatings. Coatings 6 (2016), Issue 2.
[21] ASTM INTERNATIONAL, E2109-2014: Standard Test Methods for Determining Area Percentage Porosity in Thermal Sprayed Coatings, West Conshohocken, PA, 2014.
[22] ASTM INTERNATIONAL, E384-11: Standard Test Method for Knoop and Vickers Hardness of Materials, West Conshohocken, PA, 2011.
[23] ASTM INTERNACIONAL, G32-03: Standard Test Method for Cavitation Erosion Using Vibratory Apparatus, West Conshohocken, PA, 2003.
[24] ASTM INTERNACIONAL, G59-97: Standard Test Method for Conducting Potentiodynamic Polarization Resistance Measurements, West Conshohocken, PA, 2003.
[25] ASTM INTERNACIONAL, G102-89: Standard Practice for Calculation of Corrosion Rates and related Information from Eletrochemical Measurements, West Conshohocken, PA, 1999.
[26] A. G. M. Pukasiewicz, R. F. Vaz and G. B. Sucharski: "Chemical Composition Influence on the FeMnCrSi Alloy Microstructure Deposited by HVOF", Proceedings of International Thermal Spray Conference ITSC 2015, Long Beach, May, 2015.
[27] M.H. Amushahi, F. Ashrafizadeh, M. Shamanian: Characterization of boride-rich hardfacing on carbon steel by arc spray and GMAW processes. Surface & Coatings Technology 204 (2010), pp. 2723/28.
[28] A. G. M. Pukasiewicz, A. R. Capra, R.S.C. Paredes: "Development of Arc Thermally Sprayed Fe-Mn-Cr-Si Coatings Against Cavitation Erosion", Proceedings of International Thermal Spray Conference ITSC 2011, Hamburg, September, 2011.
[29] G. Bolelli, T. Börner, A. Milanti, L. Lusvarghi, J. Laurila, H. Koivuluoto, K. Niemi, P. Vuoristo: Tribological behavior of HVOF- and HVAF-sprayed composite coatings based on Fe-Alloy + WC–12% Co. Surface & Coatings Technology 248 (2014),pp. 104/12.
[30] A. Verstak, V. Baranovski: Activated combustion HVAF: New development in solid particle spray technology, in: O. O. Popoola, N.B. Dahotre, S. Midea, H. Kopech (Eds.), Proceedings of the 1st ASM International Surface Engineering Congress, ASM International, Materials Park, OH, USA, 2003, p 685/89.
[31] L. Jacobs, M.M. Hyland, M. De Bonte: Comparative study of WC-cermet coatings sprayed via the HVOF and the HVAF Process. Journal of Thermal Spray Technology, 7 (1998), Issue 2, pp 213/18.
[32] R.K. Kumar, M. Kamaraj, S. Seetharamu, T. Pramod, P. Sampathkumaran: Effect of Spray Particle Velocity on Cavitation Erosion Resistance Characteristics of HVOF and HVAF Processed 86WC-10Co4Cr Hydro Turbine Coatings. Journal of Thermal Spray Technology 25 (2016), pp. 1217/30.
[33] G. Bolelli, L.-M. Berger, T. Börner, H. Koivuluoto, L. Lusvarghi, C. Lyphout, N. Markocsan, V. Matikainen, P. Nylén, P. Sassatelli, R. Trache, P. Vuoristo: Tribology of HVOF and HVAF-sprayed WC-10Co4Cr Hardmetal Coatings: A Comparative Assessment. Surface & Coating Technology 265 (2015), pp. 125/44.
[34] J. Nohava, B. Bonferroni, G. Bolelli, and L. Lusvarghi: Interesting Aspects of Indentation and Scratch Methods for Characterization of Thermally-Sprayed Coatings. Surface & Coating Technology 205 (2010), pp. 1127/31.
[35] T. C. Totemeier: Effect of High-Velocity Oxygen-Fuel Thermal Spraying on the Physical and Mechanical Properties of Type 316 Stainless Steel. Journal of Thermal Spray Technology 14 (2005), Issue 14, pp. 369/72.
[36] J.H .Kim, K. S. Na, G. G. Kim, J. Y. Oh, C. S. Yoon, S. J. Kim: The effects of Mn and B on the cavitation erosion resistance of austenitic Fe-base hardfacing alloys. Materials Science & Engineering A 477 (2008), pp. 204/07.
[37] S. D. Sun, D. Fabijanic, A. Ghaderi, M. Leary, J. Toton, S. Sun, M.Brandt, M. Easton: Microstructure and hardness characterisation of laser coatings produced with a mixture of AISI 420 stainless steel and Fe-C-Cr-Nb-B-Mo steel alloy powders. Surface & Coating Technology 296 (2016), pp. 76/87.
[38] G. Taillon, F. Pougoum, S. Lavigne, L. Ton-That, R. Schulz, E. Bousser, S. Savoie, L. Martinu and J-E. Klemberg-Sapieha: Cavitation erosion mechanisms in stainless steels and in composite metal–ceramic HVOF coatings. Wear 364-365 (2016), pp. 201/10.
[39] R.Q. Guo, C. Zhang, Q. Chen, Y. Yang, N. Li, L. Liu: Study of structure and corrosion resistance of Fe-based amorphous coatings prepared by HVAF and HVOF. Corrosion Science 53 (2011), pp. 2351/56.
[40] L. Gil, M.H. Staia: Influence of HVOF parameters on the corrosion resistance of NiWCrBSi coatings. Thin Solid Films 420-421 (2002), pp. 446/54.

Comparison of wear performance of thermal sprayed cermet (WC-Co) coatings from suspension and feedstock powders.

O. Ali [a], R. Ahmed [a], H. Alawadhi [b], M. Shameer [b] N.H. Faisal [c], N.M. Al-Anazi [d], M.F.A. Goosen [e]

[a] School of Engineering and Physical Sciences, Heriot-Watt University, Edinburgh, EH14 4AS, UK
[b] College of Sciences, University of Sharjah, P.O. Box: 27272, Sharjah, UAE
[c] School of Engineering, Robert Gordon University, Garthdee Road, Aberdeen, AB107GJ, UK
[d] Materials Performance Unit, Research & Development Centre, Saudi Aramco, Dhahran, 31311, Saudi Arabia
[e] Office of Research & Graduate Studies, Alfaisal University, P.O. Box 50927, Riyadh 11533, Saudi Arabia

WC-Co coatings were deposited using conventional High Velocity Oxy-Fuel Jet-Kote (HVOF-JK) and Suspension HVOF (S-HVOF) methods. Microstructural and mechanical properties along with the wear resistance of coatings were investigated. Reciprocating sliding wear tests were conducted against sintered Si_3N_4 counter-body with a normal load of 25N and total sliding distance of 500m following ASTM G133-2 standard. Coatings were characterised by Scanning Electron Microscope (SEM), X-Ray Diffraction (XRD) and nano-Indentation techniques. HVOF-JK coating showed good retention of WC whereas S-HVOF coating showed the presence of W, W_2C and amorphous/nanocrystalline phases. Nano-indentation of HVOF-JK and S-HVOF showed that the relative hardness of the HVOF-JK coating was higher but their elastic modulus was lower. The lower total wear rate was exhibited by the HVOF-JK coating. This difference in wear performance is attributed to the difference in hardness of the coatings and decarburisation of WC particles.

Keywords: Nanocomposite, sliding wear, suspension spray, tribology

1 Introduction

Thermal sprayed WC-Co coatings have been used for wear resistant applications in a number of industries such as aviation, aerospace, energy and power generation [1]. High Velocity Oxy-Fuel (HVOF) thermal spray process produces dense coatings with high hardness and good wear resistance when compared to plasma sprayed coatings. The wear resistance performance of coatings has been improving with decreasing Carbide Grain Size (CGS), due to the increase in hardness and the reduction of the mean free path which restricts crack propagation. An early investigation reported low wear resistance for nanostructured coatings (CGS was nano-scaled) due to severe decarburisation [2]. Since then nanostructured coatings have improved significantly, having better results than coatings with micro-scaled CGS, however inhomogeneous microstructure for the nano-coatings still poses problems. To produce more uniform and dense nanostructured coatings, WC-Co coatings are being produced using suspension solution by modifying existing HVOF and plasma systems for liquid feedstock. Investigations on the wear performance of Suspension HVOF (S-HVOF) have been performed previously and shown comparable wear performance to standard HVOF coatings under some tribological conditions [3-5]. In this study, a new S-HVOF coating is compared with a conventional HVOF coating in terms of wear performance, bulk mechanical and microstructural properties.

2 Experimental

2.1 Test materials

The conventional coatings are deposited using Jet Kote (HVOF-JK) thermal spray guns, details of which are given in a previous study [4]. For the S-HVOF coating a TopGun was modified for liquid feed injection. The aqueous suspension solution has 25% solid weight content. This weight content of WC-Co was based on a previous study [5]. The solid content of WC-12Co with nano-scaled carbides was supplied from GuangZhou, China. The powder was processed in a planetary ball mill before suspension solution preparation (shown in **Fig. 1.**) using the same method as used previously [3,4].

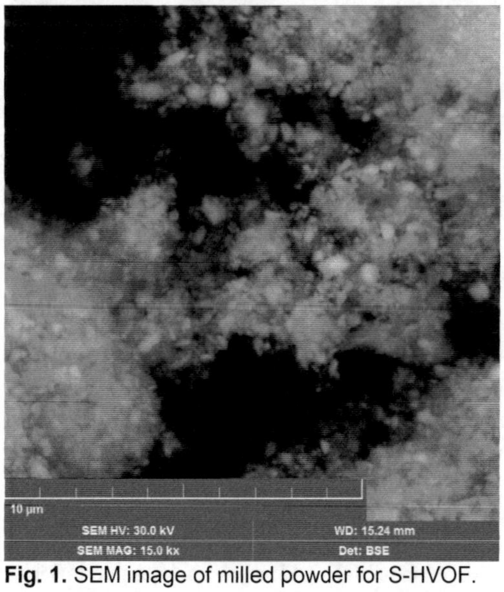

Fig. 1. SEM image of milled powder for S-HVOF.

2.2 Coating characterisation

Characterisation of coatings was conducted using Scanning Electron Microscope (SEM), X-Ray Diffraction (XRD), Energy Dispersive X-Ray Spectroscopy (EDX) and nano-indentation techniques. SEM was used to study the microstructural features of the polished cross-section of the coatings. A Bruker AXS, model D8 ADVANCE X-Ray diffractometer operating at 40kV and 40mA was used for XRD analysis, using Cu Kα radiation (wavelength λ=0.1542 nm). Nano-indentation was performed using a calibrated TriboIndenter® (Hysitron Inc., MN) equipped with Berkovich tip at the load of 50mN. Hardness and elastic modulus were calculated using the Oliver and Pharr method on coating cross-sections [6].

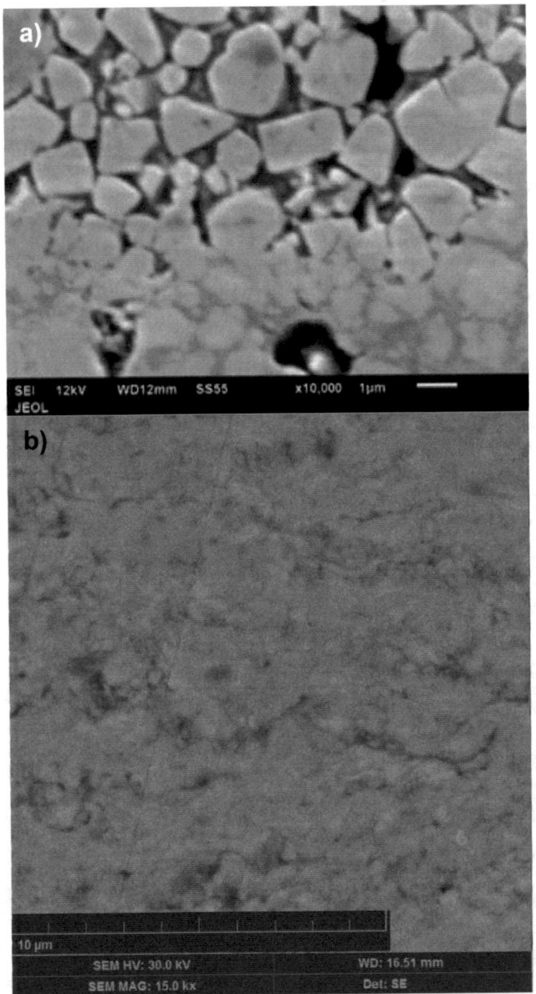

Fig. 2. SEM image of coating cross-sections of a) HVOF-JK [3] and b) S-HVOF (SE) coatings.

2.3 Sliding wear

Wear resistance of the coatings was tested under dry sliding conditions using ball-on-flat reciprocating sliding wear machine (BLR2000 M, Bud Labs, USA) at room temperature following ASTM G133-02 standard. The counter-body was sintered Si_3N_4 ball with diameter of 12.7mm (1/2 inch). The sliding distance was 500m with a normal load of 25N, reciprocating at 2Hz with a stroke length of 10mm. Sliding tests were repeated five times for each coating. Coating volume loss was measured from the average area of the wear track cross-section and the stroke length. The cross-section area was measured using interferometry (Zygo New View). Ball volume loss was measured using optical microscopy via the diameter of the wear scar on the ball. A tension-compression load cell loaded on the sliding wear machine was used to measure friction force to calculate the Coefficient of Friction (CoF). Wear tracks were analysed by use of SEM equipped with EDX, in Secondary Electron (SE) and Back Scattered Electron (BSE) modes to observe coating failure.

3. Results

3.1 Coating microstructure

Microstructural images obtained from the SEM are shown in **Fig. 2.** The HVOF-JK coating had a heterogeneous microstructure. The HVOF-JK coating had carbides which fused well with the Co matrix. The slightly rounded corners indicate that the solid carbide had slightly dissolved into the matrix [7]. S-HVOF coating had a lamella structure. Taking into consideration of thermal kinetics of suspension spraying method, some degree of decarburisation can be expected during coating deposition.

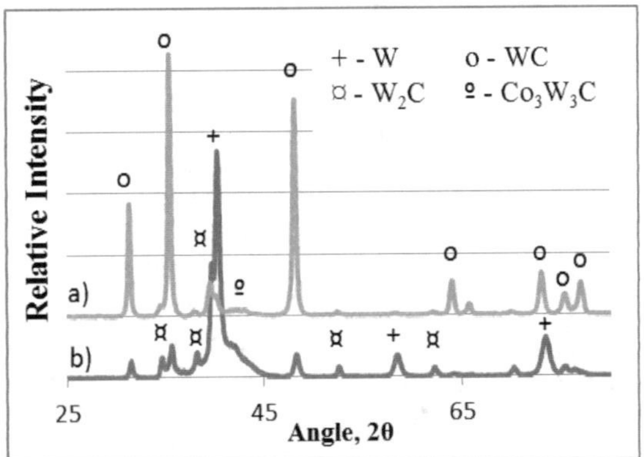

Fig. 3. XRD analysis of a) HVOF-JK [5] and b) S-HVOF coatings.

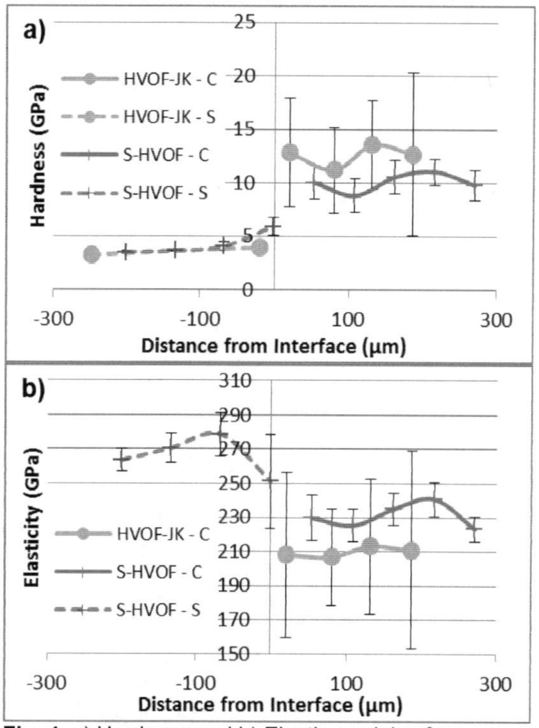

Fig. 4. a) Hardness and b) Elastic modulus from nano-indentation analysis of HVOF-JK [5] and S-HVOF coatings (C – Coating, S – Substrate).

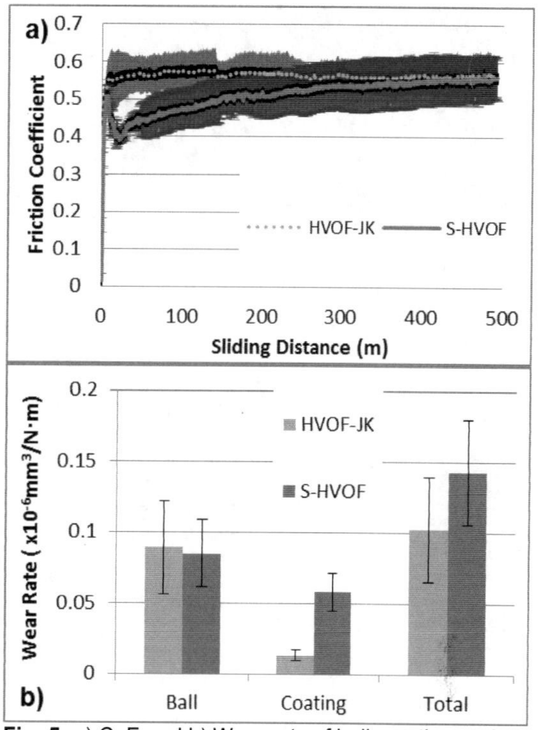

Fig. 5. a) CoF and b) Wear rate of ball, coating and total of all coatings sliding against Si_3N_4 ball for a distance of 500m.

In **Fig. 3.**, HVOF-JK coating showed limited decarburisation of WC with very limited amount of ternary η-phase (M_6C). S-HVOF coating showed relatively higher decarburisation of WC due to the presence of W and W_2C along with ternary η-phase. The hump, for values of 35-45 in 2θ, is indicative of the presence of amorphous/ nanocrystalline structures.

3.2 Nano-indentation analysis

Nano-indentation results of the coatings are compared in **Fig. 4.**, HVOF-JK coating results were performed in a previous study [5] and are compared with S-HVOF coating. HVOF-JK coating has higher hardness but lower elasticity than S-HVOF coating. The substrate for both coatings show a similar hardness.

3.3 Sliding wear test

The CoF of the sliding tests for each coating are shown in **Fig. 5a.**. The CoF for HVOF-JK were previously reported in another study [4]. The CoF of both coatings reached the same value at the end of the experiment at 500m, of roughly 0.55. The main difference between the coatings was the initial behaviour in the first 200m of the tests. HVOF-JK coating reached a stable value rapidly whereas the friction in the S-HVOF coating steadily increased with increasing sliding distance.

HVOF-JK coating showed the highest wear resistance followed by the S-HVOF coating. The total wear rate was 0.10±0.04 and 0.14±0.04 ($x10^{-6}$ $mm^3/N{\cdot}m$) for HVOF-JK and S-HVOF coatings, respectively. Ball, coating and total wear rates are shown in **Fig. 5b.** with respective standard deviation. SEM observation of the worn Si_3N_4 ball (S-HVOF test) is shown in **Fig. 6.** SEM image of the wear track of HVOF-JK is shown in **Fig. 7a.** and the wear track images of S-HVOF coating are shown in **Fig. 7b.** and **Fig. 7c.** The coatings showed no significant signs of tribofilm formation, except for the Si_3N_4 ball used against S-HVOF coating (see **Fig. 6.**). The tribofilms present on the ball contained W and Co along with increased O content. The debris collected from the sliding wear test of S-HVOF coating is shown in **Fig. 8.** The debris had flake-like morphology. The debris was mainly composed of W, O and C with varying levels of Si.

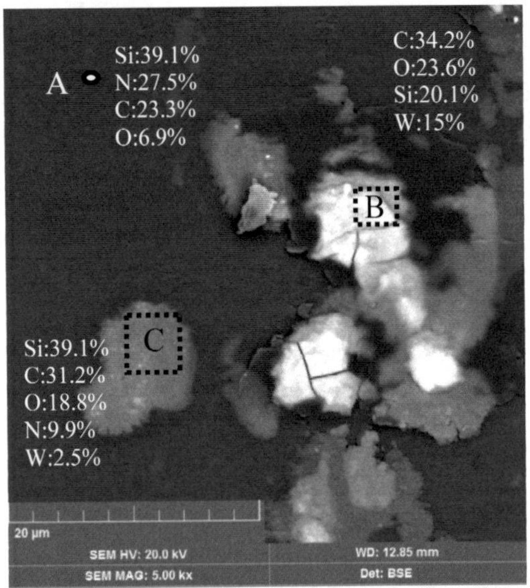

Fig. 6. Tribofilm formation on the ball for S-HVOF coating and EDX analysis of tribofilm deposited on the ball (A to C).

4. Discussion

4.1 Coating microstructure and properties

The microstructure of the HVOF-JK coating shows some signs of porous cavities while the S-HVOF has near negligible porosity. XRD analysis of HVOF-JK and S-HVOF coatings show very different microstructural phases. The larger carbides present in HVOF-JK coating are harder to melt when compared to nano-sized carbides present in the S-HVOF coating due to a relatively higher surface area to volume ratio, as described by Stewart et al. [8]. In previous studies of suspension coatings the decarburisation of WC has also been observed along with the presence of amorphous/nanocrystalline structures [3-5]. The decarburisation of WC in thermal process introduces W_2C, W and η-phase carbides in the coating microstructure. Wang et al. have suggested that this phenomenon is a combination of direct and indirect causes, the direct cause is melting of WC and indirect cause is the chemical interaction of molten Co during the deposition process [9].

4.2 Sliding wear performance

The ball wear rates of the two coatings are comparable but the respective coating wear rates had a significant difference in magnitude (see **Fig. 5b.**). The changes in the CoF are indicative of the transformation of the contact region in terms of contact area and is reflected in the coating wear rates.

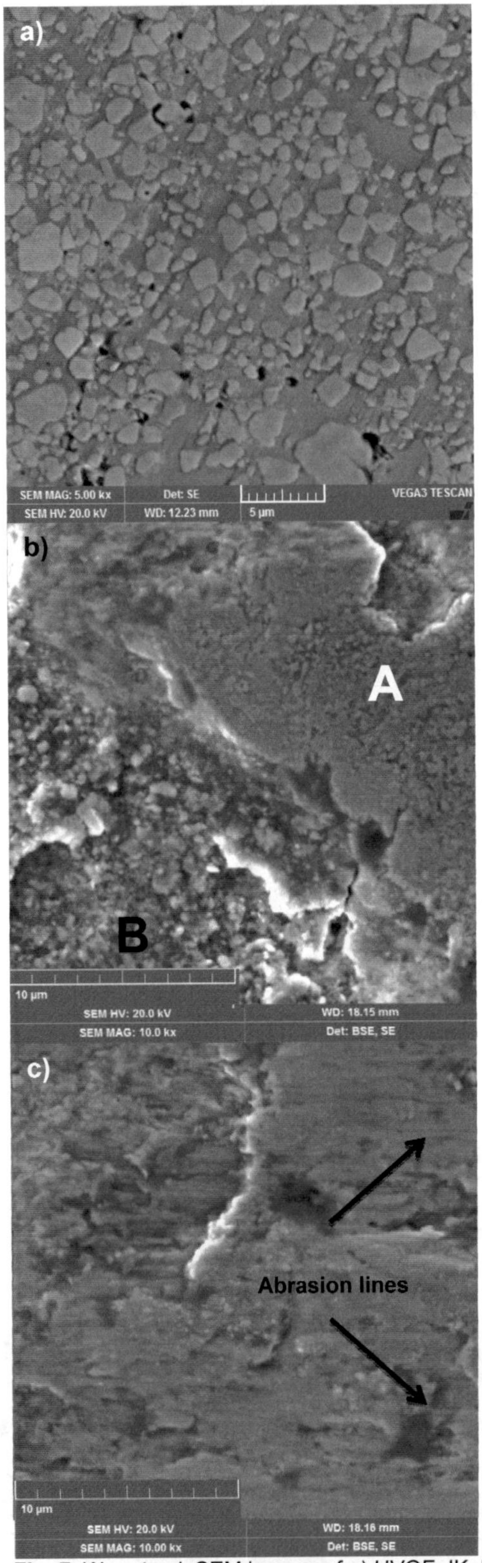

Fig. 7. Wear track SEM images of a) HVOF-JK, b) S-HVOF (A- shows matrix retention and B- shows delamination leading to exposed carbides) and c) S-HVOF coatings.

4.2.1 Coefficient of friction

The evolution of CoF with increasing sliding distance is linked to the changes in asperity interactions and the contact area of the ball. CoF for the S-HVOF coating starts off low as the asperities should be small due to the smaller carbides. Due to the higher extent of decarburisation the S-HVOF coating tends to be brittle with low toughness, especially the η-phase carbides [4]. This leads to high coating volume loss which leads to an increase in contact area causing the friction force to increase. The CoF of S-HVOF coating is plateauing at the end of the experiment, this means that the interactions are approaching a stable level. The continuous increase in contact area allows for the contact pressure to reduce. The high contact pressure combined with the brittle phases present could be a cause for the high S-HVOF coating wear rate. The conventional coating, HVOF-JK, reached stable CoF values earlier in the tests and respectively has a low coating volume loss.

4.2.2 Sliding wear observations and wear mechanisms

EDX analysis of the worn coating surfaces did not show major tribofilm formations or presence of elements originating from the ball. But a tribofilm formation was observed on the Si_3N_4 ball in the test against S-HVOF coating, which had increased content of W, Co and O. The formation and stability of this tribofilm (see **Fig. 6.**) can also be a contributing factor to plateauing of CoF curve as mentioned in the previous section. The wear track of the S-HVOF coating has signs of delamination and matrix retention, as shown in **Fig. 7b.** The cause for delamination is likely linked to the weak boundaries present in the microstructure of S-HVOF coating. In **Fig. 7c.** abrasion lines can be seen on the wear track of S-HVOF coating which is evidence that ploughing by the wear particles was taking place. The lack of tribofilm on the HVOF-JK coating has lead to matrix extrusion and carbide pull-out as the main method for volume loss (see **Fig. 7a.**). The low wear rate of the HVOF-JK coating can be attributed to the larger carbide size. Other factor contributing to the wear mechanism of both coatings included softening of the ball due to humidity in the air and tribo-chemical interactions in such an environment [10]. Ball wear rates for HVOF-JK and S-HVOF coatings are similar. In the case of S-HVOF coating the ball wear rate is high despite having tribofilm formed on the worn surface, this is attributed to the relatively lower hardness of the coating. The trapped debris contributed to 3-body abrasion and causes both the ball and coating to wear more than the HVOF-JK coating.

4.3 Role of microstructure in wear

The HVOF-JK coating has a blocky carbide microstructure, where the carbides were embedded well in the matrix and only get dislodged once the matrix has extruded sufficiently. This causes the ball wear rate to be high since no proper protective tribofilm is formed. The presence of weak boundaries in the microstructure of S-HVOF coating allows for the delamination in the coating, but the presence of region where matrix retention is occurring right alongside region of delamination indicate that once a splat has worn out, the contact pressure with relative motion causes the splat to delaminate and breakdown into fine debris (due to debris entrapment) which leads to 3-body abrasion. The maximum depth of S-HVOF coating wear tracks was on average 5.02µm compared to 1.19µm for HVOF-JK coating.

4.4 Debris morphology and role

The debris formed during the wear test of the S-HVOF coating (See **Fig. 8.**) has large flake-like morphology with some finer particles. There are two types of finer particles: ones which have an angular morphology which are likely to be formed due to the milling of the large flake-like debris and the other type of finer particles are the agglomeration of milled particles originating from the coating material. The second type of finer particles is formed due to debris entrapment and the subsequent 3-body abrasion taking place between the interacting surfaces. The flakes have one main source which is from the Si_3N_4 ball which has an increased content of Si. In **Fig.8.**, the EDX of point D has almost negligible Si content but a very high content of W. Comparing the EDX analysis of the tribofilm formed in **Fig. 6.** with the flakes in **Fig. 8.** (areas A-C in **Fig. 8.**), it can be seen that the tribofilm has higher Si content. This is due to the penetration effect of the EDX analysis, which also records the signal of the underlying material of the Si_3N_4 ball.

Fig. 8. SEM image of wear debris from sliding wear test of S-HVOF with EDX analysis.

5 Conclusion

1) Microstructural phases present in the microstructure of the S-HVOF coating were similar to the findings of previous study with the presence of η-phase ternary carbides and amorphous/nanocrystalline structures. Decarburisation of WC into W and W_2C was observed for S-HVOF coatings.
2) CoF of the coatings was almost the same for all coatings with differences in the evolution over the sliding distance.
3) Relatively lower hardness of the S-HVOF coating allows for easier removal of material causing the wear track depth to increase rapidly which conversely causes the respective ball wear rate to increase due to 3-body abrasion because of the entrapped debris.
4) Debris formed during the sliding wear test had multiple features originating from the coating and the ball. Debris formed from either origin showed distinct features.

6 References

[1] Geng, Z., Li, S., Duan, D.L. and Y. Liu: Wear behaviour of WC-Co HVOF coatings at different temperatures in air and argon. Wear 330-331 (2015), pp. 348/53.

[2] Qiao, Y, Fischer, T.E. and A. Dent: The effects of fuel chemistry and feedstock powder structure on the mechanical and tribological properties of HVOF thermal-sprayed WC-Co coatings with very fine structures. Surface and Coatings Technology 172 (2003), pp. 24/41.

[3] Ali, O., Ahmed, R., Faisal, N.H., Alanazi, N.M., Berger, L.-M., Kaiser, A., Toma, F.-L, Polychroniadis, E.K., Sall, M., Elakwah, Y.O. and M.F.A. Goosen: Influence of Post-treatment on the Microstructural and Tribomechanical Properties of Suspension Thermally Sprayed WC–12 wt%Co Nanocomposite Coatings. Tribology Letters, 65 (2017) ,Issue 2, Art. No. 33.

[4] Ahmed, R., Ali, O., Faisal, N.H., Al-Anazi, N.M., Al-Mutairi, S., Toma, F.-L., Berger, L.-M., Potthoff, A. and M.F.A. Goosen: Sliding wear investigation of suspension sprayed WC-Co nanocomposite coatings. Wear 322-323 (2015), pp. 133/50.

[5] Ahmed, R., Faisal, N.H. , Al-Anazi, N.M., Al-Mutairi, S., Toma, F.-L, Berger, L.-M., Potthoff, A., Polychroniadis, E.K., Sall, M., Chaliampalias, D. and M.F.A. Goosen: Structure Property Relationship of Suspension Thermally Sprayed WC-Co Nanocomposite Coatings. Journal of Thermal Spray Technology 24 (2014), Issue 3, pp. 357/77.

[6] Oliver, W.C. and Pharr, G.M.: An improved technique for determining hardness and elastic modulus using load and displacement sensing indentation experiments. Journal of Materials Research 7 (1992), issue 6, pp. 1564/83.

[7] Ma, N., Guo, L., Cheng, Z., Wu, H., Ye, F. and K. Zhang: Improvement on mechanical properties and wear resistance of HVOF sprayed WC-12Co coatings by optimizing feedstock structure. Applied Surface Science 320 (2014), pp. 364/71.

[8] Stewart, D.A., Shipway, P.H. and D.G. McCartney: Abrasive wear behaviour of conventional and nanocomposite HVOF-sprayed WC-Co coatings. Wear 225-229 (1999), pp. 789/98.

[9] Wang, W., Xiang, J., Chen, G., Cheng, Y., Zhao, X. and S. Zhang: Propylene flow, microstructure and performance of WC-12Co coatings using a gas-fuel HVOF spray process. Journal of Materials Processing Technology 213 (2013), pp. 1653/60.

[10] Dante, R.C. and C.K. Kajdas: A review and a fundamental theory of silicon nitride Tribochemistry. Wear 288 (2012), pp. 27/38.

Low Pressure Warm Spray of Stainless Steel Coatings. Particle Strain Localization Process.

R.Gr. Maev, V.Leshchynsky, E.Strumban, D.Dzhurinskiy, E. Maeva,
Institute for Diagnostic Imaging Research, University of Windsor, *Windsor, ON, Canada,*

Understanding the complex mechanical behavior of stainless steel based composite coatings is important for engineering applications. The focus of this research was on gaining a fundamental understanding of the structure/property relationships that exist during structure formation of the coatings made by new low pressure cold spraying with propellant gas at the temperatures of 800-1000°C (warm spray). While composition is one of the key parameter in determining the final microstructure, the specific warm spray powder shock consolidation parameters (particle velocity and gas temperature) were found to have significant effects on the development of composite structure formation and mechanical properties. Microstructural examination and modeling results revealed that the strain localization mechanism differs from that of adiabatic shear band formation that results in large differences in the ensuing microstructure of the composite coating and its properties.

1 Introduction

In recent years, coating processes based on the impact of high-velocity solid particles such as high pressure and low pressure cold spraying (CS), and aerosol deposition have been developed and attracting much industrial attention. These processes employed a supersonic nozzle into which solid particles of feedstock material were introduced and accelerated toward a substrate, either forming a thin surface coating, or being directly embedded as isolated particles in small depressions on the surface. A high pressure (up to 40 bar)technique for producing thick deposits of various metals from "cold" jets was reported by Papyrin et al.[1]. In this method, solid particles having sizes ranges from 10 to 50μm are introduced by means of a pressurized powder feeder into the high-pressure, high-temperature chamber of a converging-diverging (de Laval) nozzle and are accelerated into a supersonic stream by the gas. Recently, a CS process has been of fundamental interest for formation of metallic coatings at low temperatures. High speed (500–1200 m/s) and high strain rate (up to 10^9 /s impact of micron-scaled particles around 20-25 μm) is the main feature of this process [2]. Low pressure CS process (up to 10bar) is characterised by particle velocity (300-500m/s) and strain rate (up to 10^7 /s) [3]. Although many coatings (i.e., pure metals, alloys, polymers, composites, and bulk metallic glasses) have been obtained, the bonding mechanisms of coatings are not precisely known. A prevalent opinion is that bonding of CS coatings depends on the localization processes due to occurrence of adiabatic shear instability at the impact interface.

There are three primary flow localization processes. In the first (a typical strain-hardening material under non-adiabatic conditions) the shear stress increases due to strain hardening. For this case, material softening at certain temperature needs to be taken into account. In the second, the plastic strain energy that is dissipated as heat under adiabatic conditions, resulting in a temperature increase, in turn causing material softening whose effect overwhelms that of strain hardening. It is during this stage that the development of flow heterogeneity may be observed, indicating the onset of a weak instability process [4]. The third type of plastic flow is characterized by the marked development of strain localization, resulting in the formation of an adiabatic shear band that eventually propagates along the circumference of the specimen. The fluctuations in stress, strain, temperature or microstructure, and the inherent instability of strain softening can give rise to plastic flow (shear) localization [5]. Thus, shearing and heating (and subsequent softening) becomes highly localized, while the material strain and heating in surrounding regions remains negligible. This, in turn, causes the flow stress to rapidly decrease ("localization").

In current high pressure CS technology is of fundamental importance to create a very high speed gas that can accelerate the injected powders to velocities exceeding the critical velocity (about 500-800 m/s) at which adiabatic shear band formation occurs. The critical gas velocity that must be achieved in CS is material dependant, being affected by particle size and density, elastic-plastic properties, and structure. The low pressure air CS system possesses the particle velocities in the range of 300-500m/s that results in lack of adiabatic shear band formation effect and, consequently, low deposition efficiency of Al and Cu spraying and impossibility of stainless steel and other alloys deposition. However, adiabatic shear band strain localization may not be limited to the interface. Adiabatic shear bands may also occur in bulk particles [6] as well as in powder granular media [7]. Additionally, strain localization effects due to a particle volume temperature gradient need to be taken into account. The lack of strain localization analysis – particularly with respect to its influence on the structure of the formed coatings – limits our present understanding and further development of the low pressure CS process. The present paper is targeted at providing insight into low pressure CS particle consolidation processes where the particle significant gradients in both temperature and strain rate occur.

2 Experimental procedure

The low presure CS process was used for depositing Cu and stainless steel coatings on low carbon steel substrates. Commercially available Cu and AISI 904 powder, with particles ranging in size from 5 to 100μm has been used. The particles were accelerated to a high velocity by injecting them into a stream of high pressure carrier gas which is subsequently expanded to supersonic velocity through a converging–diverging de Laval nozzle. After exiting the nozzle, the particles are impacted onto a substrate, where the solid particles deform and create a bond with the substrate. As the process continues, particles continue to impact and form bonds with the previously consolidated material resulting in a uniform deposit with relatively little porosity and high bond strength. The low carbon steel substrates were grit blasted (blasting air pressure 0.75 MPa, stand off distance 150 mm, traverse speed 20-40 mm/s) with alumina (mesh 36) to remove the scale layer formed due to hot rolling process used to produce the substrate material. Commercially available low oxygen Cu powder (Tafa, Inc., Concord, NH, USA) and AISI 904 powder with an average diameter about 45 µm were used. To form a coating, the following CS parameters were utilized with Tessonics low pressure CS machine: propellant gas – air, gas pressure – 9-10 bar, gas temperature – 400-800°C. Optical light microscopy Leica DMI5000 M with a digital camera DFS 320 R2 was used for metallographic characterization.

3. Results and discussion

3.1. Localization vs particle temperature and velocity

The idealized representative structure of the particle with radius R_p impinged with substrate during CS will contain the particle walls of thickness ***w/2*** and particle core. The particle wall structure differs from that of the particle core because of strain localisation effects. Thus it may be assumed the walls form a topologically continuous skeleton structure within which isolated islands of the particle core are encapsulated, similar to Estrin et al [8]. If the surfaces of the deformed particles are the spheres of radius R_d, diameter of the deformed particle is $2R_f$, and its height - $2h_x$,

$$2R_d = \frac{R_f^2 + h_x^2}{h_x} \quad (1)$$

The volume fraction of inter-particle walls may be defined as

$$f = \frac{(S_d \cdot w/2)\cdot 2}{4/3\pi R_p^3} \quad (2)$$

Here S_d is the surface area of spherical segment with radius R_d and height h_x: $S_d=2\pi R_d h_x$. R_d may be defined on the base of an incompressibility rule that means volumes of the particle before and after deformation are equal.

Analysis of localization phenomenon may be made on the base of cylindrical samples impact with Taylor method described in [9,10]. The rod-on-rigid-anvil impact experiment firstly performed by Taylor in 1948 remains a popular method of investigating the mechanical response of a material to dynamic loading. In the original configuration, a cylindrically shaped specimen is impacted against a rigid anvil at velocities sufficient to induce plastic deformation without failure, where strain rates in the range of 10^3–10^6 s^{-1} are typical [10]. By comparing the shape and localisation area fraction of the the finally deformed and initial specimens for a given impact velocity, estimates of the strain localization effect may be made. Geometry of the localization area defined on the base of impact experiment results [9]. (see Fig.1)

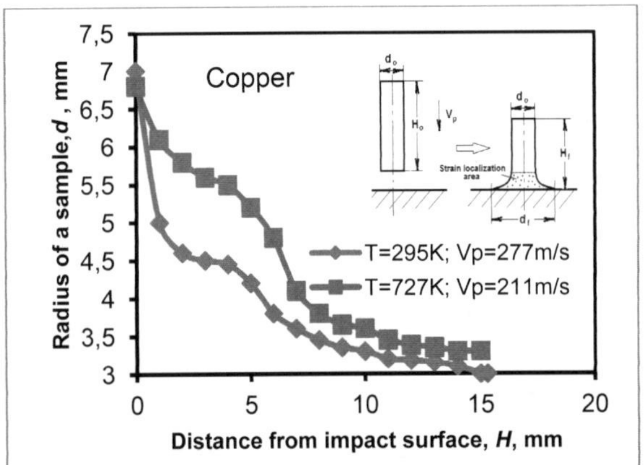

Fig.1. Geometry of localization area due to impact at ambient temperature and 727°K (the curves are calculated using Gust [9] experimental results for Copper).

One can note that strain localization volume depends on impact temperature. Increase of the temperature results in raise of the localization volume probably due to lower yield stress at higher temperature. Effect of temperature on the distorsion of the AISI 4340 samples cylindrical surface $\varepsilon_r=\Delta r/r_0$ is demonstrated on Fig.2 (Gust [9] experimental results).

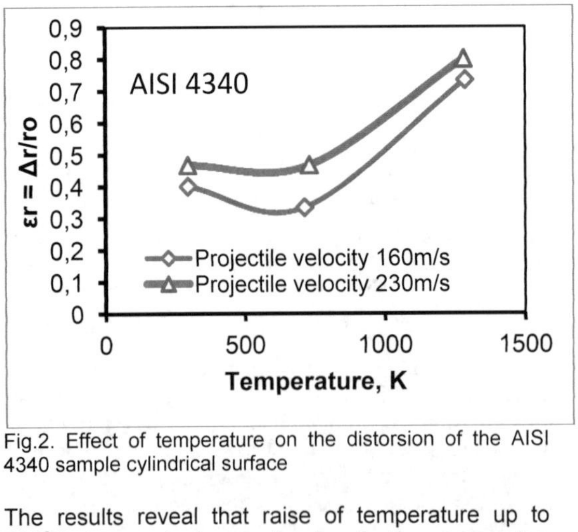

Fig.2. Effect of temperature on the distorsion of the AISI 4340 sample cylindrical surface

The results reveal that raise of temperature up to 715°K does not change the localization parameter. However, the further temperature increase results in enhancement of localization effect.

The influence of impact velocity on the strain localization area ***f*** at compression of Al and Cu cylinders is shown on Fig.3.

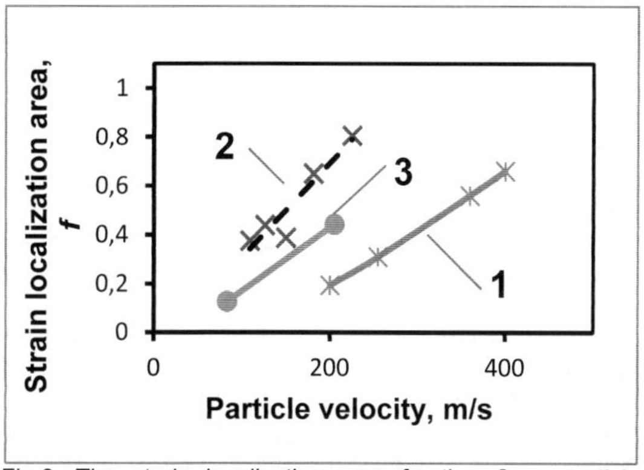

Fig.3. The strain localization area fraction ***f*** vs. particle velocity. 1 – Gust [9] experimental results of Al cylinders impact deformation; 2 – Gust [9] experimental results of Cu cylinders impact deformation; 3 – Eakins and Thadhani [10] experimental results of Cu cylinders impact deformation.

The increase of impact speed leads to enlargement of localized strain volume fraction for both Al and Cu samples in the range of 100-400m/s. The further increase of particle velocities above critical velocity thershold results in change of deformation mechanism at the interface due to adiabatic shear bands development [11-13]. So, looks like low pressure CS process allows to realise preferably the low impact velocity localization mechanism.

One of the main specific features of low pressure CS is known radial particle injection into divergent part of De Laval nozzle. The propellant gas is heated up to temperatures 300-500°C, and particles injected are heated during acceleration. The temperature distribution through the particle is function of a material thermal conductivity, particle acceleration time and surface temperature. The rough evaluation of the particle temperature gradient shows that it is possible to apply an assumption of the two layer temperature distribution for numerical simulation of particle deformation process.

3.2. Numerical simulation procedure

To study the stainless steel particle deformation during deposition with the CS process, a dynamic simulation of ~~multiple~~ impact of the AISI 316L particles iminging upon the carbon steel substrate was run, using commercial Ls-Dyna Multi Material Arbitrary Lagrangian Eulerian Method [14]. A 2-D FEM analysis was performed for 50µm particle size configurations. Fig. ~~3~~ 4 shows the scheme of the simulated model. The bottom surface of the simulated substrate was fixed and additionally the horizontal movements of the substrate in two perpendicular directions was limited by fixing the substrate sides. For 50µm AISI 316L particles impinging upon the substrate at the impact velocity V_p = 400m/s.

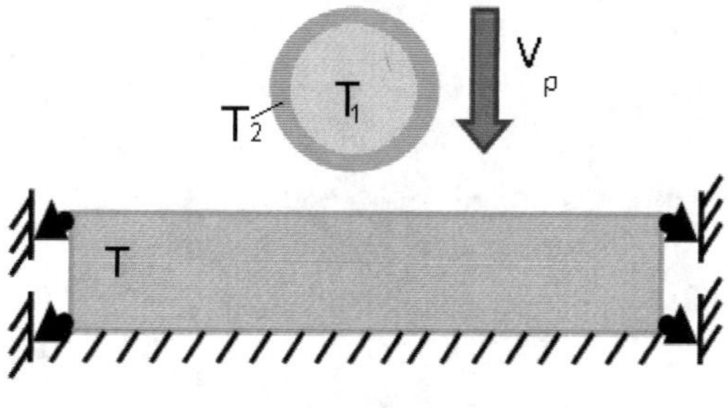

Figure 4: The schematic diagram of impact model with boundary conditions.

Flow stress of both materials was defined using Johnson-Cook model which is described by the following equation:

$$\sigma = \left(A + B\varepsilon^n\right)\left(1 + C\ln\dot{\varepsilon}^*\right)\left(1 - T^{*m}\right) \quad (3)$$

where σ is the (Von Mises) flow stress, *A* is the yield stress at reference temperature and reference strain rate, *B* is the coefficient of strain hardening, *n* is the strain hardening exponent, ε is the plastic strain, $\dot{\varepsilon}^* = \dot{\varepsilon} / \dot{\varepsilon}_0$ is the dimensionless strain rate with $\dot{\varepsilon}$ being the strain rate and $\dot{\varepsilon}_0$ the reference strain rate, and *T** is the homologous temperature and expressed as:

$$T^* = \frac{\mathrm{T} - \mathrm{T}_{\mathrm{ref}}}{\mathrm{T}_{\mathrm{m}} - T_{ref}}$$

A, B, C, m and n are Johnson-Cook constants which were taken from [11,13].

3.3.Numerical simulation results

Figure 5 a,b demonstrates the equivalent strain distribution in the particles during impact of particle with temperature gradient. It can be seen from the images in Fig. 4a that the strain of the surface layer with initial temperature 282C is about ε_i=1.1 whilst the strain of similar surface layer with initial temperture of 550C (Fig.4b) is in the range of ε_i=2.1-3.2. Thus, the preliminary modelling results reveal about intensifying localization effect due to impact deformation of the particles with temperature gradient. It is interesting to note that the temperature distribution for both cases (Fig. 6a,b) is similar in spite of difference of initial temperatures of both particle core and shell.

The numerical simulation results of stainless steel particle impact demonstrate that increase of particle initial temperature gradient allows to achieve high equivalent strains at the interface at the relatively low particle velocities without adiabatic shear band formation effect. Validation of this assumption was made by microstructure analysis of CS coatings deposited by low pressure CS

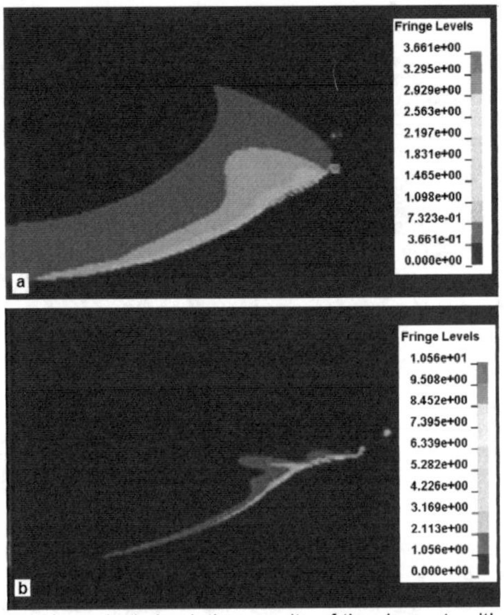

Fig. 5. Numerical simulation results of the impact with the carbon steel substrate.Equivalent strain distribution at 50µm particle impact with 400m/s. (a) - particle shell-5um, core T=25C, shell T=282C,impact time - 0.08us; (b) – particle shell-5um, core T=110C, shell T=550C,impact time - 0.08us.

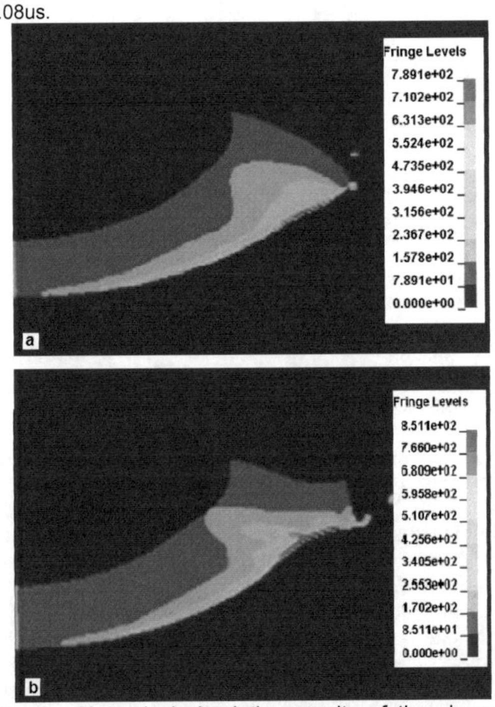

Fig. 6. Numerical simulation results of the impact with the carbon steel substrate.Temperature distribution at 50µm particle impact with 400m/s. (a) - particle shell-5um, core T=25C, shell T=282C,impact time - 0.08us; (b) – particle shell-5um, core T=110C, shell T=550C,impact time - 0.08us.

3.4.Microstructure analysis

Comparison of CS deposited copper and stainless steel coating microstructures are shown on Fig. 7 and 8. It is clearly seen on Fig. 7a that high pressure CS deposition leads to mushroom shape and jetting of the most of particles that reveals about adiabatic shear band formation at the particle-particle interface as shown in [11,12,15]. Contrary, microstructure of copper coatingsmade with low pressure CS demonstrates preferably elliptical shape of particles. Looks like adiabatic shear band localization mechanisms are not developed in the last case because the particle velocity is lower than critical velocity defined by [5,11,12]. Nevertheless, one can see the high density of low pressure CS copper coatings that reveals about occurrence of severe deformation of the copper particles with another mechanism. Based on numerical simulation results of Cu particles deformation (not shown here), one can assume that temperature gradient particle defomation mechanism may result in relatively high strains of the particle shell (up to ε_i>3.5) that leads to coating intensive densification and consolidation.

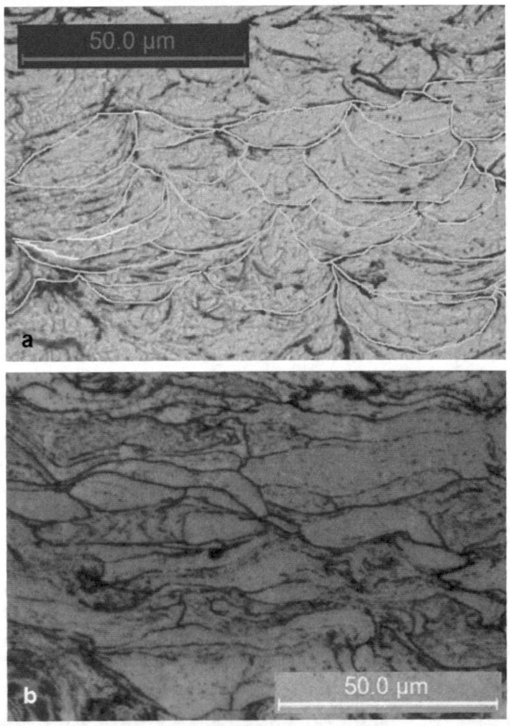

Fig. 7. Microstructure of Copper coating. (a) – high pressure CS [15], propellant gas – Nitrogen, gas pressure - 5 Mpa, gas temperature - 800°C; (b) – low pressure CS, propellant gas-air, T=500C.

Similar hyposesis of low pressure CS coating structure formation may used for stainless steel coatings. Microstructure presented on Fig.8a shows deformed particles without essential features of adiabatic shear band localization. The total strain of separate particles is high. However, some of the particles are not severely deformed. A distorsion of grains in the deformed particle allows to approximately estimate a strain distribution through the particle. It is clearly seen (circles on Fig. 8b) that the severe deformation of grains is observed in the particle surface layers that reveals about strain localization in these areas. The higher temperature of the particle surface layer (see Fig.6) leads to decrease the steel flow stress that results in strain localization during the particle impact. Contrary, the particle core with lower temperature does not deform intensively.So, this effect seems to be an evidence of presence of strain localization processes at the particle-particle interface due to deformation of the particles with the temperature gradient.

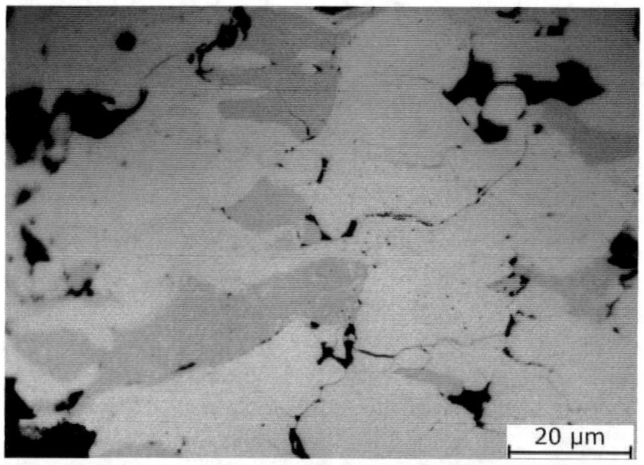

a

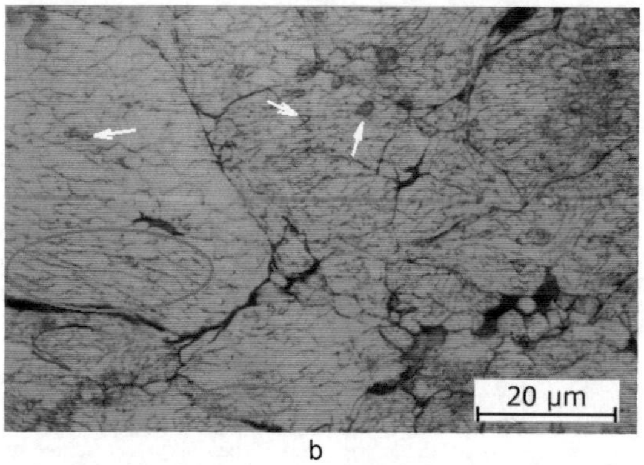

b

Fig.8. Microstructure of stainless steel (AISI 904) coating made by low pressure CS. Propellant gas-air, T=800C. (a) – unetched; (b) – after etching with Glyceregia.

One can note that some recrystallized grains are observed in the structure (white arrows on Fig.8b). It is difficult to define the origin of recrystallization, and this effect will be further studied in the ongoing work.

4. Conclusion

The effect of strain localization due low pressure cold spraying deformation of particles with temperature gradient is defined by: i) analysis of impact compression, ii) numerical simulation, and iii) analysis of the deposited Cu and stainless steel microstructure. It is found influence of particle thermal gradient on the parameters of localication at axial compression with strain rates of 10^{-3}-$10^{-7}s^{-1}$, localization areas in the stainless steel particles after deposition. The temperature gradient of the particles to be sprayed was achieved due to particle radial injection and short time of the particle acceleration in the divergent part of the De Laval nozzle.

Literature

[1] A. Papyrin, V. Kosarev, S. Klinkov, A. Alkhimov *Cold Spray Technology,* Elsevier., (2006) p 125-210

[2] V.K. Champagne, editor *The cold spray materials deposition process: Fundamentals and applications*, Woodhead Publishing Limited, Cambridge, England, (2007) p 230-312

[3] R. Gr. Maev, V. Leshchynsky, *Introduction to Low Pressure Gas Dynamic Spray*, John Willey and Son - VCH, (2007) 207p

[4] A. Molinari, Collective behaviour and spacing of adiabatic shear bands, *J. Mech. Phys. Solids,* (1997)Vol. 45, No. 9, pp. 1551-1575

[5] M.Grujicic, C.L. Zhao, W.S. DeRosset, D. Helfritch, Adiabatic shear instability based mechanism for particles/substrate bonding in the cold-gas dynamic-spray process, Materials and Design, (2004) 25, 681–688

[6] J. W.Walter, "Numerical experiments on adiabatic shear band formation in one dimension", *Int. J. Plasticity*, (1992), 8, p. 657–693

[7] N.Hu, J.F. Molinari, "Shear bands in dense metallic granular materials", *J. Mech. Phys. Solids*, (2004), 52, p.499 – 531

[8] Y. Estrin, L. S. Toth, A. Molinari and Y. Brechet, A dislocation based model for all hardenibg stages in large strain deformation, Acta mater. 46/ 15 (1998), 5509-5522

[9] W.H. Gust, High impact deformation of metal cylinders at elevated temperatures, J. Appl. Phys. 53/5 (1982) 3566-3575

[10] D. E. Eakins and N. N. Thadhani, Instrumented Taylor anvil-on-rod impact tests for validating applicability of standard strength models to transient deformation states, J. Appl. Phys (2006) 100, 073503; doi: 10.1063/1.2354326

[11] H. Assadi, F. Gartner, T.Stoltenhoff, H.Kreye, Acta Materialia, Bonding mechanism in cold spraying, Vol.51 ,2003, pp. 4379-4394

[12] H.Assadi, H.Kreye, F.Gartner, T.Klassen, Cold spraying - A materials perspective, Acta Materialia 116 (2016) 382-407

[13] T. Schmidt, F. Gartner, and H. Kreye, J., New Developments In Cold Spray Based On Higher Gas and Particle Temperatures, Therm. Spray Technol., 15 (2006), 488-494

[14] Ls-Dyna keyword user's manual, Volume I, LSTC, 2013

[15] R. Gr. Maev, V. Leshchynsky,editors, Cold Gas Dynamic Spray, 2015, CRC Press, 340p

Effect of Spraying Parameters on the Bonding Strength of Detonation Sprayed WC-Ni Coatings

Dr. L. Baldaev, Dr. B. Khamitsev, Dr. S. Baldaev, A. Ahmetgareeva
Technological Systems for Protective Coatings (TSPC), Moscow, Russia
info@tspc.ru, +7(495)7838280

Dr. M. Prokofjev,
Moscow Aviation Institute, Moscow, Russia

Abstract

Well-known detonation thermal spraying is considered to be a good method for the preparation of WC-based coating with high adhesion. The depositing parameter especial the oxygen-fuel ratio and spot frequency observably influenced the properties of the coating. The influence of the main parameters of the detonation spraying of tungsten carbide-based coatings and their mechanical properties have been investigated. Two main tasks for the development of a carbide-containing material application technology have been determined. The relationship between the phase composition of WC-based coatings and their strength, hardness and bond strength to the substrate has been identified. The technique of coating in the reductive mode has been proposed. High oxygen-fuel ratio and spot frequency decreased the decomposition of the WC-phase, which makes the WC-Ni particles harder and as a result more easily embedded in the substrate, i.e. the coating which is beneficial to enhance bonding strength.

Introduction

Composite coatings on the base of tungsten carbide have got the most widespread use among the coatings applied using the detonation method. WC-type or WC-Ni-type compositions are used particularly often. The reason for this is very good mechanical properties of the coatings obtained. Due to their significant hardness (Vickers hardness number up to 1300), these coatings feature a very high wear resistance and are widely used in aircraft industry, shipbuilding industry, manufacturing of oil-production equipment and other engineering industries [1,2]. Detonation coatings based on tungsten carbide with cobalt and nickel have the highest bond strength to the substrate among gas-thermal coatings (up to 200-250 MPa) and are often utilized in applications where the load on working surface is dynamic. Owing to their high bond strength to the substrate, WC-based compositions are also used as an intermediate layer when applying coatings from other materials.

In detonation spraying the properties of coatings are defined not only by the energy parameters of the sprayed material particles, but also, at least to the same extent, by structure and phase changes taking place in the material. These structure and phase processes start as early as during the powder particles acceleration, continue at the stage of their flying from the barrel end to the substrate surface, and end when a coating layer is formed. A good number of publications study the regularities of changes in the phase composition of coatings type WC-Co and WC-Ni. In works [3,4] detonation coatings obtained from VK-15, VK-20 and VK-8 powders in the form of a mechanical mixture of WC and Co were studied. It was found that monocarbide disintegrates $WC \rightarrow W_2C \rightarrow W$ when subjected to high temperatures and oxidizing environment. Beside that, diffraction reflections of the double carbides Co_3W_3C and $Co_3W_9C_4$ and of the intermetallic compound Co_7W_6 were detected. It should be noted that the content of these phases varied depending on the spraying process modes, especially the composition of gas explosive mixture.

In works [5,7,8], also dedicated to the study of the phase composition of coatings obtained using mechanical mixtures type VK, the intensive processes of carbide decarbonization all the way to tungsten formation are highlighted. Due to this, using mechanical mixtures as initial powders is the least practicable.

In works [1,2,11] an opinion is expressed about the presence in sintered hard alloys of a continuous skeleton of tungsten carbide crystals, in the gaps of the skeleton a continuous skeleton of binder metal being situated. Probably, this is the main peculiarity of the mechanism of the coating formation from high-melting compounds with a metal binder substance. Further, the properties of the hard alloys of the triple systems W-C-Co and W-C-Ni depend on their structure and phase composition.

In work [12], on the basis of the analysis of structure and phase transformations occurring during the spraying of powders obtained by various means, it was found that the best mechanical properties of coatings are achieved by using powders obtained through spheroidization in protective atmospheres.

In this work the influence of the most significant parameters of detonation spraying – the gas explosive mixture composition and the degree of filling of the detonation spraying system barrel – on the character and rate of structure and phase changes during coating formation and, consequently, the quality of the coatings was studied.

Experiments

The tungsten carbide-based coatings were applied using the automatic detonation spraying systems "Ob" and

"Plackart-D3". The design of the systems makes it possible to change over a wide range the composition of gas explosive mixture, the degree of filling of the system barrel with the mixture, as well as other spraying parameters. The systems are equipped with manipulators turning the sprayed part at different frequencies and moving it in a reciprocating manner at various velocities.

As initial powder material, spheroidized VN-20 (WC +20% Ni) powder. The mechanical properties of coatings obtained were assessed by the bond strength to the substrate, the strength of the coating material itself and by measuring the Vickers hardness (HV_5). The bond strength to the substrate (σ_{bond}) was determined according to the pin technique using a batch of samples consisting of 5 to 7 pcs (Fig. 1). The material of the samples: steel 45.

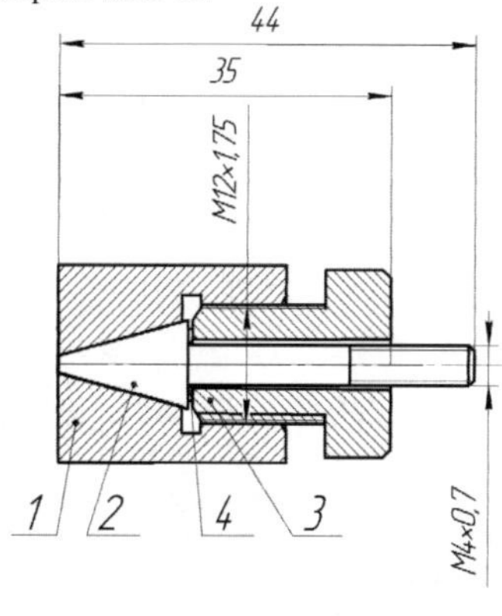

Figure 1: A pin sample for determining the bond strength of the coating to the substrate. 1 – bushing; 2 – taper pin; 3 – lock nut; 4 – washer.

The strength of the coating material itself was determined by a special technique using a batch of samples consisting of 5 to 7 pcs (Fig. 2). The material of the samples: steel 45.

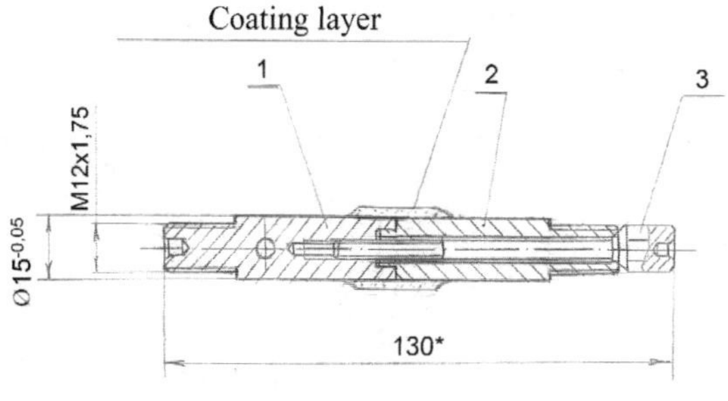

Figure 1: A sample for determining the strength of the coating material. 1 – left-hand part; 2 – right-hand part; 3 – coupling bolt.

The X-ray phase analysis (XRP analysis) of initial powders and coatings was carried out on the equipment *ALR X'TRA*, produced by Thermo-Fisher Scientific (Switzerland), with a copper anode (*CuKα*). Exposure mode: 25 mA, 40 kV (1.0 kVA). The collimating horizontal slits (restricting the vertical divergence of the X-ray beam) had a size of 1.0 and 1.0 mm. The slits on the detector of 0.9 mm and 0.5 mm restricted the registered radiation, thus ensuring a resolution of $0,11^0 2\Theta$ when registering the diffraction peaks. In the vertical direction, the beam was restricted by Soller slits with an angular divergence of 1.5 degrees. During X-ray exposures, the discrete way of the counter movement was used, the pulse collecting time being 2 s per dot.

Samples were placed into a standard cell; it was filled full. X-ray diffraction patterns were registered at a typical registration speed of 0.8 degree per minute, which provided the required sensitivity.

Results

The practice of the determination of the bond strength of WC-based coatings to the substrate using pin samples showed that usually the tearing-off occurs in the material of the coating near the end surface of the pin, and not at the boundary between the substrate and coating (cohesive tearing-off) (Fig. 3).

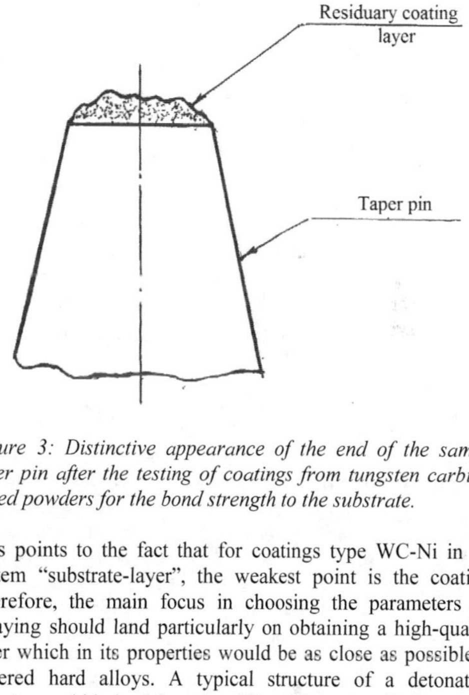

Figure 3: Distinctive appearance of the end of the sample taper pin after the testing of coatings from tungsten carbide-based powders for the bond strength to the substrate.

This points to the fact that for coatings type WC-Ni in the system "substrate-layer", the weakest point is the coating. Therefore, the main focus in choosing the parameters for spraying should land particularly on obtaining a high-quality layer which in its properties would be as close as possible to sintered hard alloys. A typical structure of a detonation tungsten carbide-based coating (Fig. 4) is pretty similar to that of hard alloys obtained by sintering. In spite of the fundamental differences in the obtainment methods, here we can also see carbide crystals distributed quite regularly in the metal constituent, the formation of solid solutions of WC in Co or Ni, although to a lesser extent than with sintering. However, the dependence of the mechanical properties on the phase composition remains.

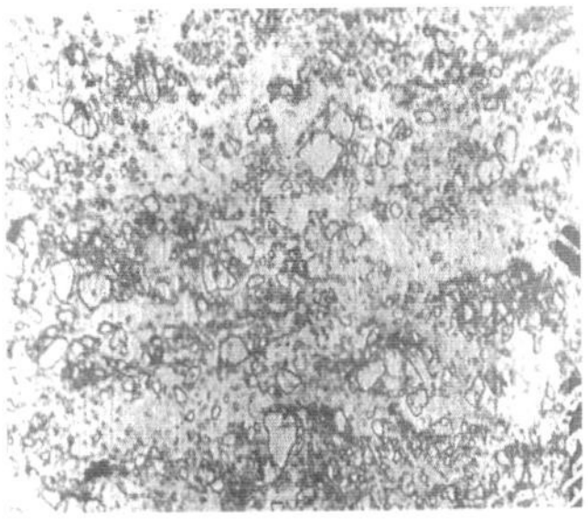

Figure 4: Typical structure of a detonation spheroidized VK-20 powder coating.

The closer the structure and phase composition of coatings are to those of sintered material the higher the strength of the layer itself, the bond strength to the sprayed surface, hardness and other operating properties must be.

It is known that for sintered hard alloys type WC-Co, from the perspective of their mechanical properties, the optimum structure is a two-phase structure consisting of a γ-phase, being a solid solution of WC in β-cobalt, and a WC phase [2,9,11].

In the conditions of detonation spraying, when heated powder particles are oxidized by detonation products and ambient air, the coating formation usually takes place with carbon deficit. At the stage of the particle acceleration and heating inside the detonation spraying system barrel, a partial decomposition of monocarbide takes place with the formation of complex carbide compounds. When the particles fly from the end of the barrel and while the coating is formed on the substrate surface, the decarbonization process continues.

In order to determine the degree to which the gas mixture composition influences the phase composition and, subsequently, the mechanical characteristics of coatings, spraying was performed at the stoichiometric (C_2H_2:O_2= 1.0:2.5) and equimolar (C_2H_2:O_2= 1.0:1.0) ratios of acetylene and oxygen, as well as at the ratio C_2H_2:O_2= 1.2:1.0 that is with the potential formation of free carbon (soot). In the latter case, a reducing environment forms in the system barrel, and the spraying mode may be nominally called the reductive mode. In the practice of the WC-based coating spraying, gas mixtures with the compositions close to equimolar (C_2H_2:O_2=1.0:1.05-1.15) are used. With such a composition, the detonation wave velocity, the detonation products flow rate and the heating degree of powder particles are close to their maximum values. A small excess of oxygen in the mixture ensures reliable operation of the electric plug in the ignition chamber of the system and prevents carbon deposition on the electrodes. The design of the detonation spraying systems used in this study allows for the application of coatings from acetylene-rich mixtures, with C_2H_2:O_2= 1.2:1.0, where Q is gas flow rate per one cycle of the system. Such an opportunity is achieved because when the barrel is filled with working gases, a mixture locally enriched with oxygen is formed. Table 1 shows the mechanical characteristics of coatings obtained from spheroidized VN-20 powder at various compositions of the gas explosive mixture.

Table 1: Mechanical characteristics of coatings from spheroidized VN-20 powder sprayed at various compositions of the gas explosive mixture.

The ratio of acetylene and oxygen flow rates	The mean bond strength of the coating to the substrate (σ_{bond}), MPa	The mean strength of the coating material (σ_B),	The mean Vickers hardness (Hv_5)
Oxidative mode, C_2H_2:O_2= 1.0:2.5	63,04±1,19	122,9±21,1	866±16,6
Neutral mode, C_2H_2:O_2= 1.0:1.05	118,68±3,8	355,8±19,3	960,6±37,8
Reductive mode, C_2H_2:O_2= 1.2:1.0	137,5±3,84	411,6±22,5	921,6±26,1

We can see from the table that the spraying of coatings using acetylene-rich gas mixtures makes it possible to significantly increase their mechanical characteristics, especially bond strength to the substrate and the strength of the coating itself.

The results of the XRP analysis of coatings obtained using gas mixtures of stoichiometric composition (oxidative mode) and acetylene-rich mixtures (reductive mode) enable to explain the growth of the strength characteristics by the corresponding structure and phase changes. Figures 5, 6 and 7 show X-ray patterns of initial powder and coatings, and Table 2 shows the results of the quantitative analysis of phase constituents and the changes in the parameter "α" of the Ni phase crystalline lattice.

Table 2: The content of the main phases in initial VN-20 powder and coatings in the oxidative and reductive spraying modes.

	Phase composition, atomic %				Parameter "α" of the Ni phase lattice
	WC	W_2C	W	Ni	
Initial powder	83,8	traces	no	16,2	3,5661
Oxidative mode, C_2H_2:O_2= 1.0:2.5	62,0	no	22,0	16,0	3,6417
Reductive mode, C_2H_2:O_2= 1.2:1.0	73,0	3,0	no	23,0	3,6059

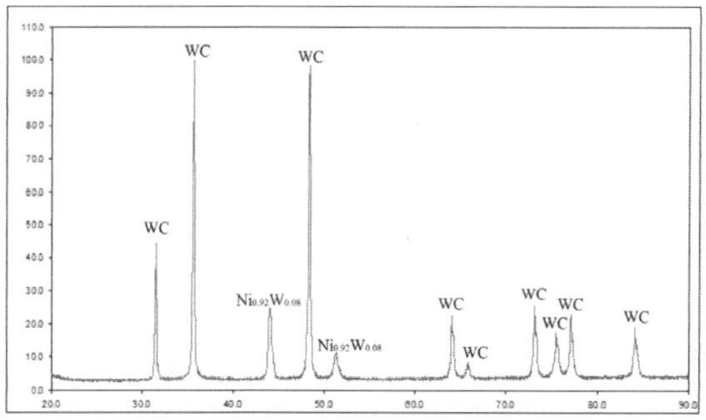

Figure 5: X-ray pattern of initial VN-20 powder.

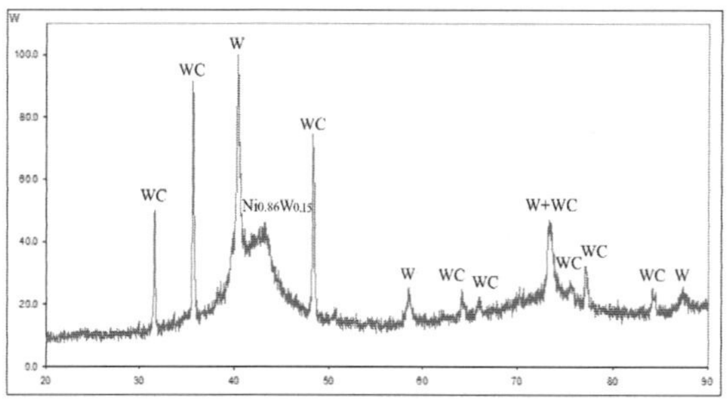

Figure 6: X-ray pattern of a coating from VN-20 powder applied in the oxidative mode.

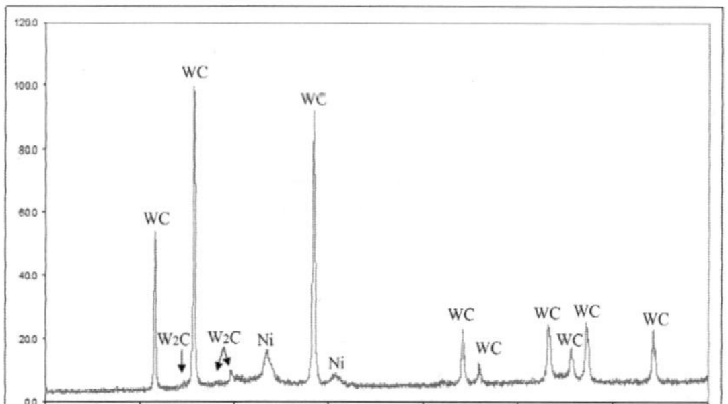

Figure 7: X-ray pattern of a coating from VN-20 powder applied in the reductive mode.

Comparing the phase compositions of the powder and coatings we can clearly observe a perceptible difference in the WC content and in the change of the parameter "α" of the nickel elementary cell. The increase in the parameter "α" is indicative of the dissolution degree of tungsten in nickel, and the difference in the WC content suggests a various oxidation rate of monocarbide.

The dissolution degree of tungsten in nickel depends on the degree of the powder particles fusion during spraying, i.e. on the degree of their heating.

Conclusions

Thus, when choosing the application modes for WC-based coatings, the optimal combination of factors related to both mutual dissolution and the decarbonization of tungsten carbide is required.

1. When detonation tungsten carbide-based coatings are applied, their mechanical properties are determined by the two main processes:

- dissolution of the carbide constituent in the catalyst metal, which contributes to the increase in the bond strength to the substrate, coating material strength and hardness;
- decarbonization of carbides when exposed to detonation products and the ambient atmosphere, which affects the coating quality in a negative way.

2. In order to decrease the rate of monocarbide decomposition, it is useful to carry out the spraying in the reductive mode i.e. using gas mixtures featuring an excess of acetylene. The presence of free carbon (soot) in detonation products can significantly increase the mechanical characteristics of coatings.

3. When developing a technology for spraying carbide-containing coatings onto specific parts, one should look for an optimal balance of the results of the processes of mutual dissolution and carbide constituent decarbonization.

References

[1] Tretiakov, V. I. Metal-Ceramic Hard Alloys. Moscow: Metallurgizdat, 1962, 470 p.

[2] Chaporova, I. N., Chernyavsky, K. S. The Structure of Sintered Hard Alloys. Moscow: Metallurgiya, 1975, 247 p.

[3] Shmyreva, T. P. The Phase Composition of Detonation Coatings from Alloys Type VK. Poroshkovaya Metallurgiya, 1982, No. 12, pp. 30-35.

[4] Kharlamov, Yu. A., Ryaboshapko, B. L., Pisklov, Yu. I., Shmyreva, T. P. The Changes in the Phase Composition of VK 8 Powder During Gas-Detonation Spraying. Poroshkovaya Metallurgiya, 1985, No. 11, pp. 28-33.

[5] Shmyreva, T. P. The Study of the Structure of Coatings Obtained by the Detonation Method. Extended abstract of a dissertation in support of candidature for a technical degree.

[6] Kharlamov, Yu. A. Obtaining Detonation Hard-Alloy Coatings from a Mechanical Mixture of Fine Powders. Poroshkovaya Metallurgiya, 1978, No. 4, pp. 41-46.

[7] Shmyreva, T. P., Vorobiev, G. M., Velichko, M. G. et al. Physical and Chemical Transformations in Powder Materials During Gas-Detonation Treatment. In "Chemical Physics of Combustion and Explosion Processes. Detonation". Chernogolovka, 1980, pp. 119-121.

[8] Alfintseva, R. A., Kadyrov, V. Kh., Fedorenko, V. K. et al. The Structural Studies of the Detonation Coatings Tungsten Carbide-Cobalt. Poroshkovaya Metallurgiya, 1982, No. 10, pp. 24-29.

[9] Chaporova, I. N., Tretiakov, D. A., Shchetilina, E. A. et al. On the Question of the Solubility of Tungsten Carbide in Cobalt and Nickel. Collection of research papers of VNIITS No. 1, "Hard Alloys". Metallurgizdat, 1959, pp. 177-190.

[10] Chaporova, I. N., Shchetilina, E. A., Serebrova, S. I. On the Influence of the Composition of Cementing Phases on Some Mechanical Properties of the Metal-Ceramic Hard Alloys WC-Co and WC-Ni. VNIITS, "Hard Alloys", Collection of research papers, 1960, No. 2, pp. 160-167.

[11] Tretiakov, V. I. The Fundamentals of General Metallurgy and Sintered Hard Alloy Production Technology. Moscow, Metallurgiya, 1976, 520 p.

[12] Khamitsev, B. G., Rubtsova, L. A., Baranova, G. M. et al. Special Aspects of the Formation of Coatings from Spheroidized VK-20 Powder. Theses of the Industry Day "Manufacturing and Restoration of Specialized Technology Products, Production Accessories and Tools Using Hardening and Protective Coatings". Zaporozhe, 1987, p. 21.

Effect of the microstructural design of plasma-sprayed thermal barrier coatings on its the thermal cycling behavior

Tao Liu, Jun-Jie Yang, Guan-Jun Yang, Chang-Jiu Li*

State Key Laboratory for Mechanical Behavior of Materials, School of Materials Science and Engineering, Xi'an Jiaotong University, Xi'an, Shaanxi 710049, China
*: Licj@mail.xjtu.edu.cn

The microstructure of thermal barrier coatings (TBC) plays an important role in the thermal cycling behavior of TBCs. In this study, ceramic coatings with different pore structures were prepared by atmospheric plasma spraying (APS). Graphite with different morphology was used as pore former to adjust the pore structure of the coatings. Then, the thermal cycling behavior of TBCs with different structure was characterized. By depositing a porous 8YSZ layer on the conventional 8YSZ layer, the thermal cycling life of TBCs can be improved. However, when the porosity of the porous layer increased to about 30%, the porous layer detached from the under layer after several cycles. An elastic energy model was applied to explain the thermal cycling behavior of TBCs

1. Introduction

Increasing gas inlet temperature is an effective approach to increase thermal efficiency of gas turbines used for aircraft propulsion and land-based power generation. To protect the superalloys operated at high temperature, thermal barrier coatings (TBCs) are widely applied. TBCs system usually consists of a ceramic top-coat of yttria stabilized zirconia (YSZ) with low thermal conductivity, a metallic bond coat (BC, MCrAlY, M= Ni/Co) and a superalloy substrate [1-3]. Currently, TBCs are usually fabricated by either electron beam-physical vapor deposition (EB-PVD) or atmospheric plasma spraying (APS). YSZ, especially zirconia containing 6-8 wt% of yttria (8YSZ) is used as the standard material of TBCs due to its low thermal conductivity and excellent mechanical properties [1].

As well known, the properties of the ceramic coating is not only related to the ceramic material, but also affected significantly by the microstructure. APS ceramic coating is formed by successive deposition of a stream of fully or partially molten particles impacting on a substrate followed by lateral flattening, rapid solidification and cooling. Subsequently, the lamellar structure with many inter-lamellar two-dimensional (2D) pores and intrasplat cracks is the typical feature of APS TBC [4]. The structure of APS ceramic coatings can be adjusted in a large range by adjusting the plasma parameter [5], preheating the substrate [6-8], powder-designing and post treatment [9,10]. In general, the thermal conductivity can be reduced by increasing the porosity of TBCs. Pore former such as polyester was always applied to increase the porosity of APS ceramic coatings [10,11]. However, the morphology of pores is difficult to control. Graphite, as a widely used pore former in ceramic sintering process and casting process [12], can generate pores with controlled shape and size in ceramic coatings. However, graphite is hard to be melted and deformed during APS process. By designing a composite powder with metal covering on the surface of graphite particles, nickel-graphite composite coating was deposited by APS [13]. Therefore, designing a ceramic/graphite composition power is a promising approach to deposit the corresponding composite coatings by APS.

In this study, 8YSZ/graphite composite powders using two different types of graphites were prepared. 8YSZ coatings with different pore structures were achieved by APS together with post-treatment. Then, the thermal cycling behavior of TBCs with different pore structure was investigated. The current results indicate that the microstructure of the ceramic coating can significantly influence the thermal cycling behavior including the cycles and the failure mode of TBCs.

2. Experimental procedures

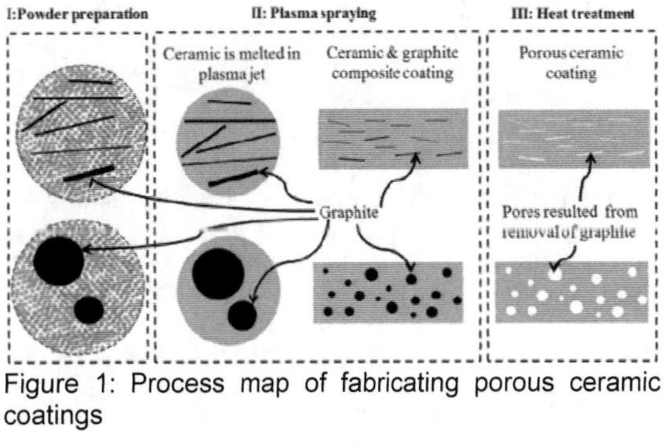

Figure 1: Process map of fabricating porous ceramic coatings

Fig. 1 shows the process map of fabricating porous ceramic coatings by using graphite as pore former. At first, composite powders with mixing nano-scaled ceramic particles with graphites in different morphologies are prepared. Then, during plasma spraying, the ceramic in composite powder is melted. When the composite particles impact on the substrate, the molten ceramic fraction acts as bindind phase and keeps the graphite in the splat to form a composite coating. According to previous study on plasma-spraying metal/graphite coatings [13], the morphology of graphite is expected to be nearly unchanged during plasma spraying. At last, the graphite is removed by oxidizing the graphite to create pores with the same morphology and locations of the corresponding graphite particles.

2.1 Powder preparation

8YSZ/graphite composite powders were prepared by spray-drying process. Nano-scaled 8YSZ particles （Tianyao Co. Ltd., QingDao） were mixed with graphite by ball milling. The mass ratio of ball to the powders was 15: 1, the mass ratio of alcohol to the powders was 2: 1. The ball milling speed was 100 rpm and the duration was 30 mins. The ball milling speed is relatively low to avoid the fracture of the graphite and keep their original morphology. 5 wt.% PVA was added as a binder in the slurry after ball milling. Spherical graphite (SG, Aoyu Co. Ltd., Ji Xi) and plated graphite (PG，Aladin Co. Ltd., Shang Hai） were used as the pore former in this study. The composite particles with a size range from 30 μm to 75 μm were used for plasma spraying.

2.1 Preparation of TBCs

Nickel-based superalloy Inconel 738 (Φ25.4 mm×3 mm) plates were used as substrates. Cold spraying (CS-2000, Xi'an Jiaotong University, China) was carried out to deposit a 100-150 μm thick bond coat with a commercial Ni23Co20Cr8.5Al4.0Ta0.6Y in spherical shape (Sulzer-Metco Amdry 997, 9~37 μm). Table 1 shows cold spray conditions. A commercial plasma spray system (GP-80, JiuJiang, China) was used to deposit YSZ coating by using Ar-H_2 plasma jet. A hollow spherical 8YSZ(Metco 204B-NS, 45~75 μm) was used to deposit an approximately 150 μm ceramic layer on the bond coat. Then, seven types of the coatings with thickness of about 150 μm were deposited on the conventional 8YSZ layer. The compositions of those TBCs are shown in table 3. Then all the coatings was put in a muffle furnace to remove the graphite. It was reported that graphite starts to be oxidized when the temperature is higher than 600 □ [12]. In this study, the oxidation condition was 800 □ for 5 h, the heating and cooling rates were set at 120 □/h. The spray powders and the coatings before and after thermal exposure at 800 □ were characterized by x-ray diffraction analysis (XRD, Rigaku D/max-2400) using Cu K_{α}□ radiation at a scanning rate of 5°/min for 2θ.

Table 1 Cold spray parameters

Parameters	Unit	Value
Accelerating gas pressure	MPa	2.0
Powders carrier gas pressure	MPa	2.3
Accelerating gas temperature	°C	730±20
Standoff distance	mm	20
Gun traverse speed over substrate	$mm{\cdot}s^{-1}$	150

2.3 Pretreatment and Thermal Cycling Test for TBCs

Pre-oxidation of the TBCs specimens was carried out in an argon atmosphere at 1000 °C for 4h and then 1080 °C for 4h to grow a uniform and dense α-Al_2O_3-based TGO as reported previously [14]. The argon atmosphere includes <10 ppm of oxygen, which could provide enough oxygen to form α-Al_2O_3 and avoid nucleation of mixed oxides prematurely. Thermal cycling test was carried out under a thermal cycling at 1120 □ with a 26-min hold and cooling down to room temperature in 4 min. The failure of the TBCs was defined as 50% of the YSZ coating delamination and/or spalling, and 3 samples at each condition were characterized. A scanning electron microscope (SEM, MIRA-III, TESCAN, Czech) was applied used to characterize the morphology of the coatings. The porosity of the coatings was determined by image analyzing method using the back scattered electron images (BEI) of the cross section of polished samples at a magnification of 1000×. At least five images were used to estimate the apparent porosity for each coating sample.

Table 2 Plasma spay parameters

Parameters	Unit	Value
Arc power	kW	42
Primary plasma gas (Ar)	slpm	50.5
Secondary plasma gas (H_2)	slpm	7.0
Spray distance	mm	90

Table 3 Components of TBCs

Coating ID	Components of spray powder for the bottom ceramic layer	Components of the spray powder for the top ceramic layer
CT	8YSZ	8YSZ
PT1	8YSZ	10vol.% PG+90 vol.% 8YSZ
PT2	8YSZ	20vol.% PG+80 vol.% 8YSZ
PT3	8YSZ	30vol.% PG+70 vol.% 8YSZ
ST1	8YSZ	10vol.% SG+90 vol.% 8YSZ
ST2	8YSZ	20vol.% SG+80 vol.% 8YSZ
ST3	8YSZ	30vol.% SG+70 vol.% 8YSZ

3. Results and discussions

3.1 Structure of the powders and the coatings

Fig. 2 shows the typical morphology of the as-prepared composite powders containing 30% graphite. It can be observed that the graphite particles either in spherical shape or planet shape were well included in the nano-scaled 8YSZ particles matrix. Apparently, composite powders containing the graphite with different morphology were achieved.

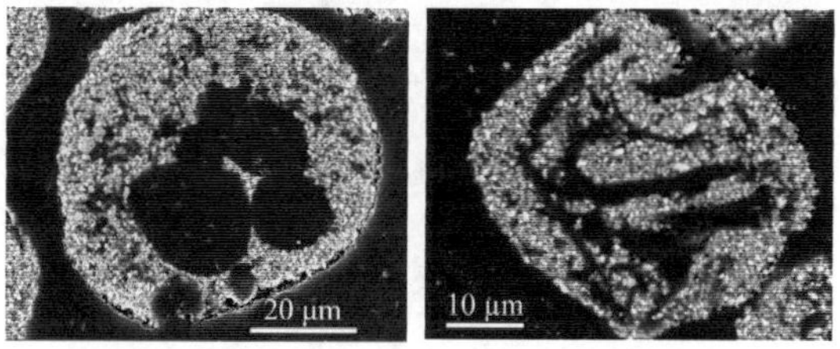

(a) 8YSZ/SG (b) 8YSZ/PG

Figure 2: Microstructure of the composite powders

Fig. 3 shows the phase structure of the spraying powder, which containing 30 vol.% SG, and the as deposited-coating before and after thermal exposure at 800 □ for 5 h. It can be observed that the graphite content in the as-sprayed coating is a little less than that in the spray powders, which may result from partial oxidation or redounding of the graphite particles during spraying process. Then, after thermal exposure at 800 □ for 5 h, graphite phase was not detected in the 8YSZ coating by XRD, and no new phase was formed. For the coating deposited with 8YSZ/PG, the results of XRD characterization is the same with the coating deposited with 8YSZ/PG powders.

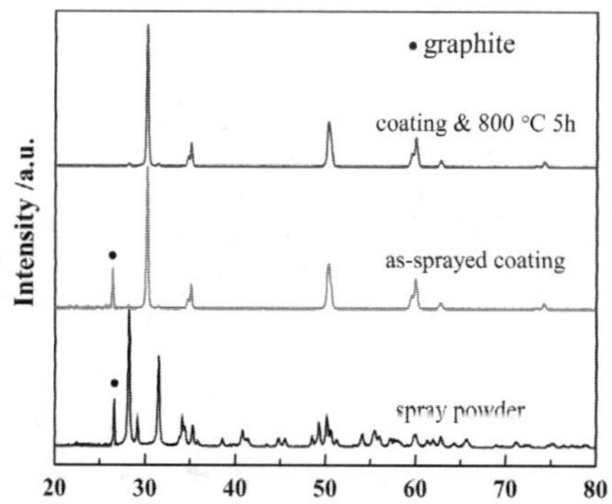

Figure 3: Phase structure of the spray powders and coatings

Fig.4 shows the cross-sectional morphology of the TBCs before thermal cycling test. In all TBCs, the thickness of the ceramic coatings is 250-300 μm. Apart form CT, which has a single ceramic layer structure of 8YSZ, the other TBCs presented a typical double-ceramic-layer structures: a conventional 8YSZ coating with thickness of about 150 μm was deposited on the bond coating, additional porous 8YSZ coatings with different pore structures were then fabricated on the conventional 8YSZ layer. **Fig. 5** shows the porosity of the top layer of the TBCs against the graphite content in the powder. As the content of graphite in the spray powders increases, the porosity of the 8YSZ coatings is increased. In addition, when the graphite contents are the same in the spray powders, using SG can yield a higher porosity in the coatings compared with PG. It means that under that experiment condition, SG is easier to be deposited in the coatings.

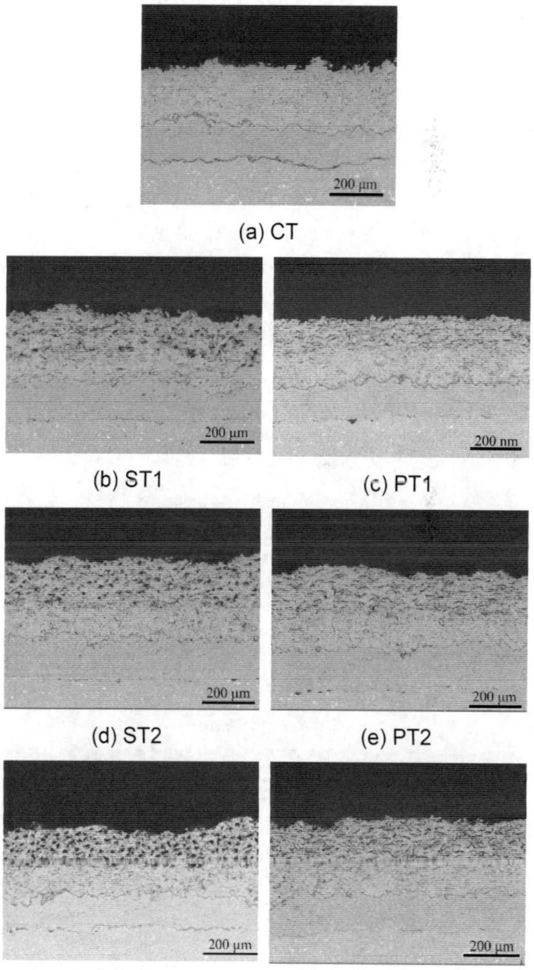

(a) CT

(b) ST1 (c) PT1

(d) ST2 (e) PT2

(f) ST3 (g) PT3

Figure 4: Microstructure of the TBCs before thermal cycling test

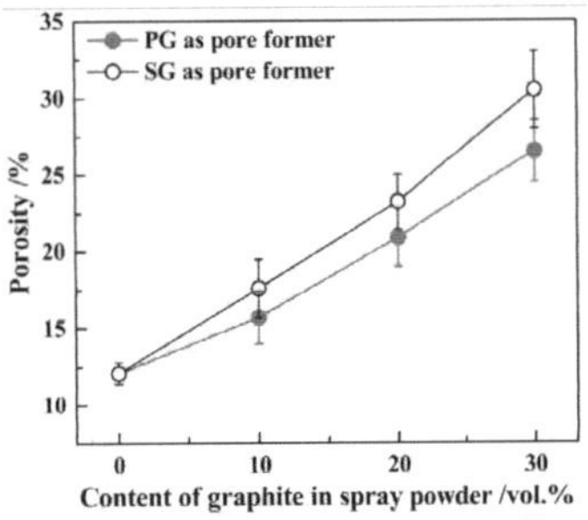

Figure 5: porosity of the coatings with different content of pore former

3.2 Thermal cycling behavior of the TBCs

Fig. 6 shows the morphology of the coatings after thermal cycling test. It was found that the failure of PT3 and ST3 TBCs occurred at the location near the interface of the porous layer and the conventional 8YSZ layer. The porous 8YSZ layer detached from the top of 8YSZ layer after only one or two cycles . For the other coatings, firstly, the thermal cycling life increased with the porosity of the porous layer as shown in Fig.7. Secondly, the failure of these TBCs occurred at the location near the interface of the conventional 8YSZ layer and TGO. From the results shown in Fig.4, it can be found that as the porosity of the top ceramic layer was increased up to 22%, the thermal cycling life of TBCs was increased by 30% in comparison of the convetional single layer TBC. However, when the porosity was increased over to 25%, the thermal cyclic lifetime decreased dramstically to several cycles.

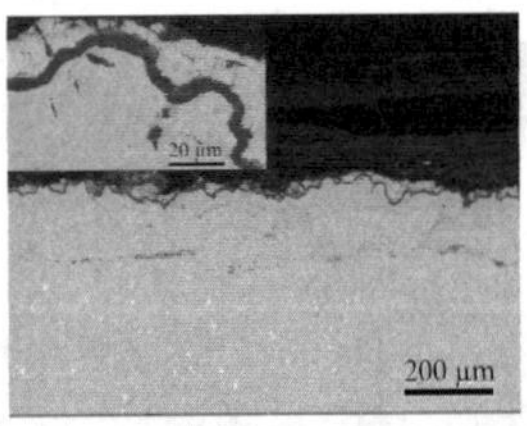

(a) CT 96 cycles

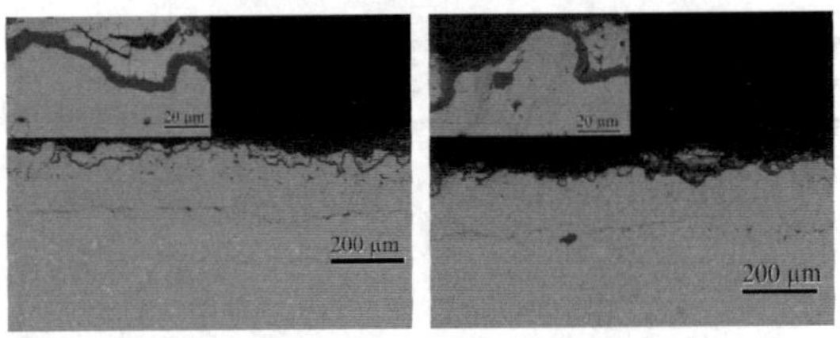

(b) ST1 107 cycles (c) PT 114 cycels

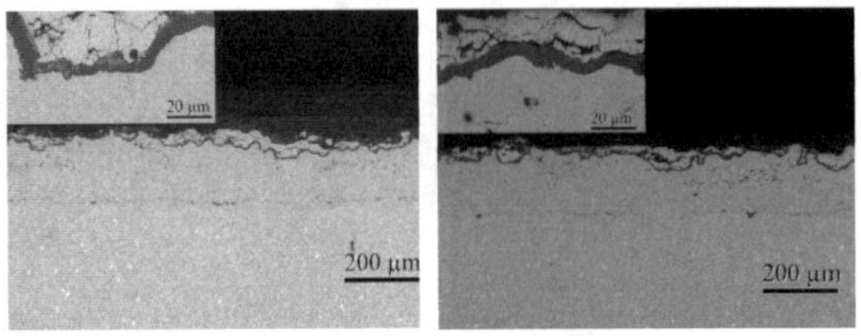

(d) ST2 127 cycles (e) PT2 135cycles

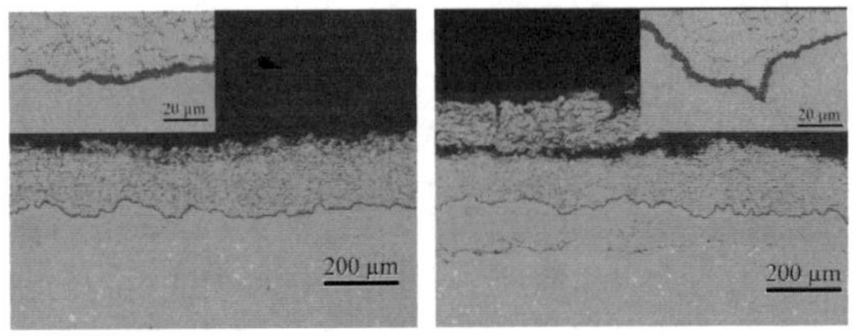

f) ST3 1 cycle (g) PT3 2 cycles

Figure 6: Microstructure of the TBCs after thermal cycling test

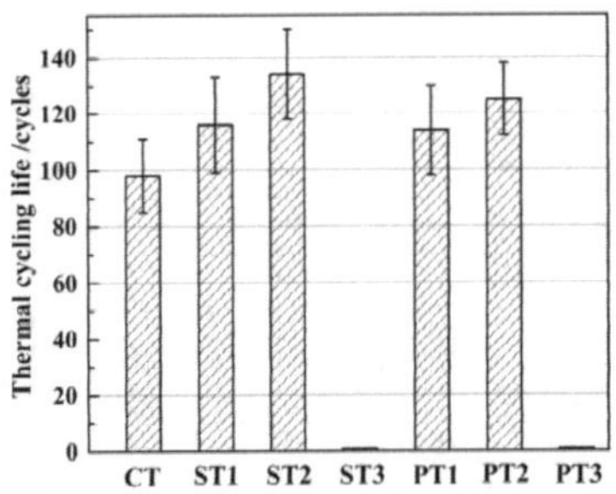

Figure 7:Thermal cycling life of the TBCs with different microstructure

3.3 Effect of elastic energy on the thermal cycling behavior of TBCs

During thermal cycling, the growing of TGO results in large local stress in the coating near TGO. Therefore, in most of situations, the cracks propagate near TGO. In addition, the thermal expansion mismatch between the ceramic coating and the metal substrate can also bring about stress that has significant influence on the thermal cycling behavior of TBCs. For the TBCs failed at the location far from TGO (*e.g.* ST3 and PT3), the influence of TGO is limited. The detaching of the ceramic layer is mainly resulted from the stress induced by the thermal expansion mismatch between the ceramic coating and the substrate.

Recently, Sampath et al. [15] developed a model to explain the effect of the structure of the ceramic coating on the thermal cycling behavior of TBCs by using the concept of elastic energy. Assuming that there is no stress in the coating when the coatings are held at high temperature condition during thermal cycling. However, when the coatings are cooled to room temperature, stress resulted from the thermal expansion mismatch arises in the coatings and the elastic energy which acts as the driving force of the propagation of cracks is accumulated from the surface of the coating in the through-thickness direction (define the surface of the ceramic layer as 0 position) as given by following equation:

$$\varepsilon_{(y)} = \alpha_{(y)}\Delta T \tag{1}$$

$$\varepsilon_s = \alpha_s \Delta T \tag{2}$$

$$\Delta\varepsilon_{(y)} = \varepsilon_{(y)} - \varepsilon_s \tag{3}$$

$$\delta_{(y)} = \frac{E_{(y)}}{1-\nu_{(y)}} \tag{4}$$

$$\sigma_{(y)} = \frac{E_{(y)}}{1-\nu_{(y)}}\Delta\varepsilon_{(y)} \tag{5}$$

$$U_{(y)} = \frac{1}{2}\int_0^y \frac{\sigma_{(y)}^2}{E_{(y)}}dy = \frac{1}{2}\int_0^y \frac{E_{(y)}}{(1-\nu_{(y)})^2}dy \tag{6}$$

Where, ΔT is the difference between the holding temperature and the ambient temperature during thermal cycling test; $\alpha_{(y)}$ is the thermal expansion coefficient of the ceramic coating at y position; $\alpha_{(s)}$ is the thermal expansion coefficient of the substrate; $\varepsilon_{(y)}$ is the strain of the ceramic coating at y position; $\varepsilon_{(s)}$ is the strain of the substrate; $\sigma_{(y)}$ is the the stress in the plane direction at y position; $\nu_{(y)}$ is the Poisson ratio of at y position; $U_{(y)}$ is the accumulated elastic energy at y position.

The elastic energy is accumulated from the surface to the internal of the coating. When the elastic energy in any position is larger than the critical strain energy, resistance against crack propagation in that place, cracking would occur. As the porosity of the coating is increased to a certain value, the fracture toughness of the coating becomes lower than the driving force of crack prorogation, TBCs would fail at that position.

For the coating failed at the position near TGO, the local stress induced by TGO is considered as the main reason [16]. Evans [16] proposed the concept of the critical thickness of TGO, that was used to evaluate the relative life of TBC:

$$h_c = 2\sqrt{\pi}[m/(m-1)](1-v^2)(K_{Ic}/E)(d^{3/2}/L) \tag{7}$$

where, h_c is the critical thickness of TGO; m is volume change ratio of the metal after oxidation; K_{IC} is fracture toughness of the ceramic coating near TGO; E is the elastic modulus of the the ceramic coating near TGO; d and L are parameters related the surface topography of the bond coating.

Apparently, the accumulated elastic energy in ceramic coatings was not took into account in this model. If considering the elastic energy, the relative critical TGO thickness can be calculated according the following formulas:

$$K_{Ic}^* = \sqrt{(G_{Ic}-U)E/(1-v^2)} \tag{8}$$

$$h_c^* = 2\sqrt{\pi}[m/(m-1)](\sqrt{(G_{Ic}-U)(1-v^2)/E}d^{3/2}/L \tag{9}$$

where, K^*_{IC} is the relative fracture toughness of the ceramic coating near TGO; h^*_c is the relative critical thickness of TGO; G_{IC} is the fracture energy release rate of the ceramic coating near TGO.

Figure 8 shows the influence of the the elastic energy on the critical thickness of TGO. As the elastic energy is increased, the critical thickness of TGO is deceased. According to formula (9), the critical thickness of TGO can be increased by reducing elastic energy or adjusting the properties of the ceramic coatings near TGO. Based on formula (7), reducing the elastic modulus of the ceramic coatings in any position can reduce the elastic energy. Therefore, for the positions (in the ceramic coating) far from the TGO surface, under the premise to resist the propagation of crack at that position, decreasing the local elastic modulus can effectively reduce the accumulation of elastic energy near TGO. For the ceramic coating near TGO, at first, the accumulated elastic energy in the ceramic coating reached the highest level. In addition, the growth of TGO would bring out higher stress in this area. As a result, increasing the resistance against crack propagation becomes an important issue to prelong the thermal cycling life time of TBCs. Depositing a dense ceramic layer with high fracture toughness between TGO and the conventional ceramic layer was proved to be an effective method to improve the thermal cycling life of TBCs [6]. However, as the elastic modulus of the dense layer is significantly higher than the conventional coating, the elastic energy per thickness in the dense layer is also much higher than the conventional coating. Therefore, as reported by Li [6], the thickness of the dense layer is a key factor to the thermal cycling life to TBCs

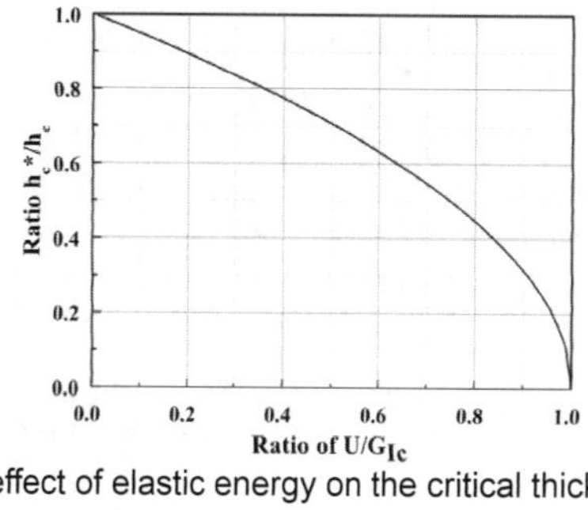

Fig. 8:effect of elastic energy on the critical thickness of TGO

4. Conclusions

The microstructure of ceramic layers has important influence on the thermal cycling behavior of the TBCs. Double-ceramic-layer TBCs were fabricated by depositing a porous 8YSZ layer on the conventional 8YSZ layer. As the porosity of the top ceramic layer was increased up to 22%, the thermal cycling life of TBCs was increased to 130 cycles compared with less than 100 cycles. When the porosity was increased over to 25%, the thermal cyclic lifetime decreased dramstically to several cycles.

An elastic energy model was applied to qualitatively explain the thermal cycling behavior of TBCs. The elastic energy is accumulated from the surface to the internal of the coating during thermal cycling. When the elastic energy in any position is larger than the local critical strain energy, cracking would occur. As the porosity of the coating was increased to 25%, it was considered that the fracture toughness of the coating became lower than the driving force of crack propagation, which resulted in spalling of porous top YSZ at very limited cycles.

Acknowledgments

The present project is supported by National Basic Research Program (Grant No. 2012CB625100)

5. Literature

[1] Padture, N.P., M. Gell, and E.H. Jordan: Thermal barrier coatings for gas-turbine engine applications. Science, 296 (2002), pp. 280/4.

[2] Hardwicke, C.U. and Y.-C. Lau: Advances in Thermal Spray Coatings for Gas Turbines and Energy Generation: A Review. Journal of Thermal Spray Technology, 22 (2013), Issue 5, pp 564/76.

[3] Feuerstein, A., et al: Technical and Economical Aspects of Current Thermal Barrier Coating Systems for Gas Turbine Engines by Thermal Spray and EBPVD: A Review. Journal of Thermal Spray Technology, 17 (2008), Issue 2, pp 199/213.

[4] Li, C.-J., G.-J. Yang, and C.-X. Li: Development of Particle Interface Bonding in Thermal Spray Coatings: A Review. Journal of Thermal Spray Technology, 22 (2012), Issue 2-3, pp. 192/206.

[5] Bakan, E., et al.: Porosity-Property Relationships of Plasma-Sprayed Gd2Zr2O7/YSZ Thermal Barrier Coatings. Journal of the American Ceramic Society, 98 (2015), Issue 8, pp. 2647/54.

[6] Li, C.-J., et al.: A Novel Plasma-Sprayed Durable Thermal Barrier Coating with a Well-Bonded YSZ Interlayer Between Porous YSZ and Bond Coat. Journal of Thermal Spray Technology, 21 (2012), Issue 3-4, pp. 383/90.

[7] Hao, S., C.-J. Li, and G.-J. Yang: Influence of Deposition Temperature on the Microstructures and Properties of Plasma-Sprayed Al2O3 Coatings. Journal of Thermal Spray Technology, 20 (2010), Issue 1-2, pp. 160/9.

[8] Xing, Y.-Z., C.-J. Li et al.: Influence of Microstructure on the Ionic Conductivity of Plasma-Sprayed Yttria-Stabilized Zirconia Deposits. Journal of the American Ceramic Society, 91 (2008), Issue 12, pp. 3931/6.

[9] Lima, R.S. and B.R. Marple: Nanostructured YSZ thermal barrier coatings engineered to counteract sintering effects. Materials Science and Engineering: A, 485 (2008), Issue 1-2, pp. 182/93.

[10] Arai, M., H. Ochiai, and T. Suidzu: A novel low-thermal-conductivity plasma-sprayed thermal barrier coating controlled by large pores. Surface and Coatings Technology, 485 (2016), pp. 120-/7.

[11] Curry, N., et al.: Next Generation Thermal Barrier Coatings for the Gas Turbine Industry. Journal of Thermal Spray Technology, 20 (2010), Issue 1-2, pp. 108/15.

[12] Sarikaya, A. and F. Dogan: Effect of various pore formers on the microstructural development of tape-cast porous ceramics. Ceramics International, 39 (2013), Issue 1, pp. 403/13.

[13] J.J. Tang, K.L., Q.Z. Yang, Y.H. Wang, P. Zhang, Y. Wang, L. Zhao, Q.Q. Fu, Z.H. Han, Y. Bai, The influence of size and distribution of graphite on the friction and wear behavior of Ni–graphite coatings. Surf Coat Technol, 252 (2014), pp. 43/55.

[14] Li Y, Li CJ, Zhang Q: Influence of TGO composition on the thermal shock lifetime of thermal barrier coatings with cold-sprayed MCrAlY bond coat[J]. J Therm Spray Technol, 19 (2010), Issue 1-2, pp. 168/77.

[15] Viswanathan, V., et al.: Multilayer, Multimaterial Thermal Barrier Coating Systems: Design, Synthesis, and Performance Assessment. Journal of the American Ceramic Society, 98 (2015), Issue 6, pp. 1769/77.

[16] A. G. Evans, D.R.M., J. W. Hutchison et al.: Mechanisms controlling the durability of thermal barrier coatings. Progress in Materials Science, 46 (2001), pp. 505/53.

Deposition mechanism of quasi-columnar YSZ coatings during plasma spray physical vapor deposition

Lihua Gao, Jiaojuan Ji, Xiaoliang Lu, Deming Zhang, Jie shen,Yueguang Yu, Beijing/CN

As a new processing technology, plasma spray physical vapor deposition (PS-PVD) is capable to deposit coatings out of the vapor phase with high deposition rate. Moreover, the resulting quasi-columnar coatings were unique, hardly deposited by other process. Due to its low thermal conductivity and excellent superior strain tolerance, quasi-columnar coatings attract much attention in the thermal barrier coatings (TBCs) field. In this paper, the morphology variation of the quasi-columnar yttria stabilized zirconia (YSZ) coatings deposited under different conditions was investigated. Combined the morphology of the initial deposits during short time spraying, the deposition mechanism of quasi-column microstructure was concluded. It can be found that the quasi-column coating was formed by co-deposition of vapor phases and solid particles. The vapor phase can develop into column, while the solid particles not only intensified the shadowing effect but also destroyed the nucleus during vapor deposition.

Keywords: Yttria stabilized zirconia (YSZ); Microstructure; Deposition mechanism; Plasma spray physical vapor deposition (PS-PVD); quasi-columnar;

1 Introduction

As a new processing technology, plasma spray physical vapor deposition (PS-PVD) makes it possible to produce columnar structured coatings out of the vapor phase using plasma spray technology. PS-PVD was developed based on the low pressure plasma spray process (LPPS, also known as vacuum plasma spraying, VPS) [1]. However, the plasma jet property of PS-PVD is greatly different from that of LPPS due to the low working pressure (1~2 mbar) and high energy plasma gun of PS-PVD [2]. The plasma plume of PS-PVD can expands to more than 2 m long and 200-400 mm in diameter, thus leading to broad deposition profile [3,4]. Moreover, the supersonic plasma plume temperatures can reach 6000-10000 K due to the high power levels (>100 kW) [5]. Therefore, the injected feedstock powders could be easily melted and builds up into a typical lamellar structure. Beyond that, when special fine grain sized powders are used, it is possible for the injected powders to be evaporated so as to deposit from gas instead of liquid phase [6,7]. Therefore, PS-PVD combines the advantages of conventional plasma spray and PVD techniques.

Recently, PS-PVD has been paid much attention, since it can deposited coatings with special designed microstructure and comply with the growing demands on modern functional coatings. As reported, different microstructures of YSZ top coat, including dense coatings with splat-like microstructure, EB-PVD-like columnar coatings, quasi-columnar coatings and mixed microstructure coating, could be formed by adjusting the feed rate, plasma gas composition and spray distance [8-10]. In order to produce these advanced microstructures with high quality, their deposition mechanisms was discussed by looking into the morphologies and phase compositions of the initial deposits during several seconds spraying. It is reported that the dense coating was stacked up mainly by liquid droplets, un-melted/semi-melted particles and vapor phases. The EB-PVD-like columnar coating was mostly deposited from vapor phases which were transported to the substrate surface in the state of vapor atoms and clusters. The hybrid coating was formed by co-deposition of vapor phases and liquid droplets [11].

However, there is little report about the deposition mechanism of quasi-columnar coatings, which attracts much attention in the thermal barrier coating field. Therefore, in order to study the deposition mechanism of quasi-columnar coatings in PS-PVD, the morphology variation of the quasi-columnar yttria stabilized zirconia (YSZ) coatings deposited under different conditions was investigated. The morphology of the initial deposits during short time spraying was also studied. Based on these experiment results, the deposition mechanism of quasi-column microstructure was concluded.

2 Experiments

All YSZ coating experiments were carried out on a PS-PVD system (Sulzer Metco, Switzerland). Fig. 1 gives an overall view of the facility. The system is composed of a vacuum chamber and a load lock allowing quick changes of the substrates without breaking the vacuum in the main chamber. The chamber was equipped with a vacuum pump unit consisted of a rotary piston pump and two roots pumps. By controlling the rotating speed, the chamber can be pumped down to a low working pressure of 100 Pa. Besides, an O3CP plasma gun with a maximum current of 3000 A and a maximum net power of 180 kW was also used. Sulzer Metco 6700 YSZ powder was used as spray feedstock as shown in Fig. 2.

Fig.1 An overall view of the PS-PVD facility

In order to study the effect of the substrate material on the quasi-columnar microstructure, polished Ni-based superalloy, graphite substrates and Ni-based superalloy coated with NiCoCrAlY bond coat were used. For the effect of substrate roughness on the microstructure, NiCoCrAlY bond coat sample without polishing were used to compare. The spray parameters were selected based on the present report of YSZ deposition, as listed in Table 1. Prior to spraying, the substrates were preheated to ~950 °C. Besides, as to the effect of substrate temperatures on the quasi-columnar, the preheating temperatures were controlled at 650 °C and 1150 °C, respectively.

Fig.2 Morphology of the Sulzer Metco 6700 YSZ powder

Table 1 Plasma spray parameters

Plasma gas	Ar 35 SLPM / He 60 SLPM
Current	2000 A
Net power	65 KW
Feed rate	10 g/min
Spray distance	900 mm
Carrier gas	Ar 16 SLPM
Substrate temperature	950 °C

SLPM: standard liters per minute

Besides, for the purpose of obtaining more information about the deposition mechanism in PS-PVD, the initial deposits were collected onto the polished substrates by controlling the spray time. The morphologies of YSZ coatings were characterized by a scanning electron microscopy (SEM, FEI Quanta 600) equipped with energy dispersive spectrometer (EDS).

3 Results and discussion

3.1 Effect of the substrate material

In order to discuss the effect of substrate material on the microstructure, a series of YSZ coatings were deposited on different substrate. Fig. 3 shows the SEM micrographs of the samples, which deposited on polished Ni-based superalloy, graphite substrate and NiCoCrAlY bond coat substrate, respectively. It can be observed that all the coating exhibited typical quasi-columnar microstructure with domed tops. Some micron-sized and nano-sized particles stick on the columnar or exist between the columnar gaps. As report, these particles may be the un-melted/semi-melted YSZ powders or re-solidified nanoparticles in the plume [9,10].

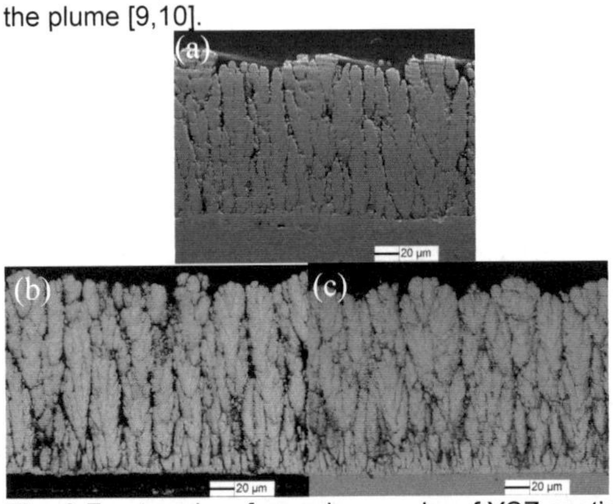

Fig. 3 Fractured surface micrographs of YSZ coatings deposited at different substrate material: (a) Ni-based superalloy, (b) graphite substrates and (c) NiCoCrAlY

3.2 Effect of the substrate temperature

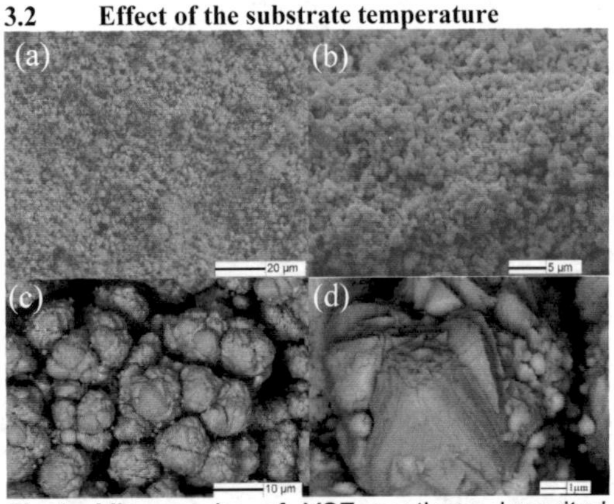

Fig.4. Micrographs of YSZ coatings deposited at different subatrste temperatures: (a) (b) 650°C and (c) (d) 1150 °C

The results of Fig. 3 show that the coating deposited at ~ 950 °C has typical quasi-columnar structure. When the substrate temperature was changed, some differences in the morphology of the coating can be observed. Fig. 4a shows the SEM micrographs of the coating deposited at 650 °C. It can be found that the no columnar was deposited. The coating is very thin, which is stacked up by many solid particles. A closer examination of the coating in Fig. 4b reveals that these particles are fine nano-particles or micron-sized particles, which was similar to that existed in quasi-columnar. This indicates that these particles result from the un-melted/semi-melted YSZ powders or re-solidified nanoparticles in the plume.

With increasing the substrate temperature to 1150 °C, the quasi-columnar structure, which is similar to that deposited at 950 °C, appears again as shown in Fig. 4c. Moreover, the top of the columnar crystal exhibits some crystallographic plane characteristics. Therefore, it can be found that the transformation mechanism of quasi-columnar coating is different from that of the EB-PVD like columnar structure. G. Mauer builds up a Structure Zone Model of pure vapor deposition in PS-PVD, which is completely consistent with the Thornton model [12]. Shadowing and surface diffusion are the basic mechanisms characterizing the growth of the columnar structure. At low substrate temperature, shadowing mainly occur and columnar with dome top would be deposited. With increasing the substrate temperature, surface diffusion would overcome the shadowing effect and faceted column would be deposited [12,13]. However, when the solid particles exist in the plume, the microstructure variation would be different, where no columnar were developed at low substrate temperature as show in Fig. 4a. One possible reason is that solid phase particles affects the formation and growth of the nucleation during vapor deposition. Detail explanation will be discussed in Section 3.4.

3.3 Effect of the substrate roughness

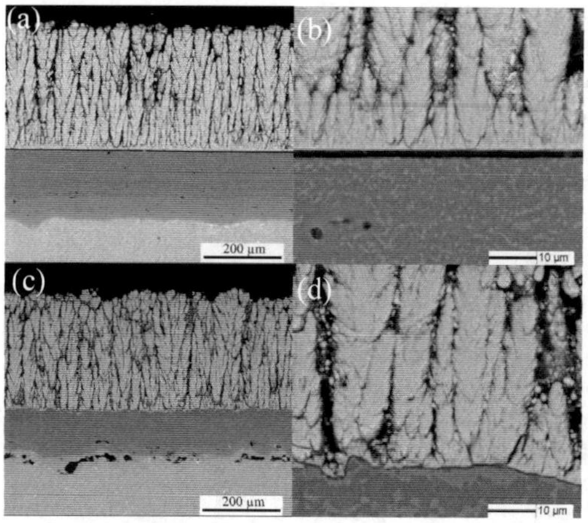

Fig.5 Fractured surface micrographs of YSZ coatings deposited at different substrate with (a)(b) and without poishing (c)(d)

Fig. 5 shows the cross-section micrograph of YSZ coating deposited on substrates with different roughness. It is obvious that both the YSZ topcoats exhibit quasi-columnar structure. However, owing to the rough surface of the bond coat, the deposited columns have larger gaps between each other than that deposited on smooth polished bond coat. Moreover, the columns are wider and random oriented. Besides, more defects can be found in the coatings as shown in Fig. 5b.

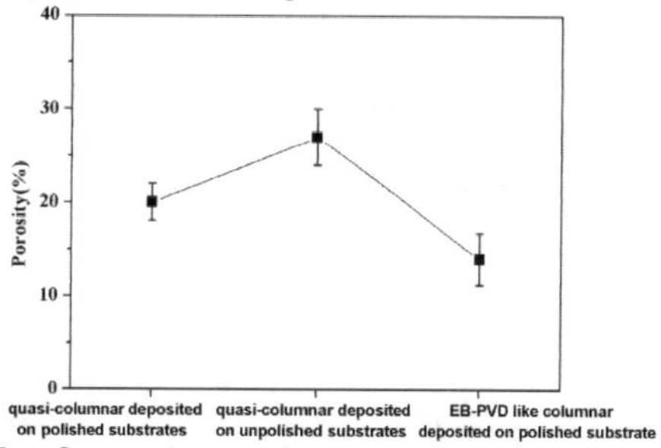

Fig.6. Comparison of the porosity for different coatings

A comparison between the porosity of these YSZ layers is depicted in Fig. 6. For the coating on smooth bond coat, the porosity is ~ 20%. In contrast to this, the porosity in the coating on as-fabricated bond coat is higher (~27%), whereas the porosity of EB-PVD like columnar is only ~14% [10]. The main mechanism for formation of these porous quasi-columnar structure is the so-called shadowing effect, which is the interaction between the roughness of the growing surface and the angular directions of the arriving particles [14,15]. Different from that on the smooth surface, the columns start to grow on the highest points of a rough surface due to the shadowing effect, hence resulting more shadowed areas and large amounts of defects and pores [16]. Besides, the solid particles may also intensify the shadowing effect.

3.4 Deposition mechanism

In order to study the deposition mechanism of quasi-column structure in PS-PVD, a series of samples with different predetermined deposition times were fabricated on the polished substrates at ~950 °C. Fig. 7 shows the SEM micrographs of the samples, corresponding to deposition times of 2 and 8 s, respectively. After 2 second deposition, fine compact columnar with domed tops are found to deposit on the substrate. As reported, the column is mostly deposited from vapor phases. The vapor atom/clusters form a nucleation center and grow into column in three-dimensional island model [11,18]. Besides, some nano-sized and micro-sized YSZ particles are found scattering on the surface. This indicates that the quasi-columnar coatings are deposited from vapor phase and solid particles. It is worthwhile to note that some micro-sized particles broken into pieces which were indicated by black arrows. In PS-PVD, solid particles, especially micro-sized particles, can be accelerated by the plasma plume and have high velocity impinging, probably getting broken as they hit the substrate. Beyond that, there are some round dents as indicated by black arrows in Fig.6a, which show the similar shape with the micro-sized particles. It can be inferred that their

formation can be probably attribute to the collision and spallation of micro-sized solid particles.
As the deposition time increased to 8 s, it is clearly observed that the columnar grow bigger and some columnar crystals are significantly greater than the other mainly due to the competitive growth. At the beginning of the columnar crystal growth, many nucleation centers are formed on the substrate. Therefore the columnar crystals are small and compact arrangement. With the deposition going on, the columnar grow competitively and some develop into columnar while some stop growing. Fig. 6c shows the fracture cross-section morphology of the YSZ coatings. The coating exhibits typical features of a quasi-columnar structure. Large amounts of nano-sized secondary columns are also clearly observed, which may be formed by secondary shadowing as reported [13,18]. It worth noticed that some columnar grow on the solid particles as marked by the white circle. This indicates that the nucleation was formed there and grew into column.

Fig. 7. SEM micrographs of samples sprayed with different deposition times: (a) 2 s and (b)(c) 8 s

Based on the above experiment results, the growth model of quasi-columnar microstructure is concluded as shown in Fig. 8. The coating is co-deposited from the mixture of vapor phase and solid particles. When the vapor atoms/clusters are absorbed on the substrate, they diffuse and nucleate on the surface of the substrate. With the continuous arriving of vapor phase, nuclei would grow into columnar crystal by the three-dimensional island growth model. While the solid particles are accelerated by the plume, they have high velocity impinging. When hitting on the substrate, some break into pieces, while some remain intact. However, they would probably spall due to the collision of the coming particles. Moreover, the growth of the columnar would be inevitably affected by the solid particles. The vapor nucleation center on the substrate would be probably destroyed due to the collision of these particles, especially micro-sized particles.

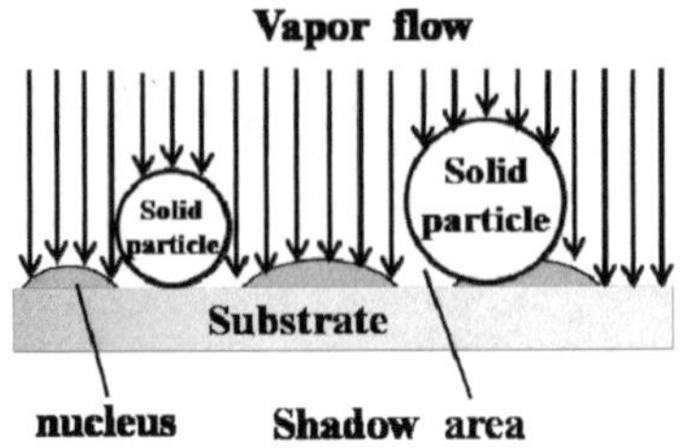

Fig. 8 Scheme of the destroy of the nucleus at the substrate and the formation of the shadow area when vapor phase and solid particle co-deposited

According to the above results, the microstructure of the coating strongly depends on the substract temperature. However, its transformation mechanism is different from that of the EB-PVD like columnar structure in PS-PVD. This may be related with the solid phase particles which affects the formation and growth of the nucleation during vapor deposition. During the vapor deposition, heterogeneous nucleation takes place on the substrate and the theoretical equation of the critical nucleation radius r* is given as follows [19,20]:

$$r^* = \frac{2\sigma_{nv} g T_m}{L_v g \Delta T}$$

Where σ_{nv} represents vapor-nucleation interface tension; L_v is latent heat of evapotation; T_m is melting point of feedstock powde and ΔT is amount of undercooling at which nucleus formed.
Nucleation rate was affected by interaction of nucleation energy and atom/cluster diffusion activation energy [20,21]:

$$N = K \exp\left(\frac{-\Delta G^*}{kT}\right) g \exp\left(\frac{-Q}{kT}\right)$$

Where ΔG^* is nucleation energy, Q is the interfacial diffusion activation energy of atom, k is Boltzmann constant. It can be found that substrate temperature is an important influencing factor for the critical nucleation radius and nucleation rate. At low substrate temperature, the critical nucleation radius decreases, resulting in lower nucleation energy. However, diffusion activation energy of vapor atom/clusters are low and their diffusion ability greatly reduced. As a consequence, the nucleation rate suffers a corresponding decrease. During the pure vapor deposition, nucleation center can grow into columnar with the arrival of the subsequent gas atoms/cluster. However, when the solid particles existed in the plume, due to their high deposition rate, solid particles would hit the substrate. When they hit the substrate, especially the nucleus, the nucleus would be probably destroyed due to the collision, leading to no columnar formed. Besides, due to the weak adhesion between the solid particles deposited, only a very thin layer of solid particles were attached on the substrate surface, as shown in Fig. 4a.

With increasing substrate temperature, according to the formula (1-2), the nucleation rate increases greatly, resulting in collision of solid particle relatively weakening and hence depositing into columnar, as shown in Fig. 4c. Moreover, vapor atom /clusters diffusivity enhance, which can diffused along the coating surface, adjust the crystals structure and finally develop into quasi-columnar with some crystallographic plane characteristics.
In contrast to the EB-PVD like columnar, the quasi-columnar coating grow many more new column in the growth stage of the columnar crystal, which is like the feature of dendrite. This may also be related with the solid particles existed in the plasma jet. When vapor phase is transported to the sample surface, it can also nucleate on solid particle, which could develop into a new column on the existing crystallites. Moreover, these solid particles can accelerate the shadowing effect during deposition and result in more shadowed areas and hence higher porosity than the EB-PVD like columnarn as shown in Fig.6. Besides, surface roughness of the substrate can also enhance shadowing effect according to the results of Fig. 5.

4 Conclusions

The morphology variation of the quasi-columnar YSZ coatings deposited under different conditions was investigated. Combined the morphology of the initial deposits during short time spraying, the deposition mechanism of quasi-column microstructure can be drawn as follows:

1) The quasi-columnar coating was formed by co-deposition of vapor phases and solid particle. The vapor phases nucleated on the substrate surface and grew into columnar crystal by the three-dimensional island growth model. While the solid particles not only destroyed the nucleus but also intensified the shadowing effect during vapor deposition.
2) The substrate temperature determined the microstructure of the coating. Especially when using low substrate temperature, the nucleus on the substrate would be probably destroyed due to the collision of the coming particles, which would not develop into column. While increasing the substrate temperature, quasi-column was deposited due to the weakening of the collision. Besides, surface roughness of the substrate can also enhance shadowing effect, thus leading to more pore and defects.

Reference

[1] T.D. Bennett, F.Yu, A nondestructive technique for determining thermal properties of thermal barrier coatings, J. Appl. Phys. 97 (2005) 1-12.

[2] K.V. Niessen, M. Gindrat, Plasma Spray-PVD: a new thermal spray process to deposit out of the vapor phase, J. Therm. Spray Technol. 20 (2011) 736-743.

[3] J.L. Dorier, M. Gindrat, Ch. Hollenstein, M. Loch, A. Refke, A. Salito, G. Barbezat, Plasma jet properties in a new process at low pressure for large area thin film deposition, International Thermal Spray Conference. Singapore (2001) 759-764.

[4] A. Refke, D. Hawley, J. Doesburg, R. Schmid, LPPS thin film technology for the application of TBC systems, International Thermal Spray Conference. Basel (2005) 438-443.

[5] K.V. Niessen, M. Gindrat, A. Refke, Vapor phase deposition using plasma spray- PVD, J. Therm. Spray Technol. 19 (2010) 502-509.

[6] G. Mauer, R. Vaßen, D. Stöver, Thin and dense ceramic coatings by plasma spraying at very low pressure, J. Therm. Spray Technol. 19(2010) 495-501.

[7] A. Hospach, G. Mauer, R. Vaßen, D. Stöver, Columnar-structured thermal barrier coatings (TBCs) by thin film low-pressure plasma spraying (LPPS-TF), J. Therm. Spray Technol. 20(2011)116-120.

[8] B.J. Harder, D. Zhu, Plasma spray-physical vapor deposition (PS-PVD) of ceramic for protective coatings, Ceram. Eng. Sci. Proc. 32 (2011) 73-84.

[9] C.Y. Li, H.B. Guo, L.H. Gao, L.L. Wei, S.K. Gong, H.B. Xu, Microstructures of yttria-stabilized zirconia coatings by plasma spray-physical vapor deposition, J. Therm. Spray Technol. 24 (2015) 534-541.

[10] L.H. Gao, H.B. Guo, L.L. Wei, C.Y. Li, S.K. Gong, H.B. Xu. Microstructure and mechanical properties of yttria stabilized zirconia coatings prepared by plasma spray physical vapor deposition, Ceram Int. 41 (2015) 8305-8311.

[11] L.H. Gao, L.L. Wei, H.B. Guo, S.K. Gong, H.B. Xu. Deposition mechanisms of yttria-stabilized zirconia coatings during plasma spray physical vapor deposition, Ceram Int. 42(2016) 5530-5536.

[12] G. Mauer, A. Hospach, N. Zotov, R. Vaßen, Process conditions and microstructures of ceramic coatings by gas phase deposition based on plasma spraying, J. Therm. Spray Technol. 22 (2013)83-89.

[13] Thornton J A. High rate thick film growth. Annu. Rev. Mater. Sci. 7(1977) 239-260

[14] A. Hospach, G. Mauer, R. Vaßen, D. Stöver, Characteristics of ceramic coatings made by thin film low pressure plasma spraying (LPPS-TF), J. Therm. Spray Technol. 21(2012)435-440.

[15] J.A. Thornton, Influence of substrate temperature and deposition rate on Structure of thick sputtered Cu coatings, J. Vac. Sci. Technol. 12(1975)830-835.

[16] A. Flores Renteria, B. Saruhan, U. Schulz, H.-J. Raetzer-Scheibe, J. Haug, A. Wiedenmann, Effect of morphology on thermal conductivity of EB-PVD PYSZ TBCs, Surf. Coat. Technol. 201 (2006) 2611-2620.

[17] S. Rezanka, G. Mauer, R. Vaßen. Improved thermal cycling durability of thermal barrier coatings manufactured by PS-PVD, J. Therm. Spray Technol. 23(2014) 182-189.
[18] K. Reichelt, Nucleation and growth of thin films, Vacuum. 38 (1988) 1083-1099
[19] X.F. Zhang, K.S. Zhou, C.M. Deng, M. Liu, Z.Q. Deng, C.G. Deng, J.B. Song, Gas-deposition mechanisms of 7YSZ coating based on plasma spray-physical vapor deposition. Journal of the European Ceramic Society,2015
[20] G. X. hu, X. Cai, Y.H. Yong, Material science foundation (third edition) [M]. Shanghai: Shang hai jiao tong university press, 2010: 229-236
[21] Z.Q.Wu, B. Wang. Thin film growth [M]. Beijing: science press, 2001:170-188.

Influence of the interface healing between the splashed particles and underlying bond coating on the cyclic oxidation behavior of LPPS MCrAlY bond coats

Bang-Yan Zhang, Guan-Jun Yang*, Cheng-Xin Li, Chang-Jiu Li, State Key Laboratory for Mechanical Behavior of Materials, School of Materials Science and Engineering, Xi'an Jiaotong University, China

The growth behavior of thermally grown oxide (TGO) on the bond coats of thermal barrier coatings (TBCs) has a significant impact on the TBCs life time. The splashed particles on the thermally sprayed MCrAlY bond coat surface have weak bonding with the underlying bulk coating, which will lead to the formation of mixed oxides and promote the failure of TBCs. In this study, the interface between the splashed particles and underlying bond coating was healed by vacuum pre-treatment (pre-diffusion). The bond coats were divided into three groups by the ways of pre-treatments before deposition of the ceramic top coating. Two groups of bond coats were subjected to two different pre-treatments (pre-diffusion-oxidation and pre-oxidation) prior to cyclic oxidation tests. The third bond coats were directly subjected to the same cyclic oxidation tests. Results show that the different TBCs present different cyclic oxidation behavior. The TBCs with pre-diffusion-oxidation exhibited the best cyclic oxidation resistance, and the ones without pre-treatment had the worst cyclic oxidation resistance. It is revealed that the healing of interface between the splashed particles and underlying bulk coating can effectively improve the cyclic oxidation resistance. This study will benefit the development of advanced TBCs with long life time and high reliability.

1 Introduction

Thermal barrier coatings (TBCs) have been widely applied to hot section parts in gas turbines for both aircraft engine and industrial electric power generation [1-4]. Thermal barrier coatings are generally composed of superalloy substrate, MCrAlY bond coat and thermal insulating yttria-stabilized zirconia (YSZ) top coating. A thermally grown oxide (TGO), which grows at the interface between the bond coat and the YSZ coating, is actually an additional part of TBCs [5]. It is reported that the TGO plays an extremely important role in determining the lifetime of TBCs [1, 6]. The preparation approach of MCrAlY bond coat and coating microstructure is closely related with the durability of the thermal insulating top coating. For atmospheric plasma sprayed (APS) YSZ top coating, a rough bond coat surface is required to create a strong interlocking adhesion. However, such interface morphology produces the out-of-plane stresses responsible for in-service failure [7], since the out-of-plane stress due to the undulating bond coat surface induces the cracking between the TGO and the bond coat at the undulation crest [8], and promotes the failure of TBCs [9].

Theoretically, a stable and slowly grown α-Al_2O_3 TGO is optimal. However, the uniformity along the interface and the constituents of TGO are significantly influenced by the surface morphology of the bond coat before YSZ deposition [7]. Besides α-Al_2O_3 scale, Cr_2O_3, NiO, and $(Ni,Co)(Cr,Al)_2O_4$ may be present in TGO. Different composition of TGO is associated with different failure mechanisms of TBCs. The formation of Cr_2O_3, being a protective scale against corrosion, needs a relatively high critical oxygen partial pressure when compared with the formation of α-Al_2O_3. However, it changes to CrO_3 at a temperature higher than 1000 °C and an oxygen partial pressure range of 0.1–10^{-4} MPa. NiO tends to grow in an atmosphere of a high oxygen partial pressure. Spinels, stoichiometrically expressed as $(Ni,Co)(Cr,Al)_2O_4$, are porous and brittle [10]. Therefore, many investigations [9, 11, 12] were conducted to aim at controlling the TGO structure.

For LPPS bond coat, the splashed particles resulting from spray droplet impact can be found everywhere on the bond coating surface. These particles are produced by the splashing of liquid droplets impacting on deposited surface with a rough topographical morphology [13]. We have found that the splashed particles are weakly bonded to the underlying coating surface, which will result in the formation of mixed oxides. We also found that the interface healing between splashed particles and underlying coating occurred after the vacuum pre-treatment (pre-diffusion), and the interface healing can avoid the generation of the mixed oxides on TGO surface. However, the influence of the interface healing between the splashed particles and the underlying bond coat on the cyclic oxidation behavior of the TBCs has not been reported yet.

In this study, the LPPS bond coat was subjected to different pre-treatments before the cyclic oxidation tests. The bond coats were divided into three groups by the ways of pre-treatments. Two groups of bond coats were respectively subjected to two different pre-treatments (pre-diffusion-oxidation and pre-oxidation) prior to cyclic oxidation tests. The third bond coats were directly subjected to the same cyclic oxidation tests. The cyclic oxidation resistance of all the bond coats were investigated.

2 Experimental procedure

The bond coats of ~150 μm in thickness were deposited by LPPS, using CoNiCrAlY powder particles in a spherical shape (Amdry 9951, Sulzer-Metco, USA). The particle size distribution (d10=10.7 μm, d90=36.9 μm) with a mean particle size of ~23.5 μm. A thin 8YSZ (8 wt% yttria-stabilized zirconia, 204B-NS, Sulzer-Metco, USA) top coating was deposited on the bond coat surface by a commercial

plasma spray system (APS, GP-80, Jiujiang, China). Nickel-based superalloy Inconel 738 (ϕ25.4 mm × 3 mm) was used as a substrate.
The LPPS spray conditions are shown in Table 2. In the LPPS process, highly purity argon (Ar ≥ 99.999%) was chosen as the main gas for spraying and the powder feeding gas, and high purity hydrogen was chosen as the auxiliary gas for spraying.

Table 1 LPPS conditions

Parameters	Value
Arc current (A)	620
Arc voltage (V)	72
Plasma gas (Ar) flow ($L \cdot min^{-1}$)	40
Plasma gas (H_2) flow ($L \cdot min^{-1}$)	8
Powder feeding gas (Ar) flow ($L \cdot min^{-1}$)	2
Chamber pressure (kPa)	15
Spray distance (mm)	280

The bond coats were divided into three groups by the ways of pre-treatments. Two groups of bond coats were respectively subjected to two different pre-treatments (pre-diffusion-oxidation and pre-oxidation) prior to cyclic oxidation tests. The third bond coats were directly subjected to the same cyclic oxidation tests. One approach, pre-oxidation, included only a pre-oxidation conducted in Ar. The other approach, pre-diffusion-oxidation, included a pre-diffusion in vacuum followed by the pre-oxidation treatment mentioned above. The detailed parameters of pre-diffusion and pre-oxidation treatments are shown in Table 2. The pre-diffusion can make small particles on the bond coating surface have a good bonding with the underlying bulk coating, which will be further shown in the results part. The pre-oxidation treatment was used to form a sufficiently thick and continuous oxide scale on the surface of the bond coats to suppress inward diffusion of oxygen and outward diffusion of cation during the subsequent high temperature oxidation.
After pre-treatments, the YSZ top coating was deposited by APS onto the bond coat surface. Table 3 shows the APS conditions. Then, thermal cyclic test of the TBCs specimen was conducted at the following thermal cycling condition: the sample was heated to 1200 °C in 2 min, held for 27 min and then rapidly cooled down to a temperature lower than 80 °C by an air blower in 4 min. The TBCs will be considered to be failed, when the peeling area of the TBCs surface is greater than or equal to 50%. The lifetime is the average of the thermal cyclic lifetimes of three TBCs specimen.

Table 2 Pre-treatment conditions

Pre-treatment	Environment	Temperature and duration
Pre-oxidation	Ar (Ar ≥ 99.99%, $P(O_2) < 3$ Pa)	4 h at 1000 °C + 4 h at 1080 °C
Pre-diffusion	High vacuum ($P(O_2) < 0.002$ Pa)	1150 °C for 3 h

Table 3 APS conditions

Parameters	Value
Arc current (A)	700
Arc voltage (V)	62
Standoff distance (mm)	100
Flow of primary gas (Ar) (SLM)	50
Flow of primary gas (H_2) (SLM)	6
Powder feeding gas (N_2) flow (SLM)	8.5

The surface morphology and cross-sectional microstructure of coatings and TGO were examined by field emission scanning electron microscope (SEM, MIRA 3 LMH, TESCAN, Czech Republic) equipped with the energy dispersive spectrometer (EDS, Aztec EDS + EBSD, Oxford Instruments, United Kingdom). Ni-electroless plating was used to deposit a layer of Ni on the pre-oxidized and pre-diffused bond coat surface, in order to protect the TGO scale from spalling off during the cutting and polishing of the samples.

3 Results and discussion

3.1 Healing of the interface between the splashed particles and underlying bond coating

Fig. 1 shows the microstructure of splashed particles on the bond coats with different pre-treatments. Lots of small splashed particles can be seen on the surface of the bond coat (**Figs. 1a, 1b**). The splashed particles have weak bond with the underlying bond coat (**Fig. 1b**) [14], which will result in a core-shell structure with a Al_2O_3 scale covering the granular metallic core (**Fig. 1c**). The core-shell structure can result in pre-generation of the mixed oxides because of insufficient diffusion effect of Al element [13, 15] between the granular metallic core and the metallic bond coat in the process of oxidation. It has been found that the interface between the splashed particle and underlying bond coat can be healed by the pre-diffusion [14] (**Fig. 1d**), which can avoid the generation of the core-shell structure (**Fig. 1e**).

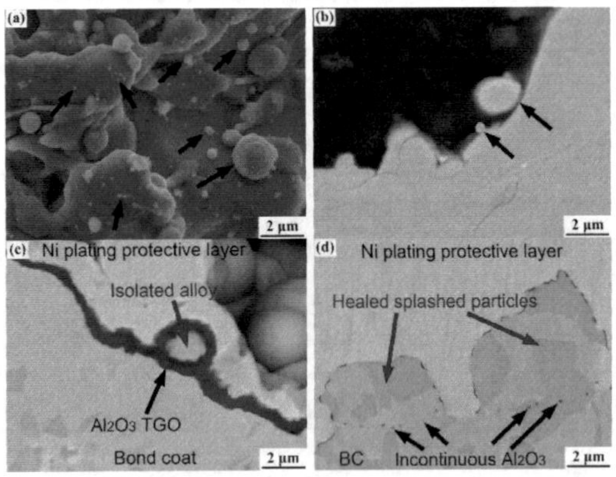

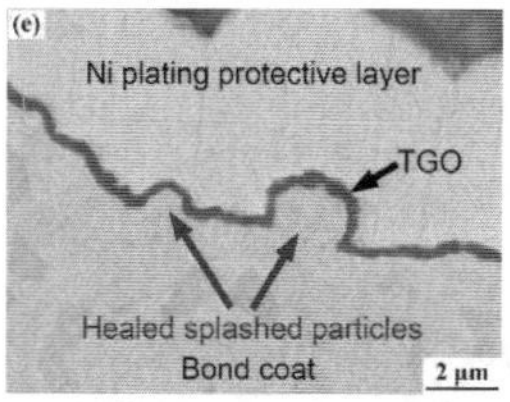

Fig. 1. Microstructure of splashed particles on the bond coats with different pre-treatments. Surface morphology (a) and cross structure (b) of as-sprayed bond coat; cross structures of the pre-oxidized bond coat (c), pre-diffused bond coat (d), and pre-diffusion-oxidation bond coat (e).

3.2 Cyclic oxidation life of the TBCs with different pre-treatments

Fig. 2 shows the cyclic oxidation life and macrographs of the TBCs with different pre-treatments (AS: As-sprayed; PO: pre-oxidation; PDO: pre-diffusion-oxidation). From the **Fig. 2**, it can be seen that the TBCs with pre-diffusion-oxidation exhibited the best cyclic oxidation resistance, and the ones without pre-treatment had the worst cyclic oxidation resistance. The TBCs with pre-diffusion-oxidation didn't have any changes when its thermal cycles are 198 which is greater than the cycles of the other samples.

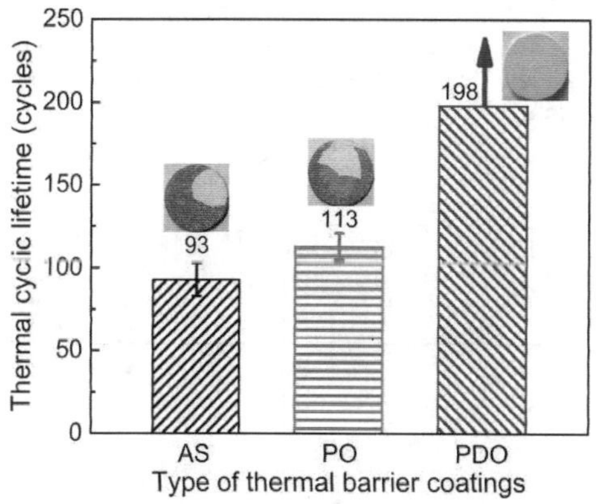

Fig. 2. Cyclic oxidation life and macrographs of the TBCs with different pre-treatments. AS: As-sprayed; PO: pre-oxidation; PDO: pre-diffusion-oxidation.

3.3 Surface morphology of the failed TBCs

Fig. 3 reveals the typical fractured surface morphology of the failed TBCs. It can be found that the area of YSZ on the failed surface is different for the two failed coatings. The area of YSZ on the failed TBCs with pre-oxidation is larger than on the failed TBCs without pre-treatment. **Figs. 4, 5** show the EDS elements mappings of the failed TBCs shown in **Figs. 3b, 3d**, respectively. It can be seen that Al_2O_3, mixed oxides and the YSZ can be found on the two kinds of the failed TBCs surface. The content of mixed oxides on the failed TBCs without pre-treatment is smaller than the other one.

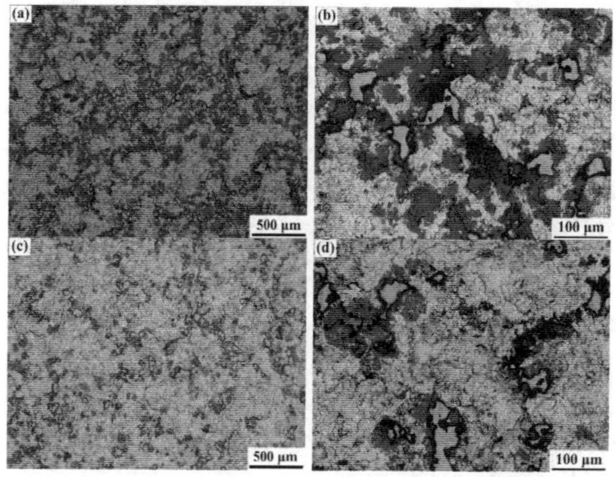

Fig. 3. Fracture surface morphology of the failed TBCs with different pre-treatments. (a) and (b) Without pre-treatment, (c) and (d) pre-oxidation.

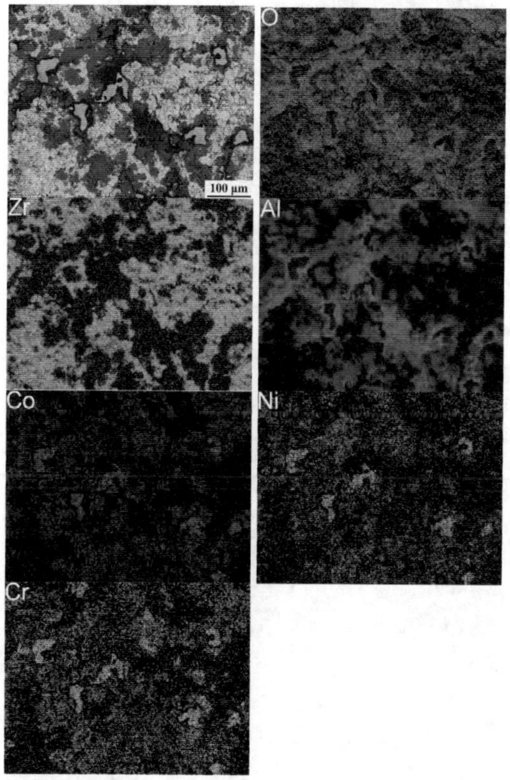

Fig. 4. EDS element mappings of the failed TBCs shown in **Fig. 3b**.

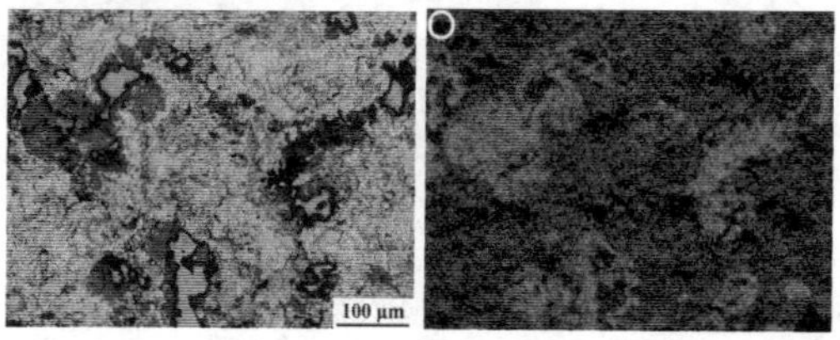

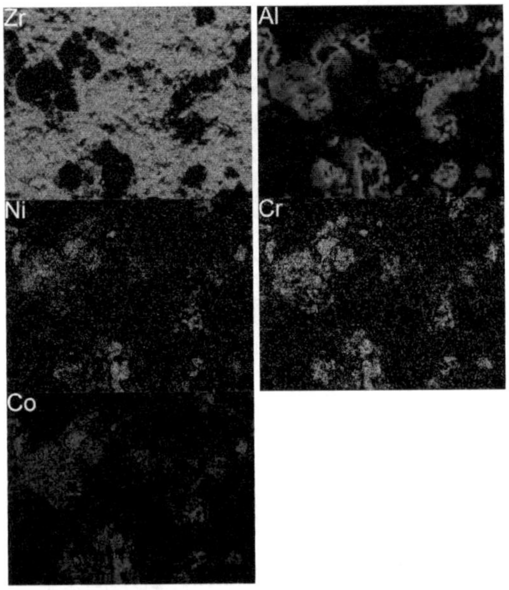

Fig. 5. EDS element mappings of the failed TBCs shown in **Fig. 3d**.

3.4 Failure mode of the failed TBCs

Fig. 6 shows the cross sectional structure of the failed TBCs with pre-oxidation). It can be found that the failing of the TBCs without pre-treatment was resulted from the mixed oxides everywhere at the interface of the bond coat and top coating, and the failing of the TBCs with pre-oxidation was mainly caused by the mixed oxides resulting from the splashed particles which had weak bonding with the underlying bond coat. From the results, it can be seen that the healing of the interface between the splashed particles and underlying bond coat can increase the thermal cycles and benefit to prolonging the lifetime of TBCs.

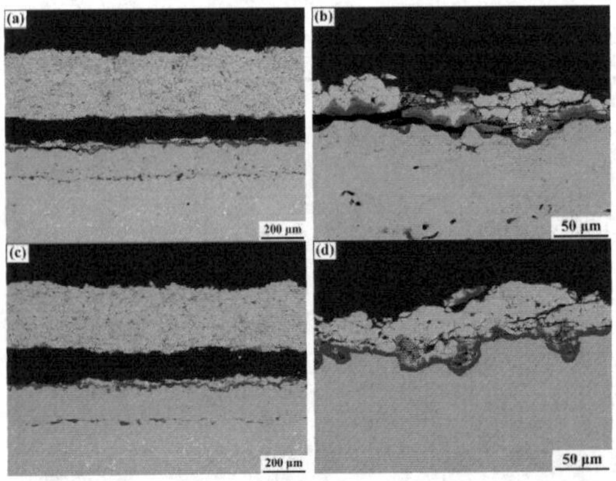

Fig. 6. Cross sectional structure of the failed TBCs. (a) and (b) Without pre-treatment, (c) and (d) pre-oxidation.

4 Summary

In this study, the interface between the splashed particles and underlying bond coat was healed by vacuum pre-treatment (pre-diffusion). The bond coats were divided into three groups by the ways of pre-treatments of samples. Two groups of bond coats were subjected to two different pre-treatments (pre-diffusion-oxidation and pre-oxidation) prior to cyclic oxidation tests. The third bond coats were directly subjected to the same cyclic oxidation tests. Results show that the different TBCs present different cyclic oxidation behavior. The TBCs with pre-diffusion-oxidation exhibited the best cyclic oxidation resistance, and the ones without pre-treatment had the worst cyclic oxidation resistance. Meanwhile, the content of mixed oxides on the surface of bond coating with pre-oxidation is smaller than on the bond coating without pre-treatment. It is revealed that the healing of interface between the splashed particles and underlying bulk coating can effectively improve the cyclic oxidation resistance. This research will benefit the development of long-life and highly reliable TBCs.

5 Acknowledgements

The present project is supported by National Basic Research Program (Grant No. 2013CB035701), the National Science Foundation of China (Grant No.: 51671159), the Fundamental Research Funds for the Central Universities and National Program for Support of Top-notch Young Professionals.

6 Literature

[1] A.G. Evans, D.R. Mumm, J.W. Hutchinson, G.H. Meier and F.S.Pettit: Mechanisms controlling the durability of thermal barrier coatings. Prog. Mater. Sci. 46 (2001), Issue 5, pp. 505/53.

[2] W.J. Brindley: Thermal barrier coatings. J. Therm. Spray Technol. 5 (1996), Issue 4, pp. 379/80.

[3] T. Strangman, D. Raybould, A. Jameel and W. Baker: Damage mechanisms, life prediction, and development of EB-PVD thermal barrier coatings for turbine airfoils. Surf. Coat. Technol. 202 (2007), Issue 4-7, pp. 658/64.

[4] Q. Zhang, C.-J. Li, Y. Li, S.-L. Zhang, X.-R. Wang, G.-J. Yang and C.-X. Li: Thermal failure of nanostructured thermal barrier coatings with cold-sprayed nanostructured NiCrAlY bond coat. J. Therm. Spray Technol. 17 (2008), Issue 5-6, pp. 838/45.

[5] S. Ahmaniemia, P. Vuoristoa, T. Mantyla, C. Gualco, A. Bonadei and R. Di Maggio: Thermal cycling resistance of modified thick thermal barrier coatings. Surf. Coat. Technol. 190 (2005), Issue 2-3, pp. 378/87.

[6] N.P. Padture, M. Gell and E.H. Jordan: Materials science - Thermal barrier coatings for gas-turbine engine applications. Science 296 (2002), Issue 5566, pp. 280/84.

[7] A. Hesnawi, H.-F. Li, Z.-H. Zhou, S.-K. Gong and H.-B. Xu: Effect of surface condition during pre-

oxidation treatment on isothermal oxidation behavior of MCrAlY bond coat prepared by EB-PVD. Surf. Coat. Technol. 201 (2007), Issue 15, pp. 6793/96.
[8] F.-H. Yuan, Z.-X. Chen, Z.-W. Huang, Z.-G. Wang and S.-J. Zhu: Oxidation behavior of thermal barrier coatings with HVOF and detonation-sprayed NiCrAlY bondcoats. Corros. Sci. 50 (2008), Issue 6, pp. 1608/17.
[9] F. Tang and J.M. Schoenung: Local accumulation of thermally grown oxide in plasma-sprayed thermal barrier coatings with rough top-coat/bond-coat interfaces. Scripta Mater. 52 (2005), Issue 9, pp. 905/09.
[10] V. Teixeira, M. Andritschky, W. Fischer, H.P. Buchkremer and D. Stover: Effects of deposition temperature and thermal cycling on residual stress state in zirconia-based thermal barrier coatings. Surf. Coat. Technol. 120(1999), pp. 103/11.
[11] A. Gil, V. Shemet, R. Vassen, M. Subanovic, J. Toscano, D. Naumenko, L. Singheiser and W.J. Quadakkers: Effect of surface condition on the oxidation behaviour of MCrAlY coatings. Surf. Coat. Technol. 201 (2006), Issue 7, pp. 3824/28.
[12] K. Ogawa, K. Ito, T. Shoji, D.W. Seo, H. Tezuka and H. Kato: Effects of Ce and Si additions to CoNiCrAlY bond coat materials on oxidation behavior and crack propagation of thermal barrier coatings. J. Therm. Spray Technol. 15 (2006), Issue 4, pp. 640/51.
[13] Y. Li, C.-J. Li, G.-J. Yang and L.-K. Xing: Thermal fatigue behavior of thermal barrier coatings with the MCrAlY bond coats by cold spraying and low-pressure plasma spraying. Surf. Coat. Technol. 205 (2010), Issue 7, pp. 2225/33.
[14] B.-Y. Zhang, J. Shi, G.-J. Yang, C.-X. Li and C.-J. Li: Healing of the interface between splashed particles and underlying bulk coating and its influence on isothermal oxidation behavior of LPPS MCrAlY bond coat. J. Therm. Spray Technol. 24(2015), Issue 4, pp. 611/21.
[15] G.-J. Yang, X.-D. Xiang, L.-K. Xing, D.-J. Li, C.-J. Li and C.-X. Li: Isothermal oxidation behavior of NiCoCrAlTaY coating deposited by high velocity air-fuel spraying. J. Therm. Spray Technol. 21 (2012), Issue 3-4, pp. 391/99.

Layer-by-layer buildup strategy for cold spray additive manufacturing

P. Vo, M. Martin, National Research Council Canada, Boucherville / CA

The use of cold spray in metal additive manufacturing (AM) offers well recognized advantages with typical commercial drivers being a rapid build rate, low process temperature, and wide range of usable alloys. Cold spray AM to date has often employed a methodology of rapid material deposition, with or without masking, into relatively simple shapes and wide tolerances that can lead to constraints in part geometries and/or significant post-spray machining. In this work, an investigation has been performed into producing more complex geometries and improving shape fidelity using a conventional AM strategy; namely, starting with a CAD drawing, slicing the CAD geometry into a layered structure, and performing a layer-by-layer build. For cold spray, technology-specific considerations must be factored into each of these steps and in particular, an effective build strategy and toolpath are critical to moving towards near-net shape parts. This requires, by extension, precise manipulation of the spray gun, or part as applicable, which was performed using industrial robot offline programming via commercially available software. Various cold spray 3d builds are used to demonstrate developments in toolpath planning and build strategy.

1 Introduction

The use of cold spray in metal additive manufacturing (AM) offers several attractive competitive advantages in comparison with laser-based additive techniques. Typical commercial drivers are rapid build rates, which can be on the order of kg/h, and low process temperatures that are well below material melting points. In addition, a wide range of alloys can be employed [1]. Several reviews of specific benefits and considerations for additive manufacturing by cold spray are available [2-4].

One significant factor when comparing cold spray with laser-based methods, however, is the difference in characteristics of a cold spray spot. The minimum spot size of typical commercial cold spray nozzles is ~4 mm and a profile is obtained across this diameter [4]. Efforts have been made to reduce spot sizes with smaller diameter 'micronozzles', although sizes are still on the order of ~1 mm [3]. The profile of deposits can also be altered on both the equipment and processing side. Xue et al. [5] developed a nozzle geometry that produced a more uniform top on spray profiles while Pattison et al. [2] sprayed on the side of a triangular deposit to produce a vertical wall.

Due in part to its inherent spray characteristics, cold spray AM to date has often employed a methodology of rapid material deposition, with or without masking, into relatively simple shapes and wide tolerances that can lead to constraints in part geometries and/or significant post-spray machining. In this work, an investigation has been performed into producing more complex geometries and improving shape fidelity through a layer-by-layer approach to build strategy and toolpath planning.

2 Methodology

2.1 Materials and Equipment

Testing was performed using powders of pure copper (Plasma Giken, Yoriimachi, Japan) or aluminum (Valimet, Stockton, USA) and substrates of 1020 steel or 6061 aluminum alloy. The powders were spherical with d_{50} values of 19 µm and 22 µm for copper and aluminum, respectively, **Fig. 1**. Deposits were produced mainly using a KM-CDS 2.2 (Inovati, Santa Barbara, USA), **Table 1**. A train wheelset (section 4.2) was produced with a Kinetiks 4000 (Oerlikon Metco, Westbury, USA) system (spray conditions not shown).

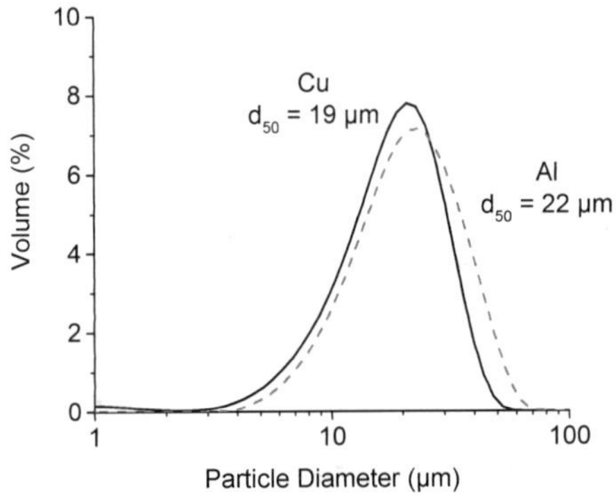

Fig. 1. Particle size distribution of Cu and Al powders.

Table. 1. Spray conditions for the KM-CDS system.

		Al	Cu
Temperature	T_g (°C)	200	480
Pressure	P_g (MPa)	0.9	0.8
Standoff	SoD (mm)	15-20	10-15

2.2 Toolpath Planning

The toolpath planning and programming required to build 3d structures with complex geometry and low (overspray) tolerances intrinsically requires precise manipulation of the spray gun, or part as applicable. This can often introduce an added layer of complexity versus that required for typical cold spray applications, which can often be performed manually by a skilled operator. Coating a surface, for example, commonly employs a straightforward x-y index for stationary substrates or a linear displacement along the axis of rotating substrates.

Toolpaths were generated using industrial robot offline programming via commercially available software. The use of software-based robot code programming provides flexibility and efficiency in toolpath planning as changes in path type, path sequencing, point density, etc. are more easily generated for evaluation. In addition, the simulation capabilities provided by commercial codes are often useful for verification and troubleshooting of build plans, e.g., collision avoidance, robot workspace, etc., **Fig. 2**.

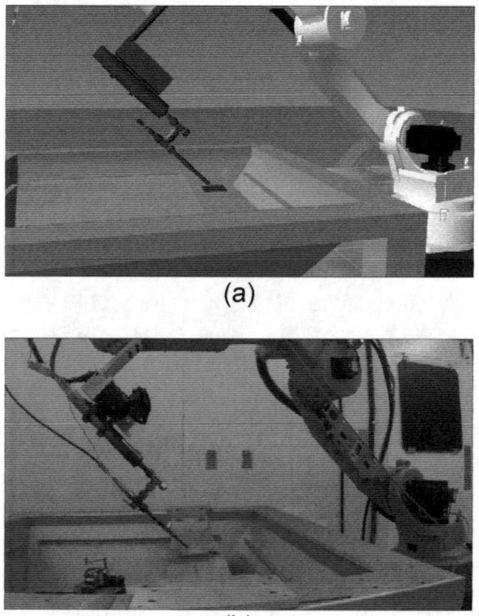
(a)

(b)

Fig. 2. (a) Simulated and (b) spray setup.

3 Build Strategy Development

3.1 Layer-by-Layer Approach

The build strategy employs a conventional AM strategy; namely, starting with a CAD drawing, slicing the CAD geometry into a layered structure, and performing a layer-by-layer build, **Fig. 3**. A slicing method suitable for cold spray builds can differ somewhat from other conventional AM techniques as cold spray AM can be used to build features on existing parts and the associated spray surface is not necessarily flat. For contoured surfaces, one slicing strategy is to produce a layer spanning the entire surface while another would be to selectively add material where needed until the entire surface becomes essentially flat before proceeding with the rest of the build.

A build within each layer is deconstructed into two sub-layer structures, which are labelled as layer filling and layer alteration in the schematic. The layer fill refers to depositing material for the body of part, which is analogous to producing cold spray AM parts by spraying oversized deposits. A difference in the proposed approach is in an alteration performed to modify the deposit edge profile at all open surfaces and in turn, improve the dimensional accuracy of the build (discussed in the following section).

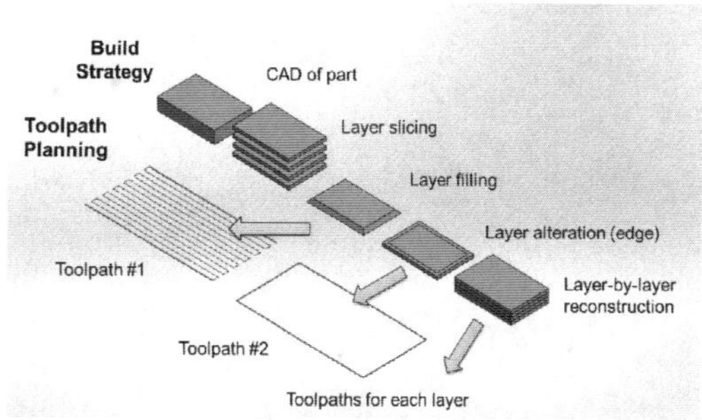

Fig. 3. Schematic of build strategy.

It is important to emphasize that although specific layering strategies are dependent on the design, the reverse also applies. A good design can greatly facilitate a cold spray build by simplifying the types of transitions and support structures that are required. The fabrication is also specific to materials, hardware, and process conditions and an integration of these various elements is critical to a well conceived, design-specific strategy.

Each sub-layer structure employs a different toolpath, with the complexity in toolpath planning heavily dependent on the geometries being produced. The toolpath accounts for the line-of-sight and proper access to the spray location. As build complexity increases and larger cold spray systems are used, this can become a significant challenge and designs become more constrained. In some cases, equipment changes, e.g., nozzles designed to produce uniform spray profiles [5], can be used to facilitate toolpath planning. The precision of a toolpath is also evaluated according to the design requirements as well as the capabilities of the manipulator (robot, gantry, etc.) and cold spray equipment (spray accuracy, process repeatability, etc.).

3.2 Profile Control

With a basic cold spray setup, variations in the gas/particle flow across the diameter of a nozzle often manifest in a well-known tapering of the sides of a deposit. A 'basic' setup here refers to a standard commercial nozzle with circular outlet geometry, no masking, and spray normal to substrate surface. A typical deposit formed after repeated gun traverses along a single line of travel showed this tapering effect, **Fig. 4a**. Beyond a threshold number of traverses, deposition no longer occurred as the impact angles became increasingly off-normal, e.g., impacting the angled sides of the triangular deposit.

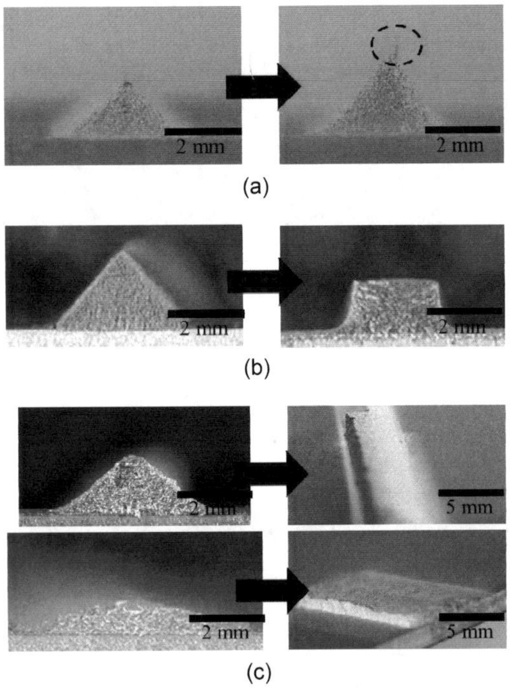

Fig. 4. Typical and altered deposit profiles: (a) Cu sprayed normal to substrate surface (circled: peak forms and deposition no longer occurs); and (b) Cu and (c) Al sprayed at multiple angles to the substrate surface.

In practice, this deposition profile effectively limits edge precision and forces spray of a significantly larger footprint than required by designs as the tapering propagates with each additional layer, i.e., tapered edge grows as deposit thickness increases. This results in drawbacks such as significant amounts of excess material sprayed for large parts and a relatively large minimum size for fine, as-sprayed features. In order to limit these drawbacks, spray angles can be changed to avoid off-normal impacts, as in the case of producing vertical walls [2].

Profile alterations for line deposits of Cu and Al were produced using a succession of spray angles, **Fig. 4b-c**. A rectangular profile was obtained for the Cu line deposit by spraying repeated passes at 90° to the substrate followed by repeated, alternating passes at 45° and 135°, **Fig. 4b**. The starting deposit profile was dependent on materials, nozzle (equipment), process conditions, etc., **Fig. 4c**. However, spray angles and/or positions were adjusted accordingly to perform profile alterations.

The development of unique toolpaths for every type of material deposit can be resource intensive, requiring extensive experimentation and/or modeling. One approach to simplify toolpath planning is to employ a common set of toolpaths for profile alteration, e.g., successive spray angles of 90°, 45°, and 135°. This method can be implemented in a straightforward manner and designs adjusted accordingly. While efficient in practice, there are potential tradeoffs in terms of build accuracy and deposit properties if starting profiles deviate significantly. Consequently, a more practical approach was taken which balanced these two extremes; the development of different sets of toolpath combinations that produced specific, predictable profile alterations from common starting deposit profiles.

In a general layer-by-layer approach, it is important to note that the toolpaths employed for layer profile alteration maintain similar spray conditions to that used in layer filling. With spray angles maintained normal to a target surface (e.g., spray at 45° to the substrate is equivalent to 90° to the triangle edge in **Fig. 4b**), deposit properties for both the layer fill and alteration should be similar. Full characterization is an area of further investigation to validate that the profile alterations do not degrade performance.

4 Cold Spray Builds

4.1 Thin-Walled Features

A layer-by-layer build strategy with profile alterations was employed to build several thin-walled (~4 mm wide) features of copper on steel substrates, **Fig. 5**. The individual layers can be clearly observed and the (vertical) profiles of each layer were retained as successive layers were deposited. The precision of the toolpath and process stability are two key factors in the uniformity of these as-sprayed features.

Adjacent vertical walls ~16 mm high and spaced apart ~3 mm are shown in **Fig. 5a**. For closely spaced features above a certain height, this method may be a suitable alternative to standard masking. Material buildup on solid metal masks is potentially an issue due to the amount of spray passes required for high height features. The equivalent structures produced by post-spray machining, without masking, would require a spray footprint with overspray area in excess of the height, e.g., area bounded by outer dimensions plus > ~16 mm for the walls shown, to account for tapering effects.

In theory, there is no limit to the feature height that can be produced as an approach for the nth layer is similar to the $(n+1)^{th}$ layer. However, process and equipment considerations can effectively add constraints. The first layer deposited is critical as sufficient substrate adhesion must be obtained to withstand spray pressures at positions far from the substrate/deposit interface. The incorporation of support structures may then become necessary for larger, complex features. The robot workspace also gives an upper limit to the size of the build. However, this is generally not a practical concern given the range of current robotic systems, particularly when compared to size constraints in other process such as powder bed AM.

(a)

(b)

(c)

Fig. 5. Thin-walled features produced by cold spray: (a) linear and (b) freeform patterns on flat substrates and (c) freeform pattern on curved substrate.

The use of a 6-axis robot to manipulate spray position provides the capability to produce freeform shapes. A circular shape ~66 mm high with ~28 mm outer diameter on a flat substrate is shown in **Fig. 5b** while a symmetric shape ~10 mm high on a curved substrate surface is shown in **Fig. 5c**. As previously mentioned, a challenge for more complex features and surfaces is the toolpath and line-of-sight. Although limits are being explored, designs can still be relatively flexible with the use of suitable fixturing and nozzle-substrate positioning during the build. In the example shown in **Fig. 5c**, each deposited layer spanned the entire feature footprint on the surface of the substrate. However, additional flexibility can be obtained by initially spraying only areas required to build out the curved surface. Once a (combined deposit and substrate) flat surface is obtained, the remainder of the feature can be built.

4.2 Hybrid Manufacturing

An additive + subtractive method integrating cold spray with machining was employed to create a 1/3rd scale train wheelset, **Fig. 6**. Each wheel was fabricated in aluminum and featured a contoured, variable thickness cross section across a ~290 mm outer diameter. A wheel axle was produced (not shown) using a typical spray configuration; a pipe substrate fixed in a rotating chuck. The spray gun was traversed linearly along the rotation axis in appropriate areas to build up the contours of the wheel axle.

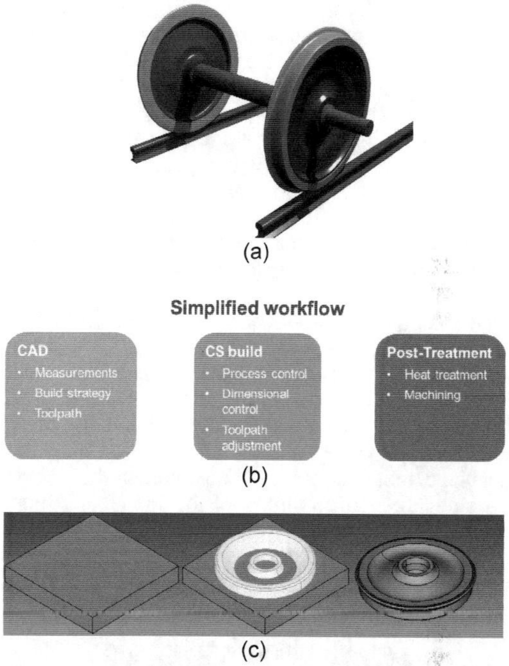

(a)

(b)

(c)

Fig. 6. Hybrid manufacturing of train wheelset: (a) CAD design, (b) simplified workflow, and (c) additive + substractive approach for wheel.

A simplified workflow for the wheel build is shown in **Fig. 6b**. The post-spray operations for this part followed typical practices for cold sprayed materials and will not be discussed. For cold spraying a part of this size, the planning based off a CAD drawing is critical to developing an efficient build, e.g., minimize spray time, overspray amount, feedstock powder usage, post-spray machining, etc.

Several build strategies were considered including radial and planar builds. For a radial build on a rotating substrate, the rotation speed should be a function of the thickness in order to obtain a consistent linear surface speed. However, the prediction and measurement of this changing radius for accurate nozzle positioning, and thus controlling the build profile, is challenging.

Based on the circular geometry, a planar build strategy with spiral toolpaths was employed. Spraying of each layer using an x-y indexing movement was not considered effective for this part due to the formation of over-thicknesses at the layer edges. These over-thicknesses were associated with continuous powder feeding and the deceleration-acceleration of the robot with sharp direction changes. Similarly, a slight over-thickness can also be expected with purely circular movements, e.g., at the transition point between two concentric circles.

The surface speed along the toolpath was determined initially using a deposition rate evaluated at the coupon-scale. However, variations between predicted and actual thicknesses were observed during the build. This was attributed in part to differences in substrate temperature (as well as feedrate and other process variations) when spraying on coupons versus the much larger plate substrate. During the cold spray build, each layer thickness was regularly measured to evaluate the actual deposition rate and confirm the transition from one layer to the next. Dimensional inspection during the build was carried out using a handheld 3d scanner (Creaform, Lévis, Canada).

While this particular wheel geometry can be produced entirely using either standalone method, the hybrid approach demonstrates a capability to minimize efforts for both cold spraying and machining, **Fig. 6c**. The strategy for manufacturing the wheel consisted of the following: (i) select a standard stock size plate that minimized material to be added/subtracted, (ii) build one side (outside of wheel) of the higher thickness hub and rim sections by cold spray, (iii) machine *both* the stock plate and deposited material to match the outside wheel contour, (iv) flip the piece and repeat steps ii-iii for the inner wheel contour.

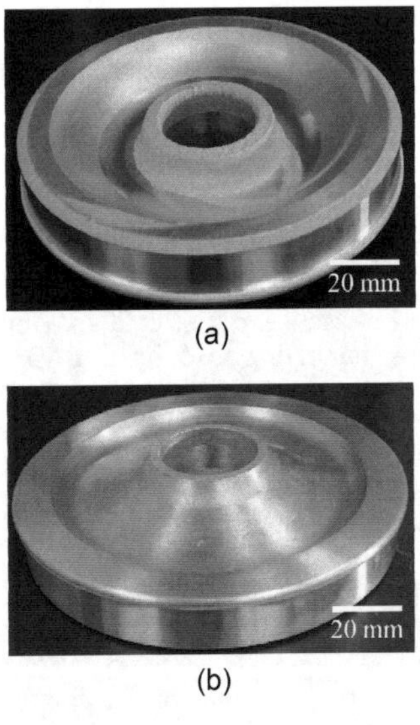

Fig. 7. (a) Partially machined wheel and (b) finished wheel (other side).

Fig. 7a shows a partially machined wheel with the hub section in an as-sprayed condition. As a practical matter, tradeoffs between build accuracy and deposition rate were often made in consideration of the total weight of material deposited. Due to the relatively low resolution and poor surface finish of the cold spray process, machining is generally necessary and considered part of the overall cold spray AM manufacturing process. **Fig. 7b** shows a finished wheel after machining.

5 Conclusions

Various cold spray 3d builds were used to demonstrate a conventional AM strategy; namely, starting with a CAD drawing, slicing the CAD geometry into a layered structure, and performing a layer-by-layer build. For cold spray, technology-specific considerations must be factored into each of these steps and in particular, an effective build strategy and toolpath are critical for producing parts with minimal overspray. A layer-by-layer cold spray AM approach can also be effectively employed in hybrid manufacturing.

6 Acknowledgements

The authors would like to acknowledge the contributions of F. Belval and M. Zeman of the National Research Council Canada.

7 References

[1] Schmidt, T., H. Assadi, F. Gartner, H. Richter, T. Stoltenhoff, H. Kreye, and T. Klassen: From particle acceleration to impact and bonding in cold spraying. J. Therm. Spray Technol. 18 (2009) pp. 794-808.

[2] Pattison, J., S. Celotto, R. Morgan, M. Bray, and W. O'Neill: Cold gas dynamic manufacturing: A non-thermal approach to freeform fabrication. Int. J. Mach. Tool. Manuf. 47 (2007) pp. 627-634.

[3] Sova, A., S. Grigoriev, A. Okunkova, and I. Smurov: Potential of cold gas dynamic spray as additive manufacturing technology. Int. J. Adv. Manuf. Technol. 69 (2013) pp. 2269-2278.

[4] Villafuerte, J.: Considering cold spray for additive manufacturing. Adv. Mater. Processes 172 (2014) pp. 50-52.

[5] Xue, L., Y. Li, and S. Wang: Direct manufacturing of net-shape functional components/test-pieces for aerospace, automotive, and other applications. J. Laser Appl. 23 (2011) pp. 042004 (1-8).

Experimental and Numerical Study of the influence of Powder Characteristics in the Cold Spraying of Al-based Alloys for Additive Manufacturing Using Low-Pressure, Medium-Pressure and High-Pressure Cold Spray Facilities

M. Bunel, F. Borit, F. Delloro, M. Jeandin, Evry / Fr
A. Bacciochini, P. Lemeille, Saint Antoine La Fôret / Fr
E. Hervé, E. Meillot, Monts / Fr
K. Roche, G. Surdon, Argenteuil / Fr

Cold spray is continuously expanding for the repair of parts made of aluminum-based alloys. Beyond repair applications, the process is now expected to be exploited efficiently for the additive manufacturing of shaped parts. However, up to now, cold spray is limited to the achievement of rather simple shapes due to a lack of basic knowledge on coating build-up mechanisms to result in dimension-controlled deposition. The objective of this work is to fill that gap through an experimental and modeling study of the coating build-up in cold spray for this specific application.
Experimentally, Al-based coatings were deposited for a large range of particle velocity due to the use of low-pressure, medium-pressure and high-pressure cold spray facilities. Particle velocity was monitored as a function of cold spray conditions. Two different types of Al 2024 (Aluminium 2024 Alloy) powders were tested. Coating porosity and microhardness were studied as a function of (both morphological and metallurgical) powder characteristics and spray conditions, primarily in the light of particle velocity. Various correlations could be exhibited.
Finite element (FE) simulations of particle impacts were developed, including particle velocity from experimental measurements. These will be used as inputs in an in-house morphological model, the first stages of which could be established successfully.

1 Introduction

Cold spray is an attractive technology to achieve thick coating with a high deposition efficiency. In this process, an inert gas like helium or nitrogen is compressed and heated to accelerate powder particles up to a supersonic speed. The particles impinge on the substrate, due to high plastic deformation to result in a dense coating with a low oxidation rate. There is no needto give more details, which can be however found in extensive recent comprehensive publications [1,2].

Cold spray is presently more and more envisaged for additive manufacturing. In first applications, cold spray technology could be used for repair [3] or direct manufacturing of preforms [4]. In 2014, Impact Innovations could achieve a titanium part of 7kg [5]. However, to apply additive manufacturing by cold spray in certain fields, e.g. the industry, properties of the coating such porosity, hardness or tensile strength must be assessed [6], which was missing very often up to now.

To try to fill the gap, this study is divided in 2 parts. First, an experimental part where the influence of the characteristics of 2 different Al 2024 powders, on coating properties like deposition efficiency (DE) and the microstructure, was shown. For given powders, the influence of spray conditions was studied extensively through the use of “high” pressure, “medium” pressure and “low” pressure cold spray facilities. Parameters, i.e. temperature,pressure and stand-off distance, could be kept in the 3 different cold spray systems to compare coating properties such as DE, porosity rate and microhardness. Then, for every cold spray facility, the influence spraying parameters was studied in the whole operating range.

In the second part, the beginning of a numerical study is presented. Elementary particles were sprayed in parallel to understand the particles interaction and how they build up.

2 Experimental

2.1 Materials

Two different Al 2024 powders were used in this study. Observation showed that the first powder was spherical with some satellites, **Fig. 1a, 1b**. This powder had been produced by water atomization by TLS (Germany). The second powder, produced by gas atomization by Chengdu Best New Materials (China) and purchased at HTC, was also spherical but with some irregular particles, **Fig. 1c, 1d**. Microhardness of the particles of the 1st and the 2nd powders was respectively 81 ± 6 HV0.001 and 88 ± 7 HV0.001. The particle size distributions of the 2 powders were taken to be very different deliberately **Fig. 1e.** The 2 powders were termed as TLS and HTC powders.

Powder was sprayed onto Alloy 2024-T3 specimens of 25x25x3 mm^3 in size.

2 Cold Spray Process and Parameters

Spraying experiments were carried out using 3 cold spray systems. High pressure processing was conducted with the 5/11 Spray System by Impact Innovations. For this facility only, the nozzle was water-cooled, which allowed to use higher temperatures. The second system, i.e. operating at a medium pressure, was the KINETICS 3000-M system by CGT-GmbH. The third system used in this study was, the SST Series-P by CenterLine, which operated at a low pressure. These were installed at Mallard / St Antoine la Forêt, MINES ParisTech – Center for spray Processing (C2P) / Evry, and Dassault Aviation /Argenteuil, respectively.

All the coatings were made with nitrogen and for a flow rate of 15 g.min^{-1}. The gas pressure used was included between 1,5 and 5 MPa and the gas temperature between 300 and 450°C.

Two types of coatings were tested, i.e. beads and fully-covered areas. Beads were obtained by moving the nozzle along a given axis whereas the nozzle scanned the substrate with a 1 mm increment step to coat large areas.

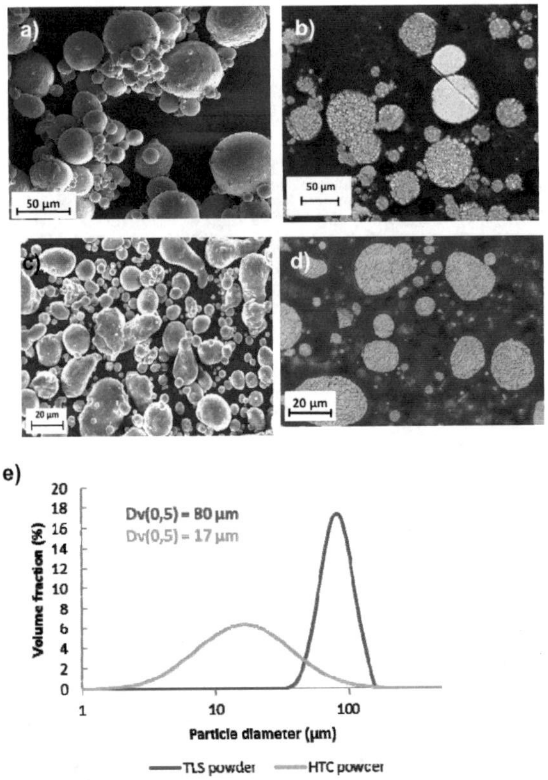

Fig. 1. Al 2024 aluminium powder: a) SEM image of loose powder, b) Optical cross-section view after "Keller's" chemical etching, Micrographs of Al 2024 powder: c) SEM image of loose powder, d) Optical cross-section view after "Keller's" chemical etching, e) particle size distribution of the 2 Al 2024 powders

3 Results

3.1 Deposition Efficiency (DE)

DE depended on many factors such as powder properties, the type of cold spray system chosen and the spraying parameters.

The influence of powder properties could be shown using the same spraying conditions for the 2 different powders. DE was higher for the TLS Al 2024 powder, i.e. 14.4% for a gas temperature of 350°C, 3 MPa Pressure and a stand-off distance of 30 mm, compared to 2.8% for the HTC Al 2024 powder. For each set of tested parameters, DE for the TLS powder was at least 5 times higher than that for the HTC powder.

A given common set of parameters (gas pressure, 1.5 MPa – gas temperature, 300°C – stand-off distance, 30 mm) could be applied to the HTC powder using the 3 cold spray facilities.This allowed to see the influence of the various spraying systems. Resulting DEs were different actually. For the low pressure system, DE was zero since no particles could adhere to the substrate. With the 2 other systems, the particles adhered to the substrate. DE was thus 0.2% and 0.6% for the middle and high pressure facilities respectively.

Gas pressure showed more influence on DE between 1.5 and 3 MPa. From 1.5 to 3 MPa, DE was at least tripled. However, beyond a certain pressure, the influence of gas pressure was no more dramatic. For a gas pressure of 3 MPa, DE was near or superior to that for 5 MPa, **Fig. 2**. The influence of gas temperature on DE was great when increasing gas temperature from 300°C to 450°C, DE was about 10 times higher whatever the gas pressure, 3 or 5 MPa [7].

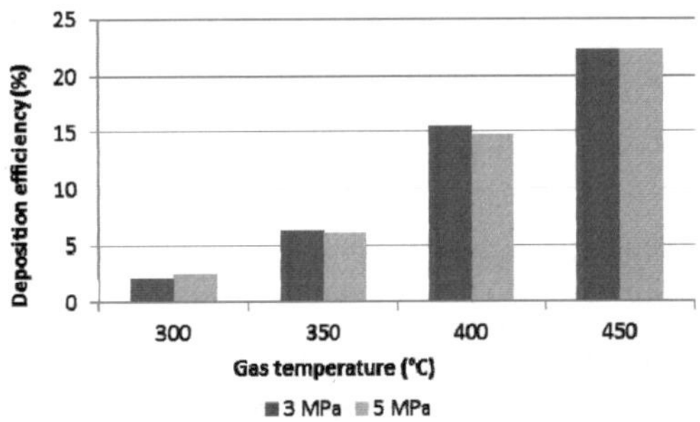

Fig. 2. Deposition efficiency for Al 2024 HTC powder

Moreover the number of passes impacted DE directly. Three coatings were tested using the same conditions 450°C, 5 MPa with the high pressure cold spray facilities. For 1 pass, DE was 27.1% and for 3 and 4 passes, DE dropped to 23.6% and 21.5% respectively.

3.2 Microstructure

After "Keller's" etching, some white particles could be seen in the microstructure of the coating when using TLS powder **Fig. 3.** These particles did not react to chemical etching unlike the other particles. Microhardness testing and EDS analysis were conducted on each type of particles. The results of those analysis showed that the white, unetched, particles was not made of Aluminum 2024 alloy but of pure aluminum. These were assumed to be due to some contamination of the starting batch of Al 2024.

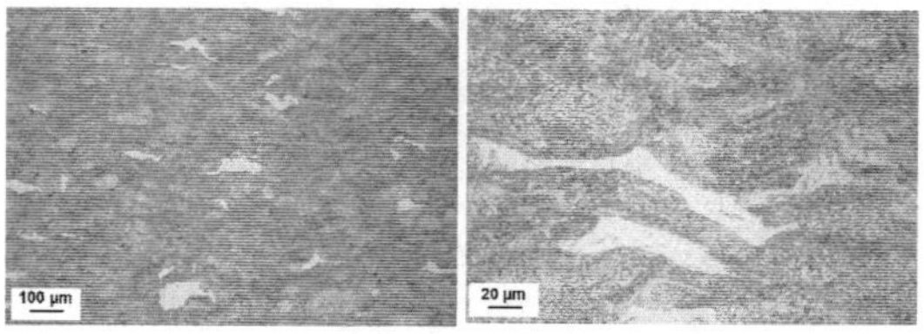

Fig. 3. Cross section optical image of cold-sprayed TLS Al 2024 powder sprayed view after "Keller's" etching

Spraying temperature had an influence on coating microstructure when using HTC powder. Cracks were seen in all the coatings, especially for gas temperatures of 300 and 350°C. Above 400°C, the cracks were sparce, **Fig.4.** In addition to these cracks, a few rather coarse porosities did exist in the microstructure.

3.3 Porosity

Porosity distribution in the beads was homogenous and porosity rate below 1%, **Fig. 5a**. The porosity rate decreased when gas pressure increased. At 350°C, the porosity rate was 0.7%, 0.5% and 0.2% for a gas pressure of 1.5, 3 and 5 MPa respectively.

In the case of fully-covered surfaces, porosity distribution in the coating was heterogeneous. The porosities concentrated between 2 overlapping passes and at the coating edges, **Fig. 5b**. These zones could show some mechanical weakness and initiate rupture. Porosity rate in this kind of coating ranged between 0.8% and 2.3%, even though it could be lower when calculated after removing of coating edges by machining.

For similar spraying conditions, the porosity rate of the multi-layered coatings was higher in comparison to that for a single bead coating [8].

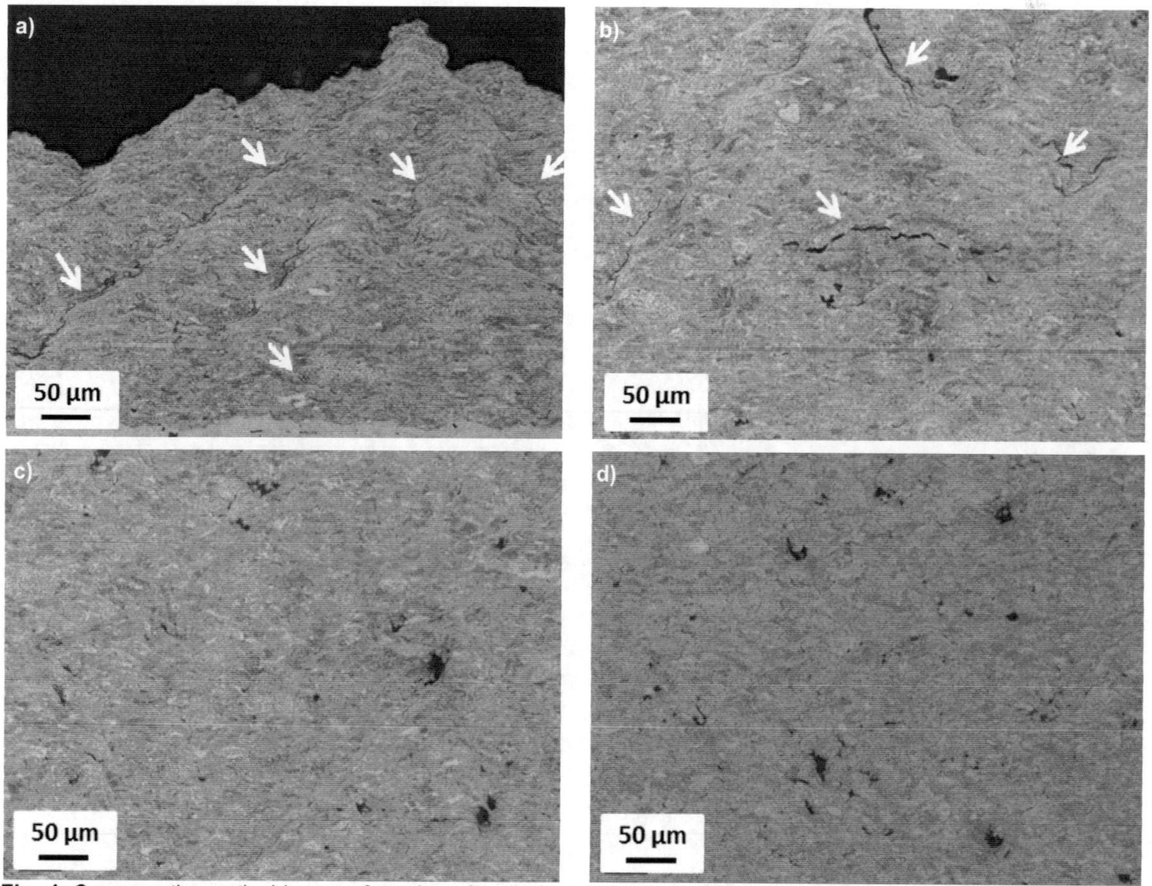

Fig. 4. Cross-section optical image of coating after "Keller's" chemical etching for a gas pressure of 3 MPa and a gas temperature of, a) 300°C, b) 350°C, c) 400°C and d) 450°C (arrows showed the cracks).

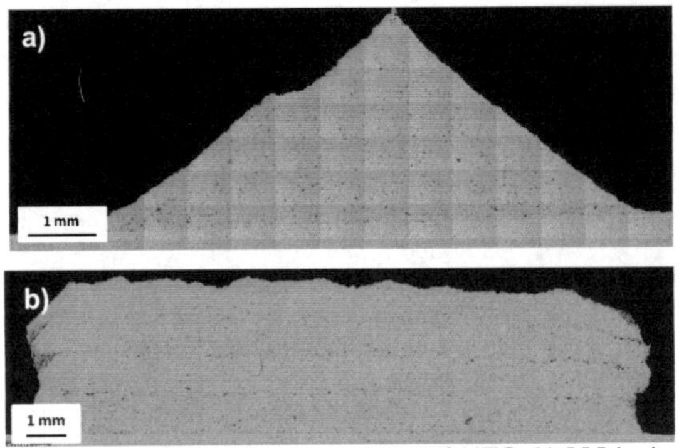

Fig. 5. Cross-section optical map of HTC Al 2024 a) bead, b) coated aera

3.4 Microhardness

Gas pressure and temperature influenced the hardness of the coating, **Fig. 6a**. Increasing gas temperature led to decreasing coating hardness. When gas temperature raised from 300°C to 450°C, microhardness felt of approximately 40 HV0.05. Unlike gas temperature, an increase in gas pressure led to a higher coating microhardness.

Microhardness in the core of the particles within the coating could be determined. The values were lower than those for the whole coating itself, **Fig. 6b**. The influence of the particles boundaries could explain the difference. Gas pressure and temperature had the same effect on intraparticular microhardness as on coating hardness.

The drop in hardness when increasing gas temperature could be explained by the decrease of the peening effect. Opposite results were obtained when considering DE [9].

3.5 Numerical Simulation

Particles and substrate were assumed to obey a "Johnson and Cook" plasticity model which included strain, strain rate hardening and thermal softening. Yield stress σ, was expressed from the Von Mises plasticity model.

$$\sigma = [A + B(\varepsilon_e{}^p)^n][1 + \ln \varepsilon^*]\left[1 - \left(\frac{T - T_0}{T - T_m}\right)^m\right] \quad \text{(Eq. 1)}$$

A, B, C, n and m were constants for the materials. The elastic behaviour followed a linear "Mie-Grüneisen" equation of state where ρ_0 was the bulk density of the material, C_0 the shock wave velocity, E the Young's Modulus and Γ_0 a material constant.

$$P = \frac{\rho_0 C_o{}^2 X\left[1 - \frac{\Gamma_0}{2}X\right]}{[1 - X]^2} + \Gamma_0 E \quad \text{(Eq. 2)}$$

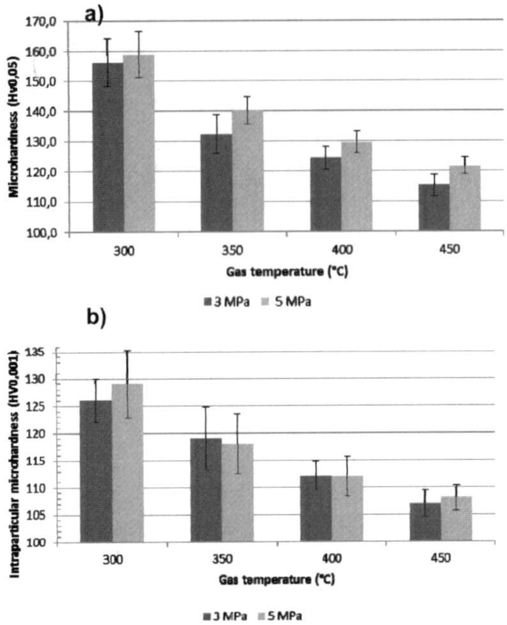

Fig. 6. Microhardness, a) of the coating, b) intraparticular within the coating

Finite element simulations were carried out using the commercial software Abaqus/Explicit in a 2D model. Deformation was locally very large, which required remeshing for adequte convergence. The ALE (Adaptive Landrangian-Eulerian) tool was used to remesh the model.

Table 1. Materials properties for simulation [10]

Coefficients	Al 2024
A (MPa)	265
B (MPa)	426
n	0.34
C	0.015
m	1.00
C_0 ($m.s^{-1}$)	5330
s	1.34
Γ_0	2.16
ρ_0 ($kg.m^{-3}$)	2790
Tm (K)	913
T_0 (K)	298
Cp ($J.kg^{-1}.K^{-1}$)	874
E (GPa)	73
ν	0.33
G (GPa)	27.6
Λ ($W.m^{-1}.K^{-1}$)	121

In a first step, the impact of single particles, which resulted in elementary splats, was investigated to create a library of shapes to feed a coating build-up model [11] on which this article does not elaborate. In this model, there was no possibility to generate porosities in the coating build up process. This will be improved in further development since some porosity was observed in the microstructure actually.

The deposition of a few particles which were distributed in the form of a pyramid was simulated to study particle impinging. Porosity formation due to particle deformation and interactions could be exhibited in particular, **Fig 7a**.

The particles of the bottom layer were more deformed compared to the other layers because of the successive impacts of the particles **Fig. 7b**.

At the end of the simulation, some porosities were present between certain particles but not between all the particles systematically **Fig 7c**. This result was in keeping with observation. Porosities could be generated in coating build-up model by introducing impact of groups of several particles in the library.

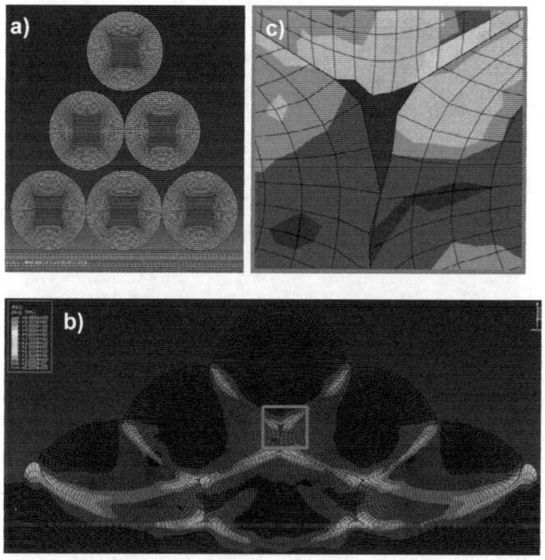

Fig. 7. Simulation of a pyramid of particles a) initial state, b) final state c) magnification on a zone between particles in b)

4 Conclusion / Discussion

As could be expected, the choice of the cold spray facility, even with similar spraying parameters, had an influence on coating. In particular, DE can change dramatically. DE was the highest for the high pressure cold spray installation.

Increasing gas pressure had different effects on the coating. DE and hardness were higher, and the porosity rate was reduced. DE increased with gas temperature, but hardness decreased. Another positive effect of gas temperature rise was the reduction of cracks and of porosities in the microstructure.

A lower hardness, including due to the presence of aluminium particle pollution (in TLS powder) improved DE, which however still remained very low compared to pure aluminium. Moreover, related mechanical properties of that for Al 2024 coating with this undeliberate aluminium addition should be lower when compared to the coating made of Al 2024 only.

Numerical simulation of a pyramidal stack of particles exhibited mecanisms for formation of interparticular porosities. This type of porosity could be observed experimentally in the microstructure.

At the moment, DE is not high enough to consider the use of cold spray for additive manufacturing of Al 2024. A solution based on the heat treatment of the starting powder batch in order to reduce hardness of the particles to increase spray efficiency is currently under investigation.

When DE is high enough to result in thick coating, mechanical properties will be assessed.

6 Acknowledgement

The authors would like to acknowledge Mr A Debray for technical assistance at MINES ParisTech / C2P.

7 Bibliography

[1] A. Vardelle, C. Moreau, J. Akedo, and al, *The 2016 thermal spray roadmap*, Journal of Thermal Spray Technology 25 (2016), Issue 8, pp.1376-1440

[2] Villafuerte, Ed., *Modern Cold Spray*, New York City: Springer International Publishing, 2015.

[3] A. Sova, S. Grigoriev, A. Okunkov, and I. Smurov, *Potential of cold gas dynamic spray as additive manufacturing technology*, The International Journal of Advanced Manufacturing Technology 69 (2013), Issue 9-12, pp.2269-2278

[4] M. Yandouzi, S. Gaydos, D. Guo, and Al, *Aircraft Skin Restauration and Evaluation*, Journal of Thermal Spray Technology 23 (2014), Issue 8, pp.1281-1290

[5] P. Richter, and L. Holzgassner, *Advence high Pressure Cold Spray Innovative Technology "Playground"*, Presented at the North America Cold Spray Conference, Bromont 2014, pp.199-220

[6] D. Macdonald, R. Fernandez, F. Delloro, and B. Jodoin, *High Deposition Efficiency Additive Manufacturing of Titanium using Low Pressure Nitrogen Cold Spray*, Presented at the International Thermal Spray Conference, Shanghai 2016, pp.12-17

[7] J. Lee, S. Shin, H. Kim, and C. Lee, *Effect of gas temperature on critical velocity and deposition characteristics in kinetic spraying*, Applied Surface Science 253 (2007), issue 7, pp.3512-3520

[8] S. Rech, A. Trentin, S. Vezzù, and Al, *Different Cold Spray Deposition Strategies: Single-*

and Multi-layers to Repair Aluminum Alloy Components, Journal of Thermal Spray Technology 23 (2014), Issue 8, pp.1237-1250.

[9] H. Lee, H. Shin, S. Lee, and K. Ko, *Effect of gas pressure on Al coatings by cold gas dynamic spray*, Materials Letters 62 (2008), Issue 10-11, pp.1579-1581

[10] Q. Blochet, F. Delloro, F. N'Guyen, and Al, *Influence of spray angle on cold spray with Al for the repair of Aircraft components*, Presented at the International Thermal Spray Conference, Barcelona 2014, pp.69-74

[11] F. Delloro, M. Faessel, H. Proudhon, and Al, *A morphological approach to the modeling of the cold spray process*, Presented at the International Thermal Spray Conference, Barcelona 2014, pp.221-226

Functional WC Cemented Carbide by the Direct Selective Laser Forming

*Junya Yamada, Hiroyuki Ibe, Kazuto Sato, Nobuaki Kato,
Thermal Spray Materials Department, Fujimi Incorporated, Kakamigahara, Gifu, Japan

Metal and polymer additive manufacturing is advancing on several applications. On the other hand, cermet materials such as WC/Co for functional structure molding by additive manufacturing are under studying. There are few reports for WC cemented carbide additive manufacturing process by forming with polymer binder then sintering. This indirect process has difficulties to make high precision functional parts due to shape control during additional sintering process. Direct forming is desired for high precision parts. However, factors and/or mechanism to achieve direct formed functional structure have been unclear in many aspects. In this study, the process conditions of the direct selective laser melting were investigated to achieve dense and hard WC cemented carbide mold parts. The optimization of laser melting conditions for WC/Co agglomerated and sintered powder was examined. In order to forming a dense and high hardness parts, the optimum conditions between powder preparation and laser energy density which related with laser power, scan speed and spot diameter were appeared by this experiments. Moldings more than 1500HV are achieved at low laser energy density. However, some of pores were observed in moldings. In addition, the dense molding could be obtained by high laser energy density. This means optimum dense functional WC cemented carbide molding is available by optimization of the molding condition. It is applicable for growing industries like automotive, aviation and cutting tool.

1 Introduction

WC cemented carbides and cermets are widely used at various applications such as cutting tools and industrial components which inquire high hardness and wear resistance. On the other hands, these components are well-known as difficult-to-machine material due to its high hardness. To make WC cemented carbides components by machining the sintered bulk take several days even if the component is small. Improvement of the productivity is desired in the manufacturing process of cemented carbides. SLM (Selective Laser Melting) process is studied to dissolve this issue. SLM is laser irradiate process on powder to build repetition layered moldings by melt-solidification. SLM is feasible approach to reduce machining process by direct manufacturing of complex shape difficult-to-machine components.
In addition resource saving, weight reduction and cost saving due to process shortage are expected.

There are few reports for WC cemented carbides direct molding in these years. However, detects of structure such as pores and crack generation [1]-[7] in the moldings are problem for practical applications.
In this study molding structure and hardness of SLM processed molding are investigated using with types of WC agglomerated and sintered powders.

2 Experimental

Process

WC/Co molding conditions were summarized on Table 1. SLM system, ProX200 (3D Systems) which prepared by Kyoei Sangyo was used. Influence of molding conditions for forming and structure were investigated by several molding pattern of laser power and scanning speed. To avoid oxidation as possible, atmosphere was replaced to N_2. 10mm squared moldings were built on SUS430 with repetition layer of 30μm thickness.

Table 1: Molding conditions

Equipment	ProX 200 (3D systems)
Laser Power	50~300W
Scanning speed	800~1600mm/sec
Layer thickness	30μm
Atmosphere	N_2 (O_2<0.2%)
Molding size	10mm×10mm×2mmt

Table 2: Powder conditions

Powder No.	1	2	3
Composition	WC/30%Co	WC/30%Co	WC/12%Co
Particle size	-20+5μm	-20+5μm	-20+5μm
Particle strength	300MPa	500MPa	300MPa

Powder

Cermet powders were prepared using with WC as ceramics and Co as metal. WC and Co composite powders were prepared by agglomerated and sintered process. Different types of WC-Co composition, particle size and particle strength are prepared. Prepared powder is summarized on Table 2.

Evaluation

Microstructure and density investiagtion of the molding was analysed by cross-sectional SEM. Molding hardness was evaluated by micro Vickers hardness tester. Crystal stracture of the molding was analysed by X-ray diffraction (XRD).

Table 3: Molding availability of each powders prepared by different laser

Scanning speed (mm/sec)		800			1200			1600		
Powder No.		1	2	3	1	2	3	1	2	3
Laser power (W)	300	△	△	○	○	○	○	○	○	○
	250	△	△	○	○	○	×	○	○	×
	200	△	○	△	○	○	△	○	×	×
	150	○	△	×	○	×	×	△	×	×
	100	△	△	×	△	×	×	△	×	×
	50	×	×	×	×	×	×	×	×	×

3 Results and discussions

Molding availability of each powders prepared by different conditions are summarized on table 3.
Circle (○) in the table means molding was available without any peeling between baseplate and molding. Triangle (△) means molding was available but a part of interface between baseplate and molding was peeled. Cross(×) means not available due to unmelt or broken due to strong residual stress. High laser power and low scan speed tends to be available to mold.
Influence of powder properties for molding availability were also remarkable. In the case of powder strength, Powder 1 which had lower particle strength than Powder 2 was available in many conditions even if low laser power and high scanning speed. It means particle strength affect for particle melting properties during laser irradiation. This tendency becomes more effective at high scanning speed. Powder 3, which included lower Co contents than Powder 1, was not molded by low laser power with each scanning speed. It means low Co contents affect for particle melting properties.

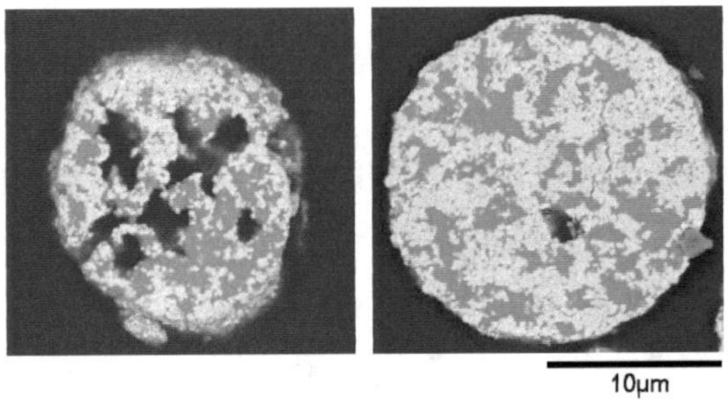

Figure 1: Cross-sectional SEM images of powders. Left : Powder 1, Right : Powder 2.

Cross-sectional SEM images and hardness of layers which were moled with different conditions of laser power and particle strength were showed in Fig2.
Microstructure and hardness were similar results regardless of particle strength when laser power was 300W. On the other hands, low particle strength leaded dense structure at 200W laser power. As shown in Fig.1, Powder 1 with low powder strength was porous. It means the defference of internal structure affect for particle melting properties.

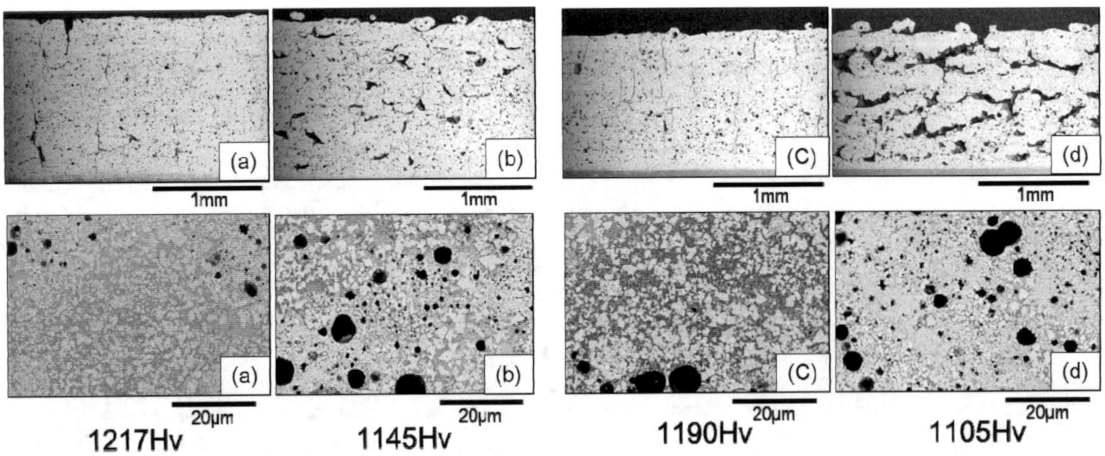

Figure 2: Cross-sectional SEM images of the molding. (a) Powder 1, Laser power 300W, scanning speed 1200mm/sec. (b) Powder 1, Laser power 200W, scanning speed 1200mm/sec. (c) Powder 2, Laser power 300W, scanning speed 1200mm/sec. (d) Powder 2, Laser power 200W, scanning speed 1200mm/sec.

Fig.3 is cross-sectional SEM images and hardness of layer which was 300W laser molded by using powder 3 with different laser scanning speed. Layer fabricated by low Co powder was hard in spite of including a lot of coarse pores. It is considered that the low metal contents of powder resulted to remain high ratio of unmelt particles and it cause detachment of particles.

Hardness of molded part of powder 3 was higher than 1500HV and it was higer than that of powder 1 and 2. Low scanning speed resulted dense microstructure. However, hardness was not high in spite of dense structure compared with several cross-sections of layer fabricated by using different powder. Grain growth of WC is observed at low speed and high laser power. It means high laser power density occur grain growth and reduce the hardness. This trend were similar to sintered bulk[8].

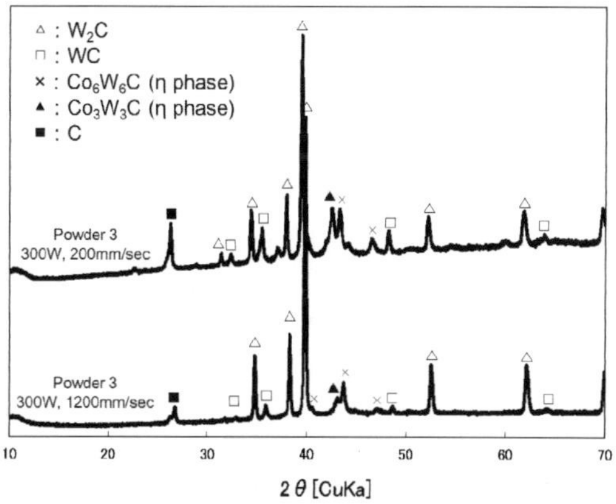

Figure 4: XRD peak patterns of modling surface using with powder 3

Fig.4 is XRD peak patterns of modling surface using with powder 3. W_2C and η phase were observed regardless of scanning speed. 200mm/s scan speed showed high ratio of η phase and free carbon phase.

4 Conclusions

In this study, molded parts were made in several laser and powder conditions. Consequently, WC/30%Co were dense molded parts with hardness of 1200Hv. However, a lot of cracks generate in the all laser conditions. The problem is asumed to be caused by rapid solidification. In addition, powder conditions which was powder strenght and composition was similer to inflence of molded parts. Therefore, it is necessary to both powder and laser conditions are optimized for the molded parts properties.

5 References

[1] Wang et al. Direct selective laser sintering of hard metal powders: Experimental study and simulation. Int. J. Adv. Manuf. Technol. 19 (2002) pp. 351/357

[2] Kruth et al. Lasers and materials in selective laser sintering. Assembly Automation, The international journal of assembly technology and management 23(2003)4, pp. 357/371.

[3] Kruth et al. Binding mechanisms in Selective Laser Sintering and Selective Laser Melting. Rapid Prototyping Journal 11(2005)1, pp. 26/36

[4] Kruth et al. Consolidation phenomena in laser and powder bed based layered manufacturing. CIRP Annals - Manufacturing technology 56(2007)2 pp. 730/759

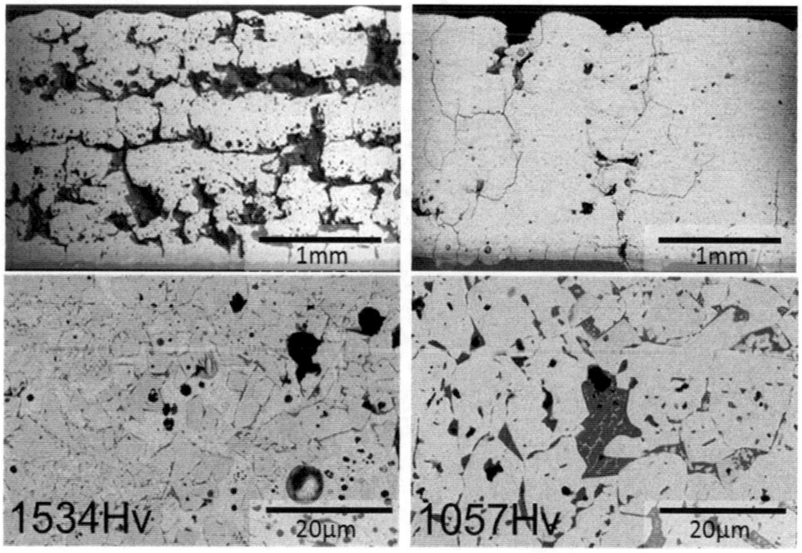

Figure 3: Cross-sectional SEM images of the molding . Left column: Powder 3, Laser power 300W, scanning speed 1200mm/sec. Right column: Powder 3, Laser power 300W, scanning speed 200mm/sec.

[5] Kumar et al. Manufacturing of WC-Co moulds using SLS machine. Journal of materials processing technology 209 (2009) pp. 3840/3848.
[6] Kang et al. Fabrication of porous structured WC-Co alloy compacts by indirect laser forming. J.Jpn.Soc.Powder Powder Metallurgy Vol.58,No.7 (2011) pp. 436/441.
[7] Khmyrov et al. Obtaining crack-free WC-Co alloys by selective laser melting. Physics Procedia. 83(2016) pp. 874/881.
[8] H. Suzuki, Cemented carbide and sintered hard material: foundation and application. Maruzen(1986).

Large size Fe-based Bulk Metallic Glass Composite prepared by Additive Manufacturing via HVOF Thermal Spraying

Wei Wang, Cheng Zhang, Lin Liu*

State Key Laboratory of Material Processing and Die & Mould Technology, School of Materials Science and Engineering, Huazhong University of Science and Technology, 430074 Wuhan, China

Abstract:

Bulk metallic glasses (BMGs) are a novel class of metallic materials with disordering atomic structure and excellent mechanical and chemical properties, and are promising for various industry applications. BMGs are usually fabricated by copper-mould casting due to the requirement of fast cooling rate for the obtainment of amorphous structure. However, this casting approach has the limitation for preparation of large size samples (the biggest Fe-based BMG obtained so far is less than 16 mm diameter). In this work, the conventional high velocity oxy-fuel (HVOF) thermal spraying technique was utilized as a novel additive manufacturing route to create large size Fe-based BMGs and BMG composites. It will be reported that a large size of 20x20x20mm BMG (Fe48Mo14Cr15Y2C15B6 (at%)) and big plate of 100×100×5 mm of Fe-based BMG composites reinforced with 50vol% 316L stainless steel powders was successfully prepared by HVOF thermal spraying. Both BMG and BMG composite showed very dense structure (porosity less than 0.4%) and good mechanical properties, Especially, BMG composite reinforced with 316 L stainless steel exhibited a yield strength of 1.8 GPa and compressive plastic strain of 2%. More importantly, this Fe-based BMG composite exhibited good toughness of K_J=21 MPa $m^{1/2}$, which is almost 4-times higher than that of as-cast BMG. This present work indicates that HVOF thermal spray can become a versatile technique for preparation of large size of bulk metallic glasses and composites with desired properties..

Keywords: Fe-based Bulk Metallic Glasses; HVOF Thermal spraying; Additive manufacturing; Mechanical Properties

Microstructure and wear behavior of Fe-based amorphous metallic coating by HVOF thermal spraying

H.H. Yao, Z. Zhou, K.Z. Tang, Z. Tan, G.H. Wang, D.Y. He
College of Materials Science and Engineering, Beijing University of Technology, Beijing, China

To manufacture a protective coating with low thermal conductivity and good frictional wear performance, a $Fe_{59}Cr_{12}Nb_5B_{20}Si_4$ was designed and produced to prepare coating by High velocity oxygen fuel (HVOF) spraying , in comparison with commercially stain steel AISI 316L coating. Both as-deposited coatings exhibit dense layered structure with the porosity below 1% and slight oxidation. The microstructure of Fe-based coating composes of amorphous matrix and some precipitated nanocrystals, resulting in the thermal conductivity of designed Fe-based coating (2.66 W/mK) is significantly lower than that of stain steel coating 316L(5.87 W/mK). Depending on the structure advantage, the Fe-based coating exhibits higher microhardness, reaching to 1258±92 HV. The friction coefficient and wear rate of Fe-based coating display an increase at 200℃ followed by a decrease at 400℃ determined by the evolution of wear mechanism at different temperature. The dominant wear mechanism of Fe-based coating at room temperature is fatigue wear accompanying with oxidative wear. At 200℃, due to the existence of "third body" abrasive wear, the wear process was accelerated. The large-area oxide layer is supposedly responsible for the decrease of friction of the coating at 400℃.

1 Introduction

In recent years, the transportation industry has been facing increasing pressure to decrease vehicle weight in order to reduce fuel consumption and emission [1]. For instance, aluminum (Al) alloy replacing the traditional steel components is an efficient way for reduction weight increasingly. It is regarded that Al alloy has lower hardness and melt point compared to steel, which is adverse to extensive application. One promising method to solve this problem is to deposit protective coatings on the surface of Al alloy parts [2-3].

Metallic glasses also perform unique combinations of mechanical, physical and chemical properties [4]; these materials are more attractive as protective coatings to withstand aggressive environments [4-7]. Up to now, amorphous metallic coatings have been produced for different purpose by using most thermal spraying methods, such as High velocity oxygen fuel (HVOF) spraying [8], Plasma spraying [9], and Arc spraying [10]. Herein, Zhou et al. [8] investigated iron based metallic glass coatings deposited by HVOF with regard to the corrosion behavior using different electrolytes. All coatings spontaneously passivated with a wide passive region and low current densities, showing excellent corrosion resistance. Zhang et al. [11] investigated the effect of different powder fractions on the tribological behavior of the coatings deposited by APS. All amorphous coatings showed better tribological performances with regard to wear resistance in comparison with the reference 316L. Arizmendi-Mirquecho et al. [12] used wire arc spraying technology to manufacture amorphous coatings and confirmed that a lower wear rate of the amorphous coating as opposed to the uncoated substrate. Except good tribological and corrosive behavior, Shin [3] and Bobzin [13] et al. investigated the thermal insulated performance of amorphous/nanocrystalline coatings. The thermal conductivity can be achieved to 2 W/mK [3] due to the high densities of scattering defects.

Fe-based amorphous metallic glasses are considered to be extremely viable candidates as surface protective coatings owing to their superior corrosion and wear resistance, good magnetic and relatively low material cost [14-17]. However, researches with respect to the heat protection and frictional wear performance of Fe-based amorphous coatings are relatively few. In this article, an amorphous alloy system of $Fe_{59}Cr_{12}Nb_5B_{20}Si_4$ was designed to prepare amorphous coatings by HVOF spraying process and explore the potential as heat protective coating in transportation field. In addition, the frictional wear performance under different temperature was also investigated in comparison with the commercially stain steel 316 coating as a reference coating under same condition.

2 Experimental Procedure

2.1 Materials Selection and preparation

A $Fe_{59}Cr_{12}Nb_5B_{20}Si_4$ powder was prepared by high pressure nitrogen gas atomization method from mixtures of pure Cr (99.1 mass %), Fe (99.1 mass %), as well as pre-alloy Fe-Nb (65 mass %), Fe-B (19 mass %) and Fe-Si (75 mass %). The as-atomized powders with particle sizes in the range of 25-45 µm were sieved and used for spraying onto the ASTM 4032 aluminum plates substrates (with dimensions of 150×100×10 mm^3). The coatings were fabricated by TAFA-JP5000 HVOF spraying system (Praxair), named H-F coatings, and the detailed spraying parameters which had been listed in **Table 1.** Two kinds of specimens were produced. The first coatings were deposited on the top surface of the ASTM 4032 aluminum with target thickness of 400-500µm. These specimens were used to investigate the coatings microstructure and sliding test. Coatings were also deposited on the s aluminum plates for nearly 1mm, aiming for easier subsequent debonding of the coatings. These free-standing coatings with a diameter of 12.7 mm were used to measure the density, thermal diffusivity and heat capacity, respectively. Before spraying all substrates were degreased by acetone, dried, and then grit-blasted. For comparison, commercially available AISI 316L powder (Cr: 18.02, Ni: 11.91, Mo: 4.27, Si: 0.95, Fe: balance, wt. %) was used to deposit reference coating under the same process conditions, named H-316 coatings.

Table 1 Spraying parameters employed in the HVOF process

Parame -ter	Kerosene flow GPH	Oxygen flow SCFH	Nitrogen flow SCFH	Feed rate, r/min	Spraying distance mm
Coating	6.2	2000	26	5.0	380

2.2 Microstructure Characterization

As The as-deposited coatings were prepared using standard metallographic procedures for microstructural investigation. X-ray diffraction (XRD) was used to determine phase composition and grain size, using D8 ADVANCE from BRUKER with Cu Kα radiation. The microstructures of the coatings were characterized using scanning electron microscopy (SEM, Quanta-200, FEI) with energy dispersive spectrometer (EDS), and transmission electron microscopy (TEM, JEM-2100, JEOL). Fifteen cross-sectional micrographs of each coating were taken with an optical microscope to evaluate the average porosity fraction of coating using an Image Pro Plus system. A Nitrogen/Oxygen determinator (G8 Galileo, BRUKER) was used to analyze the oxygen content of the free-standing coatings under highly pure Helium. The thermal stability of the coatings was examined by NETZSCH 449C differential scanning calorimetry (DSC) in a continuous heating mode at a rate of 0.33 K/s from room temperature to 1000°C under argon atmosphere.

2.3 Coating properties

The apparent density (ρ) of the sprayed coatings was measured by the Archimedes principle. Thermal diffusivity (α) measurement of the coatings was carried out at room temperature by using the laser flash thermal constant analyzer (LFA427, NETZSCH). Specific heat (C_p) of the coatings was also measured using the Differential scanning calorimetry (STA 449F3, NETZSCH) at room temperature. Then, the Thermal conductivity (k) was calculated from the relation $k= a\rho C_p$.
Vickers microhardness measurements were conducted on the cross section of the HVOF coatings with an applied load of 50 gf and indentation time of 10s, and calculated as an average of 15 measurements.
The sliding behavior of the HVOF coatings were determined using a standard SRV (Schwingungs Reibung and Verschleiss [SRV], Optimol, Instruments Prueftechnik, Germany) under dry friction and wear tester with reciprocating motion. The material of friction pair was high carbon steel GCr15 ball with a diameter of 10mm. During total duration of 30min, the steel ring with a set rotational speed of 200 rpm, was pressed onto the surface of the fixed block where the coatings were deposited, under an applied load of 40 N. And, the experimental were tested under the room temperature, 200 ℃and 400 ℃, respectively. A new frictional pair was used for the siliding test under each pre-set temperature. The friction coefficient curves were recorded automatically using a computer connected with the friction and wear tester. The wear volume loss was measured using a three dimensional (3D) non-contact surface mapping profiler (OLYMPUS 4100, Japan). The wear rate is caculted as: $K=V/SF$, where V is the wear volume loss in mm^3, S is the total siliding distance in m, F is the normal load in N.

3 Results and Discussion

3.1 Microstructure

Fig. 1 shows the SEM micrographs of self-fabricated amorphous powder $Fe_{59}Cr_{12}Nb_5B_{20}Si_4$ and commercial 316L powder. It can be seen both powders prepared by gas atomization exhibit spherical or near-spherical shape and have a size of 25-45 µm. Most of powders exhibit smooth surface as the meaning of good fluidity, which is propitious to keep the stability and uniformity during spraying process.
Fig. 2a shows the XRD patterns of the atomized Fe-based powders and as-deposited H-F coatings. It is notable that a main broad halo peak appearing in the designed Fe-based powder and coating, indicates high amorphous phase content, which is primarily because of the high GFA of $Fe_{59}Cr_{12}Nb_5B_{20}Si_4$ alloy system. The reason of the high GFA for this alloy is according to the three empirical rules for the achievement of high GFA [18]. Moreover, the high cooling rate of HVOF process also contributes to the formation of amorphous phase in the coatings. Nonetheless, compared with the powder, few peaks of α-Fe and boride are also observed in as-deposited coatings. Meanwhile the XRD patterns of reference coating H-316 is shown in **Fig. 2b**. As can be seen, the principal phase in H-316 coating is γ-(Fe, Ni), which is in agreement with 316 powder.

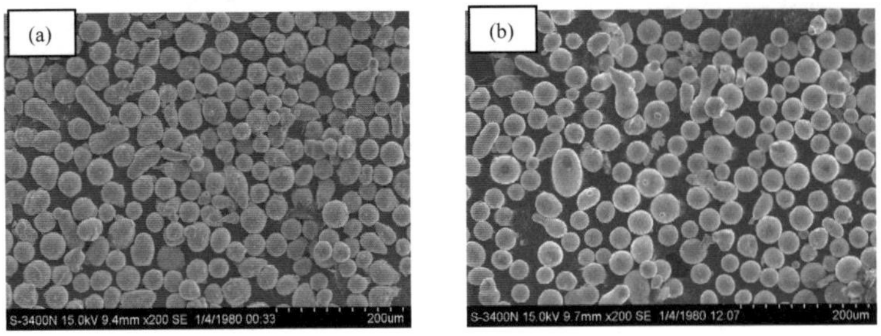

Fig. 1 A typical SEM image of spraying powders with 25-45 μm diameters: (a)$Fe_{59}Cr_{12}Nb_5B_{20}Si_4$ alloy, (b)commercial 316L.

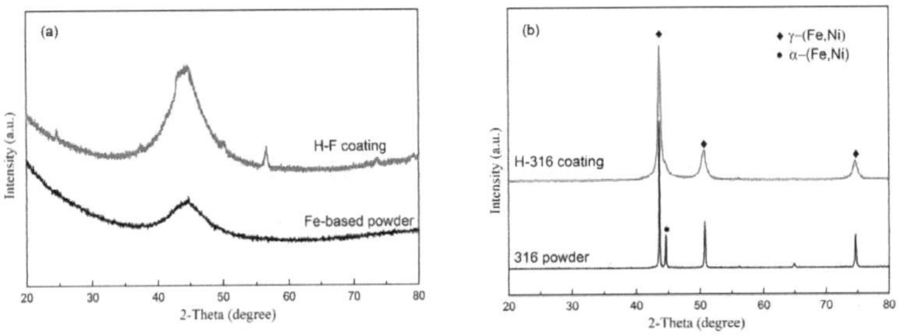

Fig. 2 XRD patterns of the atomized powders and the as-deposited coatings: (a) H-F, H-316.

Fig. 3 reveals typical SEM images from cross-sections of amorphous coating H-F, in comparison with H-316 coating. Generally, the coatings have a dense layered structure typical of HVOF thermally sprayed deposits with a good adherence to the substrate. Moreover, the coatings exhibit low oxide stringer which always comes out visibly in Fe-based and Ni-based amorphous coatings [8], although some pores exists as very dark regions can be seen from the images. The big pores located between flattened droplets are mainly caused by the loose packed layer structure or gas porosity phenomenon, while the small pores within the flattened particles originate from the shrinkage porosity [9]. To estimate the oxide level and porosity of coating, the oxygen content and porosity were measured and shown in **Table 2**. Both amorphous powder and 316 powder s express lower oxygen content than that of coatings, thus indicating slightly oxidation occurred during spraying process which is also attributes to the low temperature and high particles flying speed of HVOF. It is also note that the porosity of amorphous coating H-F, 0.95% is slightly higher than that of H-316 coating, 0.35% due to more gas or air involved in the solidification. Obviously, both H-F and H-316 coating show relatively low porosity than other sprayed coating.

Table 2 The porosity and oxygen content of powders and coatings

Sample	Fe-based powder	316L-powder	H-F coating	H-316 coating
Porosity (%)	--	--	0.95±0.06	0.35±0.0.15
Oxygen content (%)	0.017 ±0.005	0.046 ±0.005	0.16±0.01	0.223±0.01

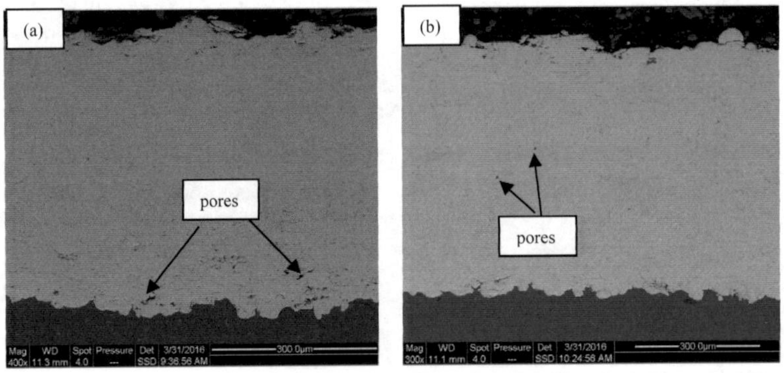

Fig. 3 SEM images of the cross-section of the as-deposited coatings:(a) H-F, (b) H-316.

TEM was undertaken to obtain more detail information on microstructure formation of H-F coating, shown in **Fig. 4**. The selected area diffraction (SAD) pattern sited at the corner of micrograph characterize that the coating are primarily composed of amorphous phase, which is according with the results of XRD examination. Nevertheless, some nanocrystalline grains, as highly faulted particles obviously ascertained in the micrograph, were embedded in amorphous matrix with grain size in the range of 100-300 nm. It validates the appearance of boride peaks in XRD results, which is regarded that the generation of these nanocrystals are principally caused by heats accumulated inside the as-deposited coatings.

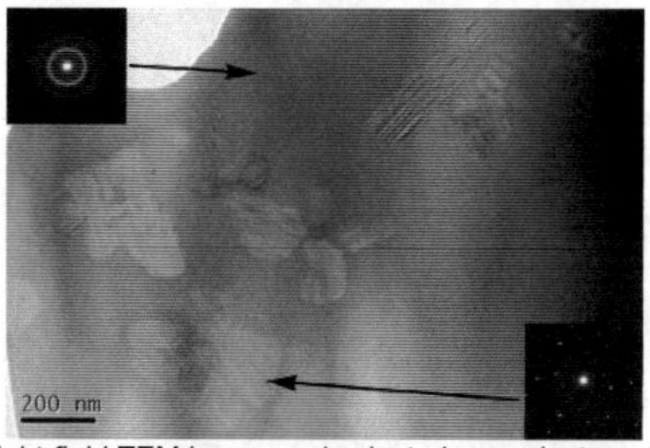

Fig. 4 Bright-field TEM image and selected area electron diffraction (SAED) patterns of the as-deposited coating H-F

3.2 Thermal Conductivity

The thermal conductivity directly indicates the ability of material to conduct heat [1]. The thermal conductivity (*k*) of the $Fe_{59}Cr_{12}Nb_5B_{20}Si_4$ coatings and commercially 316L coatings were calculated at room temperature, based on the measurement of densities (*ρ*), thermal capacities (*Cp*) and thermal diffusion coefficients (*α*) of coatings, presented in **Table 3.** Clearly, the thermal conductivity of the investigated amorphous coating H-F is 2.66 W/mK which is close to conventional YSZ TBCs, and much lower than that of H-316 coating, 5.87 W/mK. The reason could be that the amorphous structure and grain refinement are disadvantaged to transport heat of coatings [19-20].

It is well known that the thermal conductivity displays a positive correlation with the grain size due to the scattering defects[13, 20]. Since the grain size decreases into amorphous structure where lack of atomic periodicity, the amount of scattering defects increases significantly, resulting in the mean free path limited to the dimensions of the tenth of a nanometer scale. Thus the amorphous state is conducing to reduce the thermal conductivity of alloy. The H-F coating with a amorphous/nanocrystalline structure exhibits a lower thermal conductivity than H-316 coating with crystall structure, although the higher porosity exists in H-F coating which is unfavorable to the heat transfer. Therefore, the relative low thermal conductivity of H-F coating is mainly benifited from the amorphous/nanocrystalline structure.

Table 3 The densities, thermal capacity and thermal diffusion coefficient of coatings

Specimen	*ρ* (g/cm³)	C_p (J/g·K) ×10⁻¹	α (cm²/s) ×10⁻²	*K* (W/mk)
H-2	6.879	5.030	0.771	2.66
H-3	7.292	4.749	1.695	5.87

3.2 Friction and wear behavior

The microhardnesses of Fe-based amorphous metallic coating and 316L stainless steel coating prepared by HVOF spraying process have been calculated. The microhardness of H-F coating is about 1258±92 HV, and it is significantly higher than that of H-316 coating, mainly owing to the high amorphous content which has been described by microstructure analysis conducted above. In general, the dense random packed atomic configurations of the amorphous structure can effectively resist the plastic deformation arose from applied load, thus leading to the high hardness performance of the amorphous coating. In addition, the dispersion strengthening of nanocrystals precipitated in amorphous matrix, also contributes to the high hardness of H-F coating.

The variations of friction coefficient of the two kinds of coatings sliding against GCr15 ball at varied temperatures are shown in **Fig. 5**.it can be seen that H-F coating has lower friction coefficient than H-316 coating at room temperature. When sliding at 200 and 400℃, the friction coefficient of H-F is higher than that of H-316 coating. This could be because the higher microhardness of H-F coating can hinder the relative movement of the frictional pair to some extent and increasing the friciton[21-22]. Interestingly, the friction coefficient of H-F amorphous coating fistly exhibits an increase at 200℃, then followed by a decrease at 400℃. However, the friction coefficient H-316 coating decreases with the ascent of temperature.

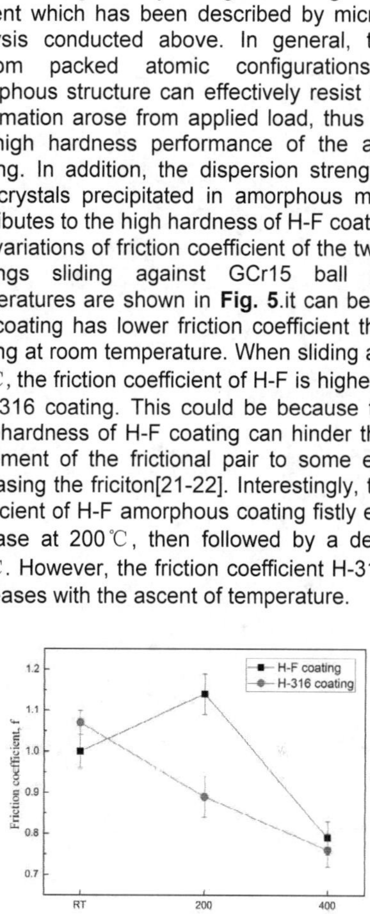

Fig. 5 Friction coefficients of the HVOF coatings sliding against GCr15 balls at various temperature

Worn surface morphologies of amorphous coating and stain steel coating were examined to clarify their wear behavior during dry frictional wear condition, as can be seen in Fig. 6. For stain steel coating sliding at room temperature, H-316, shown in **Fig. 6c**, it is visibly that there are numerous furrows formed on the surface after test, indicating a microcutting mechanism. At the same time, large amount of oxidized particles according EDS results, shown as dark area, stay on the worn surface, showing abrasive wear character, which may induce high wear rate. Besides, some traces of adhesive wear can also be observed at some of the furrow bottom. Thereby, the main wear mechanisms of 316L coating are abrasive, microcutting and oxidative wear. In contrast, **Fig. 6a** shows the contact surface of the amorphous coating H-F sliding at room temperature, few slight traces of microcutting is observed. Nevertheless, the generation and propagation of cracks could be detected apparently, implying fatigue wear occurred during sliding test. The cracks may prefer to initiate and propagate along with the defective regions on the surface of H-F coating due to the accumulation of shear stress, and leads to partial or entire flaking off of lamellae. Moreover, oxidative wear were also ascertained to occur during testing process on account of the EDS analysis of debris. It is believed that the oxygen tends to diffuse along the defects, causing the interface oxidation of coating, which accelerates the failure of fatigue wear. Therefore, the main wear mechanism of H-F coating under dry frictional wear at room temperature is fatigue wear accompanying with oxidative wear.

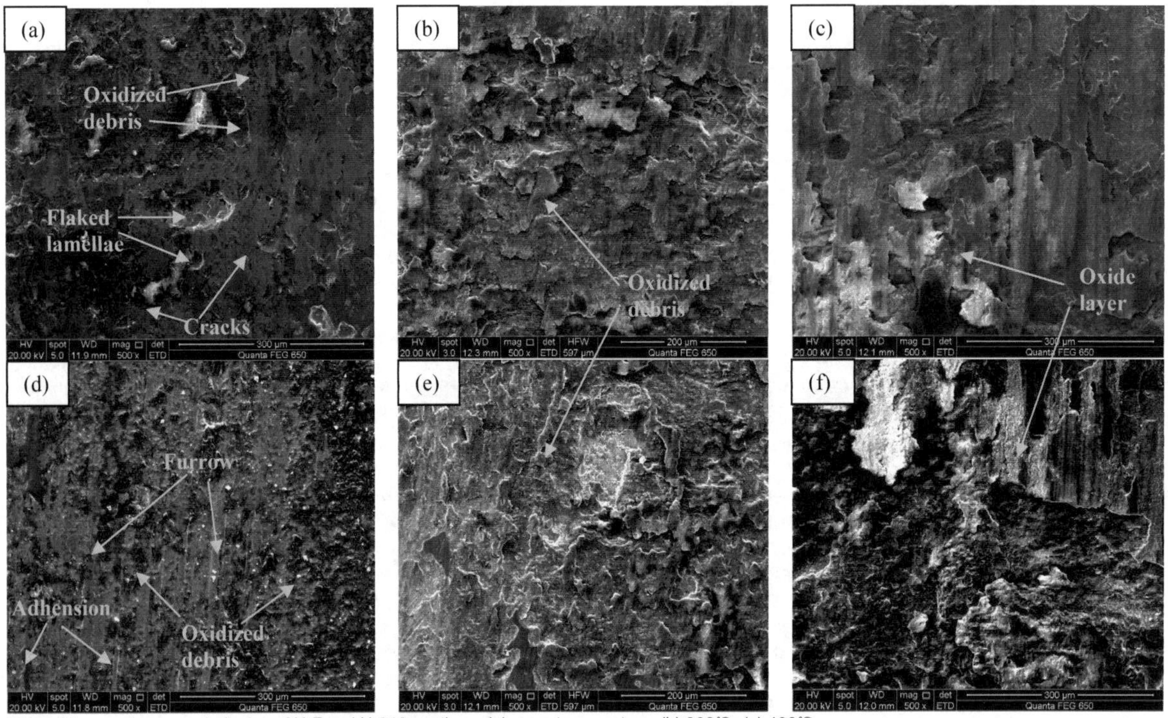

Fig. 6 Worn surface morphologies of H-F and H-316 coatings: (a) room temperature; (b) 200℃; (c) 400℃.

As the frictional wear was tested at 200℃, more oxide debris have been generated on the worn surface of coatings (**Fig. 6b, d**), because the oxidative wear was aggravated at 200 ℃. Severe interface oxidation reduces the cohesive strength of coatings, leading to the failure of fatigue wear and lamellae flaking. For the amorphous coating H-F, with the sliding time increase, some flakes crushed into particles, become a new kind of abrasive particles and participate in the frictional wear, inducing a "third body" abrasive wear. This phenomenon accelerates the wear process of H-F coating and gives rise to the increase of friction coefficient. For the H-316 coating, owing to the abrasive and microcutting wear mechanism at the beginning, a mount of oxidized debris were produced, and then those small particles proceeded to consolidate gradually on the worn surface. After the EDS analysis, the most of oxide are corresponding to iron oxides, which is benefit for decreasing wear process as a solid lubricant. Hence, the friction coefficient of H-316 displays a decrease with the ascent temperature. As the test temperature rise to 400 ℃, the oxide contents obviously increase mainly owing to the oxidation of debris caused by the friction chemical reaction, shown in **Fig. 6c, and f**. Many researches have revealed that as their contents reach a certain level, 'glazed layers' can be generated via these oxidized debris, which are intermixed, fractured into small particles and finally consolidated together onto the worn surfaces by sintering reactions [23-25]. More importantly, it is well known that these oxides are very good high-temperature solid lubricants due to softening [26-27], namely, the 'glazed layer' composed of them possesses excellent lubricating performance and can effectively decrease the friction shear strength of the contact surfaces [23, 28]. It is just the 'glazed layer' that is supposedly responsible for the decrease of friction coefficient of the coating at 400℃.

Fig. 7 illustrates the wear rate of the two kinds of coatings sliding against GCr15 balls at different temperatures. It can be seen, the wear rate of H-316 coating tend to decrease with the ascent temperature, whereas the H-F coating displays increase at 200℃ followed a decrease at 400 ℃, which is in good agreement with the variations of friction coefficient of coatings, respectively. For the amorphous H-F coating, the increase wear rate at 200℃ is because of the existence of "third body" abrasive wear at this temperature. The lower wear rate of coatings at 400℃ is arose from the oxide layer as glazed layer possessing excellent lubricating performance.

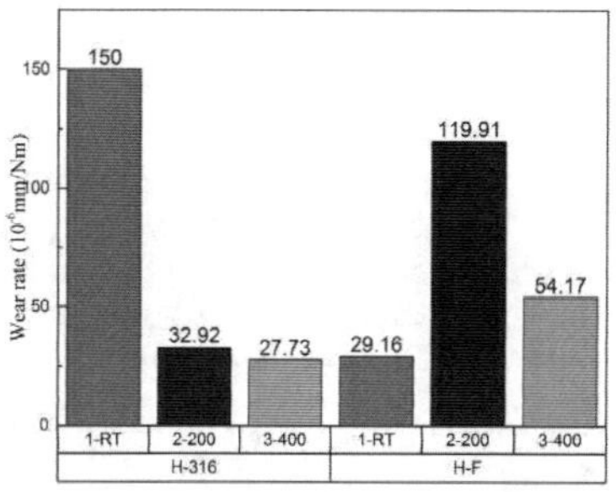

Fig. 7 Wear rates of the HVOF coatings sliding against GCr15 balls at various temperature

4 Conclusions

Gas-atomized $Fe_{59}Cr_{12}Nb_5B_{20}Si_4$ amorphous powders were deposited to form amorphous coatings by HVOF thermal spraying. The results of the present study can be summarized:

(1) The as-deposited coating exhibit dense layered structure with the porosity below 1% and slight oxidation occurs during deposition. The microstructure of Fe-based coating composes of amorphous matrix and some precipitated nanocrystals. The thermal conductivity of designed Fe-based coating is 2.66 W/mK, which is significantly lower than that of stain steel coating 316L, 5.87 W/mK, mainly benifited from the amorphous/nanocrystalline structure.
(2) Depending on the structure advantage, the designed Fe-based coating exhibits higher microhardness, reaching to 1258±92 HV. The friction coefficient and wear rate of Fe-based coating display an increase at 200℃ followed by a decrease at 400℃. The dominant wear mechanism of Fe-based coating at room temperature is fatigue wear accompanying with oxidative wear. At 200℃, due to the existence of "third body" abrasive wear, the wear process was accelerated. The large-area oxide layer is supposedly responsible for the decrease of friction of the coating at 400℃.

The microstructural characteristic and integrated performance of the $Fe_{59}Cr_{12}Nb_5B_{20}Si_4$ coating, including the thermal insulation, microhardness, and good frictional wear behavior, make it has a great significance to the actual engineering application.

5 Acknowledgements

The authors gratefully acknowledge the financial supports of the National Natural Science Foundation of China (51301009) and the Open Funds of the State Key Laboratory of Rare Earth Resource Utilization (RERU2017020).

References

[1] Poirier, D., Lamarre, J.M. and Legoux, J.G.: Thermal cycling assessment of steel-based thermal barrier coatings for Al protection. Journal of Thermal Spray Technology 24 (2014), Issue 1-2, pp. 175/84.
[2] Lugscheider, E., Dicks, R., Kowalsky, K., Cook, D., Nassenstein, K. andVerpoort, C.: A Materials System and Method of Its Application for the Wear Protection of Aluminium Engine Cylinder Bore Surfaces, Thermal Spray 2004: Advances in Technology and Applications, May 10-12, 2004, Osaka, Japan.
[3] Shin, D., Gitzhofer, F., Moreau, C.: Properties and induction plasma sprayed iron based nanostructured alloy coatings for metal based thermal barrier coatings. Journal of Thermal Spray Technology 16 (2007), Issue 1, pp. 118/27.
[4] Zhou, Z., Wang, L., He, D.Y. and etc.: Microstructure and wear resistance of Fe-based amorphous metallic coatings prepared by HVOF thermal spraying. Journal of Thermal Spray Technology 19 (2010), Issue 6, pp. 1287/93.
[5] Zhang, C., Liu, L., Chan, K.C. and etc.: Wear Behavior of HVOF-Sprayed Fe-Based Amorphous Coatings, Intermetallics 29 (2012), pp. 80/85.
[6] Yang, Y., Zhang, C., Peng, Y. and etc.: Effects of Crystallization on the Corrosion Resistance of Fe-Based Amorphous Coatings, Corrosion Science 59 (2012), pp. 10/19.
[7] Pokhmurskii V., Student M., Gvozdeckii V., Stupnyckyy T., Student O., Wielage B. and Pokhmurska H.: Arc-sprayed Iron-based coatings for erosion-corrosion protection of boiler tubes at elevated temperatures, Journal of Thermal Spray Technology 5 (2013), pp. 808/19.
[8] Zhou, Z., Wang, L., He, D.Y. and etc.: Formation and corrosion behavior of Fe-based amorphous metallic coatings by HVOF thermal spraying, Surface and Coating Technology 204 (2009), pp. 563/70.
[9] Zhou, Z., Wang, L., He, D.Y. and etc.: Microstructure and Electrochemical Behavior of Fe-Based Amorphous Metallic Coatings Fabricated by Atmospheric Plasma Spraying. Journal of Thermal Spray Technology 20 (2011), Issue1-2, pp. 344/50.
[10] Lin, J.R., Wang, Z.H., Lin, P.H. and etc.: Microstructure and cavitation erosion behavior of FeNiCrBSiNbW coating Prepared by twin wires arc spraying Process. Surface and Coatings Technology 240 (2014), Issue15, pp. 432/36.
[11] Zhang H., Xie Y., Huang L., Huang S., Zheng X., and Chen G.: Effect of Feedstock Particle Size on

Wear Resistance of Plasma Sprayed Fe-Based Amorphous Coatings, Surface and Coatings Technology, 2014, 258, pp 495-502.
[12] Arizmendi-Morquecho A., Campa-Castilla A., Leyva-Porras C., Martinez J.A.A., Gutierrez G.V., Bello K.J.M., and Lopez L.L.: Microstructural Characterization and Wear Properties of Fe-Based Amorphous-Crystalline Coating Deposited by Twin Wire Arc Spraying, Advanced Engineering Materials, 2014. doi:10.1155/2014/836739.
[13] Bobzin K., Oete M., Linke T.F., and Koenigstein T.: Process Development for Innovative Iron Alloy Metallic Glass Coating, Advanced Eengineering Materials 18 (2016), Issue 10, pp. 1833/40.
[14] Branagan, D.J., Swank, W.D., Haggard, D.C. and Fincke, J.R.: Wear-Resistant Amorphous and Nanocomposite Steel Coatings, Metallurgical and Materials Transactions A 32 (2001), Issue 10, pp. 2615/21.
[15] Otsubo, F. and Kishitake, K.: Corrosion Resistance of Fe-16%Cr-30%Mo-(C, B, P) Amorphous Coatings Sprayed by HVOF and PS Processes, Materials Transaction 46(2005), Issue 1, pp. 80/83.
[16] Guo, R.Q., Zhang, C., Chen, Q. and etc.: Study of Structure and Corrosion Resistance of Fe-Based Amorphous Coatings Prepared by HVAF and HVOF, Corrosion Science 53(2011), Issue 7, pp. 2351/56.
[17] Voyer, J.: Surface Wear Improvement of Al-Alloys by Amorphous Iron-Based Flame-Sprayed Coatings, Materials Science Forum 690(2011), pp. 405/08.
[18] Inoue, A. High strength bulk amorphous alloys with low critical cooling rates. Materials Transactions JIM 36 (1995), Issue 7, pp. 866/75.
[19] Wu, J., Guo, H.B., Zhou L.: Microstructure and Thermal Properties of Plasma Sprayed Thermal Barrier Coatings from Nanostructured YSZ. Journal of Thermal Spray Technology 19 (2010), Issue 6, pp. 1186/94.
[20] Shin, D., Gitzhofer, F., Moreau, C.: Thermal property evolution of metal based thermal barrier coatings with heat treatments. Journal of Material Science 42(2007), Issue 15, pp. 5915/23.
[21] Yaman B., Mandal H.: Wear performance of spark plasma sintered Co/WC and cBN/Co/WC composites, International Journal of Refractory Metals and Hard Materials 42 (2014), pp. 9–16.
[22] Hou G.L., An Y.L., Zhao X.Q., Chen J., Chen J.M., Zhou H.D., Liu G.: Effect of heat treatment on wear behavior of WC-(W, Cr)2CANi coating, Surface Engineering 28 (2012), pp. 786/90.
[23] Hou G.L., An Y.L., Zhao X.Q., Zhou H.D., Chen J.M.: Effect of alumina dispersion on oxidation behavior as well as friction and wear behavior of HVOF-sprayed CoCrAlYTaCSi coating at elevated temperature up to 1000 ℃, Acta Materialia, 95 (2015), pp. 164/75.
[24] Wood P.D., Evans H.E., Ponton C.B.: Investigation into the wear behaviour of stellite 6 during rotation as an unlubricated bearing at 600℃, Tribology International 44 (2011), pp. 1589/97.
[25] Inman I.A., Datta P.K., Du H.L., Burnell-Gray J.S., Pierzgalski S., Luo Q.: Studies of high temperature sliding wear of metallic dissimilar interfaces, Tribology International 38 (2005), pp. 812/23.
[26] Inman I.A., Datta S., Du H.L., Burnell-Gray J.S., luo Q.: Microscopy of glazed layers formed during high temperature sliding wear at 750 ℃, Wear 254 (2003),pp. 461/67.
[27] Peterson M.B., Murray S.F., Florek J.J.: Consideration of lubricants for temperature above 1000 F, ASLE Trans. 2 (1959) 225/34.
[28] Peterson M.B., Calabrese S.J., Li S.Z., Jiang X.X.: Friction of alloys at high temperature, J. Journal of Materials Science and Technology 10 (1994) 313/20.

Amorphous steel coatings deposited by HVOF and Cold Gas Spray processes

M. Tului[a], A. Bezzon[a], A. Marino[a], F. Marra[b], S. Matera[a], G. Pulci[b]
[a] Centro Sviluppo Materiali S.p.A., Lamezia Terme (CZ) branch office – Italy
[b] Dept. of Chemical Engineering, Materials, Environment, Sapienza University of Rome, INSTM Reference Laboratory for Engineering of Surface Treatments, Rome, Italy

In this paper it is considered the deposition of a commercial steel powder with a chemical composition that allow to obtain amorphous coatings using thermal spray techniques. In fact, these processes are characterized by high cooling speeds ($10^6 \div 10^7$ K/s) of the particles after the impact upon the substrate.
The powders were sprayed with two different processes: Cold Gas Spray (CGS) and High Velocity Oxygen Fuel (HVOF). A comparison between the samples obtained reveals that only the CGS coatings are completely amorphous while the HVOF ones present nanocrystalline phases, detected with XRD analysis and SEM micrographs. Furthermore, CGS coatings are more compact and show lower hardness, with a comparable Young modulus. An hypothesis is that amorphisation phenomenon observed is related to the plastic deformation at the impact (due to the high particles' energy), rather than to the temperature, with a mechanism that could resemble the severe plastic deformation process.
Additional thermal treatments and mechanical tests are in progress to investigate the toughness and other mechanical properties.

1 Introduction

Amorphous metals have been drawing attention in several engineering fields owing to their unique properties, such as mechanical and thermo-mechanical properties. In particular, Fe-based metallic glasses (MG) present excellent soft-magnetic properties, wear and corrosion resistance and relatively low cost [1–8].
MGs are produced by rapid quenching techniques of metals, yielding materials with a short-range order, but with hardly any ductility and with negligible plastic behaviour in compression. [9] For these reasons, any commercial application is not easily feasible yet.
Thermal sprayed coatings of amorphous metals represent a viable alternative solution to produce MGs preserving their exceptional properties and avoiding their brittleness [9].
In fact, the lack of grain boundaries, the weak points of crystalline materials, leads to a better resistance to wear and corrosion [10].
These properties allow MGs to be the starting point of new functional materials with various possible applications (e.g. protecting structural materials), putting them in the spotlight of surface engineering.
Thermal spray technologies have the suitable characteristics (in particular high cooling rate) to make depositions of amorphous metals. In last decades, researchers have been using different thermal spray techniques such as HVOF, Flame Spray (FS), Plasma Spray and CGS to obtain amorphous coatings [11–14]. Several authors have reported crystallization and oxidation in amorphous coatings when HVOF, FS and Plasma Spray are used [11,15]. Crystallization and oxidation affect mechanical properties of obtained coatings [9].
On the other hand, CGS process has arisen as alternative to the traditional thermal spray process, considering the lower temperature and the different mechanism of adhesion related to the adiabatic shear instability.

This paper focuses on the deposition of a commercial steel powder with a chemical composition that allows to obtain amorphous coatings, using two thermal spray processes: HVOF and CGS. The main purpose is to investigate the different influence of these two techniques on microstructure and consequently on mechanical properties (mainly the hardness).
The perfectly amorphous structure is only achieved with CGS and this leads to higher hardness than HVOF coatings.

2 Materials and methods

A high alloyed iron based powder, produced by Nanosteel Company Inc., Providence, Rhode Island, USA and identified by the commercial label SHS7574HVOF, was used as starting material. The producer claims that spraying such a powder, amorphous coatings can be obtained.
The chemical composition of the powder was determined by ICP-OES (Inductively Coupled Plasma Optical Emission Spectrometry) analyses.
Granulometry was determined by a laser diffraction equipment (MALVERN Mastersizer X), using Fraunhofer diffraction theory. The powder was dispersed in a deionized water/absolute ethanol mixture (50/50 by volume) to have complete dispersion in the medium.
Starting from the measured chemical composition, the phases present in the material at thermodynamic equilibrium conditions were calculated using the JMatPro® software applying database General Steel in the range from room temperature up to 1600K.
Two sets of flat samples in AISI 304 stainless steel, dimensions 50x30x3 mm, were coated; the first set using a HVOF equipment (TAFA JP 5000), the second set using a CGS equipment (IMPACT Spray System 5/11), with N_2 as process gas. Deposition parameters for the two sets are reported in Table 1.

Tab. 1. HVOF and CGS deposition parameters

HVOF			
Barrel (inch)	O_2 gas flow rate (scfh)	Kerosene flow rate (gph)	pressure in the chamber (psi)
4	1700	7.2	101
CGS			
Pressure (bar)	Temperature (°C)		Stand-off distance (mm)
40	970		20

One sample of powder, one sample coated by HVOF, and another one coated by CGS, were analysed by XRD (X-Ray Diffraction), to determine the phases, using a Siemens 500 apparatus. Cu radiation was used.

Some samples for each set were cut, embedded in resin, polished and observed on the cross section by SEM (Scanning Electron Microscopy) and EDS (Energy Dispersion Spectroscopy), using a SEM JSM 6480L apparatus.

On the cross sections of samples belonging to the first and to the second set, dynamic indentation [16] [17] measurements were carried out in order to evaluate microhardness and elastic modulus following the Oliver and Pharr [18] method (pre-load of 1,5 mN; loading rate of 20 mN/sec; 30 sec dwell at the maximum load; unloading rate of 20 mN/sec). For each sample 45 measurements were carried out with a load of 500 mN, using a Berkovich 3-sided penetrator.

3 Results

The chemical composition of the starting powder, as determined by chemical analyses, is reported in Table 2; in the same table is reported the nominal composition, as indicated in the data sheet provided by the producer.

Tab. 2. Chemical composition of the starting powder

wt%	Cr	C	B	Mo	W	Mn	Si	O	Fe
Nominal	<25	<2	<5	<15	<10	<2	<2		bal
ICPOS Analisys	18.55	0.97	3.3	12.88	6	1.6	1.57	0.018	

Figure 1 reports the calculated phase composition of the coating material, at thermodynamic equilibrium conditions, as a function of the temperature.

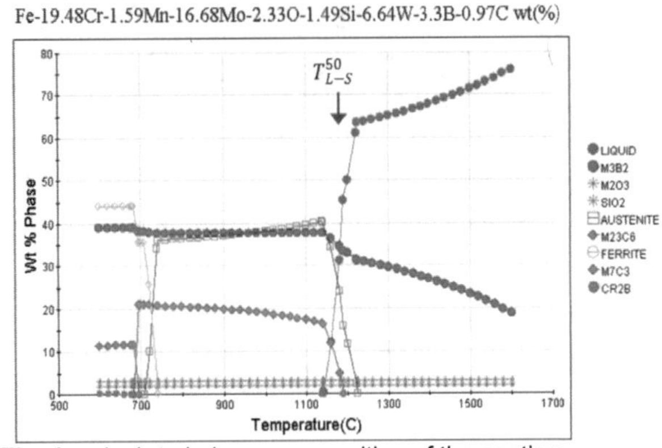

Fig. 1. calculated phase composition of the coating material as a function of the temperature

Figure 2 shows the granulometry distribution of the starting powder, a monomodal distribution can be assessed.

Figure 3 reports the XRD patterns of the starting powder, HVOF deposited coating, and CGS deposited coating, respectively.

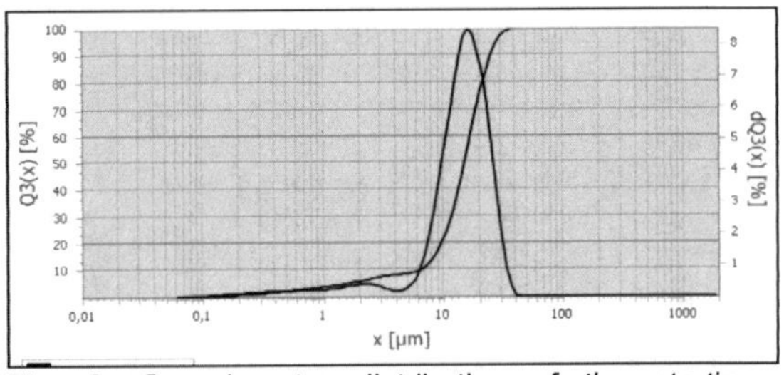

Fig. 2. Granulometry distribution of the starting powder

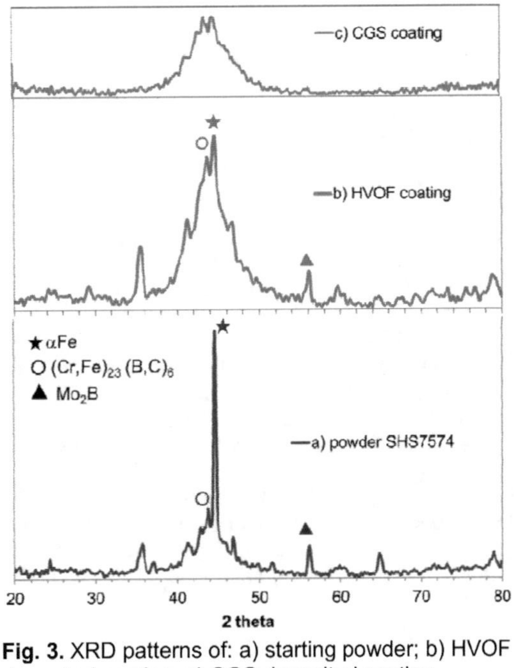

Fig. 3. XRD patterns of: a) starting powder; b) HVOF deposited coating; c) CGS deposited coating

In Figure 4 the cross sections, observed by SEM, of a HVOF deposited coating and of a CGS deposited coating, respectively, are shown. In Figure 5 the results of the instrumented indentation are reported.

4 Discussion

The chemical composition of the investigated material contains a very high amount of B and C; at the thermodynamic equilibrium, three main phases are present: metallic Fe (ferrite or austenite, depending on the temperature), borides and carbides. The first consequence of this is that the kinetics of crystallization is slow, as demonstrated by the solidus temperature (1200° C), exceptionally low for a Fe based alloy. Therefore, if cooled rapidly, this alloy can assume a nanocrystalline or even amorphous microstructure.

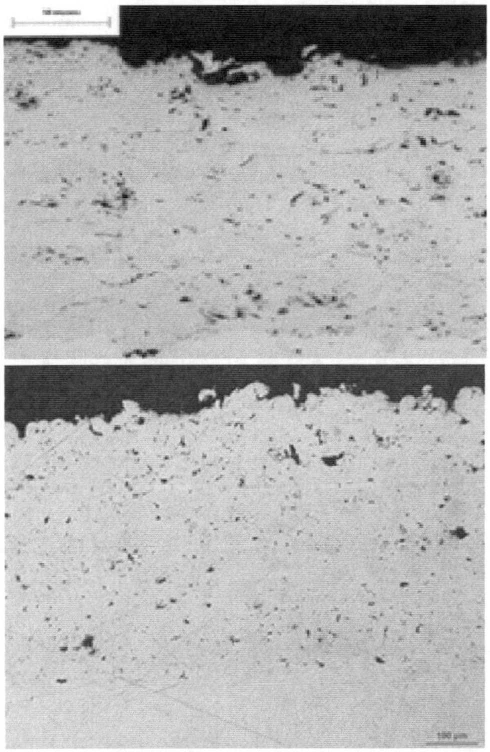

Fig. 4. Cross sections, observed by SEM, of: a) HVOF deposited coating; b) CGS deposited coating

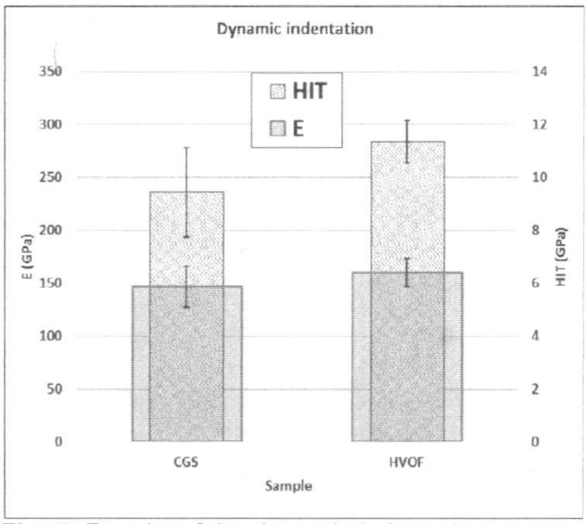

Fig. 5. Results of the dynamic indentation tests

In fact, microstructures composed by nanocrystalline stable thermodynamic phases α-Fe, $(Cr,Fe)_{23}C_6$ and M_2B have been revealed in starting powder and in HVOF coating. In both the cases, the processes caused the material to be heated up to a temperature higher than the liquidus (in the case of gas atomisation), or at least than the solidus (in the case of HVOF).
Apparently, the cooling rate is not sufficient to have a complete amorphisation, and the dominant phase is α-Fe. XRD pattern of the powder presents a higher crystalline fraction than the HVOF coating pattern. This is due to the difference in cooling speed typical of the two processes: in fact, in the case of powder the heat is extracted from particles of about 15 µm in diameter by contact with a cold gas; in the case of HVOF coating, from a few microns thick splats, by contact with a cold solid surface.
Layered completely amorphous microstructure has been observed in CGS coating. The layer morphology is given by heavily deformed sprayed metallic powder particles due to the combination of high-energy impact and severe plastic deformation [19,20,21]. The primary cause of amorphisation in the CGS deposited coating is the high-strain/strain-rate deformation with shear instability [22] (due to the high particles' energy) rather than the temperature.
Micromechanical analysis results, in Figure 5, highlight a higher hardness for the HVOF coatings, on the other hand a negligible variation can be noticed on elastic moduli. This effect can be addressed to the nucleation of nanocrystalline domains and boundary strengthening [23]. Further analysis, focusing on evolution of mechanical properties of CGS coating with thermal treatment will be carried out to investigate this behavior.

5 Conclusions

Two different techniques were used with the aim to obtain amorphous coatings. CGS technique could successfully obtain nearly complete amorphisation during deposition process. Coating obtained by CGS shows a dense microstructure, and a good adhesion with the substrate. Finally, hardness of HVOF coating is slightly higher than CGS homologous, probably due to its nanocrystalline structure.

References

[1] Yoon, S., Kim, j., Bae, G., Kim, B., Lee, C.: Formation of coating and tribological behavior of kinetic sprayed Fe-based bulk metallic glass. Journal of Alloys and Compounds 509 (2011), Issue 2, pp. 347/53

[2] Keryvin, V., Hoang, V.H., Shen, J.: Hardness, toughness, brittleness and cracking systems in an iron-based bulk metallic glass by indentation. Intermetallics 17 (2009), Issue 4, pp. 211/17.

[3] Inoe, A., Shen, B.L., Chang, C.T.: Super-high strength of over 4000 MPa for Fe-based bulk glassy alloys in $[(Fe_{1-x}Co_x)_{0.75}B_{0.2}Si_{0.05}]_{96}Nb_4$ system. Acta Materialia 52 (2004), Issue 14, pp. 4093/99.

[4] Pang, S.J., Zhang, T., Asami, K., Inoue, A.: Bulk glassy Fe–Cr–Mo–C–B alloys with high corrosion resistance. Corrosion. Science 44 (2002), Issue 8, pp. 1847/56.

[5] Makino, A., Bitoh, T., Kojima, A., Inoe, A., Masumoto, T.: Magnetic properties of zero-magnetostrictive nanocrystalline Fe–Zr–Nb–B soft magnetic alloys with high magnetic induction. Journal of Magnetism and Magnetic Materials 215–216 (2000), pp. 288/92.

[6] Trexler, M., Thadhani, N.: Mechanical properties of bulk metallic glasses. Progress in Materials Science 55 (2010), Issue 8, pp. 759–893.

[7] Kaban, I., Jovari, P., Waske, A., Stoica, M., Bednarcik, J., Beuneu, B., Mattern, N., Ecker, J.: Atomic structure and magnetic properties of Fe–Nb–B metallic glasses. Journal of Alloys and Compounds. 586 (2014), S189/93.

[8] Wang, W.H., Dong, C., Shek, C.H.: Bulk metallic glasses. Materials Science and Engineering: R: Reports 44 (2004), Issue 2, pp. 45/89.

[9] Henao, J., Concustell, A., Cano, I.G., Cinca, N., Dosta, S., Guilemany, J.M.: Influence of Cold Gas Spray process conditions on the microstructure of Fe-based amorphous coatings. Journal of Alloys and Compounds 622 (2015), pp. 995/99.

[10] Bae, G., Xiong, Y., Kumar, S., Kang, K., Lee, C.: General aspects of interface bonding in kinetic sprayed coatings. Acta Materialia 56 (2008), Issue 17, pp. 4858/68.

[11] Miura, H., Isa, S., Omuro, K.: Production of Amorphous Iron-Nickel Based Alloys by Flame-Spray Quenching and Coatings on Metal Substrates. Transactions of the Japan Institute of Metals 25 (1984), Issue 4, pp. 284/91.

[12] Voyer, J.: Surface Wear Improvement of Al-Alloys by Amorphous Iron-Based Flame-Sprayed Coatings. Materials Science Forum 690 (2011), pp. 405/8.

[13] Liu, X.Q., Zheng, Y.G., Chang, X.C., Hou, W.L., Wang, J.Q.: Influence of HVOF Thermal Spray Process on the Microstructures and Properties of Fe-based Amorphous / Nano metallic Coatings. Materials Science Forum 633–634 (2010), pp. 685/94

[14] Zhang, C., Liu, L., Chan, K.C., Chen, Q., Tang, C.Y.: Wear behavior of HVOF-sprayed Fe-based amorphous coatings. Intermetallics 29 (2012), pp. 80/5.

[15] Choi, J.H., Lee, C., Lee, D.B. Oxidation behavior of bulk amorphous $Ni_{57}Ti_{18}Zr_{20}Si_3Sn_2$ coatings between 473 and 973 K in air. Journal of Alloys and Compounds 449 (2008), Issue 1, pp. 384/88.

[16] Instrumented Indentation Test for Hardness and Materials Parameters; ISO 14577; International Organization for Standardization: Genève, Switzerland, 2007.

[17] Baiamonte, L., Marra, F., Pulci, G., Tirillò, J., Sarasini, F., Bartuli, C., Valente, T.: High temperature mechanical characterization of plasma-sprayed zirconia-yttria from conventional and nanostructured powders. Surface and Coatings Technology 277 (2015), pp.289-298.

[18] Oliver, W.C.; Pharr, G.M.: An improved technique for determining hardness and elastic modulus using load and displacement sensing indentation experiments. Journal of Materials Research 7 (1992), Issue 6, pp. 1564/83.

[19] Davis, J. R. Handbook of Thermal Spray Technology. ASM Thermal Spray, (2004), pp. 47-53, 77-84.

[20] Cavaliere, P.: Through life engineering Services – (2015), pp.175-184

[21] Hussain, T., McCartney, D.G., Shipway, P.H., Zhang, D.: Bonding Mechanisms in Cold Spraying: The Contributions of Metallurgical and Mechanical Components. Journal of Thermal Spray Technology 18 (2009), Issue 3, pp. 364/79.

[22] Lee, C., Kim, J.: Microstructure of Kinetic Spray Coatings: A Review. Journal of Thermal Spray Technology 24 (2015), Issue 4, pp. 592/610.

[23] Hansen, N.: Hall-Petch relation and boundary strengthening. Scripta Materialia 51 (2004), pp.801/6

Tribological performances of Al_2O_3/YAG amorphous ceramic coating fabricated by atmospheric plasma spraying

Kai Yang [a,*], Jian Rong [a,b], Yin Zhuang [a], Liang Wang [a], Shunyan Tao [a], Chuanxian Ding [a]

[a] Key Laboratory of Inorganic Coating Materials CAS, Shanghai Institute of Ceramics, Chinese Academy of Sciences, Shanghai 201899, P.R. China

[b] University of Chinese Academy of Sciences, Beijing 100039 , P.R. China

In this study, the spray-dried Al_2O_3/Y_2O_3 composite powder was prepared using commercially available nano-sized Al_2O_3 and Y_2O_3 feedstock. The mass ratio of Al_2O_3 to Y_2O_3 was 2:1. Atmospheric plasma spraying possesses high thermal enthalpy, large temperature gradient and rapid cooling rate. The microstructure and wear resistant performance of plasma-sprayed Al_2O_3–$Y_3Al_5O_{12}$ (YAG) coating were investigated. As-sprayed Al_2O_3–YAG coating was chiefly composed of amorphous phase, which may reveal superplastic feature in supercooled liquid region (from 503.0℃ to 906.5℃). The as-sprayed Al_2O_3–YAG coating exhibits fine plasticity and toughness. Friction and wear tests of the coatings were executed on a MMU-5GL tribological tester using a ring-on-disk arrangement. The coatings were deposited on end flat surfaces of the wear rings. The graphite disks were prepared. The wear tests were conducted at following conditions: relatively high load of 2000N; a rotational speed of 500rpm (equivalent to a sliding velocity of 0.68m/s). Friction coefficients could be obtained real-timely. The thermocouple was applied to measure the worn surface temperatures. The Al_2O_3/YAG amorphous coating/graphite pair possessed lower friction coefficient and worn surface temperature, compared with Al_2O_3 coating/graphite and Al_2O_3-Cr_2O_3 coating/graphite pairs. After wear tests, many network cracks visible to naked eye appeared the surfaces of Al_2O_3 and Al_2O_3-Cr_2O_3 coatings. However, no cracks were observed on the worn surface of the Al_2O_3/YAG amorphous coating. Therefore, plasma-sprayed Al_2O_3–YAG coating possesses excellent wear resistance under severe conditions with high pv values (p: contact pressure; v: friction velocity).

1 Introduction

Friction and wear of materials under high pv values (p: contact pressure; v: friction velocity) are commonly accompanied by high temperature, serious oxidation and strong thermal shock [1]. Oxide ceramics show high strength, high hardness, good high temperature and oxidation resistances [2]. The previous investigations were focused on the mechanical properties of the Al_2O_3 coatings [3, 4]. With respect to severe wear conditions with high pv values, the toughness, high temperature microstructure stability and thermal conduction performance of the coatings are significant.

We obtained an inspiration from Al_2O_3-based directionally solidified eutectic (DSE) oxide bulk ceramics [5]. Especially, binary Al_2O_3–YAG eutectics have been reported to shown excellent mechanical performance, thermal conduction, creep resistance and

microstructure stability at high temperatures [6]. The key preparation factors are as follows: ① high thermal enthalpy; ② large temperature gradient (10^2–10^4K·cm^{-1}); ③ rapid cooling rate. With regard to atmospheric plasma spraying, very high plasma temperatures approach 10000–20000K; the high quench rate of the molten particles during plasma spraying is about 10^6K·s^{-1}; the estimated temperature gradients in splats approach 10^4K·cm^{-1}. Rapid eutectic growth might be expected in the plasma sprayed Al_2O_3–YAG coatings.

In this present work, the as-sprayed Al_2O_3–YAG coating with dominant amorphous phases was obtained. Due to the superplastic feature of amorphous phase in supercooled liquid region,

* Corresponding author. Tel: +86 021 69906320; Fax: +86 021 69906322; E-mail address: kaiyang@mail.sic.ac.cn (K. Yang).

plasma sprayed Al_2O_3–YAG amorphous coating showed excellent wear resistance under severe conditions. The objective of this study is to evaluate wear resistance of as-sprayed Al_2O_3–YAG coating and explore the relationship between anti-wear performance and microstructure of the coating.

2 Experimental procedure

2.1 Powder preparation

Commercially available Al_2O_3 and Y_2O_3 nanopowders with the corresponding average sizes (D50) of 76.4nm and 74.2nm were utilized as raw materials. Nanostructured Al_2O_3/Y_2O_3 sprayable feedstocks were fabricated by spray drying method with the eutectic composition (82mol% Al_2O_3, 18mol% Y_2O_3).

2.2 Coating deposition

The Multicoat atmospheric plasma spraying system equipped with a F4-MB plasma gun (Sulzer Metco AG, Switzerland) was used to prepare the coating. Prior to fabricating ceramic coatings, NiCr powder was applied to deposit as a bond coating. For performance comparison, Al_2O_3 and Al_2O_3–Cr_2O_3 coatings were also deposited, which could be referred in our previous studies [7, 8].

2.3 Coating characterization

The phase composition of the as-sprayed coating was identified by X-ray diffraction (XRD) using a Rigaku D/Max2550 Diffractometer. The morphologies of the feedstock and coating were characterized by a Nova NanoSEM 230 scanning electron microscope (FEI, Hillsboro, USA). Differential thermal analysis (DTA) of the coating was performed by a thermal analysis mass spectrometer (STA449C, Netzsch, Germany). Friction and wear tests of the coating were executed on a MMU-5GL (Jinan Yihua Tribology Testing Technology Co., Ltd, China) tribological tester.

3 Results and discussion

Interestingly, the obvious amorphous scattering peak appears in the XRD pattern of the as-sprayed coating (shown in Fig. 1a), which may indicate the dominant amorphous phase in the as-sprayed coating. Simultaneously, a few α-Al_2O_3 and YAG phases were also detected. The Al_2O_3 reacted with Y_2O_3 to form YAG in the plasma spraying process. However, there are no observable two phase features in the SEM morphological image of the as-sprayed

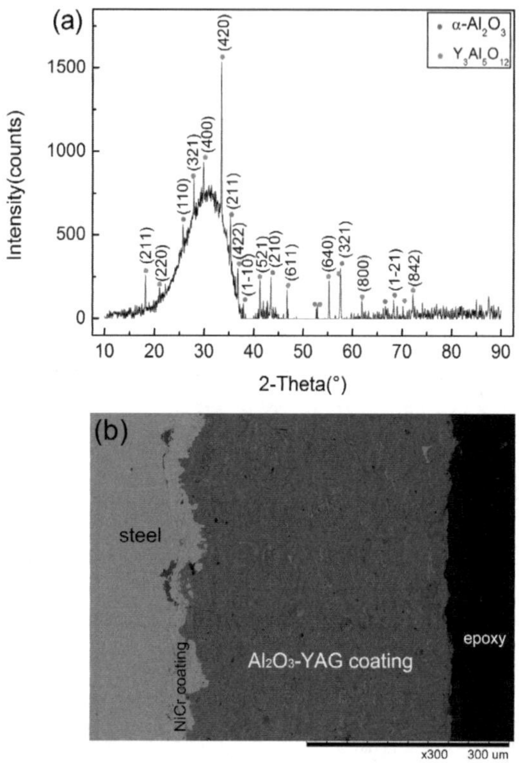

Fig.1. (a) XRD pattern of the as-sprayed Al_2O_3-YAG coating; (b) Cross-sectional morphology of the as-sprayed Al_2O_3-YAG coating.

With respect to the amorphous coating, the glass transition temperature (Tg), initial crystallization temperature (Tx) and crystallization peak temperature (Tp) are significant. According to the DTA curve of the Al_2O_3–YAG coating, the values of Tg, Tx, Tp (YAG) and Tp (α-Al_2O_3) were 503.0℃, 906.5℃, 931.3℃ and 1046.3℃, respectively (displayed in Fig. 2).

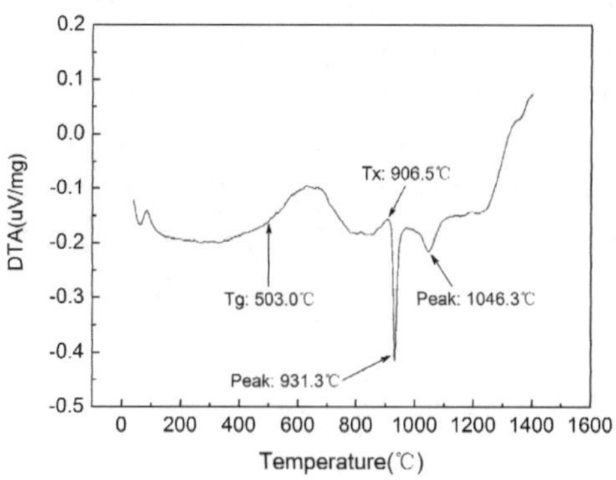

Fig.2. The DTA curve of as-sprayed coating.

Friction and wear tests were performed under the following conditions: a load of 2000N and a rotational speed of 500rpm. The dimensions of stainless steel ring and graphite disk (ISO-68, Toyo Tanso, Japan) were Φ_{outer}26mm × Φ_{inner}20mm × 12mm and Φ43mm × 3mm, respectively. The thickness of top ceramic coatings deposited onto stainless steel ring surfaces was about 300μm after grinding and polishing. The average friction coefficients of Al_2O_3 coating/graphite pair, Al_2O_3–Cr_2O_3 coating/graphite pair and Al_2O_3–YAG coating/graphite pair were 0.2225, 0.1916 and 0.1905, respectively. Moreover, the Al_2O_3–YAG coating/graphite pair possessed better friction coefficient stability (shown in Fig. 3a). Simultaneously, the thermocouple probe was set on the tribological tester for real-time monitoring frictional pair temperature. In the wear tests, the worn surface temperature follows the sequence: Al_2O_3–YAG coating/graphite pair<Al_2O_3–Cr_2O_3 coating/graphite pair<Al_2O_3 coating/graphite pair (presented in Fig. 3b).

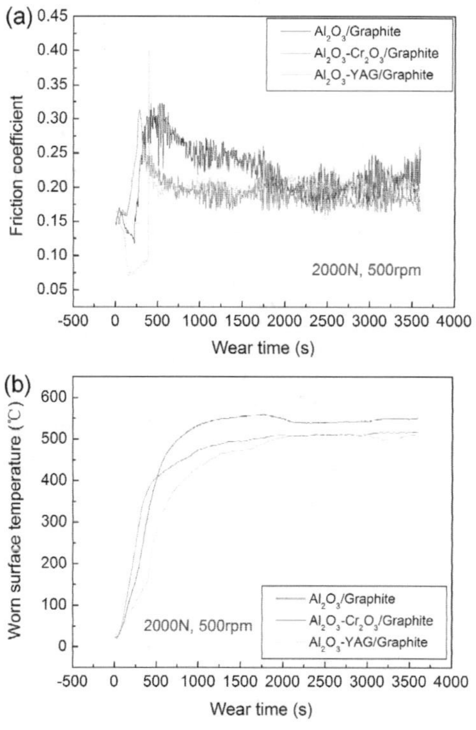

Fig.3. (a) The friction coefficients of the coating/graphite pairs (2000N, 500rpm, 1h); (b) The worn surface temperatures of the coating/graphite pairs (2000N, 500rpm, 1h).

The obvious network cracks visible to naked eye appeared on the surfaces of Al_2O_3 and Al_2O_3–Cr_2O_3 coatings under severe wear conditions (shown in Fig. 4a and 4b). However, no macroscopical cracks were observed on the Al_2O_3–YAG coating surface (shown in Fig. 4c). Furthermore, smooth and stable graphite transfer film effectively formed on the worn surface of as-sprayed Al_2O_3–YAG coating, which may be conducive to the friction coefficient stability.

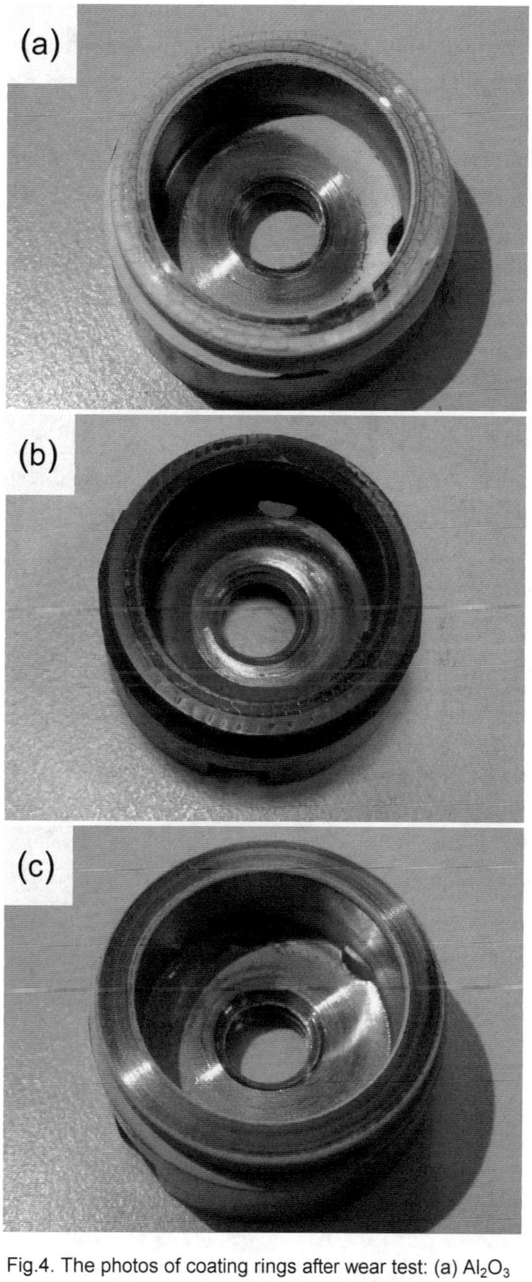

Fig.4. The photos of coating rings after wear test: (a) Al_2O_3 coating; (b) Al_2O_3–Cr_2O_3 coating; (c) Al_2O_3–YAG coating.

No solid solution in Al_2O_3–YAG eutectic system leads to rather low equilibrium distribution coefficient k, which indicates very large undercooling at the solid/liquid interface. Additionally, the high quench rate of the molten particles in plasma spraying further increases degree of undercooling. Therefore, the Al_2O_3–YAG eutectic system belongs to deep eutectic system. If the practical solidification temperature is lower than the glass-transition temperature in deep eutectic system, the eutectic crystalline phases would stop grow to form single amorphous phase. Plasma sprayed Al_2O_3–YAG coating may be just this situation. So the as-sprayed Al_2O_3–YAG coating possesses the distinct amorphous feature. According to the DTA curve (exhibited in Fig. 2), the temperature interval between Tx and Tg could be defined as supercooled liquid region, in which the amorphous materials often have the superplastic deformation behavior. In the present study, stable wear stage temperature was higher than Tg of the amorphous Al_2O_3–YAG coating. Accordingly, the as-sprayed Al_2O_3–YAG coating shows fine plasticity and toughness under severe conditions, which is greatly beneficial to relieve accumulated coating stress in the wear process and inhibit the crack generation. Accordingly, the as-sprayed Al_2O_3–YAG coating shows excellent wear resistance under severe conditions with high pv values.

4 Conclusions

In summary, the as-sprayed Al_2O_3–YAG coating is mainly comprised of amorphous phase, which may show superplastic performance in supercooled liquid region (from 503.0℃ to 906.5℃). Under the following conditions: a load of 2000N, a rotational speed of 500rpm and worn surface temperature above 500℃, Al_2O_3–YAG coating exhibits excellent wear resistance, which indicates superior coating application potential under severe wear conditions with high pv values.

Acknowledgments

The study is jointly supported by National Nature Science Foundation of China (51302299), Shanghai Nature Science Fund Project (17ZR1434700) and Youth Innovation Promotion Association, Chinese Academy of Sciences (Y65ZQ6160G). Simultaneously, we are grateful to Jinan Yihua Tribology Testing Technology Co., Ltd for providing the support for this work.

References

[1] H. Rojacz, I.A. Neacsu, L. Widder, M. Varga, J. Heiss, Thermal effects on wear and material degradation of slag pots operating in steel production, Wear 350-351 (2016) 35-45.

[2] N.P. Padture, M. Gell, E.H. Jordan, Thermal barrier coatings for gas-turbine engine applications, Science 296 (2002) 280-284.

[3] G. Sivakumar, R.O. Dusane, S.V. Joshi, A novel approach to process phase pure α-Al_2O_3 coatings by solution precursor plasma spraying, J. Eur. Ceram. Soc. 33 (2013) 2823-2829.

[4] S. Dong, B. Song, B. Hansz, H. Liao, C. Coddet, Improvement of adhesion of plasma-sprayed Al_2O_3 coatings by using dry-ice blasting, Surf. Coat. Technol. 66 (2012) 289-291.

[5] J. LLorca, V.M. Orera, Directionally solidified eutectic ceramic oxides, Prog. Mater. Sci. 51 (2006) 711-809.

[6] H.J. Su, J. Zhang, W.D. Ma, K.C. Wei, L. Liu, H.Z. Fu, S.P. Feng, A.K. Soh, In situ fabrication of highly-dense Al_2O_3/YAG nanoeutectic composite ceramics by a modified laser surface processing, J. Eur. Ceram. Soc. 34 (2014) 739-744.

[7] K. Yang, X. Zhou, H. Zhao, S. Tao, Microstructure and mechanical properties of Al_2O_3–Cr_2O_3composite coatings produced by atmospheric plasma spraying, Surf. Coat. Technol. 206 (2011) 1362-1371.

[8] K. Yang, X. Zhou, C. Liu, S. Tao, C. Ding, Sliding wear performance of plasma-sprayed Al_2O_3-Cr_2O_3 composite coatings against graphite under severe conditions, J. Therm. Spray Technol. 22 (2013) 1154-1162.

A review on the improvement of Wear, Impact and Corrosion Resistance of Iron Based Amorphous Coating by Addition of Alumina Particles

M. Yasir*[1,2], A.A.Qayyum[2],C. Zhang[1], L. Liu*[1],
[1] State Key Lab for Materials Processing and Die & Mould Technology, School of Materials Science and Engineering, Huazhong University of Science and Technology, Wuhan 430074, People's Republic of China
[2] Deaprtment of Materials Science & Engineering, Institute of Space Technology 1, Islamabad, 44000,Pakistan.

The lower wear and poor impact resistance of the amorphous coatings has been a great problem in the past few years for their use in industrial applications. Several research methods have been reported recently to overcome this issue. The present paper addresses the main work done recently on iron based amorphous composite coatings by the addition of 0-20% Alumina particles. These particles were homogeneously distributed in the amorphous matrix of the coatings which improved wear and impact resistance as compared to the monolithic coatings without any decrease in corrosion resistance. The hard alumina particles enhanced wear resistance to several times not only in air but also in salt water solution with a decrease in friction coefficient. The combined effect of wear and corrosion were also observed to become better by the alumina addition. Furthermore, the impact resistance was also improved three times by the addition of alumina particles. The hard second phase particles present in the amorphous coating matrix disperses the residual stresses generated during the impact loading. The brittle alumina particles absorb the impact energy by breaking itself which stop the initiation of cracks and also play a vital role in the crack arresting and blocking of the crack propagation.

Keywords: Al_2O_3 particle; Wear mechanism, Impact resistance, crack initiation. Finite Element Method, simulation, fracture mechanism.

1 Introduction

Iron based amorphous coatings have received huge attention because of their relatively high hardness, corrosion resistance, wear resistance and excellent hydrophobicity which makes them a very good candidate to be employed in a variety of industries such as oil and gas, marine and ship industries [1-2]. There are variety of techniques available in order to deposit amorphous coatings which includes laser cladding, kinetic spraying and thermal spraying techniques. However, among all of them High velocity Oxyfuel (HVOF) is the most attractive one because they produce amorphous coatings which have high bonding strength and low porosity [3].

Despite of higher hardness and excellent corrosion resistance, the coatings exhibit lower wear resistance than the ceramic and other commercially used coatings. Moreover, the amorphous coatings cannot be used in load-bearing applications due to poor bonding strength and lower fracture toughness as a result of intrinsic brittleness [3].

To overcome these problems, many approaches have been tried. Zhou et al. [4] tried to reinforce amorphous coatings by the help of soft stainless steel (SS) particles which resulted in higher bonding strength and better fracture toughness but the wear and corrosion resistance was decreased due introduction of these metallic reinforcements. On the other hand, Zhang et al. [5] improved the bonding strength and impact resistance to several extent by using alternative amorphous and NiCrAl layers. However these laminar structured coatings were very complex for commercial purposes.Another effort by our research group was done to to improve the wear resistance of amorphous coating by modifying feedstock powder with thin layer of Ni-W-P amorphous phase though a electroless plating process. As a result anti-oxidation Ni rich phase is formed at inters-plat region which inevitably improved wear resistance. The coatings were also not successful in improving the impact resistance of the amorphous coatings. The corrosion resistance is very important factor to consider in composite coatings. Therefore some researchers tried ceramic reinforcements such as TiN, WC-12Co and B4C particles to obtain the desired properties. As a result they showed better bonding strength, friction and wear resistance however, the problem of poor impact resistance still remains there [7].

In the recent years, we tried to add alumina particles as a reinforcement in amorphous composite coatings. It has been found that by distributing 20 wt% alumina particles homogeneously in amorphous matrix, we could develop a composite coating having reduced friction coefficient with higher wear resistance and better tribocorrosion resistance with comparison to monolithic amorphous coating. Moreover, fracture toughness on impact loading is also increased. In this paper, we will briefly highlight the recent progress in Alumina reinforced amorphous coatings [8-10].

2 Experimental

The gas-atomized amorphous powder having a composition of $Fe_{48}Mo_{14}Cr_{15}Y_2C_{15}B_6$ is blended with 20 wt.% alumina powder having the similar size of 33-55um using low energized ball milling for 5 hours. NiCrAl layer was sprayed prior to amorphous layer to achieve higher bonding strength..

*Corresponding authors: muhammadyasir85@gmail.com
lliu2000@mail.hust.edu.cn

In order to evaluate friction and wear behavior, a UMT-3 wear machine was used which have Si_3N_4 ball (4mm diameter) as a counterpart in a pin on disk mode. The same apparatus is used for tribocorrosion test with a little modification i-e a three electrode electrochemical cell was made in which Platinum is used as a counter electrode while Ag/AgCl is used as reference electrode. The electric insulation to electrochemical cell was provided by polymeric materials. The signals (current, potential and coefficient of friction) were recorded as a function of time and scanning speed is kept at 0.5mV/s for determining potentiodynamic polarization in 3.5% NaCl solution. A homemade drop-weight impact testing machine was used to measure impact resistance by varying impact energy from 1.8-21.2 J. Simulation was done to study the fracture mechanism of the coatings after impact by finite element analysis (ABAQUS software).

3 Results and Discussion

The SEM images of both coatings are shown in **Fig. 1(a, b)**. Both amorphous and amorphous composite coating have a thickness of 400 ± 20 with excellent bonding with the substrate. Enlarged SEM morphology and EDX mapping of Aluminum dispersion in amorphous matrix as shown in **Fig. 1 (c, d)** depict the homogenous distribution of alumina particles with a very low porosity.

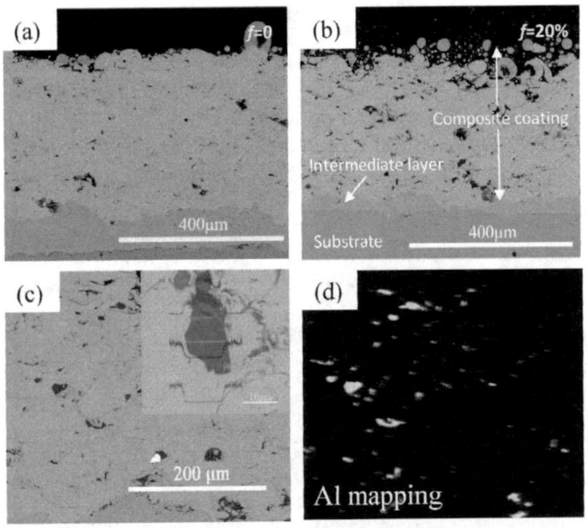

Fig. 1. Cross-sectional SEM images of both coatings with EDX Al mapping [8]

Prior to wear test, microhardness of both of these coating were observed through microindentation and it was seen that microhardness of monolithic amorphous coating was 900 ± 100 Hv while that of amorphous composite coating having 20% alumina particles as reinforcement was 1180 ± 200 Hv. Similarly, composite coatings showed higher wear resistance than pure coating. The behavior of pure amorphous coating after they were worn out under dry sliding conditions is shown in **Fig. 2(a, b)**. The edx analysis of worn surface depicts that oxidation had occurred at worn surfaces leading to high oxygen content in those surfaces is shown in **Fig. 2 (b)**. The improvement in wear resistance is observed as there is less amount of oxygen content in worn surface by adding Al_2O_3 and by the help of EDX analysis we can tell that that hard alumina particles remain part of the matrix indicated by **Fig 2 (c, d).**

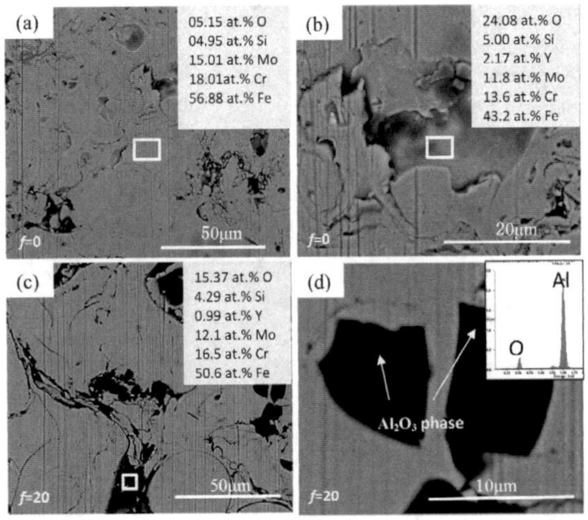

Fig. 2. Wear track of the amorphous and composite coatings with EDX analysis. [8]

Friction/ wear behavior was then observed in salt water solution(3.5% Sodium chloride) in both monolithic amorphous and composite coating, generally increase in wear rate was observed due to higher corrosion effect. However, the increase in wear

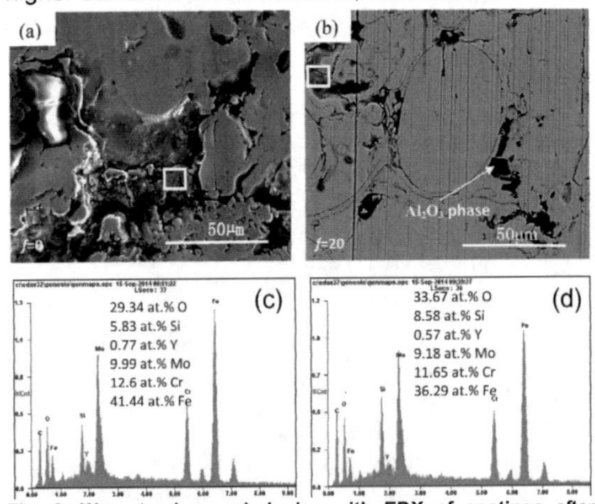

Fig. 3. Wear track morphologies with EDX of coatings after sliding in salt water solution. [8]

rate of composite amorphous coating was less than monolithic amorphous coating due to hard insulating Al_2O_3 particles. The SEM morphologies of worn

*Corresponding authors: muhammadyasir85@gmail.com lliu2000@mail.hust.edu.cn

surfaces reveals that at the intersplats regions of amorphous coatings, higher wear effect was observed as shown in **Fig. 3(a, b).** This could be because in a corrosive environment both splat delamination and pitting can occur [11, 12] .Further EDX analysis confirms that iron oxide could be the major metal oxide forming at corroded region as indicated in **Fig. 3(c, d)**.

The potentiondynamic polarization curves were drawn before and after tribocorrosion sliding for both coatings and 316L SS alloy which can be seen in **Fig. 4(a, b)**. Under no sliding conditions, it has been observed that both coatings have the same anodic and cathodic domains which then indicates that corrosion resistance of amorphous coating does not increase by introducing ceramic reinforcements (Al_2O_3). The result implies that by adding reinforcements in amorphous matrix, it may not necessarily change interfacial structure or have any other galvanic effect. However, it has been noted that stability of passive film increased for composite coating as compared to amorphous coating because they showed lower passive current density (i_{pass}). There can be two reasons for this effect in composite coating. Firstly, due to reduced porosity in composite coating and secondly, due to insulation effect provided by Al_2O_3 particles. Moreover, these two coatings have provided higher pitting potential E_{pit} (900-1000 $mV_{Ag/AgCl}$) as compared to 316L SS alloy (~300 $mV_{Ag/AgCl}$) indicating better pitting resistance. The polarization curves were measured after sliding and for 316L SS alloy, there is a negative shift in corrosion potential (E_{corr}) as well as 2-3 times increase in passive and corrosion current density as shown in **Fig 4(b).** This tells us about poor stability of passive film of 316L SS which results in decrease in corrosion resistance. The same decrease in corrosion resistance and stability of passive film is observed for amorphous coating however, this decrease is very less for composite coating. The reason for increase in passive current is that every time when a mechanical load is applied and wear occurs, a new electrochemically fresh surface is formed and this surface has less equilibrium potential. Moreover, this passive current or current density can also be increased by the presence of macroscopic galvanic corrosion as the areas which undergo wear act as anode while the and region around it act as cathode. One more thing that has also been observed was severe current oscillation during tribocorrosion sliding process as indicated in **Fig. 4(b)**. Less current oscillation is observed in amorphous coating because of their high hardness and more in 316L SS alloy which shows dynamic equilibrium between repassivation and depassivation. However, less current oscillation is observed in amorphous coatings because of high hardness more repassivation ability [12].

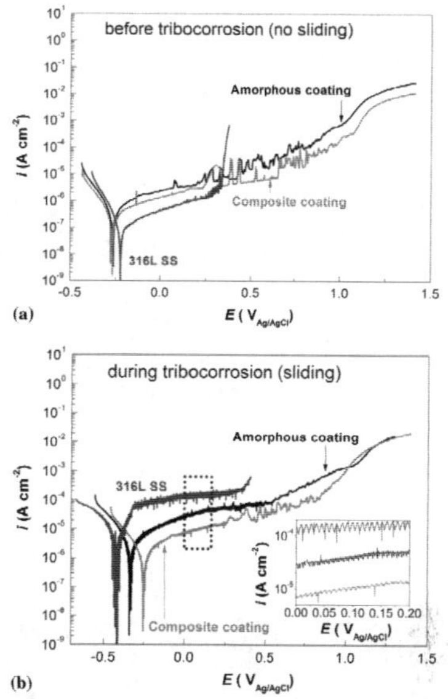

Fig. 4 Electrochemical polarization analysis of all samples with and without sliding[10]

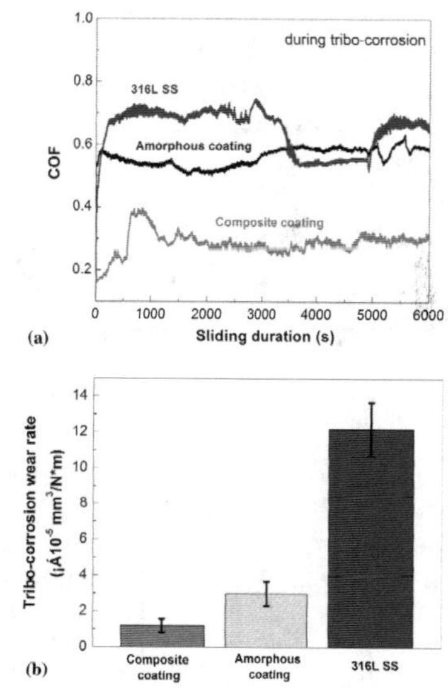

Fig. 5. COF and wear rate during tribocorrosion process for all the samples [10]

The graph of sliding duration versus coefficient of friction (COF) is shown in **Fig. 5(a)**. It depicts that COF of composite coating is least as compared to amorphous coating and 316L SS alloy. The COF for composite coating without and with potential applied in corrosive solution increases from ~ 0.15 to ~ 0.3 (under tribocorrosion condition). It is observed that

*Corresponding authors: muhammadyasir85@gmail.com
lliu2000@mail.hust.edu.cn

among these three materials, the highest wear resistance is exhibited by composite coating as shown in **Fig. 5(b)**. It is because this coating show very little wear and corrosion damage by contour materials as they are very hard particles and remain adhered to the matrix which indicates very better adhesion between these two phases while amorphous coatings mostly undergo oxidation wear mechanism.

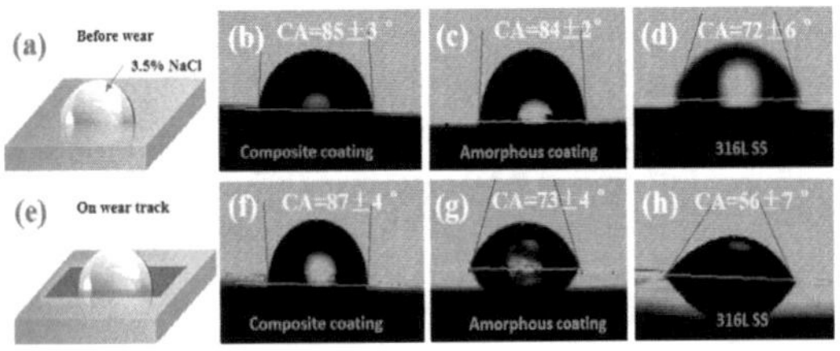

Fig. 6. Wettability measurement through CA before and after wear [10]

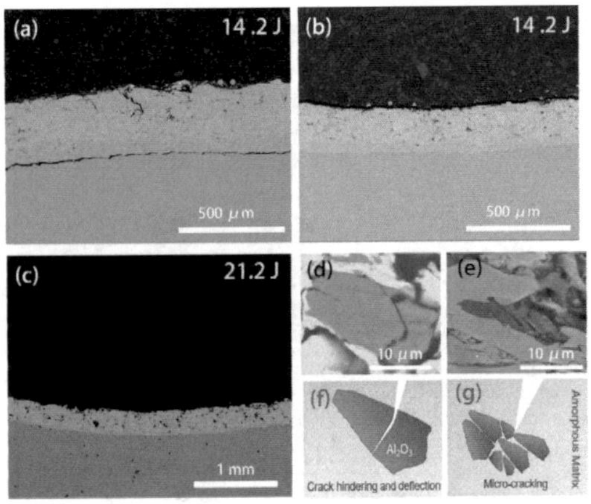

Fig. 7. Impact fracture mechanism of both the amorphous and composite coatings [9]

In order to ensure the stability of this passive layer, wettability of coatings by salt water droplet before and after wear on surface was measured as shown in **Fig. 6**. The surface energy was observed by dropping salt water droplet on wear track.The contact angles of wear tracks of composite coating before (CA ~ 85°) and after (CA > 90°) the tribocorrosion test remains same (nearly hydrophobic state) as compared to other samples whose contact angles were decreased around 10° and 16° respectively.

Impact resistance is more important for such type of coatings. To determine the impact resistance of amorphous and composite coatings, impact resistance was measured (with energies of 14.2 J and 21.2 J) and after that cross section SEM images were taken as shown in **Fig. 7**. It has been observed when amorphous coating is impacted with 14.2 J energy (80 cm height), a clear delamination occurs between the interface of substrate and NiCrAl as compared to that of composite coating. Moreover, some penetrating cracks (vertical and horizontal) were observed in amorphous coating while none of those cracks are shown in composite coating even after the energy of drop weight is increased to 21.2J.However, a large amount of plastic deformation was observed at this energy and as shown in **Fig. 7 (e,f)**. These particles might have two roles. Firstly, crack propagation was blocked by these particles and secondly, main cracks can convert themselves into microcracks when these particles absorb impact energy. The finite element modelling (FEM) was carried out which depicts that change in stress fields might occur due to alumina particles which then suppresses the interfacial delamination as shown in **Fig. 8**.

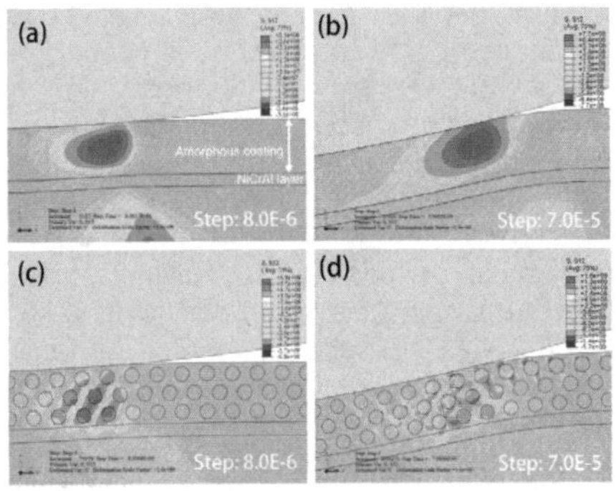

Fig. 8. Simulation analysis to explain the impact fracture mechanism of both the amorphous and composite coatings [9]

According to FEM, it has been observed that as compared to composite coating, highly significant sheer stress is created at both the edge and center of impacted region of amorphous coating which then can lead to interface delamination and formation of long penetrating cracks. Meanwhile, in case of composite coating, alumina particles were distributed throughout the matrix and the stress fields generated as a result were homogeneously distributed throughout the matrix. Also the origin of microcracks is explained by the fact that there is a huge elastic modulus difference (ΔE = 181GPa) between alumina particles and amorphous matrix; consequently, local stress concentration is formed in reinforcement and matrix interface. Due to long cracks present in amorphous coatings after impact test, corrosion resistance if measured afterwards would be less than that of composite coating.

4 Conclusion

In this paper, recent progress in Iron based amorphous coating reinforced with alumina particles were discussed. It has been shown that the composite coating show better wear rate and tribocorrosion

*Corresponding authors: muhammadyasir85@gmail.com lliu2000@mail.hust.edu.cn

resistance. The wear resistance of composite coatings increased because Al_2O_3 particles are very hard and they remain well adhere to matrix when mechanical load is applied during wear test. The stability of passive layer through potentiodynamic polarization test is increased for composiite coating because of Al_2O_3 particles are insulating in nature and reduced porosity as they distribute themselves homogenously in matrix. The contact angle measurement ensured the stability of passive layer of composite coating after wear. Furthermoe, the impact resistance was found to be increased as Al_2O_3 particles can stop the propagation of cracks and can also deflect cracks after interacting with them resulting in formation of microcracks which are less deleterious. FEM showed that highly significant sheer stress is created at both the edge and center of impacted region in amorphous coating which then can lead to interface delamination and formation of penetrating long cracks. while stress fields is distrbuted homogeneously throughout the matrix in composite coating and can result in formation of microcracks instead. It is hoped that if further investigation is made towards ceramic reinforcement in amorphous matrix, we can increase their applications in oil and gas industries, military industries and manufacturing industries.

5 References

[1] W.L. Johnson, Bulk glass-forming metallic alloys: science and technology, MRS Bull. 24 (1999) 42–56.

[2] C.A. Schuh, T.C. Hufnagel, U. Ramamurty, Mechanical behavior of amorphous alloys, Acta Mater. 55 (2007) 4067–4109.

[3] L. Liu, C. Zhang, Fe-based amorphous coatings: structures and properties, Thin Solid Films 561 (2014) 70–86.

[4] H. Zhou, C. Zhang, W. Wang, M. Yasir, L. Liu., Microstructure and mechanical properties of Fe-based amorphous composite coatings reinforced by stainless steel powders, J. Mater. Sci. Technol. 31 (2015) 43.

[5] C. Zhang, H. Zhou, L. Liu, Laminar Fe-based amorphous composite coatings with enhanced bonding strength and impact resistance, Acta Mater. 72 (2014) 239.

[6] Wang, Wei, et al. "Enhancement of oxidation and wear resistance of Fe-based amorphous coatings by surface modification of feedstock powders." Materials & Design 73 (2015) 35-41.

[7] T.Terajima, F.Takeuchi, K.Nakata, S.Adachi, K.Nakashima, T.Igarashi, Composite coating containing WC/12Co cermet and Fe-based metallic glass deposited by high-velocity oxygen fuel spraying, J.Alloy. Compd. 504 (2010) S288

[8] M. Yasir, C. Zhang, W. Wang, P. Xu, and L. Liu, Wear Behaviors of Fe-Based Amorphous Composite Coatings Reinforced by Al_2O_3 Particles in Air and in NaCl Solution, Mater.Des, 2015, 88, p 207-213

[9] M. Yasir, et al. "Enhancement of impact resistance of Fe-based amorphous coating by Al_2O_3 dispersion." Materials Letters 171 (2016) 112-116.

[10] M.Yasir, et al. "Tribocorrosion Behavior of Fe-Based Amorphous Composite Coating Reinforced by Al2O3 in 3.5% NaCl Solution." Journal of Thermal Spray Technology 25.8 (2016) 1554-1560.

[11] C. Zhang, L. Liu, K.C. Chan, Q. Chen, C.Y. Tang, Wear behavior of HVOF-sprayed Febasedamorphous coatings, Intermetallics 29 (2012) 80–85.

[12] C. Zhang, K.C. Chan, Y.Wu, L. Liu, Pitting initiation in Fe-based amorphous coatings, Acta Mater. 60 (2012) 4152–4159.

*Corresponding authors: muhammadyasir85@gmail.com lliu2000@mail.hust.edu.cn

Influence of oxidation of Ti particle during laser remelting process on the microstructure of laser remelted NiTi coatings

Yan WANG, Jiangwei LIU, Geoffrey DARUT, Thierry POIRIER, Jorge STELLA, Hanlin LIAO, Marie-Pierre PLANCHE

Commercial available Ni and Ti powder were blended together and deposited on stainless steel by atmospheric plasma spray(APS). Subsequently the as-sprayed coatings were laser remelted with a Nd -YAG pulsed laser source. Cross-sections of as-sprayed and laser-remelted coatings were characterized by scanning electron microscopy (SEM). Prior to SEM observations, the laser remelted coatings were polished and etched by Kroll etchant. Meanwhile, the energy dispersive spectrometer (EDS) was employed to analyze the chemical distribution of the coating both as-sprayed and laser remelted. The results indicated that APS sprayed NiTi coatings presented a dense microstructure with Ni splats and Ti splats distributing uniformly. Oxygen partial pressure in the argon leads to the burning of Ti splats during the laser remelting process. And Ti oxides located at the bottom of the laser molten pool because of the laser stiffness and molten flow. Moreover, the top part of the molten pool mainly involved in Ni columnar grains.

Keywords: NiTi coating, oxidation, laser remelting, microstructure

1 Introduction

Investigations have focused on NiTi intermetallic in applications for protection against cavitation and corrosion [1-6] due to its unique properties such as superelastic, high damping capacities and shape memory effect. Recently, extensive attentions have been paid to NiTi coatings instead of manufacturing large NiTi bulk components in order to reduce the material and processing costs [7-10]. And thermal spraying is an effective coating approach. Bitzer's [11] investigation has demonstrated that NiTi coatings fabricating by low-pressure plasma spraying show excellent performance against cavitation erosion due to the stress-induced martensitic phase transformation dissipating the cavitation energy. Similar results were reported by Stella [12] that after 72 h cavitation the total mass loss for vacuum plasma sprayed NiTi coating was 7.2 mg with pre-alloyed powder. In addition, mechanical alloyed NiTi powders were employed as feedstock in some studies [13] because of their low cost.

However, it was revealed that vacuum plasma sprayed element blended NiTi coatings presented less content of shape memory phase than that of pre-alloyed NiTi coatings [12]. Nevertheless, Hiraga [10] used laser plasma hybrid spraying (LPHS) to obtain NiTi coating with different mixture ratio of Ni to Ti. The results showed that NiTi phase could be formed by plasma spraying together with remelting with a high-power laser.

In the present studies, Ni and Ti powders were blended together as feedstock and fabricated by atmospheric plasma spraying. Then the as-sprayed coatings were subjected to laser remelting. The microstructure of remelted coatings was investigated. And the influence of oxidation of Ti splats on the microstructure of laser remelted NiTi coatings was discussed.

2 Experimental

2.1 Coating deposition process

Commercial Ti with irregular shape (**Fig. 1(a)**) and Ni with spherical shape (**Fig. 1(b)**) were blended together as feedstock. Carbon steel disks with specific dimensions of Ø25×9 mm were employed as substrates. The chemical composition of the steel is listed in Table 1.

Table 1 Chemical composition of carbon steel.

Ele.	Si	Mn	P	S	Cr	Mo	Ni	Fe
wt.%	<0.40	0.40 - 0.70	<0.030	<0.035	0.40	0.10	0.40	Bal.

The substrates were cleaned in an ultrasonic acetone bath and grit-blasted with alumina in order to enhance the adhesion of coatings. A Sulzer-Metco F4 plasma gun was installed on a robot to deposit NiTi coating by atmospheric plasma spraying. Plasma gas was a mixture of argon and hydrogen. Argon was also employed to carry the feedstock into plasma jet at a flow rate of 3.5 slpm. Table 2 illustrates parameters of plasma spraying.

Table 2 Parameters during plasma spraying process

Parameters	Value
Electric arc current (A)	600
Electric arc voltage (V)	60
Plasma gas (slpm) Ar	35
H2	10
Carrier gas (slpm)	3.5
Liner speed (m/s)	1.2
Stand-off distance (mm)	115

Fig. 1. Morphology of the powders Ti (a) and Ni (b).

2.2 Laser remelting process

A Nd-YAG laser source designed by Cheval company (Cheval, Pirey, France) was used to remelt the coatings. This laser can supply a maximum average power of 1100 W. And wavelength is 1064 nm with a maximum frequency of 200 Hz. The laser firing pulse duration can be adjusted between 0.5 and 10 ms. The off-set distance was fixed to 340 mm and at this focalization point the laser spot was 3 mm. During laser remelting process, Argon with a flow rate of 15 slpm was injected as protection gas (**Fig. 2**). The coatings were preheated with a laser spot of 10 mm and a scanning speed of 30 mm/s to decrease the stress between the substrate and coatings. After ten passes of preheating, the temperature of coating increases to 500 °C (**Fig. 3**). Table 3 depicted the conditions of laser treatment. A calorimeter (Labmaster 4DOPTICAL USA) was used to measure the average power of the laser spot.

Table 3 Parameters of laser remelting

Frequency (Hz)	10
Laser firing duration (ms)	5
Average power (W)	339
Scanning velocity (mm/s)	5
Off-set distance (mm)	340
Diameter of laser spot (mm)	3

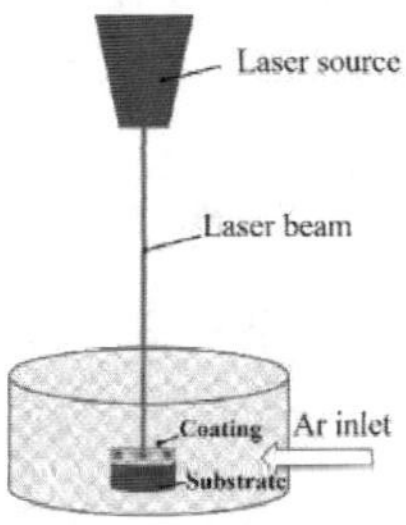

Fig. 2. Drawing of the laser remelting process.

The laser power density can be calculated by the following relationship:

$\varphi(r)=P/(2\pi R^2)\exp(-r^2/(2R^2))$ (1)

where $P/2\pi R^2$ is the power density of laser at a pulse ($W.mm^{-2}$) and R represents the radius of dispersion of the pulsed laser (1.5 mm). The energy density of the laser spot is a Gaussian distribution (equation 1)

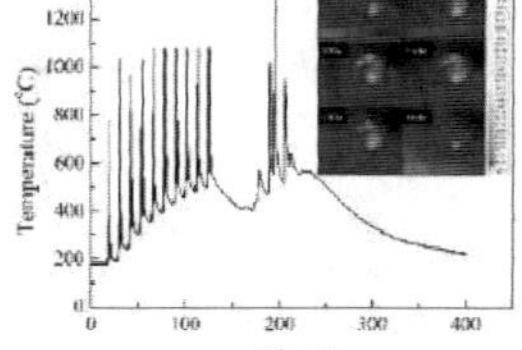

Fig. 3. Temperature evolution of coating during laser remelting.

2.3 Characterizations

Morphology of feedstock and coating were observed by scanning electron microscopy (SEM, JSW-5800LV, JOEL) and energy dispersive spectrometer (EDS). Infrared thermal imaging camera (SC 5210, FLIP) was used to measure the temperature of coating during laser remelting process.

3 Results and discusses

3.1 Microstructure of as-sprayed NiTi coatings

Fig. 4 shows the cross-sectional microstructure of the as-sprayed NiTi coatings. A dense microstructure with few pores is obtained. The coating adheres well to the substrate and no micro cracks are observed at the coating-substrate interface. The coating shows typical thermally sprayed lamellar structure. Splats boundaries are easily distinguished by different contrasts. EDS analysis (**Fig. 5**) reveals that Ni splats and Ti splats distributes uniformly in the coatings.

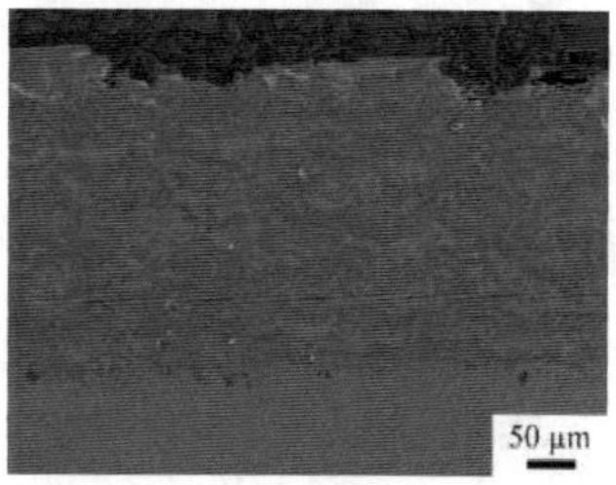

Fig. 4.. Cross-sectional microstructure of the as-sprayed NiTi coating.

Fig. 5. EDS analysis of as-sprayed NiTi coating.

3.2 Microstructure of laser remelted coatings

The laser remelted coatings consist of three distinguished parts (**Fig. 6**). Top dense layer relates to the top part of the molten pool. The middle part involving in a dark contrast and some pores is the bottom of the molten pool. And the last part shows a microstructure corresponding to the as-sprayed coating. Some microcracks are induced by the laser remelting. EDS results in **Fig. 7** indicate that Ni distributes on the top of laser remelted coatings. And Ti and O mainly locate at the bottom of the laser remelted molten pool and some of Ti and O elements are involved in the top part. This indicates that the Ti exists in the form of oxides and the oxidation of Ti splats during laser remelting. **Fig. 8** exhibits the crystalline structure of the molten pool. The top Ni exhibits columnar crystal for the reason that grains grow in favorable direction opposite to the direction of heat flow. The transmission part shows an equiaxed grains involving in some Ti oxides. And most of the Ti oxidation stacks at the bottom of molten pool.

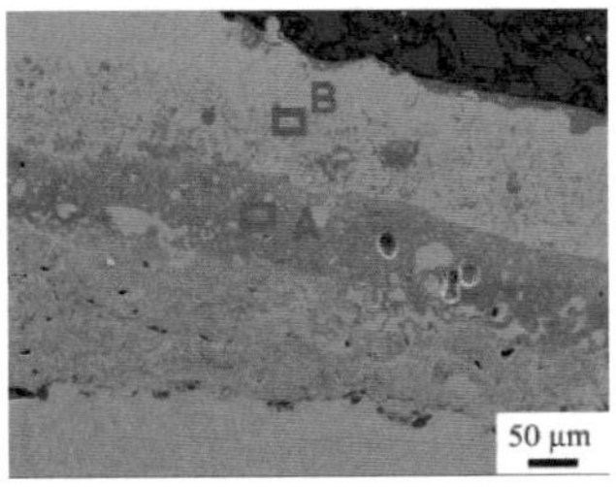

Fig. 6. Cross-sectional microstructure of the laser remelted NiTi coating.

3.3 Oxidation of splats

According to Table 4, Ti splats got severe oxidation during laser remelting process. Although argon was injected to create a protective atmosphere. There was still oxygen partial pressure in the argon. During the laser remelting process, oxygen was involved in. And Ti splats were burned due to flammability under high temperature conditions. During the laser remelting process, the coating was molten and the burned Ti splats existed in the form of Ti oxides. The laser stiffness would induce some gas and solid oxides sunk into the bottom of molten pool. Therefore, some pores were observed in **Fig. 6** and the Ti oxides located mainly at the bottom of the molten pool.

Table 4 Chemical compositions of the molten pool

		Point A			Point B		
Ele.	Series	Unn. C wt. %	Norm. C wt. %	Atom. C at. %	Unn. C wt. %	Norm. C wt. %	Atom. C at. %
O	K-series	1.49	1.67	5.50	14.45	15.05	35.50
Ti	K-series	27.75	31.08	34.18	65.25	67.99	53.59
Ni	K-series	60.60	67.25	60.32	16.28	16.69	1.091

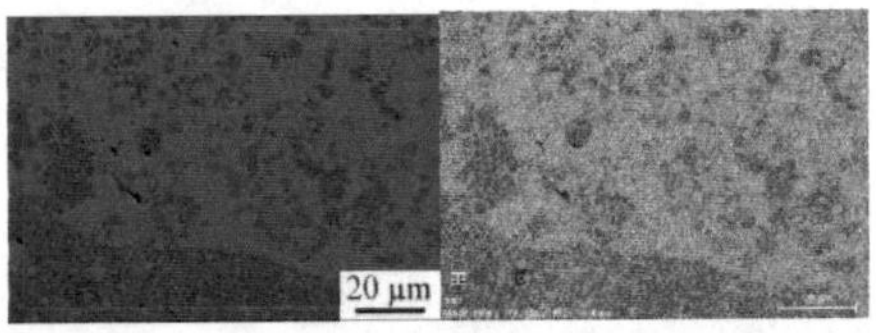

Fig. 7. EDS analysis of laser remelted NiTi coating.

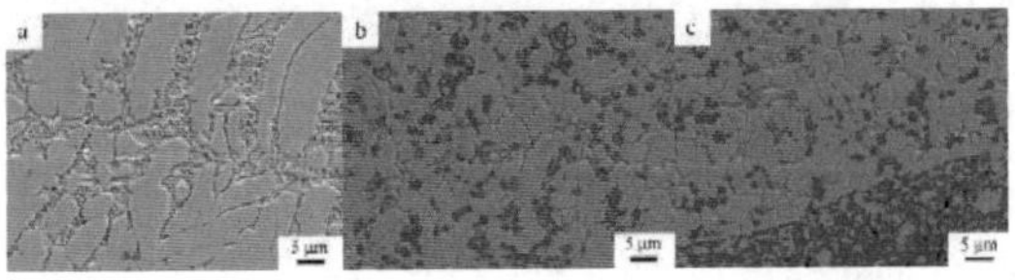

Fig. 8. Microstructure of different parts of laser remelted NiTi coatings: top (a), middle (b) and bottom (c).

4 Conclusions

NiTi coatings were fabricated by atmospheric plasma spraying with Ni and Ti blended feedstock. Dense microstructure was obtained with Ni and Ti splats distributing uniformly. And laser remelting process induces three distinguished parts of the as-sprayed coatings. Top part of the molten pool involves in Ni columnar grains. Bottom part of molten pool corresponds to Ti oxides. And the last on relates to the as-sprayed microstructure of the coating. The flammability of Ti and oxygen partial pressure in the argon contribute to the element distributions of laser remelted NiTi coatings.

5 Acknowledgements

The authors WANG Yan and LIU Jiangwei show great thanks to the China Scholarship Council for the financial supports.

6 Literature

[1] Imbeni, V. and A.N. Other: Preliminary study of micro-scale abrasive wear of a NiTi shape memory alloy. Wear (2003), Issue 12, pp. 1299/306.

[2] Richman, R.H. and A.N. Other: Cavitation erosion of NiTi explosively welded to steel. Wear (1995), pp. 80/85.

[3] Richman, R.H. and A.N. Other: Cavitation erosion of two NiTi alloys. Wear (1992), Issue 2, pp. 401/407.

[4] Abedini, M. and A.N. Other: Self-healing effect on worn surface of NiTi shape memory alloy. Materials Science and Technology (2013), Issue 3, pp. 285/288.

[5] Ng, C.H. and A.N. Other: NiTi shape memory alloy with enhanced wear performance by Laser selective area nitriding for orthopaedic applications. Surface and Coatings Technology (2017), pp. 1015/22.

[6] Pirani, C. and A.N. Other: Wear and metallographic analysis of wave one and reciproc NiTi instruments before and after three uses in root canals. scanning (2014), Issue 5, pp. 517/525.

[7] Tillmann, W. and A.N. Other: Tribological performance of near equiatomic and Ti-rich NiTi shape memory alloy thin films. Acta Materialia (2015), Issue 92, pp. 189/196.

[8] Bram, M. and A.N. Other: Vacuum plasma spraying of NiTi protection layers. Materials Letters (2002), Issue 3, pp. 647/51.

[9] Verdian, M.M. and A.N. Other:Electrochemical impedance spectroscopy of HVOF-sprayed NiTi intermetallic coatings deposited on AISI 1045 steel. Journal of Alloys and Compounds (2010), Issue 1, pp. 42/46.

[10] Hiraga, H. and A.N. Other: Cavitation erosion mechanism of NiTi coatings made by laser plasma hybrid spraying. Wear (1999), Issue 2, pp. 272/78.

[11] Bitzer, M. and A.N. Other: Cavitation-resistant NiTi coatings produced by low-pressure plasma spraying (LPPS). Wear (2015), pp. 369/77.
[12] Stella, J. and A.N. Other: Cavitation erosion of plasma-sprayed NiTi coatings. Wear (2006), Issue 9-10, pp. 1020/27.
[13] Verdian, M.M. and A.N. Other: Corrosion performance of HVOF and APS thermally sprayed NiTi intermetallic coatings in 3.5% NaCl solution. Corrosion Science (2010), Issue 3, pp. 1052/59.

On the validity of continuum computational fluid dynamics approach in very low pressure plasma spray conditions

Dmitrii Ivchenko(1), Tao Zhang(1), Gilles Mariaux(1), Armelle Vardelle(1), Simon Goutier(1), Tatiana E. Itina(2),
(1): Limoges / Fr, (2) Saint-Étienne / Fr

Plasma Spray Physical Vapor Deposition aims to substantially evaporate a powder in order to produce coatings with microstructures ranging from lamellar to columnar. This is achieved by the deposition of fine melted powder particles and nanoclusters and/or vapor condensation. The deposition process typically operates at pressure ranging between 10 and 200 Pa. In addition to experimental works, numerical works help to better understand the process and optimize the experimental conditions. However, the combination of high temperature and low pressure with the appearance of shock waves resulting from the supersonic expansion of the hot gas in the low pressure medium, makes questionable the suitability of the continuum approach for modelling such a process. This work deals with the study of (i) the effect of the pressure dependence of the thermodynamic and transport properties on the CFD predictions and (ii) the validity of the continuum approach for thermal plasma flow simulation under very low pressure conditions. It compares the flow fields predicted with a continuum approach (ANSYS Fluent CFD code) and a kinetic-based approach using a Direct Simulation Monte Carlo method (DSMC, SPARTA code). It also shows how presence of flow gradients can contribute to the errors in the results for typical PS-PVD conditions.

1 Introduction

Plasma Spray Physical Vapor Deposition (PS-PVD) is a rapidly developing technology making it possible to produce finely-structured coatings through vapor and small melted particle deposition at pressure ranging from 10 to 200 Pa [1]. The technology uses a thermal spray torch adapted to the very low pressure conditions to produce the vapor. It is aimed to fill the gap between conventional plasma spray, characterized by atmospheric operating pressure and a relatively high deposition rate (125-250 µm/min), and electron beam physical vapor deposition (EB-PVD), which produces columnar coatings, but at relatively low rates (12 µm/min) under medium vacuum conditions (<100 Pa) and with high capital costs [2, 3]. Current PS-PVD systems are capable of deposition at rates up to 100 µm/min. Typical characteristics and operating conditions are shown in **Table 1**.

The thermal plasma is produced by rapid Ohmic heating of the plasma gas in the DC-torch working close to atmospheric pressure and expands through a specifically designed nozzle into the low-pressure chamber. Due to a substantial pressure difference between the torch and coating chamber, plasma gas rapidly expands and accelerates reaching supersonic velocities. Depending on the degree of underexpansion (ratio of the pressures in the torch and chamber), pressure and Mach number, the plasma jet exhibits several shock waves, accompanied by significant temperature and velocity gradients. The gas temperature may decrease ten times or more after the nozzle exit over a 5–10 cm distance.

Table 1. Typical PS-PVD operating parameters.

Plasma spray gun	Metco F4	Metco O3CP
Electric power (kW)	45 – 55	150 – 180
Chamber pressure (Pa)	100 Pa	
Plasma gas	Ar, He, H_2 (100 slm)	
Spray distance (m)	1	2
Powder feed rate (g/min)	1-20	10-20
Particle diameter (µm)	5-15	5-25

Computational simulations may enable deeper insights into the coating process. However, as suggested by numerous studies, they might be non-trivial [4, 5] because of the rarefaction effect caused by the hot gas expansion at low pressure and enhanced by intense and numerous shocks. Therefore, the commonly used numerical descriptions of fluid motion are becoming doubtful.

The rarefaction degree is usually described by the ratio (called Knudsen number (*Kn*)) between the molecular mean free path (average distance between intermolecular collisions) and a characteristic physical length scale. When the mean free path is small enough in comparison with the macroscopic length scale of interest ($Kn<0.01$) continuum flow is observed. In this flow regime the frequent intermolecular collisions drive the distribution of molecular velocities towards a velocity equilibrium distribution (Maxwellian velocity distribution). The same can be applied for the reacting fluxes: a sufficient collision frequency leads to chemical equilibrium. When a small control volume is in local thermal equilibrium (LTE), the velocity distribution in this volume is isotropic. Thus it is possible to average the microscopic properties of gas particles (kinetic and internal energy, velocities) in macroscopic properties (pressure, temperature...) that vary continuously in space and are compatible with the scale of the system.

Under PS-PVD conditions, the rarefaction phenomenon leads to a low collision frequency of gas molecules and results in the transition from continuum flow to slip flow and eventually to free molecular flow ($Kn>0.01$). A deviation from local thermal and chemical equilibria, then, occurs (especially in the shock zones), so that the velocity distribution of the gas species deviates from the Maxwellian distribution, and is not isotropic anymore; the flow could no longer be characterized by a single translational temperature. This phenomenon is usually referred as the breakdown of the continuum fluid dynamics.

Commercially available computational fluid dynamics (CFD) software like ANSYS Fluent is commonly used to design or optimize processes involving thermal plasma jets [10-13]. The fluid dynamics is described by the set of the conservation equations [18] of mass, momentum, species, energy, constitutive equations and equation of state of the medium. However, the solution of this system under the PS-PVD conditions might be incorrect for the following reasons: i) the conservation equations assume the fluid is a continuum medium, ii) the linear relations between the stress tensor and velocity gradient and between the thermal heat flux and temperature gradient, classically used in the conservation equations, become non-linear in rarefied flows [6]. This results in the underestimation of gradient values and leads to the broadening of the shock waves. Moreover, the choice of the equation of state has a certain influence on the modelling results. E.g., the use of the ideal gas law may lead up to 300% error in the calculation of some gas properties like its density or enthalpy under low pressure conditions and high temperatures [14].

However, the effect of the aforementioned factors on predictions might be not so obvious. Regions where the continuum breaks down as well as regions where the ideal gas law produces noticeable error may be small enough to have any significant impact on the solution downstream. As the rarefaction broadens the shockwaves, the broadening caused by linear relations in constitutive equations may actually bring the numerical results closer to the real flux behavior [9]. The accuracy of the predictions is extremely dependent on the process of interest. Thereby the validity of the continuum computational fluid dynamics approach under specific PS-PVD conditions must be thoroughly investigated.

Several approaches can be used to predict more accurately the behavior of the rarefied non-equilibrium flow. They involve the use of the conservation equations that are still valid for larger deviations from the equilibrium conditions, with a correction of the stress tensor and heat flux vector to account for rarefaction effects [6-8]. Another way is to use a more general kinetic approach, which recognizes, that the gas is composed of a huge number of individual particles, involved in binary collisions with each other. This approach is based on a statistical description of positions and velocities of the particles and involves the solution of the Boltzmann equation written in terms of the phase space distribution function. However, both approaches are too complicated for practical applications and have significant computational cost.
One particular method for solving the Boltzmann equation, the so-called direct simulation Monte Carlo (DSMC) has received a wide recognition in recent years. It involves the direct simulation of statistically representative number of gas molecules. The result of the simulation is interpreted as a solution of the Boltzmann equation because it is built upon the same assumptions. Some details of the method will be highlighted in further paragraphs.

The major shortcomings of the DSMC method are its computational cost and the difficulty to simulate processes with high density variations: it is impossible to sustain good statistics in low pressure regions when they are coexisting with high pressure regions with the calculational capabilities of modern computers. These problems can be resolved by developing a numerical model combining the CFD method for the continuum regime with the DSMC method for the rarefied and non-equilibrium thermal regime [15]. This model should be capable of switching from CFD to DSMC solver when a certain criterion of continuum breakdown is fulfilled. The search for this criterion is another widely discussed topic [9, 15]. The majority of studies suggest using the Kn number, but its value corresponding to the continuum breakdown varies, depending on the formulation of the characteristic physical length and choice of the methodology to distinguish the breakdown. If diatomic molecules are present in the simulation, another criterion is the ratio of the difference between translational and rotational temperatures to translational temperature. The choice of the threshold values for these criteria may also be justified by the operating conditions and accuracy requirements for the solution downstream from the non-continuum regions.

A proper simulation of the PS-PVD process is not limited to gas dynamics and involves models of powder vaporization, vapor condensation, nucleation and coating formation. Nevertheless, the accurate simulation of the supersonic expansion of the thermal plasma jet is a first step towards this goal.

In the present study we will successively examine:
- the effect of the pressure dependence of gas thermodynamic and transport properties on the flow fields obtained by the continuum approach, as this approach is generally used with solely temperature-dependent properties at 1 atm and ideal gas law;
- the applicability of continuum approach for typical PS-PVD conditions. ANSYS Fluent was used for the CFD modelling. The predictions under low pressure conditions are compared with the predictions of a kinetic simulation performed with the SPARTA code [16]. The threshold values for the breakdown criteria, specific for the process, are estimated and the effect of the equation of state on the CFD numerical results is analyzed.

2 Models

2.1 CFD approach

The common CFD approach consists in the representation of the simulated domain as a finite set of control volumes used to approximate the fluid conservation equations of mass, momentum, species and energy (usually enthalpy) by a finite volume approach. The diffusion phenomena are considered by introducing phenomenological relations between stress tensor (heat flux) and velocity (temperature) gradients, respectively. The control volumes should be small enough to define representative local values of the fluid characteristics but large enough to respect the continuum assumption.

An equation of state, typically the ideal gas law [9], is used as a relation between the fluid density and pressure; the gas thermodynamic and transport properties are usually considered as only dependent on temperature [9, 13]. In this study a real gas model and pressure- and temperature-dependent properties were also implemented to evaluate their effect on the gas flow fields. The real gas properties, calculated by Murphy and Andre by using the Chapman-Enskog method, were used for various pressures and temperatures [14]. The calculation of these properties took into account the dissociation, recombination and ionization effects. When the ideal gas law was used, the gas properties calculated at 1 atm were used.

The discretized set of partial differential equations was solved with the Ansys Fluent 17.1 [18] CFD code. The density-based solver dedicated to high-speed compressible flows, implicit numerical scheme as well as the time derivative preconditioning algorithm (AUSM) were used. This choice of the numerical techniques was imposed by the numerical stiffness of the solved equations. The time derivative preconditioning algorithm is essential at low Mach number due to the difference between the fluid velocity and speed of sound. This algorithm modifies the time-derivative term in conservation equations, allowing to rescale the acoustic speed of the solved system, and so reducing its stiffness.

The turbulence was described by a conventional k-ω model, chosen because of its ability to control the production and dissipation of turbulent energy at low Reynolds numbers and, especially, in the viscous sublayer of near wall regions without using near-wall laws. The Shear-Stress Transport (SST) version is specifically adapted to represent laminar/turbulent transition and compressible flow.

2.2 Kinetic approach. DSMC model

Monte Carlo is a family of numerical methods for a variety of computational problems that is based on random sampling. DSMC method is the Monte Carlo method for simulation of gas flows at a molecular level. The principal limitations are the necessary dilute gas and molecular chaos assumptions. The former implies that the volume per molecule is much larger than the volume of a single molecule or species, thus the probability of triple collisions is insignificant and interaction between molecules can be considered as instantaneous. The latter assumes that the states of colliding particles are uncorrelated. Because of this, the DSMC method has the same limitations as the classical kinetic theory, which is based on the Boltzmann equation [19].

Two grids are built upon the simulated domain: collision and sampling grids. During the simulation, a statistically representative number of molecules is introduced into the system. Each molecule is characterized by its coordinates in the physical space, velocity vector and internal energy and it represents *Fn* real molecules. These molecules are involved into translational movement with velocities corresponding to the initial velocity distribution. The collision grid is chosen to ensure a sufficient amount of colliding molecules per cell. At each time step, according to the "no time counter" scheme (NTC), a number of possible collision pairs is selected in every cell according to their relative velocity and collision cross section. Collision occurs between some of the selected molecules with a probability proportional to the collision frequency. Post-collisional velocity magnitude is determined through the conservation of mass, momentum and energy. The scattering direction is chosen randomly, according to the variable soft sphere molecular model (VSS). This VSS model describes the variation of molecular collision cross section with temperature and the level of scattering anisotropy [20]. Once in several time steps the sampling of the flowfield properties occurs: the microscopic flow properties are being averaged into macroscopic in each cell of the sampling grid.

In this paper we consider both the elastic and inelastic collisions, which may occur in a reacting flow. The flow chemistry is described by the Bird's Total Collision Energy (TCE) model. The flow is composed of Ar and H_2 atoms. The dissociation reactions are taken into account but not the ionization reactions. Indeed, the light mass of electrons leads to significant velocities of the electron gas which require very short time step (10^{-10} s) for the adequate representation of possible interactions. However, under spraying conditions the temperature in the spraying chamber is low enough to ignore the ionization: near the torch outlet, the maximal ionization degree is about 30 % (10 000 Pa, 12 000 K), but due to the expansion and cooling of the gas, accompanied by rapid recombination, on the distance, about 1-2 cm from the outlet, the ionization degree falls down to $3 \cdot 10^{-5}$ % (1000 Pa, 4000 K).

The SPARTA Code predictions have been validated against the experimental data [9] for the supersonic expansion of a pure Ar jet. The code was also verified against numerous experiments and cross-verified with conventional DSMC software, like dsmcFoam and the Bird codes.

3 Breaking criteria

The following approach is used in this work. Based on an extensive numerical investigation, Boyd *et al.* [21] concluded that the gradient-length local (GLL) Knudsen number provides a reliable criterion for the continuum breakdown detection:

$$Kn_Q = (\lambda/Q)|\nabla Q|,$$

where λ is a mean free path (equation 4.77 in [19]) and Q is any flow property (temperature, density, velocity, etc.). Boyd *et al.* also suggested to take into account both viscous and thermal sources of continuum breakdown by introducing a new parameter for *Kn*:

$$Kn_{\max} = \max(Kn_D, Kn_T, Kn_V), \quad (1)$$

where *D* is the density, *T* the temperature and *V* the velocity. Breakdown occurs when this value exceeds 0.05 [9, 21].

The kinetic simulation allows to locate the thermal non-equilibrium regions using another parameters, unavailable for Navier Stokes equation solutions. The thermal non-equilibrium indicators can be defined as:

$$P_{Tne,i} = |(T_{Tr} - T_i)/T_{Tr}|, \quad (2)$$

where T_{Tr} is the translational temperature, and *i* corresponds to *x*, *r* or *R* for the axial, radial or rotational temperatures, respectively, if diatomic molecules are present in the simulation. These parameters can be useful for the verification of the threshold value for Kn_T.

4 Calculation domain and boundary conditions

The two-dimensional axisymmetric computational domain used for the CFD simulations is shown in **Fig. 1**. The domain consisted of two parts. The first part represented a part of a F4MB (Oerlikon-Metco) plasma spray torch, adapted to low pressure conditions by a convergent-divergent nozzle, with a throat section 6 mm in diameter. The second part was a conical frustum and represented the low pressure expansion chamber. Formation of the plasma jet in the torch was not modelled in this study and a preheated gas was introduced at the inlet boundary of the domain.

The plasma gas was a mixture of 80 mol% Ar and 20 mol% H_2. It has a high specific enthalpy in comparison with pure argon and thermal conductivity that favor particle heating and evaporation. The boundary conditions for the CFD simulation are the following. The preheated gas was introduced through the inlet at a constant mass flow rate equal to q_m=0.0012 kg/s (40 slm Ar/ - 8 slm H_2); the pressure was automatically adjusted to meet the mass flow rate. The temperature of the mixture was set to T = 14 200 K as this temperature corresponds to an average specific enthalpy h that satisfies to $q_m$xh(T)=P_{elec}(W)=~45 kW. The temperature of the gun walls was set to 1473 K. This temperature was adjusted to meet the actual thermal efficiency of the gun (~50%). Non-slip shear condition was applied. The low pressure chamber had free boundaries with a fixed pressure equal to 100 Pa and the backflow temperature was set to 873 K.

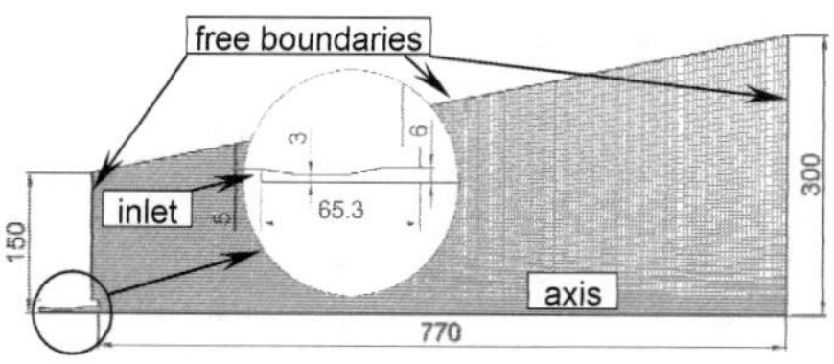

Fig. 1. Computational domain.

The structured computational grid consisted of 30 000 quadrangle cells. The mesh was refined in the regions of high-pressure (plasma gun) and in the regions where high gradients could occur (jet axis, walls). The refinement process was done iteratively until the solution became grid independent.

The computational domain used for the DSMC simulation was also 2-D and axisymmetric. To reduce the computational cost, only the low pressure chamber was simulated; it was represented as a cylinder, 770 mm in length and 300 mm in diameter. The boundary conditions (velocity, temperature and molecules number density profiles) for the torch outlet were derived from the CFD simulation. Due to the dissociation of hydrogen, the composition of the mixture was changed to 66.6 mol% Ar and 33.4 mol% H. 10^7 (Fn = $2.5\cdot10^{17}$) molecules were simulated.

A sampling grid was chosen to minimize the statistical noise in the low pressure regions. The grid was uniform and had 514 cells in axial direction and 225 cells in the radial direction, which allowed to have, at least, 20 simulated molecules per cell and led to a maximal error of about 15% for the 3σ confidence interval. In the SPARTA code the collision grid coincided with the sampling grid. The time step was chosen lower than the average residence time of the molecule in a cell (10^{-7} s).

5 Results and discussion

5.1 Effect of the gas properties on the CFD predictions

The objective consisted in evaluating the effect of the fluid properties on the predicted CFD flow fields while using (i) gas properties depending only on temperature and density calculated according to the ideal gas law ("ideal" case), (ii) the whole set of gas properties depending on temperature and pressure ("real" case). The differences in densities derived from the ideal and real gas laws are shown in **Fig. 2**.

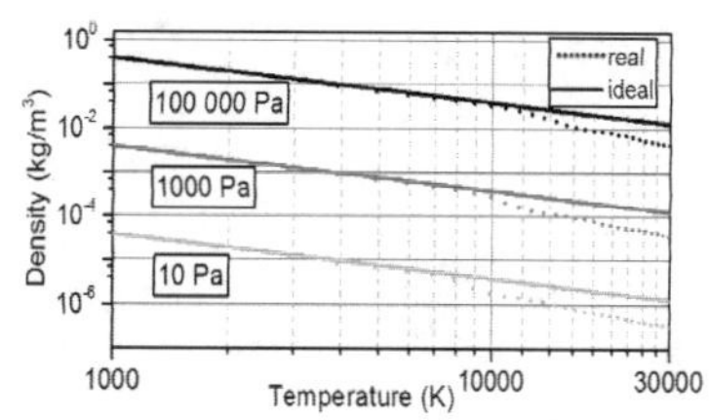

Fig. 2. Variation of gas density as a function of pressure and temperature.

The deviation from the ideal gas law becomes noticeable around 2000 – 5000 K due to the dissociation of H_2 and it is significant after 8000 – 12 000 K due to the first ionization of argon and hydrogen and may reach 100% at 12 000 K.

The real gas properties exhibit a weak pressure dependency for low ionization degrees. In particular, variation of the enthalpy with pressure before 7000 K is insignificant as it can be seen in **Fig. 3**. The peaks of thermal conductivity and heat capacity result from the dissociation of dihydrogen and ionization. These phenomena may lead to significant errors in the temperature field calculation in the region between 2000 and 4000 K if the effect of pressure on the gas properties is not considered.

A comparison of the gas axial temperature variation predicted by the CFD simulations using either ideal or real gas properties is represented in **Fig. 4**. The figure shows that the largest difference appears in the plasma jet expansion region right after the nozzle and in the first shock wave, that is in the areas where the difference between the gas properties is the most pronounced. The use of the ideal gas properties slightly narrows the first shock and decreases the temperature fall after the exit from the nozzle (in the first expansion shock). Downstream temperatures, as well as the other fields, like static pressure and velocity, are close, showing that despite the significant errors in shocks, the ideal gas law may be used to predict the flow fields downstream with deviations from the results obtained with the real gas properties reaching 20%.

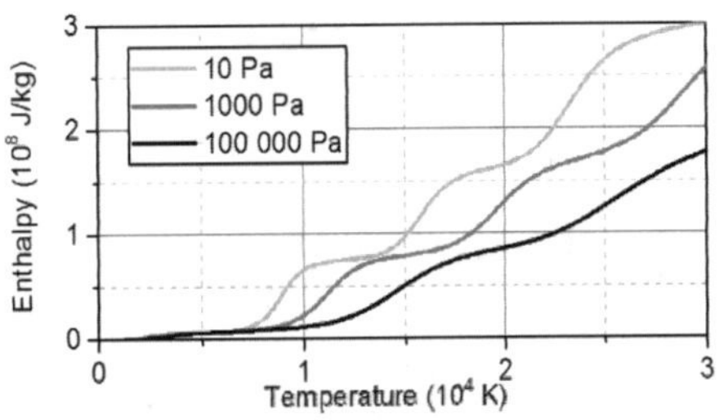

Fig. 3. Variation of gas enthalpy as a function of pressure and temperature.

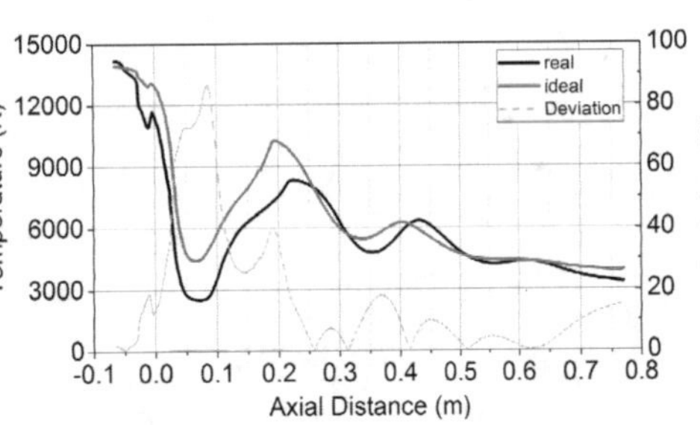

Fig. 4. Gas temperature variation and deviation from the reference case along torch axis. "0" on the horizontal axis corresponds to the torch exit.

5.1 Detection of the continuum breakdown

The results of the kinetic simulation are depicted against the CFD results in **Fig. 5**, **Fig. 6**. The Monte Carlo simulation shows the presence of a single barrel shock whereas two more additional shocks appears in the CFD one. The Monte Carlo (MC) shock has a similar length, but is located closer to the gun outlet and leads to a larger temperature drop, and smaller density drop before the shock.

The axial variation of the Knudsen number, predicted by Fluent, as well as other thermal non-equilibrium indicators, predicted by SPARTA can be seen in **Fig. 7**. Due to the steep gradients, the equilibrium breakdown occurs almost immediately after the flow issues from the torch: the deviation between the DSMC and CFD solutions is minimal in a small region about 1-2 cm long near the plasma gun outlet. The region where the deviation could no longer be ignored extends from 1-2 to 20 cm and corresponds to *Kn* higher than 0.02-0.05. This value is in agreement with the values obtained in preceding studies [9, 21]. Another reason for the breakdown can be spotted in

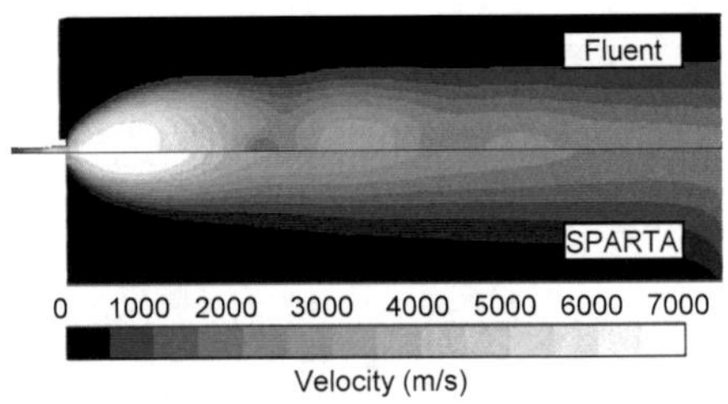

Fig. 5. Contours of velocity magnitude predicted by Ansys Fluent and SPARTA.

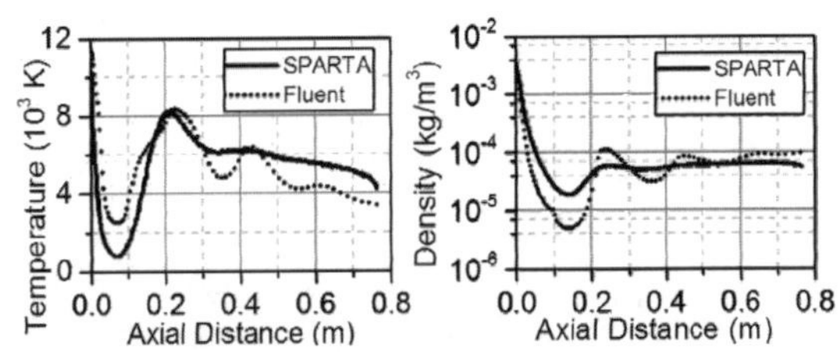

Fig. 6. Variation of gas temperature and density along torch axis.

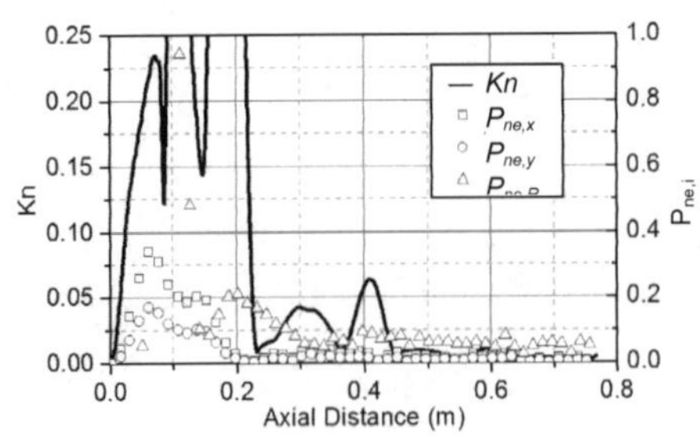

Fig. 7. Axial distribution of Kn, $P_{ne,x}$, $P_{ne,y}$, $P_{ne,R}$ (eq. 1,2).

Fig. 7. The kinetic energy of the macroscopic movement of the molecules leaving the torch in the axial direction is mostly transferred to the energy of random movement also in the axial direction. This fact leads to a significant difference between axial and radial components of temperature. According to the simulation results, if the deviation from the translational temperature is more than 3%, the results produced by Fluent are not adequate anymore. For the rotational temperature this value is about 5%.

Despite the large difference between the results in the shockwave region, the downstream fields of pressure, density and velocity are comparable. The decrease in the flow gradients leads to equilibration of the velocity distributions. The results emphasize that the CFD modelling underestimates the radius of the jet and temperature downstream. In the conditions of the study, this difference is about 1000 K and can be attributed to the CFD treatment of the non-equilibrium regions behind the shocks where, as a result of rarefaction, deviations from both thermodynamic and chemical equilibrium occur. Another possible source of discrepancy between the results is the absence of the recombination and ionization reactions in the DSMC simulation.

6 Conclusion and perspectives

The applicability of the continuum approach for modelling plasma flows under PS-PVD conditions was investigated by using the Direct Monte Carlo simulation (DSMC). The effect of the gas equation of state on CFD simulation predictions was also studied. The results show that the use of real gas properties doesn't lead to a significant improvement of the predictions downstream, but has a strong impact on the solution in the regions of rapid expansion and compression. The kinetic simulation shows the noticeable underestimation of the downstream temperature and overestimation of temperature during the expansion.

While it is possible to use the CFD approach for the estimation of the parameters of the first shock (location, width) and other flow characteristics downstream, several errors are shown to be produced in the shock. In particular, important deviations are revealed in the downstream temperature field. As the on-going studies deal with the evaporation and nucleation of the particles processed in the plasma medium, it is important to have an accurate prediction of the plasma jet temperature far from the plasma torch exit. The CFD modelling of PS-PVD processes using a lower enthalpy or a higher pressure in the expansion chamber might be possible, but under the conditions of this study, the utilization of the kinetic approach is advised.

7 Acknowledgments

The authors acknowledge Prof. Pascal Andre of Laboratoire de Physique de Clermont, France, and Dr Anthony B. Murphy of The Commonwealth Scientific and Industrial Research Organisation (CSIRO), Australia, for the calculations of the gas properties at low pressures.

8 Literature

[1] von Niessen, K. and M. Gindrat: Vapor phase deposition using a plasma spray process. Thermal Spray: Global Solutions, ITSC 2010 (Düsseldorf, Germany: DVS) e-proceedings

[2] Huang, H., Eguchi, K., Kambara, M. et al.: Ultrafast Thermal Plasma Physical Vapor Deposition of Yttria-Stabilized Zirconia for Novel Thermal Barrier Coatings. J Therm Spray Tech 15 (2006), Issue 1, pp. 83–91

[3] Jang, Byung-Koog and H. Matsubara: Influence of porosity on hardness and Young's modulus of nanoporous EB-PVD TBCs by nanoindentation. Materials Letters 59 (2005), pp. 3462-3466

[4] Josyula, E. and J. Burt: Review of Rarefied Gas Effects in Hypersonic Applications. RTO AVT/VKI Lecture Series - Models and Computational Methods for Rarefied Flows (2011)

[5] Boyd, I.D.: Predicting breakdown of the continuum equations under rarefied flow conditions. Rarefied Gas Dynamics, AIP Conference Proceedings 663 (2003), pp. 899-906

[6] Chapman, S., T.G. Cowling: Mathematical theory of non-uniform gases. Cambridge University Press (1970)

[7] Burnett, D.: Proc. London Math. Soc. 40 (1934), p. 382

[8] Zhong, X.: On numerical solutions of Burnett equations for hypersonic flow past 2-D circular blunt leading edges in continuum transition regime. AIAA 42. Fluid Dynamics Conference July 6-9, 1993, Orlando FL, AIAA 93-3092.

[9] Selezneva, S.E., M.I. Boulos et al.: Stationary supersonic plasma expansion: continuum fluid mechanics versus direct simulation Monte Carlo method. J. Phys. D: Appl. Phys. 35 (2002), Number 12, pp. 1362-1372

[10] Vautherin, B., Planche, M.-P. et al.: Vapor Deposition of Metallic Coatings by Very Low Pressure Plasma Spraying. J Therm Spray Tech 23 (2014), p. 596

[11] Jodoin, B., Gindrat, M., Dorier, J.L. et al.: Modeling and diagnostics of a supersonic DC plasma jet expanding at low pressure. Proceedings of the International Thermal Spray Conference (2002), pp. 716-720.

[12] Bolot, R., Klein, D. and C. Coddet: Design of a nozzle extension for thermal spray under very low pressure conditions. Proceedings of the International Thermal Spray Conference (2004), pp. 574-579.

[13] Bolot, R., Sokolov, D., Klein, D. and C. Coddet: Nozzle developments for thermal spray at very low pressure. J. Therm. Spray Technol. 15 (2006), pp. 827-833.

[14] Murphy, A.B.: Transport coefficients of hydrogen and argon–hydrogen plasmas. Plasma Chemistry and Plasma Processing 20 (2000), Issue 3, pp. 279-297

[15] Wu, J.-S., Lian, Y.-Y. et al.: Development and verification of a coupled DSMC–NS scheme using unstructured mesh. Journal of Computational Physics 219 (2006), Issue 2, pp. 579-607
[16] Gallis, M. A., Torczynski, J. R., Plimpton, S. J., Rader, D. J. and T. Koehler: Direct Simulation Monte Carlo: The Quest for Speed. Proceedings of the 29th Rarefied Gas Dynamics (RGD) Symposium, (2014)
[17] ANSYS, Inc.: ANSYS Fluent User's Guide. (2016)
[18] ANSYS, Inc.: ANSYS Fluent Theory Guide. (2016)
[19] Bird, G.A.: Molecular Gas Dynamics and the Direct Simulation of Gas Flows. Clarendon press, Oxford (2003)
[20] Bird, G.A.: The DSMC Method. CreateSpace (2013)
[21] Wang W.-L. and I.D. Boyd: Predicting continuum breakdown in hypersonic viscous flows. Phys. Fluids 15 (2003), Issue 1, pp. 91–100
[22] Bird, G. A.: Breakdown of Translational and Rotational Equilibrium in Gaseous Expansions. AIAA Journal 8 (1970), pp. 1998-2003
[23] Boyd, I. D., Chen, G. and G. V. Candler: Predicting Failure of the Continuum Fluid Equations in Transitional Hypersonic Flows. Physics of Fluids 7 (1995), Issue 1, pp. 210-219.

Cold spraying of WC-Co-Ni coatings using porous WC-17Co powders

S. Yin, T. Lupton, M. Meyer, R. Lupoi, Trinity College Dublin, the University of Dublin, Department of Mechanical and Manufacturing Engineering, Dublin / Ireland
E.J. Ekoi, D.P. Dowling, University College Dublin, School of Mechanical and Materials Engineering, Dublin / Ireland

Abstract: Cold spray has been successfully used to produce WC-Co coatings in recent years, showing great potential. However, due to the lack of sufficient Co matrix phase for plastic deformation, the fabrication of cold sprayed WC-Co coating naturally requires expensive propulsive gas or very high working parameters, significantly increasing the manufacturing difficulty and cost. This paper aims to use conventional high pressure cold spray to fabricate high-performance and cost-effective WC-Co-Ni coatings, and to clarify its formation mechanism. To achieve this objective, mechanically mixed porous WC-17Co and dense Ni powders were selected as the feedstock with different WC mass fraction, F1 (41.5 wt.%), F2 (64.5 wt.%) and F3 (74.7 wt.%). Working parameters were set at a moderate level (nitrogen, 3.0MPa and 350°C) to control the experimental cost. Experimental results showed that WC reinforcements had no phase transformation and were completely retained in the WC-Co-Ni coatings. The highest WC mass fraction was present in the F3 coating, reaching approximately 75% (equivalent to the feedstock), quite close to the cold sprayed WC-25Co coating. Fracture of WC-Co particles during the deposition process was found to be the reason for such high WC retainability, dominating the coating formation mechanism.

1 Introduction

WC-Co metal matrix composite (MMC) is a great wear-resistance material, commonly used in the form of thin-film coating to prevent the underlying base materials from serious wear in aggressive environments. The coating fabrication mostly relies on high-temperature thermal spray processes (primarily high-velocity oxy-fuel, HVOF) to melt the Co matrix phase in the agglomerated WC-Co powder feedstock for achieving superior adhesion [1,2]. However, high deposition temperature frequently results in decarburization, phase transformation and oxidation of hard WC reinforcements or soft Co matrix phase, significantly deteriorating the coating mechanical properties and wear-resistance performance [1–4]. Therefore, in order to eliminate the adverse effect brought by the high temperature, much effort has been devoted to explore an alternative technology. Cold spray (also known as cold gas dynamic spray, kinetic spray or super particle deposition) as a low temperature particle deposition process enables feedstock to retain its original properties, effectively preventing the coating defects in relation to high-temperature thermal spray processes. It has been found to be a promising future technology for fabricating WC-based wear-resistance coatings [5].

Recently, researchers attempted to add nickel binder phase into WC-Co powders and used low pressure cold spray (LPCS) process to fabricate WC-based wear-resistance coatings in order to reduce the experiment cost [6–9]. However, coatings fabricated in this way typically retained lower content of WC reinforcements in the coating as compared to the original feedstock and had insufficient inter-particle cohesion due to the low working parameters and deposition velocity. These shortages may significantly deteriorate the coating wear-resistance performance. Therefore, this paper aims to use conventional high pressure cold spray (HPCS) to fabricate high-cost-performance WC-Co-Ni wear-resistance coatings. To achieve this objective, porous WC-17Co powders were selected as the main feedstock because such structures were found to help the deposition of WC-Co coatings [1,10–12]. In addition, Ni powders having comparable size range with the WC-17Co powders were chosen as the binder phase because equivalent powder size facilitates the co-deposition of MMC coatings [13].

2 Experimental methodology

Nickel (-40+15 µm, Sandivik, Sweden) and porous agglomerated-sintered WC-17Co (-48+10 µm, Xinke, Wuxi, China) powders with similar spherical shape and equivalent size distribution were used as the feedstock. Fig. 1 shows the morphology of both powders observed by SEM (Carl Zeiss ULTRA, Germany). Two powders were pre-mixed at various fraction and mechanically blended before spraying. The WC mass fraction in different powders iss provided in the following: F1 (41.5 wt.%), F2 (64.5 wt.%) and F3 (74.7 wt.%). Coatings were deposited onto Al alloy substrates using a self-made HPCS system (Trinity College Dublin, Ireland). The system consists of high pressure nitrogen/helium gas from cylinders, gas heater, powder feeder, CNC working platform for controlling the substrate movement, de-Laval nozzle and computer control system. Nitrogen at the pressure of 3.0 MPa and temperature of 350°C was applied to produce the WC-Co-Ni coatings. Nozzle standoff distance and traversal speed were set as 35 mm and 15 mm/s, respectively. In addition, for understanding the coating formation mechanism, the single WC-17Co particle deposition onto a Ni substrate was also conducted at the gun traversal speed of 300 mm/s. The surface morphology of deposited single particle was then observed by SEM.

In order to examine the prospective decarburization and phase transformation of the cold sprayed WC-Co-Ni coatings during the fabrication process, the as-sprayed coatings were examined by an X-Ray diffractometer (Siemens D500, Germany) with the Cu (λ=1.54 Å) source at a current of 30 mA, voltage of 40 kV and scan step of 0.02°. Coating microstructure was studied by SEM. The measurement of WC mass fraction in the coating was performed by EDX element analysis (Oxford Instruments INCA system, UK), and also by binary image analysis. For each sample, five locations were randomly selected from the polished coating cross-section and the measured data was then averaged. To access the microstructure observation and EDX analysis, the as-sprayed coating samples were prepared using standard metallographic procedures with the final polishing applied by 0.05 µm Al_2O_3 solution. Fracture surface was also obtained by breaking the as-sprayed samples and observing by SEM.

Fig. 1. Morphology of the (a) Ni and (b) WC-17Co powders

3 Results and discussion

3.1 Element and phase transformation

Decarburization and phase transformation frequently occurred when using thermal spray processes to fabricate WC-based coatings due to the high working temperature [1,14,15]. The comparison of XRD spectra shown in Fig. 2 indicated that no decarburization and phase transformation occurred during the fabrication of cold sprayed WC-Co-Ni coatings as no other phases were detected. This fact clearly demonstrated the advantage of cold spray over other thermal spray techniques in preventing decarburization and phase transformation. In addition, the spectra presented in Fig. 2 also suggested an increasing content of WC and decreasing content of Ni and Co phases in the coating as the WC content in the feedstock increased (from F1 to F3). This phenomenon was further confirmed by the EDX and imaging element analysis shown in Fig. 3. Moreover, in Fig. 3, both measurements indicated that WC phase almost had no loss during the coating deposition process. The highest WC mass fraction in the F3 coating reached approximately 75% (equivalent to the feedstock), quite close to the cold sprayed WC-25Co coating [16]. The excellent retainability of WC phase is superior to cold sprayed WC-Co-Ni and WC-Ni coatings reported in previous works [6–9,17]. The significant reason for the excellent WC retainability will be provided in the following sections.

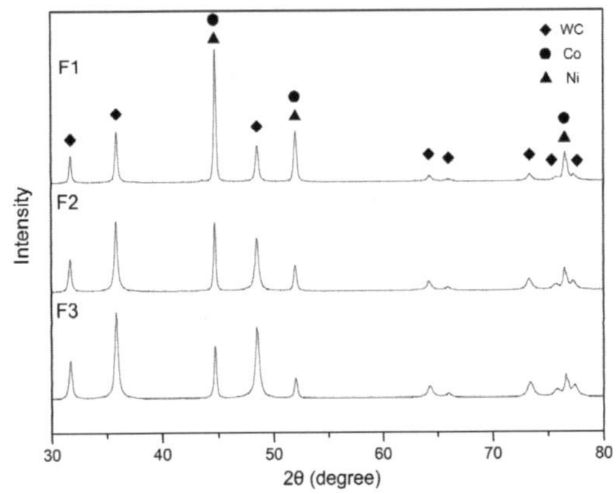

Fig. 2. Comparison of the XRD spectra of the WC-Ni-Co coatings.

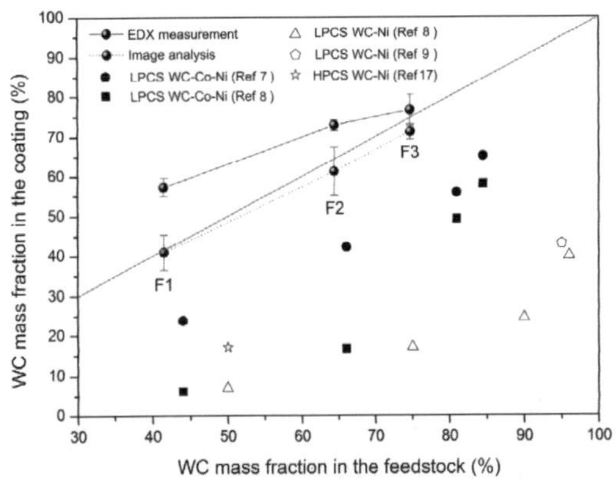

Fig. 3. Mass fraction of WC phase in the WC-CO-Ni coating as a function of that in the feedstock

3.2 Coating microstructure

Fig. 4 shows the cross-sectional SEM images of the WC-Co-Ni coatings fabricated with different feedstock. Reports indicate that a uniform distribution of the reinforcement phase in the MMC is of great importance to the final material properties [18–20]. As shown in Fig. 4a-c, for each coating, the cold sprayed WC-Co-Ni coatings fabricated in this work exhibited high-density and excellent uniformity, providing an essential condition for the high-quality material properties. The absence of cracks, observed from the coating cross-section, indicated cohesive bonding. Apart from these common features, an obvious difference in the WC content between each coating can be seen. The observed WC contents in the coatings (F3 > F2 > F1) were consistent with the EDX and imaging element analysis shown in Fig. 3. Furthermore, the magnified views of Fig. 4a-c shown in Fig. 4d-f indicated that most of the agglomerated WC-Co particles did not retain the original spherical shape, but experienced significant fracture during the co-deposition process and exhibited irregular shape in the coatings. Such fracture is a typical phenomenon that happens during the cold spray deposition of porous WC-Co powders [1,10–12]. Moreover, in the F1 and F2 coatings, inter-particle impact mostly took place between Ni and WC-Co particles. However, in the F3 coating, WC-Co particles frequently impacted onto each other due to the very high WC content in the feedstock.

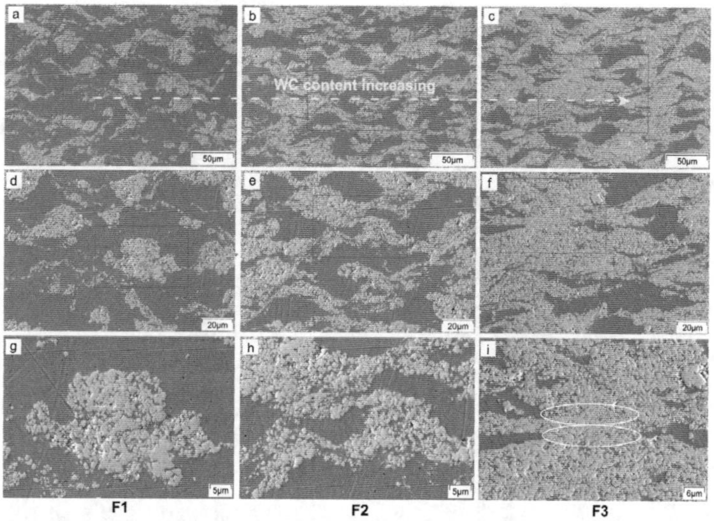

Fig. 4. Cross-sectional SEM images of the coatings fabricated with different feedstock. (a, d, g) F1 coating, (b, e, h) F2 coating and (c, f, i) F3 coating. Red boxes indicate the selected area for magnifying in the next row.

In Fig. 4g-i, the magnified views of Fig. 4d-f are provided. As can be seen from Fig. 4g and h, the inter-particle boundary between WC-Co and Ni particles could not be distinguished, which suggests good bonding between WC-Co and Ni phases in the coating. In Fig. 4i, WC-Co particles also exhibit good bonding with each other without obvious boundaries or gaps as indicated by white circles. For further understanding of the inter-particle bonding behavior in the coating, Fig. 5 gives the SEM images of fractured surfaces of the F1 and F3 coatings. As can be seen from Fig. 5a, inter-particle bonding was mainly between WC-Co and Ni particles, or between adjacent Ni particles, due to the higher content of Ni powders. Most importantly, fracture took place not only at the inter-particle interfaces but also through the WC-Co particles as marked by white circles. This phenomenon suggested that the inter-particle bonding strength in the coating was equivalent to the cohesive bonding strength of the WC-Co particles. As for the F3 coating shown in Fig. 5b, inter-particle bonding mainly happened between WC-Co particles because of the dominant volume content. The inter-particle interfaces are highlighted with white dot-lines. In this case, coating fracture mostly occurred through the WC-Co particles rather than at the inter-particle interface, which suggested that the Co-based metallic bonding between WC-Co particles was equivalent or even higher than the cohesive bonding strength of the WC-Co particle. Therefore, it could be concluded that the WC-Co-Ni coatings obtained in this work had a high-quality cohesive bonding and hence high toughness.

Fig. 5. Mass fraction of WC phase in the WC-CO-Ni coating as a function of that in the feedstock

3.3 Coating formation mechanism

It is known that hard reinforcements (e.g. BN [21], TiN [22], SiC [23], Al_2O_3 c[24], diamond [25]) in the cold sprayed MMC coatings do not undergo plastic deformation during the co-deposition process. Therefore, these hard particles do not have metallic bonding with the soft matrix phase, tending to rebound from the coating under the effect of large bulk modulus [26,27]. As a consequence, in cold sprayed MMC coatings, the hard reinforcements cannot be fully retained in the coating as compared with in the original feedstock. In this work, however, the WC phase showed hardly any loss during the co-deposition process as indicated in Fig. 3. The reason for this phenomenon could be attributed to the following two aspects. On one hand, the existing Co phase in the WC-Co powders provides the possibility for the occurrence of metallic bonding between WC-Co and Ni particles, which benefits the retainability of WC in the coating. On the other hand, the special porous structure of WC-Co powders enables the particle kinetic energy to dissipate through the fracture of particles rather than the elastic deformation that potentially causes particle rebound. The fracture phenomenon can be clearly observed from the coating cross-sectional views, as shown in Fig. 4, and the single WC-17Co deposition test (substrate material is Ni), as shown in Fig. 6 which shows cracks and splashed pieces of the WC-Co particle. Both aspects work together, resulting in the complete retention of WC phase in the coating. This fact clearly demonstrates the importance of metallic Co matrix phase in the inter-particle bonding, and the superiority of porous agglomerated WC-Co powers in fabricating cold sprayed WC-based coatings.

Fig. 6. SEM image of the surface morphology of single WC-17Co particle deposition onto a Ni substrate

Based on the above analysis, the boning mechanism of the WC-Co-Ni coatings fabricated in this work was discussed. Fig. 7 shows the schematic of the coating deposition process. As the porous WC-17Co particle impacts onto the former coating surface (Fig. 7a), the kinetic energy enables the particle to crack and fracture into small pieces (Fig. 7b). The fractured pieces retain some kinetic energy and hence splash surrounding the deposition spot and slightly embed into the coating surface (Fig. 7c). As the following particles further impact, these fractured WC-Co pieces are completely covered and retained by the following deposited materials (Fig. 7d). By continuously repeating this process, the WC-Co-Ni coating is finally formed.

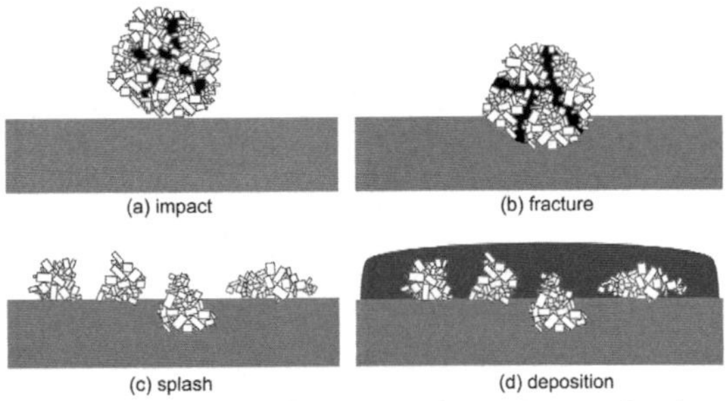

Fig. 7. Schematic of the coating formation mechanism of the WC-Co-Ni coatings

4 Conclusions

In this paper, a range of high-cost-performance WC-Co-Ni coatings were fabricated via conventional high pressure cold spray using mechanically mixed porous WC-Co powder and Ni powder. Three WC-Co/Ni feedstock with different WC mass fraction were used: F1 (41.5 wt.%), F2 (64.5 wt.%) and F3 (74.7 wt.%). Experimental results indicated that WC reinforcements had no phase transformation and completely retained in the WC-Co-Ni coatings. Such high retainability has never been achieved in previous cold sprayed WC-Co-Ni and WC-Ni coatings. The highest WC mass fraction in the F3 coating reached approximately 75% (equivalent to the feedstock), quite close to the cold sprayed WC-25Co coating. In addition, the fracture of porous WC-Co powders upon impact was found to be the coating formation mechanism, and the main reason for high WC retainability.

Acknowledgments

The authors would like to thank the CRANN Advanced Microscopy Laboratory (AML) of Trinity College Dublin and Enterprise Ireland (EI) for the support.

References

[1]. Li, C.J. and Yang, G. J. Relationships between feedstock structure, particle parameter, coating deposition, microstructure and properties for thermally sprayed conventional and nanostructured WC-Co. Int. J. Refract. Met. Hard Mater. 39 (2013), pp. 2/17

[2]. Yang, T. et al. Microstructure evolution of thermal spray WC-Co interlayer during hot filament

chemical vapor deposition of diamond thin films. J. Alloys Compd. 639 (2015), pp. 659/68.
[3]. Kumari, K., Anand, K., Bellacci, M. and Giannozzi, M. Effect of microstructure on abrasive wear behavior of thermally sprayed WC-10Co-4Cr coatings. Wear 268 (2010), pp. 1309/19.
[4]. Ryu, T., Sohn, H. Y., Hwang, K. S. and Fang, Z. Z. Plasma synthesis of tungsten carbide and cobalt nanocomposite powder. J. Alloys Compd. 481 (2009), pp. 274/7.
[5]. Papyrin, A. Cold Spray Technology. Adv. Mater. Process. 159 (2001), pp. 49/51.
[6]. Melendez, N. M., Narulkar, V. V., Fisher, G. A. and McDonald, A. G. Effect of reinforcing particles on the wear rate of low-pressure cold-sprayed WC-based MMC coatings. Wear 306 (2013), pp. 185/95.
[7]. Melendez, N. M. and McDonald, A. G. Development of WC-based metal matrix composite coatings using low-pressure cold gas dynamic spraying. Surf. Coatings Technol. 214 (2013), pp. 101/9.
[8]. Lioma, D., Sacks, N. and Botef, I. Cold gas dynamic spraying of WC–Ni cemented carbide coatings. Int. J. Refract. Met. Hard Mater. 49 (2015), pp. 365/73.
[9]. Nunthavarawong, P., Sacks, N. and Botef, I. Effect of powder feed rate on the mechanical properties of WC-5wt%Ni coatings deposited using low pressure cold spray. Int. J. Refract. Met. Hard Mater. 61 (2016), pp. 230/7.
[10]. Gao, P.-H. H., Li, C. X. C.-J. J. C.-X., Yang, G.-J. J., Li, Y.-G. G. and Li, C. X. C.-J. J. C.-X. Influence of substrate hardness on deposition behavior of single porous WC-12Co particle in cold spraying. Surf. Coatings Technol. 203 (2008), pp. 384/90.
[11]. Gao, P. H., Li, Y. G., Li, C. J. C. X., Yang, G. J. and Li, C. J. C. X. Influence of powder porous structure on the deposition behavior of cold-sprayed WC-12Co coatings. J. Therm. Spray Technol. 17 (2008), pp. 742/9.
[12]. Yang, G.-J., Gao, P.-H., Li, C.-X. and Li, C.-J. Mechanical property and wear performance dependence on processing condition for cold-sprayed WC-(nanoWC-Co). Appl. Surf. Sci. 332 (2015), pp. 80/8.
[13]. Sova, A., Papyrin, A. and Smurov, I. Influence of ceramic powder size on process of cermet coating formation by cold spray. J. Therm. Spray Technol. 18 (2009), pp. 633/41.
[14]. Couto, M., Dosta, S., Fernández, J. and Guilemany, J. M. Comparison of the Mechanical and Electrochemical Properties of WC-25Co Coatings Obtained by High Velocity Oxy-Fuel and Cold Gas Spraying. J. Therm. Spray Technol. 23 (2014), pp. 1251/8.
[15]. Couto, M., Dosta, S. and Guilemany, J. M. Comparison of the mechanical and electrochemical properties of WC-17 and 12Co coatings onto Al7075-T6 obtained by high velocity oxy-fuel and cold gas spraying. Surf. Coatings Technol. 268 (2014), pp. 180/9.
[16]. Dosta, S., Couto, M. and Guilemany, J. M. Cold spray deposition of a WC-25Co cermet onto Al7075-T6 and carbon steel substrates. Acta Mater. 61 (2013), pp. 643/52.
[17]. Alidokht, S. A., Manimunda, P., Vo, P., Yue, S. and Chromik, R. R. Cold spray deposition of Ni-WC composite coating and its dry sliding wear behavior. Surf. Coat. Technol. 308 (2016), pp. 424/34.
[18]. Chen, L. Y. et al. Novel nanoprocessing route for bulk graphene nanoplatelets reinforced metal matrix nanocomposites. Scr. Mater. 67 (2012), pp. 29/32.
[19]. Yang, N., Boselli, J. and Sinclair, I. Simulation and quantitative assessment of homogeneous and inhomogeneous particle distributions in particulate metal matrix composites. J. Microsc. 201 (2001), pp. 189/200.
[20]. Iacob, G., Ghica, V. G., Buzatu, M., Buzatu, T. and Petrescu, M. I. Studies on wear rate and micro-hardness of the Al/Al2O3/Gr hybrid composites produced via powder metallurgy. Compos. Part B Eng. 69 (2014), pp. 603/11.
[21]. Luo, X. T. and Li, C. J. Large sized cubic BN reinforced nanocomposite with improved abrasive wear resistance deposited by cold spray. Mater. Des. 83 (2015), pp. 249/56.
[22]. Li, W. Y., Yang, C. and Liao, H. Effect of vacuum heat treatment on microstructure and microhardness of cold-sprayed TiN particle-reinforced Al alloy-based composites. Mater. Des. 32 (2011), pp. 388/94.
[23]. Yu, M., Suo, X. K., Li, W. Y., Wang, Y. Y. and Liao, H. L. Microstructure, mechanical property and wear performance of cold sprayed Al5056/SiCp composite coatings: Effect of reinforcement content. Appl. Surf. Sci. 289 (2014), pp. 188/96.
[24]. Hodder, K. J., Izadi, H., McDonald, A. G. and Gerlich, A. P. Fabrication of aluminum-alumina metal matrix composites via cold gas dynamic spraying at low pressure followed by friction stir processing. Mater. Sci. Eng. A 556 (2012), pp. 114/21.
[25]. Yin, S. et al. Advanced diamond-reinforced metal matrix composites via cold spray: Properties and deposition mechanism. Compos. Part B Eng. (2017). doi:10.1016/j.compositesb.2017.01.009
[26]. Spencer, K., Fabijanic, D. M. and Zhang, M. X. The use of Al-Al2O3 cold spray coatings to improve the surface properties of magnesium alloys. Surf. Coatings Technol. 204 (2009), pp. 336/44.
[27]. Wu, J., Fang, H., Yoon, S., Kim, H. and Lee, C. The rebound phenomenon in kinetic spraying deposition. Scr. Mater. 54 (2006), pp. 665/9.

Solution heat treatment of gas atomized aluminium alloy (7075) powders: microstructural changes and resultant mechanical properties

A. Sabard[1], H. L. de Villiers Lovelock[2], P. McNutt[2], M.D.F.H. Harvey[2], T. Hussain[1]
[1]Faculty of Engineering, University of Nottingham, Nottingham, NG7 2RD, UK
[2]TWI Ltd, Granta Park, Great Abington, Cambridge, CB21 6AL, UK

Cold gas dynamic spray has significant potential for load-bearing repairs of high-value metallic components, as it is capable of producing pore and oxide-free deposits of significant thickness and with good levels of adhesion and mechanical strength. However, recently published research has shown that the rapid solidification experienced by gas atomised powders during manufacture can lead to a non-equilibrium powder microstructure, including clusters of dislocations as well as significant localised segregation of alloying elements within each particle. This paper reports on an investigation into the solution heat treatment of a precipitation hardenable aluminium alloy powder. The objective was to create a consistent and homogeneous powder phase composition and microstructure before cold spraying, with the expectation that this would also result in a more favorable heat treatment response of the cold spray deposits. Aluminium alloy 7075 gas atomized powders were solution heat treated at 450 °C for 5 hours in a sealed glass vial under vacuum and quenched in water. The powder particle microstructures were investigated using scanning electron microscopy with electron back scatter diffraction (SEM/BSE) and optical microscopy. The dendritic microstructure and solute segregation in the gas atomized powders was altered, with the heat-treated powder particles exhibiting a homogeneous distribution of solute atoms. The influence on the mechanical properties of the powder particles was studied using micro-indentation. The heat-treated powders exhibited a hardness decrease of nearly 25% compared to the as-received powders. This paper relates the behavior and the deformation of both as-received and heat-treated powders during spraying (single particle impacts), comparing the measured hardness with the deformation effect and the material jetting occurring upon impact.

1. Introduction

Cold spray is a relatively new material deposition process exploiting mainly the kinetic energy of fine powder particles to achieve a dense coating or deposit on a target substrate surface. Powder particles are introduced in a high velocity gas stream which accelerates them towards a substrate without melting them. The powder undergoes high strain rate plastic deformation upon impact and bonds with the target surface and/or with the previously deposited layers of material. The potential of cold spray processing for dimensional restoration as well as structural repair of light metallic alloys (Al, Mg and Ti mainly) has been extensively studied [1]. The specific build-up mechanisms of cold spraying, which usually imparts a high degree of work hardening to the powder particles, can lead to low-ductility coatings. In addition, defects such as porosity can adversely affect the mechanical properties of the deposits. Post deposition heat treatments have been shown to improve the properties [2].

In the case of aluminium alloys, recently published research [3], [4] has shown that the rapid solidification experienced by gas atomised powders during manufacture can lead to a non-equilibrium powder microstructures, including clusters of dislocations and significant localised segregation of alloying elements within each particle. Whilst standard industrial annealing or solution heat treatments could reduce alloying element segregation originating from the original powder and/or eliminate the work hardening resulting from cold spray deposition, such heat treatments may be undesirable in certain situations. Hence, to reduce or remove the alloying element segregation prior to spraying, solution heat treatment (SHT) of the powder has been considered here in the case of AA7075, a lightweight Al-Zn-Mg-Cu alloy widely used in the aerospace industry. The SHT homogenizes the non-equilibrium as-received microstructure of the gas atomised powder [5], dissolving any precipitates and increasing ductility for cold spraying. The supersaturated solid solution alloy is then deposited onto a substrate, which heavily deforms the powder particles.

This aim of this part of the work was to study the effect of SHT on AA7075 aluminium alloy powders before and after single-particle cold spray deposition, in terms of microstructural development and mechanical properties. In the next stage of the work, cold spray deposits will be produced and aged. The expectation is that the more consistent and homogeneous SHT powder phase composition should also result in a more predictable heat treatment response in the cold spray deposits.

2. Material and experiment

2.1 Materials

A spherical gas atomised AA7075 powder (Valimet Inc, United States), having a Dv10, Dv50 and Dv90 of 17, 34 and 60 µm respectively, was used in this study. The chemical composition of the powder was 5.72 wt% Zn, 1.69 wt% Cu, 2.30 wt% Mg, 0.20 wt % Cr, 0.17 wt % Fe, 0.03 wt % Mn, 0.13 wt % Si.

The substrates used for the cold spray experiments were 3 mm thick AA6061 (Al-Si-Mg) plates.

2.2 Heat treatment of the powder

The as-received feedstock powders were sealed in silica glass (SiO_2) vials prior to the introduction into the furnace. Once the particles were inserted in the vial, a 10 mPa vacuum was created using a diffusion pump, and the vial was then carefully closed and sealed using a flame. The dimensions of the container were as follows: outer diameter 14 mm, length ca. 100 mm and 2 mm wall thickness. Al alloy powders can potentially explode in the presence of oxygen and ignition sources [6]. Performing the heat treatment under vacuum in the tube mitigated the fire risk. Solution heat treatment of the powder was carried out using a commercial box furnace at 450 °C for 4 hours, and the quartz tubes were then quenched into cold water (5 °C) and submerged for 5 minutes after which the powder particles had reached room temperature. The quartz tube was then broken in order to collect the powder.

2.3 Microstructural analysis

The microstructure of both the powders and cold spray coatings was evaluated using optical microscopy (OM - Eclipse LV100ND) and scanning electron microscopy (SEM - JEOL JSM 64900LV). Electron probe micro-analysis (EPMA – Superprobe JEOL JXA 8200), operated at 30 kV, was performed on individual powder particles to obtain a high resolution elemental mapping using wavelength dispersive spectroscopy (WDS). OM, SEM and EPMA samples were prepared using standard metallographic techniques. Powders and cold spray deposits were cold mounted in EpoFix resin, ground with P1200 silicon carbide paper and polished to 1 micron using diamond paste. The final polish was then performed using a colloidal silica (0.06 µm) suspension. The samples were chemically etched using Keller's reagent (95 mL of H_2O, 2.5 mL HNO_3, 1.5 mL HCl, 1.0 mL HF) for 5 seconds.

2.4 Microhardness measurement

The microhardness analysis was performed using the MMT-7 Vickers Microhardness tester from Buehler. Each sample underwent 8 measurements.10 g load was applied for 10 seconds for both the mounted particles and the sprayed powder.

2.5 Cold spray deposition of individual powder particles

The cold spray experiments were performed using a custom made high pressure cold spray system at the University of Nottingham using helium as the primary accelerating gas. The rig setup has been described in detail elsewhere [7]. Deposition was carried out onto AA6061 substrates while maintaining a He gas pressure of 2.9 MPa at room temperature with a high pressure commercial powder feeder (Praxair 1264HP, Indianapolis, IN, USA). The nozzle has an area expansion ratio of 8 with a divergent length of 135 mm. Particles were sprayed using a nozzle stand-off distance of 20 mm, a powder feed rate setting of 2 rpm and a traverse speed of 600 mm/s. Swipe tests were conducted in order to observe the impact of single particles onto the substrate.

3. Results and discussion

3.1 As-received AA7075 powder

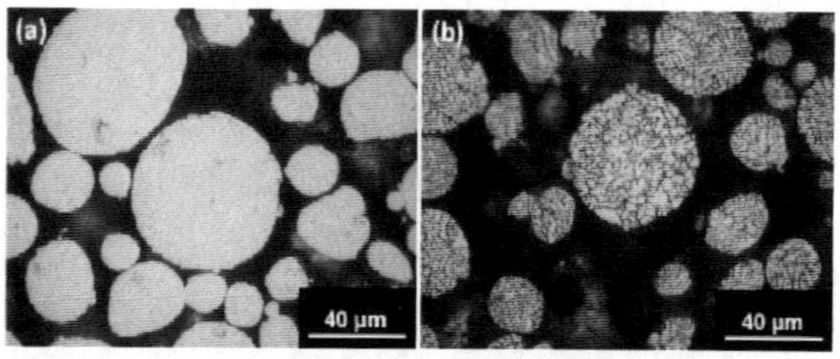

Fig. 1: Optical micrographs of the cross-section of gas atomised AA7075 powder un-etched (a) and after etching for 5 seconds with Keller's etchant (b)

When examined by optical microscopy, the gas atomized powder (Fig. 1a) shows a spherical morphology containing a mixture of both large and small particles. The microstructure of the material is revealed by Keller's etchant (Fig. 1b) and has small dendritic cells whose size varies between 1 to 4 µm. This commonly observed microstructure of gas atomised Al powder has been attributed to several phenomena occurring during the powder manufacturing process such as solidification and cooling rate, thermal equilibration and re-melting of solid particles, as well as the presence of pre-solidified micro-droplets and dendrite fragments [8]. The average cooling rate of aluminium powders prepared by gas atomising varies between 10^4 to 10^7 K/s, with the temperature of the small particles decreasing faster [9]. This wide range of cooling rates leads to different distributions and shapes of cells in the powder particles.

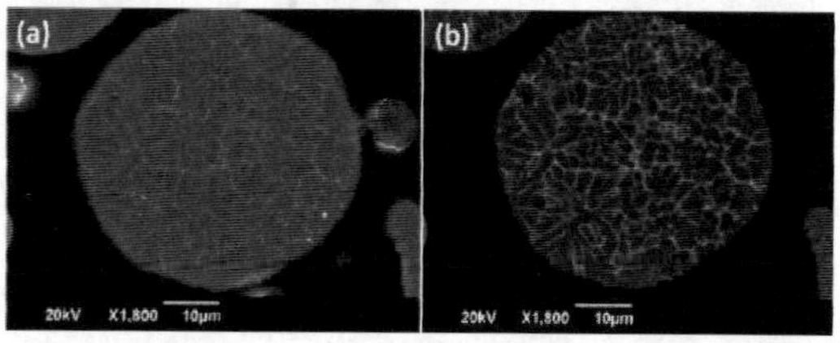

Fig. 2: SEM micrographs of an as-received AA7075 particle in cross-section. Secondary electrons (a) and back-scattered electrons (b)

When examined by SE and BSE, the as-received AA7075 powder shows a non-uniform distribution of solute elements. A typical example using a 40 µm-sized particle can be seen in Fig. 3. The BSE image reveals strong compositional variation, explained by solute segregation occurring during cooling. When the material is not cooled down below the solidus temperature quick enough, diffusion occurs during solidification leading to a migration of the solute atoms towards the grain boundaries [10]. The brightness difference between the matrix and the interdendritic regions is presumably due to the high concentration of solute atoms, especially Zn and Cu.

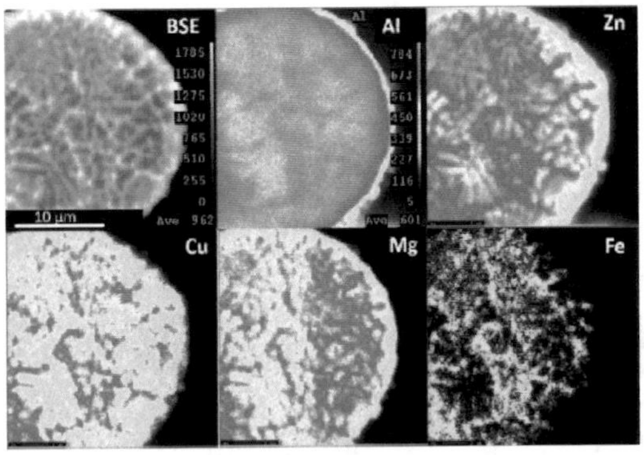

Fig. 3: EPMA-WDS map of as-received gas atomised AA7075 powder

The EPMA analysis presented in Fig. 3 reveals the distribution of the main solute atoms – i.e. Zn, Cu, Mg, Fe – in the as-received AA7075 powder. The results confirm the presence of those alloying elements at the grain boundaries. Although the grain boundaries are enriched in alloying elements, intemetallic phases such as $Mg(Zn,Cu,Al)_2$ [4], Al_2CuMg [11] or Mg_2Si [4] were not observed. Cu and Fe appear to follow the same distribution in the particle. The solute segregation occurring during gas atomisation is thus confirmed, revealing a fine dispersion of solute atoms at grain boundaries and interdendritic regions. Al seems to be homogeneously distributed throughout, despite slight variations of content between the grain boundaries and the cores of the dendritic cells.

3.2 Solution heat treated (SHT) AA7075 powder

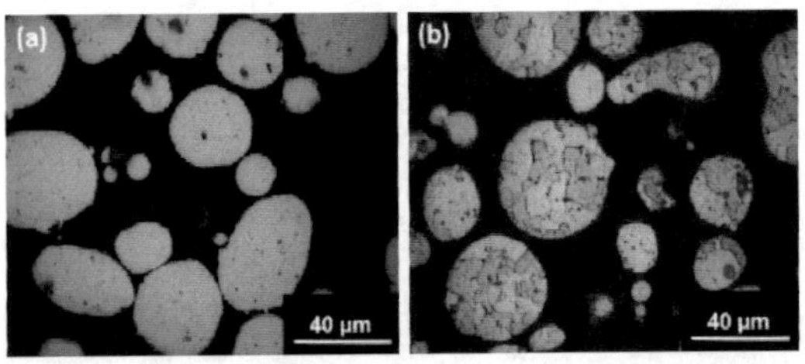

Fig. 4: Optical micrographs of the cross-section of solution heat treated AA7075 powder un-etched (a) and etched (b)

Unetched particles of AA7075 powder after solution heat treatment show no real differences with the as-received powder in cross-section (Fig. 1a, Fig. 4a), except a slight increase in terms of porosity. On the other hand, the etched powder reveals a different microstructure. Indeed, different contrasts are revealed, and the dendritic structure of the as-received powder has disappeared (Fig. 1b, Fig 4b). Differences can be observed in the powder between the smallest and the largest particles, namely the uniform white colour showed by the smallest ones. The original inconsistency of dendritic distribution observed in the as-received powder presumably led to these differences.

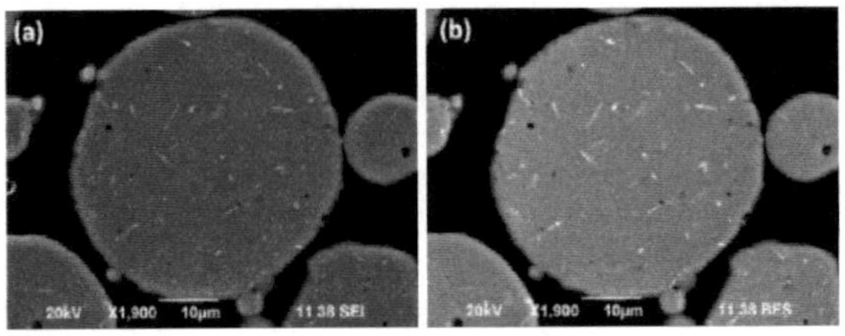

Fig. 5: SEM micrographs of a solution heat treated AA7075 article. SE (a) and BSE (b)

SEM pictures of AA7075 after solution heat treatment are shown in Fig. 5. The lack of contrasts in the BSE image reveals that the solute atoms have been dissolved into solid solution and the matrix has been homogenized. A few precipitates can also be observed on the cross-section of the particle. Needle-shaped bright precipitates are numerous in the microstructure, as well as smaller dark areas, which are potentially Mg-rich phases (e.g. Mg_2Si), based on the lower atomic number of Mg compared to Al, Zn, or Cu. As the cooling rate undergone by the powder during quenching after SHT is probably too rapid for precipitation to occur, the formation of the needle shaped precipitates presumably occurred during the heat treatment itself. This is supported by the literature which reveals that a number of phases can be expected to precipitate after SHT, due to the low solubility of some of the components in AA7075. The presence of Al_7Cu_2Fe [8] and Al_2CuMg [12] is expected, as they have been observed at the grain interiors and along the grain boundaries in wrought alloys.

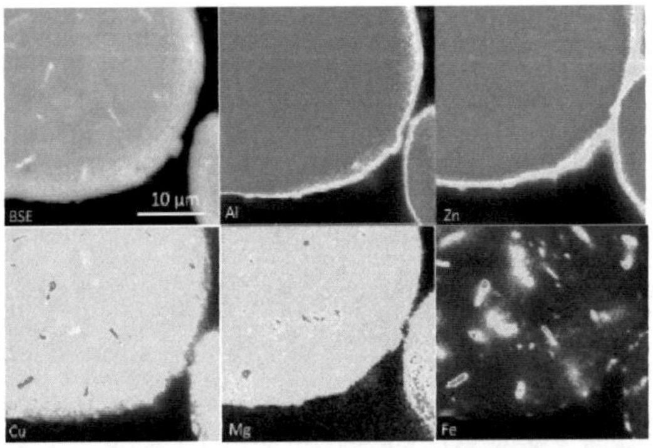

Fig. 6: EPMA-WDS map of a solution heat treated AA7075 particle

The elemental composition analysis (Fig. 6) helps identify the precipitates developed during SHT, and indicates that the bright needle-shaped precipitates are rich in Cu and Fe. The needle-like precipitates are thus presumably Al_7Cu_2Fe, forming during heat treatment from the Fe- and Cu-rich areas of the as-received powder. Those phases are commonly observed in AA7075 after solution heat treatment due mainly to the low solubility of iron [13]. The Mg-rich phases also observed on the EPMA scan can be associated with Mg_2Si, which is known to be almost insoluble in AA7075 [13].

3.3 Mechanical properties

The heat treatment of the AA7075 particles prior to spraying was performed partly to increase the ability of the powder to deform upon impact. Microhardness measurements were thus performed on the AA7075 particles, as-received and after solution heat treatment.

Table 1: Microhardness measurements of AA7075 powder

AA 7075	As-received	SHT + quenched
Hardness	101 HV	75.2 HV
Stand. dev	5.74	4.98

Typical AA7075 bulk hardness is between 165-175 HV. The as-received powder (Table 1) shows a microhardness of 101 HV ± 5.74 HV, lower than the bulk material. The SHT powder exhibits a 25% decrease in hardness, at 75.2 HV ± 4.98. This is a significant and shows that the supersaturated solid solution should be more amenable to cold spraying, despite the precipitates that form during/after SHT (as described above), possibly because these precipitates are non-coherent and therefore will not be fully effective at impeding dislocation motion.

3.4 Individual particle deformation in cold spray

3.4.1 Top view of sprayed particles

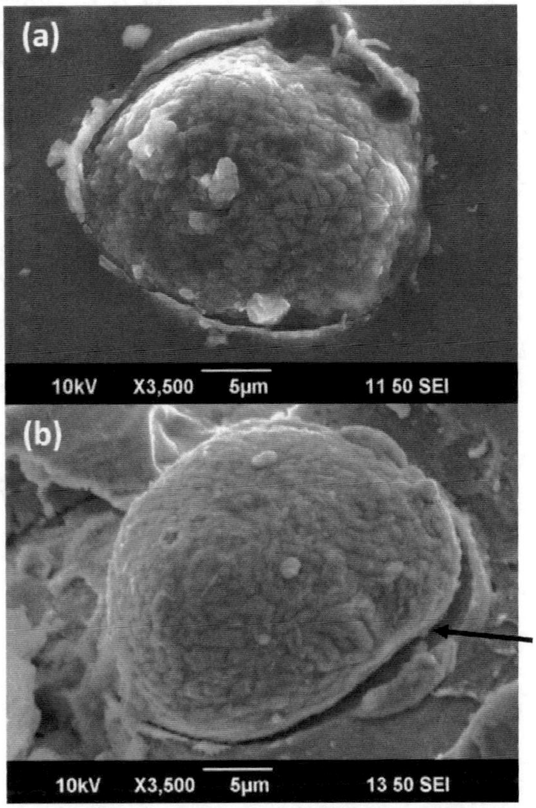

Fig. 7: SE picture of the top surface of individual as-received (a) and SHT (b) AA7075 particle after spraying

Fig. 7a shows the top view of an as-received AA7075 sprayed particle on an AA6061 substrate after swipe testing. The particle has penetrated into the substrate, leading to the material jet observed on the edge of the crater. The powder is partially embedded and most of the deformation seems to have been undergone by the substrate, with the particles retaining quite a spherical shape. This is typical of material combinations where the particle is significantly less ductile than the substrate [14].

Fig. 7b shows the top-view of a SHT AA7075 particle. The behavior observed is different compared to the as-received powder. Indeed, a ring of jet type morphology can be observed around the impact zone, as well as a gap between the particle and the substrate, as indicated by the black arrow on the picture. The particle seems to be heavily deformed and is now clearly splat-shaped, whereas the substrate has apparently undergone only a slight deformation. There seems to be the beginning of a rebounding effect, as illustrated by the gap observed at the interface.

3.4.2 Cross-section microstructure

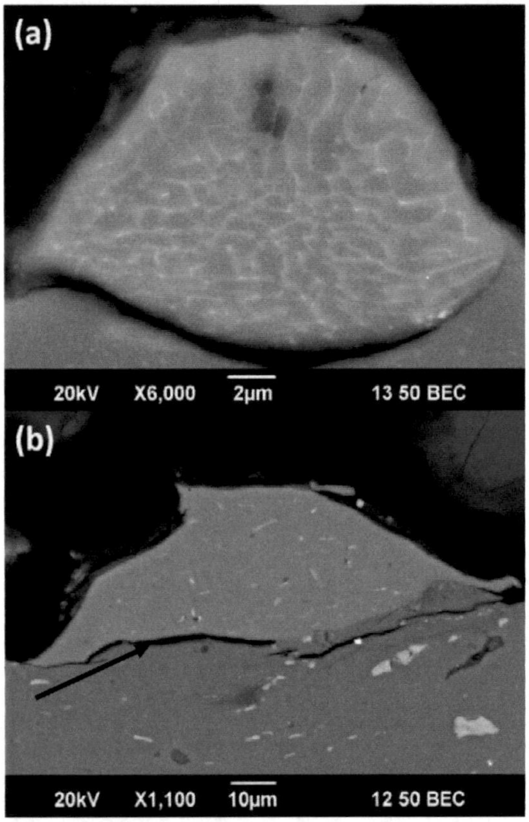

Fig. 8: BSE pictures of cross-sections of as-received (a) and SHT (b) AA7075 particle after spraying

In cross-section (Fig. 8) the differences become clearer. The dendritic structure observed previously in the as-received powder is now clearly visible as an elongated vertical shape on the upper regions of the particle. On the other hand, the lower part of the particle seems to present horizontally oriented dendritic cells, whose orientation has been presumably modified upon impact. A weak bonding is also observed between the particle and substrate, as evident by the gap at the particle-substrate interface. Only a slight material jetting is observed on the sides of the particle, illustrating a partial deformation of the powder. The high strain rate upon impact, combined with the thermal softening led to the formation of a material jet [15]. The particle kept its mostly spherical shape whilst the previously flat, polished substrate has deformated, as illustrated by the shape of the interface. For the SHT powder, the needle-shaped precipitates are still present after spraying. A hairline gap appears between the particle and the substrate, as indicated by the black arrow, showing a weak bond created upon impact. A significant jetting effect can be seen on the right side of the particle, similar to the one observed for the as-received powder. A crack-like defect is present below the particle on the right. The contrast difference there reveals that a bond has been created with the substrate, and that the particle presumably rebounded afterwards, leading to the observed gap. Overall, the deformation of the substrate is very low in the second case. The majority of the kinetic energy upon impact has been absorbed by the particle itself, leading to a heavy material jet, as well as deformation.

In summary, both powders exhibited a significant rebound effect, with gaps between the particles and the substrates, indicating weak bonding. This could potentially be related to the relative hardnesses of the powders and the substrate, which is a crucial factor in achieving good bonding in the cold spray process[14].

The more extensive deformation seen in the SHT powders is encouraging, as this should facilitate better bonding (i.e. better adhesion and mechanical strength) and lower porosity in the associated cold spray deposits. The standard ageing temperature of AA7075 is ca. 120°C [3], but this may need to be adjusted to compensate for the effect of cold work on the ageing response.

4. Conclusion

In this study, the solution heat treatment of AA7075 gas atomised powder was performed using a novel technique, and as-atomised and SHT powders were cold sprayed and the resultant particle impact behaviors compared to understand the effect of the solution heat treatment. The following conclusions can be drawn from this study:

- The non-equilibrium microstructure of gas atomised aluminium alloy powders, which is not ideal for cold spray deposition, can be homogenized using a solution heat treatment.
- The dendritic structure of the as-received powder was eliminated and its grain structure modified by SHT.
- The solute atoms present in the powder entered into a uniform solid solution, but Fe- and Mg- rich phases re-precipitated in the powder particles.
- Powder hardness was reduced by 25% due to the redistribution of the alloying elements during SHT.
- The lower hardness of the SHT particles led to greater deformation of the particles during cold spray impacts, as illustrated by material jetting and by the shapes of the splats compared to the as-received powder.
- In both untreated and SHT AA7075 powders, a weak particle-substrate interface was observed due to rebounding of the particles on the AA6061 substrates.

Future work will be focused on coating deposition, in order to see the build-up mechanisms of the SHT particles. A coating of 200 µm thickness is envisaged in order to explore the link between the lower hardness/greater ductility of the SHT powder and the coating build-up. Once deposited, aging treatments will be explored to enhance the mechanical properties. The standard ageing temperature of AA7075 is ca. 120°C [3], but this may need to be adjusted to compensate for the effect of cold work on the ageing response. The particle deposition in this work was performed 12 hours after SHT, in order to avoid the effects of natural ageing. AA7075 experiences natural ageing at room temperature, which can reduce the ductility and affect the deformation response of the particles [16]. The suppression of natural aging will also be investigated in planned future work.

References

[1] V. Champagne and D. Helfritch, "Critical Assessment 11: Structural repairs by cold spray," *Mater. Sci. Technol.*, vol. 31, no. 6, pp. 627–634, Apr. 2015.

[2] R. Huang, M. Sone, W. Ma, and H. Fukanuma, "The effects of heat treatment on the mechanical properties of cold-sprayed coatings," *Surf. Coat. Technol.*, vol. 261, pp. 278–288, 2015.

[3] H. Y. Hunsicker, "3.6 Development of Al-Zn-Mg-Cu alloys for aircraft," in *Philosophical Transactions of the Royal Society of London.*, 2016, pp. 359–376.

[4] Y. Birol, F. Birol, B. Yuksel, and O. Duygulu, "Corrosion behaviour of twin belt cast EN AW 7075 alloy," *Mater. Corros.*, vol. 64, no. 10, pp. 881–889, May 2012.

[5] M. R. Rokni, C. A. Widener, V. K. Champagne, and G. A. Crawford, "Microstructure and mechanical properties of cold sprayed 7075 deposition during non-isothermal annealing," *Surf. Coat. Technol.*, vol. 276, pp. 305–315, 2015.

[6] O. D. Neikov, I. B. Murashova, N. A. Yefimov, and S. Nabocychenko, "Handbook of Non-Ferrous Metal Powders : Technologies and Applications," pp. 554–558, 2008.

[7] T. Hussain, D. G. Mccartney, and P. H. Shipway, "Impact phenomena in cold-spraying of titanium onto various ferrous alloys," *Surf. Coat. Technol.*, vol. 205, pp. 5021–5027, 2011.

[8] W. Feng, X. Baiqing, Z. Yongan, H. Liu, and H. Xiaoqing, "Microstructural development of spray-deposited Al–Zn–Mg–Cu alloy during subsequent processing," *J. Alloys Compd.*, vol. 477, pp. 616–621, 2009.

[9] S. He, Y. Liu, and S. Guo, "Cooling Rate Calculation of Non-Equilibrium Aluminum Alloy Powders Prepared by Gas Atomization," *Rare Met. Mater. Enginerring*, vol. 38, pp. 353–356, 2009.

[10] M. R. Rokni, C. A. Widener, and V. R. Champagne, "Microstructural Evolution of 6061 Aluminum Gas-Atomized Powder and High-Pressure Cold-Sprayed Deposition," *J. Therm. Spray Technol.*, vol. 23, no. 3, pp. 514–524, 2014.

[11] D. S. MacKenzie and G. E. Totten, "Analytical Characterization of Aluminum, Steel, and Superalloys," pp. 111–125, 2005.

[12] M. R. Rokni, A. Zarei-Hanzaki, and H. R. Abedi, "Microstructure evolution and mechanical properties of back extruded 7075 aluminum alloy at elevated temperatures," *Mater. Sci. Eng. A*, vol. 532, pp. 593–600, 2011.

[13] G. E. Totten and D. S. MacKenzie, "Handbook of Aluminium, Volume 1: Physical Metallurgy and Processes," *CRC Press. New York*, vol. 7, pp. 121–123, 2003.

[14] G. Bae, Y. Xiong, S. Kumar, K. Kang, and C. Lee, "General aspects of interface bonding in kinetic sprayed coatings," *Acta Mater.*, vol. 56, pp. 4858–4868, 2008.

[15] T. Hussain, "Cold Spraying of Titanium: A Review of Bonding Mechanisms, Microstructure and Properties," *Prog. Surf. Treat. Ii*, vol. 533, pp. 53–90, 2013.

[16] A. G. Leacock, C. Howe, D. Brown, O.-G. Lademo, and A. Deering, "Evolution of mechanical properties in a 7075 Al-alloy subject to natural ageing," *Mater. Des.*, vol. 49, pp. 160–167, 2013.

Fatigue strength of Mg alloy coated with Al coatings via in-situ shot-peening assisted cold spray

Ying-Kang Wei, Xiao-Tao Luo*, Cheng-xin Li, Chang-Jiu Li*

Abstract

In this study, pure Al coatings were deposited on ZK60-T5 Mg alloy substrates via in-situ shot-peening assisted cold spray in order to study the effect of the Al coating on fatigue behavior of coated samples. Fatigue behavior of the coated and un-coated samples has been investigated through experimental tests. The size and shape distribution of powders, microstructural characteristics of coatings and fractography of fatigue test samples have been studied using scanning electron microscopy. The average microhardness of pure Al coating is higher than 70 HV_{50}. In order to obtain the fatigue S-N diagram for each set, coated and un-coated samples have been tested in a load-controlled condition. The tension-compression fatigue experiments reveal that the fatigue property of ZK60-T5 alloy coated with pure Al coatings has significantly deteriorated compared with un-coated samples.

Keywords: cold spray, aluminum coating, magnesium alloy, in-situ shot-peening, microstructure, tension-compression fatigue strength

1 Introduction

Magnesium (Mg) alloys offer a great potential application in automobile and aerospace structures and structural components in electronics and other consumer goods because of the low density, high specific strength, and excellent recyclability [1]. However, the poor corrosion resistance of Mg alloys hampers their wider applications [2]. Coating a barrier layer is an effective and economical way to prevent Mg based alloys from severe corrosion [3]. In comparison with the currently available surface treatment technologies for Mg alloys, such as chemical conversion, micro-arc oxidation, chemical vapour deposition and plating [4-7], cold spray is a relatively new and environmentally friendly technology. Unlike conventional thermal spray, cold spray process deposits coatings by plastic deformation upon impact of solid particles at a very high velocity and at a much lower temperature than the melting point of the coating material. Thus, cold spray is mostly suitable for coating on oxygen-sensitive or thermo-sensitive substrates, including magnesium alloy [8].

It was reported that cold spray process will introduce the residual compressive stress on the coating and the substrate surface [9-12]. In addition, residual stress is among the most important factors affecting coating integrity that is closely linked to the properties and service lifetime of materials and components [13]. Incorporation of big size shot-peening (SP) particles dramatically increases the density of the coating through the tamping effect upon impact of the SP particles onto the deposited particles [14-15]. Such compression may affect the residual stress of the coating and the substrate surface.

ZK60 is a high strength commercial Mg alloy which has potential applications in the load-bearing components. Meanwhile, our previous investigation revealed that the fully dense Aluminum (Al) based alloy coating deposited by in-situ shot-peening (SP) assisted cold spraying can provide completely effective protection for the magnesium material from corrosion [15]. The fatigue failure behavoir is a critical factor for the magnesium material engineering components coated with corrosion protection coatings subjected to repeated loading conditions in industry field. Thus, in present investigation, large sized stainless steel SP particles were blended into the pure Al powders to introduce an enhanced in-situ tamping effect and therefore acquire a fully dense Al coating on the ZK60-T5 magnesium alloy. Then the fatigue behavior of ZK60 alloy coated with fully dense Al coating was investigated under tension-compression loading condition to examine the effect of pure Al coating on the fatigue behavior of ZK60 alloy.

2 Materials and Methods

2.1 Materials and sample preparation

The commercial pure Al powders were purchased from Beijing Xing Rong Yuan Technology Co., Ltd., with a size range of 20-60 μm. The available stainless steel SP particles (1Cr18) have a diameter range of 200-300 μm. **Fig. 1.** shows the typical morphology of the pure Al spray powders and SP particles used in this study. Substrate for cold spraying was ZK60-T5 magnesium alloy.The chemical compositions of the Al6061 powders and the ZK60-T5 substrate are summarized in **Table 1**.

Tension-compression fatigue tests were conducted to investigate the effect of in-situ SP assisted cold spray Al coating on the fatigue properties of ZK60 alloys. After a T5 heat treatment, the ZK60 alloy rods were cut into the hourglass-shaped gauge shape (**Fig. 2.**). An approximately 700 μm thick pure Al coating was deposited on the middle of the hourglass section of the specimen. The coating was then mechanically ground and polished down to 400-500 μm thickness. For comparison, the fatigue properties of the un-coated ZK60-T5 alloy were also tested.

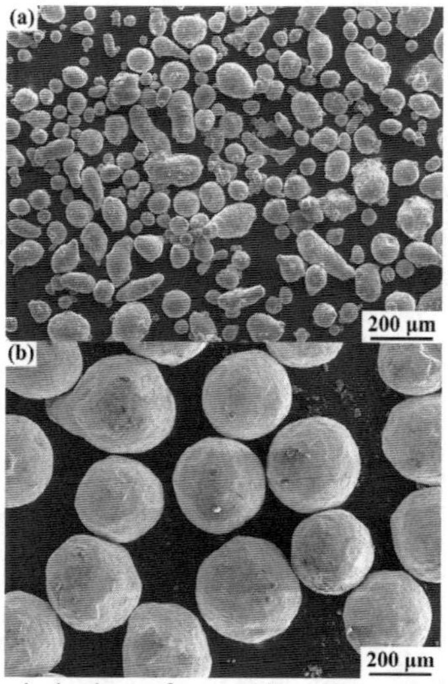

Fig. 1 Morphologies of pure Al particles and stainless steel SP particles.

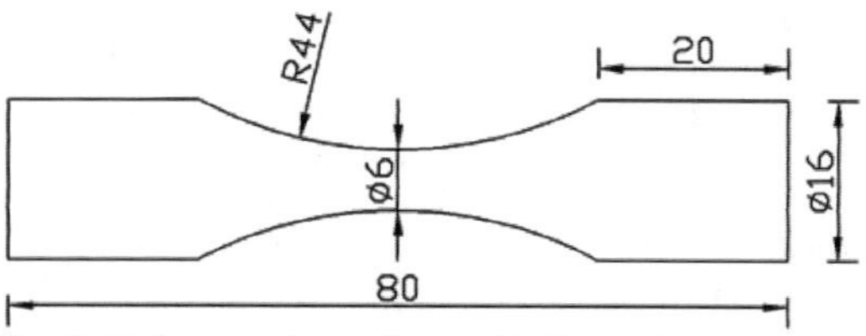

Fig. 2. Fatigue test specimen. All dimensions in millimeter.

Table 1. Chemical compositions of Al powders and ZK60 alloy substrates(wt.%)

	Al	Mg	Si	Fe	Zn	Zr
Al powders	Remainder	0.03	0.25	0.32	0.04	/
ZK60	/	Remainder	0.01	0.02	5.4	0.57

2.2 Cold spray coatings

In this study, the Al6061 coatings were prepared by a DWCS-2000 cold spraying system (Dewei Automation Company, Xi'an, China). In this system, a convergent-divergent De Laval nozzle, which is made of 2520 stainless steel, was used with a throat diameter of 2.0 mm, an outlet diameter of 6 mm and a divergent section length of 200 mm. Mixtures of Al and SP powder, at a volume fractions of SP particles (40%), were used as feedstock in in-situ SP assisted cold spray. Nitrogen was used as accelerating gas with stagnation pressure and temperature of 2 MPa and 200°C, respectively. The nitrogen was also used as the powder carrier gas at a pressure of 2.2 MPa.

For all coatings, the stand-off distance and traverse speed along the specimen central axis of the nozzle were 20 mm and 5 mm/s respectively. The rotation speed of hourglass-shaped gauge section specimen was 100 rpm. The feeding rate of feedstock powders was about 15 g/min.

In order to improve the adhesion of coating/substrate, the substrate surface was sandblasted using 150 μm alumina grits at compressed air pressure of 0.6 MPa before the deposition. After sandblasting, the substrate was cleaned in an ultrasonic acetone bath for 10 minutes.

2.3 Microstructure characterization and microhardness measurement

Cross-sectional microstructures of the coatings and fracture surface morphologies of failed specimens were examined using scanning electron microscope (SEM, MIRA3 LMH, TESCAN, Czech). The porosity of coatings was measured from the polished coating cross section by image analyzing approach. For each sample, 15 images taken at a magnification of 1000× in the back-scattered electron mode were used. Microhardness across the coating/substrate interface of the cross-sections was measured by a microhardness tester with a load of 50 g and a dwelling time of 30 s. The spacing between indentations was equal to at least five indent diagonals.

2.4 Fatigue test

High cycle fatigue experiments were conducted using a computer controlled Instron servo-hydraulic fatigue testing system, in a tension-compression configuration. Fatigue tests (stress ratio R =-1) were carried out at room temperature at a nominal frequency of 80-90 Hz.Fatigue run-out was defined as 10^7 cycles. The "brief staircase" method presented by Dixon and Massey [16]was used for performing the tests with a stress step of 10 MPa, and the Hodge–Rosenblatt [17] method was used to calculate the fatigue strength corresponding to a fatigue life of 10 million cycles. The fatigue test data has been elaborated based on the ASTM standard E739-10 [18] to obtain the S–N diagram for different cases with a failure probability of 50% on a bi-logarithmic scale.

3 Experimental Results and Discussion

3.1 Microstructure characterization

Fig. 3. shows cross-sectional microstructures of pure Al coatings deposited by in-situ SP assisted cold spraying. The coating is uniformly coated on the substrate and reveals a completely dense microstructure without obvious inter-particle cracks. Image analysis using the ImageJ software package revealed that the porosity of the pure Al coating is less than 0.3%. The SEM observation revealed that no SP particle was deposited into pure Al coating at the current spray parameters. The jagged coating/substrate interface (frame B in **Fig. 3**) may contribute to the excellent adhesive strength of coating. Meanwhile, no detectable inter-diffusion between Al and Mg across the coating/substrate interface was observed. The effect of the addition of big size SP particles into the spray feedstock on Al based alloy coating properties has been comprehensively studied previously [15].

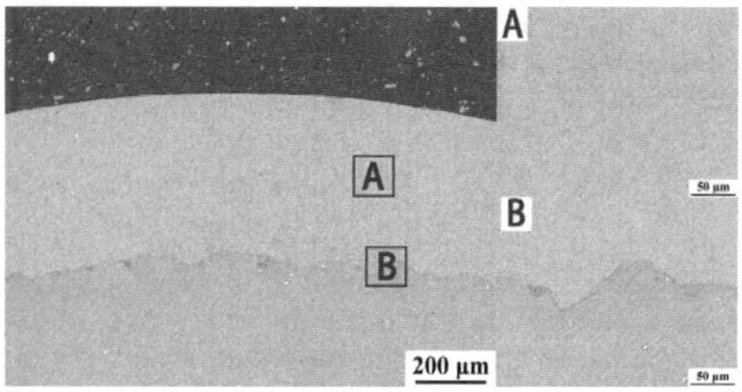

Fig. 3. Overall and localized cross-sectional microstructures of pure Al coatings deposited by in-situ SP assisted cold spraying.

3.2 Microhardness

The microhardness profile across the coating/substrate interface is plotted in **Fig. 4**. It is clear that the hardness of coating is inversely proportional to the distance away from the substrate/coating interface. Microhardness difference between cold sprayed metallic coatings and their bulk counterparts is dependent mainly on coating porosity and microstructure changes during deposition process including dislocation multiplication (work hardening) and inter-particle recrystallization (grain refinement strengthening) [8]. Generally, higher plastic deformation strain of the spraying particle results in denser coating, severer dislocation multiplication and crystal refinement and therefore higher hardness. Firstly, more addition of SP particles introduces more intensive tamping effect and thereby higher hardness. Secondly, during the spraying, the sprayed particles can be only tamped by the followed SP particle impact. This suggests the bottom coating layer near the substrate surface was hammered by more SP particles and thus experienced heavier plastic deformation than the top coating layer. Therefore, a hardness gradient as a function of the distance away from the coating/substrate interface is formed as shown in **Fig. 4.** In addition, it is noteworthy that the hardness at the substrate side (about 200 µm in depth) also increases owning to the both the grit blasting and the coating process, which is similar to the shot peening process [19]. The result means that the value of residual compressive stress of substrate surface may have been increased after cold spray process. It was reported that increasing the Mg alloy surface residual compressive stress can effectively prevent the initiation of fatigue crack and improve the fatigue life [20-21]. Hence, the tamping layer in the substrate surface may contribute to the fatigue life of ZK69-T5 alloy coated with coating.

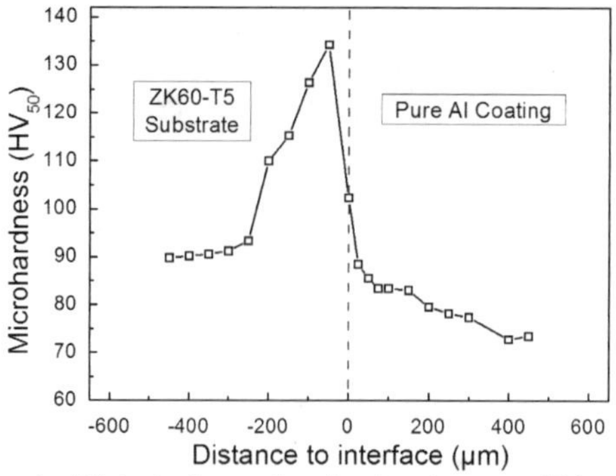

Fig. 4. Distrubution of microhardness within the substrate and coating across the substrate/coating interface of ZK60-T5 alloys with pure Al coatings deposited by in-situ SP assisted cold spraying.

3.3 Fatigue test results

Fig. 5 shows the S–N curves of the ZK60-T5 alloys with and without pure Al coatings. The fatigue run-out was achieved at 90 MPa for the ZK60-T5 alloys without coating, suggesting that the fatigue limit is between 90 and 100 MPa. However, the ZK60-T5 alloys coated with pure Al showed a significant fatigue endurance decrement. After coated with pure Al coating, all the specimen failed after less than 10^5 cycles at 90 MPa load. With the decrease of load, the number of cycles became higher and higher but still far below to 10^7. The results revealed that the tension-compression fatigue property of ZK60-T5 alloy significantly decreased after ZK60-T5 alloy was coated with pure Al coatings.

In order to further understand the fracture process, the SEM observations of specimens after failure were carried out. **Fig. 6** shows the overall and localized fractography observations of a specimen without coating, which underwent testing under stress of 130 MPa. It shows that the fatigue fractured sample was obviously divided into three regions: crack initiation region (frame A), crack propagation region (frame B) and final-fast-fracture region (frame C). The crack initiation occurred in the area of concentrated stress near the surface and rapidly expanded to the other side of the specimen (follow the direction of the red arrow) and caused the failure.

Fig. 7 presents overall images of fractured specimen with pure Al coatings, which underwent testing under stress of 80 MPa. From the rough surfaces of the fractured specimens, two crack initiation regions (maked with the blue ellipse) were clearly observed near the specimen surface and cracks extended in different directions (maked with the red arrow). Therefore, multiple locations of the initiation of cracks near the specimen surface may leads to a decline in the fatigue life of ZK60-T5 alloy coated with Al coating.

Fig. 8 shows the fracture morphology of of the coated specimen across the substrate/coating interface. It can be seen that the pure Al coating is strongly bonded to the substrate so that the interface is not distinguishable. It has been reported that the initiation of cracks is related to the critical strain and will cause the delamination of coating and substrate [22]. In addition, high residual compressive stress on the substrate surface can effectively suppress the initiation of the fatigue cracks. The above results show that the fatigue crack is not initiate in the coating/substrate interface or substrate surface. Therefore, the fatigue crack was initiated inside the Al coating near coating surface. The results reveal that although the pure Al coatings are very dense intensive peening effect of large shot-particles caused intensive hardening effect. This hardening effect may significantly decrease the ductility and toughness of pure Al coating. Therefore, although the dense coating adhered to Mg alloy excellent, lowered toughness due intensive hardening may be responsible for the significant decrease of fatigue strength of coated Mg alloy.

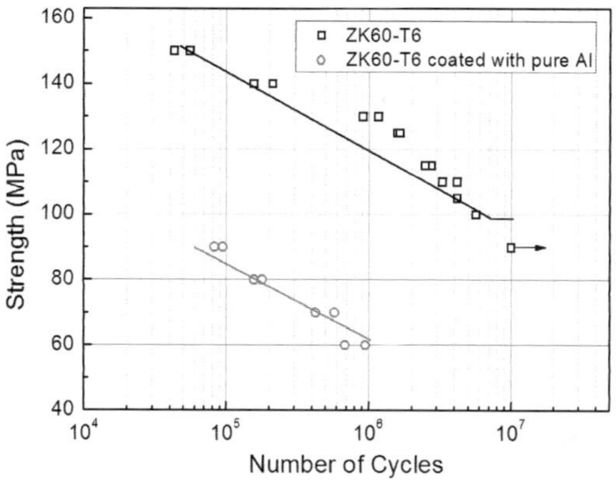

Fig. 5. The S–N diagram obtained from fatigue tests.

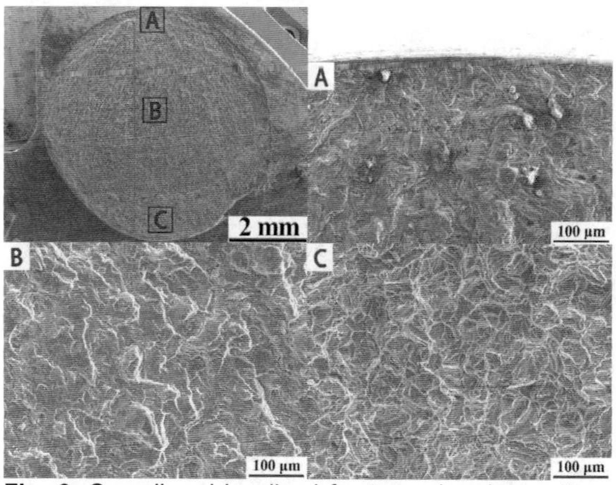

Fig. 6. Overall and localized fractography observations of ZK60-T5 alloy.

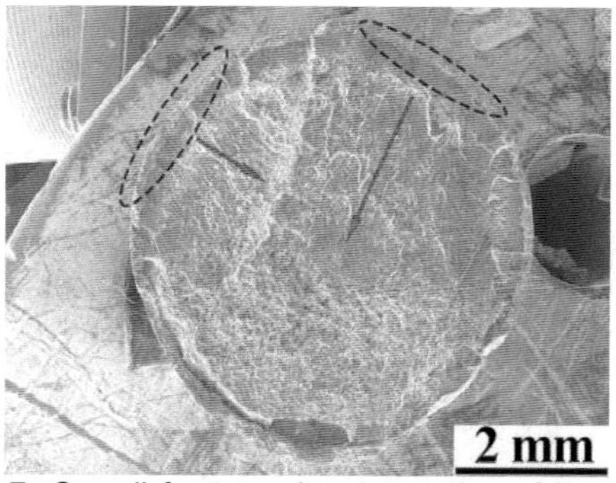

Fig. 7. Overall fractography observation of ZK60-T5 alloy coated with pure Al coting.

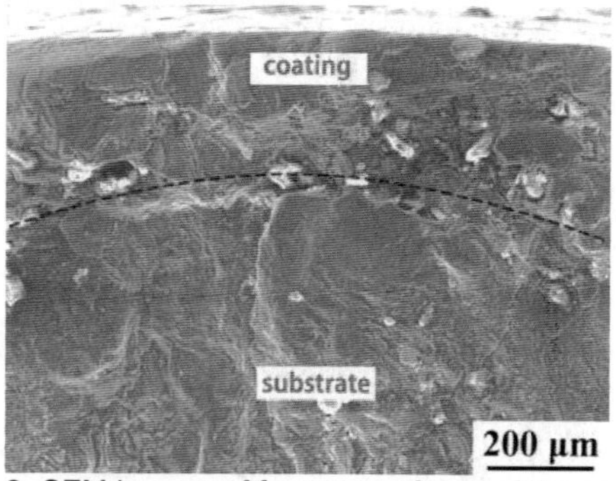

Fig. 8. SEM images of fracture surfaces of the coated specimen across the substrate/coating interface.

4 Conclusions

In this study, large sized stainless steel SP particles were blended into the pure Al powders to deposit fully dense Al coating. The pure Al coating presented a dense microstructure with porosity less than 0.3%. The hardness of coating is inversely proportional to the distance away from the substrate/coating interface and the average microhardness is higher than 70 HV_{50}. However, It was observed that the tension-compression fatigue experiments revealed that the fatigue property of ZK60-T5 alloy has significantly decreased after coated with pure Al coatings. The lowered toughness of the dense Al coating due to intensive hardening may be responsible for the low fatigue strength.

Acknowledgements

This work is supported by National Nature Science Foundation (No. 50115814) and Shaanxi Co-Innovation Projects (2015KTTSGY03-03)

5 References

[1] R.L. Edgar, In: in: K.U. Kainer (Ed.): Magnesium Alloys and their Application, Wiley-VCH, Germany, 2003, p 3.

[2] B.L Mordike, T Ebert, Magnesium: Properties-applications-potential, Mater. Sci. Eng: A. 2001, 302(1), p37-45.
[3] J.E. Gray and B. Luan, Protective Coatings on Magnesium and Its Alloys: A Critical Review, J. Alloys Compd., 2002, 336(1), p 88-113.
[4]H.F. Guo, M.Z. An, Effect of surfactants on surface morphology of ceramic coatings fabricated on magnesium alloys by micro-arc oxidation, Thin Solid Films, 2006, 500(1-2), p 186-189.
[5] A.R. Shashikala, R. Umarani, S.M. Mayanna, A.K. Sharma, Int. J. Electrochem. Sci. 2008, 3, p 993-1004.
[6] J.F. Zhang, C.W. Yan, F.H. Wang, Appl. Surf. Sci. 2009, 255, p 4926-4932.
[7] T. Ishizaki, J. Hieda, N. Saito, N. Saito, O. Takai, Electrochim. Corrosion resistance and chemical stability of super-hydrophobic film deposited on magnesium alloy AZ31 by microwave plasma-enhanced chemical vapor deposition, Acta Mater., 2010, 55, p 7094-7101.
[8] X.-T. Luo, C.-X. Li, F.-L. Shang, G.-J. Yang, Y.-Y. Wang, and C.-J. Li, High Velocity Impact Induced Microstructure Evolution during Deposition of Cold Spray Coatings: A review, Surf. Coat. Technol., 2014, 254, p 11-20.
[9] W.-B. Choi, L. Li, V. Luzin, R. Neiser, T. Gnaupel-Herold, H.J. Prask, S. Sampath, and A. Gouldstone, Integrated Characterization of Cold Sprayed Aluminum Coatings, Acta Mater., 2007, 55(3), p 857-866.
[10] Z. Arabgol, H. Assadi, T. Schmidt, F. Gartner, and T. Klassen, Analysis of Thermal History and Residual Stress in Cold-Sprayed Coatings, J. Therm. Spray Technol., 2013, 23(1–2), p 84-90.
[11] V. Luzin, K. Spencer, and M.X. Zhang, Residual Stress and Thermo-Mechanical Properties of Cold Spray Metal Coatings, Acta Mater., 2011, 59(3), p 1259-1270.
[12] K. Spencer, D.M. Fabijanic, and M.-X. Zhang, The Use of Al-Al2O3 Cold Spray Coatings to Improve the Surface Properties of Magnesium Alloys, Surf. Coat. Technol., 2009, 204(3), p 336-344.
[13] R. Ghelichi, S. Bagherifard, D. MacDonald, and I. Fernandez-Pariente, Experimental and Numerical Study Residual Stress Evolution in Cold Spray Coating, Appl. Surf. Sci., 2014, 288(1), p 26-33.
[14] X.-T. Luo, Y.-K. Wei, Y. Wang, and C.-J. Li, Microstructure and Mechanical Property of Ti and Ti6Al4V Prepared by an In-situ Shot Peening Assisted Cold Spraying, Mater. Des., 2015, 85, p 527-533.
[15] Y.-K. Wei, X.-T. Luo*, C.-X. Li, C.-J. Li*. Optimization of In-Situ Shot-Peening-Assisted Cold Spraying Parameters for Full Corrosion Protection of Mg Alloy by Fully Dense Al-Based Alloy Coating, J. Therm. Spray Technol., 2017, 26, p 173-183.
[16] Dixon W, Massey F. Introduction to statistical analysis. New York: McGraw-Hill; 1969.
[17] K.A. Brownlee, J.L. Hodges, J Am Stat Assoc 1953;48:262.
[18] ASTM standard E739-10. Standard practice for statistical analysis of linear or linearized stress life (S-N) and strain life (e–N) fatigue data; 2010. doi: http://dx.doi.org/10.1520/E0739-10.
[19] P. Zhang and J. Lindemann, Effect of Roller Burnishing on the High Cycle Fatigue Performance of the High-strength Wrought Magnesium Alloy AZ80, Scripta Mater., 2005, 52(10), p 485-490
[20] W. Wu, P.K. Liaw, K. An. Unraveling cyclic deformation mechanisms of a rolled magnesium alloy using in situ neutron diffraction, Acta Materialia, 2015, 85, p 343-353
[21] C. Potzies, K.U. Kainer. Fatigue of Magnesium Alloys, Adv. Eng.g Mater., 2004, 6, p 281-288
[22] X.-F. Zhu, B. Zhang, J. Gao, G.-P. Zhang Evaluation of the crack-initiation strain of a Cu-Ni multilayer on a flexible substrate. Scr. Mater., 2009, 60, p 178-181.

Simulation of effect of interface evolution on the bonding during the high-velocity particle impacts in cold spray by using Eulerian approach

Yu-Juan Li, Xiao-Tao Luo, Chang-Jiu Li*
State Key Laboratory for Mechanical Behavior of Materials, School of Materials Science and Engineering, Xi'an Jiao tong University, Xi'an, Shaanxi, 710049, China/ E-mail: licj@mail.xjtu.edu.cn

The bonding at the interface in cold sprayed coating is considered to be a very crucial factor determining its mechanical properties, physical and chemical behavior such as tensile strength, hardness, electrical and thermal conductivities, as well as corrosion resistance. Therefore, in this study the deformation behavior of the particle initial surface is investigated in order to reveal the evolution of free-oxide interface during the high-velocity particle impact in cold spray. The variation of the stress at the interface during the impact is also examined to evaluate the bonding between particle and substrate, and further to predict the bonding strength for the experiments. Results show that the area ratio of the free-oxide interface and the whole interface are 0.52, 0.7 and 0.76, respectively, for the case of copper particle impact at 500 m/s, 800 m/s and 1100 m/s. Moreover, the free-oxide interface in case of 800 m/s is about 3 times as much as that in case of 500 m/s while the free-oxide interface of 1100 m/s is approximately 5 times as much as that of 500 m/s. The compressive stress in the normal direction at the position where free-oxide interface occurs is higher than the yield strength of the material and during the whole impact, the tensile stress is no more than the tensile strength of the material.

1 Introduction

Many metals and its alloys have a great deal of unparalleled properties such as the superior thermal and electrical performance of Cu [1, 2], the excellent corrosion resistance of Ni [3, 4] and the high specific strength and fatigue resistance of Ti [5, 6], which make these material have been attracting wide interests due to their highly potential application in electrical engineering, corrosion environment and aeronautical industries. Compared with traditional casting process, cold spray is a more flexible and convenient coating technology. The relatively low deposition temperature of cold spray can effectively avoid a large number of defects in metal coatings, such as oxidation, phase transformations and grain growth, which always occur during the thermal-spray process [7]. The high particle velocity of cold spray makes it possible to produce dense coating with a compressive stress [8] generated in the process, which further eliminates restriction on the thickness of the coating.

It is believed that the successful bonding in coating can be obtained only when the particle velocity exceeds a critical velocity which is closely related to particle temperature and size as well for a given material [9, 10]. A large number of efforts [9, 11, 12] have been made to clarify the bonding mechanism. So far, the most widely accepted viewpoint of the bonding mechanism in cold spray is the occurrence of adiabatic shear instability, resulting from the local intensive plastic deformation at the interface. Yin et al [13, 14] proposed that at the position where adiabatic shear instability occurs, the thermal softening effect induced by adiabatic heat dominates the work-hardening effect, the metal stress starts to become stable and the plastic flow occurs, resulting in the metal behave like the viscous material. Then the viscous metal extrudes from the interface and forms a jet at the fringe contributing to clean up the break-up oxide film which exists on the initial particle and substrate surface. As a consequence, metal to metal contact and metal bonding occur.

It is generally accepted that the most dominant bonding mechanism in cold spray alloy coating is mechanical interlocking and metallurgical bonding. Interface instability resulting from the local intensive plastic deformation during the high-velocity particle impact gives rise to the material roll-ups and vortices at the interface. As a result, particle and substrate are mixed at the corresponding region and subsequently mechanical interlocking occurs [15]. Champagne and Wank et al [16, 17] observed Cu-Al and Al-Zn mixture at the interface, strongly demonstrating the occurrence of mechanical interlocking in cold-sprayed coating. Metallurgical bonding which can produce better bonding strength than any other mechanical-based bonding is normally known as the result of the atomic diffusion between different materials. Interface instability helps to clean up the oxide film at the interface resulting in the appearance of free-oxide interface. Then different atoms diffuse and the intermediate phase forms at the free-oxide interface marking the formation of metallurgical bonding [18, 19]. Lee and Bolesta et al [20, 21] have reported that there is a relatively thin intermetallic layer between particle and substrate giving a strong evidence of metallurgical bonding.

According to the bonding mechanism, metallurgical bonding which is essential to the coating strength can only occur at free-oxide interface since oxide film will block the inter-diffusion of atoms at the interface [22]. Therefore, the free-oxide interface has an important influence on the coating strength which plays a decisive role in many applications. In this study, the deformation behavior of the initial particle surface is investigated in order to reveal the evolution of free-oxide interface during the high-velocity particle impact in cold spray. The variation of the stress at the interface during the impact process is also investigated for the purpose of estimating the bonding between particle and substrate, and further expecting to predict the bonding strength.

2 Model description

2.1 Finite element method

The deformation behavior of Cu particle deposited on Cu substrate with different initial particle velocity is simulated by using an explicit finite analysis code, ABAQUS (version 6.11) with the Eulerian algorithm which has been widely used in previous works [23, 24]. The impact progress is assumed to be an adiabatic process and the dynamic explicit procedure is carried out to realize the adiabatic stress analysis. For the symmetric characteristic of the particle impact, the model is simplified as a slice with the thickness of 1 μm cut from the middle of the entity model in order to reduce the computational cost. The particle has a diameter of 28 μm and the initial surface has a thickness of 0.45 μm. **Fig.1** shows the schematic diagram of the geometric model of particle impact. The geometry is partitioned by 3D Eulerian eight-node elements with reduced integration and hourglass control (EC3D8R). The mesh size is 0.28 μm (meshing resolution of 1/100dp), except for the lower right of the model where the size is 0.875 μm (meshing resolution of 1/100dp) so as to shorten the computational time.

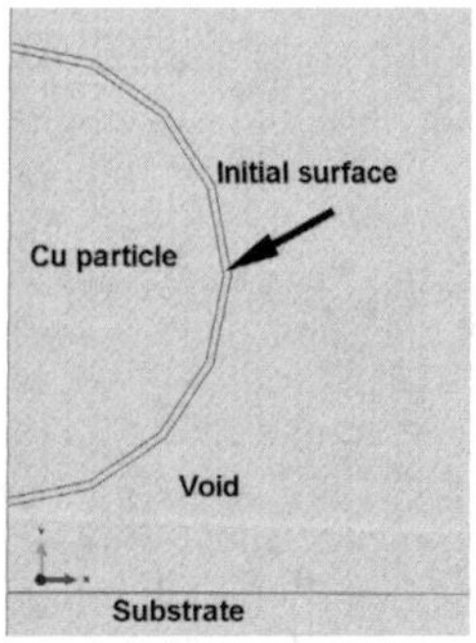

Fig. 1. Schematic diagram of the geometric model of particle impact.

2.2 Material property and other parameter settings

The material flow behavior of the particle and substrate is described as the Johnson and Cook plasticity model accounting for the strain, strain-rate hardening and thermal softening effect, which has been elaborated elsewhere [9]. A linear Mie–Gruneisen equation of state (EOS) was employed to describe the elastic behaviour [25]. All the mechanical and thermal performances are assumed to be isotropic. Previous investigation [26, 27] has shown that coating strength can be increased significantly when initial particle velocity is above 800 m/s. Therefore, in this study, the particle initial velocity is taken as 500 m/s, 800 m/s and 1100 m/s, respectively. In addition, the particle and substrate initial temperatures are set as 25 °C for all cases.

3 Results and discussion

3.1 Deformation behavior of the initial surface during the high-velocity particles impacts

Fig.2 shows the contours of the volume fraction (EVF) of a Cu particle and its initial surface, and the effective plastic strain (PEEQ) contours when it is deposited on a Cu substrate with different impact time at an initial impact velocity of 500 m/s. In order to investigate the deformation behavior of the particle initial surface, the flattened particle can be divided into two parts: the top and bottom. For the top parts, the initial surface material distributes uniformly indicating that the top part of the particle undergoes a homogeneous deformation in the whole impact process. In addition, due to the small strain at the top, the distribution of the material from the initial surface is continuous and no free-oxide interface appears during the process. While for the bottom part, the initial surface material scatters heterogeneously. At the center of the interface, it is relatively concentrated nevertheless it is extremely dispersed at the periphery of the interface. These results suggest that the deformation at the interface is heterogeneous and large deformation occurs at the rim of the interface during the impact process, which is coincidence with the distribution of the corresponding strain. This heterogeneous phenomenon gives rise to more free-oxide interface which may contribute to the bonding between particle and substrate. With the impact time increases from 20 ns to 80 ns, this heterogeneous phenomenon is increasingly apparent and therefore the free-oxide interface increases continuously. The correlation of the free-oxide interface and the bonding will be discussed in detail in the following section.

Fig.3 and **Fig.4** present the cases with an initial impact velocity of 800 m/s and 1100 m/s, respectively. The same deformation characteristic and variation tendency can be found in **Fig.3** and **Fig.4**. Noticing that as the initial particle velocity increases to 1100 m/s, the initial surface material shows a discontinuous distribution at the top from 40 ns and free-oxide interface appears. The reason is that a higher particle initial velocity brings about a more kinetic energy dissipation, then a more strain energy leading to the extension of strain from the interface to the interior of the flattened particle. It is also noticed that for the case of 800 m/s and 1100 m/s, the distribution of the particle initial surface at the interface is more dispersed and more free-oxide interface exposes resulting from the more severe plastic deformation compared with the case of 500 m/s. For the case of 1100 m/s at 80 ns, the material from the initial surface is highly dispersed at the interface which may benefit for the bonding between particle and substrate by dispersion strengthening. These reasons result in a high strength in coating produced by extremely high particle initial velocity as reported in previous investigation [27].

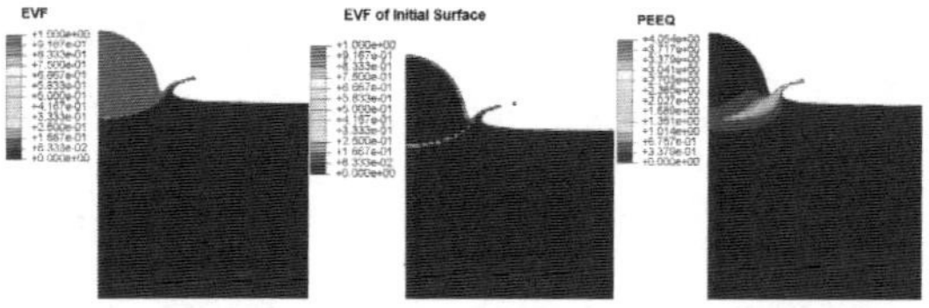

(a) 20 ns

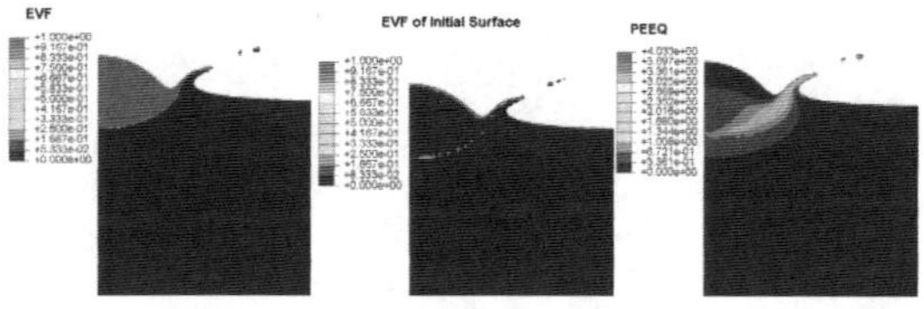

(b) 40 ns

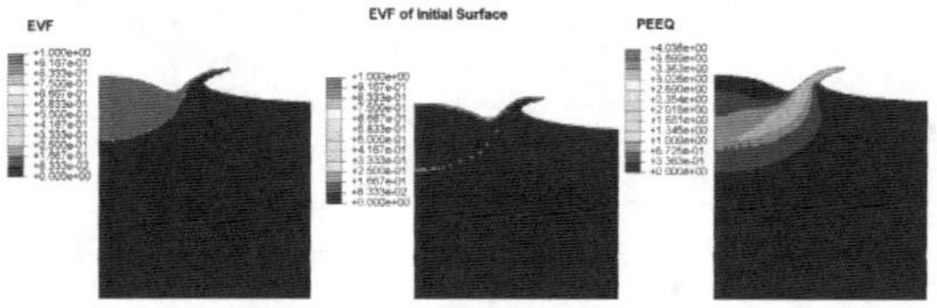

(c) 60 ns

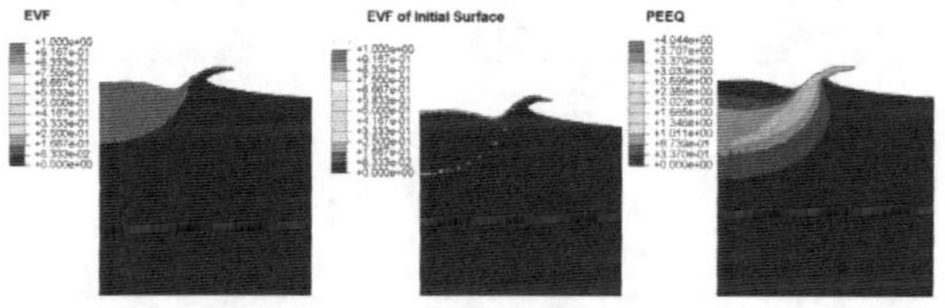

(d) 80 ns

Fig. 2. Contours of the volume fraction of a Cu particle and its surface, and its effective plastic strain contours when it is deposited on a Cu substrate at different impact times at an initial impact velocity of 500 m/s.

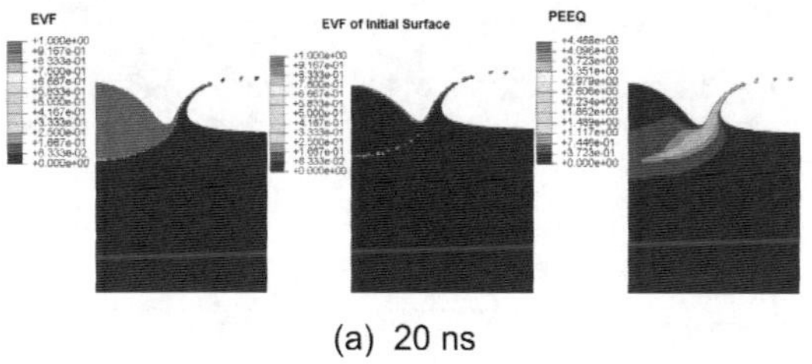

(a) 20 ns

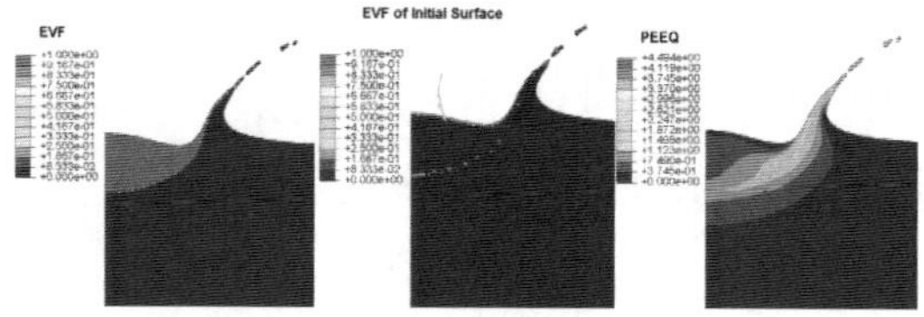

(b) 40 ns

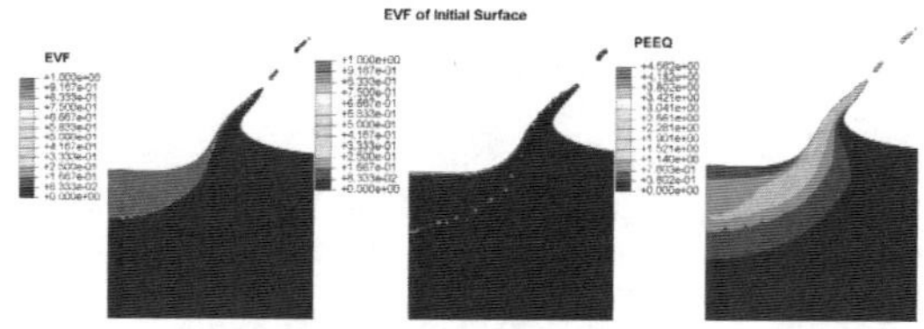

(c) 60 ns

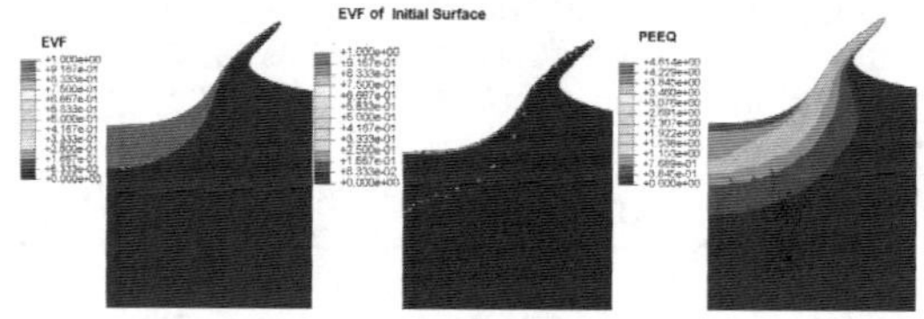

(d) 80 ns

Fig. 3. Contours of the volume fraction of a Cu particle and its surface, and its effective plastic strain contours when it is deposited on a Cu substrate at different impact times at an initial impact velocity of 800 m/s.

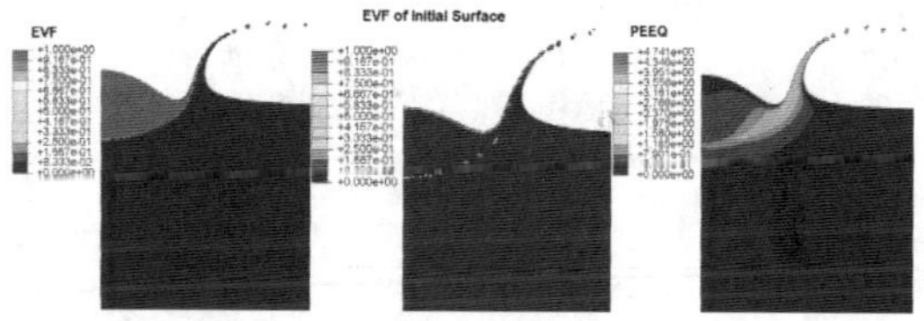

(a) 20 ns

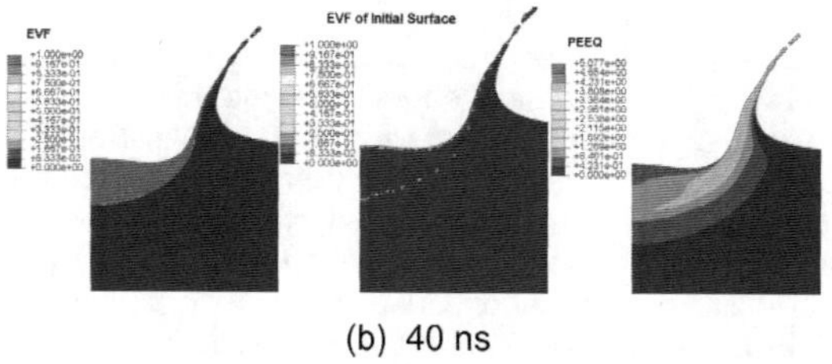

(b) 40 ns

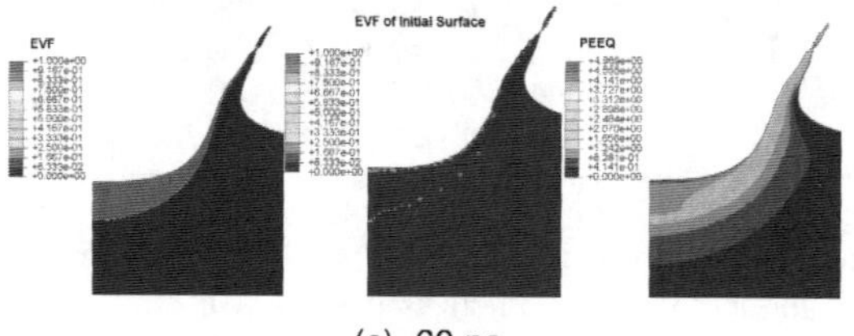

(c) 60 ns

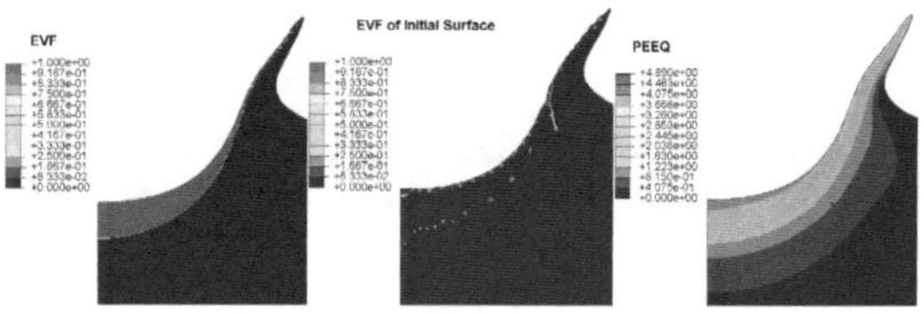

(d) 80 ns

Fig. 4. Contours of the volume fraction of a Cu particle and its surface, and its effective plastic strain contours when it is deposited on a Cu substrate at different impact times at an initial impact velocity of 1100 m/s.

In order to estimate the bonding at the interface, the interface evolutions of flattened particle at the different impact times with different initial velocities are displayed in **Fig.5**. As shown in **Fig.5**, the contour lines of the interface start at the same normalized point. Obviously, at a given particle initial velocity, the contour lines of the bottom of the interface with different impact times are almost the same which indicates that particle bottom remains little change basically during the impact process. That is the reason why the material from initial surface concentrates at the center of the interface in **Fig.2**, **3** and **4**. While the contour lines of the fringe of the interface extend outwards gradually with the increment of the impact time at the same velocity, suggesting that the intensive deformation is concentrated on the fringe of the interface and the free-oxide interface appears at that location in the process.

Table 1, **2** and **3** provides the length (L_i and L_s) and area (A_i and A_s) of the interface and the free-oxide interface during impact. In order to calculate L_i and L_s, the interface is simplified as a curve which is consist of a large number of line segment in two dimensional planar. Then, L_i and L_s is calculated according to the formula of distance. As for A_i and A_s, the interface is assumed to be a surface which is made up of a lot of little circular truncated cones in three dimensional space. A_i and A_s is calculated according to the surface area formula of circular truncated cone. As presented in **Table 1**, **2** and **3**, the length of the interface and the free-oxide interface increases with the impacting time due to the deformation at the interface. At the same impacting time, a higher particle initial velocity leads to a longer length and larger area of the interface and free-oxide interface because more kinetic energy transforms into the strain energy resulting in more severe deformation at the interface. Noting that at 80 ns, the A_s / A_i are 0.52, 0.7 and 0.76, respectively, for the case of 500 m/s, 800 m/s and 1100 m/s. These results suggesting that if bonding occurs at all the free-oxide interface, the strength of the coating is 0.52, 0.7 and 0.76 of the bulk in the corresponding work hardening state. Moreover, the A_s in case of 800 m/s is about 3 times as much as that in case of 500 m/s while the A_s of 1100 m/s is approximately 5 times as much as that of 500 m/s, indicating that the coating strength of 800 m/s and 1100 m/s is 3 and 5 times as much as that of 500 m/s.

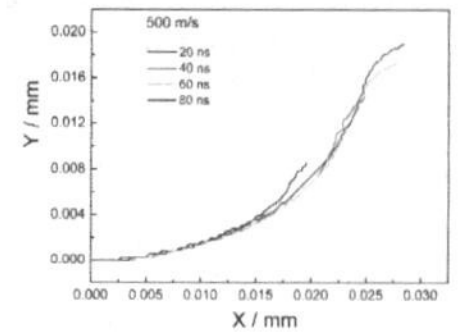

(a) 500 m/s

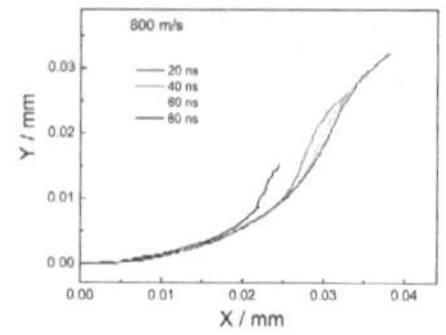

(b) 800 m/s

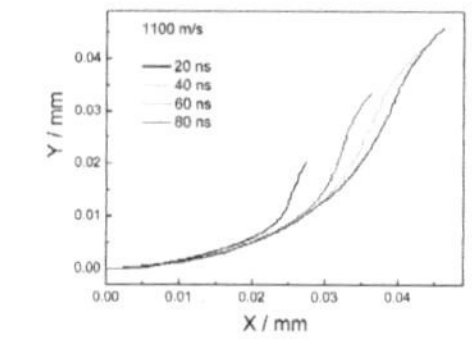

(c) 1100 m/s

Fig.5. Interface evolution of a single Cu particle deposited on a Cu substrate at different particle velocities and different impact times.

Table 1 Length and area of the interface and the free-oxide interface at the interface during impact of 500 m/s

Velocity / $m \cdot s^{-1}$	500			
Time / ns	20	40	60	80
L_i / μm	23.64	32.41	36.79	38.64
A_i / μm^2	1580	2850	3580	3880
L_s / μm	1.97	10.56	12.52	13.7
A_s / μm^2	193	1340	1770	1901

Table 2 Length and area of the interface and the free-oxide interface at the interface during impact of 800 m/s

Velocity / $m \cdot s^{-1}$	800			
Time / ns	20	40	60	80
L_i / μm	33.68	48.85	52.52	56.80
A_i / μm^2	3040	5960	6840	7870
L_s / μm	10.47	26.90	30.07	33.78
A_s / μm^2	1230	4040	4690	5520

Table 3 Length and area of the interface and the free-oxide interface at the interface during impact of 1100 m/s

Velocity / $m \cdot s^{-1}$	1100			
Time / ns	20	40	60	80
L_i / μm	39.67	58.10	69.04	74.37
A_i / μm^2	4110	8180	11200	12900
L_s / μm	21.70	36.31	43.35	44.7
A_s / μm^2	2790	6290	8660	9900

3.2 Correlation of free-oxide interface and bonding at the interface

It is known that when metal particle impacts substrate at sufficiently high velocity, adiabatic shear instability will take place at the interface between the particle and substrate due to the local extensive plastic deformation in cold spray. The occurrence of adiabatic shear instability has always been a sign of the formation of the bonding at the interface. The reason is that at the position where the adiabatic shear instability occurs, the metal material behaves like a viscous-material and extrudes the interface helping to clean up the particle initial surface and expose the free-oxide interface. As the pressure at the free-oxide interface is adequately high, necessarily more than the yield strength of the metal material, the plastic flow occurs at the normal direction, resulting in that particle and substrate contact closely, and then the atomic diffusion happens and metal to metal bonding forms at the position of the interface. Therefore, the more the free-oxide interface is, the more likely the formation of the bonding at the interface is. Furthermore, the higher the pressure at the interface is, the easier the formation of the bonding is.

In order to evaluate the possibility of the formation of bonding at the interface in cold spray when particles are deposited on substrate at different impact time with an initial impact velocity of 500 m/s, the stress in the normal direction at the interface during the impact is estimated as in **Fig.6**. As shown in **Fig.6**, when the particle is deposited on the substrate with an initial velocity of 500 m/s, the great impact force between particle and substrate induces a large compressive stress in the normal direction at the interface which keeps the materials across the interface into intimate contact. When the deformation of the material occurs, the compressive stress in the normal direction is released gradually at the interface. As the strain at the interface accumulates to a certain value with the increment of the impact time, the adiabatic shear instability is induced and the corresponding region experiences dramatic plastic deformation along the interface leading to a significantly extended interface and the occurrence of free-oxide interface. Meanwhile the compressive stress in the normal direction decreases rapidly and there even appears the tensile stress at local position due to the extremely severe plastic deformation. Thereby, adiabatic shear instability happens at 20 ns when particle is deposited on substrate at the initial velocity of 500 m/s. When the impact goes on, the region with severe plastic deformation forces the center of the interface with small deformation to deform and hence, the stress at the center decreases gradually. Due to the non-uniformity of the deformation at the fringe of the interface, the stress fluctuates within a certain range. As the impact continues, the stress at the interface tends to be stable because the deformation becomes smaller and smaller resulting from the consumption of the kinetic energy.

In the case of 800 m/s and 1100 m/s, the same deformation character and variation tendency can be observed in **Fig.7** and **Fig.8**. The difference is that due to a higher particle initial velocity, the region where the adiabatic shear instability occurs experiences more severe plastic experience resulting in a longer interface and more free-oxide interface generates. The extremely dramatic plastic deformation forces the center of the interface to deform severely and hence, there appears tensile stress at the center of the interface at 60 ns in the case of 800 m/s. At this moment, due to the extreme non-uniformity of the deformation at the whole interface, the stress fluctuates dramatically. In case of 1100 m/s at 60 and 80 ns, this unstable state is also observed and the higher the initial velocity is, the longer the duration of the unstable state. At this unstable state, the region at the center of the interface is extended and free-oxide interface may occurs, which is the reason why the material from the initial surface is more dispersed at this region compared with that in the case of 500 m/s. As the impact goes on, the stress tends to be stable.

It should be noted that the tensile stress during the whole process is no more than 200 MPa which is the tensile strength of the copper at soften state. In other words, once the bonding occurs at the interface, the bonding will remain to the end of the impact. As shown in **Fig.6**, when the particle is deposited on the substrate at 20 ns with the initial velocity, the end position of the fresh surface resulted from the adiabatic shear instability is marked by the blue line. It is found that the stress in the normal direction at this position is the compressive stress of 495 MPa which is more than the yield strength of the copper. Therefore, the bonding occurs at the moment when the free-oxide interface generates. This phenomenon is also found in the case of 800 m/s and 1100 m/s. Hence, the free-oxide interface can be defined as the effective bonding area. The effective bonding area directly determines the strength. A more effective bonding results in a higher strength in coating.

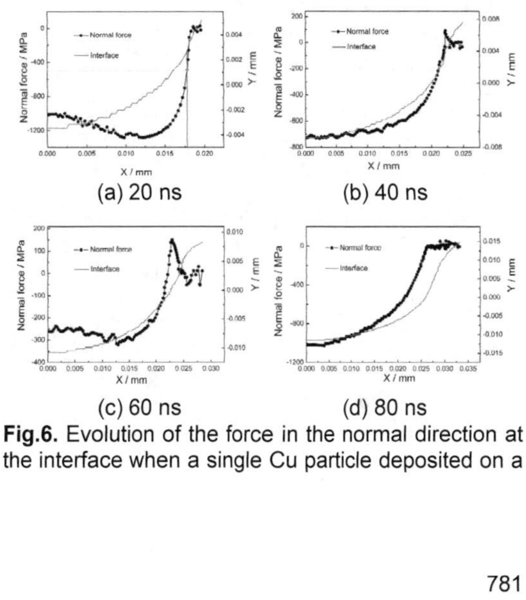

Fig.6. Evolution of the force in the normal direction at the interface when a single Cu particle deposited on a

Cu substrate at different impact times at an initial impact velocity of 500 m/s.

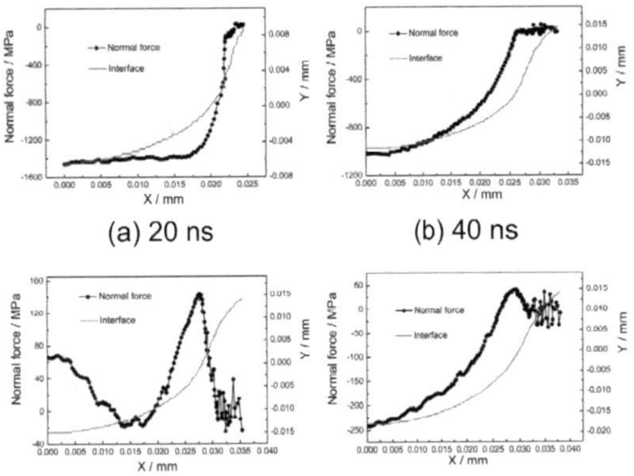

(a) 20 ns (b) 40 ns

(c) 60 ns (d) 80 ns

Fig.7. Evolution of the force in the normal direction at the interface when a single Cu particle deposited on a Cu substrate at different impacts time at an initial impact velocity of 800 m/s.

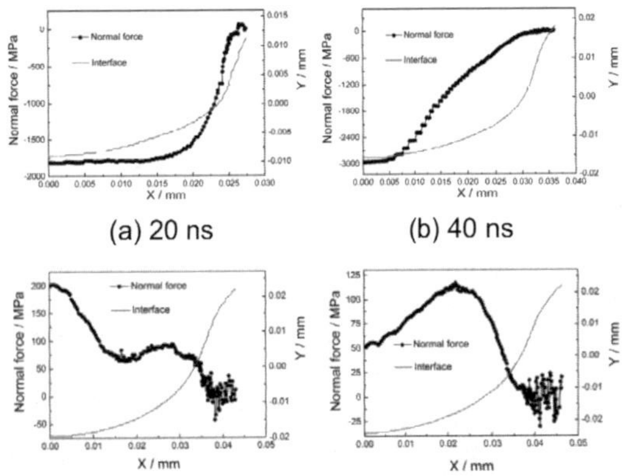

(a) 20 ns (b) 40 ns

(c) 60 ns (d) 80 ns

Fig.8. Evolution of the force in the normal direction at the interface when a single Cu particle deposited on a Cu substrate at different impact times at an initial impact velocity of 1100 m/s.

4 Conclusions

In this study, the deformation behavior of the particle initial surface is investigated during the impact at the particle initial velocity of 500 m/s, 800 m/s and 1100 m/s, respectively in order to reveal the effect of interface evolution on the bonding during the high-velocity particle impact. The conclusions can be drawn as followings. During the impact, the free-oxide interface appears due to the occurrence of the adiabatic shear instability and a higher initial velocity leads to a more free-oxide interface. According to the stress evolution in the normal direction during the impact, it is found that the bonding occurs once the free-oxide interface appears and the bonding will remain to the end of the impact. Therefore, the free-oxide interface is defined as the effective bonding area and the more the effective bonding area is, the higher the strength of the coating is. For pure copper coating the effective interface bonding area ratio is estimated to be 0.52, 0.7 and 0.76 for particle velocities of 500 m/s, 800 m/s and 1100 m/s, respectively. These results can be used to predict tthe strength of the coating supposing that the tensile strength is primarily proportional to interface bonding.

Acknowledgements

This work is supported by National Nature Science Foundation of China (No. 50115814) and Shaanxi Co-Innovation Projects (2015KTTSGY03-03).

Literature

[1] P.S. Chen, J.H. Sanders, Y.K. Liaw, F. Zimmermann, Ductility degradation of vacuum-plasma-sprayed NARloy-Z at elevated temperatures, Mater. Sci. Eng. A, 199 (1995), pp. 145-152.

[2] Koivuluoto H, Lagerbom J, Vuoristo P, Microstructural Studies of Cold Sprayed Copper, Nickel, and Nickel-30% Copper Coatings, J. Therm. Spray Technol. 16 (2007), Issue 4, pp. 488-497.

[3] D.H. Chen, N. Jin, W.W. Chen, L. Wang, S. Zhao, D. Luo, Corrosion resistance of Ni/Cu/Ni-P triple-layered coating on Mg-Li alloy, Surf. Coat. Technol. 254 (2014), pp. 440-446.

[4] X.T. Luo, Y.J. Li, C.X. Li, G.J. Yang, C.J. Li, Effect of spray conditions on deposition behavior and microstructure of cold sprayed Ni coatings sprayed with a porous electrolytic Ni powder, Surf. Coat. Technol. 289 (2016), pp. 85-93.

[5] S. Yin, P. He, H. Liao, X. Wang, Deposition Features of Ti Coating Using Irregular Powders in Cold Spray, J. Therm. Spray Technol. 23 (2014), Issue 6, pp. 984-990.

[6] D.K. Christoulis, S. Guetta, V. Guipont, M. Jeandin, The Influence of the Substrate on the Deposition of Cold-Sprayed Titanium: An Experimental and Numerical Study, J. Therm. Spray Technol. 20 (2010), Issue 3, pp. 523-533.

[7] T. Stoltenhoff, H. Kreye, H.J. Richter, An Analysis of the Cold Spray Process and Its Coatings. J. Therm. Spray Technol. 11 (2002), Issue 4, pp. 542-550.

[8] T. Suhonen, T. Varis, S. Dosta, M. Torrell, J.M. Guilemany, Residual stress development in cold sprayed Al, Cu and Ti coatings, Acta Mater. 61 (2013), Issue 17, pp. 6329-6337.

[9] H. Assadi, F. Gärtner, T. Stoltenhoff, H. Kreye. Bonding mechanism in cold gas spraying, Acta Mater. 51 (2003), Issue 15, pp. 4379-4394.

[10] C.J. Li, W.Y. Li, H. Liao, Examination of the Critical Velocity for Deposition of Particles in Cold Spraying, J. Therm. Spray Technol. 15 (2006), Issue 2, pp. 212-222.

[11] G. Bae, J.i. Jang, C. Lee, Correlation of particle impact conditions with bonding, nanocrystal formation and mechanical properties in kinetic sprayed nickel, Acta Mater. 60 (2012), Issue 8, pp.3524-3535.

[12] H. Assadi, H. Kreye, F. Gärtner, T. Klassen, Cold spraying-A materials perspective. Acta Mater. 116 (2016), pp. 382-407.
[13] S. Yin, X. Wang, X. Suo, H. Liao, Z. Guo, W. Li, C. Coddet, Deposition behavior of thermally softened copper particles in cold spraying, Acta Mater. 61 (2013), Issue 14, pp. 5105-5118.
[14] G. Bae, Y. Xiong, S. Kumar, K. Kang, C. Lee, General aspects of interface bonding in kinetic sprayed coatings, Acta Mater. 56 (2008), Issue 17, pp. 4858-4868.
[15] M. Grujicic, J.R. Saylor, D.E. Beasley, W.S. DeRosset, D. Helfritch, Computational analysis of the interfacial bonding between feed-powder particles and the substrate in the cold-gas dynamic-spray process, J. Appl. Phys. 219 (2003), Issue 3–4, pp. 211-227.
[16] V.K. Champagne, D. Helfritch, P. Leyman, S. Grendahl, B. Klotz, Interface Material Mixing Formed by the Deposition of Copper on Aluminum by Means of the Cold Spray Process, J. Therm. Spray Technol. 14 (2005), Issue 3, pp. 330-334.
[17] A. Wank, W. Bernhard, H. Podlesak, T. Grund, High-Resolution Microstructural Investigations of Interfaces Between Light Metal Alloy Substrates and Cold Gas-Sprayed Coatings, J J. Therm. Spray Technol. 15 (2006), Issue 2, pp. 280-283.
[18] F.A. Calov, A. Urena, J.M. Gomez de salazar, F. Molleda, Enhanced kinetics of diffusion coating of aluminium on copper by boundary diffusion, J. Mater. Sci. Lett. 12 (1993), pp. 920-922.
[19] I. Manna, J. Dutta Majumdar, Enhanced kinetics of diffusion coating of aluminium on copper by boundary diffusion, J. Mater. Sci. Lett. 12 (1993), pp. 920-922.
[20] H. Lee, H. Shin, K. Ko, Effects of Gas Pressure of Cold Spray on the Formation of Al-Based Intermetallic Compound, J. Therm. Spray Technol. 19 (2009), Issue 1-2, pp. 102-109.
[21] A.V. Bolestaa, V.M. Fomina, M.R. Sharafutdinovb, B.P. Tolochko. Investigation of interface boundary occurring during cold gas-dynamic spraying of metallic particles,Nucl. Instrum. Methods Phys. Res., Sect. A, 470 (2001), pp.249-252.
[22] W.Y. Li, H. Liao, C.J. Li, H.S. Bang, C. Coddet, Numerical simulation of deformation behavior of Al particles impacting on Al substrate and effect of surface oxide films on interfacial bonding in cold spraying, J. Appl. Phys. 253 (2007), Issue 11, pp. 5084-5091.
[23] W. Li, K. Yang, D. Zhang, X. Zhou, X. Guo, Interface behavior of particles upon impacting during cold spraying of Cu/Ni/Al mixture, Mater. Design, 95 (2016), pp. 237-246.
[24] M. Yu, W.Y. Li, F.F. Wang, X.K. Suo, H.L. Liao, Effect of particle and substrate preheating on particle deformation behavior in cold spraying, Surf. Coat. Technol. 220 (2013), 174-178.
[25] S. Yin, X.f. Wang, B.P. Xu, W.Y. Li, Examination on the Calculation Method for Modeling the Multi-Particle Impact Process in Cold Spraying, J. Therm. Spray Technol. 19 (2010), Issue 5, pp. 1032-1041.
[26] R. Huang, W. Ma, H. Fukanuma, Development of ultra-strong adhesive strength coatings using cold spray, Surf. Coat. Technol. 258 (2014), pp. 832-841.
[27] R. Huang, M. Sone, W. Ma, H. Fukanuma, The effects of heat treatment on the mechanical properties of cold-sprayed coatings, Surf. Coat. Technol. 261 (2015), pp. 278-288.

S/TEM and TEM investigations of thermal barrier coatings based on $Gd_2Zr_2O_7$ zirconate

G. Moskal, R. Swadźba, R. Białecki, T. Kruczek, W. Adamczyk, Gliwice / PL

Detailed investigations of thermally grown oxides (*TGO*) zone and areas adjoining to it were presented in this article. The thermal barrier coating (*TBC*) based on gadolinium zirconate (*$Gd_2Zr_2O_7$, GZ*) and deposited by air plasma spraying (*APS*) on AMS 5599 superalloy with NiCrAlY type of bond-coat was the analysed material. Analyses were made on TBC system after spraying process and after time to 500 hours of oxidation at temperature 1100°C in atmosphere of laboratory air. Ultrahigh resolution scanning-transmission (*S/TEM*) electron microscope TITAN 80-300 was used to conduct the tests. Quanta 200i Dual Beam apparatus by FEI for precise and localized focused Ga+ ion beam (*FIB*) sample preparation was used as well.The basic aim of presented investigations was related with detailed characterization of interface zone between ceramic top-coat (*TC*) of $Gd_2Zr_2O_7$ type and NiCrAlY bond-coat after spraying process and after 500 hours of deposition in oxidized atmosphere where the TGO zone was formed with the thickness ca. 5-10 µm. The special emphasis was placed on research of this zone and its morphology due to very strong influence of formed thermally grown oxides on overall life-time and destruction processes which took place on the interface between TGO and ceramic top-coat.Obtained data revealed that the most important factors influenced on destruction processes are related with the thickness of TGO zone (consisted mainly Al_2O_3, $NiAl_2O_4$ and evetually NiO) rather than formation of undesirable phases such as $GdAlO_3$ complex oxide.

1 Introduction

The rare-earth (RE) zirconates described by overall formula of $RE_2Zr_2O_7$ have a significantly lower thermal conductivity, in comparision to usually used as an insulation ceramic in thermal barrier coatings (TBC), yittria-stabilized zirconia (YSZ), and exhibit an ordered structure of pyrochlore type eventually a defected fluorite-type structure (disordered). This effect is strongly depended on the ionic size difference between RE and Zr cations [1,2]. Due to their distinctly better insulation properties, the $RE_2Zr_2O_7$ based thermal barrier coatings are very intensively analysed. These type of coatings are generally deposited by plasma spraying or physical vapour deposition processes dependely on final applications. Plasma spraying technology gives the possibilities to synthesis of $RE_2Zr_2O_7$ top-coating of TBC systems using pyrochlore feedstock powders involves an extremely high stream temperature (ca. 10^4 K) and a very fast cooling rate (ca. 10^6 K/s) [3]. TBC systems can be used mainly to provide thermal insulation for superalloys` made hot-section components of aircraft gas turbin against hot gas streams as well as to enhance operating temperature of and to improve its performance [4,5]. The operating conditions of TBC systems are extremally difficult due to high temperature of exposure, thermal shocks and aggresive environments. From overall life-time point of view the most important phenomena took places in area of interfaces between ceramic top-coat and bond-coat where thermaly growth oxides (TGO) zone is formed. During cyclic thermal exposure, a large compressive stress within the TGO zone develops as a qonsequence of the thermal expansion mismatch between the metallic NiCrAlY bond-coat and the TGO zone. This stress is strengthened by the stress arising from TGO zone growth, and this two effect is believed to be intimately involved with the failure of TBC`s. Usually the failure phenomena of TBC systems has been observed near the vicinity of TGO scale, at the TGO/bond-coat interface, within TGO and at the YSZ/TGO interface [6-10].

Detailed investiagations of these areas are usually made by "traditional" metohodology with application of scanning electron microscopy, with combination of chemical composition analysis in microareas and extremally rare with phases constituent analysis by EBSD (electron backscattered diffraction) method [11]. But the most valuable information about phenomena related with TGO zone can be obtained by transmission electron microscopy (TEM). This method has been rarely used for the detailed microstructural examination of constituents and interfaces involved in TBC`s mainly due to the difficulty in specimen preparation [12-16].

2 Research methodology

The investigation has been carried out on samples based on substrate from AMS5599 Ni-based superalloy with NiCrAlY bond-coat (thickness 125 µm) and $Gd_2Zr_2O_7$ ceramic top-coats with thickness 250 µm. Both coatings have been deposited by atmospheric plamsa spraying (APS) method. This method allows use a wide variety of powder material and for ensure high standards is fully automated. Research were carried out in as sprayed conditions and oxidation test at 1100°C. For a deeper investigation of the coating microstructure, including the identification of the phase composition, thin foils were prepared by a focused ion beam (FIB) method with Quanta 200i Dual Beam by FEI for TEM analysis. The thin foils were cut in the middle of the coatings, perpendicularly to the samples' surface. In the case of each coating, samples were cut close to the splat boundary. The samples were examined with the Ultrahigh Resolution Scanning-Transmission Electron Microscope (S/TEM) TITAN 80-300 by FEI Company, and images in bright field (BF), dark field (DF), annular dark field (ADF), high angle annular dark field (HAADF) as well as selected area diffraction patterns

(SADP) were obtained. The samples were also investigated by: X-ray phase analysis by XRD method (X-ray diffraction, Jeol JDX-7S), moreover, microstructure observation, by scanning electron microscope (*SEM*), energy dispersion spectroscopy (EDS) and electron backscatter diffraction analysis (EBSD) ware made. This investigation were made on the Hitachi 3400N scanning microscopy. Makro observation of the annealing effects was carried out on stereoscopic microscope (Nicon Eclipse MA200), whereas measurements of the observed changes in the thickness of TGO layer were made on light optical microscopy (Nikon SMZ 745T).

3 Results

The object of investigations was TBC system based on gadolinium zirconate deposited by APS method. The overall view of this TBC system with SEM/EBSD picture of microstructure and results of phases analysis was showed at fig. 1. These data showed that deposited coting is a typical for morphological point of view for plasma sprayed coatings.

The thickness of ceramic top-layer is about 250 µm and total porosity is on the level of 4,7 % measured by automati image analysis. Both XRD analysis as well as EBSD revealed that main phase structural element is gadolinium zirconate with overall formula $Gd_2Zr_2O_7$, but with fluorite type of structure rather than pyrochlore. This conclusion is a consequence of the lack of superstructure diffraction peaks expect the weakest for *d=3,19*. EBSD analysis revealed additionally that crystallite size is realtively small and its value is 0,75 µm^2 (av. area) and 0,81 µm (av.diameter). SEM analysis of bond-coat/top-coat interfeceas at as-sprayed condition was showed at fig. 2, where lack of presents of TGO zone was detected on the top and bottom of profile roughness. The cracks vertical to bond-coat surfaces small crack was only detected, especially in area of bottom profile roughness. This type of investigations don`t prejudge the presents of primary TGO zone.

The final experiment included analysis of TBC system after oxidation test at temperature 1100°C and time 500 h. The main area of interest was phenomena in TGO zone and adjacent areas. An example of TGO zone morphology characterized by SEM/EBSD methods was showed at fig. 3 where area connected to ceramic top-layer was described.

This analysis revealed presents of different type of oxides in TGO zone. $(Al,Cr)_2O_3$ was dominated oxide in the inner part of this zone, but in the area connected with ceramic layer the spinel and small amount of perovskite type of $GdAlO_3$ oxides were the main phases.

This area was the subject of detailed analysis by S/TEM method. Fig. 4 showed area between upper part of TGO zone with visible porosity in the spinel sublayer and cracks in area of ceramic layer penetrated directly from pores. Analysis of phases composition revealent presents only oxides of Al_2O_3 and $NiAl_2O_4$ type.This type of analysis don`t get occasion to confirmation of perovskite phase formation during oxidation process. Results from EBSD investigations are strong evidence of its presents in TGO zone. The thickness of this sub zone is very small, so its influence on life-time of analyzed TBC system is negligible.

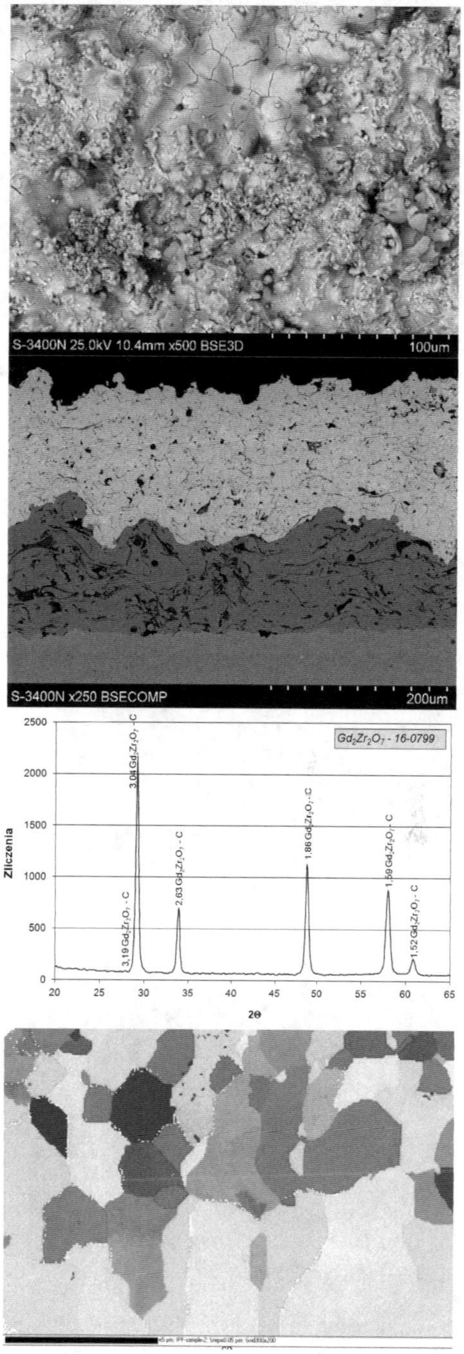

Fig.1 Overall data about investigated $Gd_2Zr_2O_7$ based TBC system.

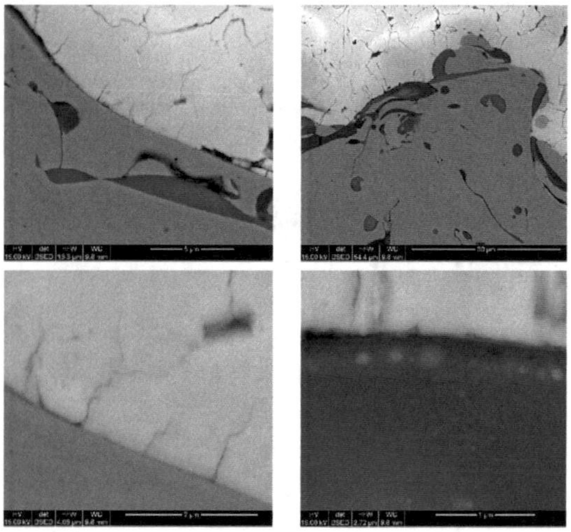

Fig.2 Detailed analysis of microstructure of bond-coat/top-coat interfaces at bottom (left) and peak (right) of profile roughness.

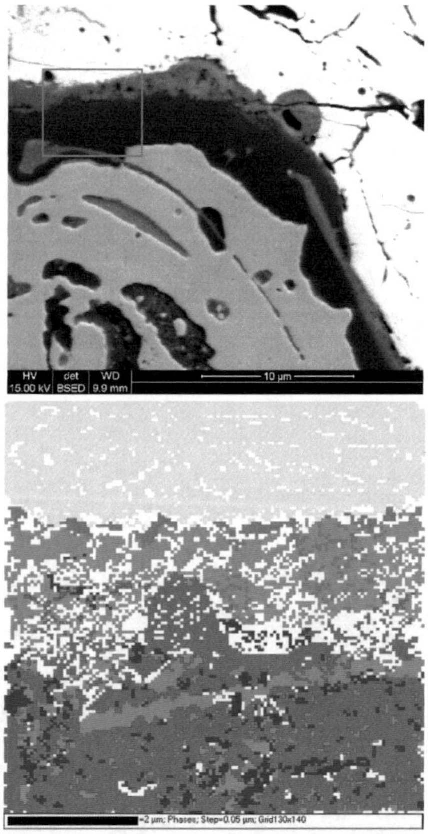

Fig.3 Morphology of TGO zone after 500 h of oxidation with EBSD analysis of phases constituent: -$Gd_2Zr_2O_7$, -$GdAlO_3$, -spinel $(Ni,Cr)Al_2O_4$, -$(Al,Cr)_2O_3$, -YAG and white regions-not indexed.

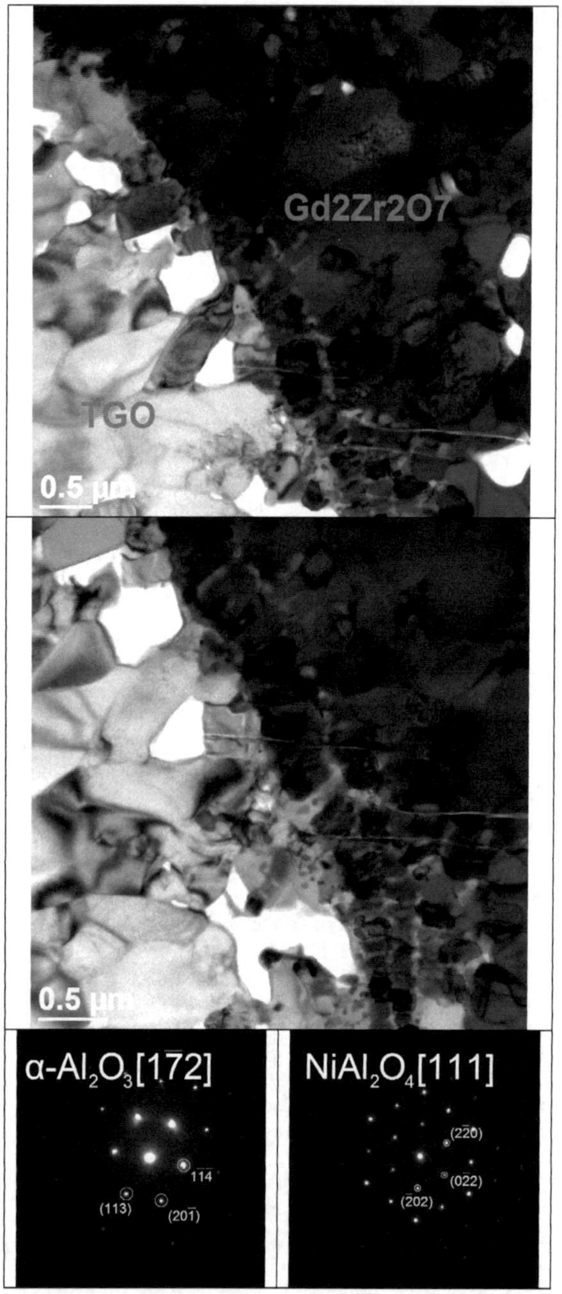

Fig.4 Morphology of TGO zone after 500 h of oxidation with TEM analysis.

S/TEM analysis of TGO zone revealed presents of columnar and smaller fraction of lower and equiaxed grains of Al_2O_3 phase.In area of spinel oxide existence the strong effect of tearing out of ceramic sublayer was detected. This the effect of voids formation and crumbling up of zirconate sublayer (fig. 5).

Much interesting possibilities gives the S/TEM analysis in the subject of chemical composition detection in small areas. The results of this analysis was showed at fig. 6, where formation of Al2O3 and NiAl2O4 oxides was clearly confirmed, as well as the effect of crumbling up of zirconate sublayer I rebuilding this area in oxides zone. From the other site very interesting observation can be made in area of zirconate layer connected with TGO zone where clearly effect of aluminium and nickel diffusion can be observed.

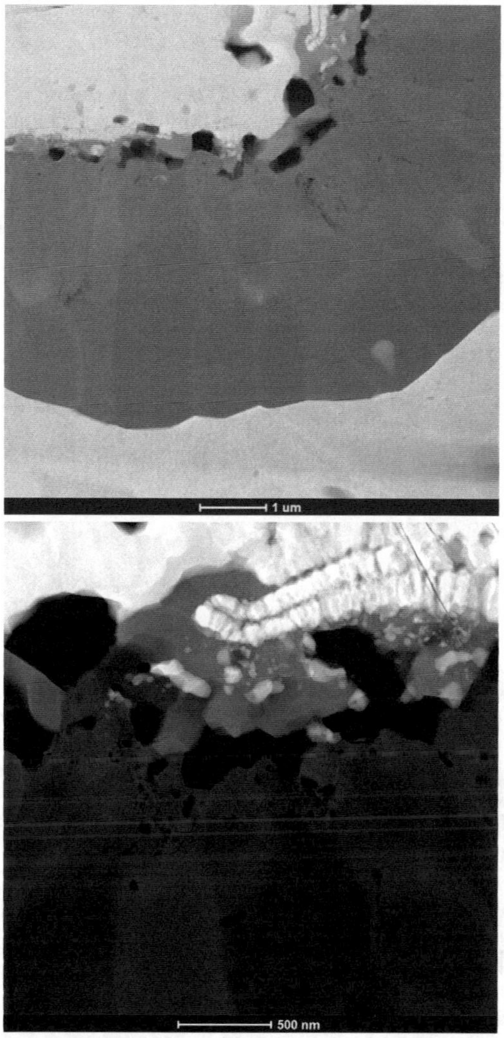

Fig.5 Morphology of TGO zone after 500 h of oxidation with S/TEM analysis.

Detalied observation of this area revealed that there are the places where higher concentration of Al and Gd is observed with lower concentration of Zr. It suggest that perovskite $GdAlO_3$ phase was formed. (see places marked by straight arrow). Thoses places not covered with places reach in Al and Ni what suggests presents of $NiAl_2O_4$ spinel oxide (see places marked by broken arrow). Analysis of Y location at fig. 5 don`t get any clearly answer what was the its role in the proces of this sublayer formation of TGO zone. The lack of Cr concentration map suggest that similar effect or rather no-effect can be suggested in the case of this element.

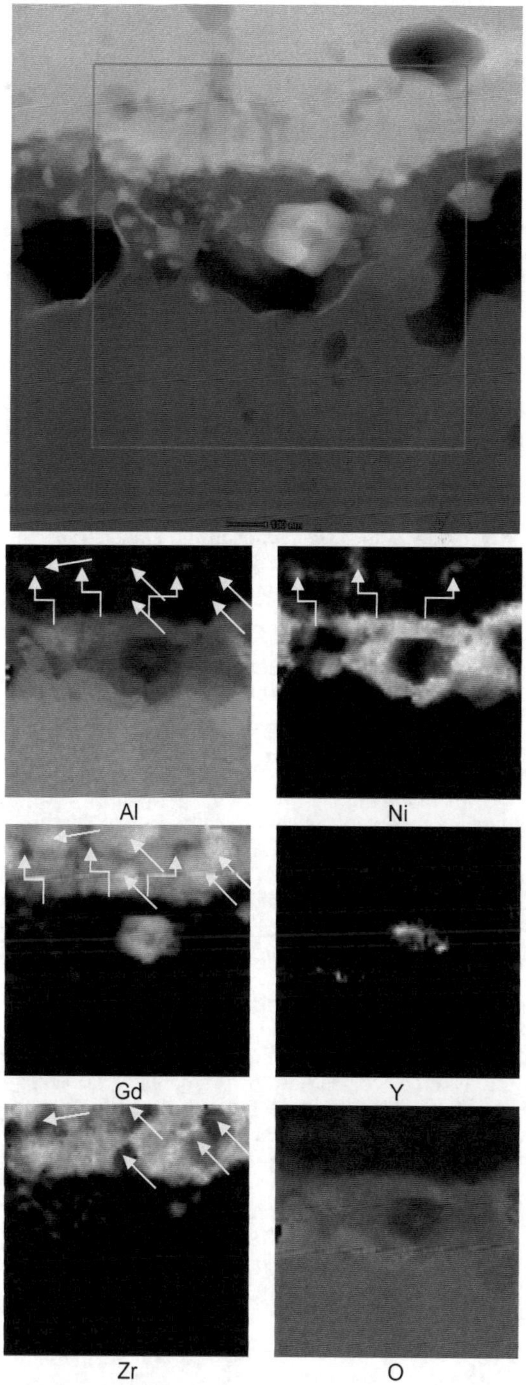

Fig.6 Morphology of TGO zone after 500 h of oxidation with element distribution in area of upper part of oxide layer connected with ceramic zirconate top-coat.

Obtained data are with good correlation with EBSD analysis exept the results where the more complexed oxide were detected such as $(Al,Cr)_2O_3$ and $(Ni,Cr)Al_2O_4$.

4 Summary and conclusions

Presented results are good example of different posibilites of electron microscopy technics of TGO zone obesrvations in TBC systems. The SEM/EBSD method of investigations get the data about basic morphological spects of TGO zone and adjacent areas with relativelly simple method of sample preparation. From the second site S/TEM and TEM methods with theirs much problematic samples preparation opening new investiagations possibilites realted with chemical and phases analysis in small areas of analysis.

Presnted investigations showed that formation of complexed oxides $REAlO_3$ with perovskite type of structure isn`t the reason of rapid destruction of rare earth zirconates based TBC systems. Observed low durability is related mianly with higher thickness of TGO where the most dominant phases element is Al_2O_3 oxide with lower fraction of spinels. Those elements are the reason of so called “white failure” of zircoantes based thermal barrier coatings.

Acknowledgments: The research has been supported by National Science Centre within Opus scheme, under contract UMO-2013/11/B/ST8/00268.

5 Literature

[1] Vassen R., Cao X., Tietz F., Basu D. and D. Stöver: Zirconates as new materials for thermal barrier coatings. Journal of American Ceramic Society 83 (2000), Issue 8, pp. 2023/28.

[2] Wu J., Wie X., Padture N. P., Klemens P. G., Gell M., García E., Miranzo P., and M. I. Osendi: Low-Thermal-Conductivity Rare-Earth Zirconates for Potential Thermal-Barrier-Coating Applications. Journal of American Ceramic Society 85 (2002), Issue 12, pp. 3031/35.

[3] Pan W., Wan C.L., Xu Q., Wang J.D. and Z.X. Qu: Thermal diffusivity of samarium–gadolinium zirconate solid solutions. Thermochimca Acta 455 (2007), Issue 1-2, pp. 16/20.

[4] Evans A.G., Clarke D.R. and Levi C.G.: The influence of oxides on the performance of advanced gas turbines Journal of European Ceramic Society. 28 (2008), Issue 28, pp. 1405/19.

[5] Bansal N. P. and Zhu D.: Effects of doping on thermal conductivity of pyrochlore oxides for advanced thermal barrier coatings. Materials Science and Engineering A 459 (2007), Isseue 1-2, pp. 192/195.

[6] R.A. Miller: Oxidation-Based Model for Thermal Barrier Coating Life. Journal of American Ceramic Society 67 (1984), Issue 8, pp. 517/21.

[7] T.E. Strangman: Thermal barrier coatings for turbine airfo. Thin Solid Films 127 (1985), Issue 3-4, pp. 93/106.

[8] Stiger M.J., Yanar N.M., Toppings M.G., Pettit F.S. and G.H. Meier: Thermal barrier coatings for the 21st century. Zeitschrift für Metallkunde 90 (1999), Issue 90, Issue 12pp. 1069/78.

[9] Evans A.G., Mumms D.R., Hutchinson J.W., G.H. Meier, Pettit F.S. and G.H. Meier: Mechanisms controlling the durability of thermal barrier coatings. Progress in Materials Science 46 (2001), Issue 5, pp. 505/56.

[10] Wright P.K. and A.G. Evans: Mechanisms governing the performance of thermal barrier coatings. Current Opinon in Solid State Mater. Sci. 4 (1999), Issue 3, pp. 255/65.

[11] Chmiela B., Sozańska M. and G. Moskal: Application of EBSD method for the investigation of microstructure and crystallographic orientation in $RE_2Zr_2O_7$ TBC, 12th European Workshop on Modern Developments in Microbeam, Analysis, IOP Conf. Series: Materials Science and Engineering 32 (2012) 012006.

[12] Chen H.C., Heberlein J. and E. Pfender: TEM characterization of plasma-sprayed thermal barrier coatings and ceramic-metal interfaces after hot isostatic pressing. Thin Solid Films 301 (1997), Issue 1-2, pp. 105/14.

[13] Unal O., Mitchell T.E. and A.H. Heuer: Microstructures of Y_2O_3-stabilized ZrO_2 electron-beam physical vapor deposition coatings on Ni-base superalloys. Journal of American Ceramic Society 77 (1994), Issue 4, pp. 984/92.

[14] Murphy K.S., More K.L. and M.J. Lance: As-Deposited mixed zone in thermally grown oxide beneath a thermal barrier coating. Surfaces and Coatings Technology 146/147 (2001), pp. 152/61.

[15] Fritscher K., Schumuecker M., Leyens C. and U. Schul: TEM investigation on the adhesion of YPSZ EB-PVD TBC`s. Materials Science Forum 251/254 (1997), pp. 965/72.

[16] Schulz U., Menzebach M., Leyens C., and Y.Q. Yang: Influence of substrate material on oxidation behavior and cyclic lifetime of EB-PVD TBC systems. Surfaces and Coatings Technology 146/147 (2001), pp. 117/23.

Effect of powder granularity in the pack on microstructure and anti-oxidation properties of SiC coatings for C/SiC composites

Fang Jia[1,2], Xiaojuan Ji[1,2], Haoran Peng[1,2], Jie Shen[1,2],Deming Zhang[1,2],Yueguang Yu[1,2]
(1.Beijing General Research Institute of Mining & Metallurgy, Beijing 100160, China;
2. Beijing Engineering Technology Research Center of Surface Strengthening and Repairing of Industry parts, Beijing 102206, China;)

Abstract：SiC coatings were prepared with pack powder in different particle sizes in a vacuum atmosphere by pack cementation technique to protect the C/SiC composites substrate from oxidation. The phase and microstructure of the coatings were investigated by XRD, SEM analyses. The relationship between powder granularity in the pack and microstructure of SiC coatings was studied. Cyclic oxidation test at 1573K in air atmosphere was performed and the effect of powder particle size in the pack on high-temperature oxidation resistance of SiC coatings was discussed in detail. It is observed that with powder granularity in the pack increasing thickness and density of SiC coatings increases, corresponding oxidation resistance of the coating is improved. Possible mechanisms related to oxidation were preliminarily discussed.

1 Introduction

C/SiC composites possess excellent high-temperature properties, such as high thermal stability and high retention of mechanical [1], so they are considered as the promising candidate materials for applications in aeronautics and astronautics fields [2]. However, C/SiC composites exhibit poor oxidation resistance in oxidative environment at high temperature. Carbon is known to react with oxygen above 400℃, property changes of a C/SiC composite during oxidation, specially the mechanical properties, are very sensitive to the oxidation of the carbon fibers and the pyrocarbon interphase [3]. Anti-oxidation coating is a logical choice to protect these composites against oxidation at high temperatures.

SiC ceramic coating has been applied widely for the anti-oxidation of C/SiC composites at high temperature owing to its excellent anti-oxidation property and good physical and chemical compatibility with C/SiC composites. The oxidation of silicon-based ceramics generates uniform and dense SiO_2 glassy film with low oxygen permeability at high temperature as well as a certain ability to seal the cracks, which provides good oxidation protection for C/SiC composites [4–6].

The preparation techniques of SiC coatings are various, such as pack cementation, CVD, slurry sintering, and laser-assisted deposition [7–10]. Chemical vapor deposition is widely used to prepare high-quality single-crystal SiC coatings, but this process requires a high working temperature. Cracks also easily form during the cooling stage, which limits chemical vapor deposition application [11]. Although the slurry sintering can coat all layers of a surface in a single process, the coatings are loose and non-uniform. The surface of SiC coatings produced by laser-assisted deposition is studded with small particles. Pack cementation is more suitable for preparing SiC coatings on C/SiC composites materials because the method is low-cost, easy to operate, and independent of the sample shape [12]. All surfaces can be covered with the coating in a single process.

Some investigations have been conducted on the oxidation behavior of C/SiC with SiC coating in air, but the effect of powder particle size in the pack on microstructure and anti-oxidation properties of SiC coatings on C/SiC composites has been seldom reported. In the present paper, the microstructure and oxidation resistance of SiC coatings prepared with pack powder of different particle size were investigated and compared. Possible mechanisms related to oxidation were preliminarily discussed. on C/SiC composites

2 Experimental procedures

C/SiC composites with the rectangle dimensions of 10 mm × 10 mm × 2 mm served as substrate materials. Prior to pack cementation, the substrate specimens were polished using a 600-grit SiC abrasive paper and then ultrasonically cleaned in alcohol and acetone and dried for 90min. The Si powder, C power, SiC powder, Al_2O_3 powder with high purity were used as starting materials to prepare the pack cementation powder.

The coatings were prepared by pack cementation and pack compositions were as follows: 40~60 wt.% Si, 20~30 wt.% SiC, 5~20 wt.% C, 7~10 wt.% Al_2O_3 during the preparation process. Two sets of experiments were conducted, and for each of experiment all of these powders have the granularity of below 325 mesh and 1250 mesh, respectively. The C/SiC composites substrates were embedded into the as-prepared mixture powders in BN crucibles with BN lid which were placed in a electric furnace. The formation of SiC coating was conducted by heating the impregnated specimens in a vacuum atmosphere at 1973K to 2173K for 2h. Following pack cementation, all specimens were removed from the crucible, then ultrasonically cleaned in the acetone for 10 min in order to remove the loosely embedded pack substance. Finally, the specimens were dried at 100℃ for 2h.

The high temperature cyclic oxidation experiment of the samples was conducted in air furnace (KSL 1700X) at 1300℃. Each cycle consisted of 50 min heating in dry and static air at 1300℃ and subsequent 10 min cooling at ambient temperature. The weight change of coated samples was recorded using an electronic balance（FA1104N） with a precision of 0.1mg. The % weigh change was calculated using the following equations (1):%weigh change = $[(m_1 - m_0)/m_0] \times 100\%$, where m_0 is the original weight of coated C/SiC composites, m_1 is the weight of the coated C/SiC composites after oxidation at high temperature. The weight change of coated samples was determined on the basis of an average value of three samples' weight changes.

The surface roughness of the coating measurement was performed on three -dimensional optical profiler (Contour GT X) equipped with 0.1 nm vertical resolution. Scanning electron microscopy (Zeiss Supra55) equipped with energy dispersive X-ray spectrum (EDXS) was used to investigate the surface and cross-sectional morphologies of the specimens. The phases of coatings were identified by X-ray diffraction (XRD, Rigaku D/max 2200PC, Japan) with Cu Kα radiation.

3 Results and discussion

Fig.1 shows X-ray patterns of as-deposited coating surfaces. It reveals that two kinds of coatings obtained with different grain powder in pack materials are both composed α-SiC and β-SiC and Si. The existence of Si could not only relieve the mismatch of the thermal expansion coefficient between SiC coating and C/SiC substrate, but also improve the oxidation resistance of the coating due to the self-sealing performance of its oxidation product SiO_2[13]. However, the surface morphologies of the as-deposited coatings shown in Fig. 2a–c have significant differences. The SiC coating prepared with coarse-grained powder in pack materials (Fig.2a) reveals relatively denser than that obtained with fine-grained powder in pack materials (Fig.2b). To notice that, no cracks appeared in the coatings shown in Fig.2.

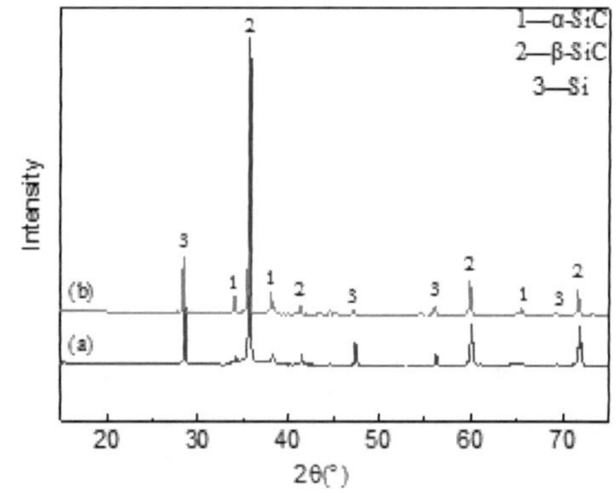

Fig. 1. XRD patterns of the as-deposited coatings: (a) fine granularity powder, (b) coarse granularity powder in pack materials.

The surface of SiC coating prepared with coarse granularity powder is coarser than that prepared with fine-grained powder due to the large size of SiC particles in the coating as shown in Fig.3.

Fig. 2. Surface morphologies of the as-deposited coatings: (a) fine granularity powder, (b) coarse granularity powder in pack materials, (c) High multiple image of (b).

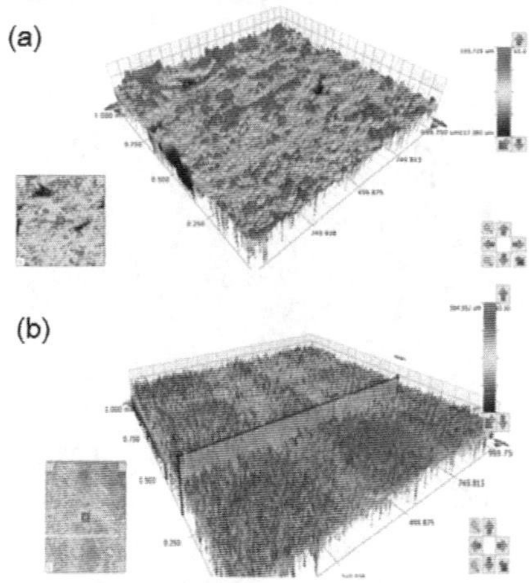

Fig. 3. Surface roughness of the as-deposited coatings: (a) fine granularity powder, (b) coarse granularity powder in pack materials.

Fig.4 exhibits the cross-section of the as-received SiC coatings. The as-received coating with a thickness of around 60 μm obtained with fine-grained powder in pack materials. From the backscattered electron image, there are some micro-cracks in the coating. The coating prepared with coarse granularity powder in pack materials is about 130 μm in thickness without any penetrable cracks or holes in the coating. Additionally, the coating reveals a dense structure with two kinds of crystalline particles characterized as light grey and dark grey in the coating. The light grey and dark grey phases can be distinguished as Si and SiC, respectively. This result is consistent with phase analysis result. The light grey phase (Si) is filled in the holes among the SiC grains in the coating.

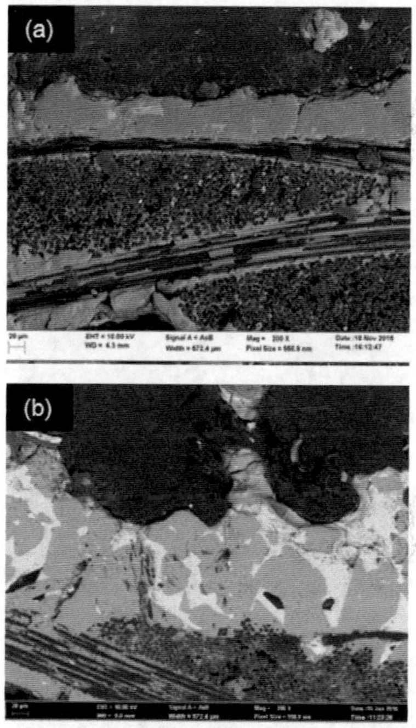

Fig.4. Cross-section morphologies of the as-deposited coatings: (a) fine granularity powder, (b) coarse granularity powder in pack materials.

To investigate the oxidation resistance of SiC coatings, the oxidation experiments were carried out at 1573K in air. Fig.5 shows the oxidation curves of the two kinds of SiC coatings. It can be seen that the weight loss of SiC coating produced with fine-grained powder reaches 15.43 % after oxidation in air for 10h. However, for the SiC coated sample prepared with coarse-grained powder, the weight loss is only 3.08 %

after oxidation in air for 10 h at 1573K. Consequently, the high temperature oxidation resistance of SiC coating prepared with coarse-grained powder is much better than that of SiC coating prepared with fine-grained powder in pack materials.

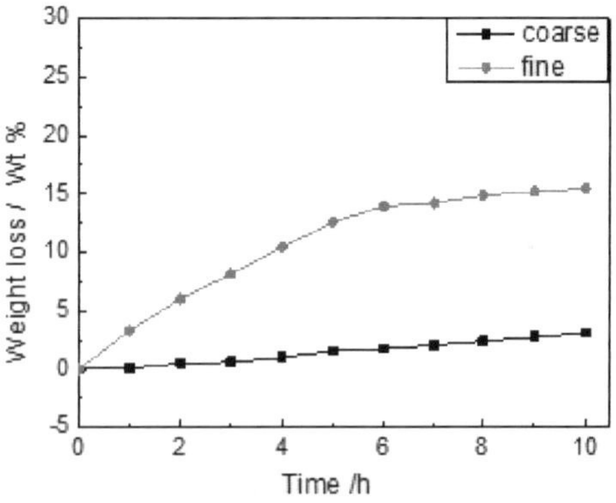

Fig. 5. Oxidation curves at 1573K of the coated samples prepared with different granularity powder in pack materials.

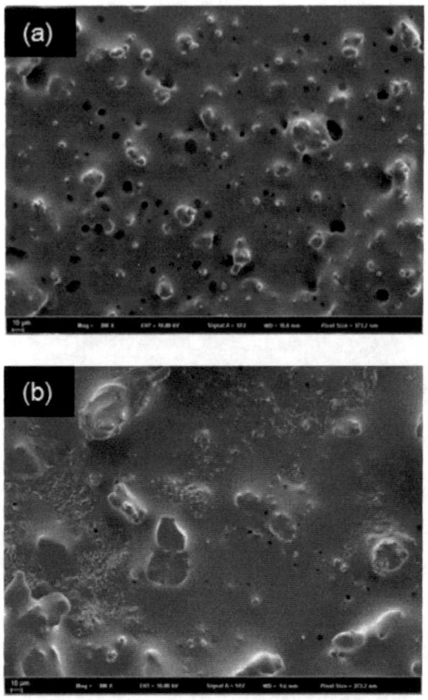

Fig. 6. Surface morphologies of coatings after 10 h oxidation at 1573K: (a) fine granularity powder, (b) coarse granularity powder in pack materials.

The surface micrographs of SiC coatings after oxidation at 1573K in air are show in Fig.6. It demonstrates that SiO_2 glass layers were formed on both SiC coatings. Dense and continuous SiO_2 oxide scale was seen on the surface of SiC coating which is prepared with coarse-grained powder shown in Fig.6b. There is no apparent holes and destruction on the scale surface, from which it can be inferred that the coating has excellent oxidation resistance. However, a large number of gas holes caused by the escape of gases can be clearly seen on the surface of SiC coating obtained with fine-grained powder. These results suggested that some defects might be generated during the oxidation process, through which the oxygen reacted with C/SiC and producing CO and CO_2 gases. The small hole of the coating could be self-cured by the molten SiO_2, but the molten SiO_2 could not fill bigger holes (Fig.6a). The formation of this kind of this hole offers entrance channels for air and results in the linear increase of weight loss rate. Therefore, the impossible cause of the weight loss of the coating prepared with fine granularity powder in pack materials at 1573K is due to the formation of holes in the coating.

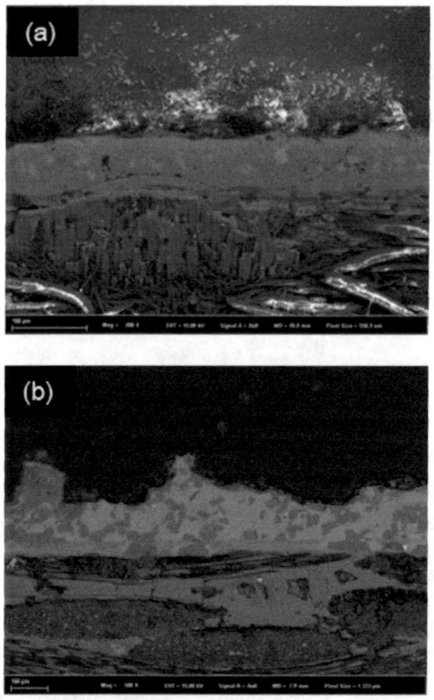

Fig. 7. Cross-section morphologies of coatings after 10 h oxidation at 1573K: (a) fine granularity powder, (b) coarse granularity powder in pack materials.

The Fig.7 shows that less carbon fiber was oxidized under SiC coating produced with coarse-grained powder, but the carbon fiber was obviously oxidized under the SiC coating with deflected crack prepared with fine-grained powder in pack materials. Additionally, micro-cracks extend along the holes in the coating forming longer continuous cracks shown in the cross-section micrographs of SiC coating produced with fine-grained powder (Fig.7a). With the extension of the oxidation time, the spallation of SiC coating might occur due to formation of the transverse long crack which would result in the loss of protection for the C/SiC substrates.

According to the above experiment results, the SiC coating produced with coarse-grained powder in pack materials exhibits much better oxidation resistance. This can be attributed to the increase in thickness and density of the coating which could protect the C/SiC substrate from oxidation. A more detailed work should be conducted in the future work.

4. Conclusions

The SiC coatings were produced by a pack cementation technique in an argon atmosphere and the oxidation behavior at 1573K was investigated. The particle size of the powder in pack materials has a great influence on the microstructure and anti-oxidation property of the coating. The coating acquired with fine-grained powder reveals a smooth surface with a lower surface roughness value, while dense SiC coating prepared with coarse-grained powder. With the increase of particle size of powder, the coating thickness is increased, and the corresponding oxidation resistance of the coating is improved.

References

[1] A.G. Capoto, W.J. Lackey. Fabrication of fiber reinforced ceramic matrix composites by chemical vapor infiltration. Ceram Eng Sci Porc 5 (1984), Issue (7-8), pp.654/67.

[2] T.M. Besmann, B.W. Sheldon, R.A. Lowden. Vapor phase fabrication and properties of continuous filament ceramic composites Science 253 (1991), Issue (6), pp. 1104/09.

[3] W.H. Glime, J.D. Cawley. Oxidation of carbon fibers and films in ceramic matrix composites: A weak link process. Carbon 33 (1995), Issue (8), pp. 1053/60.

[4] G.H. Schiroky. Oxidation behavior of chemically vapor-deposited silicon carbide. Adv Ceram Mater 2 (1987), Issue (2), pp.137/41.

[5] F. Smeacetto, M. Salvo, M. Ferraris. Oxidation protective multilayer coating for carbon/carbon composites. Carbon 40 (2002), pp. 583/87.

[6] Q.G. Fu, H.J. Li, X.H. Shi, K.Z. Li, G.D. Sun. Silicon carbide coating to protect carbon/carbon composites against oxidation. Scr. Mater 52 (2005), pp. 923/27.

[7] H.J. Li, X. Yang, Y.H. Chu, L. Li, Q.G. Fu, L.H. Qi. Oxidation protection of C/C composites with in situ bamboo-shaped SiC nanowire-toughened Si–Cr coating. Corros. Sci. 74 (2013), Issue (420), pp. 419/23.

[8] X.M. Yao, Y. Yang, X.J. Liu, Z.R. Huang, Effect of recoating slurry compositions on the microstructure and properties of SiC reticulated porous ceramics. J. Eur. Ceram. Soc. 33 (2013), Issue (15-16), pp. 2909/14.

[9] A. Udayakumar, M. Stalin, K. Venkateswarlu. Effect of CVD SiC seal coating on the mechanical properties of Cf /SiC composites generated through CVI. Surf. Coat. Technol. 219 (2014), Issue (4), pp. 75/81.

[10] F.J. Kenggo, I. Akihiko, T. Rong, G. Takashi. Laser chemical vapor deposition of SiC films with CO_2 laser. J. Alloy. Compd. 502 (2010), Issue (1), pp. 238/42.

[11] O. Knotek, F. Lofffler, L. Wolkers. Amorphous SiC PVD coatings. Diam. Relat. Mater. 2 (1993), Issue (2-4), pp. 528/30.

[12] X. Yang, Q.Z. Huang, Y.H. Zou, X. Chang, Z.A. Su, M.Y. Zhang, Z.Y. Xie. Antioxidation behavior of chemical vapor reaction SiC coatings on different carbon materials at high temperatures. Trans. NonferrousMet. Soc. China. 19 (2009). Issue (5), pp.1044/49.

[13] Q.G. Fu, H.J. Li, K.Z. Li, X.H. Shi, M. Huang. A SiC/glass oxidation protective coating for carbon/carbon composites for application at 1173 K. Carbon 45(2007), Issue (4), pp. 892/94.

Coupled thermo-mechanical analysis of the ceramic thermal barrier coatings

G. Kokot, A. Jasik, G. Moskal, R. Białecki, W. Adamczyk, T. Kruczek, Gliwice / PL

The paper is devoted to thermal and thermo-mechanical analysis of the plasma sprayed thermal barrier coatings (TBC). The temperature and stress distribution using the finite element method in the microscopic scale is presented. The finite element model is based on the microscopy scan (dimension 5mmx3.5mm) and represent three levels of barrier structure: basement AMS 5599,undercoat NiCrAlY and ceramic layers of $ZrO_2 \times 8Y_2O_3$, $Gd_2Zr_2O_7$ and DCL (double ceramic layer) $Gd_2Zr_2O_7/ZrO_2 \times 8Y_2O_3$ type. The layers material properties are fully temperature dependent and coupled thermo-mechanical analysis is performed. The numerical results will be compared with the experimental one.

1 Introduction

High efficiency is one of the most important requirements, which should be met by modern gas turbines. Increase in working temperature of critical elements of an engine, made of nickel super-alloys, is not more possible in a scale, which provides a significant progress. This limitation results from a barrier, which can be treated as drop of high temperature creep resistance of nickel super-alloys in high operation temperature [1]. This problem can be solved by application of the TBC coatings (TBC *thermal barrier coatings*) on a base of a zirconium oxide, as they are ones of the most technologically and materially advanced concepts, applied in gas turbines of engines, particularly when these coatings are provided for such elements as combustion chambers and blades. They are characterized by insulating properties, what enables to reduce temperature by approx. 170°C of elements exposed to creeping, if these elements operate in a hot section of an engine, and this reduction is possible to a scope of temperature which enables to provide long-running and safe operation [2]. The thermal barrier coatings are most often produced by the APS (*air plasma spraying*) method i. e. and by the LPPS (*low pressure plasma spraying*) method under reduced pressure or by the EB-PVD (*electron beam physical vapour deposition*) method. The barrier coatings on elements of combustion chambers and blades of a turbine are made by methods of thermal spraying with use of plasma spray guns and rotating blades are coated with use of the EB-PVD technology [3]. The TBC layers constitute a multilayer system [4]. A bond-coat is usually a coating type Ni(Co)CrAlY or a diffusive layer type (Ni,Pt)Al 75-125 µm thick. A small value of thermal conductivity, which is characteristic for ceramic materials, makes temperature be reduced in a zone of joint with a material of an bond-coat, and this material is responsible for increase in oxidation resistance and corrosion in environment comprising sulphur compounds (so called „hot corrosion"). An bond-coat should be characterized by low proneness in order to form transitory phases and high resistance for diffusion of alloy components of a layer and base [3]. A zone, where oxides occur, which are effects of surface oxidation of an bond-coat – so called TGO (thermally grown oxides), is a next significant area. An insulating layer (so called TC – top-coat) is an outer part of a system and this insulating layer is characterized by a low value of thermal conductivity. The ZrO_2 oxide, stabilized with the Y_2O_3 oxide constitutes most often the discussed insulating layer [5]. Usually, thickness of an outer ceramic layer is comprised within a scope 250-375 µm [6]. Large differences in thermal material parameters cause the high differences in the temperature and stresses fields across the coatings layers. This is one of the main reasons of forming cracks and lowering the durability of the TBC.

The determining the stresses are possible using the X-ray radiography and $\sin^2\psi$ method [7,8], but it gives only stresses on the free surface of the ceramic layer. Application of the finite element method allows to determine the temperature, stresses and strains fields in the cross section and is much more convenient to use then experimental tests.

2 Research methodology

Tests of thermal diffusivity of the ceramic layer and application of a bilayer model, which enabled to determine thermal diffusivity of a ceramic layer in a function of temperature were performed and described in [9]. This result together with data concerning specific heat and a coefficient of linear expansion (determined for charge powders [10-12]) enabled to carry out coupled thermo – mechanical analysis of the TBC layer using the finite element method. On the base of performed experimental tests the parameters presented in [9] were obtained and taken into numerical analysis. It should be underline here that full set of materials parameters required to conduct the coupled thermo-mechanical analysis was taken as the temperature dependent. Parameters which were not determined in experiment were taken from literature [13]. They were mainly mechanical properties: Young modulus and Poisson ratio.

3 Results

The first step in finite analysis is to prepare a discrete model. The discrete model was created on the base of the microscope photography presented on the Fig. 1. In order to obtain the numerical model the photography from the microscope scan was converted to CAD format (Autocad DWG format) with keeping information of layers borders using vectorization software tools.

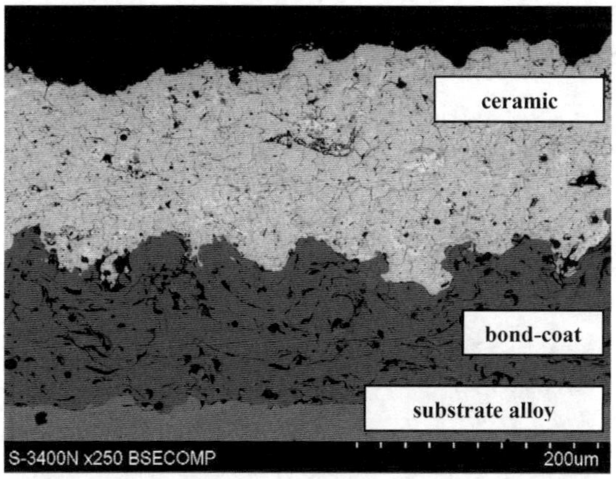

Fig. 1 The photography in the microscpic scale of TBC section (dimension 5x3.5 mm).

The vectorized CAD model is presented on the Fig .2. As the porosity of the layers have not be taken into consideration only boundaries of three layers were indentified. It is known that the porosity has influence on results. Models with the different ratio of porosity were also created and studied during the research, but those results are not included in this paper. Here calculations for the model with no porosity are presented.

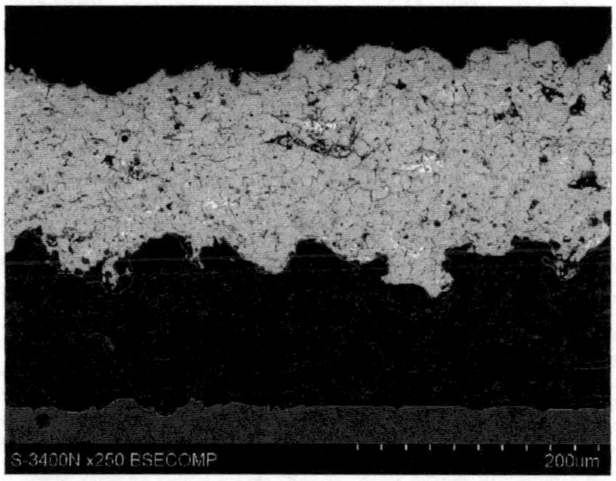

Fig. 2. The vectorized section to the DWG format with layers boundaries.

On the base of the obtained geometrical model the discrete model was prepared using MC. Patran pre-processor with the 4-noded quad element. The average size of the element is 50 μm and the fine mesh is based on the element with edge length equal to 10 μm. The model consists of about 20000 elements. In boundaries regions the fine mesh was produced.
Three types of TBC`s were analyzed - one TBC with the ceramic layer in the form of $ZrO_2{\times}8Y_2O_3$, the second TBC with the ceramic layer in the form of $Gd_2Zr_2O_7$ and the thirth type of layer of DCL morphology with internal $ZrO_2{\times}8Y_2O_3$ and oter $Gd_2Zr_2O_7$ ceramic top-layer. The thickness of both sublayer was similar. The one discrete model were prepared and the analysis were performed once for the first ceramic layer and further for the second and next one. The results are presented for the both TBC. There are three layers in the model included: first is the substarte layer of AMS 5599, second is the bond-coat NiCrAlY and third is the ceramic layer $ZrO_2{\times}8Y_2O_3$, $Gd_2Zr_2O_7$ or DCL $Gd_2Zr_2O_7/ZrO_2{\times}8Y_2O_3$. Obtained discrete FE model is presented on the Fig. 3.

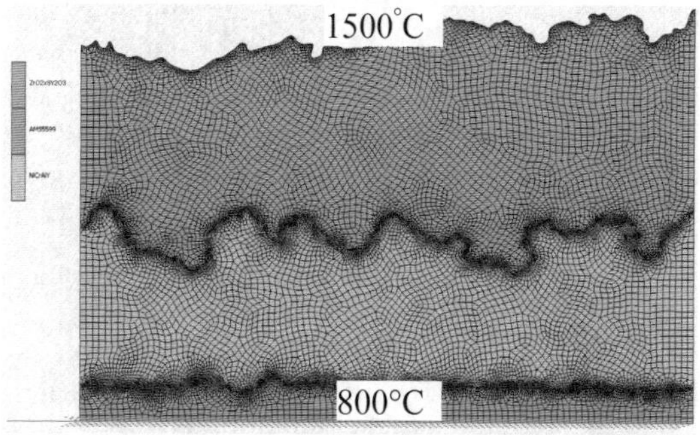

Fig. 3. The FE model of TBC layers with prescribed boundary conditions and thermal loads.

The finite element analysis have been performed as the plain strain state for microscale TBC section with dimension 5 mm × 3.2 mm presented on the Fig.1. The finite element analysis were performed in order to determine the temperature and stresses distribution in TBC layers. The coating and substrate materials were modeled as the isotropic material. Also, the model is treated as elastic body without plastic deformation, but with the temperature dependencies. The boundary conditions tried to simulate the real work situation where the temperature on the ceramic surface is about 1500°C and the substrate temperature is about 800°C. In the coupled thermo-mechanical analysis the thermal part was treated as the steady state solution with prescribed mechanical boundary conditions in the form of fixed displacements on the ground layer free edge. The problem was solved using the coupled thermo-mechanical analysis with the MSC.Marc CAE system as the solver. The obtained results in the form of temperature distribution are presented on the Fig 5. It can be seen that there is a large change in temperature values on the ceramic and substrate layers. It is caused by large differences in physical thermal parameters of this two layers materials. Ceramic layers have thermal conductivity λ = 0.8-1.5 W/mK where the substrate and ground layers have $\lambda \cong 10$ W/mK (parameters for constant temperature). But the lowest valu of temperature was obtained for DCL type of TBC system with bilayered structure of both used ceramic materials.

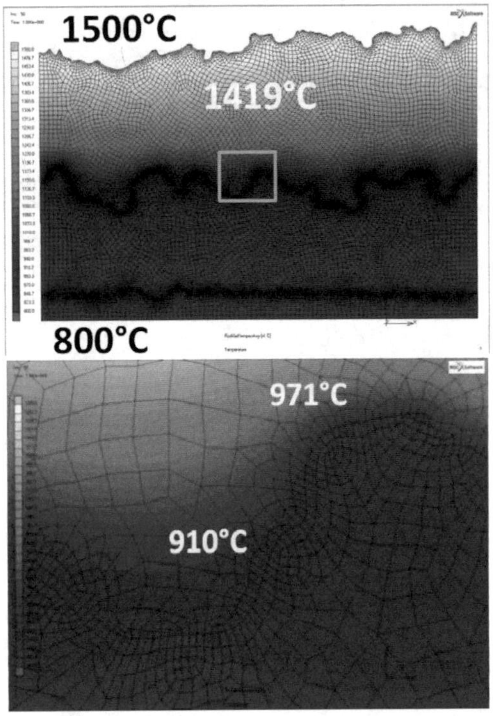

Fig. 5a. The temperature distribution in TBC with $ZrO_2 \times 8Y_2O_3$ top-coat.

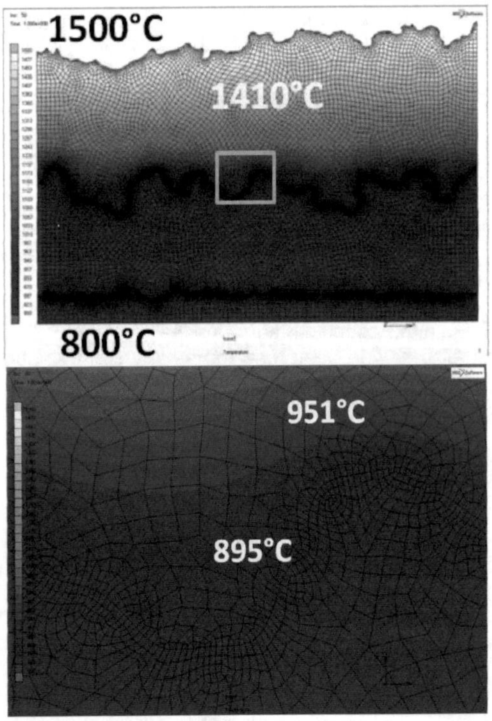

Fig. 5b. The temperature distribution in TBC with $Gd_2Zr_2O_7$ top-coat.

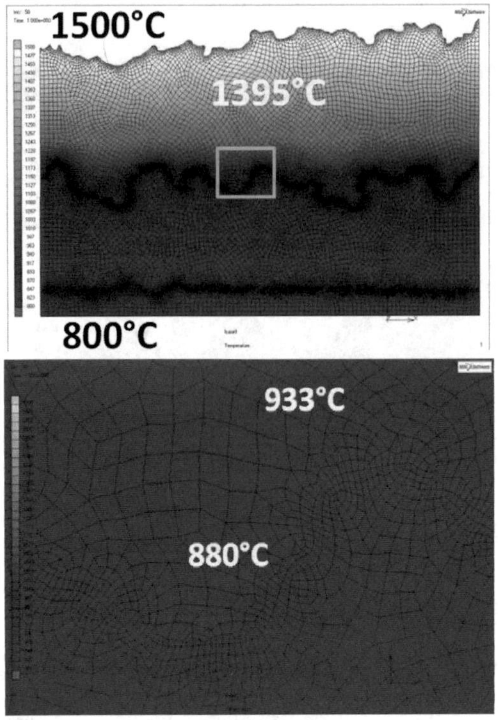

Fig.5c. The temperature distribution in TBC with DCL $Gd_2Zr_2O_7/ZrO_2 \times 8Y_2O_3$ top-coat.

Observed effect is related in consequence not only with type of used insulted materials but with the other facors as well for example with total value of intrefaces bonduaries.

On the base of experimental method of estimating the stresses using $\sin^2\psi$ method it was stated that obtained results testified to compressing nature of stresses, which occurs on the surface of a ceramic layer. This same state was observed in numerical simulation what is presented on the Fig. 6. Stresses on the free edges are respectively σ_{11}= <-1;-38> MPa for the $ZrO_2 \times 8Y_2O_3$ layer, σ_{11} = <0,-85> MPa for the $Gd_2Zr_2O_7$ layer and σ_{11}= <-5;-33> MPa for DCL TBC system. Stresses determined by $\sin^2\psi$ method are respectively σ_{11}=-11 MPa, σ_{11}=-15 MPa, and σ_{11}=-13 MPa. It means that stresses determined by experiment and numerical calculation are comparable. This is the information that numerical model is fine tuned with real situation and results especially in the rest of layers are close to real values.

A type and value of stresses in the tested TBC layers is advantageous from a durability point of view in operation conditions, however a fact should be considered that measurements are carried out on an outer area of the insulation layer. Stresses, generated near an interlayer, are decisive for durability of the layer, including stresses related to creation of a zone of the TGO oxides. However, those stresses can be fully determine using the finite element method, what was the one of the main aim of this study. Large disturbances in stress fields in the region of layers boundaries can be observed. Especially in the bottom zones of bond-coat profile roughness there are big differences in stresses values. The stresses here change their state from tension in ceramic layer to tension in substrate layer with relatively large values: σ_{11} =<-400;245> MPa and σ_{22} =<-526;281> MPa for standard $ZrO_2 \times 8Y_2O_3$ top-coat. Littlke lower values were obtained for DCL TBC system. The stresses in the second type of ceramic layer ($Gd_2Zr_2O_7$) are nearly twice higher. Additionally this is the region where the TGO oxides arise. The connection of this two types of drawbacks is the main reason to crack growth and failure (Fig. 6).

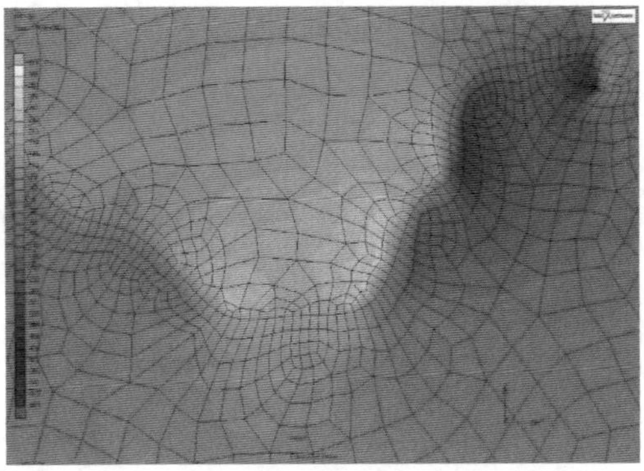

Fig. 6. The stress σ_{11} distribution in $Gd_2Zr_2O_7$ based TBC

Thanks to the finite element analysis it is possible to recognize and analyze this mechanism in details. In the further research the virtual closer crack technique (VCCT) will be implemented in the simulation for determining the crack behavior in this region.

4 Summary and conclusions

Taking into account the complicated work conditions, temperature dependent materials, full coupled thermo-mechanical process it is possible to precisely determine temperature, stresses and strain distributions what gives important information from the proper design and manufacturing process point of view. The presented research results are connected with stress distribution in TBC whit three types of ceramic layers: based on the $ZrO_2 \times 8Y_2O_3$, $Gd_2Zr_2O_7$ and DCl type. Obtained results were verified by experiments in the range were it was possible [14-16]. Performed analyses shows that differences in stress state in this three cases. Stresses in TBC with $Gd_2Zr_2O_7$ are nearly twice higher then in the TBC with $ZrO_2 \times 8Y_2O_3$ and DCL what is caused by different thermal expansions coefficient of this two ceramic layers. From thermal insulation properties point of view and stress distribution as well, the best solution should be related with doble ceramic layer systems of TBC. This type of top-layer morphology should be the best solution from working conditions point of view and overall durability of whole system.

Comparision of real conditions experiment where the cracks nucleated and increased in area the highest lewel of stress was showed at fig. 7 for gadolinium zirconate TBC system. These data confirm that area of peak of bond-coat profile roughness is the place where the inducted stress value is enough to initiation of cracking process of ceramic top-coat.

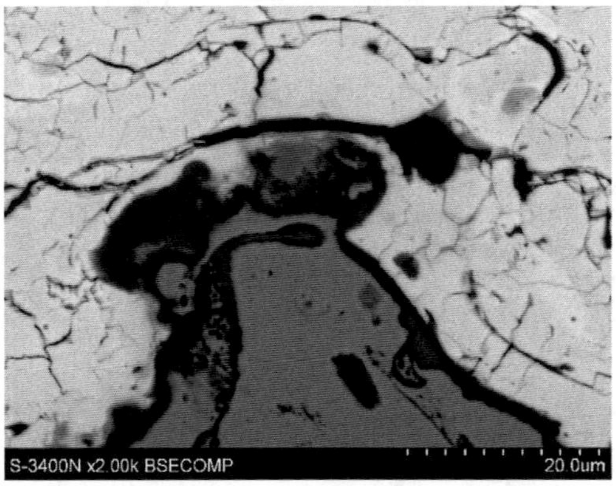

Fig. 7. The area of peak of bond-coat profile roughness with visible horizontal cracks.

Acknowledgments: The research has been supported by National Science Centre within Opus scheme, under contract UMO-2013/11/B/ST8/00268.

5 Literature

[1] Meier S.M. and D.K. Gupta: The evolution of thermal barrier coatings in gas turbine applications. Journal of Engineering for Gas Turbines and Power 116 (1994), Issue 1, pp. 250/7.

[2] Sieniawski J., Criteria and methods to evaluate materials provided for elements of turbine aircraft engines, Publishing House of Rzeszów Techical University, Rzeszów, 1995.

[3] G. Moskal: Thermal barrier coatings: Part 1 – characteristics of microstructure and properties, generation and directions of development of bond coats, Archives of Materials Science 28 (2007), Issues 1-4, pp. 100/12.

[4] Schultz U., Leyens C., Fritscher K., Peters M., Saruhan-Brings B., Lavigne O., Dorvaux J-M., Poulian M., Mevrel R. and M. Caliez: Some recent trends in research and technology of advanced thermal barrier coatings, Aerospace Science and Technology 7 (2003), Issue 1, p. 73/80.

[5] Padture N., Gell M. and E.H. Jordan: Thermal barrier coatings for gas-turbine engine applications, Science 296 (2002), pp. 280/4.

[6] R.L. Jones: Thermal barrier coatings, in Metallurgical and Ceramic Coatings, ed. by K.H. Stern, Chapman & Hall, London, 1996, p. 194.

[7] Handbook of Measurement of Residual Stresses, Ed. Lu J., Fairmont Press, Lilburn, Georgia, 1996.

[8] Noyan I.C. and J.B. Cohen: Residual stress: Measurement by Diffraction and Interpretation, Springer-Verlag, Berlin, 1987.

[9] Moskal G., Swadźba L., Hetmańczyk M., Witala B., Mendala J. and P. Sosnowy: Characterization of microstructure and thermal properties of $Gd_2Zr_2O_7$-type thermal barrier coating. Journal of European Ceramic Society 32 (2012), Issue 9, p.2025/34.

[10] Moskal G., Iwaniak A. and A. Rozmysłowska-Grund: Characterization of thermal properties of micro-sized ceramic powders for APS deposition of ceramic layers, Key Engineering Materials 484 (2011), pp.152/7.

[11] G. Moskal: Microstructural characteristics and technological properties of YSZ-type powders designed for thermal spraying of TBC, IOP Conference Series-Materials Science and Engineering 484 (2010), no 012019.

[12] G. Moskal: Microstructure and thermal diffusivity of micro- and nano- sized YSZ, Materials Science Forum 638-642 (2010), pp. 900/05.

[13] Aiguo L. and W. Yanhong: Finite element analysis of anti-spallation thermal barrier coatings, Surface and Coating Technology 165 (2003), Issue 2, pp. 154/62.

[14] Moskal G., Swadźba L., Medala B., Góral M. and M.Hetmańczyk: Degradation of the TBC system during the static oxidation test, Journal of Microscopy 237 (2010), Issue 3, pp. 450/55.

[15] Swadźba L., Moskal G., Mendala B. and T. Gancarczyk: Characterization of air plasma sprayed TBC coating during isothermal oxidation at 1100ºC. Journal of Achievements in Materials and Manufacturing Engineering 21 (2007), Issue 2, pp. 81/4.

A high bond strength interlayer for metal coating on polymer composites fabricated by super-detonation spray

Wang / C, Gao / J, Tian / H, Cui / Y, Guo / M, Zhang / H and Tang / Z, Beijing/CN

In the present work, a metal-polymer composite coating containing Al and ethylene-vinyl acetate copolymer (EVA) was prepared on the surface of a polymer matrix composite (PMC) using a detonation spraying process. The microstructure and bond strength of the as-prepared coatings were analyzed. The bonding mechanism of the coatings, especially the deposition behavior of the Al and EVA particles on the PMC surface is discussed. Results had shown that detonation spraying technique enables the deposition of metal-polymer coatings directly onto the PMC surface under precise process control. The preparation of metal-polymer composite coating on PMC via detonation spraying process presents promising application as an interlayer for the surface metallization of PMC.

1 Introduction

In recent years, polymer matrix composites (PMCs) are widely used in the aerospace industry and in the military attributed to their low density, high specific strength and stiffness, and other unique properties such as ease formation and machining, well size stability [1]. In order to increase the thrust–weight ratio, the PMCs with lighter weight and higher strength-to-weight ratio are employed to replace the metallic parts of aerocraft. PMCs in the current aircraft engines were used as the materials of fan blades, outlet guide vanes, bypass ducts, nose cone spinners, core engine fairings, variable vane rings, and cowl doors [2]. Some special characteristics such as electrical conductivity, thermal conductivity, electromagnetic shielding, erosion and radiation protection can be achieved through the surface metallization of PMCs. Therefore, the surface metallization of a PMC substrate is considered to be an effective technique for improving its surface property and extending its application [3–5].

Recently, Several conventional methods including PVD or CVD [6,7], electroplating [8–12], electroforming [13,14], and thermal spraying techniques [15–17] have been used for the surface metallization of PMCs. During these methods, PVD or CVD deposition should be in atomic dimention and workpieces should be placed in a limited cavity. High equipment and processing costs are also generated by the two techniques. Low adhesive force, poor stability in the electroplating case and the consequent pollution it generates, long production cycle and high cost of the mold for electroforming [18]. The thermal spraying technique is a flexible and cost-effective technique for manufacture coatings, which has been widely applied in the preparation of coating materials on metal or non-metal substrates. Among the thermal spraying techniques, detonation spraying has been known for about two decades and proved to be a powerful method of coating fabrication [19,20]. In detonation spraying, the coating is formed by particles heated up to their melting temperature or above and accelerated toward the substrate by the products of gaseous detonation. Computer-controlled detonation spraying (CCDS) has emerged as a promising thermal spraying method of high potential for different material systems [21,22]. One of the advantages of detonation spraying technique is the extremely low substrate temperature during spraying which is much lower than other thermal spraying methods.

However, extensive studies have been conducted on the preparation of metallic coatings on metal and alloy substrates, but few reports involve ceramics [23–28], glass [29], or PMC substrates [30–32]. Comparing with sprayed coatings on metal and alloy substrates, the melted particles with high energy could induce the degradation of the PMC substrate, or even damage, while those with low energy would cause low bonding strength between the coating and the substrate. Generally, the ceramic particles or metal powders (such as Ni) with a high melting point resulted in thermal damage of the PMC substrate surface [33]. Up to now, some researches regarding the improvement of the adhesion between the coating and PMC substrate and prevention of the thermal damage of the substrate have been performed. Zn or Al powders with a low melting point were usually used as an interlayer to protect the PMC substrate against further degradation or damage [34,35]. However, the interface between the pure Al or Zn metallic coating and PMC substrate is still mechanical bonding and the thermal expansion mismatch is still serious between the coating and substrate. The improvement of adhesion strength and thermal shock resistance of the coating is still an important problem for PMC materials.

Hot-melt adhesives (HMA) are polymer blends that are widely used for spreading over substrates in the melt state followed by solidifying after cooling. Thermoplastics, such as ethylene vinyl acetate copolymer (EVA) are often used as base polymers for HMA. EVA has attracted a great deal of theoretical and experimental attention because it is the most widely used base polymer for HMA in industry. Therefore, the main motivation of this work is to fabricate the composite coating systems with metal and polymer composition on PMCs via the detonation spraying process. The addition of EVA into the Al coating will heighten the mismatch between Al coating PMC surface. The influences on bonding strength of the coating were investigated and the possible failure factors of these coating systems were discussed.

2 Experimental

Commercially available polymer matrix composites was used as substrate. Commercially available atomized Al powders (99.9% pure) with diameters ranging from 5 to 7 μm, and ethylene-vinyl acetate copolymer (EVA) powder with diameters ranging from 10 to 20 μm were used as feedstock. In the blended spraying powder, the volume content of EVA was same as that of Al powder. Fig. 1 illustrates the nearly spherical morphologies of the Al and EVA powders. An average particle size of 25-45 μm and spherical morphology was selected to achieve the optimum flow ability of powder during spraying.

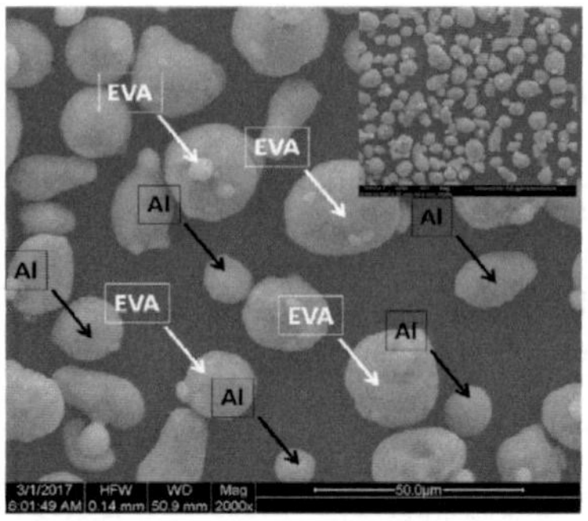

Fig. 1 SEM morphology of the Al powders blended with EVA powders

For spraying, 100×20×2 mm PMC plates were used as the substrate. Prior to coating, the plates were cleaned with acetone. The detonation spraying system used in this research was CCDS system developed by the institute of solid state chemistry and mechanochemistry SB RAS in Russia. This system consists of a1.25-m-long stainless steel detonation tube, an electrical spark located at the closed end of the tube, and a powder feeder located at a distance of 0.5 m to the nozzle exit. The D-gun spraying parameters employed in this work are listed in Table 1.The Table 2 shows the EVA/Al composite coating was deposited by different particle size (EVA), spraying passes (including EVA/Al transition composite layer and Al surface layer), heat treatment process.

Tab. 1 Spraying parameters employed during the detonation gun spraying process.

Parameter	Value
Working pressure for O_2 (MPa)	1.15
Working pressure for C_2H_2 (MPa)	1.20
Working pressure for N_2 (MPa)	1.15
Fuel ratio(O_2/C_2H_2)	70-140
Spraying distance (mm)	250
Frequency (shot/s)	4
Powder feed rate (g/s)	1.0-1.5

Tab. 2 Different process of EVA/Al composite coating

Mumble of coating	Particle size of EVA	Passes of EVA/Al transition layer	Passes of Al surface layer	Heat treatment process
No.1	20 μm	3	5	120 ℃×1h
No.2	15 μm	3	5	120 ℃×1h
No.3	5-10 μm	3	5	120 ℃×1h
No.4	5-7 μm	2	5	120 ℃×1h
No.5	None EVA			

The general set-up is sketched in Fig. 2 [36]. A channel inside gun barrel 1850–1000 mm long and 20 mm in diameter is filled with gases by a computer-controlled precision gas distribution system2; first, it is filled with a carrier gas, then with a certain portion of an explosive mixture, which results in the formation of a stratified gas medium consisting of explosive charge 3 and carrier gas 4. Then a certain amount of the feedstock powder is injected into the barrel through an orifice by a computer-controlled feeder 5 with the help of the carrier gas flow. The powder injected into the barrel forms cloud 6, which for the taken barrel diameters and injection conditions can be considered uniform in density across the barrel cross-section. The cloud extends over a distance of up to 2–3 barrel diameters from the injection point having a concentration gradient along the barrel axis. The spraying distance is measured as a distance from the exit of barrel1 to the surface of substrate7.Once the powder is injected, the computer gives a signal to initiate detonation, which is done by an electrics park. An explosive combustion of the charge occurs within a time of the order of 1 ms such that a detonation wave forms in the explosive mixture transforming into a shock wave in the carrier gas. The detonation products heated up to 3500–4500K and the carrier gas heated by the shockwave up to 1000–1500K move at a supersonic speed and exchange heat and interact with the powder during 2–5 ms. In this process, the powder particles can be heated up to the material's melting temperature and accelerated up to velocities as high as 500m/s and then deposited onto the substrate 7 to form a coating.

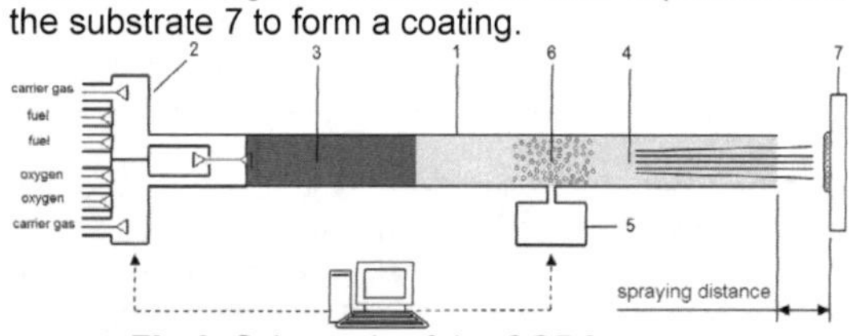

Fig.2. Schematic of the CCDS system.
(1–gun barrel,2–computer-controlled precision gas distribution system,3–explosive charge,4–carrier gas, 5–computer-controlled powder feeder,6–cloud of injected powder,7–substrate)

The surface morphology and cross-section microstructures of the coatings were analyzed by means of a FEI QUANTA-600 scanning electron microscope (SEM). At the same time, the surface of these samples was treated by spraying gold prior to the analysis. To measure the bond strength of the Al-

EVA composite coating on the substrate, the ASTM Standard C633 was followed on stretching Test System. The bond strength was calculated using maximum failure force divided by the broken cross sectional area. Three sets of specimens were prepared and tested to obtain a mean value. Then, the fracture morphology was observed by SEM.

3 Results and discussion

3.1 Surface morphology of composite coating

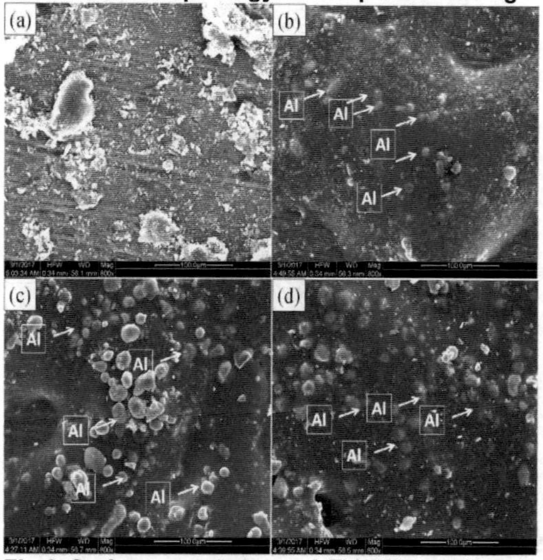

Fig.3 Surface morphology of EVA/Al transition layer (No.1-5 corresponding to a-e)

Figures 3 a-d are the SEM micrographs of transition coatings (EVA/Al) corresponding to No.1-4. Comparison of the surface morphology shows that the coating of No.1 and 2 were composed of tightly adhered irregular splats with pores and microcracks. Some particles remained unmelted or partially melted as seen in Figure 2 a and b, indicating that the larger particles size of EVA could not provides sufficient energy to melt all the particles, which results in a non-uniform microstructure in the coatings. Figure 2 c and d shows that coatings obtained within the EVA particle size of 5-10 μm exhibits less unmelted small particles than the coatings showed in Figure 2 a and b. The melted Al particles were densely embedded into the EVA softened plastic substrate with the possibility of building-up the better cohesive strength Al/EVA composite layers. However, in the case of the samples No. 3 and 4, the coating was properly obtained showing a homogeneous surface and a well-adhered morphology free of cracks.

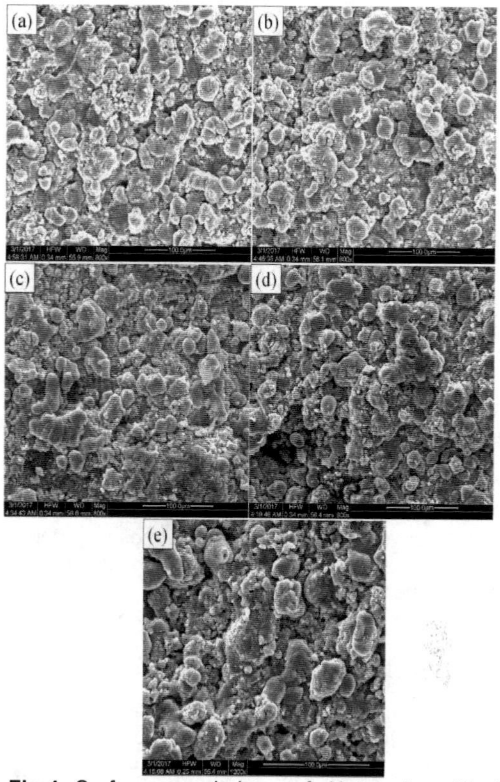

Fig.4 Surface morphology of Al top layer(No.1-5 corresponding to a-e)

Especially, the surface morphology of all the Al coatings of No. 1-5 (Figure 4 a-e) showed that almost of the particles were totally melted or super-melted, and well-flattened splats were produced. However, the uniform particles (which suggest that much melted Al particle could have sparked after they reached the surface of EVA/Al transition layer due to the prosperity energy input) can be clearly seen in Figure 3 d.

3.2 Cross-section morphology of composite coating

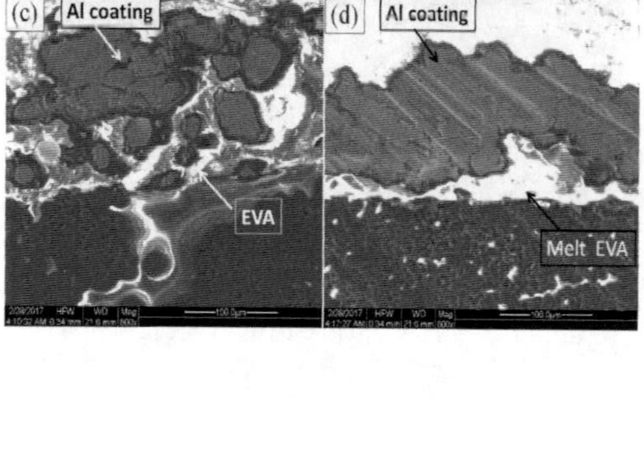

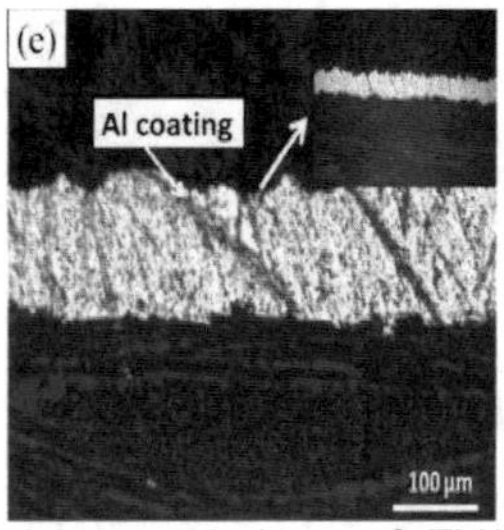

Fig.5 Cross-section morphology of EVA/Al coating (No. 1-5 corresponding to a-e)

Fig. 5 a-d show the cross-section area of the EVA/Al composite coatings. It can be seen that in the interface of PMC and transition layer, caused by the EVA softening because of the high temperature spraying stream, a certain amount of melted Al particles anchored into the EVA substrate but without degraded the PMCs matrix. This may be caused since the EVA cannot provide exorbitant heat input and erosion force to PMCs substrate in these spraying conditions for easing the plastic deformation of the particles at the impact.

Previous to this experiments it was studied the variation of the volume ratio between EVA and Al, which demonstrated that the volume ratio between 2 and 3 was the favorable result. On the other hand, spraying with lower oxygen/fuel about 75 may ensure the better wetting behavior within EVA and Al practical. In any case, it was not observed the formation of the coating when the EVA particle size higher 15 μm.

Therefore, reducing the particle size to smaller than 10 μm would avoid the non-melting of the EVA, which leads to an increase in the plastic deformation of the Al and EVA particles at the impact. Therefore, the sample No.4 that was obtained by lower oxygen/fuel spraying conditions and the appropriate particle size were satisfactorily adhered to the substrate reaching dense coating and thicknesses above 50 μm (Figure 5 d). It is also possible to observe that the superior part of the coating has certain porosity. The reason was that the coarser powder particles (20 μm) often remain unmelted or partially melted, which subsequently results in porosity and weak bonding in the coating. Nevertheless, the coating closer to PMC matrix interface is denser. This is caused by the melt EVA enhance the wettability of Al particle and PMC (Figure 3 d).

3.3 Bond strength of composite coating

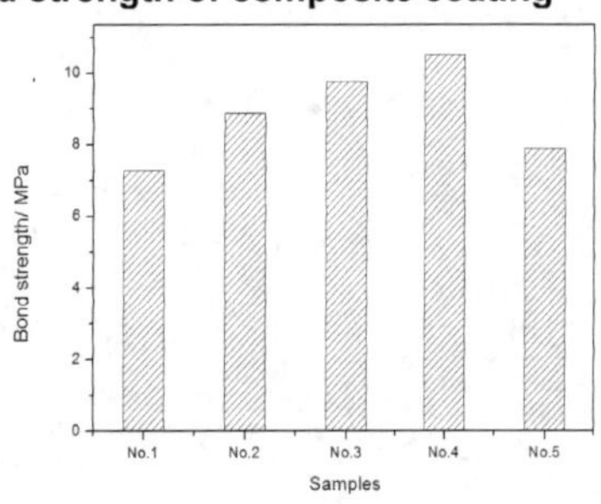

Fig.6 Bonding strength of EVA/Al composite coatings and Al coating

The bond strength of EVA/Al composite coatings and Al coating without EVA are shown in Figure 6. The values of EVA/Al composite coatings displayed in Figure 6 are measurements from averages of 8 test date. In the figure, as expected, it can be found that the bonding strength of the EVA/Al composited coatings and PMC matrix increase from 7.63 to 10.23 MPa with the decrease of EVA particle size from 400 μm to 100 μm, the relative bonding strength is significantly enhanced. The coating of No.4 had the highest bond strength, 10.23 MPa. The Al coating sprayed without contain EVA had only had the third higher bonding strength, 7.79 MPa. The coating of No.1 had a low bonding strength of 7.32 MPa. The bonding strength of Al top coatings/PMC substrate was enhanced obviously with intermediate bonding EVA/Al transition layer introduced.

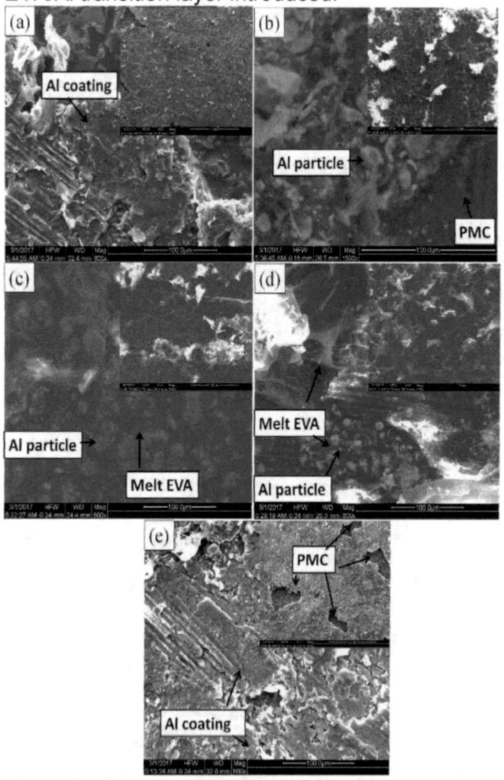

Fig.7 Surface morphology of fracture sample after bond strength test (No. 1-5 corresponding to a-e)

As the property of the coating is related to the combinative state between the coatings and substrates, many studies have been conducted on the fracture structural characterization of the coatings after bond strength test. The fracture surface areas of tensile samples after the bond strength test, all the samples of No.1-5 are presented in Figure 7 a-e. The EVA/Al transition layer shows a flattened morphology with no irregularities. On the other hand, the EVA/Al coating provides great roughnesses to the interface layer as the Al particles that reached the EVA substrate were drastically deformed. This can also be checked when observing its cross-section area in Fig. 5 d. Anyhow, to achieve a rough compatible surface is a key-point for the final application of the material because the EVA enhances the compatibility of the Al coating and the PMC matrix. Comparison of figure 6 d, the fracture of Al coating deposited without EVA showed the characterization that was smooth morphology.

3.4 TG-DTG analysis and discussion

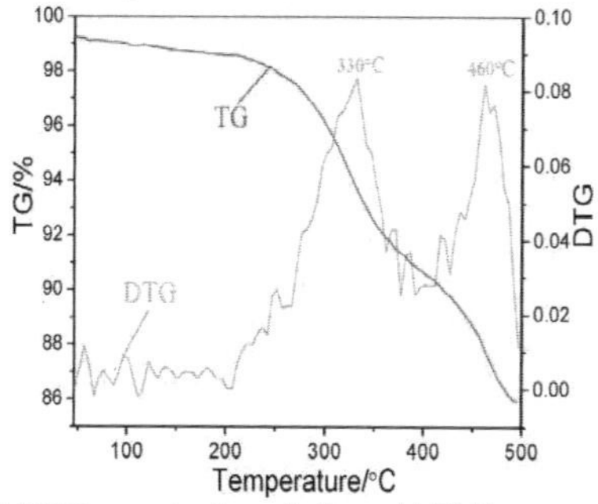

Fig.8 TG-DTG analysis of the Al-EVA composite coating

The TG-DTG analysis of the Al-EVA composite coating is shown in Figure 8. From the figure it can be seen that the Al-EVA composite coating made about 14% weight loss of the coating during test from room temperature to 500°C. From the DTG curve, it can found that there are two apparent peaks at 330°C and 460°C, respectively. The peak at 330°C is most probably the thermal decomposition of the EVA content and it brought about 8% weight loss of the coating. Because the coating specimen made for the TG test was scraped from the PMC substrate, the peak at 460°C is maybe the thermal decomposition of PMC substrate.

4 Conclusions

Densification, homogeneous and well-adhered EVA/Al composite coatings were deposited on polymer matrix composite (PMC) by Super-detonation spray. From the proposed factorial design of experiments, it is concluded that the spraying gas stream softens and melt the EVA particle which has the ability to enhance the cohesive strength the Al coating and PMC substrate. The fact of EVA particle size critically influences the deposition of the Al metallic particles and the formation of dense coating. However, it is possible to reach a balance between thermal softening of the EVA material and plastic deformation of the Al particles at the impact in orders to build-up the coating. Moreover, the mean bonding strength of the EVA/Al composite coatings was about 10 MPa. The bonding strength of Al coating and PMC substrate were enhanced obviously with EVA/Al intermediate bonding layer introduced.

5 Literature

[1] K.H. Zhang, Plastics Material Science, Northwestern Polytechnical University Press, Xi'an, 2000 in Chinese.

[2] K. Steffens, H. Wihelm, MTU Aero Engines GmbH, Munchen, Germany, 2000, Internet edition: http://www.mtu.de.

[3] F.X. Lei, Electroplating Finish. 4 (1984) 8 in Chinese.

[4] S.J. Ying, Electron. Proc. Technol. 10 (1985) 6 in Chinese.

[5] S.Q. Wang, G.S. Song, Mater. Prot. 26 (1993) 12 in Chinese.

[6] B.B. Straumal, N.F. Vershinin, A. Cantarero-Saezc, M. Friesel, P. Zieba, W. Gust, Thin Solid Films 383 (2001) 224.

[7] J. Li, S.F. Wang, T.G. Li, F.X. Tan, Electroplating Finish. 28 (2009) 18 in Chinese.

[8] X.Q. Bai, J. Li, X.P. Yan, C.H. Zhao, J. Wuhan Univ. Technol. (Transp. Sci. Eng.) 29 (2005) 947 in Chinese.

[9] Q.H. Zhou, H.W. Chen, Y. Wang, Electrochim. Acta 55 (2010) 2542.

[10] R.X. Zhu, Appl. Chem. Ind. 38 (2009) 1629 in Chinese.

[11] C. Shu, G.R. Xie, Y.L. Zhu, Electroplating Finish. 29 (2010) 5 in Chinese.

[12] M. Charbonnier, Y. Goepfert, M. Romand, Surf. Coat. Technol. 200 (2008) 5028 in Chinese.

[13] S. Li, F.X. Bai, F.Y. Kong, Fiber Reinf. Plast. Compos. 6 (2004) 37.

[14] H. Mekaru, O. Koizumi, A. Ueno, M. Takahashi, Microelectron. Eng. 87 (2010) 922.

[15] J. Voyer, P. Schulz, M. Schreiber, J. Therm. Spray Technol. 17 (2008) 818.

[16] F. Robitaille, M. Yandouzi, S. Hind, B. Jodoin, Surf. Coat. Technol. 203 (2009) 2954.

[17] F. Robitaille, M. Yandouzi, B. Jodoin, Thermal Spray 2009: Proceedings of the International Thermal Spray Conference, 2009, p. 618, doi:10.1361/cp2009itsc0618.

[18] X.L. Zhou, A.F. Chen, J.C. Liu, X.K. Wu, J.S. Zhang, Surf. Coat. Technol. 206 (2011) 132.

[19] Nikolaev YA, Vasiliev AA, Ulianitsky VYu. Comb Expl Shock Waves 39 (2003) 382.

[20] Ulianitsky V, Shtertser V, Zlobin S, Smurov I, J Thermal Spray Technol 20 (2011) 791.

[21] Lomovsky OI, Dudina DV, Ulianitsky VYu, Zlobin SB, Kosarev VF, Klinkov SV, et al. Mater. Sci. Forum 534-536 (2007) 1373.

[22] Dudina DV, Zlobin SB, Bulina NV, Bychkov AL, Korolyuk VN, Ulianitsky VYu, et al. J Eur Ceram Soc 32 (2012) 815.

[23] P.C. King, S.H. Zahiri, M. Jahedi, Acta Mater. 56 (2008) 5617.

[24] P.C. King, M. Jahedi, Appl. Surf. Sci. 256 (2010) 1735.

[25] K.H. Kim, M. Watanabe, S. Kuroda, Surf. Coat. Technol. 204 (2010) 2175.

[26] H.Y. Lee, S.H. Jung, S.Y. Lee, Y.H. You, K.H. Ko, Appl. Surf. Sci. 252 (2005) 1891.

[27 H.Y. Lee, Y.H. Yu, Y.C. Lee, Y.P. Hong, K.H. Ko, J. Therm. Spray Technol. 13 (2004) 184.

[28] D. Rafaja, T. Schucknecht, V. Klemma, A. Paul, H. Berek, Surf. Coat. Technol. 203 (2009) 3206.

[29] D. Zhang, P.H. Shipway, D.G. McCartney, J. Therm. Spray Technol. 14 (2005) 109.
[30] R. Lupoi, W. O'Neill, Surf. Coat. Technol. 205 (2010) 2167.
[31] A. Sturgeon, B. Dunn, S. Celotto, B. O'Neill, Proc. of the 10th ISMSE & the 8th ICPMSE, Collioure, France, 19–23 June 2006 (SP-616, September 2006), 2006.
[32] R.M. Tapphorn, J.A. Henness, H. Gabel, Thermal Spray 2009: Proceedings of the International Thermal Spray Conference, 2009, p. 589, doi:10.1361/cp2009itsc0589.
[33] A. Liu, M. Guo, J. Gao, and M. Zhao, Surf. Coat. Technol., 201 (2006) 2696.
[34] W. Huang, Y. Zhao, X. Fan, X. Meng, Y. Wang, X. Cai, X. Cao, Z. Wang, J Thermal Spray Technol 22 (2013) 918.
[35] L. Zhu, W. Huang, H. Cheng, X. Cao, J Thermal Spray Technol 23(2014) 1312.
[36] D. V. Dudinaa,n, I. S. Batraevb, V. Y. Ulianitskyb, M. Korchagina, Ceramics International, 40(2014)3253.

Acknowledgments

The authors express their thanks for the financial support provided by the National Natural Science Foundation of China under grant (no. 51201156, 51605455)

Corrosion Performance of WC10Co4Cr Coatings on High Strength Steel

Yang Weihua, BeiJing / China
AVIC Beijing Aeronautical Manufacturing Technology Research Institute, Science and Technology on Power Beam Processes Laboratory

Abstract: 300M steel is one of most important aerial materials, which can be used for landing gear and flap & slat track. Some surface engineering technologies are needed to be adopted on its surface, because of its bad corrosion performance. WC10Co4Cr Coatings by high velocity oxygen-fuel spray processing (HVOF) is an environmental friendly method for this protection. In this paper, WC10Co4Cr coatings were prepared on 300M by optimized HVOF processing. And their corrosion performance has been estimated by neutral salt fog test, according with ASTM B117. The results indicate that the porosity gets larger and the hardness gets higher for the dissolution of bonding phases after the test. And for the optimized coatings, there are no corrosion products in the surface and interface between the coating and 300M steel, after 2000 hours ASTM B117 test. So the coatings have a good corrosion performance.

1 Introduction

As we know, 300M steel has very high tensile strength (σ_b = 1963 MPa) and relatively good fracture toughness (K_{1C} = 70MPa•m$^{-1/2}$), so it is very important for aviation applications, such as landing gear, flap & slat track[1]. Chromium plating is the most used electrodeposited coating to obtain high levels of hardness, resistance to wear and corrosion for this material. But this processing utilizes chromium in the hexavalent state (hex-Cr), which is a known carcinogen. As a result, the environmental protection agency has issued new air emission standards for hex-Cr, which modified permissible exposure limits (PEL), from 100 micrograms/cubic meter to a range of 0.5 to 5 micrograms/cubic meter. This substantially increases the cost of chrome plating[2, 3]. So we must find some methods to alternate chromium plating.

High-velocity-oxygen-fuel (HVOF) thermal spray process is one of the most promising methods as hard chrome replacement, which has gotten to be a research hotspot for protecting 300M steel. The process has been applied in many civil aircraft, such as Boeing 737,787 and Airbus 340, 380[4, 5].

Dr. James A. Browning invented HVOF technology in 1982 and the first commercial systems were available by the late 1980's. In HVOF process, fuel and oxygen are put into the combustion chamber together, then the combustion of these gases are started by ignition spark, which produces a high temperature and high pressure in the chamber and causes the supersonic flow of the gases through the laval nozzle. The flame temperature varies in the range of 2600 °C–3000 °C, depending on the fuel, the fuel gas/oxygen ratio and the gas pressure. The powder particles get into the flame and are melted or partially melted during the flight to substrate. Finally the powder particles impact on the work piece surface where they are deformed and quenched. This process provides coatings properties that exhibit high density and uniformity[6, 7].

In this paper, WC10Co4Cr coatings were prepared on 300M by optimized HVOF processing. Their microstructure was studied by optical microscope and SEM. Their corrosion performance has been estimated by neutral salt fog test, according with ASTM B 117.

2 Experiment and Characterization

2.1 Raw materials

The substrate samples are made from 300M steel, which is machined into 20mm×10mm×5mm (type I, for morphology), 60mm×60mm×5mm(type II, for corrosion test), Φ25.4mm×7.7mm(type III, for bonding test). WC10Co4Cr powder (DG-WC10Co4Cr,10μm - 45μm , Agglomerated & Sintered, Chengdu Daguang Ltd., China) was bought as initial powder. The powder is round and made of fine WC granule. The particles separate each other, which keep it have good fluidity. In this powder, there are three phases, which are WC, Co and $Cr_{23}C_7$, no brittle W_2C phase. See more details in Fig.1.

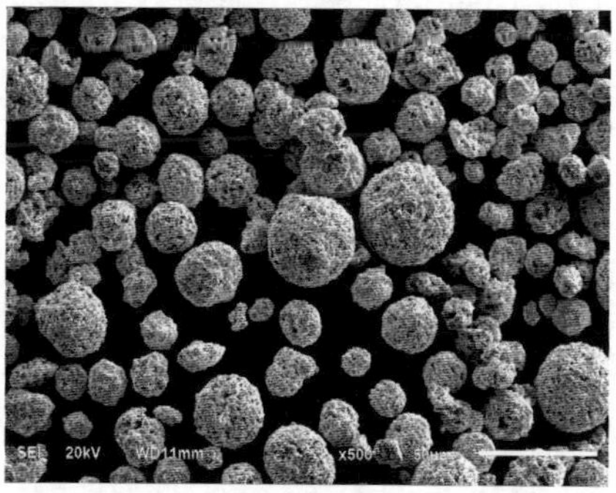

(a) Morphology of DG WC10Co4Cr

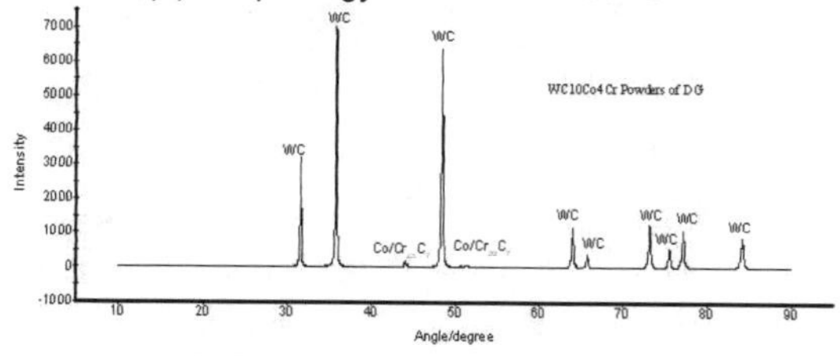

(b) Phases in DG WC10Co4Cr

Fig. 1. SEM image and XRD patterns of raw powder

2.2 Prepared of WC10Co4Cr coatings

On the samples mentioned above, WC10Co4Cr coatings were prepared by optimized HVOF processing (EvoCoat, Wokajet 440 gun, Oerlicon Metco). The parameter ranges have been listed in Table 1. And the coatings' thickness is controlled from 0.15-0.25mm for different tests. All of the substrates were blasted by using 46 mesh brown alumina under 0.4-0.6MPa, 100-120mm distance before sprayed.

Table 1 Spray parameter ranges of the coatings

parameters	Value
Kerosene flow rate(LPH)	22-24
Powder rate (g/min)	60-70
Standoff (mm)	330-340
Oxygen (LPM)	860-880
Carrier gas (dual , LPM)	10
Barrel length (mm)	100

2.3 Coatings Characterization

2.3.1 Microstructure, porosity, and microhardness

Mount samples of the type I and epoxy resin into the moulds and get them out, then grind and polish them, finally the metallographic coupons were gotten. The microstructure was observed through microscope. The porosity was automatically calculated by metallographic method (ImageJ 5.0), and the microhardness was measured by microhardness tester.

2.3.2 Phases in powder and coatings

During HVOF processing, the flame temperature is higher than 2600 ℃, which will lead WC to be decomposed. This decomposition will make the coatings more brittle, which make the coating be broken easier. So we need optimize the process by minimizing the content of W_2C (the decomposition production).

The phases in powder and coatings were analyzed by X-ray diffraction method (Rigaku D/max-rA， Cu， 40kv， 50mA). The W_2C content was calculated by the formulae as follow:

$$C_{W_2C} = \frac{I_{W_2C}}{I_{W_2C} + \frac{I_{WC}}{K_{W_2C}^{WC}}} \quad (1)$$

2.3.3 Tensile bonding Strength

The bonding strength was tested according to ASTM C633, see the scheme in fig.2. The Force will be added on the two ends, then pull the coatings off from the substrate. The boding strength will be calculated by the formula as follow:

$$\sigma_T = \frac{F}{S} \quad (2)$$

σ_T : the bonding strength of the coatings.
F: the force when the coating was pulled off.
S: the area of the section.

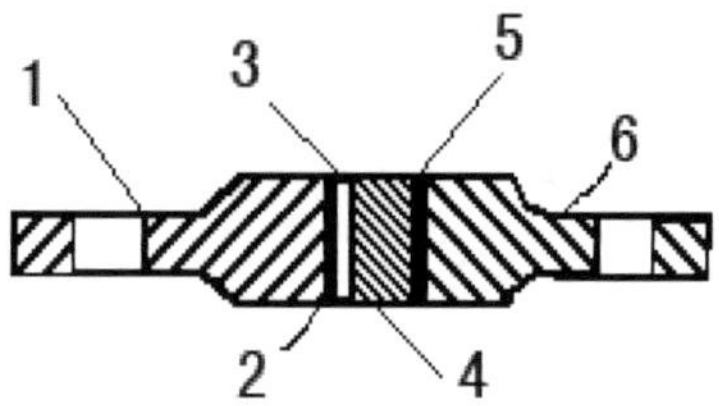

1,6:TensileHead;2,5:Adhesive;3:Coatings;4:Substrate

Fig.2. Scheme of the test method of bonding strength

2.3.4 Corrosion test

Corrosion test will be executed under these conditions as follows: 35℃, 5% NaCl solutions, 6.5-7.2 PH value, the coupons will be put in the salt fog for 2000 hours, according with ASTM B117. Then the surface and section will be observed by scan electron microscope.

3 Results and Discussion

3.1 Metallography and phase analysis

The coatings are very dense and typical lamellate structure can be seen in them. All of pores are almost round and homogeneous, which diameter is about 1 μm. The porosity is less than 1% and the average hardness is higher than $HV_{0.3}1050$. The coatings are adhered very well with the substrate; see more details in the Fig.3.

(a) optical microscope **(b) SEM**

Fig.3. Microstructure Image of the coatings

Compared with raw powders, there is a new phase (W_2C) in the coatings as sprayed because of the decomposition of WC, see it in Fig.4. The content of W_2C is 6.37%, calculated by the formula 1. The optimized parameters can keep the decomposition least, and then keep the coatings more toughness and have a longer life

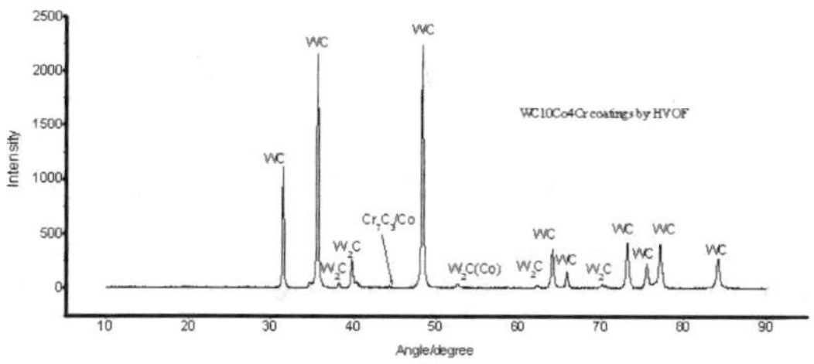

Fig.4. Microstructure of the coatings

3.2 Tensile bonding strength

The coating is attached well with the substrate, which can be seen in the Fig. 3. The tensile bonding strength test result can be seen in the Fig5. All of the tensile bonding strength is more than 70MPa. The average value is 75.2MPa.

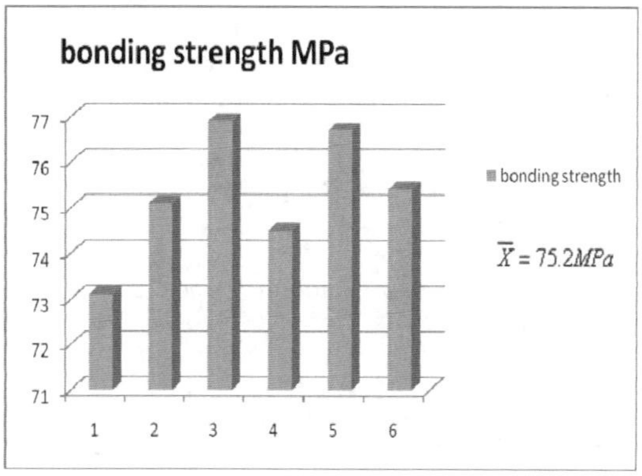

Fig.5. bonding strength of the coatings

3.3 Corrosion performance

To estimate the corrosion performance, the surface and section structures of the coatings were observed by SEM.

According to Fig.6.a and Fig.6.b, before corrosion test, the surface seems like many lamellar particles were built up together. When the raw particles were fed into the supersonic flume, the particles were heated and melted and changed into molten or semi-molten globules, then impacted and solidified onto the substrate and changed into lamellar particles. At last these lamellar particles were interlocked with Co and Cr phase and formed the coatings. After corrosion test, the surface gets more smooth and typical lamellalike characteristics are disappeared, which seems like a lot of fine granules are adhered onto the surface. And there are no featherlike corrosion products on the exterior. This indicate the cauterant did not reach into 300M steel strate, and the substrate is still protected by the coatings.

See the Fig.6.c, we can find that the coatings are very dense and pores can seldom be observed. WC10Co4Cr coatings were adhered well onto the substate. After corrosion test, the structure of the coatings is changed distinctly: the porosity gets larger and the hardness gets higher. From Fig.7., before corrosion test, the porosity is from 0.75 to 1.21 and the average value is 0.98. While, after corrosion test, the porosity is from 2.91 to 4.15 and the average value is 3.48. This is because of the dissolution of the bonding phases (Co and Cr). With this dissolution, the phases are decreased and pores gets more.. For same reasons, the hardness gets larger. The content of WC and W_2C will be increased and their hardness is higher than the bonding phases so much. So the hardness of the coatings will get higher from about $HV_{0.3}$1149.9 to about $HV_{0.3}$1256.8. See more details in Fg.7.

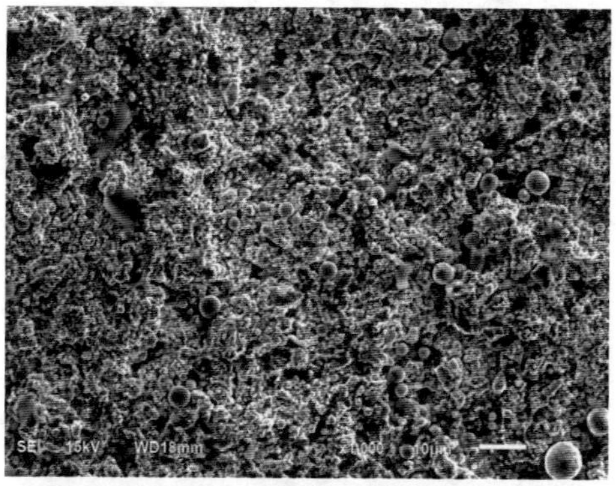

(a) Before, surface

(b) after, surface

(c) Before, section

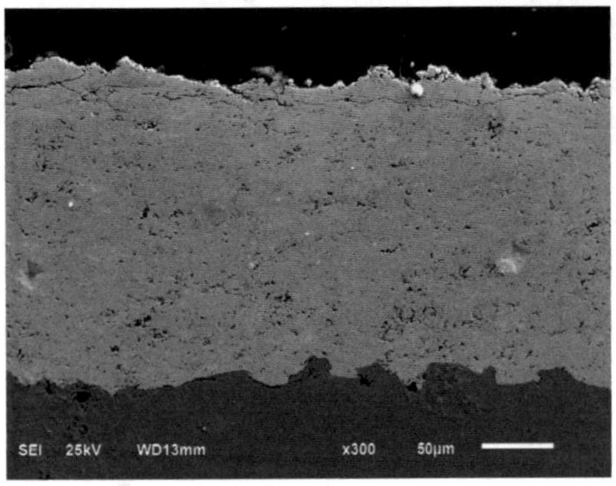

(d) after, section

Fig.6. Microstructure of the coating before and after salt fog test

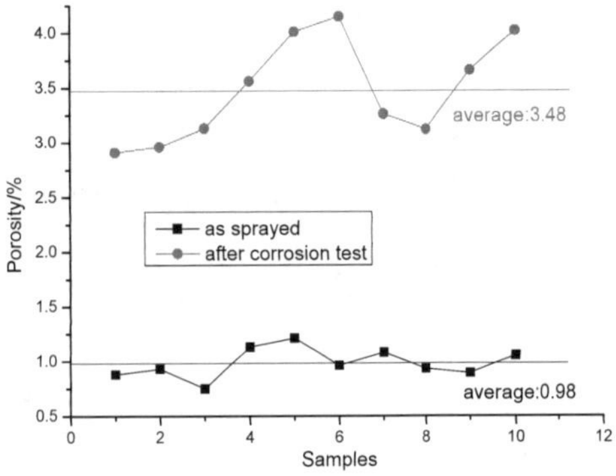

(a) Porosity of WC10Co4Cr coatings

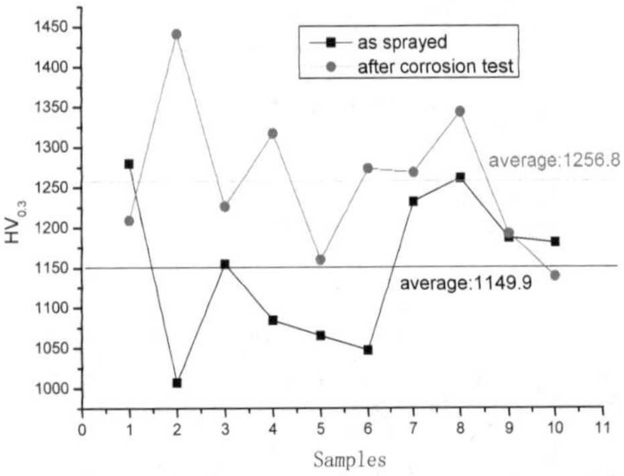

(b) Hardness of WC10Co4Cr coatings

Fig.7. Changes of the coatings before and after corrosion

4 Conclusions

1) WC10Co4Cr coatings on 300M steel can be obtained by HVOF process to protect it from corrosion conditions.

2) The optimized coatings were prepared and the coating's porosity is less than 1%. Their bonding strength is higher than 70MPa, which are according with AMS 2447.

3) There are no corrosion products in the interface between the coating and 300M steel, after 2000 hours ASTM B117 test, the optimized coatings in this paper have a good corrosion performance.

5 Literature

[1] Wu Jun, Tao Hua, Zhang Jiafu. Influence of polishing after shot peening on 300M steel. Journal of mechanical strength. 2008, 30(3), pp. 498~502

[2] Keith, Legg. Hard Chrome Replacement Options, www.rowantechnology.com

[3] Bruce Sartwell, Keith Legg, John Sauer. Validation of Thermal Spray Coatings as a Replacement for Hard Chrome Plating On Gas Turbine Engines, technical report. 2004, pp. 1-5

[4] Zhou Kesong, Song Jinbing, Liu Min. Replacement of hard chromium electroplating by thermal spraying coatings . Materials protections, 2002(12), pp. 1-4.

[5] Deng Chunming, Liu Min, Zhou Kesong. Fatigue Property of HVAF SprayedWC /17Co Coating on 300M Steel. Journal of aeronautical materials, 2007, 27(4), pp. 4-18.

[6] Wu Zi-jian, Wu Chao-jun, Wang Quan-sheng, et al. Thermal Spray Technologies and Applications, 2006, pp. 25-31

[7] W. Rusch. Comparison of Operating Characteristics for Gas and Liquid Fuel HVOF Torches [J]. Thermal Spray 2007: Global Coating Solutions, 2007. pp. 572-576.

An investigation of the corrosion and cavitation resistance of different thermally sprayed coatings in the river and marine environment

M. Fedorova, L. Baldaev, S. Baldaev, N. Baldaev, A. Akhmetgareeva, V. Martyanova
Technological systems for protective coatings Ltd, Moscow city, Russia
m.fedorova@tspc.ru

To evaluate the corrosion resistance of various coatings applied by the flame spraying method, and to compare their protective properties in sweet- and sea-water samples with seven types of coatings were made, one of them is paint, used to coat ship's bottom. SEM, Neutral spray salt, according to ASTM B117, erosion-corrosion (E-C) tests in sweet- and sea-water were carried out. The test results showed that coatings based on zinc alloys and polymer are much more effective than standard epoxy coatings and can be compared with expensive zinc-filled paints, because the latter require frequent repairs and repainting. Moreover, the developed coatings are supposed to protect against fouling. At the moment, the fouling tests are continuing.

1 Introduction

The corrosion protection of large steel structures such as bridges, pipelines, oil tanks, towers, radio and television masts, overhead walkways, and large manufacturing facilities in the metallurgical, chemical, energy, and other industries is a key issue. The protection of structures exposed to moist atmospheres and seawater such as ships, offshore platforms, seaports, is even more difficult. In most cases the surface to be protected is thousands and even tens of thousands of square meters, requiring that coating costs is competitive with those of traditional painting methods. The coating rate must be equal or higher than 10 m^2/h and if possible deposited in one unique pass. The equipment must be mobile and autonomous for operation in field conditions and must work under manual control, automation being generally difficult for largescale operations, and at last, the spray gun can be up to 30 m from its power supply and control center [1]. Flame spraying meet such requirements.

2 Experimental setup

2.1 Materials

32 samples of carbon steel were made as a base with various coatings (table 1). ZnAl15 and Zn as well-known material for anticorrosion coatings. Polymer as a good protective coating. [2] To create an anti-fouling coating, the polymer powder was mixed with silver powder and then applied to a sublayer.

According to the instructions, these paints are applied with a thickness of 450-500 μm. For comparison of corrosion and wear resistance, other coatings does not exceed standard paint in thickness. SEM cross-sections illustrating the microstructure of coatings are given in Fig. 1.

For each type of test, two samples of each type of coating were prepared. The area of each second sample was damaged to expose the underlying carbon steel.

Table 1. Values of coatings thickness

№	Coating	Thickness, mm
1	Paint	0,45-0,5
2	ZnAl15+paint	0,25-0,3
3	ZnAl15+Zn	0,2-0,3
4	ZnAl15+Zn+paint	0,35-0,5
5	ZnAl15+polymer	0,3-0,45
6	ZnAl15+Zn+ polymer	0,35-0,5
7	ZnAl15+Zn+ polymer +Ag	0,35-0,5

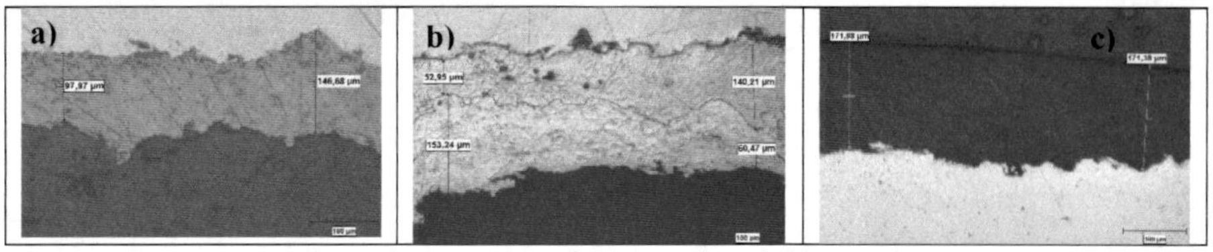

Fig. 1. SEM cross-section micrographs of (a) ZnAl and (b) ZnAl+Zn (c) polymer coatings.

2.2 Corrosion testing

The immersion corrosion test of different coatings were evaluated immersing the coatings into 3.5 wt% NaCl solution for Neutral Salt Spray (NSS) test (tab. 2), according to ASTM B117, for 360 hours, taking out the sample every 24 hours.

Table 2. NSS tests results

№	With damage	Without damage	№	With damage	Without damage
1	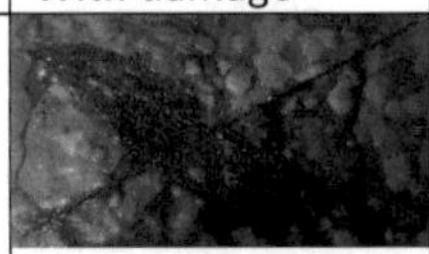		5	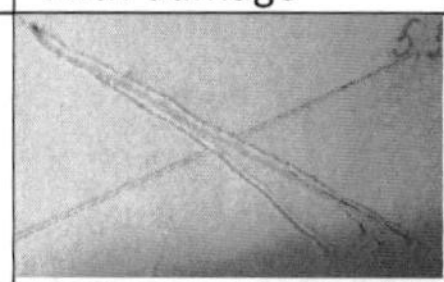	
2	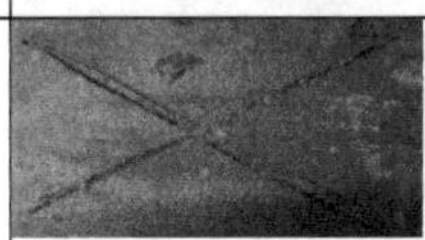		6	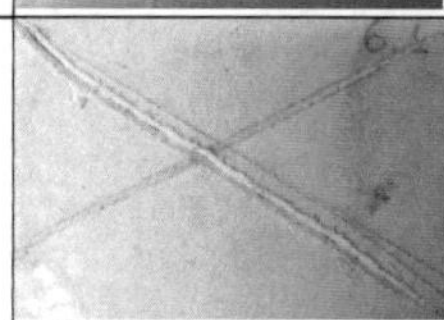	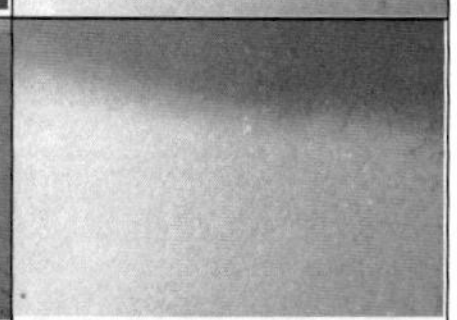
3	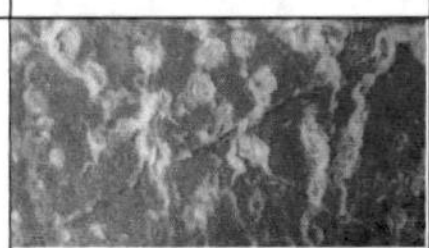	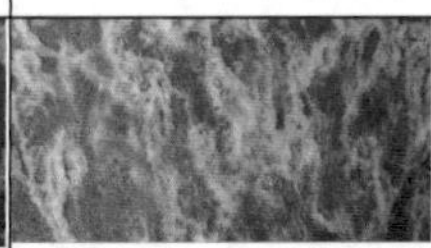	7	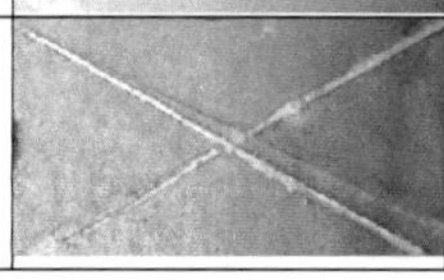	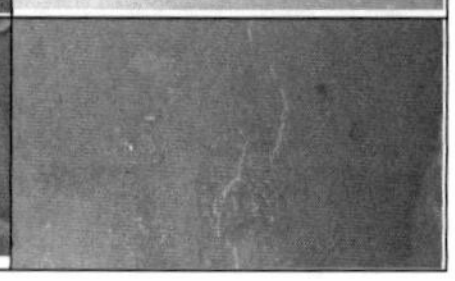
4		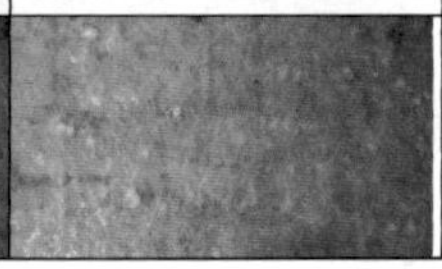			

2.3 Erosion-corrosion (E-C) testing

The aim of this test is to compare the corrosion and wear resistance of different coatings in sweet and sea-water. The coated, as well as, uncoated samples and air bubbler devices were placed in clear containers with sweet and sea-water. Test duration: 120 hours, temperature +20 °C. To assess the effect of sweet and sea-water and indirect cavitation erosion the samples were visually inspected, the coating thickness and weight were measured.

3 Results

3.1 NSS test results

The visual NSS results are presented in Fig. 4. After subjected to the salt fog for 360 h, there were rusty corrosion products on the surface of the painted sample while there was no visible rusty corrosion on the coated samples.

After testing, examination of samples with coating № 2-7 revealed the formation of calcareous deposit on the exposed steel surface with no visible steel corrosion product.

On sample with coating No. 2 without damage, there are no traces or products of corrosion. The surface is clean. On the surface of coating No. 2 with damage there are white stains. The place of the damage was overgrown with zinc and aluminum products.
On the surface of coating No. 3, a large amount of calcareous deposits was formed. This makes the appearance of the coating less presentable, but it protects the surface from biofouling. It is also noticeable that there is more deposition on the damaged sample, apparently, in this case both the sublayer and the top layer work.
Coating No. 4: the zinc sublayer affects the appearance despite the presence of paint, destroying the top layer.
Coatings No. 5-7 protect from corrosion Protect against corrosion regardless of the presence of damage. In contrast to the paint coating with a polymer top layer remain the original appearance after the tests.
The results indicate that coating based on Zink, Aluminum and polymers are highly effective for corrosion protection.

3.2 E-C test results

The visual E-C results are presented in Fig. 2, Tab. 3.

Carbon steel samples without coating after test	
Sweetwater	Seawater

Fig. 2. Appearance of carbon steel samples without coating after test.

Table 2. E-C tests results

Coating	ZnAl	ZnAl	ZnAl+Zn	ZnAl+Zn
№	1.1	2.1	3.1	4.1
Sweet-water		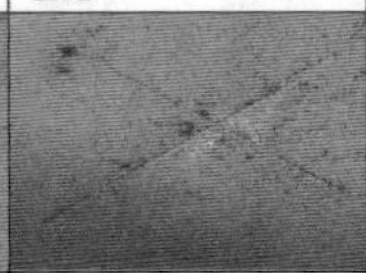		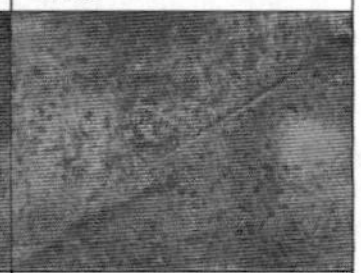
№	1.2	2.2	3.2	4.2
Sea-water		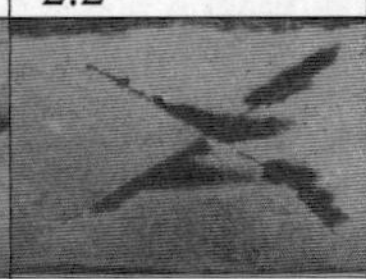	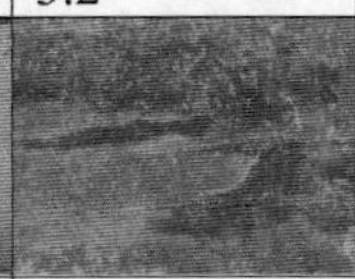	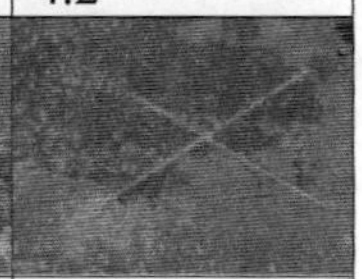
Coating	ZnAl+P	ZnAl+P	ZnAl+Zn+P	ZnAl+Zn+P
№	5.1	6.1	7.1	8.1
Sweet-water		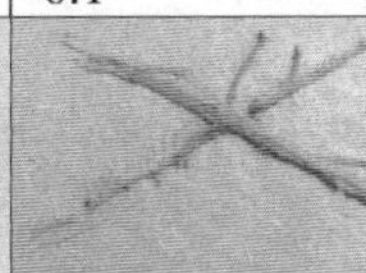		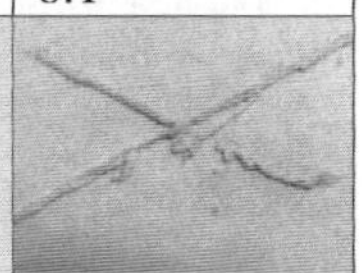
№	5.2	6.2	7.2	8.2

Sea-water	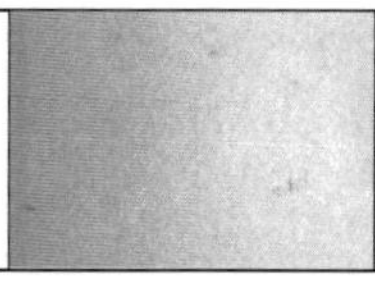			

3.2.1. E-C test in sweetwater.

After subjected to the sweetwater and air bubbling, there were rusty corrosion products on the surface of the uncoated sample.

Samples without polymer layer have white stains of calcareous deposit, visual cracks, damages, degradation of the coating. Weight gain parameter (fig. 3) for zink-coated samples is quite big, despite the fact that there is a coating erosion and as a consequence, a partial weight loss.

Samples 6-9 with polymer coating have no visual crack or erosion and in general intact. Changes in thickness and weight practically did not occur. This means that such coatings sufficiently protect the steel from corrosion and erosion, while retaining its original appearance.

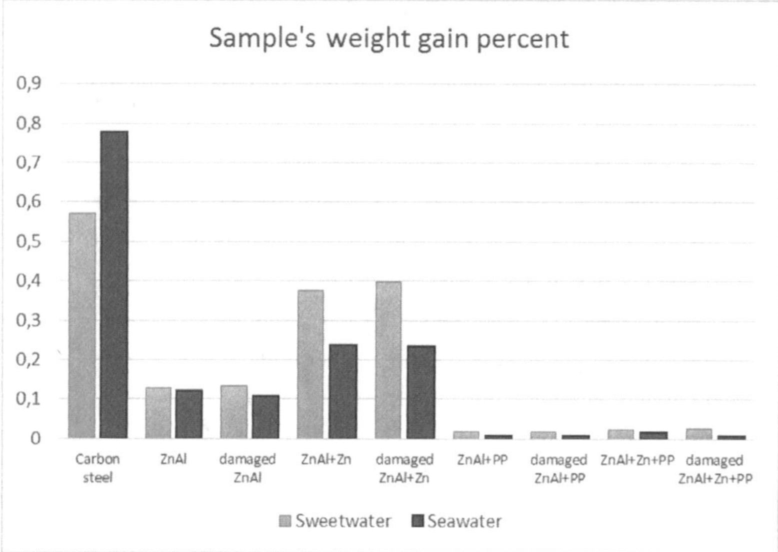

Fig. 3. Sample's weight gain percent.

3.2.2 E-C test in seawater

Uncoated sample has visual rusty corrosion and significant weight gain.

Samples without polymer layer have visual damages, and black spots where the thickness is less but there is an increase in mass. In fact, that weight gain parameter is less means more loss of coating than in the previous case.

Samples 15-18 with polymer coating have no visual crack or erosion and in general intact. Samples as well as in the previous case remained intact; damages, changes in thickness and mass are not present.

ZnAl and ZnAl+Zn coatings provide relatively less protection in the given environments as compared to coatings with polymer top layer, but it obvious that these coatings are successful to protect against corrosion of steel.

The corrosion rate of Ni-20Cr coating was found to be lower in comparison to Ni-20Cr-TiC. The XRD revealed the formation of $Cr2O3$, $Al2O3$ and $Fe2O3$ phases for the eroded-corroded scale of Ni-20Cr coating.

4 Conclusion

The formation of coating based on ZnAl and Zn with paint or polymer by flame spraying method was investigated in this research. The result revealed that the addition of a top polymer layer

increased corrosion and erosion resistance. Coatings with top layer lose less weight, it is essential for corrosion resistance.
NSS test showed that traditional paint is ineffective especially if there are some damages. In that case, all of developed systems of coatings are effective for corrosion protection, but coatings with Zink layer will provide protection against biofouling but this will have an effect on the appearance of the surface.
Coating with ZnAl sublayer and polymer top layer with a smaller coating thickness provide better protection and should be recommended for application in the river and marine environment. Biofouling resistance will be investigated in a future paper.

References

[1] Pierre L. Fauchais, Joachim V. R. Heberlein, Maher I.Boulos Thermal Spray Fundamentals (2014), pp. 1433/1435.
[2] Baldaev L. Other: Student training manual. Thermal spraying (2007), pp. 344/121.

A Review of Utilization of Niobium and Tantalum for the Enhancement in Corrosion Resistance and Biocompatibility of Bio-implants

Gurpreet Singh, Patiala, IN
Buta Singh Sidhu, Bathinda, IN
Balraj Singh, Patiala, IN

A bio-implant is a medical device used to support, replace or enhance a biological structure. Bio-implants have incalculable importance for mankind due to their crucial applications in situations like heart diseases, joint replacements, bone fixation and bone replacements etc. The materials which are used to manufacture bio-implants are generally called biomaterials. Corrosion and biocompatibility of biomaterials are the two issues, which are needed to be addressed to enhance the life span of the bio-implants. Various strategies were used to enhance the performance of biomaterials; however surface modification and alloying are the two main strategies among them. To employ these two strategies; various ceramic, metallic and composite materials were utilized as coatings and alloying elements with different biomaterials. Among all these materials, Niobium and Tantalum metals are studied by various researchers and proved to improve the corrosion resistance as well as biocompatibility of the bio-implants. This article gives a systematic overview of research work carried out in the area of bio-implants by utilizing Niobium and Tantalum for the enhancement of corrosion resistance and biocompatibility.

1 Introduction

In most of the developed and developing countries, half of the chronic diseases are either bone diseases and bone fractures including osteoporosis, bone tumor and trauma in the population over 50 years of age. According to a survey, about 10 million people in United States have osteoporosis and around 34 million are at the risk of attaining this disease [1]. The number of joint replacement procedures continues to increase significantly with over 1 million hip and knee replacement procedures taking place in the United States each year [2]. Similarly, more than 1 million stent implantation procedures are performed annually in United States [3]. So in the nearby future, these numbers will increase due to increasing of life expectancy. The field of bio-implants is not limited to the people with some kind of a disease but also cater the need of the young and dynamic people like sportspersons who need bio-implant due to an injury.

Generally, The materials which are used for the manufacturing of the bio-implants are called biomaterials. Biomaterials should possess good biocompatibility, high corrosion resistance, adequate mechanical strength and fracture toughness [4]. Biomaterials can be divided into four categories: metals, ceramics, polymers and composite materials. Metals like stainless steel (SS), titanium (Ti) alloys, nickel-titanium (NiTi) alloys and cobalt-chromium (CoCr) alloys are the most preferable material for bio-implants due to their superior mechanical properties (like strength, fracture toughness etc.). However these alloys have poor biocompatibility and release potentially harmful corrosion products (ions) in the body [5]. The corrosion products of these alloys include ions of nickel, chromium, cobalt, aluminium, vanadium etc. These ions can generate serious health problems like dermatitis, anema-B, ulcers, alzheimer disease etc. Sometimes due to excessive corrosion the fragments of the bio-implant penetrate in the surrounding tissue and a surgical procedure is necessary for the removal of metal fragments to eliminate the threat of serious infection. This re-surgical procedure adds extra cost and trauma to the patient.

Over the years, various materials had been investigated for their potential as a biomaterial. In recent times, Niobium (Nb) and Tantalum (Ta) have been attracted a lot of attention of researchers due to their potential as great biomaterials. Both these materials have excellent corrosion resistance, good biocompatibility and osteoconductivity [6]. So this paper presents a systematic overview of some of the important research studies carried out in this decade on Nb and Ta to investigate their properties as biomaterials.

2 Techniques used for investigation

For corrosion behaviour analysis, electrochemical corrosion testing was being used by most of the researchers. The output parameters which were generally stressed upon were: corrosion potential (E_{corr}), corrosion current density (I_{corr}), open circuit potential, equalibrium potential and polarization resistance (R_p) etc. Any increase/decrease in the value of these parameters is an indication of change in corrosion resistance, which is explained in the individual studies in the next section.

For biocompatibility analysis, one or more following studies used were: cell culture analysis and/or *in vivo* analysis, bioactivity analysis and antibacterial analysis. In the cell culture/antibacterial analysis, material was immersed with living cells/bacterias in body like environment clinically and after a specific time their adhesion/growth on the material was observed. *In vivo* study included surgical implantation of the specific material in the animal body and evaluation of the material after a specific time. In the bioactivity analysis, material was being immersed in simulated body fluid (SBF) and after a specific time, evaluation of the surface for the growth/bone-like apatite was done. In addition to biocompatibility/bioactivity analysis, surface wettability/hydrophilicity was also studied by some researchers in order to investigate whether surface is hydrophilic or hydrophobic. If the contact angle of liquid drop poured on surface during test was less than 90°, surface is hydrophilic otherwise hydrophobic. The degree of wettability/hydrophilicity of the surface affects biocompatibility because adhesion of bio-molecules to the surface depends upon the hydration of surface layer [7].

3 Studies on Tantalum

Various researchers investigated corrosion behaviour and biocompatibility of Ta in the form of coating and alloying element. Different coating techniques were used to obtain Ta based coatings on various materials. So in this section, some of the important studies related to Tantalum's behaviour as biomaterial are discussed.

Huang et al. [8] prepared composite coating of Ta-nitride (TaN) and silver (Ag) with different Ag contents (15 %,18 %,21 %) on pure-Ti dental implant material using magnetron sputtering method . The investigation on bacterial retention and viability revealed that all the coatings exhibited an antibacterial nature, especially the TaN coating which was found to be of hydrophobic nature (contact angle ~ 92°). Whereas, in the cell culture study, TaN coated surfaces showed better cell viability and proliferation than the uncoated material.

Chang et al. [9] produced a Ta-carbide (TaC) coating on pure-Ti substrate using magnetron sputtering method. The coating was done in two different compositions having different carbon content (~40% and ~55%). The biocompatibility along with surface wettability was evaluated. The contact angle was increased fom 47° for uncoated sample to 53° for TaC coated sample. This in turn resulted in enhanced biocompatibility determined by cell culture studies. In cell culture study, better cell viability of human cell line (MG-63) was revealed by TaC in Comparison to pure-Ti as shown in Fig 1.

Huang et al. [10] fabricated pure-Ta and Ta-oxide coatings on the Ti-substrate using magnetron sputtering method. The biocompatibility of the coatings were investigated using cell culture study and antibacterial (bacterial adhesion) analysis. The surface of pure-Ta coating was found to be hydrophobic (contact angle = 97°), while the Ta-oxide coating had a lesser contact angle (82°). In bacterial adhesion analysis, Ta-oxide coating proved to be better than pure-Ta coating. The same results were observed in cell culture studies, where tantalum oxide coatings revealed better human skin fibroblast cellular biocompatibility.

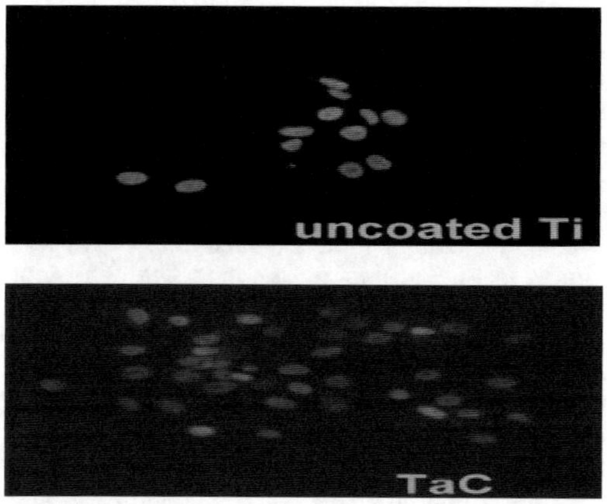

Fig. 1. Cell Cultured on Ti & TaC after 48 hours [9]

Zhang et al. [11] fabricated TaN coating on the surface of Ti-sample using magnetron sputtering method. The coating resulted in enhanced surface hardness about 12 times (2.5 GPa to 29.9 GPa). The electrochemical corrosion testing revealed the open circuit potential for coated sample had more noble level and its I_{corr} value was significantly lower (4.32 x 10^{-8} A/cm^2) than the uncoated-Ti (35.01 x 10^{-8} A/cm^2). Moreover, in bacterial adhesion testing the Ta coating remained unaffected from the bacteria. So this study concluded that the TaN coating significantly improved the corrosion resistance and proved to be of antibacterial nature.

Alves et al. [12] deposited Ta-oxide coating on the pure-Ti substrate using magnetron sputtering method. To assess the biocompatibility (osteocondutive) response of the coated material, wettability and immersion tests in SBF were performed. The results of revealed that the contact angle for coated surface was as low as 63° as compared to the uncoated surface's 74°. This study concluded that the much better apatite growth on the Ta-oxide coating substrate was attributed to the higher wettability of the surface.

Rahmati et al. [13] prepared pure-Ta coating using magnetron sputtering method on the surface of Ti-alloy (Ti-6Al-4V). Corrosion behaviour of the coating was investigated using electrochemical method and wear behaviour was also analyzed. In the corrosion testing, I_{corr} for the coated substrate (6.77 x 10^{-8} A/cm^2) was found to be significantly lower than the uncoated Ti-alloy (2.56 x 10^{-7} A/cm^2). The wear behaviour analysis revealed that the debris generation and coefficient of friction was significantly lower for the coated substrate. So this study concluded that the Ta coating enhanced the corrosion resistance as well as wear resistance of Ti-alloy.

Pham et al. [14] produced Ta coating on the CoCr-substrate using direct current sputtering. The coating improved the surface hardness about 2.5 times. Immersion test in SBF for 4 weeks revealed a healthy precipitation of apatite crystals on Ta coated sample suggested its excellent bioactivity. Moreover, the Ta coated concluded to had considerable higher biocompatibility as attachment, proliferation and differentiation of osteoblasts (MC3T3-E1) was better on it during cell culture study.

Xu et al. [15] prepared Ta-N coating on the surface of medical grade Ti-alloy (Ti-6Al-4V) using reactive sputter-deposition. The Ta-N coating was found to be highly scratch resistance thus suitable for load bearing implants. Immersion test for 14 days was performed in SBF to analyze the bioactive behaviour of the surface. In which, it was revealed that very small area with apatite layer was covered in case of Ti-alloy. However, surface of the Ta-N coated material was completely covered with bone-like apatite confirming its superior bioactivity.

Balla et al. [16] produced porous coating of Ta with varying porosities (45% & 73%) on pure-Ti using laser deposition method. Cell culture study revealed excellent cellular adherence, growth and differentiation with plentiful extracellular matrix formation on porous Ta coatings as compared to uncoated Ti. This study concluded that the improved biocompatibility was attributed to the higher wettability/hydrophillicity of the Ta surface (contact angle ~ 20°).

Roy et al. deposited [17] composite coating of Ta and magnesium oxide (MgO) on the surface of pure-Ti using Laser deposition method. The coating greatly improved the surface hardness from 185 HV (for uncoated-Ti) to 794 ± 93 HV (for coated sample). In the cell culture analysis, excellent early cellular attachment and later stage proliferation was observed on the composite coating. So this study concluded that Ta coatings had better biocompatibility than Ti and it was further improved by incorporating MgO in the coating.

Fuerst et al. [18] produced Ta coatings on Ti-alloy (Ti-6Al-4V) with different Ta content (20%, 40%, 60%, 80% & 100%) using laser deposition method. Biocompatibility analysis of the samples was done using cell culture studies. The results of this study are shown in Fig. 2, from which it was concluded that as the Ta content increased in the coating, more cell growth was observed on the samples which was evident of enhanced biocompatibility.

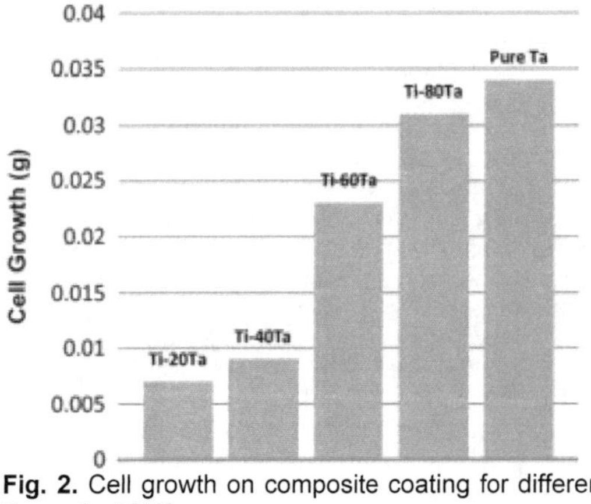

Fig. 2. Cell growth on composite coating for different Ta-content [18]

Wang et al. [19] produced Ta-oxide nanotube films on the surface of Ta sample using anodization treatment and annealing. The corrosion resistance of the surface enhanced as the I_{corr} was lesser for treated sample (10^{-9} A/cm^2) as compared to untreated sample (10^{-8} A/cm^2). Biocompatibility analysis concluded that the Ta-oxide nanotube films promoted the adhesion, proliferation, and differentiation of rabbit bone mesenchymal stem cells (rBMSCs) better than the untreated surface.

Kumar et al. [20] fabricated a Ta-oxide coating on the surface of magnesium (Mg) alloy (AZ31) using anodization method. The coating increased the surface hardness (from 80 HV to 300 HV), which enhanced the wear rate significantly. Wettability of the surface was improved incredibly as the coating had very lower contact angle (13°) than the uncoated substrate (71°). Enormous decrease in corrosion current density (I_{corr}) was observed for coated substrate (0.07 μA/cm^2) as compared to the uncoated one (25 μA/cm^2). Moreover, in cell culture studies, higher cell growth/density was observed on the Ta-oxide coating. So this study concluded that the Ta-oxide coating significantly enhanced the corrosion resistance as well the biocompatibility of the magnesium alloy.

Namur et al. [21] carried out anodization treatment on Ta-sample to produce compact Ta-oxide coating on the surface. The electrochemical corrosion analysis in artificial blood plasma revealed that the Ta-oxide layer (equalibrium potential about -60 mV) was about 10 times more stable than the pure-Ta (equalibrium potential about -600 mV). So the Ta-oxide layer was concluded to be more corrosion resistant than pute-Ta.

Arnould et al. [22] produced a Ta-oxide coating on pure-Ti substrate using sol-gel method. To investigate the bioactivity, immersion test in Simulated body Fluid (SBF) were performed for 10 days. The results revealed that an apatite layer was formed on the coated surface with a particular fiber and particle morphology, which is shown in Fig. 3. So this study concluded that the generation of the apatite layer was an indication of a biocompatible surface.

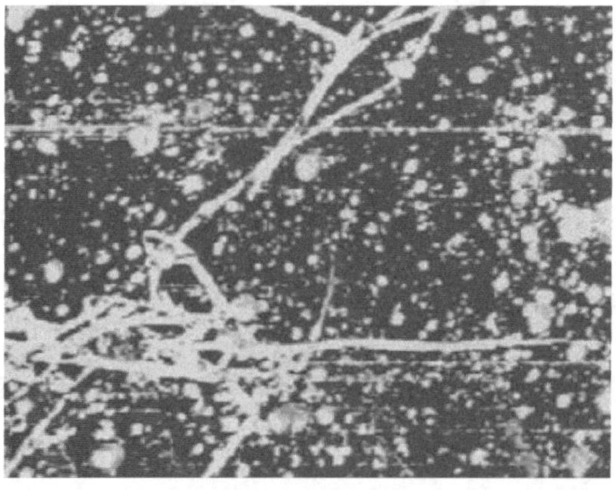

Fig. 3. Apatite formed on Ta-oxide coated sample [22]

Maho et al. [23] produced composite coating of Ta and carbon nanotubes (CNT) on the surface of pure-Ti substrate using sol gel method. The analysis for bioactivity concluded that the coated surface was covered with bone like apatite after immersion in SBF for 7 days.In another study [24], corrosion behaviour of these composite coatings was assessed. In which it was revealed that the coated surface has had significantly higher equalibrium potential and E_{corr} (45 mv & -204 mV) than the bare-Ti (-232 mV & -398 mV). In addition to having better corrosion resistance, the composite coating was found to be of hydrophilic nature (contact angle =13°).

Dong et al. [25] investigated the effect of sol-gel derived Ta-Titanium oxide composed coating on NiTi-alloy. The investigation was done on the basis of corrosion and biocompatibility testing. The contact angle was decreased as the amount of Ta was increased in the composite coating. In the corrosion testing, as the content of Ta is increased in the composite coating the corrosion current density decreased. In fact, as compared to the value for bare Niti, the corrosion current density was found to be only the half for Ta containing coating. Lastly, the platelet adhesion test was carried out to assess biocompatibility of the composite coating. The coating revealed to be excellently biocompatible and found to had acceptable hemolysis ratios for biomaterials (<5).

Li et al. [26] carried out biological evaluation of Ta coated Ti-alloy (Ti-6Al-4v). The coating was prepared using chemical vapour deposition method and both the coated and uncoated samples (scaffolds) were implanted in animal body to investigate their biocompatibility. This study concluded that the coated scaffolds presented better biocompatibility than the uncoated one as bone growth was better in case of Ta-coated scaffolds.

Yu et al. [27] fabricated Ta coating on Ti-substrate using chemical vapour deposition process. The coatings found to be of hydrophobic nature (contact angle > 100°). the corrosion behaviour of the coated surface was investigated through electrochemical corrosion testing. The results of which concluded that the coated surface had higher E_{corr} (-0.13350 V) than the uncoated surface (- 0.30546 V), hence he coating improved the corrosion resistance of Ti-substarate.

Skoog et al. [28] deposited Ta coating on the surface of CoCr-alloy (ASTM F75) using chemical vapour deposition method. The nature of the uncoated surface changed from hydrophillic (contact angle = 50°) to hydrophobic (contact angle = 93°) due to coating. Biocompatibility analysis of the Ta coating was conducted using human bone marrow-derived mesenchymal stem cells (hBMSCs) to evaluate its effect on cellular responses for orthopaedic and dental applications. The results of this study concluded that Ta coating promoted significantly better cellular interactions and growth than the uncoated CoCr-alloy.

Donkov et al. [29] produced Ta-oxide thin films on Ti-alloy (Ti-6Al-4V) using electron-beam evaporation method. The biocompatibility performance of the coating was examined using cell culture study. From surface properties analysis, it was revealed that the Ta-oxide coating lowered the contact angle, hence the wettability the surface was increased significantly. As a result, the cultured cells were well grown and attached to the coated surface which was an indication of improved biocompatibility.

Meng et al. [30] deposited Ta coating on pure-Ti using ion-implantation method. Surface of coated substrate was found to be of hydrophilic nature (Contact angle = 31°) in comparison to pure Ti substarte (contact angle = 101°). In the corrosion analysis, the coated surface exhibited higher corrosion resistance (E_{corr} = -40 V) than the bare-Ti (E_{corr} = -50 V). Moreover, after the immersion of coated sample in SBF for 28 days, bone-like apatite was formed on the surface which indicated the bioactivity of the surface. To assess biocompatibility, cell culture study was carried out. In which, it was concluded that the cells exhibited significantly higher viability and proliferation on the Ta-coating's surface.

Sun et al. [31] investigated the corrosion and biocompatibility behaviour of the Ta-oxide coated Ti. the coating was done using hydrolysis condensation process. The increase in E_{corr} (157 mV to 330 mV) was an indication of enhanced corrosion resistance. The surface wettability increased as the contact angle for coated sample was significantly lower (~ 15°) than the uncoated one (~45°). Better biocompatibility of the Ta-oxide coated surface was concluded as better cell adhesion and growth was observed on it during cell culture analysis.

Xu et al. [32] fabricated Ta-oxide coating on Ti-sample using polymer-assisted deposition. The coating found to be of hydrophillic nature (contact angle ~ 50°). In the electrochemical corrosion testing, the I_{corr} for the coated-Ti (3.08 x 10^{-7} A/cm^2) was observed much lower than the bare-Ti (2.82 x 10^{-5} A/cm^2). The cell culture study revealed that the coated surface promoted better adhesion and proliferation of the osteoblasts. So this study concluded that the Ta coated substrate fulfilled essential requirements necessary for biomaterials.

Balagna et al. [33] fabricated Ta coating on CoCr-alloy using thermal treatment in molten salts. Basically, CoCr alloys are being used as total joint substitution (hip and knee) and wear of these implants is a serious issue. So this study was done to investigate the wear behaviour of the coated Implant material. By virtue of the coating, surfcece hardness was increased from 12 GPa to 20 GPa. This study concluded that due to the coating, mean value of friction coefficient for the surface was reduced (about 40 %), so as a result the wear rate was reduced by 4 times.

Tang et al. [34] produced porous Ta coating on the surface on Ti-alloy (Ti-6Al-4V) using vaccum plasma spray method and compared its biocompatibility performance with Ti coatings deposited using the same method. The results of cell culture study demonstrated that better cell adhesion and proliferation was observed on Ta-coating, as shown in Fig. 4. The in vivo studies confirmed the results of cell culture study and concluded that better osseointegration and faster new bone formation rate was observed for Ta-coating as compared to titanium coating control.

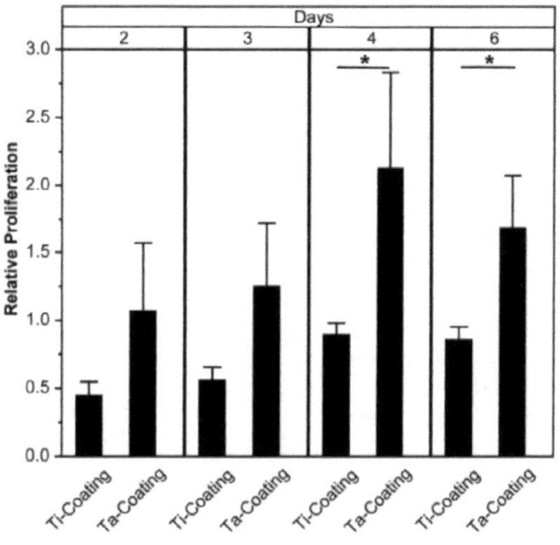

Fig. 4. Cell viability on Ti and Ta coating [34]

Wang et al. [35] evaluated the biological performance of the porous Ta. The porous Ta sample was implanted in the animal body for 12 weeks. The results of this study revealed that in the initial weeks, new bone and small blood vessels were observed at the Ta host bone interface. By the 12 week, the Ta host bone interface and pores were covered with new bone tissue while the bone trabeculae had matured and connected directly with Ta-sample.

Sagomonyants et al. [36] compared the in vitro biocompatibility of Titanium and Ta. For this purpose cell culture study using human osteoblasts (HOB) from younger (<45) and older (>60 years old) patients were compared on Ta, titanium fibermesh (TFM) and tissue culture plastic (TCP). As shown in Fig. 5. Ta greatly stimulated cell proliferation with a 4-fold and 6-fold increase relative to TFM for cells from younger and older patients, respectively, and 12- and 16-fold increase in proliferation compared to cells on TCP. So this study concluded that Ta was better in terms of cell proliferation and ability of HOB to form bone.

Cimpean et al. [37] fabricated Ti-alloy (Ti-25Ta-5Zr) with 25% content of Ta and 5% content of Zirconium (Zr) using vacuum melting technique. The in vitro corrosion and biological performance of the fabricated alloys were compared to pure-Ti. The fabricated alloys proved to be superior in terms of corrosion resistance with about two times higher impedance value (of the order of 104 Ω cm^2) as compared to pure-Ti (102 Ω cm^2). Similar trend was observed in cell culture study, cell adhesion and spreading, and enhanced cell proliferation was observed in the case of Ti-Ta alloy. So the fabricated alloy was revealed to be better than pure-Ti in terms of corrosion protection as well as in bicompatibility.

Mareci et al. [38] prepeared a Ti-alloy with 5% of Ta and 12% molybdenum (Mo) content using induction melting technique. The corrosion behaviour of Ti-alloy was compared to the pure-Ti. In which it was revealed that as compared to pure Ti, the I_{corr} decreased by about 4 times (from 8.1 µA/cm^2 to 2.1 µA/cm^2) and R_p value increased by about 2.5 times (from 1250 KΩ cm^2 to 3100 KΩ. cm^2) for Ti-Ta alloy. So it was concluded that the Ti-Ta-Mo alloy had superior corrosion resistance than the pure-Ti.

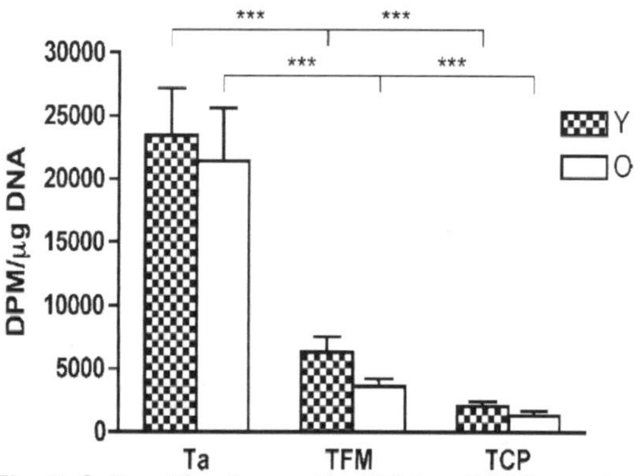

Fig. 5. Cell proliferation on Ta, TFM and TCP samples [36]

4 Studies on Niobium

Similar to Ta, various researchers investigated Nb for its utility in the field of bio-implants. Nb was investigated in the form of coating and alloying constituent. In this section, some of the important studies related to Niobium's behaviour as biomaterial are discussed.

Ramirez et al. [39] produced two coatings i.e. Nb-oxide and Nb-nitride on SS 316L substrate using magnetron sputtering method. The corrosion behaviour and biocompatibility of both the coatings were compared with each other and to medical grade SS 316L and Ti-alloy (Ti-6Al-4V). The electrochemical corrosion analysis revealed Nb-oxide and Nb-nitride had very similar I_{corr} and R_p value, as shown in Fig. 6. But as compared to SS 316L and Ti-alloy, the I_{corr} values (for both the coatings) were significantly lesser and R_p values were considerably higher. In the cell culture study, excellent cell attachment and proliferation was observed for coated surface. So it was concluded that both the Nb coatings improved the corrosion resistance as well as biocompatibility of the SS 316L.

Navarrete et al. [40] produced Nb coating on the medical grade SS 316L using magnetron sputtering method. The in vitro biological response of coated and uncoated samples were compared via cell culture study. The results of this study in form of number of cells attached on each sample surface are shown in Fig. 7, which clearly shows that significantly higher cell attachment was observed on the Nb-coated surface. So it was concluded that the Nb-coating resulted in significantly enhancing the biocompatibility of SS 316L.

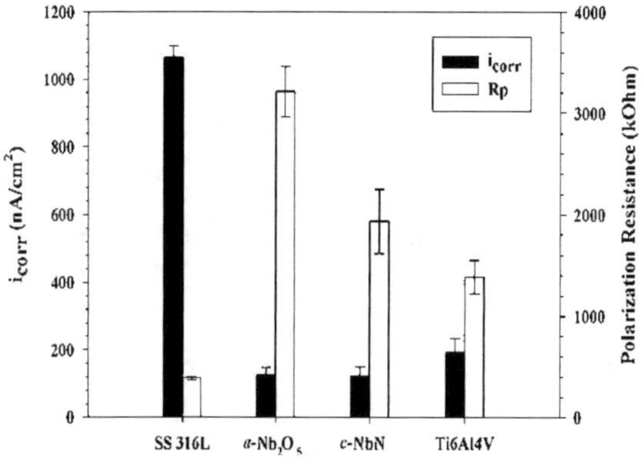

Fig. 6. Corrosion current density and polarization resistance of the surfaces [39]

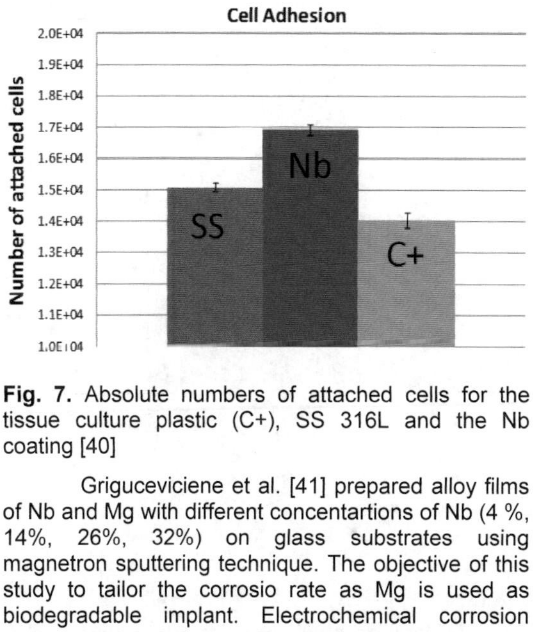

Fig. 7. Absolute numbers of attached cells for the tissue culture plastic (C+), SS 316L and the Nb coating [40]

Grigucevicíene et al. [41] prepared alloy films of Nb and Mg with different concentartions of Nb (4 %, 14%, 26%, 32%) on glass substrates using magnetron sputtering technique. The objective of this study to tailor the corrosio rate as Mg is used as biodegradable implant. Electrochemical corrosion testing revealed that as the Nb content increased, corrosion rate decreased significantly. For 4%, 14%, 26% and 32% content of Nb the degradation rate was 0.61, 0.25, 0.068 and 0.036 mm/year, respectively. So it was concluded that the corrosion rate of Mg was successfully controlled via incorporating Nb.

Huang et al. [42] produced Nb coating on polyaryl-ether-ether-ketone (PEEK) substrate using magnetron sputtering method. PEEK is considered as an alternate to mettalic implants but it suffers from insufficient biocompatibility. The Nb-coating improved the surface hardness two times which resulted in excellent wear resistance revealed by wear behaviour analysis. Moreover, the Nb-coating enhanced biocompatibility and osteogenesis of PEEK by the exhibiting excellent attachment, proliferation and differentiation of cells during cell culture study. So it was concluded that the Nb-coating could be a perfect remedy for PEEK to enhance its biocompatibility as well as mechanical properties.

Jin et al. [43] produced Nb-nitride coating over magnesium (Mg) alloy (WE43) using magnetron sputtering method. The electrochemical corrosion analysis revealed that Icorr for coated material (3.82 x 10^{-4} A/cm^2) was considerably lower than the uncoated (5.60 x 10^{-4} A/cm^2). So it was concluded that the coating successfully reduced the degradation rate of Mg-alloy.

Mazur et al. [44] compared the Nb-oxide coating and copper (Cu) reinforced Nb-oxide coating deposited on Ti-alloy (Ti-6Al-4V) alloy via magnetron sputtering method. The surface analysis of both coatings revealed that the addition of Cu in Nb-oxide did not have much effect on neither surface hardness (~8 GPa) nor wettability (contact angle ~ 75°). However, in electrochemical corrosion testing both the coatings revealed much lesser I_{corr} value (Nb-oxide = 0.24 μA/cm^2 and Cu/Nb-oxide= 0.12 μA/cm^2) than the Ti- alloy (60.50 μA/cm^2). So it was revealed that the Nb-oxide coating considerably enhanced the corrosion resistance of Ti-alloy and addition of Cu in Nb-oxide further improved it.

Pillis et al. [45] obtained Nb-oxide coating on SS 316L using magnetron sputtering method. Corrosion analysis was done using electrochemical corrosion testing and surface analysis after immersion in SBF. After immersion in SBF for 19 days, the bare SS 316L sample revealed larger pits on the surface, whereas the Nb-oxide coated sample remained almost immune from any kind of surface deterioration. The electrochemical corrosion study revealed that coated sample had about 3 times lower Icorr value (0.036 μA/cm^2) than the uncoated one (0.112 μA/cm^2). So this study concluded that the Nb-oxide coating successfully enhanced the corrosion resistance of SS 316L.

Pauline et al. [46] deposited Nb-oxide coating on the SS 316 substrate using sol gel method. To assess the corrosion behaviour and bioactivity, electrochemical corrosion analysis and immersion tests in SBF were carried out. The electrochemical corrosion testing revealed that the impedance magnitude for coated sample (about 105 Ω.cm^2) was significantly higher than that of uncoated sample (about 103 Ω.cm2). Moreover, the nanoporous morphology of the Nb-oxide coating facilitated the growth of apatite layer on the surface during bioactivity analysis. So it was concluded that the coating of Nb-oxide proved to be beneficial in enhancing the corrosion resistance and bioactivity of SS 316L.

Amarayathy et al. [47] prepared Nb-oxide coating on Mg-alloy (AZ31) using sol-gel method. The electrochemical corrosion study revealed that the value of I_{corr} for the coated alloy (0.532 x 10^{-4} A/cm^2) was about one order magnitude lesser than the coated alloy (1.194 x 10^{-4} A/cm^2).The coated surface was also found to be of hydrophillic nature and had significantly lesser contact angle (24°) than the bare alloy (65°). Moreover, the cell attachment on the coated alloy was considerably superior on the coated alloy. So it was conclude that Nb-oxide enhanced the corrosion resistance as well biocompatibility of the Mg-alloy.

Pradhan et al. [48] fabricated Nb-oxide coatings with amorphous and crystalline atomic structure on glass slides using sol gel method. Both the coatings were compared in terms of bioactivity and biocompatibility. After the immersion period of 30 days in SBF, both the coatings exhibited apatite formation ability. However, the apatite layer formed on amorphous Nb-oxide coating was more prominent. The same trend was observed in cell culture studies where better cell viability was observed in case of amorphous Nb-oxide coating. So it was concluded that both the (amorphous and crystalline) Nb-oxide coatings were proved to be bioactive and biocompatible, however, amorphous Nb-oxide coating performed better than the crystalline one.

Enayati et al. [49] produced a composite coating of Nb and hydroxyapatite (HA) on the SS 316L substrate using plasma spray method. The electrochemical corrosion testing revealed that the I_{corr} was reduced by about 12 times for Nb-HA coated sample (120 nA/cm^2) as compared to bare SS 316 L (1500 nA/cm^2). SO it was concluded that the composite coating significantly improved the corrosion resistance of SS 316L.

Nagarajan et al. [50] produced Nb-oxide coating on the surface of SS 316L using dip coating method. The electrochemical corrosion testing determined that the corrosion rate of the coated material was much lower at 0.059 x 10^{-2} mm/year as compared to uncoated one 1.8 x 10^{-2} mm/year. So this study concluded that the Nb coating significantly reduced the degradation rate of SS 316L.

Sowa et al. [51] performed anodization treatment on Nb-sample in silicate solution as electrolyte. The treatment resulted in deposition of Silicon (Si) oxide layer on the Nb surface. Surface analysis of the treated and untreated Nb revealed that the nature of the surface changed from hydrophobic (Contact angle ~ 105°)) to hydrophilic (Contact angle ~ 45°) after anodization. Results of electrochemical corrosion study showed that the Ecorr value for treated (123.6 mV) sample was significantly higher than the untreated Nb (-195.8 mV). So it was concluded that the anodization treatment enormously improved the corrosion resistance of Nb and imparted hydrophilic nature to the surface which is an indication of improved biocompatibility.

Kalisz et al. [52] prepared Nb-oxide coating on Ti-alloy (Ti-6Al-4V) using chemical vapour deposition. The coating resulted in higher surface hardness and no significant scratches were observed on the coating during wear behaviour analysis. The results of electrochemical corrosion testing revealed that the coated surface's I_{corr} (0.24 μA/cm^2) was enormously lesser than the uncoated alloy (60 μA/cm^2). So it was concluded that the Nb-oxide coating significantly improved corrosion resistance of the Ti-alloy.

Rubitschek et al. [53] fabricated a Nb-Zirconium (Zr) alloy (With 2% Zr) and compared its corrosion and corrosion fatigue behaviour with conventional medical grade Ti-alloy (Ti-6Al-4V). The electrochemical corrosion analysis revealed that the fabricated alloy and Ti-alloy had very similar E_{corr} value (~ -350 mV). In the corrosion fatigue behaviour analysis, no crack initiation behaviour was observed in SBF environment while pitting and corrosion products were also absent. So this study concluded that the fabricated Nb-Zr alloy exhibited excellent corrosion and corrosion fatigue properties comparable to conventional medical grade Ti-alloy.

Bortolome et al. [54] fabricated a composite (biocermet) of Nb and Zr and compared its biological response with pure Nb. The cell culture study revealed that the Nb sample exhibited higher cell growth than the Nb-zr composite, the results of this study are shown in Fig. 8. In vivo study of both the materials was carried out by implanting a sample material of each in an animal body. The bone response of these materials was analysed which revealed that a new bone was observed in close contact with both the material samples and no inflamed or multinucleated cells were observed nearby. So this study concluded that Nb had superior biocompatibility and osteoconductivity than the Nb-Zr composite.

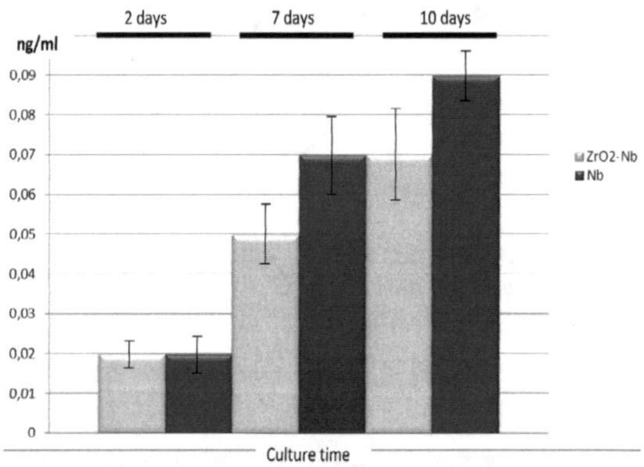

Fig. 8. Cell growth on Nb and Nb-Zr sample [54]

Mcmahon et al. [55] compared the in vitro corrosion behaviour and biological performance of Ni-Ti (Ni-50Ti) alloy and Ti-Nb (Ti-26Nb) alloy. For corrosion analysis, immersion tests were carried out for 7 days to analyze the ion release by each alloy sample in SBF. The Ni-Ti alloy released about 40 times higher ions than the Ti-Nb alloy. The results of cell culture study are shown in Fig. 9, revealed that higher number of cells were found dead on pure–Ti and Ni-Ti alloy than Ti-Nb alloy. So it was concluded that the Ti-Nb alloy proved to be more corrosion resistant and biocompatible than Ni-Ti alloy.

Dragan et al. [56] produced Ti-Nb alloy (Ti-17Nb-5Zr-5Al) using electron beam melting technique.

The corrosion behaviour of the fabricated alloy was compared to pure-Ti. The electrochemical corrosion analysis revealed that the Ti-Nb alloy had about 3 times more R_p value (~ 4.5 x 10^3 Ω.cm^2) than pure-Ti (~1.5 x 10^3 Ω.cm^2). So this study concluded that addition of Nb as alloying constituent enhanced corrosion resistance of pure-Ti.

Bai et al. [57] fabricated Ti-Nb alloy (45% Nb) by using arc melting technique. The fabricated alloy was compared to pure-Ti in terms of corrosion resistance and biocompatibility. The results of electrochemical corrosion testing revealed that I_{corr} for Ti-Nb alloy (9.48 x 10^{-8} A/cm^2) was considerably lesser than pure-Ti (15.01 x 10^{-8} A/cm^2). In vivo study revealed that in case of Ti-Nb alloy, no adverse effect on new bone formation was observed and good bone tissue compatibility was exhibited. So it was concluded that Ti-Nb alloy proved to be superior in corrosion resistance as well as biocompatibility.

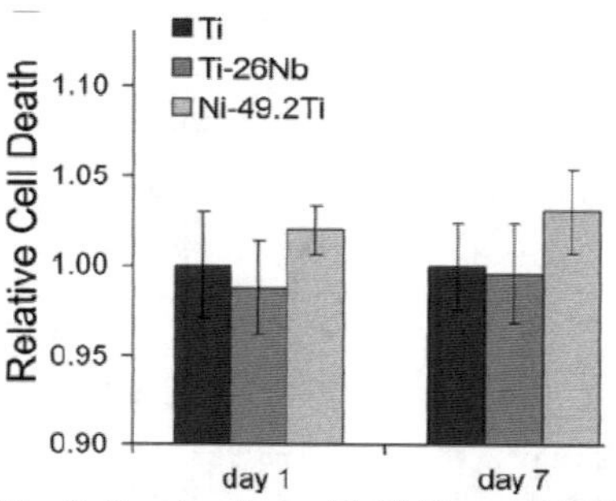

Fig. 9. Dead cells on Ti, Ti-Nb and Ni-Ti samples [55]

Weng et al. [59] prepared porous Ti-Nb alloy samples (with 25% Nb) using powder metallurgy with varying porosities (< 2%, 40%, 70%). The biological investigation of the samples was done by implating the alloy in an animal body. The results of this study revealed that as the porosity increased superior newborn bone ingrowth was observed, combined with achieving a more solid bone implant interface. So this study concluded that the porous Ti-Nb alloy implants revealed good biocompatibility as well as exhibiting a strong biological fixation between the bone and implant.

Abdi et al. [60] fabricated two Ti-alloys with and without Nb (Ti-15Si-10Zr and Ti-15Nb-15Si-10Zr) by using arc melting technique. The addition of Nb was investigated in terms of corrosion behaviour and bioactivity. The electrochemical corrosion study revealed that the incorporation of Nb imparted high pitting resistance even in extreme polarization conditions. The bioactivity analysis of the alloys revealed that after immersion in SBF for 30 days, a thicker layer of apatite was formed over the surface of Ti alloy with Nb as compared to the Ti-alloy without Nb. So this study concluded the Ti-alloy with Nb was better in terms of corrosion resistance and bioactivity as compared to Ti-alloy without Nb.

Ciurcanu et al. [61] fabricated two Ti-Nb alloys (Ti-4Mo-32Nb and Ti-10Mo-8Nb) by using cold crucible levitation melting technique. The electrochemical corrosion behaviour of the two alloys was compared. The results of this study revealed that the $I_{corr}$was at lower level (0.27 μA/cm^2 as compared to 0.49 μA/cm^2), when Nb content was increased. So it was concluded that increase in Nb content in Ti-Nb alloy improved the corrosion resistance.

Li et al. [62] fabricated a Ti-Nb alloy (Ti–25Nb– 3Zr–2Sn–3Mo) and performed a hydrothermal treatment for surface modification. The surface analysis after the treatment revealed the existence of Nb-oxide and Ti-oxide on the surface. The bioactivity analysis concluded that the formation of apatite layer took place after immersion in SBF for 15 days, which is an indication of biocompatible surface.

Anderi et al. [63] produced an alloy containing CoCr with 6% Nb using vacuum melting technique and compared its corrosion behaviour and biocompatibility with commercially available CoCr alloy (Co-28Cr-6Mo). The electrochemical corrosion study revealed that the fabricated alloy had about 20 times lesser icorr (8.29 nA/cm^2) than the commercial CoCr alloy (168.05 nA/cm^2). Similar trend was observed in cell culture study, where the cells needed more time to grow on commercial alloy and distributed randomly, but on CoCr-Nb alloy cells grew rapidly and distributed linearly. So this study concluded that the incorporation of Nb in CoCr not only significantly enhanced corrosion resistance but also revealed better biocompatibility than the commercially available CoCr-alloy.

5 Studies on Miscellaneous Ta-Nb Alloys

In some of the studies, Ta and Nb were simultaneously combined with other metals to form a new alloy to get the benefit of both these materials altogather. The important studies of this kind are discussed in this section.

Vasilescu et al. [64] fabricated an alloy of Ti-Nb-Ta (Ti-25Nb-25Ta) using cold crucible semi-levitation melting technique. Corrosion behaviour of the fabricated alloy was compared to pure-ti. The electrochemical corrosion study revealed that the Ti-Nb-Ta alloy had R_p value about 255 KΩ.cm^2, which was 23 times greater than pure-Ti (about 11 KΩ.cm^2). Moreover, the ion release rate of pure-Ti (876 ng/cm^2) in SBF was about 20 times more than the Ti-Nb-Ta alloy (43 ng/cm^2). So it was concluded that the fabricated Ti-Nb-Ta alloy was considerably better than the pure-Ti in terms of corrosion resistance.

Popa et al. [65] prepared an alloy of Ti-Nb-Zr-Ta (Ti-20Nb-10Zr-5Ta) by using vaccum melting technique. The corrosion properties of the fabricated alloy were compared to pure-Ti. Results of the electrochemical corrosion analysis revealed that the Ti-Nb-Zr-Ta alloy had about 15 times greater R_p value (192 KΩ.cm^2) than pure-Ti (13 KΩ.cm^2). Moreover, the ion release rate of pure-Ti (876 ng/cm^2) in SBF was about 5 times more than the Ti-Nb-Ta alloy (162 ng/cm^2). So this study concluded that the fabricated

Ti-Nb-Zr-Ta alloy was found to be more stable with higher corrosion resistance than pure-Ti.

Guo et al. [66] fabricated a Ti-alloy (Ti-35-Nb-3Zr-2Ta) using vacuum melting technique. The corrosion behaviour and biocompatibility of the fabricated alloy was compared to the medical grade Ti alloy (Ti-6Al-4V). The electrochemical corrosion analysis revealed that the open circuit potential for the fabricated alloy was significantly higher which revealed lower degradation rate. The fabricated alloy surface alloy exhibited higher wettability with lower contact angle (30°) as compared to Ti-6Al-4V (41°). The cell culture analysis revealed more cell adhesion and rapid proliferation on the surface of fabricated alloy. So it was concluded that the fabricated alloy seemed more suitable for bio-implant applications than Ti-6Al-4V due to its higher corrosion resistance and biocompatibility.

Barbinta et al. [67] produced two Ti-Nb-Ta alloys (Ti-25Nb-8Ta-10Zr and Ti-21Nb-15Ta-6Zr) by induction melting technique. The corrosion behaviour of both the alloys was analysed and compared to pure-Ti. The electrochemical corrosion study revealed that the polarization resistance of Ti-25Nb-8Ta-10Zr (8.4×10^{-4} $\Omega.cm^2$) and Ti-21Nb-15Ta-6Zr (9.5×10^{-4} $\Omega.cm^2$) was greater than pure-Ti (7.1×10^{-4} $\Omega.cm^2$). So it was concluded that the fabricated alloys had better corrosion resistance than pure-Ti.

Biesiekierski et al. [68] investigated corrosion behaviour and biocompatibility of fabricated Ti-Nb and Ti-Ta alloys by varying the weight percentages of Nb (0%, 12%) and Ta (0%, 7%,10%) constituents on the basis of certain electronic parameters (such as mean bond order, mean molecular d-orbital energy level and the mean electron to atom ratio). The electrochemical corrosion study revealed that the open circuit potential was on more noble side for Ti-Nb and Ti-Ta alloys than the pure-Ti and medical grade Ti-alloy (Ti-6Al-4V). Similar trend was observed in cell culture study, where excellennt cell attachment and growth was observed on fabricated Ti-Ta and Ti-Nb alloys. So it was concluded that the corrosion and biocompatibility of fabricated alloys was on par with pure-Ti and medical grade Ti-alloy.

6 Summary

In the recent times, various researchers investigated Ta and Nb for their application as biomaterials. To get the benefit of interesting properties of Ta and Nb, various other materials were combined with these metals in the form of coatings or alloying. Different coating methods were used to obtain Ta and Nb based coatings on various materials. Collaboration of Ta and Nb was also done with other malloys like Ti-alloys, Ni-Ti alloy, CoCr-alloy etc. to obtain superior combination as a new biomaterial. In almost all the cases, incorporation of Ta or/and Nb was proved to be leading to enhanced corrosion resistance and biocompatibility/bioactivity.

7 References

[1] US Department of Health and Human Services, 2004. Bone health and osteoporosis: a report of the Surgeon General. Rockville, MD: US Department of Health and Human Services, Office of the Surgeon General, 87.

[2] Kurtz, S., Ong, K., Lau, E., Mowat, F. and Halpern, M., 2007. Projections of primary and revision hip and knee arthroplasty in the United States from 2005 to 2030. J Bone Joint Surg Am, 89(4), pp.780-785.

[3] Epstein, A.J., Polsky, D., Yang, F., Yang, L. and Groeneveld, P.W., 2011. Coronary revascularization trends in the United States, 2001-2008. Jama, 305(17), pp.1769-1776.

[4] Kim, S.M., Jo, J.H., Lee, S.M., Kang, M.H., Kim, H.E., Estrin, Y., Lee, J.H., Lee, J.W. and Koh, Y.H., 2014. Hydroxyapatite-coated magnesium implants with improved in vitro and in vivo biocorrosion, biocompatibility, and bone response. Journal of Biomedical Materials Research Part A, 102(2), pp.429-441.

[5] Singh, G., Singh, H. and Sidhu, B.S., 2013. Characterization and corrosion resistance of plasma sprayed HA and HA–SiO_2 coatings on Ti–6Al–4V. Surface and Coatings Technology, 228, pp.242-247. Matsuno, H., Yokoyama, A., Watari, F., Uo, M. and Kawasaki, T., 2001. Biocompatibility and osteogenesis of refractory metal implants, titanium, hafnium, niobium, tantalum and rhenium. Biomaterials, 22(11), pp.1253-1262.

[6] Gopi, D., Murugan, N., Ramya, S. and Kavitha, L., 2014. Electrodeposition of a porous strontium-substituted hydroxyapatite/zinc oxide duplex layer on AZ91 magnesium alloy for orthopedic applications. Journal of Materials Chemistry B, 2(34), pp.5531-5540.

[7] Huang, H.L., Chang, Y.Y., Lai, M.C., Lin, C.R., Lai, C.H. and Shieh, T.M., 2010. Antibacterial TaN-Ag coatings on titanium dental implants. Surface and coatings technology, 205(5), pp.1636-1641.

[8] Chang, Y.Y., Huang, H.L., Chen, Y.C., Hsu, J.T., Shieh, T.M. and Tsai, M.T., 2014. Biological characteristics of the MG-63 human osteosarcoma cells on composite tantalum carbide/amorphous carbon films. PloS one, 9(4), p.e95590.

[9] Huang, H.L., Chang, Y.Y., Chen, H.J., Chou, Y.K., Lai, C.H. and Chen, M.Y., 2014. Antibacterial properties and cytocompatibility of tantalum oxide coatings with different silver content. Journal of Vacuum Science & Technology A: Vacuum, Surfaces, and Films, 32(2), p.02B117.

[10] Zhang, Y., Zheng, Y., Li, Y., Wang, L., Bai, Y., Zhao, Q., Xiong, X., Cheng, Y., Tang, Z., Deng, Y. and Wei, S., 2015. Tantalum nitride-decorated titanium with enhanced resistance to microbiologically induced corrosion and mechanical property for dental application. PloS one, 10(6), p.e0130774.

[11] Alves, C.A., Cavaleiro, A. and Carvalho, S., 2016. Bioactivity response of Ta_{1-x} O_x coatings deposited by reactive DC magnetron sputtering. Materials Science and Engineering: C, 58, pp.110-118.

[12] Rahmati, B., Sarhan, A.A., Basirun, W.J. and Abas, W.A.B.W., 2016. Ceramic tantalum oxide thin film coating to enhance the corrosion and wear characteristics of Ti6Al4V alloy. Journal of Alloys and Compounds, 676, pp.369-376.

[13] Pham, V.H., Lee, S.H., Li, Y., Kim, H.E., Shin, K.H. and Koh, Y.H., 2013. Utility of tantalum (Ta) coating to improve surface hardness in vitro bioactivity and biocompatibility of Co–Cr. Thin Solid Films, 536, pp.269-274.

[14] Xu, J., Liu, L., Munroe, P. and Xie, Z.H., 2015. Promoting bone-like apatite formation on titanium alloys through nanocrystalline tantalum nitride coatings. Journal of Materials Chemistry B, 3(19), pp.4082-4094.

[15] Balla, V.K., Bodhak, S., Bose, S. and Bandyopadhyay, A., 2010. Porous tantalum structures for bone implants: fabrication, mechanical and in vitro biological properties. Acta biomaterialia, 6(8), pp.3349-3359.

[16] Roy, M., Balla, V.K., Bandyopadhyay, A. and Bose, S., 2012. MgO-doped tantalum coating on Ti: microstructural study and biocompatibility evaluation. ACS applied materials & interfaces, 4(2), pp.577-580.

[17] Fuerst, J., Medlin, D., Carter, M., Sears, J. and Vander Voort, G., 2015. LASER additive manufacturing of titanium-tantalum alloy structured interfaces for modular orthopedic devices. JOM, 67(4), pp.775-780.

[18] Wang, N., Li, H., Wang, J., Chen, S., Ma, Y. and Zhang, Z., 2012. Study on the anticorrosion, biocompatibility, and osteoinductivity of tantalum decorated with tantalum oxide nanotube array films. ACS applied materials & interfaces, 4(9), pp.4516-4523.
[19] Kumar, A.M., Thangavel, E., Ramakrishna, S., Obot, I.B., Jung, H.C., Shin, K.S., Gasem, Z.M., Kim, H. and Kim, D.E., 2014. Multi-functional ceramic hybrid coatings on biodegradable AZ31 Mg implants: electrochemical, tribological and quantum chemical aspects for orthopaedic applications. RSC Advances, 4(46), pp.24272-24285.
[20] Namur, R.S., Reyes, K.M. and Marino, C.E.B., 2015. Growth and Electrochemical Stability of Compact Tantalum Oxides Obtained in Different Electrolytes for Biomedical Applications. Materials Research, 18, pp.91-97.
[21] Arnould, C., Volcke, C., Lamarque, C., Thiry, P.A., Delhalle, J. and Mekhalif, Z., 2009. Titanium modified with layer-by-layer sol–gel tantalum oxide and an organodiphosphonic acid: A coating for hydroxyapatite growth. Journal of colloid and interface science, 336(2), pp.497-503.
[22] Maho, A., Detriche, S., Delhalle, J. and Mekhalif, Z., 2013. Sol–gel synthesis of tantalum oxide and phosphonic acid-modified carbon nanotubes composite coatings on titanium surfaces. Materials Science and Engineering: C, 33(5), pp.2686-2697.
[23] Maho, A., Linden, S., Arnould, C., Detriche, S., Delhalle, J. and Mekhalif, Z., 2012. Tantalum oxide/carbon nanotubes composite coatings on titanium, and their functionalization with organophosphonic molecular films: A high quality scaffold for hydroxyapatite growth. Journal of colloid and interface science, 371(1), pp.150-158.
[24] Dong, B.H., Wu, F., Alajmi, Z., Zhang, C., Fu, T. and Ge, Y., 2014. Sol–gel derived Ta-containing TiO_2 films on surface roughened NiTi alloy. Rare Metals, 33(1), pp.21-27.
[25] Li, X., Wang, L., Yu, X., Feng, Y., Wang, C., Yang, K. and Su, D., 2013. Tantalum coating on porous Ti6Al4V scaffold using chemical vapor deposition and preliminary biological evaluation. Materials Science and Engineering: C, 33(5), pp.2987-2994.
[26] Yu, X., Tan, L., Yang, H. and Yang, K., 2015. Surface characterization and preparation of Ta coating on Ti6Al4V alloy. Journal of Alloys and Compounds, 644, pp.698-703.
[27] Skoog, S.A., Kumar, G., Goering, P.L., Williams, B., Stiglich, J. and Narayan, R.J., 2016. Biological Response of Human Bone Marrow-Derived Mesenchymal Stem Cells to Commercial Tantalum Coatings with Microscale and Nanoscale Surface Topographies. JOM, 68(6), pp.1672-1678.
[28] Donkov, N., Zykova, A., Safonov, V., Rogowska, R., Smolik, J. and Luk'yanchenko, V., 2011. Tantalum pentoxide ceramic coatings deposition on Ti4Al6V substrates for biomedical applications. Plasma Physics, 17(1), pp.131-133.
[29] Meng, F., Li, Z. and Liu, X., 2013. Synthesis of tantalum thin films on titanium by plasma immersion ion implantation and deposition. Surface and Coatings Technology, 229, pp.205-209.
[30] Sun, Y.S., Chang, J.H. and Huang, H.H., 2013. Corrosion resistance and biocompatibility of titanium surface coated with amorphous tantalum pentoxide. Thin Solid Films, 528, pp.130-135.
[31] Xu, G., Shen, X., Hu, Y., Ma, P. and Cai, K., 2015. Fabrication of tantalum oxide layers onto titanium substrates for improved corrosion resistance and cytocompatibility. Surface and Coatings Technology, 272, pp.58-65.
[32] Balagna, C., Faga, M.G. and Spriano, S., 2014. Tribological behavior of a Ta-based coating on a Co–Cr–Mo alloy. Surface and Coatings Technology, 258, pp.1159-1170.
[33] Tang, Z., Xie, Y., Yang, F., Huang, Y., Wang, C., Dai, K., Zheng, X. and Zhang, X., 2013. Porous tantalum coatings prepared by vacuum plasma spraying enhance bmscs osteogenic differentiation and bone regeneration in vitro and in vivo. PloS one, 8(6), p.e66263.
[34] Wang, Q., Zhang, H., Li, Q., Ye, L., Gan, H., Liu, Y., Wang, H. and Wang, Z., 2015. Biocompatibility and osteogenic properties of porous tantalum. Experimental and therapeutic medicine, 9(3), pp.780-786.
[35] 26. Sagomonyants, K.B., Hakim-Zargar, M., Jhaveri, A., Aronow, M.S. and Gronowicz, G., 2011. Porous tantalum stimulates the proliferation and osteogenesis of osteoblasts from elderly female patients. Journal of Orthopaedic Research, 29(4), pp.609-616.
[36] Cimpean, A., Vasilescu, E., Drob, P., Cinca, I., Vasilescu, C., Anastasescu, M., Mitran, V. and Drob, S.I., 2014. Enhancement of the electrochemical behaviour and biological performance of Ti–25Ta–5Zr alloy by thermo-mechanical processing. Materials Science and Engineering: C, 38, pp.127-142.
[37] Mareci, D., Chelariu, R., Cailean, A., Brinza, F., Bolat, G. and Gordin, D.M., 2015. Electrochemical characterization of Ti12Mo5Ta alloys in contact with saline medium. Transactions of Nonferrous Metals Society of China, 25(1), pp.345-352.
[38] Ramírez, G., Rodil, S.E., Arzate, H., Muhl, S. and Olaya, J.J., 2011. Niobium based coatings for dental implants. Applied Surface Science, 257(7), pp.2555-2559.
[39] Navarrete, R.O., Olaya, J.J., Ramírez, C. and Rodil, S.E., 2011. Biocompatibility of niobium coatings. Coatings, 1(1), pp.72-87.
[40] Grigucevičienė, A., Staišiūnas, L., Miečinskas, P., Giraitis, R., Leinartas, K. and Juzeliūnas, E., 2013. Electrochemical study of Mg-Nb alloy films in Hank's solution. chemija, 24(3), pp.190-196.
[41] Huang, J., Wan, S., Liu, B. and Xue, Q., 2014. Improved adaptability of PEEK by Nb doped graphite-like carbon composite coatings for bio-tribological applications. Surface and Coatings Technology, 247, pp.20-29.
[42] Jin, W., Wu, G., Li, P. and Chu, P.K., 2014. Improved corrosion resistance of Mg-Y-RE alloy coated with niobium nitride. Thin Solid Films, 572, pp.85-90.
[43] Mazur, M., Kalisz, M., Wojcieszak, D., Grobelny, M., Mazur, P., Kaczmarek, D. and Domaradzki, J., 2015. Determination of structural, mechanical and corrosion properties of Nb 2 O 5 and $(Nb_yCu_{1-y})O_x$ thin films deposited on Ti6Al4V alloy substrates for dental implant applications. Materials Science and Engineering: C, 47, pp.211-221
[44] Pillis, M.F., Geribola, G.A., Scheidt, G., de Araújo, E.G., de Oliveira, M.C.L. and Antunes, R.A., 2016. Corrosion of thin, magnetron sputtered Nb_2O_5 films. Corrosion Science, 102, pp.317-325.
[45] Pauline, S.A. and Rajendran, N., 2014. Biomimetic novel nanoporous niobium oxide coating for orthopaedic applications. Applied Surface Science, 290, pp.448-457.
[46] Amaravathy, P., Sowndarya, S., Sathyanarayanan, S. and Rajendran, N., 2014. Novel sol gel coating of Nb_2O_5 on magnesium alloy for biomedical applications. Surface and Coatings Technology, 244, pp.131-141.
[47] Pradhan, D., Wren, A.W., Misture, S.T. and Mellott, N.P., 2016. Investigating the structure and biocompatibility of niobium and titanium oxides as coatings for orthopedic metallic implants. Materials Science and Engineering: C, 58, pp.918-926.
[48] Enayati, M.H., Fathi, M.H. and Zomorodian, A., 2009. Characterisation and corrosion properties of novel hydroxyapatite niobium plasma sprayed coating. Surface Engineering, 25(4), pp.338-342.
[49] Nagarajan, S., Raman, V. and Rajendran, N., 2010. Synthesis and electrochemical characterization of porous niobium oxide coated 316L SS for orthopedic applications. Materials Chemistry and Physics, 119(3), pp.363-366.
[50] Sowa, M., Kazek-Kęsik, A., Krząkała, A., Socha, R.P., Dercz, G., Michalska, J. and Simka, W., 2014. Modification of niobium surfaces using plasma electrolytic oxidation in silicate solutions. Journal of Solid State Electrochemistry, 18(11), pp.3129-3142.
[51] Kalisz, M., Grobelny, M., Mazur, M., Zdrojek, M., Wojcieszak, D., Świniarski, M., Judek, J. and Kaczmarek, D., 2015. Comparison of mechanical and corrosion properties of graphene monolayer on Ti–Al–V and nanometric Nb 2 O 5 layer on Ti–Al–V alloy for dental implants applications. Thin Solid Films, 589, pp.356-363.
[52] Rubitschek, F., Niendorf, T., Karaman, I. and Maier, H.J., 2012. Corrosion fatigue behavior of a biocompatible ultrafine-grained niobium alloy in simulated body fluid. Journal of the mechanical behavior of biomedical materials, 5(1), pp.181-192.
[53] Bartolomé, J.F., Moya, J.S., Couceiro, R., Gutiérrez-González, C.F., Guitian, F. and Martinez-Insua, A., 2016. In vitro and in vivo evaluation of a new zirconia/niobium biocermet for hard tissue replacement. Biomaterials, 76, pp.313-320.
[54] McMahon, R.E., Ma, J., Verkhoturov, S.V., Munoz-Pinto, D., Karaman, I., Rubitschek, F., Maier, H.J. and Hahn, M.S., 2012. A comparative study of the cytotoxicity and corrosion resistance of nickel–titanium and titanium–niobium shape memory alloys. Acta biomaterialia, 8(7), pp.2863-2870.
[55] Dragan, L.A., Chelariu, R., Mareci, D. and Munteanu, C., 2014. Electrochemical behavior of new experimental TiNbZrAl

alloys for dental applications. Materials and Corrosion, 65(8), pp.828-836.
[56] Bai, Y., Deng, Y., Zheng, Y., Li, Y., Zhang, R., Lv, Y., Zhao, Q. and Wei, S., 2016. Characterization, corrosion behavior, cellular response and in vivo bone tissue compatibility of titanium–niobium alloy with low Young's modulus. Materials Science and Engineering: C, 59, pp.565-576.
[57] Weng, X., Yang, H., Xu, J., Li, X., Liao, Q. and Wang, J., 2016. In vivo testing of porous Ti-25Nb alloy serving as a femoral stem prosthesis in a rabbit model. Experimental and Therapeutic Medicine, 12(3), pp.1323-1330.
[58] Weng, X., Yang, H., Xu, J., Li, X., Liao, Q. and Wang, J., 2016. In vivo testing of porous Ti-25Nb alloy serving as a femoral stem prosthesis in a rabbit model. Experimental and Therapeutic Medicine, 12(3), pp.1323-1330.
[59] Abdi, S., Oswald, S., Gostin, P.F., Helth, A., Sort, J., Baró, M.D., Calin, M., Schultz, L., Eckert, J. and Gebert, A., 2016. Designing new biocompatible glass-forming Ti75-xZr10NbxSi15 (x= 0, 15) alloys: corrosion, passivity, and apatite formation. Journal of Biomedical Materials Research Part B: Applied Biomaterials, 104(1), pp.27-38.
[60] Ciurcanu, O.E., Mareci, D., Stefanescu, O.M., Trinca, L.C., Scutariu, M.M., Ilie, M. and Hritcu, L.D., 2016. Electrochemical Behaviour of TiMoNb Alloys in Hanks Balanced Salt Solution with Addition of Aminoacids Infusion Solution. REVISTA DE CHIMIE, 67(10), pp.2095-2099.
[61] Li, H., Fu, T., Li, W., Alajmi, Z. and Sun, J., 2016. Hydrothermal growth of TiO2-CaP nano-films on a Ti–Nb-based alloy in concentrated calcium phosphate solutions. Journal of Nanoparticle Research, 18(1), p.4.
[62] Andrei, M., Galateanu, B., Hudita, A., Costache, M., Osiceanu, P., Moreno, J.C., Drob, S.I. and Demetrescu, I., 2016. Electrochemical comparison and biological performance of a new CoCrNbMoZr alloy with commercial CoCrMo alloy. Materials Science and Engineering: C, 59, pp.346-355.
[63] Vasilescu, E., Drob, P., Vasilescu, C., Drob, S.I., Bertrand, E., Gordin, D.M. and Gloriant, T., 2010. Corrosion resistance of the new Ti-25Ta-25Nb alloy in severe functional conditions. Materials and Corrosion, 61(11), pp.947-954.
[64] Popa, M., Vasilescu, E., Drob, P., Raducanu, D., Moreno, J.M.C., Ivanescu, S., Vasilescu, C. and Drob, S.I., 2012. Microstructure, mechanical, and anticorrosive properties of a new Ti-20Nb-10Zr-5Ta alloy based on nontoxic and nonallergenic elements. Metals and Materials International, 18(4), pp.639-645.
[65] Guo, Y., Chen, D., Lu, W., Jia, Y., Wang, L. and Zhang, X., 2013. Corrosion resistance and in vitro response of a novel Ti35Nb2Ta3Zr alloy with a low Young's modulus. Biomedical Materials, 8(5), p.055004.
[66] Bărbînţă, A.C., Mareci, D., Chelariu, R., Bolat, G., Munteanu, C., Cho, K. and Niinomi, M., 2014. The estimation of corrosion behavior of new TiNbTaZr alloys for biomedical applications. Materials and Corrosion, 65(10), pp.1017-1023.
[67] Biesiekierski, A., Lin, J., Li, Y., Ping, D., Yamabe-Mitarai, Y. and Wen, C., 2016. Investigations into Ti–(Nb, Ta)–Fe alloys for biomedical applications. Acta biomaterialia, 32, pp.336-347.

High velocity atmospheric plasma sprayer

Yuri Chivel, Saint Etienne/FR

Abstract: Atmospheric electromagnetic plasmadynamic system and method for producing high –power plasma flows at atmospheric pressure with high frequency have been developed. In this system plasma flow velocity reaches 2-4 · 10^3 m/s with temperature up to 20 · 10^3 K. The results are presented on the use of this system for surface layer modification and its possibility for particles acceleration

1 Introduction

Atmospheric plasma generation has attracted much attention due the possibility carry uot some unique plasma technological processes in air.
Electromagnetic plasmadynamic systems (EMAPS) [1-3] belong to the most promising ones in a number of applications: surface hardening, thermal spraying, surface modifications These systems are capable of accelerating powder particles to the velocity of up to 2.5-3· 10^3 m/s [3]. Such high particle velocityy is unattainable by other spraying methods.
But EMAPS operates only at low pressures of the ambient gas (10^2- 10^3 Pa) and low frequency of operation < 0.1 Hz which presents problems as far as their industrial application is concerned. Attempts to obtain high-energy pulsed plasma flows at atmospheric pressure are confronted with difficulties of producing homogeneous discharges in the plasma accelerator channel. A high power atmospheric electromagnetic plasmadynamic system (AEMAPS) has been designed, which operates at atmospheric pressure.

2 AEMAPS principles of operation

The new method for obtaining high –energy repetitively pulsed plasma flows at atmospheric and high pressure [4,5] consists in formation a region of reduced-density gas with high degree of ionization at the beginning of the acceleration channel of the discharge device AEMAPS, where plasma current sheath is formed.

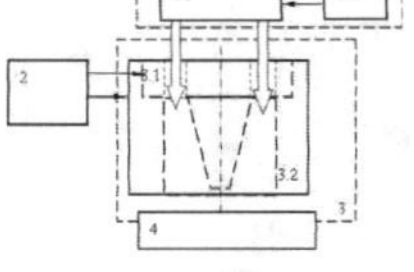
a

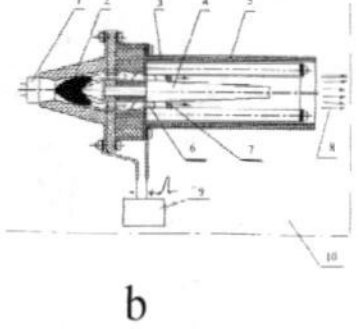
b

Fig .1. Functional scheme of AEMAPS (a): 1- DC plasma source: 1.1 - DC – torch, 1.2 – DC power supply ; 2- pulse power supply ; 3- plasma accelerator, 3.1 – cathode , 3.2 – anode ; 4- specimen.
Scheme of discharge device of AEMAPS (b,c):
1- DC plasma source , 2 –plasma torch, 3 – anode; 4 – cathode; 5 – ceramic screen

As the realization of the method a new hybrid two-stage plasmadynamic system with a first stage in the form of one or more DC plasma torch with flow temperature ~ 3000 K – 4000 K at atmospheric pressure and a second one in the form of a coaxial plasma accelerator (fig.1) has been designed [9]. In this system plasma of the DC torch fills the acceleration channel. A high-voltage impulse is applied to the accelerating gap, its narrowest zone at the beginning of the channel.

As a result, a breakdown occurs with the formation of the plasma current shell. In the area of the formation of the plasma current shell the density of the DC-induced plasma is up to 0.1 atmospheres. In accordance with Ref. [9], the degree of ionization of the DC-induced air plasma at the temperature of 3000-5000 K and atmospheric pressure is determined mainly by the associated ionization mechanism and ranges from ~ 10^{-9} at 3000 K up to ~10^{-6} at 4000 K.

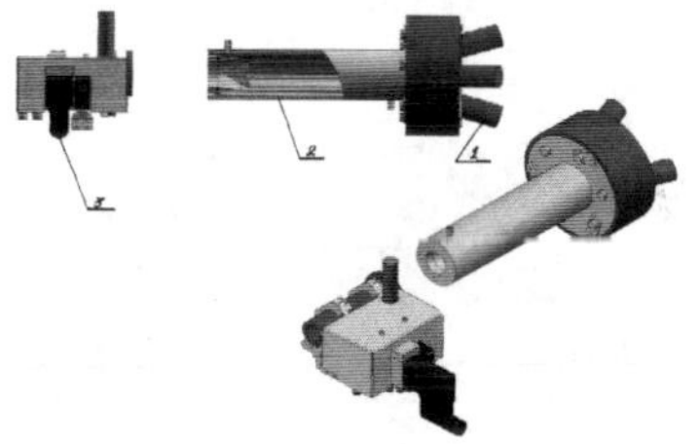

Fig. 2 New design of AEMAPS - (c) : 1- DC torch,2-acceleration channel, 3-pulsed powder feeder.

Such conditions provide for a symmetrical discharge due to a drastic decrease in the breakdown voltage and the homogeneity of the ionization along the perimeter of the accelerating channel.At the initial stage the resulting plasma current shell is accelerated by electromagnetic Lorents force F= J x B (where J presents a radial component of the current, and B — an azimuthal component of its own magnetic field) along the axis of the accelerator. In the future, with increasing density along the channel the acceleration due to gas-dynamic pressure begins to dominate. At the exit of the accelerator a large-diameter underexpanded plasma flow is formed which is characterized by high velocity and temperature.

3 Construction of AEMAPS

The AEMAPS is designed using a DC air torch with power 2 - 7kW (fig.1,2). The power supply (fig.3) provides pulse-periodic voltage of 3 - 4 kV with energy 3 - 6 kJ and frequency up to 4 Hz to anode-cathode path of the accelerator channel. The power supply is built on LC forming lines charged by high power (24 kW) inverter. The use of forming lines ensures quasi-uniform discharge current in the

Fig. 3 AEMAPS system overview-(a),

Fig 4 discharge device

4 Peculiarities of a physial processes in acceleration channel

The analysis of the physical processes in AEMAPS has been carried out. In the AEMAPS, being considered , the transition regime from vacuum to atmospheric pressure regime of plasma current shell acceleration is realized. At the early stage of acceleration after formation of a plasma current sheath, with the formation of shock wave, the process goes at rare gas and is described well in framework of plasma plunger model where plunger is accelerated mainly by the electrodynamic force. However , as plasma sheath passes into atmospheric density region, gasdynamic forses become more important. Moreover, a calculations suggest that radiative transfer has the ability to move plasma current sheath together with the shock wave.

At temperature of the gas beyond the shock wave 0.3-2 eV and pressure 30-50 bar, photons from plasma winh the energy more than 6.5 eV are absorbed by gas (Fig.4) ,heat up , ionized gas by radiation and omic heating. Electromagnetic subsonic deflagration wave (ESDW) occurs, that moves together with the shock wave. The velocity of ESDW mainly 1-4 km/s, that is observed in experiments. As this take place ,gas flows to plasma sheath with velocity 25 g/cm2 s, that causes the sheath expansion ,experimentally noted.

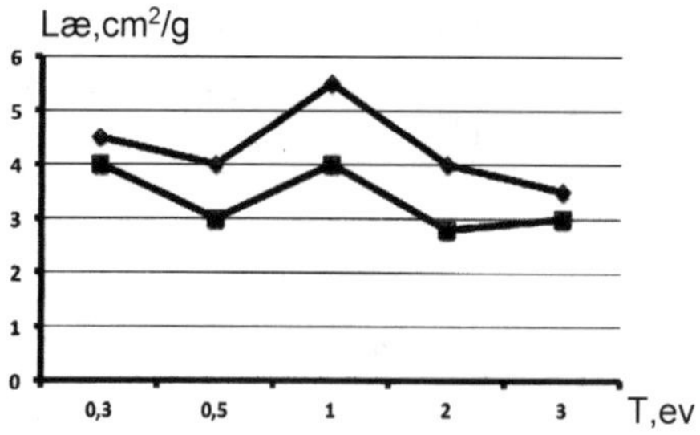

Fig. 5 Absorption of radiation by air (P=50 bar) versus air temperature. Photon energy :

◆ - 7.5 eV > E>6.5 eV: ■- E>7.7 eV. [9].

5 Investigations of the plasma flow parameters

For measurements of pulse-periodic plasma flows parameters the combined diagnostic system has been used. The velocity of plasma flow in the air was measured with the aid of a high speed streak camera GFR with linear velocity up to 10^3 m/s. The plasma brightness temperature has been measured at the wavelength 570 nm using one channel pyrometer on the basis of the Si- photodiode with time resolution 1µs. Pressure amplitude on the specimen surface was measured with a piezoceramic pressure gauge. Analysis of the results of a frame-by-frame video tape recording and high speed streak recording (fig. 4,) demonstrate that maximum plasma flow velocity may be as high as 3 – 5· 10^3 m/s (fig. 5) with the diameter of plasma flow up to 30 mm and length up to 150 -200 mm (fig.4).

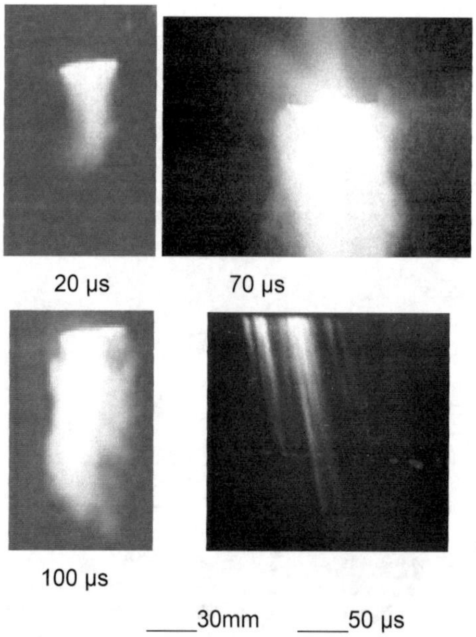

Fig. 6 Frame-by frame (a) and streak (b) recording of pulsed plasma flow in air.

The duration of the intense air plasma flow radiation varies between 150 and 300 μs and plasma brightness temperature is up to 15-20 $\cdot 10^3$K. Frequencies up to 1-4 Hz are realized at the energy of pulsed flow 3- 6 kJ.

The results of measurements of plasma flow pressure on the specimen surface (fig. 6) suggest that the pulse pressure height value may run to 5- 6 MPa during 200 μs and varies only slightly in the range 50 mm from the accelerator channel outlet.

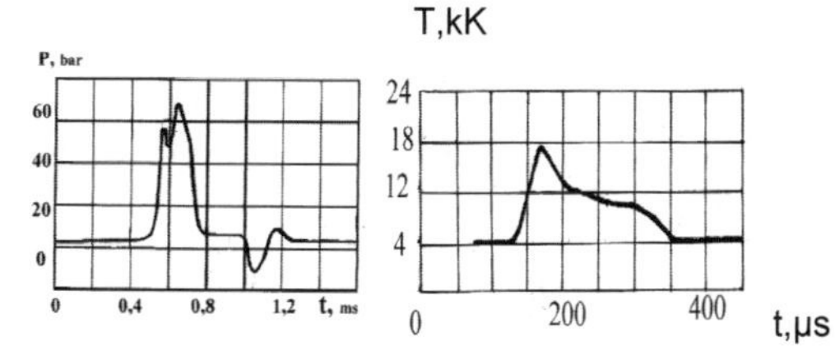

a. b.

Fig. 7 Action of pulsed plasma flow on metal specimen Plasma flow temperature –(c), plasma flow pressure pulse on the specimen surface-(b)

6 Hardening of the metal surface in air

Most commonly used technique for the surface hardening of steel is quenching or martensitic transformation. In this transformation, austenite is rapidly cooled off before the dissolved carbon atoms would escape, forming martensite, a higher hardness phase of steel. The conducted experiments demonstrate that the AEMAPS has considerable promise for steel hardening.

Interaction of pulse-periodic plasma flow with carbon steel (0, 45% C) surface (fig.4) causes formation of a modified layer with a hardness up to 10 GPa and thickness of 10 -50 μm (fig.7), depending on the number of pulses . The hardening effect is determined by fast pulse plasma heating ~ 10^6-10^7 K/s of the target surface layer above the temperature of austenite transformation followed by rapid cooling ~ 10^5-10^6 K/s caused by conduction heat transfer to the ambient media resulting in formation of fine-grained martensite structure.

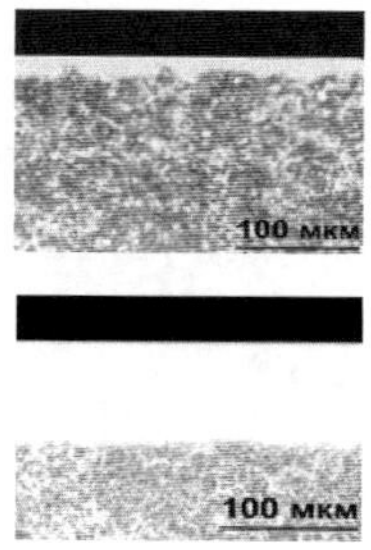

a

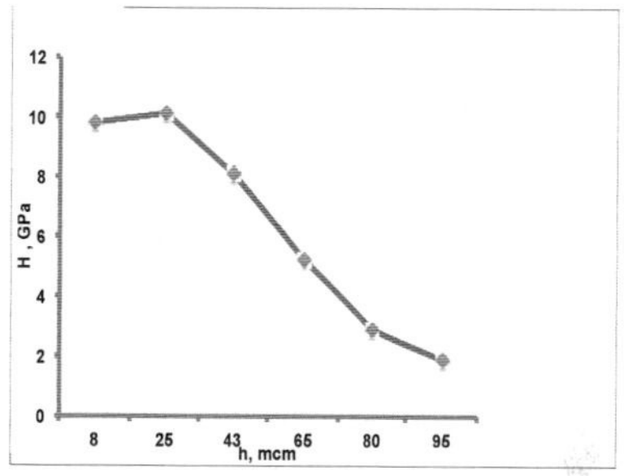

b

Fig. 8 Structure of the surface layer before and after plasma treatment –(a), distribution of microhardness –(b) . E=6kJ, 2Hz, N= 40 .

The metal surface has the highest temperature at the end of the plasma pulse:

$$T = 2\,A\cdot q\,(a\cdot \tau/\pi)^{1/2}/\lambda \qquad (4)$$

and for parameters of the heating flux it is equal to 0.87 $\cdot 10^3$ K, that may be considered as a close approximation to the steel hardening temperature T_H = 1070 K.

7 Powder particles acceleration

The possibility of pulsed periodic plasmatron toaccelerate powder particles to a high velocities up to $3\cdot 10^3$ m/s has been estimated.Plasma flow at the accelerator channel has been treated in the framework of plasma plunger model [14]. Plasma plunger consists of the leading shock wave, area of the gas compressed by shock wave and plasma current shell. At high supersonic speed the shock wave became a ionization wave and distinction between plasma current shell and shock compressed gas disappear. Particles are accelerated by dense high-temperature plasma blob with density up to 10 kg/m^3 and temperature 10- 15$\cdot 10^3$ K.

The particle acceleration in the plasma flow was calculated by a formula

$$m\cdot dVp/dt = Cx\cdot \rho\cdot S\cdot (V- Vp)^2\,;$$

where : m – particle mass, Vp – particle velocity, V – plasma flow velocity, Cx – drag coefficient , ρ – particle material density, S – cross-sectional area of the particles.
The Henderson approximation [15] for the spherical particles drag coefficient was used to calculate Cx:

$$Cx = 24\,(1 + 3.6\,Re^{0.6}) / Re + 1.85\,Re / (6251+Re);$$

where Re – particle Reynolds number : $Re = \rho \cdot d \cdot (V - Vp) / \mu$; d – particle diameter , μ - gas viscosity, calculated by the Sutherland formula

$$\mu = \mu_0\,(T/273)^{1.5}\,(273 + Ts)/(T + Ts)\ ;$$

As the calculations indicate , the drag coefficient for 10μm particle range from 1.0 to 1.9 in the particle velocity range from 0 to 3 $\cdot 10^3$m/s and plasma flow velocity range of 3 to $5 \cdot 10^3$ m/s.
As a result of integration of the equation (1) the simple formula has been worked out:

$$1/ V\text{-}Vp = 1 / V + Cx \cdot S \cdot \rho \cdot t / m$$

The steel particle with diameter 10 μm is to be accelerated up to $2{,}5 \cdot 10^3$ m/s in $t = 10^{-4}$ s, that correlates with our experimental results (Fig.8) and data from paper [16].

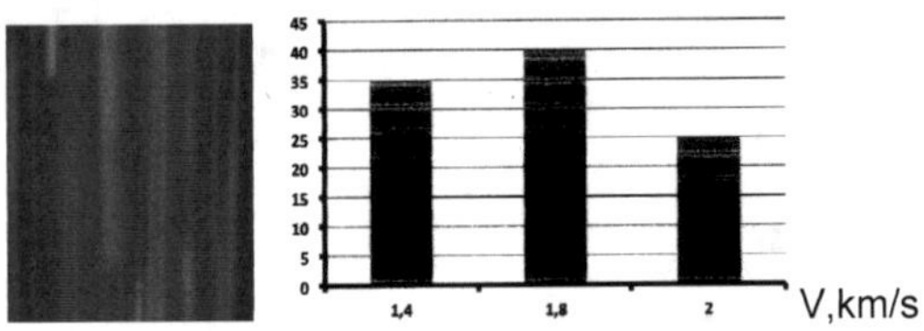

a b

Fig. 9 Particle flow (a), Particle velocity distribution – (b). Particle diameter - +10-15 μm.

8 CONCLUSION

A new method for producing high –power repetitively plasma flows at atmospheric and high pressure with high frequency has been developed. Atmospheric electromagnetic plasmadynamic system (AEMAPS) has been designed .
High-energy high frequency pulsed-periodic plasma flows at the atmospheric pressure are realized. The pulsed periodic mode allows for an increase of the exposure time and thickness of a modified layer of metals for over 50 μm. When scanning is possible the high level uniformity of the hardened layer would be provided through the high plasma flow area. Generated plasma flows with velocity $2\text{-}4 \cdot 10^3$ m/s and temperature up to $15\text{-}20 \cdot 10^3$ K make it possible to accelerate powder particles in the air up to 1,5-2,5 $\cdot 10^3$m/s that was realized [20] and to heat particles of any material to the melting temperatures. The AEMAPS offers a large potential for thermal spraying. Only electromagnetic plasmadynamic systems have the property of accelerating particles to velocities above $2\text{-}3 \cdot 10^3$ m/s. The system has an exceptional ability to scale in order to increase the plasma flow and particles parameters and the processing performance.

REFERENCES

[1] Morozov A. *Introduction to Plasmadynamics*, Nauka, Moscow, 2006, 600 pp.
[2] Tohara Hirokaz, Ando Yasutake, *Vacuum* **2008**, 83, 98 -105.
[3] Usuba S., Heimann R.B.: *J. Thermal Spray Technol.* **2006**, 15(3), 356 - 364.
[4] Chivel Yu., Method for producing high –energy pulse-periodic plasma flows at atmospheric gases and high pressure, Patent RU № 2462007, **2012**.
[5] Chivel Yu. Method for obtaining high-energy repetitively pulsed plasma flows , WO 2012025835.
[6] Raiser Yu. *Physics of Gas Discharge* . Nauka, Moscow, 1987, 588 pp.
[7] Baker K., Hwang D., Horton R. et al., *IEEE Trans. Plasma Sci.* **2002**, 30, 48 - 56.
[8] Abramovich G., *Applied Gasdynamics* ,Nauka, Moscow, 1991, 600 pp.
[9] *Optical data of the hot air* . Ed: Biberman L. , Nauka, Moscow, 1970, 285 pp.
[10] Chauveau S. , Riviere Ph. , Deron Ch. et al. *Progress in Plasma Processing of Materials* . Ed. Pierre Fauchais. Begell House,NY,2001, 447pp.
[11] Xian M., Wen-Xia P.,Cheng-Kang P., *Chin. Phys. Lett.* **2003**, 20, 685- 693.
[12] Rykalin N., Uglov A., Zuev I..,Kokora A. *Laser and Electron Beam Material Processing.* Handbook; Mir Pub.: Moscow,1988, 280 pp.
[13] Chivel Yu. : *Surf. Coat Tech.* **2010**, 205, 1088 - 1091.
[14] Thom K., Norwood J., and Jalufka N.: *Phys. Fluids* **1970**,7, S67-70 .
[15] Henderson C., *AIAA Journ.***1976**,14,707-708
[16] Uryukov V., Gasin D.: *Surf. Coating Tech.* **1991**,48 , 7- 12.

New approach in thermal spraying-laser thermal spraying

Yuri Chivel, Saint Etienne/FR

Abstract: New thermal spraying method – laser thermal spraying has been elaborated and experimentally realized. New method makes possible independent energy input to powder stream and workpiece using conical laser beams and is more efficient. New method is absolutely environmentally friendly, unlike other methods of thermal spraying.

1 Introduction

In thermal spraying, the material feedstock (commonly a powder, wire or suspension) is injected in an enthalpic source (thermal plasma or flame obtained by ionization-excitation of inert gases or combustion, resp.). The particles are melted (or partially melted), accelerated by a gas flow, and crushed on the target substrate where rapid solidification and deposit built-up occur.

For separated heating of the deposited material and surface the method of laser thermal spraying using conical laser beams has been elaborated. Initial round laser beam is divided into 2-3 circular beams with regulated distribution of the laser power throughout the circular beams. Circular beams are transformed to conical beams , which are focused separately on the surface and on the deposited material (powder stream, wire, suspension stream) for heating . A focuses of conical beams are arranged along the optical axis by which the deposited material is fed

2 Principles of method

The main idea consists in provision of laser power delivery to powder stream uniformly across the surface of stream and based on unique properties of conical laser beams [1]. Conical beams are formed on focus the laser beam by axicon – conical lens, conical mirror or conical prism (Fig.1).

In the new method [2-4] the laser energy introduction to powder stream is more efficient because heating of small particles at stationary mode of operation and a total uniform absorption of laser beam energy is possible in the dense powder stream. object at collision.

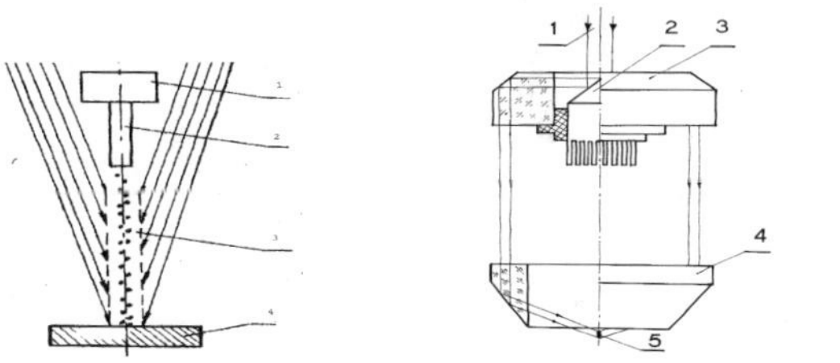

Fig.1 The scheme for delivery of laser energy to powder stream . 1 – laser beam, 2 – conical mirror, 3,4 – totally reflecting prism, 5 – focusing region

The heat build-up of surface is easy to set up through the corresponding location of surface relative focal region of conical beam. Also the particles heat the surface of object at collision.

3 Laser radiation transmission and powder heating

On powder injecting the structure of beam will be modified through laser light absorption and scattering. We consider the amplitude variations of the beam taking into account attenuation of laser light in snugpacking dispersed layer. The calculations are carried out for Ti spherical particles with diameter d=30 μm under volume density $N=10^6 cm^{-3}$ accordingly. The radius of powder beam is taking to be R= 250 μm .

For coherent light passing the exponential connection between coefficient of attenuation and layer's thickness can be used [5]:

$$T=\exp(-Q\eta) \qquad (1)$$

where $\eta = 3Rn_1/2d$ – coefficient of layer's overlap, Q – effective factor of attenuation , $n_1=0{,}74(d/\delta)^3$ - dimensionless volume density of particles ($\delta=2d$ – average distance between particles).

For layer from optically hard large particles ($\pi d/\lambda>1$, $\pi d/\lambda(n-1)>>1$) the effective factor of attenuation may be written :

$$Q=1/1{,}5n_1\ln[1+1{,}5n_1Qi\exp(1{,}5n_1Qi)] \qquad (2)$$

The transmission of powder stream has described by the relationship:

$$T=\exp[-\sigma(R-r)] \qquad (3)$$

where (R-r)- depth of laser light penetration into the powder stream in μm, σ—attenuation coefficient, Qi - factor of attenuation of insulated particle. Calculation values of attenuation coefficient for concerned powder streams are equal $\sigma= 45cm^{-1}$ (D=500 μm). Laser intensity attenuation causes the Bessel beam peak intensity to fall and the distribution of laser intensity within powder stream levelsoff

Intensity distribution of laser radiation within the powder stream in the absence of a interference excluding small centre area has described by the following expression [6]:

$$q=q_0(R/r)[e^{-\sigma(R-r)}+e^{-\sigma(R+r)}] \qquad (4)$$

The scattered laser radiation is effectively absorbed by particles under repeated scattering. The calculations of scattered radiation attenuation in the frame of transport theory [7-9] allows to determine diffuse transmission and reflection of powder stream.. For discussed above 100µm powder stream from 10µm particles Td = 0,16-0,30 for the length 50 µm , and Rd= 0,02-0,17, depending on the nature of reflection of diffuse radiation from the stream borders. The transmission of coherent light come to T=0,1 for the length 250 µm.
The potentials of the new method of laser radiation energy delivery to powder stream for production were investigated. In the model under study the dense stream ($10^6 cm^{-3}$) diameter 0,5mm from steel micropowders with particles 30 µm in size is injected into the focal region of conical beam 5-10 mm long at the rate of several m/s.Dynamics of heating and evaporation of spherical particles describe by equations:

$$4/3\pi r^3 \rho C(dT/dt)=\pi r^2 q Q_a - E - L\cdot G; \qquad (5)$$

$$dr/dt=-G/4\pi r^2 \rho; \qquad (6)$$

where r – particle radius, ρ – density of particle material, C – specific heat , q- laser radiation intensity, Qa – factor of absorption of particle, E – energy flow from particl surface, L – specific heat of evaporation, G – mass flow from surface.

Under initial step of particle heating the losses on vaporation and thermal conductivity are small and temperature dynamic may be write in view:

$$T=3q\cdot t\cdot Qa/4\rho\cdot Cr \qquad (7)$$

According the calculations at particles duration of stay in the heating region ~ 10^{-2}s , the required laser power for heating to 1500 K not exceed ~ 200 -300W.

4 Principles of laser spraying system construction

For separated heating of the deposited material and surface the new method and system of laser thermal spraying using conical laser beams [10] has been elaborated .
Space-independent and separated heating of workpiece and stream of deposited material provides optimum thermal process conditions, can significantly reduce energy costs and increase cladding accuracy
A deposited material may be a suspension, powder, aerosol or heterogeneous plasma.

The system (fig.2) comprises a laser optically coupled with an beam expander. Additionally the device includes a focusing lens 7, the conical mirrors and conical prisms One annular beam is focused by lens 7 into the surface of the workpiece 11 and other annular beam are focused by conical prism
into stream of the deposited material 10. The device operates as follows. The laser beam 1 after passing through the beam expander 2 and conical mirror is transformed into a two annular beams. One of the beams is focused by lens 7 on the product surface 11 to a spot irradiation 12 and melt surface in spot. Deposited material is fed through the tube 5 in the form of a powder stream 10. Other beam are focused by means of conical prism in the predetermined region of the stream 10 to heat powder stream.

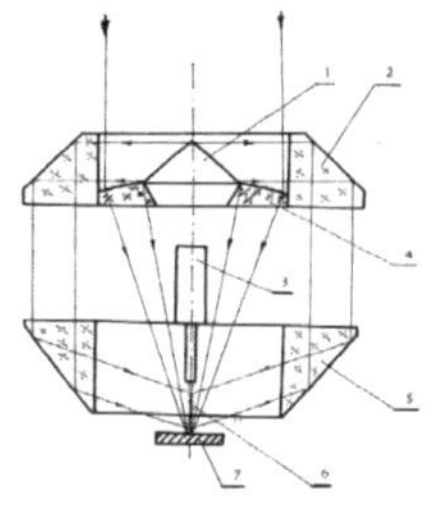

Fig. 2 The optical scheme of the laser spraying system .1 – conical mirror, 2,5 – totally reflecting conical prism, 4 – lens, 6 – focusing region in powder stream, 7- workpiece.

Variant of the system with conical reflective optics, which is important when using high-power lasers, is shown in Figure 3. The system additionally comprises a rotating mirror with an opening through which tubes for feeding gas, cooling liquid and spraying material are passed. A series of parallel annular laser beams with an adjustable distribution of laser radiation power across the annular beams is formed from an initial circular laser beam
Instead of the lens 7 (Fig. 2) is set focusing conical mirror 23. In this scheme also introduced the option of using part of the laser radiation, shaded by tube 5 . This radiation can be used for heating tube 5 and deposited material passing through it. In other variant the device (fig.3), this part of the radiation is deflected by mirror 24 to block of the laser parameters control 25.

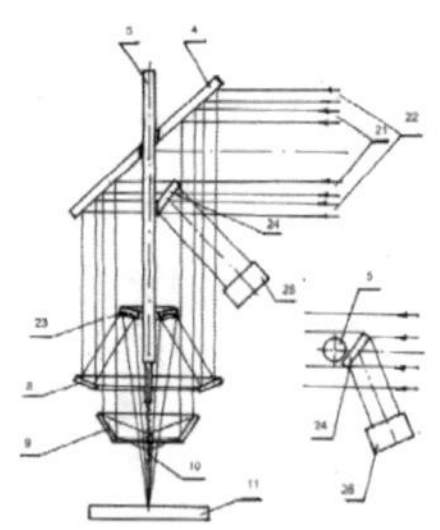

Fig. 3 Variant of the optical system with conical reflective optics.

At low absorption of laser radiation in the stream of the deposited material drops sharply efficiency of laser radiation use when a single pass through stream.For maximum use of radiation offered multipass scheme using a cylindrical mirror 36, 37 (fig. 4). It first increases the stream heating and secondly allows efficiently use of the laser radiation up to the full absorption when returning the coherent part of the laser radiation to laser resonator.

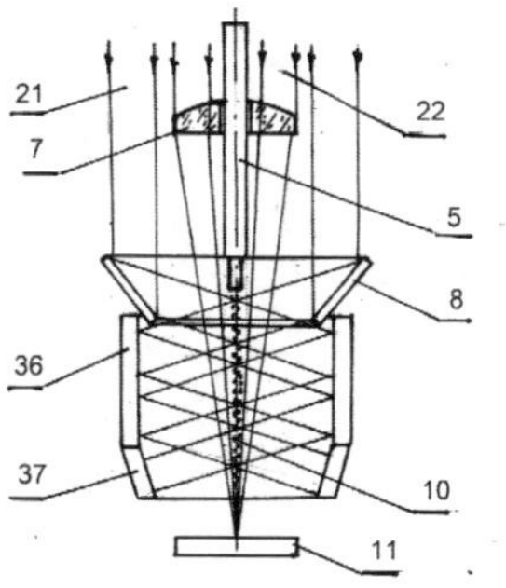

Fig. 4 Multi pass opticakl scheme of the laser spraying device .

4 Experimental

Experimental laser thermal spraying system has been elaborated and produced (Fig.5). The main optical scheme (Fig.2) has been used.

Fig 5 Photo of laser thermal spraying system

CW diode laser with power 200W was optically connected with adjustible beam expander.
Spherical INOX powder +15-30 µm in size was fed to focal region of conical beam 8 mm in lenght through tube 0.8 mm in diameter by the use oft he powder feeder at the rate ~2 m/s. According the calculation particle density in focal region was about $10^5 cm^{-3}$. As workpiece INOX plate was used.
At Fig.6 powder stream under action registered by CMOS camera are presented .

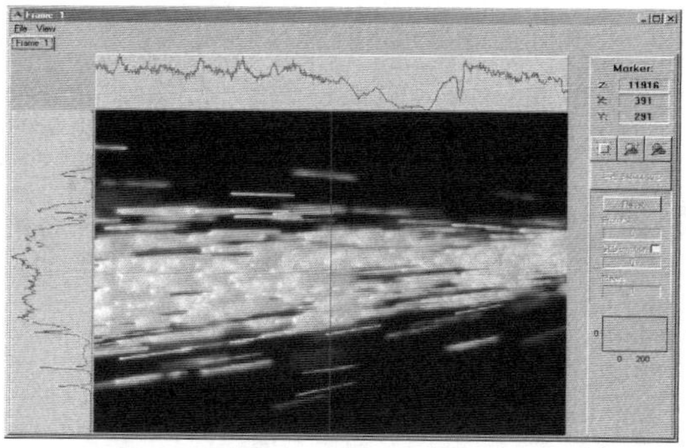

Fig.6 Powder stream image under laser heating.

Microstructure of the coating at spraying distance 12 mm is shown on Fig.7

2mm

Fig.7 Microstructure of sprayed coating

Performance evaluation of a efficiency of laser thermal spraying suggest that this process is more efficicnt that traditional metods of thermal spraying.
Specific expenditure of energy range up to η = 50-100 kW min/kg ,whereas at plasma spraying η = 1000 kW min/kg , at suspension spraying – 3000 kW min/kg.

Conclusions

By this means the harnessing of the conical beams allows to carry out a laser power delivery to powder stream uniformly across the surface of stream, resulting in gain in productivity and accuracy. The laser radiation distribution within the dense streams is sufficiently uniform, excluding a small region near the axis of stream.. The optimum heating of powder particles may be achived with high efficiency.

For separated heating of the deposited material and surface the new system of laser thermal spraying using conical laser beams has been elaborated.
Space-independent and separated heating of workpiece and stream of deposited material provides optimum thermal process conditions, can significantly reduce energy costs and increase cladding accuracy.

The calculations and modelling demonstrate that the new method of the independent energy input to powder stream and surface using conical laser beams is more efficient because heating of small particles at stationary mode of operation and a total uniform absorption of laser beam energy is possible in the dense powder stream. The optimal regimes of laser thermal spraying can be achieved using multipass optical scheme and conical beams with different wavelengths. New method also allows significantly raise parameters of the cold spray coatings and efficiency of the cold spraying. The efficiency of the laser thermal spraying can be 10 - 10^2 time higher than other thermal spray technologies. New method is absolutely environmentally friendly, unlike other methods of thermal spraying.

References

[1]. J McLeod . J.Opt.Soc.Am. v. 44. .p. 592. 1954

[2]. Yu Chivel . Proceedings SPIE , V.5399,P.228-233.2004

[3].Yu Chivel, Patent BY № 6931. 10.11.2004

[4] Chivel Yu. Cone –Shaped Beams in Selective Laser Cladding // Technical Phys.Lett. V.31. PP.1-3. 2005.

[5].A.P. Ivanov,V.A. Loiko, V.P. Dik.,The Propagation of Light in Snugpacking Dispersed Medium,.Minsk. 1988.

[6].E.P.Shafer. Appl. Phys. B.V.39. P. 1-8.1986.

[7] Chivel Yu., Khilo N., Yushkevich V. // Optical Memory and Neural Network. Alerton Press. № 4 . PP. 51-58.2007.

[8] Chivel Yu. Patent RU № 2302064 . 27.06.2007. Proc SPIE.V6334.2007..

[9] Chivel Yu. A. Patent RU № 2447979. 05.11.2010

[10] Chivel Yu. Patent RU №2580180. 06.03.2014. WO 2015132640, US Patent Application 2016.

An Alternative Die Coating Method on Low Pressure Die Casting Mould by Thermal Spraying

[1]**Ekrem ALTUNCU**, [2]Fatih USTEL, [3]Ömer Burak ÇE

[1]altuncu@sakarya.edu.tr

[1,2] Sakarya University, Thermal Spray Technology R&D Laboratory, Department of Metallurgy and Materials Engineering, Esentepe Campus, Serdivan, 54187 Sakarya, Turkey

[3]CMS Wheel and Machine Industry, AOSB, Çigli, İzmir, Turkey

Abstract

In the world, automobile rim production is carried out in a repeatable quality with low pressure die casting process. Traditional die coatings used to separate molds on casting mold surfaces cause significant losses in production due to their weak strength. The applicability of an alternative mold coating application with thermal spray coatings has been examined in this study. $MgO\text{-}ZrO_2$ based ceramic thermal barrier coatings are deposited on the mold surfaces by flexicord spray method.

Keywords: Die casting, die coating, Thermal spraying, $MgO\text{-}ZrO_2$.

1. Introduction

Low pressure die casting (LPDC) is used to manufacture complex metal components where there is a need for high structural integrity. Examples include engine blocks, wheel and key structural components of cars. Liquid metal is fed from below into the die used to form the component under a positive pressure. The die is intended to fill with uniform horizontal fronts rising vertically in each part of the die (**Fig 1**). Problems emerge when solidification fronts merge and cut off regions of still liquid metal from the new metal supply. Subsequent volume contraction then leads to the formation of void defects as the contracting metal can no longer fill the entire trapped region.In low pressure and gravity die casting the surface of each metal mould or die component, which is contacted by molten metal, is provided with a mould paint or die coating. Current die coating technology involves the use of a water-based suspension of ceramic particles containing a water-based binder, most commonly sodium silicate.

Die coatings for solidification control the heat transfer and facilitate the flow of the molten metal, thanks to just the right roughness. In this manner, they facilitate the release of the casting, extending mould life and improving the surface appearance of the finished casting. In conventional applications, the life span of the coatings is limited and varies depending on the mold type, the mold temperature, and the number of cycles [1-5]. The ability to apply commercial mold coatings is simple and cheap. However, due to the low mechanical strength of the coatings, the risk of fracture, cracking and spalling is high. This leads to the continuous renewal of the coatings leading to long mold and production stops.

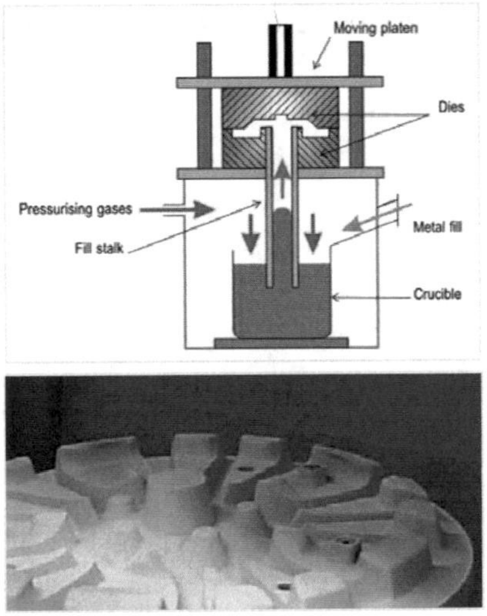

Fig. 1. Low pressure die casting process and coated die.

Inspired by the TBC (thermal barrier coating) coatings used in aircraft turbines, ceramic coating applications have been carried out with the idea of the applicability of thermal spray coatings on the mold surfaces.

In general, there are two coating applications in the TBC coatings, the bond coat and the ceramic coat. The top layer is porous and reduces the thermal conductivity of the coating. Stabilized ZrO_2-based top coatings are expected to have superior coating life compared to commercial die coatings due to their high fracture toughness. Plasma spraying is generally used in the production of TBC coatings. With plasma energy coating applications require high investment cost and operation. With the flame flexicord spray coating which is one of the new processes in the family of thermal spray coating methods, mold surfaces can be coated quickly and economically.

2.Experimetal

In experimental studies, die coatings which is provided by Foseco and thermal spray coatings were compared. There can be a number of advantages and disadvantages of a new alternative die coat application in the production process. The low pressure casting process is a series production process in which a large number of factors affect product quality. The surface quality of the part removed from the mold must have sufficient standard properties for subsequent painting operations. Besides this, casting fault is not required in the casting part and on the surface.

wt%22MgO-ZrO_2 and NiCrAl in the form of a flexible cord (diameter 4.75mm) were used as flame spray feeding material. The coatings were deposited with a master jet flame spray gun on hot work steel (H13). Pre-heating (200-300 °C) of the mold was applied before coating. After the sandblasting, the bond layer was applied to the surface and the ceramic layer was applied to the surface with the master gun and the parameters given in the **Table 1**. The preheated mold was rotated at a specific rotation speed and the coating was carried out.

Table 1. Spray parameter of flexicord flame spraying

Acetylene gas	1.2 bar
Oxygene gas	4 bar
Air	3- 4 bar
Spray distance	120-140 mm
Feed rate for BC	30-50 cm/dk
Feed rate for TC	45-70 cm/min.

The average size of the powders of FlexiCord coating materials are 38-70 µm. The powder size distribution of the fine-grained (Supra) flexicord coating is shown below in **Fig2.**

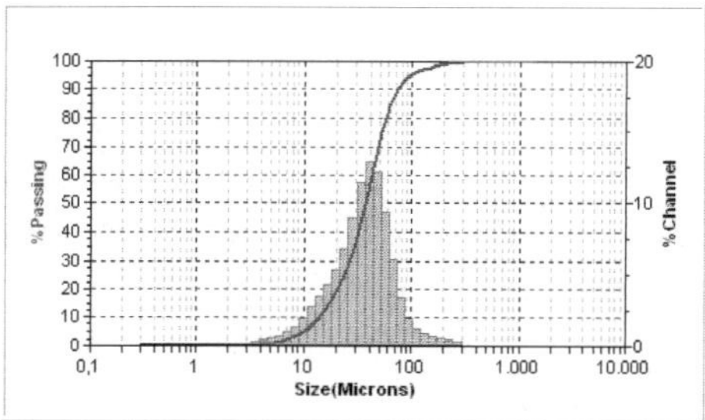

Fig2. Powder size distribution curve for ZrO_2 based feedstock materials

2. Discussion and Results

The microstructure images taken from the upper surface and cross-sectional coated specimens after the coating are shown in **Fig 3.** The coating thickness is between 100-150 microns for the bond layer and 250-750 microns for the ceramic layer. The porosity ratio in the ceramic layer is 45-60%. It can be controlled by spray parameters.

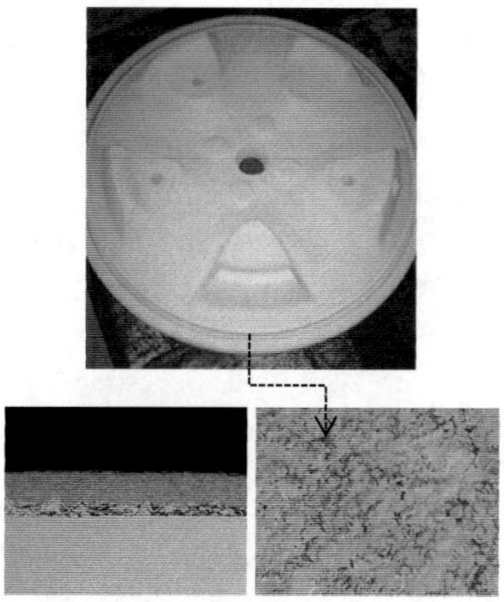

Fig 3.a.coated die, b. cross section and top surface of the TS coating.

Increasing the porosity ratio reduces the thermal conductivity of the coating but disrupts the surface quality. The increased thermal conductivity increases the problems of molten metal adhesion with the mold. The adhesion strength of coatings is 5 times higher than commercial coatings. The adhesion strength varies depending on the porosity ratio in the coating. **Table 2** compares the surface roughness of commercial die coatings and flame spray coatings.

Between commercial moldings D34 has the roughest surface roughness. It has a rough surface and low thermal conductivity due to the too much porosity of the coating structure. It is generally used together with D39 type coating solutions. When we compare the surface of the molds coated with the thermal spray method, close surface qualities can be obtained.

Table 2. Comparison of the surface roughness values of die coatings

Die coat	Commercial die coat (Foseco)			Thermals spray coating	
Roughness	D-34	D-39	D-140	ZrO_2-MgO Fine	ZrO_2-MgO Coarse
Ra (µm)	15,31	9,39	9,72	6,98	16,32
Rz (µm)	52,98	33,15	34,91	38,5	58,35

The coating surface quality and porosity ratio can be controlled by using materials having different powder sizes in coating processes and changing spray parameters. As you can see in **Figure 4**, there is a noticeable difference in porosity ratio in the top surface images.

The thermal conductivity properties vary depending on the coating thickness and porosity ratio. For this reason, thermal spray coatings were applied at different thicknesses and porosity ratios in experimental works.

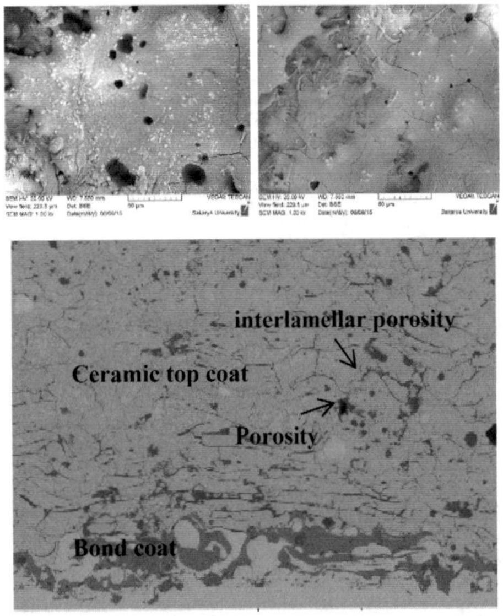

Fig.4. Porosity ratio variation on top surface with different type feedstock and spray parameters

The thermal spray coatings have a hardness variation of 500-1200 $HV_{0.3}$ and a 5 times higher abrasion resistance than commercial die coatings. Adhesion strength between 30-50 MPa could be obtained in flame spray coatings as compared to commercial coatings with poor adhesion strength (8-10MPa).

4. Conclusion

Within the scope of the study, a decrease in the consumption rate of commercial die coat consumptions was seen.However, the thermal spray coating application, a new and more technological method, increased the number of die casting (cycle) counts to an average of 5-6 times without interruption.This has resulted in significant improvements in repair and maintenance times, as well as in tool life. In this study, with the knowledge and skill gained in the operation, the firm increased the production efficiency.

Acknowlodgements

This study was financially supported by the project of 0270.STZ.2013-2 by the Ministry of Science and Technology of the Republic of Turkey. The authors present their gratitude to the Scientific Research Coordination Dept. of Sakarya University for their financial contributions through the Bapk2014-09-08-002 project.

References

[1] B. Zhang, D.M. Maijer, S.L. Cockcroft, Development of a 3-D thermal model of the low-pressure die-cast (LPDC) process of A356 aluminum alloy wheels, Materials Science and Engineering: A, Volume 464, Issues 1–2, 25 August 2007, Pages 295-305

[2]B.K. Kosec, et al., Analysis of casting die failures, Eng. Fail. Anal. 8 (4) (2001) 355–359.

[3] Peter I., Rosso M., Simone Gobber F., Study of protective coatings for aluminum die casting molds, Applied Surface Science 358 (2015) 563–571.

[4] Terek P.,Kovačević L,.Miletić A, Panjan P., Baloš S., Škorić B., Kakaš D., Effects of die core treatments and surface finishes on the sticking and galling tendency of Al–Si alloy casting during ejection, Wear, Volumes 356–357, 2016, 122-134.

[5] S. Shankar, et al., Die soldering: mechanism of the interface reaction between molten aluminum alloy and tool steel, Metall. Mater. Trans. B 33 (2002) 465–476.

Abrasive wear resistance of NiCrBSi composite claddings with different type of reinforcement

A. Gerasimov, V. Krivopusk, S. Nevezhin, S. Merchev, S. Baldaev
Technological systems for protective coatings, Russia, Moscow

Abstract

In this study abrasive wear resistance of NiCrBSi self-fluxing alloys claddings reinforced by tungsten carbide (up to 80 vol.%) and pure tungsten (up to 50 vol.%) was compared. For this purpose, metal matrix composite samples were cladded by diode laser then metallography studies and hardness measurements were provided. Finally, all specimens passed ASTM G65 abrasion test and their wear resistance was compared to alternatives claddings.

Introduction

Thermal spraying (TS) coatings from NiCrBSi self-fluxing alloys with tungsten carbide reinforcement have high wear resistance, but spraying commonly does not provide thickness up to 0.5 mm or higher that may be needed for many applications. In this respect, the use of laser cladding (LC) offers high thickness of claddings and more flexibility for their production [1].

LC of metal matrix composite (MMC) NiCrBSi+WC is one of the most effective method of protection against wear among alternative TS coatings and steel such as Hardox 400 [2, 3].

There are presented many works where Ni-WC MMC LC-s with various grades of WC hardening is compared [4]. At the same time, the wear resistance of MMC claddings with different types of reinforcements were evaluated only in some studies [5].

The aim of this work is to evaluate addition of tungsten into LC Ni-base MMC to their technological properties and wear resistance.

Experimental Methods

Two grades of NiCrBSi self-fluxing alloy (Hoganas 1035-40 and Hoganas 1560-00), tungsten carbide (TECMAT WC-125) and pure tungsten (TECMATW-25) were used as matrix and hardener in MMC claddings correspondingly (Table 1).

Table 1 Powder grades used to produce MMC claddings

	Chemical compositions in wt.%						
Powder	C	Si	B	Fe	Cr	Ni	W
Hoganas 1035-40	0.32	3.7	1.2	3.0	7.0	Bal.	-
Hoganas 1560-00	0.75	4.3	3.1	3.7	14.8	Bal.	-
TECMAT WC-125	6.1	-	-	-	-	-	Bal.
TECMAT W-25	-	-	-	-	-	-	99.9

Various MMC-s were cladded on low-carbon steel samples (75x25x5±0,5 mm) by diode laser (LDF 6000 - 100) under the following conditions: laser power - 0,8-1,2 kW, scan speed - 0,004-0,04 m/min, overlap - 1,2-1,6 mm, shielding gas flowrate - 10-20 l/min, adjusting powder feed rate was variated depending on the required volume content of the hardener in MMC claddings.
Further MMC specimens were analyzed by metallographic studies and hardness measurements by using of HVS-10 Vickers stationary digital hardness tester.
Wear resistance of claddings was determined according to G65 procedure B standard test method for measuring abrasion using the dry sand/rubber wheel apparatus.

Results and discussion

Microstructure and hardness of claddings
The soft matrix MMC claddings with different content of WC (Hoganas 1035-40+50% TECMAT WC-125 and Hoganas 1035-40+80% TECMAT WC-125) differ in hardness of the matrix and hardener (Fig. 1). In case of 50% vol. of WC in claddings they have lower hardness values of metal matrix and higher hardness values of carbides that provide crack free structure. While in 80% vol. of WC MMC-s dissolution takes place as the result there is crack formations is occurred in matrix with higher hardness.

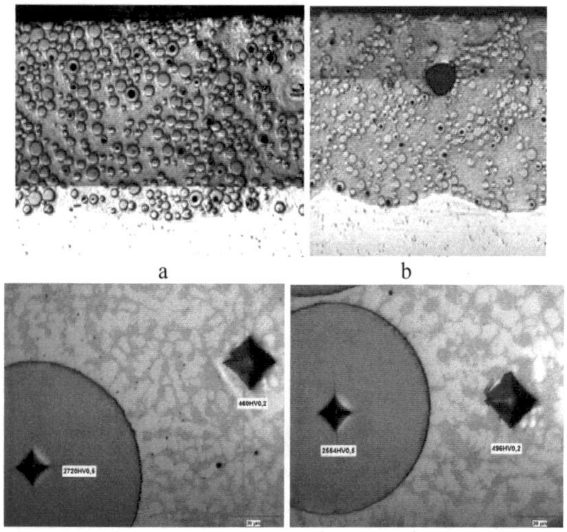

Figure 1: MMC with soft matrix: a – Hoganas 1035-40+50% TECMAT WC-125 (1); b – Hoganas 1035-40+80% TECMAT WC-125 (2); c and d – microhardness of MMC a and b correspondingly

In MMC claddings with hard matrix (Hoganas 1560-00+80% TECMAT WC-125) even when content of carbides WC exceeds 80% vol. their dissolution does not happen but formation of brittle eutectic occurs (Fig. 2). So formation of claddings without cracks is difficult.

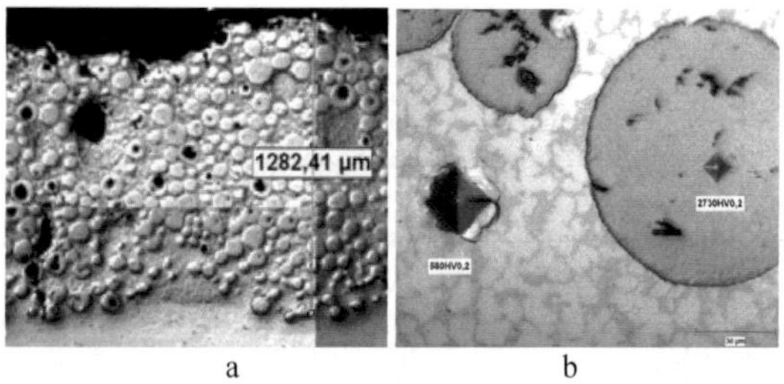

Figure 2: MMC with hard matrix: a – Hoganas 1560-00+80% TECMAT WC-125 (3); b – microhardness of MMC

One way to solve this problem is the addition of tungsten to the MMC with a hard matrix. It is observed that addition of tungsten W up to 25% vol. in MMC claddings with hard matrix prevents formation and propagation of cracks (Fig. 3). The mechanism for preventing cracking in the MMC when adding tungsten is not clear, but it can be assumed that it acts as a modifier preventing the formation of brittle eutectics in the structure of cladding [6, 7].

At the same time in case of simultaneous addition of WC and W up to 25% vol. and to 15% vol. correspondingly high wear resistance of claddings without cracks formation is possible.

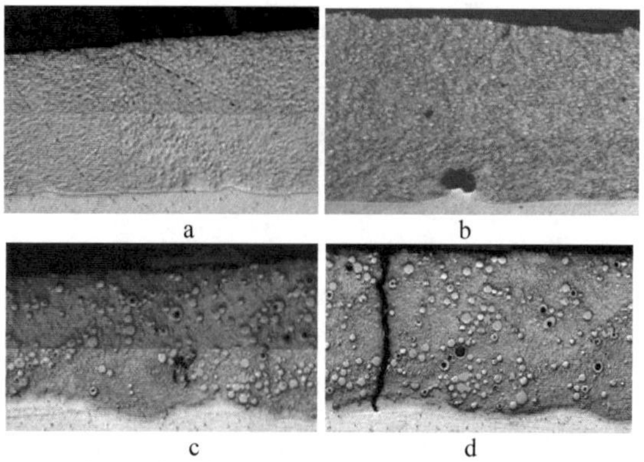

Figure 3: MMC with hard matrix and different type of reinforcement: a – Hoganas 1560-00+25% TECMAT W-25 (4); b – Hoganas 1560-00+50% TECMAT W-25 (5); c – Hoganas 1560-00+50% TECMAT WC-125+12% TECMAT W-25 (6); d – Hoganas 1560-00+30% TECMAT WC-125+14% TECMAT W-25 (7)

Wear resistance of claddings

The volume loss for the samples after abrasion tests was in range 0,002-0,005 cm^3 (Fig. 4You may see that the injection of tungsten and tungsten carbide in MMC cladding increases wear resistance more than 20% compared with the addition only tungsten.

The developed technology makes it possible to repair the MWD (measurement while drilling systems housings without dismantling directional sensors, gamma sensors whatever, whereas conventional MMC claddings do not work (Fig. 5).

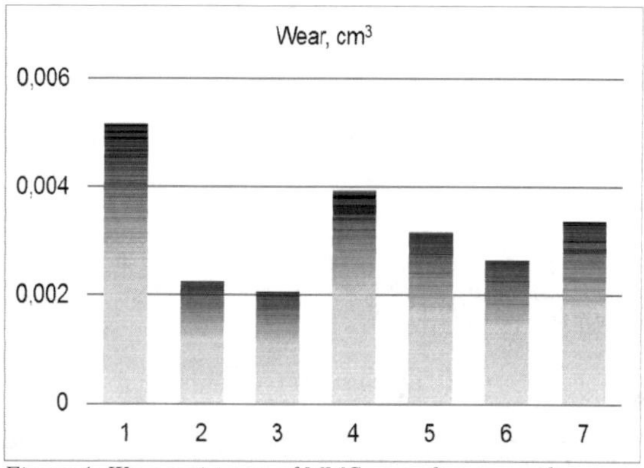

Figure 4: Wear resistance of MMC according to results of abrasion test ASTM G65 procedure B, numbers of claddings are given in accordance with the fig.1-3

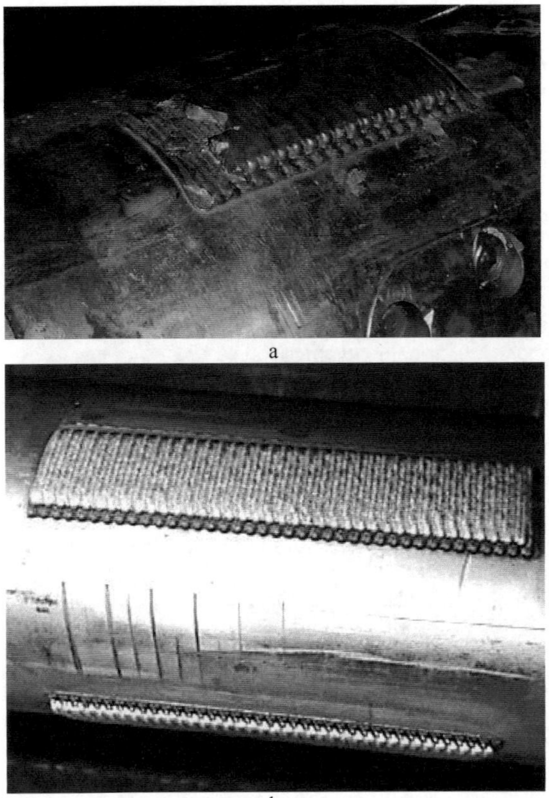

a

b

Figure 5: MWD systems housings: a – before repair; b – after LC of developed MMC

Conclusions

1) In case of using MMC with soft matrix from NiCrBSi self-fluxing alloy (below to 40 HRc) it is necessary not to exceed 50% of WC by vol. to prevent carbides dissolution and crack formation in claddings.

2) For MMC with hard matrixes (up to 60 HRC) higher content of WC (up to 80% by vol.) doesn't lead to its' dissolution but formation of brittle eutectic.

3) Addition of tungsten W up to 25% vol. in MMC with hard matrix decreases the possibility of brittle eutectic formation and cracks propagation in claddings.

4) In case of simultaneous addition of WC (up to 25%) and W (up to 15% vol.) may form high wear resistant cladding layers without cracks.

References

1. E. Toyserkani et al. *Laser Cladding*. (USA), CRC Press, 2004. - 260 p.
2. P. Kulu et al, Abrasive Wear Resistance of Recycled Hardmetal Reinforced Thick coating. *Key Engineering Materials*, 2013. - Volume 527, p. 185-190.
3. T. Liyanage. *Microstructure and Properties of Ni-Alloy and Ni-WC Composite Overlays*. (Canada), 2010. - 108 p.
4. R. Seger. *Effect of tungsten carbides properties of overlay welded WC/NiSiB composite coatings*. (Sweden), 2013. - 64 p.
5. J. Nurminen et al, Microstructure and properties of hard and wear resistant MMC coatings deposited by laser cladding. *Int. Journal of Refractory Metals & Hard Materials*, 2009, Volume 27, p 472–478.
6. I.S. Kulikov. *Metal deoxidation*. (Moscow), Metallurgy, 1975. - 504 p.
7. J. Goldstein et al. *Modification and microalloying of cast iron and steel*. (Moscow), Metallurgy, 1986. - 272 p.

Plasma processing of AB_2 alloy as negative electrode material for NiMH batteries

Ezgi ONUR ŞAHIN, Cavit EYÖVGE and Tayfur ÖZTÜRK, Middle East Technical University, Ankara/Turkey

The present work deals with a Laves phase C14 AB_2 alloy, namely $(TiZr)(VNiMnCrSn)_2$. The alloy in many ways is a good alternative to a rare earth AB5 alloy with a superior capacity reaching a value of 400 mAh/g. A drawback with this alloy is that it is difficult to activate and therefore it is desirable to develop processing techniques which would readily activate the alloy. In this study we have plasma processed the alloy so as to see if this processing would exercise a positive effect on activation. AB2 powder was therefore fed to plasma torch with 25 kW power. The powders of -325 mesh had a range of particles sizes, the finer ones were evaporated and condensed into nanoparticles less than 100 nm in size. The larger ones sphereoidized and were collected in the form of two groups of powders. We have characterized all three groups of powders both chemically and the latter two in terms of electrochemical performance.

1 Introduction

Ni/MH batteries have traditionally been based on rare earth AB5 compounds as negative electrode material [1]. However 9 fold increase in the price of rare-earth elements [2] which occurred in 2012 had an adverse effect on AB5 compounds, connected with that on NiMH batteries which has lost ground to other batteries. There are a number of alternatives to AB5. Of these AB2 alloys are particularly attractive, as they offer better capacities than AB5 [3]. A major problem in this group of alloys is that they are difficult to activate, i.e. they require a large number of cycles before they could reach their full capacity [4].

The present work is a part of an effort to improve the activation behavior of a Laves phase C14 AB2 alloy. This has been studied with a variety of methods e.g. surface modification by hot alkaline treatment, increased surface area by milling or with use of finer particles etc. The current work deals with plasma processing of active material, (TiZr) (VNiMnCrSn)2, to see if it improves its activation behavior
Please simply overwrite the text. You should only use the DVS style sheets.

2 Experimental

AB2 alloy was a Laves phase C14 alloy with the composition of (Ti0.36Zr0.64) (V0.15Ni0.58Mn0.20 Cr0.07)2. It had a particle size with a broad distribution centered on 35 µm. Discharge capacity of the alloy is given in Fig.1 where it is seen that the alloy reaches its full capacity after 12 cycles.

Plasma processing of the alloy powders was carried out in R.F. Induction plasma reactor (TEKNA), the schematic representation of the torch and the complete system are given in Fig.2. The torch was operated at a power of 25 kW. AB2 powder was fed into the torch via an injector using a carrier gas (argon, x slpm). Here small powders evaporate into the gas phase, but larger ones are subject to surface melting and as a result they just sphereoidize. The powders are collected at three locations; bottom collector, cyclone and powder collector. In the former two, the particles are relatively coarse as they could not follow the gas stream, whereas in powder collector, particles probably originate from the gas phase, i.e. from the evaporation of AB2 powders.

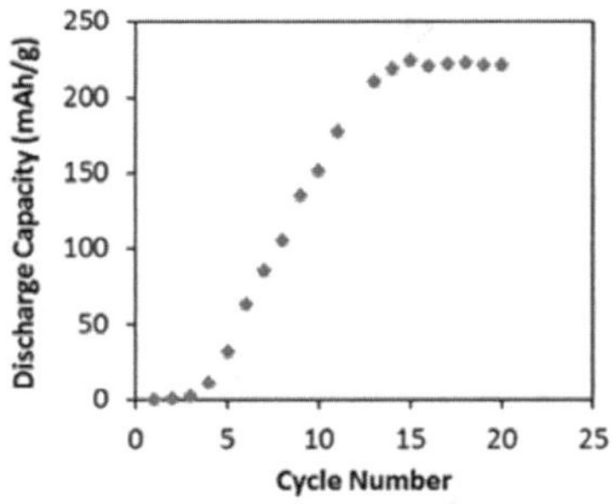

Fig. 1. Discharge capacity in AB2 alloy in as-received condition. Note that the discharge capacity reaches its full value after 12 cycles.

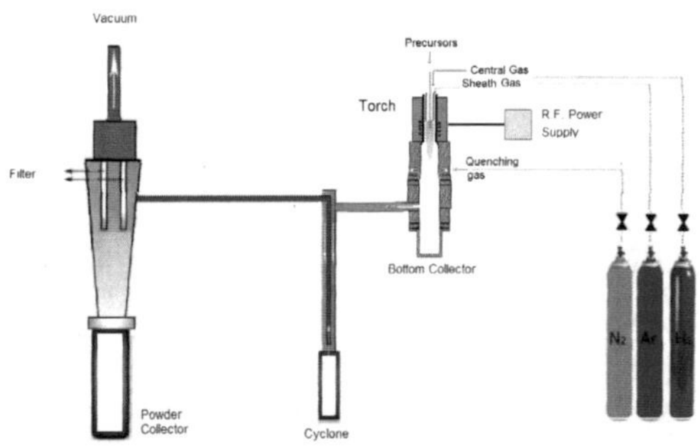

Fig. 2 Schematic representation of R.F. thermal plasma reactor. Precursors, i.e. AB2 powder, were fed to the torch and the resulting powder was collected at the bottom collector, the cyclone and the powder collector.

3 Results and discussion

A total of more than 100 g of powders were processed with a rather fast feed rate of 29 g/m. The fraction of particles collected in the bottom collector and cyclone were 0.36, 0.35 respectively and the fraction in the powder collector was only 0.05, i.e. overall yield was 76 %. Typical examples of particles collected at the bottom collector, cyclone and the powder collector are given in Fig.3. It is seen that the particle size decreases in the given order. Particles collected at the bottom collector had a spherical shape; the size measured over 100 particles yielded a value of 35 ± 15 micron. This is similar to the starting powder which when evaluated with the same method has yielded a value of 36 ± 17, μm Fig. 3(a) and (b).

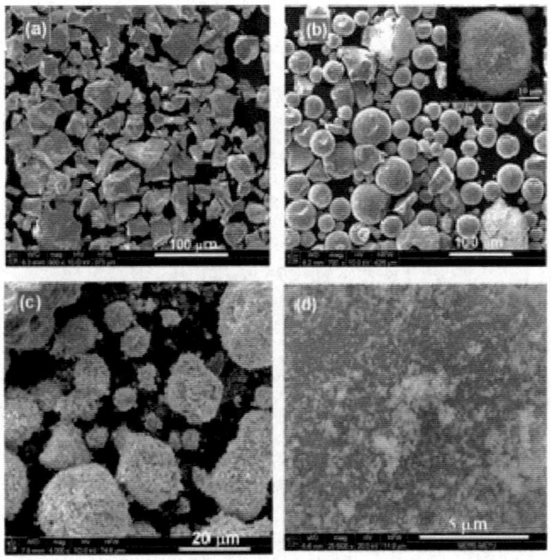

Fig. 3. SEM micrograph of powder collected from bottom collector (b), cyclone (c) and powder collector (d). a) shows the starting powder

In bottom collector, the flow makes a 90° turn, so rather heavy, i.e. large; particles are left behind and collected there. In cyclone, the flow makes 180 ° turn which would leave behind even smaller particles. As seen Fig.3(c), these particles have an average size of 17 ± 6 micron. These particles are decorated from nanosize particles that have either condensed on these particles in the reactor or attached to it while travelling to the cyclone. Particles collected in powder collector were extremely small less than 100 nm in size.

Plasma processing of powders led to drastic chemical alterations, especially in particles collected from a powder collector. Fig. 4 shows EDS spectrum recorded in the starting powder and that collected in powder collector. The initial AB2 powder that had (Ti0.36Zr0.64)(V0.15Ni0.58Mn0.20 Cr0.07)2 composition, A elements, i.e. Ti and Zr were reduced drastically; instead fine powders were richer in its Mn content. How each element varies from the starting powder to that collected in bottom to powder collector are given in Fig. 5.

XRD pattern of all three powders are given in Fig.6. It is seen that the pattern in the starting powder is compatible with C14 laves phase plus an additional minor phase. In bottom collector the pattern is

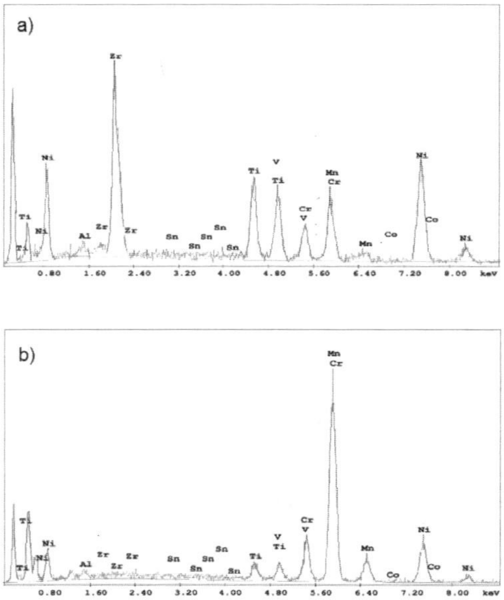

Fig. 4. EDS spectrum obtained from (a) starting powder and (b) from particles collected from powder collector. Note that plasma processed powder is richer in Mn and Cr and leaner in Zr and Ti.

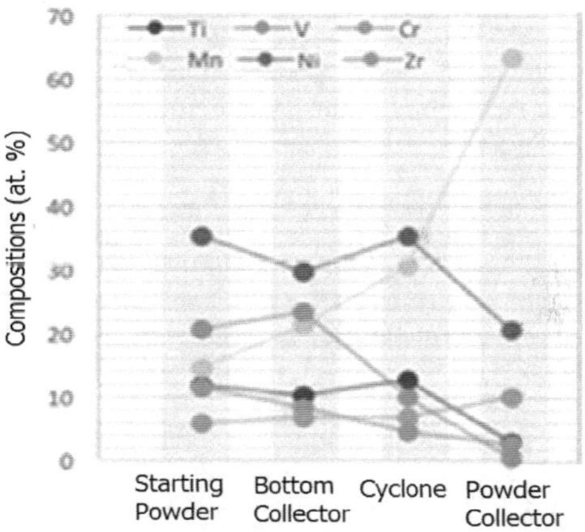

Fig. 5. Variation of elements in AB2 from starting powder to bottom, cyclone and powder collector. Note that most drastic change is in the powder collector.

identical indicating that the powders are little affected by plasma processing. Powders from cyclone are intermediate, but that of powder collector is affected quite drastically.

There is quite a pronounced broadening of the peaks in conformity with the small size of particles. In addition additional peaks could be discerned in the pattern implying that the compositional change has led to the formation of an additional phase in the material.

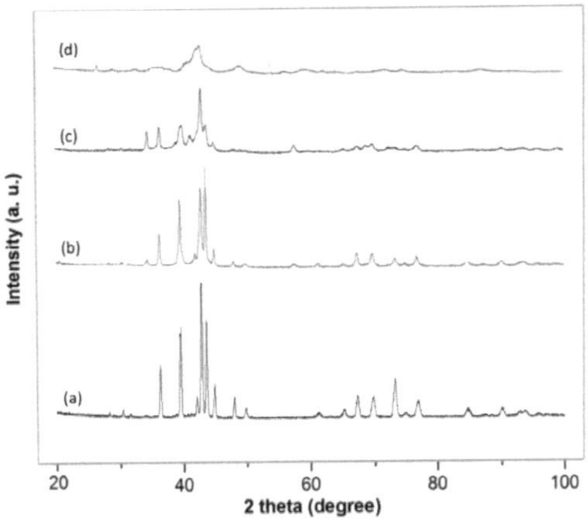

Fig. 6 XRD pattern of powders derived from C14 AB2 alloy. a) Starting powder, plasma processed powders b) from bottom collector c) Cyclone and d) powder collector.

Observations reported above are compatible with the expectations. Powders collected from bottom collector and cyclone probably derived from large size particles. The external surface of these particles may be partly evaporated, but mostly molten so that they take up spherical shape. These particles being rather large do not reach the powder collector. Small size particles, on the otherhand, evaporate and at lower down in the reactor condense into nanoparticles which can successfully follow the flow and are collected in powder collector.

With partial evaporation of the precursor, composition in the gas phase is not necessarily the same as the precursor powder. Even when it is the same, the condensation may lead to different combination of elements resulting in particles of different composition.

Powders were characterized electrochemically using pressed electrode. Discharge capacity vs. cycle number was obtained. Only the result with respect to powder collected from bottom collector and the cyclone will be reported here. Unfortunately none of these powders were active in electrochemical testing. Thus no discharge capacity could be detected for the alloys

It is known that C14 AB2 alloy when subjected to hot alkaline treatment produces Ni rich surface which help activating the alloy. Unfortunately the plasma processing did not produce such an effect. It is certain that the surface of powders, collected at the bottom collector and cyclone are modified chemically. Obviously the type of modification that occurred is not a desirable one. Further study would be required to identify the nature of surface modifications that take place in plasma processing.

Of course powders collected in the powder collector are very fine as compared to what is customary as anode material. At the time of writing this paper, the performance of these materials as negative electrode material was under evaluation and the results will be reported during the symposium.

Acknowledgement

The work reported in this paper was supported with TUBITAK (114M128) which the authors gratefully acknowledge.

References

[1] Liu Y. , Pan H., Gao M. and Wang Q., Advanced hydrogen storage alloys for Ni/MH rechargeable batteries, J. Mater. Chem., 21(2011), 4743.

[2] Li Y., Wang A., Li J., & Chen Q., A Brief Analysis of Global Rare Earth Trade Structure, Advanced Materials Research 737(2013), 3324–3331.

[3] Young K., Chao B., Bendersky L.A., Wang K.Ti12.5Zr21V10Cr8.5MnxCo1.5Ni46.5-x AB2-type metal hydrided alloys for electrochemical storage application: Part 2. Hydrogen storage and electrochemical properties. J Power Sources 218(2012), 487-494.

[4] Tan S, Shen Y, Şahin E.O., Noréus D., Öztürk T., Activation behavior of an AB2 type metal hydride alloy for NiMH batteries, International Journal of Hydrogen Energy, 41(2016), 9948–9953.

Thermal gradient cyclic lifetime of $La_2Zr_2O_7$/YSZ double ceramic layer thermal barrier coatings with equivalent thermal insulation performance

Bo-Cheng, Guan-Jun Yang*, Qiang-Zhang, Ning-Yang, Meng-Zhang, Yuming-Zhang, Cheng-Xin Li, Chang-Jiu Li
State Key Laboratory for Mechanical Behavior of Materials, School of Materials Science and Engineering, Xi'an Jiaotong University, Xi'an, Shaanxi Province 710049, PR China

Thermal insulation performance is a measurement of the thermal protection offered by the thermal barrier coatings (TBCs) to the substrate, therefore, it is essentially important to compare different double ceramic layer (DCL) TBCs on the premise of the same thermal resistance. In this study, a series of LZO/YSZ DCL-TBCs, with the equivalent thermal insulation to 500 μm thick YSZ TBCs, were prepared, and their lifetimes were evaluated by thermal gradient cyclic test at the top coat surface temperature of 1300℃. Result show that, the lifetime of DCL-TBCs was more than doubled compared to 500 μm thick YSZ TBCs, when 100μm thick YSZ coating was substituted by LZO coating. In addition, the lifetime of DCL-TBCs decreased with the increase of LZO substitutional ratio. X-ray diffraction analysis revealed that LZO maintains the pyrochlore structure after thermal cyclic test. Microstructure examination demonstrated that, with the increase of LZO substitutional ratio, the delamination position transferred from near top/bond coating interface to near LZO/YSZ interface and finally to the inside of LZO coating. Therefore, this study would shed light to further coating structure optimization towards the next generation advanced DCL-TBCs.

1 Introduction

Thermally sprayed thermal barrier coatings (TBCs) have been used successfully in hot section components to protect the metal parts from hot gases, enabling modern gas-turbine engines to operate at gas temperatures well above the melting temperature of the superalloy.[1, 2] The typical TBC is composed of a superalloy substrate, a MCrAlY (M=Ni, Co, NiCo or CoNi) bond coat (BC) to relieve the thermal mismatch between the ceramic layer and the substrate, a thermally grown oxide (TGO) inevitably grown on bond coat surface, and the outermost layer as the top ceramic coats to provide thermal insulation effect.[3, 4] The state-of-the-art top coat TBC material is 7-8 wt.% yttria-stabilized zirconia (8YSZ), primarily due to its high toughness and high coefficient of thermal expansion (CTE), and relatively low thermal conductivity.[5, 6] However, during long time exposure to high temperature, the most critical issue for YSZ is the limited operation temperature (<1200℃) for long-term application. At higher temperature, YSZ may experience phase transformations from non equilibrium tetragonal (t') phase to tetragonal (t) and cubic (c) phases, and the t-phase further to monoclinic (m) phase during cooling, accompanying volume change;[7-9] Furthermore, higher temperature induces the sintering of ceramic top coating (leading to the reduction of strain tolerance and porosity in combination with the increase of elastic modulus and thermal conductivity), which eventually would accelerate the failure of TBCs.[10] The next-generation advanced gas engines require higher turbine inlet temperature (i.e >1250℃),[11] which means higher surface temperature on TBC surface, and the conventional YSZ can not be competent to endure the harsher operating conditions. To overcome the shortcomings of YSZ and develop the higher temperature TBCs, pyrochlore structured oxides $La_2Zr_2O_7$ (LZO) are proposed as the advanced top coat material[12], due to it is phase-stable up to melting points (2300°C)[13]. What's more, $La_2Zr_2O_7$ shows lower thermal conductivity[14], and the low diffusive speed of oxygen ions in $La_2Zr_2O_7$ also decelerates the growth of TGO[15]. However, $La_2Zr_2O_7$ exhibits a low CTE and/or a poor fracture toughness, which may cause premature failure of the coating during service[14]. Given this, the concepts of functional graded TBCs (FG-TBCs)[16-17] and the double-ceramic-layered TBCs (DCL-TBCs)[13, 18] were extensively studied. It is considered that, the new architectures of TBCs provide promising solutions for TBCs applied in higher temperature with longer service time.

However, thermal insulation effect, as one of the fundamental functions, should be paid preferential attention. It is a measure of the thermal protection offered by the TBC to the substrate. In the reported literature, the design of DCL coatings is based on same total thickness of top coating while the thickness ratio of LZO to YSZ is changed. Since the thermal conductivity of LZO is much lower than that of YSZ, the thermal insulation effect of DCL coatings and the pure YSZ is evidently different from one to another. A theoretical basis must be established for the design of DCL TBCs towards a fair and comprehensive understanding of their thermal cyclic behavior.

In this paper, a series of LZO/YSZ TBCs were designed and prepared for the gradient thermal cyclic lifetime evaluation, and the microstructure, mechanical properties were characterized to investigate the thermal cyclic behavior of DCL TBCs compared with their counterpart of single ceramic layer TBCs.

2 Experimental

2.1 Preparation and Characterization of TBCs

A commercial spherical YSZ powder (Sulzer Metco 204B-NS, -75+45 μm, Westbury, NY) and LZO powder (Tian-Yao Inc., -47+12 μm, Qing Dao, China) were used to deposit the ceramic top coat by a APS system (GP-80, JiuJiang, China). The morphology of the LZO powder and YSZ powders is shown in Fig.1.

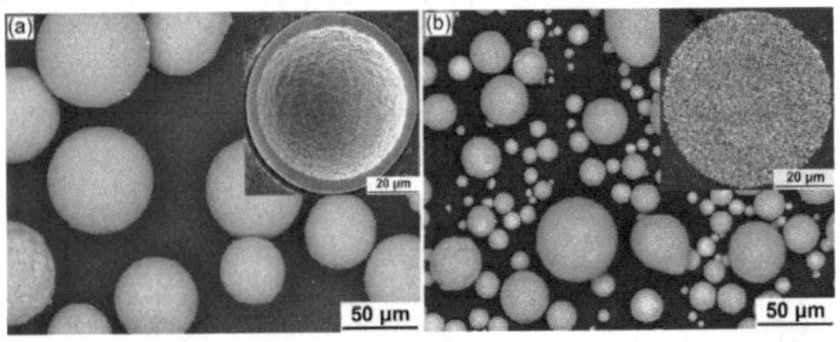

Fig. 1. Morphology of the spherical 8YSZ (a) and LZO (b) powder used for top coat. The insets show the cross section of the powder.

The plasma spray parameters for the top coat are shown in Table 1. Prior to the deposition of the top-coat, a commercially available NiCoCrAlTaY powder (Amdry 997, -37+9 μm, Sulzer Metco, Westbury, NY) was used as the bond coat, and deposited on a nickel-based superalloy (Inconel 738, Φ25.4 mm×3 mm) substrate. The bond coat was deposited to approximately 100 μm by a low pressure plasma spraying (LPPS) system.

Table 1. Parameters of air plasma spraying

Parameters	YSZ	LZO
Arc current, A	650	650
Arc voltage, V	60	60
Primary plasma gas (Ar) flow, L/min	60	50
Secondary plasma gas (H_2) flow, L/min	5.5	7
Spray distance, mm	80	80

Table 2 shows the spraying parameters of LPPS. Before the deposition of top coat, the as-prepared samples with bond coat were subjected to a preheat-treatment to forming a dense and connective Al_2O_3 TGO.

Table 2. Parameters of low-pressure plasma spray

Parameters	Value
Arc current, A	575
Arc voltage, V	150
Plasma gas (Ar) flow, L/min	40
Plasma gas (H_2) flow, L/min	6
Powder feeding gas (Ar) flow, L/min	1
Chamber pressure, kPa	15
Spray distance, mm	280
Torch traverse speed, mm/s	800

The cross-sectional microstructure of the coating was characterized using a scanning electron microscope (SEM) system (TESCAN MIRA 3 LMH, Czech Republic). Before obtaining the section of samples for a metallographic examination, the samples were infiltrated with epoxy adhesive in vacuum to protect the coatings from any artifact destruction during the sample preparation. The phase composition of YSZ coatings before and after thermal cyclic test was characterized by a x-ray diffraction (XRD, PANalytical X'pert PRO, Netherlands) system. The porosity of the different regions of top coat was determined by image analysis. The cross-sectional Vickers hardness was determined by a micro-Vickers Indenter (BUEHLER MICROMET5104, Akashi Corporation, Japan). The hardness tests were performed at a test load of 300 g and holding time of 30 s. The in-plane Young's modulus of top coat was measured by a Knoop indentation method. The parameters and the test system were the same with the measurement of microhardness.

2.2 Thermal Cyclic Test for TBCs

A gas burner test setup was designed for the thermal gradient cyclic test (Xi'an Jiaotong University, China) to simulate the actual service condition of TBCs. The preset temperature of 1300/1000℃ (1300℃ at top coating surface and 1000℃ at bond coat) was used to carry out the thermal gradient cyclic tests. One thermal cycle was defined as follow: 50 s heating, 250 s holding and 120 s cooling.

The temperatures of the specimen at surface and backside were monitored by non-contact infrared thermometers with wavelengths of 8-14 μm and 1.6 μm, respectively. The emissivity for YSZ and superalloy substrate was calibrated to be 1 and 0.91 according to their wavelength, respectively. The lifetime of TBCs was defined as the thermal cyclic lifecycles when 10% area delamination of the top coat was observed.

3 Results and Discussion

3.1 Design of coating architecture

Thermal insulation performance is one of the fundamental functions for thermal barrier coatings. The superalloy inside surface contact with air for cooling, whereas the outside surface is thermal barrier coatings and contact with hot gas, setting up a temperature gradient across the TBC.

In our previous study[1], thermal resistance is introduced to evaluate the thermal insulation performance of TBCs, and it is expressed as

$$R=\frac{\delta}{\lambda A} \qquad (1)$$

Where δ is the thickness of top coat, λ is the thermal conductivity and A is the area normal to the heat flow.

For the specific TBCs, "thermal barrier performance" is a parameter independent on the area A, hence we introduce the unit thermal resistance (UTR) as an intrinsic parameter to describe the thermal insulation performance of TBCs, it can be expressed as

$$\mathrm{UTR}=\frac{\delta}{\lambda} \qquad (2)$$

For multi-layered TBCs, the TBCs are in series system, and the UTR of the multi-layers is the sum of each layer. Hence the UTR of the DCL LZO/YSZ coating can be expressed as

$$\mathrm{UTR}=\frac{\delta_{LZO}}{\lambda_{LZO}}+\frac{\delta_{YSZ}}{\lambda_{YSZ}} \qquad (3)$$

where δ_{LZO} and δ_{YSZ} is the thickness of LZO and YSZ coatings, λ_{LZO} and λ_{YSZ} is the thermal conductivity of LZO and YSZ coatings.

The general thermal conductivities of plasma sprayed $La_2Zr_2O_7$[24] and YSZ[25] coatings are 0.74 W m^{-1} K^{-1} and 1.1 W m^{-1} K^{-1}, respectively. Therefore, if the thermal insulation effect of 100 μm thick YSZ coating is defined as 1 unit, 67μm thick LZO coating is of exactly the same thermal insulation effect of 1 unit thermal insulation effect. According the analysis above, 5 groups of TBCs with similar thermal insulation performance were designed in this study. Namely, 500μm YSZ(5Y), 400μm YSZ+ 67μm LZO (4Y/1L), 300μm YSZ+ 134μm LZO (3Y/2L), 200μm YSZ+ 201μm LZO(2Y/3L), 100μm YSZ+ 268μm LZO(1Y/4L). It can be found that the DCL-TBCs design is beneficial for reducing the top coating thickness.

3.2 Lifetime of TBCs

Figure 2 shows the lifetime of different group of TBCs under thermal gradient cyclic test. The lifetime of single YSZ ceramic layer TBCs (5Y) is 91±6 cycles. When 100μm YSZ was replaced by 67μm LZO (namely, 4Y/1L), the lifetime was more than doubled than pure YSZ (5Y). The lifetime of DCL-TBCs decreased with the further increase of LZO replace thickness. When 200μm YSZ was replaced, the lifetime of DCL-TBCs (3Y/2L) was still longer than 5Y. When 300μm and 400μm YSZ were replaced, the lifetime of DCL-TBCs (2Y/3L and 1Y/4L) decreased sharply and was much shorter than 5Y. Nevertheless, under equivalent thermal insulation performance, DCL-TBC is an effective solution to prolong the lifetime with appropriate YSZ coating thickness replaced by LZO coating.

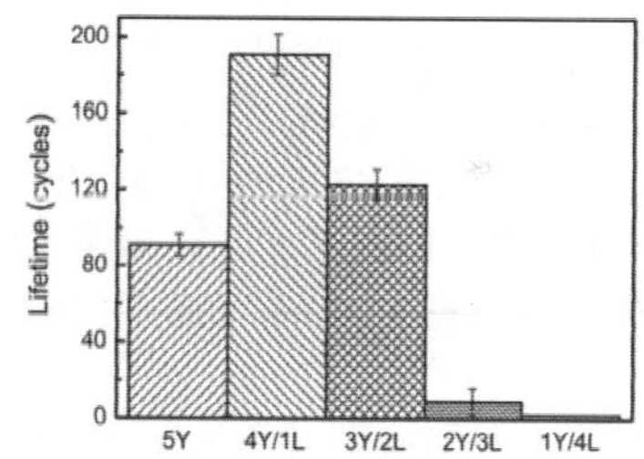

Fig. 2. Lifetime of different group of TBCs under thermal gradient cyclic test.

3.3 Failure Mode

In order to clarify the failure mode of the 5 groups of TBCs, the cross sectional SEM images of as sprayed state and after failure were examined as shown in Fig 3. For the as-sprayed TBCs, the top coats present a lamellar structure with no large cracks, and the surface shows the typical thermal sprayed features with a rough surface resulted from random stacking of splats. However, after failure, large cracks were found within top coats, and thereby the failure modes can be distinguished mainly by the location of cracks. It is found from Fig. 3(b) and 3(c) that, the failure mode of 5Y and 4Y/1L is the delamination within YSZ layers.

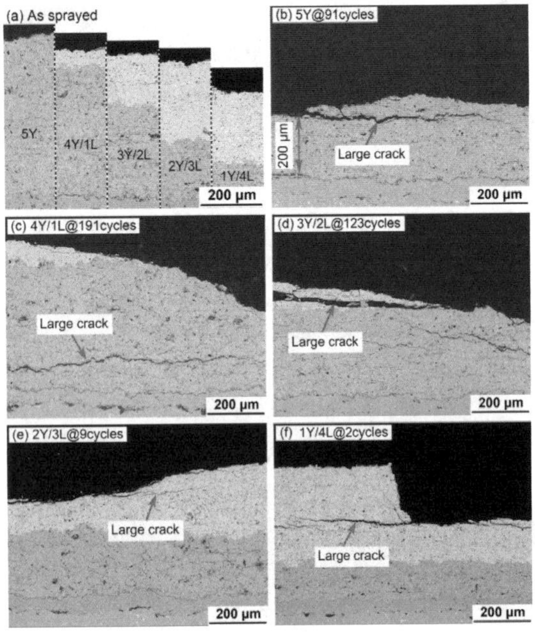

Fig. 3. The cross sectional images of different group of TBCs and their failure modes under gradient thermal cyclic test.

For 5Y, the crack depth is about 200 μm, which is less than the total YSZ coating thickness of 500 μm, and a large crack can be found as is shown in Fig. 3(b). In Fig. 3(c), a large scale crack was found within YSZ layer proved the failure mode of 4Y/1L. With the further increase of LZO replacing thickness, the delamination between LZO/YSZ interface happened, as shown in Fig. 3(d). Finally, the failure mode of 2Y/3L and 1Y/4L is coating delamination within LZO layer.

Therefore, it is reasonable to believe that with the increase of LZO substitutional ratio, the delamination position transferred from near top/bond coat interface to near LZO/YSZ interface and finally to the inside of LZO coating. The delamination of ceramic coating is essentially responsible for the failure of TBCs under thermal gradient cyclic in this study.

3.4 TGO Thickness

Generally, TGO growth is considered as one of the important factors leading to the failure of TBCs. Thus, the TGO growth behavior in the present TBC samples was examined. It was found from Fig. 3(b) to (f) that no obvious cracks were observed around the TGO, suggesting that the TGO may not be responsible for the failure of TBCs in this study. As a further step, TGO thickness was determined by the cross-sectional SEM micrographs before and after thermal cyclic test, and each value was the mean thickness of 10 different regions of TGO. Figure 4 shows the morphology of TGO for the failure samples of 2Y/3L after about 123 cycles. It can be found that the TGO is uniform and dense. TGO thickness in the as-sprayed TBCs was 0.32±0.04 μm, resulting from the pre-treatment mentioned in Section 2.1.

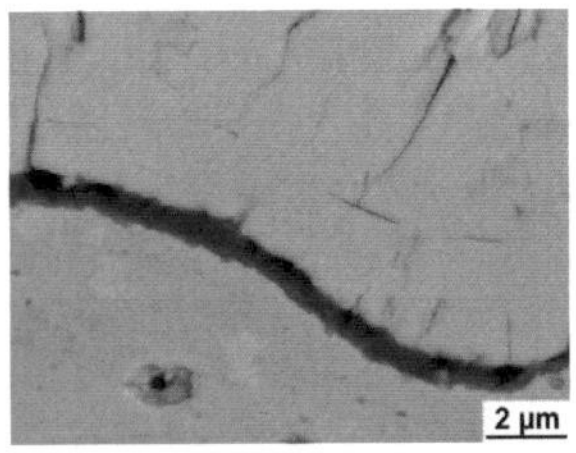

Fig. 4. The typical morphology of TGO after failure.

After thermal cyclic test to failure, the TGO slightly thickened with respect to that in the as-sprayed TBC. The thicknesses of TGO were 0.54±0.09 μm, 1.20±0.18 μm, 1.10±0.13 μm, 0.57±0.08 μm and 0.36±0.04 μm for the failure samples of 5Y, 4Y/1L, 3Y/2L, 2Y/3L and 1Y/4L, respectively.

Although the TGO thicknesses after thermal cyclic test were thicker than that in the as-prepared samples, they were significantly smaller than the reported critical value to induce a complete delamination of top coat, i.e., 5-6 μm. It is evidently proved that the slightly thickening of TGO is not the dominant factor responsible for the failure of TBCs in this study.

3.5 Phase Composition of the Top Coat

Given the fact that phase transformation might be a reason to induce the spallation of top coat, the phase composition of the top coat after TBC failure was examined, as shown in Fig. 5. It can be found from Fig.5 (a) and (b) that the phase compositions of the YSZ and LZO coatings before and after thermal cyclic test were mainly non-transformable t'-phase and pyrochlore structured LZO. What's more, it seems that neither the increase of the diffraction peaks of monoclinic phase nor non-pyrochlore structured LZO were founded after thermal gradient cyclic test.

Therefore, it evidently proved that no phase transformation of the top coat occurred. Consequently, the phase transformation is not the dominant factor responsible for the failure of TBCs in this study.

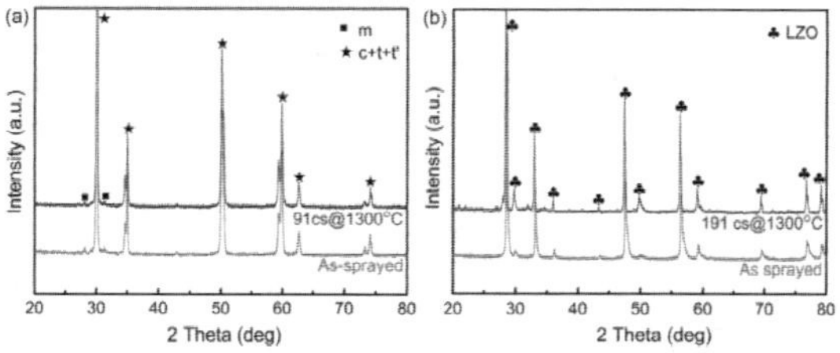

Fig. 5. XRD patterns of the top-coat before and after thermal cyclic tests: (a) YSZ and (b) LZO.

3.6 Failure Analysis

After thermal cyclic tests at different conditions, the maximum thickness of TGO is about 1.2 μm, being far less than the critical value of 5-6 μm to induce a completely delamination of the top coat. Secondly, the YSZ top coat showed no obvious phase transformation and LZO kept the pyrochlore structure the same as-sprayed state. Thirdly, dependent with the regions along through-thickness direction, high temperature exposure induced distinct changes in properties of the top coat. The significantly stiffened of top coat dominates the delamination of TBCs.

It can be concluded that, the failure of TBCs is mainly dominated by the crack propagated within top coats. The crack propagation occurs when the strain energy release rate is larger than the critical interfacial crack energy or fracture toughness. The strain energy release rate, G_h, can be obtained as follow:

$$G_h = \begin{cases} \int_0^{h_1} \frac{E_Y \varepsilon^2}{2(1-\nu)} dx & \text{for 5Y} \\ \int_0^{h_1} \frac{E_Y \varepsilon^2}{2(1-\nu)} dx + \int_{h_1}^{h_2} \frac{E_L \varepsilon^2}{2(1-\nu)} dx & \text{for 4Y} \\ \int_0^{h_2} \frac{E_Y \varepsilon^2}{2(1-\nu)} dx & \text{for 3Y/2L, 2Y/3L and 1Y/4L} \end{cases} \quad (4)$$

where h is the distance from the crack tip to top coating surface, h_1 and h_2 is the distance from the crack tip to the YSZ and LZO surface, respectively. E_Y and E_L is the local elastic modulus of YSZ and LZO at the crack tip, ε is the strain, ν is the Poisson's ratio.

As the analysis above, G_h increases with the increase of h, the coat failure happened with certain thickness. Furthermore, the sintering induced gradient distribution of mechanical properties makes the G_h increase more significantly. On the other hand, sintering caused the densification of ceramic coating will increase the cracking resistance. Therefore, whether the crack propagates mainly depends on the competition between the cracking driving force and the cracking resistance i.e., strain energy release rate and localized fracture toughness of the top coat. For 4Y/1L, the thickness of LZO is just 67 μm, the driving force in LZO layer is not enough to cause the cracking of LZO, combining with that the cracking resistance increased because of the long term high temperature exposure. The failure happens within the under YSZ layer. With the increase of LZO, the driving force increased and the failure located at LZO/YSZ interface. For 2Y/3L and 1Y/4L, since the thickness of LZO is 3 to 4 times of the 4Y/1L, G_h increased remarkably. However, the intrinsic property of LZO exhibits a low CTE and/or a poor fracture toughness. On the other hand, the contribution of sintering to the cracking resistance is limited (at most 9 cycles). Consequently, with the increase of LZO substitutional ratio, the delamination position transferred from near top/bond coating interface to near LZO/YSZ interface and finally to the inside of LZO coating.

4. Conclusions

Double ceramic layer TBCs were designed and subjected to thermal gradient tests based on the concept of equivalent thermal insulation performance proposed in this study, with the aim to reveal the failure behavior of DCL-TBCs. The following conclusions could be obtained from this work:

(1) Equivalent thermal insulation concept is introduced for the design of double ceramic coatings with different component layer ratio. TBCs with equivalent insulation performance of 500 μm YSZ coatings were designed. With the increase if LZO proportion, the total thickness of DCL TBCs decreased.

(2) The lifetime of conventional 5Y is about 91cycles under the surface/bond coat temperature of 1300/1000℃ gradient thermal cyclic test, while the lifetime of 4Y/1L DCL-TBCs showed up to 2 times higher than that of 5Y, proving that the DCL-TBCs is an effective technique for the application of TBCs in higher temperature with longer lifetime.

(3) The maximum thickness of TGO in failed TBC samples was 1.2 μm, being far less than the critical value causing the failure of TBCs through complete delamination of the top coat. In addition, TBCs showed no obvious phase change during thermal cyclic test. Therefore, the TGO thickening and phase transformation were not the essential reasons responsible for the failure of TBC in the present study.

(4) The lamellar cracking within top coating was the essential failure reason for both 5Y and DCL-TBCs. With the increase of LZO substitutional ratio, the delamination position transferred from near top/bond coating interface to near LZO/YSZ interface and finally to the inside of LZO coating.

Acknowledgments

The present project is supported by National Basic Research Program (Grant No. 2013CB035701), the National Science Foundation of China (Grant No. 51671159), the Fundamental Research Funds for the Central Universities, and the National Program for Support of Top-notch Young Professionals.

References

[1] N. P. Padture, M. Gell, and E. H. Jordan, "Thermal barrier coatings for gas-turbine engine applications," *Science,* **296**[5566] 280-4 (2002).

[2] J. Li, and C. Coddet, "Fractal Perimeters of Polishing-Induced Pull-Outs Present on Polished Cross Sections of Plasma-Sprayed Yttria-Stabilized Zirconia Coatings," *J. Am. Ceram. Soc.,* **86**[11] 1906-10 (2003).

[3] X.Q. Cao, R. Vassen, D. Stoever, " Ceramic materials for thermal barrier coatings," J. Eur. Ceram. Soc., **24**[1] 1-10 (2004).

[4] D. R. Clarke, M. Oechsner, and N. P. Padture, "Thermal-barrier coatings for more efficient gas-turbine engines," *MRS Bull.,* **37**[10] 891-98 (2012).

[5] G. W. Goward, "Progress in coatings for gas turbine airfoils," *Surf. Coat. Technol.,* **108**[1-3] 73-79 (1998).

[6] D. R. Clarke and S. R. Phillpot, "Thermal barrier coating materials," *Mater. Today.,* **8**[6] 22-29 (2005).

[7] R. W. Trice, Y. J. Su, J. R. Mawdsley, K. T. Faber, H. Wang, and W. D. Porter, "Effect of heat treatment on phase stability, microstructure, and thermal conductivity of plasma-sprayed YSZ," *J. Mater. Sci.,* **37**[11] 2359-65 (2002).

[8] J. Ilavsky and J. K. Stalick, "Phase Composition and Its Changes During Annealing of Plasma-Sprayed YSZ," *Surf. Coat. Technol.,* **127**[2] 120-29 (2000).

[9] T. Vogt, and J. Thornton, "Structural Evolution of Thermal-Sprayed Yttria-Stabilized ZrO_2 Thermal Barrier Coatings with Annealing-A Neutron Diffraction Study," *J. Am. Ceram. Soc.,* **84**[3] 678-80 (2001).

[10] S. Paul, A. Cipitria, L. Xie, M. R. Dorfman, and T. W. Clyne, "Effects of Impurity Content on the Sintering Characteristics of Plasma-Sprayed Zirconia," *J. Therm. Spray Technol.,* **16**[5] 798-803 (2007).

[11] D. Hui, X. Zhong, J. Li, Y. Zhang, M. Jian, and X. Cao, "Thermal stability of double-ceramic-layer thermal barrier coatings with various coating thickness," *Mater.Sci. Eng., A.,* **433**[1] 1-7 (2006).

[12] R. Vassen, X. Cao, F. Tietz and D. Stöver, "Zirconates as New Materials for Thermal Barrier Coatings," *J. Am. Ceram. Soc.,* **83**[8] 2023-28 (2000).

[13] X. Q. Cao, R. Vassen, and D. Stöver, "Ceramic materials for thermal barrier coatings," *J. Eur. Ceram. Soc.,* **24**[1] 1-10 (2004).

[14] H. Chen, Y. Gao and H. Luo, "Thermophysical properties of lanthanum zirconate coating prepared by plasma spraying and the influence of post-annealing," *J. Alloys. Compd.,* **486**[1-2] 391-99 (2009).

[15] A. K. Rai, M. P. Schmitt, and D. E. Wolfe, "Thermal conductivity and stability of multilayered thermal barrier coatings under high temperature annealing conditions, " *J. Eur. Ceram. Soc.,* **35**[5] 1605-12 (2014).

[16] H. Chen, Y. Liu, Y. Gao, S. Tao, and H. Luo, "Design, Preparation, and Characterization of Graded YSZ/$La_2Zr_2O_7$Thermal Barrier Coatings," *J. Am. Ceram. Soc.* **93**[6] 1732-40 (2010).

[17] K. Kokini, J. DeJonge, and B. Beardsley, "Thermal shock of functionally graded thermal barrier coatings with similar thermal resistance," *Surf. Coat. Technol.,* **154**[2-3] 223-31 (2002).

[18] T. Liu, G. J. Yang, and C. J. Li, "Properties evolution of plasma-sprayed $La_2Zr_2O_7$ coating induced by pore structure evolution during thermal exposure," *Ceram. Int.,* **42**[14] 15485-92 (2016).

[19] C. J. Li, "Propagation feature of cracks in plasma-sprayed YSZ coatings under gradient.hermal cycling," *Ceram. Int.,* **41**[3] 3481-89 (2015).

Non-parabolic isothermal oxidation kinetics of low pressure plasma sprayed MCrAlY bond coat

Bang-Yan Zhang, Guan-Jun Yang*, Cheng-Xin Li, Chang-Jiu Li, State Key Laboratory for Mechanical Behavior of Materials, School of Materials Science and Engineering, Xi'an Jiaotong University, China

The non-parabolic isothermal oxidation kinetics of low pressure plasma sprayed MCrAlY bond coat was investigated. To qualitatively explain the abnormal growth phenomenon of thermally grown oxides (TGO), the changes that occurred to their microstructure during the oxidation process were studied. Based on these observations, a modified model was developed to understand and quantitatively predict the non-parabolic oxidation and growth kinetics of TGO. This modified model, which fits well with experimental results, provides a novel method to quantitatively predict the long-term growth behaviour of TGO, and thereby benefits the development of long-life and highly reliable thermal barrier coatings.

1 Introduction

The demand to improve the efficiency in gas turbine in the recent years leads to the increased turbine inlet temperature [1-4]. In this regard, thermal barrier coatings (TBCs) have been widely applied onto the turbine parts that are exposed to the high temperatures, in order to prevent the failure of the superalloy of which they are made [2, 5, 6]. Thermal barrier coatings are generally composed of a thermal insulating ceramic top coating, a MCrAlY or NiAl-based bond coat, and a Ni-based metallic superalloy substrate [5, 7-18]. A layer of thermally grown oxides (TGO) will form at the interface between the bond coat and the top coating during thermal exposure.
It has been reported by previous studies that the TGO layer has a significant impact on the lifetime of TBCs [5, 7-9]. The TGO layer is the major source of strain which primarily derives the cracking, delamination and spallation of the top coating from the bond coat surface. During thermal exposure, the residual stress increases as the TGO layer grows, which will accelerate the spallation process of the top coating. Therefore, a deeper investigation into the growth behaviour of the TGO layer will provide an important theoretical foundation for understanding and evaluating the lifetime service and long-term reliability behaviour of TBCs.
Lowering the TGO growth rate is of great importance for prolonging the lifetime of TBCs [10-13]. The TGO layer is mainly made up of α-Al_2O_3 and mixed oxides (Cr_2O_3, NiO, and (Ni,Co)(Cr,Al)$_2$O$_4$ spinel). The fast growth of the mixed oxides in the TGO layer, which is accompanied by a large volume expansion, can cause the failure of TBCs because it induces cracks in the top coating [13, 14]. Therefore, the lifetime of TBCs can be significantly improved by suppressing or reducing the generation of mixed oxides. In this regard, a TGO layer made up of α-Al_2O_3 is preferred because the oxygen and metal ions have low diffusion coefficients, and the α-Al_2O_3 layers have a high thermal stability, which result in a low TGO growth rate. It has been proposed that the proper pre-treatment of the bond coat under a low oxygen partial pressure, before the deposition of the top coating, is an effective way to lower the TGO growth rate and suppress the growth of mixed oxides, because a thin and continuous α-Al_2O_3 TGO layer forms on the bond coat surface.
In this study, the non-parabolic oxidation kinetics of the bond coat was investigated. The TGOs oxidation behaviour can be quantitatively revealed by the growth kinetics of the TGO layer. According to the conventional Wagner oxidation theory (the parabolic law), the growth of the TGO layer is controlled by the high temperature diffusion of the elements, and the TGO growth kinetics curves obey the parabolic law. It has been reported that for a relatively short oxidation time (no more than hundreds of hours), the experimental results of the weight gain and TGO layer thickness are basically in line with the parabolic law. In addition, it has been reported by some papers that the weight gain of TBCs that underwent longer oxidation times (more than hundreds of hours) also seemed to be in line with the parabolic law. Nevertheless, weight gain changes cannot reflect variations in the TGO structure and make up during the oxidation process, so further research is necessary. The growth behaviour of the TGO layer thickness was considered to be a better way to reflect the structural changes of the TGO. The papers where changes in the TGO layer thickness (including all the oxides) were utilized to research the growth behaviour of TGO also state that the results of experiments carried out with longer oxidation times follow the parabolic law. However, after careful observations, we found that these experimental results seemed to deviate from the parabolic law, which is something that has not been observed up to now. As it has been reported that the oxygen diffusion coefficient of the outer mixed oxides is several orders of magnitude higher than that of the α-Al_2O_3 components, the research on the growth behaviour of the TGO layer would benefit if the thickness of the outside mixed oxides is neglected. In this paper, the growth kinetics of the TGO layer thickness (without the mixed oxides) for samples exposed to long oxidation times was carefully studied. Specifically, it was found that the curves of the TGO thickness growth deviated from the parabolic law. This behaviour will be discussed in this paper. Nevertheless, deep and systematic research on the mechanism of the bond coat's non-parabolic oxidation kinetics has never been reported.

In this study, low pressure plasma sprayed (LPPS) bond coat was subjected to pre-oxidation and high temperature isothermal oxidation in air. The long-term growth behaviour of the bond coat was studied, and it was found that the Al_2O_3 TGO growth curve deviated from the parabolic law during the oxidation process. The reason for the strange TGO growth behaviour, which does not fit well with the conventional Wagner oxidation theory, was examined in the perspective of the oxide grain-structure evolution. Finally, in order to further quantitatively predict the oxidation kinetics, a modified model for the oxide scale growth was developed.

2 Experimental Procedure

The bond coats of a ~150 μm thickness were deposited by LPPS using CoNiCrAlY powder in a spherical shape (Amdry 9951, Sulzer-Metco, USA). The spray powder has a mean particle size of ~25.9 μm (d10=12.9 μm, d90=38.6 μm). Nickel-based superalloy Inconel 738 was used as a substrate (Φ25.4 mm × 3 mm).

The as-sprayed bond coats were subjected to a pre-oxidation treatment conducted in Ar (4 h at 1000 °C and 4 h at 1080 °C, $P(O_2) < 3$ Pa). This pre-oxidation treatment can form a continuous α-Al_2O_3 oxide scale on the bond coat surface to suppress inward diffusion of oxygen and outward diffusion of cation during the subsequent isothermal oxidation in air. The pre-oxidation with high temperature and low oxygen partial pressure could make the deposited splats have a good metallurgical bonding through the high temperature diffusion, which would enhance elemental diffusion between each deposited splat to supply sufficient Al for the TGO growth on the bond coat surface during the subsequent oxidation [58]. After the pre-treatment, the bond coats were subjected to isothermal oxidation, for different durations, at 1000 °C in air.

The crystalline structure of the TGO layer was characterized by X-ray diffraction analysis (XRD, D8advance 3.0, Bruker, Germany). The microstructure of the coatings and the TGO layer was examined with a field emission scanning electron microscope (SEM, MIRA 3 LMH, TESCAN, Czech Republic). In order to precisely examine the grain size of the Al_2O_3 TGO layer, a bending and collecting approach was developed to obtain the exact grain size diameter at the interface between the TGO layer and the alloy matrix of the bond coat. First, the coatings (bond coat and TGO) were spalled from the substrates. Secondly, the TGO side of the coatings was pasted to a conductive tape, and then the coatings were bended into two pieces; some TGO chippings stuck on the conductive tape. Thus, the rear side of the TGO next to the bond coat could be observed by SEM.

3 Results and Discussion

3.1 TGO growth kinetics during isothermal oxidation

The cross sections of TGO layers that underwent isothermal oxidation for different durations, were observed; the images are shown in **Fig. 1**. **Fig. 1** reveals that, during isothermal oxidation, a uniform Al_2O_3 layer (dark area) remained the major component of the TGO layer found on the surface of the bond coat. In addition, **Fig. 1d** shows that after isothermal oxidation for 50 h, a small fraction of mixed oxides (light area) formed on the outside surface of the dark layer. The appearance of mixed oxides could be attributed to the unexpected oxidation of small splashed particles during the bond coat deposition. The TGO has presented a bi-layered structure (mixed oxides outside, and Al_2O_3 inside) after a 600 h isothermal oxidation (**Fig. 1f**). This might be attributed to the continuity of the small mixed oxides on the TGO surface. Nevertheless, the thickness of mixed oxides was very small, less than 1/5 the thickness of the uniform α-Al_2O_3 scale.

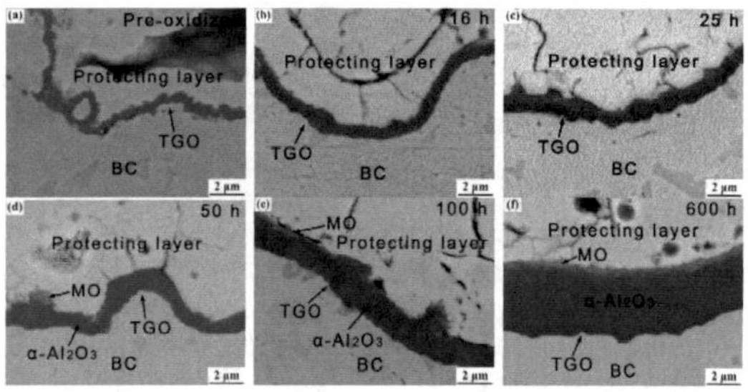

Fig. 1. Cross sectional microstructure of the TGO underwent isothermal oxidation for different durations. (a) Pre-oxidized, (b) 16 h, (c) 25 h, (d) 50 h, (e) 100 h, (f) 600 h.

To quantitatively reveal the growth kinetics of the TGO layer, the TGO layer thicknesses of bond coats oxidized for up to 100 h were obtained, as shown in **Fig. 2** (solid circles). The scattered points were well fitted with a straight line, revealing that the TGO thickness presented a good parabolic relationship with the oxidation duration. This means that the TGO growth kinetics of the coatings with short oxidation durations can be well described by the conventional Wagner theory of oxidation, using the following equation:

$$\delta^2 = 2k't \quad (1)$$

where δ is the Al_2O_3 TGO thickness, k' is the oxidation rate constant, and t is the oxidation duration. By taking the differential of Eq. 1, the Al_2O_3 TGO growth rate (oxidation rate, $\frac{d\delta}{dt}$) can be gotten, as follows:

$$\frac{d\delta}{dt} = \frac{k'}{\delta} \quad (2)$$

The above equation (Eq. 2) reveals that the Al_2O_3 TGO growth rate is inversely proportional to the Al_2O_3 TGO thickness, as k' is a constant.

According to the conventional Wagner oxidation theory, the oxidation kinetics during longer oxidation times should also follow the straight-line behaviour observed in **Fig. 2**. However, the results show that the points for longer oxidation durations (empty circles in **Fig. 2**) diverged from the straight line, and that the Al_2O_3 growth rate gradually decreased for longer oxidation times. In short, Al_2O_3 had different oxidation rate constants during an oxidation treatment of 1000 h, which is a behaviour that does not fit well with the conventional Wagner oxidation theory, i.e. parabolic function.

The dominant growth mechanism of Al_2O_3 TGO on the MCrAlY bond coat can be regarded as the diffusion of oxygen from the outside surface of the TGO, into the interface between the TGO and the alloy matrix of the bond coat. For the bi-layered TGO found on the bond coat surface of the samples that underwent long isothermal oxidation times, the oxygen diffusion to the inside went through the mixed oxides and the α-Al_2O_3 scale. The oxygen diffusion coefficient in mixed oxides is much higher than that in α-Al_2O_3 by several orders of magnitude, and the mixed oxides layer can be neglected when considering the growth of TGO via oxygen diffusion. Hence, in this study, the mixed oxides were excluded when calculating the TGO layer thickness to better understand the growth behaviour of the Al_2O_3 scale.

The k' value of the Al_2O_3 TGO layer is directly proportional to the inward diffusion rate of oxygen during the isothermal oxidation. Thus, the oxygen diffusion in the Al_2O_3 TGO layer was brought to the forefront of this research. For the oxygen diffusion in the oxide scale that covered the bond coat surface, the grain boundaries and the bulk crystalline grains are the main diffusion routes. The oxygen diffusion coefficient through the α-Al_2O_3 grain boundaries is about three orders of magnitude larger that through α-Al_2O_3 bulk crystalline grains. For the α-Al_2O_3 scales, except those with ultra-large size grains or even single crystals, the dominant diffusion route should be grain boundaries. Therefore, it is necessary to analyse the grain boundary diffusion behaviour.

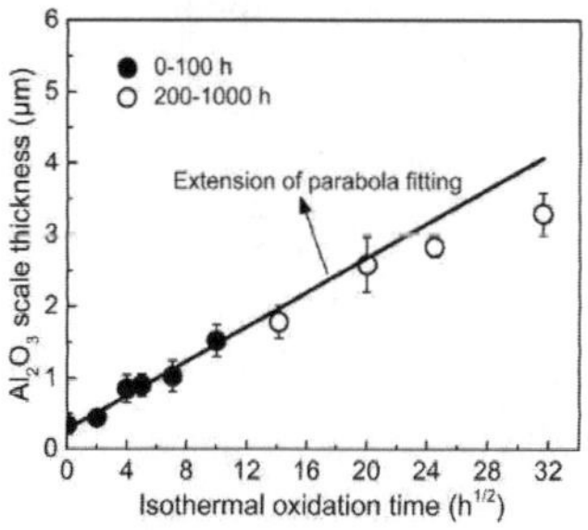

Fig. 2. Al_2O_3 TGO growth kinetics curve for long oxidation durations of up to 1000 h.

3.2 Microstructure evolution of Al_2O_3 TGO during isothermal oxidation

In order to reveal the essential reason of the non-parabolic relationship between the TGO layer thickness and oxidation duration, the grain structure of the Al_2O_3 TGO layer located on the surface of the bond coats was examined. **Fig. 3** shows the fractured cross-sectional microstructure of a TGO layer formed on the surface of a bond coat that underwent the isothermal oxidation for 50 h. From **Fig. 3**, it can be seen that the Al_2O_3 grains present columnar structures that grow inward in the direction perpendicular to the bond coat's surface. The inward growth of the Al_2O_3 grains resulted from the inward oxygen diffusion in the oxide grain boundaries. It can also be seen that the lateral size of Al_2O_3 grains at the TGO/bond coat interface increased gradually in the TGO growth direction. This change was caused by some small grains that grow slowly and merged into larger adjacent ones having faster growth rate. This will increase the grain size and decrease the grain boundary for oxygen diffusion in the TGO layer, and thereby gradually decrease the diffusion coefficient of O during the oxidation. Consequently, a lower experimental TGO thickness was observed for long-term oxidations, for example a 1000 h oxidation, than the parabola fitting of TGO thickness grain coarsening.

Fig. 3. The fractured cross-sectional microstructure of a TGO layer that formed on the surface of a bond coat that underwent isothermal oxidation for 50 h in air.

3.3 Generalized 3D model for the TGO growth behaviour

In order to further understand and quantitatively interpret the above abnormal phenomenon, a generalized 3D model of the oxide scale that forms on the surface of the bond coat was developed and a theoretical calculation was performed. In the 3D diagrammatic sketch of the α-Al_2O_3 TGO on the bond coat surface, shown in **Fig. 4**, the red lines express the grain boundaries of the α-Al_2O_3 TGO. In order to more easily understand and simplify the subsequent calculation, the Al_2O_3 scale is shown underneath the bond coat marked by dashed lines in the upper part of **Fig. 4**. From **Fig. 4** it can be seen that along the TGO growth direction (the direction of increasing z-axis), the grain boundary density decreases as the lateral grain size increases. To quantitatively evaluate the grain boundary in TGO, the grain boundary density was defined as the ratio of the grain boundary area to the unit volume of the TGO. The grain boundary density changed with the z-axis, since the grain size increased with an increase of TGO growth, as shown in **Figs. 3, 4**. The grain boundary density was not homogeneous along the z-axis direction. The oxide scale shown in **Fig. 4** is segmented to a great many of very thin and equal-thickness slices along the z-axis direction. As it was assumed that the grain boundary density of a single thin slice was homogeneous, Eq. (2) could be applied to the thin slices. The oxygen diffusion resistance per unit area in the entire Al_2O_3 scale actually was the overlapping result of the slices in series. However, to quantify the TGO growth kinetics, Eq. (2) has to be changed as follows:

$$\frac{d\delta}{dt} = \frac{1}{\frac{\delta}{k'}} \quad (3)$$

where $\frac{\delta}{k'}$ is proportional to the unit resistance of oxygen diffusion through an oxide scale with thickness of δ. For TGO layers with inhomogeneous oxygen diffusion properties along their thicknesses, due to changes in their grain size, and thereby grain boundary density, Eq. (3) has to turn into an integral:

$$\frac{d\delta}{dt} = \frac{1}{\int_0^{\delta} \frac{1}{f_{k'z}(z)} dz} \quad (4)$$

where $f_{k'z}(z)$ is the oxidation rate constant at the z local region. Then, the oxidation kinetics theory for inhomogeneous TGO can be expressed as,

$$t = \int_{\delta_0}^{\delta} \left(\int_0^{z} \frac{1}{f_{k'z}(z)} dz \right) dz \quad (5)$$

where δ_0 is the thickness of the Al_2O_3 scale on the surface of the pre-oxidized bond coat. Because $f_{k'z}(z)$ cannot be a constant, Eq. (5) will not be a parabolic function. Therefore, the Al_2O_3 TGO growth kinetics curves do not fit well into the conventional Wagner oxidation theory, which fits with the above experimental results (**Fig. 2**).

It can be found that the method of predicting the long-time TGO growth behaviour using the short-time parabola TGO growth pattern is improper, and the above discussion about the non-parabolic Al_2O_3 TGO growth behaviour should be utilized to predict the long-time TGO growth behaviour. Meanwhile, it was found that an increase of the initial Al_2O_3 lateral grain size plays an important role in retarding the thickness growth of the TGO layer. Therefore, research on this would benefit the development of long-life and highly reliable TBCs.

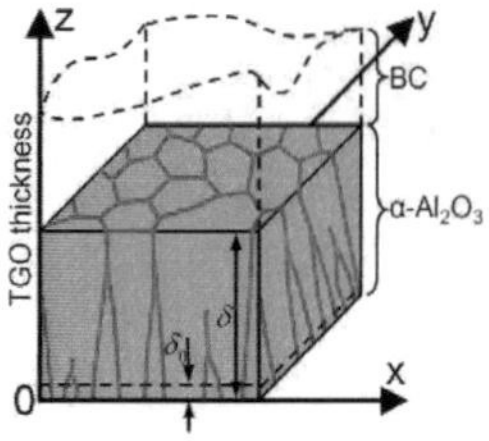

Fig. 4. 3D diagrammatic sketch of the α-Al_2O_3 TGO on the bond coat surface. The red lines express the grain boundaries of the α-Al_2O_3 TGO.

3.4 Quantitative prediction of TGO growth based on grain boundary density examinations

According to the above theoretical deduction, the oxidation rate constant, $f_{k'z}(z)$, is the key factor that affects the TGO growth behaviour. Therefore, the TGO microstructure, especially the Al_2O_3 grain boundary density at the interface of the TGO and the bond coat, is very important for the Al_2O_3 TGO growth behaviour. The grain boundary density can be obtained from the grain size because it is proportional to the reciprocal of the grain size. The Al_2O_3 grain size is a visual physical quantity, which is beneficial for understanding the above theoretical derivation. Thus, obtaining an accurate grain size is very important for analysing the Al_2O_3 TGO growth behaviour, and to verify the above theoretical calculations. However, it is extremely difficult to quantitatively obtain the grain size of this kind of thin TGO layer. It was previously reported that electron backscatter diffraction (EBSD) could be used to analyse the cross sections of TGO to examine their grain size. Although the cross sectional analysis can be realized by optimizing the sample preparation approach, the obtained results are not very precise because the cross sections of the cut grains do not always show the maximum diameter of the grains. This deviation would inevitably cause inaccuracies in the prediction. In order to precisely examine the grain size of Al_2O_3 TGO, a bending and collecting approach, which was described detailedly in the experimental section, was developed to obtain the microstructure of the rear side of the TGO layer next to the bond coat.

The microstructure of the TGO rear side (at the interface of the TGO and the bond coat), which was spalled from the samples, is partly shown in **Fig. 5**. From this figure (**Fig. 5**), it can be observed that the shape of the Al_2O_3 grains at rear side of the TGO layer is polygonal during the oxidation process, although a small fraction of the grains presents an aspect ratio larger than 2. For the purpose of having a deeper understanding of the influence that the grain structure has on the growth behaviour of the TGO layer, the changes of the average lateral Al_2O_3 grain sizes, and of the grain boundary distribution at the TGO layer/bond coat interface, were shown in **Fig. 6**. The results show that the grains at the TGO growth front, i.e. at the TGO layer/bond coat interface, increased gradually with increasing oxidation times due to grain growth and merge, which led to a reduction in the grain boundary density for oxygen diffusion. As discussed above, the decreasing grain boundary density for oxygen diffusion in the TGO will decrease the diffusion coefficient of O during the oxidation. Since the oxidation rate constant (k', Eq. (1)) is proportional to the oxygen diffusion coefficient, k' will decrease gradually as the oxidation time goes on, which would make the Al_2O_3 scale growth behaviour not fit well with the conventional Wagner oxidation theory, i.e. a parabolic function.

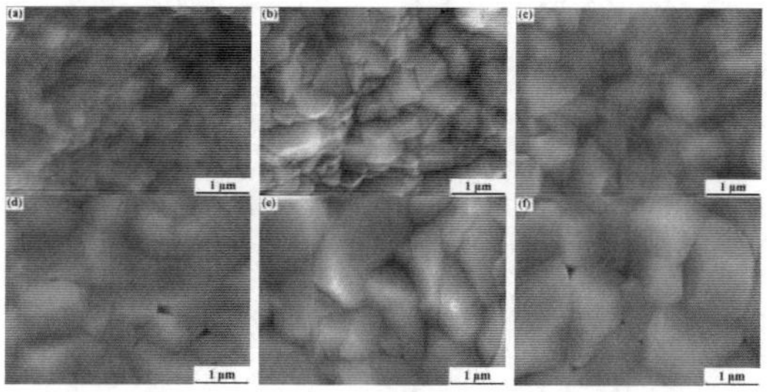

Fig. 5. Comparison between the Al_2O_3 microstructure of the TGO rear side (at the interface of the TGO and the bond coat) that underwent different isothermal oxidation durations. (a) Pre-oxidized, (b) 25 h, (c) 100 h, (d) 200 h, (e) 400 h, (f) 600 h.

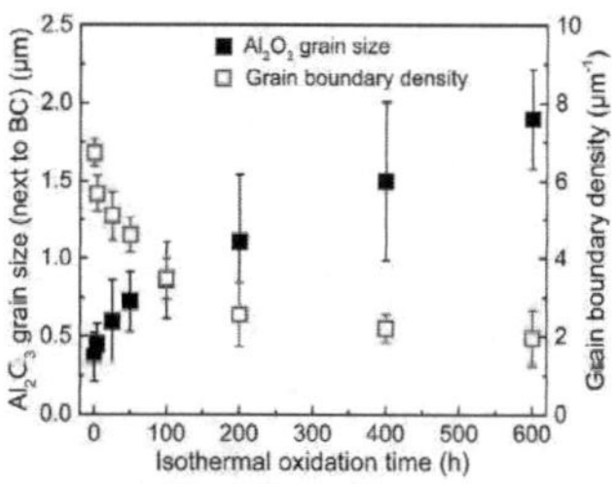

Fig. 6. Grain boundary density and Al_2O_3 grain size (next to the BC) changes of samples that underwent different isothermal oxidation durations.

The relationship between the oxidation rate constant $f_{k'z}(z)$ and the TGO thickness can be inferred from the relationship between the grain boundary density and the TGO thickness. The latter one, i.e. the relationship between the grain boundary density and the TGO thickness, can be deduced from the relationship between the oxide grain size and the TGO thickness. **Fig. 7** shows the relationship of the TGO thickness with the grain boundary density, and with the Al_2O_3 lateral grain size (next to BC) of the samples with different isothermal oxidation durations. The data of the grain sizes in this figure were fitted by the black curve shown in the figure using the following equation:

$$a(\delta) = 0.41e^{0.5\delta} \quad (6)$$

where $a(\delta)$ is the lateral size of Al_2O_3 grains at the TGO layer/bond coat interface when the TGO thickness is δ. Then, the oxidation rate constant, $f_{k'z}(z)$, can be obtained from Eq. (6), and a TGO growth kinetics theory can be acquired. However, because the oxide grains in **Fig. 5** are not round, the statistical data of the grain size obtained from **Fig. 5** inevitably have some errors, which will make the final calculated results have a bigger deviation. Therefore, in order to decrease the deviation, the statistical data of the grain boundary density are directly applied to the theoretical calculation for the TGO growth kinetics, and the grain boundary density data were fitted, shown as the red curve in **Fig. 6**, using the following equation:

$$\rho(\delta) = 7.7e^{-0.5\delta} \quad (7)$$

where $\rho(\delta)$ is the grain boundary density when the TGO thickness is δ.

Therefore, according to the above discussion, $f_{k'z}(z)$ can be calculated as follows:

$$f_{k'z}(z) = \frac{1}{C}\rho(z) = \frac{7.7e^{-0.5z}}{C} \quad (8)$$

where C is the positive constant related to the material and temperature. The thickness of the Al_2O_3 scale on the surface of the pre-oxidized bond coat is about 0.38 μm. Then, by combining Eqs. (5) and (8), the TGO thickness growth kinetics, more in tune with the practical experimental data, was obtained as follows:

$$t = \int_{0.38}^{\delta}\left(\int_{0}^{z}\frac{1}{f_{k'z}(z)}dz\right)dz = 0.52C\left(e^{0.5\delta} - 0.5\delta - 1.02\right) \quad (9)$$

According to the above experimental results, the TGO thickness is about 2.8 μm when the isothermal oxidation duration is 600 h. Then, by plugging these data ($\delta = 2.8\ \mu m$ when $t = 600\ h$) into Eq. (9), the constant C can be obtained (C=705.6). Therefore, Eqs. (8) and (9) can be expressed by the following equations:

$$f_{k'z}(z) = \frac{7.7e^{-0.5z}}{705.6} = 0.011e^{-0.5z} \quad (10)$$

$$t = 366.9\left(e^{0.5\delta} - 0.5\delta - 1.02\right) \quad (11)$$

Eq. (11) is the theoretical calculated result of the TGO thickness growth kinetics. According to Eq. (11), it can be seen that the TGO growth behaviour does not fit well with the conventional Wagner oxidation theory. In order to further verify the above discussion and theoretical calculations, **Fig. 8** shows the comparison between the experimental results and the theoretical calculated results of the thickness of the Al_2O_3 scale. It was found that the curve of the calculated Al_2O_3 TGO thickness values could fit well with the experimental data. Therefore, the above calculated theoretical results and analysis are reasonable, and the changing oxide grain structure is the key determinant that will affect the TGO growth behaviour.

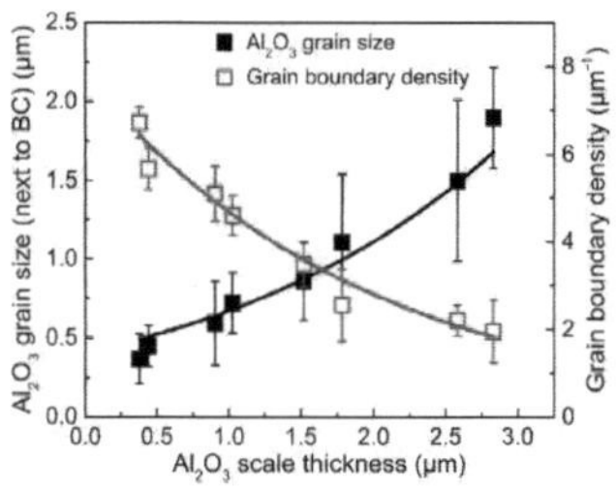

Fig. 7. The relationship of the TGO thickness with the grain boundary density, and with the Al_2O_3 lateral grain size (next to BC) of the samples with different isothermal oxidation durations.

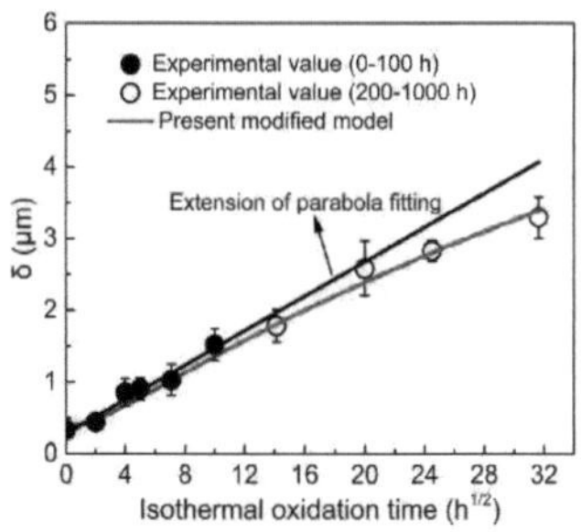

Fig. 8. Comparison between the experimental and theoretical values obtained using the model in this study.

4 Conclusions

In this study, The non-parabolic isothermal oxidation kinetics of low pressure plasma sprayed MCrAlY bond coat was investigated. To qualitatively explain the abnormal growth phenomenon of thermally grown oxides (TGO), the changes that occurred to their microstructure during the oxidation process were studied. Based on these observations, a modified model was developed to understand and quantitatively predict the non-parabolic oxidation and growth kinetics of TGO. This modified model, which fits well with experimental results, provides a novel method to quantitatively predict the long-term growth behaviour of TGO, and thereby benefits the development of long-life and highly reliable thermal barrier coatings.

5 Acknowledgements

The present project is supported by National Basic Research Program (Grant No. 2013CB035701), the National Science Foundation of China (Grant No.: 51671159), the Fundamental Research Funds for the Central Universities and National Program for Support of Top-notch Young Professionals.

6 Literature

[1] S. Ahmaniemi, M. Vippola, P. Vuoristo, T. Mantyla, F. Cernuschi and L. Lutterotti: Modified thick thermal barrier coatings: Microstructural characterization. J. European Ceram. Soc. 24 (2004),Issue 8, pp. 2247/58.
[2] S. Bose and J. DeMasiMarcin: Thermal barrier coating experience in gas turbine engines at Pratt & Whitney. J. Therm. Spray Technol. 6 (1997), Issue 1, pp. 99/104.
[3] D. Stover and C. Funke: Directions of the development of thermal barrier coatings in energy applications. J. Mater. Process. Technol. 93 (1999), pp. 195/202.
[4] J.Q. Yao, Y.D. He, D.R. Wang, H. Peng, H.B. Guo and S.K. Gong: Thermal barrier coating bonded by (Al_2O_3-Y_2O_3)/(Y_2O_3-stabilized ZrO_2) laminated composite coating prepared by two-step cyclic spray pyrolysis. Corros. Sci. 80 (2014), pp. 37/45.
[5] A.G. Evans, D.R. Mumm, J.W. Hutchinson, G.H. Meier and F.S. Pettit: Mechanisms controlling the durability of thermal barrier coatings. Prog. Mater. Sci. 46 (2001), Issue 5, pp. 505/53.
[6] Q. Zhang, C.J. Li, Y. Li, S.L. Zhang, X.R. Wang, G.J. Yang and C.X. Li: Thermal failure of nanostructured thermal barrier coatings with cold-sprayed nanostructured NiCrAlY bond coat. J. Therm. Spray Technol. 17 (2008), Issue 5-6, pp. 838/45.
[7] N.P. Padture, M. Gell and E.H. Jordan: Thermal barrier coatings for gas-turbine engine applications. Science 296 (2002), Issue 5566, pp. 280/84.
[8] J. Nicholls: Advances in coating design for high-performance gas turbines. MRS Bull. 28 (2003), Issue 9, pp. 659/70.
[9] D.R. Clarke and C.G. Levi: Materials design for the next generation thermal barrier coatings. Ann. Rev. Mater. Res. 33 (2003), pp. 383/417.
[10] D.Y. Chen, M. Gell, E.H. Jordan, E. Cao and X.Q. Ma: Thermal stability of air plasma spray and solution precursor plasma spray thermal barrier coatings. J. Am. Ceram. Soc. 90 (2007), Issue 10, pp. 3160/66.
[11] R.A. Miller and C.E. Lowell: Failure mechanisms of thermal barrier coatings exposed to elevated-temperatures. Thin Solid Films 95 (1982), Issue 3, pp. 265/73.
[12] D. Pan, M.W. Chen, P.K. Wright and K.J. Hemker: Evolution of a diffusion aluminide bond coat for thermal barrier coatings during thermal cycling. Acta Mater. 51 (2003), Issue 8, pp. 2205/17.
[13] H.B. Guo, L.D. Sun, H.F. Li and S.K. Gong: High temperature oxidation behavior of hafnium modified NiAl bond coat in EB-PVD thermal barrier coating system. Thin Solid Films 516 (2008), Issue 16, pp. 5732/35.
[14] S.W. Zeng, A.M. Zhao and H.T. Jiang: Oxidation of conventional and nanostructured 8 wt.% yttria-stabilized zirconia coating surface coatings on gamma-TiAl. Appl. Surf. Sci. 332 (2015), pp. 362/67.

Microstructural and Mechanical Investigations of Dissimilar Metal joints of T91 and T22 Ferritic Steels

Rutash Mittal[1*], Buta Singh Sidhu[2]
[1] Department of Mechanical Engineering, Malout Institute of Management & Information Technology, Malout, Dist. Sri Muktsar Sahib-152107, Punjab, India.
[2] M. R. S. Punjab Technical University, Jalandhar-Kapurthala Highway, Kapurthala, Punjab, India.
*Corresponding author, Email: rutashmittal@gmail.com

Abstract: Ferritic steels are commissioned as one of the preferred structure materials for the boilers of nuclear and thermal power plants. The mechanical characterization and microstructural evolution of shielded metal arc welded dissimilar joint between ferritic SA213T91 (T91) and ferritic SA213T22 (T22) steels has been investigated by virtue of microstructures, micro-hardness, XRD and SEM/EDS analysis. Comparative performance of distinctive zones in the weldment i.e. heat affected zone, base metal and weld pool has been explored. The T91 HAZ and T22 HAZ zones of the weldment are observed to have high micro-hardness by virtue of martensitic morphology along with the presence of delta ferrite in T91 HAZ and carbide as well as intermetalics in T22 HAZ.

1 Introduction

Welding has become as the predominant joining method in industrial practices [1]. Ferritic steels are used as the frontrunner material in the boilers of power plants by virtue of higher elevated temperature corrosion resistance and moderate strength [2]. Dissimilar metal weldments have wide applications in power plants, chemical, petrochemical and nuclear industries. Dissimilar metal weldments between different grades of ferritic steels are employed in different locations in the boilers of power plants [3]. Dissimilar metal weldment not only satisfies the service conditions, but can help to save expensive materials to be required for plant operations [4]. The adoption of dissimilar metal joint has been quoted to give flexibility in the working of system with optimized use of each material [5]. The growing availability of new materials and higher requirements being placed on materials creates a greater need for joints of dissimilar metals [6]. Normally, dissimilar metal weldments are more liable to failures and these failures are needed to be addressed. The carbon migration in the dissimilar weldments can results in the formation of a carbon extracted soft zone characterized by low micro-hardness in the low alloy ferritic steel side and a carbon enriched hard zone on the high Cr ferritic steel side of the weld interface [3]. Several studies reported the investigation of microstructural changes induced near the weld interface due to joining of different materials [5]. A detailed critique of various dissimilar metal weldments of their mechanical characterization and micro structural studies with reference to application of joints have also been covered by many researchers [9].

Dissimilar metal weldment's mechanical behavior and more importantly the performance of the heat affected zone (HAZ) of the weldment is an important issue. The microstructural alterations produced during welding can be quoted to be responsible for the changes in the mechanical properties of the weldments [8]. Micro structurally altered and heterogeneous HAZ in the weldment of 2.25Cr-1Mo steel, having the presence of various carbides has been reported, prime reason for its lower corrosion resistance [10]. The proper welding electrode/ filler material and welding process selection is an imperative aspect in the field of dissimilar weldments [5, 11]. Diffusion of carbon from low chromium side high Cr steel by virtue of carbon activity gap can be attributed to form soft and hard zones in the dissimilar metal weldment [12]. Further the authors have also investigated the sequence and chemistry of secondary phases developed in hard zones of dissimilar weldments of 9Cr-1Mo and 2.25Cr-1Mo steels [3].

Segregated precipitates with varied morphology and distribution in different zones of the weldment have also been investigated [7, 13]. The observed precipitates in HAZ were either M_7C_3 or $M_{23}C_6$ carbides. During studies on simulated microstructures in heat affected zone of 9Cr-1Mo weldment, the authors reported development of varied nature of microstructures in HAZ [14]. Cao and coworkers have observed lower tensile strength of T92 CGHAZ, by virtue of coarse tempered martensite structure during studies on T92/S304 dissimilar weldment [15]. The authors have reported that the detailed microstructural evolution and mechanical characterization of T91/347H weldment [16].

Large volume of work is reported in the area of dissimilar joints of austenitic/ferritic steels in comparison to joining of dissimilar ferritic steels [11, 15 and 17]. The parts of boilers which are having lower temperature exposure are made of low Cr ferritic steels, whereas parts having moderate to high temperature exposure, the use of high Cr ferritic steels is recommended [5]. Due to which, a dissimilar joint is required by virtue of its service conditions between two ferritic steels. The acceptance of dissimilar weldment satisfies technical as well as economical issues, as this can propel service performance and material savings [17]. Abrupt microstructural changes, higher gap in coefficient of expansion, carbide formation and comparatively higher corrosion of HAZs in ferritic steel have been reported to cause failure of the weldments [18]. Limited information is reported about the microstructures and mechanical behavior of dissimilar weldments of T91/T22. Optical and SEM microscopes, X-ray diffraction (XRD) analysis and scanning electron microscopy (SEM) /Energy dispersive spectroscopy (EDS) techniques have been used to analyze the facts.

2 Materials and Methods

2.1 Experimental Details

The base pipes of T91 and T22 were welded together by keeping V joint geometry with root gap of 2mm and 75^0 angle for the joint. Afterwards, flat strips were arranged from the dissimilar metal weldments. The individual samples of T91BM and T22BM were also machined to compare them with the weldment. The compositional details of base metals and welding electrode are listed in **Table 1**.

Table 1 Chemical composition of the base metals and welding electrode.

Base Metals/ welding electrode	C	Cr	Mn	Mo	Si	Fe
SA213T91	0.0964	8.76	0.478	1.03	0.37	Bal.
SA213T22	0.1277	2.012	0.69	0.83	0.37	Bal.
Chromotherme 2	0.06	2.4	0.8	1.10	0.44	Bal.

2.2 Weldment Characterization

The optical microstructures from all the required locations of dissimilar metal weldments were observed with an optical microscope (Leica DM4000M model) available at IIT Ropar, India. The standard metallographic procedure was used to capture the microstructures from different zones of the weldment by using an appropriate etching agent. The microstructure of T22 and its HAZ is examined by 5% Nital reagent {HNO_3 (5ml) + ethyl alcohol (100ml)} and of T91, its HAZ and weld metals were treated by fry's reagent comprising {$CuCl_2$ (5g) + HCl (40ml) +ethyl alcohol (25ml) +distilled water (30ml)}. The micro-hardness of different zones of the weldment was observed with Vickers's micro-hardness tester (Wilson instrument, model 402MVD make). The tensile testing has been done with H25K-S Hounsfield make tensile testing machine having a digital extensometer. For each observation, three samples were tested to have an average value as per ASTM E8M-04 standard [19] as depicted in "**Fig. 1**". Scanning electron microscope (SEM) of JEOL (JSM-6610LV) fitted with EDS of Oxford instruments has been employed to observe the microstructures and fractography modes. XRD facility of PAN Alytical, Model X'Pert PROMPD was operated for particular scans. Particular XRD scan of the specific zone i.e. HAZ, WM etc. was taken to find the specific knowledge of the phases [20].

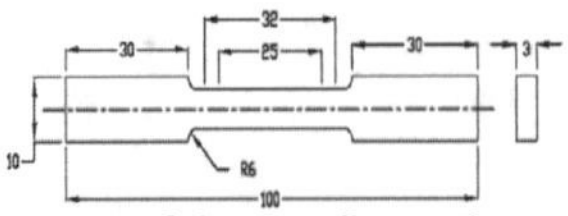

Fig. 1. Dimensions of the tensile testing specimen.

3 Results and Discussion

3.1 Visual observations and microscopic studies

The optical microstructures of different zones of T91/T22 weldment are presented in "**Fig. 2 (a-r)**". The microstructure of T91 BM consists of tempered lath martensites in which prior austenite grain boundaries as well as lath boundaries are decorated with precipitates **"(Fig. 2 a)"**. The micro mechanism responsible for the formation of martensitic structure on T91 HAZ depends on the peak temperature attained and cooling rate of that specified region. Depending on the temperature of HAZ to be above AC3 or above AC1 (i.e. between AC1 and AC3), austenitic or a mixture of ferritic and austenitic structure gets formed, which on cooling gets transformed into the martensitic structure. The T91 HAZ has a martensitic structure having presence of some delta ferrite depicted in "**Fig. 2c**". The presence of delta ferrite in the martensitic structure indicates that the temperature has reached above AC4 during welding [7]. The formation of delta-ferrite in the T91 HAZ has occurred, due to high temperature by welding. Delta ferrite is a chromium rich phase of Fe-Cr alloys and may have formed due to the depletion of Cr from the alloy matrix [13]. Depending on the peak temperature and its cooling characteristics of the thermal cycle, the phase of delta ferrite of T91 steel can be as high as 35% [13]. The presence of delta ferrite in addition to martensite structure of T91 HAZ can facilitate the depletion of Cr through precipitation and precipitation of Cr based phases affect the oxidation resistance of the T91 HAZ. It can be inferred from the microstructures of T91 HAZ that delta ferrite phase is similar in shape and size to martensite phase. The transition from T91 BM to T91 HAZ is shown at different magnification to have a clear view in "**Fig. 2 (d, e and f)**". The formation of microstructures of the HAZ is complex and depends upon the interaction of welding thermal cycles, the phase transformation and grain growth characteristics of the materials being welded. Further, it depends upon the highest temperature achieved during welding and its cooling rate with respect to its distance from the fusion boundary [5]. It results in formulating a variety of non-equilibrium microstructures, which got formed in the HAZ of T91 [13]. The effect of thermal cycles during welding can be unambiguously seen in the differentiated variation, in the microstructure of T91 BM, T91 HAZ at the interface of BM-HAZ at different magnifications in **"Fig. 2 (d-f)"**. SEM micrographs at different sections of the T91/T22 weldment are presented in "**Fig. 3 (a-d)**". A minor increase in the Cr content in the T91 HAZ can be observed w.r.t. **"Fig. 3a"** and **Table 2.**

The weld metal microstructures at different magnifications consist of a widmanstatten type structure having the presence of pearlitic structure **"Fig. 2 (j, k and l)"**. The whitish area of weld metal depicts the ferrite phase and black region is pearlite phase. The optical microstructures of T22 HAZ just adjacent to weld metal of various magnifications are presented in **"Fig. 2(m, n and o)"**. The microstructure of T22 HAZ has the bainitic matrix structure having some presence of ferrite phase **"Fig. 2o"**. Moving towards T22 BM, the T22 HAZ has inter-critical structure consisting of uniformly distributed carbides in ferritic phase, identical to the results reported [10]. The interface of T22 HAZ and T22 BM can be seen clearly in **"Fig. 2p"**. The optical microstructure of T22 BM consists of a polygonal structure having ferrite and pearlite phases **"(Fig. 2(q,r))"**. The white region of the structure is ferrite phase and a black region seems to be pearlite phase in the bainitic matrix. The SEM micrographs depicting the interfacing between weld metal /T22 HAZ and T22 HAZ/T22 BM are presented in **"Fig. 3 (c, d)"**.

The movement of Cr along with carbon from T91 BM combines with the carbon present in T91 HAZ to form the carbides of (Cr, Fe)$_7$C$_3$ and Cr$_7$C$_3$ as can be inferred from the XRD pattern of this section shown in **"Fig. 4a"**. The presence of similar type of carbides during studies of HAZs has also been reported by other researchers [3, 9, 15 and 21]. The interface of T91 HAZ and weld metal clearly show the weld interface having segregated variations in the structure in **"Fig. 2(g, h and i)"**.

The T22 HAZ has imbedded distribution of carbides of (Cr, Fe)$_7$C$_3$ and Cr$_7$C$_3$ **(Fig. 4b)**, which may be due to transfer of carbon to T22 HAZ from T22 BM **"Fig. 2(p and q)"**. The movement of elements in the interface is perhaps due to the difference in carbon activity between base metal and weld metal, which goes about as a main impetus for the dissemination of carbon [12]. Along with the carbides, XRD analysis has also identified the presence of inter-metallic of Ni-Cr-Fe, FeNi and Fe$_2$Mo **"(Fig. 4b)"**. Pilling and Ridley have suggested the precipitation of $M_{23}C_6$ and M_7C_3 (simultaneous precipitation of Cr based carbides) in the HAZ. Hence precipitation of these Cr based carbides lower the availability of free Cr and decrease the oxidation resistance of this region in as welded condition [22].

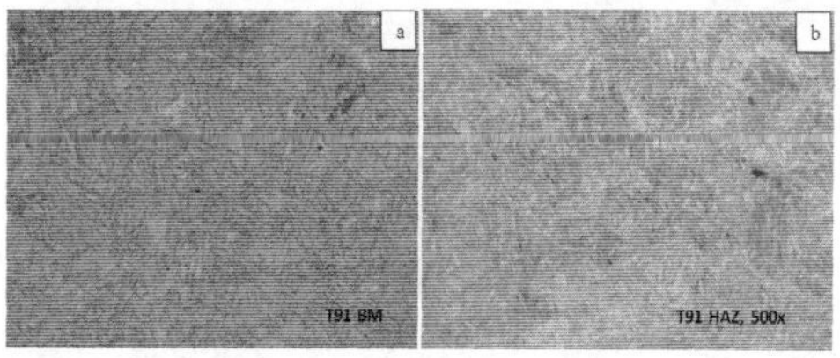

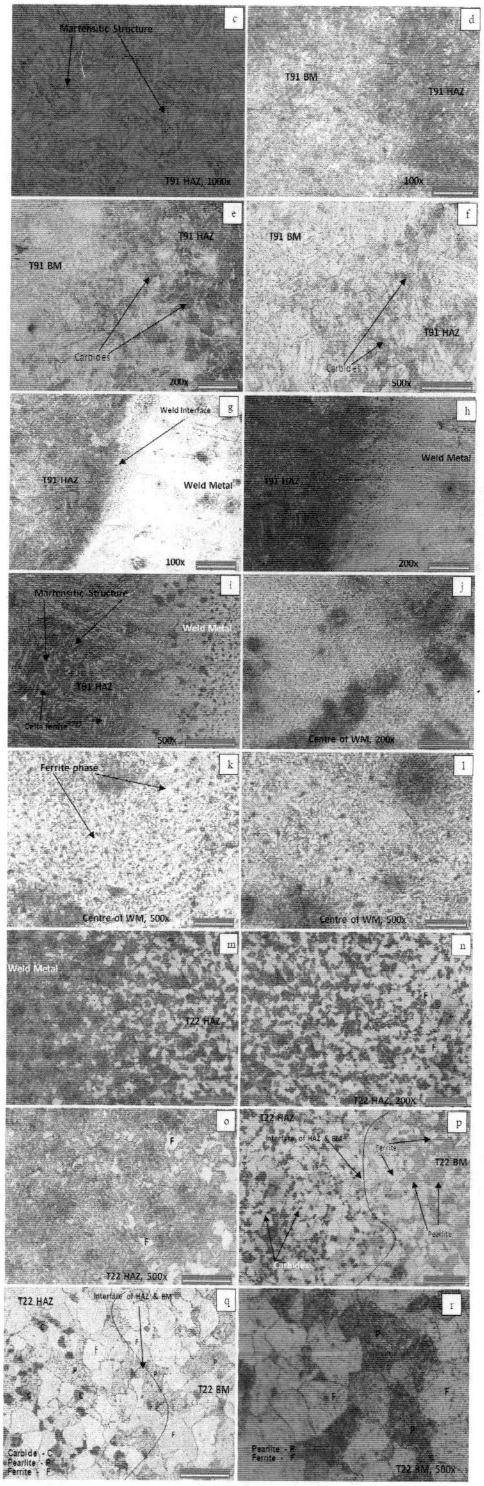

Fig. 2. Optical microstructures from different zones of T91/T22 weldment.

Table 2 Elemental variation at T91 HAZ/WM interface for T91/T22 weldment.

	T91 HAZ				WM		
Elmts	1	2	3	4	5	6	7
Cr	7.56	13.7	11.87	9.56	3.66	4.00	3.25
Fe	89.32	81.9	84.5	87.8	96.34	91.56	96.73
Mo	3.12	4.33	4.37	2.62	-	4.14	-

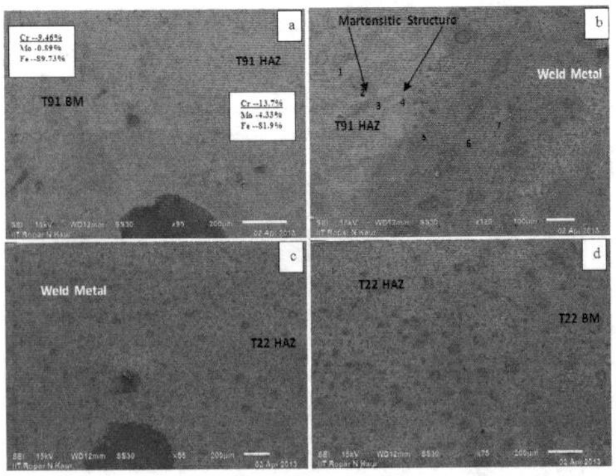

Fig. 3. SEM micrographs from different zones of T91/T22 weldment.

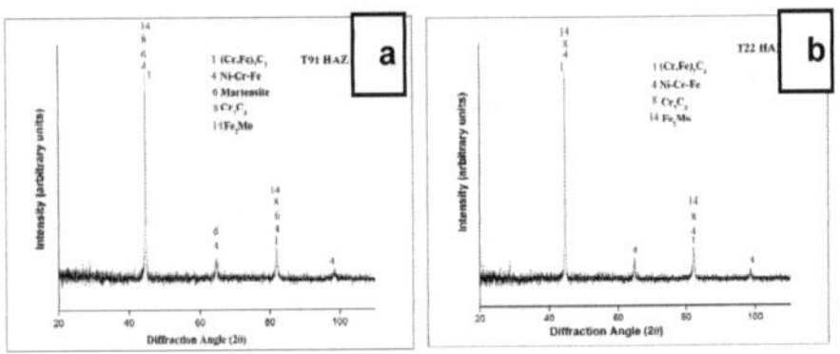

Fig. 4. X ray diffraction patterns of (a) T91 HAZ and (b) T22 HAZ of the weldment.

3.2 Micro-hardness Measurement

The phenomenon of micro-hardness measurement along the weldment is the direct manifestation of its mechanical behavior and microstructural properties. The mechanical properties will vary in accordance with the variation of micro-hardness along the weldment [23]. The micro-hardness profile of the weldment is shown in **"Fig. 5"**.

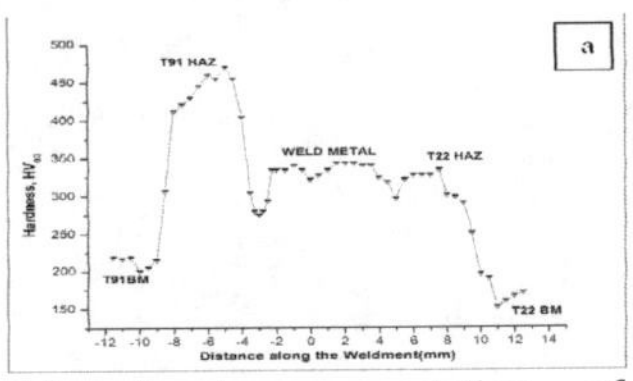

Fig. 5. Micro-hardness variation of T91/T22 weldment.

The profile of micro-hardness measurement of T91/T22 weldment depicts an increase in micro-hardness in T91 HAZ zone. The average micro-hardness of T91 HAZ is around 438Hv. The increase in micro-hardness value is observed while moving from weld metal to T91 HAZ. The observed high micro-hardness of T91 HAZ is by virtue of the presence of carbides and martensitic structure as demonstrated in **"Fig. 4a"**. A drop in micro-hardness is observed in T91 BM zone just adjacent to the T91 HAZ, which may be because of transfer of carbon from base metal to HAZ due to carbon affinity towards the weld metal [24]. While moving from T91 BM to weld metal, carbon may have not passed to weld metal, but may have combined with Cr and other elements in the T91 HAZ to form carbide precipitation phases [3, 11]. The average micro-hardness of the weld metal region is around 321Hv and distribution of micro-hardness seems to be uniform throughout the weld metal region. The observed value of micro-hardness of weld metal in uniform profile can be by virtue of widmanstaten type structure having presence of pearlitic structure. The weld metal region has crystallized form of filler material having micro alloyed presence of Cr, Mn, Ni and Mo. An increase in micro-hardness is observed in the hard zone of T22 HAZ while moving from weld metal to T22 BM having an average value to be around 294Hv. Just similar to the drop in micro-hardness value on the T91 BM of the weldment a drop in micro-hardness has also observed in the T22 BM side just adjacent to T22 HAZ. This micro-hardness drop may be due to transfer of carbon from T22 BM to T22 HAZ where it seems to form carbides in the inter-critical structure of polygonal ferrite and pearlite **"(Fig. 2n)"**, similar to the behavior as also reported by different researchers [7, 10, 12]. The formation of martensitic structure on the T91 HAZ and the presence of carbides on T91 HAZ and T22 HAZ regions also corroborate the observations of other researchers [3, 7 and 13]. The micro-hardness distribution of the weldment is asymmetrical with respect to weld center.

3.3 Tensile Testing and Fracture Analysis

The transformation and fracture processes are more complicated in welded specimens, especially in dissimilar weldments which contain different zones with diverse properties and behavior. The behavior of one zone in a joint can have an impact on its adjacent zones [25]. The tensile testing of all the base metals and welding joint have been demonstrated in **Table 3**. For dissimilar metal joints, it is common practice that mechanical properties of the dissimilar weldment should not be lower than those of the inferior base metal [4]. The T91/T22 weldment has the higher value of tensile strength (523MPa) than one of the base metals.

The fractography of the tensile tested sample is carried out using SEM. SEM/EDS fractography of both the base metals and welded joint has been shown in **"Fig. 6(a-c)"**. Dimples of varying size and shape are observed in all the fractured surfaces, which indicate the major fracturing mechanism to be ductile. Since it is known that the formation of dimples initiates at secondary phase particles, the morphology of dimples is different in each studied fracture, according to its own micro structural characteristics [26]. The fracture analysis of T91 BM presents a layered cleavage ductile fracture having a mixed mode of fracture **"(Fig. 6a)"** and fractography of T22 BM specimen depicts a ductile fracture **"(Fig. 6b)"**. The fracture has small and large shallow dimples of varying size, feature of high ductility. The fracture of T91/T22, SMAW weldment from T22 BM reveals a ductile fracture. Small and large combined dimples along with some parabolic shaped dimples have been observed in the fracture analysis of the weldment **"(Fig. 6c)"**.

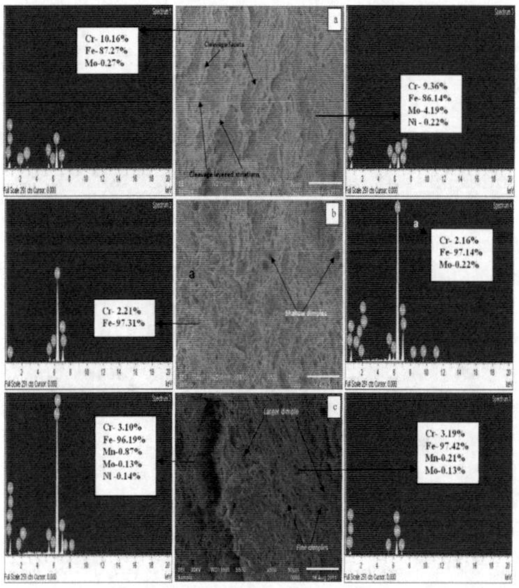

Fig. 6. Fracture Analysis of the tensile tested specimens of (a) T91 BM (b) T22 BM and (c) T91/T22 weldment.

Table 3 Mechanical properties of the base metals and weldment.

Sample	Ultimate Tensile Strength (MPa)	% Elong.	Fracture location	Fractography Details
T91/T22 Joint	523	22.50	BM of T22	Parabolic shaped dimple formation, Mixture of small and larger dimples depicting ductile.
T91 BM	624	25.71	-	Layered tearing ridge, quasi cleavage fracture
T22 BM	489	33.06	-	Ductile Fracture, small and larger shallow dimples indicative of ductility.

4 Conclusions

The major results and conclusions of the study are:-

- As the weldment has not fractured from central weld metal, it justifies the acceptability of the weldment.
- The micro-hardness of T91 HAZ is comparatively higher than other regions, by virtue of martensitic structure of that region. The presence of martensite phase along with carbides in T91 HAZ side can be another justified reason for higher micro-hardness.
- A carbon denuded soft zone is observed at the interfaces of T91 BM/ T91 HAZ and of T22 BM/ T22 HAZ in T91/T22 weldment.
- The fracture of weldments from T22 BM predicts a ductile nature.

5 Literature

[1] Bettahar, K., Bouabdallah, M., Badji, R., Gaceb, M., Kahloun, C. and B. Bacroix: Microstruture and mechanical behavior in dissimilar 13Cr/2205 stainless steel welded pipes. Materials & Design 85 (2015), pp. 221-229.

[2] Anand, R., Sudha C, Thomas Paul, V., Saroja, S. and M. Vijayalakshmi: Microstructural changes in grade 22 ferritic steel clad successively with Ni-based and 9Cr filler metals. Welding Journal 89 (2010), Issue 4, pp. 65s-73s.

[3] Sudha, C., Thomas Paul, V., Terrance, A. L. E., Saroja, S. and M. Vijayalakshmi: Microstructure and microchemistry of hard zone in dissimilar weldments of Cr-Mo steels. Welding Research Supplement 85 (2006), Issue 7, pp. 71s-80s.

[4] Han, H., Y. and Z. Sun: Welding of tube to tube joints between martensitic and austenitic stainless steels for reactor applications. International Journal of Pressure Vessels & Pipings 60 (1994), pp. 59-64.

[5] Mittal, R. and B. S. Sidhu: Microstructures and mechanical properties of dissimilar T91/347H steel weldments. Journal of Material Processing Technology 220 (2015), pp. 76-86.

[6] Satyanarayana, V. V., Madhusudan Reddy, G. and T. Mohandas: Dissimilar metal friction welding of austenitic-ferritic stainless steels. Journal of Material Processing Technology 160 (2005), pp. 128 137.

[7] Bhaduri, A. K., Venkadesan, S., Rodrigues, P. and P. G. Mukanda: Transition metal joints for steam generators- an overview. International Journal of Pressure Vessels & Pipings 58 (1994), pp. 251-265.

[8] Mortezaie, A. and M. Shamanian: An assessment of microstructure, mechanical properties and corrosion resistance of dissimilar welds between Inconel 718 and 310S austenitic stainless steel.

International Journal of Pressure Vessels & Pipings, 116 (2014), pp. 37-46.
[9] Mittal, R. and B. S. Sidhu: Studies on Microstructural Properties of GTAW Weldment of Ferritic steels. Materials Forum 38 (2016), pp. 158-166.
[10] Raman, R.K.S. and J. Gnanamoorthy: The oxidation behavior of the weld metal, heat affected zone and base metal in the weldments of 2.25Cr-1Mo Steel. Corrosion Science 34 (1993), pp. 1275-1288.
[11] Shah Hosseini, H., Shamanian, M. and A. Kermanpur: Characterization of microstructures and mechanical properties of Inconel 617/310 stainless steel dissimilar welds. Materials Characterization, 62 (2011), pp. 383-386.
[12] Sudha, C., Terrance, A. L. E., Albert, S. K. and M. Vijayalakshmi: Systematic study of formation of soft and hard zones in the dissimilar weldments of Cr-Mo steels. Journal of Nuclear Materials, 302 (2000), pp. 193-205.
[13] Raman, R. K. S. and A. K. Tyagi: Secondary ion mass spectrometry and surface profilometry characterization of oxide scales developed over various regions of weldments of 2.25Cr-1Mo steel. Material Science & Enginering A, 85 (1994), pp. 97-102.
[14] Laha, K., Chandravathi, K. S., Rao, K. B. S. and S. L. Mannan: Hot tensile properties of simulated heat affected zone microstructures of 9Cr-1Mo weldment. International Journal of Pressure Vessels & Pipings, 62 (1995), pp. 303-311.
[15] Cao. J., Gongn, Y., Yang, Z. G., Luo, X. M. and F. M. Gou: Microstructure and mechanical properties of dissimilar materials joints between T91 martensitic and S304H austenitic steel. Materials & Design 32 (2011), pp. 2763-2770.
[16] Mittal, R. and B. S. Sidhu: Microstructural and Mechanical Properties of Dissimilar Metal Weldment of Ferritic and Austenitic Steels. In Proc. of 3rd International Congress & 21st Technical Sessions on Welding and Joining Technologies held at University of Oviedo,Gijon, Spain on 17-19 May, 2016.
[17] Mittal, R. and B. S. Sidhu: Oil- ash corrosion resistance of dissimilar T22/T91 welded joint of super heater tubes. Journal of Material Engineering & Performance 24 (2015), pp. 670-682.
[18] Z. Sun: Feasibility of producing ferritic/austenitic dissimilar metal joints by high energy density laser beam process. International Journal of Pressure Vessels & Pipings, 94 (1996), pp. 153-160.
[19] ASTM International, ASTM E8-04. 2011. Standard test methods for tension testing of metallic materials. West Conshohocken, PA, United States.
[20] Arivazhagan, N., Narayanan, S., Singh, S., Prakash, S. and G. M. Reddy: High temperature corrosion studies on friction-welded low alloy steel and stainless steel in air and molten salt environment at 650^{0}C. Materials & Design 34 (2012), pp. 459-468.
[21] Mittal, R. and B. S. Sidhu: Microstructural and mechanical characterization of the different zones of the T22/T91 weldment. International Journal of Surface Engineering & Material Technology 4 (2014), pp. 45-49.
[22] Pilling, J. and N. Ridley: Tempering of 2.25Pct Cr- 1Pct Mo low carbon steels. Metallurgical Transactions A 13A (1992), pp. 587-563.
[23] Sathiya, P., Aravindan, S. and A. Noorul Haq: Some experimental investigations on friction welded stainless steel joints. Materials & Design 29 (2007), pp. 1099–1109.
[24] Sireesha, M., Albert Shaju, K., Shankar, V. and S. Sundarsan: Comparative evaluation of welding consumables for dissimilar welds between 316LN austenitic stainless steel and alloy 800. Journal of Nuclear Materials 279 (2000), pp. 65-76.
[25] Hajiannia, I., Shamanian, M. and M. Kasiri: Microstructure and mechanical properties of AISI 347 stainless steel/A335 low alloy steel dissimilar joint by gas tungsten arc welding. Materials & Design 50 (2013), pp. 566-573.
[26] Blach, J., Falat, L. and P. Serv: The Influence of hydrogen charging on the notch tensile properties and fracture behaviour of dissimilar weld joints of advanced Cr-Mo-V creep resistant steels. Engineering Failure Analysis 18(2011), pp. 485-491.

Thermal gradient cyclic lifetime of DVC and lamellar thermal barrier coatings with equivalent thermal insulation performance

Bo-Cheng, Guan-Jun Yang*, Cheng-Xin Li, Chang-Jiu Li.State Key Laboratory for Mechanical Behavior of Materials, School of Materials Science and Engineering, Xi'an Jiaotong University, Xi'an, Shaanxi Province 710049, PR China

Thermal cycle lifetime is essentially important to the application of thermal barrier coatings (TBCs) on the premise of the same thermal resistance. In this study, equivalent thermal insulation conception is introduced to the design of dense vertical crack (DVC) structured TBCs and the lamellar structured TBCs, to fairly compare the lifetime of TBCs with different structure. DVC-structured TBCs with the equivalent thermal insulation to lamellar YSZ TBCs were prepared, and their lifetimes were evaluated by thermal gradient cyclic test. Cross-sectional morphology and phase constitution before and after failure were examined by scanning electron microscope and x-ray diffraction, respectively. The failure mode was analysed. This study would shed light to further coating structure optimization.

1 Introduction

Thermal barrier coatings_(TBCs) are frequently used to provide thermal and corrosion protections for the metallic hot-section components of both aircraft engines and land-based gas turbines to achieve higher turbine inlet temperature and extended durability.[1-3] A typical TBC system consists of a superalloy substrate, a MCrAlY(M=Ni, Co, NiCo, CoNi) or NiAl-based bond coat (BC), a thermally grown oxide (TGO) inevitably grown on bond coat surface, and an air plasma-sprayed (APS) or electron-beam physical-vapor deposited (EB-PVD) ceramic top coating (TC).[4-6] APS top coatings generally presented a lamellar microstructure wherein inter-splat pores are connected with intra-splat cracks. Consequently, this unique porous structure contributes to approximately half more drop as respect to its corresponding bulk material in the through-thickness direction.[7] The state-of-the-art top coat TBC material is 7-8 wt.% yttria-stabilized zirconia (8YSZ), primarily due to its low thermal conductivity, high toughness and high coefficient of thermal expansion (CTE).[2]

However, during long time or high temperature, YSZ will experience phase transfer and inevitable sintering, which may eventually lead to the failure of the TBCs. As a multilayer system with bond coat, TGO and top-coat stacking sequentially on substrate, the delamination resistance of TBCs is highly relevant to the performance of all constituent parts. It is believed that the inducements of TBC failure are: (i) the thickening of TGO;[8-10] (ii) the phase transformation induced volume expansion of top-coat;[11-12] (iii) the sintering of top-coat.[13-16] All these three types of failure mechanisms can be attributed to the large strain and stress resulting from the meta-stable phases and microstructures mentioned above.

Vertical micro cracks are deliberately introduced in the ceramic top coat of the coating test in the present case to relieve thermal mismatch stresses, thus reducing tendency of the ceramic to crack and may extend the lifetime of TBCs, which named DVC(dense vertically-cracked) TBCs. It is reported that DVC structure extended the TBC lifetime[18-20]. However, the introduced vertical micro cracks weakened the thermal insulation effect of top coatings and on the other hand, the inevitable dense structure also increase the thermal conductivity of the top coat. Thermal insulation effect, as one of the fundamental functions, should be paid preferential attention. It is a measure of the thermal protection offered by the TBC to the substrate.

In this study, equivalent thermal insulation effect TBCs were designed with the aim to elucidate the lifetime of DVC and lamellar structured TBCs.

2 Experimental

2.1 Preparation and Characterization of TBCs

Commercially available YSZ powder (Sulzer Metco 204NS-G, -125+11 μm, Westbury, NY) was used to deposit the ceramic top coat by a APS system (GP-80, JiuJiang, China). The morphology of the YSZ powders showed in Fig.1 demonstrate a spherical morphology corresponding to the depositing powder (204NS-G). The thickness of top-coat was approximately 570 μm and 1000 μm for the traditional lamellar and DVC coatings, respectively.

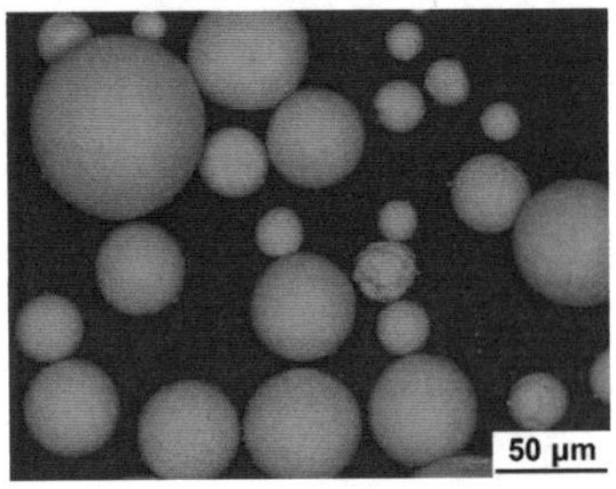

Fig. 1. Morphology of the hollow and spherical 8YSZ powder used for top-coat.

Prior to the deposition of the top-coat, a MCrAlY bond coat was deposited on a nickel-based superalloy (Inconel 738, Φ25.4 mm×3 mm). A commercially available NiCoCrAlTaY powder (Amdry 997, -37+9 μm, Sulzer Metco, Westbury, NY) was used as the bond coat material in this study. The bond coat was deposited to approximately 100 μm by a low pressure plasma spraying (LPPS) system. Table 1 shows the spraying parameters of LPPS. Before the deposition of top-coat, the as-prepared samples with bond-coat were subjected to preheat treatment by two different stages, namely, a pre-diffusion procedure (1080℃+4h, $O_2 \leqslant$ 0.01ppm) following by a pre-oxidation procedure (1080℃+4h, $O_2 \approx$ 10ppm).

Table 1. Parameters of low-pressure plasma spray

Parameters	Value
Arc current, A	575
Arc voltage, V	150
Plasma gas (Ar) flow, L/min	40
Plasma gas (H_2) flow, L/min	6
Powder feeding gas (Ar) flow, L/min	1
Chamber pressure, kPa	15
Spray distance, mm	280
Torch traverse speed, mm/s	800

The cross-sectional microstructure of the coating was investigated using a scanning electron microscope (SEM) system (TESCAN MIRA 3 LMH, Czech Republic). Before obtaining the section of samples for a metallographic examination, the samples were infiltrated with epoxy adhesive in vacuum to protect the coatings from any artifact destruction during the sample preparation. The phase composition of YSZ coatings before and after thermal cyclic test was characterized by a x-ray diffraction (XRD, PANalytical X' pert PRO, Netherlands) system.

2.2 Thermal Cyclic Test for TBCs

TBCs have been widely used to provide thermal insulation to metallic components from the hot gas in gas turbines. The top-coats are exposed to extremely high temperature to provide a major reduction on the surface temperature (approximately 100 to 300 °C dependent on various service conditions) of the underlying superalloy[1]. A gas burner test setup was designed for the gradient thermal cyclic test (Xi'an Jiaotong University, China) to simulate the actual service condition of TBCs. The combinations of pre-set temperature 1150/1000°C were used to carry out the gradient thermal cyclic tests, with the aim of elucidating the lifetime of traditional lamellar and DVC TBCs under equivalent thermal insulation effect. The temperature combination described above refers the surface temperature of top-coat and its corresponding BC temperatures.

In this study, one thermal cycle was defined as follow: 50 s heating, 70 s holding and 120 s cooling. In detail, the TBC surface was heated by a propane and oxygen flame to the pre-set temperatures in 50s, following by holding for 70s. Whilst, the backside of the TBC was cooled by compressed air to keep a thermal gradient. Then, the specimen was rapidly cooled down to room temperature in 120 s by compressed air from both the TBC surface and backside.

The temperatures of the specimen at surface and backside were monitored by a non-contact infrared thermometers with wavelengths of 8-14 μm and 1.6 μm, respectively. The emissivity for YSZ and superalloy substrate was calibrated to be 1 and 0.91 according to their wavelength, respectively. The lifetime of TBCs was defined as the thermal cyclic lifecycles when 10% area delamination of the top coat was observed.

3 Results and Discussion

3.1 Design of Coating Architecture

As mentioned above, thermal insulation effect is one of the fundamental functions for thermal barrier coatings. The superalloy inside surface contact with air for cooling, whereas the outside surface is thermal barrier coatings and contact with hot gas, setting up a temperature gradient across the TBC.

In our previous study[21], thermal resistance is introduced to evaluate the thermal insulation performance of TBCs,

$$R = \frac{\delta}{\lambda A} \quad (1)$$

Where δ is the thickness of top coat, λ is the thermal conductivity and A is the area normal to the heat flow.

For a specific TBCs, “thermal barrier performance” is a parameter independent on the area A, hence we introduce the unit thermal resistance (UTR) as an intrinsic parameter to describe the thermal insulation performance of TBCs, it can be expressed as

$$\mathrm{UTR} = \frac{\delta}{\lambda} \quad (2)$$

It is reported that the thermal conductivity of lamellar and DVC structured TBCs is about 1 W/mk[22] and 1.75 W/mk[23], respectively. Although the thermal conductivity may vary with the preparation process, it varies in a specific range. On the premise of the same thermal resistance, the thickness ratio of the lamellar and DVC structured TBCs is the ratio of the thermal conductivity value. Therefore, the top coat thickness of designed lamellar and DVC TBC is about 570 μm and 1000μm, as is shown in Fig. 2.

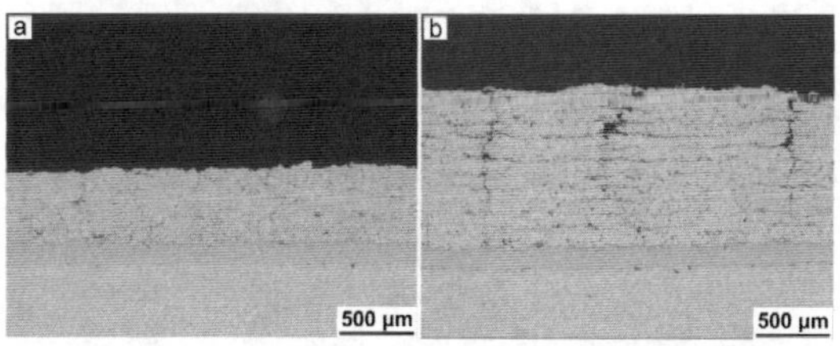

Fig. 2. The cross sectional images of the as sprayed TBCs, a. lamellar structure, b. DVC structure

It is seen from Fig. 2 that, for the as-sprayed TBCs, the top coats present a lamellar structure with no large cracks, and the surface shows the typical thermal sprayed features with a rough surface resulted from random stacking of splats. For the DVC structured TBCs, the vertical cracks extend from the surface to the bond coat, it is exactly this kind of cracks relieve thermal mismatch stresses, thus reducing tendency of the ceramic to crack and may extend the lifetime of TBCs.

3.1 Lifetime of TBCs

Figure 3 shows the lifetime of different structured TBCs under gradient thermal cyclic test. 2-3 samples were tested for each group, and the lifetime is the mean value, the error bars represent standard deviations on sets of 2-3 samples, the error bars represent standard deviations.

The average lifetime of DVC structured TBC was of 2040±103 cycles. However, the lifetime of lamellar TBC is beyond 4000 cycles, it is two times of DVC ones. Hence, it is possible to conclude that, under same thermal insulation effect, the lifetime of DVC TBCs is not longer than that of lamellar TBCs.

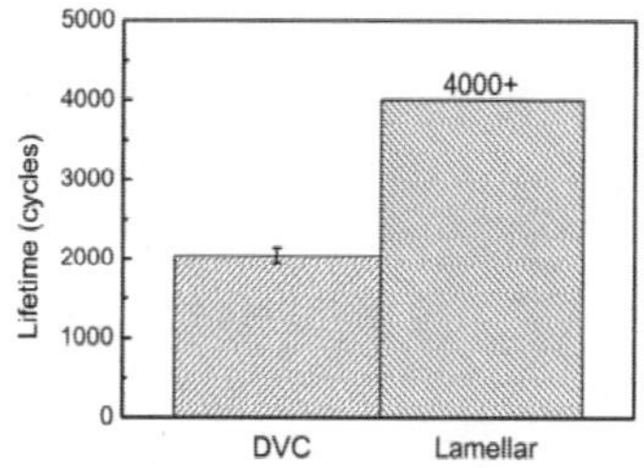

Fig. 2. The lifetime of different structured TBCs under gradient thermal cyclic test

3.2 Cross-Sectional Microstructure of TBCs After Failure

In order to clarify the failure mode of the 2 groups of TBCs, the cross sectional SEM images after failure were examined, as is shown in Fig. 3.

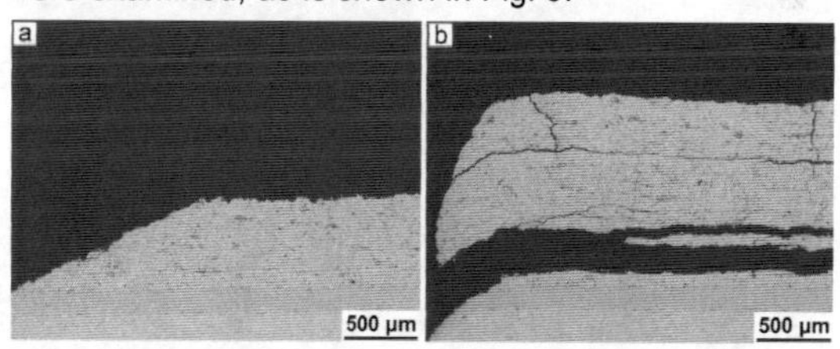

Fig. 3. The cross sectional images of the TBCs after failure, a. lamellar structure, b. DVC structure

It can be found from Fig. 3 that, for the lamellar TBCs (shows in Fig. 3a), part of top coat peeled off at the rim of the samples. However, for the DVC structured TBCs, lots of lateral and vertical cracks were found in the top coat.

3.3 TGO Thickness

Usually, TGO growth is considered as one of the dominant factors leading to the failure of TBCs. Thus, the TGO growth behaviour in the present TBC samples was examined. It was found from Fig. 3 that no obvious cracks were observed around the TGO, suggesting that the TGO may not be responsible for the failure of TBCs in this study. As a further step, TGO thickness was determined by the cross-sectional SEM micrographs before and after thermal cyclic test, and each value was the mean thickness of 10 different regions of TGO, the error bars represent standard deviations. As shown in Fig. 4, after thermal cyclic test until failure, the thicknesses of TGO were 0.49±0.07 μm and 0.64±0.07μm for the lamellar structured TBCs and DVC TBCs. They were significantly smaller than the reported critical value to induce a complete delamination of top-coat, i.e., 5-6 μm[8-9]. It is clear that the slightly thickening of TGO is not the dominant factor responsible for the failure of TBCs in this study. It is noticeable that, although the lifetime of DVC structured TBCs experience less thermal exposure, the TGO thickness is a little thicker than that of lamellar TBCs, the reason may be the vertical crack provide a passage for the transport of oxygen or the other possible reasons, it may be studied further.

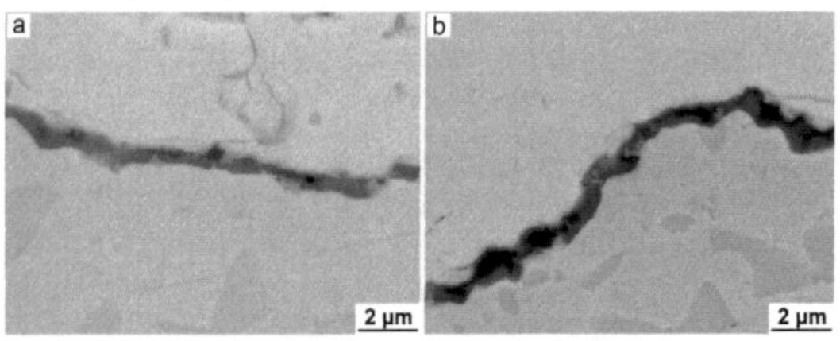

Fig. 4. The morphology of TGO after failure, a. lamellar structure, b. DVC structure

3.3 Phase Compositions of The Top Coat

Given the fact that phase transformation might be a reason to induce the spallation of top coat, the phase compositions of the top coat after TBC failure were examined, as shown in Fig. 5. It can be found that the phase compositions of the YSZ coatings before and after thermal cyclic test were mainly non-transformable t'-phase.

It seemed that there was no increase of the diffraction peaks of monoclinic phase after gradient thermal cyclic test. It was possible to observe that there was no distinct phase transformation after thermal cyclic test at different test conditions.

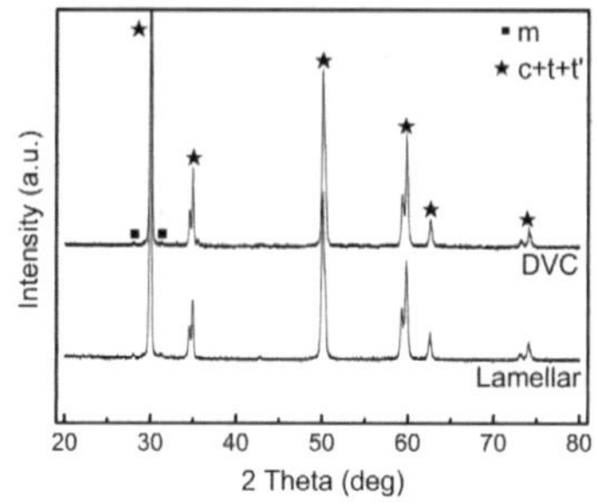

Fig. 5. XRD patterns of the top-coat after thermal cyclic tests to failure

Therefore, it evidently proved that no phase transformation of the top coat occurred. Consequently, the phase transformation may not be the dominant factor responsible for the failure of TBCs in this study.

3.4 Failure Analysis

High temperature exposure with extended periods often leads to inevitably sintering of the ceramic top coat. However, different regions of the top coat along through-thickness direction are exposed to different temperatures under gradient thermal cyclic test. The inner temperatures of YSZ coating decrease gradually from surface region to the interface between top-coat and bond coat. In our previous study, it had been confirm that, the elastic modulus of outside region (close to the flame) increased more rapidly than that of inside regions. The crack propagation occurs when the strain energy release rate is larger than the critical interfacial crack energy or fracture toughness.15 Therefore, the competition between the cracking driving force and the cracking resistance is discussed from two aspects in following parts, i.e., strain energy release rate and localized fracture toughness of the top-coat.

The strain energy release rate, G_h, can be obtained as follow:

$$G_h = \int_0^h \frac{E_x \varepsilon^2}{2(1-\nu)} dx \quad (2)$$

where h is the distance from the top coating surface to the crack tip, E_x is the local elastic modulus at the crack tip, ε is the strain, v is the Poisson's ratio.

It can be seen from equation 4 that G_h increases with the increase of h. Furthermore, the sintering induced gradient distribution of mechanical properties makes the G_h increase more significantly. The interfacial crack energy or fracture toughness (G_{hc}) is also locally enhanced by the sintering induced gradient mechanical properties. As discussed above, the elastic modulus of top coating increased because of high temperature sintering. Moreover, the elastic modulus increased faster at the regions closer to the top-coat surface due to the temperature gradient across the coating thickness. Although it is quite difficult to directly obtain the local interfacial crack energy of the relatively thinner top coat with different values across the coating thickness, it is reasonable to state that, the local fracture toughness of the top-coat increases faster at the regions closer to top-coat surface due to the temperature gradient across the coating thickness. Consequently, the cracking becomes more readily with the increase of distance from top coating surface.

For the DVC structured TBCs, at the very beginning, the large vertical cracks relieve thermal mismatch stresses between the top coat and substrate during thermal cyclic test. With the proceeding of the test, G_h increased dramatically, large lateral cracks emerged. When the lateral cracks connected with the vertical cracks, delamination happened. For the DVC structured TBCs, it may just because of the large vertical cracks accelerate the failure of TBCs after thermal cyclic test.

4 Conclusions

In this study, gradient thermal cyclic tests were conducted, with the aim to simulate and evaluate the lifetime of lamellar and DVC TBCs in laboratory conditions. The following conclusions could be obtained from this work:

(1) Equivalent thermal insulation concept is introduced for the design of thermal barrier coatings with different top coat structure. TBCs with equivalent insulation performance of 1000 μm DVC structured YSZ coatings were designed.

(2) The maximum thickness of TGO in failed TBC samples was 0.64 μm, being far less than the critical value causing the failure of TBCs through complete delamination of the top coat. In addition, there was no obvious phase transformation during thermal cyclic test. Therefore, the TGO thickening and phase transformation were not the essential reasons responsible for the failure of TBC in the present study.

(3) The lifetime of lamellar structured TBC is more than two times of that DVC TBCs with equivalent thermal insulation performance. The failure mode is the top coat delamination.

(4) Large vertical cracks connected with the lateral ones and may accelerate the failure of DVC TBCs after thermal cyclic test.

Acknowledgments

The present project is supported by National Basic Research Program (Grant No. 2013CB035701), the National Science Foundation of China (Grant No. 51671159), the Fundamental Research Funds for the Central Universities, and the National Program for Support of Top-notch Young Professionals.

References

[1] N. P. Padture, M. Gell, and E. H. Jordan, "Thermal barrier coatings for gas-turbine engine applications," *Science*, **296**[5566] 280-4 (2002).

[2] R. A. Miller, "Thermal barrier coatings for aircraft engines: History and directions," *J. Therm. Spray Technol.*, **6**[1] 35-42 (1997).

[3] J. F. Li, H. L. Liao, X. Y. Wang, and C. Coddet," Fractal Perimeters of Polishing-Induced Pull-Outs Present on Polished Cross Sections of Plasma-Sprayed Yttria-Stabilized Zirconia Coatings," *J. Am. Ceram. Soc.*, **86**[11] 1906-10 (2003).

[4] G. W. Goward, "Progress in coatings for gas turbine airfoils," *Surf. Coat. Technol.*, **108**[1-3] 73-79 (1998).

[5] X. Q. Cao, R. Vassen, W. Jungen, S. Schwartz, F. Tietz, and D. Stöver, "Thermal Stability of Lanthanum Zirconate Plasma-Sprayed Coating," **J. Am. Ceram. Soc., 84**[9] 2086-90 (2001).

[6] J. Wu, X. Wei, N. P. Padture, P. G. Klemens, M. Gell, E. Garcia, P. Miranzo, and M. I. Osendi, "Low-Thermal-Conductivity Rare-Earth Zirconates for Potential Thermal-Barrier-Coating Applications," *J. Am. Ceram. Soc.*, **85**[12] 3031-35 (2004).

[7] R. McPherson, "A Model for the Thermal-Conductivity of Plasma-Sprayed Ceramic Coatings," *J. Biol. Chem*, **112**[1] 89-95 (1984).

[8] H. Dong, G. J. Yang, C. X. Li, X. T. Luo, and C. J. Li, "Effect of TGO Thickness on Thermal Cyclic Lifetime

and Failure Mode of Plasma-Sprayed TBCs," *J. Am. Ceram. Soc.*, **97**[4] 1226-32 (2014).

[9] A. Rabiei and A. G. Evans, "Failure mechanisms associated with the thermally grown oxide in plasma-sprayed thermal barrier coatings," *Acta Mater.*, **48**[15] 3963-76 (2000).

[10] C. G. Levi, E. Sommer, S. G. Terry, A. Catanoiu, and M. Rühle "Alumina Grown during Deposition of Thermal Barrier Coatings on NiCrAlY," *J. Am. Ceram. Soc.*, **86**[4] 676-85 (2003).

[11] J. Ilavsky and J. K. Stalick, "Phase composition and its changes during annealing of plasma-sprayed YSZ," *Surf. Coat. Technol.*, **127**[2-3] 120-29 (2000).

[12] D. M. Lipkin, J. A. Krogstad, Y. Gao, C. A. Johnson, W. A. Nelson, and C. G. Levi, "Phase Evolution upon Aging of Air-Plasma Sprayed t'-Zirconia Coatings: I—Synchrotron X-Ray Diffraction," *J. Am. Ceram. Soc.*, **96**[1] 290-98 (2013).

[13] D. Zhu and R. A. Miller, "Sintering and creep behavior of plasma-sprayed zirconia- and hafnia-based thermal barrier coatings," *Surf. Coat. Technol.*, **108**[1] 114-20 (1998).

[14]. B. H. Guo, H. Murakami, and S. Kuroda, "Effects of Heat Treatment on Microstructures and Physical Properties of Segmented Thermal Barrier Coatings," *Mater. Trans.*, **46**[8] 1775-78 (2005).

[15] M. Shinozaki and T. W. Clyne, "A methodology, based on sintering-induced stiffening, for prediction of the spallation lifetime of plasma-sprayed coatings," *Acta Mater.*, **61**[2] 579-88 (2013).

[16] F. Cernuschi, L. Lorenzoni, S. Ahmaniemi, P. Vuoristo, and T. Mäntylä, "Studies of the sintering kinetics of thick thermal barrier coatings by thermal diffusivity measurements," *J. Eur. Ceram. Soc.*, **25**[4] 393-400 (2005).

[18] M. Ahrens, S. Lampenscherf, R. Vaßen, and D. Stöver. "Sintering and Creep Processes in Plasma-Sprayed Thermal Barrier Coatings," *J. Therm. Spray Technol.*, **13**[3]:432-442(2004)

[19] M. Madhwal, E. H. Jordan, M. Gell. "Failure mechanisms of dense vertically-cracked thermal barrier coatings," *Mater. Sci. Eng., A*, **384**[1-2]:151-161 (2004)

[20] E. M. Donohue, N. R. Philips, M. R. Begley, C. G. Levi. "Thermal barrier coating toughness: Measurement and identification of a bridging mechanism enabled by segmented microstructure," *Mater. Sci. Eng., A*, **564**[3]:324-330 (2013).

[21] Cheng B, Zhang Y-M,Yang N, et al. "Sintering-induced delamination of thermal barrier coatings by gradient thermal cyclic test," J Am Ceram Soc. 2017;00:1–11. doi:10.1111/jace.14713.

[22] D. M. Zhu and R. A. Miller. "Thermal Conductivity and Elastic Modulus Evolution of Thermal Barrier Coatings under High Heat Flux Conditions," *J. Therm. Spray Technol.*, **9**[2]:175-180,(2000).

[23] H. B. Guo, R. Vaßen, D Stöver. "Thermophysical properties and thermal cycling behavior of plasma sprayed thick thermal barrier coatings," *Surf. Coat. Technol.*, **192**[1]:48-56, (2005)

Heat Treatment of Cold Spray Copper on the Closure Weld Zone of Used Nuclear Fuel Containers

J.D. Giallonardo[1], D. Poirier[2], J.-G. Legoux[2], P. Vo[2], C. Hoang[2], and D. Doyle[1]
[1]Nuclear Waste Management Organization, Toronto, Ontario, Canada
[2]National Research Council of Canada, Boucherville, Quebec, Canada

The Nuclear Waste Management Organization (NWMO) has proposed the concept of a deep geological repository (DGR) for the storage of Canada's used nuclear fuel. A major engineered component is the used fuel container (UFC) consisting of a steel core coated with copper for corrosion resistance. The copper coating is required to have sufficient ductility and adhesion strength to the steel substrate for loading requirements under DGR conditions. The NWMO has identified two coating technologies for the application process: electrodeposition and cold spray. Electrodeposition is utilized to coat the bulk of the UFC components (i.e., hemi-spherical head and lower assembly). A portion of the hemi-spherical head and the lower assembly openings remain uncoated in order to facilitate the final assembly closure weld process after fuel loading. This area is then cold sprayed with copper to complete the coating on the steel. Since the cold sprayed coating is highly strained in the as-sprayed state, it requires a heat treatment to impart ductility. The ductility is assessed indirectly by measuring the hardness of the material before and after the heat treatment. A recent advancement on this front includes the implementation of an optimized band heat treatment method to prototype UFC's.

1 Introduction

The Nuclear Waste Management Organization (NWMO) was established in 2002 in accordance with the Nuclear Fuel Waste Act to assume responsibility for the long-term management of Canada's used nuclear fuel [1] by implementing Adaptive Phased Management (APM). The program's objective is to create a socially acceptable, environmentally responsible, and economically feasible solution for containment and isolation of used nuclear fuel. The technical solution to be implemented by the APM program is the creation of a Deep Geological Repository (DGR). In this concept, the used nuclear fuel will be encapsulated in long-lived Used Fuel Containers (UFC), then emplaced in a buffer box at a reference depth of 500 m underground in a suitable rock formation, and surrounded with bentonite clay [2]. The DGR method is consistent with international best practices as demonstrated by it being the approach adopted by many nuclear power producing countries around the world.

The UFC is a major engineered component of the DGR design. It is the vessel in which the used fuel is packaged before final underground emplacement. The NWMO has developed a UFC specifically designed for CANDU fuel. The use of copper coating technologies for the application of the external corrosion barrier is a unique feature of NWMO's UFC design. The choice of copper to act as a corrosion barrier was derived from evidence provided through natural analogues that include archaeological copper containing artifacts, e.g., coins, cannons, etc [3]. For the NWMO's design, a copper coating thickness of 3 mm is specified. This is based on very conservative assumptions for corrosion in a DGR environment, to which less than 1.3 mm of copper loss is expected in one million years. Furthermore, the copper must be comparable to wrought copper in order to meet corrosion performance requirements in the expected environmental conditions [4]. The characteristics of the copper are also designed accordingly based on the steel mechanical performance requirements. The copper must be ductile and well adhered in order for it to remain intact and bonded to the steel structural vessel when subjected to loading under DGR conditions [5]. Electrodeposition is utilized to coat the bulk of the UFC component surfaces (i.e., hemi-spherical head and lower assembly). The cold spray process is to be used for completing the copper coating about the closure weld zone after fuel loading and final assembly [6] in the envisioned "hot cell" or radioactive environment. Since the cold sprayed coating is highly strained in the as-sprayed state, it requires a heat treatment to impart the ductility required for the design. The current cold spray reference process employed specifically for coating the closure weld zone of the UFC has been developed with the National Research Council (Boucherville, Canada) [7].

This paper aims to summarize the recent work carried out in an effort to optimize a suitable heat treatment method to impart ductility on the cold spray copper applied to prototype UFC's. The work performed includes the verification of heat treatment conditions and optimization of a technique using representative sample geometries. Finally, the optimized heat treatment technique was applied to prototype UFC's. Plans to further advance the heat treatment technique as it would apply to the manufacturing process in a future hot cell environment will also be provided.

2 Experimental Methodology

2.1 Cold Spray Process

The reference cold spray process employed a PCS-800 system from Plasma Giken Co. (Toshima-ku, Tokyo, Japan) equipped with a de Laval type nozzle. The following parameters are defined for the cold spray process: (1) spherical low-oxygen copper powder, (2) steel surface blasted with 24 grit (alumina) to remove mill scale, (3) 100 µm bond coat, applied using helium as the carrier gas at 5 MPa and

800°C, and (4) 5 mm top coat, applied using nitrogen as the carrier gas at 5 MPa and 800°C.

2.2 Verification of Heat Treatment Conditions

Based on initial work of Vo et al. [7] further mechanical testing was performed after various heat treatments. Tensile test specimens (100 mm overall length dog bones with gauge length, width and thickness of 25, 6 and 1 mm, respectively) were electric discharge machined (EDM). These samples were extracted from the cold sprayed coatings with the tensile direction perpendicular to the spray direction. It should be noted that d_{50} = 42 μm spherical low-oxygen copper powder was used in the production of the samples. Tensile tests were performed according to ASTM E8/E8M-09 to generate engineering stress-strain curves. The heat treatments were carried out over a range of temperatures (200-600°C) and times (1-10 hrs) in an inert furnace environment. The resulting microstructures after heat treatment were characterized from polished cross-sections using an Olympus BXM53M optical microscope. Furthermore, the effect of the heat treatments on hardness were studied by performing measurements using a CLEMEX CMT MMT-X7 tester accordingly with ASTM E384-10.

2.3 Optimization of Heat Treatment Technique

Prior to applying a heat treatment method to the closure weld zone, several experiments were carried out to verify that the desired ductility was achieved on representative UFC geometries (i.e., pipe segments). The cold spray system was controlled robotically by traversing the width (axially) over the region of interest while the part is rotated. It should be noted that for this work, a commercially available spherical low-oxygen copper powder with d_{01} = 5μm or above and d_{90} = 60 μm or below spherical low-oxygen copper powder was used to produce the coatings. The pipe segments are made from seamless ASME/ASTM A106 Gr. C wiht an approximate outer diameter of 560 mm and a wall thickness of 46 mm. The lengths of the pipe were approximately 600 mm. The UFC cylindrical section was made from the same material as the pipes; however, the hemi-spherical heads are made from hot pressed ASME/ASTM A516 Gr. 70 plate having a minimum wall thickness of 30 mm. At the closure zone of the UFC, the uncoated copper section extends approximately 50 mm from the either side of the weld joint between hemi-spherical head and main UFC cylindrical section. The total width of the band of steel that requires the copper coating is approximately 100 mm. The cold sprayed copper covered this section and was over-sprayed onto the tapered electrodeposited copper (i.e., approximately 25 mm on either end).

As a baseline, one pipe segment was heat treated using a controlled inert atmosphere furnace at 350°C for 1 hr. The heat treatment was performed at Vac Aero International, Inc. (St. Laurent, Quebec, Canada). Since it is not feasible to use a furnace for heat treating UFC's that have been loaded with used nuclear fuel, a band system was employed to apply a localized heat treatment. The band system was supplied by Wattco (Lachine, Quebec, Canada) and made of stainless steel with a maximum power output of 8.4 kW. All band heat treatments were performed in a room temperature (air) environment. The samples taken from the pipe segments for analysis were initially extracted by cutting out a section. This allowed for cross-sectional (metallographical) examination of the circumferential face of the cold sprayed copper material. The resulting microstructures after heat treatment were characterized from polished cross-sections via FEG-SEM in BSEI mode (Hitachi S-4700). Vickers hardness measurements were performed on the cross-sectioned samples from the pipe segments with a Buehler Micromet II indenter according to ASTM E384-10. For each specimen, a minimum of 10 indentations were performed and the average was reported. Preliminary hardness measurements were made on the UFC using a Starrett portable Brinell tester. In this case, the surface of the copper was smoothed using 320 grit paper to produce relatively flat surface. Three measurements were taken and the average was reported.

3 Results and Discussion

3.1 Verification of Heat Treatment Conditions

It is well known that cold sprayed materials are not ductile. A heat treatment serves to impart ductility by recovery and recrystallization of the copper which has undergone substantial plastic deformation or strain hardening during the cold spray process. To verify the effects of the heat treatments, tensile tests were performed to determine the ductility (percent elongation) and select suitable conditions for further application development.

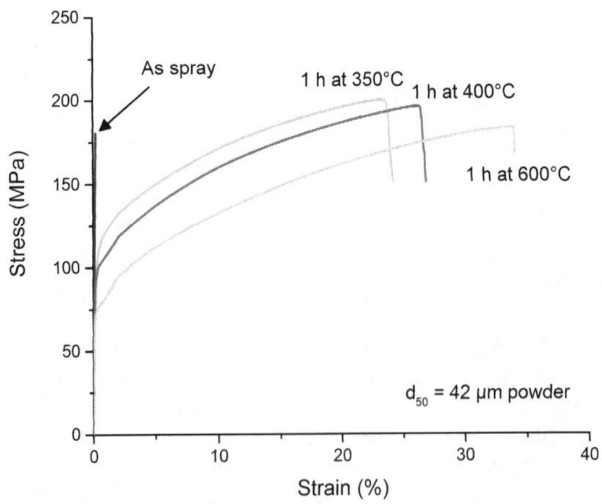

Fig 1. Stress-strain curves for select samples in the as-sprayed and selected heat treatment conditions.

Fig. 1 presents the tensile testing results obtained from cold spray coatings heat treated at various temperatures for 1 hr. The as-sprayed coating is shown to be brittle; however, the coating ductility (as measured by percent elongation) is shown to increase with heat treatment temperature reaching about 24% at 350°C for 1 hr and 34% at 600°C for 1 hr.

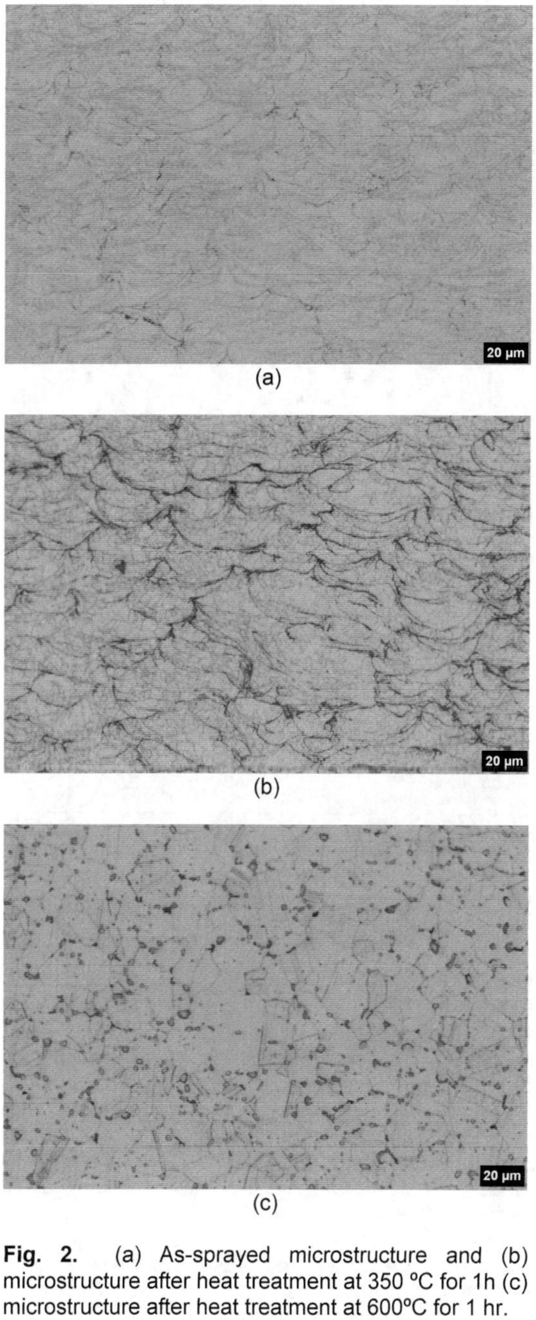

Fig. 2. (a) As-sprayed microstructure and (b) microstructure after heat treatment at 350 ºC for 1h (c) microstructure after heat treatment at 600ºC for 1 hr.

Fig. 2 compares the etched microstructures before and after heat treatment at 350°C and 600°C for 1 hr. The micrographs of the as-sprayed and heat treated (350°C for 1 hr) coatings clearly show inter-particle boundaries and heavily deformed (flattened) particles. Partially recrystallized microstructures within the deformed particles were obtained for the intermediate heat treatment temperature (Fig. 2b); however, only boundaries of equiaxed grains could be discerned in specimens heat treated at 600°C for 1 hr which is representative of full recrystallization. Also, it should be noted that in Fig. 2c, there are porosity necklaces arising from the sintering process that takes place during the heat treatment.

The increase in ductility with heat treatment temperature can be attributed to the recovery and recrystallization of the copper during heat treatment which enables the material to deform with sufficient elongation. Another factor that may have affected the strength and ductility after annealing is an increase in the amount and size of metallurgically bonded inter-particle boundaries due to diffusion during annealing, similar to sintering [8]. Metallurgical bonding reduces the number of sites for crack initiation while non-metallurgically bonded inter-particle boundaries may represent cracks exceeding the critical size, thereby promoting brittle fracture [8]. In addition to these microstructural changes, it can be noted in Fig. 1 that the yield strength of the heat treated coatings are lower than the fracture stress of the untreated samples. Upon deformation it can thus be inferred that the inter-particle bondaries in the heat treated coatings are not as stressed as the ones in the as-sprayed condition which also contributes to the increase in ductility.

When the copper coating is applied to representative geometries, the ductility after a heat treatment must be evaluated to determine if the minimum requirement is met. This may be indirectly assessed by measuring the hardness of the copper before and after heat treatment [7,9,10]. The hardeness was measured on the tensile specimens for the as-sprayed and heat treated samples (350°C and 600°C for 1 hr). The values obtained were 133 HV, 87 HV, and 51 HV, respectively. From the as-sprayed condition, this represents a 35% and 62% reduction in hardness. Since the UFC application requires a percent elongation of at least 10%, the heat treatment temperature of 350°C for 1 hr was selected for the following optimization study. That is, a reduction in hardness of approximately 30-40% is considered to be sufficient enough to suggest that ductility was imparted to the copper.

3.2 Pipe Segments

Four pipe segments were cold sprayed with copper to approximately 5 mm of thickness. The cold sprayed copper band was approximately 150 mm wide around the circumference. Fig 3. shows the pipe segment before and after the copper cold spray process. The first pipe segment was sectioned and the hardness was measured in the as-sprayed state. The remaining pieces of the pipe segment were then subjected to a heat treatment in the furnace at 350°C for 1 hr. The hardness of the heat treated copper was then measured. In the as-sprayed state, the average was 125 HV and after the heat treatment, it was 75 HV (see Table 1). The difference in hardness values represent a 40% reduction which is indicative of achieving the ductility required after heat treatment.

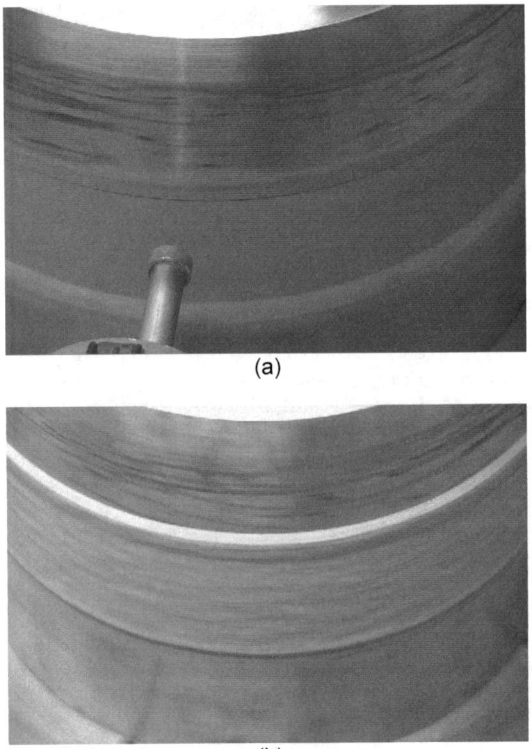

(a)

(b)

Fig. 3. Copper cold spray on a pipe segment (a) before and (b) after processing.

The remaining pipe segments were heat treated to optimize the method using the band apparatus. Fig. 4 shows a pipe segment with the band heater apparatus during a treatment cycle. It is wrapped with a thermally insulating material to minimize heat loss to the environment. For the optimization study, as-sprayed hardness values were not available since sectioning of the pipe segments beforehand would have compromised the effectiveness of the band heat treatment method. Given the results of the previous heat treatment method using the furnace, it is reasonable to take the initial as-sprayed hardness values for the three cases using the band heater to be approximately 125 HV. A summary of the optimization study results are given in Table 1.

In the first band heat treatment attempt, the controller set point was 350°C and the duration was 1 hr. The pipe segment was sectioned and the average hardness was measured to be 113 HV which represents only a 10% reduction. Although the temperature measured by the thermocouple at the surface of the copper was 350°C, the lower than expected reduction in hardness provides evidence to suggest that the copper is not reaching the desired temperature. Since the application of heat via the band is local, not only will there be heat dissipation to the steel, but there will also eventually be heat dissipation to the environment (i.e., air). Thus, it is conceivable that there is a significant temperature gradient from where the band makes contact at the surface of the copper through the thickness.

Fig. 4. Copper cold sprayed pipe segment with band heater apparatus.

Table 1. Heat treatment conditions and resulting percentage reduction in hardness for pipe segment trials.

Heat Treatment	Before	After	Reduction
Furnace, 350°C, 1 hr, vacuum	125 HV	75 HV	40 %
Band heater, 350°C, 1 hr, air	-	113 HV	*10 %
Band heater, 450°C, 2 hr, air	-	95 HV	*24 %
Band heater, 470°C, 2 hr, air	-	78 HV	*38 %

* Calculated based on the assumption that as-sprayed values are approximately 125 HV

In order to rectify this and account for the heat dissipation effect, the heat band set-point temperature and duration were increased to 450°C for 2 hrs. After sectioning, the average hardness was determined to be 95 HV representing an approximate 24% reduction. In order to further improve this, the temperature and duration were set to 470°C for 2 hrs and the resulting average hardness was determined to be 78 HV, representing a 38% reduction that is within the desired range of 30-40%. Microstructural analysis was performed on the cross-sections to observe for evidence of recrystallization and sintering. Fig. 5a shows the etched microstructure of the cold spray copper coating after heat-treatment at 470°C for 2 hrs. The dark features represent inter-particle boundaries and the deposited particles have (flattened) shapes typical of a cold sprayed material.

Fig. 5b shows the heat treated microstructure at higher magnification. The grains within the particles have sizes in the 1-10 µm range and appear to be relatively equiaxed which provides evidence of a partially recrystallized microstructure.

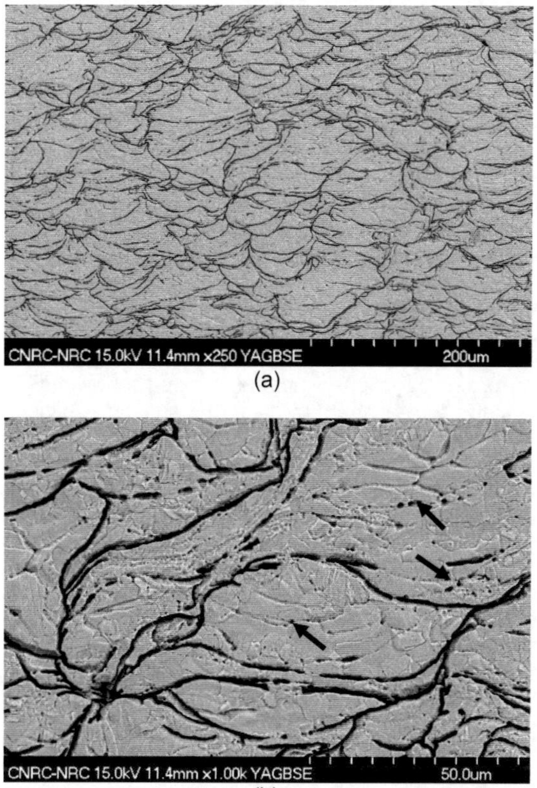

Fig. 5. Etched microstructures of band heat treated cold sprayed copper at 470°C for 2 hrs in air at (a) low magnification and (b) at high magnification.

As mentioned earlier, in other studies, the microstructures of cold sprayed copper coatings after heat treatment showed observable recrystallization that caused a reduction in hardness [9,10]. It should also be noted that in Fig. 5b, there is evidence of porosity at former inter-particle bondaries (indicated by the arrows) resulting from the sintering process which also takes place during the heat treatment.

3.3 Used Fuel Container

The UFC closure weld zone was cold sprayed with copper to a thickness of approximately 5 mm. Fig. 6 shows the UFC during the spraying process and the final copper coated closure weld zone. Two trials were performed and an assessment of the effectiveness was carried out using a non-destructive means (i.e., portable Brinell tester) for measuring the hardness of the cold sprayed copper before and after heat treatment.

Fig. 7 shows a UFC while undergoing the band heat treatment. The hardness using the portable Brinell tester was determined to be 134 HB in the before case and 96 HB after heat treatement at 470°C for 2 hrs. This represents a reduction of about 28%. In this scenario, the heat dissipation effect will be amplified by the fact that the UFC has a larger metal mass when compared to the pipe segments. In order to accommodate this, an increased set point temperature will be investigated in future work.

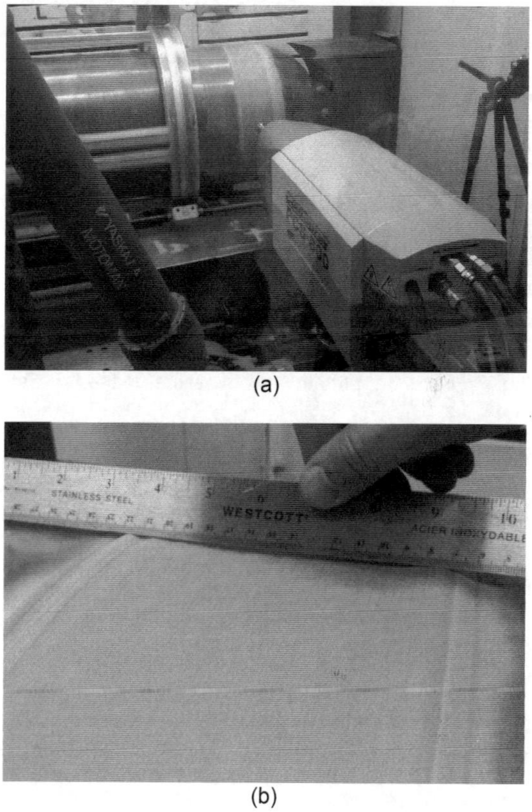

Fig. 6. Cold spray of the UFC (a) in-process and (b) at approximately 5 mm of copper coating thickness.

The band heater employed in this study is considered to be an intermediate technique and presently intended for demonstration purposes to show that the required ductility of the copper cold spray material can be imparted to the closure weld zone of prototype UFC's. In addition to this, by applying the band heat treatment technique, certain scaling issues and limitations can be anticipated or addressed ahead of implementing a final method. As mentioned in the introduction, it is envisioned that the process to cold spray copper over the closure weld zone will occur in a hot cell or radioactive environment. For this reason, other methods will be explored in the future to improve amenability to automation. Currently, there are limited solutions available that accommodate this requirement and thus, a fairly advanced technique for heat treatment is being sought for eventual implementation. For example, the use of laser assisted cold spray techniques appear to be providing some promise in addressing this challenge [11,12].

Fig. 7. Band heat treatment apparatus on the UFC at 470ºC for 2 hrs in air.

4 Conclusions

The requirement for the heat treatment arises from the fact that the as-sprayed copper ductility is not sufficient for the expected DGR conditions that the UFC experience while in service. The low ductility is a result of mainly the plastic deformation (i.e., strain hardening) that particles undergo on impact and the dominant presence of inter-particle boundaries in the as-sprayed state. In order to impart the required ductility, heat treatment conditions have been defined which effectively recrystallize the microstructure and enable sintering to eliminate inter-particle boundaries. In the work presented, heat treatment conditions were further verified by assessing the ductility via tensile testing and reduction in hardness values. Based on this, a band heat treatment technique was optimized using representative geometries (i.e., pipe segments) and finally applied to the closure weld zone of a prototype UFC. For the pipe segments, an optimized heat treatment at 470ºC for 2 hrs was shown to reduce the hardness by 38% which is within a range that is known to be consistent with achieving sufficient ductility. The band heat treatment technique was then applied to the prototype UFC and a reduction of 28% in hardness was observed. The difference in hardness reduction between the pipe segments and UFC was owed to the effect of heat dissipation since the UFC has a significantly larger metal mass. Considering this, further optimization on the application of the band heat treatment technique to the UFC will be required. Although the band heat treatment technique is considered to be an intermediate step in this development, it provides insight into the potential scaling issues that may be faced when the final method is implemented. Advanced heat treatment techniques are currently under investigation with the objective of achieving a fully automated process when in the hot cell or radioactive environment. It is anticipated that a laser assisted technique will provide some promise in achieving this.

5 References

[1] Federal Government of Canada, "Nuclear Fuel Waste Act," http://laws-lois.justice.gc.ca/eng/acts/N-27.7/ (retrieved on February 23, 2017).

[2] C. Hatton, "Development of the Canadian Used Fuel Repository Engineered Barrier System", Proceedings of the American Nuclear Society 13th International High-Level Radioactive Waste Management Conference, Charleston, April 12-13, 2015, pp. 340/345.

[3] F. King, "A natural analogue for the long-term corrosion of copper nuclear waste containers – reanalysis of a study of a bronze cannon", Applied Geochemistry, 10 (1995) pp. 477/487.

[4] P.G. Keech, P. Vo, S. Ramamurthy, J. Chen, R. Jacklin and D.W. Shoesmith, "Design and development of copper coatings for long term storage of used nuclear fuel", Corrosion Engineering, Science and Technology, 49 (2014), Issue 6, pp. 425/430.

[5] C.H. Boyle and S.A. Meguid, "Mechanical performance of integrally bonded copper coatings for the long term disposal of used nuclear fuel", Nuclear Engineering and Design, 293 (2015), pp. 403/412.

[6] P. Keech, P. Vo, D. Poirier, J.-G. Legoux, J.D. Giallonardo and D. Doyle, "Cold spray copper coating application to the closure weld zone of used nuclear fuel containers", Proceedings of the 3rd Canadian Conference on Nuclear Waste Management, Decommissioning and Environmental Restoration, Ottawa (Canada), September 11-14, 2016, In Press.

[7] P. Vo, D. Poirier, J.-G. Legoux, E. Irissou and P.G. Keech, High Pressure Cold Spray – Principles and Applications, pp. 253/276, C.M. Kay and J. Karthikeyan, Ed., ASM International, Ohio (2016).

[8] F. Gartner, T. Stoltenhoff, J. Voyer, H. Kreye, S. Riekehr and M. Kocak. "Mechanical properties of cold-sprayed and thermally sprayed copper coatings" Surface & Coatings Technology, 200 (2006), pp. 6770-6782.

[9] E. Calla, D.G. McCartney and P.H. Shipway, "Effect of deposition conditions on the properties and annealing behavior of cold-sprayed copper", Journal of Thermal Spray Technology, 15 (2006), Issue 2, pp. 255/262.

[10] T. Stoltenhoff, C. Borchers, F. Gartner and H. Kreye, "Microstructures and key properties of cold-sprayed and thermally sprayed copper coatings", Surface Coatings & Technology, 200 (2006), pp. 4947-4960.

[11] M. Perton, S. Costil, W. Wong, D. Poirier, E. Irrissou, J.-G. Legoux, A. Blouin, and S. Yue, "Effect of pulsed laser ablation and continuous laser heating on the adhesion and cohesion of cold sprayed Ti-6Al-

4V coatings“, Journal of Thermal Spray Technology, 21 (2012), Issue 6, pp.1322/1333.
[12] D. Christoulis and C. Sarafoglou, Modern Cold Spray – Materials, Process, and Applications, pp. 275/302, J. Villafuerte, Ed., Springer International, Switzerland (2015).

New Expansion of mobile Particle Shape Imaging and comparative measurements with other diagnostics

Christian Franetzky, Stephan Zimmermann and Jochen Schein, Neubiberg / D

Coatings produced by thermal spray processes are determined by particle parameters, which can be investigated using various diagnostic systems. In the first part of the paper the different PSI (Particle Shape Imaging) setup are described. Thereby a summary of former publications, presentation of measurements on different thermal spray processes and suggestions of data error evaluations with special difficulties at the particle investigation are shown. In the second part of the paper the new developed mobile experimental setup of PSI is focused. Four important particle parameters (velocity, size, shape and position) can be measured and evaluated. In this part the results and test of new different illumination sources, optical configurations and trigger units are discussed. Furthermore interesting ideas of picture processing and data evaluation are depicted and explained in more detail. First comparison of the results measured with DPV2000, SprayWatch and additional Voltage-Current-High-Speed-Camera investigations are shown. The complete paper concludes with an outlook for future combination of further established diagnostic systems (i.e. find the optimal distance between substrate and thermal spray system).

1 Introduction

2 Particle Shape Imaging

Since the importance of a high quality coating in many industries has grown a lot in the last years many people asked themselves what the best way to look on a coating process in real time would be. Due to the high velocity and temperature of the melted materials this presented itself as a difficult task. A lot of companies developing diagnostic systems saw the best chances in analysing the radiation of the particles. Using radiation laws like the Stefan-Boltzmann-law or Planck´s law they are able to determine the size and the temperature of a single particle. But by using these laws they are automatically making mistakes, which will be discussed further in this paper. The institute for plasma techniques and mathematics of the University of the federal armed forces in Munich searched for a different analysis. The goal was to see the particle how it really looks during the process. As the name of the systems already suggests the idea was to display the shadow of a particle and therefore its shape. The next chapters will describe how the system is working in detail [7, 10-11].

2.1 Particle Shape Imaging

To illuminate the particles a laser beam was divided and directed seperatly from two directions on a certain spot in the stream. Behind the process the two beams are directed on one camera chip. Here the beams show the shadows of the particles flying through the measurement area. By overlapping the two pictures the position of a particle in reference to the measuring area is determined. Are the two shadows of one particle overlapping perfectly the particle was in the center of the area. Should this be not the case, the particle was before or behind the center. But with this setup it is not possible to determine which of these two alternatives is the case (**Figure 1**).

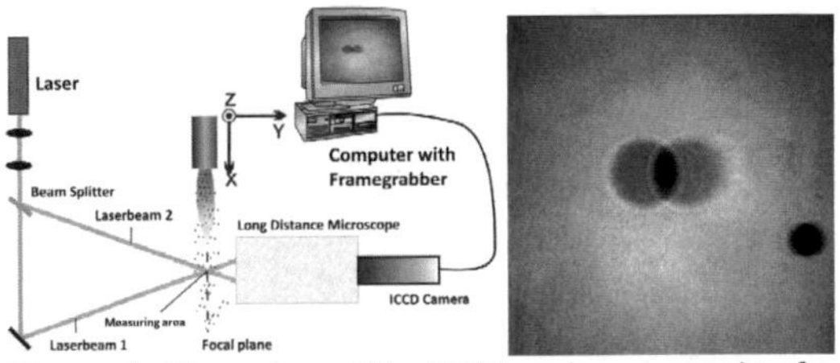

Figure 1: The setup of the PSI I and an example of a particle in the measuring area [10]

The PSI II just improves this setup by presenting the two pictures not overlapping but above each other. So by knowing which beam is the upper one it is easy to tell which one of the particles passed first and so gets displayed more to the right or to the left depending on the direction of the flying particles (**Figure 2**).

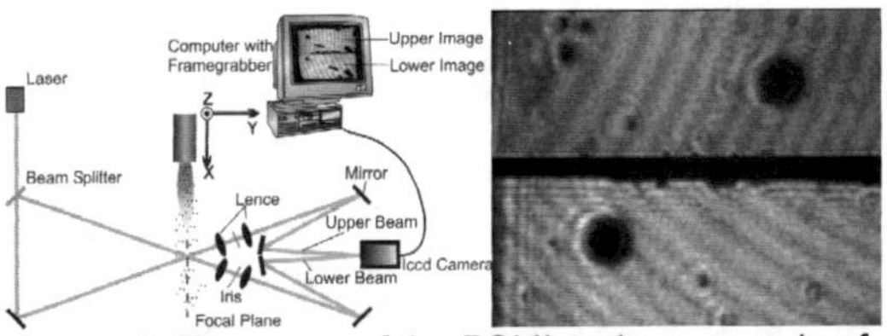

Figure 2: The setup of the PSI II and an example of a particle in the measuring area [10]

To analyse this representation of the particles a special software was developed. This software at first creates a binary picture of the original one and afterwards uses a chain code to scan the picture for particles. The number of black pixels are counted and afterwards the size gets calculated, if the scale of the representation is known. A so called particle zoo, where many black particles with certain sizes are printed on a glass, is used for estimation this scale.
The third generation of the PSI was improved by dividing the laser beam in three to get three pictures on the camera chip. By doing this the usage of a

trigger system to reduce the amount of pictures needed got easier and the redundancy got improved.
In this paper the focus is clearly on the fourth setup of the PSI. It is a mobile version, wich was invented to be used in different laboratories and for a quick relocation of the diagnostic system during a measuring campaign e.g.
This mobile version is build in a frame of "item"-profile rails and is fully movable on the ground. After many remodellings the newest setup contains two Laser each with a power of 1 W. The two beams with a wavelength of 532 nm are both focused on the measurement area and on the other side directed through a lense and on the camera chip. To enlarge the beam before the measuring area a two-lense-system is installed as shown in **Figure 3**.

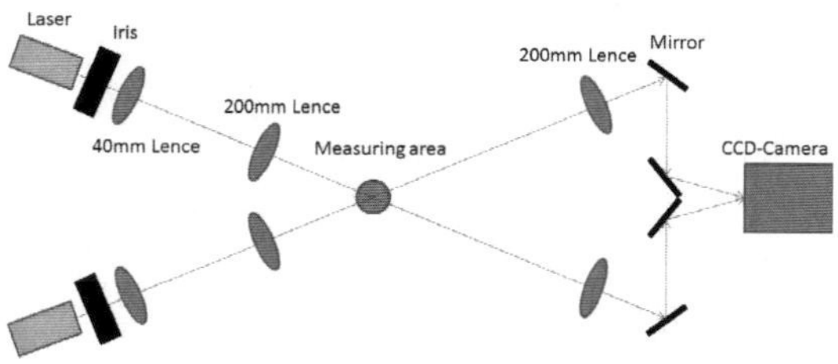

Figure 3: schematic setup of the PSI mobil

In the next chapter the PSI mobil and its diagnostic principle is explained further and more detailed.

2.2 PSI mobil

By using the profile rails it is very easy to change the setup of the PSI mobil. The Usage of a Laser with another wavelength or a change of the optical transformation, to analyse in different scales for smaller particles for example, is possible. It is also very easy to use another camera with a better performance on higher velocities e.g. Whatever setup is used, the software to analyze the pictures stays the same. This makes the PSI mobil a diagnostic system, which is capable of working in various circumstances and suits many purposes.
The best way to show what the system is capable of is by comparing it to other diagnostic systems. This will be described in the following.

3 DPV2000

The *DPV2000* developed by *Tecnar* [9] is particle diagnostic which allows to determine the velocity, the temperature and the diameter of particles in a cold or hot stream. It is capable of showing the user the data of single particles as well as showing the average data in a certain period.
Before describing the working principle of the *DPV2000* the different parts of the systems are shown. The systems contains a computer equipped with the software *DPVevolution*, which is also developed by *Tecnar*. This computer is connected with a detection module consisting of photodetectors and mirrors. Here the signals of the sensing head are processed. A glass fibre cable connects detection module and sensing head. The sensing head consists of a sensor and a mask in front of it. For placing the sensing head in the right position related to the particle stream the *DPV2000* is equipped with a two axis scanning module. With this module the user is also able to scan a two dimensional grid. The whole assembly is shown in **Figure 4**.

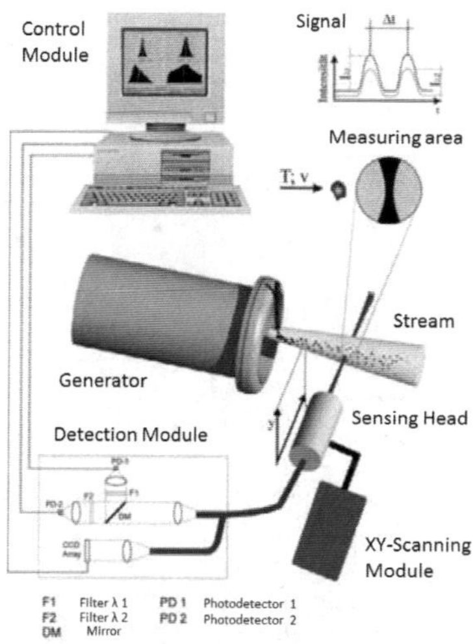

Figure 4: components of the *DPV2000* [1]

To determine the speed the system uses a mask with to gaps. By passing one gap the particle is emitting a certain emission, which gets detected by the sensing head. This gets repeated by passing the second gap and with knowledge of the distance between the gaps and the time between the two emissions the system is able to estimate the velocity of the particle [1-4].
The received emissions are used for estimating the temperature and the diameter. For this purpose the system uses the two-colour-pyrometry. The sensor observes the intensity of the emission at two different wavelengths. The wavelengths should be close to each other but not the same. By this the influence of the emission coefficient is ruled out and the determination of the temperature is within reach. With knowledge of the temperature and by using the Stefan-Boltzmann-law the *DPV2000* estimates the diameter of a spherical particle. So it expects all particles to be spherical [1-4].

4 SprayWatch

Oseir has brought their particle diagnostic on the market under the label *SprayWatch* [8]. The system is capable of estimating velocity and temperature of particles in a spray process, regardless if cold or hot. While the user is able to monitor the velocity of every single particle, the temperature is only displayed as a average temperature of the process [2].
The system consists of a camera with a CCD-chip and optical evaluation tool behind. These components

get connected with any computer and controlled with the provided software [2], **Figure 5**.

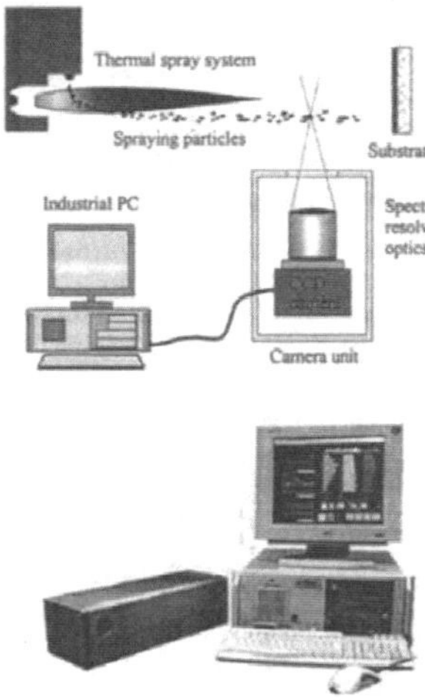

Figure 5: components of the *SprayWatch* [2]

The velocity measurement is working by taking a photo with a exposure time of 5-20 ms. In this pictures the particles produce discreet lines, which can be measured in length by the sorftware. With the knowledge of the exposure time and the length of the "lines" a calibrated *SprayWatch* can estimate quite accurate the velocity of every single particle in the pictures as long as the particle is flying at least almost horizontically through. The temperature is measured with the same principle as in the *DPV2000* and therefor it is already described in this paper. The system is able to estimate the velocity with a accuracy of 1% around the real value and the temperature with 10 degree around the real value, according to the producer [2].

5 The three systems in comparison

As the PSImobil shall be implemented as a real alternative to the established systems like DPV2000 or SprayWatch it is the first thing to do to compare the results of the three systems and find advantages and disadvantages. To do that the systems were used in a measurement of the same spraying process at the University of the Federal Armed Forces in Munich (UniBw). As the shared values measured with the three systems is the diameter and the velocity of the particles, this is the one put in the focus of this comparison. The main difference is the way these systems estimate the diameter. DPV2000 and SprayWatch use the two-colour-pyrometry, whereas the PSImobil uses a binary picture of the particles shapes. So all three systems were used in a measurement of a wire arc spraying process with a standard G3Si1-wire and of course under the same settings (5 bar air pressure, 5 m/s wire speed). The resulting histogram of the DPV2000 and the PSImobil are showing in general that the bigger part of the particles are in the same area of the histogram, **Figure 6.**

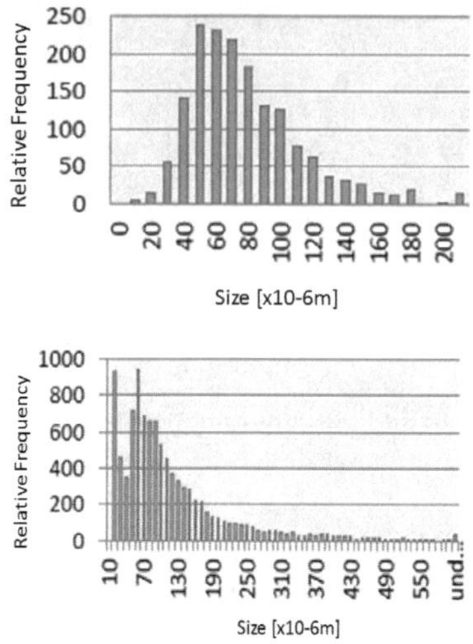

Figure 6: Diameter histograms of the DPV2000 (top) and the PSImobil (down) of the same process

A clear and smooth distribution around 60 µm Diameter is obvious in both histograms. Furthermore it is clear to see the disadvantages of both systems. The PSImobil has a high amount of detected particles below 20 µm, which is to be traced back to image noise and illustration failures. But it is capable to detect even quite big particles. The DPV2000 is on the contrary not able to detect particles bigger than 300 µm, which is a big deficit because in short distances to the spray gun bigger particles occur quite frequently. That also explains the big difference between the average values given by the DPV2000 (73,7 µm) and the PSImobil (115,19 µm). The following figure shows a picture taken with 10 mm distance to the spray gun and a obvious particle bigger than 300 µm, **Figure 7**.

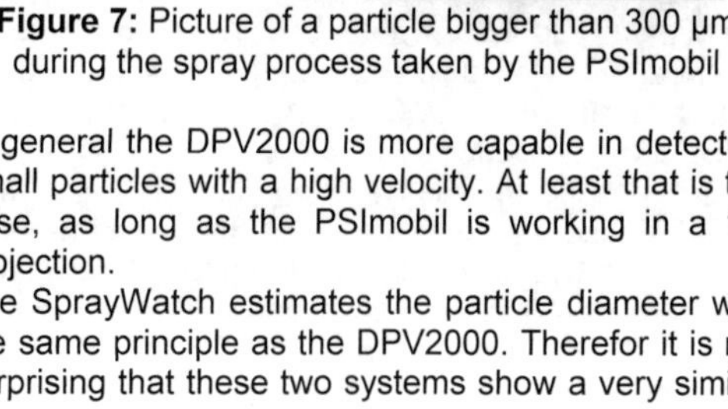

Figure 7: Picture of a particle bigger than 300 µm during the spray process taken by the PSImobil

In general the DPV2000 is more capable in detecting small particles with a high velocity. At least that is the case, as long as the PSImobil is working in a 1:1 projection.

The SprayWatch estimates the particle diameter with the same principle as the DPV2000. Therefor it is not surprising that these two systems show a very similar result in calculating the diameter. With 50 mm distance to the spray gun the DPV2000 estimates a average diameter of 68,6 µm, while the SprayWatch gives the user a value of 69,8 µm. A digaramm of the recorded process is shown in **Figure 8**.

System	Average Diameter
SprayWatch DPV2000	69,8 µm 68,8 µm

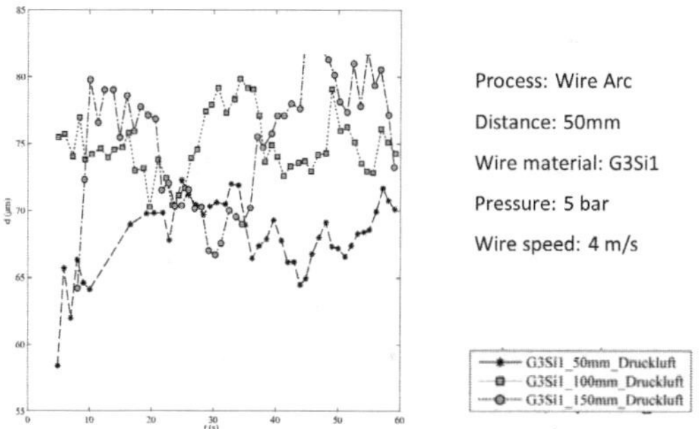

Figure 8: Diameter results of the SprayWatch system during a measurement with 50 mm distance to the spray gun

In estimating the particle velocity the DPV2000 and the SprayWatch show significant differences, which can be traced back to the different principles the systems use. While the DPV2000 calculates the velocity in the same measurement like before with 61,1 m/s, the SprayWatch calculates 82,4 m/s. Also here a diagram of the recorded process is shown (**Figure 9**).

System	Average Velocity
SprayWatch DPV2000	61,1 m/s 82,4 m/s

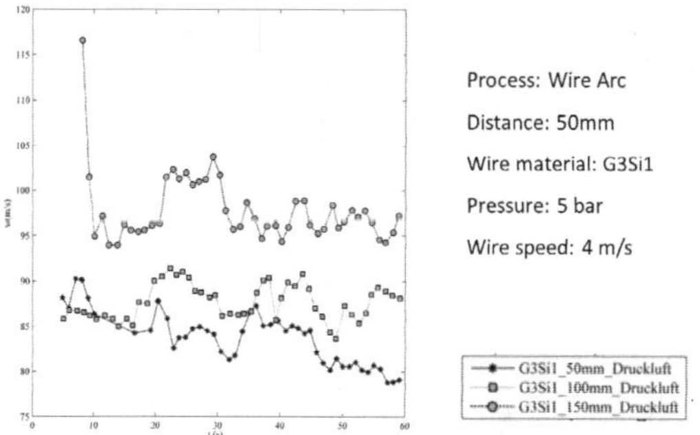

Figure 9: Velocity results of the SprayWatch sytem during a measurement with 50 mm distance to the spray gun

In conclusion it is obvious that every single one of the three system has its advantages and disadvantages. But a picture of the particle shape shows the user a pure illustration of the reality. Furthermore the user is able to review the process at a later date due to the taken pictures of the systems. Only in interpretating these pictures can be made mistakes. It is easy to imagine that particles, which have no fitting emission, are not influencing the results of the DPV2000 and the diameter results of the SprayWatch. On the other hand it is the big task to only interpretate the real particles with the PSImobil and not the noise or projection mistakes.

6 Measurement with a voltage and current monitoring

To find other possible purposes for the PSImobil the wire arc spraying process was also analyzed with a real-time voltage and current monitoring. By doing this it was the goal to gain experiences about the influences of these two parameters on the separation process. To do this multiple measuring equipment were connected to the process and to an oscilloscope. The oscilloscope and the camera were triggered external by a waveform generator [5-6], **Figure 10**.

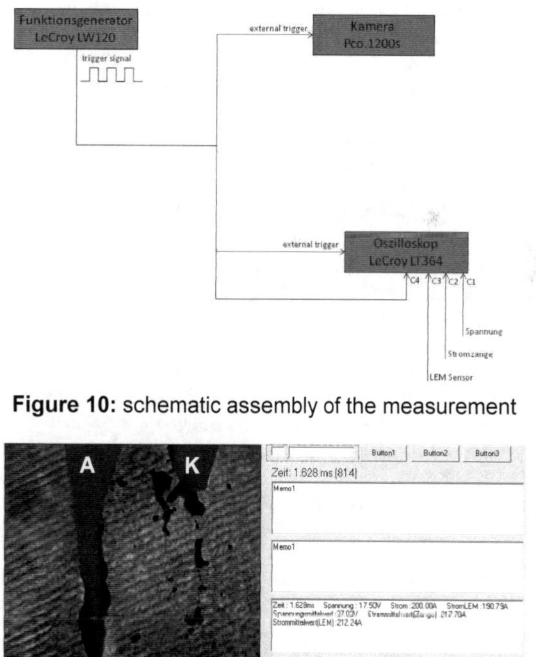

Figure 10: schematic assembly of the measurement

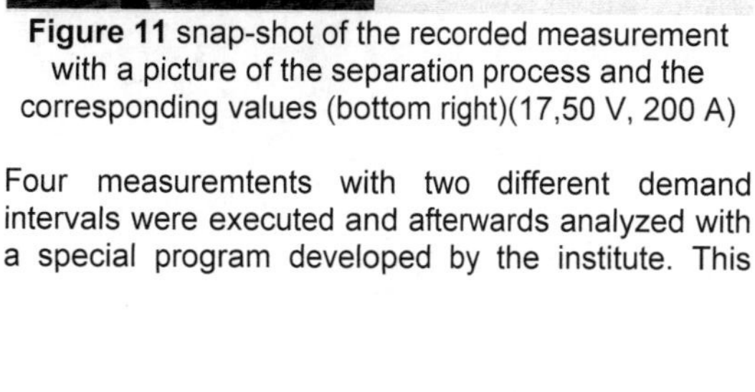

Figure 11 snap-shot of the recorded measurement with a picture of the separation process and the corresponding values (bottom right)(17,50 V, 200 A)

Four measuremtents with two different demand intervals were executed and afterwards analyzed with a special program developed by the institute. This program connects automatically the voltage and current values with the matching picture taken by the camera. The usage of the bigger high-speed camera of the institute would offer a higher frames per second, which would result in more matching pictures during a measurement. But as a first experiment it worked perfectly and the results were very promising, **Figure 11**.

7 Conclusion and perspective

The PSImobil has clearly shown its advantages during the measurements and in comparison to other systems. It gives the user a simple, pure and very flexible analysis of almost every spraying process. During the measurements a lot of experiences were gathered and influenced the setting of the PSImobil. By doing that a lot possible sources for failures or malfunctions were eliminated. The problematic illumination of the measuring area was improved with the implementation of the new 1W-Laser. The higher power and the usage of two laser result in a clearly better picture. Furthermore the pco.1200s high-spreed camera provide pictures with a much higher resolution than the 4picos-camera, which was used before.

With the rotation of the whole system of 90 degree two malfunctions were removed by one action. The enlargement of the part of the measuring area, which is perfectly in focus, and the possibility to analyse the separating process at the tip of the spray gun.

By using the pco.1200s the distances between the optical components were reduced significantly and therefore the number of possible illustration relationships got higher. By simple moving of the optical components it is possible to change from a 1:1 to a 2:1 illustration. Other illustrations are also quite easy to implement. This shows how flexible the PSImobil is in its settings.

The measurement of the wire arc spraying process showed further that the PSImobil is perfectly suitable for a analysis of the separation process.

For an analysis of the particles in higher distances to the spray gun or at impact a different high-speed camera of the institute is necessary. This camera offers a setting of a lower exposure time. But the smaller pco.1200s stays a valuable option because of its mobility.

In the near future further measurements are necessary to gain more experience with the system. In this measurements the setting of the illustration relationship should vary as well as the used camera. Also for the use in a measurement with a current and voltage monitoring the other high-speed camera would fit better because of a higher possible frames per second. Because of the flexibility and different options the PSImobil offers it is a very promising alternative to the other systems available.

8 Literature

[1] IOT Aachen, www.iot.rwth-aachen.de

[2] C. Berndt, K. Khor, E. Lugscheider: Thermal Spray 2001 – New Sur-faces for a new Millenium

[3] Maher I. Boulos, Pierre Fauchais, Emil Pfender: Thermal Plasmas, Fundamentals and Applications, Volume 1, ISBN: 0-306-44607-3

[4] Pierre L. Fauchais, Joachim V.R. Heberlein , Maher I. Boulos : Thermal Spray Fundamentals From Powder to Part, ISBN 978-0-387-28319-7

[5] M. Hugenschmidt: Lasermesstechnik, Diagnostik der Kurzzeitphysik, Springer ISBN 978-3-540-29920-2, 2007

[6] B. Jähne: Digitale Bildverarbeitung, Springer ISBN 3-540-61379-X (1997)

[7] G. Mauer, R. Vaßen, D. Stöver, S. Kirner, J.-L. Marqués, S. Zimmermann, G. Forster, J. Schein: „Improving Powder Injection in Plasma Spraying by Optical Diagnostics of the Plasma and Particle Characterization", J. Therm. Spray Technol., 2011, 20 (1-2), p 3-11

[8] http://www.oseir.com/

[9] Tecnar Automation, (DPV2000), http://www.tecnar-automation.com

[10] S. Zimmermann: „Particle Shape Imaging (PSI) – eine innovative Methode der Partikeldiagnostik bei Thermischen Beschichtungsverfahren", Dissertation, Universität der Bundeswehr München, 2007

[11] S. Zimmermann, J. Zierhut, S. Lange, J.L. Marqués, G. Forster, K. Landes und J. Schein: "Ausgewählte diagnostische Verfahren für den Plasmaspritzprozess, Selected diagnostic methods for the plasma spray process", GTS/DVS, Thermal Spray Bulletin 2/2008

The effect of splats morphology and post-treatment on the micro-structure of sprayed coating

Amirhossein Pakseresht, Armita Shahbazkhan Tehran / Iran

In the thermal spray process, particulate materials can be melted by plasma atmosphere due to its high local temperature from 8700 °C to 15,000 °C. Therefore, the material powders turn into droplets after being melted by injection into the hot flame. Molten droplets are accelerated toward a substrate and form the splats which quickly solidify; finally, the film is formed by pile-up splats. Splat morphology and post treatment can determine the microstructure, mechanical and physical properties of the coating. In this study, $BaTiO_3$ films were deposited onto a mirror polished stainless steel substrates kept at room tempuratue and 500 °C. At the elevated temperatures, the desorbtion of adsorbates and condensate at the substrate surface are the most important factor which change the morphology of the splats, from irrugula- splash morphology to disk-like shape. Splat morphology can determine deposit microstructure and improve the coating properties. The morphology of individual splats and the post treated films were studied using scanning electron microscopy. Results indicated that the porosity in the film produced at room temperature was higher than that in the film deposited on the heated substrates. Also,post heat treatment can improve coating proprties.

1 Introduction

Air plasma spraying (APS) is an important method usually used to produce thick films [1-2]. This method is used in different industries such as gas turbines, automotive and petrochemical industry and can improve engine efficiency . It has been mostly used in producing thermal barrier coatings (TBC) and anti wear films [3-4].
Because of progress in technology and the need of new coating techniques, a broad range of engineering materials are used for electrical, magnetic, optical, and medical applications using an atmospheric plasma spraying (APS) method [5]. Among them, Barium titanate, $BaTiO_3$ is so important. $BaTiO_3$ was the first material used to manufacture dielectric and ferroelectric ceramics [5-6].

$BaTiO_3$ have a perovskite structure, with the chemical formula of ABO_3 [6]. Ferroelectric properties of $BaTiO_3$ result from its three allotropic transformation [5-8]. In general, $BaTiO_3$ shows different behaviors in the form of thin film, single crystal and Bulk [5-8]. Bulk form can be synthesized by solid state reactions [6,8]; however, there are different kind of methods to process $BaTiO_3$ thin film, the most imprtant method to produce thich film was plasma spray method.
By using this method, different substrates with a variety of sizes can be used and also, the thickness of films is easily controlled. Additionally, thermal spray methods are a cost effective, quick and in-situ process [5-8].
The properties of sprayed thick films are comparable with the bulk materials [9].
In general, because of the formation of amorphous phase, cracks, voids and the non-bounded interface between splats that are usual in the plasma spray process the physical and mechanical proprties of sprayed film are less than bulk materials.[5-8]. Xing et al. [10] showed that the electrical properties of APS coatings were one-fifth to one-third of bulk materials. In APS, the physical and mechanical properties of the films are related to the morphology of the splats [9-10]. In general, in sprayed methods the film is built up by impingement of molten droplets onto substrate. Splats are created by flattening and spreading of the impacting droplets. The preheating of the powders and the substrate, roughness and oxidation state of the surface, velocity, spray condition and the angle of the particles on impact can effect on the splat morphology [6]. Among all of them, substrate preheating is an important parameter in APS; There is a transition temperature for all materials, named as (Tt), which above it , the splats morphology of materials on a substrate is changed from an splash/fragmental morphology to a disk-like shape [6].
It should be noticed that by changing the splat morphology, the physical and mechanical properties of the sprayed coating could be improved. Also, post heat-treatment can resuce the poor contact between splats results in improvement in the coating structure.
Previous investigations explained that improve in splat morphology and reduce non-bounded interface between splats play a key role in increasing the microstructure of plasma sprayed film.
In this research, the individual splats were collected on a mirror polished substrate held at two different temperatures, and microstructure of coating before and after heat treatment were studied. Also, the effect of substrate preheating on dielectric properties of plasma sprayed $BaTiO_3$ thick film was investigated.

2 Experimental

$BaTiO_3$ powders used in this investigation were prepared from Alfa Company (product No. 12348). To increase the flow ability of initial powders, they were granulated by air plasma spray method [8]. **Fig. 1.** shows the morphology of granulated powders. The average size range of the granulated powders was nominally −70 µm.
The substrates used in this study were stainless steel (AISI 316L) disks, the external diameter of the samples was 20 mm and the thickness was 0.3 mm. The substrates were grit blasted, before spraying, with alumina particles and the substrate roughness was 3.5 ± 0.5 µm.

The substrates were cleaned in an ultrasonic bath with acetone solution.

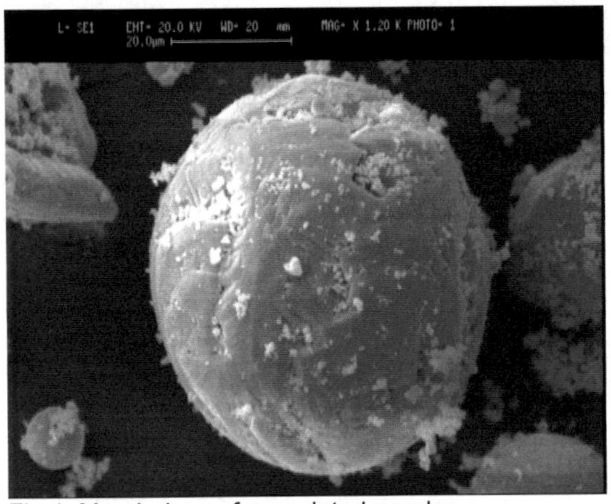

Fig.1. Morphology of granulated powder

In order to study the splat morphology, the substrates should be polished to remove surface roughness. The substrates were prepared by polishing with grit paper. The substrates were finally polished both with 0.3 μm and 0.05 μm Al_2O_3 to reach nano scale roughness.

$BaTiO_3$ film was sprayed by a Plasma Technique AG; Metco 3MB gun in air. Argon gas was the primary plasma gas and hydrogen gas was added as the secondary gas. To deposit the single splat, a stainless steel plate with holes of 1 mm in diameter to permit limited particles to reach the substrate was placed perpendicularly to the plasma jet at the distance of 20 mm ahead of the substrate.
Single splats of $BaTiO_3$ powder were collected on the polished substrates kept at room temperature, 500 °C. Before spraying, the polished substrate was preheated by plasma torch without powder. During spraying, the substrate temperature was monitored using optical pyrometer. Table 1 shows typical atmospheric plasma spray parameters and condition used for the single splat collection.

Table. 1. Experimental spraying conditions

Parameter (value)	Film	Splat collection
Current (A)	500	500
Voltage (V)	60	60
Primary gas, Ar (Scfh)	65	65
Secondary gas, H2 (Scfh)	15	15
Spray distance (mm)	80	80
Powder feed rate (g/min)	20	10

The microstructure of $BaTiO_3$ film (fracture and bottom surface) and splats was analyzed using scanning electron microscopy (VEGA//TESCAN-XMU, Russia and S360 Cambridge).
To remove the apparent amorphous phase and improve the coating properties, the films were heat treated in air at 1050 °C for 3 h with a heating rate of 10 K min–1.

3 Results and discussion

The splats morphologies are showed in **Figs. 2. and. 3.** Accordingly, the splashed / fragmented and disk-shaped splats are the morphologies related to the as-prepared and preheated substrates at 500 °C, respectively. It is clear with increasing the substrates temperature, splats morphology tends to be disk-shaped and also, splashing is decreased. It is clear that the splats morphology completely turned into disk-shape at 500 °C. In this situation, perfectly disk-shaped splats without any projection around the splat could be found on the surface of substrate. The difference in splat morphology was related to the change in surface topography of substrate at higher temperatures. Changes in the surface topology could improve thermal contact resistance between substrate and splats and wetting behaviour of droplets in impact [6]. Cedelle et al. [11] showed that the wettability of droplets could be enhanced by increasing the substrate temperature.

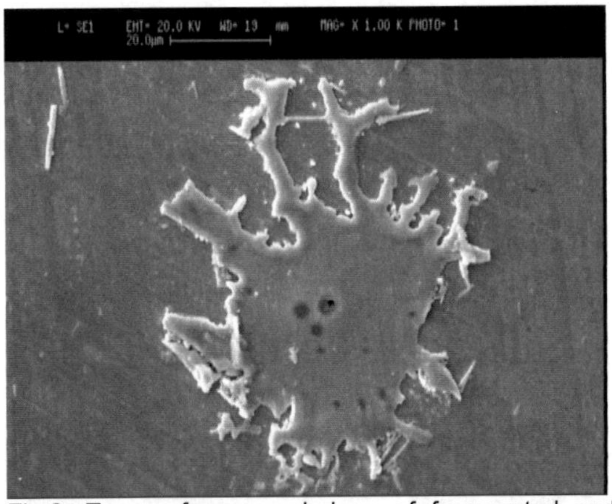

Fig.2. Top surface morphology of fragmented and splashed splat collected at room temperature

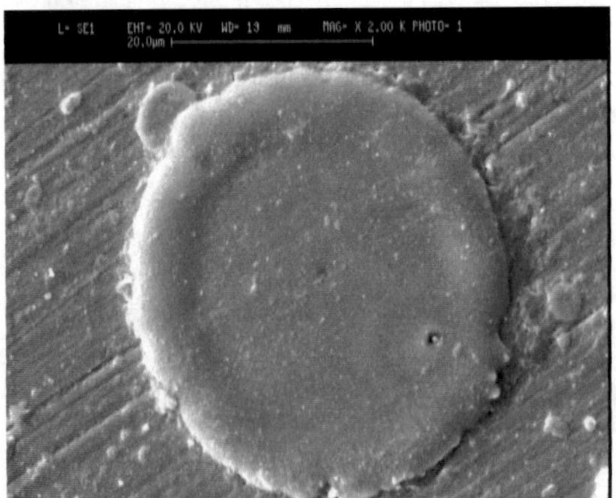

Fig.3.Top surface morphology of disk-like splat collected at high temperature

Accordingly, splashing occurred during flattening of the molten droplet at low substrate temperature because of condensates/adsorbates at the substrate surface. Actually, evaporable substances and the water were desorbed by substrate preheating; thus, the droplets impacted a clean surface and disk-shaped splats formed on substrate. However, when there are evaporable adsorbates on the surface of substrate which can evaporate after the impaction of the high temperature droplets, the fragmented droplets was formed and subsequently, the splashed splat morphology achieved [6].

3.1 Morphology of the fracture surface

Figs. **4.** and **5** show, SEM micrographs of the cross-sectional morphology of the fractured film on as-prepared and preheated substrates, respectively. In the case of conventional substrate, the micrograph showed poor contact between splats. Micro-cracks, non-bounded interfaces between splats and pores were the common defects of the sprayed coatings. Forming the pores were because of some reasons such as, evaporation of the adsorbed substances, bubble formation from the gases, extensive splashing of the droplets upon impact and poor contact between splats [12]. Measuring the porosity value at the cross-section of the coatings showed that fraction of porosity was decreased from 12.7% to 2.5% for the coating on a preheated substrate (500 °C), in comparison with the conventional one. It should be noticed that, for the preheated substrate, the adsorbed substances evaporated, caused reduce in plashing; thus, the porosity value and space between splats were decreased.

The film is formed by pile-up many individual splats and created the lamellar structure with columnar grains is obvious from **Figs.4. and 5.** In the case of pre-heated substrates, the microstructures were more uniform, showing the improved contact between the splats, and the growth of columnar grain through the splat-splat boundary [6]. This kind of microstructure causes the well- bounded lamellar layers that can create more bonding ratio between lamellae in sprayed film. Mechanical and physical properties of the sprayed coating depend on the contact of those splats with the underlying layers and substrate surface. So, coating with uniform microstructure has higher mechanical and physical properties than conventional one [5-7].

In heat treated samples the better contact between the substrate and droplets was a consequence of the evaporation of contaminants and enhanced heat transfer between the molten droplets and substrate, finally changing the splat morphology [5-7].

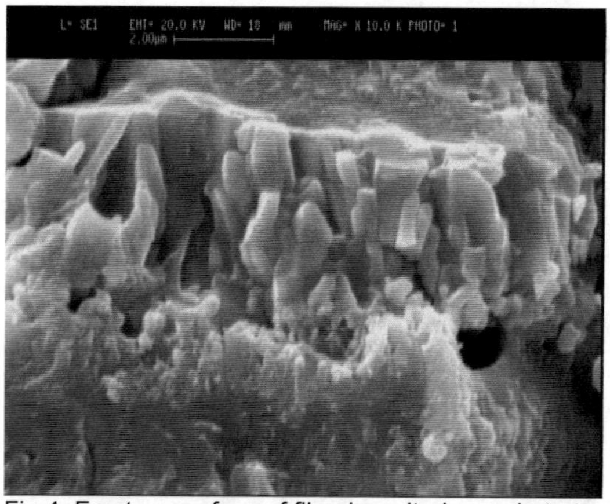

Fig.4. Fracture surface of film deposited on substrate held at room temperature

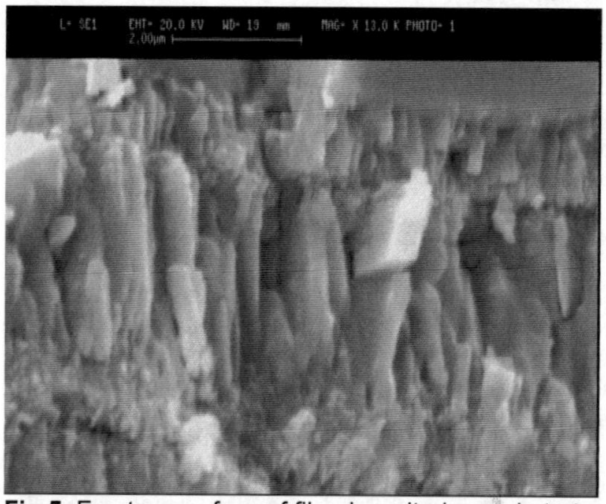

Fig.5. Fracture surface of film deposited on substrate held at high temperature

Figure.**6.** shows cross-sectional morphology of the fractured $BaTiO_3$ film after the heat treatment. As can be seen, some of the defects start to sinter and heal.

The micro-cracks, porosities, splats, and un-melted areas can be clearly seen in this Fig.**6**.

The inter-splat bonding and coherency across the splat boundary improved after heat treatment of the film. The process of heat treatment generally produces bridging interfaces between the splats in a close physical vicinity, decreases the porosity and microcrack content, and increases the contact area between the splats [5]. Moreover, some inter-splat interfaces, porosities, and large voids remain relatively unaffected in the structure.

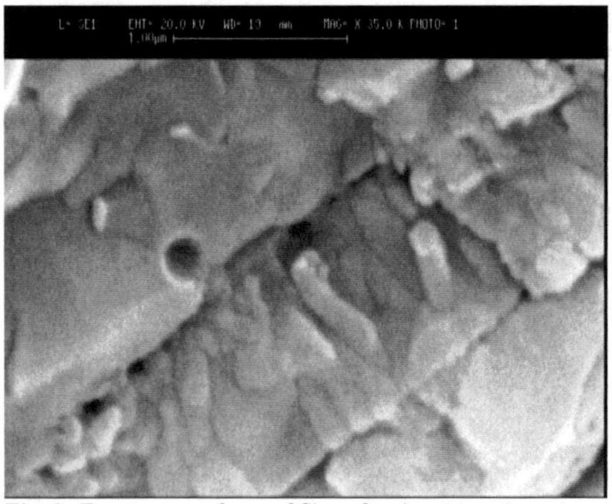

Fig.6. Fracture surface of film after heat treatment

4. Conclusion

Splat morphology can effect on final sprayed coating structure. For some materials there is a transition temperature which above this temperature splat morphology changed from splashed morphology to disk like- morphology.

Refrences:

[1] A.H. Pakseresht, and other: Evaluation of hot corrosion behavior of plasma sprayed thermal barrier coatings with graded intermediate layer and double ceramic top layer. Surface and coating technology 288 (2016), pp. 36/45.

[2] A.H. Pakseresht, and other: Spark plasma sintering of a multilayer thermal barrier coating on Inconel 738 superalloy: Microstructural development and hot corrosion behavior,. Ceramic International, 42 (2016), pp. 2770/79.

[3] A.H. Pakseresht and other: Concept of Advanced Thermal Barrier Functional Coatings in High Temperature Engineering Components." *Research Perspectives on Functional Micro-and Nanoscale Coatings*. IGI Global, (2016) , pp. 396/419.

[4] M. Nejati and other: Microstructural analysis and thermal shock behavior of CSZ, CSZ/Micro Al2O3 and CSZ/Nano Al2O3 plasma sprayed thermal barrier coatings. Surface and coating technology, 282 (2015), pp. 129/38.

[5] A.H. Pakseresht, and other: Effect of heat treatment on the microstructure and dielectric property of plasma sprayed Barium titanate films. International Journal of Materials Research. 107 (2016) pp. 28/34.

[6] A.H. Pakseresht, and other: Effect of splat morphology on microstructure and dielectric property of plasma sprayed Barium titanate films. Applied Surface Science.324 (2015) pp. 797/806.

[7] A.H. Pakseresht, and other:Effect of morphology and nonbounded interface on dielectric properties of plasma sprayed $BaTiO_3$ Coating. Journal of Advanced Materials and Processing. 2 (2014) pp. 25/32.

[8] A.H. Pakseresht, and other: Thermal plasma spheroidization and spray deposition of barium titanate powder and characterization of the sprayable powder. Materials Chemistry and Physics, 173 (2016) pp. 395/403.

[9] S. Sampath: Thermal spray applications in electronics and sensors: past, present, and future, Journal of Thermal Spray Technology 19 (2010) 921/49.

[10] Y. Xing and other: Microstructure development of plasma-sprayed yttria-stabilized zirconia and its effect on electrical conductivity, Solid State Ionics. 179 (2008), pp.1483/85.

[11] J. Cedelle and other: Influence of stainless steel substrate pre-heating on surface topography and on millimeter- and micrometer-sized splat formation. Surface and coating technology 201 (2006) pp. 1373/82.

[12] H.R. Salimijazi, and other: Measuring substrate temperature variation during application of plasma-sprayed zirconia coatings, Journal of Thermal Spray Technology. 16 (2007), pp.580/87.

Influence of bondcoat surface characteristics on lifetime in suspension plasma sprayed thermal barrier coatings

M. Gupta, N. Markocsan, University West, Trollhättan / SE
X.-H. Li, Siemens Turbomachinery, Finspång / SE
B. Kjellman, GKN Aerospace, Trollhättan / SE

Development of TBCs allowing higher combustion temperatures in gas turbines is of high commercial interest since it results in higher fuel efficiency and lower emissions. It is well known that TBCs produced by suspension plasma spraying (SPS) have lower thermal conductivity as compared to conventional systems due to their very fine porous microstructure. Moreover, columnar structured SPS TBCs are significantly cheaper to produce as compared to the conventionally used electron beam – physical vapour deposition (EB-PVD). However, SPS TBCs have not yet been commercialised due to low reliability and life expectancy of the coatings.
Lifetime of a TBC system is significantly dependent on topcoat-bondcoat interface topography. The objective of this work was to study the effect of topcoat-bondcoat interface in SPS TBCs by changing bondcoat spray parameters and bondcoat surface heat treatment. High velocity air fuel (HVAF) spraying was used for bondcoat deposition while axial-SPS was used for topcoat deposition. Same topcoat spray parameters were used for all samples. Lifetime was examined by thermal cyclic fatigue and thermal shock testing. The influence of surface roughness on lifetime has been discussed. The results show that HVAF could be a suitable process for bondcoat deposition to achieve long lifetime SPS TBCs.

1 Introduction

A thermal barrier coating (TBC) system is designed to protect gas turbines from high temperatures and harsh environments. TBCs produced by suspension plasma spraying (SPS) typically result in lower thermal conductivity than commercially used coatings produced by electron beam – physical vapour deposition (EB-PVD) or atmospheric plasma spraying (APS) [1, 2]. The columnar microstructures created by SPS are of high commercial interest as they are more porous than EB-PVD and are significantly cheaper to produce than with the EB-PVD process. However, SPS TBCs are still not favoured by industry due to low reliability and life expectancy of the coatings.
Lifetime of a TBC is significantly influenced by the topcoat-bondcoat interface topography and characteristics [3-5]. Bondcoat surface roughness is a key parameter which needs to be optimised for high TBC lifetime, especially in the case of SPS TBCs as it been shown to significantly influence the topcoat microstructure [6], which in turn affects the TBC lifetime. Preheating of the substrate surface just before spraying is a used practice when the adhesion of the coating needs to be improved. Previous studies showed that higher temperature of the bondcoat surface during spraying could significantly improve the bonding of the topcoat and its properties [7]. Therefore, to enhance the lifetime of SPS TBCs, fundamental understanding of relationships between interface characteristics, microstructure, and lifetime is essential.
The objective of this work was to study the effect of bondcoat spray parameters and bondcoat surface heat treatment on SPS TBC lifetime. The long term goal of this ongoing research work is to develop a bondcoat layer which is most suitable for SPS TBCs in order to achieve high coating lifetime.

2 Experimental

2.1 Sample preparation

In this study, TBC samples with different bondcoat spray parameters and preheating temperatures before topcoat spraying were produced. All bondcoats were sprayed by HVAF with the M3 supersonic HVAF spray gun (Uniquecoat Technologies, USA). This HVAF gun produces not only higher particle velocity, but also higher particle temperature, compared to single-nozzle HVAF spray guns. The bondcoat feedstock material used for all samples was NiCoCrAlY (AMDRY 386, Oerlikon Metco). Two HVAF nozzles and two spray distances were used to produce three set of samples as shown in Table 1, while all other spray parameters were kept the same. The 4L4 nozzle results in higher particle velocities than the 3L2 nozzle, since it is more divergent and transforms temperature to velocity in the divergent part of the de Laval nozzle. Consequently, the particle temperatures are slightly lower when using the 3L2 nozzle.

Table 1. Samples produced in this study

Sample ID	HVAF nozzle	Spray distance	Bondcoat thickness
3L2-350	3L2	350 mm	240 µm
4L4-350	4L4	350 mm	210 µm
4L4-300	4L4	300 mm	240 µm

The topcoats in this study were sprayed by SPS using the Mettech Axial III gun with NanoFeed 350 suspension feeder (Northwest Mettech Corp., Canada). The topcoat feedstock material was yttria stabilised zirconia (YSZ) suspension in alcohol with 25% solid load and D50 = 500 nm (Treibacher AG, Austria). The topcoat thickness was approximately 300 µm. Same topcoat spray parameters were used for all samples.

Just before spraying the topcoats, the bondcoat surface of the samples were preheated by the plasma gun. Two preheating temperatures were investigated in this study – around 200°C and 300°C, which were reached by five and ten strokes respectively of the plasma gun over the samples. The temperature was measured using a infrared laser thermometer which only gives a rough indication of the sample temperature. All three sets of samples showed in Table 1 were sprayed with low and high preheated bondcoats.
The substrate material used in this study was Hastelloy-X. Button shaped substrates in dimensions 25.4 mm diameter x 6 mm thickness were used for microstructure analysis and thermal shock testing, while for thermal cyclic fatigue (TCF) testing, plate substrates in dimensions 50 mm x 30 mm x 6 mm were used.

2.2 Topography

Measurement of 3D topography of the bondcoat surface is very useful as it enables better visualisation of the surface topography with high resolution and better reliability than 2D topography data.
Before spraying the topcoat layer, 3D topography of the bondcoat samples was measured by white light interferometry technique (MicroXAM, ADE Phase Shift Technology, USA) at Toponova, Sweden. The area covered by each scan was around 800 µm x 600 µm. The lateral resolution obtained with this technique was 1.1 µm along the length and 1.3 µm along the width, while the vertical resolution was 1 nm.
The captured data was analysed with the software MountainsMAP ver. 6 (DigitalSurf, France) to post-process the data as well as to compute ISO 25178 and ISO 4287 parameters.

2.3 Microstructure

One sample from each set was prepared for metallographic analysis. The samples were first cold mounted in a low viscosity epoxy resin and then cut using a diamond disc. Afterwards, they were cold mounted again in a high viscosity epoxy resin for grinding and polishing. Automatic polishing was performed using a Buehler Power Pro 5000 equipment. The failed samples after lifetime testing were also prepared in the same fashion.
Sample microstructures were analysed by scanning electron microscopy (SEM) using a Hitachi TM3000 tabletop SEM equipment.

2.4 Thermal cyclic fatigue testing

TCF testing is a common lifetime testing method performed to analyse the performance of TBCs under long exposures at high temperatures. This test allows for significant oxide growth due to bondcoat oxidation at the high temperatures. Failure in TCF testing is mainly driven by crack propagation at the topcoat-bondcoat interface due to formation of thermally grown oxide (TGO) and mismatch in thermal expansion coefficients of different layers in the TBC system. Additionally, topcoat sintering as well as the thermal fatigue of the topcoat induced by thermal cycling also play a significant role in failure.
During TCF testing, the samples were exposed to cycles of firstly heating up to dwell temperature of 1100°C for one hour, and then cooling down to around 100°C in ten minutes with compressed air. The samples were cycled in this way until failure which was determined by 20% spallation of the topcoat. A photograph of the samples was captured after each cycle to determine the failure.
Three samples of each coating set were tested by TCF testing in this study.

2.5 Thermal shock testing

Thermal shock testing is another common lifetime testing method performed to analyse the performance of TBCs under cyclic short exposures at high temperatures and then quick cooling. This test typically does not allow significant TGO growth due to the short duration of cycles. Failure in thermal shock testing is mainly driven by crack propagation near the topcoat-bondcoat interface due to the mismatch in thermal expansion coefficients of the different layers in the TBC system under the cyclic loading conditions. Strain tolerance of the topcoat is a major factor in failure as it determines the mechanical resistance of the topcoat against thermal fatigue induced by thermal cycling.
In this study, thermal shock testing was performed with the burner rig testing (BRT) equipment which has been developed in-house at GKN Aerospace, Sweden. During this test, the samples were first heated with combustion burners for 75 seconds at the front surface which results in topcoat surface temperatures up to 1300°C, and then cooled down to around 450°C with compressed air in 75 seconds. The samples are cooled down with compressed air at the back side during the heating cycle which results in a thermal gradient across the sample. The samples are cycled in this way until failure which is determined by 10% spallation of either complete topcoat or part of the topcoat layer (chipping failure). A short video of the samples is recorded after every four cycles to determine the failure.
Three samples of each coating set were tested by BRT in this study.

3 Results and discussions

3.1 Topography

The bondcoat surface profiles captured using white light interferometry technique in as-sprayed condition are shown in **Fig. 1** while two of the roughness parameters measured are shown in **Table 2**. The roughness parameter *Sa* which calculates the arithmetic mean height in 3D (corresponding to *Ra* in 2D) in Table 2 shows that the surfaces had a similar roughness with 4L4-300 showing slightly higher roughness.

(a)

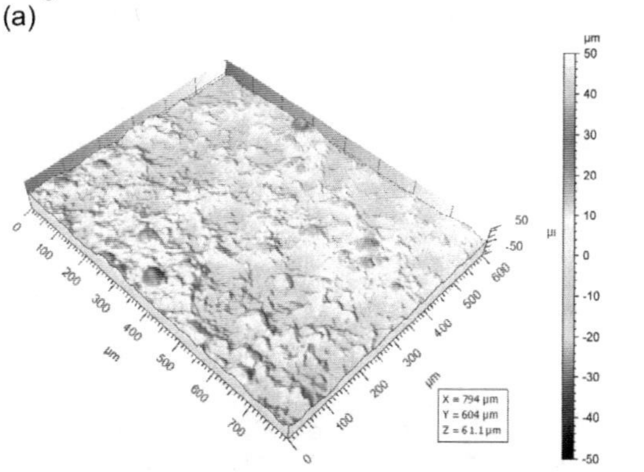

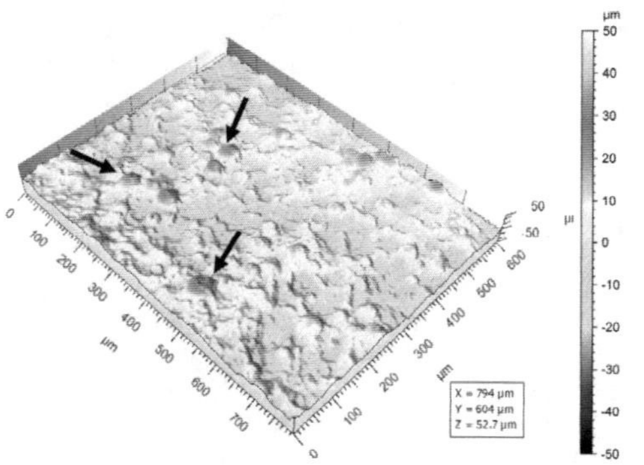

(b)

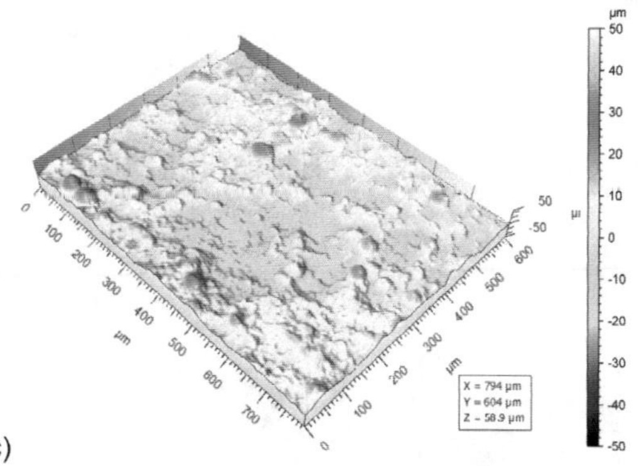

(c)

Fig. 1. Bondcoat surface profiles captured using white light interferometry technique on (a) 3L2-350, (b) 4L4-350, and (c) 4L4-300 samples in as-sprayed state.

However, differences in surface topography can be noticed in Fig. 1 when observing the surfaces carefully. The 3L2-350 bondcoat surface indicates good melting of the particles with more spread out shape and relatively few peaks. On the other hand, the 4L4-350 bondcoat surface showed several spherical shaped peaks as also indicated in Fig. 1b which are typically created due to unmelted / partially molten particles which do not deform under impact. The lateral size of these peaks was around 50-60 µm which corresponds well to the larger size particles in the AMDRY 386 powder which has a powder cut of -63+5 µm. These results corrsepond well to the HVAF spray parameters as the 3L2 nozzle creates higher particle temperature than the 4L4 nozzle resulting in better melting of the particles. The 4L4-300 bondcoat surface showed fewer peaks than the 4L4-350 bondcoat surface due to the shorter spray distance indicating that the particles did not cool down as much due to the shorter spray distance.

Table 2. Bondcoat surface roughness parameters.

Sample ID	*Sa* (µm)	*Spd* (1/mm²)
3L2-350	5.87	131
4L4-350	5.95	165
4L4-300	6.55	148

The parameter *Spd* which is a measurement of the density of peaks shown in Table 2 illustrates that 4L4 had the highest number of peaks per unit area. This quantitative result conforms to the qualitative discussion given above.

3.2 Microstructure

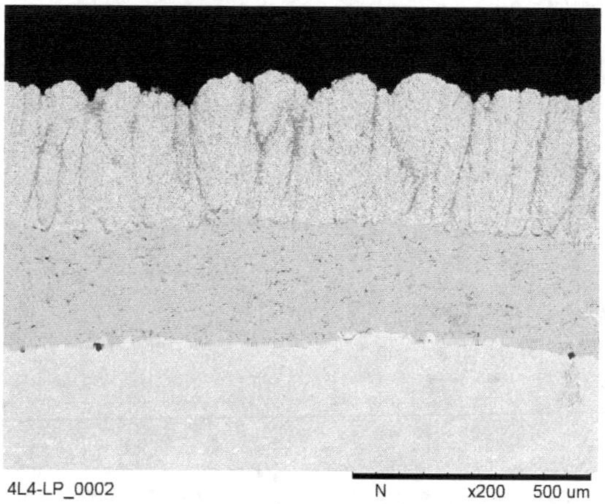

Fig. 2. Microstructure image of 4L4-350 sample with low preheated bondcoat in as-sprayed condition.

SEM microstructure image of the 4L4-350 sample with low preheated bondcoat in as-sprayed condition is shown in **Fig. 2**. The topcoat layer shows a typical columnar type microstructure created by SPS. A similar topcoat microstructure was created in other samples as well, why only one of them is shown in Fig. 2. This was expected as the topcoats were sprayed using the same spray parameters and the bondcoats had similar *Sa* roughness.

The three bondcoat micorstructures sprayed in this study are shown at a higher magnification in **Fig. 3**. All bondcoats showed a dense microstructure with similar porosity levels. Almost no significant differences could be observed among the bondcoat microstructures indicating that the change in HVAF parameters did not affect the microstructure significantly.

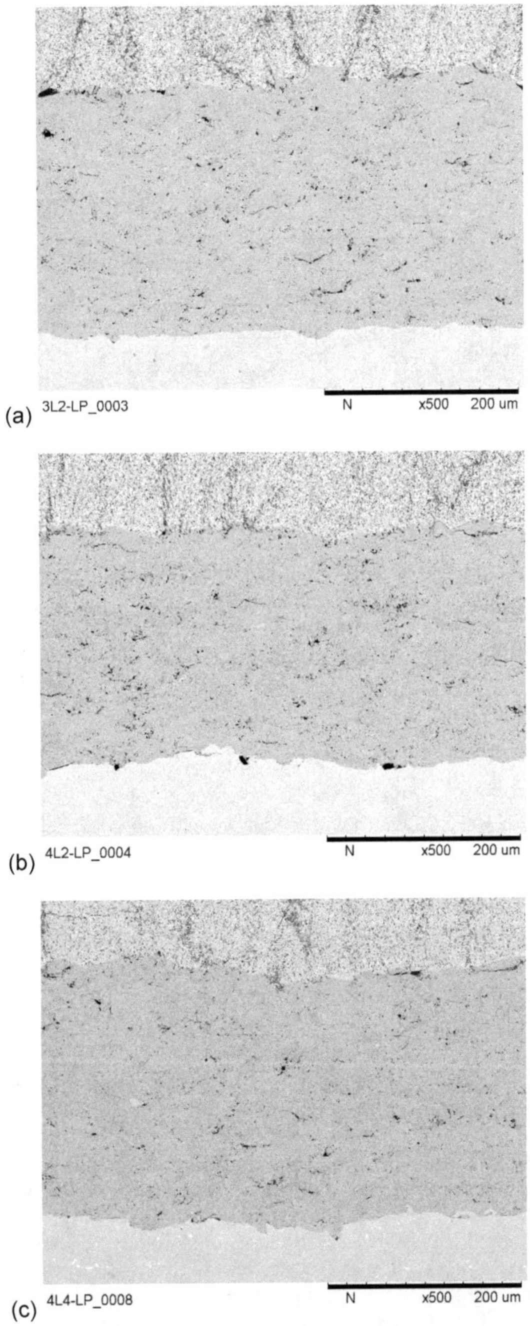

Fig. 3. Microstructure images showing the bondcoats in samples with low preheating in as-sprayed condition (a) 3L2-350, (b) 4L4-350, and (c) 4L4-300.

3.3 Thermal cyclic fatigue testing

The results from TCF testing are shown in **Fig. 4**. The samples showed good TCF lifetime results which fulfilled the industrial requirements for this type of testing. All samples with low preheated bondcoats showed a similar TCF lifetime of around 600 cycles. However, in case of high preheated bondcoats, the average lifetimes were slightly lower with 4L4-350 samples showing the lowest lifetime. These results indicate that the high preheating condition is not beneficial for TCF lifetime in this case as initially expected.

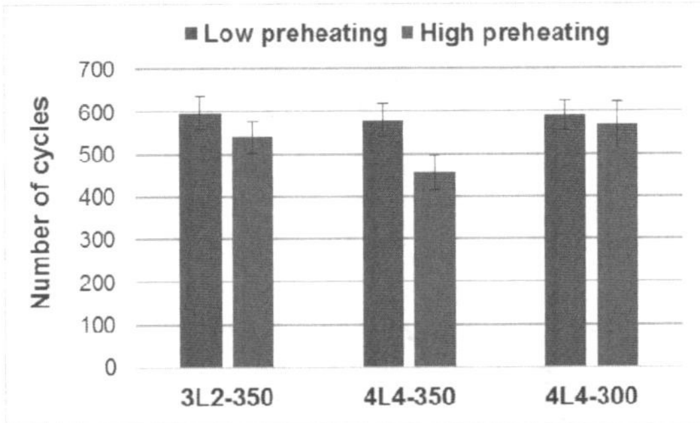

Fig. 4. Thermal cyclic fatigue lifetime results.

The reason for lower TCF lifetime of 4L4-350 could be attributed to the bondcoat surface roughness. It has shown in previous work that unmelted particles on bondcoat surface could result in lower TCF lifetime as these peaks create high stresses in the topcoat near the topcoat-bondcoat interface [5]. Since the 4L4-350 sample exhibited several spherical shaped peaks created due to unmelted particles, higher stresses could have been created in this sample resulting in the lower TCF lifetime.

3.4 Thermal shock testing

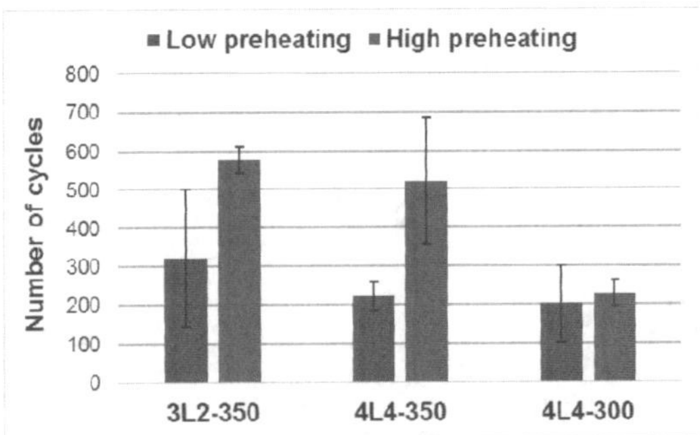

Fig. 5. Thermal shock lifetime results.

The results from thermal shock testing are shown in **Fig. 5**. The samples with low preheated bondcoats again showed a similar lifetime irrespective of the HVAF parameter setting. Among the samples with high preheated bondcoats, 4L4-300 showed the lowest lifetime. The high preheating condition seems to be beneficial for BRT lifetime in the case of 3L2-350 and 4L4-350 samples, though the high standard deviation in the results makes it difficult to draw a clear conclusion.

The general increase of the BRT lifetime after higher preheating might be explanined by the reduction of the residual stresses in the coating both in the bondcoat due to the higher heating as well as in the topcoat due the higher thermal expansion of the substrate before spraying.

4 Summary and conclusions

In this work, three NiCoCrAlY bondcoats were deposited by HVAF by varying the HVAF nozzle and spray distance. Two bondcoat surface preheating temperatures before topcoat spraying were investigated. The effect of bondcoat spray parameters and bondcoat surface treatment on lifetime of TBCs was studied.

The samples fabricated in this study showed high TCF lifetimes. The HVAF parameters created different surface roughnesses of the bondcoats which could have influenced the lifetime of the TBC samples. High preheating condition does not seem to be beneficial for TCF lifetime.

The results indicate that HVAF could be a suitable process for bondcoat deposition to achieve long lifetime SPS TBCs.

5 Acknowledgements

The authors would like to acknowledge the Knowledge Foundation for the funding of this research work. Thanks to Stefan Björklund and Kenneth Andersson at University West for the help with spraying and metallographic preparation respectively.

6 References

[1] A. Ganvir, N. Curry, S. Govindarajan, and N. Markocsan: Characterization of Thermal Barrier Coatings Produced by Various Thermal Spray Techniques Using Solid Powder, Suspension, and Solution Precursor Feedstock Material. International Journal of Applied Ceramic Technology 13 (2016), Issue 2, pp. 324/32.

[2] A. Ganvir, N. Curry, N. Markocsan, P. Nylén, S. Joshi, M. Vilemova, and Z. Pala: Influence of Microstructure on Thermal Properties of Axial Suspension Plasma-Sprayed YSZ Thermal Barrier Coatings. Journal of Thermal Spray Technology 25 (2016), Issue 1-2, pp. 202/12.

[3] A.G. Evans, D.R. Mumm, J.W. Hutchinson, G.H. Meier, and F.S. Pettit: Mechanisms controlling the durability of thermal barrier coatings. Progress in Materials Science 46 (2001), Issue 5, pp. 505/53.

[4] M.P. Taylor, W.M. Pragnell, and H.E. Evans: The influence of bond coat surface roughness on chemical failure and delamination in TBC systems. Materials and Corrosion 59 (2008), Issue 6, pp. 508/13.

[5] M. Gupta, K. Skogsberg, and P. Nylén: Influence of Topcoat-Bondcoat Interface Roughness on Stresses and Lifetime in Thermal Barrier Coatings. Journal of Thermal Spray Technology 23 (2014), Issue 1-2, pp. 170/81.

[6] N. Curry, Z. Tang, N. Markocsan, and P. Nylén: Influence of bond coat surface roughness on the structure of axial suspension plasma spray thermal barrier coatings - Thermal and lifetime performance.

[7] S. Sampath, X.Y. Jiang, J. Matejicek, A.C. Leger, and A. Vardelle: Substrate temperature effects on splat formation, microstructure development and properties of plasma sprayed coatings Part I: Case study for partially stabilized zirconia. Materials Science and Engineering A272 (1999), pp. 181/88.

Micro structure and Characterization of Hydroxyapatite coating (HAP) by Rod Flame Spray Process

Rohit Upadhyaya[1,2], Sharad Shrivastava[1], S.C.Modi[1],A.Modi[1]

[1]Birla Institute of Technology & Science, Pilani,Rajasthan, India-333031

[2]Metallizing Equipment Company Pvt Ltd, Jodhpur India -342001

Abstract

Hydroxyapatite (HAP) is available in powder form for plasma spray coating. HAP powder was fabricated indigenously in the Rod form of diameter 4.17mm. This rod was sprayed with the help of MEC make Rodojet 9810 (flame spray process). Rodojet parameter were optimised for HAP rod. Crystallinity and purity level of HAP rod was measured. XRD and SEM were used to analysed the microstructure of rod and coating. The microstructure, mechanical properties of the coating were investigated, and measure the Ca/P ratio of coating and rod. The micro-hardness and elastic modules were determined by indentation tests and bond strength was determined by tensile test. The results showed that the microstructure, mechanical properties was observed same as in plasma spray process. Porosity was observed more than 15%. The bond strength of coating was observed 20 MPa. Scratch test was done to measure the cohesive strength of coating. This new experiments play an important role for reduction the cost of the HAP powder coating.

Introduction

Different materials are used for implant application in the human body. Among all the material hydroxyapatite is sprayed by plasma spray system. The surface chemistry and surface morphology play an important role. Open and interconnected porosity allows for tissue or bone ingrowth and there should be a mechanical interlocking between the body and implant for better compatibility.Plasma spray is widely employed to enhance the osseintegration of medical implant and prostheses in the number of application, including endoprotheses such as hip,knee , glenohumeral (shoulder), ginglymus (elbow) or dental implant. The method is used in the industrial practise is plasma spray process. Coarse size of powder generally used to coat to maintain the property of the coating. The major problem in plasma spray due to high temperature of flame decomposition of powder. High flame temperature resulting incongruent melting of HAP and formation of amorphous phase.

In this paper we will study the indigenous developed rod of HAP powder. This rod was sprayed by flame spray process. The aim of the study to understand the property achieved by rob flame spray process.

2.Experimental

2.1 Sample preparation

The Rod used to spray was manufactured by indigenous method to maintained the medical condition. The sample were sprayed using Rodojet 9810 torch of MEC,India .The condition include an acetylene and oxygen mixture as a working gas and a standoff equal to 5-6 inch. The 150micron thick coating were deposited onto the steel substrate and grit blasting prior to processing.

2.2 Sample characterization

The HAP rod with fully crystallized structure was prepared by indigenous method developed by MEC,Jodhpur. Initially mean powder of size The phase analysis of rod was measured by Panalytical XRD . Ca/P ratio was also measured by XRD.

Table 1: The characteristic of rod was measure and show in the tabular form with method used to measure.

Analysis	Specifications	Results
Phase analysis **(XRD Method)**	**Mayor phase :** **Hydroxyapatite** **Purete ≥ 98%** **CaO ≤2%** **TCPα ≤2%** **TCPβ ≤2%** **TTCP ≤2%**	**Conform with JCPDS 9.432** **CaO: no detected** **TCPα: no detected** **TCPβ: no detected** **TTCP: no detected**
Ca /P **(XRD Method)**	**1.66 ≤ Ca /P ≤1.72**	**1.67 σ =0.01**
Crystallinity **(XRD Method)**	**Crystallinity >95%**	**109.9%**
Trace element	**As ≤3 ppm** **Cd ≤5 ppm** **Hg ≤5 ppm** **Pb ≤30 ppm**	**As <1.0 ppm** **Cd <1.0 ppm** **Hg <0.05 ppm** **Pb <1.0 ppm**
USP Heavy Metals	**Heavy metals ≤50 ppm**	**< 10 ppm**
Other element	**Fe <100 ppm**	**34 ppm**

	Trace metals (Cu,Ni,Zn,Co,Mn,Cr) <50 ppm	$\sum_{\text{Metal traces}}$<30 ppm
	Other metals (Fe,Cu,W,Si) <500 ppm	$\sum_{\text{Metal traces}}$<250ppm
Particle size distribution	N°101 +45 μm -150 μm ---------------- 50 μm ≤ D10 ≤ 70 μm 80 μm ≤ D50 ≤ 115 μm 100 μm ≤ D70 ≤ 125 μm 120 μm ≤ D90 ≤ 170 μm 135 μm ≤ D95 ≤ 180 μm	Camsizer particle analysis system: D10 = 60.54μm D50 = 98.22μm D70 = 120.20μm D90 = 157.94μm D95 = 177.40μm
Specific weight	N°101 0.45 g/cm³ ρ_{ap} 0.65 g/cm³	P = 0.492 g/cm³

The phase composition of the rod and coating was examined by XRD using CuKalpha irradiation at an accelerating voltage of 40kV and current of 40mA. Before the SEM observation, the coating was mounted on studs with epoxy resin and then ground and polished with 1500# SiC sand paper and a polishing cloth, respectively. The microstructure of the composite coating was analysed by Carl Zeiss EVO 18 SEM.

2.3 Mechanical property tests

The micro-hardness of the cross-sections of the composite coating were determined using Shimadzu HMV hardness instrument. The indentation were made at the middle of the cross-section of the coating as far as possible. The bond strength of coating were evaluated with the conventional tensile bond test (ASTM C-633). Two identical steel bar and a round specimen were used as a set. A high performance FM 1000 adhesive was used to join the two bars on the two surface of the round specimen. After curing , the bond strength were measured using the Instron 4302 (50 kN load cell) universal test machine at a cross head speed of 1 mm/min. Three specimen were tested, and the average was taken as the bond strength of the composite coatings.

Table 2:MICRO VICKERS'S HARDNESS -HV0.1 (ASTM-E384)

1	2	3	4	5	Average (VHN)
212	168	176	172	156	176

Table 3. COATING THICKNESS MEASUREMENT - MICROSCOPIC METHOD (EN ISO-1463)

1	2	3	4	5	Average (µm)
124.3	120.3	121.6	115.6	130.2	122.4

Table 4. Bond Strength Test (ASTM-C633)

Sample ID	Adhesion Strength (MPa)	Remarks
01.	19.40	Coating failure observed.
02.	20.59	
03.	20.71	

Figure 1. Microstructure of HAP coating with more than 12% porosity

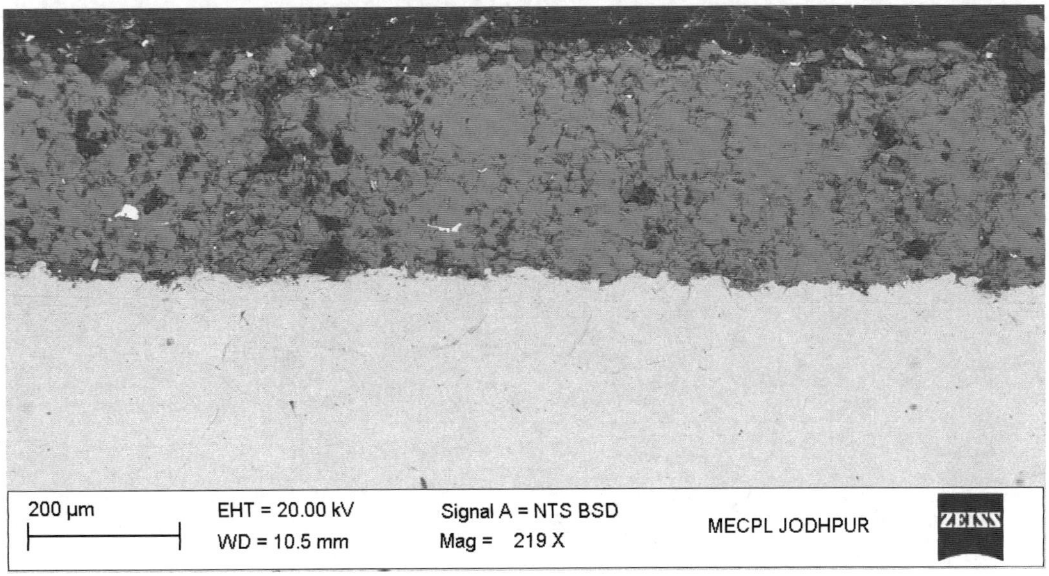

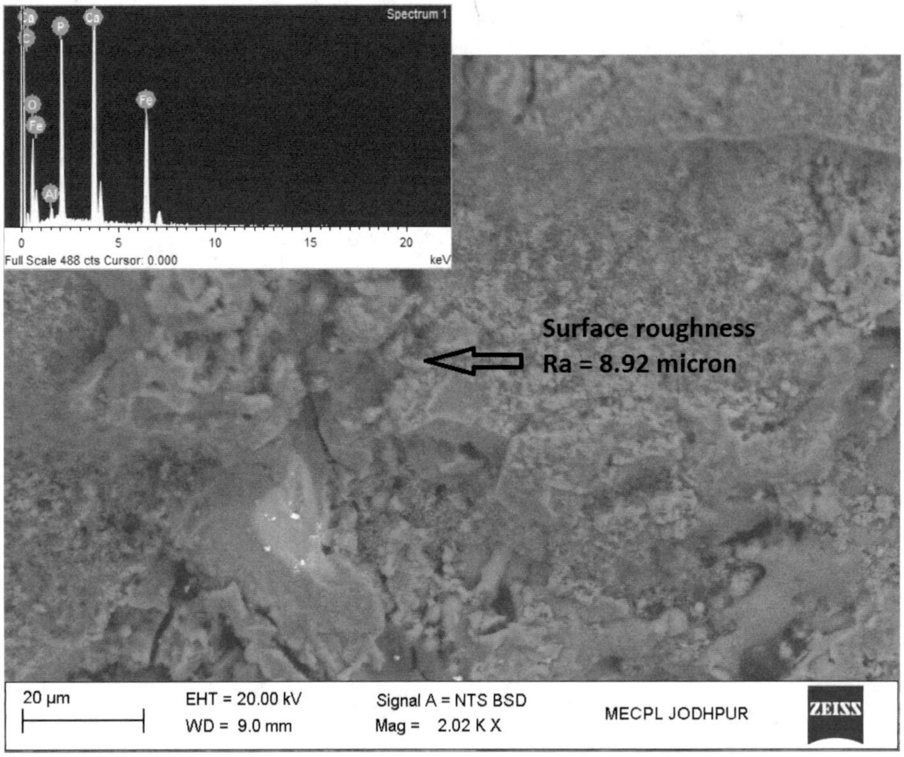

Figure 2.SEM image of as coated also with surface roughness measurement

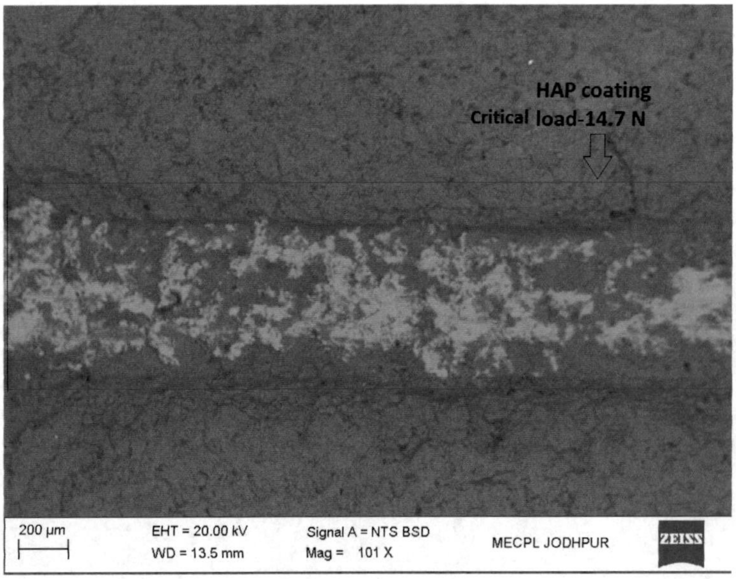

Figure 3: SEM image of scratch analysis at load 14.7 N

3.Results and Discussion

3.1 Microstructure of the coating

SEM micrograph taken from polished cross-section of the composite are displayed in Fig. 1. The thickness of the coating was 100-150 micron. Coating was free from cracks and good interface observed between the coating and the substrates. Thick lamellas coating was observed. Ca is uniformly distributed in the matrix of the coating. Some more new phase was also observed. The Ca/P ratios were very high in coating. The EDS Figure 2 showed that the Ca/P ratio were 6.7. P is nearly disappeared inside the coating, which indicate that due P was lost during the preparation of the coating.

3.2 Mechanical Properties of the coating

The value of micro-hardness and bond strength of the coating are listed in tables. For bond strength 3 sample were tested and got the average bond strength of 20.23 MPa. Scratch test was also performed to measure the check the cohesion and adhesion strength of coating. The scratch test morphologies also demonstrate after testing at 14.7 critical load. The average micro hardness was 176 HV at 300gm load as shown in table 2. The surface roughness as sprayed was Ra 8.92.

4.Conslusions

The microstructural analysis of the Rod spray HAP coating obtained by Rod spray process revealed the major phase was the crystalline HA. The Ca/p ratio was achived with the rod spray process. Same bond strength and porosity was achieved with rod spray process. The bond strength and cohesion strength was found satisfactory. Detail discussion of the result will be discussed in detail paper.

5.Reference

1. Ryan G, Pandit A, Apatsidis DP, Fabrication method of porous metal for the use in orthopaedic application, Biomaterial 2006.

2. Schneider KE, Belashchenko V, Dratwinski M, Seigmann St, Thermal spraying for the power generation components ,Weinheim

3. X.liu,C.Ding, Biomaterials 23(2002) 4065.

4. X.Zhao,X.Liu,C.Ding, J.biomed, Mater. Res A 75A (2005) 888.

5. M.P. Staiger,A.M. Pietak, j.Huadmai, G.Dias, Biomaterial 27 (2006) 1728

Study on thermal shock performance of dense vertically cracked thermal barrier coatings

Lu Xiaoliang[1,2], Ji Xiaojuan[1,2], Yu Yueguang[1,2] , Shen Jie[1,2]
1.Beijing General Research Institute of Mining & Metallurgy, Beijing 100160;
2.Beijing Engineering Technology Research Center of Surface Strengthening and Repairing of Industry Parts, Beijing 100026

Abstract: The ZrO_2-7%Y_2O_3 powder produced by domestic company was used to make dense vertically cracked (DVC) thermal barrier coatings (TBCs) via the High energy plasma spray system. The influence of preheating temperature of substrates, the thickness of single deposition and spray power on the microstructure of the coatings was studied in detail. The porosity of the thermal barrier coatings prepared by Delta gun was 2.8%-7.3% and bonding strength of the coatings was 20-30MPa. The results demonstrated that the TBCs with DVC structure had excellent thermal shock resistance, and the 1100℃ thermal shock lifetime was 206 cycles, which was obviously better than different coating structure by conventional APS (life less than 100 cycles). The mechanism of the failure of the TBCs in the thermal shock process was investigated.

1 Introduction

With the development of Aircraft engine towards high efficiency and high life, the initial temperature of the gas is also increasing, which is beyond the working temperature of the current superalloy system[1]. Thermal barrier coatings are prepared on the surface of the substrate to reduce the substrate temperature, which improves engine efficiency and extends the service life of hot-side components. The TBCs technology have wide application prospect in aviation, aerospace, shipbuilding, electricity and other fields. YSZ coatings prepared by atmospheric plasma process have been widely used in the protection of industrial hot end components. However, thermal barrier coatings in the application process meet premature failure problems. A wide range of research have been studied on this field, including the optimization of spraying process, the development of new thermal barrier coating materials, the improvement of preparation methods of coating structure. Thermal barrier coatings with vertical cracks can release thermal stress and prolong the service life of the coatings in high temperature environment. Some studies have been done on the preparation of a vertical crack structure in the coatings. for example, DU zhong[3] and GUO hongbo[4-5] used APS to prepare- DVC thermal barrier coating; Zhao yun[6] used Plaz Jet high energy plasma spray DVC coatings. But its thermal shock resistance compared to APS was not improved. M. Karger [7] used Triplex II plasma prepared DVC coatings showing horizontal cracks which are peeled off from the center of the sample during the thermal shock test which is not conducive to coating service. In this study, the DVC coating was prepared by High energy plasma spray system, the system can produce faster particle jet (about 250m / s) and higher plasma arc power (50 ~ 70kW), which can make powder melting better and coating density increased. The high plasma energy is beneficial for producing vertical cracks during the spraying process. In this paper, the relationship between the coating parameters and the thermal shock life is studied.

2 Experiment

2.1 Preparation of Coating

The substrate used in the experiment is IN718. The chemical composition of the powder was shown in Table 1. The thermal barrier coating was prepared using the ZrO_2-7% Y_2O_3 powder. The particle size distribution of the powder is 5-38μm, and the morphology of the particles is shown in Fig.1. It can be seen that the powder has a uniform particle size compact structureand are spherical. Such structure is favorable for obtaining a dense TBC. The tested powder has a loose density of 1.8972g/cm^3 and a flowability of 41.26s/50g. The bond coat material was the NiCoCrAlY powder by Beijing Institute of Mining and Metallurgy Research.

Table 1 Chemical compositions of ZrO_2-7%Y_2O_3 powder

Element	Zr	Y	Hf	Ce
wt%	67.22	5.66	1.60	<0.2

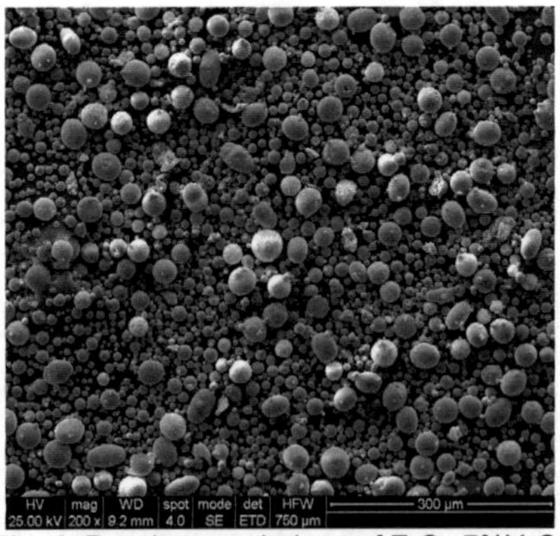

Fig. 1. Powder morphology of ZrO_2-7%Y_2O_3

2.2 Coating characterization

The microstructure of the coating was observed by scanning electron microscope (S4800). The coating porosity was determined by binarizing the cross section of the coating in the backscattering mode and calculating the proportion of the black hole. The coating bonding strength test was carried out by Chinese GB / 8642-2002 standard.

2.3 Spraying details

The bonding coating was prepared by GTV K2 fuel HVOF system. The sample was cleaned with acetone before sandblasting, then blasted with 50-70 mesh white corundum. The formation of clean rough surface enhanced the bonding strength of the coating with the substrate. The substrate material was fixed on a special stainless steel fixture for spray. In the spraying process the substrate is always cold by dry compressed air. The spray parameters are listed in Table 2.

Table 2 Spray parameter of bonding coating

Spraying gun	HOVF(K2)
Kerosene(L/h)	24
Oxygene(L/min)	850
Feed rate(g/min)	80
Spraying distance(mm)	380
Carrier gas Ar(L/min)	7

GTV DELTA plasma spraying system was used to prepare thermal barrier coating. The materials were fixed in the special stainless steel fixture. KUKA six-axis robot was used for moving the Delta plasma gun. The gun moved in S-type mobile and flat spray mode. The gun kept perpendicular to the substrate. The spraying parameters are shown in Table 3.

Table 3 Spraying parameters of the top coatings

	Power (Kw)	Preheat passes(lap)	Sediment thickness(μm/ cycle)
1#	60	4	100
2#	60	8	100
3#	60	16	100
4#	60	4	200
5#	60	4	300
6#	50	4	100
7#	55	4	100

2.4 Thermal cycling details

Thermal cycling was carried out by an automatic water-cooled thermal shock test machine designed by Beijing General Research Institute of Mining and Metallurgy. The preset temperature was 1100 ℃, holding time was 5min, water-cooling was 2min. Visually, 20% of the coating peeling area is considered coating failure

3 Results and analysis

3.1 Effect of preheating temperature on vertical crack coating

Various spraying passes (4, 6, 8) provided different substrate pre-heating temperature (230, 420, 760℃). The morphology of the coatings for the different pre-heating temperatures is shown in Fig. 2. The porosity was 2.8, 3.4, 4.9%. From the cross-sectional morphology vertical cracks can be seen at preheating temperature of 230℃, the average bond strength was 24.52MPa. When the preheating temperature was 420 ℃, the horizontal crack appeared around the vertical crack, and the bond strength was 19.05MPa. Compared to the 1 # parameter, the bonding strength was decreased. Horizontal cracks in the coating was not conducive to coating service, resulting in the service process along the horizontal cracking and peeling. When the preheating temperature was 760℃, horizontal cracks were linked up which was extremely harmful, and the bond strength was 9.05MPa. Such conduction transverse crack was not conducive to coating service.

The formation of the vertical cracks was due to the tensile stresses in top coat[6]. During the spraying process, the powder particles were heated by the plasma and impacted onto the substrate surface, and deformed and cooled to form splats. The splat-made coating shrunk when it was cooled, and horizontal tensile stresses were produced. When the tensile stress exceeded the bonding threshold of the material, vertical cracks were formed and the stresses were locally released. Thermal stresses can be produced when temperature difference existed between the substrate and the coating. Temperature change caused material property variations of thermal expansion coefficient, Young's modulus, and poison value. The thermal stress could be related to the temperature change and be influenced by the pre-heating temperature. As the temperature of the substrate increased, thermal expansion coefficient increased which caused the enhance of thermal stresses in YSZ coatings[9-10]. For Delta plasma spray, long and wide plasma stream can be obtained which was 2.5 (70 mm long) and 2 times (25 mm wide) than conventional APS. The enhanced energy density caused quicker temperature increase of the substrate and a similar temperature can be sustained as the coatings. When the substrate temperature was above 700℃, the thermal stress between the coating and the substrate became larger, resulting horizontal cracks in the coating. By the width of the vertical crack in the Fig.2(c), with the increase of coating thickness, on the one hand, due to the tensile stress of the coating, on the other hand, the deformation of the matrix also caused the crack tip to expand. So the high-energy plasma spraying process was equivalent to the thermal shock of the coating. In the coating process local temperature changes will occur in the coating, which made the coating not strong. The results showed that not only the vertical crack but also the horizontal crack was produced, and the density of the crack increased with the temperature increasing.

Through the above comparative analysis, this study was different from Guo Hongbo[5] using APS preheat temperature at 760 ℃ sprayed out of the vertical cracks coating, also different from M. Karger [8] using Triplex II to prepare a vertical cracked coating at a preheating temperature of 500 ℃. In the present study, when the preheating temperature of the substrate was 230 ℃, the coating have vertical crack and no transverse crack. So the follow-up process optimization experiment used 230 ℃ for the substrate preheating temperature.

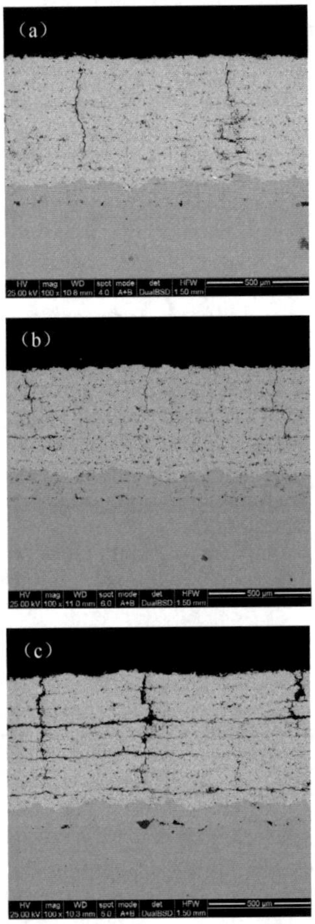

Fig.2. Cross-section of YSZ coating under different preheat temperature

3.2 Influence of single deposition thickness on vertical crack coating

Fig. 3. shows the cross-section of DVC coatings with different deposition thicknesses. Fig.4 (a), (b) and (c) are the cross-sectional SEM images of coatings with single deposition 0.1mm, 0.2mm and 0.3mm thicknesses respectively. The porosities were 2.8, 4.6 and 5.1 respectively. It can be seen that vertical cracks appeared in the coatings by three kinds of parameters. However, with the increase of the number of single deposition, the bonding strength of the coating showed a decreasing trend. The average bonding strength of the 1 #, 4 # and 5 # was 24.52MPa, 16.01MPa, 14.23MPa. Horizontal cracks occur in the coatings with the increasing number of single deposition cycles. In addition, the horizontal cracks increased and the vertical cracks were connected together with the increase of deposition thickness. For DVC coatings, tensile stress generated in the coating played a dominant role during the spraying process and the effect of thermal stress was weaken. This phenomenon also demonstrates that deposition thickness of coatings needs to be controlled in the spraying process. The temperature of coating will rise through three-anode high-energy plasma deposition which resulted in an increase of thermal stress and tensile stress of coating. Horizontal and vertical cracks in the coatings increaseed because of uneven relief of stress.

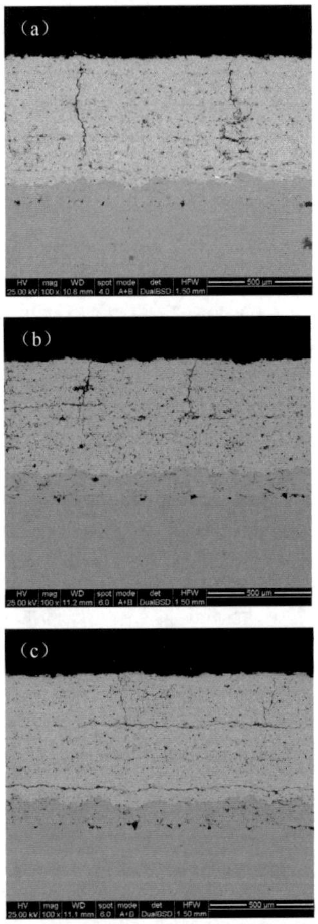

Fig.3. Cross-section of YSZ coatings with different sediment thickness

3.3 Influence of spraying power on vertical crack coating

As is shown in Fig.4.(a),(b),(c), there are the cross-sectional view of 6#,7#,1# under different spray power parameters, that was 50KW, 55KW, 60KW respectively. The porosity has been measured by microstructure gray scale method of the cross section, and the result was 7.3, 5.4, 2.8% respectively. The bonding strength was 22.1MPa, 30.03MPa, 24.52MPa and the crack density was 2 Crack/mm, 3.5 Crack/mm, 1.7 Crack/mm respectively. The vertical cracks were observed in the cross sectional under three kinds of spray parameters. The coating porosity of 50KW was obviously higher than the 60KW, this may due to the reason that the cracks can not completely penetrate the coating when the spray power was relatively lower, and the crack density in the coating was low. With the increase of spray power, the number of the crack exhibited firstly increase and then decrease and the density of the coatings were gradually improved. This confirmed the better melting state of the power when increase the spray power. But, with the much higher spray power, the number of the cracks was decreased. If the thermal stress was too high, this may affect the generation of the coating tensile stress and then reduce the coating crack density.

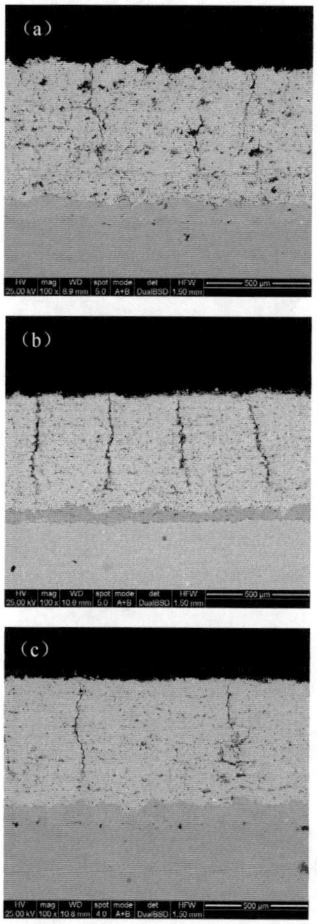

Fig.4. Cross-section of YSZ coating under different power

3.4 Thermal shock resistance of coatings

Fig. 5. shows the thermal shock lifetime of YSZ coatings deposited at different spray parameters. It could be found from Procedure 1#, 2# and 3# that the life decreased with the rising of the preheating temperature. Especially when using Procedure 3 #, the transverse cracks occurred in the coating, which leaded to earlier delamination spallation and the life decrease. Comparing the coatings deposited by procedure 1#, 4# and 5#, it can be seen that the thermal shock life of the coatings presented the decrease tendency with the increase of the thickness of single pass sprayed coating. It was chiefly because the transverse crack developed inside the coating when increasing the deposition thickness, which leaded to the coating failure. Besides, crack density of the coating showed a rising trend when increasing the spray power as shown in procedure 6#, 1# and 7#, which were 1.7 Crack/mm, 2 Crack/mm and 3.5 Crack/mm, respectively. CombiThe thermal shock life of coatings increased with the rising of crack density. All coatings retained their original appearance and the failure of the coatings mainly occurred at the TGO layer near the YSZ topcoat. The DVC existed in the coating extended vertically from the surface to the inside. Studies showed that the DVC was helpful to avoid the surface stress concentration. After 200 thermal shock test, most DVC ended at the interface between TGO and bond coat. In addition, the interface between bond coat and TGO/matrix remained intact, proving that the whole DVC coatings have excellent thermal shock resistance. Fig. 6. is for the 7 # sample thermal cycle of the macro-morphology. It can be seen that the failure process of the thermal barrier coating with vertical cracks was first formed on the surface of the substrate mesh cracks. With the increase in the number of thermal shock, coating failed from the edge to the internal point-like flake, showing excellent heat seismic performance, thermal shock frequency reached 206 times.

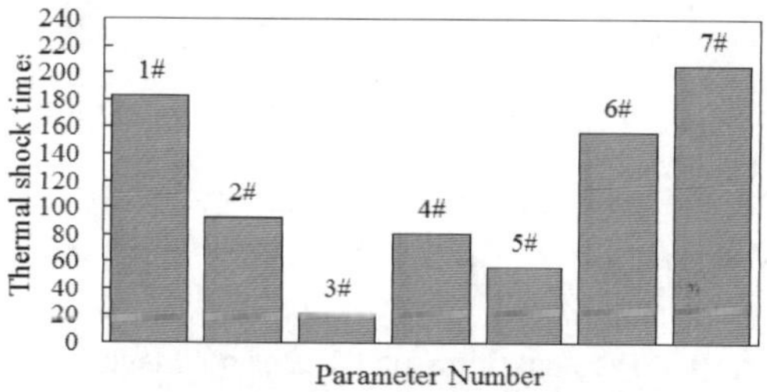

Fig.5. Different spraying parameters of YSZ coating thermal shock curve

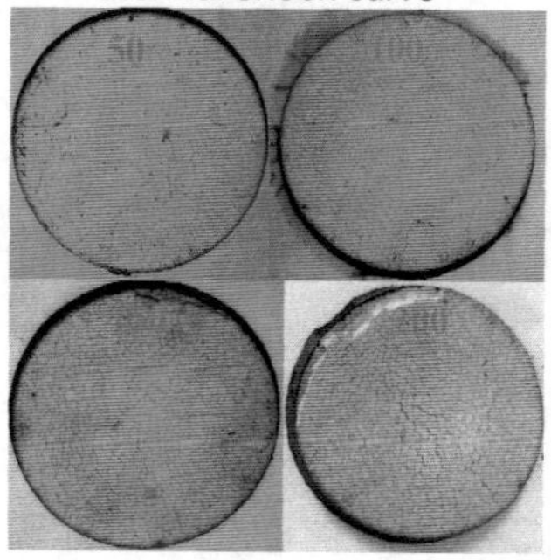

Fig.6. Macroscopic Topography of Thermal Cycles of Sample7#

4 conclusion

1) the preheating temperature of the substrate had a great influence on the structure of the coating. For the three-anode high-energy plasma spray gun, the high preheating temperature promoted the coating to produce lateral cracks to reduce the service life. Preheating temperature of 230℃ was conducive to the formation of vertical crack coating.

2) With the increase of the thickness of the single pass, the coating will produce transverse cracks, the transverse crack generation will reduce the bonding strength of the coating, but not conducive to the thermal shock life of the coating.

3) With the increase of spraying power, the number of vertical cracks in coating increased and then decreased. The thermal shock life of the coating increased with the number of vertical cracks in the coating, and the thermal shock life of the coating increased obviously.

4) The failure process of the thermal barrier coating with vertical cracks was that a network crack on the surface was first formed, and the failure of the coating was exfoliated from the edge to the interior with the increase of the number of thermal shocks. The DVC TBCs showed excellent thermal shock resistance.

References

[1] Guo Hongbo, Gong Shengkai, Xu Huibing. Research Progress of Advanced Aero-engine Thermal Barrier Coating Technology[J]. Chinese Journal of Materials Research, 2009, 28(9/10):18-26.

[2] Miller R A. Current status of the thermal barrier coatings an over view[J].Journal of Thermal Spray Technology, 1987, 21(01):22-28.

[3] R.B. Heimann, Plasma Spray Coating, 2nd Edition Wiley-VCH Verlag GmbH&Co, KGaA, Weinheim, 2008.

[4] Du Zhong. Study on Long-life Thermal Barrier Coating by Plasma Spraying[D]. Beijing Institute of Technology, 2009.

[5] GUO H B, Vaβen R, Stöver D. Atmospheric Plasma Sprayed Thick Thermal Barrier Coatings with High Segmentation Crack Density[J]. Surface and Coatings Technology, 2004, 186:353-363.

[6] GUO H B, KURODA S, MURAKAMI H. Microstructures and Properties of Plasma Sprayed Segmented Thermal Barrier Coatings[J]. Journal of American Ceramic Society, 2006, 89(4):1432-1439.

[7] Zhao Yun. Study on Vertical Barrier Thermal Barrier Coating by High Energy Plasma Spraying[D]. Beijing Institute of Technology,2015.

[8] M. Karger *, R. Vaßen, D. Stöver. Atmospheric plasma sprayed thermal barrier coatings with high segmentation crack densities: Spraying process, microstructure and thermal cycling behavior[J]. Surface and Coatings Technology, 206 (2011) 16-23.

[9] Zhao Wenguang, Wang Zhefei, Zhang Huifeng. Thermal expansion of ceria-doped yttria-stabilized zirconia[J]. Inorganic salt industry, 2011,43(3):21-26.

[10] Xiao Qiling, He Hongbo, Shao Shuying. Effect of Deposition Temperature on Residual Stress of Yttria - Stabilized Zirconia Thin Films[J]. Chinese Journal of Optics2008, 28(5):1007-1011.

Improving erosion resistance of plasma-sprayed ceramic coatings by elevating the deposition temperature based on the critical bonding temperature

Shu-Wei Yao, Guan-Jun Yang, Cheng-Xin Li, Chang-Jiu Li*
State Key Laboratory for Mechanical Behavior of Materials, School of Materials Science and Engineering, Xi'an Jiaotong University, Xi'an/China / E-mail: licj@mail.xjtu.edu.cn

Inter-lamellae bonding within thermal sprayed coatings is one of the most important factors influencing the properties and performance of coatings. It has been revealed that there exists a critical bonding temperature for a molten ceramic splat to form the bonding to the same splat surface. The erosion behaviors of thermal sprayed coatings are significantly influenced by the interface bonding between lamellae. In this study, the erosion behavior of plasma-sprayed TiO_2, Al_2O_3 and YSZ coatings deposited at different deposition temperatures was investigated. The cross section of plasma sprayed coatings was characterized by the scanning electron microscope. It was revealed that the coatings deposited at room temperature exhibit a typical lamellar structure with numerous unbonded interfaces, whereas the coatings prepared at the temperature above the critical bonding temperature present a dense structure with well bonded lamellae. The erosion rate significantly decreases with the improvement of interface bonding. In addition, the erosion mechanisms of the conventional coatings and the lamellae well bonded coatings were further discussed. The unbonded interfaces act as pre-cracks accelerating the erosion of plasma-sprayed coatings. Thus, controlling inter-lamellae bonding based on the critical bonding temperature is conducive to the improvement of erosion resistance of plasma-sprayed ceramic coatings.

1 Introduction

Thermal spray coatings are built up through the successive deposition of individual fully molten or/and semi-molten particles. These coatings have been widely used in many industrial fields, such as transport, energy, materials processing, biomedical and electronic applications [1]. A survey into the market shares of different processes in northern American for the deposition of advanced ceramic coatings revealed that about a two-third share is taken by thermal spraying [2], Thus, thermal spray plays the most important role in the production of advanced ceramic coatings.

Generally, the coatings deposited following routine approach exhibit a typical lamellar structure with numerous pores, micro cracks and unbonded lamellar interfaces [3-5]. With the systematically quantitative study on the microstructure of thermal sprayed ceramic coatings, Ohmori and Li found that the maximum bonding ratio at the interface between lamellae is about 0.32, when the substrate temperature is kept at a low level by cooling [6]. It has been clearly demonstrated that inter-lamellar bonding dominates the properties and performance of coatings, such as mechanical, thermal and electrical properties [7-9]. Thus, the coatings with different levels of inter-lamellar bonding are required to fulfil the versatile performance requirements in services. For example, thermal-sprayed YSZ coatings have been widely applied as a thermal barrier coatings (TBC) and an electrolyte layer in solid oxide fuel cell (SOFC). Generally, for a TBC, a porous microstructure with a low lamellar bonding ratio is effective to reduce thermal conductivity of and subsequently enhance the thermal barrier effect [10]. However, served as an electrolyte layer, a sufficiently dense YSZ coating with a high bonding ratio is required to acquire high ionic conductivity [11]. In addition, for most wear and corrosion protection applications, dense coatings with well boned lamellae are always expected, since increasing lamellar bonding is conducive to improving the coating performance [1]. Thus, the development of an effective approach to control the formation of interface bonding between lamellae is highly desirable for thermal spray coatings.

The velocity, size and temperature of sprayed particles are utilized to control the microstructures, adhesion and cohesion of thermal spray coatings [12]. However, it is difficult to further significantly increase the lamellar bonding ratio through adjusting the spraying parameters [13]. Gray et al. [14] developed a continuous columnar structure of 8YSZ (8 mol% Y_2O_3-ZrO_2) coating by controlling the deposition temperature. They revealed that columnar grains extend from one splat into an adjacent splat in the continuous-columnar-structured YSZ coatings. Up to now, it has been widely confirmed that increasing deposition temperature can significantly increase the inter-lamellar bonding in thermal sprayed coatings. In our previous studies, it was revealed that there exists a critical bonding temperature (T_c) for plasma-sprayed molten ceramic splats, and only when the deposition temperature is higher than T_c, can the splats form bonding to the substrate with the identical compositions to splats [15,16]. Since thermal spray coatings are built up by the successive deposition of splats, it can be considered that the plasma spray coatings with well bonded lamellae can be deposited at the critical spraying conditions for splats.

Erosion is a general phenomenon caused by the successive impact of particles at a certain speed and impact angle, which is a major life-limiting issue resulting in degradation of engineering devices. It has been reported that the erosion of the bulk materials and coatings are influenced by the fluid flow conditions and the properties of the impacting particles and target [17-19]. For thermal spray coatings, microstructure, especially the inter-lamellae bonding, is one of the most important factors influencing the erosion behavior of coatings. Li et al. [20] found that the erosion resistance of plasma-sprayed Al_2O_3 coating is proportional to the inter-lamellae bonding ratio. Levi *et al.* [21] and Davis *et al.* [22] revealed that the thermal spray TBC and protective coatings with good lamellar bonding and low porosity exhibit excellent erosion resistance. In addition, Hearley *et al.* [23], Wang *et al.* [24] and Yang *et al.* [25] demonstrated that the erosion resistance of both FeAl and NiAl coatings is improved with the improvement of lamellar bonding strength. Essentially, under the high velocity impact of erosion particles, the unbonded interfaces in thermal spray coatings act as pre-cracks and tend to propagate along the limited bonded interface sections. As a result, the erosion of thermal spray coatings occurs through the spalling of lamellae one by one. Thus, it is essentially important to enhance the bonding ratio between lamellae in order to improve the erosion resistance of thermal spray coatings.

In this study, the typical ceramic coatings, i.e. Al_2O_3, TiO_2 and YSZ coating, were prepared by plasma spray. The coatings with different microstructures, especially inter-lamellae bonding, were obtained by controlling the deposition temperatures based on the critical bonding temperature. The erosion behavior of different ceramic coatings was investigated.

2 Experimental

Typical ceramic materials, including TiO_2, Al_2O_3 and YSZ, were used to prepare coatings on the blasted Q235 steel substrate. The fuse-crushed TiO_2 powder (Sulzer-Metco, USA) with a particle size range from 10 to 45 µm was employed in this study. The fuse-crushed Al_2O_3 powder (Sulzer-Metco, USA) exhibited a particle size range from 22 to 45 µm. The fuse-crushed 8YSZ powders (Fujimi, Japan) showed a particle size between 5 and 25 µm.

A commercial plasma spray system in the 80 kW class (Jiujiang, China) was utilized to deposit ceramic coatings. The spray parameters employed in this study were displayed in Table 1. In order to precisely control the deposition temperature, a copper heating stage was assembled to preheat the substrate from its back surface, and an infrared pyrometer (RayRPM30L3U, Raytak, America) was used to monitor the surface temperature of the sample. For the coatings deposited at room temperature, a compressed air jet was used for cooling from both the back and the surface of substrate. The measurement of the substrate surface temperature by thermocouples revealed a maximum value of 100 °C during the spray process. Thus, in this case, the deposition temperature is regarded as 50 °C.

In this study, the surface morphology and the cross-section microstructure of both the as-sprayed coatings and the eroded coatings were characterized by scanning electron microscope (SEM, MIRA 3 LMH, TESCAN, Czech). Erosion test was carried out using an erosion tester, which is similar to the ACT-JP. Angular alumina particles with an average grain size of 250 µm were used as impacting particles. The pressure of the compressed air was fixed at 0.3 MPa, in order to accelerate the erosive particles. The erosion test was carried out at the impact angle of 90°, which corresponds to the maximum erosion loss for ceramic coatings. The mean erosion rate was obtained by the average value of five tests.

Table 1 Spray conditions for TiO_2 splats

Parameters	TiO_2	Al_2O_3	YSZ
Arc power (kW)	36	42	42
Arc current (A)	630	700	700
Arc voltage (V)	57	60	60
Primary plasma gas (Ar/slpm)	45	50	50
Secondary plasma gas (H_2/slpm)	4.5	6	6
Powder feed gas (N_2/slpm)	3	3	3
Spray distance (mm)	80	80	80
Traverse speed of torch (mm/s)	600	600	600
Deposition temperature (°C)	50, 150, 300, 400, 500	50, 300, 400, 500, 600	150, 650

3 Results and Discussion

3.1 Effect of deposition temperature on the microstructure of plasma spray ceramic coatings

In our previous studies, it has been revealed that the critical bonding temperatures for molten TiO_2, Al_2O_3 and YSZ particles to form bonding to the substrates with the same materials as sprayed material are 150, 300 and 650 °C, respectively [15,16]. Figs 1, 2 and 3 show the cross-sectional microstructure of plasma-sprayed TiO_2, Al_2O_3 and YSZ coatings deposited both at the room temperature and at their critical bonding temperature. As shown in Fig 1(a), the TiO_2 coating prepared at room temperature exhibiting a typical lamellar structure with numerous pores, micro cracks and unbonded interface. When the deposition temperature increases to 150 °C, all the interfaces between lamellae are well bonded together. Although micro pores still can be observed, as shown in Fig. 1(b), few unbonded interface sections can be found in the TiO_2 coating.

The similar phenomenon was observed in the deposition of Al_2O_3 and YSZ coatings, as shown in Figs. 2 and 3. With the increase of deposition temperature, the inter-lamellae bonding is significantly improved. Compared to the Al_2O_3 and YSZ coatings, TiO_2 deposit presents a better bonding between lamellae when they were deposited at room temperature. This is because TiO_2

has a low critical bonding temperature, which is very close to the maximum substrate temperature during the plasma spraying at room temperature. In addition, the YSZ coating deposited at about the critical bonding temperature exhibits more unbonded interfaces, compared to the Al_2O_3 and TiO_2 coatings. This is because that, owing to the high critical bonding temperature of YSZ splat, it is difficult to keep the substrate temperature above the critical bonding temperature all the time. Thus, it is confirmed that the bonding between lamellae in plasma-sprayed ceramic coating can be adjusted by controlling the deposition temperature based on the concept of a critical bonding temperature.

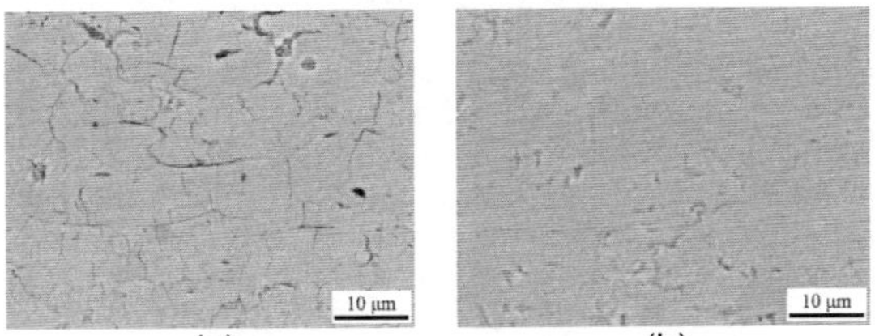

(a) (b)

Fig. 1 Cross-sectional microstructure of plasma-sprayed TiO_2 coatings at the deposition temperature of (a) 50 °C and (b) 150 °C

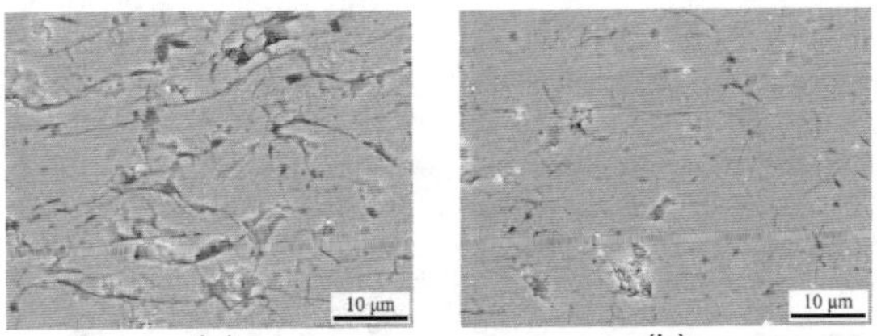

(a) (b)

Fig. 2 Cross-sectional microstructure of plasma-sprayed Al_2O_3 coatings at the deposition temperature of (a) 50 °C and (b) 300 °C

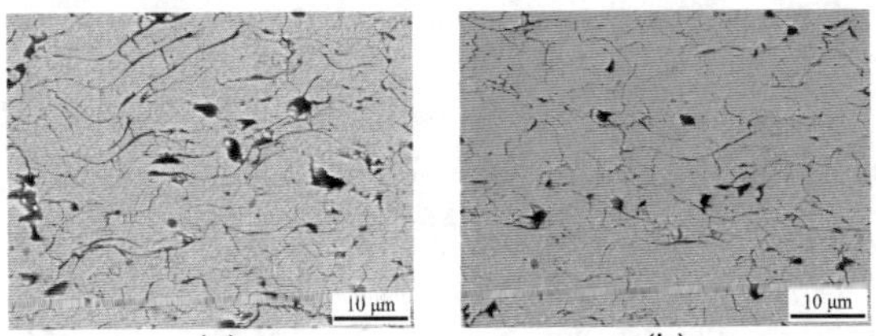

(a) (b)

Fig. 3 Cross-sectional microstructure of plasma-sprayed YSZ coatings at the deposition temperature of (a) 50 °C and (b) 650 °C

3.2 Erosion behavior of plasma spray ceramic coatings

Figure 4 shows the surface morphologies of the eroded plasma-sprayed Al_2O_3 coatings deposited at the temperate of 50 °C and 300 °C, respectively. As shown in Figs. 4(a) and (b), the coating deposited at room temperate was eroded off in the form of splat fragments, when the high velocity particles impact on it. Thus, the eroded Al_2O_3 coating exhibits a similar surface morphology to that of the as-sprayed coatings. These results were confirmed by the cross section morphology of the eroded coating. As shown in Figs. 5(a) and (b), microcracks propagate along the lamellar interface, and the erosion of coating took place through the spalling of lamellae one by one. For the Al_2O_3 coating with well bonded lamellae, the microcracks propagate across lamellae at the impact of erosion particles, as shown in Fig. 4(d). Observed from Fig. 4(e), it was found that the Al_2O_3 coating was broken into pieces, which is similar to the erosion feature of ceramic bulk.

As shown in Fig. 4(f), erosion furrows were obviously observed on the surface of well bonded Al_2O_3 coating. Generally, furrows can be formed by the impacting particles with a low impact angle. In this case, it may be caused by the secondary erosion, which will increase the loss of coatings. Since the erosion test of the two kinds of coatings are carried out at the same conditions, it was suggested that secondary erosion would also take place during the erosion test of limited bonded coatings. However, as shown in Fig. 4(a) and (b), furrows are not found on the surface of eroded coating. It may be because that the impact particles during the secondary erosion cause the spalling of splat fragments instead of furrows on the coating deposited at room temperature. These results can be confirmed by the surface morphology at the periphery of erosion crater of plasma-sprayed Al_2O_3 coatings. As shown in Fig 4(g), the Al_2O_3 coating of limited lamellar bonding exhibits a lamellar structure after the erosion test, whereas the well bonded coating presents a typical furrow morphology.

Figure 5 shows the cross-sectional microstructure of the eroded coatings. As shown in Fig. 5, cracks propagation just happens on the surface of coatings. For the coatings with limited lamellar bonding, the spalling of splat fragments takes place as the cracks propagate through the bonded interface sections. However, for the lamellae well bonded coatings, cracks propagate into splats, contributing to improve the erosion resistance. The similar phenomena were observed in the plasma spray TiO_2 and YSZ coatings. Thus, the microstructure of plasma spray coating significantly influences the erosion mechanism.

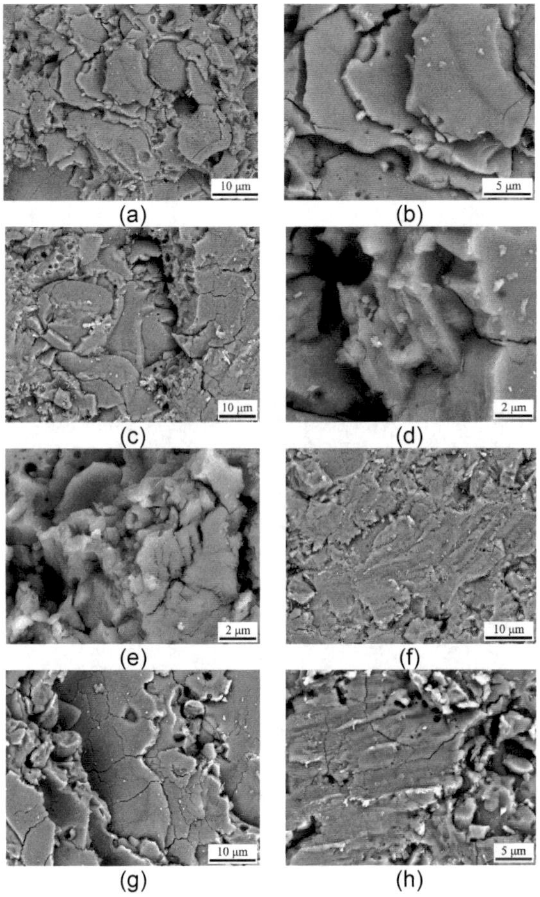

Fig. 4 Surface morphology of the eroded plasma-sprayed Al_2O_3 coatings deposited at 50 °C (a, b) and 300 °C (c, d, e, f). Surface morphology at the periphery of erosion crater of Al_2O_3 coatings deposited at 50 °C (g) and 300 °C (h)

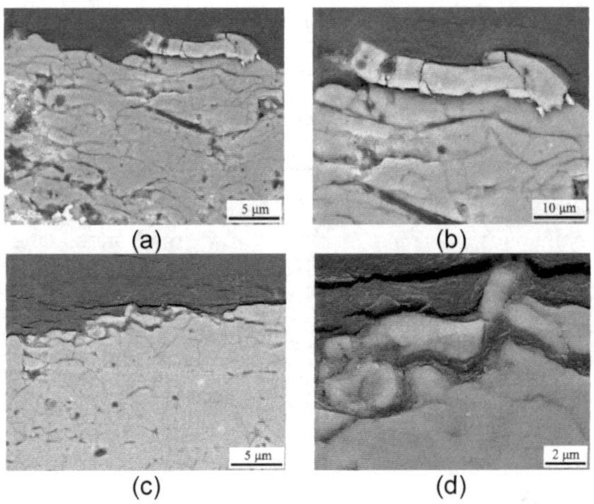

Fig. 5 Cross-sectional microstructure of the plasma-sprayed Al_2O_3 coatings deposited at 50 °C (a, b) and 300 °C (c, d) after erosion test.

Figure 6 shows the relationship between the weight of abrasives and the weight loss of plasma-sprayed ceramic coatings. It can be found that the weight loss of samples linearly increases with the increase of abrasives weight. Based on the measurement results, the erosion rates of different coatings were calculated. The effect of deposition temperate on the erosion rate is shown in Fig. 7. For TiO_2 and Al_2O_3 coatings, it can be found that the erosion rate sharply decreases with the increase of deposition temperature from room temperature to the critical bonding temperature. Then, the erosion rates decrease slightly with further increasing the deposition temperate. Although the YSZ coatings were just prepared at two temperatures since the higher deposition temperature cannot be achieved by the copper heating stage, it can be found that the erosion rate decreases when YSZ coating was deposited near its critical bonding temperature.

It has been known that increasing the deposition temperature is available approach to improve the properties and performance of plasma spray coatings. According to the erosion test results shown in Fig.7, it was suggested that there exists a critical transition point for the erosion behavior of coatings, which is equal to the critical bonding temperature. This is because that, with the increase of deposition temperature from room temperature to the critical bonding temperature, the microstructure of ceramic coatings transfers from a typical lamellar microstructure with limited interface bonding to full dense structure with well bonded interfaces. As a result, the erosion mechanism of coatings changes. With further increasing the deposition temperature, only the inter-lamellae bonding strength would be enhanced owing to bonding nature of the formation of coherency of semi-coherency interface during epitaxial growth [26,27]. Thus, the erosion resistance improves slightly with the further increase of deposition temperature.

4 Conclusions

In this study, the plasma-sprayed TiO_2, Al_2O_3 and YSZ coatings with different microstructure were prepared by controlling the deposition temperature based on the critical bonding temperature concept. The erosion behavior of different coatings was carried out. The weight loss of coating linearly increased with the increase of abrasives weight. It was revealed that the coating with different microstructure exhibited different erosion mechanisms. The coating deposited at the deposition temperature over the critical bonding temperature exhibit an excellent erosion resistance.

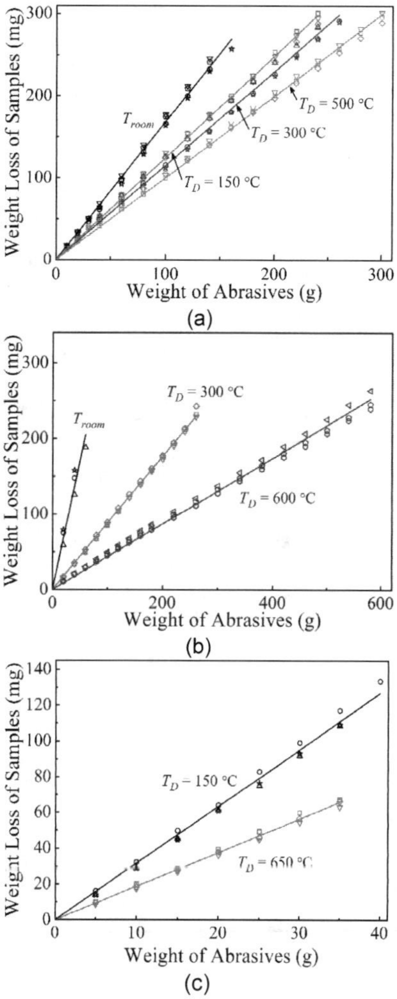

Fig. 6 Relationship between the weight of abrasives and the weight loss of plasma spray coatings: (a) TiO_2 coating (b) Al_2O_3 coating and (c) YSZ coating

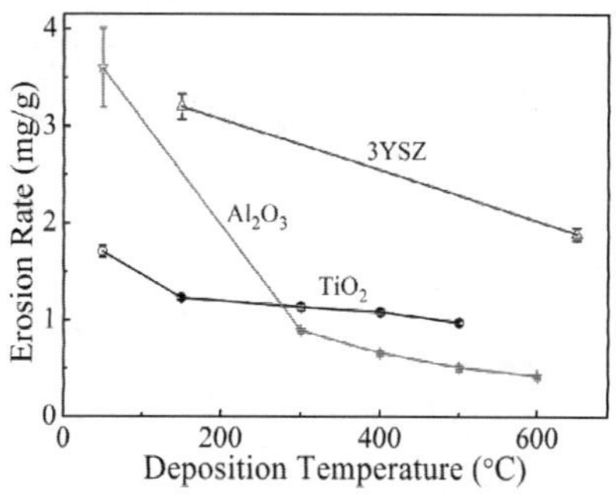

Fig. 7 Effect of deposition temperature on the erosion rates of plasma-sprayed TiO_2, Al_2O_3 and YSZ coatings.

Acknowledgments

The present project is financially supported by National Nature Science Foundation of China (No. 51171144).

References

[1] A. Vardelle, C. Moreau, J. Akedo, H. Ashrafizadeh, C.C. Berndt, J.O. Berghaus, M. Boulos, J. Brogan, A.C. Bourtsalas, A. Dolatabadi, The 2016 Thermal Spray Roadmap, J. Therm. Spray Technol., 25 (2016), 8, pp. 1376-1440.

[2] A. McWilliams, High-Performance Ceramic Coatings: Markets and Technologies, BCC Research, 1 (2016)

[3] C.J. Li, A. Ohmori, Relationships between the microstructure and properties of thermally sprayed deposits, J. Therm. Spray Technol., 11 (2002), 3, pp. 365-374.

[4] R. McPherson, B. Shafer, Interlamellar contact within plasma-sprayed coatings, Thin Solid Films, 97 (1982), 3, pp. 201-204.

[5] C. Moreau, P. Cielo, M. Lamontagne, S. Dallaire, J.C. Krapez, M. Vardelle, Temperature Evolution of Plasma-Sprayed Niobium Particles Impacting on a Substrate, Surf. Coat. Technol., 46 (1991), 2, pp. 173-187.

[6] A. Ohmori, C.-J. Li, Quantitative characterization of the structure of plasma-sprayed Al2O3 coating by using copp.er electroplating, Thin Solid Films, 201 (1991), 2, pp. 241-252.

[7] S. Kuroda, T.W. Clyne, The quenching stress in thermally sprayed coatings, Thin Solid Films, 200 (1991), 1, pp. 49-66.

[8] R. McPherson, A model for the thermal conductivity of plasma-sprayed ceramic coatings, Thin Solid Films, 112 (1984), 1, pp. 89-95.

[9] Y.Z. Xing, C.J. Li, C.X. Li, G.J. Yang, Influence of through-lamella grain growth on ionic conductivity of plasma-sprayed yttria-stabilized zirconia as an electrolyte in solid oxide fuel cells, J. Power. Sources, 176 (2008), 1, pp. 31-38.

[10] S. Boire-Lavigne, C. Moreau, R. Saint-Jacques, The relationship between the microstructure and thermal diffusivity of plasma-sprayed tungsten coatings, J. Therm. Spray Technol., 4(1995), 3, pp. 261-267.

[11] C.-J. Li, X.-J. Ning, C.-X. Li, Effect of densification processes on the properties of plasma-sprayed YSZ electrolyte coatings for solid oxide fuel cells, Surf. Coat. Technol., 190 (2005), 1, pp. 60-64.

[12] C.J. Li, G.J. Yang, C.X. Li, Development of Particle Interface Bonding in Thermal Spray Coatings: A Review, J. Therm. Spray Technol., 22 (2013), 2-3, pp. 192-206 .

[13] S.W. Yao, J.J. Tian, C.J. Li, G.J. Yang, C.X. Li, Understanding the Formation of Limited Interlamellar Bonding in Plasma Sprayed Ceramic Coatings Based on the Concept of Intrinsic Bonding Temperature, J. Therm. Spray Technol., 2016. pp. 1-14.

[14] D.M. Gray, Y.C. Lau, C.A. Johnson, M.P. Borom, W.A. Nelson, Thermal barrier coatings having an improved columnar microstructure, ed., US, 1998.
[15] G.-J. Yang, C.-X. Li, S. Hao, Y.-Z. Xing, E.-J. Yang, C.-J. Li, Critical bonding temperature for the splat bonding formation during plasma spraying of ceramic materials, Surf. Coat. Technol., 235 (2013), pp. 841-847.
[16] S.-W. Yao, C.-J. Li, J.-J. Tian, G.-J. Yang, C.-X. Li, Conditions and mechanisms for the bonding of a molten ceramic droplet to a substrate after high-speed impact, Acta Mater., 119 (2016), pp. 9-25.
[17] T. Foley, A. Levy, T. Foley, A. Levy, The erosion of heat-treated steels, Wear, 91(1983), 83, pp. 45-64
[18] A. Ninham, The effect of mechanical properties on erosion, Wear, 121 (1988), 3, pp. 307-324.
[19] Y.I. Oka, H. Ohnogi, T. Hosokawa, M. Matsumura, The impact angle dependence of erosion damage caused by solid particle impact, Wear, 203-204(96), pp. 573-579 (1997)
[20] C.J. Li, G.J. Yang, A. Ohmori, Relationship between particle erosion and lamellar microstructure for plasma-sprayed alumina coatings, Wear, 260(2006), 11-12, pp. 1166-1172.
[21] A.V. Levy, The erosion-corrosion behavior of protective coatings Surf. Coa.t Technol., 36 (1988), 1–2, pp. 387-406.
[22] A.G. Davis, D.H. Boone, A.V. Levy, Erosion of ceramic thermal barrier coatings, Wear, 110 (1986), 2, pp.101-116.
[23] J.A. Hearley, J.A. Little, A.J. Sturgeon, The erosion behaviour of NiAl intermetallic coatings produced by high velocity oxy-fuel thermal spraying, Wear, 233 (1999), 99, pp. 328-333.
[24] Y. Wang, M. Yan, The effect of CeO_2 on the erosion and abrasive wear of thermal sprayed FeAl intermetallic alloy coatings, Wear, 261 (2006), 11-12, pp. 1201-1207
[25] G.J. Yang, H.T. Wang, Effect of annealing on the microstructure and erosion performance of cold-sprayed FeAl intermetallic coatings, Surf. Coat. Technol., 205(2011), 23, pp. 5502-5509.
[26] E.-J. Yang, X.-T. Luo, G.-J. Yang, C.-X. Li, C.-J. Li, M. Takahashi, Epitaxial grain growth during 8YSZ splat formation on polycrystalline YSZ substrates by plasma spraying, Surf. Coat. Technol., 274 (2015), pp. 37-43.
[27] E.J. Yang, X.T. Luo, G.J. Yang, C.J. Li, M. Takahashi, S. Kuroda, K.H. Kim, Impact of deposition temperature on crystalline structure of plasma-sprayed Al2O3 splats revealed by FIB-HRTEM technique, Ceram. Int., 42(2016), 1, pp. 853-860.

Examination of Lamellar Pores under Thermally Sprayed Splats Using White Light Interferometry

Lin Chen, Yu-Ming Zhang, Guan-jun Yang, Cheng-Xin Li, Chang-Jiu Li

In the present study, a novel and practical method, white light interference, was proposed to characterize the lamellar pores covered by thermally sprayed YSZ and LZ splats. In this method, only an ordinary optical microscopy (OM) was employed. Colorful Newton rings and parabolic shapes of the lamellar pores were widely observed by OM. The crack spacing and the shapes of the lamellar pores captured by OM were well consistent with those by scanning electron microscopy (SEM) and focus ion beam (FIB). Besides, mechanical analyses were carried out and the results were well consistent with those by OM. Most importantly, the essential fact that the lamellar pores resulted from transverse cracking/delamination in thermal spray coatings was highly elaborated.

1 Introduction

As the case with thermally sprayed coatings which have been widely used as thermal barrier coatings (TBCs) [1,2], functional layers in solid oxide fuel cells (SOFCs) [3], and abradable seal coatings [4], numerous vertical cracks and transverse pores between lamellae are two essential features [5]. As a thermal insulation cell, TBCs are indispensable and vital components of blades in gas turbines. Every centigrade increase of thermal insulation will produce huge economic benefits. The distribution and shape of the transverse pores directly affect the thermal insulation and service life of TBCs. Therefore, it is important to explore and control the distribution of these transverse pores. However, imaging these internal transverse pores in coatings is always a problem. King [6] directly evaluated the transverse pores in a ceramic thin splat by scanning electron microscopy (SEM). But the result was not very obvious and only effective for dielectric splats on conductive substrate. Li [5] characterized the structure of Al_2O_3 coatings by copper electroplating. However, this method was only applicable to insulated coatings and of low contrast, and also not environment friendly. What's more, as an index parameter in TBCs, unfortunately, the formation mechanism of the transverse pores remains unclear up to now. It is widely believed that the transverse pores are retained during the formation process of splats for some reasons such as low impact pressure [7,8], condensates and adsorbates on the substrate surface [9,10]. All of the models mentioned above suggest that the transverse pores only exist at the edge of the splats [11]. However, Chraska and King [12-14] reported that there clearly existed transverse pores for almost each segment in both the center and periphery region of the splats. This reveals that above models are contradictory with experimental results. Compared with the large tensile stress (several GPa), residual stress in thermally sprayed coatings is usually in an order of ~100 MPa [15-17]. Resultantly, there should be a considerably large relief of the stress and strain energies. It is widely accepted that the vertical cracks is due to stress relaxation during rapid cooling [16,18]. Unfortunately, the fact that delamination also acts as an effective cracking mode in layered materials has been ignored for a long time.

In the present study, the formation mechanism of the transverse pores, delamination, is proposed. Exploring and understanding the formation mechanism as well as the shapes and distribution features of the transverse pores in TBCs are of great importance for the structure tailoring and evolution process control under a high-temperature or complex stress service condition.

2 Experiment

2.1 Materials and Splat Deposition

In this study, two kinds of splats, i.e., yttria-stabilized zirconia (YSZ) and lanthanum zirconia (LZ) were employed as demonstration materials for TBCs. The fused-crushed YSZ powder had a particle size range from 5 to 25 µm with a mean size of 17.8 µm, and LZ agglomerate powder had a particle size range from 15 to 45 µm with a mean size of 30 µm. Additionally, in order to avoid the influence of the roughness (which can result in the formation of gas pockets) and grain boundaries on substrate surface, all splats were deposited on (001) plane of single-crystal YSZ substrate with a well-polished surface (Ra <0.5 nm). All substrates were square, with a width of 10 mm and a thickness of 500 µm. The side face of the substrate was oriented along <100> direction.

The details of the experiments were reported in our former publications [11,19,20]. It was reported that deposition temperature had an obvious influence on splat/coating structures [21-23]. In addition, high substrate temperature contributed to getting rid of the adsorbates and gas pockets. In this study, the substrates were preheated to different temperatures (from 100 to 600°C with intervals of 100°C) through a copper plate heater on which the substrates were placed. To avoid the extra calefaction of the substrate by plasma arc, a shielding plate with several small holes was placed on the substrate. After splat deposition, the splat was cooled down to room temperature in atmosphere.

2.2 Characterization of Splats

We observed the splat morphologies by scanning electron microscopy (SEM, VEGA II-XMU, TESCAN, Czech Republic), and confirmed the epitaxial growth of the splats by electron back scattered diffraction (EBSD, AZTEC, OXFORD INSTRUMENTS, UK). All of the crack spacing was measured by Demo VegaTC integrated in SEM.

Most important, as for the characterization of the transverse pores covered by the splats, we proposed a simple and practical method, the white light interference. In this method, only an ordinary optical microscopy (OM, ECLIPSE MA200, NIKON, JAPAN) was employed to observe the interference fringe on the splat surfaces. The scheme of white light interference is shown in Fig. 1. A beam of light has the wave function as follows [24],

$$\vec{E}_i(x,t) = \vec{E}_0 \cos\left[2\pi f_i (t - x/u) + \varphi_i\right] \quad (1)$$

where $\vec{E}_0$, f_i and u is the base vector, frequency, and light wave velocity, respectively. When two beams of the lights meet, the interference will take place. If the two lights come from the same source (Fig. 1), the resultant interference light intensity dictates

$$I = 2I_0(1 + \cos\Delta\varphi) \quad (2)$$

where the phase difference $\Delta\varphi$ follows $\Delta\varphi = \frac{2\pi(r_2 - r_1)}{\lambda}$. When considering the half-wavelength loss, the condition of light interference can be rewritten as [24]

$$2d + \lambda = \begin{cases} (2k)\frac{\lambda}{2}, I = I_{max} \\ (2k+1)\frac{\lambda}{2}, I = I_{min} \end{cases} \quad (3)$$

where k and λ represents a random integer and the wavelength, respectively, and d refers to the lamellar width, as shown in Fig.1. When the light intensity reaches the maximum, a bright region emerges. Inversely, a dark region will be produced. Therefore, based on the positions of interference fringes, the positions and shapes of the lamellar pores (the lamellar width) can be accurately distinguished.

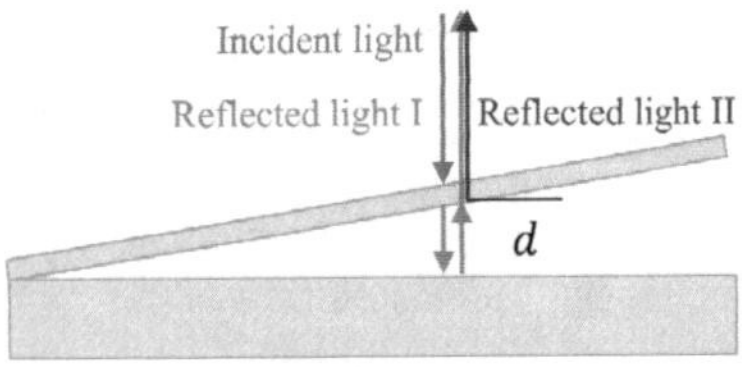

Fig. 1. The schematic diagram of white light interference. The red arrows represent the incident light. The pink and blue arrows represent the reflected lights. In principle, these light beams are overlapped (and herein separated only to easily distinguish).

3 Results

3.1 Splat morphologies by SEM

The surface morphologies of YSZ and LZ splats on single-crystal YSZ substrate at the deposition temperature from 100 to 600°C were observed by SEM, as shown in Fig. 2. The splats at all deposition temperatures presented regular and smooth crack pattern morphologies. During SEM observation, the substrate edge always kept parallel to the edge of the SEM view field. It was found that, at all deposition temperatures, vertical cracks were generally oriented at 45° angle to the substrate <100> direction (side face), which revealed <110> orientation for vertical cracks. This was consistent with the close-packed direction of face-centered-cubic (FCC) structure of YSZ and LZ. Due to regular cracking patterns, an array of nanochannels (see insets in Fig. 2) was successfully fabricated. It was found that substrate spallation occurred for all YSZ splats deposited at all different temperatures, while both interfacial and substrate spallation occurred for LZ splats deposited at temperature below 300°C, and only substrate spallation occurred at deposition temperatures above 300°C. In addition, there were many intact disk-shape splats without vertical cracks and large quantities of residual vertical cracks retained on the bare substrate (where the splat had fell off) in YSZ and LZ splats. These residual vertical cracks clearly revealed that transverse pores followed up with vertical cracking. The orientation maps by EBSD revealed that epitaxial growth along <001> orientation had readily occurred for all types of YSZ and LZ splats. Consequently, the cracks were quite straight and sharp due to the lack of intergranular cracking. The size of the square crack patterns was measured by using Demo VegaTC integrated in SEM. At least 900 patterns for every kind of splat were counted. Then, the equivalent crack spacing, the square root of the crack pattern size, was obtained, as shown in Fig. 2e. It was found that the crack pattern size changed little for epitaxial splats at the deposition temperature from 100 to 600°C. However, for every kind of splats, the crack spacing varied in an exceptionally broad range. This broad range indicated that delamination occurred substantially in many crack patterns. Epitaxial growth means the perfect bonding between the splat and substrate. This clearly indicates the lamellar pores result from transverse cracking/delamination (a new formation mechanism).

Additionally, it was found that there existed numerous residual pits or hills (corresponding to the crack patterns) on bare substrates. Apparently, these residual pits or hills were the regions where the last delamination occurred.

In brief, YSZ and LZ splats grew epitaxially on the YSZ substrate at all deposition temperatures. In the meantime, transverse pores were widely found in epitaxial splats. The crack pattern size was also counted and the result presented an exceptionally broad range, which revealed delamination had already occurred. Moreover, residual pits/hills on bare substrate were also observed. All of these revealed that the lamellar pores between the splats and substrate resulted from transverse cracking.

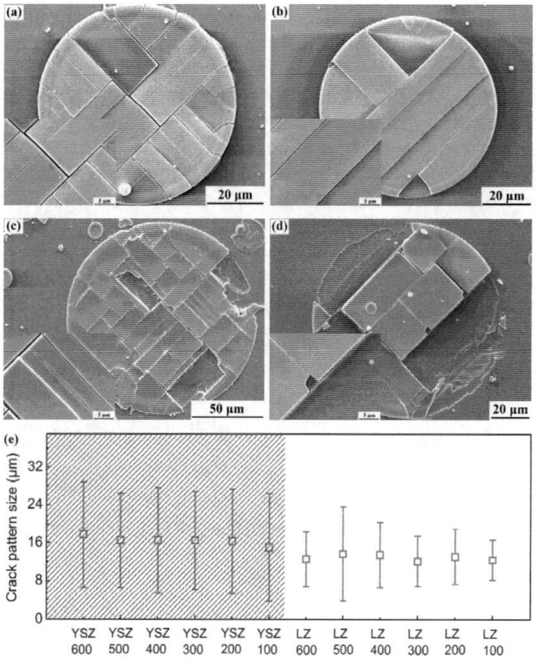

Fig. 2. The surface morphologies of YSZ (a, b) and LZ (c, d) splats on YSZ single-crystal substrate at the deposition temperature of (a, c) 300 and (b, d) 600°C. (e) The statistical crack spacing of YSZ and LZ splats from 100 to 600°C by SEM. The insets correspond to high magnification. All crack patterns present smooth and regular morphologies. All of the cracks are oriented along <110> directions.

3.2 Splat morphologies by white light interference

Most importantly, the transverse pores in epitaxial YSZ and LZ splats were characterized by using the white light interference method with an ordinary OM. As shown in Fig. 3, colorful Newton's rings clearly emerged under white light imaging by OM. As expected, almost every crack pattern had both bonded region (dark region) and unbonded region (colorful/bright region). The size of bonded regions under OM was counted by using Demo VegaTC integrated in SEM. At least 1000 patterns for every kind of splat and deposition temperature were counted. Then, the equivalent crack spacing, square root of bonded area size, was obtained, as shown in Fig. 3e. It was worth noting that the equivalent crack spacing of bonded area size captured by OM (Fig. 3e) was much smaller than this of crack patterns by SEM (Fig. 2e). This again confirmed that the lamellar pores resulted from transverse cracking.

Additionally, based on the order of interference fringes (see Fig. 3, and Eq. 3), the shape of the lamellar pores could be determined. Herein, cyan fringes (with a wavelength of 480 nm) on YSZ splat surfaces at the deposition temperature of 600°C were employed. Seventy-seven splats were counted to fit the shapes of the lamellar pores. Resultantly, a perfect parabolic shape fitting was obtained, as shown in Fig. 4.

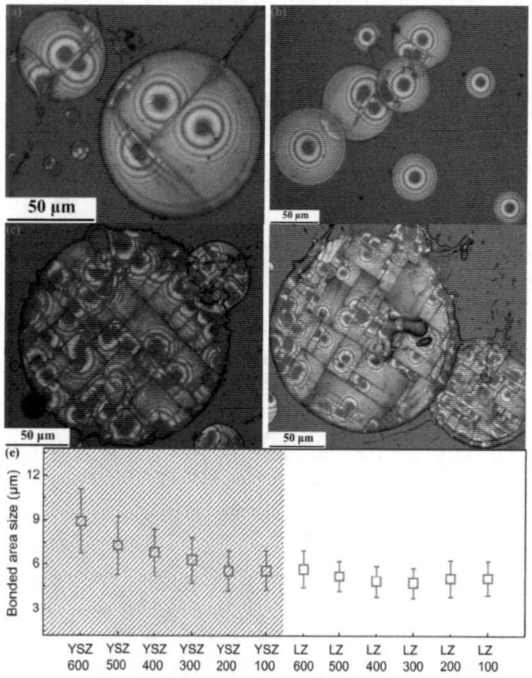

Fig. 3. The surface morphologies of YSZ (a, b) and LZ (c, d) splats at the deposition temperature of (a, c) 300 and (b, d) 600°C by OM. (e) The statistical crack spacing of YSZ and LZ splats from 100 to 600°C by OM. Colorful newton rings were clearly observed.

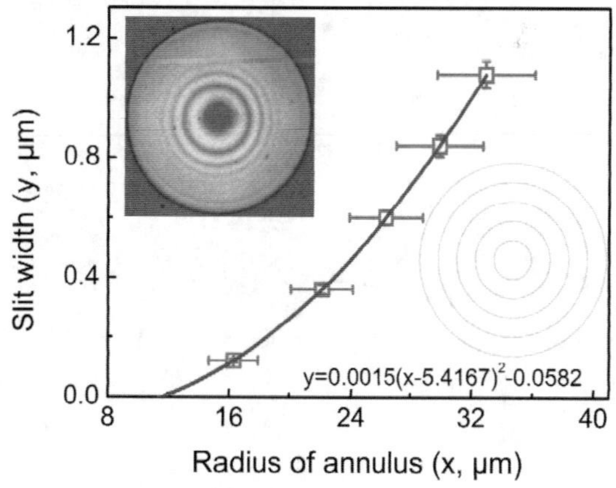

Fig. 4. The shapes of lamellar pores in YSZ splats at deposition temperature of 600°C by OM. The cyan fringes with the wavelength of 480 nm were employed. The red blocks represented the experimental results and the pink line represented the fitted result of seventy-seven splats. The perfect parabolic shape was obtained by second-order polynomial fitting.

In brief, the position and distribution of the lamellar pores were successfully observed by white light interference. Colorful Newton's rings clearly emerged under white light imaging by OM. Moreover, the bonded area size and the shape of lamellar pores were also counted under the white light interference method. Resultantly, the perfect parabolic shape of lamellar pores was successfully obtained.

3.3 Cross-sectional morphologies by SEM and FIB

To confirm the shapes of the lamellar pores covered by the fragments, the inclined cross-sectional morphologies were observed by SEM. As expected, tile and parabolic shapes were clearly observed in all splats and deposition temperatures, as shown in Figs. 5a-d. A further quantitative confirmation was also carried out by using FIB, as shown in Fig. 5e and 5f. Indeed, the perfect parabolic shapes were well consistent with the results by OM (Fig. 4).

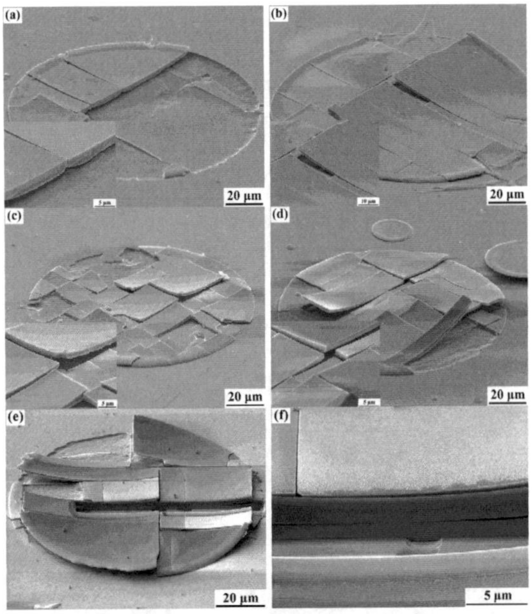

Fig. 5. The cross-sectional morphology of YSZ (a, b, e, and f) and LZ (c, d) splat at the deposition temperature of (a, c) 300, (b, d, e, and f) 600°C by SEM (a-d) and FIB (e, f). The insets corresponded to high magnifications. Tile-shaped crack patterns and parabolic-shaped lamellar pores was clearly observed.

4 Disscussion

4.1 Shapes of lamellar pores

In the present study, three aspects of deformations contribute to the shapes of lamellar pores. They are crack opening displacement (COD), the displacement of Euler buckling, and bending deformation of composite beam. The crack opening displacement is usually influenced by length scale to the crack tip. When the position is extremely close to crack tip (this region is usually called bond-breaking zone where $r_0 \leq 1$ nm for linear elastic ceramics), the COD is proportional to the distance to crack tip (r). When r is between r_0 and r_K (the length scale of fracture materials), namely the K annulus as shown in Fig. 6a, the COD is proportional to the square root of the distance to crack tip ($r^{1/2}$) and dictates [25]

$$\delta_y + i\delta_x = \frac{8(K_I + iK_{II})}{E}\sqrt{\frac{r}{2\pi}} \tag{4}$$

where K_I and K_{II} are the stress intensity factors for Mode I and II cracking, and r is the distance to the crack tip. Once the stress intensity factors are obtained, the COD can be calculated. In the case of the film and substrate having identical properties, the displacement (δ_y) follows $\sim 0.0063r^{1/2}$ when the cracking strain is of ~0.5%. Based on dimensional analysis, all of the length units in the present study are tacitly unified to μm.

The critical force for Euler buckling as shown in Fig. 6b usually dictates [26]

$$F_{cr} = \frac{\pi^2 EI_b}{(\mu l)^2}$$

(5) where l and I_b are the crack length and the bending rigidity of the composite beams. Moreover, μ is the length factor (here is equal to 2). The displacement of Euler buckling then can be expressed by

$$\nu = C_1 \sin kx + C_2 \cos kx \tag{6}$$

where k is related to the total crack length (l). In the case of the film and substrate having identical properties, the bending rigidity of composite beam (I_b) dictates

$$I_b = bd^3/12 \tag{7}$$

where b and d are the width and thickness of the composite beam. Resultantly, the critical stress for Euler buckling is about 2.82 GPa. This indicates Euler buckling can be ignored except slender stripes.

The bending deformation of the composite beam as shown in Fig. 6c usually follows [26]

$$\nu = \frac{Mx^2}{2EI} \tag{8}$$

where M is the moment of the composite beam. As shown in Figs. 4 and 5, a perfect parabolic fitting is obtained in the periphery region of crack patterns. when the film and substrate have identical properties, the moment (Fig. 6c) can be expressed by

$$M = \sigma hb \cdot d/2$$

(9) where σ_1 is residual stress in the splat.

In conjunction with Eq. (7)-(9) and the second-order coefficient of parabolic fitting (Fig. 4), the residual stress in the splat can be calculated and is about 550 MPa. It is worthwhile to note that this stress is the result after cracking and stress redistribution taking place, the total stress in the case of no cracking and redistributing must be larger than this value.

In brief, mechanical model was set up to explore the shapes of the lamellar pores. It was found that the bending deformation of the composite beam with the residual stress ~550 MPa dominated the parabolic shapes of the lamellar pores, which was well consistent with the result by OM.

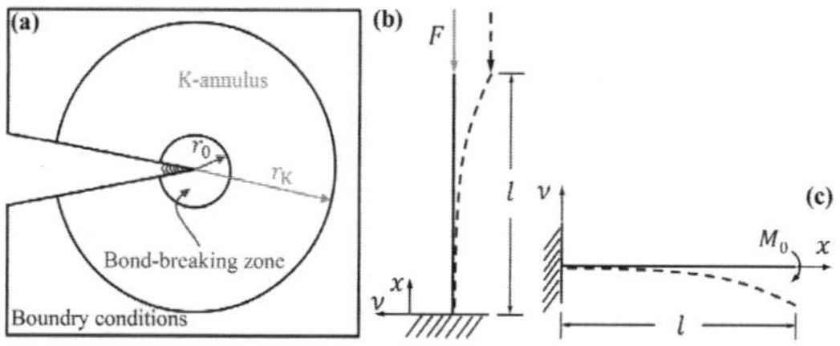

Fig. 6. The schematic diagram of (a) K annulus in linear elastic fracture mechanics, (b) Euler buckling of compressive bar, and (c) beam bending due to the moment.

4.3. Crack spacing

Based on the fracture mechanics, the crack spacing was widely investigated for various film/substrate combinations and cracking modes [27-30]. For the splat and substrate having identical properties without interfacial cracking, the crack spacing dictates [28]

$$\lambda_{equi}/h \cong 5.6\sqrt{\Gamma_f/(\bar{E}\varepsilon_0^2 h)} \quad (10)$$

where Γ_f and ε_0 is the fracture toughness and applied strain, respectively.

Because there exist both vertical and transverse cracking, the normal crack spacing has no use to sustain the mechanical equilibrium configuration. Inversely, the stress transfer from the splat to the substrate only takes place over the portion of the segment where no delamination propagates. This defines an effective length of the segments [31], which is considered as the equivalent crack spacing of bonded area size by OM as shown in Fig. 3e. Using the theory of weight function and linear fracture mechanics, the effective crack spacing dictates [32]

$$l/h = -20.83 + 59.80R + \frac{0.2811}{R_C - R} - \frac{1.074\times10^{-4}}{(R_C - R)^2} \quad (11)$$

where Rc is the critical cracking resistance number (which yields 0.6607). The cracking resistance number (R) dictates

$$R = \frac{2\gamma E}{\pi\sigma^2 h}$$

(12) where γ is the specific surface energy. Taking the epitaxial YSZ splat deposited at 600°C whose crack spacing (l/h) is about 18.14 as an example, the cracking resistance number (R) is of the value 0.56. Taking the fracture energy (2γ) as 1.37 J/m^2, the cracking stress (σ) can be caculated and is about 571 MPa, which is credible for the cracking of ceramics. Similarly, if the the cracking stress is known, the fracture energy can be caculated. It should note that the above mechanical analysis didn't condider the residual strain energies stored in the delaminated composite beams. This indicated the actual value should be larger than the present one.

5. Conclusions

In the present study, a new formation mechanism of transverse pores in thermal spray coatings, stress-driving transverse cracking/delamination, is proposed and explored. It was found that all of the YSZ and LZ splats grew epitaxially on the YSZ substrate. However, the lamellar pores were ubiquitous in almost all of the crack patterns observed by SEM and FIB. This clearly indicated that lamellar pores in thermally sprayed coatings resulted from transverse cracking. A novel and practical method, white light interference was used to quantitatively examine the distribution and shapes of lamellar pores. As a result, colorful Newton rings and parabolic shapes of the lamellar pores were observed by OM. Mechanical analyses were also carried out and the results were well consistent with those by OM. All of these strongly suggested that the lamellar pores in thermally sprayed coatings resulted from transverse cracking.

Acknoledgement

The present project is supported by National Basic Research Program (Nos. 2013CB035701), the Fundamental Research Funds for the Central Universities, and the National Program for Support of Top-notch Young Professionals

Reference

1. N.P. Padture, M. Gell, E.H. Jordan: Thermal barrier coatings for gas-turbine engine applications. Science 296 (2002), Issue 5566, pp. 280/4.
2. N.P. Padture: Advanced structural ceramics in aerospace propulsion. Nat Mater 15 (2016), Issue 8, pp. 804/9.
3. Y.Z. Xing, C.J. Li, G.J. Yang: Influence of through-lamella grain growth on ionic conductivity of plasma-sprayed yttria-stabilized zirconia as an electrolyte in solid oxide fuel cells. J Power Sources 176 (2008), Issue 1, pp. 31/8.
4. S. Wilson: Thermally Sprayed Abradable Coating Technology for Sealing in Gas Turbines. 6th International Conference on the Future of Gas Turbine Technology (2012).
5. A. Ohmori, C.J. Li: Quantitative Characterization of the Structure of Plasma-Sprayed Al_2O_3 Coating by Using Copper Electroplating. Thin Solid Splats 201 (1991), Issue 2, pp. 241/52.
6. S. Rangarajan, A.H. King: Non-destructive evaluation of delamination in ceramic thin splats on metal substrates by scanning electron microscopy. Thin Solid Splats 385 (2001), Issue 1-2, pp. 22/8.
7. V.V. Sobolev, J.M. Guilemany, J. Nutting, J.R. Miquel: Development of substrate-coating adhesion in thermal spraying. Int Mater Rev 42 (1997), Issue 3, pp. 117/36.

8. C.J. Li, J.L. Li: Transient contact pressure during flattening of thermal spray droplet and its effect on splat formation. J Therm Spray Techn 13 (2004), Issue 2, pp. 229/38.
9. X.Y. Jiang, Y.P. Wan, S. Sampath: Role of condensates and adsorbates on substrate surface on fragmentation of impinging molten droplets during thermal spray. Thin Solid Splats 385 (2001), Issue 1-2, pp. 132/41.
10. M.X. Xue, S. Chandra, J. Mostaghimi: Investigation of splat curling up in thermal spray coatings. J Therm Spray Techn 15 (2006), Issue 4, pp. 531/6.
11. L. Chen, G.J. Yang, C.X. Li: Formation of Lamellar Pores for Splats via Interfacial or Sub-interfacial Delamination at Chemically Bonded Region. J Therm Spray Techn (2017).
12. T. Chraska, A.H. King: Effect of different substrate conditions upon interface with plasma sprayed zirconia - a TEM study. Surf Coat Tech 157 (2002), Issue 2-3, pp. 238/46.
13. T. Chraska, A.H. King: Transmission electron microscopy study of rapid solidification of plasma sprayed zirconia - part I. First splat solidification. Thin Solid Splats 397 (2001), Issue 1-2, pp. 30/9.
14. T. Chraska, A.H. King: Transmission electron microscopy study of rapid solidification of plasma sprayed zirconia - part II. Interfaces and subsequent splat solidification. Thin Solid Splats 397 (2001), Issue 1-2, pp. 40/8.
15. J. Matejicek, S. Sampath, H.J. Prask: Quenching, thermal and residual stress in plasma sprayed deposits: NiCrAlY and YSZ coatings. Acta Mater, 47 (1999), Issue 2, pp. 607/17.
16. S. Kuroda, T.W. Clyne: The Quenching Stress in Thermally Sprayed Coatings. Thin Solid Splats 200 (1991), Issue 1, pp. 49/66.
17. S. Kuroda, T. Dendo, S. Kitahara: Quenching Stress in Plasma-Sprayed Coatings and Its Correlation with the Deposit Microstructure. J Therm Spray Techn 4 (1995), Issue 1, pp. 75/84.
18. S. Kuroda, T. Fukushima, S. Kitahara: Significance of quenching stress in the cohesion and adhesion of thermally sprayed coatings. J Therm Spray Techn 1 (1992), Issue 4, pp. 325/32.
19. L. Chen, G.-J. Yang, C.-X. Li, C.-J. Li: Edge Effect on Crack Patterns in Thermally Sprayed Ceramic Splats. J Therm Spray Techn (2016).
20. L. Chen, G.J. Yang, C.X. Li, C.J. Li: Hierarchical Formation of Intrasplat Cracks in Thermal Spray Ceramic Coatings. J Therm Spray Techn 25 (2016), Issue 5, pp. 959/70.
21. S. Sampath, X.Y. Jiang, J. Matejicek, A. Vardelle: Substrate temperature effects on splat formation, microstructure development and properties of plasma sprayed coatings Part I: Case study for partially stabilized zirconia. Mat Sci Eng a-Struct 272 (1999), Issue 1, pp. 181/8.
22. X. Jiang, J. Matejicek, S. Sampath: Substrate temperature effects on the splat formation, microstructure development and properties of plasma sprayed coatings Part II: case study for molybdenum. Mat Sci Eng a-Struct 272 (1999), Issue 1, pp. 189/98.
23. S. Sampath, X. Jiang: Splat formation and microstructure development during plasma spraying: deposition temperature effects. Mat Sci Eng a-Struct 304 (2001), pp. 144/50.
24. M. Born, E. Wolf: Principles of optics: electromagnetic theory of propagation, interference and diffraction of light. CUP Archive (2000).
25. H. Tada, P. Paris, G. Irwin: The analysis of cracks handbook. New York: ASME Press (2000).
26. J. Gere, S. Timoshenko: Mechanics of Materials. PWS Pub, Co., Boston (1997).
27. M.D. Thouless: Crack Spacing in Brittle Splats on Elastic Substrates. J Am Ceram Soc, 73 (1990), Issue 7, pp. 2144/6.
28. M.D. Thouless, E. Olsson, A. Gupta: Cracking of Brittle Splats on Elastic Substrates. Acta Metall Mater 40 (1992), Issue 6, pp. 1287/92.
29. Z.C. Xia, J.W. Hutchinson: Crack patterns in thin splats. J Mech Phys Solids, 48 (2000), Issue 6-7, pp. 1107/31.
30. J.W. Hutchinson, Z. Suo: Mixed-Mode Cracking in Layered Materials. Advances in Applied Mechanics 29 (1992), Issue 29, pp. 63/191.
31. R. Nahta, B. Moran: Crack spacing in brittle splats on dissimilar planar and axisymmetric elastic substrates. Engineering fracture mechanics 52 (1995), Issue 3, pp. 513/24.
32. T.Y. Zhang, M.H. Zhao: Equilibrium depth and spacing of cracks in a tensile residual stressed thin splat deposited on a brittle substrate. Engineering Fracture Mechanics 69 (2002), Issue 5, pp. 589/96.

Investigation of interfacial microstructures between LZ splats and YSZ substrate by HR-TEM

Lin Chen, Guan-Jun Yang*, Cheng-Xin Li

Thermally-sprayed LZ/YSZ double-layer coatings are promising candidate for the next generation thermal barrier coatings (TBCs) due to exceedingly low thermal conductivity and superior high-temperature phase stability. However, a delamination failure at LZ and YSZ interface were widely observed during TBCs service. Till today, the interfacial microstructure between LZ and YSZ remains unclear. In the present study, LZ splats were deposited on YSZ substrate to serve as a LZ/YSZ interface. The interfacial microstructure was explored by focused ion beam (FIB) and high-resolution transmission electron microscope (HR-TEM). The interfacial defects at splat interface were clearly observed and thoroughly discussed. These results would shed light on deeply understanding the interfacial failure of double-layer LZ/YSZ coatings.

1 Introduction

Thermally sprayed coatings have been widely used as thermal barrier coatings (TBCs) [1,2], functional layers in solid oxide fuel cells (SOFCs) [3], etc. The spray process characterizes the coatings with a large number of splats piling up layer by layer, which indicates that there exist innumerable lamellar interfaces in thermally sprayed coatings. Obviously, the quality of the interfaces determines the coating performance such as service lifetime and thermal/electric conduction. As the case with TBCs, it was widely reported that delamination at the lamellar interface such as LZ/YSZ double layer coatings was the uppermost failure mode during their service stage [4-8]. Besides, it was found that the tens/hundreds of microns long delamination had produced during the coating formation [9] which acted as the initial crack sources. These pre-existing cracks can easily coalesce each other and lead to the final failure during the services. Therefore, tailoring the qualities of these interfaces is the key to promoting the coating performances.

Apparently, the quality of lamellar interfaces were only related to the coating deposition processing. Unfortunately, how the microstructure and the interface quality is formed has been ignored for a long time. Indubitably, how to avoid the interfacial failure is related to the interface microstructure and quality tailoring. In the present study, the microstructure and quality of LZ/YSZ interface was explored to reveal the essential mechanism of interfacial failure.

2 Experiment

2.1 Film deposition

In this study, LZ splats on YSZ substrate, as double-layer TBCs, were studied. The particle size of LZ agglomerate powder ranged from 15 to 45 μm with a mean size of 30 μm. The YSZ substrates were single-crystal and square, with a width of 10 mm and a thickness of 500 μm. The side face of the substrate was oriented along <100> direction. The details of experiments were reported in our former publications [10-12]. Prior to the film deposition, we preheated the substrate to 300 and 600 °C through a copper plate heater on which the substrates were placed. The temperature was real-time monitored by a calibrated thermocouple. In addition, to avoid the extra calefaction of the substrate by plasma arc, a shielding plate with several small holes was clamped on the substrate. After deposition, the film and substrate were cooled down to room temperature in atmosphere.

2.2 Characterization of films

The splats surface morphology was observed by scanning electron microscopy (SEM, VEGA II-XMU, TESCAN, Czech Republic), and the epitaxial growth of the splats was first confirmed by electron back scattered diffraction (EBSD, AZTEC, OXFORD INSTRUMENTS, UK). Moreover, a TITAN (FEI, USA) Cs-corrected high-resolution transmission electron microscope (HR-TEM) was also employed to investigate the epitaxial interface in detail. The TEM samples were prepared by focused ion beam (FIB, FEI Quanta 200 FEG, USA). After the FIB preparation, the samples were cleaned by Nano Mill (M1040 Nano Mill) to remove the amorphous layers produced by the bombardment of FIB.

3 Results

3.1 Surface morphology of LZ splats

The LZ splats on the single crystal YSZ substrate presented regular pattern morphologies, as shown in Fig. 1. During SEM observation, the substrate edge always kept parallel to the edge of the SEM view field. It was found that vertical cracks were generally oriented at 45° angle to the substrate <100> direction (side face), which revealed <110> orientations for vertical cracks. This was consistent with the close-packed direction of face-centered-cubic (FCC) structure of LZ material. Due to regular cracking patterns, an array of nanochannels was successfully fabricated. Both interface and substrate delamination already occurred for LZ splats deposited at 300 °C (Fig. 1a), while only substrate delamination occurred at 600 °C (Fig. 1b). Residual vertical cracks were also observed on the bare substrate. The orientation maps of LZ splats were first examined by EBSD. The inverse pole figures and identical colour between film and single-crystal substrate clearly revealed the epitaxial growth had successfully occurred along the <001> direction, as shown in Figs. 1c and 1d. Notably, the interface delamination indicated a weak LZ/YSZ interface, which was incompatible with the characteristic of epitaxial growth.

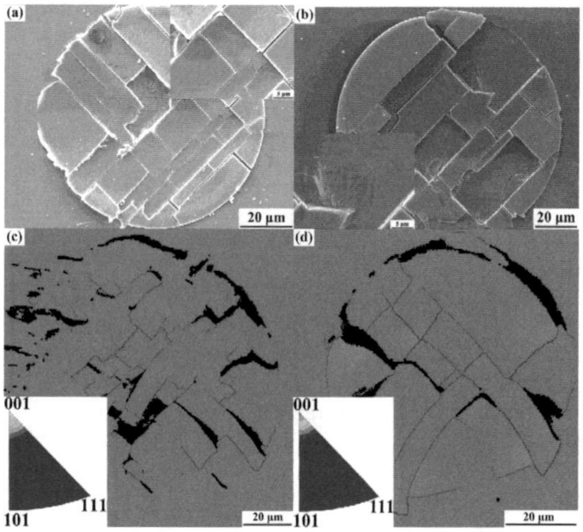

Fig. 1. The surface morphologies and orientation maps of the LZ film on the single-crystal YSZ substrate at deposition temperature of (a, c) 300 °C and (b, d) 600 °C. The insets in (a) and (b) correspond to high magnification. The insets in (c) and (d) correspond to inverse pole figures (IPFs) by EBSD. The red colour and IPFs in EBSD patterns represents s the <001> orientation. The identical colour between film and substrate clearly reveals the epitaxial growth has successfully occurred.

3.3 Dislocations at heteroepitaxial LZ interfaces

The heteroepitaxial LZ interfaces were examined by a TITAN Cs-corrected HR-TEM, as shown in Fig. 2. The film was about 800 nm thick. Under high magnification, distinct interface contrasts were observed at the heteroepitaxial interface, as shown in Figs. 2a and 2b. Generally, an opposite contrast on bright field images (BF-TEM) or dark field images (DF-TEM) was due to the existing defects. Knowing that the cracks were always along <100> direction and to associate the defects with cracking, we conducted fast Fourier transform (FFT, by DigitalMicrograph, Gatan, UK) for the HR-TEM images, and then conducted inverse fast Fourier transform (IFFT) only for the diffraction spot with g= [220]. As expected, large numbers of defects at the interface were observed from the FFT-IFFT images, as shown in Figs. 2d and 2f. Except for the stacking faults, many edge dislocations (marked by red signs) with a Burgers vector of $(a/2)\langle 110\rangle$ were present in the interfacial region (marked by two pink dashed lines). Considering that positive and negative dislocations lead to a contrary local deformation of the crystal lattice, we determined the net dislocations along the interfaces. To quantitatively evaluate and associate the dislocations with strains, the dislocation concentration [13,14] was used and defined as the ratio of the net dislocations to the total lattice numbers along the epitaxial interface. Apparently, the strain relaxed by dislocations was equal to the dislocation concentration. For the calculation, 30 high-magnified images with a total interface length of about 410 nm were used. It was found that considerable net dislocations existed in the interfacial region corresponding to the distinct interface contrasts, i.e. 6.00 ± 0.40 and 5.69 ± 0.27% for deposition temperature of 300 and 600 °C, respectively.

There are generally two types of strain components [15]. One is stress induced physical strain which obeys Hooke's law. The other is intrinsic strain caused by thermal, electric and magnetic field, and defects (such as dislocations and stacking faults), etc. This kind of intrinsic strain doesn't generate stress in free condition, such as the free thermal expansion. Based on the definition in elasticity [15], the total strain is only related to the displacement change. For single-crystal YSZ and LZ, the lattice parameter at room temperature is 0.5141 and 0.5396 nm, respectively. It can be calculated that the total strain due to the lattice mismatch between heteroepitaxial LZ film and YSZ substrate is about 4.72%, which is shared by the defects and physical elastic strain. The intrinsic strain is relaxed by two types of defects that exist in the interfacial region: stacking faults and net dislocations. The stacking faults shown in Figs. 2d and 2f are always less than 3 nm in size, and the TEM images show no stacking faults going through the whole field. Thus, we can state that all the stacking faults were local defects. Local stacking faults contribute little to the strain relaxation. As a result, the physical strain is only correlated to net dislocations, and has the value about 1.28 ± 0.40 and 0.97 ± 0.27% for deposition temperature of 300 and 600 °C, respectively.

In the conventional heteroepitaxial growth of thick film such as by molecular beam epitaxy, the lattice mismatch is almost totally relaxed by the defects, and the physical elastic strain only comes from thermal expansion mismatch. However, it is an anomalous heteroepitaxy which produce excess defects and physical strains.

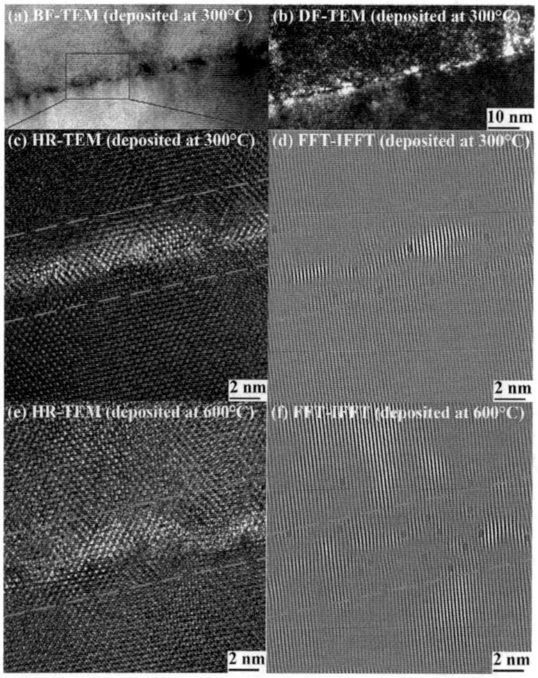

Fig. 2. Microstructure and quantitative characterization of the heteroepitaxial LZ-YSZ interface. (a) BF-TEM, (b) DF-TEM, (c) HR-TEM and (d) [220] FFT-IFFT images of the film deposited at 300 °C. (e) HR-TEM and (f) [220] FFT-IFFT images of the film deposited at 600 °C. The interface region was marked by two pink dashed lines. The dislocations in (d) and (f) were marked by the red signs.

4 Discussion

4.1 Formation mechanism of dislocations and cracking

Essentially, thermal spray is a liquid phase epitaxial process. However, excess dislocations and physical strains were produced in heteroeptaxial LZ interfacial region. Apparently, the anomalous epitaxial growths are related to the deposition processing. During the thermal spray deposition, the epitaxial film formation goes through five stages, as shown in Fig. 3. Firstly, the molten droplet heated and accelerated by plasma torch impacts the substrate with a high velocity (~150 m/s), as shown in Fig. 3I. Then, the droplet rapidly spreads on the substrate forming a liquid film, as shown in Fig. 3II. During the third stage, the solidification/epitaxial growth starts, and the liquid film transforms into a solid epitaxial film on the single-crystal substrate, as shown in Fig. 3III. After epitaxial growth, the film together with the substrate cooled down to room temperature, as shown in Fig. 3IV. Finally, an epitaxial film bonded on the single-crystal substrate was observed, as shown in Fig. 5V.

Obviously, dislocations and cracking formed during solidification and cooling process. Firstly, an emphasis was placed on the solidification. During film spreading and solidification, the lattice parameter of the film decreased with the temperature dropping. However, since the substrate was much larger than the film in all three dimensions, the substrate worked as a vast reservoir. As a result, the temperature in the whole substrate (except the region covered by the molten film) remained nearly unchanged. As expected, the localized substrate covered by molten film was significantly heated during film spreading and solidification process, but the expansion of this region was strongly constrained by its cold surroundings, as shown in Fig. 5III. As have stated above, there were two types of strains components, one was physical elastic strain and the other was intrinsic caused by thermal field, etc. Based on the definition in elasticity [15], the total strain was only related to the displacement change. Obviously, the total strain in localized substrate was nearly zero during solidification, as shown in Fig. 3III. As a result, enormous compressive physical strain was accumulated in this locally-heated substrate to negate the intrinsic thermal strain. Generally, the lattice mismatch between LZ and YSZ was 4.72%. However, the present value was much larger than this due to the constraint of locally-heated substrate. However, a liquid phase epitaxial growth on this locally-heated region would preferentially form a coherent interface. In this case, the coherent growth on the strained substrate meant that the newly grown solid film also sustained a large compressive strain (larger than 4.72%). However, any solidification process tended to form a stress-free epitaxial film. As a result, net dislocations (ideally, ~4.72%) and excess net dislocations (related to the degree of the constraint of the locally-heated substrate) were produced in the interface layer to negate the lattice mismatch at the film-substrate interface. It was interesting to realize that cracking in the epitaxial film would not happen during epitaxial growth since enormous physical compressive strain was accumulated in the epitaxial film.

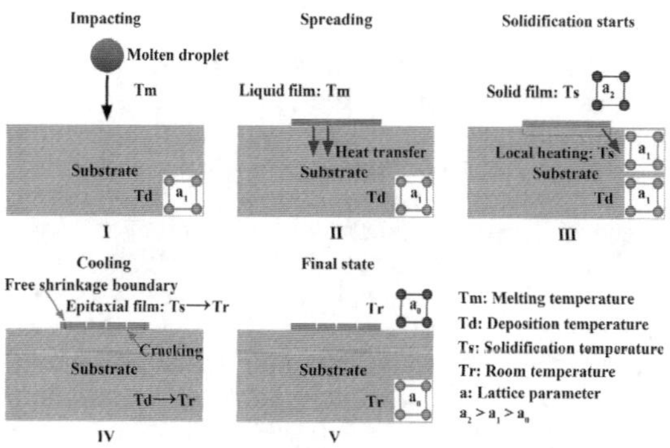

Fig. 3. The detailed film formation process by thermal spray technology. A molten droplet with high-speed impacts on single-crystal substrate surface (I). The droplet quickly spreads on the substrate forming a liquid film (II). The liquid film starts solidification on substrate forming a solid epitaxial film (III). During the

subsequent cooling, the epitaxial film would shrink with the temperature from Ts to Tr (IV). An epitaxial film/substrate system was formed at room temperature (V). The blue dashed line in epitaxial film in (IV) denotes the position in the case of free shrinkage. Additionally, the sign, a, is the theoretical lattice parameter of the epitaxial film or single-crystal substrate.

After epitaxial growth, the film together with the substrate would cool down to room temperature, as shown in Fig. 3IV. However, the degree of thermal shrinkage between the film and substrate was significantly different. The temperature change in epitaxial film during cooling was from Ts to Tr (where Ts and Tr was solidification and room temperature, respectively), while it was actually only from Td to Tr (where Td was deposition temperature, 300 or 600 °C) for the substrate, as shown in Fig. 3IV. Compared with free shrinkage (marked by blue dashed lines in Fig. 5IV), a large tensile stress in epitaxial film was produced due to the constraints of the substrate. Once the stress in film exceed its tensile strength, cracking would take place. Taking the elastic modulus of 175 GPa and solidification temperature of 1800 °C into account, the maximum tensile stress in LZ splats could be as high as ~2.36 GPa. Therefore, vertical cracks inevitably occurred to relax this large tensile stress. Furthermore, the cracking direction could be predicted to be along the <110> directions, since the <110> directions had the severest lattice distortion for this FCC structure. The prediction was well consistent with the cracking behaviour shown in Figs. 1a and 1b.
In brief, large quantities of dislocations and extra dislocations were produced during solidification process, while cracking occurred during cooling process. Both of the extra dislocations and cracking were due to the constraint effect of cold surroundings in the locally-heated substrate. The locally-heated substrate with unchanged lattice parameter is the essential difference between conventional and present liquid phase epitaxial growth.

4.2 Relaxation mechanism of strain by dislocations in interface region

The strain relaxation mechanisms observed in highly strained epitaxial films generally include relaxation by plastic deformation (such as by dislocations and stacking faults), cracking, and surface roughening during the conventional epitaxial growth process [16]. Whether or not any mechanism occurs in a given system is determined by a number of factors, among which the lattice mismatch at the film-substrate interface is a crucial one. First, strain relaxation takes place by defects such as the dislocations and stacking faults along the film-substrate interface [17]. In this case, the defects have to go through the whole interface and reach the exterior of this crystalline region at least in one direction. Considering that all stacking faults are local and have a small size of <3 nm, thus only dislocations should be considered. Second, relaxation by cracking is commonly observed in highly strained heterostructures in tension [18]. This mechanism can be ignored in this study, because the strain in the film was compressive during the epitaxial growth process, as shown in Fig. 6III. Third, the relaxation of purely elastic stresses results in the development of surface undulations [19]. This mechanism also can be ignored in this study, since the films presented a very smooth surface profile, see Fig. 1. In brief, forming dislocations is the only effective mechanism to compensate the apparent lattice mismatch at the film-substrate interface during the solidification process.
Generally, the critical strain required to induce dislocations at the film-substrate interface is about 0.30% [16]. In present study, this critical strain was nearly the same as the error bar of the strain relaxed by dislocations. Consequently, the dislocation concentration data can reveal the true strain at the film-substrate interface. As a result, we can also quantitatively deduce the interface/solidification temperature (Ts) during the epitaxy using the strain relaxed by dislocations as follows:

$$Ts=Td+(\varepsilon_d-4.72\%)/\alpha \quad (1)$$

where ε_d is the strain relaxed by dislocations, and α is the coefficient of thermal expansion (CTE, 9.1×10-6 /K). Therefore, the epitaxial temperature of LZ films were 1543 ± 388 and 1542 ± 262 °C at the deposition temperature of 300 and 600 °C, respectively. Up to now, it is impossible to experimentally measure the epitaxial temperature (or the solidification temperature of the liquid) in liquid phase epitaxy mainly because of the remarkably short solidification time (generally microseconds). However, the interface dislocation density calculated in this study potentially opens a door to determine the interface epitaxial/solidification temperature.
Note that dislocation initiation was a thermal activation process because of large energy barriers. The activation energy of a perfect dislocation in this study was about 29 eV, which was calculated on the basis of the literatures [20]. The free energy available from thermal fluctuation at the typical solidification temperature (about 1600 °C) was about 10 eV, i.e. of the same magnitude as the activation energy. As a result, the formation of dislocations became quite easy with the additional energy provided by the local strain caused by local substrate heating. Therefore, the formation of many positive and negative dislocations could be interpreted as follows.
Severe thermal fluctuations result in the formation of larger quantities of dislocations. Additionally, note that the epitaxial growth by thermally sprayed technique is a rapid non-equilibrium process. The cooling rate of the films can be as large as 10^4–10^6 K/s [21]. In this case, the thermal energy may spread in the form of thermal waves. As a result, the molten films are turbulent because of thermal fluctuations. These wave-like thermal fluctuations result in the formation of high-density dislocations including both positive and negative dislocations. Concomitantly, a fair number of stacking faults arise to alleviate the local high strains

between the dislocation pairs. In addition, rapid cooling probably prohibits the annihilation and climb of these supersaturated dislocations. As a result, super-high-density dislocations are retained and observed at room temperature. In brief, both large lattice mismatch and sufficient thermal fluctuation create high-density dislocations.

The high-density dislocations lead to a weak-bonded interface. This may be the reason for the interface delamination of LZ splats regardless of its epitaxial growth at deposition temperature of 300 °C. It should be noted that these high-density dislocations arise in single-crystal/epitaxial films in the present study. The densities in general polycrystalline coatings must be much higher. These interfacial defects significantly contribute to interface cracking and may be the essential formation mechanism of lamellar pores in thermally sprayed coatings.

5 Conclusion

In the present study, the interfacial microstructures of thermally sprayed heteroepitaxial LZ splats were identified. An anomalous heteroepitaxial growth with super-high dislocation densities and cracking strains in LZ splats was observed. The super-high dislocations in heteroepitaxial LZ splats were attributed to three aspects, namely, the intrinsic mismatch between LZ and YSZ (the value ~4.72%), the extra mismatch due to the constraint of locally-heated substrate, and the high cooling rate and severe thermal fluctuations leading to retaining dislocation pairs. The large cracking strains (the value ~1.5%) resulted from the different shrinkage between the epitaxial film and substrate which was essentially due to the the constraint of locally-heated substrate. The constraint of locally-heated substrate was the essential difference between conventional and present epitaxial growth.

The super-high-density dislocations lead to a weakly-bonded interface and a weak cracking resistance. The large cracking strains lead to a strong cracking motivation. Both of them become the essential disadvantage of LZ/YSZ double-layer TBCs and result in easy delamination at LZ/YSZ interfaces during TBCs service.

Acknowledgements

This work is supported by National Basic Research Program (Nos. 2013CB035701), the Fundamental Research Funds for the Central Universities, and the National Program for Support of Top-notch Young Professionals. Authors would like to thank Engineer Chuan-sheng Ma for his help with manipulation of TITAN.

Reference

1. N.P. Padture, M. Gell, E.H. Jordan: Thermal barrier coatings for gas-turbine engine applications. Science 296 (2002), Issue 5566, pp. 280/4.
2. N.P. Padture: Advanced structural ceramics in aerospace propulsion. Nat Mater 15 (2016), Issue 8, pp. 804/9.
3. Y.Z. Xing, C.J. Li, G.J. Yang: Influence of through-lamella grain growth on ionic conductivity of plasma-sprayed yttria-stabilized zirconia as an electrolyte in solid oxide fuel cells. J Power Sources 176 (2008), Issue 1, pp. 31/8.
4. H. Dong, G.J. Yang, H.N. Cai, C.J. Li: Propagation feature of cracks in plasma-sprayed YSZ coatings under gradient thermal cycling. Ceram Int 41 (2015), Issue 3, pp. 3481/9.
5. C.J. Li, Y. Li, G.J. Yang, C.X. Li: Evolution of Lamellar Interface Cracks During Isothermal Cyclic Test of Plasma-Sprayed 8YSZ Coating with a Columnar-Structured YSZ Interlayer. J Therm Spray Techn 22 (2013), Issue 8, pp. 1374/82.
6. S. Ahmadian, E.H. Jordan: Explanation of the effect of rapid cycling on oxidation, rumpling, microcracking and lifetime of air plasma sprayed thermal barrier coatings. Surf Coat Tech 244 (2014), pp. 109/16.
7. W. Ma, S. Gong, H. Li, H. Xu: Novel thermal barrier coatings based on $La_2Ce_2O_7$/8YSZ double-ceramic-layer systems deposited by electron beam physical vapor deposition. Surf Coat Tech 202 (2008), Issue 12, pp. 2704/08 .
8. L. Wang, Y. Wang, X.G. Sun, J.Q. He, Z.Y. Pan, C.H. Wang: Thermal shock behavior of 8YSZ and double-ceramic-layer $La_2Zr_2O_7$ /8YSZ thermal barrier coatings fabricated by atmospheric plasma spraying. Ceram Int 38 (2012), Issue 5, pp. 3595/606.
9. Z.Y.-M. Cheng B, Yang N, et al: Sintering-induced delamination of thermal barrier coatings by gradient thermal cyclic test. J Am Ceram Soc. (2017).
10. L. Chen, G.J. Yang, C.X. Li, C.J. Li: Hierarchical Formation of Intrasplat Cracks in Thermal Spray Ceramic Coatings. J Therm Spray Techn 25 (2016), Issue 5, pp. 959/70.
11. L. Chen, G.-J. Yang, C.-X. Li, C.-J. Li: Edge Effect on Crack Patterns in Thermally Sprayed Ceramic Splats. J Therm Spray Techn (2016).
12. L. Chen, G.J. Yang, C.X. Li: Formation of Lamellar Pores for Splats via Interfacial or Sub-interfacial Delamination at Chemically Bonded Region. J Therm Spray Techn (2017).
13. S. Kaiser, H. Preis, W. Gebhardt, M. Stutzmann, A. Rosenauer, D. Gerthsen: Quantitative Transmission Electrorn Microscopy Investigation of the Relaxation by Misfit Dislocations Confined at the Interface of GaN/Al_2O_3(0001). Japanese Journal of Applied Physics 37 (1999), Issue 1, pp. 84/9.

14. J.H. Joo, K.J. Greenberg, M. Baram, D.R. Clarke, E.L. Hu: Aqueous Epitaxial Growth of ZnO on Single Crystalline Au Microplates. Cryst Growth Des 13 (2013), Issue 3, pp. 986/91.
15. S.P.T. Goodier, S.P. Timoshenko, J.N. Goodier: Theory of elasticity. MacGraw Hill International Book (1970).
16. Y. Li, G.C. Weatherlyy, M. Niewczas: TEM studies of stress relaxation in GaAsN and GaP thin films. Philosophical Magazine 85 (2005), Issue 26-27, pp. 3073/90.
17. L.B. Freund: The Mechanics of Dislocations in Strained-Layer Semiconductor-Materials. Adv Appl Mech 30 (1994), pp. 1/66.
18. M.D. Thouless, E. Olsson, A. Gupta: Cracking of Brittle Splats on Elastic Substrates. Acta Metall Mater 40 (1992), Issue 6, pp. 1287/92.
19. X. Wu, G.C. Weatherly: Composition modulations in tensile strained In1-xGaxAsyP1-y films grown on (100) InP substrates. Journal of Crystal Growth 233 (2001), Issue 1-2, pp. 88/98.
20. J.P. Hirth, J. Lothe, T. Mura: Theory of Dislocations (2nd ed.). JOHN WILEY & SONS, New York (1982).
21. M. Vardelle, A. Vardelle, A.C. Leger, P. Fauchais, D. Gobin: Influence of Particle Parameters at Impact on Splat Formation and Solidification in Plasma Spraying Processes. J Therm Spray Techn 4 (1995), Issue 1, pp. 50/8.

Morphology and durability of $Sm_2Zr_2O_7$+8YSZ TBC systems during oxidation test at temperature 1100°C

S. Jucha, G. Moskal, M. Mikuśkiewicz, Gliwice / PL

The zirconates of rare earth elements, such as $Sm_2Zr_2O_7$ ceramic, can be an alternative material, for commercially and widely used for insulation layers in thermal barrier coatings (TBC) systems, zirconia modified by yttria (8YSZ). Presented article described the internal morphology of composite TBC coatings of $Sm_2Zr_2O_7$+8YSZ type with different ratio of both used to coatings deposition powders (25/75, 50/50 and 75/25) as well as the TBC of double ceramic layer (DCL) type with a 8YSZ internal layer and the outer layer of $Sm_2Zr_2O_7$ type. The main subject of presented investigations is related with the oxidation resistance of TBC systems during static oxidation test at temperature 1100°C. In this case, the special emphasis was took on characterization of thermally grown oxides (TGO) zone thickness where the most important phenomena related with overall live-time of TBC system usually take place.

1 Introduction

Thermal barrier coatings (TBC) have been widely used in production of high-temperature and corrosion-resistant coatings. Their aim is protection substrate materials against thermal exposition as well as corrosion and oxidation processes. This type of coatings is know from 40 years but still is exist a continuing need to improve the material and technological solutions for obtain new type TBC`s systems with better insulation properties and durability during real condition working [1,2]. Materials usually used for TBC deposition can be safety classified as the multifunctional systems. They have to fulfil many criteria, where do not fulfil any of them, exclude this material. These criteria include among others: low value of thermal conductivity (below 2,5 W/mK), high coefficient of thermal expansion (above 10^{-5}°C^{-1}), high working temperature (up to 1600°C), the ability to experiencing deformation under thermal expansion stresses which are associated with cyclic heating and cooling as well as the compatibility of deformation all materials including in TBC system. Each material has its own thermal expansion coefficient, the differences between them can not be too large, to avoid the delamination of coat. Moreover the desired material should be characterized by the absence of the phase transformations in the range of operating temperature, the erosion resistance and ability to work in repeated of thermal cycles [3-5]. Properties which are the subject of current research include: high-temperature oxidation resistant, required due to the operation conditions of this coating under air environment. The secon subject is thermodynamic compatibility of the coating material with thermally grown oxides (TGO) zone, consisting mainly of Al_2O_3, formed by oxidation the bond coat (BC). Usually BC is made from Ni(Co)CrAlY type of alloys or aluminides.

The main parameter that characterizes the usefulness of the TBC layer to work under oxidative conditions is its ability to protect TGO layer against their growth i.e. blocking oxygen diffusion to the substrate. With the increase the TGO layer is followed decrease of strength TBC coating and increases the probability of delamination [3, 6-7]. Especially unfavorable regarding the relevant strength of the coating to delamination, as result of work in oxidizing environment is TGO growth on the tops of NiCrAlY rougness profile, because that in these points comes the most often to cracking the ceramic top-coat and delamination. Fig. 1 is a diagram illustrating described model of destruction the TBC coating.

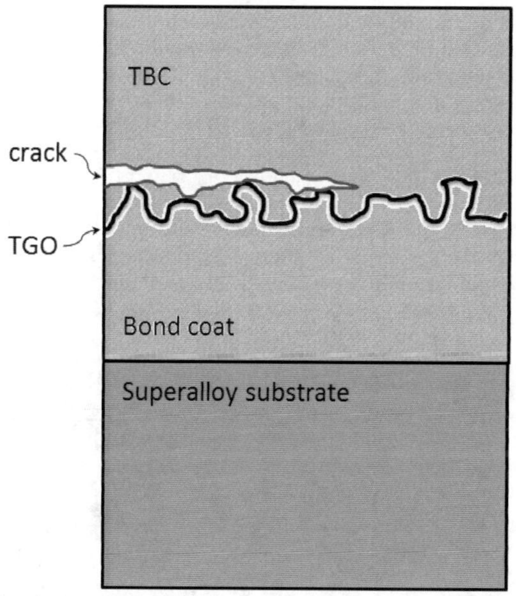

Fig.1. Schematic illustartion of cracking the TBC coating.

2 Research methodology

The investigation has been carried out on samples based on compounds from AMS5599 Ni-based superalloy with NiCrAlY bond-coat and different types of ceramic top-coats. 8YSZ and $Sm_2Zr_2O_7$ feedstock powders in various proportions (75%:25%; 50%:50%; 25%:75%), marked as Zr75Sm25, Zr50Sm50, Zr25Sm75 and the sample covered with a DCL TBC system 8YSZ/$Sm_2Zr_2O_7$ marked as (100/100 double layer) were used. Coatings have been depsoited by atmospheric thermal spraying (APS) method. This method allows use a wide variety of powder material and for ensure high standards is fully automated. Research were carried out in oxidation environment at 1100°C by 2, 10, 48, 175 and 500 h.

The samples after annealing in oxidation condition, were respectively investigated by: X-ray phase analysis by XRD method (X-ray diffraction, Jeol JDX-7S), moreover, microstructure observation, by scanning electron microscope (*SEM*), energy dispersion spectroscopy (EDS) and electron backscatter diffraction analysis (EBSD) ware made. This investigation were made on the Hitachi 3400N scanning microscopy. Makro observation of the annealing effects was carried out on stereoscopic microscope (Nicon Eclipse MA200), whereas measurements of the observed changes in the thickness of TGO layer were made on light optical microscopy (Nikon SMZ 745T).

3 Results

First step of investigation was realted with TGO thickness measurement on each samples after 2, 10, 48, 175 and 500 h of deposition. Based on the changes of TGO zone thickness, were created graphs of kinetics of their growth in time. It was important to identify the best checking up material in oxidation environment. Because the surface area between BC and TBC is characterized by significant level od rougness (technological requirement), and in consequence the strong differencess in TGO thickness was observed, the measurements of TGO thicknes were performed in two places i.e. at the vertices of BC layer and the bottom. The places of measurement is presented at fig. 2 and the final results of this analysis was showed at fig. 3.

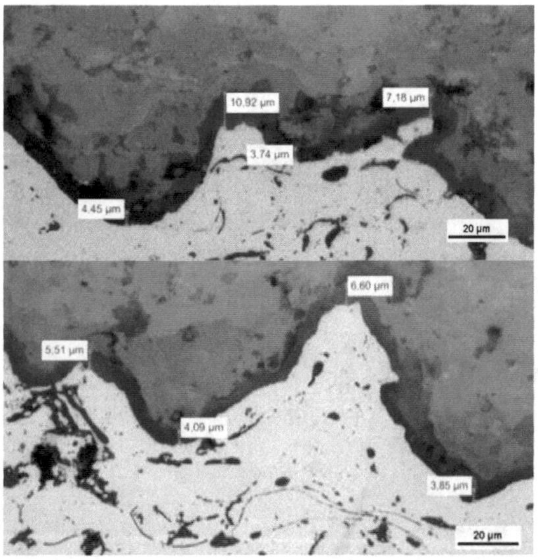

Fig.2. Examples of the measurements places of the TGO thickness.

Among the tested materials, those having a gradient 100/100 coatings exhibit the lowest value of TGO thickness in the case of vertices places of measurement. The secon place was dedicated to TBC system of Zr75Sm25 type where the volume of 8YSZ ceramic was realtivelly high. The others types of coating exhibite much higher level of oxide zone gowth, what was revealed at fig. 3. The vertices of bond-coat profile rougness is critical place from final durability of TBC systems point of view, beceause in this places the process of cracking nucleation and growth is started. These are the places to which oxygen arrives the fastest and therefore observed the most rapid growth of TGO layer and in the next step the highest level of tensile strength can be observed. The best behaviour from this point of view revealed TBC system of 100/100 type which average thickness after 48, 175 and 500h in oxidation environment at a temperature 1100°C amounted to 2,36 µm, 3,73 µm and 4,63 µm. Good protection provided also the coating marked as Zr75Sm25, with an average thickness measured at the same time intervals and places (vertices) at 3,15 µm, 4,98 µm and finally 5,37 µm. Despite the initial high-speed growth TGO layer, that after 100h it were shown, like in the case double layer coating, the logarithmic course of the oxidation kinetics. In the case of coatings containing lager amounts of $Sm_2Zr_2O_7$ ceramic, that althought initially the logarithmic character of the oxidation kinetics course, already after 48 hour it is able to observe change the kinetics course in the direction of linear course. TGO thickness measurements into the samples Zr50Sm50 and Zr25Sm75, had after 500 h of annealingin oxidation environment, respectively average thickness 7,23 µm and 7,08 µm respectively. Observed differences are related with phases constituent of TGO zone in this area, where the formation of more complexed oxides than alumina was usually observed for example oxides of spinel type.

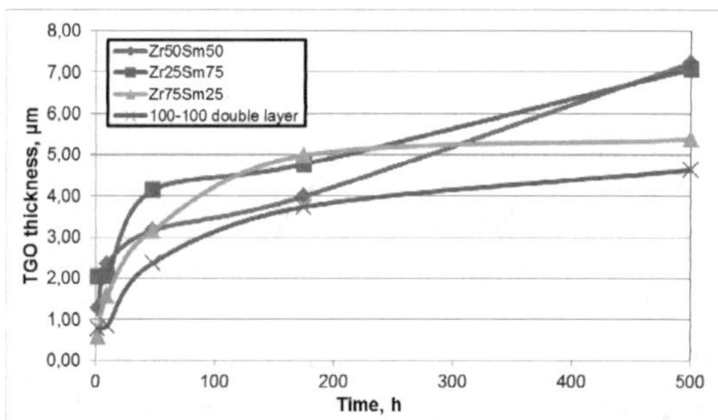

Fig.3. TGO zone growth, measurements carried out at the vertices of BC.

The situation is look somewhat different in the case of measurements carried out at the bottom of bond-coat profile roughness as was showed at fig. 4. The course of the TGO layer growth in all tested samples have similar to each other logarithmic character of the oxidation kinetics course. Differences between measurments of TGO thickness are within the range of the standard deviation. It can`t indicate the superiority any of the materials. Measurements carried out after 48, 175 and 500h are within the range of respectively 1,77-2,13 µm, 2,66-3,12 µm and 3,75-3,97 µm. This is the effect of similar phase constituent of TGO zone in this area, where the Al_2O_3 oxide is usually typical structural elements. In consequence the local thickness of this zone is very similar.

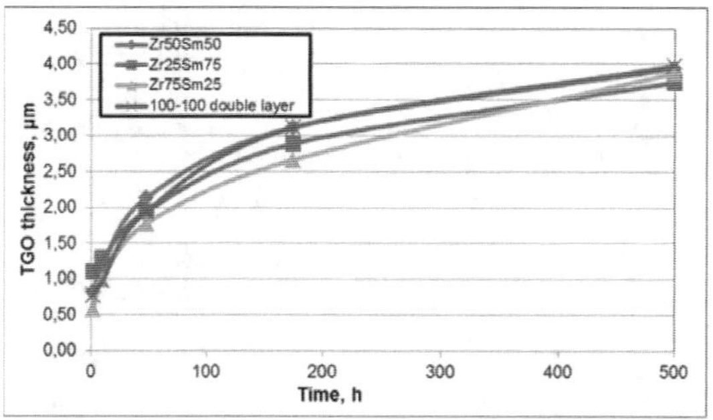

Fig. 4. TGO layer growth, measurements carried out at the buttom of BC.

Presented analysis showed that application of $Sm_2Zr_2O_7$ as a structural elements of deposited thermal barrier coatings has a strong influence on overall TGO thickness increasing in area of vertices of bond-coat profile roughness. From the other site high level of zirconate ceramic in top-coat due to spallation effect of TBC system as was presented at fig. 5.

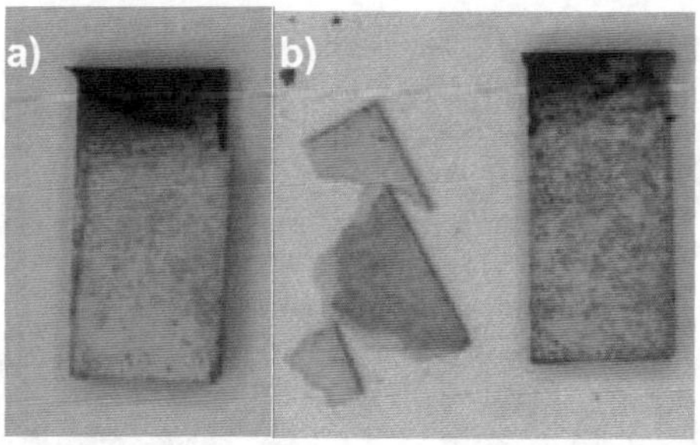

Fig.5. Macrophotography of Zr25Sm75 samples, annealing in 1100°C in oxidizing environment by a) 45h; b) 500h.

This negative effect of spallation can be related with lower compatibility of this compound with Al_2O_3, and formation in consequence complexed oxides of rare elements with alumina. Additionally in the ceramic, near the TGO layer occure yttrium diffusion from 8YSZ, therefore it is able to observe, with using EBSD locally occurring tetragonal and monoclinic varieties of zirconium oxide. In unstabilized zirconium oxide, occure phase transformation which in the temperature range of operation, causes a significant decrease their durability. As given that this is a martensitic transformation it involves with change a volume of zirconium oxide, which results in formation a significant stress inside the material [8]. Incompatibility most clearly highlighted in the case of Zr25Sm75 samples. After only 48h in 1100°C in oxidizing environment, comes to completely breakdown of the coating. Another element related with crack formations are phenomena on the top surfaces of TBC systems. Fig.6. a) and b) presents typical surface morphology of the layer made by thermal spraying technology.

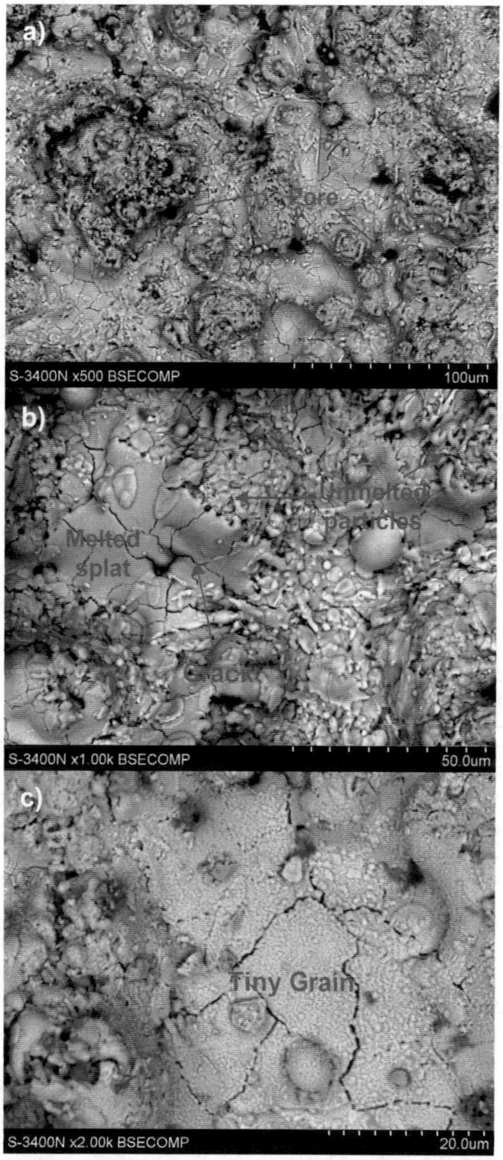

Fig.6. Microphotography of 100/100 double layer samples a) material after spraying, 500x; b) material after spraying, 1000x; c) material after annealing in 1100°C in oxidizing environment by 175h, mag. 2000x.

The coating consists of melted splats, partially melted and unmelted particles. On the sprayed coating surface, can be observed lot of fine cracks arising from the rapid heating and cooling during spraying. The coating has also many pores. Those pores and fine cracks does not constitute the disadvantage of the layer, but rather an advantage. Thanks them, comes to the stress relaxation inside material, resulting from cycle heating operation. As a result of operation at high temperature development and growth of larger cracks can be observed in ceramic top-coat materials. As long as, those cracks are vertical, that allow them o reduce stress during heating and cooling without much impact on the reduction of thermal barrer effect. The situation changes when it starts to development of horizontal cracks. Their propagation leads to delamination of coating which in turn has directly affects at the rapid degradation of superalloy substrate. Fig. 6 c) presents coating after annealing in 1100°C in oxidizing environment, by 175h. It could be observed additionall effect of high-temperature exposure in the form numerous, small grains which were created as a result of recrystallization on the coating surface [9-10]. The cross-section micrographs of the 100/100 double layer TBC coating after annealing in 1100°C by 500h in oxidising environment are shown at Fig. 7. Althought after oxidation proces in double ceramic layer was observed growth of TGO, that there were no alarming signs of damage the coatings. The TBC layer has a compact character i.e. there were no there were no spallation areas as well as there was no notice the presence of large horizontal cracks. Both the $Sm_2Zr_2O_7$ layer and 8YSZ layer, characterized by the presence of porosity and tiny cracks Fig. 7a. In larger pores can be observed fine grine zone Fig. 7b.
Zr50Sm50 and Zr25Sm75 samples didn't show such high resistant to working at 1100°C. Already at the samples annealed by 48h it can be observed clearly visible horizontal cracks, leading to final delamination (fig. 8). Among the coating prepared by spraying mixtures of two ceramics, the best high-temperature corrosion resistant in oxidizing environment has demonstrated Zr75Sm25 coating (fig. 8), where after 500 h of exposure no significant effects of destruction were observed.

4 Summary and conclusions

Results of the investigations on high-temperature corrosion resistant in oxidizing environment of different types of TBC systems based on samarium zirconates ceramic showed the superiority double layered TBC coating over the mixtures of two ceramics (8YSZ and $Sm_2Zr_2O_7$). The DCL gradient coatings due to it construction have lot of advantages. They combine the advantages of two types ceramics: 8YSZ by excellent protection against high-temperature oxidation and thus reducing the speed of the TGO layer growth and in consequence it thikness, as well as keeping superior thermochemical resistance, and $Sm_2Zr_2O_7$ which has much lower thermal conductivity, and beter resistance to AVA, CMAS or liquid salt deposits. These results confirm other studies, as well as literature data e.g. [11-21]. Not all kind of ceramics can be combinated. The limiting factor is differences between thermal expansion coefficient. Too large difference leads to cracking and as the results, delamination or spallation of the coating.

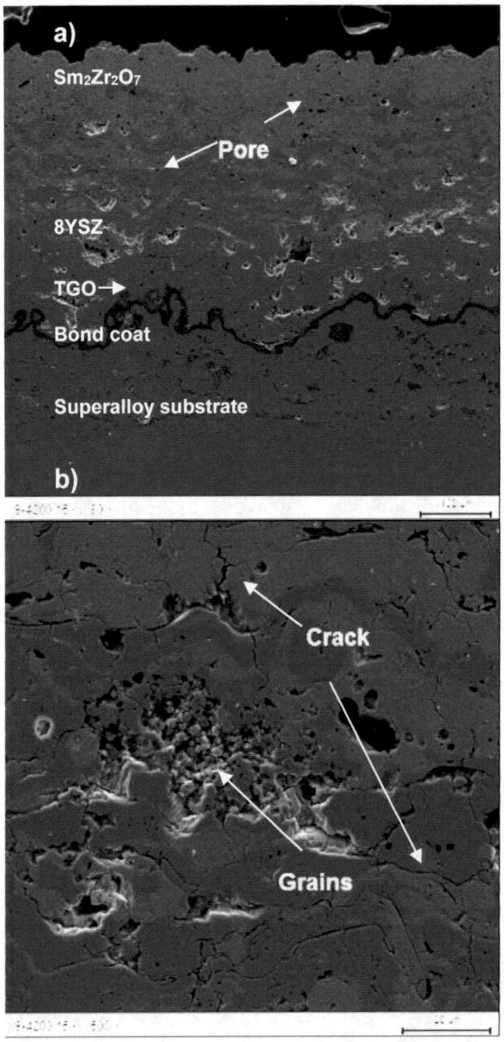

Fig.7. The cross-section micrographs of the 100-100 double layer TBC coating after annealing in 1100°C by 500h a) an entire layer view, x80 b) areas with visible micro-cracks, pores and fine grine zone, x500.

The sample covered with a gradient 8YSZ at the substrate and the outer layer $Sm_2Zr_2O_7$, demonstrated the smallest increase of TGO layer. The result was achieved thanks good thermochemical compatibility of 8YSZ with Al_2O_3 and better blockade of oxygen diffusion by yttria stabilized zirconia layer. In the case of coatings with samarium zirconate this efect was much lower due to high level of internall defects promoted strong effect of oxygen diffusion (but lowering thermal conductivity). In consequence higher thickness of TGO zone was observed especially in the area of vertices of bond-coat profile roughness.

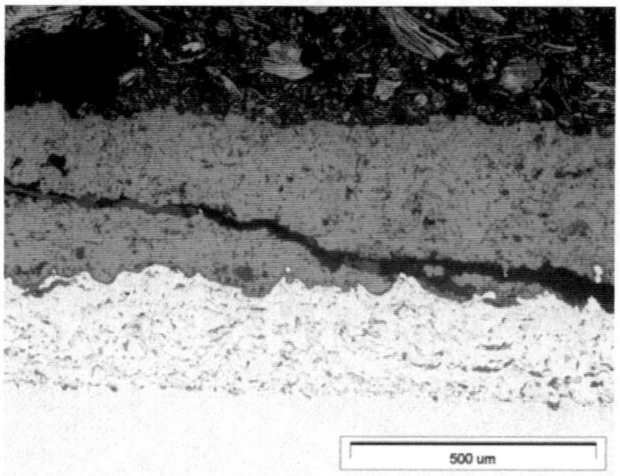

Fig.8. The cross-section micrographs of the Zr50Sm50 TBC system after oxidation test in 1100°C by 48h.

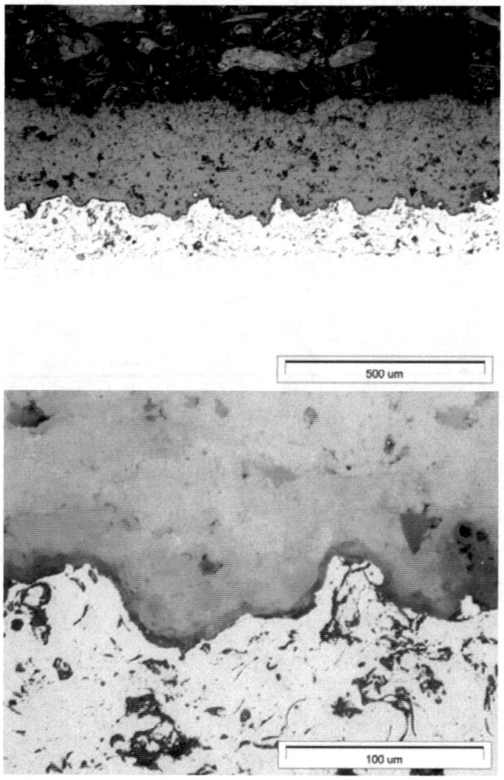

Fig.9. The cross-section micrographs of the Zr75Sm25 TBC system after oxidation test in 1100°C by 500 h.

Although, from the point of view the resistance to high-temperature oxidation, the samarium zirconate does not bring any positive changes, that the main conclusion flowing from this research is that, it does not also no changes affecting adversely. It allows to use $Sm_2Zr_2O_7$ in double layer coatings as outer layer. This material has many better properties than 8YSZ, including: lower thermal conductivity, much beter resistance to AVA, CMAS or salty seawater.

Good connection between 8YSZ and $Sm_2Zr_2O_7$ confirm the possibility to use this kind double layer coatings in place of, currently widely used 8YSZ TBC coatings.

Acknowledgments: The research has been supported by National Science Centre within Opus scheme, under contract UMO-2016/21/D/ST8/01687.

5 Literature

[1] Sohn Y.H., Lee E.Y., Nagaraj B.A., Biederman R.R. and R.D Sisson Jr.: Microstructural characterization of thermal barrier coatings on high pressure turbine blades. Surface and Coatings Technology. 146-147 (2001), pp. 132/39;

[2] Hejwowski T. and A. Weroński: The effect of thermal barrier coatings on diesel engine performance. Vacuum 65 (2202), Issues 3-4, pp. 427/32.

[3] Clarke D.R. and S.R. Phillpot: Thermal barrier coating materials. Materials Today 8 (2005), Issue 6, pp. 22/9.

[4] Moskal G., Rozmysłowska A., Gazda A. and M. Homa: Wybrane termofizyczne właściwości proszków cyrkonianowych na bazie pierwiastków ziem rzadkich typu $RE_2Zr_2O_7$ (RE - Gd, La, Sm, Nd) przeznaczonych do natryskiwania cieplnego powłokowych warstw barierowych, Prace Instytutu Odlewnictwa XLIX (2009), Zeszyt 4, pp. 15/26;

[5] Moskal G., Swadźba L. and A. Rozmysłowska: Kierunki rozwoju powłokowych barier cieplnych, XXXVI Szkoła Inżynierii Materiałowej Kraków – Krynica 23-26 IX 2008.

[6] Zhao H., Begley M.R., Heuer A.,Sharghi-Moshtaghin R., H.N.G Wadley: Reaction, transformation and delamination of samarium zirconate thermal barrier coatings, Surface and Coatings Technology 205 (2011), Issue 19, p. 4355/65.

[7] Zhao H., Yu Z. and H.N.G. Wadley: The influence of coating compliance on the delamination of thermal barrier coatings. Surface and Coatings Technology 204 (2010), Issue 15, pp. 2432/41.

[8] Ch. Wang: Experimental and Computational phase studies of the ZrO_2-based systems for thermal barrier coatings, Dissertation an der Universität Stuttgart, September 2006.

[9] Zhao S., Gu L., Zhao Y., Huang W., Zhu L., Fan X., Zou B and X. Cao: Thermal cycling behavior and failure mechanism of $La_2(Zr_{0.7}Ce_{0.3})_2O_7/Eu^{3+}$ - doped 8YSZ thermal barrier coating prepared by atmospheric plasma spraying. Journal of Alloys and Compounds 580 (2013), pp. 101/07.

[10] Kawasaki A. and R. Watanabe: Thermal fracture behavior of metal/ceramic functionally graded materials. Engineering Fracture Mechanics 69 (2002), Issues 14-16, pp. 1713/28.
[11] Poerschke D.L. and C.G. Levi: Effects of cation substitution and temperature on the interaction between thermal barrier oxides and molten CMAS. Journal of the European Ceramic Society 35 (2015), Issue 2, pp. 681-691.
[12] A. Aygun: Novel thermal barrier coatings (TBC) that are resistant to high temperature attack by CaO-Mgo-Al_2O_3-SiO_2 (CMAS) glassy deposits, Dissertation presented in partial fulfillment of the requirements for the degree Doctor of Philosophy in the Graduate School of The Ohio State University, 2008
[13] J.M. Drexler: Thermal barrier coatings resistant to glassy deposits, Dissertation presented in partial fulfillment of the requirements for the degree Doctor of Philosophy in the Graduate School of The Ohio State University, 2011
[14] Schulz U. and W. Braue: Degradation of La2Zr2O7 and other novel EB-PVD thermal barrier coatings by CMAS (CaO–MgO–Al2O3–SiO2) and volcanic ash deposits. Surface and Coatings Technology 235 (2013), pp. 165/73.
[15] Drexler J.M., Ortiz A.L. and N.P. Padture: Composition effects of thermal barrier coating ceramics on their interaction with molten Ca–Mg–Al-silicate (CMAS) glass. Acta Materiallia 60 (2012), Issue 15, pp. 5437/47.
[16] Krämer S., Yang J. and C.G. Levi: Infiltration-inhibiting reaction of gadolinium zirconate thermal barrier coatings with CMAS melts. Journal of American Ceramic Society 91 (2008), Issue 2, pp. 576/83.
[17] Levi C.G., Hutchinson J.W., Vidal-Sétif M-H, and C.A. Johnson: Environmental degradation of thermal-barrier coatings by molten deposits. MRS Bulletin 37 (2012), Issue 10, pp. 932/41.
[18] Drexler J.M., Gledhill A.D., Shinoda K., et all, Jet Engine Coatings for Resisting Volcanic Ash Damage, Adv.Mater., (2011), Issue 23, pp. 2419/24;
[19] Gledhill A.D., Reddy K.M., Drexler J.M., et all, Mitigation of damage from molten fly ash to air-plasma-sprayed thermal barier coatings, Materials Science and Engineering A 528, (2011), pp. 7214/21;
[20] Xu Z., He L., Mu R., et all, Hot corrosion behavior of rare earth zirconates and yttria partially stabilized zirconia thermal barrier coatings, Surface and Coatings Technology, Vol. 204, Issues 21–22, August (2010), pp. 3652/61
[21] Habibi M.H., Hot corrosion behaviour of new candidates for thermal barrier coatings application in turbine simulated environments, A Dissertation Submitted to the Graduate Faculty of the Louisiana State University and Agricultural and Mechanical College in partial fulfillment of the requirements for the degree of Doctor of Philosophy in The Department of Mechanical and Industrial Engineering, May 2014.

Development of methodology of cohesive strength evaluation of thermally sprayed coatings

J. Schubert, Z. Česánek, Pilsen / CZ

Measuring the cohesive strength of thermally sprayed coatings is relatively difficult matter, which can be accessed in many directions. This issue is nowadays solved by use of Scratch test method. This method is not completely sufficient for the cohesive strength testing because the coating is under load of combined stresses during the Scratch test. The reason to develop this method was need for exact measurement of tensile cohesion toughness of thermally sprayed coatings, which could provide results as close to a classic tensile test as possible. Another reason for development of this method was the impossibility of direct comparison with results obtained by other methods. Tested coatings were prepared using HP / HVOF (Stellite 6, NiCrBSi, CrC-NiCr and Hastelloy C-276). These coatings were selected as commonly used in commercial sector and on because of rising customer demand for ability to provide such coating characteristics. The tested coatings were evaluated in terms of cohesive strength (method based on tensile strength test). Final fractures were evaluated by scanning electron microscopy and EDS analysis. As expected higher cohesive strength showed metallic coatings with top results of coating Stellite 6. Carbide coatings showed approximately third of the cohesion strength in comparison with metal based coating.

1 Introduction

This paper is based on research which was conducted with aim to develop simple and fast method allowing to measure cohesive strength of coatings. Another condition was that the method has to be able provide reputable results with low costs. The study is part of more complex research focused on influence of coating on mechanical properties of substrate material (mainly fatigue life). Reason for development of such method was demand of customers which asks for precise specification of mechanical properties of coatings used commercially. The method described further in this paper proved to fulfil all requirement and can provide all necessary data together with sufficient repeatability. There are few drawbacks related to measurement of carbide based coatings which leads to a slightly higher volatility of measured results. The reasons for this deviation are described father in this paper. This issue is nowadays solved by use of Scratch test method [1] which is not completely sufficient for such kind of testing. The demand was for test as close to standard tensile testing as possible. As it was mentioned above, this method will be used with combination with other mechanical properties evaluation processes. This complete overview of properties is necessary to determine the influence of coatings on fatigue life of coating/substrate system. As it is clear, there are many factors affecting fatigue life. One of them is residual stress mainly located around interface between coating and substrate material. This factor can also play role in affecting nucleation of cracks in case of cohesive (tensile) stress test [2,3,4].
Furthermore, cohesive strength is one of at the most influencing factor determining wear mechanism of coated parts and it is possible to find correlation between them [5].

2 Experiment

2.1 Coating deposition

Commercially available powders were used for sample preparation: Amperit 588.074 (Cr_3C_2-25%NiCr) with a particle size distribution suitable for HVOF (-45+15 mm); Alloy-6 (Stellite 6, 484.33) with a particle size distribution suitable for HVOF (-45+15 mm); M-771.33 (NiCrBSi) with a particle size distribution suitable for HVOF (-53+20 mm) and M-341.33 (Alloy-276) with a particle size distribution suitable for HVOF (-53+20 mm). All coatings were deposited by HP/HVOF technology with JP-5000 torch from the company TAFA Incorporated. Material of substrate samples was 316L stainless steel. Before coating process samples were degreased and grit blasted with brown corundum (with particle distribution F22). Previously optimized spray parameters were used for each coating. Coatings were deposited on substrate material and their surface was further metallographically grinded to the surface roughness Ra_{max}=1 µm.

2.2 Experimental equipment

Selected coatings were tested on cohesion strength properties using newly constructed device. The method is based two cylindric arts inserted in to each other (shown in **Fig. 1**). During the spray process both parts of the sample are tightened together by threaded bolt. Dividing plane is coated with appropriate coating in 40 mm width and with 400 µm thickness. Afterwards the sample is subjected to tensile test (in this case provided by device ZWICK Z250 in laboratories of VZÚ Plzeň). Design of this sample together with appropriate sample mounting should provide elimination of shear force during the test procedure.
After the tensile tests the samples were further examined. First of all, the microstructure of each coating was captured and evaluated by scanning electron microscopy (SEM – FEI Quanta 200).

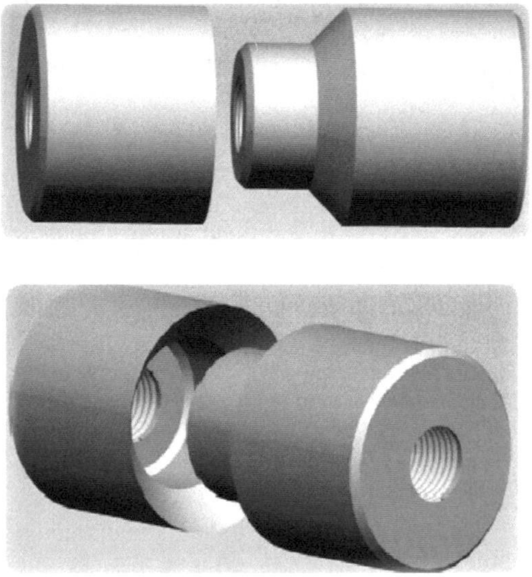

Fig. 1 Design of test sample

2.3 Measurement conditions

The measurement conditions during tensile test were as follows:

- Preload 10 MPa
- Load sped 0,0001 1/s

2.4 Results

High temperature corrosion test was based on corrosive aggressive action of Na_2SO_4 and V_2O_5 salt mixture at the temperature of 750 °C. The salt mixture in ratio of 40/60 (mass percentage) was mixed with alcohol and applied on specimen surface in an amount of 3-4 g/cm^2. Prior to salt mixture application, specimen surface was rinsed with alcohol and heated in an oven at 250 °C in order to ensure sufficient adhesion of salt layer. After the application of salt layer, specimens were left 3 hours in the oven at 100 °C in order to evaporate ethyl alcohol and to dry the salt layer. The test contains 50 cycles of heating at 750 °C for 1 hour in the silicon carbide oven and subsequent cooling for 20 minutes at room temperature. Specimens were weighed with a precision of 0.0001 g before and during the test.

3 Results

3.1 Scanning electron microscopy evaluation

The reason to analyses microstructure was to prove that the samples were prepared under same conditions as commercially produced coatings, and to prove that there is no difference between test sample microstructure and commercially produced microstructure of coating. Difference in microstructure (pores, particle distribution, etc.) would lead to not corresponding results. All tested coatings prepared for this experiment shown satisfactory comparable microstructure with our standards. Examples of individual microstructure in cross-section is shown in following figures (**Fig. 2-5**).

3.1.1 Cr3C2-NiCr coating

Commercially available carbide based powder (FST 588.071) was used for deposition of this coating. Following figures (**Fig. 2**) represent microstructure captured by SEM microscopy. The resulting microstructure was acknowledged as suitable and identical to commercially produced coatings.

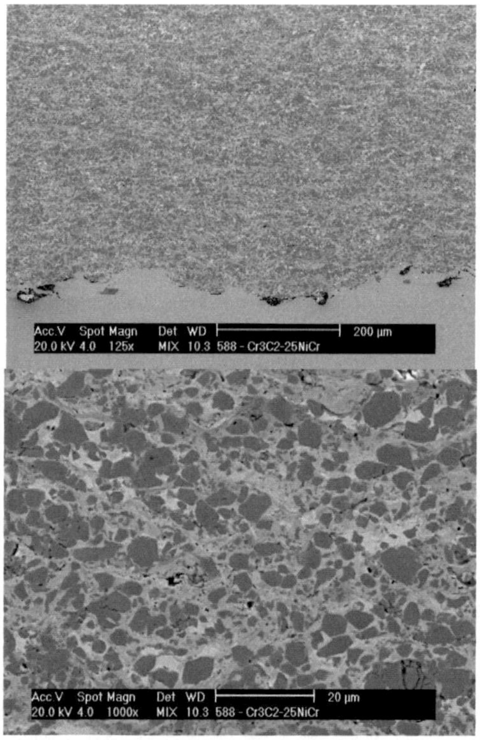

Fig. 2 CrC-NiCr coating microstructure cross-section; b) 125 magnification, a) 1000 magnification

3.1.2 Alloy-C276

Commercially available alloy based powder (FST 341.33) was used for deposition of this coating. Following figures (**Fig. 3**) represent microstructure captured by SEM microscopy. The resulting microstructure was acknowledged as suitable and identical to commercially produced coatings.

Fig. 3 Alloy C-276 coating microstructure cross-section; b) 125 magnification, a) 1000 magnification

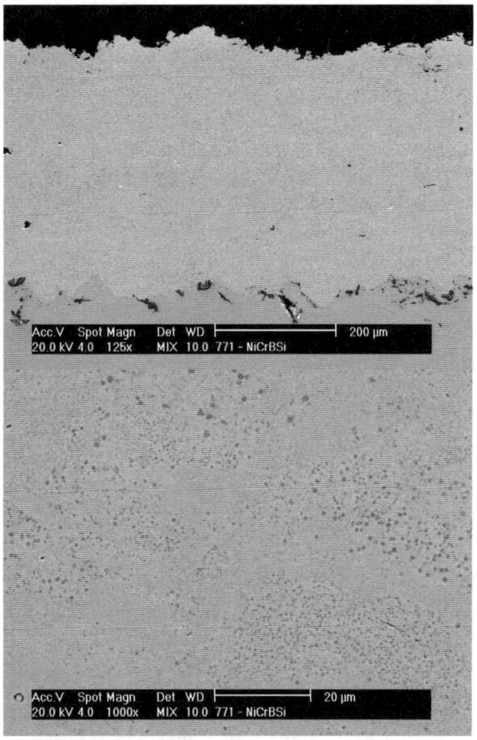

Fig. 4 NiCrBSi coating microstructure cross-section; b) 125 magnification, a) 1000 magnification

3.1.3 NiCrBSi coating

Commercially available alloy based powder (FST 771.33) was used for deposition of this coating. Following figures (**Fig. 4**) represent microstructure captured by SEM microscopy. The resulting microstructure was acknowledged as suitable and identical to commercially produced coatings.

3.1.4 Stellite 6 coating

Commercially available alloy based powder (FST 484.33) was used for deposition of this coating. Following figures (**Fig. 5**) represent microstructure captured by SEM microscopy. The resulting microstructure was acknowledged as suitable and identical to commercially produced coatings.

Fig. 5 Stellite 6 coating microstructure cross-section; b) 125 magnification, a) 1000 magnification

3.2 Cohesion strength test

Following section of this paper represents results of cohesion strength testing conducted according to previously mentioned parameters. **Fig. 6-9** represents results of cohesion strength testing of each selected coating separately. Results of average values for each coating are provided in **Tab. 1** and **Fig. 10**. Each coating was tested seven times to ensure relevant results. The lowest and the highest value obtained for each and singe coating was excluded. This procedure should ensure more accurate results. Results of coating CrC-NiCr (see **Fig. 6**) were the most volatile from all measurements. Also, this coating shown lowest cohesive strength from all tested materials. The reason of this results is the most inhomogeneous microstructure of all tested coatings (as shown in **Fig. 2**). Carbide particles embedded in metallic matrix acts as tension concentrator and serves as initiator of fracture cracking.

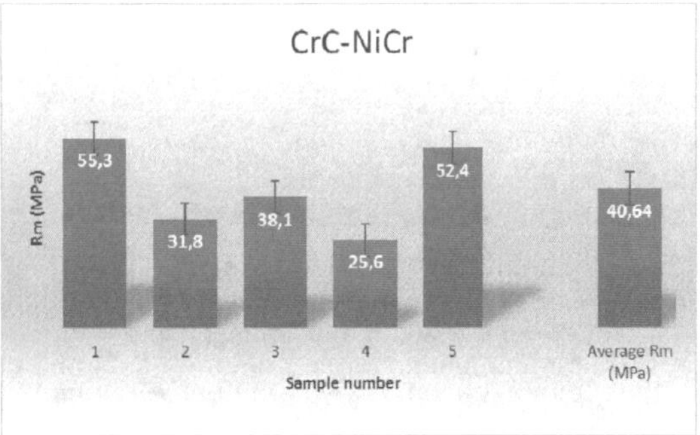

Fig. 6 Cohesive strength results of coating C3r2C-NiCr.

Coating Hastelloy C-276 (results shown in **Fig. 7**) provided lower results than expected but with low deviation. Expectations ranged to be somewhere around Stellite 6 results (displayed further in this paper). The reason for this behaviour is still unclear because even the microstructure (**Fig. 3**) is fine and homogenous. Causes of this behaviour will be further examined.

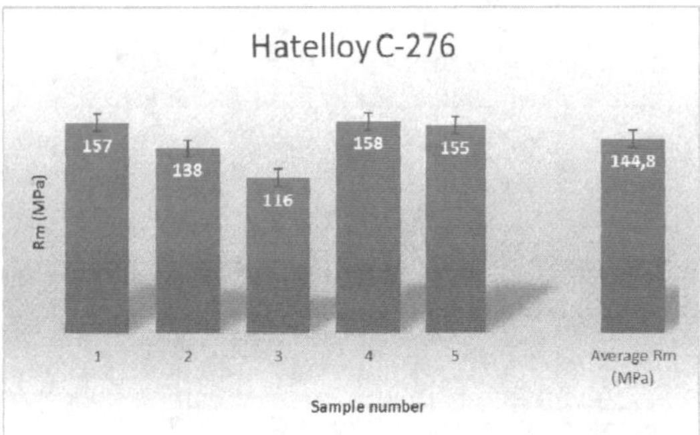

Fig. 7 Cohesive strength results of coating Alloy C-276.

Alloy based coating NiCrBSi (**Fig. 8**) showed average values between all tested coatings with decent result deviation. It is expected that reason for lower values of this coating is formation of new structures in microstructure (**Fig. 4**).

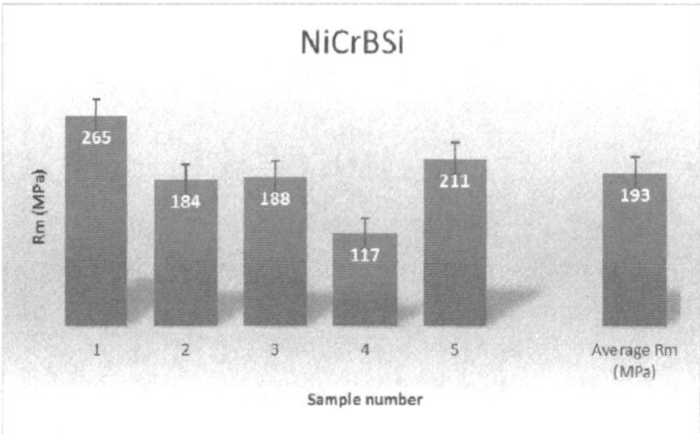

Fig 8 Cohesive strength results of coating NiCrBSi.

Last but not least alloy based coating Stellite 6 (**Fig. 9**) provided best results according to expectance. Low deviation of values ensures reproducibility of the test. This results also points at well done optimisation of thermal spray process ensuring fine microstructure with low share of flaws and defects.

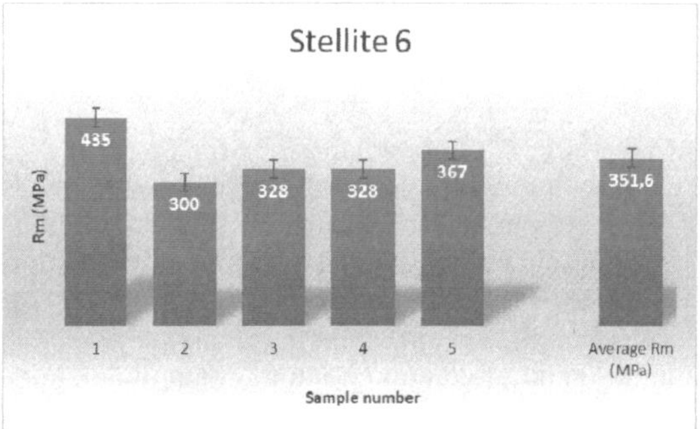

Fig. 9 Cohesive strength results of coating Stellite 6.

Tab. 1 Cohesive tensile strength average values

Coating	Rm (MPa)	Standard deviation
NiCrBSi	193	23,8
Stellite 6	351,6	23,4
CrC-NiCr	40,6	5,7
Hastelloy C-276	144,8	8,0

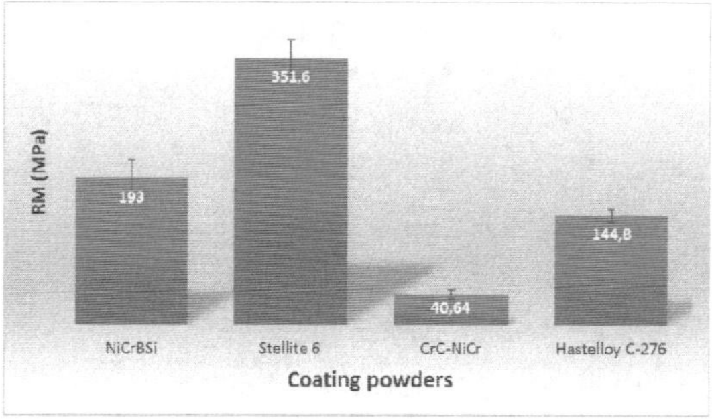

Fig. 10 Comparison of average cohesive strength of tested coatings

5 Conclusion

Results obtained by this research proved, that there are still new ways how to evaluate mechanical properties of thermally sprayed coatings. Quick, cheap and reliable determination of cohesive strength of coatings is important from many perspectives. Excluding the customers demand to know exact properties of delivered coating there are other useful applications for this method.

It is possible to correlate cohesive strength with mechanism of wear which allows to predict the behaviour of specific coating under defined stress conditions. On the other hand, this method can be used to correlate fatigue properties of coated system.

Results indicate that this method is primarily suitable for alloy based coatings where it provides replicable measurements. Alloy based coatings exhibit low volatility in measured values with low standard deviation in proportion to measured values. For this reason, this method seems to be reliable for such use.

Results obtained measuring carbide based coatings ale less convincing but still the 12.5% deviation is acceptable.

The research will continue as mentioned above and will focus on other carbide based coatings to prove or disproof usefulness of such method for this type of coatings. Furthermore, the correlation between cohesive strength and wear mechanism will be examined together with influence of residual stress influence.

Acknowledgement

The paper was prepared thanks to the project of Technology Agency of the Czech Republic no. TE01020068.

References

[1] Vencl A., Arostegui S., Favaro G., Zivic F., Mrdak M., Mitrović S., Popovic V., Evaluation of adhesion/cohesion bond strength of the thick plasma spray coatings by scratch testing on coatings cross-sections, Tribology International/Elsevier, 2011, pp 1281-1288

[2] Yang Y. C., Chang E., Influence of residual stress on bonding strength and fracture of plasma-sprayed hydroxyapatite coatings on Ti–6Al–4V substrate, Biomaterials / Elsevier, 2008, pp 1827–1836

[3] Kadolkar P., Dahotre N. B., Effect of processing parameters on the cohesive strength of laser surface engineered ceramic coatings on aluminum alloys, Materials Science and Engineering, Elsevir, 2003, pp 183–191

[4] Ivosevic M., Knight R., Kalidindi S. R., Palmese G. R., Sutter K. L., Adhesive/cohesive properties of thermally sprayed functionally graded coatings for polymer matrix composites, Journal of Thermal Spray Technology, 2005, pp 45–51

[5] Kim H. J., Kweon Y. G., Chang R. W., Wear and erosion behavior of plasma-sprayed WC-Co coatings, Journal of Thermal Spray Technology, 1194, pp 169–178

Cavitation wear characteristics of Al_2O_3-ZrO_2-ceramic coatings deposited by APS and HVOF -processes

J. Kiilakoski[1], F. Lukac[2], H. Koivuluoto[1], P. Vuoristo[1]
1: Laboratory of Materials Science, Tampere University of Technology, Tampere/FIN
2: Department of Materials Engineering, Institute of Plasma Physics CAS, Prague/CZ

Thermally sprayed ceramic coatings are used in environments requiring good wear- and corrosion resistance among others. However, a typical issue with ceramic coatings is their low impact resistance and tendency to fail catastrophically by cracking. In bulk ceramics, the Al_2O_3-ZrO_2–composition has been of interest for long since already small additions of ZrO_2 into Al_2O_3 have shown improvements in fracture toughness compared to pure Al_2O_3. Efforts are being made to induce this increased resistance to fracturing in thermally sprayed coatings as well, resulting in higher wear resistance due to a more predictable behavior and damage-tolerance. In this work, Al_2O_3-ZrO_2-coatings have been deposited by atmospheric plasma spray (APS) and high-velocity oxy-fuel spray (HVOF) processes. The wear characteristics of the coatings were evaluated with cavitation erosion, delving into the mechanics of the erosion and the resulting microstructural changes in the coatings. Evidence of phase transformation of t-ZrO_2 to m-ZrO_2 was found during the erosion. The HVOF-sprayed coating exhibited greater wear resistance against the cavitating bubbles due to its finer microstructure.

1 Introduction

Ceramic thermally sprayed coatings are often used in tribological applications, such as center press rolls of paper machines, process valves, mechanical seals etc. [1] The coatings are used for their combined corrosion and wear resistance, but often susceptibility to microscale and impact wear due to their brittleness and interlamellar cracking hinder their usage. [2]
In technical ceramics research bringing the long-known excellent toughness properties of ZrO_2 to Al_2O_3 has been in focus for some time and already small additions of ZrO_2 into Al_2O_3 have proven to give large increases in fracture toughness. [3–5] The toughness of ZrO_2 is attributed to phase transformation of tetragonal to monoclinic and the following volume change as well as the ferroelastic domain switching in tetragonal ZrO_2. [6,7]
Thermally sprayed Al_2O_3-coatings are used in some tribological applications, often due to their relatively low price. However, their tribological attributes are often inferior to other options, eg. Cr_2O_3-coatings. [8,9] Therefore, the possibility of improved properties and new applications for thermally sprayed Al_2O_3-based coatings has led to research on achieving the desired phase transformation ability of ZrO_2 also into Al_2O_3-ZrO_2 –coatings. Oberste-Berghaus et al. [10] have found increased crack propagation resistance of Al_2O_3-ZrO_2 sprayed with HVOF as compared to APS due to the retention of small ZrO_2 particles in a Al_2O_3 and Al_2O_3-ZrO_2 matrix that enables crack arresting by phase transformation of ZrO_2.
In this study, HVOF and APS –sprayed Al_2O_3-40ZrO_2 –coatings were investigated by cavitation erosion and resulting surfaces and detached particles along with the virgin coatings were investigated with SEM, XRD and optical profilometry.

2 Experimental

The coatings were prepared from commercially available feedstock, fused & crushed powders with two different particle size distributions (Ceram GmbH, Germany). The APS coating was sprayed using an F4MB atmospheric plasma torch (Oerlikon Surface Solutions AG, Switzerland) and a Plasma Technik A-3000S 4/2 plasma spray system. The HVOF coating was sprayed with a TopGun (GTV Verschleißschutz GmbH, Germany) with ethene as the fuel gas. The coatings were deposited on AISI 5120 low carbon steel and AISI 316L stainless steel plates which were grit-blasted with alumina (grit 36) prior to spraying. The deposition parameters are listed in **Table 1**.
The coating microstructures were characterized by scanning electron microscopy (SEM, Zeiss Evo 15, Zeiss GmbH, Germany) and X-ray diffraction with quantitative Rietveld refinement phase analysis performed by TOPAS v5 software (XRD, Bruker, Karlsruhe, Cu-Kα radiation, Germany) and qualitative analysis of the cavitation dust (XRD, Empyrean, PANAlytical, Cu-Kα radiation, The Netherlands). The microhardness of the coatings was measured from ten indentations using a Vickers hardness tester (MMT-X7, Matsuzawa, Japan) and a load of 300 gf ($HV_{0.3}$).

Table 1 **Deposition parameters of the coatings**

Sample name	APS Al_2O_3-ZrO_2	HVOF Al_2O_3-ZrO_2
Process	APS	HVOF
Powder Name	Ceram Al_2O_3-40ZrO_2	Ceram Al_2O_3-40ZrO_2
Nominal Chemical Composition [wt.%]	Al_2O_3-40ZrO_2	Al_2O_3-40ZrO_2
Powder Size [µm]	-51+20	-30+10
Current [A]	610	-
Voltage [V]	70	-
Power [kW]	42.7	-
C_2H_4 [slpm]	-	90
O_2 [slpm]	-	257
Ar [slpm]	41	-
H_2 [slpm]	13	-
Standoff distance [mm]	140	150
Surface speed [m/min]	87	179
Offset [mm/pass]	7	2.9
Coating thickness [µm]	300	330
Thickness per pass [µm]	8.3	9.2

The cavitation erosion test was performed with an ultrasonic transducer (VCX-750, Sonics & Materials, USA), according to the ASTM G32-10 standard for indirect cavitation erosion. The vibration tip was an alloy of Ti-6Al-4V. The sample surfaces were ground flat and polished with a polishing cloth and diamond suspension (3 µm). Samples were cleaned in an ultrasonic bath with ethanol and weighed after drying. Samples were attached on a stationary sample holder and the head of the ultrasonic transducer was placed at a distance of 0.5 mm. Samples were weighed after 15, 30, 60 and 90 min. After the test, detached particles were collected from the water bath. They were characterized with XRD and the worn surfaces with SEM (XL30, Philips, The Netherlands) and Alicona Infinitefocus G5 (Alicon Imaging GmbH, Austria).

3 Results and discussion

3.1 Coating microstructures

The coating cross-sections are presented in **Fig. 1**. Both coatings exhibit a high melting degree, with the HVOF-coating having slightly more unmolten ZrO_2 (white) particles embedded in the composite matrix consisting of Al_2O_3 and ZrO_2. This is due to the lower temperature and dwell time of the particles in the flame of the HVOF that easily leads to a lower melting degree when compared with APS. The APS-coating exhibits a substantial amount of microcracks in the structure, likely from the heating/cooling cycles of the substrates combined with the high pass thickness. [11] The HVOF-coating has fewer microcracks, even though the pass thickness was essentially the same or more than in the APS-coating. This leads to believe that the cracking is a result of the larger particle size and temperature gradient upon cooling of the plasma-sprayed particles. The coating mircohardnesses were 796±59 $HV_{0.3}$ for the APS coating and 917±48 $HV_{0.3}$ for the HVOF coating.

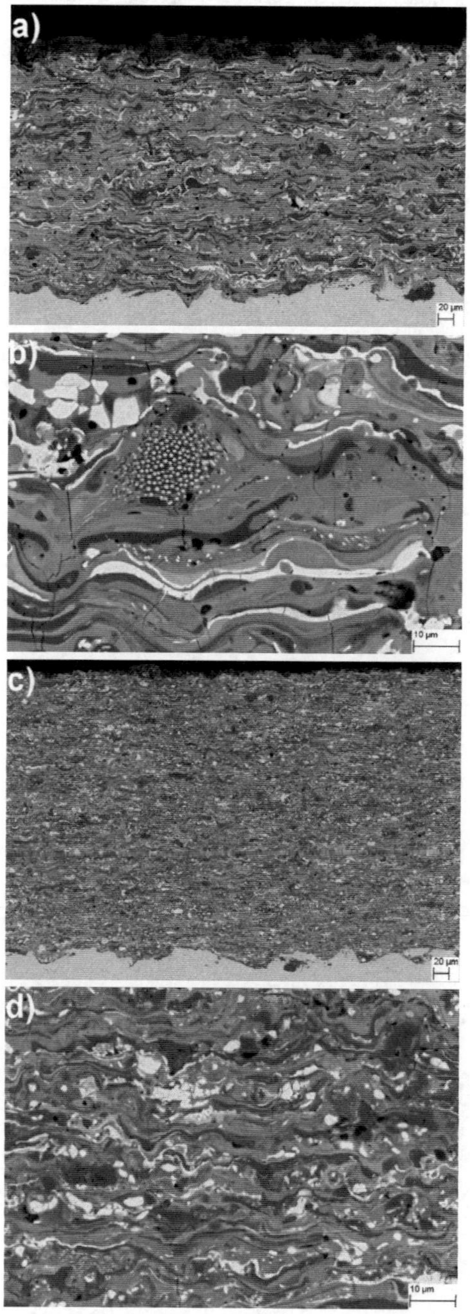

Fig. 1. SEM-images of the cross-sections of the Al_2O_3-40ZrO_2 –coatings deposited with a),b) APS and c),d) HVOF.

3.3 X-ray diffraction analyses

Quantitative Rietveld refinement analysis was performed on the XRD profiles of the samples in order to determine the exact phase compositions of the coatings. The whole diffraction pattern fits are presented in **Fig. 2**. Both coatings consisted of α- and γ-Al_2O_3 and monoclinic and tetragonal ZrO_2. The HVOF-coating had a larger portion of tetragonal ZrO_2 and α-Al_2O_3 than the APS-coating. This is likely due to the lower temperature of the HVOF-process leading to the temperature of the coating dropping under the transition temperature faster which enables higher retention of the tetragonal ZrO_2 Both coatings had a substantial amount of amorphous phase, the crystallinity was 45% for the APS coating and 46% for the HVOF coating. The feedstock powders had also all phases present, but mainly contained α-Al_2O_3 and monoclininc ZrO_2 phases. The phase compositions are presented in **Table 2**.

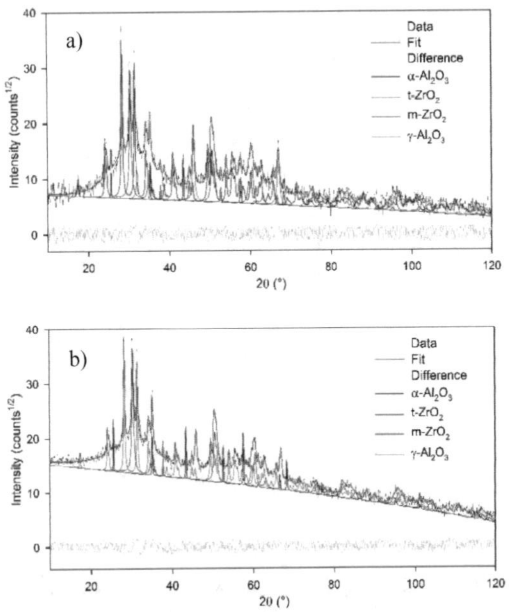

Fig. 2. Rietveld-fits of XRD-graphs of the a) APS Al_2O_3-40ZrO_2 b) HVOF Al_2O_3-40ZrO_2.

Table 2 Phase compositions of the coatings as calculated from the Rietveld analysis.

	APS	HVOF
Crystallinity	45	46
α-Al_2O_3	17,3±0,5 %	27,5±0,5 %
γ-Al_2O_3	24±1 %	11±1 %
t-ZrO_2	18±1 %	25±1 %
m-ZrO_2	40±1 %	36±1 %

The c/a√2 –ratio of the t-ZrO_2 –phase was measured to find out if it is transformable or non-transformable (t'), as is convention [12–15]. The ratios were 1.02 and 1.019 for the APS- and HVOF-coatings, respectively, meaning the t-ZrO_2 in both is most likely transformable.

The theoretical densities were calculated using the exact phase compositions of the coatings. The resulting densities were 4.90 mg/mm^3 for the APS-coating and 5.11 mg/mm^3 for the HVOF-coating. The difference stems largely from the higher amount of the denser α-Al_2O_3 in the HVOF-coating.

3.2 Cavitation erosion

Cavitation erosion of the coatings showed a clear difference between the coatings, with the APS-coating wearing 27.9±1.9 mg and the HVOF-coating 6.6±0.5 mg during the 90 minute test. The results are presented in **Fig. 3** and SEM-images of the cavitated surfaces in
Fig. 4.

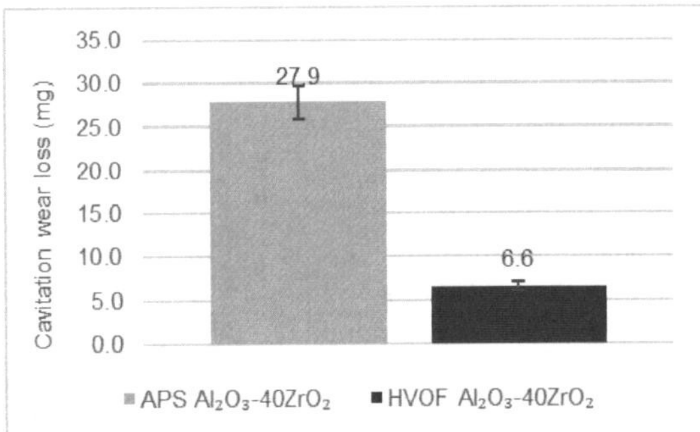

Fig. 3. Cavitation wear mass loss of the coatings.

From the SEM-images of the cavitated surfaces, it is evident that the scale of wear is different for these two coatings. In the APS-coating craters of roughly 200 µm can be seen. The wear mechanism in cavitation erosion is such that the cavitating bubbles prefer to initiate on cavities of the surface and when a cavity wears, the probability of a new bubble in this site increases due to the increase in surface area. Additionally, the "drilling effect" of cavitation is one of the main reasons for the erosion of the coatings, where weak sites are preferentially attacked. A cavity of >10 µm the probability of nucleation of a bubble increases drastically. [16] Due to the smaller splat size of the HVOF-coating this phenomenon is not evident, leading to a more even wear and higher resistance against cavitation.

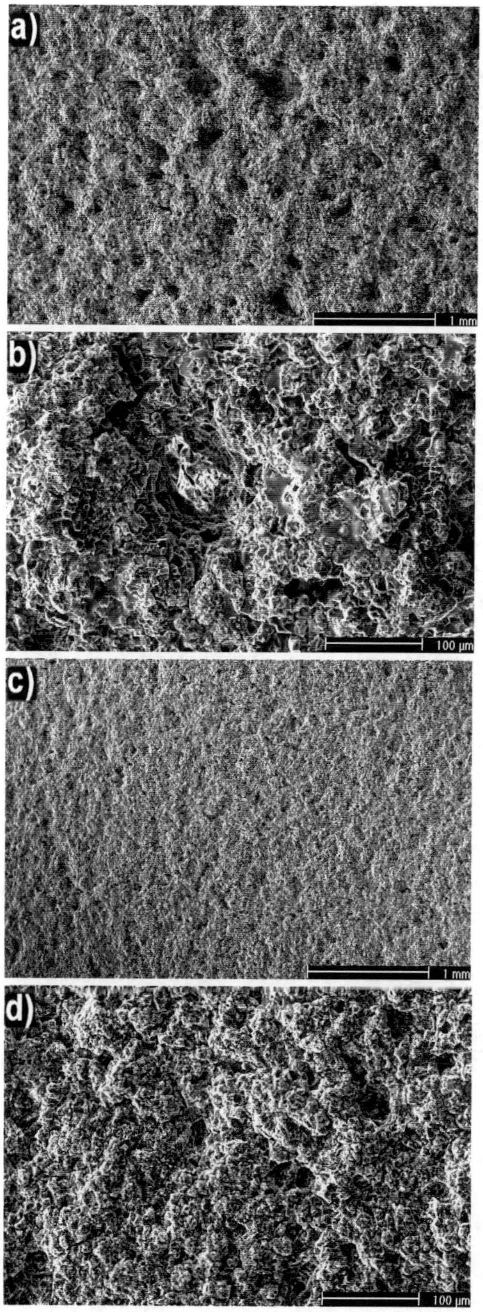

Fig. 3. SEM-images of the cavitation wear area of the Al_2O_3-40ZrO_2 –coatings deposited with a), b) APS and c), d) HVOF.

The coating surfaces were investigated with an optical profilometer after testing in order to determine the true volume loss and porosity of the samples. The profilometer image of the APS-sample is presented in **Fig. 4**. In these cases the mass losses were 28.9 mg and 7 mg for the APS- and HVOF samples, respectively, leading to volume losses of 5.91 mm^3 and 0.74 mm^3. From these values the calculated densities are 4.89 and 9.45 mg/mm^3. The values are slightly surprising, as porosity in the coatings should lead to a reduction in the density measured this way as compared to the theoretical value calculated in section 3.1. Obviously, the low detected volume of the HVOF-sample was a result of undetected damage possibly in the form of small, deep cavities or damage in smaller scale than the resolution of the instrument. Additionally, the deepest hole seen with the profilometer was 222 µm for the APS-coating and 55 µm for the HVOF-coating, underlining the more pronounced drilling effect seen in the SEM-images of the APS-coatings. The following conclusions can be drawn from the results: 1) The coatings are relatively dense as the APS sample gave the exact value of the theoretical density and 2) the damage occurred (ie. splat size) on the HVOF sample is smaller than 14.07 µm, which was the lateral resolution of the profilometer.

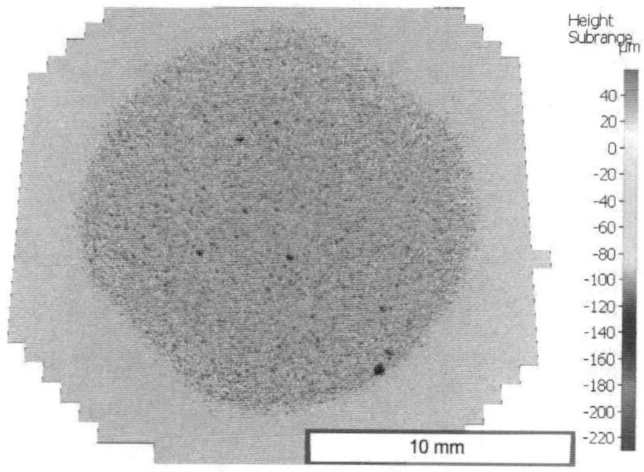

Fig. 4. Topographical image of the APS-sample taken with an optical profilometer.

XRD-analyses were performed on some cavitated coating dust collected from the deionized water after the test in order to see evidence of possible phase change of t-ZrO_2 to m-ZrO_2 during wear. As presented in **Fig. 5** some evidence of such transformation was found. The peak at 44.7° that is visible in the dust of the APS-coating is more than likely due to iron detached from the cooling system of the test and not to any material of interest.

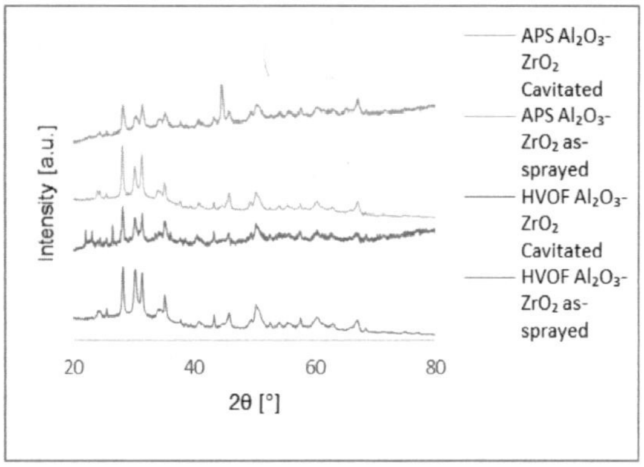

Fig. 5. XRD-graphs of the coatings compared with cavitation dust collected after the test.

The relative amounts of m-ZrO_2 and t-ZrO_2 are presented in **Table 3**. The phase amounts were calculated from the respective net areas of the characteristic peaks of the phases, at 28.2° for m-ZrO_2 and at 30.2° for t-ZrO_2. As is evident, 39 % (APS) and 44 % (HVOF) of the t-ZrO_2 does go through the phase transformartion to m-ZrO_2. Since both coatings exhibit roughly the same amount of transformation, it is unlikely to lead to significant differences in wear resistance when compared with other factors, such as structural integrity and the scale of the microscopical features of the coating.

Table 3 Relative amount of m-ZrO_2 and t-ZrO_2 before and after cavitation erosion.

		m-ZrO_2	t-ZrO_2	Reduction of t-ZrO_2 [%]
APS	Initial	1	0.659	39
	After cavitation		0.405	
HVOF	Initial		1.144	44
	After cavitation		0.642	

4. Conclusion

In this study, the behaviour of APS- and HVOF-sprayed Al_2O_3-40ZrO_2 –coatings in cavitation erosion was studied. Both coatings exhibited a typically dense structure with clear segregation of phases. The HVOF-coating consisted of a greater amount of α-Al_2O_3 and t-ZrO_2 probably due to the lower process temperature. The HVOF-coating was 5 times more resistant to cavitation than its plasma-sprayed counterpart. The SEM-analysis of the worn areas revealed a difference in the scale of the phenomenon in these two coatings. XRD analyses were performed before and after cavitation, where transformable tetragonal ZrO_2 was found in both coatings, and evidence of transformation to monoclinic ZrO_2 was found. However, both coatings exhibited roughly the same amount of phase transformation leaving its actual effect on wear resistance still in doubt since other factors like structural integrity and the scale of microscopical features seemed to be have a more prominent effect.

Acknowledgements

The work has been done within DIMECC Ltd and its HYBRIDS programme. The authors gratefully acknowledge the financial support from Tekes (Finnish Funding Agency for Technology and Innovation) and the participating companies. The authors would like to thank Mr. Mikko Kylmälahti, of Tampere University of Technology for spraying the coatings and Dr. Radek Musalek from the Department of Materials Engineering, Institute of Plasma Physics, Czech Academy of Sciences for the SEM-imaging of the cross-sections.

References

[1] P. Vuoristo, Thermal Spray Coating Processes, Comprehensive Materials. Processing Vol. 4 Films and Coatings: Technology and Recent Development, S. Hashmi (Ed.), Elsevier, The Netherlands, 2014, pp. 229–276.

[2] L. Pawlowski, The Science and Engineering of Thermal Spray Coatings, 2nd edition, John Wiley & Sons, England, 2008, 656 p.

[3] J. Chevalier, S. Deville, G. Fantozzi, J.F. Bartolomé, C. Pecharroman, J.S. Moya, L.A. Diaz, R. Torrecillas, Nanostructured ceramic oxides with a slow crack growth resistance close to covalent materials, Nano Lett. 5 (2005), Issue 7, pp. 1297–1301.

[4] J. Chevalier, A.H. De Aza, G. Fantozzi, M. Schehl, R. Torrecillas, Extending the lifetime of ceramic orthopaedic implants, Adv. Mater. 12 (2000), Issue 21, pp. 1619–1621.

[5] E. Kannisto, M.E. Cura, E. Levänen, S.P. Hannula, Mechanical Properties of Alumina Based Nanocomposites, Key Eng. Mater. 527 (2012), pp. 101–106.

[6] A. Evans, Perspective on the Development of High-Toughness Ceramics, J. Am. Ceram. Soc. 73 (1990), Issue 2, pp. 187–206.

[7] G. V. Srinivasan, J.F. Jue, S.Y. Kuo, A. V. Virkar, Ferroelastic domain switching in polydomain tetragonal zirconia single crystals, J. Am. Ceram. Soc. 72 (1989), Issue 11, pp. 2098–2103.

[8] G. Bolelli, V. Cannillo, L. Lusvarghi, T. Manfredini, Wear behaviour of thermally sprayed ceramic oxide coatings, Wear 261 (2006), Issues 11–12, pp. 1298–1315.

[9] R.B. Heimann, Applications of plasma-sprayed ceramic coatings, Adv. Ceram. Mater. 122-124 (1996), pp. 399–441.

[10] J. Oberste Berghaus, J.-G. Legoux, C. Moreau, F. Tarasi, T. Chráska, Mechanical and Thermal Transport Properties of Suspension Thermal-Sprayed Alumina-Zirconia Composite Coatings, J. Therm. Spray Technol. 17 (2007), Issue 1, p 91–104.

[11] P.L. Fauchais, J.V.R. Heberlein, M.I. Boulos, Thermal Spray Fundamentals, Springer US, USA, 2014, 1566 p.

[12] H.G. Scott, Phase relationships in the zirconia-yttria system, J. Mater. Sci. 10 (1975), Issue 9, pp. 1527–1535.

[13] P.D. Harmsworth, R. Stevens, Phase composition and properties of plasma-sprayed zirconia thermal barrier coatings, J. Mater. Sci. 27 (1992), Issue 3, pp. 611–615.

[14] M. Leoni, R.. Jones, P. Scardi, Phase stability of scandia–yttria-stabilized zirconia TBCs, Surf. Coat. Technol. 108-109 (1998), pp. 107–113.

[15] C. Viazzi, J.-P. Bonino, F. Ansart, A. Barnabé, Structural study of metastable tetragonal YSZ powders produced via a sol–gel route, J. Alloys Compd. 452 (2008), Issue 2, pp. 377–383.

[16] C.E. Brennen, *Cavitation and bubble dynamics*, Oxford University Press Inc, USA, 1995, 282 p.

Reactive Plasma Sprayed Nitride Coatings with Improved Conductivity

Mohammed Shahien[1,2], Motohiro Yamada[1], Masahiro Fukumoto[1]
[1] Toyohashi University of Technology, TUT, Toyohashi/JP
[2] Central Metallurgical R & D Institute, CMRDI, Cairo/EG

Decomposition of the nitride ceramic particles like aluminium nitride (AlN) during conventional thermal spray process prevents their deposition on the substrate. Reactive plasma spraying (RPS) is a promising solution to fabricate AlN coatings. It is based on reaction and deposition of molten particles in active nitrogen ambient to form the AlN phase. Several thick AlN based coatings were fabricate successfully by reactive plasma spraying of aluminium and/or alumina particles. This study shows our recent achievements of fabrication of AlN coatings with improved conductivity.
It was possible to fabricate AlN based coatings through reactive spraying of fine alumina particles mixed with fine AlN additives. Using small particle size powders improved the particles melting, surface area, therefore nitriding conversion and the AlN content. The fabricated AlN based coating contains several of oxide phases, with low density and high porosity, therefore its thermal conductivity was very low (about 2.6 W/m.K). To fabricate AlN coatings with high thermal conductivity, a liquid phase promoting additive (yttria) was added to the feedstock powder. It assists the formation of the yttrium aluminate (Y-Al-O) phase and therefore the sintering of the coatings during heat treatment. Finally, AlN coating with improved thermal conductivity (above 90 W/m.K) was developed.

1 Introduction

The low thermal conductivity of alumina limits its applicability in advanced electronic packaging devices. Among non-oxide ceramics, aluminium nitride (AlN) had intensely considered for electronic ceramic packaging. It is because its novel properties of high thermal conductivity, appropriate thermal expansion coefficient, high resistance to molten metals, wear corrosion and oxidation, and nontoxic nature [1, 2]. Development of thick coatings of AlN ceramics is considered as a promising solution for structural and power devices applications [3].

Plasma spray is one of the most widely developed technology for thick coatings deposition in industry, due to it high deposition rates. The process is characterized with the high temperature and specific energy densities of the thermal plasmas, which easily melt any form of materials (with stable molten/semi –molten phase). The deposition in plasma spraying includes the heating and propelling of the molten, semi-molten or heat-softened material toward the substrate at high velocity. Various materials with high melting points such as ceramics and/or non-metallics had been developed by plasma spray process [4].
On the other hand, fabrication of the nitride ceramic coatings like AlN directly by the conventional thermal spray was impossible. It is due to, the thermal decomposition of the AlN particles without any stable melting/semi-melting phases in the high temperature conditions. This prevents the AlN particles deposition during plasma spraying.

In order to, overcome this problem using the reactivity of the plasma (Reactive plasma spraying; RPS), which is responsible for melting the particles and accelerate it to the substrate was a promising way to fabricate the AlN plasma sprayed coatings [5]. It is basically depending on spraying of molten or semi-molten materials under a reactive plasma condition to assist the reaction between the feedstock material and the surrounding active species in the plasma [5]. In the last decade, we successfully developed several AlN based coatings through reactive plasma nitriding of several Al, Al based and Al_2O_3 powders [6-14] in the atmospheric N_2-H_2 plasma. The particle size of the starting feedstock powder is among the most important and controlling factors during reactive plasma spraying process. Thus, it significantly affects the nitriding formation, coating microstructure and therefore properties.
This study will show our recent achievements of fabrication of AlN coatings with improved conductivity, through reactive plasma spraying of fine particles. Furthermore, the influence of feedstock materials on the coating formation, microstructure and properties will be investigated.

2 Materials and Experimental Set-up

2.1 Materials and Feedstock Preparation

The primary powders are fine particles of commercial Al_2O_3 powder (1 μm; Kojundo Chemical Lab Co., Ltd., Saitama, Japan) with fine commercial AlN powder (1.14 μm; Tokuyama Co., Tokyo, Japan). Since such fine particles could not be used as direct sprayable feedstocks, the powders were granulated to sprayable micron-sized powders by spray drying. A commercial L-8i spray dryer (Ohkawara Kakohki Co., Ltd., Yokohama, Japan) was used. The microstructures of the primary fine particles as well as the spray dried mixture are shown in **Fig. 1**. It is shows the characteristic spherical or donut shape of the spray dried particles. Furthermore, the SEM images show that the particle size of the spray dried mixture is about 70 μm, which is suitable for plasma spraying. **Figure 2** shows the particle size distribution of the spray dried mixture. For some experiments the fine yttrium oxide (Y_2O_3: 0.5 μm) was added to the initial feedstock during spray drying.

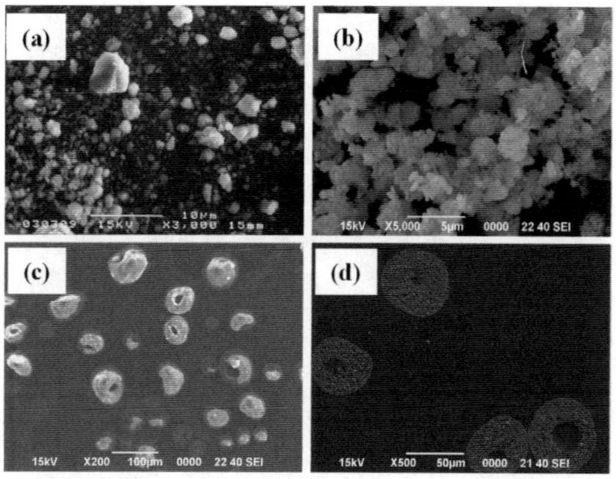

Fig. 1. SEM microstructure of: (a) primary AlN particles, (b) primary Al_2O_3 particles, (c) spray dried feed stock mixture, and (d) feedstock mixture cross-section microstructure.

2.2 Methodology and Plasma Spraying System

The coatings were plasma sprayed using the atmospheric DC plasma spraying system (APS: 9MB, Oerlikon Metco (previously Sulzer Metco), Switzerland). The system is based on a stick-type cathode and uses two plasma gases that may include Ar-H_2, N_2-H_2, Ar-He, and sometimes plasma gases mixtures of Ar-H_2-He. In our study, for reactive plasma spraying the N_2 gas was used as the primary gas and the H_2 gas as secondary gas. **Figure 3** shows schematic illustration of the APS system. The system and RPS process had been reported in details previously [6-14]. Table 1 shows the operational spraying parameters. The feedstock powder was supplied into the plasma stream with a carrier gas of N_2. Grid blasted mild steel plates (50mm × 50 mm × 5mm) were prepared as the substrate material.

2.3 Coating Characterizations

The coatings phase composition was identified using the x-ray diffraction (XRD, RINT-2500; Rigaku, Tokyo, Japan) analysis using Cu Ka radiation. The morphology of the feedstock powder and the coatings cross-sections were observed using a scanning electron microscope (SEM, JSM-6390; JEOL, Tokyo, Japan). The porosity of coatings was estimated by image analysis of SEM cross-section images.

Table 1 Operational spraying parameters

Primary gas: Pressure (kPa) : Flow rate (SCFH)	N_2: 330.9 100
Secondary gas: Pressure (kPa) : Flow rate (SCFH	H_2: 344.7 5
Spray distance (mm)	100
Carrier gas:	N_2
Substrate materials	SS400/Cu

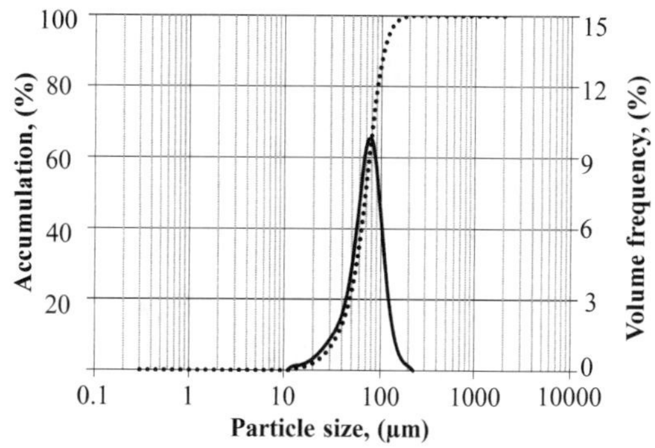

Fig. 2. The particle size distribution profile of the feedstock mixture.

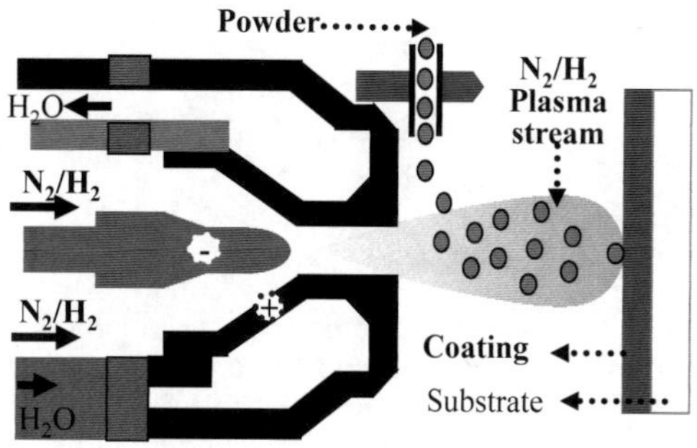

Fig. 3. Schematic diagram of the reactive atmospheric plasma spraying system.

Thermal conductivity of the coatings was calculated by measuring the thermal diffusivity, density and specific heat of the freestanding coatings by the following equation:

$$\lambda = \alpha \rho C_p$$

where:

λ = thermal conductivity (W/m.K)
α= thermal diffusivity (m^2/s).
ρ= density (kg/m^3)
Cp = specific heat (J/kg.K).

The thermal conductivity measurement was done for the free-standing coatings with a disk shape of 10 mm in diameter and about 0.5 mm of thickness. Both front and back sides of the samples were polished for surface uniformity. Thermal diffusivity of the samples was measured by Laser Flash method using the LFA 447 Nanoflash® instrument (NETZSCH, Inc., Germany) at 25 °C. The through-plane diffusivity had been deduced from back side temperature increase measurement and using the following equation:

$$\alpha = \frac{0.1388\, L^2}{t_{1/2}} * K$$

where:

$t_{1/2}$ = half maximum temperature increase (is the time required for the back-side surface temperature to reach half of the maximum temperature rise, the constant value of approximately 0.139 is known as the Fourier number, K = the energy loss constant and L = sample thickness (m)

Differential scanning calorimeter measurement was used to measure the specific heat capacity of the coating at 25 °C by as follow:

$$C_p = \frac{Q}{m\Delta T}$$

where:
Cp = specific heat (J/kg.K), Q = Quantity of required energy, m = mass and ΔT = change in the temperature.

Coating bulk density was calculated from the sample volume and mass by using Shimadzu Co. electronic balance (AuW120D) at 25 °C:

$$\rho = \frac{m}{S * L}$$

where:
ρ= density (kg/m^3), m = mass (kg), S = surface area (m^2), and L = thickness (m).

3 Results and Discussion

Figure 4 shows the phase composition of the primary powders and the spray dried mixture. It reveals that, the spray dried mixture is composed of AlN and α-Al_2O_3 without any impurities. With spraying the spray dried mixture thick AlN based coating was fabricated. The coating consists of *h*-AlN, *c*-AlN, Al_5O_6N, γ-Al_2O_3, and a small amount of α-Al_2O_3, as shown in the XRD spectra of as sprayed coatings in **Fig. 5**.
Formation of the *c*-AlN phase is attributed to the reactive plasma nitriding of the fine Al_2O_3 particles in the N_2/H_2 plasma, which agree with the previously reported study [14]. On the other hand, formation of the *h*-AlN is attributed to the deposition of AlN particles from the feedstock mixture.

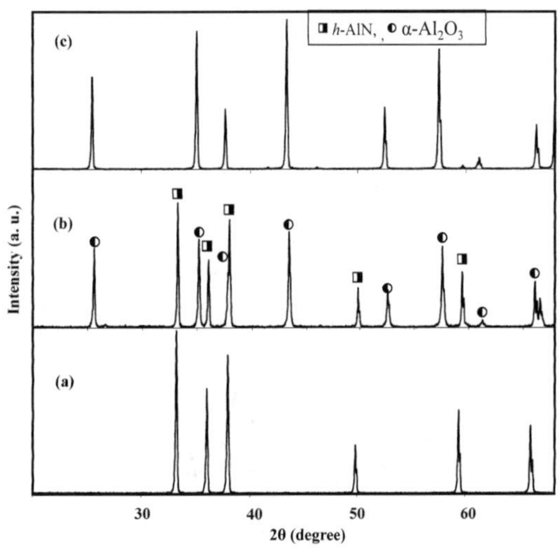

Fig. 4. XRD spectra of: (a) primary AlN, (b) spray dried mixture, and (c) primary Al_2O_3 powder.

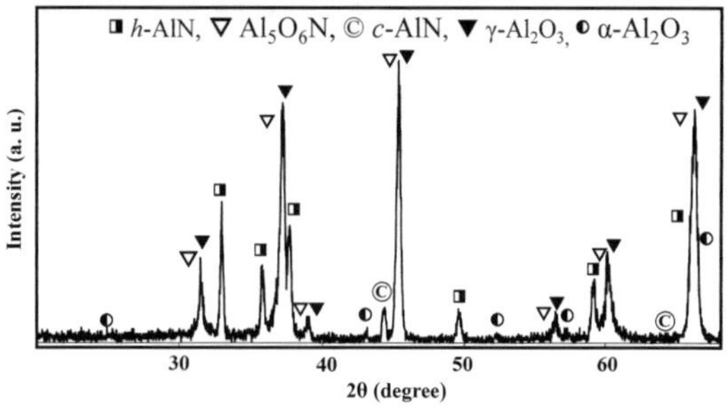

Fig. 5. XRD spectrum of the fabricated coating using the spray dried mixture.

Besides that, the formation of oxynitride phase is due to the reaction between the Al_2O_3 particles with the surrounding nitrogen species. Moreover, formation of the γ-Al_2O_3 phase in the coatings is due to the transformation from the α-Al_2O_3 of the feedstock powder. Thus, the feedstock α-Al_2O_3 melted, and after that the molten Al_2O_3 solidified on the cold substrate surface in the form of γ-Al_2O_3 due to the rapid quenching of the molten Al_2O_3 and the rapid solidification rate in the plasma spray process. This transformation occurs easily under the high temperature and rapid solidification rate of the plasma spray process [12, 15].

The XRD spectrum of the fabricated coatings also reveals the high content of γ-Al_2O_3 and very small content of the α-Al_2O_3 phase. It is strongly related to the effect of using the fine particle size on the melting behaviour of the particle. Thus, fine particles heat capacity is much smaller than that of the large particles; therefore, their temperature will increase faster than large one. Furthermore, at constant weight the heat conduction and decomposition rates of the fine particles are faster than the large ones. In addition, the smaller particles attain higher temperature than the large particles. It is due to the larger heat absorbed per unit volume of the smaller particles under the same plasma temperature.

It indicates that, with using the fine Al_2O_3, the particles are fully melted and therefore transformed to γ-Al_2O_3 phase in the plasma sprayed coatings. It is also worth to mention that, using the fine particles improved its melting and therefore its nitriding. Thus, the nitriding reaction starts from the surface of the particles. Thus, the surface area of a large particle increases by four times when compared with a particle of half the diameter. Thereby, with decreasing the particle size the surface area (so called reaction start point) increased, and thereby improves the nitriding reaction. Therefore, the high content of γ-Al_2O_3 phase with using the fine particles is related to the high proportion of melting. The small amount of α-Al_2O_3 phase in the fabricated coating is due to the unmelted ingredient in the feedstock droplets during deposition. The fabricated coating properties and phase composition are shown in Table 2.

Table 2 The properties and the phase content of the fabricated coating.

Thermal Conductivity λ, (W/m.K)	Density, P (g/cm³)	Porosity (%)	Av Hardness	The phase content in the coating, (%), *				
				γ-Al_2O_3	Al_5O_6N	*h*-AlN	*c*-AlN	α- Al_2O_3
2.6	2.17	18	990	31.5	23	27	17	1.5
* Phase content calculated by comparing the peak intensities from the XRD data								

It reveals that, the coating thermal conductivity is about 2.5 W/m.K, which is very low for AlN sintered bodies. Besides that, the coating has density of about 2.17 g/cm^3, which is lower than the AlN density (3.16 g/cm^3). Furthermore, the coating has a porosity of 18 %, as estimated from the SEM microstructure by the image analysis. The coating has an average Vickers hardness of 990. The coating porosity and hardness are lower than the values with using large Al_2O_3 particles [13]. Thus, using the fine particle improved the melting and improved the formation of disk-like splats and decreased the porosity. Decreasing the hardness is strongly related to increase the AlN content of the lower hardness in the coatings.

The lower thermal conductivity of the fabricated coatings can be explained as follow:
As shown in Table 2, the coating contains high amounts of oxide phases in the form of γ-Al_2O_3, Al_5O_6N, as well as α-Al_2O_3. The high content of Al_2O_3 phases lowered the coating conductivity so much due to their low conductivity values compered to AlN. Moreover, the higher content of γ-Al_2O_3, strongly suppress the coating conductivity, it is due to its lower density compared to the α-Al_2O_3 phase. In addition: the γ-phase has the cubic spinel structure (it has higher Al (III) vacancies) compared to α-phase (trigonal structure). Furthermore, the relative disorder of the γ-Al_2O_3 metastable phase decreased the mean free path for the scattering of phonons and therefore suppressed the diffusivity so much. Such factors lead to further decrease in the coating conductivity due to higher content of γ-phase than the α-phase. In addition, the large amount of the oxynitride phase leads to further decrease in the conductivity due to their lower conductivity compered to Al_2O_3 values.

On the other hand, the lower value of the coating density strongly suppresses the coating conductivity. On the same time, the higher porosity level of the plasma sprayed coating as shown in table 2 and in **Figure 6** (SEM microstructure of the coatings), lead to further decrease in its conductivity. These pores in the coating are filled with air with low conductivity.
Besides that, the anisotropic layer structure of the plasma sprayed coatings with individual splats oriented parallel to the substrate surface lowers its conductivity. It is related to the interfacial thermal contact by pores or secondary phases. Moreover, the interfaces due to the thermal or impact stresses reduces the heat transfer and suppressed the thermal conductivity of the coatings.

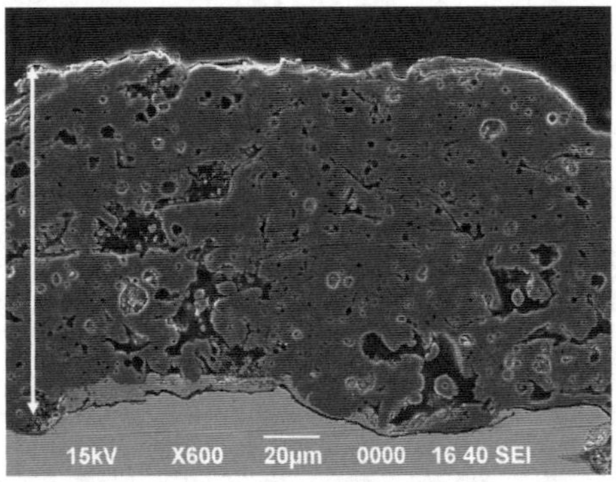

Fig. 6. SEM cross-section microstructure of the fabricated coating using the spray dried mixture.

It is concluded that, it was possible to fabricate AlN based coatings with homogenous microstructure and relatively high nitride content through reactive plasma spraying of fine Al_2O_3/AlN powders. However, the coating conductivity is much lower than the sintered AlN bodies. The lower conductivity is strongly related to the phase composition of the coating, higher oxygen content, higher porosity level and lower density. To improve the conductivity decreasing the dissolved oxygen and improving the microstructure are required factors. Therefore, the next section will discuss the feasibility of using some sintering additives on the feedstock power.

The AlN manufacturing showed that, oxygen is the main issue of this material. Thus, the AlN particles contain more than 1wt% of oxygen. This oxygen has been dissolved in AlN lattice is essentially in the form of Al_2O_3 dissolved in the lattice [16, 17]. The oxygen in the AlN lattice occupies nitrogen sites and creates aluminium vacancies to balance the electric charge as follow [16, 17]:

$$Al_2O_3 \rightarrow 2Al_{Al} + 3O_N + V_{Al}$$

Where: The V_{Al} represents the Al vacancies.
These Al vacant sites assist the mass and strain misfits and therefore increasing the scattering cross section of

phonons, and thereby decreasing the thermal conductivity [18].

In order to reach higher AlN ceramic material, using rare earth oxide additives like (yttrium oxide: Y_2O_3) can be used as sintering promoting additive. Thus, with the presence of Al_2O_3 phase it will act as liquid phase promoting additive and form yttrium aluminate Y-Al-O phase as wetting agent. **Figure 7** shows the XRD spectra of the coatings after using the Y_2O_3 additives.

The phase composition of the coatings showed the formation of the aluminum yttrium oxide (Y-Al-O) phase in the fabricated coating, besides the *c*-AlN, *h*-AlN, Al_5O_6N, γ-Al_2O_3 and the α-Al_2O_3 phases. Formation of the YAG phase is well known features of the Al_2O_3-AlN-Y_2O_3 system under high temperatures. Therefore, it was possible to introduce the Y_2O_3 additive in the AlN/Al_2O_3 coating system and formation of the Y-Al-O phase as liquid phase promoting agent.

Besides that, with the using the Y_2O_3 additives the AlN content improved and the Al_5O_6N and γ-Al_2O_3 content decreased in the coating. However, the α-Al_2O_3 amount increased with the Y_2O_3 addition. It ireveals that, the transformation of the α-Al_2O_3 to the γ-Al_2O_3 phase suppressed with Y_2O_3.

Although, the Y-Al-O was formed on the fabricated coatings, and slightly improved the phase composition of the coating as shown in XRD spectrum. The coating still has the porous microstructure as shown in **Fig. 8**, with porosity level about 11 %. Furthermore, the coating conductivity was about 4 W/m.K, which still far from the AlN sintered compacts thermal conductivity. The lower value of the coating conductivity after using the Y_2O_3 is due to the large amount of oxide phases still appear in the coating, higher porosity, and low density of the coatings. Furthermore, it indicates that, the main role of the Y-Al-O formation on the AlN sintering was not achieved completely yet.

Based on our experience and thermal behaviour of the as sprayed coatings, the effect of heat treatment was investigated. After heat treatment the oxygen content significantly decrease (Al_2O_3 phases mainly disappeared and very small amount of the oxynitride phase appeared) and the fabricated coating consist of AlN phase and small amount of the Y-Al-O phase as shown in the XRD spectrum in **Fig. 9**.
During the heat treatment (sintering) process, the yttrium aluminate (Y-Al-O) on the as-sprayed coating react with the alumina dissolved in the AlN based mixture.

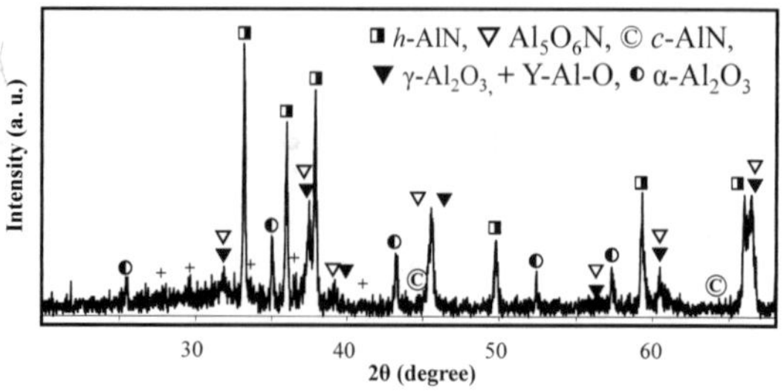

Fig. 7. XRD spectrum of the fabricated coating with using the yttria additives.

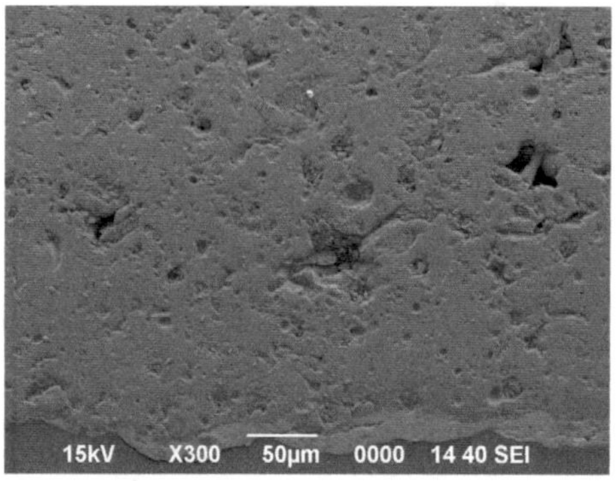

Fig. 8. SEM cross-section microstructure of the fabricated coating using yttria additives.

Besides that, this chemical reaction is acting as a driving force for the oxygen diffusion out slowly from the AlN grains itself. With decreasing the alumina content on the mixture and once the concentration on the surface of the AlN grains decreased enough, the de-Wetting reaction occurs. This leads to migration and collection of the Y-Al-O phase on the triple points in the microstructure.

Figure 10 shows the SEM microstructure of the fabricated coatings after heat treatment. The microstructure significantly shows the particles sintering and the nicked-shape sintered structure. Furthermore, the coating porosity decreased so much (less than 2 %) and the density improved with the heat treatment.

Figure 11 also shows the effect of heat treatment on the coating thermal conductivity. It clearly shows the improved thermal conductivity of the coatings (above 90 W/m.K) after heat treatment.

Therefore, in order to achieve high conductivity AlN plasma sprayed coatings using a liquid phase promoting additive like Y_2O_3, formation of the yttrium aluminate Y-Al-O phase and followed by heat treatment is a promising solution during reactive plasma spraying process.

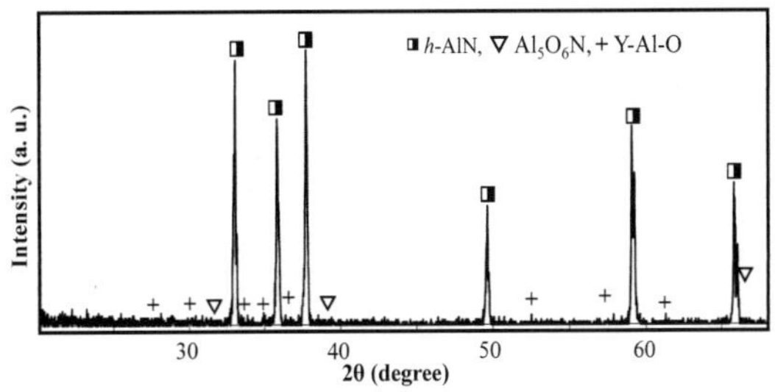

Fig. 9. XRD spectrum of the fabricated coating after the heat treatment.

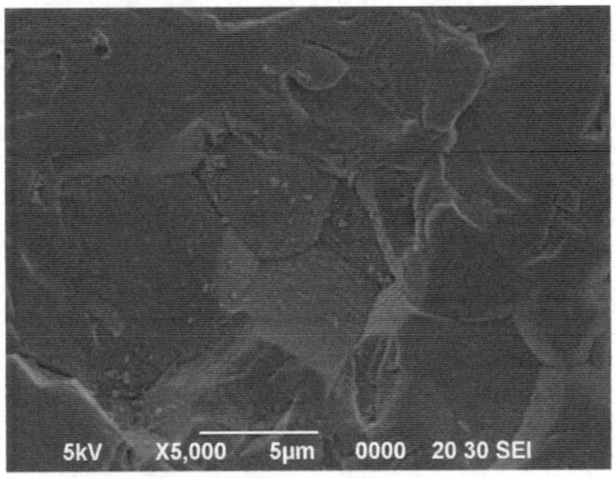

Fig. 10. SEM cross-section microstructure of the coating after heat treatment.

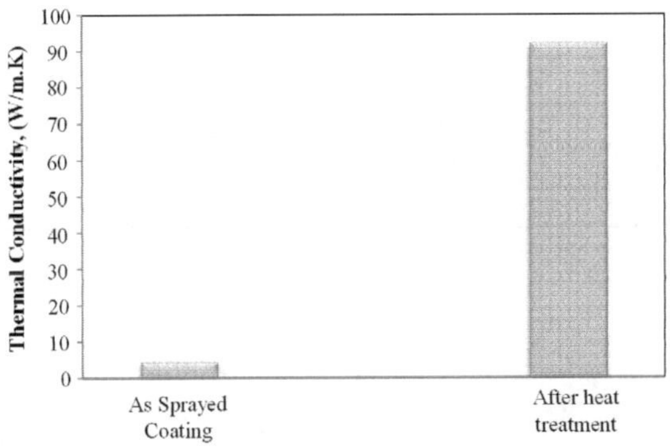

Fig. 11. The thermal conductivity of the coating before and after heat treatment.

4 Conclusion

Reactive plasma spraying of fine particles was investigated upon fabrication of AlN coatings in N_2/H_2 plasma. Spraying of fine particles enhanced the melting of the particles as well as the nitriding conversion (by increasing the surface area). Furthermore, using an AlN additive in the feedstock powder improved the nitride content in the coating. However, the large amount of oxide phases, lower density and higher porosity of the fabricated coatings significantly suppressed the coating thermal conductivity (less than 5 W/m.K).

Homogenous AlN coatings with high thermal conductivity more than 90 W/m.K, were successfully fabricated through addition of Y_2O_3 powder as liquid phase promoting additive and followed by heat treatment of the as sprayed coating.

5 References

[1] Wemer, A.W.: Carbide, Nitride and Boride Materials Synthesis and Processing, Chapman & Hall, London, 1997, pp. 6-68.

[2] Pierson, H.O.: Handbook of Refractory Carbides and Nitrides, Noyes, NJ, 1996, pp. 237-239.

[3] Krishna, L.R., Sen, D., Rao, Y.S., Rao, G.V.N., and Sundararajan, G.: Thermal Spray Coating of Aluminum Nitride Utilizing the Detonation Spray Technique, J. Mater. Res., 2002, 17(10), pp. 2514-2523.

[4] Fauchais, P., Vardelle, A. and Dussoubs, B.: Quo vadis thermal spraying?, Journal of Thermal Spray Technology, 10-1 (2001), pp. 44-66.

[5] Yamada, M., Yasui, T., Fukumoto, M. and Takahashi, K.: Nitridation of Aluminum Particles and Formation Process of Aluminum Nitride Coatings by Reactive RF Plasma Spraying, Thin Solid Films, 2007, 515(9), pp. 4166-4171.

[6] Shahien, M., Yamada, M., Yasui, T. and Fukumoto, M.: Reactive Atmospheric Plasma Spraying of AlN Coatings: Influence of Aluminum Feedstock Particle Size, J. Therm. Spray Technol., 2011, 20(3), pp. 580-589.

[7] Shahien, M., Yamada, M., Yasui, T. and Fukumoto, M.: Influence of NH_4Cl Powder Addition for Fabrication of Aluminum Nitride Coating in Reactive Atmospheric Plasma Spray Process, J. Therm. Spray Technol., 2011, 20(1-2), pp. 205-212.

[8] Shahien, M., Yamada, M., Yasui, T. and Fukumoto, M.: Cubic Aluminum Nitride Coating Through Atmospheric Reactive Plasma Nitriding, J. Therm. Spray Technol., 2010, 19, pp. 635-641.

[9] Shahien, M., Yamada, M., Yasui, T. and Fukumoto, M.: Influence of Plasma Gases on Fabrication of AlN Coatings Through Atmospheric Plasma Nitriding Process, Ind. Appl. Plasma Process., 2010, 3, pp. 1-10.

[10] Shahien, M., Yamada, M., Yasui, T. and Fukumoto, M.: Fabrication of AlN Coatings by Reactive Atmospheric Plasma Spray Nitriding of Al Powders, Mater. Trans., 2010, 51, pp. 957-961.

[11] Shahien, M., Yamada, M., Yasui, T. and Fukumoto, M.: In Situ Fabrication of AlN Coating by Reactive Plasma Spraying of Al/AlN Powder, Coatings, 2011, 1-2, pp. 88-107.

[12] Shahien, M., Yamada, M., Yasui, T. and Fukumoto, M.: Synthesis of Cubic Aluminum Nitride Coating from Al_2O_3 Powder in Reactive Plasma Spray process, Mater. Trans., 2013, 54(2), pp. 207-214.

[13] Shahien, M., Yamada, M., Yasui, T. and Fukumoto, M.: N_2 and H_2 Plasma Gases Effects in Reactive Plasma Spraying of Al_2O_3 Powder, Surf. Coat. Technol., 2013, 216, pp. 308-317.
[14] Shahien, M., Yamada, M., Yasui, T. and Fukumoto, M.: Reactive Plasma Spraying of Fine Al_2O_3/AlN Feedstock Powder, J. Therm. Spray Technol., 2013, 22(8), pp. 1283-1293.
[15] Zhao, L., Seemann, K., Fischer, A. and Lugscheider, E.: Study on Atmospheric Plasma Spraying of Al_2O_3 Using On-Line Particle Monitoring, Surf. Coat. Technol., 2003, 168, pp. 186-190.
[16] Slack, G. A.: Nonmetallic crystals with high thermal conductivity, J. Phys. Chem. Solids, 1973, 34(2), pp. 321-335.
[17] Slack, G. A., Tanzilli, R. A., Pohl, R. O. and Vandersande, J. W.: The intrinsic thermal-conductivity of AlN, J. Phys. Chem. Solids, 1987, 48(7), pp. 641-647.
[18] Virkar, A. V., Jackson, T. B. and Cutler, R. A.: Thermodynamic and kinetic effects of oxygen removal on the thermal conductivity of aluminum nitride, J. Am. Ceram. Soc., 1989, 72(11), pp. 2031-2042.

Corrosion Behavior of HVAF-Sprayed Bi-Layer Coatings; Effect of Intermediate Layer's Microstructure and Chemical Composition

Esmaeil Sadeghimeresht*, Nicolaie Markocsan, Per Nylén

[1] Department of Engineering Science, University West, Trollhättan, 46153, Sweden

Abstract: In a variety of engineering applications, components are exposed to corrosive/erosive environment. Protective coatings are essential to improve the functional performances and/or extend the lifetime of the components. Thermal spraying as a cost-effective coating deposition technique offers high flexibility in coatings' chemistry/morphology/microstructure design. However, the pores formed during spraying inherently restrict the use of coatings for corrosion protection. In view of the above gap to have a high quality coating, bi-layer coatings have been developed to boost the corrosion perfromance of the coatings. In a bi-layer coating, an intermediate layer is deposited on the substrate before spraying the coating. The electrochemical behavior of each layer is critical to ensure a good corrosion protection. The corrosion behavior of the layers strongly depends on coating composition and microstructure, which are affected by feedstock material and spraying process. In the present work, Cr_3C_2-NiCr top layer with different intermediate layers (i. e., Fe-, and Ni-based) were sprayed by high-velocity air fuel (HVAF) process. Microstructure analysis, as well as electrochemical tests, e.g., open-circuit potential (OCP) and polarization were performed. The results showed a direct link between the OCP of each layer in a bi-layer coating and corrosion mechanisms. It was found that the higher corrosion resistance of Ni-based intermediate layers than Fe-based coatings was due to higher OCP of the coating in the galvanic couple with top layers. Splat boundaries and interconnected pores reduced the corrosion resistance of the intermediate layers, however a sufficient reservoir of protective scale-forming elements (such as Cr or Al) improved the corrosion behavior.

1 Introduction

Cr_3C_2-NiCr coatings are thermally sprayed on components in wear/corrosion applications. However, the applications in which the hard coatings are used normally involve environments that are intrinsically corrosive [1], [2]. Several thermal spray techniques, e.g., high-velocity oxy fuel (HVOF), high-velocity air fuel (HVAF) and atmospheric plasma spraying (APS) have been used to deposit a defect-free Cr_3C_2-NiCr coating, however the formation of pores is inevitable.

The pores, especially when they are interconnected, adversely affect the corrosion resistance by providing preferential paths for corrosive ions to penetrate through the coatings [3], [4]. Furthermore, NiCr metal matrix acts as anodes in connection with Cr_3C_2 and forms some localized micro-galvanic cells which can help the diffusion of corrosive ions [5].

In many industrial applications, in order to maintain corrosion protection of the coatings, very thick coatings are deposited which increase both the price and the weight of the components [6]. Therefore, to integrate the good wear resistance of the Cr_3C_2-NiCr coating with the appropriate electrochemical properties of an anti-corrosion coating at low cost, bi-layer coating systems were introduced by adding an intermediate layer of a dense metallic coating between the Cr_3C_2-NiCr top layer and the substrate [7].

The advantages of bi-layer coatings over the single-layer ones have been studied in several works [3], [4], [7]–[10]. The intermediate layer was shown to be beneficial for impeding the penetration paths of corrosive ions to the metallic substrate by adding a layer with more positive corrosion potential (E_{corr}), which eventually led the intermediate layer to be cathode in the top layer/ intermediate layer galvanic couple [3]. The rate of attack by corrosive ions could be effectively controlled through the bi-layer coating systems.

The corrosion performance of the bi-layer coating was found to be dependent on the intermediate layer microstructure, as a galvanic couple could be formed not only among different sites of a coating layer, but also among the top layer/intermediate layer and intermediate layer/substrate couples [4]. Therefore, a good control over the microstructure of these coatings is important in order to obtain satisfactory corrosion protection.

It was proposed that Cr and Al in NiCr and NiAl coatings respectively could form a corresponding thin passive layer of chromia or alumina to control the corrosion [11], [12]. However, the synergistic effects of Cr and Al in two-phase NiCoCrAlY microstructure, which contains microgalvanic cells, have not been yet studied.

Coatings produced by HVAF process, which has been recently developed, is capable of producing very dense coatings with a minimum number of defects such as *in situ* oxides. It ensures that the beneficial elements (e.g., Cr or Al) are preserved for the corrosion protection and not depleted while spraying [13].

Very few attempts have been studied the role of intermediate layer chemical composition and microstructure in corrosion behaviour of bi-layer coatings. In this study, various single layer and bi-layer coatings were deposited by HVAF, HVOF, and APS processes. Corrosion mechanisms were studied through salt spray, open-circuit potential (OCP) and potentiodynamic polarization techniques in 3.5 wt% NaCl solution along with SEM analysis.

2 Experimental Details

2.1 Specimen preparation

Domex 355 carbon steel (0.10C-0.03Si-1.50Mn-0.02P-0.01S-0.20V-0.15Ti wt%) substrates of 50x30x3 mm dimension were grit blasted using alumina particles (220 µm mesh size) to create a roughness profile of around 6 µm on R_a scale. The powder composition is given in Table I. The powders were deposited on the grit blasted substrates using the HVAF process (Uniquecoat M3[TM], Oilville, VA, USA).

Table I. Chemical composition of the powders

Powder	Chemical composition (wt %)							
	Ni	Al	Cr	O	C	Co	Mo	Fe
Cr_3C_2-NiCr	20.5		Bal.	0.3	8.9	-	-	-
Fe-based 1	11.0	-	30	0.2	0.6	-	-	Bal.
Fe-based 2	17.0	-	28	0.1	1.8	-	4.5	Bal.
Fe-based 3	4.0	-	10	0,2	2.0	-	5.5	Bal.
Ni	Bal.	-	-	0.1	-	-	-	-
NiCr	Bal.	-	21.1	0.1	-	-	-	-
NiAl	Bal.	5.5	-	0.1	-	-	-	-
NiCoCrAlY	Bal.	8.5	20	0.2	-	23.1	-	-

Several spray parameter sets were initially employed to produce optimized coatings in term of minimum pore content. The APS, HVOF, and HVAF spraying process parameters for different powders are given in Table II, III and IV respectively. Details of the spray processes for HVAF, HVOF and APS are described in previous works [6], [7], [14]–[16].

Table II. Spraying process parameters, a) APS, b) HVOF, c) HVAF

a)

Coatings	Current (A)	Voltage (V)	Primary gas (Ar) flow (l/min)	Secondary gas (H_2) flow (l/min)	Powder carrier gas (Ar) flow (l/min)	Feed rate (g/min)	SoD (mm)
Ni	600	65	55	10	3.5	50	140
NiCr	600	68	65	14	2.7	60	140
NiAl	600	66	45	11	2.5	50	140

b)

Coatings	P(O_2) (MPa)	P(Fuel) (MPa)	P(Air) (MPa)	O_2 flow (l/min)	Fuel flow (l/min)	Air flow (l/min)	N_2 flow (l/min)	Feed rate (g/min)	SoD (mm)
Ni	1.17	0.97	0.69	214	635	344	12.5	40	225
NiCr	1.17	0.97	0.69	214	635	344	12.5	40	225
NiAl	1.17	0.97	0.69	214	635	344	12.5	40	225

c)

Coatings	Nozzle	P(Air) (MPa)	P(Fuel 1) (propane) (MPa)	P(Fuel 2) (propane) (MPa)	Powder carrier gas (N_2) flow (l/min)	Feed rate (g/min)	SoD (mm)
Fe-based 1	3L2G	0.8	0.69	0.69	60	100	400
Fe-based 2	3L2G	0.8	0.69	0.69	60	100	400
Fe-based 3	3L2G	0.8	0.69	0.69	60	100	400
Ni	4L4	0.82	0.7	0.7	60	150	350
NiCr	4L4	0.82	0.69	0.69	60	150	300
NiAl	3L2G	0.82	0.69	0.69	60	150	300
Cr_3C_2-NiCr	4L2	0.8	0.7	0.7	60	150	350
NiCoCrAlY	3L2	0.8	0.66	0.7	60	150	350

2.2 Coating characterization

The cross-section of the as-sprayed and corroded single- and bi-layer coatings were analyzed by SEM (Hitachi TM 3000, Tokyo, Japan) equipped with an EDS. The phase composition of top surface of as-sprayed coatings was analyzed using X'Pert PRO diffractometer with Cu K_α (λ=0.154 nm) radiation. Porosity measurements were done using "ImageJ" software [17]. Details of the procedure are discussed in our previous work [7]. Hardness measurements (Shimadzu HMV-2, Tokyo, Japan) were performed on the polished cross-section of the coatings with a Vickers indenter at the load of 300 g and dwell time of 15 s [18]. The surface roughness (R_a) of the coatings was measured using a stylus-based profilometer (Surftest 301, Mitutoyo, Japan).

The coated samples were located in the salt spray chamber at the temperature of 25°C under an aqueous spray (5 wt% NaCl solution with the pH between 6.5 and 7.2) at the rate of 1-2 ml/h.80 cm^2 (ASTM standard B 117). The samples were taken after 240 hours and the corrosion of the coatings was studied by EDS/SEM. A three-electrode cell was used to perform the electrochemical measurements consisted of as-sprayed sample as the working electrode (WE), saturated calomel electrode (SCE) as the reference electrode, and platinum foil as the counter electrode. The surface area exposed to the electrolyte was 0.2 cm^2. The samples were immersed in 3.5 wt% NaCl and the OCP was monitored for 3 h. An IVIUMSTAT instrument (IVIUM, Eindhoven, Netherlands) was used to apply a voltage scan from −250 to +2000 V (vs. OCP) at a scan rate of 0.2 mV/s. Prior to each potentiodynamic polarization measurement, the as-sprayed samples were immersed in the electrolyte at the OCP for 1 h to obtain a stable state.

3 Results & Discussion

3.1 Microstructure analysis of corroded Cr_3C_2-NiCr coating

As already discussed, the corrosion behavior of Cr_3C_2-NiCr is affected by (a) porosity, (b) galvanic effect between the Cr_3C_2 and the NiCr metallic binder, and (c) galvanic effect between the matrix zones of higher Cr_3C_2 concentration (subsequently lower NiCr concentration) and the matrix of higher NiCr concentration (subsequently lower Cr_3C_2 concentration) [7], as can be seen in Fig. 1. The NiCr was more anodic than Cr_3C, as NiCr corrded in the couple. Therefore, the dissolution of NiCr in the galvanic couple was expected to be more. Conversely, the dissolution of the Cr_3C_2 was expected to be hindered owing to cathodic protection by the NiCr. It means that even there were only closed pores in the HVAF-sprayed Cr_3C_2-NiCr coating, corrosive ions could find their paths by corroding the NiCr metallic binder and making the pores interconnected. In this context, the bi-layer coatings were developed by adding an intermediate layer beneath the Cr_3C_2-NiCr coating to address the above-mentioned corrosion challenges.

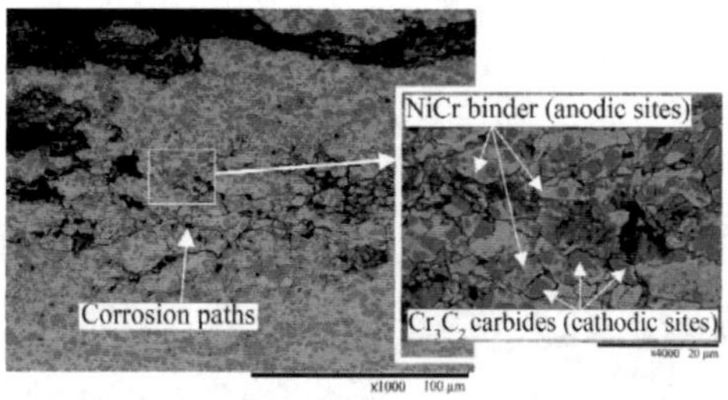

Fig. 1. Corrosive ion penetration paths, cathodic and anodic sites within the single-layer Cr_3C_2-NiCr coating after salt spray test at 24°C for 240 h.

Once the electrolyte penetrates through the top layer, the corrosion resistance is variably associated with the electrochemical properties of the intermediate layer mainly defined by its chemical composition and microstructure. As already discussed, the corrosion of the bi-layer coating could be controlled if the corrosion potential (or open-circuit potential) of the intermediate layer would be higher than the OCP of the top layer.

3.2 Microstructural and compositional effect of intermediate layer on corrosion behavior

In this part, the electrochemical behavior of single-layer coatings such as Fe-based and Ni-based were studied in 3.5 wt% NaCl at room temperature. The experiments were performed in order to address how a single-layer coating with different chemistries shows different corrosion properties, in particular OCP.

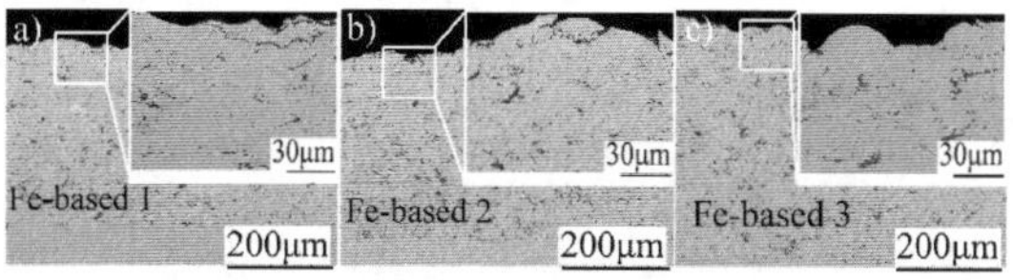

Fig. 2. Back-scattered SEM micrographs of cross sections of as-sprayed (a) Fe-based 1, (b) Fe-based 2 and (c) Fe-based 3.

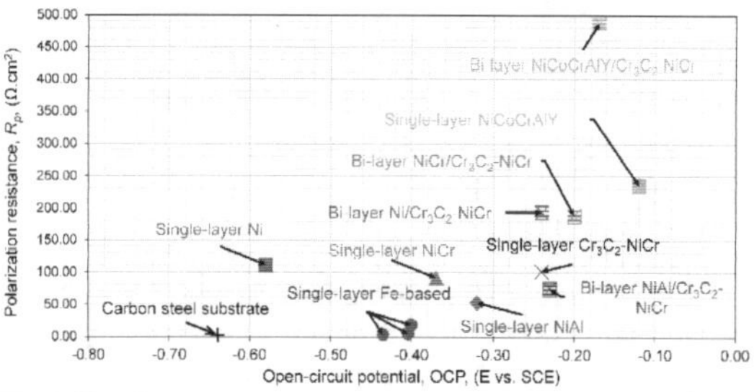

Fig. 3. Polarization resistance (R_p) vs. open-circuit potential (OCP) of various HVAF single- and bi-layer coatings in 3.5 wt% NaCl solution.

The typical microstructures of the Fe-based coatings sprayed by HVAF process are shown in Fig. 2. The coatings presented different microstructural features such as porosity, inter-lamellar boundaries, etc. which adversely affect the corrosion behavior. The high i_{corr} and low E_{corr} values of both coatings from the polarization test confirmed that these chemistries are unable to provide good corrosion protection in the corrosive environment, as seen in Fig. 3. The single-layer Ni-based coatings were studied as an alternative to Fe-based coatings in order to obtain a better corrosion perfromance.

The microstructure of APS-, HVOF-, and HVAF-, sprayed Ni, NiCr, NiAl and NiCoCrAlY coatings are shown in Fig. 4. As it can be seen, the least and most porous microstructures were deposited using the HVAF and APS processes respectively. The highest density of the HVAF coatings is attributed to the low temperature and high velocity of in-flight particles impacting the substrate [19].

The HVAF-sprayed Ni-based coatings presented the best corrosion behavior, revealing that they are superior to the other tested coatings from both feedstock composition and microstructure points of view. It was concluded that these coatings will be used as the intermediate layer in the next investigation on the bi-layer coatings.

As seen in Fig. 3, the higher OCP value of the NiCoCrAlY layer than that of the top layer's, impedes the electrolyte to pass through the top layer and reach the top layer/intermediate layer interface, whereas in the Ni and NiCr intermediate layers, the OCP was lower than that of the top layer's. In the former situation, the intermediate layer would be sacrificially protected by the top layer, thus in case of failure in the top layer, the intermediate layer still serves as a corrosion barrier. In the latter, once the electrolyte meets the intermediate layer/top layer interface, the intermediate layer corrodes which eventually permits the electrolyte to further penetrate and reach the substrate. The figure also shows that the pure Ni coating cannot properly fulfil the requirements required for intermediate layers to protect the bi-layer coating against corrosion.

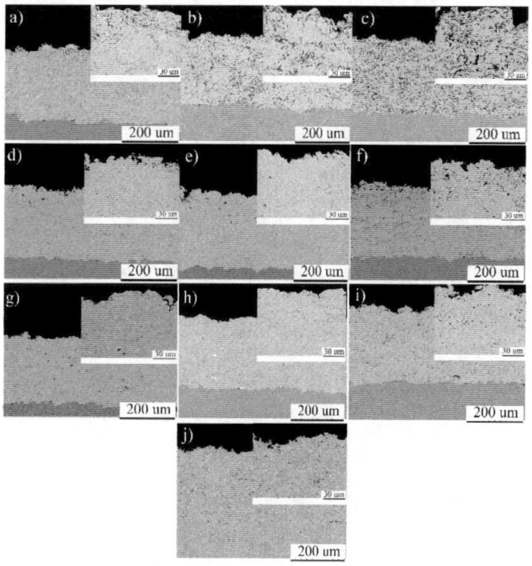

Fig. 4. Back-scattered SEM micrographs of cross sections of as-sprayed a) APS Ni, b) APS NiCr, c) APS NiAl, d) HVOF Ni, e) HVOF NiCr, f) HVOF NiAl, g) HVAF Ni, h) HVAF NiCr, i) HVAF NiAl, and j) HVAF NiCoCrAlY.

Regarding the polarization values (R_p) in Fig. 3, the lowest i_{corr} was reported for bi-layer Cr_3C_2-NiCr/NiCoCrAlY coating. Considering the values of E_{corr} and i_{corr}, the Cr_3C_2-NiCr/NiCoCrAlY system showed superior corrosion resistance than the other coatings. The SEM results in Fig. 5 (a-c) show that the electrolyte penetrated through the interconnected pores of the Cr_3C_2-NiCr top layer and reached the top/intermediate layer interface. The corrosive ions penetrated through the NiCr and NiAl intermediate layers, as these intermediate layers provided lower OCP than that of the Cr_3C_2-NiCr top layer's.

It was found that although the Ni intermediate layer presented a better inter-lamellar cohesion, and lower porosity compared to the other intermediate layers (Fig. 4), the Ni itself was not inherently able to last in 3.5 wt% NaCl. The NiCoCrAlY intermediate layers did not show a perfect coating in term of density compared to the Ni coating. However, the systems containing this intermediate layer presented better corrosion resistance than the Cr_3C_2-NiCr/NiCr or Cr_3C_2-NiCr/NiAl systems. As the top layer/intermediate layer interface acted as an additional barrier to avoid the corrosion of the substrate, the electrolyte failed to penetrate easily through the splat boundaries of the intermediate layer. Formation of galvanic cells between the substrate and intermediate layers was also facilitated, as the Cr_3C_2-NiCr was more anodic than, for instance, the NiCoCrAlY intermediate layer (based on OCP measurements shown in Fig. 3). Therefore, if the electrolyte meets the intermediate layer through the defects of the top layer, the intermediate layer will be sacrificially protected by the top layer.

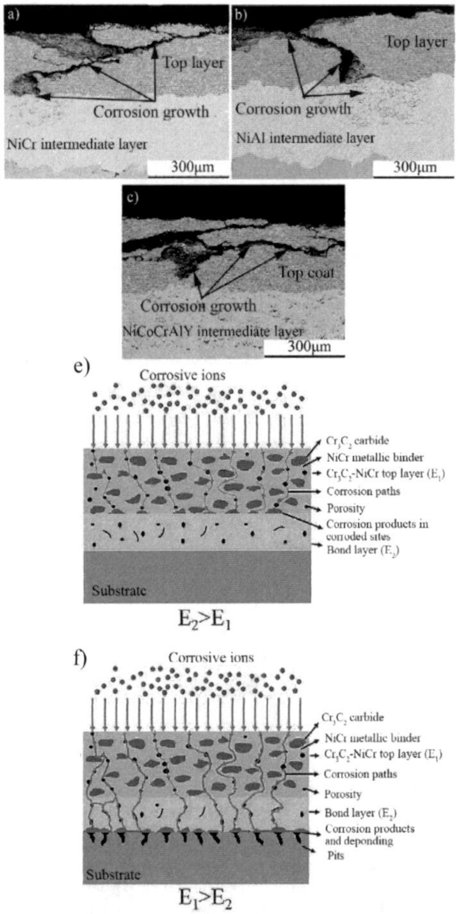

Fig. 5. Microstructure of corroded samples, a) Cr_3C_2-NiCr/Ni, b) Cr_3C_2-NiCr/NiCr, c) Cr_3C_2-NiCr/NiCoCrAlY and d) Cr_3C_2-NiCr/CoNiCrAlY (BSE images), e) Schematic view of bi-layer coatings with different configurations, a) E_{corr} of the intermediate layer is higher than OCP of the top layer, f) OCP of the top layer is higher than that of the intermediate layer.

Other works also suggested that Cr and Al could form a thin passive layer to control the corrosion [20]. A less porous coating as obtained in this study using the HVAF process, allows the Cr and Al to easily diffuse further to the corroded sites where the passive film is supposed to form. Consistent with the OCP results, the simultaneous addition of Cr and Al to Ni-based coatings provided a cathodic intermediate layer, which was highly beneficial for the bi-layer coating with the Cr_3C_2-NiCr top layer. The composition of NiCoCrAlY coating encouraged the formation of the γ-NiCoCr and β-NiAl phases and also defects within the coating (which develop the pitting). However, a full support of the passive layer by Cr and Al together compromised the effect of coating's defects and undesired galvanic couples.

Fig. 5 (e & f) show schematic views of bi-layer coatings with different settings in terms of galvanic coupling between the top and intermediate layers. It shows that the OCP (or E_{corr}) of the intermediate layer affects the corrosion behavior of the bi-layer coatings. The severity of corrosion depends on the magnitude of the potential differential ($\Delta E = E_{corr}^{intermediate\ layer} - E_{corr}^{top\ layer}$) between the top layer and intermediate layer. Higher $E_{corr}^{intermediate\ layer}$ than $E_{corr}^{top\ layer}$ ensures that the intermediate layer is sacrificially protected by the top layer, therefore the lifetime of the coating system will be longer.

According to Fig. 3, the chromia forming coating of NiCr showed more positive value of OCP compared to alumina forming coating of NiAl, thus the former revealed a better protection of the substrate, thermodynamically. The high corrosion protection of the NiCoCrAlY coating was due to the homogeneous alloy composition without significant forming of macro-galvanic cells.

Corrosion growth mechanisms in different single-layer coatings can be seen in Fig. 6. The figure shows that while corrosive ions can only penetrate through the inter-lamellar boundaries surrounding one NiCoCrAlY particle, corrosion can grow through both the lamellae boundary and inside of the particles in NiCr and NiAl coatings. The Ni phase in NiCr and NiAl was unable to tolerate the corrosive environment not only because of the low corrosion resistance of Ni, but also due to a great number of microgalvanic cells that formed between Ni and alloying elements like Cr or Al in the corresponding coatings. In the NiCoCrAlY coating, the presence of two phases (γ-NiCoCr and β-NiAl) which most probably have E_{corr} close to each other resulted in corrosion propagation through the splat boundaries around a single NiCoCrAlY particle. Therefore, the high corrosion protection by the NiCoCrAlY coating could be due to the homogeneous alloy composition without forming of macro-galvanic cells [21].

Fig. 6. Backscattered cross-sectional SEM micrographs of corroded samples in 3.5 wt% NaCl at 25°C, a) single-layer NiCr, b) single-layer NiAl and c) single-layer NiCoCrAlY coatings.

4 Summary

1) While high-alloyed Fe-based coatings a dense structure, the corrosion results of all three coatings showed low corrosion protection. As further microstructure optimization failed in reaching a very high density of these coatings (as a requirement for better corrosion protection), the study was directed

towards Ni-based coatings, as this type of coatings could provide less porous microstructure.

2) Three thermal spray techniques and feedstocks with different chemistries were employed to spray Ni-based coatings (Ni, NiCr, and NiAl) with different microstructures. APS and HVOF coatings showed large defects in the coating microstructure, in particular high porosity. The pores facilitated channels forming for electrolyte that reached the substrate quite quickly. HVOF-sprayed coatings exhibited a wider range of corrosion behavior. The corrosion behavior was directly related to the evolution of microstructure and porosity level. This in turn was dependent on the processing parameters selected for spraying. Presence of defects undermined the corrosion behavior of the coating by decreasing the OCP and increasing the i_{corr} values. HVAF-sprayed coatings showed the least porous coatings compared to the other spraying techniques. Although in a low amount, the presence of the pores still diminished the corrosion resistance of the coating. HVAF process optimization was carried out by studying the effect of different spray parameters on porosity.

3) The OCP of the intermediate layer determines the corrosion behavior of the bi-layer coating. Higher OCP of the intermediate layer than that of the top layer leads to a galvanic couple which favors the intermediate layer to be sacrificially protected. In this situation, corrosion diffusion is stopped at the intermediate layer/ top layer interface.

4) Both microstructure and chemical composition of the intermediate layer affected the corrosion behavior of the bi-layer coatings. The chemical composition had higher proportion to the corrosion protection; however, the effect of defects in thermal spray coatings cannot be neglected.

5) The defects reduced the diffusion rate of beneficial elements like Cr to form the passive layer, helped the corrosive agents to diffuse rapidly, provided galvanic couples, and caused Cr-depletion, etc.

6) A desired the intermediate layer microstructure (less defects and higher density) enabled to focus on the next important step of the study, which was chemical composition of the intermediate layers.

7) Addition of alloying elements such as Cr or Al improved the corrosion behavior of the coating. The NiCr or NiCoCrAlY coatings presented higher content of defects but better corrosion behavior than the denser Ni coating. The passivation by Cr was shown to be better than Al in room temperature.

8) The most promising coating system to be used for industrial applications is Cr_3C_2-NiCr top layer with an intermediate layer of NiCoCrAlY coating.

Acknowledgements

The authors would like to thank Mr. Jonas Olsson and Mr. Stefan Björklund for spraying the coatings.

References

[1] J. A. Picas, A. Forn, A. Igartua, and G. Mendoza, "Mechanical and tribological properties of high velocity oxy-fuel thermal sprayed nanocrystalline CrC NiCr coatings," *Surf. Coat. Technol.*, vol. 174–175, pp. 1095–1100, Sep. 2003.

[2] J. K. N. Murthy and B. Venkataraman, "Abrasive wear behaviour of WC–CoCr and Cr3C2–20(NiCr) deposited by HVOF and detonation spray processes," *Surf. Coat. Technol.*, vol. 200, no. 8, pp. 2642–2652, Jan. 2006.

[3] P. Poza, C. J. Múnez, M. A. Garrido-Maneiro, S. Vezzù, S. Rech, and A. Trentin, "Mechanical properties of Inconel 625 cold-sprayed coatings after laser remelting. Depth sensing indentation analysis," *Surf. Coat. Technol.*, vol. 243, pp. 51–57, Mar. 2014.

[4] S. García-Rodríguez, A. J. López, B. Torres, and J. Rams, "316L stainless steel coatings on ZE41 magnesium alloy using HVOF thermal spray for corrosion protection," *Surf. Coat. Technol.*, vol. 287, pp. 9–19, Feb. 2016.

[5] A. Lekatou, D. Zois, A. E. Karantzalis, and D. Grimanelis, "Electrochemical behaviour of cermet coatings with a bond coat on Al7075: Pseudopassivity, localized corrosion and galvanic effect considerations in a saline environment," *Corros. Sci.*, vol. 52, no. 8, pp. 2616–2635, 2010.

[6] E. Sadeghimeresht, N. Markocsan, and P. Nylen, "Microstructural and electrochemical characterization of Ni-based bilayer coatings produced by HVAF process," *Surf. Coat. Technol.*, 2016.

[7] E. Sadeghimeresht, N. Markocsan, P. Nylén, and S. Björklund, "Corrosion performance of bi-layer Ni/Cr2C3–NiCr HVAF thermal spray coating," *Appl. Surf. Sci.*, vol. 369, pp. 470–481, Apr. 2016.

[8] C. Godoy, M. M. Lima, M. M. R. Castro, and J. C. Avelar-Batista, "Structural changes in high-velocity oxy-fuel thermally sprayed WC–Co coatings for improved corrosion resistance," *Surf. Coat. Technol.*, vol. 188, pp. 1–6, 2004.

[9] G. Bolelli *et al.*, "Functionally graded WC–Co/NiAl HVOF coatings for damage tolerance, wear and corrosion protection," *Surf. Coat. Technol.*, vol. 206, no. 8, pp. 2585–2601, 2012.

[10] S. Kuroda, J. Kawakita, T. Fukushima, and S. Tobe, "Importance of the adhesion of HVOF sprayed coatings for aqueous corrosion resistance," *Mater. Trans.*, vol. 44, no. 3, pp. 381–388, 2003.

[11] Z. Bergant, U. Trdan, and J. Grum, "Effect of high-temperature furnace treatment on the microstructure and corrosion behavior of NiCrBSi flame-sprayed coatings," *Corros. Sci.*, vol. 88, pp. 372–386, Nov. 2014.

[12] Z. B. Bao, Q. M. Wang, W. Z. Li, J. Gong, T. Y. Xiong, and C. Sun, "Corrosion behaviour of AIP NiCoCrAlYSiB coating in salt spray tests," *Corros. Sci.*, vol. 50, pp. 847–855, 2008.

[13] G. Bolelli *et al.*, "Tribological behavior of HVOF- and HVAF-sprayed composite coatings based on Fe-Alloy + WC–12% Co," *Surf. Coat. Technol.*, vol. 248, pp. 104–112, Jun. 2014.

[14] E. Sadeghimeresht, N. Markocsan, and P. Nylén, "Microstructure Effect of Intermediate Coat Layer on Corrosion Behavior of HVAF-Sprayed Bi-Layer Coatings," *J. Therm. Spray Technol.*, vol. 26, no. 1–2, pp. 243–253, Jan. 2017.

[15] E. Sadeghimeresht, N. Markocsan, and P. Nylén, "A Comparative Study of Corrosion Resistance for HVAF-Sprayed Fe- and Co-Based Coatings," *Coatings*, vol. 6, p. 16, Mar. 2016.

[16] E. Sadeghimeresht, N. Markocsan, and P. Nylen, "A comparative study on Ni-based coatings prepared by HVAF, HVOF, and APS methods for corrosion protection applications," *J Therm Spray Technol*, 2016.

[17] C. A. Schneider, W. S. Rasband, and K. W. Eliceiri, "NIH Image to ImageJ: 25 years of image analysis," *Nat. Methods*, vol. 9, no. 7, pp. 671–675, Jul. 2012.

[18] "ASTM E384-11e1, Standard Test Method for Knoop and Vickers Hardness of Materials , ASTM International, West Conshohocken, PA, 2011, www.astm.org." .

[19] E. Sadeghimeresht, H. Hooshyar, N. Markocsan, S. Joshi, and P. Nylén, "Oxidation Behavior of HVAF-Sprayed NiCoCrAlY Coating in H2–H2O Environment," *Oxid. Met.*, pp. 1–16, Jul. 2016.

[20] J. M. Guilemany, N. Cinca, S. Dosta, and A. V. Benedetti, "Corrosion behaviour of thermal sprayed nitinol coatings," *Corros. Sci.*, vol. 51, no. 1, pp. 171–180, 2009.

[21] X. Fan *et al.*, "Investigation of the bond coats for thermal barrier coatings on Mg alloy," *Appl. Surf. Sci.*, vol. 265, pp. 264–273, Jan. 2013.

KCl-induced corrosion behavior of HVAF-sprayed Ni-based coatings in ambient air

R. Jafari[a], E. Sadeghimeresht[b], T. Shahrabi Farahani[a], N.Markocsan[b], S. Joshi[b]
a Department of Material Science and Engineering, Tarbiat Modares University, Tehran, 14115, Iran
b Department of Engineering Science, University West, Trollhättan, 46153, Sweden
Corresponding author(s): esmaeil.sadeghimeresht@hv.se, tshahrabi34@modares.ac.ir

Abstract: It is well known that the presence of KCl deposited on superheater tubes in biomass- and waste-fired boilers leads to a severe corrosion and premature damage. In order to protect such critical components which are routinely exposed to aggressive environments, thermal spray coatings are frequently proposed as a potential solution. By virtue of the techno-commercial benefits that provides as a direct outcome of its ability to cost-effectively deposit coatings virtually free of porosity and *in situ* formed oxides, the high velocity air-fuel (HVAF) process offers a particularly attractive approach. In the present work, the influence of KCl on the oxidation behavior of four HVAF-sprayed Ni-based coatings (Ni21Cr, Ni5Al, Ni21Cr7Al1Y, and Ni21Cr9Mo) has been investigated. The coatings were deposited onto specimens of 16Mo3 steel, a widely used boiler tube material. High temeparture corrosion tests were carried out in ambient air at 600°C, with 0.1 mg/cm^2 KCl being sprayed onto the samples prior to the exposure. Uncoated substrates and an identical test environment without KCl were used as reference. SEM/EDS and XRD techniques were utilized to characterize the as-sprayed and exposed samples. The results showed that the small addition of KCl significantly accelerated damage to the coatings. It was further revealed that the alumina-forming NiAl coating was capable of forming a more protective oxide scale compared to other chromia and mixed-oxide scale forming coatings. In general, the oxidation resistance of the coatings based on the kinetic studies had the following ranking (from the best to the worst): NiAl >NiCr> NiCrAlY> NiCrMo.
Keywords: High Velocity Air-Fuel (HVAF), Thermal Spray Coating, Corrosion protection, Biomass Boiler, Cl-induced corrosion

1 Introduction

In recent decades, biomass and waste fuels are being increasingly used as renewable and CO_2 free resources of energy in power generation industry [1]. High temperature flow gases from the combustion of such fuels mainly contain chlorine compounds which may deposit on degradation-prone parts of boilers, in particular the superheater tubes. This process leads to form a harsh environment that appears highly corrosive to the tube materials [2]. Demands for higher efficiency through increasing the operational temperature makes the situation extremely aggressive to even expensive corrosion resistant alloys which have been developed over the last decades [3]. Alkali chloride/chlorine accelerates corrsoion by reacting with protective metal oxid scale resulting in active oxidation, material wastage, internal attack, defect formation, and non-adherent scale loss [4], [5]. Despite limitation of literature concerning the effect of biomass fuels on oxidation mechanism, KCl is found as a dominant component of deposit near the metal surface; indicating its key role in Cl-induced corrosion problems of many power plants [5], [6]. In order to control the oxidation damages, high alloy Fe- and Ni-based components have been widely used instead of traditional low alloy steels [7]. However, developing more noble materials is not always economically favorable. Flexibility of coating technologies allows to manufacture dense and high adherent coatings which can meet the critical protection requirements of such a boiler environment. The rapid development of thermal spray processes offers attractive and cost effective routes to avoid degradation of low alloy tubes by selecting appropriate coating's chemistry and manufacturing process [8], [9]. A variety of spray methods has been successfully used to fabricate corossion resistance coatings [10], [11]. High velocity air-fuel (HVAF), known as a novel and promising method to prouduce dense and adherent coating works based on high velocity impact of molten or semi-molten particles which could be used to apply a wide range of coating materials including Ni-based powders [12]–[15]. Recent advances in design and development of new HVAF guns reduce the cost and enhance the quality of product in collation with typical high velocity oxy-fuel (HVOF) method resulting in replacement and also booming markets for HVAF technology [16]–[18]. The coatings produced by HVAF characterized with lower porosity, less oxide content and more interlamellar cohesion, are capable of controlling the diffusion of aggressive agents such as chlorine to substrate/coating interface [19], [20]. The few studies performed on HVAF Ni-based coatings show that the coatings present excellent oxidation behavior in oxidizing environments [21]. However, the corrosion mechanisms of HVAF Ni-based coatings in oxidizing-chloridizing enviroenment needs to be studied further in terms of comprehensively investigating both thermodynamic and kinetic aspects. In the present study, high temperature corosion behavior of four commercially available Ni-based coatings –Ni21Cr, Ni5Al, Ni21Cr9Mo, Ni21Cr7Al1Y- sprayed by HVAF was investigated in a simplified oxidizing-chloridizing boiler environment concerning the effect of KCl. The selection of the feedstocks was based on presence of beneficial alloying elements, i.e., Al and/or Cr which play key roles in formation of protective oxide scale in aggressive chlorine environment [22].

2 Experimental Details

2.1 Materials

Commercially available Ni-based powders Ni21Cr - Amperit 251.001, Ni5Al -Amperit 281.003, Ni21Cr7Al1Y -Amperit 413.001 sourced from H.C Starck Co. and Ni21Cr9Mo supplied by M.H. Engineering Co. used as feedstock materials. 16Mo3, a low alloy ferritic steel commonly used in boiler parts was selected as substrate of coatings and also as reference material. 304L and Sanicro25 were also used due to their high temperature recommended applications as reference materials. The KCl salt with particle size of 45-155 µm was used as chlorine containing deposit.

2.2 Spray method and sample preparation

A HVAF gun (Uniquecoat M3^{TM}, Oilville, VA, USA) was utilized to spray the coatings. The 16Mo3 rods (length=500 mm) were fixed in a horizontal rotating mandrel for applying the coating on the cylindrical surface. After spraying, the rod was cut into pieces of 5 mm length and coatings were applied on the flat faces to ensure that the specimens were coated on all sides. The substrates were grit blasted by alumina grits (mesh 220) prior to spraying for cleaning and roughing the surfaces. Several combinations of thermal spray parameters were initially tried and the resulting coatings characterized to determine spray conditions that yielded least porosity. The selected process parameters are presented in Table I. All coatings were sprayed to a thickness of around 250 µm, and then polished in order to unify the effect of surface roughness for all coatings.

Table I. HVAF spray process parameters

Variables	NiCr	NiAl	NiCrMo	NiCrAlY
Nozzle type	4L2G	4L4G	3L2G	4L2G
Fuel 1 (Propane) pressure, (Mpa)	0.7	0.7	0.7	0.7
Fuel 2 (Propane) pressure, (Mpa)	0.7	0.7	0.8	0.7
Carrier gas(N_2) pressure, (Mpa)	0.4	0.4	0.4	0.4
Feed rate, g/min	150	150	150	150

2.3 High temperature corrosion test

KCl dissolved in a solution (ethanol/water mixture) sprayed on the samples to form 0.1 mg/cm^2 deposit. Isothermal oxidation test was performed at 600 °C in an air furnace for 168h. To diminish the shielding effect, nearby sample placed in 50mm distance. Corrosion performance of coatings is discussed in relation to microstructure characterization, coating cross section investigation andmass change, in comparison with reference materials in the same condition.

2.4 Characterisation

The powders, coatings' morphologies and cross sections were studied by scanning electron microscope (SEM) equipped with energy dispersive spectroscopy system (EDS). Mass changes after exposure was recorded to study the corrosion kinetics of the coatings in comparison with the substrate and reference materials. The phase composition of the as-received powders, as-sprayed, and exposed coatings was determined by a Philips- PW1730 X-ray diffractometer (XRD; Cu) over a 2θ range of 10–80°. Samples cold mounted in a low shrinkage epoxy were cut by a low speed saw prior to the metallographic preparations. Cross-sectional analysis performed to investigate the coatings' features and corrosion performance. ImageJ, an open access image analyzer software was used to measure the coating porosity of the precise prepared metallographic cross-sections.

3 Results & discussion

3.1 Characterization of coatings and feedstocks

Fig. 1 shows spherical or near-spherical shape morphology of the feedstock powders.

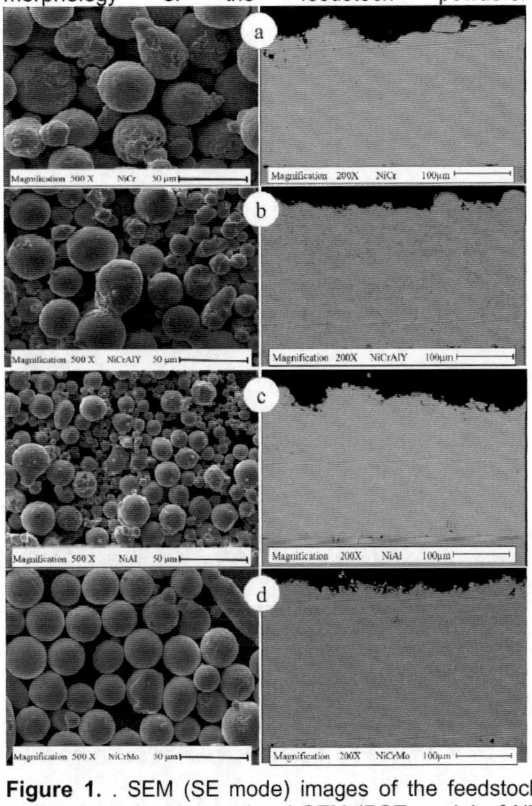

Figure 1. . SEM (SE mode) images of the feedstock materials and cross sectional SEM (BSE mode) of Ni-based HVAF as-sprayed coatings, a) NiCr, b) NiCrAlY, c) NiAl, and d) NiCrMo.

Cross-sectional SEM images taken in BSE mode (Fig. 1) exhibited almost dense and defect-free coatings for all compositions. The semi-molten particles deformed sufficiently upon impact to substrate or other particles. The splat boundaries are barely visible near coating surfaces, which is due to lack of hammering effect by subsequent particle impact. Good interlocking at substrate/coating interface, well-adhered splats and low porosity content are evidence of proper parameter selection and HVAF potential for manufacturing dense and uniform coatings [19]. Table II shows the summery of the feedstock features and coatings' properties.

Table II. Summery of coating characterization and feedstock materials feature

Materials	Powder size (µm)	Coating thickness (µm)*	Coating roughness, R_a(µm)	Coating Porosity (vol.%)
NiCr	16 to 52	221 ± 21	<0.1	0.2±0.02
NiAl	3 to 32	235 ± 12	<0.1	0.2±0.07
NiCrAlY	5 to 38	236±24	<0.1	0.3±0.10
NiCrMo	19 to 30	218±7	<0.1	0.3±0.09

*Average of 20 individual measurements.

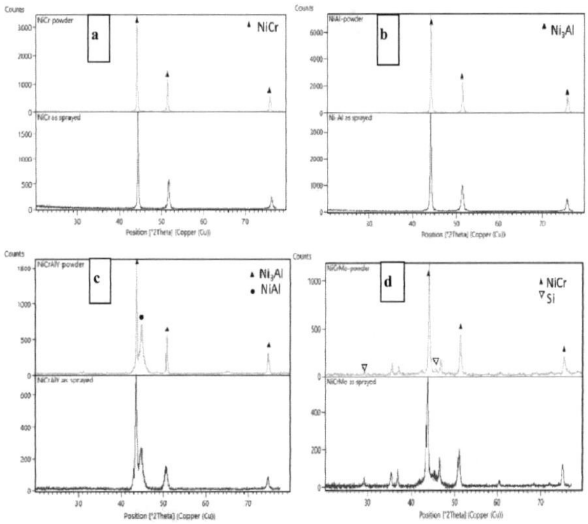

Figure 2. XRD pattern of feedstocks (green) and as-sprayed coatings(blue) limited to 20-80 degree. a) NiCr, b) NiAl, c) NiCrAlY, and d) NiCrMo.

Minimum phase change between the as-sprayed coatings and feedstocks in XRD patterns is illustrated in Fig. 2. No feedstock oxidation and loss related to the spray method was observed. Thus, the good interlamellar adhesion discussed in section 3.1 could be also explained with lower oxidation state of splashed particles during spray process.

3.2 Corrosion test results

3.2.1 Mass changes

In order to evaluate the corrosion resistance of the coatings, the weight changes of the samples were measured after removing the residual deposited salts. NiCr and NiAl showed less change in weight as an indication of better performance in comparison with the other coatings and refrence materials (see Fig. 3a).

Fig. 3b also reveals lower corrosion degradation of coatings by comparing median value of formed scale

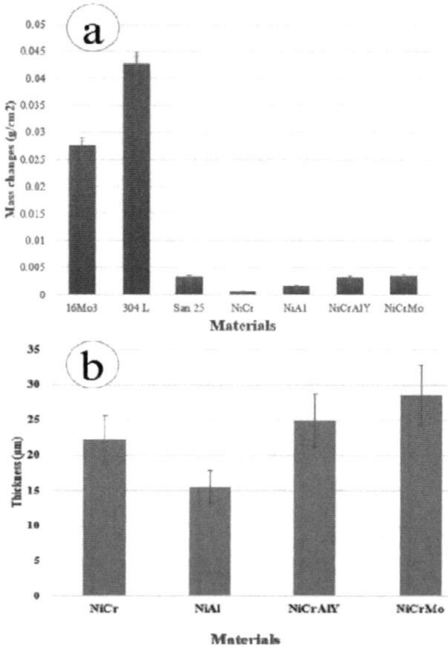

Figure 3. a)Mass change of exposed sample to 0.1mg /cm^2 KCl deposit at 600 °C for 168h, b) Thickness of formed scale on HVAF Ni-based coatings (average of 10 measurments).

thickness on each coating cross section. Note that more protective behavior is expected from dense and slow growing oxide scale [23].

3.2.2 Phase and microstructure characterization of the exposed samples

Dark gray as-sprayed samples turned to shiny metallic color after grinding/polishing. The obvious change in the specimen appearances including color change and peeled off spot (especially in NiCrMo) observed after 168h exposure at 600°C to KCl deposit (Fig. 4). The reference materials showed noticeable changes including spallation and swelling of scales verifying lower protection of the formed oxide scales.

Figure 4. Test samples appearance and coatings' morphologies after the exposure at 600 °C for 168h, a) NiCr, b) NiAl, c) NiCrAlY, d) NiCrMo, and e) Reference materials; half disc:16Mo3, arc shape: Sanicro25, rectangular: 304L.

Fig. 4 also shows the typical surface morphologies of the Ni-based coatings exposed to KCl deposit at 600°C which were markedly different from the polished coatings. Fig. 4 (a) and (b) show more compacted scales in comparison to Fig. 4 (c) and (d). In particular, some cracks found in the oxide scale on the NiCrMo sample (d), could act as penetration path for corrosive agents. Non-protective and flake shape scale on NiCrAlY outer surface could give an explanation of the coating degradation under the test condition.

EDS images in Fig. 5 shows that the Cl penetration occured in the coatings with different compositions. However, the growth of the formed oxide layer varied in size could be responsible for controlling the Cl penetration through the coating. There were seemingly more cracks in the NiCrMo and NiCrAlY coatings after exposure, compared to the other two coatings in Fig. 5. Consequently, oxygen and Cl could diffuse faster through those cracks. Moderate oxide scale formed on NiCr, which is rich in Cr and O in upper oxide layer and less Cr content at the lower oxide layer confimed the formation of Cr_2O_3. For NiAl, less oxygen and Cl diffusion to substrate observed due to the thin protective oxide scale formed on the surface. Clearly, Al migrated to top surface to form alumina. Furthermore, Al easily diffused to cracks and porosity appeared after the exposure. On the contrary to the NiAl coating, NiCrMo (Fig. 5 (d)) had a thick oxide scale including Ni, Cr and K, resulted in the spinel formation However, there is no clear evidence for Cl penetration path through the coating. The Cr depletion in Fig. 5 (c) occurred around a wide crack. It seems that no continues protective scale could formed on NiCrAlY and the coating encountered internal oxidation, despite presence of Al and Cr. The XRD patterns of corroded sample shown in Fig. 6 represent oxides and spinel structures, e.g., $NiCr_2O_4$ - that formed in a reaction between NiO and Cr_2O_3 [24]. No sign of transition metal chlorides, e.g., $NiCl_2$, $NiCl_3$ or $CrCl_3$ was seen in Fig. 6 (a), and (b) due to higher stability of the protective oxides, e.g., Cr_2O_3 and Al_2O_3 at high oxygen partial pressure. Formation of $NiCl_2$ is more possible among other metallic chlorides as supplied by (c), and (d) in Fig. 61. which were unable to form a continuous and protective scale. Besides, K remained at NiCrMo surface, followed by non-protective oxide spinel formation which grew faster than protective oxide scales.

Chlorine is able to penetrate through oxide scales along cracks and react with coating's or substrates' elements such as Cr to form metallic chlorides where there is a poor access to oxygen (low oxygen partial pressure). Formation of metallic chlorides to volatile chlorides are thermodynamically possible in these locations. The volatile chlorides tend to transfer outward through the diffusion paths up to the coating surface. These chlorides change to oxides where oxygen is more available (higher oxygen partial pressure) which result in porous oxide scale [23].

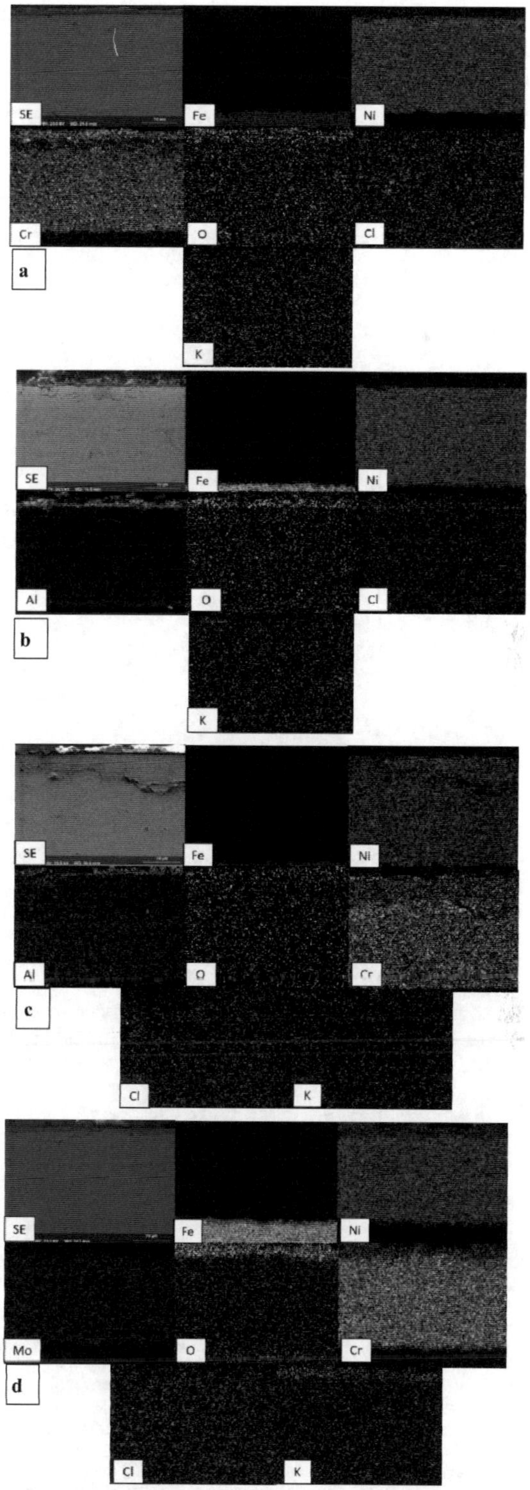

Figure 5. SEM/EDS map of major present elements in coating cross section after the exposure at 600 °C for 168h. a) NiCr, b) NiAl, c) NiCrAlY, d) NiCrMo.

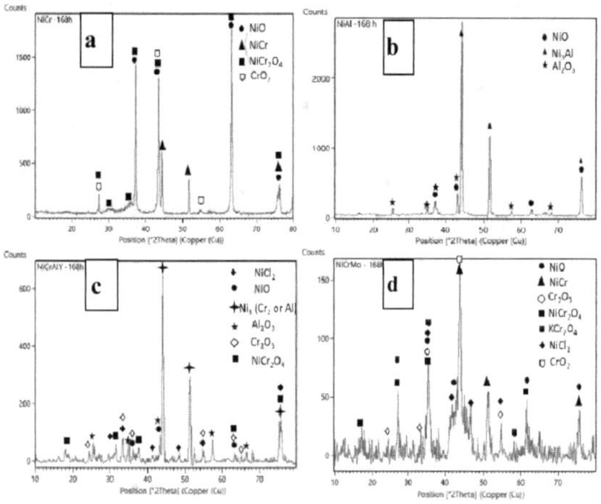

Figure 6. XRD pattern of coatings after HTC. a)NiCr, b)NiAl, c)NiCrAlY, d)NiCrMo.

In oxidizing-chloridizing environment, the competition between Cl and O depends on alloying elements. Rather the slow growth of the oxide layer formed on the NiAl enhanced the corrosion protection (see Fig. 3). In contrast, higher thickness of scales for the other coatings refer to a much higher degree of corrosion.

4 Conclusion

The HVAF Ni-based coatings produced in this study were capable of corrosion control in alkali chloride-containing environement. In all coatings which had different microstructure and density, the oxide scales covered the sample surfaces after the exposure to KCl for a week at 600°C. More protective layers formed on the NiAl and NiCr coatings in which less alloying element depletion and mass gain observed in comparison with NiCrMo and NiCrAlY. By the way, all coatings improved high temperature corrosion resistance in collation to reference materials. The coatins' corrosion resistance could be ranked as (from best to worst) NiAl>NiCr>NiCrAlY>NiCrMo. As a result, dense and well-adhered Ni-based coatings with lamellar structure are able to control the corrodants' penetration and protect the substarte.

5 References

[1] R. Saidur, E. A. Abdelaziz, A. Demirbas, M. S. Hossain, and S. Mekhilef, "A review on biomass as a fuel for boilers," *Renew. Sustain. Energy Rev.*, vol. 15, no. 5, pp. 2262–2289, 2011.

[2] N. Otsuka, "A thermodynamic approach on vapor-condensation of corrosive salts from flue gas on boiler tubes in waste incinerators," vol. 50, pp. 1627–1636, 2008.

[3] Y. Kawahara, M. Nakamura, H. Tsuboi, and K. Yukawa, "Evaluation of New Corrosion-Resistant Superheater Tubing in High-Efficiency Waste-to-Energy Plants," *Corrosion*, vol. 54, no. 7, pp. 576–589, 1998.

[4] H. P. Nielsen, F. J. Frandsen, K. Dam-Johansen, and L. L. Baxter, "Implications of chlorine-associated corrosion on the operation of biomass-fired boilers," *Prog. Energy Combust. Sci.*, vol. 26, no. 3, pp. 283–298, 2000.

[5] T. Blomberg, "Which are the right test conditions for the simulation of high temperature alkali corrosion in biomass combustion?," *Mater. Corros.*, vol. 57, no. 2, pp. 170–175, 2006.

[6] C. Berlanga and J. A. Ruiz, "Study of corrosion in a biomass boiler," *J. Chem.*, vol. 2013, 2013.

[7] Y. Kawahara, "Application of high temperature corrosion-resistant materials and coatings under severe corrosive environment in waste-to-energy boilers," *J. Therm. Spray Technol.*, vol. 16, no. 2, pp. 202–213, 2007.

[8] L. Pawlowski, *The science and engineering of thermal spray coatings*, vol. 34. 1995.

[9] D. L. Singbeil, "Superheater Corrosion Produced by Biomass Fuels," no. 1.

[10] S. Tuurna, T. Varis, K. Penttil??, K. Ruusuvuori, S. Holmstr??m, and S. Yli-Olli, "Optimised selection of new protective coatings for biofuel boiler applications," *Mater. Corros.*, vol. 62, no. 7, pp. 642–649, 2011.

[11] T. Varis, D. Bankiewicz, P. Yrjas, M. Oksa, T. Suhonen, S. Tuurna, K. Ruusuvuori, and S. Holmström, "High temperature corrosion of thermally sprayed NiCr and FeCr coatings covered with a KCl-K2SO4 salt mixture," *Surf. Coatings Technol.*, vol. 265, pp. 235–243, 2015.

[12] G. Bolelli, T. Börner, A. Milanti, L. Lusvarghi, J. Laurila, H. Koivuluoto, K. Niemi, and P. Vuoristo, "Tribological behavior of HVOF- and HVAF-sprayed composite coatings based on Fe-Alloy + WC-12% Co," *Surf. Coatings Technol.*, vol. 248, pp. 104–112, 2014.

[13] A. P. Wang, T. Zhang, and J. Q. Wang, "Formation and properties of Ni-based amorphous metallic coating produced by HVAF thermal spraying," *Mater. Trans.*, vol. 46, no. 5, pp. 1010–1015, 2005.

[14] A. Verstak and V. Baranovski, "Activated Combustion HVAF Coatings for Protection against Wear and High Temperature Corrosion," *Therm. Spray 2003 Adv. Sci. Appl. Technol.*, pp. 535–541, 2003.

[15] E. Sadeghimeresht, N. Markocsan, P. Nylén, and S. Björklund, "Corrosion performance of bi-layer Ni/Cr2C3-NiCr HVAF thermal spray coating," *Appl. Surf. Sci.*, vol. 369, pp. 470–481, 2016.

[16] A. Verstak, "High Velocity Air-Fuel Spraying and Its Applications in Oil and Gas Industry," no. Ref 5, pp. 529–534, 2012.

[17] C. Lyphout, K. Sato, S. Houdkova, E. Smazalova, L. Lusvarghi, G. Bolelli, and P. Sassatelli, "Tribological Properties of Hard Metal Coatings Sprayed by High-Velocity Air Fuel Process," *J. Therm. Spray Technol.*, vol. 25, no. 1–2, pp. 331–345, 2016.

[18] O. Andreas, "Thermal spraying by HVAF as an environmentally friendly alternative to electrolytic hard chrome plating of piston rods." .

[19] E. Sadeghimeresht, H. Hooshyar, N. Markocsan, S. Joshi, and P. Nylén, "Oxidation Behavior of HVAF-Sprayed NiCoCrAlY Coating in H2–H2O Environment," *Oxid. Met.*, no. July, 2016.

[20] G. Wang, Z. Huang, P. Xiao, and X. Zhu, "Spraying of Fe-based amorphous coating with high corrosion resistance by HVAF," *J. Manuf. Process.*, vol. 22, pp. 34–38, 2016.

[21] E. Sadeghimeresht, N. Markocsan, M. Huhtakangas, and S. Joshi, "Isothermal oxidation of HVAF-sprayed Ni-based chromia, alumina and mixed-oxide scale forming coatings in ambient air," *Surf. Coatings Technol.*

[22] R. Bender and M. Sch??tze, "The role of alloying elements in commercial alloys for corrosion resistance in oxidizing-chloridizing atmospheres part I: Literature evaluation and thermodynamic calculations on phase stabilities," *Mater. Corros.*, vol. 54, no. 8, pp. 567–586, 2003.

[23] M. A. Uusitalo, P. M. J. Vuoristo, and T. A. Mäntylä, "High temperature corrosion of coatings and boiler steels below chlorine-containing salt deposits," *Corros. Sci.*, vol. 46, no. 6, pp. 1311–1331, 2004.

[24] T. S. Sidhu, A. Malik, S. Prakash, and R. D. Agrawal, "Cyclic oxidation behavior of Ni- and Fe-based superalloys in air and Na 2 SO 4 -25 % NaCl molten salt environment at 800 o C," *Int. J. Phys. Sci.*, vol. 1, no. September, pp. 27–33, 2006.

Comparison of oxidation resistance and TGO zone morphology of NiCrAlY plasma sprayed coating and sintered alloy

D. Niemiec, G. Moskal, R. Białecki, W. Adamczyk, T. Kruczek, K. Szymański, Gliwice / PL

The basic aim of presented article is related with comparison of internal morphology of atmospheric plasma sprayed NiCrAlY coating and sintered alloy obtained from the same feedstock powder. This part of investigations consisted characterization of phases composition and internal morphology characterization by SEM/EDS method. The second level of this comparing process is characterization of oxidation process in both cases. In order to properly perform these tests the top surfaces of sintered sample was sand blasted to the roughness level adequate to surface condition after plasma spraying process. The oxidation tests were performed at temperature 1000 and 1100°C during the time to 1000h of exposure in air. The specimens after determined intervals were moving out from the furnace and characterized by visual inspection of top surface as well as by SEM/EDS method and XRD analysis to characterization of phases generated during oxidation. The special emphasis was took on the characterization of TGO zone on cross sectioned samples where detailed assessment of oxides zone morphology was made by SEM/EDS method. The oxide thickness was measured as well with dividing of overall oxides zone thickness on the sublayers related with alumina and spinels formation.

1 Introduction

High efficiency is one of the most important requirements to be met by modern aircraft gas turbines. This parameters can be improved by increasing of operating temperature. In consequence, the exploitation temperature of critical engine components made of nickel based superalloys is increasing as well. This is the base limitation which stems from the barrier, which is a drop in nickel superalloys creep resistance at high temperature [1]. This problem can be solved by using the thermal barrier coating (TBC) systems based on ceramic top-layer with strong insulation properties, Ni alloys based bond-coat with excellence oxidation resistance and basic substrate alloy. The most important phenomena critical for TBC durability are related with processes took places at interfaces zone between bond-coat and ceramic top-coat. In this zone the different types of oxides are formed as a result of bond-coat oxidation process. The most important parameters described this zone called as TGO (*thermal grown oxide)* zone are the thickness and phases constituent [2].

The usually used bond-coat materials are Ni(Co)CrAlY type of alloys (for APS - *atmospheric plasma spraying* TBC systems), that protects base substrate alloy against oxidation and corrosion destruction. There are two mechanisms of degradation. The first mechanism is so called *wedging*, which occurs with thicker and more defected layers of TGO zone oxides but also characterized by high adhesion to substrate. The second is *buckling* mechanism involving on bending of the oxide layer are TGO. This is characteristic mechanism for the thin and dense oxide layer with low adhesion to the surface [3]. Bond-coat obtained by APS method characteristic in high roughness level of surface, what is related with necessity to good level of adhesion occurring. During cooling process of APS coats the compressive stress in the bottom of bond-coat profile of roughness is generated. This kind of stress generated in the next step the micro-cracks in area of the peak of bond-coat roughness and in consequence delamination process of the ceramic top-layer of TBC system is starting. The value of the stress generated is dependent on the thickness of the TGO zone [4,5]. Additionally the most important role of the bond-coat is corrosion protection in an environment containing the sulphur compounds. The low tendency to brittle transitions and high resistance to diffusion of alloying elements to the substrate is another requirements towards bond-coat materials [6-9]. The difference in corrosion performance is generally attributed to inhomogeneous coatings structure primarily caused by pores and oxides formed during different technologies of coating deposition, as well as an elemental segregation arising from the rapidly solidified structure [10].

The aim of this work is a more detailed understanding of how specific features of the microstructure can affect the resistance and corrosion by the morphology of the resulting oxides in the case of NiCrAlY. Study was undertaken to examine the relationship between variations of the bond-coat microstructure and properties of resistance to oxidation for the NiCrAlY samples obtained by APS and sintering processes from the same feedstock materials.

2 Experimental description

Material used for testing was a nickel based superalloy Inconel 625 with Ni-22Cr-10Al-Y (wt.%) coating deposited by APS method. The same NiCrAlY powder used for APS spraying was also used to produce the sintered samples. Sinters were made using the Degussa high-temperature press. The powder was sintered at 950°C for 30 minutes in vacuum 3×10^{-6} MPa and in an actual pressure of 15 MPa. Sinters been sand blasted in order to bring to the surface roughness comparable to the NiCrAlY coating. Firstly short characterization of basic materials was showed. It included characterization of mictrostructure and phases constituent of coating and sinter. In the second part the characterization of observed phenomena in the microstructure of coating and sinter in area of TGO zone were described by XRD phases analysis (diffractometr JEOL JDX-7s) of oxidized at 1000 and 1100°C top-surfaces.

Additionally the investigations of TGO oxide layer morphology was performed with a SEM/EDS method (Hitachi 3400N).

3 Results

The first analysed material was obtained by the sintering method. As a feedstock material the powder of Ni–22Cr–10Al–Y was used. Observed material has a multi-phase internal morphology with two-phase matrix and at least two different lighter and darker participations. Average chemical composition of sinter is very similar to that observed for feedstock powder (fig. 1). Generally, matrix has been built from $(Ni,Cr)_3Al$, and Ni(Cr) phases with small amount of Al. In the case of participation, Ni(Cr)–Y with different levels of Y were detected. Phase analysis by XRD method revealed that detected diffraction peaks correspond to Ni_3Al and Ni phases. Additionally Ni–Y compounds of NiY and Ni_7Y type were confirmed. Probably, tertiary compounds from Ni–Y–Al system can exist as well [11].

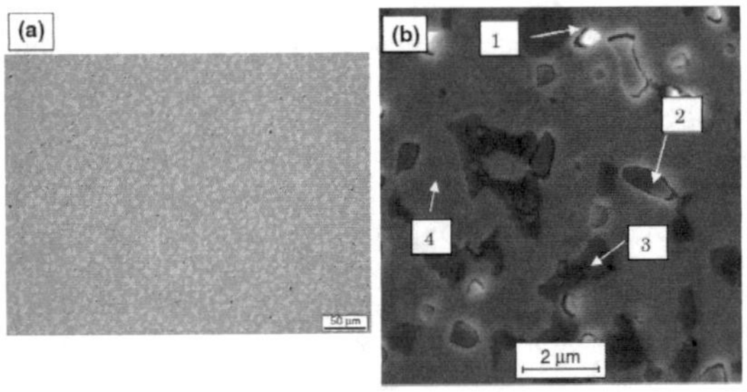

Mass%	Al	Cr	Ni	Y
Pt. 1	7.37	14.65	47.81	29.24
Pt. 2	4.83	8.14	45.26	41.15
Pt. 3	16.65	6.75	76.60	
Pt. 4	5.29	31.06	63.65	
av. comp.	8.63	22.61	67.58	1.17

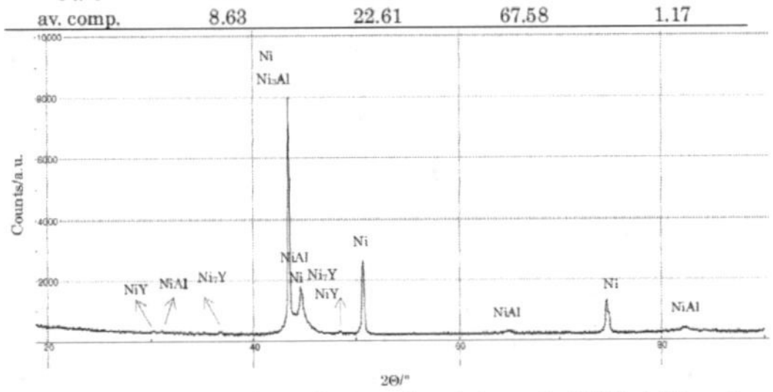

Fig. 1 The main view of sintered NiCrAlY alloy morphology [11].

The second analysed morphological form of initial powder was plasma-sprayed coating deposited during standard process of this type of coating deposition. The thickness of coating was 125 Im. Its internal morphology is relatively most complicated. Observed differences are related with high temperature in plasma beam during the thermal spraying process. OM picture of coating microstructure is shown in Fig. 3 In this case, the brightness matrix consisted darker areas of other phases such as oxides (mainly Al_2O_3) or contaminations are visible. Though internal microstructure of coating was different, its phase constituent is very similar to this observed after sintering. Additionally, presence of oxides should be detected on diffraction pattern in this case [11].

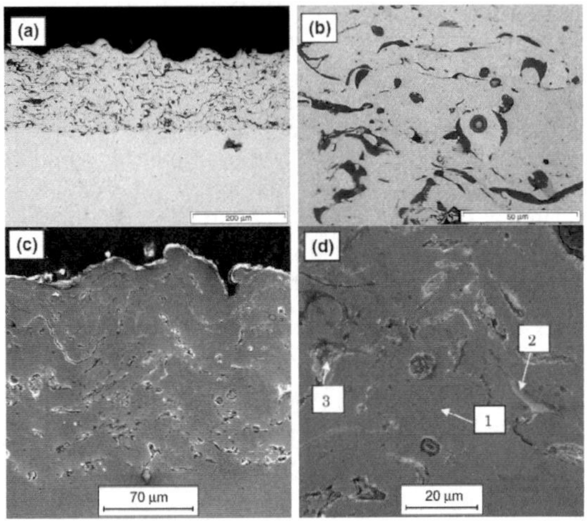

Mass%	Al	Cr	Ni	Y
Pt. 1	8.2	22.3	69.2	0.3
Pt. 2	0.8	17.1	82.1	-
Pt. 3	40.1	31.7	25.1	3.1
av. comp.	5.00	22.55	71.54	0.91

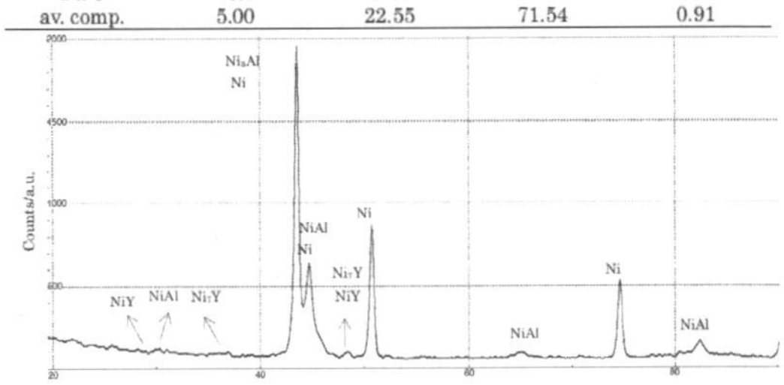

Fig. 2 The main view of NiCrAlY coating morphology [11].

The comparosion the surfaces roughness in as-sprayed condition and for the sample after sand blasting process is showed in fig. 3, where 3D visualization of investigated surcaes can be observed. Quantitative data revealed that rougness of both specimens are comparable.

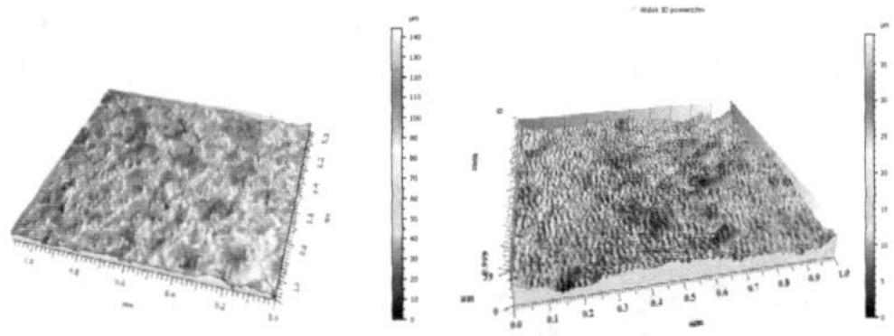

Parameter [µm]	Surface as-sprayed	Surface sand blasted
Rz	10,23	8,08
Ra	2,45	1,22
Rq	2,21	1,59

Fig. 3 Top-surfaces characterization of topography of analysed samples.

Sintered NiCrAlY specimen and coating were oxidized in a furnace at a temperature of 1000°C and 1100°C in laboratory air. Then, they were drawn out of the furnace after the successively 25, 300, 500, 750 and 1000 hours, as is typical when conducting this type of tests. In first step of presented investigation, results of phase's analysis from top surfaces of specimens after oxidation test by 1000h were showed at fig 4.

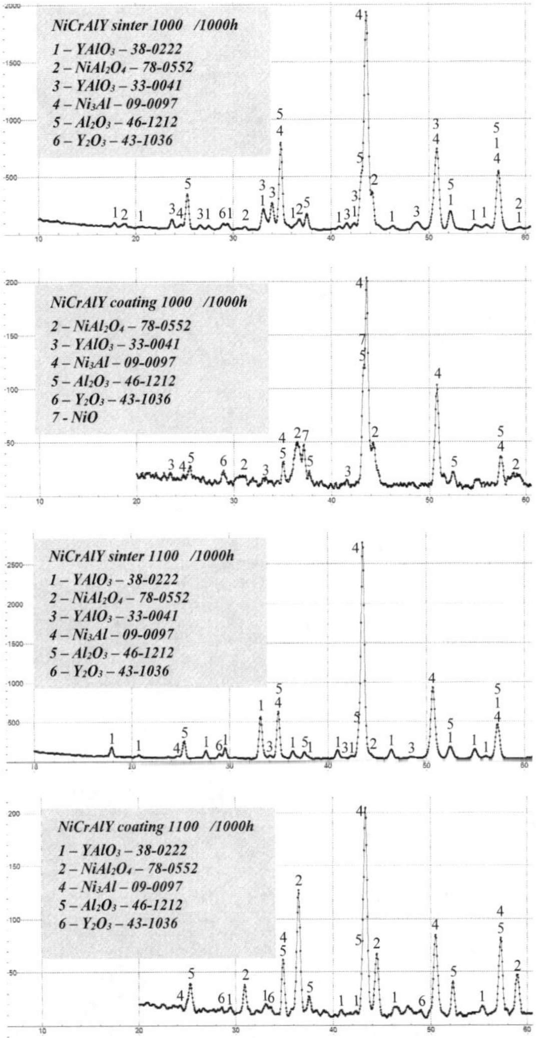

Fig. 4 XRD patterns of TGO zone of oxidized NiCrAlY alloys in the form of as sprayed and sintered.

Data obtained during XRD phases analysis from top-surfaces of oxidized specimens revealed significant differences in oxides form generated during performed test. The main difference is related with presents of NiO oxide as a main phase element in APS sprayed NiCrAlY coatings tested at 1000 and 1100 after 1000 h of exposure. In the case of sintered alloys this phase was not observed. In all analysed specimens spinel oxide of $NiAl_2O_4$ type was detected as well as Al_2O_3, but the intensity of diffraction peaks has an inversely type. In the case of sintered NiCrAlY alloy the strongest peaks were observed for alumina, and for NiCrAlY coatings the same effect was observed for spinel diffraction peaks. Next observed difference is related with formation of complexed oxide of Y-Al-O type and formula $YAlO_3$. Relatively strong diffraction peaks were observed only in the case of sintered specimens.

These observations were confirmed during morphological analysis of TGO zone after 1000h of exposure. The example of oxide layer morphology after test at 1100 was showed at fig. 5.

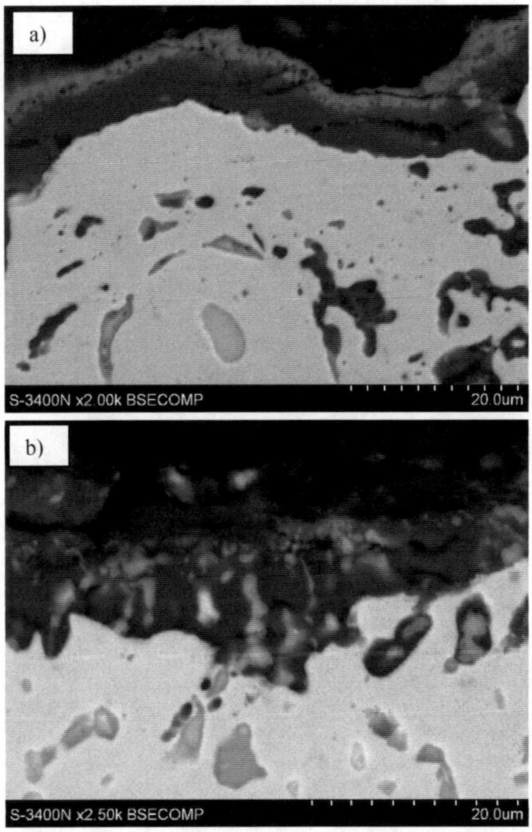

Fig. 5 TGO zone morphology of as-sprayed specimen (a) and NiCrAlY alloy in as-sintered condition (b).

As is visible at this picture the TGO zone` morphology generated on NiCrAlY material in the as-coated condition revealed presents of two different areas adequate to outer spinel sub-layer with high level of porosity, and dens internal sub-layer consisted mainly alumina as a base oxide. In the case of as sintered specimen there is no clearly highlighted in the TGO zone.

Figures 6 and 7 present main element surfaces distribution in area of TGO zone. Presented maps confirmed data obtained during XRD and SEM analysis especially in the case of lack of spinel sub-layer on the surface of NiCrAlY sintered specimens after oxidation at temperature 1100 .

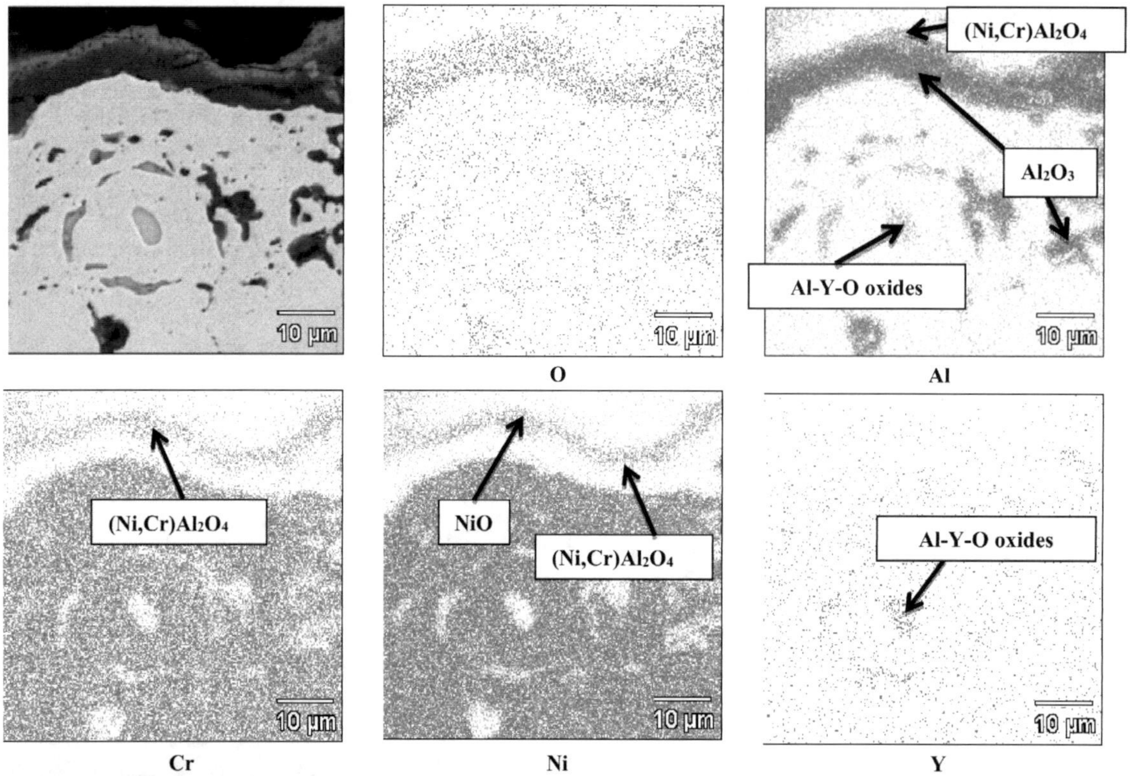

Fig.6 Elements distribution on cross-section of as-sprayed sample after 1000 hours at a temperture 1100°C.

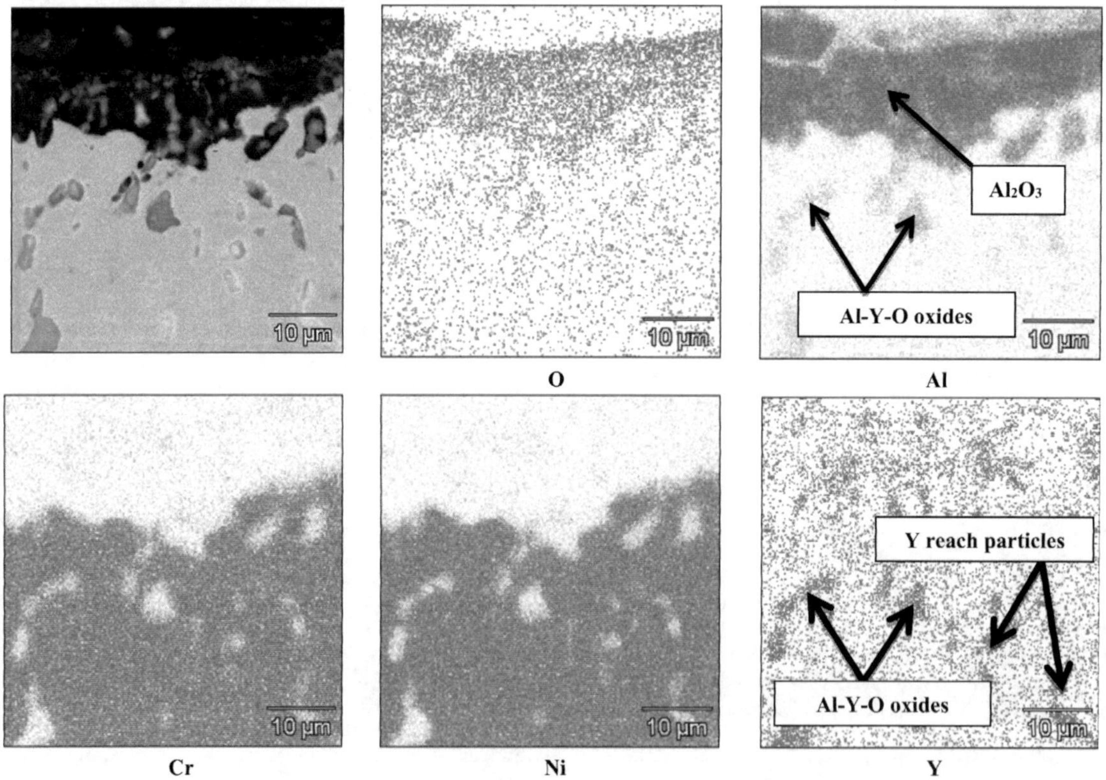

Fig. 7 Elements distribution on cross-section of sintered sample after 1000 hours at a temperture 1100°C.

Data presented at those figures showed additionally that internal effect of oxidation was obtained in the case of as-sprayed specimen, what was not in this scale for sample after sintering process. This internal oxidation revealed as an alumina and $YAlO_3$ (or Y_2O_3) island formation in core of the coating. The deepness of internal oxidation is much higher too in comparison to sintered specimen where only very weak effect of $YAlO_3$ (or Y_2O_3) formation was observed, especially in area of TGO zone. For the other site the presents of Y reach particles were observed as well.

4 Summary

The results show a significant influence of the initial structure of analysed materials in the form of NiCrAlY sinter and coating on the oxidation process, especially related with oxide zone formation and its morphology. In the case of as-sprayed condition main phase detected in TGO zone are $NiAl_2O_4$ or rather $(Ni,Cr)Al_2O_4$ spinel and NiO oxide. The alumina fraction is relatively lower and exist in the form of internal dens layer. Different morphology of TGO zone was observed for as-sintered NiCrAlY material, where main detected phase in this place was alumina with relatively small amount of spinel oxide. Additionally the presents of $YAlO_3$ oxide in oxide zone was detected.
Obtained data strongly suggest that oxidation mechanism in both cases was different and was related with internal structure of basic material. This internal microstructure was created during manufacturing process such as atmospheric plasma spraying and sintering. Both process are totally different especially due the temperature level, much higher in plasma stream. In consequence in as-sprayed condition stronger effect of internal oxidation was observed and as a consequence the depletion of β-NiAl phases in Al should be observed [12], but this phenomena was not subject of this article.

Acknowledgments: The research has been supported by National Science Centre within Opus scheme, under contract UMO_2013/11/B/ST8/00268.

5 Literature

[1] Quadakkers W. J., Tyagi A. K., Clemens D., Anton R. and I. Singheiser: The significance of bond coat oxidation for the life of TBC coatings, Elevated temperature coatings: surface and technology III, The minerals, Metals & Materials Society (1999), pp. 119/30.
[2] Swadźba L., Moskal G., Mendala B. and M. Hetmańczyk: Characterization of microstructure and properties of TBC systems with gradient of chemical composition and porosity, Archives of Metallurgy and Materials 53 (2008), Issue 3, pp. 945/54.
[3] Naumenko D., Shermet V., Singheiser L. and W.J. Quadakkers: Failure mechanism of thermal barrier coatings on MCrAlY-type bondcoats associated with the formation of the thermally grown oxide, Journal of Materialas Science 44 (2009), Issue 7, pp. 1687/703.
[4] Evans A.G., Crumley G.B. and R. E. Demaray, The mechanical behavior of brittle coatings and layers, Oxidation of Metals 20 (1983), Issue 5, pp. 193/216.
[5] Hsueh C.-H., Haynes J.A., Lance M.J., Becher P.F., Ferber M.K., Fuller E.R., Langer S.A., Carter W.C. and W.R. Cannon: Effects of interface roughness on residual stresses in thermal barrier coatings, Journal of American Ceramic Society 82 (1999), Issue 4, pp. 1073/75.
[6] Tang F. and J. Schoenung: Local accumulation of thermally grown oxide in plasma-sprayed thermal barrier coatings with rough top-coat/bond-coat interfaces, Scripta Materiallia 52 (2205), Issue 9, pp. 905/9.
[7] Swadźba L., Moskal G., Mendala B. and M. Hetmańczyk: Characterization of microstructure and properties of TBC systems with gradient of chemical composition and porosity, Archives of Metallurgy and Materials 53 (2008), Issue 3, pp. 945/54.
[8] Zielińska M., Sieniawski J., Zagula-Yavorska M. and M. Motyka: Influence of chemical composition of nickel based superalloy on the formation of aluminide coatings, Archives of Metallurgy and Materials 1 (2011), Issue 1, pp. 193/7.
[9] Zagula-Yavorska M., Sieniawski J. and M. Zielińska: Functional properties aluminide layer deposited by CVD method on Inconel 713 LC Ni-base superalloy, Archives of Metallurgy and Materials 1 (2011), Issue 1, pp. 187/92.
[10] Ahmed N., Bakare M.S., McCartney D.G. and K.T. Voisey: The effects of microstructural features on the performance gap in corrosion resistance between bulk and HVOF sprayed Inconel 625, Surface and Coatings Technology 204 (2010), Issue 14, pp. 2294/301.
[11] Moskal G. and A. Jasik: Thermal diffusivity characterization of bond-coat materials used for thermal barrier coatings, Journal of Thermal Analysis and Calorimetry 126 (2016), Issue 1, pp. 9/17.
[12] Chen H. and D.G. McCartney: Some aspects on modelling of the β-phase depletion behaviour under different oxide growth kinetics in HVOF CoNiCrAlY coatings, Surface and Cotings Technology 313 (2017), pp. 107/14.

The influence of deposition temperature and thermal conductivity of substrate on the splat formation

Jun Wang, Chang-Jiu Li
State Key Laboratory for Mechanical Behavior of Materials, School of Materials Science and Engineering, Xi'an Jiaotong University, China / E-mail: licj@mail.xjtu.edu.cn

Ni20Cr splats were sprayed onto polished substrates at different preheating temperatures in an argon atmosphere by Low Pressure Plasma Spray to reveal the dominating factor on the effective interface bonding formation. The splat morphology, microstructure and splat-substrate interface bonding were characterized by SEM and EBSD. The interface for examination of typical splats was prepared by FIB. Disk splats were obtained on AISI 304 stainless steel substrates preheated to temperatures of 100 °C (cooling from 350 °C), 350 and 550 °C. Moreover, typical distinct two-zone microstructure feature was observed on the splat surface by SEM and EBSD, including central coarse grain and marginal fine grain. When the preheating temperature was higher than 350 °C, effective bonding formed only in the entire central coarse zone, whereas no effective bonding was observed in the fine grain zone. By using glass, copper, nickel and 304 SS as substrates, it was found that increasing thermal conductivity of metallic substrates has little effect on splat diameter and morphology and however decreased the area fraction of central coarse grain zone. It was revealed that the melt/substrate interface temperature plays a crucial role on the interface bonding formation.

1 Introduction

In last decades, as one of the important coating deposition technologies, thermal spraying has made great progress and been applied to many industrial fields to improve the corrosion resistance, wear resistance and high temperature resistance of engineering components. A thermal spray coating is formed by successive impacts of a stream of molten droplets followed by spreading and rapid solidification and cooling. The independent impact and solidification process results in a typical lamellar microstructure with limited lamellae interface bonding [1], which limits the performance of the coating. Therefore, it is necessary to study the bonding formation mechanism to enhance the properties of a coating.

The interface bonding in a coating mainly contributes to the mechanical, chemical-metallurgical and physical adhesion and cohesion of coating. The formation of chemical-metallurgical bonding at the splat interface is of critical importance to ensure adequate cohesion.

It has been reported that the substrate melting occurs upon droplet impact when the temperature of a molten droplet is sufficiently higher than the melting point of the substrate. The formation of metallurgical bonding was confirmed by experiments when Mo or W particles were sprayed onto steel substrate [2-4]. It was also well known that the melting of steel substrate may be caused by impact of high temperature of Ni-Al droplet resulting from the exothermic reaction of Ni-Al composite powders during spraying process [5]. Furthermore, Tian et al. [6] proposed a new novel shell-core powder particle design by cladding spherical Ni20Cr powders with refractory Mo as alloying element to raise particle temperature significantly high enough to cause impact melting. Until now, except for Brossard et al. [7] who reported that localized substrate melting occurred for NiCr particles sprayed onto steel substrate, few researchers reported the evidence of the substrate melting resulting from the impact of a molten particle with a melting point similar to that of the substrate. Therefore, results suggest that substrate melting occurs only when the molten particle could supply sufficient thermal energy to substrate upon impact.

Taking account of the fact that the interface temperature depends on the droplet and substrate property, increasing substrate temperature may promote the interface bonding formation. Recent investigations confirmed that deposition temperature dominated the splat bonding formation for plasma spraying of ceramic coating and proposed the concept of the critical bonding temperature [8]. In addition, Yao et al. [9, 10] proposed a model for the bonding mechanism at the melt/solid interface to explain the bonding formation reasonably. Thus, the interface bonding formation is benefited from the increase of substrate temperature. Similarly, the deposition temperature has significant effect on the performance of metallic coatings. Jiang et al. [11] observed that the fracture characteristics of Mo coatings changed from interlamellar to translamellar mode, illustrating enhanced cohesion resulting from increasing deposition temperature. Apparently, the increase of deposition temperature may promote the interface bonding formation for metallic coatings. However, since increasing deposition temperature may induce additional oxidation of splat surface which inhibits the formation of a metallurgical bonding. Thus at present, the influence of deposition temperature on the interface bonding formation of metallic coatings was not clear yet due to lack of systematical investigation. Moreover, the thermal properties of substrate itself, especially the thermal conductivity, may have significant influence on the interface bonding, since it influences the melt/substrate interface temperature. In this study, the influence of deposition temperature and thermal conductivity of the substrate on the splat morphology and interface bonding formation was investigated to reveal the interface bonding formation mechanism.

2 Experimental procedures

Spherical Ni20Cr powder with a size range from 50 to 57 μm was plasma-sprayed in an argon atmosphere at 50 kPa by low pressure plasma spraying (LPPS) to eliminate the oxidation of substrate and splats during spraying. To reveal the influence of substrate temperature on the bonding formation, Ni20Cr splats were deposited on polished 304 stainless steel substrates at the deposition temperatures of 100, 350 and 550 °C using a special heater. In order to eliminate the re-adsorption of absorbates and condensates on the substrate surface, the highest temperature during the preheating process should exceed the transition temperature. Thus, the preheating to the deposition temperature of 100 °C was implemented by firstly heating the substrate up to 350 °C and then cooling down to 100 °C by argon flow. All other substrate samples were directly heated up to the designed deposition temperature at a rate of 100 °C/min. Moreover, besides 304 stainless steel, Ni20Cr splats were also deposited on glass, mirror-polished nickel, and copper substrates at 350 °C. To maintain the consistency of characteristics of the in-flight particles, the same spraying conditions were adopted.

Splats deposited at various conditions were examined by SEM. Splat diameter was measured by the image-pro plus software. EBSD analysis was also performed to reveal the crystallographic orientation on the splat surface. To examine the bonding formation at the splat-substrate interface by SEM, the cross-sections of splats were prepared by FIB cutting at different regions.

3 Results

3.1 Effect of deposition temperature on splat morphology and interface bonding formation

3.1.1 Splat morphology and microstructure

Fig. 1 shows the morphologies of typical splats deposited on 304 SS substrates at 100, 350 and 550 °C. It was observed that all splats at different deposition temperatures presented a typical disk-shape, which confirmed that the surface of substrates after preheating was enough clean to eliminate splashing based on the evaporated-gas-induced splashing model [12]. When the preheating temperature was increased to 550 °C, few short projections at the periphery and a new cellular-skeleton structure in the central zone were observed on the splat.

Fig. 2 shows the typical microstructure on the surface of the splat deposited at 350 °C by SEM. The top surface can be divided into two parts including central zone and rim zone, due to the difference of microstructure. The central zone contained coarse columnar grains perpendicular to substrate, and rim zone exhibited fine grains. Fig. 3 presents the crystal orientation images of splats deposited at 100, 350 and 550 °C by electron backscatter diffraction, which evidently shows the presence of two types of distinct microstructures in the two zones in the splat. Furthermore, with the increase of the deposition temperature, no clear difference was observed in the microstructure distribution. However, the grain size at the central zone close to the periphery decreased for splats deposited at 550 °C compared with 350 °C, which may be result from the measurement deviation of EBSD technique due to surface unevenness.

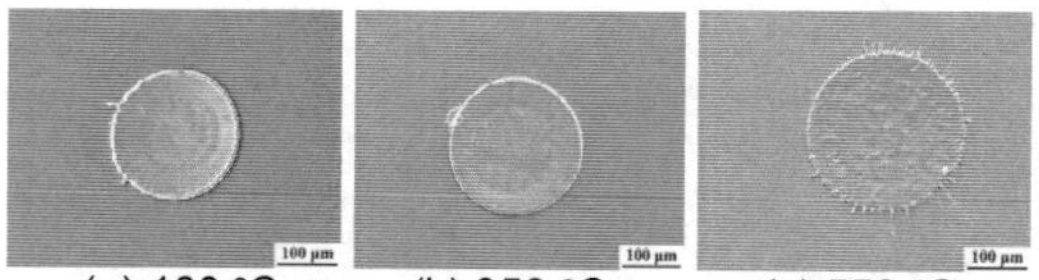

(a) 100 °C (b) 350 °C (c) 550 °C

Fig.1. Splat morphology at different deposition temperatures

(a) (b) (c)

Fig. 2. Microstructure of (a) whole splat, (b) central zone and (c) rim zone marked in (a)

(a) 100 °C (b) 350 °C (c) 550 °C

Fig. 3. Grain orientation image of splats at different deposition temperatures

3.1.2 Interface bonding formation

The cross sections revealing the bonding state were prepared by focus ion beam (FIB). As mentioned above, the splat presented distinct microstructures in two zone. In order to reveal the relationship between the bonding state and microstructure, two cross sections were prepared on the splat surface. One is located at the central coarse grain zone and the other one was across the border of the central coarse grain zone and marginal fine grain zone. The morphology of cross section and corresponding grain distribution images for splats deposited at 100 °C, 350 °C and 550 °C were observed by SEM and EBSD, as shown in Fig. 4, Fig. 5 and Fig. 6, repectively.

The examination into the interface bonding from the splat cross sections in Fig. 4 revealed that the bonding formed at the interface marked A, locating at the inner part of the splat. Nevertheless, clearly no effective bonding formed for entire interface located at the periphery marked as B. The grain size distribution image in Fig. 4c shows that the region B was across the central coarse grain zone and marginal fine grain zone. The fact indicates that the effective bonding forms only in part of the central coarse grain zone at a deposition temperature of 100 °C. When the deposition temperature increased to 350 °C, effective bonding formed in the whole interface in region A, and part of the interface in region B exhibited a clear splat-substrate interface, as shown in Fig. 5b. Furthermore, by comparing the cross section B in Fig. 5b and grain size distribution image in Fig. 5c, it can be speculated that the border of bonded interface and unbonded interface is close to the border of central coarse grain zone and marginal fine grain zone. The cross-sections of the splat deposited at 550 °C in Fig. 6 also showed similar bonding state with that at 350 °C. The proportion of the bonded interface occupied 2/3 part of the splat diameter from the splat center.

Evidently, the above results showed that the interface bonding formation should be dependent on the microstructure. When the deposition temperature exceeds 350 °C, the effective bonded interface is limited in the central coarse grain zone, occupying 2/3 part of the splat diameter.

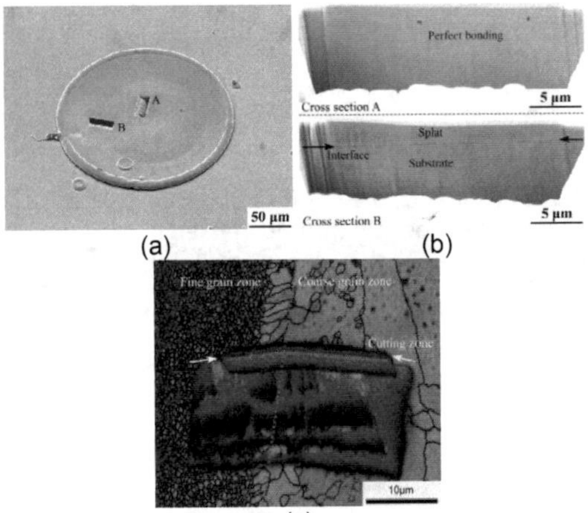

Fig. 4. FIB sampling positions for typical splat interface at a deposition temperature of 100 °C (a), (b) cross sections of the splat at typical splat regions marked as A and B in (a), (c) grain distribution image for the region marked as B in (a)

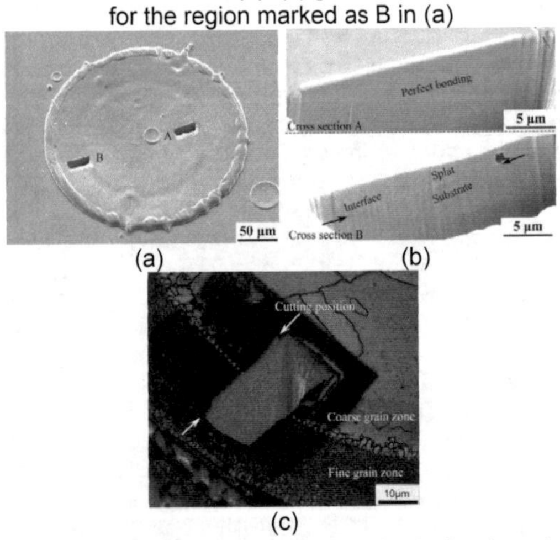

Fig. 5. FIB sampling positions for typical splat interface at a deposition temperature of 350 °C (a), (b) cross sections of the splat at typical splat regions marked as A and B in (a), (c) grain distribution image for the region marked as B in (a)

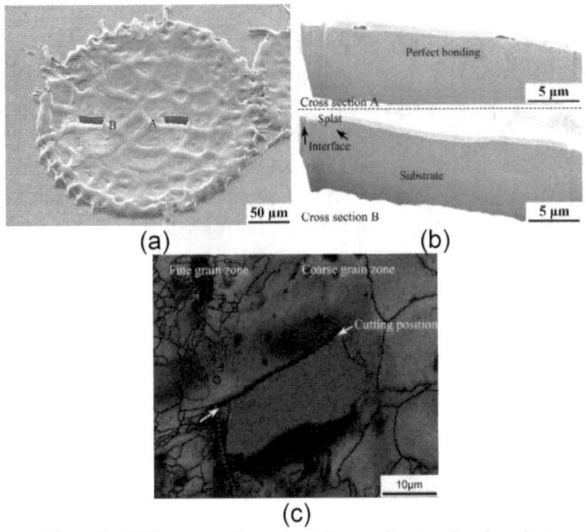

Fig. 6. FIB sampling positions for typical splat interface at a deposition temperature of 550 °C (a), (b) cross sections of the splat at typical splat regions marked as A and B in (a), (c) grain distribution image for the region marked as B in (a)

3.2 Effect of substrate conductivity on splat morphology and interface bonding formation

As discussed in section 3.1, the interface temperature plays an important role in determining the bonding formation, which depends on the substrate conductivity. Hence, Ni20Cr particles were sprayed onto four kinds of substrates with different thermal conductivity at a deposition temperature of 350 °C to reveal the bonding formation regime. The thermal conductivities of the substrates are presented in Fig. 7, and the morphology of splats deposited on individual substrates was exhibited in Fig. 8. Perfect disk-shape splats were deposited on all metallic substrates, whereas a few spheres in the center and a solid ring surrounding it were observed on the glass substrate.

Fig. 9 shows the grain orientation images of splats deposited on 304 stainless steel, Ni and Cu substrates. The splats deposited on the Ni substrate also presented distinct two-zone microstructures, being comparable with that on 304 stainless steel substrate. However, the border of the two-zone microstructures is indistinct for the splat deposited on Cu substrate. The grains in the central zone became finer and some coarse grains observed in the central zone were surrounded by fine grains.

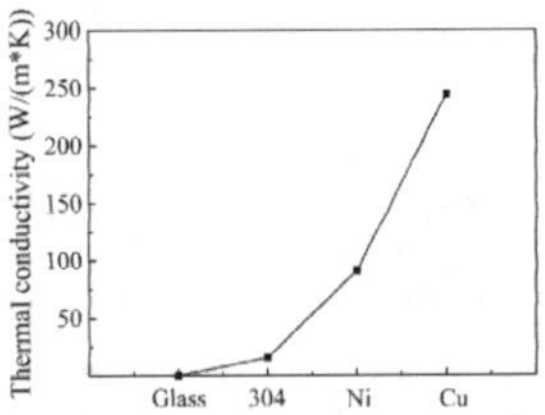

Fig. 7. Thermal conductivity of four kinds of substrates

The relationship between substrate thermal conductivity and average splat diameter/area percentage of central zone is illustrated in Fig. 10. The flattening ratio of splats usually depends on the Reynolds number of impacting droplet and the solidification process. The Reynolds numbers of molten droplets for splats on these different substrates were identical due to the same thermal spray parameters and particles. It was found that the diameter of splat on glass substrate was 1.5 times that on metallic substrates as shown in Fig. 10, indicating that higher thermal conductivity had a great influence on the spreading process of droplet on a substrate and subsequently solidification process. Increasing the substrate thermal conductivity may increase splat cooling rate and subsequently lead to the increase of viscosity of spreading melt, which further limits the flattening of droplet. However, the splat diameter on metallic substrates decreased only slightly with the increase of substrate thermal conductivity in the magnitude by one order, suggesting that the heat transfer depends on the interface thermal resistance instead of thermal resistance of the substrate itself. In addition, the area percentage of central zone decreased with the increase of thermal conductivity of substrate, as shown in Fig. 10. As discussed in section 3.1.2, the interface bonding formation is associated with the microstructure. Hence, it can be speculated that the effective bonding ratio decreases with increasing substrate thermal conductivity.

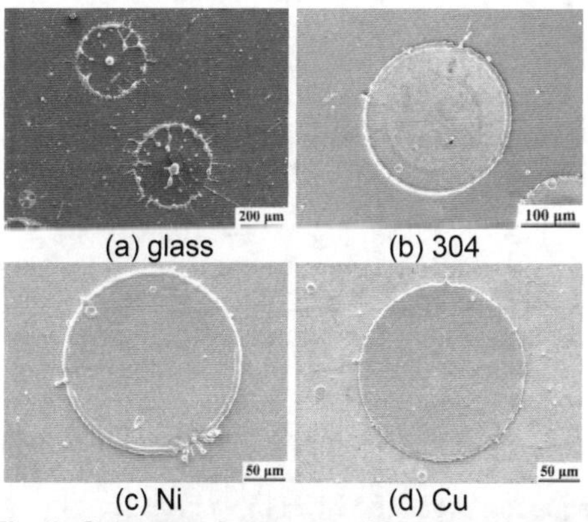

Fig. 8. Splat morphology on different substrates

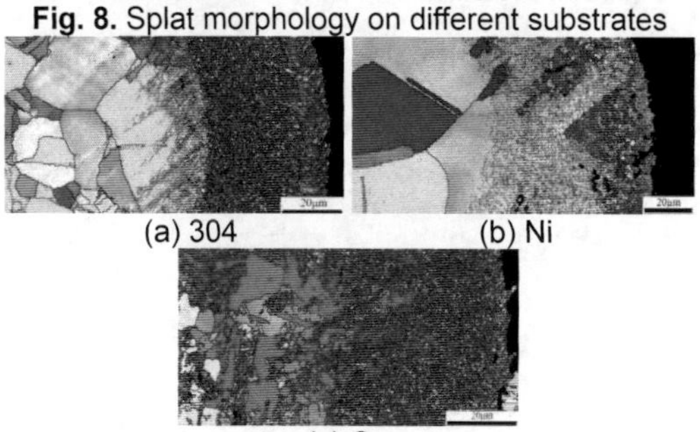

Fig. 9. Grain orientation images of splats on different substrates by EBSD

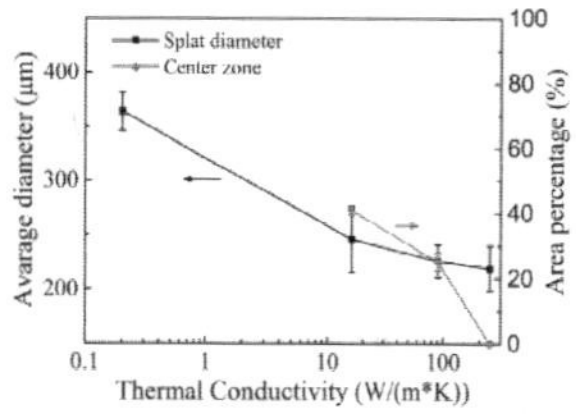

Fig. 10. The relationship between substrate thermal conductivity and average splat diameter/area percentage of central zone

4 Discussions

4.1 Effect of deposition temperature and substrate thermal conductivity on splat morphology

In three last decades, numerous investigations have shown that the splat morphology can be changed from fragment irregular shape to disk-shape when increasing the substrate temperature during plasma spraying for certain droplet/substrate combinations. Several classic models [13, 14] have been proposed to explain the splashing mechanisms, including adsorbed gas induced splashing model, substrate melting-induced splashing model, Sommerfeld number model and solidification parameter model.

In this paper, all substrates were heated to 350 °C during the preheating process, which made the substrate surface clear of absorbates and condensates. Therefore, although a substrate was cooled down to 100 °C during a short time, near disk-shape splats were also observed on all substrates. However, few short fine fingers were observed at the periphery of the splat deposited at 550 °C. With continuing increase of substrate temperature, the Sommerfeld number becomes no significant change due to adopting similar spraying parameters. It is also found [2] that substrate melting will not happen when the melting point of droplet is close to that of substrate. As a result, only the solidification parameter model can be concerned to explain the splashing phenomena of splat deposited at 550 °C. A solid layer may form rapidly after a droplet impacts the substrate, which will restrain the splat from spreading too far. If the higher deposition temperature decreases the freezing rate of the splat, the top surface of splat will have enough time to spread over the edge of the solidifying layer and result in splashing. Allen et al. [15] suggested that fingers formed around the periphery of a spreading droplet due to a Rayleigh-Taylor instability.

In addition, a new cellular-skeleton structure appeared on the top surface of splat deposited at 550 °C. Since the contact angle may be larger than 90° during spreading of melt, the hydrophobic surface of substrate transfers to a weak hydrophilic or even hydrophobic surface [10], which leads to the contraction of liquid under the surface tension and lifting up of periphery region. This fact lowers cooling rate and thus jetting leading to fingering happens.

Moreover, the lower thermal conductivity of glass reduces significantly heat-sink effect and maintain spread melt film at liquid state longer to lead to splashing to small spherical droplets by contraction due to surface tension effect. (Fig. 8a).

4.2 Effect of deposition temperature and substrate thermal conductivity on splat microstructure

The typical lamellar microstructure is usually observed in thermal spray coating. The microstructure of a splat, being the basic unit of the coating, will be a key factor determining the mechanical performance of coatings. Typical columnar grains perpendicular to the substrate was found in the YSZ splat presented by Tomas Chraska et al. [16]. A complex structure containing coarse columnar grains and a narrow ring-shaped zone surrounding the central zone with smaller grain size was observed in Cu splat by Yang [17]. Whereas, Jiang [18] presented equiaxed grains in the central portion and dendrite grains align along the splat rim towards the center for molybdenum splat on glass substrate. In Fig. 2 and Fig. 3, a distinct two-zone microstructure including central coarse grain zone and marginal fine grain zone was found from splat surface.

In general, grain size is determined by both the nucleation and growth rates during solidification. The nucleation rate increases much rapid than growth rate for metallic solidification when increasing the undercooling. Therefore, the grain size is inversely proportional to the degree of undercooling. During the splat spreading process, the contact pressure is the highest at the center of the contact area and decreases with spreading distance over the substrate surface [19], which results in close contact between splat and substrate in a limited central zone. Generally, at the close contact zone the melt has a higher cooling rate and large undercooling [20]. Therefore, the grain size in the central zone should be smaller than that at the edge during spreading process. However, the two-zone microstructure feature in Fig. 9 is not consistent with the above theoretical hypothesis. Then the first question is why the grain size at the edge is much smaller than that in the center.

As observed in this study, the splat diameter decreased greatly when metallic substrate was used instead of glass substrate. It indicates that the cooling effect of substrate has a great influence on splat spreading and could not be neglected. When a droplet impacts onto the substrate, a thinner liquid layer [21] injects from the contact point along the surface with the maximum spreading velocity reaching to 3 times of droplet impacting velocity. The higher thermal conductivity of metallic substrate decreases the liquid temperature at the bottom of the droplet. The longer the spreading distance is, the lower the liquid temperature is. Thus, the undercooling at the edge should be much higher than that in the center of the splat, which will result in finer grains at the edge zone. It is also observed in Fig. 9 that the area percentage of central coarse zone decreases with the increase of substrate thermal conductivity, which is consistent with this hypothesis.

4.3 Effect of deposition temperature and substrate thermal conductivity on interface bonding formation

Numerous investigations have shown that the interface bonding formation is one of the key factors dominating the coating property [22]. At present, few papers have been reported to show the interface bonding directly, due to the difficulty of preparing and observing cross section of splat-substrate interface. It was found that the formation of chemical bonds between the ceramic splat and metal substrate benefits from the presence of a thin oxide layer [16]. Yang [8] proved that splat interface bonding was significantly influenced by the deposition temperature for ceramic splat. Afterwards, Yao [10] proposed the intrinsic bonding temperature concept to explain the formation mechanism. Then the second question is how temperature influences the interface bonding formation. The present study was designed to alter substrate temperature and thermal conductivity to investigate the effect of interface temperature on interface bonding formation.

Results clearly showed that when the deposition is carried out at the temperature of 350 °C, the bonding is present at the interface over the central coarse grain zone of the splat whereas no effective bonding is observed at the interface located at the marginal fine grain zone. Therefore, it is clear that the microstructure is correlated to the splat bonding formation. Based on the significant difference of grain size, it can be considered that the temperature of bottom surface in the center is much higher than that at the edge during splatting. Hence, the effective bonding forms in the central zone instead of the marginal zone. According to heat transfer theory, the interface temperature increases as the deposition temperature increases and decreases as the thermal conductivity of substrate increases. The results in Fig. 9 and Fig. 10 indicates that the increase of the substrate thermal conductivity rather than deposition temperature has a great effect on interface bonding formation. The interface temperature calculated from the equation proposed by Li [2] is increased by 114 K if increasing the deposition temperature by 200 K, and decreased by 483 K as the Cu substrate is replaced by 304 substrate. The limited increase of deposition temperature may be not enough to increase the interface temperature to form effective bonding at the rim zone. Whereas, the increase of the substrate thermal conductivity not only decreases the cooling rate during spreading, but also decreases the interface temperature, which benefits to the formation of an effective interface bonding. Therefore, the altering interface temperature due to the high thermal conductivity plays a critical influence on the interface bonding formation for metallic splat.

At last, the bonded interface decreased in the central coarse zone in Fig. 4, although the microstructure distribution of the splat deposited at 100 °C is similar with that at 350 °C. Excluding the factor of interface temperature, the interface contact condition is responsible for the unbonded formation. During the cooling process of substrate from 350 °C to 100 °C, the slight adsorption of adsorbates and condensates may result in the formation of weak interface, which will rupture because of quenching stress.

5 Conclusion

In the present study, the influence of deposition temperature on splat microstructure and splat-substrate interface bonding were investigated. Results indicated that typical distinct two-zone microstructure feature was observed on splat surface, including central coarse grain and marginal fine grain. The interface bonding formation was closely related to the microstructure distribution. Effective bonding formed in the interface of whole central coarse grain zone, accounting 2/3 part of whole interface, and no bonding formed in the marginal fine grain zone, when the deposition temperature was higher than 350 °C. Increasing the deposition temperature has little effect on the interface bonding. However, weakly bonded interface resulting from slight adsorption of adsorbates and condensates may form, when the deposition temperature was decreased to 100 °C.

In addition, substrates with different thermal conductivity were used to investigate the influence of melt/substrate interface temperature during splatting on the interface bonding formation. It was found that the average diameter of splats decreased a little with increasing the thermal conductivity of metallic substrates. Whereas, the fraction of central coarse grain zone decreased rapidly, implying that the fraction of the bonded interface area decreased due to the above mentioned relationship between microstructure and interface bonding formation. Therefore, the control of the interface temperature during splatting may play a crucial role on the interface bonding formation.

Acknowledgments

The present project is financially supported by National Nature Science Foundation of China (No. 51474171).

Literature

[1] A. Ohmori, and C.-J. Li, Quantitative characterization of the structure of plasma-sprayed Al_2O_3 coating by using copper electroplating, Thin Solid Films 201(1991), Issue 2, pp. 241-252

[2] L. Li, X.Y. Wang, G. Wei, A. Vaidya, H. Zhang, and S. Sampath, Substrate melting during thermal spray splat quenching, Thin Solid Films 468(2004), Issue 1–2, pp. 113-119

[3] S. Kitahara, and A. Hasui, A study of the bonding mechanism of sprayed coatings, Journal of Vacuum Science and Technology 11(1974), Issue 4, pp. 747-753

[4] C.-J. Li, C.-X. Li, G.-J. Yang, and Y.-Y. Wang, Examination of Substrate Surface Melting-Induced Splashing During Splat Formation in Plasma Spraying, Journal of Thermal Spray Technology 15(2006), Issue 4, pp. 717-724

[5] H.S.I. Jr., Bonding of flame - sprayed Ni–Al, Journal of Vacuum Science and Technology 12(1975), Issue 4, pp. 773-776

[6] J.-J. Tian, S.-W. Yao, X.-T. Luo, C.-X. Li, and C.-J. Li, An effective approach for creating metallurgical self-bonding in plasma-spraying of NiCr-Mo coating by designing shell-core-structured powders, Acta Materialia 110(2016), pp. 19-30

[7] S. Brossard, P.R. Munroe, A.T. Tran, and M.M. Hyland, Study of the Splat Microstructure and the Effects of Substrate Heating on the Splat Formation for Ni-Cr Particles Plasma Sprayed onto Stainless Steel Substrates, Journal of Thermal Spray Technology 19(2010), Issue 5, pp. 1100-1114

[8] G.-J. Yang, C.-X. Li, S. Hao, Y.-Z. Xing, E.-J. Yang, and C.-J. Li, Critical bonding temperature for the splat bonding formation during plasma spraying of ceramic materials, Surface and Coatings Technology 235(2013),pp. 841-847

[9] S.-W. Yao, J.-J. Tian, C.-J. Li, G.-J. Yang, and C.-X. Li, Understanding the Formation of Limited Interlamellar Bonding in Plasma Sprayed Ceramic Coatings Based on the Concept of Intrinsic Bonding Temperature, Journal of Thermal Spray Technology 25(2016), 8, pp. 1617-1630

[10] S.-W. Yao, C.-J. Li, J.-J. Tian, G.-J. Yang, and C.-X. Li, Conditions and mechanisms for the bonding of a molten ceramic droplet to a substrate after high-speed impact, Acta Materialia 119(2016),pp. 9-25

[11] X. Jiang, J. Matejicek, and S. Sampath, Substrate temperature effects on the splat formation, microstructure development and properties of plasma sprayed coatings: Part II: case study for molybdenum, Materials Science and Engineering: A 272(1999), Issue 1, pp. 189-198

[12] C.-J. Li, and J.-L. Li, Evaporated-gas-induced Splashing Model for Splat Formation during Plasma Spraying, Surface and Coatings Technology 184(2004), Issue 1, pp. 13-23

[13] S. Sampath, and X. Jiang, Splat formation and microstructure development during plasma spraying: deposition temperature effects, Materials Science and Engineering: A 304–306(2001), pp. 144-150

[14] K. Yang, M. Liu, K. Zhou, and C. Deng, Recent Developments in the Research of Splat Formation Process in Thermal Spraying, Journal of Materials 2013(2012), pp. 1-14

[15] R.F. Allen, The role of surface tension in splashing, Journal of Colloid and Interface Science 51(1975),Issue 2, pp. 350-351

[16] T. Chraska, and A.H. King, Effect of different substrate conditions upon interface with plasma sprayed zirconia—a TEM study, Surface and Coatings Technology 157(2002), Issue 2–3, pp. 238-246

[17] K. Yang, M. Fukumoto, T. Yasui, and M. Yamada, Role of substrate temperature on

microstructure formation in plasma-sprayed splats, Surface and Coatings Technology 214(2013), pp. 138-143
[18] X. Jiang, S. Sampath, and H. Herman, Grain morphology of molybdenum splats plasma-sprayed on glass substrates, Materials Science and Engineering: A 299(2001), Issue 1–2, pp. 235-240
[19] C.-J. Li, and J.-L. Li, Transient Contact Pressure During Flattening of Thermal Spray Droplet and Its Effect on Splat Formation, Journal of Thermal Spray Technology 13(2004), Issue2, pp. 229-238
[20] E.-J. Yang, X.-T. Luo, G.-J. Yang, C.-X. Li, and C.-J. Li, A TEM Study of the Microstructure of Plasma-Sprayed YSZ Near Inter-splat Interfaces, Journal of Thermal Spray Technology 24(2015),Issue 6, pp. 907-914
[21] J. Cedelle, M. Vardelle, B. Pateyron, and P. Fauchais, Investigation of plasma sprayed coatings formation by visualization of droplet impact and splashing on a smooth substrate, IEEE Transactions on Plasma Science 33(2005), Issue 2 I, pp. 414-415
[22] R. McPherson, and B.V. Shafer, Interlamellar contact within plasma-sprayed coatings, Thin Solid Films 97(1982),Issue 3, pp. 201-204

Performance of CoCrAlSiY Alloy Powder

Shuting ZHANG[1,2,3], Kaiping DU[1,2,3], Ru Lin Peng[4] Ji SHI[1,2,3], Jianming LIU[1,2,3,]

(1. Beijing General Institute of Mining and Metallurgy, Beijing 100070;
2. Beijing Key Laboratory of Special Coating Materials and Technology, Beijing 102206;
3. Beijing Industrial Parts Surface Strengthening and Repairing Engineering Technology Research Center, Beijing 102206;
4. Linköping University, Sweden)

Abstract: The CoCrAlSiY alloy powder with Si mass concentrations of 0, 2% and 5% was prepared in this work. The oxidation kinetics curves of all three kinds of powders after 300 h oxidation at 1000 °C were plotted. In addition, the phase constitution of alloy powder and the distribution of β phase were analyzed by SEM and EDS. Furthermore, the effect of Si-addition on the melting temperature and oxidation resistance of the alloy powder were investigated by DSC-TG from the room temperature to 1400 °C. And the element concentrations at the grain boundary of alloy powder with Si addition of 2% were also analyzed. The results show that the melting temperature of alloy powder decreases as increasing Si content, which indicates that adding Si element could influence on the selective oxidation of Al and Cr elements in the alloy system, and improve the oxidation resistance of CoCrAlY powder. In addition, the weight gain of powder with Si addition of 2% is lowest. And Si element has a enrich tendency in the grain boundary. Therefore, the higher Si content would have a negative effect on the high temperature oxidation resistance of powder.

Key Words: MCrAlY; Si; powder; oxidation; grain boundary

1. Introduction

MCrAlY coatings (M = Ni, Co or their combination), one of the most important coating systems, are the oxidation and corrosion resistant coatings of metal materials at the high temperature. They have been widely used in the aerospace, energy and petrochemical [1-3]. Recently, adding modifying elements (Re, Zr, Hf, Ta, Si, etc.) to MCrAlY coating system is a hotspot to further improve the high temperature performance of MCrAlY coating. MCrAlY coating with added Si element is an important kind of high temperature protective coating, which is applied to protect hot corrosion of equipment in the marine environment[4-5]. Clemens[6] et al. investigated the effect of Si on the oxidation behavior of NiCrAlY alloy.They found that adding 1-2% of Si could obviously improve the anti-spalling of the alloy oxide film, which enhanced its oxidation resistance. Toto Sudiro[7]. proved that Si element could significantly improve the corrosion resistant of coatings. However, the current study only focus on the coating performance, the work related to the effect of Si element on the alloy performance is rather limited. Therefore, the effect of Si element on its distribution in the alloy powder, the melting temperature of alloy and the oxidation resistance of alloy are discussed in this work, which provide guidance for practical application.

2. Experiment

Three kinds of CoCrAlY alloy powders with Si mass concentrations of 0%, 2% and 5% were designed and prepared by VIGA-16 Germany vacuum atomization furnace in the study. Thus their nominal composition is Co29Cr6Al0.5Y, Co29Cr6Al2Si0.5Y and Co29Cr6Al5Si0.5Y respectively. A constant oxidation test at 1000 °C was performed in a QM-ISPZ-CL type muffle furnace. Then these powders were taken out and weighed every 25 h. The thermal performance of

International cooperation project(2015DFA51530)
Corresponding author: Shuting ZHANG (1978-), Ph.D, Email: shutingzhang007@163.com

alloy powder were analyzed by STC445F3 type differential scanning calorimetry (DSC). In this process, the samples were heated up by 10 °C/min in an alumina crucible with 60 ml/min argon gas protection. The microstructure of powder were observed by Quanta600 type scanning electron microscopy (SEM). In addition, the distribution of Si element in the powder was studied by focusing ion beam (FIB) and projection electron microscopy (TEM).

3. Results and Discussion

3.1 Phase constitution and microstructure of CoCrAlSiY alloy powder

The phase constitution of MCrAlY powder mainly includes β phase (NiAl or CoAl) and Ni-based or Co-based matrix phase. In order to investigate the effect of Si element on the performance of alloy system, the phase constitution and microstructure of the powder with different Si contents were analyzed. The previous work[8] showed that all three kinds of powders consist of the same phases, they are a CoAl β phase and Co_3Al as the matrix on it. It seems that the addition of Si does not have a significant influence on phase constitution of the powders in this work. Scanning electron micrographs of the three types of alloy powders are shown in Figure 1. It can be seen that the element distribution in the different powders is uniform. The powders with Si mass concentrations of 2% and 5% clearly have a two-phase microstructure, in which the phase showing light contrast is the matrix, and that the β phase shows a dark contrast. In addition, the size of β phase was gradually refined, while its volume content was enhanced as increasing Si content.

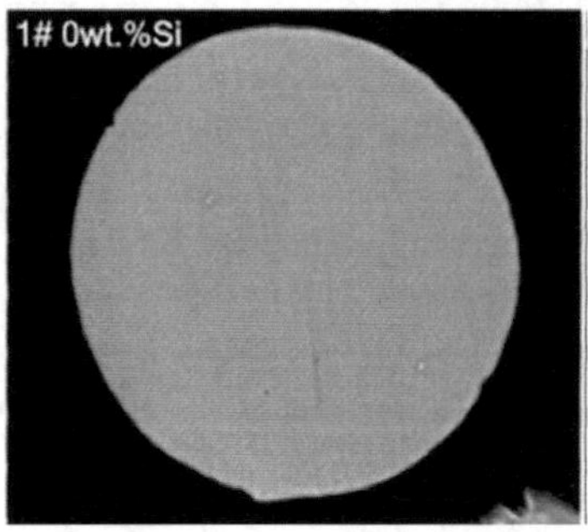

a) 0 Si

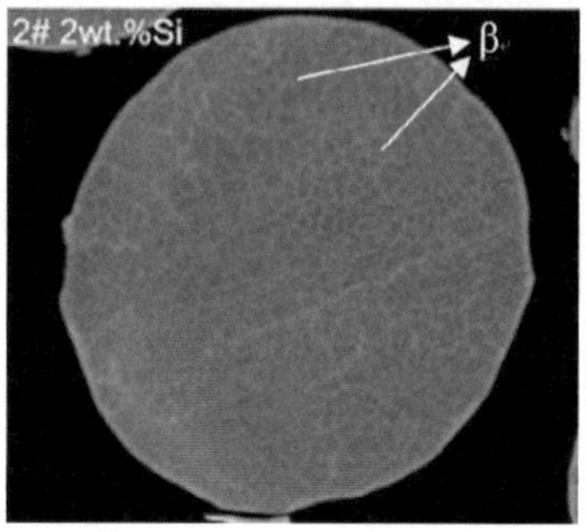

b) 2% Si

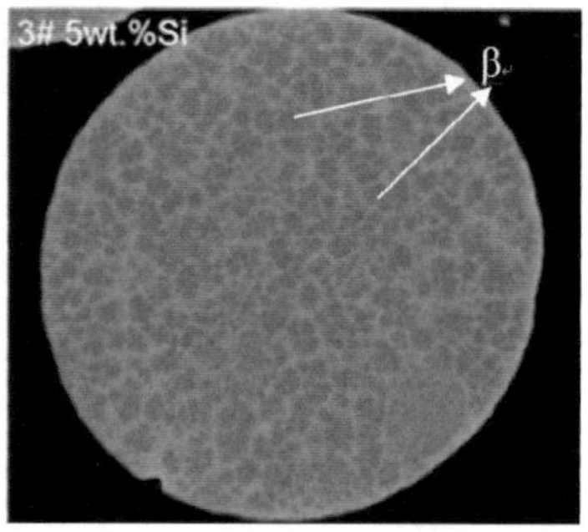

c) 5% Si

Figure. 1. Scanning electron micrograph of three different testing alloy powders (×3000)

Scanning electron micrograph of alloy powder after 300 h oxidation at 1000 °C is shown in Figure 2. Compared with Figure 1, the powder with a Si mass concentration of 2% only include a single phase, while that with a Si mass concentration of 5% is still obvious two-phase structure. EDS analysis of alloy powder after 300 h oxidation at 1000 °C are summarized in Table 1. It can be seen that the powder with Si addition of 2% after the oxidation mainly consists of Co, Cr and Si. And the mass concentration of Cr and Si element is almost equal to the nominal composition of $Co29Cr6Al2Si0.5Y$. The matrix phase (C point) and β phase (D point) of the powder with Si addition of 5% after the oxidation not only includes Co, Cr and Si, but also contains Al element. However, the Al content in the matrix phase (1.17%) is obviously lower than that in the β phase (9.77%). While the Cr content in the β phase decreases by 10% compared with the nominal composition of Co29Cr6Al2Si0.5Y. According to the above analysis, after 300 h oxidation at 1000 °C, the Al element, the key oxidation resistant element, in the powder with Si addition of 2% element have been consumed, which results in the disappearing of β phase. In addition, for the powder with Si addition of 5%, the difference of both Al and Cr in the matrix and β phase is obvious, which indicates that the Si addition of 5% has a significant effect on the selective oxidation of Al and Cr elements in the MCrAlY alloy system.

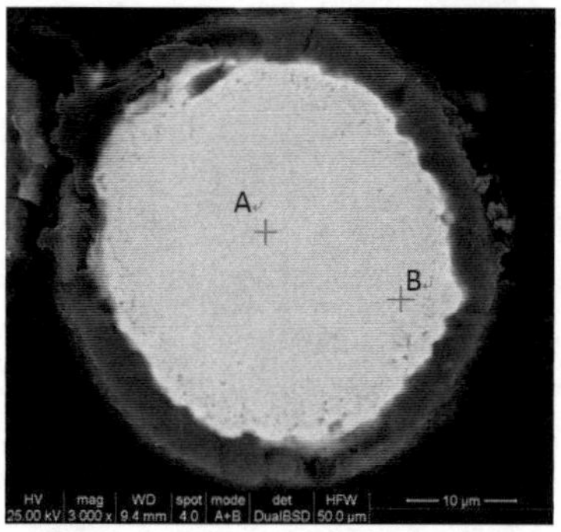

a) 2% Si

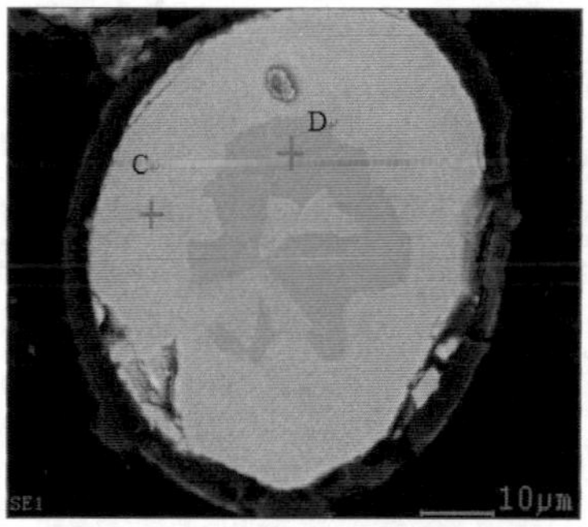

b) 5% Si

Figure. 2. Scanning electron micrograph of alloy powder after 300 h oxidation at 1000 °C

Table. 1. EDS analysis of alloy powder after 300 h oxidation at 1000 °C

Wt / Element	2%Si		5%Si	
	A	B	C	D
AlK	—	—	01.17	09.77
SiK	02.61	02.45	07.91	06.16
CrK	31.33	31.62	34.41	18.57
CoK	66.06	65.93	56.51	65.50
Matrix	Correction			

3.2 DSC-TG analysis of CoCrAlSiY alloy powder

DSC curves of three kinds of alloy powders is shown in Figure 3. It can be seen that the initial melting temperature of the alloy powders with Si mass concentrations of 0%, 2% and 5% is 1380 °C, 1281 °C and 1258 °C respectively. For the alloy powders with Si addition of 2% and 5%, a phase of low melting point occurs around the melting point. In addition, the melting temperature of CoCrAlY powder without Si element is highest, and that with Si addition of 2% and 5% decreases by about 100 °C and 130 °C respectively. The melting latent heat of the alloy powder with Si mass concentrations of 0%, 2% and 5% is 113.7 J/g, 125.7 J/g and 171.4 J/g respectively. Furthermore, the temperature range of endothermic peak for the alloy powder with a lower Si content is narrow, which shows a characteristics of coherent melting. As increasing Si content, the temperature range of endothermic peak widens, which means that the solid-liquid coexistence temperature range enlarges. Thus the formation probability of dendrites in the solidification process is high. Therefore, the alloy material with a high Si content has a negative influence on the uniformity of the coating structure and structure. The TG curve shows that the oxidation weight gain of alloy powder without Si element begins at about 700 °C, and it reaches to about 2% at 1400 °C. While the oxidation weight gain of alloy powder with Si addition of 2% and 5% is not obvious, which indicates that the addition of Si element could improve the oxidation resistant of powder.

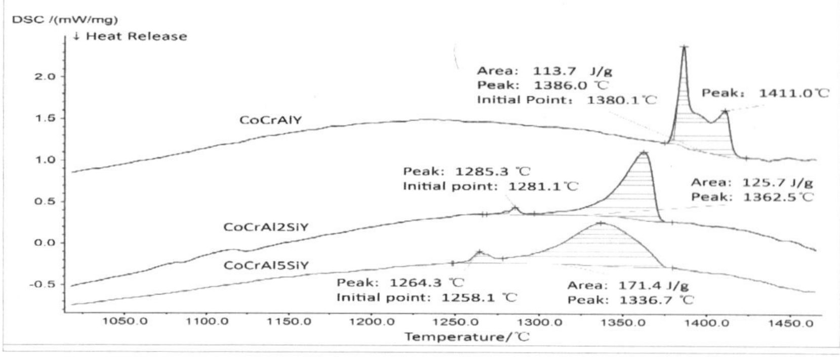

Figure 3. DSC curves of three kinds of alloy powders

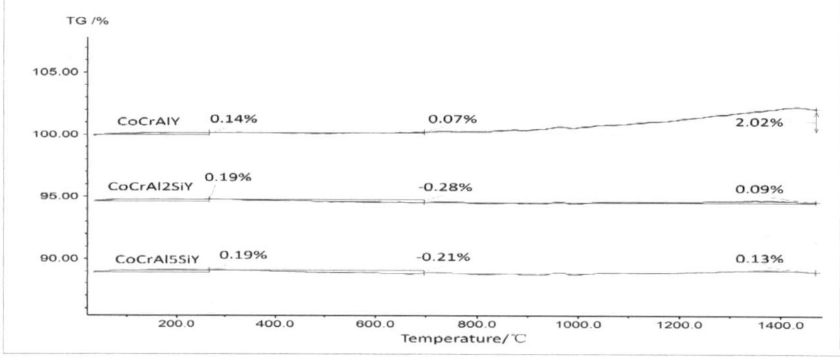

Figure 4. TG curves of three kinds of alloy powders

3.3 Oxidation resistance of CoCrAlSiY alloy powder

The oxidation kinetics curves of the three types of alloy after 300 h oxidation at 1000 °C is given in Figure 5. It can be seen that the weight of powders increases continuously with the increasing of the oxidation time, which indicate that the oxidation of powder continues to occur. In addition, the weight gain of powder without Si element is the largest, followed by that with a Si mass concentration of 5%, and that with a Si mass concentration of 2% is the lowest. In order to investigate the effect of Si element on the performance of MCrAlY alloy powder, the CoCrAlSiY alloy with a Si mass concentration of 2% was prepared by FIB and observed by S-TEM, the concentration of alloying elements in the different positions was analyzed by EDS. As shown in Figure 6 and 7. The concentration of Si element in the grain boundary is higher than that in the matrix and the precipitated phase, which shows that the segregation phenomena occurs in the grain boundary for CoCrAlSiY alloy system with a Si mass concentration of 2%. In the CoCrAlSiY alloy system, the atomic radius of Si element is 110 pm, while that of Co and Cr is about 135 pm. Thus Si element exists in the form of replacement type solid solution. The effect of Si element on XRD analysis shows that adding Si element could not produce a new phase, but it could inhibit the grain growth. According to **Harrison**[11] equation at high temperature, the effective diffusion coefficient Deff in the mean medium is the weighted average of the grain boundary diffusion coefficient Db and the bulk diffusion coefficient D:

$$D_{eff} = fD_b + (1 - f) D$$

Where f is the volume fraction of the grain boundary in the crystal. f =q*δ/d, where q is numerical factor of

International cooperation project(2015DFA51530)
Corresponding author: Shuting ZHANG (1978-), Ph.D, Email: shutingzhang007@163.com

the grain shape and δ is grain boundary width. If the diffusion atoms aggregates in the grain boundary, f should multiply by the segregation factor (s)[12], thus f =q*s*δ/d. When Si element aggregates in the grain boundary, the volume fraction of the grain boundary (f) and grain boundary diffusion content increase. Therefore, after adding Si element to the CoCrAlSiY system, the alloy grain boundary is a rapid diffusion channel of the film-forming element or a short-circuit diffusion channel. It is favorable for the formation of protective outer oxide film of alloy, which could inhibit the matrix oxidation and improve the oxidation resistance of the alloy[13]. The above phenomenon is the positive effect of grain boundary diffusion. However, as the Si content increases from 2% to 5%, the volume fraction of the grain boundary (f) increases and the grain boundary area enlarges. It increases the supply flux of cation and decreases the number of interfacial voids. At the same time, the increasing grain boundary improves the diffusion channel of the oxidation reaction, which increases the brittleness of the CoCrAlSiY alloy[14-15]. Therefore, the effect of Si element on the high temperature oxidation resistance of the powder is the negative effect of grain boundary diffusion. It is the reason that the oxidation resistance of alloy powder with a Si mass concentration of 2% is better.

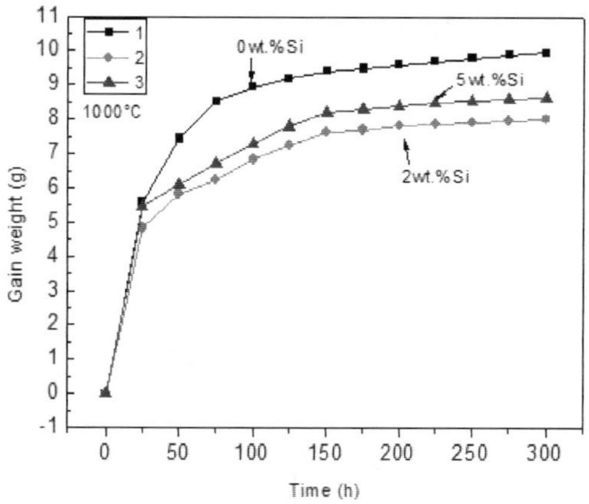

Figure. 5. Weight gain curve of powders after 300h oxidation at 1000°C

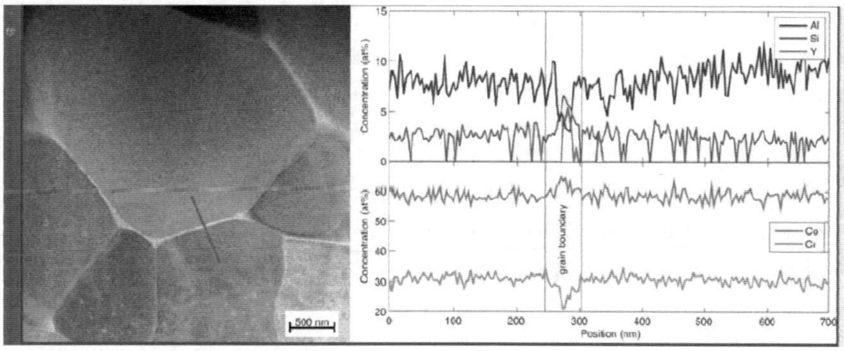

Figure. 6. EDS analysis of powder with a Si mass concentration of 2% by S-TEM

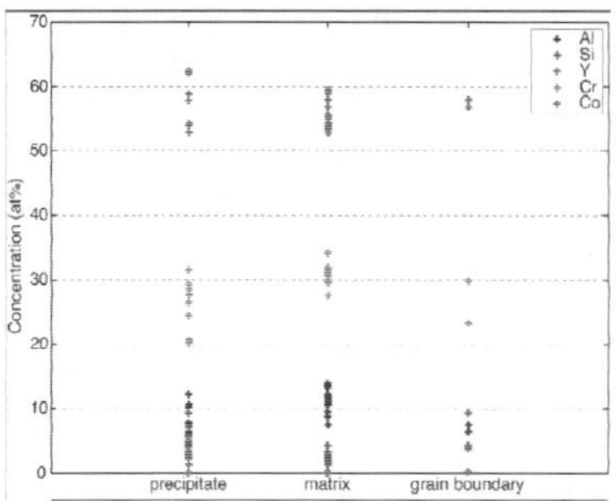

Figure. 7. Atom concentration statistical diagram of powder with a Si mass concentration of 2%

4. Conclusions

1) Si element impacts the melting point of the alloy powder and the latent heat of phase change. With the increase of Si content, the alloy melting point decreases. The melting point of the CoCrAlY powder without Si element is about 1380 °C, while that with Si mass concentrations of 2% and 5% decreases by 100

and 130 °C respectively. With the increase of Si mass concentration in the alloy powder, the latent heat of phase change of three kinds of alloy powder is 113.7J/g, 125.7 J/g and 171.4 J/g respectively.

2) Adding Si element could influence on the selective oxidation of Al and Cr elements in the alloy system, which improves the oxidation resistance of COCrAlY powder. The oxidation kinetics curve of powder after 300 h oxidation at 1000 °C shows that the powder weight gain increases with the increasing of the oxidation time, which indicates that the powder oxidation occur continuously. The weight gain of the powder without Si element is the largest, followed by that with a Si mass concentration of 5%, and that with a Si mass concentration of 2% is the lowest.

3) Si element has a enrich tendency in the grain boundary of CoCrAlY alloy powder. The EDX analysis of the alloying elements in the different positions shows that the concentration of Si element in the grain boundary is higher than that in the matrix and the precipitated phase. According to Harrison Equation, when the Si content increases from 2% to 5%, the effect of Si element on the high temperature oxidation resistance of powder is the negative effect of the grain boundary diffusion, which weakens the oxidation resistance of the powder.

Acknowledgement

The authors would like to thank Prof. Changhai LI from Chalmers University for the guidance on the powder performance analysis and Doctor Olivier Messe from Cambridge University for the assistance on the TEM testing.

References

[1] WANG X ,XIN L , Wei H , et al. Progress of High-temperature Protective Coatings[J],Corrosion Science & Protection Technology[J], 2013, 25(3):175-183

[2] PADTURE N P, GELL M, JORDAN E H. Thermal barrier coatings for gas-turbine engine applications [J]. Science, 2002, 296(5566): 280-4

[3] Shahzad Salam, Zhang Y D, Wang H H, et al. Influence of Re on High-Temperature Oxidation Behavior of MCrAlY Type Coating Alloys[J], Rare Metal Materials and Engineering, 2015, 44(S1): 316-319.

[4] SALAM S, HOU P Y, ZHANG Y D, et al. Compositional effects on the high-temperature oxidation lifetime of MCrAlY type coating alloys[J]. Corrosion Science, 2015, 95:143-151.

[5] Beele W,Czech N,Quadakkers W J, et al. Long-term oxidation tests on a Re-containing MCrAlY coating[J].Surface and Coatings Technology,1997,94:41-45.

[6] Clemens D,Vosberg V,Hobbs W,et al.TEM and SNMS studies of protective alumina scales on NiCrAlY-alloys[J].Fresenius' journal of analytical chemistry,1996,355(5-6):703-706.

[7] Sudiro T,Sano T,Kyo S,et al. High Temperature Corrosion of CoNiCrAlY-Si Alloys in an Air-Na2SO4-NaCl Gas Atmosphere[J].Materials transactions,2011,52(3):433-438.

[8] ZHANG S T,GUO D H,WANG H,et al. Research on the antioxidant property of MCrAlY alloy powders, Rare Metal Materials and Engineering [J],2014,43(1): 2580-2583.

[9] N Curry, K Vanevery, T Snyder, N Markocsan. Thermal Conductivity Analysis and Lifetime Testing of Suspension Plasma-Sprayed Thermal Barrier Coatings. Coatings, 2014, 4(3):630-650

[10] HD Scheckenbach, CW Smith. Thermal spray powder incorporating a particular high temperature polymer[P].US, 2002

[11] Mishin Y, Herzig Chr, Bernodini J, et al .Grain boundary diffution: fundamental to recent developments[J]. Inter. Mates. Res. ,1997 ,42 (4) :155

[12] Abe F ,Araki H ,Yoshida H, et al . The effect of grain size on the corrosion behavior of inconel 600 in high temperature steam[J] . Corros. Sci. ,1981 ,21 (12) :819

[13] Singh Raman R K, Khanna A S ,Tiwari R K, et al. Influence of grain size on the oxidation resistance of 2.25Cr21Mo steel[J] . Oxid Met . ,1992 ,37 (1/ 2) :1

[14] Singh Raman R K,Gnananroothy J B. Influence of prioraustenite grain size on oxidation behavior of 9Cr21Mo steel[J] . Oxid. Met . ,1992 ,38 (5/ 6) :483

[15] Singh Raman R K,Gnanamootloy J B ,Roy S K. Synergistics influence of alloy grain size and silicon content on oxidation behavior of 9Cr21Mo steels[J] . Oxid. Met . ,1994 ,42 (5/ 6) :33

Liquid flame spray construction of polyimide-copper layers for marine antifouling applications

Yi Liu, Xinkun Suo, Zhe Wang, Yongfeng Gong, Xin Wang, Hua Li*
Key Laboratory of Marine Materials and Related Technologies, Ningbo Institute of Materials Technology and Engineering, Chinese Academy of Sciences, Ningbo 315201, China. *: lihua@nimte.ac.cn.

Polyimide-copper layers consisting of individual capsule-like splats were one-step fabricated by solution precursor flame spray through controlling the reaction between dianhydride and diamine dissolved in copper nanoparticles-containing dimethylformamide solvent. The polyimide splat exhibited hollow structure with an inner pore of 10-15 μm and a tiny hole of 1-5 μm on its top surface. Transversal cut by focused ion beam milling of the individual splats and scanning electron microscopy characterization further revealed unique dispersion of the copper nanoparticles inside the polyimide shell. After 1000 h exposure to the testing synthetic seawater, continuous release of copper from the coatings containing up to 30wt.%Cu kept remarkable. Antifouling performances of the constructed layers were assessed by examining colonization behaviors of typical bacteria *Bacillus sp.* and marine algae *Phaeodactylum tricornutum* and *Chlorella* on their surfaces. Distribution of the inorganic nanoparticles endows the polyimide coatings with special capsule structure and exciting hydrophobicity and antifouling performances. The liquid flame spray route and the encapsulated structure of the polyimide-Cu coatings would open a new window for designing and constructing environment-friendly marine antifouling layers for long-term applications.

1 Introduction

Any surface immersed in seawater is prone to the settlement of marine organisms, such as protein, bacteria, algae, or mollusks [1]. As a consequence, artificial marine infrastructures usually suffer from biofouling attacks throughout their services in the marine environment [2, 3]. Among the measures taken so far to solve the abovementioned problems, construction of an antifouling layer has been proven to be effective in offering long-term antifouling performances. Exciting research progresses have been made in recent years towards developing novel environmental friendly antifouling materials and coatings, some of which have shown promising performances.

The antifouling technique based on the use of biocides has been the most widespread approach in modern maritime industries [4]. During the last decades, however, growing awareness of environmental issues pertaining to the use of biocides have triggered booming research efforts in searching green alternatives [5]. Copper has been the most important alternative biocide [6-9] since organotin compounds were banned. Regardless of the role copper plays as the essential element for marine organisms, it is still a big concern about its tolerable concentration, above which unexpected deterioration usually happens to marine species. Design and fabrication of smart coatings with appropriate structure for sustainable release of copper ions for long-term antifouling performances are therefore essentially required.

To achieve constrained release of copper ions without deteriorating antifouling efficiency, the measures taken to incorporate copper into polymeric matrices are to be further explored. Due to their inert nature, polymers as surface coatings on metal surface could act as barrier resisting invasion of corrosive species. Among the polymeric materials developed for marine applications, aromatic polyimide attracted extensive attention owing to its excellent mechanical properties, superior thermal stability, good chemical resistance and processability [10, 11].

In this paper, the liquid flame spray route was proposed for constructing copper-containing polymer coatings. Nano copper particles were added in polyimide precursor for subsequent synthesis of polyimide and coating fabrication. Microstructural features of individual polyimide-copper splats and coatings were characterized by focused ion beam technique and antifouling performances of the coatings were assessed by examining colonization behaviors of Bacillus sp. bacteria and Phaeodactylum tricornutum and Chlorella. This research provides a new processing route for large-scale fabrication of capsules with polymer as shell and inorganic particles as filler.

2 Materials and methods

2.1 Synthesis of polyimide and polyimide-Cu precursors

Kapton-type aromatic polyimide precursor solution was prepared by mixing monomer pyromellitic dianhydride and 4,4'-oxydianiline in DMF solvent. For the polyimide-Cu suspension preparation, Cu particles with the size of ~300 nm were dispersed in DMF-ODA solutions and PMDA was then added with an equivalent molar ratio of ODA.

2.2 Fabrication of coatings

After ultrasonic treatment, the precursor suspensions were atomized by a home-made spray atomizer. The precursor feedstock injector with a diameter of 1.5 mm was positioned just next to the flame torch, and the angle between the injector and flame was 30 degrees. Pressure of the atomizing air was 0.7 MPa. Prior to coating deposition, the grit-blasted carbon steel Q235 plates and polished stainless steel substrates were rinsed in distilled water and ultrasonically cleaned in acetone. The spraying was carried out using the Castodyn DS 8000 system (Castolin Eutectic, Germany). For the liquid flame spraying, acetylene was used as the fuel gas with the flow rate of 1.5 Nm^3/hr and working pressure of 0.1 MPa. Pressure and flow rate of oxygen were 0.5 MPa and 2.5 Nm^3/hr respectively. The precursor feed rate was 40 ml/min and the spray distance was 200 mm.

2.3 Characterization of microstructure

Microstructure of the splats and the coatings was characterized by field emission scanning electron microscopy (S4800, Hitachi, Japan). For further examination, focused ion beam (FIB) milling of the samples was performed in a FIB/scanning electron microscope dual beam system (Helios Nanolab 660, FEI, USA). Transversal cut of the samples was carried out at a stage tilt of 52°. A focused beam of Ga+ ions was rastered across the specimen surface. Larger beam currents quickly milled away sections of material from the sample surface, while lower currents were used for polishing the milled surfaces. The samples were also analyzed by fourier transform infrared spectroscopy (Nicolet 6700, Thermo Fisher Scientific, USA) with the resolution of 4 cm^{-1} and a scan range of 4000-400 cm^{-1}. Release rate of copper ions from the coatings as a function of immersion time was also examined through measuring the concentration of copper ions in testing solution using inductively coupled plasma mass spectrometer (NexION 300, USA).

2.4 Antifouling testing

Antifouling performances of the coatings were assessed by examining formation of bacterial biofilm and colonization of algae on their surfaces. Marine *Bacillus sp.* (MCCC No. 1A00791) was used in this study. *Bacillus sp.* bacteria were cultured in 2216E (CM 0471) media. Diatoms *Phaeodactylum tricornutum* and green algae *Chlorella sp.* were used in this study to further clarify the antifouling performances of the coatings. Adhesion of the algae on the surfaces of the samples for 1 week was inspected. After the fixation by 2.5% glutaraldehyde for 1 h, the samples were observed by confocal laser scanning microscopy (CLSM, TCS SP5, Leica, Germany).

3 Results and discussion

Polyimide and copper-containing polyimide coatings were successfully fabricated by the liquid flame spray deposition. Cross-sectional view of the coatings shows typical lamellar morphological feature of thermal sprayed coatings and porous microstructure (**Fig. 1 a**). Surprisingly, it is noted that the copper-containing polyimide coatings exhibit a typical capsule structure with nano copper particles being enwrapped by polyimide (**Fig. 1 b**). The coating comprises individual micro-capsules and close interconnection among them is seen (**Fig. 1 a**), leaving no obvious through-thickness flaws. This feature offers the probability of release of the encapsulated antifoulants in a controllable manner. The microcapsules are small particles with biocides as core materials being surrounded by polymer shell. It is anticipated that this structure might also give rise to long-term survival of the coatings against biofouling in harsh marine environment. Enlarged inner views of the coatings clearly show dispersed single Cu particles and agglomerated Cu clusters that are encapsulated by polyimide layer (**Fig. 1 c and d**). Micro-capsules are uniformly distributed in the coatings and copper nanoparticles exhibit an entrapped state, in turn avoiding undesirable interaction between active Cu and seawater for spontaneous leakage. This is exciting since polyimide is physicochemically stable and performs well in seawater. The polymeric shell controls diffusive release of the biocide to the surrounding medium for marine antifouling (**Fig. 1 e**) while protecting the remaining encapsulated biocide from possible corrosion.

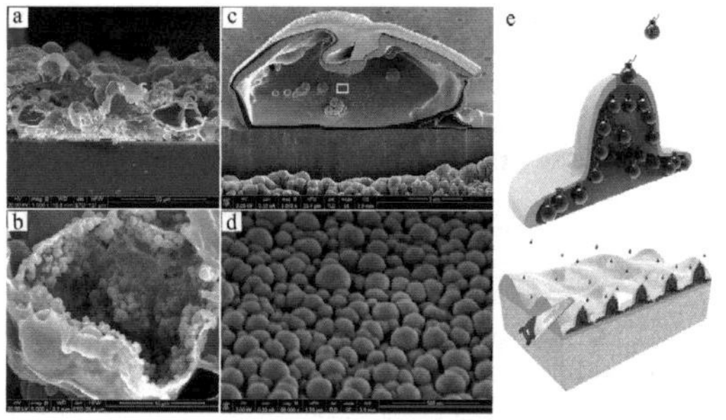

Fig. 1. Microstructure of the polyimide-30wt.%Cu coatings and splats.

It is well established that thermal sprayed coatings are built up by the accumulation of individual splats, which usually involves impingement of molten/partilly molten particles and subsequent quenching. The particle parameters such as velocity, temperature, and degree of solidification, and substrate variables affect the splatting behaviors [12]. Poly (amic acid) (PAA) and PAA-Cu precursor solutions were initially synthesized via in situ polymerization and following thermal imidization by flame spraying. During the spraying, combustion products of DMF solvent are carbon oxides (CO, CO_2), nitrogen oxides (NO, NO_2), etc. Consequently, high temperature heating results in solvent evaporation and water loss in PAA macromolecules [13], leading to generation of gas in particle core. This presumably accounts mainly for the microsized hollow polyimide spheres formed in the coatings. The quenching-induced solidification of polyimide obviously results in constrained escape of the gas generated during the imine formation.

To further elucidate the formation mechanisms of the structure, polyimide particles were collected by spraying the precursor directly into distilled water contained in a stainless steel container and futher dried. Spherical polyimide particles with perfect capsule structure are attained (**Fig. 2 a**) and they have the sizes consistent to the spheres realized in the coatings. It is noted that the spheres have smooth surfaces and no open holes are seen. Further characterization of individual polyimide splats collected on stainless steel substrate suggests their well-flattened morphology (**Fig. 2 b**) and trace of gas escape during flattening stage (**Fig. 2 c**). These phenoma are schematically depicted in Fig. 2 d. Accumulation formation of thermal sprayed coating on fresh substrate surface is a complex dynamic process. For hollow droplet like polyimide droplet in this case, upon its impact on substrate, liquid jets out from the bottom of the droplet and spreads in the radial direction (jetting-out region B in **Fig. 2 d**), but solidification is usually fast enough so that solidification of the portion of the droplet in contact with substrate would have completed (bonded region A in **Fig. 2 d**). High-velocity impact would result in sudden increase in inner pressure of the hollow sphere, in turn likely triggering breaking along the direction opposite to the impact to form a tiny hole (top hole zone C in **Fig. 2 d**).

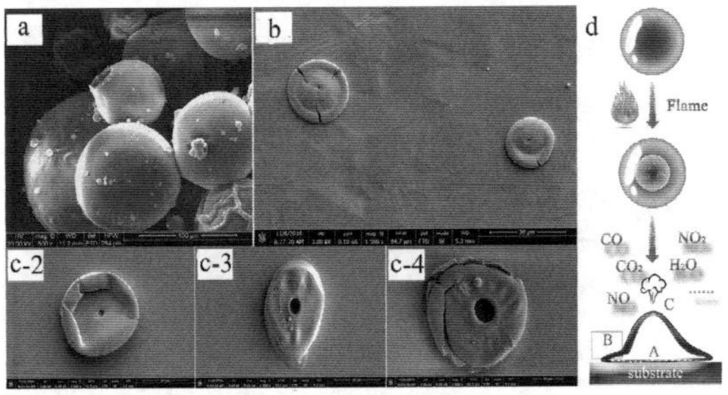

Fig. 2. Morphology of the as-sprayed powder and splats.

The structural features of the flattened polyimide-based particles were further examined by SEM through milling the splats with Ga+ beam by FIB. Prior to dissection of the splats, a protective strip of platinum was deposited (inset in **Fig. 3 a**) over the site of interest (**Fig. 3 a**) and appeared as a lighter layer covering the sectioned splat (**Fig. 3 b and c**). The cross-section of the hat-like polyimide splat clearly shows the hollow structure (**Fig. 3 b and c**). The thickness of the thin bottom layer, the lateral, and the upper shell varies showing the value of ~0.2 μm, ~1.5 μm, and ~4 μm, respectively. The inner surfaces of the hollow splat are smooth and it is interesting to note that the polyimide splat bonds very well with the substrate and the intimately contacting polyimide layer has the thickness of 0.1~0.3 μm (**Fig. 3 b and c**). This is important since the well-bonded polyimide-substrate interface provides the coatings with favorable adhesive strength. It should be noted that prior to the deposition, the substrate was already pre-heated by the flame spray gun. Surface pre-heating causes removal of absorbed gas and formation of Fe_2O_3 layer on the surface, which probably increases the contact resistance and further promotes spreading and bonding of the sprayed material. To gain further information about the exceptional structure, subsequently impacted droplet on the pre-flattened splat was also characterized by FIB approach (**Fig. 3 d**). Excellent cross-linking is recognized between the adjacent two splats (**Fig. 3 d and e**). Interestingly, sectioned view shows unnoticable interface in between the splats at their bottom layers (**Fig. 3 f**). It should be noted that the splats have isolated inner pores of ~12 μm and top holes of 1-5 μm. This special capsule strucure is promising for controlled release of packed biocides for potential antifouling functions. In addition, the flawed morphology at the center of the top layer of the splat (**Fig. 3 a and d**) further proves our speculation (as depicted in **Fig. 2 d**) that the possible gas leakage as a result of sudden increase in inner pressure of the hollow sphere easily results in formtion of the through-thickness hole.

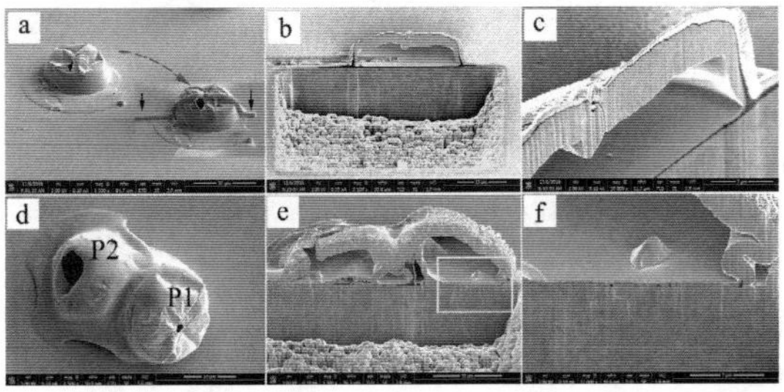

Fig. 3. Morphologies of the as-deposited polyimide splats as revealed by FESEM characterization after FIB dissection.

Thermal spray technique already showed great potential for synthesis of new materials [14, 15]. It seems clear that flame spray has accomplished complete synthesis of polyimide by thermal imidization during the coating deposition. Degree of imidization was examined by FTIR (**Fig. 4 a**). IR spectra of the polyimide coatings display the characteristic peaks at 1720 cm^{-1} for symmetric stretching vibration of C=O, 1780 cm^{-1} for asymmetric stretching vibration of C=O, 1380 cm^{-1} for CN stretching vibrations, and 725 cm-1 for bending vibration of C=O. For all the as-deposited coatings, full imidization has been achieved. This is not surprising since liquid droplets of polyimide precursor solution injected into the flame are subject to a relative controllable temperature range which is suitable for thermal imidization.

Further electrochemical testing was carried out for the coatings, since corrosion resistance is one of the essential requirements for the coatings for marine applications. Results show that E_{oc} of the polyimide and polyimide-Cu coatings in artificial seawater is more positive and corrosion current density is obviously lower than that of carbon steel Q235 (**Fig. 4 b**), indicating better anti-corrosion performances of the coatings. This nevertheless suggests that the polyimide coating alone already significantly enhanced the anti-corrosion properties of the steel substrate. The addition of copper even with increased content does not deteriorate the electrochemical properties of the polyimide coatings, which is likely attributed to the special capsule structure and the well-bonded interconnecting state of adjacent capsules.

It is known that the effectiveness of biocides differs depending on their concentration (release rate) and exposure duration. Therefore, controlling the release of copper ions from the capsules is of top importance. Morphologies of the polyimide splats already showed exciting hollow structures and copper nanoparticles are distributed inner polyimide shell. The tiny top hole of the polyimide splat would facilitate sustained release of copper ions. Further assessment of the releasing of copper ions shows that the polyimide-Cu coatings immersed in artificial seawater for 48 h exhibit the formation of many flowerlike particles on their surfaces (**Fig. 4 c**). The individual particles have nearly spherical shape with an average size of ~3 μm. This structure is similar to a previously reported 3D nanostructure of CuO as obtained by immersing copper foil into hot NaOH solutions [16]. Release rate testing of copper from the coatings suggests continuous yet relatively stable release (**Fig. 4 d**). It is not surprising that higher content of copper in the coatings brings about higher release rate. Compared to the biocide-encapsulated coating, mixture of polyimide and Cu powder without encapsulated structure shows 11 times faster release rate. Polyimide as the shell in the coatings protects efficiently quick consumption of copper by the seawater, which is promising for long-term antifouling functions.

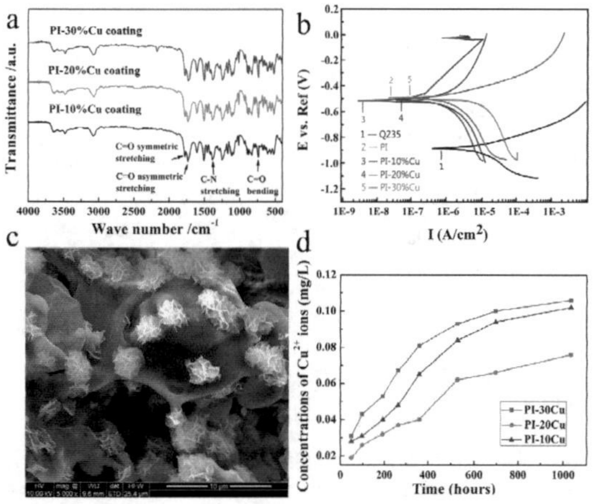

Fig. 4. Characteristics of the coatings and their electrochemical and copper release performances

.

In addition, antifouling properties of the Cu-free and the Cu-containing polyimide coatings against marine bacteria were examined. For marine applications, surface characteristics in particular wettability and surface energy of the layers play important roles in deciding their biocide-release, antifouling, and anti-corrosion performances. It is clear that the addition of Cu particles altered the surface wettability of the coatings (the inset in **Fig. 5 a**). The coatings turn to be hydrophobic after addition of Cu particles and increase in Cu content brought about significantly increased water contact angles. The angle increase from ~68o for the pure polyimide coating to ~131° for the layer comprising 30wt.%Cu. Hydrophobicity is generally explained as a result of reduced surface energy due to the combination of micro-hollow polyimide capsule and Cu nano-particles. Besides, this hierarchically micro-nano-structured capsules give rise to air-trapping below the droplet on the coating surface, forming Cassie's state [17]. This hydrophobicity likely promotes anti-corrosion performances of the coatings and constrains release of copper from the coatings. In fact, design of antifouling structure based on hydrophobic surface and encapsulated biocides is a promising method towards preventing biofouling in modern maritime industries.

Excellent sterilization performances were further revealed for the Cu-containing coatings (**Fig. 5 a**). For the coatings containing 20wt.%Cu and 30wt.%Cu, disappearance of the bacterial colonies in the Petri dishes indicates that most of the bacteria were killed by copper ions. After 12 h exposure, ~100% bacteria were already extinguished (data not shown). Further examination of bacterial adhesion shows that after being immersed in the bacteria-containing seawater for 2 days, the Cu-containing polyimide coatings exhibit remarkable bacterial removal and increased Cu content in the coatings gives rise to enhanced removal (**Fig. 5 b, c, and d**). For the polyimide-30 wt.%Cu coating, there are a large number of spores instead of bacteria seen on its surface (**Fig. 5 d**). Copper typically exists as Cu (II) outside bacterial cell and Cu (I) inside the cell. It was evidenced that binding of heavy metals to the DNA of bacteria and bacterial spores occurred. The DNA in dormant spores of Bacillus subtilis as well as other Bacillus species is extremely well protected against damage resulting from treatments such as desiccation, heat, oxidizing agents, or UV and γ radiation. This high degree of DNA protection is a major factor for species survival when subjected to environmental stresses. Moreover, it is observed that the biocides release is promoted against Bacillus sp., and obvious CuO layer is seen on the surface area (**Fig. 5 c and d**), possibly indicating more adhered bacteria, which in turn facilitates secreted extracellular polymeric substances (EPS) and biofilm-triggered corrosion of Cu.

To gain further insight into antifouling performances of the coatings with the capsule structure, adhesion of typical algae on the polyimide-based layers were examined. After 1 week of exposure, algae nouniformly adhered on the areas where the capsules are seen on the surface of the polyimide-Cu coating (**Fig. 5 e-1 and f-1**). SLCM observation shows remarkably inhibited adhesion of Phaeodactylum tricornutum and Chlorella on the copper-containing polyimide coatings. More Cu in the coatings results in less recruited algae (**Fig. 5 e and f**). This is attrituted presumably to continuous release of copper ions from the Cu-encapsulated coatings. The coating with the addition of 20 wt.%Cu performs the best among the samples in terms of removal of the bacteria and the algae, suggesting possible synergestic effect of the distinct composite structure and release of copper.

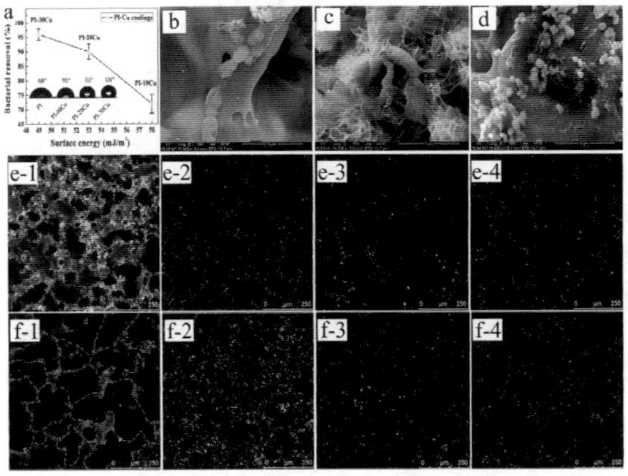

Fig. 5. Antifouling properties of the coatings.

4 Conclusions

Polyimide-copper coatings have been fabricated by liquid flame spray and the coatings have typical capsule structure with nano copper particles being enwrapped by polyimide shell. These structural characteristics facilitate constrained release of copper for antifouling functions and yet provide excellent anti-corrosion performances. The Cu-containing polyimide coatings show remarkable capability of resisting adhesion and colonization of Bacillus sp. bacteria and algae Phaeodactylum tricornutum and Chlorella. The newly constructed polyimide-Cu coatings show encouraging promises as anti-corrosion/fouling layers for marine structures. The thermal spray processing route might open a new window for making organic-inorganic composites or their coatings with capsule structure.

5 Acknowledgements

This work was supported by National Natural Science Foundation of China (grant # 31500772 and 41476064), Key Research and Development Program of Zhejiang Province (grant # 2017C01003), Ningbo Natural Science Foundation (grant # 2015A610019), and China Postdoctoral Science Foundation (grant # 2014M561800 and 2016T90554).

6 References

[1] Callow, J.A.and Callow, M.E.: Trends in the Development of Environmentally Friendly Fouling-Resistant Marine Coatings. Nat. Commun. 2 (2011) , pp. 244.

[2] Callow, M.E. and Callow, J.A. :Marine Biofouling: A Sticky Problem. Biologist 49 (2002) , Issue 13, pp.1/5.

[3] Fitridge, I.; Dempster, T.; Guenther, J. and de Nys, R. The Impact and Control of Biofouling in Marine Aquaculture: A Review. Biofouling 28 (2012) , Issue 7, pp. 649/69.

[4] Thouvenin, M.; Peron, J.J.; Charreteur, C.; Guerin, P.; Langlois. J.Y. and Vallee-Rehel, K.: A Study of the Biocide Release from Antifouling Paints. Prog. Org. Coa. 44 (2002) , Issue 2, pp.75/83.

[5] Diego, M.Y.; Søren, K. and Kim, D.J. Antifouling Technology-Past, Present and Future Steps Towards Efficient and Environmentally Friendly Antifouling Coatings: Review. Prog. Org. Coa. 50 (2004) , Issue 2, pp.75-104.

[6] Jia, B.; Mei, Y.; Cheng, L.; Zhou, J.P. and Zhang, L.N. Preparation of Copper Nanoparticles Coated Cellulose Films with Antibacterial Properties through One-Step Reduction. ACS Appl. Mater. Inter. 4 (2012) , Issue 6, pp.2897/02.

[7] Wei, X.; Yang, Z.; Wang, Y.; Tay, S.L. and Gao, W. Polymer Antimicrobial Coatings with Embedded Fine Cu and Cu Salt Particles. Appl. Microbiol. Biot. 98 (2014) , Issue 6, pp.6265/74.

[8] Jiang, S.; Sreethawong, T.; Lee, S.S.C.; Low, M.B.J.; Win, K.Y.; Brzozowska, A.M.; Teo, S. L.M.; Vancso, G.J.; Janczewski, D. and Han, M.Y.: Fabrication of Copper Nanowire Films and Their Incorporation into Polymer Matrices for Antibacterial and Marine Antifouling Applications. Adv. Mater. Interfaces 1400483 (2015) , Issue 3, pp.1/8.

[9] Anyaogu, K.C.; Fedorov, A.V.; Neckers, D.C.: Synthesis, Characterization, and Antifouling Potential of Functionalized Copper Nanoparticles. Langmuir 24 (2008) , Issue 8, pp.4340/46.

[10] Vanherck, K.; Koeckelberghs, G. and Vankelecom, I.F.J.: Crosslinking Polyimides for Membrane Applications: A Review. Prog. Polym. Sci. 38 (2013) , Issue 6, pp.874/96.

[11] Liaw, D.J.; Wang, K.L.; Huang, Y.C.; Lee, K.R.; Lai, J.Y. and Ha, C.S.: Advanced Polyimide Materials: Syntheses, Physical Properties and Applications. Prog. Polym. Sci. 37 (2012) , Issue 7, pp.907/74.

[12] Pasandideh-Fard, M.; Pershin, V.; Chandra, S. and Mostaghimi, J.: Splat Shapes in a Thermal Spray Coating Process: Simulations and Experiments. J. Therm. Spray Techno. 11(2002) , Issue 2, pp.206/17.

[13] Li, Y.; Pei, X.; Shen, B.; Zhai, W.; Zhang, L. and Zheng, W.: Polyimide/Graphene Composite Foam Sheets with Ultrahigh Thermostability for Electromagnetic Interference Shielding. RSC Adv. 5 (2015) , Issue 31, pp.24342/51.

[14] Fauchais, P.; Vardelle and A.; Denoirjean, A.: Reactive Thermal Plasmas: Ultrafine Particle Synthesis and Coating Deposition. Surf. Coat. Tech. 97 (1997) , Issue 1-3, pp.66/78.

[15] Liu, H. and Huang, J.: Reactive Thermal Spraying of TiC-Fe Composite Coating by Using Asphalt as Carbonaceous Precursor. J. Mater. Sci. 40 (2005), Issue 15, pp.4149/51.

[16] Liu, J.; Huang, X.; Li, Y.; Sulieman, K.M.; He, X. and Sun, F.: Hierarchical Nanostructures of Cupric Oxide on a Copper Substrate: Controllable Morphology and Wettability. J. Mater. Chem. 16 (2006), Issue 45, pp.4427/34.
[17] Bhushan, B.; Jung, Y.C. and Koch, K.: Micro-, Nano-and Hierarchical Structures for Superhydrophobicity, Self-Cleaning and Low Adhesion. Philos. Trans. R. Soc. London, Ser. A 367 (2009) , pp.1631/72.

Investigation of pretreatment and coating of vulcanized fiber by means of thermal spraying

W. Tillmann, A. Brinkhoff, I. Baumann, K. Schmidt, R. Zielke, Dortmund / D

Abstract – In light weight constructions, researches focus more and more on ecological aspects of materials. In this way vulcanized fiber relive a renaissance because it is produced from renewable raw materials. Vulcanized fiber is a composite material, which was approximately discovered in 1855. It is manufactured by parchmentizing of pulp with the use of a zinc chloride solution. This material is well known for its good strength to weight relationship, but it is susceptible to the contact with aqueous fluids and also shows only a low resistance to wear. Therefore, a functionalization of the surface is needed.
The functionalization can be realized by thermal spray processes to increase the potential of vulcanized fiber. Especially the pretreatment of the specimen to enable the coating application is one of the key topics of this paper. Vulcanized fiber is a thermal sensitive material. The impact of hot spray materials can cause undesired reactions such as the combustion of fibers on the substrate's surface. To reduce this effect, twin wire arc spraying and low-pressure cold gas spraying processes were used to apply low-melting materials (copper, zinc) onto the specimen. Thereby, the influence of the coating on the vulcanized fiber surface was investigated. In addition, non-destructive testing methods were evaluated to gain information about the vulcanized fiber.

1 Introduction

Light weight constructions play an increasingly important role in many areas of mechanical engineering. In addition to weight savings, research is more and more focusing on ecological aspects of materials. Due to their biodegradability and production from renewable raw materials, vulcanized fibers nowadays represent an interesting ecological alternative [1].
The production process of vulcanized fiber based on cotton was already patented in 1859 [6]. Since then, numerous industrial applications, such as cases, protective clothing, seals or building materials, have been developed for this material. Suitable raw materials for vulcanized fiber production are natural fibers with a very high cellulose content.
Vulcanized fiber is industrially produced either in a discontinuous process (winding process) or continuously on a production line. An important step in the production process is parchmenting. For thin vulcanized fibers, sulfuric acid and for thicker fibers a zinc chloride acid can be used as a reactant. The acid is absorbed by the paper and dissolves the surfaces of the fibers. Multiple sugar molecules are formed and the fibers become twisted and shrink. During this time, it is possible to connect individual layers of paper to a larger homogenous composite by means of a slight contact pressure. The acid is then removed again in order to prevent a final decomposition of the fibers and to obtain a safe end product [2].
Vulcanized fiber features light weight, high strength and good machinability (e.g. bending, drilling, milling, grinding, punching) [1]. Despite the diverse positive properties of vulcanized fiber, the requirements placed on today's construction materials are not adequately covered. This mainly affects the limited resistence against sliding, abrasive and erosive wear, a moderate temperature resistance to approx. 200°C as well as susceptibility to aqueous or aggressive solutions and dirt [6]. In order to counteract these problems, the possibilities and limitations of the coating of vulcanized fiber by means of thermal spraying are investigated. The goal is to develop a comprehensive concept for the production of novel surfaces for the protection as well as to expand the application possibilities of construction components and / or products from vulcanized fiber. This includes the improvement of the biochemical and thermomechanical protection properties (hydrophobing, dirt repellent, increased abrasion resistance, heat protection).
The focus of this research work is the pretreatment of the surface of vulcanized fiber by means of various strategies. Within the scope of the investigations carried out here, boundary surface phenomena, which occur between the coating and the substrate material during pretreatment and during coating represent central scientific questions. These include, in particular, the structural influence of the vulcanized fiber surface or the fiber structure on the bonding of the coating.
One of the most common preparation methods for thermal spraying is the grit blasting process. The surface to be coated is roughened by means of corundum particles, which enables the coating to adhere to the surface.
Within the scope of this study the influence of different preparation parameters on the substrate material as well as its influence on the coating development were investigated. Subsequently the surfaces of the vulcanized fibers were examined. Due to the thermal sensitivity of the vulcanized fibers, thermal spraying processes with lower thermal influence on the substrate material were used. These include, among others, twin wire arc spraying (TWAS) and low-pressure cold-gas spraying (LP-CS). Low-melting metals (such as copper and zinc) were used as caoting materials and adapted spray parameter settings were applied.
Finally, non-destructive testing methods were tested and scrutinized in order to detect any defects in the coating (such as delaminations or inclusions). Therefore, different test methods such as the X-ray

method or thermography, which are commonly used for this purpose [3,4,5], were employed.

2 Experimental

2.1 Materials and preparation

For this study, rectangular specimen with edge length of 50 mm were used as substrates. The substrates were made from vulcanized fiber with thicknesses of 2 mm and 8 mm. To prepare the surface by grit blasting, corundum with different grain sizes EKF 14 (1,180 – 1,700 µm) and EKF 80 (150 - 212 µm) was used. The blasting parameters were varied in two pressure stages for each grain size during the experiments. These stages are 2 bar (Stage 1) and 4 bar (Stage 2). The distance of the blasting nozzle was set to 100 mm, the angle of the blasting nozzle was 45°.
Beside the blasted substrates, also unblasted substrates were employed for the coating experiments. The following variations of the pretreatment parameters result from the varied paramters:

- Variation 1: untreated
- Varaition 2: EKF 14, 4 bar
- Variation 3: EKF 14, 2 bar
- Variation 4: EKF 80, 4 bar
- Variation 5: EKF 80, 2 bar

2.2 Thermal spray processes

The spray equipment used for the experiments was the twin wire arc spraying device Smart arc by Oerlikon Metco (Pfäffikon, Switzerland). Furthermore, a low-pressure cold-gas device by Dycomet Europe (Akkrum, Nederland) was used for feasibility studies of alternative coating tests. Due to the thermal sensitivity of the vulcanized fibers, low-melting spray materials and adapted spray parameter settings, such as zinc-aluminum (ZnAl2), were initially used as feedstock for the coating tests by means of TWAS. Optimal parameters were used for these investigations, which have been proven to be optimal in previous studies. The parameters are summarized in **Table 1**.

Table 1. Spray parameters used in this study.

Parameter	Value
Stand-off distance	150 mm
Track pitch	5 mm
Primery pressure	2.4 bar
Current	220 A
Voltage	26 V
Gun movement	12,000 mm/min
Number of passes	1

For the coating experiments with the low-pressure cold gas device, copper was used as feedstock. Optimal parameters were also investigated in other studies. The spray distance was set to 10 mm. Further, a pressure of 6 bar and a temperature stage 5 were used. The handling of the spraying device was performed by a indurstrial Robot IRB 4600 by ABB Automation GmbH (Friedberg, Germany). The handling parametres are a gun movement of 2,000 mm/min and a track pitch of 4 mm.

2.3 Surface analysis

One of the main objectives of this study is the analysis of the influence of different preparation parameters on the substrate material with a special focus on the influence of the coating on the vulcanized fiber. Therefore, the substrate topography was investigated in the SEM (scanning electron microscope) after the preparation process as well as after the coating process. For this study, a SEM type JSM 70 1F (Jeol GmbH, Freising, Germany) was employed.

2.4 Non destructive analysis

Eddy current test

The eddy current method was used to examine the coated vulcanized fiber samples. A transmitting coil, which is flowed through by an alternating current, generates a magnetic alternating field and thereby induces eddy currents in electrically conductive materials. These eddy currents themselves generate a magnetic field. The superposition of the two fields is measured by means of a receive coil. Since the coating applied to the vulcanized fiber is electrically conductive in contrast to the vulcanized fiber itself, the layer thickness could be tested well with this method. It was also possible to detect defects such as cracks in the coating moprhology.

Thermography

For the thermography tests, the samples were heated from the backside and the temperature distribution on the surface was recorded by a thermal camera. If delaminations are present in the layer, the heat is transferred worse to the surface of the coating. This allows to detect delamination or inclusions in the microstructure.

3 Results and discussion

3.1 Pretreatment parameter and variations

In order to obtain information on the pretreatment methods and the influence of the coating process on the vulcanized fiber, the specimen were coated by various methods. First, the coating of vulcanized fibers was examined by means of TWAS. Initially, untreated samples with a thickness of 8 mm were used to find out whether it is necessary to pretreat the substrates for the coating process. On these specimens, however, after the arc spraying process had been carried out, no coating adhesion to the entire surface could be realized. **Figure 1** shows a delaminated coating on the untreated substrate. The image shows that detachment during the coating tests occurs mainly in the center of the sample. Only in the edge areas minimal adhesion of the coating materials could be found.

Fig. 1. Delaminated coating of the untreated substrate

An explanation for this phenomenon lies in the separation process of the substrate. The specimen were mechanically cut into their shape. The edge areas of the substrates had slightly protruding fibers caused by the separation process. In contrast, the surface of the sample was not pretreated. This did not reveal any exposed fibers. The surface texture in the edge areas as well as the centre of the specimen leads to the conclusion that a roughened surface is better suited for the bonding of the coating. However, a pretreatment of the surface of the vulcanized fibers results in a partial exposure of individual fibers, as the coating on the substrate can mechanically clamp. Therefore, a pretreatment of the samples is indispensable.
To analyse functional relationships between the surface roughness and the coating adhesion as well as to ensure or even to improve the latter one, the grit blasting parameter settings were varied. During the blasting experiments a special focus was also placed on the influence of the basting procedure on the condition of the vulcanized fibers. Besides a good adhesion an undesired damage of the fiber material should be prevented by the grit blasting procedure.

The differently pretreated samples were coated by means of TWAS. It was found that the layer adhesion was significantly higher for grit blasted samples in comparison to the untreated ones. The former revealed no detachments. However, macroscopically no significant differences of the coatings produced by different pretreatment parameters are recognizable (**Figure 2**).

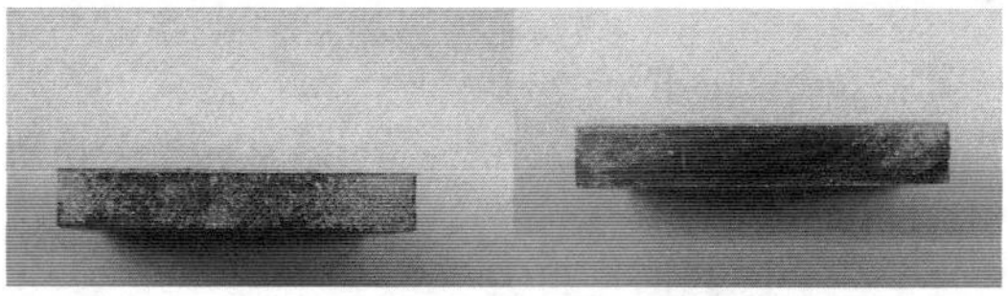

Fig. 2. Coated vulcanized fiber specimen with different pretreatment parameter settings (left: Variation 4; right: Variation 5)

In order to get more information on the bonding mechanism and interface phenomena between coating and substrate material with regard to the roughness profile, the samples were examined by means of scanning electron microscopy (SEM) prior to the coating operation. It was found that a pretreatment with large corundum particles and high pressures result in a severe surface damage, whereby individual fibers of the substrate material protrude from the surface area into the applied layer. On the one hand, this supports the bonding and adhesion of the layer, but on the other hand, the erosion of the surface can lead to a weakening of the substrate.

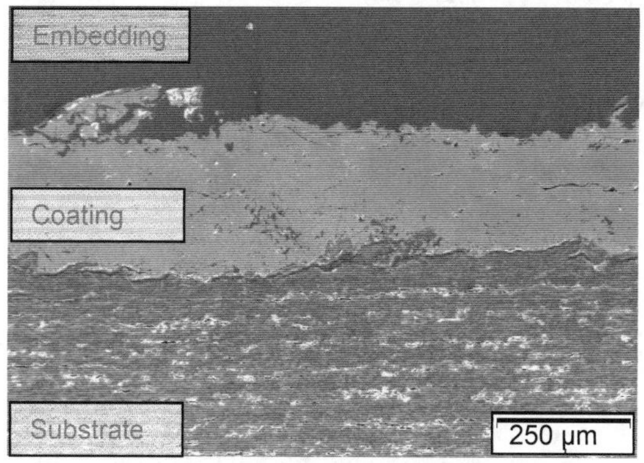

Fig. 3. Coated vulcanized fiber by use of variation 4

Figure 3 shows that a pretreatment with parameter variation 4 (EKF 14, 4 bar) leads to a strong erosion of the surface combined with a breakage of the fibers. It can be seen that by the pretreatment process many fibers were protruded. However, at the same time a good interface bonding between coating and vulcanized fiber surface can be achieved.

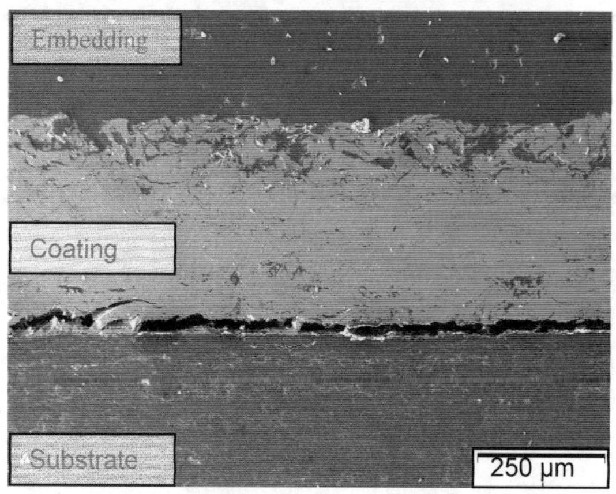

Fig. 4. Coated vulcanized fiber by use of variation 5

Figure 4 shows the interface between coating and substrate materials, pretreatment with the blasting parameters of variation 5 (EKF 80, 2 bar). In contrast to pretreatment parameters of variation 4, the

parameters of variation 5 do not provide an uniformly adhering coating. Furthermore, it can be recognized that a pretreatment with the parameter variation 5 did not protrude as much fibers as the variation 5 and did not destroy the surface.
From this it can be concluded that a certain surface pretreatment is required to ensure a good bonding of the layer. On the other hand, the parametersetting of the pretreatment process should not be too strong that the surface of the substrate is destroyed or weakened. These effects occur, for example, with coarse corundum and high pressures of the pretreatment process.
In order to achieve a continuous connection, the grit blasting parameters were further varied. On the one hand by increasing the blasting pressure in terms of using smaller corundum fractions (Variation 2) and, on the other hand, reducing the pressure by use of corundum with a larger grain size (Variation 1). Both variations provide a good connection between coating and substrate.
Due to the increased surface destruction and the associated exposure of individual fibers from the substrate material, it was found that a pretreatment with EKF 80 and a pressure of 4 bar ensured a sufficient adhesionof the coating with minimal surface damage.

After the coating tests by means of TWAS and the examination of the pretreatment parameters, further coating tests were carried out by means of LP-CS spraying. The optimum pretreatment parameters (EKF 80, 2bar) previously determined were used.
The coating experiments using LP-CS spraying have shown interesting results regarding to the thermal sensitivity of vulcanized fibers. Various thermally induced phenomena were observed. During the spray experiments no continuous layers could be applied. However, several alterations of the substrate's surface topography could be observed. On the one hand, the presence of carbon black on the substrate surface. On the other hand, partial bubble formation was found on the coated surface.

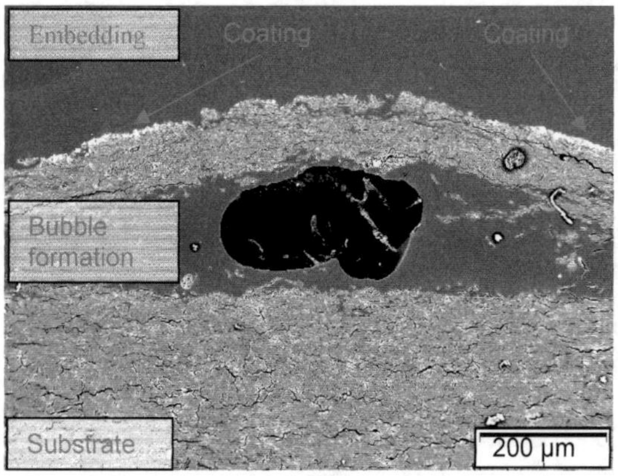

Fig. 5. Copper coated specimen with delamination in the vulcanized fiber

Figure 5 shows a SEM image of the coated specimen with a bubble formation on the surface. A minimal layer adhesion on the outside of the bubble is visible. A continuous layer could not be created.
Furthermore, it is found that the bubble is located in the substrate material. The individual layers of the vulcanized fiber have partially detached. One possible explanation for the bubble formation is the moisture which is present in the substrate material. Due to the heat influence of the coating process, parts of the residual moisture evaporate and have formed the shown bubble by the associated expansion.
Possible approaches to prevent the formation of bubbles are for example the reduction of the heat input into the sample. In order to achieve this, the spray or handling parameter settings as the spray distance or gun movement should be adapted. A further option is a drying process of the vulcanized fibers ahead of the coating process.

To conclude the investigations, the influence of the substrate thickness on the coating ability of the vulcanized fibers was investigated. In addition to the 8 mm samples, further ones were coated with 2 mm thickness. In these tests as well, the previously determined optimal preparation parameters (EKF 80, 4 bar) were used.

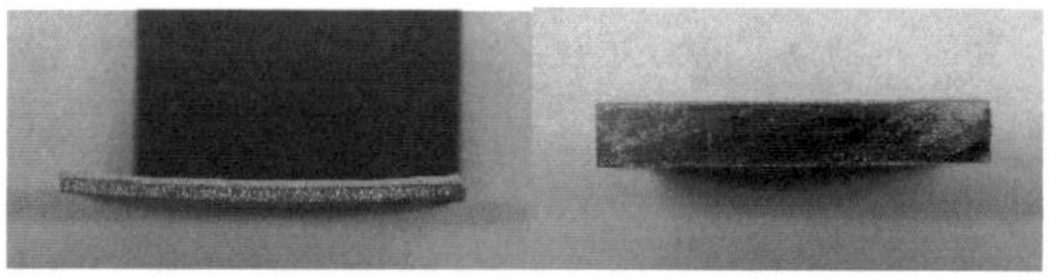

Fig. 6. Comparison of the coated specimen with different thickness (left: 2 mm; right: 8 mm)

Figure 6 shows a comparison between the coated specimen with different thickness. A deformation of the 2 mm sample caused by the coating process is noticeable. A variation of the layer thickness by increasing of the number of passes showed a worsening of the deformation. From this it can be concluded that by applying the coatings, thermal stresses are induced in the component, which can lead to deformations from a lesser thickness of the component. The stresses on the surface of the vulcanized fibers result of the different thermal mismatch in the expansion of the materials. The different thickness leads to the fact that in case of the 2 mm specimen, smaller forces are required to deform the material.
Furthermore, the vulcanized fiber is a polymer which has temperature-dependent strengths. The heat input reduces the strength of the material, which means that small forces are sufficient to deform the material.

3.2 Non-destructive analysis

Following the investigation of the influence of the different pretreatment parameters on the layer bonding of the ZnAl2 layer to the vulcanized fibers, non-destructive methods were applied to analyse the layer. For this purpose, the eddy current test method was used, which could provide reliable information on the layer thickness due to the electrical conductivity of the applied layer [5]. In addition to the eddy current test, thermography tests were continued. **Figure 7** shows the evaluation of a coated vulcanized fiber sample.

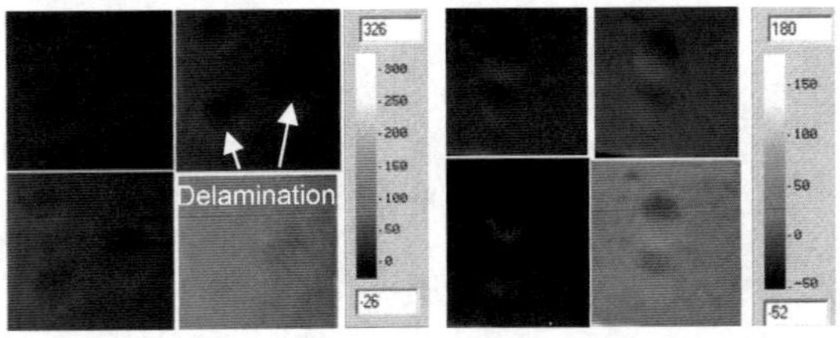

Fig. 7. Thermography analysis of coated vulcanized fiber samples showing delamination [5]

It was found that the thermography method is suitable for measuring on the one hand the layer thicknesses and on the other hand detecting defects in the coating. In the context of other research, defects in the coating, such as bubbles or inclusions, were detected in the surface [5].

4 Conclusion

The coating experiments showed that it is possible to apply coatings by mean of TWAS process onto the thermally sensitive vulcanized fiber. Due to the functionalisability of the surface, the vulcanized fiber represents an ecological alternative as construction material in the field of lightweight construction. Within the scope of this work, it was shown that a corresponding pretreatment of the substrate is required in order to ensure an adequate connection of the coating to the substrate. However, extreme pretreatment parameters which provide an intense roughness profile (e.g. coarse corundum and high pressures), lead to a severe damage to the specimen surface. A pretreatment with fine corundum (EKF 80) and pressures of 4 bar have proven to be suitable parameters in order to enable a good bonding of the coating and, on the other hand, to minimize the damages of the substrates surface.

Furthermore, the coating by means of LP-CS spraying has shown that the heat input can have an influence on the dimensional stability of the component and a certain thickness of the material is necessary in order to counteract any stresses which are formed during coating. It was found that in addition to the pretreatment parameters, the residual moisture in the material is of great importance to avoid delaminations and bubbles on the components surface. In addition, it has been observed that the coating process results in stresses in the component surface, which can lead to deformation if the thickness is too low.

To obtain more information about the coatability of vulcanized fibers, further experiments by means of other thermal spray processes, such as HVOF spraying process (high-velocity oxygen fuel) should be performed. The aim of these investigations is to apply an application-oriented coating (for example a wear-resistant coating). The thermal influence on the substrate surface will also be of great importance with these methods or materials.

5 Literature

[1] M. Joseph: Vulkanfiber, Renaissance eines alten Werkstoffes. naro.tech, Internationales Symposium, Werkstoffe aus Nachwachsenden Rohstoffen, 3, International Congress (2001)

[2] Mittendorf, R.: Entwicklung einer Richtlinie für die Auslegung von Direktschraubverbindungen in Vulkanfiber. Dissertation, TU-Dortmund, 2016

[3] Sievers, N., Tillmann, W., Becker, S., Rademacher, H.-G. and R. Zielke: 3D-Terahertz - Eine Möglichkeit zur zerstörungsfreien Prüfung von Vulkanfiber. DGZfP-Jahrestagung, Berichtsband 141, Dresden (2013)

[4] Tillmann, W., Sievers, N., Zielke, R., Rademacher, H.-G., Becker, S., Steinhausen, R. and M. Kiel: Welche Verfahren sich zur Prüfung von Vulkanfiber eignen – Anspruchsvolle Analyse. Carl Hanser Verlag, München QZ Jahrgang 59 (2014), pp. 64-67

[5] Zielke, R. Tillmann, W. Kuck, M. Rademacher, H.-G. Nellesen, J. Brinkhoff A. and I. Baumann: Charakterisierung von Vulkanfiberbauteilen mit Hilfe zerstörungsfreier Prüfverfahren. Tagung Werkstoffprüfung, Neu-Ulm (2016), pp. 201-206

[6] DIN 7737: Schichtpreßstoff-Erzeugnisse; Vulkanfiber, Typen. Beuth Verlag, Berlin (1959)

[7] Penning, B., Walther, F., Dumke, D. and B. Künne: Einfluss der Verformungsgeschwindigkeit und des Feuchtegehaltes auf das quasistatische Verformungsverhalten technischer Vulkanfiber. Materials Testing 55 (2013), No. 4, pp. 276-284

The influence of post treatment presence on high temperature corrosion resistance of YSZ thermal barrier coatings

Mrs. Alsu Akhmetgareeva, TSPC Ltd., Moscow, Russia

Dr. Sergey Baldaev, TSPC Ltd., Moscow, Russia

Dr. Ivan Mazilin, TSPC Ltd., Moscow, Russia

Dr. Lev Baldaev, TSPC Ltd., Moscow, Russia

Dr. Tat'yana Il'inkova, Kazan National Research Technical University named after A.N.Tupolev - KAI (KNRTU-KAI), Kazan, Russia

Abstract.

Well known that high temperature corrosion resistance of thermal barrier coatings (TBC) may depend not only on chemical composition, thickness, the rate of formation of TGO layer, but also on the presence of post treatment.

In this work we studied the influence of presence of post treatment on properties of TBC particularly on high temperature corrosion. TBC system were sprayed on Ni-based substrate and consists of HVOF NiCo-based bond coat (BC) and APS yttria-stabilized zirconia (YSZ) top coat. Thermal cycling test in the presence of corrosive environment was used to determine effect of post treatment presence. Before and after Thermal cycling test cross-section of specimens were studied by SEM and EDX analyzes. Preliminary results demonstrated that high temperature corrosion resistance of TBC with APS-top layer and HVOF-bond layer can be improved by applying post treatment.

Investigation of Microstructure and Mechanical Properties of Alumina-Titania Coatings by Flexicord Flame Spraying

[1]**Ekrem ALTUNCU**, [2]Fatih USTEL

[1]altuncu@sakarya.edu.tr

[1,2] Sakarya University, Thermal Spray Technology R&D Laboratory, Department of Metallurgy and Materials Engineering, Esentepe Campus, Serdivan, 54187 Sakarya, Turkey

Abstract:

Thermally sprayed Al_2O_3-TiO_2 coatings were in the area of interest over the last decade because they showed improved wear properties over conventional coatings. In this study, flexicord flame spray gun was used to deposit Al_2O_3-TiO_2 coatings at different spray parameters. The microstructural morphology variation and phase transformation of coatings were investigated. In addition, as one of the most important properties for ceramic coatings, hardness, solid particle abrasive wear resistance of coatings were measured before and after heat treated condition. Test results show that the higher mechanical properties and wear resistance by the heat treatment at elevated temperatures.

Keywords: Thermal Spray, Alumina-Titania, Flexicord feed stock, Wear

1. Introduction

Thermal spray coatings are versatile in today's industrial applications. Coating materials in the form of powder or wire can be deposited at the surface of the metallic substrate materials at high speed with different spray guns. Coating quality varies depending on spray parameters and coating material properties. The production of ceramic based coatings obtained by the plasma spray coating method is costly and requires high investment. Coatings obtained with the HVOF (high velocity oxy fuel) process have higher adhesion strength than plasma spray coatings, although they have lower porosity, but considerable gas consumption during deposition. The melting behavior of powder particles during the coating process significantly affects the coating properties and microstructure. Powder size, size distribution, powder shape and production method, as well as the deposition parameters controlling the deposition efficiency. Thermal spray flexicords are an alternative solution to produce high quality ceramic, carbide and metallic thermal sprayed coatings. Flexicord spraying technology does not compete against plasma spraying or HVOF and this complementary thermal spray technology provides different coating benefits with faster setup of the spray operation.

Solid particle erosion/abrasion at extreme conditions (high hardness and strength of abrasives and materials to be ground, high velocity and pressure, cyclic impact load, elevated temperatures, etc.) is a serious problem for industrial equipment, for example milling and mixing devices. To reduce the damaging effects of erosion, various surface modification techniques are utilized. These techniques provide low cost solution to deal with such degradation. Among these techniques, thermal spraying has attracted considerable attention worldwide owing to its versatile nature. A relationship between erosion wear and coating characteristics is not always found, but in general, one expects a decrease of erosion resistance with the increase in porosity and discontinuities. Thermally-sprayed alumina based materials, e.g., alumina-titania (Al_2O_3-TiO_2), are commonly applied as wear resistant coatings in industrial applications. alumina-titania (Al_2O_3-TiO_2) materials are widely used as thermally-sprayed protective coatings against abrasion, erosion, and cavitation erosion wear, e.g., in textile, pulp and paper, and pump industries. The melting temperature of (Al_2O3-TiO_2) is approximately 2000 °C, which makes it possible to be sprayed also by combustion based spray processes, e.g., flame spraying, HVOF spraying and plasma spraying [1-4]. In this study flexicord thermal sprayed (**Fig1**.) alumina-titania coatings wear behaviour with heat treated conditions are investigated.

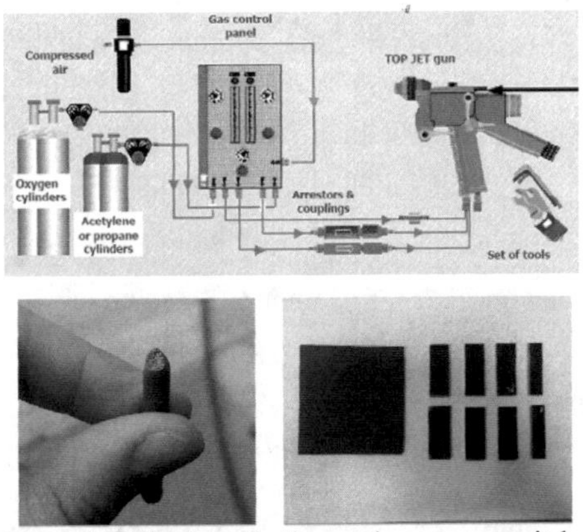

Fig. 1.a.Flexicord thermal spray unit.b. feedstock cord, c. samples

2. Experimental Techniques

Flame spray master jet equipment (spray gun) is very versatile and ideal for high quality coatings. Master jet (Saint Gobain) allows ceramic and carbide based coatings for small and large part surfaces.

4.75mm diameter flexicord material (AT40:%60 Alumina-%40 Titania, AT3: Alumina-%3 Titania) was used in this study. Alumina-Titania fine powders are compressed into a polymer sheath. During spraying, the polymer sheath melts and the powders perform a homogeneous melting. The powder scattering during spraying has been substantially reduced. A double layer coating layer was obtained. After the bonding surface, a bond coat (NiCr) is applied to the base surface (304 stainless steel palte 3mm thickness) and then the ceramic layer is coated.

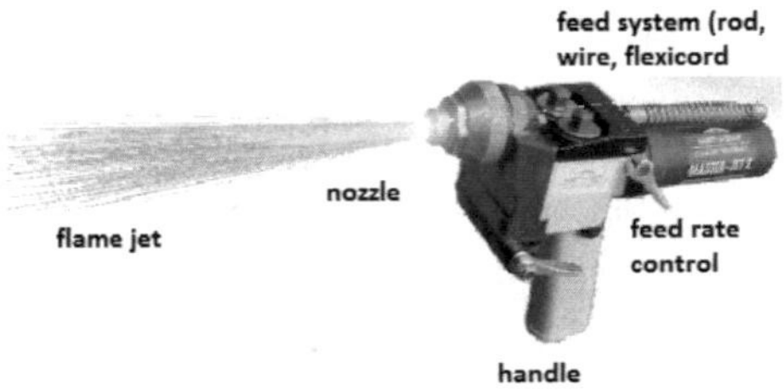

Fig. 2. Masterjet gun for flexicord spraying

The basic spray parameters in the FlexiCord flame spray process are the gas pressure, feed rate, distance to the surface. Spray parameters of supplier company were used for coating parameters. The spray gun parameters are shown in **Table 1.**

Table 1. Masterjet spray gun parameters

Parameters	Value
Asetilen/Oksijen; bar/bar	1.4./ 4
Spray distance; mm	100
Feed rate; kg/hr	1,25
Cord speed; cm/min	60

Wear test was performed on the produced coating samples according to ASTMG65 test standard (**Fig.3.**) at room temperature and at high temperatures after heat treatment. Weight loss and wear rate were calculated. Top surface was observed before and after wear with electron microscopy (SEM). The hardness changes of the coatings were determined by the Vickers hardness ($HV_{0,3}$) method depending on the heat treatment condition (at 400,600 °C for 24 hr). The average value was calculated by taking 6 measurements for each sample.

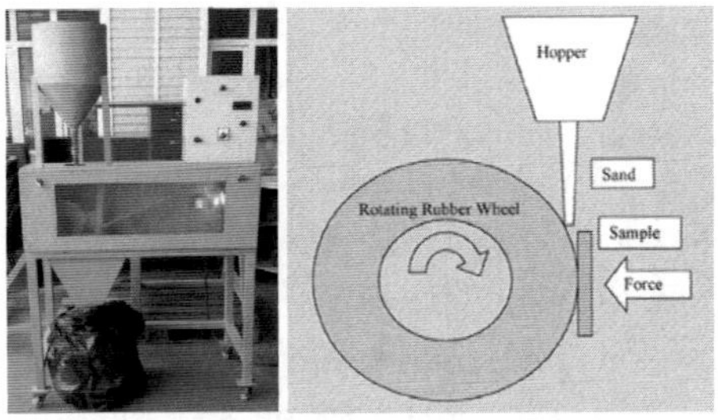

Fig.3. ASTMG65 test system

3. Results and Discussion

The coatings are produced manually and the total coating thicknesses are 150±25 µm. Deposition efficiency is %45. The microstructural features of the resulting coatings are lamellar and heterogeneous as observed in typical thermal spray coatings. There are non-melted particles, micro cracks and pores in the coating section (**Fig4.**). Surface roughness Ra about 5.5 µm. Microhardness of the coating is at the 900±30 $HV_{0.3}$ level.

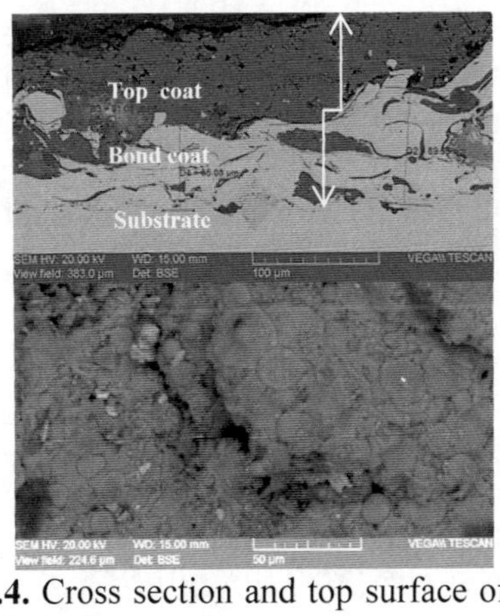

Fig.4. Cross section and top surface of the coating micrograph

Figure 5 shows the wear track of the coatings after wear test at ambient temperature.

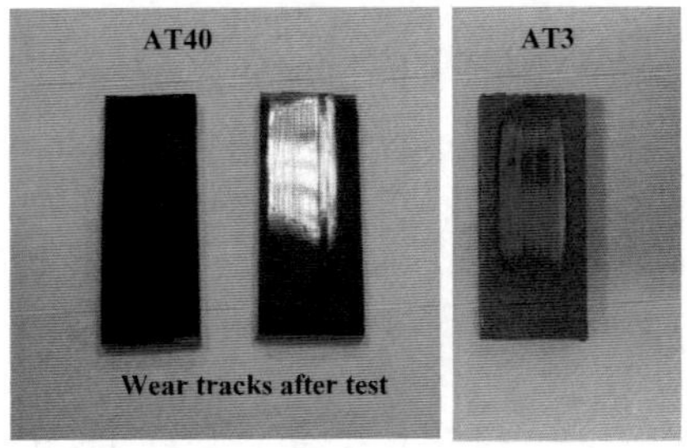

Fig.5. Wear tracks at room temperature

The wear performance results of the coated specimens subjected to different heat treatment conditions are presented in **Fig. 6.** It is observed that the wear resistance of the coating increases as compared to the base material. As a result of the heat treatment, there is a marked increase in the wear resistance of the coating as a result of the decrease in the microstructural discontinuities (as porosity) of the coating and the increase in microhardness.

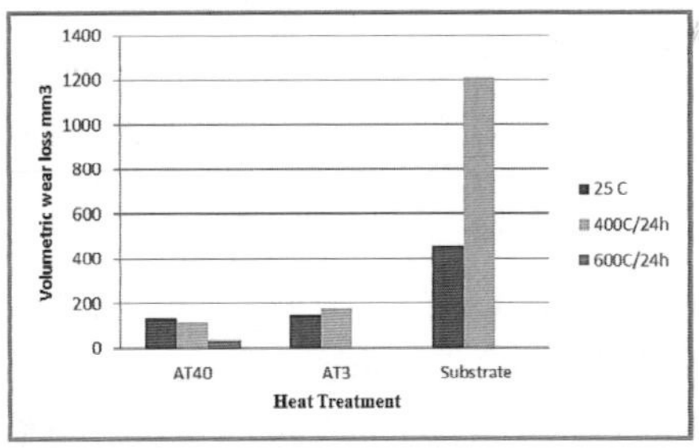

Fig. 6. Volumetric wear loss comparison at different HT condition

With the increase in heat treatment temperature, the coating microhardness increased from 900±30 HV to 1250±50 HV0.3. Porosity determination with image analysis has been observed in post-heat treatment coatings in a considerably reduced porosity. The porosity ratio decreased from 7% to 3%. XRD analysis was performed before and after the heat treatment in the material characterization phase (**Fig 7**). It has been found that Al_2TiO_5 hard phases are formed as a result of the phase transformation in the coating microstructure due to the heat treatment applied to the coatings. The molten Al_2O_3 and TiO_2 phase combined into the solid solution $Al_2O3 \cdot TiO_2$ (Al_2TiO_5) phase under high temperature during spraying or heat treatement. The higher the content of TiO_2, the larger the amount of Al_2TiO_5 will be generated. The presence of these phases played an important role in the increase of wear resistance. The formation of the Al_2TiO_5 phase causes the wear resistance to improve for coatings operating at high temperatures.

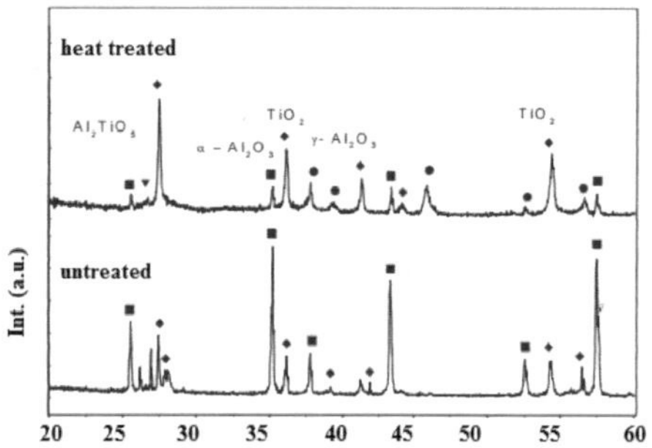

Fig.7. XRD pattern of coatings

Aluminium titanate Al_2TiO_5, is known to be a promising candidate material in refractory and engine components because of its low thermal expansion, excellent thermal shock resistance, and low thermal conductivity.

4. Conclusions

Flexicord flame spray coatings are outperformed all other coatings in the wear tests. Higher velocity and small particle size resulted in dense and fine microstructure, which provided the coatings with high wear resistance. The effect of porosity and phase type are significant and needs to be accounted for in erosion wear. When there is high porosity content at intersplat surfaces, its effect is superimposed in the cohesion effect. Wear test results indicated that wear resistance of the coatings at elevated temperature is better than that in ambient conditions.

Acknowlodgements

The authors thank Sakarya University Scientific Project (BAP project number: 2013-09-08-003) Department for financial support of this work. Authors would also like to thank TESLAB workers for helping with the sample preparation and testing. he formation of the Al_2TiO_5 phase during the coating production and the subsequent increase in the heat treatment have significantly affected the wear resistance of the coating. For this reason the AT40 wear resistance is higher than AT3.

References

[1]Ahn, J.; Hwang, B.; Song, E.P.; Lee, S.; Kim, N.J. Correlation of microstructure and wear resistance of Al_2O_3-TiO_2 coatings plasma sprayed with nanopowders. Met. Mater. Trans. A **2006**, 37, 1851–1861.

[2]Wang, M.; Shaw, L.L. Effects of the powder manufacturing method on microstructure and wear performance of plasma sprayed alumina-titania coatings. Surf. Coatings Technol. **2007**, 202, 34–44.

[3] Lima, R.S.; Moreau, C.; Marple, B.R. HVOF-sprayed coatings engineered from mixtures of nanostructured and submicron Al_2O_3-TiO_2 powders: An enhanced wear performance. J. Therm. Spray Technol. **2007**, 16, 866–872.

[4] Matikaine V., Niemi K., Koivuluoto H. Vuoristo P., Abrasion, Erosion and Cavitation Erosion Wear Properties of Thermally Sprayed Alumina Based Coatings, Coatings **2014**, 4, 18-36.

Evaluation of Low-pressure Cold Spray MMC Coatings by Acoustic Emission-coupled Four-point Flexural Testing

Y.T.R. Lee[a], M. Shibly[b], T. Hussain[b], G. Fisher[c], A. McDonald[a]
[a]University of Alberta, Edmonton / Alberta / Canada
[b]University of Nottingham, Nottinghan / United Kingdom
[c]Innotech Alberta, Edmonton / Alberta / Canada

Low-pressure cold spray has been used as an innovative method to deposit metal matrix composite (MMC) coatings: boron carbide-nickel (B_4C-Ni) and tungsten carbide-cobalt-nickel (WC-Co-Ni) composites. The coatings were studied using scanning electron microscopy, X-ray diffraction with Rietveld refinement, and acoustic emission-coupled four-point flexural test. Indentation fracture toughness tests were performed on the WC-Co-Ni coatings, only. The results showed that the composites had reinforcing particle volume fractions of 45.8 ± 0.3 vol.% and 22.7 ± 0.1 vol.% for the WC-Co-Ni and B_4C-Ni MMC coatings, respectively. Flexural tests were used to evaluate the fracture strain of the composites. In these tests, the WC-Co-Ni composite failed by brittle facture at approximately 0.5% nominal strain. The B_4C-Ni composite showed flexural behaviour similar to that of an unreinforced Ni matrix. These results suggest that there was insufficient B_4C within the coating to affect significantly the ductile failure mode of Ni matrix. Post bending fracture analysis showed the presence of straight, continuous cracks on the WC-Co-Ni surface and the indentation fracture toughness of WC-Co-Ni was found to be 1.2 ± 0.2 MPa·$m^{0.5}$. Discontinuous, random cracks were observed on the B_4C-Ni surface. The quantification of these properties is essential in evaluating the performance of the low-pressure cold spray coatings to determine their potential applications.

1 Introduction

Metal matrix composites (MMCs) are advanced engineered materials with tailorable properties to serve in specific industrial applications. One of the most common uses of MMCs is as wear resistant coatings [1 – 4]. The combination of a tough ductile matrix and the high hardness of the reinforcing particles results in high resistance to wear for the MMC coatings [1, 4]. Conventional deposition techniques for these MMC coatings typically involves thermal spraying. Techniques such as high-velocity oxy-fuel (HVOF) spraying and plasma spraying have been successful in the deposition of WC-based MMC coatings [5, 6]. However, these processes typically require high temperatures, where decarburization or phase change of the material can occur [7, 8]. Decarburization of the deposited material can reduce the performance of these coatings, specifically reducing their wear resistance [8, 9].

Low-pressure cold spraying has been explored as an alternative deposition technique for MMC coatings. In low-pressure cold spraying, small particles (1 - 50 µm) are accelerated to impact upon a surface at high velocities (300 – 1200 m/s) [10]. The high velocity and momentum of the particles induces plastic deformation upon impact, causing mechanical interlocking of the impacting particles with the substrate. As such, cold spraying can be used to fabricate coatings at temperatures that are well below the melting temperature of the feedstock materials. Thus, undesirable phase changes can be minimized or eliminated [11].

Melendez, et al. [1, 11] successfully created WC-Co-Ni MMC coatings using low-pressure cold spraying. X-ray diffraction (XRD) analysis showed that the cold-sprayed WC-Co-Ni coatings experienced no decarburization. Measurement of the hardness and wear resistance indicated that the cold-sprayed WC-Co-Ni coating had lower hardness than the HVOF-sprayed WC-based coatings, but had comparable abrasive wear resistance. Lee et al. [12] also successfully created MMC coatings using low-pressure cold spraying. The coatings created were B_4C-based. The B_4C-Ni coatings were found to have slightly lower hardness than that of the cold-sprayed WC-Co-Ni coatings, but had much higher wear rates than the WC-Co-Ni coatings [12]. These differences between the two coatings can be partially attributed to the difference in carbide content between the coatings, but further analysis of the fracture behaviour of these coatings was not conducted.

The objectives of this study were to: a) investigate the fracture behaviour of low-pressure cold-sprayed WC-Co-Ni and B_4C-Ni coatings by using acoustic emission-coupled four-point flexural testing and b) compare the fracture behaviour of the low-pressure cold-sprayed coatings to that of HVOF-sprayed WC-based coatings that are currently in use in industry.

2 Experimental Method

2.1 Powder and Substrate

Three powders were used in this study. The B_4C powder (Panadyne, Montgomeryville, PA, USA) was fused and crushed, resulting in a highly angular structure. The WC-Co powder (72F-NS, Sulzer Metco, Westbury, NY, USA) was a sintered and crushed powder with 12 wt.% Co and 88 wt.% WC. The Ni powder (N5001, Centerline, Windsor, ON, Canada) was specially designed for cold spray and had a highly dendritic structure to promote adhesion. The morphology of these powders is shown in Fig. 1. The Ni powder was mechanically blended with the carbide powders to create a composite powder feed stock. The composition of the mechanical blends are shown in Table 1. These powder blends were cold-sprayed onto low carbon steel substrates for characterization and on titanium alloy (Ti6Al4V) substrates for flexural testing. The titanium alloy substrate was selected to maximize the nominal strain before plastic deformation of the substrate. The substrates for flexural testing were 120

Fig. 1. Morphology of (a) Ni, (b) B_4C, and (c) WC-Co powders

Table 1. Compositions of the powder blends.

Powder blend	Powder (wt.%)		
	B_4C	WC-Co	Ni
Ni	-	-	100
92 wt.% B_4C + 8 wt.% Ni	92	-	8
96 wt.% WC-Co + 4 wt.% Ni	-	96	4

mm x 19 mm x 7 mm. All substrates were grit-blasted with alumina grit #24 at 551.6 kPa (80 psig) to promote adhesion of the coating.

2.2 Cold Spray Deposition

A low-pressure cold spray nozzle (SST series P, CenterLine Ltd., Windsor, ON, Canada) was used to fabricate the MMC coatings. An industrial robot (HP-20, Motoman, Yaskawa Electric Corp., Waukegan, IL, USA) was used for automatic control of the deposition process. Compressed air at a pressure of 634 kPa (92 psig) and 550°C was used as the working fluid. The stand-off-distance was set to 5 mm and the traverse speed of the nozzle was 5 mm/s. For the powder blend containing B_4C, the traverse speed of the nozzle was decreased to 2.5 mm/s to improve deposition.

2.3 Coating Characterization

The as-sprayed coatings were cut into cross sections and cold-mounted into epoxy resin. The coatings were then ground and polished for microscopy and image analysis. Cross sections of the coatings after flexural testing were also analysed. The cross sections were taken from the area that was subjected to the maximum bending moment, then the samples were mounted, ground, and polished.

Optical microscopy (OM) and scanning electron microscopy (SEM) were used to investigate the overall microstructure of the coating. Image analysis (ImagePro, Media Cybernetics, Bethesda, MD, USA) was conducted on a minimum of five images, with an approximate cross sectional area of 300 µm x 200 µm, to investigate the carbide loading within the coating. OM and SEM images were also captured of the coating after bending. These images were used to investigate crack propagation within the coatings.

X-ray diffraction (XRD) was used to identify any phase change that had occurred during the cold spraying process. XRD was preformed on the fabricated coatings prior to flexural testing. The coatings were characterized using a copper anode with a point detector arranged in a Bragg-Brentano geometry and a nickel-beta filter was used to minimize the copper K-beta wavelengths. The current was fixed at 25 mA and a voltage of 40 kV. The 2θ angle was changed from 10° to 90° at a rate of 0.75°/min. Rietveld refinement was also conducted on the coating XRD patterns to determine the weight percentage of the phases present in the materials.

2.4 Vickers Microhardness

A Vickers microhardness tester (Buehler, Lake Buff, IL, USA) was used to measure the microhardness of the coatings. A total of ten indentations were conducted at 300 gf with a dwell time of ten seconds. A spacing equivalent to four indentation diagonals was placed between each indentation.

2.5 Acoustic Emission-coupled Flexural Testing

Four-point bending was conducted on an universal testing system (Instron 5969, Norwood, MA, USA). The four-point fixture had an outer span of 64 mm and an inner span of 32 mm. The as-sprayed coatings were prepared for these flexural testing by grinding the coating down to an approximate coating thickness of 200 µm. The coatings were then loaded in tension in the four-point bending fixture. The mid-point deflection was measured with a linear variable differential transformer (LVDT). The percentage nominal strain was limited to values lower than 0.8% to prevent plastic deformation in the substrate.

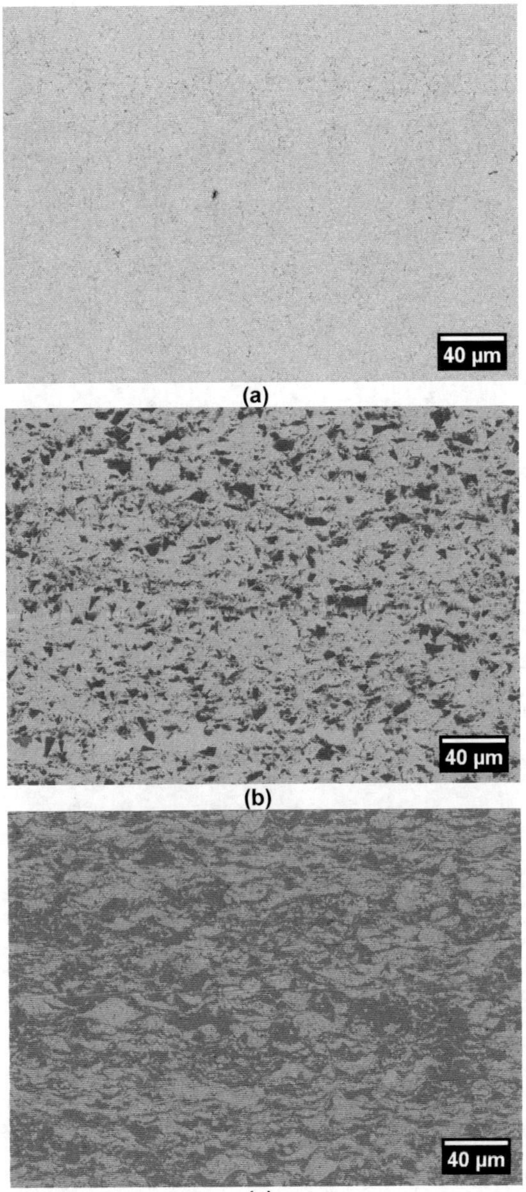

Fig. 2. Microstructure of the fabricated coatings of (a) Ni, (b) B_4C-Ni, and (c) WC-Co-Ni

Table 2. Content of B_4C and WC particles in the fabricated coatings

	Powder Blend		Coating	
	vol.%	wt.%	vol.%	wt.%
B_4C-Ni	97.6	92	22.7 ± 0.1	7.7 ± 0.04
WC-Co-Ni	93.4	96	45.8 ± 0.3	58.7 ± 0.3

Two acoustic transducers (Physical Acoustics Corporation, Princeton Junction, NJ, USA) were used to gather the acoustic emission during fracture of the coatings. The two acoustic sensors were set to different decibel ranges in order to gather as much data as possible. The first sensor was set between 45 and 100 dB, while the second sensor was set between 70 and 140 dB. The data showed significantly more detail for the first sensor and thus the analysis was conducted using only the data from that first sensor.

2.6 Indentation Fracture Toughness

The indentation fracture toughness of the coatings was investigated using the Evan and Wilshaw model [13]. This model relies on the applied load and crack dimensions from a Vickers macroindentation. The equation from the Evan and Wilshaw model is as follows:

$$K_C = 0.079\left(\frac{P}{a^{1.5}}\right)\log\left(\frac{4.5a}{c}\right), \qquad (1)$$

where, K_C is the indentation fracture toughness (in $MPa \cdot m^{0.5}$), P is the applied load (in N), a is the half diagonal of the indent (in µm), and c is the crack length from the centre of the indent (in µm). Ten indents were produced from a 10 kgf load. The cracks that were formed parallel to the coating-substrate interface were measured.

3 Results and Discussion

3.1 Coating Carbide Content

Image analysis was used to estimate the volume fraction of carbide particles within the coatings. Typical images that were used for carbide content analysis are shown in Fig. 2. Table 2 shows the coating carbide content for both B_4C and WC-Co. The standard error of the mean is shown. The carbide content was converted from volume percent to weight percent using the following densities: 8.92 g/cm^3 for Ni; 2.52 g/cm^3 for B_4C; and 14.97 g/cm^3 for WC-Co. The results in Table 2 show that despite starting with similar carbide volume fractions, the WC-Co coatings have a much higher carbide retention within the coating.

The higher carbide content in the WC-Co-based coating indicates a higher deposition efficiency for the WC-Co powder. The higher deposition efficiency may be due to the composite nature of the WC-Co powder, which allows it to have higher deformability than the pure ceramic B_4C powder. Evidence of the higher deformability of the WC-Co particle can be seen in Fig. 2c. The WC-Co particles appear to be flattened, suggesting that particle deformation occurred upon impact. Since cold spray deposition relies on the deformation of the particles and the substrate, WC-Co has a higher deposition efficiency.

3.2 X-ray Diffraction

X-ray diffraction (XRD) was used to confirm that no phase change occurred during the cold spray deposition process. Figure 3 shows the XRD patterns of the coatings. In the figure, the blue curve is the data that was obtained from the XRD and the red curve is the XRD profile based on the Rietveld refinement. The difference between these two curves is the grey curve at the bottom of the graph. Additionally, the weight percentages of the phases based on Rietveld refinement are shown on the left of the profiles.

There was minimal phase change during the cold spray deposition process. Figure 3a shows that marginal oxidation occurred within the Ni coating. The B_4C-Ni coatings show only B_4C phases and Ni phases, indicating that there was no significant phase change. Additionally, the WC-Co-Ni coating also only showed WC and Ni phases. The broad peak at about 37° may be attributed to some contamination on the sample.

The quantitative phase analysis from Reitveld refinement shows a different carbide content from that obtained from image analysis. The variation may be due to the orientation of the analysed surface. In the case of the Reitveld refinement, the analysed surface was parallel to the substrate surface, while for the image analysis, the study was conducted on a cross section of the coating. This difference was more prominent in the WC-Ci-Ni coatings. The WC content, as obtained from Reitveld refinement (87 wt.%), was much higher in the WC content obtained from image analysis (58 wt.%). The deformation of the WC-Co particles upon impact created splats that greatly changed the aspect ratio of the WC-Co particles within the coating. As such, the impacting surface and the cross section of the WC-Co-Ni coating have different coverage areas of WC. Grinding the top surface down to 200 µm for the flexural test may also have increased the apparent WC content in the coating as the abrasive may have preferentially removed Ni from the surface. The B_4C particles did not deform and did not form splats within the MMC coating; therefore, the difference between the impacting surface and the cross section was less significant. Thus, the B_4C content, as obtained from the two measurement techniques (7.7 wt.% and 9.7 wt.%), was close.

3.3 Vickers Microhardness

The higher carbide content in the WC-Co-based coating can also have an impact on the microhardness of the coatings. The WC-Co coatings have the highest hardness, as shown in Table 3. The results show that the carbide particles have a large reinforcing effect on the coating and greatly increase the hardness of the composite. Overall, the high volume fraction of carbide particles in the WC-Co-based coating results in the MMC with the highest microhardness. This high microhardness is due to the close proximity of the WC particles, or low mean free path between reinforcing particles, which allows the reinforcing particles to load share and increase the hardness of the coating [11, 12, 14].

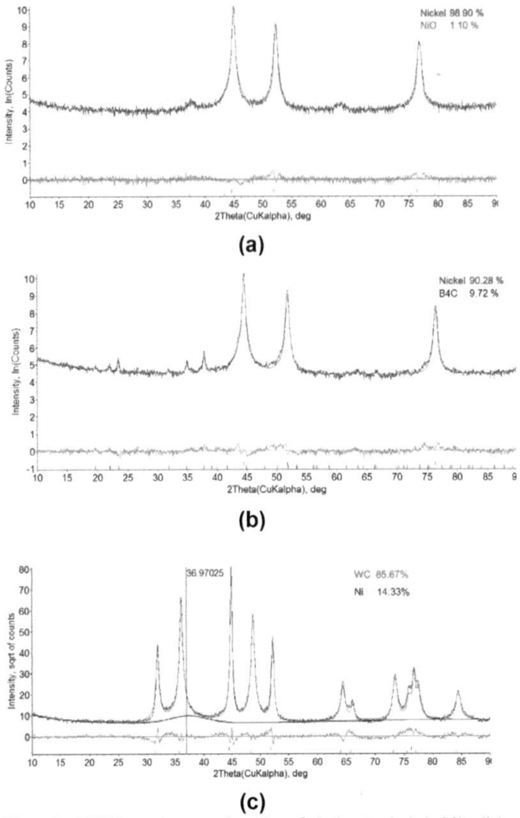

Fig. 3. XRD patterns for the fabricated (a) Ni, (b) B_4C-Ni, and (c) WC-Co-Ni coatings

Table 3. Microhardness of the coatings

Coating	Microhardness (kgf/mm^2)
Ni	116.3 ± 3.5
B_4C-Ni	377.0 ± 8.4
WC-Co-Ni	556.7 ± 9.0

3.4 Acoustic Emission-coupled Flexural Testing

Four-point bending was used to investigate the fracture behaviour of the coatings. The acoustic emission data from the tests was collected and compiled in order to calculate a cumulative absolute energy. Figure 4 shows the total acoustic energy that was released during the flexural tests. The maximum energy for both the Ni and B_4C-Ni coatings were on the order of 10^4 aJ, the same order of magnitude as the uncoated substrate. This similarity suggests that the B_4C-Ni coatings have a fracture behaviour similar to that of the unreinforced coatings and are not significantly different from the uncoated substrate. The small amount of energy that was measured could be the result of the support cylinders settling on the samples.

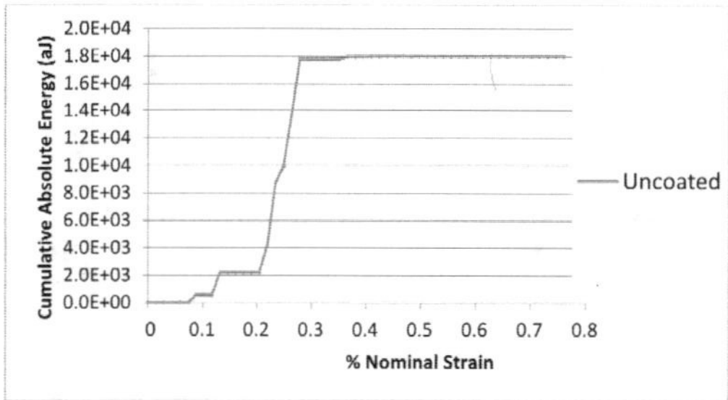

(a)

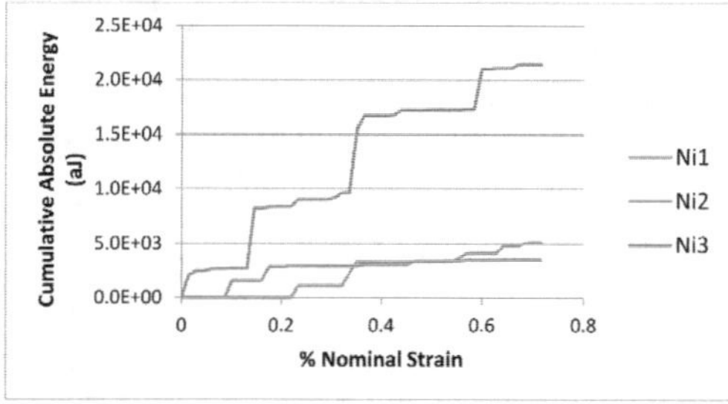

(b)

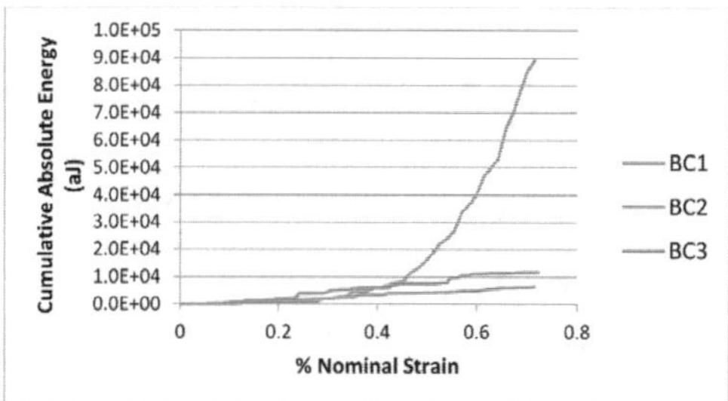

(c)

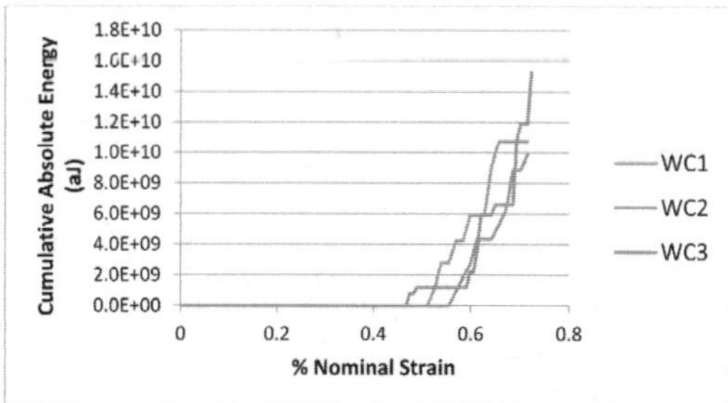

(d)

Fig. 4. Cumulative acoustic energy released during four-point bending of (a) Uncoated, (b) Ni-, (c) B_4C-Ni-, and (d) WC-Co-Ni-coated substrates

The WC-Co-Ni coatings exhibited the highest cumulative energy, and it was on the order of 10^{10} aJ (see Fig. 4d). This higher energy is evidence of brittle fracture behaviour within the coating. In a brittle coating under a tensile load, stress is concentrated at defects. The stress accumulates until it is suddenly released and a crack is formed. The sudden release of stress also produces the acoustic energy that was arrested. This reduced the stress concentration and prevented further failure until additional strain was applied.

(a)

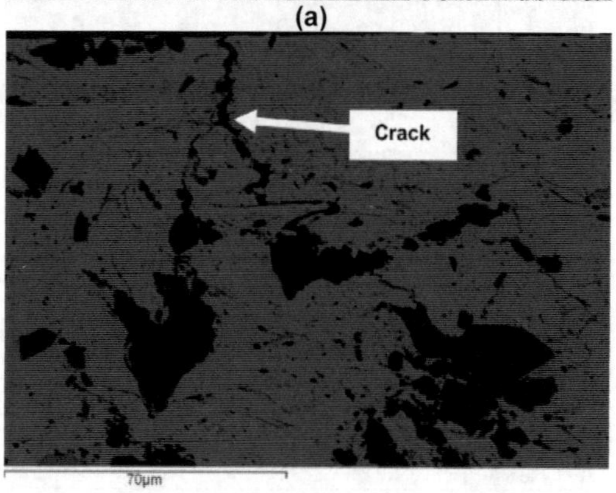

(b)

Fig. 5. Image of the cracks within the (a) top surface and (b) cross-section of the B_4C-Ni coating

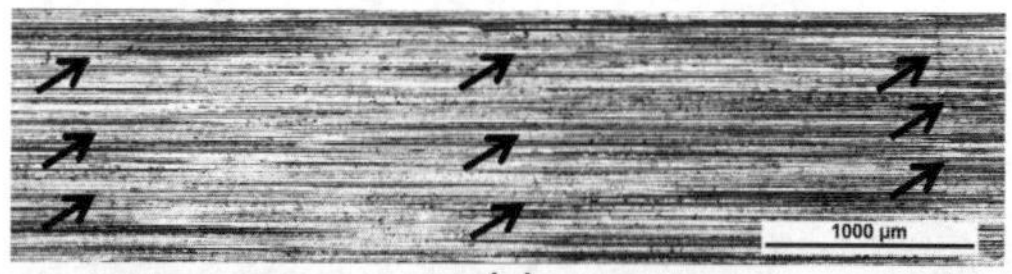

(a)

(b)

Fig. 6. Image of the cracks within the (a) top surface and (b) cross-section of the WC-Co-Ni coatings

The strain to failure for the WC-Co-Ni coatings was determined to be ~0.5% (see Fig. 4d). This strain to failure is comparable to that of HVOF-sprayed WC-based coatings, which range from 0.6% - 1.12% [15]. The HVOF-sprayed coatings may have higher strain to failure since they typically have higher WC density.

An analysis of the cracks that were formed in the coatings provides additional information about the failure behaviour. Figures 5 and 6 show the top surface of the coating and cross sections of the cracks. The cracks in the B_4C-Ni coating were discontinuous and were randomly spaced along the top of the coating (see Fig. 5a). Discontinuous cracks may be due to the ductile nature of the B_4C-Ni coatings, which allowed the ductile regions of the coating to deform with the substrate and cause internal decohesion between the splats, resulting in cracks that follow the splat boundaries (see Fig. 5b). Conversely, the cracks in the WC-Co-Ni coatings were continuous along the width of the coating (see Fig. 6a). This was similar to the cracks observed in the WC-based HVOF coatings [16]. The cracks in the cold-sprayed WC-Co-Ni coatings did not penetrate through the coating to the substrate-coating interface (see Fig. 6b), unlike those of the WC-based HVOF coatings. The shallow depth of these cracks may be due to the presence of Ni, which increased the ductility of the MMC coating.

3.5 Indentation Fracture Toughness

Indentation fracture toughness tests were conducted to provide a direct measurement of the fracture toughness of the coatings. Only the WC-Co-Ni coatings formed cracks during the tests and only in the direction parallel to the coating-substrate interface. The indentation fracture toughness along this direction was found to be 1.21 ± 0.2 MPa·m$^{0.5}$. This value is comparable to the fracture toughness of WC-based HVOF coatings, 0.65 – 1.85 MPa·m$^{0.5}$, where minimal decarburization occurs [15]. No cracks were formed on the B_4C-Ni coatings, which was likely due to the dominant behaviour of the ductile Ni matrix. No cracks were formed perpendicular to the coating-substrate interface, suggesting that the coatings are anisotropic. A further study will be conducted to investigate the fracture toughness along this orientation.

4 Conclusions

This study has shown that low-pressure cold spraying can be an effective and viable method of producing MMC coatings. In fact, the cold-sprayed WC-Co-Ni coatings have comparable four-point bending behaviour to that of conventional WC-based HVOF coatings, with a nominal strain to failure of ~0.5%. Additionally, the fracture toughness of the cold-sprayed WC-Co-Ni coatings, 1.21 ± 0.2 MPa·m$^{0.5}$, was also similar to the WC-based HVOF coatings. It was concluded that the B_4C-Ni coatings had insufficient B_4C content and was therefore dominated by the ductile Ni matrix.

5 References

[1] N. M. Melendez, V. V. Narulkar, G. A. Fisher, and A. G. McDonald. Effect of reinforcing particles on the wear rate of low-pressure cold-sprayed WC-based MMC coatings. Wear, vol. 306 (2013), no. 1–2, pp. 185–195.

[2] I. M. Hutchings. Tribological properties of metal matrix composites. Mater. Sci. Technol., vol. 10 (1994), no. 6, pp. 513–517.

[3] K.-H. Zum Gahr. Microstructure and wear of materials. Elsevier (1987).

[4] I. A. Ibrahim, F. A. Mohamed, and E. J. Lavernia. Particulate reinforced metal matrix composites - A review. J. Mater. Sci., vol. 26 (1991), no. 5, pp. 1137–1156.

[5] M. A. Zavareh, A. A. D. M. Sarhan, P. A. Zavareh, B. B. A. Razak, W. J. Basirun, and M. B. C. Ismail. Development and protection evaluation of two new, advanced ceramic composite thermal spray coatings, Al_2O_3–$40TiO_2$ and Cr_3C_2–20NiCr on carbon steel petroleum oil piping. Ceram. Int., vol. 42 (2016) no. 4, pp. 5203–5210.

[6] A. Al-Hamed, H. Y. Al-Fadhli, S. Al-Mutairi, B. S. Yilbas, M. S. J. Hashmi, and J. Stokes. Investigation of HVOF thermal sprayed nanostructured WC-12Co mixed with Inconel-625 coatings for oil/gas applications. Surface Effects and Contact Mechanics XI, vol. 78 (2013), pp. 215–225.

[7] J. M. Guilemany, J. M. de Paco, J. R. Miguel, and J. Nutting. Characterization of the W_2C phase formed during the high velocity oxygen fuel spraying of a WC + 12 pct Co powder. Metall. Mater. Trans. A, vol. 30 (1999), no. 8, pp. 1913–1921.

[8] H. Wang, X. Song, X. Liu, C. Wei, Y. Gao, and J. Fu. Effect of heat-treatment of spray-dried powder on properties of ultrafine-structured WC–Co coating. Surf. Coat. Technol., vol. 207 (2012), pp. 117–122.

[9] H. Chen and I. Hutchings. Abrasive wear resistance of plasma-sprayed tungsten carbide–cobalt coatings. Surf. Coat. Technol., vol. 107 (1998), no. 2–3, pp. 106–114.

[10] A. Papyrin. Cold spray technology. Elsevier (2007).

[11] N. M. Melendez and A. G. McDonald. Development of WC-based metal matrix composite coatings using low-pressure cold gas dynamic spraying. Surf. Coat. Technol., vol. 214 (2013), pp. 101–109.

[12] Y. T. R. Lee, H. Ashrafizadeh, G. Fisher, and A. McDonald. Effect of content and type of reinforcing particles on the hardness and wear rate of low-pressure cold-sprayed TiC and B4C-based metal matrix composite coatings. International Thermal Spray Conference (ITSC) 2016 (2016), pp. 310–315.

[13] A. G. Evans and T. R. Wilshaw. Quasi-static solid particle damage in brittle solids—I. Observations analysis and implications. Acta Metall., vol. 24 (1976]), no. 10, pp. 939–956.

[14] K. J. Hodder, J. A. Nychka, and A. G. McDonald. Comparison of 10 µm and 20 nm Al-Al2O3 metal matrix composite coatings fabricated by low-pressure cold gas dynamic spraying. J. Therm. Spray Technol. vol. 23 (2014), no. 5, pp. 839–848.

[15] Lynsey Driver. HVOF spraying of WC-Co & MCrAlY coatings for aeroengine components. University of Nottingham (2004).

[16] S. Bouaricha, J.-G. Legoux, and P. Marcoux. Bending behavior of HVOF produced WC-17Co coating: Investigated by acoustic emission. J. Therm. Spray Technol. vol. 13 (2004), no. 3, pp. 405–414.

Characterization of CrMoBW- Fe base in-flight particles and splats fabricated by nano-structured cored wire arc spraying

Man Tuiprae[1,a*], Kiattisak Meekhanthong[2,b], Komsanti Chokethawai[3,c], Sittichai Wirojanupatump[3,d]
[1] Faculty of Engineering, Rajamangala University of Technology lanna, Chiang Mai 50200, Thailand.
[2] Advanced Surface Technology Co., Ltd. Thailand.
[3] Faculty of Science, Chiang Mai University, Chiang Mai 50200, Thailand.
[a]Tman@rmutl.ac.th, [b]kiattisak.ast@gmail.com, [c]komsanti@gmail.com, [d]wish- 2000@hotmail.com

Key words: cored wire arc spray, in-flight particle, splat

Abstract

Properties and performance of the coating fabricated by wire arc spray process strongly depend on various factors including spray parameters and characteristics of both in-flight particle and splat. In this study, relationship between in-flight particle, splat formation and microstructure of the CrMoBW-Fe based nanocomposite coating fabricated by wire arc spray process were investigated. The Focus ion beam (FIB) milling was used to prepare a transversal cross-section of the in-flight particle and splat to be characterized by scanning electron microscopy (SEM) along with an energy dispersive x-ray spectroscopy (SEM-EDS). In addition, phase compositions were also analyzed by using x-ray diffractometer (XRD). The results revealed that size of the in-flight particles and splat morphology greatly influenced microstructure and properties of the coatings. In-flight particles were revealed a spherical-shape with a broad size distribution ranging from 0.8 to 115 μm at the average of 26 μm particle diameter. In-flight particles were well flattened on impact giving flower-shape splat with the average diameter of 100 μm. Flattening degree was found to be 3.8. Arc sprayed CrMoBW-Fe base nanocomposite coating showed very thin splat with fine lamella structure which implied that particles were fully molten upon impact onto substrate.

1 Introduction

Serviceable engineering components depend not only on their bulk properties but also on surface characteristics. Twin wire arc spraying has been widely used as a quick and convenient process for producing thick coatings. It has been widely used in thermal spray industry as being an economical spraying process which combines low equipment cost and operating cost along with very high deposition efficiencies [1-4].

Arc spray process can be applied with a wide range of electrical conductive materials including metal, alloy and metal matrix composites or cermet known as cored wire. Properties of the arc spray coatings strongly affected by spray parameters and relate with droplet formation, size of in-flight particle and particle velocity.[5-9]. Many researchers reported that substrate parameters such as temperature, surface profiles and their interactions on impact influence the splat formation and coating properties [10-11]. However, understanding of the mechanisms of the splat formation is still limited. In this work, the splat formation of the arc sprayed CrMoBW-Fe base cored wire onto a polished stainless steel substrate was investigated.

2 Experimental procedures

2.1 Characterization of cored wire

CrMoBW-Fe base cored wire was studied. Characterization of the as recieved cored wire including morphology, particle size, microstructure, chemical compositions and phase compositions. Nominal compositions of the cored wire materials provided by supplier was shown in Table1. Analysis of the filler particles taken from the cored wire was prepared by putting particles on a carbon tape followed by gold sputtering just prior to SE-SEM, BSE-SEM investigation along with EDS-SEM for chemical analysis (JEOL JSM-5910, Japan). Size distribution of filler particles was analyzed by laser particle size analyzer. X-ray diffractometer (Rigaku, TTRAX III) was employed for phase analysis.

2.2 Collection of in-flight particle and Splat

CrMoBW-Fe base cored wire was sprayed by arc spray system (TAFA Arcjet 9000, USA). Arc spray parameters were shown in Table 2. In-flight particles were collected by spraying through Ø 3 mm holes of stainless steel plate into distilled water (Fig.1a). The in-flight particles were collected and cleaned by alcohol follow by air drying immediately. Splat collection was done by spraying onto polished surface of stainless steel plate (Fig. 1b).

2.3 Characterization of in-flight particle

Morphology and microstructure of the in-flight particls were revealed by SEM along with EDS-SEM for chemical analysis. Cross-sectioning of in-flight particles was prepared by FIB technique. Size distribution was also analyzed by laser particle size analyzer. Phase analysis of the in-flight particls was analyzed by X-Ray diffraction technique. The XRD was operated at 50 kV and 30 mA using Cu K_α radiation (1.54 Å). The goniometer was set with a step-width of 0.02° and a scan rate of 2.4 degree per second over a 2θ ranging from 20° to 40°. Phase compositions were identified by using JCPDS.

2.4 Characterization of a single splat

Characteristics of splat produced by CrMoBW-Fe base cored wire were investigated by OM and SEM. FIB technique was used for microstructure characterization of a single splat cross-section. Measurement of perimeter and area of individual splat were performed by using image analysis with SCENTIS software. Measurement was randomly performed on 100 splats in order to obtain the average value. Splat diameter, d_s was calculated by equation (1) using a measured splat area. The degree of flattening is the ratio of the splat diameter (d_s) to the in-flight particle diameter (d_i) as shown in equation (2), while the degree of splashing is the ratio of calculated splat area ($P^2/4\pi$) to actual splat area (A) measured by image analysis as shown in equation (3) [5].

$$d_s = (4A/\pi)^{1/2} \qquad (1)$$

$$DF = d_s/d_i \qquad (2)$$

$$DS = (P^2/4\pi) \times (1/A) \qquad (3)$$

When P is the perimeter of splat, and A is the actual (measured) area of splat.

Table 1 Chemical compositions of studied cored wires as nominated by supplier.

Designated code	Chemical Compositions (wt%)							
	Si	B	Mn	Cr	Mo	WC	Nb	Fe
CrMoBW - Fe base	2	5	3	25	6	16	12	Bal.

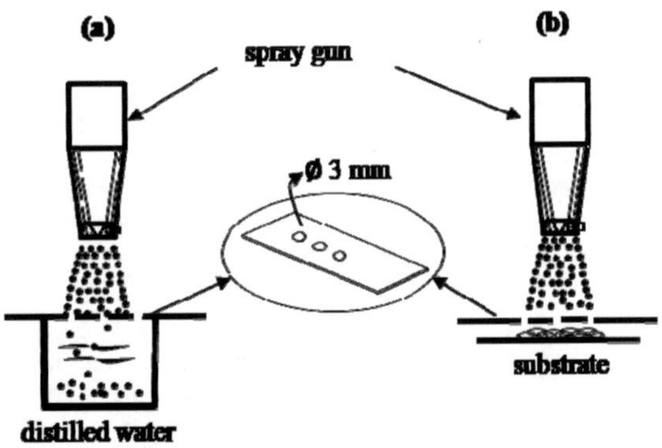

Figure 1 Schematic diagram shows collection methods (a) in-flight particle collection and (b) splat collection.

Table 2 Arc spraying parameters.

Arc Voltage (V)	35
Arc current (A)	200
Air pressure (bar)	4.2
Wire feed rate (g/min)	95
Spray distance (mm)	150

3 Results and discussion

3.1 Cored wire characterization

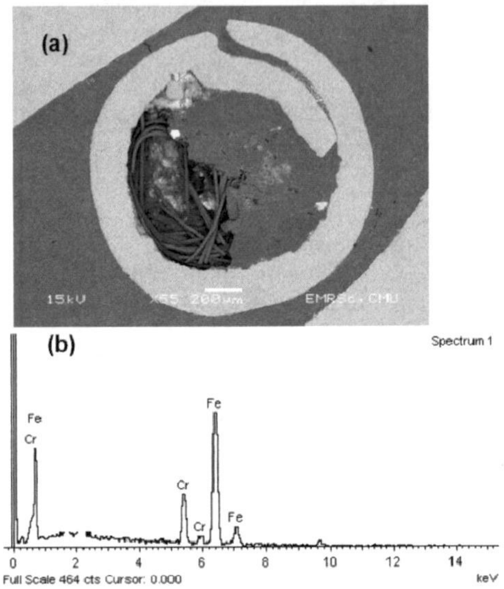

Figure 2 (a) BSE-SEM micrograph and (b) EDS spectrum of CrMoBW- Fe base shell.

Fig. 2(a) shows a cross-section of CrMoBW-Fe base cored wire illustrating shell and filler component of cored wire. The shell part was about 200 μm thick with a diameter of 1.6 mm. Chemical analysis of CrMoBW-Fe base cored wire showed Fe and Cr as can be seen in Fig. 2(b). Morphology of the filler particles showed as irregular shape as shown in Fig. 3. Filler particles had broad size distribution ranging from 0.2 to 125 μm with a mean diameter of 52 μm (Fig. 4). Phases found in CrMoBW-Fe base cored wires filler including W_2C, WC, B_4C, FeNb, Fe_2O_4, and FeCr as shown in Fig. 5.

3.2 In-flight characterization

Analysis of in-flight particles from CrMoBW- Fe base cored wire by SEM indicated that in-flight particles are spherical shape with rather smooth particle surface (Fig. 6 (a)). The average diameter of CrMoBW- Fe base in-flight particles was 26.7 μm. with particle size distribution ranging from 0.8 to 115 μm as shown in Fig. 6(b). SEM cross-section image of CrMoBW- Fe base in-flight particle (Fig. 7(a)) shows a good homogeneity of the eutectic and dendrite structure. EDS-SEM analysis of chemical compositions taken from mapping mode was shown in Fig. 7 (b). It can be observed from the EDS spectrum that all elements including Fe, Cr, Nb, Mo, W, C, Si and Mn were fully melted during arc spraying process.

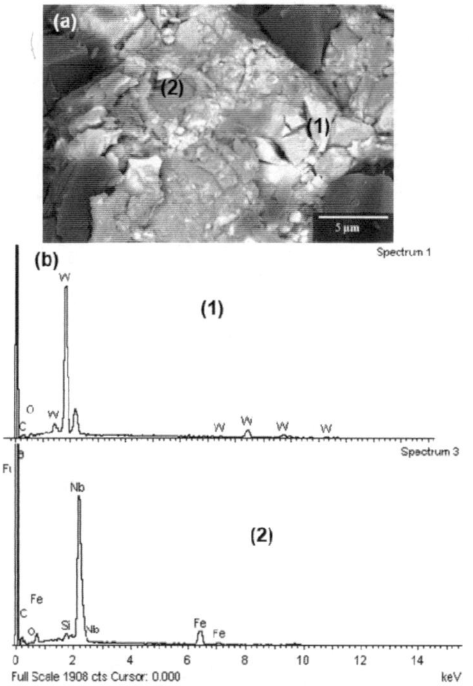

Figure 3 (a) SE-SEM micrographs show morphology of CrMoBW- Fe base filler (b) EDS spectrums of area 1 and 2 of CrMoBW- Fe base filler

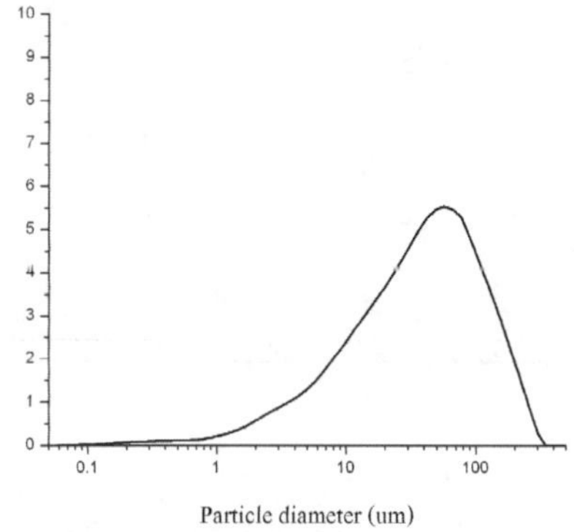

Figure 4 Size distribution of filler particle of CrMoBW- Fe base cored wire.

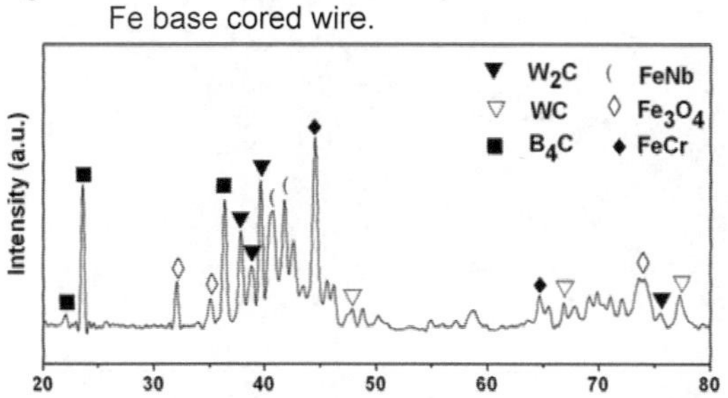

Figure 5 X-ray diffraction patterns of the filler of CrMoBW- Fe base.

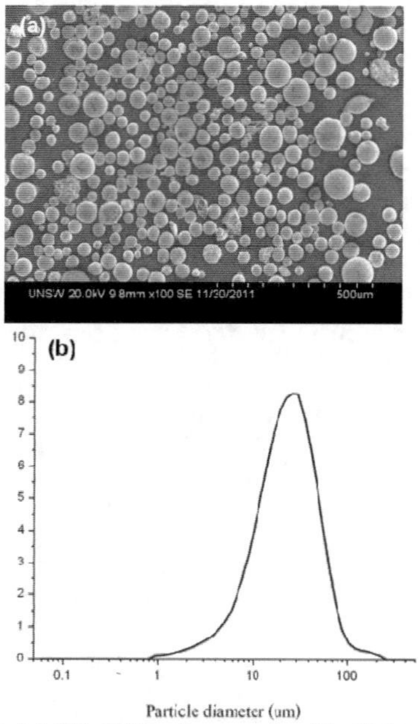

Figure 6 (a) SE-SEM micrograph of in-flight particles (b) Size distribution of in-flight particles from CrMoBW- Fe base core wire.

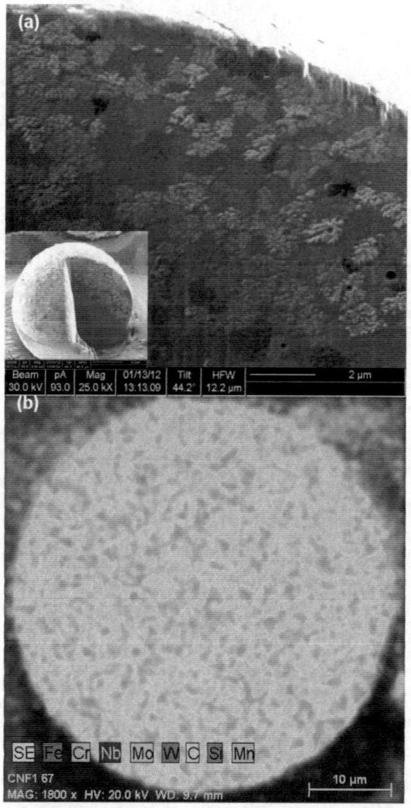

Figure 7 (a) SEM-FIB image of in-flight particles of CrMoBW- Fe base (b) EDS spectrum mapping analysis of the in-flight particle

In Fig. 8 also shows peaks of FeCr as a predominant phase formed along with carbide and oxide including W_2C, Fe_3O_4 and Fe_3C.

3.3 Splat characterization

Fig. 9 shows SEM images of the CrMoBW- Fe base splats with a disk shape like (Fig.9a) and a flower shape like (Fig.9b). The average diameter of the splat was calculated to be 100 μm. The degree of flattening and degree of splashing were 3.8 and 3.3, respectively (Table 3).

Fig.9(c) shows well bonded interface between splat and substrate with a thin splat layer (2 μm). It also revealed a typical void inside splat structure as shown in Fig. 9(d). SEM images of the coating cross-section revealed a thin splat and fine lamella structure (Fig. 10). This implied that particles were fully molten upon impact. The micro pores can also be observed.

Table 3 Characteristics of in-flight particle and splat of CrMoBW- Fe base.

Mean size (μm)			Degree of flattening	Degree of splashing
Filler particle	In-flight particle	Splat		
52	26	100	3.8	3.3

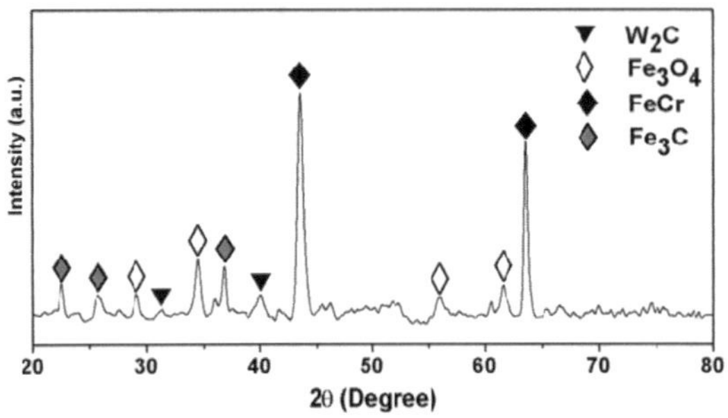

Figure 8 X-ray diffraction patterns of the in-flight particles of CrMoBW- Fe base cored wire

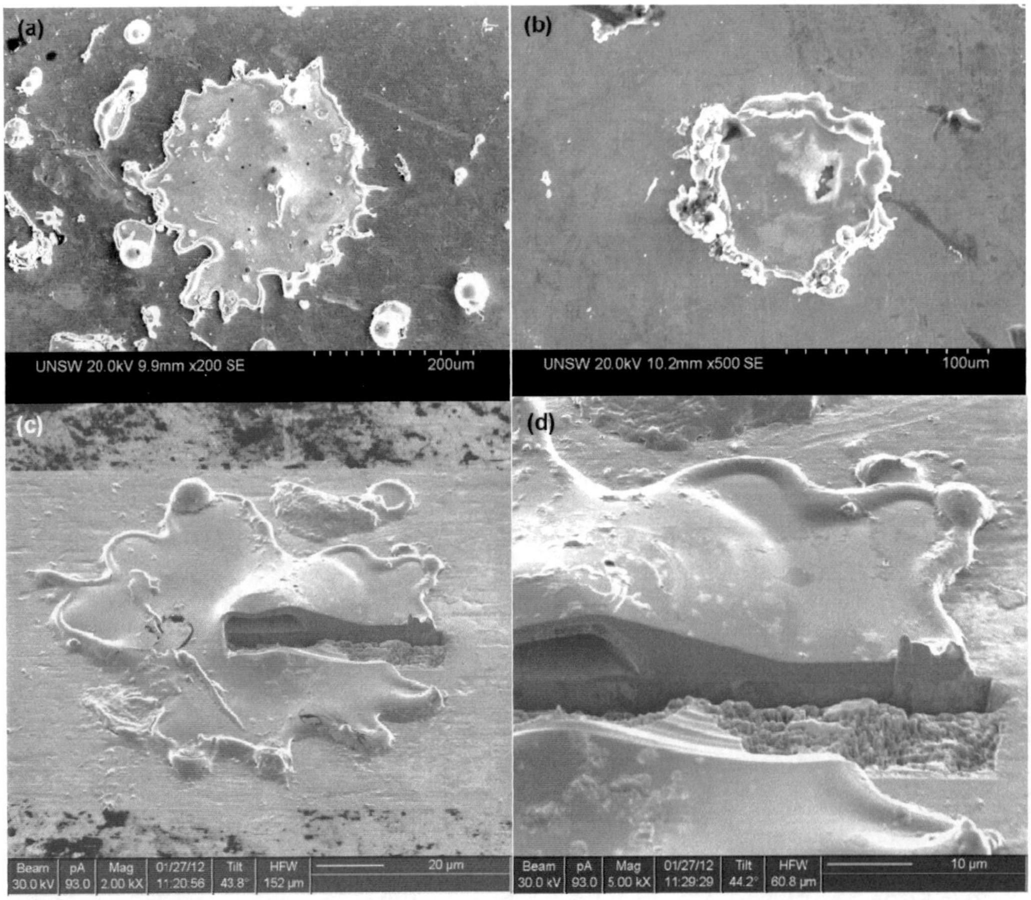

Figure 9 SE-SEM micrographs and SEM-FIB micrographs of CrMoBW- Fe base splats particles

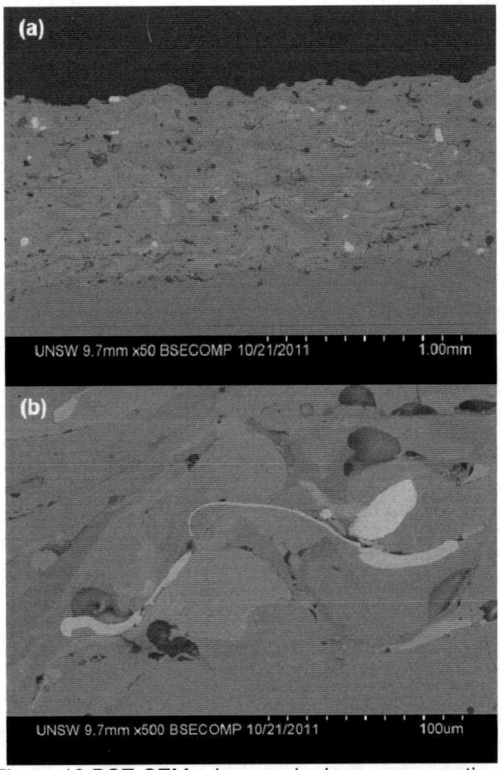

Figure 10 BSE-SEM micrograph show cross-sections of CrMoBW- Fe base coating at (a) 50X and (b) 500X magnification.

4 Conclusions

A spherical shape like of the in-flight particles of CrMoBW-Fe base nanocomposite with a narrow size distribution were successfully obtained by cored wire arc spray process. Upon impact onto the stainless steel substrate, particles were well flattened showing flower-shape splat with the average diameter of 100 μm. Microstructure of the arc sprayed CrMoBW-Fe base nanocomposite coating exhibited a thin splat and fine lamella structure. This implied that particles were fully molten.

5 Acknowledgements

The authors would like to acknowledge the financial support from Rajamangala University of technology Lanna (Hands-on Researcher Development) and Chiang Mai University. The authors are also gratefully acknowledge Advance Surface Technology Co., Ltd., for facilitating spraying process.

6 Reference

[1] I. Gedzevicius, A.V. Valiulis, Analysis of wire arc spraying process variables on coatings properties, J. Materials Processing Technology 175 (2006)206-211.

[2] M.P. Planche, H. Liao, C. Oddet: Relationships between in-flight particle characteristics and coating microstructure with a twin wire arc spray process and different working conditions, Surface & Coatings Technology 182(2004)215-226.

[3] C.M. Fernandes, V. Popovich, M. Matos, A.M.R. Senos, M.T. Vieira, Carbide phases formed in WC-M (M=Fe/Ni/Cr) systems, Ceramics International 35(2009)369-372.

[4] V. Stoica, R. Ahmed, T. Itsukaichi, Influence of heat-treatment on the sliding wear of thermal spray cermet coating, Surface & Coating Technology 199 (2005)7-21.

[5] A. Pourmousa, J. Mostaghimi, A. Aabedini, S. Chandra, Particle size distribution in a wire-arc spraying system, J. Thermal Spray Technology. 14(2005)502-510.

[6] K. Cooke, G. Oliver, V. Buchanan, N. Palmer, Optimisation of the electric wire arc-spraying process for improved wear resistance of sugar mill roller shells, Surface & Coating Technology 202(2007)185-188.

[7] I. Gedzevicius, A.V. Valiulis, Analysis of wire arc spraying process variables on coating properties, Journal of Materials Processing Technology 175 (2006)206-211.

[8] M.P. Planche, H. Liao, C. Coddet, Relationships betaween in-flight particle characteristics and coating microsturcture with a twin wire arc spray process and different working conditions, Surface & Coating Technology 182(2004)215-226.

[9] K. Venugopal, M. Agrawal, Evalution of arc sprayed coating for erosion protection of tubes in atmosheric fluidsed bes combustion (AFBC) boilers, Wear 264(2008)139-145.

[10] S. Brossard, P.R. Munroe, A.T.T. Tran, and M.M. Hyland, Effects of Substrate Roughness on Splat Formation for Ni-Cr Particles Plasma Sprayed onto Aluminum Substrates, J. Thermal Spray Technology 19(5)(2010)1131-1141.

[11] S. Brossard, P.R. Munroe, A.T. Tran, and M.M. Hyland, Study of the Splat Microstructure and the Effects of Substrate Heating on the Splat Formation for Ni-Cr Particles Plasma Sprayed onto Stainless Steel Substrates, J. Thermal Spray Technology 19(5)(2010)1100-1114.

The oxidation behavior and mechanism of plasma spraying ZrB_2 ceramic coating with SiC addition

Xiaoqiang Feng, Xing Wang, Longjiang Deng, Jiangliang Xie
University of Electronic Science and Technology of China, Chengdu, CHN

Plasma spraying ZrB_2 ceramic coating is considered as potential candidate method of Thermal Protective System. While the application of plasma sprayed ZrB_2 coating is restricted due to its oxidation. Therefore, it is important to study the oxidation behavior of the ZrB_2 material. In this research, oxidation behavior of the ZrB_2 ceramic, which achieve different density by using SPS and pressureless sintering, is studied to explain the oxidation behavior of plasma spraying ZrB_2 ceramic coating.
The oxidation behavior of ZrB_2 ceramics is investigated using SEM, XRD and EDS. The ZrB_2–based ceramic coating is gravely oxidized at 600°C, but the block ZrB_2–based ceramics also possess excellent oxidation resistance above 1000°C. The density of ZrB_2 ceramics significantly increase when changing the sintering method from pressureless sintering to SPS. The high density has beneficial effect to improve the oxidation resistance of ZrB_2 ceramic, for there are few open pores channel in high density ceramics. The oxygen can not diffuse to the inner through pores, as a result, the high density ceramics can only be oxidized from outside to inside progressively, unlike low density ceramics, whose surface and inner is oxidized simultaneously.

1 Introduction

Zirconium diboride with a very high melting points (> 3000°C), is one of the most promising materials for working in high temperature environment. It belongs to a class of materials called ultra high temperature ceramics (UHTCs). Other borides, nitrides and carbides of transition metals such as HfB_2, TiB_2, TiC, ZrC, TaN and ZrN are also considered as potential candidate material for high temperature application, but ZrB_2 is the most investigated UHTCs, because it has a relatively lower density (6.09g/cm^3), higher thermal conductivity (58.2W(mK)$^{-1}$) and higher electrical conductivity (10.3×10^4S/cm).[1-4]
However, the application of ZrB_2 is restricted due to its weaknesses. One of them is that the ZrB_2 would be oxidized at high temperatures [5,6].
The oxidation behavior of ZrB_2 is studied in numerous investigations, and researchers find a lot of methods to improve the oxidation resistance of ZrB_2[7-13]. But the oxidation of plasma spraying ZrB_2-based ceramic coating in depth study is relatively small and most researchers ignore the influence of density on oxidation behavior.
The objectives of the present study are to study the oxidation behavior of plasma spraying ZrB_2-based ceramic coating and to understand the effects of density on the oxidation resistance of ZrB_2.

2 Experimental procedure

2.1 Powder preparation and characterization

Commercially available ZrB_2 was used to prepare the materials for this study. SiC powder was added as second phase. The average particle sizes of the raw powders were calculated by laser granularity (MASTERSIZER 3000, Malvern). The particle size, purity, and suppliers are summarized for each powder in **Table 1**.

Table 1. Raw Material Characteristics

Powder	Particle Size(μm)	Purity(wt%)	Supplier
ZrB_2	10-15	99.5	Eno Material, China
SiC	1.6-1.7	99.5	Eno Material, China

The powders were ball-milled in ethanol for 10 h. These powders were then dried in air in a free convection at 80°C. The particle size of the powder after ball-milled is 1.4~1.6 μm. By using steps of slurry forming, spray drying and heat treatment, the powder was reprocessed into spherical agglomerates for plasma spraying.

2.2 Sample preparation

The ZrB_2 ceramic coating were prepared by the plasma spraying. Two different process parameters were used, the spray power of one is lower and the spray power of another is higher.

2.3 Oxidation testing

The oxidation process of ZrB_2 ceramic coating was conducted in a tube furnace in air at 400°C, 500°C, 600°C, 1000°C and 1200°C. The heating rate of the furnace up to the oxidation temperature was 6°C /min. After the dwell time at the oxidation temperature, the furnace was shut off, therefore, the cooling rate was quite fast down to room temperature. The furnace temperature was controlled by a programmable temperature controller.

2.4 Sample characterization

Crystalline phases of the spherical agglomerates and the ZrB_2 ceramic coating before and after oxidation were determined by X-ray diffraction (XRD, XRD-

7000, Shimadzu). The morphology and microstructures of the spherical agglomerates and the ZrB_2 ceramic coating before and after oxidation were observed using scanning electron microscopy (SEM; JEOL, JSM-7600F).

3 Results and discussion

3.1 Microstructure of processed spherical agglomerates

Fig.1 is the SEM images of ZrB_2/SiC agglomerates. As show in **Fig.1(a)**, almost all of the agglomerates are regular sphere.
The good sphericity is obtained and no perforation cracks appear on the surface as show in **Fig.1(b)**. But the surface is coarse and of lax structure. This is because the raw powders which constitute the whole spherical agglomerate have anomalistic shapes and sharp corners. In addition, the rapid evaporation of the water inside of the spherical agglomerates would make some stoma on the surface of spherical agglomerates. So some defects appear on the surface of spherical agglomerates. As seen from **Fig.1(c)**, the size of the spherical agglomerates is in the range from 40 microns to 120 microns, meeting the requirements of plasma spraying for powders.

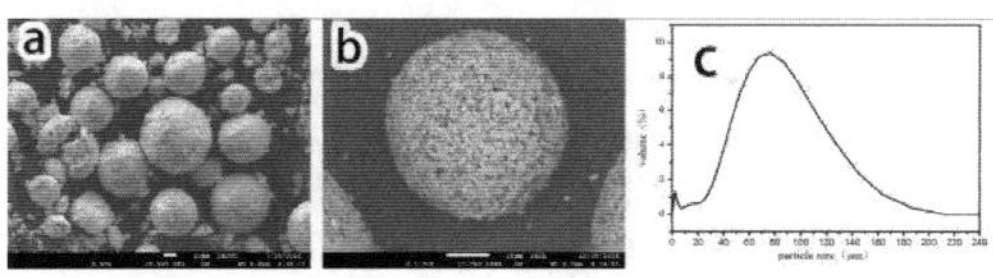

Fig.1. SEM images of ZrB_2/SiC agglomerates: (a) spherical agglomerates, (b) high magnification of (a), (c) The particle size distribution.

3.2 Microstructure of the ZrB_2 ceramic coating

Fig.2 is the SEM images of the plasma spraying ZrB_2 ceramic coating. The surface of the coatings has two kinds of microscopic morphology when the spray power is lower. A part of the ZrB_2/SiC spherical agglomerates are melted adequately, then the morphology is an unbroken cohesive whole. Another part of the ZrB_2/SiC spherical agglomerates are not melted adequately, so the agglomerates is broken to fine raw powders and the powders coat the surface of the substrate. The surface of the coatings is oxidized when the spray power is higher. And many spherical particles are distributed on the surface of the higher spraying power coating because of the spattering of the molten ZrB_2 droplets when they impact the surface of the substrate.
The crystalline phase composition of the plasma spraying ZrB_2 ceramic coating was verified by analysis of the XRD patterns at room temperature. The XRD patterns of the plasma spraying ZrB_2 ceramic coating in **Fig.2(f)** show the ZrB_2 as the dominant phase. In addition, some low-intensity additional peaks were observed in the higher spraying power ZrB_2 coating showed the surface of the coating is mild oxidized when the spray power is higher.

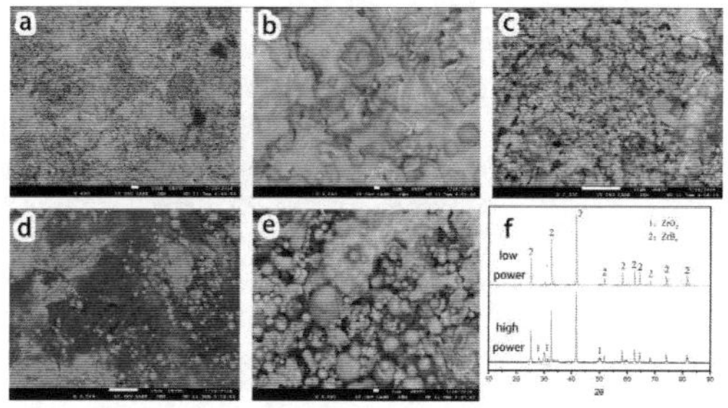

Fig.2. SEM images of (a) lower spraying power ZrB_2–based ceramic coatings, (b) surface consist of adequately melted agglomerates of (a), (c) surface consist of un-melted agglomerates of (a), (d) higher spraying power ZrB_2–based ceramic coatings, (e) high magnification of (d), (f) XRD patterns of the ZrB_2 coatings.

3.3 Oxidation behavior

The oxidized plasma spraying ZrB_2 ceramic coating was studied using XRD, SEM and EDS.
The XRD patterns of the oxidized plasma spraying ZrB_2 ceramic coating is showed in **Fig.3**. The hexagonal ZrB_2 is also the dominant phase after oxidized at 400°C and 500°C. Some new peaks are observed in plasma spraying ZrB_2 ceramic coating after oxidized at 600°C, which to be indexed to ZrO_2. And the intensity of the ZrO_2 peaks is increased as the oxidized time increased.
Fig.4 is the SEM images of plasma spraying ZrB2 ceramic coating after oxidation. The plasma spraying ZrB_2 ceramic coating is barely oxidized when it is oxidized at 400°C. But small amount of elements oxygen is detected on the granular surface using EDS, as shown in **Fig.5(a)**. It may be because these regions have large specific surface area and strong adsorption, make the oxidation more available.
Some tiny cracks appear on the surface of the plasma spraying ZrB_2 ceramic coating after oxidized at 500°C for 9 hours , which indicates the occurrence of the slow oxidation.
As shown in **Fig.4** and **Fig.5(b)**, the plasma spraying ZrB_2 ceramic coating is oxidized at 600°C. Glass phase B_2O_3 appears on the surface of the coating. Evenly, the coatings were oxidized completely when the oxidation time is increased to 9 hours, and the coatings have become white ZrO_2 powders.
The plasma spraying ZrB_2 ceramic coating cracked, flaked off and even swelled after oxidized above 1000°C, as shown in **Fig.6**. This show the substrates are also oxidized.

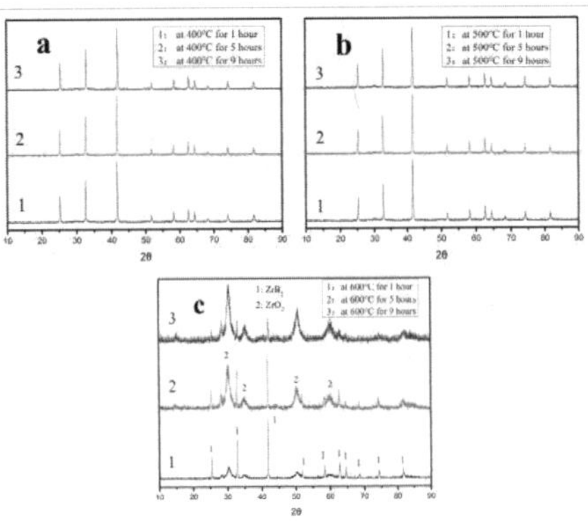

Fig.3. XRD patterns of the plasma spraying ZrB_2-based ceramic coatings after oxidized: (a) at 400°C, (b) at 500°C, (c) at 600°C.

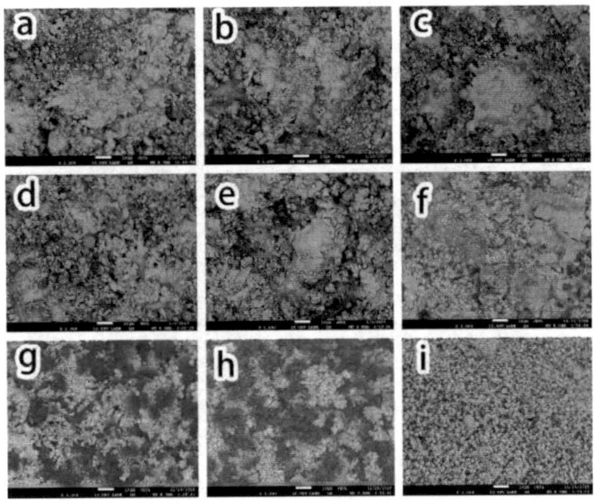

Fig.4. SEM images of plasma spraying ZrB_2 ceramic coating after oxidation: (a) under 400°C for 1 hour, (b) under 400°C for 5 hours, (c) under 400°C for 9 hours, (d) under 500°C for 1 hour, (e) under 500°C for 5 hour, (f) under 500°C for 9 hour, (g) under 600°C for 1 hour, (h) under 600°C for 5 hour, (i) under 600°C for 9 hour.

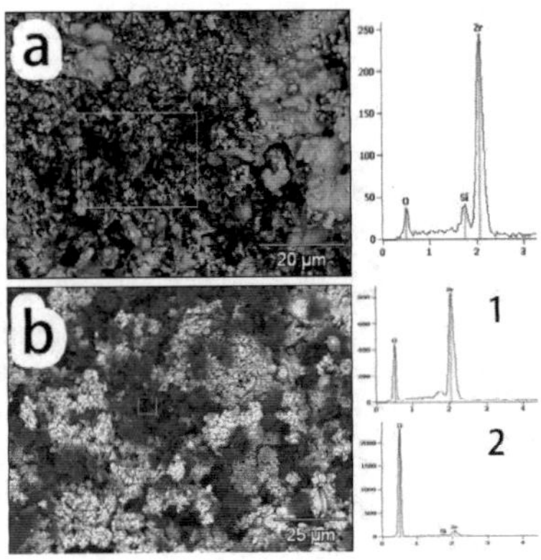

Fig.5. Elemental analysis results to the surface of ZrB_2 coating after oxidized: (a) oxidized under 400°C for 9 hours, (b),oxidized under 600℃ for 5 hours.

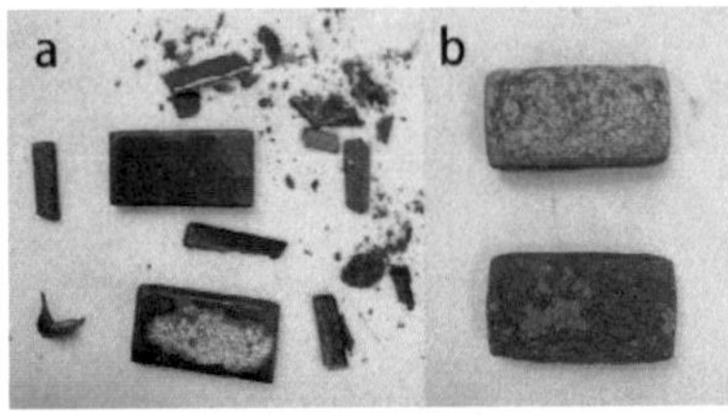

Fig.6. Images of plasma spraying ZrB_2 ceramic coating after oxidation for 3 hours. (a) at 1000°C, (b) at 1200°C.

3.4 Oxidation resistance analysis

The study of the oxidation mechanism of ZrB_2 ceramics can provide theoretical guidance to improve the oxidation resistance of ZrB_2 ceramics, estimate the service life of ZrB_2 ceramics, and predict the failure behaviors of the ZrB_2 ceramics.

The block ZrB_2-based ceramics also have excellent oxidation resistance above 1000°C [15]. But the plasma spraying ZrB_2-based ceramic coatings have been oxidized completely to ZrO_2 at 600°C. The different density between two forms ZrB_2-based ceramic maybe the major factor.

Two sets of block ZrB_2-based ceramics with different density were sintered by using spark plasma sintering (SPS) and pressureless sintering. The powder mixtures were ball-milled in ethanol for 1 h. These powders were then dried in air in a free convection at 80°C. The powders were densified by pressureless sintering at 1800°C for 2 hours under argon atmosphere and spark plasma sintering at 1800°C in a vacuum environment, respectively. There green bodies densified by spark plasma sintering were sintered using a graphite mould for 5 min under 40MPa of uniaxial pressure. The density of ZrB_2 ceramics significantly increase when using SPS. The relative density of pressureless sintering ZrB_2 ceramics and SPS ZrB_2 ceramics is 80% and 95% respectively.

The oxidation process was conducted in a tabular furnace with molybdenum disilicide ($MoSi_2$) heating elements capable of operating at 1800°C. **Fig.7** is the SEM images of the cross section of the block ZrB_2-based ceramics after oxidized in air at 1200°C for 3 hours.

As seen from **Fig.7(a)** and **Fig.7(b)**, there is no layering in the pressureless sintering ZrB_2-based ceramics after oxidized, the interior and the exterior were oxidized simultaneously. The reason is that there are many pore channels for oxygen in the low-density ZrB_2-based ceramics, the oxygen could diffuse to the inner of the ZrB_2-based ceramics through pores and react with the inner ZrB_2, similarly to the oxidation behavior of ZrB_2 powders[14].

As seen from **Fig.7(c)**, a glass phase protective layer is observed on the surface of the SPS ZrB_2-based ceramics after oxidized, and the interior is not oxidized. Because the oxygen would be blocked off by the dense surface of the high-density ZrB_2-based

ceramics. So the oxygen can only react with the outermost layer of ZrB_2 at the beginning and produce a glass phase layer to obstruct the diffusion of oxygen.
So the density and the porosity would be one of the major factors which led to the poor oxidation resistance of ZrB_2-based ceramics coating compare with the oxidation resistance of block ZrB_2-based ceramics.

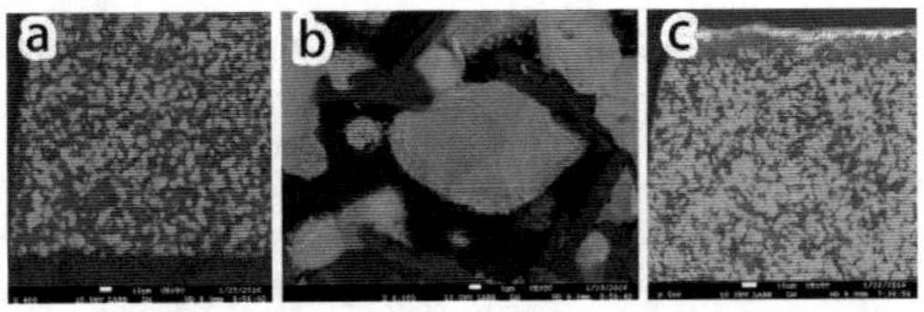

Fig.7. Cross-section images for block ZrB_2-based ceramics after oxidation at 1200°C for 3h: (a) low-density block ZrB_2-based ceramics, (b) high magnification of (a), (c) high-density block ZrB_2-based ceramics.

4 Conclusion

ZrB_2-based ceramic coatings were prepared by plasma spraying method, and the oxidation behavior and mechanism of the ZrB_2-based ceramic coatings was analyzed according to experimental results. The spray power can influence the microscopic morphology of the ZrB_2-based coating. The ZrB_2 agglomerates can not melted adequately with lower spray power, but the coating will be broken with higher spray power. The ZrB_2-based ceramic coatings is obviously oxidized at 600°C in air, indicate the oxidation resistance of the ZrB_2-based ceramic coatings is worse than block ZrB_2-based ceramics. The relative density of pressureless sintering ZrB_2 ceramics and SPS ZrB_2 ceramics is 80% and 95% respectively. The high-density ceramics can only be oxidized from outside to inside progressively, but in low-density ceramics, the surface and inner is oxidized simultaneously, as the oxygen can diffuse to the inner through pores. So the high density is beneficial to improve the oxidation resistance of ZrB_2 ceramic, for there are few open pores channel in high density ceramics.

Acknowledgements

This work was supported by the Science and Technology Project of Sichuan Province: Applied Basic Research Project (2015JY0179) and the Fundamental Research Funds for the Central Universities (106142014027).

References

[1] Kuwabara, K., Some characteristics and applications of ZrB2 ceramics, Bull. Ceram. Soc. Jpn., 37(2002), Issue 4, pp. 267/271

[2] Dong Gao, Yue Zhang, Oxidation kinetics of hot-pressed ZrB2-SiC ceramic matrix composites, Ceramics International 39 (2013), pp. 3113/3119

[3] Matthew J. Gasch, Donald T. Ellerby, Sylvia M. Johnson, Ultra High Temperature Ceramic Composites, Handbook of ceramic composites.

[4] Ping Hu, Wang Guolin, Zhi Wang, Oxidation mechanism and resistance of ZrB2-SiC composites, Corrosion Science 51 (2009), pp. 2724/2732.

[5] R. F. Voitovich, E. A. Pugach, L. A. Menshikova, High-temperature oxidation of zirconium diboride, Poroshkovaya Metallurgiya 6 (1967), pp. 44/48

[6] Shi C. Zhang, Greg E. Hilmas, Oxidation of zirconium diboride with tungsten carbide additions, J. Am. Ceram. Soc., 94(2011), Issue 4, pp. 1198/1205

[7] Shi C. Zhang, Greg E. Hilmas, Improved oxidation resistance of zirconium diboride by tungsten carbide additions, J. Am. Ceram. Soc., 91(2008), Issue 11, pp. 3530/3535.

[8] F. Monteverde, A. Bellosi, Oxidation of ZrB2-based ceramics in dry air, Journal of The Electrochemical Society 150(2003), Issue 11, pp. B552/B559.

[9] Laura Silvestroni, Elena Landi, Oxidation behavior and kinetics of ZrB2 containing SiC chopped fibers, Journal of European Ceramic Society 35(2015), pp. 4377/4387

[10] M. Kazamzadeh Dehdashti, W. G. Fahrenholtz, G. E. Hilmas, Effects of transition metals on the oxidation behavior of ZrB2 ceramics, Corrosion Science 9 (2015), pp. 224/231

[11] Laura Silverstroni, Giacomo Meriffi, Diletta Sciti, Oxidation behavior of ZrB2 composites doped with various transition metal silicides, Corrosion Science 83(2014), pp. 281/291

[12] Peter A. Williams, Ridwan Sakidja, Oxidation of ZrB2-SiC ultra-high temperature composites over a wide range of SiC content, Journal of the European Ceramic Society 32(2012), pp. 3875/3883

[13] Diletta Sciti, Mylene Brach, Oxidation behavior of a pressureless sinteren ZrB2-Mosi2 ceramic composite, J. Mater. Res., 20, (2005), Issue 4, pp. 922/930

[14] Wei-Ming Guo, Guo-Jun Zhang, oxidation of ZrB2 powder in the temperature range of 650-800℃, Journal of Alloys and Compounds 471 (2009), pp. 502/506

[15] P. Sarin, P.E. Driemeyer, In situ studies of oxidation of ZrB2 and ZrB2-SiC composites at high temperatures, Journal of European Ceramic Society 30 (2010), pp. 2375/2386

Effect of Thermal Aging on Properties and Microstructure of AlSi-hBN Coating

Jianming Liu[1, 2], Yueguang Yu[2*,1], Tong Liu[2], Suyuan Yang[3], Xingwang Cheng[3]

1. Northeastern University, Shenyang, 110819, China
2. Beijing General Research Institutes of Mining and Metallurgy, Beijing, 100160, China
3. Beijing Institute of Technology, Beijing 100081, China

Abstract: A commercial abradable coating AlSi-hBN was fabricated by atmospheric plasma spraying and was thermal aged at 450 °C in the atmospheric environment for 1000h. Thermal aging effect on coating abradablity, hardness, bond strength and microstructure were evaluated. It was found that coating abradability increased obviously after 2h thermal aging and slightly decreased for extended thermal aging time. A sharp decrease in both coating hardness and bond strength were found after 2h thermal aging and a slow increase occurred with the extended aging time. Relationships between coating properties and microstructure were studied. Decomposition of organic binder and sintering of AlSi matrix metal generated by thermal aging were found to be reasons for coating properties changes.

Key Words: Abradable coating, AlSi-hBN, Thermal aging , Microstructure

1 Introduction

Thermal spray abradable coatings have been widely applied in the aero-engine for over 50 years[1]. Abradable coatings can be easily rubbed in by the blades and labyrinth teeth to protect the rotating parts from wearing and keep the clearance between rotor and stator at the minimums to maintain engine efficiency. Abradable coatings serve in extreme environments with high velocities up to 650 m/s and high temperature up to 1200 ºC[2]. To meet the high reliability standard of aero engines the stability of abradable coatings in the serious environments is very important. Long time service under high temperature is one of the working conditions of abradable coatings. There for, thermal aging effect on abradable coatings is necessary to be studied. While most of the previous studies mainly focus on coating initial properties[3] ~ [9] and rare papers published concerning coating properties after long time service except those of post engine service investigation[10]. In the work, a commercial abradable coating AlSi-hBN which is widely used in the engine compressor sections was thermal aged at 450°C for 1000h. Coating properties including the abradability at high temperature and high velocity conditions and microstructure of different aging time were investigated. Relationships between coating properties and microstructure were studied.

2 Experiment

2.1 Material

AlSi-hBN coating samples were atmospheric plasma sprayed using a mechanically cladded composite powder supplied by Beijing General Research Institute of Mining and Metallurgy (Beijing, China). Fig. 1 gives the section morphology of powder. AlSi-hBN feedstock powder mainly consists of particles of coarse hBN cladded by fine AlSi and has a nominal composition of 20wt% hBN and 8wt% Si, balanced with aluminum. Coating samples were sprayed onto Ti6Al4V substrates by a F6 plasma torch (GTV Verschleiss-schutz GmbH, Germany) and the substrate was pre-sprayed by a Ni-Al bond coating powder (KF-6, Beijing General Research Institute of Mining and Metallurgy, Beijing, China) before the AlSi-hBN top coating was applied. Sample dimensions for hardness, bond strength, abradability tests and microstructure evaluation are ø40mm×10mm, ø25mm×5mm, 40mm×60mm×8mm and 40mm×80mm×2mm, respectively.

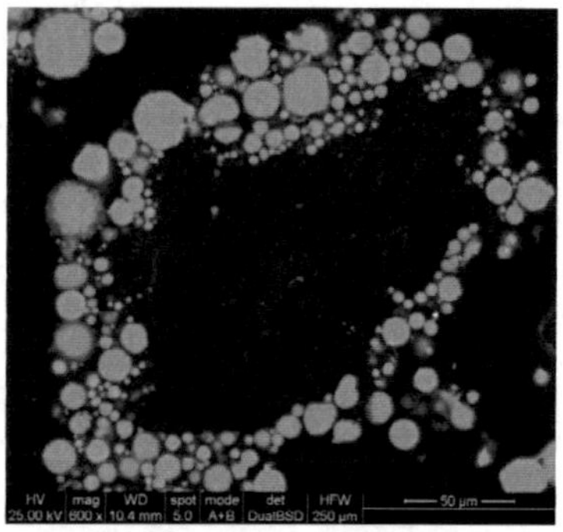

Fig. 1 Cross section morphology of AlSi-hBN particle.

2.2 Thermal aging procedure

Thermal aging experiment was carried out at 450 ºC in the static air by a SX-12-10 model resistance furnace (Beijing resistance furnace factory, Ltd., China). The hardness samples were taken out from the furnace periodically to make hardness test and when obvious coating hardness changes occurred the aging time was marked and 1 abradability sample, 3 bond strength samples and 1 microstructure evaluation were taken out and tested.

2.3 Properties and microstructure evaluation

The abradability tests were carried out on a high temperature and high speed abradability test rig (BKV-HVT 300/800model, Beijing General Research Institute of Mining and Metallurgy, China). In the work, parameters for abradability tests were set constant for all samples and are as follows: blade velocity is 300m/s, incursion rate is 50μm/s, set incursion depth is 500μm and temperature is 450 °C. And all coating samples were rubbed by Ti6Al4V dummy blades with a tip dimension of 0.7mm×10mm×3mm. Abradability was quantitatively evaluated by the ratio of the blade wear to coating incursion depth (IDR) which is created by Metco[7]. IDR is calculated according to the percentage of blade height change to the precise incursion depth into the coating (wear depth). The IDR was close to zero mean the improved abradability of the seal system. The ratio was positive when blade wear was predominant, otherwise negative value mean that adhesive transferred from the coating to the tip.

Coating harness was measured using Rockwell R15Y hardness scale (600MRD-S, WOLPERT, China) and the average values were calculated by 5 indentations for each coating. A WDW-100A mole tensile machine (Huaheng test Ltd., China) was using to test coating bond strength and the coating samples were bonded to the fixture using FM1000 films (Cytec Industries Inc, USA). The average bond strength values were calculated by 3 as tested results. Samples for microstructure evaluation were sectioned by a wire-electrode cutting machine and mounted using a vacuum impregnation system (Cast N 'Vac 1000 , BUEHLER, Illinois, United States) with epoxy resin. The specimens were grounded off more than 1.5mm with 80 # sand paper to get rid of the cutting effected zones before fine grounding and polishing. A FEI Quanta 200 model scanning electron microscope (SEM, FEI, Oregon, United States) was used to observe microstructure of coating. The average area fraction value of metallic phase in the coating section SEM picture was calculated based on five fields of view chosen randomly at 100x magnification by using a software ImageJ (National Institutes of Health, US). To determine the decomposition temperature of residual organic binder in the coating, a piece of as sprayed freestanding AlSi-hBN coating was subjected to differential thermal analysis (Thermo Plus EVO , RIGAKU , Japan).

3 Results and Discussion

Coating hardness values of samples aged by different time were plotted to the aging time in Fig. 2. A sharp decrease in coating hardness occurred after thermal aged for 2 hours at 450 °C. Coating hardness increased slowly till 300 hours and then became steady till 1000 hours. A significant reduction in coating hardness (from 65 R15Y to 50 R15Y) was found after the AlSi-hBN coating was thermal aged for 1000 hours at 450 °C. According to hardness test results, samples aged for 2h, 50h, 300h and 1000h were selected for abradability, bond strength tests and microstructure evaluation.

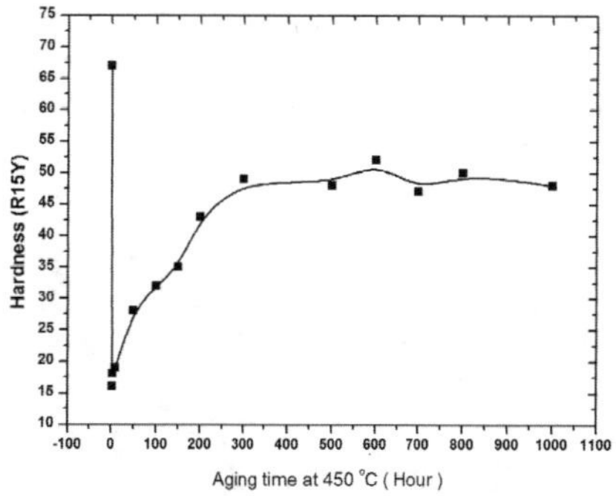

Fig.2 Hardness variation curve of coatings aged at 450°C for 1000h

Figure 3 illustrated the IDR values of coating aged for different hours. To help making intuitive comparison the bond strength was also plotted in Fig.3. It can be seen that IDR values and coating bond strength exhibited similar trend to coating hardness as the thermal aging time increasing. The IDR value decreased to 0 after being aged for 2 and 50 hours and then increased till 1000 hours. An obvious reduction in IDR value which means an improvement in abradability was made after being thermal aged for 1000 hours.

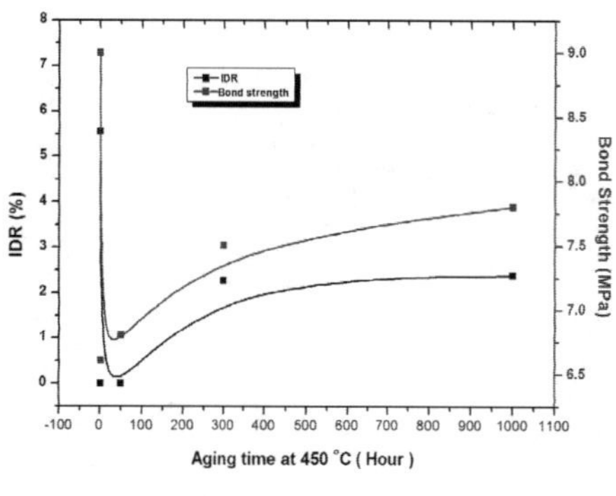

Fig.3 Transformation curve of IDR and bond strength with the increase of thermal aging time at 450°C

Images of AlSi-hBN coating microstructure for different aging time were given in Fig.4. AlSi-hBN coating consists mainly of AlSi matrix and hBN secondary filler, and pores and oxides are not distinguishable due to their similar contrast to hBN. It can be seen that with the elongation of aging time, hBN morphology and the metallic phase area fraction changed. For the first 50 hours hBN secondary filler tends to be coarser and exhibited more irregular shape, and the area fraction of metallic phase decreased compared with the as sprayed coating. The secondary filler became finer and shown a shape similar to the as sprayed coating as the increase of aging time. The metallic phase area fractions of coatings aged for different time were plotted in Fig.5 and a similar trend to coating hardness as the thermal aging time increasing is obvious.

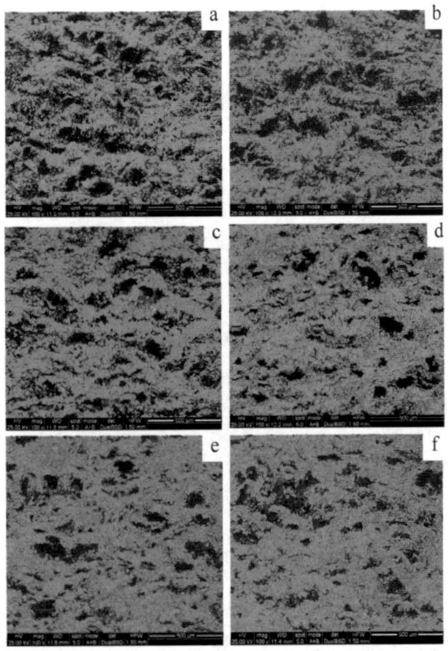

Fig.4 Images of coating microstructure thermal aged at 450°C for (a)0h, (b)2h, (c) 50h, (d) 300h, (e) 700h, (f) 1000h

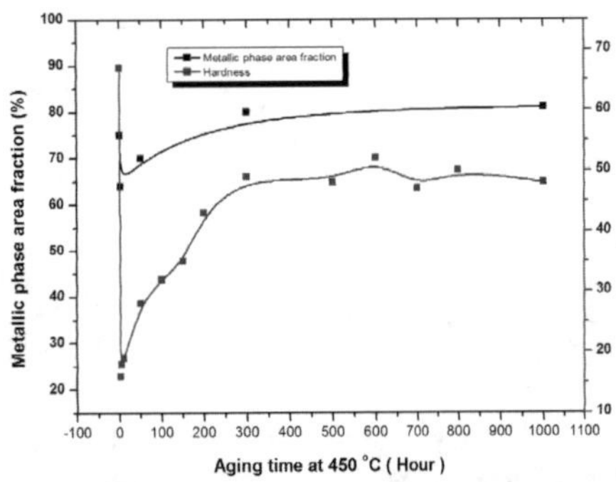

Fig.5 Transformation curve of metallic phase area fraction and hardness with the increase of thermal aging time at 450°C

AlSi-hBN feedstock powder consists of Al-12Si alloy, hBN filler and organic binder. Previous study revealed that a certain amount of organic binder will be left in the as sprayed coating. The as sprayed coating was subjected to a differential thermal analysis and an endothermic peak was found at 395 °C which indicated the decomposition temperature of organic binder in the coating. Clearly, as the coating was thermal aged for 2 hours at 450 °C the decomposition of organic binder occurred. Meanwhile, residual stress in the as sprayed coating should be also relieved. The decomposition of organic binder and the relief of residual stress should be the reason for sharp decrease of coating hardness after being aged for 2 hours. As the decomposition of organic binder the bonding effect of coating microstructure is weakened and a decrement in coating bond strength occurred. A loose microstructure morphology should be also concluded as the result of organic binder decomposition. Al-12Si eutectic alloy has a melting point of 577°C and sintering effect of AlSi metallic phase was protruding when coating was thermal aged at 450 °C for a prolonged period. And sintering effect of AlSi should be the reason for the slowly increase in coating hardness, bond strength and area fraction of metallic phase for prolonged aging hours. Sintering effect of metallic phase did not contribute to coating hardness increase as the aging time reached 300 hours and a contribution to bond strength continued till 1000 hours.

Rubbing at high velocity and high temperature is a complex process involving compress, tensile and shear between coating material and blades. There for abradability is a comprehensive coating property related to hardness, bond strength and microstructure. Clearly, the decrease of coating hardness, bond strength and metallic phase area fraction made it easier for blade rubbing into coating, hence abradability was increased as in the initial aging stage. In contrast, the abradability decreased as the increase of coating hardness, bond strength and metallic phase area fraction for prolonged aging period.

4 Conclusions

Thermal aging effect on properties and microstructure of AlSi-hBN coating was experimentally studied. AlSi-hBN coating was thermal aged at 450°C for 1000h in the atmosphere environment, and transformation in coating abradability, hardness, bond strength and microstructure with the increase of aged hours were investigated. It was found that the IDR value, hardness, bond strength and metallic phase area fraction all decreased in the first 2 hours and then grown up as the aging time increased. Coating hardness reached a stable state when aged for 300 hours and the other properties values and metallic phase area fraction risen slowly till 1000 hours. An improvement in abradability rubbed by Ti6Al4V dummy blades at 300m/s, 50µm/s and 450 °C was measured as the AlSi-hBN coating was thermal aged at 450°C for up to 1000 hours. The decomposition of organic binder and relief of residual stress in the first 2 aging hours were found to be responsible for the improvement of abradability, and decrease of hardness, bond strength and metallic phase area fraction. The sintering effect of AlSi matrix at 450 °C is the reason for slowly improvements in IDR values, coating hardness, bond strength and metallic phase area fraction with the increase of aging time.

5 Literature

[1] Majumdar S, Sengupta P, Kale G B et al. Surface and Coatings Technology, 2006, 200: 3713

[2] W.D. Marscher and C. Osborne, Abradable Turbine Shroud Development, Final Report, 1983

[3] X. Ma and A. Matthews, Surface and Coatings Technology, 2007, 202: 1214

[4] R.E. Johnston and W.J. Evans, Surface and Coatings Technology, 2007, 202: 725

[5] R. E. Johnston, Journal of Thermal Spray Technology, 2009, 18: 1004

[6] Rodolphe Bolot, Jean-Louis Seichepine, Journal of Thermal Spray Technology, 2011, 20: 39

[7] Richard. K. Schmid, New High Temperature Abradables for Gas Turbines, 1997, 83

[8] W.H. Xue, S.Y. Gao, D.L. Duan, Wear, 215, 322: 76-90

[9] M. Bounazef, S. Guessasma, and B.A. Saadi, Material Letters, 2004, 58: 3375

[10] M.O. Borel, A.R. Nicoll, H.W. Schläpfer et al. Surface and Coatings Technology, 1989, 39: 117

Research on Abradability of Different Abradables Rubbed against Ti_2AlNb Blades

Xuying Cheng1,2, Jianming Liu1,2, Yueguang Yu1*,2, Tong Liu1, Deming Zhang1, Jie Shen1
1. Beijing General Research Institutes of Mining and Metallurgy, Beijing, 100160, China
2. Northeastern University, Shenyang, 110819, China

Abstract: Ti_2AlNb Blade is one of the best choices for high pressure compressor of aero-engine to improve the thrust force and reduce the weight. A series of abradability tests were carried out on the CuAlNi-Graphite, NiCrAl-Graphite, NiCrAl-Bentonite, and NiCrFeAl-hBN abradable coatings rubbed against Ti_2AlNb dummy blades with the maximum blade-tip velocity up to 300 m/s at 500 ℃. In consideration of the effect of engine's working condition, some tests with incursion rate as the single variable were carried out. The scratched surfaces of the samples were observed by the stereoscopic optical camera, and a ratio of the blade wear to shroud incursion depth (IDR) was introduced to characterize the abradability of coatings. The results show that NiCrAl-Graphite and NiCrFeAl-hBN abradable coatings perform very well rubbed against the Ti_2AlNb blade, and the blade-tip wear is not obvious after abradability tests.

1 Introduction

As aero engines are developing to higher thrust force and lower weight, titanium based materials are more and more inclined to be the choice of engine structural materials. Ti_2AlNb alloy is a new type of Ti-Al intermetallic developed in recent years. Compared with the conventional high temperature titanium alloy, it's of many advantages such as high strength, excellent creep resistance, outstanding specific strength and very good ability of oxidation resistance at higher temperature. Ti_2AlNb alloy has been an ideal candidate for lightweight high-temperature structural materials [1, 2]. After research and development for decades, the key technology of Ti_2AlNb alloy processing has been mastered, and this alloy has been successfully applied in the engine rotor and stator [2].
The abradable coatings are widely used on the inner face of the stator in the aero-engines to control the clearance between the rotor and stator, which made the engine to be assembled as tightly as possible. This technique could effectively improve the aero-engine efficiency and reduce the fuel consumption [3]. A lot of researches about abradability evaluation for abradable seal coatings rubbed against Ni-based superalloy and stainless steel rotor under high temperature and high speed simulated conditions have been carried out. However, the alternative change for material choice of the rotor structure needs more advanced seal coating to match. For instance, the wear resistance of Ti_2AlNb alloy blade is inferior to that of nickel based superalloy blade, so better abradability of seal coating is required.
In this work, special experiment apparatus was employed to simulate the high velocity and temperature operating condition for rotor rubbed against seal coating. The abradability of CuAlNi-Graphite, NiCrAl-Graphite, NiCrAl-Bentonite, and NiCrFeAl-hBN abradable coatings rubbed against Ti_2AlNb dummy blades were investigated. This work can support effective experiment data for the engine choosing appropriate seal coating.

2 Experiments

2.1 Material

This series of tests were performed using four thermally sprayed abradable coatings, which are commonly used in the compressor section of the aero-engine, and **Table 1** shows the main information of coatings. The coatings generally consist of two components: metal matrix and solid lubricant. The importance of the metallic matrix is to give the coating its oxidation resistance and interparticle cohesive strength, while solid lubricant helps to minimize the coating's transfer to the blade material. All coatings were sprayed to the thickness of 1.9~2.0 mm and the hardness of 50~80 HR15Y, whose surfaces were grinded to the roughness below 10 μm with 400 grit SiO_2 grinding paper. **Fig. 1** shows the SEM pictures of coating microstructures.

Table 1. Main composition of the abradable coatings

Coating name	Naminal composition	The maxim service temperature
CuAlNi-Graphite	CuAlNi-14%Graphite	550℃
NiCrAl-Graphite	NiCrAl-13%Graphite	650℃
NiCrAl-Bentonite	NiCrAl-21%Bentonite	650℃
NiCrFeAl-hBN	NiCrFeAl-6.5%hBN	750℃

2.2 Testing Rig and Testing Parameters

To simulate the actual operating conditions of aero turbine engines, abradability tests were carried out on the BGRIMM BKY-HVT 300/800 abradablity test rig [4]. This rig is capable of operating at up to 800 °C. Blade tip speeds can be adjusted from 100 to 350 m/s with corresponding incursion rate settings ranging from 5 to 1000 μm/s and thus representing a wide range of possible engine rub conditions. The testing parameters are listed in **Table 2**. The abradability testings were done at 500 ℃ and the maximum blade tip velocity up to 300 m/s. Taking the effect of engine's working condition into consideration, the incursion rate varied between different tests. All coatings were rubbed by the Ti_2AlNb dummy blades with a tip thickness of 0.7 mm.

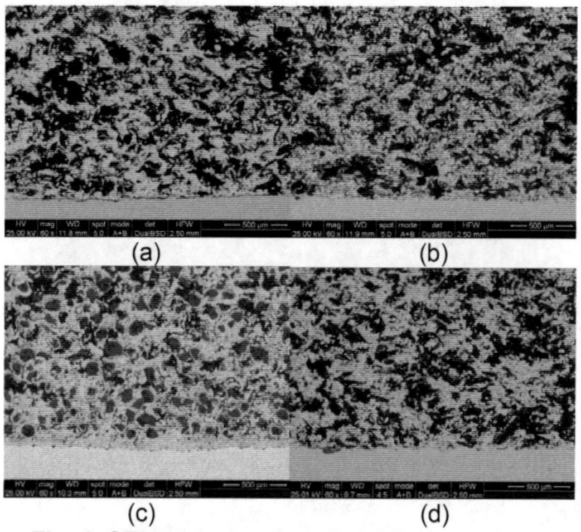

Fig. 1. SEM pictures of coating microstructures, a:CuAlNi-Graphite, b:NiCrAl-Graphite, c:NiCrAl-Bentonite, d:NiCrFeAl-hBN.

The blade height is measured before and after the test to calculate the height change. The coating scratch depth is calculated geometrically using the scratch length. And a ratio of the blade wear to shroud incursion depth (IDR) [5] is defined as the percentage of the blade height change to the coating scratch depth, which is used to indicate the abradability of coatings. If the blade height is reduced, the IDR is marked as positive. If coating material adhesion was observed on the blade tip causing the increase of the blade height, the IDR is marked as minus.

Table 2. The selected experimental parameters of abradablitytestings

Test number	Temperature (℃)	Blade tip velocity (m/s)	Incursion rate (μm/s)	Incursion depth (μm)
1	500	300	5	500
2	500	300	50	500
3	500	300	480	500

3 Results and Discussion

Fig. 2. shows the calculated IDRs of four coatings at the different incursion rate. It can be found that all of Ti_2AlNb dummy blades experienced different degrees of wear because all IDRs were positive. For a fixed incursion depth, the closer to zero the IDR is, the better the abradable seal system is. The NiCrAl-Graphite and NiCrFeAl-hBN coatings exhibit obviously good abradability and as the incursion rate decreases the IDR is reduced. This indicates that the coating wear plays the dominant role in the rubbing process, while severer blade-tip wear is only done at a great speed of incursion. The Ti_2AlNb blade-tip experienced the severest wear against the CuAlNi-Graphite coating and the less incursion rate resulted in the higher IDR. The counterpart wear was mainly focused on the blade-tip material loss and the wear degree became even worse with the incursion rate decreased. The performance of NiCrAl-Bentonite coating is little better than that of CuAlNi-Graphite coating but still worse than those of the other two coatings.

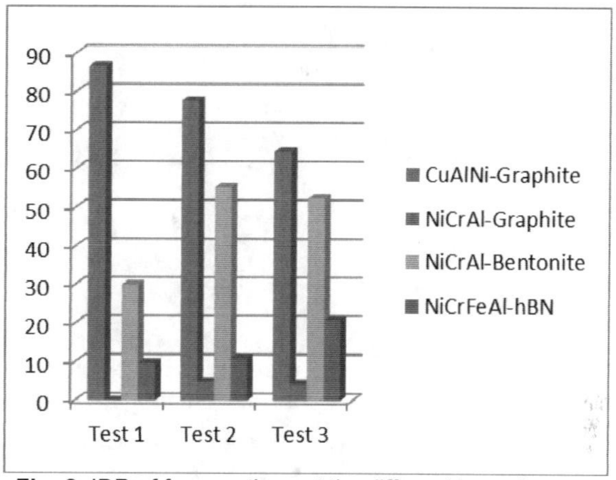

Fig. 2. IDR of four coatings at the different incursion rate.

All samples were photographed by stereo microscope after testing, see Fig. 3 ~ 6. The stereoscopic optical pictures display the wear surface morphology of the blade and the coating under the rubbing condition that the incursion rate is controlled as the only variable. The surface morphology of CuAlNi-Graphite coatings and its corresponding blades revealed that the blades suffered severe wear and the coating scratching surface was very short within some clearly grooving (seen in Fig. 3). It should be pointed that the surface morphology of the CuAlNi-Graphite coating sample for test condition 1 also shows trails caused by local over-heating that resulted from a low incursion rate and a longtime rubbing. As the incursion rate increased, the grooving become less obvious and some pocks appeared which were formed by micro-rupture. Apparently, the incursion rate is a sensitive factor to the wear mechanism of CuAlNi-Graphite abradable coating. The low incursion rates induced significant grooving and over-heating, while the main wear mechanism is micro-rupture and grooving at high incursion rate.

A similar morphology of blade and substrate was also observed for NiCrAl-Bentonite coatings (Fig. 4). Grooving also occurred in the coating which is unfavorable for the abradability and the abraded area on the substrate showed that a lot of pock marked or pitted due to the break-out of packets of wear particles, together with ploughing of the blade tip, which should resulted in much higher IDRs than others. It may indicate that grooving and micro-rupture are the predominant wear mechanisms for NiCrAl-

Bentonite coating rubbed against the Ti_2AlNb blade at 500 ℃ with the blade tip velocity of 300 m/s.

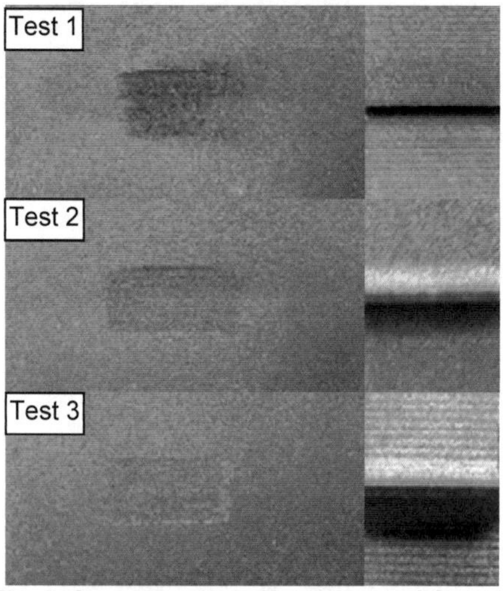

Fig. 3. Stereoscopic optical pictures of CuAlNi-Graphite coatings and Ti_2AlNb dummy blades.

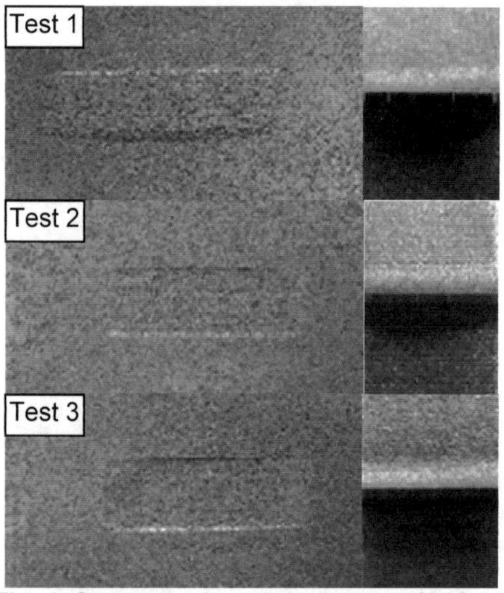

Fig. 4. Stereoscopic optical pictures of NiCrAl-Bentonite coatings and Ti_2AlNb dummy blades.

For the tests of NiCrAl-Graphite and NiCrFeAl- hBN coatings, a relatively long and uniform abraded area on the substrate and much smoother blade tip profile were observed. The phenomenon indicated that these two coatings were cut by Ti_2AlNb blades, together with the less change of the blades. The tests exhibited a desired cutting wear mechanism of the NiCrAl-Graphite and NiCrFeAl- hBN coating. So the coatings performed satisfying abradability when they were rubbed against the Ti_2AlNb blade under the rubbing condition of the high temperature of 500 ℃ associated with the high blade tip velocity of 300 m/s.

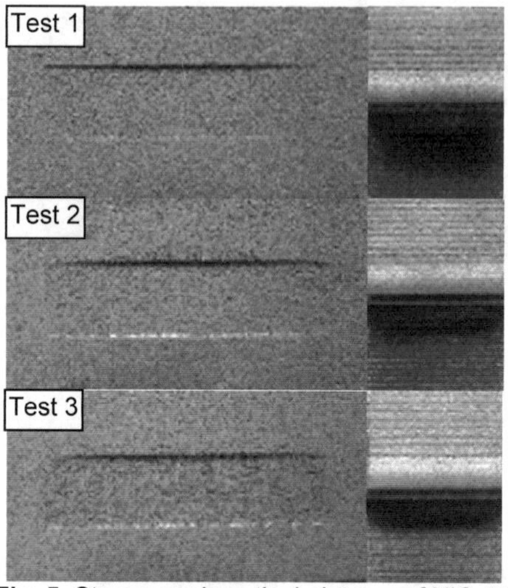

Fig. 5. Stereoscopic optical pictures of NiCrAl-Graphite coatings and Ti_2AlNb dummy blades.

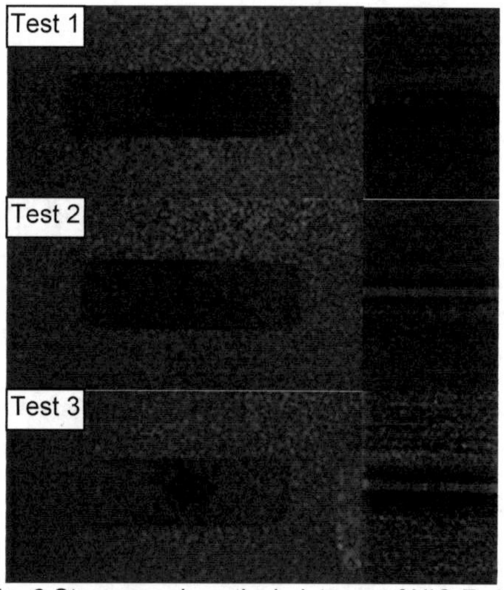

Fig. 6.Stereoscopic optical pictures of NiCrFeAl-hBNcoatings and Ti_2AlNb dummy blades.

4 Conclusions

A series of abradability tests were carried out on the CuAlNi-Graphite, NiCrAl-Graphite, NiCrAl- Bentonite, and NiCrFeAl-hBN abradable coatings rubbed against Ti_2AlNb dummy blades with the maximum blade-tip velocity up to 300 m/s at 500 ℃. The single variable tests of the incursion rate (5 μm/s、50 μm/s and 480 μm/s) were done. The results show that NiCrAl-Graphite and NiCrFeAl-hBN coatings performed good abradability and exhibited a cutting morphology under the different incursion rates. As to CuAlNi-Graphite coating, the incursion rate has a significant effect on the wear mechanism, which can be transformed from grooving/over-heating to grooving/micro-rupture with the incursion rate increased. For NiCrAl-Bentonite coating, the wear mechanism may be the synergism of the grooving and micro-rupture effects. The NiCrAl-

Graphite and NiCrFeAl-hBN coating show a desired cutting wear mechanism. So both the two performed satisfying abradability.

In a conclusion, the NiCrAl-Graphite and NiCrFeAl-hBN coating show better performance of abradability rubbed against the Ti_2AlNb blade than those of the CuAlNi-Graphite and NiCrAl- Bentonite coating. The NiCrAl-Graphite and NiCrFeAl-hBN coating are very promising as the abradable seal coating when Ti_2AlNb alloy were applied in the engine rotor.

5 Literature

[1] A. K. Gogia, T. K. Nandy, D. Banerjee et al. Microstructure and mechanical properties of orthorhombic alloys in Ti-Al-Nb system. Intermetallics, Vol. 6 (1998), Issue 7-8, pp. 741~748.

[2] R. G. Rowe. Ti_2AlNb based alloys outperform conventional titanium aluminides. Advanced Materials and Processes, 1992, Issue 3, pp. 33~35.

[3] F. Ghasripoor, R. Schmid, and M. Dorfman. Abradable coatings increase gas turbine efficiency. Materials World, Vol. 5 (1997), Issue 6, pp. 328-330.

[4] Q. Y. Lu, J. Shen, Y. G. Yu et al. Progress on facilities and methodology to evaluate abradable seal coatings. Advanced Materials Research, Vols. 690-693 (2013), pp. 1992-1998.

[5] R. K. Schmid. New high temperature abradables for gas turbines. Diss. ETH No. 1223 (1997), pp. 83-89.

Effect of Mo content on tribological properties of atmospheric plasma-sprayed Mo-NiCrBSi composite coatings under dry and oil-lubricated conditions

H.F. Xu, L.M. Liu, J.K. Xiao, C. Zhang, Yangzhou/CN, G. Zhang, Qingdao/CN, H. Liao, Belfort/F

NiCrBSi is a material generally used in wear-resistant coatings. In order to improve the tribological properties of atmospheric plasma-sprayed NiCrBSi coatings, Molybdenum (Mo) was incorporated into the NiCrBSi coatings to reduce the friction coefficient and wear rate under dry and oil-lubricated conditions. In this paper, Mo-NiCrBSi composite coatings with Mo content of 5, 10, 20 and 30 wt.% were deposited on stainless steel substrates respectively by atmospheric plasma spray. X-ray diffraction, optical microscopy and scanning electron microscopy equipped with energy dispersive spectroscopy were utilized to investigate the phase structure and surface morphology of the composite coatings. Reciprocating friction tests were conducted to measure the friction coefficients and 3D optical microscopy was used to depict the wear track profiles. The results showed that the 30 wt.% Mo-NiCrBSi coating exhibits the best tribological performance. In addition, MoO_2 and MoS_2 films were formed in the friction process under dry condition and oil-lubricated condition respectively.

1 Introduction

In the past decades, thermal sprayed coatings gradually became an alternative of electroplating chrome coatings since electroplating produces hexavalent chromium ions having a long-term toxic effects on human body and natural environment [1]. NiCrBSi feedstock powder is heated and accelerated to a high velocity by thermal spray gun, and then impacted on the substrate. The melted or partially melted powder solidifies and cools rapidly after spreads on the substrate surface, which ultimately forms splats. Therefore, thermal-sprayed NiCrBSi coatings usually have a structure of lamellar splats with good wear performance. Planche et al. [2] prepared NiCrBSi coatings by flame spray (FS), atmospheric plasma spray (APS) and high-velocity oxygen-fuel spray (HVOF). The in-flight velocity of powder increased from 39, 137 to 439 m/s, and the thickness of the lamella in the coatings decreased correspondingly. They found that high velocity is favorable for the particle flattening and formation of dense coating. The hardness of NiCrBSi coating is generally about 700 - 1000 HV, which is one of the most important reasons that NiCrBSi coatings have good wear-resistant performance [3]. The hardness of NiCrBSi coating is directly related with γ-Ni phase (matrix phase) and precipitate compound phase (enhanced phase). With respect to the effects of elements, the presence of Cr improves the hardness and corrosion resistance of the alloy. B and Si reduce the melting point and improve the self-fluxing properties. B also forms hard phases with Ni, such as Ni_3B, thereby enhancing the hardness of sprayed coatings [4, 5]. The wear resistance of NiCrBSi coatings are proved to be good, but in the aspect of self-lubrication, it still needs more research [6, 7].

Mo has a high melting point of 2620°C and a low hardness of 170 HV. Under dry friction condition in air atmosphere, MoO_2 layer can be formed on the contacting surface. MoO_2 has a high hardness, meanwhile, as a solid lubricant, MoO_2 significantly reduces the friction coefficient between coating and counterpart [8]. Therefore, Mo has been successfully used in many applications, such as automobile synchronous ring and piston ring [9, 10].

In certain conditions, high hardness and low friction coefficient are simultaneously needed. Mo and NiCrBSi can be used to obtain a composite material. The incorporation of Mo into NiCrBSi coating also improves the anti-occlusion performance of coatings [11]. During the friction process, element Mo on the surface can be oxidized by oxygen in the air and then form a MoO2 film rapidly [12]. The MoO_2 film has good lubricating performance, and also prevents the direct adhesion on the contact surface of friction pair, which plays an important role in friction. On the other hand, in boundary or oil-lubricated conditions, Mo may be reacted into MoS_2 [13]. MoS_2 is a lubrication material with good lubrication and anti-corrosion abilities. After adding Mo into NiCrBSi coatings, Mo can be easily sulfured under high temperature and pressure and form MoS_2 film, even when oil is exhausted of oil. Niranatlumpong et al. [14] studied the influence of Mo content on the friction and wear properties of Mo-NiCrBSi coating. They found that 25wt.% Mo-NiCrBSi coating has a lower hardness, a higher porosity and a better wear performance than pure NiCrBSi coating. Excessive Mo reduces the interlayer adhesion and results in coating failure [15, 16]. With regard to the performance under oil-lubricated condition, there are few reports on the Mo-NiCrBSi coatings when different contents of Mo are incorporated.

2 Experimental

2.1 Material preparation

The substrates were 304 stainless steel with size of 40 mm×60 mm×2.5 mm. Prior to spray, the samples were grit blasted with 24-mesh corundum sand and the pressure of the compressed air was 0.4 MPa. The size of NiCrBSi powder was in range of 40-60 μm. The particle size of Mo powder was 44-74 μm and the purity was higher than 99.9%. The fraction of Mo in the composite powders were fixed as 0, 5, 10, 20 and 30 wt.% respectively, and the composited powders was mixed at 200 r/min for two hours by a ball-milling machine. Prior to spray, composite powders were dried in an oven at 100°C for two hours to increase the liquidity.

2.2 Coating deposition

The atmospheric plasma spray F4MB gun (Oerlikon Metco, Switzerland) was used for coating deposition. The six-axis robot arm (ABB, Sweden) was used to achieve a uniform and reproducible deposition. During the spraying, the substrate was cooled with 0.2 MPa compressed air and the coating thickness was controlled between 100-150 μm. Other spray parameters were listed in Table 1.

Table 1: Plasma spraying parameters.

Parameters	Value
Current /A	565
Voltage /V	66
Ar flow /L·min^{-1}	50
H_2 flow / L·min^{-1}	9
Powder feed rate / g·min^{-1}	30
Spray distance /mm	120
Torch traverse speed /mm·s^{-1}	100

2.3 Wear tests and characterization

The cross-sectional morphology of the coatings was observed by Leica optical microscopy. Phase structure were analyzed using an X-ray diffractometer (Bruker D8 Advance) with Cu-Kα radiation. Vickers hardness were measured using a load of 100 grams and a load time of 10 seconds. 15 indentations on different areas of the coatings were performed. Then the average and variance values were calculated.

The reciprocating friction tests were performed on a UMT-2 tribometers under dry and oil (SAE 5W-40) lubricated conditions. Si_3N_4 balls with a diameter of 4 mm were used as counterpart. The load was fixed at 10 N and sliding speed was 10 m/s. Stroke length was 8 mm and the total friction length was 60 m. The friction coefficient was recorded automatically by friction force/load. The track profile was observed by a three-dimensional optical microscopy (Contour GT-K Bruker). Wear surface was inspected by field emission-scanning electron microscopy (FE-SEM) and optical microscopy.

3 Results and discussion

3.1 Coating microstructure

Fig. 1 shows the cross-sectional images of Mo-NiCrBSi coating. The thickness of the four coatings was about 100-150 μm. It can be clearly seen that Mo (light gray) and NiCrBSi (dark gray) are distributed well in the coatings, indicating that the composite powders are melted and spread well during spraying. In addition, black void can be observed, which may be attributed to the presence of pores and peeling off of oxide particles during metallographic polishing.

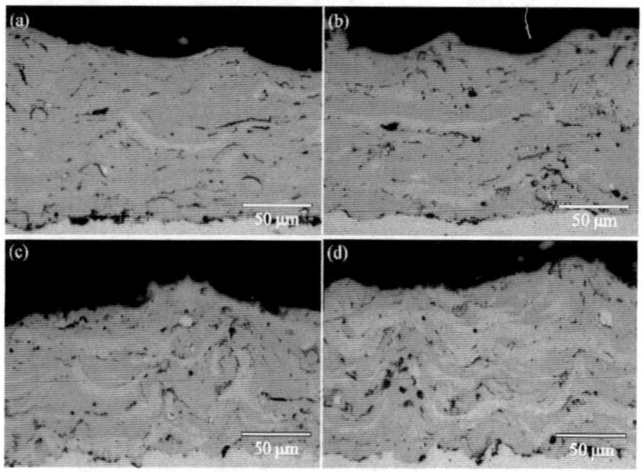

Fig. 1: Cross-sectional microstructure of Mo-NiCrBSi coating (a) 5%Mo-NiCrBSi, (b) 10%Mo-NiCrBSi-, (c) 20%Mo-NiCrBSi, (d) 30%Mo-NiCrBSi.

Fig. 2 is XRD patterns of the five coatings. It can be seen that the main peak of these coatings are Mo at 40.54° and γ-Ni at 44.38° respectively, as well as hard phases such as $Cr_{13}Ni_5Si_2$, Ni_3B, Cr_2B, Cr_3C_2. The hard phase in the coating has a positive performance on the hardness. Furthermore, it can be clearly seen that the intensity of Mo peak become more obvious with the increase of Mo content.

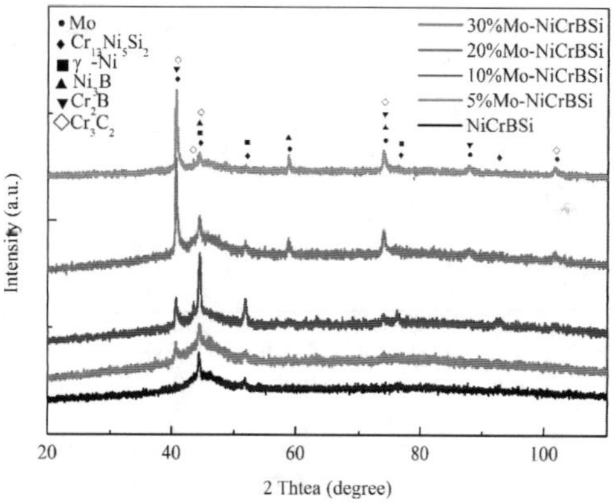

Fig. 2: XRD patterns of the NiCrBSi and Mo-NiCrBSi composite coatings.

3.2 Hardness of Mo-NiCrBSi coatings

The micro hardness of the coating is an important index to evaluate the coating quality, which influences the strength, wear resistance and service life of the coatings. In this work, the microhardness was measured on the polished cross-sectional surface and the distance between two measuring points was 3 times longer than the diagonal of indentation. The hardness of pure NiCrBSi coating is 903.6±59.7 $HV_{0.1}$. With the increase of Mo content, the hardness decreases gradually, from 765.1 ± 106.0 $HV_{0.1}$ for

5%Mo-NiCrBSi to 683.1 ± 63.9 $HV_{0.1}$ for 30%Mo-NiCrBSi. It can be deduced that the reduction of coating hardness should be attributed to the low hardness of Mo.

3.3 Tribological performance

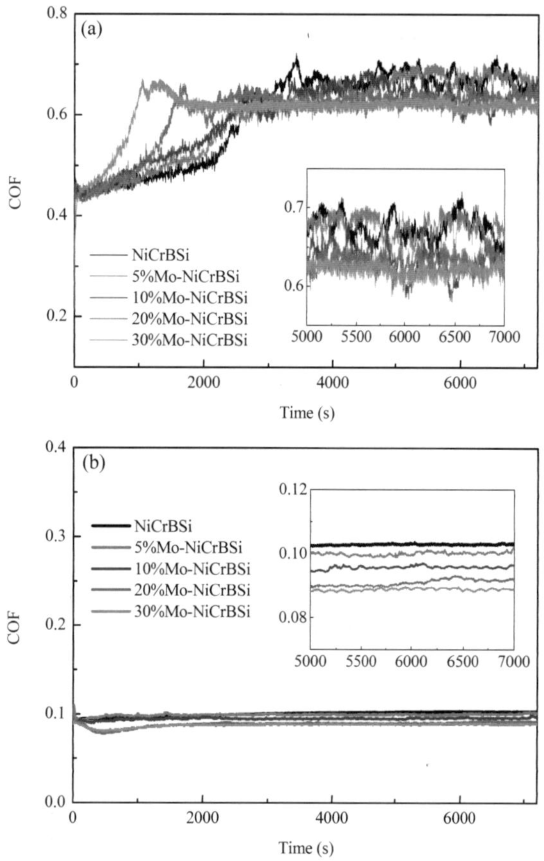

Fig. 3: Friction coefficients of the Mo-NiCrBSi coatings under (a) dry friction (b) oil lubricated friction.

As shown in **Fig. 3a**, the friction coefficients of the coatings are increased in the initial few minutes and then get stable. This change should be related with the polished coating surface prior to friction. It can be further found that time needed for the unstable period has a direct relationship with Mo content. When Mo content is high, the lubricating MoO_2 makes the coating to reach a stable stage quickly. The friction coefficients vary between 0.6 and 0.7, and it decreased with the increment of Mo content, which shows that the addition of Mo reduces the friction coefficients of the composite coatings slightly.

The friction coefficients of the composite coatings under oil-lubricated conditions were given in **Fig. 3b**. The coefficients reduced significantly under oil lubricated condition, and are stabilized at about 0.09-0.10. Similar to dry condition, Mo content has a slight and positive influence on friction coefficients. The fluctuations of coefficients under oil-lubricated condition are much smaller than that under dry condition. This should be attributed to that the dry friction are greatly affected by the irregular coating surface whereas an oil film forms between coating and counterpart ball during oil-lubricated friction.

The volume loss was measured using three-dimensional optical microscopy. Under dry friction condition, the NiCrBSi coating was 4359 μm³/N·m and it was decreased to 1422 μm³/N·m when the Mo content was 30 wt.%. The results show that Mo has a good anti-wear performance. Similarly, Mo also shows a positive effect on the wear rates when under the oil lubricated condition. The wear rate under oil-lubricated condition were decreased to c.a. 1/20 of those under dry condition. The depth of the track and wear rate of the coatings are inversely proportional. The deepest wear track is pure NiCrBSi coating. Wear track depth decreases with the increment of Mo content. The wear profiles after oil lubricated conditions show an obviously improvement compared with those after dry friction. Table 2 lists the wear track parameters. The values indicated that the Mo enhances the wear resistance.

MoO_2 is well-known to play an important role in the dry friction. In order to confirm the existence of MoO_2 on the coating surface, EDS scanning is performed before and after dry friction, results were shown in Table 3. Mass fraction of oxygen after the friction test increased from 0% to 1.1%, which indicates that Mo in the coating surface reacts with O_2 and form a MoO_2 layer during the friction process. When the experiments were conducted under oil-lubricated condition, the friction coefficients of the five coatings were in an extremely low and stable state. The porosity in the coatings play a role of oil storage making the friction pair unveiled in the oil during the whole process.

Table 2: Track parameters for the five coatings.

	Dry friction		Oil lubricated friction	
	volume loss /μm³	depth /μm	volume loss /μm³	depth /μm
NiCrBSi	1.50×10^8	10.25	1.08×10^6	1.43
5%Mo-NiCrBSi	1.40×10^8	7.43	8.01×10^5	0.96
10%Mo-NiCrBSi	1.21×10^8	6.57	2.05×10^5	0.36
20%Mo-NiCrBSi	0.95×10^8	5.15	0.89×10^5	0.15
30%Mo-NiCrBSi	0.48×10^8	3.26	0.41×10^5	0.08

Table 3: EDS compositions of the 30%Mo-NiCrBSi coating before and after wear test (wt.%).

Dry friction	O	Si	Cr	Fe	Ni	Mo
Before wear	0.0	3.0	14.7	4.3	47.6	30.3
After wear	1.1	2.2	9.3	3.0	33.3	51.0
oil lubricated friction	Si	C	Cr	Fe	Ni	Mo
Before wear	2.4	0.0	18.1	4.7	56.6	17.2
After wear	3.5	0.3	11.2	4.0	39.7	42.3

As a result of elimination of most oxygen in air, Mo reacted with S in the oil to generate a better lubricant MoS_2. The elemental mass change before and after oil lubricated process are analyzed by EDS shown in Table 3. There is 0.34 wt% S in the coating surface even after cleaned by acetone for 5 hours in ultrasonic cleaning machine. In contrast, the S content in the non-frictional part under the same condition is 0 wt%, which suggests that S has been in the state of compound (MoS_2) in the wear surface.

3.4 Wear mechanisms

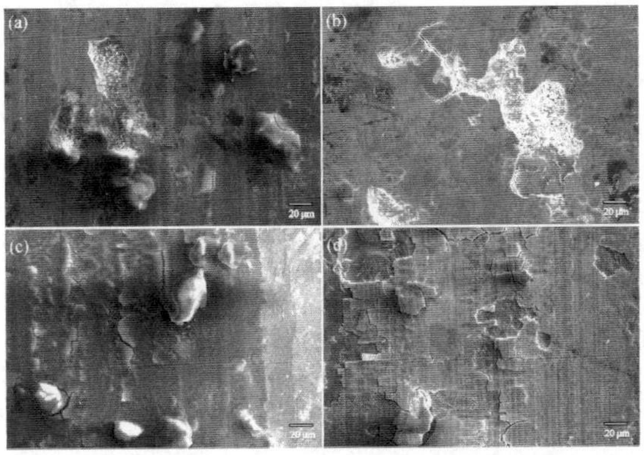

Fig. 4: FE-SEM graphs of (a) 5%Mo-NiCrBSi, (b) 10%Mo-NiCrBSi, (c) 20%Mo-NiCrBSi and (d) 30%Mo-NiCrBSi coatings after dry friction.

There are many peelings in the pure NiCrBSi coating after the dry friction, which should be ascribed to the deformation caused by compressive stress. Cracks and pores are the inherent properties of atmospheric plasma-sprayed coatings, after a long-time friction, stresses accumulate at the interlayer, the unbounded areas increase and eventually caused particles falling off and form abrasive debris. In addition, there are many grooves along the sliding direction, which should be ascribed to the plowing effect caused by the falling off particles between friction pair. Therefore, under dry friction condition, the wear mechanisms involved in the NiCrBSi coating mainly are abrasive wear and adhesive wear.

Fig. 4 shows worn surfaces of 5, 10, 20 and 30 wt% Mo-NiCrBSi coatings after dry friction respectively. The friction coefficients and track profile data showed that 30% Mo had a best wear-resistant performance. When the Mo content is 5 wt%, as shown in **Fig. 4a**, the worn surface still has similar peeling off particles and grooves as the pure NiCrBSi coating. When the Mo content increases, the number of peeling off particles decreases while a layer structure began to appear due to the form of MoO_2 with a good lubrication properties at high temperature and repeated extrusion. During the friction process, MoO_2 formed a lubricating layer between the coating and the ball, acting as a lubricating film and effectively reduced the friction coefficient and avoided the formation of abrasive particles. Therefore, the main wear mechanism in the Mo-NiCrBSi coatings should be adhesive wear.

More information can be observed in the SEM graphs. From **Fig. 4c** and **4d**, it can be observed that the lubricating film have fish-scale cracks, due to the high flash temperature in the friction process. MoO_2 film were continuous with cracks on the surface. After the friction process, the temperature decreased rapidly and cracks have been produced, because of the difference in thermal expansion coefficient between MoO_2 and NiCrBSi [17, 18], and eventually formed these cracks.

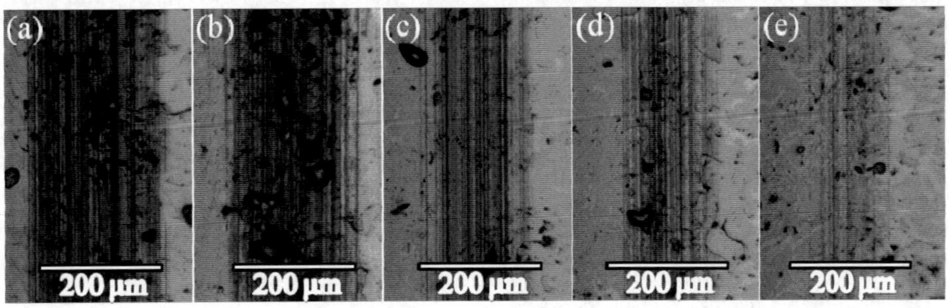

Fig. 5: Images of wear scars (a) NiCrBSi, (b) 5%Mo-NiCrBSi, (c) 10%Mo-NiCrBSi, (d) 20%Mo-NiCrBSi and (e) 30%Mo-NiCrBSi coatings after oil-lubricated friction.

Under oil-lubricated conditions, the friction behavior is simplified, since oil can form an oil film and effectively decrease the friction coefficient. At the same time, Mo reacted with S in the oil and generated MoS_2 film in the surface of Mo-NiCrBSi coating. The MoS_2 film further reduces the wear rate effectively. It is clearly shown in **Fig. 5** that the friction mechanisms in the oil-lubricated conditions are plowing, which should be attributed to the self-lubrication ability of Mo and its sulfide in the coatings. In sample (a) and (b), a large number of peeling is observed, which is similar to dry friction condition. The lack of MoS_2 film may be the reason. With the increment of Mo content. In addition, the depth of wear track is decreased with the increment of Mo content, from 1.43 μm in the NiCrBSi coating to 0.08 μm in the 30%Mo-NiCrBSi coating.

4 Conclusions

In this paper, NiCrBSi coatings and 5, 10, 20 and 30wt.% Mo-NiCrBSi composite coatings were deposited by atmospheric plasma spray. These coatings were characterized and performed under dry and oil-lubricated conditions. Some conclusions can be drawn:

a) Atmospheric plasma-sprayed Mo-NiCrBSi coatings have a good particle flattening.

b) Among the five coatings prepared, 30%Mo-NiCrBSi coating shows the best friction and wear properties.

c) With the increment of Mo content from 0 to 30%, the wear mechanism changes from coexistence of abrasive wear and adhesive wear to adhesive wear.

Mo forms MoO_2 and MoS_2 films under dry and oil lubricated conditions, effectively reducing the friction coefficients and wear rates.

5 Literature

[1] Kocberber, N. and Donmez, G.: Chromium (VI) bioaccumulation capacities of adapted mixed cultures isolated from industrial saline wastewater. Bioresource Technology 98 (2007), pp. 2178-2183.

[2] Planche, M.P., Liao, H., Normand, B. and Coddet, C.: Relationships between NiCrBSi particle characteristics and corresponding coating properties using different thermal spraying processes. Surface and Coatings Technology, 200 (2005), pp 2465-2473.

[3] Miguel, J.M., Guilemany, J.M., and Vizcaino, S.: Tribological study of NiCrBSi coating obtained by different processes. Tribology International, 36 (2003), pp 181-187.

[4] Verné, E., Ferraris, M., Ventrella A., Paracchini, L. Krajewski, A. and Ravaglioli, A.: Sintering and plasma spray deposition of bioactive glass-matrix composite for medical application. Journal of European Ceramic Society, 18 (1998), pp 363-372.

[5] Navas, C., Colaço, R., de Damborenea, J. and Vilar, R.: Abrasive wear behaviour of laser clad and flame sprayed-melted NiCrBSi coatings. Surface and Coatings Technology, 200 (2006), pp 6854-6862.

[6] Krishnaveni, K., Sankara, T.S.N., Narayanan, and Seshadri, S.K.: Electroless Ni–B coatings: preparation and evaluation of hardness and wear resistance. Surface and Coatings Technology, 190 (2005), pp 115-121.

[7] Houdková, Š., Smazalová, E., Vostřák, M. and Schubert, J.: Properties of NiCrBSi coating, as sprayed and remelted by different technologies. Surface and Coatings Technology, 253 (2014), pp 14-267.

[8] Serres, N., Hlawka, F., Costil, S., Langlade, C., Machi, F. and Cornet, A.: Dry coatings and ecodesign Part. 2 - Tribological performances. Surface and Coatings Technology, 204 (2009), pp 197-204.

[9] Laribi, M., Mesrati, N., Vannes, A.B. and Treheux, D.: Adhesion and residual stresses determination of thermally sprayed molybdenum on steel. Surface and Coatings Technology, 166 (2003), pp 206-212.

[10] Hwang, B., Lee, S., and Ahn, J.: Correlation of microstructure and wear resistance of molybdenum blend coatings fabricated by atmospheric plasma spraying. Materials Science and Engineering: A, 366 (2004), pp 152-163.

[11] Tura, J.M., Traveria, A., de Castellar, M.D., Pujadas, J., Blouet, J., Gras, R., Magham, H.G., Belair, P., Hanau, T. and Romero, A.: Frictional properties and wear of a molybdenum coating and a bronze (Cu–10%Sn) with friction modifier fillers. Wear, 189 (1995), pp 70-76.

[12] Laribi, M., Mesrati, N., Vannes, A.B. and Treheux, D.: Adhesion and residual stresses determination of thermally sprayed molybdenum on steel. Surface and Coatings Technology, 166 (2003), pp 206-211.

[13] Wayne, S.F., Sampath, S. and Anand, V.: Wear mechanisms in thermally-sprayed Mo based coatings. Tribology Transactions, 37 (1994), pp 636-640.

[14] Niranatlumpong, P. and Koiprasert, H.: The effect of Mo content in plasma-sprayed Mo-NiCrBSi coating on the tribological behavior. Surface and Coatings Technology, 205 (2010), pp 483-489.

[15] Wayne, F., Sampath, S. and Anand, V.: Wear mechanisms in thermally-sprayed Mo based coatings. Tribology Transactions, 37 (1994), pp 636.

[16] Usmani, S. and Sampath, S.: Time-dependent friction response of plasma-sprayed molybdenum. Wear, 225-229 (1999), pp 1131-1136.

[17] Alfonso, J.E., Garzo´n, R. and Moreno, L.C.: Behavior of the thermal expansion coefficient of a-MoO_3 as a function of the concentration of the Nd^{3+} ion. Physica B, 407 (2012), pp 4001–4004.

[18] Xuan, H.F., Wang, Q.Y., Bai, S.L., Liu, Z.D., Sun, H.G. and Yan, P.C.: A study on microstructure and flame erosion mechanism of a graded Ni–Cr–B–Si coating prepared by laser cladding. Surface and Coatings Technology, 244 (2014), pp 203–209.

Direct Metal laser sintering (DMLS) technique for prospective wear related applications

G. Grewal[1], G.Kaushal[2] B Krishan[2]
[1] *Research Scholar*, Punjabi University, Patiala, Punjab, India

[2]*Assistant Professor*, Mechanical Engineering Section, Yadavindra College of Engineering, Punjabi University, Guru Kashi Campus, Talwandi Sabo, Punjab, India

Direct Metal Laser Sintering (DMLS) technique is one of the technologies which is generally used to built prototypes and tooling applications. DMLS uses powder bed fusion to bond particles together by laser energy. A new powder layer is spread on top of the previous layers and the process is repeated up to required shape of part can be produced.
This review paper presents development, current status and challenges of the DMLS technique with emphasises on material processed by DMLS and is aimed to understand influence of density, microstructure, micro-hardness, tensile strength and wear behaviour of built-up parts. It also highlights the process through proofs based on classical results in terms of advantages and applications.

Keywords- Direct Metal Laser Sintering (DMLS), Wear, and Linear Laser Energy Density (LED)

1. INTRODUCTION

Direct Metal Laser Sintering (DMLS) technique is one among the recent technologies used to build prototypes and tooling applications. DMLS produces complex geometry, low volume and strong individualization parts directly from three-dimensional (3D) computer-aided design (CAD) data in a short amount of time without the need for tooling. [1, 2] The fundamental of the technique is explained by following diagram as shown in figure 1. The setup of DMLS process, with major components as

i) CAD system
ii) Powder handling system
iii) Scanning system

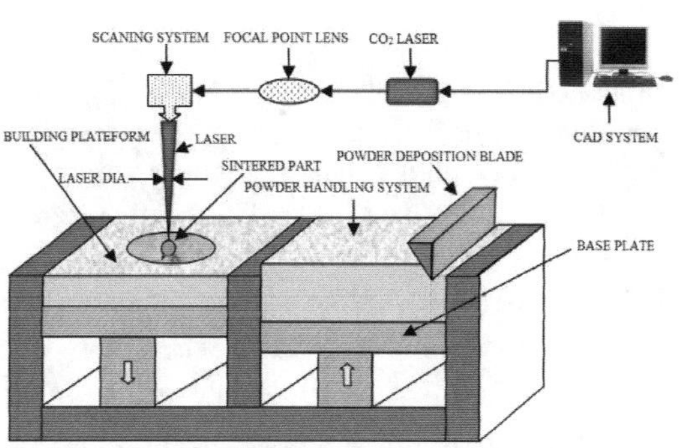

Figure 1. Set-up of the DMLS Technique

CAD system is used to convert 3D CAD models of functional parts and tooling inserts into triangulated surface models in the standard STL format. The process computer slices the standard STL format to 50 µm thick horizontal layers, representing the part into a Stack of thin slices. Powder handling system is used which deposits a layer of powder on a steel base plate allowing the laser beam to sinter in liquid phase and levelled. Here, a low melting point metal, called the binder, is melted by laser radiation in the matrix and wetted to a high melting point metal powder which serves as structural sintered part. As the binder melts it fills the void between the structural particles by capillary action and consolidates the matrix on cooling. [3] Then, a powder layer (about 50 µm in thickness) is spread on the base plate using a powder deposition blade. The scanning system scans the laser beam on the powder surface. The laser energy causes the powder particles to bond together. After laser scanning, the building platform is lowered by 50 µm, and a fresh layer of powder is distributed on top of the previous layers, the process is repeated, and by altering the shape of each scan layer, a part of uniformed shape can be produced. [1]

2. SIGNIFICANT PROCESS PARAMETERS OF DMLS TECHNIQUE

DMLS technique is a complex process where a large number of parameters can influence the quality of the final part. Figures 2 show the various parameters and the mechanisms taking place in the designing process of

DMLS technique. The significant process parameters are Laser Energy Density, Hatch Spacing, Powder Properties, Scanning Pattern Style and Thickness of Layer in DMLS technique. [4,5,6]

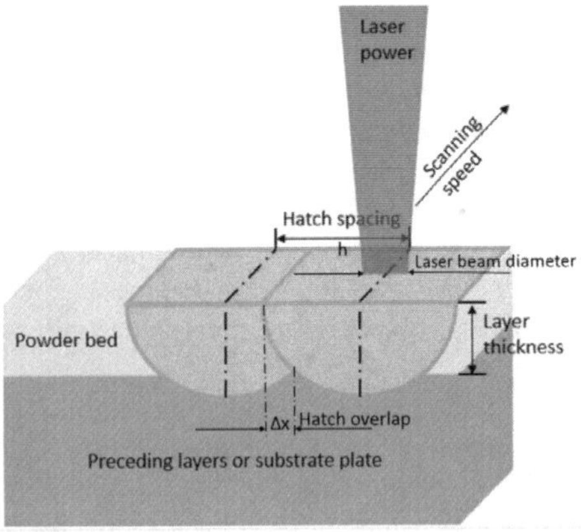

Figure 2 Process parameters of DMLS technique [4].

2.1 Laser Energy Density

The laser energy density is the amount of average useful energy to volume of objects during scanning of layer. The laser energy density is a key factor that affects final part's quality in the (DMLS) technique to measure input of energy. An integrated parameter (LED) is used to assess the combined effect of laser power (P) and scan speed (v) involved in individual line scanning and in order to estimate the laser energy in put to the powder layer being melted. Linear energy density is defined as the ratio of laser power to scan speed with the amount of Joule/Millimetre (J/mm). The combined effect of these parameters is evaluated by another factor terms as Volumetric Laser Energy Density (VED) with unit of Joule/millimeter^3 which further controls the (DMLS) technique integrally. VED is defined as Laser Power (P) divided by scan speed (v), hatch spacing (h) and Layer thickness (t)

$$\text{VED} = \frac{P}{V.h.t} \tag{1}$$

Where units of, P = watt (w)
V = millimetre / second (mm/s)
h = mm (millimetre)
t = millimetre (mm)

The equation 1 exhibits precisely that energy density depend upon four factors like

i) The incident laser power (P)
ii) Laser scan speed (v)
iii) Hatch distance (h)
iv) Laser beam thickness (t) [7]

Krishnan et al. performed experiment for the influence of Laser Energy Density to determine the process ranges from 1.2-1.8 J/mm2 of energy density for AlSi10Mg in DMLS technique and concluded that, the combination of 195W laser power, 700 mm/s scanning speed and 0.17 mm hatching distance yields highest density.[8]

2.2 Hatch Spacing

Hatch Spacing is the space between two nearest hatch lines. Hatch spacing is generally less than laser beam diameter. Figure 2 illustrate that one DMLS technique layer is made up of many hatches separated by hatch distance *"h"*. To get a sample of better quality, condition should be set in such manner so as to have overlap between two hatches which is called as hatch overlap, denoted as Δx. To have continuing among tracks, it is necessary to have overlap in DMLS parts which leads to a solid sample. In order to have better quality samples, an overlap of minimum 20% is maintained in most of DMLS technique. [9]. *Thijs et al.* reported that the hatch spacing considerably effect of hardness and concluded hardness improved by decreasing hatch spacing distance. [10]

2.3 Powder Properties

In DMLS technique, a significant role is played by powder properties such that size, shape and distribution of powder grains. Particle size distribution represents the flow ability of the matter an important powder property and is a mathematical task that defines particle relative quality by mass according to their size- ranges. As shape decides powder processing possibility therefore, selected powder must be spherical. [11]. In DMLS technique, the powder is deposited through a system of valves due to gravity. If the deposition of powder is irregular, it will strike in the valves and resulting in not creating a smooth layer as shown in figure 3 [5,12]. Large amount of small particles reduces the energy required to melt the material and improves the surface roughness of DMLS parts, thereby, reducing layer. [13].

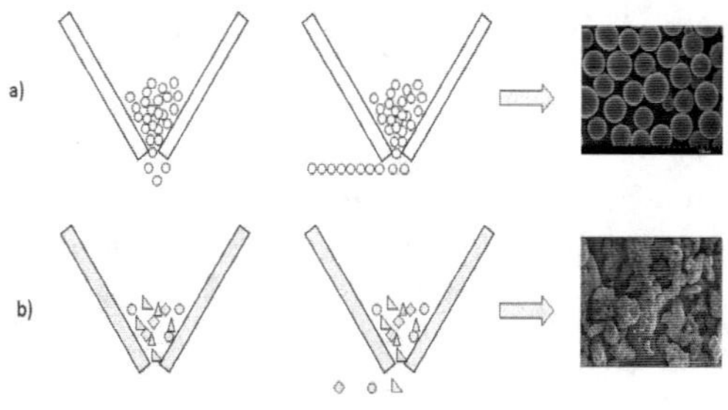

Figure 3. Powder Layer deposition (a) Spherical powder, and (b) Irregular powder. [5]

Lumley et al. found that the usage of fine additive powders resulted in a really reduced extent of liquid phase formed whereas coarse powders increased the amount of liquid phase that formed and prolonged its existence. They concluded that where there was no mutual solid solubility, particle size is unimportant in liquid development for improving the densification mechanisms of parts. [14]. However, *Olakanmi et al* demonstrated a key role of powder particle size and size distribution in the selective laser sintering densification mechanism given that chemical composition had a important effect on the propensity of the resulting alloy to density, and an appropriate component ratio of the powder blend is needed for a range of particle sizes, shapes, and surface morphology in the laser sintering of blended Al–Si powders. Theoretically, there ought to exist a relationship between the sintered properties and the particle size distribution of a candidate powder for the selective laser sintering process. A comparison of the selective laser sintering process with the conventional powder metallurgy sintering process, suggests the following features of such a relationship:
(1) Smaller sized powder particles tend to sinter more quickly since the sintering stress is inversely related to particle diameter.
(2) Powder particle densification should take place through neck formation at points of contact between particles. The consequence of this is that with a higher packing density of similarly sized particles in a powder, the tendency should be for a faster sintering rate to be obtained. [15]

2.4 Scanning Pattern Style

The scanning pattern is defined as the design/ pattern where in the hatches are oriented within and between the layers. The scanning pattern may be varied by Specific requirement of DMLS part and User's creativity. Basic hatch styles examples are exhibited in the figure 4. These scanning patterns may be repeated every layer with or without the presence of scanning pattern rotations between the layers. The rotation of the scanning pattern between the layers is carried out to have a better bonding between the layers. [9]. *Thijs et al.* prepared rectangular samples with a width, length and height of 5, 10 and 5 mm, respectively, were produced. The three types of scanning strategy that were used are illustrated in figure 4 and found to have a negligible effect on the relative density of the part, using the optimum values, a 'zig-zag' pattern had a relative density of 99.6%, a unidirectional pattern had a relative density of 99.35%, and a cross-hatch pattern had a density of 99.9%. This cross-hatching also produced pores that were more spheriodized than the other strategies, which, according to the paper, is less detrimental to the parts mechanical properties. [10]

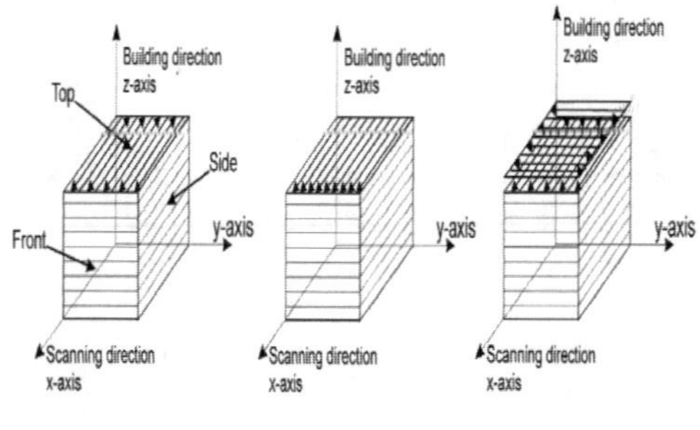

Figure 4 Different scanning patterns styles: (a) zigzag, (b) unidirectional, and (c) cross-hatching [10]

2.5 Thickness of Layer

An essential requirement for layer manufacturing in a thin layer of Powder because the its difficult to fuse the bond between conductive layers by the pre-placed powder layer and to make a strong fusion bond, the underlying solidified layer needs to be re-melted. The transmitted energies through the Powder layer decides determine the degree of re-melting. Hence the bond between layers will get increase by a smaller layer thickness, resulting in higher density components. [16]. *Gu et al.* studied, powder layer thicknesses were generally less than 0.40 mm, *i.e.*, within the thermal penetration depth of the laser beam, the particulate dispersion states in the sintered structures vary considerably with the used layer thickness and pointed out that although the shape and size of the melt pools cannot be observed directly in the solidified structures. In other words, a proper reduced in the powder layer thickness facilitates an enhancement in the thermodynamic and kinetic quality of both molten liquid and solid particles in the melt pool during sintering, so as to obtain a more homogeneous dispersion of the reinforcing particulates in the matrix metal on cooling.[17]

3. SOME RELEVANT IMPORTANT STUDIES RELATED TO DMLS TECHNIQUE

Thijs et al. [10] investigated the control of scanning speed, hatching spacing and scanning strategy on the properties of material produced by SLM and used powder of 5-50 μm particle size. The spot diameter was 52 μm. The optimized laser power, scanning velocity, and hatching spacing are 42 W, 200 m/s and 75 μm respectively. Rectangular samples with 5 mm width, 10 mm length and 5 mm were produced. Three types of scanning strategy used are zig-zag, unidirectional and cross-hatching. They found that the hatch spacing and scanning velocity significantly affect the hardness and melt pool width such a way that hardness increased by decreasing hatch spacing and melt pool width increased by reducing scanning velocity and melt pool is less stable at very low scanning velocities.

Zhu et al. [11] investigated the influence of the powder density on the final product for selective laser sintering. Different copper powders were used, and powder packing models were considered. They noted that there exists an influence of surface roughness and shape of the powder on the packing density. The final density of the object increases with the increase of apparent density. The apparent density can be enhanced by mixing different size powders. If two different powder sizes were used, the limiting density can be computed using:

$$\rho = \rho l + \frac{\rho s}{1 - \rho l}$$

where ρ is limiting density, ρL is the density of large powder diameter, and ρS is the density of small powder diameter.

Simchi et al [18] proposed fractional density of investigated powders as a function of P/V as illustrated in Figure 5, it indicated that the density is linearly proportional to the ratio of laser power to scan rate in semi-log scale. One may also notice that the density depends on the layer thickness (w) and scan line spacing (h). They also observed the effect of these parameters on the sintered density of iron powder with

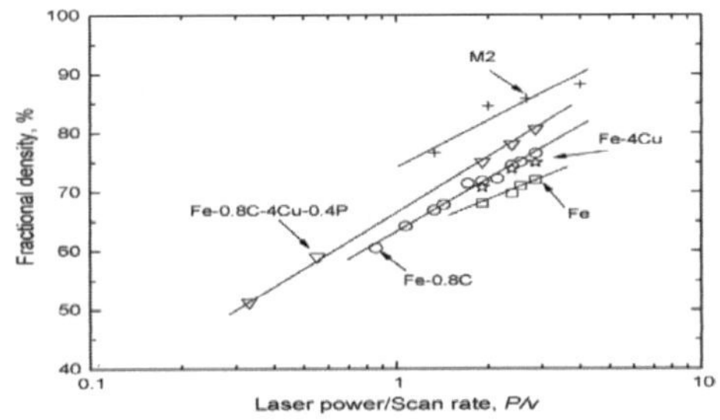

Figure 5 Fractional density of laser sintered powders as a function of the ratio of laser power to scan rate. [18]

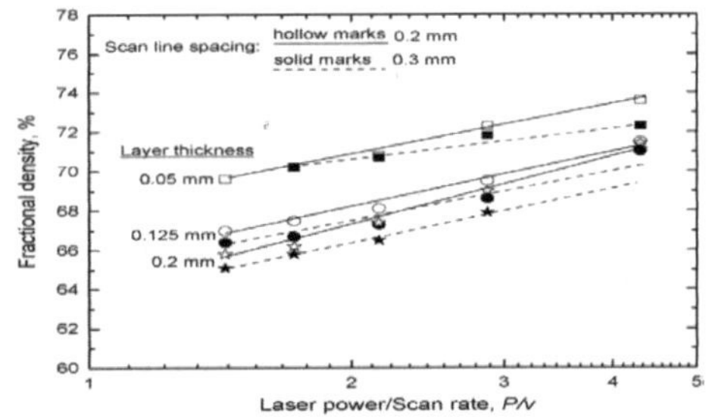

Figure 6. Effect of layer thickness and scan line spacing on the fractional density of sintered iron [18]

D50 =51μm in Figure 6. It was seen; with raising the layer thickness lower density was achieved. Also, with declining the line scan spacing higher density is obtained. [18]

Dongdong et al. [19] studied processing conditions and microstructural characteristics of direct laser sintered porous 316L stainless steel parts and defined Linear Energy Density (LED), by tailoring a partial melting mechanism which gave the ratio of laser power to scan speed. The mixture of a linear energy density of ˷3400–˷6000 J/m and a lower scan speed less than 0.06 m/s contributed to the occurrence of a reasonable sintering mechanism, and obtained porosities of ~21–~55%, in laser sintered part, under optimal processing conditions with favorable stress–strain response, *i.e.*, the tensile strength of 152 MPa.

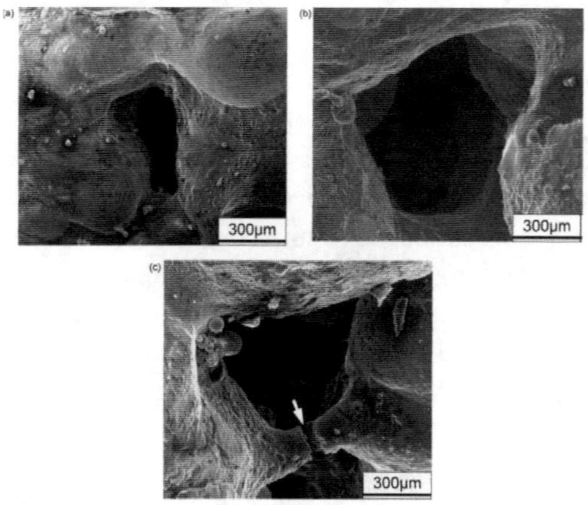

Figure 7. SEM images showing characteristic pore features in laser sintered samples prepared [19]

They determined a reason in a suitable sintering mechanism, reduce laser power or enhance in scan speed, scan line spacing, and powder layer thickness generally LED to a higher sintered porosity. The characteristic morphologies of pores on surfaces of laser sintered samples are provided in Figure 7. The pores were existed in laser sintered porous metals due to:

(i) The formation of liquid bridges between partially melted particles during laser irradiation.

(ii) The presence and growth of sintering necks during solidification, leaving residual pores between solidified metallic agglomerates.

Mumtaz et al. [20] investigated the behaviour of a commercial nickel super alloy. They measured the contact angle and bead geometry, respectively, and analyzed the bonding behaviour using metallographic micro sections. Different parameters were changed, e.g. pulse width, percentage overlap, hatch strategy and scan strategy, etc. They depicted the results in a process map, displaying pulse width against specific energy. It was shown that higher levels of porosity were produced at lower pulse energies, and the formation of these pores predominantly exists around layer boundaries.

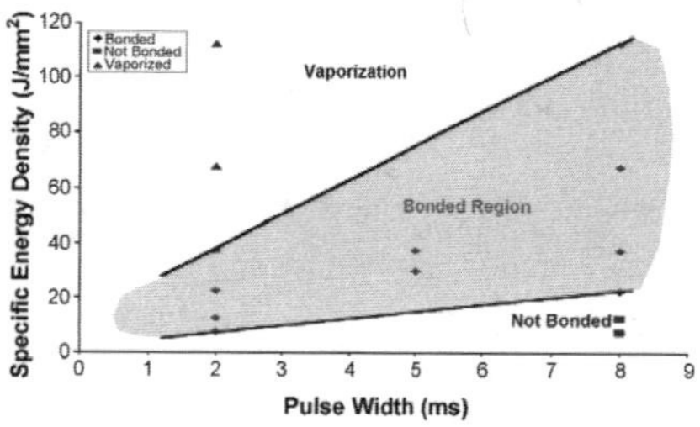

Figure 8. Specific Energy Density against pulse width [20]

Figure 8 showed result of greater heat intensities being generated at lower pulse widths. Vaporization of material is more likely to occur at lower pulse widths and therefore requires less specific energy to fully fuse the powder. As the pulse width increases the specific energy required for complete bonding.

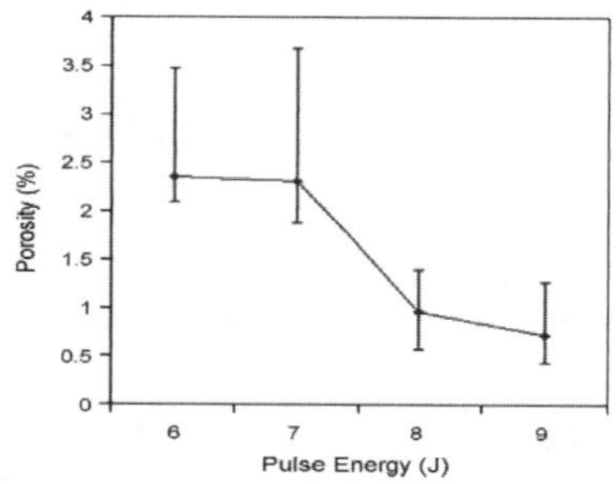

Figure 9. Layered specimen, porosity against pulse energy [20]

Figure 9 showed the quantity of porosity within layered specimens produced at a pulse width of 5ms, repetition rate 10 Hz, scan speed 168mm/min and pulse energies between 6 and 9 J. It was observe that reduction of porosity with increment pulse energy, low pulse of energies cannot produce sufficient heat to completely melt powder and due to this higher porosity was present.

Childs et al. [21] compared theoretical results with experiments for different tool steels and stainless steels. A CO_2 laser source with 10 to 200 W power, scan speeds up to 50 mm/s and an argon shielding gas was used. The tests were concluded that the quality of the molten pool is affected by scanning speed and laser power. A higher scanning velocity speed and a lower laser power cause the non-continuous molten pool. Too large scanning space will make the molten pools are not close enough to each other.

4. WEAR AND ITS IMPLICATIONS

Modern industries are concerned about losses incurring due to wear related problems and subsequent cost involved in replacements. Wear is a deformation of material due to mechanical action such as contact and relative motion in between solid surface. There are various ways to address this problem such as by applying thermal spray coatings or by using super alloy materials, but economic considerations have restricted this use to a select few applications. DMLS technique is an additive manufacturing process and it is proposed study to prevent surface degradation by DMLS technique.

4.1 Recent advances of DMLS Technique in Wear Based Applications

Aerospace industry: Mostly component are made of costly and time consuming. Nickel superalloys, Titanium alloys high temperature ceramics and these generally have complex geometries. Moreover their manufacturing is also difficult. Mostly, DMLS technique is used in the production of turbine blades, jet engines and rib web structural components. [[22]]

Automotive Industry: The developments of new products in automotive industries are very costly and time-consuming. Industries are used the DMLS as an important tool in design and development of component and functional parts such as drive shaft, gear box components engine exhausts, piston and wheels. [22]

Biomedical Industries: DMLS has become as a life-saving development in the medical sector. DMLS gave recent developments in biomedicine and biomaterial. This technique is used in biomedical field to produce products such as orthopedic implants, dental applications, artificial organs, tissue scaffolds, medical devices, artificial bladders, bio-printing, painted organs, microvasculature networks and biologic chips. [[23]]

5. CONCLUDING REMARKS

Direct Metal Laser Sintering (DMLS) technique is an upcoming process where, a low melting point metal, called the binder, is melted by laser radiation in the matrix and wetted to a high melting point metal powder which serves as structural sintered part. It has already grown its root in Aerospace, Automotive and Biomaterial Industries for to resist wear based problems and may benefit in other spheres where wear based problems result in substantial losses such as turbine blades of hydro power plants.

Among the significant process parameter of DMLS technique, Laser Energy Density followed by Hatch Spacing and Powder Properties play an important role.

Surface roughness of DMLS part is improved by use of large amount of small particles which reduce the energy required to melt the material.

Homogenous dispersion of reinforcing particulates in matrix metal on cooling can be obtained by reduction in powder layer thickness which, helps in the thermodynamic and kinetic quality of both molten liquid and solid particles in the melt pool during sintering.

6. REFERNCES

[1]. A. Amanov, S. Sasaki, I. S. Cho, Y. Suzuki, H. J. Kim, D.E. Kim (2013), "An investigation of the tribological and nano-scratch behaviours of Fe–Ni–Cr alloy sintered by direct metal laser sintering", *Materials and Design,* 47, 386–394.

[2]. A. Sova, D. Pervushin, I. Smurov (2010), "Development of multimaterial coatings by cold spray and gas detonation spraying", *Surface & Coatings Technology,* 205, 1108–1114.

[3]. K. Gaiser, P. Erickson, P. Stroeve, J. P. Delplanque (2016), "An experimental investigation of design parameters for pico-hydro Turgo turbines using a response surface methodology", *Renewable Energy,* 85, 406-418

[4]. C. Y. Yap, C. K. Chua, Z. L. Dong, Z. H. Liu, D. Q. Zhang, L. E. Loh, S. L. Sing (2015), "Review of selective laser melting: Materials and applications",

Applied Physics Reviews Appl. Phys. Rev., *2*(4), 041101.

[5]. T. Kurzynowski, E. Chlebus, B. Kuźnicka, J. Reiner (2012), "Parameters in selective laser melting for processing metallic powders. High Power Laser Materials Processing: Lasers, Beam Delivery, Diagnostics, and Applications.

[6]. A. Stwora, G. Skrabalak (2013), "Influence of selected parameters of Selective Laser Sintering process on properties of sintered materials", *Journal of Achievements in Materials and Manufacturing Engineering, 61*(2). Retrieved May 19, 2016

[7]. *Laser Additive Manufacturing of High-Performance Materials.* (2015). Berlin: Springer-Verlag Berlin and Heidelberg GmbH & KG.

[8]. M. Krishnan, E. Atzeni, R. Canali, L.Iuliano, (2014), "On the effect of process parameter properties of $AlSi_{10}Mg$ part produced by DMLS", *Rapid Prototyping Journal*, DOI: 10.1108 / RPJ-03-2013-0028

[9]. K. G. Prashanth, (2014), "Selective laser melting of Al-12Si", *Doctoral dissertation, Dresden, Techn. Univ., Fak. Maschinenwesen, Diss*

[10]. L. Thijs, F. Verhaeghe, T. Craeghs, J. V. Humbeeck, J. Kruth (2010), "A study of the microstructural evolution during selective laser melting of Ti–6Al–4V", *Acta Materialia, 58*(9), 3303-3312

[11]. H. Zhu, J. Fuh, L. Lu (2007), "The influence of powder apparent density on the density in direct laser-sintered metallic parts" *International Journal of Machine Tools and Manufacture, 47*(2), 294-298.

[12]. A. Simchi (2004), "The role of particle size on the laser sintering of iron powder", *Metall and Materi Trans B Metallurgical and Materials Transactions B, 35*(5), 937-948

[13]. N. Karapatis, J. V. Griethuysen , R. Glardon (1998), "Direct rapid tooling: A review of current research", *Rapid Prototyping Journal, 4*(2), 77-89.

[14]. R. N. Lumley, G.B Schaffer (1996), "The effect of solubility and particle size on liquid phase sinterin", *Scripta Mater 35(5), 589–95*

[15]. E. O. Olakanmi, K. W. Dalgarno, R. F. Cochrane (2012), *"Laser sintering of blended Al–Si powders"*, Rapid Prototyp J, 18, 109–19

[16]. M. Agarwala, D. Bourell, J. Beaman, H. Marcus, J. Barlow (1995), "Direct selective laser sintering of metals", *Rapid Prototyping Journal, 1*(1), 26-36

[17]. D. Gu, Y. Shen (2006), "Processing and microstructure of submicron WC–Co particulate reinforced Cu matrix composites prepared by direct laser sintering", *Mater Sci Eng A 435–436, 54–61*

[18]. A. Simchi (2006), " Direct laser sintering of metal powders: Mechanism, kinetics and microstructural features", *Materials Science and Engineering A 428 (2006) 148–158*

[19]. G. Dongdong, S. Yifu (2008), "Processing conditions and microstructural features of porous 316L stainless steel components by DMLS", *Applied Surface Science* 255 (2008) 1880–1887

[20]. K. A. Mumtaz, P. Erasenthiran, N. Hopkinson (2008), "High density selective laser melting of Waspaloy", *Journal of Materials Processing Technology, 195*(1-3) 77-87

[21]. T. Childs, C. Hauser, M. Badrossamay, (2004), "Mapping and Modelling Single Scan Track Formation in Direct Metal Selective Laser Melting", *CIRP Annals - Manufacturing Technology, 53*(1), 191-194. doi:10.1016/s0007-8506(07)60676-3

[22]. N. Guo, M. C. Leu M. C. (2013), "Additive manufacturing: Technology, applications and research needs", *Frontiers of Mechanical Engineering Front. Mech. Eng., 8*(3), 215-243

[23]. 3D Printing Electronics Laser Additive Manufacturing Systems. (n.d.) Retrieved May 19, 2016, from http://www.optomec.com/

Deposition of PVD thin films on thermal barrier coatings for a wear resistant thermal insulation

W. Tillmann, M. Dildrop, Dortmund / D

Abstract

Due to the superposed thermal and mechanical stress profile, thermo-mechanically coupled forming processes require tools and machine components which meet high demands. High forming forces and process temperatures in the contact zone between the tool and the workpiece limit the life span of these tools. A promising approach for protecting such tools is a combination of thermally sprayed coatings and physical vapor deposited layers. This coating system combines the characteristics of the individual layers and leads to superior mechanical, tribological as well as thermal properties under the mentioned coupled stresses.

In this study thermally sprayed alumina (Al_2O_3) and yttria-stabilized zirconia (ZrO_2) coatings were produced by atmospheric plasma spraying. Therefor different coating porosities were adjusted in order to varied the effect of thermal insulation for the substrate made of AISI H11 (1.2343). After the coating process the surface roughness of the thermal barrier coatings (TBC) were reduced by polishing process in preparation for the PVD top layer. Subsequently, wear and heat resistant hard TiAlSiN and CrAlSiN coatings were deposited on top of the polished TBCs by using magnetron sputtering process. As a reference the PVD coatings were also applied on a nitrided steel samples. Titanium and chromium interlayers were applied by PVD technique in different coating thicknesses (50 – 150 µm) between PVD and thermally sprayed coatings. Afterwards the influence of these metallic interlayers on coating adhesion of PVD coatings were analyzed by performing scratch tests. Hardness and young's modulus of PV coatings were investigated by means of nanoindentation. The morphology and topography of the coatings were analyzed by scanning electron microscopy, light microscopy and optical three-dimensional surface analyzer. EDX analyses and X-ray diffraction were used to determine the chemical composition of the PVD coatings. Finally the wear resistant of the PVD top layers were determined at different temperatures (20°C, 500°) by using a high temperature tribometer.

1 Introduction

Plasma sprayed zirconia or alumina coatings are often used as thermal barrier coatings (TBC) for high-temperature parts like engines of automobiles, rockets and planes because of the porous microstructure and low thermal conductivity. The high surface roughness is one principal characteristic of these coatings and leads to high friction coefficients. Due to this high dynamic stress loads appear on the surface of the coatings during running cycles of coated parts [1, 2].

Recently, intensive research is devoted to develop coating system which combine the positive properties of different technologies. This combination could be a solution for reducing the surface roughness and thus the friction coefficient of thermally barrier coatings [3-5]. Different thermally sprayed layers and CVD or PVD coatings can be combined to develop deverse multilayer layer architectures. Above all, the combination of thermal sprayed and PVD coatings show promising approaches [3-13].

By means of a thermally sprayed coating between the steel substrate and the PVD layer abrupt property changes can be reduced. Thus, similarly to a plasma nitriding of conventional steel substrates, the thermally sprayed coating can effect a reduction of the hardness gradient between substrate and PVD coating. This could increase the load carrying capacity and thus the wear resistance and adhesion of the PVD layer [4, 6, 7].

Furthermore studies also show that the combination of PVD and thermally sprayed layers effect an improvement in the oxidation resistance of the bond coat. Due to this the TGO film growth in the bond coat is reduced [13].

In this study combined coating system, consisting of plasma sprayed thermal barrier coatings and magnetron sputtered PVD layers, are developed. It's the aim to develop a coating system which withstand both high thermal and mechanical loads. The main challenge is to produce a sufficient adhesion between PVD and thermally sprayed coating.

2 Experimentals

2.1 Sample preparation

First of all both the thermally sprayed and the PVD coatings were applied on steel substrates. Therefore the hot work steel AISI H11 (1.2343) was used as a substrate material for the coating experiments. For the application of the thermally sprayed coatings, the samples with a diameter of 40 mm and a thickness of 4 mm, were grit blasted and cleaned in an ultrasonic ethanol bad. Corundum (EKF 24) with a grain size of 600 to 850 µm were used as grit blasting material to increase the surface roughness of the steel substrates.

After the production of alumina and zirconia coatings they were grinded and polished to get a smooth surface for deposition of the PVD coatings on top. CrAlSiN and TiAlSiN PVD coatings were deposited on both steel substrates and the polished plasma sprayed coatings in order to compare the properties of both coating system. **Fig. 1** shows the schematic diagram of the combined coating systems.

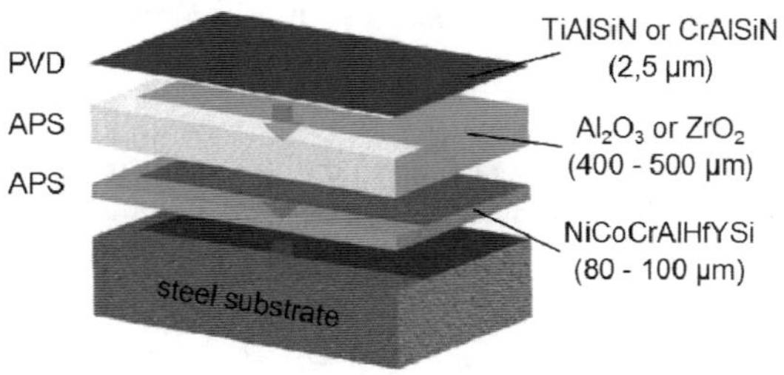

Fig. 1. Combined coating system

In preparation for the PVD coatings the steel substrates were polished and plasma nitrided to increase the hardness and thus the carrying load of the coating/substrate system, avoiding a shell egg effect during the analysis. The nitriding time was 8 h and the substrate temperature was kept constant at about 650°C. Plasma nitriding was performed with a pressure of 2 mbar and a bias voltage of U = -650V while the atmosphere composition was H_2:N_2 = 3:1. After nitriding process a surface roughness of R_Z = 0.8 was measured. The hardness of the substrate could be increased from 3 GPa up to 11 GPa.

2.2 Thermal Spraying process

Zirconia and alumina thermal barrier coatings were manufactured by atmospheric plasma spraying (Plasma Gun F4-MB, Oerlikon Metco, Switzerland). For both, zirconia and alumina, two different grain-sized feedstocks were used to varied the porosities of the coatings. Furthermore an high temperature oxidation resistant layer (NiCoCrAlHfYSi) was applied on the substrate to increase the adhesion of the TBCs. **Tab. 1** shows the alumina and yttrium stabilized zirconia feedstocks from two different manufactures (Oerlikon Metco, Switzerland and Durum Verschleißschutz, Germany). The powder feeding system which were used (Twin-120-AH) consisting of two powder containers. One of them was used for the spray powder for adhesion layer and the other one for the spray powders of the TBCs. The powders were injected radially into the plasma flame in an angle of 90° using argon as carrier gas.

Tab. 1. Feedstocks for producing TBCs

Type	Material	Grain size
Metco 204F	ZrO_2-8Y_2O_3	-45 + 15 µm
Metco 200C-NS	ZrO_2-8Y_2O_3	-140 + 45 µm
Metco 6051	Al_2O_3	-31 + 5,5 µm
Durum 603	Al_2O_3	-90 + 45 µm
Amdry 997	NiCoCrAlHfYSi	-38 + 5 µm

2.3 PVD coating process

The PVD thin films were deposited using a magnetron sputtering device (CC800/9 sinox ML, Cemecon, Germany). The PVD device is equipped with four unbalanced planar magnetron sources. For deposition of TiAlSiN and CrAlSiN one pair of TiAl60- or AlCr24-targets, one single Ti- or Cr-target and one single Si-target were used, as shown in in **Fig. 2**. The TiAl60-targets were made of titanium with 60 cylindrical aluminum plugs and the AlCr24-targets consist of aluminium with 24 chromium plugs.

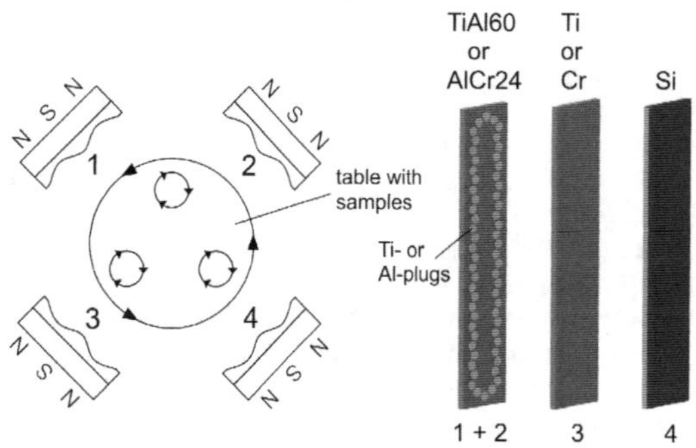

Fig. 2. Target configuration for deposition of PVD coatings

The other targets consist of pure silicon and pure titanium or chromium. This target configuration makes it possible to varied the silicon content by keeping the content of the other coating elements constant. The power of the Si target (0 – 4 kW) was varied to obtain different Si contents inside the TiAlSiN and CrAlSiN coatings. The target configuration, the power of the other targets and the flows of the plasma gases argon and krypton were kept constant for each coating system. Bias voltage was set to U = -100 V for TiAlSiN or rather U = -120 V for the CrAlSiN coatings. The process gas nitrogen was regulated by a nitrogen gas controller that was adjusted to a total chamber pressure of 500 or 580 mPa. The coatings were deposited with a thickness of 4.5 or 2.5 microns. **Tab. 2** gives an overview of the PVD process parameters.

Tab. 2. Process parameter for deposition of PVD coatings

	CrAlSiN	TiAlSiN
2 x TiAl60 [kW]	-	9.5
2 x AlCr24 [kW]	5.0	-
Ti [kW]]	-	2.0
Cr [kW]]	1.0	-
Si [kW]]	0 - 3	0 - 4
DC Bias [V]	-100.0	-200.0
Ar [mln]	295.0	295.0
Kr [mln]	120.0	200.0
N_2 controller [mPa]	580.0	500.0
Coating time [h]	5.2/ 2.7	2.2/ 1.4
Process temperature [°C]	~ 600	~ 700

2.4 Analytics

The surface roughness of the PVD and thermally sprayed coatings were measured by an optical 3D-surface analyzer (Infinite Focus Alicona, Austria). The morphology and structure of the coatings were examined using scanning electron microscopy (JSM-7001F, Jeol, Germany), energy dispersive scattering and a light microscope (type Axiophot, Zeiss, Germany). For analyses of the coating morphology the samples were prepared by grinding and polishing them with diamond suspension (9 µm, 6 µm, 3 µm, and 1 µm). These cross-section samples were also used to determine the porosity and thickness of the TBCs by means of image analysis using the software Axiovision Autmess (Zeiss) and Stream Motion (Olympus).
The Hardness of the thermally sprayed coating were measured by Vickers test, applying a test load of 0,1 kp (HV 0.1). To determine the micohardness of the PVD layers on steel substrate and TBCs nanoindenation tests were performed. Therefor a Berkovich diamond penetrates the PVD coating to determine the mechanical properties in the nanoscale range. In this study, the maximal depth of the indents is t_{max} = 2 µm.
The adhesion between the PVD and thermally sprayed coatings were determined by scratch tests. The scratch tests, which were performed according to DIN 1071-3 with a tip radius of r = 200 mm resulted partially in no spallation of the coatings. Due to this a sharper Rockwell Diamond tip (r = 100 µm) were used to scratch across the surface of the coatings. The normal force was increased linearly from F_N = 0 N to F_N = 100 N over a distance of l = 10 mm, with a velocity of v = 10 mm/min. The scratch tracks were analysed by scanning electron microscopy to determine the critical loads.
In order to investigate the tribological properties of the combined coatings at elevated temperatures, ball-on-disc tests were performed at two different temperature levels (20°C, 500°C). For this, a high temperature tribometer (CSM Instruments, Switzerland) was used. The balls used as counterparts for the rotating disc (coated samples) consist of alumina (Al_2O_3) and have a diameter of 6 mm. The velocity of the rotating disc (v = 0.4 m/s) and the normal force (F_N = 2 N) were kept constant during the tests. The tribometer tests were stopped at 4000 rotations and no external lubricants were employed. The optical 3D-surface analyzer (Infinite Focus Alicona, Austria) was used to measure the wear volume of the wear tracks of the samples. Afterwards, the wear coefficient was calculated.

3 Results and discussion

3.1 Coating morphology

The morphology of the four thermally sprayed coating systems is very different, depending on the grain size of the feedstocks, as shown in **Fig. 3**. The spray powders of Al_2O_3 lead to coating porosities of about 9 % and 16 %. In contrast the coarser spray powders of zirconia lead to higher coating porosities of about 12 % and 28 %. In all cases the coating structures are homogeneous and the thicknesses of the TBCs are in a range of 520 to 600 microns. Studies show that low coating thicknesses (< 210 µm) can lead to flaking of PVD coatings on top [3]. Furthermore, the porosity of the adhesion layer is about 5 % and the coating thickness about 90 microns.

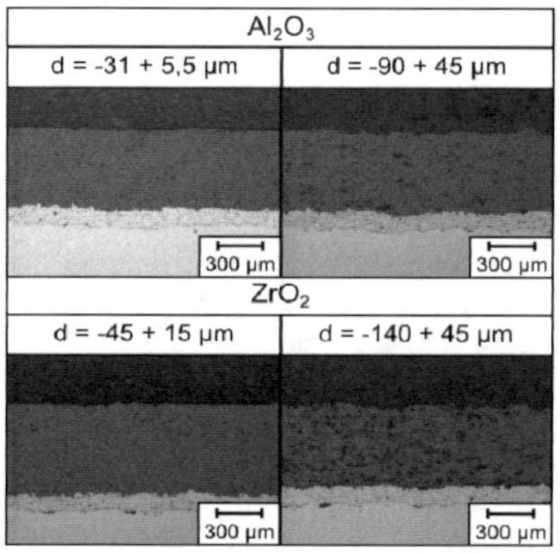

Fig. 3. Light microscope pictures of the TBCs

3.3 Surface roughness

The manufactured TBCs were grinded and polished after coating process to get a smooth surface for deposition of the PVD coatings on top. It was possible to decrease the surface roughness of the alumina and zirconia coatings, which were produced with the fine feedstocks, from R_Z = 24,8 µm to R_Z = 1 µm, as shown in **Fig. 4 a**. The surface roughness of the porous TBCs could be decreased from R_Z = 37,4 µm to R_Z = 2,83 µm (Al_2O_3) or rather from R_Z = 51,8 µm to R_Z = 6,7 µm (ZrO_2). By polishing process the pores of the coatings were opened and led to craters on the surface, as shown in **Fig. 4 b**. The craters of the porous coatings are very large and are responsible for the high surface roughness after polishing process. Due to the deep of the craters (up to 6 microns) the polished porous alumina and zirconia coatings aren't suitable for the application of PVD layers on top. The craters would disturb the PVD film growth and inhibit the application of a continuous PVD coating with a thickness of a few microns. As a consequence, only the two fine powder sprayed coatings were used for application of the PVD layers on top.

3.3 Coating hardness

3.3.1 Thermal barrier coatings

For the dense zirconia coatings a Vickers hardness of about 838 HV0.1 were measured. Due to the rougher microstructure the porous zirconia coating show a

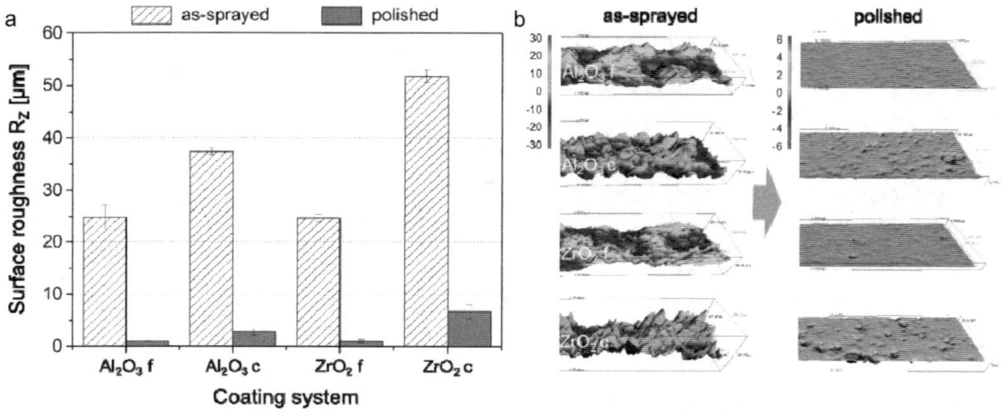

Fig. 4. Surface roughness of TBC before and after polishing process; a) Surface roughness R_Z; b) 3D Topography

lower Vickers hardness of about 788 HV0.1. As expected the hardness of alumina coatings is about 100 HV0.1 higher than the hardness of the zirconia coatings. Here the porous coatings show also a lower hardness (912 HV0.1) than the fine powder sprayed coatings (972 HV0.1). The bonding layer has a Vickers hardness of about 337 HV0.1.

3.3.2 PVD coatings

Fig. 5 visualizes the influence of the Si content on hardness and elastic modulus of magnetron sputtered TiAlSiN and CrAlSiN coatings on steel substrate. The incorporation of silicon leads to a decrease of the grain size and thus an increase in the strength and hardness of the coating. Al/Ti ratio and Al/Cr ratio was kept constant at about 1.16 and the nitrogen content is about 52 at% for all coatings. Starting from a silicon content of about 7 at% (TiAlSiN) or 4 at% (CrAlSiN), the coating structure changes from cubic to a more hexagonal structure. An amorphous Si_3N_4 matrix is formed around Ti(AlSi)N crystallites. The result is a nanocomposite structure consisting of a crystalline Ti(AlSi)N phase and an amorphous Si_3N_4 matrix. A silicon content of 6 - 8 at% shows the ideal ratio between amorphous phases and nanocrystals where the highest binding energies occurred. A further increase of the silicon content leads to the formation of increasingly more Si_3N_4 phases. Thus, the properties of the nanocomposite increasingly approach the matrix properties, which results in a reduction of coating hardness [14-16].

Pretests show a very good tribomechanical behaviour for $TiAlSi_{7.9}N$ and $CrAlSi_{7.5}N$ at high temperatures up to 750°C. Due to this it is the aim within the framework of this work to deposit these PVD coatings on top of the TBCs.

These selected PVD coatings show different coating properties depending on the substrate, as shown in **Fig. 6**. Both, $TiAlSi_{7.9}N$ and $CrAlSi_{7.5}N$, show highest hardness on the nitrided steel substrates and the lowest hardness on the polished zirconia coatings.

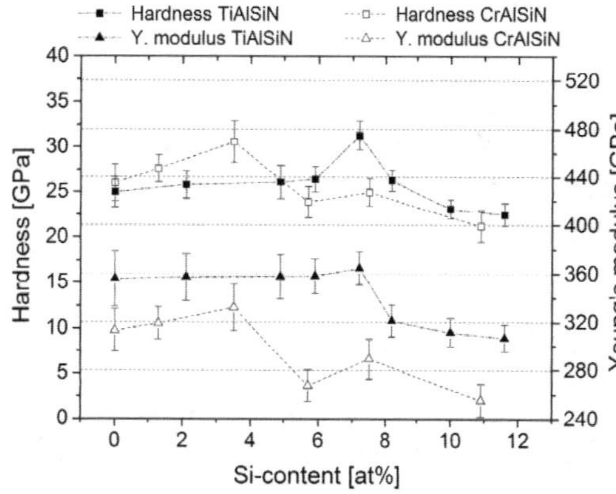

Fig. 5. Influence of Si content on Hardness of TiAlSiN and CrAlSiN on nitrided steel substrate

During the PVD coating process the targets (coating material) are bombared with energetic argon ions, whereby atoms are sputtered out of the targets and enter the gas phase. After that the coating material is transported through the plasma and condensing on the substrate. In addition a voltage is given to the

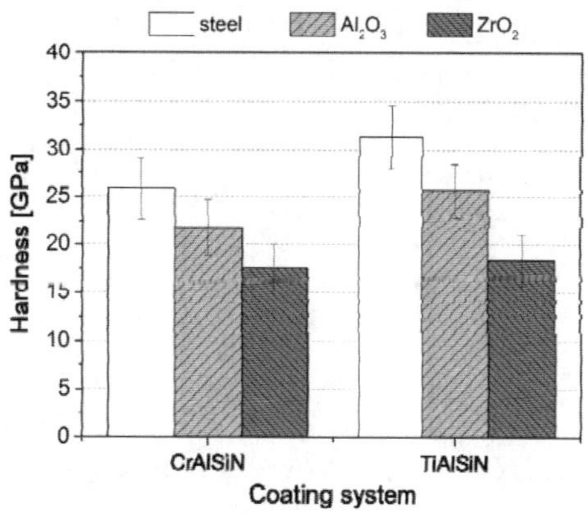

Fig. 6. Microhardness of $CrAlSi_{7.5}N$ and $TiAlSi_{7.9}N$ on steel substrate and TBCs

substrate (bias voltage) to compress the layer structure by ion bombardment of noble gas ions. Thereby growth faults are eliminated and the grain size is reduced by continuous nucleation and leads to an increase of the coating hardness [17].
The different physical properties of the substrate materials lead various film growth and hardness of the PVD coatings. Due to the low electrical conductivity of the TBCs in comparison to the steel substrate the coatings are less compressed by the ion bombardment.

3.4 Coating adhesion

First of all the PVD coatings were applied on polished and nitrided steel substrates and polished TBCs with a coating thickness of about 4.5 microns. Optional a metal ion etching process with ions of TiAl for $TiAlSi_{7.9}N$ or ions of AlCr for $CrAlSi_{7.5}N$ was performed prior to the PVD coating process. By metal ion etching for about one hour the surface of the substrates is cleaned, activated and the surface roughness can be increased a little bit. In these experiments the surface roughness of the substrate is nearly the same before and after ion etching process ($R_a \approx 0,16$ µm).
In the next step the PVD coatings were deposited without metal ion etching and the thickness of the coatings were reduced from 4.5 to 2.5 microns to analyse the influence on the coating adhesion. Furthermore, metallic interlayers of Ti or Cr were applied between PVD and thermally sprayed coatings to increase the coating adhesion. **Fig. 7** shows an overview of the critical loads of the PVD coatings on steel substrate and TBCs. The critical load Lc_3 indicates the total delamination from the PVD coating of the substrate material.

3.4.1 Al_2O_3

Metal ion etching prior to the PVD coating process has a bad effect on both, the adhesion of $TiAlSi_{7.9}N$ and $CrAlSi_{7.5}N$ on of top of the polished Al_2O_3 coatings. Both the reduction of the coating thickness from 4.5 µm to 2.5 µm and leaving off metal ion etching process leads to an improved adhesion of $TiAlSi_{7.9}N$ (Lc_3 = 8 N to Lc_3 = 36 N) and $CrAlSi_{7.5}N$ (Lc_3 = 0 N to Lc_3 = 32 N) coatings on Al_2O_3. The critical load of $TiAlSi_{7.9}N$ could be further increased by applying a titanium interlayer (50-150 nm) between the two different coatings. A maximum adhesion (Lc_3 = 61 N) between $TiAlSi_{7.9}N$ and alumina could be achieved by an titanium interlayer of 100 nm. The metallic interlayer avoid an abrupt change of properties between PVD coating and TBC and the hardness gradient is reduced. Due to the lower shear strength of thicker metallic interlayers above 100 µm, the coating adhesion is decreased [18].
In contrast to the $TiAlSi_{7.9}N$ coating, the adhesion between $CrAlSi_{7.5}N$ and Al_2O_3 could not be increased by incorporating metallic interlayer. The applied chrome interlayers (50-150 nm) lead to a decrease of the adhesion between $CrAlSi_{7.5}N$ and alumina (Lc_3 = 32 N to Lc_3 = 10 N). In the case of the thick chrome interlayers (100 nm, 150 nm) the $CrAlSi_{7.5}N$ coatings even delaminated from Al_2O_3.

3.4.2 ZrO_2

In contrast to the PVD coatings on the alumina coatings they show a really poor adhesion to zirconia. The reduction of coating thickness from 4.5 µm to 2.5 µm enable only a very low adhesion (Lc_3 = 4 N) of $TiAlSi_{7.9}N$ on top of the thermally sprayed zirconia coating. Leaving off metal ion etching prior to the PVD coating process and a titanium interlayer (50-150 nm) show no improvement of the adhesion of $TiAlSi_{7.9}N$ PVD layers on top of the zirconia coatings.
In case of the CrAlSiN PVD coatings only the contribution of chromium interlayers (50-150 nm) enable a poor adhesion to zirconia. Independet of the thickness of the interlayers critical loads of Lc_3 = 7 N were measured. To achieve a more precise analysis of the adhesion between PVD and thermally sprayed

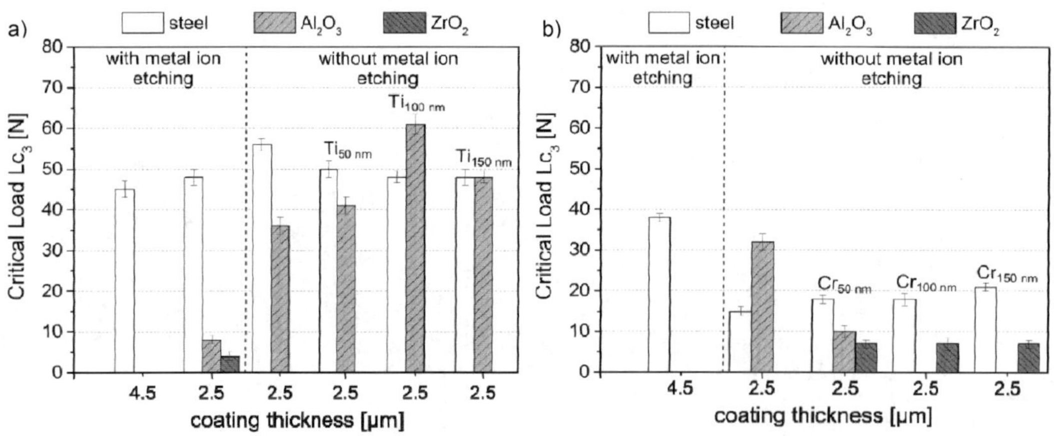

Fig. 7. Coating adhesion of PVD coatings on TBCs and steel substrate; a) $TiAlSi_{7.9}N$; b) $CrAlSi_{7.5}N$

Coatings SEM pictures of the scratch tracks were made, as shown in **Fig. 8**. A poor adhesion of the PVD coatings to the thermally sprayed coatings is characterized by widescaled near-surface spallations of the PVD coatings (see fig. 8 a, f, h, i, j and k). This is an indication for high residual stresses inside the PVD coatings. The scratch tracks of the strongly adherent PVD coatings on alumina (fig. 8 b, c, d, e and g) show deep trenches of the thermally sprayed coating outside the track, instead of widescaled near-surface spallations of the PVD layers. This effect occurs due to the high forces which are introduced into the alumina coating by the diamant tip during the scratch test.

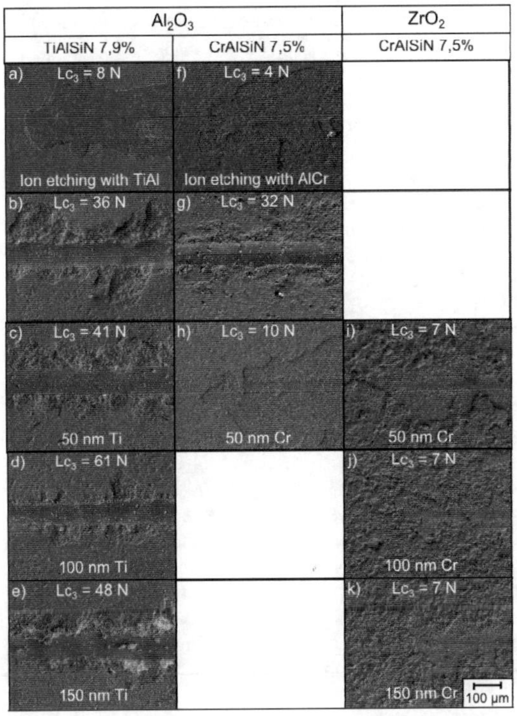

Fig. 8. SEM pictures of the scratch tracks of the PVD coatings on top of the TBCs (range of the critical load Lc_3)

3.5 Tribometer test

The $TiAlSi_{7.9}N$ (Lc_3 = 61 N) and $CrAlSi_{7.5}N$ (Lc_3 = 32 N) coatings which show the best adhesion to the TBCs (see Fig. 8 d and 8 g) were choosen for the tribometer tests. Due to the poor coating adhesion, it was not possible to perform tribometer tests with the PVD coatings which were applied on top of the zirconia coatings. The tests show that the wear resistance of thermally sprayed alumina coatings could be increased by a PVD top layer, as shown in **Fig. 9**. At room temperature the wear coefficients of the combined coating systems are nearly the same than the Al_2O_3 coating. But at elevated temperatures of 500°C the wear resistance of alumina could be increased by a $TiAlSi_{7.9}N$ top layer. Due to the low adhesion (Lc_3 = 32 N) $CrAlSi_{7.5}N$ partly delaminated from Al_2O_3 during the tribometer test at 500°C. The delamination of the PVD coating led to an increased abrasive wear of the thermally sprayed alumina coating. Subsequently the wear coefficient of this combined coating system (Al_2O_3 + $CrAlSi_{7.5}N$) is on a higher level than the wear coefficient of the Al_2O_3 reference sample.

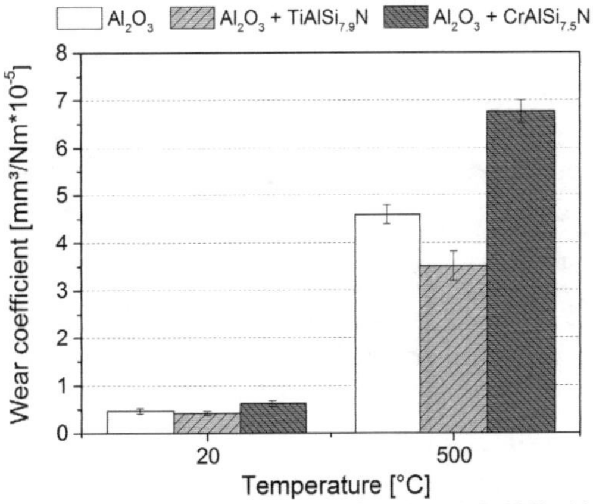

Fig. 9. Wear resistance of $TiAlSi_{7.9}N$ and $CrAlSi_{7.9}N$ on Al_2O_3 by using an alumina counterpart with F_N = 2 N

4 Conclusion and Outlook

Zirconia and alumina coatings were produced by atmospheric plasma spraying and were polished in preparation for the application of the PVD coatings. By polishing process the pores of the coatings were opened and led to craters on the surface. Due to the deep of the craters (up to 6 microns) porous TBCs (P > 12 %) aren't suitable for the application of PVD layers on top.

$TiAlSi_{7.9}N$ (H = 32 GPa) and $CrAlSi_{7.5}N$ (H = 25 GPa) PVD coatings were applied on polished thermally sprayed alumina (H = 972 HV0.1) and zirconia coatings (H = 838 HV0.1). TiAlSiN PVD coatings show a good adhesion to the alumina coatings. The best adhesion (up to Lc_3 = 61 N) were achieved by applying $TiAlSi_{7.9}N$ without metal ion etching on polished alumina coatings and contributing a titanium interlayer (100 nm) between PVD and thermally sprayed coating. CrAlSiN PVD coatings show a worse adhesion to the alumina coating in comparison to the TiAlSiN coatings. Metal ion etching process with AlCr ions and metallic interlayers of chrome decrease the adhesion between $CrAlSi_{7.5}N$ and Al_2O_3. Critical loads up to Lc_3 = 32 N were achieved. All PVD coatings show a very poor adhesion to the thermally sprayed zirconia coatings. Obviously, there are no chemical or physical reactions and diffusion processes between ZrO_2 and PVD coating elements at the coating process temperatures of 600 to 700°C.

The tribometer tests show that the wear resistance of the thermally sprayed alumina coating could be increased by an PVD top layer ($TiAlSi_{7.9}N$) at elevated

temperatures (500°C). In comparison to the alumina coating the combined coating system (Al_2O_3 + $TiAlSi_{7.9}N$) show a decrease of the wear coefficient up to about 30 %.
For the future it would be interesting to analyse the influence of the PVD process temperatures (> 700°C) and different interlayers on the adhesion between TBCs and PVD coatings.

Acknowledgement

The authors thank the DFG (German Science Foundation) within the projects TI343/102-1 (Untersuchung der tribo-mechanischen Eigenschaften von TiAlSiN- und CrAlSiN-Nanokompositschichten unter Hochtemperatureinfluss) and HO2356/11-1 (Herstellung und Applikation von gradierten Schichtgefügestrukturen für den Prozess des Reib-Drückens unter Einsatz von PVD- und Thermischen Spritzverfahren) for their financial support.

5 Literature

[1] Cao, X.Q.; Vassen, R.; Stoever, D.: Cermics materials for thermal barrier coatings. Journal of thermal barrier coatings 24 (1) (2004), pp. 1-10

[2] Vassen, R.; Cao, X.; Tietz, F.; Bas, D.; Stöver, D.: Zirconates as new materials for thermal barrier coatings. Journal of the American Cermaic Society 83 (8) (2000), pp. 2023-2028

[3] Kobayashi, A.; Yatsuzuka, M.: Bonding strength of DLC film on zirconia coating prepared by gas tunnel type plasma spraying. Surface and Coatings Technology 202 (24) (2008), pp. 5914–5918

[4] Bemporad, E.; Sebastiani, M.; Felicis, D.; Carassiti, F.; Valle, R.; Casadei, F.: Production and characterization of duplex coatings (HVOF and PVD) on Ti–6Al–4V substrate. International conference on surfaces coatings and nanostructured materials nanoSMat 2005 515 (1) (2006), pp. 186–194

[5] Bolelli, G.; Lusvarghi, L.; Pighetti Mantini, F.; Pitacco, F.; Volz, H.: Enhanced tribological properties of PECVD DLC coated thermally sprayed coatings. 3 Rencontres Internationales de la Projection Thermique3-RIPT 20073 Rencontres Internationales de la Projection Thermique 202 (18) (2008), pp. 4382–4386

[6] Casadei, F.; Tului, M.: Combining thermal spraying and PVD technologies: A new approach of duplex surface engineering for Ti alloys. Proceedings of the 40th International Conference on Metallurgical Coatings and Thin Films (ICMCTF) 237 (2013), pp. 415–420

[7] Bemporad, E.; Sebastiani, M.; Staia, M. H.; Puchi Cabrera, E.: Tribological studies on PVD/HVOF duplex coatings on Ti6Al4V substrate. Proceedings of the 35th International Conference on Metallurgical Coatings and Thin Films (ICMCTF) 203 (5–7) (2008), pp. 566–571

[8] Bolelli, G.; Lusvarghi, L.; Montecchi, M.; Pighetti Mantini, F.; Pitacco, F.; Volz, H.; Barletta, M.: HVOF-sprayed WC-Co as hard interlayer for DLC films. Proceedings of the 35th International Conference on Metallurgical Coatings and Thin Films (ICMCTF) 203 (5-7) (2008), pp. 699–703

[9] Bemporad, E.; Sebastiani, M.; Carassiti, F.; Casadei, F.; Valle, R.: Development of a Duplex Coating Procedure (HVOF and PVD) on TI-6AL-4V Substrate for Automotive Applications. Advanced Ceramic Coatings and Interfaces II: Ceramic and Engineering Science Proceedings, Volume 28, Issue 3 (eds U. Schulz, H.-T. Lin, J. Salem and D. Zhu), John Wiley & Sons, Inc., Hoboken, NJ, USA (2007)

[10] Bemporad, E.; Sebastiani, M.; Felicis, D.; Carassiti, F.; Valle, R.; Casadei, F.: Production and characterization of duplex coatings (HVOF and PVD) on Ti–6Al–4V substrate. International conference on surfaces coatings and nanostructured materials nanoSMat 2005 515 (1) (2006), pp. 186–194

[11] Gadow, R.; Scherer, D.: Composite coatings with dry lubrication ability on light metal substrtes. Proceedings of Symposium C on Prtoective Coating and Thin Films 151–152 (2002), pp. 471–477

[12] Hübsch, C.; Dellinger, P.; Maier, H. J.; Stemme, F.; Bruns, M.; Stiesch, M.; Borchers, L: Protection of yttria-stabilized zirconia for dental applications by oxidic PVD coating. Acta Biomaterialia 11 (2015), pp. 488–493

[13] Lima, C. R. C.; Cinca, N.; Guilemany, J.M.: Study of the high temperature oxidation performance of Thermal Barrier Coatings with HVOF sprayed bond coat and incorporating a PVD ceramic interlayer. Ceramics International 38 (8) (2012), pp. 6423–6429

[14] P. J. Martin, A. Bendavid, J. M. Cairney, M. Hoffman, Nanocomposite Ti–Si–N, Zr–Si–N, Ti–Al–Si–N, Ti–Al–V–Si–N thin film coatings deposited by vacuum arc deposition, Surface and Coatings Technology, 200 (2005), pp. 2228–2235.

[15] Y. Tanaka, N. Ichimiya, Y. Onishi, Y. Yamada, Structure and properties of Al–Ti–Si–N coatings prepared by the cathodic arc ion plating method for high speed cutting applications, Proceedings of the 28th International Conference on Metallurgic Coatings and Thin Films, 146–147 (2001), pp. 215–221.

[16] I.-W. Park, S. R. Choi, J. H. Suh, C.-G. Park, K. H. Kim, Deposition and mechanical evaluation of superhard Ti–Al–Si–N nanocomposite films by a hybrid coating system, Proceedings of the 30th International Conference on Metallurgical Coatings and Thin Films 447–448 (2004), pp. 443–448.

[17] Tillmann, W.; Sprute, T.; Hoffmann, F.; Chang, Y.-Y., Tsai, C.-Y.: Influence of bias voltage on residual stresses and tribological properties of TiAlVN-coatings at elevated temperatures. Surface and Coatings Technology 231 (2011), pp. 122-125

[18] Bouzakis, K.-D.; Makrimallakis, S.; Katirzoglou, G.; Skordaris, G.; Gerardis, S.; Bouzakis, E.; Leyendecker, T. et al.: Adaption of graded Cr/CrN-interlayer thickness to cemented carbide substrates' roughness for improving the adhesion of HPPMS PVD films and the cutting performance. Surface and Coatings Technology 205 (2010), pp. 1564-1570

Production advanteges using laser cladding as an additive manufacturing method

U.D. Bihlet, M. Honoré, P.T. Nielsen, FORCE Technology Denmark

The potential of additive manufacturing has reached a point where the techniques are considered highly relevant for production purposes. In general, the manufacturing industry greets the new approach with enthusiasm, as it offers innovative designs and potentially reduced production costs. However, questions arise concerning the durability of additively manufactured components.

This paper describes industrial trials with laser cladding and precipitation hardening heat treatment of thinwalled structures with the 17-4 PH stainless steel alloy. Due to the great relevance of the AM production methods for the aviation industry, the mechanical strength of the alloy given by the MMPDS document is used as a baseline. In order to improve the properties of the produced specimens, hot isostatic pressing was applied. The results show that a post processing treatment consisting of a HIP cycle and a conventional precipitation hardening, vastly improves the mechanical strength and elongation values of printed specimens, causing them to exceed the specified values.

1 Introduction

In these years, additive manufacturing (AM) is receiving a massive amount of attention. The term covers a wide range of production technologies, which substantially differ in their suitable application area **[1]**.

One of these technologies is laser cladding (LC). As an AM method, LC has an outstanding build rate and is suitable for fabricating additions to existing complex structures as it has a large work envelope. In practice, the main difference between LC and the other wide spread method, powder bed fusion, is constituted by their powder delivery methods.

Currently, LC is being used industrially for reconditioning of turbine blades, and to build large free standing structures or to add features to large castings or forgings, as well as in the aviation, dental and space industries.

The material properties are important to the application, and in the case of the aviation industry, the Metallic Materials Properties Development and Standardization (MMPDS)**[2]** provides design strength values suitable for aircraft construction with most metallic materials. These design values are tabulated for the alloys according to their production method, of which additive manufacturing is not included as a possibility. That raises the question of how to make a safe design with AM parts.

A common post-treatment which is routinely used on cast turbine components and on powder metallurgical parts is Hot Isostatic Pressing (HIP). This treatment causes a significant improvement of the mechanical properties by closing and diffusion bonding internal pores and defects. This is accomplished at a high external pressure in the range of 800-1000 bar and at temperatures close to the melting point of the alloy.

17-4 PH (AISI 630) is a precipitation hardening stainless steel alloy, which is industrially applied with heat treatments that results in a martensitic microstructure. Standardized heat treatments for this alloy consists of three steps, i.e. solution heating, quenching and ageing. During the ageing treatment, which is normally carried out in the temperature interval of 482-621 °C, coherent Cu-rich precipitates form in the microstructure**[3]**. The size and number of the precipitates, which determines the strength and ductility of the alloy, is affected by the production method and by the heat treatment. Thereby, by selecting a suitable precipitation hardening heat treatment, the strength level can be engineered to suit the application.

A number of works have measured the hardness and tensile strengths of AM 17-4 PH[4-6]. These earlier works used powder bed fusion methods and focused on the phase composition of the microstructure.

In the present work, the mechanical properties of AM 17-4 PH stainless steel produced by laser cladding is tested both before and after HIP'ing and the results are compared to MMPDS values.

2 Experimental

2.1 Laser cladding process and equipment

The AM 17-4 PH samples are manufactured using purpose-built and commercially available laser cladding system, based on a high-power diode-laser, powder supply and an integrated co-axial nozzle, capable of delivering the laserbeam, powder and protection gasses co-axially at the process point.

The powder entering the laser focal point is melted and deposited on the sample surface. By relative displacement of the sample and the laser beam, 3-

dimensional features may be built to measure, via a layer-by-layer deposition.

Process monitoring is applied in order to maintain a constant melt pool temperature and flow, and consequently a stable build-up quality. Laser powers in the range 1.4 kW to 2.1 kW were applied.

2.2 17-4 PH for additive manufacturing

Prealloyed 17-4 PH powder was used, with a particle size distribution of which 93% fell between 10-270 µm.

The heat treaments shown in **Table 1** were used. Hot isostatic pressing was performed prior to the precipitation heat treatment at 1130-1160°C x 1000 bar and solution treatment and aging was performed in a vacuum furnace at a pressure of 10^{-2} hPa.

Table 1. Heat treatments.

Treatment	HIP	Solution treatment	Ageing
As fabricated	-	-	-
H900	-	1030°C x ½ hour	482°C x 1 hour
HIP+H900	X	1030°C x ½ hour	482°C x 1 hour
HIP+H1100	X	1030°C x ½ hour	593°C x 4 hours

2.3 Trials and tests

The chemical composition was measured using optical emission spectroscopy, as described by ASTM E 415.

The mechanical properties were quantified by tensile testing of flat specimens which were machined on all sides. Except for the HIP'd specimens, which were single runs, each result is presented as an average of two specimens. The test was conducted as described by ISO EN 6892-1. Samples were oriented in the build direction as well as perpendicular to the buid direction. The fracture surfaces of seleced specimens were inspected using scanning electron microscopy.

Unstrained specimens were sectioned transverse to the long axis to expose the microstructure. Metallurgical specimens were prepared by embedding the cross sections in epoxy and grinding and polishing them with conventional metallurgical methods to a finish of 1 µm diamond paste. To determine the phase composition, specimens were electrolytically etched with oxalic acid The specimens were inspected with a light optical Zeiss microscope.

3 Results and discussion

Free standing structures with varying heights and a wall thickness of approximately 4 mm were printed on a mild steel base plate, **Fig. 2**. A single structure with a larger wall thickness was also printed as a first run. This structure separated from the build plate, indicating a high level of tensile residual stress in the fabricated material.

Fig. 2. Printed structures.

3.1 Chemical composition

The chemical composition of the fabricated material is compared in **Table 2** with ASTM A564 limits and with the values given on the powder supplier certificate. The Cr content of the fabricated material drops slightly compared to that of the powder material, and only just clears the A564 minimum content. The impurity levels of C, P and S increase very little or not at all as a consequence of the fabrication process.

Table 2. Chemical composition in wt%. Fe bal.

	C	Mn	P	S	Si	Cr	Ni	Nb	Cu
Powder certificate	0.01	0.41	0.036	0.012	0.31	15.34	4.50	0.23	3.24
As fabricated	0.011	0.49	0.038	0.032	0.26	15.0	4.72	0.24	3.72
ASTM A564 min	-	-	-	-	-	15.00	3.00	-	3.00
ASTM A564 max	0.07	1.00	0.040	0.030	1.00	17.00	5.00	-	5.00

3.2 Mechanical properties

The results of tensile testing compared with MMPDS minimum values for sheet, strip, plate and rod is shown in **Fig. 3.** There is little correlation between the strength of specimens and their orientation with regard to the build direction. In the fabricated state, the strength is uniformly low, and less than that obtained with a H1100 heat treatment. The large difference between yield and tensile strength indicates a high degree of strain hardening. The fabricated material responds readily to the H900 heat treatment, and the highest strength level is obtained with a combination of hot isostatic pressing and a H900 treatment. Without the HIP treatment, the H900 specimens comply with the minimum MMPDS values, but with the HIP treatment, the strength values as well as the difference between the yield and ultimate tensile strength increases further. Similarly, the HIP+H1100 heat treatment state clears the minimum MMPDS values by a substantial margin.

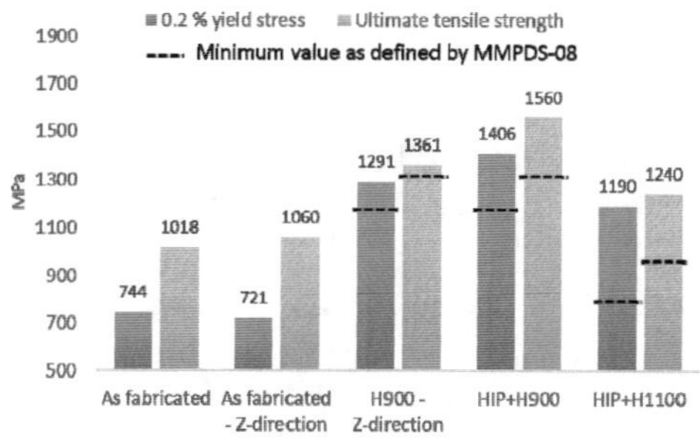

Fig. 3. Tensile strength of specimens.

The elongations are compared with MMPDS values for sheet, strip, plate and rod in **Fig. 4.** The elongation of the as fabricated material is lower in the build direction. Due to the layerwise production method, this is not unexpected. A further drop in elongation is observed after the H900 heat treatment. Again, this is to be expected as the significant increase in strength will cause a stronger sensitivity towards internal defects in the microstructure. By subjecting specimens to a HIP treatment prior to their precipitation hardening treatments, the elongations increase dramatically, and are brought to comply with the MMPDS minimum values.

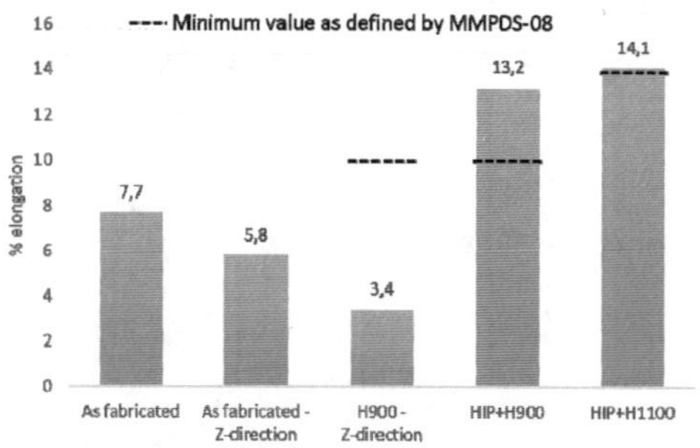

Fig. 4. Elongation of tensile specimens.

The fracture surface of the vertically oriented as fabricated specimen is shown in **Fig. 5**. Even though the majority of the fracture surface is dimpled, indicating a ductile fracture, numerous cleaved areas (dark in the figure) are riddled throughout the fracture surface. Each of these cleaved areas has one or more spherical cavities placed approximately in the center. It would appear that the fracture initiated from these internal defects, which have the appearance of gas porosities.

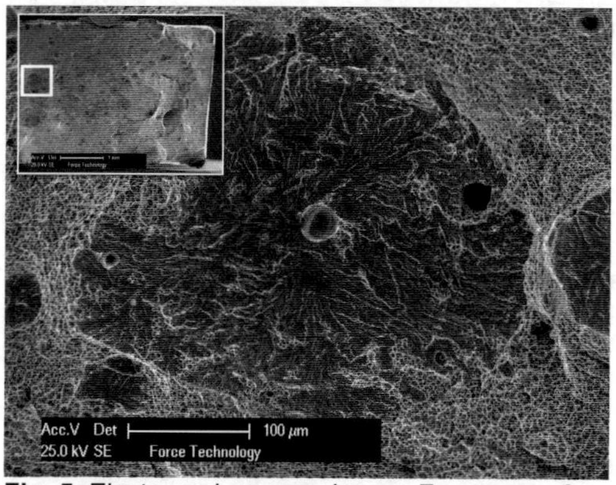

Fig. 5. Electron microscopy image. Fracture surface of vertically oriented tensile specimen in the as fabricated condition.

3.2 Microstructure observations

The microstructure of the as fabricated material is compared with that of the HIP+H900 material in **Fig. 6.** It is apparent that the HIP treatment removes a large number of spherical porosities and oxide inclusions of sizes of up to approximately 20 µm present in the as fabricated microstructure. The sizes and spatial distribution of these pores corresponds with the cleavage areas on the fracture surface, shown in **Fig. 5.**

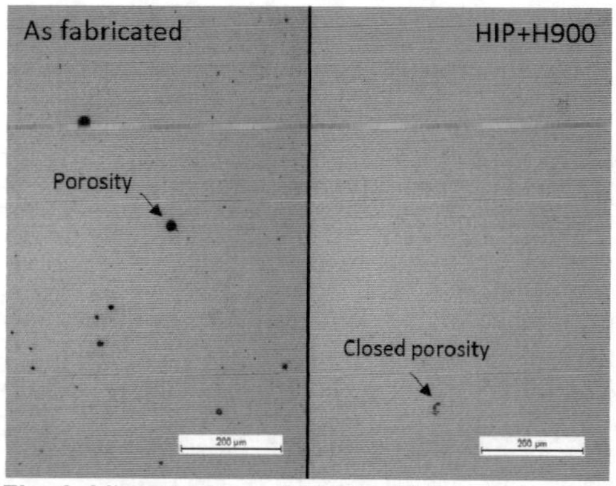

Fig. 6. Microstructure of AM 17-4 PH before and after the HIP treatment.

3.3 Interpretation of results

The strength of laser cladded AM 17-4 PH stainless steel is, in the as fabricated condition, inferior to that of conventional 17-4 PH subjected to standard heat treatments. The strength and elongation levels are similar to those found by Mower et al[4] who used the powder bed fusion method instead of laser cladding. Work by LeBrun et al[5] found higher elongation values but similar strength and work hardening levels, also for as fabricated powder bed fusion material.

There, the cause of the work hardening was found to be a consequence of transformation of retained austenite, formed in the build process, into strain induced martensite. The H900 heat treatment causes the yield strength and ultimate tensile strength to exceed the MMPDS minimum values. The elongation is however much lower. Inspection of the fracture surface and the microstructure indicates that the cause of low strength and elongation is a large content of inclusions and pores, which act as stress concentrators during tensile testing. These pores are effectively eliminated by the HIP treatment, with significant increases in both strength and ductility levels.

3.4 Evaluation of production potential

At first glance the inferior mechanical properties of as-fabricated additively manufactured parts might call for a dismissal of the laser cladding method completely. The present investigations indicate, that such dismissal would be premature, seeing that a simple HIP'ing of the parts is sufficient to improve the mechanical properties sufficiently. Obviously the HIP process adds an additional process step to the fabrication of the parts, which must be considered.

For parts additively manufactured from high-cost powders and alloys, the savings in material cost alone may suffice to justify the HIP process. In other applications the specific benefits of AM relative to traditional machining, in terms of freedom of design, offering the possibility of manufacturing components with internal cooling channels, light-weigth hollow structures etc., may be the economic enabler for the HIP.

4 Summary and conclusions

The present work treats the fabrication of 17-4 PH stainless steel from an applied perspective, and uses the MMPDS strength values as the benchmark standard. This work documents that when faced with a porous microstructure, an industry standard HIP treatment works just as well on AM parts as it does on a casting with micro shrinkages. To summarize:

- The fabricated AM 17-4 PH responds readily to conventional H900 and H1100 heat treatments.
- Poor elongation caused by a porous initial microstructure can be improved by a HIP treatment.
- With the HIP treatment, the AM 17-4 PH complies with MMPDS minimum mechanical strength and elongation values.

5 Outlook

It would be highly relevant for the application of 17-4 PH to the aviation industry to verify whether the fatigue strength properties are also comparable to those of conventionally produced 17-4 PH.

6 Literature

[1] Gu, D.D., Meiners, W., Wissenbach, K., Poprawe, R.: Laser additive manufacturing of metallic components: materials, processes and mechanisms, Intl. Materials Rev., Vol 57, No. 3, (2012), pp. 133/184.

[2] Batelle Memorial Institute: Metallic Materials Properties Development and Standardization, Federal Aviation Administration, (2013)

[3] Murayama, M., Hono, K., Katayama, Y.: Microstructural evolution in a 17-4 PH stainless steel after aging at 400 °C. Metall. Mater. Trans. A 30 (2) (1999), pp. 345/353

[4] Mower, T.M., Long, M.J.: Mechanical behaviour of additive manufactures, powder-bed laser-fused materials, Mat. Science & Engineering A 651, (2016), pp. 198-213.

[5] Lebrun, T., Nakamoto, T., Horikawa, K., Kobayashi, H.: Effect of retained austenite on subsequent thermal processing and resultant mechanical properties of selective laser melted 17-4 PH stainless steel, Mat. and Design 81, (2015), pp. 44-53.

[6] Murr, E.L., Martinez, E., Hernandez, J., Collins, S., Amato, K.N., Gaytan, S.M., Shindo, P.W.: Microstructures and properties of 17-4 PH stainless steel fabricated by selective laser melting, J. Mater. Res. Technol. 1(3), (2012), pp. 167-177.

THE INFLUENCE OF SIZE AND SPEED OF DROPS ON THE STRUCTURE AND PROPERTIES OF ELECTRIC-ARC SPRAYED COATINGS

H. Pokhmurs'ka, M. Student, V. Hvozdetskyi, T. Stupnytskyi,
V. Posuvailo, Lviv / Ua

Increase of the pressure of air flow from 0.6 to 1.2 MPa provides the growth of its speed from 300 to 600 m/s. The increase of air flow speed reduces the flight time of drop from the arc to the substrate and them against the prepared surface at higher temperatures. Increase of the pressure of air provides reduction the size of drops and oxides between lamellas as well as improves the mechanical characteristics.

1 Introduction

Among the methods of thermal spray coatings application, arc spraying (AS) is technologically the most simple and cheap method. The recent use of electrode materials in the form of special cored wires (CW) has allowed us to expand the scope of the method of arc spraying and to get new regenerative and protective coatings of various functional purposes with high-performance [1-3]. However the electric arc coatings are characterized by high porosity – 7-10 %, high residual tensile stresses of 60-70 MPa, 80-90 MPa, low cohesion and adhesion – 8-10 MPa (compared with coatings deposited by other thermal spray methods). Recently, significant investigations are carried out in the world to improve the physicomechanical characteristics of the electric arc coatings (AS method). For this purpose in a number of works a compressed air is replaced by a mixture of combustible gases and oxygen [3, 4]. This method allows, by increasing the velocity of the molten droplets formed on the coating, to improve significantly the mechanical properties and to reduce the coating porosity. However, in this case the design of the equipment is more complicated and the energy costs and as well as the cost of the coating deposition increase greatly. In this paper, to improve the mechanical characteristics of electric arc coatings the effect of the increased flight speed of the molten droplets in the spray coating due to the higher air pressure in the spray-gun from 0.6 to 1.2 MPa is investigated.

2 Experimental methods

The electric arc coating 1 mm thick was applied by a metallizer GT-1 produced at Physico-Mechanical Institute of the NAS of Ukraine. For coating spraying the following electrode materials were used: wire of solid cross-section Sv 08 and powder wires of type 1 – 140Cr14; 2 – 140Cr14V; 3 – 140Cr14Ni2TiAl; 4 – 140Cr16B3Ti2Ni2; 5 – 140Cr16B3; 6 – Cr6B3Al3Mg2. Coating was sprayed under the following conditions: the voltage of arc U – 32 V, current I – 120 A, the air pressure 0.6 and 1.2 MPa for air flow of 1.3 m3/min, the distance from the arc to the sprayed surface in all cases was 120 mm. Before spraying samples were grit-blasted with corundum. Microhardness was determined by the device PMT-3 under loading of 200 g.

To obtain supersonic speed of the air flow by the Mach number 2 the Laval nozzle with a vertical place¬ment of two air channels in the nozzle having a critical diameter of each hole of 2.2 mm was used. (Fig. 1).

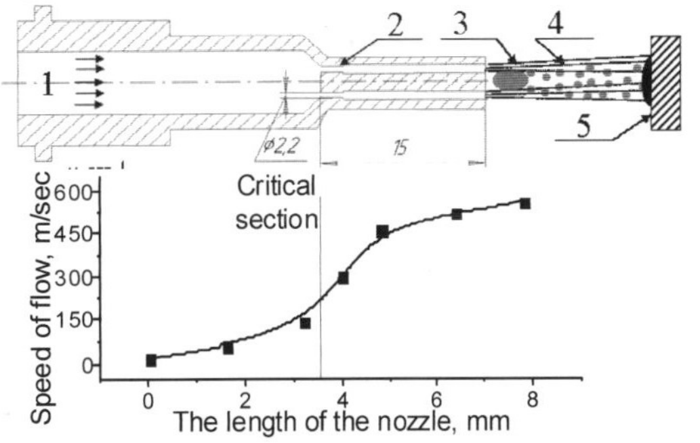

Fig. 1. Scheme of the nozzle for supersonic flow: 1 – air flow; 2 – a critical nozzle section; 3 – melt of electrode materials; 4 – metal-air flow; 5 – steel substrate with deposited coating

The length of the supersonic section was 15 mm. The Laval nozzle was calculated by the formula:

a) cross-section area:

$$f_i = G/\rho_i w_i$$

b) diameter:

$$d_i = ((4/\pi) \cdot f_i)^{0.5};$$

c) the length of the nozzle:

subsonic section

$$l_1 = (d_i - d_{kp})/2tg(\alpha_2/2);$$

2) supersonic section

$$l_2 = (d_i - d_{kp})/2tg(\alpha_1/2\ [5].$$

The distance from the nozzle to the arc is 10 mm.

The air flow rate was determined by the formula [5], where w is speed of the air flow in the nozzle, k is adiabatic index k = Cp/Cv, R is gas constant of air, T0 and Ti is temperature of the air flow at subsonic and supersonic section of the nozzle. Speed of the droplets dispersed by the air flow was determined experimentally using two rotating discs.

In order to determine the surface temperature of the sprayed surface under spraying at a pressure of 0.6 and 1.2 MPa, the coating was deposited using an unmoving metallizer for 30 sec on a plate of thickness 5 mm, on the reverse side of which a thermocouple was fixed in the hole at a depth of 4 mm. In the

process of the coating spraying from the electrode wire of the weight is 300 g, the temperature on the element surface in the spraying spot varied parabolically. At the air pressure of 0.6 MPa the maximum temperature of 450 °C was recorded, and at a pressure of 1.2 MPa – 600 °C.
Cohesive strength of the coatings was determined on tubular specimens (Fig. 2) consisting of two halves of a total length of 100 mm. The halves of the samples were connected by a bolt and a nut. The surface of the samples was coated with coatings with a thickness of 1 mm. The samples were then released from the mandrel and subjected to tension on a tensile testing machine FPZ-10, the force at the fracture moment being recorded. Cohesive strength was determined by the well-known formula:

$$\sigma = \frac{P}{S}$$

where P is fixed force at the fracture moment; S is the area of sprayed coating.

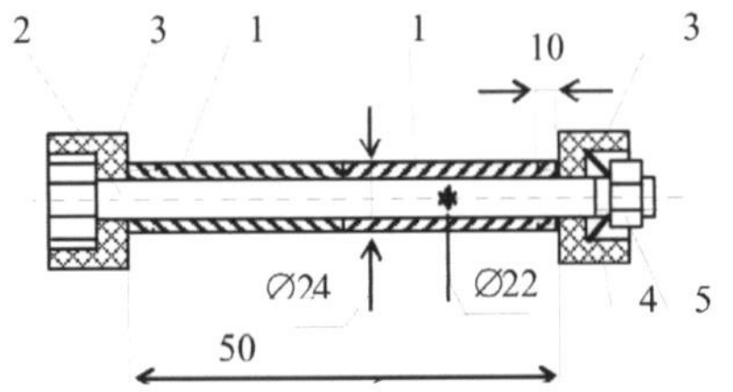

Fig. 2. Scheme for coating samples to determine the value of its cohesive strength:
1 – sample surface for spraying coatings; 2 – pin; 3 – protecting washer; 4 – spring; 5 – nut

Stresses were calculated by the method developed for bimetallic rings (diameter 60 mm, height 20 mm, thickness 4 mm) with a longitudinal section, in to which the ratio of the radii of the center line to the radial thickness of the ring is not less than 10. On the outer surface of the rings the AS of thickness 1 mm was applied. In order to determine the stress the modulus of elasticity of the steel ring of the substrate and the sprayed layer, which is 70 MPa, is taken into account [6]. The difference was measured with a caliper ШЦ-I-125-0.05, State Standard (GOST) 166-89.

3 Results and discussion

Calculation of the air flow velocity by the above formula demonstrates that the increase in the air flow pressure from 0.6 to 1.2 MPa, causes two-fold increase in the air speed from 300 to 600 m/s (Fig. 1). In this case the speed of the droplets, dispersed by air flow under spraying the cored wire electrodes, increases from 60...90 m/s to 160...220 m/s (Fig. 3), and their size decreases. Reduction of the flight time of the dispersed droplets from the arc to the sprayed surface provides their high temperature when striking the sprayed surface.

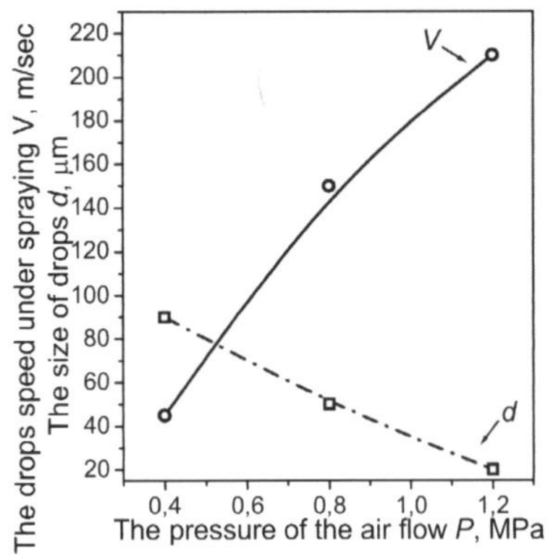

Fig. 3. Change of the drops speed and droplets of diameter depending on the pressure of the air jet

Increasing pressure of the air flow from 0.6 to 1.2 MPa promotes an intense dispersion of the melt of providing the electrode wire, the reduction in the droplets size which formed the coating. Under air pressure of 0.6 MPa a droplet of size from 30 to 100 µm (Fig. 4a), and at air pressure of 1.2 MPa – 10...30 µm or less (Fig. 4b) is formed. Metallographic studies confirm the results of fractionl distribution of droplets obtained under spraying with a pressure of 1.2 MPa (Fig. 5a).

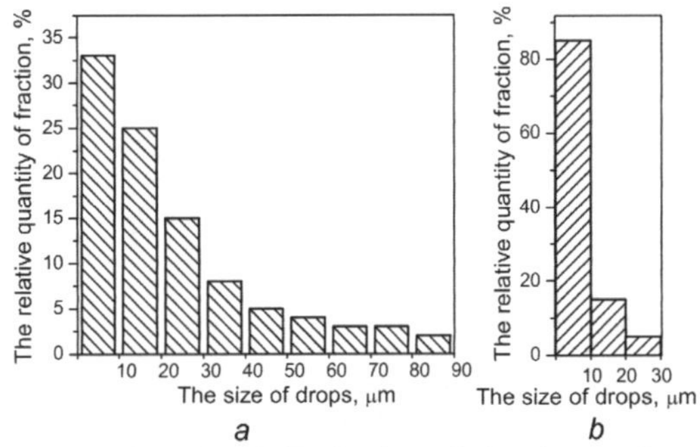

Fig. 4. Fractional distribution of the drops of cored wire Cr6B3Al3Mg2: *a* – 0.6 MPa; *b* – 1.2 MPa

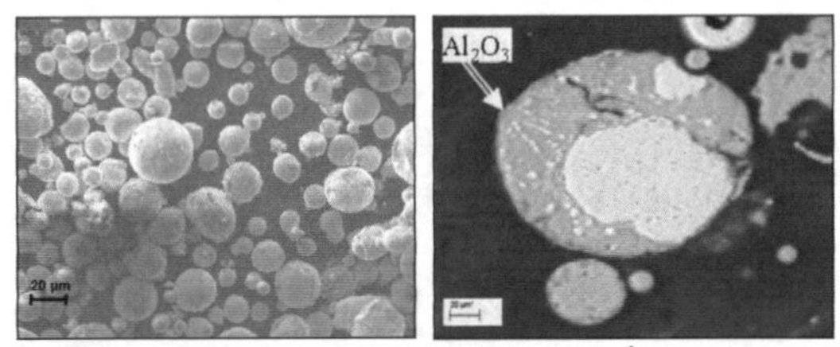

a *b*

Fig. 5. The atomized drops of the melt metal from the CW140Cr14Ni2TiAl:
a – general view of drops from the surface;
b – transverse microsection sprayed in snow drops target: full and partial oxidation of drops from the surface

On the surface of dispersed droplets on the surface of oxide films of various compositions are formed.

Reduc¬tion of the droplet sizes provides an increased area of contact with air and consequently their partial or total oxidation (Fig. 5b), which increases the amount of oxides and the formation of the composite coating structure (Fig. 6).

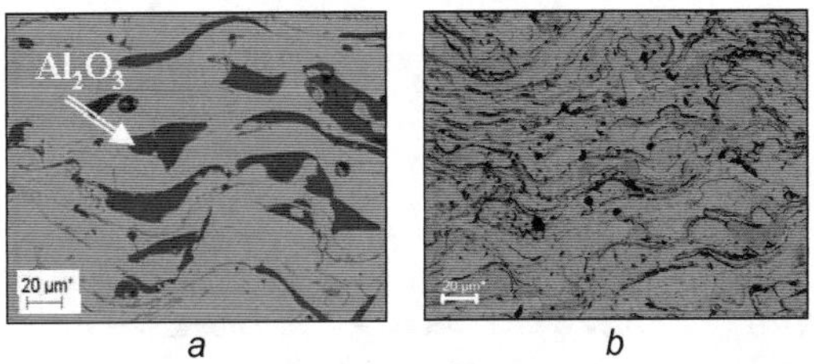

Fig. 6. Placing of oxide films in the structure of CW140Cr14Ni2TiAl: *a* – 0.6 MPa; *b* – 1.2 MPa

Under spraying pressure of 0.6 MPa, the amount of oxide phase in the coating structure from the CW140Cr14Ni2TiAl is 15...20 % vol., and at a pressure of 1.2 MPa the oxide content increases by 1.3...1.5 times and is 20...30 % vol. The oxide phase is usually located between the coating lamellae (Fig. 6a), though in some cases it is formed by the oxide lamellae of the completely oxidized drops (Fig. 6b).
The analysis of the obtained experimental data shows that the maximum content of oxides in the coating is observed when ferrophosphorus, ferrotitanium and ferrochromium are added to the charge (Fig. 7), and in this case the oxides of Ti, Cr and Fe or their mixtures are mainly formed in the coating.

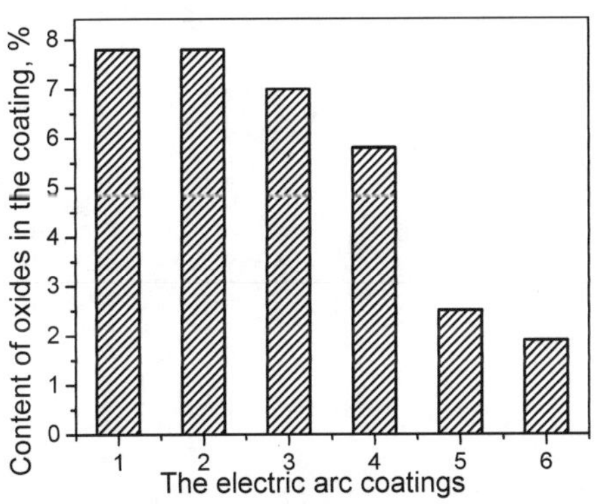

Fig. 7. Effect of charge components cored wires on oxygen content in the coatings with CW:
1 – 140Cr14; 2 – 140Cr14PP; 3 – 140Cr14Ni2TiAl; 4 – 140Cr16B3Ti2Ni2; 5 – 140Cr16B3; 6 – Cr6B3Al3Mg2

Addition of the CW charge of the elements like B, Al and Mg significantly reduces the content of oxides in the coating, and in this case their chemical composition is changed with advantageous formation of Mg or Al oxides in the coating due to reduction other oxides to the metal.
The thickness of the lamellar oxide films in the coating under a spraying pressure of 1.2 MPa is 1...3 mm (Fig. 6b), at the same time the thickness of oxide films at a pressure of 0.6 MPa can reach 20 m (Fig. 6a).

The microhardness of the iron oxide inclusions – magnetite, hematite, and wustite in the coating sprayed from the unalloyed steel wire is within 700-800 HV. The microhardness of the coating formed of non-alloyed wire Sv 08 is 200...300 HV at a pressure of 0.6 MPa. The spraying of the same coating at a pressure of 1.2 MPa reduces the size of the lamellas and oxides and their volume content increases, providing an increase in the microhardness up to 320...450 HV. The sprayed CWCr6B3Al6Mg2 coating at a pressure of 0.6 MPa and 1.2 MPa demonstrate an increase in microhardness from 600...650 to 850...900 HV, respectively. This is ensured by a higher content of aluminum oxide (Al2O3 microhardness is 2000 HV) and a more uniform distribution of FeCrB2 borides in the coating structure (Fig. 8).

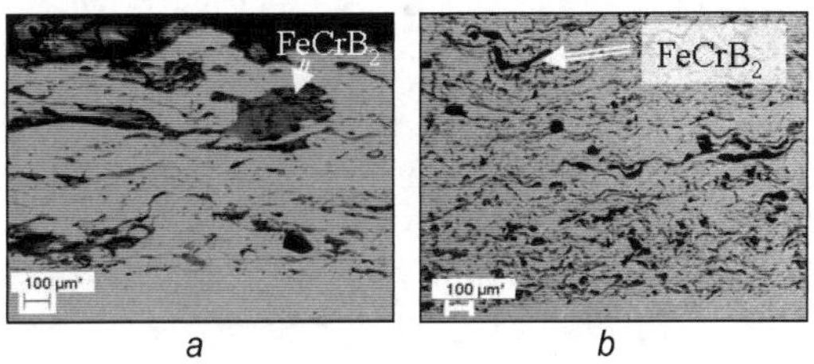

Fig. 8. Microstructure of coatings with CWCr6B3Al3Mg2: a – 0.6 MPa; b – 1.2 Mpa

It is found that the air flow pressure greatly affects the coating porosity. On an example of CW140Cr14Ni2TiAl coating it is shown that with increasing pressure of the air flow from 0.4 MPa to 1.2 MPa not only the total porosity decreases from 8.0 to 2.0 %, but also the geometrical dimensions of its pores from 30 to 3 m (Table 1).

Table 1. Effect of pressure air jet on porosity and geometric dimensions of pores in the coating of CW140Cr14Ni2TiAl

The pressure of the air flow *P*, MPa	The volume open porosity, %	The average thickness of pore, µm
0.4	8.0	30
0.6	4.4	10
0.8	3.0	5
1.2	2.0	3

With increasing pressure of air flow from 0.6 to 1.2 MPa the level of residual tensile stresses of the first kind decreases (Fig. 9). Due to the strong connection between the lamellas and the formation of thin oxide films of great length as a reinforcing component of the coating, the cohesive strength of the sprayed CW coating at a pressure of 1.2 MPa increases twice. The increase of cohesion strength at simultaneous reduction of residual tensile stresses leads to enhancing of abrasive wear resistance at 1.4 times (fig. 10).

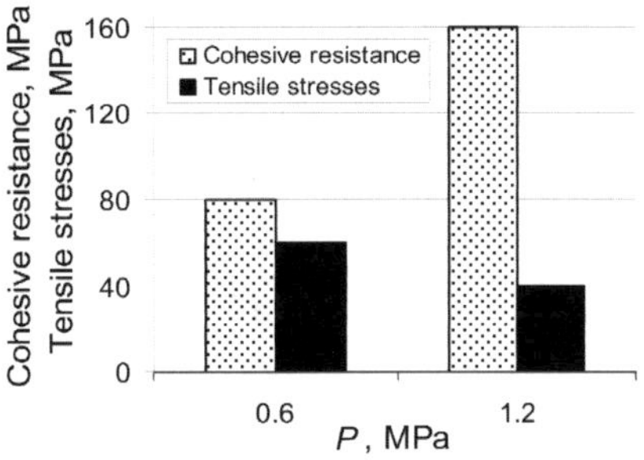

Fig. 9. Cohesive resistance and residual tensile stresses of the coating from the cored wire Cr6B3Al3Mg2

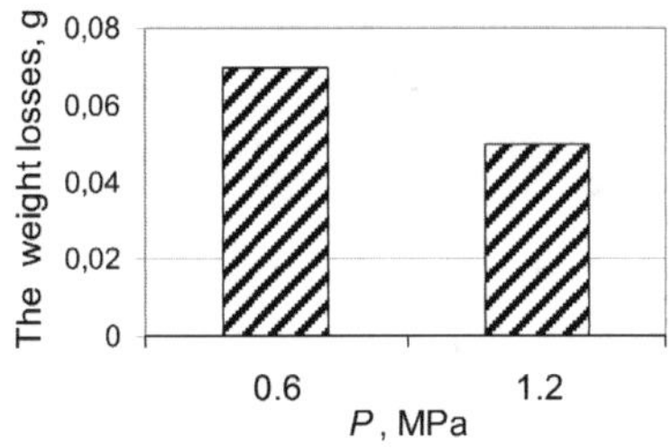

Fig. 10. The abrasive wear resistance coating from the cored wire Cr6B3Al3Mg2 sprayed at pressure 0.6 and 1.2 MPa

4 Conclusions

1. It is established that an increase in the pressure of air flow from 0.6 to 1.2 MPa ensures the growth of the air flow speed from 300 to 600 m/s, and the dispersed droplets speed from 120 to 220 m/s.
2. Increase of the pressure of the air flow from 0.6 to 1.2 MPa leads to the decrease in the coating thickness of the lamellas, formation of a greater number of oxide phases in the coating, in this case the hardness, cohesive strength and abrasive wear resistance increases and the level of residual stresses of the first circular kind in the coating decrease.

5 Literature

[1] Wielage B., Pokhmurska H., Student M., Gvozdeckii V., Stupnytckyi T., Pokhmurskii V. Iron-based coatings arc-sprayed with cored wires for applications at elevated temperatures. Surface and coating technology (2013), №220, pp. 27/35

[2] Pokhmurskyi V. I., Student M.M., Pokhmurska H.V., Student O.Z, Hvozdecky V.M., Stupnytskyi T.R. Arc-sprayed iron-based coatings for erosion-corrosion protection of boiler tubes at elevated temperatures. Journal of Thermal Spray Technology (2013), Issue. 22. pp. xx-yy

[3] Student M.M., Pokhmurs'ka H.V., Hvozdets'kyi V.M., Holovchuk M.Ya, Romaniv M.S. Effect of high-temperature corrosion on the gas-abrasive resistance of electric-arc coatings. Materials Science 45 (2009), Issue 4, pp. 481/89.

[4] Student M., Dzioba Yu., Hvozdets'kyi V., Pokhmurska H., Wielage B., Grund T. High-temperature corrosion of electric-arc coatings sprayed from powder core wires based on the Fe–Cr–B–Al system. Materials Science 44 (2008), Issue 5, 693/99.

[5] Ванєєв С.М. Методичні вказівки до індивідуальних занять та виконання індивідуальної роботи «Розрахунок сопла Лаваля» з курсів «Спецрозділи теплофізики» та "Розширювальні і компресорні турбомашини для газової та нафтової промисловості" СумДУ (2009), 22с.

[6] Бабичев М.А. Методы определения внутренних напряжений в деталях машин. М.: Изд-во АН СССР (1955), 132.

Cold sprayed alumina-silicon dioxide composite coatings on AZ31 magnesium alloy

L.J. Fang, J.G. Gong, J. Huang, Z.M. Jia, X.K. Suo, H. Li / CN

An alumina-silicon dioxide composite coating was fabricated by cold spraying on AZ31 magnesium alloy. The microstructure, mechanical properties (microhardness, bonding strength and tribological behaviour) and anti-corrosion property of the coating as a function of the ceramic volume were investigated. The results show that the composite coating presents higher bonding strength and microhardness. Addition of silicon dioxide significantly enhances the anti-wear and anti-corrosion performances of AZ31 magnesium alloy.

1 Introduction

Magnesium alloys are promising lightweight materials applied in automotive, aerospace and electronics industries due to their combination of excellent special strength. However, the relatively low corrosion and wear resistance of magnesium alloys limit their potential industry applications [1]. One of the most effective solutions to prevent corrosion is to coat an anti-corrosion layer on magnesium alloys. Many surface techniques have been developed and used to improve the anti-corrosion of magnesium alloys, such as thermal spraying, chemical vapour deposition (CVD), sol-gel, plating, anodizing oxidation, microarc oxidation, etc. [2-7].

Cold spraying (CS) differs from the conventional thermal spray methods. In CS process, micron-sized particles are accelerated by an inert gas stream with a high velocity. The feedstock remains in the solid state throughout the entire process. It can avoid some problems associated with high temperature, such as chemical reaction, phase transition, oxidation and unfavourable structural changes. Therefore, CS is particularly suitable for the materials sensitive to heat or oxidation during spraying. In particular, cold spray coatings are denser than the thermal spray coatings [8], contributing to better mechanical properties and corrosion resistance.

Ceramic materials have many advantages, such as high hardness, high melting point, good thermal stability and good abrasion resistance. Addition of ceramic materials to a coating can be effectively improve wear resistance, corrosion resistance or temperature oxidation resistance of the coating [9, 10]. Lee [11, 12], Shkodkin [13] and Irissou [9] reported that incorporating ceramic particles into metallic coatings can not only improve coating quality by reducing the porosity but also increase bonding strength. Porosity reduction of the coating contributes to the improvement of the corrosion resistance [9]. Addition of hard ceramic particles into cold sprayed coatings also improves the hardness and wear resistance [13, 14].

Different ceramic such as Al_2O_3 [11, 15], SiC [16-19] and TiN [20, 21] were investigated as the ceramic phrase in metallic matrix ceramic composites (MMCs). SiO_2, as a ceramic with good chemical stability, wear resistance and low cost, has not been studied. Therefore, in this paper, Al-SiO_2 composite coatings were deposited by CS on AZ31 magnesium alloy. The effects of SiO_2 content on the microstructure, mechanical properties and anti-corrosion property were investigated systematically.

2 Materials and methods

2.1 Materials and coatings preparation

Spherical commercial pure aluminium powder (Beijing General Research Institute of Mining & Metallurgy, China) with a mean size of 25 μm and angular SiO_2 powder (Beijing General Research Institute of Mining & Metallurgy, China) with a mean size of 16 μm were used in this study. Al and SiO_2 powders were mixed by mechanically blending. The composite coatings with different SiO_2 content were fabricated by cold spraying (CGT KINETIKS 4000, Germany). Nitrogen was used as acceleration gas and carrier gas with the temperature of 350 ℃ and the pressure of 3 MPa. The spraying distance was 30 mm, and the traverse speed of the gun was 200 mm/s. The composite coatings with 10 vol.%, 20 vol.% and 40 vol.% SiO_2 particles were named as Coating10, Coating20 and Coating 40, respectively. AZ31 magnesium alloy (AZ31) plates with a dimension of 20 mm×20 mm×3 mm were used as substrates. Prior to spraying, the substrates were surface grit blasted using 60 mesh black fused alumina sands and subsequently degreased by sonication cleaning in acetone.

2.2 Characterization of the coatings

Phase structures of the powder and the coatings were characterized by X-ray diffraction (XRD, D8 Advance, Bruker AXS, Germany) at a scanning rate of 0.02 °/s over a 2θ range of 10° ~ 90° using Cu Kα radiation operated at 35 mA and 40 kV. Topography and cross-sectional morphology of the coatings were examined by field emission scanning electron microscope (FESEM, Hitachi S-550N, Japan). Element analyses were carried out by using energy dispersive X-ray spectra (EDX) equipped with FESEM. The porosity of the coating was measured by image analysis and at least ten images per coating were taken. The microhardness of the coating was measured using a Vickers hardness tester with 100 g load with dwelling time of 10 s. Five points per sample were acquired randomly to get a reliable data. Bonding strength of the coating was tested following the ASTM standard C633-01.

The tribological properties of the coatings were evaluated using a reciprocating-type ball-on-disc tribometer (JLTB-02, J&L Tech Co. Ltd, Korea). The tests were performed under a load of 3 N for 30 mins at 18.6°C with a relative humidity of 70%. Steel balls (GCr15) with a diameter of 6 mm were used as the counterparts. The coatings and balls were ultrasonically cleaned in acetone prior to the tests, and a new ball or a new position of the ball was used for each friction test. The friction coefficients and sliding time were automatically recorded during the tests. For each sample, three measurements were repeated. Electrochemical impedance spectroscopy (EIS) measurements were performed using a Solartron Modulab system (2100A, UK). The tests were conducted in 3.5 wt.% NaCl solution at room temperature. AC signal of 10 mV and the frequency ranging from 100 kHz to 0.01 Hz were employed. Before the electrochemical measurement, the coating samples were immersed in an aerobic chamber containing 3.5 wt.% NaCl solution for 30min. Each measurement was repeated three times.

3 Results and discussion

3.1 Microstructure of the coatings

The topographical morphology of the coating was examined, and the result is shown in Fig. 1. Different topographical morphologies of the Al coating and the composite coatings could be found (Fig.1a vs Fig.1b-d). In Fig. 1a, several shallow and smooth craters as marked by white arrows can be found, which might be created by rebounded Al particles during deposition. Conversely, several deep craters with edges and corners could be observed in the composite coatings which is due to the rebounded hard ceramic particles (Fig.1b-d). The typical cross-sectional microstructures of the coatings were shown in Fig. 1e and 1f. The composite coating present fewer cracks and pores than the Al coating. The quantitative of the defects and ceramic contents of the coatings were measured, and the results are shown in Table 1. It is found that addition of SiO_2 effectively improves the density of the coating. The SiO_2 content in the composite coating is much lower than that in the feedstock, especially for Coating20 and Coating40. It suggests that most of ceramic particles rebounded during deposition, which contributes to the decrease of coating porosity due to the more severe deformation of the deposited Al particles by the rebounded ceramic particles.

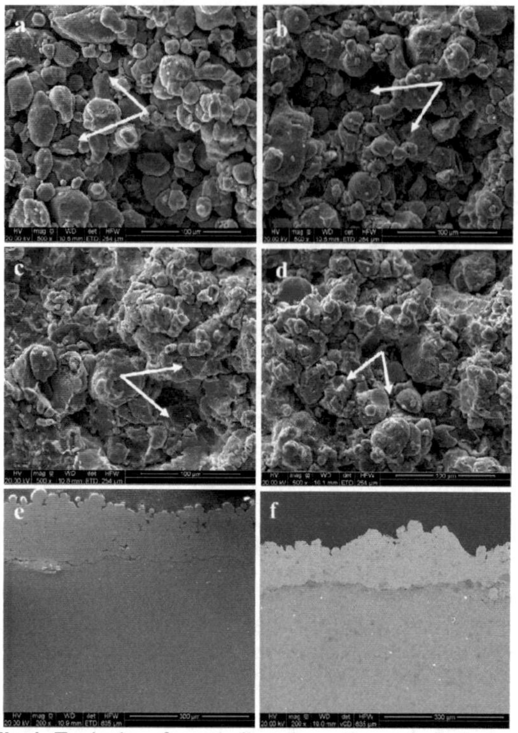

Fig.1. Typical surface and cross section morphologies of the as-sprayed (a) pure Al coating, (b) coating10, (c) coating20, (d) coating40, (e) pure Al coating and (f) coating40.

Table 1 Porosity and SiO_2 content of the coatings

Coating	Porosity (%)	SiO_2 (%)
Pure Al coating	5.11 ± 1.39	0
Coating10	1.23 ± 0.45	8.98 ± 0.97
Coating20	1.33 ± 0.70	11.46 ± 1.85
Coating40	0.77 ± 0.19	15.75 ± 1.37

The typical EDS mapping of SiO_2 distribution and the XRD spectra of the coatings are showed in Fig.2. The results suggest that no oxidation is found during spraying (Fig. 2a). Moreover, SiO_2 peaks could not be detected in the composite coatings due to their relative few content.

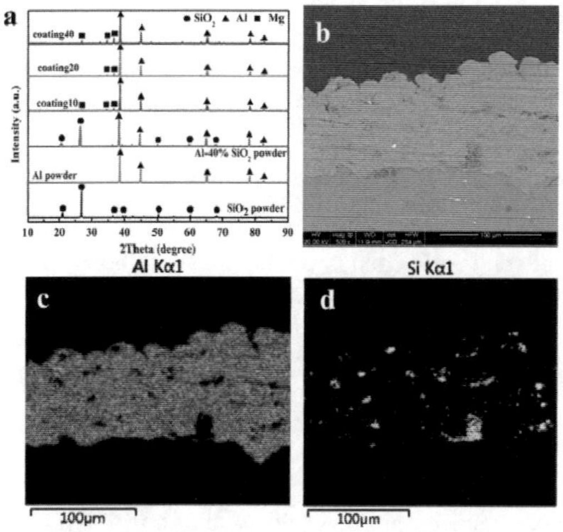

Fig.2. (a) XRD spectra of the powders and the coatings, (b-d) typical EDS mapping analysis of the composite coatings.

The microhardness and bonding strength of the coatings are also measured, and the results are shown in Fig.3. Moderate increase of the microhardness and the bonding strength of the composite coating is found as SiO_2 content increases, which could be benifited from the increase of the coating density and ceramic content. The decrease of the defects or pores and the increased of ceramic content limit the deformation of the surface under compression load, which results in the increase of the microhardness of the coating. The decrease of the defects and pores also increases the bonding aera between particles. Meanwhile, the ceramic particles play as anchor between particles under traction load, therefore the bonding strength of the coating increases.

The tribological behaviour of the coating was also investigated, and the result is shown in Fig.4. The friction coefficients of the coatings increase compared to substrate (Fig.4a), which may be due to the interior defects and ceramic particles [22-24]. The mass loss of the Al coating is higher than that of the substrate, which is due to the interior defect in the coating. For the composite coating, the mass loss significantly decreases as the SiO_2 content increase (Fig.4b). The wear morphology of the coatings are shown in Fig.5. A typical characteristic of adhesive wear is observed for the substrate and the Al coating (Fig.5a and 5b). On the surface of the composite coating, the plough and extrusion characterization is found, suggesting the abrasive wear mechanism.

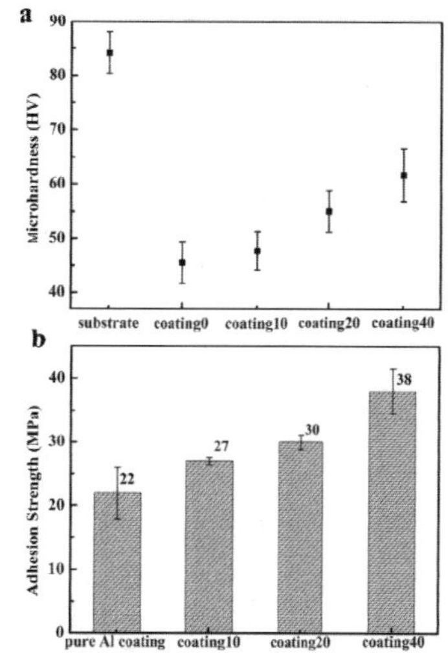

Fig.3. (a) Microhardness and (b) bonding strength of the coatings.

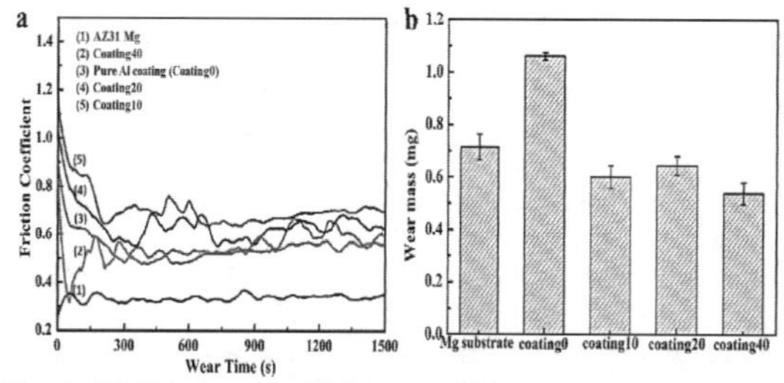

Fig.4. (a) Friction coefficient and (b) Wear mass of the coatings and substrates.

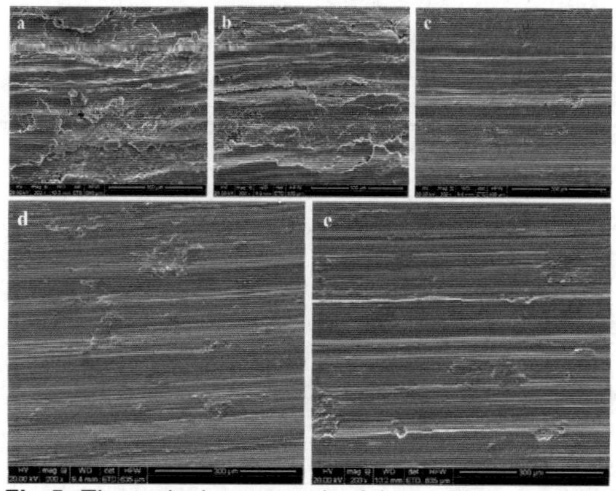

Fig.5. The typical wear track of the coating. (a) AZ31 magnesium alloy sunstrate, (b) pure Al coating, (c) coating10, (d) coating20 and (e) coating40.

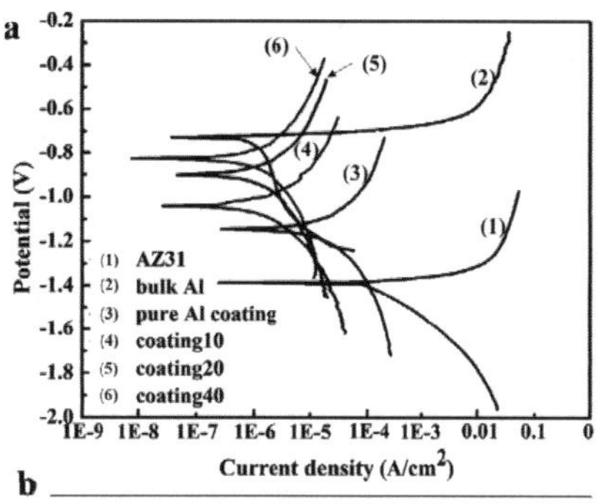

b

Samples	Corrosion potential (E_{corr}) (V/SCE)	Corrosion current density (I_{corr}) (A/cm²)
AZ31 substrate	-1.387	8.796×10^{-5}
Bulk pure Al	-0.7276	7.335×10^{-7}
Pure Al coating	-1.146	5.79×10^{-6}
Coating10	-1.04	6.66×10^{-7}
Coating20	-0.9	6.04×10^{-7}
Coating40	-0.826	4.87×10^{-7}

Fig.6. Potentiodynamic polarization curves and Electrochemical characteristics of the coatings, substrates and bulk pure Al.

The corrosion behaviour of the composite coating was evaluated by potentiodynamic polarization technique and the results are showed in Fig.6. The polarization curves show that the current density of the composite coatings is much lower than that of the Al coating and the substrate, suggesting better corrosion resistance.

4 Conclusions

Al/SiO_2 composite coatings were deposited on AZ31 magnesium alloy substrate by cold spraying. The composite coatings show lower porosity, higher microhardness, higher bonding strength, higher wear and corrosion resistance than the Al coating or the AZ31 substrate. With the addition of SiO_2 particles, the wear mechanism transfers from adhesive wear to abrasive wear.

5 Literature

[1] Christoglou, C., Voudouris, N., Angelopoulos, G.N., Pant, M. and Dahl, W: Deposition of aluminium on magnesium by a CVD process. Surface & Coatings Technology 184 (2004), Issue 2-3, pp. 149/55.

[2] Lin, C.S. and Li, W.J: Corrosion resistance of cerium-conversion coated AZ31 magnesium alloys in cerium nitrate solutions. Materials Transactions 47 (2006), Issue 4, pp. 1020/25.

[3] Shao, Z., Cai, Z. and Shi, J: Preparation and Performance of Electroless Nickel on AZ91D Magnesium Alloy. Materials & Manufacturing Processes 31 (2015), Issue 9, pp. 1238/45.

[4] Zhang, R.F. and Zhang, S.F: Formation of micro-arc oxidation coatings on AZ91HP magnesium alloys. Corrosion Science 51 (2009), Issue 12, pp. 2820/25.

[5] Zhong, X., Li, Q., Chen, B., Wang, J., Hu, J. and Hu, W: Effect of sintering temperature on corrosion properties of sol–gel based Al_2O_3 coatings on pre-treated AZ91D magnesium alloy. Corrosion Science 51 (2009), Issue 12, pp. 2950/58.

[6] Altun, H. and Sen, S: The effect of DC magnetron sputtering AlN coatings on the corrosion behaviour of magnesium alloys. Surface and Coatings Technology 197 (2005), Issue 2-3, pp. 193/00.

[7] Carboneras, M., López, M.D., Rodrigo, P., Campo, M., Torres, B., Otero, E. and Rams, J: Corrosion behaviour of thermally sprayed Al and Al/SiCp composite coatings on ZE41 magnesium alloy in chloride medium. Corrosion Science 52 (2010), Issue 3, pp. 761/68.

[8] Van Steenkiste, T.H., Smith, J.R. and Teets, R.E: Aluminum coatings via kinetic spray with relatively large powder particles. Surface and Coatings Technology 154 (2002), pp. 237/52.

[9] Irissou, E., Legoux, J.G., Arsenault, B. and Moreau, C: Investigation of Al-Al_2O_3 Cold Spray Coating Formation and Properties. Journal of Thermal Spray Technology 16 (2007), Issue 5-6, pp. 661/68.

[10] Tao, Y., Xiong, T., Sun, C., Jin, H., Du, H. and Li, T: Effect of α-Al2O3 on the properties of cold sprayed Al/α-Al_2O_3 composite coatings on AZ91D magnesium alloy. Applied Surface Science 256 (2009), Issue 1, pp. 261/66.

[11] Lee, H.Y., Jung, S.H., Lee, S.Y., You, Y.H. and Ko, K.H: Correlation between Al2O3 particles and interface of Al-Al_2O_3 coatings by cold spray. Applied Surface Science 252 (2005), Issue 5, pp. 1891/98.

[12] Lee, H.Y., Yu, Y.H., Lee, Y.C., Hong, Y.P. and Ko, K.H: Cold Spray of SiC and Al_2O_3 with Soft Metal Incorporation: A Technical Contribution. Journal of Thermal Spray Technology 13 (2004), Issue 2, pp. 184/89.

[13] Shkodkin, A., Kashirin, A., Klyuev, O. and Buzdygar, T: Metal Particle Deposition Stimulation by Surface Abrasive Treatment in Gas Dynamic Spraying. Journal of Thermal Spray Technology 15 (2006), Issue 3, pp. 382/86.

[14] Irissou, E., Legoux, J.G., Ryabinin, A.N., Jodoin, B. and Moreau, C: Review on Cold Spray Process and Technology: Part I-Intellectual Property, Journal of Thermal Spray Technology 17 (2008), Issue 4, pp. 495/16.

[15] Wang, Q., Spencer, K., Birbilis, N. and Zhang, M.X: The influence of ceramic particles on bond strength of cold spray composite coatings on AZ91 alloy substrate. Surface and Coatings Technology 205 (2010), Issue 1, pp. 50/56.

[16] Sansoucy, E., Marcoux, P., Ajdelsztajn, L. and Jodoin, B: Properties of SiC-reinforced aluminum alloy coatings produced by the cold gas dynamic spraying process. Surface and Coatings Technology 202 (2008), Issue 16, pp. 3988/96.

[17] Eesley, G., Elmoursi, A. and Patel,N: Thermal properties of kinetic spray Al-SiC metal-matrix

composite. Journal of Materials Research 18 (2003), Issue 4, pp. 855/60.
[18] Yu, M., Suo, X.K., Li, W.Y., Wang, Y.Y. and Liao, H.L: Microstructure, mechanical property and wear performance of cold sprayed Al5056/SiCp composite coatings: Effect of reinforcement content. Applied Surface Science 289 (2014), pp. 188/96.
[19] Lee, H. and Ko, K: Effect of SiC particle size on cold sprayed Al-SiC composite coatings. Surface Engineering 25 (2013), Issue 8, pp. 606/11.
[20] Li, W.Y., Zhang, G., Liao, H.L. and Coddet, C: Characterizations of cold sprayed TiN particle reinforced Al2319 composite coating. Journal of Materials Processing Technology 202 (2008), Issue 1-3, pp. 508/13.
[21] Li, W.Y., Zhang, G., Zhang, C., Elkedim, O., Liao, H. and Coddet, C: Effect of Ball Milling of Feedstock Powder on Microstructure and Properties of TiN Particle-Reinforced Al Alloy-Based Composites Fabricated by Cold Spraying. Journal of Thermal Spray Technology 17 (2008), Issue 3, pp. 316/22.
[22] Zoo, Y.S., An, J.W., Lim, D.P. and Lim, D.S: Effect of Carbon Nanotube Addition on Tribological Behavior of UHMWPE. Tribology Letters 16 (2004), Issue 4, pp. 305/09.
[23] Kelly, R.G., Scully, J.R., Shoesmith, D.W. and Buchheit, R.G: Electrochemical techniques in corrosion science and engineering, Chemphyschem 4 (2003), pp. 296/98.
[24] Aruna, S.T., Ezhil Selvi, V., William Grips, V.K. and Rajam, K.S: Corrosion- and wear-resistant properties of Ni-Al_2O_3 composite coatings containing various forms of alumina. Journal of Applied Electrochemistry 41 (2011), Issue 4, pp. 461/68.

Studies on the Cold Sprayability of Mixed 316L/Fe Powders

Xin Chu, Phuong Vo, Stephen Yue, Montreal/CDN

Abstract: In this work, single component 316L and Fe coatings, as well as mixed 316L/Fe coatings with a dual powder feeder to obtain various feedstock compositions, were deposited to measure the deposition efficiency (DE). Individual particle impact tests were performed on single component and composite coatings to understand the particle impact behaviors during deposition. Bond ratio (BR) were determined for the impact tests to correlate with the DE. Results show that the 316L powder has a better DE than Fe, whereas the DE of the mixed 316L/Fe powders increases with increasing feedstock Fe content. The BR results correspond well with the DE of single component powders and mixed powders. The BR of 316L impacts onto composite coatings decreases with increasing Fe content, while the BR of Fe impacts plateaus at a high value regardless of composite coating composition, which leads to the increase of overall mixture DE.

1. Introduction

Cold spray of mixed metal powders has been increasingly studied by researchers. It has been found to offer similar benefits as ceramic additions to metal powders in improving cold sprayability of the component powder (e.g. increase deposition efficiency and decrease porosity) [1]. In ceramic-metal system, the impinging ceramic powders can not be deposited alone as they only contribute to tamping and roughening to the surface layers [2]. While in metal-metal system, information regarding the behaviors of the mixed powders upon codeposition is still lacking, thus the "unpredicted" situations in its cold sprayability are often observed.

The cold spray coating deposition can be mainly divided into two distinct steps: (1) breeding layer formation; and (2) coating build-up. Due to the ill-defined nature of the bonding mechanisms involved in cold spray, the quantitative prediction of deposition efficiency (DE) is difficult. To evaluate the cold sprayability of the spray material, a few researchers have used the individual particle impact test as an indication of DE. However, the impact test is commonly generated by low feed rate spraying of powders onto the substrate, thus it corresponds to the step of breeding layer formation. Quantitatively speaking, the breeding layer formation only has limited contribution to DE at the initial deposition stages, once the first layer has been formed, the coating deposition then shift to particle-particle bonding. Therefore, using the common approach of impact test one can often mispredict the DE result of a feedstock, especially the deformability of the powder and substrate materials is markedly dissimilar.

In this work, the DE of mixed 316L stainless steel and commercial purity Fe powders was investigated as part of an overall effort to produce metal-metal composites with controlled corrosion properties for biodegradable implants [3]. The individual particle impact tests were performed on the as-polished coatings to reflect the particle behaviors during coating build-up. The results offer an important mechanistic insight towards the various deposition mechanisms of 316L/Fe mixed powders and provide directions for the future endeavors to fully explain the DE characteristics.

2. Methods

Commercially available 316L stainless steel powder and commercial purity Fe powder were used as the feedstock. SEM images of the feedstock powders are shown in Fig. 1. Information regarding the feedstock are shown in Table 1.

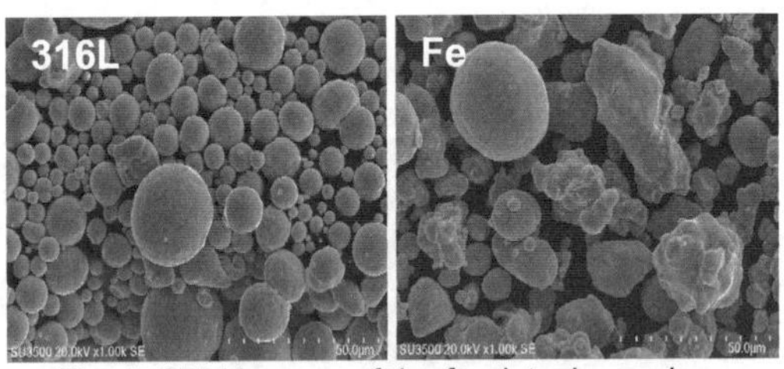

Fig. 1: SEM images of the feedstock powders.

Table 1: Information regarding the feedstock powders.

Powder	Supplier	Shape	Size	HV
316L	Sandvik	Spherical	29	262
Fe	QMP	Spherical+ irregular	22	141

Spraying was performed at the McGill-NRC cold spray facility, located at the National Research

Council Canada, Boucherville, using a PLASMA GIKEN PCS-800 cold spray system (Plasma Giken, Tokyo, Japan) with a PNFC2-010-30S carbide nozzle. Nitrogen was used as the propellant gas. The spraying was performed at a condition of 700 ℃ gas temperature, 4 MPa gas pressure and 80 mm stand-off distance. Two deposition modes were used: high feed rate with low gun travel speed (300 m/s) for coating deposition, and low feed rate and high gun speed (1000 m/s) for individual particle impact. The detailed powder feed rates of two deposition modes are shown in Tables 2 and 3.

Table 2: Process parameters for coating deposition of 316L/Fe mixed powders.

Powder	Gun speed (mm/s)	Powder feed rate (g/min)	
		316L	Fe
316L	300	23.8	-
10Fe	300	18.3	2.4
50Fe	300	7.8	10.9
80Fe	300	3.9	18.1
90Fe	300	3.1	21.7
Fe	300	-	21.7

Table 3: Process parameters for individual particle impact tests.

Impact test	Gun speed (mm/s)	Powder feed rate (g/min)	
		316L	Fe
316L on 316L	1000	8.5	-
316L on Fe	1000	8.5	-
316L on 50Fe	1000	8.5	-
Fe on 316L	1000	-	11
Fe on Fe	1000	-	11
Fe on 50Fe	1000	-	11

For coating deposition, a dual feeder was used to separately feed the 316L and Fe powders to obtain the desired feedstock compositions. The substrates used were mild steel plates with dimensions of 76.2 mm×76.2 mm×6.4 mm, blasted to promote adhesion. After spraying, DE was measured as the mass gain of the substrate divided by the total mass consumption of spray material during the time that the gun was over the substrate. The coating cross-sections (transverse to the gun travel direction) were observed using a light optical microscope (LOM). For individual particle impact, the particles were sprayed onto the above-deposited coatings. Before use as substrates for impact tests, the as-sprayed coatings were polished to a mirror finish, in order to clearly reveal craters and deposits. After impacts, the surface and micrographs of the craters and deposits were characterized using a Hitachi SU3500 SEM (Hitachi, Tokyo, Japan). The bond ratio (BR), defined as the fraction of the number of bonded particles (deposits) to the total incident particles (craters+deposits), was determined by an image analysis method.

3. Results

3.1. Deposition efficiency

The deposition efficiency results are shown in Fig. 2. For single component powders, 316L powder exhibits a slightly superior DE to Fe (50% vs 45%). However, for the mixed powders, it is found that the DE increases with increasing feedstock Fe. Particularly, a small addition of 316L to Fe, i.e. 90Fe, is seen to "stimulate" the mixture DE to exhibit an even higher value than pure 316L (55% vs 50%). In turn, a small addition of Fe to 316L (10Fe), gives the lowest DE, which is in fact much lower than the Fe DE (38% vs 45%). Moreover, also seen in the Fig. 2, if compared the experimental mixture DE with the expected DE calculated by the rule of mixtures (ROM), $i.e. DE = DE_{316L} \times f_{316L} + DE_{Fe} \times f_{Fe}$, the trends of DE results with increasing feedstock Fe are totally opposite to each other. This finding attracts the interest since the addition of a poorer DE component can instead facilitate the deposition of a better DE component, detailed study is carried out as follows.

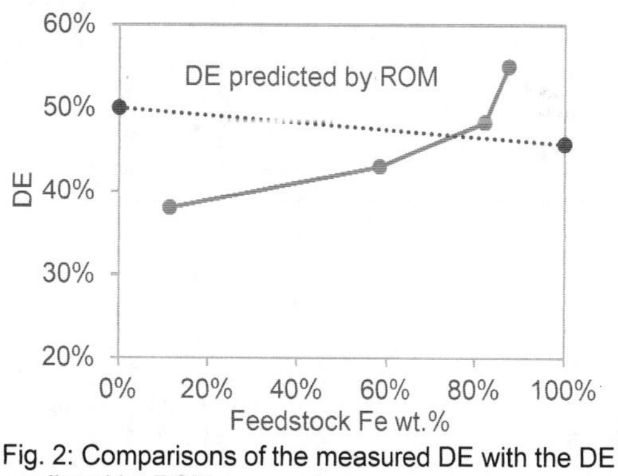

Fig. 2: Comparisons of the measured DE with the DE predicted by ROM.

3.2. Coating microstructure

The cross-sectional microstructure close to the coating surfaces of two DE extremes: 10Fe (lowest DE) and 90Fe (highest DE), is shown in Figs. 3 (a) and (b). The images are oriented such that the impact direction is as indicated by the white arrows. The dark regions in the images are Fe, the light regions are 316L and the black spots are pores. Defects are also identified along the interparticle boundaries,

indicating either poor interparticle bonding or strong rebound of the impacted particles.
For composite coatings, the addition of a second component powder has introduced the mixed 316L/Fe interfaces. The mixed 316L/Fe interfaces in 10Fe and 90Fe coatings are compared and very distinct characteristics can be seen. For instance, 10Fe, as shown in Fig. 3 (a), there are far more defects present at the mixed interfaces compared to 90Fe in Fig. 3 (b). In particular, a few crack-like defects in Fig. 3 (a) are identified to be laterally distributed, transverse to the impact direction. These defects distributing along the particle impact interfaces are considered as indicators of a strong rebound of the impacted particles. Also it is noticed that these lateral defects mostly locate on the "top" of Fe and the "bottom" of 316L, as marked by red arrows. This finding implies a stronger rebound of 316L particles impacting on the previously deposited Fe layer compared with Fe particles on 316L layer during coating build-up. However, for 90Fe, as shown in Fig. 3 (b), there are no signs of any lateral defects and all the mixed interfaces exhibit very close contact with much better optical characteristics than the Fe/Fe interfaces, which appears to correlate with the highest DE.

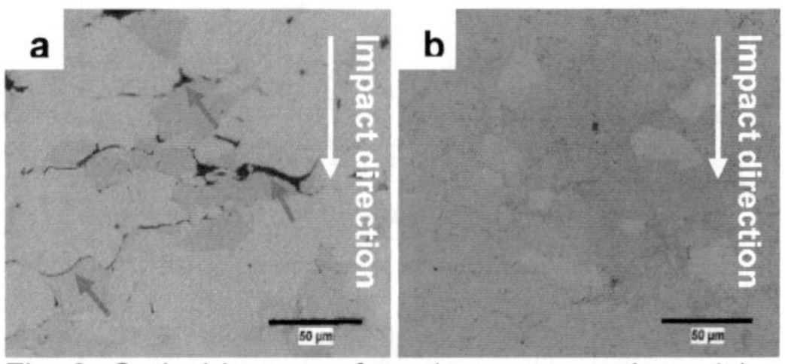

Fig. 3: Optical images of coating cross-sections: (a) 10Fe and (b) 90Fe. Red arrows: lateral defects at the mixed 316L/Fe interfaces

3.3. Individual particle impacts

The SEM surface morphology of the single component coatings after individual particle impact tests are shown in Fig. 4 (a) to (d). The corresponding bond ratio (BR) results determined for each impact tests are tabulated in Table 4. Note that ultrafine 316L deposits were not included, since these particles having negligible masses barely contribute to the DE results. It is seen that the impact of 316L particles on 316L coating has a higher BR compared with Fe on Fe. This is consistent with the observation that 316L single component powder has higher DE than Fe. For the two types of impacts between 316L and Fe, it is seen that Fe on 316L has a much higher BR than 316L on Fe and is by far highest of all the combinations tested.

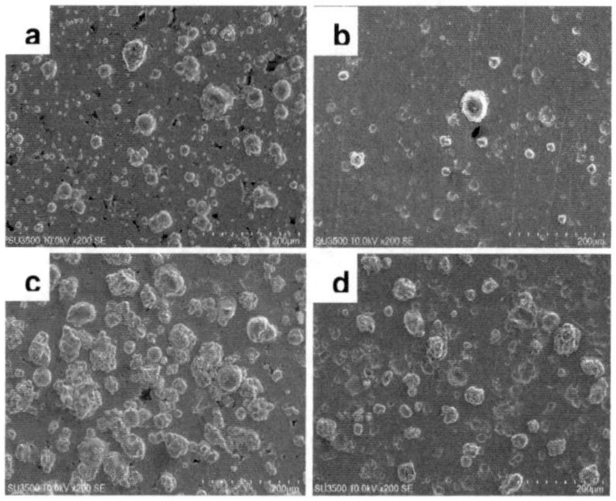

Fig. 4: SEM surface morphology after individual particle impact tests of: (a) 316L on 316L, (b) 316L on Fe, (c) Fe on 316L, and (d) Fe on Fe.

Table 4: Bond ratio (BR) of individual particle impacts onto single component coatings.

Coating	316L BR, %	Fe BR, %
316L	35.8±0.8%	87.3±2.3%
Fe	8.3±1.1%	25.4±1.5%

As shown in Fig. 5 (a) to (d) are the magnified images of deposits. Comparing the two types of impacts between 316L and Fe, the 316L deposits can be seen to lift off the Fe coating at the edges (Fig. 5 (b)); whereas the Fe deposits show a close contact to the 316L coating (Fig. 5 (c)). This result is consistent with the finding of preferred location of lateral defects at the mixed 316L/Fe interfaces, showing that the 316L impacts exhibit a stronger rebound on Fe layer than Fe on 316L. Comparing the two types of single component impacts, it appears that the 316L deposits exhibit a closer contact to 316L surface than Fe on Fe, which correlates with the higher BR of 316L on 316L than Fe on Fe, but the difference is not that significant compared with that of the former two impacts.

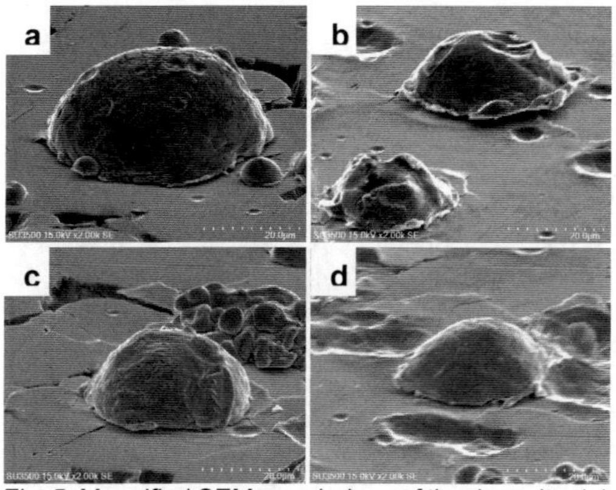

Fig. 5: Magnified SEM morphology of the deposits: (a) 316L on 316L, (b) 316L on Fe, (c) Fe on 316L and (d) Fe on Fe.

Similarly, the BR results from impact tests of 316L or Fe onto as-polished composite coatings are tabulated in Table 4. It can be seen from Table 5 that the BR results of 316L and Fe particles vary distinctly on different composite coatings. With increasing the Fe fraction in the composite coatings, the BR result of the 316L particle monotonically decreases from 30.1% to 6.9%, while that of the Fe somehow plateaus at a high value of about 55%. The stable BR of Fe is consistent with the observation of that high Fe mixed coating, e.g. 90Fe, exhibits a dense coating microstructure.

Table 5: Bond ratio (BR) of individual particle impacts onto composite coatings.

Coating	316L BR, %	Fe BR, %
10Fe	30.1±1.7%	59.5±3.8%
50Fe	16.0±1.1%	54.8±1.9%
80Fe	12.3±4.0%	54.3±2.5%
90Fe	7.9±0.4%	57.0±2.0%

4. Discussions

The mild steel substrate used in this study is a commonly used material on which particles can be easily deposited to form the first layer. Therefore, the particle-particle cohesion during coating build-up that should mainly determine the DE characteristics. As the individual impact tests performed onto as-polished coatings well reflect the particle impact behaviors during this process, the BR results were compared with the measured DE results. For single component powder deposition, the DE results of 316L and Fe were compared with the BR results of 316L on 316L and Fe on Fe, respectively. For the mixed powders, assuming that no interactions between mixed powders have occurred during flight, the mixed powder BR is determined as the averaged contribution of measured BR results for 316L or Fe on composite layers, weighted by their initial feedstock compositions:

$$BR = f_{Fe} \times BR_{Fe} + (1 - f_{Fe}) \times BR_{316L} \quad \text{(Eq 1)}$$

Taking the BR data from Tables 4 and 5, the mixed powders BR was calculated and a dotted line was drawn to compare with the measured DE results. As shown in Fig. 6, the BR results are seen to exhibit good correlation with the DE results, regarding both single component and mixed powders deposition. In indicates that using the impact tests onto as-polished coatings can well reflect the DE of feedstock materials. It is understood that the future model optimization the effects such as roughness and tamping needs to be improved to be considered, however, it appears in this study, those effects are less significant for the feedstock both with certain cold sprayability and relatively high density.

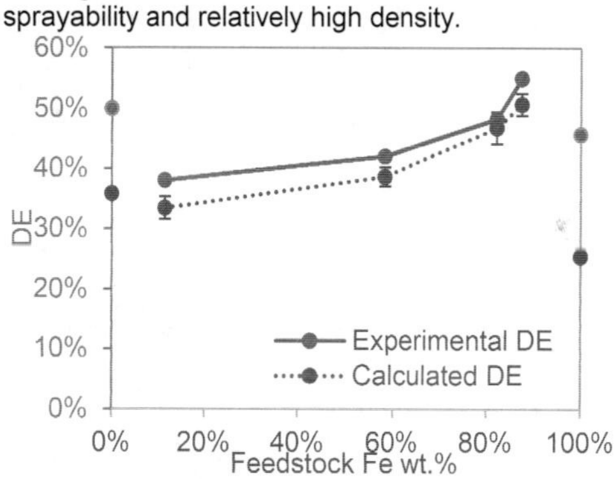

Fig. 6: Comparisons between the results of measured DE and calculated BR.

For BR expectations onto composite coatings, both 316L and Fe impacts should have a decreasing BR with increasing Fe feedstock fraction, since the Fe layer has already exhibited a more violent rebound of the impacting particles compared to 316L, as seen in Table 5. However, only the 316L impacts exhibit a good correlation with the expected BR, while the Fe BR appears to be independent of the composite composition.

The high plateau of Fe BR regardless of composite composition well explains the elevated BR at high Fe mixed regions, however, the impact behaviors of Fe onto composite layers itself is not clear. Intuitively, the particle impact behavior appears to only relate to the first previously deposited layer, considering in this case, the particle impact types are all "solid on solid" which means no embedding has occurred. However, this study indicates that there is a "substrate" effect for Fe impacts onto composite coatings, e.g. the surrounding and underlying 316L phases provide a cushioning effect to the Fe matrix to retain Fe impacts. Therefore, to validate this hypothesis, replacing the 316L to a harder component with similar density will be a plausible approach. Moreover, it is also not sure that the two composite composition extremes, i.e. 10Fe and 90Fe, could exhibit similar behaviors in terms of retaining Fe impacts. Considering the particle deposition being a competitive process between adhesion and rebound, the adhesion strength/energy of the bulk composite as well as the single deposits should also be determined in the future.

5. Conclusions

In summary, single component 316L and Fe powders, as well as mixed 316L/Fe powders were deposited in this study. The DE of the mixed powders is found to increase with increasing fraction of feedstock Fe despite the single component Fe powder having an inferior DE to the 316L. The BR results determined from individual particle impact tests onto as-polished coatings have good correlation with the DE of both single component powders and the mixed powders. The BR of 316L impacts onto composite 316L/Fe layers can be well expected from the BR of 316L impacts onto each single component layer; while the BR of Fe impacts plateau at a high level regardless of composite layer composition, which leads to the increase of overall mixture DE.

Acknowledgements

The authors would like to acknowledge the financial support of Natural Sciences and Engineering Research Council of Canada (NSERC).

References

[1] S. Yue, W. Wong, H. Aydin, R. Mongrain, R. Barua, P. Vo, R. Dolbec, Improving cold sprayability: mixed metal powders, Proceedings from the International Thermal Spray Conference, 2015, pp. 473-478.

[2] E. Irissou, J.-G. Legoux, B. Arsenault, C. Moreau, Investigation of Al-Al2O3 cold spray coating formation and properties, J. Therm. Spray Technol., 16 (2007) 661-668.

[3] J. Frattolin, R. Barua, H. Aydin, S. Rajagopalan, L. Gottellini, R. Leask, S. Yue, D. Frost, O.F. Bertrand, R. Mongrain, Development of a Novel Biodegradable Metallic Stent Based on Microgalvanic Effect, Ann. Biomed. Eng., 44 (2016) 404-418.

A novel way to analyze the residual strain and stress in cold-sprayed coatings

Yun Ye, Renzhong Huang*, Jian Huang, Guangdong Institute of New Materials/CN
Hirotaka Fukanuma, Plasma Giken Co., Ltd/JP
Takahisa Shobu, Japan Atomic Energy Agency/JP

Residual stress and adhesive/cohesive strength of cold-sprayed coatings are extremely important factors and their banance determines the coating failure, such as delamination or crack. Dominated residual stress to bonding stress should lead to coating peel off from its substrate. Up to now, it is still difficult to measure the residual stress of cold-sprayed coating especially inside it. In the present study, a high-energy X-rayed beam was utilized to penetrate the cold-sprayed coatings and the change of its diffraction angle can be detected. This gave a way to measure the residual strain inside the cold-sprayed coatings. With the scanning of the X-rayed beam and the detector at different locations, the strain of cold-sprayed coating at 2 directions can be obtained. Moreover, the residual stress of cold-sprayed coating can be calculated with the measured strain.

1 Introduction

Cold spraying is a versatile and efficient method for deposition of relatively thick metallic coatings. So far, cold spray has been used to spray dutile material such as Cu[1], Al[2], and Ni-based alloys[3] but also metal matrix composites[4], cermets[5]. Compared to the traditional themal spary, cold spray has lower processing temperature, and the coating formed based on the intense plastic deformation with the feedstock material and its substrate. It is widely received that the amounts of compressive stress exists in the cold-sprayed coatings due to the kinetic effects generated during the severe impact deformation [6-7]. However, the opposite opinion that the thermal history of cold-sprayed particles during deposition still generates some residual tensile stress have been reported[8]. The exitence of residual tensile or compress stress in cold-sprayed coating gravely effect its properties, such as decreasing machical properties or deforming the workpiece. This limits the cold-sprayed coating application on big-size structure parts due to existence of its residual strain. It is essential to detect the residual stress in the cold-sprayed coating. Up to now, there is no better way to measure residual stress inside cold-sprayed coatings. In this study, a high-energy X-rayed beam was employed to beam and penetrate the cold-sprayed coatings, and then detected the change of its diffraction angle. The residual strain inside the cold-sprayed coatings can be measured based on Bragg's law. With the scanning of the X-rayed beam and the detector at different locations, the strain of cold-sprayed coating at 2 directions can be obtained. With the measured strain, the residual stress of cold-sprayed coating at two different directions can be calculated.

2 Experiment procedures

2.1 Specimen preparation

Comercially available stainless steel 304(called SS304)and Inconel 718(called In718) powders are used in the study. The SS304 powder shows near-spheric shape and the In718 powder shows perfect-spheric shape as shown in Fig. 1. The average diameter of SS304 is 41 µm, and 23 µm for In718 powder.

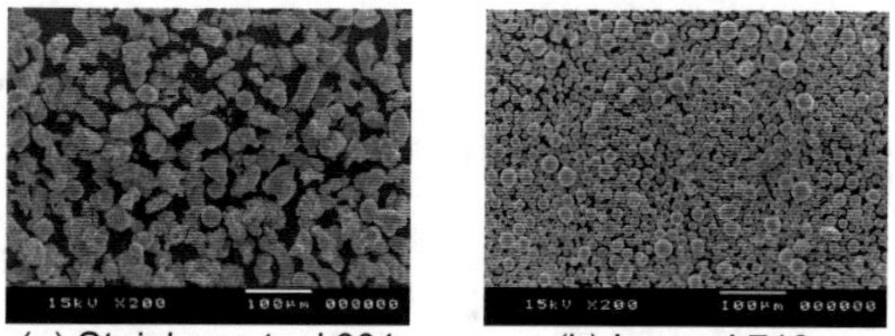

(a) Stainless steel 304 (b) Inconel 718

Fig. 1. Morphlogies of powders

A commercial cold spray system, model number PCS-1000 designed by PLASMA GIKEN CO. LTD., was used to prepare the coatings. A convergent-divergent(De-Laval) nozzle was configured to accelerate the working gas to supersonic speed.Nitrogen and helium was used as the propelant gas in this test respectively. In order to observe the influence of spray conditions to coating residual stress, coatings were prepared under different conditions as shown in the table1,the coating thickness is 1~3mm. stainless steel 304 plates with the demension of 50 mm X 50 mm X 10 mm were used as substrates.

Table1 : Spray conditions

Materials	SS304		In718	
Gas	N_2	He	N_2	He
Gas pressure(MPa)	5	3	5	3
Gas temperature(°C)	1000			
Substrate	SS304			

The residual strain&stress specimen was cut from the cold-sprayed plate by wire-cut as shown in Fig2 (a), and the dimension of the cut specimen is shown in Fig. 2 (b).

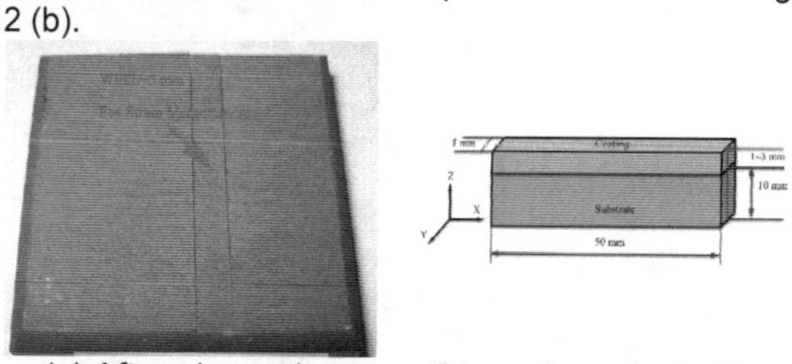

(a) After wire cutting (b) specimen for testing

Fig. 2. Specimen preparation

2.2 Measurement theory of strain and stress

Acccording to Bragg equation, the strain of the coating ε can be derived by the following equation when coating diffraction angel θ and ideal diffraction angel θ_0 was detected :

:

$$\varepsilon = \frac{d - d_0}{d_0} = \frac{\sin\theta_0 - \sin\theta}{\sin\theta} \qquad (2.1)$$

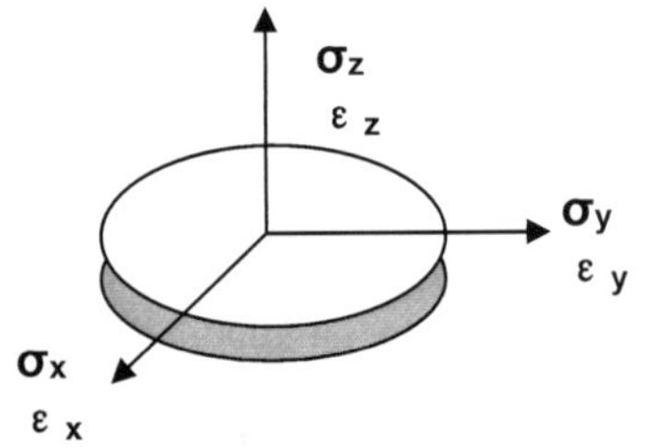

Fig.3.Coordinate system for strain ans stress

The relationship of main strain σ_x, σ_y, σ_z and main stress ε_x, ε_y, ε_z as fowling

$$\varepsilon_x = \frac{1}{E}[\sigma_x - \nu(\sigma_y + \sigma_z)]$$

$$\varepsilon_y = \frac{1}{E}[\sigma_y - \nu(\sigma_z + \sigma_x)] \qquad (2.2)$$

$$\varepsilon_z = \frac{1}{E}[\sigma_z - \nu(\sigma_x + \sigma_y)]$$

E: Young's modulus, v Poisson's ratio

Then,if θ and θ_0 are known, ε strain can be calculated by (2.1) ,then stress σ will be calculated by (2.2) in a three-demension space. In this study we considered that Y direction stress was released as testing specimen is very thin (5mm) in the direction Y, that is $\sigma_y=0$, the equaition (2.2)would be deduced as following:

$$\sigma_x = \frac{E}{1-\nu^2}[\varepsilon_x + \nu\varepsilon_z)] \qquad (2.3)$$

$$\sigma_z = \frac{E}{1-\nu^2}[\varepsilon_z + \nu\varepsilon_x)]$$

2.3 Strain scanning equipment and parameters

In this study, a high energy white X-ray (70 keV) obtained from B22XU in Spring-8 (Fig4(a)) which belongs to Japan Atomic Energy Agency, was used for the strain scanning, the scanning process is as Fig4(b). Wavelength is 0.1783Å, Monochromatic crystal is Si (111), Analyzer is Ge (111). Two slits were set for high-accuracy of energy collecting in the space. Stress in y direction could be considered as 0, and the strain of x and z direction were detected during our experiment.

For different coating, the diffraction lattice plane is different. No strain in the powder can be considered in this study, therefore, θ_0 is intead by x-ray diffraction result of spraying powder as Fig.5, and E and v is shown as Table 2:

Table 2: Parameters used in the X-ray analysis

Material	SS 304	In718
Diffraction 2 θ_0	9.3949341	9.375506
Young's modulus(GPa)	222	200
Poisson's ratio v	0.266	0.29

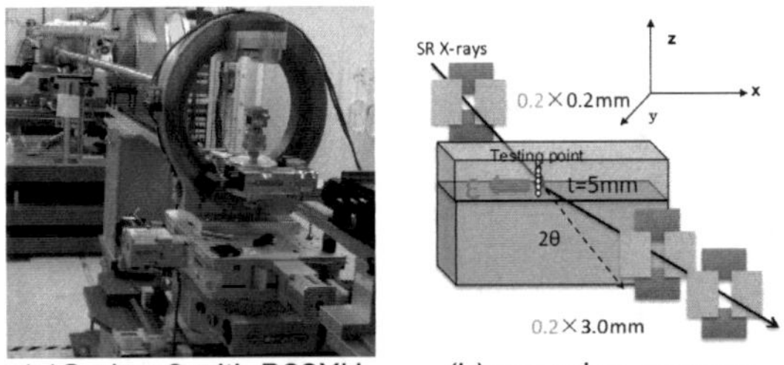

(a)Spring-8 with B22XU (b) scanning process

Fig. 4. Scanning equipment and process

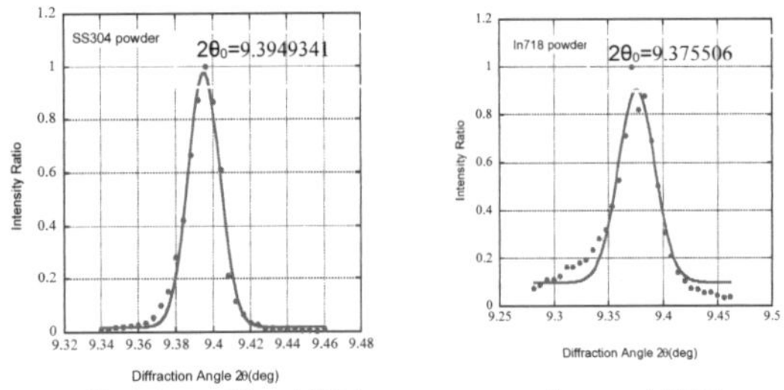

(a) Stainless Steel 304 (b) Inconel 718

Fig. 5. Spaying powder diffraction

3 Result and discussion

3.1 Microstructure of coatings

Fig.6. shows the microstructure of SSteel 304 coatings. Compared to traditional thermal spray coatings, fewer oxides can be observed inside the cold-sprayed coatings due to its lower processing temperature.

Comparing (a) with (b), it can be seen obviously the coating prapared with He gas looks denser and less defects than the one with N_2 gas even the gas pressure is higher with N_2 gas. The particles sprayed with He gas are flatter, as the particle velocity accelerated by He gas is higher than the one of N_2 gas. Morover, the enthalpy of He gas is higher than N_2 gas and this lead to higher temperature at the depoists surface, consequently, substrate and particles deformed more easily.

The microstructures of cold-sprayed In718 coatings are shown in Fig.7. Compared to SS304 coatings under same spraying condition respectively, In718 coatings are denser and more homogeneous. This is due to the fine powder of In718 and leads to higher particle velocity of In718 particles than the one of SS 304 particles. Dense In718 coating even prepared by N_2 gas can be obtained.

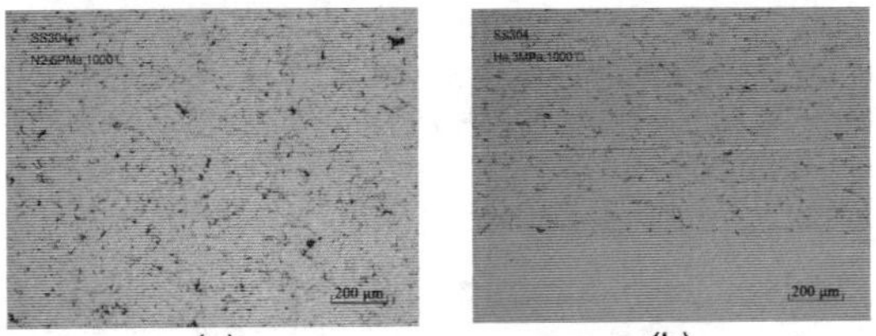

(a) (b)

Fig. 6. Microstucture of SS304 coatings

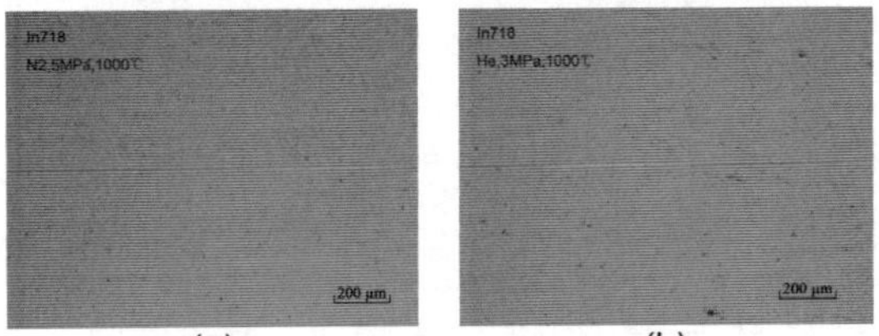

(a) (b)

Fig. 7.Microstucture of In718 coatings

3.2 Residual strain and stress

The cold-sprayed SS304 coatings residual strain and stress from the interface of coating and substrate to the surface of the coating are detected. The results are shown in Fig.8 and 9.

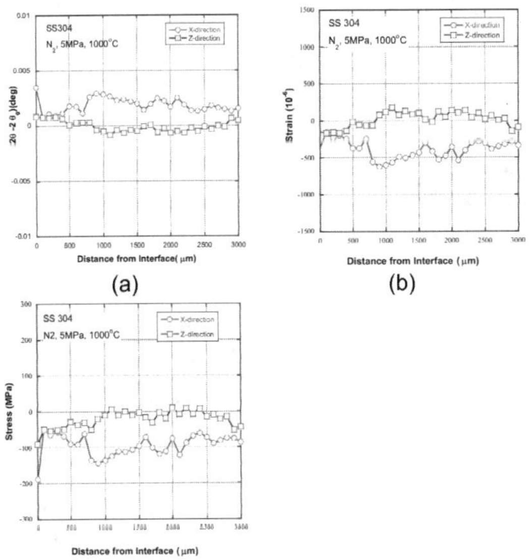

(a) (b)

(c)

Fig. 8. Scanning result of N_2 SS304 coating

It can be seen that 2θ change and strain of SS304 coating under N_2 gas is very stable from interface to surface at both X-direction and Z-direction. On contrast, the strain of SS304 coating prepared by He gas vibrates a lot along with normal direction of coating surface, and reachs limit value at the distance of 300 and 600 μm from the substrate/coating interface.

The calculated residual stress of cold-sprayed SS304 coating prepared by N_2 gas reached more than 100 MPa somewhere of inside the coating, and more than 200 MPa for He gas. It seems that the higher particle velocity generates more intense deformation with He gas, consequently higher residual strain and stress. Generately, the only compressive stress can be observate in the coating sprayed by N_2 gas, and tensile stress can be observated in the coating prepared by He gas.

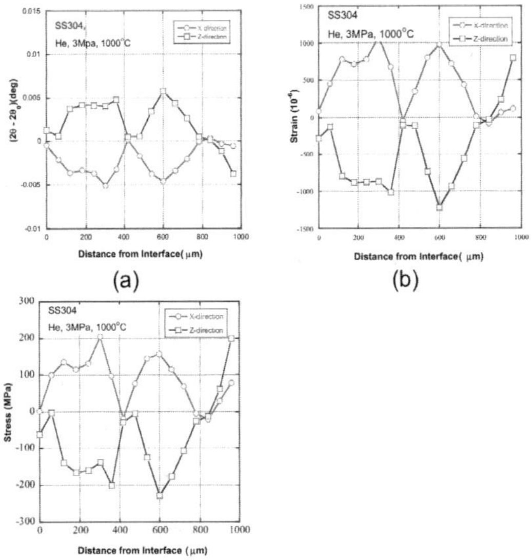

(a) (b)

(c)

Fig. 9. Scanning results of He SS304 coatings

The cold-sprayed In718 coatings residual strain and stress from the interface of coating and substrate to the surface of the coating are detected. The results are shown in Fig.10 and 11.

Compared with Fig. 10 and 11, the strain and stress of Inconel 718 coating sprayed by He gas is more fluctuant than that sprayed with N_2 gas. Similar to the results of SS304, the strain and stress of He coating is higher than the N_2 coating due to the same reason. The calculated residual stress of cold-sprayed Inconel 718 coating prepared by N_2 gas reached more than 100 MPa somewhere of inside the coating, and more than 200 MPa for He gas. Different from the cold-sprayed SS304 coatings, the Inconel 718 coating exhibits compressive strain and stress at X direction and tensile strain and stress at Z direction irrelevant to the propelant gas type.

4 Conclusion

In this study,cold sprayed SS304 and In718 coatings' residual strain was detected by using a high energy white X-ray scanning. The spraying condition influence to the stress was discussed.

1) The coating residual strain can be detected and residual stress can be calculated out. The type of the stress (pressure or tensile) can be studied.
2) The strain type lies on the spraying material, that will make different coatings having different deformation.

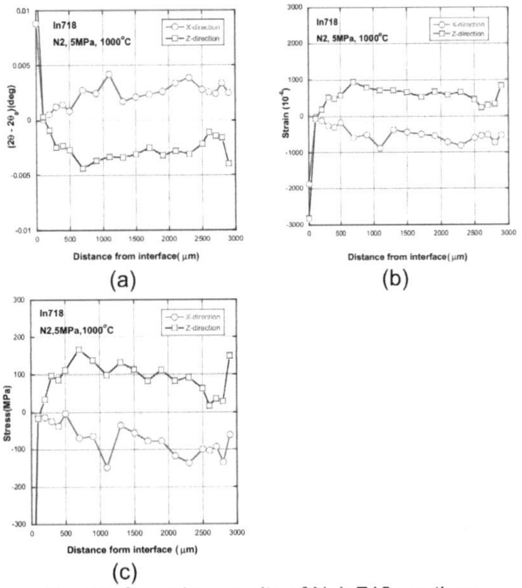

Fig. 10. Scanning results of N_2 In718 coatings

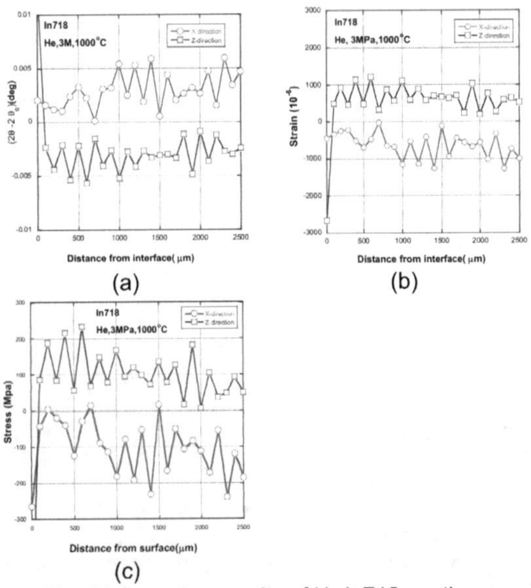

Fig. 11 Scanning results of He In718 coatings

3) No matter the SS304 or In718, He spraying coating is denser, meanwhile residual strain is also greater than N_2 coating. Compressive and tensile stress exist in the cold-sprayed coatings sitmultaneously.

Acknowledgements

The authors would like to acknowledge the financial support provided by the High-level Leading Talent Introduction Program of GDAS (Grants No. 2016GDASRC-0204), the Laboratory Apparatus Upgradation Program of Guangdong science and Technology Department (No. 2013B061800053), and Openning atomic facility useage project supported by Japan Atomic Energy Agency (No. 2012B-E19).

5 Literature

[1] E.Calla,D.G. McCarthney,P.H. Shipway,in: D.von Hofe(Ed.) ASM International,Osaka, Japan,May10-12 2004,pp.382-387

[2] R.Morgan,P.Fox,J.Pattison,C.Sutcliffe,W.OÖNeil, Mater.Lett.58(7-8)(2004)1317-1320

[3] F.Raletz,G.Ezoo,M.Vardelle,M.Ducos,in:D.von Hofe(Ed.)ASM International,Osaka,Japan,May 10-12 2004,pp.344-349

[4]D.Morelli,A.Elmoursi,T.Vansteenksite,D.Gorkiewicz ,B.Gillispie,in:B.R. Marple.C.Moreau(Eds.),2,ASM International,Orlando,FL,2003,pp.85-90

[5] J.Karthikeyan,C.M.Kay,J.Lindemann,in:C.C. Berndt,K.A.Khor,E.F.Lugscheider(Eds.),ASM International,Singapore,May 28-302001,pp.146-152

[6]R.Ghelichi,S.Bagherifard,D.MacDonald,I.Fernandez-Pariente,B.Jodoin,M.Guagliano,Appl.Surf.Sci.288(1) (2014)26-33

[7] K.Spencer,V.Luzin,N.Matthews, M.-X.Zhang, Surf.Coat.Technol.206(19-20)(2012)4249-4255

[8]Z.Arabgol,H.Assadi,T.Schmidt,F.Gãrtner,T.Klassen ,J.Therm.Spray Technol.23(1-2)(2014)84-90

Comparison of microstructure and tribological behavior of WC reinforced maraging steel 300 composites prepared by cold spraying and selective laser melting

X.C. Yan, C.J. Huang, R. Bolot, D. Lucas, H.L. Liao, IRTES-LERMPS, ICB UMR 6303, CNRS, Univ. Bourgogne Franche-Comté, UTBM, F-90010 Belfort, France
R.Z. Huang, W.Y. Ma, M. Liu*, Guangdong Institute of New Materials, 363 Changxing Road, Guangzhou 510651, P.R. China

WC is a well-known material used to increase the wear resistance of iron-based composite materials which show a favorable wettability with iron alloy particles. In this work, two different additive manufacturing technologies, i.e. the cold-spraying additive-manufacturing (CSAM) and the selective laser melting (SLM), were used to fabricate WC/maraging steel 300 (WC/MS300) composites. A comparional investigation on their microstructure and tribological behavior was carried out. In addition, the evolution of the reinforment phase during these two processes was calssified by the methods of SEM and EDS.

Keywords: CSAM, SLM, WC/MS300, Microstructure, Wear

1 Introduction

Maraging steel 300 (MS300) possesses many applications in aircraft, aerospace, and tooling fields due to their ultra-high strength [1]. However, a relatively poor wear resistance of maraging steels limits their applications [2]. Nevertheless, their wear resistance could be improved substantially by joining the WC particles and maraging steels to form the iron-based metal matrix composites (MMCs) [3-4]. Therefore, these MMCs are of great significance for industrial applications owing to their excellent combination of higher specific strength and improved wear resistance, compared with the iron-based alloys.

Cold spraying (CS), has been widely investigated as a solid-state deposition process, due to its high deposition efficiency and volume production of coatings or parts. In this process, the deposition of particles occurs through an intensive plastic deformation upon impact in a solid-state at a temperature well below the melting point of sprayed materials [5]. Consequently, the deleterious effects of oxidation, phase transformation and other problems inherent to the conventional thermal spraying can be minimized or eliminated. Recently, it has been recognized as a useful and powerful tool and developed into cold-spraying additive-manufacturing (CSAM). For example, a gear has been made by coordinating the spray nozzle motion and the external motor-drive axis [7], but its related research works just started.

Different to CSAM, another high-performance AM process, selective laser melting (SLM) is regarded as an additive melting technique that allows the production of complex-shaped parts by successive deposition of melted metallic material according to a pattern conceived by computer aided design. The gradual accumulation of layers on the top of each other is carried out by using the thermal energy supplied by a focused laser beam to melt powders. Recently, many MMCs, e.g., Fe-based alloys [1, 6], Al-based alloys [8, 9], Ti-based alloys [10, 11], have been successfully prepared by SLM.

Therefore, this present work mainly focuses on studying the microstructure and tribological behavior of WC/maraging steel 300 (WC/MS300) composites fabricated by CSAM and SLM, and the corresponding characteristics of the microstructure, mechanical properties and tribological behavior are also compared and discussed.

2 Materials and experimental procedures

2.1 Materials

The spherically gas atomized MS300 powder (EOS GmbH, German) with a size range of 33-40 μm and the irregular WC powder (XIAMENJINGLU, China) with a size range of 1-6 μm were used as starting materials to make a powder mixture at a weight ratio of 85/15. The blended powders were mechanically mixed in a tumbling mixer for 200 min before being processed. The morphology of the mixed powders is shown in Fig. 1. The chemical composition of MS300 is presented in Table. 1.

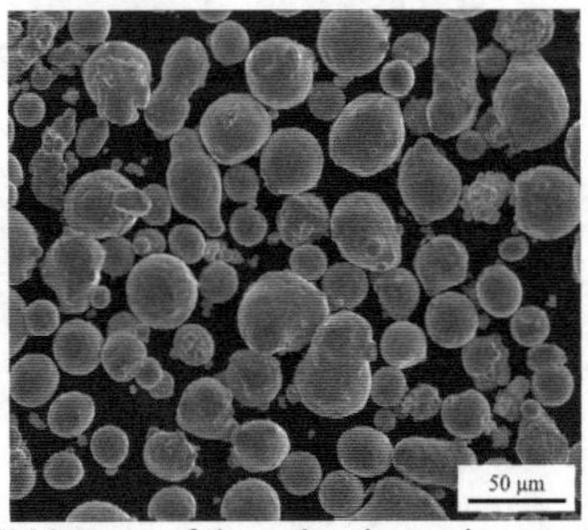

Fig. 1. SEM image of the mixed powder morphology.

Table 1. Chemical compostion of MS300

EL	Fe	Ni	Co	Mo	Ti	Al	Cr,Cu	C	Mn,Si	P,S
wt%	balance	17-19	8.5-9.5	4.5-5.2	0.6-0.8	0.05-0.15	≤0.5	≤0.03	≤0.1	≤0.01

2.2 Experimental procedures

A WC/MS300 composite was produced using a PCS-1000 cold spraying system. Nitrogen was used as the driving gas, at a pressure of 3.0 MPa at a temperature of 500 °C. The standoff distance from the nozzle exit to the substrate surface was 30 mm. A grit-blasted 304 stainless steel plate was used as the substrate to deposit.

Another WC/MS300 composite was processed with an EOS M290 system (EOS GmbH) equipped with an Yb-Fiber laser (1064 nm) with a maximum power of 400 W. The laser beam focus spot (100 µm), the power (285 W), the layer thickness (50 µm), the hatch distance (50 µm) and the laser beam moving speed (0.96 m s^{-1}) were kept constant during this work. The whole experimental campaign was operated in a chamber filled with nitrogen and a residual oxygen content of 1.3 % in volume. Cubic SLM specimens with dimensions of 15mm×15mm×8mm were prepared.

The polished composites were finally etched at room temperature in a solution of 25 ml HNO_3, 50 ml HCl and 1 g $CuCl_2$ in distilled water for 20 s. The samples were prepared according to standard procedure and the densification of the specimens was characterized by an optical microscope (OM). The microstructure was then observed by a scanning electron microscope (SEM) equipped with an energy-dispersive spectroscopy (EDS) unit (JSM5800LV, JEOL, Japan). The cross-section density of the samples was calculated by image analysis (NIH Image J, Software, USA) and 15 analyses were carried out for each sample. The Vickers hardness was measured using a micro-hardness tester (Leiz-Wetzlar, Germany) at a load of 200g and an indentation time of 25 s. Vickers hardness measurements were carried out at different locations on the cross section of the sample, and the final average hardness value is the mean value for 15 measurements.

Dry sliding wear tests were conducted at air temperature with CSEM tribometer implement (Switzerland). Before testing, the roughness *Ra* of the sample is less than 0.15 µm. The counterpart material is a CO/WC ball with 6 mm diameter which was polished and cleaned with ethanol before testing. The friction test load was 500g and rotated at 10 cm/s for 100 m with 3 mm radius. The coefficient of friction (COF) was recorded during sliding. After the friction test, the worn surfaces were observed using the SEM and analysed by EDS. The cross-sections of the worn tracks were measured by an Altisurf 500 profilometer (France). The worn volumes of the samples were calculated via the product of the cross-sectional areas and the length of worn tracks.

3 Results and discussion

3.1 Microstructure and formation mechanisms

The etched cross-section morphologies of WC/MS300 composites fabricated by CSAM and SLM are presented in Fig. 2. As shown in Fig. 2a, the phase of MS300 deforms significantly, acting as matrix, while the hard phase of WC particles embedded in the matrix, acts as reinforcement. An intimate interface was formed between matrix and WC, and severe plastic deformation of MS300 led to grains elongation, while no trace of deformation is observed for the WC phase. During the deposition, MS300 particles are deformed through both the impact of the other MS300 particles and the impact of the hard and heavy WC particles. It is clearly seen that the CSAM composite presents certain obvious pores and cracks. In Fig. 2b, dendritic solidification patterns distributed on the cross-section and perpendicular to the layer addition direction can be found. Compared with Fig. 2a, less WC phase appears in the observation region. This is due to the high energy input from the laser source during SLM processing, which results in the partial melting of the WC phase. The lower melting point MS300 particles were fully liquefied and welded the WC phase in the matrix, thus showing higher densification.

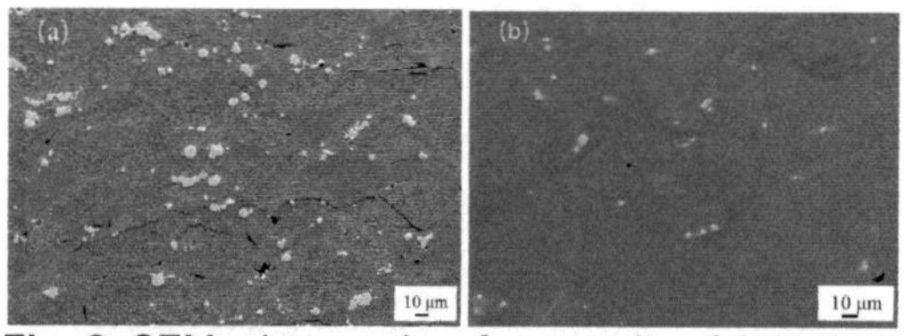

Fig. 2. SEM micrographs of composites fabricated via CSAM (a) and SLM (b) .

A dense cross-section with a relative density of 99.6% (Fig. 3) is shown in Fig. 2b. During the SLM process, the composite with nearly 100% was obtained by the high energy laser source. Due to the lower melting temperature comparing with WC phase, the MS300 phase were completely liquified. The temperature and concentration gradient create a molten pool flow (Marangoni flow) which can accelerate the movement and harmonize the distribution of WC phase. The favorable wettability of WC particles with MS300 matrix is also advantageous to form the dense specimen with 0.4 % porosity, which can be seen in Fig. 3. In addition, it is found that the composite prepared by CSAM appears less dense than that fabricated via SLM. This is due to the fact that an incomplete compaction results in a higher porosity, whereas poor bonding will result in low mechanical property of the coating.

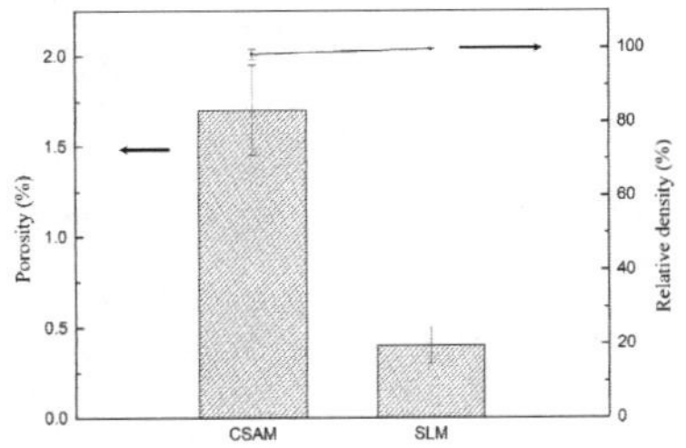

Fig. 3. Porosity and relative density of the composites manufactured by CSAM and SLM.

For describing the formation mechanisms of composites manufactured by CSAM and SLM, the tillustration is presented in Fig. 4. During CSAM as shown in Fig. 4a, before MS300 particles deposition, the grain boundaries are homogeneously distributed in the particles with a low dislocation density. However, with the deformation of the particles increased, the dislocation density increased. A further deformation leads to the formation of elongated dislocation cells and subgrains. Then, the deformation of the particle breaks the subgrains, and the rotation of the broken subgrains complete the recrystallized structure formation in the shear localization area, which has experienced the severe plastic deformation. Additionaly, the tamping effect of reinforcements promotes the deformation of metal phase and enhances the coating density and hardness. Different to MS300 particles in the deposition, the WC particles do not plastically deform and just deposit on the soft matrix and become surrounded by the subsequent matrix materials. However, the breakup and detachment of the embedded WC particles may occur when they are impacted by subsequent WC particles.

Fig. 4b demonstrates the formation mechanism of the SLM-processed WC/MS300 composites. Within the mixed powder, the MS300 particles encountered sufficient laser energy which can amply melt the relatively low melting point powder. WC particles were hardely melted due to its comparatively high melting temperature and gathered due to the Marangoni flow in the melted pool. However, in the molten pool partially melted WC phases can release W and C atoms which can contribute to the reaction with the metallic substrate. In Fig. 4b, high energy laser beam accelerated the melting of the matrix meanwhile the moving WC particles at the top of the molten pool dissolved ending with smaller diameter, which can be proved by Fig. 3b. On the other hand, comparatively large WC particles with partially melted interface were emerged at the bottom of the molten pool.

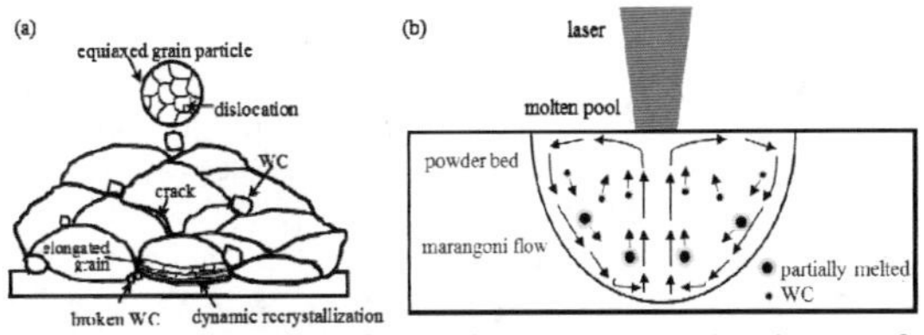

Fig. 4. Illustration of the formation mechanisms of composites manufactured by CSAM (a) and SLM (b).

3.2 WC evolution

For examining the evolution of WC particles during the two AM processes, the SEM micrographs with higher magnification of the CSAM and SLM composites are conducted in Fig. 5. As presented in Fig. 5a, the WC phase is substantially intact and distributed around the MS300 particles. In addition, it has been stated that the deposition has no effect on the WC deformation. They were just distributed among the deposited MS300 particles and lodged in the coating during CSAM process. When the embedded WC particles collided with the subsequent WC particles, they would break as shown in Fig. 5a and even be detached from the coating leaving a crater. During SLM, all the MS300 particles were completely melted and partially WC particles can hardly be observed, due to the forceful heat radiations from laser source at the top of the molten pool. The WC particles are present in a variety of molten states that can be seen at the different locations of the internal molten pool in Fig. 5b. However, there exists slight WC specks with a relative sizable concentration gradient area in the middle of the pool. It is due to the heat conduction energy being insufficient to entirely melt WC particles. At the bottom of the molten pool, with the heat conduction energy attenuation, partially melted and relatively complete WC particles were discovered. It can be concluded that WC particles were considerably or partially melted and their combination with the matrix in SLM composite made it exceedingly tough, which is ascribable to the high energy laser source.

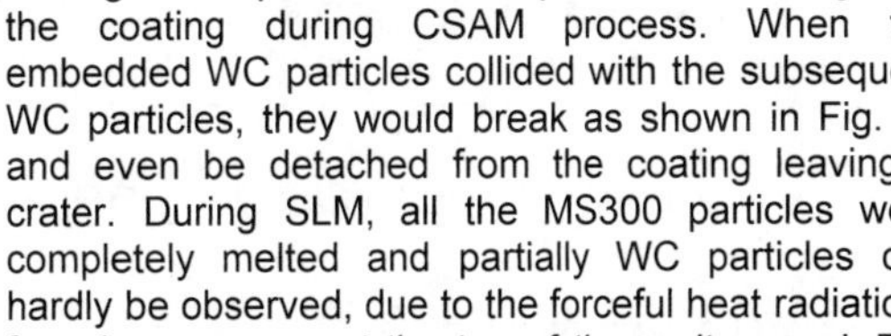

Fig. 5. SEM micrographic mages of the CSAM (a) and SLM (b) processed composites.

To further specify the chemical composition of the WC phase distributed in the CSAM and SLM-processed parts, EDX linear analysis was carried out and showed in Fig. 6. In Fig. 6a, it can seen that an intimate interface is formed between the two phases, MS300 and WC. The results of linear EDS as shown in Fig. 6b, it was conducted across the WC particles and the interface. The main elements of the matrix: Fe and Ni, and the abundant elements of particles: W and C, were selected. The elements changed sharply which demonstates that there is no obvious diffusion between them. In Fig. 6c, it can be observed that indistinct contour existed around the partially melted WC phase. To evaluate the contour elements distribution, linear EDS was conducted across the WC particles and the affected region. Fig. 6d, at the interface, the concentration of W element gradually increased while the density of Fe and Ni atoms reduced slowly. The fluctuation of C element is not apparent because of its relatively low content. This indicated that the high laser energy input provided the activation energy necessary to diffuse. Besides, W and C atoms from the WC particles surface diffused mutually with Fe and Ni atoms released from the matrix. During the mutual diffusion, the chemical diffusion process ended in the formation of carbides of different atomic ratios [3]. Since WC powders possess higher activation energy, partially melted WC phase can not liberate adequate quantities of W and C atoms which limits further diffusion. Therefore, the concentration gradient interface can only appear within a certain distance from a WC particle surface. The future study will focus on the new phase formed in the diffusion layer.

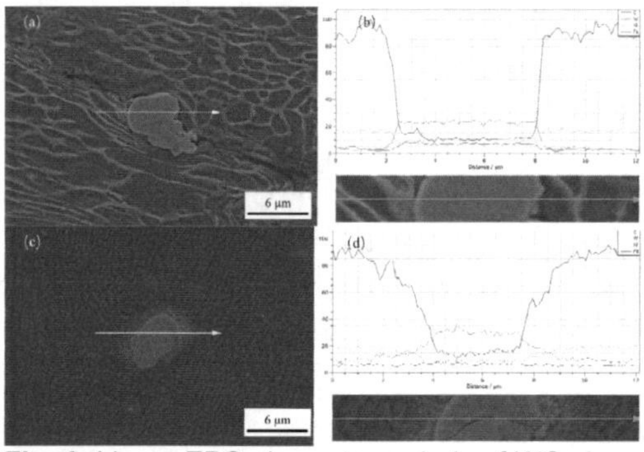

Fig. 6. Linear EDS elements analysis of WC phase in CSAM and SLM composites.

3.3 Micro-hardness

Fig. 7 shows the micro-hardness of two WC/MS300 composites, the CSAM-processed average value is about 393.8 Hv and SLM-processed about 385 Hv. It should be noted that in order to gauge relatively accurate values during the micro-hardness test, the testing probe was located so as to avoid the hard WC phase. The micro-hardness of original SLMed MS300 parts is 371 Hv [1]. Compared with the WC/MS300, the enhancement of micro-hardness was considered to be caused by the reinforcement particles WC, which can release C and W elements into the metal matrix melted pool and compose new hard metal compounds. On the other hand, WC particles were uniformly distributed in the denser MS300 matrix in Fig. 2b. This can also contribute to the improvememt of the micro-hardness. It can be concluded that WC particles play a major role in the development of the microhardness of the composite samples. However, the CSAM part shows a different strengthening mechanism, its hardness is higher than that of the SLM composite. It is due to the severe plastic deformation of the MS300 matrix and the tamping effect of hard phase (WC). In addition, the uniformly dispersed ceramic particles of WC also contribute to the increase of the overall hardness by restricting the deformation of the soft MS300 matrix.

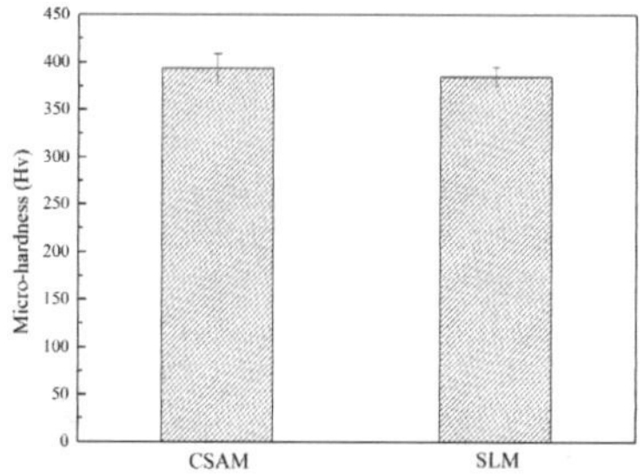

Fig. 7. Micro-hardness of CASM and SLM-processed composites.

3.4 Tribological behaviours

Fig. 8. shows two curves depicting the evolution of the coefficient of friction (COF) for the two different AM methods. Compared with the CSAMed sample, the SLM composite part reveals a higher average COF, which is 0.52 versus 0.49. Obviously, the COF of the CSAM composite experienced rapid increase and slow decline before 2000 laps, then steadily stabilizes at a certain level.

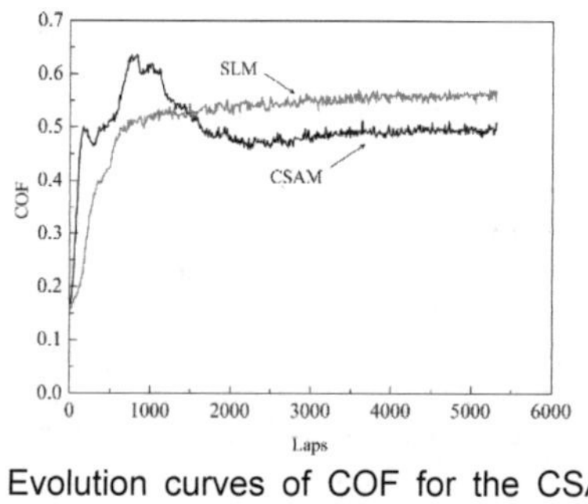

Fig. 8. Evolution curves of COF for the CSAM and SLM-processed WC/MS300 composites.

For a better understanding of the changes of COF, a comprehensive relationship between the wear resistance and the WC content was investigated. At first, the different contents and distributions of WC phase have a significant effect on the friction. The calculated relative contents of WC phase in CSAM and SLM composites are 3.9 and 0.6 respectively. Due to its higher content of WC particles and lower relative densification, the COF of CSAM composite experienced significant fluctuations. However, the COF of SLM composite with lower content of WC phase increased steadily to 0.52. Then the high energy input from the SLM process forms a consistant bonding between WC particles and the matrix. Thereby, even though the SLM composite possesses a higher COF than the CSAM composite, the wear rate of the SLM-processed composite is much lower than that of the CSAM, due to the strong cohesion between the particles which can be illustrated in Fig. 9.

The wear rates and wear tracks on cross-sections of the CSAM and SLM-processed composites are displayed in Fig. 9. Apparently, the wear rate of the CSAMed part is 4.96 $\times$ 10^{-5} mm^3 $N^{-1}m^{-1}$, which is evidently more than the SLM composite : 4.6 $\times$ 10^{-6} mm^3 N^{-1} m^{-1}. It is reasonable to believe that in the SLMed WC/MS300 part, the bonding strength of WC particles with matrix is much higher due to grain boundaries diffusion reaction.

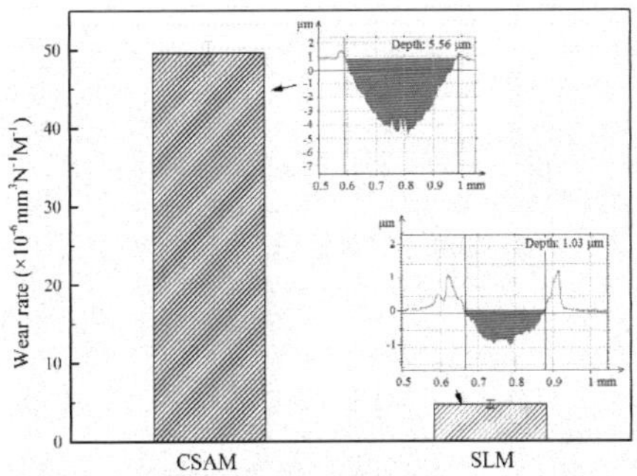

Fig. 9. Comparison of wear rates and wear cross-sections of the CSAM and SLM-processed composites

To investigate the wear mechanism, SEM images of the worn surfaces of the two processes were displayed in Fig. 10. Almost no bumps and grooves are found on the wear tracks of parts, as shown in Figs. 10a and c. On the contrary, a mass of oxide layer exists on the worn surface of the CSAM composite, as shown in Fig. 10a. The EDS results from the inset of Fig. 11b demonstrate that the main composition of this layer is iron oxides. Therefore, the oxides would be more easily abraded by the counterpart and well-bonded WC particles during the adhesion wear process. Then, the surface oxidation layers would be crushed into the wear interface under the shearing force of friction. This shows a typical feature of the oxidation wear process. However, the iron oxide particles distribution on the worn tracks will contribute to a third-body abrasion in the sliding process and give rise to heterogeneous deformation, which could significantly decrease the friction coefficient and prevent serious ploughing of the worn tracks. Consequently, the adhesion between the sliding pairs is no longer a major factor of the tribological behavior.

However, when comparing with Fig. 10a, the smoother and clearer friction marks can be found in Fig. 10b, which means less deformation happened during the same sliding procedure. In Fig. 11d, the elements of Fe, Ni, Mo were released from the maraging steel matrix, so were C and W from WC particles. Oxygen element was detected on the slightly worn scrap thus attesting that oxidation occured during the heat accumulation of friction process.

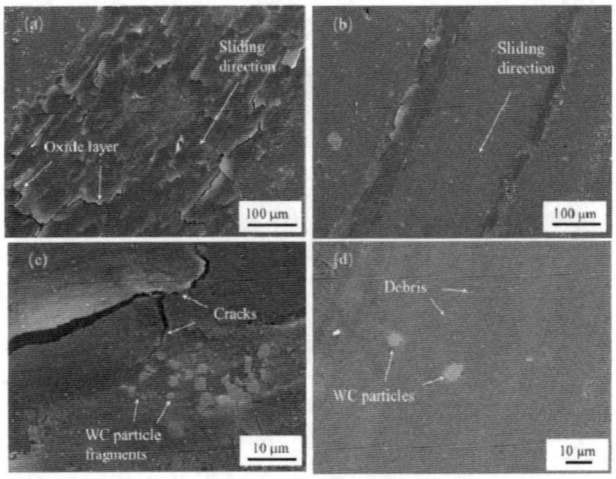

Fig. 10. SEM images of worn surface morphologies and worn tracks at high magnification, for the CSAM (a,c) and SLM-processed (b,d) parts.

The wear surface of the remaining embedded WC particles indicate that after experiencing a huge shear stress it still did not flake. In other words, the particle cohesion in the SLM-processed composite is stronger than the mechanical occlusion in the CSAM composite. As a result, the wear rate of the SLM composite was much lower in contrast with CSAM composite, as can be seen in Fig. 9. Also, the SLM composite possesses a better wear resistance ability. The main wear mechanism is abrasive. In Fig. 10d, the high magnification SEM image presents WC particles with a semi-polished surface and almost no obvious scratches can be found, revealing the relatively stable changes of the friction coefficient and wear behavior. Besides, relatively fine debris appeared on the friction path of the WC particles, it is thus reasonable to consider that adhesive wear was the main wear mechanism for the SLM composite.

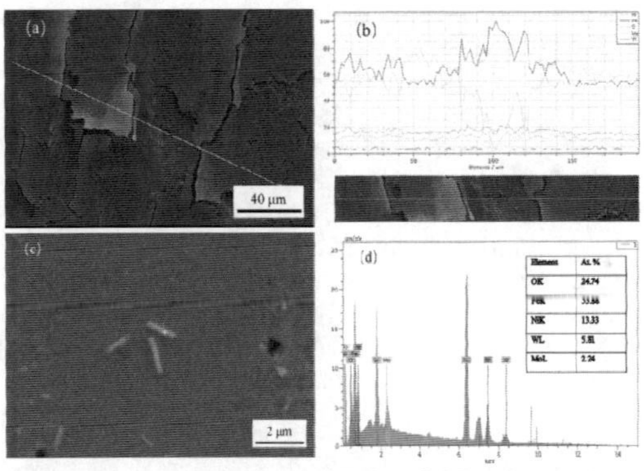

Fig. 11. Linear element analysis of CSAM sample (a) and corresponding EDS element distribution (b); point element analysis of SLM worn debris (c) and corresponding chemical composition (d)

4 Conclusion

In this study, two AM technologies (CSAM and SLM) were used to fabricate the relatively dense WC/MS300 composites. After a series of investigations were done on microstructure and tribological behavior. The results were presented and discussed.

(1) the CSAM processed composite is featured by the plastically deformed partices with uniform WC particles embedding at the interfaces between MS300 particles. However, the microstructure of SLM processed composite revealed dendritic solidification patterns distributed perpendicular to the powder bed plane.

(2) The WC particle in CSAM was bonded due to the mechanical interlocking. However, the completely or partially melted WC particles were uniform distributed and firmly welded with the metal matrix, due to the high energy caused molten flow in SLM composite.

(3) The CSAM and SLM processed composites present a different wear behavior and the wear rate of the CSAM is evidently more than the SLM.

5 Acknowledgements

One of the authors, X.C YAN, is grateful for the financial supports provided by the China Scholarship Council (NO. 201504490031), Guangdong Provincial Science and Technology Department, International Cooperation Base Construction Project China-France Modern Material Surface Engineering Technology International Cooperation Base Construction (NO, 2014B050502008), Guangzhou Science and Technology Innovation Committee, Production and Research Collaborative Innovation Major Projects of Foreign Cooperation Projects, Research and Application of Laser 3D Printing Die-Casting Mold Technology (NO, 2016201604030028), Guangdong Provincial Science and Technology Department, Provincial Scientific Research Institutions Reform and Innovation Projects, Laser 3D Printing Technology and Application of Innovative Research Team Building (NO, 2016B070701020), Zhongshan Science and Technology Bureau, Zhongshan City-Guangdong Province Science Institute of Technical Transfer Special Funds Project, Study on Industrial Application of Laser 3D Printing Type Cooling Injection Mold (NO, 2016G1FC0004), Guangdong Provincial Science and Technology Department, Research Platform Environment and Capacity Building Special Fund Project, China-UK Joint Manufacturing Technology Research Center-3D Printing Technology Innovation Platform (NO, 2016GDASPT-03102), Guangdong Provincial Science and Technology Department Research Platform Environment and Capacity Building Projects of Special Funds, "Modern Materials Surface Engineering Technology National Engineering Laboratory" Scientific Research Platform to Enhance Capacity (NO, 2016GDSPT-0206).

6 Literature

1. Kempen, K. and E. Yasa. Other: Microstructure and mechanical properties of Selective Laser Melted 18Ni-300 steel. Physics Procedia 12 (2011), Issue, pp. 255-63.
2. Hermann Becker, T. and D. Dimitrov: The achievable mechanical properties of SLM produced Maraging Steel 300 components. Rapid Prototyping Journal 22 (2016), Issue 3, pp. 487-94.
3. Rong, T. and D. Gu. Other: Effects of tailored gradient interface on wear properties of WC/Inconel 718 composites using selective laser melting. Surface and Coatings Technology 307, Part A (2016), Issue, pp. 418-27.
4. Gu, D. and Y. Shen: Direct laser sintered WC-10Co/Cu nanocomposites. Applied Surface Science 254 (2008), Issue 13, pp. 3971-78.
5. Huang, C. and W. Li. Other: Microstructural evolution and mechanical properties enhancement of a cold-sprayed CuZn alloy coating with friction stir processing. Materials Characterization 125 (2017), Issue, pp. 76-82.
6. Riemer, A. and S. Leuders. Other: On the fatigue crack growth behavior in 316L stainless steel manufactured by selective laser melting. Engineering Fracture Mechanics 120 (2014), Issue, pp. 15-25.
7. P. Richter, Athens, Greece. (2014).
8. LIU, G. and G. LI. Other, The influence of Strontium addition on wear properties of Al―20 wt% Si alloys under dry reciprocating sliding condition. Vol. 32, Kidlington, ROYAUME-UNI, Elsevier. 2011, 6.
9. Mazur, M. and M. Leary. Other, 5 - Mechanical properties of Ti6Al4V and AlSi12Mg lattice structures manufactured by Selective Laser Melting (SLM), in Laser Additive Manufacturing. 2017, Woodhead Publishing. p. 119-61.
10. Vrancken, B. and L. Thijs. Other: Heat treatment of Ti6Al4V produced by Selective Laser Melting: Microstructure and mechanical properties. Journal of Alloys and Compounds 541 (2012), Issue, pp. 177-85.
11. Wysocki, B. and P. Maj. Other: Microstructure and mechanical properties investigation of CP titanium processed by selective laser melting (SLM). Journal of Materials Processing Technology 241 (2017), Issue, pp. 13-23.

Post Deposition Heat Treatment of Cold Sprayed C355 Deposits for Repair: Microstructure and Mechanical Properties

M.V. Zuccoli[1,2], A. Sabard[1], M. Guagliano[2], H.L. de Villiers Lovelock[3], T. Hussain[1]
[1]Faculty of Engineering, University of Nottingham, Nottingham, NG7 2RD, UK
[2]Politecnico di Milano, 20156, Milano, Italy
[3]TWI Ltd, Granta Park, Great Abington, Cambridge, CB21 6AL, UK

Cold gas dynamic spray is increasingly used for dimensional repair in the aerospace sector as it is capable of producing dense, oxide-free deposits of significant thickness and with good levels of adhesion and inherent mechanical strength. There is significant interest in extending the application of cold spray deposits to include structural, load-bearing repairs. However, particularly for high strength aluminium alloys, cold spray deposits can exhibit high levels of porosity and micro-cracks, leading to mechanical properties that are inadequate for most load bearing applications. In this work, heat treatment was investigated as a potential means of improving the properties of a cold sprayed Al alloy C355 deposit. C355 alloy deposits were produced using two process gas temperatures (350°C and 500°C) and three gas pressures (40, 50 and 60 bar) using a commercially available HPCS system. Microstructural analysis of the coatings revealed that the optimal microstructure (ca. 1% porosity) was obtained at 500°C and 60 bar. Therefore, coatings produced with process conditions of 500°C and 60 bar were heat treated at 175, 200, 225, 250°C for 4h in air and the evolution of the microstructure and microhardness was analysed. The results show that heat treatment at 225°C can decrease porosity (<0.2%) and retain high hardness (105 $HV_{0.05}$ vs 130 $HV_{0.05}$ as-sprayed). Further investigation was performed on as-sprayed and 225°C heat treated deposits. The results show that this heat treatment can halve residual stress (-50 MPa vs -100 MPa as-sprayed), and improve tensile properties (UTS). Therefore, this work has demonstrated that the heat treatment of C355 cold sprayed deposits at 225°C can significantly improve their properties.

1 Introduction

Cold Spray (CS) is a relatively recent solid-state deposition process that exploits kinetic energy, rather than thermal energy, for depositing 5-50 micrometer - sized powders onto a substrate. Its main advantage is that CS is able to produce ultra-thick deposits avoiding the deleterious effects of high temperature that other thermal spray processes imply. Extensive literature exists on this process [1], since it is a promising technology for repair of aeronautical compoments of light alloys (Aluminium, Magnesium and Titanium alloys) that have experienced wear and corrosion. At the current moment, the quality of Al alloy CS deposits is enough to satisfy the requirements for geometrical repair of Al and Mg components. However, Al alloy CS coatings are usually featured by high strength and stiffness, but nearly no ductility [2]. Furthermore, the presence of porosity and other deposition defects can adversely affect mechanical properties of the repairs [1]. Finally, if the compressive residual stress state that is typical of CS deposits is excessive, it can lead to decohesive failure of the deposits [1]. These conditions prevent CS technology from qualifying for structural repair of Al alloys components.

In order to overcome these problems, post deposition heat treatment of CS deposits can be performed. Some studies exist on heat treatment of pure aluminium [2], but less work is present in literature on CS Al alloys.

In this work, post-deposition heat treatment of C355 Al alloy CS deposits has been carried out in order to improve microstructure and mechanical properties of the as-sprayed coatings. C355 Al alloy (Si 4.5-5.5%, Cu 1-1.5%, Mg 0.4-0.6%) has been chosen in this study since it is a quaternary alloy (Al-Si-Cu-Mg) which is widely used for structural aerospace applications [3].

2 Experimental Methods

2.1 Cold Spray Deposition and Heat Treatment Plan

C355 Al alloy coatings were produced using gas-atomised C355 Al powder (LPW Technology, UK), -65+45 μm in size. The deposits were made using a commercial high pressure CS system (Impact 5/11 at TWI, Camrbidge, UK) equipped with a SiC watercooled nozzle and nitrogen process gas. Various process gas temperatures (350-500°C), pressures (40,50 and 60 bar), and stand-off distances – SOD – (20-50 mm) were employed in order to investigate their effect on microstructure and mechanical properties. All the coatings have been produced using 3 rpm powder feed rate, 90° spray angle, 1 mm track spacing, and 100 mm/s traverse speed. AA6082-T6 (Si 0.7-1.3%, Mg 0.6-1.2%, Mn 0.4-1.0%) substrate grit blasted with 120 grit was used. Finally, a total deposition thickness of 400 μm was achieved in this study with two passes.

Stress-relief heat treatment at different temperatures (175, 200, 225, 250°C) was carried out on the coating that showed the least porosity (500°C, 60 bar, 50 mm). The dwell time was 4 hours for all the treatments. All the heat treatments were performed in air using commercial furnaces, each one pre-heated to the desired temperature. After 4 hours, the samples were removed from the furnaces and left in air for cooling.

2.2 Microstructural Analysis

As-sprayed and heat treated deposits were prepared for metallographic analysis according to established standards. The samples were sectioned, hot mounted, ground, polished up to 1 μm diamond finish and followed by 0.06 μm colloidal silica finish. Powders were cold mounted and finished down to 0.06 μm colloidal silica. All the samples were etched using modified Keller's reagent (2 ml HF, 3 ml HCl, 20 ml HNO_3, 175 ml water). Microstructural analysis was conducted via optical microscopy (OM – Eclipse LV100ND, Nikon), scanning electron microscopy (SEM – JEOL JSM 64900LV), electron probe micro-analysis (EPMA – Superprobe JEOL JXA 8200), and X-ray diffractometry (XRD – D500 Diffractometer, Siemens). Image analysis for porosity estimation was performed using ImageJ software. A minimum of ten images (dimensions 300x240 μm each) were used for calculating the average level of porosity and the standard deviation. SEM operating at 20 kV was used to perform energy dispersive spectroscopy (EDS) for phase identification. Superprobe operating at 30 kV was used for high resolution elemental mapping via wavelength dispersive spectroscopy (WDS). XRD scans were run using Cu*Kα* radiation in the range 2θ 18-120°, with 0.02° step size and 10s per step. X'Pert HighScore with access to PDF-2 database was used for phase identification. Furthermore, Rietveld refinement [4] was performed in TOPAS V5 software package for quantitative phase composition and lattice parameter refinement. Finally, the Whole Powder Pattern Modelling (WPPM) [5] was employed for determining the average sizes of coherently scattering domains (CDD) and microstrains. The instrumental broadening was accounted for by using the so called fundamental parameters approach, emplying known instrumental parameters. The relation $\rho=2\sqrt{3}\varepsilon d^{-1}b^{-1}$ was used in order to calculate dislocation density ρ [m^{-2}], where ε [%] is the value of the microstrain, d is the grain size (estimated using the measured CDD) [nm], and b=0.286 nm is the magnitude of the Burgers vector for an FCC Al alloy [6]. This relation is commonly used when studying aluminium alloys subjected to severe plastic deformation, thus it has been rational to use it to estimate the dislocation density in the deposits, both as-sprayed and heat treated, but not for the feedstock powder.

2.3 Mechanical Characterisation

Microhardness measurement was performed using a Vickers Microhardness tester (MMT-7, Buehler) equipped with a standard diamond pyramid indenter. 50 gf load was used for all CS deposits, and 10 gf for powders. For each sample, a minimum of 15 measurements were acquired, with spacing according to ASTM standard (E384-16). Nanohardness measurement was performed using a 10 mN load. A total of 25 and 100 measurements were performed for powders and deposits respectively. The measured nanohardness in GPa has been converted into Vickers Hardness number for comparison with microhardness results (ASTM E92-16). Micro-flat tensile (MFT) testing was performed at room temperature using a tensile testing system (5969, Instron) with a cross-head speed of 0.5 mm/min. Elongation was measured with a non-contacting extensometer. MFT specimens were obtained from 0.4 mm thick free-standing coatings and cut using micro-EDM. Samples' geometry and dimensions (10 mm gauge length, 3 mm gauge width, 0.4 thickness) were chosen according to ASTM E8/E8M-16a and other studies [7]. The yield strength was determined using 0.2% offset method. Finally, first order residual stress profile was measured using a portable X-ray diffractometer (AST X-Stress 3000) emitting Cr*Kα* radiation at 30 kV and 6.5 mA. Measurements were performed with diffraction angle 2θ calibrated at 139.3° for Al, irradiating an area of 4 mm diameter. A total of 11 Chi tilts in the range of -45° to 45° along three rotations (0°, 45°, 90°) were run. The residual stress profile was calculated using the $\sin^2(\psi)$ method [8]. Seven meausurements were performed from top surface to interface, by removing each time 0.06 mm material layer through electro-polishing with a solution of 94% acetic acid - 6% perchloric acid. MFT test and residual stress measuement have been performed on AS deposits produced at 500°C-60 bar, and on the same samples after HT at 225°C.

3 Results and Discussion

3.1 Microstructural Analysis

3.1.1 Feedstock Powder

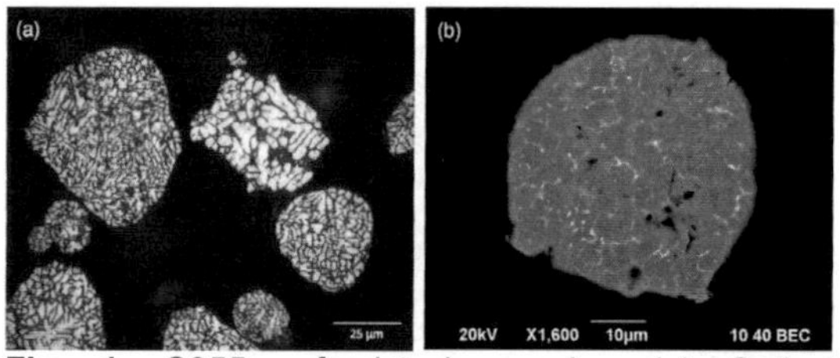

Fig. 1. C355 feedstock powder. (a) Optical micrograph of etched cross-section of particles. (b) Backscattered SEM image of powder cross-section.

Fig. 1.a shows the morphology of powders and evidences their higly irregular structures. The same image reveals the dendritic/ cellular-like structure of powders constituted by grains. The grain size is mostly homogeneous within a single particle, although it varies significantly among different particles (1 to 15 μm). The Backscattered SEM image in **Fig. 1.b** confirms the dendritic structure of the powders, and shows significant solute segregation at grain boundaries (GBs), as the non uniform distribution of the solute elements is evidenced by the sharp atomic number contrast in **Fig. 1.b**. The phenomenon of solute segregation in Al alloys gas-atomised powders has been extensively studied by Rokni et al. (2015) [6]. EDS and EPMA-WDS analyses are further proof that the alloying elements present – i.e. Si, Cu, Mg –

are mainly located at GBs, suggesting that typical precipitates of C355 Al – i.e. θ-Al_2Cu, Q-$Al_5Cu_2Mg_8Si_6$, and β-Mg_2Si – may be present there [3]. XRD pattern in **Fig. 2.** shows the typical peaks of aluminium, silicon and θ-Al_2Cu phases, thus confirming the presence of θ precipitates at GBs. Though, the presumed amount of Q and β phases presenti s below the detection range of XRD (1-2%), preventing their peaks from being identified.

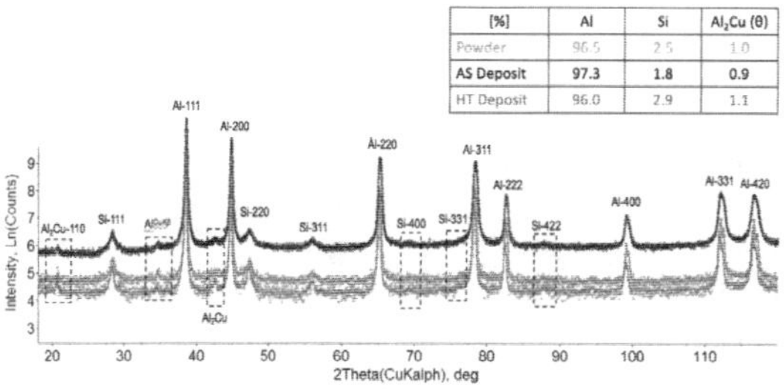

Fig. 2. XRD patterns of powder, as-sprayed and HT-225°C deposits.

3.1.2 As-sprayed Deposits

Optical micrographs of as-sprayed (AS) deposits produced at 500°C and 60 bar are shown in **Fig. 3**. The coatings appear as subsequent compact layers of splats (see **Fig. 3.a**), and reveal extensive degree of particle deformation due to the high kinetic energy employed by the process (see **Fig. 3.b**). 500°C and 60 bar spray parameters have been selected in this study since they produced the lowest level of porosity (~ 1%). This result can be explained by the fact that increased gas temperature and/or pressure allow for higher particle velocity and/or temperature which, in turn, result in enhanced particle deformation, small pores, thin inter-splat cracks, and reduced overall level of porosity. From **Fig. 3.a** it is possible to see that porosity is mainly present in the form of isolated pores. In **Fig. 4.a** and **Fig. 4.c** triple-point and quadruple-point voids are visible. These defects are due to imperfect deformation of the particles which prevents from filling all the voids. Furthermore, also inter-splat cracks are visible in **Fig. 3.c**, and are typically more pronounced near the pores.

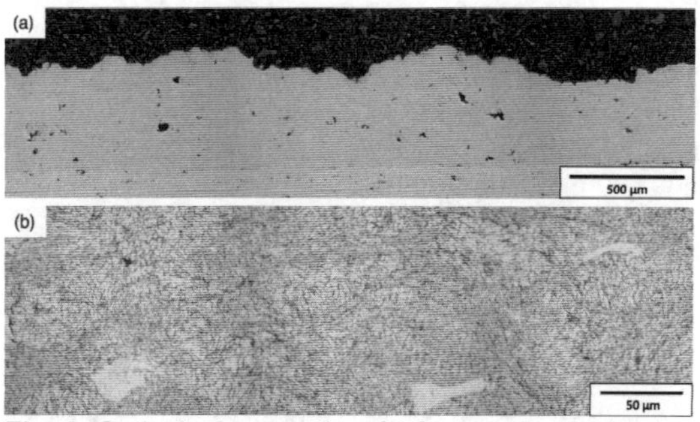

Fig. 3. Optical micrographs of AS deposit produced at 500°C and 60 bar (a) unetched, and (b) etched.

Overall, the microstructure shown in **Fig. 3.b** and **Fig. 4.a** presents the same dendritic features that were found in the feedstock powder (**Fig. 1.a**), even though the particles and grains have been heavily deformed after the process, as it is demonstrated by the fact that the dendrites are heavily deformed in the coatings (see details in **Fig. 4.a**). This suggests that, at a macro-scale, no major change has happened during deposition from feedstock powder to deposited splats. Furhtermore, EDS, and EPMA-WDS analyses (**Fig. 4.b**), and XRD pattern (**Fig. 2.**) confirm that the phases present after deposition are the same (Al, Si, θ) as those in the feedstock powder, and in the same virtual amount (Table in **Fig. 2.**).

Observing the optical micrograph in **Fig. 4.a**, two different microstructural features can be distinguished. The first one are the nearly undeformed inner particle regions (indicated with white arrows) that retain their original structure (see **Fig. 1.a**). The second one are the higly deformed regions and localised jetting at the particle interface, indicated with black arrows. The latter phenomenon is attributed to adiabatic shear instabilities, which can cause local modification of the microstructure through grain refinement, recovery, recrystallization, and precipitation [1]. However, these severely localized plastic deformation areas are highly limited to interfacial regions, thus involving a small portion of the overall deposits, as it appears by observing **Fig. 3.b** and **Fig. 4.a**.

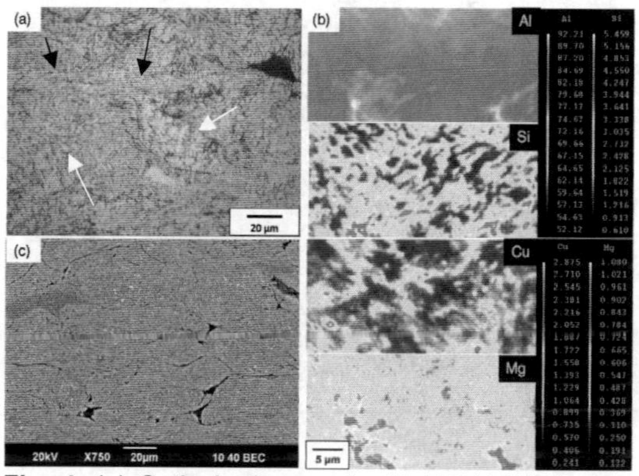

Fig. 4. (a) Optical micrograph, (b) EPMA-WDS map, and (c) BSE SEM image of AS deposit produced using 500°C and 60 bar spray parameters.

3.1.3 Heat Treated Deposits

Optical micrographs of Heat Treated (HT) deposits are shown in **Fig. 5**. From the observation of the micrographs in **Fig. 5.**, it is visible at first sight that the level of porosity in the HT deposits decreases for increasing treatment temperature ranging from 175°C to 250°C, with a reduction by at least 40% for each step increment of 25°C. In particular, HT at 225°C (**Fig. 5.d**) allows to reduce porosity by 90% compared to the AS state (**Fig. 5.a**). Furhtermore, the large pores present before HT (**Fig. 5.a**) have reduced in size, becoming nano-sized (**Fig. 5.d**) or vanishing completely (**Fig. 5.e**). Also the inter-splat cracks seem to have become thinner, as the contact surface among the particles is improved with respect to the

AS state, leading in turn to enhanced particle cohesion (compare **Fig. 4.c** and **Fig. 6.c**). However, the morphology of the pores is not modified by the heat treatment since the same pore shapes can be found after heat treatment at 175 and 200°C (see **Fig. 5.a**, **Fig. 5.b** and **Fig. 5.c**).

The reason for the decrease in porosity after HT, and for the improved cohesion at the particle interfaces, can be explained by the diffusion mechanism that takes place in Al alloys at temperatures around 200°C [2]. The energy provided to the atoms during the treatment in form of heat allows them to move and rearrange to lower energy level configurations. This phenomenon is enhanced by increasing the heat treatment temperature [2], as the higher thermal input provides energy to more atoms to move and/or to travel longer distances. At 175°C (**Fig. 5.b**) and 200°C (**Fig. 5.c**) diffusion is low, and its effect becomes more pronounced for sufficiently high temperatures. For treatment temperatures above 225°C (**Fig. 5.d** and **Fig. 5.e**), some particle boundaries start disappearing and additional chemical/metallurgical bonding is created where the interfaces were bonded soley by mechanical interlocking. Also, when the interface of the particles becomes obscure at high treating temperatures (225, 250°C), improvement in the bond strength at the particle interface can be expected. As a result, the adhesive and cohesive bonding is highly enhanced [1]. However, the temperatures employed in this study are not enough to initiate recrystallization, as explained before, or to produce any other major modification in the grains.

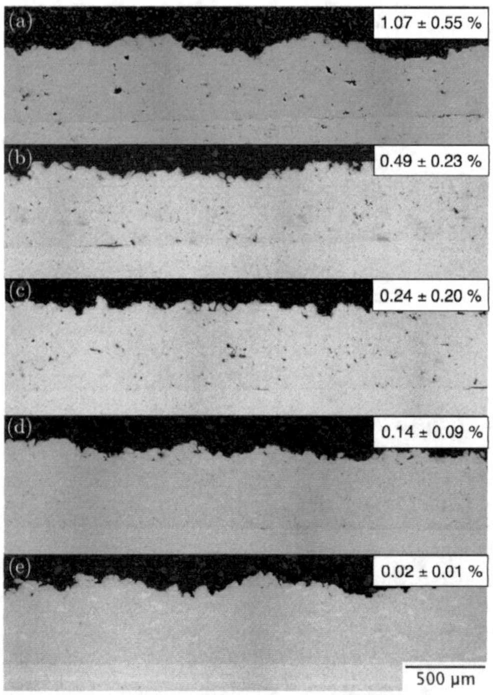

Fig. 5. Deposits produced at 500°C-60bar (a) as-sprayed, and after heat treatment at (b) 175°C, (c) 200°C, (d) 225°C and (e) 250°C for 4h in air.

In addition to this, the relatively low temperatures employed in this study prevent from completely removing the defects present in the AS state (**Fig. 5.a**), as in **Fig. 5.d** and **Fig. 5.e** some minor pores can still be detected.

Fig. 6.a shows the microstructure of the coating after post-deposition HT at 225°C. It presents the same dendritic structure that features the as-sprayed deposits (**Fig. 3.a**), as well as the feedstock powder (**Fig. 1.a**). Furthermore, EDS, EPMA-WDS (**Fig. 6.b**), and XRD phase identification and quantification (**Fig. 2.**) detected no phase change or major modification in the distribution of the elements after HT at 225°C, and no recrystallization. Comparing **Fig. 6.b** with **Fig. 4.b**, it is possible to see in detail that the HT did not have major effect on the distribution of the elements, which further suggests that the heat treatment did not modify the dendritic structure of the AS deposits. In addition, it can be noticed that the dendrites are mainly constituted by the alloying elements (Si, Cu, Mg) in both the AS and HT deposits, as visible in **Fig. 4.b** and **Fig. 6.b**. These results were expected since the treatment teperatures chosen, 225°C in particular, are well below the recrystallization temperature of pure aluminium, i.e. 300°C [2], and thus of aluminium alloys.

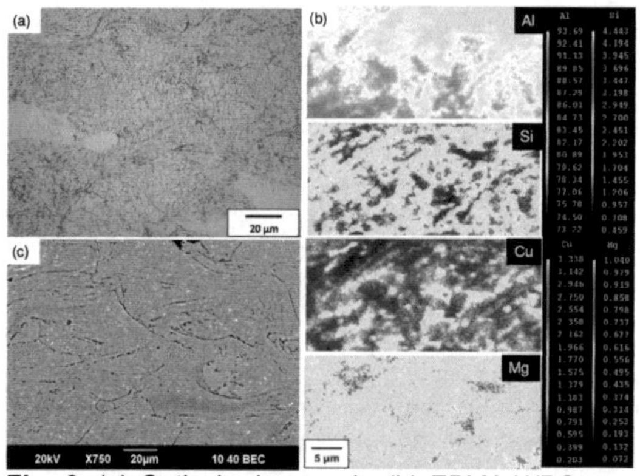

Fig. 6. (a) Optical micrograph, (b) EPMA-WDS map, and (c) BSE SEM image of HT deposit at 225°C.

3.2 Mechanical Characterisation

3.2.1 Microhardness

Fig. 7. shows the results of Vickers microhardness and nanohardness measurements. Regarding Vickers microhardness, the AS deposits have the same hardness value of 130 ± 5 HV$_{0.05}$, regardless the spray parameters, and do not present a significant trend along the thickness. The hardness of AS deposits, as expected, is much higher than that of the feedstock powder (83±8 HV$_{0.01}$), consistently with other works [9] that report an increase in hardness of CS coatings by 50% compared to feedstock powder. Furthermore, the hardness of AS deposits is 25% higher than that of C355 Al bulk in its peak aged (T6) condition (104 ± 10 HV$_{0.05}$).

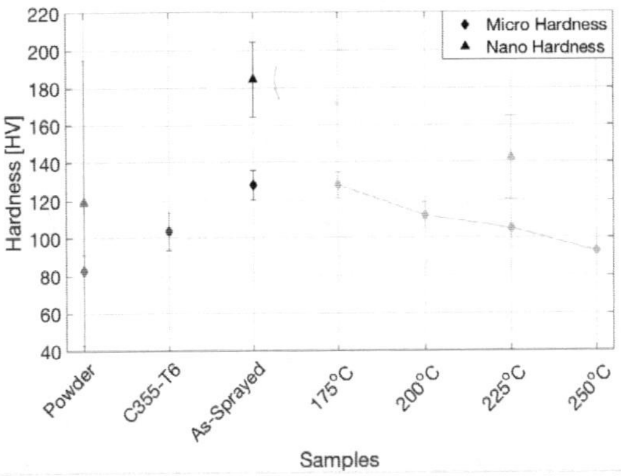

Fig. 7. Vickers microhardness of powder, C355-T6 bulk, as-sprayed and stress relieved deposits. Nanohardness of powder, as-sprayed and HT-225°C deposits.

The high hardness of the coatings is attributed to the high-strain and high-strain-rate plastic deformation that the deposited particles have undergone during deposition. The predominant strengthening mechanism is work hardening due to the formation of a great number of dislocations during spraying, as the value of dislocation density calculated in the AS deposits (0.8 $10^{15}m^{-2}$) suggests. The high dislocation density is mainly due to the increased average mircrostrain (from 0.34% to 0.73%) and decreased CDD (from >200 nm to 118 nm) from powder to AS coating. This latter phenomenon is indication that also grain refinement is responsible for strengthening, as the Hall-Petch relation [10] states that hardness is inversely proportional to the square root of average grain size, i.e. CDD.
The effect of post-deposition stress relief heat treatment at various temperatures on hardness is shown in **Fig. 7**. Hardness decreases quasi-linearly with increasing the treatment temperature, from a value comparable to the AS deposits (128±7 $HV_{0.05}$ at 175°C) to 93±8 $HV_{0.05}$ at 250°C. Medium temperatures – 200, 225°C – yield hardness of 112±7 $HV_{0.05}$ and 105±6 $HV_{0.05}$ respectively. Overall, HT at 225°C allows to retain a significant amount of hardness, comparable to that of C355 bulk-T6. The decrease in hardness evidences that work hardening is gradually released during heat treatment due to rearrangement and reduction of dislocations, i.e. dislocation density decreases from 0.8 $10^{15}m^{-2}$ to 0.3 $10^{15}m^{-2}$ after HT at 225°C, as microstrains decrease from 0. 73% to 0.39%, and CDD increases from 118 nm to 147 nm. As in the AS state, also the grain size plays a role in the decrease of hardness, according to Hall-Petch theory.
Regarding nanohardness measurement, the results converted to Vickers Hardness number are indicated in **Fig. 7**. As it is visible, the trend is in accordance with that of microhardness, even though the value is slightly higher. The difference in the results is expected due to the difference in area considered, as in nanoindentation hardness the projected contact area calculated from penetration depth is used, whereas in Vickers hardness the whole contact area is considered.

3.2.2 Tensile Properties

The results of MFT test are presented in **Fig. 8.** AS deposits show highly brittle behaviour beyond the elastic limit, without any evidence of necking prior to failure. In the AS state, UTS, Young's Modulus and elongation are 199 MPa, 33.9 GPa and 0.67% respectively. The results obtained mainly indicate two aspects: (1) the presence of defects, i.e. porosity and micro-cracks, badly affects the mechanical properties [1], to the point that its negative effect overcomes the positive contribution to strengthening coming from high particle deformation and the related cold working; (2) failure mode is primarily de-cohesive, happening at the interface between particles prior to experiencing plastic deformation. The presence of defects is the primary reason why UTS, Young's Modulus and elongation of the AS deposits are lower than C355 bulk-T6 (199 MPa vs 248 MPa, 33.9 GPa vs 69.6 GPa, 0.67% vs 2.5% respectively) [11].
The HT-225°C deposits show improved behaviour comapared to the AS state, in terms of UTS (230 MPa) and elongation (0.97%), in particular in the plastic region. Furhtermore, HT coatings have higher Yield Strength compared to C355 bulk-T6 (228 MPa vs 172 MPa) [11]. Within the elastic limit, the behaviour is very similar, as both AS and HT samples yield comparable values of Young's Modulus (**Fig. 8.**). The improvement in tensile properties is primarly due to: (1) additional chemical/metallurgical bonding and (2) decreased level of porosity and defects which enhance particle cohesion, and (3) partial release of severe strain hardening state. The first two factors lead to higher UTS, while the third one is the main responsible for improved ductility and reduced brittle behaviour at rupture.

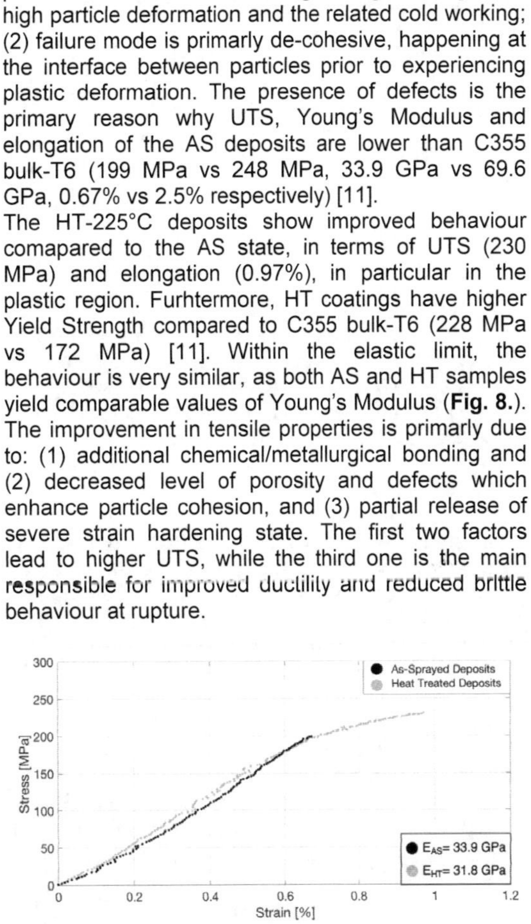

Fig. 8. Results of MFT test of 500°C-60 bar as-sprayed and stress relieved (225°C) deposits.

3.2.2 Residual Stress

Fig. 9. compares the residual stress profile along the thickness of the AS and HT-225°C deposits. In the AS state, the stress is compressive along the whole thickness, being slightly reduced on the very top surface. This relaxation is due to the reduced tamping effect experienced by the last deposited layers. On the contrary, the increased stress approaching the interface is evidence of the peening effect induced on the substrate by cold spraying, which adds up to the compressive stress caused by grit blasting.

After HT at 225°C, the stress state is still compressive but halved in magnitude, and presents the same relaxation on the surface. This finding is in agreement with halving of the microstrains discussed in **3.2.1**. The partial release of stress after HT is beneficial since, even though compressive state is beneficial for fatigue life and stress-corrosion cracking, excessive compression tends to cause adhesive and/or cohesive failure.

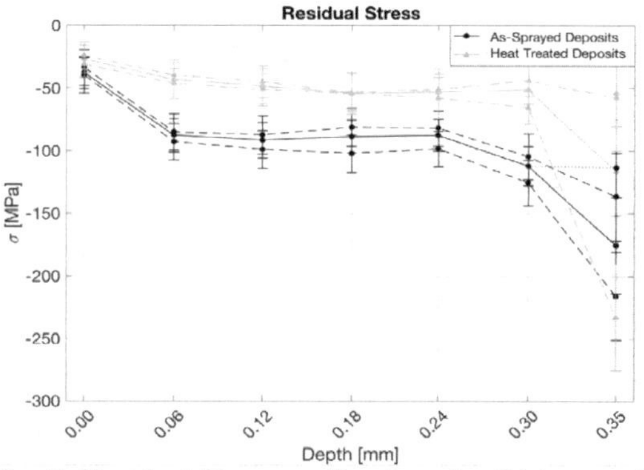

Fig. 9. Residual Stress profile along the thickness of AS and HT-225°C deposits. The solid lines represent the average along three rotations (0°, 45°, 90°), and the dashed lines are the maximum and minimum principal stresses.

4 Conclusion

In this study, dense C355 Al deposits were produced using a commercial HPCS system. Optimal spray parameters were selected upon microstructural and mechanical investigation of the coatings. Finally, stress-relief heat treatment at different temperatures - i.e. 175, 200, 225, 250°C - was performed, and the most promising condition that allows to enhance both microstructural and mechanical properties was identified and reported here. The main findings of the study are:

- CS temperature of 500°C and pressure of 60 bar yield a level of porosity of ~1%, harndess value of 130 ± 5 $HV_{0.05}$, and compressive residual stress state along the thickness; however, the AS deposits show poor ductility;
- Stress-relief heat treatment of C355 Al CS deposits can reduce porosity due to diffusion mechanism (<0.2%), being more effective for increasing treatment temperature;
- The best combination of decrease in porosity and hardness retention is obtained for stress-relief treatment temperature of 225°C, i.e. <0.2% porosity, 105 ± 6 $HV_{0.05}$;
- Stress-relief heat treatment at 225°C is beneficial to tensile properties, as it increase ductility by ~30% and strength by ~15%, without affecting the elastic properties; furthermore it halves the residual stress state, which is beneficial for preventing cohesive/adhesive failure.

5 References

[1] C. Lee, J. Kim, C. Lee, J. Kim, and K. Spray, "Microstructure of Kinetic Spray Coatings: A Review," *J. Therm. Spray Technol.*, vol. 24, no. April, pp. 592–610, 2015.

[2] R. Huang, M. Sone, W. Ma, and H. Fukanuma, "The effects of heat treatment on the mechanical properties of cold-sprayed coatings," *Surf. Coat. Technol.*, vol. 261, pp. 278–288, 2015.

[3] G. Mrowka-Nowotnik and J. Sieniawski, "Microstructure and Mechanical Properties of C355.0 Cast Aluminium Alloy," *Arch. Mater. Sci. Eng.*, vol. 47, no. 2, pp. 85–94, 2011.

[4] H. M. Rietveld, "Line profiles of neutron powder-diffraction peaks for structure refinement," *Acta Crystallogr. Sect. A*, no. 22, 1967.

[5] P. Scardi and M. Leoni, "Whole powder pattern modelling," *Acta Crystallogr. Sect. A Found. Crystallogr.*, vol. 58, no. 2, pp. 190–200, 2002.

[6] M. R. Rokni, C. A. Widener, V. K. Champagne, and G. A. Crawford, "Microstructure and Mechanical Properties of Cold Sprayed 7075 Deposition During Non-Isothermal Annealing," *Surf. Coat. Technol.*, vol. 276, pp. 305–315, 2015.

[7] T. Schmidt, F. Gaertner, and H. Kreye, "New Developments in Cold Spray Based on Higher Gas and Particle Temperatures," *J. Therm. Spray Technol.*, vol. 15, no. December, pp. 488–494, 2006.

[8] M. Guagliano, "L 'applicazione della diffrattometria dei raggi X per l'analisi del cedimento dei componenti meccanici," *Frat. ed Integrità Strutt.*, vol. 2, pp. 25–35, 2007.

[9] S. Rech, A. Trentin, S. Vezzù, J. Legoux, E. Irissou, and M. Guagliano, "Influence of Pre-Heated Al 6061 Substrate Temperature on the Residual Stresses of Multipass Al Coatings Deposited by Cold Spray," *J. Therm. Spray Technol.*, vol. 20, no. January, pp. 243–251, 2011.

[10] K. Ma, H. Wen, T. Hu, T. D. Topping, D. Isheim, D. N. Seidman, E. J. Lavernia, and J. M. Schoenung, "Mechanical behavior and strengthening mechanisms in ultrafine grain precipitation-strengthened aluminum alloy," *Acta Mater.*, vol. 62, pp. 141–155, 2014.

[11] *Metals Handbook, Vol.2 - Properties and Selection: Nonferrous Alloys and Special-Purpose Materials.* ASM International, 1990.

DEVELOPMENT OF HIGH-PERFORMANCE COLD-SPRAYED NANOSTRUCTURED NI-20Cr COATINGS FOR HARSH ENVIRONMENT OF POWER PLANT BOILERS

M. Kumar[1], H. Singh[2], N. Singh[3], N. M. Chavan[4], S. Kumar [5], S. V. Joshi [6]

1) Department of Mechanical Engineering, CGC College of Engineering, Landran, Mohali, Punjab, India
2) School of Mechanical, Materials & Energy Engineering, Indian Institute of Technology Ropar, Rupnagar, Punjab, India
3) Department of Chemistry, Indian Institute of Technology Ropar, Rupnagar, Punjab, India
4,5,6) International Advanced Research Centre for Powder Metallurgy & New Materials (ARCI), Hyderabad 500005, India.

In the current work, Ni-20Cr coatings have been developed for potential use in harsh environments of power plant boilers. A pre-synthesized Ni-20Cr nanocrystalline powder was deposited on T22 boiler steel using cold-spray process. The high temperature oxidation behavior of the coating was investigated under cyclic conditions at 900° C for 50 cycles, so as to understand the kinetics of oxidation. Moreover, high temperature erosion-corrosion (E-C) behaviour of the coating was ascertained under cyclic conditions in an actual boiler at 740 ± 10°C for 1500 hours. The oxidized and eroded-corroded samples were characterized using X-ray Diffraction (XRD), Scanning Electron Microscopy/Energy Dispersive Spectroscopy (SEM/EDS) analyses. The microhardness, oxidation and E-C data for the developed coating was compared with an earlier reported cold-spray Ni-20Cr coating, which was developed by using a commercially available micron-sized Ni-20Cr powder. The results showed that the developed coating was found to have 33% high microhardness in comparison with the microstructured Ni-20Cr coating. The oxidation and E-C rates of the steel were found to decrease significantly after the application of the developed coating by 89% and 68% respectively. Moreover the nanostructured coating outperformed the corresponding micro-structured Ni-20Cr coating with regard to high temperature oxidation and E-C resistance to boiler steel by a significant fraction. The investigated coating was found to have oxidation protective oxides such as Cr_2O_3 and NiO in its oxide scale and was found to be spallation-free.

1 Introduction

Materials degradation under high temperature environment in boilers, gas turbines, waste incinerators, diesel engines, coal gasification plants, chemical plants and other energy generation systems, poses serious threat to the efficiency of systems [1], which in turn results in significant economic losses. It has been reported that erosion and corrosion are responsible for multi-million loss every year to the boiler industry in India [2]. The coal fired boilers are frequently used in power generation industry in India and the share of coal based power generation is increasing continuously [3]. The boilers tubes are generally made of iron based steels, which often face the problem of rupturing due to corrosion as-well-as erosion under high temperature and aggressive environment [4]. Such failures contribute to 50% - 75% arrest time of the boilers [5-6]. Therefore, it is important to understand the phenomenon of material degradation under high temperatures [7-10]. In order to address this issue of hot corrosion and erosion, several surface modification techniques have been studied [11-14]. Among the different surface modification techniques thermal spraying is one of most technically and industrially viable techniques [15]. It has been concluded from the literature that a relatively younger thermal spraying process, named cold spraying has not been explored much for high temperature applications [16-18]. This process is a solid state process, which uses high velocity (Mach > 2) of the powder carrier gas (H_2, N_2 or air) with operating temperatures less than the melting point of feedstock powder, and thus avoids many deleterious high temperature reactions [19-20]. During spraying, the powder particles deform plastically upon striking the substrate surface and form interlinking splats, resulting into the coating. Due to these salient features, the cold-spray process is a promising technique to develop nanostructured coatings [20]. Demand for nanostructured coatings has increased recently owing to their relatively superior properties such as high hardness, ductility, high corrosion resistance and high wear resistance in comparison with their microstructure counterparts [21-23].

Nickel-based alloys, having high chromium content, are widely used as corrosion- resistant coating materials, since they offer good resistance to oxidation/corrosion and have good thermal conductivity [24-27]. Furthermore, when these alloys are used as nanostructured materials, they offer superior resistance to oxidation/corrosion, due to formation of high-density Cr_2O_3 oxide layer by selective oxidation. Cr_2O_3 is a thermodynamically stable phase upto very high temperatures due to its very high melting point and forms a dense, continuous and adherent layer that inhibits interaction of oxygen with the underlying coating/substrate [28-29]. In the present study, the erosion-corrosion (E-C) behavior of a cold-spray Ni-20Cr nanostructured coating on T22 boiler steel was examined in the low-temperature superheater zone (740 ± 10°C) of an actual power plant boiler for 1500 hours under cyclic conditions. The authors have already reported the detailed deposition process of this coating powder on SA 516 steel elsewhere [30]. The outcome of the study shall be useful to explore the possibilities of use of the developed coatings for enhancing the life of boiler tubes of coal-fired power plant boilers.

2 Experimental Procedures

2.1 Development of Coatings

Substrate Material

The substrate material selected for the present study was SAE213-T22 (here after denoted as T22), which is a commonly used boiler tube material in the North India's power plant boilers. The chemical composition (wt%) of steel is Carbon 0.15, Manganese 0.3-0.6, Silicon 0.5, Sulphur 0.03, Phosphorous 0.03, Chromium 1.9-2.6, Molybdenum 0.87-1.13 and balance iron. The test specimens of dimensions 20 mm × 15 mm × 5 mm were cut from the boiler tubes and polished down to 220 grit size of SiC emery paper.

Coating Formulation

The Ni-20Cr nanocrystalline powder was synthesized by blending two types of powders (Table 1) (Ni-powder-74 μm, Cr nano-particles-60 nm) in a planetary ball mill (Fritsch, P-7 premium line, Germany) for 20 h [30]. These two powders were mixed so as to achieve overall composition (wt%) of Ni:80% and Cr:20%. The parameters used in ball milling and complete synthesis process of the powder has been reported elsewhere [30]. The nanocrystalline Ni-20Cr powder was coated on the steel specimens using an in-house cold-spray system at International Advanced Research Centre for Powder Metallurgy and New Materials (ARCI), Hyderabad, India [30, 32].

Table 1 Composition of Ni-20Cr coating powder (wt%) for cold-sprayed coating [30]

Type of Powder	**Blended powders (Wt%)**	
	Micron-sized	Nano-sized
Name of powder	Ni (CA*)	Cr (BM*)
Composition (%)	80%	20%
Average particle size	74 μm	60 nm

*CA- Commercial available, BM- Ball milled

2.2 Characterization of as-sprayed Coatings

Phase analysis of the coated samples was done using XRD machine (PANalytical, X'Pert-Pro model with Cu-Kα radiation, Netherlands). Scanning electron microscopy/energy dispersive spectroscopy (SEM/EDS) (JEOL, JSM-6610LV, Japan/ Oxford INCAx-act, 51-ADD0013, Germany) was done to study the surface and cross-sectional morphology/composition of the coating. The surface images of the polished sample were obtained for measuring the porosity by an Optical Microscope (Leica DM4000M) fitted with Metallurgical software (Leica Microsystems, LMW V 3.6.6). The apparent porosity measurement was made with an image analyzer system having software of MicroCAM 4.1 software (Radical, Ambala, India). Each reported value of the porosity is the average of ten readings taken at different micrographs of a sample. After surface characterization, the samples were sectioned using a high precision slow-speed diamond cutter (Scientific, Precimet-1, India) and then mounted in the transoptic powder using a mounting press (Metco, Bainmount, India). Further, these mounted samples were mirror polished using the standard metallurgical procedure to study the cross-sectional morphology and elemental analysis of as-sprayed samples by SEM and X-ray mappings analysis [31,33].

The surface roughness of the coated sample was measured using a contact type surface profilometer (SurfCom, 130A, USA). Microhardness of the coating along the cross-section was measured using a microhardness tester (Wilson, 402MVD, Illinois) at a load of 100 g with a dwell period of 10 s [33]. The fracture toughness of the coating was calculated on the basis of indentation method using a microhardness tester (Wilson, 402MVD, USA). The indent was applied in the sectioned coating area using Vickers indenter at a load of 200 gf for a 10 s dwell time. Fracture toughness (K_{IC}) was determined with the help of expression given by the Evans and Wilshaw formula [34].

$$K_{IC} = 0.079\ (P/a)^{3/2} \log (4.5a/c)$$

where, P is indentation load in mN, 'c' is the indentation half diagonal (μm) and 'a' is crack length measured from the indentation centre (μm). Ten indentations were performed at different locations and the average of ten values was reported.

The adhesion strength of the coating was measured in terms of critical load requirred to peel-off the coating. The test was performed according to ASTM standard C1624-05 under progressive loading condition, with the help of a scratch testing machine (Ducom, TR-101, India) (Rockwell indenter with the tip of 120° cone angle and 200 μm radius) [35]. Three consecutive scratches of length 5 mm, with loading rate of 5 N/mm and indenter velocity of 0.2 mm/s were made for the cold-spray coating. The test was repeated thrice to establish the repeatability [32].

2.3 High Temperature Air Oxidation Study

The cyclic isothermal air oxidation studies were conducted at 900°C for 50 cycles in a silicon carbide tube furnace (TGIS, Roorkee, India) fitted with a PID controller and calibrated to ±1°C. The details of high temperature air oxidation study have been reported elsewhere [33]. The kinetics of oxidation was established using weight change data. After the exposure for 50 cycles, the oxidized samples were characterized by using XRD and SEM/EDS for the surface as well as cross-sectional analysis as per the already mentioned procedure under section characterisation of as-sprayed coatings.

2.4 E-C Studies in Actual Environment

E-C studies of coated and uncoated samples were performed at Guru Gobind Singh Super Thermal Power Plant (GGSSTPP), Ropar (India) under cyclic conditions for 15 cycles (1500 h). The samples were exposed to low temperature superheater (LTSH) zone. The temperature of the flue gases was measured with the help of thermocouple (Chromel Alumel Type K) and was found to be 740 ± 10°C. This particular location was chosen based on the feedback (severe failure zone due to E-C) from the power plant authorities. The detailed procedure involved in study has been mentioned elsewhere [31]. The chemical analysis of the Indian coal and that of ash and flue gas inside the boiler is reported elsewhere [31]. After every cycle, the samples were cleaned with a soft brush and washed with acetone. Subsequently, the weight and thickness changes were measured for each sample. The extent of E-C was established based on the thickness loss data.

3 RESULTS

3.1 Characterisation of As-sprayed Coatings

Average values of mechanical properties of the investigated coating are reported in Table 2.

Table 2 Properties of investigated cold-spray Ni-20Cr coating on T22 steel

Coating/ Substrate	Average apparent porosity	Average roughness R_a (μm)	Average microhardness of coating (Hv)	Fracture toughness (MPa√m)	Critical load (N)
Ni-20Cr / T22	1.52%	6.04	356 ± 16	0.72	55

The XRD analysis (inset of Fig. 1a) reveals the presence of γ-Ni as the main phase in the coated steel. No oxide phase was observed, which shows that oxidation of feedstock powder has not taken place during coating process, which is a positive attribute. The grain size for the coating was calculated using Scherrer's formula and was found to be 18 nm. This value endorses the formation of a nano-structured coating [36-38].

The microstructure (Fig. 1a) of cold-sprayed coating consists of flattened particles. The interlocking of the particles is also seen in the SEI image. The EDS analysis at selected points revealed the presence of Ni as the primary element and Cr as the secondary element. No O is seen, which predicts the formation of oxide-free coating, which is a positive attribute from the oxidation resistance point-of-view. The average coating thickness as measured from cross-sectional SEM image (Fig. 1b) is found to be 250 ± 12 μm.

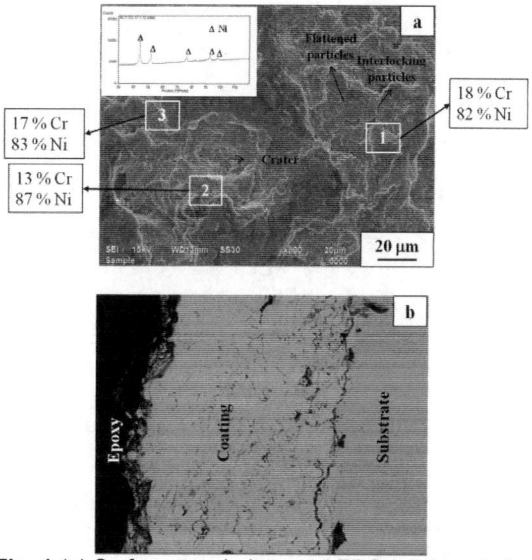

Fig. 1 (a) Surface morphology and EDS point analysis along with X-ray diffraction profile (inset) (b) SEM morphology and variation of elemental composition across the cross section of the cold-sprayed Ni-20Cr coating on T22 boiler steel

3.2 Cyclic Oxidation Studies

The weight gain for the uncoated and coated T22 steels was measured as 193.66 mg/cm^2 and 20.32 mg/cm^2 respectively. The cold-sprayed Ni-20Cr coatings could reduce the weight gain of T22 steel by 89%, which depicts that the necessary protection against oxidation has been provided by the Ni-20Cr coating. Furthermore the plots between (weight change/area)^2 versus number of cycles are shown in Fig. 2, which reveal that the coated steel samples show conformance with the parabolic rate law of oxidation. This indicates that the coating has shown tendency to act as diffusion barrier to the oxidizing species. On the other hand the bare steel has shown relatively higher deviation from the parabolic rate law. The parabolic rate constants for the uncoated and coated boiler steels calculated on the basis of 50 cycle's exposure data are 2212 x 10^{-10} g^2/cm^4/s and 20 x 10^{-10} g^2/cm^4/s respectively.

The XRD analysis of oxidised samples reveals that the oxide scale of bare steel contains Fe_2O_3 as the main phase. On the other hand, the analysis of coated steel indicates the presence of NiO and Cr_2O_3 as the main phases, along with some nascent Ni peaks.

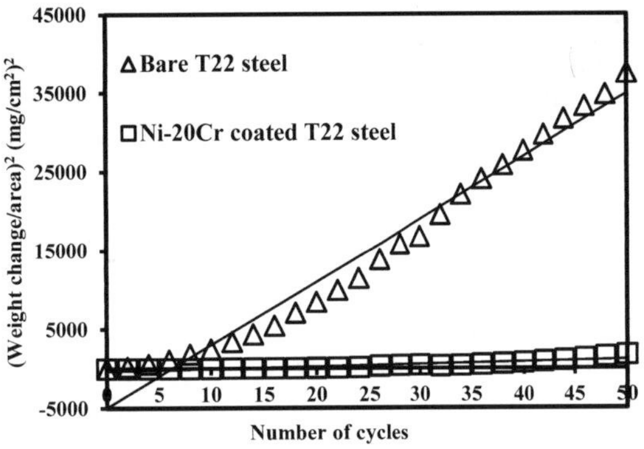

Fig. 2 (Weight gain per unit area)^2 vs. number of cycles' plots for bare [32] and cold-spray Ni-20Cr coated T22 boiler steel subjected to cyclic oxidation in air at 900°C for 50 cycles.

SEM/EDS analysis of the uncoated T22 and coated steel samples subjected to cyclic air oxidation at 900°C for 50 cycles is shown in Fig. 3. The SEM analysis of the bare T22 (Fig. 3a) indicates the presence of a granular scale, whereas the EDS analysis reveals that the scale is rich in Fe and O throughout the composition of the scale. This indicates the possibility of formation of Fe_2O_3 phase in the oxide scale, as has also been revealed by XRD analysis. The analysis of coated steel reveals a scale with a typical pyramidal, crystalline structure, which is a typical NiO scale (Fig. 3b). The EDS analysis indicates a significant concentration of Ni and O in the oxide scale along with small amounts of Cr, which predicts the possibility of formation of oxides of Ni and/or Cr.

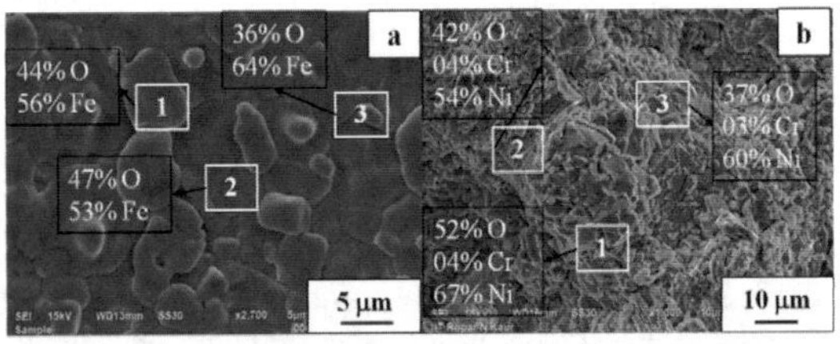

Fig. 3 Surface SEM/EDS analysis for (a) bare T22 steel [32] (b) cold-sprayed T22 steels subjected to cyclic oxidation in air at 900°C for 50 cycles.

The SEM micrograph of oxidized coating on T22 steel reveals that the coating has retained its identity (Fig. 4). The average oxide scale thickness is found to be 250 ± 12 µm. Further the EDS analysis reveals the presence of significant amount of Ni, Cr and O (line 1 and 2), which may be responsible for the presence of NiO and Cr_2O_3 phases in the oxide scale, as has been indicated by XRD analysis. Near the interface of scale and substrate, the presence of a large amount of Fe and O is revealed along with small amount of Ni, Cr, Si and Mo (line 3). This indicates the possible diffusion of Fe from the base steel to the coating area. An enlarged SEM image of the coating area represents densely packed oxidized splat boundaries.

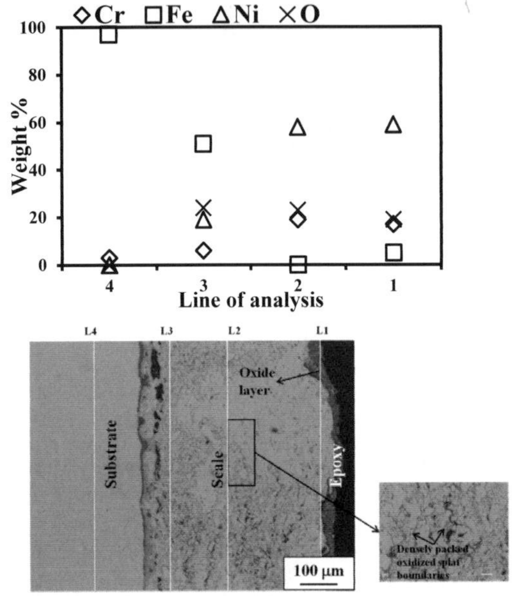

Fig. 4 Oxide scale morphology and variation of elemental composition across the cross-section of cold-spray Ni-20Cr T22 boiler steel subjected to cyclic oxidation in air at 900°C for 50 cycles.

3.3 E-C Studies in Actual Boiler

The corrosion rate (mil per year) calculated based on thickness loss data, for uncoated and coated steel was found to be 108 mpy and 34 mpy respectively after exposure to harsh boiler conditions for 1500 hours. It is clear from erosion-corrosion rates that the investigated coating has reduced the corrosion rate of base steel by 68%.

X-ray diffraction analysis of the cold-spray coated T22 steel, reveals the presence of NiO, Cr_2O_3, SiO_2 and Al_2O_3 phases. Out of these phases, Al_2O_3 belongs to the ash constituents, where Cr_2O_3 might have formed by selective oxidation of Cr present in the coating.

The SEM analysis (Fig.5) for the eroded-corroded cold-sprayed Ni-20Cr coating on T22 steel shows that the coating is still intact with the substrate steel and has retained its original morphology. An enlarged view of the scale indicates that the original splat-like morphology of the coating is still retained. Line scan analysis along the oxide scale (line 1 and 2) revealed the presence of rich concentrations of Ni with significant amounts of Cr and O. S and Si were also seen along the lines. Near the interface, line analysis (line 3) revealed the presence of Ni, Cr, O and Fe.

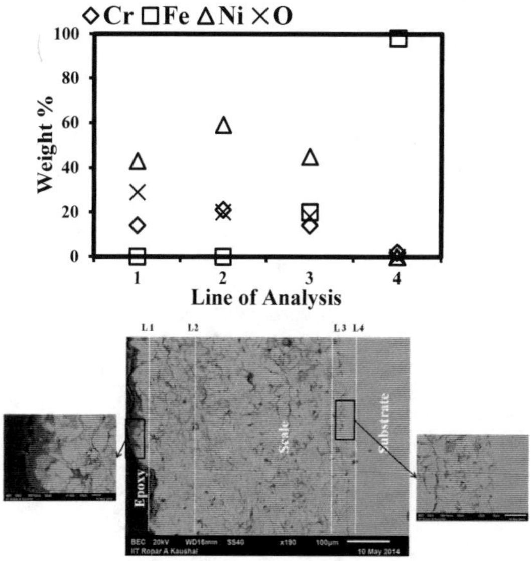

Fig. 5 Oxide scale morphology and variation of elemental composition across the cross-section of cold-sprayed T22 boiler steel subjected to actual boiler environment at 740 ± 10°C for 15 cycles.

CONCLUSIONS

Ni-20Cr nanocrystalline powder was successfully formulated and deposited on T22 boiler steels by the cold-spray technique at the given process parameters. The coating was found to be nano-structured. The coating was found to have dense morphology. The investigated coating was found to have higher hardness in comparison with its conventional microstructured counterpart, owing to its finer grain structure. This higher hardness is useful to enhance the erosion-corrosion resistance of the coatings. The developed coating was found to be useful to control the high temperature oxidation and erosion-corrosion of T22 steel by a substantial magnitude. This could be attributed to the presence of Cr_2O_3 and NiO phases in the exposed coatings. The coating was found to be spallation-free in both the environments of study. Moreover, the high temperature performance of the coating was observed to be better than its microstructured counterpart. which could be attributed to its finer grain structure.

REFERENCES

[1]. A.M. Beltran, and D.A. Shores: Ch. 11: Hot Corrosion," in 'The Superalloys,' Eds. C.T. Sims, and Acta Hagel, 44 (24) (1972), 4289-4296.

[2]. http://www.financialexpress.com/news/-cost-of-corrosion-in-india-at-rs-36000-cr-/118088/ accessed on 27 October 2012

[3]. www.powermin.nic.in

[4]. M.A. Uusitalo, P.M.J. Vuoristo and T.A. Mantyla, High Temperature Corrosion of Coatings and Boiler Steels in Oxidizing Chlorine-Containing Atmosphere, *Material Science and Engineering A*, 2003, 346(1-2),p 168-177.

[5]. R. A. Rapp and Y. S. Zhang, Hot Corrosion of Materials: Fundamental Studies, *JOM*, 1994, 46, p 47-55.

[6]. V. H. Hidalgo, F. J. B. Varela, and E. F. Rico: Erosion Wear and Mechanical Properties of Plasma-Sprayed Nickel- and Iron-Based Coatings Subjected to Service Conditions in Boilers, *Tribology International*, 1997, 30, p 641-649.

[7]. J. Stringer and I. G. Wright: Current Limitations of High-Temperature Alloys in Practical Applications, *Oxidation of Metals*, 1995, 44(1), p 265-308.

[8]. N. J. Simms, *Environmental Degradation of Boiler Components, Power Plant Life Management and Performance Improvement*, 2011, J.E. Oakey, Ed., Woodhead Publishing.

[9]. N. J. Simms, P. J. Kilgallon, and J. E. Oakey: Fireside Issues in Advanced Power Generation Systems, *Energy Materials: Materials Science and Engineering for Energy Systems*, 2007, 2(3), p 154-160.

[10]. N. J. Simms, P. J. Kilgallon, and J. E. Oakley: Degradation of Heat Exchanger Materials under Biomass Co-firing Conditions, *Materials at High Temperatures*, 2007, 24(4), p 333-342.

[11]. K. L. Choy: Chemical Vapor Deposition of Coatings, *Progresses in Material Science*, 2003, 48, p 57-170.

[12]. M. A. Young: Health Hazards of Electroplating, *Journal of Occupational and Environmental Medicine*, 1965, 7, p 348-352.

[13]. J. A. Koropchak and S.B. Roychowdhury: Evidence for Aerosol Ionic Redistribution within Aerosols produced by Chrome Electroplating, *Environmental Science and Technology,* 1990, 24, p 1861-1863.

[14]. W. T. Tsai, H. P. Chen and W. Y. Hsien: A Review of uses, Environmental Hazards and Recovery/Recycle Technologies of Perfluorocarbons (PFCs) Emissions from the Semiconductor Manufacturing Processes, *Journal of Loss Prevention in the Process Industry*, 2002, 15, p 65-75.

[15]. R. C. McCune, A. N. Papyrin, J. N. Hall, W. L. Riggs and P. H. Zajchowski: *Advances in Thermal Spray Science and Technology*, 1995, C. C. Berndt and S. Sampath, eds., ASM International, Materials Park, OH.

[16]. A. Papyrin, V. Kosarev, S. Klinkov, A. Alkimov, V. Fomin, *Cold Spray Technology,* 2007, Elsevier.

[17]. L. Pawlowski, *The Science and Engineering of Thermal Spray Coatings*, 2008, John Wiley and Sons, Ltd..

[18]. J. R. Davis, *Handbook of Thermal Spray Technology*, 2004, TSS/ASM International, Materials Park, OH.

[19]. J. Karthikeyan, *Cold Spray Technology: International Status and USA Efforts*, 2004.

[20]. R. S. Lima, J. Karthikeyan, C. M. Kay, J. Lindemann, C. C. Berndt: Microstructural Characteristics of Cold-sprayed Nanostructured WC–Co Coatings, *Thin Solid Films,* 2002, 416, p 129-135.

[21]. H. Gleiter, Nanostructured Materials: Basic Concepts and Microstructure, *Acta Materilia*, 2000, 48, p 1-29.

[22]. G. Cao, Nanostructures and Nanomaterials – Synthesis. Properties and Applications, 2004, Imperial College Press, London.

[23]. B. R. Marple, J. Voyer and J. F. Bisson, C. Moreau: Thermal Spraying of Nanostructured Cermet Coatings, *Journal of Materials Processing Technology*, 2001, 117, p 418-423.

[24]. T.S. Sidhu, S. Prakash, R.D. Agrawal: Curr. Sci., 2006, 90 (1), p 41–47.

[25]. J. He, M. Julie and J. M. Schoenung: Nanocrystalline Ni Coatings Strengthened with Ultrafine Particles, *Metallurgical and Materials Transactions A*, 2003, 34, p 673-683.

[26]. S. Zhao, X. Xie, G. D. Smith and S. J. Patel, The Corrosion of INCONEL Alloy 740 in Simulated Environments for Pulverized Coal-fired Boiler, *Materials Chemistry and Physics*, 2005, 90, p 275-281.

[27]. S. J. Matthews, *Erosion-Corrosion of Cr_3C_2-NiCr High Velocity Thermal Spray Coatings,* 2004, Ph.D. Thesis, The University of Auckland.

[28]. B. Wang, J. Gong, A. Y. Wang, C. Sun, R. F. Huang and L. S. Wen, Oxidation Behaviour of NiCrAlY Coatings on Ni-based Superalloy, *Surface and Coatings Technology*, 2002, 149, p 70-75.

[29]. T. H. Steenkiste, J. R. Smith, R. E. Teets, J. J. Moleski: Aluminum Coating via Kinetic Spray with Relatively Large Powder Particles, *Surface and Coatings Technology,* 2002, 154(2-3), p 237–252.

[30]. M. Kumar, H. Singh and N. Singh: Synthesis and Deposition of Ni-20Cr Powder using Cold Spraying, *Surface Engineering,* 2013, 29(6), p 419-426.

[31]. M. Kumar, H. Singh and N. Singh: Fire Side Erosion-corrosion Protection of Boiler Tubes by Nano-structured Coatings, *Materials and Corrosion,* DOI: 10.1002/maco. 201407954.

[32]. N. Kaur, M. Kumar, S. K. Sharma, D. Y. Kim, S. Kumar, N. M. Chavan, S. V.Joshi, N. Singh, and H. Singh: Study of Mechanical Properties and High Temperature Oxidation Behavior of a Novel Cold-spray Ni-20Cr Coating on Boiler Steels, *Applied Surface Science*, 2015, 328, p 13-25

[33]. M. Kumar, H. Singh and N. Singh: Production of a Nano-Crystalline Ni-20Cr Coatings for High-Temperature Applications, *Journal of Thermal Spray Technology*, 2013, 23(4), p 692-707.

[34]. A.G. Evans and T.R. Wilshaw, Quasi-static Solid Particle Damage in Brittle Solids-I. Observations Analysis and Implications. *Acta Metallurgica,* 1976, 24, p 939-56.

[35]. J. Takadoum, Materials and Surface Engineering in Tribology, 2008, Wiley-ISTE.

[36]. H.P. Klug, L.E. Alexander, X-ray Diffraction Procedures, 1974, Second ed. Wiley, NY, USA.

[37]. B.E. Warren, X-ray Diffraction, Dover Publications, Inc., 1990, NY, USA.

[38]. H.G. Jiang, M. Ruhle, E.J. Lavernia: On the Applicability of the X-Ray Diffraction Line Profile Analysis in Extracting Grain Size and Microstrain in Nanocrystalline Materials, *Journal of Material Research,* 1999, 14 (2), p 549–559.

Design of Fe-based Amorphous Composite Coatings

Cheng Zhang*, Wei Wang, Peng Xu, M. Yasir, Lin Liu*

State Key Laboratory of Material Processing and Die & Mould Technology, School of Materials Science and Engineering, Huazhong University of Science and Technology, 430074 Wuhan, China

Abstract:

The practical application of bulk metallic glasses (BMGs) as structural materials is still restricted due to the intrinsic brittleness of BMGs although they have high hardness and strength and even pretty good toughness. As an alternative form of metallic glasses, amorphous coatings based on BMG systems can exactly overcome the drawback of BMGs, but carry forward the superiority in corrosion and wear resistance, thus exhibiting promising applications in surface engineering. However, the monolithic amorphous coatings faces new challenges, such as low adhesion strength and impact toughness. In this presentation, we present some new findings on the design of novel Fe-based amorphous composite coatings by HVOF technique. These include the add stainless steel ductile phase, hard ceramic particles and in-situ carbon phases into the amorphous coating, in addition to form laminar structural coatings. It will be shown that the second phase and their interface structure with amorphous matrix play important role in the resultant mechanical properties of the coatings. The combination of the good mechanical, physical and chemical properties warrants amorphous composite coatings to have extensive applications in industry in near future.

High temperature hot corrosion behavior of selected thermally sprayed coatings in an aggressive environment

Z. Česánek, Š. Houdková, J. Schubert, Pilsen / CZ

Cr_3C_2-25%NiCr, Stellite 6, NiCrBSi and Hastelloy C-276 coatings were deposited on substrate material P91 by HP/HVOF (High Pressure / High Velocity Oxygen Fuel) thermal spraying technology. The resistance against high temperature corrosion was evaluated exposition of coatings to corrosive-aggressive environment in the form of molten salts mixture with composition of 60% V_2O_5 and 40% Na_2SO_4 at temperature of 750 °C. Further, coatings were exposed to cyclic conditions.
After the corrosion tests, all coatings were analyzed using scanning electron microscope (SEM), and analysis of elemental composition (EDX). Alloys-based coatings showed very similar corrosion mechanism in the selected aggressive environment and the same can be stated about cermet coatings. The obtained results prove that HVOF deposited coatings can replace current surface protection of components in power equipment such as nitriding.

1 Introduction

The construction of energy facilities working under high temperature puts high demands on materials used. Besides very high temperatures causing oxidation, the situation is often more complicated by the presence of corrosive aggressive media. Due to decreasing reserves of conventional fuels, residual fuel oils are being nowadays used as a fuel in boilers, refinery furnaces and gas turbines. The elements contained in oils, especially sodium, vanadium and sulfur, produce during the combustion process compounds (e.g. Na_2SO_4 (melting point at 884 °C) and V_2O_5 (melting point at 670 °C) [1] and [2]. The resulting compounds, generally known as ash, are deposited on component surfaces and are responsible for acceleration of material oxidation, so called hot corrosion. Certain salts are at operating temperatures in liquid state or form complex salt mixtures which have lower melting point than individual components of these mixtures separately. The corrosion under such conditions is therefore called corrosion in molten salts. The principle and mechanism of high temperature corrosion in molten salts has been described many times [3].
The corrosion rate reduction can be achieved by appropriately chosen corrosion inhibitors, e.g. MgO, CaO, CeO_2, MnO_2 etc. However, their use in real environment is limited by the dosage problems with fuel. Another possibility is to control the process parameters of boilers and gas turbines, such as air/fuel ratio, temperature, pressure etc., but it is useful only to some extent because these parameters can only be controlled within certain limits. To limited extent, the addition of significant amount of Al, Cr and small amount of Y, Zr and Hf can be used to increase the material corrosion resistance of power equipment components. However, higher concentrations of these elements negatively affect the mechanical properties of alloys. The application of corrosion resistant alloy and cermet based protective coatings using thermal spraying technology appears to be the most reliable and economical method to achieve desired corrosion resistance [4].
Generally, the coating properties determine its behavior in operation. The simulation of real future operating conditions is therefore the best quality test. Operational tests, however, are often impossible to implement due to time demands. For this reason, coatings are characterized by means of laboratory tests evaluating their physical and mechanical properties. The attention is usually focused on the evaluation of mechanical and physical properties and their dependence on temperature, corrosive environments, procedures and conditions (such as the composition of furnace atmosphere etc.) [5].
HVOF technology is one of the most suitable technologies for application of corrosion resistant coating. It enables to produce coatings from alloy and cermet based materials having a very low porosity. The resistance of HVOF coatings against high temperature corrosive aggressive environment has been addressed by many authors [4], [9] and [10]. It was shown in their studies that certain coatings produced by HP/HVOF technology provide better protection to components in high temperature corrosive environment in comparison with bare materials. The resistance of Stellite 6, NiCrBSi, Hastelloy C-276 and Cr_3C_2-25% NiCr coatings was evaluated in this paper. It is possible to find enough information in literature about high temperature corrosion of all evaluated coatings besides Hastelloy C-276. This coating has a very good potential to succeed in the environment of high temperature corrosion. This paper deals with the comparison of Hastelloy C-276 with other evaluated coatings in the environment simulating conditions in boilers for solid and liquid fuels.

2 Experiment

2.1 Coating deposition

Commercially available powders were used for sample preparation: Amperit 588.074 (Cr_3C_2-25%NiCr) with a particle size distribution suitable for HVOF (-45+15 mm); M-484.33 (CoCrWC) with a particle size distribution suitable for HVOF (-53+20 mm); M-771.33 (NiCrBSi) with a particle size distribution suitable for HVOF (-53+20 mm) and M-341.33 (Alloy-276) with a particle size distribution suitable for HVOF (-53+20 mm). All coatings were deposited by HP/HVOF technology with JP-5000 torch from the company TAFA Incorporated. Previously optimized spray parameters were used for each coating, see **Tab 1.** Coatings were deposited on substrate material 1.4923 (P91) which is in nitrided state used as an effective constructional material for components in power equipment. Flat specimens with dimensions of 10x10x5 mm were sprayed, and their surface was further metalographically grinded to the surface roughness Ra_{max}=1µm.

Tab. 1. Spraying parameters

Coating	Hastelloy C-276	NiCrBSi
Equivalent ratio:	0,9	0,75
Oxygen:	857 l/m (1824 scfh)	990 l/m (2106 scfh)
Kerosin:	22,7 l/h	23 l/h
Deposition distance:	380 mm	360 mm
Barel:	4"	4"
Coating	**Cr_3C_2-25%NiCr**	**Stellite 6**
Equivalent ratio:	1,07	0,93
Oxygen:	872 l/m (1855 scfh)	996 l/m (1751 scfh)
Kerosin:	27,2 l/h	27 l/h
Deposition distance:	360 mm	360 mm
Barel:	4"	6 "

2.2 Microstructure evaluation

After the corrosion tests, specimens were cut and poured in order to conduct cross-sectional metallographic comparison. Metallographic evaluation was conducted using scanning electron microscope FEI Quanta 200. Exposed specimens were further evaluated by elemental composition change using EDS analysis. It is important to mention that the specimen preparation for metallographic evaluation is not quite standard process after the corrosion tests, but it is an essential part of microstructure study of thermally sprayed coatings. Separation method was used to make cross sections because this method enables to obtain more precise information for the evaluation of thermally sprayed coatings microstructure after the corrosion testing. It is a method of semi-automatic preparation. The equipment used for metallographic specimen preparation was LECO Spectrum System, 2000.

Thermally sprayed coatings require special care when separating a part out of the total specimen. Pure alcohol was used as a coolant. The separated part was further rinsed again with alcohol. Lateral specimen sides and beveled sharp edges of coating were grounded. It was also highly suitable to grind the bottom edge of substrate and to remove impurities in order to prevent possible oxidation. So processed specimen was further rinsed with alcohol and dried in warm air.

Cold mounting recommended for thermally sprayed coatings was used for all coatings. This procedure is used to prevent the coating exposure to high temperatures, which could affect the resulting coating microstructure.

Several grinding papers attached in LECO equipment were used for grinding. The length of every grinding process did not exceed 1 minute.

A holder with nine attachment holes was used to attach specimens. The attachment of specimens into the holder was made on a special rotating plate. Coating orientation towards the direction of rotation is shown in **Fig. 1.** The holder with attached specimens was then inserted into the rotating head of grinding machine. After each grinding process, all specimens were cleaned with alcohol and dried with compressed air. The next step was polishing. The grinding and subsequent polishing operations included the 50 s steps on abrasives with grain size 320, 800, 1200, 3 µm and 1 µm, finished with DP-Spray 1µm + MD Nap suspension. Alcohol was used as a coolant in all cases. Rotation speed was 300 m/min and specimen loading was 15 N.

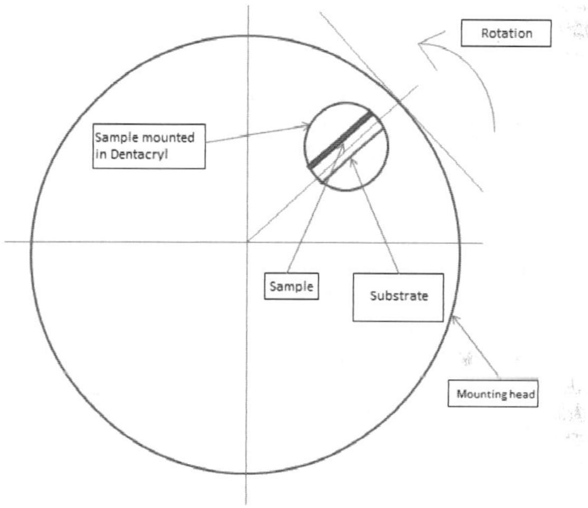

Fig. 1 Scheme of coating orientation towards the holder rotation

2.3 Microhardness measurements using $HV_{0.3}$

Microhardness of all evaluated specimens was measured before and after the corrosion tests with LECO DM-400A Hardness Tester using $HV_{0.3}$ method in the company VZÚ Plzen s.r.o. Seven indentations under 300 g load for 10 s of indentor intrusion into the coating were always conducted. The resulting values are the average of above mentioned 7 measurements. Coating microhardness was evaluated using Vickers hardness test in coating cross-section.

2.4 High temperature corrosion test

High temperature corrosion test was based on corrosive aggressive action of Na_2SO_4 and V_2O_5 salt mixture at the temperature of 750 °C. The salt mixture in ratio of 40/60 (mass percentage) was mixed with alcohol and applied on specimen surface in an amount of 3-4 g/cm^2. Prior to salt mixture application, specimen surface was rinsed with alcohol and heated in an oven at 250 °C in order to ensure sufficient adhesion of salt layer. After the application of salt layer, specimens were left 3 hours in the oven at 100 °C in order to evaporate ethyl alcohol and to dry the salt layer. The test contains 50 cycles of heating at 750 °C for 1 hour in the silicon carbide oven and subsequent cooling for 20 minutes at room temperature. Specimens were weighed with a precision of 0.0001 g before and during the test.

3 Results

3.1 Corrosion kinetics

The graph in **Fig. 2** shows how the evaluated coatings responded to the aggressive environment of high temperature corrosion. The coating with designation 771.33 (NiCrBSi) was totally destroyed in the eleventh cycle, and therefore no further values are included. The graph shows that besides NiCrBSi coating all other evaluated coatings exhibited corrosion resistance. Carbide coating showed higher weight gains and alloy based coatings lower weight gains. These results were largely consistent with foreign experiments which have been conducted under very similar conditions.

In case of Cr_3C_2-25% NiCr carbide coating, Sidhu et al. proved the formation of nickel and chromium oxides and nickel and chromium spinels. According to the author, this coating is resistant up to 900 °C in very similar corrosive aggressive environment [8].

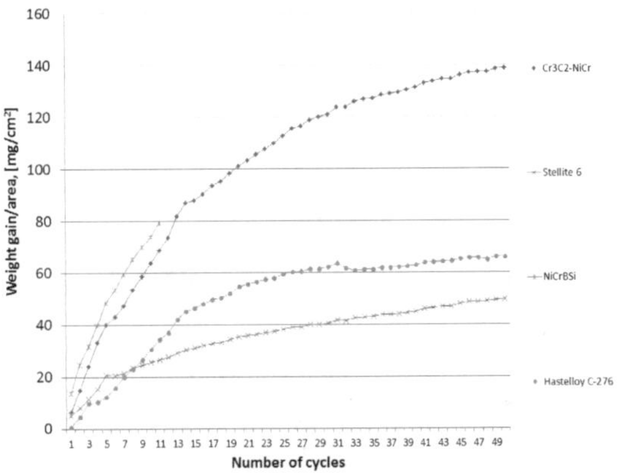

Fig. 2 Weight gains depending on the number of cycles in the environment of Na_2SO_4 - 60% V_2O_5 at 750 °C

3.2 SEM/EDS analysis

This chapter contains photographs of scanning electron microscope (SEM) with elemental composition determined by EDX analysis in tables. **Fig. 3** illustrates the microstructure of Cr_3C_2-25% NiCr coating in cross section after the exposure to corrosive environment. The photograph apparently shows the mechanism of corrosion damage, namely coating disruption processing from the surface. There was formed an oxide layer which was further attacked by molten salts. This coating lost its protective layer resulting in progressive breaking from the surface.

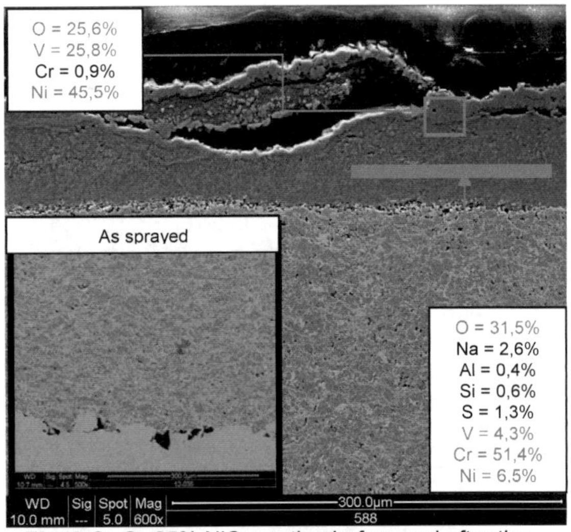

Fig. 3 Cr_3C_2-25% NiCr coating before and after the exposure at 750 °C in corrosive environment of molten salts Na_2SO_4 - 60% V_2O_5 after 50 cycles

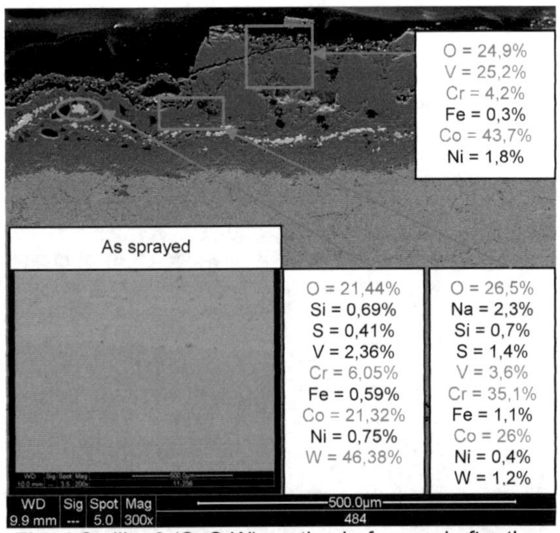

Fig. 4 Stellite 6 (CoCrW) coating before and after the exposure at 750 °C in corrosive environment of molten salts Na_2SO_4 - 60% V_2O_5 after 50 cycles

Fig. 4 shows the microstructure of Stellite 6 coating in cross section after the exposure to corrosive environment. The photograph apparently shows the mechanism of corrosion damage - creation of oxides on coating surface. These oxides reacted with salt mixture resulting in the attack on individual coating splats. However, the corrosion attack appears to be negligible in terms of coating functionality.

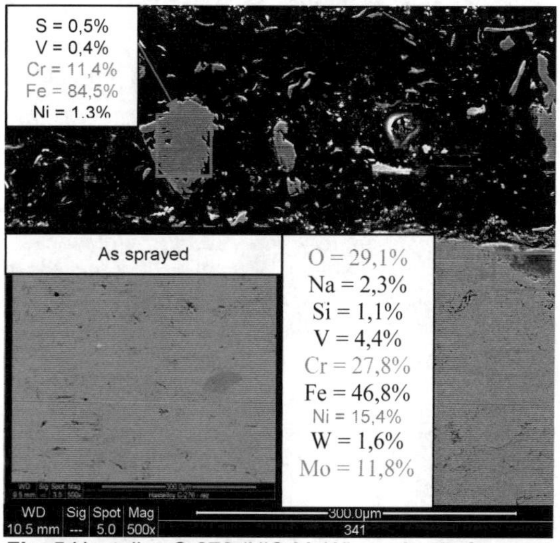

Fig. 5 Hastelloy C-276 (NiCrMoW) coating before and after the exposure at 750 °C in corrosive environment of molten salts Na_2SO_4 - 60% V_2O_5 after 50 cycles

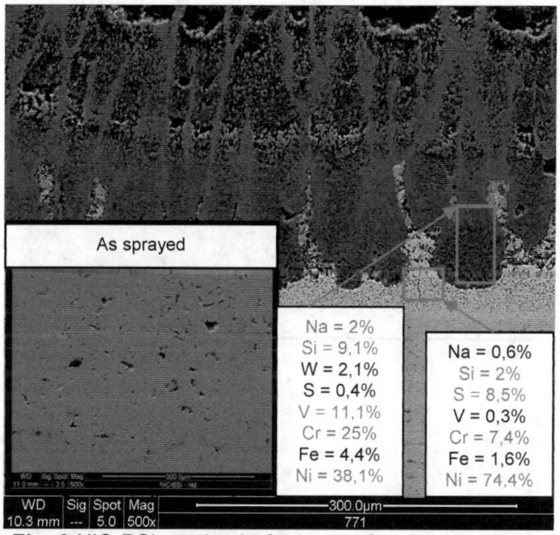

Fig. 6 NiCrBSi coating before and after the exposure at 750 °C in corrosive environment of molten salts Na_2SO_4 - 60% V_2O_5 after 50 cycles

Fig. 5 shows the microstructure of Hastelloy C-276 coating in cross section after the exposure in corrosive environment. The photograph shows identical mechanism of corrosion damage like Stellite 6 coating. It proves that these coatings exhibit very similar corrosion mechanism. **Fig. 6** shows the microstructure of NiCrBSi coating in cross section after the exposure to corrosive environment. This coating suffered from very aggressive attack. Thin oxide layer was damaged and it resulted in coating destruction. Based on the results obtained, this coating is not suitable for this type of corrosive environment.

3.3 Microhardness $HV_{0.3}$

Microhardness $HV_{0.3}$ values of evaluated coatings before and after the corrosion tests are given in **Tab. 2.** The resulting coating microhardness is an arithmetic average of seven measurements. Measurements were conducted according to DIN EN ISO 6507-1 in the metallographic laboratory of VZÚ Plzen s.r.o

Tab. 2 Measured microhardness $HV_{0.3}$ values of evaluated coatings before and after corrosion tests

Coating	Before test	After test
Cr_3C_2-25%NiCr	847±36	846±43
Hastelloy C-276	477±52	472±16
Stellite 6	628±57	599±26
NiCrBSi	815±59	592±81

4 Discussion

Stellite 6 is a coating based on CoCrW. Stellite alloys have an excellent hardness and toughness, and also are generally very corrosion resistant. Stellite alloys often have a very high melting point due to the content of cobalt and chromium [6]. The high temperature corrosion resistance of CoCrW coating and the tendency to behave as a diffusion barrier for degradation media are mainly attributed to the formation of chromium oxide on splats boundaries rich in Co and to the formation of surface oxides, particularly cobalt and chromium oxides, together with cobalt-chromium and nickel-chromium spinels. The formation of $CoCr_2O_4$ spinel blocks the diffusion using cobalt oxide (CoO) resulting in further suppressing of cobalt oxide formation. Increasing growth of $CoCr_2O_4$ and Cr_2O_3 phases together with the formation of CoO and Co_3O_4 increases the corrosion resistance of this alloy. Splats rich in Ni and Co remain mostly in un-oxidized state [4].

NiCrMo (Hastelloy) based coatings generally exhibit good corrosion resistance. Hastelloy coating in sprayed state ("as sprayed"), as well as polished or thermally processed (reflowed) versions perform with excellent corrosion resistance, as reported in studies [7].

Cr_3C_2-25% NiCr coating is currently used in a wide range of industrial applications, such as gas turbines, automotive industry, manufacturing equipment, boiler superheaters or piping systems. The structure of thermally sprayed Cr_3C_2-25% NiCr coating is mainly composed of nickel solid solution γ together with Cr_7C_3 and Cr_2O_3 phases. The coating has porosity lower than 1.5 %. Inclusions, unmelted and partially melted powder particles also occur in the coating structure. Cr_3C_2-25% NiCr coating is characterized by high resistance to high temperature corrosion due to the formation of nickel and chromium oxides, as well as nickel-chromium spinel. Thermally sprayed Cr_xC_y-NiCr coatings can be used in high temperature applications due to the excellent corrosion resistance, oxidation of nickel-chromium alloy and acceptable wear resistance of chromium carbides at the temperatures close to 900 °C. One of the most important properties of coatings resistant to high temperature corrosion in aggressive environments is porosity. Porosity in coating structure creates corrosion paths through which corrosive media may penetrate though the coating to the substrate resulting in rapid corrosion attack. Coatings with lower porosity generally provide better corrosion resistance than porous ones [8]. This coating exhibits excellent corrosion resistance due to selective oxidation, which occurs along splat boundaries rich in nickel resulting in the formation of chromium oxides (Cr_2O_3). This oxide blocks paths in pores and enables the improvement of coating resistance to penetration and propagation of corrosive media. Moreover, very low porosity and flat splat structure contribute to the development of resistance to high temperature corrosion. The corrosion usually spreads along splat boundaries and through pores and cavities. Because of dense splat coatings structure, the distance from the coating surface to coating-substrate interphase along spalt boundaries rapidly increases and results in coating resistance to hot corrosion.

Regarding the weight gains, NiCrBSi coating exhibited better resistance to hot corrosion than e.g. CoCrWCSi (Stellite 6) coating. NiCrBSi coating exhibited only the oxidation of upper coating part of 100-110 microns from the surface and the remaining part has from visual point of view similar structure like before the corrosion testing. Two thick oxide layers were formed in the coating. The top layer is rich in silica and the bottom layer is rich in chromium oxide. Such layer distribution contributes to better resistance to high temperature corrosion. Furthermore, NiCrBSi coating contains Cr_2O_3 and SiO_2 as main phases together with $NiCr_2O_4$ spinel in the surface after the exposure to high temperature corrosion. The top layer of NiCrBSi coating is dense, compact and exhibits no cracking [4]. Wang et al. [11] reported that addition of Si and B may increase selective oxidation, which leads to the formation of protective layer in initial corrosion stages and improves coating-substrate adhesion in further process of high temperature corrosion.

5 Conclusion

Stellite 6:

- Formation of Cr oxides on splat boundaries was observed by Stellite 6 coating.
- Further, surface oxides of Cr and Co were formed in the coating resulting in improved resistance against high temperature corrosion.
- The graph on **Fig. 2** apparently shows that Stellite 6 coating reached the lowest weight changes when exposed to corrosive environment, which is a sign of good resistance of this surface protection.
- This fact was further confirmed by microhardness $HV_{0.3}$ measurements. The coating did not show any significant change in microhardness values after the exposure to corrosive environment. The differences in resulting values are in allowable tolerance range.

Hastelloy C-276:

- Hastelloy C-276 coating is known for its corrosion resistance.
- The graph on **Fig. 2** shows that the coating reached low weight changes when exposed to corrosive environment, which is a sign of good resistance of this surface protection.
- This fact was further confirmed by microhardness $HV_{0.3}$ measurements. The coating did not show any significant change in microhardness values after the exposure to corrosive environment. The differences in resulting values are in allowable tolerance range.

Cr_3C_2-25%NiCr:

- Ni and Cr oxides were formed in the coating.
- The coating generally shows higher porosity than it is usual by alloy based coatings.
- The graph on **Fig. 2** shows big weight changes after the exposure to corrosive environment, but the coating withstood in this environment.
- This fact was further confirmed by microhardness $HV_{0.3}$ measurements. The coating did not show any significant change in microhardness values after the exposure to corrosive environment. The differences in resulting values are in allowable tolerance range.

NiCrBSi:

- This coating exhibited the lowest corrosion resistance out of all evaluated coatings.
- This fact was further confirmed by microhardness $HV_{0.3}$ measurements which revealed significant change after the exposure to corrosive environment.
- It may be concluded that this coating is not suitable for component protection in this type of corrosive environment.

Acknowledgement
The paper was prepared thanks to the project of Technology Agency of the Czech Republic no. TE01020068.

References

[1] Reid W. T., „External Corrosion and Deposits - Boilers and Gas Turbines, "New York, Elsevier, 1971, p. 115..

[2] Alexander, P.A., Marsden, R.A., London, Conference on Mechanism of Corrosion by Fuel Impurities, 1963, p. 542.

[3] Khanna, A.S., High Temperature oxidation and Corrosion, Delhi: ASM International, 2002.

[4] Sidhu, T.S., Prakash, S., Agrawal, R.D., A Comparative Study of Hot Corrosion Resistance of HVOF Sprayed NiCrBSi and Stellite-6 Coated Ni-Base Super alloy at 900°C, Proceedings of the Journal of Thermal Spray Technology, 2006, pp. 210-218.

[5] Tsui, Y. C., Clyne, T. W., R. C. Reed „On the Change in Stress State Associated with Bond Coat Oxidation during Heat Treatment of a Thermal Barrier Coating System," Thermal Spray 1997: A United Forum for Scientific and Technological Advances (ASM International), 1998, p. 267-276.

[6] Guilemany, J.M., Lorenzana, C., Miguel, J.M., Delgado, J., A Study of the Microstrukture, Wear and Corrosion Behaviour of High Velocity Oxygen Fuel (HVOF) Thermally Sprayed Coatings and Self-Standing Forms, Obtained with Blendsof Cr3C2-NiCr – NiCrBSi, Thermal Spray 2002: International Thermal Spray Conference (DVS-ASM), 2002, p. 884-889.

[7] Kuroda, S., Fukushima, T., Kawakita, J., Kodama, T., „Marine Corrosion Performance of HVOF Sprayed Alloy Coatings: Comparison of Field Exposure Test and Laboratory Test," Thermal Spray 2002: International Thermal Spray Conference (DVS-ASM), p. 819-824, 2002.

[8] Sidhu, T. S., Prakash, S., Agrawal, R. D., Charakterization and Hot Corrosion Resistence of Cr_3C_2-NiCr Coating on Ni-base Superalloys in an Aggresive Enviroment, Journal of Thermal Spray Technology, vol. 15, 2006.

[9] Chatha, S. S., Sidhu, H. S., Sidhu, B. S., High temperature hot corrosion behaviour of NiCr and Cr_3C_2-NiCr coatings on T91 boiler steel in an aggressive environment at 750 °C, Surface & Coatings Technology 206, 2012, p. 3839-3850.

[10] Tuominen, J., Honkanen, M., Uusitalo, M. Ahmaniemi, S., Vuoristo, P., Mantyla, T., Hot Corrosion resistance Laser Coatings in Diesel Engine, Proceedings of the Thermal Spray, 2007.

[11] Wang, Q.M., Wu, Y.U., Ke, P.L. Cao, H.T. Gong, J., Sun, C., Wen, L.S., „Hot Corrosion Behaviour of AIP NiCoCrAlY (SiB) Coatings on Nickel Base Superalloys," Surface nad Coating Technology, vol. 186 (3), 2004, pp. 389-397.

Effect of particle and carbide grain sizes on a HVOAF WC-Co-Cr coating for the future application on internal surfaces: microstructure and wear

J. Pulsford[1], S. Kamnis[2], J. Murray[1], M. Bai[1], T. Hussain[1]

[1]Department of Mechanical, Materials and Manufacturing Engineering, University of Nottingham, Nottingham, UK
[2]Castolin Eutectic -Monitor Coatings Ltd, North Shields, UK

The use of nanoscale WC grain or finer feedstock particles are two possible methods of improving the properties and performance of WC-Co-Cr coatings. Finer powders are being pursued for the development of coating internal surfaces, as less thermal energy is required to melt the finer powder compared to coarse powders, permitting spraying at smaller stand-off distances. Three WC-10Co-4Cr coatings, with two different particle sizes and two different carbide grain sizes, were sprayed using a high velocity oxy-air fuel (HVOAF) thermal spray system developed by Monitor Coatings Ltd. in the UK. The powder and coating microstructure were characterised using XRD and SEM. Fracture toughness and dry sliding wear performance were investigated using a ball-on-disc tribometer with a WC counter-body. It was found that the finer powder had a higher hardness but relatively lower fracture toughness. When performing sliding wear testing at the lower 96N load the nanostructured coating performed best; however at 240N this coating was displayed the highest specific wear rates, with the other two powders performing to a similar, better standard.

1 Introduction

In recent years, failure of critical components in high value manufacturing industries have been observed due to extreme sliding wear and one of the most extensively used types of wear resistant coatings are industrial hard chromium or electrolytic hard chrome (EHC) [1]. However, due to the recent introduction of restrictions regarding the use of hexavalent chromium, due to REACH legislations in the EU, the need for industry to replace this process has risen. High velocity oxy fuel (HVOF) thermal sprayed WC-Co-Cr cermet coatings are currently a promotion deposition technology that possibly can replace EHC coatings; however, HVOF processes are limited by their line of sight nature. Internal diameter (ID) thermal spraying is a modern technology operating at lower power levels but at the same time retaining high kinetic energy, allowing a greater range of internal surfaces to be effectively coated. ID thermal spraying is a field that has been gaining the attention of researchers, with Lyphout and Björklund successfully applying a WC-Co-Cr coating on an internal surface of a pipe with a 200 mm ID using a newly developed ID HVAF system produced by UniqueCoat Technologies, USA. The resulting coating had a hardness of 900 (HV0.3) and a porosity of approximately 1% [2].
In HVOF thermal spraying, the temperature of the in-flight particles is strongly dependant on the length of the barrel, as a longer barrel will result in the powder particles being exposed to the high temperature combustion product gases for a longer time [3]. Therefore, powders used in traditional HVOF systems may not work effectively in ID HVOF systems due to the shorter residence time in the barrel. Powders with a reduced particle size range can be used to mitigate the effect of the reduced barrel length, as the reduced size of the overall powder will result in the powders being more sensitive to temperature effects. Alongside these materials, some researchers have reported that using nanostructured WC-Co-Cr feedstock powders result in coatings with better performance in sliding wear resistance than their coarse carbide grain counterparts, with the reason behind this being that the carbide grain size is reduced for a given binder phase content, resulting in the mean free path of matrix being reduced [4]. However, both finer and nanostructured WC-Co-Cr powder feedstocks have been previously shown to be more vulnerable to decomposition of the WC phase, due to the finer particles being more susceptible to overheating due to the increased surface area to volume ratio of the nano sized WC grains in the nanostructured powders [5].
In this study, an HVOAF (High velocity oxy-air fuel) spray system developed by Monitor Coatings Ltd. (UK) was used to spray three separate WC-Co-Cr powders with the same composition but differing particle size ranges and carbide grain sizes. The spray parameters used in this study are from a first trial of the spray gun and as a result are unoptimised. The microstructures and the associated mechanical properties were studied. Unlubricated sliding wear testing was used to determine if changing the particle size range or carbide size has an impact on the coating's sliding wear resistance and as a means to test the performance of the three coatings.

2 Experimental

2.1 Materials and coating deposition

Three separate powders (H.C Starck Ltd., Germany) with the same composition (WC-10%Co-4%Cr) but differing particle size distributions and carbide grain sizes were sprayed (60mm SOD) onto 60 x 25 x 2 mm plates of mild carbon steel (0.05-0.25% C, < 0.4% Mn) for general characterisation and onto stainless steel 316 discs with a diameter of 38.1mm and thickness of 6 mm for wear testing.
A summary of the powders used in this study is shown in Table 1. All substrates were grit blasted and ultrasonically cleaned in ethanol prior to coating deposition.

Commercial Designation	Powder Code	Processing Method	Powder Size (µm)	Carbide Grain Size
AMPERIT® 557.059	MC-30	Agglomerated and sintered	-30 + 5	Medium (1-2 µm)
AMPERIT® 507.059	NC-30	Agglomerated and sintered	-30 + 5	Nanometric (200-400nm)
AMPERIT® 554.067	MC-15	Sintered and crushed	-15 + 5	Medium (1-2 µm)

Table 1 - Powder Specifications from manufacturer (H.C. Starck Ltd. Germany)

2.2. Characterisation methods

X-ray diffraction was performed on the feedstock powders and as sprayed coatings using a Siemens D500 with Cu Kα radiation (1.5406Å) in the $20° \leq 2\theta \leq 90°$ range to identify the phases present in the powders and coatings. A step size of 0.02° and dwell time of 3s was used. The coating cross sections were cut from the bulk samples using a diamond tipped precision cutting wheel. Powder and coating cross sections were obtained by mounting in a conductive mounting compound (Metprep Ltd, UK), ground with silica papers (down to 15.3 µm) then polished to a 1 µm diamond finish. The powder morphologies and coating microstructures were investigated using a scanning electron microscope (SEM, Jeol 6490LV) operated in high vacuum mode, utilising both secondary electron imaging (SEI) and backscattered electron imaging (BSE). The average size of the carbide grains of the starting powders was measured by using an image processing program (Image J) (NIH, USA), with the average of 10 measurements used.
The microhardness of the coatings was measured on polished cross sections by using a Vickers indenter (Buehler microhardness tester) at a load of 300 gf (≈3 N) and a dwell time of 10s; for each sample 6 impressions were made in the centre of the coatings running parallel with the substrate. Fracture toughness of the coatings was measured by indenting on polished cross sections with a 2.5 kgf (≈25 N) load and measuring the crack lengths parallel to the substrate using optical microscopy; cracks propagating from the left and right tips of the indent were considered (see Figure 1). The fracture toughness K_c was determined using the method described by Evans and Wilshaw [6], shown in Eq. 1 below:

$$K_c = 0.079\left(\frac{P}{a^{3/2}}\right)\log\left(\frac{4.5a}{c}\right) \qquad \text{Eq. 1}$$

Valid when $0.6 \leq c/a \leq 4.5$
Where P = applied indentation load (N)
a = indentation half diagonal (m)
c = crack length from indent centre (m)

2.3 Wear Testing

The sample surface was prepared for the wear testing by diamond grinding down to a grit size of 15.3 µm, followed by polishing down to a 1 µm finish. Unlubricated sliding wear testing at room temperature was undertaken using a ball-on-disc tribometer, with loads of 240 N and 96 N, a sliding distance of 500 m and a sliding speed of 0.5 m/s. The counterbody used was a sintered WC ball measuring 9.6 mm in diameter, with the ball being replaced for each test. These conditions were chosen in order to simulate a typical conditions for components in the aerospace industry under sliding wear. The friction coefficient was monitored during the test and the mass of the sample was measured before and after testing. The volume loss of each disc was determined using profilometry (Advanced Metrology System Ltd.); four line traces were made across the wear track and from each of these the cross sectional area of material loss could be determined. These were averaged and multiplied by the total track length to calculate the total volume of material lost. The specific wear rate could then be calculated by dividing the total volume of material lost by the product of the total distance slid and the load used. The morphology of the wear scar and fracture mechanisms were then examined using SEM (Jeol 6490LV), with optical microscopy (Nikon Eclipse LV100ND) used to image the WC counterbody.

3 Results

3.1 Microstructure of powders and coatings

Fig. 1 shows the morphology and cross-section of the three powder feedstocks. The MC 30 and NC 30 powders both exhibited a spherical shape with the internal structure being porous (Fig.1). The cross sectional image of the MC-30 reveals a structure with large interconnected pores, while the pores seen in the cross section of the NC-30 powder are much smaller in comparison. However in contrast the MC-15 powder had an angular blocky shape with very little porosity (Fig.1). These differences can be attributed to the different manufacturing methods used to produce the powders, with the MC-30 and NC-30 powders being produced by the agglomerated and sintering method and the MC-15 powder being sintered and crushed. The average carbide size was measured to be approximately 1.47µm, 1.26µm and 0.32µm for the MC-30, MC-15 and NC-30 powders respectively. Figure 2 shows low magnification images of the coatings in the top row, the second row shows BSE images of the coating cross sections, third row shows the SE images taken from the same site and finally he bottom row displays the XRD results of the coatings. Low magnification SEM images (Fig. 2) show that each coating has a macroscopically dense structure. Coating thickness varied between the three coatings,

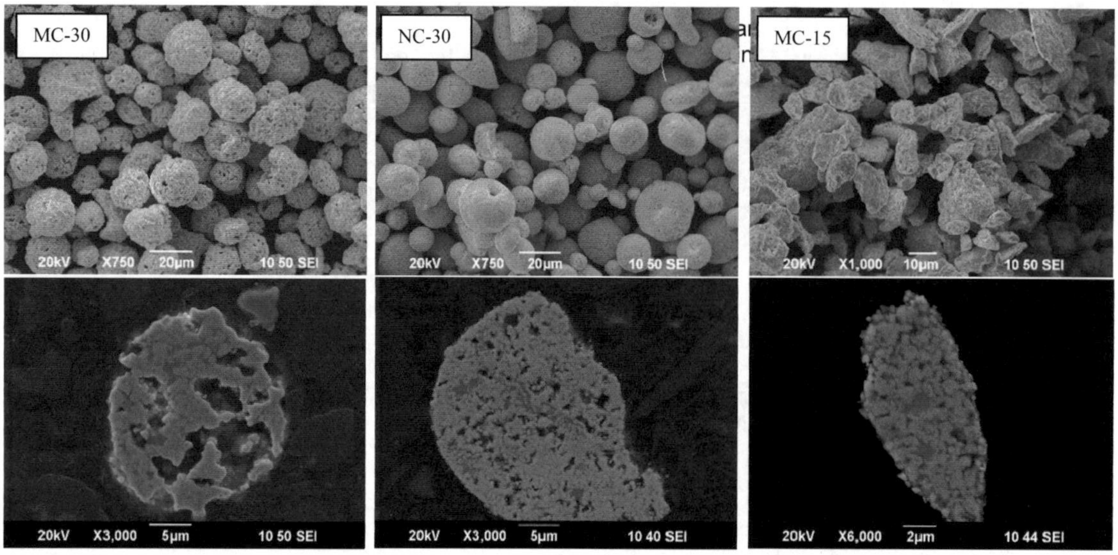

Figure 1 – Top row: SEM images showing powder morphology. Bottom row: SEM images of powder cross section

with the thickness of the MC-30, NC-30 and MC-15 coatings being approximately 350 µm, 480 µm and 200 µm respectively. The MC-30 and NC-30 coatings both appear well bonded to the substrate, however the MC-15 coating shows evidence of particles from grit blasting trapped inside the coating-substrate interface. Higher magnification images in Fig 2 (2nd-3rd rows) show a small amount of porosity, with the microstructure of the coatings consisting of blocky grains of WC with rounded edges and the metallic binder, which exhibits three distinct levels of contrast under BSE (Fig. 2). The brightest areas are a result of WC dissolving into the molten binder during spraying, areas which appear grey are Co rich and the areas that appear black are not porosity (confirmed by the SEI images in Fig. 2c, g and k) but are Cr rich regions, as also observed by other researchers; an example region is marked on Figure 2 [7]. The XRD analysis (Fig. 2) reveals that the 3 powders consist of the primary WC phase, metallic Co f.c.c and Co_3W_3C, although the peak intensity of the latter two phases is lower in the NC-30 powder. The phase composition changed during spraying, as the Co f.c.c and Co_3W_3C phases were no longer detectable in the coatings; all 3 coatings showed that some W_2C had formed and this was the only other phase detectable aside from WC. The W_2C peak intensities were larger in relation to the WC peaks in the MC-15 coating. The presence of this phase can be seen in the BSE SEM images of the polished coating cross sections in Fig. 2; some of the blocky WC grains are surrounded by a phase that appears brighter on the BSE images, indicating some W_2C formation; these regions are circled on Figure 2 for clarity. In the NC-30 coating, some carbide grains appear to have significantly decarburised; these grains appear as small spots of the brightest contrast

The normalised microhardness of the MC-30, NC-30 and MC-15 coatings are 0.86, 0.81 and 1 respectively, with the fracture toughness being measured at 5.59 ± 0.31, 5.15 ± 0.12 and 4.89 ± 0.12 MPa.m$^{0.5}$. The finer powder feedstock produced a coating with the highest hardness but the lowest fracture toughness while reducing the carbide grain size resulted in the coating with the lowest microhardness.

3.2 Sliding wear behaviour

The specific wear rates of the coatings and their respective counter bodies tested under the two applied loads are displayed in Fig. 4. At 96N the MC-30 and MC-15 coatings perform similarly, with the coating specific wear rates being higher than the counterbody specific wear rate. However at this load the NC-30 showed negligible wear, with the counterbody wear not being measurable. At 240N the specific wear rates of the MC-30 and MC-15 are again very similar, except the counterbody specific wear rate is noticeably higher for the MC-15 coating. The specific wear rate of the NC-30 is significantly higher at 240N, indicating a different mechanism of wear may be taking place. To attempt to gather a greater understanding of the wear mechanisms taking place the worn surfaces were imaged using SEM imaging. In the MC-30 and MC-15 coatings at both loads extensive cracking can be seen in areas with little to no WC coverage, indicating that these areas are paths for crack propagation. A tribofilm may be present in both MC-30 and MC-15 coatings, although further investigation is required to confirm this. At both loads carbide grain pullout can be observed, with this effect being greater at 240N than 96N. No cracking of the carbide grains themselves was observed at either load.

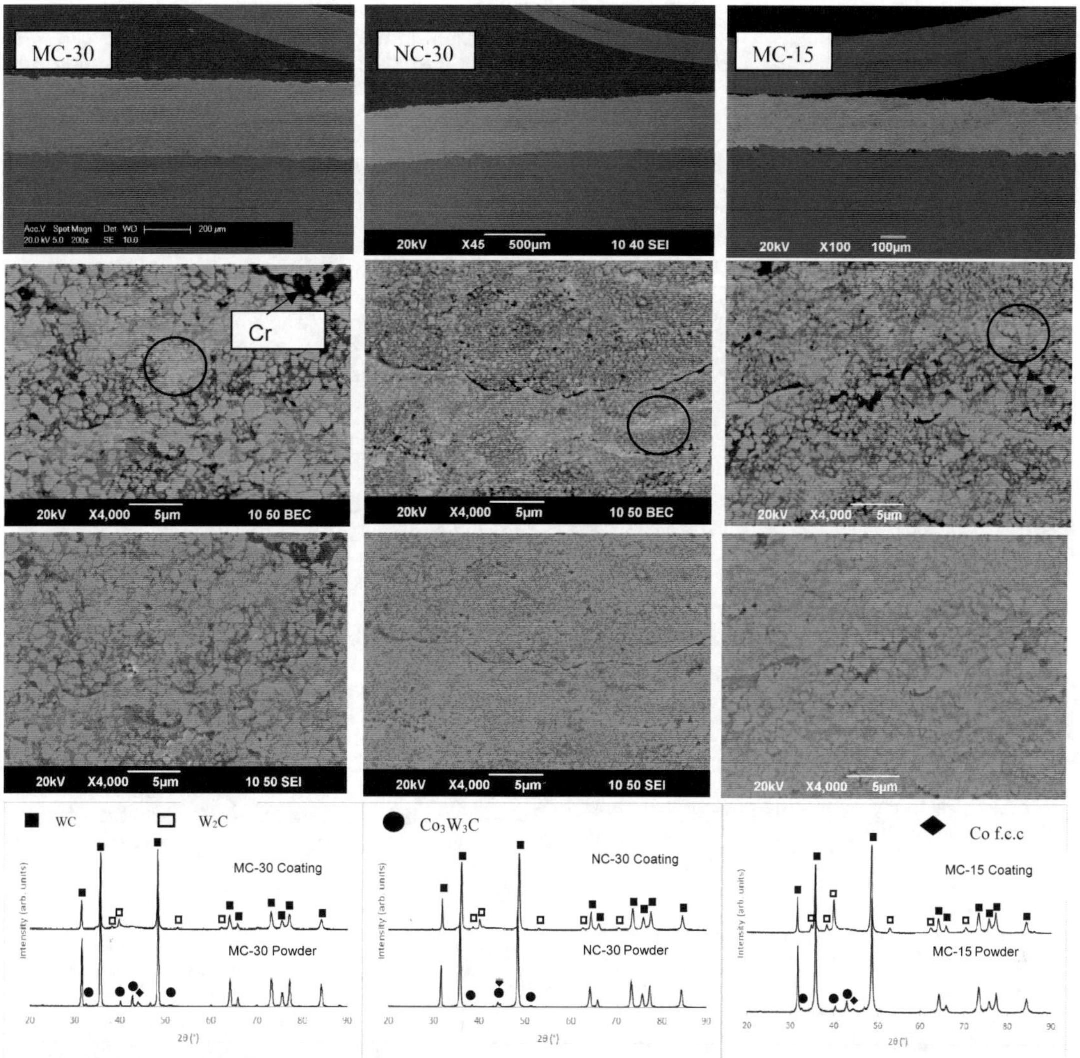

Figure 2 – Top row: Low magnification SE SEM image of coating, 2nd row: BSE image of coating cross section, 3rd row: SE image of coating cross section and bottom row: XRD spectra of powder and coating

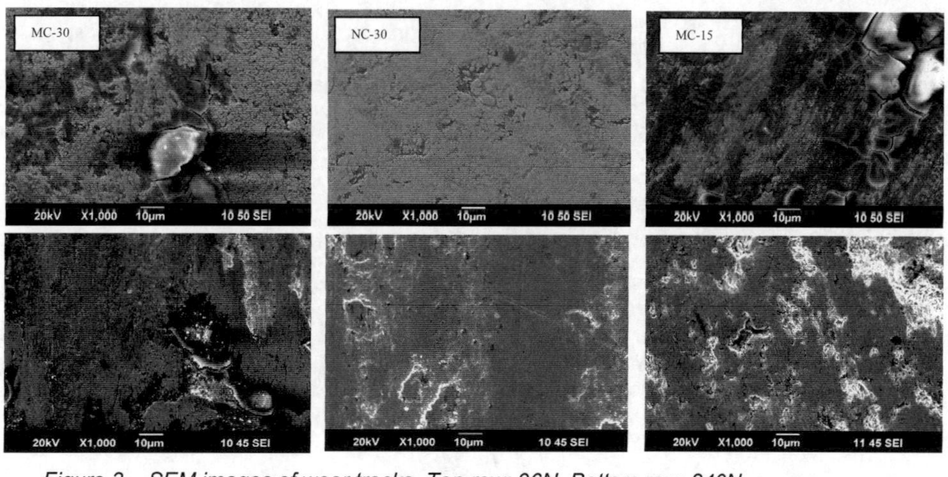

Figure 3 – SEM images of wear tracks, Top row: 96N. Bottom row: 240N

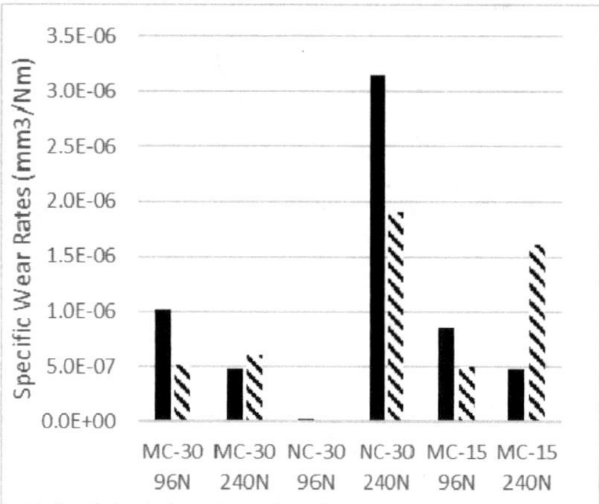

Figure 4 - Specific wear rates of coatings (solid black) and sintered WC counterbody (striped bars) during sliding wear testing

4 Discussion

The XRD spectra in Fig. 2 indicate that WC is main ceramic peak and small amount of W_2C has formed in all three coatings. However a different ratio of WC to W_2C peak intensity is observed for the MC-15 coating, indicating that a greater amount of W_2C is present in relation to WC. It has been reported that reducing the carbide grain size to the nanoscale greatly increased the amount of W_2C in the coating [4] [7]; however, the results of this study show that when spraying using these parameters the peak intensity ratio between the WC and W_2C peaks in the MC-30 and NC-30 coatings is very similar, indicating that reducing the carbide size had little effect on W_2C formation in this case. These results can be explained by the fact that this effect is more highly dependent on the particle size range rather than the carbide size, as the finer cut powder will reach a greater temperature during spraying.

Previous researchers have stated that nanostructured WC-Co-Cr powder feedstocks result in coatings with a much higher hardness; their reasoning being that reducing the carbide size results in increased hardness and fracture toughness due to the Hall-Petch effect [7]. This is because decreasing the carbide grain size and binder mean free path results in increased suppression of dislocation propagation and this jamming of slip therefore means fracture toughness is increased [5]. However, the hardness of the nanostructured NC-30 coating measured in this study contradicts these reports as this coating was measured to have the lowest microhardness. The fracture toughness measurements showed that the MC-30 coating had the highest fracture toughness, with the MC-15 coating having the lowest fracture toughness and the highest hardness; this is possibly because of the higher amount of W_2C, as this phase can increase hardness but reduce fracture toughness [9].

The MC-30 and MC-15 coatings show similar patterns in regards to the sliding wear, with the specific wear rate of the coatings reducing and counterbody wear rates increasing when the load is increased from 96N to 240N. These results may indicate that a similar mechanism of wear is taking place for both coatings at the low and high loads and that the increased load has a greater effect on the counterbody specific wear rate than the coating specific wear rate for these 2 coatings. SEM images of the coating cross sections reveal that the microstructure of the MC-30 and MC-15 is also very similar, indicating that they may perform in a similar way. Areas lacking WC hard phase coverage appear to be sites of crack propagation, possibly due to the reduced hardness of the binder phase.

The NC-30 coating appears to behave in a much different way under wear testing, with negligible wear detected at 96N for the coating and counter-body and the highest recorded specific wear rates for the coating and counter-body at 240N. In Figure 3 it can be observed that multiple circular areas are cracking, with the pullout of WC grains occurring simultaneously.

The results of the wear testing appear to show that the NC-30 coating performed best at 96N and the MC-30 and MC-15 coatings offered very similar performance at 240N. Figure 3c shows a coating surface with good uniform WC coverage meaning less of the soft binder phase is exposed; as the other two coatings show clear evidence of cracking in these areas, the better coverage of WC could be a factor in the better performance at 96N. However the overall lower hardness costs the NC-30 coating at the 240N load, negating this effect. At the higher load the two harder coatings MC-30 and MC-15 perform better than the nanostructured coating; the higher hardness of the coatings resulted in a much more significant increase in the counterbody specific wear rate, rather than the coating specific wear rate. In this case, although the finer powder coating decarburised more during spraying, the effects of this may be beneficial to coating performance, as the higher hardness may outweigh the lower fracture toughness. Previous work has also argued that a certain degree of decarburisation in WC cermet coatings may be beneficial as too little decarburisation leads to a large amount of material being removed from brittle island boundaries, while too much decarburisation results in the composition of WC falling to a level at which the soft binder phase is too exposed [10].

For this particular torch the results of this study show that the MC-15 powder is a promising candidate for spraying complex geometries internally, as the wear testing showed it compared well with the conventional MC-30 powder. Although the nanostructured NC-30 powder proved itself at the lower load, its poor wear performance at higher loads show that while this material may offer good performance at low loads, applications that require wear resistance at higher loads may not be suitable for this material.

It must be noted that further work is required in order to validate the results of this study; sliding wear testing at a number of other loads and more SEM investigations on the wear track cross sections would uncover more information to help improve the understanding of the tribological mechanisms taking place.

5 Conclusion

Three WC-10%Co-4%Cr based powders with varying particle size range and carbide grain sizes were deposited using an HVOAF system developed by Castolin Eutectic-Monitor Coatings Ltd. (UK). The microstructure of the powders and coatings were examined by SEM and XRD, followed by performance testing of the coatings by unlubricated sliding wear testing. From the results of this study the following conclusions can be made:

- Characterisation of the powder and coating microstructures using SEM and XRD revealed that the MC-15 powder decarburised during spraying to a greater extent compared to the MC-30 and NC-30 powders, possibly due to the finer powder reaching a higher temperature during spraying.
- The higher particle kinetic energy for the MC-15 powder resulted in it having a higher hardness but reduced fracture toughness compared to the other coatings tested. The reduction in fracture toughness may also be attributed to the relatively higher decarburisation compared to MC-30 and NC-30.
- In the sliding wear testing at 96N the nanostructured coating performed best, making it the most suitable for low load applications. However at 240N this coating displayed the highest specific wear rates; with the MC-30 and MC-15 coatings both outperforming it by a similar degree; therefore the medium carbide coatings should be chosen for higher load applications based on this data.

6 Literature

[1] M. Bielewski, Replacing Cadmium and Chromium, the Research and Technology Organisation and NATO, AG-AVT-140 (2011), Chp. 23, p 1/22

[2] Lyphout, C. and S. Björklund. "Internal Diameter HVAF Spraying For Wear And Corrosion Applications". *Journal of Thermal Spray Technology* (2014)

[3] Hackett, C. (1996). *The Gas Dynamics of High-Velocity-Oxy-Fuel Thermal Sprays* (PhD Thesis). Dept. of Mechanical and Nuclear Engineering, Pennsylvania State University.

[4] Stewart, D., Shipway, P., & McCartney, D. (1999). Abrasive wear behaviour of conventional and nanocomposite HVOF-sprayed WC–Co coatings. *Wear*, *225-229*, 789-798.

[5] K. Jia and T.E. Fischer, Sliding Wear of Conventional and Nanostructured Cemented Carbides, Wear, 1997, 203–204, p 310–318

[6] Evans, A. G., and To R. Wilshaw. "Quasi-static solid particle damage in brittle solids—I. Observations analysis and implications." *Acta Metallurgica* 24.10 (1976): 939-956.

[7] Thakur, L. & Arora, N. (2012). Sliding and Abrasive Wear Behavior of WC-CoCr Coatings with Different Carbide Sizes. *J. Of Materi Eng And Perform*, *22*(2), 574-583.

[8] Kear, B., Skandan, G., & Sadangi, R. (2001). Factors controlling decarburization in HVOF sprayed nano-WC/Co hardcoatings. *Scripta Materialia*, *44*(8-9), 1703-1707.

[9] Sudaprasert, T. (2002*). An Investigation of Microstructure and Sliding Wear in Thermal Sprayed WC-Co Coatings*, Ph.D thesis, The University of Nottingham, UK

[10] Verdon, C., Karimi, A., & Martin, J. (1997). Microstructural and analytical study of thermally sprayed WC-Co coatings in connection with their wear resistance. *Materials Science And Engineering*: A, 234-236, 731-734

The influence of process parameters on structure and functional properties of laser clad WC-NiCrBSi coating

Vostřák Marek, Houdková Šárka, Zdeněk Česánek, University of West Bohemia, Univerzitní 8, Plzeň, 306 14, Czech Republic, mvostrak@ntc.zcu.zc, tel: +420377634827

In this study, the influence of main laser cladding parameters on the final microstructure of cladded WC-NiCrBSi coating is discussed. As the used powder is the blend of 60% tungsten carbide spherical particles and 40 % NiCrBSi matrix, the carbides' morphology and distribution in the final coating are dependent on laser cladding parameters. The variability of coating's microstructure strongly influenced their wear resistance. In the paper, the influence of laser processing parameters on final coating microstructure is studied. The functional properties of cladded coatings are evaluated by dry sand rubber wheel test ASTM G-65. The wear resistance of the coating is dependent on the amount of undissolved carbides particle in the coating.

1 Introduction

Laser cladding is a technology capable of producing compact functional coatings from a variety of materials. For example, it is widely used for the production of wear resistant layer. In the laser cladding, the functional properties and structure of final coating are strongly dependent on the right selection of process parameters [1].
The metal matrix composites (MMCs) are materials which excel in their high wear resistance due to the tough matrix and hard reinforcing particles [2]. Among them, the combination of Ni-based alloy and tungsten carbide is widely used.
In the case of laser cladding of such complex material, the process parameters will play a crucial role in the final appearance of carbides in the coating and thus on its final functional properties.
In various studies, the role powder composition – the percentage and the shape of the carbides on final coating hardness and wear resistance was tested [3] [4] [5]. The influence of process parameters on the microstructure of WC-reinforced Ni-based matrix single clads was studied previously [6] [7] [8].
In the work of Tobar et al [6], the processing window, enabling to reach the most appropriate carbide dissolution level of WC-NiCrBSi was described. The high laser energy density produced a dissolution of WC particles in the matrix and high level of dilution. During solidification, the mixed carbides were produced, resulting in carbides blocks distributed in the coating. In the case of low laser energy density, the boundary of WC particles was not melted, resulted in poor bonding between particles and matrix.
The studies focusing on the single laser clads cannot fully describe the microstructure and properties of the coating, where the overlapping plays its role in further heat input resulting in microstructure changes.
In the paper, the processability, microstructure, microhardness and abrasive wear properties of WC-40%NiCrBSi coatings deposited by 5kW solid state disc laser Trumpf TruDisk is studied in dependence on laser processing parameters.

2 Experimental procedure

2.1 Feedstock powder

The commercially available feedstock powder from Oerlicon Metco company designated as MetcoClad 52052 was used for cladding. The powder is composed of 60% of WC spherical particles blended with 40% of gas atomized NiCrBSi (8.0Cr3.5Si1.6B0.3C Ni bal.) particles. The particle size distribution lays between 45-106 µm. The WC was manufactured in a unique way, resulting in non-acicular shape with higher hardness than conventional fused and crushed WC [9].

2.2 Laser cladding process and parameters

Laser cladding was realized by the Solid state laser Trumpf TruDisk 8002 with Precitec coaxial 4-way cladding head YC52. The experiment was designed to evaluate the influence of laser deposition parameters on the coatings microstructure and related wear resistance.
Two series of the experiment were performed. First, the process speed S was changing, while powder feed rate F (g/min) remains constant. Therefore the amount of powder per line - F/S [g/m] was decreasing with increasing speed S. As a result, the thickness of the coating differs for different speed (the thickest coating for the slowest speed S). Power P was also constant and thus the Fluence [J/mm^2] was decreasing with increasing speed S. In the second series, the amount of powder per line – F/S [g/m] remain constant, as well as the Fluence [J/mm^2]. The power was changing accordingly. In this case, it was intended that the thickness of the cladded coating will remain same for all process speed. The P/F [J/g] ratio – the energy per gram of the powder is nearly constant in both series. The process parameters are summarized in **Table 1**.
Sheets of carbon steel (EN10083-2: C45) with dimensions 100 × 100 × 20 mm were used as a substrate to ensure sufficient heat dissipation. A pre-heating to 350 °C was used to avoid cracking of the coating due to fast cooling.

Table 1. Laser processing parameters

No.	Power P [W]	Process speed S [mm/s]	Feed rate F [g/min]	F/S [g/m]	Fluence H [J/mm²]
1_1	3000	2.5	28.5	19	333
1_2		5.0		9.5	200
1_3		8.3		5.7	150
1_4		12.5		3.8	109
1_5		16.7		2.85	90
2_1	900	2.5	8.42	5.6	150
2_2	1800	5.0	16.8		
2_3	3000	8.3	28.5		
2_4	4500	12.5	42.57		

2.3 Characterization

The microstructure of the coatings was evaluated on the cross sections (grinded and polished by an automatic Leco grinding and polishing equipment) by optical microscope Nikon Epiphot 200, digital optical 3D microscope Hirox KH7700, and SEM Quanta 200 from FEI equipped by EDAX NEWXL-30 Silicon doped by Lithium detector. The thickness of the coating was measured on the cross section in the areas of individual tracks overlaps (the minimal thickness). It was averaged at least for 10 measurements.
The microhardness HV1 was evaluated on the coatings cross section. Due to the non-homogenous coatings microstructure, the microhardness was measured in the sub-surface region, relevant for wear response. Six measurements were done in similar distance from the surface, the average values are reported.
The abrasive wear resistance was evaluated by Dry Sand Rubber Wheel Test according to ASTM G-65. Test parameters were as follows: 22 N Load; Al_2O_3 abrasive media; AG 29 rubber counterpart; 718 m total abrasive distance. The samples were weighted after each 143 m and the cumulative mass loss [g] was calculated. The wear rate [g/m] was determined from two independent measurements. The surface of the coating was left in the as-received state, not to exclude the upper coatings layers from evaluation.

3 Results and discussion

3.1 Microstructure

As it can be seen in **Fig. 1**, the thickness of the cladded coating differs significantly for the series 1 (due to different F/S ratio) and even for the series 2, where the F/S ratio was constant. Even if the calculated ratio P/F (energy per gram of the powder) is constant, the efficiency of the powder melting is dependent on laser power [1]. Thus in the series 2, the thickness of the coating is increasing with increasing laser power, as more of the powder is utilized. It is in agreement with Bartkowski [10] who shows that with increasing laser power, the thickness of the MMC coatings is increasing, but the wear resistance is decreasing.
The cross sections of cladded coatings are presented in **Fig. 2.** The OM images are arranged as in the coating, the top layer is on the top line, the middle section of the coating in middle line and the boundary are in the bottom line (In the case of thinner coating, only two images are sufficient to display whole thickness of the coating). In the coatings cladded by the slowest process speed, there is a high percentage of not dissolved WC spherical particles and with increasing speed, their amount is decreasing. It is with contradiction with our assumption. Such observation is in contradiction with the assumption, that the slow process speed, related to high heat input, will lead to higher carbide dissolution, as it was reported in the literature [6] [7] [11]. Similarly, in the cladding of stainless steel, higher solidification rate at higher cladding speeds decreases the cell spacing. Similarly, as in cladding of stainless steel, higher solidification rate at higher cladding speeds decreases the cell spacing [12].

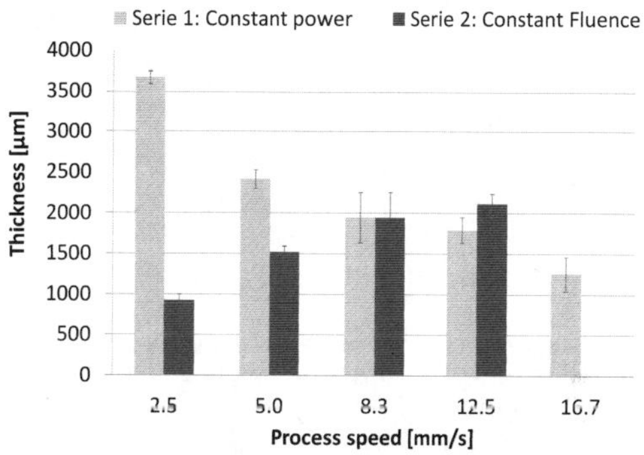

Fig. 1. The measured thickness of the final coatings.

In this study, the complex relation between laser power, scanning velocity and powder feed rate was found regarding the carbide dissolution.
In the first series, where the laser power is constant, the decisive parameter, related to carbide particles dissolution, is the amount of powder per line (F/S).
In the second series, the amount of dissolved carbides can be connected with laser power. With increasing power, the efficiency of powder melting is growing and more carbide particles are dissolved.
It should be noted that all of the prepared coatings contain some level of pores. By variating of process parameters, we were not able to fully eliminate them. The reason of the pores origins can be in the initial powder composition (it contains 60% WC), it was presented, that higher amount than 45% of tungsten carbide can lead to the formation of pores and cracks [6] [13]. The decomposition of WC, producing the free carbon, can contribute to the rise of porosity in the coating [8].

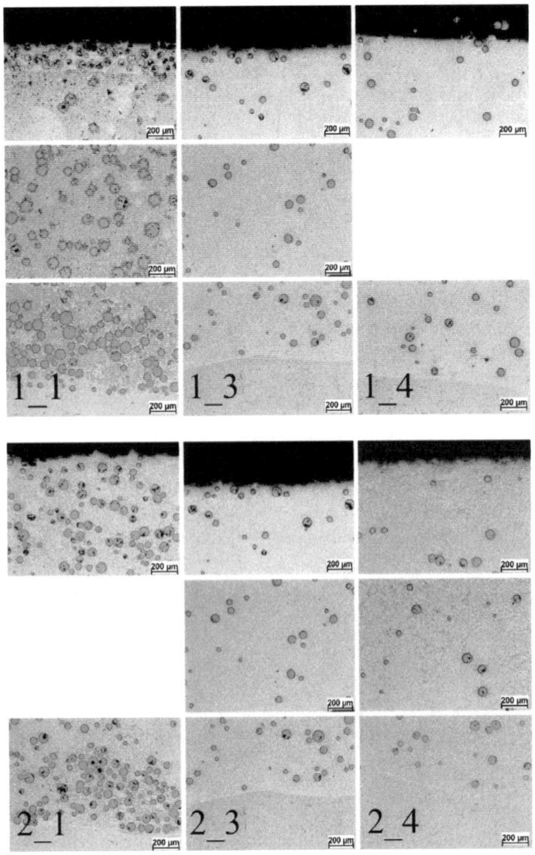

Fig. 2. OM of selected coatings.

The microstructure of the coatings is highly non-homogenous (**Fig. 3**). Under the coating surface, the high amount of WC spherical particles is preserved (1). In the coating, the majority of the carbide particles were dissolved during cladding and solidify again in different shapes and sizes, in dependence on the location. Both needle-like and blocks-shape WC/W2C carbides morphology was observed, similarly to the shapes reported in [11].
In the literature, the distribution of carbides in the laser bead is often reported. Most often, their sinking-in the bottom of the melt pool due to their higher density was observed [4] [14] [15] [16].
In [7] the primary spherical particles was preserved on the single clads borders with surroundings due to the fast cooling by convection. Also in [17], the higher concentration of carbides was recorded in the upper part of the coating, related to rapid solidification of the melt.
In the clad coatings microstructure, the significant influence of overlapping was identified: the subsequent melting of previous laser bead leads to a sharp needle-like carbide morphology (2). In the coatings cross sections, the regions with the different chemical composition can be also observed (3).
EDS analyses showed a high amount of Fe in this regions. During melting, the molten substrate material penetrates into the material of the coating not only through homogenous dilution but also in form of bounded regions. Such regions can be caused by circulation of the molten material inside the melt pool as a result of Marangoni convection force [18] [19]. The overlaying results also in the non-homogenous distribution of spherical particles in the subsurface area. In the overlapping region, the particles are dissolved and the carbides solidify again in smaller precipitates.

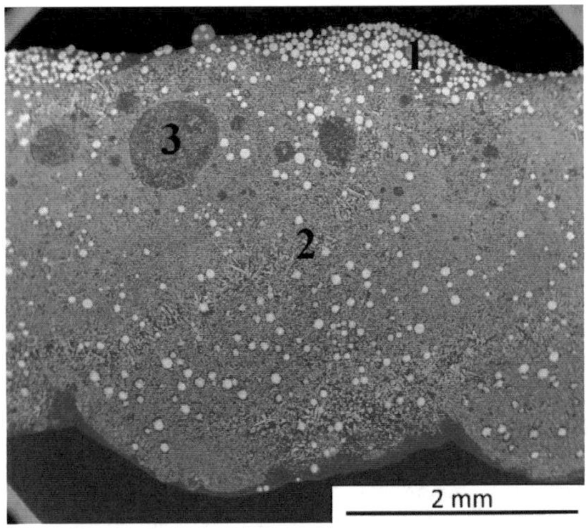

Fig. 3. BSE of coating, deposited by parameters No.1_1

3.2 Hardness and wear resistance

The results of dry sand rubber wheel test (**Fig. 4**) showed a strong dependence on the laser clad microstructure. The higher the amount of preserved spherical WC particles in the subsurface region, the higher was the abrasive wear resistance. Most pronounced it can be seen for the coatings with highest F/S, but a clear dependence was found also for the coatings with constant F/S, but variable laser power (**Fig. 4b**).
The higher wear resistance is related with the microhardness HV1, measured in the subsurface region (**Fig. 5**). The high content of WC spherical particles in the subsurface region caused the higher measured microhardness. The observation is valid for both series (variable F/S and constant F/S resp.). In agreement with the abrasive wear test, the coatings deposited by parameters 1_3 - 1_5 does not show a significant difference in subsurface microhardness. Their microstructure is more similar, both the level of carbide dissolution as well as the amount of Fe dilution, measured by EDS.
The negative effect of high level of carbide dissolution on the abrasive wear resistance was also previously mentioned by [17].
As the coatings microstructure is not homogenous and the distribution of spherical particles in the subsurface region follows the direction of cladding

with stripes of needle-like carbide precipitates, the wear of the coating was not homogenous either. In the **Fig. 6**, the wear scar on the surface of the 1_1 coating can be seen. In **Fig. 5b**, the detail made by BSE confirmed the non-homogenous surface structure and presence of stripes with a higher amount of spherical WC particles.

The negative effect of overlapping during the laser hardening is well known [20] [21]. Similar for laser cladding, the overlapping modified the microstructure of MMC laser clads in a negative way. Such effect was not observed for Co-based alloy laser clads [22] and was not found in the literature considering MMC laser cladding either.

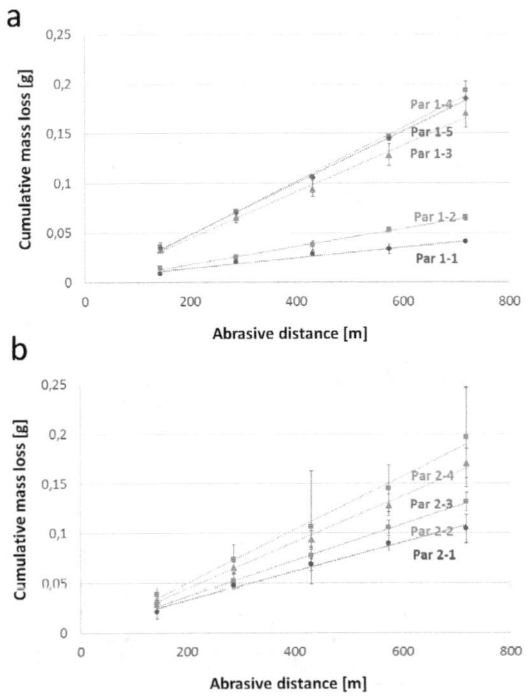

Fig. 4. Abrasive wear in dependence on laser processing parameters

4 Conclusion

In the paper, the commercially available powder with a high amount of WC spherical particles (WC-40%NiCrBSi) was deposited onto the steel substrates by Solid state laser, varying the laser process parameters. The used laser parameters had an influence on coating microstructure and related wear properties:

- No single laser parameter, responsible for the coating microstructure was found. In the case of constant laser power, the amount of powder per line (F/S) was decisive. In the case of constant F/S, the laser power is important for the effectivity of powder melting.
- The level of WC decomposition depends on the laser parameters. The lowest WC carbide particle decomposition was found for the highest F/S values or lowest P.
- The non-homogenous microstructure is caused not only by the different thermal gradient during solidification, but also due to the strong influence of overlapping. The morphology of secondary carbides was further changed by the remelting.

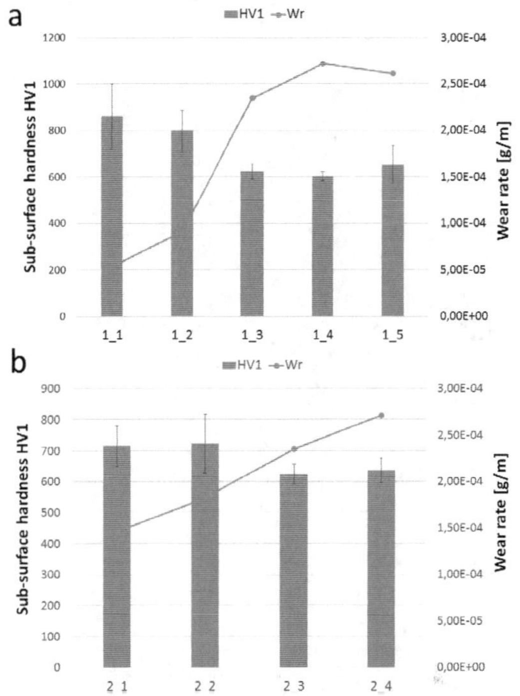

Fig. 5. Sub-surface microhardness HV1 in dependence on the laser processing parameters

Acknowledgements

The research outcome was developed within the CENTEM project, reg. no. CZ.1.05/2.1.00/03.0088, co-funded by the ERDF as part of the Ministry of Education, Youth and Sports OP RDI programme and, in the follow-up sustainability stage, supported through CENTEM PLUS (LO1402) by financial resources from the Ministry of Education, Youth and Sports under the "National Sustainability Programme I" and project SGS-2016-005.

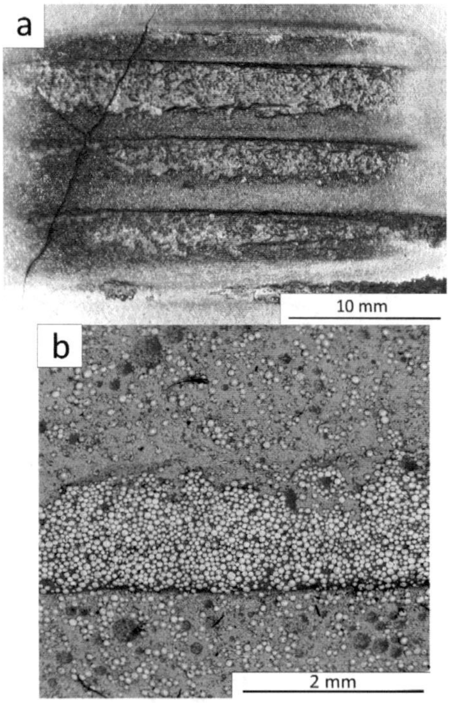

Fig. 6. Wear scar in the coating, deposited by parameters No. 1_1

Literature

[1] de Oliveira, U., V. Ocelík and J.T.M. De Hosson:, Analysis of coaxial laser cladding processing conditions. Surface and Coatings Technology 197 (2005) Issue 2–3, pp.127–36

[2] Nurminen, J., J. Näkki and P. Vuoristo:, Microstructure and properties of hard and wear resistant MMC coatings deposited by laser cladding. International Journal of Refractory Metals and Hard Materials 27 (2009) Issue 2, pp.472–8

[3] Luo, X., J. Li and G.J. Li:, Effect of NiCrBSi content on microstructural evolution, cracking susceptibility and wear behaviors of laser cladding WC/Ni-NiCrBSi composite coatings. Journal of Alloys and Compounds 626 (2015) pp.102–11

[4] Fernández, M.R., A. García, J.M. Cuetos, et al.:, Effect of actual WC content on the reciprocating wear of a laser cladding NiCrBSi alloy reinforced with WC. Wear 324–325 (2015) pp.80–9

[5] Deschuyteneer, D., F. Petit, M. Gonon and F. Cambier:, Processing and characterization of laser clad NiCrBSi/WC composite coatings - Influence of microstructure on hardness and wear. Surface and Coatings Technology 283 (2015) pp.162–71

[6] Tobar, M.J., C. Álvarez, J.M. Amado, et al.:, Morphology and characterization of laser clad composite NiCrBSi-WC coatings on stainless steel. Surface and Coatings Technology 200 (2006) Issue 22–23 SPEC. ISS., pp.6313–7

[7] Amado, J.M., M.J. Tobar, J.C. Alvarez, et al.:, Laser cladding of tungsten carbides (Spherotene®) hardfacing alloys for the mining and mineral industry. Applied Surface Science 255 (2009) Issue 10, pp.5553–6

[8] Amado, J.M., M.J. Tobar, A. Yáñez, et al.:, Crack Free Tungsten Carbide Reinforced Ni (Cr) Layers obtained by Laser Cladding. Physics Procedia 12 (2011) pp.338–44

[9] Oerlikon Metco:, Material Product Data Sheet Tungsten Carbide – Nickel Alloy Powder Blend for Laser Cladding, 2014

[10] Bartkowski, D. and G. Kinal:, Microstructure and wear resistance of Stellite-6/WC MMC coatings produced by laser cladding using Yb:YAG disk laser. International Journal of Refractory Metals and Hard Materials 58 (2016) pp.157–64

[11] Vespa, P., P.T. Pinard, R. Gauvin and M. Brochu:, Analysis of WC / Ni-Based Coatings Deposited by Controlled Short-Circuit MIG Welding. Journal of Materials Engineering and Performance 21 (2012) Issue 6, pp.865–76

[12] Hemmati, I., V. Ocelík and J.T.M. De Hosson:, The effect of cladding speed on phase constitution and properties of AISI 431 stainless steel laser deposited coatings. Surface and Coatings Technology 205 (2011) Issue 21, pp.5235–9

[13] Van Acker, K., D. Vanhoyweghen, R. Persoons and J. Vangrunderbeek:, Influence of tungsten carbide particle size and distribution on the wear resistance of laser clad WC/Ni coatings. Wear 258 (2005) Issue 1–4 SPEC. ISS., pp.194–202

[14] Garía, A., M.R. Fernández, J.M. Cuetos, et al.:, Study of the Sliding Wear and Friction Behavior of WC + NiCrBSi Laser Cladding Coatings as a Function of Actual Concentration of WC Reinforcement Particles in Ball-on-Disk Test. Tribology Letters 63:41 (2016)

[15] Guo, C., J. Zhou, J. Chen, et al.:, High temperature wear resistance of laser cladding NiCrBSi and NiCrBSi/WC-Ni composite coatings. Wear 270 (2011) Issue 7–8, pp.492–8

[16] Zhou, S., X. Zeng, Q. Hu and Y. Huang:, Analysis of crack behavior for Ni-based WC composite coatings by laser cladding and crack-free realization. Applied Surface Science 255 (2008) Issue 5, pp.1646–53

[17] Huang, S.W., M. Samandi and M. Brandt:, Abrasive wear performance and microstructure of laser clad WC/Ni layers. Wear 256 (2004) Issue 11–12, pp.1095–105

[18] Goodarzi, D.M., J. Pekkarinen and A. Salminen:, Effect of process parameters in

laser cladding on substrate melted areas and the substrate melted shape. Journal of Laser Applications 27 (2015) Issue 52,

[19] Limmaneevichitr, C. and S. Kou:, Experiments to Simulate Effect of Marangoni Convection on Weld Pool Shape. Welding Research Supplement 79 (2000) Issue 5, pp.126–35

[20] Cordovilla, F., Á. García-Beltrán, P. Sancho, et al.:, Numerical/experimental analysis of the laser surface hardening with overlapped tracks to design the configuration of the process for Cr-Mo steels. Materials & Design 102 (2016) pp.225–37

[21] Hruška, M., M. Vostřák, E. Smazalová and M. Švantner:, Standard and scanning laser hardening procedure, in: Met. 2013 - 22nd Int. Conf. Metall. Mater. Conf. Proc., TANGER LTD, 2013: pp. 1009–14

[22] Houdková, Š., Z. Pala, E. Smazalová, et al.:, Microstructure and sliding wear properties of HVOF sprayed, laser remelted and laser clad Stellite 6 coatings. Surface and Coatings Technology (2016)

Transient coating build-up and thermal analysis of cold spray process by finite element modelling

C. CHEN, Y. XIE, C. VERDY, H. LIAO, S. DENG*, IRTES – LERMPS, ICB UMR 6303, CNRS, Univ. Bourgogne Franche-Comté, UTBM, F-90010 Belfort, France
Z. REN, Shanghai University & State Key Laboratory of Advanced Special Steel, 149 Yanchang Road, Shanghai 200072, P.R. China

Since cold spray is widely considered as an additive manufacturing and damage repair technology, it is crucial to understand the coating build-up process and the temperature evolution. In this work, a 3D numerical model was developed to simulate the transient coating build-up process as well as the heat transfer in cold spray. By coupling the heat transfer with the ALE (Arbitrary Lagrangian–Eulerian) moving mesh and coating thickness model, this 3D model is able to investigate the temperature evolution of a coating which simultaneously grows according to the nozzle trajectory. The nozzle trajectory that represents the heat source and mass flux of particle impact is generated and simulated in the offline programming software RobotStudio™. By assigning the results of coating thickness distribution, the simultaneous build-up of coating computational domain is achieved by ALE moving mesh method. The validation of the FEA (finite element analysis) model was carried out by measuring the coating surface temperature via an infrared imaging camera. With the proposed model, it is able to study the actual coating build-up process as well as the heat transfer phenomena, which may provide more insights for the application in additive manufacturing and damage repair.

1 Introduction

Cold spray that is also called cold gas dynamic spray (CGDS) or kinematic spray (KS) has drawn more and more attention due to its unique 'cold' characteristics [1-3]. In this process, the particles are accelerated to a high velocity (300 m/s - 1200 m/s) by a supersonic protective nitrogen, helium gas or air at a relatively low temperature, and deposited onto the substrate. With its feature of low temperature and high impacting velocity, a dense and thick coating with low porosity and less oxidation can be achieved. Recently, cold spray has been widely applied in the domains of additive manufacturing [4, 5], damaged component restoration [6], protective coating building [7].

In order to better control the additive manufacturing by cold spray and improve the sample quality, it is of great importance to study the thermal distribution during the coating build-up process. Numerous studies have been made to understand the heat exchange [8] and temperature distribution by different approaches. By methods of thermocouple [9] and infrared thermal camera [10], it is able to measure the transient temperature variation during spray process. Meanwhile, numerical simulation is also used to provide more details about temperature distribution in cold spray process. However, the difficulty stays with the transient coating build-up process by the simultaneous nozzle movement. In most of early studies [9, 11], coating was partitioned as a part of predefined geometry without accounting the transient and gradual coating deposition process. Obviously, such treatment has significant difference from the actual coating deposition process, where the thickness and topology of coating changes along with nozzle movement. Lately, different kinds of numerical approaches have been developed to investigate such transient coating build-up process.

In order to simulate the temperature evolution with transient coating build-up process, a 3D model based on ALE (Arbitrary Lagrangian–Eulerian) moving mesh method was established by accounting the transient coating build-up as real coating deposition process. By coupling between the heat transfer, moving mesh and coating thickness simulation, the proposed model enables the simulation of the transient temperature distribution as the simultaneous nozzle movement in cold spray. Based on the results of coating thickness model, the transient coating build-up is achieved by assigning the coating thickness distribution to the mesh deformation value of coating surface, which is namely the coating build-up. This work aims to study the transient temperature evolution in the cold spray process. Experimental validation was also performed by measuring the transient temperature history on the coating surface by imaging infrared camera. Furthermore, the effects of nozzle traverse speed on heat transfer as well as coating thickness are investigated by the proposed model.

2 Numerical model

2.1 Governing equation and boundary conditions

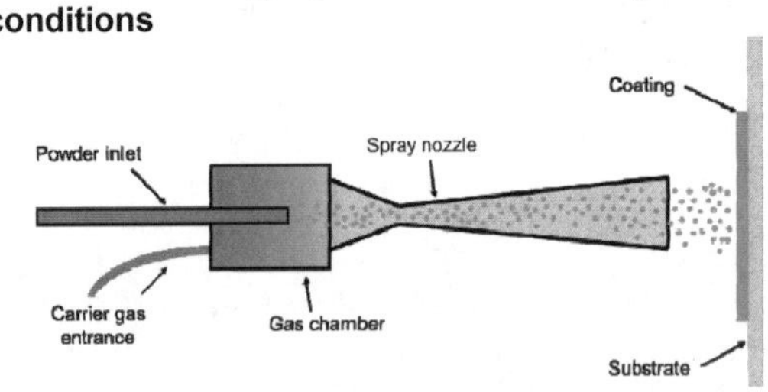

Fig. 1. Schematic of the cold spray system.

As the schematic is shown in Fig. 1, dissimilar to traditional thermal spray process, cold sprayed coating is formed by the deposition of solid-state particle. Particles are carried by heated gas and then deposit onto substrate. According to the finite element analysis (FEA) of particle acceleration and heating in the De Laval nozzle used in cold spray [12], the particle in cold spray before impacting onto substrate is found to be slightly heated. Thus, the heat transfer by the impact of particle can be neglected in this process. Similar with the related work [13], only the heat transfer between the gas and coating-substrate structure is taken into account in this model. The governing transient heat transfer equation for the coating and the substrate in three dimensional is written as below.

$$\rho \cdot C_p \frac{\partial T}{\partial t} = \nabla \cdot (k \nabla T) + Q \qquad (1)$$

Where ρ is the density C_p is the heat capacity, k is the thermal conductivity, Q is the additional heat source. At the coating surface, a third kind of boundary condition was used to describe the heat transfer on substrate surface by heat flux from the nozzle, which is given in Eq. (2).

$$-n \cdot (-k \nabla T) = h \cdot (T_{ext} - T) \qquad (2)$$

Where, n is the normal vector to the surface, h is the heat transfer coefficient and T_{ext} is the temperature of the external fluid far from the boundary. The term Text was defined as the temperature of gas prior to the impact onto the substrate, which was obtained by the simulation in ANSYS-Fluent 14.1 based on the powder and working conditions in the experiment. The details of the ANSYS-Fluent model can be referred to the work [14]. Based on the Gaussian distribution of heat flux out of nozzle that was frequently applied in the numerical studies [15-17], the heat transfer coefficient between the gas and substrate was set as the Eq. (3).

$$h = h_0 \cdot \frac{A}{\sigma \sqrt{2\pi}} (-(\frac{(x-x_0)^2 + (y-y_0)^2}{2\sigma^2})) \qquad (3)$$

Where the h_0 is the heat transfer coefficient obtained by the equation of external forced convection [18]. The definition of the heat transfer coefficient by external force convection is given as below. The σ is the standard deviation of the Gaussian distribution, which means that the diameter of the jet impact is 6σ. The (x_0, y_0) is the center coordinate of the jet impact out of nozzle, which is defined through the nozzle trajectory. The details about the nozzle trajectory generation will be discussed in the following section.

$$h = \begin{cases} 2\frac{k}{L} \frac{0.3387 Pr^{1/3} Re_L^{1/2}}{(1+(0.0468/Pr)^{2/3})^{1/4}} & if\ Re_L \le 5 \times 10^5 \\ 2\frac{k}{L} Pr^{1/3} (0.037 Re_L^{4/5} - 871) & if\ Re_L > 5 \times 10^5 \end{cases} \qquad (4)$$

where $Pr = \mu C_p / k$ and $Re_L = \rho U_{ext} L / \mu$, k is the thermal conductivity of the gas. The plate length of L is the length of the substrate. The exterior velocity of Uext is the gas velocity prior to the impact upon substrate, which is calculated by ANSYS-Fluent 14.1 simulation as mentioned above. The heat transfer due to the surrounding air circulation in workspace was considered as the heat loss by cooling effect. Similar to the heat accumulation, the heat transfers by cooling effect is also treated as the third kind of boundary condition in Eq. (2). Under the consideration of perfect bonding between coating and substrate, the heat transfer between them is regarded as thermal conduction. The numerical model is built in the software COMSOL and solved by coupling the problems of heat transfer, moving mesh, coating thickness distribution, which will be discussed in the following sections.

Table 1. Material properties.

Material	Temperature (K)	Density (kg/m^3)	Thermal conductivity (W/(m·K))	Specific heat (J/(kg·K))
Ni	295	8900	90.6	444.3
	350	8880.6	85	463.6
	400	8862	80.2	482.3
	500	8822.4	72.2	530
	600	8780.5	65.5	593.7
Al	-	2700	160	900

2.2 Description of coating build-up model

According to related work, a proper coating formation model has dominant effects on the finite element analysis results [19]. It is reported that the stochastic deposition model can provide a more objective result of the transient heat transfer of the coating/substrate than that of the layer-by-layer model [19]. However, these methods lack accuracies to account the transient coating build-up process in a 3D model. Therefore, it is of great importance to propose this coating build-up model in 3D that is based on the transient coating thickness distribution model. In this model, the coating build-up process is achieved by the ALE moving mesh method based on the transient coating thickness distribution. It allows the evolution of computational domain according to the coating formation as the spray nozzle trajectory.

As the overall model is shown in Fig. 2(a), mesh deformation of the geometry is achieved based on the control of boundary condition. Because the generation of a new computational domain is not allowed, it is necessary to give a preliminary computational domain of coating. As indicated in the left of Fig. 2(a), the preliminary computational domain is modeled allowing the successive build-up of coating on the substrate surface. The initial coating thickness is controlled at such a small value of 50 µm (approximately the thickness of a single particle layer). Thus, an assumption is made here that the preliminary computational domain of coating has little effect on the heat transfer with the substrate. As the moving mesh method description shown in Fig. 2(b), the mesh displacement value of the coating surface is zero in the directions of x and y, which means no horizontal coating grow-up is allowed and the coating growth on the substrate surface is restricted vertically.

A variable that represents the coating thickness distribution on the substrate surface is assigned as the coating growth value in the direction of axis-z. The transient distribution of coating thickness is obtained by the results of the coating thickness distribution model, which will be introduced in the next section. The mesh deformation in the computational domain of substrate is fixed as zero, which means no deformation is assigned to substrate.

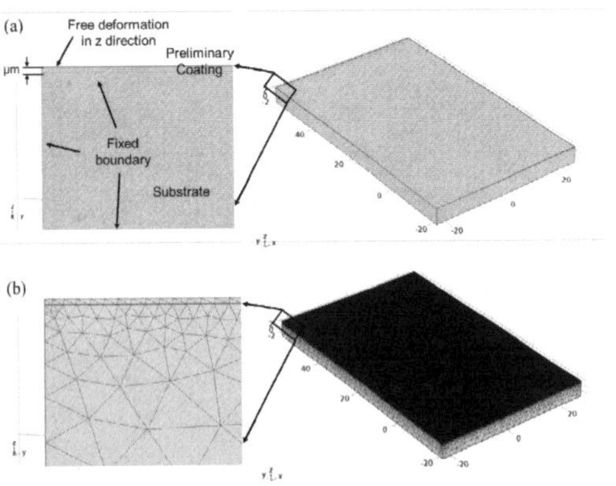

Fig. 2. 3D numerical model based on deformed mesh method: (a) overall geometry and detailed description of ALE moving mesh method; (b) meshing.

2.3 Coating thickness distribution

According to the related studies [20], the operating parameters in spray process can directly affect the coating quality as well as the coating thickness. However, the previous work on the finite element analysis in the coating formation process seldom takes into account the effects of operating parameters on coating thickness. The computational domain of coating is usually treated as a static as-sprayed one, instead of transient coating formation process. In this study, the coating thickness distribution is simulated according to the coating thickness model proposed in the reference [21]. It takes into account the effects of operating parameters in spray process, such as nozzle traverse speed, spray angle, scanning step between two successive passes of trajectory. The coating thickness model is given as an ODE (ordinary differential equation) in equation below, which is solved prior to the heat transfer problem during the simulation to define the transient coating thickness growth.

$$\varphi = \zeta(\theta)\int_0^T \left(\int \frac{A}{\sigma\sqrt{2\pi}} e^{-\left(\frac{(x-\mu_x)^2}{2\sigma^2}+\frac{(y-\mu_y)^2}{2\sigma^2}\right)} dxdy\right)dt \quad (5)$$

Where ξ(θ) is the coefficient to account the influence of spray angle on coating thickness, A is the amplitude factor obtained from experimental data, σ is the standard deviation of the Gaussian distribution. The coating profile deposited at the duration of time step dt is considered as a Gaussian distribution, which is widely applied in the simulation of coating thickness [22]. The variables in the Gaussian distribution equation are obtained by measurement of coating profile by single nozzle pass in experiment. By assigning the transient coating thickness distribution to the mesh displacement of the coating surface, the coating build-up process can be realized. It allows the simultaneous calculation of thermo-mechanical effects during the coating build-up process.

2.4 Robot kinematics

The nozzle trajectory represents the nozzle movement in relative to the substrate during spray process, which is also the movement of heat source and particle source. Manipulated by robot, the nozzle trajectory in real spray process is found to be directly affected by the robot kinematics [23, 24]. For example, an undesired robot performance can result in the fluctuation of nozzle traverse speed [24], which can cause local over-heating [15], uneven coating thickness, and arise of residual stress. Thus, it is of great significance to include the robot kinematics into the simulation of cold spray process. As is shown in Fig. 3, a round-trip trajectory was generated in this study by an offline programming software RobotStudio™. The predefined TCP (tool center point) speed is 150 mm/s. The trajectory has a full coverage of the substrate surface with an offset of 5 mm at the edge of the substrate, which is used to maintain the stability of TCP speed at the predefined value after each change of movement direction. By using the virtual robot system in RobotStudio™, the position and the speed of TCP that is the representation of the movement of particle impact point were obtained through trajectory simulation. The TCP speed at each point in the trajectory are shown in Fig. 3, with the color indicating its speed value. It can be found that the nozzle speed within the substrate area is well maintained at the value of 150 mm/s. By importing the position and the speed of TCP in the model as a form of interpolation, the nozzle movement was well defined by accounting its speed variation along the trajectory. Thus, the nozzle trajectory by the simulation in RobotStudio™ can be used to account the movement of heat source and mass flux of particle impact in cold spray.

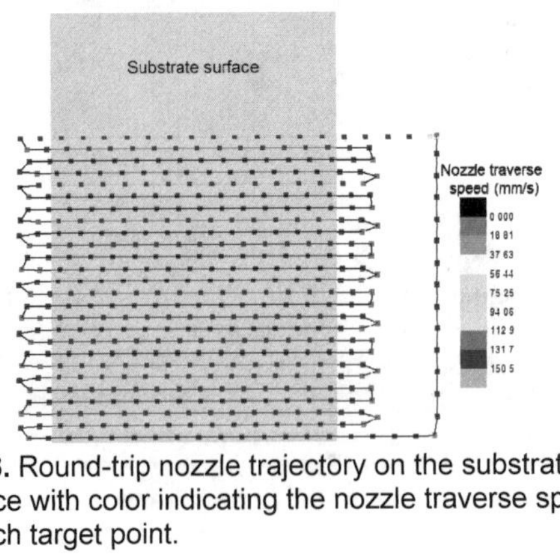

Fig. 3. Round-trip nozzle trajectory on the substrate surface with color indicating the nozzle traverse speed of each target point.

3 Experimental details

In order to validate the finite element model, experimental validation was carried out by measuring the temperature history during the coating deposition process in cold spray. Coatings were fabricated by using a CGT K3000 system (LERMPS, UTBM, France). Compressed air was used as the driving gas with the pressure and the temperature at the inlet of 2.4 MPa and 360 °C, respectively. The standoff distance from the nozzle exit to the substrate surface was set as 30 mm. The round-trip nozzle trajectory as well as the nozzle traverse speed were defined as the description in previous section of nozzle trajectory. Under such conditions, the coating was obtained by repeating the nozzle trajectory 5 times. The gas-atomized pure Ni powder (Praxair Surface Technologies, Inc., U.S.A) with spherical morphology was chosen as the feedstock, while aluminum was chosen as the substrate. The powder morphology by scanning electron microscopy (SEM) as well as its size distribution are shown in Fig. 4. Microstructures of coatings were examined by optical microscope (OM, Nikon, Japan). In order to evaluate the temperature history during the coating build-up process, the surface temperature of coating was measured and simultaneously recorded by a thermal imaging camera (SC5000, FLIR Systems, U.S.A). The temperature variation at the middle point of coating surface was continuously recorded by measurement software, which enables the investigation of temperature evolution on the coating surface. The camera operating at a frequency of 10 Hz and a waveband of 2.5-5.1 µm was fixed on a tripod with a distance of about 1.0 m to the sample.

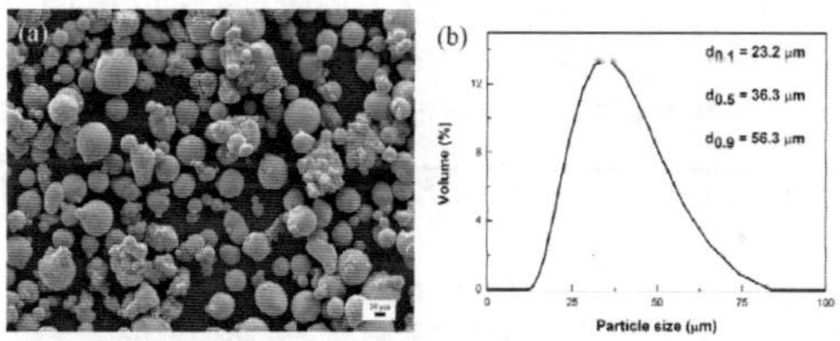

Fig. 4. (a) SEM morphology and (b) size distribution of Ni particles.

4 Results and discussion

Before experimental validation, a numerical validation was made to study the influence of preliminary computational domain of coating on the heat transfer with the substrate. For such purpose, two finite element analyses of the heat transfer process that does not account the transient coating build-up process was made respectively with the computational domain that includes and excludes the preliminary computational domain of coating. The trajectory of the heat source of gas flow out of nozzle exit was represented by the nozzle trajectory simulated by RobotStudio as given in the section 2.4. The comparison of temperature history on the center of coating surface in the case with preliminary coating domain and substrate surface in the case without it was shown in Fig. 5. It can be seen that the preliminary coating domain has little effect on the numerical simulation of the temperature evolution and heat transfer between the substrate during cold spray. The average deviation of surface temperature between these two results is 3.47 °C. Thus, it can be confirmed that this preliminary coating domain with thickness of 50 µm that is nearly the size of particle size used in this work has little effect on the heat transfer between the gas flow and the substrate domain.

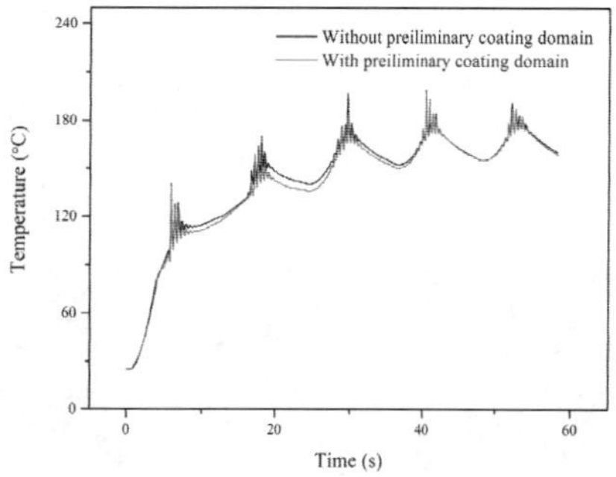

Fig. 5. Validation of the dependence of the preliminary coating domain on the heat transfer simulation without accounting transient coating build-up.

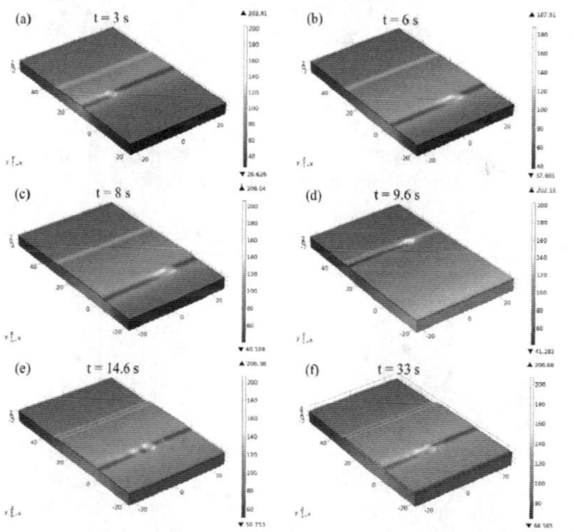

Fig. 6. Transient coating build-up process during the nozzle movement with contour indicating the temperature magnitude.

The simulation results of the transient coating build-up process on substrate is shown in Fig. 6, where the contour indicates the temperature magnitude. The predefined nozzle traverse speed in this cases is 150 mm/s. It can be found that the coating formation continues gradually from the top-end of substrate to the bottom with the round-trip trajectory of nozzle movement, which presents the heat source and the mass flux of the particle impact. In order to better observe the coating build-up process, a detail image of coating build-up process t=0.1 s is given in Fig. 7. It can be seen that the computational domain of coating is gradually deformed along with the movement of nozzle, which represents the build-up process of coating. The Gaussian distribution of coating thickness as nozzle movement can also be noticed. At the meantime, a Gaussian distributed temperature can be found at the end of as-sprayed coating, which is the heat source by the nozzle impacting onto the coating surface. The scanning step that is the separation between two successive nozzle pass is 2 mm in this simulation. As shown in Fig. 6, the coating deposited by each single nozzle pass can be clear observed. At the same time, with the movement of heat source and the formation of coating, the temperature at coating surface is gradually increasing. After the execution of first pass of trajectory, a thin coating can be observed in Fig. 6(d).

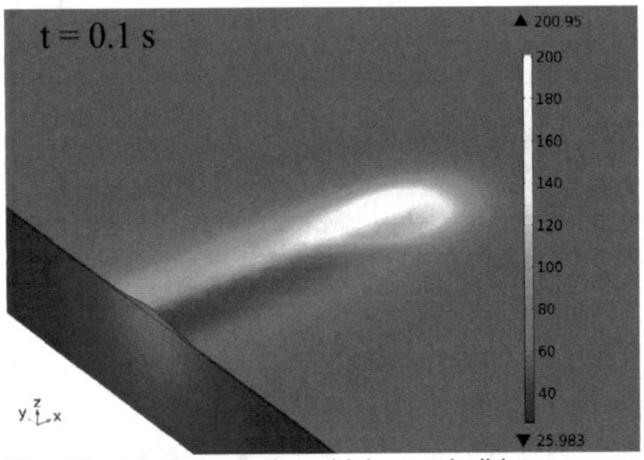

Fig. 7. detailed coating thickness build-up process with the nozzle movement at t = 0.1 s and the nozzle traverse speed of 150 mm/s.

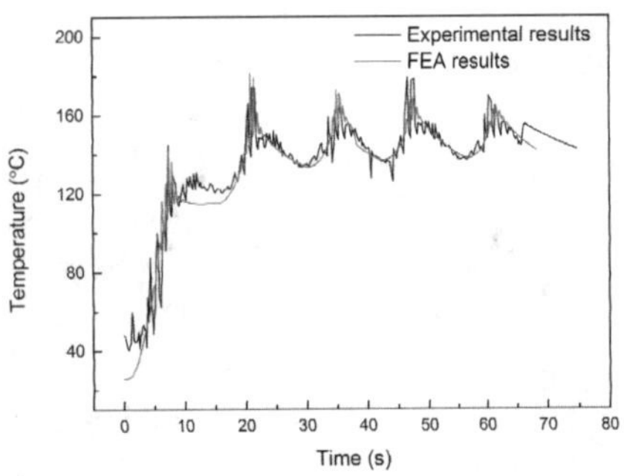

Fig. 8. Comparison of temperature history on the coating surface between numerical simulation and experimental measurement by imaging infrared camera.

As shown in Fig. 8, the temperature history on the central position of coating surface during its build-up obtained by numerical simulation is compared with experimental measurement by imaging infrared camera. Due to the existence of measurement lag by the thermocouple [9], the imaging infrared camera was used to collect the temperature history instead of thermocouple. The temperature evolution from simulation is simultaneously obtained at the center of coating surface as the coating build-up process. A good agreement is found to be obtained between the numerical model and experimental validation. Thus, it can be convinced that the proposed coating build-up model is well validated by the experimental results. The temperature increases rapidly from room temperature as the execution of nozzle trajectory. 5 temperature peaks can be observed in Fig. 9, which indicates that the nozzle is traversing by the central point of coating surface. Coating surface temperature decreases between two successive passes of nozzle trajectory, which is caused by the cooling effects by the ambient temperature. As the continuation of nozzle trajectory, coating surface temperature maintains around 150 °C and no further increment of temperature can be observed after three passes of trajectory.

5 Conclusion

In this work, a 3D FEA model was developed to simulate the transient coating build-up process and the heat transfer in cold spray. The proposed model enables the study of coating thickness as well as the temperature distribution under the influence of robot trajectory and different operating parameters. Meanwhile it can also achieve the investigation of additive manufacturing process of samples with complex geometry by CS.

By coupling the heat transfer with the ALE moving mesh and coating thickness modeling, this 3D model is able to investigate the temperature evolution of a coating which is transiently building up according to the nozzle trajectory. By assigning the results of coating thickness modeling, the simultaneous build-up of coating computational domain is achieved by ALE moving mesh method. Thus, it enables the heat transfer modeling of a transient coating build-up process. The validation of the FEA model was carried out by measuring the coating surface temperature via an infrared imaging camera during a cold spray experiment. The satisfactory agreement was found between the FEA modeling result and the experimental measurement.

Acknowledgments

The authors would like to acknowledge the support by China Scholarship Council (CSC) and Marie Curie FP7-IPACT-268696(EU).

6 Literature

[1] Pawlowski, L., The science and engineering of thermal spray coatings. John Wiley & Sons. 2008.

[2] Papyrin, A.: Cold spray technology. Adv. Mater. Process. 159 (2001), Issue 9, pp. 49-51.

[3] Li, W. and D. Zhang. Other: Modelling of impact behaviour of cold spray particles: review. Surf Eng 30 (2014), Issue 5, pp. 299-308.
[4] Pattison, J. and S. Celotto. Other: Cold gas dynamic manufacturing: A non-thermal approach to freeform fabrication. Int J Mach Tool Manu 47 (2007), Issue 3, pp. 627-34.
[5] Dupuis, P. and Y. Cormier. Other: Flow structure identification and analysis in fin arrays produced by cold spray additive manufacturing. Int J Heat Mass Tran 93 (2016), Issue, pp. 301-13.
[6] Champagne, V.K.: The repair of magnesium rotorcraft components by cold spray. Journal of Failure Analysis and Prevention 8 (2008), Issue 2, pp. 164-75.
[7] Dzhurinskiy, D. and V. Leshchinsky. Other: Protective coatings for mechanical aluminum-magnesium joints. Surf Eng 31 (2015), Issue 10, pp. 740-46.
[8] Ryabinin, A.N. and E. Irissou. Other: Simulation of gas-substrate heat exchange during cold-gas dynamic spraying. Int J Therm Sci 56 (2012), Issue, pp. 12-18.
[9] Arabgol, Z. and H. Assadi. Other: Analysis of Thermal History and Residual Stress in Cold-Sprayed Coatings. J. Therm. Spray. Technol. 23 (2014), Issue 1-2, pp. 84-90.
[10] Yin, S. and X. Wang. Other: Deposition behavior of thermally softened copper particles in cold spraying. Acta Mater. 61 (2013), Issue 14, pp. 5105-18.
[11] Zhang, T. and Y. Bao. Other: Effect of a moving flame on the temperature of polymer coatings and substrates. Prog Org Coat 70 (2011), Issue 1, pp. 45-51.
[12] Yin, S. and M. Meyer. Other: Gas Flow, Particle Acceleration, and Heat Transfer in Cold Spray: A review. J. Therm. Spray. Technol. (2016), Issue, pp. 1-23.
[13] Yin, S. and Y. Sun. Other: Effect of Spray Angle on Temperature Distribution within the Metallic Substrate in Cold Spraying. J. Therm. Spray. Technol. 22 (2013), Issue 6, pp. 983-91.
[14] Xie, Y. and M.-P. Planche. Other: Investigation on the influence of particle preheating temperature on bonding of cold-sprayed nickel coatings. Surf. Coat. Technol. Issue
[15] Bolot, R. and S. Deng. Other: A Coupled Model Between Robot Trajectories and Thermal History of the Workpiece During Thermal Spray Operation. J. Therm. Spray. Technol. 23 (2014), Issue 3, pp. 296-303.
[16] Liu, J.W. and R. Bolot. Other: Transient thermal and mechanical analysis of NiCrBSi coatings manufactured by hybrid plasma spray process with in-situ laser remelting. Surf. Coat. Technol. 292 (2016), Issue, pp. 132-43.
[17] Liu, J.W. and R. Bolot. Other: Residual stresses and final deformation of an alumina coating: Modeling and measurement. Surf. Coat. Technol. 268 (2015), Issue, pp. 241-46.
[18] Bergman, T.L. and F.P. Incropera. Other, Fundamentals of heat and mass transfer. John Wiley & Sons. 2011.
[19] Chen, Y. and X. Liang. Other: Finite element modeling of coating formation and transient heat transfer in the electric arc spray process. Int J Heat Mass Tran 53 (2010), Issue 9–10, pp. 2012-21.
[20] Cai, Z. and S. Deng. Other: The Effect of Spray Distance and Scanning Step on the Coating Thickness Uniformity in Cold Spray Process. J. Therm. Spray. Technol. 23 (2014), Issue 3, pp. 354-62.
[21] Chen, C. and Y. Xie. Other: Modelling of coating thickness distribution and its application in offline programming software. Surf. Coat. Technol. (2016), Issue
[22] Leigh, S. and C. Berndt: Evaluation of off-angle thermal spray. Surf. Coat. Technol. 89 (1997), Issue 3, pp. 213-24.
[23] Deng, S. and H. Liang. Other: Kinematic Optimization of Robot Trajectories for Thermal Spray Coating Application. J. Therm. Spray. Technol. 23 (2014), Issue 8, pp. 1382-89.
[24] Chen, C. and H. Liao. Other: Nozzle Mounting Method Optimization Based on Robot Kinematic Analysis. J. Therm. Spray. Technol. 25 (2016), Issue 6, pp. 1138-48.

Cold-gas sprayed deposition of metallic coatings onto ceramic substrates using laser surface texturing pre-treatment

R. Kromer*, Y.Danlos, S. Costil

ICB-LERMPS, UMR 6303, CNRS, Univ. Bourgogne Franche-Comté, UTBM, F-90010 Belfort, France

*E-mail: robin.kromer@utbm.fr

Abstract

Cold spraying consists in depositing a variety of metals as dense coatings onto metal surfaces. Indeed, copper, stainless steel, nickel, chromium, aluminum, cobalt, titanium, niobium and other metals can all be deposited, as well as metal alloys according to these base-metals and braze powders. The particle-substrate contact time, contact temperature and contact area upon impact are parameters influencing physico-chemical and mechanical bonds. The resultant bonding arose from plastic deformation and temperature at the interface which illustrates why metal coating cannot be sprayed onto rough ceramic substrates. Laser surface texturing has been used as prior treatment to create specific topography. Metal-ceramic has demonstrated a non-deformation of the substrate minimizing intimate bonds. Particle compressive states indicate anchoring mechanisms for laser textured surfaces. Consequently, cold spray coating parameters depend on the target material and a methodology can be established with particle parameters (diameters, velocities, temperatures) and particle/substrate properties to adapt the surface topography. Mechanical adhesion is a key issue in cold spray process. As a result, laser surface texturing is a promising tool to adapt the surface to improve adhesion. Metalization process can be achieved.

Keywords: laser surface texturing, cold spray, ceramics, adhesion, in-contact area, anchoring mechanisms

1 Introduction

Recently cold spraying has been developed for many industrial applications due to its high rate deposition and material property conservation [1]. Particles are accelerated in a supersonic jet of compressed gas to high velocities to promote adhesion to substrate [2]. After impact and deformation, high contact pressure and temperature create intimate bonds [3]. Besides, some phenomena have been commonly observed such as: jet formation by particle penetration into soft substrate, removal of the surface oxide, grain refinement, dislocation formation and intermixing between feedstock and substrate (roll-up vortex formation)[4]. Several studies on those mechanisms and their reliability depending on the materials were reported [5]. However, the mechanical deformability is limited for ceramic substrates [6]. Those adhesion mechanisms become more complex due to the deformation processes, including crater formation, interfacial jet and vortex-like feature. In this case, intimate bonding hence is so difficult.

Thus, initial surface conditions (including surface topography) have been studied. Papyrin showed that particle adhered to the surface only after some delay, when the surface topography state was changed by impinging particles [7]. The first particles rebound, thus, modifying the surface and the coating starts to build-up when substrate topography was patterned [8]. The particles activated the surface by oxide removal and concave shape might promote direct attachment zones [9]. Consequently, several studies have evaluated the benefits of prior-surface treatments [10]. Substrate micro-structure near interface and impacted area shape can affect particle deformation and metallurgical bond[11].

Among surface pre-treatments, degreasing and grit-blasting are the most widespread preparations. Degreasing agent leads to remove organic of the surface and grit-blasting modifies the surface topography [12]. Initial roughness before spraying can lead to mechanical anchoring mechanisms during coating deposition, but their applications are limited especially on ceramics. Surface treatments by pulsed laser beam have been used to provide specific surface modifications such as cleaning state, topography morphology, heating level, etc [13]. Laser system in manufacturing process have several advantages such as easy automation, localized treatment and three dimensional treatments. Laser surface texturing were studied to promote higher mechanical anchoring of thick coatings [14]. Furthermore, surface topography can be adapted to spray process and materials [15]. The textured substrate surface has an influence on deposition efficiency (DE) and adhesion strength (AD) [16].

This paper investigates laser surface texturing enhancement for aluminum and copper coating onto alumina and titanium coating onto silicon carbide. DE and AD have been evaluated for both textured substrates with two pattern morphologies. Adhesion mechanisms have been discussed to propose a new strategy for metallization of ceramics by cold spray process.

2 Methods and Materials

2.1 Materials

The powders used for cold spraying were pure aluminum (99.5 wt.% Al) and copper (99.5 wt.% Cu) powders (fabricated at LERMPS Laboratory, France) and a commercially pure titanium powder (min 99 wt.% Ti) (supplied by Active Metals Ltd., Sheffield, UK. Powder). The particle size distributions are in the range of −55/+15 µm, −45/+12 µm and -65/+8 µm respectively (analysis with a Mastersize 2000 laser, Malvern Instruments Ltd, UK). Two ceramic substrates were used: Al_2O_3 (GoodFellow, France) and SiC (SCERAM Ceramics, France). The substrates were 10 mm thick buttons with a diameter of 25 mm.

2.2 Cold spray process

In this study, 300µm thick coatings were produced by cold gas dynamic spraying using a Kinetics 3000 gun (CGT GmbH, Germany). The spray gun is a converging-diverging "de Laval" nozzle with an expansion of 5.6 and a divergence zone length of 150 mm, mounted on a XYZ robot (ABB robot - Affolternstrasse 44 CH-8050 Zurich Switzerland). The gas conditions were selected to obtain two cases of particle velocities. Parameters of particles temperature and speed are studied using the Software Kinetic Spray Solutions (Kinetic Spray Solutions GmbH, Holmer Weg, 13 21244 Buchholz). The gas parameters as well as the spray conditions are given in Table 1.

Table 1: The cold spray parameters

GAS FLOW RATE [SLPM]	Al/Al_2O_3	Ti/SiC	Cu/Al_2O_3
Spray Distance[mm]	30	30	30
Gas Pressure [Bars]	22	30	30

Gas Temperature [°C]	300	600	500
Step-Size [mm]	2	2	2
Powder Feed rate [g/min]	3	3	3
Gun Traverse Speed [mm/s]	150	150	150

2.3 Surface Treatments

Conventionally, substrates are prepared in two steps: cleaning and roughening. Degreasing agents were used in order to remove organic substances of the substrates. Laser has been used as an alternative technique to modify the substrate morphology with a reduced environmental impact. In this study, the laser interaction with the substrate generates a temperature rise to melted and vaporization point. Hole pattern can then be created by multiplying laser impacts to increase mechanical interlock as "root" effect. Regarding laser treatment, experiments were conducted with a pulsed fiber laser (Laseo, Ylia M20, Quantel France) (Figure 1-A). The laser operates at a nominal wavelength of 1.06 μm with pulse duration of 100 ns, a maximum power of 20 W and frequencies varying from 20 to 100 kHz. Laser beam is circular with a diameter of 60 μm at the focal point and Gaussian energy distribution. The laser patterning consists of series of equidistant lines covered with a number of patterns (several laser pulses generate a pattern) (Figure 1-B). In this study, a hole pattern was chosen. The textured substrate has an adhesion area depending on the shape, height, orientation, distribution and density of holes (Figure 1-C). Two specific pattern morphologies have been chosen corresponding to numerical impact model results.

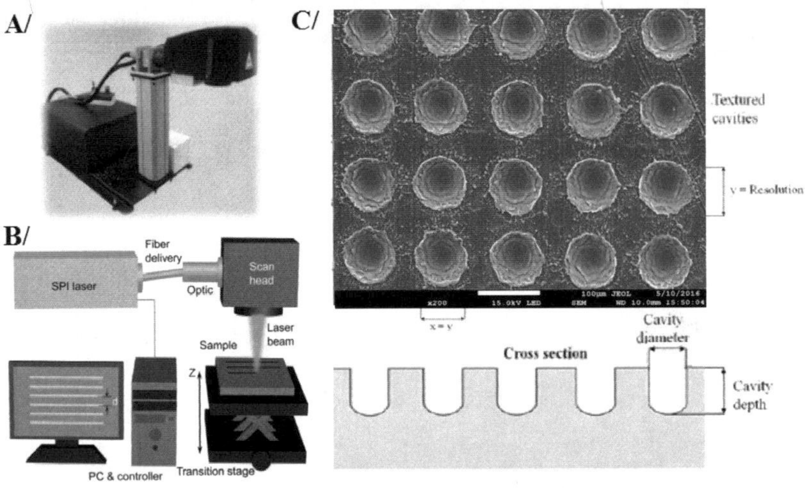

Figure 1: A/ Laser Laseo Ylia 20, B/ system schematic of laser surface texturing, C/ Pattern examples on aluminum surfaces

For cold spray prior-treatments, a pattern has been defined so that its morphology was equal or 25% larger to half mean particle size. T1 and T2 correspond to two hole patterns with the same pattern density (100 pattern/mm²)(Figure 2). The distance between patterns was fixed to 100 µm. The diameter opening was larger than the particle mean size. A half-ellipsoid diameter opening was equivalent to 40 and 60 µm thus depth of 40 and 60 µm, respectively. The same patterns were created for Al_2O_3 and SiC substrates. Laser parameters have been optimized and defined specifically for both materials to create a pattern.

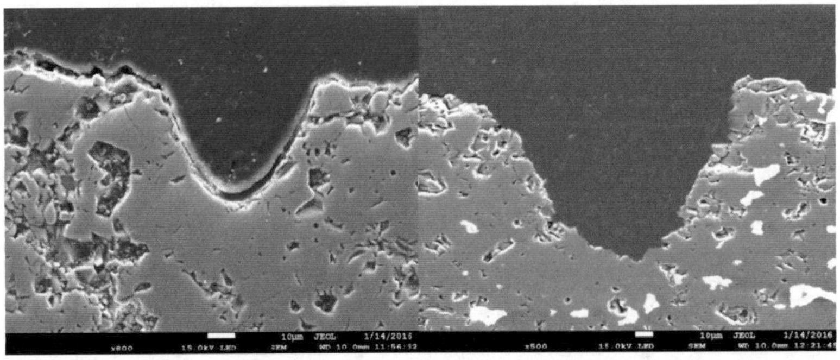

Figure 2: Cross-sections of T1 and T2 patterns

2.4 Coating and interface characterizations

Surface topography with and without coating was analyzed by light scanning electron microscopy (JEOL JSM 7800F SEM). One of the most important characteristics of the process is the deposition efficiency, which is defined as the ratio of the mass gain of the substrate during exposure time to the flow with the proper of parameter set and the decrease in powder mass during the same time.

Also, 300µm coatings thickness have been achieved to carry out adhesion bond strength evaluation for raw and textured surfaces. Coating adhesion is usually tested by pull-off tests (described in DIN EN 582 - ASTM Standard C633). Testing buttons samples are joined with cylindrical counterparts using an adhesive film (FM1000). A constant displacement rate (1.026 mm/s) is applied with a tensile test. As-sprayed coatings were tested at room temperature.

3 Results and Discussions

3.1 Impact and deposition efficiency

In general, high speed impact creates plastic deformation of the particle and the substrate. The incoming particles modify particle and substrate extreme surface microstructure. The substrate deformation is correlated to the impacted zone morphology, particle temperature and velocity. Kumar et al. have observed that particle adhesion is related to target shape (concave/convex) [17]. Particle and pattern sizes must be similar to maximize the contact-time and contact zone.

Initially, the first aluminum bonded particles have been evaluated for raw and textured ceramic substrates. Figure 3 shows the particle bonds on raw substrate. A particle flattening behavior was observed and no deformation occurred to the substrate. The particle tied on peaks.

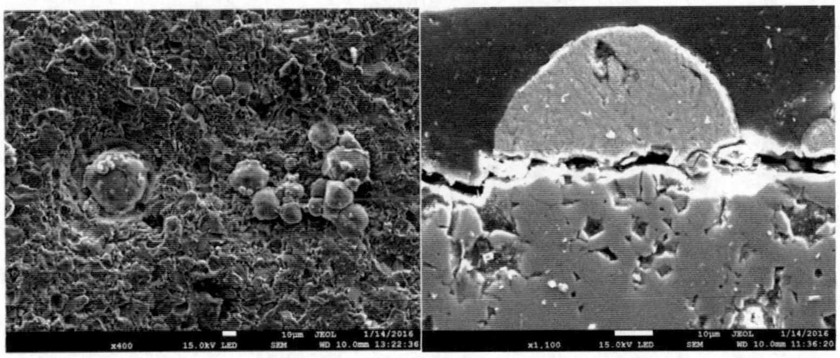

Figure 3: Top view and cross-section of aluminum particles on raw alumina substrates

Inversely, as observed in Figure 4, particles were attached in concave zones (created by laser surface texturing) and blocked inside holes. Few particles were observed on plateau. The particles were deformed and a jet of matter can be observed on the particle edges. The particle-substrate contacts were larger in this case. Particles were anchored in the patterns.

Figure 4: Top view and cross-section of aluminum impacted particles on textured substrates

After impact, the residual stress was in compression in the direction of the surface in front of any obstacles of the spreading [18]. As proposed by the authors [19], a numerical evaluation of the stress distribution confirmed such behavior (Figure 5). On raw surfaces, compressive stresses appear on extreme layer substrate. Large compressive states could be evaluated in cavities providing anchoring.

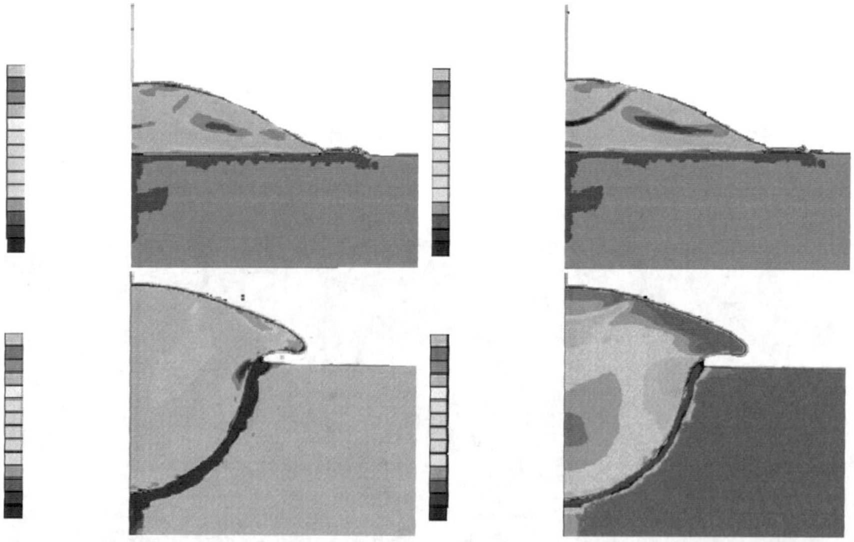

Figure 5: Maximal principal and Sxx stress distributions for aluminum particles impacted on ceramic raw and textured substrates [19]

The same observations could be drawn for the three configurations. Then, DE has been evaluated for the different couples (Figure 6). The deposition efficiency has been measured after the complete 300 μm thick coating build-up which requires 3 layers. The first layer corresponds to soft/hard impact mechanism whereas the second and third layers are particle/coating build-up mechanism. Regarding the sprayed materials, the deposition efficiency has been evaluated to 75 - 90 and 55% for Al, Cu and Ti respectively for particles impacted on the build-up coating. Regarding the first layer corresponding to the particle impacts onto the substrates, DE is obviously influenced by the particle but also by the substrate morphology (Figure 6).

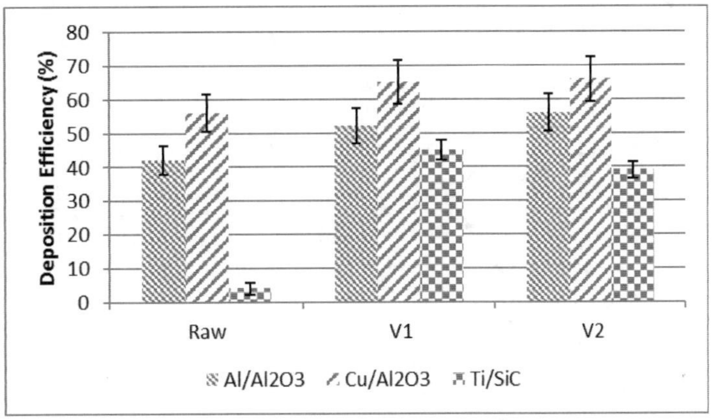

Figure 6: Deposition efficiency for raw and textured (V1 and V2) specimens

As presented on the graph, an increase of the deposition efficiency can be noticed particularly using laser surface texturing. Compared to raw surfaces, DE improvements occurred systematically whatever the materials considered which confirms the strong effect of the first layer and the pattern efficiency for the direct bonding in cavities. In addition, titanium coatings were built on raw substrate (only two samples on fives). Only few particles stayed attached to the surface and were debonded during cold spray process. Cavities created zones for direct bonds and can be defined as roots for coating build-up. The large in-contact area and the compressive stress state enabled direct bonding for the three configurations.

In addition, the pattern shape had an effect on DE. V2 was larger than V1 for Al and Cu coatings. The opposite could be observed for Ti coating. The particle distribution had to be in accordance with the pattern sizes. For instance, the particle distributions were Gaussian-like for aluminum and copper, but it was shifted (Figure 7). The number of particle statistically enabled to get in patterns and had the conditions to adhere were linked to the particle-pattern size ratios and the particle mechanical properties. The conditions were more efficient for the case V2 for aluminum and copper and for the case V1 for titanium. Pattern volume and concavity ratio could be estimated. The volumes have been estimated to 14000 μm^3 and 33000 μm^3, respectively and the patterns were half-ellipsoid-like shape. Besides, the mean particle volumes were equal to 23000, 28000 and 19000 μm^3 for aluminum, copper and titanium, respectively. The volume should be equal to maximize adhesion. However, cavity shape effect has not been taken into account. For instance, deep patterns would be difficult to full-fill and large patterns would be flat-like and no anchoring mechanisms could occur. So, pattern optimization is in progress.

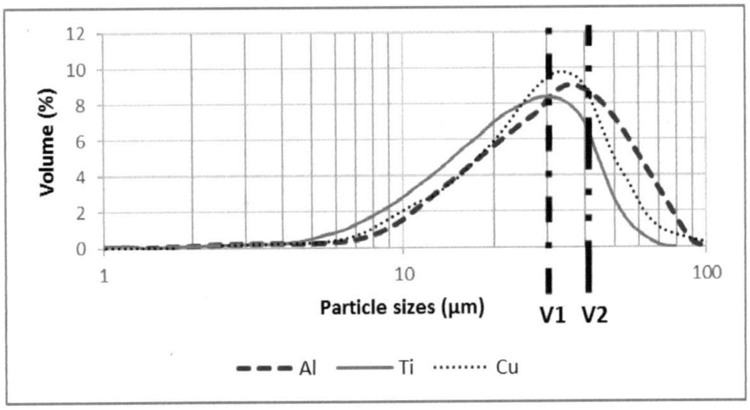

Figure 7: Powder granulometry of aluminum, copper and titanium powders

3.2 Coating and adhesion strengths

Adhesion bond strength has been evaluated to observe laser prior-treatment enhancement. First, the interfaces have been studied after complete coating build-up. Figure 8 shows the coating structures on raw surface. A dense coating has been elaborated, but the interface seems irregular where cracks can be observed at the interface.

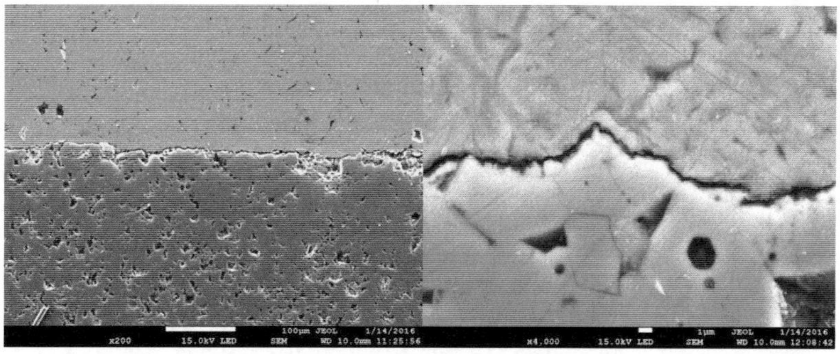

Figure 8: Cross-sections of aluminum coating on raw substrate - large and fine magnitude

In comparison, dense coatings have also been observed on textured samples and as in Figure 8, the interface may suit the requirements and an obvious intimate contact can be observed on the edges. Micro-roughness could be noticed which could create friction forces, in other wordsanchoring forces.

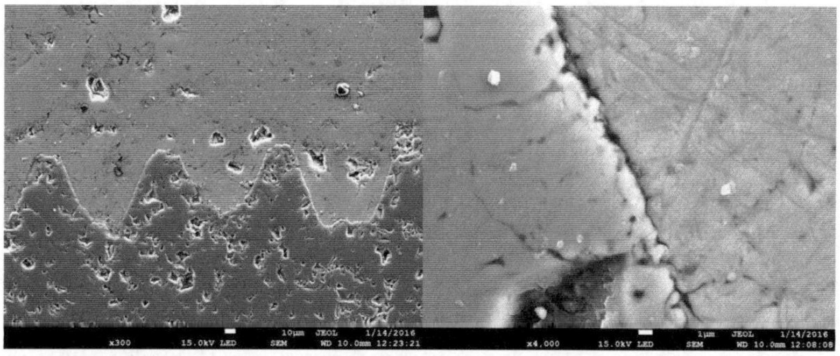

Figure 9: Cross-sections of aluminum coating on textured substrate - large and fine magnitude

Besides, similar behaviors occurred for the other configurations. The first particle bonding enabled a complete coating build-up. However, the coating interface quality was different. For instance, in Figure 10, cracks could be observed on the edges and defaults were distinguished on plateaus.

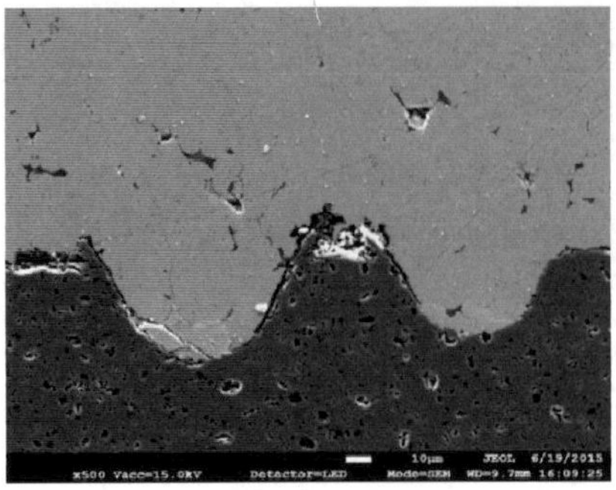

Figure 10: Cross-section of titanium on textured substrate

Finally, adhesion strengths have been evaluated (Figure 11). In any cases, adhesion strengths are limited for raw substrates (Al: 7 MPa Cu: 5 MPa and Ti: 2 MPa). In addition, laser surface texturing has shown an enhancement by a factor of three or four for the different configurations. The in-contact areas have been increased. Thus, the total energy to break the interface can be assumed larger. Besides, the compressive states created obstacles for interface crack propagation [20]. Adhesion strength strongly depends on anchoring effects . V1 and V2 patterns increased the adhesion up to 14 and 20 MPa for every application.

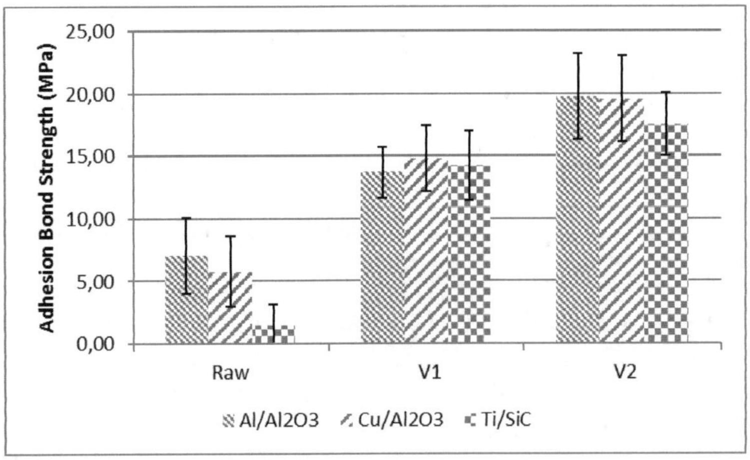

Figure 11: Adhesion bond strengths for Al, Cu and Ti coatings on raw and textured substrates (V1 and V2)

Crack propagation mechanisms can be distinguished as demonstrated by the specimens studied after pull-off tests. As illustrated in Figure 12, the cracks kinked out the interface near hole patterns and the coatings were locked in. The adhesion strength can then be increased due to those obstacles.

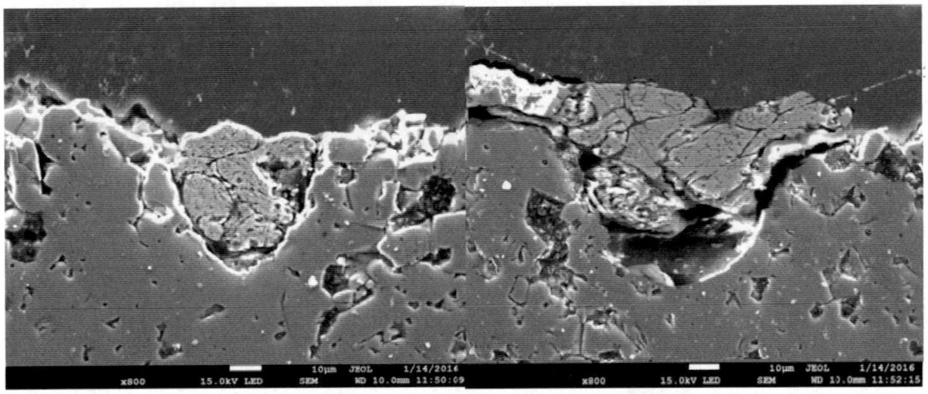

Figure 12: Crack propagation through the coating and the interface

Adhesion strength has been improved by anchoring mechanisms. Particles were locked in the substrates; thus the coating structure is crucial. The pattern filling needs to be maximized to amplify these mixed-mode failure effects. Those "roots" are beneficial and can increase coating-substrate configuration in cold spray process. Laser surface texturing enables more thermal spray application in the future.

4 Conclusions

Laser surface texturing has shown interesting aspects for cold spray coating applications. It promotes larger adhesion area and adapts the surface to particle. The deposition efficiency was increased for adapted cavities and it enables coating build-up for specific coating-substrate. Adhesion strengths have been increased in every case. Coating compressive stresses have been evaluated in patterns which promote anchoring mechanisms. It has been demonstrated that specific pattern shape can also create mixed-mode failures. Cracks propagate into the coating instead of following the interface. Laser surface texturing can create infinite surface topography as such adapted to particle size distribution and materials.

Acknowledgement

This work was performed in connection with the ANR ARCOLE (12-BS09-0009) project.

References

[1] R. C. Dykhuizen and M. F. Smith, "Gas dynamic principles of cold spray," *J. Therm. Spray Technol.*, vol. 7, no. 2, pp. 205–212, 1998.

[2] H. Assadi *et al.*, "On Parameter Selection in Cold Spraying," *J. Therm. Spray Technol.*, vol. 20, no. 6, pp. 1161–1176, Dec. 2011.

[3] T. Hussain, D. G. McCartney, P. H. Shipway, and D. Zhang, "Bonding Mechanisms in Cold Spraying: The Contributions of Metallurgical and Mechanical Components," *J. Therm. Spray Technol.*, vol. 18, no. 3, pp. 364–379, Sep. 2009.

[4] V. K. Champagne, "The cold spray materials deposition process: fundamentals and applications," *Boca Raton Woodhead*, 2007.

[5] W. Y. Li, M. Yu, F. F. Wang, and H.-L. Liao, "A generalized critical velocity window based on material property for cold spraying by eulerian method," *J. Therm. Spray Technol.*, vol. 23, no. 3, pp. 557–566, 2014.

[6] K. H. Ko, J. O. Choi, and H. Lee, "The interfacial restructuring to amorphous: A new adhesion mechanism of cold-sprayed coatings," *Mater. Lett.*, vol. 175, pp. 13–15, Jul. 2016.

[7] A. Papyrin, V. Kosarev, S. Klinkov, A. Alkhimov, and V. M. Fomin, *Cold spray technology*. Elsevier, 2006.

[8] J. Wu, H. Fang, S. Yoon, C. Lee, and H. Kim, "Critical velocities for high speed particle deposition in kinetic spraying," *Mater. Trans.*, vol. 47, no. 7, pp. 1723–1727, 2006.

[9] W.-Y. Li, H. Liao, C.-J. Li, H.-S. Bang, and C. Coddet, "Numerical simulation of deformation behavior of Al particles impacting on Al substrate and effect of surface oxide films on interfacial bonding in cold spraying," *Appl. Surf. Sci.*, vol. 253, no. 11, pp. 5084–5091, Mar. 2007.

[10] Y. Danlos, S. Costil, X. Guo, H. Liao, and C. Coddet, "Ablation laser and heating laser combined to cold spraying," *Surf. Coat. Technol.*, vol. 205, no. 4, pp. 1055–1059, Nov. 2010.

[11] T. Goyal, R. S. Walia, and T. S. Sidhu, "Surface roughness optimization of cold-sprayed coatings using Taguchi method," *Int. J. Adv. Manuf. Technol.*, vol. 60, no. 5–8, pp. 611–623, May 2012.

[12] L. Sluatineanu, S. Potarniche, M. Coteata, I. Grigoras, L. Gherman, and F. Negoescu, "Surface roughness at aluminium parts sand blasting," *Proc. Manuf. Syst.*, vol. 6, no. 2, 2011.

[13] D. Garcia-Alonso, N. Serres, C. Demian, S. Costil, C. Langlade, and C. Coddet, "Pre-/During-/Post-Laser Processes to Enhance the Adhesion and Mechanical Properties of Thermal-Sprayed Coatings with a Reduced Environmental Impact," *J. Therm. Spray Technol.*, vol. 20, no. 4, pp. 719–735, Mar. 2011.

[14] R. Kromer *et al.*, "Laser surface patterning to enhance adhesion of plasma sprayed coatings," *Surf. Coat. Technol.*, vol. 278, pp. 171–182, Sep. 2015.

[15] R. Kromer, J. Cormier, and S. Costil, "Role of Powder Granulometry and Substrate Topography in Adhesion Strength of Thermal Spray Coatings," *J. Therm. Spray Technol.*, May 2016.

[16] R. Kromer, S. Costil, J. Cormier, L. Berthe, P. Peyre, and D. Courapied, "Laser Patterning Pretreatment before Thermal Spraying: A Technique to Adapt and Control the Surface Topography to Thermomechanical Loading and Materials," *J. Therm. Spray Technol.*, vol. 25, no. 3, pp. 401–410, Feb. 2016.

[17] S. Kumar, G. Bae, and C. Lee, "Deposition characteristics of copper particles on roughnened substrates htrough kinetic spraying," *Appl. Surf. Sci.*, vol. 265, pp. 3472–3479, 2009.

[18] Z. Arabgol, H. Assadi, T. Schmidt, F. Gärtner, and T. Klassen, "Analysis of Thermal History and Residual Stress in Cold-Sprayed Coatings," *J. Therm. Spray Technol.*, vol. 23, no. 1–2, pp. 84–90, Jan. 2014.

[19] R. Kromer, Y. Danlos, S. Costil, D. Moskal, and S. Houdkova, "Coating adehsion enhancement by laser surface texturing - deposition characteristics of metallic particles on different classes of substrates in cold spraying process," *XXX Int. Conf. Surf. Modif. Technol.*, 2016.

[20] R. Kromer, S. Gojon, C. Verdy, and H. L. Liao, "Laser patterning prior-treatments in cold spraying to enhance adhesion bond strength by anchoring mechanisms," *7th RIPT Conf.*, 2015.

Effects of momentum and heat transfer between plasma and suspensions on an axial injection plasma spraying

T. Suzuki, H. Saito, T. Fujino, and M. Suzuki, Tsukuba / JPN

The paper reports numerical simulation results of a direct current (DC) suspension plasma spray with an axial injection system. In the numerical modelling, a two-way coupled Eulerian-Lagrangian approach was employed to simulate plasma flow and suspension behavior. As effects of two-way interaction, momentum transfer and energy transfer were chosen. The plasma spray was assumed to have a rotationally symmetric, two-dimensional shape around the injection axis of suspensions (the central axis), and therefore the axisymmetric two-dimensional approximation was adopted in the numerical modelling. Working gas was pure argon and was supplied from both the anode-side and the cathode-side ports. A total volume flow rate of the working gas from anode and cathode was set to 46.5 slm. A feed rate of suspension was parametrically set to 0, 15, 25, or 35 g/min. Numerical results indicate that the temperature of a plasma hot-core region and the velocity of a plasma jet around the central axis drop more with increasing feed rate of suspension mainly because of a decrease in Joule heating with increasing it. The numerical results also suggest that the increase in feed rate of suspension leads to practically lengthening flight distance of suspensions required for completing evaporation process of disperse medium.

1 Introduction

Suspension plasma spraying (SPS) process employs liquid droplets which contain micron and submicron powders [1]. Spraying of such powders forms small splats on the substrate, leading to finely structured coatings. The SPS processes are generally operated with the radial injection of feedstock material, although it is difficult to penetrate suspensions into a hot core region of plasma plume.

To avoid that issue, a plasma spray with axial injection system of the suspension is employed [2, 3]. The plasma spray can inject suspensions directly into the hot spot of the plasma jet and it can recognize efficient transport of momentum and energy from the plasma jet to the suspensions [4]. While an axial feeding system has these merits, more complicated plasma-suspension interaction occurs. To deeply understand the plasma-suspension interaction is important for controlling the SPS process with axial injection system. Numerical simulation is one of approaches to obtain the findings of the plasma-suspension interaction and its effects. Other researchers [5, 6] conducted numerical simulation focusing on the plasma-suspension interaction in the SPS process with a radial injection plasma spray. They reported that operating conditions such as the flow rate of the gas, suspension injector angles, positions, and injection velocities affect the heating and the accelerating processes of suspension. Numerical simulation for SPS processes with axial injection plasma spray considering the plasma-suspension interaction, however, has yet to be reported. In this work, therefore, we conducted numerical modelling of the plasma-suspension interaction in the SPS process with an axial injection plasma spray. In the modelling, momentum transfer and heat transfer between plasma and suspension were chosen as effects of two-way interaction.

The present paper reports the outlines of the numerical modelling and numerical results of two-way plasma-suspension interaction in the SPS process with the axial injection plasma spray.

2 Numerical modelling

2.1 Governing equations for fluid dynamics

Governing equations for fluid dynamics are written as follows:

- Mass conservation equation

$$\frac{\partial \rho}{\partial t} + \boldsymbol{\nabla} \cdot (\rho \boldsymbol{u}) = 0 \tag{1}$$

- Momentum conservation equations

$$\frac{\partial (\rho \boldsymbol{u})}{\partial t} + \boldsymbol{\nabla} \cdot (\rho \boldsymbol{u}\boldsymbol{u}) = \boldsymbol{\nabla} \cdot \bar{\bar{\tau}} - \boldsymbol{\nabla} p - \boldsymbol{S}_M \tag{2}$$

- Energy conservation equation

$$\frac{\partial (\rho E_{flu})}{\partial t} + \boldsymbol{\nabla} \cdot (\rho H_{flu} \boldsymbol{u}) = \boldsymbol{\nabla} \cdot (\kappa \boldsymbol{\nabla} T) + \bar{\bar{\tau}} : \boldsymbol{\nabla} \boldsymbol{u} + \boldsymbol{J} \cdot \boldsymbol{E} - \dot{q}_{rad} - S_E \tag{3}$$

- Ideal gas law

$$p = \rho R T \tag{4}$$

where t is the time, ρ the mass density, $\boldsymbol{u}$ the velocity of the gas. $\bar{\bar{\tau}}$ the viscous stress tensor, p the pressure. E_{flu}, H_{flu} and κ are the total internal energy per unit mass, the total enthalpy per unit mass, and the thermal conductivity, respectively. $\boldsymbol{J}$, $\boldsymbol{E}$ and $\dot{q}_{rad}$ are the current density, the electric field and the radiation heat loss, respectively. R is the specific gas constant. $\boldsymbol{S}_M$ is the momentum transfer from the gas to suspensions, S_E the energy transfer from the gas to suspensions. The numerical modes for $\boldsymbol{S}_M$ and S_E will be given later.

Ar gas is adopted as operating gas. An arc and a surrounding gas are assumed to be in locally thermal and chemical equilibrium (LTCE) state. Under the assumption, thermodynamic and transport properties of Ar gas are calculated with standard statistical thermodynamics and the well-known Chapman-Enskog method, respectively. The radiation heat loss is calculated under optically-thin assumption.

Although a detailed description on a numerical solver of Eqs. (1) – (4) is omitted due to space limitation, we employ a finite volume method for discretization coupled with the AUSM-DV scheme [7] for evaluating

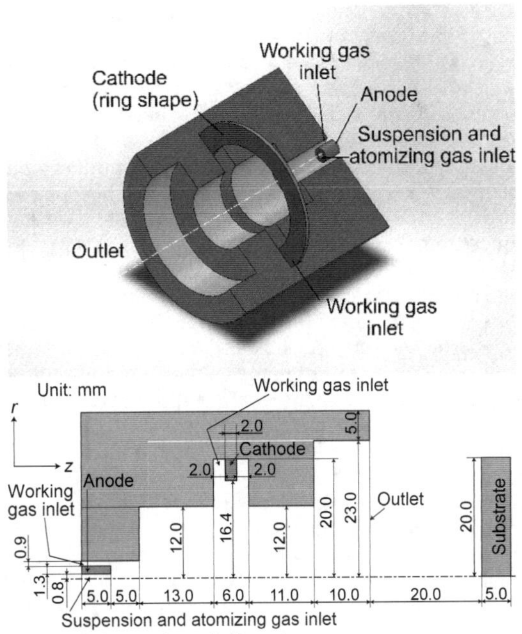

Fig. 1. Schematics of plasma spray.

numerical flux of convective terms and the 4-step Runge-Kutta scheme for time integration.

2.2 Governing equations for electrodynamics

Governing equations for electrodynamics are written as follows:

$$\boldsymbol{E} = -\boldsymbol{\nabla}\phi \tag{5}$$

$$\boldsymbol{\nabla}\cdot\boldsymbol{J} = 0 \tag{6}$$

$$\boldsymbol{J} = \sigma\boldsymbol{E} \tag{7}$$

where ϕ and σ are the scalar potential and the electric conductivity, respectively. An elliptic type partial differential equation of ϕ is derived from Eqs. (5) – (7), and then it is discretized by the Galerkin-FEM.

2.3 Governing equations for suspensions

In this work, it is assumed that suspensions and powder particles are spherical [8] and no collision or breakup happens. The suspensions and the powder particles are tracked by the Lagrangian approach. The governing equations for them are written as follows.

- Dynamic equations of particle motion:

$$\frac{d\boldsymbol{u}_p}{dt} = \frac{3\rho C_D}{8\rho_p r_p}|\boldsymbol{u}-\boldsymbol{u}_p|(\boldsymbol{u}-\boldsymbol{u}_p) \tag{8}$$

- Thermal energy transport equation:

$$\frac{dT_p}{dt} = \frac{\dot{Q}_{conv}-\dot{Q}_{vap}-\dot{Q}_{rad}}{mc_p} \tag{9}$$

- Equation of particle radius:

$$\frac{dr_p}{dt} = \frac{\dot{m}}{4\pi\rho_p r_p^2} \tag{10}$$

- Momentum transfer and energy transfer

$$\boldsymbol{S}_M = \frac{1}{V_{cell}}\sum_i \left(m\frac{d\boldsymbol{u}_p}{dt}\right)_i \tag{11}$$

Table 1. Operating conditions.

Gas type	Ar
Current [A]	200
Flow rate of working gas from anode side [slm]	17
Flow rate of atomizing gas [slm]	1.3
Flow rate of working gas from cathode side [slm]	29.5
Substrate distance from spray outlet [mm]	20
Substrate radius [mm]	20
Feedstock powder	ZrO_2
Disperse medium	C_2H_5OH
Concentration of ZrO2 in suspension [wt.%]	10
Powder size [μm]	5.0
Suspension size [μm]	80
Suspension feed rate [g/min]	0, 15, 25, 35
Initial injection position [mm]	$(r, z) = (0.5, 4.5)$
Initial velocity [m/s]	$(u_{p,r}, u_{p,z}) = (0, 10)$

$$S_E = \frac{1}{V_{cell}}\sum_i \left(\frac{1}{2}m\frac{d|\boldsymbol{u}_p|^2}{dt} + mc_p\frac{dT_p}{dt}\right) \tag{12}$$

where $\boldsymbol{u}_p$ is the particle velocity, r_p, ρ_p, and C_D are the particle radius, the particle density, and the drag coefficient, respectively. T_p, m, c_p $\dot{Q}_{conv}$, $\dot{Q}_{vap}$, and $\dot{Q}_{rad}$ are the particle temperature, the particle mass, the particle specific heat, the convective heat flux, the evaporative heat flux, and the radiative heat flux, respectively [9]. When T_p reaches the evaporation temperature of disperse medium $T_{p,d}$, the temperature of suspensions keeps $T_{p,d}$ until disperse medium gets enough latent heat for vaporization. When T_p reaches the melting temperature of powder particles $T_{p,m}$, the temperature of powder particles keeps $T_{p,m}$ until they get enough latent heat for melting. $\dot{m}$ is the evaporation rate of the particle. V_{cell} is the volume of the computational cell.

2.4 Numerical domain and condition

Figure 1 illustrates the schematic drawings of a twin-cathode type DC plasma spray with an axial injection system analyzed in this work, which has a rotationally symmetric shape. The basic structure of the spray is designed by referring the plasma spray used by Shahien et al [10], although their spray has three-dimensional shape.

The authors adopt axisymmetric two-dimensional approximation for numerical modelling for the plasma torch shown in Fig. 1. Table 1 summarizes operating conditions. In this work, only a feed rate of suspension is parametrically varied. The radial position of suspension's initial injection has normal distribution whose standard deviation is 0.2 mm.

On the wall surface boundary, non-slip wall condition is applied and wall temperature is fixed at 300 K. On the boundary sufficiently apart from the plasma spray, the pressure is set to 1 atm and the temperature and the

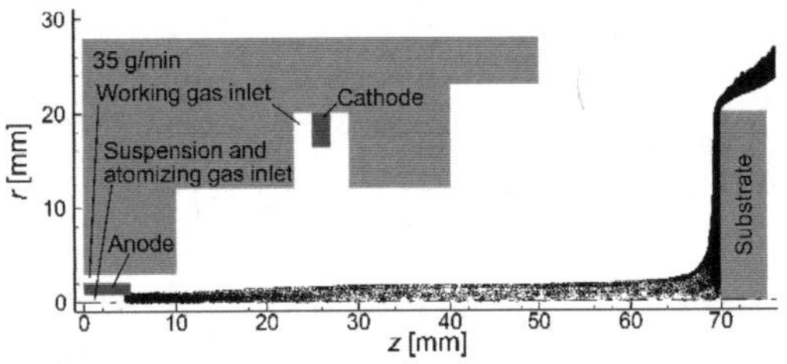

Fig. 2. Space distribution of suspensions at an instant of time for a feed rate of suspension of 35 g/min.

pressure are decided using non-gradient condition in the direction normal to boundary surface. The electrical potential at the cathode is set to 0 V, while the one on the anode is determined so as to satisfy an electric current of 200 A between the anode and the cathode.

3 Results and discussion

3.1 Flight characteristics of suspensions in plasma jet

Figure 2 shows the space distribution of suspensions at an instant of time when a feed rate of suspension is 35 g/min. Suspensions injected from the inlet port fly toward the substrate inside the spray without largely expanding in the radial direction. Consequently, momentum and heat transfer around the central axis are particularly popular. For other feed rates of suspension, almost same suspensions' distributions as the one in Fig. 2 were obtained.

Figures 3(a) and 3(b) depict the histories of temperature and velocity, respectively, of a single suspension with the initial radial position of $r = 0.3$ mm for three different feed rates of suspension: 15, 25, and 35 g/min. Figures 3(a) and 3(b) also show those histories obtained by using one-way coupling manner where the momentum and heat transfer from plasma to suspension are taken into account. The temperature of suspension injected from the inlet port increases to the boiling point of disperse medium with 351 K, and then the suspension's temperature is kept at the boiling point until the disperse medium's evaporation process is finished. After that, an emerged powder particle is heated and its temperature rapidly increases to 2950 K, which corresponds to the melting point. Thereafter, the particle with keeping molten state flies toward the substrate. In the vicinity of the substrate, the particle's temperature decreases suddenly. This is due to the existence of thermal boundary layer of fluid flow near the substrate. Comparing temperature histories of different suspension's feed rates, the plateau temperature region related to evaporation of disperse medium extends more downstream with increasing feed rate of suspension. This indicates that an amount of heat transfer from plasma jet to a single suspension decreases with increasing feed rate of suspension. However, since powder particles melt in any condition as shown in Fig. 3(a), the heating ability of the plasma spray is sufficiently high. From Fig. 3(b), the suspension is drastically accelerated after the evaporation of disperse medium. Thereafter, its velocity takes a peak value and then decreases as it goes toward the substrate. In the vicinity of the substrate, the particle in molten state has a very small velocity. This is because it has small inertia, so that it is susceptible to rapidly decelerated plasma jet near the substrate. Figure 3(b) also shows that the increase in feed rate of suspensions leads to decreasing suspension's velocity over the whole region between the injection point and the substrate, indicating that momentum transfer from plasma jet to a single suspension decreases with increasing suspension's feed rate. We can forecast from this result that a collision velocity of the particle in molten state with the substrate gets lower by increasing a feed rate of suspension.

In Fig.3 we can recognize substantial differences of temperature and velocity history between one-way and two-way coupling manners, and also those differences become large as a feed rate of suspension increases. The one-way coupling manner underestimates a flight distance of suspension required for completing evaporation process of disperse medium and overestimates a flight velocity of particle in both solid and molten states. The one-way coupling manner is incapable of simulating a reduction of plasma jet's temperature and velocity due to the interaction

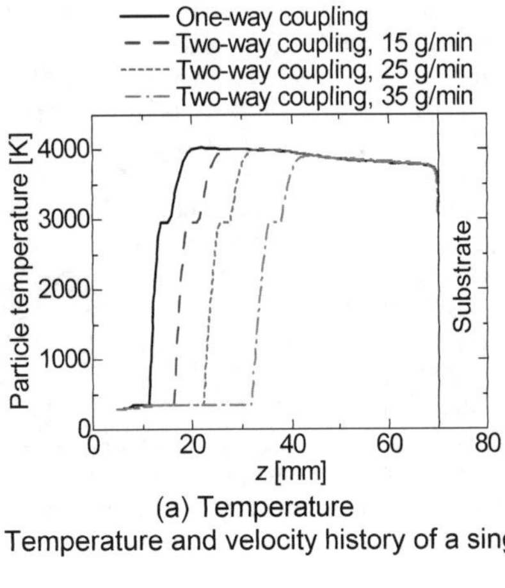

(a) Temperature

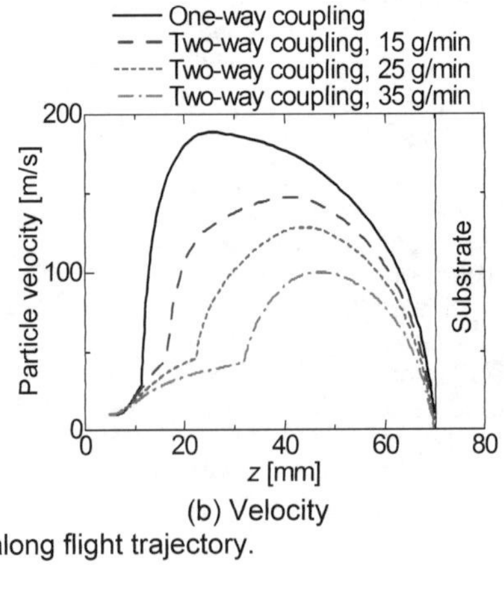

(b) Velocity

Fig. 3. Temperature and velocity history of a single suspension along flight trajectory.

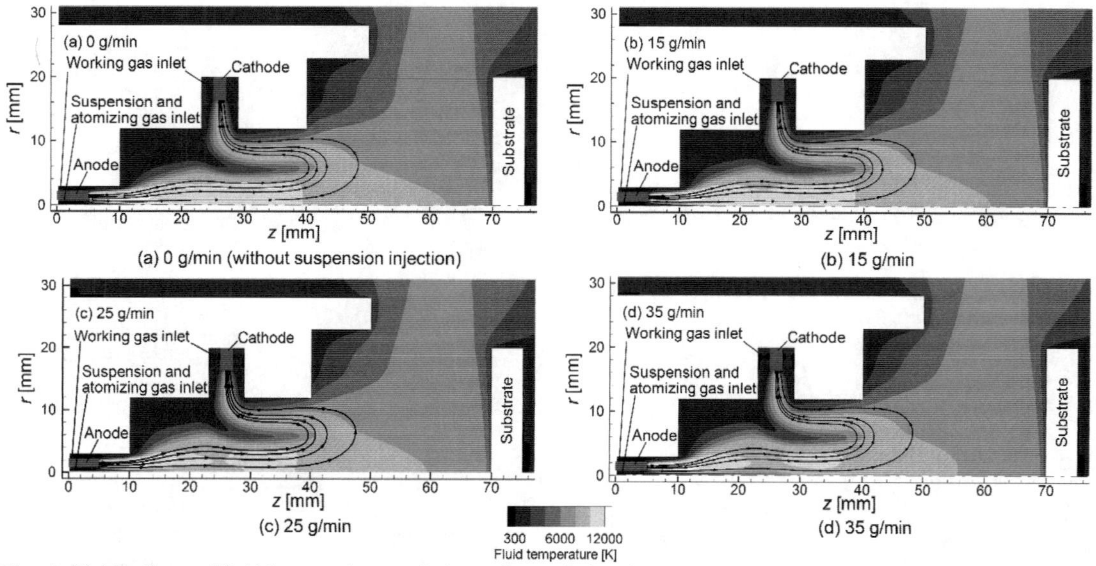

Fig. 4. Distributions of fluid temperature and streamlines of electric current.

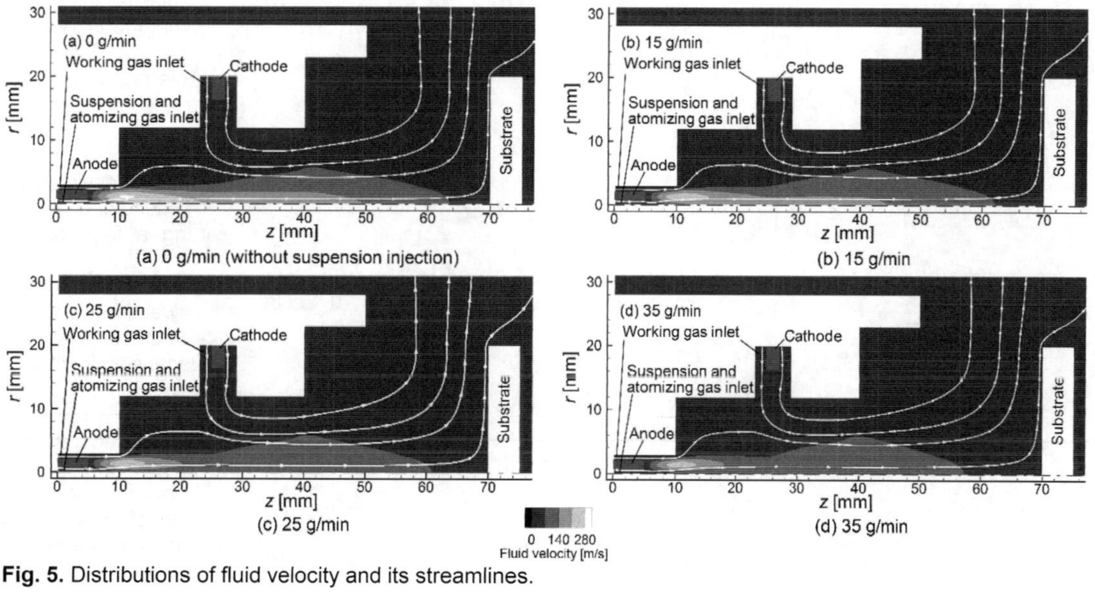

Fig. 5. Distributions of fluid velocity and its streamlines.

between plasma jet and suspensions, as will be discussed in the following section.

3.2 Influences of suspensions on plasma jet's characteristics

Figure 4 shows distributions of fluid temperature and streamlines of electric current for the following four different feed rates of suspension: 0, 15, 25, and 35 g/min. The electric current extracted from the anode passes through a core of arc jet around the center axis, and then it is largely bent toward the cathode near the plasma spray's outlet (z = 50 mm). Arc core's temperature around the central axis drops by increasing a feed rate of suspension. This is because the increase in its feed rate leads to an increase in heat transport amount from plasma jet to suspensions.

Figure 5 depicts distributions of fluid velocity and its streamlines for the four different feed rates of suspension, showing that arc core's velocity as well as its temperature near the central axis decreases as the feed rate of suspension increases.

To examine the effect of suspensions on plasma jet in detail, the distributions of temperature, velocity, and Joule heating in plasma jet along the line r = 0.5 mm are shown in Fig. 6. The drop in arc temperature due to injection of suspensions occurs in the whole region between the injection point and the substrate,

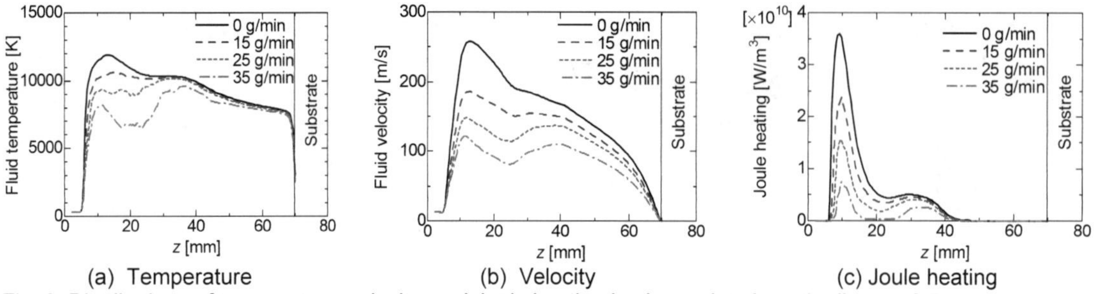

Fig. 6. Distributions of temperature, velocity, and Joule heating in plasma jet along the line r = 0.5 mm.

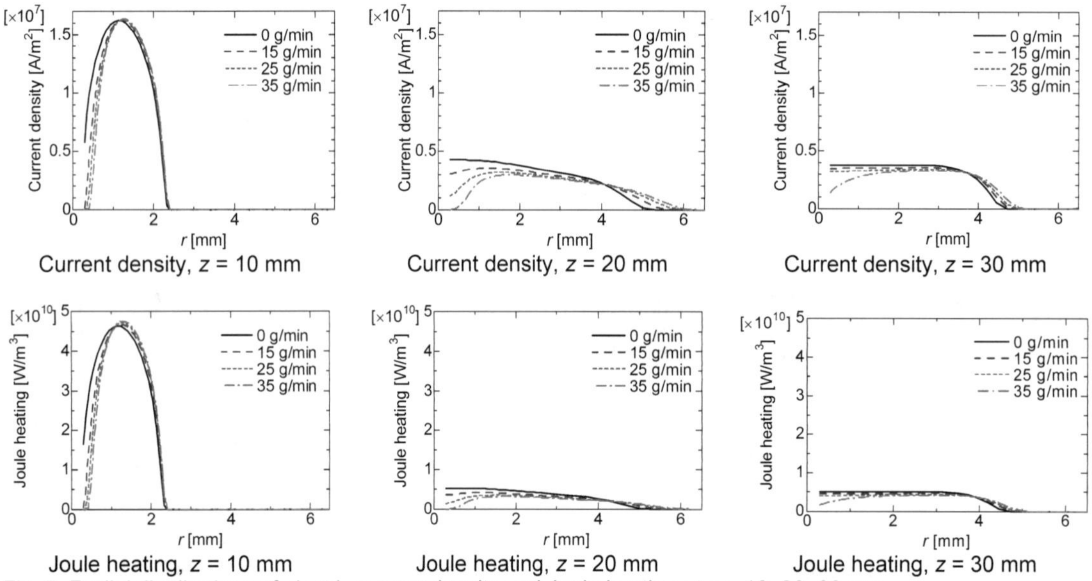

Fig. 7. Radial distributions of electric current density and Joule heating at z = 10, 20, 30 mm.

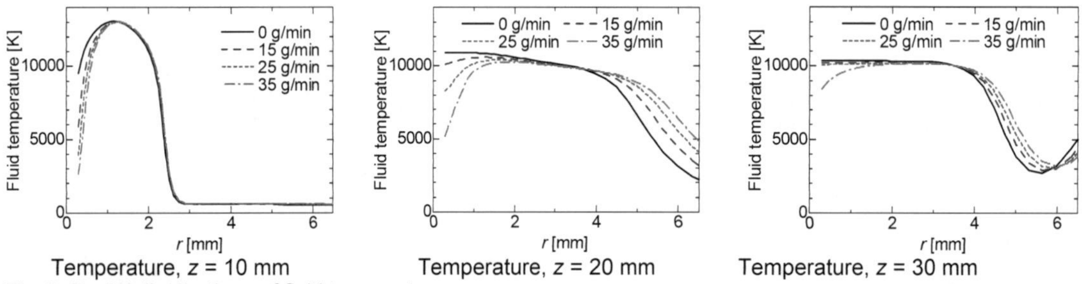

Fig. 8. Radial distributions of fluid temperature.

particularly in the region from z = 5 mm to z = 30 mm, as shown in Fig. 6(a). Furthermore, an amount of temperature drop grows as the suspension's feed rate increases. The velocity of arc jet decreases with the increase of the feed rate, as shown in Fig. 6(b). Figure 6(c) indicates that the Joule heating, which yields directly the rise in arc jet's temperature and indirectly the increase in its velocity, is strong in the two regions: around z = 10 mm and z = 30 mm for all of the feed rates of suspensions. Furthermore, we can recognize from Figs. 6(a) – 6(c) a strongly positive correlation between Joule heating and arc jet's temperature and the one between Joule heating and arc jet's velocity, although there is a slightly lag of variations of temperature and velocity from variation of Joule heating along the line.

Figure 7 depicts the radial distributions of electric current density and Joule heating at z = 10, 20, and 30 mm. The increase in feed rate of suspension leads to a drop in both electric current density and Joule heating near the central axis (r = 0 mm), while it leads to a slight inflation of the area with electric current to the radial direction except for the region near the injection port. The trends in the distributions of current density and Joule heating generally agree.

Figure 8 illustrates the radial distributions of fluid temperature at z = 10, 20, and 30 mm. Corresponding to the decrease in Joule heating around the central axis with increasing feed rate of suspension, the fluid temperature around the central axis drops with increasing it. Although figures are omitted, this leads to a drop in electrical conductivity around the central axis. Consequently, the electric current density around the central axis decreases with increasing feed rate of suspension, inevitably requiring further expansion of the area with electric current under constant operation current condition (200 A in this work).

4 Conclusions

The authors conducted axisymmetric two-dimensional numerical simulations of a direct-current (DC) suspension plasma spray with an axial injection system, in which momentum and heat transfer between plasma jet and suspensions were taken into account. A feed rate of suspension was parametrically varied. Main results obtained by this work are summarized as follows:

(1) Momentum and heat transfer from plasma jet to suspension practically influence the fluid characteristics of a plasma jet. Therefore, numerical modelling of the axial injection DC plasma spray should require a two-way coupling manner between plasma jet and suspensions, not a one-way coupling manner from only plasma jet to suspensions.
(2) Increase in feed rate of suspension results in decreasing plasma jet's temperature and velocity, lengthening flight distance of suspensions required for completing evaporation process of disperse medium, and decreasing flight velocity of particles in molten state.
(3) Electric current density and Joule heating in the region around the central axis, through which the suspensions pass, drop by increasing feed rate of suspension, while the area with electric current slightly inflates to the radial direction under constant operation current condition.

Literature

[1] Fazilleau, J., Delbos, C., Rat, V., Coudert, J. F., Fauchais, P., and Pateyron, B. : Phenomena Involved in Suspension Plasma Spraying Part 1: Suspension Injection and Behavior. Plasma Chemistry and Plasma Processing 26 (2006), Issue 4, pp. 371/391.

[2] Berghaus, J. O., Marple, B., and Moreau, C. : Suspension Plasma Spraying of Nanostructured WC-12Co Coatings. Journal of Thermal Spray Technology 15 (2006), Issue 4, pp. 676/681.

[3] Tarasi, F., Medraj, M., Dolatabadi, A., Berghaus, J. O., and Moreau, C. : Effective Parameters in Axial Injection Suspension Plasma Spray Process of Alumina-Zirconia Ceramics. Journal of Thermal Spray Technology 17 (2008), Issue 5, pp. 685/691.

[4] Waldbillig, D. and Kesler, O. : Characterization of Metal-supported Axial Injection Plasma Sprayed Solid Oxide Fuel Cells with Aqueous Suspension Plasma Sprayed Electrolyte Layers. Journal of Power Sources 191 (2009), Issue 2, pp. 320/329.

[5] Jabbari, F., Jadidi, M., Wuthrich, R., and Dolatabadi, A. : A Numerical Study of Suspension Injection in Plasma-Spraying Process. Journal of Thermal Spray Technology 23 (2014), Issue 1, pp. 3/13.

[6] Farrokhpanah, A., Coyle, T. W., Mostaghimi, J. : Numerical Study of Suspension Plasma Spraying. Journal of Thermal Spray Technology 26 (2017), Issue 1, pp. 12/36.

[7] Yamamoto, S., Kano, S., and Daiguji, H. : An Efficient CFD Approach for Simulating Unsteady Hypersonic Shock-shock Interference Flows. Computers & Fluids 27 (1998), Issue 5-6, pp. 571/580.

[8] Xiong, HB. and Lin, JZ. : Nanoparticles Modelling in Axially Injection Suspension Plasma Spray of Zirconia and Alumina Ceramics. Journal of Thermal Spray Technology 18 (2009), Issue 5-6, pp. 887/895.

[9] Wan, Y. P., Prasad, V., Wang, G.-X., Sampath, S., and Fincke, J. R. : Model and Powder Particle Heating, Melting, Resolidification, and Evaporation in Plasma Spraying Processes. Journal of Heat Transfer 121 (1999), Issue 3, pp. 691/699.

[10] Shahien, M. and Suzuki, M. : Low Power Consumption Suspension Plasma Spray System for Ceramic Coating Deposition. Surface & Coatings Technology (2016), 6 pages. http://dx.doi.org/10.1016/j.surfcoat.2016.07.040

Submodeling of alumina coatings by APS

Martínez García, V., Gadow, R., Killinger, A., Moreno Nicolás, J. A., Düsseldorf / D

Traditional low-cost bulk materials are unable to fulfil the increasing requirements of actual technology and functional coatings – e.g. APS alumina coatings on aluminium substrates for tribological properties – are a suitable alternative. The development of new thermal spray coatings is usually based in experimental procedures which involve long development times and high costs. Nowadays, numerical simulation allows the researcher a better understanding of thermal spray processes as well as reducing the time and cost for the optimization of processes, but it requires a deep insight into the physics of the phenomenon. The submodelling approach allows the researchers to work with a local model, a thin layer – just a few microns – on the metal substrate surface, and more realistic boundary conditions, which requires a little specialised knowledge. The current study involves the workflow to manage the modelling at piece scale, and the transfer and interpolation of boundary conditions in each scale. The coating is divided into several layers, which represent the successive splats deposited during the process, and the heat flow from the torch is modelled by radiation and convection. A code is implemented in order to generate the routine needed by the FEM software, in which the results are processed and interpolated for the subsequent submodel. Furthermore, the material plasticity is considered and several tests are performed in order to check the simulation results.

1 Introduction

Lightweight is one of the great factor for vehicles and machines desing, where light materials, e.g. aluminium, with poor mechanical properties at high temperatures as well as poor tribological properties are widely used. Therefore, coating application by thermal spraying is a sophisticated and quasi-monolithic solution to improve superficial properties of parts, with high flexibility in terms of component geometry and layer composite material combination. Hence, it can be easily integrated into existing production lines, coating materials are environmentally inert, and recycling is easy and uncritical [1].
Due to the dynamic heat and mass transfer processes during thermal spraying and subsequent coating cooling, the component is exposed to a variety of different thermal situations with great temperature gradients. Temperature compensation between coating and substrate depends on the thermal conductivities of the materials, while different thermophysical properties of the metal or ceramic coating and the light metal alloy substrate cause different expansion gradients, with emphasis on the coefficient of thermal expansion. Therefore, changes in the thermal situation lead to thermal stresses. During the cooling period of the component, the high contraction of the light metal substrate compared to the coating causes compressive thermal stresses in the coating [1].
A previous step to the analysis of the residual stress generation mechanisms in the whole component is the temperature field analysis. Besides, the understanding of the physical phenomenon must be applied in a scale close to the droplet size, but the boundary conditions are applied in the piece wall. Thus, the necessary resolution implies a droplet scale, while the available data come from the piece scale. Numerical methods, such as the finite element method (FEM), are useful tools solving this kind of problems, but the design of a mesh fine enough to match with the local region leads to an excessive number of nodes and degrees of freedom, which can burden the available computing resources. An alternative approach is to selectively refine the region of interest and to maintain a coarser mesh elsewhere [2].
Previous works used FEM to carry out a one-way coupled temperature stress field analysis, including mass transfer, and estimate the stress generation during thermal spraying, but present deficiencies in the connection between coating and substrate [1, 3, 4]. Further models coupled numerical simulation and off-line programming to adapt process kinematics in order to control the substrate temperature, and consequently, to influence the development of residual stresses in the coating composite, but encompass the macroscopic phenomena without microscopic resolution. [5, 6]. On the other hand, models which estimate the residual stress distribution in coatings has been developed with 3-D stochastic models for a small volume of the coating and substrate, based on microstructure data from scanning electron microscope (SEM) [7].
There are several techniques for local mesh refinement. One is to use multigrid techniques, in which a coarse grid is solved and then used as an initial guess for an iterative procedure for reducing the stiffness matrices. This technique naturally combines with adaptive refinement, where the finite element mesh changes from a coarse mesh in the far field to a finer mesh around the region of interest. Another approach is substructuring, where symmetries of the mesh are exploited to reduce the size of the stiffness matrices. A further approach is submodelling, in which a global grid is solved and the region of interest is isolated as a submodel and analysed separately with a finer mesh. The initial and boundary conditions for the submodel are taken from the solution of the global mesh. This technique is also occasionally known as substructuring (in the past), or as zooming global-local analysis [8].
Submodeling is based on St. Venant's principle, which states that if an actual distribution of forces is replaced by a statically equivalent system, the

distribution of stress and strain is altered only near the regions of load application. This implies that stress concentration effects are localized around the concentration; therefore, if the boundaries of the submodel are far enough away from the stress concentration, reasonably accurate results can be calculated in the submodel [9]. Therefore, the same can be applied for thermal analysis.
For load-history dependent problems, the boundary conditions must be adapted from the coarse model at multiple substeps to the fine-mesh model analysis at equivalent substeps. The more boundary conditions are being read and implemented, the more accurate are the results of the fine-mesh model analysis [9].
In this work, the thermal and stress analysis using submodeling has been validated by experimental results.

2 Methods

2.1 Governing equations

The necessary equations to study the coating process can be divided into:
- The equation of heat flow balance (inside the body and over its surfaces) [10]:

$$-\frac{\partial}{\partial x}\left(k_x \cdot \frac{\partial T}{\partial x}\right) - \frac{\partial}{\partial y}\left(k_y \cdot \frac{\partial T}{\partial y}\right) - \frac{\partial}{\partial z}\left(k_z \cdot \frac{\partial T}{\partial z}\right) + q_V + q_S = \rho \cdot C \cdot \frac{\partial T}{\partial t} \quad (1)$$

where the boundary conditions are the heat flux from the APS torch, and the heat loss by convection and radiation.
- The equilibrium equation or Navier equation [11]:

$$\mu \nabla^2 \mathbf{u} + (\lambda + \mu)\nabla(\nabla \cdot \mathbf{u}) - \alpha(3\lambda + 2\mu)\nabla T + \mathbf{f} = \mathbf{0} \quad (2)$$

with the temperature field as boundary condition for the calculation of thermal elongation. In the finite element model described above, equation 2 is solved in its weak (integral) form with implicit time integration [12].

2.2 Submodeling

In this section, the necessity of using submodeling and the selection of the parameters for submodeling are described.
One of the difficulties found in this study is the big difference between the scales used in the analysis of the problem. The boundary conditions must be applied to a component with a size of 115×115×10 mm^3 (the standard size for experimental testing). In comparison, the region of interest to understand the effects of the temperature gradient consists of a very small agglomeration of splats. To be more precise, four adjacent columns, each with a thickness of ten splats, resulting in a volume of 120×120×50 μm^3, are defined. This means a volume ratio of 1.837×10^8, which will be accounted for by applying submodeling techniques. Due to the large volume ratio between the global (coarse mesh) model and the submodel, a medium submodel region with a volume of 600×600×500 μm^3 will be considered additionally, which allows the study of the interaction between coating and metal layer. Thus, the volume ratio between the component and the first (intermediate) submodel is 7.347×10^5, and the volume ratio between the first submodel and the second one is 250.
The incoming heat flux area that directly receives the heat flux from the moving APS torch, i.e. surface nodes, must follow the torch movement, leading to a time-dependent change in the heat transfer conditions of the surface, and thus, large temperature gradients. However, inside the material, the temperature change due to heat conduction is lower, and logically, the gradient is lower too. To reduce the number of elements used in the model of the whole component, element size grows with distance from the surface. **Fig. 1** shows the mesh used for the whole component. The mesh is refined towards the surface, with the smallest elements on the surface representing the coating/substrate interface. The global component mesh only includes the substrate, but no coating volume elements. The coating is modelled by shell elements on the surface in order to allocate coating material properties for convective and radiative heat transfer after coating deposition.

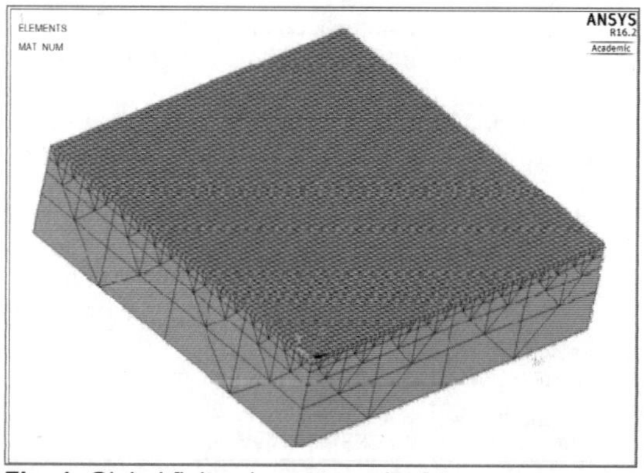

Fig. 1. Global finite element mesh of the component

As shown in the first paragraph, the heat flux from the spray torch is defined as forced convection. This heat flux is applied on the surface of 12×12 elements in the global model, equivalent to 7.2×7.2 mm^2. In each load step of the calculation, a row of 12 elements perpendicular to the torch movement path start to receive the heat flux, and 12 elements leave the area covered by the torch. A very important parameter to de define the heat interchanged is the film coefficient, h, which is defined by the Nusselt number, Nu [13]. Other parameters which characterized the process are the APS nozzle diameter, D, which is 6 mm, and the thermal conductivity of the spray jet gases, k [13].
The time step is related to the torch velocity, vflame, 0.5 m/s, and the width of the elements, Δx, as follows:

$$t_{step} = \frac{\Delta x}{v_{flame}} \quad (3)$$

The first submodel is built with the size corresponding to one surface element of the global component model. This element is subdivided in 10×10×10 elements, and a coating element layer is added, see **Fig. 2**. This time the coating element is equivalent in size to the underlying substrate one.

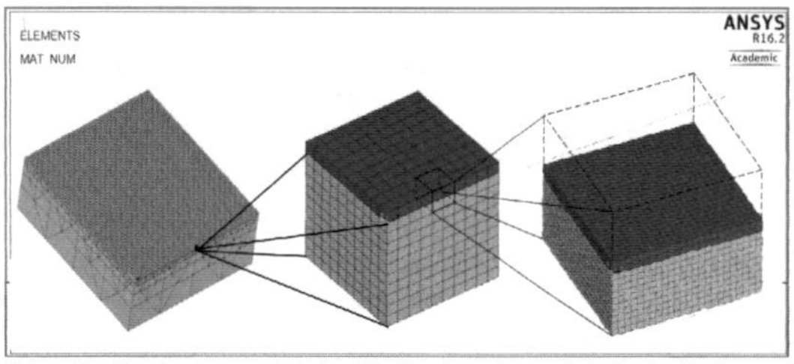

Fig. 2. Global model (left), first submodel (middle) and second submodel (right); second step, corresponding to the deposition of two splats in the second submodel.

The boundary conditions on the nodes on the bottom face of the first submodel are adopted from the node value interpolation of the respective, neighbouring element in the global element model at the end of each step. The same procedure is applicable to the lateral faces, with the exception of the nodes on the edge of the top face.

Modelling of the coating process and heat flux on the top face involves the following steps as well as initial and boundary conditions:

- Convection and radiation of the uncoated surface before the torch reaches the element are represented by metal surface properties. Coating elements are deactivated.
- On the torch position, the coating layer still does not exist and the boundary condition is represented by forced convection on the substrate surface nodes.
- Convection and radiation on the coated surface are represented by coating material properties. The coating elements are now activated.

The second submodel is built from four adjacent elements in the metal layer of the first submodel, subdividing each one in 12×12×10 elements. When the flame position was allocated on the metal surface, the coating elements grow on this position, 24×24×1 per step (corresponding to one splat in volume), see **Fig. 2**. When the process is finished, there are four 24×24×10 coating elements.

The boundary conditions in the nodes on the bottom face of the second submodel (metal material) come from the node value interpolation of the selected element in the first submodel. The same procedure is applicable to the lateral faces, with the exceptions of the nodes on the corner of the top face.

Modelling of the splat growing process and heat flux by convection on the top face involves the following steps as well as initial and boundary conditions:

- The initial temperatures are interpolated from the results of the prior step in the first submodel,
- Convection and radiation of the uncoated surface before the torch reaches the element are represented by metal surface properties. Coating elements not yet deposited are deactivated,
- Ten coating layers appear progressively on the top surface - element activation - during torch loading, each one in a separate step. The boundary condition is represented by forced convection,
- The initial temperature of the nodes on the top face of the coating layer is of 1900 K,
- Convection and radiation on the surface after the coating are represented by coating material properties.

Due to the modelling procedure in this second submodel, that each of the ten coating layers must appear incrementally, the element birth and death technique is not applicable []. Therefore, each layer deposition (each step, respectively) requires a new model including the next, deposited layer (corresponding to one splat), adopting initial conditions from the previous calculation step, which is automated with a routine programming, as shown in **Fig. 3** for the files flowchart in the thermal analysis between the gobal model and first submodel.

The boundary conditions are represented by long dash dot lines, initial conditions by round dot lines and solutions by solid line. The file extension .inp correponds to ANSYS® input files and the file extension .rth corresponds to thermal results files.

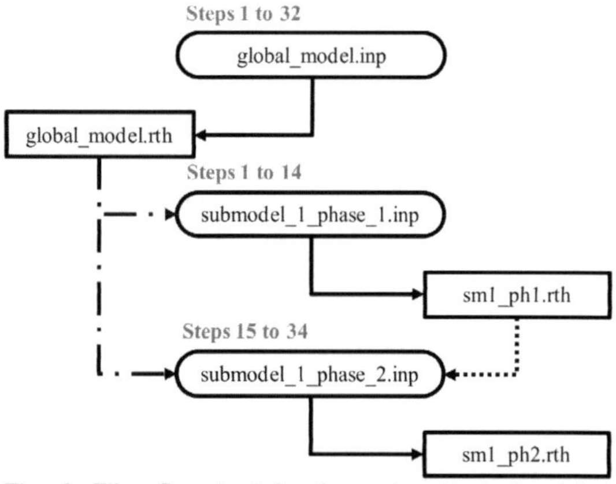

Fig. 3. Files flowchart for thermal analysis in global and first submodel.

3 Simulation results

A sample plate of aluminium and an APS thermal spray coating process have been modelled. The coating material is alumina, and the material properties of substrate and coating are obtained from a technical handbook [14]. Uncoupled thermal stress calculations of the component coating process were conducted. In this section will be presented only the results of the thermal and tensional analysis of the first submodel. The temperature/depth profile and its evolution is studied for eleven nodes of the substrate, which are located in the same vertical on the centre of this model (element), and two upper nodes corresponding to the coating, as shown in **Fig. 4**. The distance between two consecutive nodes is 50 μm.

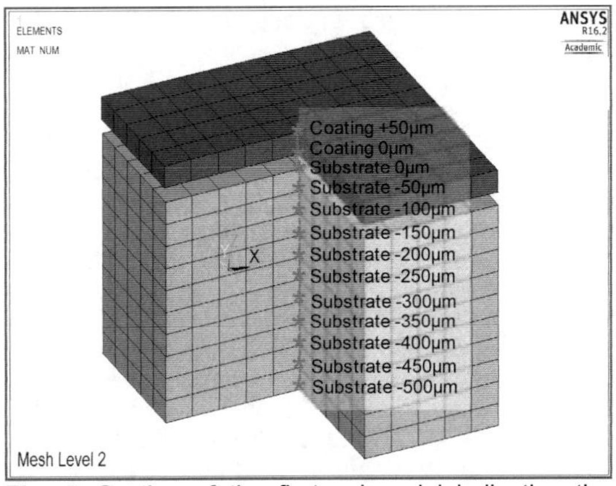

Fig. 4. Section of the first submodel indicating the nodes selection for the temperature/depth analysis.

Fig. 5 and **Fig. 6** show the temperature evolution and the thermal stresses in x-direction during the coating deposition at different nodes, for the first submodel scale, respectively. Thermal stress calculations were made considering plastic behaviour. The dash line represents the instant in which the coating is deposited in the surface, which corresponds with a time of 16,8 ms for the selected. The temperature and stress evolutions are represented until 40,8 ms in order to obtain a zoom of the first instants after the coating deposition.

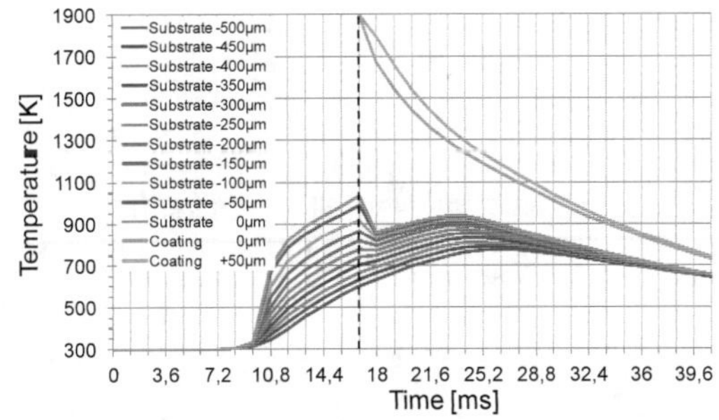

Fig. 5. Temperature evolution in first submodel.

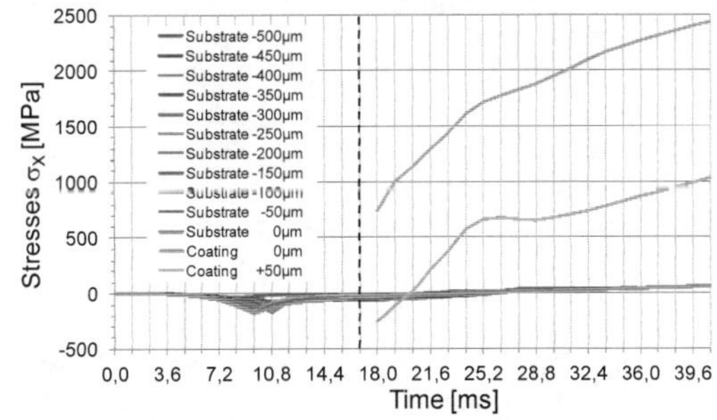

Fig. 6. Stress evolution (σ_x) in first submodel.

It is remarkable the presence of a high temperature diference in the interface during the first instant after the coating deposition, which is explained by the definition of a contact resistance between the metal susbtrate and the ceramic coating. The characteristics of the interface is of great importance for the thermal analysis but more in the tensional analysis, where the stresses of the coating are to high at the beginning of the process, with high mismatch between the coating and substrate strains.In this stage, the cooling down of the lastly deposited coating layer results in its contraction and simultaneous expansion of the adjacent material. This cooling is so-called first cooling and generates the quenching stresses, also called first stresses intrinsic or deposition stresses [15]. As the temperature compensation between the substrate and coating occurs, the so-called second cooling takes place and the stresses in the coating increase due to the generation of the so-called thermal stresses [16].

Fig. 7 and **Fig. 8** show the temperature evolution and the thermal stresses in x-direction during the coating deposition of different splats at different nodes, for the second submodel scale, respectively. Each splat is deposited every 1,2 ms after the previous splat and the first splat is deposited at the instant 14,4 ms, as shown by the dash line in **Fig. 7**. The temperature and stress evolutions are represented for the upper and lower substrate nodes as well as for all the coating nodes.

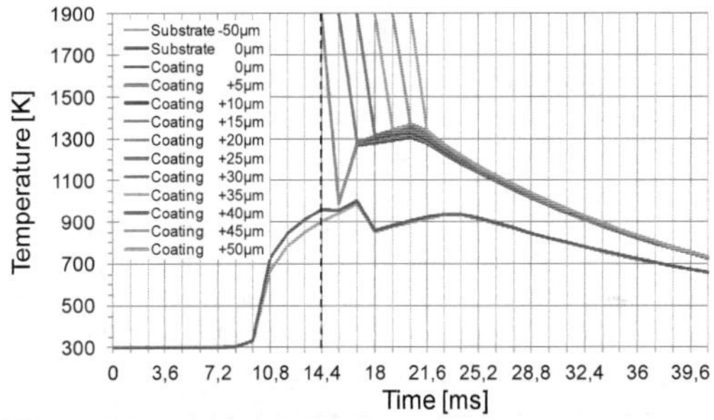

Fig. 7. Temperature evolution in second submodel.

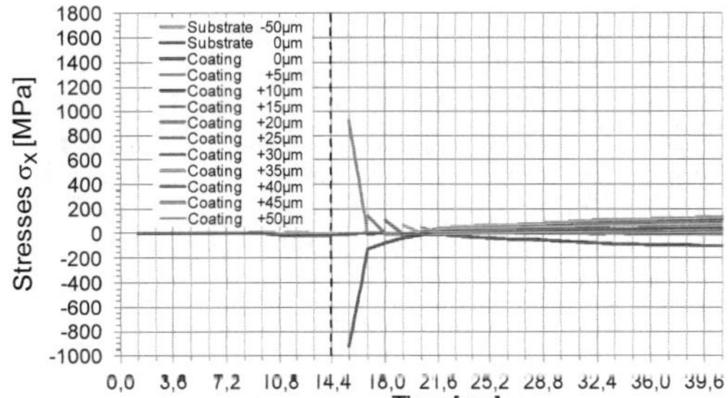

Fig. 8. Stress evolution (σ_x) in second submodel.

4 Experimental validation

A test plate is prepared, where a alumina coating is deposited by atmospheric plasma spraying technique on aluminum substrate. Measurements of the residual stress are performed by the incremental-step, drilling and circular milling method, are performed, as described in [17], [18]. The coating thickness is of approximately 50 µm in order to be compared with the simulation results. From the experimental analysis, the stresses in the coating present a compressive stress value of 130 MPa in the region near the coating surface, which decrease to a compressive stress value of around 25 MPa in the interface, continuing decreasing with the depth on the substrate, as shown in **Fig. 9**.

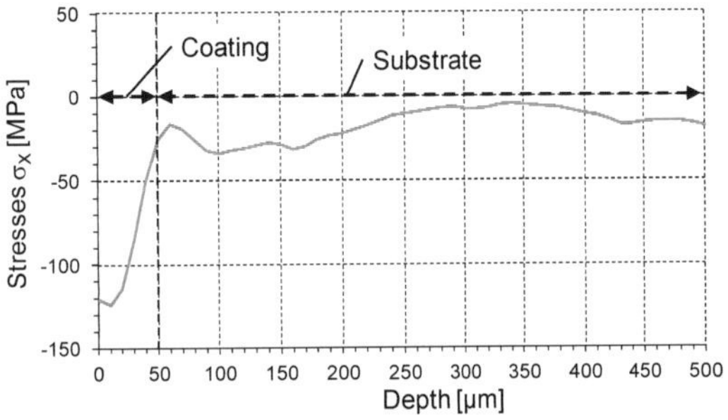

Fig. 9. Residual stress profiles in x-direction (σ_x) obtained by experimental measurement.

5 Summary and conclusions

Aiming to overcome the limitations of previous works, a realistic modelling of coating build-up and related heat flux, and thus, allows an insight into the thermal stress evolution in the coating during the process is presented. However, the success of the model depends on the assumption of elastic deformation of the material (coating and substrate), and a reliable material properties model will be of great importance to achieve a better correlation between the calculated values and experimental data. Moreover, a right definition of the coating/susbtrate interface is of great importance to describe the coating process.

In comparison to other studies where the coating deposition is considered [1, 3, 4], i.g. mass transfer, the stress results obtained with the presented model are improved due to a better definition of the interface between the coating and substrate, which results in a more real stress generation during the coatring process. Othe studies are basically centered only on the process kinematics in order to control heat transfer, and consequently, the substrate temperature ans its influence in the development of residual stresses in the coating composite, but with without enough microscopic resolution. [5, 6]. Finally, models which estimate the residual stress distribution in coatings has been developed for a small volume of the coating and substrate, based on microstructure data from scanning electron microscope (SEM) and without a good connection with a global model of the coating process [7].

The final redidual stress level obtained by simulation of the second submodel tends to a compressive stress value of 100 MPa in the interface for the coating and tensile stress value close to 150 MPa in the coating surface. These values are in the order of magnitude of the maximum stress level of 150 MPa from a reference study [4]. The final redidual stress level obtained by simulation of the first submodel tend to tensile stress values close to 75 MPa in the interface for the substrate, which is in the order of magnitude of the maximum stress level of 75 MPa obtained from the same reference. From the experimental results, a compressive stress value close to 125 MPa is obtained in the coating surface, which is the same order of magnitude as the obtained by simulation. The difference in the stress sign, i.g. compressive or tensile stresses, is highly influenced by the modelling of the convective heat flux on the surface during the coating process, and therefore a better implementation of the convective film coefficient and process kinematic is nececesary. Furthermore, a reliable material properties model is necessary for ceramic materials and take into account the stress relaxation mechanisms presented during the coiating process.

6 Acknowledgment

The authors are grateful to the Deutsche Forschungsgemeinschaft DFG (German Research Foundation) for supporting this research, project number GA 589/10-2.

7 Literature

[1] Wenzelburger, M.; Escribano, M.; Gadow, R.: Modeling of thermally sprayed coatings on light metal substrates: - Layer growth and residual stress formation, Surface and Coatings Technology, Vol. 180 -181 (2004), pp. 429–435.

[2] Voleti, S.R.; Chandra, N.; Miller, J.R.: Global-Local Analysis of Large-Scale Composite Structure Using Finite Element Methods, Computers & Structures, Vol. 58. No. 3 (1996), pp. 453–464.

[3] Buchmann, M.: Herstellung und Optimierung von thermisch gespritzten, tribologisch wirksamen Schichtsystemen auf Leichtmetall-Zylinderlaufflächen, Thesis, University of Stuttgart, Faculty for Mechanical Engineering (2002), Shaker Verlag, Aachen. ISBN 978-3-8322-0885-1

[4] Buchmann, M.; Gadow, R.; Tabellion, J.: Experimental and numerical residual stress analysis of layer coated composites, Materials Science and Engineering: A, Vol. 288 (2000), pp. 154–159.

[5] Candel, A.: Entwicklung von thermisch gespritzten Schichtverbundwerkstoffen durch Optimierung robotergestützter Brennerbewegungs-abläufe, Thesis, University of Stuttgart, Faculty for Mechanical Engineering (2009), Shaker Verlag, Aachen. ISBN 978-3-8322-8377-3.

[6] Floristán Zubieta, M.: Off-line generation of robot trajectories for Thermal Spray processes with respect to the heat and mass transfer, Thesis, University of Stuttgart, Faculty for Mechanical Engineering (2013), Shaker Verlag, Aachen. ISBN 978-3-8440-181-3.

[7] Mostaghimi, J.: Ghafouri-Azar, R.; Chandra ,S.: Modeling development of residual stresses in thermal spray coatings, Computational Materials Science, Vol. 35 (2006), pp. 13-26.
[8] Cormier, N.G.; Smallwood, B.S.; Sinclair, G.B.; Meda, G.: Aggressive submodelling of stress concentrations, Int. J. Numer. Meth. Engng., Vol. 46 (1999), pp. 889–909.
[9] Advanced Analysis Guide - Chapter 9: Submodeling, Release 12.1, ASNYS Inc., Canonsburg, USA (2009).
[10] Siegel, R.; Howell, J.R.: Thermal radiation heat transfer. Hemisphere Publising Corporation, 1992, Hemisphere Publishing Corporation, Washington, USA (1992).
[11] Morales, J.L.; Moreno, J.A.; Alhama. F.: Numerical solution of 2D elastostatic problems formulated by potential functions. Applied Mathematical Modelling, Vol. 37 (2013), pp. 6339-6353.
[12] Zienkiewicz, O.C.; Taylor, RL.: The Finite Element Method, Butterworth-Heinemann, Burlington, USA (2000).
[13] VDI-Wärmeatlas, Springer, Berlin-Heidelberg, Germany (1997).
[14] Touloukian, Y. S.: Thermophysical properties of high temperature solid materials: Vol. 2. Nonferrous alloys & Vol. 4. Oxides and their solutions and mixtures. Macmillan, New York, USA (1967).
[15] Berreth, K.; Buchmann, M. ; Gadow, R.; Tabellion, J.: Evaluation of Residual Stresses in Thermal Sprayed Coatings, In: Conference Proceedings UTSC´99, United Thermal Spray Conference and Exposition, Beschichten für die Praxis, E. Lugscheider, P. A. Kammer (Eds.), DVS-Verlag GmbH, Düsseldorf (1999), ISBN 3-87155-653-X, pp. 670-675.
[16] Buchmann, M ; Gadow, R. ; Tabellion, J.: Residual Stress Optimization of Thermally Sprayed Ceramic Coatings on Light Metal Components, In: R. Gadow (ed.), Advanced Ceramics and Composites / 6th Interregional European Colloquium on Ceramics and Composites, expert, Renningen (2000), ISBN 3-8169-1830-1, pp. 268-281.
[17] Wenzelburger, M.; López, D.; Gadow, R.: Methods and application of residual stress analysis on thermally sprayed coatings and layer composites, Surf. Coat. Technol., Vol. 201, Issue 5 (2006), pp 1995-2001.
[18] Gadow, R.; Riegert-Escribano, M.J.; Buchmann, M.: Residual Stress Analysis in Thermally Sprayed Layer Composites, Using the Hole Milling and Drilling Method, J. Therm. Spray Technol., Vol. 14, Nr. 1 (2005), pp 100-108.

Modeling of Heating the Particles of Powders In the New Detonation System

A. Rigin, A. Maximov, O. Maximova, S. Parygina,
Cherepovets State University, Russia
thermalspray@protonmail.com, +7(921)480-54-38

Abstract

Detonation spraying is used for applying metal, cermet, polymer and ceramic coatings. The advantages of this technology include good adhesion and density of coatings. Minimal residual stresses in the spray process give the possibility to apply coatings to thin-walled parts and foil. Numerical modeling methods for energy parameters of powder particles give the possibility to predict the interaction of powder particles with the gas flow in and properties of coatings depending on spray parameters. In many cases, existing advantages such as high quality of coatings, little heating of the substrate, efficient and economical use of the powder, make the detonation spray process more attractive in comparison with other spray methods. The new design of D-Gun and software gives the possibility to overcome these limitations and significantly increase the productivity. This result was obtained due to a new design of the valve head, new computer control system, electronics and software that allows the firing rate frequency to be increased, ensure the reliability and extend the maintenance-free lifetime. The paper discusses the results of modeling of acceleration and heating the particles of powders in the new detonation system and also methods of computer simulation of the formation of adhesive contact between powder particles and metal surface.

Introduction

Detonation spraying is a pulsed process, where gas mixture explosive energy is used to accelerate and heat up the particles of sprayed material. The detonation systems improvement is intended to extend their technological potential. Significant lightening of the equipment enables to install the D-Gun on a machine datum with programmed numerical control or onto a robot. It opens up new possibilities to use the detonation spray unit as a system for additive manufacturing of simply-shaped parts made of powder without any melting processes, which has a number of advantages, such as low residual stresses due to the lack of metal melting. For example, in SLS systems, the laser melting of powder is used to produce parts with additive manufacturing, which enables to use only homogeneous metallic material; The use of metal ceramics and ceramics is problematic or even impossible. Detonation spraying with one and the same unit allows spraying several different materials, for example, solid-state metals or ceramics like zirconium dioxide. It allows manufacturing of multimaterial parts (it is possible to install up to six powder feeders with different materials on one D-Gun of a new construction). It allows manufacturing of sandwich-type parts and coatings with 3D gradient structure, which can be created by setting up the D-Gun barrel movement relatively to the part. To solve a problem of obtaining multimaterial coatings, it is crucial to develop mathematical models to calculate adhesion and cohesion of sprayed material, which was done by the author. In this article we provide one of the developed mathematical models that is used to assess impact of powder particles temperature on adhesion. The mathematical model will be taken as a basis of the program that allows calculating the detonation spraying parameters to ensure the defined coating properties. This software will make it possible to calculate the bulk data of outcome variables of the process (the powder particles temperature, velocity and energy) with different input parameters of processing methods (ratio and amount of combustion gas, reflux gas and oxygen, barrel length, etc.). Using the mathematical statistics methods enables to determine the best technological parameters of the detonation spraying process to apply a particular material.

Mathematical modeling of this problem comprise several parts, including the modeling of gas mixture detonation process and detonation wave propagation; Modeling of particles behavior in the detonation product stream, including the calculation of temperature, state and velocity of the particles. In the first section, we provide fundamental formulas used for the mathematical model of heating the powder particles in the detonation system. These processes are thoroughly studied. The theoretical foundations for solving this problem were laid by a distinguished Russian physicist, Zeldovich I.B [1], who proposed the model of detonation wave propagation in gas: The compression wave front adiabatically pressurize the gas to the temperature, at which the chemical reactions of burning start, which, in their turn, promote stable propagation of compression wave. The further development of detonation spraying systems and mathematical methods of modeling, as well as investigational studies were carried out by the scientists of Lavrentyev Institute of Hydrodynamics in Novosibirsk.

However, the questions of modeling the processes occurring in coating formation, prediction of adhesion, cohesion and other qualities of the coatings so far are ill-defined and require solving a lot of theoretical problems. The second section describes the solution of one of these problems. It is a new mathematical model that allows assessing the relationship between the coating adhesion and temperature of the particles for such materials as metals and ceramics (for example, aluminium, chrome and zirconium oxides), carbides and polymers. The particle temperature is not the only factor that has impact on coating adhesion. Adhesion is extremely sensitive to material and qualities of the base surface, as well as to velocity and state of the powder particles; But these problems can be solved within certain models. The calculation model of the relationship between adhesion and the particle temperature allows determining the optimal temperature of the powder particles during the coating application. It is very important to solve this problem to limit the particles heating. It is required to prevent chemical decomposition of the particles (for example, WC and other carbides) and parts overheating due to the excessive heat input, as well as to raise the ratio of powder use (powder that is heated insufficiently bounces off the surface) and to ensure its total adhesion, which is very important during the AM of parts.

Modeling of Heating the Powder Particles

Detonation spraying is marked by a significant quantity of technological parameters. The main of them are:

- the depth of powder feeding, that is the distance between the delivery point of the powder into the detonation system barrel and its cross section;
- the ratio of flow rate of gases – combustion gas, oxygen and reflux gas for dilution (composition of working mixture);
- the total flow rate of explosive gas mixture that determines the occupation density of the barrel (the ratio between the total flow rate of gases in one cycle and the total volume of the barrel and the mixing chamber);
- the thickness of a sprayed layer in one cycle;
- the distance of spraying is a distance between the barrel cross section and the sprayed surface.

The basic data for the particles heating model are the variables obtained as the result of a modeling of a detonation process that leads to detonation wave generation. These input variables are used to simulate the particles acceleration and heating process. Figure 1 displays the D-Gun diagram and the main points and parameters for setting objectives of mathematical modeling.

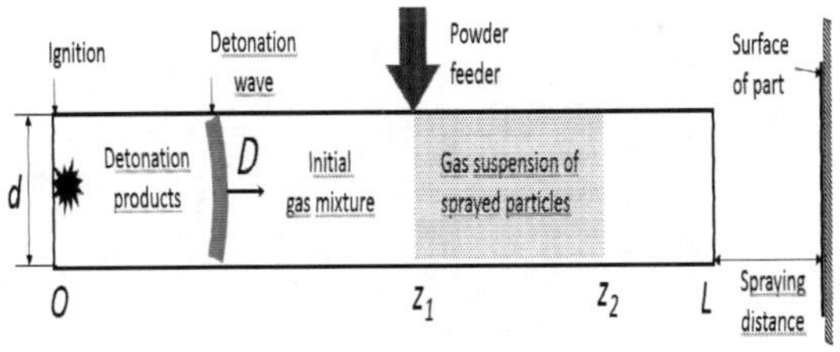

Fig. 1. D-gun barrel diagram for the model.

The detonation system has a D-Gun, L in length, with cylindrical form and inner diameter d, that is partly or fully filled with detonating gas mixture with initial pressure of P_0, density of ρ_0 and temperature of T_0. Inside the barrel, between points z_1, z_2 ($0 < z_1 < z_2 < L$) there is a gas suspension of hard sprayed spherical particlesd_p in diameter. With the ignition, the detonation wave forms in the left close end of the barrel and expands in the gas mixture at a velocity of D (Fig. 1). The detonation wave and the following expansion wave accelerate and transfer the suspended powder particles to outlet of the barrel. When the detonation wave reaches the right open end of the D-Gun barrel, the efflux of detonation products and particles starts with the following spraying on the part surface.

In the work [1], the theoretical foundations for gas detonation processes are developed. In works on detonation spraying [2-4], one-dimensional accelerating of the disperse mixtures in the detonation system barrel is considered. The impact of the zone out of barrel on the flow parameters is highlighted in the work [6]. In these works, the following assumptions are made: the pressure is made only by gas, an impact of the powder particles is neglected; viscosity and heat conduction of phases is taken into consideration only in processes. Thus, the set of equations of two-dimensional axially symmetric nonstationary motion of the gas suspension is given by [9]:

$$\frac{\partial\rho_i}{\partial t}+\frac{1}{r}\frac{\partial(r\rho_i v_i)}{\partial r}+\frac{\partial(\rho_i u_i)}{\partial z}=0,$$
$$\frac{\partial(\rho_i v_i)}{\partial t}+\frac{1}{r}\frac{\partial(r\rho_i v_i^2)}{\partial r}+\frac{\partial(\rho_i v_i u_i)}{\partial z}+\varepsilon_i\frac{\partial p}{\partial r}=(-1)^i f_r n;$$
$$\frac{\partial(\rho_i u_i)}{\partial t}+\frac{1}{r}\frac{\partial(r\rho_i v_i u_i)}{\partial r}+\frac{\partial(\rho_i u_i^2)}{\partial z}+\varepsilon_i\frac{\partial p}{\partial z}=(-1)^i f_z n; \qquad (1)$$
$$\frac{\partial(\rho_2 e_2)}{\partial t}+\frac{1}{r}\frac{\partial(r\rho_2 e_2 v_2)}{\partial r}+\frac{\partial(\rho_2 e_2 u_2)}{\partial z}=qn;$$
$$\sum_{i=1}^{2}\left[\frac{\partial(\rho_i E_i)}{\partial t}+\frac{1}{r}\frac{\partial(r\rho_i v_i E_i)}{\partial r}+\frac{1}{r}\frac{\partial(r\varepsilon_i p v_i)}{\partial r}+\frac{\partial(\rho_i u_i E_i)}{\partial z}+\frac{\partial(\varepsilon_i p u_i)}{\partial z}\right]=0.$$

$$\rho_i=\varepsilon_i\rho_i^0, E_i=e_i+(v_i^2+u_i^2)/2,$$
$$n=6\varepsilon_2/\pi d_p^3, \varepsilon_1+\varepsilon_2=1, \rho_2^0=\mathrm{const}, i=1,2,$$

where v_i and u_i. are velocity components in radial r and axial z directions, respectively; ei, Ei are specific internal and total energy of i-phase, respectively; f_z, f_r are the components of force interaction from gas to the dispersed particle in cylindrical coordinates; q is the intensity of the heat inflow to the surface of a certain particle; n is the quantity of dispersed particles in one unit volume of the mixture. The mixture volume fraction, occupied by i-phase is marked by its cubical content ε_i. Each point of the mixture volume is placed in correspondence with an average density of phases t that characterizes the phase mass in one unit volume, and a real density of ρ_i, that characterizes the density of comprising elements. Indices i = 1 belong to gas phases and indices i = 2 belong to disperse phase. To define pressure, the ideal gas law is used:

$$p=\rho_1^0 RT_1/\mu_1, \qquad (2a)$$

where R is a universal gas constant; μ_1 is a molecular mass of detonation products, T_1 is a temperature of gas phase. Before the detonation products of reacting gas flow start efflux in the D-Gun barrel, we use equations of internal gas energy $e_1(T_1\mu_1)$ and chemical equilibrium $\mu_1(\rho 1, T_1)$ [7, 8]; After starting the detonation products efflux for internal gas energy, another equation is used:

$$e_1 = \int_{T_0}^{T_i} c_1(T)dt, \quad (2b)$$

where $c_1(T)$ is specific gas heat capacity with its constant volume. The set of equations for internal energy [7, 8] takes into consideration the phase transformation (particle melting):

$$e_2 = \begin{cases} \int_{T_0}^{T_2} c_2(T)dT, \; if \; T_2 < T_{2m}, \\ \int_{T_0}^{T_2} c_2(T)dT + (1 - m/m_2)\Delta h, \text{if } T_2 = T_{2m} \\ \int_{T_0}^{T_2} c_2(T)dT + \Delta h, \; if \; T_2 > T_{2m}, \end{cases} \quad (3)$$

where $c_1(T)$ is specific particle heat capacity, T_2 is the temperature of the disperse phase; T_{2m} is the melting temperature of the disperse phase; m/m_2 is a relative fraction of a layer melted by the mass m in the particle mass m_2, $m_2 = (\pi d^3_p \rho_2^0)/6$; Δh is the specific melting heat. Solutions of simultaneous equations(1) - (3) are used to define an interphase force and thermal interaction between gas and disperse phases[5]:

$$\begin{aligned} f_z &= \frac{1}{2} C_d \rho_1^0 \Delta v(u_1 - u_2)\frac{\pi d_p^2}{4}; \\ f_r &= \frac{1}{2} C_d \rho_1^0 \Delta v(v_1 - v_2)\frac{\pi d_p^2}{4}; \\ q &= \pi d_p \lambda_1 \mathrm{Nu}(T_1 - T_2); \\ \Delta v &= \sqrt{(v_1 - v_2)^2 + (u_1 - u_2)^2}, \end{aligned} \quad (4)$$

where C_d *(Re, M)* is a resistance coefficient of spherical particles d_p *in diameter; Re* is the Reynolds number, *M* is the Mach number; Nu is the Nusselt number; A is a gas mixture thermal conductivity; Δv is the modulus of the vector of relative velocity of gas and disperse phases.

The data obtained as the result of solution of simultaneous equations (1.1 – 1.4) are used to analyse the dynamics and particle behavior [2], [4] in a nonstationary flow of detonation product:

$$\begin{aligned} m_2\left(\frac{\partial u_2}{\partial t} + \frac{1}{r}\frac{\partial(r u_2 v_2)}{\partial r} + \frac{\partial(u_2^2)}{\partial z}\right) &= f_z; \\ m_2\left(\frac{\partial v_2}{\partial t} + \frac{1}{r}\frac{\partial(r v_2^2)}{\partial r} + \frac{\partial(u_2 v_2)}{\partial z}\right) &= f_r; \\ m_2\left(\frac{\partial e_2}{\partial t} + \frac{1}{r}\frac{\partial(r e_2 v_2)}{\partial r} + \frac{\partial(e_2 u_2)}{\partial z}\right) &= q. \end{aligned} \quad (5)$$

The calculation of thermal field in spherical particle that is moving in the detonation-gas camera is made with a nonstationary equation of thermal conductivity:

$$\rho_2 c_2(T)\frac{\partial T_p}{\partial t} = \frac{1}{r_p^2}\frac{\partial}{\partial r_p}\left(r_p^2 \lambda_2(T)\frac{\partial T_p}{\partial r_p}\right), \quad (6)$$

For the purposes of use in a computer program, this set of equations is solved with a mesh method. The mesh method is used to solve differential and integral equations and allows narrowing the solution down to an algorithm that can be used in a computer modeling program. In this method, the range of continuous change of arguments, where the equation solution is sought, is replaced by discrete set of points. Instead of the function of continuous argument, the functions of a discrete argument are considered in the mesh points. Derivatives that appear in the equation, initial and boundary conditions, are approximated by difference coefficients; Integrals are approximated by quadrature formulae, here, initial integral equation is replaced by a set of finite-difference equations, for differential equations – with difference scheme. The use of this method allows a reasonably accurate prediction of the particle temperature during its transfer by detonation products.

2. Simulation of the formation of adhesive contact.

Molecular theory treats the adhesion as the result of molecular interaction forces between contacting molecules of adhesive and substrate. Therefore, it is important that the adhesive and the substrate molecules possess polar functional groups capable of interaction [11]. Thermodynamic properties of the coating can be examined if it is considered, for example, as a set of dipoles interacting with a solid surface and with each other. When the particle faces the surface as a result of thermal motion, it can be mirrored from the substrate, or be firmly bound to the surface, that is, to become adsorbed one. The opposite phenomenon of desorption is also possible when the adsorbed particle having sufficient kinetic energy can leave the surface. Fig. 2 schematically shows the potential energy of the particle versus the distance to the flat surface [12]. In case of physical adsorption, the particle binds to the surface by means of weak van der Waals forces, which are characterized by the energy E_p. These bonds are not accompanied by charge transfer from substrate to particle or v.v. (physical absorption). At higher temperatures, the exchange of electrons between adsorbed powder particles and the metal surface atoms is possible, which leads to rather strong chemical bonds between them, and then such powder particle is considered as chemisorbed one [12]. To get into this well, the particle must overcome an energy barrier E_b. Then, it is in a much deeper potential well E_c. The value E_c+ E_b defines the work function *W*.

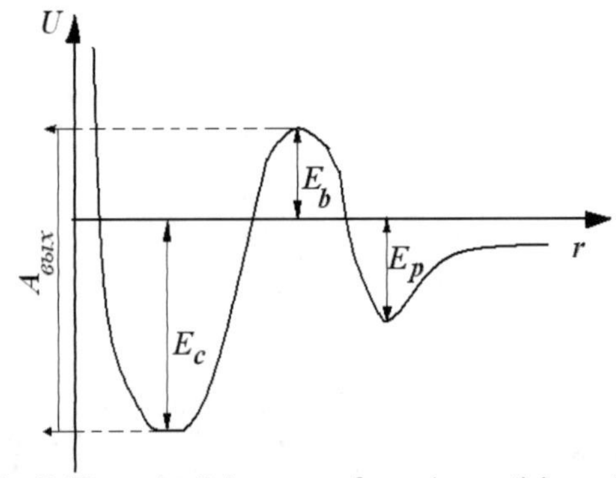

Fig. 2. The potential energy of powder particle vs. its distance to metal surface [12].

The modified Langmuir adsorption model may be used for calculations of the particles sticking degree to the substrate. This model is based on the following assumptions: 1) all adsorptive sites of the lattice are equivalent; 2) only one particle can occupy an adsorptive site; 3) influence of adsorption forces is restricted by thickness of the monolayer coating. At these conditions, the adsorption layer corresponds to two-dimension model of lattice gas [11].

At computer simulation by Monte Carlo method, the metal surface is represented as two-dimensional rectangular lattice containing N sites (adsorption points) along axis X, and M sites – along axis Y. The position of the lattice site is defined by two numbers i and j, and its employment – by the number n_{ij}, which is equal to zero, if the adsorption point is free, and to unity – otherwise [13]. In this model, at the initial moment (t = 0), it is assumed that all the values of the occupied lattice sites n_{ij} = 0; i.e. there is no adsorption.

The random selection of a lattice site and the calculation of the energy of the particle located near this site are carried out as follows. If the corresponding site is free ($n_{i,j}$ = 0), the particle of the adsorbent is not captured by the surface (non-adsorbed one), its energy is equal to ε_T. The value of ε_T is a random number determined by Maxwell distribution at the corresponding temperature of particle of the powder adsorbent. The adsorbent particle with probability P_1 is captured by the substrate surface, if it can overcome the energy barrier (Fig. 2), i.e. if $\varepsilon_T > E_b$, then the site is full, and $n_{i,j}$ = 1. If the corresponding site is full ($n_{i,j}$ = 1), then the particle can leave the metal surface with probability P_2, if its energy $\varepsilon > W$. The considered operations are repeated for a certain time proportional to the number of Monte Carlo steps, which is set by the lattice size and the temperature of metal surface.

Relative degree of filling of the metal surface θ (degree of sticking) by powder particles is defined as the ratio of the number of adsorbed sites to the total number of lattice sites:

$$\theta = \frac{1}{NM} \sum_{i,j=1}^{N,M} n_{i,j} \tag{7}$$

Obviously, the adhesion strength of coating to metal proportional to the degree of adhesion θ. Fig. 3 shows the results of the calculations of degree of sticking θ vs. the renormalized temperature T/T_0 for the lattice 100x100 at periodic boundary conditions (T_0 is the optimal temperature for the powder coating).

It turns out that at low temperatures the degree of sticking of powder particles to metal surface is low. For dielectric materials, there is a threshold at which there is no interaction, in fact, there is abrasion treatment of the metal surface. At higher temperatures there is the adhesion of the powder particles to the metal surface. Reheating the particles over the optimal temperature T_0 does not make sense, and can only lead to a deterioration of the quality of the coating; i.e. the coating can keep up with the metal surface, which is the cause of the defects. The optimal temperature value is calculated based on the physical properties of the material of the powder and the metal substrate.

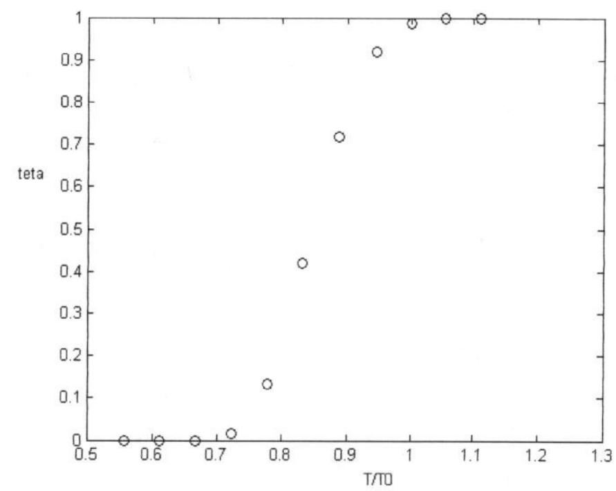

Fig. 3. The degree of sticking θ of the powder coating vs. the renormalized temperature T/T_0.

This graph shows a generalized temperature dependence of the particle for the mathematical model. As seen from the graph at low temperature, the degree of adhesion is small. There is barrier at which no interaction, in fact this is abrasion of the metal surface. At higher temperatures there is the adhesion of the particles of the adhesive to the metal surface. Reheating the particles over the optimum temperature T_0 is not meaningful, and can only lead to a deterioration of the quality of the coating; coating can keep up with the metal surface, which is the cause of the defects. In the transition to individual cases, the optimum temperature value is calculated based on the physical properties of the material of the particles and the substrate.

Conclusion

On the basis of the proposed models can calculate the optimum temperature of the particles, and parameters to control detonation unit (mainly, a detonating composition of the gas mixture) so as to minimize the temperature and to obtain better coating quality. Excessive overheating of the powder particles usually leads to a deterioration in the quality of the coating of heat into the substrate, oxidation or decay of the coating material.

For example, WC falls into W_2C when heated in an oxidizing environment, and Co_3W_3C и $Co_3W_9C_4$ and Co_7W_6 carbides are also formed. Preliminary calculations in the model allows to avoid it. The model given in the article is one of the components to the overall coating adhesion theoretical model that takes into account other important factors (speed and state of the particles, the material and surface condition). The model can be used in the software to calculate the optimal process parameters of the coating process of the various materials. Numerical modeling methods for energy parameters of powder particles allow to predict the interaction of powder particles with the gas flow in and to improve properties of coatings. The software, based on this model allows to expand the technological capabilities of the method of detonation spraying. Pulsed nature of the process provides minimum residual stresses that together with the low heat level eliminates structural changes, buckling and distortion of parts. Minimal residual stresses in the spray process give the possibility to apply coatings to thin-walled parts and foil. Therefore, detonation spraying is widely used in the manufacture and repair of blades and other parts of aircraft engines, restoring repair of crankshafts and camshafts of internal combustion engines, face seals of shafts, rollers of shafts in metal processing and metallurgical production. Its use is also efficient in chemical engineering for hardening parts, instruments (rollers, knives, cutters, injection molds), etc. Expenses of powder for detonation spraying several times smaller than HVOF for small parts. It is also good for the spraying of small local areas (e.g. threaded surfaces for pump and compressor pipes [10]). In many cases, existing advantages such as high quality of coatings, little heating of the substrate, efficient and economical use of the powder, make the detonation spray process more attractive in comparison with other spray methods. Minimizing temperature enables application detonation spraying for application of coatings of substantial thickness and Additive Manufacturing composite parts from ceramics, polymers, metals and alloys without melting.

References

[1] Zeldovich Y.B., Kompaneets A. Detonation theory. - M .: State Publishing House technical and theoretical literature, 1955. - 268 p.

[2]. Investigation of the dynamics of acceleration and heating of metal particles of a detonation wave / V.M. Boyko, V. V. Grigoriev, S.A. Zhdan et al. // Physics of combustion and explosion - 1983. - № 4. - S. 133-136.

[3] A.I. Dolmatov, P.D. Zhemanyuk, Numerical modeling of two-phase flow in the barrel of the detonation installation // Technologist system. - 2001. - № 4. - S. 12-16.

[4] Gladilin A. M., Karpilovskaya E.I., Kornev A.D. Calculation of the two-phase parameters of the medium in the barrel of the detonation apparatus used for coating // Physics of combustion and explosion. - 1978. - № 1. - S. 123-128.

[5] Ivandaev A.I., Kutushev A.G., Nigmatulin R.I. Gas dynamics of multiphase media. Shock and detonation waves in gas suspensions. - M .: VINITI, 1981. - V. 16. - P. 209-287. - (Results of Science and Technology, Fluid Dynamics Series).

[6]. Kantor L.A., Kantor S.A., Strongin MP Calculation process of detonation-gas protective coatings // Physics of combustion and explosion. - 1987. - № 4. - S. 131-135.

[7] Nikolaev Y.A. Model of the kinetics of chemical reactions at high temperatures // Physics of combustion and explosion. - 1978. - № 4. - S. 73-76.

[8] Zhdan S.A., Fedenyuk V.I. Equilibrium parameters of the gas flow in the barrel of the detonation installation // Physics of combustion and explosion. - 1982. - № 6. - S. 103-107.

[9] [9] Nigmatulin RI Fundamentals of mechanics of heterogeneous media. - M .: Nauka, 1978 - 336 p.

[10] L. Baldaev, S. Mankovsky, D. Ishmukhametov, A. Rigin "Protecting Threaded Surfaces in Pump and Compressor Pipes Used in the Oil and Gas Industry", Thermal Spray 2015, Long Beach, CA, USA, Proceedings of the ITSC 2015, p. 1134

[11] V.I. Mamonov, M.V. Mamonov, Surface physics. Theoretical models and experimental metods, M.: FIZMATLIT 2011.

[12] Pratton M, Introduction to the physics of surface, Izhevsk: NITs RHD 2000.

[13] Proutorov E.V., Maksimova O.G., Maksimov A.V, Journal of Physics "Simulation of the adhesion contact on the interface of polymer-metal": Conference Series, vol. 633 (2 015), no. 012044.

Adhesive-Corrosion Resistance of High Velocity Metal Droplet on Atmospheric Plasma Sprayed Ni-WSe_2-$BaF_2 \cdot CaF_2$-Y-Ag-hBN Solid Lubricant Coatings

XIAO-JING YUAN*, XIAO-HU CHEN, BAI-LIN ZHA, CHUNJIANG YAO

(Xi'an Research Institute of Hi-Tech, Xi'an, Shannxi 710025, China)

Corresponded author: yxj2003@263.net

Abstract: Adhesive-corrosion resistance is an important issue for the application life of solid lubricant coatings. With abundant metal droplets deposited on the surface of a solid self-lubricating coating, the coefficient of friction of the coating changes and results in adhesive wear. In this paper, a new method for evaluating the adhesive resistance for solid self-lubricating Ni-WSe2-BaF2·CaF2-Y-Ag-hBN coatings was reported. The microstructures and anti-adhesive characteristics under different angles of aluminum metal droplets obtained with an arc supersonic nozzle were investigated for high temperature solid self-lubricating coatings produced by plasma spray. The results demonstrate that the friction coefficient of Ni-WSe2-BaF2·CaF2-Y-Ag-hBN solid self-lubricant coatings is distributed between 0.086 and 0.299 at 25-800℃. The effect of molten metal drops on the coating adhesive-corrosion rate increases with the deposition angle. At 90°, the deposition rate of metal droplets on the coating and substrate surface is maximized, and the hexagonal Boron Nitride (hBN) self-lubricating coating deposition rate is only 58 mg/(cm^2·s). hBN can effectively decrease the adhesive layer of the coating by the rate of the polishing and stripping.

Keywords: composite solid lubrication coating; anti-adhesive; friction and wear; plasma spray

1 Introduction

Solid self-lubricating coatings that exhibit such properties as being resistant to load, having a wide temperature range and long life, being free of dust and rain water and being non-perishable are suitable for a tribo-pair, such as the rail and adapter. These coatings should carry a large overload and be clean. Recently, NASA (National Aeronautics and Space Administration) and AFRL (Air Force Research Laboratory) introduced a series of solid self-lubricating composite coatings, such as the PS (Plasma Spray) coating series (PS100-PS400) and adaptive coating (such as YSZ-Ag-Mo/TiN adaptive multilayer coating) [1-3]. Sliney et al. [4] developed a nickel-based solid self-lubricating coating prepared by plasma spray, which contains silver, eutectic barium fluoride and calcium fluoride, and has good mechanical properties and lubricating properties. The LICP (Lanzhou Institute of Chemical Physics, Chinese Academy of Sciences)[5] found that the average friction factor of Ni_3Al-$BaF_2 \cdot CaF_2$-Ag-Cr composite coating can reach 0.3-0.36 in the temperature range of 20-800℃. Huang C. et al [6] studied the NiCr/Cr_3C_2-NiCr/hBN composite over a wide operating temperature range of the coefficient of friction. Tungsten diselenide (WSe_2) is a semiconductor and part of the transition metal dichalcogenide (TMDC) family.[7] Pauschitz A.[8] found that WSe_x ($1.97<x<2$) has an obvious advantage over other transition metal dichalcogenides because of its narrow stoichiometric range, making this phase more resistant to oxygen penetration. Our group developed the NiCr-WSe_2-BaF_2·CaF_2-Y-Ag-hBN solid lubricant coating using plasma spray, which exhibits excellent lubricating properties over a wide temperature range, and the CrO_4^- ceramic glaze layer generated on the coating's surface at high temperature can effectively reduce the friction coefficient [9, 10].

Solid self-lubricating coatings are needed for special circumstances, such as in a high energy gas-solid stream, where the instantaneous combustion engine generates temperatures of 2000 to 3000℃ and gas speeds of 5 to 7 Ma, resulting in high-speed hot melt residue erosion, adhesive-residue, and produces cavitation damage. Recent research showed that thermal erosion damage is more prominent than ablation damage in a high aluminum engine, and the air flow contains a large amount of viscous substances, including Al, residues and condensed phase products of the combustion process. These adhesive-corrosion residue deposited on the surface of the rail will change the mechanical and tribological properties, causing serious oxidation wear, affecting the service life of the coating and leading to launching system fault, reducing the equipment service life. [11] Therefore, metal droplet adhesion, erosion, oxidation and so on, as well as the cleaning of the coating surface, must be thoroughly studied [12, 13].

However, there is little published research on how to quantitatively characterize the adhesive-corrosion performance of solid self-lubricant coatings at high temperature. In this paper, the performance of a series of adhesive-corrosion resistance solid self-lubricating coatings were systematically analyzed and characterized on design-related test equipment using plasma sprayed Ni-WSe_2-BaF_2·CaF_2-Y-Ag-hBN over a wide temperature range.

2 Experimental Procedures

2.1 Coating material

Table 1 shows the design ratio of Ni-WSe_2-BaF_2·CaF_2-Y-Ag-hBN powder for high temperature solid self-lubricating coatings. Polyvinyl alcohol was added to the mixed powder, and the solution was heated and stirred continuously. Solution drying and heating continued for 2 h. Afterwards, the resulting solid was crushed, ground into a mortar, and screened to obtain a composite solid lubricating powder. The Ni-WSe_2-CaF_2·BaF_2-Ag-Y particles were mixed and heated at 500°C for 2 h to form the composite (Ni:WSe_2: BaF_2·CaF_2:Ag:Y=80:8:7:2:3 in weight, named NWY coating). Next, the particles were crushed into a mortar and screened with a standard mesh. The Ni-WSe_2-CaF_2·BaF_2-Ag-Y-hBN (Ni: WSe_2: BaF_2·CaF_2: Ag: Y: hBN=60:16:15:2:3:4 in weight, named NWB coating) composite powder was prepared using the same methods as the Ni-WSe_2-CaF_2·BaF_2-Ag-Y powders.

Tab.1 Ni-WSe_2-BaF_2·CaF_2-Y-Ag-hBN lubricant composite powders

NO.	Ni(W.%)	BaF_2·CaF_2(W.%)	WSe_2(W.%)	Ag(W.%)	Y(W.%)	hBN(W.%)
NWB	60. 0	15	15. 0	2. 0	3. 0	4. 0
NWY	80. 0	7	8. 0	2. 0	3. 0	-

2.2 Plasma spraying

Plasma spraying is the most versatile of the thermal spray processes. Plasma is capable of spraying all materials that are reviewed suitable. In plasma spray, an arc is created in between two electrodes in a plasma forming gas, which usually consists of either argon/hydrogen or argon/helium. As the plasma gas is heated by the arc, it expands and is accelerated through a shaped nozzle. Temperatures in the arc zone approach 36,000°F (20,000°K). Temperatures in the plasma jet are still 18,000°F (10,000°K) at the exit of the nozzle. In this experiment, using the Metco-9M plasma spraying system, composite solid self-lubricating powder was deposited on GH4145 steel substrates before experimental pretreatment of the substrate surface, removing surface oil, oxides and other impurities by the ion spraying optimized parameters indicated in Tab.2. The preparation of the coating thickness ranges from 0.3-0.5 mm.

Tab.2 Technical parameters of the atmospheric plasma spraying process

Current/A	Voltage/V	N_2 flowrate/(scfm)	H_2 flowrate/(scfm)	Feedstock rate /(ml/h)	Spraying distance/mm
500	80	100	10	200	105

2.3 High-temperature self-lubricating coating characterization

The cross-section microstructures of the coating were characterized by TESCAN VEGA II XMU scanning electron microscope. An X-ray diffractometer (XRD, X'Pert Powder PANAlytical with Cu K radiation) was utilized to analyze the constituent phases in feedstock powder and plasma sprayed coating.

The Vickers micro-hardness of the APS-sprayed coatings was measured on the polished cross-section using a MICROMET 5100SERIES Automatic Microhardness System (USA BUEHLER Company) with 1.96 N and 15 s. An average of five measurements were taken at different locations on the transverse section of the coating.

The lubricant coatings were tested using a CSM high-temperature tribometer instrument (CSM Co. Rue de la Gare4 Galileo Center Ch-2034 Peseux, Switzerland). A commercial Si_3N_4 ceramic ball (hardness 1600 H_V, roughness Ra (0.2-0.6) μm), which was 10 mm in diameter, was also used. The test was performed with a normal load of 30 N and a sliding velocity of 0.24 m/s for approximately 60 min. The standard sample was a small cylinder 30 mm × 8 mm in dimension. The test temperature was from 25°C up to 800°C. During the test, the friction force was directly measured with a sensor, fed into a computer, and then divided by the load to obtain the friction coefficient.

2.4 Adhesive-corrosion test of metal droplet on High-temperature self-lubricating coating

The adhesive-corrosion resistance of high-temperature self-lubricating coatings is an important indicator of the anti-adhesive performance of solid self-lubricating coatings. At present, several reported evaluation methods, such as thermal simulation of materials (no erosion of oxyacetylene ablation test) and cold simulation (high speed blow sand erosion test) [11], obviously cannot test the actual properties, such as the slide surface of a high-temperature and high speed particle. In our work, a test frame was designed for erosion resistance of high-temperature self-lubrication coatings (Fig. 1) [14]. Fig. 1 (a) shows the principle diagram of the Al metal molten droplet corroding the coating surface, including where the high speed Al melted droplets were formed through the supersonic arc nozzle. The adhesive-corrosion angles of specimens were set at 0°, 30°, 45°, 60°, and 90°. After the test of adhesive-corrosion, the weight of metal particle deposited on the sample surface was evaluated.

Then, in order to evaluate the cleaning ability of the coatings, the adhesive-corrosion products on the coatings surface was worn off with grinding and polishing equipment, shown in Fig. 1 (b). The weight of the sample was measured at a given polishing speed and predetermined time intervals.

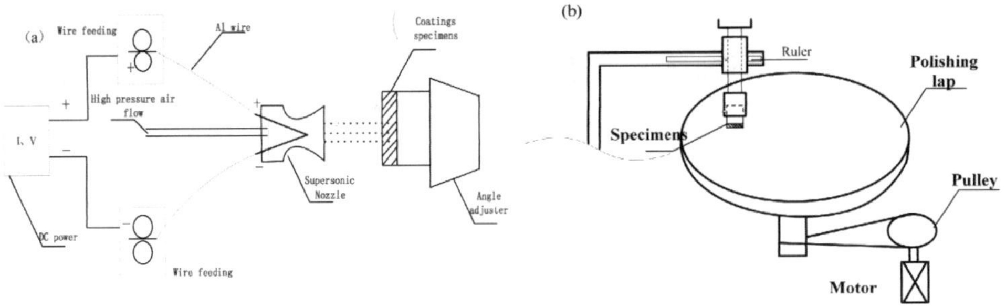

Fig. 1 Schematic diagram of the anti-adhesive test of high temperature self-lubricating coating

(a) Metal droplet adhesive-corrosion principle; (b) Peeling off the surface of the coating

3 Results and Discussion

3.1 Microstructure of Ni-WSe_2-BaF_2·CaF_2-Y-Ag-hBN solid self-lubricant coating

Fig. 2 shows the SEM images of plasma sprayed Ni-WSe_2-BaF_2·CaF_2-Y-Ag-hBN high temperature solid self-lubricating coating. In this experiment, the cross-section morphology of NWY (Fig. 2 (a)) and NWB (Fig. 2 (b)) coatings were obtained. Both the NWY and NWB coatings take on a lamellar structure and a non-uniform structure due to the plasma spray. Furthermore, there is low porosity in the coatings. Because of the high temperature and low velocity of plasma flow, plenty of particles were melted, and the flattening ratio of the particles is lower in the impact matrix. In addition, the hBN changes the interfacial properties of the flattened particles, which can make the stress concentrated in the coatings, and then influence the binding degree of the particles and the matrix (as showed in Fig. 2 (b)). The XRD patterns of plasma sprayed NWY and NWB solid lubricant coatings are shown in Fig. 2 (c). Here, the coatings basically maintain the structure of the spraying powder, otherwise the small amount of NiO.

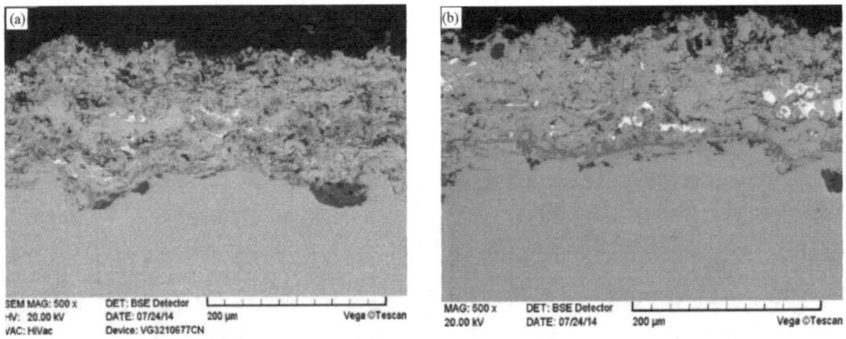

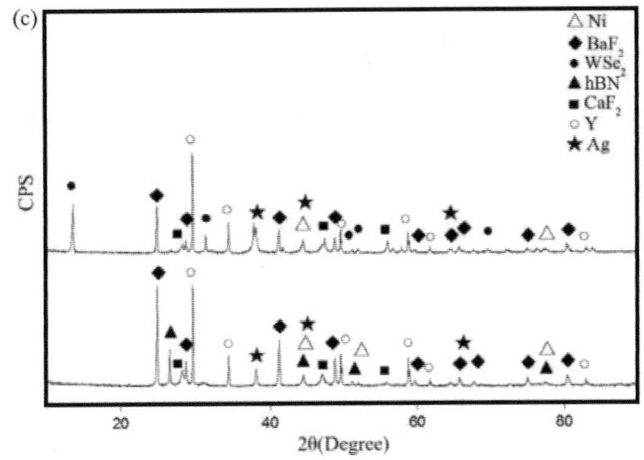

Fig. 2 Microstructure of plasma sprayed Ni-WSe_2-BaF_2·CaF_2-Y-Ag-hBN high temperature solid self-lubricant coating

(a) NWY; (b)NWB; (c)XRD

3.2 Friction characteristics of Ni-WSe_2-BaF_2·CaF_2-Y-Ag-hBN solid self-lubricant coating in wide temperature range

Fig. 3(a) shows the friction coefficient for plasma sprayed Ni-WSe_2-BaF_2·CaF_2-Y-Ag-hBN high temperature solid self-lubricant coating over a wide temperature range (25-800℃). The friction coefficient of NWY and NWB solid self-lubricant coating is distributed from 0.142 to 0.427 and 0.086 to 0.299, respectively. At the experimental temperature of 500℃, NWY lubricant coating has a higher friction coefficient than the other lubricant coatings. For the NWB lubricant coating, the friction coefficient increases with the friction temperature and reaches 0.299. As a result, the composition of NWB lubricant coating contains some high temperature solid lubricating phase of hBN. Therefore, the friction coefficient of NWB coating is less than that of NWY coating. The average friction coefficient of NWB coating is 0.2224 over a wide temperature range (25-800℃).

Obviously, the addition of hBN in the NWB lubricant coating promotes the synergistic lubrication of each lubricating phase (e.g., BaF_2·CaF_2, WSe_2, Ag) at different temperatures. Compared with NWY lubricant coatings, the friction coefficient of NWB lubricant coating changes smoothly at 500℃.

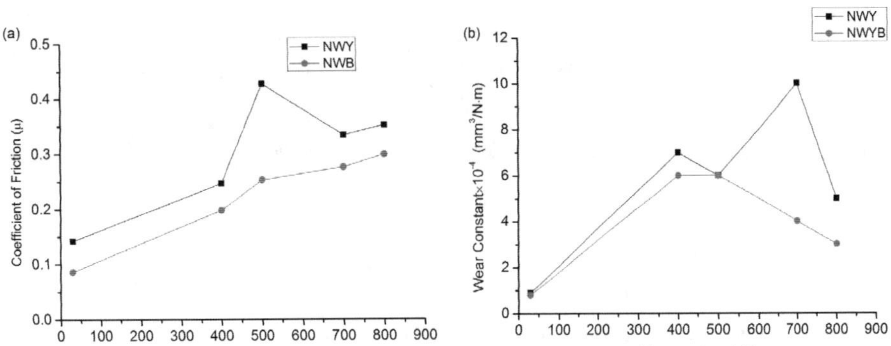

Fig. 3 Friction coefficient of solid lubricant coating by plasma spray: (a) the friction coefficients and (b) the wear constants of the coatings at temperatures from 25°C to 800°C.

As shown in Fig. 3 (b), the wear constants of the coatings increase with temperature. The wear rate of the Ni-WSe_2-$BaF_2 \cdot CaF_2$-Y and Ni-WSe_2-$BaF_2 \cdot CaF_2$-Y-hBN coatings is 0.9×10^{-4} - 10×10^{-4} $mm^3\ N^{-1}\cdot m^{-1}$ and 0.8×10^{-4} - 6×10^{-4} $mm^3\cdot N^{-1}m^{-1}$, respectively. In the figure, the wear rate for the Ni40Cr-WSe_2-$BaF_2 \cdot CaF_2$-Y-hBN coating increases sharply with temperature. Above 500°C, the friction coefficients of the coatings decrease, and the wear rates of both coatings has a significant increase. It is clearly revealed that the incorporation of a relatively small amount of $BaF_2 \cdot CaF_2$ high-temperature solid lubricant phase into the coating causes surprising enhancements in self-lubricating properties and, especially, wear resisting properties under high-temperature sliding wear test conditions. Conversely, the interfacial bond between the matrix and the reinforcement particle plays a significant role in the wear process. In short, the friction coefficients and wear constants of the Ni-WSe_2-$BaF_2 \cdot CaF_2$-Y-hBN counterpart are low and stable at 300-800°C. The wear rate of the Ni-WSe_2-$BaF_2 \cdot CaF_2$-Y-hBN coating is lower than that of the Ni-WSe_2-$BaF_2 \cdot CaF_2$-Y coating under the same evaluated conditions.

As shown in Fig. 4, from the friction marks of NWY and NWB coating at 500℃ after the friction test, it is apparent that the wear scar surface of NWB lubricant coatings are relatively wide and shallow, while that of NWY lubricant coating is narrow and deep. The microhardness of the NWY and NWB lubricant coating was measured by MICROMET 5100SERIES Automatic Microhardness System (USA BUEHLER Company). The average hardness of NWY and NWB was 151.1 $HV_{0.2}$ and 147.2 $HV_{0.2}$, respectively. Therefore, the wear rate of NWB coating is higher than that of NWY coating at this temperature.

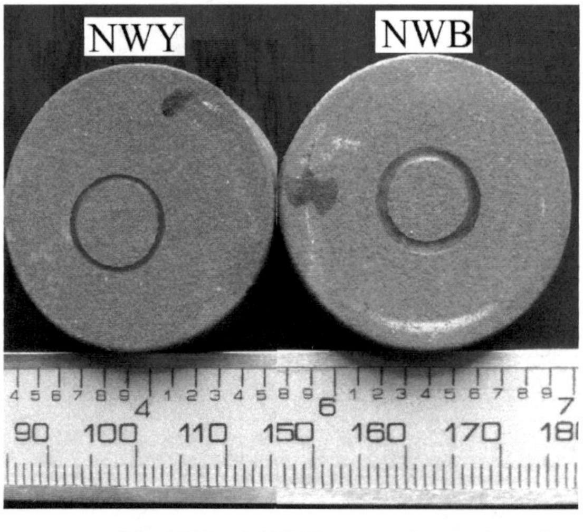

Fig.4 Wear scar of the NWY and NWB composite coatings（500℃）

3.3 Adhesive-corrosion characteristics of metal droplets in high temperature solid self-lubrication coatings

In order to examine and evaluate the anti-adhesive properties of NWY and NWB solid lubricant coatings, the bonding process of the coatings was monitored with the experimental set-up demonstrated in Fig. 1 (a). The rate of adhesive-corrosion of metal droplets on the coating surface can be calculated according to Formula (1).

$$Z = \frac{g_i - g_0}{\Delta t A} \qquad (1)$$

In this equation, Z is the average rate of adhesive-corrosion of metal droplets; g_i, g_0 are the weight of the tested sample and the reference weight, respectively; Δt is the depositing time of the metal droplet; A is the tested surface area for the sample.

The relationship between the coatings sample and the adhesive-corrosion angle is shown in Fig. 5. The adhesive-corrosion rate of Al metal droplets was lowest on the NWB coating surface containing 60 wt.% Ni. Additionally, the adhesive-corrosion rate of Al metal droplets deposited on the coatings sample surface increases with the increase of the metal deposition angle.

When the adhesive-corrosion angle is 90°, the depositing rate of Al metal droplets on the surface of the coating and the substrate reaches a maximum value, and the average rate on GH4145 steel substrates, NWY coating and NWB coating are 176.22 mg/(cm^2·s), 64.4 mg/(cm^2·s), and 58 mg/(cm^2·s), respectively.

This shows that hBN has a significant effect on improving the adhesive-corrosion resistance of the coatings. The NWY coating and NWB coatings contain Ni and, according to surface wetting behavior [15-16], the gradient for Al metal droplets increases with increasing Ni content. When the etching angle is less than 60°, the amount of Al on the substrate decreases, and the NWY and NWB coating are obviously increased.

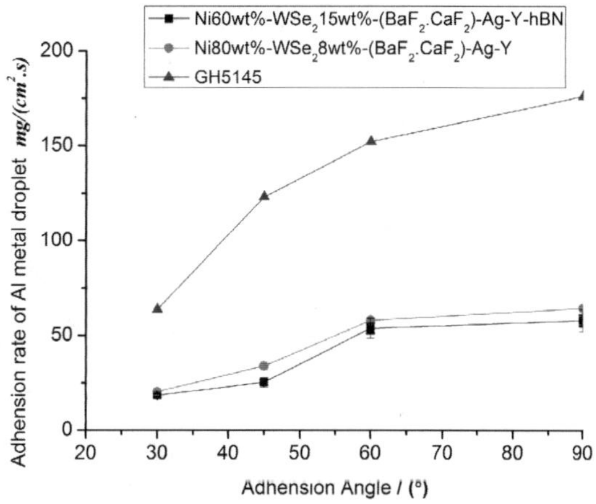

Fig.5 the relationship between the adhesive-corrosion angle and the coating

3.4 Self-cleaning ability of the coating surface for Al droplet

Coating will be applied as the friction pairs, the Al metal droplet deposited on the coating surface should be cleaned by the coating self. So, another index will be characterized for evaluating peeling capacity of the Al adhesive on the coating surface, named as the self-cleaning ability of the solid lubricant coating.

According to the relevant requirements of the design of the experiment, some parameters were set as sandpaper 1000 mesh, polishing rate was 700r/min, and the peeling weight of sample were measured every 2 minutes with an analytic balance, FA2104 electrical balance made by HANGPING Co., having an accuracy of 0.1 mg.

The weight of the metal peeling from the surface of the high temperature self-lubrication coating is shown in Tab.3. Obviously, with the increase of time, the weight of Al on the surface of each sample shows a decreasing trend, but the rate of the adhesive of the GH4145 substrate is relatively small.

Tab.3 the peeling capacity of Al layer on the surface of high temperature self-lubrication coating

angle (°)	NWY (g)			NWB (g)			GH4145 (g)		
90	0.10	0.14	0.04	0.45	0.09	0.05	0.06	0.05	0.07
60	0.21	0.05	0.06	0.32	0.47	0.02	0.05	0.10	0.09
45	0.09	0.14	0.12	0.15	1.06	0.03	0.12	0.09	0.05
30	0.10	0.10	0.05	0.57	0.11	0.09	0.06	0.04	0.05

4 Conclusions

In this paper, the anti-adhesive properties of plasma deposited Ni-WSe_2-$BaF_2 \cdot CaF_2$-Y-Ag-hBN high temperature solid self-lubricant coating are shown as follows:

- Plasma deposited Ni-WSe_2-$BaF_2 \cdot CaF_2$-Y-Ag-hBN high temperature solid self-lubricating coating has a friction coefficient ranging from 0.086 to 0.299 between 25 and 800°C.
- The adhesive-corrosion rate of Al metal droplet on the coating increases with the deposition angle. The maximum rate is taken on at 90°, where the average deposition rate on the substrate and the NWB coating are 176.22 mg/(cm^2·s) and 58 mg/(cm^2·s), respectively.

ACKNOWLEDGMENTS

The authors acknowledge the financial support of the National Natural Science Foundation of China (NSFC) (grant No. 51405497).

REFERENCES

[1] DellaCorte C, Fellenstein J A. The effect of compositional tailoring on the thermal expansion and tribological properties of PS300: a solid lubricant composite coating[J]. Tribol. Trans, 1997, 40: 639-642.

[2] Blanchet T A, Kim J H, Calabrese S J, DellaCorte C. Thrustwasher evaluation of self-lubricating PS304 composite coatings in high temperature sliding contact[J]. Tribol. Trans, 2002, 45: 491-498.

[3] C. Muratore , A.A. Voevodin, J.J. Hu, J.S. Zabinski. Multilayered YSZ-Ag-Mo/TiN adaptive tribological nanocomposite coatings. Tribology Letters, 2006. 24(3): 201-206

[4] Sliney, H. E.. Wide temperature spectrum self-lubricating coating prepared by plasma spraying. Thin Solid Films, (1979) 64, 211–217 .

[5] Zhu S Y, Bi Q L, Yang J, et al. influence of fluoride content on friction and wear performance of ni3al matrix high temperature self-lubricating composites[J].tribology letters,2011,43(3):341-349.

[6] Huang C B, Zhang W G, etal. Preparation and wear performance of NiCr/Cr3C2-NiCr/hBN plasma sprayed composite coating[J]. Surface and Coatings Technology, 2011, 205: 3722-3728.

[7] Daniel Gerchman, Annelise Kopp Alves, Solution-processable exfoliation and suspension of atomically thin WSe2, Journal of Colloid and Interface Science 468 (2016) 247–252

[8] Pauschitz, A., Badischa, E. Roy, M. and Shtansky, D.V., —On the scratch behavior of self-lubricating WSe 2 films, Wear. 267, (2009), pp. 1909–1914.

[9] Xia Jie, Chen Xiaohu, Yuan Xiaojing, etc. Fabrication and High Temperature Tribological Properties of Plasma Sprayed NiCoCrAlY/BaF_2/CaF_2/C/Y Self-lubricating Coatings, China Surface Engineering. 2014, 27(6):67-74(in chinese)

[10] Xiao-Jing Yuan, Xiao-Hu Chen, Bai-Lin Zha & Zhi-Hang Yu, Structural and Tribological Performance of solidNiCr-WSe2 -BaF2·CaF2-Y-hBN and NiCr-WSe2 -BaF2·CaF2-Y lubricant coatings produced by Atmospheric Plasma Spray, Tribology Transactions, DOI: 10.1080/10402004.2016.1158337

[11] Xiao Jun, Zhou Huidi, Li Tiehu, etc. Protection of Ablative Adhesion and Erosion of Slide Track of Missile Launcher by High-Temperature Two-Phase Exhaust Using MoS2-Based Bonded Solid Lubricant Film, TRIBOLOGY. 2003(09):435-449(in chinese)

[12] Finnie, I. Erosion of surfaces by solid particles [J]. Wear, 1960, V3(4):87-103

[13] Wang W C. Application of a high temperature self-lubricating composite coating on steam turbine components. Surf. & Coat Technol. 2004, 177-178: 12-17.

[14] Yuan Xiaojing, Chen Xiaohu, etc. an Detection method for anti-adhesion property of solid self-lubricating coating, National invention patent in China：CN 104502257 A

[15] Wang Zhenting, Chen Huahui. Microstructure and Friction and Wear Behavior of Induction Clad Ni-Based Composite Coating Reinforced with Micro- and Nano-WC Particulates [J]. TRIBOLOGY. 2005(03):203-206

[16] M. Esmaeilpour, B. Niroumand, A. Monshi, B. Ramezanzadeh, E. Salahi, The role of surface energy reducing agent in the formation of self-induced nanoscale surface features and wetting behavior of polyurethane coatings. Progress in Organic Coatings, Volume 90, January 2016, Pages:317-323

Investigating influences of suspension plasma spray parameters on photocatalytic activity of TiO_2 coatings

T.P. Vu[1], N. Otto[2], A. Vogel[1], F. Kern[1], A. Killinger[1], R. Gadow[1]
[1] IFKB, University of Stuttgart, Allmandring 7b, 70569, Stuttgart, Germany
[2] ISWA, University of Stuttgart, Bandtäle 2, 70569, Stuttgart, Germany

Abstract: In recent studies, the suspension plasma spray technology (SPS) was shown as a promising method for deposition of photocatalytic TiO_2 coatings on glass substrates. However, only little information about the effects of SPS process parameters on the resulting photocatalytic activity is available. In this study, several suspensions were prepared from different powders and various dispersion mediums successively sprayed. Four titania powders with a proven photocatalytic track record, Evonik P25, Kronoclean 7050, Hombicat UV100 and Sigma Aldrich were used to study the influence of starting powder type. Furthermore, compositions of the dispersing medium were varied in order to investigate their influence on coating structure, adhesion to substrate and photocatalytic activity. Degradation of methylene blue was tested as a marker for the photocatalytic activity. At chosen torch parameters, alcohol and water in a ratio of 1:3 was identified as the best composition. Suspensions sprayed with higher alcohol content suffered from the reduction in anatase phase; higher water content reduced the coating adhesion to the substrate. Regarding the initial powders, films made of Evonik P25 and Sigma Aldrich TiO_2 had higher photocatalytic activity than others. Moreover, the photocatalytic activity of doped TiO_2 coatings illuminated by both an artificial light and sunlight were investigated.
Keywords: titania, suspension plasma spray, photocatalysis, water treatment.

1 Introduction

Photocatalytic reactions using titania as an active photocatalyst irradiated with ultraviolet light date back at least to 1973 [1] [2]. Since then thousands of publications on the subject have followed, in which the effect was used for synthesis reactions such as the oxidation of isobutene to acetone [3], alkylation of ammonia on Pt/TiO_2 or the production of hydrogen [4] [5]. Later the photocatalytic decomposition of pollutants on TiO_2 surfaces was one of the preferred target applications somewhat shifting the focus from chemical reaction engineering to environmental technologies [6] [7]. Quite early it was discovered that anatase phase has higher photocatalytic activity than rutile giving rise to the manufacturing of suitable powders anatase phase [8]. Technical applications have been reported using TiO_2 dispersions e.g. for the cleaning of effluents from textile mills, fresh and wastewater treatment and lately food safety applications [9] [10] [11]. It was even proposed to use titania coating on building materials to destroy air pollutants [11]. Recently research in titania photocatalysts focuses on broadening the exploitable frequency range and also applies visible light instead of UV by introduction of various dopants [12] [13] [14] [15].
The possible way for technical reasons is to use the titania not as a powder, which is impossible to completely separate from suspension media after using, but as a coating. Numerous studies have been carried out investigating the feasibility of photochemical reactions with the heterogeneous catalysis. There are several methods to coat titanium dioxide with various substrates such as sol-gel [16] [17], thermal sprayings, etc [18] [19] [20] [21] [22] [23] [24] [25] [26] [27]. Among those methods, suspension plasma spraying is a promising one. Since the SPS coatings have many advantages such as good adhesion, easy to scale up, reasonable cost [27].
In this study it investigated how the parameters of the SPS process affect to its coating properties, especially the photocatalytic activity of the active layer. The titania materials used for the study were four nanoscale anatase-rich powders including TiO_2 Evonik P25, Hombicat UV100, Kronoclean 7050, and Sigma Aldrich. In which the first two are frequently used in studies over the last years.

Table 1: Some important properties of the titanium dioxide powders used to make SPS suspensions

	Evonik P25	Hombicat UV 100
crystal composition	85% anatase 15% rutil	100% anatase
primary particle	20-40 nm	<10 nm
particle size in deionized water	$D_{(v,0.1/0.5/0.9)}$ 3.2/ 105/ 5320 nm	$D_{(v,0.1/0.5/0.9)}$ 523/ 920/ 1610 nm
specific surface area (BET)	50 [m²/g] [28]	300 [m²/g] [28]
	Sigma-Aldrich	**Kronocl. 7050**
crystal composition	100% anatase	100% anatase
primary particle	<10 nm	<10 nm
particle size in deionized water	$D_{(v,0.1/0.5/0.9)}$ 230/447/1200 nm	$D_{(v,0.1/0.5/0.9)}$ 1450/2760/4880 nm

2 Materials and Methods

2.1 Materials

Regarding titania powders, there are four different commercial products were used. They included TiO_2 Evonik P25, Hombicat UV100, Kronoclean 7050, and Sigma Aldrich. TiO_2 Evonik P25 and Hombicat UV 100 were reported often in many studies. Some general properties of those powders are mentioned in table 1 and **Fig.1**. Deionized water (DW) having conductivity of 5 µS/cm was used to make suspension.

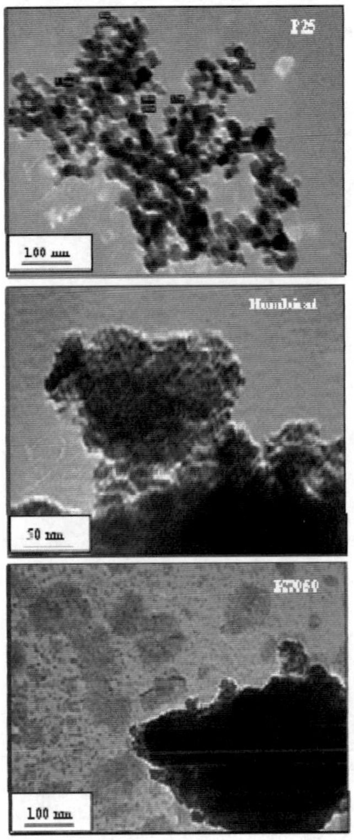

Fig. 1: TEM images of TiO_2 Evonik P25 (upper), Hombicat UV100 (middle), and Kronoclean 7050 (lower). As the images shown, P25 has the biggest crystallite size about 20 nm of anatase and 40 nm of rutil. The primary size of Hombicat and Kronoclean is much smaller in comparison with those of P25. In addition, the last two powders show the hard-agglomeration behavior, while P25 secondary particles are not.

2.2 Methods

As regards suspension preparation, an attritor and 0.75 kg of ZrO_2 milling-balls 0.5-0.9 mm were used. 60 g of TiO_2 added into 540 ml DW were milled for 1h at 400 rpm. Regarding plasma torch parameters, they are listed in table 2. Substrates were made of glass ceramic.

Table 2: Spraying parameters of SPS process with a F6 gun of Oerlikon Metco, GTV GmbH

SPS parameters	value
arc current	600 A
H_2	8 L/min
Ar	40 L/min
spraying gap	70 mm
torch travel speed	500 mm/s
meander gap	3 mm

In order to test the photocatalytic activity of those coatings, methylene blue (MB) was used as a marker at 0.5 L of a concentration 10 µmol/L. Coatings of 10 x 10 cm were tested under the radiation of a home-made LED system having wavelength between UVA-blue light. The methylene blue solution was kept in contact with a coating in darkness for 2 hours so as to reach adsorption equilibrium before starting the radiation.

Regarding methods for characterizing coatings and testing of the photocatalytic activity, XRD, light microscope, TEM, SEM, UV-Vis spectrometer, and helium pycnometry were used.

3. Results and discussion

3.1 Cross section

As shown in **Fig. 2**, the four initial powders caused the different thickness of the coatings. The difference comes from their primary and secondary particle size, which affect to the melting behavior of the particle during SPS. However, all those coatings are similar in the sense of porosity and contain both melting and un-melting zones. Since the difference of thickness might affect to their photocatalytic activity. We have coated three coatings of P25 (5, 15, 30 µm) and tested their activity. The results show the photocatalytic activity was not affected by the thickness in the tested-thickness range of 5-30 µm

3.2 XRD

In general crystal phases of photcatalyst play an important role in its photocatalytic activity. Though anatase has larger direct bandgap in comparison with rutil, it is believed that anatase is more active than rutil. Since its indirect bandgap is narrower than the direct one [29]. **Fig.3** shows both quantification and qualification of the phases of those coatings. The P25 coating has the lowest percent of anatase, because its powder contains only 85% of anatase, while the others consist of 100% anatase. Depending on the anatase parameter, it is predicted that the P25 coating has lowest photocatalytic activity. However, section 3.3 shows the inverse result to the prediction. It does not mean that the anatase percentage is not important. It means besides the anatase amount, there are other parameters affecting the activity, such as crystallite size, impurity, specific surface area, etc [30]. As shown in table 3, the difference of the initial powders causes the dissimilarity in the crystallite size of TiO_2 in those coatings.

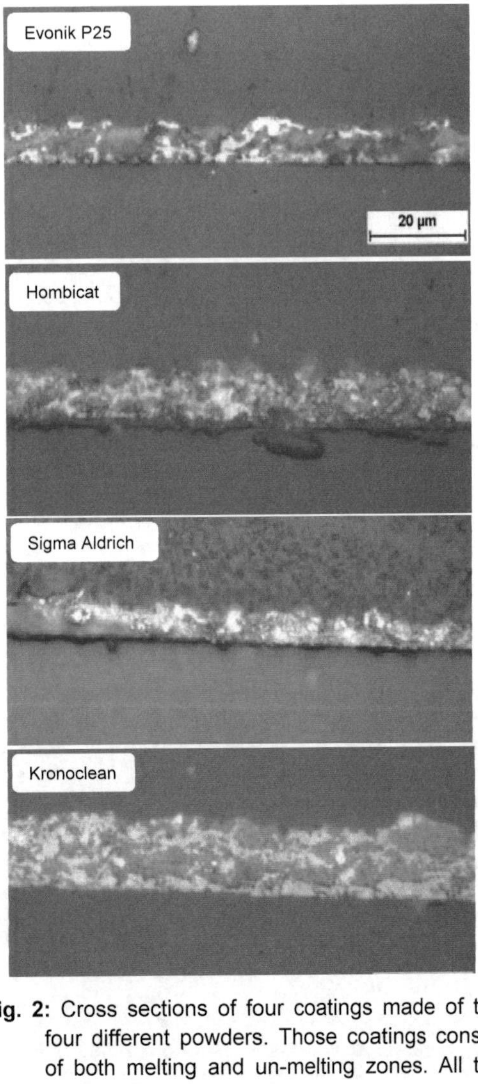

Fig. 2: Cross sections of four coatings made of the four different powders. Those coatings consist of both melting and un-melting zones. All the coatings have porous structures. However, they are different in thickness. Since the powders have differences in both primary and secondary particles size.

Table 3: Crystallite size of the powders and the coatings. The primary size is calculated by Scherrer equation. Regarding P25, size of anatase did not change too much, while the anatase in the coating of Sigma Aldrich grew several times bigger than the others.

	in powders		in coatings	
	anatase [nm]	rutil [nm]	anatase [nm]	rutil [nm]
TiO2 Evonik P25	22	41	25	92
Hombicat UV100	<10	-	43	78
Kronoclean 7050	<10	-	18	105
Sigma Aldrich	<10	-	170	87

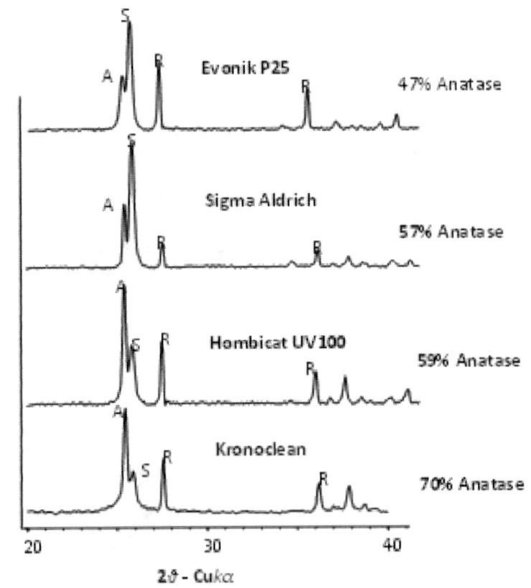

Fig. 3: XRD of SPS coatings made of different TiO_2 powders. Two phases of TiO_2 found are anatase (**A**) and rutil (**R**). Besides the two crystal phases of the titania, a peak of the glass ceramic substrate (**S**) also appeared there. Since the thickness of coatings are relatively thin. In tendency more anatase was found in coatings of the three pure anatase powders in comparison with TiO_2 Evonik P25.

3.3 Photocatalytic activity

Fig. 4 shows the result of the photocatalytic test. It is reconfirmed again that P25 has highest activity, though its coating contains smaller percentage of anatase. The second position is Kronoclean 7050.

The third one is Sigma Aldrich and the last one is Hombicat UV100. The highly active behavior of TiO_2 P25 was reported by many studies over the last years. Even though testing conditions of those studies were not similar.

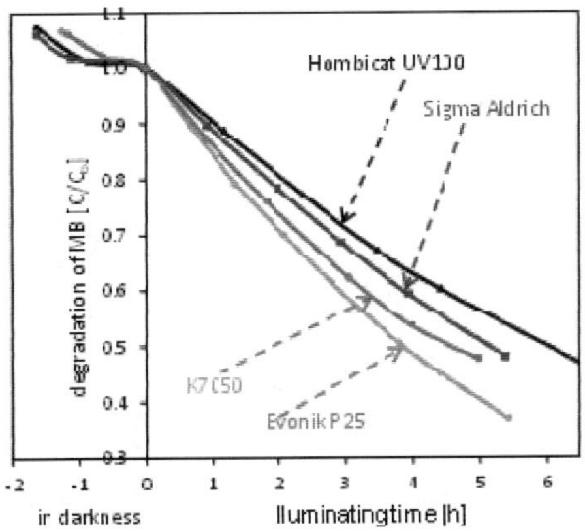

Fig. 4: Photocatalytic test results of the four coatings. The solution of MB is put in darkness for 2 hours before starting the illumination of 6 hours. The change in concentration of MB was measured every hour by a UV-Vis spectrometer. Though the coating of TiO_2 Evonik P25 contains less anatase percentage, its photocatalytic activity is higher than the others.

4. Conclusions

Results in this study shows that SPS is a good option to produce an active photocatalytic coating for many environmental cleaning applications such as water and wastewater treatments. The starting powder plays an important role in the activity of the coatings. TiO_2 P25 with our testing conditions is the best candidate for SPS, though its anatase percentage is lower than those of other coatings. The noble powder P25 showed the highly activity in hundreds of publications, which tested with many different conditions. Anatase amount is an important parameter, but it is not everything. Other parameters such as crystallite size, specific surface area, impurity also have effects on the resulting coatings. In case of P25, its best performance can come from the harmony of all the parameters [30].

Reference

[1] R. Bickley, "Photoadsorption and photocatalysis at rutile surfaces I. Photoadsorption of oxygen," *J. Catal.*, vol. 31, no. 3, pp. 389–397, Dec. 1973.

[2] R. Bickley, "Photoadsorption and photocatalysis at rutile surfaces II. Photocatalytic oxidation of isopropanol," *J. Catal.*, vol. 31, no. 3, pp. 398–407, Dec. 1973.

[3] J. Herrmann, "Heterogeneous photocatalysis: In situ photoconductivity study of TiO2 during oxidation of isobutane into acetone," *J. Catal.*, vol. 60, no. 3, pp. 369–377, Dec. 1979.

[4] B. Ohtani, H. Osaki, S. Nishimoto, and T. Kagiya, "A redox combined photocatalysis: New method of N-alkylation of ammonia by TiO2/Pt suspended in alcohols," *Tetrahedron Lett.*, vol. 27, no. 18, pp. 2019–2022, Jan. 1986.

[5] J. Kiwi and M. Gratzel, "Heterogeneous photocatalysis: enhanced dihydrogen production in titanium dioxide dispersions under irradiation. The effect of magnesium promoter at the semiconductor interface," *J. Phys. Chem.*, vol. 90, no. 4, pp. 637–640, Feb. 1986.

[6] H. Al-Ekabi and N. Serpone, "Kinetics studies in heterogeneous photocatalysis. I. Photocatalytic degradation of chlorinated phenols in aerated aqueous solutions over titania supported on a glass matrix," *J. Phys. Chem.*, vol. 92, no. 20, pp. 5726–5731, Oct. 1988.

[7] V. Augugliaro *et al.*, "Photocatalytic degradation of nitrophenols in aqueous titanium dioxide dispersion," *Appl. Catal.*, vol. 69, no. 1, pp. 323–340, Jan. 1991.

[8] A. Wold, "Photocatalytic properties of titanium dioxide (TiO2)," *Chem. Mater.*, vol. 5, no. 3, pp. 280–283, Mar. 1993.

[9] P. A. Pekakis, N. P. Xekoukoulotakis, and D. Mantzavinos, "Treatment of textile dyehouse wastewater by TiO2 photocatalysis," *Water Res.*, vol. 40, no. 6, pp. 1276–1286, Mar. 2006.

[10] M. Mehrjouei, S. Müller, and D. Möller, "A review on photocatalytic ozonation used for the treatment of water and wastewater," *Chem. Eng. J.*, vol. 263, pp. 209–219, Mar. 2015.

[11] V. K. Yemmireddy and Y.-C. Hung, "Selection of photocatalytic bactericidal titanium dioxide (TiO2) nanoparticles for food safety applications," *LWT - Food Sci. Technol.*, vol. 61, no. 1, pp. 1–6, Apr. 2015.

[12] W. Choi, A. Termin, and M. R. Hoffmann, "The Role of Metal Ion Dopants in Quantum-Sized TiO2: Correlation between Photoreactivity and Charge Carrier Recombination Dynamics," *J. Phys. Chem.*, vol. 98, no. 51, pp. 13669–13679, Dec. 1994.

[13] A.-W. Xu, Y. Gao, and H.-Q. Liu, "The Preparation, Characterization, and their Photocatalytic Activities of Rare-Earth-Doped TiO2 Nanoparticles," *J. Catal.*, vol. 207, no. 2, pp. 151–157, Apr. 2002.

[14] W. Raza, M. M. Haque, M. Muneer, M. Fleisch, A. Hakki, and D. Bahnemann, "Photocatalytic degradation of different chromophoric dyes in

aqueous phase using La and Mo doped TiO2 hybrid carbon spheres," *J. Alloys Compd.*, vol. 632, pp. 837–844, May 2015.

[15] M. Senna *et al.*, "Modification of titania nanoparticles for photocatalytic antibacterial activity via a colloidal route with glycine and subsequent annealing," *J. Mater. Res.*, vol. 28, no. 03, pp. 354–361, Feb. 2013.

[16] J. Schneider *et al.*, "Understanding TiO $_2$ Photocatalysis: Mechanisms and Materials," *Chem. Rev.*, vol. 114, no. 19, pp. 9919–9986, Oct. 2014.

[17] A. Kubacka, M. Fernández-García, and G. Colón, "Advanced Nanoarchitectures for Solar Photocatalytic Applications," *Chem. Rev.*, vol. 112, no. 3, pp. 1555–1614, Mar. 2012.

[18] T. Kanazawa and A. Ohmori, "Behavior of TiO2 coating formation on PET plate by plasma spraying and evaluation of coating's photocatalytic activity," *Surf. Coat. Technol.*, vol. 197, no. 1, pp. 45–50, Jul. 2005.

[19] P. Ctibor *et al.*, "Titanium Dioxide Coatings Sprayed by a Water-Stabilized Plasma Gun (WSP) with Argon and Nitrogen as the Powder Feeding Gas: Differences in Structural, Mechanical and Photocatalytic Behavior," *J. Therm. Spray Technol.*, vol. 21, no. 3–4, pp. 425–434, Jun. 2012.

[20] F.-L. Toma *et al.*, "Comparative study on the photocatalytic behaviour of titanium oxide thermal sprayed coatings from powders and suspensions," *Surf. Coat. Technol.*, vol. 203, no. 15, pp. 2150–2156, May 2009.

[21] M. Yamada, H. Isago, H. Nakano, and M. Fukumoto, "Cold Spraying of TiO2 Photocatalyst Coating With Nitrogen Process Gas," *J. Therm. Spray Technol.*, vol. 19, no. 6, pp. 1218–1223, Dec. 2010.

[22] M. Gardon *et al.*, "Photocatalytic Activity of Nanostructured Anatase Coatings Obtained by Cold Gas Spray," *J. Therm. Spray Technol.*, vol. 23, no. 7, pp. 1135–1141, Oct. 2014.

[23] F.-L. Toma, G. Bertrand, S. O. Chwa, C. Meunier, D. Klein, and C. Coddet, "Comparative study on the photocatalytic decomposition of nitrogen oxides using TiO2 coatings prepared by conventional plasma spraying and suspension plasma spraying," *Surf. Coat. Technol.*, vol. 200, no. 20–21, pp. 5855–5862, May 2006.

[24] I. Burlacov, J. Jirkovský, M. Müller, and R. B. Heimann, "Induction plasma-sprayed photocatalytically active titania coatings and their characterisation by micro-Raman spectroscopy," *Surf. Coat. Technol.*, vol. 201, no. 1–2, pp. 255–264, Sep. 2006.

[25] G. Bolelli, V. Cannillo, R. Gadow, A. Killinger, L. Lusvarghi, and J. Rauch, "Properties of High Velocity Suspension Flame Sprayed (HVSFS) TiO2 coatings," *Surf. Coat. Technol.*, vol. 203, no. 12, pp. 1722–1732, Mar. 2009.

[26] A. Killinger, R. Gadow, G. Mauer, A. Guignard, R. Vaßen, and D. Stöver, "Review of New Developments in Suspension and Solution Precursor Thermal Spray Processes," *J. Therm. Spray Technol.*, vol. 20, no. 4, pp. 677–695, Jun. 2011.

[27] C. Lee, H. Choi, C. Lee, and H. Kim, "Photocatalytic properties of nano-structured TiO2 plasma sprayed coating," *Surf. Coat. Technol.*, vol. 173, no. 2–3, pp. 192–200, Aug. 2003.

[28] J. Kirchnerova, M.-L. Herrera Cohen, C. Guy, and D. Klvana, "Photocatalytic oxidation of n-butanol under fluorescent visible light lamp over commercial TiO2 (Hombicat UV100 and Degussa P25)," *Appl. Catal. Gen.*, vol. 282, no. 1–2, pp. 321–332, Mar. 2005.

[29] T. Luttrell, S. Halpegamage, J. Tao, A. Kramer, E. Sutter, and M. Batzill, "Why is anatase a better photocatalyst than rutile? - Model studies on epitaxial TiO2 films," *Sci. Rep.*, vol. 4, Feb. 2014.

[30] P. Pichat, Ed., *Photocatalysis and water purification: from fundamentals to recent applications.* Weinheim: Wiley-VCH, 2013.

Influence of substrate properties on the formation of suspension plasma sprayed coatings

Yongli Zhao[a], Zexin Yu[a], Marie-Pierre Planche[a], Audrey Lasalle[b] ,Alain Allimant [b], Ghislain Montavon[a], Hanlin Liao[a]

[a] ICB UMR 6303, CNRS, Université de Bourgogne Franche-Comté, UTBM, F-90010 Belfort, France
[b] Saint-Gobain C.R.E.E., 84300 Cavaillon, France

In this study, YSZ coatings were deposited on different substrate materials (stainless steel and aluminum) using suspension plasma spray (SPS) technique. The effects of substrate properties (material, surface topology, temperature, and thickness) on the formation of coatings were investigated. The results showed that, with the identical spray parameters, the porosity is higher for the coatings deposited on aluminum than that on stainless steel due to the high thermal transfer ability of the former substrate material. The SEM results revealed that the microstructure of as-prepared coatings could be tailored from the vertical cracked structure to the columnar structure by increasing the substrate surface roughness and their formation mechanisms were discussed. The substrate preheating temperature has an influence on the microstructure of the coatings, especially in the interfacial region and increasing the substrate temperature is an effective means for reducing the interface defects in the coatings. With the increase of the substrate thickness, the quantity of the vertical cracks in the coatings is reduced and their width becomes narrower.

1 Introduction

Finely structured or even nanostructured coatings have better mechanical, thermal and chemical properties in comparison with micro-structured coatings for numerous technical and industrial applications [1-3]. To form finely structured coatings by thermal spraying technique, a feasible solution consists of reducing the size of the precursors to a sub-micrometric or nanometric scale. However, these smaller particles are difficult to feed using conventional equipment since they have poor flowability and tend to form larger aggregates that can cause blocking in the injection path. It is the reason why injection of such submicrometric or nanometric size powders into a high temperature plasma or flame can be achieved with the help of liquid carriers [4-6]. In the last decade, suspension plasma spray (SPS) has continuously gained increasing interest in the scientific world [7]. Nowadays it is considered as an emerging process to form the coating by the piling up of molten particles with a size ranging from a few tens of nanometers to a few micrometers at impact on the substrate [1].
The coating properties have a close relationship with its microstructures. A certain favorable coating microstructure should be considered depending on applications. For instance, the columnar microstructure in thermal barrier coatings (TBCs) tends to show good thermomechanical properties [8], and the fine-grained lamellar microstructure is more efficient to reduce the thermal conductivity [9]. In the field of solid oxide fuel cells (SOFCs), a thin and dense coating is needed for the electrolytes [10], on the contrary, the cathodes should exhibit finely porous structures [11]. However, for the wear-resistant and corrosion-resistant coatings, porous structures could reduce its anti-wear and anti-corrosion performance. As the conventional plasma spraying (APS), the microstructures of the SPS coatings also could be tailored by controlling the spray conditions. However, SPS is more complicated than APS hence it is more difficult to optimize the process parameters. [6, 12]. The substrate properties are also important process parameters for thermal spray coatings. Lots of researches have been done to investigate the thermal sprayed coatings deposited on different substrates [13-15] . However, there is still no systematic study reported the role of the substrate properties on the SPS coatings. Thus, the investigation of the influence of substrate properties on the formation of suspension plasma sprayed coatings is significantly meaningful from the practical viewpoint.

This work focuses on preparing YSZ coatings using SPS technique. Different coating microstructures induced by different substrate conditions were analyzed. The effects of the substrate properties on porosity and the mechanical performance of the coatings were investigated in order to get a better understanding of the influence of substrate properties on the formation of SPS coatings.

2 Experimental procedure

2.1 Substrate preparation

Stainless steel (304L) and aluminum (1050) of 25 mm diameter and three different thicknesses (5mm,10mm, and 20mm) were used as substrates. They were polished or grit-blasted to obtain different surface topologies. All of the substrates were cleaned up in the ultrasonic bath of ethanol prior to spraying. The 3D profiles of substrates topographies and the surface roughness were made by a 3D profilometer (Altisurf 500, France) and are presented in **Fig.1**.

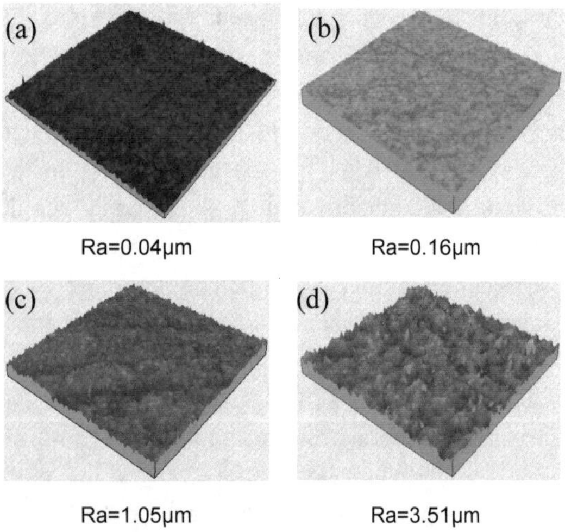

Fig.1. 3D profile of the substrate surface and the surface roughness

2.2 Coating fabrication

A Commercial feedstock YSZ powder (Saint-Gobain, France, 8 wt.% yttria, 3 wt.% alumina, SSA =7m^2/g, D_{50} = 360 nm) was sprayed in this study. A solid mass load of 25 wt.% powder in absolute ethanol was chosen to prepare the suspension.

The SPS coatings were produced using the atmospheric plasma ProPlasma torch (Saint-Gobain Coating Solutions, Avignon, France) equipped with a twin-fluid atomizer system. The plasma torch, with a 6.5 mm diameter anode worked on the "high performance" mode was equipped. Plasma conditions used a mixture of 50 SLPM argon and 10 SLPM hydrogen as process gas is corresponding to the net power of 41 kW. The torch was attached to a 6-axis robotic arm. The atomizer is positioned perpendicularly to the torch axis, at 10 mm from the torch axis and 6 mm downstream the anode face.

During the spray process, suspension was stored in a pressurized tank, under magnetic stirring to avoid sedimentation issues, and was delivered through the twin-fluid atomize supplied with argon as atomizing gas. Substrates were fixed to a rotating samples holder. The linear speed was 1000 mm/s and the spray distance 40 mm. The temperature of the substrates was also monitored using an infrared thermometer.

2.3 Characterizations

The microstructural analyses of the as-sprayed SPS coatings are performed on the cross sections and the top surfaces using the field emission scanning electron microscopy (JSM-7800F, JEOL, Japan). Specimens for cross-section analysis were sectioned using a diamond cutting blade and mounted in resin. Samples were subsequently grinded with emery papers, polished with diamond paste, and cleaned with water and ethanol in an ultra-sonic device. All observed samples were sputtered with a thin gold layer to allow the ceramic layers to be observed clearly.

Microhardness measurements were performed on the polished cross-section of coatings using a Vickers hardness indenter (Leitz, RZD-DO, Germany) with a load of 300 g and a holding time of 30 s. Ten indentation measurements were taken randomly within different regions of each sample.

The porosity was finally calculated with the use of the software ImageJ. For each sample, ten SEM images at a magnification of 1000X were collected randomly across the entire cross-section in regions not close to the edges. It should be noted that this method is limited to detect the pores with the dimension less than 100 nm at this magnification.

3 Results and discussion

3.1 Effect of the substrate material

Fig.2 shows the top surface and cross-section SEM images of the coatings deposited respectively on stainless steel and aluminum using identical spray parameters. It can be seen that: 1) The as-prepared coatings exhibit a typical granular structure, which is composed of flattened lamellae and spherical particles. 2) The diameters of the spherical particles are on the same order of that of the original powder (around 300 nm). and 3) There are more spherical particles on the coating deposited on aluminum than that on stainless steel.

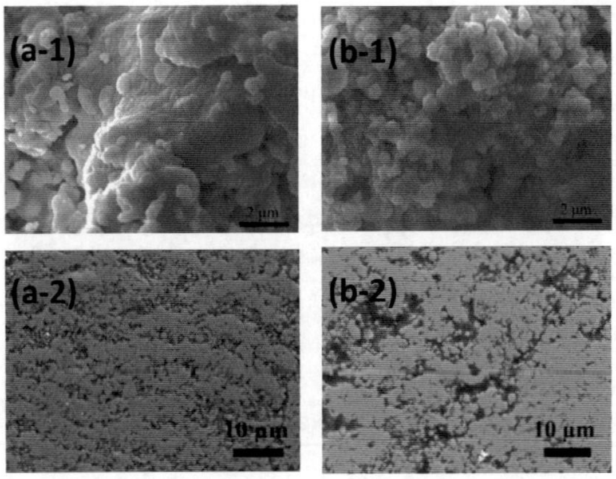

Fig.2 SEM images of the coatings deposited on different substrate materials: a) SS-304L, b) Al-1050

Considering that the two types of substrates have the same shape (button, diameter 25 mm, thickness 10 mm), these phenomena may result from the different thermal properties between stainless steel and aluminum. The specific heat capacity and thermal conductivity of aluminum are both higher than those of stainless steel (around 25°C, c(ss304)=502 J·Kg-1·K-1, c(Al1050)=897 J·Kg-1·K-1, k(ss304)=12.1 W·m-1·K-1, k(Al1050)=227 W·m-1·K-1). This way of the heat transfer to the environment is higher for aluminum than for stainless steel. The particles deposited on the aluminum solidify more rapidly than those projected on the stainless steel and this rapid solidification on aluminum (quenching) allows the particles to solidify in spherical form and possibly induce porosity in the coatings. On the opposite, the melted particles on the stainless steel can remain in the form of splats and splashes. The shapes of the single splat projected on the two substrates agree with what have been proposed on the role of the thermal behavior of the substrates, as shown in **Fig.3**. The measurements of the coating porosity were carried out on the as-sprayed samples, and the results also confirmed the inference mentioned above, that is to say, the average porosity is 25.9 % and 30.5 % for the coatings deposited on stainless steel and on aluminum respectively.

One should consider that when the coating is deposited up to a certain thickness, the thermal transfer from the molten particles to the substrate will be reduced by the low thermal conductivity of YSZ material and the interface defects. Correlatively, the solidification rate of the molten particles should therefore decrease with the increase of the coating thickness. So the same denser microstructure could be formed in the top part of the coating regardless of the substrate material. Our works is in progress to prove this hypothesis.

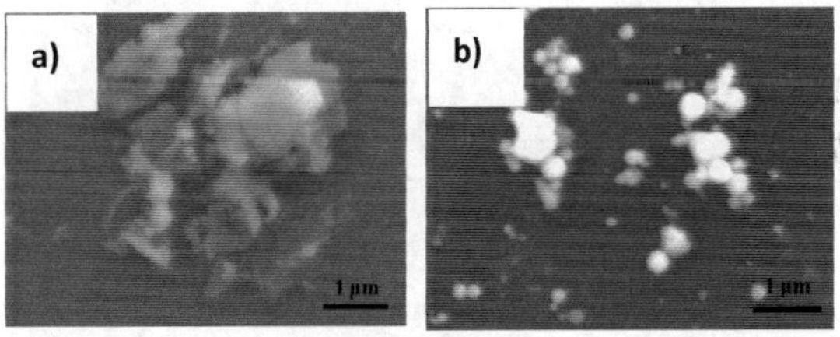

Fig.3. YSZ single splat deposited on a) SS-304L, b) Al-1050

3.2 Effect of the substrate surface topology

Fig.4 presents the SEM cross-section images of the as-prepared coatings deposited on 304L substrates with four different surface topologies (see **Fig.1**). It can be observed that the substrate surface condition has a significant influence on the coating microstructure. With the increase of the substrate roughness, the coating microstructure changes from slight vertical cracks to obvious columnar structure. Both of the coatings in Fig.4 (a) and (b) exhibit vertical-crack microstructure, but they display a subtle difference between them. There are fewer cracks in the coating (a) than that of coating (b), and it seems that the average width of the cracks in the latter is larger. Furthermore, there is almost no crack in the interfacial region of the coating (a), however, as the arrows showed in Fig.4(b), small cracks can be found clearly in the same region of the coating (b). This tendency is enhanced obviously as shown in the coating (c). Finally, on a substrate with the highest roughness, the vertical cracks are very distinctly, and some of them widen to columnar porosity bands, as shown in Fig.4(d). A schematic of the coating microstructure deposited on different substrates was summarized in **Fig.5**. These findings suggest that it is possible to tailor the coating microstructures by changing the substrate surface topology keeping the same spray conditions.

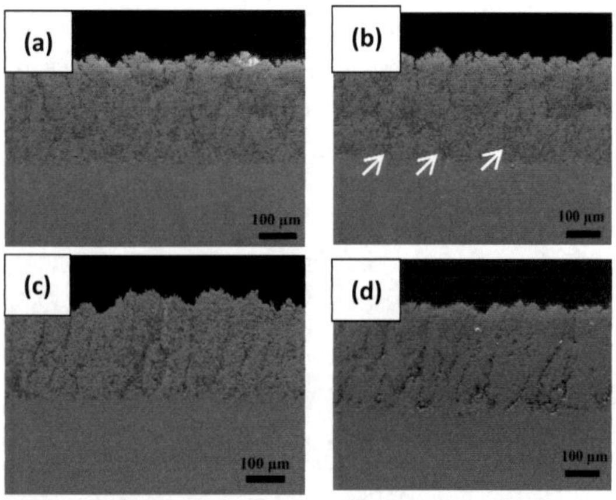

Fig.4. SEM cross-section images of coatings deposited on stainless steel substrates with different surface topologies: a) Ra=0.04μm, b) Ra=0.16μm, c) Ra=1.05 μm, and d) Ra=3.51μm

The microstructure differences between the coatings given in Fig.4 may certainly result from the substrate surface topology. In this study, the axis of plasma jet is set perpendicular to the substrate surface. When a big particle impacts the substrate, its movement direction remains perpendicular to the substrate surface. However, for the small particles, that have relatively low momentum, they might be influenced by the drag force of the hot gas swirls especially in the boundary layer close to the substrate. This drag force causes them to deviate from their original spray direction leading to impact the substrate at a shallow angle. Numerical computations pointed out that zirconia particles of the size less than 5 μm could be affected by the drag force component parallel to the substrate [16]. According to the results in section 3.1 of this paper, the flattened lamellae and spherical particles of the as-sprayed coatings have a size of several hundred nanometers. So the trajectory of these particular particles must be strongly dependent on the drag force. When the particles were projected on a rough surface, shadowing effect could be employed for explaining the mechanism of columnar growth [5, 17]. It suggested that columnar microstructures are resulted from the deposition of particles having a significant velocity component parallel to the substrate near the asperities on the substrate surface. In contrast, on a smooth surface, the movement of the particles is almost unhindered in the direction parallel to the substrate surface, the particles should be deposited in a relative large range. Therefore, the shadow effect does not act because there are no asperities on the substrate surface and a relative dense coating without vertical cracks could be formed. Subsequently, the single particles rooted on the smooth substrate will become as the new asperities, which will hinder the movement of the following particles. The accumulation of these following particles leads to the new asperities becoming bigger and higher. So the coating surface roughness will increase with the spraying process. Until the roughness reaches a certain critical value, the cracks will appear. The corresponding image of the coating structure is shown in Fig.5(1). However, on a moderate rough surface, the accumulation of particles on the asperities starts from the beginning of the coating formation. This is the reason why the vertical cracks could appear more early in the coatings nearer to the substrate with a higher surface roughness (see Fig.5(2)). As for the highest rough substrate surface, maybe the particles are confined among the high substrate peaks, the depositions sprayed by different spray passes could not overlap very well each other. So the large cracks or columnar porosity bands may be created on this kind of substrates.

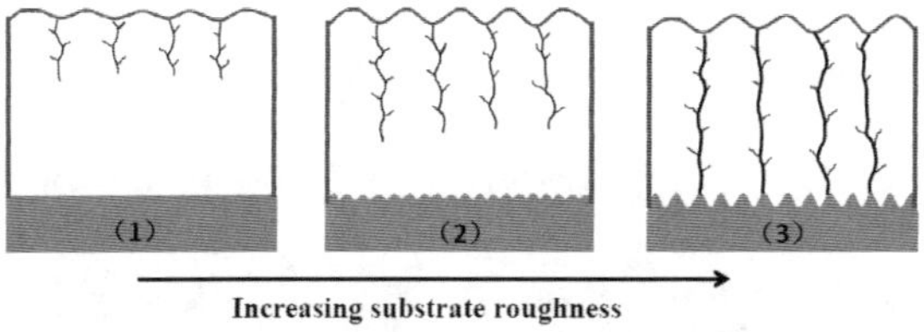

Fig.5. Microstructure Schematic of coatings deposited on substrates with different roughness

The porosity and vickers hardness of these four case coatings are exhibited in **Fig.6**. With the increase of the substrate surface roughness, the average porosity of the coating is increased from 25.2% to 31.3%. According to the results, one can draw a conclusion that it is more difficult to fabricate a dense coating on a high rough surface. It should be noted that the porosity standard deviation of the coating on mirror polished substrate is much smaller than those on other substrates. This means that this coating is with the best uniformity. The substrate surface topology also has a great effect on the coating hardness as shown in **Fig.6**, the hardness of coating is decreasing with the increase of the substrate surface roughness. These changes of the coating microstructure will inevitably influence the thermal, mechanical and chemical properties of the coating.

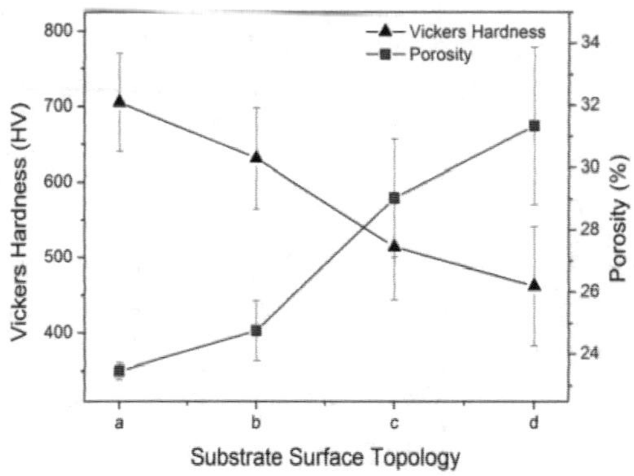

Fig.6. Vickers Hardness and porosity of coatings deposited on SS304L with different surface roughness: a) Ra=0.04μm, b) Ra=0.16μm, c) Ra=1.05 μm, and d) Ra=3.51μm

3.3 Effect of the substrate temperature

The cross section SEM images of the coatings deposited on SS304L substrate preheated at different temperatures are displayed in **Fig.7**. It shows that the preheating temperature influences the microstructure of the coatings, especially in the interfacial region. More obvious defects in the coating deposited on the substrate without preheating are visible. One can observe clearly that the interface is greatly improved with the increase of the substrate preheating temperature. So controlling the substrate temperature is an effective means for reducing the interface defects in the coatings. It should be noted there are more obvious vertical cracks which run through the coating thickness appeared with the increasing of the substrate preheating temperature.

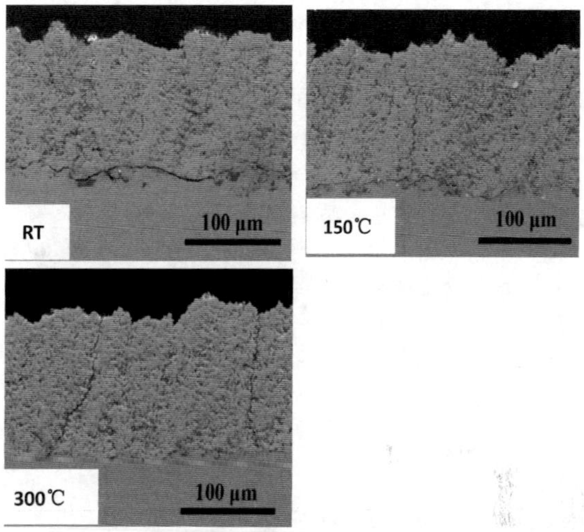

Fig.7. SEM images of the coatings deposited on SS304L substrate preheated at: a) without preheating, b) 150 °C, and c) 300 °C

The wettability, spreading, and interfacial phenomena of the molten particles could be varied by the substrate temperature. Some studies on APS coatings reported that the splat morphology changes with the increase of substrate temperature from fragmented to more continuous shape [18]. The splat size of the SPS coatings is smaller, but these splats have the similar morphology as the APS coatings [3, 19]. On a low temperature substrate, the wettability between the molten particles and the substrate is lower. The molten particles thus splashed extensively after impact and had very irregular contours. Furthermore, low temperature results in a high temperature gradient at the splat/ substrate interface. This makes the molten particles solidify very rapidly. So they have no enough time to flow into surface cavities hence to form good contact with the substrate or previous splats. On the other hand, the irregular contours present voids at the coating-substrate interface, which decrease the adhesion of the coating. On the contrary, a higher preheating temperature could reduce the cooling rate of particles impinging on the surface. The slow solidification permits these particles to have a good flowability, which might promote the formation of an intimate contact between the atoms in the melt and the solid substrate surface [20]. Another possible reason is that high temperature could evaporate moisture and other volatile contaminants on the surface, allowing also better contact between impacting particles and the substrate [18].

The formation of the vertical cracks in the coatings can be explained from the view of the release of thermal stress. The stress derived from the mismatch of thermal expansion between the coating and the substrate could be released adequately at the default regions. This is the reason why fewer cracks appear in the coating with obvious defaults in the interfacial region. Conversely, a good adhesion between the coating and the substrate compel to develop cracks in the coating.

3.4 Effect of substrate thickness

As it can be seen from **Fig.8**, with the increase of the substrate thickness, the porosity of the coating decreases slightly. The porosity changes from 27.9% to 25.7% average values when the thickness of the substrate increases from 5 mm to 20 mm. In addition, there are fewer vertical cracks in the coatings deposited on a thicker substrate, and their widths are obviously much narrower.

As the heat transfer is slower on the thicker substrates, the coating temperature is higher and consequently the solidification of the impacted particle is slower as well. This leads to denser coatings. A slower heat transfer means that the temperature gradient is smaller in the coating and thereby induces the thermal stress to be released more slowly. Thus, a denser coating with fewer cracks could be formed, and the width of these cracks is also narrower.

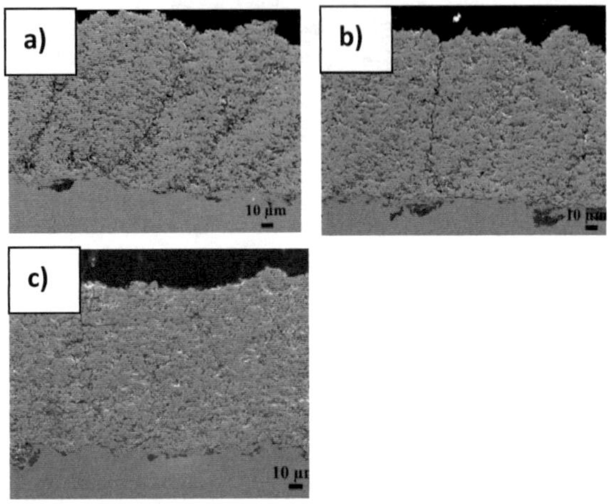

Fig.8 SEM cross-section images of the coatings deposited on SS304L substrates with different thicknesses: a) 5mm, b) 10mm, and c) 20mm

4 Conclusion

YSZ coatings are successfully produced using suspension plasma spraying. This study demonstrates that the formation of SPS coating is influenced significantly by the substrate properties. The results are summarized as follows:

- Higher thermal transfer ability of the substrate can speed up the particle's cooling rate and makes the deposited particles more spherical, which could form a coating with higher porosity. A higher porosity will further reduce the mechanical properties of the coating.
- There is a direct correlation between the cracks formation in the coating and the substrate surface morphology. Microstructure of the SPS coating could be tailored from the vertical cracks to the columnar structure by increasing the substrate surface roughness.
- Substrate temperature has an important effect on the adhesion of the coating. Increasing the substrate preheating temperature is an effective means for reducing the interface defects in the coatings.
- The substrate thickness has a slight effect on the microstructure of the coating. The coating

deposited on a thicker substrate has a lower porosity and smaller micro-cracks. This is due to the lower heat transfer ability of the thicker substrate.

5 Literature

[1] Armelle Vardelle, C.M., Jun Akedo et al, The 2016 Thermal Spray Roadmap. Journal of Thermal Spray Technology, 2016. 25(3): p. 1376-1440.

[2] Fauchais, P., G. Montavon, and B.R. Marple, Engineering a new class of thermal spray nano-based microstructures from agglomerated nanostructured particles, suspensions and solutions : an invited review. Journal of Physics D: Applied Physics, 2011. 44: p. 93001-1-93001-53.

[3] Joulia, A., et al., Tailoring the Spray Conditions for Suspension Plasma Spraying. Journal of Thermal Spray Technology, 2015. 24(1): p. 24-29.

[4] Tarasi, F., et al., Effective Parameters in Axial Injection Suspension Plasma Spray Process of Alumina-Zirconia Ceramics. Journal of Thermal Spray Technology, 2008. 17(5): p. 685-691.

[5] VanEvery, K., et al., Column Formation in Suspension Plasma-Sprayed Coatings and Resultant Thermal Properties. Journal of Thermal Spray Technology, 2011. 20(4): p. 817-828.

[6] Fauchais, P., et al., Operating parameters for suspension and solution plasma-spray coatings. Surface and Coatings Technology, 2008. 202(18): p. 4309-4317.

[7] Toma, F.-L., et al., Demands, Potentials, and Economic Aspects of Thermal Spraying with Suspensions: A Critical Review. Journal of Thermal Spray Technology, 2015. 24(7): p. 1143-1152.

[8] Curry, N., et al., Thermal Conductivity Analysis and Lifetime Testing of Suspension Plasma-Sprayed Thermal Barrier Coatings. Coatings, 2014. 4(3): p. 630-650.

[9] Nait-Ali, B., et al., Thermal conductivity of highly porous zirconia. Journal of the European Ceramic Society, 2006. 26(16): p. 3567-3574.

[10] Stöver, D., et al., Plasma sprayed components for SOFC applications. Surface and Coatings Technology, 2006. 201(5): p. 2002-2005.

[11] Zhao, C., et al., Fabrication and characterization of a cathode-supported tubular solid oxide fuel cell. Journal of Power Sources, 2009. 192(2): p. 552-555.

[12] Pawlowski, L., Suspension and solution thermal spray coatings. Surface and Coatings Technology, 2009. 203(19): p. 2807-2829.

[13] Fukumoto, M., H. Nagai, and T. Yasui, Influence of surface character change of substrate due to heating on flattening behavior of thermal sprayed particles. Journal of Thermal Spray Technology, 2006. 15(4): p. 759-764.

[14] Sokołowski, P., et al., The key process parameters influencing formation of columnar microstructure in suspension plasma sprayed zirconia coatings. Surface and Coatings Technology, 2014. 260: p. 97-106.

[15] Curry, N., et al., Influence of bond coat surface roughness on the structure of axial suspension plasma spray thermal barrier coatings — Thermal and lifetime performance. Surface and Coatings Technology, 2015. 268: p. 15-23.

[16] J. Oberste Berghaus, S.B., J.G. Lagroux, C. Moreau, Injection conditions and in-flight particles states in suspension plasma spraying of alumina and zirconia nano-ceramics. Thermal Spray 2005 Conference, 2005: p. 512 - 518.

[17] Sokołowski, P., et al., Advanced Microscopic Study of Suspension Plasma-Sprayed Zirconia Coatings with Different Microstructures. Journal of Thermal Spray Technology, 2016. 25(1): p. 94-104.

[18] Pershin, V., et al., Effect of substrate temperature on adhesion strength of plasma-sprayed nickel coatings. Journal of Thermal Spray Technology, 2003. 12(3): p. 370-376.

[19] Delbos, C., et al., Phenomena Involved in Suspension Plasma Spraying Part 2: Zirconia Particle Treatment and Coating Formation. Plasma Chemistry and Plasma Processing, 2006. 26(4): p. 393-414.

[20] Yao, S.-W., et al., Conditions and mechanisms for the bonding of a molten ceramic droplet to a substrate after high-speed impact. Acta Materialia, 2016. 119: p. 9-25.

Numerical analyzing the in-flight particle temperature and velocity in plasma spray

Wenhua Ma, Guangdong Industrial Analysis and Testing Center/China
Renzhong Huang, Jian Huang, Guangdong Institute of New Materials/China

The in-flight particle temperature and velocity are very important factors that determine the deposition characteristics and coating qualities in plasma spray. Therefore, predicting the in-flight particle temperature and velocity with spray conditions is benefit to control the prepared coating qualities. In the present study, a steady-state 3D model was used to calculate the gas flow field of a plasma jet. The plasma gas temperature and velocity profiles applied to the plasma jet as the boundary conditions were obtained with our previous studies of plasma arc behaviors inside plasma torch. After obtained the gas temperature and velocity of the plasma jet, the in-flight particle trajectories, temperature, and velocity were investigated with numerical calculation method. Also the in-flight particle temperature and velocity were verified by experimental measurements.

1 Introduction

The two key parameters that determine the properties of coatings produced by plasma spraying are the particle velocity and temperature in the plasma jet. Consequently, many studies have investigated plasma jets in plasma sprays. In particular, plasma jet characteristics and the behavior of injected particles have been extensively studied numerically.

Borisov et al. [1] developed a model for simulating subsonic turbulent plasma jets generated by plasma spray torches and the behavior of sprayed particles in such jets. The calculated results showed reasonable agreement with measurements. Bolot et al. [2] modeled plasma jets using two commercial CFD packages, PHOENICS and FLUENT. They describe some differences regarding the ways in which these two codes solve this type of problem. Wang et al. [3] described the mole fractions of different species, the temperature, and the velocity in a multicomponent plasma jet using the eddy dissipation model and a realizable k–ε turbulence mathematical representation. Their results revealed that the eddy dissipation model is useful for predicting the various chemical reactions that occur in a multicomponent plasma gas, including recombination of dissociative atoms and ions, charge exchange reactions, and nitrogen and oxygen dissociation. Ramachandran and Nishiyama [4] developed a fully coupled 3D model to clarify two-way interactions between plasma and particles in terms of energy, momentum, and turbulence. Plasma–particle two-way interactions were modeled by adopting a Lagrangian approach for particle behavior and an Eulerian approach for plasma flow. The effects of each two-way interaction on energy and momentum transfer and turbulence modulation were investigated by numerical simulations. Particles in the plasma jet cause local deformations of the plasma jet kinetic energy and its dissipation rate fields. Kang et al. [5] developed numerical models using the CFD program FLUENT and used them to investigate the effect of a substrate on the behavior of the plasma flow fields and in-flight particles. Their results revealed that although a perpendicular or inclined substrate significantly affects plasma flow fields near the substrate, the particle behavior was relatively unaffected. The insignificant effect of the substrate on particle behavior was qualitatively verified by experimental observations using SprayWatch imaging diagnostic system. Xiong et al. [6] used a 3D computational model, LAVA-P-3D, to investigate air entrainment in plasma jets and metal particle oxidation by entrained oxygen. They obtained good agreement with experimental data. Li and Chen [7] developed a 3D model for a turbulent thermal plasma jet with a transversely injected carrier gas and metal particles at atmospheric pressure. They employed the standard k–ε model together with momentum and energy continuity equations to numerically simulate turbulent plasma flow. An improved particle stochastic-trajectory model was used in the calculations to predict the motion of injected particles in the turbulent flow field. The heating histories of the injected particles during their movement were also calculated. The modeling results revealed it is very important to include the effect of the carrier gas on the jet and particle behaviors. Transverse injection of the carrier gas deflected the plasma jet from its geometrical axis; the particle trajectories were also appreciably altered by carrier gas injection. The particles dispersed about their average trajectories in the turbulent flow field. Furthermore, Li and Pfender [8] realized more accurate prediction of 3D temperature and flow fields inside and outside of a DC arc plasma torch for transverse particle and carrier gas injection into the plasma jet. The 3D trajectories and heating histories of the particles were predicted from the computed plasma temperature and flow fields.

Most of the previous studies of plasma jets focused on the gas flow field of a plasma jet and the behavior of the injected particles. The boundary conditions for the plasma jet inlet are almost always derived using an empirical formula or experimental results. In our previous studies [9-10], an improved LTE (local thermodynamic equilibrium) mode can predict the plasma arc behavioes inside the torch with a relatively high accuracy. However, the plasma jet conditions still not report based on the results calculated by the improved LTE mode.

In this research, a steady-state 3D model is used to calculate the gas flow field of a plasma jet. The boundary conditions of plasma gas temperature and velocity profiles applying to the plasma jet are obtained form the results calculated with the improved LTE mode based on our previous studies. After obtained the gas temperature and velocity of the plasma jet, the in-flight particle trajectories, temperature, and velocity are investigated with numerical calculation method. The calculated particle temperature and velocity were also compared with the experimentally measured results.

2 Mathematical Model and Experimental procedure

2.1 Model Assumptions

The gas flow fields of a plasma jet outside plasma torch and the in-flight particle behaviors were calculated based on the following main assumptions in this study.

(1) The plasma jet flow is quasi-steady and turbulent, and has temperature-dependent properties.
(2) The plasma jet is in local thermodynamic equilibrium (i.e., the temperatures of the gas atoms, ions, and electrons at a point are the same) allowing it to be characterized by a single temperature.
(3) The plasma jet flow is in an optically thin, compressible, and steady system.
(4) An argon plasma jet is injected into argon gas (rather than air) at atmospheric pressure.
(5) The effect of ionization reactions of atomic argon resulting from dissociative reactions in the gas field of the plasma jet can be ignored.
(6) The effect of the carrier gas on the plasma jet can be neglected.
(7) The particles in the plasma jet are spherical. The temperature distributions within the particles are uniform. The latent heat of the particles can be neglected.
(8) The injected particles have negligible effects on the plasma jet characteristics, and particle–particle interactions are negligible; this is reasonable for a low-particle loading rate.
(9) The effect of gravity on the particles is negligible.

2.2 Governing Equations

Based on the above assumptions, the governing equations for a 3D steady gas flow of a plasma jet can be written as:

Conservation of mass:

$$\nabla \cdot (\rho \vec{V}) = 0$$

Conservation of momentum:

$$\rho \vec{V}.\nabla\vec{V} = -\nabla\left[P + \tfrac{2}{3}\mu(\nabla \cdot \vec{V})\right] + 2\nabla \cdot (\mu\vec{S})$$

Conservation of energy:

$$\rho c_p \vec{V}.\nabla T = \nabla \cdot (\lambda \nabla T) - S_r$$

Where ρ is gas mass density, $\vec{V}$ gas velocity, P gas pressure, μ dynamic viscosity, $\vec{S}$ strain rate tensor, c_p specific heat at constant pressure, T gas temperature, S_r volumetric net radiation losses, and λ gas thermal conductivity.

Lagrangian equations of motion and heat balance are used to simulate particle behavior in plasma jet. The governing equations for a particle in plasma jet can be written as follows:

Particle velocity:

$$m_p \frac{d\vec{V_p}}{dt} = \frac{1}{2} C_d \rho |\vec{V} - \vec{V_p}|(\vec{V} - \vec{V_p}) \cdot \frac{\pi}{4} d_p^2$$

Particle temperature:

$$m_p c_{pp} \frac{dT_p}{dt} = h(T - T_p) \cdot \pi d_p^2$$

where m_p is particle mass, $\vec{V_p}$ particle velocity, C_d drag coefficient, d_p particle diameter, c_{pp} specific heat of particle, T_p particle temperature, and h heat transfer coefficient.

The drag coefficient is given by [11]:

$$C_d = \alpha_1 + \frac{\alpha_2}{Re} + \frac{\alpha_3}{Re^2}$$

Where *Re* is the relative Reynolds number, which is defined as:

$$Re = \frac{\rho d_p |\vec{V} - \vec{V_p}|}{\mu}$$

α_1, α_2, and α_3 are constants that given in Ref. 11.

The heat transfer coefficient *h* is deduced from the Nusselt number N_u, which is evaluated from the well-known Ranz–Marshall correlation [12].

$$N_u = \frac{hd_p}{\lambda} = 2.0 + 0.6Re^{0.5}Pr^{0.33}$$

where *Pr* is the Prandtl number of the gas.

2.3 Computational domain and boundary conditions

Figure 1 shows the computational domain of the plasma jet used in the study. It is discretized using structured hexahedral cells with 133,285 nodes and 129,360 cells.

To allow the boundary conditions to be specified, the computational domain is divided into four faces: gas inlet, nozzle wall, atmospheric boundary, and gas outlet as shown in Fig. 1..

Fluent 6.3 was used to calculate the gas flow field of the plasma jet. The temperature and velocity profiles of gas inlet boundary are obtained from the simulation results referenced Ref. 9 and 10 at the plasma torch exit. The nozzle wall is considered to be an adiabatic boundary. The gas outlet is regarded as a free outlet to the atmosphere. In addition, the atmospheric boundary is defined as an open boundary through which gas flow can pass depending the gas pressure. The k–ε model is utilized in the current simulation and the governing equations are solved by the SIMPLE algorithm.

The particle injection position is shown in Fig. 1 fed from the edge of the plasma torch with only an initial radial velocity. Acceleration and heating of the in-flight particles are computed using the discrete phase model (DPM) in Fluent [13].

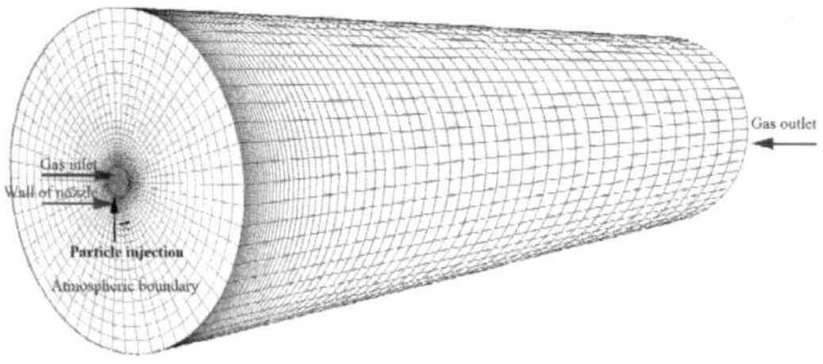

Fig. 1. Computational domain and mesh of plasma jet

2.4 Spray conditions

The time-averaged gas temperature and velocity at the torch exit calculated by the improved LTE mode in Ref. 9 were used as the gas inlet boundary conditions for the domain shown in Fig. 1. Table 1 lists the spray conditions for the plasma jet simulations. Electrical currents of 400, 500, and 600 A were applied. Figure 2 shows the corresponding boundary conditions of the gas inlet based on the results form the improved LTE mode. The thermodynamic and transport properties of the plasma gas are taken from Ref. 14 and 15.

Table 1: Spray conditions

Plasma gas	Argon
Torch	SG-100
Gas flow rate (SLM)	50
Applied Current (A)	400, 500, 600

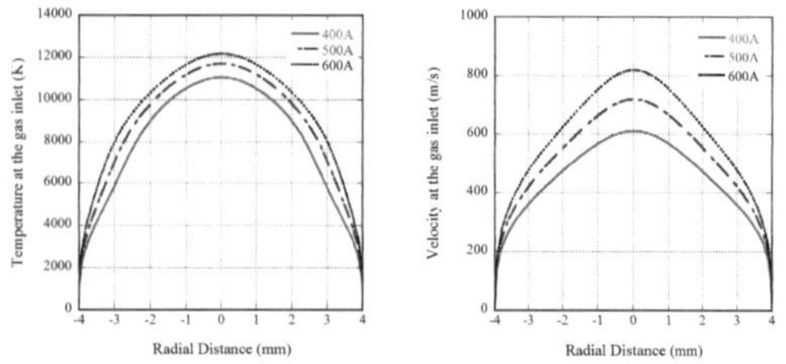

(a) (b)

Fig. 2. Gas temperature (a) and velocity (b) at the gas inlet

Alumina particles were used as the feedstock to calculate the in-flight particle temperature and velocity. Table 2 lists the properties of alumina. Particles with diameters in the range 20 to 40 µm are fed from the particle injection position. The initial radial velocities of particles were set to be 5, 10, and 20 m/s, while the axial velocity was set to zero.

Table 2: Properties of particles utilized in simulation

Particle	Al_2O_3
Density (kg/m^3)	3900
Specific heat (J.kg/K)	1340
Thermal conductivity [w/(m.K)]	6.55

2.5 Experimental Methods

An 80 kW automatic plasma spray system developed by Plasmadyne (a division of Geotel Inc.) was used to spray the commercially available oxide ceramic powders of PC.AT13T (Al_2O_3-TiO_2), supplied by Saint-Gobain Ceramic Materials Co. Ltd. The powder morphology is shown in Fig. 2 and it shows irregular shape. The in-flight particle temperature and velocity were measured by DPV-2000 and SprayWatch systems to verify the numerical calculated results.

Fig. 2. Morphology of PC.AT13T powder

3 Results and Discussions

3.1 Gas flow of plasma jet

Figure 3 shows the calculated gas temperature distributions of the plasma jet for the conditions described above. It shows that the plasma jet temperature and length increase with increasing applied current. The gas temperature decreases with increasing distance from the gas inlet at all the three conditions of different applied current due to the involvement of the surrounding gas. The gas temperature of the plasma jet tends to increase with increasing applied current due to the increase in the input electrical power.

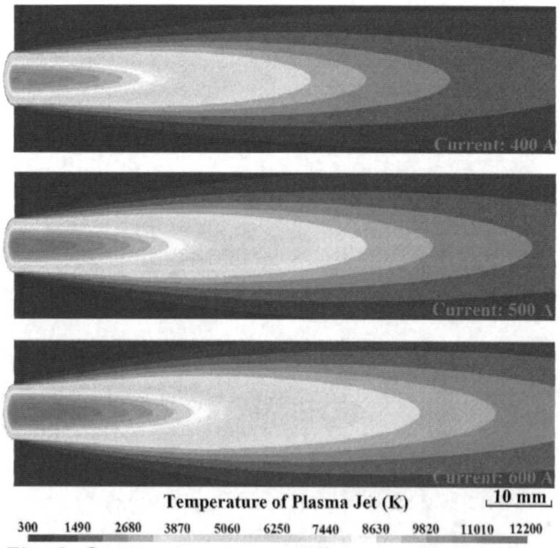

Fig. 3. Gas temperature at different applied current

Figure 4 shows the calculated gas velocity distributions of the plasma jet at different applied currents. Similar to the gas temperature, the input electrical power greatly affects the gas velocity because the gas velocity increases with increasing applied current. The gas velocity also decreases with increasing standoff distance due to the involvement of the surrounding gas.

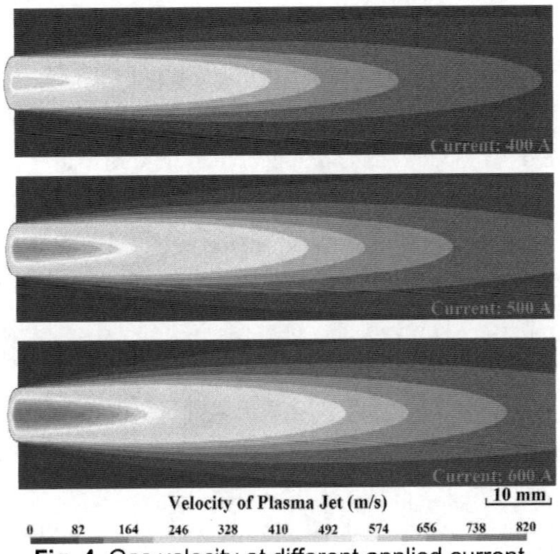

Fig. 4. Gas velocity at different applied current

3.2 Calculated Particle temperature and velocity

The particle trajectories for different particle diameters and initial velocities at the applied current of 500 A are superimposed on the gas temperature distribution of the plasma jet in Fig. 5. It shows that the particles travel in different regions in the plasma jet depended on their particle diameter and initial velocity; consequently, the particles are accelerated and heated by different amounts.

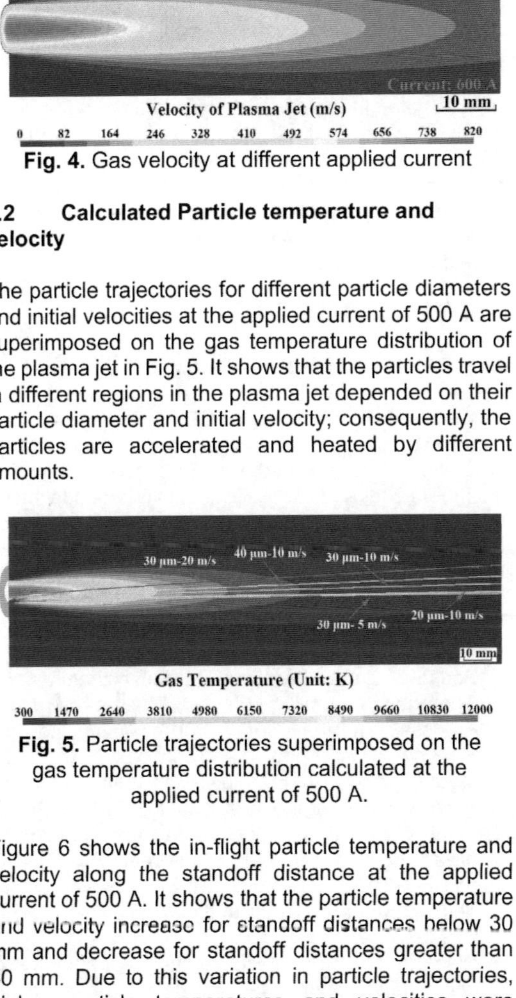

Fig. 5. Particle trajectories superimposed on the gas temperature distribution calculated at the applied current of 500 A.

Figure 6 shows the in-flight particle temperature and velocity along the standoff distance at the applied current of 500 A. It shows that the particle temperature and velocity increase for standoff distances below 30 mm and decrease for standoff distances greater than 30 mm. Due to this variation in particle trajectories, higher particle temperatures and velocities were obtained at certain initial radial velocities (e.g., 20 m/s); this is attributed to particles residing longer in the gas field of high temperature and velocity at these initial radial velocities, giving them higher temperatures and velocities. During their flight in the plasma jet, larger particles are heated to higher temperatures, whereas smaller particles are accelerated to higher velocities near the torch exit and decelerated much more rapidly when the particle velocity exceeds the gas velocity due to their lower inertia.

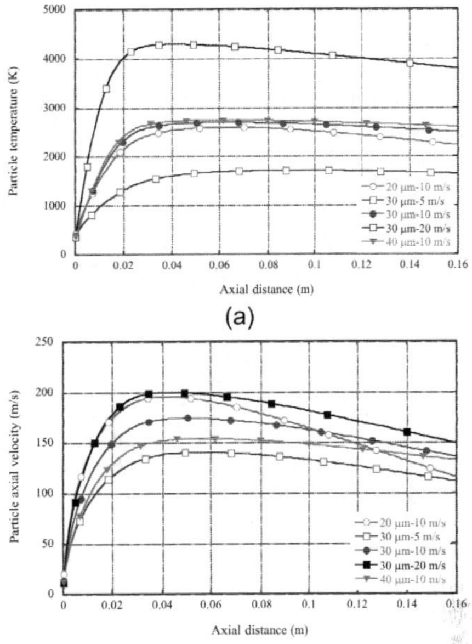

Fig. 6. In-flight particle temperature (a) and velocity (b) calculated at the applied current of 500 A

Figure 7 shows the particle temperature and velocity for particles with a diameter of 30 μm and an initial radial velocity of 10 m/s for different applied currents at a spray distance of 50 mm. It shows that both the particle temperature and velocity increase with increasing applied current due to increases in the gas temperature and velocity. Therefore, a high applied current is required to spray refractory materials such as high-melting-point metals and ceramics.

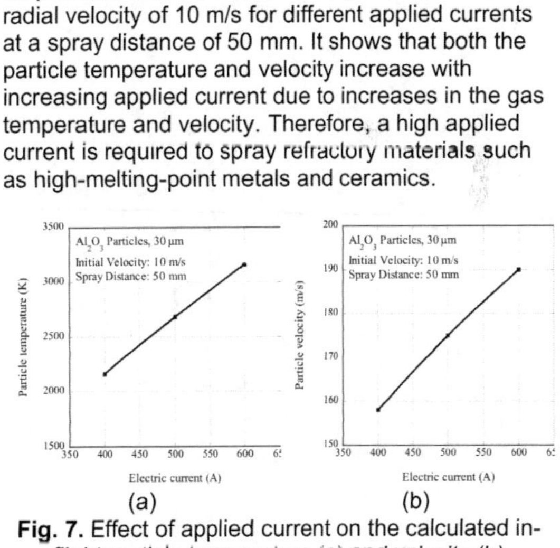

Fig. 7. Effect of applied current on the calculated in-flight particle temperature (a) and velocity (b)

3.3 Comparison with experimental results

Based on the flow rate of the powder-feeding gas and the diameter of the orifice from which the powder is injected, in the in-flight particle diagnostic experiments, the particles are estimated to have an initial radial velocity of 10 m/s. Therefore, the calculation results for an initial radial velocity of 10 m/s were compared with the experimental particle temperature and velocity measured by DPV-2000 and SprayWatch systems. The main differences between the simulation and the experiment are the powder feeding position and the powder composition. In the experiment, the powder feeding position is located inside the plasma torch and 9 mm away from the torch exit, whereas it is located at the torch exit in the simulations. In addition, the powder used in the experiment is 85% alumina and 15% titania rather than pure alumina. Figure 8 compares the calculated particle temperature with the measured one for an applied current of 500 A. It shows that the calculated particle temperature agrees well with the experimental one. Figure 9 compares the calculated particle velocity with the measured one for an applied current of 500 A. The measured particle velocity is slightly higher than the calculated one. This discrepancy may be caused by the different powder feeding positions. Because the particle is accelerated inside the plasma torch, the measured particle velocity is higher than the calculated one. There is another reason related to the discrepancy, the drag coefficient. The actual drag coefficient of solid particle at the initial phase is some higher than the one in the simulations because the drag coefficient of spherical particle was adopted, whereas the particles are not spherical before their melt. The results imply that plasma jet simulations can effectively predict the steady-state particle temperature and velocity.

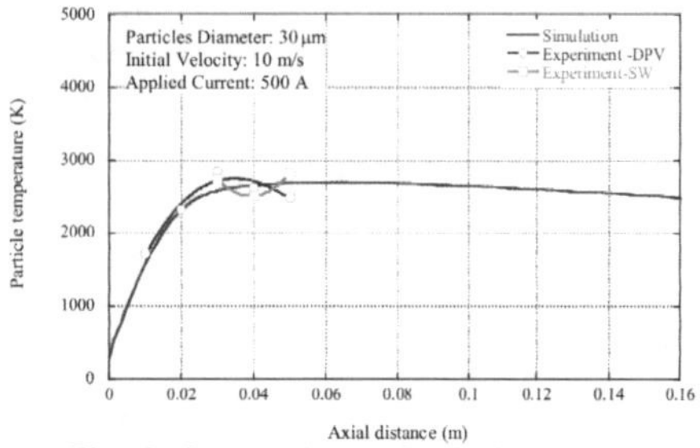

Fig. 8. Comparison of calculated and experimentally measured particle temperatures

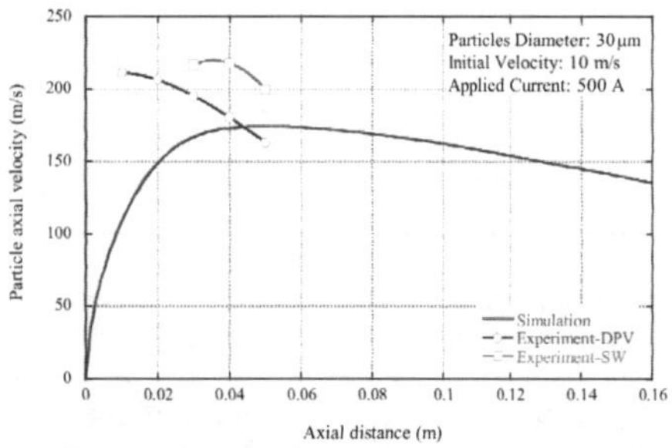

Fig. 9. Comparison of calculated and experimentally measured particle velocities

4 Conclusions

In this study, the in-flight particle temperature and velocity in a plasma jet were predicted using a 3D steady-state model based on the arc behaviour in our previous researches. Gas flow fields for the plasma jets outside the plasma torch in the steady state were obtained by numerical calculation method. The particle trajectory, temperature, and velocity were determined from the gas temperature and velocity fields of plasma jets. The numerical calculation results revealed that the gas temperature and velocity increase with increasing applied current due to the increase in the input electrical energy. The in-flight particle trajectory was influenced by the initial particle velocity and diameter. Different trajectories cause particles to experience different heating rates and accelerations. The calculated in-flight particle temperature and velocity generally increased as the standoff distance increased to about 30 mm and they subsequently decreased. Similar to the gas temperature and velocity, the in-flight particle temperature and velocity also increased with increasing applied current.
The calculated in-flight particle temperature and velocity are very similar to the measured ones with DPV-2000 and SprayWatch system. This demonstrates that it is an effective way to predict the plasma spraying jet behaviors using the boundary conditions calculated from the improved LTE mode inside the plasma torch.

Acknowledgements

The authors would like to acknowledge the financial support provided by the High-level Leading Talent Introduction Program of GDAS (Grants No. 2016GDASRC-0204), Innovation Ability Construction Foundation of Key Institute in Guangdong Province (No. [2014] 208) and Ability Construction Foundation of Quality Supervision Test Center for National Mineral and Regeneration Metal (No. 2016GDASPT-03, 2016GDASPT-0201).

5 Literature

[1] Borisov Yu., Krivtsun I., Muzhichenko A., Computer Modelling of the Plasma Spraying Process, Thermal Spray 2003: Advancing the Science & Applying the Technology, (Ed.) C. Moreau and B. Marple, Published by ASM International, Materials Park, Ohio, USA, 2003.

[2] Bolot R., Li J. and Coddet C., Modeling of thermal plasma jets: a comparison between PHOENICS and FLUENT, Thermal Spray 2004: Advancing in Technology and Application, ITSC, May 2004, Osaka, Japan.

[3] Wang Fuchi, Fan Qunbo, Wang Lu and Wang Quansheng, Numerical simulation of a multi-component reacting plasma jet in plasma spraying, Thermal Spray 2004: Advancing in Technology and Application, ITSC, May 2004, Osaka, Japan.

[4] Ramachandran Kandasamy and Nishiyama Hideya, Fully coupled 3D modeling of plasma–particle interactions in a plasma jet, Thin Solid Films, 457 (2004), 158–167.

[5] Kang C. W., Ng H. W., Yu S. C. M., Comparative Study of Plasma Spray Flow Fields and Particle Behavior Near to Flat Inclined Substrates, Plasma Chem Plasma Process (2006) 26:149–175.
[6] Xiong Hong-bing, Zhu Ze-fei and Lin Jiang-zhong, Three-dimensional simulation of air entrainment in plasma jet and particle oxidation, Journal of Zhejiang University (Engineering Science), Vol. 40 No. 4, Apr. 2006.
[7] Li He-Ping, Chen Xi, Three-dimensional simulation of a plasma jet with transverse particle and carrier gas injection, Thin Solid Films, 390 (2001), 175-180.
[8] Li He-Ping and Pfender E., Three Dimensional Modeling of the Plasma Spray Process, Journal of Thermal Spray Technology, Volume 16(2) June 2007, 245–260.
[9] Huang R., Fukanuma H., Uesugi Y., Tanaka Y.; An Improved Local Thermal Equilibrium Model of DC Arc Plasma Torch; IEEE Transactions on Plasma Science; Volume: 39, Issue: 10, p 1974 – 1982
[10] Huang R., Fukanuma H., Uesugi Y., Tanaka Y.; Comparisons of Two Models for the Simulation of a DC Arc Plasma Torch; Journal of Thermal Spray Technology, Vol. 22, Issue. 2-3, 2013, p183-191.
[11] Morsi S.A. and Alexander A.J., An investigation of particle trajectories in two-phase flow systems, Journal of Fluid Mechanics 55 (1972) 193–208.
[12] Ranz W.E. and Marshall W.R., Evaporation from drops, Part I, Chemical Engineering Progress 48 (1952).
[13] Ansys Inc., Fluent 6.3 User's Guider, NH, 2006.
[14] Maher I. Boulos, Pierre Fauchais, Emil Pfender, Thermal plasmas: fundamentals and applications, Springer, 1994
[15] V. Colombo, E. Ghedini*, P. Sanibondi, Thermodynamic and transport properties in non-equilibrium argon, oxygen and nitrogen thermal plasmas, Progress in Nuclear Energy, V50 (2008), p921–933

Pulsed current and pulsed powder operation of the One-Cathode-One-Anode-Plasma-Generator (OCOAPG)

Stephan Zimmermann, Alexander Atzberger and Jochen Schein, Neubiberg / D
Michal Szulc and Jochen Zierhut, Neubiberg / D

In most plasma spraying SMEs a One-Cathode-One-Anode-Plasma-Generator (OCOAPG) is used due to its cost-effectiveness. To achieve high deposition rates, the highest possible fraction of the injected powder has to be melted and accelerated towards the substrate. Adequate to the amount and size of the particles, a sufficiently long and reproducible residence time in plasma is therefore needed. This can be achieved by a long plasma jet with little or no temporal variation in length and temperature. In OCOAPG an arc is operated between a cathode and a central tubular anode, which causes different instabilities in the effluent plasma jet. Due to the instable interaction between the plasma jet and the carrier gas jet continuously incorporating the powder, fluctuations occur resulting in reduced coating quality. Coating systems with a higher amount of electrodes (and hence using several arcs) show higher stability and therefore can provide higher coating quality. However, due to their complexity and cost intensity, the investment hurdle for SMEs usually turns out to be too high. Recently, research steps to improve the plasma spraying process with OCOAPG have been undertaken by using a controllable current source to create a uniform particle gas jet interaction. As the movement of the anodic arc attachment point can be actively controlled by current pulses, the plasma jet can be lengthened and shortened at periodic intervals. Furthermore, by pulsing the particle delivery synchronously to the power modulation an improved particle penetration and consistent residence times can be achieved. First diagnostic results, including coatings, are presented and discussed within the paper.

1 Introduction

In OCOAPG an arc is operated between a cathode and a central tubular anode, which causes different instabilities in the effluent plasma jet. Due to the instable interaction between the plasma jet and the carrier gas jet continuously incorporating the powder, fluctuations occur resulting in reduced coating quality, **Fig. 1** [1-4].

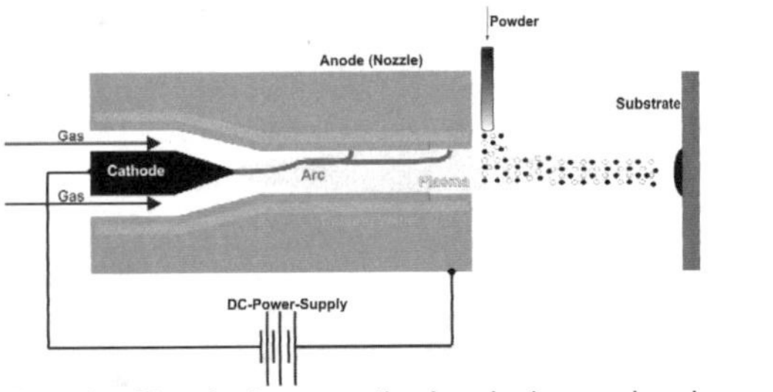

Fig. 1: Classical one-cathode single-anode plasma generator (F4/F6)

To improve the coating quality during plasma spraying by means of a fixing of the anode batch has led to plasma generator concepts to a stationary anode attachment by dividing the entire plasma generator current into several partial streams or single-cell arcs. A commercial implementation of these concepts the plasma generator systems "Triplex" (three cathodes, one anode) and "delta" (one cathode, three anodes), wherein in each case one total arc is divided into three partial arcs. Both systems allow a much more uniform interaction between gasjet and introduced spray powder than is the case with OCOAPG. In return the purchase of these multi-arc systems is also significantly more cost-intensive, since besides the pure plasma generator cost also more expensive costs for the further plant technology required for operation. The aim of the OCOAPG research is to measure/control the fluctuations of the gasjet length, **Fig. 2**. The idea is to pulsed and synchronized the current with the gasjet length and the spray powder input. As the movement of the anodic arc attachment point can be actively controlled by current pulses, the plasma jet can be lengthened and shortened at periodic intervals. Furthermore, by pulsing the particle delivery synchronously to the power modulation an improved particle penetration and consistent residence times can be achieved.

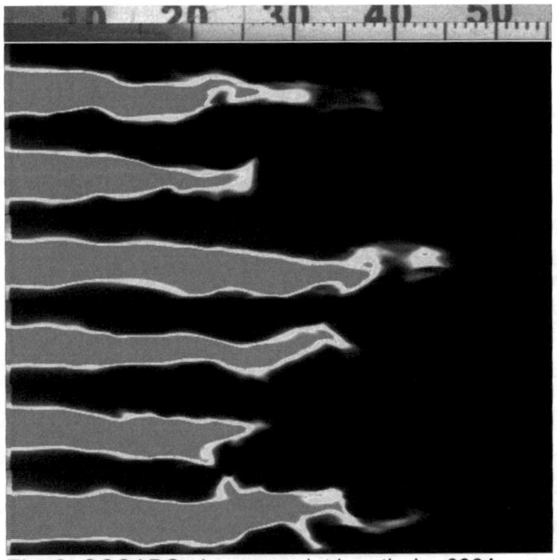

Fig. 2: OCOAPG plasma gasjet length; I = 600A, Φ_{Gas} = 50SLPM Argon / 10SLPM Hydrogen, dc-operation; high-speed camera: pco.dimax HS4: 2000x302Pixels, 13923fps, images: Δt = 71.823µs, exposure time: 1µs

2 Experimental Setup

In a first step the characteristic arc root displacement frequencies for different OCOAPG parameters were investiged and calculated. In these experiments, a classical one-cathode single-anode plasma generator (F4/F6) was used, **Table 1**. In this OCOAPG the anodic arc root point is moving depending on the parameter. For the measurements of the characteristic arc root displacement at dc operation the following **Fig. 3** shows the diagnostic tool. Thereby the sound, voltage, current, emission (high speed camera) are measured and the characteristic arc root displacement frequencies are calculated and analyzed by a self programmed Delphi/Matlab FFT-tool and a image postprocessing tool [1-4].

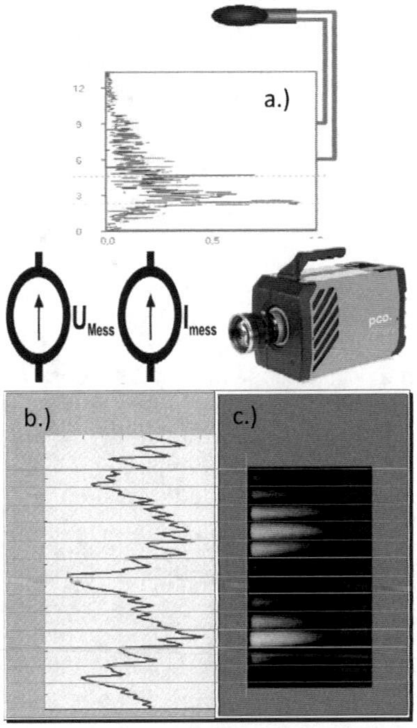

Fig. 3: Diagnostic tools for determination of characteristic arc root displacement frequencies during dc plasma operation (F4/F6)

nitrogen parameter of the experiments			
Parameter_Nr.	nozzle	current(A)	Ar (SLPM)/ N_2 (SLPM)
I	6mm	350	40/6
II	6mm	500	40/6
III	6mm	150	40/6
IV	8mm new	350	40/6
V	8mm old	350	40/6
hydrogen parameter of the experiments			
Parameter_Nr.	nozzle	current(A)	Ar (SLPM)/ H_2 (SLPM)
1	6mm	600	50/10
2	6mm	600	45/15
3	6mm	500	40/6
4	6mm	350	40/6
5	6mm	200	40/6
6	6mm	150	40/6

Table 1: Parameter list of the experiments with the classical one-cathode single-anode plasma generator (F4/F6)

For the investigation of the pulsed plasma operation process the following electrical setup are used, **Fig. 4**.

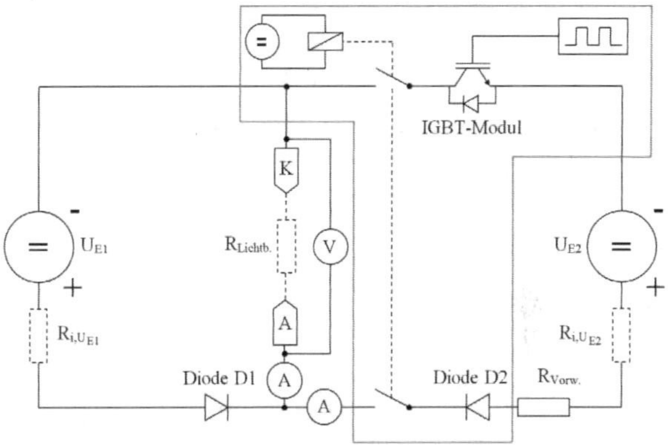

Fig. 4: Electrical setup for the determination of the characteristic behavior of the arc and plasma during pulsed operation (F4/F6)

3 Results and Discussion

For the pulsed operation it is an necessary condition to determine the characteristic arc root displacement frequencies during dc plasma operation. In the following **Table 2** the calculated frequencies under argon/nitrogen and argon/hydrogen gas mixture condition are shown.

characteristic arc root displacement frequencies (N_2)	
Parameter I	3113 Hz
Parameter II	3500 Hz
Parameter III	2044 Hz
Parameter IV	2200 Hz
Parameter V	2300 Hz
characteristic arc root displacement frequencies (H_2)	
Parameter 1	4249 Hz
Parameter 2	4499 Hz
Parameter 3	3999 Hz
Parameter 4	3625 Hz
Parameter 5	2875 Hz
Parameter 6	2156 Hz

Table 2: Calculated characteristic arc root displacement frequencies for the different parameters

In **Fig. 5** an first example for the current and voltage detection during pulsed operation is depicted. Thereby the gas mixture is only argon.

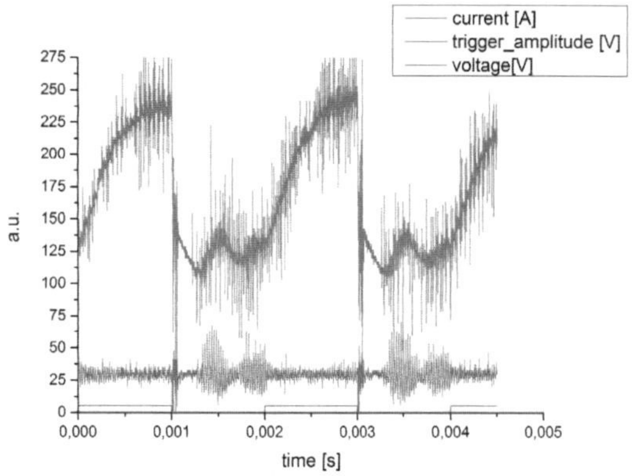

Fig. 5: Example for the current and voltage detection during pulsed operation (F4/F6); I(Primary current)=120A, I(secondary current)=42A, Φ_{Gas} = 40SLPM Argon, f(pulsed frequency)=500Hz

In the next **Fig. 6** the comparison of the pulsed parameter III is shown.

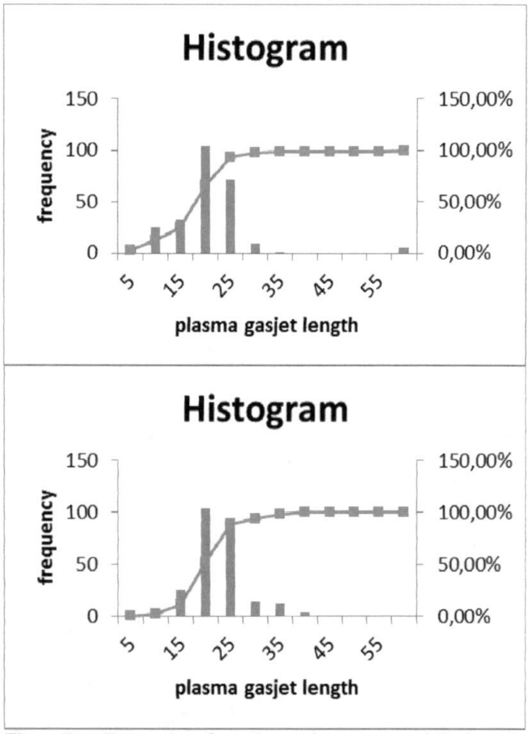

Fig. 6: Example for the plasma gasjet length determine/calculated (256 high speed images), F4/F6, I(Primary current)=127A, I(secondary current)=22,5A, Φ_{Gas} = 40SLPM Argon and 6SLPM nitrogen, top: f(pulsed frequency)=2000Hz (pulsed operation), bottom: f(pulsed frequency)=0Hz (dc operation)

A first example of the high speed image processing to determine and visualisate the pulsed plasma gasjet behavior. In addition a first idea for the synchronous power modulation/injection are calculated with these results, **Fig. 7**.

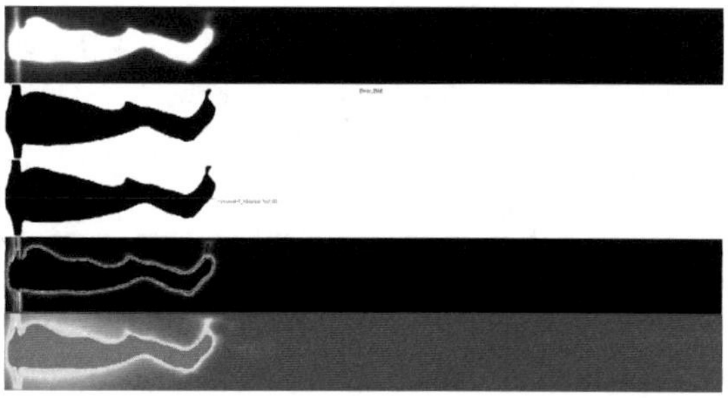

Fig. 6: Example for the high speed image processing, F4/F6, I(Primary current)=127A, I(secondary current)=22,5A, Φ_{Gas} = 40SLPM Argon and 6SLPM nitrogen, f(pulsed frequency)=2000Hz, top to bottom: original-, binary- (threshold: 64), plasma gasjet length- (threshold: 64), sobel-, false colour image

4 Outlook

In future measurements the pulsed frequency and secondary current will be changed to investigate the maximum of this pulsed parameter. The goal is to reach the parameter 1 and parameter 2 (industry). Also coating determination and comparisons will be carried out to find an optimum for the pulsed OCOAPG operation (synchronous power modulation).

5 Acknowledgements

The authors gratefully acknowledge the financial support of the AiF/DVS within the project IGF 18.963N / DVS 02.099.

6 Literature

[1] J.-L. Marqués, M. Fuchs, F. Dschung, S. Lange, J. Zierhut und J. Schein, Fluctuations in one-cathode plasma torches: characterization and modelling

[2] S. Schmettlach, Akustische Emissionsanalyse von Thermischen Spritzprozessen, 2014

[3] S. Zimmermann, K. Kreher, P. Weiland, L. Schnauber, J.L. Marqués, K. Landes und J. Schein, Entwicklungen im Bereich der DC Plasmageneratoren und deren mögliche Applikationsfelder, 2012

[4] P.L. Fauchais, J.V.R. Heberlein , M. Boulos : Thermal Spray Fundamentals From Powder to Part, ISBN 978-0-387-28319-7

Self-lubricating cobalt-based composite coatings deposited by Plasma Transferred Arc

Lidia Baiamonte*, Giovanni Pulci*, Edoardo Hlede+, Cecilia Bartuli*, Teodoro Valente*, Francesco Marra*

* Dept. of Chemical Engineering Materials Environment, University of Rome "La Sapienza", INSTM Reference Laboratory for Engineering of Surface Treatments, Via Eudossiana 18, Roma, 00184, RM, Italy.

+ Wärtsilä Italy S.p.A., Bagnoli della Rosandra 334, San Dorligo della Valle, 34018, TS, Italy

Wear-resistant cobalt–based alloy (Stellite 12) coatings deposited by plasma transferred arc (PTA), commonly used to protect critical mechanical components in harsh environments, were modified by addition of hard ceramic particles (TiC) and solid lubricant compounds (MoS_2 and CaF_2) to improve the overall tribological performance.
In this preliminary study, microstructural, microhardness and tribological analyses were carried out to assess: a) the feasibility of PTA deposition of thermally sensitive phases characterised by very low density; b) the effect of the addition of a mixture of soft and hard phases on the coating hardness; c) the effect of the modified composition in terms of wear resistance; d) the effect of the addition in terms of lubrication (friction coefficient and produced heat).
Results showed that: a) an appropriate pre-consolidation of feedstock materials can be effective in preserving the heat-sensitive phases within the microstructure of PTA deposits; b) the addition of a total amount of 5% wt. of solid lubricants and reinforcing carbides produced a limited decrease in the coating hardness (about 13%) and an evident improvement in terms of friction coefficient but, on the other hand, a remarkable reduction (about 30%) in wear resistance.
Further investigation will be addressed to optimize the composition of modified feedstock to counteract the softening effect of lubricant phases without depressing the self-lubrication behaviour.

1 Introduction

Plasma Transferred Arc (PTA) weld surface is a widely used deposition technique to produce wear and corrosion resistant metallic coatings, due to high deposition rate, excellent process stability, wide range of selection of materials and very good adhesion provided by the metallurgical bonding created between the deposited layer and the substrate. [1] [2]
Cobalt-based alloys, such as those belonging to the Stellite family, are commonly selected as hardfacing coatings for their excellent wear and corrosion resistance over a wide range of temperatures. [3] [4]
To improve the coatings, wear behaviour solid lubricating compounds, mostly sulphides or fluorides, can be added, often in combination with carbides used as a reinforcing medium to improve the hardness. Lubricating phases, in fact, may affect the alloy's microhardness so that the addition of ceramic particles could be of help in maintaining, or improving, the hardness of the material. [5] [6] [7]
In this paper MoS_2, CaF_2 and TiC were added in the form of powders to the Stellite 12 feedstock. While MoS_2 was selected for its consolidated use as a solid lubricant, CaF_2 is known to be chemically stable other than providing good lubricating properties.
The purpose of this study is, at first, to understand whether the additive compounds survive the complete fusion provided by the PTA deposition process. Then, tribological behaviour, hardness and microstructure of standard and modified Stellite 12 hardfacing layer will be compared, to assess whether and to what extent the solid lubricating compounds affect the coating tribological behaviour in terms of wear resistance and friction coefficient.

2 Materials and methods

Plain Stellite alloy 12 (Kennametal, Latrobe, PA, USA) thick coatings were deposited onto AISI 420 martensitic steel discs by a plasma transferred arc plant (ROBO 9, Commersald SpA, Modena, Italy) located at Centro Sviluppo Materiali SpA (Rome, Italy). The same facility was used to deposit a Stellite 12 coating modified by the addition of TiC, MoS_2 and CaF_2 (Alfa Aesar GmbH & Co.KG, Karlsruhe, Germany) with an overall amount of 5 wt%. Additives were agglomerated in a composite powder by spray drying (NIRO Atomizer Mobile Minor™, GEA Process Engineering, Soeborg, Denmark) before deposition.
Operating parameters for the deposition of metallic and modified composite coatings are reported in **Table 1**.

Table 1: PTA operating parameters for the deposition of Stellite 12 and Stellite 12 + solid lubricants coatings

Sample	Current (A)	Feed rate (rpm)
Stellite 12	130	12
Stellite 12 + solid lubricants	130	15

Scanning electron microscopy (Philips SEM XL40), assisted by energy dispersion spectrometry (EDS) microanalysis (EDAX Octane SDD) was used to investigate the microstructure of coated samples.
X-ray diffraction analysis (XRD) was performed on the modified coatings (Philips X'Pert3, PANalytical B.V., Almelo, The Netherlands), to investigate whether the additive compounds survived the deposition process. The XRD device was operated at 40 kV and 40 mA with Cu Kα radiation, with a scan range of 10–80° (2θ), step size of 0.02° and a counting time of 2 s for each step.
Vickers microhardness measurements were performed on polished cross section of samples at 0.98, 9.8 and 19.6 N (0.1, 1, and 2 kgf respectively; maximum load was reached in 10 s and then maintained for 15 s), followed by optical evaluation of

the indentation size (ASTM E-384-89) [8] using a Leica VMHT Microhardness Tester (Leica Microsystems GmbH, Wetzlar, Germany). Fifteen indentations were carried out for each load.
Five cubic-shaped (~13 mm) samples per coating were cut from the coated disc for tribological characterization; they were ultrasonically cleaned in acetone, dried in furnace at 105°C and weighted. Wear tests were performed at room temperature with a PLINT TE53 Multi Purpose Friction and Wear Tester tribometer in inverted block-on-ring configuration (ASTM G77) [9] [10] [11] [12].
The samples were tested for 5000 seconds against a 100Cr6 steel ring rotating at 318 rpm (~1 m/s), with a 200 N normal force applied. The tested specimens were then cleaned and weighted, and wear tracks were measured by profilometric analysis using a Talyscan 150 profilometer (Taylor Hobson, Leicester, UK) equipped with a laser probe [13]. Both weight and volume losses were calculated considering for each wear track the 5 mm wide central zone, in order to avoid geometrical and experimental errors (small misalignments, relative position of the block over the ring, different sample dimensions, etc.).

3 Results and discussion

SEM micrographs of cross sections of standard and modified Stellite 12 coatings are shown in **Fig. 1.**
The coatings show the typical dendritic microstructure usually exhibited by PTA deposits of Stellite alloys [4]; moreover, the addition of solid lubricant and reinforcing phases does not seem to affect the density of the base-alloy.
An EDS analysis was carried out on the cross section of the modified coating to investigate the distribution of the base-alloy elements and to confirm the presence of the solid lubricant phases. An elemental map is shown in **Fig. 2**.

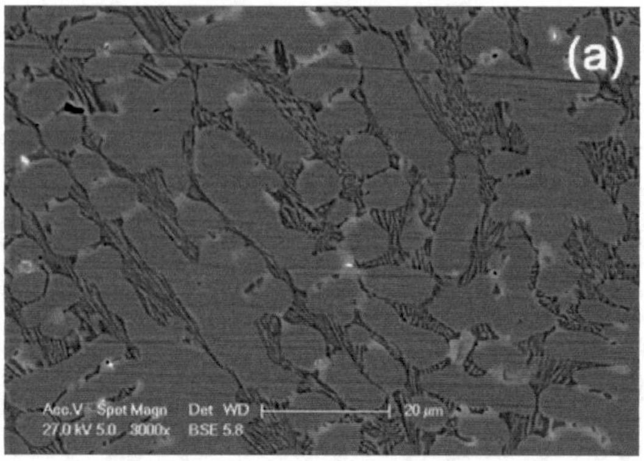

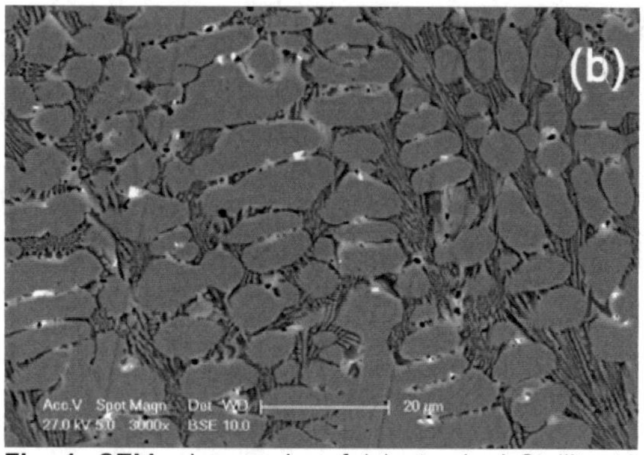

Fig. 1: SEM micrographs of (a) standard Stellite 12 and (b) Stellite 12 + solid lubricants coatings cross sections (3000x, BSE detector).

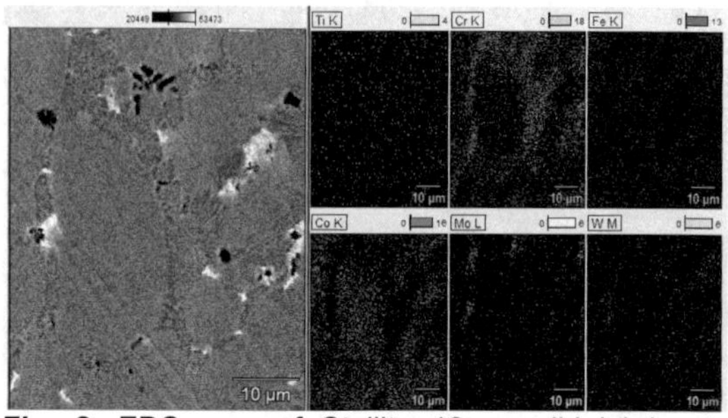

Fig. 2: EDS map of Stellite 12 + solid lubricants coating.

It was confirmed that the dendritic structure is mainly made of cobalt and iron, while chromium is concentrated in the interdendritic regions.
The dark spots, also visible as rounded small particles between the branches in **Fig.1**, are found to be especially rich in molybdenum, thus confirming the permanence of this element in the coating after the deposition process and evidencing its homogeneous distribution throughout the section.
The EDS analysis (**Fig. 3**) of the general cross section confirmed the presence of Mo and evidenced unquantified amounts of Ca and Ti; F and C remain below detectable limits, while S peak is probably hidden by Mo peak as they generally overlap.

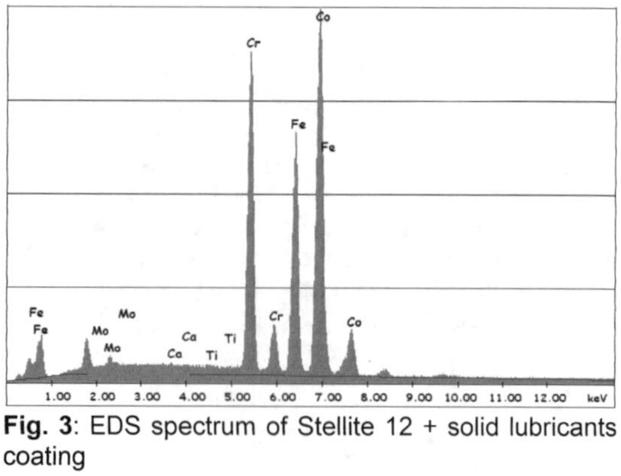

Fig. 3: EDS spectrum of Stellite 12 + solid lubricants coating

Results of the XRD analysis of modified Stellite coatings are shown in **Fig. 4**.
The spectrum clearly indicates the presence of CaF_2, confirming that this solid lubricant phase has survived the deposition process.
XRD analysis could not identify the peaks corresponding to the other dispersed phases (MoS_2 and TiC).

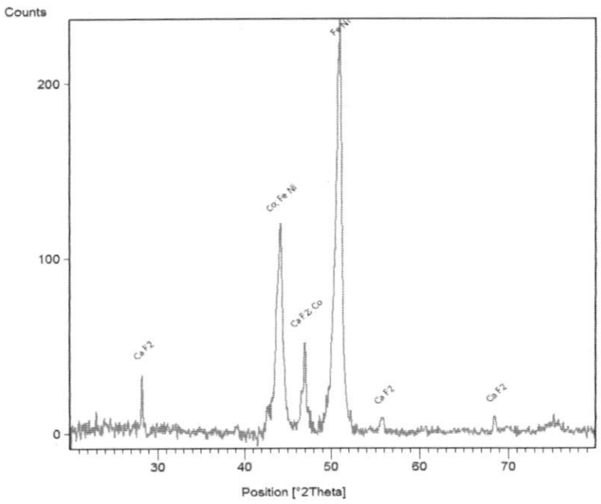

Fig. 4: XRD spectrum Stellite 12 + solid lubricants coating.

Results for Vickers microhardness measurement (average values and standard deviations) are shown in **Fig. 5**. An average decrease in hardness of about 13% can be observed, and it is independent on the applied normal load.
The decrease of the microhardness in the modified coatings is an expected result, as all the added dispersed phases act as inclusions and defects in the homogeneous Stellite matrix. More specifically it is known [6] that the solid lubricant phases, CaF_2 and MoS_2, are compound characterized by low values of hardness.
On the other hand, TiC certainly acts as a reinforcing agent [7], and is thus intended to mitigate the softening action of the solid lubricant compounds.

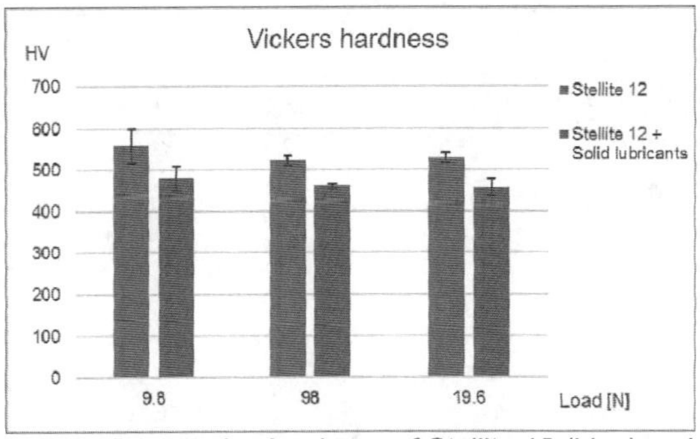

Fig. 5: Vickers microhardness of Stellite 12 (blue) and Stellite 12 + solid lubricants (red) coatings, as measured for increasing applied normal load.

Weight and volume losses are represented in the histograms in **Fig. 6** and **Fig. 7**, respectively:

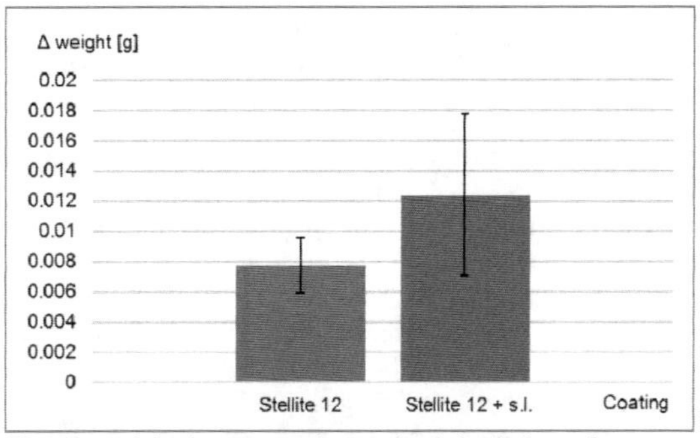

Fig. 6: weight losses and standard deviations for Stellite 12 (blue) and Stellite 12 + solid lubricants (red) after tribological test.

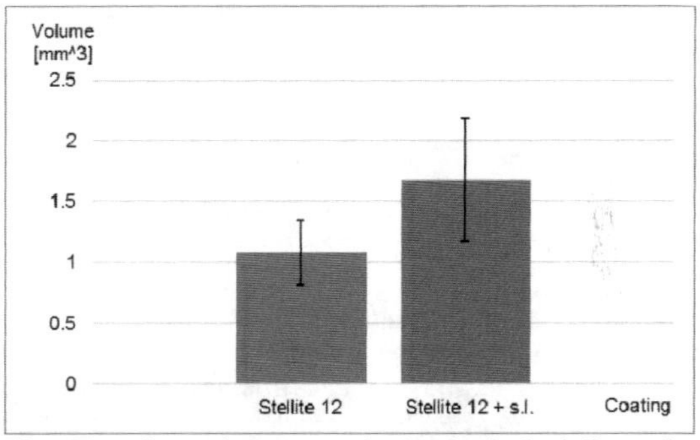

Fig. 7: volume losses and standard deviations for Stellite 12 (blue) and Stellite 12 + solid lubricants (red) after tribological test.

The decreasing in microhardness experienced by Stellite 12 modified with solid lubricating compounds reflects on weight and volume losses [10]. The modified coating, in fact, loses more material than the conventional coating after wear tests.
Results from tribological tests, in terms of friction coefficient and block temperature over the test time, are displayed in **Fig. 8**.

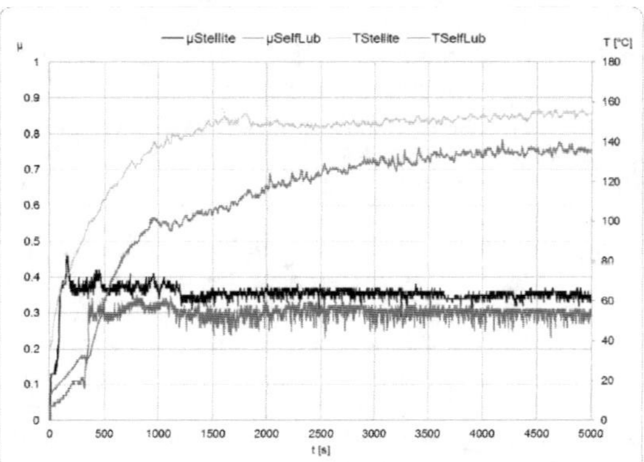

Fig. 8: comparison between friction coefficients and block temperatures of Stellite 12 (black and yellow

lines) and Stellite 12 + solid lubricants (green and red lines).

Conventional Stellite 12 coating friction coefficient features a rapid transient state, after which a stable value of ~0.35 is maintained. Block temperature reaches ~155°C.
Compared to plain Stellite 12, the modified coating exhibits an improved behaviour: after a slower transient state, the friction coefficient reaches a value of ~0.3, while the block temperature attains a lower constant value of ~135°C.
Thus, the modification of the composition considerably affects the wear behaviour of the hardfacing alloy, providing an effective lubricating action and lowering both friction coefficient and block temperature. However, the consequent reduction in surface mechanical stress is not sufficient to overcome the softening effect of dispersed phases, so that an overall reduction in wear resistance is observed.

4 Conclusions

A comparison between PTA-deposited Stellite 12 and Stellite 12 + 5wt% solid-lubricating compounds was carried out in the present study. CaF_2 and MoS_2 were selected for their lubricating properties, while TiC was selected as a reinforcing medium.
Compositional and microstructural analyses, performed on the Stellite 12 + solid lubricants coating, were carried out to determine the persistence of the added phases within the metallic matrix. XRD revealed the presence of CaF_2 peaks, while EDS confirmed that Mo was present in the coating, even if maps could not exactly locate it.
The modified Stellite 12 coating experienced a drop in the hardness values, as expected because of the presence of the solid-lubricants; TiC content in this formulation does not seem to be enough to counteract the hardness decrease, through the material reinforcement.
The effect of the presence of solid lubricating compounds affect heavily the mass and volume losses measured after the tribological tests: the modified coating experiences in fact more material loss than the conventional one.
On the other hand, both friction coefficient and block temperature are lower than those registered for plain Stellite 12, meaning that the added compounds actively counteract the effects of friction.
Further activities will be addressed on the assessment of the effect of reinforcement (TiC) and lubricant (CaF_2, MoS_2) phases, with the aim to maintain the self-lubricating ability without an important reduction of hardness and, in turn, wear resistance.
The present study, however, has a preliminary intent, as it aims to assess the feasibility to produce a self-lubricating coating by PTA technique by just modifying commercial powders. Furthermore, the purpose is to determine the effects of the addition of phases by a direct comparison in terms of tribological properties. More attention on the microstructural aspects of this compositional modification is required, as it is also important to investigate how and to what extent the deposition parameters and the overall amount of solid lubricating compounds affect the mechanical and tribological properties of the coating.

References

[1] Q. Y. Hou, J. S. Gao, and F. Zhou, "Microstructure and wear characteristics of cobalt-based alloy deposited by plasma transferred arc weld surfacing," Surf. & Coat. Technol. vol. 194, pp. 238–243, 2005.

[2] V. V. Díaz, J. C. Dutra, A. Sofia, and C. D. Oliveira, "Hardfacing by Plasma Transferred Arc Process", Arc Welding, Prof. Wladislav Sudnik (Ed.), InTech, pp. 3–21, 2008.

[3] A. Motallebzadeh, E. Atar, and H. Cimenoglu, "Sliding wear characteristics of molybdenum containing Stellite 12 coating at elevated temperatures," Tribology Int., vol. 91, pp. 40–47, 2015.

[4] Y. J. Kim, "Wear and corrosion resistance of PTA weld surfaced Ni and Co based alloy layers", Surf. Eng., vol. 15, n. 6, pp. 495–501, 1999.

[5] J. Shin, J. Doh, J. Yoon, D. Lee, and J. Kim, "Effect of molybdenum on the microstructure and wear resistance of cobalt-base Stellite hardfacing alloys", Surf. Coat. Technol., vol. 166, n. 2-3, pp. 117–126, 2003.

[6] H. Yan, J. Zhang, P. Zhang, Z. Yu, C. Li, P. Xu, Y. Lu, "Laser cladding of Co-based alloy/TiC/CaF_2 self-lubricating composite coatings on copper for continuous casting mold", Surf. Coat. Technol., vol. 232, pp. 362–369, 2013.

[7] P. Skarvelis and G. D. Papadimitriou, "Plasma transferred arc composite coatings with self lubricating properties, based on Fe and Ti sulfides: microstructure and tribological behavior", Surf. Coat. Technol., vol. 203, n. 10–11, pp. 1384–1394, 2009.

[8] ASTM, E384-11e1, Standard Test Method for Knoop and Vickers Hardness of Materials, ASTM International, West Conshohocken, PA, 2011.

[9] ASTM, G77, Standard Test Method for Ranking Resistance of Materials to Sliding Wear Using Block-on-Ring Wear Test. ASTM International, West Conshohocken, PA, 2010.

[10] F. Cipri, F. Marra, G. Pulci, J. Tirillò, C. Bartuli, T. Valente, "Plasma sprayed composite coatings obtained by liquid injection of secondary phases", Surf. Coat. Technol., vol. 203, n. 15, pp. 2116–2124, 2009.

[11] G. Di Girolamo, L. Pilloni, G. Pulci, F. Marra, "Tribological Characterization of WC–Co Plasma Sprayed Coatings", J. Am. Ceram. Soc., vol. 92, n. 5, pp. 1118–1124, 2009.

[12] G. Di Girolamo, F. Marra, G. Pulci, J. Tirillò, T. Valente, L. Pilloni, "Microstructure and wear behaviour of plasma-sprayed nanostructured WC-Co coatings", Int. J. Appl. Ceram. Tec., vol. 10, no. 1, pp. 60–71, 2013.

[13] L. Baiamonte, G. Pulci, E. Hlede, F. Marra, and C. Bartuli, "Thermal spray coatings for corrosion and wear protection of naval Diesel engines components", Metall. Ital., vol. 106, n. 6, pp. 9–13, 2014.

Development of bondcoat layer for long lifetime suspension plasma sprayed thermal barrier coatings

M. Gupta, N. Markocsan, University West, Trollhättan / SE
X.-H. Li, Siemens Turbomachinery, Finspång / SE
L. Östergren, GKN Aerospace, Trollhättan / SE

Development of thermal barrier coatings (TBCs) manufactured by suspension plasma spraying (SPS) is of high commercial interest as SPS has been shown capable to produce columnar microstructures similar to the conventionally used electron beam – physical vapour deposition (EB-PVD) process. Moreover, SPS is a significantly cheaper process and can also produce more porous coatings than EB-PVD. However, lifetime of SPS coatings needs to be improved further for them to be applicable in commercial applications.
The bondcoat microstructure as well as topcoat-bondcoat interface topography affect the TBC lifetime significantly. The objective of this work was to investigate the feasibility of different bondcoat deposition process for SPS TBCs. In this work, a NiCoCrAlY bondcoat deposited by high velocity air fuel (HVAF) was compared to commercial NiCoCrAlY and PtAl bondcoats. All bondcoat variations were prepared with and without grit blasting the bondcoat surface. SPS was used to deposit the topcoats on all samples using the same spray parameters. Lifetime of these samples was examined by thermal cyclic fatigue and thermal shock testing. The effect of bondcoat deposition process and interface topography on lifetime in each case has been discussed. The results show that HVAF could be a suitable process for bondcoat deposition in SPS TBCs.

1 Introduction

A thermal barrier coating (TBC) system is designed to protect a gas turbine component from high temperatures. TBCs produced by suspension plasma spraying (SPS) can significantly improve the thermal insulation ability due to a very fine porous microstructure [1]. SPS TBCs are of significant commercial interest as the technique is considerably cheaper than commercially used Electron Beam Physical Vapour Deposition (EB-PVD), both in terms of equipment cost as well as running cost. In addition, the SPS process has higher deposition rates and the layers are porous so that the thermal conductivity is significantly reduced [2]. However, SPS TBCs have not yet been commercially introduced due to low reliability and life expectancy of the coatings. Improvements in reliability and lifetime of SPS TBCs which enables an alternative to the EB-PVD process would result in huge cost savings as well as lower emissions.
Lifetime of a TBC system is strongly dependent on bondcoat microstructure, fabrication process as well as topcoat-bondcoat interface topography [3-5]. The bondcoat layer in EB-PVD TBCs is typically deposited by vacuum plasma spraying (VPS) in case of NiCoCrAlY bondcoats and diffusion process in case of platinum modified alumunide (PtAl) bondcoats, while in case of atmospheric plasma sprayed TBCs, the NiCoCrAlY bondcoats are typically deposited by High Velocity Oxy-Fuel (HVOF) spraying. It is not fully understood yet which bondcoat deposition process would be most suitable for SPS TBCs as the deposition process affects the bondcoat microstructure and interface topography, which in turn affects TBC lifetime.
In previous work, high velocity air fuel (HVAF) has shown promising results as bondcoat deposition method in SPS TBCs [6]. HVAF is a cheap and energy efficient process as it uses compressed air as oxidant instead of oxygen as in HVOF and does not require a special vacuum chamber for spraying as in VPS. HVAF could be a more suitable process than HVOF for SPS TBCs since it creates smoother and denser bondcoat layers which were shown to be beneficial for good adhesion of SPS topcoats [7].
In this work, HVAF bondcoat was compared to commercial VPS and diffusion bondcoats for application in SPS TBCs. The research objective of this work was to investigate the relationships between bondcoat fabrication process, topcoat-bondcoat interface topography, and lifetime in SPS TBCs. The long term goal of this ongoing research work is to develop a bondcoat layer which is most suitable for SPS TBCs in order to achieve high coating lifetime.

2 Experimental

2.1 Sample preparation

Three bondcoat processes were investigated in this study – HVAF, VPS, and diffusion coating. The material which was used for HVAF and VPS was NiCoCrAlY (AMDRY 386-2, Oerlikon Metco). These two bondcoats were compared to the PtAl bondcoat. The HVAF bondcoat was sprayed with the M3 supersonic HVAF spray gun (Uniquecoat Technologies, USA) using the spray parameters based on prior experience with the process. The VPS and PtAl bondcoats were ordered from commercial suppliers. The thickness of the HVAF bondcoat was around 200 µm, VPS bondcoat was around 160 µm and PtAl bondcoat was around 75 µm in as-sprayed condition.
As the influence of grit blasting on the samples lifetime was also of interest in this study, from each set of samples some were grit blasted prior the topcoat deposition. A light grit blasting with a robotized pressure feed grit blasting gun was performed over some of the samples of each bondcoat after the bondcoat spraying. The selected grit blasting parameters would normally result in a

roughness of approximately *Ra* = 3 µm on plain uncoated steel samples.

After the grit blasting step, both grit blasted as well as non grit blasted samples were sprayed with topcoats using the same spray parameters. The topcoats were sprayed by SPS using the Mettech Axial III gun with NanoFeed 350 suspension feeder (Northwest Mettech Corp., Canada). The topcoat material was yttria stabilised zirconia (YSZ) suspension in alcohol with 25% solid load. The samples were preheated with the plasma gun prior to topcoat spraying. The topcoat thickness was around 260 µm.

Substrates in dimensions 25.4 mm diameter x 6 mm thickness (coupons) were used for microstructure analysis and thermal shock testing, while for thermal cyclic fatigue (TCF) testing, substrates in dimensions 50 mm x 30 mm x 6 mm (plates) were used. The substrate material for all TCF plates and HVAF bondcoat coupons was Hastelloy-X, while for VPS and PtAl bondcoat coupons, substrate material was Inconel 792.

2.2 Topography

3D topography of bondcoat samples in as-sprayed condition (before spraying topcoat layer) was measured by white light interferometry technique (MicroXAM, ADE Phase Shift Technology, USA) at Toponova, Sweden. Each scan covered an area of around 800 µm x 600 µm. The lateral resolution obtained with this technique was 1.1 µm along the length and 1.3 µm along the width, while the vertical resolution was 1 nm.

The captured data was analysed with the software MountainsMAP ver. 6 (DigitalSurf, France) to post-process the data as well as to compute ISO 25178 and ISO 4287 parameters.

2D topography of the bondcoat samples before and after grit blasting was measured by contact profilometer. Six measurements were done by the profilometer on each sample.

2.3 Microstructure

After topcoat spraying, one sample from each set was prepared for metallography analysis. The samples were first cold mounted in low viscosity epoxy resin, then cut using a diamond cutting disc, and then cold mounted again in high viscosity epoxy resin for grinding and polishing. The polishing was performed automatically using a Buehler Power Pro 5000 equipment. The failed samples after lifetime testing were also prepared in the same way.

Microstructures of both as-sprayed and failed samples were analysed by scanning electron microscopy (SEM) using a Hitachi TM3000 tabletop SEM equipment.

2.4 Thermal cylic fatigue testing

TCF testing is one of the common lifetime tests performed to analyse the performance of TBCs under cyclic heating and cooling under long term exposures at high temperatures. Due to the long exposure, this test allows for significant oxide growth due to bondcoat oxidation. Swelling of thermally grown oxide (TGO) and mismatch in thermal expansion coefficient of different layers in the TBC system during cycling are the main driving mechanisms leading to failure in TCF testing. Sintering of topcoat during the heating cycle as well as thermal fatigue of the coating due to thermal cycling also play a significant role in failure.

In this study, TCF testing was performed at Siemens Industrial Turbomachinery, Sweden. During this test, the samples were first heated in a furnace at 1100°C for one hour, and then cooled down to approximately 100°C in ten minutes by blowing compressed air on the coating surface of the samples. The heating and cooling cycles were repeated in this fashion until failure, which was determined by 20% spallation of the topcoat. The samples were photographed after each cycle to determine the failure.

Two samples of each coating set were tested by TCF testing in this study.

2.5 Thermal shock testing

Thermal shock testing is another common lifetime test performed to analyse the performance of TBCs under cyclic heating and cooling under short term exposures at high temperatures and then quick quenching. Due to the short exposure, this test does not allow significant TGO growth but samples are exposed to high thermal gradients and transients. Thermal expansion mismatch along with thermal fatigue of the topcoat are the major driving mechanisms leading to failure in thermal shock testing. Strain tolerance of the topcoat and its sintering resistance play a key role in failure as they determine the resistance of topcoat against thermal fatigue induced by cyclic loading conditions.

In this study, thermal shock testing was performed using the burner rig testing (BRT) equipment developed in-house at GKN Aerospace, Sweden. During this test, the samples were first heated up to topcoat surface temperatures of 1300°C with combustion burners in 75 seconds and then cooled down from the back side to around 450°C with compressed air in 75 seconds. During the heating cycle, the samples are also exposed to compressed air from the back side which enforces a thermal gradient across the sample. The heating and cooling cycles are repeated in this fashion until failure which is determined by 10% spallation of topcoat either through its entire thickness or part of its thickness (chipping failure). The samples are video recorded after every four cycles to determine the failure.

Four samples of each coating set were tested by BRT in this study.

3 Results and discussions

3.1 Bondcoat roughness

It is well known that topcoat-bondcoat interface topography affects the lifetime of TBCs significantly [7,

8]. Therefore, precise measurement of bondcoat surface topography and careful assessment of the measurement data is essential to understand TBC failure mechanisms and subsequently enhance their lifetime. Measurement of 3D topography of samples is highly beneficial to analyse the topography of different samples as it provides high resolution data with much better visualisation and reliability than 2D topography data.

The bondcoat surface roughness data of different samples measured by contact profilometer before and after grit blasting is shown in **Table 1**. Light grit blasting of bondcoat samples was performed to remove high protrusions on the bondcoat surface as these features could induce high stresses during thermal cycling and be detrimental to TBC lifetime [8]. It can be seen in Table 1 that the HVAF bondcoat showed the highest roughness both before and after grit blasting, while VPS bondcoat showed similar roughness with a bit lower value. Grit blasting reduced the surface roughness in both HVAF and VPS bondcoats suggesting that the high protrusions on the bondcoat surface were at least partially removed. On the other hand, the PtAl showed the lowest surface roughness by far among all bondcoats which was slightly increased after grit blasting as it would normally do on a smooth surface.

Table 1. Bondcoat surface roughness measured by profilometer before and after grit blasting.

Sample	Bondcoat roughness *Ra* (μm)	
	Before	After
HVAF	8.11 ± 1.23	7.17 ± 0.68
VPS	6.77 ± 0.75	6.19 ± 0.52
PtAl	1.13 ± 0.29	1.35 ± 0.28

The bondcoat surface profiles captured using white light interferometry technique in as-sprayed condition are shown in **Fig. 1**. It can be clearly noticed that all three surfaces have significantly different surface features. The HVAF bondcoat surface consists of mainly spherical hills evenly spread out along with a few high hills as indicated in Fig. 1a. The reason for this type of surface profile is believed to be the low process temperature in HVAF spraying as well as the a bit wide particle size distribution of the powder feedstock, resulting in incomplete melting of the particles. These semi-molten particles then adhere on the surface and form the spherical hills. As the larger particles in the feedstock material would be only partially molten, these particles could result in the high hills present on the HVAF bondcoat surface. The lateral size of the high hills in Fig. 1a, around 50-80 μm, correspond well to the larger size particles in the AMDRY 386-2 powder which has a powder cut of -88+16 μm.

The VPS bondcoat surface shown in Fig 1b consists of several closely spaced sharp hills as also indicated in the figure along with smaller hills evenly spread out. The reason for this type of surface profile is believed to be the high process temperature in VPS resulting in completely molten particles which could splash on the surface and create the small steep hills. The sharp hills could have been created by the larger particles in the feedstock powder which could not have been completely molten.

The PtAl bondcoat surface shown in Fig 1c consisted of a very smooth profile (please note the different scale bar) with evenly distributed small hills and valleys. The low roughness of PtAl bondcoat resulted from the diffusion process adapted to EB-PVD topcoats which require very smooth bondcoat surfaces.

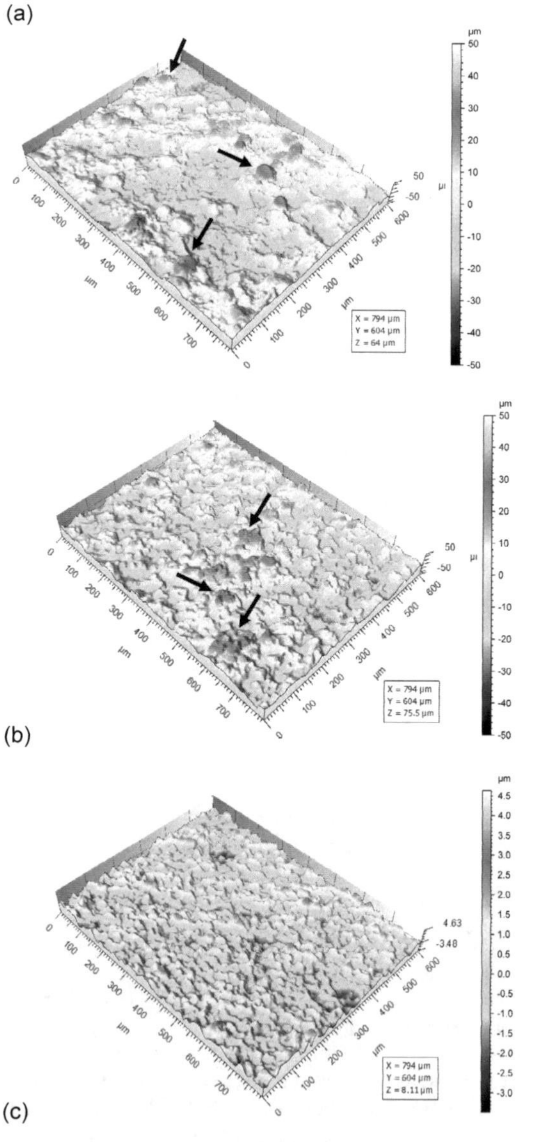

Fig. 1. Bondcoat surface profiles captured using white light interferometry technique on (a) HVAF, (b) VPS, and (c) diffusion samples in as-sprayed state.

3.2 Microstructure

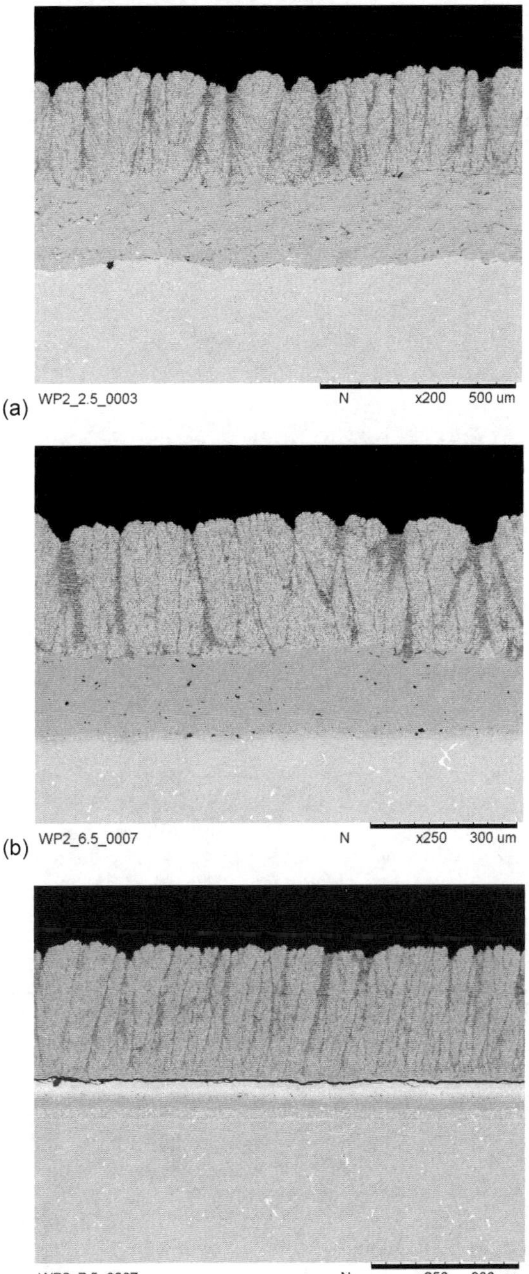

Fig. 2. Microstructure images of samples in as-sprayed condition with grit blasted bondcoats (a) HVAF, (b) VPS, and (c) diffusion.

The microstructure images of the samples in as-sprayed condition with grit blasted bondcoats are shown in **Fig. 2**. It can be observed that both HVAF and VPS bondcoats had a dense microstructure with low porosity. Some splat boundaries (delaminations) but fewer and smaller globular voids could be observed in the HVAF bondcoat (which create the dark porous areas). The topcoats showed a columnar structure typically produced by SPS on these bondcoats with quite similar column density and porosity. This was expected as the topcoats were sprayed using the same spray parameters and the bondcoats had a close *Ra* roughness.

The PtAl bondcoat showed a typical microstructure created by diffusion coating. Due to the low roughness of the bondcoat, very thin columns were created in the topcoat even though same spray parameters were used for all samples. This phenomenon has also been observed in previous work where low bondcoat roughness resulted in thin columns due to higher number on initiation peaks for column formation [7]. This topcoat microstructure looked very similar to a EB-PVD topcoat which usually exhibits high strain tolerance. However, due to the very low roughness of the PtAl bondcoat (~1 µm), the adhesion of the SPS topcoats was very poor due to poor mechanical bonding unlike EB-PVD topcoats where chemical bonding is the major bonding mechanism. This resulted in complete spallation of the topcoat during metallography preparation as can be noticed in Fig. 2c. This also indicates that a higher roughness is necessary for good adhesion of SPS topcoats which would imply wider columns with perhaps lower strain tolerance than a microstructure with high strain tolerance.

3.3 Thermal cyclic fatigue testing

The results from TCF testing are shown in **Fig. 3**. It should be pointed out that tested samples showed good TCF results, fulfilling the industrial requirements for this type of test. Between the HVAF and VPS samples which were sprayed with the same feedstock material, HVAF bondcoat showed around 20% higher TCF lifetime on average over grit blasted and non grit blasted VPS samples. Both HVAF and VPS samples showed a slighlty lower TCF lifetime with grit blasted bondcoats indicating that grit blasting did not improve the lifetime as expected. This could be due to the roughness of bondcoats which could have made the effect of grit blasting insignificant. It must be noted that HVAF bondcoat had around 25% higher thickness than VPS bondcoat which could have also influenced the TCF lifetime results. However, it was observed that the beta-phase was still present in both bondcoats after failure in TCF testing showing that there was still an aluminium reservoir left in the bondcoat (though smaller in the VPS bondcoat) and failure did not occur due to chemical failure of the bondcoat. Nevertheless, the result shows that HVAF could be a suitable process for bondcoat deposition in SPS TBCs, especially given that it is a much cheaper and less sophisticated process than VPS.

The PtAl bondcoat showed the highest TCF lifetime among all bondcoats. The reason for high TCF lifetime of PtAl bondcoats is deemed to be due to a different chemical composition of the bondcoat which results in slower TGO growth without any detrimental oxides such as chromia, spinel, nickel oxide, etc., as well as the high strain tolerance of the topcoat microstructure with very thin columns similar to a EB-PVD topcoat. The TCF lifetime of PtAl bondcoat samples was further increased with grit blasted bondcoats indicating better adhesion of the SPS topcoat in case of a low initial roughness of the bondcoat.

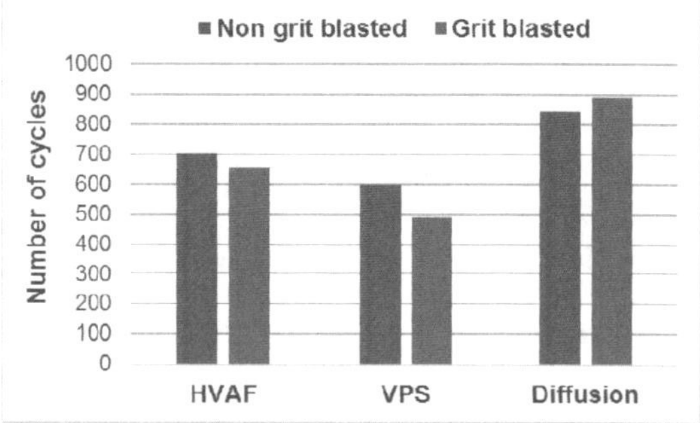

Fig. 3. Thermal cyclic fatigue lifetime results.

3.4 Thermal shock tesing

The results from thermal shock tesing are shown in **Fig. 4**. Between the HVAF and VPS samples sprayed with same feedstock material, similar to the TCF lifetime results, HVAF bondcoat showed around 50% higher BRT lifetime on average over grit blasted and non grit blasted VPS samples. VPS samples showed a slightly lower BRT lifetime with grit blasted bondcoats while HVAF showed a similar BRT lifetime when considering the standard deviation, again showing that grit blasting did not improve the lifetime in this case as expected.

The reason for lower BRT lifetime of VPS samples is believed to be the sharp hills on the VPS bondcoat surface as shown in Fig. 1b. It has been shown in previous work that for a given lateral dimension of the hill on bondcoat surface, the stresses in topcoat near the topcoat-bondcoat surface increase significantly with the amplitude of the hill, and vice versa [9]. This implies that a sharper hill as in VPS bondcoat would induce much higher stresses in the topcoat than a flatter hill as in HVAF bondcoat. These induced stresses are particularly high close to the topcoat-bondcoat surface where failure usually occurs during BRT [9]. Lower bondcoat thickness in VPS samples as well as a different substrate material (Inconel 792 instead of Hastelloy-X) could have also affected the BRT lifetime.

Due to poor adhesion of the topcoat layer on the non grit blasted PtAl bondcoats, the samples spalled off immediately after a few cycles in the harsh conditions during BRT and therefore could not be tested. In case of grit blasted PtAl bondcoats, two out of four samples spalled off immediately while other two showed a low BRT lifetime creating a very high standard deviation why its not shown in Fig. 4. Clearly, the roughness of the PtAl needs to be increased to apply SPS topcoats for better adhesion.

The poor adhesion of SPS topcoats on PtAl bondcoats did not seem to affect the TCF lifetime of these samples due to much lower heating and cooling rates in TCF testing than BRT and thus lower induced stresses. Additionally, it is believed that the initial residual stresses in the coating in as-sprayed condition are typically relaxed in a few hours during TCF testing due to high creep rates in the ceramic coatings [10]. Stress relaxtion is not possible in BRT due to much shorter cycles with significanlty higher heating and cooling rates which in turn create stress build-up in the topcoat during subsequent cycling and thus could lead to early failure.

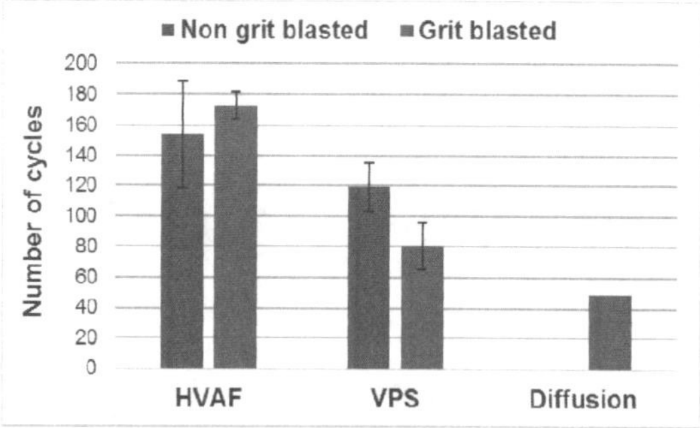

Fig. 4. Thermal shock lifetime results.

4 Summary and conclusions

In this work, NiCoCrAlY bondcoats deposited by HVAF were compared with commercial NiCoCrAlY bondcoat deposited by VPS and PtAl bondcoat. The YSZ topcoats were deposited by SPS using the same spray parameters in all cases. The effect of topcoat-bondcoat interface topography on topcoat microstructure and lifetime were discussed.

The results show that the bondcoat topography can significantly vary the column formation in topcoat microstructure. The smooth PtAl bondcoat resulted in a thin columnar structure while the rough HVAF and VPS bondcoats resulted in wider columns even though same topcoat spray parameters were used.

The HVAF bondcoat showed better lifetime results than the VPS bondcoat, even though both of them were deposited using the same feedstock material. The PtAl bondcoat showed high TCF lifetime due to different bondcoat chemistry but a low BRT lifetime due to poor adhesion of the topcoat on the smooth bondcoat. Light grit blasting of the bondcoat surface did not seem to influence the lifetime significantly, probably due to the high roughness of HVAF and VPS bondcoats making grit blasting insignificant.

The results indicate that HVAF could be a more suitable process for bondcoat deposition in SPS TBCs as compared to VPS, especially when considering that HVAF is a significantly cheaper and less sophisticated process than VPS.

5 Acknowledgements

The authors would like to acknowledge the Knowledge Foundation for the funding of this research work. Thanks to Stefan Björklund and Kenneth Andersson at University West for the help with spraying and metallographic preparation respectively.

6 References

[1] A. Ganvir, N. Curry, S. Govindarajan, and N. Markocsan: Characterization of Thermal Barrier Coatings Produced by Various Thermal Spray Techniques Using Solid Powder, Suspension, and Solution Precursor Feedstock Material. International Journal of Applied Ceramic Technology 13 (2016), Issue 2, pp. 324/32.

[2] A. Ganvir, N. Curry, N. Markocsan, P. Nylén, S. Joshi, M. Vilemova, and Z. Pala: Influence of Microstructure on Thermal Properties of Axial Suspension Plasma-Sprayed YSZ Thermal Barrier Coatings. Journal of Thermal Spray Technology 25 (2016), Issue 1-2, pp. 202/12.

[3] A.G. Evans, D.R. Mumm, J.W. Hutchinson, G.H. Meier, and F.S. Pettit: Mechanisms controlling the durability of thermal barrier coatings. Progress in Materials Science 46 (2001), Issue 5, pp. 505/53.

[4] M.P. Taylor, W.M. Pragnell, and H.E. Evans: The influence of bond coat surface roughness on chemical failure and delamination in TBC systems. Materials and Corrosion 59 (2008), Issue 6, pp. 508/13.

[5] A. Scrivani, U. Bardi, L. Carrafiello, A. Lavacchi, F. Niccolai, and G. Rizzi: A Comparative Study of High Velocity Oxygen Fuel, Vacuum Plasma Spray, and Axial Plasma Spray for the Deposition of CoNiCrAlY Bond Coat Alloy. Journal of Thermal Spray Technology 12 (2003), Issue 4, pp. 504/07.

[6] N. Curry, K. VanEvery, T. Snyder, and N. Markocsan: Thermal Conductivity Analysis and Lifetime Testing of Suspension Plasma-Sprayed Thermal Barrier Coatings. Coatings 4 (2014), pp. 630/50.

[7] N. Curry, Z. Tang, N. Markocsan, and P. Nylén: Influence of bond coat surface roughness on the structure of axial suspension plasma spray thermal barrier coatings - Thermal and lifetime performance. Surface & Coatings Technology 268 (2015), pp. 15/23.

[8] M. Gupta, K. Skogsberg, and P. Nylén: Influence of Topcoat-Bondcoat Interface Roughness on Stresses and Lifetime in Thermal Barrier Coatings. Journal of Thermal Spray Technology 23 (2014), Issue 1-2, pp. 170/81.

[9] M. Ahrens, R. Vaßen, and D. Stöver: Stress distributions in plasma-sprayed thermal barrier coatings as a function of interface roughness and oxide scale thickness. Surface and Coatings Technology 161 (2002), pp. 26/35.

[10] R. Vaßen, S. Giesen, and D. Stöver: Lifetime of Plasma-Sprayed Thermal Barrier Coatings: Comparison of Numerical and Experimental Results. Journal of Thermal Spray Technology 18 (2009), Issue 5-6, pp. 835/45.

Axial Feeding Features of Low Power Suspension Plasma Spraying

Mohammed Shahien[1,2], Masato Suzuki[1]
[1] National Institute of Advanced Industrial Science and Technology, AIST, Tsukuba/JP
[2] Central Metallurgical R & D Institute, CMRDI, Cairo/EG

Recently we successfully developed a new suspension plasma spray (SPS) system by using a twin-cathode plasma spray gun. The system consists of three plasma torches (central main torch with cathode nozzle and anode, and two sub-torches on the sides with anode nozzle and cathode). The system is characterized by using only argon (Ar) as the plasma gas and low input power, which makes it a gas and power consumption SPS system, compared to the other available SPS systems.

This study will discuss the unique features of the axial injection spraying and its effect on the coating formation by the new system. The axial injection showed a significant potential during suspension feeding. It keeps the liquid suspension on the plasma core zone; therefore, the liquid encountered the highest heat and momentum transfers. This enhances the precious control of the coating microstructure through controlling of the droplet size and splat size. It was possible to control the droplet size and therefore the coating microstructure through axia injection and controlling the solid load content on the suspensions. Furthermore, gas atomization control and adjustment showed a strong potential on controlling the mist distribution on the plasma plume and coating formation during axial injection suspension plasma spraying.

1 Introduction

Plasma spraying is recognised as one of the most outstanding coating technology in industry. It is due to the unique features of easy to use, high deposition rates, wide flexibility, highly efficient and economically industrial applicability. However, it was difficult to use the nano and/or few micron size powders and take the benefits of the superior mechanical properties of nano-structured coatings directly in conventional plasma spray [1]. It is due to, the low flowability of the fine particles, the aggregation and clogging of the feeding lines. Furthermore, the poor rheological properties of the fine particles leads to non-uniform powder transporting or injection in the core of plasma plume (high enthalpy flow).

Since the late of 1990s, the most recent innovation on plasma spray technology is promoting the application of nano and/or few micron size feedstock particles directly [2-8]. Recently two plasma spraying technologies have been developed: the suspension plasma spray (SPS) processes and solution precursor plasma spray (SPPS). These new innovations are based on using a liquid carrier for transporting the nano-structured powders into the plasma jet in the form of suspension (SPS) and/or solution precursors (SPSS). This assists the formation of thinner lamella with small diameters, almost no inter-lamellar cracks and lower residual stresses. Therefore, formation of nano-structured coatings and microstructure control.

The first patent on suspension spraying was granted in 1997 by F. Gitzhofer et. al., [2]. The process is basically based on dispersing the nano-particle (20-100 nm) into the liquid phase (water or alcohol) by means of dispersants. Followed by injection of this liquid suspension into the plasma stream. The liquid stream rapidly fragmented upon penetration the plasma stream, due to flow shear forces and vaporized. Then the liquid phase evaporates and forms a single particle or aggregates of a few particles. Followed by melting of the particle, formation of liquid drop, impact, solidification and formation of small flattened lamellae.

The feed particle size in SPS process is much smaller than particle size in conventional plasma spray and it may lose the absorbed heat and momentum faster. Therefore, the interaction between the in-flight particle and the plasma jet in SPS is important factor and has a significant effect on the coating formation and therefore its quality.

In SPS there are two ways for suspension injection:

- Radial or external injection by a feeding nozzle as shown in **Fig. 1**. It is the most used technique for SPS [9-10]. During the radial injection, the drops firstly meet the plasma gases flowing on the fringes of the plasma, thus, although the high velocities are sufficient for fragmentation and liquid vaporization, it is not enough for adequate heating of the suspended particles. These poorly heated/treated particles form stacking defects. Moreover, the arc fluctuations, the angle and the distance of the injection nozzle significantly affect the liquid droplet penetration and fragmentation during radial injection.

- Axial or internal injection, it not so common for conventional/ordinary plasma spray guns used for SPS. During the axial injection the drops are directly injected in hot gases zone of the plasma (the highest heat and momentum transfers are between the plasma and feedstock suspension)) and are rapidly fragmented in tiny droplets that are then vaporized. Although, the axial injection is more favourable for suspension spraying, for the best of our knowledge few torches with specific designs enable the axial injection. The recent system from Mettech, Vancouver, Canada, (Mettech axial III [11, 12]) torch, and hollow cathode radio frequency (RF) plasma torches [13], are using axial injection for the suspensions.

Furthermore, recently we developed a new plasma spray torch with axial injection function [8]. It is low power and gas consumption. It is based on using three plasma torches, the electrode of the main torch has hollowed anode and therefore enables the axial injection of the suspensions. Furthermore, the system requires only argon as the plasma gas and works on low power level compered to available spray systems, which makes it a promising SPS system. This study will show the axial feeding features of the new SPS system, and its effect on the coating microstructure.

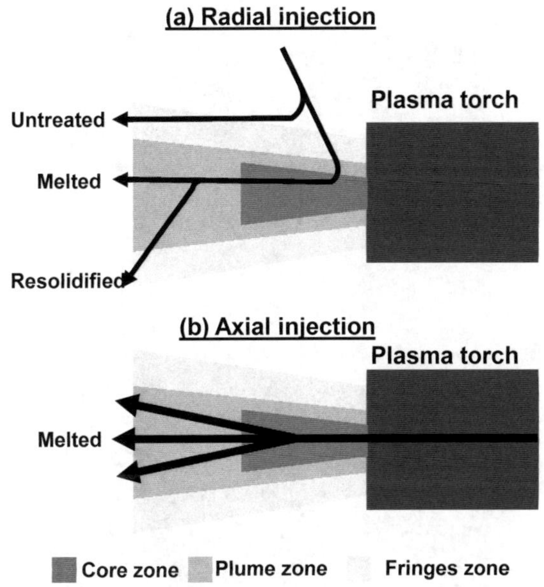

Fig. 1. Schematic expiration of the radial and axial injection plasma spraying.

2 Experimental Set-up

2.1 Plasma Spraying and Injection Systems

Figure 2 shows the real image and schematic diagram of the spraying system. It is a twin-cathode type plasma spray gun (Aeroplasma Corporation, Japan). The plasma torch is composed of three plasma torches. The main torch (P-torch; +), is located on the center of the gun, it has a reversed polarity (an anode and a cathode nozzle). Two sub-torches (N-torches: -), one on the right side and one on the left side of the main torch, and both have normal polarity (a cathode and an anode nozzle).

It is clear that, the stream of the plasma is working on the same gas flow of the main plasma torch. During ignition of the plasma, the process starts by plasma generation between the two cathodes of the sub-torches and the anode of the main-torch. The system is working with a low power level (using high voltage and low current) compared to other spray systems. Besides that, it is gas consumption system, thus only the argon (Ar) gas is enough for plasma generation.

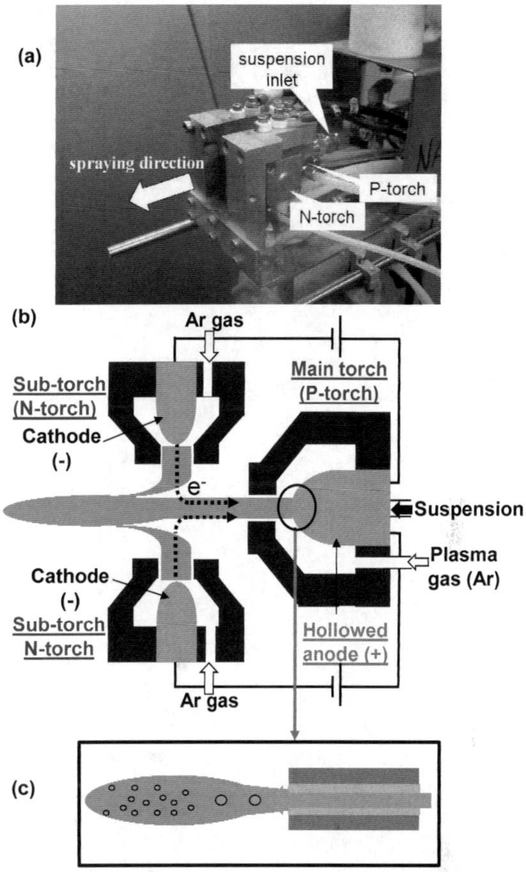

Fig. 2. (a) The real view of the spraying system, (b) schematic cross-section of the plasma torch and (c) the schematic view of the suspension injection point.

Thus, enthalpy of plasma jet is high enough (the voltage: 150~190 V) with Ar for melting the ceramic particles with this system. Therefore, it is recognized as a cost reduction spraying system.

On the other hand, the anode of the main-torch is a hollowed electrode; therefore it enables the axial injection of the suspension in the plasma jet. For suspension injection, a commercial dual tube nozzle (HM-6L: Fuso Seiki Co., Ltd.) was used. It consists of two tubes as simply shown in **Fig. 2 (c)**. The inner tube is used for suspension injection to the plasma jet. The outer tube is used for the atomizing gas (Ar gas) around the suspension.

2.2 Material, Methodology and Characterization

The suspension was prepared using the ethanol as solvent and using the Poly-Ethlene-Imine (PEI) as stabilizer in the lab. The mixture of the suspension was ball milled with using the ZrO2 balls (Φ10 mm). The suspensions were prepared with various concentrations (5~15wt%). The feedstock suspensions were injected in the plasma jet using the pneumatic system (Ar gas). All the typical spray parameters are shown in table1. The grid-blasted stainless steel (SUS304) plates were used as substrates. **Table 1**, shows the operational spraying parameters.

Table 1 Operational spraying parameters

Plasma Gas Flow Rate	P-torch: 40 L/min N-torch: 4 L/min
Current	90 A x 2
Power input	27 kW
Voltage	150 V
Atomizing Gas Flow Rate (Ar)	5 l/min
Spray distance (mm)	10-45
Solvent	Ethanol
Substrate	SS 400
Suspension feed rate	35 ml/min
Suspension concentration	5-15 wt%

The morphology of the feedstock powder, the coatings cross-sections and surface were observed using a scanning electron microscope (FE-SEM: JSM-6300F; JEOL, Tokyo, Japan). The porosity of coatings was estimated by image analysis of SEM cross-section images. The coatings phase composition was identified using the x-ray diffraction (XRD: MXP3A, Bruker AXS K.K., Japan) analysis using Cu Ka radiation.

3 Results and Discussion

3.1 Coating fabrication by the new SPS system

Figure 3 shows the SEM cross-section microstructure of the fabricated coatings by the new system and applying the axial injection. Several ceramic coatings with different melting points including: Al_2O_3, Y_2O_3, TiO_2, and YSZ were successfully fabricated by the new system. It reveals that, the enthalpy of the plasma is enough with using only Ar gas to melt these different oxide ceramics. The coatings shown in **Fig 3**, were fabricated with using different suspensions and different solid content as follow: 5 wt.% of Al_2O_3 powder at spray distance of 15 mm, 5 wt.% of YSZ powder at spray distance of 40 mm, 15 wt.% of TiO_2 powder at spraying distance of 15 mm and 10 wt.% Y_2O_3 suspension at 15 mm.

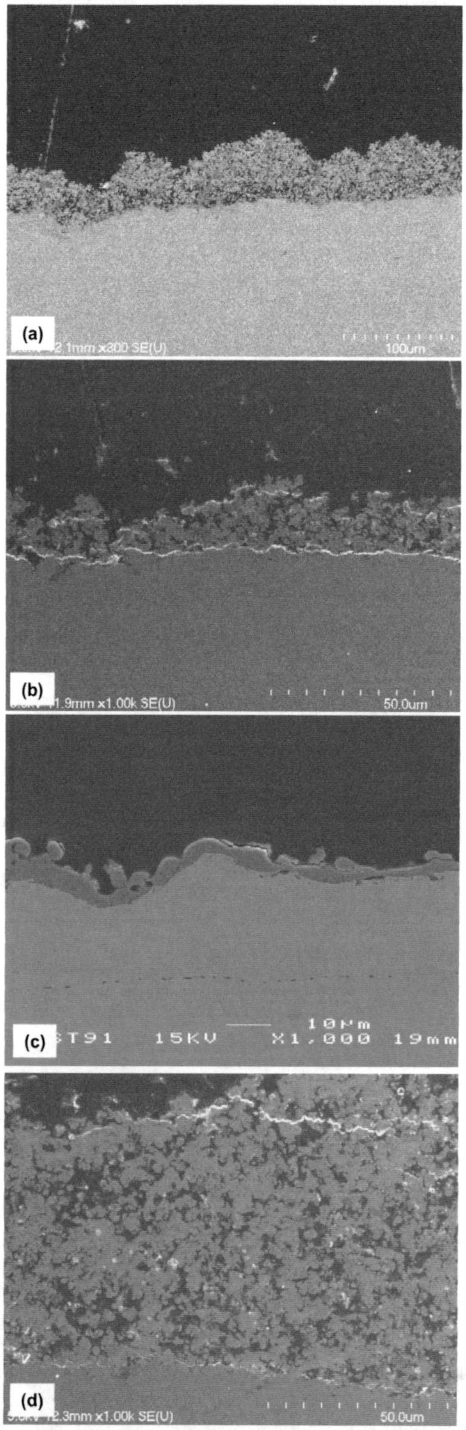

Fig. 3. The SEM cross sections of the fabricated coatings with using suspension of: (a) Al_2O_3, (b) YSZ, (c) TiO_2 and (d) Y_2O_3 powders.

It indicates that, the coating microstructure is strongly affected with the spray parameters of spray distance and the solid content in the suspension. Furthermore, it reveals that, the new SPS system is will strongly enables controlling the microstructure of the suspension plasma spray coatings in wide range.

3.2 Axial spraying novel features

During suspension plasma spray, after injecting the suspension in the plasma plume it get accelerated in the plume and follow the sequence of mist atomization, vaporization of the solvent and particle aggregation and finally melting and then deposition on the substrate. The injection way and position of the particles in the plasma jet is important factor to control the splat formation and the coating deposition is SPS process. Thus, it will affect the liquid penetration on the plasma plume. The most important condition for deposition of the coating and is the penetration of the liquid in the hot temperature (central zone of the plasma). With using the axial injection it is relatively easy to and faceplate the feasibility of controlling the microstructure through controlling and optimization of the process operational parameters.

In order to clarify the effect and benefits of using the axial inject and changing the solid content of suspension, the suspensions were supplied into the inner tube and injected vertically downward into static atmospheric air. In which the ethanol and/or suspensions were injected to the p-torch through the dual-tube nozzle without using plasma. Thus, suspension is supplied to the atomizers from pressurized liquid container. Furthermore, the injection pressure was also controlled.

The metal halide lamp was used for observation and photograph. The laser diffraction method by using LDSA1500A (Tohnichi computer applications co. ltd.), using the histogram mode and the focus length of lens is 300 mm were applied. These experiments were investigated due to difficulty of observing the details of the suspension species in the plasma during the normal plasma spray process due to the high temperature of the plasma.

The influence of suspension concentration was confirmed during investigating the atomization parameters. The real observation of the spray in case of using only ethanol and suspensions with different solid load content is shown in **Fig. 4**. It reveals that, with increasing the suspension concentration (increase the solid load) gradually, the amount of the injected suspension improved. In addition, the suspensions directly follow the gas stream flow (plasma gas flow during the real experiments), this can clearly reveals the advantages of the axial injection spraying.

Furthermore, it reveals that, during spray and the axial injection, the suspended particles will stay on the core of the plasma jet with the highest heat and momentum transfer. Therefore, the droplet of slurry in plasma in will follow a homogenies temperature regime during its deposition sequence:

- Aerodynamic break down
- Evaporation of liquid phase
- Sintering of some fine solids
- Melting of fine solids and of agglomerates
- Evaporation of liquid material from melt
- Impact on the substrate surface

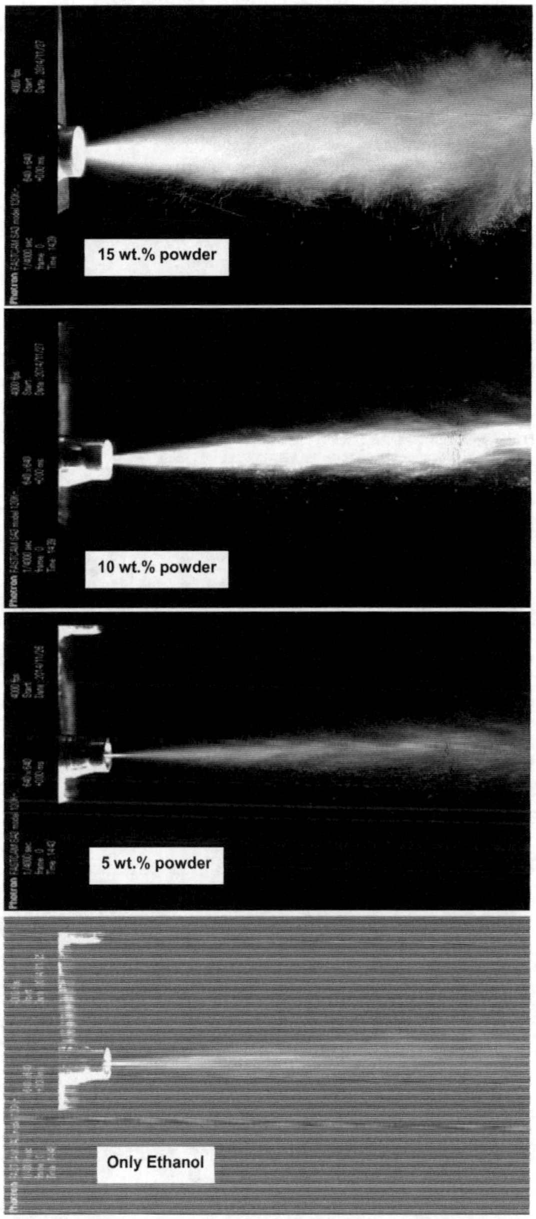

Fig. 4. The influence of solid load content during supplying only suspension with the axial injection.

The above observations revealed that, with using axial injection, all the different suspensions stayed in the plasma and this will significantly assist the feasibility of controlling the microstructure.

Figure 5 shows the surface morphology of the fabricated coatings by using different ceramic solid load content during axial injection suspension spraying. It shows that, the morphology of the coatings and splat size is significantly controlled by changing the solid content on the suspensions. Thus, using low solid content on suspension leads to formation of coatings consisting of high amount of spherical splats and small amount of flatten splats. However, using high solid content the suspensions leads to formation of coatings consisting of high amount of flatten splats and small spherical splats.

Thus, in the case of suspension with low solid content, during spraying after the solvent vaporization the mist tends to aggregate and forms small (light) droplet which deposit on the substrate as spherical splats and forms porous structure. On the other side, in case of suspension with high solid content, during spraying after the solvent vaporization the mist tends to aggregate and forms large (heavy) droplet which flatten on the substrate and form dense coatings.

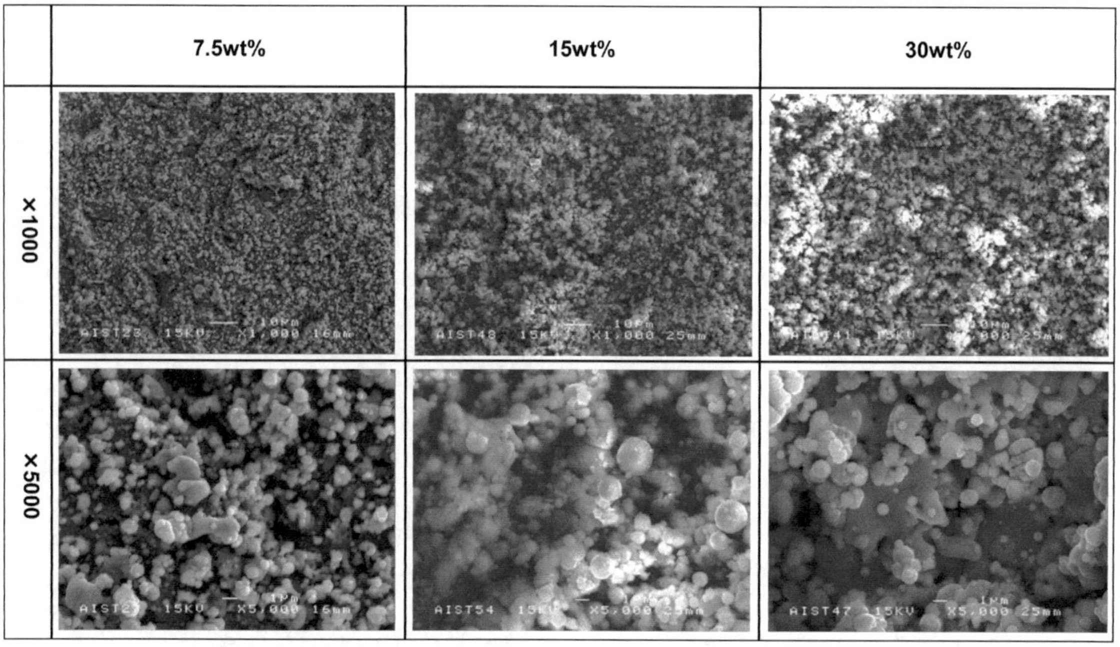

Fig. 5. The SEM microstructure of the surface of the fabricated coatings with using different Y_2O_3 suspension with different solid content of powders.

The formation mechanism of dense coatings with using high solid load suspension and formation of thin coatings with low solid content under axial injection can be explained as follow:
The light mist forms spherical splats and forms porous coatings and the heavier droplets forms flatten splats and therefore dense coatings is related to the momentum and movement of the droplet near the substrate before deposition. The gas flow decrease gradually on the direction of the substrate and leads to decreasing the plasma jet velocity and therefore decreasing the particles velocity. The decreasing tendency of the smaller (lighter) particles is more significant than the larger ones (heavier).

Therefore, the small (light) droplet will have very low velocity and very low momentum and therefore it will stick on the substrate as it is without flattening and forming porous coatings. On the other hand, large (heavy) droplet will have higher velocity (compared to the smaller ones) and higher momentum and therefore it will hit the substrate strongly, flatten and forming dense coatings.

It reveals that, with using axial injection the coating microstructure can be easily controlled by controlling the spraying conditions like solid content on the suspensions. For example, with using low suspension concentration, the small (light) spherical droplet (with low velocity and small momentum), will stick on the substrate as it is and lead to formation of porous coatings. On the other hand, with using high suspension concentration the large (heavy) droplet (with higher velocity and large momentum), will hit the substrate hardly and flatten and leads to formation of dense coatings.
In addition, during the axial injection the atomizing gas flow rate is showing a significant effect on during spray, coating formation as well as microstructure. **Figure 6** shows some of our current investigations on this issue. It is clear that, the atomizing gas flow affects the breakout point of the suspension, which

will affect the mist distribution on the plasma plume and therefore the melting tendency and coating microstructure.

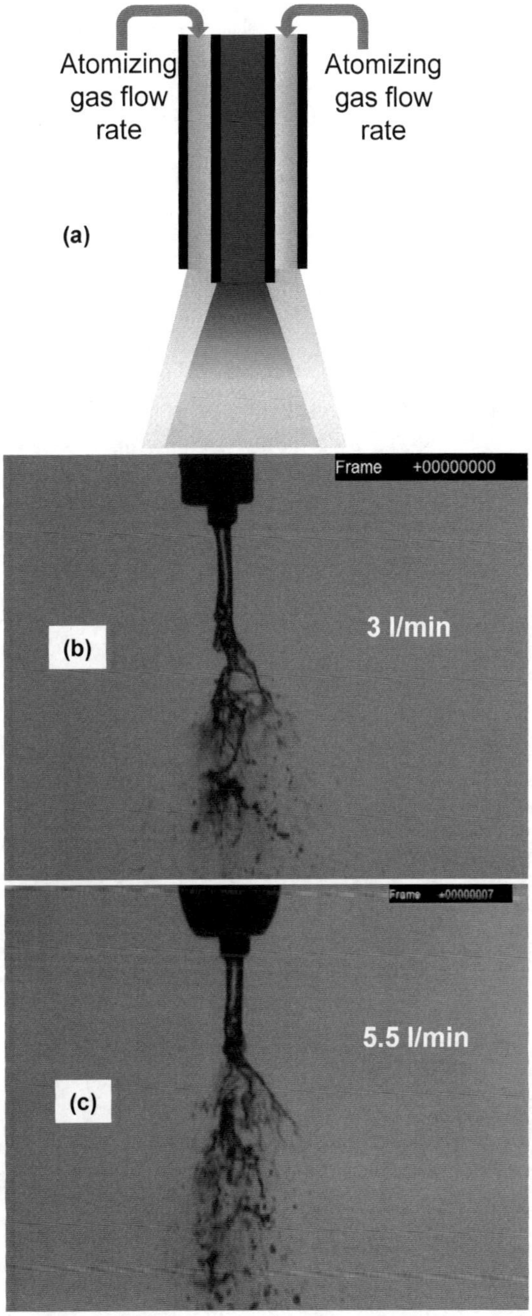

Fig. 6. Schmatic illustraion of the atomizing gas effect experiments and the real image of the experiment with different atomizing gas flow.

4 Conclusion

Applying the axial injection on suspension plasma spraying is recognised as one of the most outstanding issues to control the process and improve its performance. A new suspension plasma spraying system with an axial feeding was successfully developed. Upon axial injection it was possible to inject and keep the suspensions on the core zone of the plasma even with changing the solid content of the ceramic particles. It keeps the suspension droplets on homogeneous conditions of the highest heat and momentum transfers from the plasma.
Furthermore, this enables the precious control of the droplet size and therefore the splat size and coating microstructure, by changing the processing parameters like the solid load content. On the other hand, atomization of the suspension mist during axial injection is an important factor to control the mist distribution on the plasma plume and thereby coating build up structure.

5 References

[1] Karthikeyan, J., Berndt, C.C., Tikkanen, J., Wang, J.Y., King, A.H. and Herman, H.: Preparation of nanophase materials by thermal spray processing of liquid precursors, NanoStructured Materials, (1997), 8 (1), pp. 61-74.

[2] Gitzhofer, F., Bouyer, E. and Boulos, M.I.: Suspension plasma spraying. US Patent 5 609 921, 3 November 1997.

[3] Fauchais, P., Montavon, G., Lima, R. and Marple, B.: Engineering a new class of thermal spray nano-based microstructures from agglomerated nanostructured particles, suspensions and solutions: an invited review, J. Phys. D: Appl. Phys., (2011), 44 (no. 093001), pp. 1-53.

[4] Pawlowski, L.: Suspension and solution thermal spray coatings, Surf. Coat. Technol., (2003), 202 (19), pp. 2807-2829.

[5] Fauchais, P. and Vardelle, A.. Innovative and emerging processes in plasma spraying: from micro- to nano-structured coatings, J. Phys. D: Appl. Phys., (2011), 44 (no. 194011), pp. 1-4.

[6] Pawlowski, L.: Finely grained nanometric and submicrometric coatings by thermal spraying: a review, Surf. Coat. Technol., (2008), 202 (18), pp. 4318-4328.

[7] Viswanathan, V., Laha, T., Balani, K., Agarwal, A. and Seal, S.: Challenges and Advances in Nanocomposite Processing Techniques, Mat. Sc. and Eng. R, (2006) Vol. 54, pp. 121–285.

[8] Shahien, M. and Suzuki, M.: Low power consumption suspension plasma spray system for ceramic coating deposition, Surf. Coat. Technol.,(2016) http://dx.doi.org/10.1016/j.surfcoat.2016.07.040

[9] Fauchais, P., Rat, V., Coudert, J.-F., Etchart-Salas, R. and Montavon G.: Operating parameters for suspension and solution plasma-spray coatings, Surf. Coat. Technol. (2008), 202, pp. 4309–4431.

[10] Kassner, H., Siegert, R., Hathiramani, D., Vassen, R. and Stoever, D.: Application of Suspension Plasma Spraying (SPS) for

manufacture of ceramic coatings, J. Therm. Spray Technol., (2008), 17–1, pp. 115–123.
[11] Oberste-Berghaus, J., Marple, B. and Moreau, C.: Suspension plasma spraying of nanostructured WC-12Co coatings, J. Therm. Spray Technol. (2006), 15–4, pp. 676–681.
[12] Oberste-Berghaus, J., Legoux, J.-G., Moreau, C., Tarasi, F. and Chraska, T.: Mechanical and thermal transport properties of suspension thermal-sprayed alumia-zirconia composite coatings, J. Therm. Spray Technol., (2008), 17–1, pp. 91–104.
[13] Fauchais, P.: Understanding plasma spraying: an invited review, J. Phys. D. Appl. Phys., (2004), 37, R86–R108.

The possibility of using ultrasonic waves to support Suspension Plasma Spraying process

Paweł Sokołowski, Monika Gałafin, Stefan Kozerski, Marcin Korzeniowski

Faculty of Mechanical Engineering, Wroclaw University of Technology, Łukasiewicza 5, 50-371 Wrocław, Poland

Suspension Plasma Spraying is one of the newest thermal spray processes. It uses liquid feedstock instead of dry powders, as it is in case of conventional Powder Plasma Spraying. The suspension preparation process should be optimized individually depending on the powder material to avoid poor suspension stabilization, fast sedimentation of solids or too high viscosity. However, regardless of the liquid feedstock preparation, each kind of suspension causes some technological problems during spraying. In this study the possibility of using high power ultrasonic waves to support the suspension feeding and injection was investigated. The idea was to break the powder agglomerates before the suspension goes through the injector (in the pipeline system). The in-line ultrasonic dispersion system was designed. The geometry of all important components (transducer, booster and horn) was analyzed and calculated depending on the frequency and power of ultrasounds. By using modal analysis and Finite Element Method (FEM) the propagation of ultrasonic wave was simulated for different sonotrode geometries, called: (i) stepped, (ii) exponential, (iii) catenoidal and (iv) tapered. The in-house system was initially tested in order to see if all elements were designed and manufactured properly. Finally, the ultrasonic cavitation was observed inside the feeding pipeline system, which should allow dispersing the suspension prior to its injection into the plasma jet.

Key words: suspension plasma spraying, suspension preparation, suspension feeding, ultrasonic cavitation

1 Introduction

The use of suspensions or solutions is going to be more and more popular in thermal spray technology. One of the most important advantage of using liquid feedstocks over the dry powders is the possibility to obtain finely-grained coatings [1]–[3]. When using conventional powder spraying methods, the injection of fine solids into the high-energy stream of gas is very difficult, e.g. due to their low mass [4]. In such processes the coarse, micrometer sized, powders are usually used. Thus, this results in the formation of coatings showing micrometer-sizes features [5]. The application of spray materials in a form of liquids (suspensions or solutions) allows to avoid such problem. The liquid feedstock is injected in a form of continuous stream or atomized droplets and then its fragmentation occurs. In case of suspension spraying, the fine powder particles reach the high-energy stream of gas just after the evaporation of solvent (suspension is a mixture of submicrometer- or nanometer-sized powders and solvent) [6]. During solution spraying, the fine grained solids are in-flight formed from the precursor being injected into the high-energy stream of gas [7].
Especially Suspension Plasma Spraying (SPS) is of great interest now for both research and industry. The technology transfer of SPS can be observed already. One of the main issues in the development of SPS technology is still the evaluation and control of process repeatability. It was found that one of the key elements influencing the process dynamics is the formulation of suspension [8], [9]. The stable and homogeneous suspension is required to obtain e.g. consistent feeding and injection of feedstock. Thus, the suspension formulation process should be optimized individually for each kind of solid/solvent material to avoid poor suspension stabilization, fast sedimentation of solids etc. However, the own experience shows that even a good suspension can causes many different problems during spraying. Many different aspects can be considered here:

- feeder: coagulation/flocculation of powder particles; sedimentation of suspension; inconsistent feeding due to the agitation or change of suspension characteristics,
- supply lines: suspension pipe flow characteristics depending on the solid morphology/content, solvent type, feeding parameters and supply system design; settling of solid particles inside the narrow pipes and pipeline components, powder buildup in so-called stagnation points,
- injector: clogging of injector nozzle; injector degradation; powder buildup inside the torch nozzle and its degradation in case of internal injection; incorrect feedstock penetration and fragmentation due to the wrong selection of injection point and injection parameters in case of external injection.

The above causes that some improvements should be proposed in order to support feeding and injection stages in SPS process. In this work the idea of using the in-line suspension dispersion system is proposed.

2 General idea of the in-line ultrasonic suspension dispersion

The proper preparation and feeding of the liquid feedstock is a relevant part of Suspension Plasma Spraying process. It is already well-known also that the ultrasonic waves can be very useful in the colloid chemistry. The high-intensity ultrasounds are very often used to homogenize liquid-solid systems due to the phenomena called ultrasonic cavitation [10]. When the colloid is subjected to high-intensity ultrasound then the nucleation of bubbles, e.g inside the powder aggregates, occur. The compression and expansion of the ultrasonic wave cause the pulsing and continuous growth of the bubbles. After reaching of a very high local temperature and pressure the bubbles collapse. The intense shock waves emanate on collapse and allows breaking of powder clusters. This results in the formulation of well-dispersed colloid [11]. This phenomena is presented in Fig. 1 [12].

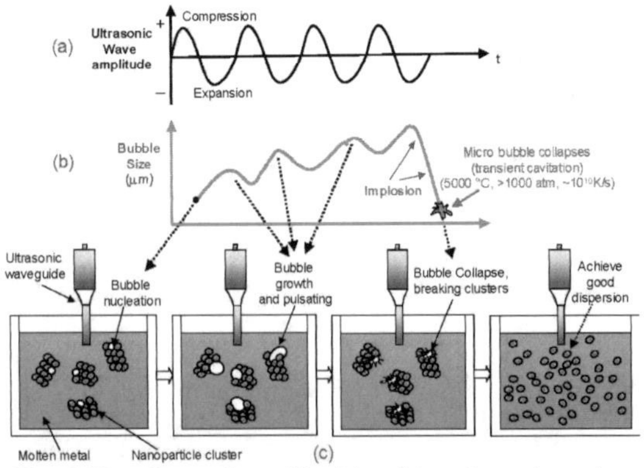

Fig. 1 The dispersion of liquid-solid systems by using the ultrasonic cavitation phenomena [12]

One of the important part of the suspension formulation process is the homogenization stage. It is performed usually by using conventional tip sonicators and takes place a long before the spraying process occurs. In this study, the possibility of breaking the powder agglomerates directly during the suspension feeding is presented. The in-house designed sonicator was implemented in the feeding pipeline system. This would allow carrying out a short homogenization stage of liquid feedstock just prior to its injection into the plasma jet. Furthermore, the settled powder particles can be activated and suspended in the solvent again. This can also prevent the small diameter supply lines against the powder buildup inside and the clogging of feeding system. The schematic illustration of the proposed idea is shown in Fig. 2

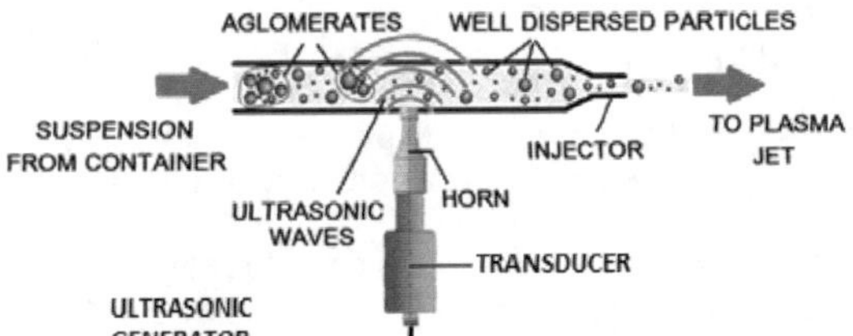

Fig. 2 The ultrasonic suspension dispersion system implemented in the feeding pipeline system

In order to ensure a proper operation of the in-house made ultrasonic system, it was necessary to properly select (or design) all components presented in Fig. 2. The set-up consisted of: ultrasonic generator, transducer (converter), the waveguide (booster) and the ultrasonic working tool (horn/sonotrode). The high-energy ultrasonic generator with a set frequency of 25 kHz and a power output in the range of 250 - 500 Watt was applied. Then the high-efficiency sandwich ultrasonic transducer with a screw-on permanent booster were selected. A working element, supplying ultrasonic energy to the feeding line and feedstock material, was an ultrasonic sonotrode. The role of sonotrode is to amplify the vibration amplitude and transmit the ultrasonic wave supplied by the ultrasonic generator. It was designed in a way to reach the maximum vibration amplitude of the ultrasonic wave at the sonotrode tip. This should allow to promote the ultrasonic cavitation phenomena inside the suspension supply line.

3. Design of the working element

As described in previous paragraph, the sonotrode was found to be a key component of the system. But the design and manufacturing of the horn had to be performed very carefully. If the sonotrode does not fit to the system then its efficiency is very limited or, in some cases, the ultrasonic generator can be even damaged.

The dimensions of sonotrode (diameter and length) were calculated depending on the frequency and wavelength of ultrasounds. In general, the length of the horn can be assumed as a half of the length of the longitudinal wave (λ/2). But this is the true only if the shape of the sonotrode is very simple, e.g. cylindrical. For more complex horn geometries different aspects should considered, e.g. the length coefficient, which is influenced by the distribution of internal stresses in the sonotrode. Usually, the sonotrodes with circular cross-section are used. In this study the four different horn shapes were considered: stepped, exponential, catenoidal and tapered. The analyzed horns are presented in Fig. 3. The input sonotrode diameter D_o was 60 mm and it was influenced by the booster geometry. The output diameter, labelled as *d,* was influenced by diameter of the suspension supply line (which had a diameter of 6 mm). The lengths of the sonotrodes l_k were calculated by using the equations given by A. I. Markov [13], which were as follows:

- stepped sonotrode:

$$l_k = \frac{\lambda}{2} = \frac{c_l}{2f}$$

- exponential sonotrode:

$$l_k = \frac{c_l}{2f} \cdot \sqrt{1 + \left(\frac{\ln(N)}{\pi}\right)^2}$$

- catenoidal sonotrode:

$$l_k = \frac{\lambda}{2} \cdot \sqrt{\frac{(kl)^2 + (ArchN)^2}{\pi^2}}$$

- tapered sonotrode:

$$l_k = \frac{\lambda}{2} \cdot \frac{(\alpha l_k)}{\pi}$$

where:
λ - wavelength [m],
c_l - ultrasonic wave propagation velocity [m/s],
f - frequency [Hz],
N - ratio of input sonotrode diameter / output sonotrode diameter (D_o/d) [-],
k'l - length coefficient of catenoidal sonotrode [-],
αl_k - length coefficient of tapered sonotrode [-].

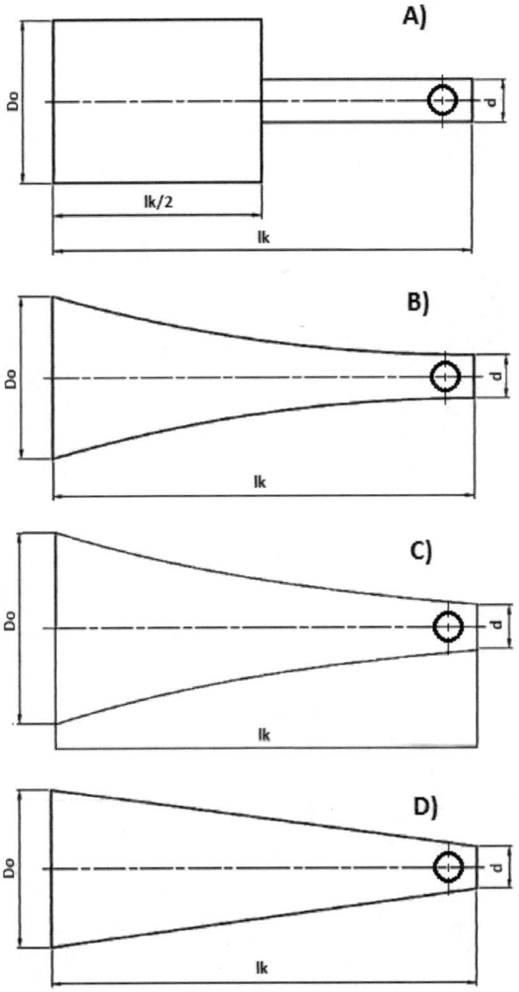

Fig. 3 The different horns analyzed by FEM: (a) stepped, (b) exponential, (c) catenoidal and (d) tapered

The sonotrode should be made of material characterized by: (i) sufficient acoustic properties (to efficiently transmit the wave energy), (ii) high strength (due to high internal stresses) and (iii) being machinable. In this study the two different materials were considered: AlCuMg2 aluminum alloy and C45 non-alloy quality steel. The calculated lengths of sonotrodes in accordance to the mentioned methodology are summarized in Table 1.

Table 1 The calculated lengths of sonotrodes depending on the material and sonotrode shape

Sonotrode	Stepped	Exponential	Catenoidal	Tapered
Length of Al alloy sonotrode [mm]	102±2	120±2	104±2	116±2
Length of steel sonotrode [mm]	105±1	124±1	107±1	120±1

4 Results and discussion

4.1 Numerical analysis

The modal analysis of ultrasonic wave propagation was performed by using *Autodesk Inventor* software (Fig. 4). The surface fixed to the booster was highlighted in light blue. The model included also the hole for suspension supply line. The FEM model was built using the quadrilateral elements (not presented in Fig. 4). Each sonotrode model was subjected to ultrasonic vibrations in the range of 20 to 30 kHz (equivalent to the frequency of the ultrasonic generator control unit). Then, the natural frequency of each sonotrode was analyzed. The position of the vibration nodes and the propagation of ultrasonic wave was also investigated. The natural frequency should be similar to that of ultrasonic generator (=25 kHz) and the greatest vibrations should occur near the hole that was designed for the suspension supply line.

Fig. 4 The exemplary 3D model used for FEM analysis - stepped sonotrode

The best values of natural frequencies obtained by the FEM analysis for each kind of sonotrode are summarized in Table 2. It can be noticed that the values are lower than 25 kHz, so the frequency given by the ultrasonic generator. This was caused by several reasons. The given equations for the sonotrode length calculation assume that the resonance take place at the end of the sonotrode. But for the effective suspension dispersion it is needed to introduce the resonance to the liquid feedstock inside the feeding pipe. So, the hole was implemented to the sonotrode structure and it changes significantly the distribution of ultrasonic wave in the sonotrode tip.

Table 2 The best natural frequencies obtained by the FEM analysis

Sonotrode	Material	Best frequency [Hz]
Stepped	Al Alloy	23987
	Steel	23747
Exponential	Al Alloy	21681
	Steel	21727
Catenoidal	Al Alloy	23571
	Steel	23723
Tapered	Al Alloy	22469
	Steel	22578

The propagation of the ultrasonic wave was observed to be very different depending on the geometry and the natural frequency of the sonotrode. The most useful results were obtained for tapered and catenoidal sonotrodes (Fig. 5). The greatest ultrasonic vibrations were located near to the hole designed for the suspension supply line. This was were necessary in order to induce the ultrasonic cavitation inside the feeding pipeline system.

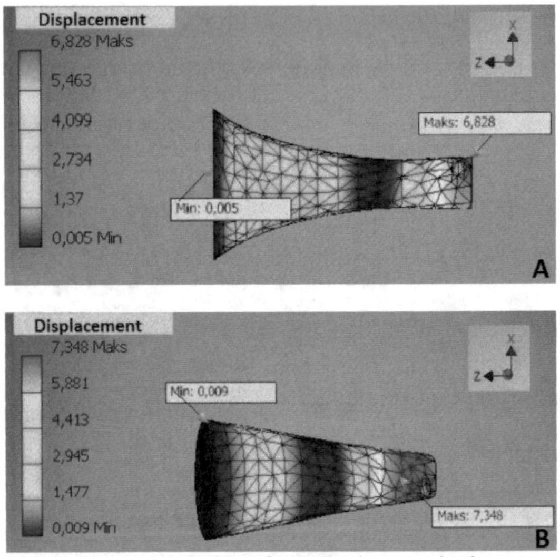

Fig. 5 The propagation of ultrasonic wave in the case of: (a) catenoidal and (b) tapered sonotrodes

4.2 The initial test of in-line suspension dispersion system

The performed calculations and FEM analysis allowed to design the prototype of ultrasonic dispersion system. The preliminary tests were performed in order to see if the phenomena of ultrasonic cavitation can be induced inside the feeding line. The catenoidal sonotrode was used. As can be seen in Fig. 6 the ultrasonic waves can be sufficiently transmitted to the feedstock material. The air bubbles can be observed inside the pipe (please, see red rectangle in Fig. 6a).

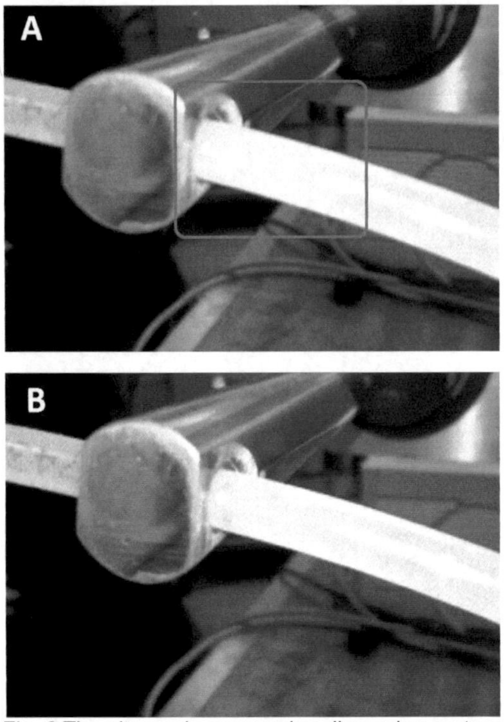

Fig. 6 The ultrasonic suspension dispersion system; (a) UT "on" and (b) UT "off"

The ultrasonic cavitation takes place in a limited length of the pipeline (a few cm). The time needed for a nucleation, growth and collapse of the air bubble is very short. There should be no risk that the suspension flow and deposition process will be disturbed by the air bubbles introduced to the feedstock material. But this will be verified during the plasma spraying tests.

5 Conclusions

In this study the idea of using the ultrasonic waves to support feeding and injection of suspension during Suspension Plasma Spraying process was investigated. The in-line ultrasonic dispersion system was analyzed and designed. The proper components (ultrasonic generator, transducer, booster) were selected. The shape and length of working tool (sonotrode) were considered depending on the frequency and power of ultrasounds. By Finite Element Method (FEM) the propagation of ultrasonic wave through the sonotrode was simulated. The results showed the use of tapered and catenoidal sonotrodes allows obtaining the required displacement distribution. In accordance to the numerical simulations, the greatest ultrasonic vibration should occur near the hole designed for suspension supply line. Finally, the system basing on catenoidal sonotrode was built and preliminary tested. The ultrasonic waves were sufficiently transmitted to the suspension supply line and the ultrasonic cavitation was induced inside the line.

The further research are carried out in order to see how the ultrasonic dispersion system operates during the deposition process. The influence of the ultrasonic dispersion system on the pipeline system lifetime, spray process dynamics and coating microstructure is under consideration now.

6 Acknowledgements

The work was financed by the research project Preludium 2014/15/N/ST8/02660, awarded to the Faculty of Mechanical Engineering at the Wroclaw University of Science and Technology by the National Science Centre (Poland). The help of Ewa Harapińska from Wrocław University of Science and Technology is also acknowledged.

6 References

[1] L. Pawlowski, "Suspension and solution thermal spray coatings," *Surf. Coat. Technol.*, vol. 203, no. 19, pp. 2807–2829, Jun. 2009.

[2] J. Karthikeyan, C. C. Berndt, J. Tikkanen, J. Y. Wang, A. H. King, and H. Herman, "Preparation of nanophase materials by thermal spray processing of liquid precursors," *Nanostructured Mater.*, vol. 9, no. 1–8, pp. 137–140, 1997.

[3] P. Fauchais and G. Montavon, "Latest Developments in Suspension and Liquid Precursor Thermal Spraying," *J. Therm. Spray Technol.*, vol. 19, no. 1–2, pp. 226–239, Jan. 2010.

[4] P. Fauchais and A. Vardelle, "Solution and Suspension Plasma Spraying of Nanostructure Coatings," in *Advanced Plasma Spray Applications*, H. Salimi Jazi, Ed. InTech, 2012.

[5] S. Sampath and H. Herman, "Rapid solidification and microstructure development during plasma spray deposition," *J. Therm. Spray Technol.*, vol. 5, no. 4, pp. 445–456, Dec. 1996.

[6] J. Fazilleau, C. Delbos, V. Rat, J. F. Coudert, P. Fauchais, and B. Pateyron, "Phenomena Involved in Suspension Plasma Spraying Part 1: Suspension Injection and Behavior," *Plasma Chem. Plasma Process.*, vol. 26, no. 4, pp. 371–391, Apr. 2006.

[7] S. Govindarajan, R. O. Dusane, and S. V. Joshi, "In situ Particle Generation and Splat Formation During Solution Precursor Plasma Spraying of Yttria-Stabilized Zirconia Coatings," *J. Am. Ceram. Soc.*, vol. 94, no. 12, pp. 4191–4199, Dec. 2011.

[8] R. Rampon, O. Marchand, C. Filiatre, and G. Bertrand, "Influence of suspension characteristics on coatings microstructure obtained by suspension plasma spraying," *Surf. Coat. Technol.*, vol. 202, no. 18, pp. 4337–4342, Jun. 2008.

[9] F.-L. Toma, A. Potthoff, L.-M. Berger, and C. Leyens, "Demands, Potentials, and Economic Aspects of Thermal Spraying with Suspensions: A Critical Review," *J. Therm. Spray Technol.*, vol. 24, no. 7, pp. 1143–1152, Oct. 2015.

[10] J. H. Bang and K. S. Suslick, "Applications of Ultrasound to the Synthesis of Nanostructured Materials," *Adv. Mater.*, vol. 22, no. 10, pp. 1039–1059, Mar. 2010.

[11] K. S. Suslick and G. J. Price, "Applications of Ultrasound to Materials Chemistry," *Annu. Rev. Mater. Sci.*, vol. 29, no. 1, pp. 295–326, 1999.

[12] S. Nimityongskul *et al.*, "Characterization of Hot Extruded Mg/SiC Nanocomposites Fabricated by Casting," in *Magnesium Technology 2011*, W. H. Sillekens, S. R. Agnew, N. R. Neelameggham, and S. N. Mathaudhu, Eds. Springer International Publishing, 2011, pp. 443–446.

[13] A. I. Markov, *Ultrasonic Machining of Intractable Materials [by] A.I. Markov; Edited by E.A. Neppiras, Translated [from the Russian] by Scripta Technica, Ltd.* 1966.

AUTHOR INDEX

AUTHOR INDEX

AUTHOR INDEX

AUTHOR INDEX

AUTHOR INDEX

AUTHOR INDEX